TEACHER'S EDITION

HOLT
Pre-Algebra

Jennie M. Bennett

David J. Chard

Audrey Jackson

Jim Milgram

Janet K. Scheer

Bert K. Waits

HOLT, RINEHART AND WINSTON

A Harcourt Education Company

Orlando • **Austin** • New York • San Diego • Toronto • London

STAFF CREDITS

Editorial

Lila Nissen, *Vice President*
Robin Blakely, *Associate Director*
Joseph Achacoso, *Assistant Managing Editor*
Threasa Boyar, *Editor*

Student Edition

Glenn Worthman, *Senior Editor*
Tessa Henry, *Editor*
Kristi Smith, *Associate Editor*

Teacher's Edition

Kelli Flanagan, *Senior Editor*
Thomas Hamilton, *Editor*

Ancillaries

Mary Fraser, *Executive Editor*
Higinio Dominguez, *Associate Editor*

Technology Resources

John Kerwin, *Executive Editor*
Robyn Setzen, *Senior Editor*
Patricia Platt, *Senior Technology Editor*
Manda Reid, *Technology Editor*

Copyediting

Denise Nowotny, *Copyediting Supervisor*
Patrick Ricci, *Copyeditor*

Support

Jill Lawson, *Senior Administrative Assistant*

Design

Book Design

Marc Cooper, *Design Director*
Tim Hovde, *Senior Designer*
Lisa Woods, *Designer*
Teresa Carrera-Paprota, *Designer*
Bruce Albrecht, *Design Associate*
Ruth Limon, *Design Associate*
Holly Whittaker, *Senior Traffic Coordinator*

Teacher's Edition

José Garza, *Designer*
Charlie Taliaferro, *Design Associate*

Cover Design

Pronk & Associates

Image Acquisition

Curtis Riker, *Director*
Tim Taylor, *Photo Research Supervisor*
David Knowles, *Photo Researcher*
Elaine Tate, *Art Buyer Supervisor*
Nicole McLeod, *Art Buyer*
Sam Dudgeon, *Senior Staff Photographer*
Victoria Smith, *Staff Photographer*
Lauren Eischen, *Photo Specialist*

New Media Design

Ed Blake, *Design Director*

Media Design

Dick Metzger, *Design Director*
Chris Smith, *Senior Designer*

Graphic Services

Kristen Darby, *Director*
Eric Rupprath, *Ancillary Designer*
Linda Wilbourn, *Image Designer*

Prepress and Manufacturing

Mimi Stockdell, *Senior Production Manager*
Susan Mussey, *Production Supervisor*
Rose Degollado, *Senior Production Coordinator*
Sara Downs, *Production Coordinator*
Jevara Jackson, *Senior Manufacturing Coordinator*
Ivania Lee, *Inventory Analyst*
Wilonda Ieans, *Manufacturing Coordinator*

Holt Pre-Algebra

Teacher's Edition Contents

REVIEWERS

Audrey Jackson
Principal
Claymont Elementary School
Ballwin, MO

RESEARCH

Glickman, C. 2002.
Leadership for learning.
Alexander, VA: Association for
Supervision and Curriculum
Development.

**National Council of Teachers of
Mathematics. 2000.**
*Principles and standards for school
mathematics.*
Reston, VA: National Council of
Teachers of Mathematics.

Senge, P. 1994.
The fifth discipline fieldbook.
New York, NY: Doubleday.

Tomlinson, C. 1995.
*How to differentiate instruction in
mixed-ability classrooms.*
Alexander, VA: Association for
Supervision and Curriculum
Development.

Wiggins, G., and J. McTighe. 1998.
Understanding by design.
Alexander, VA: Association for
Supervision and Curriculum
Development.

Classroom and Learning Environment

"Effective mathematics teaching requires understanding what students know and need to learn and then challenging and supporting them to learn it well."

National Council of Teachers of Mathematics 2000

The fundamental goal of *Holt Middle School Math* is to provide teachers with the necessary tools and understanding of school mathematics to ensure student success at all levels.

Differentiating Instruction

Holt Middle School Math enables teachers to easily differentiate instruction. By implementing the process of scaffolding, the program provides continuous support for challenging work. The section planner in *Holt Middle School Math* assists the teacher in planning and pacing for all students at all levels, while the lesson plans provided with the program help the teacher determine which resources to use to differentiate instruction. The exercises ensure that students have ample opportunities, with guidance, to master the skills taught in the lessons and to then apply these skills with critical thinking. Each lesson includes multiple examples and opportunities to reach all learners through extensions, journal activities, the use of manipulatives, and home connections.

Fostering Successful Instructional Strategies

Holt Middle School Math promotes successful learning by supporting numerous teaching strategies, including direct instruction and cooperative learning. The Hands-On and Technology Labs are ideal for cooperative learning in heterogeneous groups, and the Explorations are designed for discovery learning. The Focus on Problem Solving and Think and Discuss features in each chapter are intended to stimulate student interaction. The exercise sets at the end of each lesson are tied to specific examples to encourage students to direct their own learning and to foster parental help on assignments. Thus, *Holt Middle School Math* can be used to accommodate various styles of the teacher as well as the students.

Creating a Community of Learners

The way we think about our classrooms might be different today, but our goal is the same—mathematics success for all. This program strives to assist teachers in creating the positive environment necessary to build a community of competent and confident learners in a mathematics class. Imagine a classroom where diversity in learning is the norm and the teacher responds to the learners' needs with flexible strategies, open dialogue, and ongoing assessment. Students will learn best when learning opportunities are natural and when connections can be easily made. *Holt Middle School Math* aids teachers in maximizing the capacity of each student every day.

David J. Chard, Ph.D.
*Assistant Professor and
Director of Graduate Studies
in Special Education*
University of Oregon
Eugene, OR

RESEARCH

Bransford, J. D., A. L. Brown, and R. R. Cocking, eds. 2000.
How people learn: Brain, mind, experience, and school.
Washington, DC: National Research Council.

Gersten, R., D. J. Chard, and S. Baker. 2002.
A meta-analysis of research on mathematics instruction for students with learning disabilities.
Eugene, OR: Eugene Research Institute.

Mathematics Learning Study Committee. 2001.
Adding it up: Helping children learn mathematics.
Washington, DC: National Academy Press.

National Council of Teachers of Mathematics. 2000.
Principles and standards for school mathematics.
Reston, VA: National Council of Teachers of Mathematics.

Vygotsky, L. 1962.
Thought and Language.
Cambridge, MA: MIT Press.

Accessibility for All Learners

"Students exhibit different talents, abilities, achievements, needs, and interests in mathematics. Nevertheless, all students must have access to the highest-quality mathematics instructional programs."

National Council of Teachers of Mathematics 2000

One of the primary goals of *Holt Middle School Math* is to provide teachers with a resource for teaching students new skills and strategies important for developing their comprehension of mathematics.

Coherent Pedagogical Approach

This program was designed with instructional features that represent a coherent pedagogical approach to mathematics instruction. Each lesson begins with carefully wrought examples of all of the skills, concepts, and strategies addressed. Additionally, the program's examples and counter-examples assist students in understanding the distinct boundaries that exist within each concept and the context in which particular skills and strategies are useful (Bransford, Brown, and Cocking 2000).

Procedural Fluency Development

A second goal is to develop procedural fluency in specific mathematical skills (Mathematics Learning Study Committee 2001). For many students with cognitive disabilities, insufficient practice hampers their ability to develop this fluency. *Holt Middle School Math* provides students with ample opportunities to practice specific computation and problem-solving procedures. Once fluent, students will then have ready access to these tools for use in more sophisticated mathematics.

Sufficient Scaffolding

Key to any instructional program is sufficient scaffolding to support student learning (Vygotsky 1962). In a typical middle school math classroom, some students will require substantial assistance in developing strategies for solving problems. Still others will already have the knowledge necessary to solve problems with little support. In this program, the instructional framework builds the background knowledge essential for ensuring that students are able to solve increasingly complex problems. Scaffolding is utilized in a number of ways throughout the program, from graduated difficulty of new content and applications to frequent opportunities for review, substantive reteaching lessons, and additional examples for extended instruction.

Jim Milgram, Ph.D.
Professor of Mathematics
Stanford University
Stanford, CA

RESEARCH

Morris, Anne K., and Vladimir M. Stoutsky. 1998.
Understanding of logical necessity: Developmental antecedents and cognitive consequences.
Child Development 69 (3): 721–41.

Mathematics Learning Study Committee. 2001.
Adding it up: Helping children learn mathematics.
Washington, DC: National Academy Press.

Schmidt, William, Richard Houang, and Leland Cogan. 2002.
A coherent curriculum: The case of mathematics.
American Educator 26 (2): 1-17.

Wu, H. H. 2001.
How to prepare students for algebra.
American Educator 25 (2): 10-17.

Transition to Advanced Mathematics

"...throughout the grades from pre-K through 8 all students can and should ... understand mathematical ideas, compute fluently, solve problems, and engage in logical reasoning."

Mathematics Learning Study Committee 2001

Middle school mathematics instruction occurs at a critical time in students' development and must address the needs specific to this period of learning.

From Foundation Skills to Advanced Topics

When students enter the middle grades, they must prepare for the transition to more advanced mathematical topics such as algebra and geometry while enhancing their basic arithmetic knowledge. It is crucial that they develop abstract reasoning and symbolic manipulation skills. In *Holt Middle School Math* these areas are carefully developed using methods aligned with standard best practices. The program addresses national and state standards while recognizing that some mathematical topics require more sophisticated instruction. All instructional materials, including the vocabulary lists, examples, and reference materials, reflect accurate mathematics. The integrity of the math represented in the program is strictly maintained so that the instructional design contributes positively to students' understanding of the discipline.

Instructional Sequencing

While the introduction of advanced topics requires that students broaden their understanding of mathematical ideas, it also reflects the hierarchical and sequential nature of mathematics as a discipline (Schmidt, Houang, and Cogan 2002). Students need to see the relationships between the math they are learning and real-world scenarios. To foster the development of these connections, the sequence of instruction within each grade and across this program accounts for the elements of mathematics that should be taught first in order to prepare students for later insights. The presentation of mathematical concepts in this program is aligned with that of the most successful international programs.

Enhancing the Role of the Teacher

Middle school mathematics instruction must be supported by materials that assist teachers in helping students successfully learn and do more complex mathematics. Care has been taken to ensure that each instructional lesson develops enough background information so that teachers can demonstrate to their students how the concepts they learn today will tie in to their later mathematical education. Teachers can use this foundation material as a resource when relaying information to their classes. In this way, *Holt Middle School Math* is an asset for teachers as well as for their students.

Jennie M. Bennett, Ed.D.
Instructional Mathematics Supervisor
Houston Independent School District
Houston, TX

RESEARCH

Artzt, Alice F., and Shirel Yaloz-Femia. 1999.
Mathematical reasoning during small-group problem solving. In Developing mathematical reasoning in grades K–12.
Reston, VA: National Council of Teachers of Mathematics.

Jensen, Eric. 1998.
Teaching with the brain in mind.
Alexandria: VA: Association for Supervision and Curriculum Development.

Krulik, Stephen, and Jesse A. Rudnick. 1999.
Innovation tasks to improve critical- and creative-thinking skills. In Developing mathematical reasoning in grades K–12.
Reston, VA: National Council of Teachers of Mathematics.

Levine, Mel. 2002.
A Mind at a Time.
New York, NY: Simon & Schuster.

Schell, Vicki J. 1981.
Learning partners: Reading and mathematics.
Paper presented at 14th annual meeting of the Missouri State Council of the International Reading Association. 13-17 May, at Columbia, MO.

Sullivan, Peter, and David Clarke. 1991.
Catering to all abilities through 'good' questions.
Arithmetic Teacher 39 (2): 14-18.

Strategic Problem Solving

"The single best way to grow a better brain is through challenging problem solving."

Eric Jensen, 1998

Unlike simple numeric computation problems, word problems present unique challenges to some students. One of the goals of this program is to teach students strategies to comprehend and solve word problems.

Development of Critical Thinking Skills

Holt Middle School Math ensures that students have the necessary tools to approach word problems strategically by teaching problem solving as a planned step-by-step process (Levine 2002). Using Polya's method for solving problems (understand the problem, create a plan, carry out the plan, and look back) activates students' critical thinking skills and engages students in making decisions and thinking logically. This program provides students with in-school systemic problem-solving experiences in which these critical skills are developed. Students generate different strategies for solving a problem, select the most feasible one, and arrive at a reasonable solution. These problem-solving skills help students understand how to approach real-world problems strategically both inside and outside the classroom.

Reading Connections

Reading comprehension is necessary to all subjects at all levels, including mathematics (Schell 1981). When students read word problems, they must synthesize or integrate their ideas and determine the operations to use. Reading is therefore a pivotal partner in problem solving; it sets the stage for understanding the problem itself. In this program students are asked to state the details of a problem, identify the necessary information, restate the problem in their own words, and demonstrate knowledge of mathematical vocabulary.

Asking Good Questions

Good questions engage the student in a more active role in learning. When the teacher asks such questions, learning is student-centered rather than teacher-centered. Good questions stimulate and activate communication between the teacher and students and allow students to respond in their own way when solving math problems. These questions can guide students at all levels to experience success with problem solving. Another goal of *Holt Middle School Math* is to assist teachers by suggesting good questions through features such as Focus on Problem Solving and Reaching All Learners.

Concrete Understanding

Janet K. Scheer, Ph.D.
Executive Director
Create A Vision™
Foster City, CA

RESEARCH

Bohan, Harry J., and Peggy Bohan Shawaker. 1994.
Using manipulatives effectively: A drive down rounding road.
Arithmetic Teacher 41 (5): 246-48.

National Council of Teachers of Mathematics. 2000.
Principles and standards for school mathematics.
Reston, VA: National Council of Teachers of Mathematics.

Stein, Mary Kay, and Jane W. Bovalino. 2001.
Manipulatives: One piece of the puzzle.
Mathematics Teaching in the Middle School 6 (6): 356-59.

Threadgill-Sowder, Judith, and Patricia Juilfs. 1980.
Manipulative versus symbolic approaches to teaching logical connectives in junior high school: An aptitude x treatment interaction study.
Journal for Research in Mathematics Education 11 (5): 367-74.

"When students gain access to mathematical representations and the ideas they represent, they have a set of tools that significantly expand their capacity to think mathematically."

National Council of Teachers of Mathematics 2000

Holt Middle School Math makes use of mathematical modeling and provides many options for the use of manipulatives to enhance student understanding of abstract concepts.

Manipulatives for Concrete Understanding

Educational research demonstrates the effectiveness of hands-on learning in supplementing understanding of mathematical ideas for some students (Threadgill-Sowder and Juilfs 1980). This is especially important in the middle grades, when students are exposed to increasingly abstract concepts. While some middle school students are ready to embrace these abstract topics, others still need the concrete foundation that manipulatives can supply. This program utilizes algebra tiles, pattern blocks, and two-color counters in Hands-On Labs to model topics such as fraction operations and grouping of terms in algebraic expressions. Additionally, the Reaching All Learners features provide teachers with concrete methods for presenting selected topics.

Bridging: Concrete to Symbolic Understanding

Most theories of developmental learning support the use of physical tools to establish a foundation for abstract thought. For this approach to be successful, students must make connections between the manipulatives with which they are working and the abstract mathematical concepts the materials represent; in this way, concrete action is transferred to symbolic understanding (Bohan and Shawaker 1994). It is essential that sufficient context and introduction to a manipulative lesson be provided so that students are able to form connections to symbolic representations in a guided manner (Stein and Bovalino 2001). The modeling activities within this program are intentionally placed after foundation skills have been developed and before symbolic computation is emphasized.

Opportunities to Expand Knowledge

Another advantage of manipulative lessons rooted in foundational skills is that students are given the opportunity to discover new mathematical concepts. The discovery-based knowledge that is developed in this program's Hands-On Labs is solidified by lessons that formalize the mathematical rules and symbolic representations of the concept. In this way, concrete understanding facilitates application of the learned mathematical concepts.

Bert K. Waits, Ph.D.
Professor Emeritus of Mathematics
The Ohio State University
Columbus, OH

RESEARCH

Graham, A. T., and M.O.J. Thomas. 2000.
Building a versatile understanding of algebraic variables with a graphic calculator.
Educational Studies in Mathematics 41 (3): 265-82.

Hollar, Jeannie C., and Karen Norwood. 1999.
The effects of a graphing-approach intermediate algebra curriculum on students' understanding of function.
Journal for Research in Mathematics Education 30 (2): 220-26.

National Commission on Mathematics and Science Teaching for the 21st Century. 2000.
Before it's too late:
The Glenn Commission report.

National Council of Teachers of Mathematics. 2000.
Principles and standards for school mathematics.
Reston, VA: National Council of Teachers of Mathematics.

Technology to Enhance Learning

"Technology is essential in teaching and learning mathematics; it influences the mathematics that is taught and enhances students' learning."

National Council of Teachers of Mathematics 2000

A wide array of technological tools is available for use in mathematics classrooms, including graphing calculators, spreadsheet programs, and geometry software. *Holt Middle School Math* makes use of these tools to reinforce student learning.

Integrated Use of Technology

Research has demonstrated that technology, when used appropriately, can improve students' mathematical understanding and problem-solving skills (Hollar and Norwood 1999). Similarly, technological tools can help teachers challenge students to use and understand mathematics in real-world scenarios. Through the use of integrated technology labs, this program gives students a solid foundation for understanding how to use technology appropriately to learn mathematics. The program puts the NCTM Technology Principle into action in every chapter.

Balanced Curriculum

Holt Middle School Math acknowledges that students must utilize all available tools in the mathematics-learning process. Traditional paper-and-pencil skills are emphasized throughout the program but are supplemented by technology components. Proficiency in mental math computation is stressed as well. Thus, technology is presented not as an end in itself but rather as a means for understanding and application. Current research supports this use of technology. The many technology labs in the program are devoted to the pedagogical use of spreadsheets and dynamic geometry software. Students also use graphing calculators, which are powerful, portable computers with built-in software for graphing, data analysis, and statistics.

Professional Development

This program helps teachers achieve one of the Glenn Commission's major goals in its report to the nation on the crisis in mathematics and science education today. The report states that in order to "inform efforts to promote higher student achievement, teachers should actively work to improve [their] knowledge and skills to incorporate educational technology into [their] learning and teaching" (National Commission on Mathematics and Science Teaching 2000).

Problem Solving Handbook

CHAPTER 1

Algebra Toolbox

Interdisciplinary LINKS

Life Science 7, 37
Earth Science 47
Physical Science 5, 41
Entertainment 7, 12, 17
Sports 12, 27, 31
Social Studies 16, 17, 22
Money 20
Business 27, 31, 37
Retail 35
History 37
Music 41

Student Help

Remember 4, 18, 24, 29
Helpful Hint 9, 10, 14, 23, 28, 34, 35, 38
Test Taking Tip 57

internet connect

Homework Help
Online
6, 11, 16, 21, 26, 36, 40, 45

KEYWORD: MP4 HWHelp

Algebra *Indicates algebra included in lesson development*

Integers and Exponents

CHAPTER 3

Rational and Real Numbers

Student Help

Remember 113, 131, 136, 140, 147
Helpful Hint 121, 122, 146, 156, 161
Test Taking Tip 171

🔲 internet connect   **go.hrw.com**
Homework Help Online
115, 119, 124, 129, 133, 138, 142, 148, 152, 158
KEYWORD: MP4 HWHelp

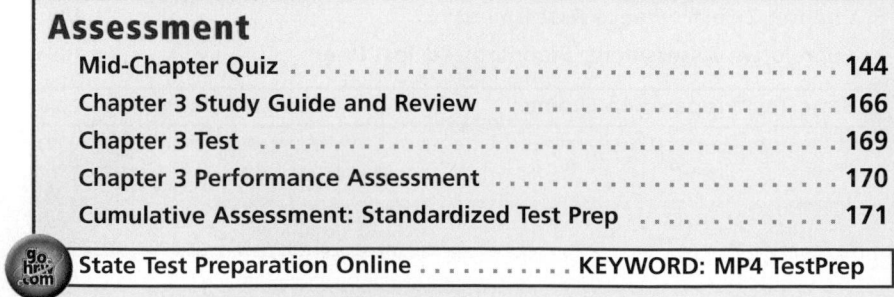

Algebra   *Indicates algebra included in lesson development*

Collecting, Displaying, and Analyzing Data

Interdisciplinary LINKS

Life Science 177, 207
Earth Science 191
Business 177
Money 177
Language Arts 183
Astronomy 185, 187
Geography 192

Student Help

Helpful Hint 197, 205
Test Taking Tip 219

 internet connect

Homework Help
Online

176, 181, 186, 190, 198,
202, 206
KEYWORD: MP4 HWHelp

Plane Geometry

Interdisciplinary LINKS

Earth Science 243
Physical Science 226, 231
Art 231, 257, 267
Social Studies 238, 262

Student Help

Reading Math 223, 254
Remember 228, 265
Writing Math 229
Helpful Hint 241, 245, 255, 259
Test Taking Tip 277

internet connect (go. hrw. com)
Homework Help
Online

224, 230, 236, 241, 246, 252, 256, 261, 266

KEYWORD: MP4 HWHelp

Algebra *Indicates algebra included in lesson development*

Perimeter, Area, and Volume

Interdisciplinary LINKS

Life Science 311, 321, 327
Earth Science 323
Physical Science 288
Social Studies 284, 293, 311, 315, 323
Construction 293, 309
Transportation 295, 306, 315
Entertainment 297, 311
Food 297
Sports 297, 319
Technology 306
History 313
Architecture 315
Career 315
Art 317

Student Help

Helpful Hint 280, 282, 290, 307
Reading Math 286
Remember 294, 307
Test Taking Tip 339

internet connect
Homework Help Online
283, 287, 292, 296, 304, 310, 314, 318, 322, 326
KEYWORD: MP4 HWHelp

CHAPTER 7

Ratios and Similarity

Interdisciplinary LINKS

Life Science 354, 373, 377, 378
Earth Science 343
Physical Science 352, 357, 358, 371, 385
Business 344, 349, 379, 383
Transportation 344, 352, 354
Computers 345
Entertainment 345, 346, 349, 379
Hobbies 345
Communications 349
Sports 354
Health 359
Photography 365
Art 371, 385
Architecture 375, 379

Student Help

Reading Math 342, 372
Helpful Hint 350, 356, 362, 382
Remember 369
Test Taking Tip 397

internet connect (go.hrw.com)
Homework Help Online
344, 348, 353, 358, 364, 370, 374, 378, 384
KEYWORD: MP4 HWHelp

Algebra *Indicates algebra included in lesson development*

Percents

CHAPTER 8

Interdisciplinary LINKS

Life Science 408, 411, 416, 419

Earth Science 408, 419, 423

Physical Science 403, 410, 423

Language Arts 408

Social Studies 401, 408, 413

Sports 423

Economics 427

Student Help

Remember 400
Reading Math 400
Helpful Hint 406, 420
Test Taking Tip 443

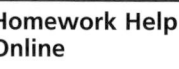 internet connect

Homework Help
Online

402, 407, 412, 418
KEYWORD: MP4 HWHelp

Probability

Interdisciplinary LINKS

Life Science 459, 466, 475
Earth Science 454
Business 450, 485
Entertainment 450
Safety 452
Art 475
Sports 475
Games 481

Student Help

Helpful Hint 457, 472
Reading Math 471
Test Taking Tip 495

internet connect go.com

Homework Help
Online

449, 453, 458, 465, 469, 474, 480, 484

KEYWORD: MP4 HWHelp

Algebra   *Indicates algebra included
in lesson development*

More Equations and Inequalities

CHAPTER **10**

Interdisciplinary LINKS

Life Science 501
Earth Science 511
Physical Science 505,
  511, 522
Money 503
Sports 505, 509, 518
Business 516
Economics 518
Entertainment 517, 527

Student Help

Remember 503, 520
Helpful Hint 508, 509,
  520, 524, 525
Test Taking Tip 537

 internet connect
Homework Help
Online
500, 504, 509, 517, 521,
525
KEYWORD: MP4 HWHelp

CHAPTER 11

Graphing Lines

Interdisciplinary LiNKS

Life Science 554, 559, 566
Earth Science 559, 571
Physical Science 543, 564, 566
Sports 542, 571, 573
Business 544, 571
Entertainment 544, 552
Safety 549
Medical 557
Cooking 566
Economics 575

Student Help

Remember 545, 547, 572
Reading Math 540
Helpful Hint 546, 551, 562, 567, 568, 569, 573
Test Taking Tip 587

☑ internet connect
Homework Help Online
543, 548, 553, 558, 565, 570, 574
KEYWORD: MP4 HWHelp

 Algebra Indicates algebra included in lesson development

Sequences and Functions

CHAPTER 12

Interdisciplinary LINKS

Life Science 599, 614, 616
Physical Science 599, 618, 625, 631
Travel 592
Business 594, 612, 616, 625
Recreation 594, 616
Money 597
Finance 631
Music 605, 629
Home Economics 612
Sports 612
Health 620
Astronomy 623
Hobbies 625
Economics 599, 616

Student Help

Helpful Hint 590, 618, 628
Writing Math 591
Reading Math 609
Remember 622
Test Taking Tip 641

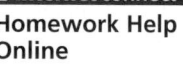
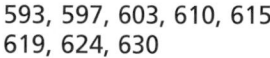
Homework Help Online
593, 597, 603, 610, 615, 619, 624, 630

KEYWORD: MP4 HWHelp

Assessment

CHAPTER 13

Polynomials

Life Science 652, 673
Art 653, 657
Business 651, 659, 661, 663
Health 667
Physics 645
Sports 671
Transportation 647, 659

Introduction to Polynomials *Algebra*

Polynomial Operations *Algebra*

Student Help

Remember 674
Helpful Hint 670
Test Taking Tip 685

🔌 **internet** connect ═══ go.hrw.com
Homework Help Online
646, 652, 658, 662, 666, 672
Keyword: MP4 HWHelp

Assessment

State Test Preparation Online . . . . **Keyword: MP4 TestPrep**

Algebra *Indicates algebra included in lesson development*

Set Theory and Discrete Math

CHAPTER 14

internet connect

Homework Help Online

690, 694, 698, 704, 710, 714, 718

Keyword: MP4 HWHelp

Student Handbook

Math Skills for Life Science

Math Skill	Where Taught
Balance equations (change on one side requires an equivalent change on the other).	Lessons 1-3, 1-4, 2-4, 3-6, 10-1, 10-2, 10-3
Calculate slope, graph a line with a given slope, and understand slope as a rate of change.	Lessons 5-5, 11-1, 11-2, 11-3, 11-4; Technology Lab 11A
Combinations	Lesson 9-6
Convert between standard and metric units.	Lesson 7-3
Convert between temperature units.	Lesson 2-5; Skills Bank p. 781
Convert between units in one dimension, two dimensions, or three dimensions.	Lessons 6-6, 6-7, 6-8, 6-9; Skills Bank p. 780
Display and read data in bar graphs, line graphs, and circle graphs.	Lesson 4-5, 4-6; Hands-On Lab 8A
Find angle measures.	Lesson 5-1
Find areas.	Lessons 6-1, 6-2, 6-4
Find average (mean).	Lesson 4-3
Find perimeter.	Lessons 6-1, 6-2
Find probability.	Lessons 9-1, 9-2, 9-3, 9-4, 9-5, 9-6, 9-7
Find surface area.	Lessons 6-8, 6-9, 6-10
Find surface area to volume ratios.	Lesson 7-9
Find volume.	Lessons 6-6, 6-7, 6-10
Graph on a coordinate plane.	Lessons 1-7, 1-8, 1-9, 5-5
Operate with exponents.	Lessons 2-6, 2-7, 2-8, 3-8; Hands-On Lab 3B
Operate with fractions.	Lessons 3-2, 3-3, 3-4, 3-5
Operate with large whole numbers.	Lessons 2-6, 2-9
Operate with percents.	Lessons 8-1, 8-2, 8-3, 8-4, 8-5
Operate with powers of ten.	Lessons 2-6, 2-7, 2-8, 2-9
Operate with units of rate and convert to compatible units.	Lessons 7-2, 7-3
Read metersticks and inch rulers.	Skills Bank p. 782
Round decimal and whole numbers.	Skills Bank pp. 760, 783
Set up and solve proportions.	Lessons 7-1, 7-4; Hands-On Lab 7A
Simplify fractions.	Lesson 3-1
Solve equations.	Lessons 1-3, 1-4, 2-4, 3-6, 10-1, 10-2, 10-3
Solve literal equations (rewrite formulas).	Lesson 10-5
Translate words to equations or algebraic expressions.	Lessons 1-1, 1-2
Understand exponential growth behavior.	Lesson 12-6
Understand nonlinear relationships, such as half-life.	Lesson 12-6
Use and operate with scientific notation.	Lesson 2-9
Use appropriate one-, two-, and three-dimensional measurement units.	Lessons 6-1, 6-2, 6-3, 6-4, 6-5, 6-6, 6-7, 6-8, 6-9, 6-10
Use scale models.	Lessons 7-7, 7-8, 7-9; Hands-On Lab 7C
Write equivalent percents, fractions, and decimals.	Lessons 8-1, 8-2, 8-3, 8-4, 8-5
Write equivalent ratios and ratios in simplest form.	Lessons 7-1, 7-2; Hands-On Lab 7A

Math Skills for **Earth Science**

Math Skill	Where Taught
Calculate slope, graph a line with a given slope, and understand slope as a rate of change.	Lessons 5-5, 11-1, 11-2, 11-3, 11-4; Technology Lab 11A
Convert between standard and metric units.	Lesson 7-3
Convert between temperature units.	Lesson 2-5; Skills Bank p. 781
Display and read data in bar graphs, line graphs, and circle graphs.	Lessons 4-5, 4-6; Hands-On Lab 8A
Find angle measures.	Lesson 5-1
Find areas.	Lessons 6-1, 6-2, 6-4
Find average (mean).	Lesson 4-3
Find surface area.	Lessons 6-8, 6-9, 6-10
Find volume.	Lessons 6-6, 6-7, 6-10
Graph on a coordinate plane.	Lessons 1-7, 1-8, 1-9, 5-5
Operate with exponents.	Lessons 2-6, 2-7, 2-8, 3-8; Hands-On Lab 3B
Operate with fractions.	Lessons 3-2, 3-3, 3-4, 3-5
Operate with large whole numbers.	Lessons 2-6, 2-9
Operate with percents.	Lessons 8-1, 8-2, 8-3, 8-4, 8-5
Operate with powers of ten.	Lessons 2-6, 2-7, 2-8, 2-9
Operate with units of rate and convert to compatible units.	Lessons 7-2, 7-3
Read metersticks and inch rulers.	Skills Bank p. 782
Round decimal and whole numbers.	Skills Bank pp. 760, 783
Set up and solve proprotions.	Lessons 7-1, 7-4; Hands-On Lab 7A
Simplify fractions.	Lesson 3-1
Solve equations.	Lessons 1-3, 1-4, 2-4, 3-6, 10-1, 10-2, 10-3
Solve literal equations (rewrite formulas).	Lesson 10-5
Translate words to equations or algebraic expressions.	Lessons 1-1, 1-2
Understand nonlinear relationships, such as Richter scale, and half-life.	Lesson 12-6; Skills Bank pp. 784, 785
Use and operate with scientific notation.	Lesson 2-9
Use appropriate one-, two-, and three-dimensional measurement units.	Lessons 6-1, 6-2, 6-3, 6-4, 6-5, 6-6, 6-7, 6-8, 6-9, 6-10
Use scale models.	Lessons 7-7, 7-8, 7-9; Hands-On Lab 7C
Write equivalent percents, fractions, and decimals.	Lessons 8-1, 8-2, 8-3, 8-4, 8-5
Write equivalent ratios and ratios in simplest form.	Lessons 7-1, 7-2; Hands-On Lab 7A

Math Skills for *Physical Science*

Math Skill	Where Taught
Balance equations (change on one side requires an equivalent change on the other).	Lessons 1-3, 1-4, 2-4, 3-6, 10-1, 10-2, 10-3
Calculate slope, graph a line with a given slope, and understand slope as a rate of change.	Lessons 5-5, 11-1, 11-2, 11-3, 11-4; Technology Lab 11A
Convert between standard and metric units.	Lesson 7-3
Convert between temperature units.	Lesson 2-5; Skills Bank p. 781
Convert between units in one dimension, two dimensions, or three dimensions.	Lessons 6-6, 6-7, 6-8, 6-9; Skills Bank p. 780
Display and read data in bar graphs, line graphs, and circle graphs.	Lessons 4-5, 4-6; Hands-On Lab 8A
Find angle measures.	Lesson 5-1
Find areas.	Lessons 6-1, 6-2, 6-4
Find average (mean).	Lesson 4-3
Find volume.	Lessons 6-6, 6-7, 6-10
Graph on a coordinate plane.	Lessons 1-7, 1-8, 1-9, 5-5
Operate with exponents.	Lessons 2-6, 2-7, 2-8, 3-8; Hands-On Lab 3B
Operate with fractions.	Lessons 3-2, 3-3, 3-4, 3-5
Operate with integers.	Lessons 2-1, 2-2, 2-3; Hands-On Lab 2A
Operate with large whole numbers.	Lessons 2-6, 2-9
Operate with percents.	Lessons 8-1, 8-2, 8-3, 8-4, 8-5
Operate with powers of ten.	Lessons 2-6, 2-7, 2-8, 2-9
Operate with units of rate and convert to compatible units.	Lessons 7-2, 7-3
Read metersticks and inch rulers.	Skills Bank p. 782
Round decimal and whole numbers.	Skills Bank pp. 760, 783
Set up and solve proportions.	Lessons 7-1, 7-4; Hands-On Lab 7A
Simplify fractions.	Lesson 3-1
Solve equations.	Lessons 1-3, 1-4, 2-4, 3-6, 10-1, 10-2, 10-3
Solve literal equations (rewrite formulas).	Lesson 10-5
Translate words to equations or algebraic expressions.	Lessons 1-1, 1-2
Understand inverse relationships.	Lesson 12-8
Understand linear relationships.	Lessons 11-1, 11-5, 12-5
Understand nonlinear relationships, such as acceleration, pH scale, and half-life.	Lessons 12-6, 12-7; Skills Bank p. 784
Understand parallel and perpendicular line relationships.	Lessons 5-5, 11-2
Use and operate with scientific notation.	Lesson 2-9
Use and understand binary numbers.	Chapter 3 Extension
Use appropriate one-, two-, and three-dimensional measurement units.	Lessons 6-1, 6-2, 6-3, 6-4, 6-5, 6-6, 6-7, 6-8, 6-9, 6-10
Use scale models.	Lessons 7-7, 7-8, 7-9; Hands-On Lab 7C
Write equivalent percents, fractions, and decimals.	Lessons 8-1, 8-2, 8-3, 8-4, 8-5
Write equivalent ratios and ratios in simplest form.	Lessons 7-1, 7-2; Hands-On Lab 7A

NCTM Standards For Grades 6–8

Number and Operations

• Understand numbers, ways of representing numbers, relationships among numbers, and number systems

COURSE 1

Lessons 1-1, 1-3, 3-1, 3-5, 4-1, 4-2, 4-3, 4-4, 4-5, 4-6, 4-7, 5-5, 8-1, 8-2, 8-7, 8-8, 8-9, 8-10, 9-1, 9-2; Hands-On Labs 3A, 3B, 4A, 4B, 8A; Chapters 1, 4, 8, 9 Extensions

COURSE 2

Lessons 2-1, 2-2, 2-4, 2-5, 2-6, 2-7, 2-8, 2-9, 3-1, 3-7, 3-8, 3-9, 3-10, 5-1, 5-2, 6-1, 6-3, 6-5; Hands-On Labs 2B, 8C; Chapter 3 Extension

PRE-ALGEBRA

Lessons 2-6, 2-9, 3-1, 3-10, 7-1, 8-1, 8-2, 8-3, 8-4, 8-5, 8-6, 8-7; Hands-On Lab 7A; Technology Labs 3A, 8B; Chapters 3, 8 Extensions

• Understand meanings of operations and how they relate to one another

Lessons 1-4, 1-5, 2-4, 2-5, 2-6, 2-7, 3-9, 3-10, 5-1, 5-2, 5-3, 5-4, 5-7, 5-8, 5-9, 5-10, 9-8; Hands-On Labs 3B, 3D, 5A, 5B, 5C, 5D, 9A, 9B, 9C; Technology Lab 1A

Lessons 2-3, 2-7, 2-8, 2-11, 2-12, 3-3, 3-4, 3-5, 4-2, 4-3, 4-4, 4-5, 4-6, 4-7, 4-8, 4-10, 4-11, 4-12, 6-1, 6-4, 8-2, 8-7, 8-8; Hands-On Labs 2B, 3A, 3B, 4A, 4B, 4C, 6A, 8B; Technology Labs 2A, 2, 4; Chapter 3 Extension

Lessons 1-1, 1-2, 1-3, 1-4, 1-5, 1-6, 2-1, 2-2, 2-3, 2-7, 3-2, 3-4, 3-5, 3-6, 3-7, 3-8, 3-9, 10-1, 10-2, 10-4; Hands-On Labs 2A, 3B, 10A; Technology Lab 12B; Chapter 3 Extension

• Compute fluently and make reasonable estimates

Lessons 1-2, 1-3, 1-4, 1-6, 3-2, 3-3, 3-6, 3-7, 3-8, 4-8, 4-9, 5-6, 6-2, 8-4, 8-5, 8-6, 8-7, 8-8, 8-9, 9-4, 9-5, 9-6, 9-7, 11-4; Hands-On Lab 11B

Lessons 3-3, 3-4, 3-5, 4-1, 4-2, 4-3, 4-4, 4-5, 4-7, 4-8, 4-9, 4-10, 4-11, 5-2, 5-3, 5-6, 5-7, 6-2, 6-3, 6-5, 6-6; Hands-On Lab 6A; Chapter 10 Extension

Lessons 1-6, 2-1, 2-2, 2-3, 2-6, 2-7, 2-8, 3-2, 3-3, 3-4, 3-5, 3-9, 4-3, 7-2, 7-3, 8-1, 8-2, 8-3, 8-4, 8-5, 8-6, 8-7, 9-6; Hands-On Lab 3B; Chapter 8 Extension

Algebra

• Understand patterns, relations, and functions

COURSE 1

Lessons 1-7, 6-6, 12-1, 12-2; Hands-On Lab 12A

COURSE 2

Lessons 1-7, 1-8, 12-1, 12-2, 12-3, 12-4, 12-5, 12-6, 12-7; Hands-On Labs 1C, 5A, 7F, 10A, 12A; Technology Lab 12; Chapters 7, 8, 11 Math-Ables

PRE-ALGEBRA

Lessons 1-7, 1-8, 1-9, 2-8, 11-1, 11-2, 11-5, 11-7, 12-1, 12-2, 12-3, 12-4, 12-5, 12-6, 12-7, 12-8; Hands-On Labs 9B, 12A; Technology Lab 1A

• Represent and analyze mathematical situations and structures using algebraic symbols

Lessons 2-1, 2-2, 2-3, 2-4, 2-5, 2-6, 2-7, 3-10, 5-4, 5-10, 6-6, 9-3, 9-8; Hands-On Labs 3B, 3D, 5A, 5B, 5C, 5D, 9A, 9B, 9C; Chapter 2 Extension

Lessons 2-7, 2-10, 2-11, 2-12, 3-6, 4-6, 4-12, 6-4, 11-1, 11-2, 11-3, 11-4, 11-5, 11-6, 11-7, 12-5, 12-8; Hands-On Labs 2C, 3C, 11A; Chapters 11, 12 Extensions

Lessons 1-1, 1-2, 1-6, 2-4, 2-5, 3-7, 7-4, 10-4, 10-5, 10-6, 11-3, 11-4, 13-1, 13-2, 13-3, 13-4, 13-5, 13-6; Hands-On Labs 2A, 13A, 13B; Technology Lab 11A; Chapter 13 Extension

Problem Solving

● Build new mathematical knowledge through problem solving

Appears throughout each course in features such as the following:

Problem Solving on Location, Math-Ables, and *Performance Assessment* at the end of each chapter

Problem Solving Applications within each chapter

For example:

COURSE 1

Pages 13, 36–37, 78–79, 86, 128, 140–141, 200, 206, 227, 262, 312, 318

COURSE 2

Pages 48–49, 83, 120, 182, 248–249, 294, 404, 458–459, 502, 592–593, 600, 644

PRE-ALGEBRA

Pages 48–50, 75, 108, 128, 150–151, 162–164, 218, 338, 351, 434–436, 442, 536, 578–580, 676–677, 722

● Solve problems that arise in mathematics and in other contexts

Appears throughout each course in features such as the following:

Problem Solving on Location and *Math-Ables* at the end of each chapter

Interdisciplinary exercise sets in each chapter and *Problem Solving Applications* within each chapter

For example:

Pages 11, 15, 23, 70, 161, 198–199, 260–261, 285, 310–311, 382, 440, 490

Pages 50, 153, 180–181, 250, 292–293, 399, 402–403, 493, 502, 548, 594, 642–643

Pages 22, 67, 100–102, 134, 226, 264, 268–270, 328, 330–332, 413, 448, 501, 528–530, 575, 620, 678, 684, 720–721, 728

● Apply and adapt a variety of appropriate strategies to solve problems

Appears throughout each course in features such as the following:

Problem Solving Skill lessons, *Focus on Problem Solving,* and *Problem Solving Applications* within each chapter

Problem Solving Handbook at the front of the book.

For example:

Pages xviii–xxix, 28–29, 31–32, 52–53, 113, 165, 187, 353, 406, 455

Pages xviii–xxix, 19, 161, 192–195, 219, 230–233, 277, 387, 485, 565, 608–611

Pages xx–xxxi, 8–12, 25, 33, 92–95, 141, 145, 249, 350–354, 361, 415, 421, 456, 498, 513, 607, 665, 701

● Monitor and reflect on the process of mathematical problem solving

Appears throughout each course in features such as the following:

Focus on Problem Solving within each chapter and *Problem Solving Applications* at the end of each chapter

Problem Solving Handbook at the front of the book

For example:

Pages xviii–xxix, 36–37, 78–79, 140–141, 198–199, 231, 283, 343, 438–439, 521

Pages xviii–xxix, 48–49, 248–249, 402–403, 443, 500–501, 529, 573, 592–593

Pages xx–xxxi, 48–49, 56, 83, 100–101, 170, 195, 210–211, 276, 299, 388–389, 396, 415, 461, 486–487, 494, 561, 632–633, 640, 655, 717

Reasoning and Proof

● **Recognize reasoning and proof as fundamental aspects of mathematics**

Appears throughout each course in features such as the following:
Think and Discuss in every lesson and lab
Hands-On Labs and *Technology Labs* in every chapter
Math-Ables at the end of each chapter
Additional examples include:

COURSE 1
Lessons 1-6, 1-7, 2-2, 3-9, 6-8, 7-6, 7-7, 7-8, 8-2, 8-3, 10-4, 11-4; Chapter 4 Extension

COURSE 2
Lessons 2-1, 7-7, 7-9, 10-6, 10-7, 12-2, 12-3, 12-8

PRE-ALGEBRA
Lessons 5-1, 5-2, 5-3, 5-4, 5-5, 12-1, 12-2, 12-3, 14-1, 14-2, 14-3, 14-6, 14-7

● **Make and investigate mathematical conjectures**

Lessons 1-7, 7-8; Hands-On Labs 4C, 7B, 8A, 10A, 11B, 12A; Technology Labs 1, 7, 10, 12; Chapters 1, 9 Extensions

Lessons 7-3, 7-5, 10-1, 10-2, 10-4, 10-5; Hands-On Labs 3A, 3B, 4C, 5A, 7D, 12A; Technology Labs 2, 7, 8, 10; Chapters 8, 9, 10 Extensions

Lessons 2-8, 4-7, 6-6, 6-7, 6-8, 6-9, 9-3, 9-7, 9-8, 11-7, 12-1, 12-2, 12-3; Hands-On Labs 3B, 6A, 7B, 12A; Technology Lab 9A

● **Develop and evaluate mathematical arguments and proofs**

Appears throughout each course in features such as the following:
Think and Discuss in every lesson and lab
Hands-On Labs and *Technology Labs* in every chapter
Write About It in the exercise sets
Additional examples include:

Lessons 2-4, 2-5, 2-6, 2-7, 3-10, 4-1, 5-4, 5-10, 7-5, 7-6, 9-8, 11-4; Chapters 4, 9 Extensions

Lessons 2-2, 2-9, 2-10, 3-3, 3-4, 5-5, 7-8, 8-4, 8-5, 8-6, 9-2, 12-6, 12-7

Lessons 4-6, 5-1, 5-2, 5-3, 5-6, 6-1, 6-2, 6-3, 6-4, 6-6, 6-7, 6-8, 6-9, 6-10, 7-6, 7-7, 14-4, 14-5; Technology Lab 14A; Chapter 7 Extension

● **Select and use various types of reasoning and methods of proof**

Appears throughout each course in features such as the following:
Think and Discuss in every lesson and lab
Focus on Problem Solving and *Problem Solving Applications* within each chapter
Performance Assessment at the end of each chapter
Problem Solving Handbook at the front of the book
Choose a Strategy in the exercise sets
Additional examples include:

Lessons 1-7, 2-4, 2-5, 2-6, 2-7, 3-10, 4-8, 4-9, 5-1, 5-4, 5-7, 5-10, 7-6, 7-8, 8-2, 8-3, 9-8, 11-1, 11-2, 11-3; Hands-On Labs 4B, 8A, 10A, 10B, 10C

Lessons 2-10, 7-8, 9-4, 12-6; Hands-On Labs 2C, 7C, 10A, 11A; Technology Labs 5, 7, 8, 9; Chapters 11, 12 Extensions; Chapters 1, 5, 10, 12 Math-Ables

Lessons 2-8, 4-7, 5-1, 5-2, 5-3, 5-4, 6-6, 6-7, 6-8, 6-9, 9-3, 9-7, 9-8, 11-7, 12-1, 12-2, 12-3, 14-5; Hands-On Labs 3B, 4A, 5C, 6A, 7B, 12A; Technology Lab 9A

Stepping into the Future

All the Ways You Teach

Being ready for the future means making sure no student is left behind. Throughout the *Student Edition, Teacher Edition, Premier Online Edition*, and program resources for *Holt Pre-Algebra,* you'll find the help you need to teach, assist, and assess every ability level in your class without spending hours of extra time assigning special assignments and grading extra worksheets.

All the Ways They Learn

The perfect pre-algebra curriculum for the future ensures that students can master math concepts found in state tests, they are able to learn at their own speed and style, and that they are able to manipulate information to use it in everyday terms with everyday technology.

We're With Them Every Step of the Way

Meeting State Standards and Assessment

This program reflects the increased expectation of what students should "know and be able to do" by offering test preparation at the lesson and chapter levels and online through **go.hrw.com**.

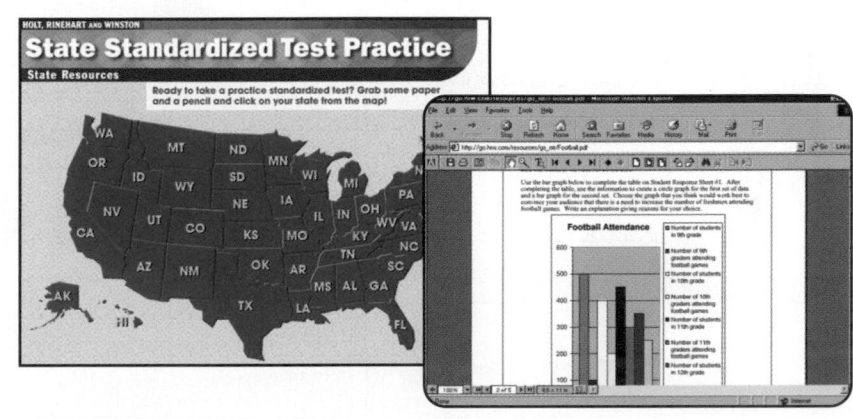

Differentiating Instruction

By offering early intervention and examples that allow students to get help early and practice at their own level of learning, this program helps you reach all learners, including students who need extra help and advanced learners.

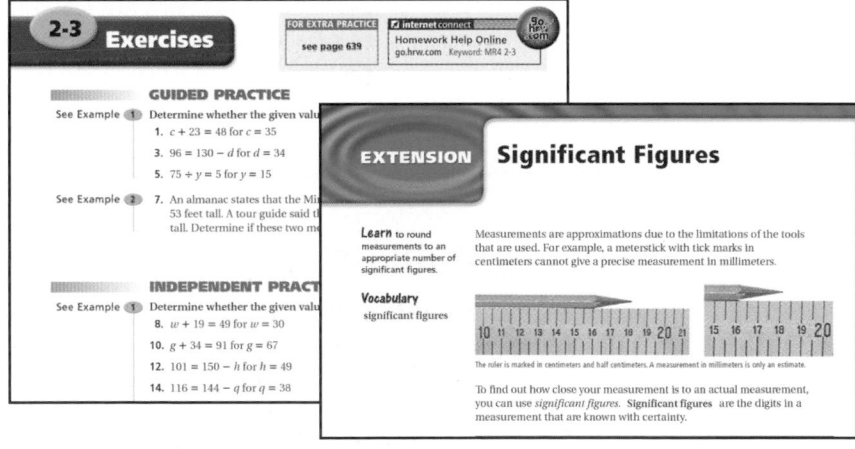

Integrating Technology

This program offers you state-of-the-art technology integrated with your curriculum saving you time and increasing your efficiency with the new test and practice generator on the *One-Stop Planner* ® *CD-ROM*, the *Are You Ready? Intervention CD-ROM,* and the *Holt Pre-Algebra Premier Online Edition.*

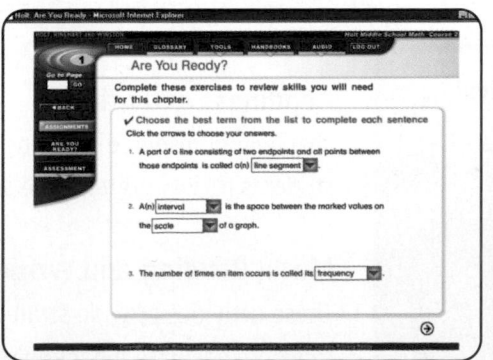

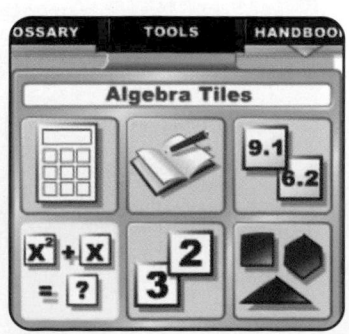

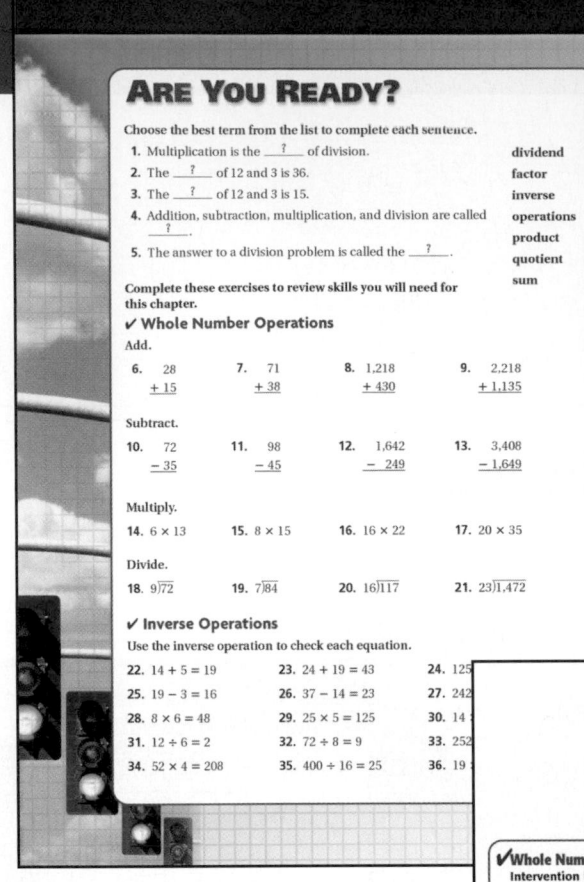

ARE YOU READY?

Choose the best term from the list to complete each sentence.

1. Multiplication is the ___?___ of division.
2. The ___?___ of 12 and 3 is 36.
3. The ___?___ of 12 and 3 is 15.
4. Addition, subtraction, multiplication, and division are called ___?___.
5. The answer to a division problem is called the ___?___.

dividend
factor
inverse
operations
product
quotient
sum

Complete these exercises to review skills you will need for this chapter.

✔ **Whole Number Operations**

Add.

6. 28   7. 71   8. 1,218   9. 2,218
+ 15   + 38   + 430   + 1,135

Subtract.

10. 72   11. 98   12. 1,642   13. 3,408
− 35   − 45   − 249   − 1,649

Multiply.

14. 6×13 15. 8×15 16. 16×22 17. 20×35

Divide.

18. $9\overline{)72}$ 19. $7\overline{)84}$ 20. $16\overline{)117}$ 21. $23\overline{)1,472}$

✔ **Inverse Operations**

Use the inverse operation to check each equation.

22. $14 + 5 = 19$ 23. $24 + 19 = 43$ 24. 125
25. $19 - 3 = 16$ 26. $37 - 14 = 23$ 27. 242
28. $8 \times 6 = 48$ 29. $25 \times 5 = 125$ 30. 14
31. $12 \div 6 = 2$ 32. $72 \div 8 = 9$ 33. 252
34. $52 \times 4 = 208$ 35. $400 \div 16 = 25$ 36. 19

Assessing Prior Knowledge

INTERVENTION ◀▷
Diagnose and Prescribe

Evaluate your students' performance on this page to determine whether intervention is necessary or whether enrichment is appropriate. Options that provide instruction, practice, and a check are listed below.

Resources for Are You Ready?

■ *Are You Ready? Intervention and Enrichment*

■ **Recording Sheet for Are You Ready?**
Chapter 2 Resource Book . p. 00

💿 *Are You Ready? Intervention CD-ROM*

Diagnose and Prescribe

Holt Pre-Algebra includes intervention strategies that diagnose students' difficulties with mathematics while providing intervention resources that will bring success to every learner.

ARE YOU READY?
Were students successful with Are You Ready?

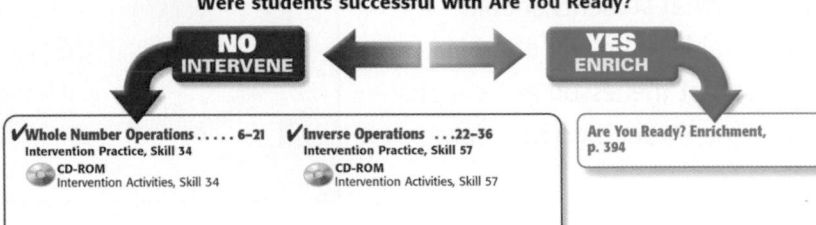

NO INTERVENE ◀▷ YES ENRICH

✔ **Whole Number Operations** 6–21
Intervention Practice, Skill 34
💿 CD-ROM
Intervention Activities, Skill 34

✔ **Inverse Operations** . . . 22–36
Intervention Practice, Skill 57
💿 CD-ROM
Intervention Activities, Skill 57

Are You Ready? Enrichment, p. 394

Intervention Tools

Each chapter begins with **Are You Ready?** assessing students' knowledge of prerequisite skills, and helping you assign either more help or enrichment depending on where students stand.

Intervention Components

Are You Ready? Intervention and Enrichment

This workbook provides additional help for students who have difficulty with a particular math concept through direct instruction, conceptual models, and scaffolded practice. Enrichment masters for every lesson enhance critical-thinking skills, as well as extend lesson objectives. Also available: *Are You Ready? Intervention CD-ROM*.

Readiness Activities

This helpful resource contains activities for every chapter and helps students review prerequisite skills needed to complete each lesson.

Math: Reading and Writing in the Content Area

These activities provide strategies for students to help organize their thinking, navigate a page in a math book, and master vocabulary.

Assessment that Gets Results

Throughout each chapter, students have access to a series of assessment resources, including **Chapter Review, Chapter Test, Performance Assessment,** and **Standardized Test Prep.**

Show What You Know

Students are asked to select work from the recently completed chapter—including section reviews, homework assignments, etc.—and create a portfolio from those pieces.

Short Response

Students are asked to execute a series of tasks, including creating their own problems with explanations of how they arrived at their conclusions and collecting magazine or newspaper articles that mention mathematical concepts.

Extended Problem Solving

Students are asked to solve a set of problems using a graph, table, etc. Students are able to use the strategy of their choice to complete the series.

Assessment Resources

- Inventory Assessment
- Section Quizzes
- Chapter Tests
- Performance Assessment
- Cumulative Tests
- End-of-year Test

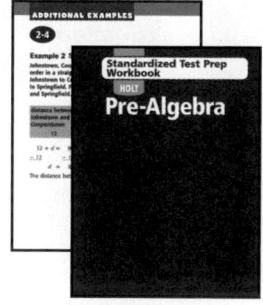

Test Prep Tool Kit

Countdown to Testing Transparencies

Multiple-choice and critical-thinking questions are featured on these transparencies, preparing students for state assessment by building problem-solving skills.

Standardized Test Prep Workbook

This resource includes a two-page test for every chapter as well as two state-specific tests, a diagnostic test and **Test-Taking Tips.**

Standardized Test Prep CD-ROM

This convenient assessment tool provides an easy way to create practice worksheets and tests that correlate to your state assessment.

Standardized Test Prep Video

This video gives visual demonstrations of math problems correlated to the book and to state standards. Hints and suggestions guide students toward solutions, while timed practice gives students experience solving multiple-choice and short-answer problems.

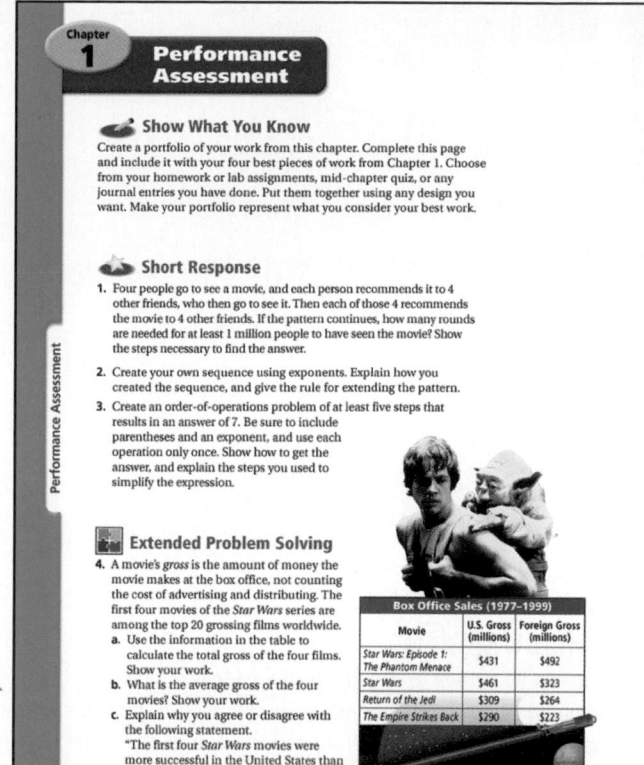

Focus on Problem Solving

To support problem-solving skills needed in state testing, this feature gives students a real-world scenario with the steps, examples, and practice problems needed to fully master the concept.

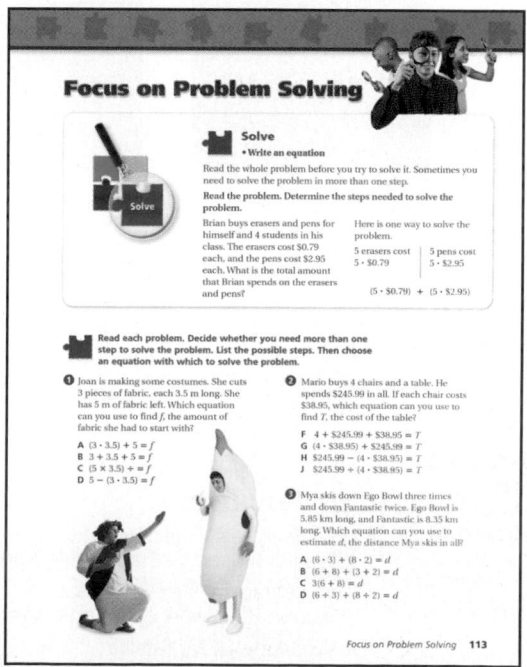

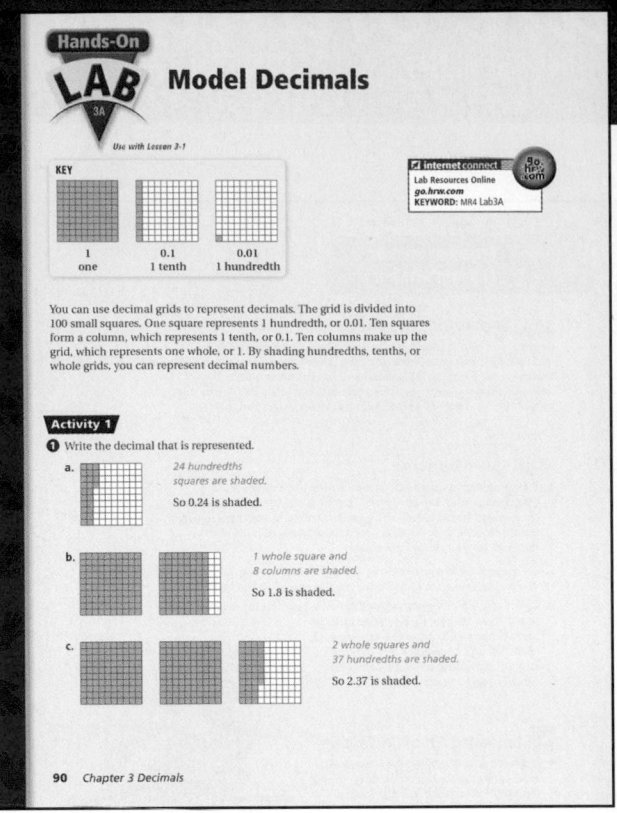

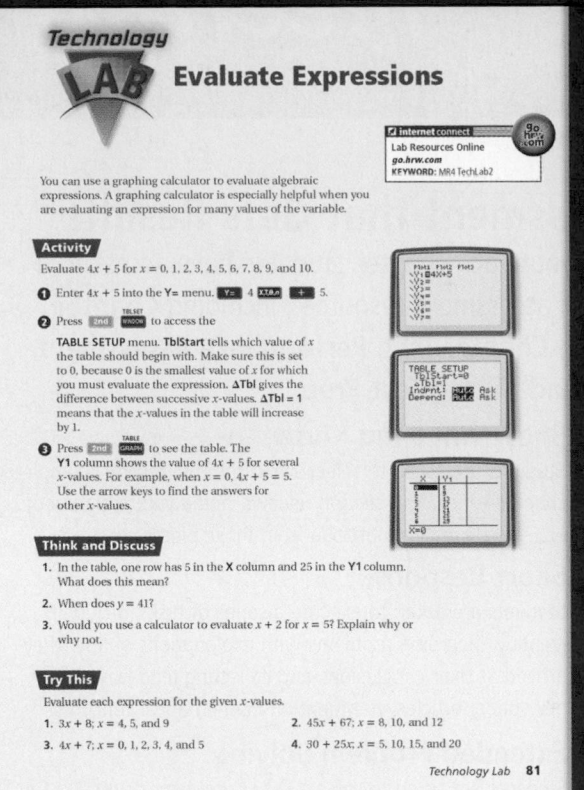

Kinesthetic Models

Give students an activity to step up learning. Models, algebra tiles, graphs, pictures that represent real-world math, and other visual representations are all incorporated into **Hands-On Lab** giving the reinforcement needed for rigorous content. **Lab Resources** also available online.

Visual Appeal

A graphing calculator can give learners the support they need to illustrate concepts and help retention of skills. **Technology Lab** walks students through a set of problems using their calculators. **Lab Resources** also available online.

 ## Reaching All Learners

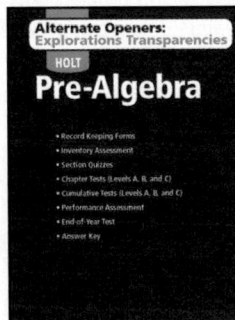

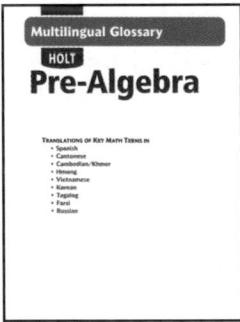

Alternate Openers: Exploration Transparencies

This notebook of transparencies provides an alternate way to teach each lesson.

Interactive Problem Solving

Blackline masters enable students to work problems for the lesson with step-by-step prompts.

Success for English Language Learners

These masters present the same concepts as the student lesson, using fewer words and more visuals. Also includes teacher support with suggested activities and teaching tips.

Multilingual Glossary

This glossary contains translations of key mathematical terms in Spanish, Cantonese, Cambodian/Khmer, Hmong, Vietnamese, Korean, Tagalog, Farsi, and Russian.

SPANISH STUDENT EDITIONS AVAILABLE!

Spanish Resources

- Libro de Trabajo: Guía Interactiva de Estudio (Spanish Interactive Study Guide Workbook)
- Libro de Trabajo: Tarea y Práctica (Spanish Homework and Practice Workbook)
- Activades de Apoyo Familia (Family Involvement Activities)

Internet Connect:

Students can link to homework help online directly related to chapter content with this in-text feature.

Homework Help Online
go.hrw.com   Keyword: MR4 2-7

Differentiating Instruction

The risk of death in a house fire can be reduced by up to 50% if the home has a working smoke alarm.

Spark Interest
Show your students that math is needed in the real world.

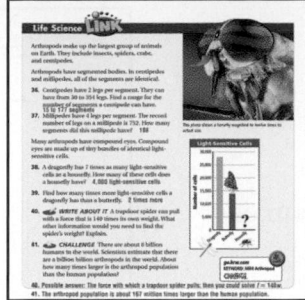

Interdisciplinary Link and Internet Activities
Interdisciplinary Links are featured throughout the program, generating interest and showing students how important math is in our everyday lives. Internet activities online at **go.hrw.com**.

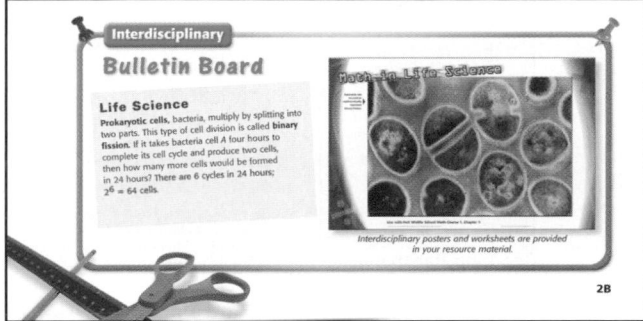

Interdisciplinary Bulletin Board
This feature gives you suggestions about how to update your class visually to complement the section you're about to teach.

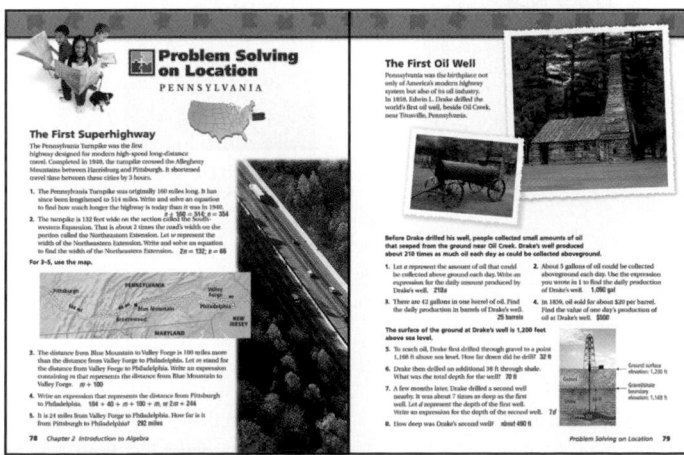

Problem Solving on Location
This feature takes students on a journey to places across the nation. Students become familiar with geography and also learn a math lesson specific to that location, helping them sharpen their critical-thinking skills needed for state testing.

Teacher to Teacher

What is Differentiated Instruction?
Differentiated instruction answers the question of how to teach every student in a classroom—regardless of ability level or background—in a way that is equal, effective, and efficient.

Differentiated instruction is more and more becoming the alternative to a "one-size-fits-all classroom." Instead of only offering core information and basic instruction of skills, a curriculum centered around differentiated instruction offers information tailored to a student's readiness, interest, and profile (the way they learn).

How Do I Reach All of My Students?
Holt Pre-Algebra ensures differentiated instruction by offering content, resources, and technology that speak to all types of learners simultaneously. A visual learner, an auditory learner, a kinesthetic learner, and an English-language learner can all benefit from features and resources such as:
- **Hands-On Lab**
- **Technology Lab**
- **Career Link**
- **Interdisciplinary Links**
- **Internet Activities**
- **Interdisciplinary Bulletin Board**
- **Problem Solving on Location**

What better way to ensure understanding than through a program that helps you do all of this without having to maintain a separate agenda for each student? Differentiated instruction gives each student what he or she needs while they are learning with everyone else in the class. State and national requirements are also taught throughout the program. The differentiated instruction curriculum in *Holt Pre-Algebra* takes care of the teacher as well, making everything streamlined and easy to manage.

Background information on Differentiated Instruction was researched from Carol Ann Tomlinson's book *How to Differentiate Instruction in Mixed-Ability Classrooms* (Alexandria, VA: Association for Supervision and Curriculum Development, 1995). Tomlinson is Professor of Educational Leadership, Foundations, and Policy at the Curry School of Education, UVA. ASCD (Assoc. for Supervision and Curriculum Development) publishes much of her work.

Holt offers a comprehensive and systematic training program to complement *Holt Pre-Algebra,* providing high-quality and accessible professional learning opportunities designed to relate to the unique needs of the educator. For more information on professional development services provided by Holt, email us at **holtinfo@hrw.com**.

HOLT Professional Development

For Teachers

Chapter 1 Resource Book

HOLT
Pre-Algebra

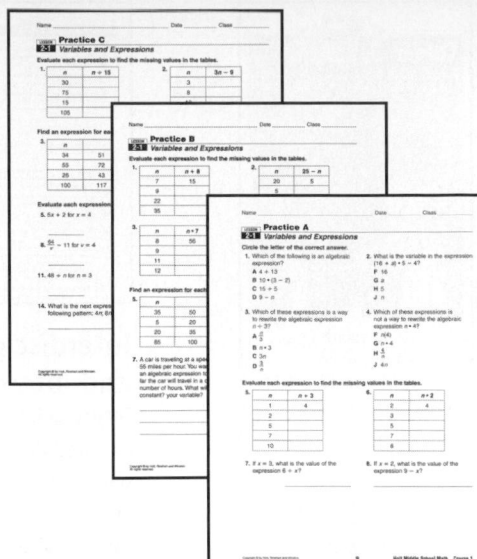

FOR THE CHAPTER:
- Parent Letter
- Are You Ready?
  Recording Sheet
- Chapter Planning and
  Pacing Guides
- Section Planning Guides
- Chapter Review
- Teacher Tools

FOR EVERY LESSON:
- Exploration Transparency
  Reproduction
- Exploration Reading Sheets
- Practice Levels A, B, C
- Reteach
- Challenge
- Problem Solving
- Puzzles, Twisters & Teasers

TRANSPARENCIES:
- Daily Transparencies
- Teaching Transparencies
- Additional Examples
  Transparencies

ANSWERS:
- Worksheet Answer Key
- Transparency Answer Key

Chapter Resources Booklet

These comprehensive books for each of the fourteen chapters include all of the items needed to extend and reinforce the students' and teacher's books. Blackline masters, transparencies, and more are found all in one place making it convenient to plan your lesson.

- Practice A, B, and C
- Reteach Masters
- Challenge Masters
- Problem Solving Masters
- Puzzles, Twisters & Teasers
- and More!

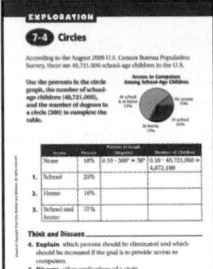

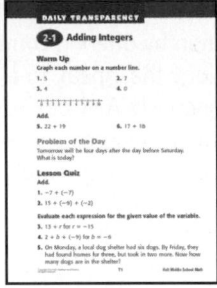

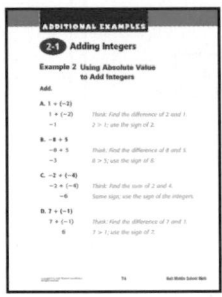

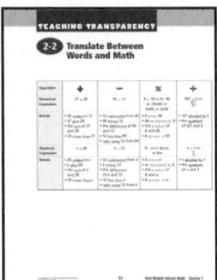

Exploration Transparency

Daily Transparency

Additional Examples Transparency

Teaching Transparencies

Assessment and Everyday Teaching Resources

- Assessment Resources
  - Inventory Assessment
  - Section Quizzes
  - Chapter Tests
  - Performance Assessment
  - Cumulative tests
  - End-of-year Test
- Test Prep Tool Kit
- Lesson Plans
- Solution Key
- Answer Transparencies

Homework and Practice Workbook

Interactive Study Guide Workbook

Standardized Test Prep Workbook

Reaching All Learners

- Alternate Openers:
  Explorations Transparencies
- Are You Ready? Intervention and Enrichment
- Consumer and Career Math
- Family Involvement Activities
- Actividades De Apoyo Familiar
  (Family Involvement Activities)
- Hands-On Lab Activities

- Interdisciplinary Posters and Worksheets
- Interactive Problem Solving
- Math: Reading and Writing
  in the Content Area
- Multilingual Glossary
- Technology Lab Activities
- Success for English Language Learners
- Readiness Activities

- Interactive Study Guide Workbook
- Standardized Test Prep Workbook
- Homework and Practice Workbook
- Libro de Trabajo: Guía Interactiva de Estudio
  (Spanish Interactive Study Guide Workbook)
- Libro de Trabajo: Tarea y Práctica
  (Spanish Homework and Practice Workbook)

Technology that Engages and Expands Learning

Holt Pre-Algebra offers an array of technology products that promote mathematics teaching and learning.

One-Stop Planner® CD-ROM with New Test and Practice Generator

This convenient tool for planning and managing lessons contains all the print-based teaching resources plus customizable lesson plans. You'll also be able to create tests and quizzes that correlate to your state assessment with the new test generator. All of these resources are accessible with the click of the mouse!

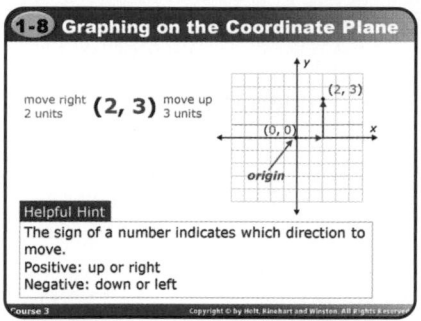

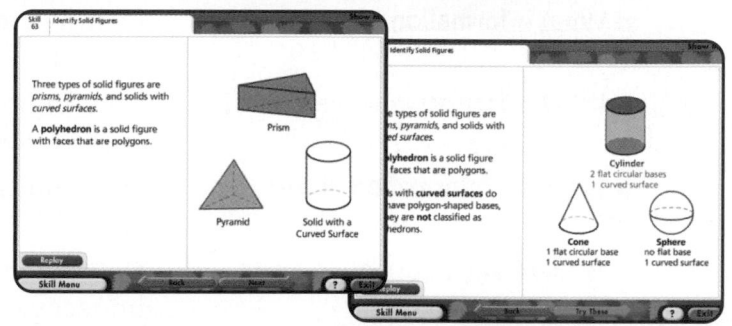

Lesson Presentations CD-ROM

This resource contains colorful, animated electronic lesson presentations—a convenient alternative to the blackboard or overhead projector! This program can also be used by individual students as a tutorial.

Are You Ready? Intervention CD-ROM

This CD-ROM provides an easy method of evaluating students' knowledge and administering additional help with prerequisite skills necessary for success. Students can work independently with computer-guided instruction and practice and can have their skills checked by computer-administered testing.

Electronic Textbooks Lighten the Load

Textbooks from **Holt Online Learning** are portable, expandable, interactive, and yet weigh nothing at all. You'll find interactive exercises and feedback, homework help, presentation materials, and much more.

Premier Online Edition

The *Premier Online Edition* features content and assessment correlated to state standards.

Student Edition CD-ROM
Teacher Edition CD-ROM

Internet Connect

Students can link to homework help online—directly related to chapter content—with this in-text feature.

internet connect
Homework Help Online
go.hrw.com   Keyword: MR4 2-7

Problem Solving Handbook

The Problem Solving Plan

In order to be a good problem solver, you need to use a good problem-solving plan. The plan used in this book is detailed below. If you have another plan that you like to use, you can use it as well.

UNDERSTAND the Problem

- **What are you asked to find?** Restate the question in your own words.

- **What information is given?** Identify the important facts in the problem.

- **What information do you need?** Determine which facts are needed to answer the question.

- **Is all the information given?** Determine whether all the facts are given.

- **Is there any information given that you will not use?** Determine which facts, if any, are unnecessary to solve the problem.

Make a PLAN

- **Have you ever solved a similar problem?** Think about other problems like this that you successfully solved.

- **What strategy or strategies can you use?** Determine a strategy that you can use and how you will use it.

SOLVE

- Follow your plan. Show the steps in your solution. Write your answer as a complete sentence.

LOOK BACK

- **Have you answered the question?** Be sure that you answered the question that is being asked.

- **Is your answer reasonable?** Your answer should make sense in the context of the problem.

- **Is there another strategy you could use?** Solving the problem using another strategy is a good way to check your work.

- **Did you learn anything that could help you solve similar problems in the future?** Try to remember the problems you have solved and the strategies you used to solve them.

Using the Problem Solving Plan

Roy has a rectangular piece of land that he wants to put a fence around. He will place a post every 9 ft along the perimeter. Each post is 5 ft tall. The land is 63 ft long and 45 ft wide. How many posts does Roy need?

UNDERSTAND the Problem

Roy has a piece of land that is 63 ft by 45 ft. He wants a post every 9 ft along the perimeter. You must find out how many posts Roy needs.

Make a PLAN

You can **draw a diagram** to show how many posts Roy needs for his fence.

SOLVE

Draw a rectangle that is similar to Roy's land. Place marks along the perimeter of the rectangle to represent the posts to be placed every 9 ft.

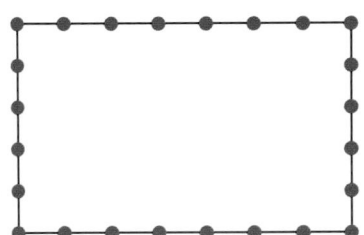

Count the number of marks you placed around the rectangle. Each corner should only have one mark.

Roy needs 24 posts for his fence.

LOOK BACK

The perimeter is 63 + 45 + 63 + 45 = 216 ft. If a post is placed every 9 ft, there will be 216 ÷ 9 = 24 posts. The answer is reasonable.

Problem Solving Handbook

Draw a Diagram

When problems involve objects, distances, or places, drawing a diagram can make the problem clearer. You can **draw a diagram** to help understand the problem and to solve the problem.

Problem Solving Strategies

Draw a Diagram	Make a Table
Make a Model	Solve a Simpler Problem
Guess and Test	Use Logical Reasoning
Work Backward	Use a Venn Diagram
Find a Pattern	Make an Organized List

June is moving her cat, dog, and goldfish to her new apartment. She can only take 1 pet with her on each trip. She cannot leave the cat and the dog or the cat and the goldfish alone together. How can she get all of her pets safely to her new apartment?

 **Understand the Problem**

The answer will be the description of the trips to her new apartment. At no time can the cat be alone with the dog or the goldfish.

Make a Plan   **Draw a diagram** to represent each trip to and from the apartment.

Solve   In the beginning, the cat, dog, and goldfish are all at her old apartment.

Old Apartment		New Apartment	
June, Cat, Dog, Fish	June, Cat →	June, Cat	Trip 1: She takes the cat and returns alone.
June, Dog, Fish	← June	Cat	
June, Dog, Fish	June, Dog →	June, Dog, Cat	Trip 2: She takes the dog and returns with the cat.
June, Cat, Fish	← June, Cat	Dog	
June, Cat, Fish	June, Fish →	June, Dog, Fish	Trip 3: She takes the fish and returns alone.
June, Cat	← June	Dog, Fish	
June, Cat	June, Cat →	June, Cat, Dog, Fish	Trip 4: She takes the cat.

 **Look Back**   Check to make sure that the cat is never alone with either the fish or the dog.

PRACTICE

1. There are 8 flags evenly spaced around a circular track. It takes Ling 15 s to run from the first flag to the third flag. At this pace, how long will it take her to run around the track twice?   **120 s, or 2 min**

2. A frog is climbing a 22-foot tree. Every 5 minutes, it climbs up 3 feet, but slips back down 1 foot. How long will it take it to climb the tree? **55 min**

Make a Model

A problem that involves objects may be solved by making a model out of similar items. **Make a model** to help you understand the problem and find the solution.

Problem Solving Strategies

Draw a Diagram	Make a Table
Make a Model	Solve a Simpler Problem
Guess and Test	Use Logical Reasoning
Work Backward	Use a Venn Diagram
Find a Pattern	Make an Organized List

The volume of a rectangular prism can be found by using the formula $V = \ell wh$, where ℓ is the length, w is the width, and h is the height of the prism. Find all possible rectangular prisms with a volume of 16 cubic units and dimensions that are all whole numbers.

Understand the Problem
You need to find the different possible prisms. The length, width, and height will be whole numbers whose product is 16.

Make a Plan
You can use unit cubes to make a model of every possible rectangular prism. Work in a systematic way to find all possible answers.

Solve
Begin with a $16 \times 1 \times 1$ prism.

$16 \times 1 \times 1$

Keeping the height of the prism the same, explore what happens to the length as you change the width. Then try a height of 2. Notice that an $8 \times 2 \times 1$ prism is the same as an $8 \times 1 \times 2$ prism turned on its side.

$8 \times 2 \times 1$ **Not a rectangular prism**   **$4 \times 4 \times 1$**   **$4 \times 2 \times 2$**

The possible dimensions are $16 \times 1 \times 1$, $8 \times 2 \times 1$, $4 \times 4 \times 1$, and $4 \times 2 \times 2$.

Look Back
The product of the length, width, and height must be 16. Look at the prime factorization of the volume: $16 = 2 \cdot 2 \cdot 2 \cdot 2$. Possible dimensions:

$1 \cdot 1 \cdot (2 \cdot 2 \cdot 2 \cdot 2) = 1 \cdot 1 \cdot 16$ $1 \cdot 2 \cdot (2 \cdot 2 \cdot 2) = 1 \cdot 2 \cdot 8$

$1 \cdot (2 \cdot 2) \cdot (2 \cdot 2) = 1 \cdot 4 \cdot 4$ $2 \cdot 2 \cdot (2 \cdot 2) = 2 \cdot 2 \cdot 4$

PRACTICE

1. Four unit squares are arranged so that each square shares a side with another square. How many different arrangements are possible?   7

2. Four triangles are formed by cutting a rectangle along its diagonals. What possible shapes can be formed by arranging these triangles?

2. Isosceles triangle, parallelogram, rectangle, kite, and other shapes

Problem Solving Handbook

Guess and Test

When you think that guessing may help you solve a problem, you can use **guess and test.** Using clues to make guesses can narrow your choices for the solution. Test whether your guess solves the problem, and continue guessing until you find the solution.

 Problem Solving Strategies

Draw a Diagram	Make a Table
Make a Model	Solve a Simpler Problem
Guess and Test	Use Logical Reasoning
Work Backward	Use a Venn Diagram
Find a Pattern	Make an Organized List

North Middle School is planning to raise $1200 by sponsoring a car wash. They are going to charge $4 for each car and $8 for each minivan. How many vehicles would have to be washed to raise $1200 if they plan to wash twice as many cars as minivans?

Understand the Problem

You must determine the number of cars and the number of minivans that need to be washed to make $1200. You know the charge for each vehicle.

Make a Plan

You can **guess and test** to find the number of cars and minivans. Guess the number of cars, and then divide it by 2 to find the number of minivans.

Solve

You can organize your guesses in a table.

	Cars	Minivans	Money Raised	
First guess	200	100	$4(200) + $8(100) = $1600	*Too high*
Second guess	100	50	$4(100) + $8(50) = $800	*Too low*
Third guess	150	75	$4(150) + $8(75) = $1200	

They should wash 150 cars and 75 minivans, or 225 vehicles.

Look Back

The total raised is $4(150) + $8(75) = $1200, and the number of cars is twice the number of minivans. The answer is reasonable.

PRACTICE

1. At a baseball game, adult tickets cost $15 and children's tickets cost $8. Twice as many children attended as adults, and the total ticket sales were $2480. How many people attended the game?
   80 adults and 160 children = 240 people

2. Angie is making friendship bracelets and pins. It takes her 6 minutes to make a bracelet and 4 minutes to make a pin. If she wants to make three times as many pins as bracelets, how many pins and bracelets can she make in 3 hours? **30 pins and 10 bracelets**

Work Backward

To solve a problem that asks for an initial value that follows a series of steps, you may want to **work backward**.

 Problem Solving Strategies

Draw a Diagram	Make a Table
Make a Model	Solve a Simpler Problem
Guess and Test	Use Logical Reasoning
Work Backward	Use a Venn Diagram
Find a Pattern	Make an Organized List

Tyrone has two clocks and a watch. If the power goes off during the day, the following happens:

- **Clock A stops and then continues when the power comes back on.**
- **Clock B stops and then resets to 12:00 A.M. when the power comes back on.**

When Tyrone gets home, his watch reads 4:27 P.M., clock B reads 5:21 A.M., and clock A reads 3:39 P.M. What time did the power go off, and for how long was it off?

 **Understand the Problem**

You need to find the time that the power went off and how long it was off. You know how each clock works.

 **Make a Plan**

Work backward to the time that the power went off. Subtract from the correct time of 4:27, the time on Tyrone's watch.

 **Solve**

The difference between the correct time and the time on clock A is the length of time the power was off.

$$4:27 - 3:39 = 0:48 \qquad \textit{The power was off for 48 minutes.}$$

Clock B reset to 12:00 when the power went on.

$$5:21 - 12:00 = 5:21 \qquad \textit{The power came on 5 hours and 21 minutes ago.}$$

Subtract 5:21 from the correct time to find when the power came on.

$$4:27 - 5:21 = 11:06 \qquad \textit{The power came on at 11:06 A.M.}$$

Subtract 48 minutes from 11:06 to find when the power went off.

$$11:06 - 0:48 = 10:18$$

The power went off at 10:18 A.M. and was off for 48 minutes.

 **Look Back**

If the power went off at about 10 A.M. for about an hour, it would come on at about 11 A.M., and each clock would run for about $5\frac{1}{2}$ hours.

PRACTICE

1. Jackie is 4 years younger than Roger. Roger is $2\frac{1}{2}$ years older than Jade. Jade is 14 years old. How old is Jackie? **$12\frac{1}{2}$ years old**

2. Becca is directing a play that starts at 8:15 P.M. She wants the cast ready 10 minutes before the play starts. The cast needs 45 minutes to put on make-up, 15 minutes for a director's meeting, and then 35 minutes to get in costume. What time should the cast arrive? **6:30 P.M.**

Problem Solving Handbook

Find a Pattern

If a problem involves numbers, shapes, or even codes, noticing a pattern can often help you solve it. To solve a problem that involves patterns, you need to use small steps that will help you **find a pattern**.

Problem Solving Strategies

Draw a Diagram	Make a Table
Make a Model	Solve a Simpler Problem
Guess and Test	Use Logical Reasoning
Work Backward	Use a Venn Diagram
Find a Pattern	Make an Organized List

Gil is trying to decode the following sentence, which may have been encoded using a pattern. What does the coded sentence say?

QEB NRFZH YOLTK CLU GRJMP LSBO QEB IXWV ALD.

Understand the Problem

You need to find whether there was a pattern used to encode the sentence and then extend the pattern to decode the sentence.

Make a Plan

Find a pattern. Try to decode one of the words first. Notice that *QEB* appears twice in the sentence.

Solve

Gil thinks that *QEB* is probably the word *THE*. If *QEB* stands for *THE*, a pattern emerges with respect to the letters and their position in the alphabet.

Q: 17th letter T: 20th letter   *+ 3 letters*

E: 5th letter H: 8th letter   *+ 3 letters*

B: 2nd letter E: 5th letter   *+ 3 letters*

Continue the pattern. Although there is no 27th, 28th, or 29th letter of the alphabet, the remaining letters should be obvious ($27 = 1 = A$, $28 = 2 = B$, and $29 = 3 = C$).

QEB NRFZH YOLTK CLU GRJMP LSBO QEB IXWV ALD.

THE QUICK BROWN FOX JUMPS OVER THE LAZY DOG.

Look Back

The sentence makes sense, so the pattern fits.

PRACTICE

Decode each sentence.

1. RFC DGTC ZMVGLE UGXYPBQ HSKN OSGAIJW.

($RFC = THE$) **THE FIVE BOXING WIZARDS JUMP QUICKLY.**

2. U PYLS VUX KOUWE GCABN DCHR TCJJS ZIQF.

($U = A$) **A VERY BAD QUACK MIGHT JINX ZIPPY FOWLS.**

Make a Table

To solve a problem that involves a relationship between two sets of numbers, you can **make a table.** A table can be used to organize data so that you can look at relationships and find the solution.

Problem Solving Strategies

Draw a Diagram
Make a Model
Guess and Test
Work Backward
Find a Pattern

Make a Table
Solve a Simpler Problem
Use Logical Reasoning
Use a Venn Diagram
Make an Organized List

Jill has 12 pieces of 2 ft long decorative edging. She wants to use the edging to enclose a garden with the greatest possible area against the back of her house. What is the largest garden she can make?

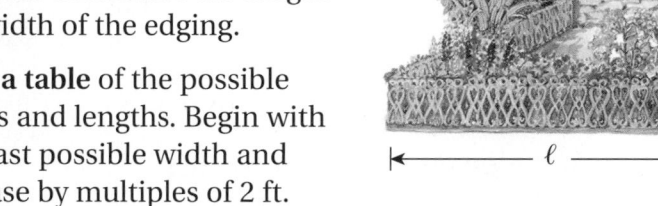

 **Understand the Problem**

You must determine the length and width of the edging.

 **Make a Plan**

Make a table of the possible widths and lengths. Begin with the least possible width and increase by multiples of 2 ft. Remember that the width is the same on two sides.

 **Solve**

Use the table to solve.

Width (ft)	Length (ft)	Garden Area (ft²)
2	20	40
4	16	64
6	12	72
8	8	64
10	4	40

The maximum area that the garden can be is 72 ft², with a width of 6 ft and a length of 12 ft.

 **Look Back**

She can use 3 pieces of edging for the first side, 6 pieces for the second side, and another 3 pieces for the third side.

3 + 6 + 3 = 12 pieces
6 ft + 12 ft + 6 ft = 24 ft

PRACTICE

1. Suppose Jill decided not to use the house as one side of the garden. What is the greatest area that she could enclose?   **36 ft²**

four 3-packs and one 2-pack   **2.** A store sells batteries in packs of 3 for $3.99 and 2 for $2.99. Barry got 14 batteries total for $18.95. How many of each package did he buy?

Problem Solving Handbook

Solve a Simpler Problem

If a problem contains large numbers or requires many steps, try to **solve a simpler problem** first. Look for similarities between the problems, and use them to solve the original problem.

Problem Solving Strategies

Draw a Diagram	Make a Table
Make a Model	**Solve a Simpler Problem**
Guess and Test	Use Logical Reasoning
Work Backward	Use a Venn Diagram
Find a Pattern	Make an Organized List

Noemi heard that 10 computers in her school would be connected to each other. She thought that there would be a cable connecting each computer to every other computer. How many cables would be needed if this were true?

Understand the Problem You know that there are 10 computers and that each computer would require a separate cable to connect to every other computer. You need to find the total number of cables.

Make a Plan Start by **solving a simpler problem** with fewer computers.

Solve The simplest problem starts with 2 computers.

2 computers
1 connection

3 computers
3 connections
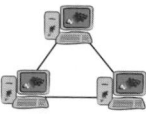
4 computers
6 connections
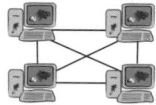

Organize the data in a table to help you find a pattern.

Number of Computers	Number of Connections
2	1
3	1 + 2 = 3
4	1 + 2 + 3 = 6
5	1 + 2 + 3 + 4 = 10
10	1 + 2 + 3 + 4 + 5 + 6 + 7 + 8 + 9 = 45

So if a separate cable were needed to connect each of 10 computers to every other one, 45 cables would be required.

Look Back Extend the number of computers to check that the pattern continues.

PRACTICE

1. A banquet table seats 2 people on each side and 1 at each end. If 6 tables are placed end to end, how many seats can there be? **26**

2. How many diagonals are there in a dodecagon (a 12-sided polygon)? **54**

Use Logical Reasoning

Sometimes a problem may provide clues and facts to help you find a solution. You can **use logical reasoning** to help solve this kind of problem.

 Problem Solving Strategies

Draw a Diagram	Make a Table
Make a Model	Solve a Simpler Problem
Guess and Test	**Use Logical Reasoning**
Work Backward	Use a Venn Diagram
Find a Pattern	Make an Organized List

Kim, Lily, and Suki take ballet, tap, and jazz classes (but not in that order). Kim is the sister of the person who takes ballet. Lily takes tap.

Understand the Problem

You want to determine which person is in which dance class. You know that there are three people and that each person takes only one dance class.

Make a Plan

Use logical reasoning to make a table of the facts from the problem.

Solve

List the types of dance and the people's names. Write *Yes* or *No* when you are sure of an answer. Lily takes tap.

	Ballet	Tap	Jazz
Kim		No	
Lily	No	**Yes**	No
Suki		No	

The person taking ballet is Kim's sister, so Kim does not take ballet. Suki must be the one taking ballet.

	Ballet	Tap	Jazz
Kim		No	
Lily	No	**Yes**	No
Suki	**Yes**	No	No

← *Kim must be the one taking jazz.*

Kim takes jazz, Lily takes tap, and Suki takes ballet.

Look Back

Make sure none of your conclusions conflict with the clues.

PRACTICE

1. Patrick, John, and Vanessa have a snake, a cat and a rabbit. Patrick's pet does not have fur. Vanessa does not have a cat. Match the owners with their pets. **Patrick: snake, John: cat, Vanessa: rabbit**

2. Keifer: sixth, Dylan: seventh, Isabella: eighth, Chrissy: ninth

2. Isabella, Keifer, Dylan, and Chrissy are in the sixth, seventh, eighth, and ninth grades. Isabella is not in seventh grade. The sixth-grader has band with Dylan and lunch with Isabella. Chrissy is in the ninth grade. Match the students with their grades.

Use a Venn Diagram

You can **use a Venn diagram** to display relationships among sets in a problem. Use ovals, circles, or other shapes to represent individual sets.

Problem Solving Strategies

Draw a Diagram	Make a Table
Make a Model	Solve a Simpler Problem
Guess and Test	Use Logical Reasoning
Work Backward	**Use a Venn Diagram**
Find a Pattern	Make an Organized List

Patricia took a poll of 100 students. She wrote down that 32 play basketball, 45 run track, and 19 do both. Mrs. Thornton wants to know how many of the students polled only play basketball.

Understand the Problem

You know that 100 students were polled, 32 play basketball, 45 run track, and 19 play basketball *and* run track.

The answer is the number of students who only play basketball.

Make a Plan

Use a Venn diagram to show the sets of students who play basketball, students who run track, and students who do both.

Solve

Draw and label two overlapping circles in a rectangle. Work from the inside out. Write 19 in the area where the two circles overlap. This represents the number of students who play basketball and run track.

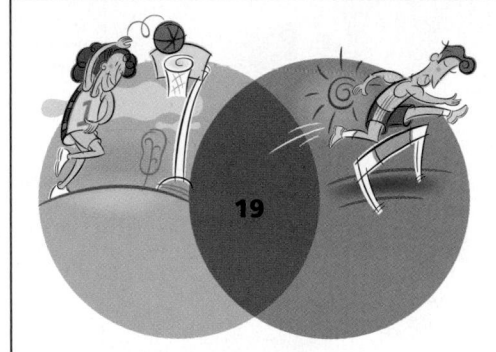

Use the information in the problem to complete the diagram. You know that 32 students play basketball, and 19 of those students run track.

So 13 students only play basketball.

Look Back

When your Venn diagram is complete, check it carefully against the information in the problem to make sure it agrees with the facts given.

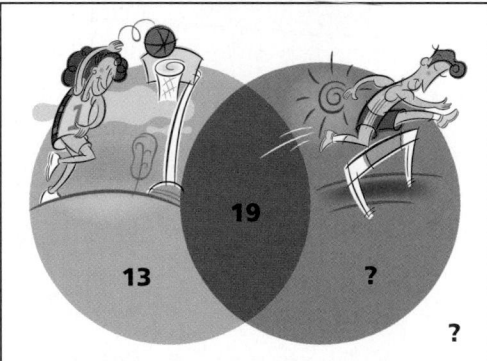

PRACTICE

1. How many of the students only run track?   **26**

2. How many of the students do not play basketball or run track?   **42**

Make an Organized List

Sometimes a problem involves many possible ways in which something can be done. To find a solution to this kind of problem, you need to **make an organized list.** This will help you to organize and count all the possible outcomes.

Problem Solving Strategies

Draw a Diagram	Make a Table
Make a Model	Solve a Simpler Problem
Guess and Test	Use Logical Reasoning
Work Backward	Use a Venn Diagram
Find a Pattern	**Make an Organized List**

What is the greatest amount of money you can have in coins (quarters, dimes, nickels, and pennies) without being able to make change for a dollar?

 **Understand the Problem**

You are looking for an amount of money. You cannot have any combinations of coins that make a dollar, such as 4 quarters or 3 quarters, 2 dimes, and a nickel.

 Make a Plan

Make an organized list, starting with the maximum possible number of each type of coin. Consider all the ways you can add other types of coins without making exactly one dollar.

Solve

List the maximum number of each kind of coin you can have.

3 quarters = 75¢   9 dimes = 90¢   19 nickels = 95¢   99 pennies = 99¢

Next, list all the possible combinations of two kinds of coins.

3 quarters and 4 dimes = 115¢	9 dimes and 1 quarter = 115¢
3 quarters and 4 nickels = 95¢	9 dimes and 1 nickel = 95¢
3 quarters and 24 pennies = 99¢	9 dimes and 9 pennies = 99¢

19 nickels and 4 pennies = 99¢

Look for any combinations from this list that you could add another kind of coin to without making exactly one dollar.

3 quarters, 4 dimes, and 4 pennies = 119¢
3 quarters, 4 nickels, and 4 pennies = 99¢
9 dimes, 1 quarter, and 4 pennies = 119¢
9 dimes, 1 nickel, and 4 pennies = 99¢

The largest amount you can have is 119¢, or $1.19.

 **Look Back**

Try adding one of any type of coin to either combination that makes $1.19, and then see if you could make change for a dollar.

PRACTICE

1. How can you arrange the numbers 2, 6, 7, and 12 with the symbols +, ×, and ÷ to create the expression with the greatest value?
$12 \times 7 + 6 \div 2 = 87$

2. How many ways are there to arrange 24 desks in 3 or more equal rows if each row must have at least 2 desks? **5**

Algebra Toolbox

Section 1A	Section 1B
Equations and Inequalities	**Graphing**
Lesson 1-1 Variables and Expressions	**Lesson 1-7** Ordered Pairs
Lesson 1-2 Write Algebraic Expressions	**Lesson 1-8** Graphing on a Coordinate Plane
Lesson 1-3 Solving Equations by Adding or Subtracting	**Technology Lab 1A** Create a Table of Solutions
Lesson 1-4 Solving Equations by Multiplying or Dividing	**Lesson 1-9** Interpreting Graphs and Tables
Lesson 1-5 Solving Simple Inequalities	
Lesson 1-6 Combining Like Terms	

Pacing Guide for 45-Minute Classes

Chapter 1

DAY 1	DAY 2	DAY 3	DAY 4	DAY 5
Lesson 1-1	Lesson 1-2	Lesson 1-3	Lesson 1-4	Lesson 1-5

DAY 6	DAY 7	DAY 8	DAY 9	DAY 10
Lesson 1-6	**Mid-Chapter Quiz** Lesson 1-7	Lesson 1-8	**Technology Lab 1A**	Lesson 1-9

DAY 11	DAY 12
Chapter 1 Review	**Chapter 1** **Assessment**

Pacing Guide for 90-Minute Classes

Chapter 1

DAY 1	DAY 2	DAY 3	DAY 4	DAY 5
Lesson 1-1 Lesson 1-2	Lesson 1-3 Lesson 1-4	Lesson 1-5 Lesson 1-6	**Mid-Chapter Quiz** Lesson 1-7 Lesson 1-8	**Technology Lab 1A** Lesson 1-9

DAY 6	DAY 7
Chapter 1 Review Lesson 2-1 Lesson 2-2	**Chapter 1** **Assessment** Lesson 2-3

COURSE 1
- Write and evaluate algebraic expressions.
- Identify solutions of and solve one-step equations.
- Solve one-step whole number inequalities.
- Make connections between linear functions and graphs.

COURSE 2
- Write and evaluate algebraic expressions.
- Combine like terms.
- Solve one-step equations.
- Solve and graph solutions of one-step inequalities.
- Make and interpret graphs and tables.
- Make connections between linear functions and graphs.

PRE-ALGEBRA
- Write and evaluate algebraic expressions.
- Combine like terms.
- Solve one-step equations.
- Solve and graph solutions of one-step inequalities.
- Make and interpret graphs and tables.
- Make connections between linear functions or two-variable equations and graphs.

Across the Curriculum

LANGUAGE ARTS

Math: Reading and Writing in the Content Area . . . . pp. 1–9

Focus on Problem Solving
  Solve . . . . SE p. 33

Journal . . . . TE, last page of each lesson

Write About It. . . . . SE pp. 7, 12, 17, 27, 31, 37, 41

SOCIAL STUDIES LINK

History . . . . SE p. 37

Social Studies . . . . SE pp. 16, 17, 22

Geography . . . . SE p. 15

SCIENCE LINK

Life Science . . . . SE pp. 7, 37

Earth Science . . . . SE p. 47

Physical Science. . . . . SE pp. 5, 41

TE = *Teacher's Edition*   **SE** = *Student Edition*

Interdisciplinary

Bulletin Board

Earth Science

As a rocket moves away from Earth, it gains speed due to fuel loss and the decrease of gravitational pull and air resistance. The speed of a space shuttle in low Earth orbit is about 17,500 miles an hour. Write and solve an equation to approximate the distance traveled by the shuttle in 8 hours.
$d = rt = 17{,}500(8) = 140{,}000$ mi

Interdisciplinary posters and worksheets are provided in your resource material.

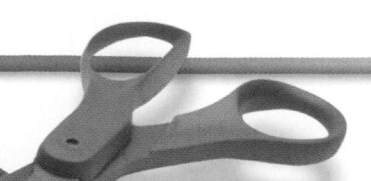

Resource Options

Chapter 1 Resource Book

Student Resources

Practice (Levels A, B, C)..... pp. 9–11, 18–20, 27–29, 36–38, 45–47, 54–56, 64–66, 73–75, 82–84

Reteach .... pp. 12, 21, 30, 39, 48, 57, 67, 76, 85–86

Challenge .... pp. 13, 22, 31, 40, 49, 58, 68, 77, 87

Problem Solving .... pp. 14, 23, 32, 41, 50, 59, 69, 78, 88

Puzzles, Twisters & Teasers .... pp. 15, 24, 33, 42, 51, 60, 70, 79, 89

Recording Sheets .... pp. 3–4, 8, 17, 26, 35, 44, 53, 63, 72, 81, 93, 96, 101–102, 105

Chapter Review .... pp. 90–92

Teacher and Parent Resources

Chapter Planning and Pacing Guide. .... p. 5

Section Planning Guides .... pp. 6, 61

Parent Letter .... pp. 1–2

Teaching Tools .... pp. 96–105

Teacher Support for Chapter Project .... p. 94

Transparencies .... pp. T1–T40

• Daily Transparencies

• Additional Examples Transparencies

• Teaching Transparencies

Reaching All Learners

English Language Learners

Success for English Language Learners .... pp. 1–18

Math: Reading and Writing in the Content Area .... pp. 1–9

Spanish Homework and Practice .... pp. 1–9

Spanish Interactive Study Guide .... pp. 1–9

Spanish Family Involvement Activities. .... pp. 1–8

Multilingual Glossary

Individual Needs

Are You Ready? Intervention and Enrichment .. pp. 29–32, 145–148, 213–216, 225–228, 237–240, 405–406

Alternate Openers: Explorations. .... pp. 1–9

Family Involvement Activities .... pp. 1–8

Interactive Problem Solving. .... pp. 1–9

Interactive Study Guide .... pp. 1–9

Readiness Activities .... pp. 1–2

Math: Reading and Writing in the Content Area .... pp. 1–9

Challenge ..... CRB pp. 13, 22, 31, 40, 49, 58, 68, 77, 87

Hands-On

Hands-On Lab Activities .... pp. 1–10

Technology Lab Activities .... pp. 1–7

Alternate Openers: Explorations. .... pp. 1–9

Family Involvement Activities .... pp. 1–8

Applications and Connections

Consumer and Career Math .... pp. 1–4

Interdisciplinary Posters .... Poster 1, TE p. 2B

Interdisciplinary Poster Worksheets. .... pp. 1–3

Transparencies

Alternate Openers: Explorations .... pp. 1–9

Exercise Answers Transparencies

Chapter 1 Resource Book .... pp. T1–T40

• Daily Transparencies

• Additional Examples Transparencies

• Teaching Transparencies

Technology

Teacher Resources

Lesson Presentations CD-ROM. .... Chapter 1

Test and Practice Generator CD-ROM .... Chapter 1

One-Stop Planner CD-ROM .... Chapter 1

Student Resources

Are You Ready? Intervention CD-ROM
Skills 5, 34, 51, 54, 57

▪ internet connect   **go.hrw.com**

Homework Help Online	KEYWORD: MP4 HWHelp1
Math Tools Online	KEYWORD: MP4 Tools
Glossary Online	KEYWORD: MP4 Glossary
Chapter Project Online	KEYWORD: MP4 PSProject1
Chapter Opener Online	KEYWORD: MP4 Ch1

CNN student News™   KEYWORD: MP4 CNN1

SE = *Student Edition*   **TE** = *Teacher's Edition*   **AR** = *Assessment Resources*   **CRB** = *Chapter Resource Book*   **MK** = *Manipulatives Kit*

Assessment Options

Assessing Prior Knowledge

Determine whether students have the required prerequisite concepts and skills.

Are You Ready?....................................... SE p. 3
Inventory Test.................................... AR pp. 1–4

Test Preparation

Provide review and practice for chapter and standardized tests.

Standardized Test Prep........................... SE p. 57
Spiral Review with Test Prep ..... SE, last page of each lesson
Study Guide and Review .... SE pp. 52–54
Test Prep Tool Kit

Technology

 Test and Practice Generator CD-ROM

internet connect

State-Specific Test Practice Online KEYWORD: MP4 TestPrep

Performance Assessment

Assess students' understanding of chapter concepts and combined problem-solving skills.

Performance Assessment .... SE p. 56
  Includes scoring rubric in TE
Performance Assessment .... AR p. 118
Performance Assessment Teacher Support......... AR p. 117

Portfolio

Portfolio opportunities appear throughout the Student and Teacher's Editions.

Suggested work samples:

Problem Solving Project .... TE p. 2
Performance Assessment .... SE p. 56
Portfolio Guide.............................. AR p. xxxvii
Journal....................... TE, last page of each lesson
Write About It .... SE pp. 7, 12, 17, 27, 31, 37, 41

Daily Assessment

Obtain daily feedback on students' understanding of concepts.

Spiral Review and Test Prep ...... SE, last page of each lesson

Also Available on Transparency In Chapter 1 Resource Book

Warm Up..................... TE, first page of each lesson
Problem of the Day............. TE, first page of each lesson
Lesson Quiz................... TE, last page of each lesson

Student Self-Assessment

Have students evaluate their own work.

Group Project Evaluation .... AR p. xxxiv
Individual Group Member Evaluation .... AR p. xxxv
Portfolio Guide............................... AR p. xxxvii
Journal....................... TE, last page of each lesson

Formal Assessment

Assess students' mastery of concepts and skills.

Section Quizzes .... AR pp. 5–6
Mid-Chapter Quiz............................. SE p. 32
Chapter Test .... SE p. 55
Chapter Tests (Levels A, B, C) .... AR pp. 33–38
Cumulative Tests (Levels A, B, C)........... AR pp. 145–156
Standardized Test Prep
  Cumulative Assessment .... SE p. 57
End-of-Year Test......................... AR pp. 313–316

Technology

 Test and Practice Generator CD-ROM

Make tests electronically. This software includes:
- Dynamic practice for Chapter 1
- Customizable tests
- Multiple-choice items for each objective
- Free-response items for each objective
- Teacher management system

SE = *Student Edition*   **TE** = *Teacher's Edition*   **AR** = *Assessment Resources*   **CRB** = *Chapter Resource Book*   **MK** = *Manipulatives Kit*

Chapter 1 Tests

Three levels (A,B,C) of tests are available for each chapter in the *Assessment Resources.*

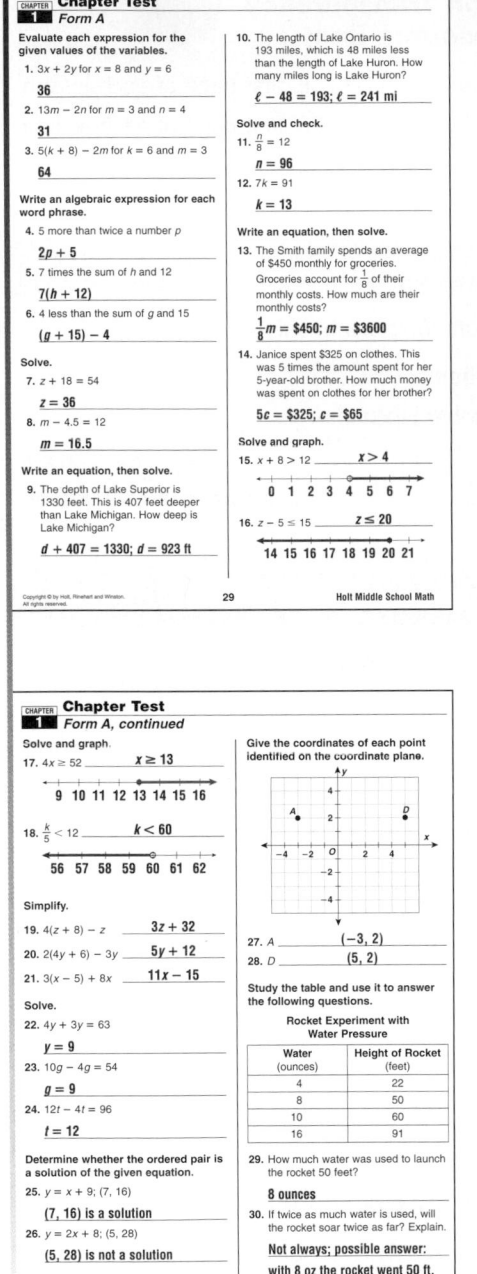

LEVEL A

CHAPTER 1 **Chapter Test**
Form A

Evaluate each expression for the given values of the variables.

1. $3x + 2y$ for $x = 8$ and $y = 6$

   36

2. $13m - 2n$ for $m = 3$ and $n = 4$

   31

3. $5(k + 8) - 2m$ for $k = 6$ and $m = 3$

   64

Write an algebraic expression for each word phrase.

4. 5 more than twice a number p

   $2p + 5$

5. 7 times the sum of h and 12

   $7(h + 12)$

6. 4 less than the sum of g and 15

   $(g + 15) - 4$

Solve.

7. $z + 18 = 54$

   $z = 36$

8. $m - 4.5 = 12$

   $m = 16.5$

Write an equation, then solve.

9. The depth of Lake Superior is 1330 feet. This is 407 feet deeper than Lake Michigan. How deep is Lake Michigan?

   $d + 407 = 1330; d = 923$ ft

10. The length of Lake Ontario is 193 miles, which is 48 miles less than the length of Lake Huron. How many miles long is Lake Huron?

    $\ell - 48 = 193; \ell = 241$ mi

Solve and check.

11. $\frac{n}{8} = 12$

    $n = 96$

12. $7k = 91$

    $k = 13$

Write an equation, then solve.

13. The Smith family spends an average of $450 monthly for groceries. Groceries account for $\frac{1}{8}$ of their monthly costs. How much are their monthly costs?

    $\frac{1}{8}m = \$450; m = \$3600$

14. Janice spent $325 on clothes. This was 5 times the amount spent for her 5-year-old brother. How much money was spent on clothes for her brother?

    $5c = \$325; c = \$65$

Solve and graph.

15. $x + 8 > 12$ **$x > 4$**

    0 1 2 3 4 5 6 7

16. $z - 5 \le 15$ **$z \le 20$**

    14 15 16 17 18 19 20 21

29   Holt Middle School Math

CHAPTER 1 **Chapter Test**
Form A, continued

Solve and graph.

17. $4x \ge 52$ **$x \ge 13$**

    9 10 11 12 13 14 15 16

18. $\frac{k}{5} < 12$ **$k < 60$**

    56 57 58 59 60 61 62

Simplify.

19. $4(z + 8) - z$ **$3z + 32$**

20. $2(4y + 6) - 3y$ **$5y + 12$**

21. $3(x - 5) + 8x$ **$11x - 15$**

Solve.

22. $4y + 3y = 63$

    $y = 9$

23. $10g - 4g = 54$

    $g = 9$

24. $12t - 4t = 96$

    $t = 12$

Determine whether the ordered pair is a solution of the given equation.

25. $y = x + 9$; (7, 16)

    (7, 16) is a solution

26. $y = 2x + 8$; (5, 28)

    (5, 28) is not a solution

Give the coordinates of each point identified on the coordinate plane.

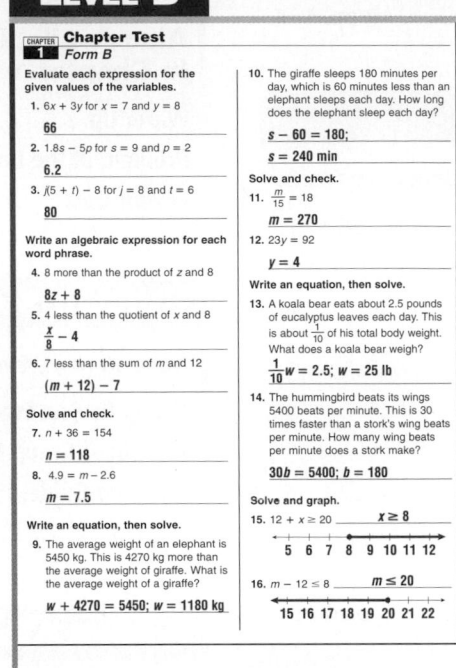

27. A   **$(-3, 2)$**

28. D   **$(5, 2)$**

Study the table and use it to answer the following questions.

Rocket Experiment with Water Pressure

Water (ounces)	Height of Rocket (feet)
4	22
8	50
10	60
16	91

29. How much water was used to launch the rocket 50 feet?

    8 ounces

30. If twice as much water is used, will the rocket soar twice as far? Explain.

    Not always; possible answer: with 8 oz the rocket went 50 ft, but with 16 oz it did not go 100 ft.

LEVEL B

CHAPTER 1 **Chapter Test**
Form B

Evaluate each expression for the given values of the variables.

1. $6x + 3y$ for $x = 7$ and $y = 8$

   66

2. $1.8s - 5p$ for $s = 9$ and $p = 2$

   6.2

3. $j(5 + t) - 8$ for $j = 8$ and $t = 6$

   80

Write an algebraic expression for each word phrase.

4. 8 more than the product of z and 8

   $8z + 8$

5. 4 less than the quotient of x and 8

   $\frac{x}{8} - 4$

6. 7 less than the sum of m and 12

   $(m + 12) - 7$

Solve and check.

7. $n + 36 = 154$

   $n = 118$

8. $4.9 = m - 2.6$

   $m = 7.5$

Write an equation, then solve.

9. The average weight of an elephant is 5450 kg. This is 4270 kg more than the average weight of giraffe. What is the average weight of a giraffe?

   $w + 4270 = 5450; w = 1180$ kg

10. The giraffe sleeps 180 minutes per day, which is 60 minutes less than an elephant sleeps each day. How long does the elephant sleep each day?

    $s - 60 = 180;$
    $s = 240$ min

Solve and check.

11. $\frac{m}{15} = 18$

    $m = 270$

12. $23y = 92$

    $y = 4$

Write an equation, then solve.

13. A koala bear eats about 2.5 pounds of eucalyptus leaves each day. This is about $\frac{1}{10}$ of his total body weight. What does a koala bear weigh?

    $\frac{1}{10}w = 2.5; w = 25$ lb

14. The hummingbird beats its wings 5400 beats per minute. This is 30 times faster than a stork's wing beats per minute. How many wing beats per minute does a stork make?

    $30b = 5400; b = 180$

Solve and graph.

15. $12 + x \ge 20$ **$x \ge 8$**

    5 6 7 8 9 10 11 12

16. $m - 12 \le 8$ **$m \le 20$**

    15 16 17 18 19 20 21 22

CHAPTER 1 **Chapter Test**
Form B, continued

Solve and graph.

17. $13y < 104$ **$y \le 8$**

    4 5 6 7 8 9 10 11

18. $6 \ge \frac{x}{2}$ **$12 \ge x$**

    7 8 9 10 11 12 13 14

Simplify.

19. $7(2b - 8)$ **$14b - 56$**

20. $4(4x - 4) + 2x$ **$18x - 16$**

21. $3(4b + 3) - 5$ **$12b + 4$**

Solve.

22. $24k - 7k = 51$ **$k = 3$**

23. $45 = 3y + 6y$ **$y = 5$**

24. $\frac{z}{12} = 9$ **$z = 108$**

Determine whether the ordered pair is a solution of the given equation.

25. $y = 7x + 7$; (4, 30)

    (4, 30) is not a solution

26. $y = 4x + 12$; (3, 24)

    (3, 24) is a solution

Give the coordinates of each point identified on the coordinate plane.

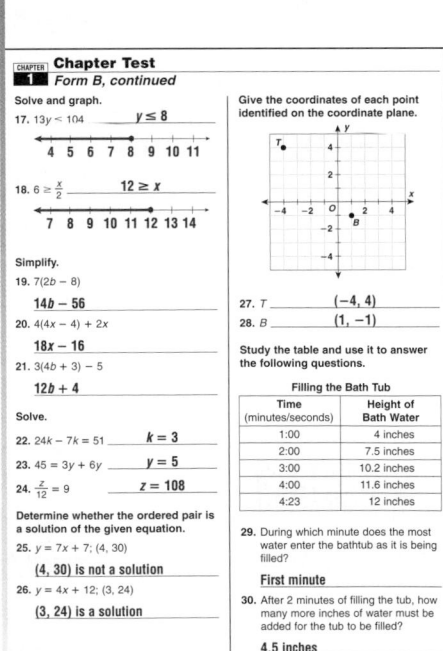

27. T   **$(-4, 4)$**

28. B   **$(1, -1)$**

Study the table and use it to answer the following questions.

Filling the Bath Tub

Time (minutes/seconds)	Height of Bath Water
1:00	4 inches
2:00	7.5 inches
3:00	10.2 inches
4:00	11.6 inches
4:23	12 inches

29. During which minute does the most water enter the bathtub as it is being filled?

    First minute

30. After 2 minutes of filling the tub, how many more inches of water must be added for the tub to be filled?

    4.5 inches

LEVEL C

CHAPTER 1 **Chapter Test**
Form C

Evaluate each expression for $x = 2.5$, $y = 12$, and $z = 4$.

1. $2y - 2x$

   19

2. $z(4 + x) + 8$

   34

3. $5xyz$

   600

Write an algebraic expression for each word phrase.

4. 5 less than the product of 8 and k

   $8k - 5$

5. twice the sum of k and 8

   $2(k + 8)$

6. half the sum of 12 and t

   $\frac{12 + t}{2}$ or $\frac{1}{2}(12 + t)$

Solve and check.

7. $2.8 + t = 9.4$

   $t = 6.6$

8. $18 = d - 5$

   $d = 23$

Write an equation, then solve.

9. The world's largest cherry pie weighs 37,740 pounds. This is 7625 pounds heavier than the largest apple pie. How heavy is the largest apple pie?

   $c + 7625 = 37,740;$
   $c = 30,115$ lb

10. The world's largest lollipop weighs 2220 pounds, which is 10,126 pounds lighter than the largest popsicle. What is the weight of the largest popsicle?

    $p - 10,126 = 2,220;$
    $p = 12,346$ lb

Solve and check.

11. $\frac{n}{6} - 4 = 8$

    $n = 72$

12. $13x + 14 = 40$

    $x = 2$

Write an equation, then solve.

13. The population of Nevada is close to 1.6 million, which is about $\frac{1}{5}$ of Michigan's population. What is the estimated population of Michigan?

    $\frac{1}{5}m = 1.6; m = 8$ million people

14. There are about 1.2 million people in the U.S. who speak Chinese. This is about 3 times the number of people who speak Greek. How many people speak Greek?

    $3c = 1.2; c = 0.4$ million or 400,000 people

Solve and graph.

15. $3.2 + x \ge 9$ **$x \ge 5.8$**

    5.5 5.6 5.7 5.8 5.9 6.0 6.1

16. $8 < x - 3$ **$x > 11$**

    8 9 10 11 12 13 14 15

CHAPTER 1 **Chapter Test**
Form C, continued

Solve and graph.

17. $3x + 8 \ge 41$ **$x \ge 11$**

    8 9 10 11 12 13 14 15

18. $6 < \frac{a}{4}$ **$a > 24$**

    20 21 22 23 24 25 26 27

Simplify.

19. $6(3h + 9) - 4h$ **$14h + 54$**

20. $6(5y - 7) + 8y$ **$38y - 42$**

21. $4(2y - 3) + 5y$ **$13y - 12$**

Solve.

22. $12g + 13g = 450$ **$g = 18$**

23. $45 = 3y + 6y$ **$y = 5$**

24. $42 = 9k - 2k$ **$k = 6$**

Determine whether the ordered pair is a solution of the given equation.

25. $y = 17x - 3$; (2, 31)

    (2, 31) is a solution

26. $y = 4x - 12$; (8, 4)

    (8, 4) is not a solution

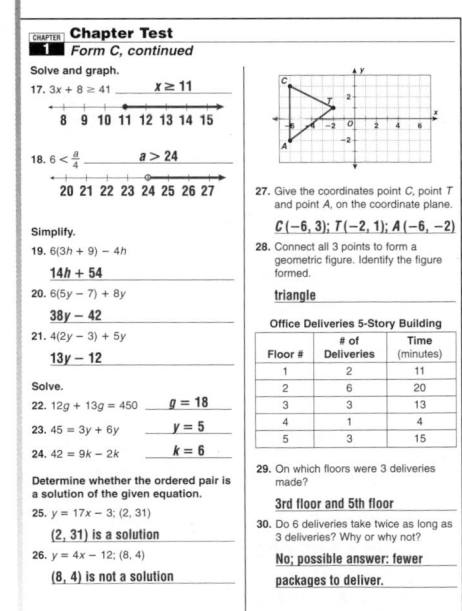

27. Give the coordinates point C, point T and point A, on the coordinate plane.

    $C(-6, 3); T(-2, 1); A(-6, -2)$

28. Connect all 3 points to form a geometric figure. Identify the figure formed.

    triangle

Office Deliveries 5-Story Building

Floor #	# of Deliveries	Time (minutes)
1	2	11
2	6	20
3	3	13
4	1	4
5	3	15

29. On which floors were 3 deliveries made?

    3rd floor and 5th floor

30. Do 6 deliveries take twice as long as 3 deliveries? Why or why not?

    No; possible answer: fewer packages to deliver.

Test and Practice Generator
CD-ROM

Create and customize multiple versions of the same tests with corresponding answers for any chosen chapter objectives.

Chapter 1 State and Standardized Test Preparation

Test Taking Skill Builder and Standardized Test Practice
are provided for each chapter in the *Test Prep Tool Kit.*

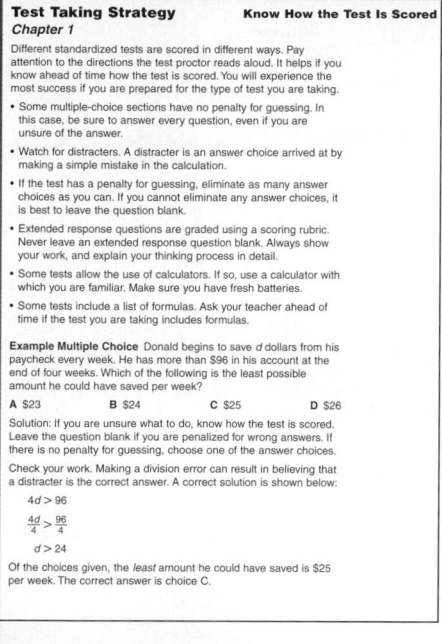

TEST TAKING SKILL BUILDER

Test Taking Strategy Know How the Test Is Scored
Chapter 1

Different standardized tests are scored in different ways. Pay
attention to the directions the test proctor reads aloud. It helps if you
know ahead of time how the test is scored. You will experience the
most success if you are prepared for the type of test you are taking.

- Some multiple-choice sections have no penalty for guessing. In
  this case, be sure to answer every question, even if you are
  unsure of the answer.
- Watch for distracters. A distracter is an answer choice arrived at by
  making a simple mistake in the calculation.
- If the test has a penalty for guessing, eliminate as many answer
  choices as you can. If you cannot eliminate any answer choices, it
  is best to leave the question blank.
- Extended response questions are graded using a scoring rubric.
  Never leave an extended response question blank. Always show
  your work, and explain your thinking process in detail.
- Some tests allow the use of calculators. If so, use a calculator with
  which you are familiar. Make sure you have fresh batteries.
- Some tests include a list of formulas. Ask your teacher ahead of
  time if the test you are taking includes formulas.

Example Multiple Choice Donald begins to save d dollars from his
paycheck every week. He has more than $96 in his account at the
end of four weeks. Which of the following is the least possible
amount he could have saved per week?

A $23   B $24   C $25   D $26

Solution: If you are unsure what to do, know how the test is scored.
Leave the question blank if you are penalized for wrong answers. If
there is no penalty for guessing, choose one of the answer choices.

Check your work. Making a division error can result in believing that
a distracter is the correct answer. A correct solution is shown below:

$$4d > 96$$
$$\frac{4d}{4} > \frac{96}{4}$$
$$d > 24$$

Of the choices given, the *least* amount he could have saved is $25
per week. The correct answer is choice C.

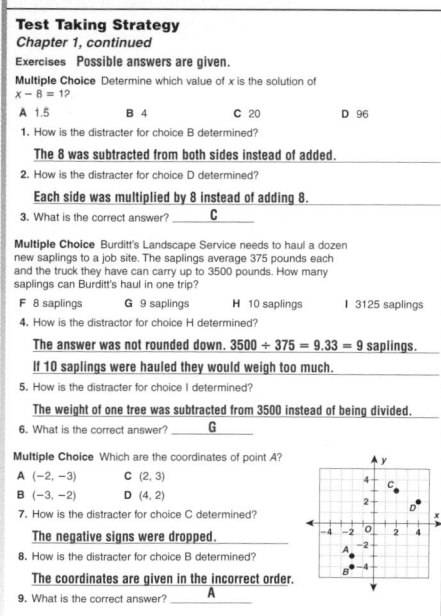

Test Taking Strategy
Chapter 1, continued
Exercises Possible answers are given.

Multiple Choice Determine which value of x is the solution of
$x - 8 = 1$?

A 1.5   B 4   C 20   D 96

1. How is the distracter for choice B determined?

   <u>The 8 was subtracted from both sides instead of added.</u>

2. How is the distracter for choice D determined?

   <u>Each side was multiplied by 8 instead of adding 8.</u>

3. What is the correct answer? ___<u>C</u>___

Multiple Choice Burditt's Landscape Service needs to haul a dozen
new saplings to a job site. The saplings average 375 pounds each
and the truck they have can carry up to 3500 pounds. How many
saplings can Burditt's haul in one trip?

F 8 saplings   G 9 saplings   H 10 saplings   I 3125 saplings

4. How is the distracter for choice H determined?

   <u>The answer was not rounded down. 3500 ÷ 375 = 9.33 = 9 saplings.</u>

   <u>If 10 saplings were hauled they would weigh too much.</u>

5. How is the distracter for choice I determined?

   <u>The weight of one tree was subtracted from 3500 instead of being divided.</u>

6. What is the correct answer? ___<u>G</u>___

Multiple Choice Which are the coordinates of point A?

A (−2, −3)   C (2, 3)
B (−3, −2)   D (4, 2)

7. How is the distracter for choice C determined?

   <u>The negative signs were dropped.</u>

8. How is the distracter for choice B determined?

   <u>The coordinates are given in the incorrect order.</u>

9. What is the correct answer? ___<u>A</u>___

STANDARDIZED TEST PRACTICE

Standardized Test Practice
Chapter 1
Select the best answer for Questions 1–8.

1. Simplify. $4(2x + 5) - 3x$

   A $11x + 4$
   B $5x + 5$
   C $5x + 20$
   D $11x + 20$

2. Which algebraic equation represents
   the word sentence "3 more than the
   product of 5 and a number y"?

   F $3(5) + y$
   G $3 + 5 + y$
   H $3y + 5$
   I $3 + 5y$

3. If c is the price of a small CD player,
   the expression $c + 0.07c$ can be
   used to find the total cost of the CD
   player including 7% sales tax. What
   would be the total cost of a CD
   player that is on sale for $45?

   A $48.60
   B $49.51
   C $48.15
   D $48.51

4. Don wants a $300 bicycle and he
   has saved $\frac{1}{4}$ of the money. Which
   equation could be used to determine
   the amount he has saved?

   F $\frac{1}{4}x = 300$
   G $(\frac{1}{4})(300) = x$
   H $\frac{1}{4} + x = 300$
   I $\frac{1}{4} = 300 - x$

5. The table shows how many bricks
   3 people moved during a 4-day period.
   Which person corresponds to moving
   more bricks on Day 3 than on Day 1?

Day	1	2	3	4
Person 1	250	130	350	0
Person 2	280	185	100	0
Person 3	280	85	200	290

   A Person 1
   B Person 2
   C Person 3
   D No one

6. Solve for m. $m + 3 = 22$

   F $m = 19$
   G $m = 25$
   H $m = 66$
   I $m = 128$

7. The home football team was losing
   35 to 10 at half time. Which inequality
   could be used to show how many
   points the home team needs to score
   to win the game? Let p represent the
   number of points needed.

   A $35 + 10 > p$
   B $10 + p > 35$
   C $p - 35 < 10$
   D $10 + p < 35$

8. Which inequality is represented by
   this graph?

   F $n \le 4$ H $n \ge 4$
   G $n > 4$ I $n < 4$

Standardized Test Practice
Chapter 1, continued
Gridded Response
Solve the problems. Use the answer
sheet to write and grid in your answer.

9. Solve for t to determine the total time
   it takes Sal to back up his computer.
   $4t = 16$.

10. Marian is buying a new car and will
    pay a total of $12,600 on her loan.
    Her monthly payment will be $360.
    For how many months did she take
    out the loan?

Short Response
Use the coordinate grid for Exercises 11
and 12.

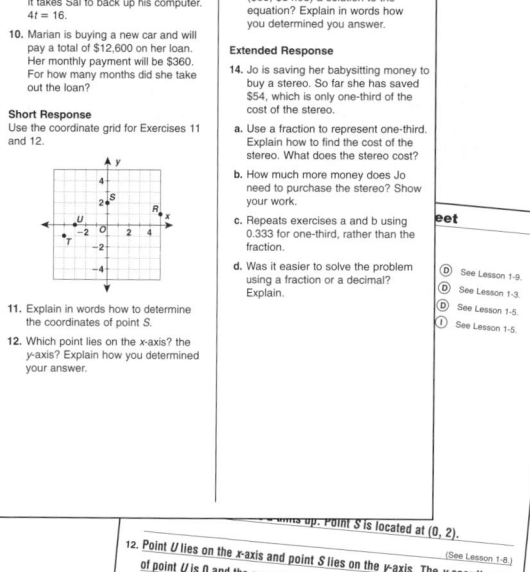

11. Explain in words how to determine
    the coordinates of point S.

12. Which point lies on the x-axis? the
    y-axis? Explain how you determined
    your answer.

13. Franklin is buying a coat for $33. The
    equation for the cost is $c = 1.06p$,
    where p is the price of the coat
    before tax. Is the ordered pair
    ($33, $34.98) a solution to the
    equation? Explain in words how
    you determined your answer.

Extended Response

14. Jo is saving her babysitting money to
    buy a stereo. So far she has saved
    $54, which is only one-third of the
    cost of the stereo.

    a. Use a fraction to represent one-third.
       Explain how to find the cost of the
       stereo. What does the stereo cost?

    b. How much more money does Jo
       need to purchase the stereo? Show
       your work.

    c. Repeats exercises a and b using
       0.333 for one-third, rather than the
       fraction.

    d. Was it easier to solve the problem
       using a fraction or a decimal?
       Explain.

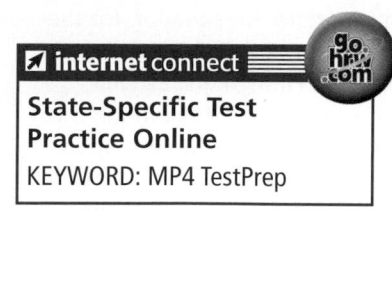

**State-Specific Test
Practice Online**
KEYWORD: MP4 TestPrep

Test Prep Tool Kit

- Standardized Test Prep Workbook
- Countdown to Testing transparencies
- State Test Prep CD-ROM
- Standardized Test Prep Video

Customized answer sheets give
students realistic practice for
actual standardized tests.

(answer sheet excerpt)

D	See Lesson 1-9.
D	See Lesson 1-3.
D	See Lesson 1-5.
I	See Lesson 1-5.

...ints up. Point S is located at (0, 2).

12. <u>Point U lies on the x-axis and point S lies on the y-axis. The x-coordinate</u> (See Lesson 1-8.)
    <u>of point U is 0 and the x-coordinate of point S is 0.</u>

13. <u>Yes, the ordered pair is a solution. I substituted the values ($33, $34.98)</u> (See Lesson 1-8.)
    <u>into the equation and checked to see if it was a true statement.</u>

    $c = 1.06p$
    $34.98 = 1.06(33)$
    $34.98 = 34.98$

Extended Response (See Lesson 1-7.)
Write your answers for Problem 14 on the back of this paper.
See Lesson 1-4.

Algebra Toolbox

Why Learn This?

Tell students that fire fighting is an example of a field in which different variables combine to create unique situations. A fire in which plastic is burning creates a different situation for the firefighter than one in which a refrigerant is burning. Firefighters, as well as those who make their equipment, must take into account many factors when determining the best method of extinguishing the fire and protecting the people around the fire.

Using Data

To begin the study of this chapter, have students:

- Identify whether hydrogen chloride is toxic at a concentration of 40 ppm. No, it is toxic at 50 ppm or more.

- Identify which two gases have the same danger level. Hydrogen chloride and hydrogen cyanide have the same danger level.

- Compare the danger level of HCl to that of $COCl_2$. The danger level of HCl is 25 times the danger level of $COCl_2$.

Algebra Toolbox

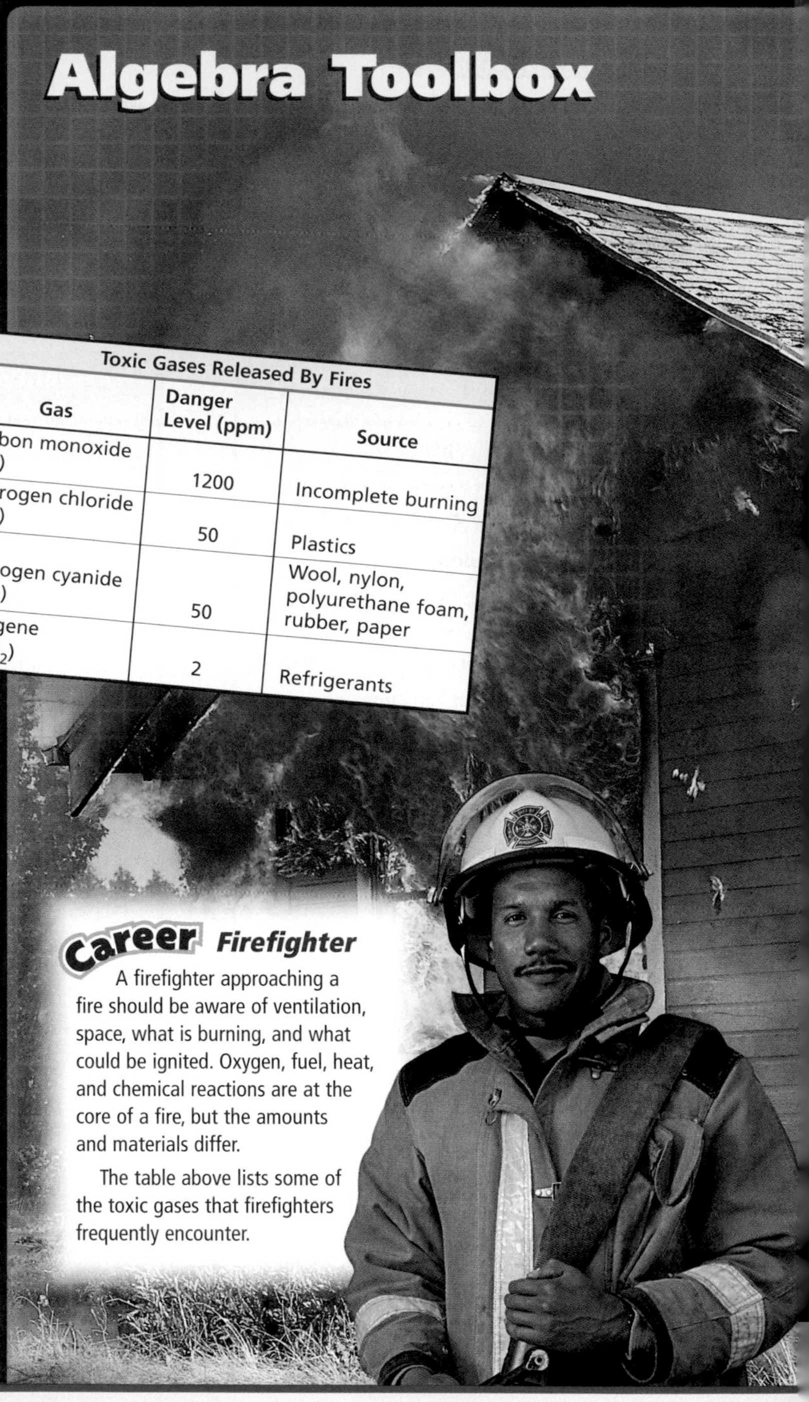

Toxic Gases Released By Fires		
Gas	Danger Level (ppm)	Source
Carbon monoxide (CO)	1200	Incomplete burning
Hydrogen chloride (HCl)	50	Plastics
Hydrogen cyanide (HCN)	50	Wool, nylon, polyurethane foam, rubber, paper
Phosgene ($COCl_2$)	2	Refrigerants

internet connect

Chapter Opener Online
go.hrw.com
KEYWORD: MP4 Ch1

Career *Firefighter*

A firefighter approaching a fire should be aware of ventilation, space, what is burning, and what could be ignited. Oxygen, fuel, heat, and chemical reactions are at the core of a fire, but the amounts and materials differ.

The table above lists some of the toxic gases that firefighters frequently encounter.

Problem Solving Project

Physical Science Connection

Purpose: To create algebraic expressions and graphs to solve problems

Materials: Firefighter worksheet

Understand, Plan, Solve, and Look Back

Have students:

✔ Complete the Firefighter worksheet to discover the relationships among toxic gases.

✔ Determine an algebraic expression that relates phosgene, carbon monoxide, hydrogen chloride, and hydrogen cyanide.

✔ Create a graph comparing the danger levels of the gases.

✔ Research the effects of toxic gases on the human body.

✔ Check students' work.

internet connect

Chapter Project Online: *go.hrw.com*
KEYWORD: MP4 PSProject1

ARE YOU READY?

Choose the best term from the list to complete each sentence.

1. __?__ is the __?__ of addition. subtraction; opposite operation
2. The expressions 3 · 4 and 4 · 3 are equal by the __?__. Commutative Property
3. The expressions 1 + (2 + 3) and (1 + 2) + 3 are equal by the __?__. Associative Property
4. Multiplication and __?__ are opposite operations. division
5. __?__ and __?__ are commutative. addition; multiplication

- addition
- Associative Property
- Commutative Property
- division
- opposite operation
- multiplication
- subtraction

Complete these exercises to review skills you will need for this chapter.

✔ Whole Number Operations

Simplify each expression.

6. 8 + 116 + 43   7. 2431 − 187   8. 204 · 38   9. 6447 ÷ 21
   167   2244   7752   307

✔ Compare and Order Whole Numbers

Order each sequence of numbers from least to greatest.

10. 1050; 11,500; 105; 150   11. 503; 53; 5300; 5030   12. 44,400; 40,040; 40,400; 44,040
   105; 150; 1050; 11,500   53; 503; 5030; 5300   40,040; 40,400; 44,040; 44,400

✔ Inverse Operations

Rewrite each expression using the inverse operation.

13. 72 + 18 = 90   14. 12 · 9 = 108   15. 100 − 34 = 66   16. 56 ÷ 8 = 7
   90 − 18 = 72   108 ÷ 9 = 12   66 + 34 = 100   7 · 8 = 56

✔ Order of Operations

Simplify each expression.

17. 2 + 3 · 4   14   18. 50 − 2 · 5   40   19. 6 · 3 · 3 − 3   51   20. (5 + 2)(5 − 2)   21
21. 5 − 6 ÷ 2   2   22. 16 ÷ 4 + 2 · 3   23. (8 − 3)(8 + 3)   24. 12 ÷ 3 ÷ 2 + 5   7
                           10   55

✔ Evaluate Expressions

Determine whether the given expressions are equal.

25. (4 · 7) · 2 and 4 · (7 · 2)   yes   26. (2 · 4) ÷ 2 and 2 · (4 ÷ 2)   yes   27. 2 · (3 − 3) · 2 and (2 · 3) − 3   no   28. 5 · (50 − 44) and 5 · 50 − 44   no
29. 9 − (4 · 2) and (9 − 4) · 2   no   30. 2 · 3 + 2 · 4 and 2 · (3 + 4)   yes   31. (16 ÷ 4) + 4 and 16 ÷ (4 + 4)   no   32. 5 + (2 · 3) and (5 + 2) · 3   no

---

Assessing Prior Knowledge

INTERVENTION

Diagnose and Prescribe

Evaluate your students' performance on this page to determine whether intervention is necessary or whether enrichment is appropriate. Options that provide instruction, practice, and a check are listed below.

Resources for Are You Ready?

- *Are You Ready? Intervention and Enrichment*
- **Recording Sheet for Are You Ready?** *Chapter 1 Resource Book* . .pp. 3–4
- *Are You Ready? Intervention* **CD-ROM**

ARE YOU READY?

Were students successful with Are You Ready?

NO INTERVENE ← → **YES** ENRICH

✔Whole Number
Operations 6–9
Intervention Practice, Skill 34
 CD-ROM
Intervention Activities, Skill 34

✔Compare and Order Whole
Numbers 10–12
Intervention Practice, Skill 5
CD-ROM
Intervention Activities, Skill 5

✔Inverse Operations 13–16
Intervention Practice, Skill 57
CD-ROM
Intervention Activities, Skill 57

✔Order of Operations . . . 17–24
Intervention Practice, Skill 51
CD-ROM
Intervention Activities, Skill 51

✔Evaluate Expressions . . 25–32
Intervention Practice, Skill 54
CD-ROM
Intervention Activities, Skill 54

Are You Ready? Enrichment, pp. 405–406

Equations and Inequalities

One-Minute Section Planner

Lesson	Materials	Resources
Lesson 1-1 Variables and Expressions **NCTM:** Number and Operations, Algebra, Communication, Connections **NAEP:** Algebra 3b ☑ SAT-9 ☑ SAT-10 ☑ ITBS ☑ CTBS ☑ MAT ☑ CAT	**Optional** Recording Sheet for Reaching All Learners *(CRB, p. 96)*	• *Chapter 1 Resource Book,* pp. 7–15 • Daily Transparency T1, CRB • Additional Examples Transparencies T2–T4, CRB • *Alternate Openers: Explorations,* p. 1
Lesson 1-2 Write Algebraic Expressions **NCTM:** Number and Operations, Algebra, Communication **NAEP:** Algebra 3a ☑ SAT-9 ☑ SAT-10 ☑ ITBS ☑ CTBS ☑ MAT ☑ CAT	**Optional** Teaching Transparency T6 *(CRB)* Flash cards for Reaching All Learners *(CRB, pp. 97–100)*	• *Chapter 1 Resource Book,* pp. 16–24 • Daily Transparency T5, CRB • Additional Examples Transparencies T7–T9, CRB • *Alternate Openers: Explorations,* p. 2
Lesson 1-3 Solving Equations by Adding or Subtracting **NCTM:** Number and Operations, Communication, Connections **NAEP:** Algebra 4a ☑ SAT-9 ☑ SAT-10 ☑ ITBS ☑ CTBS ☑ MAT ☑ CAT	**Optional** Balance scale Teaching Transparency T11 *(CRB)* Recording Sheet for Reaching All Learners *(CRB, pp. 101–102)*	• *Chapter 1 Resource Book,* pp. 25–33 • Daily Transparency T10, CRB • Additional Examples Transparencies T12–T14, CRB • *Alternate Openers: Explorations,* p. 3
Lesson 1-4 Solving Equations by Multiplying or Dividing **NCTM:** Number and Operations, Communication, Connections **NAEP:** Algebra 4a ☑ SAT-9 ☑ SAT-10 ☑ ITBS ☑ CTBS ☑ MAT ☑ CAT	**Optional** Egg carton Teaching Transparency T16 *(CRB)* Media advertisements	• *Chapter 1 Resource Book,* pp. 34–42 • Daily Transparency T15, CRB • Additional Examples Transparencies T17–T19, CRB • *Alternate Openers: Explorations,* p. 4
Lesson 1-5 Solving Simple Inequalities **NCTM:** Number and Operations, Problem Solving, Communication **NAEP:** Algebra 3b ☑ SAT-9 ☑ SAT-10 ☑ ITBS ☐ CTBS ☑ MAT ☑ CAT	**Optional** Teaching Transparency T21 *(CRB)* Index cards	• *Chapter 1 Resource Book,* pp. 43–51 • Daily Transparency T20, CRB • Additional Examples Transparencies T22–T25, CRB • *Alternate Openers: Explorations,* p. 5
Lesson 1-6 Combining Like Terms **NCTM:** Number and Operations, Algebra, Communication, Connections **NAEP:** Algebra 4a ☑ SAT-9 ☑ SAT-10 ☑ ITBS ☑ CTBS ☑ MAT ☑ CAT	**Optional** Cutout shapes for Reaching All Learners *(CRB, pp. 103–104)*	• *Chapter 1 Resource Book,* pp. 52–60 • Daily Transparency T26, CRB • Additional Examples Transparencies T27–T28, CRB • *Alternate Openers: Explorations,* p. 6
Section 1A Assessment		• Mid-Chapter Quiz, SE p. 32 • Section 1A Quiz, AR p. 5 • *Test and Practice Generator* CD-ROM

SAT = *Stanford Achievement Tests*   **ITBS** = *Iowa Test of Basic Skills*   **CTBS** = *Comprehensive Test of Basic Skills/Terra Nova*
MAT = *Metropolitan Achievement Test*   **CAT** = *California Achievement Test*
NCTM = Complete standards can be found on pages T29–T35.   **NAEP** = Complete standards can be found on pages A54–A58.
SE = *Student Edition*   **TE** = *Teacher's Edition*   **AR** = *Assessment Resources*   **CRB** = *Chapter Resource Book*   **MK** = *Manipulatives Kit*

Section Overview

Writing and Evaluating Algebraic Expressions

Lessons 1-1, 1-2

Why? Formulas are written using algebraic expressions that show relationships between quantities. When you evaluate these algebraic expressions, you get a value for the formula.

> The **perimeter** of a rectangle is the **sum** of **twice the length** and **twice the width**
> $P = 2l + 2w$

Find the perimeter of a rectangle with length 3 inches and width 4 inches.

$P = 2l + 2w$
$\quad = 2(3) + 2(4)$
$\quad = 6 + 8$
$\quad = 14$

The perimeter is 14 inches.

> To **evaluate** an algebraic expression, **substitute** a given number for the variable, and find the value of the resulting numerical expression.

Solving Equations and Inequalities

Lessons 1-3, 1-4, 1-5

Why? Many real-world problems may be solved using equations and inequalities.

To solve an equation or inequality, *undo* whatever operation is indicated on the variable.

$$25 = r - 15$$
$$\underline{+\ 15 = \ +\ 15}$$
$$40 = r$$

> In the equation, subtracting 15 is indicated. To undo subtracting 15, add 15 to both sides of the equation.

$$4x > 24$$
$$\frac{4x}{4} > \frac{24}{4}$$
$$x > 6$$

> In the inequality, multiplying by 4 is indicated. To undo multiplying by 4, divide both sides of the inequality by 4.

Combining Like Terms

Lesson 1-6

Why? Combining like terms helps to simplify algebraic expressions.

> **Like terms** can be grouped together because they have the same variable raised to the same power.

Simplify: $7(a + b) - 4a + 6 - 5b + 8$
$$7a + 7b - 4a + 6 - 5b + 8$$
$$7a + 7b - 4a + 6 - 5b + 8$$
$$3a + 2b + 14$$

> Combine Coefficients:
> $7 - 4 = 3$
> $7 - 5 = 2$
> $6 + 8 = 14$

> Identify like terms.

> The Distributive Property states that $a(b + c) = ab + ac$ for all a, b, and c.

Warm Up

Evaluate.

1. $21 - 2(3)$ **15**
2. $4 + 3 \cdot 9$ **31**
3. $2(9) + (3)$ **21**
4. $6(1.4) + 12$ **20.4**
5. $7(2.9) - 5$ **15.3**

Problem of the Day

Miss Smith obtained the prices below from the landscape company. She plans to buy five birch, two elm, one dogwood, and two oak trees. The landscape company will charge her $15 to plant each tree. How much will it cost? **$337.90**

Trees	Price	Trees	Price
Maple	$22.99	Dogwood	$23.99
Elm	$16.99	Crab apple	$26.99
Oak	$19.99	Birch	$17.99

Available on Daily Transparency in CRB

1-1 Variables and Expressions

Learn to evaluate algebraic expressions.

Vocabulary
variable
coefficient
algebraic expression
constant
evaluate
substitute

The nautilus is a sea creature whose shell has a series of chambers. Every lunar month (about 30 days), the nautilus creates and moves into a new chamber of the shell.

Let n be the number of chambers in the shell. You can approximate the age, in days, of the nautilus using the following expression:

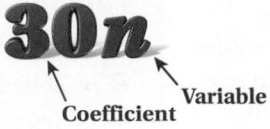

Coefficient   Variable

This nautilus shell has about 34 chambers. Using this information, you can determine its approximate age.

A **variable** is a letter that represents a value that can change or vary. The **coefficient** is the number multiplied by the variable. An **algebraic expression** has one or more variables.

In the algebraic expression $x + 6$, 6 is a **constant** because it does not change. To **evaluate** an algebraic expression, **substitute** a given number for the variable, and find the value of the resulting numerical expression.

EXAMPLE 1   Evaluating Algebraic Expressions with One Variable

Evaluate each expression for the given value of the variable.

A $x + 6$ for $x = 13$

$13 + 6$ *Substitute 13 for x.*
19 *Add.*

> **Remember!**
>
> Order of Operations
> PEMDAS:
> 1. Parentheses
> 2. Exponents
> 3. Multiply and Divide from left to right.
> 4. Add and Subtract from left to right.

B $2a + 3$ for $a = 4$

$2(4) + 3$ *Substitute 4 for a.*
$8 + 3$ *Multiply.*
11 *Add.*

C $3(5 + n) - 1$ for $n = 0, 1, 2$

n	Substitute	Parentheses	Multiply	Subtract
0	$3(5 + 0) - 1$	$3(5) - 1$	$15 - 1$	**14**
1	$3(5 + 1) - 1$	$3(6) - 1$	$18 - 1$	**17**
2	$3(5 + 2) - 1$	$3(7) - 1$	$21 - 1$	**20**

1 Introduce

Alternate Opener

EXPLORATION

1-1 Variables and Expressions

Catherine's dance team is planning a spring trip to the coast. Catherine is saving money in a bank account to pay for the trip. Her parents started her account with $100. She sells Christmas plants and adds $2.50 to her account for each plant she sells.

How much will be in her account if she sells 50 plants?

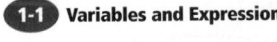
(initial amount)   (Price × number of plants)

$100 + 2.5n$

Evaluate the expression $100 + 2.5n$ by substituting 50 for n.

$100 + 2.5(50)$
$100 + 125$
225

There will be $225.00 in Catherine's account if she sells 50 plants.

Evaluate the expression $100 + 2.5n$ for each value of n.

1. $n = 10$ $100 + 2.5(10)$
2. $n = 25$ $100 + 2.5(\ )$
3. $n = 75$ $100 + 2.5(\ )$

Think and Discuss

4. **Explain** why n is called the *variable* in the expression $100 + 2.5n$.
5. **Describe** how to evaluate an expression.

Motivate

Ask students for the meanings of the words *variable* and *constant* in a context such as the following: "The air temperature in the desert was quite *variable* yesterday; it was cold overnight and warm during the day. The temperature at the equator was *constant* for 24 hours." Explain that the words have the same meaning in mathematics. A constant is a value that does not change, such as the number 5, and a variable is a symbol for a quantity that is not fixed, such as x.

Exploration worksheet and answers on Chapter 1 Resource Book pp. 8 and 106

2 Teach

Lesson Presentation

Guided Instruction

In this lesson, students learn to evaluate algebraic expressions. Explain that to evaluate an algebraic expression, students must replace variables with given numbers and then evaluate the resulting expression by using the order of operations.

> **Teaching Tip**   Remind students that when there is a coefficient in front of a variable, multiplication is indicated. Therefore, when they replace the variable with a value, they need to insert parentheses. For example, the expression $3x + 4$ should be written $3(5) + 4$ if x is replaced with 5.

EXAMPLE 2 **Evaluating Algebraic Expressions with Two Variables**

Evaluate each expression for the given values of the variables.

A $2x + 3y$ for $x = 15$ and $y = 12$

$2(15) + 3(12)$	*Substitute 15 for x and 12 for y.*
$30 + 36$	*Multiply.*
66	*Add.*

B $1.5p - 2q$ for $p = 18$ and $q = 7.5$

$1.5(18) - 2(7.5)$	*Substitute 18 for p and 7.5 for q.*
$27 - 15$	*Multiply.*
12	*Subtract.*

EXAMPLE 3 *Physical Science Application*

If c is a temperature in degrees Celsius, then $1.8c + 32$ can be used to find the temperature in degrees Fahrenheit. Convert each temperature from degrees Celsius to degrees Fahrenheit.

A freezing point of water: 0°C

$1.8c + 32$	
$1.8(0) + 32$	*Substitute 0 for c.*
$0 + 32$	*Multiply.*
32	*Add.*
$0°C = 32°F$	

Water freezes at 32°F.

B world's highest recorded temperature (El Azizia, Libya): 58°C

$1.8c + 32$	
$1.8(58) + 32$	*Substitute 58 for c.*
$104.4 + 32$	*Multiply.*
136.4	*Add.*
$58°C = 136.4°F$	

The highest recorded temperature in the world is 136.4°F.

Think and Discuss

1. **Give an example** of an expression that is algebraic and of an expression that is not algebraic.

2. **Tell** the steps for evaluating an algebraic expression for a given value.

3. **Explain** why you cannot find a numerical value for the expression $4x - 5y$ for $x = 3$.

Example 1

Evaluate each expression for the given value of the variable.

A. $x - 5$ for $x = 12$ **7**

B. $2y + 1$ for $y = 4$ **9**

C. $6(n + 2) - 4$ for $n = 5, 6, 7$
$6(5 + 2) - 4$ **38**
$6(6 + 2) - 4$ **44**
$6(7 + 2) - 4$ **50**

Example 2

Evaluate each expression for the given values of the variables.

A. $4x + 3y$ for $x = 2$ and $y = 1$ **11**

B. $9r - 2p$ for $r = 3$ and $p = 5$ **17**

Example 3

Use the expression $1.8c + 32$ to convert each boiling point temperature from degrees Celsius to degrees Fahrenheit.

A. Boiling point of water at sea level: 100°C   **212°F**

B. Boiling point of water at an altitude of 4400 meters: 85°C   **185°F**

---

3 **Close**

Reaching All Learners
Through Grouping Strategies

Give each pair of students a recording sheet (Chapter 1 Resource Book p. 96) with columns of exercises like the one below.

	A	B
1. $x = 3$	$3x - 1$	$x + 5$

Have one student in each pair evaluate the expression in column A and the other evaluate the expression in column B. Have them compare the values to determine whether the expressions have the same value for the given value of the variable.

Summarize

Ask the students to decide which of the following is a constant and which is a variable: your age (variable); the year in which you were born (constant).

Have the students suggest additional examples of variables and constants.

Ask the students if an algebraic expression contains one or more variables (yes). Ask the students if they would expect to see any variables in an expression after evaluating that expression. Why or why not?

Possible answer: No; because to evaluate an algebraic expression, you must replace the variables with numbers.

Answers to Think and Discuss

Possible answers:

1. algebraic: $x + 1$; not algebraic: $25 + 9$

2. Substitute the given value for the variable, and then find the value of the resulting numerical expression.

3. The y-value is not given.

1-1 Exercises

FOR EXTRA PRACTICE
see page 732

internet connect
Homework Help Online
go.hrw.com Keyword: MP4 1-1

Students may want to refer back to the lesson examples.

Assignment Guide

If you finished Example **1** assign:
Core 1–3, 10–12, 19–29 odd, 57–62
Enriched 19–30, 57–62

If you finished Example **2** assign:
Core 1–5, 10–14, 19–30, 31–49 odd, 57–62
Enriched 19–50, 56–62

If you finished Example **3** assign:
Core 1–18, 19–30, 31–49 odd, 51–53, 57–62
Enriched 18–62

Answers

47. 24

48. 70

GUIDED PRACTICE

See Example **1** Evaluate each expression for the given value of the variable.

1. $x + 5$ for $x = 12$ **17**   **2.** $3a + 5$ for $a = 6$ **23**   **3.** $2(4 + n) - 5$ for $n = 0$ **3**

See Example **2** Evaluate each expression for the given values of the variables.

4. $3x + 2y$ for $x = 8$ and $y = 10$ **44**   **5.** $1.2p - 2q$ for $p = 3.5$ and $q = 1.2$ **1.8**

See Example **3** You can make cornstarch slime by mixing $\frac{1}{2}$ as many tablespoons of water as cornstarch. How many tablespoons of water do you need for each number of tablespoons of cornstarch?

6. 10 tbsp **5 tbsp**   **7.** 16 tbsp **8 tbsp**   **8.** 23 tbsp **11.5 tbsp**   **9.** 34 tbsp **17 tbsp**

INDEPENDENT PRACTICE

See Example **1** Evaluate each expression for the given value of the variable.

10. $x + 7$ for $x = 23$ **30**   **11.** $5t + 3$ for $t = 6$ **33**   **12.** $6(2 + k) - 5$ for $k = 0$ **7**

See Example **2** Evaluate each expression for the given values of the variables.

13. $5x + 4y$ for $x = 7$ and $y = 8$ **67**   **14.** $4m - 2n$ for $m = 25$ and $n = 2.5$ **95**

See Example **3** If q is the number of quarts, then $\frac{1}{4}q$ can be used to find the number of gallons. Find the number of gallons for each of the following.

15. 16 quarts **4 gal**   **16.** 24 quarts **6 gal**   **17.** 8 quarts **2 gal**   **18.** 32 quarts **8 gal**

PRACTICE AND PROBLEM SOLVING

Evaluate each expression for the given value of the variable.

19. $12d$ for $d = 0$ **0**   **20.** $x + 3.2$ for $x = 5$ **8.2**   **21.** $30 - n$ for $n = 8$ **22**

22. $5t + 5$ for $t = 1$ **10**   **23.** $2a - 5$ for $a = 7$ **9**   **24.** $3 + 5b$ for $b = 1.2$ **9**

25. $12 - 2m$ for $m = 3$ **6**   **26.** $3g + 8$ for $g = 14$ **50**   **27.** $x + 7.5$ for $x = 2.5$ **10**

28. $15 - 5y$ for $y = 3$ **0**   **29.** $4y + 2$ for $y = 3.5$ **16**   **30.** $2(z + 8)$ for $z = 5$ **26**

Evaluate each expression for $t = 0$, $x = 1.5$, $y = 6$, and $z = 23$.

31. $y + 5$ **11**   **32.** $2y + 7$ **19**   **33.** $z - 2x$ **20**   **34.** $3z - 3y$ **51**

35. $2z - 2y$ **34**   **36.** xy **9**   **37.** $2.6y - 2x$ **12.6**   **38.** $1.2z - y$ **21.6**

39. $4(y - x)$ **18**   **40.** $3(4 + y)$ **30**   **41.** $4(2 + z) + 5$ **105**   **42.** $2(y - 6) + 3$ **3**

43. $3(6 + t) - 1$ **17**   **44.** $y(4 + t) - 5$ **19**   **45.** $x + y + z$ **30.5**   **46.** $10x + z - y$ **32**

47. $3y + 4(x + t)$ **48.** $3(z - 2t) + 1$ **49.** $7tyz$ **0**   **50.** $z - 2xy$ **5**

Math Background

In Lessons 1-8 and 11-1 through 11-4, students will learn to graph linear equations. An equation of a line that is not vertical is often written in the form $y = mx + b$. In this context, the letters x and y are variables, but the letters m and b are constants. For example, if m is the constant 2 and b is the constant 3, then the equation is $y = 2x + 3$. The graph of this equation is the set of all points (x, y) that make the equation true. The letters x and y are variables because their values vary.

RETEACH 1-1

LESSON 1-1 Reteach
Variables and Expressions

An *algebraic expression* uses at least one letter, or *variable*, which represents a value that can change.
A number that multiplies a variable is its *coefficient*.
A *constant* is a specific number, whose value does not change.

Algebraic Expression
$$4x + 7$$
Coefficient — Variable — Constant

To *evaluate* an algebraic expression, *substitute* a given number for a variable, and find the value of the resulting numerical expression.

Follow the order of operations:
1. Parentheses
2. Multiply or Divide
3. Add or Subtract

Evaluate $5(m + 1) + 8n$ for $m = 10$ and $n = 2$.
$5(m + 1) + 8n$
$5(10 + 1) + 8(2)$ Substitute 10 for m and 2 for n.
$5(11) + 8(2)$ Parentheses, simplify inside.
$55 + 16$ Multiply, from left to right.
71 Add.

Complete to evaluate each expression.

1. $9 + 7z$ for $z = 3$
$9 + 7 \cdot$ __3__
$9 +$ __21__
__30__

2. $5(q - 8)$ for $q = 17$
$5 \cdot ($ __17__ $- 8)$
$5 \cdot ($ __9__ $)$
__45__

3. $25 - 2x$ for $x = 8$
$25 - 2 \cdot ($ __8__ $)$
$25 -$ __16__
__9__

4. $2(x + 6) + 4$ for $x = 9$
$2($ __9__ $+ 6) + 4$
$2($ __15__ $) + 4$
__30__ $+ 4$
__34__

5. $42 - 3(x + 1)$ for $x = 3$
$42 - 3 \cdot ($ __3__ $+ 1)$
$42 - 3 \cdot ($ __4__ $)$
$42 -$ __12__
__30__

6. $22 + 5(2z)$ for $z = 4$
$22 + 5 \cdot (2 \cdot$ __4__ $)$
$22 + 5 \cdot ($ __8__ $)$
$22 +$ __40__
__62__

PRACTICE 1-1

LESSON 1-1 Practice B
Variables and Expressions

Evaluate each expression for the given value of the variable.
1. $6x + 2$ for $x = 3$ __20__
2. $18 - a$ for $a = 13$ __5__
3. $\frac{1}{4}y$ for $y = 16$ __4__
4. $9 - 2b$ for $b = 3$ __3__
5. $44 - 12n$ for $n = 3$ __8__
6. $7.2 + 8k$ for $k = 2$ __23.2__
7. $20(b - 15)$ for $b = 19$ __80__
8. $n(18 - 5)$ for $n = 4$ __52__

Evaluate each expression for the given values of the variables.
9. $2x + y$ for $x = 7$ and $y = 11$ __25__
10. $4j - k$ for $j = 4$ and $k = 10$ __6__
11. $9a - 6b$ for $a = 6$ and $b = 2$ __42__
12. $5s + 5t$ for $s = 15$ and $t = 12$ __135__
13. $7(n - m)$ for $m = 4$ and $n = 15$ __77__
14. $w(14 - y)$ for $w = 8$ and $y = 5$ __72__

A lemonade mix calls for $\frac{1}{4}$ cup of mix for one quart of lemonade. How much mix is needed to make each amount of lemonade?
15. 2 quarts $\frac{1}{2}$ cup
16. 8 quarts   2 cups
17. 12 quarts   3 cups
18. 18 quarts $4\frac{1}{2}$ cups

19. For each ride, Bill's Taxi Company charges $2 plus $0.35 for each minute.
 a. Write an expression for the cost of a taxi ride of x minutes.   __$2 + 0.35x$__
 b. How much will it cost for a 12-min taxi ride?   __$6.20__

51. LIFE SCIENCE Measuring your heart rate is one way to check the intensity of exercise. Studies show that a person's maximum heart rate depends on his or her age. The expression $220 - a$ approximates a person's maximum heart rate in beats per minute, where a is the person's age. Find your maximum heart rate.

52. LIFE SCIENCE In the Karvonen Formula, a person's resting heart rate r, age a, and desired intensity I are used to find the number of beats per minute the person's heart rate should be during training.

training heart rate (THR) = $I(220 - a - r) + r$

What is the THR of a person who is 45 years old, and who has a resting heart rate of 85 and a desired intensity of 0.5?
130 beats per minute

53. ENTERTAINMENT There are 24 frames, or still shots, in one second of movie footage.

a. Write an expression to determine the number of frames in a movie.

b. Using the running time of *E.T. the Extra-Terrestrial*, determine how many frames are in the movie.

E.T. the Extra-Terrestrial (1982) has a running time of 115 minutes, or 6900 seconds.

 54. CHOOSE A STRATEGY A baseball league has 192 players and 12 teams, with an equal number of players on each team. If the number of teams were reduced by four but the total number of players remained the same, there would be _____ players per team. **D**

A four more   B eight fewer   C four fewer   D eight more

 55. WRITE ABOUT IT A student says that for any value of x the expression $5x + 1$ will always give the same result as $1 + 5x$. Is the student correct? Explain.

 56. CHALLENGE Can the expressions $2x$ and $x + 2$ ever have the same value? If so, what must the value of x be?
Yes; when $x = 2$, both expressions have a value of 4.

Spiral Review

Identify the odd number(s) in each list of numbers. (Previous course)   **101, 411, 117,**
57. 15, 18, 22, 34, 21, 62, 71, 100   **15, 21, 71**   **58.** 101, 114, 122, 411, 117, 121   **121**

59. 4, 6, 8, 16, 18, 20, 49, 81, 32   **49, 81**   **60.** 9, 15, 31, 47, 65, 93, 1, 3, 43
All are odd numbers.

61. TEST PREP Which is **not** a multiple of 21?   **62. TEST PREP** Which is a factor of 12?
(Previous course)   **C**   (Previous course)   **F**

A 21   C 7   F 4   H 8
B 42   D 105   G 24   J 36

Journal
Have students write at least three examples of variables (quantities that change value) and constants (numbers that stay the same) from their everyday lives.

Test Prep Doctor
For Exercises 61 and 62, remind students of how the words *factor* and *multiple* are related to one another. Give them an equation such as $3 \times 4 = 12$. Ask them to determine which numbers are the factors and which number is the multiple. If students chose choice **A** for Exercise 61, they may have forgotten that any number is a multiple of itself (by 1).

CHALLENGE 1-1

LESSON 1-1 Challenge
Etaulave: Evaluate Backwards

Expression	Value for Variable	Substitution	Value of Expression
$2x + 5$	1	$2(1) + 5$	7
$2x + 5$	2	$2(2) + 5$	9
$2x + 5$	3	$2(3) + 5$	11

In the table above you use the values of the variable to evaluate the given expression. What if you are given the values of the expression and the values of the variable? How can you work backward to determine the expression?

Expression	Value for Variable	Substitution	Value of Expression
	1		5
	2		8
	3		11

Complete the following statements.

1. As the values of the variable increase by 1, the values of the expression increase _____ **by 3**.

2. Because the values of the expression depend on the values of the variables, your answer to Question 1 tells you the **coefficient** _____ of the variable.

3. Using the coefficient and the variable x, you know that **3x** _____ is part of the expression.

4. After each value of the variable is multiplied by _____ **3**, you still need to _____ **add 2** to get the value of the expression.

Write the expression given the values of x and their corresponding values of the expression.

5. The values of the expression are 12, 14, 16, 18, 20 when $x = 10, 11, 12, 13$, and 14.
2x − 8

6. The values of the expression are 50, 45, 40, 35, 30 when $x = 2$, 3, 4, 5, and 6.
60 − 5x

PROBLEM SOLVING 1-1

LESSON 1-1 Problem Solving
Variables and Expressions

Write the correct answer.

1. If l is the length of a room and w is the width, then lw can be used to find the area of the room. Find the area of a room with $l = 15$ ft and $w = 10$ ft.
150 square feet

2. If l is the length of a room and w is the width, $2l + 2w$ can be used to find the perimeter of a room. Find the perimeter of a room with $l = 16$ ft and $w = 12$ ft.
56 feet

3. Jaime earns 20% commission on her sales. If s is her total sales, then $0.2s$ can be used to find the amount she earns in commission. Find her commission if her sales are $1200.
$240

4. If p is the regular hourly rate of pay, then $1.5p$ can be used to find the overtime rate of pay. Find the overtime rate of pay if the regular hourly rate of pay is $6.00 per hour.
$9.00 per hour

Choose the letter for the best answer.

5. A plumber charges a fee of $75 per service call plus $15 per hour. If h is the number of hours the plumber works, then $75 + 15h$ can be used to find the total charges. Find the total charges if the plumber works 2.5 hours.
A $37.50
B $112.50
C $225
D $1127.50

6. Tickets to the movies cost $4 for students and $6 for adults. If s is the number of students and a is the number of adults, $4s + 6a$ can be used to find the cost of the tickets. Find the cost of the tickets for 3 students and 2 adults.
F $15
G $17
H $24
J $26

7. If c is the number of cricket chirps in a minute, then the expression $\frac{c}{4} + 20$ can be used to estimate the temperature in degrees Fahrenheit. If there are 92 cricket chirps in a minute, find the temperature.
A 43 degrees
B 33 degrees
C 102 degrees
D 75 degrees

8. Flowers are sold in flats of 6 plants each. If f is the number of flats, then $6f$ can be used to find the number of flowers. Find the number of flowers in 18 flats.
F 3 flowers
G 108 flowers
H 24 flowers
J 12 flowers

Lesson Quiz

Evaluate each expression for the given values of the variables.

1. $6x + 9$ for $x = 3$ **27**
2. $x + 14$ for $x = 8$ **22**
3. $4x + 3y$ for $x = 2, y = 3$ **17**
4. $1.6x - 2.9y$ for $x = 19, y = 6$ **13**
5. If n is the amount of money in a savings account, then the expression $n + 0.03n$ can be used to find the amount in the account after it has earned interest for one year. Find the total in the account after one year if $500 is the initial amount.   **$515**

Available on Daily Transparency in CRB

Warm Up

**Evaluate each expression for the
given values of the variables.**

1. $9y - 13$ for $y = 4$ **23**

2. $6n + 2p$ for $n = 2$ and $p = 3$ **18**

3. $3x - y$ for $x = 1$ and $y = 2$ **1**

**Which operation symbol goes
with each word?**

4. Sum $+$ **5.** Product $\times$

6. Quotient $\div$ **7.** Difference $-$

Problem of the Day

Find a pair of numbers that fits the
description. Their product is 221
and their sum is 30.   **13, 17**

Available on Daily Transparency in CRB

It was impossible to know what sort of
mood the equation was in; it had a
variable expression.

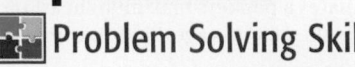

Write Algebraic Expressions

 Problem Solving Skill

Learn to write
algebraic expressions.

Each 30-second block of
commercial time during
Super Bowl XXXV cost an
average of $2.2 million.

This information can be
used to write an algebraic
expression to determine
how much a given
number of 30-second
blocks would have cost.

Eighty-three commercials aired during the 2002 Super Bowl.

	Word Phrases	Expression
+	• a number plus 5 • add 5 to a number • sum of a number and 5 • 5 more than a number • a number increased by 5	$n + 5$
−	• a number minus 11 • subtract 11 from a number • difference of a number and 11 • 11 less than a number • a number decreased by 11	$x - 11$
✖	• 3 times a number • 3 multiplied by a number • product of 3 and a number	$3m$
÷	• a number divided by 7 • 7 divided into a number • quotient of a number and 7	$\frac{a}{7}$ or $a \div 7$

EXAMPLE 1 **Translating Word Phrases into Math Expressions**

Write an algebraic expression for each word phrase.

A a number n decreased by 11

n decreased by 11

n $-$ 11

$n - 11$

---

1 Introduce

Alternate Opener

EXPLORATION

1-2 Write Algebraic Expressions

The table shows the prices for different activities at a park.

Activity	Price
Park ride	$1.25 per ride
Jet ski rental	$25.00 + $5.00 per hour
Scooter rental	$10.00 + $2.50 per hour
Bike rental	$5.00 + $2.00 per hour
Bay cruise	$25.00 per person

An algebraic
expression for this
cost is 1.25r.

Use a variable to write an expression for each cost.

	Activity	Cost	Variable	Expression
	Park ride	$1.25 per ride	r (rides)	1.25r
1.	Jet ski rental	$25.00 + $5.00 per hour	h (hours)	
2.	Scooter rental	$10.00 + $2.50 per hour	h (hours)	
3.	Bike rental	$5.00 + $2.00 per hour	h (hours)	
4.	Bay cruise	$25.00 per person	p (people)	

To find the cost of renting a jet ski for 5 hours, evaluate the
expression $25 + 5h$ by substituting 5 for h.

$25 + 5(5) = 25 + 25 = 50$

5. Use your expressions from above to evaluate the cost
of a 5-hour scooter rental and of a 5-hour bike rental.

Think and Discuss

6. Describe a real-world situation in which you might use the
expression $50 + 2x$.

7. Explain what you need to write an algebraic expression.

Motivate

Ask the students if any of them speak a
language other than English. Ask volunteers
to translate simple expressions, such as
"hello" or "how are you?" into another language. Explain that in this lesson, they will
learn how to "translate" words into algebraic
expressions.

***Exploration worksheet and answers on
Chapter 1 Resource Book pp. 17 and 108***

---

2 Teach

**Lesson
Presentation**

Guided Instruction

In this lesson, students learn to
write algebraic expressions. Review the table
of phrases and expressions with students
(available on Teaching Transparency T6 in
the Chapter 1 Resource Book). Point out that
there are several different words for each
operation.

While reviewing the word problems in
Examples 2 and 3, make sure students can
identify the key words that determine the
operation(s) that will be used to solve the
problem. You may want to spend extra time
to review the additional examples as well.

Write an algebraic expression for each word phrase.

B the quotient of 3 and a number *h*

quotient of   3 and *h*

3   ÷   *h*

$\frac{3}{h}$

Helpful Hint

In Example 1C parentheses are not needed because multiplication is performed first by the order of operations.

C 1 more than the product of 12 and *p*

1   more than   the   product of   12 and *p*

1   +   (12   ·   *p*)

1 + 12*p*

D 3 times the sum of *q* and 1

3   times   the   sum of   *q* and 1

3   ·   (*q*   +   1)

3(*q* + 1)

To solve a word problem, you must first interpret the action you need to perform and then choose the correct operation for that action. When a word problem involves groups of equal size, use multiplication or division. Otherwise, use addition or subtraction. The table gives more information to help you decide which operation to use to solve a word problem.

Action	Operation	Possible Question Clues
Combine	Add	How many altogether?
Combine equal groups	Multiply	How many altogether?
Separate	Subtract	How many more? How many less?
Separate into equal groups	Divide	How many equal groups?

EXAMPLE **Interpreting Which Operation to Use in Word Problems**

A Monica got a 200-minute calling card and called her brother at college. After talking with him for *t* minutes, she had *t* less than 200 minutes remaining on her card. Write an expression to determine the number of minutes remaining on the calling card.

200 − *t*   *Separate t minutes from the original 200.*

B If Monica talked with her brother for 55 minutes, how many minutes does she have left on her calling card?

200 − 55 = 145   *Evaluate the expression for t = 55.*

There are 145 minutes remaining on her calling card.

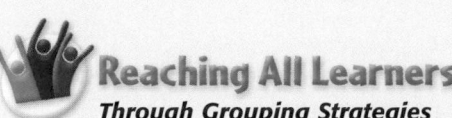

Reaching All Learners

Through Grouping Strategies

Give each pair of students a few flash cards with a word phrase on one side and the corresponding algebraic expression on the other (Chapter 1 Resource Book pp. 97–100). Have them place the cards, word phrase side up, between them. Students then take turns writing an algebraic expression for the word phrase. If the expression is correct, the student keeps the card. If not, the card is returned to the bottom of the deck. When the entire deck is gone, or when time is called, the student with the most cards wins.

Additional Examples

Example 1

Write an algebraic expression for each word phrase.

A. the product of 8 and a number *n*
8*n*

B. 9 less than a number *w*
w − 9

C. 3 increased by the difference of *p* and 5
3 + (*p* − 5)

D. 24 decreased by the product of 6 and *q*
24 − 6*q*

Example 2

A. Jared worked for *h* hours at the pay rate of $5 each hour. Write an expression to determine how much money Jared earned.
5*h*

B. How much money will he earn if he works a total of 18 hours?
$90

Example 3

Write an algebraic expression to evaluate each word problem.

A. Amed bought a new sweater for $27 plus sales tax *t*. If the tax was $1.76, what was the total cost of the sweater?   27 + *t*; $28.76

B. The cost to rent a banquet hall is $240. If the cost will be shared equally among all of the people who attend the event, how much will it cost each person if 12, 15, 16, or 20 people attend?

Number of People *n*	240 ÷ *n*	Cost per Person
12	240 ÷ 12	$20
15	240 ÷ 15	$16
16	240 ÷ 16	$15
20	240 ÷ 20	$12

C. An airplane was flying at an altitude of 20,000 feet when it began its descent at 9:00 P.M. After 10 minutes, it had descended *a* feet. If the plane descended 8500 feet during the ten minutes, what was its altitude at 9:10?

20,000 − *a*; 11,500 feet

EXAMPLE 3 Writing and Evaluating Expressions in Word Problems

Write an algebraic expression to evaluate each word problem.

A Rob and his friends buy a set of baseball season tickets. The 81 tickets are to be divided equally among p people. If he divides them among 9 people, how many tickets does each person get?

$81 \div p$ *Separate the tickets into p equal groups.*

$81 \div 9 = 9$ *Evaluate for p = 9.*

Each person gets 9 tickets.

Helpful Hint

Some word problems give more numbers than are necessary to find the answer. In Example 3B, 30 seconds describes the length of a commercial, and the number is not needed to solve the problem.

B A company airs its 30-second commercial n times during Super Bowl XXXV at a cost of $2.2 million each time. What will the cost be if the commercial is aired 2, 3, 4, and 5 times?

$2.2 \text{ million} \cdot n$ *Combine n equal amounts of $2.2 million.*

$2.2n$ *In millions of dollars*

n	$2.2n$	Cost
2	2.2(2)	$4.4 million
3	2.2(3)	$6.6 million
4	2.2(4)	$8.8 million
5	2.2(5)	$11 million

Evaluate for n = 2, 3, 4, and 5.

C Before Benny took his road trip, his car odometer read 14,917 miles. After the trip, his odometer read m miles more than 14,917. If he traveled 633 miles on the trip, what did the odometer read after his trip?

$14,917 + m$ *Combine 14,917 miles and m miles.*

$14,917 + 633 = 15,550$ *Evaluate for m = 633.*

The odometer read 15,550 miles after the trip.

Think and Discuss

1. Give two words or phrases that can be used to express each operation: addition, subtraction, multiplication, and division.

2. Express $5 + 7n$ in words in at least two different ways.

Math Background

The Egyptians had different symbols for addition and subtraction than those that are used today. The symbol for addition was two feet walking from right to left, which was the direction in which the Egyptians wrote. The symbol for subtraction was a pair of legs walking in the opposite direction, left to right.

3 Close

Summarize

Write the four operational symbols ($+$, $-$, $\times$, $\div$) on the board, and ask students to think of as many words as they can to represent each one. Remind students that "translating" will help them solve many types of word problems.

Possible answers: addition: sum, added, more than, total, increased, plus; subtraction: difference, less than, minus, decrease, take away; multiplication: times, product, multiplied, each; division: divided, split, quotient, separated

Answers to Think and Discuss

Possible answers:

1. addition: increased by, more than; subtraction: decreased by, less than; multiplication: product, times; division: quotient, divide

2. 5 plus the product of 7 and a number n; the product of 7 and a number n, increased by 5

FOR EXTRA PRACTICE
see page 732

internet connect
Homework Help Online
go.hrw.com Keyword: MP4 1-2

go.hrw.com

1-2 PRACTICE & ASSESS

GUIDED PRACTICE

See Example 1 Write an algebraic expression for each word phrase.

1. the quotient of 6 and a number t $6 \div t$
2. a number y decreased by 25 $y - 25$
3. 7 times the sum of m and 6 $7(m + 6)$
4. the sum of 7 times m and 6 $7m + 6$

See Example 2
5. a. Carl walked n miles for charity at a rate of \$8 per mile. Write an expression to find out how much money Carl raised. $8n$
   b. How much money would Carl have raised if he had walked 23 miles? $8(23) = \$184$

See Example 3 Write an algebraic expression to evaluate the word problem.

6. Cheryl and her friends buy a pizza for \$15.00 plus a delivery charge of d dollars. If the delivery charge is \$2.50, what is the total cost? $15 + d;\ \$17.50$

INDEPENDENT PRACTICE

See Example 1 Write an algebraic expression for each word phrase.

7. a number k increased by 34 $k + 34$
8. the quotient of 12 and a number h $\frac{12}{h}$ or $12 \div h$
9. 5 plus the product of 5 and z $5 + 5z$
10. 6 times the difference of x and 4 $6(x - 4)$

See Example 2
11. a. Mr. Gimble's class is going to a play. The 42 students will be seated equally among p rows. Write an expression to determine how many people will be seated in each row. $42 \div p$
    b. If there are 6 rows, how many students will be in each row?   **7 students**

See Example 3 Write an algebraic expression and evaluate each word problem.

12. Julie bought a card good for 35 visits to a health club and began a workout routine. After y visits, she had y fewer than 35 visits remaining on her card. After 18 visits, how many visits did she have left? $35 - y;\ $ **17 visits**

13. Myron bought n dozen eggs for \$1.75 per dozen. If he bought 8 dozen eggs, how much did they cost? $1.75n;\ \$14$

PRACTICE AND PROBLEM SOLVING

Write an algebraic expression for each word phrase.

14. 7 more than a number y $y + 7$
15. 6 times the sum of 4 and y $6(4 + y)$
16. 11 less than a number t $t - 11$
17. half the sum of m and 5 $\frac{1}{2}(m + 5)$
18. 9 more than the product of 6 and a number y $6y + 9$
19. 6 less than the product of 13 and a number y $13y - 6$
20. 2 less than a number m divided by 8 $\frac{m}{8} - 2$
21. twice the quotient of a number m and 35 $2\left(\frac{m}{35}\right)$

Students may want to refer back to the lesson examples.

Assignment Guide

If you finished Example **1** assign:
  Core 1–4, 7–10, 15–21 odd, 31–40
  Enriched 7, 9, 14–24, 31–40

If you finished Example **2** assign:
  Core 1–5, 7–11, 15–21 odd, 22–24, 31–40
  Enriched 5, 11, 14–24, 28–40

If you finished Example **3** assign:
  Core 1–13, 15–27 odd, 31–40
  Enriched 7–13 odd, 14–40

RETEACH 1-2

LESSON 1-2 Reteach
Writing Algebraic Expressions

What words tell you to add, subtract, multiply, or divide?

Add $n + 6$	Subtract $n - 6$	Multiply $6n$	Divide $n \div 6$
n plus 6	n minus 6	n times 6	n divided by 6
the sum of n and 6	the difference between n and 6	the product of n and 6	the quotient of n and 6
6 added to n	6 subtracted from n	6 multiplied by n	6 equal shares of n
n increased by 6	6 less than n		
	n decreased by 6		

Give the algebraic expression that represents the word expression.

1. the quotient of n and 3 $n \div 3$
2. 3 more than n $n + 3$
3. 3 multiplied by n $3n$
4. n decreased by 3 $n - 3$
5. 3 less than n $n - 3$
6. the product of n and 3 $3n$
7. n increased by 3 $n + 3$
8. one-third times n $\frac{1}{3}n$ or $n \div 3$

9. Michelle had \$142 in her savings account. Write an expression to determine the number of dollars left in her account after she withdrew d dollars. $142 - d$

10. The Mortons are donating \$1000 to buy gifts for the children spending Thanksgiving in a hospital ward. Write an expression to determine the amount to be spent on each of c children if they are to share equally in the donated money. $1000 \div c$

11. The circumference of the earth is about 25,000 miles. Write an expression to determine the number of miles traveled after an airplane pilot travels x times around the earth. $25,000x$

PRACTICE 1-2

LESSON 1-2 Practice B
Writing Algebraic Expressions

Write an expression for each word phrase.

1. a number n divided by 5 $\frac{n}{5}$
2. the sum of 4 and a number y $4 + y$
3. a number c decreased by 11 $c - 11$
4. the product of a number n and 9 $9n$
5. 6 less than a number x $x - 6$
6. the quotient of 21 and a number b $\frac{21}{b}$
7. 3 times the sum of a number b and 5 $3(b + 5)$
8. 10 times the difference of a number d and 3 $10(d - 3)$
9. the sum of 11 times a number s and 3 $11s + 3$
10. 7 minus the product of 2 and a number x $7 - 2x$

11. a. Madeline worked x hours at a rate of \$7 per hour. Write an expression to find out how much money Madeline earned. $7x$
    b. How much money would Madeline have earned if she worked 18 hours?   \$126

12. a. Jonah wants to divide his baseball card collection of c cards equally among 6 friends. Write an expression to determine the number of cards he will give each friend. $\frac{c}{6}$
    b. If Jonah has 144 cards, how many will he give each friend?   24 cards

Write an algebraic expression and use it to solve each word problem.

13. Mike bought 45 stamps at the post office. After he used x stamps, he had x less than 45 stamps remaining. How many stamps did he have left after using 33 stamps? $45 - x;\ $ 12 stamps

14. At the deli Jackie bought y pounds of ham salad for \$4.19 per pound. If she buys 3 pounds, how much will it cost? $4.19y;\ \$12.57$

Answers

22. Possible answer: 3 less than the product of 4 and a number *b*

23. Possible answer: the sum of 12 and a number *t*

24. Possible answer: 3 times the sum of a number *m* and 4

28. Possible answer: The student did not apply the Distributive Property correctly: $3(n - 5) = 3n - 15$.

29. Possible answer: Multiplication is repeated addition, so either operation can be used to solve the problem. Adding the cost 5 times or multiplying the cost by 5 will give the correct answer.

30. $1 + 2n$; yes; twice an odd number is always an even number, and adding 1 to an even number always results in an odd number.

Journal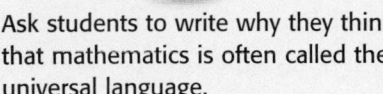

Ask students to write why they think that mathematics is often called the universal language.

Test Prep Doctor

Remind students that when they are answering questions, they must be careful to read every word. In Exercise 39 for example, if they skip the word *not,* they may know the mathematical concept but still answer the question incorrectly.

Lesson Quiz

Write an algebraic expression for each word phrase.

1. 5 less than a number *k* $k - 5$

2. a number *x* divided by 11 $\frac{x}{11}$

3. 4 times the sum of *n* and 5 $4(n + 5)$

Write an algebraic expression to evaluate the word problem.

4. Karen buys *n* raffle tickets for $0.50 each. If she buys 13 of them, how much will they cost? $0.50n$; $6.50

Available on Daily Transparency in CRB

Translate each algebraic expression into words.

22. $4b - 3$ **23.** $t + 12$ **24.** $3(m + 4)$

25. ENTERTAINMENT Ron bought two comic books on sale. Each comic book was discounted $1 off the regular price *r*. Write an expression to find what Ron paid before taxes. If each comic book was regularly $2.50, what was the total cost before taxes? $2(r - 1); 2(2.50 - 1) = 3

26. SPORTS In basketball, players score 2 points for each field goal, 3 points for each three-point shot, and 1 point for each free throw made. Write an expression for the total score for a team that makes *g* field goals, *t* three-point shots, and *f* free throws. Find the total score for a team that scores 23 field goals, 6 three-pointers, and 11 free throws. $2g + 3t + f$; 75 points

27. At age 2, a cat or dog is considered 24 "human" years old. Each year after age 2 is equivalent to 4 "human" years. Fill in the expression [24 + ▢ (a − 2)] so that it represents the age of a cat or dog in human years. Copy the chart and use your expression to complete it. $24 + 4(a - 2)$

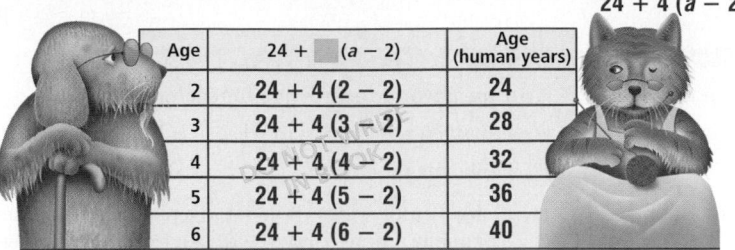

Age	24 + ▢ (a − 2)	Age (human years)
2	24 + 4 (2 − 2)	24
3	24 + 4 (3 − 2)	28
4	24 + 4 (4 − 2)	32
5	24 + 4 (5 − 2)	36
6	24 + 4 (6 − 2)	40

 28. WHAT'S THE ERROR? A student says $3(n - 5)$ is equal to $3n - 5$. What's the error?

 29. WRITE ABOUT IT Paul used addition to solve a word problem about the weekly cost of commuting by toll road for $1.50 each day. Fran solved the same problem by multiplying. They both had the correct answer. How is this possible?

 30. CHALLENGE Write an expression for the sum of 1 and twice a number *n*. If you let *n* be any odd number, will the result always be an odd number?

Spiral Review

Find each sum, difference, product, or quotient. (Previous course)

31. $200 + 2$ **202**   **32.** $200 \div 2$ **100**   **33.** $200 \cdot 2$ **400**   **34.** $200 - 2$ **198**

35. $200 + 0.2$ **200.2**   **36.** $200 \div 0.2$ **1000**   **37.** $200 \cdot 0.2$ **40**   **38.** $200 - 0.2$ **199.8**

39. TEST PREP Which is **not** a factor of 24?   **C**   **40. TEST PREP** Which is a multiple of 15?   **J**
(Previous course)   (Previous course)

A 24	C 48	F 1	H 3
B 8	D 12	G 5	J 15

CHALLENGE 1-2

LESSON 1-2 Challenge
Amazing Math

Write an algebraic expression for each word phrase on the board. Evaluate each expression for *x* = 2.

Then find a path from the top row to the bottom row that gives a total of 22.

3 times *x*	1 less than twice *x*	6 more than *x*	*x* increased by 3	the quotient of twice *x* and 2
$3x = 6$	$2x - 1 = 3$	$x + 6 = 8$	$x + 3 = 5$	$\frac{2x}{2} = 2$
1 more than *x*	the product of 3 and *x*	*x* decreased by 1	half of *x*	twice *x* increased by 3
$x + 1 = 3$	$3x = 6$	$x - 1 = 1$	$\frac{x}{2} = 1$	$2x + 3 = 7$
1 less than 3 times *x*	the difference between 3 and *x*	the difference between 2 and *x*	the product of 4 and *x*	the sum of 6 and twice *x*
$3x - 1 = 5$	$3 - x = 1$	$2 - x = 0$	$4x = 8$	$6 + 2x = 10$
twice *x*	the difference between *x* and 1	the sum of *x* and 5	1 more than half of *x*	the product of 4 and 3 times *x*
$2x = 4$	$x - 1 = 1$	$x + 5 = 7$	$\frac{x}{2} + 1 = 2$	$4(3x) = 24$
x increased by 2	the quotient of *x* and 2	7 increased by *x*	the quotient of 6 and *x*	5 times *x* divided by 2
$x + 2 = 4$	$\frac{x}{2} = 1$	$7 + x = 9$	$\frac{6}{x} = 3$	$\frac{5x}{2} = 5$
22				

PROBLEM SOLVING 1-2

LESSON 1-2 Problem Solving
Writing Algebraic Expressions

Write the correct answer.

1. Morton bought 15 new books to add to his collection of books *b*. Write an algebraic expression that represents the total number of books in Morton's collection now.

$b + 15$

2. Paul exercises *m* minutes per day 5 days a week. Write an algebraic expression that represents how many minutes Paul exercises in two weeks.

$2(5m)$

3. Helen bought 3 shirts that each cost *s* dollars and 4 pairs of pants that each cost *p* dollars. Write an algebraic expression that represents the total cost of the shirts and pants, not including tax.

$3s + 4p$

4. Bernice made *c* cookies to divide evenly among four friends and herself. Write an algebraic expression that represents the number of cookies each person will receive.

$\frac{c}{5}$

Choose the letter for the best answer.

5. Jonas collects baseball cards. He has 245 cards in his collection. For his birthday, he received *r* more cards, then he gave his brother *g* cards. Which algebraic expression represents the total number of cards he now has in his collection?
   A $245 + r + g$
   B $245 - r - g$
   C $245 + r - g$
   D $r + g - 245$

6. Monique is saving money for a computer. She has *m* dollars saved. For her birthday, her dad doubled her money, but then she spent *s* dollars on a shirt. Which algebraic expression represents the amount of money she has now saved for her computer?
   F $m + 2 - s$
   G $2m - s$
   H $2m + s$
   J $m + 2s$

7. Which algebraic expression represents the number of years in *m* months?
   A $12m$
   B $\frac{m}{12}$
   C $12 + m$
   D $12 - m$

8. Which algebraic expression represents how many minutes are in *h* hours?
   F $60h$
   G $\frac{h}{60}$
   H $h + 60$
   J $h - 60$

1-3 Solving Equations by Adding or Subtracting

Learn to solve equations using addition and subtraction.

Vocabulary

equation

solve

solution

inverse operation

isolate the variable

Addition Property of Equality

Subtraction Property of Equality

Mexico City is built on top of a large underground water source. Over the 100 years between 1900 and 2000, as the water was drained, the city sank as much as 30 feet in some areas.

If you know the altitude of Mexico City in 2000 was 7350 feet above sea level, you can use an *equation* to estimate the altitude in 1900.

In 1910, the Monumento a la Independencia was built at ground level. It now requires 23 steps to reach the base because the ground around the monument has sunk.

An **equation** uses an equal sign to show that two expressions are equal. All of these are equations.

$$3 + 8 = 11 \qquad r + 6 = 14 \qquad 24 = x - 7 \qquad 9n = 27 \qquad \frac{100}{2} = 50$$

To **solve** an equation that contains a variable, find the value of the variable that makes the equation true. This value of the variable is called the **solution** of the equation.

EXAMPLE 1 **Determining Whether a Number Is a Solution of an Equation**

Determine which value of x is a solution of the equation.

$x - 4 = 16; x = 12, 20,$ or 21

Substitute each value for x in the equation.

$x - 4 = 16$
$12 - 4 \overset{?}{=} 16$ *Substitute 12 for x.*
$8 \overset{?}{=} 16$ ✗

So 12 **is not** a solution.

$x - 4 = 16$
$20 - 4 \overset{?}{=} 16$ *Substitute 20 for x.*
$16 \overset{?}{=} 16$ ✔

So 20 **is** a solution.

$x - 4 = 16$
$21 - 4 \overset{?}{=} 16$ *Substitute 21 for x.*
$17 \overset{?}{=} 16$ ✗

So 21 **is not** a solution.

1-3 Organizer

Pacing: Traditional 1 day
Block $\frac{1}{2}$ day

Objective: Students solve equations using addition and subtraction.

Warm Up

Write an algebraic expression for each word phrase.

1. a number x decreased by 9 $x - 9$
2. 5 times the sum of p and 6
       $5(p + 6)$
3. 2 plus the product of 8 and n
       $2 + 8n$
4. the quotient of 4 and a number c
       $\frac{4}{c}$

Problem of the Day

Janie's horse refused to do 5 jumps today and cleared 14 jumps. Yesterday, the horse cleared 9 more jumps than today. How many jumps did the horse clear in the two-day jumping event?   37

Available on Daily Transparency in CRB

Math Humor

After he put his algebra homework in a jar of hot water, the student explained, "I thought you told us to *dissolve* the equations!"

---

1 Introduce

Alternate Opener

EXPLORATION

1-3 Solving Equations by Adding or Subtracting

Evaluate the expressions for each given value of x.

1.

x	$x + 1$
0	
1	
2	
3	

2.

x	$x - 2$
3	
4	
5	
6	

Each expression below has been evaluated. Find the value of x.

3.

x	$x + 2$
	3
	4
	5
	6

4.

x	$x - 5$
	0
	1
	2
	3

Think and Discuss

5. **Explain** how you evaluated the expressions in numbers 1 and 2.
6. **Explain** how you found the values of x in numbers 3 and 4.

Exploration worksheet and answers on Chapter 1 Resource Book pp. 26 and 110

Motivate

Place a balance scale with equal weights on each side where all the students can see it. Ask the students what will happen if you add or subtract an equal amount of weight on both sides. Demonstrate each. Tell the students that equations are like balance scales; we must always keep the sides equal.

---

2 Teach

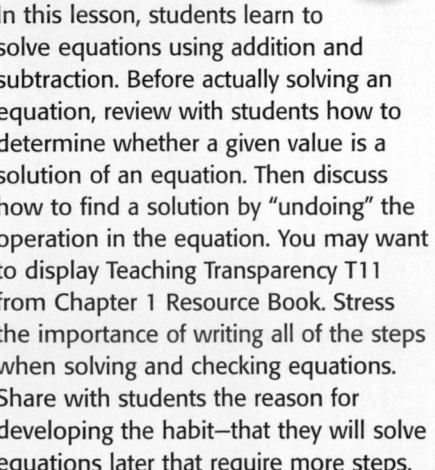

Lesson Presentation

Guided Instruction

In this lesson, students learn to solve equations using addition and subtraction. Before actually solving an equation, review with students how to determine whether a given value is a solution of an equation. Then discuss how to find a solution by "undoing" the operation in the equation. You may want to display Teaching Transparency T11 from Chapter 1 Resource Book. Stress the importance of writing all of the steps when solving and checking equations. Share with students the reason for developing the habit—that they will solve equations later that require more steps.

Addition and subtraction are **inverse operations**, which means they "undo" each other. To solve an equation, use inverse operations to **isolate the variable**. In other words, get the variable alone on one side of the equal sign.

To solve a subtraction equation, like $y - 15 = 7$, you would use the **Addition Property of Equality**.

ADDITION PROPERTY OF EQUALITY		
Words	**Numbers**	**Algebra**
You can add the same number to both sides of an equation, and the statement will still be true.	$\begin{array}{r} 2 + 3 = 5 \\ +4 \quad +4 \\ \hline 2 + 7 = 9 \end{array}$	$x = y$ $x + z = y + z$

There is a similar property for solving addition equations, like $x + 9 = 11$. It is called the **Subtraction Property of Equality**.

SUBTRACTION PROPERTY OF EQUALITY		
Words	**Numbers**	**Algebra**
You can subtract the same number from both sides of an equation, and the statement will still be true.	$\begin{array}{r} 4 + 7 = 11 \\ -3 \quad -3 \\ \hline 4 + 4 = 8 \end{array}$	$x = y$ $x - z = y - z$

EXAMPLE 2 **Solving Equations Using Addition and Subtraction Properties**

Solve.

Ⓐ $3 + t = 11$

$$\begin{array}{r} 3 + t = 11 \\ -3 \qquad -3 \\ \hline 0 + t = 8 \end{array}$$ *Subtract 3 from both sides.*

$t = 8$ *Identity Property of Zero: $0 + t = t$*

Check

$3 + t = 11$

$3 + 8 \overset{?}{=} 11$ *Substitute 8 for t.*

$11 \overset{?}{=} 11 ✔$

Teach

Reaching All Learners
Through Grouping Strategies

Have students work in pairs. Each student in the pair receives a different recording sheet with addition and subtraction equations (Chapter 1 Resource Book, pp. 101–102). Each student solves the first equation and then gives the paper to his or her partner to check the solution. If both solutions are correct, they exchange again to get their original papers and solve the next equations. If a solution is not correct, the two students work together to find the correct solution.

Solve.

B $m - 7 = 11$

$$m - 7 = 11$$
$$\underline{+7 \quad +7} \qquad \text{Add 7 to both sides.}$$
$$m + 0 = 18$$
$$m = 18$$

C $15 = w + 14$

$$15 = w + 14$$
$$15 - 14 = w + 14 - 14 \qquad \text{Subtract 14 from both sides.}$$
$$1 = w + 0$$
$$1 = w$$
$$w = 1 \qquad \text{Definition of Equality}$$

EXAMPLE 3 *Geography Applications*

A The altitude of Mexico City in 2000 was about 7350 ft above sea level. What was the approximate altitude of Mexico City in 1900 if it sank 30 ft during the 100-year period?

beginning altitude	−	altitude sank	=	altitude in 2000

Solve: $\qquad x \qquad - \qquad 30 \qquad = \qquad 7350$

$$x - 30 = 7350$$
$$\underline{+30 \quad +30} \qquad \text{Add 30 to both sides.}$$
$$x + 0 = 7380$$
$$x = 7380$$

In 1900, Mexico City was at an altitude of 7380 ft.

B From 1954 to 1999, shifting plates increased the height of Mount Everest from 29,028 ft to 29,035 ft. By how many feet did Mount Everest's altitude increase during the 45-year period?

Solve: $29{,}028 \text{ ft} + h = 29{,}035 \text{ ft}$

$$29{,}028 + h = 29{,}035$$
$$\underline{-29{,}028 \qquad -29{,}028} \qquad \text{Subtract 29,028 from both sides.}$$
$$0 + h = 7$$
$$h = 7$$

Mount Everest's altitude increased 7 ft between 1954 and 1999.

Geography LINK

Mexico City, above, sank 19 inches in one year while Venice, Italy, possibly the most famous sinking city, has sunk only 9 inches in the last century.

go.hrw.com
KEYWORD: MP4 Sinking

CNN Student News

Think and Discuss

1. **Explain** whether you would use addition or subtraction to solve $x - 9 = 25$.

2. **Explain** what it means to isolate the variable.

---

3 Close

Summarize

Show students a group of several addition equations. For example:

$$x + 4 = 9 \qquad 12 + y = 15$$
$$10 = n + 3 \qquad 16 = 4 + z$$

Point out the plus sign in each of the equations. Ask students what operation they would use to solve each of these equations (subtraction). Show the solutions and point out the subtraction step in each one. Repeat the procedure with a group of subtraction equations.

Answers to Think and Discuss

Possible answers:

1. Addition; the equation uses subtraction, and addition is the inverse operation of subtraction.

2. To isolate the variable means to get the variable by itself on one side of the equation.

Additional Examples

Example 1

Determine which value of x is a solution of the equation.

$x + 8 = 15$; $x = 5$, 7, or 23
7 is a solution.

Example 2

Solve.

A. $10 + n = 18$ $\qquad n = 8$

B. $p - 8 = 9$ $\qquad p = 17$

C. $22 = y - 11$ $\qquad y = 33$

Example 3

A. Jan took a 34-mile trip in her car, and the odometer showed 16,550 miles at the end of the trip. What was the original odometer reading?
The original odometer reading was 16,516 miles.

B. From 1980 to 2000, the population of a town increased from 895 residents to 1125 residents. What was the increase in population during that 20-year period?
The increase in population was 230.

Example 3 note: For word problems, encourage students to clearly label the variable before they begin working (e.g., "Let $x =$ the beginning altitude").

FOR EXTRA PRACTICE
see page 732

internet connect
Homework Help Online
go.hrw.com   Keyword: MP4 1-3

Students may want to refer back to the lesson examples.

Assignment Guide

If you finished Example **1** assign:
Core 1, 2, 7, 8, 13–19 odd, 43–46
Enriched 3–20, 43–46

If you finished Example **2** assign:
Core 1-5, 7-11, 13–35 odd, 43–46
Enriched 17–35, 41–46

If you finished Example **3** assign:
Core 1–12, 13–33 odd, 34–39, 43–46
Enriched 13–46

Notes

GUIDED PRACTICE

See Example **1** Determine which value of x is a solution of each equation.

1. $x + 9 = 14$; $x = 2, 5,$ or 23 **5**   **2.** $x - 7 = 14$; $x = 2, 7,$ or 21 **21**

See Example **2** Solve.

3. $m - 9 = 23$ $m = 32$ **4.** $8 + t = 13$ $t = 5$ **5.** $13 = w - 4$ $w = 17$

See Example **3** **6.** At what altitude did a climbing team start if it descended 3600 feet to a camp at an altitude of 12,035 feet? **15,635 feet**

INDEPENDENT PRACTICE

See Example **1** Determine which value of x is a solution of each equation.

7. $x - 14 = 8$; $x = 6, 22,$ or 32 **22**   **8.** $x + 7 = 35$; $x = 5, 28,$ or 42 **28**

See Example **2** Solve.

9. $9 = w + 8$ $w = 1$ **10.** $m - 11 = 33$ $m = 44$ **11.** $4 + t = 16$ $t = 12$

See Example **3** **12.** If a team camps at an altitude of 18,450 feet, how far must it ascend to reach the summit of Mount Everest at an altitude of 29,035 feet?
10,585 feet

PRACTICE AND PROBLEM SOLVING

Determine which value of the variable is a solution of the equation.

13. $d + 4 = 24$; $d = 6, 20,$ or 28 **20** **14.** $m - 2 = 13$; $m = 11, 15,$ or 16 **15**

15. $y - 7 = 23$; $y = 30, 26,$ or 16 **30** **16.** $k + 3 = 4$; $k = 1, 7,$ or 17 **1**

17. $12 + n = 19$; $n = 7, 26,$ or 31 **7** **18.** $z - 15 = 15$; $z = 0, 15,$ or 30 **30**

19. $x + 48 = 48$; $x = 0, 48,$ or 96 **0** **20.** $p - 2.5 = 6$; $p = 3.1, 3.5,$ or 8.5 **8.5**

Solve the equation and check the solution.

21. $7 + t = 12$ $t = 5$ **22.** $h - 21 = 52$ $h = 73$ **23.** $15 = m - 9$ $m = 24$

24. $m - 5 = 10$ $m = 15$ **25.** $h + 8 = 11$ $h = 3$ **26.** $6 + t = 14$ $t = 8$

27. $1785 = t - 836$ $t = 2621$ **28.** $m + 35 = 172$ $m = 137$ **29.** $x - 29 = 81$ $x = 110$

30. $p + 8 = 23$ $p = 15$ **31.** $n - 14 = 31$ $n = 45$ **32.** $20 = 8 + w$ $w = 12$

33. $0.8 + t = 1.3$ $t = 0.5$ **34.** $5.7 = c - 2.8$ $c = 8.5$ **35.** $9.87 = w + 7.97$ $w = 1.9$

36. $73,142 + n = 81,607$; 8465 people

36. *SOCIAL STUDIES* In 1990, the population of Cheyenne, Wyoming, was 73,142. By 2000, the population had increased to 81,607. Write and solve an equation to find n, the increase in Cheyenne's population from 1990 to 2000.

Math Background

An equation with a variable is sometimes called an open sentence. To find a solution to an equation is to find a number that makes an open sentence a true sentence. For example, $x + 5 = 7$ is an open sentence. If you substitute the solution 2 for x, you get the true sentence $2 + 5 = 7$.

Two properties used in solving equations in this lesson are the Addition and Subtraction Properties of Equality. After discussing negative numbers in Chapter 2, the Addition Property of Equality alone can serve the purpose of both properties. To solve the equation $x + 5 = 7$, you can either subtract 5 from both sides or add -5 to both sides.

RETEACH 1-3

LESSON 1-3 Reteach
Solving Equations by Adding or Subtracting

To solve an addition equation, use subtraction.	To solve a subtraction equation, use addition.
Solve $x + 5 = 12$.	Solve $9 = w - 3$.
$x + 5 = 12$ Subtract the number added to the variable. $\underline{-5 \quad -5}$ $x = 7$	$9 = w - 3$ Add the number subtracted from the variable. $\underline{+3 \quad +3}$ $12 = w$

Tell what you would add or subtract to solve the equation.

1. $a + 2 = 7$ **2.** $x - 2 = 4$ **3.** $6 + y = 9$ **4.** $34 = b + 13$
  subtract 2   add 2   subtract 6   subtract 13

5. $21 = z - 9$ **6.** $14 + r = 20$ **7.** $18 = d - 11$ **8.** $6 = 5 + p$
  add 9   subtract 14   add 11   subtract 5

Complete to solve the equation. In Exercises 9–11 check the solution.

9. $x - 2 = 15$ $\underline{+2 \quad +2}$ $x - \underline{0} = \underline{17}$ $x = \underline{17}$
Check: $x - 2 = 15$ $\underline{17} - 2 = 15$ $\underline{15} = 15$ ✔

10. $z + 6 = 14$ $\underline{-6 \quad -6}$ $z + \underline{0} = \underline{8}$ $z = \underline{8}$
Check: $z + 6 = 14$ $\underline{8} + 6 = 14$ $\underline{14} = 14$ ✔

11. $7 = 4 + n$ $\underline{-4 \quad -4}$ $\underline{3} = \underline{0} + n$ $\underline{3} = \underline{7}$ ✔
Check: $7 = 4 + n$ $7 = 4 + \underline{3}$ $7 = \underline{7}$ ✔

12. $t + 5 = 16$ $\underline{-5 \quad -5}$ $t + \underline{0} = \underline{11}$ $t = \underline{11}$

13. $a - 7 = 13$ $\underline{+7 \quad +7}$ $a - \underline{0} = \underline{20}$ $a = \underline{20}$

14. $28 = 9 + b$ $\underline{-9 \quad -9}$ $\underline{19} = \underline{0} + b$ $\underline{19} = b$

PRACTICE 1-3

LESSON 1-3 Practice B
Solving Equations by Adding or Subtracting

Determine which value is a solution of the equation.

1. $d - 6 = 12$; $d = 6, 8,$ or 18 **2.** $9 + b = 17$; $b = 6, 8,$ or 26
  $d = 18$ $b = 8$

3. $k - 12 = 26$; $k = 14, 38,$ or 40 **4.** $a + 18 = 59$; $a = 37, 41,$ or 77
  $k = 38$ $a = 41$

Solve.

5. $n - 8 = 11$ **6.** $9 + g = 13$ **7.** $m + 10 = 37$
  $n = 19$ $g = 4$ $m = 27$

8. $b - 12 = 42$ **9.** $16 + h = 23$ **10.** $11 + x = 11$
  $b = 54$ $h = 7$ $x = 0$

11. $a + 35 = 51$ **12.** $k - 22 = 0$ **13.** $52 = j + 45$
  $a = 16$ $k = 22$ $j = 7$

14. $7.5 + c = 10.6$ **15.** $y - 1.7 = 0.6$ **16.** $m - 2.25 = 4.50$
  $c = 3.1$ $y = 2.3$ $m = 6.75$

17. Two sisters, Jenny and Penny, play on the same basketball team. Last season they scored a combined total of 458 points. Jenny scored 192 of the points. Write and solve an equation to find the number of points Penny scored.
  $192 + p = 458$; $p = 266$

18. After his payment, Mr. Weber's credit card balance was $245.76. His payment was for $75.00. Write and solve an equation to find the amount of his credit card bill.
  $x - 75.00 = 245.76$; $x = 320.76$

37. SOCIAL STUDIES In 1804, explorers Lewis and Clark began their journey to the Pacific Ocean at the mouth of the Missouri River. Use the map to determine the following distances.

   a. from Blackbird Hill, Nebraska, to Great Falls, Montana

   b. from the meeting point, or confluence, of the Missouri and Yellowstone Rivers to Great Falls, Montana

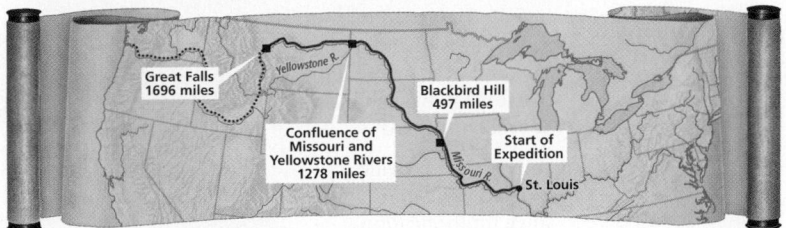

Great Falls 1696 miles
Yellowstone R.
Blackbird Hill 497 miles
Confluence of Missouri and Yellowstone Rivers 1278 miles
Start of Expedition
Missouri R.
St. Louis

38. SOCIAL STUDIES The United States flag had 15 stars in 1795. How many stars have been added since then to make our present-day flag with 50 stars? Write and solve an equation to find s, the number of stars that have been added to the United States flag since 1795. $s + 15 = 50$; $s = 35$ **stars**

39. ENTERTAINMENT Use the bar graph about movie admission costs to write and solve an equation for each of the following.

   a. Find c, the increase in cost of a movie ticket from 1940 to 1990.

   b. The cost c of a movie ticket in 1950 was \$3.82 less than in 1995. Find the cost of a movie ticket in 1995.

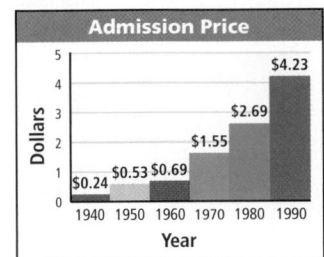

Admission Price

Dollars — Year
$0.24 (1940), $0.53 (1950), $0.69 (1960), $1.55 (1970), $2.69 (1980), $4.23 (1990)

 40. WRITE A PROBLEM Write a subtraction problem using the graph about admission costs. Explain your solution.

 41. WRITE ABOUT IT Write a set of rules to use when solving addition and subtraction equations.

 42. CHALLENGE Explain how you could solve for h in the equation $14 - h = 8$ using algebra. Then find the value of h.

Spiral Review

Evaluate each expression for the given value of the variable. (Lesson 1-1)

43. $x + 9$ for $x = 13$ **22**   **44.** $x - 8$ for $x = 18$ **10**   **45.** $14 + x$ for $x = 12$ **26**

46. TEST PREP Which is "3 times the difference of y and 4"? (Lesson 1-2)   **C**

   A $3 \cdot y - 4$ **B** $3 \cdot (y + 4)$ **C** $3 \cdot (y - 4)$ **D** $3 - (y - 4)$

Answers

37. a. $497 + m = 1696$; 1199 miles

    b. $1278 + m = 1696$; 418 miles

39. a. $0.24 + c = 4.23$; \$3.99

    b. $c - 3.82 = 0.53$; \$4.35

40. Possible answer: Find the difference between the cost of a movie ticket in 1980 and the cost in 1950; $2.69 - 0.53 = 2.16$; The difference is \$2.16.

41. Possible answer: If the equation has a plus sign in it, subtract the number after the sign from both sides. If the equation has a minus sign in it, add the number following the sign to both sides.

42. Add h to both sides and then subtract 8 from both sides; $h = 6$.

Journal

Have students explain why they think addition and subtraction equations are often used to find the amount of increase or decrease in a problem.

Test Prep Doctor

For Exercise 46, remind students that the word *times* indicates multiplication and *difference* indicates subtraction. The only choices that have both operations are **A** and **C**. Choice **C** is correct because the difference must be found first.

CHALLENGE 1-3

LESSON 1-3 Challenge
Equation Bingo

The card shown below is a Bingo card for the game Equation Bingo. The numbers called out will be solutions to the list of equations. The numbers will be called from solving the equations in alphabetical order.

Determine which column, row, or diagonal gives a Bingo first and at which lettered equation it occurs.

List of Equations

a. $x + 4 = 17$ 13   **b.** $y - 12 = 4$ 16
c. $6 + z = 10$ 4   **d.** $24 = r + 10$ 14
e. $15 = b - 9$ 24   **f.** $4 + s = 22$ 18
g. $28 = v + 11$ 17   **h.** $8 = 6 + k$ 2
i. $9 + d = 17$ 8   **j.** $g + 14 = 20$ 6
k. $13 = h + 10$ 3   **l.** $m - 9 = 3$ 12
m. $10 = 5 + w$ 5   **n.** $23 = e + 12$ 11
o. $18 = j - 9$ 27   **p.** $26 + t = 200$ 174

Bingo is made with letter j and is the diagonal from the upper right to the lower left.

Equation Bingo

5	16	27	24
17	13	6	3
11	4	12	8
2	174	14	18

PROBLEM SOLVING 1-3

LESSON 1-3 Problem Solving
Solving Equations by Adding or Subtracting

Write the correct answer.

1. Lisa sold her old bike for \$140 less than she paid for it. She sold the bike for \$65. Write an equation you would use to determine how much Lisa paid for her bike.

$b - 140 = 65$

2. Wesley had 479 stamps in his stamp collection. Now his collection has 563 stamps. Write an equation to determine how many stamps Wesley has added to his collection.

$479 + s = 563$

3. From December 1998 to August 2000, the number of Internet users ages 9–17 increased from 15,396 to 19,579. Write an equation and solve to determine the increase in the number of Internet users ages 9–17.

$15,396 + i = 19,579$; $i = 4183$

4. The number of forests in the National Forest System in 2000 was 155. This is an increase of 117 forests from the number in 1900. Write an equation and solve to determine the number of forests in the National Forest System in 1900.

$f + 117 = 155$; $f = 38$

Choose the letter for the best answer.

5. In the women's Olympic 50-meter freestyle, the 1988 winner swam in 25.49 seconds, while in 2000, the winner swam in 24.32 seconds. By how much did the winning time decrease?

  A 1.17 sec   **C** 2.38 sec
  B 4.81 sec   **D** 49.81 sec

6. In the men's Olympic discus throw, the 2000 winner threw the discus 131.7 feet farther than the 1896 winner. If the 2000 winner threw the discus 227.3 ft, how far did the 1896 winner throw?

  F 359 ft   **H** 228.6 ft
  G 183.4 ft   **J** 95.6 ft

7. Naples, FL, is the second fastest growing metropolitan area. In 1990, the population was 152,099. By 2000, the population was 251,377. How much did the population grow?

  A 403,476   **C** 152,099
  B 99,278   **D** 87,354

8. Female teens watch an average of 2.5 hours of TV per week less than male teens. If female teens watch 18.3 hours per week, how many hours a week do male teens watch TV?

  F 7.3 h   **H** 15.8 h
  G 18.3 h   **J** 20.8 h

Lesson Quiz

Determine which value of x is a solution of the equation.

1. $x + 9 = 17$; $x = 6, 8,$ or 26 **8**

2. $x - 3 = 18$; $x = 15, 18,$ or 21 **21**

Solve.

3. $a + 4 = 22$ $a = 18$

4. $n - 6 = 39$ $n = 45$

5. The price of your favorite cereal is now \$4.25. In prior weeks the price was \$3.69. Write and solve an equation to find n, the increase in the price of the cereal.

$3.69 + n = 4.25$; \$0.56

Available on Daily Transparency in CRB

Pacing: Traditional 1 day
Block $\frac{1}{2}$ day

Objective: Students solve equations using multiplication and division.

Warm Up

Solve.

1. $x - 3 = 9$ $x = 12$
2. $16 = n + 9$ $n = 7$
3. $5 + k = 6$ $k = 1$
4. $47 = t - 19$ $t = 66$
5. $15 = x + 2$ $x = 13$

Problem of the Day

Jackie went shopping for a new wardrobe. She bought seven new shirts and four new pairs of pants. She made sure that they can all be worn together in combinations. How many outfits can she put together? **28**

Available on Daily Transparency in CRB

Teacher: Evaluate $x + 4$ if $x = 7$.

Student: Is this a trick question? Yesterday you told us that $x = 2$!

Learn to solve equations using multiplication and division.

Vocabulary

Division Property of Equality

Multiplication Property of Equality

In 1912, Wilbur Scoville invented a way to measure the hotness of chili peppers. The unit of measurement became known as the Scoville unit.

You can use Scoville units to write and solve multiplication equations for substituting one kind of pepper for another in a recipe.

You can solve a multiplication equation using the **Division Property of Equality** .

A Thai pepper is 90,000 Scoville units. This means it takes 90,000 cups of sugar water to neutralize the hotness of one cup of Thai peppers.

DIVISION PROPERTY OF EQUALITY		
Words	**Numbers**	**Algebra**
You can divide both sides of an equation by the same nonzero number, and the statement will still be true.	$4 \cdot 3 = 12$ $\frac{4 \cdot 3}{2} = \frac{12}{2}$ $\frac{12}{2} = 6$	$x = y$ $\frac{x}{z} = \frac{y}{z}$

EXAMPLE 1 **Solving Equations Using Division**

Solve $7x = 35$.

$7x = 35$

$\frac{7x}{7} = \frac{35}{7}$ *Divide both sides by 7.*

$1x = 5$ $1 \cdot x = x$

$x = 5$

Check

$7x = 35$

$7(5) \overset{?}{=} 35$ *Substitute 5 for x.*

$35 \overset{?}{=} 35$ ✔

Remember!

Multiplication and division are inverse operations.

$\frac{8 \cdot 3}{3} = 8$

1 Introduce

Alternate Opener

 EXPLORATION

 **1-4** Solving Equations by Multiplying or Dividing

Evaluate the expressions for each given value of x.

1.	x	$2x$
	0	
	1	
	2	
	3	

2.	x	$\frac{x}{2}$
	0	
	2	
	4	
	6	

Each expression below has been evaluated. Find the value of x.

3.	x	$3x$
		3
		6
		9
		12

4.	x	$\frac{x}{4}$
		0
		1
		2
		3

Think and Discuss

5. **Explain** how you evaluated the expressions in numbers 1 and 2.

6. **Explain** how you found the values of x in numbers 3 and 4.

Motivate

Show the students an egg carton for a dozen eggs. Tell them, "I spent $1.20 for this carton of eggs. I wonder how much one egg cost." Write the equation $12c = \$1.20$. Say to the students, "If I solve this equation to find the value of c, I will know the cost of one egg."

Exploration worksheet and answers on Chapter 1 Resource Book pp. 35 and 112

2 Teach

Lesson Presentation

Guided Instruction

In this lesson, students learn to solve equations using multiplication and division. Explain that the principles in this lesson are very similar to those in Lesson 1-3. Point out that multiplication and division are also inverse operations, so one "undoes" the other. You may want to use Teaching Transparency T16 from Chapter 1 Resource Book to review the properties in the lesson.

To solve two-step equations, "undo" the addition or subtraction first, and then "undo" the multiplication or division. This is in the *reverse order* of the steps you would use to evaluate an expression.

You can solve division equations using the
Multiplication Property of Equality .

MULTIPLICATION PROPERTY OF EQUALITY		
Words	**Numbers**	**Algebra**
You can multiply both sides of an equation by the same number, and the statement will still be true.	$2 \cdot 3 = 6$ $4 \cdot 2 \cdot 3 = 4 \cdot 6$ $8 \cdot 3 = 24$	$x = y$ $zx = zy$

 EXAMPLE 2 Solving Equations Using Multiplication

Solve $\frac{h}{3} = 6$.

$\frac{h}{3} = 6$

$3 \cdot \frac{h}{3} = 3 \cdot 6$ *Multiply both sides by 3.*

$h = 18$

 EXAMPLE 3 *Food Application*

A recipe calls for 1 tabasco pepper, but Jennifer wants to use jalapeño peppers. How many jalapeño peppers should she substitute in the dish to equal the Scoville units of 1 tabasco pepper?

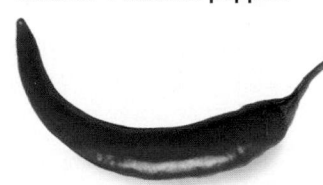

Scoville Units of Selected Peppers	
Pepper	**Scoville Units**
Ancho (Poblano)	1,500
Bell	100
Cayenne	30,000
Habanero	360,000
Jalapeño	5,000
Serrano	10,000
Tabasco	30,000
Thai	90,000

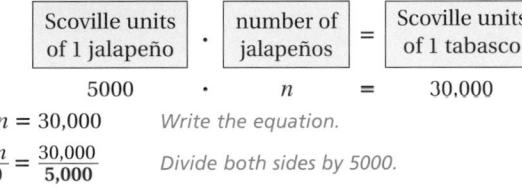

Scoville units of 1 jalapeño	$\cdot$	number of jalapeños	$=$	Scoville units of 1 tabasco
5000	$\cdot$	n	$=$	30,000

$5{,}000n = 30{,}000$ *Write the equation.*

$\frac{5{,}000n}{5{,}000} = \frac{30{,}000}{5{,}000}$ *Divide both sides by 5000.*

$n = 6$

Six jalapeños are about as hot as one tabasco pepper. Jennifer should substitute 6 jalapeños for the tabasco pepper in her recipe.

Additional Examples

Example 1

Solve $8x = 32$. $x = 4$

Example 2

Solve $\frac{n}{5} = 7$. $n = 35$

Example 3

Joe has enough flour to bake one sheet cake but would rather make cookies. How many dozen cookies can he make?

Dessert	**Cups of Flour**
Bread pudding	5
Cookies (1 doz.)	2
Sheet cake	8

Joe can make 4 dozen cookies.

Example 4

Meg has saved $50, which is one-fourth of the amount she needs for a school trip. What is the total amount she needs?   Meg needs $200 total.

Example 5

Solve $3y - 7 = 20$. $y = 9$

Reaching All Learners
Through Critical Thinking

Give students copies of a circular or of a newspaper ad. Have them find examples of prices, such as two for 88 cents or six for $1.50. For each of the examples they find, have students write and solve an equation to find the cost of a single item. When students have completed this activity, give them a multiplication equation and have them write a sample ad for the equation.

EXAMPLE *Money Application*

Helene's band needs money to go to a national competition. So far, band members have raised $560, which is only one-third of what they need. What is the total amount needed?

fraction of total amount raised so far	·	total amount needed	=	amount raised so far
$\frac{1}{3}$	·	x	=	$560

$$\frac{1}{3}x = 560 \qquad \textit{Write the equation.}$$

$$3 \cdot \frac{1}{3}x = 3 \cdot 560 \qquad \textit{Multiply both sides by 3.}$$

$$x = 1680$$

The band needs to raise a total of $1680.

Sometimes it is necessary to solve equations by using two inverse operations. For instance, the equation $6x - 2 = 10$ has multiplication and subtraction.

Variable term

Multiplication ⟶ $6x - 2 = 10$

Subtraction

To solve this equation, add to isolate the term with the variable in it. Then divide to solve.

EXAMPLE **Solving a Simple Two-Step Equation**

Solve $2x + 1 = 7$.

Step 1:

$$\begin{array}{rl} 2x + 1 = & 7 \\ -1 = & -1 \\ \hline 2x = & 6 \end{array}$$

Subtract 1 from both sides to isolate the term with x in it.

Step 2:

$$\frac{2x}{2} = \frac{6}{2} \qquad \textit{Divide both sides by 2.}$$

$$x = 3$$

Think and Discuss

1. **Explain** what property you would use to solve $\frac{k}{2.5} = 6$.

2. **Give** the equation you would solve to figure out how many ancho peppers are as hot as one cayenne pepper.

Math Background

The equations $\frac{1}{6}x = 3$ and $\frac{x}{6} = 3$ can be solved the same way because the expressions $\frac{1}{6}x$ and $\frac{x}{6}$ are equivalent. Students may need an explanation to help them understand that these expressions are equivalent. Illustrate the following:

$$\frac{1}{6}x = \frac{1}{6} \cdot x = \frac{1}{6} \cdot \frac{x}{1} = \frac{1 \cdot x}{6 \cdot 1} = \frac{x}{6}$$

3 Close

Summarize

Ask students to explain each of the four properties of equality. Show the class some one-step equations (e.g., $5x = 35$ and $y - 8 = 13$) and ask them to identify which property should be used to solve each equation. Then show them some two-step equations (e.g., $\frac{x}{3} + 2 = 10$ and $4x - 3 = 13$) and ask them which property they should apply first.

Possible answers: Students should explain that each property applies the operation to both sides of the equation so that the equation remains true. Answers to the samples are the Division, Addition, Subtraction, and Addition Properties of Equality, respectively.

Answers to Think and Discuss

Possible answers:

1. Because this is a division equation, you would use the Multiplication Property of Equality.

2. $1500a = 30,000$

FOR EXTRA PRACTICE
see page 732

internet connect
Homework Help Online
go.hrw.com Keyword: MP4 1-4

1-4 PRACTICE & ASSESS

GUIDED PRACTICE

See Example ① Solve.

1. $4x = 28$ $x = 7$ 2. $7t = 49$ $t = 7$ 3. $3y = 42$ $y = 14$ 4. $2w = 26$ $w = 13$

See Example ② 5. $\frac{l}{15} = 4$ $l = 60$ 6. $\frac{k}{8} = 9$ $k = 72$ 7. $\frac{h}{19} = 3$ $h = 57$ 8. $\frac{m}{6} = 1$ $m = 6$

See Example ③ 9. One serving of milk contains 8 grams of protein, and one serving of steak contains 32 grams of protein. Write and solve an equation to find the number of servings of milk n needed to get the same amount of protein as there is in one serving of steak. $8n = 32$; $n = 4$ **servings**

See Example ④ 10. Gary needs to buy a suit to go to a formal dance. Using a coupon, he can save \$60, which is only one-fourth of the cost of the suit. Write and solve an equation to determine the cost c of the suit. $\frac{1}{4}c = 60$; $c = \$240$

See Example ⑤ Solve.

11. $3x + 2 = 23$ $x = 7$ 12. $\frac{k}{5} - 1 = 7$ $k = 40$ 13. $3y - 8 = 1$ $y = 3$ 14. $\frac{m}{6} + 4 = 10$ $m = 36$

INDEPENDENT PRACTICE

See Example ① Solve.
15. $3d = 57$ $d = 19$ 16. $7x = 105$ $x = 15$ 17. $4g = 40$ $g = 10$ 18. $16y = 112$ $y = 7$

See Example ② 19. $\frac{n}{9} = 63$ $n = 567$ 20. $\frac{h}{27} = 2$ $h = 54$ 21. $\frac{a}{6} = 102$ $a = 612$ 22. $\frac{j}{8} = 12$ $j = 96$

See Example ③ 23. An orange contains about 80 milligrams of vitamin C, which is 10 times as much as an apple contains. Write and solve an equation to find n, the number of milligrams of vitamin C in an apple. $10n = 80$; $n = 8$ mg

See Example ④ 24. Fred gathered 150 eggs on his family's farm today. This is one-third the number he usually gathers. Write and solve an equation to determine the number n that he usually gathers. $\frac{1}{3}n = 150$; $n = 450$

See Example ⑤ Solve.

25. $6x - 5 = 7$ $x = 2$ 26. $\frac{n}{3} - 4 = 1$ $n = 15$ 27. $2y + 5 = 9$ $y = 2$ 28. $\frac{h}{7} + 2 = 2$ $h = 0$

PRACTICE AND PROBLEM SOLVING

Solve.

29. $2x = 14$ $x = 7$ 30. $4y = 80$ $y = 20$ 31. $6y = 12$ $y = 2$ 32. $9m = 9$ $m = 1$

33. $\frac{k}{8} = 7$ $k = 56$ 34. $\frac{1}{5}x = 121$ $x = 605$ 35. $\frac{b}{6} = 12$ $b = 72$ 36. $\frac{n}{15} = 1$ $n = 15$

37. $3x = 51$ $x = 17$ 38. $15g = 75$ $g = 5$ 39. $16y + 18 = 66$ $y = 3$ 40. $3z - 14 = 58$ $z = 24$

42. $m = 576$

43. $n = 35$

44. $a = 12$

41. $\frac{b}{4} = 12$ $b = 48$ 42. $\frac{m}{24} = 24$ 43. $\frac{n}{5} - 3 = 4$ 44. $\frac{a}{2} + 8 = 14$

Students may want to refer back to the lesson examples.

Assignment Guide

If you finished Example ① assign:
Core 1–4, 15–18, 29–32, 49–54
Enriched 1–4, 15–18, 29–32, 49–54

If you finished Example ② assign:
Core 1–8, 15–22, 29, 33, 49–54
Enriched 15–22, 29–38, 49–54

If you finished Example ③ assign:
Core 1–9, 15–23, 29–35 odd, 45, 49–54
Enriched 15–23, 29–38, 45–54

If you finished Example ④ assign:
Core 1–10, 15–24, 29–35 odd, 45, 49–54
Enriched 15–24, 29–38, 45–54

If you finished Example ⑤ assign:
Core 1–28, 37–47 odd, 49–54
Enriched 15–54

Notes

RETEACH 1-4

Reteach
1-4 *Solving Equations by Multiplying or Dividing*

To solve a multiplication equation, use division.

Solve $3x = 24$.

$3x = 24$
$\frac{3x}{3} = \frac{24}{3}$
$x = 8$

To solve a division equation, use multiplication.

Solve $\frac{x}{4} = 20$.

$\frac{x}{4} = 20$
$4 \cdot \frac{x}{4} = 20 \cdot 4$
$x = 80$

When an equation has two operations, undo addition or subtraction first. Then undo multiplication or division.

Solve $2x + 11 = 35$.

$2x + 11 = 35$
$\underline{-11 \quad -11}$ Undo the addition.
$2x = 24$
$\frac{2x}{2} = \frac{24}{2}$ Undo the multiplication.
$x = 12$

Tell what number you would multiply or divide by to solve the equation.

1. $5a = 60$ **Divide by 5.**   2. $\frac{x}{6} = 12$ **Multiply by 6.**   3. $144 = 12f$ **Divide by 12.**

Solve.

4. $6x = 42$
$\frac{6x}{6} = \frac{42}{6}$
$x = 7$

5. $\frac{a}{3} = 9$
$3 \cdot \frac{a}{3} = 9 \cdot 3$
$a = 27$

6. $25 = \frac{k}{5}$
$5 \cdot 25 = \frac{k}{5} \cdot 5$
$125 = k$

7. $2x + 3 = 11$
$\underline{-3 \quad -3}$
$\frac{2x}{2} = \frac{8}{2}$
$x = 4$

8. $\frac{b}{4} + 5 = 6$
$\underline{-5 \quad -5}$
$4 \cdot \frac{b}{4} = 1 \cdot 4$
$b = 4$

9. $5t - 9 = 36$
$\underline{+9 \quad +9}$
$\frac{5t}{5} = \frac{45}{5}$
$t = 9$

PRACTICE 1-4

Practice B
1-4 *Solving Equations by Multiplying or Dividing*

Solve.

1. $4w = 48$ $w = 12$ 2. $8y = 56$ $y = 7$ 3. $4b = 64$ $b = 16$

4. $\frac{x}{4} = 9$ $x = 36$ 5. $\frac{v}{6} = 14$ $v = 84$ 6. $\frac{n}{21} = 3$ $n = 63$

7. $5a = 75$ $a = 15$ 8. $54 = 3q$ $q = 18$ 9. $23b = 161$ $b = 7$

10. $\frac{k}{21} = 15$ $k = 315$ 11. $\frac{w}{17} = 17$ $w = 289$ 12. $11 = \frac{r}{34}$ $r = 374$

13. $672 = 24b$ $b = 28$ 14. $\frac{u}{25} = 13$ $u = 325$ 15. $42m = 966$ $m = 23$

16. $3x + 7 = 16$ $x = 3$ 17. $\frac{t}{5} + 8 = 10$ $t = 10$ 18. $5 = 2n - 3$ $n = 4$

19. Alex scored 13 points in the basketball game. This was $\frac{1}{5}$ of the total points the team scored. Write and solve an equation to determine the total points the team scored. $\frac{t}{5} = 13$; $t = 65$

20. Jar candles at the Candle Co. cost \$4. Nikki spent \$92 buying jar candles for party favors. Write and solve an equation to determine how many jar candles Nikki bought at the Candle Co. $4c = 92$; $c = 23$

Interdisciplinary LINK

Social Studies

Exercises 45–48 focus on solving problems about interstate highways in the United States. The construction of the highways was a major accomplishment and was initiated by President Eisenhower in 1956.

Answers

45. $16m = 42{,}000$; $m = 2625$ miles

Journal

Ask students to think about inverse operations as operations that "undo" each other. Have students write about everyday things that "undo" each other, such as accelerator and brake pedals or locking and unlocking a door.

Test Prep Doctor

For Exercise 53, remind students that "prime factorization of 72" requires that two conditions be met. The factors must give 72 as the product, and the factors must all be prime numbers. In choice **B**, the product is not 72. In choices **C** and **D**, the product is 72, but not all of the factors are prime numbers. Only choice **A** meets both conditions.

Lesson Quiz

Solve.

1. $\dfrac{k}{6} = 3$ $k = 18$

2. $16t = 112$ $t = 7$

3. $\dfrac{b}{9} = 14$ $b = 126$

4. $3n + 17 = 29$ $n = 4$

5. Joan is making a scrapbook of her pictures. She wants to put 3 pictures on each page. She has 63 photos. Write and solve an equation to find how many pages she will need for all of the photos. $3x = 63$; 21 pages

Available on Daily Transparency in CRB

Social Studies LINK

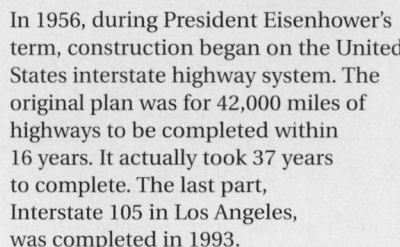

In 1956, during President Eisenhower's term, construction began on the United States interstate highway system. The original plan was for 42,000 miles of highways to be completed within 16 years. It actually took 37 years to complete. The last part, Interstate 105 in Los Angeles, was completed in 1993.

45. Write and solve an equation to show how many miles m needed to be completed per year for 42,000 miles of highways to be built in 16 years.

46. Interstate 35 runs north and south from Laredo, Texas, to Duluth, Minnesota, covering 1568 miles. There are 505 miles of I-35 in Texas and 262 miles in Minnesota. Write and solve an equation to find m, the number of miles of I-35 that are not in either state.
$m + (505 + 262) = 1568$; $m = 801$ miles

47. A portion of I-476 in Pennsylvania, known as the Blue Route, is about 22 miles long. The length of the Blue Route is about one-sixth the total length of I-476. Write and solve an equation to calculate the length of I-476 in miles m. $\dfrac{1}{6}m = 22$; $m = 132$ miles

48. **CHALLENGE** Interstate 80 extends from California to New Jersey. At right are the number of miles of Interstate 80 in each state the highway passes through.

   a. _?_ has 134 more miles than _?_.   **Iowa; Indiana**

   b. _?_ has 174 fewer miles than _?_.   **Ohio; Nevada**

Number of I-80 Miles	
State	Miles
California	195 mi
Nevada	410 mi
Utah	197 mi
Wyoming	401 mi
Nebraska	455 mi
Iowa	301 mi
Illinois	163 mi
Indiana	167 mi
Ohio	236 mi
Pennsylvania	314 mi
New Jersey	68 mi

Spiral Review

Solve. (Lesson 1-3)

49. $3 + x = 11$ $x = 8$ **50.** $y - 6 = 8$ $y = 14$ **51.** $13 = w + 11$ $w = 2$ **52.** $5.6 = b - 4$ $b = 9.6$

53. **TEST PREP** Which is the prime factorization of 72? (Previous Course)   **A**

   A $3 \cdot 3 \cdot 2 \cdot 2 \cdot 2$ **C** $3 \cdot 2 \cdot 2 \cdot 6$

   B $3^3 \cdot 2^2$ **D** $3^2 \cdot 4 \cdot 2$

54. **TEST PREP** What is the value of the expression $3x + 4$ for $x = 2$? (Lesson 1-1)   **J**

   F 4   **H** 9

   G 6   **J** 10

CHALLENGE 1-4

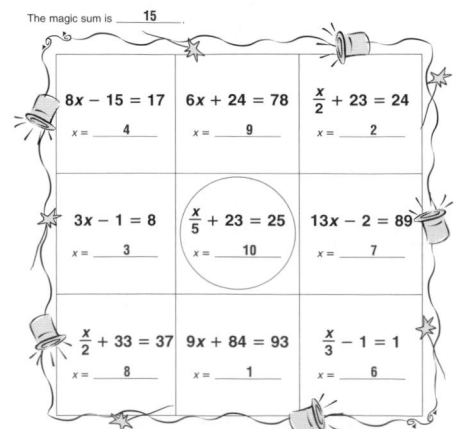

LESSON 1-4 Challenge

Which Square is Magic?

A *magic square* has the same sum for every row, column, and diagonal. In the magic square below, the solution of each equation is a number of the magic square. But, one of the equations has a solution that does not fit in this magic square.

Solve each equation. Find the magic sum. Circle the equation that does not fit in this magic square. Write a new equation with a solution that completes the magic square. **Replacement for the equation in the center box will vary, but its solution should be 5. Possible replacement:** $2x + 15 = 25$.

The magic sum is   **15**

$8x - 15 = 17$	$6x + 24 = 78$	$\dfrac{x}{2} + 23 = 24$
$x =$ **4**	$x =$ **9**	$x =$ **2**
$3x - 1 = 8$	$\dfrac{x}{5} + 23 = 25$	$13x - 2 = 89$
$x =$ **3**	$x =$ **10**	$x =$ **7**
$\dfrac{x}{2} + 33 = 37$	$9x + 84 = 93$	$\dfrac{x}{3} - 1 = 1$
$x =$ **8**	$x =$ **1**	$x =$ **6**

PROBLEM SOLVING 1-4

LESSON 1-4 Problem Solving

Solving Equations by Multiplying or Dividing

Write the correct answer.

1. Brett is preparing to participate in a 250-kilometer bike race. He rides a course near his house that is 1.25 km long. Write an equation to determine how many laps he must ride to equal the distance of the race.

   $1.25l = 250$

2. The average life span of a duck is 10 years, which is one year longer than three times the average life span of a guinea pig. Write and solve an equation to determine the lifespan of a guinea pig.

   $3y + 1 = 10$; $y = 3$

3. The speed of a garden snail is one-fifth that of a three-toed sloth. If a garden snail can travel at 0.03 mi/h, what is the speed of a three-toed sloth? Write an equation and solve.

   $\left(\dfrac{1}{5}\right)s = 0.03$; $s = 0.15$ mi/h

4. In 2002, the movie with the highest box office sales was *Titanic*, which made about 2.74 times the box office sales of the number 25 movie, *Mrs. Doubtfire*. If *Titanic* made $600.8 million, about how much did *Mrs. Doubtfire* make? Write and solve an equation.

   $2.74m = 600.8$; $m = \$219.3$ million

Choose the letter for the best answer.

5. An inch is 2.54 times bigger than a centimeter. If a pencil is 18 cm long, how many inches long is it?

   A 45.72 in.   C 15.45 in.

   B 20.54 in.   **D** 7.1 in.

6. When Maria doubles a recipe, she uses 8 cups of flour. How many cups of flour are in the original recipe?

   F 2 cups   H 8 cups

   G 4 cups   J 16 cups

7. The depth of water is often measured in fathoms. A fathom is one-sixth of a foot. If the average depth of the Gulf of Mexico is 29,244 fathoms, what is the average depth in feet?

   A 4874 ft   C 29,250 ft

   B 98,867 ft   D 175,464 ft

8. Four times as many pet birds have lived in the White House as pet goats. Sixteen pet birds have lived in the White House. How many pet goats have there been?

   F 4   H 12

   G 20   J 64

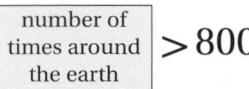

1-5 Solving Simple Inequalities

Learn to solve and graph inequalities.

Vocabulary
inequality

algebraic inequality

solution of an inequality

solution set

Laid end to end, the paper used by personal computer printers each year would circle the earth *more than* 800 times.

$$\boxed{\begin{array}{c}\text{number of}\\\text{times around}\\\text{the earth}\end{array}} > 800$$

An **inequality** compares two quantities and typically uses one of these symbols:

is less than   *is greater than*   *is less than or equal to*   *is greater than or equal to*

EXAMPLE 1 Completing an Inequality

Compare. Write < or >.

A $12 - 7 \boxed{} 6$

$5 \boxed{} 6$

$5 < 6$

B $3(8) \boxed{} 16$

$24 \boxed{} 16$

$24 > 16$

Helpful Hint

An open circle means that the corresponding value is not a solution. A solid circle means that the value is part of the solution set.

An inequality that contains a variable is an **algebraic inequality** . A number that makes an inequality true is a **solution of the inequality** .

The set of all solutions is called the **solution set** . The solution set can be shown by graphing it on a number line.

Word Phrase	Inequality	Sample Solutions	Solution Set
x is less than 5	$x < 5$	$x = 4$ $4 < 5$ $x = 2.1$ $2.1 < 5$	0 1 2 3 4 5 6 7
a is greater than 0 a is more than 0	$a > 0$	$a = 7$ $7 > 0$ $a = 25$ $25 > 0$	−3 −2 −1 0 1 2 3
y is less than or equal to 2 y is at most 2	$y \le 2$	$y = 0$ $0 \le 2$ $y = 1.5$ $1.5 \le 2$	−3 −2 −1 0 1 2 3 4 5
m is greater than or equal to 3 m is at least 3	$m \ge 3$	$m = 17$ $17 \ge 3$ $m = 3$ $3 \ge 3$	−1 0 1 2 3 4 5 6

---

1 Introduce

Alternate Opener

EXPLORATION

 Solving Simple Inequalities

To raise money for a school trip, Angela sells cookies for $2.50 a box. She has to raise at least $100 from the cookie sales.

1. Which of the numbers of boxes of cookies below gives Angela

 a. less than her goal of $100?

 b. exactly $100?

 c. more than her goal of $100?

   30 boxes

You can use the inequality $2.5b \ge 100$, which means "2.5 times a number *b* is greater than or equal to 100," to model the problem.

Give a real-world situation for each inequality and solve each inequality.

2. $x + 5 < 12$

3. $x - 6 < 12$

4. $\frac{x}{5} \le 7$

5. $25x \ge 500$

< less than
≤ less than or equal to
> greater than
≥ greater than or equal to
≠ not equal

Think and Discuss

6. **Discuss** the strategies you used for solving the inequalities in numbers 1–5.

7. **Explain** what is meant by the phrase "solution of an inequality."

Motivate

Make some statements like "Tonight you will have fewer than fifteen homework questions to answer" and "Tomorrow the temperature is going to be greater than 60 degrees." Ask students what the phrases "fewer than" and "greater than" mean. Have them give specific numbers that would make the statements true.

Exploration worksheet and answers on
Chapter 1 Resource Book pp. 44 and 114

1-5 Organizer

Pacing: Traditional 1 day
Block $\frac{1}{2}$ day

Objective: Students solve and graph inequalities.

Warm Up

Solve.

1. $x + 6 = 13$ ⟶ $x = 7$

2. $8n = 48$ ⟶ $n = 6$

3. $t - 2 = 56$ ⟶ $t = 58$

4. $6 = \frac{z}{6}$ ⟶ $z = 36$

Problem of the Day

Bill and Brad are taking drivers education. Bill drives with his instructor for one and a half hours three times a week. He needs a total of 27 hours. Brad drives two times a week, two hours each time. He needs 26 hours. Who will finish his hours first?   Bill

Available on Daily Transparency in CRB

Math Fact

To completely escape Earth's gravity, a rocket or other object must travel at a speed equal to or greater than about 25,000 miles per hour. To simply go into orbit, the rocket can travel slower— at a speed greater than or equal to about 10,000 miles per hour.

2 Teach

Lesson Presentation

Guided Instruction

In this lesson, students learn to solve and graph inequalities. Begin by reminding students that the symbols <, >, ≤, and ≥ are used in inequalities. The ≠ symbol may also be considered an inequality. Explain why statements like 6 > 2 and 4 ≥ 4 are true, while statements like 7 < 5 and 4.1 ≤ 4 are false. Review the table of examples (available on Teaching Transparency T21 in Chapter 1 Resource Book).

Show students that the processes previously used to solve equations are the same processes that are used to solve inequalities.

Most inequalities can be solved the same way equations are solved. Use inverse operations on both sides of the inequality to isolate the variable. (There are special rules when multiplying or dividing by a negative number, which you will learn in the next chapter.)

EXAMPLE 2 Solving and Graphing Inequalities

Solve and graph each inequality.

A $x + 7.5 < 10$

$\quad\quad \underline{-7.5 \quad -7.5}$ *Subtract 7.5 from both sides.*

$\quad\quad\quad x < 2.5$

Remember!

The inequality symbol opens to the side with the greater number.

$2 < 10$

According to the graph, 2.4 should be a solution, since $2.4 < 2.5$, and 3 should not be a solution because $3 > 2.5$.

Check $x + 7.5 < 10$

$\quad\quad 2.4 + 7.5 \overset{?}{<} 10$ *Substitute 2.4 for x.*

$\quad\quad\quad\quad 9.9 \overset{?}{<} 10$ ✔

So 2.4 is a solution.

Check $x + 7.5 < 10$

$\quad\quad 3 + 7.5 \overset{?}{<} 10$ *Substitute 3 for x.*

$\quad\quad\quad 10.5 \overset{?}{<} 10$ ✘

And 3 is not a solution.

B $6n \geq 18$

$\quad \dfrac{6n}{6} \geq \dfrac{18}{6}$ *Divide both sides by 6.*

$\quad\quad n \geq 3$

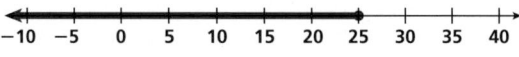

C $t - 3 \leq 22$

$\quad\quad \underline{+3 \quad +3}$ *Add 3 to both sides.*

$\quad\quad\quad t \leq 25$

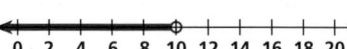

D $5 > \dfrac{w}{2}$

$\quad 2 \cdot 5 > 2 \cdot \dfrac{w}{2}$ *Multiply both sides by 2.*

$\quad 10 > w$ *$10 > w$ is the same as $w < 10$.*

Teach

Reaching All Learners
Through Grouping Strategies

Have students work in pairs. Have each pair of students create two sets of index cards: one set with basic inequalities (e.g., $x \geq 2$ and $x < 5$) and another set with the corresponding graphs of the solution sets. Then have the students shuffle each set and trade both sets with another pair of students. Students then can take turns matching each inequality with its graph.

EXAMPLE 3 **PROBLEM SOLVING APPLICATION**

PROBLEM SOLVING

If all of the sheets of paper used by personal computer printers each year were laid end to end, they would circle the earth more than 800 times. The earth's circumference is about 25,120 mi (1,591,603,200 in.), and one sheet of paper is 11 in. long. How many sheets of paper are used each year?

1 **Understand the Problem**

The **answer** is the number of sheets of paper used by personal computer printers in one year. **List the important information:**

• The amount of paper would circle the earth *more than* 800 times.

• Once around the earth is 1,591,603,200 in.

• One sheet of paper is 11 in. long.

Show the relationship of the information:

| the number of sheets of paper | · | the length of one sheet | > | 800 | · | the distance around the earth |

2 **Make a Plan**

Use the relationship to *write an inequality*. Let *x* represent the number of sheets of paper.

$$x \cdot 11 \text{ in.} > 800 \cdot 1{,}591{,}603{,}200 \text{ in.}$$

3 **Solve**

$11x > 800 \cdot 1{,}591{,}603{,}200$

$11x > 1{,}273{,}282{,}560{,}000$ *Multiply.*

$\dfrac{11x}{11} > \dfrac{1{,}273{,}282{,}560{,}000}{11}$ *Divide both sides by 11.*

$x > 115{,}752{,}960{,}000$

More than 115,752,960,000 sheets of paper are used by personal computer printers in one year.

4 **Look Back**

To circle the earth once takes $\frac{1{,}591{,}603{,}200}{11} = 144{,}691{,}200$ sheets of paper; to circle it 800 times would take $800 \cdot 144{,}691{,}200 = 115{,}752{,}960{,}000$ sheets.

Think and Discuss

1. Give all the symbols that make $5 + 8 \ \blacksquare\ 13$ true. Explain.

2. Explain which symbols make $3x \ \blacksquare\ 9$ false if $x = 3$.

---

3 **Close**

Summarize

Tell students that one way to remember the meaning of each of the symbols < and > is to think that the small end of each symbol "points" to the lesser number. Ask students if they have any other suggestions for remembering the meanings of the symbols < and >.

Possible answer: The "mouth" of the sign opens to eat the greater number.

Answers to Think and Discuss

1. =, ≤, ≥; Since both sides are equal, it can be any symbol that includes equality.

2. Since both sides are equal, either the < or > (or ≠) would make the statement false.

Additional Examples

Example 1

Compare. Write < or >.

A. $23 - 14 \ \blacksquare\ 6$

>

B. $5(12) \ \blacksquare\ 70$

<

Example 2

Solve and graph each inequality.

A. $x + 2.5 \leq 8$

$x \leq 5.5$

0 1 2 3 4 5 6 7 8 9 10

B. $5t > 15$

$t > 3$

0 1 2 3 4 5 6 7 8 9 10

C. $w - 1 < 8$

$w < 9$

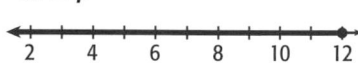
0 1 2 3 4 5 6 7 8 9 10

D. $3 \geq \dfrac{p}{4}$

$12 \geq p$

2   4   6   8   10   12

Example 3

An interior designer is planning to place a wallpaper border along the edges of all four walls of a room. The total distance around the room is 88 feet. The border comes in packages of 16 feet. What is the least number of packages that must be purchased to be sure that there is enough border to complete the room?

Because whole rolls must be purchased, at least 6 rolls of border must be purchased to ensure that there is enough to complete the room.

1-5 PRACTICE & ASSESS

1-5 Exercises

FOR EXTRA PRACTICE
see page 732

internet connect
Homework Help Online
go.hrw.com Keyword: MP4 1-5

GUIDED PRACTICE

See Example **1** Compare. Write $<$ or $>$.

1. $4 + 8$ ▨ 13 $<$
2. $4(2)$ ▨ 7 $>$
3. $27 - 13$ ▨ 11 $>$
4. $5(9)$ ▨ 42 $>$
5. $9 + 2$ ▨ 10 $>$
6. $3(8)$ ▨ 27 $<$
7. $52 - 37$ ▨ 14 $>$
8. $8(7)$ ▨ 54 $>$

See Example **2** Solve and graph each inequality. **For graphs, see margin.**

9. $x + 3 < 4$ $x < 1$
10. $4b \geq 20$ $b \geq 5$
11. $m - 4 \leq 28$ $m \leq 32$
12. $5 > \frac{x}{3}$ $15 > x$
13. $y + 8 \geq 25$ $y \geq 17$
14. $6f < 30$ $f < 5$
15. $z - 8 > 13$ $z > 21$
16. $7 \leq \frac{x}{2}$ $14 \leq x$

See Example **3** 17. For a field trip to the museum, the science club can purchase individual tickets for $4 each or a group pass for $160. How many club members are necessary for it to be cheaper to buy a group pass than to buy individual tickets? Write and solve an inequality to answer the question.
$m > 40$; more than 40 members

INDEPENDENT PRACTICE

See Example **1** Compare. Write $<$ or $>$.

18. $4 + 7$ ▨ 12 $<$
19. $6(4)$ ▨ 25 $<$
20. $15 - 9$ ▨ 4 $>$
21. $7(6)$ ▨ 40 $>$
22. $13 + 5$ ▨ 17 $>$
23. $5(2.3)$ ▨ 12 $<$
24. 7 ▨ $19 - 13$ $>$
25. 12 ▨ $3(4.2)$ $<$

See Example **2** Solve and graph each inequality. **For graphs, see margin.**

26. $b + 4 < 8$ $b < 4$
27. $7x \geq 49$ $x \geq 7$
28. $h - 2 \geq 3$ $h \geq 5$
29. $1 < \frac{t}{4}$ $4 < t$
30. $6 + a > 9$ $a > 3$
31. $3x \geq 12$ $x \geq 4$
32. $f - 9 \leq 2$ $f \leq 11$
33. $2 < \frac{a}{3}$ $6 < a$

See Example **3** 34. There are 88 keys on a new piano. If there are 12 pianos in a room of broken pianos, some of which may have missing keys, how many piano keys could be on the pianos? Write and solve an inequality to answer the question. $k \leq 1056$; not more than 1056 keys

PRACTICE AND PROBLEM SOLVING

Write the inequality shown by each graph.

35.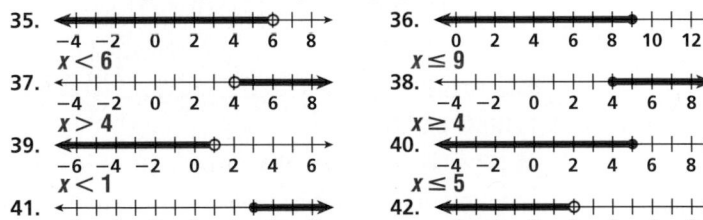
$x < 6$
36.
$x \leq 9$
37.
$x > 4$
38.
$x \geq 4$
39.
$x < 1$
40.
$x \leq 5$
41. $x \geq 5$
42. $x < 0$
41.
42.

Students may want to refer back to the lesson examples.

Assignment Guide

If you finished Example **1** assign:
Core 1–8, 18–25, 50–57
Enriched 1–8, 18–25, 50–57

If you finished Example **2** assign:
Core 1–16, 18–33, 35–41 odd, 50–57
Enriched 1–15 odd, 18–33, 35–42, 45, 48–57

If you finished Example **3** assign:
Core 1–34, 35–45 odd, 46, 50–57
Enriched 1–17 odd, 18–57

Answers

9.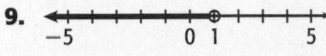
10.
11.
12.
13.
14.
15.
16.

Math Background

In this lesson, the graphs of inequalities are placed on number lines. Graphs of the same inequalities may be placed on coordinate planes. For example, the graph of $x > 2$ below shows the solution set consisting of all ordered pairs (x, y) that have x-values that are greater than 2.

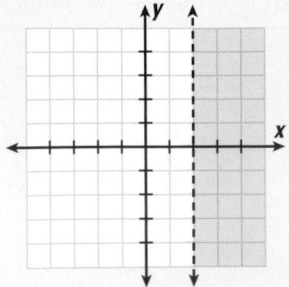

RETEACH 1-5

LESSON
1-5 Reteach
Solving Simple Inequalities

A **solution of an inequality** is a number that makes the inequality true. An inequality usually has more than one solution. All the solutions are contained in the **solution set**.
As with equations, solve a simple inequality by using inverse operations to isolate the variable.

Solve and graph $x + 4 > 9$.

$$x + 4 > 9$$
$$\underline{-4 \quad -4} \quad \text{Subtract 4.}$$
$$x > 5$$

Draw an open circle at 5 to show that 5 is not included in the solution set.
Draw an arrow to the right of 5 to show that all numbers greater than 5 are included in the solutions.

According to the graph, 6 should be a solution and 4 should not be a solution.
Check:
$x + 4 > 9$ $x + 4 > 9$
$6 + 4 \overset{?}{>} 9$ $4 + 4 \overset{?}{>} 9$
$10 > 9$ $8 \not> 9$
So, 6 *is* in the solution set and 4 *is not* in the solution set.
Thus, the solution set for the inequality $x + 4 > 9$ is $x > 5$.

Write true or false.

1. $7 < 4$ **false**
2. $0 \leq 9$ **true**
3. $1.9 > 19$ **false**

Using the variable n, write the inequality shown by each graph.

4. $n \geq 3$
5. $n < 6$

Complete. Is the given value in the solution set? Answer *is* or *is not*.

6. 3 **is not** in the solution set of $x - 1 > 5$.
$x - 1 > 5$
$3 - 1 \overset{?}{>} 5$
$2 > 5$

7. 0 **is** in the solution set of $4 \geq z + 4$.
$4 \geq z + 4$
$4 \overset{?}{\geq} 0 + 4$
$4 \geq 4$

8. 25 **is** in the solution set of $10 \leq \frac{x}{2}$.
$10 \leq \frac{x}{2}$
$10 \overset{?}{\leq} \frac{25}{2}$
$10 \leq 12.5$

PRACTICE 1-5

LESSON
1-5 Practice B
Solving Simple Inequalities

Use $<$ or $>$ to compare each inequality.

1. $7 + 10$ $\boxed{>}$ 16
2. 21 $\boxed{>}$ $4(5)$
3. $25 - 7$ $\boxed{<}$ 19
4. 58 $\boxed{>}$ $7(8)$
5. $43 - 18$ $\boxed{>}$ 23
6. 60 $\boxed{<}$ $9(7)$

Solve and graph each inequality.

7. $x + 4 > 9$ $x > 5$
8. $c - 6 \leq 1$ $c \leq 7$
9. $3y \geq 12$ $y \geq 4$
10. $\frac{v}{3} < 3$ $v < 9$
11. $7 + x \leq 10$ $x \leq 3$
12. $5s < 25$ $s < 5$
13. $b - 2 \leq 5$ $b \leq 7$
14. $\frac{n}{7} > 2$ $n > 14$
15. $\frac{r}{2} \geq 6$ $r \geq 12$
16. $9w < 54$ $w < 6$
17. $14 + k > 25$ $k > 11$
18. $120 \geq 8z$ $z \leq 15$

19. A piano has 88 keys. A piano repair shop has 4 broken pianos in its shop, some with keys missing. How many piano keys are in the shop? Write and solve and inequality to answer the question.
$\frac{1}{4}p \leq 88$; $p < 352$; less than 352

20. Gabe has an SUV that can tow up to 10,000 pounds. He needs to haul some steel. Steel weighs about 500 pounds per cubic foot. How many cubic feet of steel would be too much for Gabe to tow? Write and solve and inequality to answer the question.
$500x > 10,000$; $x > 20$; more than 20 ft^3

Sports LINK

The BT Global Challenge 2000 began on September 10, 2000, and ended June 30, 2001. Of that almost 10-month period, the crews spent about 177 days, 19 hours, and 20 minutes at sea.

43. Reginald's cement truck can carry up to 2200 pounds of cargo. He needs to haul 50 bags of cement that weigh 50 pounds each. Write and solve an inequality to determine whether Reginald will be able to carry all of the cement in one trip. **50(50) > 2200; 2500 > 2200; no**

44. **SPORTS** There were 7 legs of the BT Global Challenge 2000 yacht race. If the crew of the winning boat, the *LG Flatron*, sailed at a rate of at least 6 knots (6 nautical miles per hour) continuously, how many hours would it have taken them to sail the leg from Cape Town, South Africa, to La Rochelle, France?

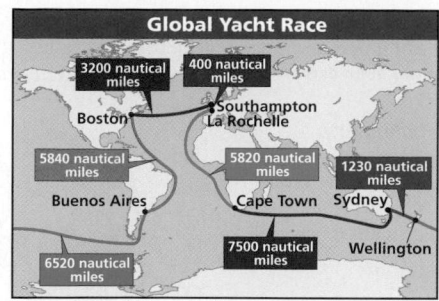

6x ≤ 5820; x ≤ 970; at most 970 hours

45. Suly earned an 87 on her first test. She needs a total of 140 points on her first two tests to pass the class. What score must Suly make on her second test to ensure that she passes the class? **x ≥ 53**

46. **BUSINESS** A rule of thumb for electronic signs is that a sign with letters *n* inches tall is readable from up to 50*n* feet away. How tall should the letters be to be readable from 900 feet away?

at least 18 in. tall

47. **WRITE A PROBLEM** The weight limit for an elevator is 2500 pounds. Write, solve, and graph a problem about the elevator and the number of 185-pound passengers it can safely carry.

48. **WRITE ABOUT IT** In mathematics, the conventional way to write an inequality is with the variable on the left, such as $x > 5$. Explain how to rewrite the inequality $4 ≤ x$ in the conventional manner.

49. **CHALLENGE** $3 ≤ x < 5$ means both $3 ≤ x$ and $x < 5$ are true at the same time. Solve and graph $6 < x ≤ 12$. **x > 6 and x ≤ 12.**

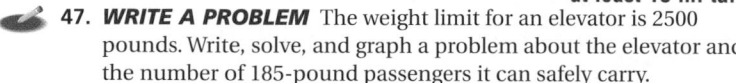

Spiral Review

Evaluate each expression for the given values of the variable. (Lesson 1-1)

50. $2(4 + x) - 3$ for $x = 0, 1, 2, 3$ **5; 7; 9; 11**
51. $3(8 - x) - 2$ for $x = 0, 1, 2, 3$ **22; 19; 16; 13**
52. $5(x - 1) - 1$ for $x = 5, 6, 7, 8$ **19; 24; 29; 34**
53. $4(x + 2) - 3$ for $x = 2, 4, 6, 8$ **13; 21; 29; 37**
54. $3(7 + x) + 4$ for $x = 2, 4, 6, 8$ **31; 37; 43; 49**
55. $2(9 - x) + 3$ for $x = 3, 4, 5, 6$ **15; 13; 11; 9**

56. **TEST PREP** A company prints *n* books at a cost of \$9 per book. What is the total cost of the books? (Lesson 1-2) **D**

   A \$9 − *n*
   B *n* + \$9
   C $\frac{n}{\$9}$
   D \$9*n*

57. **TEST PREP** Which value of *x* is the solution of the equation $x - 5 = 8$? (Lesson 1-3) **H**

   F 3
   G 11
   H 13
   J 15

Answers

26.  (number line: −5, 0, 4, 5)
27. (number line: 0, 5, 7, 10)
28. (number line: −5, 0, 5, 10)
29. (number line: −5, 0, 4, 10)
30. (number line: −5, 0, 3, 5)
31. (number line: −5, 0, 4, 5)
32. (number line: 3, 5, 7, 9, 11)
33. (number line: −5, 0, 6, 10)

47–48. See p. A1.

Journal

At an amusement park there is a sign that reads, "You must be at least 4 feet tall to ride this roller coaster." Ask students to write about any other real-world situation in which an inequality is represented.

Test Prep Doctor

For Exercise 56, suggest to students that they choose a test number to represent the number of books to be printed, use it to decide what operation should be used, and then replace their number with the variable *n*.

CHALLENGE 1-5

LESSON 1-5 Challenge
You Make the Call

Sometimes, an inequality is expressed with words like *no* or *not*. But, the algebraic inequality may be clearer if you avoid those words.

Example
The fire regulation says that this restaurant may seat no more than 350 people.
• the inequality as stated: no more than 350
• the equivalent without *no*: less than or equal to 350
• algebraic inequality: Let *x* = the number of diners allowed. $x ≤ 350$

Sometimes, two conditions of inequality can be expressed as a single inequality.

Example
$x > 3$ and $x < 9$ means that *x* is between 3 and 9, and can be written as $3 < x < 9$.

Write an algebraic inequality, identifying what the variable represents.

1. The recipe calls for not less than 15 oz of butter.
   Let *x* = oz of butter; $x ≥ 15$

2. Mr. Valdez says he cannot contribute more than \$500.
   Let *x* = money contributed; $x ≤ 500$

3. On his typing test, Philip can have no more than 4 errors to pass.
   Let *x* = errors allowed; $x ≤ 4$

4. The team will have to score no fewer than 20 points to win.
   Let *x* = points needed; $x ≥ 20$

5. This canister can hold at most 5 lb of rice.
   Let *x* = number of lb; $x ≤ 5$

6. Ken needs to save at least \$200 to pay for his computer upgrade.
   Let *x* = money saved; $x ≥ 200$

If *x* is a whole number, write the solution of each inequality.

7. $7 ≤ x < 11$ **7; 8; 9; 10**
8. $15 > x ≥ 9$ **14; 13; 12; 11; 10; 9**
9. $2.4 < x < 7.7$ **3; 4; 5; 6; 7**
10. $\frac{1}{3} < \frac{x}{9} < \frac{2}{3}$ **3; 4; 5**

PROBLEM SOLVING 1-5

LESSON 1-5 Problem Solving
Solving Simple Inequalities

Use the table.

1. Write an inequality that compares the population *p* of Los Angeles to the population of New York.
   $p < 8,008,278$

2. Write an inequality that compares the population *p* of Los Angeles to the population of Chicago.
   $p > 2,896,016$

Top 3 U.S. Cities by Population 2000

Rank	City	Population
1	New York	8,008,278
2	Los Angeles	*p*
3	Chicago	2,896,016

Write the correct answer.

3. The sum of any two sides of a triangle is greater than the third side. In an isosceles triangle, two sides are equal. If the equal sides of an isosceles triangle have length *a*, and the third side is 12 in, find *a*.
   $2a > 12; a > 6$

4. To avoid a service charge, Jose must keep more than \$500 in his account. His current balance is \$536.24, but he plans to write a check for \$157.35. Find out how much he must deposit to avoid a service charge.
   $536.24 - 157.35 + d > 500;$
   $d > \$121.11$

Choose the letter for the best answer.

5. It costs \$0.25 per word to place an ad in the paper. Mia wants to spend no more than \$20 on the ad. How many words *w* can Mia's ad be?
   A $w > 80$
   B $w < 80$
   C $w < 5$
   D $w > 5$

6. An auto shop estimates parts and labor for a repair will cost less than \$200. Parts will cost \$58.49. Find the maximum cost *c* of the labor.
   F $c < \$141.51$
   G $c < \$258.49$
   H $c > \$141.51$
   J $c > \$258.49$

7. To advance to the next level of a competition, Rachel must earn at least 180 points. She has already earned 145 points. Find the minimum number of points *p* she needs to advance to the next level of the competition.
   A $p < 35$
   B $p < 325$
   C $p > 35$
   D $p > 325$

8. Hurricane Andrew that hit Florida in 1992 did more than 8 times the damage of Hurricane Fran in North Carolina in 1996. If Andrew did \$26.5 billion in damage, about how much damage *d* did Fran do?
   F $d > 212$ billion
   G $d > 3.3$ billion
   H $d < 212$ billion
   J $d < 3.3$ billion

Lesson Quiz

Compare. Write < or >.

1. $13 > 5(2)$
2. $14 - 2 > 11$

Solve and graph each inequality.

3. $k + 9 < 12$ $k < 3$
   (number line: −5, 0, 3, 5)

4. $3 ≤ \frac{m}{2}$ $6 ≤ m$
   (number line: −5, 0, 6, 10)

5. A school bus can hold 64 passengers. Three classes would like to use the bus for a field trip. Each class has 21 students. Write and solve an inequality to determine whether all three classes will fit on the bus. $3(21) \overset{?}{\leq} 64; 63 ≤ 64;$ **yes**

Available on Daily Transparency in CRB

1-5 Solving Simple Inequalities   **27**

1-6 Organizer

Pacing: Traditional 1 day
Block $\frac{1}{2}$ day

Objective: Students combine like terms in an expression.

Warm Up

Simplify.

1. $9 + 13 - 5 + 3$ 20
2. $16 - 8 + 4 - 1$ 11
3. $6 + 9 - 10 + 3$ 8
4. $17 + 8 - 20 - 2$ 3

Problem of the Day

Ray and Katrina are wandering through the wildlife preserve. They observe and count a total of 15 wild turkeys and deer and a total of 46 legs. How many of each did they see?   7 turkeys, 8 deer

Available on Daily Transparency in CRB

Math Fact !!

From 1939 to 1970, there were more than 30 volumes of mathematics published under the name Nicolas Bourbaki. In fact, there was no such person. The name was used by a group of mathematicians whose membership varied over the years.

1-6 Combining Like Terms

Learn to combine like terms in an expression.

Vocabulary

term

like term

equivalent expression

simplify

The district choir festival combines choirs from all three high schools in the district. The festival director has received the following rosters from each choir.

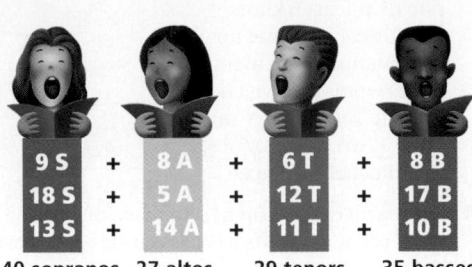

9 S +	8 A +	6 T +	8 B	Johnson High 31 members
18 S +	5 A +	12 T +	17 B	Kennedy High 52 members
13 S +	14 A +	11 T +	10 B	Filmore High 48 members

40 sopranos   27 altos   29 tenors   35 basses

To find the total number in each section, the director groups together like parts from each school. Students from different schools who sing in the same section are similar to *like terms* in an expression.

Terms in an expression are separated by plus or minus signs.

Helpful Hint

Constants such as 4, 0.75, and 11 are like terms because none of them have a variable.

$$7x + 5 - 3y + 2x$$

Like terms can be grouped together because they have the same variable raised to the same power. Often, like terms have different coefficients. When you combine like terms, you change the way an expression looks, but not the value of the expression. **Equivalent expressions** have the same value for all values of the variables.

EXAMPLE 1   **Combining Like Terms to Simplify**

Combine like terms.

A $\boxed{5x} + \boxed{3x}$ *Identify like terms.*

$8x$ *Combine coefficients: 5 + 3 = 8*

B $\boxed{5m} - \boxed{2m} + \boxed{8} - \boxed{3m} + \boxed{6}$ *Identify like terms.*

$0m + 14$ *Combine coefficients: 5 − 2 − 3 = 0*

14 *and 8 + 6 = 14*

---

1 Introduce

Alternate Opener

EXPLORATION

1-6   **Combining Like Terms**

Like terms are combined separately from unlike terms. Look at how the like terms are combined in the diagram.

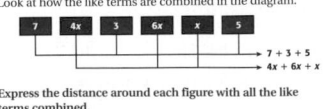

Express the distance around each figure with all the like terms combined.

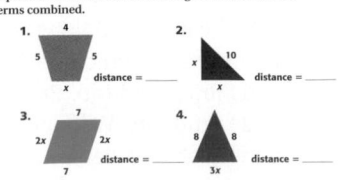

Think and Discuss

5. **Explain** what it means to combine like terms.
6. **Describe** a real-world situation in which you combine like terms.

Motivate

Ask the students if they have ever heard the phrase "You are trying to compare apples and oranges." Have them tell you what they think it means. Explain that apples and oranges cannot be compared because they are unlike objects.

Exploration worksheet and answers on Chapter 1 Resource Book pp. 53 and 116

---

2 Teach

Lesson Presentation

Guided Instruction

In this lesson, students learn to combine like terms in an expression. Make sure that students understand that all constants are like terms and that terms with variables are like terms only if they have the same variable raised to the same power. Point out that unlike terms remain separate in a simplified expression.

Teaching Tip    Students may use one of the following methods to be sure they have included all terms:

- color-coded underlining
- drawing circles, squares, or triangles
- lightly crossing out terms

EXAMPLE **Combining Like Terms in Two-Variable Expressions**

Combine like terms.

A $6a + 8a + 4b + 7$

$\boxed{6a} + \boxed{8a} + \left(4b\right) + \left(7\right)$ *Identify like terms.*

$14a + 4b + 7$ *Combine coefficients: 6 + 8 = 14*

B $k + 3n - 2n + 4k$

$\boxed{1k} + \left(3n\right) - \left(2n\right) + \boxed{4k}$ *Identify like terms; the coefficient of k is 1, because 1k = k.*

$5k + n$ *Combine coefficients.*

C $4f - 12g + 16$

$\boxed{4f} - \left(12g\right) + \left(16\right)$ *No like terms*

To **simplify** an expression, perform all possible operations, including combining like terms.

EXAMPLE **Simplifying Algebraic Expressions by Combining Like Terms**

Remember!

The Distributive Property states that $a(b + c) = ab + ac$ for all *a*, *b*, and *c*. For instance, $2(3 + 5) = 2(3) + 2(5)$.

Simplify $4(y + 9) - 3y$.

$4(y + 9) - 3y$

$4(y) + 4(9) - 3y$ *Distributive Property*

$4y + 36 - 3y$ *4y and 3y are like terms.*

$1y + 36$ *Combine coefficients: 4 − 3 = 1*

$y + 36$

EXAMPLE **Solving Algebraic Equations by Combining Like Terms**

Solve $8x - x = 112$.

$8x - x = 112$ *Identify like terms. The coefficient of x is 1.*

$7x = 112$ *Combine coefficients: 8 − 1 = 7*

$\dfrac{7x}{7} = \dfrac{112}{7}$ *Divide both sides by 7.*

$x = 16$

Think and Discuss

1. Describe the first step in simplifying the expression $2 + 8(3y + 5) - y$.

2. Tell how many sets of like terms are in the expression in Example 1B. What are they?

3. Explain why $8x + 8y + 8$ is already simplified.

COMMON ERROR ALERT

Students sometimes will combine terms incorrectly in an expression like $4x + 3 + x$. They may overlook the fact that *x* has the same value as $1x$. Encourage students to write the coefficient 1 in front of any variable without a coefficient in front of it.

Additional Examples

Example 1

Combine like terms.

A. $14a - 5a$
$9a$

B. $7y + 8 - 3y - 1 + y$
$5y + 7$

Example 2

Combine like terms.

A. $9x + 3y - 2x + 5$
$7x + 3y + 5$

B. $5t + 7p - 3p - 2t$
$3t + 4p$

C. $4m + 9n - 2$
no like terms

Example 3

Simplify $6(5 + n) - 2n$.
$30 + 4n$

Example 4

Solve $x + 3x = 48$.
$x = 12$

---

3 Close

Reaching All Learners

Through Concrete Manipulatives

Provide students with cut-out shapes like the following: squares labeled with *x*'s, circles labeled with *y*'s, and triangles labeled with 1's. Use the same shapes but different colors for subtracted expressions (e.g., "minus *x*"). Photocopy masters for these shapes are available in the Chapter 1 Resource Book, pp. 103–104. Have students use the cut-outs to simplify expressions, such as $2x + 1 + 3y + 5 + 4x$ (**6x + 3y + 6**) and $5x + 6y - 3 - 2x - 3y$ (**3x + 3y − 3**). Remind students that a pair of cut-outs having the same shape but different colors equals zero.

Summarize

Show students the expression $2a + 5b + 5 - a + 3$. Ask students how many terms are in the expression. Ask students to identify the like terms. Ask students to simplify the expression by combining the like terms. Remind students that simplifying like terms is another important step in simplifying expressions and solving equations.

Possible answers: 5; 2a and a, 5 and 3; $a + 5b + 8$

Answers to Think and Discuss

Possible answers:

1. Use the Distributive Property to simplify $8(3y + 5)$ to $24y + 40$.

2. There are two sets of like terms: 5m, 2m, and 3m; and 8 and 6.

3. There are no like terms.

1-6 Combining Like Terms   **29**

FOR EXTRA PRACTICE

see page 732

internet connect

Homework Help Online

go.hrw.com Keyword: MP4 1-6

Students may want to refer back to the lesson examples.

Assignment Guide

If you finished Example **1** assign:
Core 1–6, 19–27, 56–66
Enriched 1–5 odd, 19–27, 47, 50, 54, 56–66

If you finished Example **2** assign:
Core 1–12, 19–33, 51, 56–66
Enriched 1–11 odd, 19–33, 46, 47, 50, 51, 53, 54, 56–66

If you finished Example **3** assign:
Core 1–15, 19–39, 51, 56–66
Enriched 1–15 odd, 19–39, 46, 47, 50, 51, 53, 54, 56–66

If you finished Example **4** assign:
Core 1–45, 50, 51, 56–66
Enriched 1–17 odd, 19–66

GUIDED PRACTICE

See Example **1** Combine like terms.

1. $7x - 3x$ **4x**

2. $2z + 5 + 3z$ **5z + 5**

3. $4f + 2 - 2f + 6 + 6f$ **8f + 8**

4. $9g + 8g$ **17g**

5. $5p - 8 - p$ **4p − 8**

6. $2x + 7 - x + 5 + 3x$ **4x + 12**

See Example **2**

7. $4x + 3y - x + 2y$ **3x + 5y**

8. $5x + 2y - y + 4x$ **9x + y**

9. $3x + 4y + 2x - 3y$ **5x + y**

10. $7p + 2p + 5z - 2z$ **9p + 3z**

11. $7g + 5h - 12$ **7g + 5h − 12**

12. $2h + 3m + 8h - 3m$ **10h**

See Example **3** Simplify.

13. $3(r + 2) - 2r$ **r + 6**

14. $5(2 + x) + 3x$ **10 + 8x**

15. $7(t + 8) - 5t$ **2t + 56**

See Example **4** Solve.

16. $4n - 2n = 84$ **n = 42**

17. $y + 3y = 96$ **y = 24**

18. $5p - 2p = 51$ **p = 17**

INDEPENDENT PRACTICE

See Example **1** Combine like terms.

19. $8y + 5y$ **13y**

20. $5z - 6 - 3z$ **2z − 6**

21. $2a + 4 - a + 7 + 6a$ **7a + 11**

22. $4z - z$ **3z**

23. $8x + 2 - 5x$ **3x + 2**

24. $9b + 6 - 3b - 3 - b$ **5b + 3**

25. $12p - 7p$ **5p**

26. $7a + 8 - 3a$ **4a + 8**

27. $2x + 8 + 2x - 5 + 5x$ **9x + 3**

See Example **2**

28. $2z + 5z + b - 7$ **7z + b − 7**

29. $4a + a + 3z - 2z$ **5a + z**

30. $9x + 8y + 2x - 8 - 4y$ **11x + 4y − 8**

31. $5x + 3 + 2x + 5q$ **7x + 5q + 3**

32. $7d - d + 3e + 12$ **6d + 3e + 12**

33. $15a + 6c + 3 - 6a + c$ **9a + 7c + 3**

See Example **3** Simplify.

34. $5(y + 2) - y$ **4y + 10**

35. $3(4y - 6) + 8y$ **20y − 18**

36. $4(x + 8) + 9x$ **13x + 32**

37. $2(3y + 4) + 9$ **6y + 17**

38. $6(2x + 8) - 9x$ **3x + 48**

39. $3(3x - 3) + 2x$ **11x − 9**

See Example **4** Solve.

40. $5x - x = 48$ **x = 12**

41. $8p - 3p = 25$ **p = 5**

42. $p + 2p = 18$ **p = 6**

43. $3y + 5y = 64$ **y = 8**

44. $a + 5a = 72$ **a = 12**

45. $9x - 5x = 56$ **x = 14**

PRACTICE AND PROBLEM SOLVING

Simplify.

46. $7(3l + 5k) - 14l + 12$ **7l + 35k + 12**

47. $6d + 8 + 5d - 3d - 7$ **8d + 1**

Solve.

48. $13(g + 2) = 78$ **g = 4**

49. $7x - 12 = x + 2 + 2x - 3x$ **x = 2**

Math Background

The Distributive Property is usually written in the form $a(b + c) = ab + ac$.

Another way to write it is in the form $ba + ca = (b + c)a$. When you combine like terms that have a variable, you are using the Distributive Property.

$5x + 3x = (5 + 3)x = 8x$
$ba + ca = (b + c)a$

So when you simplify expressions like $2(y + 9) + 3y$, you are actually using the Distributive Property twice.

$2(y + 9) + 3y = 2(y) + 2(9) + 3y$
$= 2y + 18 + 3y$
$= 2y + 3y + 18$
$= (2 + 3)y + 18$
$= 5y + 18$

RETEACH 1-6

LESSON 1-6 Reteach
Combining Like Terms

The parts of an expression separated by plus or minus signs are called **terms**.

$5a + 7b - 3a + 6a^2$

The expression shown has four terms. You can combine two of these terms to **simplify** the expression.

Like terms have the same variable raised to the same power

$\begin{aligned} 5a + 7b - 3a + 6a^2 \\ 5a - 3a + 7b + 6a^2 \\ 2a + 7b + 6a^2 \end{aligned}$

Equivalent expressions have the same value for all values of the variables.

Some algebraic equations can be solved by first combining like terms.
Solve $4w - w = 24$.

$4w - w = 24$ Identify like terms; w is $1w$.
$3w = 24$ Combine coefficients of like terms.
$\frac{3w}{3} = \frac{24}{3}$ Divide both sides by 3.
$w = 8$

Complete to combine like terms.

1. $9z + 4z$

$(9 + \underline{4})z$

$\underline{13} z$

2. $9r + 5q - 2r$

$(9 - \underline{2})r + 5q$

$\underline{7} r + 5q$

3. $5t + 12f - 1t - 3f$

$(5 - \underline{1})t + (12 - \underline{3})f$

$\underline{4} t + \underline{9} f$

Simplify.

4. $7m + 3n - m + 2n$

6m + 5n

5. $15r + 4 - 3r - 2$

12r + 2

6. $6x + 3z - y$

6x + 3z − y

Complete to solve.

7. $5h + 2h = 21$

$\underline{7} h = 21$

$\frac{7h}{7} = \frac{21}{7}$

$h = \underline{3}$

8. $16w - 5w = 44$

$\underline{11} w = 44$

$\frac{11w}{11} = \frac{44}{11}$

$w = \underline{4}$

9. $48 = 13x - x$

$48 = \underline{12} x$

$\frac{48}{12} = \frac{12x}{12}$

$\underline{4} = x$

PRACTICE 1-6

LESSON 1-6 Practice B
Combining Like Terms

Combine like terms.

1. $8a - 5a$ **3a**

2. $12g + 7g$ **19g**

3. $4a + 7a + 6$ **11a + 6**

4. $6x + 3y + 5x$ **11x + 3y**

5. $10k - 3k + 5h$ **7k + 5h**

6. $3p - 7q + 14p$ **17p − 7q**

7. $3k + 7k + 5k$ **15k**

8. $5c + 12d - 6$ **5c + 12d − 6**

9. $13 + 4b + 6b - 5$ **8 + 10b**

10. $4f + 6 + 7f - 2$ **11f + 4**

11. $x + y + 3x + 7y$ **4x + 8y**

12. $9n + 13 - 8n - 6$ **n + 7**

Simplify.

13. $4(x + 3) - 5$ **4x + 7**

14. $6(7 + x) + 5x$ **42 + 11x**

15. $3(5 + 3x) - 4x$ **15 + 5x**

Solve.

16. $6y + 2y = 16$ **y = 2**

17. $14b - 9b = 35$ **b = 7**

18. $3q + 9q = 48$ **q = 4**

19. Gregg has q quarters and p pennies. His brother has 4 times as many quarters and 8 times as many pennies as Gregg has. Write the sum of the number of coins they have, and then combine like terms.

q + p + 4q + 8p; 5q + 9p

20. If Gregg has 6 quarters and 15 pennies, how many total coins do Gregg and his brother have?

165 coins

Write and simplify an expression for each situation.

50. **BUSINESS** A museum charges $5 for each adult ticket, plus an additional $1 per ticket for tax. What is the total cost of x tickets? $5x + 1x = 6x$

51. **SPORTS** Use the information below to find how many medals of each kind were won by the four countries in the 2000 Summer Olympics. $52g + 41s + 49b$

United States	Great Britain	Brazil	Lithuania
39 Gold	11 Gold	0 Gold	2 Gold
25 Silver	10 Silver	6 Silver	0 Silver
33 Bronze	7 Bronze	6 Bronze	3 Bronze

Write and solve an equation for each situation.

52. **BUSINESS** The accounting department ordered 12 cases of paper, and the marketing department ordered 20 cases of paper. If the total cost of the combined order was $896 before taxes, what is the price of each case of paper? $12x + 20x = 896; x = 28; \$28$

53. **WHAT'S THE ERROR?** A student said that $2x + 3y$ can be simplified to $5xy$ by combining like terms. What error did the student make?

54. **WRITE ABOUT IT** Write an expression that can be simplified by combining like terms. Then write an expression that cannot be simplified, and explain why it is already in simplest form.

55. **CHALLENGE** Simplify and solve $2(7x + 5 - 3x) + 4(2x - 2) = 50$. $16x + 2 = 50; x = 3$

Spiral Review

Solve each equation. (Lesson 1-3)

56. $4 + x = 13$ $x = 9$
57. $x - 4 = 9$ $x = 13$
58. $17 = x + 9$ $x = 8$
59. $19 = x + 11$ $x = 8$
60. $5 + x = 22$ $x = 17$
61. $x - 24 = 8$ $x = 32$
62. $x - 7 = 31$ $x = 38$
63. $41 = x + 25$ $x = 16$
64. $x + 8 = 15$ $x = 7$

65. **TEST PREP** Determine which value of x is a solution of the equation $3x + 2 = 11$. (Lesson 1-3) **B**

A $x = 2.2$
B $x = 3$
C $x = 4.3$
D $x = 3.6$

66. **TEST PREP** Determine which value of x is a solution of the equation $4x - 3 = 13$. (Lesson 1-3) **J**

F $x = 3$
G $x = 3.5$
H $x = 2.5$
J $x = 4$

CHALLENGE 1-6

LESSON 1-6 Challenge
Mission Operation

You can create your own operation. Take a symbol, such as ▲, and tell what that symbol means using the standard operations of +, −, ×, and ÷.

Example
If x ▲ $y = x + 2xy$, find 5 ▲ 3.

x ▲ $y = x + 2xy$ Use the operation as defined.
5 ▲ 3 = 5 + 2(5)(3)   Substitute 5 for x and 3 for y.
       = 5 + 30   Carry out the standard operations.
       = 35

Apply the definition of ▲. First, show how to substitute the values or expressions that replace x and y. Then, carry out the standard operations to simplify completely.

1. 7 ▲ 2
$$\frac{7 + 2(7)(2)}{35}$$

2. 2 ▲ 7
$$\frac{2 + 2(2)(7)}{30}$$

3. $(a \blacktriangle b) + (b \blacktriangle a)$
$$\frac{a + 2ab + b + 2ba}{a + 4ab + b}$$

4. $[a \blacktriangle (4b)] + [(2a) \blacktriangle b]$
$$\frac{a + 2a(4b) + 2a + 2(2a)(b)}{3a + 12ab}$$

Let $a \not\approx b = 2a + \frac{b}{2}$. Apply the definition of ✳. Simplify the resulting expression.

5. $(9 ✳ 4) + (3 ✳ 8)$
$$\frac{2(9) + \frac{4}{2} + 2(3) + \frac{8}{2}}{30}$$

6. $(r ✳ 4t) + (4r ✳ 2t)$
$$\frac{\left(2r + \frac{4t}{2}\right) + 2(4r) + \frac{2t}{2}}{10r + 3t}$$

PROBLEM SOLVING 1-6

LESSON 1-6 Problem Solving
Combining Like Terms

Write the correct answer.

1. An item costs x dollars. The tax rate is 5% of the cost of the item, or 0.05x. Write and simplify an expression to find the total cost of the item with tax.
$x + 0.05x = 1.05x$

2. A sweater costs d dollars at regular price. The sweater is reduced by 20%, or 0.2d. Write and simplify an expression to find the cost of the sweater before tax.
$d - 0.2d = 0.8d$

3. Consecutive integers are integers that differ by one. You can represent consecutive integers as x, $x + 1$, $x + 2$ and so on. Write an equation and solve to find three consecutive integers whose sum is 33.
10, 11, 12

4. Consecutive even integers can be represented by x, $x + 2$, $x + 4$ and so on. Write an equation and solve to find three consecutive even integers whose sum is 54.
16, 18, 20

Choose the letter for the best answer.

5. In Super Bowl XXXV, the total number of points scored was 41. The winning team outscored the losing team by 27 points. What was the final score of the game?
A 33 to 8
B 34 to 7
C 22 to 2
D 18 to 6

6. A high school basketball court is 34 feet longer than it is wide. If the perimeter of the court is 268, what are the dimensions of the court?
F 234 ft by 34 ft
G 67 ft by 67 ft
H 70 ft by 36 ft
J 84 ft by 50 ft

7. Julia ordered 2 hamburgers and Steven ordered 3 hamburgers. If their total bill before tax was $7.50, how much did each hamburger cost?
A $1.50
B $1.25
C $1.15
D $1.02

8. On three tests, a student scored a total of 258 points. If the student improved his performance on each test by 5 points, what was the score on each test?
F 81, 86, 91
G 80, 85, 90
H 75, 80, 85
J 70, 75, 80

Chapter
1 **Mid-Chapter Quiz**

Purpose: *To assess students' mastery of concepts and skills in Lessons 1-1 through 1-6*

Assessment Resources

Section 1A Quiz
Assessment Resources . . . . p. 5

 Test and Practice Generator CD-ROM

Additional mid-chapter assessment items in both multiple-choice and free-response format may be generated for any objective in Lessons 1-1 through 1-6.

Answers

22.
9   9.5   9.7   10

23.
0 1 2 3 4 5 6 7 8 9 10

24.
6.5   7.1   7.5

25.
0   4   8   10   12   16

Mid-Chapter Quiz

LESSON **1-1** (pp. 4–7)

Evaluate each expression for the given values of the variables.

1. $4x + 7y$ **63**
for $x = 7$ and $y = 5$

2. $5(r - 8t)$ **340**
for $r = 100$ and $t = 4$

3. $2(3m + 7n)$ **190**
for $m = 13$ and $n = 8$

LESSON **1-2** (pp. 8–12)

Write an algebraic expression for each word phrase.

4. 12 more than twice a number n **$2n + 12$**

5. 5 less than 3 times a number b **$3b - 5$**

6. 6 times the sum of p and 3   **$6(p + 3)$**

7. 10 plus the product of 16 and m **$10 + 16m$**

Write an algebraic expression to represent the problem situation.

8. Sami has a calendar with 365 pages of cartoons. After she tears off p pages, how many pages of cartoons remain?   **$365 - p$**

LESSON **1-3** (pp. 13–17)

Solve.

9. $5 + x = 26$ **$x = 21$**

10. $p - 8 = 16$ **$p = 24$**

11. $32 = h + 21$ **$h = 11$**

12. $60 = k - 33$ **$k = 93$**

Write and solve an algebraic equation for the word problem.

13. The deepest location in Lake Superior is 1333 feet, which is 1123 feet deeper than the deepest location in Lake Erie. What is the deepest location in Lake Erie?   **Let e = depth of Lake Erie; $e + 1123 = 1333$; $e = 210$; The deepest location in Lake Erie is 210 ft.**

LESSON **1-4** (pp. 18–22)

Solve.

14. $4m = 88$ **$m = 22$**

15. $\frac{w}{50} = 50$ **$w = 2500$**

16. $100y = 50$ **$y = \frac{1}{2}$**

17. $\frac{1}{2}x = 16$ **$x = 32$**

18. $3x + 4 = 10$ **$x = 2$**

19. $4z - 1 = 11$ **$z = 3$**

20. $\frac{1}{3}y - 2 = 7$ **$y = 27$**

21. $16 = 10 + 2m$ **$m = 3$**

LESSON **1-5** (pp. 23–27)   **For graphs, see margin.**

Solve and graph each inequality.

22. $x + 2.3 < 12$ **$x < 9.7$**

23. $3n > 15$ **$n > 5$**

24. $y - 4.1 \geq 3$ **$y \geq 7.1$**

25. $6 \leq \frac{z}{2}$ **$z \geq 12$**

LESSON **1-6** (pp. 28–31)

Solve.

26. $7y - 4y = 6$ **$y = 2$**

27. $\frac{5x + 3x}{2} = 20$ **$x = 5$**

28. $2(t + 5t) = 48$ **$t = 4$**

Focus on Problem Solving

Solve

• **Choose an operation: Addition or Subtraction**

To decide whether to add or subtract, you need to determine what action is taking place in the problem. If you are combining numbers or putting numbers together, you need to add. If you are taking away or finding out how far apart two numbers are, you need to subtract.

Action	Operation	Illustration
Combining or putting together	Add	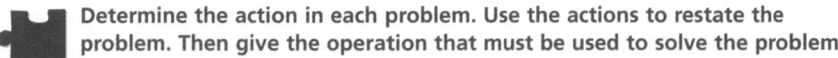
Removing or taking away	Subtract	
Finding the difference	Subtract	

Jan has 10 red marbles. Joe gives her 3 more. How many marbles does Jan have now? The action is combining marbles. Add 10 and 3.

Determine the action in each problem. Use the actions to restate the problem. Then give the operation that must be used to solve the problem.

❶ The state of Michigan is made up of two parts, the Lower Peninsula and the Upper Peninsula. The Upper Peninsula has an area of about 16,400 mi², and the Lower Peninsula has an area of about 40,400 mi². Estimate the area of the state.

❷ The average temperature in Homer, Alaska, is 53.4°F in July and 24.3°F in December. Find the difference between the average temperature in Homer in July and in December.

❸ Einar has $18 to spend on his friend's birthday presents. He buys one present that costs $12.35. How much does he have left to spend?

❹ Dinah got 87 points on her first test and 93 points on her second test. What is her combined point total for the first two tests?

Answers

1. 16,400 + 40,400 = 56,800 mi²

2. 53.4 − 24.3 = 29.1°F

3. 18 − 12.35 = $5.65

4. 87 + 93 = 180 points

Focus on Problem Solving

Purpose: *To focus on choosing an operation: addition or subtraction*

Problem Solving Resources

Interactive Problem Solving . . . . pp. 1–9

Math: Reading and Writing in the Content Area . . . . pp. 1–9

Problem Solving Process

This page focuses on the third step of the problem-solving process: **Solve**

Discuss

Have students discuss the action that is taking place in each exercise and the operation that they chose to perform that action.

Possible answers:

1. Combining; combine the area of the Lower Peninsula of Michigan with the area of the Upper Peninsula of Michigan: 16,400 + 40,400.

2. Finding the difference; find the temperature difference in Homer between July and December: 53.4 − 24.3.

3. Taking away; if Einar took away $12.35 from $18.00, find how much he has left: 18 − 12.35.

4. Combining; combine the point totals 87 and 93 to find the total points for the first two tests: 87 + 93.

Lesson	Materials	Resources
Lesson 1-7 Ordered Pairs **NCTM:** Algebra, Geometry, Communication, Representation **NAEP:** Algebra 2c ☑ SAT-9 ☑ SAT-10 ☑ ITBS ☑ CTBS ☑ MAT ☑ CAT		• *Chapter 1 Resource Book,* pp. 62–70 • Daily Transparency T29, CRB • Additional Examples Transparencies T30–T32, CRB • *Alternate Openers: Explorations,* p. 7
Lesson 1-8 Graphing on a Coordinate Plane **NCTM:** Algebra, Geometry, Communication, Representation **NAEP:** Algebra 2c ☑ SAT-9 ☑ SAT-10 ☑ ITBS ☑ CTBS ☑ MAT ☑ CAT	**Optional** Street map Teaching Transparency T34 (CRB) Recording Sheet for Reaching All Learners (CRB, p. 105)	• *Chapter 1 Resource Book,* pp. 71–79 • Daily Transparency T33, CRB • Additional Examples Transparencies T35–T37, CRB • *Alternate Openers: Explorations,* p. 8
Technology Lab 1A Create a Table of Solutions **NCTM:** Algebra, Representation **NAEP:** Algebra 2a ☑ SAT-9 ☑ SAT-10 ☐ ITBS ☐ CTBS ☑ MAT ☑ CAT	**Required** Graphing calculators	• *Technology Lab Activities,* p. 1
Lesson 1-9 Interpreting Graphs and Tables **NCTM:** Algebra, Communication, Representation **NAEP:** Data Analysis and Probability 1a ☐ SAT-9 ☑ SAT-10 ☐ ITBS ☑ CTBS ☑ MAT ☑ CAT	**Optional** Graphs from newspapers or magazines	• *Chapter 1 Resource Book,* pp. 80–89 • Daily Transparency T38, CRB • Additional Examples Transparencies T39–T40, CRB • *Alternate Openers: Explorations,* p. 9
Section 1B Assessment		• Section 1B Quiz, AR p. 6 • *Test and Practice Generator* CD-ROM

SAT = *Stanford Achievement Tests*   **ITBS** = *Iowa Test of Basic Skills*   **CTBS** = *Comprehensive Test of Basic Skills/Terra Nova*

MAT = *Metropolitan Achievement Test*   **CAT** = *California Achievement Test*

NCTM = Complete standards can be found on pages T29–T35.   **NAEP** = Complete standards can be found on pages A54–A58.

SE = *Student Edition*   **TE** = *Teacher's Edition*   **AR** = *Assessment Resources*   **CRB** = *Chapter Resource Book*   **MK** = *Manipulatives Kit*

Section Overview

Ordered Pair Solutions of Equations in Two Variables

Lesson 1-7, Technology Lab 1A

Why? Functional relationships between two quantities can be expressed using ordered pairs of numbers.

A solution of a two-variable equation is written as an **ordered pair.**

When the numbers in the ordered pair are substituted in the equation, they make it true.

Find solutions of $y = x + 3$.

$(1, 4)$ is a solution. $\rightarrow 4 = 1 + 3$

Make a table of solutions for $y = x + 3$.

x	$y = x + 3$	(x, y)
1	4	$(1, 4)$
2	5	$(2, 5)$
4	7	$(4, 7)$

Graphing an Equation

Lesson 1-8

Why? The graph of an equation displays relationships and patterns between two variables.

Graph the equation $y = x + 3$.

From the table above, $(1, 4)$, $(2, 5)$, and $(4, 7)$ are points of the equation.

Plot these points.

Draw the line through the points. The line represents all possible solutions.

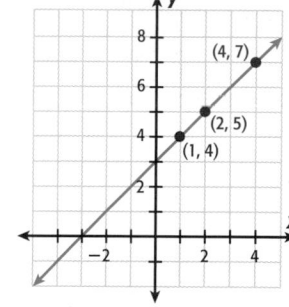

The graph of an equation is the set of all ordered pairs that are solutions of the equation.

Interpreting Graphs and Tables

Lesson 1-9

Why? Real-world situations may be described by tables or graphs.

Sally walks to the end of the street and then quickly runs down a steep hill. At the bottom of the hill, Sally stops to tie her shoe. Then she walks over to Amy's house and stops.

Time	Speed (mi/h)
4:00	4
4:01	7
4:02	0
4:03	4
4:04	0

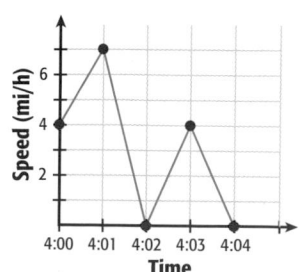

Pacing: Traditional 1 day
Block $\frac{1}{2}$ day

Objective: Students write solutions of equations in two variables as ordered pairs.

Warm Up
Solve.

1. $x - 8 = 19$ $x = 27$

2. $5 = a - 2$ $a = 7$

3. $7 + n = 24$ $n = 17$

4. $3c - 7 = 32$ $c = 13$

5. $17y + 7 = 58$ $y = 3$

Problem of the Day
A moving van travels 50 miles per hour. Use the equation $y = 50x$. How far will the van travel in 4.5 hours? **225 miles**

Available on Daily Transparency in CRB

Math Humor

Can you believe a student actually tried to cheat on his graphing test? Some people just don't know where to draw the line.

1-7 Ordered Pairs

Learn to write solutions of equations in two variables as ordered pairs.

Vocabulary
ordered pair

A sign at the store reads "Birthday Banners $8. Personalize for $1 per letter."

Cecelia has 7 letters in her name, and Dowen has 5 letters in his. Figure out how much it will cost to get a personalized birthday banner for each of them.

| Price of banner | = | $8 | + | $1 | · | Number of letters in name |

Let y be the price of the banner and x be the number of letters in the name; the equation for the price of a banner is $y = 8 + x$.

For Cecelia's banner: $x = 7$, $y = 8 + 7$ or $y = 15$
For Dowen's banner: $x = 5$, $y = 8 + 5$ or $y = 13$

A solution of a two-variable equation is written as an **ordered pair**. When the numbers in the ordered pair are substituted in the equation, the equation is true.

Ordered pair
$$(x, y)$$

$(7, 15)$ is a solution → $15 = 8 + 7$
$(5, 13)$ is a solution → $13 = 8 + 5$

EXAMPLE 1 **Deciding Whether an Ordered Pair Is a Solution of an Equation**

Determine whether each ordered pair is a solution of $y = 3x + 2$.

A $(1, 4)$

$y = 3x + 2$
$4 \overset{?}{=} 3(1) + 2$ *Substitute 1 for x and 4 for y.*
$4 \overset{?}{=} 5$ ✗

$(1, 4)$ is *not* a solution.

Helpful Hint

The order in which a solution is written is important. The first variable is called the *independent variable,* and the second variable is called the *dependent variable.*

B $(2, 8)$

$y = 3x + 2$
$8 \overset{?}{=} 3(2) + 2$ *Substitute 2 for x and 8 for y.*
$8 \overset{?}{=} 8$ ✔ *A solution since 8 = 8*

$(2, 8)$ is a solution.

C $(16, 50)$

$y = 3x + 2$
$50 \overset{?}{=} 3(16) + 2$ *Substitute 16 for x and 50 for y.*
$50 \overset{?}{=} 50$ ✔

$(16, 50)$ is a solution.

---

1 Introduce

Alternate Opener

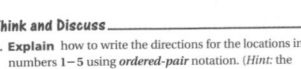

EXPLORATION

1-7 Ordered Pairs

You are about to give directions to the locations labeled on the grid. First find where you are on the map. The only restrictions are the following:

- You can move only horizontally (sideways) or vertically (up and down). Use the directions *left, right, west,* or *east* for each first move, and *up, down, north,* or *south* for each second move.
- Your first move should be horizontal, and your second move should be vertical.

Give directions to go to each location.

1. the museum
2. the park
3. the mall
4. the zoo
5. the planetarium

Think and Discuss

6. **Explain** how to write the directions for the locations in numbers 1–5 using *ordered-pair* notation. (*Hint:* the museum is at $(-4, 6)$.)

Motivate

Show the students a simple two-variable equation, such as $x + y = 10$. Ask students to supply pairs of numbers whose sum is 10. Organize their answers in a table using one value of each pair for x and the other in the same pair for y. Explain to students that because there are two variables, each solution is a pair of numbers.

Exploration worksheet and answers on Chapter 1 Resource Book pp. 63 and 118

2 Teach

Lesson Presentation

Guided Instruction

In this lesson, students learn to write solutions of equations in two variables as ordered pairs. To see if an ordered pair is a solution of a two-variable equation, substitute the first number in the ordered pair for x and the second number in the ordered pair for y. If the result is true, then the ordered pair is a solution of the equation. To find a solution of a two-variable equation, substitute any number for x in the equation, and then solve the equation for y.

EXAMPLE 2 **Creating a Table of Ordered Pair Solutions**

Use the given values to make a table of solutions.

A $y = 3x$ for $x = 1, 2, 3, 4$

Helpful Hint

A table of solutions can be set up vertically or horizontally.

x	$3x$	y	(x, y)
1	3(1)	3	(1, 3)
2	3(2)	6	(2, 6)
3	3(3)	9	(3, 9)
4	3(4)	12	(4, 12)

B $n = 4m - 3$ for $m = 1, 2, 3, 4$

m	1	2	3	4
$4m - 3$	4(1) − 3	4(2) − 3	4(3) − 3	4(4) − 3
n	1	5	9	13
(m, n)	(1, 1)	(2, 5)	(3, 9)	(4, 13)

EXAMPLE 3 **Retail Application**

In most states, the price of each item is not the total cost. Sales tax must be added. If sales tax is 8 percent, the equation for total cost is $c = 1.08p$, where p is the price before tax.

A How much will Cecelia's $15 banner cost after sales tax?

$c = 1.08(15)$ *The price of Cecelia's banner before tax is $15.*

$c = 16.2$

Cecelia's banner is $15.00, and after tax it will cost $16.20, so (15, 16.20) is a solution of the equation.

B How much will Dowen's $13 banner cost after sales tax?

$c = 1.08(13)$ *The price of Dowen's banner before tax is $13.*

$c = 14.04$

Dowen's banner is $13.00, and after tax it will cost $14.04, so (13.00, 14.04) is a solution of the equation.

Think and Discuss

1. Describe how to find a solution of a two-variable equation.

2. Explain why an equation with two variables has an infinite number of solutions.

3. Give two equations using x and y that have (1, 2) as a solution.

Additional Example

Example 1

Determine whether each ordered pair is a solution of $y = 4x - 1$.

A. (3, 11) **B.** (10, 3) **C.** (11, 43)
   yes   no   yes

Example 2

Use the given values to make a table of solutions.

A. $y = x + 3$ for $x = 1, 2, 3, 4$

x	$x + 3$	y	(x, y)
1	1 + 3	4	(1, 4)
2	2 + 3	5	(2, 5)
3	3 + 3	6	(3, 6)
4	4 + 3	7	(4, 7)

B. $n = 6m - 5$ for $m = 1, 2, 3$

m	1	2	3
$6m - 5$	6(1) − 5	6(2) − 5	6(3) − 5
n	1	7	13
(m, n)	(1, 1)	(2, 7)	(3, 13)

Example 3

A salesman marks up the price of everything he sells by 20%. The equation for the sales price p is $p = 1.2w$, where w is wholesale cost.

A. What will be the sales price of a sweater with a wholesale cost of $48? **$57.60**

B. What will be the sales price of a jacket with a wholesale cost of $85? **$102**

---

3 Close

Reaching All Learners

Through Number Sense

Have students make a table of solutions for $y = 2x$ using $x = 1, 2, 3,$ and 4. When they have the solutions (1, 2), (2, 4), (3, 6), and (4, 8), have them describe any patterns they notice. Help students see that the y-value is always twice the x-value. Have them find y when x is 1000 ($y = 2000$). Then have them find x when y is 1000 ($x = 500$).

Summarize

Discuss the similarities and differences between one-variable equations and two-variable equations. Ask students to define an ordered pair. Make sure students understand that two-variable equations have an infinite number of solutions.

Possible answers: Both types of equations have variables, and solutions can be checked by substituting values and solving. One-variable equations usually have one solution that is a single number, but two-variable equations have many solutions that are ordered pairs of numbers.

Answers to Think and Discuss

Possible answers:

1. Substitute a value for one variable into the equation and solve to find the value of the other variable.

2. In general, there is no limit to the number of values you can substitute into the equation. So there is an unlimited number of ordered pairs that make the equation true.

3. $y = x + 1$ and $y = 2x$

FOR EXTRA PRACTICE
see page 733

🔲 internet connect 🔲
Homework Help Online
go.hrw.com Keyword: MP4 1-7

Students may want to refer back to the lesson examples.

GUIDED PRACTICE

See Example **1** Determine whether each ordered pair is a solution of $y = 2x - 4$.

1. $(2, 1)$ no **2.** $(4, 4)$ yes **3.** $(6, 8)$ yes **4.** $(5, 5)$ no

See Example **2** Use the given values to make a table of solutions.

5. $y = 2x$ for $x = 1, 2, 3, 4, 5, 6$ **6.** $y = 3x - 2$ for $x = 1, 2, 3, 4, 5, 6$

See Example **3** **7.** The cost of mailing a letter is $0.23 per ounce plus $0.14. The equation that gives the total cost c of mailing a letter is $c = 23w + 14$, where w is the weight in ounces. What is the cost of mailing a 5-ounce letter? **$1.29**

INDEPENDENT PRACTICE

See Example **1** Determine whether each ordered pair is a solution of $y = 4x + 3$.

8. $(1, 7)$ yes **9.** $(4, 20)$ no **10.** $(2, 11)$ yes **11.** $(6, 25)$ no

See Example **2** Use the given values to make a table of solutions.

12. $y = 4x - 1$ for $x = 1, 2, 3, 4, 5, 6$ **13.** $y = 2x + 8$ for $x = 1, 2, 3, 4, 5, 6$

14. $y = 2x - 3$ for $x = 2, 4, 6, 8, 10$ **15.** $y = 3x - 4$ for $x = 2, 4, 6, 8, 10$

See Example **3** **16.** The fine for speeding in one town is $75 plus $6 for every mile over the speed limit. The equation that gives the total cost c of a speeding ticket is $c = 75 + 6m$, where m is the number of miles over the posted speed limit. Terry was issued a ticket for going 71 mi/h in a 55 mi/h zone. What was the total cost of the ticket? **$171**

PRACTICE AND PROBLEM SOLVING

Determine whether each ordered pair is a solution of $y = x + 3$.

17. $(2, 5)$ yes **18.** $(4, 6)$ no **19.** $(5, 8)$ yes **20.** $(3, 7)$ no

Determine whether each ordered pair is a solution of $y = 3x - 5$.

21. $(2, 2)$ no **22.** $(4, 7)$ yes **23.** $(6, 13)$ yes **24.** $(5, 10)$ yes

Use the given values to make a table of solutions.

25. $y = 4x - 3$ for $x = 1, 2, 3, 4, 5, 6$ **26.** $y = 3x - 1$ for $x = 1, 2, 3, 4, 5, 6$

27. $y = x + 8$ for $x = 1, 2, 3, 4, 5, 6$ **28.** $y = 2x + 1$ for $x = 2, 4, 6, 8, 10$

29. $y = 2x + 4$ for $x = 2, 4, 6, 8, 10$ **30.** $y = 2x - 3$ for $x = 3, 6, 9, 12, 15$

31. Name an equation with solutions $(1, 1)$, $(2, 2)$, $(3, 3)$, and (n, n) for all values of n. **Possible answer:** $x = y$

Assignment Guide

If you finished Example **1** assign:
Core 1–4, 8–11, 17–23 odd, 39–46
Enriched 8–11, 17–24, 39–46

If you finished Example **2** assign:
Core 1–6, 8–15, 17–29 odd, 39–46
Enriched 8–15 odd, 17–31, 34, 36, 37, 39–46

If you finished Example **3** assign:
Core 1–16, 17–29 odd, 32–35, 39–46
Enriched 8–15 odd, 16–46

Answers

5–6, 12–15. Complete answers on p. A1.

5. (1, 2); (2, 4); (3, 6); (4, 8); (5, 10); (6, 12)

6. (1, 1); (2, 4); (3, 7); (4, 10); (5, 13); (6, 16)

12. (1, 3); (2, 7); (3, 11); (4, 15); (5, 19); (6, 23)

13. (1, 10); (2, 12); (3, 14); (4, 16); (5, 18); (6, 20)

14. (2, 1); (4, 5); (6, 9); (8, 13); (10, 17)

15. (2, 2); (4, 8); (6, 14); (8, 20); (10, 26)

Math Background

In Think and Discuss problem 2, it is noted that a two-variable equation generally has an infinite number of solutions. It is easy to see that some solutions of the equation $y = x + 2$ are (1, 3), (2, 4), (3, 5), (4, 6), (5, 7), etc. What may not be as obvious is that solutions can contain negative values and non-integer values. Some other solutions of the same equation are $(-1, 1)$, $(1.4, 3.4)$, and $\left(\frac{3}{8}, 2\frac{3}{8}\right)$.

It is also important to note that to find a solution of a two-variable equation, you can start by picking a value for y. For example, to find a solution of $y = x + 2$, you can let $y = 10$; then $x = 8$.

RETEACH 1-7

LESSON 1-7 Reteach
Ordered Pairs

An **ordered pair** can be used to write a solution for a two-variable equation. For the equation $y = x + 5$, a solution is (0, 5). When the x-value is 0, the y-value is 5.

(0, 5)
x-value y-value

Which of the ordered pairs (5, 3) or (3, 5) is a solution of $y = 2x - 1$?

$y = 2x - 1$
$3 \stackrel{?}{=} 2(5) - 1$ — Substitute 5 for x and 3 for y.
$3 \stackrel{?}{=} 10 - 1$
$3 \neq 9$
So, (5, 3) *is not* a solution of $y = 2x - 1$.

$y = 2x - 1$
$5 \stackrel{?}{=} 2(3) - 1$ — Substitute 3 for x and 5 for y.
$5 \stackrel{?}{=} 6 - 1$
$5 = 5$
So, (3, 5) *is* a solution of $y = 2x - 1$.

Determine whether each ordered pair is a solution of the given equation. Write *is* or *is not*.

1. $y = 4x + 3$; (1, 6) — is not
2. $y = 4x + 3$; (0, 3) — is
3. $y = x + 3$; (3, 0) — is not
4. $y = 5x$; (3, 15) — is
5. $y = 3x - 4$; (5, 3) — is not
6. $y = 6 - x$; (4, 2) — is

A two-variable equation has infinitely many solutions. Use a table to find and record some solutions to a given equation. Use $x = 1, 2,$ and 3, for example, to make a table of values for $y = 5x - 1$. Substitute each given value of x in the expression for x. Evaluate the expression to find the value of y that completes the ordered pair.

x	$5x - 1 = y$	(x, y)
1	$5(1) - 1 = 4$	(1, 4)
2	$5(2) - 1 = 9$	(2, 9)
3	$5(3) - 1 = 14$	(3, 14)

Complete each table.

7. $y = 4x$

x	$4x = y$	(x, y)
0	$4(0) = 0$	(0, 0)
1	$4(1) = 4$	(1, 4)
2	$4(2) = 8$	(2, 8)

8. $y = 5x - 3$

x	$5x - 3 = y$	(x, y)
1	$5(1) - 3 = 2$	(1, 2)
2	$5(2) - 3 = 7$	(2, 7)
3	$5(3) - 3 = 12$	(3, 12)

PRACTICE 1-7

LESSON 1-7 Practice B
Ordered Pairs

Determine whether each ordered pair is a solution of $y = 4 + 2x$.

1. (1, 1) — no **2.** (2, 8) — yes **3.** (0, 4) — yes **4.** (8, 2) — no

Determine whether each ordered pair is a solution of $y = 3x - 2$.

5. (1, 1) — yes **6.** (3, 7) — yes **7.** (5, 15) — no **8.** (6, 16) — yes

Use the given values to complete the table of solutions.

9. $y = x + 5$ for $x = 0, 1, 2, 3, 4$

x	$x + 5$	y	(x, y)
0	$0 + 5$	5	(0, 5)
1	$1 + 5$	6	(1, 6)
2	$2 + 5$	7	(2, 7)
3	$3 + 5$	8	(3, 8)
4	$4 + 5$	9	(4, 9)

10. $y = 3x + 1$ for $x = 1, 2, 3, 4, 5$

x	$3x + 1$	y	(x, y)
1	$3(1) + 1$	4	(1, 4)
2	$3(2) + 1$	7	(2, 7)
3	$3(3) + 1$	10	(3, 10)
4	$3(4) + 1$	13	(4, 13)
5	$3(5) + 1$	16	(5, 16)

11. $y = 2x + 6$ for $x = 0, 1, 2, 3, 4$

x	$2x + 6$	y	(x, y)
0	$2(0) + 6$	6	(0, 6)
1	$2(1) + 6$	8	(1, 8)
2	$2(2) + 6$	10	(2, 10)
3	$2(3) + 6$	12	(3, 12)
4	$2(4) + 6$	14	(4, 14)

12. $y = 4x - 2$ for $x = 2, 4, 6, 8, 10$

x	$4x - 2$	y	(x, y)
2	$4(2) - 2$	6	(2, 6)
4	$4(4) - 2$	14	(4, 14)
6	$4(6) - 2$	22	(6, 22)
8	$4(8) - 2$	30	(8, 30)
10	$4(10) - 2$	38	(10, 38)

13. Alexis opened a savings account with a $120 deposit. Each week she will put $20 into the account. The equation that gives the total amount t in her account is $t = 120 + 20w$, where w is the number of weeks since she opened the account. How much money will Alexis have in her savings account after 5 weeks?

$220

32. BUSINESS The manager of a pizza restaurant finds that its daily food cost is $60 plus $3 per pizza. Write an equation for food cost c in terms of the number of pizzas sold p. Then solve the equation to find the daily food cost on a day when 113 pizzas were sold. Write your answer as an ordered pair. **$c = 3p + 60$; (113, 399)**

33. GEOMETRY The perimeter P of a square is four times the length of one side s, which can be expressed as $P = 4s$. Is (13, 51) a solution of this equation? If not, find a solution that uses one or the other of the given values. **No; (13, 52) or (12.75, 51)**

34. Given the equation $y = 2x - 8$, find the ordered-pair solution when $x = 4$ and the ordered-pair solution when $y = 4$. **(4, 0) and (6, 4)**

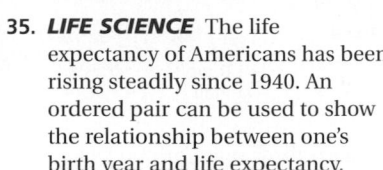

In 1513, Ponce de León went in search of the legendary Fountain of Youth, which people believed would give them eternal youth. While searching, he discovered Florida, which he named Pascua de Florida.

35. LIFE SCIENCE The life expectancy of Americans has been rising steadily since 1940. An ordered pair can be used to show the relationship between one's birth year and life expectancy.

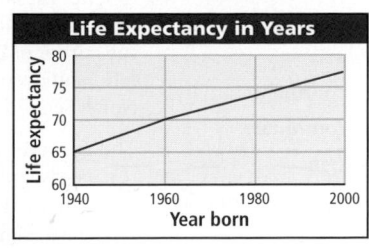

a. Write an ordered pair that shows the approximate life expectancy of an American born in 1980. **(1980, 74)**

b. The data on the chart can be approximated by the equation $L = 0.2n - 323$, where L is the life expectancy and n is the year of birth. Use the equation to find an ordered pair that shows the approximate life expectancy for an American born in 2020. **(2020, 81)**

36. WHAT'S THE ERROR? A table of solutions shows that (4, 10) is a solution to the equation $y = \frac{x}{2} - 1$. What's the error?

37. WRITE ABOUT IT Write an equation that has (3, 5) as a solution. Explain how you found the equation.

38. CHALLENGE In football, a touchdown is worth 6 points and a field goal is worth 3 points. If x equals the number of touchdowns scored, and y equals the number of field goals scored, find the possible solutions of the equation $54 = 6x + 3y$. **(0, 18), (1, 16), (2, 14), (3, 12), (4, 10), (5, 8), (6, 6), (7, 4), (8, 2), and (9, 0)**

Spiral Review

Evaluate each expression for the given value of the variable. (Lesson 1-1)

39. $x - 4$ for $x = 11$ **7**

40. $2x + 3$ for $x = 9$ **21**

41. $3x - 2$ for $x = 2$ **4**

42. $4(x + 1)$ for $x = 8$ **36**

43. $3(x - 1)$ for $x = 5$ **12**

44. $2(x + 4)$ for $x = 3$ **14**

45. TEST PREP Determine which value of x is the solution of $x + 3 = 14$. (Lesson 1-3) **B**

A 9
B 11
C 17
D 21

46. TEST PREP Determine which value of x is the solution of $x - 4 = 3$. (Lesson 1-3) **J**

F 1
G 12
H −1
J 7

Journal

Give students examples like the following: The more apples I buy, the more money it is going to cost me. The longer I ride my bicycle, the more distance I cover. Ask students to write about any real-world situation in which one quantity is directly affected by the other.

Test Prep Doctor

When answering multiple-choice questions, students sometimes stop looking at the other choices when they think that they have found the correct one. For Exercises 45 and 46, students should substitute all choices into the equation to make sure that their choice is the only correct one.

CHALLENGE 1-7

LESSON 1-7 Challenge

1-2-3, x-y-z

The solution of a three-variable equation is written as an ordered triple. For example, (x, y, z) represents a solution of the equation $3x + 2y + 7z = 32$. You can verify that $(1, 4, 3)$ is a solution of $3x + 2y + 7z = 32$ by substituting $x = 1$, $y = 4$, and $z = 3$.

Arrange the numbers of the given ordered triple so that the result is a solution of the given equation.

1. (0, 0, 8)   **2.** (7, 3, 0)   **3.** (2, 3, 10)

$5x + 8y - 3z = 64$ $4x + 5y - z = 8$ $7x - y + 2z = 10$

__(0, 8, 0)__   __(0, 3, 7)__   __(2, 10, 3)__

4. $\left(\frac{1}{2}, \frac{1}{3}, \frac{1}{4}\right)$ **5.** $\left(4, 0, \frac{1}{5}\right)$ **6.** $\left(2, \frac{1}{3}, 1\right)$

$10x + 9y + 8z = 10$ $15x - 21y + z = 7$ $3x + 12y - 2z = 9$

$\left(\frac{1}{2}, \frac{1}{3}, \frac{1}{4}\right)$ $\left(\frac{1}{5}, 0, 4\right)$ $\left(\frac{1}{3}, 1, 2\right)$

Determine the value of the indicated variable(s) so that the resulting ordered triple is a solution of the given equation.

7. $(x, 0, 5)$ **8.** $(2, y, 0)$ **9.** $(2, 3, z)$

$2x - 0y + z = 13$ $5x - y + 3z = 9$ $3x + 5y + 8z = 37$

__(4, 0, 5)__   __(2, 1, 0)__   __(2, 3, 2)__

10. $(x, y, 0)$ **11.** $(x, 8, z)$ **12.** $(0, y, z)$

$7x + 5y - 4z = 25$ $x + 3y - 5z = 15$ $3x - 4y - z = 0$

__(0, 5, 0)__   __(1, 8, 2)__   __(0, 0, 0)__

Possible answers are shown for Exercises 10–12.

PROBLEM SOLVING 1-7

LESSON 1-7 Problem Solving

Ordered Pairs

Use the table at the right for Exercises 1–2.

1. Write the ordered pair that shows the average miles per gallon in 1990.

__(1990, 20.2)__

2. The data can be approximated by the equation $m = 0.30887x - 595$ where m is the average miles per gallon and x is the year. Use the equation to find an ordered pair (x, m) that shows the estimated miles per gallon in the year 2020.

__(2020, 28.9)__

Year	Miles per Gallon
1970	13.5
1980	15.9
1990	20.2
1995	21.1
1996	21.2
1997	21.5

Average Miles per Gallon

For Exercises 3–4 use the equation $F = 1.8C + 32$, which relates Fahrenheit temperatures F to Celsius temperatures C.

3. Write ordered pair (C, F) that shows the Celsius equivalent of 86°F.

__(30, 86)__

4. Write ordered pair (C, F) that shows the Fahrenheit equivalent of 22°C.

__(22, 71.6)__

Choose the letter for the best answer.

5. A taxi charges a $2.50 flat fee plus $0.30 per mile. Use an equation for taxi fare t in terms of miles m. Which ordered pair (m, t) shows the taxi fare for a 23-mile cab ride?
A (23, 6.90)   C (23, 9.40)
B (23, 18.50)   D (23, 64.40)

6. The perimeter p of a square is four times the length of a side s, or $p = 4s$. Which ordered pair (s, p) shows the perimeter for a square that has sides that are 5 in.?
F (5, 1.25)   H (5, 9)
G (5, 20)   J (5, 25)

7. Maria pays a monthly fee of $3.95 plus $0.10 per minute for long distance calls. Use an equation for the phone bill p in terms of the number of minutes m. Which ordered pair (m, p) shows the phone bill for 120 minutes?
A (120, 15.95)   C (120, 28.30)
B (120, 474.10)   D (120, 486.00)

8. Tickets to a baseball game cost $12 each, plus $2 each for transportation. Use an equation for the cost c of going to the game in terms of the number of people p. Which ordered pair (p, c) shows the cost for 6 people?
F (6, 74)   H (6, 84)
G (6, 96)   J (6, 102)

Lesson Quiz

Determine whether each ordered pair is a solution of $y = 4x - 7$.

1. (2, 15) no   **2.** (4, 9) yes

3. Use the given values to make a table of solutions.

$y = 4x - 6$ for $x = 2, 4, 6, 8,$ and 10

x	$4x - 6$	y	(x, y)
2	$4(2) - 6$	2	(2, 2)
4	$4(4) - 6$	10	(4, 10)
6	$4(6) - 6$	18	(6, 18)
8	$4(8) - 6$	26	(8, 26)
10	$4(10) - 6$	34	(10, 34)

Available on Daily Transparency in CRB

Pacing: Traditional 1 day
Block $\frac{1}{2}$ day

Objective: Students graph points and lines on the coordinate plane.

Warm Up

Find the values for *y* by substituting 1, 3, 5, and 7 for *x*.

1. $y = 2x - 3$ −1, 3, 7, 11

2. $y = 4x + 1$ 5, 13, 21, 29

3. $y = 5x - 5$ 0, 10, 20, 30

Problem of the Day

Tomorrow will be four days after the day before Saturday. What is today?
Monday

Available on Daily Transparency in CRB

Additional Example

Example 1

Give the coordinates of each point.

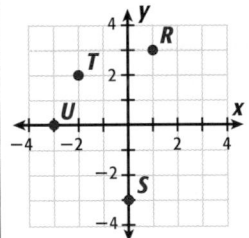

$R\,(1, 3)$
$S\,(0, -3)$
$T\,(-2, 2)$
$U\,(-3, 0)$

1-8 Graphing on a Coordinate Plane

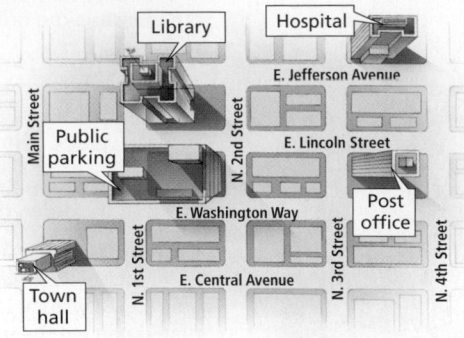

Learn to graph points and lines on the coordinate plane.

Vocabulary

coordinate plane

x-axis

y-axis

x-coordinate

y-coordinate

origin

graph of an equation

Kim left a message for José that read, "Meet me on Second Street."

But José did not know where on Second Street. A better message would have been "Meet me at the corner of East Jefferson Avenue and North Second Street."

The **coordinate plane** is like a map formed by two number lines, the **x-axis** and the **y-axis**, that intersect at right angles. Ordered pairs are the locations, or points, on the map. The **x-coordinate** and **y-coordinate** of an ordered pair tell the direction and number of units to move.

x-coordinate
move right or left **(x, y)** *y*-coordinate
move up or down

Helpful Hint

The sign of a number indicates which direction to move.
Positive: up or right
Negative: down or left

To plot an ordered pair, begin at the **origin**, the point (0, 0), which is the intersection of the *x*-axis and the *y*-axis. The first coordinate tells how many units to move left or right; the second coordinate tells how many units to move up or down.

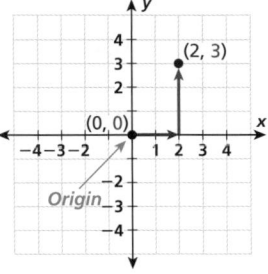

*move right
2 units* **(2, 3)** *move up
3 units*

EXAMPLE 1   Finding the Coordinates of Points on a Plane

Give the coordinates of each point.

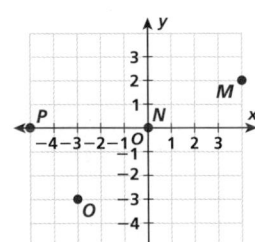

Point *M* is (4, 2).
  4 units right, 2 units up
Point *N* is (0, 0).
  0 units right, 0 units up
Point *O* is (−3, −3).
  3 units left, 3 units down
Point *P* is (−5, 0).
  5 units left, 0 units up

---

1 Introduce

Alternate Opener

EXPLORATION

1-8 Graphing on a Coordinate Plane

Any point on a *coordinate plane* is an ordered pair that can be expressed as the following:

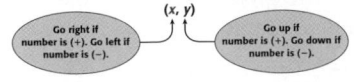

(x, y)

Go right if number is (+). Go left if number is (−).

Go up if number is (+). Go down if number is (−).

1. Plot the points in the table on a coordinate plane. The first point has been plotted for you.

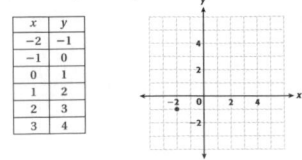

2. Use a straight line to connect the points you plotted in number 1.

Think and Discuss

3. **Explain** how to plot ordered pairs on a coordinate plane.
4. **Describe** real-world items that resemble planes and points.

Motivate

Discuss some situations in which locations are determined by a grid. Examples may include street maps, latitude and longitude, or games like computer chess or Battleship®. Show the students a map with a grid. Demonstrate how to use the labels on the top, bottom, and sides to find a particular area on the map. Explain that a similar procedure will be used in the lesson.

Exploration worksheet and answers on Chapter 1 Resource Book pp. 72 and 120

2 Teach

Lesson Presentation

Guided Instruction

In this lesson, students learn to graph points and lines on a coordinate plane. To begin, show students how to create a coordinate plane by using graph paper. Explain that each axis is a number line, and show them how to label and number the axes. You may want to display Teaching Transparency T34 from the Chapter 1 Resource Book. Use your sample to define the vocabulary terms. Show students how to identify the location of a point by giving its coordinates, and then show them how to graph points and lines.

EXAMPLE 2 Graphing Points on a Coordinate Plane

Graph each point on a coordinate plane. Label the points *A–D*.

A (2, 5)
right 2, up 5

B (0, 4)
right 0, up 4

C (−1, 2)
left 1, up 2

D (2, −3)
right 2, down 3

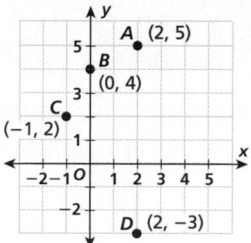

The **graph of an equation** is the set of all ordered pairs that are solutions of the equation.

EXAMPLE 3 Graphing an Equation

Complete each table of ordered pairs. Graph each equation on a coordinate plane.

A $y = 2x$

x	2x	y	(x, y)
1	2(1)	2	(1, 2)
2	2(2)	4	(2, 4)
3	2(3)	6	(3, 6)
4	2(4)	8	(4, 8)

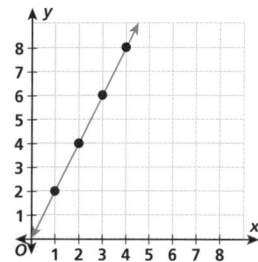

The points of each equation are on a straight line. Draw the line through the points to represent all possible solutions.

B $y = 3x − 2$

x	3x − 2	y	(x, y)
0	3(0) − 2	−2	(0, −2)
1	3(1) − 2	1	(1, 1)
2	3(2) − 2	4	(2, 4)
3	3(3) − 2	7	(3, 7)

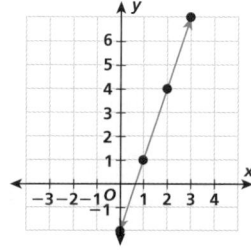

Think and Discuss

1. Give the coordinates of a point on the *x*-axis and a point on the *y*-axis.

2. Give the missing *y*-coordinates for the solutions to $y = 5x + 2$: $(1, y), (3, y), (10, y)$.

Additional Examples

Example 2

Graph each point on a coordinate plane. Label the points *A–D*.

A. (3, 4)   **B.** (4, 0)

C. (−4, 4)   **D.** (−1, −3)

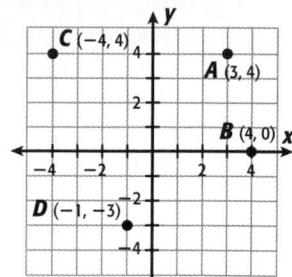

Example 3

Complete the table of ordered pairs. Graph the equation on a coordinate plane.

$y = 2x − 3$

x	2x − 3	y	(x, y)
0	2(0) − 3	−3	(0, −3)
1	2(1) − 3	−1	(1, −1)
2	2(2) − 3	1	(2, 1)
3	2(3) − 3	3	(3, 3)

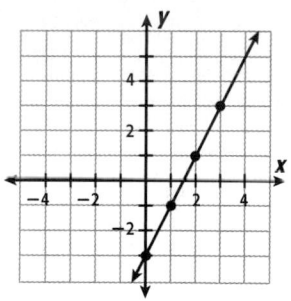

---

3 Close

Reaching All Learners
Through Critical Thinking

Give each student a recording sheet (Chapter 1 Resource Book, p. 105) that contains some two-variable equations and corresponding tables of ordered pairs. For each set, have the students graph each of the four points to determine which one is not a solution to the equation (i.e., the point not in line with the other three). Then have them check their work algebraically.

Summarize

Draw an unlabeled coordinate plane on the board and plot a point with the coordinates labeled (?, ?). Ask students to identify and label any important parts of the graph.

Possible answers: *x*-axis, *y*-axis, origin, point, (3, 4)

Remind students that when graphing points on a coordinate plane, order is important. Show them that the points that correspond to ordered pairs like (2, 3) and (3, 2) are in different locations on the coordinate plane.

Answers to Think and Discuss

1. Possible answer: *x*-axis: (3, 0); *y*-axis: (0, 4)

2. $y = 7$; $y = 17$; $y = 52$

FOR EXTRA PRACTICE
see page 733

internet connect
Homework Help Online
go.hrw.com   Keyword: MP4 1-8

go.hrw.com

Students may want to refer back to the lesson examples.

Assignment Guide

If you finished Example 1 assign:
Core 1–6, 13–18, 25–31 odd, 39–44
Enriched 1, 2, 13–18, 25–32, 39–44

If you finished Example 2 assign:
Core 1–10, 13–22, 25–31 odd, 39–44
Enriched 1–9 odd, 13–22, 25–32, 37, 39–44

If you finished Example 3 assign:
Core 1–24, 25–31 odd, 34, 39–44
Enriched 1–11 odd, 12, 13–23 odd, 24–44

Answers

7–10.

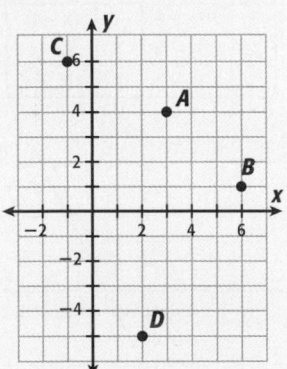

11–12, 19–24. See p. A1.

GUIDED PRACTICE

See Example **1** Give the coordinates of each point.

1. A (−2, 3)   2. B (3, 5)   3. C (2, −3)
4. D (5, −1)   5. E (5, 5)   6. F (−3, −4)

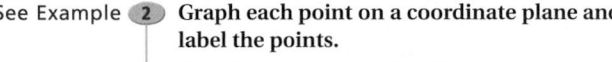

See Example **2** Graph each point on a coordinate plane and label the points.

7. $A(3, 4)$ 8. $B(6, 1)$
9. $C(−1, 6)$ 10. $D(2, −5)$

See Example **3** Complete each table of ordered pairs. Graph each equation on a coordinate plane.

11. $y = x + 1$

x	x + 1	y	(x, y)
0	▩	▩	(0, 1)
1	▩	▩	(1, 2)
2	▩	▩	(2, 3)

12. $y = 2x − 1$

x	2x − 1	y	(x, y)
0	▩	▩	(0, −1)
1	▩	▩	(1, 1)
2	▩	▩	(2, 3)

INDEPENDENT PRACTICE

See Example **1** Give the coordinates of each point.

13. G (0, 3)   14. H (−1, 5) 15. J (2, −4)
16. K (3, 2)   17. L (−2, 5) 18. M (−4, −3)

See Example **2** Graph each point on a coordinate plane and label the points.

19. $A(2, 6)$ 20. $B(0, 4)$
21. $C(−3, −7)$ 22. $D(−3, 0)$

See Example **3** Complete each table of ordered pairs. Graph each equation on a coordinate plane.

23. $y = 3x$

x	3x	y	(x, y)
0	▩	▩	(0, 0)
1	▩	▩	(1, 3)
2	▩	▩	(2, 6)

24. $y = 2x + 1$

x	2x + 1	y	(x, y)
0	▩	▩	(0, 1)
1	▩	▩	(1, 3)
2	▩	▩	(2, 5)

Math Background

Knowing how coordinate graphs are used in real-world situations can make the topic more meaningful for students. Cartography, the science of map making, is an application of graphing on a coordinate plane. Cartographers map a region of the surface of Earth onto part of a plane. Because Earth is nearly spherical and the coordinate plane is flat, there are different methods for mapping Earth's surface onto a flat plane. The different methods are called *projections*.

RETEACH 1-8

LESSON Reteach
1-8 Graphing on the Coordinate Plane

Point A is described by the ordered pair (3, −2). The first number, 3, is the **x-coordinate** and the second number, −2, is the **y-coordinate**.

Using a grid to represent a **coordinate plane**, graph point A by starting at the **origin**, where the **x-axis** and **y-axis** intersect. The x-coordinate, 3, tells you to go *right* 3. The y-coordinate, −2, tells you to go *down* 2.

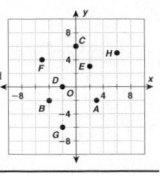

Write an ordered pair to describe each point on the coordinate plane above.

1. point C ___(0, 6)___   2. point F ___(−5, 4)___   3. point G ___(−2, −6)___

Graph each ordered pair.

4. $J(2, 1)$ 5. $K(−3, 0)$
6. $L(5, −4)$ 7. $M(−4, −4)$
8. $N(−2, 6)$ 9. $P(3, −2)$

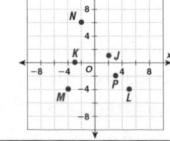

The graph of an equation is the set of all ordered pairs that are solutions of the equation.

Complete the table. Then graph the equation.

10. $y = 4x$

x	4x = y	(x, y)
0	4(0) = 0	(0, 0)
1	4(1) = 4	(1, 4)
2	4(2) = 8	(2, 8)

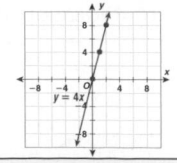

PRACTICE 1-8

LESSON Practice B
1-8 Graphing on the Coordinate Plane

Give the coordinates of each point.

1. F 2. X
___(−2, −3)___   ___(−6, 3)___
3. T 4. B
___(8, −3)___   ___(5, 0)___
5. D 6. R
___(−4, 4)___   ___(1, −8)___
7. H 8. Y
___(3, 8)___   ___(−5, 0)___

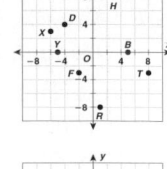

Graph each point on a coordinate plane. Label points A–F.

9. $A(5, 6)$ 10. $B(0, 4)$
11. $C(2, −8)$ 12. $D(−2, 2)$
13. $E(−7, 0)$ 14. $F(−5, −3)$

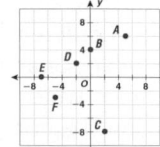

Complete the table of ordered pairs. Graph the equation on a coordinate plane.

15. $y = 3x$

x	3x	y	(x, y)
0	3(0)	0	(0, 0)
1	3(1)	3	(1, 3)
2	3(2)	6	(2, 6)

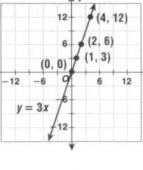

16. A barber can cut 3 heads of hair per hour. Use the equation and coordinate grid in Exercise 15 to graph the point for x = 4. The y-value represents how many heads of hair that the barber can cut in 4 hours. How many can the barber cut in a 4-hour day?

___12___

Music LINK

Some dance rhythms, such as *dununba*, from Guinea, are played on drums at high speed for hours at a time. *Dununba* may be played at 350 beats per minute.

For each ordered pair, list two other ordered pairs that have the same *y*-coordinates. **Possible answers are given.**

25. (4, 0) **(1, 0),** **(2, 0)**   **26.** (6, 2) **(5, 2),** **(4, 2)**   **27.** (3, 7) **(2, 7),** **(4, 7)**   **28.** (1, 4) **(2, 4),** **(3, 4)**

For each ordered pair, list two other ordered pairs that have the same *x*-coordinates. **Possible answers are given.**

29. (4, 7) **(4, 3),** **(4, 5)**   **30.** (2, 5) **(2, 4),** **(2, 3)**   **31.** (0, 3) **(0, 4),** **(0, 5)**   **32.** (6, 1) **(6, 2),** **(6, 3)**

33. *MUSIC* One drumming pattern originating in Ghana can be played at 2.5 beats per second. To find the number of beats played in *s* seconds, use the equation $b = 2.5s$. How many beats are played in 30 seconds? **75 beats**

34. *PHYSICAL SCIENCE* A car travels at 60 miles per hour. To find the distance traveled in *x* hours, use the equation $y = 60x$. Make a table of ordered pairs and graph the solution. How far will the car travel in 3.5 hours? **210 miles**

35. *CONSTRUCTION* To build house walls, carpenters place a stud, or board, every 16 inches unless there are doors or windows. Use the equation $y = \frac{x}{16} + 1$ to determine the number of studs in a wall of length *x* inches with no doors or windows. Make a table of ordered pairs and graph the solution. How many studs should be placed in a wall 8 feet long? (*Hint:* There are 12 inches in a foot.) **7 studs**

 36. *WRITE A PROBLEM* Write a problem whose solution is a geometric shape on the coordinate plane.

 37. *WRITE ABOUT IT* Assume you are in a city that is arranged in square blocks, much like a coordinate grid, and you are looking at a map. Explain how to get from point (4, 6) to point (1, 2).

38. *CHALLENGE* Find the missing number in the equation shown by using the table of ordered pairs. Graph the equation.

x	5*x* + ▦	*y*	(*x*, *y*)
1	5(1) + ▦	8	(1, 8)
2	5(2) + ▦	13	(2, 13)

Spiral Review

Write an algebraic expression for each word phrase. (Lesson 1-2)

39. the difference of a number and 13 $x - 13$

40. a number divided by 6 $\frac{x}{6}$ or $x \div 6$

41. the sum of a number and 31 $x + 31$

42. 8 divided into a number $\frac{x}{8}$ or $x \div 8$

43. **TEST PREP** Solve the equation $7x = 42$. (Lesson 1-4) **C**

  A $x = 35$ **C** $x = 6$
  B $x = 294$ **D** $x = 49$

44. **TEST PREP** Solve the equation $\frac{x}{3} = 7$. (Lesson 1-4) **F**

  F $x = 21$ **H** $x = 10$
  G $x = 4$ **J** $x = \frac{7}{3}$

Answers

34–35, 38. Complete answers on p. A1

36. Possible answer: Graph the points (1, 3), (7, 3), (4, 0), and (4, 6). Connect the points and identify the figure. (square)

37. Possible answer: Go 3 blocks to the left and 4 blocks down.

38. $y = 5x + 3$

Journal

Ask students to draw a map of the streets around the school, their houses, or any other areas that interest them. Have students describe the location of one of the buildings in the area.

Test Prep Doctor ✚

For Exercises 43 and 44, remind students of the methods used for "undoing" multiplication and division equations.

Lesson Quiz

Give the coordinates of each point.

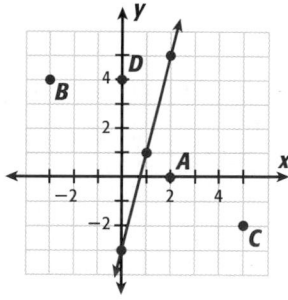

1. *A* **(2, 0)**   **2.** *B* **(−3, 4)**

Graph each point on a coordinate plane. Label points C, D.

3. *C*(5, −2)   **4.** *D*(0, 4)

5. Complete the table for $y = 4x - 3$. Graph the equation on a coordinate plane.

x	4*x* − 3	*y*	(*x*, *y*)
0	4(0) − 3	−3	(0, −3)
1	4(1) − 3	1	(1, 1)
2	4(2) − 3	5	(2, 5)

Answers on graph above
Available on Daily Transparency in CRB

CHALLENGE 1-8

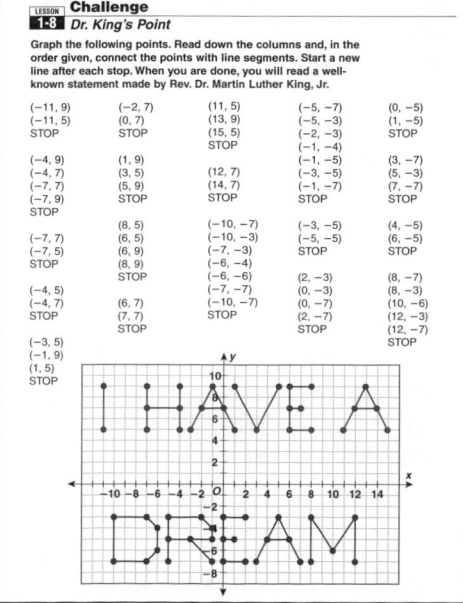

LESSON 1-8 **Challenge**
Dr. King's Point

Graph the following points. Read down the columns and, in the order given, connect the points with line segments. Start a new line after each stop. When you are done, you will read a well-known statement made by Rev. Dr. Martin Luther King, Jr.

(−11, 9)   (−2, 7)   (11, 5)   (−5, −7)   (0, −5)
(−11, 5)   (0, 7)   (13, 9)   (−5, −3)   (1, −5)
STOP   STOP   (15, 5)   (−2, −3)   STOP
                     STOP   (−1, −4)
(−4, 9)   (1, 9)   (−1, −5)   (3, −7)
(−4, 7)   (3, 5)   (12, 7)   (−3, −5)   (5, −3)
(−7, 7)   (5, 9)   (−3, −5) (−1, −5)   (7, −7)
(−7, 9)   STOP   (14, 7)   STOP   STOP
STOP   STOP
           (8, 5)   (−3, −5)   (4, −5)
(−7, 7)   (6, 5)   (−10, −7) (−5, −5)   (6, −5)
(−7, 5)   (6, 9)   (−10, −3) STOP   STOP
STOP   (8, 9)   (−7, −3)
           STOP   (−6, −4)
(−4, 5)   (−6, −6) (2, −3)   (8, −7)
(−4, 7)   (6, 7)   (−7, −7) (0, −3)   (8, −3)
STOP   (7, 7)   (−10, −7) (0, −7)   (10, −6)
           STOP   STOP   (2, −7)   (12, −7)
(−3, 5)   STOP   (12, −7)
(−1, 9)   STOP
(1, 5)
STOP

PROBLEM SOLVING 1-8

LESSON 1-8 **Problem Solving**
Graphing on the Coordinate Plane

Complete the table of ordered pairs. Graph each solution. Answer the question.

1. John earns $150 per week plus 5% of his computer software sales. John's weekly pay *y* in terms of his sales *x* is $y = 150 + 0.05x$. Complete the table. How much does John get paid for $200 in sales?

$160

x	*y*	(*x*, *y*)
0	150	0, 150
10	150.5	10, 150.5
25	151.25	25, 151.25
50	152.5	50, 152.5
100	155	100, 155

2. Margarite starts out with $100. Each week, she spends $6 to go to the movies. The amount of money *y* Margarite has left each week *x*, is $y = 100 − 6x$. How much money does she have left after 11 weeks?

$34

x	*y*	(*x*, *y*)
0	100	0, 100
1	94	1, 94
2	88	2, 88
3	82	3, 82
4	76	4, 76

The graph at the right represents the miles traveled *y* in *x* hours. Use the graph to choose the best let...

3. Which of the ordered pairs below represents a solution?
  Ⓐ (1, 65)   **C** (5, 120)
  B (2, 70)   **D** (7, 300)

4. The graph represents a car traveling how fast?
  F 40 mi/h   **H** 70 mi/h
  Ⓖ 65 mi/h   **J** 75 mi/h

Technology LAB 1A
Create a Table of Solutions

Pacing: Traditional 1 day
Block $\frac{1}{2}$ day

Objective: To use a graphing calculator to make a table of values

Materials: Graphing calculator

Lab Resources

Technology Lab Activities . . . . p. 1

Using the Page

This technology activity shows students how to make a table of values on any graphing calculator. Specific keystrokes may vary, depending on the make and model of the graphing calculator used. The keystrokes given are for a TI-83 model. For keystrokes to other models, visit go.hrw.com.

The Think and Discuss problem can be used to assess students' understanding of the technology activity. While Try This problems 1–2 can be done without a graphing calculator, they are meant to help students become familiar with using a graphing calculator to generate tables.

Assessment

1. Make a table of solutions to the equation $y = 3x + 2$ using **TblStart** = 0 and **ΔTbl** = 1.

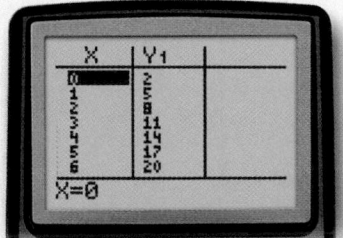

2. What are the first three x- and y-values in the table? **(0, 2), (1, 5), and (2, 8)**

3. Find the value of y when $x = 21$. **65**

Create a Table of Solutions

Use with Lesson 1-8

The *Table* feature on a graphing calculator can help you make a table of values quickly.

📶 **internet** connect
Lab Resources Online
go.hrw.com
KEYWORD: MP4 Lab1A

Activity

1 Make a table of solutions of the equation $y = 2x - 3$. Then find the value of y when $x = 29$.

To enter the equation, press the [Y=] key. Then press 2 [X,T,θ,n] [—] 3.

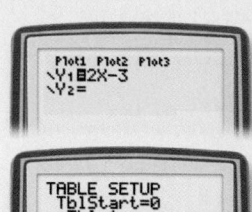

Press [2nd] [WINDOW] to go to the Table Setup menu. In this menu, **TblStart** shows the starting x-value, and **ΔTbl** shows how the x-values increase. If you need to change these values, use the arrow keys to highlight the number you want to change and then type a new number.

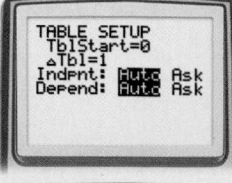

Press [2nd] [GRAPH] to see the table of values.

On this screen, you can see that $y = 7$ when $x = 5$.

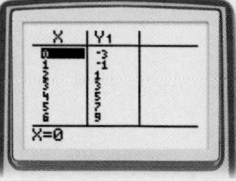

Use the arrow keys to scroll down the list. You can see that $y = 55$ when $x = 29$.

To check, substitute 29 into $y = 2x - 3$.

$y = 2x - 3$
$= 2(29) - 3 = 58 - 3 = 55$

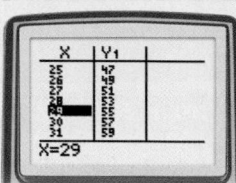

Think and Discuss

1. On an Internet site, pencils can be purchased for 17¢ each, but they only come in boxes of 12. You decide to make a table to compare x, the number of pencils, to y, the total cost of the pencils. What **TblStart** and **ΔTbl** values will you use? Explain.

Try This

For each equation, use a table to find the y-values for the given x-values. Give the **TblStart** and **ΔTbl** values you used.

1. $y = 3x + 6$ for $x = 1, 3,$ and 7   **2.** $y = \frac{x}{4}$ for $x = 5, 10, 15,$ and 20

Answers

Think and Discuss

1. Possible answer: Because pencils come in boxes of 12, and you are only interested in the cost of positive numbers of pencils, use **TblStart** = 12 and **ΔTbl** = 12.

Try This

1. $y = 9$ when $x = 1$; $y = 15$ when $x = 3$; $y = 27$ when $x = 7$; possible answers: **TblStart** = 1; **ΔTbl** = 2

2. $y = 1.25$ when $x = 5$; $y = 2.5$ when $x = 10$; $y = 3.75$ when $x = 15$; $y = 5$ when $x = 20$; possible answers: **TblStart** = 5; **ΔTbl** = 5

1-9 Interpreting Graphs and Tables

Learn to interpret information given in a graph or table and to make a graph to solve problems.

The table below shows how quickly the temperature can increase in a car that is left parked during an afternoon of errands when the outside temperature is 93°F.

Location	Temperature on Arrival	Temperature on Departure
Home	—	140° at 1:05
Cleaners	75° at 1:15	95° at 1:25
Mall	72° at 1:45	165° at 3:45
Market	80° at 4:00	125° at 4:20

EXAMPLE 1   Matching Situations to Tables

The table gives the speeds of three dogs in mi/h at given times. Tell which dog corresponds to each situation described below.

Time	12:00	12:01	12:02	12:03	12:04
Dog 1	8	8	20	3	0
Dog 2	0	10	0	7	0
Dog 3	0	4	4	0	12

A David's dog chews on a toy, then runs to the backyard, then sits and barks, and then runs back to the toy and sits.

Dog 2—The dog's speed is 0 to start, while he sits and barks, and when he gets back to the toy. It is positive while he is running.

B Kareem's dog runs with him and then chases a cat until Kareem calls for him to come back. The dog returns to his side and sits.

Dog 1—The dog is running at the start, so his speed is positive. His speed increases while he chases the cat and then decreases to 0 when he sits.

C Janelle's dog sits on top of a pool slide, slides into the swimming pool, and swims to the ladder. He gets out of the pool and shakes and then runs around the pool.

Dog 3—The dog's speed is 0 at the top of the slide, 4 while swimming, and 12 while he runs around the pool.

1-9 Organizer

Pacing: Traditional 1 day
Block $\frac{1}{2}$ day

Objective: Students learn to interpret information given in a graph or a table and to make a graph to solve problems.

Warm Up

Use the table for problems 1 and 2.

Week	1	2	3	4	5
Savings	$3	$5	$10	$2	$5

1. In which week did Alicia earn the most?   week 3

2. What is the average amount Alicia earned in one week? $5

Problem of the Day

The late fee at a video store is $1 for the first day and $2.50 for each day after that. There is also a $1 fee if the movie is not rewound. What is the total fine if the movie is 4 days late and not rewound? **$9.50**

Available on Daily Transparency in CRB

---

1 Introduce

Alternate Opener

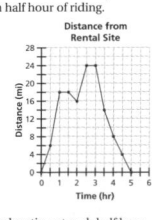

EXPLORATION

1-9 Interpreting Graphs and Tables

Graphs and tables can be used to represent many different situations.

Carmen rented a bicycle. The graph shows how far away she is from the rental site after each half hour of riding.

1. Use the graph to describe Carmen's trip. You can start the description like this: "In the first half hour, Carmen rode the bike for 6 miles. In the second half hour, she increased her speed because…"

2. Complete the table to show her location at each half hour.

Time (hr)	0	0.5	1	1.5	2	2.5	3	3.5	4	4.5	5
Location											

Think and Discuss

3. Explain how the graph shows how fast Carmen was riding.

4. Determine during which half hour Carmen was riding fastest.

Motivate

Find some simple line graphs and tables that were published in recent newspapers or magazines, and show them to the students. Explain that the graphs and tables can show important information if you know how to read them. Point out that tables and graphs are two ways to visually organize information.

Exploration worksheet and answers on Chapter 1 Resource Book pp. 81 and 122

2 Teach

Lesson Presentation

Guided Instruction

In this lesson, students learn to interpret information given in a graph or a table and to make a graph to help solve problems. Show students a simple table and explain how to read it. Emphasize the importance of headings for columns and rows. Illustrate how to find a particular piece of information in the table by finding the intersection of the appropriate row and column (similar to the coordinates of a point on a plane).

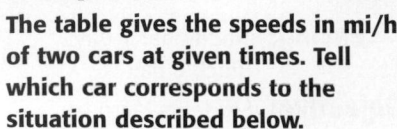

Additional Examples

Example 1

The table gives the speeds in mi/h of two cars at given times. Tell which car corresponds to the situation described below.

Time	1:00	1:05	1:10	1:15	1:20
Car 1	50	50	30	25	0
Car 2	55	10	0	0	55

Mr. Lee is traveling on the highway. He pulls over, stops, and then gets back onto the highway.   car 2

Example 2

Tell which graph corresponds to the situation described above.

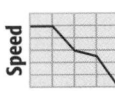   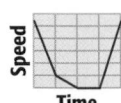

Graph 1   Graph 2

Graph 2

Example 3

Create a graph that illustrates the temperature (°F) inside the car.

Location	Arrival	Departure
Work	68° at 12:30	42° at 4:30
Cleaners	65° at 4:50	60° at 5:00
Market	65° at 5:10	49° at 5:40

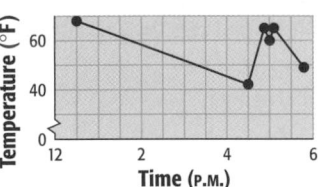

EXAMPLE 2 Matching Situations to Graphs

Tell which graph corresponds to each situation described in Example 1.

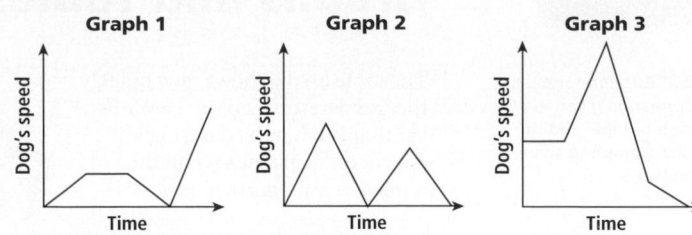

Graph 1   Graph 2   Graph 3

A David's dog

Graph 2—The dog's speed is 0 when the graph is on the *x*-axis.

B Kareem's dog

Graph 3—The dog's speed is not 0 when the graph starts.

C Janelle's dog

Graph 1—The dog is running at the end, so his speed is not 0.

EXAMPLE 3 Creating a Graph of a Situation

The temperature inside a car can get dangerously high. Create a graph that illustrates the temperature inside a car.

Location	Temperature (°F)	
	On Arrival	On Departure
Home	—	140° at 1:00
Cleaners	75° at 1:10	95° at 1:20
Mall	72° at 1:40	165° at 3:40
Market	80° at 3:55	125° at 4:15

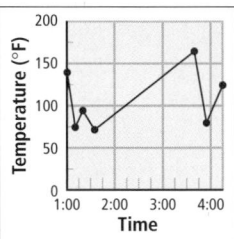

Think and Discuss

1. Describe what it means when a graph of speed starts at (0, 0).

2. Give a situation that, when graphed, would include a horizontal segment.

Teach

Reaching All Learners
Through Curriculum Integration

History Have each student research the biography of a historical figure. Have them create a table that summarizes important events in the person's life (similar to a timeline). Column headings might include the date, place, and description of the event. Then ask students questions about the person's life, and have them use their tables to answer them.

3 Close

Summarize

Discuss different ways of presenting information, including written paragraphs, tables, and graphs. Ask students for the pros and cons of each. Ask if they have an opinion about which of these methods is the most efficient.

Possible answers: Paragraphs are clear but they take a long time to read. Tables are good for numbers and finding specific information quickly, but sometimes they leave out details. Graphs give you a visual idea of changes in data, but they are not as specific or detailed as the other ways.

Answers to Think and Discuss

Possible answers:

1. If a graph starts at (0, 0), the object was not moving at the start.

2. Gina rides her bike to the park, stops and talks to a friend, and then rides home.

FOR EXTRA PRACTICE
see page 733

internet connect
Homework Help Online
go.hrw.com Keyword: MP4 1-9

go.hrw.com

1-9 PRACTICE & ASSESS

GUIDED PRACTICE

See Example ① 1. Tell which table corresponds to the situation described below.

Jerry rides his bike to the end of the street and then rides quickly down a steep hill. At the bottom of the hill, Jerry stops to talk to Ryan. After a few minutes, Jerry rides over to Reggie's house and stops.   **Table 2**

Table 1	
Time	Speed (mi/h)
3:00	0
3:05	8
3:10	0
3:15	5
3:20	3

Table 2	
Time	Speed (mi/h)
3:00	5
3:05	12
3:10	0
3:15	5
3:20	0

Table 3	
Time	Speed (mi/h)
3:00	6
3:05	3
3:10	2
3:15	0
3:20	5

See Example ② 2. Tell which table from Exercise 1, if any, corresponds to each graph.

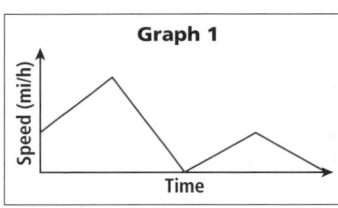

Graph 1

Table 2

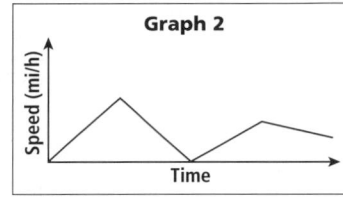

Graph 2

Table 1

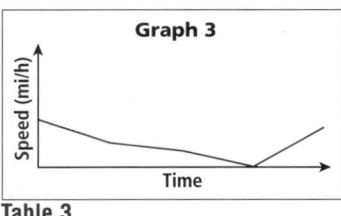

Graph 3

Table 3

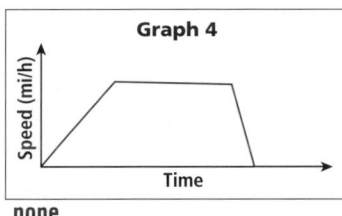

Graph 4

See Example ③ 3. Create a graph that illustrates the information in the table about a ride at an amusement park.

Time	3:20	3:21	3:22	3:23	3:24	3:25
Speed (mi/h)	0	14	41	62	8	0

Students may want to refer back to the lesson examples.

Assignment Guide

If you finished Example ① assign:
Core 1, 4, 11–16
Enriched 1, 4, 11–16

If you finished Example ② assign:
Core 1, 2, 4, 5, 11–16
Enriched 1, 2, 4, 5, 9, 11–16

If you finished Example ③ assign:
Core 1–7, 9, 11–16
Enriched 4–16

Answers

3.

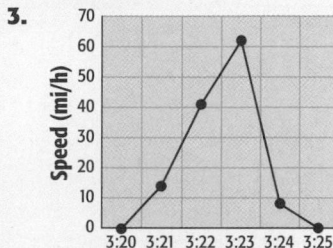

Answers

6.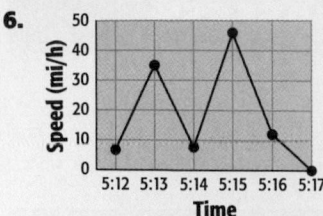

7.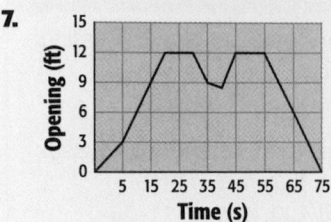

8. Possible answer: The geyser erupts abruptly, stays near the same level for about 5 minutes, and then stops very quickly.

9. a. Old Faithful
b. Riverside

In 1736, Leonhard Euler (pronounced "oiler") published a famous paper in which he solved a popular puzzle, called the Königsberg bridge problem. The problem was based on determining whether someone could cross each of the seven bridges of Königsberg, Prussia, once and only once. Euler's solution relied on a diagram that he called a *graph*. The graph was composed of points, called *vertices*, and segments joining the points, called *edges*. Euler's solution introduced a new branch of mathematics known as graph theory (See Lesson 14-6).

INDEPENDENT PRACTICE

See Example ① **4.** Tell which table corresponds to the situation. **Table 3**

An airplane sits at the gate while the passengers get on. Then the airplane taxis away from the gate and out to the runway. The plane waits at the end of the runway for clearance to take off. Then the plane takes off and continues to accelerate as it ascends.

Table 1	
Time	Speed (mi/h)
6:00	0
6:10	20
6:20	40
6:30	0
6:40	80

Table 2	
Time	Speed (mi/h)
6:00	20
6:10	0
6:20	10
6:30	80
6:40	300

Table 3	
Time	Speed (mi/h)
6:00	0
6:10	10
6:20	0
6:30	80
6:40	350

See Example ② **5.** Tell which graph corresponds to each table described in Exercise 4.

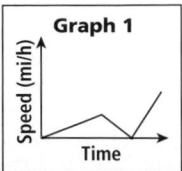
Table 1

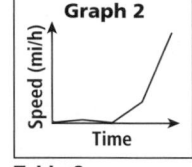
Table 3

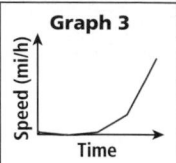
Table 2

See Example ③ **6.** Create a graph that illustrates the information in the table about Mr. Schwartz's commute from work to home.

Time	Speed (mi/h)	Time	Speed (mi/h)
5:12	7	5:15	46
5:13	35	5:16	12
5:14	8	5:17	0

PRACTICE AND PROBLEM SOLVING

7. Use the table to graph the movement of an electronic security gate.

Time (s)	0	5	10	15	20	25	30	35
Gate Opening (ft)	0	3	6	9	12	12	12	9

Time (s)	40	45	50	55	60	65	70	75
Gate Opening (ft)	8	12	12	12	9	6	3	0

RETEACH 1-9

LESSON 1-9 Reteach
Interpreting Graphs and Tables

Tell which table best corresponds to the given situation.
Situation: Mr. Savoy begins math lessons by going over yesterday's homework assignment. The class discussion about the new topic is lengthy. The remaining class time is spent in group work.

Table 1	
Activity	Minutes
Review	12
New topic	24
Group work	24

Table 2	
Activity	Minutes
Review	12
New topic	36
Group work	12

Table 3	
Activity	Minutes
Review	12
New topic	12
Group work	33

Focus on review. Since all three tables indicate the same time for review, no tables can be eliminated. More time is spent on the new topic than on group work. Eliminate Table 1 since equal time is spent on each. Eliminate Table 3 since more time is spent on group work than on the new topic. So, Table 2 best corresponds to the given situation.

Which table best corresponds to the given situation?

At home today, Tim spent two hours on his schoolwork.
He had no science homework today.
His French assignment was to go over the errors made on yesterday's test.
Tim had only one minor error on that test.
Tomorrow is a math quiz and Tim is confident about the current topic.
Tim needed to do some research for his civics report, due tomorrow.

Table 1	
Subject	Minutes
Civics	30
French	60
Math	30
Science	0

Table 2	
Subject	Minutes
Civics	20
French	5
Math	100
Science	0

Table 3	
Subject	Minutes
Civics	90
French	5
Math	30
Science	0

1. No table can be eliminated based on science time, since ___they all list 0 min___

2. The French review should not have taken long, so eliminate Table ___1___

3. Given the circumstance, Tim would have spent more time on ___Civics___ than on ___Math___

4. So, the table that best corresponds is Table ___3___

PRACTICE 1-9

LESSON 1-9 Practice B
Interpreting Graphs and Tables

The table gives the speed of three dogs in mi/h at the given times. Tell which dog corresponds to each situation described below.

Time	5:00	5:01	5:02	5:03	5:04
Dog 1	0	1	12	0	0
Dog 2	5	23	4	0	0
Dog 3	14	0	18	2	9

1. Leshaan walks his dog. Then he lets the dog off the leash and it runs around the yard. Then they go into the house and the dog stands eating from his dog dish and drinking from his water bowl. ___Dog 2___

2. Luke's dog is chasing its tail. Then it stops and pants. The dog then runs to the backyard fence and walks along the fence, barking at a neighbor. Then it runs to Luke at the back door. ___Dog 3___

Tell which graph corresponds to each situation in Exercises 1–2.

3.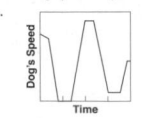
Exercise 2 (Dog 3)

4.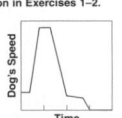
Exercise 1 (Dog 2)

5. Create a graph that illustrates the temperature inside the car.

Location	Temperature on Arrival	Temperature on Departure
Home	—	74° at 8:30
Summer job	77° at 9:00	128° at 12:05
Pool	92° at 12:15	136° at 2:30
Library	95° at 2:40	77° at 5:10

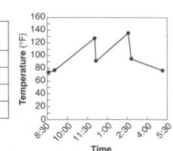

Earth Science LINK

Geyser is an Icelandic word meaning "to gush or rush forth." Geysers erupt because underground water begins to boil. Pressure builds as the temperature rises until the geyser erupts as a fountain of steam and water.

8. Explain what the data tells about Beehive geyser. Make a graph.

Average Water Height of Beehive Geyser

Time	1:00	1:01	1:02	1:03	1:04	1:05	1:06	1:07
Average Height (ft)	0	147	153	155	152	148	0	0

9. Use the chart to choose the correct geyser name to label each graph.

Yellowstone National Park Geysers

Geyser Name	Old Faithful	Grand	Riverside	Pink Cone
Duration (min)	1.5 to 5	10	20	80

a.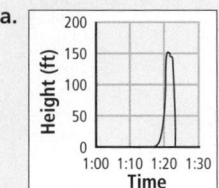

b.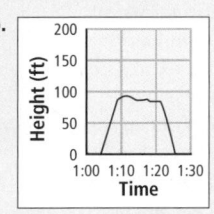

Old Faithful Eruption Information

Duration	Time Until Next Eruption
1.5 min	48 min
2 min	55 min
2.5 min	70 min
3 min	72 min
3.5 min	74 min
4 min	82 min
4.5 min	93 min
5 min	100 min

10. ⭐ **CHALLENGE** Old Faithful erupts to heights between 105 ft and 184 ft. It erupted at 7:34 A.M. for 4.5 minutes. Later it erupted for 2.5 minutes. It then erupted a third time for 3 minutes. Use the table to determine how many minutes followed each of the three eruptions. Sketch a possible graph.

Old Faithful is the most famous geyser at Yellowstone National Park.

go.hrw.com
KEYWORD: MP4 Geyser

Spiral Review

Solve. (Lesson 1-3)

11. $4 + x = 13$ **x = 9**
12. $13 = 9 + x$ **x = 4**
13. $x - 9 = 2$ **x = 11**
14. $x - 2 = 5$ **x = 7**

15. **TEST PREP** Solve $x + 7 < 15$. (Lesson 1-5) **D**

    A $x > 22$
    B $x > 8$
    C $x < 22$
    D $x < 8$

16. **TEST PREP** Solve $4 \leq \frac{x}{2}$. (Lesson 1-5) **H**

    F $x \leq 8$
    G $x \geq 2$
    H $x \geq 8$
    J $x \leq 2$

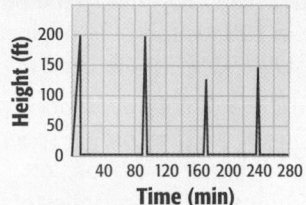

CHALLENGE 1-9

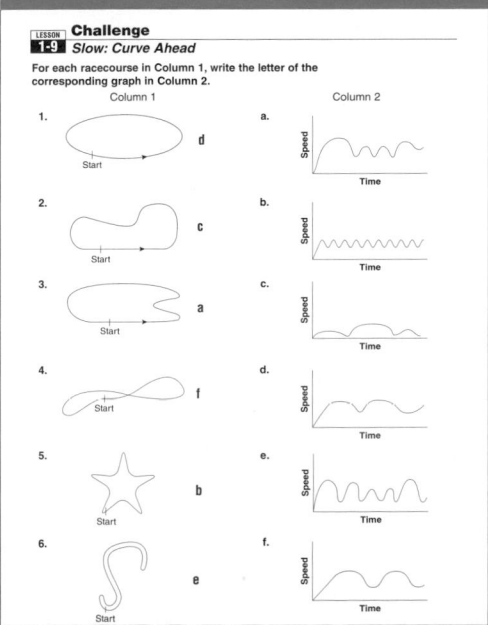

PROBLEM SOLVING 1-9

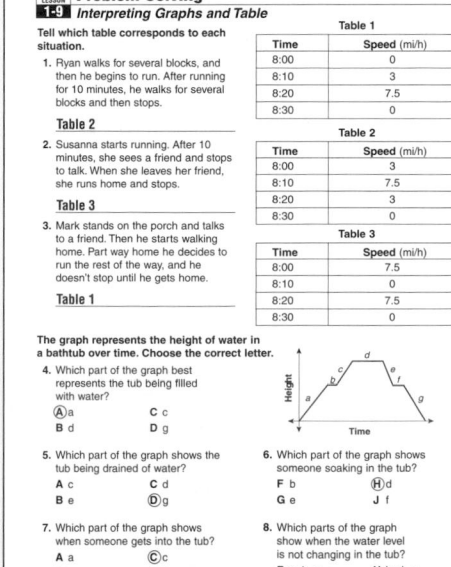

Problem Solving on Location

Illinois

Purpose: *To provide additional practice for problem-solving skills in Chapter 1*

Hometown of Superman

- After problem 2, have students consider the following problem: Write and solve an equation to determine how many feet taller the statue purchased in 1993 is than the original 7-foot statue. **7 + *t* = 15; *t* = 8**

- After problem 3, have students determine at what recording speed (SP = 2 hours, LP = 4 hours, or EP = 6 hours) all the episodes of the serial could have been recorded on a VCR tape. 244 minutes is more than 4 hours. Therefore, the episodes would have to be recorded at the 6 hour speed.

Extension Have students make a sketch to compare their own height with the heights of both the 7-foot and 15-foot Superman statues. Based on their sketches, ask the students to estimate the minimum number of life-size statues of themselves they would need to stack to exceed the height of each Superman statue.
Check students' work.

Problem Solving on Location

ILLINOIS

Hometown of Superman

The town of Metropolis, Illinois, was founded in 1839. It is a small town located in the southern part of the state on the Ohio River. Metropolis has an approximate population of 6500.

1. **a.** In the first *Superman* comic book, Superman arrives in a town named Metropolis. This comic book appeared on shelves *n* years after the 1839 founding of Metropolis, Illinois. Write an algebraic expression to represent the year the first Superman comic appeared. **1839 + *n***

   b. Metropolis was officially declared the hometown of Superman (*n* − 65) years after the first comic appeared. (This is the same *n* from part **a**.) Use your expression from part **a** to write a new algebraic expression representing the year that Metropolis was declared Superman's hometown. Simplify your expression by combining like terms. **1839 + *n* + *n* − 65; 1774 + 2*n***

   c. Metropolis, Illinois, was declared Superman's hometown in 1972. Use this information and your expression from part **b** to solve for *n*. **1774 + 2*n* = 1972; *n* = 99**

Metropolis, Illinois, holds a Superman celebration every year.

2. A 7-foot-tall statue of Superman was purchased in 1986. In 1993, a 15-foot-tall bronze statue costing about $100,000 replaced it. The 15-foot statue cost 100 times more than the original statue. Write and solve an equation to find the cost of the original statue. **100*c* = 100,000; *c* = $1000**

3. In 1948 a movie serial called *The Adventures of Superman* appeared. It had *e* episodes that were each about 16 minutes long. The total running time of the serial was 244 minutes. Write and solve an equation to find the number of episodes in the serial. (Round to the nearest whole number.) **16*e* = 244; *e* = 15.25 ≈ 15 episodes**

Chicago Skyline

The Sears Tower, in Chicago, is the tallest building in the world in several categories.

Chicago's three tallest buildings are, from left, the John Hancock Center, the Sears Tower, the Aon Center.

Category	Height (ft)
Height of highest occupied floor	1431
Height to top of roof	1450
Height to top of spire or antenna	1730

1. The height of the Sears Tower from the main entrance on the east side of the building to the top of the roof is 1450 ft. But the height from the west side of the tower to the roof is 1454 ft. This is because the street on the west side of the tower is y feet lower than the street on the east side. Write and solve an equation to find the difference in feet in the levels of the two streets. $y + 1450 = 1454; y = 4$ ft

On a clear day you can see four states from the sky deck of the Sears Tower: Illinois, Michigan, Indiana, and Wisconsin.

2. Write and solve an equation to find the distance d from the top of the roof to the top of the antenna. $1450 + d = 1730; d = 280$ ft

3. Assume that all of the 110 stories of the Sears Tower are the same height, for a total height of 1450 ft. Write and solve an equation to find the height of each floor to the nearest foot. $110h = 1450, h \approx 13$ ft

The Sears Tower isn't the only tall building in Chicago. The John Hancock Center and the Aon Center are also among the world's tallest buildings.

4. The highest indoor swimming pool in the United States is on the 44th floor of the Hancock Center. Assume that each floor of the Hancock Center is h ft tall, and give the expression that tells the height at which the pool is located. $44h$

5. The Hancock Center has 46 residential floors, which is fewer than the total number of floors divided by 2. Using j for the total number of floors in the Hancock Center, write and solve an inequality that expresses the possible number of stories the Hancock Center has. $46 < \dfrac{j}{2}; j > 92$ stories

Chicago Skyline

- After problem 3, discuss the following: The Sears Tower has more than 2230 steps. If the stairs were equally divided among the stories, approximately how many steps would there be per story? > 20 steps

- After problem 3, discuss why it is important to assume that all the 110 stories are the same height to write and solve an equation. To solve the equation $110h = 1450$, you will divide. Division assumes breaking into equal groups.

Extension The Superman statue in Metropolis is 15 feet tall. The roof of the Sears Tower in Chicago is 1450 feet from the ground. Write and solve an equation to find how many Superman statues, stacked on top of one another, it would take to reach the roof of the Sears Tower. $15s = 1450; s \approx 96.7$; it would take 97 statues to reach the roof of the Sears Tower.

Game Resources

Puzzles, Twisters & Teasers
Chapter 1 Resource Book

Math Magic

Purpose: *To apply the problem-solving skill of translating words into math to perform a fun trick*

Discuss: Ask students to explain how the trick works. What does the variable *n* represent? How does this trick use combining like terms?
Possible answer: Operations are performed on the variable in such a way that the result is a constant. The variable *n* represents whatever number a person begins with. In the last step, when like terms are combined, the variable disappears because the like terms are opposites.

Extend: Challenge students to create their own math magic tricks. Have them explain how their tricks work by using variables and combining like terms.
Possible answer: Think of a number. Multiply the number by 10. Divide the result by 5. Add 7. Subtract 2 times the original number. Your answer is 7. The algebraic representation is as follows: n, $10n$, $\frac{10n}{5}$, $2n + 7$, $2n + 7 - 2n = 7$.

Crazy Cubes

Purpose: *To apply the problem-solving skill of guess and check to a classic brainteaser*

Discuss: Discuss some strategies that students can use to begin the game.
Possible answer: Set the first cube, and then try to place the other cubes, one at a time, so that no number is repeated along the front, back, top, and bottom.
Ask students why this game is a challenge. Possible answer: because you can't see all the sides at once and because the order of the cubes might change as you go along

Extend: Have students explore the number of possible ways to position a cube. There are 24 possible ways to position the cube (6 faces to serve as base, with 4 options for the front).

MATH-ABLES

Math Magic

You can guess what your friends are thinking by learning to "operate" your way into their minds! For example, try this math magic trick.

Think of a number. Multiply the number by 8, divide by 2, add 5, and then subtract 4 times the original number.

No matter what number you choose, the answer will always be 5. Try another number and see. You can use what you know about variables to prove it. Here's how:

	What you say:	What the person thinks:	What the math is:
Step 1:	Pick any number.	6 (for example)	n
Step 2:	Multiply by **8**.	$8(6) = 48$	$8n$
Step 3:	Divide by **2**.	$48 \div 2 = 24$	$8n \div 2 = 4n$
Step 4:	Add **5**.	$24 + 5 = 29$	$4n + 5$
Step 5:	Subtract **4** times the original number.	$29 - 4(6) = 29 - 24 = 5$	$4n + 5 - 4n = 5$

Invent your own math magic trick that has at least five steps. Show an example using numbers and variables. Try it on a friend!

Crazy Cubes

This game, called The Great Tantalizer around 1900, was reintroduced in the 1960s as "Instant Insanity™." The goal is to line up four cubes so that each row of faces has four different sides showing. Make four cubes with paper and tape, numbering each side as shown.

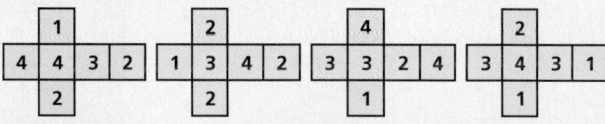

Line up the cubes so that 1, 2, 3, and 4 can be seen along the top, bottom, front, and back of the row of cubes. They can be in any order, and the numbers do not have to be right-side up.

Technology LAB — Graph Points

internet connect
Lab Resources Online
go.hrw.com
KEYWORD: MP4 TechLab1

On a graphing calculator, the **WINDOW** menu settings determine which points you see and the spacing between those points. In the standard viewing window, the x- and y-values each go from -10 to 10, and the tick marks are one unit apart. The boundaries are set by **Xmin, Xmax, Ymin**, and **Ymax. Xscl** and **Yscl** give the distance between the tick marks.

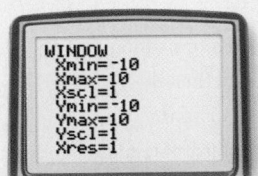

Activity

1 Plot the points $(2, 5)$, $(-2, 3)$, $(-\frac{3}{2}, 4)$, and $(1.75, -2)$ in the standard window. Then change the minimum and maximum x- and y-values of the window to -5 and 5.

Press **WINDOW** to check that you have the standard window settings.

To plot $(2, 5)$, press **2nd** **PRGM** *DRAW* **POINTS** **ENTER**.

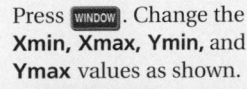

Then press 2 **,** 5 **ENTER**. After you see the grid with a point at $(2, 5)$, press **2nd** **MODE** *QUIT* to quit. Repeat the steps above to graph $(-2, 3)$, $(-\frac{3}{2}, 4)$, and $(1.75, -2)$.

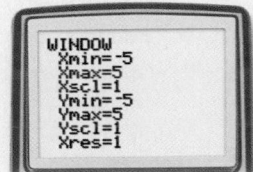

This is the graph in the standard window.

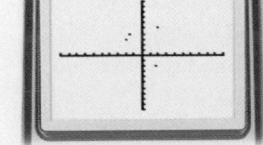

Press **WINDOW**. Change the **Xmin, Xmax, Ymin,** and **Ymax** values as shown.

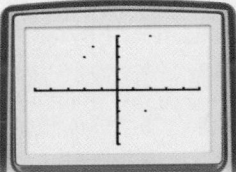

Repeat the steps above to graph the points in the new window.

Think and Discuss

1. Compare the two graphs above. Describe and explain any differences you see.

Try This

Graph the points $(-4, -8)$, $(1, 2)$, $(2.5, 7)$, $(3, 8)$, and $(-4.5, 12)$ in each window.

1. standard window   **2. Xmin** $= -5$; **Xmax** $= 5$; **Ymin** $= -20$; **Ymax** $= 20$; **Yscl** $= 5$

Answers

Think and Discuss

1. Possible answer: On the second graph, the points appear to be farther apart. Using the new settings, the window displays a smaller part of the coordinate plane on the same calculator screen, so the distance between tick marks is wider.

Try This

1.

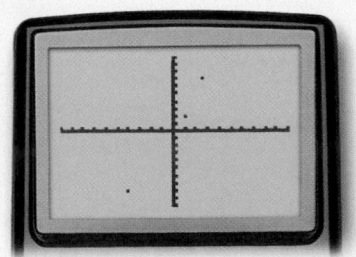

Note that $(-4.5, 12)$ is not visible in this window.

2.

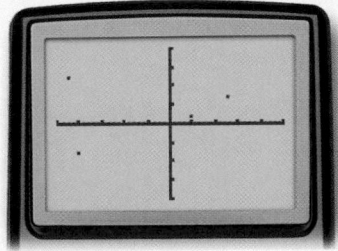

It is best to turn off any equation in **Y=** which may obscure points.

Technology LAB — Graph Points

Objective: To use a graphing calculator to plot points described by ordered pairs and to adjust the viewing window

Materials: Graphing calculator

Lab Resources

Technology Lab Activities . . . . p. 7

Using the Page

This technology activity shows students how to plot points, which can be done on any graphing calculator. Specific keystrokes may vary, depending on the make and model of the graphing calculator used. The keystrokes given are for a TI-83 model. For keystrokes to other models, visit go.hrw.com.

The Think and Discuss problem can be used to assess students' understanding of window settings on the calculator. Try This problems 1 and 2 are meant to reinforce students' understanding of the window settings and to encourage students to think about how a change in these settings can cause the position of the points to appear different.

Assessment

1. How could you set the calculator to show the x- and y-axes from -100 to $+100$ by tens?

Press WINDOW.
Enter the following settings:
Xmin $= -100$
Xmax $= 100$
Xscl $= 10$
Ymin $= -100$
Ymax $= 100$
Yscl $= 10$

2. If you plot the points $(4, 6)$ and $(5, 8)$ in the window described above, how will they appear? Explain. Possible answer: The points will appear to be very close together because the view is so large.

Purpose: *To help students review and practice concepts and skills presented in Chapter 1*

Assessment Resources

Chapter Review
Chapter 1 Resource Book . . . pp. 90–92

 Test and Practice Generator CD-ROM

Additional review assessment items in both multiple-choice and free-response format may be generated for any objective in Chapter 1.

Answers

1. ordered pair; *x*-coordinate; *y*-coordinate

2. solution set; inequality

3. 147

4. 152

5. 278

6. $2(k + 4)$

7. $4t + 5$

Vocabulary

Addition Property of Equality 14	inequality 23	solution set 23
algebraic expression 4	inverse operation 14	solve 13
algebraic inequality 23	isolate the variable 14	substitute 4
coefficient 4	like term 28	Subtraction Property of Equality ...14
constant 4	Multiplication Property of Equality 19	term 28
coordinate plane 38	ordered pair 34	variable 4
Division Property of Equality 18	origin 38	*x*-axis 38
equation 13	simplify 29	*x*-coordinate 38
equivalent expression ...28	solution of an equation 13	*y*-axis 38
evaluate 4	solution of an inequality 23	*y*-coordinate 38
graph of an equation 39		

Complete the sentences below with vocabulary words from the list above. Words may be used more than once.

1. In the ___?___ (4, 9), 4 is the ___?___ and 9 is the ___?___.

2. $x < 3$ is the ___?___ to the ___?___ $x + 5 < 8$.

1-1 Variables and Expressions (pp. 4–7)

EXAMPLE

■ Evaluate $4x + 9y$ for $x = 2$ and $y = 5$.

$4x + 9y$
$4(2) + 9(5)$ *Substitute 2 for x and 5 for y.*
$8 + 45$ *Multiply.*
53 *Add.*

EXERCISES

Evaluate each expression.

3. $9a + 7b$ for $a = 7$ and $b = 12$

4. $17m - 3n$ for $m = 10$ and $n = 6$

5. $1.5r + 19s$ for $r = 8$ and $s = 14$

1-2 Writing Algebraic Expressions (pp. 8–12)

EXAMPLE

■ Write an algebraic expression for the word phrase "2 less than a number *n*."

$n - 2$ *Write as a subtraction.*

EXERCISES

Write an algebraic expression for each phrase.

6. twice the sum of *k* and 4

7. 5 more than the product of 4 and *t*

1-3 Solving Equations by Adding or Subtracting (pp. 13–17)

EXAMPLE

Solve.

■ $x + 7 = 12$
$$\begin{array}{l} \underline{-7 \quad -7} \\ x + 0 = 5 \\ \quad\quad x = 5 \end{array}$$
    Subtract 7 from each side.

    Identity Property of Zero

■ $y - 3 = 1.5$
$$\begin{array}{l} \underline{+3 \quad +3} \\ y + 0 = 4.5 \\ \quad\quad y = 4.5 \end{array}$$
    Add 3 to each side.

    Identity Property of Zero

EXERCISES

Solve and check.

8. $z - 9 = 14$ **9.** $t + 3 = 11$
10. $6 + k = 21$ **11.** $x + 2 = 13$

Write an equation and solve.

12. A polar bear weighs 715 lb, which is 585 lb less than the a sea cow. How much does the sea cow weigh?

13. The Mojave Desert, at 15,000 mi², is 11,700 mi² larger than Death Valley. What is the area of Death Valley?

1-4 Solving Equations by Multiplying or Dividing (pp. 18–22)

EXAMPLE

Solve.

■ $4h = 24$
$$\frac{4h}{4} = \frac{24}{4}$$ *Divide each side by 4.*
$1h = 6$ $4 \div 4 = 1$
$h = 6$ $1 \cdot h = h$

■ $\frac{t}{4} = 16$
$4 \cdot \frac{t}{4} = 4 \cdot 16$ *Multiply each side by 4.*
$1t = 64$ $4 \div 4 = 1$
$t = 64$ $1 \cdot t = t$

EXERCISES

Solve and check.

14. $7g = 56$ **15.** $108 = 12k$ **16.** $0.1p = 8$
17. $\frac{w}{4} = 12$ **18.** $20 = \frac{y}{2}$ **19.** $\frac{z}{2.4} = 8$

Write an equation to solve.

20. The Lewis family drove 235 mi toward their destination. This was $\frac{2}{3}$ of the total distance. What was the total distance?

21. Luz will pay a total of $9360 on her car loan. Her monthly payment is $390. For how many months is the loan?

1-5 Solving Simple Inequalities (pp. 23–27)

EXAMPLE

Solve and graph.

■ $x + 5 \le 8$
$$\begin{array}{l} \underline{-5 \quad -5} \\ x \le 3 \end{array}$$
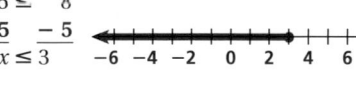

■ $3w > 18$
$$\frac{3w}{3} > \frac{18}{3}$$
$w > 6$
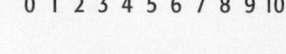

EXERCISES

Solve and graph.

22. $h + 3 < 7$ **23.** $y - 2 > 5$ **24.** $2x \ge 8$
25. $4p < 2$ **26.** $2m > 4.6$ **27.** $3q \le 0$
28. $\frac{w}{2} \ge 4$ **29.** $\frac{x}{3} \le 1$ **30.** $\frac{y}{4} > 4$
31. $4 < x + 1$ **32.** $2 < y - 4$ **33.** $8 \ge 4x$

Answers

8. $z = 23$

9. $t = 8$

10. $k = 15$

11. $x = 11$

12. Let c = weight of sea cow; $c - 585 = 715$; $c = 1300$; a sea cow weighs 1300 lb.

13. Let d = area of Death Valley; $d + 11,700 = 15,000$; $d = 3300$; the area of Death Valley is 3300 mi².

14. $g = 8$

15. $k = 9$

16. $p = 80$

17. $w = 48$

18. $y = 40$

19. $z = 19.2$

20. Let d = total distance; $\frac{2}{3}d = 235$; $d = 352.5$; the total distance was 352.5 mi.

21. Let m = number of months; $390m = 9360$; $m = 24$; the loan is for 24 months.

Study Guide and Review

Answers

22. $h < 4$;

23. $y > 7$;

24. $x \ge 4$;

25. $p < \frac{1}{2}$;

26. $m > 2.3$;
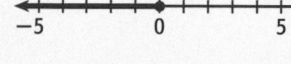

27. $q \le 0$;

28. $w \ge 8$;

29. $x \le 3$;

30. $y > 16$;

31. $x > 3$;

32. $y > 6$;

33. $x \le 2$;

Answers

34. $11m - 4$

35. $14w + 6$

36. $y = 5$

37. $z = 8$

38. yes

39. no

40.

x	3x + 2	y	(x, y)
0	3(0) + 2	2	(0, 2)
1	3(1) + 2	5	(1, 5)
2	3(2) + 2	8	(2, 8)
3	3(3) + 2	11	(3, 11)
4	3(4) + 2	14	(4, 14)

41.–46.

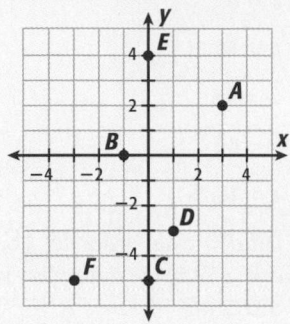

47. 5

48. 8

49. 20

50. Oven E

Study Guide and Review

1-6 Combining Like Terms (pp. 28–31)

EXAMPLE

■ Simplify.

$3(z - 6) + 2z$

$3z - 3(6) + 2z$ *Distributive Property*

$3z - 18 + 2z$ *3z and 2z are like terms.*

$5z - 18$ *Combine coefficients.*

EXERCISES

Simplify.

34. $4(2m - 1) + 3m$ **35.** $12w + 2(w + 3)$

Solve.

36. $6y + y = 35$ **37.** $9z - 3z = 48$

1-7 Ordered Pairs (pp. 34–37)

EXAMPLE

■ Determine whether (8, 3) is a solution of the equation $y = x - 6$.

$y = x - 6$

$3 \overset{?}{=} 8 - 6$

$3 \overset{?}{=} 2$ ✗

(8, 3) is not a solution.

EXERCISES

Determine whether the ordered pair is a solution of the given equation.

38. $(27, 0); y = 81 - 3x$ **39.** $(4, 5); y = 5x$

Use the values to make a table of solutions.

40. $y = 3x + 2$ for $x = 0, 1, 2, 3, 4$

1-8 Graphing on a Coordinate Plane (pp. 38–41)

EXAMPLE

■ Graph $A(3, -1)$, $B(0, 4)$, $C(-2, -3)$, and $D(1, 0)$ on a coordinate plane.

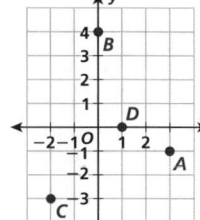

EXERCISES

Graph each point on a coordinate plane.

41. $A(3, 2)$ **42.** $B(-1, 0)$ **43.** $C(0, -5)$

44. $D(1, -3)$ **45.** $E(0, 4)$ **46.** $F(-3, -5)$

Give the missing coordinate for the solutions of $y = 3x + 5$.

47. $(0, y)$ **48.** $(1, y)$ **49.** $(5, y)$

1-9 Interpreting Graphs and Tables (pp. 43–47)

EXAMPLE

■ Which car has the faster acceleration?

Acceleration	Car A (s)	Car B (s)
0 to 30 mi/h	1.8	3.2
0 to 40 mi/h	2.8	4.7
0 to 50 mi/h	3.9	6.4
0 to 60 mi/h	5.1	8.8

Car A accelerates from 0 to each measured speed in fewer seconds than car B.

EXERCISES

50. Which oven had not been preheated?

Time (min)	Oven D (°F)	Oven E (°F)
0	450°	70°
1	435°	220°
2	445°	450°
3	455°	440°
4	450°	450°

Notes

Evaluate each expression for the given values of the variables.

1. $4x + 5y$ for $x = 9$ and $y = 7$ **71**
2. $5k(6 - 6m)$ for $k = 2$ and $m = \frac{1}{2}$ **30**

Write an algebraic expression for each word phrase.

3. 3 more than twice p **$2p + 3$**
4. 4 times the sum of t and 5
   $4(t + 5)$
5. 6 less than half of n
   $\frac{1}{2}n - 6$

Solve.   **$m = 10$**

6. $m + 15 = 25$
7. $4d = 144$ **$d = 36$**
8. $50 = h - 3$ **$h = 53$**
9. $\frac{x}{3} = 18$ **$x = 54$**
10. $y - 4 \geq 1.1$
    $y \geq 5.1$
11. $\frac{x}{3} < 6$ **$x < 18$**
12. $w + 1 < 4.5$ **$w < 3\frac{1}{2}$**
13. $2p > 15$
    $p > 7.5$

Graph each inequality.

14. $x > 4$
15. $y \leq 8$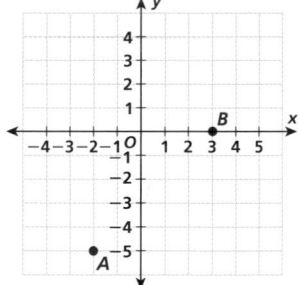

Write and solve an equation for each problem.

16. Acme Sporting Products manufactures 3216 tennis balls a day. Each container holds 3 balls. How many tennis ball containers are needed daily?

17. In the 1996 presidential election, Bill Clinton received 2,459,683 votes in Texas. This was 177,868 more votes than he had received in 1992 in Texas. How many votes did Bill Clinton get in Texas in 1992?

Solve.

18. $4x + 3 = 19$
    $x = 4$
19. $\frac{y}{2} - 5 = 1$
    $y = 12$
20. $10z + 2z = 108$
    $z = 9$
21. $26 = 3f + 10f$
    $f = 2$

Determine whether the ordered pair is a solution of the given equation.

22. $(6, 5)$; $y = 5x - 25$ **yes**

Give the coordinates of each point shown on the coordinate plane.

23. A **$(-2, -5)$**

24. B **$(3, 0)$**

25. Use the table to graph the speed of the car over time.

Time (s)	0	5	10	15
Speed (mi/h)	0	20	30	35

Purpose: *To assess students' mastery of concepts and skills in Chapter 1*

Assessment Resources

Chapter 1 Tests (Levels A, B, C)
Assessment Resources . . . . . . pp. 33–38

 ***Test and Practice Generator* CD-ROM**

Additional assessment items in both multiple-choice and free-response format may be generated for any objective in Chapter 1.

Answers

16. Let c = the number of containers; $3c = 3216$; $c = 1072$; 1072 containers are needed each day.

17. Let v = the number of votes in 1992; $v + 177{,}868 = 2{,}459{,}683$; $v = 2{,}281{,}815$; Bill Clinton got 2,281,815 votes in Texas in 1992.

25.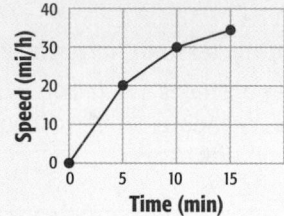

Purpose: *To assess students' under-standing of concepts in Chapter 1 and combined problem-solving skills*

Assessment Resources

Performance Assessment
Assessment Resources . . . . p. 118

Performance Assessment Teacher Support
Assessment Resources . . . . p. 117

Answers

1-2. See p. A2.

3. See Level 3 work sample below.

Scoring Rubric for Problem Solving Item 3

Level 3
Accomplishes the purposes of the task.

Student gives clear explanations, shows understanding of mathematical ideas and processes, and computes accurately.

Level 2
Purposes of the task not fully achieved.

Student demonstrates satisfactory but limited understanding of the mathematical ideas and processes.

Level 1
Purposes of the task not accomplished.

Student shows little evidence of under-standing the mathematical ideas and processes and makes computational and/or procedural errors.

Performance Assessment

Show What You Know

Create a portfolio of your work from this chapter. Complete this page and include it with your four best pieces of work from Chapter 1. Choose from your homework or lab assignments, mid-chapter quiz, or any journal entries you have done. Put them together using any design you want. Make your portfolio represent what you consider your best work.

Short Response

1. Find the solution set for the equation $x + 6 = 10$. Find the solution set for the inequality $2x \geq 8$. Explain what the solution sets have in common and then explain why they are different.

2. The average rise and fall of the tides in Eastport, Maine, is 5 ft 10 in. more than twice the average in Philadelphia, Pennsylvania. Write an algebraic expression you can use to find the measurement for Eastport, Maine. Then find that measurement.

Average Rise and Fall of Tides		
Place	ft	in.
Boston, MA	10	4
Charleston, SC	5	10
Eastport, ME		
Fort Pulaski, GA	7	6
Key West, FL	1	10
Philadelphia, PA	6	9

Extended Problem Solving

Choose any strategy to solve each problem.

3. A soft-drink company is running a contest. A whole number less than 100 is printed on each bottle cap. If you collect a set of caps with a sum of exactly 100, you win a prize. Below are some typical bottle caps.

  a. Write the prime factorization of each of the numbers on the bottle caps.

  b. What do all of the numbers on the bottle caps have in common?

  c. Do the bottle caps contain a winning set? Explain.

48

15

75

33

60

93

27

6

Student Work Samples for Item 3

Level 3

3a. 60 = 2 · $\overset{2}{\underset{30}{\cancel{15}}}$ 3 · 5
  75 = 3 · $\cancel{25}$ 5 · 5   b. They all have 3 as a prime factor.
  93 = 3 · 31
  33 = 3 · 11
  48 = 2 · $\cancel{24}$ 2 · $\cancel{12}$ 2 · $\cancel{6}$ 2 · 3
  27 = 3 · 3 · 3
  15 = 3 · 5
  6 = 2 · 3

  c. No. By the Distributive Prop., the sum of any bottle cap numbers would have to be ÷ by 3, and 100 is not ÷ by 3.

The student demonstrated an understand-ing of prime factors, identified the common factor, and understood why there was no winning set.

Level 2

3a. 60 = 2 · 3 · 2 · 5
  75 = 3 · 5 · 5
  93 = 31 · 3   b. They all have a prime factor of 3.
  33 = 11 · 3
  48 = 2 · 2 · 2 · 2 · 3
  27 = 3 · 3 · 3
  15 = 5 · 3
  6 = 3 · 2

  c. 60 + 33 + 6 = 99   48 + 27 + 15 + 6 = 96
  75 + 27 = 102
  93 + 6 = 99   There is no winning set.
  33 + 60 + 6 = 99
  33 + 27 + 15 + 6 = 81
  33 + 48 + 15 = 96

The student demonstrated an understand-ing of prime factors, identified the common factor, but could not apply this knowledge to the problem.

Level 1

3a. 60 = 2 × 30
  75 = 3 × 25
  93 Prime
  33 = 3 × 11
  48 = 2 × 24
  15 = 5 × 3
  6 = 2 × 3

  b. Nothing in common
  c. They don't have one.

The student showed little evidence of understanding prime factorization, made errors, and was unable to identify common factors or a strategy to solve the problem.

Cumulative Assessment, Chapter 1

1. Which algebraic equation represents the word sentence "15 less than the number of computers c is 32"?   **D**

(A) $\frac{c}{15} = 32$ (C) $15 - c = 32$

(B) $15c = 32$ (D) $c - 15 = 32$

2. Which inequality is represented by this graph?   **J**

2   4   6   8   10   12   14

(F) $x < 7$ (H) $7 < x$

(G) $x \le 7$ (J) $7 \le x$

3. Bill is 3 years older than his cat. The sum of their ages is 25. If c represents the cat's age, which equation could be used to find c?   **D**

(A) $c + 25 = c + 3$ (C) $c + 3c = 25$

(B) $c + 25 = 3c$ (D) $c + (c + 3) = 25$

4. The solution of $k + 3(k - 2) = 34$ is   **F**

(F) $k = 10$ (H) $k = 8$

(G) $k = 9$ (J) $k = 7$

5. Jamal brings $20 to a pizza restaurant where a plain slice costs $2.25, including tax. Which inequality can he use to find the number of plain slices he can buy?   **C**

(A) $2.25 + s \le 20$ (C) $2.25s \le 20$

(B) $2.25 + s \ge 20$ (D) $2.25s \ge 20$

6. When twice a number is decreased by 4, the result is 236. What is the number?   **J**

(F) 29.5   (H) 116

(G) 59   (J) 120

7. A number n is increased by 5 and the result is multiplied by 5. This result is decreased by 5. What is the final result?   **D**

(A) $5n$ (C) $5n + 10$

(B) $5n + 5$ (D) $5n + 20$

8. Which has the greatest value?   **H**

(F) $(2 + 3)(2 + 3)$ (H) $(2 \cdot 3)(2 \cdot 3)$

(G) $2 + 3 \cdot 3$ (J) $2 \cdot 2 + 3 \cdot 3$

TEST TAKING TIP!
To convert from a larger unit of measure to a smaller unit, multiply by the conversion factor. To convert from a smaller unit of measure to a larger unit, divide by the conversion factor.

9. **SHORT RESPONSE** Jo has 197 fund-raising posters. She decides to use four 5-inch strips of tape to hang each poster. Each roll of tape is 250 feet long. Estimate the number of whole rolls Jo will need to hang all of the posters. Explain in words how you determined your estimate. (*Hint:* 12 in. = 1 ft)

10. **SHORT RESPONSE** Mrs. Morton recorded the lengths of the telephone calls she made this week.

Length of call (min)	2	5	7	12	15
Number of calls	7	x	2	2	3

The number of calls shorter than 6 minutes is equal to the number of calls longer than 6 minutes. Write an equation that could be used to determine the number of 5-minute calls Mrs. Morton made. Solve your equation.

Purpose: *To provide review and practice for Chapter 1 and standardized tests*

Assessment Resources

Cumulative Tests (Levels A, B, C)
Assessment Resources. . . . pp. 145–156

State-Specific Test Practice Online
KEYWORD: MP4 TestPrep

Test Prep Doctor

To help students with questions involving a variable in an inequality, as in item 2, encourage them to read the variable first. The inequality $7 \le x$ can be read "x is greater than or equal to 7."

Supplement the test-taking tip given for item 8 by encouraging students to create a table of values, such as the following:

Tape Left on Roll (ft)	Number n of Posters Mounted
300	0
298	1
296	2

By substituting the values of n from column 2 into the expressions given in the answer choices, the student can verify that the correct expression is $300 - 2n$.

Answers

9. 2 rolls; It takes 20 inches of tape to hang 1 poster. One roll of tape is $12 \cdot 250 = 3000$ inches long, so $3000 \div 20 = 150$ posters can be hung with a roll of tape. Jo has 197 posters, so she needs 2 rolls of tape.

10. $x = 0$

$$7 + x = 2 + 2 + 3$$
$$7 + x = 7$$
$$\underline{-7 \qquad -7}$$
$$x = 0$$

Resource Options

Chapter 2 Resource Book

Student Resources

Practice (Levels A, B, C). . . . . pp. 9–11, 18–20, 27–29,
36–38, 45–47, 55–57, 64–66, 73–75, 82–84
Reteach . . . . pp. 12, 21, 30, 39, 48, 58, 67, 76, 85
Challenge . . . . pp. 13, 22, 31, 40, 49, 59, 68, 77, 86
Problem Solving . . . . pp. 14, 23, 32, 41, 50, 60, 69, 78, 87
Puzzles, Twisters & Teasers . . . . pp. 15, 24, 33, 42, 51, 61,
70, 79, 88
Recording Sheets . . . . pp. 3–4, 8, 17, 26, 35, 44, 54, 63,
72, 81, 92, 97
Chapter Review . . . . pp. 89–91

Teacher and Parent Resources

Chapter Planning and Pacing Guide. . . . . p. 5
Section Planning Guides . . . . pp. 6, 52
Parent Letter . . . . pp. 1–2
Teaching Tools . . . . pp. 95–97
Teacher Support for Chapter Project . . . . p. 93
Transparencies . . . . pp. T1–T44
• Daily Transparencies
• Additional Examples Transparencies
• Teaching Transparencies

Reaching All Learners

English Language Learners

Success for English Language Learners . . . . pp. 19–36
*Math: Reading and Writing
in the Content Area* . . . . pp. 10–18
Spanish Homework and Practice . . . . pp. 10–18
Spanish Interactive Study Guide . . . . pp. 10–18
Spanish Family Involvement Activities. . . . . pp. 9–16
Multilingual Glossary

Individual Needs

Are You Ready? Intervention and Enrichment. pp. 157–160,
213–216, 241–244, 261–264, 407–408
Alternate Openers: Explorations . . . . pp. 10–18
Family Involvement Activities . . . . pp. 9–16
Interactive Problem Solving. . . . . pp. 10–18
Interactive Study Guide . . . . pp. 10–18
Readiness Activities . . . . pp. 3–4
*Math: Reading and Writing
in the Content Area* . . . . pp. 10–18
Challenge . . . . . CRB pp. 13, 22, 31, 40, 49, 59, 68, 77, 86

Hands-On

Hands-On Lab Activities . . . . pp. 11–18
Technology Lab Activities . . . . pp. 8–13
Alternate Openers: Explorations . . . . pp. 10–18
Family Involvement Activities . . . . pp. 9–16

Applications and Connections

Consumer and Career Math. . . . . pp. 5–8
Interdisciplinary Posters . . . . Poster 2, TE p. 58B
Interdisciplinary Poster Worksheets . . . . pp. 4–6

Transparencies

Alternate Openers: Explorations . . . . pp. 10–18
Exercise Answers Transparencies
Chapter 2 Resource Book . . . . pp. T1–T44
• Daily Transparencies
• Additional Examples Transparencies
• Teaching Transparencies

Technology

Teacher Resources

 Lesson Presentations CD-ROM. . . . . Chapter 2

 Test and Practice Generator CD-ROM . . . . Chapter 2

 One-Stop Planner CD-ROM . . . . Chapter 2

Student Resources

 Are You Ready? Intervention CD-ROM
Skills 37, 51, 58, 63

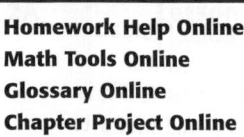

☑ internet connect

Homework Help Online	**KEYWORD:** MP4 HWHelp2
Math Tools Online	**KEYWORD:** MP4 Tools
Glossary Online	**KEYWORD:** MP4 Glossary
Chapter Project Online	**KEYWORD:** MP4 PSProject2
Chapter Opener Online	**KEYWORD:** MP4 Ch2

 **KEYWORD:** MP4 CNN2

SE = *Student Edition*   **TE** = *Teacher's Edition*   **AR** = *Assessment Resources*   **CRB** = *Chapter Resource Book*   **MK** = *Manipulatives Kit*

Assessment Options

Assessing Prior Knowledge

Determine whether students have the required prerequisite concepts and skills.

Are You Ready? . . . . SE p. 59
Inventory Test . . . . AR pp. 1–4

Test Preparation

Provide review and practice for chapter and standardized tests.

Standardized Test Prep . . . . SE p. 109
Spiral Review with Test Prep . . . . . SE, last page of each lesson
Study Guide and Review . . . . SE pp. 104–106
Test Prep Tool Kit

Technology

 ***Test and Practice Generator* CD-ROM**

☐ internet connect

State-Specific Test Practice Online KEYWORD: MP4 TestPrep

Performance Assessment

Assess students' understanding of chapter concepts and combined problem-solving skills.

Performance Assessment . . . . SE p. 108
   Includes scoring rubric in TE
Performance Assessment . . . . AR p. 120
Performance Assessment Teacher Support . . . . AR p. 119

Portfolio

Portfolio opportunities appear throughout the Student and Teacher's Editions.

Suggested work samples:

Problem Solving Project . . . . TE p. 58
Performance Assessment . . . . SE p. 108
Portfolio Guide . . . . AR p. xxxvii
Journal . . . . TE, last page of each lesson
Write About It . . . . SE, pp. 63, 67, 71, 77. 81, 87, 91, 99

Daily Assessment

Obtain daily feedback on students' understanding of concepts.

Spiral Review and Test Prep . . . . . . SE, last page of each lesson

Also Available on Transparency In Chapter 2 Resource Book

Warm Up . . . . TE, first page of each lesson
Problem of the Day . . . . TE, first page of each lesson
Lesson Quiz . . . . TE, last page of each lesson

Student Self-Assessment

Have students evaluate their own work.

Group Project Evaluation . . . . AR p. xxxiv
Individual Group Member Evaluation . . . . AR p. xxxv
Portfolio Guide . . . . AR p. xxxvii
Journal . . . . TE, last page of each lesson

Formal Assessment

Assess students' mastery of concepts and skills.

Section Quizzes . . . . AR pp. 7–8
Mid-Chapter Quizzes . . . . SE p. 82
Chapter Test . . . . SE p. 107
Chapter Tests (Levels A, B, C) . . . . AR pp. 39–44
Cumulative Tests (Levels A, B, C) . . . . AR pp. 157–168
Standardized Test Prep
   Cumulative Assessment . . . . SE p. 109
End-of-Year Test . . . . AR pp. 313–316

Technology

 ***Test and Practice Generator* CD-ROM**

   Make tests electronically. This software includes:

   • Dynamic practice for Chapter 2
   • Customizable tests
   • Multiple-choice items for each objective
   • Free-response items for each objective
   • Teacher management system

SE = *Student Edition*   **TE** = *Teacher's Edition*   **AR** = *Assessment Resources*   **CRB** = *Chapter Resource Book*   **MK** = *Manipulatives Kit*

Chapter 2 Tests

Three levels (A,B,C) of tests are available for each chapter in the *Assessment Resources*.

LEVEL A

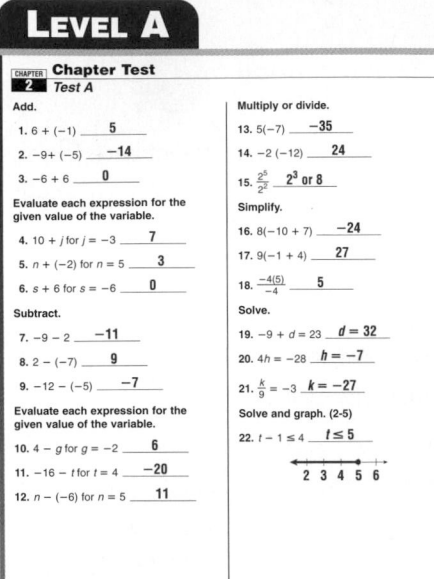

LEVEL B

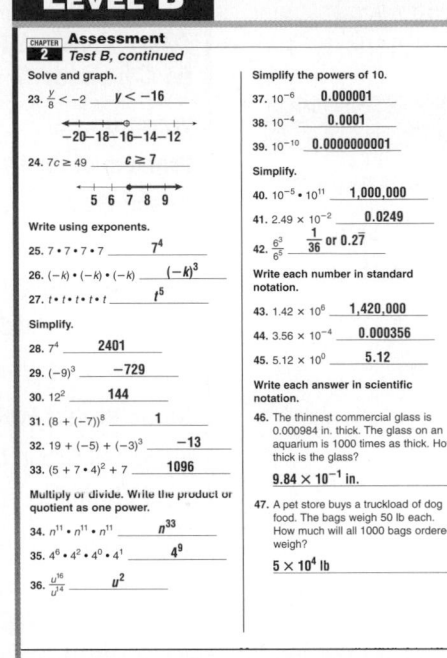

LEVEL C

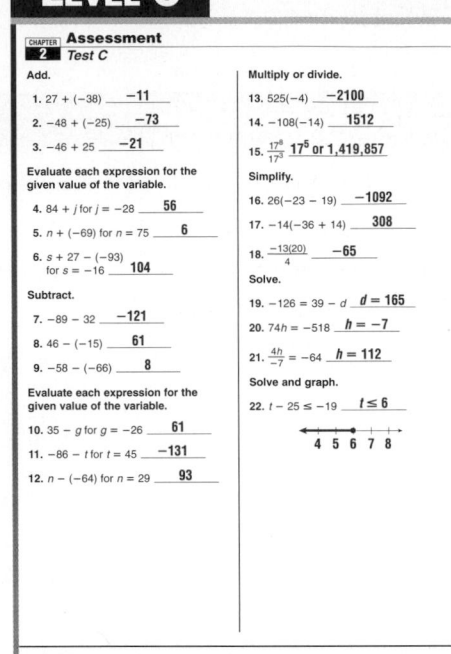

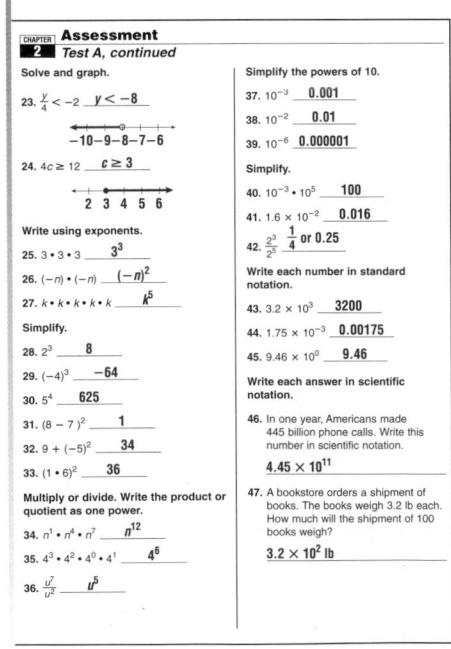

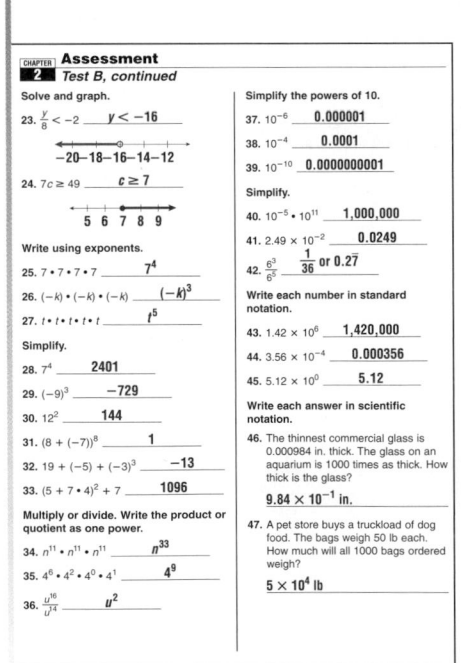

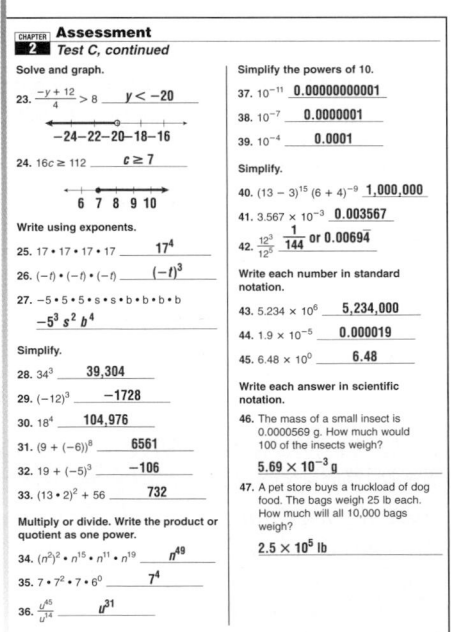

Test and Practice Generator
CD-ROM

Create and customize multiple versions of the same tests with corresponding answers for any chosen chapter objectives.

Chapter 2 State and Standardized Test Preparation

Test Taking Skill Builder and Standardized Test Practice
are provided for each chapter in the *Test Prep Tool Kit.*

TEST TAKING SKILL BUILDER

Test Taking Strategy | **Bubble**
Chapter 2

Even if you solve every problem on a test correctly, you will not score well unless you can enter the answers correctly on the answer sheet.

- Be sure to fill in the entire bubble.
- Make your mark dark, but not so dark that you rip the paper.
- Be sure to choose only one answer for each question.

Bubble Grids typically look like this:

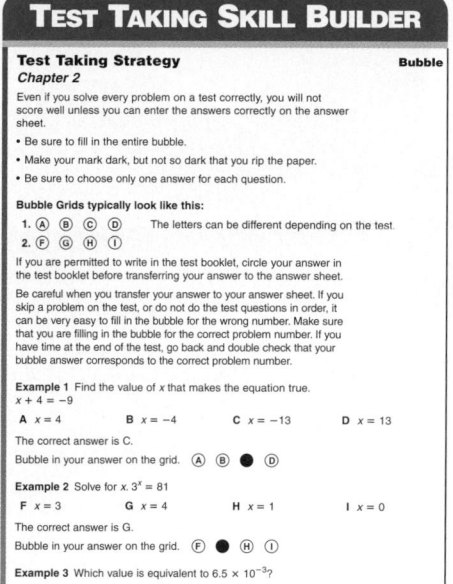

1. (A) (B) (C) (D)   The letters can be different depending on the test.
2. (F) (G) (H) (I)

If you are permitted to write in the test booklet, circle your answer in the test booklet before transferring your answer to the answer sheet.

Be careful when you transfer your answer to your answer sheet. If you skip a problem on the test, or do not do the test questions in order, it can be very easy to fill in the bubble for the wrong number. Make sure that you are filling in the bubble for the correct problem number. If you have time at the end of the test, go back and double check that your bubble answer corresponds to the correct problem number.

Example 1 Find the value of x that makes the equation true.
$x + 4 = -9$

A $x = 4$ B $x = -4$ C $x = -13$ D $x = 13$

The correct answer is C.
Bubble in your answer on the grid. (A) (B) ● (D)

Example 2 Solve for x. $3^x = 81$

F $x = 3$ G $x = 4$ H $x = 1$ I $x = 0$

The correct answer is G.
Bubble in your answer on the grid. (F) ● (H) (I)

Example 3 Which value is equivalent to 6.5×10^{-3}?

A 0.0065   B 0.065   C 0.65   D 6500

The correct answer is A.
Bubble in your answer on the grid. ● (B) (C) (D)

Test Taking Strategy
Chapter 2, continued

Exercises

Tell what error, if any, was made in each bubble grid response below.

1. Which number is equivalent to 3^{-3}?

A $\frac{1}{3}$ B $\frac{1}{9}$ C $\frac{1}{27}$ D 27

2. Which is $4 \times 4 \times 4 \times 5 \times 5$ expressed in exponential form?

F 4×5 G 4×5^2 H $3^4 \times 2^5$ I $4^3 \times 5^2$

3. Solve. $w + 5 \leq -6$

A $w \leq -1$ B $w \leq -11$ C $w \geq -1$ D $w \geq -11$

4. Evaluate the expression for the given value of the variable.
$c + 21$ for $c = -12$.

F 9   G -9 H -33 I 33

Bubble Grid | **Possible answers are given.**

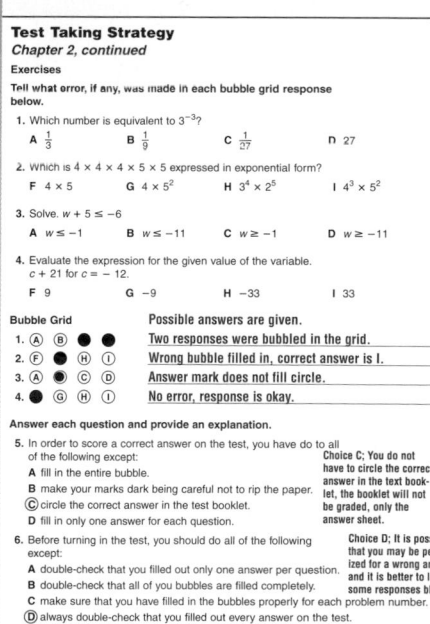

1. (A) (B) ● ●   Two responses were bubbled in the grid.
2. (F) ● (H) (I)   Wrong bubble filled in, correct answer is I.
3. (A) ● (C) (D)   Answer mark does not fill circle.
4. ● (G) (H) (I)   No error, response is okay.

Answer each question and provide an explanation.

5. In order to score a correct answer on the test, you have do to all of the following except:

A fill in the entire bubble.
B make your marks dark being careful not to rip the paper.
C circle the correct answer in the test booklet.
D fill in only one answer for each question.

Choice C; You do not have to circle the correct answer in the text booklet, the booklet will not be graded, only the answer sheet.

6. Before turning in the test, you should do all of the following except:

A double-check that you filled out only one answer per question.
B double-check that all of you bubbles are filled completely.
C make sure that you have filled in the bubbles properly for each problem number.
D always double-check that you filled out every answer on the test.

Choice D; It is possible that you may be penalized for a wrong answer and it is better to leave some responses blank

---

STANDARDIZED TEST PRACTICE

Standardized Test Practice
Chapter 2

Select the best answer for Questions 1–9.

1. Which number is equivalent to $7^4 \cdot 7^5$?

A 7^{20} C 7^8
B 7^9 D 7^6

2. The temperature dropped from $-5°F$ to $-28°F$ in 4 hours. How much did the temperature change?

F 22°   H 23°
G 33°   I 37°

3. The Wildcats took possession of the football on their own 20-yd line. They gained 16 yd, lost 8 yd, gained 5 yd, and then lost 4 yd. On what yard line were the Wildcats after the 4 plays?

A 11 yard line
B 19 yd line
C 29 yard line
D 41 yard line

4. A diver was 68 ft below the surface. She then rose to 50 ft below the surface. How many feet did the diver ascend?

F 118 ft
G 18 ft
H 28 ft
I 68 ft

5. Simplify. $\frac{(2)(-3)(6)}{(-4)}$

A -9 C -12
B 9   D 12

6. Solve. $\frac{a}{3} = -12$

F -4
G 4
H -36
I 6

7. Which value is a solution to $-6x \geq 24$?

A -6
B -2
C 0
D 4

8. Mike would like to save $10 a month for the next 100 months. Which equation can be used determine how much money he will save?

F $10 \cdot 10^3 = m$
G $10 \cdot 10^2 = m$
H $10 \cdot 10^4 = m$
I $10 \cdot 10^1 = m$

9. When Martin opened a savings account, he opened it with $500. A week later, he withdrew $30 for a class field trip. The next week, he deposited $45 in the bank. Which expression can be used to determine how much money he now has in the bank?

A $500 - $30 + $45
B $500 + $30 - $45
C $500 + ($30 + $45)
D $500 + ($30 - $45)

Standardized Test Practice
Chapter 2, continued

Gridded Response
Solve the problems. Use the answer sheet to write and grid-in your answer.

10. What is the exponent on 6 when you simplify $6^{-4} \cdot 6^6$?

11. Simplify. $(3 \cdot 7^4) - 19$

12. During a school fundraiser, Travis sold $28 in magazines. The next day he sold another magazine for $7. At the end of the third day he had sold a total of $56. What were his sales for the third day?

Short Response
Solve the problems. Use the answer sheet to write your answers.

13. The mass of a very small insect weighs 0.00000497 grams. Explain the benefit of expressing this number in scientific notation and provide the number.

14. Explain why the scientific notation for 1,070,000 is NOT 107×10^4.

Extended Response

15. The following chart shows the low temperatures for a week in Alaska during the winter months.

Day	Low Temperature
Mon.	$-16°F$
Tues.	$-12°F$
Wed.	$-22°F$
Thurs.	$-17°F$
Fri.	$-8°F$
Sat.	$-13°F$
Sun.	$4°F$

a. Write an equation to find the average low temperature for the weekdays, and then solve the equation.

b. What was the average low temperature for the weekend? Explain, in words, how you came up with your answer.

c. What was the average low temperature for the entire week? Explain how your process was different from part b.

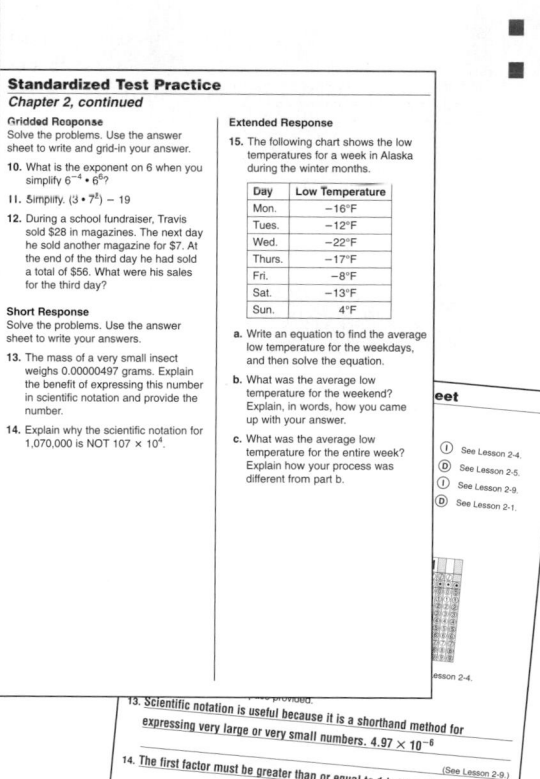

...eet
(I) See Lesson 2-4.
(D) See Lesson 2-5.
(I) See Lesson 2-9.
(D) See Lesson 2-1.

...esson 2-4.

13. Scientific notation is useful because it is a shorthand method for expressing very large or very small numbers. 4.97×10^{-6}

14. The first factor must be greater than or equal to 1 but less than 10. (See Lesson 2-9.)

Extended Response   (See Lesson 2-9.)
Write your answers for Problem 15 on the back of this paper.
See Lesson 2-4.

State-Specific Test Practice Online
KEYWORD: MP4 TestPrep

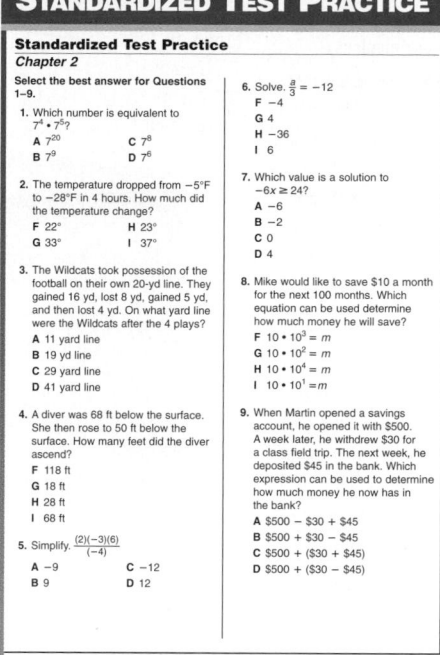

Test Prep Tool Kit

- Standardized Test Prep Workbook
- Countdown to Testing transparencies
- State Test Prep CD-ROM
- Standardized Test Prep Video

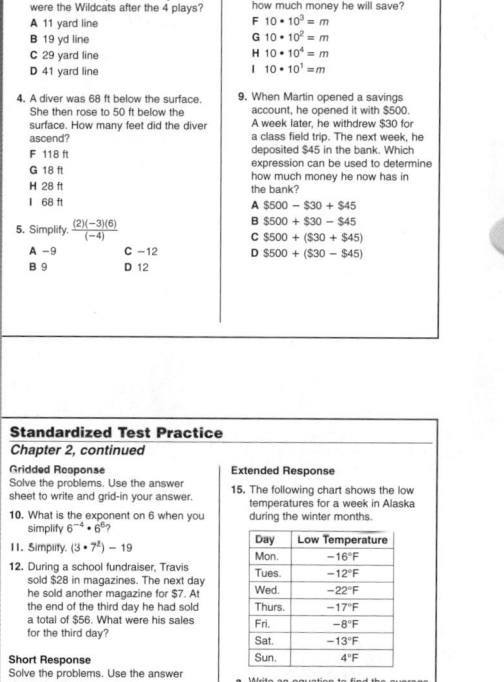

Customized answer sheets give students realistic practice for actual standardized tests.

Chapter 2

Integers and Exponents

Why Learn This?

Tell students that equations are used in science to determine the value of unknown quantities. Also, exponents are used to help express very large or very small numbers. For example, the independent life span of a muon is expressed as 2.2×10^{-6} seconds instead of 0.0000022 seconds because it is easier to understand and use numbers written in scientific notation.

Using Data

To begin the study of this chapter, have students:

- Write a number sentence using <, >, or = to compare the independent life span of a neutron to that of a muon. **$920 > 2.2 \times 10^{-6}$**
- Determine how long the independent life span of a neutron is in minutes. **15 minutes and 20 seconds** Write 920 as 9.2 times a power of 10. **9.2×10^2**

Chapter 2

TABLE OF CONTENTS

internet connect
Chapter Opener Online
go.hrw.com
KEYWORD: MP4 Ch2

Integers and Exponents

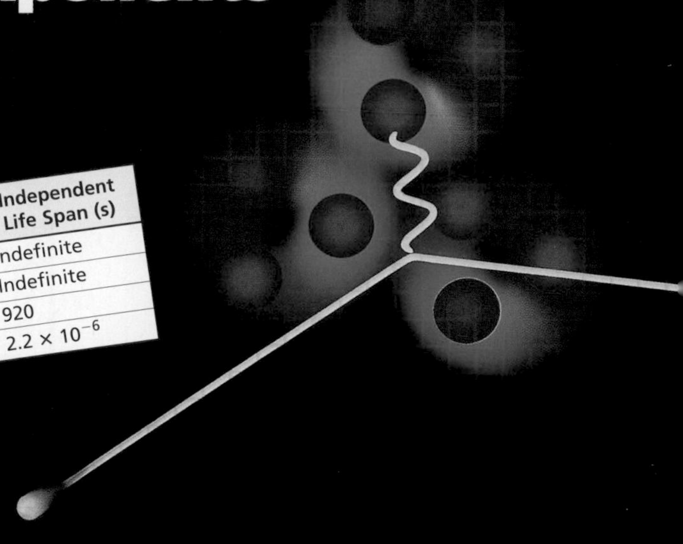

Atomic Particle	Independent Life Span (s)
Electron	Indefinite
Proton	Indefinite
Neutron	920
Muon	2.2×10^{-6}

Career ***Nuclear Physicist***

The atom was defined by the ancient Greeks as the smallest particle of matter. We now know that atoms are made up of many smaller particles.

Nuclear physicists study these particles using large machines—such as linear accelerators, synchrotrons, and cyclotrons—that can smash atoms to uncover their component parts.

Nuclear physicists use mathematics along with the data they discover to create models of the atom and the structure of matter.

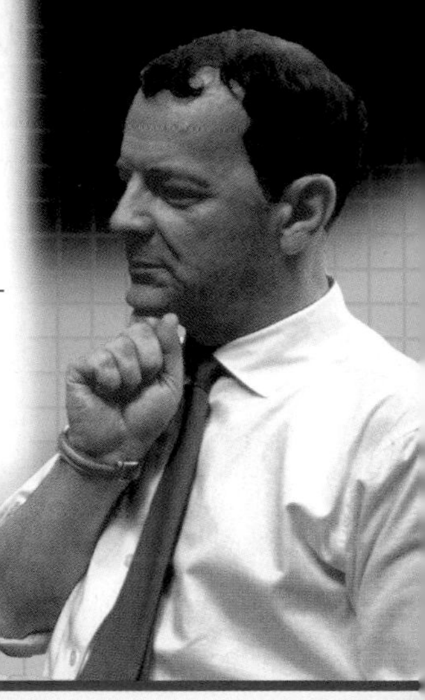

PROBLEM SOLVING

Problem Solving Project

Physical Science Connection

Purpose: To use exponents to compare time spans

Materials: The Lives of Particles worksheet, modeling materials

internet connect
Chapter Project Online: ***go.hrw.com***
KEYWORD: MP4 PSProject2

Understand, Plan, Solve, and Look Back

Have students:

- ✓ Complete The Lives of Particles worksheet to learn about the independent life spans of some atomic particles.
- ✓ Construct a drawing or model of the atom and its component parts.
- ✓ Check students' work.

ARE YOU READY?

Choose the best term from the list to complete each sentence.

1. According to the __?__, you must multiply or divide before you add or subtract when simplifying a numerical __?__. **order of operations; expression**

2. An algebraic expression is a mathematical sentence that has at least one __?__. **variable**

3. In a(n) __?__, an equal sign is used to show that two quantites are the same. **equation**

4. You use a(n) __?__ to show that one quantity is greater than another quantity. **inequality**

expression

inequality

order of operations

variable

equation

Complete these exercises to review skills you will need for this chapter.

✔ Order of Operations

Simplify by using the order of operations.

5. $(12) + 4(2)$ **20**   6. $12 + 8 \div 4$ **14**   7. $15(14 - 4)$ **150**

8. $(23 - 5) - 36 \div 2$ **0**   9. $12 \div 2 + 10 \div 5$ **8**   10. $40 \div 2 \cdot 4$ **80**

✔ Equations

Solve.

11. $x + 9 = 21$ **$x = 12$**   12. $3z = 42$ **$z = 14$**   13. $\frac{w}{4} = 16$ **$w = 64$**

14. $24 + t = 24$ **$t = 0$**   15. $p - 7 = 23$ **$p = 30$**   16. $12m = 0$ **$m = 0$**

✔ Match a Number Line to an Inequality

Write an inequality that describes the set of points shown on each number line.

17.  $x < 4$
$-6\ -4\ -2\ \ 0\ \ 2\ \ 4\ \ 6$

18. 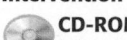 $x \le 6$
$-6\ -4\ -2\ \ 0\ \ 2\ \ 4\ \ 6$

19. $x \ge 2$
$-4\ -2\ \ 0\ \ 2\ \ 4\ \ 6\ \ 8$

20. $x > 5$
$-4\ -2\ \ 0\ \ 2\ \ 4\ \ 6\ \ 8$

✔ Multiply and Divide by Powers of Ten

Multiply or divide.

21. $358(10)$ **3580**   22. $358(1000)$ **358,000**   23. $358(100,000)$ **35,800,000**

24. $\frac{358}{10}$ **35.8**   25. $\frac{358}{1000}$ **0.358**   26. $\frac{358}{100,000}$ **0.00358**

Assessing Prior Knowledge

INTERVENTION

Diagnose and Prescribe

Evaluate your students' performance on this page to determine whether intervention is necessary or whether enrichment is appropriate. Options that provide instruction, practice, and a check are listed below.

Resources for Are You Ready?

- *Are You Ready? Intervention and Enrichment*

- **Recording Sheet for Are You Ready?**
  Chapter 2 Resource Book pp. 3–4

   Are You Ready? Intervention CD-ROM

ARE YOU READY?
Were students successful with Are You Ready?

NO INTERVENE ⟵ ⟶ **YES ENRICH**

✔ **Order of Operations** 5–10
Intervention Practice, Skill 51
  CD-ROM Intervention Activities, Skill 51

✔ **Equations** 11–16
Intervention Practice, Skill 58
  CD-ROM Intervention Activities, Skill 58

✔ **Match a Number Line to an Inequality** 17–20
Intervention Practice, Skill 63
  CD-ROM Intervention Activities, Skill 63

✔ **Multiply and Divide by Powers of Ten** 21–26
Intervention Practice, Skill 37
  CD-ROM Intervention Activities, Skill 37

Are You Ready? Enrichment, pp. 407–408

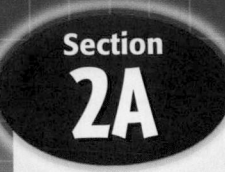
 **One-Minute Section Planner**

Lesson	Materials	Resources
Lesson 2-1 Adding Integers **NCTM:** Number and Operations, Communication **NAEP:** Number Properties 3a ☑ SAT-9 ☑ SAT-10 ☑ ITBS ☑ CTBS ☑ MAT ☑ CAT	**Optional** Teaching Transparency T2 *(CRB)* Two-color counters *(MK)*	• *Chapter 2 Resource Book*, pp. 7–15 • Daily Transparency T1, CRB • Additional Examples Transparencies, T3–T5 CRB • *Alternate Openers: Explorations*, p. 10
Lesson 2-2 Subtracting Integers **NCTM:** Number and Operations, Communication, Connections **NAEP:** Number Properties 3a ☐ SAI-9 ☑ SAT-10 ☐ ITBS ☑ CTBS ☑ MAT ☑ CAT	**Optional** Teaching Transparency T7 *(CRB)*	• *Chapter 2 Resource Book*, pp. 16–24 • Daily Transparency T6, CRB • Additional Examples Transparencies T8–T10, CRB • *Alternate Openers: Explorations*, p. 11
Lesson 2-3 Multiplying and Dividing Integers **NCTM:** Number and Operations, Communication, Connections **NAEP:** Number Properties 3a ☐ SAT-9 ☑ SAT-10 ☐ ITBS ☑ CTBS ☑ MAT ☑ CAT	**Optional** Teaching Transparency T12 *(CRB)*	• *Chapter 2 Resource Book*, pp. 25–33 • Daily Transparency T11, CRB • Additional Examples Transparencies T13–T15, CRB • *Alternate Openers: Explorations*, p. 12
Hands-On Lab 2A Model Solving Equations **NCTM:** Number and Operations, Algebra, Representation **NAEP:** Algebra 4a ☐ SAT-9 ☐ SAT-10 ☐ ITBS ☐ CTBS ☐ MAT ☐ CAT	**Required** Algebra tiles *(MK)*	• *Hands-On Lab Activities*, pp. 14–16, 93
Lesson 2-4 Solving Equations Containing Integers **NCTM:** Algebra, Problem Solving, Communication, Connections **NAEP:** Algebra 4a ☐ SAT-9 ☑ SAT-10 ☐ ITBS ☑ CTBS ☑ MAT ☑ CAT	**Optional** Squares and circles for Reaching All Learners *(CRB, p. 95)*	• *Chapter 2 Resource Book*, pp. 34–42 • Daily Transparency T16, CRB • Additional Examples Transparencies T17–T19, CRB • *Alternate Openers: Explorations*, p. 13
Lesson 2-5 Solving Inequalities Containing Integers **NCTM:** Algebra, Communication, Connections **NAEP:** Algebra 4a ☐ SAT-9 ☐ SAT-10 ☐ ITBS ☐ CTBS ☐ MAT ☐ CAT	**Optional** Teaching Transparency T21 *(CRB)* Number lines for Reaching All Learners *(CRB, p. 96)*	• *Chapter 2 Resource Book*, pp. 43–51 • Daily Transparency T20, CRB • Additional Examples Transparencies T22–T25, CRB • *Alternate Openers: Explorations*, p. 14
Section 2A Assessment		• Mid-Chapter Quiz, SE p. 82 • Section 2A Quiz, AR p. 7 • *Test and Practice Generator* CD-ROM

SAT = *Stanford Achievement Tests*   **ITBS** = *Iowa Test of Basic Skills*   **CTBS** = *Comprehensive Test of Basic Skills/Terra Nova*
MAT = *Metropolitan Achievement Test*   **CAT** = *California Achievement Test*
NCTM = Complete standards can be found on pages T29–T35.   **NAEP** = Complete standards can be found on pages A54–A58.
SE = *Student Edition*   **TE** = *Teacher's Edition*   **AR** = *Assessment Resources*   **CRB** = *Chapter Resource Book*   **MK** = *Manipulatives Kit*

Section Overview

Adding and Subtracting Integers

Lessons 2-1, 2-2

 Why? To evaluate expressions and solve equations and inequalities, we need to be able to add and subtract integers. ← The set of integers includes the set of whole numbers and their **opposites**.

Adding Integers		Subtracting Integers
If the signs are the same . . .	*If the signs are different . . .*	To subtract an integer, add its opposite.
Add the absolute values. The sign for the sum will be the same as the sign of the integers you are adding. $5 + 2 = 7$ $-5 + (-2) = -7$	Subtract the absolute values. The sign for the difference will be the same as that of the integer with the larger absolute value. $5 + (-2) = 3$ $-5 + 2 = -3$	$5 - 2 = 5 + (-2) = 3$ $5 - (-2) = 5 + 2 = 7$

Multiplying and Dividing Integers

Lesson 2-3

Why? To evaluate expressions and solve equations and inequalities, we need to be able to multiply and divide integers as well as add and subtract integers.

Multiplying and Dividing Integers			
If the signs are the same . . .		*If the signs are different . . .*	
The sign of the product or quotient will be positive.		The sign of the product or quotient will be negative.	
$6(2) = 12$ $-6(-2) = 12$	$\frac{6}{2} = 3$ $\frac{-6}{-2} = 3$	$6(-2) = -12$ $-6(2) = -12$	$\frac{6}{-2} = -3$ $\frac{-6}{2} = -3$

Solving Equations and Inequalities Containing Integers

Hands-On Lab 2A, Lessons 2-4, 2-5

Why? Solving equations and inequalities is necessary in many problem-solving situations.

To solve an equation or inequality, *undo* whatever operation is indicated on the variable.

Remember: If you multiply or divide both sides of an inequality by a negative number, reverse the inequality symbol.

$$310 = C + 273$$
$$\underline{-\ 273 = \quad -\ 273}$$
$$37 = C$$

$$-3x > 18$$
$$\frac{-3x}{-3} < \frac{18}{-3}$$
$$x < -6$$

In the equation, adding +273 is indicated. To undo adding +273, add −273 to both sides of the equation.

In the inequality, multiplying by −3 is indicated. To undo multiplying by −3, divide both sides of the inequality by −3.

Pacing: Traditional $\frac{1}{2}$ day
Block $\frac{1}{3}$ day

Objective: Students add integers.

Warm Up

Graph each number on a number line.

1. 5   **2.** 7

3. 4   **4.** 0

Add.

5. 22 + 19 **41**   **6.** 17 + 18 **35**

Problem of the Day

Start at (0, 0). Walk 3 blocks west and 2 blocks north. Then walk 4 blocks south and 1 block east. What are your coordinates? (−2, −2)

Available on Daily Transparency in CRB

Math Humor

No wonder his company went bankrupt. He tried to make larger negative profits because they had a greater absolute value.

2-1 Adding Integers

Learn to add integers.

Vocabulary
integer
opposite
absolute value

Katrina keeps a health journal. She knows that when she eats she adds calories and when she exercises she subtracts calories. So she uses *integers* to find her daily total.

Integers are the set of whole numbers, including 0, and their **opposites**. The sum of two opposite integers is zero.

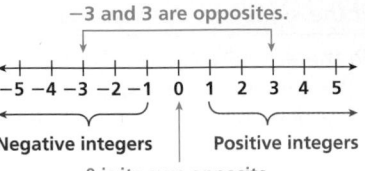

−3 and 3 are opposites.

−5 −4 −3 −2 −1 0 1 2 3 4 5

Negative integers   Positive integers

0 is its own opposite.

EXAMPLE 1 Using a Number Line to Add Integers

Use a number line to find the sum.

$4 + (-6)$

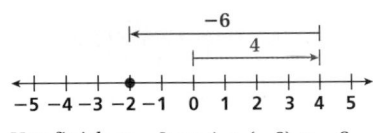

Move right 4 units. From 4, move left 6 units.

Helpful Hint

To add a **positive** number, move to the **right**. To add a **negative** number, move to the **left**.

You finish at −2, so $4 + (-6) = -2$.

Another way to add integers is to use absolute value. The **absolute value** of a number is its distance from 0. The absolute value of −4, written as $|-4|$, is 4; and the absolute value of 5 is 5.

ADDING INTEGERS	
If the signs are the same...	**If the signs are different...**
find the sum of the absolute values. Use the same sign as the integers.	find the difference of the absolute values. Use the sign of the integer with the larger absolute value.

---

1 Introduce

Alternate Opener

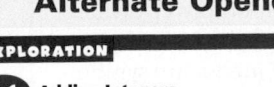

EXPLORATION

2-1 Adding Integers

You can use a thermometer to model addition of integers.

1. Suppose the temperature starts at −50°F and increases 40° during the day. Complete the addition statement to show the new temperature.
   $-50° + 40° = $ ___

2. Suppose the temperature starts at 40°F and drops 70° overnight. Complete the addition statement to show the new temperature.
   $40° + (-70°) = $ ___

Complete the addition statement modeled by each number line.

3. $-4 + $ ___ $= 5$
   −6 −4 −2 0 2 4 6

4. $-1 + ($ ___ $) = -8$
   −10 −8 −6 −4 −2 0 2

Think and Discuss

5. **Explain** how to add integers on a number line.

Motivate

To introduce students to adding integers, ask for examples of real-world situations that could be represented by integers (e.g., receiving $2 for allowance could be represented by +2; giving away three cookies could be represented by −3). Suggest a set of integers such as {−3, 1, −5, 2}. Ask students to order the integers from least to greatest. You may want to use Teaching Transparency T2 (CRB) to review integers and opposites on a number line.

Exploration worksheet and answers on Chapter 2 Resource Book pp. 8 and 98

---

2 Teach

Lesson Presentation

Guided Instruction

In this lesson, students learn to add integers. Show students how to use a number line to add. Starting at zero, move to the first number in the addition expression. From there, move the number of spaces indicated by the second number, moving left for a negative number or right for a positive number. After students have mastered this method, show them that the same results are obtained by using the rules.

Teaching Tip
Suggest that students first determine the sign of the sum, which is always the sign of the number with the greatest absolute value.

EXAMPLE 2 Using Absolute Value to Add Integers

Add.

A −3 + (−5)

$$-3 + (-5)$$
$$-8 \qquad \textit{Same sign; use the sign of the integers.}$$
Think: Find the sum of 3 and 5.

B 4 + (−7)

$$4 + (-7) \qquad \textit{Think: Find the difference of 7 and 4.}$$
$$-3 \qquad \textit{7 > 4; use the sign of 7.}$$

C −3 + 6

$$-3 + 6 \qquad \textit{Think: Find the difference of 6 and 3.}$$
$$3 \qquad \textit{6 > 3; use the sign of 6.}$$

EXAMPLE 3 Evaluating Expressions with Integers

Evaluate *b* + 12 for *b* = −5.

$$b + 12$$
$$(-5) + 12 \qquad \textit{Replace b with −5.}$$
Think: Find the difference of 12 and 5.
$$-5 + 12 = 7 \qquad \textit{12 > 5; use the sign of 12.}$$

EXAMPLE 4 *Health Application*

Katrina wants to check her calorie count after breakfast and exercise. Use information from the journal entry to find her total.

$$145 + 62 + 111 + (-110) + (-40)$$
Use a positive sign for calories and a negative sign for calories burned.

$$(145 + 62 + 111) + (-110 + -40)$$
Group integers with same signs.

$$318 + (-150)$$
Add integers within each group.

$$168$$
318 > 150; use the sign of 318.

Katrina's calorie count after breakfast and exercise is 168 calories.

Monday Morning

Calories

Oatmeal	145
Toast w/jam	62
8 fl oz juice	111

Calories burned

Walked six laps	110
Swam six laps	40

Think and Discuss

1. Compare the sums 10 + (−22) and −10 + 22.

2. Explain whether an absolute value is ever negative.

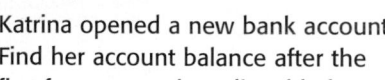
Additional Examples

Example 1

Use a number line to find the sum.

(−6) + 2
−4

Example 2

Add.

A. 1 + (−2)
−1

B. (−8) + 5
−3

C. (−2) + (−4)
−6

D. 7 + (−1)
6

Example 3

Evaluate *c* + 4 for *c* = −8.
−4

Example 4

Katrina opened a new bank account. Find her account balance after the first four transactions, listed below.

Deposits: $200, $20

Withdrawals: $166, $38

Katrina's account balance after the first four transactions is $16.

---

3 Close

Reaching All Learners
Through Concrete Manipulatives

Have students complete the following addition problems using two-color counters (in Manipulatives Kit) to represent the numbers. Have students use yellow for positive numbers and red for negative numbers. (Pennies can also be used, with heads representing positive and tails representing negative.) Remind students that a positive counter and a negative counter are opposites and their sum is zero.

1. −3 + 2   −1

2. 5 + (−4)   1

3. (−1) + (−5)   −6

Summarize

Have students write in their own words the rules for adding integers. Discuss how students can check their answers.

Possible answers: If the signs are the same, add the numbers and place the same sign in the answer. If the signs are different, determine which number is greater without the signs. The sign of the greater number (ignoring signs) will be the sign for the answer. Subtract the numbers. You can check your answer using a number line.

Answers to Think and Discuss

Possible answers:

1. 10 + (−22) = −12; −10 + 22 = 12; the sums are opposites.

2. No; an absolute value is a distance, and distance cannot be negative.

2-1 Exercises

FOR EXTRA PRACTICE
see page 734

internet connect
Homework Help Online
go.hrw.com   Keyword: MP4 2-1

GUIDED PRACTICE

Students may want to refer back to the lesson examples.

See Example **1** Use a number line to find each sum.

1. $3 + 2$ **5** **2.** $6 + (-4)$ **2** **3.** $-6 + 10$ **4** **4.** $-4 + (-2)$ **−6**

See Example **2** Add.

5. $-11 + 3$ **−8** **6.** $8 + (-2)$ **6** **7.** $-12 + 15$ **3** **8.** $-7 + (-9)$ **−16**

See Example **3** Evaluate each expression for the given value of the variable. **−8**

9. $t + 16$ for $t = -5$ **11** **10.** $m + 8$ for $m = -4$ **4** **11.** $p + (-4)$ for $p = -4$

See Example **4** **12.** Ron is balancing his checkbook. Use the information at right to find the difference in his checking account. Note that checks represent account withdrawals. **$297**

Checks	Deposits
$128	$500
$46	$175
$204	

INDEPENDENT PRACTICE

See Example **1** Use a number line to find each sum.

13. $5 + (-7)$ **−2** **14.** $-5 + 5$ **0** **15.** $5 + (-8)$ **−3** **16.** $-4 + 7$ **3**

See Example **2** Add.

17. $9 + 12$ **21** **18.** $-7 + (-8)$ **−15** **19.** $-9 + (-9)$ **−18** **20.** $16 + (-4)$ **12**

See Example **3** Evaluate each expression for the given value of the variable.

21. $q + 10$ for $q = 12$ **22** **22.** $x + 16$ for $x = -6$ **10** **23.** $z + (-7)$ for $z = 16$ **9**

See Example **4** **24.** A hospital clerk is checking her records. Use the information at right to find the net change in the number of patients for the week. **50**

	Admissions	Discharges
Monday	14	8
Tuesday	25	4
Wednesday	13	11
Thursday	17	0
Friday	9	5

PRACTICE AND PROBLEM SOLVING

Write an addition equation for each number line diagram.

25.
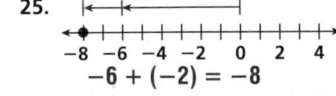
$-6 + (-2) = -8$

26.
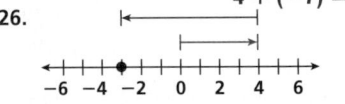
$4 + (-7) = -3$

Assignment Guide

If you finished Example **1** assign:
  Core 1–4, 13–16, 25–34, 50–54
Enriched 13–16, 25–38, 50–54

If you finished Example **2** assign:
  Core 1–8, 13–20, 25–37 odd, 50–54
Enriched 13–20, 25–38, 48–54

If you finished Example **3** assign:
  Core 1–11, 13–23, 25, 26, 27–45 odd, 50–54
Enriched 13–23, 25–45, 48–54

If you finished Example **4** assign:
  Core 1–26, 27–45 odd, 46, 50–54
Enriched 13–54

Answers

1–4, 13–16. Complete answers on p. A2

Math Background

To find the sum of any two integers, exactly one of the following cases will apply:

Case 1 Add two numbers with the same sign. This requires two jumps in the same direction on the number line. Because both jumps are in the same direction, you add absolute values to get the final distance from zero.

Case 2 Add two numbers with different signs. This requires two jumps in opposite directions on the number line. Because the jumps are in opposite directions, you subtract absolute values to get the final distance from zero.

RETEACH 2-1

LESSON 2-1 Reteach
Adding Integers

You can model integer addition using two-color counters. Use the yellow side for 1 and the red side for –1. Remember that one yellow counter and one red counter are opposites, so their sum is zero.

$7 + 5 = $ ⓎⓎⓎⓎⓎⓎⓎ ⓎⓎⓎⓎⓎ = 12 yellow counters = 12

$7 + (-5) = $ ⓎⓎⓎⓎⓎⓎⓎ ⓇⓇⓇⓇⓇ = 2 yellow counters = 2

$-7 + (-5) = $ ⓇⓇⓇⓇⓇⓇⓇ ⓇⓇⓇⓇⓇ = 12 red counters = −12

$-7 + 5 = $ ⓇⓇⓇⓇⓇⓇⓇ ⓎⓎⓎⓎⓎ = 2 red counters = −2

If the given integers were added, state whether the result would be positive or negative.

1. $-4 + (-6)$ negative
2. $-3 + 8$ positive
3. $-5 + 2$ negative

Notice if the counters are the same color, you add the absolute values of the integers. The answer is the sign of the integers. If the counters are both colors, you subtract the absolute values of the integers. Use the sign of the integer with the greater absolute value.

To add the given integers, state whether you need to add or subtract absolute values.

4. $8 + 3$ add
5. $-4 + (-1)$ add
6. $3 + (-6)$ subtract

Complete to find each sum.   **7.** $5 + (-9) = ?$ **8.** $-6 + (-4) = ?$

Are the signs the same or different? different   same
Which sign will you use for the sum? negative   negative
Will you add or subtract absolute values? subtract   add
Write the sum. −4   −10

PRACTICE 2-1

LESSON 2-1 Practice B
Adding Integers

Use a number line to find each sum.

1. $3 + 1$
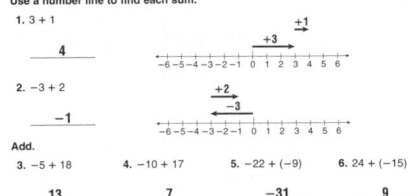
4

2. $-3 + 2$
−1

Add.

3. $-5 + 18$ 13
4. $-10 + 17$ 7
5. $-22 + (-9)$ −31
6. $24 + (-15)$ 9

Evaluate each expression for the given value of the variables.

7. $r + 7$ for $r = 3$ 10
8. $m + 5$ for $m = 9$ 14
9. $x + 9$ for $x = 4$ 13

10. $-6 + t$ for $t = -8$ −14
11. $-7 + y$ for $y = -4$ −11
12. $x + 9$ for $x = -8$ 1

13. $-5 + d$ for $d = -2$ −7
14. $x + (-4)$ for $x = -4$ −8
15. $k + (-3)$ for $k = -5$ −8

16. $-8 + b$ for $b = 13$ 5
17. $-10 + d$ for $d = -2$ −12
18. $t + (-3)$ for $t = 3$ 0

19. Joleen has 2560 trading cards in her collection. She buys 165 new cards for the collection. How many trading cards does she have now?
2725 trading cards

20. The running back for the Bears carries the ball twice in the first quarter. The first run he gained fifteen yards and the second run he lost eight yards. How many yards did the two runs total?
7 yards

Use a number line to find each sum.

27. $-8 + (-5)$ **−13**   **28.** $16 + (-22)$ **−6**   **29.** $-36 + 18$ **−18**

30. $55 + 27$ **82**   **31.** $57 + (-59)$ **−2**   **32.** $-14 + 85$ **71**

33. $52 + (-9)$ **43**   **34.** $-26 + (-26)$ **−52**   **35.** $-41 + 41$ **0**

36. $-7 + 9 + (-8)$ **−6**   **37.** $-11 + (-6) + (-2)$ **38.** $32 + (-4) + (-15)$ **13**
                                    −19

Evaluate each expression for the given value of the variable.

39. $c + 16$ for $c = -8$ **8**   **40.** $k + (-12)$ for $k = 4$ **−8**

41. $b + (-3)$ for $b = -17$ **−20**   **42.** $15 + r$ for $r = -18$ **−3**

43. $-9 + w$ for $w = -6$ **−15**   **44.** $1 + n + (-7)$ for $n = 6$ **0**

45. Evaluate $2 + x + y$ for $x = 7$ and $y = -4$. **5**

46. ECONOMICS Refer to the data below about U.S. international trade for the year 2000. Consider values of exports as positive quantities and values of imports as negative quantities.

	Exports	Imports
Goods	$772,210,000,000	$1,224,417,000,000
Services	$293,492,000,000	$217,024,000,000

Source: 2000 U.S. Census

The number one port for foreign trade by water in the United States is the Port of Houston. In 2000, the port recorded 6801 vessel calls totaling over 175 million tons of cargo.

a. What was the total of U.S. exports in 2000? **$1,065,702,000,000**

b. What was the total of U.S. imports in 2000? **−$1,441,441,000,000**

c. The sum of exports and imports is called the *balance of trade*. Write an addition equation to show the 2000 U.S. balance of trade. **$1,065,702,000,000 + (−$1,441,441,000,000) = −$375,739,000,000**

 47. WHAT'S THE ERROR? A student evaluating $-3 + f$ for $f = -4$ gave an answer of 1. What could be wrong?

 48. WRITE ABOUT IT Explain the different ways it is possible to add two integers and get a negative answer.

49. CHALLENGE What is the sum of $1 + (-1) + 1 + (-1) + \ldots$ when there are 12 terms? 17 terms? 20 terms? 23 terms? Explain any patterns that you find. **0; 1; 0; 1; When there are an odd number of terms the sum is 1; when there are an even number of terms the sum is 0.**

Spiral Review

Solve. (Lessons 1-3 and 1-4)

50. $p - 8 = 12$ **p = 20**   **51.** $f + 9 = 15$ **f = 6**   **52.** $\frac{m}{4} = 16$ **m = 64**   **53.** $7q = 42$ **q = 6**

54. TEST PREP Which number below is **not** a solution of $n - 7 < 1$? (Lesson 1-5) **D**

A 2   B 4   C 6   D 8

Journal

Have students write a story involving one of these topics: a football team gaining and losing yards, a submarine diving and rising in the water, or the rising and falling value of a stock.

Test Prep Doctor

For Exercise 54, encourage students to substitute each answer choice into the inequality. Have students focus on the word **not** in the question. Remind students that they are looking for the one choice that does *not* make the sentence true. The correct choice is **D.**

CHALLENGE 2-1

Challenge
2-1 *Presto, Chango!*

A magic square has the same sum for every row, column, and diagonal.

4	−3	2
−1	1	3
0	5	−2

Magic Square

Magic Square A has a magic sum of __3__.

Add 2 to each integer in Magic Square A to create Magic Square B.

6	−1	4
1	3	5
2	7	0

Magic Square B
magic sum is __9__

Add 5 to each integer in Magic Square A to create Magic Square C.

9	2	7
4	6	8
5	10	3

Magic Square C
magic sum is __18__

Add 23 to each integer in Magic Square A to create Magic Square D.

27	20	25
22	24	26
23	28	21

Magic Square D
magic sum is __72__

Describe the relationship between the magic sum of Magic Square A and the magic sum of a new magic square created by adding any integer n to each integer in Magic Square A.

When *n* is added to each integer of Magic Square A,

then 3*n* is added to its magic sum.

Use the relationship you just described to predict the magic sum of Magic Square E, the magic square you would get if you added −6 to each integer in Magic Square A.

My prediction for the magic sum of magic square E is: __3 + (−18) = −15__

Verify your prediction by creating Magic Square E, and calculating its magic sum.

−2	−9	−4
−7	−5	−3
−6	−1	−8

Magic Square E

Magic Square E has a magic sum of __−15__.

PROBLEM SOLVING 2-1

Problem Solving
2-1 *Adding Integers*

Use the following information for Exercises 1–3. In golf, par 73 means that a golfer should take 73 strokes to finish 18 holes. A score of 68 is 5 under par, or −5. A score of 77 is 4 over par, or +4.

1. Use integers to write Tiger Woods's score for each round as over or under par.
   −5, +1, +1, −8

2. Add the integers to find Tiger Woods's overall score.
   −11

3. Was Tiger Woods's overall score over or under par?
   11 under par

Tiger Woods's Scores
Mercedes Championship
January 6, 2002
Par 73 course

Round	Score
1	68
2	74
3	74
4	65

Choose the letter for the best answer.

4. At 9:00 A.M., the temperature was −15°. An hour later, the temperature had risen 7°. What is the temperature now?
   A −22°   C −8°
   B 8°   D 22°

5. Sandra is reviewing her savings account statement. She withdrew amounts of $35, $20, and $15. She deposited $65. If her starting balance was $657, find the new balance.
   F $652   H $662
   G $522   J $507

6. During a possession in a football game, the Vikings gained 22 yards, lost 15 yards, gained 3 yards, gained 20 yards and lost 5 yards. At the end of the possession, how many yards had they lost or gained?
   A gained 43 yards
   B lost 43 yards
   C lost 25 yards
   D gained 25 yards

7. A submarine is cruising at 40 m below sea level. The submarine ascends 18 m. What is the submarine's new location?
   F 58 m below sea level
   G 22 m below sea level
   H 18 m below sea level
   J 12 m below sea level

Lesson Quiz

Add.

1. $-7 + (-7)$ **−14**

2. $15 + (-9) + (-2)$ **4**

Evaluate each expression for the given value of the variable.

3. $13 + r$ for $r = -15$ **−2**

4. $2 + b + (-9)$ for $b = -6$ **−13**

5. On Monday, a local dog shelter had six dogs. By Friday, they had found homes for three, but took in two more. Then how many dogs were in the shelter? **5 dogs**

Available on Daily Transparency in CRB

Pacing: Traditional $\frac{1}{2}$ day
Block $\frac{1}{3}$ day

Objective: Students subtract integers.

Warm Up

Add.

1. $-7 + 2$ -5

2. $-12 + (-9)$ -21

3. $32 + (-19)$ 13

4. $-6 + (-28)$ -34

5. $104 + (-87)$ 17

6. $-18 + (-24)$ -42

Problem of the Day

Copy and complete the magic square. The magic sum is 0.

+5	−8	+3
−2	0	+2
−3	+8	−5

Available on Daily Transparency in CRB

Math Fact

A number and its opposite are called *additive inverses* of each other. The sum of additive inverses is always zero.

2-2 Subtracting Integers

Learn to subtract integers.

Some roller coasters have maximum drops that are greater than their heights.

Riders enter underground tunnels at speeds of up to 85 miles per hour. The underground depths of the rides can be represented by negative integers.

Subtracting a smaller number from a larger number is the same as finding how far apart the two numbers are on a number line. Subtracting an integer is the same as adding its opposite.

SUBTRACTING INTEGERS

Words	Numbers	Algebra
Change the subtraction sign to an addition sign and change the sign of the second number.	$2 - 3 = 2 + (-3)$ $4 - (-5) = 4 + 5$	$a - b = a + (-b)$ $a - (-b) = a + b$

EXAMPLE 1 **Subtracting Integers**

Subtract.

A $-5 - 5$

$-5 - 5 = -5 + (-5)$ *Add the opposite of 5.*

$\qquad\qquad = -10$ *Same sign; use the sign of the integers.*

B $2 - (-4)$

$2 - (-4) = 2 + 4$ *Add the opposite of −4.*

$\qquad\qquad = 6$ *Same signs; use the sign of the integers.*

C $-11 - (-8)$

$-11 - (-8) = -11 + 8$ *Add the opposite of −8.*

$\qquad\qquad = -3$ *$11 > 8$; use the sign of 11.*

---

1 Introduce

Alternate Opener

EXPLORATION

2-2 Subtracting Integers

You can use a number line to model subtracting integers.

To subtract 10 from 50, begin at the number being subtracted, 10, and count the number of units to the number 50.

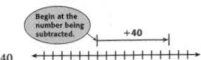

$50 - 10 = 40$

The direction is **right**, so the difference is **positive**.

To subtract −10 from −70, begin at the number being subtracted, −10, and count the number of units to the number −70.

$-70 - (-10) = -60$

The direction is **left**, so the difference is **negative**.

Use the number line to complete each subtraction statement.

1. $5 - (-4) =$ ___   **2.** $-8 - (-1) =$ ___

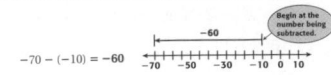

Think and Discuss

3. Discuss a different strategy for subtracting integers.

Motivate

To introduce students to subtraction of integers, review with them how to determine the opposite of a number. You may want to use a number line to demonstrate this (Teaching Transparency T7 in Chapter 2 Resource Book).

Review with students how to add integers. Have students recite the rules and apply them to several addition problems.

Exploration worksheet and answers on Chapter 2 Resource Book pp. 17 and 100

2 Teach

Lesson Presentation

Guided Instruction

In this lesson, students learn to subtract integers. Explain to students that a subtraction sign can be read as "plus the opposite of." For example, $5 - 2$ can be read "5 minus 2" or "5 plus the opposite of 2." Stress that the answer is the same for either interpretation. You may want to use Teaching Transparency T00 from the Chapter 2 Resource Book.

Teaching Tip Point out that subtraction is not commutative. In fact, if the order in a subtraction expression is reversed, the value of the new expression is the opposite of the original value. For example, $5 - 2 = 3$, and $2 - 5 = -3$.

EXAMPLE 2 Evaluating Expressions with Integers

Evaluate each expression for the given value of the variable.

A $4 - t$ for $t = -3$.

$4 - t$

$4 - (-3)$ *Substitute −3 for t.*

$= 4 + 3$ *Add the opposite of −3.*

$= 7$ *Same sign; use the sign of the integers.*

B $-5 - s$ for $s = -7$.

$-5 - s$

$-5 - (-7)$ *Substitute −7 for s.*

$= -5 + 7$ *Add the opposite of −7.*

$= 2$ *7 > 5; use the sign of 7.*

C $-1 - x$ for $x = 8$.

$-1 - x$

$-1 - 8$ *Substitute 8 for x.*

$= -1 + (-8)$ *Add the opposite of 8.*

$= -9$ *Same sign; use the sign of the integers.*

EXAMPLE 3 Architecture Application

Desperado Roller Coaster

209 ft

225 ft

Ground level, 0 ft

? ft

The roller coaster Desperado has a maximum height of 209 ft and maximum drop of 225 ft. How far underground does the roller coaster go?

$209 - 225$ *Subtract the drop from the height.*

$209 + (-225)$ *Add the opposite of 225.*

$= -16$ *225 > 209; use the sign of 225.*

Desperado goes 16 ft underground.

Think and Discuss

1. **Explain** why $10 - (-10)$ does not equal $-10 - 10$.

2. **Describe** the answer that you get when you subtract a larger number from a smaller number.

COMMON ERROR ALERT

Some students might change the subtraction symbol to an addition symbol without changing the sign of the second number. Caution students that they need to change both the operation symbol and the sign of the second number.

Additional Examples

Example 1

Subtract.

A. $-7 - 4$
     -11

B. $8 - (-5)$
     13

C. $-6 - (-3)$
     -3

Example 2

Evaluate each expression for the given value of the variable.

A. $8 - j$ for $j = -6$
     14

B. $-9 - y$ for $y = -4$
     -5

C. $n - 6$ for $n = -2$
     -8

Example 3

The top of the Sears Tower, in Chicago, is 1454 feet above street level, while the lowest level is 43 feet below street level. How far is it from the lowest level to the top?

It is 1497 feet from bottom to top.

---

3 Close

Reaching All Learners
Through Critical Thinking

Have students compare the following expressions and determine which pairs have the same value.

$3 + 3$ $3 - 3$

$3 + (-3)$ $3 - (-3)$

Both $3 + 3$ and $3 - (-3)$ equal 6.
Both $3 - 3$ and $3 + (-3)$ equal 0.

Summarize

Have the students describe how to express a subtraction as an addition. Then have them state the rules for addition. Emphasize the importance of the rules for adding and subtracting integers.

Possible answers: To write a subtraction as an addition, change the subtraction sign to addition and change the sign of the second number. To add two integers with the same sign, add the absolute values of the integers and use the same sign. To add two integers with different signs, subtract the absolute values of the integers and use the sign of the number with the greater absolute value.

Answers to Think and Discuss

Possible answers:

1. The differences are not the same because $10 - (-10)$ is positive and $-10 - 10$ is negative.

2. The answer is negative. If both numbers are positive, when you subtract the larger number, you move left past zero, and the answer is negative. If the smaller number is negative, then you move even farther left of zero. If both numbers are negative, then the smaller number is farther from zero. When you subtract, you move to the right, but not far enough to reach zero.

FOR EXTRA PRACTICE
see page 734

internet connect
Homework Help Online
go.hrw.com   Keyword: MP4 2-2

go.hrw.com

> Students may want to refer back to the lesson examples.

Assignment Guide

If you finished Example **1** assign:
  Core 1–4, 9–16, 25–31 odd, 44–48
Enriched 9–16, 24–31, 44–48

If you finished Example **2** assign:
  Core 1–7, 9–22, 25–37 odd, 44–48
Enriched 9–22, 24–37, 44–48

If you finished Example **3** assign:
  Core 1–23, 25–41 odd, 44–48
Enriched 9–48

Answers

24. $-6 - (-4) = -2$

25. $5 - 8 = -3$

GUIDED PRACTICE

See Example **1**   Subtract.

1. $-7 - 8$ **−15**   2. $-7 - (-4)$ **−3**   3. $9 - (-5)$ **14**   4. $-10 - (-3)$ **−7**

See Example **2**   Evaluate each expression for the given value of the variable.

5. $7 - h$ for $h = -6$ **13**   6. $-8 - m$ for $m = -2$ **−6**   7. $-3 - k$ for $k = 12$ **−15**

See Example **3**   8. The temperature rose from $-4°F$ to $45°F$ in Spearfish, South Dakota, on January 22, 1943, in only 2 minutes! By how many degrees did the temperature change? *Source: The Weather Book*, Random House, Inc. **49°F**

INDEPENDENT PRACTICE

See Example **1**   Subtract.

9. $-2 - 9$ **−11**   10. $12 - (-7)$ **19**   11. $11 - (-6)$ **17**   12. $-9 - (-3)$ **−6**

13. $-8 - (-11)$ **3**   14. $-14 - 8$ **−22**   15. $-5 - (-9)$ **4**   16. $30 - (-12)$ **42**

See Example **2**   Evaluate each expression for the given value of the variable.

17. $12 - b$ for $b = -4$ **16**   18. $-9 - q$ for $q = -12$ **3**   19. $-7 - f$ for $f = 10$ **−17**

20. $7 - d$ for $d = 16$ **−9**   21. $-7 - w$ for $w = 7$ **−14**   22. $-3 - p$ for $p = -3$ **0**

See Example **3**   23. A submarine cruising at 25 m below sea level, or -25 m, descends 15 m. What is its new depth?   **40 m below sea level, or −40 m**

PRACTICE AND PROBLEM SOLVING

Write a subtraction equation for each number line diagram.

24. 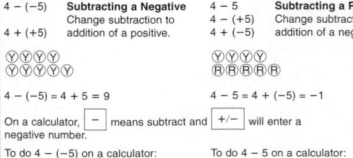   25.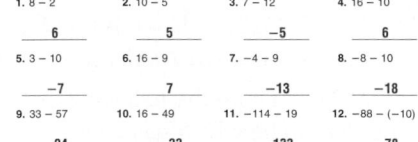

Perform the given operations.

26. $-7 - (-10)$ **3**   27. $24 - (-27)$ **51**   28. $-31 - 11$ **−42**

29. $-31 - 31$ **−62**   30. $-12 - 9 + (-4)$ **−25**   31. $-13 - (-5) + (-8)$ **−16**

Evaluate each expression for the given value of the variable.

32. $x - 15$ for $x = -3$ **−18**   33. $6 - t$ for $t = -7$ **13**   34. $-14 - y$ for $y = 9$ **−23**

35. $s - (-21)$ for $s = -19$ **2**   36. $1 - r - (-2)$ for $r = 5$ **−2**   37. $-3 - w + 3$ for $w = 42$ **−42**

Math Background

The ancient Egyptians used the following symbols for the given numbers:

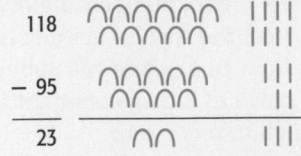
1,000   100   10   1

The number 118 would look like this:

If a subtraction required regrouping, the Egyptians replaced one symbol with ten symbols. The subtraction problem 118 − 95 = 23, written with Egyptian symbols, might look like this:

RETEACH 2-2

LESSON **2-2** Reteach
Subtracting Integers

To subtract one integer from another, rewrite the subtraction as the addition of an opposite. Then use the rules for adding integers.

$4 - (-5)$ **Subtracting a Negative** $4 - 5$ **Subtracting a Positive**
        Change subtraction to   Change subtraction to
$4 + (+5)$ addition of a positive. $4 + (-5)$ addition of a negative.

$4 - (-5) = 4 + 5 = 9$ $4 - 5 = 4 + (-5) = -1$

On a calculator, $-$ means subtract and $+/-$ will enter a negative number.

To do $4 - (-5)$ on a calculator: To do $4 - 5$ on a calculator:
Input: 4 $-$ 5 $+/-$ $=$ Input: 4 $-$ 5 $=$
Display: 9   Display: -1

Complete to find the difference. Remember to change two signs.

1. $7 - (-6)$ is the same as 7 $+$ $+$ 6 = **13**
2. $-4 - 3$ is the same as -4 $+$ $-$ 3 = **−7**
3. $-2 - (-9)$ is the same as -2 $+$ $+$ 9 = **7**
4. $14 - 16$ is the same as 14 $+$ $-$ 16 = **−2**
5. $7 - (-10)$ is the same as 7 $+$ $+$ 10 = **17**
6. $-8 - (-19)$ is the same as -8 $+$ $+$ 19 = **11**
7. $-5 - 12$ is the same as -5 $+$ $-$ 12 = **−17**

Find each difference. Use a calculator to check.

8. $7 - 12 =$ **−5**   9. $-3 - 8 =$ **−11**   10. $17 - (-4) =$ **21**

11. $-14 - (-3) =$ **−11**   12. $5 - 8 =$ **−3**   13. $-6 - 4 =$ **−10**

PRACTICE 2-2

LESSON **2-2** Practice B
Subtracting Integers

Subtract.

1. $8 - 2$ **6**   2. $10 - 5$ **5**   3. $7 - 12$ **−5**   4. $16 - 10$ **6**

5. $3 - 10$ **−7**   6. $16 - 9$ **7**   7. $-4 - 9$ **−13**   8. $-8 - 10$ **−18**

9. $33 - 57$ **−24**   10. $16 - 49$ **−33**   11. $-114 - 19$ **−133**   12. $-88 - (-10)$ **−78**

Evaluate each expression for the given value of the variables.

13. $x - 8$ for $x = 10$ **2**   14. $w - 10$ for $w = 15$ **5**   15. $15 - w$ for $w = 8$ **7**

16. $12 - t$ for $t = -8$ **20**   17. $15 - x$ for $x = -12$ **27**   18. $w - 20$ for $w = -15$ **−35**

19. $-15 - x$ for $x = -10$ **−5**   20. $-9 - x$ for $x = -20$ **11**   21. $-11 - d$ for $d = -15$ **4**

22. $y - (-10)$ for $y = -10$ **0**   23. $x - (-15)$ for $x = -5$ **10**   24. $a - (-12)$ for $a = 10$ **22**

25. The altitude of Mt. Blackburn in Alaska is 16,390 feet. The altitude of Mt. Elbert in Colorado is 14,433 feet. What is the difference in the altitudes of the two mountains?

   1957 feet

26. In January, Jesse weighed 230 pounds. By November, he weighed 185 pounds. How much did Jesse's weight change?

   −45 pounds

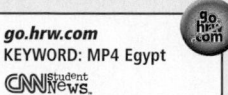

Use the timeline to answer the questions. Use negative numbers for years B.C. Assume that there was a year 0 (there wasn't) and that there have been no major changes to the calendar (there have been).

go.hrw.com
KEYWORD: MP4 Egypt
CNN Student News.

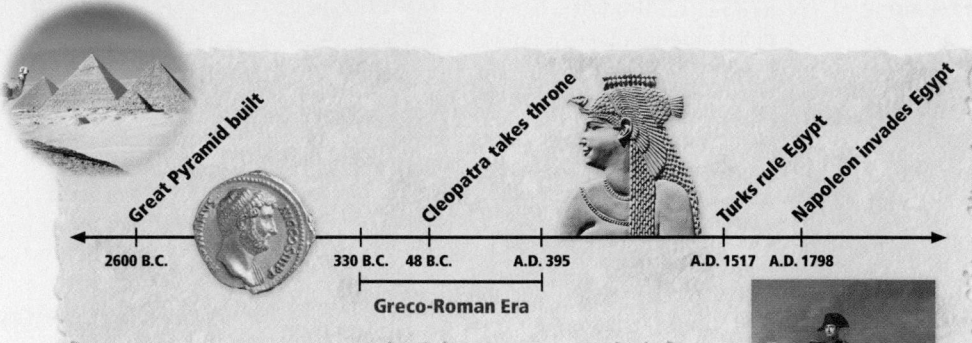

Great Pyramid built   Cleopatra takes throne   Turks rule Egypt   Napoleon invades Egypt

2600 B.C.   330 B.C.   48 B.C.   A.D. 395   A.D. 1517   A.D. 1798

Greco-Roman Era

38. How long was the Greco-Roman era, when Greece and Rome ruled Egypt? **725 years**

39. Which was a longer period of time: from the Great Pyramid to Cleopatra, or from Cleopatra to the present? By how many years?
Great Pyramid to Cleopatra; about 500 years

40. Queen Neferteri ruled Egypt about 2900 years before the Turks ruled. In what year did she rule? **1383 B.C.**

41. There are 1846 years between which two events on this timeline?
Cleopatra takes throne and Napoleon invades Egypt.

42. **WRITE ABOUT IT** What is it about years B.C. that makes negative numbers a good choice for representing them?

43. **CHALLENGE** How would your calculations differ if you took into account the fact that there was no year 0?

Spiral Review

Combine like terms. (Lesson 1-6)

44. $9m + 8 - 4m + 7 - 5m$ **15** **45.** $6t + 3k - 15$ **no like terms** **46.** $5a + 3 - b + 1$ **$5a - b + 4$**

47. TEST PREP Which of the following is **not** a solution of $y = 5x + 1$? (Lesson 1-7) **C**

  A (0, 1) **B** (1, 6) **C** (21, 4) **D** (22, 111)

48. TEST PREP Which of the following is the value of $-7 + 3h$ when $h = 5$? (Lesson 2-1) **H**

  F −8 **G** −22 **H** 8 **J** 22

CHALLENGE 2-2

LESSON 2-2 Challenge
Teeter Totter

Joel wants his math average for 6 tests to be at least 85. So far, his grades on the first 5 tests were 82, 91, 73, 83, and 88.

Complete the following table to help Joel figure out what grade he has to get on the 6th test in order to achieve his goal.

Grade	73	82	83	88	91
Grade − Average	73 − 85 =	82 − 85 =	83 − 85 =	88 − 85 =	91 − 85 =
	−12	−3	−2	3	6

After calculating the differences between the existing grades and the desired average of 85, notice that some of the differences are negative and some are positive. Add the positive and negative differences separately.

1. Sum of negative differences
$-12 + (-3) + (-2) = -17$

2. Sum of positive differences
$3 + 6 = 9$

Joel thinks that since the average is the "middle" grade, the differences below the average should balance the differences above the average. If Joel's reasoning is correct:

3. Should the 6th grade be higher or lower than the average? **higher**

4. By how many points? **8 points**

5. What must the 6th grade be in order to achieve an average of at least 85? **93**

6. Verify your answer by using it as the 6th grade. Find the average by your usual method: add the 6 grades and divide by 6.

$\frac{73 + 82 + 83 + 88 + 91 + 93}{6} = 85$

Use Joel's method to find the necessary 6th grade for each set of grades if the given average is to be achieved.

7. 80, 93, 75, 82, 85; average to be 85
6th grade should be **95**

8. 85, 80, 90, 100, 80; average to be 88
6th grade should be **93**

PROBLEM SOLVING 2-2

LESSON 2-2 Problem Solving
Subtracting Integers

Write the correct answer.

1. In Fairbanks, Alaska, the average January temperature is −13°F, while the average April temperature is 30°F. What is the difference between the average temperatures?
43°F

2. The highest point in North America is Mt. McKinley, Alaska, at 20,320 ft above sea level. The lowest point is Death Valley, California, at 282 ft below sea level. What is the difference in elevations?
20,602 ft

3. The temperature fell from 44°F to −56°F in 24 hours in Browning, Montana, on January 23−24, 1916. By how many degrees did the temperature change?
100°F

4. The boiling point of chlorine is −102°C, while the melting point is −34°C. What is the difference between the melting and boiling points of chlorine?
68°C

Use the table below to answer Exercises 5−7. The table shows the first and fifth place finishers in a golf tournament. In golf, the winner has the lowest total for all five rounds. Choose the letter for the best answer.

5. By how many points did Mickelson beat Kelly in Round 2?
  Ⓐ 2 C 5
  B 3 D 8

6. By how many points did Kelly beat Mickelson in Round 3?
  F 2 Ⓗ 5
  G 3 J 9

7. Who won the Bob Hope Chrysler Classic and how many points difference was there between first and fifth place?
  A Kelly; 4 C Kelly; 3
  B Mickelson; 4 Ⓓ Mickelson; 3

Bob Hope Chrysler Classic
January 20, 2002

Round	J. Kelly	P. Mickelson
1	−8	−8
2	−3	−5
3	−7	−2
4	−4	−7
5	−5	−8

Interdisciplinary

Social Studies

Exercises 38–43 involve reading a timeline. Reading and interpreting timelines is a prerequisite skill for middle school social studies courses, such as Holt, Rinehart & Winston's *People, Places, and Change.*

Answers

42. Possible answer: The number of the year expressed in B.C. decreases as time goes forward. This is what happens with the negative part of the number line. As you move to the right and get closer to zero, the absolute value of the numbers decreases.

43. Possible answer: The calculations that involve an A.D. and B.C. year would be one less than originally calculated. But any calculations strictly in A.D. or B.C. would remain the same.

Journal

Have students write about a situation involving a vertical change, such as traveling on an elevator. Have them also write about another situation that might involve subtraction of integers.

Test Prep Doctor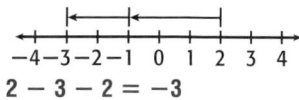

For Exercise 47, students should remember that the first number in the ordered pair is the *x*-value and the second number is the *y*-value. If students find that only answer choice **C** makes the equation true, they may have reversed this order.

Lesson Quiz

1. Write a subtraction equation for the number line diagram.

−4 −3 −2 −1 0 1 2 3 4

$2 - 3 - 2 = -3$

Perform the given operations.

2. $-6 - (-4)$ −2

3. $-9 - 4 + (-3)$ −16

Evaluate each expression for the given value of the variable.

4. $9 - s$ for $s = -5$ 14

5. $-4 - w + 5$ for $w = 21$ −20

Available on Daily Transparency in CRB

Objective: Students multiply and divide integers.

Warm Up

Multiply or divide.

1. 5(8) **40**
2. 6(12) **72**
3. $\frac{36}{9}$ **4**
4. $\frac{49}{7}$ **7**
5. 18(7) **126**
6. $\frac{192}{16}$ **12**

Problem of the Day

Complete the pyramid by filling in the missing numbers. Each number is the sum of the numbers in the two boxes below it.

Available on Daily Transparency in CRB

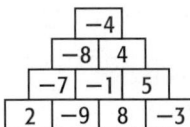

Who invented algebra? a clever X-pert

2-3 Multiplying and Dividing Integers

Learn to multiply and divide integers.

On *Jeopardy! Teen Tournament,* a correct answer is worth the dollar amount of the question, and an incorrect answer is worth the opposite of the dollar amount of the question. If a contestant answered three $200 questions incorrectly, what would the score be?

A positive number multiplied by an integer can be written as repeated addition.

$$3(-200) = -200 + (-200) + (-200) = -600$$

From what you know about adding integers, you can see that a positive integer times a negative integer is negative.

You know that multiplying two positive integers together gives you a positive answer. Look for a pattern in the integer multiplication at right to understand the rules for multiplying two negative integers.

$$3(-200) = -600 \quad \Big\} +200$$
$$2(-200) = -400 \quad \Big\} +200$$
$$1(-200) = -200 \quad \Big\} +200$$
$$0(-200) = 0 \quad \Big\} +200$$
$$-1(-200) = 200$$
$$-2(-200) = 400$$
$$-3(-200) = 600$$

The product of two negative integers is a positive integer.

MULTIPLYING AND DIVIDING TWO INTEGERS

If the signs are the same, the sign of the answer is **positive**.

If the signs are different, the sign of the answer is **negative**.

EXAMPLE 1 Multiplying and Dividing Integers

Multiply or divide.

A 6(−7)   *Signs are different.*
    −42   *Answer is **negative**.*

B $\frac{-45}{9}$ *Signs are different.*
    −5   *Answer is **negative**.*

C −12(−4)   *Signs are the same.*
    48   *Answer is **positive**.*

D $\frac{18}{-6}$ *Signs are different.*
    −3   *Answer is **negative**.*

---

1 Introduce

Alternate Opener

2-3 Multiplying and Dividing Integers

Imagine a person walking on a number line. If the person faced a **positive** direction, it would be **to the right**. If the person faced a **negative** direction, it would be **to the left**.

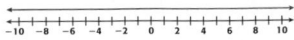

Suppose each step is 2 units long.

1. A person who is standing at 0 and facing a **positive** direction takes 3 steps **backward.** Complete the multiplication statement to find the person's location.

    3 · −2 = ___

2. A person who is standing at 0 and facing a **negative** direction takes 4 steps **forward.** Complete the multiplication statement to find the person's location.

    −4 · 2 = ___

3. A person who is standing at 0 and facing a **negative** direction takes 5 steps **backward.** Complete the multiplication statement to find the person's location.

    −5 · −2 = ___

Think and Discuss

4. **Describe** a situation for the multiplication statement 6 · −3 = −18 by using a number line.

Motivate

To introduce students to multiplication and division of integers, point out that multiplication of whole numbers is repeated addition. Give examples of repeated addition, such as 6 + 6 + 6 = 3 · 6. Review the rules for adding integers.

Exploration worksheet and answers on Chapter 2 Resource Book pp. 26 and 102

2 Teach

Lesson Presentation

Guided Instruction

In this lesson, students learn to multiply and divide integers. Review with students the pattern given in the lesson. They should be able to determine the sign rules for multiplication and division (Teaching Transparency T12, CRB). Point out that the sign rules are the same for both operations.

Teaching Tip
Remind students of the order of operations. Ask students to explain each step in Example 2, referring to both the rules for multiplication and division of integers and the order of operations.

EXAMPLE 2 Using the Order of Operations with Integers

Simplify.

A $-2(3 - 9)$

$-2(3 - 9)$	*Subtract inside the parentheses.*
$= -2(-6)$	*Think: The signs are the same.*
$= 12$	*The answer is positive.*

B $4(-7 - 2)$

$4(-7 - 2)$	*Subtract inside the parentheses.*
$= 4(-9)$	*Think: The signs are different.*
$= -36$	*The answer is negative.*

C $-3(16 - 8)$

$-3(16 - 8)$	*Subtract inside the parentheses.*
$= -3(8)$	*Think: The signs are different.*
$= -24$	*The answer is negative.*

Remember!

Order of Operations
1. Parentheses
2. Exponents
3. Multiply and divide from left to right.
4. Add and subtract from left to right.

The order of operations can be used to find ordered pair solutions of integer equations. Substitute an integer value for one variable to find the value of the other variable in each ordered pair.

EXAMPLE 3 Plotting Integer Solutions of Equations

Complete a table of solutions for $y = -2x - 1$ for $x = -2$, -1, 0, 1, and 2. Plot the points on a coordinate plane.

x	−2x − 1	y	(x, y)
−2	−2(−2) − 1	3	(−2, 3)
−1	−2(−1) − 1	1	(−1, 1)
0	−2(0) − 1	−1	(0, −1)
1	−2(1) − 1	−3	(1, −3)
2	−2(2) − 1	−5	(2, −5)

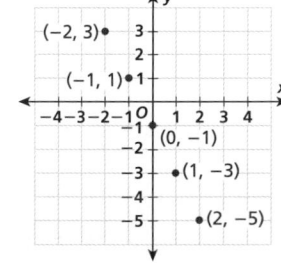

Think and Discuss

1. **List** all possible multiplication and division statements for the integers with absolute values of 5, 6, and 30. For example, $5 \cdot 6 = 30$.

2. **Compare** the sign of the product of two negative integers with the sign of the sum of two negative integers.

Additional Examples

Example 1

Multiply or divide.

A. $-6(4)$ **B.** $-8(-5)$
   -24 40

C. $\dfrac{-18}{2}$ **D.** $\dfrac{-25}{-5}$
   -9 5

Example 2

Simplify.

A. $3(-6 - 12)$ **B.** $-5(-5 + 2)$
   -54 15

C. $-2(14 - 5)$
   -18

Example 3

Complete a table of solutions for $y = 3x - 1$ for $x = -2, -1, 0, 1,$ and 2. Plot the points on a coordinate plane.

x	3x − 1	y	(x,y)
−2	3(−2) − 1	−7	(−2, −7)
−1	3(−1) − 1	−4	(−1, −4)
0	3(0) − 1	−1	(0, −1)
1	3(1) − 1	2	(1, 2)
2	3(2) − 1	5	(2, 5)

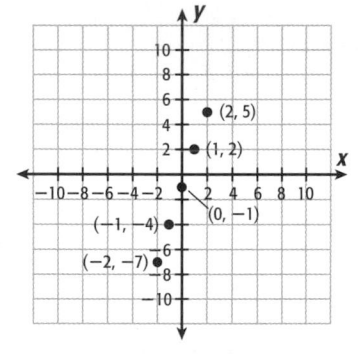

---

❸ Close

Reaching All Learners
Through Critical Thinking

Have students answer True or False to the following statements, and then provide an example and an explanation for each.

If a product is negative, then there must be an odd number of negative factors in the expression. True; $-3(-2)(-1) = -6$. Because every two negative factors make a positive, an odd number results in one extra negative.

If a quotient is negative, then there must be an even number of positive factors in the numerator and denominator combined. False; $\dfrac{-3(6)(2)}{(4)} = -9$. The sign of the quotient is determined only by the number of negative factors.

Summarize

Review the rules for the four operations. Discuss why the rules for multiplication and division are the same, while the rules for addition and subtraction are different.

Possible answers: The rules for multiplication and division are the same because one can rewrite division as multiplication without changing any of the signs. The rules for addition and subtraction are different because to change subtraction into addition, the sign of the second number must change.

Answers to Think and Discuss

Possible answers:

1. $5 \cdot 6 = 30$, $(-5) \cdot 6 = -30$,
   $5 \cdot (-6) = -30$, $(-5) \cdot (-6) = 30$,
   $6 \cdot 5 = 30$, $(-6) \cdot 5 = -30$,
   $6 \cdot (-5) = 30$, $(-6) \cdot (-5) = 30$,
   $30 \div 5 = 6$, $30 \div 6 = 5$,
   $(-30) \div 5 = -6$,
   $(-30) \div (-6) = 5$,
   $30 \div (-5) = -6$,
   $30 \div (-6) = -5$,
   $(-30) \div (-5) = 6$,
   $(-30) \div 6 = -5$

2. The product of two negative numbers is positive, while the sum of two negative numbers is negative.

2-3 **Exercises**

FOR EXTRA PRACTICE
see page 734

internet connect
Homework Help Online
go.hrw.com Keyword: MP4 2-3

go.hrw.com

GUIDED PRACTICE

See Example **1** Multiply or divide.

1. $9(-3)$ **−27** **2.** $\frac{-56}{7}$ **−8** **3.** $-6(-5)$ **30** **4.** $\frac{32}{-8}$ **−4**

See Example **2** Simplify.

5. $-7(5-12)$ **49** **6.** $7(-3-8)$ **−77** **7.** $-6(-5+9)$ **8.** $12(-8+2)$ **−72**
−24

See Example **3** Complete a table of solutions for each equation for $x = -2, -1, 0, 1,$ and 2. Plot the points on a coordinate plane.

9. $y = 3x + 1$ **10.** $y = -3x - 1$ **11.** $y = 2x + 2$

INDEPENDENT PRACTICE

See Example **1** Multiply or divide.

12. $-4(-9)$ **36** **13.** $\frac{77}{-7}$ **−11** **14.** $12(-7)$ **−84** **15.** $\frac{-42}{6}$ **−7**

See Example **2** Simplify.

16. $10(7-15)$ **17.** $-13(-2-8)$ **18.** $15(9-12)$ **19.** $10+4(5-8)$
−80 **130** **−45** **−2**

See Example **3** Complete a table of solutions for each equation for $x = -2, -1, 0, 1,$ and 2. Plot the points on a coordinate plane.

20. $y = -2x$ **21.** $y = -2x + 1$ **22.** $y = -x - 3$

PRACTICE AND PROBLEM SOLVING

Perform the given operations.

23. $-9(5)$ **−45** **24.** $\frac{-121}{11}$ **−11** **25.** $-6(-6)$ **36**

26. $\frac{100}{-25}$ **−4** **27.** $3(-4)(-2)$ **24** **28.** $\frac{-96}{-12}$ **8**

29. $12(3)(-2)$ **−72** **30.** $\frac{-15(3)}{-5}$ **9** **31.** $-10(-1)(-8)$ **−80**

32. $\frac{3(-8)}{2}$ **−12** **33.** $-9(2-9)$ **63** **34.** $\frac{-12(-6)}{-2}$ **−36**

Evaluate the expressions for the given value of the variable.

35. $-3t - 4$ for $t = 5$ **36.** $-x + 2$ for $x = -9$ **37.** $-7(s + 8)$ for $s = -10$
−19 **11** **14**
38. $\frac{-r}{7}$ for $r = 49$ **−7** **39.** $\frac{-27}{t}$ for $t = -9$ **3** **40.** $\frac{y - 10}{-3}$ for $y = 37$ **−9**

Complete a table of solutions for each equation for $x = -2, -1, 0, 1,$ and 2. Plot the points on a coordinate plane.

41. $y = 2x + 4$ **42.** $y = 5 - 4x$ **43.** $y = 1 + 3x$

Students may want to refer back to the lesson examples.

Assignment Guide

If you finished Example **1** assign:
Core 1–4, 12–15, 23–26, 28, 45, 50–55
Enriched 12–15, 23–28, 44, 45, 47, 48, 50–55

If you finished Example **2** assign:
Core 1–8, 12–19, 23–39 odd, 44, 45, 50–55
Enriched 12–19, 23–40, 44, 45, 50–55

If you finished Example **3** assign:
Core 1–34, 35–45 odd, 50–55
Enriched 12–55

Answers
9.

x	−2	−1	0	1	2
y	−5	−2	1	4	7

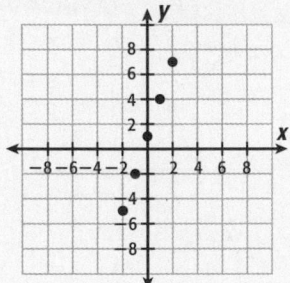

10–11, 20–22, 41–43. See pp. A2–A3.

Math Background

One situation that illustrates multiplication of integers is as follows:

Imagine a road that is laid out like a number line. Mile markers to the north are positive, and to the south negative. A car moving south at 30 mi/h can be represented by −30 mi/h. If the car starts at mile marker 0, its position after 2 hours would be at mile marker −60 ($2 \times -30 = -60$). The car's position 2 hours before it reached mile marker 0 would have been at mile marker 60 ($-2 \times -30 = 60$). Similar examples can be created using a northbound car traveling at +30 mi/h.

RETEACH 2-3

LESSON 2-3 Reteach
Multiplying and Dividing Integers

Since multiplication is a shortcut for addition, a pattern becomes apparent when you multiply or divide integers. Look at the two multiplication problems written as repeated addition.

$8 \times 3 = 8 + 8 + 8 = 24$ $-8 \times 3 = -8 + -8 + -8 = -24$
$8 \times 3 = 24$ $-8 \times 3 = -24$
$8 \times 2 = 16$ *Notice the product* $-8 \times 2 = -16$ *Notice the product*
$8 \times 1 = 8$ *is decreasing by 8.* $-8 \times 1 = -8$ *is increasing by 8.*
$8 \times 0 = 0$ $-8 \times 0 = 0$
$8 \times -1 = -8$ *Keep decreasing* $-8 \times -1 = 8$ *Keep increasing by*
$8 \times -2 = -16$ *by 8 even though* $-8 \times -2 = 16$ *8 even though both*
$8 \times -3 = -24$ *one factor is negative.* $-8 \times -3 = 24$ *factors are negative.*

Multiplication and division have the same rules for multiplying integers.

Positive × Positive = Positive   Positive × Negative = Negative
Negative × Positive = Negative   Negative × Negative = Positive

Tell if the signs are the same or different and if the result will be positive or negative.

1. $9 \times (-5)$
signs are **different**
product is **negative**

2. $-14 \div (-2)$
signs are **the same**
quotient is **positive**

3. $27 \div 3$
signs are **the same**
quotient is **positive**

4. $50 \div (-2)$
signs are **different**
quotient is **negative**

Enter ÷ or − for the product or quotient.

5. $(-14) \div (-2) = $ **+** 7
6. $8 \times 7 = $ **+** 56
7. $(-3) \times 5 = $ **−** 15
8. $16 \times (-3) = $ **−** 48
9. $25 \div (-5) = $ **−** 5
10. $(-100) \div 50 = $ **−** 2

Find the product or quotient.

11. $7 \times (-2) = $ **−14**
12. $-99 \div (-11) = $ **9**
13. $8 \times (-6) = $ **−48**
14. $64 \div 16 = $ **4**
15. $-12 \times (-3) = $ **36**
16. $-48 \div 12 = $ **−4**

PRACTICE 2-3

LESSON 2-3 Practice B
Multiplying and Dividing Integers

Multiply or divide.

1. $6 \cdot 7$ 2. $\frac{-15}{5}$ 3. $-7 \cdot 3$ 4. $\frac{20}{-4}$
   42   **−3**   **−21**   **−5**
5. $\frac{-36}{-4}$ 6. $-8(-9)$ 7. $\frac{-48}{-6}$ 8. $7(-7)$
   9   **72**   **8**   **−49**

Simplify.

9. $-5(3 + 7)$ 10. $10(8 - 2)$ 11. $-4(12 - 3)$ 12. $9(15 - 8)$
   −50   **60**   **−36**   **63**
13. $12(-9 + 4)$ 14. $-11(7 - 13)$ 15. $15(-12 + 8)$ 16. $-10(-8 - 6)$
   −60   **66**   **−60**   **140**

Complete the table for the equation $y = 3x - 1$. Then plot the points on the coordinate plane.

x	3x − 1	y	(x, y)
17. | 2 | 3(2) − 1 | 5 | (2, 5) |
18. | 1 | 3(1) − 1 | 2 | (1, 2) |
19. | 0 | 3(0) − 1 | −1 | (0, −1) |
20. | −1 | 3(−1) − 1 | −4 | (−1, −4) |
21. | −2 | 3(−2) − 1 | −7 | (−2, −7) |

22. Kristin and her three friends buy a pizza with twelve slices and split it equally. How many slices will each person receive?
3 slices

23. The temperature was −1°F, −5°F, 8°F, and −6°F on four consecutive days. What was the average temperature for those days?
−1°F

Science **LINK**

Anoplogaster cornuta, often called a fangtooth or ogrefish, is a predatory fish that reaches a maximum length of 15 cm. It can be found in tropical and temperate waters at −16,000 ft.

44. **EARTH SCIENCE** The ocean floor is extremely uneven. It includes underwater mountains, ridges, and extremely deep areas called *trenches*. To the nearest foot, find the average depth of the trenches shown. **−32,148 ft**

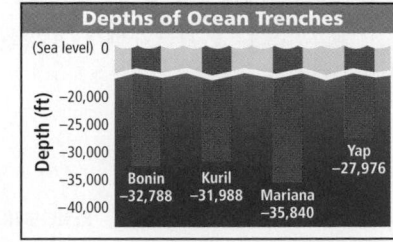

Depths of Ocean Trenches

45. **BUSINESS** A leak in a commercial water tank changes the amount of water in the tank each day by −6 gallons. When the total change is −192 gallons, the pump will stop working. How many days will it take from the time the tank is full until the pump fails? **32 days**

46. **EARTH SCIENCE** Ocean tides are the result of the gravitational force between the sun, the moon and the earth. When ocean tides occur, the earth also moves. This is called an earth tide. The formula for the height of an earth tide is $y = \frac{x}{3}$, where x is the height of the ocean tide. Fill in the table and plot the points on a coordinate plane.

Ocean Tide (x)		$\frac{x}{3}$	Earth Tide (y)
High:	12		
Low:	−9		
High:	6		
Low:	−12		

47. **CHOOSE A STRATEGY** P is the set of positive factors of 20, and Q is the set of negative factors of 12. If x is a member of set P and y is a member of set Q, what is the greatest possible value of $x \cdot y$? **D**

   A   220   B   212   C   210   D   −1

48. **WRITE ABOUT IT** If you know that the product of two integers is negative, what can you say about the two integers? Give examples.

49. **CHALLENGE** Complete a table of solutions of $x + y = 10$ for $x = -2, -1, 0, 1,$ and 2. Plot the points on a coordinate plane.

48. Possible answer: One integer must be negative, and the other must be positive. For instance, if the product is −6, the integers could be 3 and −2, 2 and −3, 6 and −1, or 1 and −6. They could not be 3 and 2 because the product of two positive numbers is positive, and they could not be −3 and −2 because the product of two negative numbers is positive.

Spiral Review

Solve. (Lessons 1-3 and 1-4)

50. $z - 13 = 5$ $z = 18$ 51. $8 + w = 19$ $w = 11$ 52. $\frac{x}{5} = 25$ $x = 125$ 53. $3h = 0$ $h = 0$

54. **TEST PREP** Which ordered pair is a soluton of $2y - 3x = 8$? (Lesson 1-7) **A**

   A   (6, 13)   B   (19, 4)   C   (10, 4)   D   (4, 0)

55. **TEST PREP** Which of the following is equivalent to $|7 - (-3)|$? (Lesson 2-2) **G**

   F $|7| - |-3|$ G $|7| + |-3|$ H   −10   K   4

Answers

46.

$\frac{x}{3}$	$\frac{12}{3}$	$\frac{-9}{3}$	$\frac{6}{3}$	$\frac{-12}{3}$
y	4	−3	2	−4

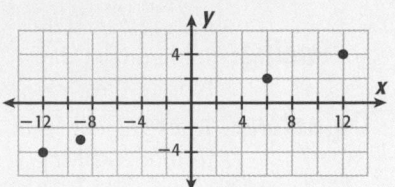

49.

x	−2	−1	0	1	2
y	12	11	10	9	8

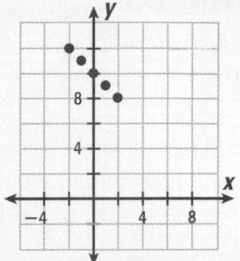

CHALLENGE 2-3

LESSON Challenge
2-3 Pearls of Wisdom

For security, bead bracelets are strung with knots between beads.

Starting to the right of the equals-sign clasp and moving clockwise, write × or ÷ as the knots between numbered beads to make the expression around the bracelet true.

Example   **Possible answers given.**

③ ÷ 1 × 4 ÷ −2 ÷ −2 × 8 × −1 ÷ −4 ÷ 2 = ③

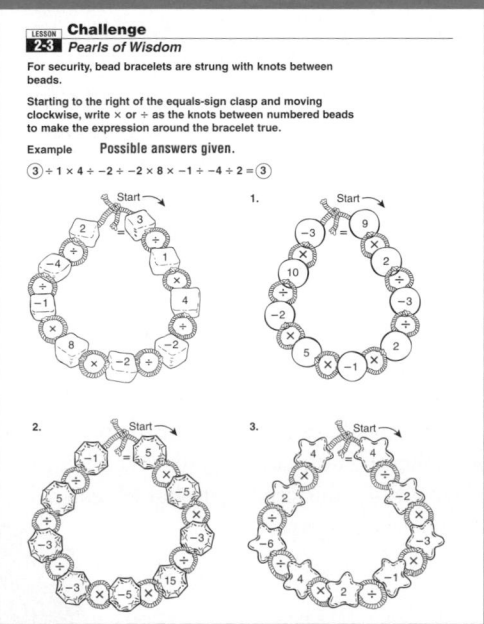

PROBLEM SOLVING 2-3

LESSON Problem Solving
2-3 Multiplying and Dividing Integers

Write the correct answer.

1. A submersible started at the surface of the water and was moving down at −12 meters per minute toward the ocean floor. The submersible traveled at this rate for 32 minutes before coming to rest on the ocean floor. What is the depth of the ocean floor?

   −384 m

2. For the first week in January, the daily high temperatures in Bismarck, North Dakota, were 7°F, −10°F, −10°F, −7°F, 8°F, 12°F, and 14°F. What was the average daily high temperature for the week?

   2°F

3. Sally went golfing and recorded her scores as −2 on the first hole, −2 on the second hole, and 1 on the third hole. What is her average for the first three holes?

   −1

4. The ocean floor is at −96 m. Tom has reached −15 m. If he continues to move down at −3 m per minute, how far will he be from the ocean floor after 7 minutes?

   60 m

Use the table below to answer Exercises 5–7. Choose the letter for the best answer.

5. What is the caloric impact of 2 hours of in-line skating?

   A −477 Cal   C −583 Cal
   B −479 Cal   **D** −954 Cal

6. What is the caloric impact of eating a hamburger and then playing Frisbee for 3 hours?

   F 220 Cal   H 190 Cal
   G −190 Cal   J −220 Cal

7. Tim plays basketball for 1 hour, skates for 5 hours, and plays Frisbee for 4 hours. What is the average amount of calories Tim burns per hour?

   A −375 Cal   C −545 Cal
   B −1250 Cal   D −409 Cal

Calories Consumed or Burned	
Food or Exercise	Calories
Apple	125
Pepperoni pizza (slice)	181
Hamburger	425
Basketball (1hr)	−545
In-line skating (1 hr)	−477
Frisbee (1 hr)	−205

Lesson Quiz

Perform the given operations.

1. $-8(4)$ **−32**

2. $\frac{-12(5)}{-10}$ **6**

Evaluate the expressions for the given value of the variable.

3. $-4t - 9$ for $t = -6$ **15**

4. $\frac{-36}{t}$ for $t = 9$ **−4**

5. Complete a table of solutions for $y = 4x + 1$ for $x = -3, -1, 1,$ and 3.

x	−3	−1	1	3
y	−11	−3	5	13

Available on Daily Transparency in CRB

2A Model Solving Equations

Pacing: Traditional 1 day
Block $\frac{1}{2}$ day

Objective: To use algebra tiles to model solving equations

Materials: Algebra tiles

Lab Resources

Hands-On Lab Activities.. pp. 14–16, 93

Using the Pages

Discuss with students what each algebra tile represents and how to represent each side of an equation using algebra tiles.

Use algebra tiles to model and solve each equation.

1. $x + 1 = 4$

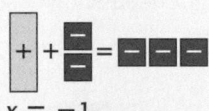

$x = 3$

2. $x + (-2) = -3$

$x = -1$

Model Solving Equations

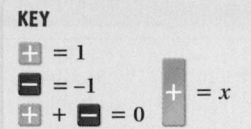

Use with Lesson 2-4

internet connect
Lab Resources Online
go.hrw.com
KEYWORD: MP4 Lab2A

go.hrw.com

KEY

$\boxed{+} = 1$

$\boxed{-} = -1$

$\boxed{+} + \boxed{-} = 0$ $\boxed{+} = x$

REMEMBER
It will not change the value of an expression if you add or remove zero.

You can use algebra tiles to help you solve equations.

Activity

To solve the equation $x + 3 = 5$, you need to get x alone on one side of the equal sign. You can add or remove tiles as long as you add the same amount or remove the same amount on both sides.

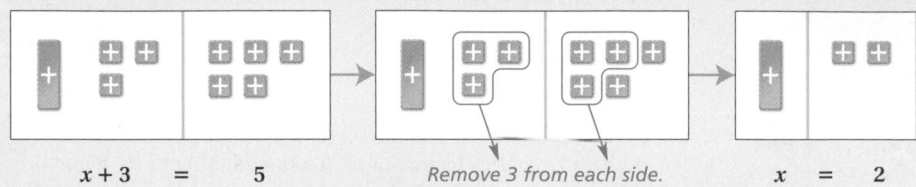

$x + 3 \quad = \quad 5$ *Remove 3 from each side.* $x \quad = \quad 2$

1 Use algebra tiles to model and solve each equation.

    a. $x + 1 = 2$ **b.** $x + 2 = 7$ **c.** $x + (-6) = -9$ **d.** $x + 4 = 4$

The equation $x + 4 = 2$ is more difficult to solve because there are not enough yellow tiles on the right side. You can use the fact that $1 + (-1) = 0$ to help you solve the equation.

$x + 4 \quad = \quad 2$ *Add zero.*

Remove 4 from each side. $x \quad = \quad -2$

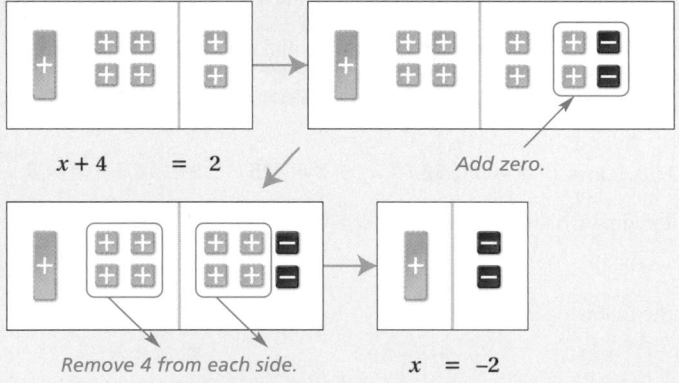

Karen Smith
Duxbury, Massachusetts

Teacher to Teacher

Instead of using algebra tiles, I use *cups and chips*. I provide each student with a few yellow and red disposable cups and several red and yellow chips.

To start the activity, I model an equation with cups representing variables and chips representing numbers (e.g., for the equation $x - 4 = 2$, I would place a yellow cup and four red chips on one side and two yellow chips on the other). I explain to the students that we want to find out how much the cup represents so we must get the cup alone.

I ask the students how we can do this, and then we solve the equation together by adding or removing the appropriate number of chips to both sides of the equation.

❷ Use algebra tiles to model and solve each equation.

a. $x + 3 = 7$ b. $x + 9 = 2$ c. $x + (-3) = -1$ d. $x + (-11) = -4$

Modeling $x - 4 = 2$ is similar to modeling $x + 4 = 2$. Remember that you can add zero to an equation and the equation's value does not change.

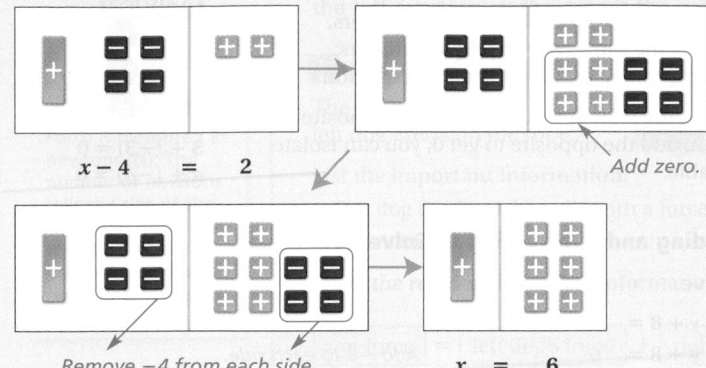

$x - 4 = 2$ Add zero.

Remove −4 from each side. $x = 6$

❸ Use algebra tiles to model and solve each equation.

a. $x - 1 = 2$ b. $x - 2 = 5$ c. $x - 4 = -3$ d. $x - 7 = 4$

Think and Discuss

1. When you add zero to an equation, how do you know the numbers of yellow square tiles and red square tiles that you need to represent the addition?

2. When you remove tiles, what operation are you representing? When you add tiles, what operation are you representing?

3. How can you use the original model to check your solution?

4. Give an example of an equation with a negative solution that would require your adding 2 red square tiles and 2 yellow square tiles to model and solve it.

5. Give an example of an equation with a positive solution that would require your adding 2 red square tiles and 2 yellow square tiles to model and solve it.

Try This

Use algebra tiles to model and solve each equation.

1. $x - 7 = 10$ **X = 17** 2. $x + 5 = -8$ **X = −13** 3. $x + 3 = 4$ **X = 1** 4. $x + 2 = -1$ **X = −3**
   X = 12
5. $x + (-4) = 8$ **X = 12** 6. $x - 6 = 2$ **X = 8** 7. $x + (-1) = -9$ **X = −8** 8. $x - 7 = -6$ **X = 1**

Think and Discuss

Possible answers:

1. Determine how many tiles you need to remove from each side. Add pairs of positive and negative tiles until you have enough tiles to remove the correct number.

2. subtraction; addition

3. To check, replace the x-tile with the number of unit tiles from the answer and see if the equation balances.

4. $x + 2 = -5$

5. $x + (-2) = 4$

1. a.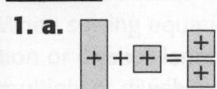

   $x = 1$

   b.

   $x = 5$

   c.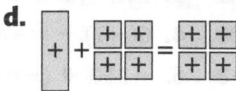

   $x = -3$

   d.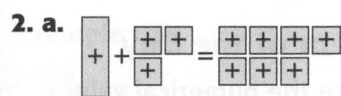

   $x = 0$

2. a.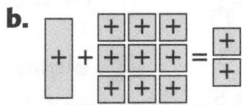

   $x = 4$

   b.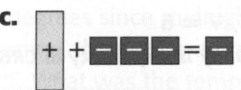

   $x = -7$

   c.

   $x = 2$

   d.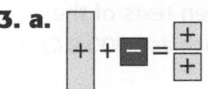

   $x = 7$

3. a.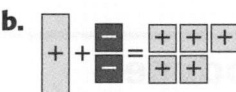

   $x = 3$

   b.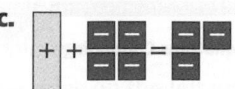

   $x = 7$

   c.

   $x = 1$

   d.

   $x = 11$

Chapter 2 — Mid-Chapter Quiz

Purpose: To assess students' mastery of concepts and skills in Lessons 2-1 through 2-5

Assessment Resources

Section 2A Quiz
Assessment Resources .... p. 7

Test and Practice Generator CD-ROM

Additional mid-chapter assessment items in both multiple-choice and free-response format may be generated for any objective in Lessons 2-1 through 2-5.

Answers

20. $m \geq -3$

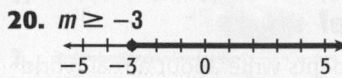

21. $t < 2$

22. $r \leq -8$

23. $k \geq -5$

ηηηηηηηηηηηηηη

Notes

Chapter 2 — Mid-Chapter Quiz

LESSON 2-1 (pp. 60–63)

Evaluate each expression for the given value of the variable.

1. $p + 12$ for $p = -5$ **7** **2.** $w + (-9)$ for $w = -4$ **−13** **3.** $t + (-14)$ for $t = 8$ **−6**

4. In a 12-hour time period in Granville, North Dakota, on Feb. 21, 1918, the temperature increased 83°F. If the beginning temperature was −33°F, what was the temperature 12 hours later? *(Source: Time Almanac 2000)* **50°F**

LESSON 2-2 (pp. 64–67)

Subtract.

5. $12 - (-8)$ **20** **6.** $-9 - (-3)$ **−6** **7.** $-5 - (-16)$ **11** **8.** $-20 - 7$ **−27**

9. The approximate surface temperature of Pluto, the coldest planet, is −391°F, while the approximate surface temperature of Venus, the hottest planet, is 864°F. How much hotter is Venus than Pluto? **1255°F**

LESSON 2-3 (pp. 68–71)

Multiply or divide.

10. $(-8)(-6)$ **48** **11.** $\frac{-21}{3}$ **−7** **12.** $\frac{39}{-3}$ **−13** **13.** $(-4)(-7)(-3)$ **−84**

14. In a *magic square*, all sums—horizontal, vertical, and diagonal—are the same.

Start with magic square A and create magic square B by dividing each entry of A by 2. What is the magic sum of B? **3**

8	−6	4
−2	2	6
0	10	−4

Magic square A

LESSON 2-4 (pp. 74–77)

Solve.

15. $t - 12 = -4$ $t = 8$ **16.** $\frac{x}{-2} = -16$ $x = 32$ **17.** $7x = -91$ $x = -13$ **18.** $10 + y = 24$ $y = 14$

19. After balancing her checkbook Barbara had exactly $0. Her bank said her balance was −$18. She realized she had not been recording her daily $2 debit charge for cups of coffee. For how many days had she forgotten to record her coffee purchases? **9 days**

LESSON 2-5 (pp. 78–81)

Solve and graph.

20. $m + 1 \geq -2$ **21.** $t - 5 < -3$ **22.** $\frac{r}{-2} \geq 4$ **23.** $-3k \leq 15$

Mid-Chapter Quiz

82 Chapter 2 Integers and Exponents

Focus on Problem Solving

Look Back
• **Is your answer reasonable?**

After you solve a word problem, ask yourself if your answer makes sense. You can round the numbers in the problem and estimate to find a reasonable answer. It may also help to write your answer in sentence form.

Read the problems below and tell which answer is most reasonable.

1 Tonia makes $1836 per month. Her total expenses are $1005 per month. How much money does she have left each month? **C**
- **A.** about −$800 per month
- **B.** about $1000 per month
- **C.** about $800 per month
- **D.** about −$1000 per month

2 The Qin Dynasty in China began about 2170 years before the People's Republic of China was formed in 1949. When did the Qin Dynasty begin? **A**
- **A.** before 200 B.C.
- **B.** between 200 B.C. and A.D. 200
- **C.** between A.D. 200 and A.D. 1949
- **D.** after A.D 1949

3 On Mercury, the coldest temperature is about 600°C below the hottest temperature of 430°C. What is the coldest temperature on the planet? **C**
- **A.** about 1030°C
- **B.** about −1030°C
- **C.** about −170°C
- **D.** about 170°C

4 Julie is balancing her checkbook. Her beginning balance is $325.46, her deposits add up to $285.38, and her withdrawals add up to $683.27. What is her ending balance? **A**
- **A.** about −$70
- **B.** about −$600
- **C.** about $700
- **D.** about $1300

Purpose: *To focus on determining whether an answer is reasonable*

Problem Solving Resources

Interactive Problem Solving. . pp. 10–18

Math: Reading and Writing in the Content Area . . . . pp. 10–18

Problem Solving Process

This page focuses on the last step of the problem-solving process: **Look Back**

Discuss

Have students discuss how they determined which answer was the most reasonable. How does estimation help eliminate some of the answer choices?

Possible answer: Estimation can help you decide whether the answer should be positive or negative and approximately what the answer should be. Answer choices that are not close to the estimate can be eliminated.

1. $1800 − 1000 = 800$; the negative choices can be eliminated since her income is more than her rent.

2. $1950 − 2200 = −250$; since the difference is negative, the positive (A.D.) choices can be eliminated.

3. $400 − 600 = −200$; −170 is the only answer choice close to −200.

4. $300 + 300 − 700 = −100$; C and D are positive and can be eliminated; B is too low of a balance.

Section 2B
Exponents and Scientific Notation

One-Minute Section Planner

Lesson	Materials	Resources
Lesson 2-6 Exponents **NCTM:** Number and Operations, Communication **NAEP:** Number Properties 3a ☑ SAT-9 ☑ SAT-10 ☑ ITBS ☑ CTBS ☑ MAT ☑ CAT	**Optional** Calculators Counters for Reaching All Learners *(MK)*	• *Chapter 2 Resource Book,* pp. 53–61 • Daily Transparency T26, CRB • Additional Examples Transparencies T27–T29, CRB • *Alternate Openers: Explorations,* p. 15
Lesson 2-7 Properties of Exponents **NCTM:** Number and Operations, Communication **NAEP:** Algebra 3b ☐ SAT-9 ☑ SAT-10 ☐ ITBS ☑ CTBS ☑ MAT ☑ CAT	**Optional** Teaching Transparency T31 *(CRB)*	• *Chapter 2 Resource Book,* pp. 62–70 • Daily Transparency T30, CRB • Additional Examples Transparencies T32–T34, CRB • *Alternate Openers: Explorations,* p. 16
Lesson 2-8 Look for a Pattern in Integer Exponents **NCTM:** Number and Operations, Algebra, Problem Solving, Reasoning and Proof, Communication, Connections **NAEP:** Algebra 1a ☐ SAT-9 ☐ SAT-10 ☐ ITBS ☐ CTBS ☐ MAT ☐ CAT	**Optional** Teaching Transparencies T36–T37 *(CRB)*	• *Chapter 2 Resource Book,* pp. 71–79 • Daily Transparency T35, CRB • Additional Examples Transparencies T38–T40, CRB • *Alternate Openers: Explorations,* p. 17
Lesson 2-9 Scientific Notation **NCTM:** Number and Operations, Communication, Representation **NAEP:** Number Properties 1f ☑ SAT-9 ☑ SAT-10 ☑ ITBS ☐ CTBS ☑ MAT ☑ CAT	**Optional** Newspapers or magazines	• *Chapter 2 Resource Book,* pp. 80–88 • Daily Transparency T41, CRB • Additional Examples Transparencies T42–T44, CRB • *Alternate Openers: Explorations,* p. 18
Section 2B Assessment		• Section 2B Quiz, AR p. 8 • *Test and Practice Generator* CD-ROM

SAT = *Stanford Achievement Tests*   **ITBS** = *Iowa Test of Basic Skills*   **CTBS** = *Comprehensive Test of Basic Skills/Terra Nova*
MAT = *Metropolitan Achievement Test*   **CAT** = *California Achievement Test*

NCTM = Complete standards can be found on pages T29–T35.   **NAEP** = Complete standards can be found on pages A54–A58.

SE = *Student Edition*   **TE** = *Teacher's Edition*   **AR** = *Assessment Resources*   **CRB** = *Chapter Resource Book*   **MK** = *Manipulatives Kit*

Section Overview

Exponents

Lesson 2-6

Why? Exponents are used to write multiplication expressions that have repeated factors.

Simplify $3 \cdot (-10)^3$.

$3 \cdot (-10)^3$

$= 3 \cdot (-10) \cdot (-10) \cdot (-10)$

$= 3 \cdot (-1000)$

$= -3000$

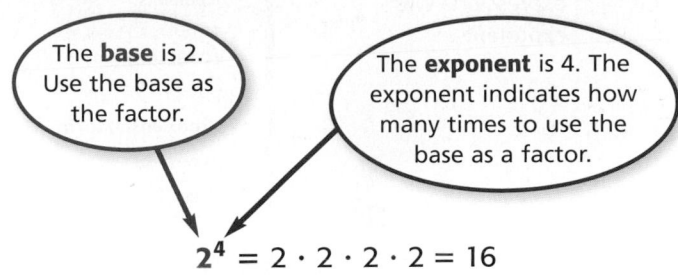

The **base** is 2. Use the base as the factor.

The **exponent** is 4. The exponent indicates how many times to use the base as a factor.

$2^4 = 2 \cdot 2 \cdot 2 \cdot 2 = 16$

2^4 is read "two to the fourth power."

Using Properties of Exponents

Lesson 2-7

Why? Properties of exponents allow us to simplify expressions that have exponents.

Properties	Examples
$b^m \cdot b^n = b^{m+n}$	$5^3 \cdot 5^4 = 5^{3+4} = 5^7$
$\dfrac{b^m}{b^n} = b^{m-n}$, if $b \neq 0$.	$\dfrac{3^6}{3^2} = 3^{6-2} = 3^4$
$a^0 = 1$, if $a \neq 0$.	$9^0 = 1$, $(-4)^0 = 1$, 0^0 is not defined.

Negative Exponents and Scientific Notation

Lessons 2-8, 2-9

Why? Scientific notation is a useful way to express very large or very small numbers. To express very small numbers in scientific notation, we need negative exponents.

If the values $m = 2$ and $n = 5$ are used in the property $\dfrac{b^m}{b^n} = b^{m-n}$, we have $\dfrac{b^2}{b^5} = b^{2-5} = b^{-3}$. Cases such as this suggest another reason to have a definition for negative exponents.

Negative Exponents

Definition	Examples
$b^{-n} = \dfrac{1}{b^n}$, if $b \neq 0$.	$3^{-2} = \dfrac{1}{3^2} = \dfrac{1}{3 \cdot 3} = \dfrac{1}{9}$
	$(-2)^{-4} = \dfrac{1}{(-2)^4} = \dfrac{1}{(-2) \cdot (-2) \cdot (-2) \cdot (-2)} = \dfrac{1}{16}$

Scientific Notation			Standard Notation
2.3×10^3	$= 2.3 \times 1000 =$		2300
9.05×10^{-2}	$= 9.05 \times 0.01 =$		0.0905

Use a number that is at least one but less than ten here.

Use a power of 10 here.

Pacing: Traditional 1 day
Block $\frac{1}{2}$ day

Objective: Students evaluate expressions with exponents.

Warm Up

Find the product.

1. $5 \cdot 5 \cdot 5 \cdot 5$ 625

2. $3 \cdot 3 \cdot 3$ 27

3. $(-7)(-7)(-7)$ -343

4. $9 \cdot 9$ 81

Problem of the Day

What two positive integers when multiplied together also equal the sum of the same two numbers?
2 and 2

Available on Daily Transparency in CRB

Math Fact

The modern notation for exponents is believed to have originated with René Descartes around 1637.

2-6 Exponents

Learn to evaluate expressions with exponents.

Vocabulary
power
exponential form
exponent
base

Fold a piece of $8\frac{1}{2}$-by-11-inch paper in half. If you fold it in half again, the paper is 4 sheets thick. After the third fold in half, the paper is 8 sheets thick. How many sheets thick is the paper after 7 folds?

With each fold the number of sheets doubles.

$$2 \cdot 2 \cdot 2 \cdot 2 \cdot 2 \cdot 2 \cdot 2 = 128 \text{ sheets thick after 7 folds.}$$

This multiplication problem can also be written in *exponential form*.

$$2 \cdot 2 \cdot 2 \cdot 2 \cdot 2 \cdot 2 \cdot 2 = 2^7$$

The number 2 is a factor 7 times.

The term 2^7 is called a **power**. If a number is in **exponential form**, the **exponent** represents how many times the **base** is to be used as a factor.

Base   Exponent

EXAMPLE 1   Writing Exponents

Write in exponential form.

A $3 \cdot 3 \cdot 3 \cdot 3 \cdot 3 \cdot 3$
$3 \cdot 3 \cdot 3 \cdot 3 \cdot 3 \cdot 3 = 3^6$ *Identify how many times 3 is a factor.*

Reading Math

Read 3^6 as "3 to the 6th power."

B $(-2) \cdot (-2) \cdot (-2) \cdot (-2)$
$(-2) \cdot (-2) \cdot (-2) \cdot (-2) = (-2)^4$ *Identify how many times -2 is a factor.*

C $n \cdot n \cdot n \cdot n \cdot n$
$n \cdot n \cdot n \cdot n \cdot n = n^5$ *Identify how many times n is a factor.*

D 12
$12 = 12^1$ *12 is used as a factor 1 time, so $12 = 12^1$.*

EXAMPLE 2   Evaluating Powers

Helpful Hint

Always use parentheses to raise a negative number to a power.
$(-8)^2 = (-8) \cdot (-8)$
$\quad = 64$
$-8^2 = -(8 \cdot 8)$
$\quad = -64$

Evaluate.

A 2^6
$2^6 = 2 \cdot 2 \cdot 2 \cdot 2 \cdot 2 \cdot 2$ *Find the product of six 2's.*
$\quad = 64$

B $(-8)^2$
$(-8)^2 = (-8) \cdot (-8)$ *Find the product of two -8's.*
$\quad = 64$

---

1 Introduce

Alternate Opener

EXPLORATION

2-6 Exponents

You can multiply $(-5) \cdot (-5) \cdot (-5) \cdot (-5) \cdot (-5) \cdot (-5)$ using exponents and a calculator.

The number -5 is a factor 6 times, so you can write it as $(-5)^6$.

The expressions are equivalent because they have the same value.

$(-5) \cdot (-5) \cdot (-5) \cdot (-5) \cdot (-5) \cdot (-5) = 15{,}625$ and $(-5)^6 = 15{,}625$

Guess the value of the variable. Use a calculator after each guess to check your answer.

1. $b^5 = -243$ 2. $b^5 = -7776$
$b = \_\_\_$ $b = \_\_\_$

3. $(-2)^x = 1024$ 4. $(-2)^x = -512$
$x = \_\_\_$ $x = \_\_\_$

5. $(-3)^x = 81$ 6. $(-2)^x = 64$
$x = \_\_\_$ $x = \_\_\_$

Think and Discuss

7. **Describe** the strategies you used to find the values of the variables.

Motivate

To introduce the concept of exponents, ask students if there is a simpler expression for the sum $5 + 5 + 5 + 5$. They might suggest the multiplication expression $4 \cdot 5$. Explain that just as repeated addition can be simplified by multiplication, repeated multiplication can be simplified by using exponents. For example, $5 \cdot 5 \cdot 5 \cdot 5 = 5^4$.

Exploration worksheet and answers on Chapter 2 Resource Book pp. 54 and 108

2 Teach

Lesson Presentation

Guided Instruction

In this lesson, students learn to evaluate expressions with exponents. Show students an example of a power, such as 3^4. Point out the *base,* the *exponent,* and the *power.* Show them how to evaluate the power ($3 \cdot 3 \cdot 3 \cdot 3 = 81$). You may want to show that the same result can be obtained by working the multiplications in any order.

Teaching Tip Point out that the sign rules for multiplication still apply. Work the examples containing negative bases step by step so students will understand why the result is sometimes positive (when the exponent is even) and sometimes negative (when the exponent is odd).

Evaluate.

C $(-5)^3$

$(-5)^3 = (-5) \cdot (-5) \cdot (-5)$ *Find the product of three −5's.*

$\qquad = -125$

EXAMPLE **Simplifying Expressions Containing Powers**

Simplify $50 - 2(3 \cdot 2^3)$.

$50 - 2(3 \cdot 2^3)$

$= 50 - 2(3 \cdot 8)$ *Evaluate the exponent.*

$= 50 - 2(24)$ *Multiply inside the parentheses.*

$= 50 - 48$ *Multiply from left to right.*

$= 2$ *Subtract from left to right.*

EXAMPLE **4** *Geometry Application*

The number of diagonals of an *n*-sided figure is $\frac{1}{2}(n^2 - 3n)$. Use the formula to find the number of diagonals for a 5-sided figure.

$\frac{1}{2}(n^2 - 3n)$

$\frac{1}{2}(5^2 - 3 \cdot 5)$ *Substitute the number of sides for n.*

$\frac{1}{2}(25 - 3 \cdot 5)$ *Evaluate the exponent.*

$\frac{1}{2}(25 - 15)$ *Multiply inside the parentheses.*

$\frac{1}{2}(10)$ *Subtract inside the parentheses.*

5 diagonals *Multiply.*

Verify your answer by sketching the diagonals.

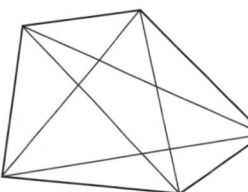

Think and Discuss

1. Describe a rule for finding the sign of a negative number raised to a whole number power.

2. Compare $3 \cdot 2$, 3^2, and 2^3.

3. Show that $(4 - 11)^2$ is not equal to $4^2 - 11^2$.

COMMON ERROR
/// **ALERT** \\\

Some students may multiply the exponent and the base (e.g., $3^4 = 12$). Encourage students to expand the power before multiplying (e.g., $3^4 = 3 \cdot 3 \cdot 3 \cdot 3 = 81$).

Additional Examples

Example **1**

Write in exponential form.

A. $4 \cdot 4 \cdot 4 \cdot 4$
   4^4

B. $d \cdot d \cdot d \cdot d \cdot d$
   d^5

C. $(-6) \cdot (-6) \cdot (-6)$
   $(-6)^3$

D. $5 \cdot 5$
   5^2

Example **2**

Evaluate.

A. 3^5 **B.** $(-3)^5$
   243   −243

C. $(-4)^4$ **D.** 2^8
   256   256

Example **3**

Simplify $(2^5 - 3^2) + 6(4)$.
47

Example **4**

Use the formula $\frac{1}{2}(n^2 - 3n)$ to find the number of diagonals in a 7-sided figure.

14 diagonals

---

3 **Close**

 **Reaching All Learners**

Through Concrete Manipulatives

Give each student or group of students a set of 16 counters (provided in the Manipulatives Kit). Ask students to divide the counters into 4 equal groups and to write an addition expression that represents the grouping $(4 + 4 + 4 + 4)$. Then have them write a multiplication expression that represents the same grouping $(4 \cdot 4)$ and then an expression with an exponent (4^2). After simplifying each expression to get 16, they should begin to see how addition, multiplication, and powers are related.

Summarize

Review the vocabulary terms *power, base,* and *exponent.* Discuss how each term is related to multiplication. You may wish to have students identify the base, exponent, and power in the expression 2^3.

Possible answers: A power is an expression where a number is multiplied by itself a certain number of times. A base is the factor in a power. An exponent is the number of times the factor is used. In the expression 2^3, 2 is the base, 3 is the exponent, and 2^3 is the power.

Answers to Think and Discuss

Possible answers:

1. A negative number raised to an odd power is negative, and a negative number raised to an even power is positive.

2. $3 \cdot 2 = 6$; $3^2 = 3 \cdot 3 = 9$; $2^3 = 2 \cdot 2 \cdot 2 = 8$

3. $(4 - 11)^2 = (-7)^2 = 49$; $4^2 - 11^2 = 16 - 121 = -105$; $49 \neq -105$. Expressed another way, the square of any nonzero integer is positive, so $(4 - 11)^2$ is positive. But since $11 > 4$ and $11^2 > 4^2$, $4^2 - 11^2$ is negative. Therefore, the two expressions cannot be equal.

FOR EXTRA PRACTICE
see page 735

internet connect

Homework Help Online
go.hrw.com Keyword: MP4 2-6

Students may want to refer back to the lesson examples.

Assignment Guide

If you finished Example **1** assign:
  Core 1–4, 14–17, 27–30, 58–64
  Enriched 1–4, 14–17, 27–30, 58–64

If you finished Example **2** assign:
  Core 1–8, 14–21, 27–37 odd, 58–64
  Enriched 14–21, 27–38, 47, 48, 56, 58–64

If you finished Example **3** assign:
  Core 1–12, 14–25, 27–53 odd, 58–64
  Enriched 14–25, 27–53, 58–64

If you finished Example **4** assign:
  Core 1–26, 27–53 odd, 54, 58–64
  Enriched 14–64

Notes

GUIDED PRACTICE

See Example **1** Write in exponential form.
1. 14 14^1 2. $15 \cdot 15$ 15^2 3. $b \cdot b \cdot b \cdot b$ b^4 4. $(-1) \cdot (-1) \cdot (-1)$ $(-1)^3$

See Example **2** Evaluate.
5. 3^4 **81**   6. $(-5)^2$ **25**   7. $(-3)^5$ **−243**   8. 7^4 **2401**

See Example **3** Simplify.
9. $(3 - 6^2)$ **−33** 10. $42 + (3 \cdot 4^2)$ **90** 11. $(8 - 5^3)$ **−117** 12. $61 - (4 \cdot 3^3)$ **−47**

See Example **4** 13. The sum of the first n positive integers is $\frac{1}{2}(n^2 + n)$. Check the formula for the first four positive integers. Then use the formula to find the sum of the first 12 positive integers. **78**

INDEPENDENT PRACTICE

See Example **1** Write in exponential form.
14. $6 \cdot 6 \cdot 6 \cdot 6 \cdot 6 \cdot 6 \cdot 6$ 6^7 15. $(-7) \cdot (-7) \cdot (-7)$ $(-7)^3$
16. -6 $(-6)^1$ 17. $c \cdot c \cdot c \cdot c \cdot c$ c^5

See Example **2** Evaluate.
18. 6^6 **46,656**   19. $(-4)^4$ **256**   20. 8^4 **4096**   21. $(-2)^9$ **−512**

See Example **3** Simplify.
22. $(1 - 7^2)$ **−48**   23. $27 + (2 \cdot 5^2)$ **77**
24. $(8 - 10^3)$ **−992**   25. $45 - (5 \cdot 3^4)$ **−360**

See Example **4** 26. A circle can be divided by n lines into a maximum of $\frac{1}{2}(n^2 + n) + 1$ regions. Use the formula to find the maximum number of regions for 7 lines. **29**

3 lines → 7 regions

PRACTICE AND PROBLEM SOLVING

Write in exponential form.
27. $(-2) \cdot (-2) \cdot (-2)$ $(-2)^3$ 28. $h \cdot h \cdot h \cdot h$ h^4
29. $4 \cdot 4 \cdot 4 \cdot 4$ 4^4 30. $(5)(5)(5)(5)(5)$ 5^5

Evaluate.
31. 7^3 **343**   32. 8^2 **64**   33. $(-12)^3$ **−1728** 34. $(-6)^5$ **−7776**
35. $(-3)^6$ **729**   36. $(-9)^3$ **−729** 37. 4^1 **4**   38. 2^9 **512**

Math Background

Exponents are necessary in many topics in this course, including scientific notation, area, volume, and right triangles. Exponents are also at the heart of place value notation.

The second power of a number is commonly called the *square* of the number. This name comes from the fact that the area of a square is given by the formula $A = s^2$, where s is the length of a side of the square.

The third power of a number is called the *cube* of the number because the volume of a cube is given by the formula $V = s^3$, where s is the length of an edge of the cube.

RETEACH 2-6

LESSON **2-6** **Reteach** *Exponents*

The fifth power of 3 $3^5 = 3 \cdot 3 \cdot 3 \cdot 3 \cdot 3$
base / exponent
3 used as a factor 5 times

Complete to write each expression using an exponent. State the power.
1. $5 \cdot 5 \cdot 5 \cdot 5 = 5^{\underline{4}}$ the __fourth__ power of 5
2. $(-7) \cdot (-7) \cdot (-7) = (-7)^{\underline{3}}$ the __third__ power of −7

Complete to evaluate each expression.
3. $(-2)^3 = (-2)(-2)(-2) = \underline{-8}$
4. $10^4 = \underline{10} \cdot \underline{10} \cdot \underline{10} \cdot \underline{10} = \underline{10,000}$
5. $(-5)^4 = (\underline{-5})(\underline{-5})(\underline{-5})(\underline{-5}) = \underline{625}$

When an expression is a product that includes a power, you simplify the power first.
$3 \cdot 2^3 = 3 \cdot (2 \cdot 2 \cdot 2) = 3 \cdot 8 = 24$

Complete to simplify each expression.
6. $4 \cdot (-2)^3 = 4(\underline{-2})(\underline{-2})(\underline{-2}) = \underline{-32}$
7. $5 \cdot 3^3 = \underline{5} \cdot \underline{3} \cdot \underline{3} \cdot \underline{3} = \underline{135}$
8. $(3 \cdot 2)^3 = 6^3 = \underline{6} \cdot \underline{6} \cdot \underline{6} = \underline{216}$
9. $-4(-2)^3 = (\underline{8})^3 = (\underline{8})(\underline{8})(\underline{8}) = \underline{512}$
10. $25 - 3(4 \cdot 3^2)$
  $= 25 - 3(4 \cdot \underline{9})$
  $= 25 - 3(\underline{36})$
  $= 25 - \underline{108}$
  $= \underline{-83}$
11. $-100 - 2(3 \cdot 4)^2$
  $= -100 - 2(\underline{12})^2$
  $= -100 - 2(\underline{144})$
  $= -100 - \underline{288}$
  $= \underline{-388}$
12. $15 - 4(3 + 3^2)$
  $= 15 - 4(3 + \underline{9})$
  $= 15 - 4(\underline{12})$
  $= 15 - \underline{48}$
  $= \underline{-33}$

PRACTICE 2-6

LESSON **2-6** **Practice B** *Exponents*

Write using exponents.
1. $6 \cdot 6 \cdot 6 \cdot 6 \cdot 6 \cdot 6$ 6^6 2. $7 \cdot 7 \cdot 7 \cdot 7$ 7^4
3. $(-8) \cdot (-8) \cdot (-8) \cdot (-8)$ $(-8)^4$

Evaluate.
4. 10^2 **100**   5. $(-6)^2$ **36**   6. 8^2 **64**   7. $(-7)^2$ **49**
8. $(-5)^3$ **−125**   9. 12^2 **144**   10. $(-9)^2$ **81**   11. $(-4)^3$ **−64**

Simplify.
12. $5^3 - 10$ **115**   13. $50 - 7^2$ **1**   14. $(-9)^2 + 19$ **100**   15. $6^2 - 2^4$ **20**
16. $7^2 - 2^3 + 1$ **42**   17. $10 + 3^3 \cdot 2$ **64**   18. $8^2 + 4 \cdot 5^2$ **164**   19. $2(5^3 + 10^2)$ **450**

Evaluate for the given value of the variable.
20. n^3 for $n = 4$ **64**   21. $4x^2$ for $x = 2$ **16**   22. $y^4 - 10$ for $y = 3$ **71**

23. Write an expression for five times a number that is used as a factor three times. $5x^3$
24. Find the volume of a regular cube if the length of a side is 10 cm. (Hint: $V = l^3$.) **1000 cm³**

Simplify.

39. $(9 - 5^3)$ **−116**   40. $(18 - 7^3)$ **−325**   41. $42 + (8 - 6^3)$ **−166**   42. $16 + (2 + 8^3)$ **530**

43. $32 - (4 \cdot 3^2)$ **−4**   44. $(5 + 5^5)$ **3130**   45. $(5 - 6^1)$ **−1**   46. $86 - [6 - (-2)^5]$ **48**

Evaluate each expression for the given value of the variable.

47. a^3 for $a = 6$ **216**   48. x^7 for $x = -1$ **−1**   49. $n^4 + 1$ for $n = 4$ **257**   50. $1 - y^5$ for $y = 2$ **−31**

51. **LIFE SCIENCE** Bacteria can divide every 20 minutes, so one bacterium can multiply to 2 in 20 minutes, 4 in 40 minutes, 8 in 1 hour, and so on. How many bacteria will there be in 6 hours? Write your answer using exponents, and then evaluate. $2^{18} = 262,144$ **bacteria**

Most bacteria reproduce by a type of simple cell division known as binary fission. Each species reproduces best at a specific temperature and moisture level.

52. Make a table with the column headings n, n^2, and $2n$. Complete the table for $n = -5, -4, -3, -2, -1, 0, 1, 2, 3, 4,$ and 5.

53. For any whole number n, $5^n - 1$ is divisible by 4. Verify this for $n = 3$ and $n = 5$.

54. The chart shows Han's genealogy. Each generation consists of twice as many people as the generation after it.

a. Write the number of Han's great-grandparents using an exponent. 2^3

b. How many ancestors were in the fifth generation back from Han? **32**

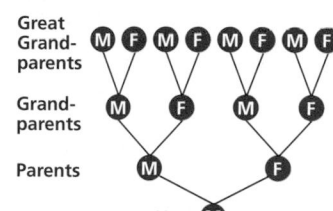

55. **CHOOSE A STRATEGY** Place the numbers 1, 2, 3, 4, and 5 in the boxes to make a true statement: $\square \cdot \square^3 = \square^2 - \square$ $3 \cdot 2^3 = 5^2 - 1^4$

56. **WRITE ABOUT IT** Compare 10^3 and 3^{10}. For any two numbers, which usually gives the greater number, using the larger number as the base or as the exponent? Give at least one exception.

57. **CHALLENGE** Write $(3^2)^3$ using a single exponent.
$(3^2)^3 = (3 \cdot 3)(3 \cdot 3)(3 \cdot 3) = 3^6$

Spiral Review

Multiply or divide. (Lesson 2-3)

58. $7(-8)$ **−56**   59. $\dfrac{-63}{-7}$ **9**   60. $\dfrac{38}{-19}$ **−2**   61. $-8(-13)$ **104**   62. $-6(15)$ **−90**

63. **TEST PREP** Which represents the phrase *the difference of a number and 32*? (Lesson 1-2) **C**

A $n + 32$ C $n - 32$

B $n \times 32$ D $32 \div n$

64. **TEST PREP** Which of the values -2, -1, and 0 are solutions of $x - 2 > -3$? (Lesson 2-5) **G**

F -1 and 0   H -2 and -1

G only 0   J $-2, -1,$ and 0

Answers

52.

n	n^2	$2n$
−5	25	−10
−4	16	−8
−3	9	−6
−2	4	−4
−1	1	−2
0	0	0
1	1	2
2	4	4
3	9	6
4	16	8
5	25	10

53. $5^3 - 1 = 125 - 1 = 124$, and $124 \div 4 = 31$; $5^5 - 1 = 3125 - 1 = 3124$, and $3124 \div 4 = 781$

56. Possible answer: $10^3 = 1000$ and $3^{10} = 59,049$, so $10^3 < 3^{10}$; using the larger number as the exponent generally gives the greater number; exception: $3^2 > 2^3$.

Journal

Have students imagine that they have invested $100 in an account and that their money will double every five years. Have them write about what they would be able to buy with the money after 20 years.

Test Prep Doctor

For Exercise 64, encourage students to test each possible solution independently. Have them record each result on scrap paper and then use those results to choose the correct response.

CHALLENGE 2-6

LESSON 2-6 Challenge
Check This Out

Imagine an 8 × 8 checkerboard.
- Put 1 penny on the first square.
- Stack 2 pennies on an adjacent square.
- Stack 4 pennies on an adjacent square.
- Stack 8 pennies on an adjacent square.

Assume the pattern continues so that each square has double the number of pennies as the previous square.

Complete the table.

Square	1	2	3	4	5	6	7	8
1. Number of Pennies	1	2	4	8	16	32	64	128
2. Exponent Form	2^0	2^1	2^2	2^3	2^4	2^5	2^6	2^7

3. Look for a pattern in the table. How many pennies are on the 12th square? the 15th square?

square $n = 2^{n-1}$ pennies; square $12 = 2^{11}$ or 2048 pennies;
square $15 = 2^{14}$ or 16,384 pennies

4. Find the number of pennies on the 25th square. How much money is this? Estimate the height of the stack.

2^{24} or 16,777,216 pennies = $167,772.16;
Possible answer: About 1 million in., or over 15 mi

5. Find the number of pennies on the 10th square. Find the total number of pennies on the first nine squares. Which number is greater?

512 pennies on 10th square; 511 pennies on the first nine squares.
512 > 511

6. Explain why it is impossible to stack pennies on every square in the manner described at the top of the page.

Possible answer: By the 25th square, the stack is already over
15 mi high, and there are 39 more squares to stack.

PROBLEM SOLVING 2-6

LESSON 2-6 Problem Solving
Exponents

Write the correct answer.

1. The formula for the volume of a cube is $V = e^3$ where e is the length of a side of the cube. Find the volume of a cube with side length 6 cm.

216 cm³

2. The distance in feet traveled by a falling object is given by the formula $d = 16t^2$ where t is the time in seconds. Find the distance an object falls in 4 seconds.

256 feet

3. The surface area of a cube can be found using the formula $S = 6e^2$ where e is the length of a side of the cube. Find the surface area of a cube with side length 6 cm.

216 cm²

4. John's father offers to pay him 1 cent for doing the dishes the first night, 2 cents for doing the dishes the second, 4 cents for the third, and so on, doubling each night. Write an expression using exponents for the amount John will get paid on the tenth night.

2^9 cents

Use the table below for Exercises 5–7, which shows the number of e-mails forwarded at each level if each person continues a chain by forwarding an e-mail to 10 friends. Choose the letter for the best answer.

5. How many e-mails were forwarded at level 5 alone?

A 5^{10} C 2^{10}
B 2^5 D 10^5

6. How many e-mails were forwarded at level 6 alone?

F 100,000   H 10,000,000
G 1,000,000   J 100,000,000

Forwarded E-mails	
Level	E-mails forwarded
1	10
2	100
3	1000
4	10,000

7. Forwarding chain e-mails can create problems for e-mail servers. Find out how many total e-mails have been forwarded after 6 levels.

A 1,111,110   C 1,000,000
B 6,000,000   D 100,000,000

Lesson Quiz

Write in exponential form.

1. $n \cdot n \cdot n \cdot n$ n^4

2. $(-8)(-8)(-8)$ $(-8)^3$

3. Evaluate $(-4)^4$.   **256**

4. Simplify $99 - 3(4 \cdot 2^3)$.   **3**

5. A population of bacteria doubles in size every minute. The number of bacteria after 5 minutes is $15 \cdot 2^5$. How many are there after 5 minutes?   **480**

Available on Daily Transparency in CRB

Pacing: Traditional 1 day
Block $\frac{1}{2}$ day

Objective: Students apply the properties of exponents and evaluate the zero exponent.

Learn to apply the properties of exponents and to evaluate the zero exponent.

The factors of a power, such as 7^4, can be grouped in different ways. Notice the relationship of the exponents in each product.

$$7 \cdot 7 \cdot 7 \cdot 7 = 7^4$$
$$(7 \cdot 7 \cdot 7) \cdot 7 = 7^3 \cdot 7^1 = 7^4$$
$$(7 \cdot 7) \cdot (7 \cdot 7) = 7^2 \cdot 7^2 = 7^4$$

MULTIPLYING POWERS WITH THE SAME BASE		
Words	**Numbers**	**Algebra**
To multiply powers with the same base, keep the base and add the exponents.	$3^5 \cdot 3^8 = 3^{5+8} = 3^{13}$	$b^m \cdot b^n = b^{m+n}$

Warm Up

Evaluate.

1. 3^3 27

2. $4 \cdot 4 \cdot 4 \cdot 4$ 256

3. b^2 for $b = 4$ 16

4. $n^2 r$ for $n = 3$ and $r = 2$ 18

EXAMPLE 1 Multiplying Powers with the Same Base

Multiply. Write the product as one power.

A $3^5 \cdot 3^2$
$3^5 \cdot 3^2$
3^{5+2} *Add exponents.*
3^7

B $a^{10} \cdot a^{10}$
$a^{10} \cdot a^{10}$
a^{10+10} *Add exponents.*
a^{20}

C $16 \cdot 16^7$
$16 \cdot 16^7$
$16^1 \cdot 16^7$ *Think: $16 = 16^1$*
16^{1+7} *Add exponents.*
16^8

D $6^4 \cdot 4^4$
$6^4 \cdot 4^4$ *Cannot combine; the bases are not the same.*

Problem of the Day

Calculate 6 to the fourth power minus 56. **1240**

Available on Daily Transparency in CRB

Notice what occurs when you divide powers with the same base.

$$\frac{5^5}{5^3} = \frac{5 \cdot 5 \cdot 5 \cdot 5 \cdot 5}{5 \cdot 5 \cdot 5} = \frac{\cancel{5} \cdot \cancel{5} \cdot \cancel{5} \cdot 5 \cdot 5}{\cancel{5} \cdot \cancel{5} \cdot \cancel{5}} = 5 \cdot 5 = 5^2$$

Math Humor

Teacher: What is a number to the zero power?

Student: a number that didn't pay its electric bill

DIVIDING POWERS WITH THE SAME BASE		
Words	**Numbers**	**Algebra**
To divide powers with the same base, keep the base and subtract the exponents.	$\frac{6^9}{6^4} = 6^{9-4} = 6^5$	$\frac{b^m}{b^n} = b^{m-n}$

1 Introduce

Alternate Opener

EXPLORATION

2-7 Properties of Exponents

- To multiply powers with the same base, add the exponents.
$$4^2 \cdot 4^3 = 4^{2+3} = 4^5$$
- To divide powers with the same base, keep the base and subtract the exponents.
$$\frac{3^4}{3^2} = 3^{4-2} = 3^2$$

Complete each table.

	Product of Powers	Add Exponents	Resulting Power
1.	$2^3 \cdot 2^4$		
2.	$3^2 \cdot 3^5$		
3.	$5^3 \cdot 5^4$		

	Quotient of Powers	Subtract Exponents	Resulting Power
4.	$\frac{2^5}{2^3}$		
5.	$\frac{3^7}{3^3}$		
6.	$\frac{10^6}{10^5}$		

Think and Discuss

7. **Describe** how to multiply two powers that have the same base.

8. **Describe** how to divide two powers that have the same base.

Motivate

Before introducing students to the properties of exponents, review what an exponent is. Write an expression such as 3^5 on the chalkboard. Have students identify the base and the exponent. Ask them for a step-by-step explanation of how to simplify the expression using multiplication. Encourage students to use the proper vocabulary.

Exploration worksheet and answers on Chapter 2 Resource Book pp. 63 and 110

2 Teach

Lesson Presentation

Guided Instruction

In this lesson, students learn to apply the properties of exponents and to evaluate the zero exponent. Help students discover the property $b^m \cdot b^n = b^{m+n}$ by having them write several expressions in expanded form, such as $2^3 \cdot 2^4 = (2 \cdot 2 \cdot 2)(2 \cdot 2 \cdot 2 \cdot 2) = 2^7$. Use a similar process for the property $\frac{b^m}{b^n} = b^{m-n}$. You may want to display Teaching Transparency T31 (CRB). To discover the property $a^0 = 1$ if $a \neq 0$, have students write several statements, such as $\frac{5^2}{5^2}$, and simplify them using both the expansion method and the properties of exponents.

EXAMPLE 2 Dividing Powers with the Same Base

Divide. Write the quotient as one power.

A $\dfrac{100^9}{100^3}$

$\dfrac{100^9}{100^3}$

100^{9-3} *Subtract exponents.*

100^6

B $\dfrac{x^8}{y^5}$

$\dfrac{x^8}{y^5}$ *Cannot combine; the bases are not the same.*

Helpful Hint

0^0 does not exist because 0^0 represents a quotient of the form
$$\dfrac{0^n}{0^n}.$$
But the denominator of this quotient is 0, which is impossible, since you cannot divide by 0.

When the numerator and denominator of a fraction have the same base and exponent, subtracting the exponents results in a **0** exponent.

$$1 = \frac{4^2}{4^2} = 4^{2-2} = 4^0 = 1$$

This result can be confirmed by writing out the factors.

$$\frac{4^2}{4^2} = \frac{(4 \cdot 4)}{(4 \cdot 4)} = \frac{(\cancel{4} \cdot \cancel{4})}{(\cancel{4} \cdot \cancel{4})} = \frac{1}{1} = 1$$

THE ZERO POWER		
Words	Numbers	Algebra
The zero power of any number except 0 equals 1.	$100^0 = 1$ $(-7)^0 = 1$	$a^0 = 1$, if $a \neq 0$

EXAMPLE 3 *Physical Science Application*

There are about 10^{25} molecules in a cubic meter of air at sea level, but only 10^{23} molecules at a high altitude (33 km). How many times more molecules are there at sea level than at 33 km?

You want to find the number that you must multiply by 10^{23} to get 10^{25}. Set up and solve an equation. Use x as your variable.

$(10^{23})x = 10^{25}$ *"10^{23} times some number x equals 10^{25}."*

$\dfrac{(10^{23})x}{10^{23}} = \dfrac{10^{25}}{10^{23}}$ *Divide both sides by 10^{23}.*

$x = 10^{25-23}$ *Subtract the exponents.*

$x = 10^2$

There are 10^2 times more molecules per cubic meter of air at sea level than at 33 km.

Think and Discuss

1. Explain why the exponents cannot be added in the product $14^3 \cdot 18^3$.

2. List two ways to express 4^5 as a product of powers.

Additional Examples

Example 1

Multiply. Write the product as one power.

A. $6^6 \cdot 6^3$
6^9

B. $n^5 \cdot n^7$
n^{12}

C. $2^5 \cdot 2$
2^6

D. $24^4 \cdot 24^4$
24^8

Example 2

Divide. Write the quotient as one power.

A. $\dfrac{7^5}{7^3}$
7^2

B. $\dfrac{x^{10}}{x^9}$
x^1 or x

Example 3

A light-year, or the distance light travels in one year, is almost 10^{18} centimeters. To convert this number to kilometers, you must divide by 10^5. How many kilometers is a light-year?
A light-year is almost 10^{13} km.

3 Close

Reaching All Learners
Through Critical Thinking

To reinforce the concepts that $x = x^1$ and $x^0 = 1$ for $x \neq 0$, ask students to simplify the following expressions using both the expansion method and the properties of exponents: $4 \cdot 4^2$ and $4^0 \cdot 4^2$. Have students compare the results.

Possible answers:
$4 \cdot 4^2 = (4)(4 \cdot 4) = 4^3$
$4 \cdot 4^2 = 4^1 \cdot 4^2 = 4^{1+2} = 4^3$
$4^0 \cdot 4^2 = 4^{0+2} = 4^2$
$4^0 \cdot 4^2 = (1)(4 \cdot 4) = 4^2$

Summarize

Review the rules by showing an example for each and then showing the same example in expanded form. Remind students that if they have trouble remembering the rules, they can use the expansion method to relearn the rules. Ask students which method would be better for solving a problem with large exponents, such as $\dfrac{7^{23}}{7^{19}}$, and have them explain their answers.

Possible answer: The rules would be better because the expansion would take a lot of time and space to work out.

Answers to Think and Discuss

Possible answers:

1. The exponents cannot be added because the bases are not the same. However, the bases could be multiplied together under the same exponent, for example, $(14 \cdot 18)^3$.

2. $4^5 \cdot 4^0$, $4^3 \cdot 4^2$, $2^5 \cdot 2^5$, $2^8 \cdot 2^2$

2-7 Exercises

FOR EXTRA PRACTICE
see page 735

internet connect
Homework Help Online
go.hrw.com Keyword: MP4 2-7

Students may want to refer back to the lesson examples.

Assignment Guide

If you finished Example **1** assign:
Core 1–4, 10–13, 23–26, 31–34, 51–59
Enriched 10–13, 23–26, 31–34, 41, 42, 46, 48, 51–59

If you finished Example **2** assign:
Core 1–8, 10–17, 19–41 odd, 51–59
Enriched 10–17, 23–42, 48, 51–59

If you finished Example **3** assign:
Core 1–35, 37–47 odd, 51–59
Enriched 10–59

Notes

GUIDED PRACTICE

See Example **1** Multiply. Write the product as one power.

1. $3^4 \cdot 3^7$ **3^{11}**   2. $12^3 \cdot 12^2$ **12^5**   3. $m \cdot m^5$ **m^6**   4. $14^5 \cdot 8^5$
cannot combine

See Example **2** Divide. Write the quotient as one power.

5. $\frac{8^7}{8^5}$ **8^2**   6. $\frac{a^9}{a^1}$ **a^8**   7. $\frac{12^5}{12^5}$ **$12^0 = 1$**   8. $\frac{7^{18}}{7^6}$ **7^{12}**

See Example **3** 9. A scientist estimates that a sweet corn plant produces 10^8 grains of pollen. If there are 10^{10} grains of pollen, how many plants are there?
10^2 plants

INDEPENDENT PRACTICE

See Example **1** Multiply. Write the product as one power.

10. $10^{10} \cdot 10^7$**10^{17}** 11. $2^3 \cdot 2^3$ **2^6**   12. $r^5 \cdot r^4$ **r^9**   13. $16 \cdot 16^3$ **16^4**

See Example **2** Divide. Write the quotient as one power.

14. $\frac{7^{12}}{7^8}$ **7^4**   15. $\frac{m^{10}}{d^3}$ **cannot combine**   16. $\frac{t^8}{t^5}$ **t^3**   17. $\frac{10^8}{10^8}$ **$10^0 = 1$**

See Example **3** 18. There are 8^2 small squares on a standard chessboard, but 8^3 small squares on a 3-D chessboard. How many times more squares are on the 3-D chessboard? **8^1, or 8 times as many**

PRACTICE AND PROBLEM SOLVING

Multiply or divide. Write the product or quotient as one power.

19. $\frac{6^8}{6^5}$ **6^3**   20. $7^9 \cdot 7^1$ **7^{10}**   21. $\frac{a^3}{a^2}$ **a**   22. $\frac{10^{18}}{10^9}$ **10^9**

23. $x^3 \cdot x^7$ **x^{10}**   24. $a^7 \cdot b^8$**cannot combine**   25. $6^4 \cdot 6^2$ **6^6**   26. $4 \cdot 4^2$ **4^3**

27. $\frac{12^5}{6^3}$ **cannot combine**   28. $\frac{11^7}{11^6}$ **11^1 or 11**   29. $\frac{y^9}{y^9}$ **$y^0 = 1$**   30. $\frac{2^9}{2^3}$ **2^6**

31. $x^5 \cdot x^3$ **x^8**   32. $c^9 \cdot d^3$**cannot combine**   33. $4^4 \cdot 4^2$ **4^6**   34. $9^2 \cdot 9^2$ **9^4**

35. $10^5 \cdot 10^9$ **10^{14}**   36. $\frac{k^6}{p^2}$ **cannot combine**   37. $n^8 \cdot n^8$ **n^{16}**   38. $\frac{9^{11}}{9^6}$ **9^5**

39. $4^9 \div 4^5$ **4^4**   40. $2^{12} \div 2^6$ **2^6**   41. $6^2 \cdot 6^3 \cdot 6^4$ **6^9**   42. $5^3 \cdot 5^6 \cdot 5^0$ **5^9**

43. There are 26^3 ways to make a 3-letter "word" (from *aaa* to *zzz*) and 26^5 ways to make a 5-letter word. How many times more ways are there to make a 5-letter word than a 3-letter word? **26^2, or 676**

44. 10^{27} tons

44. **ASTRONOMY** The mass of the known universe is about 10^{23} solar masses, which is 10^{50} metric tons. How many metric tons is one solar mass?

Math Background

Students may have trouble understanding the rule $a^0 = 1$, if $a \neq 0$. Recall that an exponent indicates the number of times that the base is used as a factor. An exponent of zero indicates that the base is not used as a factor at all. In this case, the only factor is 1. This approach can also be demonstrated in the following pattern:

$2^4 = 1 \cdot 2^4 = 1 \cdot 2 \cdot 2 \cdot 2 \cdot 2$
$2^3 = 1 \cdot 2^3 = 1 \cdot 2 \cdot 2 \cdot 2$
$2^2 = 1 \cdot 2^2 = 1 \cdot 2 \cdot 2$
$2^1 = 1 \cdot 2^1 = 1 \cdot 2$
$2^0 = 1 \cdot 2^0 = 1$

RETEACH 2-7

LESSON 2-7 Reteach
Properties of Exponents

To multiply powers with the same base, keep the base and add exponents.

$x^a \cdot x^b = x^{a+b}$
$4^5 \cdot 4^2 = 4^{5+2} = 4^7$
$8^3 \cdot 8 = 8^{3+1} = 8^4$

To divide powers with the same base, keep the base and subtract exponents.

$x^a \div x^b = x^{a-b}$
$4^5 \div 4^2 = 4^{5-2} = 4^3$
$8^3 \div 8 = 8^{3-1} = 8^2$

Any nonzero number raised to the zero power equals 1.

$x^0 = 1$ with $x \neq 0$
$17^0 = 1$

Complete to see why the rules for exponents work.

1. $4^5 \cdot 4^2 = (\underline{4})(\underline{4})(\underline{4})(\underline{4})(\underline{4}) \cdot (\underline{4})(\underline{4}) = 4^{\underline{7}}$

2. $8^3 \cdot 8 = (\underline{8})(\underline{8})(\underline{8}) \cdot (\underline{8}) = 8^{\underline{4}}$

3. $4^5 \div 4^2 = \frac{4^5}{4^2} = \frac{4 \cdot 4 \cdot 4 \cdot 4 \cdot 4}{4 \cdot 4} = 4^{\underline{3}}$

4. $8^3 \div 8 = \frac{8^3}{8} = \frac{8 \cdot 8 \cdot 8}{8} = 8^{\underline{2}}$

5. $\frac{6^3}{6^3} = 6^{3-3} = 6^{\underline{0}}$ Also, $\frac{6^3}{6^3} = \frac{6 \cdot 6 \cdot 6}{6 \cdot 6 \cdot 6} = \underline{1}$ So, $6^0 = \underline{1}$

Complete to write each product or quotient as one power.

6. $12^3 \cdot 12^2 = 12^{3+2} = 12^{\underline{5}}$ 7. $9^4 \cdot 9^3 = 9^{4+3} = 9^{\underline{7}}$

8. $\frac{7^6}{7^2} = 7^{6-2} = 7^{\underline{4}}$ 9. $\frac{12^6}{12^4} = 12^{6-4} = 12^{\underline{2}}$

Write each product or quotient as one power.

10. $10^4 \cdot 10^6 = \underline{10^{10}}$ 11. $5^5 \cdot 5 = \underline{5^6}$ 12. $4^5 \cdot 4 \cdot 4^3 = \underline{4^9}$

13. $\frac{15^6}{15^2} = \underline{15^4}$ 14. $\frac{9^9}{9} = \underline{9^4}$ 15. $\frac{2^{10}}{2^2} = \underline{2^8}$

Complete to evaluate each expression.

16. $2 \cdot 5^0 = 2 \cdot \underline{1} = \underline{2}$ 17. $6 + 8^0 = 6 + \underline{1} = \underline{7}$ 18. $(5 + 1)^0 = (\underline{6})^0 = \underline{1}$

PRACTICE 2-7

LESSON 2-7 Practice B
Properties of Exponents

Multiply. Write the product as one power.

1. $10^5 \cdot 10^7$ 2. $x^9 \cdot x^8$ 3. $14^7 \cdot 14^9$ 4. $12^6 \cdot 12^8$
$\underline{10^{12}}$ $\underline{x^{17}}$ $\underline{14^{16}}$ $\underline{12^{14}}$

5. $y^{12} \cdot y^{10}$ 6. $15^9 \cdot 15^{14}$ 7. $(-11)^{20} \cdot (-11)^{10}$ 8. $(-a)^6 \cdot (-a)^7$
$\underline{y^{22}}$ $\underline{15^{23}}$ $\underline{(-11)^{30}}$ $\underline{(-a)^{13}}$

Divide. Write the quotient as one power.

9. $\frac{12^9}{12^2}$ 10. $\frac{(-11)^{12}}{(-11)^8}$ 11. $\frac{x^{10}}{x^5}$ 12. $\frac{16^{10}}{16^2}$
$\underline{12^7}$ $\underline{(-11)^4}$ $\underline{x^5}$ $\underline{16^8}$

13. $\frac{17^{19}}{17^2}$ 14. $\frac{14^{16}}{14^{13}}$ 15. $\frac{23^{17}}{23^9}$ 16. $\frac{(-a)^{12}}{(-a)^7}$
$\underline{17^{17}}$ $\underline{14^2}$ $\underline{23^8}$ $\underline{(-a)^5}$

Write the product or quotient as one power.

17. $\frac{22^8}{22^2}$ 18. $d^8 \cdot d^5$ 19. $(-15)^5 \cdot (-15)^{10}$
$\underline{22^6}$ $\underline{d^{13}}$ $\underline{(-15)^{15}}$

20. $\frac{w^{12}}{w^3}$ 21. $31^{16} \cdot 31^8$ 22. $\frac{25^{20}}{25^{10}}$
$\underline{w^9}$ $\underline{31^{24}}$ $\underline{25^{10}}$

23. $(-x)^{18} \cdot (-x)^9$ 24. $\frac{r^{18}}{r^7}$
$\underline{(-x)^{27}}$ $\underline{r^{11}}$

25. Jefferson High School has a student body of 6^4 students. Each class has approximately 6^2 students. How many classes does the school have? Write the answer as one power.
$\underline{6^2}$

26. Write the expression for a number used as a factor fifteen times being multiplied by the same number used as a factor ten times. Then, write the product as one power.
$\underline{x^{15} \cdot x^{10} = x^{25}}$

45. **BUSINESS** Using the manufacturing terms below, tell how many dozen are in a great gross. How many gross are in a great gross? **12^2; 12^1**

1 dozen	$= 12^1$ items
1 gross	$= 12^2$ items
1 great gross	$= 12^3$ items

46. A googol is the number 1 followed by 100 zeros.
   a. What is a googol written as a power? **10^{100}**
   b. What is a googol times a googol written as a power? **10^{200}**

Peanuts © Charles Schulz. Dist. by Universal Press Syndicate. Reprinted with Permission. All rights reserved.

47. **ASTRONOMY** The distance from Earth to the moon is about 22^4 miles. The distance from Earth to Neptune is about 22^7 miles. How many one-way trips from Earth to the moon are about equal to one trip from Earth to Neptune? **22^3 trips**

 48. **WHAT'S THE ERROR?** A student said that $\frac{4^7}{8^7}$ is the same as $\frac{1}{2}$. What mistake has the student made?

 49. **WRITE ABOUT IT** Why do you add exponents when multiplying powers with the same base?

50. **CHALLENGE** A number to the 10th power divided by the same number to the 7th power equals 125. What is the number? **5**

Spiral Review

Evaluate each expression for $m = -3$. (Lesson 2-1)
51. $m + 6$ **3**   52. $m + -5$ **−8**   53. $-9 + m$ **−12**   54. $m + 3$ **0**

Subtract. (Lesson 2-2)
55. $-8 - 8$ **−16**   56. $-3 - (-7)$ **4**   57. $-10 - 2$ **−12**   58. $11 - (-9)$ **20**

59. **TEST PREP** Which is **not** a solution to $-3x > 15$? (Lesson 2-5) **D**

A -20 B -100 C -6 D -5

Answers
48. Possible answer: The student did not consider the exponents.
49. Possible answer: When multiplying powers with the same base, you add exponents because the sum of the exponents is the total number of factors in the product.

Journal

Remind students of the physical science application example (Example 3) in the lesson. Ask students to write about a topic in science or any other subject that seems likely to make use of large numbers.

Test Prep Doctor

In Exercise 59, the term **not** may be especially troublesome in a question about an inequality with a negative number. Encourage students to be sure to find the three choices that *are* solutions. This is a way to verify that the remaining choice is not a solution.

CHALLENGE 2-7

Challenge
2-7 Square Dance

Study these patterns.

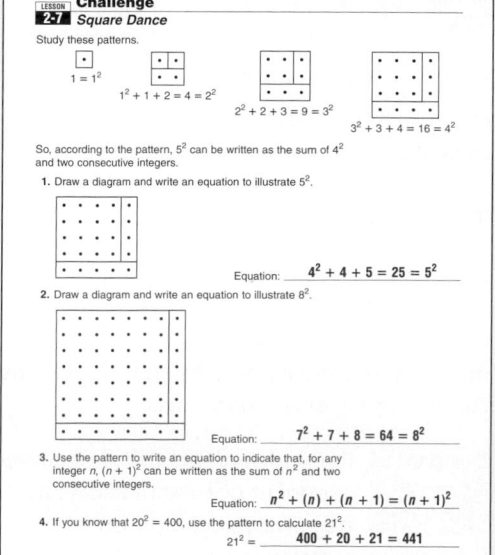

$1 = 1^2$
$1^2 + 1 + 2 = 4 = 2^2$
$2^2 + 2 + 3 = 9 = 3^2$
$3^2 + 3 + 4 = 16 = 4^2$

So, according to the pattern, 5^2 can be written as the sum of 4^2 and two consecutive integers.

1. Draw a diagram and write an equation to illustrate 5^2.

Equation: **$4^2 + 4 + 5 = 25 = 5^2$**

2. Draw a diagram and write an equation to illustrate 8^2.

Equation: **$7^2 + 7 + 8 = 64 = 8^2$**

3. Use the pattern to write an equation to indicate that, for any integer n, $(n + 1)^2$ can be written as the sum of n^2 and two consecutive integers.

Equation: **$n^2 + (n) + (n + 1) = (n + 1)^2$**

4. If you know that $20^2 = 400$, use the pattern to calculate 21^2.

$21^2 =$ **$400 + 20 + 21 = 441$**

PROBLEM SOLVING 2-7

Problem Solving
2-7 Properties of Exponents

Write each answer as a power.

1. Cindy separated her fruit flies into equal groups. She estimates that there are 2^{10} fruit flies in each of 2^2 jars. How many fruit flies does Cindy have in all?

2^{12} fruit flies

2. Suppose a researcher tests a new method of pasteurization on a strain of bacteria in his laboratory. If the bacteria are killed at a rate of 8^9 per sec, how many bacteria would be killed after 8^2 sec?

8^{11} bacteria

3. A satellite orbits the earth at about 13^4 km per hour. How long would it take to complete 24 orbits, which is a distance of about 13^5 km?

13 hr

4. A scientist estimates that a certain plant produces 10^7 grains of pollen. If there are 10^4 plants, how many grains of pollen do they produce?

10^{11} grains of pollen

Use the table to answer Exercises 5–6. The table describes the number of people involved at each level of a pyramid scheme. In a pyramid scheme each individual recruits so many others to participate who in turn recruit others, and so on. Choose the letter of the best answer.

5. Using exponents, how many people will be involved at level 6?
   A 6^6 C 5^5
   B 6^5 Ⓓ 5^6

6. How many times more people will be involved at level 6 than at level 2?
   Ⓕ 5^4 H 5^5
   G 5^3 J 5^6

Pyramid Scheme Each person recruits 5 others.	
Level	Total Number of People
1	5
2	5^2
3	5^3
4	5^4

7. There are 10^3 ways to make a 3-digit combination, but there are 10^6 ways to make a 6-digit combination. How many times more ways are there to make 6-digit combination than a 3-digit combination?
   A 5^{10} C 2^6
   B 2^{10} Ⓓ 10^3

Lesson Quiz

Write the product or quotient as one power.

1. $n^3 \times n^4$ n^7
2. $8 \cdot 8^8$ 8^9
3. $\dfrac{10^9}{10^5}$ 10^4
4. $\dfrac{t^9}{t^7}$ t^2
5. $3^2 \cdot 3^3 \cdot 3^5$ 3^{10}

6. A school would like to purchase new globes. They can get six dozen for $705.80 from Company A. From Company B, they can buy a half gross for $725.10. Which company should they buy from? (1 gross = 12^2 items) Company A

Available on Daily Transparency in CRB

Warm Up

Evaluate.

1. 10^3 **1000**

2. 10^0 **1**

3. $10^2 \cdot 10^2$ **10,000**

4. $\dfrac{10^7}{10^4}$ **1000**

5. $\dfrac{10^6}{10^6}$ **1**

Problem of the Day

Find two different numbers for the values of x and y that will make x^y and y^x equal.   **2 and 4**

Available on Daily Transparency in CRB

Math Humor

Why did the integer get a bad evaluation at work? He had a negative attitude.

2-8 Look for a Pattern in Integer Exponents

 Problem Solving Skill

Learn to evaluate expressions with negative exponents.

The nanoguitar is the smallest guitar in the world. It is no larger than a single cell, at about 10^{-5} meters long. Can you imagine 10^{-5} meters?

Look for a pattern in the table to extend what you know about exponents to include negative exponents. Start with what you know about positive and zero exponents.

The nanoguitar is carved from crystalline silicon. It has 6 strings that are each about 100 atoms wide.

10^2	10^1	10^0	10^{-1}	10^{-2}
$10 \cdot 10$	10	1	$\frac{1}{10}$	$\frac{1}{10 \cdot 10}$
100	10	1	$\frac{1}{10} = 0.1$	$\frac{1}{100} = 0.01$

$\div 10 \qquad \div 10 \qquad \div 10 \qquad \div 10$

EXAMPLE 1   **Using a Pattern to Evaluate Negative Exponents**

Evaluate the powers of 10.

A 10^{-3}

$10^{-3} = \dfrac{1}{10 \cdot 10 \cdot 10}$ *Extend the pattern from the table.*

$10^{-3} = \dfrac{1}{1000} = 0.001$

B 10^{-4}

$10^{-4} = \dfrac{1}{10 \cdot 10 \cdot 10 \cdot 10}$ *Extend the pattern from Example 1A.*

$10^{-4} = \dfrac{1}{10,000} = 0.0001$

C 10^{-5}

$10^{-5} = \dfrac{1}{10 \cdot 10 \cdot 10 \cdot 10 \cdot 10}$ *Extend the pattern from Example 1B.*

$10^{-5} = \dfrac{1}{100,000} = 0.00001$

So how long is 10^{-5} meters?

10^{-5} m $= \dfrac{1}{100,000}$ m $\longrightarrow$ "one hundred-thousandth of a meter"

---

1 Introduce

Alternate Opener

EXPLORATION

2-8 Look for Patterns in Integer Exponents

Suppose the height of a magic plant doubles every hour, beginning with a height of 1 inch at the "zero hour."

Use the bar graph to think about how tall the plant was 1 hour before (-1) and 2 hours before (-2) the zero hour.

1. Draw and label a bar on the bar graph to show the height at hour -1 and hour -2. (*Hint:* The height doubles each hour.)

2. Use the graph and the pattern in the table to find the value of each exponential expression.

Time, x	Plant Height, 2^x
-3	$2^{-3} = \frac{1}{2^3} = \frac{1}{8}$
-2	
-1	
0	
1	
2	
3	

Think and Discuss
3. **Describe** the pattern in the bar graph.
4. **Describe** the pattern in the table.

Motivate

Prior to beginning this lesson, ask students to evaluate the following expressions:

A. 4^5 **1024**

B. 5^0 **1**

C. 7^1 **7**

After they have evaluated them, ask them to describe in their own words how they got their answers.

Exploration worksheet and answers on Chapter 2 Resource Book pp. 72 and 112

2 Teach

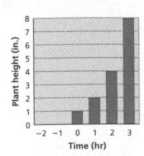 **Lesson Presentation**

Guided Instruction

In this lesson, students learn to evaluate expressions with negative exponents. Begin by using the pattern shown in the table (Teaching Transparency T36, CRB) to provide a meaningful definition of an expression with a negative exponent (Teaching Transparency T37, CRB). Then give students an opportunity to apply the meaning to evaluate powers and expressions containing negative exponents.

 **Teaching Tip** Remind students that a negative exponent does not indicate a negative value, but instead a fractional expression.

NEGATIVE EXPONENTS		
Words	**Numbers**	**Algebra**
A power with a negative exponent equals 1 divided by that power with its opposite exponent.	$5^{-3} = \frac{1}{5^3} = \frac{1}{125}$	$b^{-n} = \frac{1}{b^n}$

EXAMPLE **2** **Evaluating Negative Exponents**

Evaluate $(-2)^{-3}$.

$(-2)^{-3}$

$\frac{1}{(-2)^3}$ *Write the reciprocal; change the sign of the exponent.*

$\frac{1}{(-2)(-2)(-2)}$

$-\frac{1}{8}$

Remember!

The reciprocal of a number is 1 divided by that number.

EXAMPLE **3** **Evaluating Products and Quotients of Negative Exponents**

Evaluate.

A $10^3 \cdot 10^{-3}$

$10^3 \cdot 10^{-3}$

$10^{3 + (-3)}$ *Bases are the same, so add the exponents.*

$10^0 = 1$ *Check* $10^3 \cdot 10^{-3} = 10^3 \cdot \frac{1}{10^3} = \frac{10^3}{10^3} = \frac{10 \cdot 10 \cdot 10}{10 \cdot 10 \cdot 10} = 1$

B $\frac{2^4}{2^7}$

$\frac{2^4}{2^7}$

2^{4-7} *Bases are the same, so subtract the exponents.*

2^{-3}

$\frac{1}{2^3}$ *Write the reciprocal; change the sign of the exponent.*

$\frac{1}{8}$ *Check* $\frac{2^4}{2^7} = \frac{2 \cdot 2 \cdot 2 \cdot 2}{2 \cdot 2 \cdot 2 \cdot 2 \cdot 2 \cdot 2 \cdot 2} = \frac{1}{8}$

Think and Discuss

1. Express $\frac{1}{2}$ using an exponent.

2. Tell whether the statement is true or false: If a power has a negative exponent, then the power is negative. Justify your answer.

3. Tell whether an integer raised to a negative exponent can ever be greater than 1.

Additional Examples

Example **1**

Evaluate the powers of 10.

A. 10^{-2} $\frac{1}{100}$ or 0.01

B. 10^{-1} $\frac{1}{10}$ or 0.1

C. 10^{-6} $\frac{1}{1,000,000}$ or 0.000001

Example **2**

Evaluate.

A. 5^{-3} $\frac{1}{125}$

B. $(-10)^{-3}$ $-\frac{1}{1,000} = -0.001$

Example **3**

Evaluate.

A. $2^{-5} \cdot 2^3$ $\frac{1}{4}$

B. $\frac{6^5}{6^8}$ $\frac{1}{216}$

C. $3^{-4} \cdot 3^5$ 3

Example 3 note: For Example 3A, remind students that any nonzero number to the zero power is one.

3 Close

Reaching All Learners
Through Grouping Strategies

Have the class work in pairs. Show students a series of equivalent expressions, such as $\frac{5^2}{5^4} = 5^{-2} = \frac{1}{5^2} = \frac{1}{25}$. Explain that all of the expressions are equal. Have each student create two expressions involving exponents that can be simplified. Encourage students to use negative exponents. Then have the students in each pair exchange problems and solve each other's work.

Possible answers:

Simplify. **1.** $(-4)^{-2}$ **2.** $\frac{6^2}{6^5}$

Answers: **1.** $\frac{1}{16}$ **2.** $\frac{1}{216}$

Summarize

Have students describe how to evaluate an expression with a positive exponent and then how to evaluate an expression with a negative exponent. Ask students to provide examples of each.

Possible answers: To evaluate an expression with a positive exponent, multiply, using the base as a factor the number of times indicated by the exponent. To evaluate an expression with a negative exponent, place the expression in the denominator of a fraction with 1 as the numerator and replace the negative exponent with its opposite. Then evaluate the new expression.

Answers to Think and Discuss

Possible answers:

1. 2^{-1}

2. False. A negative exponent does not make a power negative. For example, 2^{-1} is not a negative number.

3. No, but 1^{-1} is 1.

FOR EXTRA PRACTICE
see page 735

✐ internet connect
Homework Help Online
go.hrw.com Keyword: MP4 2-8

go.hrw.com

Students may want to refer back to the lesson examples.

GUIDED PRACTICE

See Example **1** Evaluate the powers of 10.

1. 10^{-7} **0.0000001** 2. 10^{-3} **0.001** 3. 10^{-6} **0.000001** 4. 10^{-1} **0.1**

See Example **2** Evaluate.

5. $(-2)^{-4}$ $\frac{1}{16}$ 6. $(-3)^{-2}$ $\frac{1}{9}$ 7. 2^{-3} $\frac{1}{8}$ 8. $(-2)^{-5}$ $-\frac{1}{32}$

See Example **3** 9. $10^7 \cdot 10^{-4}$ **1000** 10. $3^5 \cdot 3^{-7}$ $\frac{1}{9}$ 11. $\frac{6^8}{6^5}$ **216** 12. $\frac{3^6}{3^9}$ $\frac{1}{27}$

INDEPENDENT PRACTICE

See Example **1** Evaluate the powers of 10.

13. 10^{-2} **0.01** 14. 10^{-9} **0.000000001** 15. 10^{-5} **0.00001** 16. 10^{-11} **0.00000000001**

See Example **2** Evaluate.

17. $(-4)^{-3}$ $-\frac{1}{64}$ 18. 3^{-2} $\frac{1}{9}$ 19. $(-10)^{-4}$ **0.0001** 20. $(-2)^{-1}$ $-\frac{1}{2}$

See Example **3** 21. $10^5 \cdot 10^{-1}$ **10,000** 22. $\frac{2^3}{2^5}$ $\frac{1}{4}$ 23. $\frac{5^2}{5^2}$ **1** 24. $\frac{3^7}{3^2}$ **243**

25. $\frac{2^1}{2^4}$ $\frac{1}{8}$ 26. $4^2 \cdot 4^{-3}$ $\frac{1}{4}$ 27. $10^3 \cdot 10^{-6}$ **0.001** 28. $6^4 \cdot 6^{-2}$ **36**

PRACTICE AND PROBLEM SOLVING

Evaluate.

29. 2^7 **128** 30. $\frac{5^7}{5^5}$ **25** 31. $\frac{m^9}{m^2}$ m^7

32. $x^{-5} \cdot x^7$ x^2 33. $\frac{(-3)^2}{(-3)^4}$ $\frac{1}{9}$ 34. $8^4 \cdot 8^{-4}$ **1**

35. $4^9 \cdot 4^{-4}$ **1024** 36. $\frac{7^2}{8^6}$ **cannot combine** 37. $2^{-2} \cdot 2^{-2} \cdot 2^3$ $\frac{1}{2}$

38. $\frac{(7-3)^3}{(5-1)^6}$ $\frac{1}{64}$ 39. $(5-3)^{-7} \cdot (7-5)^5$ $\frac{1}{4}$ 40. $\frac{(4-11)^5}{(1-8)^2}$ **−343**

41. $(2 \cdot 6)^{-5} \cdot (4 \cdot 3)^3$ $\frac{1}{144}$ 42. $\frac{(3+2)^4}{5(7-2)^3}$ **1** 43. $(2+2)^{-5} \cdot (1+3)^6$ **4**

44. **COMPUTER SCIENCE** Computer files are measured in bytes. One byte contains approximately 1 character of text.

	Byte	Kilobyte (KB)	Megabyte (MB)	Gigabyte (GB)
Value (bytes)	$2^0 = 1$	2^{10}	2^{20}	2^{30}

a. If a hard drive on a computer holds 2^{35} bytes of data, how many gigabytes does the hard drive hold?

b. A Zip® disk holds about 2^8 MB of data. How many bytes is that?

Assignment Guide

If you finished Example **1** assign:
Core 1–4, 13–16, 50–54
Enriched 1–4, 13–16, 50–54

If you finished Example **2** assign:
Core 1–8, 13–20, 29, 50–54
Enriched 1–8, 13–20, 45, 50–54

If you finished Example **3** assign:
Core 1–28, 29–39 odd, 45, 47, 50–54
Enriched 13–54

Answers

44. a. $2^{35} \div 2^{30} = 2^{35-30} = 2^5$ GB
b. $2^8 \cdot 2^{20} = 2^{8+20} = 2^{28}$ bytes

Math Background

The pattern of powers of ten shown at the beginning of this lesson is an effective introduction to scientific notation, which will be investigated further in the next lesson.

Because scientists must use very large and small numbers frequently, they use powers of ten to keep the numbers simple. Positive exponents are used for very large numbers, such as those used in astronomy and physics. Negative exponents are used for very small numbers, such as those used in chemistry and biology.

RETEACH 2-8

LESSON **2-8** **Reteach**
Look for Patterns in Integer Exponents

To rewrite a negative exponent, move the power to the denominator of a unit fraction. $5^{-2} = \frac{1}{5^2}$

Complete to rewrite each power with a positive exponent.

1. $7^{-3} = \frac{1}{7^3}$ 2. $9^{-5} = \frac{1}{9^5}$ 3. $13^{-4} = \frac{1}{13^4}$

Complete each pattern.

4. $10^{-1} = \frac{1}{10} = 0.1$
$10^{-2} = \frac{1}{10^2} = \frac{1}{100} = 0.01$
$10^{-3} = \frac{1}{10^3} = \frac{1}{1000} = 0.001$

5. $5^{-1} = \frac{1}{5}$
$5^{-2} = \frac{1}{5^2} = \frac{1}{5 \cdot 5} = \frac{1}{25}$
$5^{-3} = \frac{1}{5^3} = \frac{1}{5 \cdot 5 \cdot 5} = \frac{1}{125}$

6. $3^{-1} = \frac{1}{3}$
$3^{-2} = \frac{1}{3^2} = \frac{1}{3 \cdot 3} = \frac{1}{9}$
$3^{-3} = \frac{1}{3^3} = \frac{1}{3 \cdot 3 \cdot 3} = \frac{1}{27}$

7. $4^{-3} = \frac{1}{4^3} = \frac{1}{4 \cdot 4 \cdot 4} = \frac{1}{64}$
$4^{-4} = \frac{1}{4^4} = \frac{1}{4 \cdot 4 \cdot 4 \cdot 4} = \frac{1}{256}$

Complete to evaluate.

8. $4^5 \cdot 4^{-2} = 4^{\underline{3}} = \underline{64}$
9. $9^4 \cdot 9^{-2} = 9^{\underline{2}} = \underline{81}$
10. $10^{-3} \cdot 10^6 = 10^{\underline{3}} = \underline{1000}$
11. $5^{-3} \cdot 5^3 = 5^{\underline{0}} = \underline{1}$
12. $7^2 \cdot 7^{-2} = 7^{\underline{0}} = \underline{1}$
13. $8^3 \cdot 8^{-3} \cdot 8 = 8^{\underline{1}} = \underline{8}$

14. $(2 \cdot 12)^{-3} \cdot (8 \cdot 3)^3$
$(\underline{24})^{-3} \cdot (\underline{24})^3$
$= \underline{24^0}$
$= \underline{1}$

15. $(6 \cdot 2)^{-4} \cdot (3 \cdot 4)^2$
$(\underline{12})^{-4} \cdot (\underline{12})^2$
$= \underline{12^{-2}}$
$= \frac{1}{144}$

16. $(25 \cdot 4)^3 \cdot (50 \cdot 2)^{-5}$
$(\underline{100})^3 \cdot (\underline{100})^{-5}$
$= \underline{100^{-2}}$
$= \underline{0.0001}$

PRACTICE 2-8

LESSON **2-8** **Practice B**
Look for Patterns in Integer Exponents

Evaluate the powers of 10.

1. 10^{-3} **0.001** 2. 10^3 **1000** 3. 10^{-5} **0.00001** 4. 10^{-2} **0.01**

5. 10^0 **1** 6. 10^4 **10,000** 7. 10^1 **10** 8. 10^5 **100,000**

Evaluate.

9. $(-6)^{-2}$ $\frac{1}{36}$ 10. $\frac{12^4}{12^6}$ $\frac{1}{144}$ 11. $\frac{x^7}{x^{12}}$ $\frac{1}{x^5}$

12. $(-9)^{-3}$ $-\frac{1}{729}$ 13. $8^{-4} \cdot 8^{-1}$ $\frac{1}{32,768}$ 14. $\frac{4^7}{4^{11}}$ $\frac{1}{256}$

15. $12^{-3} \cdot 12^6$ **1728** 16. $\frac{17^{10}}{17^{12}}$ $\frac{1}{289}$ 17. $\frac{(-13)^2}{(-13)^5}$ $-\frac{1}{2197}$

18. $(-15)^{-2} \cdot (-15)^{-1}$ $-\frac{1}{3375}$ 19. $\frac{20^4}{20^6}$ $\frac{1}{400}$ 20. $(18)^{-2} \cdot (18)^{-1}$ $\frac{1}{5832}$

Express the answer using powers of 10 and negative numbers.

21. 1 milliliter $= \frac{1}{1000}$ liter $= \underline{10^{-3}}$ liters.

22. Find the volume of a cube with a side that measures 0.1 cm. (Hint: $V = s^3$.)
$\underline{10^{-3}}$ cm^3

Prefixes for the International System of Units

Factor	10^3	10^2	10^1	10^{-1}	10^{-2}	10^{-3}	10^{-6}	10^{-9}	10^{-12}	10^{-15}
Prefix	kilo-	hecto-	deca-	deci-	centi-	milli-	micro-	nano-	pico-	femto-
Symbol	k	h	da	d	c	m	μ	n	p	f

45. The sperm whale is the deepest diving whale. It can dive to depths greater than 10^{12} nanometers. How many kilometers is that?

46. The greatest known depth of the Arctic ocean is about 10^6 millimeters. How many hectometers is that?

47. The primary food in the blue whale's diet is a crustacean called a krill. One krill weighs approximately 10^{-5} kg.

   a. How many grams does one krill weigh?

   b. If a blue whale eats 10^7 krill, how many grams of krill is that?

   c. How many decagrams do 10^7 krill weigh?

48. The tropical brittlestar is a sea creature that lives in the coral reef. It is covered with 20,000 crystal eyes that are each about 100 micrometers wide.

   a. How many meters wide is one crystal eye?

   b. How long would a row of 10^5 crystal eyes be in meters?

49. ⭐ **CHALLENGE** A cubic centimeter is the same as 1 mL. If a humpback whale has more than 1 kL of blood, how many cubic centimeters of blood does the humpback whale have?

Krill may be up to 2 in. long, nearly $\frac{1}{288}$ of the length of a humpback whale, pictured above.

Science

Exercises 45–49 involve using integer exponents to describe characteristics of some sea creatures. Integer exponents are used throughout middle school science programs such as *Holt Science & Technology*.

Answers

45. 1 kilometer

46. 10 hectometers

47. a. $10^{-5} \cdot 10^3 = 10^{-2}$ g
   b. $10^{-2} \cdot 10^7 = 10^5$ g
   c. $10^5 \div 10^1 = 10^{5-1} = 10^4$; 10^4 decagrams

48. a. $10^2 \cdot 10^{-6} = 10^{-4}$; 10^{-4} m
   b. $10^5 \cdot 10^{-4} = 10$; 10 m

49. 1 kL $= 10^3$ L $= 10^6$ mL $= 10^6$ cm^3; A humpback whale has more than 10^6 cm^3 of blood.

Journal

Have students write about something that is very small. Have them describe the measurement using negative exponents. Examples do not need to be factual but should be realistic. For example: The flea was 2^{-4} inches long.

Test Prep Doctor

For Exercise 54, encourage students to eliminate the negative choices first. Students can then choose from answers **B** and **C** by comparing them with 1.00000. A number line may be useful for the comparison.

Spiral Review

Evaluate each expression for the given values of the variables. (Lesson 1-1)

50. $2x - 3y$ for $x = 8$ and $y = 4$ **4**

51. $6s - t$ for $s = 7$ and $t = 12$ **30**

52. $7w + 2z$ for $w = 3$ and $z = 0$ **21**

53. $5x + 4y$ for $x = 9$ and $y = 10$ **85**

54. TEST PREP Which of the following numbers is greater than 1? (previous course) **B**

   A -235 **B** 1.000008   **C** 0.99999   **D** -5.88

CHALLENGE 2-8

LESSON 2-8 Challenge
Stuff It!

$9^{\frac{1}{2}}$ means $\sqrt[2]{9^1}$.

   To find the value, first evaluate the root: $\sqrt[2]{9} = 3$.
   Then, raise the result to the indicated power: $3^1 = 3$.
   So, $9^{\frac{1}{2}}$ is $\sqrt[2]{9^1} = 3^1 = 3$.

In general, here's the way to rewrite a term with a fractional exponent:

$$x^{\frac{a}{b}} = \sqrt[b]{x^a}$$

Evaluate $8^{\frac{2}{3}}$.

   $8^{\frac{2}{3}} = \sqrt[3]{8^2}$ Rewrite using radical form.
   $= 2^2$ Evaluate the root; $\sqrt[3]{8} = 2$ since $2 \cdot 2 \cdot 2 = 8$.
   $= 4$ Evaluate the power.

Rewrite each term using radical form. Evaluate the root. Evaluate the power.

1. $64^{\frac{1}{2}} = \underline{\sqrt[2]{64^1}}$
   $= \underline{8^1}$
   $= \underline{8}$

2. $100^{\frac{1}{2}} = \underline{\sqrt[2]{100^1}}$
   $= \underline{10^1}$
   $= \underline{10}$

3. $400^{\frac{1}{2}} = \underline{\sqrt[2]{400^1}}$
   $= \underline{20^1}$
   $= \underline{20}$

4. $64^{\frac{2}{3}} = \underline{\sqrt[3]{64^2}}$
   $= \underline{4^2}$
   $= \underline{16}$

5. $216^{\frac{2}{3}} = \underline{\sqrt[3]{216^2}}$
   $= \underline{6^2}$
   $= \underline{36}$

6. $1000^{\frac{2}{3}} = \underline{\sqrt[3]{1000^2}}$
   $= \underline{10^2}$
   $= \underline{100}$

7. $625^{\frac{3}{4}} = \underline{\sqrt[4]{625^3}}$
   $= \underline{5^3}$
   $= \underline{125}$

8. $32^{\frac{2}{5}} = \underline{\sqrt[5]{32^2}}$
   $= \underline{2^2}$
   $= \underline{4}$

9. $10,000^{\frac{1}{4}} = \underline{\sqrt[4]{10,000^5}}$
   $= \underline{10^5}$
   $= \underline{100,000}$

PROBLEM SOLVING 2-8

LESSON 2-8 Problem Solving
Look for Patterns in Integer Exponents

Write the correct answer.

1. The weight of 10^7 dust particles is 1 gram. Evaluate 10^7.

   10,000,000

2. If the weight of 10^7 dust particles is 1 gram, what is the weight written as a power of 10 of one dust particle?

   10^{-7} g

3. As of 2001, only 10^6 rural homes in the United States had broadband Internet access. Evaluate 10^6.

   1,000,000

4. Atomic clocks measure time in microseconds. A microsecond is 0.000001 second. Write this number using a power of 10.

   10^{-6} sec

Use the table to answer Exercises 6–7. Choose the best answer.

Prefixes for the Interational System of Units

Factor	10^3	10^2	10^1	10^{-1}	10^{-2}	10^{-3}	10^{-6}	10^{-9}	10^{-12}	10^{-15}
Prefix	kilo-	hecto-	deka-	deci-	centi-	milli-	micro-	nano-	pico-	femto-
Symbol	l	h	da	d	c	m	μ	n	p	f

5. The diameter of the nucleus of an atom is about 10^{-15} m. Evaluate 10^{-15}.

   A 0.0000000000001
   B 0.00000000000001
   C 0.000000000000001
   (D) 0.0000000000000001

6. The diameter of an atom is about 10^{-11} m. How many nanometers is the diameter of an atom?

   F 10^{-1}
   (G) 10^{-2}
   H 10^{-9}
   J 10^{-20}

7. The diameter of the nucleus of an atom is about 10^{-15} m. How many nanometers is the diameter of the nucleus of an atom?

   (A) 10^{-6}
   B 10^{-9}
   C 10^{-15}
   D 10^{-24}

8. How many times larger is the diameter of an atom (10^{-11}) than the diameter of the nucleus of the atom (10^{-15})?

   F 100
   G 1,000
   (H) 10,000
   J 100,000

Lesson Quiz

Evaluate the powers of 10.

1. 10^{-3} 0.001

2. 10^{-7} 0.0000001

Evaluate.

3. $(-6)^{-2}$ $\frac{1}{36}$

4. $7^4 \cdot 7^{-4}$ 1

5. $\dfrac{9^2}{9^5}$ $\frac{1}{729}$

6. In engineering notation, a *tera* is equal to 10^{12}, and a *mega* is equal to 10^6. How many megas are equal to a tera? 10^6

Available on Daily Transparency in CRB

Objective: Students express large and small numbers in scientific notation.

Warm Up

Order each set of numbers from least to greatest.

1. 10^4, 10^{-2}, 10^0, 10^{-1}
10^{-2}, 10^{-1}, 10^0, 10^4

2. 8^2, 8^{-2}, 8^3, 8^0
8^{-2}, 8^0, 8^2, 8^3

3. 2^3, 2^{-6}, 2^{-4}, 2^1
2^{-6}, 2^{-4}, 2^1, 2^3

4. 5.2^2, 5.2^9, 5.2^{-1}, 5.2^{-2}
5.2^{-2}, 5.2^{-1}, 5.2^2, 5.2^9

Problem of the Day

Order the powers from least to greatest: 2^6, 6^2, 36^0, 16^1.
36^0, 16^1, 6^2, 2^6

Available on Daily Transparency in CRB

Math Fact

The mass of Earth is about 5,980,000,000,000,000,000,000,000 kilograms. This can be rewritten as 5.98×10^{24} kg.

2-9 Scientific Notation

Learn to express large and small numbers in scientific notation.

Vocabulary
scientific notation

An ordinary penny contains about 20,000,000,000,000,000,000,000 atoms. The average size of an atom is about 0.00000003 centimeter across.

The length of these numbers in standard notation makes them awkward to work with.

Scientific notation is a shorthand way of writing such numbers.

$$1.8 \times 10^4$$

In scientific notation the number of atoms in a penny is 2.0×10^{22}, and the size of each atom is 3.0×10^{-8} centimeters across.

EXAMPLE 1 Translating Scientific Notation to Standard Notation

Write each number in standard notation.

Helpful Hint

The sign of the exponent tells which direction to move the decimal. A positive exponent means move the decimal to the right, and a negative exponent means move the decimal to the left.

A 2.64×10^7
2.64×10^7
$2.64 \times 10,000,000$ $10^7 = 10,000,000$
$26,400,000$ *Think: Move the decimal right 7 places.*

B 1.35×10^{-4}
1.35×10^{-4}
$1.35 \times \frac{1}{10,000}$ $10^{-4} = \frac{1}{10,000}$
$1.35 \div 10,000$ *Divide by the reciprocal.*
0.000135 *Think: Move the decimal left 4 places.*

C -5.8×10^6
-5.8×10^6
$-5.8 \times 1,000,000$ $10^6 = 1,000,000$
$-5,800,000$ *Think: Move the decimal right 6 places.*

---

1 Introduce
Alternate Opener

EXPLORATION

2-9 Scientific Notation

1. Complete the table of values for the powers of ten.

Exponent	Power
−6	$10^{-6} =$
−5	$10^{-5} =$
−4	$10^{-4} =$
−3	$10^{-3} =$
−2	$10^{-2} = \frac{1}{10^2} = \frac{1}{10 \times 10} = 0.01$
−1	$10^{-1} = \frac{1}{10^1} = \frac{1}{10} = 0.1$
0	$10^0 = 1$
1	$10^1 = 10$
2	$10^2 = 10 \times 10 = 100$
3	$10^3 =$
4	$10^4 =$
5	$10^5 =$
6	$10^6 =$

Think and Discuss

2. Discuss the pattern you see in the table of values for powers of ten.
3. Explain how you know that $10^{-9} = 0.000000001$.

Motivate

Ask students how they would simplify the following expressions: 2×10^1, 3.5×10^4, and 3.5×10^{15}. Show them how to multiply by a power of ten by moving the decimal point to the right the number of spaces equal to the exponent and then adding zeros. Show some examples and explain that this lesson is about using powers of ten to simplify working with large and small numbers.

Exploration worksheet and answers on Chapter 2 Resource Book pp. 81 and 114

2 Teach

Lesson Presentation

Guided Instruction

In this lesson, students learn to express large and small numbers in scientific notation. First, show students how to convert scientific notation to standard notation. Review how to multiply by a power of ten by moving the decimal point to the left or right. Move the decimal point to the right for a positive exponent and to the left for a negative exponent.

Next, discuss how to write numbers in standard notation in scientific notation. Show students how to move the decimal so that there is only one digit in front of the decimal point. Then write the correct power of ten by counting the number of decimal spaces moved.

EXAMPLE 2 **Translating Standard Notation to Scientific Notation**

Write 0.000002 in scientific notation.

0.000002

2   *Move the decimal to get a number between 1 and 10.*

$2 \times 10^{\blacksquare}$ *Set up scientific notation.*

            Think: The decimal needs to move left to change 2 to 0.000002, so the exponent will be negative.

            Think: The decimal needs to move 6 places.

So 0.000002 written in scientific notation is 2×10^{-6}.

Check $2 \times 10^{-6} = 2 \times 0.000001 = 0.000002$

EXAMPLE 3 **Money Application**

Suppose you have a million dollars in pennies. A penny is 1.55 mm thick. How tall would a stack of all your pennies be? Write the answer in scientific notation.

$1.00 = 100$ pennies, so $1,000,000 = 100,000,000$ pennies.

1.55 mm $\times$ 100,000,000   *Find the total height.*

155,000,000 mm

$1.55 \times 10^{\blacksquare}$ *Set up scientific notation.*

           Think: The decimal needs to move right to change 1.55 to 155,000,000, so the exponent will be positive.

           Think: The decimal needs to move 8 places.

In scientific notation the total height of one million dollars in stacked pennies is 1.55×10^8 mm. This is about 96 miles tall.

Think and Discuss

1. **Explain** the benefit of writing numbers in scientific notation.

2. **Describe** how to write 2.977×10^6 in standard notation.

3. **Determine** which measurement would be least helpful in scientific notation: size of bacteria, speed of a car, or number of stars.

COMMON ERROR ALERT

Students might be confused about the relationship between positive and negative exponents and moving the decimal left or right. Encourage students to remember that when written in scientific notation, a number smaller than 1 will have a negative exponent and a number larger than 1 will have a positive exponent.

Additional Examples

Example 1

Write each number in standard notation.

A. 1.35×10^5
135,000

B. 2.7×10^{-3}
0.0027

C. -2.01×10^4
$-20,100$

Example 2

Write 0.00709 in scientific notation.
7.09×10^{-3}

Example 3

A pencil is 18.7 cm long. If you were to lay 10,000 pencils end to end, how many millimeters long would they be? Write the answer in scientific notation.
The 10,000 pencils would be 1.87×10^6 mm long, laid end to end.

---

③ Close

Reaching All Learners
Through Home Connection

Have students work with an adult to find some real-world examples of very large or very small numbers. They may find these examples in newspapers, books, or magazines, or they may use an example from the adult's workplace. Have students record five numbers and write them in both standard notation and scientific notation.

Check students' work.

Summarize

Discuss with students the meaning and uses of scientific notation. Ask students how many zeros are in a million dollars. Then ask how they would write one million in scientific notation. Repeat the process with one billion dollars. Challenge the class to write one cent in dollars in scientific notation.

Possible answers: Scientific notation is a way to write very large or very small numbers without including a lot of zeros. Scientific notation can make it easier to compare numbers. One million has six zeros, so it can be written as 1×10^6. One billion has nine zeros, so it can be written as 1×10^9. One cent can be written as 1×10^{-2} dollars.

Answers to Think and Discuss

Possible answers:

1. It makes extremely large or extremely small numbers less awkward to work with.

2. The exponent is 6, so move the decimal six places. The sign of the exponent is positive, so move the decimal to the right.

3. The speed of a car, because it is usually a 1-, 2-, or 3-digit number that is easy to write in standard notation

2-9 Exercises

FOR EXTRA PRACTICE
see page 735

internet connect
Homework Help Online
go.hrw.com   Keyword: MP4 2-9

go.hrw.com

Assignment Guide

If you finished Example **1** assign:
Core 1–4, 10–13, 19–30, 45,
50–58
Enriched 10–13, 19–30, 45, 47–58

If you finished Example **2** assign:
Core 1–8, 10–17, 19–45 odd,
50–58
Enriched 10–17, 22–39, 47–58

If you finished Example **3** assign:
Core 1–38, 43, 45, 50–58
Enriched 10–58

Answers

39. 9×10^2

40. 5×10^{-6}

41. 6×10^6

42. 9.5678×10^{-3}

Students may want to refer back to the lesson examples.

GUIDED PRACTICE

See Example **1**   Write each number in standard notation.

1. 3.15×10^3 **2.** 1.25×10^{-7} **3.** 4.1×10^5 **4.** 3.9×10^{-4}
  3150   **0.000000125**   **410,000**   **0.00039**

See Example **2**   Write each number in scientific notation. 4.89×10^6 1.4×10^{-7}
5. 0.000057   **6.** 0.0003   **7.** 4,890,000   **8.** 0.00000014
  5.7×10^{-5} 3×10^{-4}

See Example **3**   **9.** The temperature on the Sun's surface is about 5500°C. Scientists believe that the temperature at the center of the Sun is 270 times hotter. What is the temperature at the center of the Sun? Write the answer in scientific notation. $(1.485 \times 10^6)°C$

INDEPENDENT PRACTICE

See Example **1**   Write each number in standard notation.   **0.0021**   **63,700,000**

10. 8.3×10^5 **11.** 6.7×10^{-4} **12.** 2.1×10^{-3} **13.** 6.37×10^7
  830,000   **0.00067**

See Example **2**   Write each number in scientific notation. 1×10^9 3×10^{-8}

14. 0.000009   **15.** 7,800,000   **16.** 1,000,000,000   **17.** 0.00000003
  9×10^{-6} 7.8×10^6

See Example **3**   **18.** Protons and neutrons make up the nucleus of an atom and are the most massive particles in the atom. In fact, if a nucleus were the size of an average grape, it would have a mass greater than 9 million metric tons. A metric ton is 1000 kg. What would the mass of a grape-size nucleus be in kilograms? Write your answer in scientific notation. 9×10^9 **kg**

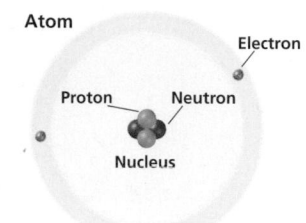

Atom
Electron
Proton   Neutron
Nucleus

PRACTICE AND PROBLEM SOLVING

Write each number in standard notation.   **0.00000013**

19. 1.3×10^4 **20.** 4.45×10^{-2} **21.** 5.6×10^1 **56**   **22.** 1.3×10^{-7}
  13,000   **0.0445**
23. 5.3×10^{-8} **24.** 9.567×10^{-5} **25.** 8.58×10^6 **26.** 7.1×10^3
  0.000000053   **0.00009567**   **8,580,000**   **7100**
27. 9.112×10^6 **28.** 3.4×10^{-1} **29.** 2.9×10^{-4} **30.** 6.8×10^2
  9,112,000   **0.34**   **0.00029**   **680**

Write each number in scientific notation.   **5.6 × 10⁷**   **8.079 × 10⁹**

31. 0.00467   **32.** 0.00000059   **33.** 56,000,000   **34.** 8,079,000,000
  4.67×10^{-3} 5.9×10^{-7}
35. 0.0076   **36.** 0.0000000002   **37.** 3500   **38.** 0.0000000091
  7.6×10^{-3} 2×10^{-10} 3.5×10^3 9.1×10^{-9}
39. 900   **40.** 0.000005   **41.** 6,000,000   **42.** 0.0095678

Math Background

Although people often think of zero as having no value, it can make a number much greater or much smaller when used as a placeholder. For instance, by writing 11 zeros to the right of the number 2, you get the approximate number of stars in the Andromeda Galaxy, 200,000,000,000.

Writing numbers with so many zeros can be very cumbersome. Scientific notation was developed to alleviate this problem. Scientific notation can be used with very large and very small numbers, and it makes computation easier and tables more readable.

RETEACH 2-9

LESSON 2-9 Reteach
Scientific Notation

Standard Notation	Scientific Notation	
	1st factor is between 1 and 10.	2nd factor is an integer power of 10.
430,000	4.3×10^5	positive integer for large number
0.0000057	5.7×10^{-6}	negative integer for small number

To convert from scientific notation, look at the power of 10 to tell how many places and which way to move the decimal point.

Complete to write each in standard notation.

1. 4.12×10^6 **2.** 3.4×10^{-5}

Is the exponent positive or negative?   **positive**   **negative**
Move the decimal point right or left? How many places?   **right 6**   **left 5**
Write the number in standard notation.   **4,120,000**   **0.000034**

Write each number in standard notation.

3. 8×10^5 **4.** 7.1×10^{-4} **5.** 3.14×10^8
  800,000   **0.00071**   **314,000,000**

To convert to scientific notation, determine the factor between 1 and 10. Then determine the power of 10 by counting from the decimal point in the first factor to the decimal point in the given number.

Complete to write each in scientific notation.

6. 32,000,000   **7.** 0.0000000712

What is the first factor?   **3.2**   **7.12**
From its location in the first factor, which way must the decimal move to its location in the given number? How many places?   **right 7**   **left 8**
Write the number in scientific notation. 3.2×10^7 7.12×10^{-8}

Write each number in scientific notation.

8. 41,000,000   **9.** 0.0000000643   **10.** 1,370,000,000
  4.1×10^7 6.43×10^{-8} 1.37×10^9

PRACTICE 2-9

LESSON 2-9 Practice B
Scientific Notation

Write each number in standard notation.

1. 2.54×10^2 **2.** 6.7×10^{-2} **3.** 1.14×10^3 **4.** 3.8×10^{-1}
  254   **0.067**   **1140**   **0.38**
5. 7.53×10^{-3} **6.** 5.6×10^4 **7.** 9.1×10^5 **8.** 6.08×10^{-4}
  0.00753   **56,000**   **910,000**   **0.000608**
9. 8.59×10^5 **10.** 3.331×10^6 **11.** 7.21×10^{-3} **12.** 5.88×10^{-4}
  859,000   **3,331,000**   **0.00721**   **0.000588**

Write each number in scientific notation.

13. 75,000,000   **14.** 208   **15.** 907,100
  7.5×10^7 2.08×10^2 9.071×10^5
16. 56   **17.** 0.093   **18.** 0.00006
  5.6×10^1 9.3×10^{-2} 6.0×10^{-5}
19. 0.00852   **20.** 0.0505   **21.** 0.003007
  8.52×10^{-3} 5.05×10^{-2} 3.007×10^{-3}
22. 5226   **23.** 0.04   **24.** 98,856
  5.226×10^3 4.0×10^{-2} 9.8856×10^4

25. Jupiter is about 778,120,000 kilometers from the Sun. Write this number in scientific notation.
  7.7812×10^8

26. The size of a bacterium that sours milk is approximately 1.6×10^{-11}. Write this number in standard notation.
  0.000000000016

43. SOCIAL STUDIES

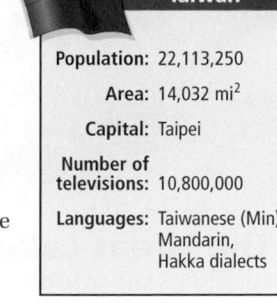

a. Express the population and area of Taiwan in scientific notation.

b. Divide the number of square miles by the population to find the number of square miles per person in Taiwan. Express your answer in scientific notation.

Taiwan	
Population:	22,113,250
Area:	14,032 mi²
Capital:	Taipei
Number of televisions:	10,800,000
Languages:	Taiwanese (Min), Mandarin, Hakka dialects

Life Science LINK

This frog is covered with duckweed plants. Duckweed plants can grow both in sunlight and in shade. They produce tiny white flowers that are nearly invisible to the human eye.

44. LIFE SCIENCE Duckweed plants live on the surface of calm ponds and are the smallest flowering plants in the world. They weigh about 0.00015 g.

a. Write this number in scientific notation. **1.5×10^{-4} g**

b. If left unchecked, one duckweed plant, which reproduces every 30–36 hours, could produce 1×10^{30} (a nonillion) plants in four months. How much would one nonillion duckweed plants weigh? **1.5×10^{26} g**

45. LIFE SCIENCE The size of a bacterium that sours milk is approximately 7.8×10^{-5} in. Write this number in standard notation. **0.000078 in.**

46. PHYSICAL SCIENCE The *atomic mass* of an element is the mass, in grams, of one *mole* (mol), or 6.02×10^{23} atoms. **$3.5(6.02 \times 10^{23}) =$**

a. How many atoms are there in 3.5 mol of carbon? **2.107×10^{24} atoms**

b. If you know that 3.5 mol of carbon weighs 42 grams, what is the atomic mass of carbon? **$42 \div 3.5 = 12$ g**

c. Using your answer from part **b**, find the approximate mass of one atom of carbon. **$12 \div (6.02 \times 10^{23}) \approx 1.99 \times 10^{-23}$ g**

47. WRITE A PROBLEM A proton has a mass of about 1.7×10^{-24} g. Use this information to write a problem.

48. WRITE ABOUT IT Two numbers are written in scientific notation. How can you tell which number is greater?

49. CHALLENGE Where on a number line does the value of a positive number in scientific notation with a negative exponent lie?

Spiral Review

Simplify. (Lesson 2-3)

50. $-3(6 - 8)$ **6**   **51.** $4(-3 - 2)$ **-20**   **52.** $-5(3 + 2)$ **-25**   **53.** $-3(1 - 8)$ **21**

Solve. (Lesson 2-4)

54. $m - 2 = 7$ **$m = 9$**   **55.** $8 + t = -1$ **$t = -9$**   **56.** $y - 24 = -19$ **$y = 5$**   **57.** $b + 4 = -23$ **$b = -27$**

58. TEST PREP Which number is equivalent to -64? (Lesson 2-6) **A**

A $(-4)^3$ B $(-4)^{-3}$ C 4^3 D 4^{-3}

CHALLENGE 2-9

LESSON 2-9 Challenge
The Wild Blue Yonder

Astronomers measure distances within our solar system in *astronomical units* (AU).
1 AU ≈ 92,956,000 mi or 149,600,000 km (the distance from Earth to the Sun)

Mean Distance From the Sun

Planet	km	Scientific Notation	AU
Mercury	57,900,000	5.79×10^7	0.4
Venus	108,200,000	1.082×10^8	0.7
Earth	149,600,000	1.496×10^8	1.0
Mars	227,900,000	2.279×10^8	1.5
Jupiter	778,400,000	7.784×10^8	5.2
Saturn	1,429,400,000	1.4294×10^9	9.6
Uranus	2,875,000,000	2.875×10^9	19.2
Neptune	4,504,300,000	4.5043×10^9	30.1
Pluto	5,900,100,000	5.9001×10^9	39.4

1. The table gives each planet's mean distance from the Sun in kilometers. Write these distances in scientific notation.

2. Convert to AUs by dividing each planet's mean distance from the Sun by 1.496×10^8. Use scientific notation. Round your answers to the nearest tenth of an AU.

Example $\frac{5.79 \times 10^7}{1.496 \times 10^8} = \frac{5.79}{1.496} \times \frac{10^7}{10^8} = 3.87 \times 10^{-1} = 0.387 \approx 0.4$ AU

3. Approximately how many times greater is Saturn's distance from the Sun than is Earth's distance? Answer to the nearest tenth.

about 9.6 times

4. Approximately how many times greater is the distance of the farthest planet from the Sun than is the distance of the closest planet to the Sun? Answer to the nearest tenth.

Pluto's distance from the Sun is about 101.9 times that of Mercury.

PROBLEM SOLVING 2-9

LESSON 2-9 Problem Solving
Scientific Notation

Write the correct answer.

1. In June 2001, the Intel Corporation announced that they could produce a silicon transistor that could switch on and off 1.5 trillion times a second. Express the speed of the transistor in scientific notation.

1.5×10^{12}

2. With this transistor, computers will be able to do 1×10^9 calculations in the time it takes to blink your eye. Express the number of calculations using standard notation.

1,000,000,000

3. The elements in this fast transistor are 20 nanometers long. A nanometer is one-billionth of a meter. Express the length of an element in the transistor in meters using scientific notation.

2×10^{-8} m

4. The length of the elements in the transistor can also be compared to the width of a human hair. The length of an element is 2×10^{-5} times smaller than the width of a human hair. Express 2×10^{-5} in standard notation.

0.00002

Use the table to answer Exercises 5–9. Choose the best answer.

5. Express a light-year in miles using scientific notation.
   A 58.8×10^{11} C 588×10^{10}
   B 5.88×10^{12} D 5.88×10^{-13}

Distance From Earth To Stars
Light-Year = 5,880,000,000,000 mi.

Star	Constellation	Distance (light-years)
Sirius	Canis Major	8
Canopus	Carin	650
Alpha Centauri	Cantaurus	4
Vega	Lyra	23

6. How many miles is it from Earth to the star Sirius?
   F 4.705×10^{12} H 7.35×10^{12}
   G 4.704×10^{13} J 7.35×10^{11}

7. How many miles is it from Earth to the star Canopus?
   A 3.822×10^{15} C 3.822×10^{14}
   B 1.230×10^{15} D 1.230×10^{14}

8. How many miles is it from Earth to the star Alpha Centauri?
   F 2.352×10^{13} H 2.352×10^{14}
   G 5.92×10^{13} J 5.92×10^{14}

9. How many miles is it from Earth to the star Vega?
   A 6.11×10^{13} C 6.11×10^{14}
   B 1.3524×10^{13} **D** 1.3524×10^{14}

Lesson Quiz

Write each number in standard notation.

1. 1.72×10^4 **17,200**

2. 6.9×10^{-3} **0.0069**

Write each number in scientific notation.

3. 0.0053   **5.3×10^{-3}**

4. 57,000,000   **5.7×10^7**

5. A human body contains about 5.6×10^6 microliters of blood. Write this number in standard notation. **5,600,000**

Available on Daily Transparency in CRB

Problem Solving on Location

Michigan

Purpose: *To provide additional practice for problem-solving skills in Chapters 1–2*

The Great Lakes State

- After problem 1, have students consider the following problems: Which lake's average depth is closest to the mean maximum depth? Lake Huron

- After problem 5, have students explain how they translated the word problem into an equation. Possible answer: Translate the words *minus* and *is equal to* into symbols. Substitute −923 for the maximum depth of Lake Michigan. Discuss how to solve an equation in which x is subtracted. Possible answer: Isolate the −x by adding 923 to both sides. Then divide both sides by −1.

Extension Have students research the Great Lakes to determine the area of each lake's surface and the volume of each lake. Have them rank the Great Lakes according to several different criteria. Check students' work.

Problem Solving on Location

MICHIGAN

The Great Lakes State

Michigan is known as the Great Lakes State because the shores of the state's two peninsulas touch four of the five Great Lakes.

For 1–6, use the table.

Great Lakes Bordering Michigan	
Lake	Maximum Depth (ft)
Lake Erie	−210
Lake Huron	−750
Lake Michigan	−923
Lake Superior	−1333

A person standing anywhere in Michigan is within 85 miles of at least one of the Great Lakes.

1. Find the average maximum depth to the nearest foot of the Great Lakes bordering Michigan. **−796 ft**

2. Which lake has the shallowest maximum depth? Which lake has the deepest maximum depth? What is the difference in the maximum depths of these two lakes? **Lake Erie; Lake Superior; 1123 ft**

3. Write and solve an equation to determine the depth d of the lake that has a maximum depth 540 ft shallower than Lake Huron. $d - 540 = -750$; $d = -210$ (Lake Erie)

4. If Lake Erie were 6 times as deep as it is, would any Great Lake be deeper? Explain. **Lake Superior; −1333 < −1260**

5. The maximum depth of Lake Michigan minus the maximum depth x of which lake is equal to −173? $x = -750$; Lake Huron

6. Tahquamenon Falls, in Michigan's Upper Peninsula, is one of the largest waterfalls west of the Mississippi River. Its drop d is only about $\frac{1}{27}$ of the maximum depth of Lake Superior. How many feet does Tahquamenon Falls drop? **about −49 ft**

7. Including Lake Ontario (the only Great Lake that does not touch Michigan), the Great Lakes and their connecting channels contain a total of 6,000,000,000,000,000 gallons of water. Write this number in scientific notation. 6×10^{15}

Ship crews in the Great Lakes refer to the *J. W. Westcott II* mail deliveries as "mail by the pail."

J. W. Westcott Company

Since 1895 the J. W. Westcott Company has been making mid-river mail deliveries by tugboat to the riverboats and lake boats of the Great Lakes area. The company runs the only floating post office in the world, and the Westcott Boat Station is the only boat station in the United States with its own postal code. The mail boat makes approximately 30 trips per day, 275 days per season. In 1968 the mail boat delivered nearly a million pieces of mail. Now it delivers closer to 400,000 pieces.

1. The *J. W. Westcott II* has a 220-horsepower diesel engine. One horsepower is the power needed to raise 550 pounds through a height of 1 foot in 1 second. So 1 horsepower (hp) is equal to 550 foot pounds force per second (550 ft lb f/s). How many ft lb f/s does the engine of the *J. W. Westcott II* have? What is this number estimated to the nearest power of ten? **121,000; 10^5**

2. Suppose the J. W. Westcott Company delivered an average of 750,000 pieces of mail per year over all the years it has been in operation. Approximately how many pieces of mail would have been delivered by the 100th anniversary in 1995? Write your answer in scientific notation. **7.5×10^7 pieces**

3. The J. W. Westcott Company provides mail service 24 hours a day from April through December. That is 275 days.

   a. Find the number of hours the mail boat offers service. Write your answer in scientific notation. **6.6×10^3**

   b. Write and solve an equation to find the number of eight-hour shifts from April through December. **$8n = 6600$; $n = 825$ shifts**

4. The mail boat delivers to freighters that weigh more than 250 tons. One ton is 2000 pounds. How many pounds do these freighters weigh? Write your answer in scientific notation. **more than 5×10^5 lb**

Game Resources

Puzzles, Twisters & Teasers
Chapter 2 Resource Book

Magic Squares

Purpose: *To apply the skill of writing and solving equations to completing a magic square*

Discuss: Ask students to explain what a magic square is. How can writing and solving equations help you find the missing numbers in a magic square?

Possible answer: In a magic square, the sum of the numbers in any row, column, or diagonal is the same. To find the missing numbers, assign a variable to each. Then write equations and solve them to find the value of one variable. Use that value to find the sum of one row, column, or diagonal. Then use the sum to find the values of the other variables.

Extend: Challenge students to create a magic square using the numbers 1–9.

Possible answer:

8	1	6
3	5	7
4	9	2

Answers

2. Possible answer:

−1	−2	3
4	0	−4
−3	2	1

MATH-ABLES

Equation Bingo

Purpose: *To practice solving equations in a game format*

Discuss: When students get "bingo," have them demonstrate for the class how the winning solution was obtained.

Extend: Have students create new equation cards for each solution on their bingo cards. Use the new equation cards to play again.

MATH-ABLES

Magic Squares

A *magic square* is a square with numbers arranged so that the sums of the numbers in each row, column, and diagonal are the same.

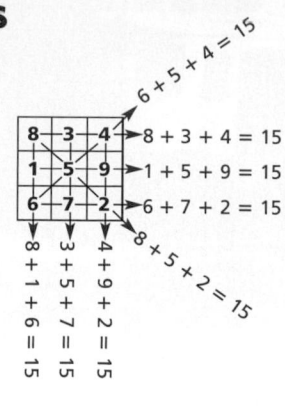

6 + 5 + 4 = 15

8	3	4
1	5	9
6	7	2

8 + 3 + 4 = 15
1 + 5 + 9 = 15
6 + 7 + 2 = 15
8 + 5 + 2 = 15

8 + 1 + 6 = 15
3 + 5 + 7 = 15
4 + 9 + 2 = 15

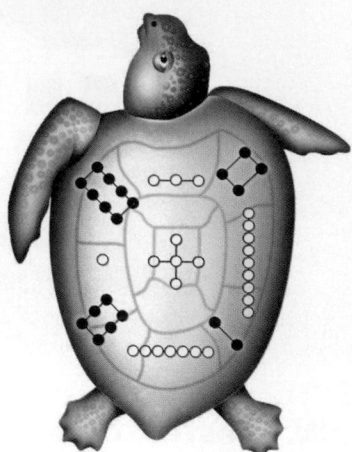

According to an ancient Chinese legend, a tortoise from the Lo river had the pattern of this magic square on its shell.

1. Complete each magic square below.

−1

6		4
1	3		5
	7		0

2

1

	−6	−1
−4		0	−2
−3	2		−5

0

7

−7		6	−4
4	−2		1	−1
	2	3	−3
5	−5	−6		8

2. Use the numbers −4, −3, −2, −1, 0, 1, 2, 3, and 4 to make a magic square with row, column, and diagonal sums of 0.

Equation Bingo

Each bingo card has numbers on it. The caller has a collection of equations. The caller reads an equation, and then the players solve the equation for the variable. If players have the solution on their cards, they place a chip on it. The winner is the first player with a row of chips either down, across, or diagonally.

internet connect
go.hrw.com

For a complete set of rules and cards, visit **go.hrw.com**
KEYWORD: MP4 Game2

Technology LAB

Evaluate Expressions

Use with Lesson 2-8

✔ **internet** connect
Lab Resources Online
go.hrw.com
KEYWORD: MP4 Lab2

A graphing calculator can be used to evaluate expressions that have negative exponents.

Activity

1 Use the **STO►** button to evaluate x^{-3} for $x = 2$. View the answer as a decimal and as a fraction.

 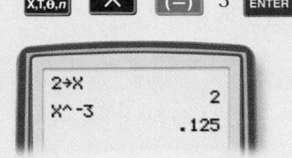 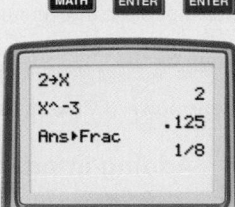

Notice that $2^{-3} = 0.125$, which is equivalent to $\frac{1}{2^3}$, or $\frac{1}{8}$.

2 Use the **TABLE** feature to evaluate 2^{-x} for several x-values. Match the settings shown.

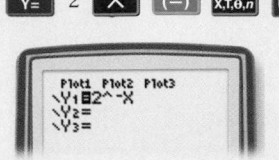 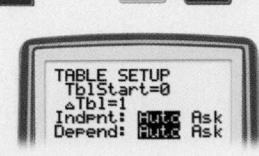 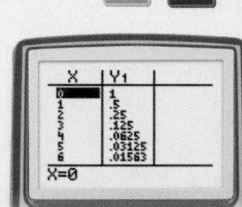

The **Y1** list shows the value of 2^{-x} for several x-values.

Think and Discuss

1. When you evaluated 2^{-3} in Activity 1, the result was not a negative number. Is this surprising? Why or why not?

Try This

Evaluate each expression for the given x-value(s). Give your answers as fractions and as decimals rounded to the nearest hundredth.

1. 4^{-x}; $x = 2$ **2.** 3^{-x}; $x = 1, 2$ **3.** x^{-2}; $x = 1, 2, 5$

Answers

Think and Discuss

1. Possible answer: No, it's not surprising because negative exponents indicate fractions not negative values.

Try This

1. $\frac{1}{16} \approx 0.06$

2. $\frac{1}{3} \approx 0.33$; $\frac{1}{9} \approx 0.11$

3. $\frac{1}{1} = 1$; $\frac{1}{4} = 0.25$; $\frac{1}{25} = 0.04$

Technology LAB

Evaluate Expressions

Objective: To use a graphing calculator to evaluate an expression

Materials: Graphing calculator

Lab Resources

Technology Lab Activities . . . . p. 13

Using the Page

This technology activity shows students how to evaluate expressions with negative exponents for several values of x, which can be done on any graphing calculator. Specific keystrokes may vary, depending on the make and model of the graphing calculator used. The keystrokes given are for a TI-83 model. For keystrokes to other models, visit go.hrw.com.

The Think and Discuss problem is meant to reinforce that negative exponents do not necessarily simplify to negative numbers. Try This problems 1–3 are meant to help students become familiar with using a graphing calculator to evaluate expressions.

Assessment

How could you use the calculator to evaluate the expression x^{-3} for $x = 2$, 3, and 4? Enter

Y= **X,T,θ,n** **∧** **(−)** 3

and then **2nd** **WINDOW** **TblStart** = 2

and **ΔTbl** = 1. Enter **2nd** **GRAPH** and read the values in the table: 0.125, 0.03704, 0.01563.

Purpose: *To help students review and practice concepts and skills presented in Chapter 2*

Assessment Resources

Chapter Review
Chapter 2 Resource Book . . . pp. 89–91

🔵 **Test and Practice Generator**
CD-ROM

Additional review assessment items in both multiple-choice and free-response format may be generated for any objective in Chapter 2.

Answers

1. opposite

2. scientific notation; power

3. exponent, base

4. -2

5. -12

6. -3

7. 1

8. -24

9. 8

10. -8

11. -16

12. 17

13. 3

14. 15

15. -22

16. -4

17. 16

18. -5

Vocabulary

absolute value 60	exponential form 84	power 84
base 84	integer 60	scientific notation 96
exponent 84	opposite 60	

Complete the sentences below with vocabulary words from the list above. Words may be used more than once.

1. The sum of an integer and its ___?___ is 0.

2. A number in ___?___ is a number from 1 to 10 times a(n) ___?___ of 10.

3. In the power 3^5, the 5 is the ___?___ and the 3 is the ___?___ .

2-1 Adding Integers (pp. 60–63)

EXAMPLE

■ Add.

$-8 + 2$ *Find the difference of 8 and 2.*
-6 *8 > 2; use the sign of the 8.*

■ Evaluate.

$-4 + a$ for $a = -7$
$-4 + (-7)$ *Substitute.*
-11 *Same sign*

EXERCISES

Add.

4. $-6 + 4$
5. $-3 + (-9)$
6. $4 + (-7)$
7. $4 + (-3)$
8. $-11 + (-5) + (-8)$

Evaluate.

9. $k + 11$ for $k = -3$
10. $-6 + m$ for $m = -2$

2-2 Subtracting Integers (pp. 64–67)

EXAMPLE

■ Subtract.

$-3 - (-5)$
$-3 + 5$ *Add the opposite of −5.*
2 *5 > 3; use the sign of the 5.*

■ Evaluate.

$-9 - d$ for $d = 2$
$-9 - 2$ *Substitute.*
$-9 + (-2)$ *Add the opposite of 2.*
-11 *Same sign*

EXERCISES

Subtract.

11. $-7 - 9$
12. $8 - (-9)$
13. $-2 - (-5)$
14. $13 - (-2)$
15. $-5 - 17$
16. $16 - 20$

Evaluate.

17. $9 - h$ for $h = -7$
18. $12 - z$ for $z = 17$

Study Guide and Review

2-3 Multiplying and Dividing Integers (pp. 68–71)

EXAMPLE

Multiply or divide.

- $4(-9)$ The signs are **different**.
  -36 The answer is **negative**.

- $\dfrac{-33}{-11}$ The signs are the **same**.
  3 The answer is **positive**.

EXERCISES

Multiply or divide.

19. $7(-5)$ **20.** $\dfrac{72}{-4}$ **21.** $-4(-13)$

22. $\dfrac{-100}{-4}$ **23.** $8(-3)(-5)$ **24.** $\dfrac{10(-5)}{-25}$

2-4 Solving Equations with Integers (pp. 74–77)

EXAMPLE

Solve.

- $\begin{array}{r} x - 9 = -12 \\ +9 = +9 \\ \hline x = -3 \end{array}$ ■ $\begin{array}{r} y + 4 = -11 \\ -4 = -4 \\ \hline y = -15 \end{array}$

- $\begin{array}{c} 4m = 20 \\ \dfrac{4m}{4} = \dfrac{20}{4} \\ m = 5 \end{array}$ ■ $\begin{array}{c} \dfrac{t}{-2} = 10 \\ (-2) \cdot \dfrac{t}{-2} = (-2) \cdot 10 \\ t = -20 \end{array}$

EXERCISES

Solve.

25. $p - 8 = 1$ **26.** $t + 4 = 7$

27. $6 + k = 9$ **28.** $-7g = 42$

29. $\dfrac{w}{-4} = 20$ **30.** $10 = \dfrac{b}{-2}$

31. $8 = -2a$ **32.** $-13 = \dfrac{h}{7}$

33. $-15 + s = 23$

2-5 Solving Inequalities with Integers (pp. 78–81)

EXAMPLE

Solve and graph.

- $\begin{array}{r} x + 5 \leq -1 \\ -5 \quad -5 \\ \hline x \leq -6 \end{array}$

- $\begin{array}{c} -3q > 21 \\ \dfrac{-3q}{-3} > \dfrac{21}{-3} \\ q < -7 \end{array}$

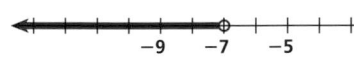

EXERCISES

Solve and graph.

34. $b + 3 < 1$ **35.** $r - 2 > 4$

36. $2m \geq 6$ **37.** $4p < -8$

38. $-2z > 10$ **39.** $-3q \leq -9$

40. $\dfrac{m}{2} \geq 2$ **41.** $\dfrac{x}{-3} < 1$

42. $\dfrac{y}{-1} > -4$ **43.** $4 + x > 1$

44. $-3b \geq 0$ **45.** $-2 + y < 4$

Answers

19. -35

20. -18

21. 52

22. 25

23. 120

24. 2

25. $p = 9$

26. $t = 3$

27. $k = 3$

28. $g = -6$

29. $w = -80$

30. $b = -20$

31. $a = -4$

32. $h = -91$

33. $s = 38$

34. $b < -2$

35. $r > 6$

36. $m \geq 3$

37. $p < -2$

38. $z < -5$

39. $q \geq 3$

40. $m \geq 4$

41. $x > -3$

42. $y < 4$

43. $x > -3$

44. $b \leq 0$

45. $y < 6$

Answers

46. 7^3

47. $(-3)^2$

48. k^4

49. 625

50. −32

51. −1

52. 4^7

53. 9^6

54. p^4

55. 8^3

56. 9^2

57. m^5

58. 5^3

59. y^5

60. k^0

61. $\frac{1}{125}$

62. $-\frac{1}{64}$

63. $\frac{1}{11}$

64. 1

65. 1

66. 1

67. $\frac{1}{8}$

68. $-\frac{1}{27}$

69. 1620

70. 0.00162

71. 910,000

72. 0.000091

73. 8.0×10^{-9}

74. 7.3×10^7

75. 9.6×10^{-6}

76. 5.64×10^{10}

Study Guide and Review

2-6 Exponents (pp. 84–87)

EXAMPLE

■ Write in exponential form.

$4 \cdot 4 \cdot 4$

4^3

■ Evaluate the power.

$(-2)^3$

$(-2) \cdot (-2) \cdot (-2)$

-8

EXERCISES

Write in exponential form.

46. $7 \cdot 7 \cdot 7$ **47.** $(-3) \cdot (-3)$

48. $k \cdot k \cdot k \cdot k$

Evaluate each power.

49. 5^4 **50.** $(-2)^5$ **51.** $(-1)^9$

2-7 Properties of Exponents (pp. 88–91)

EXAMPLE

Write the product or quotient as one power.

■ $2^5 \cdot 2^3$

2^{5+3}

2^8

■ $\frac{10^9}{10^2}$

10^{9-2}

10^7

EXERCISES

Write the product or quotient as one power.

52. $4^2 \cdot 4^5$ **53.** $9^2 \cdot 9^4$ **54.** $p \cdot p^3$

55. $\frac{8^5}{8^2}$ **56.** $\frac{9^3}{9}$ **57.** $\frac{m^7}{m^2}$

58. $5^0 \cdot 5^3$ **59.** $y^6 \div y$ **60.** $k^4 \div k^4$

2-8 Looking for a Pattern in Integer Exponents (pp. 92–95)

EXAMPLE

Evaluate.

■ $(-3)^{-2}$

$\frac{1}{(-3)^2}$

$\frac{1}{9}$

■ $\frac{2^5}{2^5}$

2^{5-5}

2^0

1

EXERCISES

Evaluate.

61. 5^{-3} **62.** $(-4)^{-3}$

63. 11^{-1} **64.** $\frac{7^4}{7^4}$

65. $\frac{5^7}{5^7}$ **66.** $\frac{x^3}{x^3}$

67. $(9-7)^{-3}$ **68.** $(6-9)^{-3}$

2-9 Scientific Notation (pp. 96–99)

EXAMPLE

Write in standard notation.

■ 3.58×10^4

$3.58 \times 10,000$

$35,800$

■ 3.58×10^{-4}

$3.58 \times \frac{1}{10,000}$

$3.58 \div 10,000$

0.000358

Write in scientific notation.

■ $0.000007 = 7 \times 10^{-6}$ ■ $62,500 = 6.25 \times 10^4$

EXERCISES

Write in standard notation.

69. 1.62×10^3 **70.** 1.62×10^{-3}

71. 9.1×10^5 **72.** 9.1×10^{-5}

Write in scientific notation.

73. 0.000000008   **74.** 73,000,000

75. 0.0000096   **76.** 56,400,000,000

Perform the given operations.

1. $-9 + (-12)$ **−21**
2. $11 - 17$ **−6**
3. $6(-22)$ **−132**
4. $(-20) \div (-4)$ **5**
5. $42 - (-5)$ **47**
6. $-18 \div 3$ **−6**
7. $-9 - (-13)$ **4**
8. $12 - (-6) + (-5)$ **13**
9. $-2(-21 - 17)$ **76**
10. $(-15 + 3) \div (-4)$ **3**
11. $(54 \div 6) - (-1)$ **10**
12. $-(16 + 4) - 20$ **−40**

13. The temperature on a winter day increased 37°F. If the beginning temperature was −9°F, what was the temperature after the increase? **28°F**

Evaluate each expression for the given value of the variable.

14. $16 - p$ for $p = -12$ **28**
15. $t - 7$ for $t = -14$ **−21**
16. $13 - x + (-2)$ for $x = 4$ **7**
17. $-8y + 27$ for $y = -9$ **99**

Solve.

18. $y + 19 = 9$ **$y = -10$**
19. $4z = -32$ **$z = -8$**
20. $52 = p - 3$ **$p = 55$**
21. $\frac{w}{3} = 9$ **$w = 27$**
22. $t + 1 < 7$ **$t < 6$**
23. $z - 4 \geq 7$ **$z \geq 11$**
24. $\frac{m}{-2} \leq 6$ **$m \geq -12$**
25. $-3q > 15$ **$q < -5$**

Graph each inequality.

26. $x > -4$![number line from −8 to 8]
27. $n \leq 3$![number line from −4 to 4]

Evaluate each power.

28. 4^3 **64**
29. $(-5)^4$ **625**
30. $(-3)^5$ **−243**

Multiply or divide. Write the product or quotient as one power.

31. $7^4 \cdot 7^5$ **7^9**
32. $\frac{12^5}{12^2}$ **12^3**
33. $x \cdot x^3$ **x^4**

Evaluate.

34. $(12 - 3)^2$ **81**
35. $40 + 5^3$ **165**
36. $\frac{3^4}{3^7}$ **$\frac{1}{27}$**
37. $10^4 \cdot 10^{-4}$ **1**

Write each number in standard notation.

38. 3×10^6 **3,000,000**
39. 3.1×10^{-6} **0.0000031**
40. 4.52×10^5 **452,000**

Write each number in scientific notation.

41. 3000 **3×10^3**
42. 42,000,000 **4.2×10^7**
43. 0.00000092 **9.2×10^{-7}**

44. A sack of cocoa beans weighs about 132 lb. How much would one thousand sacks of cocoa beans weigh? Write the answer in scientific notation. **1.32×10^5 lb**

Chapter Test Chapter **2**

Purpose: To assess students' mastery of concepts and skills in Chapter 2

Assessment Resources

Chapter 2 Tests (Levels A, B, C)
Assessment Resources . . . . pp. 39–44

Test and Practice Generator CD-ROM

Additional assessment items in both multiple-choice and free-response format may be generated for any objective in Chapter 2.

Purpose: *To assess students' under-standing of concepts in Chapter 2 and combined problem-solving skills*

Assessment Resources ✓

Performance Assessment
Assessment Resources . . . . p. 120

Performance Assessment Teacher Support
Assessment Resources . . . . p. 119

Answers

1–3. See p. A3.

4. See Level 3 work sample below.

Scoring Rubric for Problem Solving Item 4

Level 3
Accomplishes the purposes of the task.

Student gives clear explanations, shows understanding of mathematical ideas and processes, and computes accurately.

Level 2
Purposes of the task not fully achieved.

Student demonstrates satisfactory but limited understanding of the mathematical ideas and processes.

Level 1
Purposes of the task not accomplished.

Student shows little evidence of under-standing the mathematical ideas and processes and makes computational and/or procedural errors.

Performance Assessment

Show What You Know

Create a portfolio of your work from this chapter. Complete this page and include it with your four best pieces of work from Chapter 2. Choose from your homework or lab assignments, mid-chapter quiz, or any journal entries you have done. Put them together using any design you want. Make your portfolio represent what you consider your best work.

⭐ Short Response

1. **a.** Complete the following rules for operations involving odd and even numbers:

   even + even = __?__   odd + odd = __?__   odd + even = __?__
   even · even = __?__   odd · odd = __?__   even · odd = __?__

   b. Compare the rules from part **a** with the rules for finding the sign when multiplying two integers.

2. Write the subtraction equation $4 - 6 = -2$ as an addition equation. Draw a number-line diagram to illustrate the addition equation.

3. Consider the statement "Half of a number is less than or equal to -2." Write an inequality for this word sentence and solve it. Show your work.

Extended Problem Solving

4. The formula for converting degrees Celsius (°C) to degrees Fahrenheit (°F) is $F = \frac{9}{5}C + 32$. A way to estimate the temperature in degrees Fahrenheit is to double the temperature in degrees Celsius and add 30.

   a. Write the way of estimating as a formula.

   b. Compare the results for the exact formula and the estimate formula for $-10°C$, $0°C$, $30°C$, and $100°C$.

   c. For which of the values was the estimate closest to the exact answer? Find a temperature in degrees Celsius for which the estimate and the exact answer are the same. Show your work.

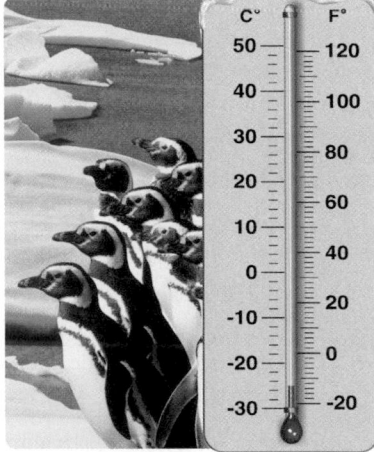

Student Work Samples for Item 4

Level 3

4q. F = 2C + 30

b. C -10° 0° 30° 100°
$\frac{9}{5}$C | $\frac{-90}{5}$=-18 0 54 180
+32 | (14 32 86 212)
2C | -20 0 60 200
+30 | (10 30 90 230)

c. For 0°C, the estimate was closest. The temperature is between 0 and 30, and is closer to 0.
$\frac{9}{5}(10)=18$ 18 + 32 = 50
2(10) = 20   20 + 30 = 50
The temperature is 10°C

Level 2

4q. F = 2C + 30

b. 2 · -10 + 30 = 10
   2 · 0 + 30 = 30
$\frac{9}{5}$ · -10 + 32 = 14
$\frac{9}{5}$ · 0 + 32 = 32
2 · 30 + 30 = 90
$\frac{9}{5}$ · 30 + 32 = 86
2 · 100 + 30 = 230
$\frac{9}{5}$ · 100 + 32 = 212

c. 20 degrees

Level 1

4q. F = 2 + Temp × 30

b. 2 + -10 × 30
   = -8 × 30
$\frac{9}{5}$ = 2 + Temp × 30

c. ?

The student correctly wrote the formula, and found the values for which both for-mulas give equal results.

The student correctly wrote the formula and found values but was unable to proceed beyond this point. No work was shown.

The student was unable to write the formula or attempt to find a value for which the formulas give equal results.

internet connect

State-Specific Test Practice Online
go.hrw.com   Keyword: MP4 TestPrep

Standardized Test Prep

Chapter
2

Standardized Test Prep

Chapter
2

Cumulative Assessment, Chapters 1–2

1. If $(n + 3)(9 - 5) = 16$, then what does n equal?   **A**

Ⓐ 1   Ⓒ 4
Ⓑ 7   Ⓓ 9

2. If $x = -\frac{1}{4}$, which is least?   **J**

Ⓕ $1 - x$ Ⓗ x
Ⓖ $x - 1$ Ⓙ $1 \div x$

TEST TAKING TIP!

Make comparisons: Express quantities in a common number base.

3. Which ratio compares the value of a hundred \$1000 bills with the value of a thousand \$100 bills?   **B**

Ⓐ 1 to 10   Ⓒ 5 to 1
Ⓑ 1 to 1   Ⓓ 10 to 1

4. Which is $3 \times 3 \times 3 \times 3 \times 11 \times 11 \times 11$ expressed in exponential form?   **G**

Ⓕ $4^3 \times 3^{11}$ Ⓗ 33^7
Ⓖ $3^4 \times 11^3$ Ⓙ 33^3

5. Which number is equivalent to 2^{-5}?   **B**

Ⓐ $\frac{1}{10}$ Ⓒ $-\frac{1}{10}$
Ⓑ $\frac{1}{32}$ Ⓓ $-\frac{1}{32}$

6. Which is 8.1×10^{-5}?   **H**

Ⓕ 8,100,000   Ⓗ 0.000081
Ⓖ 810,000   Ⓙ 0.0000081

7. Which power is equivalent to $5^{12} \div 5^4$?   **D**

Ⓐ 1^3 Ⓒ 5^3
Ⓑ 1^8 Ⓓ 5^8

8. The bar graph shows the average daily temperatures in Sturges, Michigan, for five months. Between which two months did the average temperature change by the greatest amount?   **G**

Ⓕ January and February
Ⓖ February and March
Ⓗ March and April
Ⓙ April and May

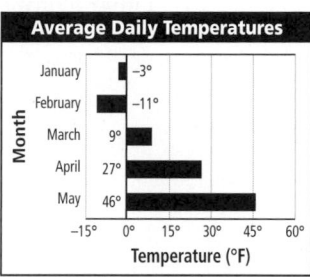

Average Daily Temperatures

9. SHORT RESPONSE Linda takes her grandson Colin to the ice cream parlor every Wednesday and spends \$6.50. During a 30-day month that began on a Monday, how much money did Linda spend at the ice cream parlor? Explain how you found your answer.

10. SHORT RESPONSE An elevator begins 7 floors above ground level and descends to a floor that is 2 floors below ground level. Each floor is 12 feet high. Draw a diagram to determine the number of feet the elevator traveled.

Standardized Test Prep

Chapter
2

Purpose: *To provide review and practice for Chapters 1–2 and standardized tests*

Assessment Resources

Cumulative Tests (Levels A, B, C)
Assessment Resources . . . . . pp. 157–168

State-Specific Test Practice Online
KEYWORD: MP4 TestPrep

Test Prep Doctor ➕

Remind students that substituting answer choices is often the easiest way to arrive at the correct answer. In item 1, the student could either solve the equation or simply substitute the answer choices in the equation and see which choice makes the equation true.

Expand on the test-taking tip given for item 3 by reviewing with students how to express a power of ten in exponential form. The values, in dollars, are:

$100 \times 1000 = 10^2 \times 10^3 = 10^5 = 10,000$

$1000 \times 100 = 10^3 \times 10^2 = 10^5 = 10,000$

Answers

9. \$26; The first Wednesday of the month would be the 3rd day of the month. To find the dates for the other Wednesdays, add 7: 3, 10, 17, 24. There are 4 Wednesdays, so she spent $4 \cdot 6.50 = 26$ dollars.

10. 108 ft

12	7
12	6
12	5
12	4
12	3
12	2
12	1
12	Ground floor
12	−1
	−2

Rational and Real Numbers

Pacing Guide for 45-Minute Classes

Chapter 3

DAY 24	DAY 25	DAY 26	DAY 27	DAY 28
Lesson 3-1	Lesson 3-2	Lesson 3-3	Lesson 3-4	Lesson 3-5

DAY 29	DAY 30	DAY 31	DAY 32	DAY 33
Technology Lab 3A	Lesson 3-6	Lesson 3-7	Mid-Chapter Quiz Lesson 3-8	Lesson 3-9

DAY 34	DAY 35	DAY 36	DAY 37	DAY 38
Hands-On Lab 3B	Lesson 3-10	Extension	Chapter 3 Review	Chapter 3 Assessment

Pacing Guide for 90-Minute Classes

Chapter 3

DAY 12	DAY 13	DAY 14	DAY 15	DAY 16
Chapter 2 Assessment Lesson 3-1	Lesson 3-2 Lesson 3-3	Lesson 3-4 Lesson 3-5	Technology Lab 3A Lesson 3-6	Lesson 3-7 Lesson 3-8

DAY 17	DAY 18	DAY 19	DAY 20	
Mid-Chapter Quiz Lesson 3-9 Hands-On Lab 3B	Lesson 3-10 Extension	Chapter 3 Review Lesson 4-1 Hands-On Lab 4A	Chapter 3 Assessment Lesson 4-2	

COURSE 1

- Identify, compare, order, and write equivalent rational numbers.
- Estimate sums and differences of rational numbers.
- Add, subtract, multiply, and divide rational numbers.
- Solve equations containing rational numbers.

COURSE 2

- Identify, compare, order, and write equivalent rational numbers.
- Estimate sums and differences of rational numbers.
- Add, subtract, multiply, and divide rational numbers.
- Solve equations containing rational numbers.

PRE-ALGEBRA

- Write equivalent rational numbers.
- Add, subtract, multiply, and divide decimals and rational numbers.
- Solve equations and inequalities containing rational numbers.
- Estimate and find square roots, and solve problems involving square roots.
- Determine whether a number is rational or irrational.
- Convert between bases.

Across the Curriculum

LANGUAGE ARTS LINK

SOCIAL STUDIES LINK

SCIENCE LINK

TE = *Teacher's Edition*   **SE** = *Student Edition*

Interdisciplinary

Bulletin Board

Physical Education

A marathon runner can determine her average minutes per mile rate by dividing her total recorded race time in minutes by the number of miles she ran. If the runner ran a 26.2-mile marathon in 3.29 hours (197.4 minutes), what was her average minutes per mile rate?   about 7.53 minutes per mile

Math in Physical Education

Use with Holt Middle School Math, Chapter 3

Interdisciplinary posters and worksheets are provided in your resource material.

Resource Options

Chapter 3 Resource Book

Student Resources

Teacher and Parent Resources

- Daily Transparencies
- Additional Examples Transparencies
- Teaching Transparencies

Reaching All Learners

English Language Learners

Individual Needs

Hands-On

Applications and Connections

Transparencies

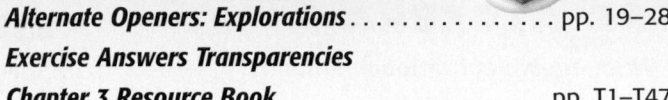

- Daily Transparencies
- Additional Examples Transparencies
- Teaching Transparencies

Technology

Teacher Resources

Student Resources

Are You Ready? Intervention CD-ROM
Skills 18, 21, 22, 24

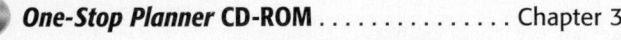

internet connect

Homework Help Online	KEYWORD: MP4 HWHelp3
Math Tools Online	KEYWORD: MP4 Tools
Glossary Online	KEYWORD: MP4 Glossary
Chapter Project Online	KEYWORD: MP4 PSProject3
Chapter Opener Online	KEYWORD: MP4 Ch3

KEYWORD: MP4 CNN3

SE = *Student Edition*   **TE** = *Teacher's Edition*   **AR** = *Assessment Resources*   **CRB** = *Chapter Resource Book*   **MK** = *Manipulatives Kit*

Assessing Prior Knowledge

Determine whether students have the required prerequisite concepts and skills.

Are You Ready?.................................. SE p. 111
Inventory Test................................. AR pp. 1–4

Test Preparation

Provide review and practice for chapter and standardized tests.

Standardized Test Prep........................... SE p. 171
Spiral Review with Test Prep..... SE, last page of each lesson
Study Guide and Review.................. SE pp. 166–168
Test Prep Tool Kit

Technology

 Test and Practice Generator CD-ROM

🔗 **internet** connect

State-Specific Test Practice Online KEYWORD: MP4 TestPrep

Performance Assessment

Assess students' understanding of chapter concepts and combined problem-solving skills.

Performance Assessment...................... SE p. 170
  Includes scoring rubric in TE
Performance Assessment...................... AR p. 122
Performance Assessment Teacher Support......... AR p. 121

Portfolio

Portfolio opportunities appear throughout the Student and Teacher's Editions.

Suggested work samples:

Problem Solving Project...................... TE p. 110
Performance Assessment...................... SE p. 170
Portfolio Guide............................. AR p. xxxvii
Journal....................... TE, last page of each lesson
Write About It .. SE pp. 116, 120, 125, 130, 139, 143, 149, 159

Daily Assessment

Obtain daily feedback on students' understanding of concepts.

Spiral Review and Test Prep ...... SE, last page of each lesson

Also Available on Transparency In Chapter 3 Resource Book

Warm Up...................... TE, first page of each lesson
Problem of the Day............ TE, first page of each lesson
Lesson Quiz................... TE, last page of each lesson

Student Self-Assessment

Have students evaluate their own work.

Group Project Evaluation .... AR p. xxxiv
Individual Group Member Evaluation .... AR p. xxxv
Portfolio Guide.............................. AR p. xxxvii
Journal....................... TE, last page of each lesson

Formal Assessment

Assess students' mastery of concepts and skills.

Section Quizzes .... AR pp. 9–10
Mid-Chapter Quiz.......................... SE p. 144
Chapter Test .... SE p. 169
Chapter Tests (Levels A, B, C) .... AR pp. 45–50
Cumulative Tests (Levels A, B, C) .... AR pp. 169–180
Standardized Test Prep
  Cumulative Assessment .... SE p. 171
End-of-Year Test.......................... AR pp. 313–316

Technology

 Test and Practice Generator CD-ROM

Make tests electronically. This software includes:

• Dynamic practice for Chapter 3
• Customizable tests
• Multiple-choice items for each objective
• Free-response items for each objective
• Teacher management system

SE = *Student Edition*   **TE** = *Teacher's Edition*   **AR** = *Assessment Resources*   **CRB** = *Chapter Resource Book*   **MK** = *Manipulatives Kit*

Chapter 3 Tests

Three levels (A,B,C) of tests are available for each chapter in the *Assessment Resources*.

LEVEL A

CHAPTER 3 Chapter Test
Form A

Simplify.

1. $\frac{6}{9}$ $\frac{2}{3}$

2. $-\frac{25}{75}$ $-\frac{1}{3}$

3. $\frac{12}{28}$ $\frac{3}{7}$

Write each decimal as a fraction in simplest form.

4. 0.39 $\frac{39}{100}$

5. 0.24 $\frac{6}{25}$

Write each fraction as a decimal.

6. $\frac{7}{10}$ 0.7

7. $\frac{1}{5}$ 0.2

Add or subtract.

8. $\frac{3}{5} - \frac{1}{5}$ $\frac{2}{5}$

9. $-1.4 + 1.9$ 0.5

Evaluate each expression for the given value of the variable.

10. $2.9 + j$ for $j = 1.3$ 4.2

11. $4 + n$ for $n = -\frac{2}{3}$ $3\frac{1}{3}$

Multiply. Write each answer in simplest form.

12. $\frac{3}{5}\left(\frac{1}{2}\right)$ $\frac{3}{10}$

13. $5.3(4.1)$ 21.73

14. $-0.2(2.7)$ -0.54

Evaluate $\frac{1}{6}y$ for each value of y.

15. $y = 4$ $\frac{2}{3}$

16. $y = \frac{1}{9}$ $\frac{1}{54}$

17. $y = \frac{2}{13}$ $\frac{1}{39}$

Divide. Write each answer in simplest form.

18. $\frac{9}{16} \div \frac{1}{2}$ $1\frac{1}{8}$

19. $9 \div \frac{3}{4}$ 12

Divide.

20. $0.36 \div 0.24$ 1.5

21. $9.12 \div 0.5$ 18.24

CHAPTER 3 Chapter Test
Form A, continued

Evaluate $\frac{2}{x}$ for each value of x.

22. $x = 0.4$ 5

23. $x = 0.2$ 10

Add or subtract.

24. $\frac{1}{2} + \frac{1}{6}$ $\frac{2}{3}$

25. $3\frac{1}{6} + 1\frac{7}{12}$ $4\frac{3}{4}$

Evaluate each expression for the given value of the variable.

26. $\frac{1}{8} + x$ for $x = \frac{1}{4}$ $\frac{3}{8}$

27. $x - \frac{4}{5}$ for $x = \frac{1}{5}$ $-\frac{3}{5}$

Solve.

28. $m - 1.6 = 5.2$ $m = 6.8$

29. $\frac{2}{9}m = \frac{2}{3}$ $m = 3$

30. $3.0h = 12.0$ $h = 4$

Solve.

31. $y - \frac{1}{8} \geq \frac{5}{8}$ $y \geq \frac{3}{4}$

32. $7 + y > 8.2$ $y > 1.2$

33. $3a < \frac{1}{6}$ $a < \frac{1}{18}$

Simplify each expression.

34. $\sqrt{2} + 2$ 2

35. $\sqrt{36} - \sqrt{4}$ 4

36. $\sqrt{16} + 2$ 6

Use a calculator to find each value. Round to the nearest tenth.

37. $\sqrt{196}$ 14.0

38. $\sqrt{12}$ 3.5

30. $-\sqrt{29}$ -5.4

State if the number is rational, irrational, or not a real number.

40. 0.91   rational

41. -5 rational

42. $0.\overline{3}$ rational

43. Jessie bought 6 carnations that were \$0.39 each. How much did she spend?   \$2.34

44. There is a fourth of a pie left. Your mom says you can have half of it. How much of the pie is that? $\frac{1}{8}$

LEVEL B

CHAPTER 3 Chapter Test
Form B

Simplify.

1. $\frac{21}{30}$ $\frac{7}{10}$

2. $-\frac{45}{120}$ $-\frac{3}{8}$

3. $\frac{15}{17}$ already simplified

Write each decimal as a fraction in simplest form.

4. -3.46 $-3\frac{23}{50}$

5. -0.07 $-\frac{7}{100}$

Write each fraction as a decimal.

6. $\frac{7}{25}$ 0.28

7. $-\frac{25}{5}$ -5

Add or subtract.

8. $\frac{7}{23} - \frac{9}{23}$ $-\frac{2}{23}$

9. $-1.4 + 0.72$ -0.68

Evaluate each expression for the given value of the variable.

10. $8 + n$ for $n = -\frac{1}{4}$ $7\frac{3}{4}$

11. $\frac{4}{12} + f$ for $f = -\frac{9}{12}$ $-\frac{5}{12}$

Multiply. Write each answer in simplest form.

12. $\frac{3}{4}\left(\frac{1}{8}\right)$ $\frac{3}{32}$

13. $-5\left(1\frac{5}{6}\right)$ $-9\frac{1}{6}$

14. $4.73(3.1)$ 14.663

Evaluate $5\frac{2}{3}y$ for each value of y.

15. $y = 4$ $22\frac{2}{3}$

16. $y = -\frac{1}{9}$ $-\frac{17}{27}$

17. $y = \frac{7}{17}$ $2\frac{1}{3}$

Divide. Write each answer in simplest form.

18. $\frac{5}{6} \div \frac{9}{16}$ $1\frac{13}{27}$

19. $6\frac{1}{3} \div 3\frac{1}{2}$ $1\frac{17}{21}$

Divide.

20. $1.9 \div 0.05$ 38

21. $7.15 \div 1.3$ 5.5

CHAPTER 3 Chapter Test
Form B, continued

Evaluate $\frac{8}{x}$ for each value of x.

22. $x = 2.5$ 3.2

23. $x = 0.04$ 200

Add or subtract.

24. $\frac{2}{3} + \frac{4}{5}$ $1\frac{7}{15}$

25. $2\frac{5}{6} - 1\frac{3}{10}$ $1\frac{8}{15}$

Evaluate each expression for the given value of the variable.

26. $\frac{1}{8} + x$ for $x = -4\frac{1}{8}$ -4

27. $n - 3\frac{5}{6}$ for $n = -\frac{1}{15}$ $-3\frac{9}{10}$

Solve.

28. $m - 5.6 = -0.9$ $m = 4.7$

29. $\frac{3}{7}m = -\frac{6}{7}$ $m = -2$

30. $3.7h = 0.74$ $h = 0.2$

Solve.

31. $y - \frac{1}{8} \geq \frac{3}{4}$ $y \geq \frac{7}{8}$

32. $5.5 + y > 2.3$ $y > -3.2$

33. $6a < -\frac{5}{8}$ $a < -\frac{5}{48}$

Simplify each expression.

34. $\sqrt{7} + 9$ 4

35. $\sqrt{36} - \sqrt{81}$ -3

36. $\sqrt{25} + 24$ 29

Use a calculator to find each value. Round to the nearest tenth.

37. $\sqrt{136}$ 11.7

38. $-\sqrt{34}$ -5.8

39. $\sqrt{82.9}$ 9.1

State if the number is rational, irrational, or not a real number.

40. $\sqrt{27}$ irrational

41. $-\sqrt{\frac{81}{16}}$ rational

42. $\sqrt{\frac{1}{9}}$ rational

43. If you buy 2.5 pounds of hamburger and it costs \$2.10 per pound, how much does the package cost?   \$5.25

44. A book of stamps contains 20 stamps. If you used one fourth of them, how many did you use?   5

LEVEL C

CHAPTER 3 Chapter Test
Form C

Simplify.

1. $\frac{75}{204}$ $\frac{25}{68}$

2. $-\frac{108}{320}$ $-\frac{27}{80}$

3. $\frac{14}{53}$ already simplified

Write each decimal as a fraction in simplest form.

4. -4.3125 $-4\frac{5}{16}$

5. 0.03 $\frac{1}{30}$

Write each fraction as a decimal.

6. $\frac{9}{16}$ 0.5625

7. $-\frac{19}{40}$ -0.475

Add or subtract.

8. $\frac{5}{136} - \frac{9}{136}$ $-\frac{1}{34}$

9. $-17.67 + 25.13$ 7.46

Evaluate each expression for the given value of the variable.

10. $54 + n$ for $n = -\frac{7}{16}$ $53\frac{9}{16}$

11. $\frac{2}{9} + f$ for $f = -\frac{8}{9}$ $-\frac{2}{3}$

Multiply. Write each answer in simplest form.

12. $13\frac{4}{7}\left(\frac{3}{5}\right)$ $8\frac{1}{7}$

13. $11\left(4\frac{5}{12}\right)$ $48\frac{7}{12}$

14. $-0.47(92.8)$ -43.616

Evaluate $16\frac{2}{5}y$ for each value of y.

15. $y = 4$ $65\frac{3}{5}$

16. $y = \frac{1}{16}$ $1\frac{1}{40}$

17. $y = \frac{5}{29}$ $2\frac{24}{29}$

Divide. Write each answer in simplest form.

18. $13\frac{3}{4} \div 1\frac{2}{3}$ $8\frac{1}{4}$

19. $6\frac{2}{3} \div 7\frac{1}{2}$ $\frac{8}{9}$

Divide.

20. $0.512 \div 0.08$ 6.4

21. $2002 \div 5.2$ 385

CHAPTER 3 Chapter Test
Form C, continued

Evaluate $\frac{24}{x}$ for each value of x.

22. $x = 0.9$ 26.6

23. $x = 0.012$ 2000

Add or subtract. Write each answer in simplest form.

24. $\frac{4}{5} + \frac{3}{8}$ $1\frac{7}{40}$

25. $6\frac{5}{8} - 1\frac{9}{10}$ $4\frac{29}{40}$

Evaluate each expression for the given value of the variable.

26. $\frac{2}{3} + x$ for $x = \frac{7}{12}$ $1\frac{1}{4}$

27. $n - 5\frac{5}{6}$ for $n = -4\frac{7}{8}$ $-10\frac{17}{24}$

Solve.

28. $m + 25.6 = -0.19$ $m = -25.79$

29. $\frac{4}{17}m = -\frac{8}{17}$ $m = -2$

30. $23.7h = 16.59$ $h = 0.7$

Solve.

31. $y - \frac{1}{2} \geq \frac{7}{8}$ $y \geq 1\frac{3}{8}$

32. $43.6 + y > 32.3$ $y > -11.3$

33. $-3a < -13\frac{4}{5}$ $a > 4\frac{3}{5}$

Simplify each expression.

34. $\sqrt{460} + 24$ 22

35. $\sqrt{121} - \sqrt{196}$ -3

36. $\sqrt{324} + 324$ 342

Use a calculator to find each value. Round to the nearest tenth.

37. $\sqrt{1436}$ 37.9

38. $-\sqrt{284}$ -16.9

39. $\sqrt{982.9}$ 31.4

State if the number is rational, irrational, or not a real number.

40. $\sqrt{11}$ irrational

41. $\sqrt{576}$ rational

42. $\sqrt{5\frac{1}{16}}$ rational

43. The area of a rectangle is its length times its width. Find the area if the length is 2.3 feet and the width is 0.7 foot.   1.61 feet^2

44. Find the area of a rectangle if the length is $2\frac{2}{5}$ inches and the width is $\frac{4}{9}$ inches. $1\frac{5}{27}$ in^2

Test and Practice Generator
CD-ROM

Create and customize multiple versions of the same tests with corresponding answers for any chosen chapter objectives.

Chapter 3 State and Standardized Test Preparation

Test Taking Skill Builder and Standardized Test Practice
are provided for each chapter in the *Test Prep Tool Kit*.

TEST TAKING SKILL BUILDER

Test Taking Strategy   Gridded Response
Chapter 3

Gridded response questions require that you fill in your answer
on the grid provided on the answer sheet.

Response Grids have these parts:

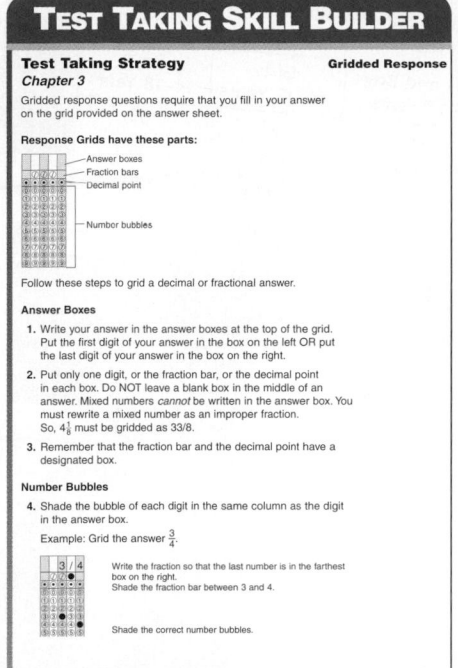

— Answer boxes
— Fraction bars
— Decimal point
— Number bubbles

Follow these steps to grid a decimal or fractional answer.

Answer Boxes

1. Write your answer in the answer boxes at the top of the grid.
   Put the first digit of your answer in the box on the left OR put
   the last digit of your answer in the box on the right.

2. Put only one digit, or the fraction bar, or the decimal point
   in each box. Do NOT leave a blank box in the middle of an
   answer. Mixed numbers *cannot* be written in the answer box. You
   must rewrite a mixed number as an improper fraction.
   So, $4\frac{1}{8}$ must be gridded as 33/8.

3. Remember that the fraction bar and the decimal point have a
   designated box.

Number Bubbles

4. Shade the bubble of each digit in the same column as the digit
   in the answer box.

   Example: Grid the answer $\frac{3}{4}$.

   Write the fraction so that the last number is in the farthest
   box on the right.
   Shade the fraction bar between 3 and 4.

   Shade the correct number bubbles.

Test Taking Strategy
Chapter 3, continued

Exercises
What should go in the first box on the left for each gridded
response answer?

1. $\frac{5}{6}$ **5**   2. $8\frac{1}{2}$ **1**   3. $\frac{7}{12}$ **7**   4. $18\frac{1}{2}$ **3**

You place the first digit of your answer in the far left column.
Which column should the fraction bar go in for each gridded
response answer?

5. $\frac{3}{5}$ **second**   6. $12\frac{3}{4}$ **third**   7. $\frac{12}{7}$ **third**   8. $\frac{2}{7}$ **second**

Tell what error was made in each gridded response below.

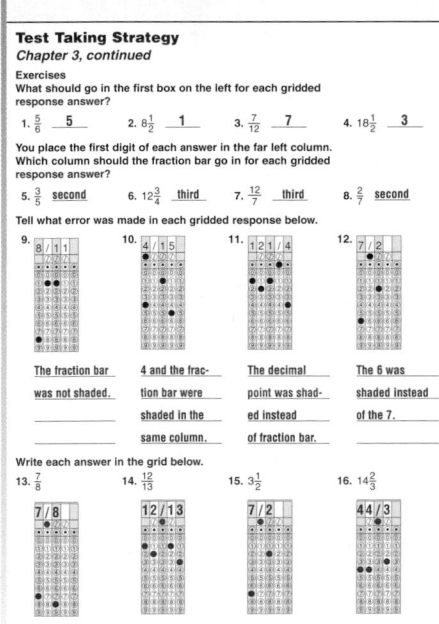

9.   10.   11.   12.

The fraction bar   4 and the frac-   The decimal   The 6 was
was not shaded.   tion bar were   point was shad-   shaded instead
   shaded in the   ed instead   of the 7.
   same column.   of fraction bar.

Write each answer in the grid below.

13. $\frac{7}{8}$ 14. $\frac{12}{13}$ 15. $3\frac{1}{2}$ 16. $14\frac{2}{3}$

STANDARDIZED TEST PRACTICE

Standardized Test Practice
Chapter 3

Select the best answer for Questions 1–9.

1. The value of $\sqrt{32}$ is between what
   two numbers?
   A 2 and 8
   B 5 and 6
   C 30 and 40
   D 16 and 10

2. What number goes in the box
   to make the equation true?
   $$\frac{7}{12} - \frac{?}{12} = \frac{1}{3}$$
   F 6   H 4
   G 5   I 3

3. Which decimal gives the best
   estimate of the amount of the
   circle that is shaded?

   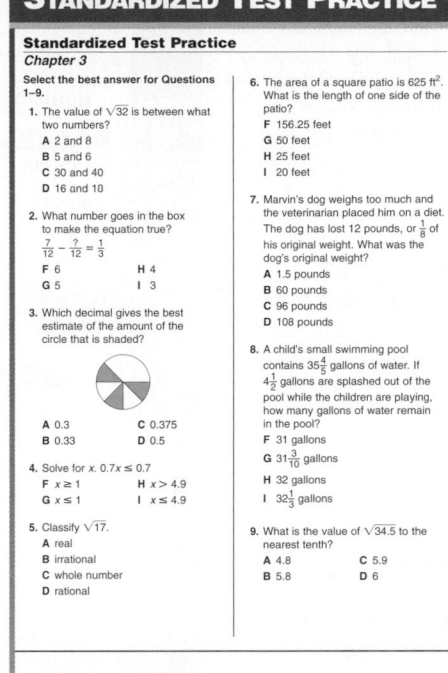

   A 0.3   C 0.375
   B 0.33   D 0.5

4. Solve for x. $0.7x \le 0.7$
   F $x \ge 1$ H $x > 4.9$
   G $x \le 1$ I $x \le 4.9$

5. Classify $\sqrt{17}$.
   A real
   B irrational
   C whole number
   D rational

6. The area of a square patio is 625 ft^2.
   What is the length of one side of the
   patio?
   F 156.25 feet
   G 50 feet
   H 25 feet
   I 20 feet

7. Marvin's dog weighs too much and
   the veterinarian placed him on a diet.
   The dog has lost 12 pounds, or $\frac{1}{8}$ of
   his original weight. What was the
   dog's original weight?
   A 1.5 pounds
   B 60 pounds
   C 96 pounds
   D 108 pounds

8. A child's small swimming pool
   contains $35\frac{4}{5}$ gallons of water. If
   $4\frac{1}{2}$ gallons are splashed out of the
   pool while the children are playing,
   how many gallons of water remain
   in the pool?
   F 31 gallons
   G $31\frac{3}{10}$ gallons
   H 32 gallons
   I $32\frac{1}{3}$ gallons

9. What is the value of $\sqrt{34.5}$ to the
   nearest tenth?
   A 4.8   C 5.9
   B 5.8   D 6

Standardized Test Practice
Chapter 3, continued

Gridded Response
Solve the problems. Use the answer
sheet to write and grid-in your answer.

10. The body and the head of a pigmy
    shrew is 1.74 inches long. If the total
    length of the shrew from the tip of its
    nose to the end of its tail is 3.1 inches
    long, how many inches is the tail of
    the shrew?

11. Schmidt's Used Car Lot has 68
    vehicles for sale. If $\frac{1}{4}$ of the vehicles
    are sports utility vehicles, how many
    vehicles on the lot are SUVs?

12. Your grandmother's punch bowl
    holds 48 cups of punch. The
    matching glasses hold 6 ounces,
    which is $\frac{3}{4}$ of a cup. How many
    glasses of punch will you be able to
    serve from the bowl?

Short Response
Solve the problems. Use the answer
sheet to write your answers.

13. To finish a shirt, Brenda needs $16\frac{3}{4}$
    inches of lace. If she has $18\frac{7}{8}$ inches
    of lace, how much will she have left
    for another project? Explain in words
    how you determined your answer.

14. While shopping, Mrs. Lylson looks at
    a square rug for her dining room. The
    rug has an area of 256 ft^2. What is
    the length of each side of the rug?
    How did you determine your answer.

Extended Response

15. For a dinner banquet, the Crestwood
    High School decorating committee
    has set up 8 ft tables. It is
    determined that $9\frac{1}{2}$ ft of table
    covering is needed to make a table
    cloth for each table. A roll of table
    covering comes 68 ft to a roll.

    a. How many tables can be covered
       with one roll of covering? Write and
       solve an equation.

    b. How much table covering will be
       wasted per roll? Explain in words
       how you determined your answer.

    c. How many rolls of table covering
       will they need to cover 25 tables
       without piecing any of the tablecloths
       together. Explain in words how you
       determined your answer.

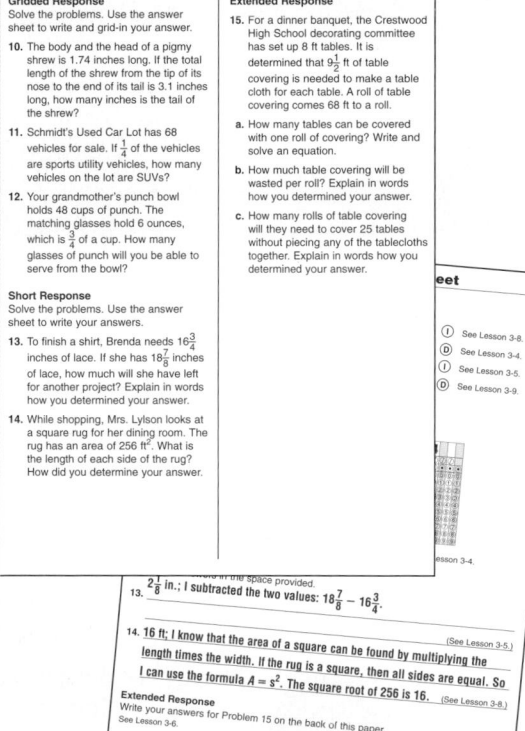

...eet

ⓘ See Lesson 3-8.
ⓓ See Lesson 3-4.
ⓘ See Lesson 3-5.
ⓓ See Lesson 3-9.

...son 3-4.

13. $2\frac{1}{8}$ in.; I subtracted the two values: $18\frac{7}{8} - 16\frac{3}{4}$.

14. 16 ft; I know that the area of a square can be found by multiplying the
    length times the width. If the rug is a square, then all sides are equal. So
    I can use the formula $A = s^2$. The square root of 256 is 16.   (See Lesson 3-8.)
                                           (See Lesson 3-5.)

    Extended Response
    Write your answers for Problem 15 on the back of this paper.
    See Lesson 3-6.

internet connect

**State-Specific Test
Practice Online**
KEYWORD: MP4 TestPrep

Test Prep Tool Kit

- Standardized Test Prep Workbook
- Countdown to Testing transparencies
- State Test Prep CD-ROM
- Standardized Test Prep Video

Customized answer sheets give
students realistic practice for
actual standardized tests.

Rational and Real Numbers

Why Learn This?

Tell students that the set of real numbers includes all numbers they have studied so far—whole numbers, fractions, decimals, square roots, etc. Tell them that numbers that are not real are called *imaginary numbers* and that they will study them in a few years. *Fractions* will now be referred to as *rational numbers.* Point out to students that all the numbers in the table can be classified as real numbers and as rational numbers.

Using Data

To begin the study of this chapter, have students:

- Give the daily iron requirement for a 16-year-old boy. **11 mg**

- Calculate how many more calories are needed by a 15-year-old boy than by a 15-year-old girl. **800**

- Calculate how much more protein a boy needs when he is 17 than when he is 10. **20 g**

internet connect

Chapter Opener Online
go.hrw.com
KEYWORD: MP4 Ch3

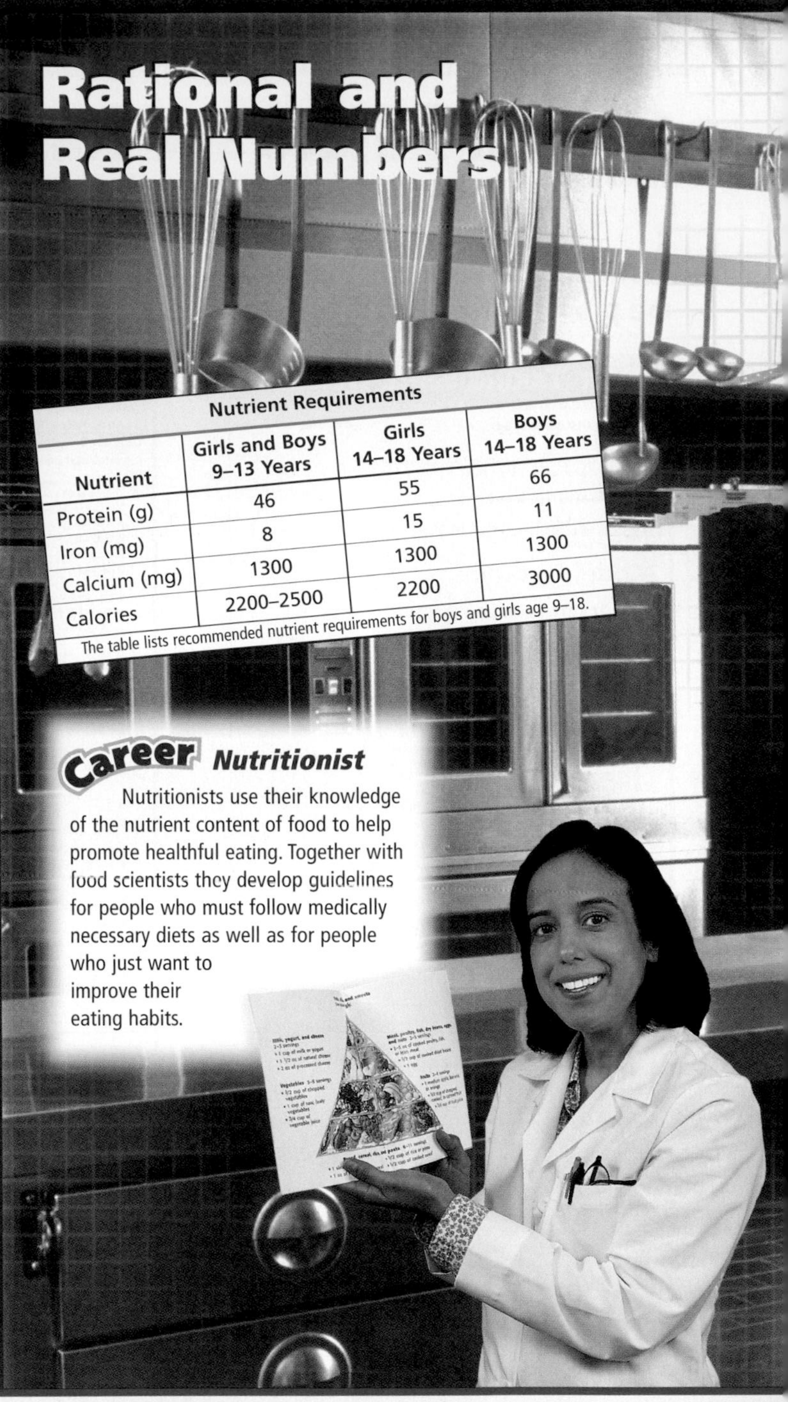

Rational and Real Numbers

Nutrient Requirements			
Nutrient	**Girls and Boys 9–13 Years**	**Girls 14–18 Years**	**Boys 14–18 Years**
Protein (g)	46	55	66
Iron (mg)	8	15	11
Calcium (mg)	1300	1300	1300
Calories	2200–2500	2200	3000

The table lists recommended nutrient requirements for boys and girls age 9–18.

Career *Nutritionist*

Nutritionists use their knowledge of the nutrient content of food to help promote healthful eating. Together with food scientists they develop guidelines for people who must follow medically necessary diets as well as for people who just want to improve their eating habits.

Problem Solving Project

Life Science and Health Connections

Purpose: To solve problems using fractions, decimals, and mixed numbers

Materials: Food for Thought worksheet

internet connect

Chapter Project Online: *go.hrw.com*
KEYWORD: MP4 PSProject3

Understand, Plan, Solve, and Look Back

Have students:

✔ Complete the Food for Thought worksheet to learn how foods can provide varying amounts of the daily-required nutrients.

✔ Research why the human body needs the nutrients in the table. What happens when a body is deficient in one of them?

✔ Bring in some labels from food products. Compare the nutrients on the labels with those in the tables and calculate the fraction of the daily requirements a serving of each food contains.

✔ Check students' work.

ARE YOU READY?

Choose the best term from the list to complete each sentence.

equivalent fraction
fraction
improper fraction
mixed number
proper fraction

1. A number that consists of a whole number and a fraction is called a(n) __?__. **mixed number**

2. A(n) __?__ is a number that represents a part of a whole. **fraction**

3. A fraction whose absolute value is greater than 1 is called a(n) __?__, and a fraction whose absolute value is between 0 and 1 is called a(n) __?__. **improper fraction; proper fraction**

4. A(n) __?__ names the same value. **equivalent fraction**

Complete these exercises to review skills you will need for this chapter.

✔ Model Fractions

Write a fraction to represent the shaded portion of each diagram.

5. 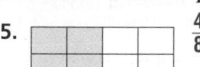 $\frac{4}{8}$

6. $\frac{4}{10}$

7.  $\frac{5}{4}$

8. 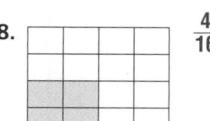 $\frac{4}{16}$

✔ Write a Fraction as a Mixed Number

Write each improper fraction as a mixed number.

9. $\frac{22}{7}$ $3\frac{1}{7}$ 10. $\frac{18}{5}$ $3\frac{3}{5}$ 11. $\frac{104}{25}$ $4\frac{4}{25}$ 12. $\frac{65}{9}$ $7\frac{2}{9}$

✔ Write a Mixed Number as a Fraction

Write each mixed number as an improper fraction.

13. $7\frac{1}{4}$ $\frac{29}{4}$ 14. $10\frac{3}{7}$ $\frac{73}{7}$ 15. $5\frac{3}{8}$ $\frac{43}{8}$ 16. $11\frac{1}{11}$ $\frac{122}{11}$

✔ Write Equivalent Fractions

Supply the missing information.

17. $\frac{3}{8} = \frac{\blacksquare}{24}$ **9**   18. $\frac{5}{13} = \frac{\blacksquare}{52}$ **20**   19. $\frac{7}{12} = \frac{\blacksquare}{36}$ **21**   20. $\frac{8}{15} = \frac{\blacksquare}{45}$ **24**

Assessing Prior Knowledge

INTERVENTION
Diagnose and Prescribe

Evaluate your students' performance on this page to determine whether intervention is necessary or whether enrichment is appropriate. Options that provide instruction, practice, and a check are listed below.

Resources for Are You Ready?

• *Are You Ready? Intervention and Enrichment*

• *Recording Sheet for Are You Ready?*
  Chapter 3 Resource Book . . . . . . p. 3

  Are You Ready? Intervention CD-ROM

ARE YOU READY?
Were students successful with Are You Ready?

NO INTERVENE ← → **YES** ENRICH

✔ **Model Fractions** 5–8
Intervention Practice, Skill 18
 **CD-ROM**
Intervention Activities, Skill 18

✔ **Write a Fraction as a Mixed Number** 9–12
Intervention Practice, Skill 21
 **CD-ROM**
Intervention Activities, Skill 21

✔ **Write a Mixed Number as a Fraction** 13–16
Intervention Practice, Skill 22
 CD-ROM
Intervention Activities, Skill 22

✔ **Write Equivalent Fractions** 17–20
Intervention Practice, Skill 24
 CD-ROM
Intervention Activities, Skill 24

Are You Ready? Enrichment, pp. 409–410

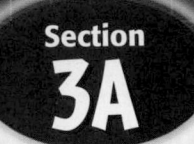

Section 3A

Rational Numbers and Operations

One-Minute Section Planner

Lesson	Materials	Resources
Lesson 3-1 Rational Numbers **NCTM:** Number and Operations, Communication **NAEP:** Number Properties 1d ☑ SAT-9 ☑ SAT-10 ☑ ITBS ☑ CTBS ☑ MAT ☑ CAT	**Optional** Teaching Transparency T2 *(CRB)* Recording Sheet for Reaching All Learners *(CRB, p. 105)*	• *Chapter 3 Resource Book,* pp. 6–15 • Daily Transparency T1, CRB • Additional Examples Transparencies T3–T5, CRB • *Alternate Openers: Explorations,* p. 19
Lesson 3-2 Adding and Subtracting Real Numbers **NCTM:** Number and Operations, Communication, Connections **NAEP:** Number Properties 3a ☑ SAT-9 ☑ SAT-10 ☑ ITBS ☑ CTBS ☑ MAT ☑ CAT	**Optional** Recording Sheet for Reaching All Learners *(CRB, p. 106)*	• *Chapter 3 Resource Book,* pp. 16–24 • Daily Transparency T6, CRB • Additional Examples Transparencies T7–T9, CRB • *Alternate Openers: Explorations,* p. 20
Lesson 3-3 Multiplying Rational Numbers **NCTM:** Number and Operations, Communication, Connections **NAEP:** Number Properties 3a ☑ SAT-9 ☑ SAT-10 ☑ ITBS ☑ CTBS ☑ MAT ☑ CAT	**Optional** Teaching Transparency T11 *(CRB)*	• *Chapter 3 Resource Book,* pp. 25–33 • Daily Transparency T10, CRB • Additional Examples Transparencies T12–T15, CRB • *Alternate Openers: Explorations,* p. 21
Lesson 3-4 Dividing Rational Numbers **NCTM:** Number and Operations, Problem Solving, Communication **NAEP:** Number Properties 3a ☑ SAT-9 ☑ SAT-10 ☑ ITBS ☑ CTBS ☑ MAT ☑ CAT	**Optional** Teaching Transparency T17 *(CRB)* Recording Sheet for Reaching All Learners *(CRB, p. 107)*	• *Chapter 3 Resource Book,* pp. 34–42 • Daily Transparency T16, CRB • Additional Examples Transparencies T18–T20, CRB • *Alternate Openers: Explorations,* p. 22
Lesson 3-5 Adding and Subtracting with Unlike Denominators **NCTM:** Number and Operations, Communication **NAEP:** Number Properties 3a ☑ SAT-9 ☑ SAT-10 ☑ ITBS ☑ CTBS ☑ MAT ☑ CAT	**Optional** Cutouts and Recording Sheet for Reaching All Learners *(CRB, pp. 108–109)*	• *Chapter 3 Resource Book,* pp. 43–52 • Daily Transparency T21, CRB • Additional Examples Transparencies T22–T24, CRB • *Alternate Openers: Explorations,* p. 23
Technology Lab 3A Explore Repeating Decimals **NCTM:** Number and Operations **NAEP:** Number Properties 1d ☐ SAT-9 ☐ SAT-10 ☐ ITBS ☐ CTBS ☐ MAT ☐ CAT	**Required** Graphing calculators	• *Technology Lab Activities,* pp. 14–15
Lesson 3-6 Solving Equations with Rational Numbers **NCTM:** Number and Operations, Communication, Connections **NAEP:** Algebra 4c ☐ SAT-9 ☑ SAT-10 ☐ ITBS ☑ CTBS ☐ MAT ☑ CAT	**Optional** Recording Sheet for Reaching All Learners *(CRB, p. 110)*	• *Chapter 3 Resource Book,* pp. 53–61 • Daily Transparency T25, CRB • Additional Examples Transparencies T26–T28, CRB • *Alternate Openers: Explorations,* p. 24
Lesson 3-7 Solving Inequalities with Rational Numbers **NCTM:** Number and Operations, Algebra, Problem Solving, Communication, Connections **NAEP:** Algebra 4c ☐ SAT-9 ☐ SAT-10 ☐ ITBS ☐ CTBS ☐ MAT ☐ CAT	**Optional** Recording Sheet for Reaching All Learners *(CRB, p. 111)*	• *Chapter 3 Resource Book,* pp. 62–70 • Daily Transparency T29, CRB • Additional Examples Transparencies T30–T33, CRB • *Alternate Openers: Explorations,* p. 25
Section 3A Assessment		• Mid-Chapter Quiz, SE p. 144 • Section 3A Quiz, AR p. 9 • *Test and Practice Generator* CD-ROM

SAT = *Stanford Achievement Tests*   **ITBS** = *Iowa Test of Basic Skills*   **CTBS** = *Comprehensive Test of Basic Skills/Terra Nova*
MAT = *Metropolitan Achievement Test*   **CAT** = *California Achievement Test*
NCTM = Complete standards can be found on pages T29–T35.   **NAEP** = Complete standards can be found on pages A54–A58.
SE = *Student Edition*   **TE** = *Teacher's Edition*   **AR** = *Assessment Resources*   **CRB** = *Chapter Resource Book*   **MK** = *Manipulatives Kit*

Section Overview

Operations with Rational Numbers   *Lessons 3-1 through 3-5, Technology Lab 3A*

Why? Many quantities and measurements are expressed with rational numbers.

A **rational number** is a number that can be expressed as a *ratio* (fraction) in the form $\frac{n}{d}$, where n and d are integers, and $d \neq 0$.

Examples:

$$5 = \frac{5}{1} \quad \frac{3}{4} \quad 1.59 = 1\frac{59}{100} = \frac{159}{100}$$

Suppose that a recipe requires $\frac{3}{4}$ cup of sugar and that sugar costs \$1.59 for 5 pounds.

5, $\frac{3}{4}$, and 1.59 are all rational numbers, but $\frac{3}{4}$ and 1.59 are not integers.

Addition

$$\frac{3}{8} + \frac{7}{8}$$
$$= \frac{3 + 7}{8}$$
$$= \frac{10}{8}$$
$$= \frac{5}{4}$$
$$= 1\frac{1}{4}$$

Add numerators and keep the common denominator.

Subtraction

$$\frac{3}{4} - \frac{1}{6}$$
$$= \frac{3}{4}\left(\frac{3}{3}\right) - \frac{1}{6}\left(\frac{2}{2}\right)$$
$$= \frac{9}{12} - \frac{2}{12}$$
$$= \frac{7}{12}$$

When necessary, multiply the fractions by **a form of 1** to obtain common denominators.

Multiplication

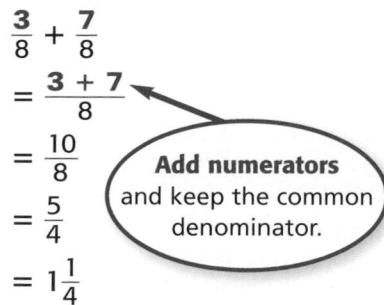

$$1\frac{3}{8}\left(-\frac{2}{3}\right)$$
$$= \frac{11}{8}\left(-\frac{2}{3}\right)$$
$$= \frac{(11)(-2)}{(8)(3)}$$
$$= \frac{-22}{24}$$
$$= -\frac{11}{12}$$

Write the mixed number $1\frac{3}{8}$ as an improper fraction, $\frac{11}{8}$.

Division

$$\frac{8}{9} \div \frac{2}{3}$$
$$= \frac{8}{9} \cdot \frac{3}{2}$$
$$= \frac{24}{18}$$
$$= \frac{4}{3}$$
$$= 1\frac{1}{3}$$

To divide by $\frac{2}{3}$, multiply by its reciprocal, $\frac{3}{2}$.

Algebra with Rational Numbers   *Lessons 3-6, 3-7*

Why? Most real-world applications involve rational numbers.

Solving Equations

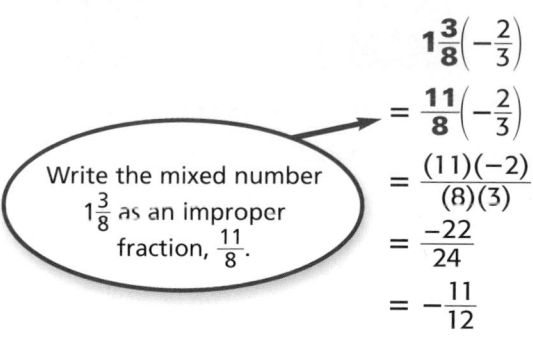

$$x - \frac{1}{3} = \frac{1}{2}$$
$$+ \frac{1}{3} = + \frac{1}{3}$$
$$x = \frac{1}{2} + \frac{1}{3}$$
$$x = \frac{3}{6} + \frac{2}{6}$$
$$x = \frac{5}{6}$$

$$1.25m = 40$$
$$\frac{1.25m}{1.25} = \frac{40}{1.25}$$
$$m = 32$$

$$\frac{2}{3}x = -\frac{1}{6}$$
$$\frac{2}{3}x \cdot \frac{3}{2} = -\frac{1}{6} \cdot \frac{3}{2}$$
$$x = -\frac{3}{12}$$
$$x = -\frac{1}{4}$$

Solving Inequalities

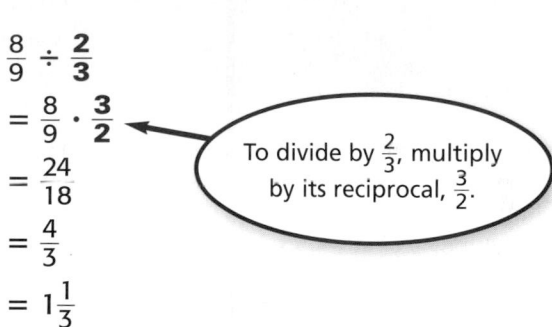

$$-2\frac{1}{2}y < 5$$
$$-\frac{5}{2}y < 5$$
$$\left(-\frac{2}{5}\right)\left(-\frac{5}{2}\right)y > \left(-\frac{2}{5}\right)5$$
$$y > -2$$

Reverse the inequality symbol when you multiply or divide by a negative number.

Pacing: Traditional 1 day
Block $\frac{1}{2}$ day

Objective: Students write rational numbers in equivalent forms.

Warm Up

Divide.

1. $36 \div 3$ **12** **2.** $144 \div 6$ **24**

3. $68 \div 17$ **4** **4.** $345 \div 115$ **3**

5. $1024 \div 64$ **16**

Problem of the Day

An ice cream parlor has 6 flavors of ice cream. A dish with two scoops can have any two flavors, including the same flavor twice. How many different double-scoop combinations are possible? **21**

Available on Daily Transparency in CRB

Math Humor

Numbers like $\frac{1}{6}$ and $\frac{2}{3}$ in decimal form are like people who make sense but say the same thing over and over. They're rational, but they repeat!

3-1 Rational Numbers

Learn to write rational numbers in equivalent forms.

Vocabulary

rational number

relatively prime

In 2001, the Wimbledon tennis tournament increased the number of "seeds" from 16 to 32. Since Wimbledon has 128 total players, $\frac{32}{128}$ of the players are seeded.

A **rational number** is any number that can be written as a fraction $\frac{n}{d}$, where n and d are integers and $d \neq 0$.

Decimals that terminate or repeat are rational numbers.

Some ranked players, like Venus and Serena Williams, are "seeded" so that they will not meet until late in a tournament.

Numerator

$$\frac{n}{d}$$

Denominator

Rational Number	Description	Written as a Fraction
-1.5	Terminating decimal	$\frac{-15}{10}$
$0.8\overline{3}$	Repeating decimal	$\frac{5}{6}$

The goal of simplifying fractions is to make the numerator and the denominator *relatively prime*. **Relatively prime** numbers have no common factors other than 1.

You can often simplify fractions by dividing both the numerator and denominator by the same nonzero integer. You can simplify the fraction $\frac{12}{15}$ to $\frac{4}{5}$ by dividing both the numerator and denominator by 3.

12 of the 15 boxes are shaded.  $\frac{12}{15} = \frac{4}{5}$ 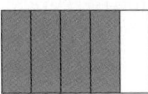 *4 of the 5 boxes are shaded.*

The same total area is shaded.

EXAMPLE 1 **Simplifying Fractions**

Simplify.

A $\frac{6}{9}$

$\frac{6}{9} = \frac{6 \div 3}{9 \div 3}$ $6 = 2 \cdot 3$
$9 = 3 \cdot 3$; 3 is a common factor.

$= \frac{2}{3}$ *Divide the numerator and denominator by 3.*

---

1 Introduce

Alternate Opener

EXPLORATION

3-1 Rational Numbers

Comparing numbers with $\frac{1}{2}$ is useful in many situations.

Phil surveyed 317 voters and found that 156 supported his reelection. He reasoned the following way to determine whether he had a majority.
$156 \times 2 = 150 \times 2 + 6 \times 2$

$= 300 + 12$

$= 312$

Phil concluded that he did not have a majority. Look at how Phil checked his estimate with a calculator.

Double the numerator and compare it with the denominator of each fraction to determine whether the fraction is greater than or less than $\frac{1}{2}$. Check your work with a calculator.

	Fraction	$> \frac{1}{2}$?	$< \frac{1}{2}$?
1.	$\frac{51}{101}$		
2.	$\frac{221}{425}$		
3.	$\frac{260}{513}$		
4.	$\frac{578}{1152}$		

Think and Discuss

5. Explain how you checked your work with a calculator.

6. Name two fractions with the denominator 437 that are close in value to $\frac{1}{2}$.

Motivate

Show students this diagram:

Ask students: "What fraction of each bar is shaded? Is there a way to write one of these fractions as a decimal?" The responses should illustrate that there are three equivalent forms of the same rational number: $\frac{2}{5}$, $\frac{4}{10}$, and 0.4.

Exploration worksheet and answers on Chapter 3 Resource Book pp. 7 and 116

2 Teach

Lesson Presentation

Guided Instruction

In this lesson, students learn to write rational numbers in equivalent forms. Review the definition of *rational number* (Teaching Transparency T2, CRB). Discuss how to simplify a fraction and the meaning of *relatively prime*. Review the place-value chart (Teaching Transparency T2, CRB) to prepare students for writing decimals as fractions. Explain that the fraction bar indicates division of the numerator by the denominator. Work Example 3B out a few extra places so that students can see the pattern developing. Point out that many fractions will have a repeating pattern of digits when written as decimals.

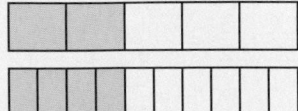

Simplify.

B $\frac{21}{25}$ $21 = 3 \cdot 7$
$25 = 5 \cdot 5$; there are no common factors.

$\frac{21}{25} = \frac{21}{25}$ 21 and 25 are relatively prime.

C $\frac{-24}{32}$

$\frac{-24}{32} = \frac{-24 \div 8}{32 \div 8}$ $24 = \boxed{2 \cdot 2 \cdot 2} \cdot 3$
$32 = \boxed{2 \cdot 2 \cdot 2} \cdot 2 \cdot 2$ 8 is a common factor.

$= \frac{-3}{4}$, or $-\frac{3}{4}$ Divide the numerator and denominator by 8.

Remember!

$\frac{0}{a} = 0$ for $a \neq 0$

$\frac{a}{a} = 1$ for $a \neq 0$

$\frac{-3}{4} = \frac{3}{-4} = -\frac{3}{4}$

To write a finite decimal as a fraction, identify the place value of the digit farthest to the right. Then write all of the digits after the decimal point as the numerator with the place value as the denominator.

Place Value

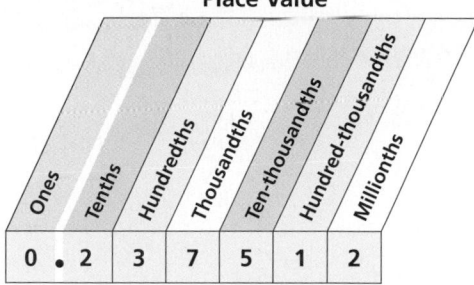

0	.	2	3	7	5	1	2

EXAMPLE **2** **Writing Decimals as Fractions**

Write each decimal as a fraction in simplest form.

A 0.5

0.5

$= \frac{5}{10}$ 5 is in the tenths place.

$= \frac{1}{2}$ Simplify by dividing by the common factor 5.

B −2.37

−2.37

$= -2\frac{37}{100}$ 7 is in the hundredths place.

C 0.8716

0.8716

$= \frac{8716}{10,000}$ 6 is in the ten-thousandths place.

$= \frac{2179}{2500}$ Simplify by dividing by the common factor 4.

COMMON ERROR
ALERT

Some students might partially simplify a fraction but not write it in simplest form. A fraction is not in simplest form until the numerator and denominator are relatively prime.

$\frac{24 \div 4}{32 \div 4} = \frac{6}{8}, \frac{6 \div 2}{8 \div 2} = \frac{3}{4}$

Additional Examples

Example 1

Simplify.

A. $\frac{5}{10}$ $\frac{1}{2}$ **B.** $\frac{16}{80}$ $\frac{1}{5}$

C. $\frac{-18}{29}$ $-\frac{18}{29}$

Example 2

Write each decimal as a fraction in simplest form.

A. −0.8 $-\frac{4}{5}$ **B.** 5.37 $5\frac{37}{100}$

C. 0.622 $\frac{311}{500}$

Example 3

Write each fraction as a decimal.

A. $\frac{11}{9}$ $1.\overline{2}$ **B.** $\frac{7}{20}$ 0.35

Reaching All Learners
Through Critical Thinking

Give each student a recording sheet (Chapter 3 Resource Book p. 105) with sets of fractions (e.g., $\frac{1}{2}, \frac{2}{3}, \frac{3}{4}, \frac{4}{5}$) in random order. Have students write each fraction as a decimal and use the values to write the fractions in order from least to greatest. Have students analyze the pattern, and then see if they can put other fractions in a set in ascending order without writing them as decimals.

y

3-1 Rational Numbers   **113**

To write a fraction as a decimal, divide the numerator by the denominator. You can use long division.

When writing a long division problem from a fraction, put the numerator inside the "box," or division symbol. It may help to write the numerator first and then say "divided by" to yourself as you write the division symbol.

$$\frac{\text{numerator}}{\text{denominator}} \rightarrow \text{denominator} \overline{)\text{numerator}}$$

EXAMPLE 3 **Writing Fractions as Decimals**

Write each fraction as a decimal.

A $\frac{5}{4}$

$$
\begin{array}{r}
1.25 \\
4\overline{)5.00} \\
-4\downarrow \\
\hline
1\,0 \\
-8\downarrow \\
\hline
20 \\
-20 \\
\hline
0
\end{array}
$$

The remainder is 0. This is a terminating decimal.

The fraction $\frac{5}{4}$ is equivalent to the decimal 1.25.

B $\frac{1}{6}$

$$
\begin{array}{r}
0.1\overline{6} \\
6\overline{)1.000} \\
-6\downarrow \\
\hline
40 \\
-36\downarrow \\
\hline
40
\end{array}
$$

The pattern repeats, so draw a bar over the 6 to indicate that this is a repeating decimal.

The fraction $\frac{1}{6}$ is equivalent to the decimal $0.1\overline{6}$.

Think and Discuss

1. **Explain** how you can be sure that a fraction is simplified.

2. **Give** the sign of a fraction in which the numerator is negative and the denominator is negative.

3 Close

Summarize

Remind students that a fraction, such as $\frac{3}{4}$, is another way of showing a division problem (3 divided by 4). To write a fraction as a decimal, solve the division problem. Point out that fractions and decimals are two ways to show numbers that are smaller than 1. Ask students to write the value of 0.20 in as many ways as they can, and write their responses on the chalkboard. Point out that all correct responses have the same value.

Possible answers:

0.2; 0.200; $\frac{20}{100}$; $\frac{2}{10}$; $\frac{1}{5}$; 20 ÷ 100; 100$\overline{)20}$; twenty-hundredths; two-tenths

Answers to Think and Discuss

1. Possible answer: When the numerator and denominator have no common factors other than 1, the fraction is in simplest form.

2. Positive; a negative number divided by a negative number gives a positive quotient.

3-1 Exercises

FOR EXTRA PRACTICE
see page 736

internet connect
Homework Help Online
go.hrw.com   Keyword: MP4 3-1

go.hrw.com

3-1 PRACTICE & ASSESS

GUIDED PRACTICE

See Example **1** Simplify.

1. $\frac{12}{15}$ **$\frac{4}{5}$**   2. $\frac{6}{10}$ **$\frac{3}{5}$**   3. $-\frac{16}{24}$ **$-\frac{2}{3}$**   4. $\frac{11}{27}$ **$\frac{11}{27}$**

5. $\frac{57}{69}$ **$\frac{19}{23}$**   6. $-\frac{20}{24}$ **$-\frac{5}{6}$**   7. $-\frac{7}{27}$ **$-\frac{7}{27}$**   8. $\frac{49}{112}$ **$\frac{7}{16}$**

See Example **2** Write each decimal as a fraction in simplest form.

9. 0.75 **$\frac{3}{4}$**   10. 1.125 **$1\frac{1}{8}$**   11. 0.431 **$\frac{431}{1000}$**   12. 0.8 **$\frac{4}{5}$**

13. −2.2 **$-2\frac{1}{5}$**   14. 0.625 **$\frac{5}{8}$**   15. 3.21 **$3\frac{21}{100}$**   16. −0.3878 **$-\frac{1939}{5000}$**

See Example **3** Write each fraction as a decimal.

17. $\frac{7}{8}$ **0.875**   18. $\frac{3}{5}$ **0.6**   19. $\frac{5}{12}$ **$0.41\overline{6}$**   20. $\frac{3}{4}$ **0.75**

21. $\frac{16}{4}$ **4.0**   22. $\frac{1}{8}$ **0.125**   23. $\frac{12}{5}$ **2.4**   24. $\frac{9}{4}$ **2.25**

INDEPENDENT PRACTICE

See Example **1** Simplify.

25. $\frac{21}{28}$ **$\frac{3}{4}$**   26. $\frac{25}{60}$ **$\frac{5}{12}$**   27. $-\frac{17}{34}$ **$-\frac{1}{2}$**   28. $-\frac{18}{21}$ **$-\frac{6}{7}$**

29. $\frac{13}{17}$ **$\frac{13}{17}$**   30. $\frac{22}{35}$ **$\frac{22}{35}$**   31. $\frac{64}{76}$ **$\frac{16}{19}$**   32. $-\frac{78}{126}$ **$-\frac{13}{21}$**

See Example **2** Write each decimal as a fraction in simplest form.

33. 0.4 **$\frac{2}{5}$**   34. 3.5 **$3\frac{1}{2}$**   35. 0.71 **$\frac{71}{100}$**   36. −0.183 **$-\frac{183}{1000}$**

37. 1.377 **$1\frac{377}{1000}$** 38. 1.450 **$1\frac{9}{20}$** 39. −1.4 **$-1\frac{2}{5}$** 40. −2.9 **$-2\frac{9}{10}$**

See Example **3** Write each fraction as a decimal.

41. $\frac{3}{8}$ **0.375**   42. $\frac{11}{12}$ **$0.91\overline{6}$**   43. $\frac{7}{5}$ **1.4**   44. $\frac{9}{20}$ **0.45**

45. $\frac{34}{50}$ **0.68**   46. $\frac{23}{5}$ **4.6**   47. $\frac{29}{25}$ **1.16**   48. $\frac{7}{3}$ **$2.\overline{3}$**

PRACTICE AND PROBLEM SOLVING

49. Make up a fraction that cannot be simplified that has 36 as its denominator. **Possible answer: $\frac{25}{36}$**

50. Make up a fraction that cannot be simplified that has 27 as its denominator. **Possible answer: $\frac{2}{27}$**

Students may want to refer back to the lesson examples.

Assignment Guide

If you finished Example **1** assign:
Core 1–8, 25–32, 49, 50, 58–63
Enriched 1–7 odd, 25–32, 49, 50, 53, 55, 57–63

If you finished Example **2** assign:
Core 1–16, 25–40, 49, 50, 58–63
Enriched 9–15 odd, 25–40, 49, 50, 53–55, 57–63

If you finished Example **3** assign:
Core 1–48, 58–63
Enriched 1–23 odd, 25–63

Notes

RETEACH 3-1

LESSON 3-1 Reteach
Rational Numbers

A rational number is a *ratio* of two integers.

Rational Number = $\frac{\text{Integer}}{\text{Integer}}$ ← Numerator ← Denominator

Rational Numbers
| Integers | Fractions |
| Repeating Decimals | Terminating Decimals |

The set of rational numbers contains:
all integers
all fractions
decimals that repeat, such as $0.4\overline{6}$
decimals that terminate, such as 3.5

To simplify a fraction, divide numerator and denominator by the highest common factor.

$\frac{5}{15} = \frac{5 \div 5}{15 \div 5} = \frac{1}{3}$

Complete to simplify each fraction.

1. $\frac{8}{16} = \frac{8 \div 8}{16 \div 8} = \frac{1}{2}$ 2. $\frac{15}{45} = \frac{15 \div 15}{45 \div 15} = \frac{1}{3}$

3. $\frac{12}{30} = \frac{12 \div 6}{30 \div 6} = \frac{2}{5}$ 4. $\frac{12}{24} = \frac{12 \div 12}{24 \div 12} = \frac{1}{2}$

5. $\frac{5}{35} = \frac{5 \div 5}{35 \div 5} = \frac{1}{7}$ 6. $\frac{14}{49} = \frac{14 \div 7}{49 \div 7} = \frac{2}{7}$

Simply each fraction.

7. $\frac{8}{56}$ **$\frac{1}{7}$**   8. $\frac{15}{50}$ **$\frac{3}{10}$**   9. $\frac{8}{36}$ **$\frac{2}{9}$**

To write a decimal as a fraction, use the number of decimal places to get the denominator. Then simplify.

$0.4 = \frac{4}{10} = \frac{4 \div 2}{10 \div 2} = \frac{2}{5}$

Complete to write each decimal as a fraction in simplest form.

10. $0.25 = \frac{25}{100} = \frac{25 \div 25}{100 \div 25} = \frac{1}{4}$ 11. $0.375 = \frac{375}{1000} = \frac{375 \div 125}{1000 \div 125} = \frac{3}{8}$

Write each decimal as a fraction in simplest form.

12. 0.55 **$\frac{11}{20}$**   13. 0.32 **$\frac{8}{25}$**

PRACTICE 3-1

LESSON 3-1 Practice B
Rational Numbers

Simplify.

1. $\frac{6}{9}$ **$\frac{2}{3}$**   2. $\frac{8}{64}$ **$\frac{1}{8}$**   3. $\frac{9}{36}$ **$\frac{1}{4}$**   4. $-\frac{7}{28}$ **$-\frac{1}{4}$**

5. $\frac{15}{40}$ **$\frac{3}{8}$**   6. $-\frac{4}{48}$ **$-\frac{1}{12}$**   7. $-\frac{14}{63}$ **$-\frac{2}{9}$**   8. $\frac{12}{72}$ **$\frac{1}{6}$**

Write each decimal as a fraction in simplest form.

9. 0.72 **$\frac{18}{25}$**   10. 0.058 **$\frac{29}{500}$**   11. −1.65 **$-1\frac{13}{20}$**   12. 2.1 **$2\frac{1}{10}$**

13. 0.036 **$\frac{9}{250}$**   14. −4.06 **$-4\frac{3}{50}$**   15. 2.305 **$2\frac{61}{200}$**   16. 0.0064 **$\frac{4}{625}$**

17. −0.60 **$-\frac{3}{5}$**   18. 6.95 **$6\frac{19}{20}$**   19. 0.016 **$\frac{2}{125}$**   20. 0.0005 **$\frac{1}{2000}$**

Write each fraction as a decimal.

21. $\frac{1}{8}$ **0.125**   22. $\frac{8}{3}$ **$2.66\overline{6}$**   23. $\frac{14}{15}$ **$0.93\overline{3}$**   24. $\frac{16}{5}$ **3.2**

25. $\frac{11}{16}$ **0.6875**   26. $\frac{7}{9}$ **$0.77\overline{7}$**   27. $\frac{4}{5}$ **0.8**   28. $\frac{31}{25}$ **1.24**

29. Make up a fraction that cannot be simplified that has 24 as its denominator.
sample answer: $\frac{5}{24}$

Math Background

A fraction can be written in simplest form as either a mixed number or an improper fraction. Students may prefer a mixed number because they will probably have a clearer concept of its value if it can be easily compared to integers. For example, students may understand that $5\frac{3}{7}$ is between 5 and 6, but they may have difficulty making the same conclusion about the equivalent improper fraction $\frac{38}{7}$. For this reason, the answers to the Exercises have been given as mixed numbers. In more advanced courses, however, improper fractions may prove more useful.

Answers

51. a. $\frac{3}{4}, \frac{1}{6}, \frac{5}{9}, \frac{17}{20}, \frac{13}{32}, \frac{11}{25}, \frac{19}{24}, \frac{8}{15}$

    b. 2×2; 2×3; 3×3; $2 \times 2 \times 5$; $2 \times 2 \times 2 \times 2 \times 2$; 5×5; $2 \times 2 \times 2 \times 3$; 3×5

    c. 0.75 terminating; $0.1\overline{6}$ repeating; $0.\overline{5}$ repeating; 0.85 terminating; 0.40625 terminating; 0.44 terminating; $0.791\overline{6}$ repeating; $0.5\overline{3}$ repeating

53. GCF = 4; $\frac{12}{19}$; No, the fraction cannot be further simplified because the numerator and denominator are relatively prime.

55. Possible answer: A negative number divided by a negative number results in a positive. The simplified fraction should be positive.

56. Possible answer: If the prime factors of the denominator are 2's and 5's, the fraction is equivalent to a terminating decimal. If there are any other prime factors, the fraction is equivalent to a repeating decimal.

Journal

Have students write some real-world situations that involve fractions and/or decimals. Examples might include cooking recipes, distance measurements, and monetary calculations.

Test Prep Doctor

For Exercise 62, some students might think that answer choices **A** and **B** are the same answer. Have these students choose some numbers that would satisfy the inequalities to see that they are actually different.

Lesson Quiz

Simplify.

1. $\frac{18}{42}$ $\frac{3}{7}$ **2.** $\frac{15}{21}$ $\frac{5}{7}$

Write each decimal as a fraction in simplest form.

3. 0.27 $\frac{27}{100}$ **4.** -0.625 $-\frac{5}{8}$

5. Write $\frac{13}{6}$ as a decimal. $2.1\overline{6}$

6. Tommy had 13 hits in 40 at bats for his baseball team. What is his batting average? (Batting average is the number of hits divided by the number of at bats, expressed as a decimal.)   **0.325**

Available on Daily Transparency in CRB

---

51. a. Simplify each fraction below.

$$\frac{9}{12} \qquad \frac{5}{30} \qquad \frac{15}{27} \qquad \frac{68}{80}$$

$$\frac{39}{96} \qquad \frac{22}{50} \qquad \frac{57}{72} \qquad \frac{32}{60}$$

    b. Write the denominator of each simplified fraction as the product of prime factors.

    c. Write each simplified fraction as a decimal. Label each as a terminating or repeating decimal.

52. The ruler is marked at every $\frac{1}{16}$ in. Do the labeled measurements convert to terminating or repeating decimals? **terminating**

53. Remember that the greatest common factor, GCF, is the largest common factor of two or more given numbers. Find and remove the GCF of 48 and 76 from the fraction $\frac{48}{76}$. Can the resulting fraction be further simplified? Explain.

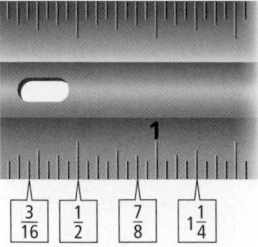

$$\frac{3}{16} \quad \frac{1}{2} \quad \frac{7}{8} \quad 1\frac{1}{4}$$

54. Prices on one stock market are shown using decimal equivalents for fractions or mixed numbers. Write the stock price 13.625 as a mixed number. $13\frac{5}{8}$

 55. WHAT'S THE ERROR? A student simplified a fraction in this manner: $\frac{-12}{-18} = -\frac{2}{3}$. What error did the student make?

 56. WRITE ABOUT IT Using your answers to Exercise 51, examine the prime factors in the denominators of the simplified fractions that are equivalent to terminating decimals. Then examine the prime factors in the denominators of the simplified fractions that are equivalent to repeating decimals. What pattern do you see?

   57. CHALLENGE A student simplified a fraction to $-\frac{3}{7}$ by removing the common factors, which were 3 and 7. What was the original fraction?

$$-\frac{63}{147}$$

Spiral Review

Evaluate each expression for the given values of the variable. (Lesson 1-1)

58. $3x + 5$ for $x = 2$ and $x = 3$ **11; 14**   **59.** $4(x + 1)$ for $x = 6$ and $x = 11$ **28; 48**

60. $2x - 4$ for $x = 5$ and $x = 7$ **6; 10**   **61.** $7(3x + 2)$ for $x = 1$ and $x = 0$ **35; 14**

62. TEST PREP Solve the inequality $7 > \frac{x}{3}$. (Lesson 1-5)   **B**

   A $21 < x$ **B** $x < 21$ **C** $2.333 > x$ **D** $\frac{7}{3} > x$

63. TEST PREP Solve the inequality $8x \le 24$. (Lesson 1-5)   **H**

   F $x \le 32$ **G** $x < 3$ **H** $x \le 3$ **J** $x \le 16$

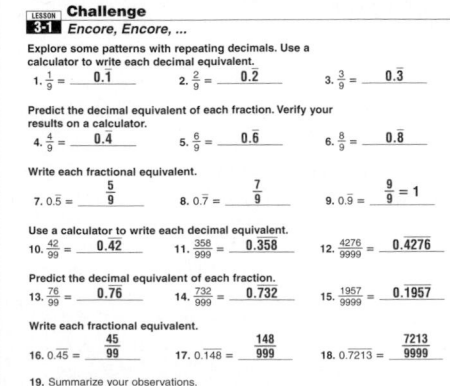

CHALLENGE 3-1

LESSON 3-1 Challenge
Encore, Encore, ...

Explore some patterns with repeating decimals. Use a calculator to write each decimal equivalent.

1. $\frac{1}{9} = $ **0.1**   **2.** $\frac{2}{9} = $ **0.2**   **3.** $\frac{3}{9} = $ **0.3**

Predict the decimal equivalent of each fraction. Verify your results on a calculator.

4. $\frac{4}{9} = $ **0.4**   **5.** $\frac{6}{9} = $ **0.6**   **6.** $\frac{8}{9} = $ **0.8**

Write each fractional equivalent.

7. $0.\overline{5} = $ $\frac{5}{9}$ **8.** $0.\overline{7} = $ $\frac{7}{9}$ **9.** $0.\overline{9} = $ $\frac{9}{9} = 1$

Use a calculator to write each decimal equivalent.

10. $\frac{42}{99} = $ **0.42**   **11.** $\frac{358}{999} = $ **0.358**   **12.** $\frac{4276}{9999} = $ **0.4276**

Predict the decimal equivalent of each fraction.

13. $\frac{76}{99} = $ **0.76**   **14.** $\frac{732}{999} = $ **0.732**   **15.** $\frac{1957}{9999} = $ **0.1957**

Write each fractional equivalent.

16. $0.\overline{45} = $ $\frac{45}{99}$ **17.** $0.\overline{148} = $ $\frac{148}{999}$ **18.** $0.\overline{7213} = $ $\frac{7213}{9999}$

19. Summarize your observations.

A single digit repeating decimal equates to a fraction with denominator 9, a 2-digit repeating decimal to denominator 99, and so on.

PROBLEM SOLVING 3-1

LESSON 3-1 Problem Solving
Rational Numbers

Write the correct answer.

1. Fill in the table below which shows the sizes of drill bits in a set.

2. Do the drill bit sizes convert to repeating or terminating decimals?

Terminating decimals

13-Piece Drill Bit Set

Fraction	Decimal	Fraction	Decimal	Fraction	Decimal
$\frac{1}{4}$"	0.25	$\frac{11}{64}$"	0.171875	$\frac{3}{32}$"	0.09375
$\frac{15}{64}$"	0.234375	$\frac{5}{32}$"	0.15625	$\frac{5}{64}$"	0.078125
$\frac{7}{32}$"	0.21875	$\frac{9}{64}$"	0.140625	$\frac{1}{16}$"	0.0625
$\frac{13}{64}$"	0.203125	$\frac{1}{8}$"	0.125		
$\frac{3}{16}$"	0.1875	$\frac{7}{64}$"	0.109375		

Use the table at the right that lists the world's smallest nations. Choose the letter for the best answer.

3. What is the area of Vatican City expressed as a fraction in simplest form?

   A $\frac{8}{50}$ **C** $\frac{17}{1000}$

   B $\frac{4}{25}$ ⓓ $\frac{17}{100}$

World's Smallest Nations

Nation	Area (square miles)
Vatican City	0.17
Monaco	0.75
Nauru	8.2

4. What is the area of Monaco expressed as a fraction in simplest form?

   F $\frac{75}{100}$ ⓗ $\frac{3}{4}$

   G $\frac{15}{20}$ **J** $\frac{2}{3}$

5. What is the area of Nauru expressed as a mixed number?

   A $8\frac{1}{50}$ **C** $8\frac{2}{100}$

   B $8\frac{2}{50}$ ⓓ $8\frac{1}{5}$

6. The average annual precipitation in Miami, FL is 57.55 inches. Express 57.55 as a mixed number.

   ⓕ $57\frac{11}{20}$ **H** $57\frac{5}{100}$

   G $57\frac{55}{1000}$ **J** $57\frac{1}{20}$

7. The average annual precipitation in Norfolk, VA is 45.22 inches. Express 45.22 as a mixed number.

   ⓐ $45\frac{11}{50}$ **C** $45\frac{11}{20}$

   B $45\frac{22}{1000}$ **D** $45\frac{1}{5}$

3-2 Adding and Subtracting Rational Numbers

Learn to add and subtract decimals and rational numbers with like denominators.

The 100-meter dash is measured in thousandths of a second, so runners must react quickly to the starter pistol.

If you subtract a runner's reaction time from the total race time, you can find the amount of time the runner took to run the actual 100-meter distance.

Pressurized pads in the starting blocks ensure that a runner does not "jump the gun."

EXAMPLE 1 Sports Application

In the 2001 World Championships 100-meter dash, it took Maurice Green 0.132 seconds to react to the starter pistol. His total race time, including this reaction time, was 9.82 seconds. How long did it take him to run the actual 100 meters?

$$9.820 \longleftarrow \textit{Add a zero so the decimals align.}$$
$$\underline{-\ 0.132}$$
$$9.688$$

The time he spent running the actual distance was 9.688 seconds.

EXAMPLE 2 Using a Number Line to Add Rational Numbers

Use a number line to find each sum.

A $-0.4 + 1.3$

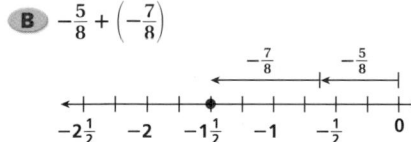

Move left 0.4 units. From -0.4, move right 1.3 units.

You finish at 0.9, so $-0.4 + 1.3 = 0.9$.

B $-\frac{5}{8} + \left(-\frac{7}{8}\right)$

Move left $\frac{5}{8}$ units. From $-\frac{5}{8}$, move left $\frac{7}{8}$ units.

You finish at $-\frac{12}{8}$, which simplifies to $-\frac{3}{2} = -1\frac{1}{2}$.

---

1 Introduce
Alternate Opener

EXPLORATION

3-2 Adding and Subtracting Rational Numbers

Beckie has $454.96 in a checking account. She needs to pay bills in the amounts $25.95, $313.00, $45.76, and $87.95.

Beckie estimates the following:

Account: $454.96 ≈ $455.00

Bills: $25.95 ≈ $ 25.00
        $313.00 = $310.00
        $45.76 ≈ $ 45.00
        $87.95 ≈ $ 90.00
                 $470.00

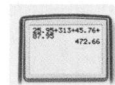

Beckie determines that she does not have enough money in her account to pay her bills. She then checks her estimate with a calculator.

Estimate the solution to each expression in the table. Then use a calculator to solve.

		Estimate	Actual
1.	120 − 9.8		
2.	45 − 17.8 + 15.9 + 16.1 − 1.07		
3.	88.10 + 109.85		
4.	34.12 − 18.30 + 65.25		

Think and Discuss
5. **Describe** the estimation strategies you used.

Motivate

Use the following examples to review the rules for adding signed numbers (Lesson 2-1):

Same signs: $5 + 2 = 7$, $-5 + (-2) = -7$

Opposite signs: $5 + (-2) = 3$, $-5 + 2 = -3$

Review how to use a number line to perform these additions. Remind students that to subtract a number, you add its opposite (Lesson 2-2).

Exploration worksheet and answers on Chapter 3 Resource Book pp. 17 and 118

3-2 Organizer

Pacing: Traditional 1 day
Block $\frac{1}{2}$ day

Objective: Students add and subtract decimals and rational numbers with like denominators.

Warm Up
Simplify.
1. $\frac{21}{14}$ $1\frac{1}{2}$ 2. $\frac{12}{30}$ $\frac{2}{5}$
3. $\frac{24}{56}$ $\frac{3}{7}$

Write each decimal as a fraction in simplest form.
4. 1.15 $1\frac{3}{20}$ 5. -0.22 $-\frac{11}{50}$

Problem of the Day

Four sprinters run a race. In how many different ways can they arrive at the finish line, assuming there are no ties? **24**

Available on Daily Transparency in CRB

Math Humor

How do we know that the fractions $\frac{3}{c}$, $\frac{6}{c}$, and $\frac{8}{c}$ are not from the United States? Because their numerators are all over c's!

2 Teach *Lesson Presentation*

Guided Instruction

In this lesson, students learn to add and subtract decimals and rational numbers with like denominators. Review with students how to align decimals for addition and subtraction. Explain that you can't add or subtract digits that are in different places because they don't share the same value. Review Example 2 to show students how to add and subtract fractions using a number line, and then point out that the same results are obtained by adding or subtracting the numerators and keeping the same denominator. Remind students to write their answers in simplest form.

Additional Examples

Example 1

In August 2001 at the World University Games in Beijing, China, Jimyria Hicks ran the 200-meter dash in 24.08 seconds. Her best time at the U.S. Senior National Meet in June of the same year was 23.35 seconds. How much faster did she run in June?

She ran 0.73 second faster in June.

Example 2

Use a number line to find each sum.

A. $0.3 + (-1.2)$ -0.9 **B.** $\frac{1}{5} + \frac{2}{5}$ $\frac{3}{5}$

Example 3

Add or subtract.

A. $-\frac{2}{9} - \frac{5}{9}$ $-\frac{7}{9}$ **B.** $\frac{6}{7} + \left(-\frac{3}{7}\right)$ $\frac{3}{7}$

Example 4

Evaluate each expression for the given value of the variable.

A. $12.1 - x$ for $x = -0.1$ 12.2

B. $\frac{7}{10} + m$ for $m = 3\frac{1}{10}$ $3\frac{4}{5}$

ADDING AND SUBTRACTING WITH LIKE DENOMINATORS		
Words	Numbers	Algebra
To add or subtract rational numbers with the same denominator, add or subtract the numerators and keep the denominator.	$\frac{2}{7} + -\frac{4}{7} = \frac{2 + (-4)}{7}$ $= \frac{-2}{7}$, or $-\frac{2}{7}$	$\frac{a}{d} + \frac{b}{d} = \frac{a+b}{d}$

EXAMPLE 3 Adding and Subtracting Fractions with Like Denominators

Add or subtract.

A $\frac{6}{11} + \frac{9}{11}$

$\frac{6}{11} + \frac{9}{11} = \frac{6+9}{11}$ *Add numerators. Keep the denominator.*

$= \frac{15}{11}$, or $1\frac{4}{11}$

B $-\frac{3}{8} - \frac{5}{8}$

$-\frac{3}{8} - \frac{5}{8} = \frac{-3}{8} + \frac{-5}{8}$ $-\frac{5}{8}$ *can be written as* $\frac{-5}{8}$.

$= \frac{-3 + (-5)}{8} = \frac{-8}{8} = -1$

EXAMPLE 4 Evaluating Expressions with Rational Numbers

Evaluate each expression for the given value of the variable.

A $23.8 + x$ for $x = -41.3$

$23.8 + (-41.3)$ *Substitute −41.3 for x.*

-17.5 *Think: 41.3 − 23.8. 41.3 > 23.8. Use sign of 41.3.*

B $-\frac{1}{8} + t$ for $t = 2\frac{5}{8}$

$-\frac{1}{8} + 2\frac{5}{8}$ *Substitute* $2\frac{5}{8}$ *for t.*

$= \frac{-1}{8} + \frac{21}{8}$ $2\frac{5}{8} = \frac{2(8) + 5}{8} = \frac{21}{8}$

$= \frac{-1 + 21}{8} = \frac{20}{8}$ *Add numerators. Keep the denominator.*

$= \frac{5}{2}$, or $2\frac{1}{2}$

Think and Discuss

1. **Give an example** of an addition problem that involves simplifying an improper fraction in the final step.

2. **Explain** why $\frac{7}{9} + \frac{7}{9}$ does not equal $\frac{14}{18}$.

---

3 Close

Reaching All Learners
Through Critical Thinking

Give each pair or group of students a recording sheet (Chapter 3 Resource Book p. 106) with sets of related exercises such as these:

1. $2.51 + 4.2$ **6.71**

2. $-2.51 + (-4.2)$ **−6.71**

3. $-2.51 + 4.2$ **1.69**

4. $2.51 + (-4.2)$ **−1.69**

Have students complete the worksheet and compare the answers for each set. Then have each student create a similar set of problems, exchange with his or her partner, and solve the problems.

Summarize

Remind students that decimal points must be aligned when adding or subtracting decimals. Remind students that to add or subtract fractions with like denominators, they should add or subtract the numerators and keep the same denominator. Ask students if it's possible to add a decimal and a fraction, and ask them how they would do it.

Possible answer: Yes; you can write the fraction as a decimal and then add the decimals together, or you can write the decimal as a fraction and add the fractions together.

Answers to Think and Discuss

Possible answers:

1. $\frac{5}{8} + \frac{5}{8} = \frac{10}{8} = \frac{5}{4} = 1\frac{1}{4}$

2. To add fractions with like denominators, add the numerators, but keep the same denominator. $\frac{7}{9} + \frac{7}{9} = \frac{14}{9} = 1\frac{5}{9}$

3-2 Exercises

FOR EXTRA PRACTICE
see page 736

☐ internet connect
Homework Help Online
go.hrw.com   Keyword: MP4 3-2

3-2 PRACTICE & ASSESS

GUIDED PRACTICE

See Example ① 1. In the World Championships for the 100-meter dash in Edmonton, Alberta, Canada, on August 5, 2001, Tim Montgomery had a reaction time of 0.157 seconds. His total race time was 9.85 seconds. How long did it take him to run the actual distance? **9.693 seconds**

> Students may want to refer back to the lesson examples.

See Example ② Use a number line to find each sum.

2. $-0.7 + 2.1$ **1.4**   3. $-\frac{3}{4} + \left(-\frac{5}{4}\right)$ **−2**   4. $-1.3 + 0.9$ **−0.4**   5. $-\frac{1}{2} + \left(-\frac{4}{2}\right)$ **$-2\frac{1}{2}$**

6. $-1.8 + 0.3$ **−1.5**   7. $-\frac{1}{9} + \left(-\frac{4}{9}\right)$ **$-\frac{5}{9}$**   8. $-3.6 + 1.7$ **−1.9**   9. $-\frac{2}{3} + \left(-\frac{7}{3}\right)$ **−3**

See Example ③ Add or subtract.

10. $\frac{4}{9} - \frac{7}{9}$ **$-\frac{1}{3}$**   11. $-\frac{5}{12} - \frac{11}{12}$ **$-1\frac{1}{3}$**   12. $\frac{1}{10} + \frac{7}{10}$ **$\frac{4}{5}$**   13. $-\frac{3}{20} + \frac{11}{20}$ **$\frac{2}{5}$**

14. $\frac{5}{8} - \frac{1}{8}$ **$\frac{1}{2}$**   15. $-\frac{4}{17} + \frac{9}{17}$ **$\frac{5}{17}$**   16. $\frac{13}{5} + \frac{8}{5}$ **$4\frac{1}{5}$**   17. $-\frac{17}{18} - \frac{29}{18}$ **$-2\frac{5}{9}$**

See Example ④ Evaluate each expression for the given value of the variable.

18. $17.3 + x$ for $x = -13.1$ **4.2**   19. $-\frac{1}{5} + x$ for $x = \frac{3}{5}$ **$\frac{2}{5}$**

20. $35.3 + x$ for $x = -13.9$ **21.4**   21. $-\frac{3}{5} + x$ for $x = 1$ **$\frac{2}{5}$**

Assignment Guide

If you finished Example ① assign:
  Core 1, 22, 49–54
  Enriched 1, 22, 49–54

If you finished Example ② assign:
  Core 1–9, 22–30, 49–54
  Enriched 1–9, 22–30, 49–54

If you finished Example ③ assign:
  Core 1–17, 22–38, 49–54
  Enriched 1–17 odd, 22–38, 43–54

If you finished Example ④ assign:
  Core 1–42, 49–54
  Enriched 1–17 odd, 18–54

INDEPENDENT PRACTICE

See Example ① 22. In the men's 5000 m short-track speed-skating relay in the 2002 Olympics, the Canadian team won the gold medal with a time of 411.579 seconds, defeating the second place Italian team by 4.748 seconds. How long did it take the Italian team to finish the race? **416.327 seconds**

See Example ② Use a number line to find each sum.

23. $-3.4 + 1.8$ **−1.6**   24. $-\frac{3}{4} + \left(-\frac{3}{4}\right)$ **$-1\frac{1}{2}$**   25. $-0.9 + 2.5$ **1.6**   26. $-\frac{1}{12} + \left(-\frac{7}{12}\right)$ **$-\frac{2}{3}$**

27. $-1.7 + 3.6$ **1.9**   28. $-\frac{7}{10} + \left(-\frac{3}{10}\right)$ **−1**   29. $-4 + 1.3$ **−2.7**   30. $-\frac{15}{16} + \left(-\frac{9}{16}\right)$ **$-1\frac{1}{2}$**

See Example ③ Add or subtract.

31. $\frac{8}{11} - \frac{3}{11}$ **$\frac{5}{11}$**   32. $-\frac{4}{13} - \frac{8}{13}$ **$-\frac{12}{13}$**   33. $\frac{9}{17} + \frac{16}{17}$ **$1\frac{8}{17}$**   34. $-\frac{19}{25} + \frac{13}{25}$ **$-\frac{6}{25}$**

35. $\frac{11}{32} - \frac{27}{32}$ **$-\frac{1}{2}$**   36. $-\frac{1}{15} + \frac{13}{15}$ **$\frac{4}{5}$**   37. $\frac{8}{21} + \frac{15}{21}$ **$1\frac{2}{21}$**   38. $-\frac{31}{57} - \frac{49}{57}$ **$-1\frac{23}{57}$**

See Example ④ Evaluate each expression for the given value of the variable.

39. $47.3 + x$ for $x = -18.6$ **28.7**   40. $-\frac{9}{10} + x$ for $x = \frac{3}{10}$ **$-\frac{3}{5}$**

41. $13.95 + x$ for $x = -30.29$ **−16.34**   42. $-\frac{16}{23} + x$ for $x = \frac{11}{23}$ **$-\frac{5}{23}$**

RETEACH 3-2

Reteach
3-2 Adding and Subtracting Rational Numbers

To add fractions that have the same denominator:
• Use the common denominator for the sum.
• Add the numerators to get the numerator of the sum.
• Write the sum in simplest form.

$\frac{1}{8} + \frac{3}{8} = \frac{1+3}{8} = \frac{4}{8} = \frac{1}{2}$

To subtract fractions that have the same denominator:
• Use the common denominator for the difference.
• Subtract the numerators.
  Subtraction is addition of an opposite.
• Write the difference in simplest form.

$\frac{3}{6} - \left(-\frac{1}{6}\right) = \frac{3+1}{6} = \frac{4}{6} = \frac{2}{3}$

Complete to add the fractions.

1. $\frac{3}{14} + \frac{4}{14} = \frac{7}{14} = \frac{1}{2}$ 2. $\frac{2}{10} + \left(-\frac{4}{10}\right) = \frac{-2}{10} = -\frac{1}{5}$

3. $-\frac{5}{12} + \left(-\frac{3}{12}\right) = \frac{-8}{12} = -\frac{2}{3}$

Complete to subtract the fractions.

4. $\frac{8}{9} - \frac{2}{9} = \frac{6}{9} = \frac{2}{3}$ 5. $\frac{9}{15} - \left(-\frac{3}{15}\right) = \frac{12}{15} = \frac{4}{5}$

6. $-\frac{10}{24} - \left(-\frac{2}{24}\right) = \frac{-8}{24} = -\frac{1}{3}$

To add or subtract decimals, line up the decimal points and then add or subtract from right to left as usual.

```
  12.83        35.78
 +24.17       −14.55
  37.00        21.23
```

Complete to add the decimals.

7. $14.23 + 3.56 = $ __17.79__   8. $44.02 + 8.07 = $ __52.09__

9. $1.39 + 13.6 = $ __14.99__

Complete to subtract the decimals.

10. $124.33 - 13.16 = $ __111.17__   11. $33.47 - 0.6 = $ __32.87__

12. $25.15 - 25.06 = $ __0.09__

PRACTICE 3-2

Practice B
3-2 Adding and Subtracting Rational Numbers

1. Gretchen bought a sweater for $23.89. In addition, she had to pay $1.43 in sales tax. She gave the sales clerk $30. How much change did Gretchen receive from her total purchase?

__$4.68__

2. Jacob is replacing the molding around two sides of a picture frame. The measurements of the sides of the frame are $4\frac{3}{16}$ in. and $2\frac{5}{16}$ in. What length of molding will Jacob need?

__$6\frac{1}{2}$ in.__

Use a number line to find each sum.

3. $-0.5 + 0.4$

__−0.1__

4. $-\frac{2}{7} + \frac{6}{7}$

__$\frac{4}{7}$__

Add or subtract. Write answers in simplest form.

5. $\frac{3}{8} + \frac{1}{8}$ __$\frac{1}{2}$__   6. $-\frac{6}{10} + \frac{7}{10}$ __$\frac{3}{5}$__   7. $\frac{5}{14} - \frac{3}{14}$ __$\frac{1}{7}$__   8. $\frac{4}{15} + \frac{7}{15}$ __$\frac{11}{15}$__

9. $\frac{5}{18} - \frac{7}{18}$ __$-\frac{1}{9}$__   10. $-\frac{2}{17} + \frac{6}{17}$ __$\frac{10}{17}$__   11. $-\frac{1}{16} + \frac{5}{16}$ __$\frac{1}{4}$__   12. $\frac{3}{20} + \frac{1}{20}$ __$\frac{1}{5}$__

Evaluate each expression for the given value of the variable.

13. $38.1 + x$ for $x = -6.1$ __32__   14. $18.7 + x$ for $x = 8.5$ __27.2__   15. $\frac{8}{15} + x$ for $x = -\frac{4}{15}$ __$\frac{4}{15}$__

Math Background

The Commutative Property of Addition states that $a + b = b + a$. It holds for all real numbers a and b (e.g., $0.21 + 3.4 = 3.4 + 0.21 = 3.61$). There is no commutative property for subtraction, because $a - b = b - a$ is *not* true for all real numbers a and b (e.g., $\frac{2}{3} - \frac{1}{3} = \frac{1}{3}$, but $\frac{1}{3} - \frac{2}{3} = -\frac{1}{3}$). The Associative Property of Addition states that $a + (b + c) = (a + b) + c$. It holds for all real numbers a, b, and c. That is, the answer is the same no matter how you *associate* (group) the numbers. However, there is no associative property for subtraction (e.g., $1.2 - (0.8 - 0.3) = 0.7$, but $(1.2 - 0.8) - 0.3 = 0.1$).

Answers

Answers

46. Possible answer: Ed had a board 8.75 feet in length. He sawed off a piece that was $3\frac{1}{4}$ feet long. What is the length of the remaining board? Answer: $5\frac{1}{2}$ feet

47. Possible answer: The denominators are not added because the size of each part of the whole does not change; only the number of parts changes.

Journal

Have students write a word problem that requires addition or subtraction of rational numbers for the solution.

Test Prep Doctor ✚

For Exercises 53 and 54, remind students that to subtract a number, they need to add the opposite of the number.

Lesson Quiz

Add or subtract.

1. $-1.2 + 8.4$ **7.2**

2. $2.5 + (-2.8)$ -0.3

3. $\frac{3}{4} + -\frac{5}{4}$ $-\frac{1}{2}$

4. Evaluate $62.1 + x$ for $x = -127.0$
   -64.9

5. Sarah's best broad jump is 1.6 meters, and Jill's best is 1.47 meters. How much farther can Sarah jump than Jill?   0.13 m

Available on Daily Transparency in CRB

PRACTICE AND PROBLEM SOLVING

43. *DESIGN* In a mechanical drawing, a hidden line is represented by dashes $\frac{4}{32}$ inch long with $\frac{1}{32}$-inch spaces between them. Without measuring, how long is each set of dashes?

a. - - - - - $\frac{29}{32}$ in. b. - - - - - - - - $1\frac{7}{32}$ in. c. - - - - $\frac{19}{32}$ in.

44. *SPORTS* A college football must be between $10\frac{14}{16}$ inches and $11\frac{7}{16}$ inches long. What is the greatest possible difference in length between two college footballs that meet these standards? $\frac{9}{16}$ in.

45a.
3.63 quadrillion Btu

45. *ENERGY* The circle graph shows the sources of renewable energy and their use in the United States in British thermal units (Btu).

a. How many quadrillion Btu's created by hydroelectric, solar, and wind methods combined were used?

b. How many more Btu's created by wood and waste were used than those created by geothermal, solar, and wind sources combined?   **2.717 quadrillion Btu**

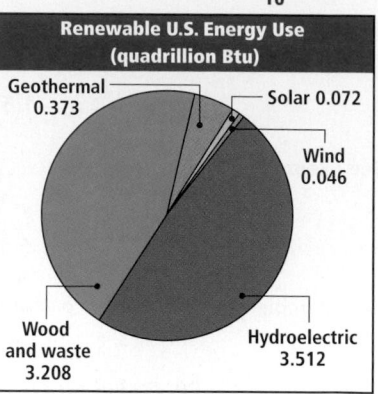

Renewable U.S. Energy Use (quadrillion Btu)

Geothermal 0.373
Solar 0.072
Wind 0.046
Wood and waste 3.208
Hydroelectric 3.512

46. *WRITE A PROBLEM* Write a problem that requires a decimal to be converted to a fraction and that also involves addition or subtraction of fractions.

47. *WRITE ABOUT IT* When a student was adding fractions, the denominators were not added. Explain why.

48. *CHALLENGE* The gutter of a bowling lane measures $9\frac{5}{16}$ inches wide. This is $\frac{3}{16}$ inch less than the widest gutter permitted and $\frac{5}{16}$ inch greater than the narrowest gutter permitted. What is the greatest possible difference in the width of two gutters? $\frac{1}{2}$ in.

Spiral Review

Combine like terms. (Lesson 1-6)

49. $7x - 5y + 18$ $7x - 5y + 18$

50. $3x + y + 5y - 2x$ $x + 6y$

51. $34x + 17y + 3 - 18x + 5y + 8$
   $16x + 22y + 11$

52. $48x + 23y + 5x + 6 - 3y + 15$
   $53x + 20y + 21$

53. **TEST PREP** Subtract $-4 - (-12)$.
(Lesson 2-2)   **A**

A 8   B -16 C -8 D 16

54. **TEST PREP** Subtract $-15 - (-8)$.
(Lesson 2-2)   **G**

F 7   G -7 H 23   J -23

CHALLENGE 3-2

Challenge
3-2 Number Code

Each sum is the code for a letter. As you find a sum, write its letter code in the message below. Write the sum in simplest form. Some letters appear more than once. An example is done for you.

$4.5 + (-6.5)$	**-2** , T	
2. $\frac{7}{8} + \left(-1\frac{3}{8}\right)$	**$-\frac{1}{2}$** , M	
4. $-1.05 + 0.85$	**-0.2** , I	
6. $-7.08 + (-12.02)$	**-19.1** , S	
8. $\frac{-4}{5} + 1$	**$\frac{1}{5}$** , E	
10. $1\frac{2}{4} + \left(\frac{-3}{4}\right)$	**$\frac{3}{4}$** , P	
12. $8 + (-6.4)$	**1.6** , Y	
14. $\frac{7}{8} + \left(-1\frac{7}{8}\right)$	**-1** , L	
16. $-62.3 + 23.9$	**-38.4** , A	
18. $-2.9 + 0.85$	**-2.05** , O	

1. $14.56 + (-10.09)$	**4.47** , V	
3. $\frac{6}{8} + \left(-\frac{3}{8}\right)$	**$\frac{3}{8}$** , N	
5. $\frac{-2}{4} + \left(\frac{-3}{4}\right)$	**$-1\frac{1}{4}$** , U	
7. $-9.5 + 3.1$	**-6.4** , E	
9. $-1\frac{1}{2} + \left(-1\frac{1}{2}\right)$	**-3** , H	
11. $5 + \left(-4\frac{1}{10}\right)$	**$\frac{9}{10}$** , I	
13. $-3\frac{1}{4} + 3\frac{1}{4}$	**0** , S	
15. $6.52 + (-5)$	**1.52** , Z	
17. $9\frac{1}{8} + (-10)$	**$-\frac{7}{8}$** , R	
19. $2.7 + (-0.9)$	**1.8** , O	

S	U	R	E		Y	O	U		A	R	E
-19.1	$-1\frac{1}{4}$	$-\frac{7}{8}$	-6.4		1.6	-2.05	$-1\frac{1}{4}$		-38.4	$-\frac{7}{8}$	$\frac{1}{5}$

N	O	T		L	E	S	S		T	H	A	N
$\frac{3}{8}$	1.8	-2		-1	$\frac{1}{5}$	-19.1	0		-2	-3	-38.4	$\frac{3}{8}$

Z	E	R	O	?
1.52	-6.4	$-\frac{7}{8}$	-2.05	

I'	M		P	O	S	I	I	I	V	E	!
-0.2	$-\frac{1}{2}$		$\frac{3}{4}$	-2.05	0	$\frac{9}{10}$	-2	-0.2	4.47	$\frac{1}{5}$	

PROBLEM SOLVING 3-2

Problem Solving
3-2 Adding and Subtracting Rational Numbers

Write the correct answer.

1. In 1928, Elizabeth Robinson of the United States won the Olympic Gold in the 100-m dash with a time of 12.20 seconds. In 2000, American Marion Jones won the 100-m dash with a time of 10.75 seconds. How many seconds faster did Marion Jones run the 100-m dash?

1.45 sec

2. The snowfall in Rochester, NY in the winter of 1999–2000 was 91.5 inches. Normal snowfall is about 76 inches per winter. How much more snow fell in the winter of 1999–2000 than is normal?

15.5 inches

3. In a survey, $\frac{76}{100}$ people indicated that they check their e-mail daily, while $\frac{23}{100}$ check their e-mail weekly, and $\frac{1}{100}$ check their e-mail less than once a week. What fraction of people check their e-mail at least once a week?

$\frac{99}{100}$

4. To make a small amount of play dough, you can mix the following ingredients: 1 cup of flour, $\frac{1}{2}$ cup of salt and $\frac{1}{2}$ cup of water. What is the total amount of ingredients added to make the play dough?

2 cups

Choose the letter for the best answer.

5. How much more expensive is it to buy a ticket in Boston than in Minnesota?

A $32.51
B $9.14
C $6.45
D $15.59

6. How much more expensive is it to buy a ticket in Boston than the league average?

F $32.51
G $9.14
H $6.45
J $15.59

Baseball Ticket Prices

Location	Average Price
Minnesota	$8.46
League Average	$14.91
Boston	$24.05

7. What is the total cost of a ticket in Boston and a ticket in Minnesota?

A $32.51
B $23.37
C $38.96
D $48.10

3-3 Multiplying Rational Numbers

Learn to multiply fractions, mixed numbers and decimals.

Kendall invited 36 people to a party. She needs to triple the recipe for a dip, or multiply the amount of each ingredient by 3. Remember that multiplication by a whole number can be written as repeated addition.

Favorite Vegetable Dip
1 c sour cream
1/2 c mayonnaise
1 envelope dry Italian dressing mix
1/2 tsp thyme
1/4 tsp curry powder
Mix and chill 24 hours. Serves 12.

Repeated addition	Multiplication
$\frac{1}{4} + \frac{1}{4} + \frac{1}{4} = \frac{3}{4}$	$3\left(\frac{1}{4}\right) = \frac{3 \cdot 1}{4} = \frac{3}{4}$

Notice that multiplying a fraction by a whole number is the same as multiplying the whole number by just the numerator of the fraction and keeping the same denominator.

> ### RULES FOR MULTIPLYING TWO RATIONAL NUMBERS
>
> **If the signs of the factors are the same, the product is positive.**
>
> $(+) \cdot (+) = (+)$ or $(-) \cdot (-) = (+)$
>
> **If the signs of the factors are different, the product is negative.**
>
> $(+) \cdot (-) = (-) \cdot (+) = (-)$

EXAMPLE 1 Multiplying a Fraction and an Integer

Multiply. Write each answer in simplest form.

Helpful Hint

To write $\frac{12}{5}$ as a mixed number, divide:

$\frac{12}{5} = 2$ R2

$= 2\frac{2}{5}$

A $6\left(\frac{2}{3}\right)$

$6\left(\frac{2}{3}\right)$

$= \frac{6 \cdot 2}{3}$

$= \frac{12}{3}$

$= 4$

B $-4\left(2\frac{3}{5}\right)$

$-4\left(2\frac{3}{5}\right)$

$= -4\left(\frac{13}{5}\right)$ $\quad 2\frac{3}{5} = \frac{2(5) + 3}{5} = \frac{13}{5}$

$= -\frac{52}{5}$ *Multiply.*

$= -10\frac{2}{5}$ *Simplify.*

---

1 Introduce

Alternate Opener

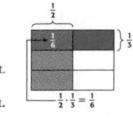

EXPLORATION

3-3 Multiplying Rational Numbers

To find the product of $\frac{1}{2} \cdot \frac{1}{3}$, use one color to color in $\frac{1}{2}$ of a square vertically. Use a second color to color in $\frac{1}{3}$ of the square horizontally. The product is represented by the area where the two colors intersect.

Use a similar model to find each product.

1. $\frac{1}{2} \cdot \frac{4}{6}$ **2.** $\frac{2}{3} \cdot \frac{3}{4}$

To multiply $2 \cdot 1\frac{1}{2}$, color in 2 squares across and $1\frac{1}{2}$ down.

$2 \cdot 1\frac{1}{2} = 3$

Use a similar model to find each product.

3. $3 \cdot 4\frac{1}{2}$ **4.** $8\frac{1}{4} \cdot 4$

Think and Discuss

5. Explain how the model shows that $2.5 \cdot 2.5 = 6.25$.

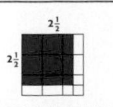

Motivate

Ask students if they have ever mixed ingredients for a recipe. Ask them if they have had to use fractions to measure the ingredients. Ask them if they would have known how much of each ingredient to use to make a double batch, a half batch, or a batch and a half. Tell the students that it is helpful to be able to multiply rational numbers for such purposes.

Exploration worksheet and answers on Chapter 3 Resource Book pp. 26 and 120

3-3 Organizer

Pacing: Traditional 1 day
Block $\frac{1}{2}$ day

Objective: Students multiply fractions and mixed numbers.

Warm Up

Write each number as an improper fraction.

1. $2\frac{1}{3}$ $\frac{7}{3}$ **2.** $1\frac{7}{8}$ $\frac{15}{8}$ **3.** $3\frac{2}{5}$ $\frac{17}{5}$

4. $6\frac{2}{3}$ $\frac{20}{3}$ **5.** $5\frac{3}{8}$ $\frac{43}{8}$

Problem of the Day

The sum of three consecutive integers is 168. What are the three integers?   55, 56, and 57

Available on Daily Transparency in CRB

Math Humor

5 out of 4 people have trouble with fractions.

2 Teach

Lesson Presentation

Guided Instruction

In this lesson, students learn to multiply fractions and mixed numbers. Review the concept of multiplication as repeated addition and the rules for multiplying signed numbers (Teaching Transparency T11, CRB). Point out the model for multiplying fractions, and then review Example 2. Point out that students may "cancel out" common factors between numerators and denominators. For multiplying decimals, remind students that the number of decimal places in the product should be the total number of decimal places in the two factors.

Example 1

Multiply. Write each answer in simplest form.

A. $-8\left(\frac{6}{7}\right)$ $-6\frac{6}{7}$ **B.** $2\left(5\frac{1}{3}\right)$ $10\frac{2}{3}$

Example 2

Multiply. Write each answer in simplest form.

A. $\frac{1}{8}\left(\frac{6}{7}\right)$ $\frac{3}{28}$ **B.** $-\frac{2}{3}\left(\frac{9}{2}\right)$ -3

C. $4\frac{3}{7}\left(\frac{1}{2}\right)$ $2\frac{3}{14}$

Example 3

Multiply.

A. $2(-0.51)$ -1.02

B. $(-0.4)(-3.75)$ 1.5

Example 4

Evaluate $-3\frac{1}{8}x$ for each value of x.

A. $x = 5$ $-15\frac{5}{8}$

B. $x = \frac{2}{7}$ $-\frac{25}{28}$

Example 2 note: Students may assume that multiplying two numbers results in a product that is greater than both of its factors (e.g., $2 \cdot 3 = 6$). Explain to students that because fractions involve division, multiplying two fractions does not necessarily result in a greater product (e.g., $\frac{1}{3} \cdot \frac{1}{2} = \frac{1}{6}$). Because decimals represent fractions, this also applies to decimals ($0.2 \cdot 0.3 = 0.06$).

A model of $\frac{3}{5} \cdot \frac{2}{3}$ is shown. Notice that to multiply fractions, you multiply the numerators and multiply the denominators.

If you place the first rectangle on top of the second, the number of green squares represents the numerator, and the number of total squares represents the denominator.

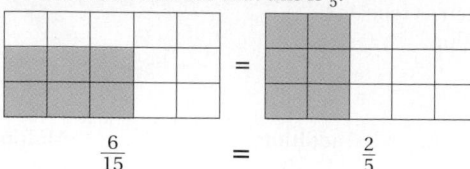

To simplify the product, rearrange the six green squares into the first two columns. You can see that this is $\frac{2}{5}$.

$$\frac{6}{15} = \frac{2}{5}$$

EXAMPLE 2 · Multiplying Fractions

Multiply. Write each answer in simplest form.

A $-\frac{1}{2}\left(-\frac{3}{5}\right)$

$$-\frac{1}{2}\left(-\frac{3}{5}\right) = \frac{-1}{2}\left(\frac{-3}{5}\right)$$

$$= \frac{(-1)(-3)}{2(5)}$$ *Multiply numerators.*
Multiply denominators.

$$= \frac{3}{10}$$ *Simplest form*

B $\frac{5}{12}\left(-\frac{12}{5}\right)$

$$\frac{5}{12}\left(-\frac{12}{5}\right) = \frac{5}{12}\left(\frac{-12}{5}\right)$$

$$= \frac{\overset{1}{\cancel{5}}(\overset{-1}{\cancel{-12}})}{\underset{1}{\cancel{12}}(\underset{1}{\cancel{5}})}$$ *Look for common factors: 12, 5.*

$$= \frac{-1}{1} = -1$$ *Simplest form*

C $6\frac{2}{3}\left(\frac{7}{20}\right)$

$$6\frac{2}{3}\left(\frac{7}{20}\right) = \frac{20}{3}\left(\frac{7}{20}\right)$$ *Write as an improper fraction.*

$$= \frac{\overset{1}{\cancel{20}}(7)}{3(\underset{1}{\cancel{20}})}$$ *Look for common factors: 20.*

$$= \frac{7}{3}, \text{ or } 2\frac{1}{3}$$ *7 ÷ 3 = 2 R1*

Helpful Hint

A fraction is in lowest terms, or simplest form, when the numerator and denominator have no common factors.

 **Reaching All Learners**
Through Critical Thinking

Show students two ways to evaluate $\frac{1}{2}(0.7)$:

$$\frac{1}{2}(0.7) = 0.5 \cdot 0.7 = 0.35$$

$$\frac{1}{2}(0.7) = \frac{1}{2} \cdot \frac{7}{10} = \frac{7}{20}$$

Show that $\frac{7}{20} = 0.35$ by dividing:

$$20\overline{)7.00}\quad 0.35$$. Have students find each of the following products both ways and show that the results are equal in each case.

1. $\frac{1}{2}(0.3)$ $\frac{3}{20}$, 0.15   **2.** $\frac{1}{5}(0.4)$ $\frac{2}{25}$, 0.08

3. $\frac{3}{4}(0.1)$ $\frac{3}{40}$, 0.075   **4.** $1\frac{1}{2}(0.2)$ $\frac{3}{10}$, 0.3

EXAMPLE 3 **Multiplying Decimals**

Multiply.

A $-2.5(-8)$

$-2.5 \cdot (-8) = 20.0$ *Product is positive with 1 decimal place.*

You can drop the zero after the decimal point.

$= 20$

B $-0.07(4.6)$

$-0.07 \cdot 4.6 = -0.322$ *Product is negative with 3 decimal places.*

COMMON ERROR ALERT

When multiplying a mixed number by a whole number, some students may multiply the whole numbers and leave the fraction unchanged. Remind students to write the mixed number as an improper fraction before multiplying.

EXAMPLE 4 **Evaluating Expressions with Rational Numbers**

Evaluate $-5\frac{1}{2}t$ for each value of t.

A $t = -\frac{2}{3}$

$-5\frac{1}{2}t$

$= -5\frac{1}{2}\left(-\frac{2}{3}\right)$ *Substitute $-\frac{2}{3}$ for t.*

$= -\frac{11}{2}\left(-\frac{2}{3}\right)$ *Write as an improper fraction.*

$= \frac{11 \cdot \overset{1}{\cancel{2}}}{\cancel{2} \cdot 3}$ *The product of 2 negative numbers is positive.*

$= \frac{11}{3}$, or $3\frac{2}{3}$ *$11 \div 3 = 3$ R2*

B $t = 8$

$-5\frac{1}{2}t$

$= -\frac{11}{2}(8)$ *Substitute 8 for t.*

$= -\frac{88}{2}$

$= -44$

Think and Discuss

1. **Name** the number of decimal places in the product of 5.625 and 2.75.

2. **Explain** why products of fractions are like products of integers.

3. **Give an example** of two fractions whose product is an integer due to common factors.

Close

Summarize

Ask students to explain how to multiply each of the following:

1) a fraction and a mixed number
2) two decimals
3) a decimal and a fraction

Possible answers:

1. Change the mixed number to an improper fraction, multiply numerators, multiply denominators, and write in simplest form.

2. Multiply the numbers and count the total number of decimal places in the two factors.

3. Either write the decimal as a fraction and multiply the fractions, or write the fraction as a decimal and multiply the decimals.

Answers to Think and Discuss

1. 5 decimal places; the number of decimal places is the sum of the decimal places in the factors.

2. Possible answer: Products of fractions are actually products of two sets of integers, the numerators and the denominators.

3. Possible answer: $\frac{3}{2} \cdot \frac{4}{3} = 2$

FOR EXTRA PRACTICE
see page 736

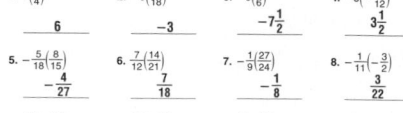
internet connect
Homework Help Online
go.hrw.com   Keyword: MP4 3-3

GUIDED PRACTICE

Students may want to refer back to the lesson examples.

See Example **1** Multiply. Write each answer in simplest form.

1. $4\left(\frac{1}{3}\right)$ $1\frac{1}{3}$
2. $-6\left(2\frac{2}{5}\right)$ $-14\frac{2}{5}$
3. $3\left(\frac{5}{8}\right)$ $1\frac{7}{8}$
4. $-2\left(1\frac{9}{10}\right)$ $-3\frac{4}{5}$

5. $7\left(\frac{4}{9}\right)$ $3\frac{1}{9}$
6. $-5\left(1\frac{8}{11}\right)$ $-8\frac{7}{11}$
7. $9\left(\frac{3}{4}\right)$ $6\frac{3}{4}$
8. $3\left(2\frac{1}{8}\right)$ $6\frac{3}{8}$

See Example **2** Multiply. Write each answer in simplest form.

9. $-\frac{1}{3}\left(-\frac{4}{7}\right)$ $\frac{4}{21}$
10. $\frac{3}{8}\left(-\frac{7}{10}\right)$ $-\frac{21}{80}$
11. $6\frac{2}{5}\left(\frac{5}{9}\right)$ $3\frac{5}{9}$
12. $-\frac{2}{3}\left(-\frac{3}{8}\right)$ $\frac{1}{4}$

13. $\frac{5}{13}\left(-\frac{5}{6}\right)$ $-\frac{25}{78}$
14. $4\frac{7}{8}\left(\frac{5}{12}\right)$ $2\frac{1}{32}$
15. $-\frac{7}{8}\left(-\frac{2}{3}\right)$ $\frac{7}{12}$
16. $\frac{5}{12}\left(-\frac{11}{16}\right)$ $-\frac{55}{192}$

See Example **3** Multiply.

17. $-3.1(-4)$ **12.4**
18. $0.04(3.6)$ **0.144**
19. $-7.3(-5)$ **36.5**
20. $-0.15(2.8)$ **−0.42**

21. $-5.9(-7)$ **41.3**
22. $0.5(7.3)$ **3.65**
23. $-4.7(-3)$ **14.1**
24. $-0.08(5.2)$ **−0.416**

See Example **4** Evaluate $3\frac{2}{7}x$ for each value of x.

25. $x = 4$ $13\frac{1}{7}$
26. $x = 1\frac{3}{4}$ $5\frac{3}{4}$
27. $x = -2$ $-6\frac{4}{7}$
28. $x = -\frac{3}{7}$ $-1\frac{20}{49}$

29. $x = 7$ **23**
30. $x = 2\frac{1}{3}$ $7\frac{2}{3}$
31. $x = -3$ $-9\frac{6}{7}$
32. $x = -\frac{3}{10}$ $-\frac{69}{70}$

INDEPENDENT PRACTICE

See Example **1** Multiply. Write each answer in simplest form.

33. $3\left(\frac{1}{5}\right)$ $\frac{3}{5}$
34. $-4\left(1\frac{5}{8}\right)$ $-6\frac{1}{2}$
35. $2\left(\frac{9}{16}\right)$ $1\frac{1}{8}$
36. $-5\left(1\frac{3}{4}\right)$ $-8\frac{3}{4}$

37. $9\left(\frac{14}{15}\right)$ $8\frac{2}{5}$
38. $-2\left(4\frac{7}{8}\right)$ $-9\frac{3}{4}$
39. $6\left(\frac{2}{3}\right)$ **4**
40. $-7\left(3\frac{1}{5}\right)$ $-22\frac{2}{5}$

See Example **2** Multiply. Write each answer in simplest form.

41. $-\frac{2}{3}\left(-\frac{5}{6}\right)$ $\frac{5}{9}$
42. $\frac{2}{5}\left(-\frac{9}{10}\right)$ $-\frac{9}{25}$
43. $2\frac{5}{7}\left(\frac{2}{9}\right)$ $\frac{38}{63}$
44. $-\frac{1}{2}\left(-\frac{11}{12}\right)$ $\frac{11}{24}$

45. $\frac{4}{5}\left(-\frac{3}{8}\right)$ $-\frac{3}{10}$
46. $5\frac{1}{3}\left(\frac{13}{16}\right)$ $4\frac{1}{3}$
47. $-\frac{3}{4}\left(-\frac{1}{8}\right)$ $\frac{3}{32}$
48. $\frac{7}{8}\left(\frac{3}{5}\right)$ $\frac{21}{40}$

See Example **3** Multiply.

49. $-2.9(-3)$ **8.7**
50. $-0.02(5.9)$ **−0.118**
51. $-6.2(-7)$ **43.4**
52. $-0.25(3.5)$ **−0.875**

53. $-4.8(-7)$ **33.6**
54. $-0.07(4.8)$ **−0.336**
55. $-3.6(-8)$ **28.8**
56. $-0.04(9.2)$ **−0.368**

See Example **4** Evaluate $2\frac{3}{4}x$ for each value of x.

57. $x = 6$ $16\frac{1}{2}$
58. $x = 2\frac{1}{3}$ $6\frac{5}{12}$
59. $x = -4$ -11
60. $x = -\frac{3}{8}$ $-1\frac{1}{32}$

61. $x = 3$ $8\frac{1}{4}$
62. $x = 4\frac{7}{8}$ $13\frac{13}{32}$
63. $x = -7$ $-19\frac{1}{4}$
64. $x = -\frac{7}{9}$ $-2\frac{5}{36}$

Assignment Guide

If you finished Example **1** assign:
Core 1–8, 33–40, 65, 72–79
Enriched 1–8, 33–40, 65, 72–79

If you finished Example **2** assign:
Core 1–16, 33–48, 65, 72–79
Enriched 1–15 odd, 33–48, 65–79

If you finished Example **3** assign:
Core 1–24, 33–56, 65, 72–79
Enriched 9–24, 33–56, 65–79

If you finished Example **4** assign:
Core 1–65, 72–79
Enriched 9–79

Notes

Math Background

Models can be useful for demonstrating multiplication of rational numbers and the Commutative Property. The expression $3 \cdot \frac{1}{2}$ is read "three times one-half," and it means $\frac{1}{2} + \frac{1}{2} + \frac{1}{2}$.

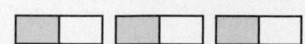

The expression $\frac{1}{2} \cdot 3$ is read "one-half times three," and it means "one-half of three."

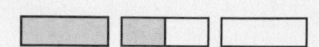

By the Commutative Property, $3 \cdot \frac{1}{2} = \frac{1}{2} \cdot 3$. This equality is evident on the models; the shaded portions are equal.

RETEACH 3-3

Reteach
3-3 *Multiplying Rational Numbers*

To model $\frac{1}{3} \times \frac{3}{4}$:

Divide a square into 4 equal parts. Lightly shade 3 of the 4.

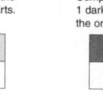
Darken 1 of the 3 shaded parts.

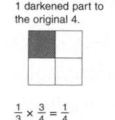
Compare the 1 darkened part to the original 4.

$\frac{1}{3} \times \frac{3}{4} = \frac{1}{4}$

Model each multiplication. Write the result. Possible models are shown.

1. $\frac{1}{2} \times \frac{2}{4} = \frac{1}{4}$
2. 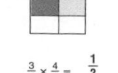 $\frac{3}{4} \times \frac{4}{6} = \frac{1}{2}$
3. $\frac{2}{3} \times \frac{3}{9} = \frac{2}{9}$

To multiply fractions:
• Cancel common factors, one in a numerator and the other in a denominator.
• Multiply the remaining factors in the numerator and in the denominator.
• If the signs of the factors are the same, the product is positive. If the signs of the factors are different, the product is negative.

$\frac{3}{4} \times \frac{8}{9} = \frac{1 \times 2}{1 \times 3} = \frac{2}{3}$

Multiply. Answer in simplest form.

4. $\frac{1}{2} \times \frac{4}{9} = \frac{2}{9}$
5. $\frac{2}{3} \times \frac{6}{7} = \frac{4}{7}$
6. $\frac{3}{5} \times \frac{15}{17} = \frac{9}{17}$
7. $\frac{2}{3} \times \left(-\frac{9}{10}\right) = -\frac{3}{5}$
8. $\left(-\frac{2}{9}\right) \times \frac{27}{40} = -\frac{3}{20}$
9. $\left(-\frac{4}{7}\right) \times \left(-\frac{21}{8}\right) = 1\frac{1}{2}$

PRACTICE 3-3

Practice B
3-3 *Multiplying Rational Numbers*

Multiply. Write each answer in simplest form.

1. $8\left(\frac{3}{4}\right)$ 6
2. $-6\left(\frac{9}{18}\right)$ −3
3. $-9\left(\frac{5}{6}\right)$ $-7\frac{1}{2}$
4. $-6\left(-\frac{7}{12}\right)$ $3\frac{1}{2}$

5. $-\frac{5}{18}\left(\frac{8}{15}\right)$ $-\frac{4}{27}$
6. $\frac{7}{12}\left(\frac{14}{21}\right)$ $\frac{7}{18}$
7. $-\frac{7}{9}\left(\frac{27}{24}\right)$ $-\frac{1}{8}$
8. $-\frac{8}{11}\left(-\frac{3}{4}\right)$ $\frac{3}{22}$

9. $\frac{7}{20}\left(-\frac{15}{28}\right)$ $-\frac{3}{16}$
10. $\frac{16}{25}\left(-\frac{18}{32}\right)$ $-\frac{9}{25}$
11. $\frac{1}{9}\left(-\frac{18}{17}\right)$ $-\frac{2}{17}$
12. $\frac{17}{20}\left(-\frac{12}{34}\right)$ $-\frac{3}{10}$

Multiply.

13. $-2(-5.2)$ **10.4**
14. $0.53(0.04)$ **0.0212**
15. $(-7)(-3.9)$ **27.3**
16. $-2(8.13)$ **−16.26**

17. $0.02(-4.62)$ **−0.0924**
18. $0.5(-7.8)$ **−3.9**
19. $(-0.41)(-8.5)$ **3.485**
20. $-8(6.3)$ **−50.4**

21. $15(-0.05)$ **−0.75**
22. $(-3.04)(-1.7)$ **5.168**
23. $10(-0.09)$ **−0.9**
24. $(-0.8)(-0.15)$ **0.12**

Evaluate $3\frac{1}{4}x$ for each value of x.

25. $x = -3$ $-9\frac{3}{4}$
26. $x = -\frac{2}{3}$ $-2\frac{1}{6}$
27. $x = 2$ $6\frac{1}{2}$
28. $x = \frac{1}{2}$ $1\frac{5}{8}$

29. Travis painted for $6\frac{2}{3}$ hours. He received $27 an hour for his work. How much was Travis paid for doing this painting job?
$180

PRACTICE AND PROBLEM SOLVING

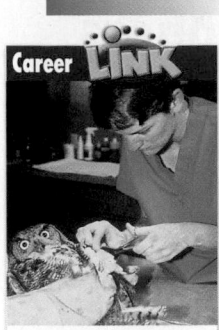

Career LINK

Becoming a veterinarian requires at least two years at an undergraduate college and four years at a veterinary college. There are fewer than 30 veterinary colleges in the United States.

65. *HEALTH* As a rule of thumb, people should drink $\frac{1}{2}$ ounce of water for each pound of body weight per day. How much water should a 145-pound person drink per day? **$72\frac{1}{2}$ ounces**

66. People who are physically active should increase the daily amount of water they drink to $\frac{2}{3}$ ounce per pound of body weight. How much water should a 245-pound football player drink per day? **$163\frac{1}{3}$ ounces**

67. *ANIMALS* The label on a bottle of pet vitamins lists dosage guidelines. What dosage would you give to each of these animals?

Do-Good Pet Vitamins

- **Adult dogs:**
  $\frac{1}{2}$ tsp per 20 lb body weight
- **Puppies, pregnant dogs, or nursing dogs:**
  $\frac{1}{2}$ tsp per 10 lb body weight
- **Cats:**
  $\frac{1}{4}$ tsp per 2 lb body weight

a. a 50 lb adult dog   **$1\frac{1}{4}$ tsp**

b. a 12 lb cat   **$1\frac{1}{2}$ tsp**

c. a 40 lb pregnant dog   **2 tsp**

68. *CONSUMER ECONOMICS* At a clothing store, the ticketed price of a sweater is $\frac{1}{2}$ the original price. You have a discount coupon for $\frac{1}{2}$ off the ticketed price. What fraction of the original price is the additional discount? **$\frac{1}{4}$**

69. *WHAT'S THE ERROR?* A student multiplied two mixed numbers in the following fashion: $3\frac{3}{8} \cdot 4\frac{1}{3} = 12\frac{1}{8}$. What's the error?

70. *WRITE ABOUT IT* In the pattern $\frac{1}{3} + \frac{1}{4} + \frac{1}{5} + \dots$, which fraction makes the sum greater than 1? Explain.

71. *CHALLENGE* On January 20, 2001, George W. Bush was inaugurated as the forty-third president of the United States. Of the 42 presidents before him, $\frac{1}{3}$ had served as vice-president. Of those previous vice-presidents, $\frac{3}{7}$ served as president for more than four years. What fraction of the first 42 presidents were former vice-presidents who also served more than four years as president?
$\frac{1}{3} \times \frac{3}{7} = \frac{1}{7}$

Spiral Review

Solve. (Lesson 1-3)

72. $7 + x = 13$ **$x = 6$**   **73.** $x - 5 = 7$ **$x = 12$**   **74.** $x + 8 = 19$ **$x = 11$**

75. $12 + x = 46$ **$x = 34$**   **76.** $x - 27 = 54$ **$x = 81$**   **77.** $x + 31 = 75$ **$x = 44$**

78. TEST PREP Solve the inequality $-3a \geq 24$. (Lesson 2-5)   **D**

A $a \geq -8$ B $a > 8$ C $a < 8$ D $a \leq -8$

79. TEST PREP Solve the inequality $\frac{a}{2} < -22$. (Lesson 2-5)   **F**

F $a < -44$ G $a > -44$ H $a > -11$ J $a < 11$

Answers

69. Possible answer: The student multiplied the whole numbers together and then multiplied the fractions together instead of writing both mixed numbers as improper fractions and then multiplying.
$3\frac{3}{8} \cdot 4\frac{1}{3} = \frac{27}{8} \cdot \frac{13}{3} = \frac{117}{8} = 14\frac{5}{8}$

70. Possible answer: $\frac{1}{7}$; The sum through $\frac{1}{6}$ is $\frac{19}{20}$, which is less than 1. Adding the $\frac{1}{7}$ makes the sum $\frac{153}{140}$, which is greater than 1.

CHALLENGE 3-3

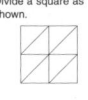 **LESSON 3-3 Challenge**
Curtains

Variations of modern long multiplication were introduced into Europe by a 13th-century Italian, Leonardo of Pisa (Fibonacci). Many multiplication techniques can be traced to a book called *Lilaviti*, written by Bhaskara for his daughter in 12th-century India.

Here's how to do multiplication by the **Gelosia Method**, named after *jalousie*, the iron grill Italians placed over the windows. The method is also called the **Lattice Method of Multiplication**.

Consider: 38 × 56

Divide a square as shown.

Align the factors. Insert the individual products.

Sum each diagonal; begin with lower right. As needed, carry numbers into next diagonal sum.

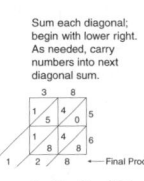

So, 38 × 56 = 2128.

Use the Gelosia Method to multiply.
Verify results by your usual method.

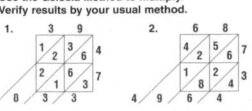

39 × 47 = **1833**   68 × 73 = **4964**   358 × 64 = **22,912**

PROBLEM SOLVING 3-3

LESSON 3-3 Problem Solving
Multiplying Rational Numbers

Use the table at the right.

1. What was the average number of births per minute in 2001?

252 births

Average World Births and Deaths per Second in 2001	
Births	$4\frac{4}{5}$
Deaths	1.7

2. What was the average number of deaths per hour in 2001?

6,120 deaths

3. What was the average number of births per day in 2001?

362,880 births

4. What was the average number of births in $\frac{1}{2}$ of a second in 2001?

$2\frac{1}{10}$ births

5. What was the average number of births in $\frac{1}{4}$ of a second in 2001?

$1\frac{1}{20}$ births

Use the table below. During exercise, the target heart rate is 50–75% of the maximum heart rate. The lower and upper ends of the target rate are found by simplifying the expression $0.50m$ or $0.75m$ where m is the maximum heart rate.

6. What is the target heart rate range for a 14 year old?

A 7–10.5
B 103–154.5
C 145–166
D 206–255

Age	Maximum Heart Rate
13	207
14	206
15	205
20	200
25	195

Source: American Heart Association

7. What is the target heart rate range for a 20 year old?

F 100–150
G 125–175
H 150–200
J 200–250

8. What is the target heart rate range for a 25 year old?

A 25–75
B 85–125
C 97.5–146.25
D 195–250

Lesson Quiz

Multiply.

1. $9\left(\frac{1}{7}\right)$ $1\frac{2}{7}$

2. $\frac{2}{3}\left(-\frac{5}{8}\right)$ $-\frac{5}{12}$

3. $-0.47(2.2)$ -1.034

4. Evaluate $2\frac{1}{2}(x)$ for $x = \frac{4}{5}$.   2

5. Teri is shopping for new shoes. Her mom has agreed to pay half the cost (and all the sales tax). The shoes that Teri likes are normally $30 a pair but are on sale for $\frac{1}{3}$ off. How much money does Teri need to buy the shoes?   **$10**

Available on Daily Transparency in CRB

Warm Up

Multiply.

1. $-3\left(\frac{5}{6}\right)$ $-2\frac{1}{2}$

2. $-15\left(-\frac{2}{3}\right)$ 10

3. $0.05(2.8)$ 0.14

4. $-0.9(16.1)$ -14.49

Problem of the Day

Katie made a bookshelf that is 5 feet long. The first 6 books she put on it took up 8 inches of shelf space. About how many books should fit on the shelf? 45 books

Available on Daily Transparency in CRB

Math Humor

Teacher: Do you know how many quarters go into a half?

Student: No, but I know how many go into a video game!

3-4 Dividing Rational Numbers

Learn to divide fractions and decimals.

Vocabulary
reciprocal

A number and its **reciprocal** have a product of 1. To find the reciprocal of a fraction, exchange the numerator and the denominator. Remember that an integer can be written as a fraction with a denominator of 1.

Number	Reciprocal	Product
$\frac{3}{4}$	$\frac{4}{3}$	$\frac{3}{4}\left(\frac{4}{3}\right) = 1$
$-\frac{5}{12}$	$-\frac{12}{5}$	$-\frac{5}{12}\left(-\frac{12}{5}\right) = 1$
6	$\frac{1}{6}$	$6\left(\frac{1}{6}\right) = 1$

Multiplication and division are inverse operations. They undo each other.

$$\frac{1}{3}\left(\frac{2}{5}\right) = \frac{2}{15} \longrightarrow \frac{2}{15} \div \frac{2}{5} = \frac{1}{3}$$

Notice that multiplying by the reciprocal gives the same result as dividing.

$$\left(\frac{2}{15}\right)\left(\frac{5}{2}\right) = \frac{2 \cdot 5}{15 \cdot 2} = \frac{1}{3}$$

DIVIDING RATIONAL NUMBERS IN FRACTION FORM		
Words	**Numbers**	**Algebra**
To divide by a fraction, multiply by the reciprocal.	$\frac{1}{5} \div \frac{2}{3} = \frac{1}{5} \cdot \frac{3}{2} = \frac{3}{10}$	$\frac{a}{b} \div \frac{c}{d} = \frac{a}{b} \cdot \frac{d}{c} = \frac{ad}{bc}$

EXAMPLE 1 **Dividing Fractions**

Divide. Write each answer in simplest form.

Ⓐ $\frac{7}{12} \div \frac{2}{3}$

$\frac{7}{12} \div \frac{2}{3} = \frac{7}{12} \cdot \frac{3}{2}$ *Multiply by the reciprocal.*

$= \frac{7 \cdot \overset{1}{\cancel{3}}}{\underset{4}{\cancel{12}} \cdot 2}$ *Reduce common factors.*

$= \frac{7}{8}$ *Simplest form*

1 Introduce

Alternate Opener

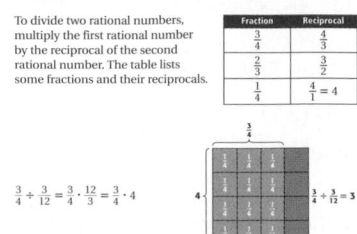

3-4 Dividing Rational Numbers

To divide two rational numbers, multiply the first rational number by the reciprocal of the second rational number. The table lists some fractions and their reciprocals.

Fraction	Reciprocal
$\frac{3}{4}$	$\frac{4}{3}$
$\frac{2}{3}$	$\frac{3}{2}$
$\frac{1}{4}$	$\frac{4}{1} = 4$

$\frac{3}{4} \div \frac{3}{12} = \frac{3}{4} \cdot \frac{12}{3} = \frac{3}{4} \cdot 4$ $\frac{3}{4} \div \frac{3}{12} = 3$

Use reciprocals and modeling to divide the rational numbers.

1. $1\frac{1}{2} \div \frac{3}{4}$ **2.** $\frac{1}{2} \div \frac{3}{4}$

3. $\frac{5}{12} \div \frac{1}{6}$ **4.** $\frac{11}{12} \div \frac{1}{3}$

Think and Discuss

5. Discuss whether you could multiply by the reciprocal of the first rational number instead of the second rational number when you divide.

6. Explain how you could start a division problem when the numbers you are dividing are decimals and fractions.

Motivate

Write the following problems on the board: $12 \div 3 = ?$ and $12 \cdot \frac{1}{3} = ?$. Ask for volunteers to solve both problems and to show the solutions on the board. Both equal 4. Point out that 3 can be expressed as $\frac{3}{1}$. Ask students to describe the relationship between $\frac{3}{1}$ and $\frac{1}{3}$. Explain that these numbers are called *reciprocals*. Tell students that in this lesson, they will see that division is the same as multiplication by the reciprocal.

Exploration worksheet and answers on Chapter 3 Resource Book pp. 35 and 122

2 Teach

Lesson Presentation

Guided Instruction

In this lesson, students learn to divide fractions. Define and discuss *reciprocals* (Teaching Transparency T17, CRB). Show that dividing by a number and multiplying by the reciprocal of the number give the same result. Remind students that they may "cancel out" common factors once they have changed the division into multiplication by a reciprocal. For Example 2, show students how to choose which power of 10 to use to clear the decimal from the denominator. You may want to review place value (Lesson 3-1 and Teaching Transparency T2, CRB).

Divide. Write each answer in simplest form.

B $3\frac{1}{4} \div 4$

$$3\frac{1}{4} \div 4 = \frac{13}{4} \div \frac{4}{1} \qquad \textit{Write as improper fractions.}$$

$$= \frac{13}{4}\left(\frac{1}{4}\right) \qquad \textit{Multiply by the reciprocal.}$$

$$= \frac{13 \cdot 1}{4 \cdot 4} \qquad \textit{No common factors.}$$

$$= \frac{13}{16} \qquad \textit{Simplest form}$$

When dividing a decimal by a decimal, multiply both numbers by a power of 10 so you can divide by a whole number. To decide which power of 10 to divide by, look at the denominator. The number of decimal places is the number of zeros to write after the 1.

$$\frac{1.32}{0.4} = \frac{1.32}{0.4}\left(\frac{10}{10}\right) = \frac{13.2}{4}$$

1 decimal place   *1 zero*

...Decimals

$$= \frac{2.92}{0.4}\left(\frac{10}{10}\right) = \frac{29.2}{4}$$

$$= 7.3 \qquad \textit{Divide.}$$

...g Expressions with Fractions and Decimals

...each expression for the given value of the variable.

...or $n = 0.24$

$$\frac{}{} = \frac{7.2}{0.24}\left(\frac{100}{100}\right) \qquad \textit{0.24 has 2 decimal places, so use } \frac{100}{100}.$$

$$= \frac{720}{24} \qquad \textit{Divide.}$$

$$= 30$$

When $n = 0.24$, $\frac{7.2}{n} = 30$.

$m \div \frac{3}{8}$ for $m = 7\frac{1}{2}$

$$7\frac{1}{2} \div \frac{3}{8} = \frac{15}{2} \cdot \frac{8}{3}$$

$$= \frac{\overset{5}{\cancel{15}} \cdot \overset{4}{\cancel{8}}}{\underset{1}{\cancel{2}} \cdot \underset{1}{\cancel{3}}} = \frac{20}{1} = 20$$

When $m = 7\frac{1}{2}$, $m \div \frac{3}{8} = 20$.

Additional Examples

Example 1

Divide. Write each answer in simplest form.

A. $\frac{5}{11} \div \frac{1}{2}$ $\frac{10}{11}$ **B.** $2\frac{3}{8} \div 2$ $1\frac{3}{16}$

Example 2

Divide.

$0.384 \div 0.24$ **1.6**

Example 3

Evaluate each expression for the given value of the variable.

A. $\frac{5.25}{n}$ for $n = 0.15$ **35**

B. $k \div \frac{4}{5}$ for $k = 5$ $6\frac{1}{4}$

Example 4

A cookie recipe calls for $\frac{1}{2}$ cup of oats. You have $\frac{3}{4}$ cup of oats. How many batches of the cookies can you bake?

You can bake $1\frac{1}{2}$ batches of the cookies.

...g All Learners

...ritical Thinking

...recording sheet (Chapter ...107). Explain to students ...: 32 oz. Ask students to ...ervings are in a quart ...3 oz. 4 servings Ask ...how they found the ...e students find out how ...uart contains for the fol-

2. 4 oz   8

4. 1 oz   32

6. $\frac{1}{2}$ oz   64

Example 4 note: You may want to replace the fractions with whole numbers first to show how to set up the division. For example, "If you pour 8 cups into a container and the serving size is 2 cups, how many servings did you pour?"

8	÷	2	=	4
amount poured	÷	serving size	=	number of servings

EXAMPLE 4 PROBLEM SOLVING

You pour $\frac{2}{3}$ cup of a sports drink into a glass. The serving size is 6 ounces, or $\frac{3}{4}$ cup. How many servings will you consume? How many calories will you consume?

 **Understand the Problem**

The number of calories you consume is the number of calories in the fraction of a serving.

List the **important information:**
- The amount you plan to drink is $\frac{2}{3}$ cup.
- The amount of a full serving is $\frac{3}{4}$ cup.
- The number of calories in one serving is 50.

2 Make a Plan

Set up an equation to find the numb servings you will drink.

| amount you drink | ÷ | serv

Using the number of serving

| number of servings | · | c

 **Solve**

Let n = number of servings.

Servings: $\frac{2}{3} \div \frac{3}{4} = n$

$\frac{2}{3} \cdot \frac{4}{3} = n$

$\frac{8}{9} = n$

You will drink $\frac{8}{9}$ of a serving, which

4 Look Back

You did not pour a full serving, so $\frac{8}{9}$ i than 1, and 44.4 calories is less than t

Think and Discuss

1. **Tell** what happens when you divide a frac you are correct using multiplication by the

2. **Model** the product of $\frac{2}{3}$ and $\frac{1}{4}$.

3 Close

Summarize

Ask students to explain how multiplication and division of fractions are the same and how they are different. Review the process for dividing decimals. Ask the students what power of 10 they would use to simplify the expression 22.5 ÷ 1.125.

Possible answer: Division of fractions is the same as multiplication except that in division you multiply by the reciprocal; 1000.

Answers to Think a

Possible answers:

1. When you divide a frac you multiply it by its re the answer is 1.

2.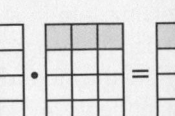

$\frac{2}{3}$ · $\frac{1}{4}$ = $\frac{2}{12}$

EXAMPLE **2** **Dividing**

Divide.

$2.92 \div 0.$

$2.92 \div 0.4$

EXAMPLE **3** **Evaluatin**

Evaluate

 **A** $\dfrac{7.2}{n}$

$\dfrac{7.2}{0.2}$

B

 **Reachin**
Through

Give each student
3 Resource Book
that a quart conta
find out how mar
if the serving size
students to descri
answer. division H
many servings one
lowing serving size

1. 5 oz   6.4

3. 3 oz $10\frac{2}{3}$

5. $\frac{2}{3}$ oz   48

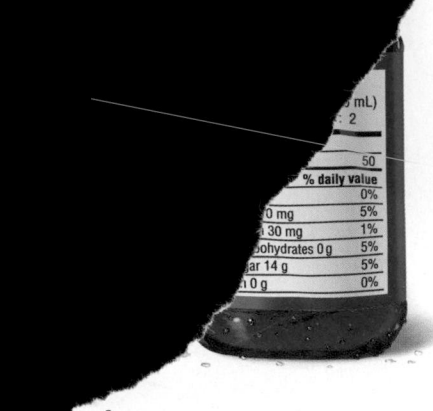

		% daily value
		0%
0 mg		5%
30 mg		1%
ohydrates 0 g		5%
ar 14 g		5%
0 g		0%

er of

| g size | = | number of servings |

you can find the calories consumed.

| ories per serving | = | total calories |

Let c = total calories.

Calories: $\frac{8}{9} \cdot 50 = c$

$\frac{8 \cdot 50}{9} = c$

$\frac{400}{9} = c \approx 44.4$

is about 44.4 calories.

s a reasonable answer. It is less
he calories in a full serving, 50.

ction by itself. Show that
e reciprocal.

nd Discuss

ion by itself,
iprocal, and

or $\frac{1}{6}$

FOR EXTRA PRACTICE
see page 736

internet connect
Homework Help Online
go.hrw.com   Keyword: MP4 3-4

go.hrw.com

3-4 PRACTICE & ASSESS

GUIDED PRACTICE

See Example 1 Divide. Write each answer in simplest form.

1. $\frac{2}{3} \div \frac{5}{6}$ $\frac{4}{5}$
2. $2\frac{1}{4} \div 3\frac{2}{5}$ $\frac{45}{68}$
3. $-\frac{6}{7} \div 3$ $-\frac{2}{7}$
4. $\frac{7}{8} \div \frac{3}{10}$ $2\frac{11}{12}$

5. $3\frac{3}{16} \div 2\frac{5}{8}$ $1\frac{3}{14}$
6. $-\frac{5}{9} \div 6$ $-\frac{5}{54}$
7. $\frac{9}{10} \div \frac{3}{5}$ $1\frac{1}{2}$
8. $2\frac{5}{12} \div \frac{5}{6}$ $2\frac{9}{10}$

See Example 2 Divide.

9. $3.72 \div 0.3$ **12.4**
10. $3.4 \div 0.05$ **68**
11. $10.71 \div 0.7$ **15.3**
12. $3.44 \div 0.4$ **8.6**

13. $3.46 \div 0.9$ **3.84**
14. $14.08 \div 0.8$ **17.6**
15. $7.86 \div 0.006$ **1310**
16. $2.76 \div 0.3$ **9.2**

See Example 3 Evaluate each expression for the given value of the variable.

17. $\frac{4.5}{x}$ for $x = 0.2$ **22.5**
18. $\frac{8.4}{x}$ for $x = 0.4$ **21**
19. $\frac{40.5}{x}$ for $x = 0.9$ **45**

20. $\frac{9.2}{x}$ for $x = 2.3$ **4**
21. $\frac{20.8}{x}$ for $x = 1.6$ **13**
22. $\frac{21.6}{x}$ for $x = 0.08$ **270**

See Example 4 23. You drink $\frac{3}{4}$ pint of spring water. One serving of the water is $\frac{7}{8}$ pint. How much of a serving did you drink? $\frac{6}{7}$ **serving**

INDEPENDENT PRACTICE

See Example 1 Divide. Write each answer in simplest form.

24. $\frac{1}{8} \div \frac{2}{5}$ $\frac{5}{16}$
25. $3\frac{1}{2} \div 1\frac{7}{8}$ $1\frac{13}{15}$
26. $-\frac{5}{12} \div \frac{2}{3}$ $-\frac{5}{8}$
27. $\frac{9}{10} \div \frac{1}{4}$ $3\frac{3}{5}$

28. $1\frac{3}{4} \div 4\frac{1}{8}$ $\frac{14}{33}$
29. $-\frac{2}{9} \div \frac{7}{12}$ $-\frac{8}{21}$
30. $\frac{2}{5} \div \frac{5}{16}$ $1\frac{7}{25}$
31. $2\frac{3}{8} \div 1\frac{1}{6}$ $2\frac{1}{28}$

32. $-\frac{3}{11} \div \frac{4}{7}$ $-\frac{21}{44}$
33. $\frac{3}{16} \div \frac{3}{4}$ $\frac{1}{4}$
34. $3\frac{11}{12} \div 2\frac{1}{4}$ $1\frac{20}{27}$
35. $-\frac{3}{4} \div \frac{1}{6}$ $-4\frac{1}{2}$

See Example 2 Divide.

36. $10.86 \div 0.6$ **18.1**
37. $1.94 \div 0.02$ **97**
38. $9.76 \div 0.8$ **12.2**
39. $8.55 \div 0.5$ **17.1**

40. $6.52 \div 0.004$ **1630**
41. $24.66 \div 0.9$ **27.4**
42. $9.36 \div 0.03$ **312**
43. $17.78 \div 0.7$ **25.4**

44. $11.128 \div 0.52$ **21.4**
45. $24 \div 0.75$ **32**
46. $13.608 \div 0.81$ **16.8**
47. $3.6864 \div 0.64$ **5.76**

See Example 3 Evaluate each expression for the given value of the variable.

48. $\frac{6.3}{x}$ for $x = 0.3$ **21**
49. $\frac{9.1}{x}$ for $x = 0.7$ **13**
50. $\frac{12}{x}$ for $x = 0.02$ **600**

51. $\frac{15.4}{x}$ for $x = 1.4$ **11**
52. $\frac{3.69}{x}$ for $x = 0.9$ **4.1**
53. $\frac{22.2}{x}$ for $x = 0.06$ **370**

54. $\frac{1.6}{x}$ for $x = 3.2$ **0.5**
55. $\frac{0.56}{x}$ for $x = 0.8$ **0.7**
56. $\frac{94.05}{x}$ for $x = 28.5$ **3.3**

See Example 4 57. The platform on the school stage is $8\frac{3}{4}$ feet wide. Each chair is $1\frac{5}{12}$ feet wide. How many chairs will fit across the platform? **6 chairs**

Students may want to refer back to the lesson examples.

Assignment Guide

If you finished Example 1 assign:
Core 1–8, 24–35, 65–72
Enriched 1–8, 24–35, 65–72

If you finished Example 2 assign:
Core 1–16, 24–47, 65–72
Enriched 1–16, 24–47, 65–72

If you finished Example 3 assign:
Core 1–22, 24–56, 65–72
Enriched 1–22, 24–56, 65–72

If you finished Example 4 assign:
Core 1–59, 65–72
Enriched 1–15 odd, 17–72

Notes

RETEACH 3-4

LESSON 3-4 Reteach
Dividing Rational Numbers

To write the **reciprocal** of a fraction, interchange the numerator and denominator.

$\frac{2}{3} \quad\rightleftarrows\quad \frac{3}{2}$
Fraction   Reciprocal

The product of a number and its reciprocal is 1.

$\frac{2}{3} \times \frac{3}{2} = 1$

Write the reciprocal of each rational number.

1. The reciprocal of $\frac{3}{5}$ is: $\frac{5}{3}$
2. The reciprocal of 6 is: $\frac{1}{6}$
3. The reciprocal of $2\frac{1}{3}$ is: $\frac{3}{7}$

To divide by a fraction, multiply by its reciprocal.

$6 \div \frac{2}{3}$ $\frac{3}{5} \div \frac{9}{10}$
$6 \times \frac{3}{2}$ $\frac{3}{5} \times \frac{10}{9}$
$\frac{6 \times 3}{2} = 9$ $\frac{3 \times 10}{5 \times 9} = \frac{2}{3}$

Complete to divide and simplify.

4. $12 \div \frac{3}{8} = 12 \times \frac{8}{3} = \frac{96}{3} = 32$
5. $16 \div \frac{4}{3} = \frac{16 \times \left(\frac{3}{4}\right)}{4} = \frac{48}{4} = 12$

6. $\frac{5}{7} \div \frac{20}{21} = \frac{5}{7} \times \frac{21}{20} = \frac{3}{4}$
7. $-\frac{3}{4} \div \left(-\frac{9}{8}\right) = -\frac{3}{4} \times \left(-\frac{8}{9}\right) = \frac{2}{3}$

Change a decimal divisor to a whole number. Using the number of places in the divisor, move the decimal point to the right in both the divisor and the dividend.

$0.7\overline{)4.34} \rightarrow 0.7\overline{)4.3.4} \rightarrow 7\overline{)43.4}$ $\quad\quad 6.2$

Rewrite each division with a whole-number divisor. Then, do the division.

8. $1.6\overline{)11.52} \rightarrow 16\overline{)115.2} = 7.2$
9. $0.3\overline{)45} \rightarrow 3\overline{)450} = 150$
10. $1.32\overline{)71.28} \rightarrow 132\overline{)7128} = 54$
11. $3.58\overline{)5728} \rightarrow 358\overline{)572,800} = 1600$

PRACTICE 3-4

LESSON 3-4 Practice B
Dividing Rational Numbers

Divide. Write each answer in simplest form.

1. $\frac{1}{5} \div \frac{3}{10}$ $\frac{2}{3}$
2. $-\frac{5}{8} \div \frac{3}{4}$ $-\frac{5}{6}$
3. $\frac{1}{4} \div \frac{1}{8}$ 2
4. $-\frac{2}{3} \div \frac{4}{15}$ $-2\frac{1}{2}$

5. $\frac{2}{9} \div \left(-\frac{2}{3}\right)$ $-\frac{1}{3}$
6. $-\frac{7}{10} \div \left(-\frac{2}{5}\right)$ $1\frac{3}{4}$
7. $\frac{6}{11} \div \frac{3}{22}$ 4
8. $\frac{4}{9} \div \left(-\frac{8}{15}\right)$ $-\frac{5}{6}$

9. $-15 \div \frac{3}{8}$ -40
10. $\frac{5}{6} \div 12$ $\frac{5}{72}$
11. $6 \div \frac{3}{8}$ 16
12. $\frac{9}{10} \div (-6)$ $-\frac{3}{20}$

Divide.

13. $20 \div 0.5$ 40
14. $2.16 \div 0.4$ 5.4
15. $-3.16 \div (-0.02)$ 158
16. $0.702 \div 0.3$ 2.34

17. $8.736 \div 1.3$ 6.72
18. $79.36 \div (-6.2)$ -12.8
19. $42 \div 0.21$ 200
20. $63.81 \div 9$ 7.09

21. $-121 \div 1.1$ -110
22. $6.006 \div 0.22$ 27.3
23. $21.12 \div 1.5$ 14.08
24. $8.068 \div 4$ 2.017

Evaluate each expression for the given value of the variable.

25. $\frac{-18}{x}$ for $x = 0.12$ -150
26. $\frac{10.8}{x}$ for $x = 0.03$ 360
27. $\frac{-9.018}{x}$ for $x = -1.2$ 7.515

28. A can of fruit contains $3\frac{1}{2}$ cups of fruit. The suggested serving size is $\frac{1}{2}$ cup. How many servings are in the can of fruit?
7 servings

Math Background

A division expression containing fractions can be written as a *complex fraction*.

$\frac{1}{2} \div \frac{3}{4}$ can be written as $\dfrac{\frac{1}{2}}{\frac{3}{4}}$.

To find the quotient, multiply the numerator and denominator of the complex fraction by the reciprocal of the fraction in the denominator. The denominator then becomes one and the numerator is a multiplication expression that can be simplified:

$$\frac{\frac{1}{2}}{\frac{3}{4}} = \frac{\frac{1}{2} \cdot \frac{4}{3}}{\frac{3}{4} \cdot \frac{4}{3}} = \frac{\frac{2}{3}}{1} = \frac{2}{3}$$

Answers

61. Yes; $29 \div \frac{5}{8} = 46\frac{2}{5}$, which is greater than the number of DVDs.

62. Possible answer: The student should have multiplied by $\frac{1}{3}$ instead of dividing.
$\frac{7}{8} \cdot \frac{1}{3} = \frac{7}{24}$ cup of rice

63. Possible answer: The quotient will have an even denominator. When the division is changed to multiplication by the reciprocal, the factor 3 will be in the numerator. This factor can be divided out with the 6 in the denominator, leaving the factor 2 in the denominator. Because the denominator of the quotient will have a factor of 2, it must be even.

Journal

Ask students to write about a situation from their everyday lives in which they might need to divide by a fraction.

Test Prep Doctor ✚

For Exercise 71, remind students of the meaning of a negative exponent. Show two ways to evaluate the expression.
$$3^4 \cdot 3^{-2} = \frac{3^4}{1} \cdot \frac{1}{3^2} = \frac{3^4}{3^2} = \frac{81}{9} = 9$$
$$3^4 \cdot 3^{-2} = 3^{4-2} = 3^2 = 9$$

Choice **C** is correct.

Lesson Quiz

Divide.

1. $2\frac{5}{6} \div \left(-1\frac{1}{2}\right)$ $-1\frac{8}{9}$

2. $-14 \div 1.25$ -11.2

3. $3.9 \div 0.65$ **6**

4. Evaluate $\frac{112}{x}$ for $x = 6.3$.   **17.7̄**

5. A penny weighs 2.5 grams. How many pennies would it take to equal one pound (453.6 grams)?
  181

Available on Daily Transparency in CRB

PRACTICE AND PROBLEM SOLVING

58. Reba is eating her favorite cereal. There are $3\frac{2}{3}$ servings remaining in the box. Reba pours only $\frac{1}{3}$ of a serving into her bowl at a time. How many more bowls can Reba have before the box is empty?   **11 bowls**

59. The thickest vinyl floor tiles available are $\frac{1}{8}$ inch thick. The thinnest tiles are $\frac{1}{20}$ inch thick. How many thin tiles would equal the thickness of one of the thick tiles? $2\frac{1}{2}$ **tiles**

60. Nesting dolls called *matrushkas* are a well-known type of Russian folk art. Use the information in the picture to find the height of the largest doll.

$3\frac{31}{48}$ in.

$\frac{6}{25}x = \frac{7}{8}$ in. x in.

61. Cal has 41 DVDs in cases that are each $\frac{5}{8}$ inch thick. Can he put all the DVDs on a shelf that is 29 inches long?

62. **WHAT'S THE ERROR?** A student had a recipe that called for $\frac{7}{8}$ cup of rice. Since he wanted to make only $\frac{1}{3}$ of the whole recipe, he calculated the amount of rice he would need: $\frac{7}{8} \div \frac{1}{3} = \frac{7}{8} \cdot \frac{3}{1} = 2\frac{5}{8}$ cups of rice. What was his error?

63. **WRITE ABOUT IT** A proper fraction with denominator 6 is divided by a proper fraction with denominator 3. Will the denominator of the quotient be odd or even? Explain.

64. **CHALLENGE** According to the 2000 U.S. census, about $\frac{1}{30}$ of the U.S. population resides in Los Angeles County. About $\frac{1}{8}$ of the U.S. population resides in California. What fraction of the California population resides in Los Angeles County? $\frac{4}{15}$

Spiral Review

Solve each equation. (Lesson 1-4)

65. $7x = 45.5$ $x = 6.5$ **66.** $\frac{x}{6} = 11.2$ $x = 67.2$ **67.** $1032 = 129x$ $x = 8$

68. $\frac{x}{5} = 16.25$ $x = 81.25$ **69.** $13x = 58.5$ $x = 4.5$ **70.** $\frac{x}{2} = 1.38$ $x = 2.76$

71. **TEST PREP** Evaluate $3^4 \cdot 3^{-2}$. (Lesson 2-8)   **C**

   A $\frac{1}{9}$ **B** 72   **C** 9   **D** 6

72. **TEST PREP** Evaluate $\frac{2^5}{2^9}$. (Lesson 2-8)   **F**

   F $\frac{1}{16}$ **G** 16   **H** 8   **J** -16

CHALLENGE 3-4

LESSON 3-4 Challenge
A New License to Operate

You can invent new operations based on the familiar operations of addition, subtraction, multiplication, and division, and the familiar order of operations.

If $a \triangle b = \frac{a+b}{2}$ where a and b represent any rational numbers,

then $3 \triangle 5 = \frac{3+5}{2} = 4$.

Use the given definition of operation $\triangle$ to evaluate each expression.

1. $\frac{1}{2} \triangle (-10) =$ __-4.75__   **2.** $\frac{100 \triangle (-10)}{10} =$ __4.5__

3. $4 \triangle 6 \triangle 3 =$ __4__   **4.** $[5.5 \triangle (-6)] + [-6 \triangle 5.5] =$ __-0.5__

Use the operation shown to answer each question. $\begin{vmatrix} a & b \\ c & d \end{vmatrix} = ac - bd$

5. $\begin{vmatrix} 1 & 8 \\ 3 & 4 \end{vmatrix} =$ __-29__   **6.** $\begin{vmatrix} -2 & 3 \\ 3 & -2 \end{vmatrix} =$ __0__

7. If $\begin{vmatrix} 1 & 3 \\ x & 2 \end{vmatrix} = 18$, then $x =$ __24__   **8.** If $\begin{vmatrix} 6 & 2 \\ x & x \end{vmatrix} = 12$, then $x =$ __3__

Use the operation shown to answer each question. $a \searrow b = \frac{a^2}{b^2}$

9. $(3 \searrow 5) =$ __$\frac{9}{25}$__   **10.** $(1 \searrow 8) - (5 \searrow 8) =$ __$-\frac{3}{8}$__

11. $(1 \searrow 3) \times (3 \searrow 6) =$ __$\frac{1}{36}$__   **12.** $(1 \searrow 10)^2 =$ __$\frac{1}{10,000}$__

If $\lrcorner n \lrcorner$ means 1 less than the number of digits in the integer n, then, for example, $\lrcorner 77 \lrcorner = 1$ since 77 has 2 digits.

Use the definition of $\lrcorner n \lrcorner$ to answer each question.

13. If n is a positive integer less than 100, what is the greatest value for $\lrcorner n \lrcorner$? __1__

14. If n is a positive integer less than 1001, what is the greatest value for $\lrcorner n \lrcorner$? __3__

15. If n has 100 digits, what is the value of $\lrcorner \lrcorner n \lrcorner \lrcorner$? Explain.

By definition, $\lrcorner$100-digit number$\lrcorner = 99$. Then, since 99 has 2 digits, $\lrcorner 99 \lrcorner = 1$.

PROBLEM SOLVING 3-4

LESSON 3-4 Problem Solving
Dividing Rational Numbers

Use the table at the right that shows the maximum speed over a quarter of a mile of different animals. Find the time is takes each animal to travel one-quarter mile at top speed. Round to the nearest thousandth.

1. Quarter horse
   0.005 hours

2. Greyhound
   0.006 hours

3. Human
   0.009 hours

4. Giant tortoise
   1.471 hours

5. Three-toed sloth
   1.667 hours

Maximum Speeds of Animals	
Animal	Speed (mph)
Quarter Horse	47.50
Greyhound	39.35
Human	27.89
Giant Tortoise	0.17
Three-toed Sloth	0.15

Choose the letter for the best answer.

6. A piece of ribbon is $1\frac{7}{8}$ inches long. If the ribbon is going to be divided into 15 pieces, how long should each piece be?

   (A) $\frac{1}{8}$ in.
   B $\frac{1}{15}$ in.
   C $\frac{2}{3}$ in.
   D $28\frac{1}{8}$ in.

7. The recorded rainfall for each day of a week was 0 in., $\frac{1}{4}$ in., $\frac{3}{4}$ in., 1 in., 0 in., $1\frac{1}{2}$ in., $1\frac{1}{4}$ in. What was the average rainfall per day?

   F $\frac{9}{10}$ in.
   (G) $\frac{9}{14}$ in.
   H $\frac{7}{8}$ in.
   J $4\frac{1}{2}$ in.

8. A drill bit that is $\frac{7}{32}$ in. means that the hole the bit makes has a diameter of $\frac{7}{32}$ in. Since the radius is half of the diameter, what is the radius of a hole drilled by a $\frac{7}{32}$ in. bit?

   A $\frac{14}{32}$ in.
   C $\frac{9}{16}$ in.
   B $\frac{7}{32}$ in.
   (D) $\frac{7}{64}$ in.

9. A serving of a certain kind of cereal is $\frac{2}{3}$ cup. There are 12 cups of cereal in the box. How many servings of cereal are in the box?

   (F) 18
   G 15
   H 8
   J 6

Adding and Subtracting with Unlike Denominators

Learn to add and subtract fractions with unlike denominators.

Vocabulary

least common denominator (LCD)

A pattern for a double-circle skirt requires $9\frac{1}{3}$ yards of 45-inch-wide material. To add a ruffle takes another $2\frac{2}{5}$ yards. If the total amount of material for the skirt and ruffle are cut from a bolt of fabric $15\frac{1}{2}$ yards long, how much fabric is left?

To solve this problem, add and subtract rational numbers with unlike denominators. First find a common denominator using one of the following methods:

Method 1   Find a common denominator by multiplying one denominator by the other denominator.

Method 2   Find the **least common denominator (LCD)** ; the least common multiple of the denominators.

EXAMPLE 1   Adding and Subtracting Fractions with Unlike Denominators

Remember!

The least common multiple of two numbers is the smallest number other than zero that is a multiple of the two numbers.

Add or subtract.

A $\frac{2}{3} + \frac{1}{5}$

Method 1: $\frac{2}{3} + \frac{1}{5}$ *Find a common denominator: 3(5) = 15.*

$= \frac{2}{3}\left(\frac{5}{5}\right) + \frac{1}{5}\left(\frac{3}{3}\right)$ *Multiply by fractions equal to 1.*

$= \frac{10}{15} + \frac{3}{15}$ *Rewrite with a common denominator.*

$= \frac{13}{15}$ *Simplify.*

B $3\frac{2}{5} + \left(-3\frac{1}{2}\right)$

Method 2: $3\frac{2}{5} + \left(-3\frac{1}{2}\right)$

$= \frac{17}{5} + \left(-\frac{7}{2}\right)$ *Write as improper fractions.*

Multiples of 5: 5, ⑩, 15, 20, . . .   *List the multiples of each*
Multiples of 2: 2, 4, 6, 8, ⑩, . . .   *denominator and find the LCD.*

$= \frac{17}{5}\left(\frac{2}{2}\right) + \left(-\frac{7}{2}\right)\left(\frac{5}{5}\right)$ *Multiply by fractions equal to 1.*

$= \frac{34}{10} + \left(-\frac{35}{10}\right)$ *Rewrite with a common denominator.*

$= -\frac{1}{10}$ *Simplify.*

---

1 Introduce

Alternate Opener

3-5 Adding and Subtracting with Unlike Denominators

You can use models to show addition and subtraction of fractions with unlike denominators. Look at the models for $\frac{1}{2} + \frac{1}{3}$ and $\frac{1}{2} - \frac{1}{3}$.

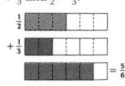 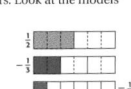

1. Draw a picture to show that $\frac{1}{2} + \frac{4}{5} = 1\frac{3}{10} = 1.3$.
2. Draw a picture to show that $\frac{1}{2} + \frac{2}{3} = \frac{7}{6} = 1\frac{1}{6}$.
3. Complete each addition problem. Simplify your answers.
   a. $\frac{7}{10} + \frac{4}{5}$ b. $\frac{3}{5} + \frac{3}{4}$

Think and Discuss

4. **Explain** how to add and subtract fractions with unlike denominators.
5. **Discuss** why the fraction that results from adding or subtracting fractions with unlike denominators has a different denominator than the fractions.

Motivate

To introduce adding and subtracting with unlike denominators, ask students to list the multiples of 5 and the multiples of 2. Then ask the students to identify the *least common* multiple. Show them that although there are many common multiples (10, 20, 30, etc.), there is only one least common multiple (10).

5, ⑩, 15, 20, 25, 30, . . .

2, 4, 6, 8, ⑩, 12, 14, 16, 18, 20, 22, 24, 26, 28, 30, . . .

Exploration worksheet and answers on Chapter 3 Resource Book pp. 44 and 124

3-5 Organizer

Pacing: Traditional 1 day
          Block $\frac{1}{2}$ day

Objective: Students add and subtract fractions with unlike denominators.

Warm Up

Add or subtract.

1. $1\frac{3}{5} + \left(-\frac{2}{5}\right)$ $1\frac{1}{5}$
2. $3\frac{11}{12} - 2\frac{7}{12}$ $1\frac{1}{3}$
3. $6.5 + -1.2$ 5.3
4. $3.4 - 0.9$ 2.5

Problem of the Day

A school installed a new phone system with four-digit extensions. How many extensions are possible? all numbers from 0000 to 9999, or 10,000

Available on Daily Transparency in CRB

Math Humor

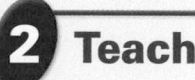

Clerk: That $2.50 pen is on sale for $2.00.

Customer: But the sign says, "Half off everything."

Clerk: That's right. It was $2\frac{1}{2}$ dollars, so I took off the $\frac{1}{2}$.

---

2 Teach

Lesson Presentation

Guided Instruction

In this lesson, students learn to add and subtract fractions with unlike denominators. Discuss the two methods for finding a common denominator. To help students understand why fractions must have common denominators before adding across the numerators, you may want to compare the denominators of fractions with units of measure. For example, just as you cannot add 2 inches and 3 centimeters without first writing them in terms of the same unit, you cannot add $\frac{2}{5}$ and $\frac{3}{10}$ without first writing them in terms of the same denominator.

EXAMPLE 2 **Evaluating Expressions with Rational Numbers**

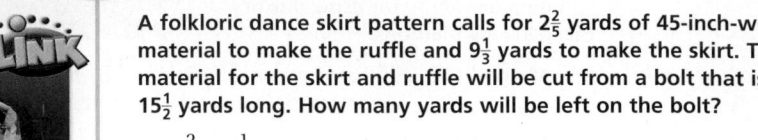

Evaluate $n - \frac{11}{16}$ for $n = -\frac{1}{3}$.

$$n - \frac{11}{16}$$

$$= \left(-\frac{1}{3}\right) - \frac{11}{16} \qquad \text{Substitute } -\frac{1}{3} \text{ for } n.$$

$$= \left(-\frac{1}{3}\right)\left(\frac{16}{16}\right) - \frac{11}{16}\left(\frac{3}{3}\right) \qquad \text{Multiply by fractions equal to 1.}$$

$$= -\frac{16}{48} - \frac{33}{48} \qquad \text{Rewrite with a common denominator: } 3(16) = 48.$$

$$= -\frac{49}{48}, \text{ or } -1\frac{1}{48} \qquad \text{Simplify.}$$

EXAMPLE 3 **Consumer Application**

Social Studies **LINK**

There are three categories of folkloric dance in Mexico: *danza*, *mestizo*, and *bailes regionales*. These dances are performed by traveling groups such as the Ballet Folklorico.

A folkloric dance skirt pattern calls for $2\frac{2}{5}$ yards of 45-inch-wide material to make the ruffle and $9\frac{1}{3}$ yards to make the skirt. The material for the skirt and ruffle will be cut from a bolt that is $15\frac{1}{2}$ yards long. How many yards will be left on the bolt?

$$2\frac{2}{5} + 9\frac{1}{3} \qquad \text{Add to find length needed for the skirt and ruffle.}$$

$$= \frac{12}{5} + \frac{28}{3} \qquad \text{Write as improper fractions. The LCD is 15.}$$

$$= \frac{36}{15} + \frac{140}{15} \qquad \text{Rewrite with a common denominator.}$$

$$= \frac{176}{15}$$

The amount needed for the skirt and ruffle is $\frac{176}{15}$, or $11\frac{11}{15}$ yards. Now find the number of yards remaining.

$$15\frac{1}{2} - 11\frac{11}{15} \qquad \text{Subtract amount needed from bolt length.}$$

$$= \frac{31}{2} - \frac{176}{15} \qquad \text{Write as improper fractions. The LCD is 30.}$$

$$= \frac{465}{30} - \frac{352}{30} \qquad \text{Rewrite with a common denominator.}$$

$$= \frac{113}{30}, \text{ or } 3\frac{23}{30} \qquad \text{Simplify.}$$

There will be $3\frac{23}{30}$ yards left on the bolt.

Think and Discuss

1. Give an example of two denominators with no common factors.

2. Tell if $-2\frac{1}{5} - \left(-2\frac{3}{16}\right)$ is positive or negative. Explain.

3. Explain how to add $2\frac{2}{5} + 9\frac{1}{3}$ without first writing them as improper fractions.

Teach

Reaching All Learners
Through Concrete Manipulatives

Give each student (or group of students) a recording sheet and a set of manipulative cutouts that represent $\frac{1}{2}, \frac{1}{3}, \frac{1}{4}, \frac{1}{6},$ and $\frac{1}{12}$. (Chapter 3 Resource Book pp. 108–109).

Have students use the cutouts to complete the exercises.

3 **Close**

Summarize

Review both methods of finding common denominators. Remind students that fractions can only be added or subtracted when the denominators are the same. Ask them if this is true for multiplication or division. no Ask the students to find the sum of $-1\frac{1}{4}$ and $\frac{5}{6}$ by both of the methods presented in the lesson. Remind students to give the answer in simplest form.

$$-1\frac{1}{4} + \frac{5}{6} = -\frac{5}{4} + \frac{5}{6} = -\frac{30}{24} + \frac{20}{24}$$
$$= -\frac{10}{24} = -\frac{5}{12}$$

$$-1\frac{1}{4} + \frac{5}{6} = -\frac{5}{4} + \frac{5}{6} = -\frac{15}{12} + \frac{10}{12} = -\frac{5}{12}$$

Answers to Think and Discuss

Possible answers:

1. 3 and 7; 4 and 9; 27 and 16

2. Negative; $\left|-2\frac{1}{5}\right| = 2\frac{3}{15}$, and $\left|-2\frac{3}{16}\right| = 2\frac{3}{16}$; since $2\frac{3}{15} > 2\frac{3}{16}$, the answer takes the sign of $-2\frac{1}{5}$.

3. Using the Commutative Property, add the whole number portions, add the fractional portions by finding a common denominator, and combine the two results.

$$2\frac{2}{5} + 9\frac{1}{3} = 2 + 9 + \frac{2}{5} + \frac{1}{3} = 11\frac{11}{15}$$

FOR EXTRA PRACTICE
see page 736

internet connect
Homework Help Online
go.hrw.com   Keyword: MP4 3-5

go.hrw.com

3-5 **PRACTICE & ASSESS**

GUIDED PRACTICE

See Example **1** Add or subtract.

1. $\frac{5}{8} + \frac{1}{6}$ $\ \mathbf{\frac{19}{24}}$ 2. $\frac{5}{16} + \frac{2}{7}$ $\ \mathbf{\frac{67}{112}}$ 3. $\frac{1}{3} - \frac{7}{9}$ $\ \mathbf{-\frac{4}{9}}$ 4. $\frac{3}{4} - \frac{5}{16}$ $\ \mathbf{\frac{7}{16}}$

5. $2\frac{1}{5} + \left(-5\frac{2}{3}\right)$ $\ \mathbf{-3\frac{7}{15}}$ 6. $4\frac{11}{12} + \left(-7\frac{3}{8}\right)$ $\ \mathbf{-2\frac{11}{24}}$ 7. $3\frac{7}{12} + \left(-2\frac{4}{5}\right)$ $\ \mathbf{\frac{47}{60}}$ 8. $5\frac{3}{5} - 3\frac{7}{8}$ $\ \mathbf{1\frac{29}{40}}$

See Example **2** Evaluate each expression for the given value of the variable.

9. $2\frac{3}{5} + x$ for $x = -1\frac{1}{8}$ $\ \mathbf{1\frac{19}{40}}$ 10. $n - \frac{4}{7}$ for $n = -\frac{5}{9}$ $\ \mathbf{-1\frac{8}{63}}$

11. $3\frac{1}{2} + x$ for $x = -2\frac{7}{8}$ $\ \mathbf{\frac{5}{8}}$ 12. $n - \frac{7}{16}$ for $n = -\frac{1}{3}$ $\ \mathbf{-\frac{37}{48}}$

See Example **3** 13. A $2\frac{1}{4}$-foot-long piece of wood is needed to replace a window sill. If this amount is cut from a piece of wood $8\frac{7}{8}$ feet long, how much remains? $\mathbf{6\frac{5}{8}}$ **ft**

INDEPENDENT PRACTICE

See Example **1** Add or subtract.

14. $\frac{5}{12} + \frac{3}{7}$ $\ \mathbf{\frac{71}{84}}$ 15. $\frac{1}{5} + \frac{7}{9}$ $\ \mathbf{\frac{44}{45}}$ 16. $\frac{15}{16} - \frac{9}{10}$ $\ \mathbf{\frac{3}{80}}$ 17. $\frac{1}{3} + \frac{11}{12}$ $\ \mathbf{1\frac{1}{4}}$

18. $5\frac{4}{5} + \left(-3\frac{2}{7}\right) \mathbf{2\frac{18}{35}}$ 19. $\frac{5}{7} - \frac{13}{16}$ $\ \mathbf{-\frac{11}{112}}$ 20. $1\frac{2}{3} - 4\frac{5}{8} \mathbf{-2\frac{23}{24}}$ 21. $\frac{1}{5} + \frac{8}{9}$ $\ \mathbf{1\frac{4}{45}}$

See Example **2** Evaluate each expression for the given value of the variable.

22. $1\frac{7}{8} + x$ for $x = -2\frac{5}{6}$ $\ \mathbf{-\frac{23}{24}}$ 23. $n - \frac{2}{3}$ for $n = \frac{9}{16}$ $\ \mathbf{-\frac{5}{48}}$

24. $2\frac{5}{8} + x$ for $x = -1\frac{9}{10}$ $\ \mathbf{\frac{29}{40}}$ 25. $n - \frac{13}{15}$ for $n = \frac{3}{4}$ $\ \mathbf{-\frac{7}{60}}$

See Example **3** 26. A DVD contains a movie that takes up $4\frac{1}{3}$ gigabytes of space. If the DVD can hold $9\frac{2}{5}$ gigabytes, how much space on the disk is unused? $\mathbf{5\frac{1}{15}}$ **GB**

PRACTICE AND PROBLEM SOLVING

27. **MEASUREMENT** Bernard I. Pietsch measured the sides of the base of the Washington Monument. The north side measured $661\frac{3}{8}$ inches, the west side measured 661 inches, the south side measured $660\frac{13}{25}$ inches, and the east side measured 661 inches. Find the average side length. $\mathbf{660\frac{779}{800}}$ **in.**

28. **CONSTRUCTION** A water pipe has an outside diameter of $1\frac{1}{4}$ inches and a wall thickness of $\frac{5}{16}$ inch. What is the inside diameter of the pipe? $\mathbf{\frac{5}{8}}$ **in.**

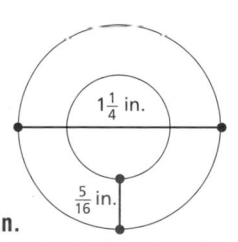

$1\frac{1}{4}$ in.

$\frac{5}{16}$ in.

Students may want to refer back to the lesson examples.

Assignment Guide

If you finished Example **1** assign:
Core 1–8, 14–21, 34–41
Enriched 1–8, 14–21, 34–41

If you finished Example **2** assign:
Core 1–12, 14–25, 34–41
Enriched 1–12, 14–25, 34–41

If you finished Example **3** assign:
Core 1–26, 29, 34–41
Enriched 1–13 odd, 14–41

Notes

RETEACH 3-5

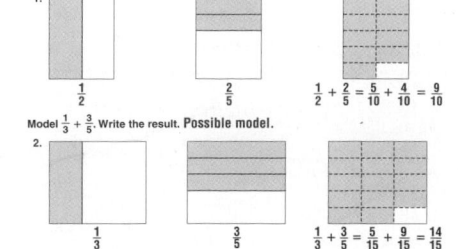

Reteach
3-5 *Adding and Subtracting with Unlike Denominators*
To model $\frac{1}{2} + \frac{1}{3}$, use two rectangles of the same size and shape.

A. 1st rectangle: Shade $\frac{1}{2}$ vertically.   B. 2nd rectangle: Shade $\frac{1}{3}$ horizontally.

$\frac{1}{2}$ $\frac{1}{3}$

C. Separate the shaded portions into parts of equal size.   D. Use a new rectangle to show the sum.

$\frac{1}{2} = \frac{3}{6}$ $\frac{1}{3} = \frac{2}{6}$ $\frac{1}{2} + \frac{1}{3} = \frac{3}{6} + \frac{2}{6} = \frac{5}{6}$

Model $\frac{1}{2} + \frac{2}{5}$. Write the result. **Possible model.**
1.

$\frac{1}{2}$ $\frac{2}{5}$ $\frac{1}{2} + \frac{2}{5} = \frac{5}{10} + \frac{4}{10} = \frac{9}{10}$

Model $\frac{1}{3} + \frac{3}{5}$. Write the result. **Possible model.**
2.

$\frac{1}{3}$ $\frac{3}{5}$ $\frac{1}{3} + \frac{3}{5} = \frac{5}{15} + \frac{9}{15} = \frac{14}{15}$

PRACTICE 3-5

Practice B
3-5 *Adding and Subtracting with Unlike Denominators*
Add or subtract. Write answer in simplest form.

1. $\frac{2}{3} + \frac{1}{2}$ $\mathbf{1\frac{1}{6}}$ 2. $\frac{3}{5} + \frac{1}{3}$ $\mathbf{\frac{14}{15}}$ 3. $\frac{3}{4} - \frac{1}{3}$ $\mathbf{\frac{5}{12}}$ 4. $-\frac{5}{9} + \frac{1}{2}$ $\mathbf{-\frac{1}{18}}$

5. $\frac{5}{16} + \left(-\frac{5}{6}\right)$ $\mathbf{-\frac{5}{16}}$ 6. $\frac{7}{9} - \left(-\frac{5}{6}\right)$ $\mathbf{1\frac{11}{18}}$ 7. $\frac{7}{8} - \frac{1}{4}$ $\mathbf{\frac{5}{8}}$ 8. $-\frac{3}{8} - \frac{5}{6}$ $\mathbf{-1\frac{5}{24}}$

9. $\frac{7}{8} + \frac{5}{12}$ $\mathbf{1\frac{7}{24}}$ 10. $1\frac{2}{9} + 2\frac{1}{18}$ $\mathbf{3\frac{5}{18}}$ 11. $2\frac{2}{3} - \frac{3}{5}$ $\mathbf{2\frac{1}{15}}$ 12. $-2\frac{3}{4} - 1\frac{5}{6}$ $\mathbf{-4\frac{7}{12}}$

13. $8 - 3\frac{5}{9}$ $\mathbf{4\frac{4}{9}}$ 14. $5\frac{1}{3} + 1\frac{11}{12}$ $\mathbf{6\frac{1}{4}}$ 15. $-12 + \left(-\frac{5}{12}\right)$ $\mathbf{-12\frac{5}{12}}$ 16. $5\frac{3}{10} - 14$ $\mathbf{-8\frac{7}{10}}$

Evaluate each expression for the given value of the variable.

17. $2\frac{3}{8} + x$ for $x = 1\frac{5}{6}$ $\mathbf{4\frac{5}{24}}$ 18. $-2\frac{5}{9} + x$ for $x = 1\frac{1}{3}$ $\mathbf{-1\frac{1}{15}}$ 19. $-3\frac{5}{7} - x$ for $x = -1\frac{3}{5}$ $\mathbf{-1\frac{29}{35}}$

20. $-1\frac{5}{8} + x$ for $x = 2\frac{1}{6}$ $\mathbf{\frac{13}{24}}$ 21. $-2\frac{5}{8} - x$ for $x = -2\frac{1}{4}$ $\mathbf{-\frac{7}{20}}$ 22. $-3\frac{1}{10} - x$ for $x = -4\frac{4}{5}$ $\mathbf{2\frac{9}{10}}$

23. Ana worked $6\frac{1}{2}$ h on Monday, $5\frac{3}{4}$ h on Tuesday and $7\frac{1}{6}$ h on Friday. How many total hours did she work these three days? $\mathbf{19\frac{5}{12}}$ **h**

Math Background

Another method for finding the least common multiple of a set of numbers involves prime factors. To find the LCM of 60 and 45, factor each number into its prime factors, using exponents: $60 = 2^2 \cdot 3 \cdot 5$ and $45 = 3^2 \cdot 5$. The LCM is the product of the greatest powers of all the prime factors: LCM $= 2^2 \cdot 3^2 \cdot 5 = 180$. This method is useful for rational expressions as well. To find the sum $\frac{1}{6ab^2} + \frac{1}{4a^2b}$, you would need the LCM of the denominators.

$6ab^2 = 2 \cdot 3 \cdot a \cdot b^2$

$4a^2b = 2^2 \cdot a^2 \cdot b$

LCM $= 2^2 \cdot 3 \cdot a^2 \cdot b^2 = 12\,a^2 b^2$

Earth Science

Exercises 29–33 involve using facts about the erosion of Niagara Falls. Water flow and erosion are studied in middle school Earth science programs, such as *Holt Science & Technology*.

Journal

Ask students to consider the two methods for finding a common denominator, multiplying the denominators or finding the least common denominator. Ask them to describe situations in which it would be better to use one method over the other.

Test Prep Doctor

For Exercise 41, remind students of the order of operations. The students who chose **F** may have subtracted 2 from 8 before applying the exponent. The students who chose **G** probably neglected to add 41, and the students who chose **J** may have added the numbers inside the parentheses instead of subtracting.

Earth Science

Niagara Falls, on the border of Canada and the United States, has two major falls, Horseshoe Falls on the Canadian side and American Falls on the U.S. side. Surveys of the erosion of the falls began in 1842. From 1842 to 1905, Horseshoe Falls eroded $239\frac{2}{5}$ feet.

29. In 1986, Thomas Martin noted that American Falls eroded $7\frac{1}{2}$ inches and Horseshoe Falls eroded $2\frac{4}{25}$ feet. What is the difference between the two measurements? $18\frac{21}{50}$ in.

30. From 1842 to 1875, the actual yearly erosion of Horseshoe Falls varied from a minimum of $\frac{61}{100}$ meter to a maximum of $1\frac{17}{50}$ meters. By how much did these rates of erosion differ? $\frac{73}{100}$ meter

31. In the 48 years between 1842 and 1890, the average rate of erosion at Horseshoe Falls was $\frac{33}{50}$ meter per year. In the 22 years between 1905 and 1927, the rate of erosion was $\frac{7}{10}$ meter per year. Approximately how much total erosion occurred during these two time periods? $47\frac{2}{25}$ meters

32. Lake Erie, which feeds Niagara Falls, has a six-month average precipitation rate of $48\frac{1}{2}$ centimeters. From September 1999 to February 2000, the precipitation was $40\frac{1}{5}$ centimeters. How far below the average was precipitation during this period? $8\frac{3}{10}$ cm

33. **CHALLENGE** Rates of erosion of American Falls have been recorded as $\frac{23}{100}$ meter per year for 33 years, $\frac{9}{40}$ meter per year for 48 years, and $\frac{1}{5}$ meter per year for 4 years. What is the total amount of erosion during these three time spans? $19\frac{19}{100}$ meters

Spiral Review

Simplify. (Lesson 2-3)

34. $-4(6-8)$ **8**

35. $3(-5-4)$ **−27**

36. $-2(4-9)$ **10**

37. $-8(-5-6)$ **88**

38. $7(2-5)$ **−21**

39. $-3(-3-3)$ **18**

40. **TEST PREP** Simplify the expression $100 - 2(4 \cdot 3^2)$. (Lesson 2-6) **B**

  A 104   B 28   C 14,112   D −188

41. **TEST PREP** Simplify the expression $41 + 3(8 - 2^3)$. (Lesson 2-6) **H**

  F 689   G 0   H 41   J 89

Lesson Quiz

Add or subtract.

1. $\frac{5}{14} + \frac{1}{7}$ $\frac{1}{2}$

2. $8\frac{2}{3} - 1\frac{1}{2}$ $7\frac{1}{6}$

3. $\frac{3}{5} + \left(-2\frac{2}{3}\right)$ $-2\frac{1}{15}$

4. Evaluate $1\frac{3}{8} - n$ for $n = \frac{9}{16}$. $\frac{13}{16}$

5. Robert is 5 feet $6\frac{1}{2}$ inches tall. Judy is 5 feet $3\frac{3}{4}$ inches tall. How much taller is Robert than Judy? $2\frac{3}{4}$ in.

Available on Daily Transparency in CRB

CHALLENGE 3-5

LESSON **3-5** **Challenge**
Please Repeat That.

A decimal that repeats one digit is equivalent to a fraction with denominator 9.
$0.\overline{1} = \frac{1}{9}$ $0.\overline{2} = \frac{2}{9}$ $0.\overline{5} = \frac{5}{9}$

A decimal that repeats two digits is equivalent to a fraction with denominator 99.
$0.\overline{43} = \frac{43}{99}$ $0.\overline{61} = \frac{61}{99}$ $0.\overline{38} = \frac{38}{99}$

The pattern continues so that $0.\overline{681} = \frac{681}{999}$ and $0.\overline{24793} = \frac{24,793}{99,999}$.

Use a calculator to write each decimal equivalent.

1. $\frac{1}{90} =$ **$0.0\overline{1}$**

2. $\frac{21}{990} =$ **$0.0\overline{21}$**

3. $\frac{358}{9990} =$ **$0.0\overline{358}$**

Predict the decimal equivalent of each fraction. Verify your results on a calculator.

4. $\frac{4}{90} =$ **$0.0\overline{4}$**

5. $\frac{62}{990} =$ **$0.0\overline{62}$**

6. $\frac{617}{9990} =$ **$0.0\overline{617}$**

Write each fractional equivalent and simplify.

7. $0.\overline{7} =$
$\frac{7}{9}$

8. $0.0\overline{8} =$
$\frac{8}{90} = \frac{4}{45}$

9. $0.00\overline{24} =$
$\frac{24}{9900} = \frac{2}{825}$

When one digit repeats but does not begin in the first decimal place, and the digit in the first place is other than 0, you must add fractions.
$0.3\overline{7} = 0.3 + 0.0\overline{7}$
$= \frac{3}{10} + \frac{7}{90}$
$= \frac{27}{90} + \frac{7}{90} = \frac{34}{90} = \frac{17}{45}$

Write each repeating decimal as the sum of two fractions. Find the sum and simplify. Verify.

10. $0.2\overline{8} =$
$\frac{2}{10} + \frac{8}{90} = \frac{26}{90} = \frac{13}{45}$

11. $0.25\overline{32} =$
$\frac{25}{100} + \frac{32}{9900} = \frac{2507}{9900}$

12. $0.12\overline{7} =$
$\frac{1}{10} + \frac{27}{990} = \frac{126}{990} = \frac{7}{55}$

13. $0.75\overline{483} =$
$\frac{75}{100} + \frac{483}{99,900} = \frac{75,408}{99,900} = \frac{6284}{8325}$

PROBLEM SOLVING 3-5

LESSON **3-5** **Problem Solving**
Adding and Subtracting with Unlike Denominators

Write the correct answer.

1. Nick Hysong of the United States won the Olympic gold medal in the pole vault in 2000 with a jump of 19 ft $4\frac{1}{4}$ inches, or $232\frac{1}{4}$ inches. In 1900, Irving Baxter of the United States won the pole vault with a jump of 10 ft $9\frac{7}{8}$ inches, or $129\frac{7}{8}$ inches. How much higher did Hysong vault than Baxter?
$102\frac{3}{8}$ inches

2. In the 2000 Summer Olympics, Ivan Pedroso of Cuba won the Long jump with a jump of 28 ft $3\frac{3}{4}$ inches, or $336\frac{3}{4}$ inches. Alvin Kraenzlein of the Unites States won the long jump in 1900 with a jump of 23 ft $6\frac{7}{8}$ inches, or $282\frac{7}{8}$ inches. How much farther did Pedroso jump than Kraenzlein?
$53\frac{7}{8}$ inches

3. A recipe calls for $\frac{1}{8}$ cup of sugar and $\frac{3}{4}$ cup of brown sugar. How much total sugar is added to the recipe?
$\frac{7}{8}$ cup

4. The average snowfall in Norfolk, VA for January is $2\frac{3}{5}$ inches, February $2\frac{9}{10}$ inches, March 1 inch, and December $\frac{9}{10}$ inch. If these are the only months it typically snows, what is the average snowfall per year?
$7\frac{2}{5}$ inches

Use the table at the right that shows the average snowfall per month in Vail, Colorado.

5. What is the average annual snowfall in Vail, Colorado?

  A $15\frac{13}{20}$ in.   C $187\frac{7}{10}$ in.
  B 153 in.   Ⓓ $187\frac{4}{5}$ in.

6. The peak of the skiing season is from December through March. What is the average snowfall for this period?

  F $30\frac{18}{20}$ in.   Ⓗ $123\frac{4}{5}$ in.
  G $123\frac{3}{5}$ in.   J 127 in.

Average Snowfall in Vail, CO

Month	Snowfall (in.)	Month	Snowfall (in.)
Jan	$36\frac{7}{10}$	July	0
Feb	$35\frac{7}{10}$	August	0
March	$25\frac{2}{5}$	Sept	1
April	$21\frac{1}{5}$	Oct	$7\frac{4}{5}$
May	4	Nov	$29\frac{7}{10}$
June	$\frac{3}{10}$	Dec	26

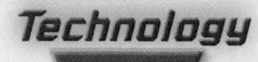

Technology LAB 3A

Explore Repeating Decimals

Use with Lesson 3-5

internet connect
Lab Resources Online
go.hrw.com
KEYWORD: MP4 Lab3A

You can divide to display decimal equivalents of fractions using your graphing calculator. To display decimals as fractions, use the MATH key.

You can also use the MATH key to find fractions equivalent to repeating decimals.

Activity

❶ Use a graphing calculator to find the decimal equivalent of each fraction. Look for patterns in the fraction and decimal forms.

$$\frac{1}{9} \quad \frac{4}{9} \quad \frac{23}{99} \quad \frac{47}{99} \quad \frac{461}{999} \quad \frac{703}{999}$$

For example, type 1 ÷ 9, and press ENTER .

```
1/9
       .1111111111
```

The decimal equivalent is a repeating decimal, $0.\overline{1}$.

Notice that $0.5\overline{3}$ can be written as a sum of a repeating and a terminating decimal.

$$\begin{array}{r} 0.3333\ldots \\ + \ 0.2 \\ \hline 0.5333\ldots \end{array}$$

❷ Find the fraction for $0.5\overline{3}$.

To find the fraction for $0.5\overline{3}$, write the decimals as fractions and add.

$$\begin{array}{r} 0.3333\ldots \\ + \ 0.2 \\ \hline 0.5333\ldots \end{array} \qquad \begin{array}{r} \frac{1}{3} = \frac{10}{30} \\ + \frac{2}{10} = \frac{6}{30} \\ \hline = \frac{16}{30} = \frac{8}{15} \end{array}$$

Think and Discuss

1. Based on the pattern you found in ❶, how would you write the repeating decimal $0.\overline{3726}$ as a fraction? Divide to check your answer.

Write the four digits over 9999, and simplify if necessary:

Try This $\frac{3726}{9999} = \frac{414}{1111}$. $414 \div 1111 = 0.37263726\ldots$

Write each decimal as a sum or difference of a repeating and a terminating decimal. Then write the repeating decimals as a fraction. Check by dividing.

1. $0.1\overline{5}$ 2. $0.1\overline{3}$ 3. $0.6\overline{51}$ 4. $0.9\overline{15}$ 5. $0.4\overline{532}$

$0.1 + 0.0\overline{5}; \frac{7}{45}$ $0.\overline{3} - 0.2; \frac{2}{15}$ $0.\overline{15} + 0.5; \frac{43}{66}$ $0.\overline{51} + 0.4; \frac{151}{165}$ $0.\overline{253} + 0.2; \frac{2264}{4995}$

Answers

Activity

1. $\frac{1}{9} = 0.\overline{1}$

$\frac{4}{9} = 0.\overline{4}$

$\frac{23}{99} = 0.\overline{23}$

$\frac{47}{99} = 0.\overline{47}$

$\frac{461}{999} = 0.\overline{461}$

$\frac{703}{999} = 0.\overline{703}$

Technology LAB 3A

Explore Repeating Decimals

Pacing: Traditional 1 day
Block $\frac{1}{2}$ day

Objective: To use a graphing calculator to explore repeating decimals

Materials: Graphing calculator

Lab Resources

Technology Lab Activities. . . . pp. 14–15

Using the Page

This technology activity shows students how to write fractions as repeating decimals, and repeating decimals as fractions, both of which can be done on any graphing calculator. Specific keystrokes may vary, depending on the make and model of the graphing calculator used. The keystrokes given are for a TI-83 model. For keystrokes to other models, visit go.hrw.com.

The Think and Discuss problem can be used to assess students' understanding of the technology activity. While Try This problems 1–5 can be done without a graphing calculator, they are meant to help students become familiar with using a graphing calculator to convert repeating decimals to fractions.

Assessment

1. What pattern do you see for fractions that have a numerator from 1 to 8 and a denominator of 9?
   The equivalent decimal is formed by repeating the numerator after the decimal point (for example, $\frac{1}{9} = 0.\overline{1}$; $\frac{2}{9} = 0.\overline{2}$).

2. How could you write $0.3\overline{1}$ as the sum of two fractions? $\frac{1}{5} + \frac{1}{9}$

Warm Up

Add or subtract.

1. $\frac{7}{10} + \frac{5}{10}$ $1\frac{1}{5}$ **2.** $2\frac{3}{8} - 1\frac{5}{16}$ $1\frac{1}{16}$

3. $4.8 + 3.6$ **8.4** **4.** $2.4 - 0.05$ **2.35**

Problem of the Day

A computer word is made of strings of 0's and 1's. How many different words can be formed using 8 characters? (An example is 01010101.) **256**

Available on Daily Transparency in CRB

Math teachers never die; they just reduce to lowest terms.

3-6 Solving Equations with Rational Numbers

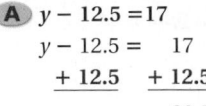

Learn to solve equations with rational numbers.

One of the world's most famous jewels is the Hope diamond. The roughly cut Hope diamond was sold to King Louis XIV of France in 1668. When the king's jeweler cut it, the diamond was reduced by $45\frac{1}{16}$ carats to a steely blue $67\frac{1}{8}$ carats.

You can write an equation using these fractions and solve for the weight of the roughly cut diamond.

EXAMPLE 1 Solving Equations with Decimals

Solve.

A $y - 12.5 = 17$

$$y - 12.5 = 17$$
$$\underline{+12.5 \quad +12.5}$$
$$y = 29.5$$

Add 12.5 to both sides.

> **Remember!**
> Once you have solved an equation, it is a good idea to check your answer. To check your answer, substitute your answer for the variable in the original equation.

B $-2.7p = 10.8$

$$-2.7p = 10.8$$
$$\frac{-2.7}{-2.7}p = \frac{10.8}{-2.7}$$
$$p = -4$$

Divide both sides by -2.7.

C $\frac{t}{7.5} = 4$

$$\frac{t}{7.5} = 4$$
$$7.5 \cdot \frac{t}{7.5} = 7.5 \cdot 4$$
$$t = 30$$

Multiply both sides by 7.5.

EXAMPLE 2 Solving Equations with Fractions

Solve.

A $x + \frac{1}{5} = -\frac{2}{5}$

$$x + \frac{1}{5} = -\frac{2}{5}$$
$$x + \frac{1}{5} - \frac{1}{5} = -\frac{2}{5} - \frac{1}{5}$$
$$x = -\frac{3}{5}$$

Subtract $\frac{1}{5}$ from both sides.

---

1 Introduce

Alternate Opener

EXPLORATION

3-6 Solving Equations with Rational Numbers

A box of cookies has 6 servings. Each serving contains 40.5 calories. How many calories are in the whole box?

40.5	40.5	= 81
40.5	40.5	= 81
40.5	40.5	= 81
		243

Equation
$40.5 \cdot 6 = c$
$243 = c$

Estimate the solution for each equation. Then use a calculator to solve.

	Equation	Estimate	Actual
1.	$124.75 - x = 50$		
2.	$x + 16.9 = 15.5$		
3.	$0.6x = 15$		
4.	$\frac{x}{1.25} = 8$		

5. Write a real-world situation for each equation in numbers 1–4.

Think and Discuss

6. Describe how you estimated the solutions for the equations in numbers 1–4.

7. Discuss whether it is easier to estimate the solutions of some equations than the solutions of others.

Motivate

Show the students equations such as $n + 4 = 9$, $3x = 12$, and $\frac{y}{5} = 4$.

Remind students that they learned how to solve these equations in Chapter 1. Have the students solve and check these equations. Tell the students that they will solve equations in the new lesson by the same methods.

Exploration worksheet and answers on Chapter 3 Resource Book pp. 54 and 126

2 Teach

Lesson Presentation

Guided Instruction

In this lesson, students learn to solve equations with rational numbers. To begin, remind students of the methods they used to solve equations with whole numbers (Lessons 1-3 and 1-4) and integers (Lesson 2-4). Demonstrate the similarity between solving those equations and solving equations in which the numbers are decimals and fractions. Emphasize that the algebra procedures are the same as those used to solve whole number and integer equations.

Solve.

B
$$x - \frac{1}{4} = \frac{3}{8}$$

$$x - \frac{1}{4} = \frac{3}{8}$$

$$x - \frac{1}{4} + \frac{1}{4} = \frac{3}{8} + \frac{1}{4} \qquad \textit{Add } \frac{1}{4} \textit{ to both sides of the equation.}$$

$$x = \frac{3}{8} + \frac{2}{8} \qquad \textit{Find a common denominator, 8.}$$

$$x = \frac{5}{8}$$

C
$$\frac{3}{5}w = \frac{3}{16}$$

$$\frac{3}{5}w = \frac{3}{16}$$

$$\frac{5}{3} \cdot \frac{3}{5}w = \frac{5}{3} \cdot \frac{3}{16} \qquad \textit{Multiply both sides by } \frac{5}{3}. \textit{ Simplify.}$$

$$w = \frac{5}{16}$$

E X A M P L E **3** **Solving Word Problems Using Equations**

In 1668 the Hope diamond was reduced from its original weight by $45\frac{1}{16}$ carats to a diamond weighing $67\frac{1}{8}$ carats. How many carats was the original diamond?

Convert fractions:

$$45\frac{1}{16} = \frac{45(16) + 1}{16} = \frac{721}{16} \qquad 67\frac{1}{8} = \frac{67(8) + 1}{8} = \frac{537}{8}$$

Write an equation:

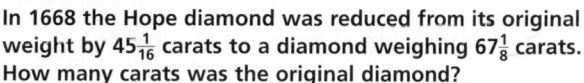

Original weight	−	Amount cut	=	Weight after cut
w	−	$\frac{721}{16}$	=	$\frac{537}{8}$

$$w - \frac{721}{16} = \frac{537}{8}$$

$$w - \frac{721}{16} + \frac{721}{16} = \frac{537}{8} + \frac{721}{16} \qquad \textit{Add } \frac{721}{16} \textit{ to both sides.}$$

$$w = \frac{1074}{16} + \frac{721}{16} \qquad \textit{Find a common denominator, 16.}$$

$$w = \frac{1795}{16}, \text{ or } 112\frac{3}{16} \qquad \textit{Simplify.}$$

The original Hope diamond was $112\frac{3}{16}$ carats.

Think and Discuss

1. Explain the first step in solving an addition equation with fractions having *like* denominators.

2. Explain the first step in solving an addition equation with fractions having *unlike* denominators.

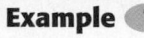

Additional Examples

Example **1**

Solve.

A. $m + 4.6 = 9$ $\qquad m = 4.4$

B. $8.2p = -32.8$ $\qquad p = -4$

C. $\frac{x}{1.2} = 15$ $\qquad x = 18$

Example **2**

Solve.

A. $n + \frac{2}{7} = -\frac{3}{7}$ $\qquad n = -\frac{5}{7}$

B. $y - \frac{1}{6} = \frac{2}{3}$ $\qquad y = \frac{5}{6}$

C. $\frac{5}{6}x = \frac{5}{8}$ $\qquad x = \frac{3}{4}$

Example **3**

Mr. Rios wants to prepare a casserole that requires $2\frac{1}{2}$ cups of milk. If he makes the casserole, he will have only $\frac{3}{4}$ cup of milk left for his breakfast cereal. How much milk does Mr. Rios have?

Mr. Rios has $3\frac{1}{4}$ cups of milk.

---

3 **Close**

Reaching All Learners
Through Critical Thinking

Have students work in pairs. Give each student a recording sheet (Chapter 3 Resource Book p. 110) with a set of incomplete one-step equations, such as $x + \_\_\_ = \_\_\_$; 3.4, 1.2. Tell students to replace the blanks with the given constant values in each equation and then solve the equation. Tell students to reverse the constant values to get a different equation and then solve that equation. Have them repeat the procedure for the remaining equations and then to try to identify a pattern.

Summarize

Remind students that one-step equations are solved the same way whether the constant numbers in the equations are whole numbers, integers, or rational numbers. Ask students what steps they would take to solve the equations that follow: $x + 8 = 12$, $x + 6 = -3$, $x + 2.4 = 1.5$, and $x + \frac{3}{4} = \frac{7}{8}$.

Possible answers: Subtract 8 from both sides; subtract 6 from both sides; subtract 2.4 from both sides; subtract $\frac{3}{4}$ from both sides.

Point out that these equations are all addition equations and that they all should be solved by subtraction.

Answers to Think and Discuss

Possible answers:

1. Subtract the same fraction from both sides of the equation to isolate the variable on one side of the equation.

2. Find the least common denominator, or subtract the same fraction from both sides of the equation to isolate the variable on one side of the equation.

FOR EXTRA PRACTICE
see page 736

internet connect
Homework Help Online
go.hrw.com   Keyword: MP4 3-6

> Students may want to refer back to the lesson examples.

Assignment Guide

If you finished Example **1** assign:
Core 1–6, 14–19, 31, 33, 35, 48–54
Enriched 14–19, 33, 35, 38–41, 48–54

If you finished Example **2** assign:
Core 1–12, 14–25, 27–32, 48–54
Enriched 14–25, 27–41, 46, 48–54

If you finished Example **3** assign:
Core 1–32, 44, 48–54
Enriched 14–54

Answers

39. $y = 64.1$

40. $f = -6.7$

41. $m = -2.8$

GUIDED PRACTICE

See Example **1** Solve.

1. $y + 23.4 = -52$ $y = -75.4$
2. $-6.3f = 44.1$ $f = -7$
3. $\frac{m}{3.2} = -6$ $m = -19.2$
4. $r - 17.9 = 36.8$ $r = 54.7$
5. $\frac{s}{13.21} = 5.2$ $s = 68.692$
6. $0.04g = 0.252$ $g = 6.3$

See Example **2** Solve.

7. $x + \frac{1}{7} = -\frac{3}{7}$ $x = -\frac{4}{7}$
8. $-\frac{2}{9} + k = -\frac{5}{9}$ $k = -\frac{1}{3}$
9. $\frac{3}{5}w = -\frac{7}{15}$ $w = -\frac{7}{9}$
10. $m - \frac{4}{3} = -\frac{4}{3}$ $m = 0$
11. $\frac{7}{19}y = -\frac{63}{19}$ $y = -9$
12. $t + \frac{4}{13} = \frac{12}{39}$ $t = 0$

See Example **3** **13.** The Hope diamond has a width of $21\frac{39}{50}$ millimeters. Its width is equal to its length plus $3\frac{41}{50}$ millimeters. How many millimeters long is the Hope diamond? $17\frac{24}{25}$ **mm**

INDEPENDENT PRACTICE

See Example **1** Solve.

14. $y + 16.7 = -49$ $y = -65.7$
15. $5.8m = -52.2$ $m = -9$
16. $-\frac{h}{6.7} = 3$ $h = -20.1$
17. $k - 2.1 = -4.5$ $k = -2.4$
18. $\frac{z}{10.7} = 4$ $z = 42.8$
19. $c + 2.94 = 8.1$ $c = 5.16$

See Example **2** Solve.

20. $j + \frac{1}{3} = \frac{3}{4}$ $j = \frac{5}{12}$
21. $\frac{5}{8}d = \frac{6}{18}$ $d = \frac{8}{15}$
22. $6h = \frac{12}{37}$ $h = \frac{2}{37}$
23. $x - \frac{1}{12} = \frac{5}{12}$ $x = \frac{1}{2}$
24. $r + \frac{5}{9} = -\frac{1}{9}$ $r = -\frac{2}{3}$
25. $\frac{5}{6}c = \frac{7}{24}$ $c = \frac{7}{20}$

See Example **3** **26.** Among all minerals, sapphires rank second to diamonds in hardness. One of the largest blue star sapphires, the Star of India, weighs 563 carats. How much more does the Star of India weigh than the original Hope diamond, which weighed $112\frac{3}{16}$ carats? $450\frac{13}{16}$ **carats**

PRACTICE AND PROBLEM SOLVING

Solve.

27. $z - \frac{5}{9} = \frac{1}{9}$ $z = \frac{2}{3}$
28. $-5f = -1.5$ $f = 0.3$
29. $\frac{j}{8.1} = -4$ $j = -32.4$
30. $t - \frac{3}{4} = 6\frac{1}{4}$ $t = 7$
31. $-2.9g = -26.1$ $g = 9$
32. $\frac{4}{9}d = -\frac{2}{9}$ $d = -\frac{1}{2}$
33. $\frac{v}{5.5} = -5.5$ $v = -30.25$
34. $r + \frac{5}{8} = -2\frac{3}{8}$ $r = -3$
35. $y + 3.8 = -1.6$ $y = -5.4$
36. $-\frac{1}{12} + r = \frac{3}{4}$ $r = \frac{5}{6}$
37. $-5c = \frac{5}{24}$ $c = -\frac{1}{24}$
38. $m - 2.34 = 8.2$ $m = 10.54$
39. $y - 68 = -3.9$
40. $-14 = -7.3 + f$
41. $\frac{2m}{0.7} = -8$

Math Background

It is important for students to be able to solve equations that contain fractions and decimals as well as integers. Help students understand that the properties of equality that were introduced in Chapter 1 apply to all of these types of numbers. In fact, the properties apply to all real numbers. It is helpful if students begin to recognize equations by the included operation(s) rather than by the types of numbers they contain. For example, $x + \frac{3}{4} = -\frac{1}{2}$ should be recognized as an addition equation rather than a fraction problem. When teaching the Guided Practice Exercises, you may want to have students first identify the type of equation before they solve it.

RETEACH 3-6

LESSON **3-6** **Reteach**
Algebra: Solving Equations with Rational Numbers

Solving equations with rational numbers is basically the same as solving equations with integers or whole numbers:
Use inverse operations to isolate the variable.

$\frac{1}{4}z = -16$ Multiply each side by 4.
$4 \cdot \frac{1}{4}z = -16 \cdot 4$
$z = -64$

$y - \frac{3}{8} = \frac{7}{8}$ Add $\frac{3}{8}$ to each side.
$+ \frac{3}{8} \quad + \frac{3}{8}$
$y = \frac{10}{8} = 1\frac{2}{8} = 1\frac{1}{4}$

$x + 3.5 = -17.42$ Subtract 3.5 from each side.
$- 3.5 \quad - 3.5$
$x = -20.92$

$-26t = 317.2$ Divide each side by −26.
$\frac{-26t}{-26} = \frac{317.2}{-26}$
$t = -12.2$

Tell what you would do to isolate the variable.
1. $x - 1.4 = 7.82$ **add 1.4**
2. $\frac{1}{4} + y = \frac{7}{4}$ **subtract $\frac{1}{4}$**
3. $3z = 5$ **divide by 3**

Solve each equation.
4. $14x = -129.5$ $x = 9.25$
5. $\frac{1}{3}y = 27$ $y = 81$
6. $265.2 = \frac{z}{22.1}$ $5860.92 = z$
7. $x + 53.8 = -1.2$ $x = -55$
8. $25 = \frac{1}{5}k$ $125 = k$
9. $m - \frac{2}{3} = \frac{3}{5}$ $m = 1\frac{4}{15}$

PRACTICE 3-6

LESSON **3-6** **Practice B**
Algebra: Solving Equations with Rational Numbers

Solve.
1. $x + 6.8 = 12.19$ **5.39**
2. $y - 10.24 = 5.3$ **15.54**
3. $0.05w = 6.25$ **125**
4. $\frac{a}{9.05} = 8.2$ **74.21**
5. $-12.41 + x = -0.06$ **12.35**
6. $\frac{d}{-8.4} = -10.2$ **85.68**
7. $-2.89 = 1.7m$ **−1.7**
8. $n - 8.09 = -11.65$ **−3.56**
9. $\frac{x}{5.4} = -7.18$ **−38.772**

Solve. Write each answer in simplest form.
10. $\frac{7}{9} + x = 1\frac{1}{9}$ $\frac{1}{3}$
11. $\frac{6}{11}y = -\frac{18}{22}$ $-1\frac{1}{2}$
12. $\frac{7}{10}d = \frac{21}{20}$ $1\frac{1}{2}$
13. $x - \left(-\frac{9}{14}\right) = \frac{5}{7}$ $\frac{1}{14}$
14. $x - \frac{15}{21} = 2\frac{6}{7}$ $3\frac{4}{7}$
15. $\frac{8}{15}a = \frac{9}{10}$ $-1\frac{11}{16}$

16. A recipe calls for $2\frac{1}{3}$ cups of flour and $1\frac{1}{4}$ cups of sugar. If the recipe is tripled, how much flour and sugar will be needed?
7 cups of flour and $3\frac{3}{4}$ cups of sugar

17. Daniel filled the gas tank in his car with 14.6 gal of gas. He then drove 284.7 mi before needing to fill his tank with gas again. How many miles did the car get to a gallon of gasoline?
19.5 mi

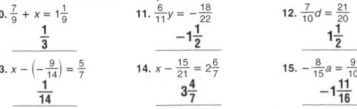

Earth Science LINK

Diamonds are found in several forms: the well-known gemstone, bort, ballas, and carbonado. Carbonado, ballas, and bort are used to cut stone and for the cutting edges of drills and other cutting tools.

42. **EARTH SCIENCE** The largest of all known diamonds, the Cullinan diamond, weighed 3106 carats before it was cut into 105 gems. The largest cut, Cullinan I, or the Great Star of Africa, weighs $530\frac{1}{5}$ carats. Another cut, Cullinan II, weighs $317\frac{2}{5}$ carats. Cullinan III weighs $94\frac{2}{5}$ carats, and Cullinan IV weighs $63\frac{3}{5}$ carats.

  a. How many carats of the original Cullinan diamond were left after the Great Star of Africa and Cullinan II were cut? **$2258\frac{2}{5}$ carats**

  b. How much more does Cullinan II weigh than Cullinan IV? **$253\frac{4}{5}$ carats**

  c. Which diamond weighs 223 carats less than Cullinan II? **Cullinan III**

43. Jack is tiling along the walls of the rectangular kitchen with the tile shown. The kitchen has a length of $243\frac{3}{4}$ inches and a width of $146\frac{1}{4}$ inches.

  a. How many tiles will fit along the length of the room? **15 tiles**

  b. How many tiles will fit along its width? **9 tiles**

  c. If Jack needs 48 tiles to tile around all four walls of the kitchen, how many boxes of ten tiles must he buy? (*Hint:* He must buy whole boxes of tile.) **5 boxes**

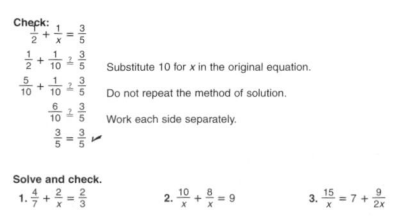
KITCHEN FLOOR PLAN
$16\frac{1}{4}$ in
$16\frac{1}{4}$ in

44. **LIFE SCIENCE** Each tablet in a box of allergy medicine weighs 0.3 gram. The total weight of all the tablets is 15 grams. How many tablets are in the box? **50 tablets**

45. **WHAT'S THE ERROR?** Ann's CD writer burns 0.6 megabytes of information per second. A computer salesperson said that if Ann had 28.8 megabytes of information to burn, she could burn it in a little more than 15 seconds with this writer. What was the error?

46. **WRITE ABOUT IT** If a is $\frac{1}{3}$ of b, is it correct to say $\frac{1}{3}a = b$? Explain.

47. **CHALLENGE** A 150-carat diamond was cut into two equal pieces to form two diamonds. One of the diamonds was cut again, reducing it by $\frac{1}{3}$ its weight. In a final cut, it was reduced by $\frac{1}{4}$ its new weight. How many carats remained after the final cut? **$37\frac{1}{2}$ carats**

Spiral Review

Evaluate each expression for the given values of the variables. (Lesson 1-1)

48. $4x + 5y$ for $x = 3$ and $y = 9$ **57**

49. $7m - 2n$ for $m = 5$ and $n = 7$ **21**

Write each number in scientific notation. (Lesson 2-9)

50. -0.000348
-3.48×10^{-4}

51. 0.00000524
5.24×10^{-6}

52. $-4,870,000,000$
-4.87×10^9

53. $64,000,000,000$
6.4×10^{10}

54. **TEST PREP** If $x + y = 6$, then $x + y - 2 = \underline{?}$. (Lesson 1-5) **A**

  A 4   B 8   C 3   D −4

Answers

45. Possible answer: The salesperson multiplied 28.8 by 0.6 instead of dividing by 0.6 to get 48 seconds.

46. Possible answer: No, a is $\frac{1}{3}$ of b describes the equation $a = \frac{1}{3}b$, but $\frac{1}{3}a = b$ means that b is $\frac{1}{3}$ of a.

Journal

Remind students that they solved one-step equations in Chapters 1 (whole numbers), 2 (integers), and 3 (rational numbers). Have students write about how the equations in these three chapters are the same and how they are different.

Test Prep Doctor

For Exercise 54, explain to students that because they know that the value of $(x + y)$ is 6, they can replace the entire expression with 6. That way, $x + y - 2$ can be viewed as $(x + y) - 2$, or $(6) - 2$, which equals 4.

CHALLENGE 3-6

Challenge
3-6 Location, Location, Location

An equation that has a variable in the denominator of one or more of its terms is called a **fractional equation**.

One method of solution is to clear the equation of fractions by multiplying each side of the equation by the LCD.

$\frac{1}{2} + \frac{1}{x} = \frac{3}{5}$ The LCD of 2, x, and 5 is 10x, with $x \neq 0$.

$10x\left(\frac{1}{2} + \frac{1}{x}\right) = 10x\left(\frac{3}{5}\right)$ Multiply each side by 10x.

$10x \cdot \frac{1}{2} + 10x \cdot \frac{1}{x} = 10x \cdot \frac{3}{5}$ Distributive Property

$5x + 10 = 6x$ Simplify.

$5x - 5x + 10 = 6x - 5x$ Subtract 5x from each side.

$10 = x$

Check:
$\frac{1}{2} + \frac{1}{x} = \frac{3}{5}$

$\frac{1}{2} + \frac{1}{10} = \frac{3}{5}$ Substitute 10 for x in the original equation.

$\frac{5}{10} + \frac{1}{10} = \frac{3}{5}$ Do not repeat the method of solution.

$\frac{6}{10} = \frac{3}{5}$ Work each side separately.

$\frac{3}{5} = \frac{3}{5}$ ✓

Solve and check.

1. $\frac{4}{7} + \frac{2}{x} = \frac{2}{3}$
   $x = 21$

2. $\frac{10}{x} + \frac{8}{x} = 9$
   $x = 2$

3. $\frac{15}{x} = 7 + \frac{9}{2x}$
   $x = 1\frac{1}{2}$

PROBLEM SOLVING 3-6

Problem Solving
3-6 Algebra: Solving Equations with Rational Numbers

Write the correct answer.

1. In the last 150 years, the average height of people in industrialized nations has increased by $\frac{1}{3}$ foot. Today, American men have an average height of $5\frac{7}{12}$ feet. What was the average height of American men 150 years ago?
   $5\frac{1}{4}$ feet

2. Jaime has a length of ribbon that is $23\frac{1}{2}$ in. long. If she plans to cut the ribbon into pieces that are $\frac{3}{4}$ in. long, into how many pieces can she cut the ribbon? (She cannot use partial pieces.)
   31 pieces

3. Todd's restaurant bill for dinner was $15.55. After he left a tip, he spent a total of $18.00 on dinner. How much money did Todd leave for a tip?
   $2.45

4. The difference between the boiling point and melting point of Hydrogen is 6.47°C. The melting point of Hydrogen is −259.34°C. What is the boiling point of Hydrogen?
   −252.87°C

Choose the letter for the best answer.

5. Maurice Green won the Olympic gold in the 100-m dash in 2000 with a time of 9.87 seconds. His time was 0.93 seconds faster than Francis Jarvis who won the 100-m dash in 1900. What was Jarvis' time in 1900?
   A 8.94 seconds
   B 10.61 seconds
   C 10.80 seconds
   D 11.23 seconds

6. The balance in Susan's checking account was $245.35. After the bank deposited interest into the account, her balance went to $248.02. How much interest did the bank pay Susan?
   F $1.01
   G $2.67
   H $3.95
   J $493.37

7. After a morning shower, there was $\frac{17}{100}$ in. of rain in the rain gauge. It rained again an hour later and the rain gauge showed $\frac{1}{4}$ in. of rain. How much did it rain the second time?
   A $\frac{2}{25}$ in.
   B $\frac{1}{6}$ in.
   C $\frac{21}{50}$ in.
   D $\frac{3}{8}$ in.

8. Two-third of John's savings account is being saved for his college education. If $2500 of his savings is for his college education, how much money in total is in his savings account?
   F $1666.67
   G $3750
   H $4250.83
   J $5000

Lesson Quiz

Solve.

1. $x - 23.3 = 17.8$ $x = 41.1$

2. $j + \frac{2}{3} = -14\frac{3}{4}$ $j = -15\frac{5}{12}$

3. $9y = \frac{3}{5}$ $y = \frac{1}{15}$

4. $\frac{d}{4} = 2\frac{3}{8}$ $d = 9\frac{1}{2}$

5. Tamara had 6 bags of mulch for her garden. Each bag contained $8\frac{1}{3}$ lb of mulch. What was the total weight of the mulch? **50 lb**

Available on Daily Transparency in CRB

Pacing: Traditional 1 day
Block $\frac{1}{2}$ day

Objective: Students solve inequalities with rational numbers.

Learn to solve inequalities with rational numbers.

The minimum size for a piece of first-class mail is 5 inches long, $3\frac{1}{2}$ inches wide, and 0.007 inch thick. For a piece of mail, the combined length of the longest side and the distance around the thickest part may not exceed 108 inches. Many inequalities are used in determining postal rates.

Warm Up

Solve.

1. $t + 8.7 = -12.4$ $t = -21.1$
2. $r + 4\frac{4}{9} = 11\frac{1}{3}$ $r = 6\frac{8}{9}$
3. $3.2x = 14.4$ $x = 4.5$
4. $\frac{P}{8.6} = 5.4$ $p = 46.44$

EXAMPLE 1 Solving Inequalities with Decimals

Solve.

A $0.5x \geq 0.5$

$0.5x \geq 0.5$

$\frac{0.5}{0.5}x \geq \frac{0.5}{0.5}$ *Divide both sides by 0.5.*

$x \geq 1$

B $t - 7.5 > 30$

$t - 7.5 > 30$

$t - 7.5 + 7.5 > 30 + 7.5$ *Add 7.5 to both sides of the equation.*

$t > 37.5$

Problem of the Day

Arnie built a fence around a rectangular field that measures 120 feet by 90 feet. He put a post in each corner and every 6 feet along all four sides. How many fence posts did he use? **70**

Available on Daily Transparency in CRB

EXAMPLE 2 Solving Inequalities with Fractions

Solve.

A $x + \frac{1}{2} < 1$

$x + \frac{1}{2} < 1$

$x + \frac{1}{2} - \frac{1}{2} < 1 - \frac{1}{2}$ *Subtract $\frac{1}{2}$ from both sides.*

$x < \frac{1}{2}$

B $-3\frac{1}{3}y \geq 10$

$-3\frac{1}{3}y \geq 10$

$-\frac{10}{3}y \geq 10$ *Rewrite $-3\frac{1}{3}$ as the improper fraction $-\frac{10}{3}$.*

$\left(-\frac{3}{10}\right)\left(-\frac{10}{3}\right)y \leq \left(-\frac{3}{10}\right)10$ *Multiply both sides by $-\frac{3}{10}$. Change $\geq$ to $\leq$.*

$y \leq -3$

Remember!

When multiplying or dividing an inequality by a *negative* number, reverse the inequality symbol.

Math Humor

The left side of the inequality was losing badly when one term said, "We just need to believe in ourselves." The other term said, "Forget that positive thinking, we need some negativity to turn this thing around!"

---

1 Introduce
Alternate Opener

EXPLORATION

3-7 Solving Inequalities with Rational Numbers

Beckie has $124.75 in her account. If she wants to maintain a $50 balance, how much can she spend?

Can She Spend ...	$124.75 − x	Yes	No
$60.00?	124.75 − 60.00 = 64.75	✓	
$70.00?	124.75 − 70.00 = 54.75	✓	
$75.00?	124.75 − 75.00 = 49.75		✓

This problem can be represented with the inequality $124.75 − x \geq 50$. Look at the last row in the table. To maintain a $50.00 balance, Beckie should spend $0.25 less than $75.00, or $74.75. This solution is written as $x \leq 74.75$.

Estimate the solution for each inequality, and then use a calculator to solve.

	Equation	Estimate	Actual
1.	$124.75 − x \geq 50$		
2.	$x + 16.9 \leq 50$		
3.	$14.5x \leq 100$		
4.	$\frac{x}{2.5} \geq 50$		

Think and Discuss

5. **Describe** how you estimated the solutions for the inequalities in numbers 1−4.

Motivate

Show students an example of a one-step inequality with rational numbers (e.g., $x + \frac{3}{4} \leq \frac{5}{16}$). Ask if they recognize any connections to any other work that they have done so far. Help students see the connections to one-step inequalities with real numbers or integers (Lessons 1-5 and 2-5) and equations with rational numbers (Lesson 3-6).

Exploration worksheet and answers on Chapter 3 Resource Book pp. 63 and 128

---

2 Teach

Lesson Presentation

Guided Instruction

In this lesson, students learn to solve inequalities with rational numbers. Review the procedures for solving inequalities (Lessons 1-5 and 2-5). Emphasize that the procedures used to solve inequalities with whole numbers or integers are the same for rational numbers. Remind students to reverse the inequality symbol when they multiply or divide both sides by a negative number.

EXAMPLE 3 PROBLEM SOLVING APPLICATION

With first-class mail, there is an extra charge in any of these cases:

- The length is greater than $11\frac{1}{2}$ in.
- The height is greater than $6\frac{1}{8}$ in.
- The thickness is greater than $\frac{1}{4}$ in.
- The length divided by the height is less than 1.3 or greater than 2.5.

The height of an envelope is 4.5 inches. What are the minimum and maximum lengths to avoid an extra charge?

1 Understand the Problem

The **answer** is the minimum and maximum lengths for an envelope to avoid an extra charge. List the **important information:**

- The height of the piece of mail is 4.5 inches.
- If the length divided by the height is between 1.3 and 2.5, there *will not be* an extra charge.

Show the **relationship** of the information:

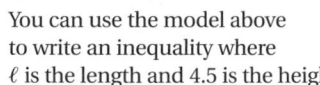

$$1.3 \leq \frac{length}{height} \leq 2.5$$

2 Make a Plan

You can use the model above to write an inequality where ℓ is the length and 4.5 is the height.

$$1.3 \leq \frac{\ell}{4.5} \leq 2.5$$

3 Solve

$1.3 \leq \frac{\ell}{4.5}$ and $\frac{\ell}{4.5} \leq 2.5$

$4.5 \cdot 1.3 \leq \ell$ and $\ell \leq 4.5 \cdot 2.5$ *Multiply both sides of each inequality by 4.5.*

$\ell \geq 5.85$ and $\ell \leq 11.25$ *Simplify.*

4 Look Back

The length of the envelope must be between 5.85 in. and 11.25 in.

Think and Discuss

1. Explain the first steps in solving $0.5x > 7$ and solving $\frac{3}{5}x > 3$.

2. Give an example of an inequality with a fraction in which the sign changes during solving.

Some students may forget to reverse the inequality symbol when reversing is necessary. Remind students to reverse the inequality symbol when they multiply or divide by a negative number and also when they rewrite an inequality in reverse order. For example, $3 > n$ can be rewritten $n < 3$.

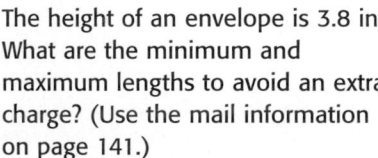

Additional Examples

Example 1

Solve.

A. $0.4x \leq 0.8$ **B.** $y - 3.8 < 11$

   $x \leq 2$ $y < 14.8$

Example 2

Solve.

A. $x + \frac{2}{3} > 2$ **B.** $-2\frac{1}{4}n \leq 9$

   $x > 1\frac{1}{3}$ $n \geq -4$

Example 3

The height of an envelope is 3.8 in. What are the minimum and maximum lengths to avoid an extra charge? (Use the mail information on page 141.)

The length of the envelope must be between 4.94 inches and 9.5 inches to avoid extra charges.

---

3 Close

 Reaching All Learners
Through Critical Thinking

Give each student a recording sheet (Chapter 3 Resource Book p. 111) containing inequalities with rational numbers that are solved *incorrectly*. The solution for each inequality contains one error. Examples of possible errors are: using the incorrect operation, failing to reverse the inequality symbol, omitting a negative sign, and making a computation error. Ask students to identify each error and supply the correct solution.

Summarize

Discuss the similarities between solving equations and solving inequalities. Help students see that the procedures are almost identical. Ask students how the type of number in an *equation* affects the steps you would take to solve it. Help students see that the type of number does not affect the steps you would take, but only affects the way the calculations are performed (i.e., the arithmetic). Ask how the type of number in an *inequality* affects the steps you would take to solve it. Remind students that they must reverse the inequality symbol when multiplying or dividing by a negative number.

Answers to Think and Discuss

Possible answers:

1. Divide both sides of the inequality by 0.5; multiply both sides by $\frac{5}{3}$.
2. $-\frac{3}{4}x \leq -6$

3-7 **Exercises**

FOR EXTRA PRACTICE
see page 736

internet connect
Homework Help Online
go.hrw.com   Keyword: MP4 3-7

GUIDED PRACTICE

Students may want to refer back to the lesson examples.

Assignment Guide

If you finished Example **1** assign:
Core 1–6, 14–19, 27, 29, 47–56
Enriched 14–19, 27, 29, 33–35, 37, 38, 45–56

If you finished Example **2** assign:
Core 1–12, 14–25, 27–30, 47–56
Enriched 14–25, 27–41, 45–56

If you finished Example **3** assign:
Core 1–30, 43, 47–56
Enriched 14–56

Notes

See Example **1** Solve.

1. $0.3x \geq 0.6$ $x \geq 2$
2. $k - 7.2 > 2.1$ $k > 9.3$
3. $\frac{g}{-0.5} \geq -\frac{7}{0.5}$ $g \leq 7$
4. $h + 0.79 < 1.58$ $h < 0.79$
5. $6.07w \leq 1.4568$ $w \leq 0.24$
6. $z - 0.75 > -0.75$ $z > 0$

See Example **2** Solve.

7. $k - \frac{2}{5} > \frac{3}{15}$ $k > \frac{3}{5}$
8. $y + \frac{7}{9} \geq \frac{56}{72}$ $y \geq 0$
9. $13q \leq -\frac{1}{13}$ $q \leq -\frac{1}{169}$
10. $x + \frac{1}{3} < 2$ $x < 1\frac{2}{3}$
11. $-3f < -\frac{4}{5}$ $f > \frac{4}{15}$
12. $3\frac{1}{4}m \geq 13$ $m \geq 4$

See Example **3** 13. Timothy is driving from Sampson to Williamsbery, a distance of 366.5 miles. If he averages between 45 mi/h and 55 mi/h, how long will it take him to get to Williamsbery to the nearest tenth of an hour, assuming he does not stop? **between 6.7 and 8.1 hours**

INDEPENDENT PRACTICE

See Example **1** Solve.

14. $0.6 + y \geq -0.72$ $y \geq -1.32$
15. $m - 5.8 \leq -5.87$ $m \leq -0.07$
16. $-0.8x \geq -0.56$ $x \leq 0.7$
17. $\frac{g}{-2.7} \geq 9$ $g \leq -24.3$
18. $c + 11.7 < 6$ $c < -5.7$
19. $\frac{w}{-0.4} \geq \frac{3}{0.8}$ $w \leq -1.5$

See Example **2** Solve.

20. $\frac{5}{9} + n \leq \frac{9}{5}$ $n \leq 1\frac{11}{45}$
21. $2\frac{2}{5}k \geq 1\frac{2}{3}$ $k \geq \frac{25}{36}$
22. $-\frac{2}{7} + x < 3$ $x < 3\frac{2}{7}$
23. $x + \frac{2}{5} \geq 5$ $x \geq 4\frac{3}{5}$
24. $7t < -\frac{14}{15}$ $t < -\frac{2}{15}$
25. $-6\frac{1}{8}m \geq 7$ $m \leq -1\frac{1}{7}$

See Example **3** 26. It takes an elevator 2 seconds to go from floor to floor. Each passenger takes 1.5 to 2.0 seconds to board or exit. Twelve passengers board on the first floor and exit on the fourth floor. How long will you wait if you are on the seventh floor and ring for the elevator just as it leaves the first floor? **between 30 and 36 seconds**

PRACTICE AND PROBLEM SOLVING

Solve.

27. $-0.5d \geq 1.5$ $d \leq -3$
28. $-3\frac{3}{4}m \geq 7\frac{1}{2}$ $m \leq -2$
29. $\frac{2g}{0.5} \geq -\frac{4}{0.5}$ $g \geq -2$
30. $x + \frac{2}{5} \geq 3$ $x \geq 2\frac{3}{5}$
31. $-4t < -\frac{12}{13}$ $t > \frac{3}{13}$
32. $r + 9.3 > 4.2$ $r > -5.1$
33. $-1.6y \leq 12.8$ $y \geq -8$
34. $c - 15.3 < 61.7$ $c < 77$
35. $\frac{w}{-1.6} \geq \frac{1}{4.8}$ $w \leq -\frac{1}{3}$
36. $6f > -\frac{4}{9}$ $f > -\frac{2}{27}$
37. $5 < c + 1.9$ $c > 3.1$
38. $\frac{2}{-0.4} \geq -\frac{r}{0.8}$ $r \geq 4$
39. $2 > c - 1\frac{1}{3}$ $c < 3\frac{1}{3}$
40. $-f < \frac{6}{7}$ $f > -\frac{6}{7}$
41. $3\frac{1}{4}t \leq 19.5$ $t \leq 6$

Math Background

The inequality used in Example 3, $1.3 \leq \frac{\ell}{4.5} \leq 2.5$, is a shortened version of the following compound inequality:

$1.3 \leq \frac{\ell}{4.5}$ AND $\frac{\ell}{4.5} \leq 2.5$. This type of compound inequality is called a *conjunction*, and the solution set is the *intersection* of the solution sets for the individual inequalities (See Lesson 14-2).

Another type of compound inequality, such as $x < 3$ OR $x > 4.5$, is called a *disjunction*, and its solution set is the *union* of the solution sets for the individual inequalities.

RETEACH 3-7

Reteach
3-7 *Algebra: Solving Inequalities with Rational Numbers*

Solving inequalities with rational numbers is basically the same as solving inequalities with integers or whole numbers:

Use inverse operations to isolate the variable.
When multiplying or dividing by a negative number, reverse the inequality symbol.

$-\frac{1}{2}z > -8$

$-2 \cdot \left(-\frac{1}{2}\right)z > -8 \cdot (-2)$ Multiply by −2.

$z < 16$ Change > to <.

$2.6w < -23.4$
$\frac{2.6w}{2.6} < \frac{-23.4}{2.6}$ Divide by 2.6.
$\frac{2.6w}{2.6} < \frac{-234}{26}$ whole-number divisor
$w < -9$ < symbol remains

$x - \frac{2}{5} \leq \frac{3}{5}$
$+ \frac{2}{3} \quad + \frac{2}{3}$ Add $\frac{2}{3}$ to each side.
$x \leq \frac{3}{5} + \frac{2}{3}$ LCD = 15
$x \leq \frac{9}{15} + \frac{10}{15}$ Rewrite fractions.
$x \leq \frac{19}{15}$ Add fractions.
$x \leq 1\frac{4}{15}$ Rewrite fraction.

Tell how to isolate the variable.
Tell if the symbol changes or stays the same.
1. $-1.4x \geq 28$ 2. $-\frac{1}{7} + z < -\frac{3}{14}$ 3. $-\frac{5}{8}t > 24$
divide by −1.4   **add $\frac{1}{7}$**   **multiply by $-\frac{8}{5}$**
symbol changes   **same symbol**   **symbol changes**

Complete to solve each inequality.
4. $7.3x < -87.6$
$\frac{7.3x}{7.3} < \frac{-87.6}{7.3}$
$x < \frac{-876}{73}$
$x < -12$

5. $\frac{2}{3} + w \geq -\frac{5}{6}$
$-\frac{2}{3} \quad -\frac{2}{3}$
$w \geq -\frac{5}{6} - \frac{4}{6}$
$w \geq -\frac{9}{6}$
$w \geq -1\frac{1}{2}$

PRACTICE 3-7

Practice B
3-7 *Algebra: Solving Inequalities with Rational Numbers*

Solve.
1. $-12.7 + x > -15.1$ 2. $4.8y \geq 10.56$ 3. $-\frac{m}{1.8} < -4.15$
x > −2.4   **y ≥ 2.2**   **m < −7.47**
4. $-0.4k \leq 6.08$ 5. $\frac{d}{-1.2} > 4.05$ 6. $-8.6 + r \geq -6.8$
k ≥ −15.2   **d < −4.86**   **r ≥ 1.8**

Solve. Write answers in simplest form.
7. $x + \frac{5}{16} \geq \frac{11}{16}$ 8. $y - \frac{7}{12} < \frac{5}{6}$ 9. $1\frac{5}{9}k > -\frac{7}{18}$
x ≥ $\frac{3}{8}$ **y < $1\frac{5}{12}$**   **k > $-\frac{1}{4}$**
10. $\frac{x}{-8} \leq \frac{9}{16}$ 11. $x - \frac{7}{15} < -3\frac{4}{5}$ 12. $-2\frac{1}{8}w \geq -4\frac{3}{4}$
x ≥ $-4\frac{1}{2}$ **x < $-3\frac{1}{3}$**   **w ≤ $2\frac{4}{17}$**
13. $x + \left(-2\frac{3}{10}\right) < -3\frac{3}{4}$ 14. $5\frac{5}{9}y \leq -3\frac{4}{27}$ 15. $d - \left(-3\frac{5}{18}\right) > -2\frac{5}{6}$
x < $-1\frac{9}{20}$ **y ≤ $-\frac{17}{30}$**   **d > $-6\frac{1}{9}$**

16. Sarah bought a 55-yard bolt of ribbon. If she wants to cut pieces that are $1\frac{1}{3}$-yard long, what is the maximum number of ribbons she can cut from the bolt?
41 pieces

42. Use the information in the box to explain whether each piece of mail with the given measures is subject to an extra charge. Explain your answers.

Additional Postage Required

Length greater than $11\frac{1}{2}$ in.

Height greater than $6\frac{1}{8}$ in.

Thickness greater than $\frac{1}{4}$ in.

Length divided by height is less than 1.3 or greater than 2.5

a. Length $10\frac{3}{4}$ in., height $4\frac{1}{4}$ in., thickness $\frac{3}{16}$ in.

b. Length $11\frac{1}{4}$ in., height $4\frac{1}{2}$ in., thickness $\frac{5}{16}$ in.

c. Length 11 in., height $4\frac{5}{8}$ in., thickness $\frac{7}{32}$ in.

43. *LIFE SCIENCE* There are over 2000 species of jellyfish. The bell of the largest jellyfish, the lion's mane, can be up to 72 inches across. The smallest jellyfish measures only one-quarter of an inch across. Suppose 50 different species of jellyfish were lined up next to each other. What is the minimum and maximum width their bells would cover?

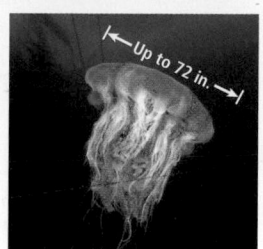

|← Up to 72 in. →|

44. *WHAT'S THE QUESTION?* Pens at the Pen Station cost between $0.39 and $5.59 each. If the answer is *at least $7.02 but no more than $100.62*, what might the question be if it related to the cost of a certain number of pens?

45. *WRITE ABOUT IT* If $0.3 + 0.7 > y$, is y greater than 1? Explain.

46. *CHALLENGE* Asbestos fiber can be as thin as 2×10^{-5} mm in diameter. Wool fiber is 2.5×10^{-2} mm in diameter. How many times as thick is wool fiber than asbestos fiber? **Up to 1250 times as thick**

Answers

42. a. yes; because the length divided by the height is $10\frac{3}{4} \div 4\frac{1}{4} = \frac{43}{17} \approx 2.53$ in., which is greater than 2.5 in.

b. Yes; the thickness is $\frac{5}{16}$, which is greater than $\frac{1}{4}$.

c. No; the dimensions are all within the limits. The length divided by the height is $11 \div 4\frac{5}{8} = \frac{88}{37} \approx 2.38$ in., which is between 1.3 and 2.5 in.

43. at least 12.5 in. but not more than 3600 in.

44. Possible answer: How much could 18 pens cost?

45. Possible answer: No; $0.3 + 0.7 > y$ simplifies to $1 > y$ or $y < 1$.

Journal

Have students write about whether they find inequalities easy to solve. Have students give a reason for their answers.

Spiral Review

Add or subtract. (Lesson 3-2)

47. $-0.4 + 0.7$ **0.3**   **48.** $1.35 - 5.6$ **−4.25**   **49.** $-0.01 - 0.25$ **−0.26**   **50.** $-0.65 + -0.12$ **−0.77**

Multiply. (Lesson 3-3)

51. $-2.4(-7)$ **16.8**   **52.** $3.2(-1.7)$ **−5.44**   **53.** $-0.03(8.6)$ **−0.258**   **54.** $-1.07(-0.6)$ **0.642**

55. TEST PREP If $y = -\frac{3}{9}$, which is not equal to y? (Lesson 3-1) **C**

A $\frac{-1}{3}$ **B** $-\frac{1}{3}$ **C** $-\left(\frac{-1}{3}\right)$ **D** $-\left(\frac{-1}{-3}\right)$

56. TEST PREP If $24x = 2.4$, what is $12x$? (Lesson 3-6) **G**

F 0.2   **G** 1.2   **H** 4.8   **J** 2

Test Prep Doctor

One way to solve Exercise 56 is to solve for x and then multiply that value by 12. A quicker way is to divide each side of the equation by 2. Dividing by 2 gives the answer in one step: $12x = 1.2$. The correct answer is choice **G**.

CHALLENGE 3-7

Challenge
3-7 Look Closely!

Think about the solution sets of some different inequalities.

Consider the inequality $x + 3 > x$.

1. Simplify the inequality. Write the result. __$3 > 0$__

2. Explain why the simplified inequality does not show a solution set.

__The variable has been eliminated.__

3. Tell which rational numbers you think are in the solution set. Explain.

__all the rational numbers, since $3 > 0$ is a true statement.__

4. Draw a conclusion about an inequality that leads to a statement that is always true.

__The inequality is true for all rational numbers.__

Consider the inequality $x + 5 + 3x > 4x + 9$.

5. Simplify the inequality. Write the result. __$5 > 9$__

6. Explain why the simplified inequality does not show a solution set.

__The variable has been eliminated.__

7. Tell which rational numbers you think are in the solution set. Explain.

__no rational numbers, since $5 > 9$ is a false statement.__

8. Draw a conclusion about an inequality that leads to a statement that is never true.

__The inequality is not true for any rational number.__

Tell which rational numbers you think are in the solution set of each inequality.

9. $|x| \geq 0$ __all__   10. $|x| < 0$ __none__

11. $|x| \leq 0$ __0__   12. $|x| > -3$ __all__

13. $|x| < -3$ __none__   14. $|x| \leq -3$ __none__

15. $|x| > 3$ __$x > 3$ or $x < -3$__   16. $|x + 3| \leq 6$ __$x \leq 3$ and $x \geq -9$__

PROBLEM SOLVING 3-7

Problem Solving
3-7 Solving Inequalities with Rational Numbers

Write the correct answer.

1. The cost of an ad is $1.25 for every 5 words. If Maria wants to spend no more than $15 on her ad, what is w the maximum number of words she can use?

__$w \leq 60$__

2. John has written too many checks and his bank balance is −$53.23. The bank will charge him an overdraft fee of $24.75. How much does he need to deposit so that his balance will be at least $50.00?

__$d \geq $127.98__

3. A brighter star has a lower magnitude. The difference between the magnitude the star Canopus and the magnitude of Sirius is less than 1. Sirius is the brightest star and has a magnitude of −1.6. Find the magnitude m of the star Canopus.

__$m < -0.6$__

4. Susanna has a job that pays $6.00 per hour. She is told that she will receive an annual raise of at least $0.45 per hour. How many years y must she work before her hourly wage is at least $7.35 per hour?

__$y \leq 3$__

In 1932, the United States Golf Association set standards for golf balls. The maximum weight of a golf ball is 1.620 oz. The diameter of a golf ball cannot be less than 1.680 in. Choose the letter for the best answer.

5. A newly designed golf ball is too heavy. If the weight of the ball is reduced by 0.17 oz. it will meet the standard. What is the current weight of the new ball?

A $w > 1.450$ oz   C $w > 1.790$ oz
B $w \leq 1.450$ oz   (D) $w \leq 1.790$ oz

6. How much can a golf ball weigh and be no more than 0.03 ounces less than the standard?

(F) 1.590 oz   H 1.710 oz
G 1.650 oz   J 1.617 oz

7. Write an inequality that represents the standard for the diameter d of a golf ball.

A $d > 1.680$ in.   C $d \leq 1.680$ in.
(B) $d \geq 1.680$ in.   D $d < 1.680$ in.

8. If 200 golf balls are lined up, what is the length ℓ the balls would cover?

F $\ell \leq 28$ ft   H $\ell \geq 336$ ft
(G) $\ell \geq 28$ ft   J $-\ell < 336$ in.

Lesson Quiz

Solve.

1. $w + 1.25 \leq 5.12$ $w \leq 3.87$
2. $f - 1\frac{3}{8} \geq 4\frac{1}{4}$ $f \geq 5\frac{5}{8}$
3. $1.2x > 17.04$ $x > 14.2$
4. $\frac{h}{0.7} < 5.5$ $h < 3.85$
5. Rosa's car gets between 20 and 21 mi/gal on the highway. She knows that her gas tank holds at least 18 gallons. What is the minimum distance Rosa could drive her car on the highway between fill-ups?   **360 miles**

Available on Daily Transparency in CRB

Chapter 3 Mid-Chapter Quiz

Purpose: To assess students' mastery of concepts and skills in Lessons 3-1 through 3-7

Assessment Resources

Section 3A Quiz
Assessment Resources . . . . p. 9

Test and Practice Generator CD-ROM

Additional mid-chapter assessment items in both multiple-choice and free-response format may be generated for any objective in Lessons 3-1 through 3-7.

Mid-Chapter Quiz

LESSON 3-1 (pp. 112–116)

Simplify.

1. $\frac{12}{36}$ $\frac{1}{3}$
2. $\frac{18}{45}$ $\frac{2}{5}$
3. $\frac{27}{63}$ $\frac{3}{7}$
4. $\frac{55}{121}$ $\frac{5}{11}$

Write each decimal as a fraction in simplest form.

5. 0.4 $\frac{2}{5}$
6. 0.75 $\frac{3}{4}$
7. 0.18 $\frac{9}{50}$
8. 0.825 $\frac{33}{40}$

LESSON 3-2 (pp. 117–120)

Evaluate each expression for the given value of the variable.

9. $72.9 - x$ for $x = 31.31$ **41.59**
10. $-\frac{2}{5} + z$ for $z = 5\frac{3}{5}$ $5\frac{1}{5}$
11. $\frac{3}{4} + y$ for $y = -3\frac{1}{4}$ $-2\frac{1}{2}$

LESSON 3-3 (pp. 121–125)

Multiply. Write each answer in simplest form.

12. $3\left(5\frac{3}{4}\right)$ $17\frac{1}{4}$
13. $2\frac{3}{4}\left(\frac{7}{22}\right)$ $\frac{7}{8}$
14. $\frac{2}{5}\left(\frac{-5}{6}\right)$ $-\frac{1}{3}$
15. $\frac{-1}{5}\left(\frac{-2}{3}\right)$ $\frac{2}{15}$

LESSON 3-4 (pp. 126–130)

Divide. Write each answer in simplest form.

16. $\frac{3}{5} \div \frac{4}{15}$ $2\frac{1}{4}$
17. $\frac{3}{5} \div 5$ $\frac{3}{25}$
18. $-\frac{3}{4} \div 1$ $-\frac{3}{4}$
19. $-6\frac{7}{8} \div 1\frac{2}{3}$ $-4\frac{1}{8}$

LESSON 3-5 (pp. 131–134)

Add or subtract.

20. $\frac{3}{8} + \frac{1}{3}$ $\frac{17}{24}$
21. $2\frac{1}{2} + 3\frac{7}{10}$ $6\frac{1}{5}$
22. $7\frac{5}{8} - 2\frac{1}{6}$ $5\frac{11}{24}$
23. $3\frac{1}{6} - 1\frac{3}{4}$ $1\frac{5}{12}$

LESSON 3-6 (pp. 136–139)

Solve.

24. $x + \frac{1}{5} = -\frac{1}{5}$ $x = -\frac{2}{5}$
25. $y - \frac{8}{9} = \frac{1}{9}$ $y = 1$
26. $10 = \frac{7}{2}m$ $m = 2\frac{6}{7}$
27. $\frac{2}{3}d = \frac{1}{6}$ $w = \frac{1}{4}$

28. A basketball team has 87 points after three-fourths of a game. How many points will the team finish with if the players keep the same pace? **116 points**

LESSON 3-7 (pp. 140–143)

Solve.

29. $x + \frac{1}{10} \geq 5$ $x \geq 4\frac{9}{10}$
30. $-2t > 1$ $t < -\frac{1}{2}$
31. $-6 \leq \frac{-g}{3}$ $g \leq 18$
32. $m + 1.3 \leq 0.5$ $m \leq -0.8$

33. A charity has met less than $\frac{2}{5}$ of its donation goal with one week to go in its pledge drive. It has received $7400 in pledges. How much is its goal? **At least $18,500**

Focus on Problem Solving

Solve
- **Choose an operation**

To decide whether to add or subtract to solve a problem, you need to determine the action taking place in the problem.

Action	Operation
Combining numbers, or putting numbers together	Addition
Taking away or finding out how far apart two numbers are	Subtraction
Combining equal groups	Multiplication
Splitting things into equal groups or finding how many equal groups you can make	Division

Determine the action for each problem. Write the problem using the actions. Then show what operation you used to get the answer.

1 Mary is making a string of beads. If each bead is 0.7 cm wide, how many beads does she need to make a string that is 35 cm long?

2 A cake recipe calls for $2\frac{1}{2}$ cups of sugar for the cake and $1\frac{1}{2}$ cups of sugar for the icing. How much sugar do you need to make the cake?

3 Suppose $\frac{1}{3}$ of the fish in a lake are considered game fish. Of these, $\frac{2}{5}$ meet the legal minimum size requirement. What fraction of the fish in the lake are game fish that meet the legal minimum size requirement?

4 Part of a checkbook register is shown below. Find the amount in the account after the transactions shown.

RECORD ALL CHARGES OR CREDITS THAT AFFECT YOUR ACCOUNT

TRANSACTION	DATE	DESCRIPTION	AMOUNT	FEE	DEPOSITS	BALANCE	$287.34
Withdrawal	11/16	autodebit for phone bill	$43.16				$43.16
Check 1256	11/18	groceries	$27.56				$27.56
Check 1257	11/23	new clothes	$74.23				$74.23
Withdrawal	11/27	ATM withdrawal	$40.00	$1.25			$41.25

Answers
1. 50
2. 4 cups
3. $\frac{2}{15}$
4. $101.14

Focus on Problem Solving

Purpose: *To focus on choosing an operation*

Problem Solving Resources
Interactive Problem Solving. . pp. 19–28

Math: Reading and Writing in the Content Area . . . . pp. 19–28

Problem Solving Process
This page focuses on the third step of the problem-solving process: **Solve**

Discuss
Have students discuss which action is taking place in each problem, rewrite the problem using action words, and then indicate the operation used to get the answer.

Possible answers:

1. Finding how many equal groups you can make; how many equal lengths of 0.7 cm are there in 35 cm?; division.

2. Combining numbers; what is the combined amount of sugar needed?; addition.

3. Splitting things into equal groups (multiplying by a fraction is the same as dividing); what is $\frac{1}{3}$ of $\frac{2}{5}$?; multiplication.

4. Taking away; what remains of $287.34 after $43.16, $27.56, $74.23, and $41.25 have been taken away?; subtraction.

Focus on Problem Solving **145**

Real Numbers

One-Minute Section Planner

Lesson	Materials	Resources
Lesson 3-8 Squares and Square Roots **NCTM:** Number and Operations, Communication **NAEP:** Algebra 3b ☑ SAT-9 ☑ SAT-10 ☐ ITBS ☑ CTBS ☑ MAT ☑ CAT	**Optional** Calculators Number Cards for Reaching All Learners *(CRB, p. 112)*	• *Chapter 3 Resource Book,* pp. 72–80 • *Daily Transparency T34, CRB* • *Additional Examples Transparencies T35–T36, CRB* • *Alternate Openers: Explorations,* p. 26
Lesson 3-9 Finding Square Roots **NCTM:** Number and Operations, Problem Solving, Communication **NAEP:** Number Properties 2d ☑ SAT-9 ☑ SAT-10 ☐ ITBS ☑ CTBS ☑ MAT ☑ CAT	**Required** Calculators **Optional** Number Cards for Reaching All Learners *(CRB, p. 113)*	• *Chapter 3 Resource Book,* pp. 81–89 • *Daily Transparency T37, CRB* • *Additional Examples Transparencies T38–T41, CRB* • *Alternate Openers: Explorations,* p. 27
Hands-On Lab 3B Explore Cubes and Cube Roots **NCTM:** Number and Operations, Reasoning and Proof, Connections, Representation **NAEP:** Number Properties 2d ☐ SAT-9 ☐ SAT-10 ☐ ITBS ☐ CTBS ☐ MAT ☐ CAT	**Required** Centimeter cubes *(MK)*	• *Hands-On Lab Activities,* pp. 21–24
Lesson 3-10 The Real Numbers **NCTM:** Number and Operations, Communication **NAEP:** Number Properties 1e ☐ SAT-9 ☐ SAT-10 ☐ ITBS ☐ CTBS ☐ MAT ☐ CAT	**Optional** Calculators Teaching Transparency T43 *(CRB)* Recording Sheet for Reaching All Learners *(CRB, p. 115)*	• *Chapter 3 Resource Book,* pp. 90–98 • *Daily Transparency T42, CRB* • *Additional Examples Transparencies T44–T46, CRB* • *Alternate Openers: Explorations,* p. 28
Extension Other Number Systems **NCTM:** Number and Operations **NAEP:** Number Properties 1a ☐ SAT-9 ☐ SAT-10 ☐ ITBS ☐ CTBS ☐ MAT ☐ CAT	**Optional** Table of powers *(CRB, p. 114)*	• *Additional Examples Transparency T47, CRB*
Section 3B Assessment		• *Section 3B Quiz, AR p. 10* • *Test and Practice Generator CD-ROM*

SAT = *Stanford Achievement Tests*   **ITBS** = *Iowa Test of Basic Skills*   **CTBS** = *Comprehensive Test of Basic Skills/Terra Nova*
MAT = *Metropolitan Achievement Test*   **CAT** = *California Achievement Test*
NCTM = Complete standards can be found on pages T29–T35.   **NAEP** = Complete standards can be found on pages A54–A58.
SE = *Student Edition*   **TE** = *Teacher's Edition*   **AR** = *Assessment Resources*   **CRB** = *Chapter Resource Book*   **MK** = *Manipulatives Kit*

Section Overview

Squares and Square Roots
Lessons 3-8 and 3-9

Why? Squares and square roots are important and necessary concepts in algebra, geometry, and higher levels of mathematics.

Squares

The **square** of both 6 and −6 is 36.
$$6^2 = 36$$
$$(-6)^2 = 36$$

Square Roots

The **positive square root** of 36 is 6.
$$6 = \sqrt{36}$$
The **negative square root** of 36 is −6.
$$-6 = -\sqrt{36}$$

Perfect Square

• A number whose principal square root is an integer
• Examples: 0, 1, 4, 9, 16, 25, . . .

Principal Square Root

• Positive square root or square root of zero
• Indicated by the symbol $\sqrt{}$

Estimate $\sqrt{27}$ to the nearest tenth.

Step 1:
$$\sqrt{25} = 5 \text{ and } \sqrt{36} = 6$$
So $\sqrt{27}$ is between 5 and 6, closer to 5.

Step 2:
$$5.1^2 = 26.01 \text{ (too low) and}$$
$$5.2^2 = 27.04 \text{ (too high)}$$
To the nearest tenth, $\sqrt{27} \approx 5.2$.

The Real Numbers
Hands-On Lab 3B, Lesson 3-10

Why? The set of real numbers includes rational and irrational numbers.

Real Numbers

A **rational number** can be written as a quotient of two integers. Every rational number can be written as a decimal that either terminates or repeats.

$$3\frac{4}{5} = 3.8 \qquad -3 = -3.0 \qquad \frac{2}{3} = 0.\overline{6}$$
$$\sqrt{1.44} = 1.2 \qquad \sqrt{\frac{4}{25}} = \frac{2}{5} = 0.4 \qquad \frac{0}{2} = 0$$

An **irrational number** cannot be written as a quotient of two integers. There is no exact decimal representation for an irrational number.

$$\sqrt{7} \approx 2.646 \qquad \sqrt{2.8} \approx 1.673$$
$$\sqrt{\frac{3}{8}} \approx 0.612 \qquad \pi \approx 3.14159 \approx \frac{22}{7}$$

> A fraction with a zero denominator, such as $\frac{3}{0}$, is **undefined**.

> **Density Property:** Between any two real numbers is another real number.

Warm Up

Simplify.

1. 5^2 25   2. 8^2 64

3. 12^2 144   4. 15^2 225

5. 20^2 400

Problem of the Day

A Shakespearean sonnet is a poem made up of 3 quatrains (4 lines each) and a couplet (2 lines). Each line is in iambic pentameter (which means it has 5 iambic feet). So how many iambic feet long is a Shake-spearean sonnet? **70**

Available on Daily Transparency in CRB

Math Humor

Why wouldn't the tree fit in the round pot? It had square roots!

3-8 Squares and Square Roots

Learn to find square roots.

Vocabulary

principal square root

perfect square

Think about the relationship between the area of a square and the length of one of its sides.

area = 36 square units
side length = $\sqrt{36}$ = 6 units

Taking the square root of a number is the inverse of squaring the number.

$$6^2 = 36 \qquad \sqrt{36} = 6$$

Every positive number has two square roots, one positive and one negative. One square root of 16 is 4, since $4 \cdot 4 = 16$. The other square root of 16 is -4, since $(-4)(-4)$ is also 16. You can write the square roots of 16 as ± 4, meaning "plus or minus" 4.

Quilts are often pieced together from small squares to form a large design.

Helpful Hint

$\sqrt{-49}$ is not the same as $-\sqrt{49}$. A negative number has no real square roots.

When you press the $\sqrt{}$ key on a calculator, only the nonnegative square root appears. This is called the **principal square root** of the number.

$$+\sqrt{16} = 4 \qquad\qquad -\sqrt{16} = -4$$

The numbers 16, 36, and 49 are examples of perfect squares. A **perfect square** is a number that has integers as its square roots. Other perfect squares include 1, 4, 9, 25, 64, and 81.

EXAMPLE 1 Finding the Positive and Negative Square Roots of a Number

Find the two square roots of each number.

A 64
$$\sqrt{64} = 8 \qquad\qquad \text{8 is a square root, since } 8 \cdot 8 = 64.$$
$$-\sqrt{64} = -8 \qquad\qquad -8 \text{ is also a square root, since } -8 \cdot -8 = 64.$$

B 1
$$\sqrt{1} = 1 \qquad\qquad \text{1 is a square root, since } 1 \cdot 1 = 1.$$
$$-\sqrt{1} = -1 \qquad\qquad -1 \text{ is also a square root, since } -1 \cdot -1 = 1.$$

C 121
$$\sqrt{121} = 11 \qquad\qquad \text{11 is a square root, since } 11 \cdot 11 = 121.$$
$$-\sqrt{121} = -11 \qquad\qquad -11 \text{ is also a square root, since } -11 \cdot -11 = 121.$$

---

1 Introduce

Alternate Opener

EXPLORATION

3-8 Squares and Square Roots

The sequence shows the square numbers 1, 4, 9, and 16.

$$1^2 = 1 \quad 2^2 = 4 \quad 3^2 = 9 \quad 4^2 = 16$$
$$3 \qquad 1+3 \qquad 1+3+5 \qquad 1+3+5+7$$

1. Draw a picture to show that $5^2 = 1 + 3 + 5 + 7 + 9$.
2. Add the odd numbers $1 + 3 + 5 + 7 + 9 + \cdots + 17 + 19$. What square number do you get?
3. The table starts with $11^2 = 1 + 3 + 5 + 7 + 9 + 11 + 13 + 15 + 17 + 19 + 21 = 121$. Complete the table by adding the next odd number to this sum.

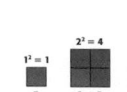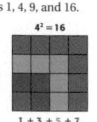

11^2	12^2	13^2	14^2	15^2	16^2	17^2	18^2	19^2	20^2
121									

Think and Discuss

4. **Explain** how you can determine square numbers using sums of odd numbers.
5. **Demonstrate** that the value of 22^2 can be determined by adding odd numbers.

Motivate

Use a calculator to demonstrate the use of the square-root button. Use the calculator to find $\sqrt{9}$, $\sqrt{16}$, and $\sqrt{25}$. Ask students what number they think will be the output if you press $\sqrt{36}$. Ask students to describe in their own words the function of the square-root key.

Exploration worksheet and answers on Chapter 3 Resource Book pp. 73 and 130

---

2 Teach

Lesson Presentation

Guided Instruction

In this lesson, students learn to find square roots. Remind students that they are familiar with inverse operations, such as addition and subtraction. Tell students that they will learn about another pair of inverse operations: squaring and finding a square root. Explain that 4 and -4 are the two square roots of 16 because $4^2 = 16$ and $(-4)^2 = 16$. Discuss the fact that the $\sqrt{}$ symbol indicates *principal square root,* which is always either positive or 0. Emphasize that the opposite of a square root is a real number, but that a negative number has no real square roots. For example, $-\sqrt{49} = -7$, but $\sqrt{-49}$ is not a real number.

EXAMPLE 2 *Computer Application*

The square computer icon contains 676 pixels. How many pixels tall is the icon?

Find the square root of 676 to find the length of the side. Use the positive square root; a negative length has no meaning.

$$26^2 = 676$$

So $\sqrt{676} = 26$.

The icon is 26 pixels tall.

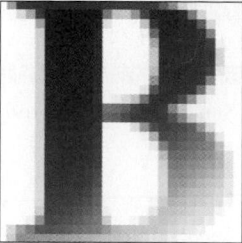

The square computer icon contains 676 colored dots that make up the picture. These dots are called *pixels*.

Remember!

The area of a square is s^2, where s is the length of a side.

In the order of operations, a square root symbol is like an exponent. Everything under the square root symbol is treated as if it were in parentheses.

$$\sqrt{5 - 3} = \sqrt{(5 - 3)}$$

EXAMPLE 3 **Evaluating Expressions Involving Square Roots**

Evaluate each expression.

A $2\sqrt{16} + 5$

$2\sqrt{16} + 5 = 2(4) + 5$	*Evaluate the square root.*
$= 8 + 5$	*Multiply.*
$= 13$	*Add.*

B $\sqrt{9 + 16} + 7$

$\sqrt{9 + 16} + 7 = \sqrt{25} + 7$	*Evaluate expression under square root symbol.*
$= 5 + 7$	*Evaluate the square root.*
$= 12$	*Add.*

Think and Discuss

1. Describe what is meant by a perfect square. Give an example.

2. Explain how many square roots a positive number can have. How are these square roots different?

3. Decide how many square roots 0 has. Tell what you know about square roots of negative numbers.

Show students...**COMMON ERROR ALERT**

Some students may try to apply a property that is not true. Show that $\sqrt{9 + 16} \neq \sqrt{9} + \sqrt{16}$ by evaluating each expression.

$\sqrt{9 + 16} = \sqrt{25} = 5$, but

$\sqrt{9} + \sqrt{16} = 3 + 4 = 7$

Additional Examples

Example 1

Find the two square roots of each number.

A. 49 ±7   **B.** 100 ±10

C. 225 ±15

Example 2

A square window has an area of 169 square inches. How wide is the window? The window is 13 inches wide.

Example 3

Evaluate each expression.

A. $3\sqrt{36} + 7$ 25

B. $\sqrt{21 - 5} + 9$ 13

---

3 Close

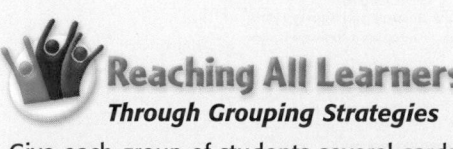 **Reaching All Learners**

Through Grouping Strategies

Give each group of students several cards containing integers from −5 to 5 (Chapter 3 Resource Book p. 112). Give the class problems involving square roots that have those integers as answers. For example, some questions could be as follows:

• Find $\sqrt{9} + 1$. **4**

• Find $-\sqrt{25}$. **−5**

• Find a square root of 1. **±1**

• What number has exactly one square root? **0**

Have groups solve each problem and hold up the card with the correct answer.

Summarize

Remind students that every positive number has two square roots, one positive and one negative. The statement that gives the positive, or principal, square root of 9 is $\sqrt{9} = 3$. The statement that gives the negative square root of 9 is $-\sqrt{9} = -3$. Remind students that the square root of a negative number is not a real number. Ask students to help you create a list of the first 15 perfect squares.

1, 4, 9, 16, 25, 36, 49, 64, 81, 100, 121, 144, 169, 196, and 225

Answers to Think and Discuss

Possible answers:

1. A perfect square is a number that has integers as its square roots. An example is 25.

2. Each positive number has two square roots. One of these roots is positive; the other is negative and is the opposite of the first.

3. The number zero has one square root, 0. Square roots of negative numbers do not exist in the real number system because the product of two negative numbers or two positive numbers is always a positive number.

3-8 Exercises

FOR EXTRA PRACTICE
see page 737

internet connect
Homework Help Online
go.hrw.com   Keyword: MP4 3-8

GUIDED PRACTICE

See Example 1   **Find the two square roots of each number.**

1. 25 ± 5 2. 144 ± 12 3. 4 ± 2 4. 400 ± 20

5. 1 ± 1 6. 81 ± 9 7. 9 ± 3 8. 16 ± 4

See Example 2   9. A square court for playing the game four-square has an area of 256 ft². How long is one side of the court?   **16 ft**

Area = 256 ft²

See Example 3   **Evaluate each expression.**

10. $\sqrt{9 + 16}$ **5**   11. $\frac{\sqrt{64}}{4}$ **2**

12. $2\sqrt{100} - 75$ **−55**   13. $-\left(\sqrt{169} - \sqrt{144}\right)$ **−1**

INDEPENDENT PRACTICE

See Example 1   **Find the two square roots of each number.**

14. 121 ± 11 15. 225 ± 15 16. 484 ± 22 17. 169 ± 13

18. 196 ± 14 19. 441 ± 21 20. 64 ± 8 21. 361 ± 19

See Example 2   22. Roger found a square digital relief map on a Web site. The map contained 160,000 pixels. How many pixels high is the map?   **400 pixels**

See Example 3   **Evaluate each expression.**

23. $\sqrt{16} - 7$ **−3**   24. $\sqrt{\frac{64}{4}}$ **4**   25. $-\left(\sqrt{25}\sqrt{16}\right)$ **−20**   26. $10(\sqrt{400} - 15)$ **50**

PRACTICE AND PROBLEM SOLVING

Find the two square roots of each number.

27. 49 ± 7 28. 100 ± 10 29. 289 ± 17 30. 576 ± 24

31. 900 ± 30 32. 36 ± 6 33. 529 ± 23 34. 324 ± 18

You can find the square root of a fraction that does not reduce to a whole number by using the method shown:

$$\sqrt{\frac{9}{4}} = \frac{\sqrt{9}}{\sqrt{4}} = \frac{3}{2}$$

Find the two square roots of each number.

35. $\frac{1}{4}$ $\pm\frac{1}{2}$ 36. $\frac{1}{100}$ $\pm\frac{1}{10}$ 37. $\frac{25}{4}$ $\pm\frac{5}{2}$ 38. $\frac{81}{16}$ $\pm\frac{9}{4}$

39. $\frac{9}{4}$ $\pm\frac{3}{2}$ 40. $\frac{256}{64}$ ± 2 41. $\frac{100}{10,000}$ $\pm\frac{1}{10}$ 42. $\frac{121}{484}$ $\pm\frac{1}{2}$

43. **SPORTS** A karate match is held on a square mat that has an area of 676 ft². What is the length of the mat?   **26 ft**

Notes

Math Background

By agreement among mathematicians, the $\sqrt{}$ symbol means *principal square root*, which is nonnegative. Therefore, $\sqrt{36}$ represents just one number, 6. The expression $\pm\sqrt{36}$ represents both square roots of 36.

An important property of square roots is used in Practice and Problem Solving Exercises 35–42.

Property	Example
$\sqrt{\frac{a}{b}} = \frac{\sqrt{a}}{\sqrt{b}}, b \neq 0$	$\sqrt{\frac{1}{4}} = \frac{\sqrt{1}}{\sqrt{4}} = \frac{1}{2}$

In the property, *a* and *b* represent nonnegative real numbers.

RETEACH 3-8

Reteach
3-8 Squares and Square Roots

A perfect square has two identical factors.
$25 = 5 \times 5 = 5^2$ or $25 = (-5) \times (-5) = (-5)^2$ then 25 is a perfect square.

Tell if the number is a perfect square. If yes, write its identical factors.

1. 121 11^2 or $(-11)^2$ 2. 200 not a perfect square

3. 400 20^2 or $(-20)^2$

Since $5^2 = 25$ and also $(-5)^2 = 25$, both 5 and −5 are **square roots** of 25.
$\sqrt{25} = 5$ and $-\sqrt{25} = -5$
The **principal square root** of 25 is 5: $\sqrt{25} = 5$

Write the two square roots of each number.

4. $\sqrt{81} = $ __9__   5. $\sqrt{625} = $ __25__   6. $\sqrt{169} = $ __13__
$-\sqrt{81} = $ __−9__ $-\sqrt{625} = $ __−25__ $-\sqrt{169} = $ __−13__

Write the principal square root of each number.

7. $\sqrt{144}$ __12__   8. $\sqrt{6400}$ __80__   9. $\sqrt{10,000}$ __100__

Use the principal square root when evaluating an expression. For the order of operations, do square root first, as you would an exponent.

$5\sqrt{100} - 3$
$5(10) - 3$
$50 - 3$
47

Complete to evaluate each expression.

10. $3\sqrt{144} - 20$ 11. $\sqrt{25 + 144} + 13$ 12. $\frac{\sqrt{100}}{20} + \frac{1}{2}$

$3 \times $ __12__ $- 20$ $\sqrt{169} + 13$ $\frac{10}{20} + \frac{1}{2}$

__36__ $- 20$ __13__ $+ 13$ $\frac{1}{2} + \frac{1}{2}$

__16__   __26__   __1__

PRACTICE 3-8

Practice B
3-8 Squares and Square Roots

Find the two square roots for each number.

1. 36   2. 81   3. 49   4. 100
6, −6   9, −9   7, −7   10, −10

5. 64   6. 121   7. 25   8. 144
8, −8   11, −11   5, −5   12, −12

Evaluate each expression.

9. $\sqrt{32 + 17}$ 10. $\sqrt{100 - 19}$ 11. $\sqrt{64 + 36}$ 12. $\sqrt{73 - 48}$
7   9   10   5

13. $2\sqrt{64} + 10$ 14. $36 - \sqrt{36}$ 15. $\sqrt{100} - \sqrt{25}$ 16. $\sqrt{121} + 16$
26   30   5   27

17. $\frac{\sqrt{49}}{4}$ 18. $-3.5\sqrt{16}$ 19. $\frac{\sqrt{81}}{\sqrt{9}}$ 20. $\frac{\sqrt{144}}{-6}$
$1\frac{3}{4}$ −14   3   −2

The Pyramids of Egypt are often called the first wonder of the world. This group of pyramids consists of Menkaura, Khufu, and Khafra. The largest of these is Khufu, sometimes called Cheops. During this time in history, each monarch had his own pyramid built to bury his mummified body. Cheops was a king of Egypt in the early 26th century B.C. His pyramid's original height is estimated to have been 482 ft. It is now approximately 450 ft. The estimated completion date of this structure was 2660 B.C.

21. If the area of the base of Cheops' pyramid is 570,025 ft², what is the length of one of the sides of the ancient structure?
(Hint: $s = \sqrt{A}$)
755 ft

22. If a replica of the pyramid were built with a base area of 625 in², what would be the length of each side?
(Hint: $s = \sqrt{A}$)
25 in.

44. LANGUAGE ARTS *Crelle's Journal* is the oldest mathematics periodical in existence. Zacharias Dase's incredible calculating skills were made famous by *Crelle's Journal* in 1844. Dase produced a table of factors of all numbers between 7,000,000 and 10,000,000. He listed 7,022,500 as a perfect square. What is the square root of 7,022,500? **2650**

45. SOCIAL STUDIES Zerah Colburn was born in Vermont in 1804. At the age of 8, he could calculate the square root of 106,929 mentally. What is the square root of 106,929? **327**

46. RECREATION A chessboard contains 32 black and 32 white squares. How many squares are along each side of the game board? **8**

47. INDUSTRIAL ARTS A carpenter wants to use as many of his 82 small wood squares as possible to make a large square inlaid box lid.

a. How many squares can the carpenter use? How many squares would he have left? **81; 1**

b. How many more small wood squares would the carpenter need to make the next larger possible square box lid? **18**

 48. WHAT'S THE ERROR? A student said that since the square roots of a certain number are 2.5 and −2.5, the number must be their product, −6.25. What error did the student make?

 49. WRITE ABOUT IT Explain how you know whether $\sqrt{29}$ is closer to 5 or 6 without using a calculator.

 50. CHALLENGE The square root of a number is five less than six times four. What is the number? **361**

In 1997, Deep Blue became the first computer to win a match against an international chess grand master when it defeated world champion Garry Kasparov.

go.hrw.com
KEYWORD:
MP4 Chess

Spiral Review

Solve. (Lesson 1-3)

51. $9 + t = 18$ $t = 9$ **52.** $t - 2 = 6$ $t = 8$ **53.** $10 + t = 32$ $t = 22$ **54.** $t + 7 = 7$ $t = 0$

Evaluate. (Lesson 2-8)

55. $(-3)^{-2}$ $\frac{1}{9}$ **56.** $(-2)^{-3}$ $-\frac{1}{8}$ **57.** $(1)^{-3}$ **1**   **58.** $\frac{6^8}{6^8}$ **1**

59. TEST PREP If a number is divisible by 15, then it is also divisible by   ? . (Previous course) **D**

   A 10   **B** 30   **C** 5 and 10   **D** 3 and 5

60. TEST PREP If a number is divisible by 5 and by 8, it is also divisible by   ? . (Previous course) **F**

   F 40   **G** 3   **H** 13   **J** 24

Answers

48. Possible answer: The student multiplied the two square roots together instead of squaring one of them. The number should be 2.5^2 or $(-2.5)^2$, which equal 6.25.

49. Possible answer: $\sqrt{29}$ is closer to 5 than 6 because 29 is closer to 25 than it is to 36.

Journal

Ask students to list as many perfect squares as they can remember from the lesson.

Test Prep Doctor

For Exercise 59, if a number is divisible by 15, it must also be divisible by numbers that divide evenly into 15. Only answer choice **D** gives two numbers that divide evenly into 15.

CHALLENGE 3-8

LESSON 3-8 **Challenge**

Dig It!

Find the **digital root** of a number by adding its digits, adding the digits of the result, and so on, until the result is a single digit.

$358 \rightarrow 3 + 5 + 8 = 16 \rightarrow 1 + 6 = 7$ The digital root of 358 is 7.

1. Complete the table to find the digital roots of the squares of 1–17.

		Digital Root		
Number	Square	Calculation		
1	1		=	1
2	4		=	4
3	9		=	9
4	16	1 + 6	=	7
5	25	2 + 5	=	7
6	36	3 + 6 = 9	=	9
7	49	4 + 9 = 13 → 1 + 3	=	4
8	64	6 + 4 = 10 → 1 + 0	=	1
9	81	8 + 1	=	9
10	100	1 + 0 + 0	=	1
11	121	1 + 2 + 1	=	4
12	144	1 + 4 + 4	=	9
13	169	1 + 6 + 9 = 16 → 1 + 6	=	7
14	196	1 + 9 + 6 = 16 → 1 + 6	=	7
15	225	2 + 2 + 5	=	9
16	256	2 + 5 + 6 = 13 → 1 + 3	=	4
17	289	2 + 8 + 9 = 19 → 1 + 9 = 10 → 1 + 0	=	1

2. Make an observation about the results. **Possible answers:**

The only results are 1, 4, 7, or 9.

3. Make a conjecture about the digital root of any whole-number perfect square. Verify your conjecture by using at least three more perfect squares.

The result is one of the numbers 1, 4, 7, or 9. Choices vary.

4. A **palindrome** is a number that is the same when read forward or backward, such as 14741. Find two palindromes in the table.

The digital roots of the squares of the numbers 1–8 and then 1–17.

PROBLEM SOLVING 3-8

LESSON 3-8 **Problem Solving**

Squares and Square Roots

Write the correct answer.

1. For college wrestling competitions, the NCAA requires that the wrestling mat be a square with an area of 1764 square feet. What is the length of each side of the wrestling mat?

42 feet

2. For high school wrestling competitions, the wrestling mat must be a square with an area of 1444 square feet. What is the length of each side of the wrestling mat?

38 feet

3. The Japanese art of origami requires folding square pieces of paper. Elena begins with a large sheet of square paper that is 169 square inches. How many squares can she cut out of the paper that are 4 inches on each side?

9 squares

4. When the James family moved into a new house they had a square area rug that was 132 square feet. In their new house, there are three bedrooms. Bedroom one is 11 feet by 11 feet. Bedroom two is 10 feet by 12 feet, and bedroom three is 13 feet by 13 feet. In which bedroom will the rug fit?

Bedroom three

Choose the letter for the best answer.

5. A square picture frame measures 36 inches on each side. The actual wood trim is 2 inches wide. The photograph in the frame is surrounded by a bronze mat that measures 5 inches. What is the maximum area of the photograph?

  A 841 sq. inches   B 900 sq. inches
  C 1156 sq. inches   **D** 484 sq. inches

6. To create a square patchwork quilt wall hanging, square pieces of material are sewn together to form a larger square. Which number of smaller squares can be used to create a square patchwork quilt wall hanging?

  F 35 squares   **G** 64 squares
  H 84 squares   J 125 squares

7. A can of paint claims that one can will cover 400 square feet. If you painted a square with the can of paint, how long would it be on each side?

  A 200 feet   B 65 feet
  C 25 feet   **D** 20 feet

8. A box of tile contains 12 tiles. If you tile a square area using whole tiles, how many tiles will you have left from the box?

  F 9   G 6
  H 3   J 0

Lesson Quiz

Find the two square roots of each number.

1. 81 ±9

2. 2500 ±50

Evaluate each expression.

3. $3\sqrt{16} + 1$ **13**

4. $7\sqrt{9} - 2\sqrt{49}$ **7**

5. Ms. Estefan wants to put a fence around 3 sides of a square garden that has an area of 225 ft². How much fencing does she need? **45 ft**

Available on Daily Transparency in CRB

Objective: Students estimate square roots to a given number of decimal places and solve problems using square roots.

Warm Up

Find the two square roots of each number.

1. 144   ±12   **2.** 256   ±16

Evaluate each expression.

3. $8 + \sqrt{144}$ 20   **4.** $7\sqrt{289}$ 119

Problem of the Day

A pyramid of blocks is built in layers. The bottom layer has 6^2, or 36, blocks. The next layer has 5^2 blocks, and so on until the top layer has 1 block. How many blocks are there in all? 91 blocks

Available on Daily Transparency in CRB

Math Fact

The sum of any number of consecutive odd whole numbers, beginning with 1, is a perfect square (e.g., $1 + 3 = 4$, $1 + 3 + 5 = 9$, $1 + 3 + 5 + 7 = 16$, etc.).

3-9 Finding Square Roots

Learn to estimate square roots to a given number of decimal places and solve problems using square roots.

A museum director wants to install a skylight to illuminate an unusual piece of art. It must be square and have an area of 300 square inches, with wood trim around it. Can you calculate the length of trim that you need? You can do this using your knowledge of squares and square roots.

EXAMPLE 1   Estimating Square Roots of Numbers

Each square root is between two integers. Name the integers.

A $\sqrt{30}$

Think: What are perfect squares close to 30?

$5^2 = 25$ $25 < 30$
$6^2 = 36$ $36 > 30$ $5 < \sqrt{30} < 6$

$\sqrt{30}$ is between 5 and 6.

B $-\sqrt{150}$

Think: What are perfect squares close to 150?

$(-12)^2 = 144$ $144 < 150$
$(-13)^2 = 169$ $169 > 150$

$-\sqrt{150}$ is between -12 and -13. $-13 < -\sqrt{150} < -12$

EXAMPLE 2   PROBLEM SOLVING APPLICATION

You want to install a square skylight that has an area of 300 square inches. Calculate the length of each side and the length of trim you will need, to the nearest tenth of an inch.

1 **Understand the Problem**

First find the length of a side. Then you can use the length of a side to find the *perimeter*, the length of the trim around the skylight.

2 **Make a Plan**

The length of a side, in inches, is the number that you multiply by itself to get 300. To be accurate, find this number to the nearest tenth.

If you do not know a step-by-step method for finding $\sqrt{300}$, use guess and check.

---

1 Introduce

Alternate Opener

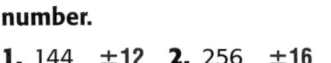

EXPLORATION

3-9 **Finding Square Roots**

Knowing the square numbers can help you estimate square roots.

1. Complete the table of squares.

1^2	2^2	3^2	4^2	5^2	6^2	7^2	8^2	9^2	10^2
1									

11^2	12^2	13^2	14^2	15^2	16^2	17^2	18^2	19^2	20^2
121									

Use the table of squares above to help you estimate each square root to the nearest tenth. Use a calculator to check your estimates. Round to two decimal places.

	Square Root	Estimate	Calculator
2.	$\sqrt{10}$		
3.	$\sqrt{20}$		
4.	$\sqrt{200}$		
5.	$\sqrt{300}$		
6.	$\sqrt{57}$		
7.	$\sqrt{130}$		

Think and Discuss

8. Discuss your strategy for estimating square roots.

9. Explain how you could estimate $\sqrt{1000}$.

Motivate

Ask students to find $\sqrt{4}$ and $\sqrt{9}$. Show the number line diagram:

$\sqrt{4}$ $\sqrt{5}$ $\sqrt{8}$ $\sqrt{9}$
────┼───┼──────────┼───┼────
   2 2.24   2.83   3

Ask students where they think $\sqrt{8}$ should be placed on the diagram. If they say "between 2 and 3," ask whether it should be closer to 2 or to 3. Then use a calculator to find the approximation $\sqrt{8} \approx 2.83$. Then try the same process with $\sqrt{5}$ (≈ 2.24).

Exploration worksheet and answers on Chapter 3 Resource Book pp. 82 and 132

2 Teach

Lesson Presentation

Guided Instruction

In this lesson, students learn to estimate square roots to a given number of decimal places and solve problems using square roots. Tell students that all positive numbers have square roots, but most of those square roots are not integers. If students have calculators, you may want to ask them to enter $\sqrt{7}$ to see an example. Discuss with students how to identify the two integers that a given square root is between. Show students how to get an approximation of a square root to the nearest tenth by repeated use of the guess-and-check method.

3 Solve

Because 300 is between 17^2 (289) and 18^2 (324), the square root of 300 is between 17 and 18.

Guess 17.5	Guess 17.2	Guess 17.4	Guess 17.3
$17.5^2 = 306.25$	$17.2^2 = 295.84$	$17.4^2 = 302.76$	$17.3^2 = 299.29$
Too high	Too low	Too high	Too low
Square root is between 17 and 17.5.	Square root is between 17.2 and 17.5.	Square root is between 17.2 and 17.4.	Square root is between 17.3 and 17.4.

The square root is between 17.3 and 17.4. To round to the nearest tenth, look at the next decimal place. Consider **17.35**.

$17.35^2 = 301.0225$ *Too high*

The square root must be *less than* 17.35, so you can round *down*. To the nearest tenth, $\sqrt{300}$ is about 17.3.

The length of a side of the skylight is **17.3** inches, to the nearest tenth of an inch. Now estimate the length around the skylight.

$4 \cdot 17.3 = 69.2$ *Perimeter = 4 · side*

The trim is about 69.2 inches long.

4 Look Back

The length 70 inches divided by 4 is 17.5 inches. A 17.5-inch square has an area of 306.25 square inches, which is close to 300, so the answers are reasonable.

EXAMPLE 3 **Using a Calculator to Estimate the Value of a Square Root**

Use a calculator to find $\sqrt{300}$. Round to the nearest tenth.
Using a calculator, $\sqrt{300} \approx 17.32050808...$. Rounded, $\sqrt{300}$ is 17.3.

Think and Discuss

1. Discuss whether 9.5 is a good first guess for $\sqrt{75}$.

2. Determine which square root or roots would have 7.5 as a good first guess.

3 Close

Reaching All Learners
Through Number Sense

Place papers showing the integers 1 to 10 (one per sheet) around the room in order. Give each student a card showing the square root of a positive number between 1 and 100, such as $\sqrt{28}$ (Chapter 3 Resource Book p. 113). Have each student place his or her card between the appropriate pair of integers. For example, $\sqrt{28}$ should be placed between 5 and 6. Continue until all the cards are placed. Then have the class determine if all the placements are correct.

Summarize

Remind students that sometimes it is sufficient to approximate a square root by naming the two integers it is between. At other times, they may need to approximate a square root to a given number of decimal places. Emphasize the value of knowing at least the first ten perfect squares: 1, 4, 9, 16, 25, 36, 49, 64, 81, and 100.

Before students begin the Exercises, you may want to clarify the procedure for solving multistep problems that involve rounding. Answers are given in the most accurate form with rounding as the final step.

Answers to Think and Discuss

Possible answers:

1. No; $9^2 = 81$, so $\sqrt{75}$ must be less than 9; 8.5 is a better first guess.

2. Two good choices would be $\sqrt{56}$ and $\sqrt{57}$. $7^2 = 49$ and $8^2 = 64$, so a number that has a square root close to 7.5 is about halfway between 49 and 64. In fact, $7.5^2 = 56.25$.

FOR EXTRA PRACTICE
see page 737

internet connect
Homework Help Online
go.hrw.com   Keyword: MP4 3-9

Students may want to refer back to the lesson examples.

Assignment Guide

If you finished Example **1** assign:
Core 1–4, 10–13, 19–23 odd, 43–55
Enriched 10–13, 19–24, 43–55

If you finished Example **2** assign:
Core 1–5, 10–14, 19–27 odd, 43–55
Enriched 10–14, 19–28, 43–55

If you finished Example **3** assign:
Core 1–18, 19–39 odd, 43–55
Enriched 10, 14, 15, 19–55

Answers

35. ±5.20

36. ±7.35

37. ±317.02

38. ±60.10

GUIDED PRACTICE

See Example **1** Each square root is between two integers. Name the integers.

1. $\sqrt{40}$ **2.** $-\sqrt{72}$ **3.** $\sqrt{200}$ **4.** $-\sqrt{340}$
6 and 7   **−8 and −9**   **14 and 15**   **−18 and −19**

See Example **2** **5.** A square table has a top that has an area of 11 square feet. To the nearest hundredth, what length of edging is needed to go around all edges of the tabletop?   **≈13.27 ft**

See Example **3** Use a calculator to find each value. Round to the nearest tenth.

6. $\sqrt{83}$ **9.1**   **7.** $\sqrt{42.3}$ **6.5**   **8.** $\sqrt{2500}$ **50**   **9.** $\sqrt{190}$ **13.8**

INDEPENDENT PRACTICE

See Example **1** Each square root is between two integers. Name the integers.

10. $-\sqrt{50}$ **11.** $\sqrt{3}$ **12.** $\sqrt{610}$ **13.** $-\sqrt{1000}$
−7 and −8   **1 and 2**   **24 and 25**   **−31 and −32**

See Example **2** **14.** Each square on Laura's chessboard is 13 square centimeters. A chessboard has 8 squares on each side. To the nearest hundredth, what is the width of Laura's chessboard?   **≈28.84 cm**

See Example **3** Use a calculator to find each value. Round to the nearest tenth.

15. $\sqrt{69}$ **8.3**   **16.** $\sqrt{91.5}$ **9.6**   **17.** $\sqrt{650}$ **25.5**   **18.** $\sqrt{200}$ **14.1**

PRACTICE AND PROBLEM SOLVING

Write the letter that identifies the position of each square root.

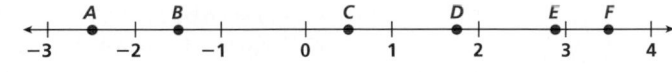

19. $-\sqrt{2}$ **B**   **20.** $\sqrt{3}$ **D**   **21.** $\sqrt{8}$ **E**

22. $-\sqrt{6}$ **A**   **23.** $\sqrt{12}$ **F**   **24.** $\sqrt{0.25}$ **C**

Use guess and check to estimate each square root to two decimal places.

25. $\sqrt{51}$ **7.14**   **26.** $-\sqrt{80}$ **−8.94**   **27.** $\sqrt{135}$ **11.62**   **28.** $\sqrt{930}$ **30.50**

Find each product to two decimal places.

29. $\sqrt{51} \cdot \sqrt{36}$ **42.85**   **30.** $-\sqrt{80} \cdot \sqrt{25}$ **−44.72**   **31.** $\sqrt{135} \cdot (-\sqrt{1})$ **−11.62**

32. $-\sqrt{164} \cdot \sqrt{4}$ **−25.61**   **33.** $\sqrt{22} \cdot (-\sqrt{49})$ **−32.83**   **34.** $\sqrt{260} \cdot \sqrt{144}$ **193.49**

Find each number to two decimal places.

35. What number squared is 27?   **36.** What number squared is 54?

37. What number squared is 100,500?   **38.** What number squared is 3612?

Math Background

In this lesson, students use guess-and-check to estimate square roots. A more reliable method for estimating square roots involves repeated division. For example, to find $\sqrt{28}$, choose the integer whose perfect square is closest to 28 ($5^2 = 25$).

Divide 28 by that integer: $5\overline{)28.0}$ = 5.6

Find the average of the quotient and the divisor: $(5 + 5.6) \div 2 = 5.3$.

Check the result: $5.3^2 = 28.09$.

You can then repeat the process with your new estimate and continue repeating the algorithm until your estimate has the desired accuracy.

RETEACH 3-9

LESSON Reteach
3-9 *Finding Square Roots*

To locate a square root between two integers, refer to the table.

Number	1	2	3	4	5	6	7	8	9	10
Square	1	4	9	16	25	36	49	64	81	100
Number	11	12	13	14	15	16	17	18	19	20
Square	121	144	169	196	225	256	289	324	361	400

Locate $\sqrt{260}$ between two integers.
260 is between the perfect squares 256 and 289:
256 < 260 < 289
So: $\sqrt{256} < \sqrt{260} < \sqrt{289}$
And: $16 < \sqrt{260} < 17$

Use the table to complete the statements.

1. __36__ < 39 < __49__   2. __121__ < 130 < __144__
   $\sqrt{36} < \sqrt{39} < \sqrt{49}$ $\sqrt{121} < \sqrt{130} < \sqrt{144}$
   __6__ < $\sqrt{39}$ < __7__   __11__ < $\sqrt{130}$ < __12__

After locating a square root between two integers, you can determine which of the two integers the square root is closer to.
27 is between the perfect squares 25 and 36:   25 < 27 < 36
So: $\sqrt{25} < \sqrt{27} < \sqrt{36}$
And: 5 < $\sqrt{27}$ < 6

The difference between 27 and 25 is 2;   25 < 27 < 36
the difference between 36 and 27 is 9.
So, $\sqrt{27}$, is closer to 5.   2   9

Complete the statements.

4. 100 < 106 < 121   5. __225__ < 250 < __256__
   $\sqrt{100} < \sqrt{106} < \sqrt{121}$ $\sqrt{225} < \sqrt{250} < \sqrt{256}$
   __10__ < $\sqrt{106}$ < __11__   __15__ < $\sqrt{250}$ < __16__
   106 − 100 = __6__   250 − __225__ = __25__
   121 − 106 = __15__   __256__ − 250 = __6__
   $\sqrt{106}$ is closer to __10__ than __11__ $\sqrt{250}$ is closer to __16__ than __15__

PRACTICE 3-9

LESSON Practice B
3-9 *Finding Square Roots*

Each square root is between two integers. Name the integers.

1. $\sqrt{6}$ 2. $\sqrt{20}$ 3. $\sqrt{28}$ 4. $\sqrt{44}$
   __2 < $\sqrt{6}$ < 3__ __4 < $\sqrt{20}$ < 5__ __5 < $\sqrt{28}$ < 6__ __6 < $\sqrt{44}$ < 7__

5. $\sqrt{34}$ 6. $\sqrt{17}$ 7. $\sqrt{31}$ 8. $\sqrt{52}$
   __5 < $\sqrt{34}$ < 6__ __4 < $\sqrt{17}$ < 5__ __5 < $\sqrt{31}$ < 6__ __7 < $\sqrt{52}$ < 8__

Use a calculator to find each value. Round to the nearest tenth.

9. $\sqrt{14}$ 10. $\sqrt{42}$ 11. $\sqrt{21}$ 12. $\sqrt{47}$
   __3.7__   __6.5__   __4.6__   __6.9__

13. $\sqrt{58}$ 14. $\sqrt{60}$ 15. $\sqrt{35}$ 16. $\sqrt{75}$
   __7.6__   __7.7__   __5.9__   __8.7__

17. $\sqrt{66}$ 18. $\sqrt{55}$ 19. $\sqrt{67}$ 20. $\sqrt{80}$
   __8.1__   __7.4__   __8.2__   __8.9__

Police use the formula $r = 2\sqrt{5L}$ to approximate the rate of speed in miles per hours of a vehicle from its skid marks, where L is the length of the skid marks in feet.

21. About how fast is a car going that leaves skid marks of 80 ft?
   __40 mi/h__

22. About how fast is a car going that leaves skid marks of 245 ft?
   __70 mi/h__

23. If the formula for finding the length of the skid marks is $L = \frac{r^2}{20}$, what would be the length of the skid marks from a vehicle traveling 80 mi/h?
   __320 ft__

Tsunamis, sometimes called tidal waves, move across deep oceans at high speeds with barely a ripple on the water surface. It is only when tsunamis hit shallow water that their energy moves them upward into a mammoth destructive force.

Tsunamis can be caused by earthquakes, volcanoes, landslides, or meteorites.

39. The rate of speed of a tsunami, in feet per second, can be found by the formula $r = \sqrt{32d}$, where d is the water depth in feet. Suppose the water depth is 20,000 ft. How fast is the tsunami moving?

40. The speed of a tsunami in miles per hour can be found using $r = \sqrt{14.88d}$, where d is the water depth in feet. Suppose the water depth is 25,000 ft.

  a. How fast is the tsunami moving in miles per hour?

  b. How long would it take a tsunami to travel 3000 miles if the water depth were a consistent 10,000 ft?

41.  **WHAT'S THE ERROR?** Ashley found the speed of a tsunami, in feet per second, by taking the square root of 32 and multiplying by the depth, in feet. What was her error?

42. **CHALLENGE** Find the depth of the water if a tsunami's speed is 400 miles per hour. **approximately 10,753 ft**

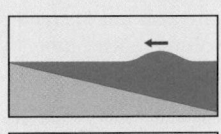
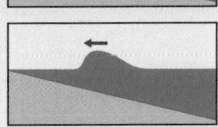
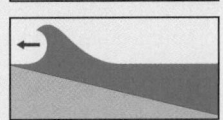

As the wave approaches the beach, it slows, builds in height, and crashes on shore.

go.hrw.com
KEYWORD: MP4 Wave
CNN Student News.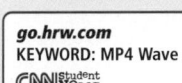

Spiral Review

Solve. (Lesson 3-6)

43. $y - 27.6 = -32$
$y = -4.4$

44. $-5.3f = 74.2$
$f = -14$

45. $\frac{m}{3.2} = -8$
$m = -25.6$

46. $x + \frac{1}{8} = -\frac{5}{8}$
$x = -\frac{3}{4}$

Evaluate. (Lesson 3-7)

47. $x + \frac{1}{3} < 6$ $x < 5\frac{2}{3}$

48. $-7f < -\frac{4}{5}$ $f > \frac{4}{35}$

49. $3\frac{1}{4}m \geq 26$ $m \geq 8$

50. $0.7x \geq -1.4$
$x \geq -2$

Find the square roots of each number. (Lesson 3-8)

51. 16 **4 and −4**

52. 81 **9 and −9**

53. 100 **10 and −10**

54. 1 **1 and −1**

55. TEST PREP It took Tina 6 minutes to saw a board into 3 equal pieces. How long would it have taken her to saw it into 9 equal pieces? (*Hint:* Think about the number of cuts she must make.) (Lesson 1-4) **D**

  A 2 min   **B** 18 min   **C** 21 min   **D** 24 min

Interdisciplinary

Science

Exercises 39–42 involve using formulas with square roots to determine the speeds of tsunamis. Tsunamis are studied in middle school science programs such as *Holt Science & Technology*.

Answers

39. 800 ft/s

40. a. approximately 610 mi/h

  b. approximately 7.8 hr

41. Possible answer: The student should have multiplied before taking the square root.

Journal

Ask students to describe how they would try to estimate the square root of 1000.

Test Prep Doctor

For Exercise 55, remind students that to cut the board into 3 equal pieces, only 2 cuts were made. If 2 cuts took 6 minutes, then each cut took 3 minutes. To cut the board into 9 pieces, she would have needed to make 8 cuts. Eight cuts at 3 minutes each would have taken 24 minutes, choice **D**.

CHALLENGE 3-9

LESSON 3-9 Challenge *Dig Deeper!*

The **digital root** of a number is found by adding its digits, adding the digits of the result, and so on, until the result is a single digit. $918 \rightarrow 9 + 1 + 8 = 18 \rightarrow 1 + 8 = 9$ The digital root of 918 is 9.

1. Complete the table to display numbers and their digital roots and to determine if they are divisible by 3 (remainder = 0). Make an observation about the results.

A number is divisible by 3 if its digital root is divisible by 3.

Number	Divisible by 3?	Digital Root Calculation		Divisible by 3?
81	yes	8 + 1	= 9	yes
92	no	9 + 2 = 11 → 1 + 1	= 2	no
226	no	2 + 2 + 6 = 10 → 1 + 0	= 1	no
315	yes	3 + 1 + 5	= 9	yes
659	no	6 + 5 + 9 = 20 → 2 + 0	= 2	no
704	no	7 + 0 + 4 = 11 → 1 + 1	= 2	no
1064	no	1 + 0 + 6 + 4 = 11 → 1 + 1	= 2	no

2. Complete the table to display the products of numbers and the products of their digital roots. Make an observation about the results.

The digital root of a product of whole numbers equals the product of their digital roots.

Product	Digital Root of Factor	Digital Root of Factor	Product of Digital Roots of Factors	Digital Root of Product
24 × 32 = 768	2 + 4 = 6	3 + 2 = 5	6 × 5 = 30 → 3 + 0 = 3	7 + 6 + 8 = 21 → 2 + 1 = 3
11 × 17 = 187	1 + 1 = 2	1 + 7 = 8	2 × 8 = 16 → 1 + 6 = 7	1 + 8 + 7 = 16 → 1 + 6 = 7
121 × 42 = 5082	1 + 2 + 1 = 4	4 + 2 = 6	4 × 6 = 24 → 2 + 4 = 6	5 + 0 + 8 + 2 = 15 → 1 + 5 = 6
243 × 35 = 8505	2 + 4 + 3 = 9	3 + 5 = 8	9 × 8 = 72 → 7 + 2 = 9	8 + 5 + 0 + 5 = 18 → 1 + 8 = 9
81 × 72 = 5832	8 + 1 = 9	7 + 2 = 9	9 × 9 = 81 → 8 + 1 = 9	5 + 8 + 3 + 2 = 18 → 1 + 8 = 9
360 × 54 = 19,440	3 + 6 + 0 = 9	5 + 4 = 9	9 × 9 = 81 → 8 + 1 = 9	1 + 9 + 4 + 4 + 0 = 18 → 1 + 8 = 9

PROBLEM SOLVING 3-9

LESSON 3-9 Problem Solving *Finding Square Roots*

The distance to the horizon can be found using the formula $d = 112.88\sqrt{h}$ where d is the distance in kilometers and h is the number of kilometers from the ground. Round your answer to the nearest kilometer.

1. How far is it to the horizon when you are standing on the top of Mt. Everest, a height of 8.85 km?

336 km

2. Find the distance to the horizon from the top of Mt. McKinley, Alaska, a height of 6.194 km.

281 km

3. How far is it to the horizon if you are standing on the ground and your eyes are 2 m above the ground?

5 km

4. Mauna Kea is an extinct volcano on Hawaii that is about 4 km tall. You should be able to see the top of Mauna Kea when you are how far away?

at most 226 km

You can find the approximate speed of a vehicle that leaves skid marks before it stops. The formulas $S = 5.5\sqrt{0.7L}$ and $S = 5.5\sqrt{0.8L}$, where S is the speed in miles per hour and L is the length of the skid marks in feet, will give the minimum and maximum speeds that the vehicle was traveling before the brakes were applied. Round to the nearest mile per hour.

5. A vehicle leaves a skid mark of 40 feet before stopping. What was the approximate speed of the vehicle before it stopped?

  A 25–35 mph   **C** 29–31 mph
  B 28–32 mph   **D** 68–70 mph

6. A vehicle leaves a skid mark of 100 feet before stopping. What was the approximate speed of the vehicle before it stopped?

  F 46–49 mph   **H** 62–64 mph
  G 50–55 mph   **J** 70–73 mph

7. A vehicle leaves a skid mark of 150 feet before stopping. What was the approximate speed of the vehicle before it stopped?

  A 50–55 mph   **C** 55–70 mph
  B 53–58 mph   **D** 56–60 mph

8. A vehicle leaves a skid mark of 200 feet before stopping. What was the approximate speed of the vehicle before it stopped?

  F 60–63 mph   **G** 65–70 mph
  H 72–78 mph   **J** 80–90 mph

Lesson Quiz

Each square root is between two integers. Name the two integers.

1. $\sqrt{27}$ **5 and 6**

2. $-\sqrt{456}$ **−22 and −21**

Use a calculator to find each value. Round to the nearest tenth.

3. $\sqrt{89}$ **9.4**

4. $\sqrt{1223}$ **35.0**

5. A square field has an area of 2000 square feet. To the nearest foot, how much fencing would be needed to enclose the field?

179 ft

Available on Daily Transparency in CRB

Hands-On LAB

**3B
Explore
Cubes and
Cube Roots**

Pacing: Traditional 1 day
Block $\frac{1}{2}$ day

Objective: To use base-ten blocks
to explore cubes and
cube roots

Materials: Base-ten blocks

Lab Resources

Hands-On Lab Activities .... pp. 21–24

Using the Pages

Discuss with students what is represented by each base-ten block.

1. What is the volume of 1 block?
1 cubic unit

2. What is the volume of the solid?

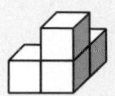 4 cubic units

3. How many blocks are needed to model a solid with volume 9 cubic units? **9**

Hands-On LAB 3B

Explore Cubes and Cube Roots

Use with Lesson 3-9

WHAT YOU NEED:
Smallest base-10 blocks
(Rainbow cubes or centimeter
cubes will also work.)

REMEMBER
• All edges of a cube are the same length.
• Volume is the number of cubic units needed to fill the space of a solid.

internet connect
Lab Resources Online
go.hrw.com
KEYWORD: MP4 Lab3B

The number of small unit blocks it takes to construct a cube is equal to the volume of the cube. By building a cube with edge length x and counting the number of unit blocks needed to build the cube, you can find x^3 (x-cubed), the volume.

Activity 1

1 Find 2^3.

You need to build a cube with an edge length of 2.

Build 3 edges of length 2.

Fill in the rest of the cube.

Count the number of unit cubes you needed to build a cube with an edge length of 2.

To make a cube with edge length 2, you need 8 unit blocks. So $2^3 = 8$.

Think and Discuss Possible answers:

1. Why would it be difficult to model 2^4? **Space is only 3-dimensional. We can build in only 3 directions: left/right, up/down, back/front. It would take 4 dimensions to model 2^4.**

2. How can you find the value of a number squared from the model of that number cubed?
The number of blocks that make up the base of the cube with side length x is equal to x^2.

Try This

Model the following. How many blocks do you need to model each?

1. 1^3 **1**

2. 3^3 **27**

3. 4^3 **64**

Teacher to Teacher

As an introduction to cubes and cube roots, I divided the class into pairs and gave each pair of students a random number of centimeter cubes. I asked my students if they could create a larger cube using all of their centimeter cubes. After a few minutes, we created a class table with two columns, one for the number of cubes and one indicating whether or not that number of cubes could create a larger cube with no small cubes left over. When a larger cube could be made, the class counted to see how many centimeter cubes made up the length, width, and height of the larger cube. The students immediately noticed that all of the measurements were the same.

*Stacie Tarbet
Prince William County, Virginia*

You can determine whether any number x is a perfect cube by trying to build a cube out of x unit blocks. If you can build a cube with the given number of blocks, then that number is a perfect cube. Its *cube root* will be the length of one edge of the cube that is formed.

Activity 2

❶ Try to build a cube using 27 unit blocks. Is 27 a perfect cube? If so what is its cube root?

Start by building a cube with an edge length of 2, since $1^3 = 1$ and $27 > 1$.

You still have 19 unit blocks left over. So try building a cube with an edge length of 3. Remember that when you add 1 unit cube to any edge you must do the same to all three edges to keep the cube shape.

A cube with edges of length 3 can be made with 27 blocks.
length = 3
width = 3
height = 3

You can make a cube with edges of length 3 by using 27 small blocks. So 27 is a perfect cube. Its cube root is 3. We write $\sqrt[3]{27} = 3$.

Think and Discuss

1. Is 100 a perfect cube? Why or why not?

2. How would you estimate the cube root of 100?

3. $\sqrt[3]{125} = 5$. Does $\sqrt[3]{2(125)} = \sqrt[3]{250} = 2(\sqrt[3]{125}) = 10$? Why or why not?

4. Use blocks to model a solid with a length of 3, a height of 2, and a width of 2. How many blocks did you use? Is this a perfect cube?

Try This

Model to find whether each number is a perfect cube. If the number is a perfect cube, find its cube root. If not, find the whole numbers that the cube root is between.

1. 64   **4**   2. 75   **between 4 and 5**   3. 125   **5**   4. 200   **between 5 and 6**

Answers

Activity 2

Think and Discuss

Possible answers:

1. No; you cannot model a cube with exactly 100 blocks.

2. It is between 4 and 5 because 4^3 is 64 and 5^3 is 125. It is closer to 5 than to 4.

3. No; 10^3 is $10 \times 10 \times 10 = 1000$. Also, $6^3 = 216$ and $7^3 = 343$, so $\sqrt[3]{250}$ is between 6 and 7.

4. 12; no

3-10 PRACTICE & ASSESS

3-10 Exercises

Students may want to refer back to the lesson examples.

internet connect
Homework Help Online
go.hrw.com   Keyword: MP4 3-10

GUIDED PRACTICE

See Example **1** Write all names that apply to each number.

1. $\sqrt{12}$ 2. $\sqrt{49}$ 3. 0.15 4. $-\dfrac{\sqrt{25}}{2}$

See Example **2** State if the number is rational, irrational, or not a real number.

5. $\sqrt{4}$ **rational** 6. $\sqrt{\dfrac{4}{25}}$ **rational** 7. $\sqrt{72}$ **irrational** 8. $-\sqrt{-2}$ **not real**

9. $-\sqrt{36}$ **rational** 10. $\sqrt{-4}$ **not real** 11. $\sqrt{\dfrac{16}{-25}}$ **not real** 12. $\dfrac{0}{0}$ **not real**

See Example **3** Find a real number between each pair of numbers.

13. $5\frac{1}{6}$ and $5\frac{2}{6}$ $5\frac{1}{4}$ 14. 3.14 and $\frac{22}{7}$ $\frac{2199}{700}$ 15. $\frac{1}{8}$ and $\frac{1}{4}$ $\frac{3}{16}$

INDEPENDENT PRACTICE

See Example **1** Write all names that apply to each number.

16. $\sqrt{35}$ **irrational, real**   17. $\dfrac{7}{9}$ **rational, real**   18. 2 **whole, integer, rational, real**   19. $\dfrac{\sqrt{100}}{-5}$ **integer, rational, real**

See Example **2** State if the number is rational, irrational, or not a real number.

20. $\dfrac{-\sqrt{25}}{-5}$ 21. $-\sqrt{\dfrac{0}{9}}$ 22. $\sqrt{-12(-3)}$ 23. $-\sqrt{3}$

24. $\dfrac{\sqrt{16}}{5}$ 25. $\sqrt{18}$ 26. $\sqrt{-\dfrac{1}{4}}$ 27. $-\sqrt{\dfrac{9}{0}}$

See Example **3** Find a real number between each pair of numbers.

28. $3\frac{2}{5}$ and $3\frac{3}{5}$ $3\frac{1}{2}$ 29. $-\frac{1}{100}$ and 0 $-\frac{1}{200}$ 30. 3 and $\sqrt{4}$ 2.5

PRACTICE AND PROBLEM SOLVING

Write all names that apply to each number.

31. 8 32. $-\sqrt{36}$ 33. $\sqrt{20}$ 34. $\dfrac{2}{3}$

35. $\sqrt{3.24}$ 36. $\sqrt{25}+5$ 37. $0.\overline{15}$ 38. $\dfrac{\sqrt{100}}{20}$

39. -6.5356 **rational, real**   40. $\sqrt{4.5}$ **irrational, real**   41. -122 **integer, rational, real**   42. $\dfrac{0}{5}$ **whole, integer, rational, real**

Give an example of each type of number.

43. an irrational number that is less than -5 **Possible answer:** $-\sqrt{50}$

44. a rational number that is less than 0.5 **Possible answer:** 0.4

45. a real number between $\frac{5}{9}$ and $\frac{6}{9}$ **Possible answer:** $\frac{11}{18}$

46. a real number between $-5\frac{4}{7}$ and $-5\frac{5}{7}$ **Possible answer:** $-5\frac{9}{14}$

Assignment Guide

If you finished Example **1** assign:
 Core 1–4, 16–19, 31–41 odd, 62–75
 Enriched 16–19, 31–42, 62–75

If you finished Example **2** assign:
 Core 1–12, 16–27, 31–43 odd, 62–75
 Enriched 16–27, 31–44, 60–75

If you finished Example **3** assign:
 Core 1–30, 31–57 odd, 62–75
 Enriched 16–75

Answers

1. irrational, real
2. whole, integer, rational, real
3. rational, real
4. rational, real
20. rational
21. rational
22. rational
23. irrational
24. rational
25. irrational
26. not real
27. not real
31. whole, integer, rational, real
32. integer, rational, real
33. irrational, real

Math Background

The only numbers that students at this level have encountered are real numbers. The use of the word *real* implies that there are numbers that are not real numbers.

In fact, there are *imaginary* numbers. Imaginary numbers are useful for specific purposes in science and engineering. The imaginary unit is $\sqrt{-1}$; it is represented by the letter i. An example of an imaginary number is $\sqrt{-9}$, which is simplified as follows:

$\sqrt{-9} = \sqrt{9} \cdot \sqrt{-1} = 3i.$

RETEACH 3-10

Reteach
3-10 *The Real Numbers*

The set of **rational numbers** contains all integers, all fractions, and decimals that end or repeat.

[Real Numbers → Rational Numbers, Irrational Numbers]

Irrational numbers can only be written as decimals that do not end or repeat.

Together, the rational numbers and the irrational numbers form the set of **real numbers**.

Square roots of numbers that are perfect squares are rational.

Square roots of numbers that are not perfect squares are irrational.

$\sqrt{25} = 5$ $\sqrt{3} = 1.732050807\ldots$

Tell if each number is rational or irrational.

1. $\sqrt{7}$ 2. $\sqrt{81}$ 3. $\sqrt{169}$ 4. $\sqrt{101}$

irrational   rational   rational   irrational

The square of a nonzero number is positive. $3^2 = 9$ and $(-3)^2 = 9$
So, the square root of a negative number is not a real number.
$\sqrt{-9}$ is not a real number.

Tell if each number is real or not real.

5. -8 6. $-\sqrt{8}$ 7. $\sqrt{-8}$ 8. $\sqrt{-25}$

real   real   not real   not real

Between any two real numbers, there is always another real number. One way to find a number between is to find the number halfway between.

To find a real number between $7\frac{1}{5}$ and $7\frac{2}{5}$,
divide their sum by 2: $7\frac{1}{5} + 7\frac{2}{5} = \left(14\frac{3}{5}\right) \div 2 = 7\frac{3}{10}$.

Find a real number between each pair. **Possible answers are shown.**

9. $8\frac{3}{7}$ and $8\frac{4}{7}$ 10. -1.6 and -1.7 11. $-3\frac{7}{9}$ and $-3\frac{2}{9}$ 12. $6\frac{1}{2}$ and $6\frac{3}{4}$

$8\frac{1}{2}$ -1.65 $-3\frac{1}{2}$ $6\frac{5}{8}$

PRACTICE 3-10

Practice B
3-10 *The Real Numbers*

Write all names that apply to each number.

1. $-\frac{7}{8}$ 2. $\sqrt{0.15}$ 3. $\sqrt{\frac{18}{2}}$

rational; real   irrational; real   whole; integer; rational; real

4. $\sqrt{45}$ 5. -25 6. -6.75

irrational; real   integer; rational; real   rational; real

State if the number is rational, irrational, or not a real number.

7. $\sqrt{14}$ 8. $\sqrt{-16}$ 9. $\frac{6.2}{0}$ 10. $\sqrt{49}$

irrational   not real   not real   rational

11. $\frac{7}{20}$ 12. $-\sqrt{81}$ 13. $\sqrt{\frac{7}{9}}$ 14. -1.3

rational   rational   irrational   rational

Find a real number between each pair of numbers.

15. $7\frac{3}{5}$ and $7\frac{4}{5}$ 16. 6.45 and $\frac{13}{2}$ 17. $\frac{7}{8}$ and $\frac{9}{10}$

sample answer: $7\frac{7}{10}$ sample answer: 6.48 sample answer: $\frac{22}{25}$

18. Give an example of a rational number between $-\sqrt{4}$ and $\sqrt{4}$.
   sample answer: 0

19. Give an example of an irrational number less than 0.
   sample answer: $-\frac{\sqrt{22}}{7}$

20. Give an example of a number that is not real.
   sample answer: $\frac{2}{0}$

47. Find a rational number between $\sqrt{\frac{1}{4}}$ and $\sqrt{1}$. **Possible answer:** $\frac{3}{4}$

48. Find a real number between $\sqrt{2}$ and $\sqrt{3}$. **Possible answer: 1.5**

49. Find a real number between $\sqrt{5}$ and $\sqrt{11}$. **Possible answer: 3**

50. Find a real number between $\sqrt{70}$ and $\sqrt{75}$. **Possible answer: 8.5**

51. Find a real number between $-\sqrt{20}$ and $-\sqrt{17}$. **Possible answer: -4.25**

52. a. Find a real number between 1 and $\sqrt{2}$.

    b. Find a real number between 1 and your answer to part a.

    c. Find a real number between 1 and your answer to part b.

For what values of x is the value of each expression a real number?

53. $\sqrt{x}$ $x \geq 0$ 54. $5 - \sqrt{x}$ $x \geq 0$ 55. $\sqrt{x + 3}$ $x \geq -3$

56. $\sqrt{2x - 4}$ $x \geq 2$ 57. $\sqrt{5x + 2}$ $x \geq -\frac{2}{5}$ 58. $\sqrt{1 - \frac{x}{3}}$ $x \leq 3$

 59. **WHAT'S THE ERROR?** A student said that the Density Property is true for integers because between the integers 2 and 4 is another integer, 3. Explain why the student's argument does not show that the Density Property is true for integers.

60. **WRITE ABOUT IT** Can you ever use a calculator to determine if a number is rational or irrational? Explain.

61. **CHALLENGE** The circumference of a circle divided by its diameter is an irrational number, represented by the Greek letter π (*pi*). Could a circle with a diameter of 2 have a circumference of 6? Why or why not?

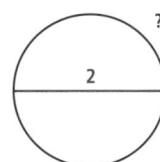

Spiral Review

Estimate each square root to two decimal places. (Lesson 3-9)

62. $\sqrt{30}$ **5.48**   63. $\sqrt{40}$ **6.32**   64. $\sqrt{50}$ **7.07**   65. $\sqrt{60}$ **7.75**

66. $\sqrt{1.8}$ **1.34**   67. $-\sqrt{17}$ **-4.12**   68. $\sqrt{12}$ **3.46**   69. $2 \cdot \sqrt{3}$ **3.46**

Write each number in scientific notation. (Lesson 2-9)

70. 1,970,000,000   **1.97×10^9**   71. 2,500,000   **2.5×10^6**

72. 31,400   **3.14×10^4**   73. 5,680,000,000,000,000   **5.68×10^{15}**

74. **TEST PREP** If $20 \cdot 4000 = 8 \cdot 10^x$, then $x = \underline{\ ?\ }$. (Lesson 2-7)   **A**

     A 4   **B** 1000   **C** 3   **D** 10

75. **TEST PREP** If $\frac{12}{36} = 2w$, what is w? (Lesson 3-6)   **J**

     F $\frac{24}{72}$ **G** $\frac{1}{3}$ **H** $\frac{24}{36}$ **J** $\frac{1}{6}$

CHALLENGE 3-10

LESSON Challenge
3-10 *Searching for Perfection*

Numbers that are equal to the sum of all their factors (not including the number itself) are called **perfect numbers**.

$6 = 1 + 2 + 3$ 6 is the smallest perfect number.

1. Which of the numbers 24 or 28 is a perfect number? Explain.

24 is not perfect since $1 + 2 + 3 + 4 + 6 + 8 + 12 \neq 24$.

28 is perfect since $1 + 2 + 4 + 7 + 14 = 28$.

The ancient Greek mathematician Euclid devised a method for computing perfect numbers.

• Begin with the number 1 and keep adding powers of 2 until you get a sum that is a *prime number* (only factors are itself and 1).

• Multiply this sum by the last power of 2.

2. Complete the table to write the first three perfect numbers.

	Sum	Prime?	Euclid's Method	Perfect Number
$1 + 2$	$= 3$	yes	2×3	6
$1 + 2 + 4$	$= 7$	yes	4×7	28
$1 + 2 + 4 + 8$	$= 15$	no		
$1 + 2 + 4 + 8 + 16$	$= 31$	yes	16×31	496

So the first three perfect numbers are   **6, 28, 496**.

The next perfect number is tedious to calculate in this manner. If, however, the calculations are written with exponents, a new pattern emerges.

3. Complete the table to write the sums using exponents.

Series	Sum
$1 + 2^1$	$= 2^2 - 1$
$1 + 2^1 + 2^2$	$= 2^3 - 1$
$1 + 2^1 + 2^2 + 2^3$	$= 2^4 - 1$
$1 + 2^1 + 2^2 + 2^3 + 2^4$	$= 2^5 - 1$

Incorporating this information, Euclid proved that whenever a prime number of the form $2^n - 1$ is found, a perfect number can be written.

If $2^n - 1$ is prime, then $2^{n-1}(2^n - 1)$ is a perfect number.

4. Find the fourth perfect number. $\underline{2^6(2^7 - 1) = 8128}$

5. Find the fifth perfect number. $\underline{2^{12}(2^{13} - 1) = 33,550,336}$

PROBLEM SOLVING 3-10

LESSON Problem Solving
3-10 *The Real Numbers*

Write the correct answer.

1. Twin primes are prime numbers that differ by 2. Find an irrational number between twin primes 5 and 7.

     Possible answer: $\underline{\sqrt{31}}$

2. Rounded to the nearest ten-thousandth, $\pi = 3.1416$. Find a rational number between 3 and π.

     Possible answer: $\underline{\frac{31}{10}}$

3. One famous irrational number is e. Rounded to the nearest ten-thousandth $e \approx 2.7823$. Find a rational number that is between 2 and e.

     Possible answer: $\underline{\frac{5}{2}}$

4. Perfect numbers are those that the divisors of the number sum to the number itself. The number 6 is a perfect number because $1 + 2 + 3 = 6$. The number 28 is also a perfect number. Find an irrational number between 6 and 28.

     Possible answer: $\underline{\sqrt{43}}$

Choose the letter for the best answer.

5. Which is a rational number?

     A the length of a side of a square with area 2 cm²

     B the length of a side of a square with area 4 cm²

     C a non-terminating decimal

     D the square root of a prime number

6. Which is an irrational number?

     F a number that can be expressed as a fraction

     G the length of a side of a square with area 4 cm²

     H the length of a side of a square with area 2 cm²

     J the square root of a negative number

7. Which is an integer?

     A the number half-way between 6 and 7

     B the average rainfall for the week if it rained 0.5 in., 2.3 in., 0 in., 0 in., 0 in., 0.2 in., 0.75 in. during the week

     C the money in an account if the balance was $213.00 and $21.87 was deposited

     D the net yardage after plays that resulted in a 15 yard loss, 10 yard gain, 6 yard gain and 5 yard loss

8. Which is a whole number?

     F the number half-way between 6 and 7

     G the total amount of sugar in a recipe that calls for $\frac{1}{4}$ cup of brown sugar and $\frac{3}{4}$ cup of granulated sugar

     H the money in an account if the balance was $213.00 and $21.87 was deposited

     J the net yardage after plays that resulted in a 15 yard loss, 10 yard gain, 6 yard gain and 5 yard loss

---

Answers

34. rational, real

35. rational, real

36. whole, integer, rational, real

37. rational, real

38. rational, real

52. a. Possible answer: 1.3
     b. Possible answer: 1.2
     c. Possible answer: 1.1

59. Possible answer: The student's example works for nonconsecutive integers like 2 and 4, but not for consecutive integers like 2 and 3.

60. Possible answer: If the calculator shows a terminating decimal, then the number is rational. If the decimal does not terminate, you cannot tell from the display on the calculator whether it repeats or not because you see only a limited number of digits.

61. Possible answer: no, because the circumference divided by the diameter would be 3, which is rational

Journal

Have students write about what they think it means when a number is called real. Have them write about what kinds of numbers might not be real.

Test Prep Doctor

For Exercise 74, suggest to students that they rewrite $20 \cdot 4000$ as follows:

$$20 \cdot 4000 = (2 \cdot 10) \cdot (4 \cdot 1000)$$
$$= (2 \cdot 4) \cdot (10 \cdot 1000)$$
$$= (8) \cdot (10,000)$$
$$= 8 \cdot 10^4$$

So $x = 4$, choice **A**.

Lesson Quiz

Write all names that apply to each number.

1. $\sqrt{2}$ real, irrational

2. $-\frac{\sqrt{16}}{2}$ real, integer, rational

State if the number is rational, irrational, or not a real number.

3. $\frac{\sqrt{25}}{0}$ not a real number

4. $\sqrt{4} \cdot \sqrt{9}$ rational

5. Find a real number between $-2\frac{3}{4}$ and $-2\frac{3}{8}$. Possible answer: $-2\frac{5}{8}$

Available on Daily Transparency in CRB

Pacing: Traditional 1 day
Block $\frac{1}{2}$ day

Objective: Students convert between bases.

Using the Pages

In Lesson 3-10, students learned that the set of whole numbers is a subset of the set of rational numbers. In this extension, students will learn how to convert whole numbers from one base to another base. These skills can be important, especially when working with computers. Computer programmers use binary (2-digit) and hexadecimal (16-digit) number systems.

EXTENSION # Other Number Systems

Learn to convert between bases.

Vocabulary
octal
binary

We use the base 10, or *decimal* number system, because we have ten fingers, or digits. Most cartoon characters have only eight fingers because cartoonists need to reduce detail. Cartoon characters could use the base 8, or **octal**, system.

Base 10

• Place values are powers of 10.
• Digits are 0, 1, 2, 3, 4, 5, 6, 7, 8, 9.

$4316_{decimal}$ = 4 thousands · 3 hundreds · 1 ten · 6 ones

10^3	10^2	10^1	$10^0 = 1$
4	3	1	6

$4 \times 10^3 + 3 \times 10^2 + 1 \times 10 + 6 \times 1$

Base 8

• Place values are powers of 8.
• Digits are 0, 1, 2, 3, 4, 5, 6, 7.

4316_{octal} = 4 five hundred twelves · 3 sixty-fours · 1 eight · 6 ones

8^3	8^2	8^1	$8^0 = 1$
4	3	1	6

$4 \times 8^3 + 3 \times 8^2 + 1 \times 8 + 6 \times 1$

$= 2048 + 192 + 8 + 6$

$= 2254_{decimal}$

EXAMPLE 1 **Changing from Base 8 to Base 10**

Change 271_{octal} to base 10.

8^2	8^1	$8^0 = 1$
2	7	1

$2 \times 8^2 + 7 \times 8^1 + 1 \times 8^0$

$= 128 + 56 + 1$

$= 185$

$271_{octal} = 185_{decimal}$

---

❶ Introduce

Motivate

Ask students if they have heard of the word *binary* and if they know what it means. Discuss the fact that *binary* means *consisting of two parts.* Point out that computers are programmed entirely in a binary, or base 2, number system.

Ask students to name the largest digit that we use in our decimal number system (9). Tell students that they will learn about number systems in base 8 and base 2. Ask students what they think are the largest digits used in number systems with base 8 (7) and base 2 (1).

❷ Teach

Lesson Presentation

Guided Instruction

In this extension, students learn how to convert between bases. Discuss how to represent a base 10 number using exponents, and then discuss how to represent a base 8 number using exponents.

Explain that to change $185_{decimal}$ to base 8, you need to know that $8^2 = 64$ is the greatest power of 8 that is contained in $185_{decimal}$. Another way to say the same thing is to say that $185_{decimal}$ is between $8^2 = 64$ and $8^3 = 512$.

EXAMPLE 2 **Changing from Base 10 to Base 8**

Helpful Hint

There is at least one multiple of 8^2 in 185, but no multiples of $8^3, 8^4, 8^5, \ldots$, since these are all greater than 185.

Change $185_{decimal}$ to base 8.

185 is between $8^2 = 64$ and $8^3 = 512$.
Do repeated divisions, by 8^2, 8^1, and finally 8^0.

$185 \div 8^2 = 2$ remainder 57
$\quad 57 \div 8^1 = 7$ remainder 1
$\quad\quad 1 \div 8^0 = 1$ remainder 0

$185_{decimal} = 271_{octal}$

Check
$2 \times 8^2 + 7 \times 8^1 + 1 \times 8^0 = 185$

EXTENSION Exercises

Change each number in base 8 to base 10.

1. 63_{octal} **51**$_{decimal}$ 2. 357_{octal} **239**$_{decimal}$ 3. 1042_{octal} **546**$_{decimal}$

Change each number in base 10 to base 8.

4. $74_{decimal}$ **112**$_{octal}$ 5. $229_{decimal}$ **345**$_{octal}$ 6. $3339_{decimal}$ **6413**$_{octal}$

Base 2, or the **binary** system, is the number system used by computers. The binary system works in the same way as base 10 and base 8, except the place values are powers of 2 and the only digits are 0 and 1.

Change each number in base 2 to base 10.

7. 11_{binary} **3**$_{decimal}$ 8. 1010_{binary} **10**$_{decimal}$ 9. 111010_{binary}
$\quad\quad\quad\quad\quad\quad\quad\quad\quad\quad\quad\quad\quad\quad\quad\quad\quad\quad\quad\quad\quad$**58**$_{decimal}$

Change each number in base 10 to base 2.

10. $13_{decimal}$ **1101**$_{binary}$ 11. $222_{decimal}$ 12. $1024_{decimal}$
$\quad\quad\quad\quad\quad\quad\quad\quad\quad\quad$**11011110**$_{binary}$ **10000000000**$_{binary}$

The binary system can be used in a code to represent symbols such as letters, numbers, and punctuation. There are four possible two-digit codes.

Possible Two-Digit Codes
00, 01, 10, 11

13. **a.** Write the possible binary three-digit codes.
   b. Write the possible binary four-digit codes.

14. ***WHAT'S THE ERROR?*** The binary number 1010110_{binary} is supposed to equal $78_{decimal}$. Correct the mistake in the binary number.

15. ***CHALLENGE*** What would be the digits for base 5? for base n?

COMMON ERROR ALERT

Students may interpret the rightmost place of a number in base 8 as representing 8^1 instead of 8^0. Tell them that when they are writing 271_{octal} in base 10, $(2 \times 8^2) + (7 \times 8^1) + (1 \times 8^0)$ is correct.

Additional Examples

Example 1
Change 1046_{octal} to base 10.
$550_{decimal}$

Example 2
Change $302_{decimal}$ to base 8.
456_{octal}

Answers

13. **a.** 000, 001, 010, 011, 100, 101, 110, 111

   b. 0000, 0001, 0010, 0011, 0100, 0101, 0110, 0111, 1000, 1001, 1010, 1011, 1100, 1101, 1110, 1111

14. Possible answer: $78_{decimal} = 1(64) + 0(32) + 0(16) + 1(8) + 1(4) + 1(2) + 0(1) = 1001110_{binary}$

15. 0, 1, 2, 3, 4; 0, 1, 2, 3, $\ldots$, $n - 1$

3 Close

Summarize

Teaching Tip

Display the following table of powers (Chapter 3 Resource Book p. 114).

Powers of 10	Powers of 2	Powers of 8
$10^0 = 1$	$2^0 = 1$	$8^0 = 1$
$10^1 = 10$	$2^1 = 2$	$8^1 = 8$
$10^2 = 100$	$2^2 = 4$	$8^2 = 64$
$10^3 = 1000$	$2^3 = 8$	$8^3 = 512$
$10^4 = 10,000$	$2^4 = 16$	$8^4 = 4096$
	$2^5 = 32$	
	$2^6 = 64$	

Use an example such as the following to review base 10 and base 8 number systems.

$128_{dec.} = 1 \times 10^2 + 2 \times 10^1 + 8 \times 10^0$
$\quad\quad\quad = 100 + 20 + 8$

$127_{octal} = 1 \times 8^2 + 2 \times 8^1 + 7 \times 8^0$
$\quad\quad\quad = 64 + 16 + 7$

Emphasize that *all* the numbers to the right of the equal signs are base 10 numbers. Any time a base is not specified, base 10 is assumed. Ask the students why we could not write 128_{octal}. Ask students for the greatest possible value of a 2-digit base 8 number.

Possible answers: The greatest digit in base 8 is 7; $77_{octal} = 7 \times 8^1 + 7 \times 8^0 = 63$.

Problem Solving on Location

New York

Purpose: To provide additional practice for problem-solving skills in Chapters 1–3

Adirondack Park

- After problem 2, ask students to describe all the steps needed to solve the problem. Write an equation and solve it. Let x be the number of trails in the unknown region. Then $139 + x = \frac{67}{147}$ (588).

  $139 + x = 268$

  $\underline{-139 \qquad -139}$

  $x = 129$

- After problem 4, have students write an equation that could be used to model the problem. $8.5 + 12.2 + m = 31.15$ What steps are needed to solve the equation? Combine like terms ($8.5 + 12.2 = 20.7$). Subtract 20.7 from each side to get $m = 10.45$.

Extension Have students use problems 1–3 as models for making up their own problems. Have them exchange papers and solve.

Check students' work.

Problem Solving on Location

NEW YORK

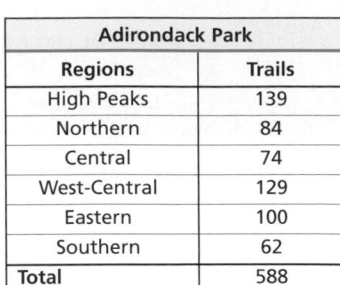

Adirondack Park

Adirondack Park covers one-fifth of the state of New York, making the park larger than Connecticut, Delaware, Hawaii, New Jersey, or Rhode Island. The park can be broken up into six regions plus the Northville–Lake Placid trail. There are 589 trails, totaling over 2000 miles in length. The 135-mile Northville–Lake Placid trail is the longest.

For 1–3, use the table. Simplify your answers.

1. a. Express the number of trails in the northern region as a fraction of all of the trails in the table. $\frac{1}{7}$

   b. Express the number of trails in the central region as a fraction of all of the trails in the table. $\frac{37}{294}$

   c. What fraction of the trails in the table do the High Peaks region and the central region combined make up? $\frac{71}{196}$

2. The High Peaks region combined with which other region contain $\frac{67}{147}$ of the trails? **West Central Region**

3. The northern region has $\frac{21}{25}$ as many trails as which region? **Eastern Region**

Adirondack Park	
Regions	Trails
High Peaks	139
Northern	84
Central	74
West-Central	129
Eastern	100
Southern	62
Total	**588**

4. Bradley walked one trail each day on a three-day trip to Adirondack Park. On the first day, he walked the Black Mountain trail, which is 8.5 miles. On the second day, he walked the Dead Creek Flow trail, which is a round-trip distance of 12.2 miles. And on day three, he walked the Mount Marcy–Elk Lake trail. If his three-day total mileage was 31.15 miles, how long was the Mount Marcy–Elk Lake trail? **10.45 miles**

Zoos of New York City

- After problem 1, have students find the decimal equivalent of the fraction they found. $0.0\overline{18}$
- After problem 3, have students discuss how to make an estimate for this problem. Possible answer: Round 6.5 to 7; round 400 to 420; $7 \div 420 = \frac{1}{60}$.

Extension Have students research the number of animals and species represented at a local zoo. Have them make a table to compare their data with the data given in this feature. Check students' work.

Answers

4. a comparison of the species

Zoos of New York City

The Bronx Zoo, located in New York City, has been operating since 1899. It opened with 22 exhibits and 843 animals. Currently, workers at the Bronx Zoo system care for more than 15,000 animals at five facilities in New York City, including Central Park Zoo in Manhattan.

Unusual species at the Bronx Zoo include snow leopards, lowland gorillas, Mauritius pink pigeons, and Chinese alligators.

The Congo Gorilla Forest is the zoo's 6.5 acre African rain forest habitat. It features 400 animals and 55 species. The exhibit's 23 lowland gorillas make up one of the largest breeding groups in the United States.

1. What fraction of the species in the African rain-forest habitat are lowland gorillas? $\frac{1}{55}$

2. What fraction of the animals in the African rain-forest habitat are lowland gorillas? $\frac{23}{400}$

3. If the 6.5-acre African rain-forest habitat were divided equally, about what fraction of an acre would each of the 400 animals have? $\frac{13}{800}$

4. Central Park Zoo has 1400 animals and 130 species. The Queens Zoo has 400 animals and 70 species. Which represents a larger fraction, a comparison of the species at the Queens Zoo with those at Central Park Zoo, or a comparison of the number of animals at each zoo?

Game Resources

Puzzles, Twisters & Teasers
Chapter 3 Resource Book

Egyptian Fractions

Purpose: *To apply operations with fractions to a historical activity*

Discuss: Ask students: What fraction would you subtract first in order to express the fraction $\frac{3}{8}$ as the Egyptians would have? Explain. You subtract $\frac{1}{3}$ from $\frac{3}{8}$ first, because it is the largest unit fraction that is less than $\frac{3}{8}$. How do you know when to stop? when the difference is a unit fraction

Extend: Have students determine how many fractions can be written using the fractions $\frac{1}{2}$, $\frac{1}{3}$, $\frac{1}{4}$, and/or $\frac{1}{5}$ as addends a maximum of one time each. Have them write all the fractions.

11 fractions:
$$\frac{1}{2} + \frac{1}{3} = \frac{5}{6}$$
$$\frac{1}{2} + \frac{1}{4} = \frac{3}{4}$$
$$\frac{1}{2} + \frac{1}{5} = \frac{7}{10}$$
$$\frac{1}{3} + \frac{1}{4} = \frac{7}{12}$$
$$\frac{1}{3} + \frac{1}{5} = \frac{8}{15}$$
$$\frac{1}{4} + \frac{1}{5} = \frac{9}{20}$$
$$\frac{1}{2} + \frac{1}{3} + \frac{1}{4} = \frac{13}{12}$$
$$\frac{1}{2} + \frac{1}{3} + \frac{1}{5} = \frac{31}{30}$$
$$\frac{1}{2} + \frac{1}{4} + \frac{1}{5} = \frac{19}{20}$$
$$\frac{1}{3} + \frac{1}{4} + \frac{1}{5} = \frac{47}{60}$$
$$\frac{1}{2} + \frac{1}{3} + \frac{1}{4} + \frac{1}{5} = \frac{77}{60}$$

Egg Fractions

Purpose: *To practice adding fractions in a game format*

Discuss: Have students give a combination of fractions with a sum of $\frac{2}{3}$.
Possible answer: $\frac{7}{12}$ and $\frac{1}{12}$

Extend: Have students play the game again, but this time they must make each sum using 3 tokens instead of 2. (Students should replace $\frac{1}{6}$ with $\frac{11}{12}$.)

MATH-ABLES

Egyptian Fractions

If you were to divide 9 loaves of bread among 10 people, you would give each person $\frac{9}{10}$ of a loaf. The answer was different on the ancient Egyptian Ahmes papyrus, because ancient Egyptians used only *unit fractions*, which have a numerator of 1. All other fractions were written as sums of different unit fractions. So $\frac{5}{6}$ could be written as $\frac{1}{2} + \frac{1}{3}$, but not as $\frac{1}{6} + \frac{1}{6} + \frac{1}{6} + \frac{1}{6} + \frac{1}{6}$.

Method	Example	
Suppose you want to write a fraction as a sum of different unit fractions.	$\frac{9}{10}$	
Step 1. Choose the largest fraction of the form $\frac{1}{n}$ that is less than the fraction you want.	(number line: 0 $\frac{1}{5}$ $\frac{1}{4}$ $\frac{1}{3}$ $\frac{1}{2}$ $\frac{9}{10}$ $\frac{1}{1}$)	
Step 2. Subtract $\frac{1}{n}$ from the fraction you want.	$\frac{9}{10} - \frac{1}{2} = \frac{2}{5}$ remaining	
Step 3. Repeat steps 1 and 2 using the difference of the fractions until the result is a unit fraction.	(number line: 0 $\frac{1}{5}$ $\frac{1}{4}$ $\frac{2}{5}$ $\frac{1}{2}$ $\frac{1}{1}$) $\frac{2}{5} - \frac{1}{3} = \frac{1}{15}$ remaining	
Step 4. Write the fraction you want as the sum of the unit fractions.	$\frac{9}{10} = \frac{1}{2} + \frac{1}{3} + \frac{1}{15}$	

Write each fraction as a sum of different unit fractions. **Possible answers:**

1. $\frac{3}{4}$ $\frac{1}{2} + \frac{1}{4}$ **2.** $\frac{5}{8}$ $\frac{1}{2} + \frac{1}{8}$ **3.** $\frac{11}{12}$ $\frac{1}{2} + \frac{1}{4} + \frac{1}{6}$ **4.** $\frac{3}{7}$ $\frac{1}{4} + \frac{1}{7} + \frac{1}{28}$ **5.** $\frac{7}{5}$ $1 + \frac{1}{3} + \frac{1}{15}$

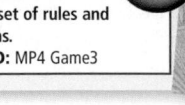

This game is played with an empty egg carton. Each compartment represents a fraction with a denominator of 12. The goal is to place tokens in compartments with a given sum.

internet connect
Go to *go.hrw.com* for a complete set of rules and instructions.
KEYWORD: MP4 Game3

You can add and subtract fractions using your graphing calculator. To display decimals as fractions, use the **MATH** key.

internet connect
Lab Resources Online
go.hrw.com
KEYWORD: MP4 TechLab3

Activity

1 Use a graphing calculator to add $\frac{7}{12} + \frac{3}{8}$. Write the sum as a fraction.

Type 7 **÷** 12 and press **ENTER** .

You can see that the decimal equivalent is a repeating decimal, $0.58\overline{3}$.

Type **+** 3 **÷** 8 **ENTER** . The decimal form of the sum is displayed.

Press **MATH** **ENTER** **ENTER** .

The fraction form of the sum, $\frac{23}{24}$, is displayed as 23/24.

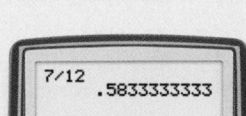

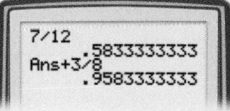

2 Use a graphing calculator to subtract $\frac{3}{5} - \frac{2}{3}$. Write the difference as a fraction.

Type 3 **÷** 5 **—** 2 **÷** 3 **MATH** **ENTER** **ENTER** .

The answer is $-\frac{1}{15}$.

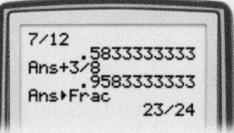

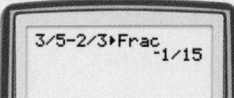

Think and Discuss

1. Why is the difference in **2** negative?

2. Type 0.33333... (pressing 3 at least twelve times). Press **MATH** **ENTER** **ENTER** to write $0.\overline{3}$ as a fraction. Now do the same for $0.\overline{9}$. What happens to $0.\overline{9}$? How does the fraction for $0.\overline{3}$ help to explain this result?

Try This

Use a calculator to add or subtract. Write each result as a fraction.

1. $\frac{1}{2} + \frac{2}{5}$ $\frac{9}{10}$
2. $\frac{7}{8} - \frac{2}{3}$ $\frac{5}{24}$
3. $\frac{7}{17} + \frac{1}{10}$ $\frac{87}{170}$
4. $\frac{1}{3} - \frac{5}{7}$ $\frac{-8}{21}$
5. $\frac{5}{32} + \frac{2}{11}$ $\frac{119}{352}$
6. $\frac{33}{101} - \frac{3}{7}$ $\frac{-72}{707}$
7. $\frac{4}{15} + \frac{7}{16}$ $\frac{169}{240}$
8. $\frac{1}{35} - \frac{1}{37}$ $\frac{2}{1295}$

Answers

Think and Discuss

Possible answers:

1. The difference is negative because $\frac{2}{3} > \frac{3}{5}$. A greater number is being subtracted from a lesser number.

2. 0.99999... becomes 1 (if enough 9's are used) when converted to a fraction on the graphing calculator. If $0.\overline{3}$ is $\frac{1}{3}$, then $0.\overline{9}$ is displayed as 1 because it is 3 times greater. $0.\overline{3} \times 3 = 0.\overline{9}$; $\frac{1}{3} \times 3 = 1$.

Technology

LAB Add and Subtract Fractions

Objective: To use a graphing calculator to add and subtract fractions

Materials: Graphing calculator

Lab Resources

Technology Lab Activities . . . . p. 18

Using the Page

This technology activity shows students how to add and subtract fractions on a graphing calculator. Specific keystrokes may vary, depending on the make and model of the graphing calculator used. The keystrokes given are for a TI-83 model. For keystrokes to other models, visit go.hrw.com.

The Think and Discuss problems can be used to assess students' understanding of the technology activity. While Try This problems 1–8 can be done without a graphing calculator, they are meant to help students become familiar with using a graphing calculator to add and subtract fractions.

Assessment

1. How could you use the graphing calculator to add $\frac{1}{2} + \frac{2}{3}$?

Enter 1 **÷** 2 **+** 2 **÷** 3

MATH **ENTER** **ENTER** .

2. What is the sum in simplest form?
   $\frac{7}{6}$

Chapter 3 Study Guide and Review

Purpose: *To help students review and practice concepts and skills presented in Chapter 3*

Assessment Resources

Chapter Review
Chapter 3 Resource Book .. pp. 99–101

Test and Practice Generator CD-ROM

Additional review assessment items in both multiple-choice and free-response format may be generated for any objective in Chapter 3.

Answers

1. rational number

2. real numbers; irrational numbers

3. relatively prime

4. principal square root

5. perfect square

6. $\frac{3}{5}$

7. $\frac{1}{4}$

8. $\frac{21}{40}$

9. $\frac{2}{3}$

10. $\frac{2}{3}$

11. $\frac{3}{4}$

12. $\frac{-6}{13}$

13. $1\frac{2}{5}$

14. $\frac{5}{9}$

15. $\frac{1}{6}$

Vocabulary

Density Property 157
irrational number 156
perfect square 146
principal square root 146

rational number 112
real number 156
reciprocal 126
relatively prime 112

Complete the sentences below with vocabulary words from the list above. Words may be used more than once.

1. Any number that can be written as a fraction $\frac{n}{d}$ (where n and d are integers and $d \neq 0$) is called a ___?___.

2. The set of ___?___ is made up of the set of rational numbers and the set of ___?___.

3. Integers that have no common factors other than 1 are ___?___.

4. The nonnegative square root of a number is called the ___?___ of the number.

5. A number that has rational numbers as its square roots is a ___?___.

3-1 Rational Numbers (pp. 112–116)

EXAMPLE

■ Write the decimal as a fraction.

$0.8 = \frac{8}{10}$ *8 is in the tenths place.*

$= \frac{8 \div 2}{10 \div 2}$ *Divide numerator and denominator by 2.*

$= \frac{4}{5}$

EXERCISES

Write each decimal as a fraction.

6. 0.6   7. 0.25   8. 0.525

Simplify.

9. $\frac{14}{21}$ 10. $\frac{22}{33}$ 11. $\frac{75}{100}$

3-2 Adding and Subtracting Rational Numbers (pp. 117–120)

EXAMPLE

■ Add or subtract.

$\frac{3}{7} + \frac{4}{7} = \frac{3+4}{7} = \frac{7}{7} = 1$

$\frac{8}{11} - \left(\frac{-2}{11}\right) = \frac{8-(-2)}{11} = \frac{8+2}{11} = \frac{10}{11}$

EXERCISES

Add or subtract.

12. $\frac{-8}{13} + \frac{2}{13}$ 13. $\frac{3}{5} - \left(\frac{-4}{5}\right)$

14. $\frac{-2}{9} + \frac{7}{9}$ 15. $\frac{-5}{12} - \left(\frac{-7}{12}\right)$

Study Guide and Review

3-3 Multiplying Rational Numbers (pp. 121–125)

EXAMPLE

■ Multiply. Write the answer in simplest form.

$$5\left(3\frac{1}{4}\right) = \left(\frac{5}{1}\right)\left(\frac{3(4)+1}{4}\right)$$

$$= \left(\frac{5}{1}\right)\left(\frac{13}{4}\right) \quad \text{Write as improper fractions.}$$

$$= \frac{65}{4} \quad \text{Multiply.}$$

$$= 16\frac{1}{4} \quad \text{Write in simplest form.}$$

EXERCISES

Multiply. Write each answer in simplest form.

16. $3\left(-\frac{2}{5}\right)$ **17.** $2\left(3\frac{4}{5}\right)$

18. $\frac{-2}{3}\left(\frac{-4}{5}\right)$ **19.** $\frac{8}{11}\left(\frac{-22}{4}\right)$

20. $5\frac{1}{4}\left(\frac{3}{7}\right)$ **21.** $2\frac{1}{2}\left(1\frac{3}{10}\right)$

3-4 Dividing Rational Numbers (pp. 126–130)

EXAMPLE

■ Divide. Write the answer in simplest form.

$$\frac{7}{8} \div \frac{3}{4} = \frac{7}{8} \cdot \frac{4}{3} \quad \begin{array}{l}\text{Multiply by the reciprocal.} \\ \text{Write as one fraction.}\end{array}$$

$$= \frac{7 \cdot 4}{8 \cdot 3}$$

$$\frac{7 \cdot \overset{1}{\cancel{4}}}{\underset{2}{\cancel{8}} \cdot 3} = \frac{7 \cdot 1}{2 \cdot 3} \quad \begin{array}{l}\text{Divide by common factor, 4.}\end{array}$$

$$\frac{7}{6} = 1\frac{1}{6}$$

EXERCISES

Divide. Write each answer in simplest form.

22. $\frac{3}{4} \div \frac{1}{8}$ **23.** $\frac{3}{10} \div \frac{4}{5}$

24. $\frac{2}{3} \div 3$ **25.** $4 \div \frac{-1}{4}$

26. $3\frac{3}{4} \div 3$ **27.** $1\frac{1}{3} \div \frac{2}{3}$

3-5 Adding and Subtracting with Unlike Denominators (pp. 131–134)

EXAMPLE

■ Add.

$$\frac{3}{4} + \frac{2}{5} \quad \text{Multiply denominators, } 4 \cdot 5 = 20.$$

$$\frac{3 \cdot 5}{4 \cdot 5} = \frac{15}{20} \quad \frac{2 \cdot 4}{5 \cdot 4} = \frac{8}{20} \quad \begin{array}{l}\text{Rename fractions with the LCD 20.}\end{array}$$

$$\frac{15}{20} + \frac{8}{20} = \frac{15+8}{20} = \frac{23}{20} = 1\frac{3}{20} \quad \begin{array}{l}\text{Add and simplify.}\end{array}$$

EXERCISES

Add or subtract.

28. $\frac{5}{6} + \frac{1}{3}$ **29.** $\frac{5}{6} - \frac{5}{9}$

30. $3\frac{1}{2} + 7\frac{4}{5}$ **31.** $7\frac{1}{10} - 2\frac{3}{4}$

3-6 Solving Equations with Rational Numbers (pp. 136–139)

EXAMPLE

■ Solve.

$$\begin{array}{l} x - 13.7 = -22 \\ \underline{+13.7} \quad \underline{+13.7} \quad \text{Add 13.7 to each side.} \\ x = -8.3 \end{array}$$

EXERCISES

Solve.

32. $y + 7.8 = -14$ **33.** $2.9z = -52.2$

34. $w + \frac{3}{4} = \frac{1}{8}$ **35.** $\frac{3}{8}p = \frac{3}{4}$

Answers

16. $-1\frac{1}{5}$

17. $7\frac{3}{5}$

18. $\frac{8}{15}$

19. -4

20. $2\frac{1}{4}$

21. $3\frac{1}{4}$

22. 6

23. $\frac{3}{8}$

24. $\frac{2}{9}$

25. -16

26. $1\frac{1}{4}$

27. 2

28. $1\frac{1}{6}$

29. $\frac{5}{18}$

30. $11\frac{3}{10}$

31. $4\frac{7}{20}$

32. $y = -21.8$

33. $z = -18$

34. $w = -\frac{5}{8}$

35. $p = 2$

Answers

36. $m > -\frac{1}{12}$

37. $t \geq -12$

38. $y \leq -3\frac{1}{4}$

39. $x > -\frac{1}{2}$

40. ± 4

41. ± 30

42. ± 26

43. 5

44. $\frac{1}{2}$

45. 9

46. 89.4 in.

47. 167.3 cm

48. rational

49. irrational

50. not real

51. irrational

52. rational

53. not real

3-7 Solving Inequalities with Rational Numbers (pp. 140–143)

EXAMPLE

■ Solve.

$-3x > \frac{6}{7}$

$-\frac{1}{3}(-3x) > -\frac{1}{3}\left(\frac{6}{7}\right)$ *Multiply each side by $-\frac{1}{3}$.*

$x < -\frac{2}{7}$ *Change > to <, since you multiplied by a negative.*

EXERCISES

Solve.

36. $4m > -\frac{1}{3}$ **37.** $-2.7t \leq 32.4$

38. $7\frac{1}{2} - y \geq 10\frac{3}{4}$ **39.** $x + \frac{4}{5} > \frac{3}{10}$

3-8 Squares and Square Roots (pp. 146–149)

EXAMPLE

■ Find the two square roots of 400.

$20 \cdot 20 = 400$

$(-20) \cdot (-20) = 400$

The square roots are 20 and -20.

EXERCISES

Find the two square roots of each number.

40. 16   **41.** 900   **42.** 676

Evaluate each expression.

43. $\sqrt{4 + 21}$ **44.** $\frac{\sqrt{100}}{20}$ **45.** $\sqrt{3^4}$

3-9 Finding Square Roots (pp. 150–153)

EXAMPLE

■ Find the side length of a square with area 359 ft² to one decimal place. Then find the distance around the square.

$18^2 = 324$, $19^2 = 361$

Side $= \sqrt{359} \approx 18.9$

Distance around $\approx 4(18.9) \approx 75.6$ feet

EXERCISES

Find the distance around each square with the area given. Answer to the nearest tenth.

46. Area of square ABCD is 500 in².

47. Area of square MNOP is 1750 cm².

3-10 The Real Numbers (pp. 156–159)

EXAMPLE

■ State if the number is rational, irrational, or not a real number.

$-\sqrt{2}$ real, irrational   *The decimal equivalent does not repeat or end.*

$\sqrt{-4}$ not real   *Square roots of negative numbers are not real.*

EXERCISES

State if the number is rational, irrational, or not a real number.

48. $\sqrt{81}$ **49.** $\sqrt{122}$ **50.** $\sqrt{-16}$

51. $-\sqrt{5}$ **52.** $\frac{0}{-4}$ **53.** $\frac{7}{0}$

Simplify.

1. $\frac{36}{72}$ **$\frac{1}{2}$**
2. $\frac{21}{35}$ **$\frac{3}{5}$**
3. $\frac{16}{88}$ **$\frac{2}{11}$**
4. $\frac{18}{25}$ **$\frac{18}{25}$**

Write each decimal as a fraction in simplest form.

5. 0.225 **$\frac{9}{40}$**
6. 0.04 **$\frac{1}{25}$**
7. 0.101 **$\frac{101}{1000}$**
8. 0.875 **$\frac{7}{8}$**

Write each fraction as a decimal.

9. $\frac{7}{8}$ **0.875**
10. $\frac{13}{25}$ **0.52**
11. $\frac{5}{12}$ **0.41$\overline{6}$**
12. $\frac{4}{33}$ **0.$\overline{12}$**

Add or subtract. Write each answer in simplest form.

13. $\frac{-3}{11} - \left(\frac{-4}{11}\right)$ **$\frac{1}{11}$**
14. $7\frac{1}{4} - 2\frac{3}{4}$ **$4\frac{1}{2}$**
15. $\frac{5}{6} + \frac{7}{18}$ **$1\frac{2}{9}$**
16. $\frac{5}{6} - \frac{8}{9}$ **$-\frac{1}{18}$**
17. $4\frac{1}{2} + 5\frac{7}{8}$ **$10\frac{3}{8}$**
18. $8\frac{1}{5} - 1\frac{2}{3}$ **$6\frac{8}{15}$**

Multiply or divide. Write each answer in simplest form.

19. $9\left(\frac{-2}{27}\right)$ **$-\frac{2}{3}$**
20. $\frac{7}{8} \div \frac{5}{24}$ **$4\frac{1}{5}$**
21. $\frac{2}{3}\left(\frac{-9}{20}\right)$ **$-\frac{3}{10}$**
22. $3\frac{3}{7}\left(1\frac{5}{16}\right)$ **$4\frac{1}{2}$**
23. $34 \div 3\frac{2}{5}$ **10**
24. $-4\frac{2}{3} \div 1\frac{1}{6}$ **-4**

Solve.

25. $x - \frac{1}{4} = -\frac{3}{8}$ **$x = -\frac{1}{8}$**
26. $-3.14y = 53.38$ **$y = -17$**
27. $-2k < \frac{1}{4}$ **$k > -\frac{1}{8}$**
28. $h - 3.24 \leq -1.1$ **$h \leq 2.14$**

Find the two square roots of each number.

29. 196 **±14**
30. 1 **±1**
31. 0.25 **±0.5**
32. 6.25 **±2.5**

Each square root is between two integers. Name the integers.

33. $\sqrt{230}$ **15 and 16**
34. $\sqrt{125}$ **11 and 12**
35. $\sqrt{89}$ **9 and 10**
36. $-\sqrt{60}$ **-8 and -7**

State whether the number is rational, irrational, or not real.

37. $-\sqrt{121}$ **real, rational**
38. $-1.\overline{7}$ **real, rational**
39. $\sqrt{-9}$ **not real**

Solve.

40. Michelle wants to put a fence along one side of her square-shaped vegetable garden. The area of the garden is 1250 ft². How much fencing should she buy, to the nearest foot? **36 ft**

Chapter Test

Purpose: *To assess students' mastery of concepts and skills in Chapter 3*

Assessment Resources

Chapter 3 Tests (Levels A, B, C)
Assessment Resources . . . . pp. 45–50

 Test and Practice Generator CD-ROM

Additional assessment items in both multiple-choice and free-response format may be generated for any objective in Chapter 3.

Chapter
3

Performance Assessment

Purpose: *To assess students' under- standing of concepts in Chapter 3 and combined problem-solving skills*

Assessment Resources

Performance Assessment
Assessment Resources .... p. 122

Performance Assessment Teacher Support
Assessment Resources .... p. 121

Answers
1–3. See p. A3.

4. See Level 3 work sample below.

Scoring Rubric for Problem Solving Item 4

Level 3
Accomplishes the purposes of the task.

Student gives clear explanations, shows understanding of mathematical ideas and processes, and computes accurately.

Level 2
Purposes of the task not fully achieved.

Student demonstrates satisfactory but limited understanding of the mathemat- ical ideas and processes.

Level 1
Purposes of the task not accomplished.

Student shows little evidence of under- standing the mathematical ideas and processes and makes computional and/or procedural errors.

Performance Assessment

Show What You Know

Create a portfolio of your work from this chapter. Complete this page and include it with your four best pieces of work from Chapter 3. Choose from your homework or lab assignments, mid-chapter quiz, or any journal entries you have done. Put them together using any design you want. Make your portfolio represent what you consider your best work.

Short Response

1. A square chessboard is made up of 64 squares. If you placed a knight in each of the squares around the edge of the board, how many knight pieces would you need? Show or explain how you determined your answer.

2. In a mechanical drawing, a hidden line is usually represented by dashes $\frac{1}{8}$ in. long with $\frac{1}{32}$ in. spaces between them. How long is a line represented by 26 dashes? Show or explain how you determined your answer.

3. Write the multiplication equation $\frac{3}{4} \cdot \frac{5}{7} = \frac{15}{28}$ as a division equation. Use your result to explain why dividing by a fraction is the same as multiplying by the fraction's reciprocal.

Extended Problem Solving

4. Use a diagram to model multiplication of fractions.

   a. Draw a diagram to model the fraction $\frac{5}{6}$.

   b. Shade $\frac{2}{5}$ of the part of your diagram that represents $\frac{5}{6}$. What product does this shaded area represent?

   c. Use your diagram to write the product in simplest form.

Student Work Samples for Item 4

Level 3

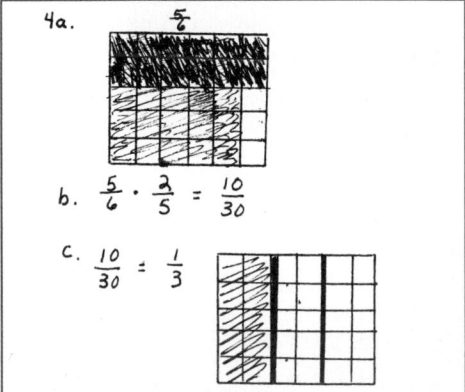

The student correctly divided and shaded the region and modeled the product in simplest form.

Level 2

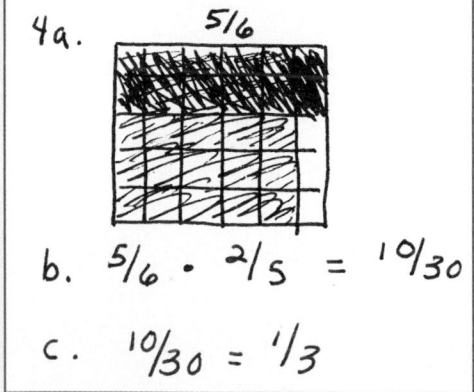

The student correctly divided and shaded the region but did not model the product in simplest form.

Level 1

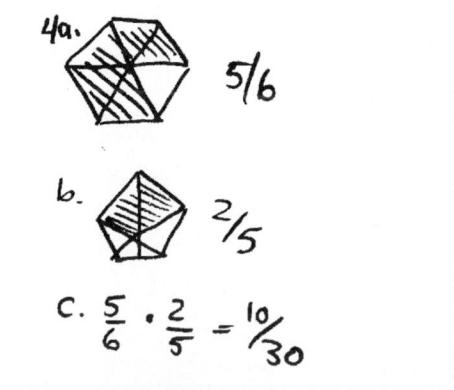

The student showed little understanding of the computational or modeling process.

Cumulative Assessment, Chapters 1–3

1. Which ordered pair lies on the negative portion of the y-axis? **B**
   (A) $(-4, -4)$ (C) $(4, -4)$
   (B) $(0, -4)$ (D) $(-4, 0)$

2. The sum of two numbers that differ by 1 is x. In terms of x, what is the value of the greater of the two numbers? **H**
   (F) $\frac{x-1}{2}$ (H) $\frac{x+1}{2}$
   (G) $\frac{x}{2}$ (J) $\frac{x}{2}+1$

3. If the sum of the consecutive integers from -22 through x is 72, what is the value of x? **B**
   (A) 23   (C) 50
   (B) 25   (D) 75

4. If $xy + y = x + 2z$, what is the value of y when $x = 2$ and $z = 3$? **G**
   (F) $\sqrt{8}$ (H) $\sqrt[3]{8}$
   (G) $\frac{8}{3}$ (J) 24

5. A local library association has posted the results of community contributions to the building fund.

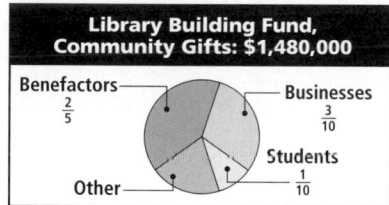

Library Building Fund, Community Gifts: $1,480,000

Benefactors $\frac{2}{5}$ — Businesses $\frac{3}{10}$ — Students $\frac{1}{10}$ — Other

   How much money does "Other" represent? **B**
   (A) $148,000   (C) $444,000
   (B) $296,000   (D) $592,000

6. Which number is equivalent to 3^{-3}? **G**
   (F) $\frac{1}{9}$ (H) $-\frac{1}{9}$
   (G) $\frac{1}{27}$ (J) $-\frac{1}{27}$

7. What is the value of $10 - 2 \cdot 3^2$? **B**
   (A) -26 (C) 72
   (B) -8 (D) 576

TEST TAKING TIP!

Making comparisons: When assigning test values, try different kinds of numbers, such as negatives and fractions.

8. If x is any real number, then which statement **must** be true? **J**
   (F) $x^2 > x$
   (G) $x^3 > x$
   (H) $x^3 > x^2$
   (J) No relationship can be determined.

9. **SHORT RESPONSE** The total weight of Sam and his son Dan is 250 pounds. Sam's weight is 10 pounds more than 3 times Dan's weight. Write an equation that could be used to determine Dan's weight. Solve your equation.

10. **SHORT RESPONSE** There was $1000 in the bank teller's drawer when the bank opened. After the first customer's withdrawal, the drawer still had greater than $900 in it, and it had an equal number of $1, $5, $10, $20, $50, and $100 bills in it. How much money did the first customer withdraw? Show or explain how you found your answer.

Standardized Test Prep

Purpose: *To provide review and practice for Chapters 1–3 and standardized tests*

Assessment Resources

Cumulative Tests (Levels A, B, C)
Assessment Resources . . . . . pp. 169–180

State-Specific Test Practice Online
KEYWORD: MP4 TestPrep

Test Prep Doctor ✚

Suggest to students that for item 2 they represent the unknown numbers as follows:

greater number: y
lesser number: $y - 1$.

Then students can solve the following equation to find the greater number y:

$$y + y - 1 = x$$
$$2y - 1 = x$$
$$2y = x + 1$$
$$y = \frac{x+1}{2}$$

For item 3, point out that the sum of consecutive integers from -22 through 22 is zero. Students can then begin with 23 and add consecutive integers to get the sum of 72: $23 + 24 + 25 = 72$. The last integer, x, is 25.

Answers

9. 60 pounds;

$$10 + 3d + d = 250$$
$$10 + 4d = 250$$
$$\underline{-10 \qquad\qquad -10}$$
$$4d = 240$$
$$\frac{4d}{4} = \frac{240}{4}$$
$$d = 60$$

10. $70;

Let x be the number of each type of bill in the drawer. Then
$$1x + 5x + 10x + 20x + 50x + 100x > 900.$$
$$\frac{186x}{186} > \frac{900}{186}$$
$$x > 4.8$$

Since x must be a whole number, $x \geq 5$. If $x = 5$, there is $186(5) = 930$ in the drawer, so the first customer withdrew $70.

Collecting, Displaying, and Analyzing Data

Pacing Guide for 45-Minute Classes

Chapter 4

DAY 39	DAY 40	DAY 41	DAY 42	DAY 43
Lesson 4-1 Hands-On Lab 4A	Lesson 4-2	Lesson 4-3	Lesson 4-4	Technology Lab 4B

DAY 44	DAY 45	DAY 46	DAY 47	DAY 48
Mid-Chapter Quiz Lesson 4-5	Lesson 4-6	Lesson 4-7	Extension	Chapter 4 Review

DAY 49				
Chapter 4 Assessment				

Pacing Guide for 90-Minute Classes

Chapter 4

DAY 19	DAY 20	DAY 21	DAY 22	DAY 23
Chapter 3 Review Lesson 4-1 Hands-On Lab 4A	Chapter 3 Assessment Lesson 4-2	Lesson 4-3 Lesson 4-4	Technology Lab 4B Lesson 4-5	Mid-Chapter Quiz Lesson 4-6 Lesson 4-7

DAY 24	DAY 25			
Extension Chapter 4 Review	Chapter 4 Assessment Lesson 5-1			

COURSE 1

- Organize data in tables, stem-and-leaf plots, bar graphs, histograms, box-and-whisker plots, and line graphs.
- Find measures of central tendency.
- Recognize misleading graphs.
- Graph ordered pairs on a coordinate grid.

COURSE 2

- Identify populations and samples.
- Organize data in tables, stem-and-leaf plots, bar graphs, histograms, box-and-whisker plots, and line graphs.
- Find measures of central tendency.
- Read and interpret circle graphs.
- Recognize misleading graphs and statistics.
- Create and interpret scatter plots.

PRE-ALGEBRA

- **Recognize biased samples and identify sampling methods.**
- **Organize data in tables, stem-and-leaf plots, bar graphs, histograms, box-and-whisker plots, and line graphs.**
- **Find appropriate measures of central tendency and variability.**
- **Recognize misleading graphs and statistics.**
- **Create and interpret scatter plots.**

LANGUAGE ARTS LINK

Math: Reading and Writing in the Content Area . . . . pp. 29–35
Focus on Problem Solving
  Make a Plan . . . . SE p. 195
Language Arts . . . . SE p. 183
Journal . . . . TE, last page of each lesson
Write About It . . . . SE pp. 177, 187, 192, 199, 203

SOCIAL STUDIES LINK

Geography . . . . SE p. 192
Social Studies . . . . SE p. 181

SCIENCE LINK

Life Science . . . . SE p. 207
Earth Science . . . . SE p. 191
Astronomy . . . . SE pp. 185, 187

TE = *Teacher's Edition*   **SE** = *Student Edition*

Interdisciplinary

Bulletin Board

Social Studies

The mean area per country for 15 countries in Southwest Asia is 152,470 square miles. This calculation does not take into account the 304,426 square miles of Turkey. How does the mean change if you include the area of Turkey? It increases by about 9497 square miles.

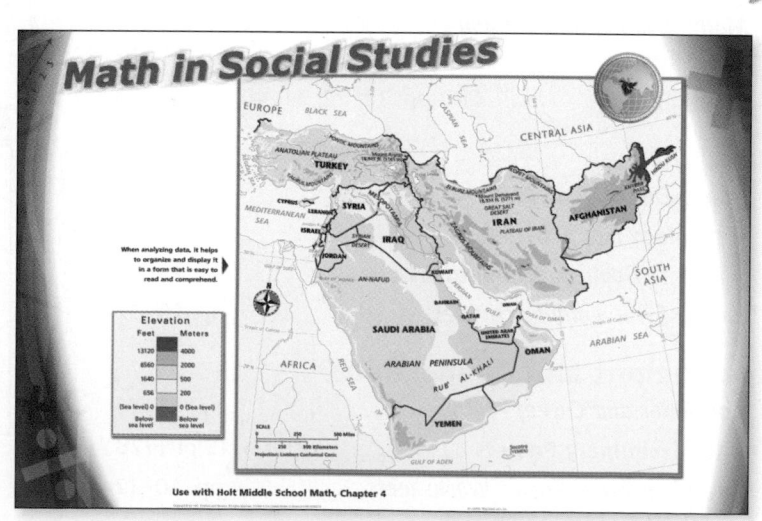

Use with Holt Middle School Math, Chapter 4

Interdisciplinary posters and worksheets are provided in your resource material.

Resource Options

Chapter 4 Resource Book

Student Resources

Practice (Levels A , B, C) . . . . pp. 8–10, 17–19, 27–29, 36–38, 46–48, 56–58, 65–67

Reteach. . . . . pp. 11, 20–21, 30, 39, 49–50, 59, 68

Challenge . . . . pp. 12, 22, 31, 40, 51, 60, 69

Problem Solving . . . . pp. 13, 23, 32, 41, 52, 61, 70

Puzzles, Twisters & Teasers. . . . . pp. 14, 24, 33, 42, 53, 62, 71

Recording Sheets . . . . pp. 3, 7, 16, 26, 35, 45, 55, 64, 75, 78–79

Chapter Review . . . . pp. 72–74

Teacher and Parent Resources

Chapter Planning and Pacing Guide. . . . . p. 4

Section Planning Guides . . . . pp. 5, 43

Parent Letter . . . . pp. 1–2

Teaching Tools . . . . pp. 78–79

Teacher Support for Chapter Project . . . . p. 76

Transparencies . . . . pp. T1–T34

• Daily Transparencies
• Additional Examples Transparencies
• Teaching Transparencies

Reaching All Learners

English Language Learners

Success for English Language Learners . . . . pp. 57–70

Math: Reading and Writing in the Content Area . . . . pp. 29–35

Spanish Homework and Practice . . . . pp. 29–35

Spanish Interactive Study Guide . . . . pp. 29–35

Spanish Family Involvement Activities . . . . pp. 25–32

Multilingual Glossary

Individual Needs

Are You Ready? Intervention and Enrichment. . . pp. 13–16, 73–80, 369–372, 411–412

Alternate Openers: Explorations . . . . pp. 29–35

Family Involvement Activities . . . . pp. 25–32

Interactive Problem Solving. . . . . pp. 29–35

Interactive Study Guide . . . . pp. 29–35

Readiness Activities . . . . pp. 7–8

Math: Reading and Writing in the Content Area . . . . pp. 29–35

Challenge . . . . CRB pp. 12, 22, 31, 40, 51, 60, 69

Hands-On

Hands-On Lab Activities. . . . . pp. 25–29

Technology Lab Activities. . . . . pp. 19–23

Alternate Openers: Explorations . . . . pp. 29–35

Family Involvement Activities . . . . pp. 25–32

Applications and Connections

Consumer and Career Math. . . . . pp. 13–16

Interdisciplinary Posters . . . . Poster 4, TE p. 172B

Interdisciplinary Poster Worksheets. . . . . pp. 10–12

Transparencies

Alternate Openers: Explorations . . . . pp. 29–35

Exercise Answers Transparencies

Chapter 4 Resource Book . . . . pp. T1–T34

• Daily Transparencies
• Additional Examples Transparencies
• Teaching Transparencies

Technology

Teacher Resources

Lesson Presentations CD-ROM. . . . . Chapter 4

Test and Practice Generator CD-ROM . . . . Chapter 4

One-Stop Planner CD-ROM . . . . Chapter 4

Student Resources

Are You Ready? Intervention CD-ROM
Skills 1, 16, 17, 90

internet connect

Homework Help Online	KEYWORD: MP4 HWHelp4
Math Tools Online	KEYWORD: MP4 Tools
Glossary Online	KEYWORD: MP4 Glossary
Chapter Project Online	KEYWORD: MP4 PSProject4
Chapter Opener Online	KEYWORD: MP4 Ch4

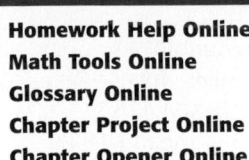  KEYWORD: MP4 CNN4

SE = Student Edition   TE = Teacher's Edition   AR = Assessment Resources   CRB = Chapter Resource Book   MK = Manipulatives Kit

172C Chapter 4 Collecting, Displaying, and Analyzing Data

Assessment Options

Assessing Prior Knowledge

Determine whether students have the required prerequisite concepts and skills.

Are You Ready?.................................. SE p. 173
Inventory Test.................................. AR pp. 1–4

Test Preparation

Provide review and practice for chapter and standardized tests.

Standardized Test Prep........................... SE p. 219
Spiral Review with Test Prep ..... SE, last page of each lesson
Study Guide and Review .... SE pp. 214–216
Test Prep Tool Kit

Technology

 Test and Practice Generator CD-ROM

internet connect

State-Specific Test Practice Online KEYWORD: MP4 TestPrep

Performance Assessment

Assess students' understanding of chapter concepts and combined problem-solving skills.

Performance Assessment .... SE p. 218
  Includes scoring rubric in TE
Performance Assessment .... AR p. 124
Performance Assessment Teacher Support......... AR p. 123

Portfolio

Portfolio opportunities appear throughout the Student and Teacher's Editions.

Suggested work samples:

Problem Solving Project .... TE p. 172
Performance Assessment .... SE p. 218
Portfolio Guide................................ AR p. xxxvii
Journal....................... TE, last page of each lesson
Write About It.............. SE pp. 177, 187, 192, 199, 203

Daily Assessment

Obtain daily feedback on students' understanding of concepts.

Spiral Review and Test Prep ...... SE, last page of each lesson

Also Available on Transparency In Chapter 4 Resource Book

Warm Up...................... TE, first page of each lesson
Problem of the Day ...... TE pp. 174, 179, 184, 188, 200, 204
Lesson Quiz.................... TE, last page of each lesson

Student Self-Assessment

Have students evaluate their own work.

Group Project Evaluation .... AR p. xxxiv
Individual Group Member Evaluation .... AR p. xxxv
Portfolio Guide................................ AR p. xxxvii
Journal....................... TE, last page of each lesson

Formal Assessment

Assess students' mastery of concepts and skills.

Section Quizzes .... AR pp. 11–12
Mid-Chapter Quiz............................. SE p. 194
Chapter Test .... SE p. 217
Chapter Tests (Levels A, B, C) .... AR pp. 51–56
Cumulative Tests (Levels A, B, C)........... AR pp. 181–192
Standardized Test Prep
  Cumulative Assessment .... SE p. 219
End-of-Year Test........................... AR pp. 313–316

Technology

 Test and Practice Generator CD-ROM

Make tests electronically. This software includes:

- Dynamic practice for Chapter 4
- Customizable tests
- Multiple-choice items for each objective
- Free-response items for each objective
- Teacher management system

SE = *Student Edition*   **TE** = *Teacher's Edition*   **AR** = *Assessment Resources*   **CRB** = *Chapter Resource Book*   **MK** = *Manipulatives Kit*

Three levels (A,B,C) of tests are available for each chapter in the *Assessment Resources.*

LEVEL A

CHAPTER 4 **Chapter Test**
Form A

Identify the population, sample, and sampling method.

1. Every 15th student eating in the school cafeteria was asked his or her favorite dessert.

   All students eating in the school cafeteria; every 15th student; systematic.

2. 300 households in Sylvania were selected by random digit dialing to respond to a survey.

   All households in Sylvania with telephones; 300 households called; random.

3. Ten video stores in a large city were randomly selected and 50 customers were randomly selected from each store and asked what was their favorite video.

   People going to the video stores in the city; those asked the questions; stratified.

Organize the data to make a stem-and-leaf plot.

4. The following are test scores for a math class.

Test Scores				Stem	Leaves
98	85	70	93	7	0 4 8
85	74	98	81	8	1 5 5 5 9
89	78	100	85	9	3 8 8
				10	0

Use the data. 10, 25, 18, 15, 21, 20, 12 Round to the nearest tenth.

5. Find the mean. **17.3**
6. Find the median. **18**
7. Find the mode. **none**
8. Find the range. **15**
9. Find the third quartile. **21**
10. Find the first quartile. **12**
11. Make a box-and-whisker plot.

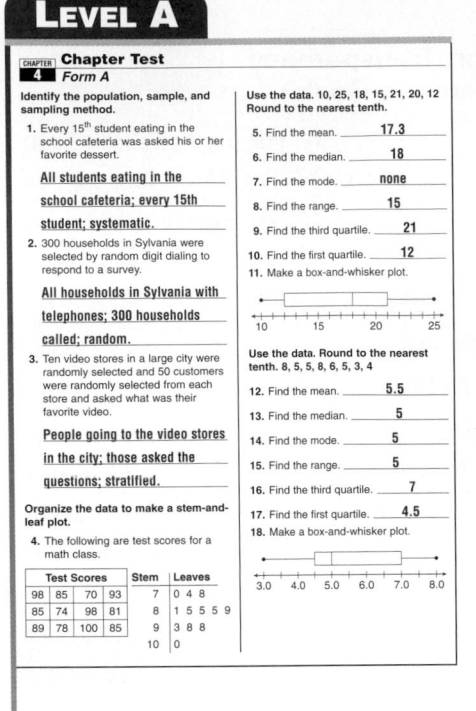

Use the data. Round to the nearest tenth. 8, 5, 5, 8, 6, 5, 3, 4

12. Find the mean. **5.5**
13. Find the median. **5**
14. Find the mode. **5**
15. Find the range. **5**
16. Find the third quartile. **7**
17. Find the first quartile. **4.5**
18. Make a box-and-whisker plot.

CHAPTER 4 **Chapter Test**
Form A, continued

Use the data.
21, 21, 19, 22, 24, 19, 21, 24, 21, 24, 22, 19, 22, 21, 19, 24, 21

19. Make a frequency table.

Scores	Frequency
19	4
21	6
22	3
24	5

20. Make a bar graph.

Explain why the graph is misleading.

21.

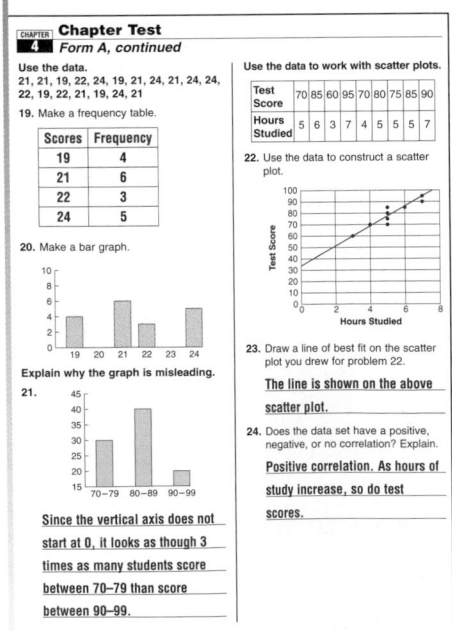

Since the vertical axis does not start at 0, it looks as though 3 times as many students score between 70–79 than score between 90–99.

Use the data to work with scatter plots.

Test Score	70	85	60	95	70	80	75	85	90
Hours Studied	5	6	3	7	4	4	5	5	7

22. Use the data to construct a scatter plot.

23. Draw a line of best fit on the scatter plot you drew for problem 22.

   The line is shown on the above scatter plot.

24. Does the data set have a positive, negative, or no correlation? Explain.

   Positive correlation. As hours of study increase, so do test scores.

LEVEL B

CHAPTER 4 **Chapter Test**
Form B

Identify the population, sample, and sampling method.

1. 250 students in an elementary school were randomly selected and asked what was their favorite color.

   All students in the school; 250 students; random.

2. Every twentieth visitor to the local zoo was asked a series of questions about the animal exhibits.

   All visitors to the zoo; those asked the questions; systematic.

3. Five grocery stores in a city were randomly selected and 100 customers were randomly surveyed from each store as to their favorite flavor of ice cream.

   Grocery store customers; those surveyed; stratified.

4. Make a back-to-back stem-and-leaf plot.

Test Scores for 2 Math Classes

Class #1				Class #2			
98	85	70	93	88	94	82	100
85	74	98	81	88	100	84	100
89	78	100	85	88	94	94	71
89	93	85	81	94	82	82	94

Leaves	Stem	Leaves
0 4 8	7	1
1 1 5 5 5 5 9 9	8	2 2 2 4 8 8 8
3 3 8 8	9	4 4 4 4 4
0	10	0 0 0

Use the data. 98, 95, 89, 85, 89, 90

5. Find the mean. **91**
6. Find the median. **89.5**
7. Find the mode. **89**
8. Find the range. **13**
9. Find the third quartile. **95**
10. Find the first quartile. **89**
11. Make a box-and-whisker plot.

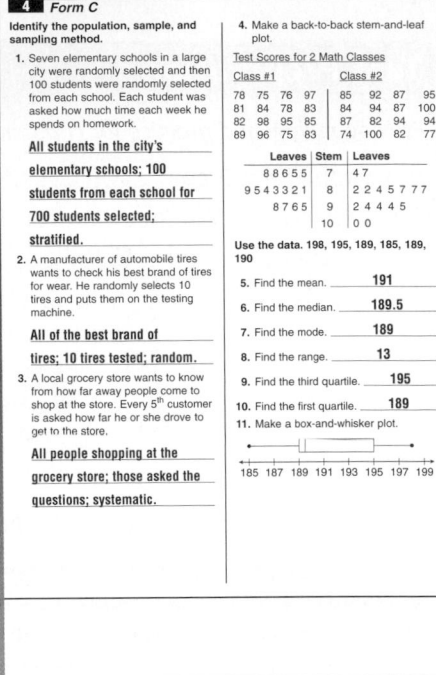

Use the data. Round to the nearest tenth. 28, 25, 28, 22, 28, 29, 23, 24

12. Find the mean. **25.9**
13. Find the median. **26.5**
14. Find the mode. **28**
15. Find the range. **7.0**
16. Find the third quartile. **28.0**
17. Find the first quartile. **23.5**
18. Make a box-and-whisker plot.

CHAPTER 4 **Chapter Test**
Form B, continued

Use the test score data to answer the questions.
38,15,18,17,28,29,22,24,34,20,35,31,25, 33,14

19. Make a frequency table with an interval of 10.

Scores	Frequency
10–19	4
20–29	6
30–39	5

20. Make a histogram.

Explain why the graph is misleading.

21. **Bookings**

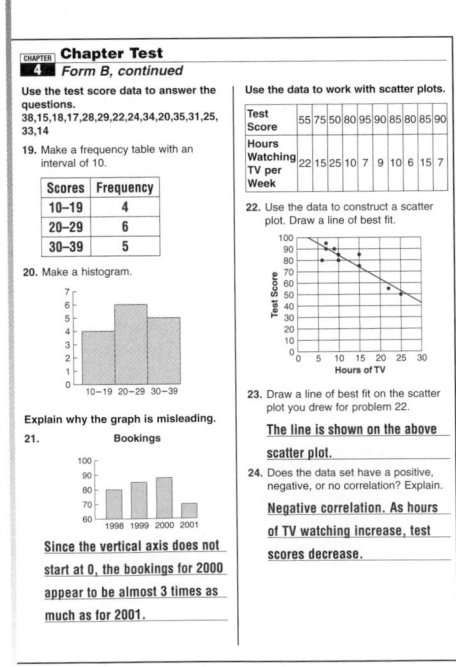

Since the vertical axis does not start at 0, the bookings for 2000 appear to be almost 3 times as much as for 2001.

Use the data to work with scatter plots.

Test Score	55	75	50	80	95	90	85	80	85	90
Hours Watching TV per Week	22	15	25	10	7	9	10	6	15	7

22. Use the data to construct a scatter plot. Draw a line of best fit.

23. Draw a line of best fit on the scatter plot you drew for problem 22.

   The line is shown on the above scatter plot.

24. Does the data set have a positive, negative, or no correlation? Explain.

   Negative correlation. As hours of TV watching increase, test scores decrease.

LEVEL C

CHAPTER 4 **Chapter Test**
Form C

Identify the population, sample, and sampling method.

1. Seven elementary schools in a large city were randomly selected and then 100 students were randomly selected from each school. Each student was asked how much time each week he spends on homework.

   All students in the city's elementary schools; 100 students from each school for 700 students selected; stratified.

2. A manufacturer of automobile tires wants to check his best brand of tires for wear. He randomly selects 10 tires and puts them on the testing machine.

   All of the best brand of tires; 10 tires tested; random.

3. A local grocery store wants to know from how far away people come to shop at the store. Every 5th customer is asked how far he or she drove to get to the store.

   All people shopping at the grocery store; those asked the questions; systematic.

4. Make a back-to-back stem-and-leaf plot.

Test Scores for 2 Math Classes

Class #1				Class #2			
78	75	76	97	85	92	87	95
81	84	78	83	84	94	87	100
82	98	95	85	87	82	94	94
89	96	75	83	74	100	82	77

Leaves	Stem	Leaves
8 8 6 5 5	7	4 7
9 5 4 3 3 2 1	8	2 2 4 5 7 7 7
8 7 6 5	9	2 4 4 4 5
	10	0 0

Use the data. 198, 195, 189, 185, 189, 190

5. Find the mean. **191**
6. Find the median. **189.5**
7. Find the mode. **189**
8. Find the range. **13**
9. Find the third quartile. **195**
10. Find the first quartile. **189**
11. Make a box-and-whisker plot.

CHAPTER 4 **Chapter Test**
Form C, continued

Use the data. 1128, 507, 1634, 989, 1350, 1275, 1647, 1301, 1035

12. Find the mean. **1207.3**
13. Find the median. **1275**
14. Find the mode. **none**
15. Find the range. **1140**
16. Find the third quartile. **1492**
17. Find the first quartile. **1012**
18. Make a box-and-whisker plot.

Use the test score data.
138, 115, 128, 117, 128, 129, 122, 124, 134, 120, 135, 131, 125, 133, 114

19. Make a frequency table with an interval of 10.

Scores	Frequency
110–119	3
120–129	7
130–139	5

20. Make a histogram

Test Scores

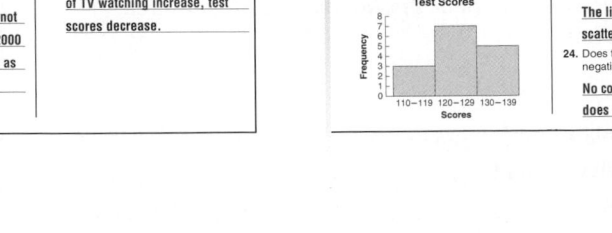

21. The graph shows the number of cars and trucks on a highway. Explain why the graph is misleading.

= 50 cars
= 50 trucks

100 cars   150 trucks

The size of the cars versus the size of the trucks distorts the comparison.

Test Scores	85	80	67	89	74	91	81	79	96	79
Hours of Exercise	3.0	8.0	3.0	8.5	1.0	8.0	5.5	9.5	2.0	2.5

22. Use the data above to construct a scatter plot. Draw a line of best fit.

23. Draw a line of best fit on the scatter plot you drew for problem 22.

   The line is shown on the above scatter plot.

24. Does the data set have a positive, negative, or no correlation? Explain.

   No correlation. Hours of exercise does not affect test scores.

Test and Practice Generator
CD-ROM

Create and customize multiple versions of the same tests with corresponding answers for any chosen chapter objectives.

Chapter 4 State and Standardized Test Preparation

Test Taking Skill Builder and Standardized Test Practice
are provided for each chapter in the *Test Prep Tool Kit.*

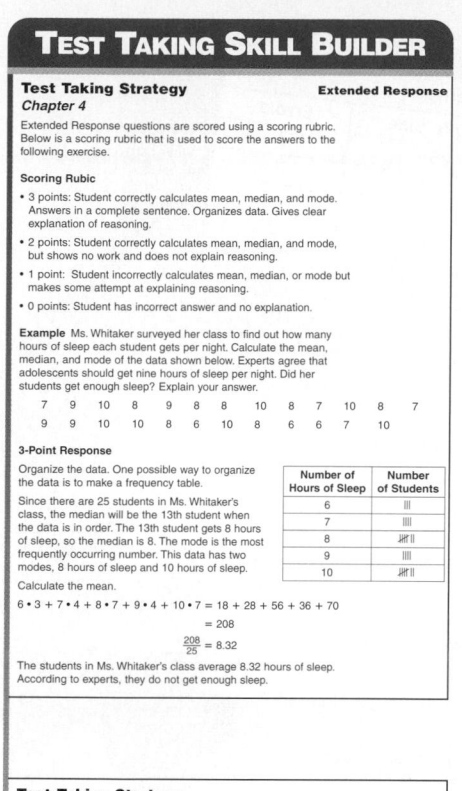

TEST TAKING SKILL BUILDER

Test Taking Strategy
Chapter 4   Extended Response

Extended Response questions are scored using a scoring rubric.
Below is a scoring rubric that is used to score the answers to the
following exercise.

Scoring Rubic

- 3 points: Student correctly calculates mean, median, and mode.
  Answers in a complete sentence. Organizes data. Gives clear
  explanation of reasoning.

- 2 points: Student correctly calculates mean, median, and mode,
  but shows no work and does not explain reasoning.

- 1 point: Student incorrectly calculates mean, median, or mode but
  makes some attempt at explaining reasoning.

- 0 points: Student has incorrect answer and no explanation.

Example Ms. Whitaker surveyed her class to find out how many
hours of sleep each student gets per night. Calculate the mean,
median, and mode of the data shown below. Experts agree that
adolescents should get nine hours of sleep per night. Did her
students get enough sleep? Explain your answer.

7   9   10   8   9   8   8   10   8   7   5
9   9   10   10   8   6   10   8   6   6   7   10

3-Point Response

Organize the data. One possible way to organize
the data is to make a frequency table.

Since there are 25 students in Ms. Whitaker's
class, the median will be the 13th student when
the data is in order. The 13th student gets 8 hours
of sleep, so the median is 8. The mode is the most
frequently occurring number. This data has two
modes, 8 hours of sleep and 10 hours of sleep.

Number of Hours of Sleep	Number of Students
6	III
7	IIII
8	JHH II
9	IIII
10	JHH II

Calculate the mean.

$6 \cdot 3 + 7 \cdot 4 + 8 \cdot 7 + 9 \cdot 4 + 10 \cdot 7 = 18 + 28 + 56 + 36 + 70$

$= 208$

$\frac{208}{25} = 8.32$

The students in Ms. Whitaker's class average 8.32 hours of sleep.
According to experts, they do not get enough sleep.

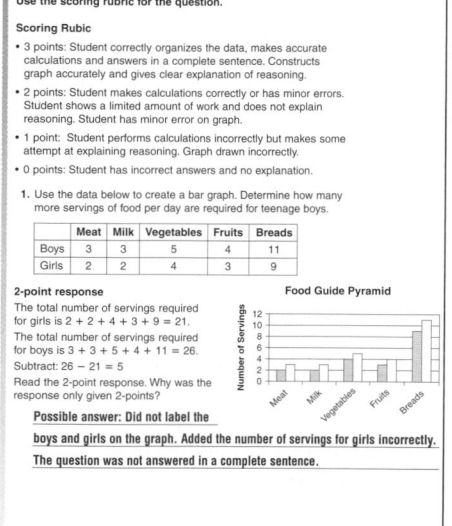

Test Taking Strategy
Chapter 4, continued

Exercises
Use the scoring rubric for the question.

Scoring Rubic

- 3 points: Student correctly organizes the data, makes accurate
  calculations and answers in a complete sentence. Constructs
  graph accurately and gives clear explanation of reasoning.

- 2 points: Student makes calculations correctly or has minor errors.
  Student shows a limited amount of work and does not explain
  reasoning. Student has minor error on graph.

- 1 point: Student performs calculations incorrectly but makes some
  attempt at explaining reasoning. Graph drawn incorrectly.

- 0 points: Student has incorrect answers and no explanation.

1. Use the data below to create a bar graph. Determine how many
   more servings of food per day are required for teenage boys.

	Meat	Milk	Vegetables	Fruits	Breads
Boys	3	3	5	4	11
Girls	2	2	4	3	9

2-point response

The total number of servings required
for girls is $2 + 2 + 4 + 3 + 9 = 21$.
The total number of servings required
for boys is $3 + 3 + 5 + 4 + 11 = 26$.
Subtract: $26 - 21 = 5$
Read the 2-point response. Why was the
response only given 2-points?

Food Guide Pyramid

Possible answer: Did not label the
boys and girls on the graph. Added the number of servings for girls incorrectly.
The question was not answered in a complete sentence.

STANDARDIZED TEST PRACTICE

Standardized Test Practice
Chapter 4

Select the best answer for Questions 1–8.

1. Every fifth customer entering the mall
   at the south entrance was surveyed.
   Identify the sampling method used.
   A Systematic   C Stratified
   B Random   D None

Use the table to answer questions 2 through 3.

	Population	Altitude (in feet)
Swanton	6170	213
Winters	20,140	4480
Margaretta	15,643	3280
Fremont	7210	1489

2. If you went from Winters to Fremont,
   how much would the altitude
   decrease?
   F 1200 feet   H 4267 feet
   G 2991 feet   I 5969 feet

3. Which is the best estimate of the
   average population of the four cities?
   A 10,000   C 20,000
   B 12,000   D 25,000

4. Using the following data, which
   statement is true? 7, 8, 6, 8, 3, 4
   F median = mean
   G median < mode
   H median < mean
   I median = mode

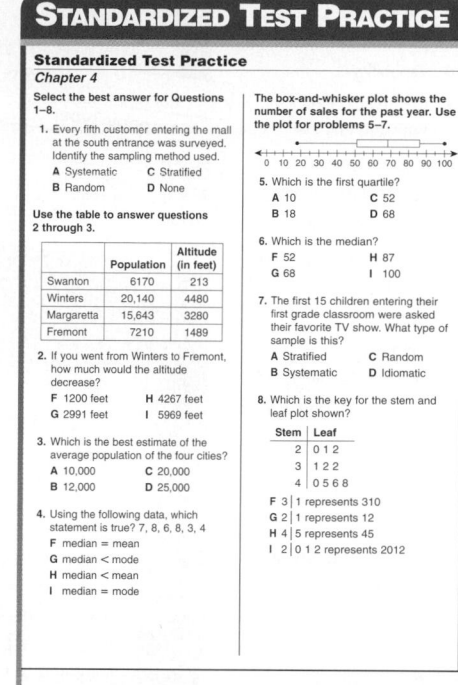

The box-and-whisker plot shows the
number of sales for the past year. Use
the plot for problems 5–7.

0 10 20 30 40 50 60 70 80 90 100

5. Which is the first quartile?
   A 10   C 52
   B 18   D 68

6. Which is the median?
   F 52   H 87
   G 68   I 100

7. The first 15 children entering their
   first grade classroom were asked
   their favorite TV show. What type of
   sample is this?
   A Stratified   C Random
   B Systematic   D Idiomatic

8. Which is the key for the stem and
   leaf plot shown?

Stem	Leaf
2	0 1 2
3	1 2 2
4	0 5 6 8

   F 3 | 1 represents 310
   G 2 | 1 represents 12
   H 4 | 5 represents 45
   I 2 | 0 1 2 represents 2012

Standardized Test Practice
Chapter 4, continued

Gridded Response
Solve the problems. Use the answer
sheet to write and grid-in your answer.

9. According to the stem-and-leaf plot,
   how many students are on the bus
   more than 15 minutes?

   Time on the School Bus

Stem	Leaf
1	1 2 2 4 4 5 6 7 8 9
2	0 5 6 8
3	2 2 4 7
4	0

10. Use the stem-and-leaf plot from
    Question 9. How many students
    are on the bus for more than
    30 minutes?

Short Response
Solve the problems. Use the answer
sheet to write your answers.

11. A scatter plot shows a random
    display of data. Describe in words
    the correlation, if any.

12. Explain in words the correlation, if
    any, between the cost of a pair of
    boots and the size of the boots.

13. Explain in words why it can be
    beneficial to draw a misleading
    graph.

Extended Response

14. The following data shows the
    percentage of employee attendance
    each month at a factory.

   Percentage of Employee Attendance

Jan.	86.4	July	86.7
Feb.	92.3	Aug.	98.5
March	94.6	Sept.	92.5
April	76.5	Oct.	98.7
May	89.1	Nov.	93.4
June	68.7	Dec.	87.6

   a. Construct a line graph based on the
      data in the table.

   b. Explain what, if any, pattern you see.

   c. Give an explanation for the pattern
      you described in part b.

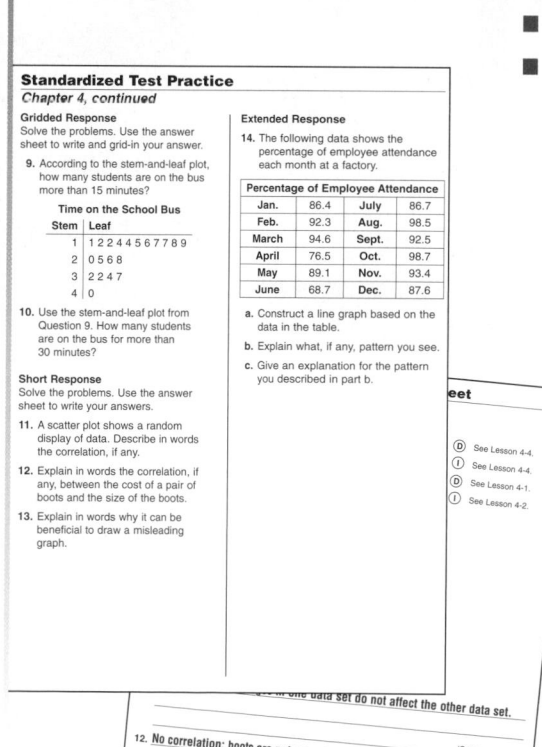

...one data set do not affect the other data set.

12. No correlation; boots are not priced by size.   (See Lesson 4-7.)

13. A misleading graph can be beneficial if you are trying to persuade
    someone into a certain viewpoint.   (See Lesson 4-7.)

Extended Response   (See Lesson 4-6.)
Write your answers for Problem 14 on the back of this paper.
See Lesson 4-2.

...eet

   (D) See Lesson 4-4.
   (I) See Lesson 4-4.
   (D) See Lesson 4-1.
   (I) See Lesson 4-2.

✓ internet connect
**State-Specific Test
Practice Online**
KEYWORD: MP4 TestPrep

Test Prep Tool Kit

- Standardized Test Prep Workbook
- Countdown to Testing transparencies
- State Test Prep CD-ROM
- Standardized Test Prep Video

Customized answer sheets give
students realistic practice for
actual standardized tests.

Collecting, Displaying, and Analyzing Data

Why Learn This?

Tell students that quality assurance is just one of many fields in which information from a small sample can be used to derive information about a large population. For example, a quality assurance specialist can use the information in the table to predict the number of errors in an entire month's production of software.

Using Data

To begin the study of this chapter, have students:

- Identify the company type with the largest number of items sampled. **stoneworks**

- Identify the company type with the fewest number of errors found. **software**

- Predict the number of errors that would be found if 100 items were sampled at a tool company. **8**

☑ internet connect
Chapter Opener Online
go.hrw.com
KEYWORD: MP4 CH4

Collecting, Displaying, and Analyzing Data

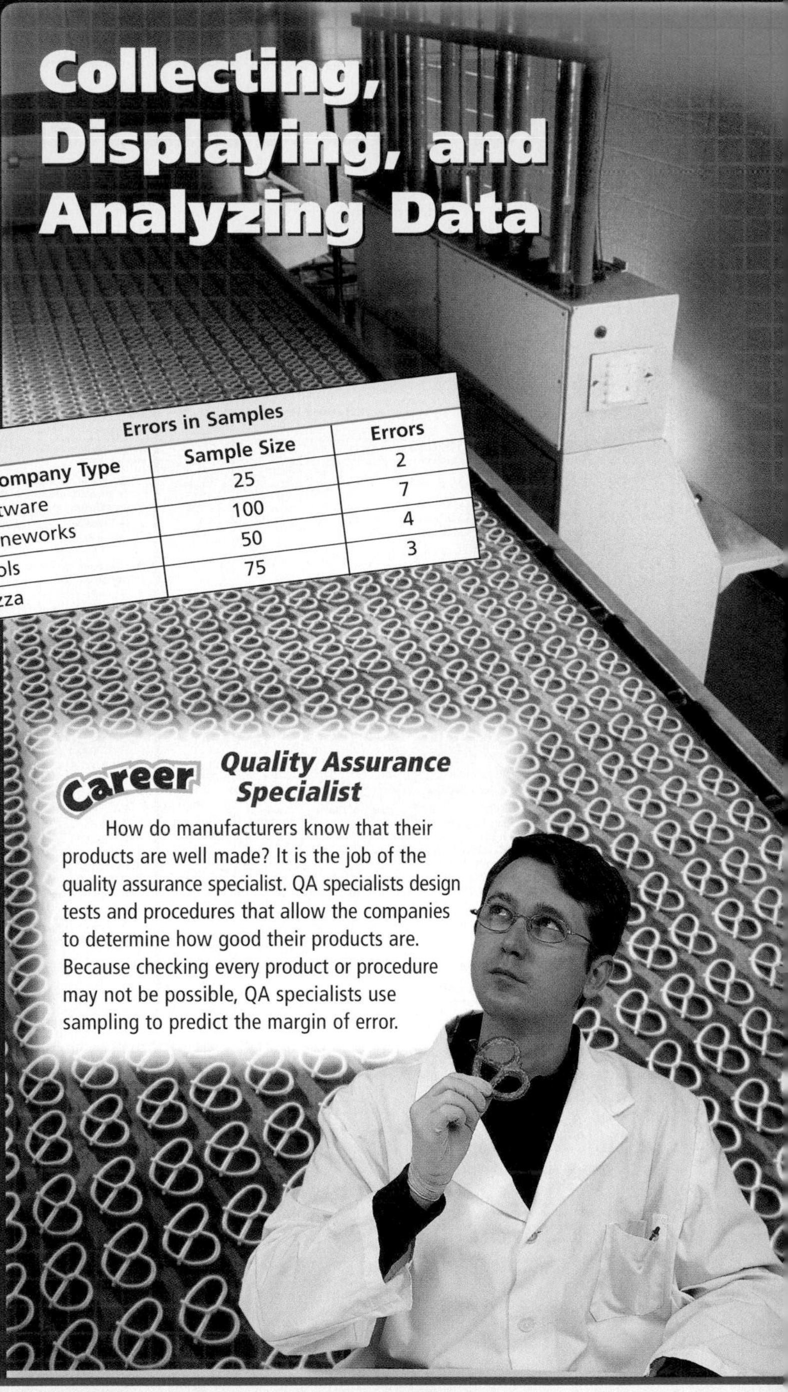

Errors in Samples		
Company Type	Sample Size	Errors
	25	2
Software	100	7
Stoneworks	50	4
Tools	75	3
Pizza		

Career *Quality Assurance Specialist*

How do manufacturers know that their products are well made? It is the job of the quality assurance specialist. QA specialists design tests and procedures that allow the companies to determine how good their products are. Because checking every product or procedure may not be possible, QA specialists use sampling to predict the margin of error.

Problem Solving Project

Social Studies Connection

Purpose: To use samples to estimate larger populations to solve problems

Materials: Making Quality Products worksheet

☑ internet connect
Chapter Project Online: *go.hrw.com*
KEYWORD: MP4 PSProject4

Understand, Plan, Solve, and Look Back

Have students:

✔ Complete the Making Quality Products worksheet to practice organizing data to describe results.

✔ Select a company and create a graph. Have them compare their company to the companies of other students.

✔ Discuss the role of quality assurance in companies. Why do they think this work is important?

✔ Research companies that have gotten into trouble because of their errors. What did they do to correct their problem?

✔ Check students' work.

ARE YOU READY?

Choose the best term from the list to complete each sentence.

1. A __?__ is a uniform measure where equal distances are marked to represent equal amounts. **scale**

2. __?__ is the process of approximating to a given __?__. **Rounding; place value**

3. Ordered pairs of numbers are graphed on a __?__. **coordinate grid**

coordinate grid
place value
rounding
scale

Complete these exercises to review skills you will need for this chapter.

✔ Round Decimals

Round each number to the indicated place value.

4. 34.7826; nearest tenth **34.8**

5. 137.5842; nearest whole number **138**

6. 287.2872; nearest thousandth **287.287**

7. 362.6238; nearest hundred **400**

✔ Compare and Order Decimals

Order each sequence of numbers from greatest to least.

8. 3.005, 3.05, 0.35, 3.5 **3.5; 3.05; 3.005; 0.35**

9. 0.048, 0.408, 0.0408, 0.48 **0.48; 0.408; 0.048; 0.0408**

10. 5.01, 5.1, 5.011, 5.11 **5.11; 5.1; 5.011; 5.01**

11. 1.007, 0.017, 1.7, 0.107 **1.7; 1.007; 0.107; 0.017**

✔ Place Value of Whole Numbers

Write each number in standard form.

12. 1.3 million **1,300,000**

13. 7.59 million **7,590,000**

14. 4.6 billion **4,600,000,000**

15. 2.83 billion **2,830,000,000**

✔ Read a Table

Use the table for problems 16–18.

16. Which activity experienced the greatest change in participation from 2000 to 2001? **basketball**

17. Which activity experienced the greatest positive change in participation from 2000 to 2001? **soccer**

18. Which activity experienced the least change in participation from 2000 to 2001? **softball**

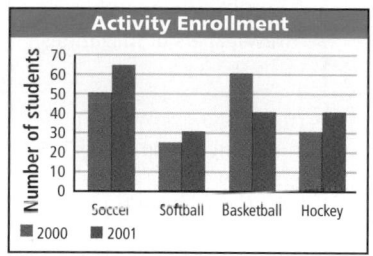

Activity Enrollment

Number of students — Soccer, Softball, Basketball, Hockey
■ 2000 ■ 2001

Assessing Prior Knowledge

INTERVENTION

Diagnose and Prescribe

Evaluate your students' performance on this page to determine whether intervention is necessary or whether enrichment is appropriate. Options that provide instruction, practice, and a check are listed below.

Resources for Are You Ready?

- **Are You Ready? Intervention and Enrichment**
- **Recording Sheet for Are You Ready?** Chapter 4 Resource Book . . . . . p. 3

 Are You Ready? Intervention CD-ROM

ARE YOU READY?

Were students successful with Are You Ready?

 **NO INTERVENE**

 **YES ENRICH**

 Round Decimals **4–7**
Intervention Practice, Skill 16
CD-ROM Intervention Activities, Skill 16

 Compare and Order Decimals **8–11**
Intervention Practice, Skill 17
CD-ROM Intervention Activities, Skill 17

 Place Value of Whole Numbers **12–15**
Intervention Practice, Skill 1
CD-ROM Intervention Activities, Skill 1

✔ **Read a Table** **16–18**
Intervention Practice, Skill 90
CD-ROM Intervention Activities, Skill 90

Are You Ready? Enrichment, pp. 411–412

Collecting and Describing Data

One-Minute Section Planner

Lesson	Materials	Resources
Lesson 4-1 Samples and Surveys **NCTM:** Data Analysis and Probability, Communication, Representation **NAEP:** Data Analysis and Probability 3a ☐ SAT-9 ☐ SAT-10 ☐ ITBS ☑ CTBS ☐ MAT ☐ CAT	**Optional** Teaching Transparency T2 (CRB)	• *Chapter 4 Resource Book,* pp. 6–14 • *Daily Transparency T1, CRB* • *Additional Examples Transparencies T3–T4, CRB* • *Alternate Openers: Explorations,* p. 29
Hands-On Lab 4A Explore Sampling **NCTM:** Data Analysis and Probability, Reasoning and Proof, Connections, Representation **NAEP:** Data Analysis and Probability 3d ☐ SAT-9 ☐ SAT-10 ☐ ITBS ☑ CTBS ☐ MAT ☐ CAT		• *Hands-On Lab Activities,* pp. 25–26
Lesson 4-2 Organizing Data **NCTM:** Algebra, Data Analysis and Probability, Communication, Representation **NAEP:** Data Analysis and Probability 1a ☑ SAT-9 ☑ SAT-10 ☑ ITBS ☑ CTBS ☑ MAT ☑ CAT	**Optional** Tape measures *(MK)*	• *Chapter 4 Resource Book,* pp. 15–24 • *Daily Transparency T5, CRB* • *Additional Examples Transparencies T6–T8, CRB* • *Alternate Openers: Explorations,* p. 30
Lesson 4-3 Measures of Central Tendency **NCTM:** Number and Operations, Data Analysis and Probability, Communication **NAEP:** Data Analysis and Probability 2a ☑ SAT-9 ☑ SAT-10 ☑ ITBS ☑ CTBS ☑ MAT ☑ CAT	**Optional** Teaching Transparency T10 (CRB) Newspapers or magazines	• *Chapter 4 Resource Book,* pp. 25–33 • *Daily Transparency T9, CRB* • *Additional Examples Transparencies T11–T13, CRB* • *Alternate Openers: Explorations,* p. 31
Lesson 4-4 Variability **NCTM:** Data Analysis and Probability, Communication **NAEP:** Data Analysis and Probability 2d ☐ SAT-9 ☐ SAT-10 ☐ ITBS ☐ CTBS ☐ MAT ☑ CAT	**Optional** Teaching Transparency T15 (CRB) Recording Sheet for Reaching All Learners (CRB, p. 78)	• *Chapter 4 Resource Book,* pp. 34–42 • *Daily Transparency T14, CRB* • *Additional Examples Transparencies T16–T18, CRB* • *Alternate Openers: Explorations,* p. 32
Technology Lab 4B Create Box-and-Whisker Plots **NCTM:** Data Analysis and Probability, Representation **NAEP:** Data Analysis and Probability 2b ☐ SAT-9 ☐ SAT-10 ☐ ITBS ☐ CTBS ☐ MAT ☐ CAT	**Required** Graphing calculators	• *Technology Lab Activities,* p. 22
Section 4A Assessment		• Mid-Chapter Quiz, SE p. 194 • Section 4A Quiz, AR p. 11 • *Test and Practice Generator* CD-ROM

SAT = *Stanford Achievement Tests*　　**ITBS** = *Iowa Test of Basic Skills*　　**CTBS** = *Comprehensive Test of Basic Skills/Terra Nova*
MAT = *Metropolitan Achievement Test*　　**CAT** = *California Achievement Test*
NCTM = Complete standards can be found on pages T29–T35.　　**NAEP** = Complete standards can be found on pages A54–A58.
SE = *Student Edition*　　**TE** = *Teacher's Edition*　　**AR** = *Assessment Resources*　　**CRB** = *Chapter Resource Book*　　**MK** = *Manipulatives Kit*

Section Overview

Sampling Methods and Biases

Hands-On Lab 4A, Lesson 4-1

Why? The value and usefulness of data depend on how it is obtained.

For most surveys, a sample of a population is polled to get the data. For the data to have as little bias as possible, appropriate sampling methods are required.

Sampling Method	How Members Are Chosen
Random	By chance
Systematic	According to a rule or formula
Stratified	At random from randomly chosen subgroups

Organizing Data

Lesson 4-2

Why? To be able to analyze data, it is necessary to organize it. Tables and stem-and-leaf plots provide two methods of organizing data.

Table

Tire Comparison			
Brand	Traction	Temperature	Tread Wear
Grabber	A	C	A
TigerPaw	B	B	A
TuffTire	B	A	A

A table is often a useful way to compare and contrast data.

Stem-and-Leaf Plot

Quiz Scores

Stems	Leaves
7	0 0
8	1 2 5 8 8
9	0 3 5
10	0

Key: 8|1 means 81

The 11 scores in this stem-and-leaf plot are 70, 70, 81, 82, 85, 88, 88, 90, 93, 95, and 100.

Central Tendency and Variability

Technology Lab 4B, Lessons 4-3, 4-4

Why? To analyze a set of numerical data, we need ways to describe it. Central tendency is a way to describe a data set using one number. Variability (variation) of a data set can be described by several methods.

Measures of Central Tendency

Quiz Scores:
70 70 81 82
85 88 88 90
93 95 100

Mean $= \dfrac{70 + 70 + 81 + 82 + 85 + 88 + 88 + 90 + 93 + 95 + 100}{11} \approx 85.6$

Median (the middle value, or average of two middle values): 88

Mode (the value or values that occur most often): 70 and 88

Measures of variability can be shown using a **box-and-whisker plot.**

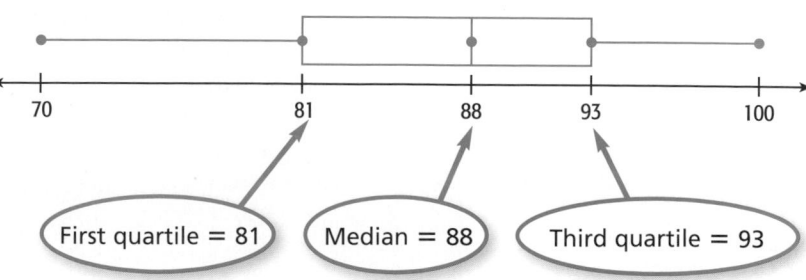

Range = largest value − smallest value

= 100 − 70 = 30

First quartile = 81   Median = 88   Third quartile = 93

Warm Up

Evaluate.

1. $t + 15$ for $t = -5$ 10

2. $n + (-13)$ for $n = 7$ -6

Solve for x.

3. $\frac{x}{-3} = -21$ $x = 63$

4. $7x = -98$ $x = -14$

Problem of the Day

Mr. Gray's 29 students will be sitting in the gym to watch a play. There are two rows of five chairs, three rows of four chairs, and four rows of two chairs. Is there enough room for all of the students to sit? yes

Available on Daily Transparency in CRB

Math Humor

Did you hear the news that everyone in the world now has a telephone? In a recent survey, 100% of the people who were called said that they had one.

4-1 Samples and Surveys

Learn to recognize biased samples and to identify sampling methods.

Vocabulary

population

sample

biased sample

random sample

systematic sample

stratified sample

A fitness magazine printed a readers' survey. Statements 1, 2, and 3 are interpretations of the results. Which do you think the magazine would use?

1. The average American exercises 3 times a week.
2. The average reader of this magazine exercises 3 times a week.
3. The average reader who responded to the survey exercises 3 times a week.

The **population** is the entire group being studied. The **sample** is the part of the population being surveyed.

For statement 1, the population is all Americans and the sample is readers of the fitness magazine who chose to respond. This is a **biased sample** because it is not a good representation of the population.

People who read fitness magazines are likely to be interested in exercise. This could make the sample biased in favor of people who exercise more times per week.

EXAMPLE **1** **Identifying Biased Samples**

Identify the population and sample. Give a reason why the sample could be biased.

A A radio station manager chooses 1500 people from the local phone book to survey about their listening habits.

Population	Sample	Possible Bias
People in the local area	Up to 1500 people who take the survey	Not all people are in the phone book.

B An advice columnist asks her readers to write in with their opinions about how to hang the toilet paper on the roll.

Population	Sample	Possible Bias
Readers of the column	Readers who write in	Only readers with strong opinions write in.

---

1 Introduce

Alternate Opener

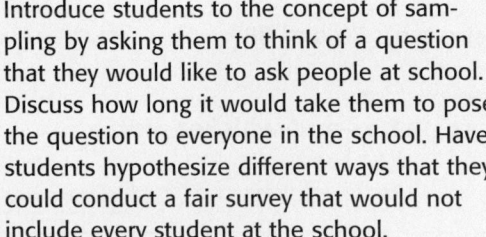

EXPLORATION

4-1 Samples and Surveys

A survey is being conducted to determine what items should be offered at the school snack bar. Instead of including every student in the survey, a method of selecting a fair representation of all students must be used.

1. Determine which method of selecting students might provide a sample that fairly represents all of the students in the school.

   Method 1: selecting students whose last names begin with A through M

   Method 2: selecting students who purchased snack bar items in the past week

   Method 3: selecting all eighth-grade students

   Method 4: selecting one-fourth of all students at random

   Method 5: selecting the first 100 students to come to the office to take the survey

2. Explain one way to make the method you chose in number 1 even more likely to give a representative sample.

Think and Discuss

3. **Explain** why the methods you did not choose in number 1 do not provide a representative sample of students.

Motivate

Introduce students to the concept of sampling by asking them to think of a question that they would like to ask people at school. Discuss how long it would take them to pose the question to everyone in the school. Have students hypothesize different ways that they could conduct a fair survey that would not include every student at the school.

Exploration worksheet and answers on Chapter 4 Resource Book pp. 7 and 80

---

2 Teach

Guided Instruction

In this lesson, students learn to recognize biased samples and identify sampling methods. Discuss the definitions of *population* and *sample* with them. While reviewing the examples (Teaching Transparency T2, CRB), discuss bias with the students. For Example 1A, you might ask students which groups of people might not be listed in the phone book (celebrities, homeless people, etc.) and how this might affect survey results.

Teaching Tip Students might have difficulty understanding the different sampling methods. Use the class as the population and create different samples using each of the methods given.

Identify the population and sample. Give a reason why the sample could be biased.

C Surveyors in a mall choose shoppers to ask about product preferences.

Population	Sample	Possible Bias
All shoppers in the mall	The people who are polled	Surveyors are more likely to approach shoppers who look agreeable.

To get accurate information, it is important to use a good sampling method. In a **random sample**, every member of the population has an equal chance of being chosen. A random sample is best, but other methods are often used for convenience.

Sampling Method	How Members Are Chosen
Random	By chance
Systematic	According to a rule or formula
Stratified	At random from randomly chosen subgroups

EXAMPLE **Identifying Sampling Methods**

Identify the sampling method used.

A An exit poll is taken of every tenth voter.

systematic   *The rule is to question every tenth voter.*

B In a statewide survey, five counties are randomly chosen and 100 people are randomly chosen from each county.

stratified   *The five counties are the random subgroups. People are chosen randomly from within the counties.*

C Students in a class write their names on strips of paper and put them in a hat. The teacher draws five names.

random   *Names are chosen by chance.*

Think and Discuss

1. Describe ways to eliminate the possible bias in Example 1C.

2. Decide which sampling method would be best to find the number of times a week the average student in your school exercises.

3 Close

Reaching All Learners
Through Home Connection

Have students ask a family member or neighbor to help them develop a survey. The survey can be about the neighborhood, something at school, or anything of interest to the student. Have students get the family member or neighbor to help them develop a survey question, determine a population, and decide on a sampling method that will be unbiased.

Possible answer: Question: What is your favorite type of music?; population: the entire school; method: systematic; survey every third person who walks into school at the beginning of the day.

Summarize

Review the vocabulary words from the lesson. Discuss biased samples and ask students why someone might intentionally use a biased sample. Have students give an example of each of the sampling methods discussed in the lesson.

Possible answers: Advertisers might use biased samples to help sell their products. Random: Five students' names are drawn out of hat. Systematic: A store owner surveys every third person who enters the store. Stratified: A phone surveyor chooses 5 random letters and then calls 20 random people whose last names begin with each of the letters.

Answers to Think and Discuss

Possible answers:

1. The surveyor could use the systematic method and poll every tenth person who walks by.

2. Stratified; randomly choose 10 classes and survey 10 people randomly chosen from each class.

4-1 Exercises

FOR EXTRA PRACTICE
see page 738

internet connect
Homework Help Online
go.hrw.com   Keyword: MP4 4-1

Students may want to refer back to the lesson examples.

GUIDED PRACTICE

See Example **1** Identify the population and sample. Give a reason why the sample could be biased.

1. A pet store owner surveys 100 customers to find out what brand of dog food is purchased most frequently.

See Example **2** Identify the sampling method used.

2. People with a house number ending with 1 are polled.   **systematic**

3. A surveyor flips through the phone book and selects 30 names. **random**

INDEPENDENT PRACTICE

See Example **1** Identify the population and sample. Give a reason why the sample could be biased.

4. A deli owner asks Sunday's customers to choose a favorite mustard.

See Example **2** Identify the sampling method used.

5. People at the theater seated in an even-numbered row and an odd-numbered seat are surveyed.   **systematic**

6. Ten study groups at the library are chosen, and one person is selected at random from each group.   **stratified**

PRACTICE AND PROBLEM SOLVING

Identify the population and sample. Give a reason why the sample could be biased.

7. A cafeteria worker asks students who buy the entrée if they like the food in the cafeteria.

8. A theater manager asks the last ten people to leave a movie, "Did you like the movie?"

9. A chef asks the first four customers who order the new cheese sauce if they like it.

10. A biologist studying trees samples blossoms of trees along the river.

Identify the sampling method used.

11. Every fifth name is called from a list of voters.   **systematic**

12. Students each write a question on a slip of paper and put it in a box. The teacher draws one question to discuss.   **random**

13. One hundred shoppers are chosen by chance from four randomly chosen computer stores.   **stratified**

Assignment Guide

If you finished Example **1** assign:
Core 1, 4, 7–10, 22–28
Enriched 1, 4, 7, 9, 18, 22–28

If you finished Example **2** assign:
Core 1–13, 22–28
Enriched 1–15 odd, 17–28

Answers

1. Population: pet store customers; sample: 100 customers; possible bias: not all customers have dogs.

4. Population: deli customers; sample: Sunday's customers; possible bias: the sample does not include week-day customers.

7. Population: students; sample: students who buy the entrée; possible bias: the students who buy the entrée may be the people who like the food in the cafeteria.

8. Population: theater customers; sample: last ten customers to leave; possible bias: the last ten people to leave may have enjoyed the movie more than those who left early.

9. Poplulation: restaurant customers; sample: first four customers who order the cheese sauce; possible bias: if the customers ordered cheese sauce, then they probably like cheese.

Math Background

The United States Census Bureau conducts a survey every ten years to determine the population of the United States. The first U.S. census taken was in 1790. Two hundred men rode on horseback to count the number of people in each household, and data was recorded on small scraps of paper. This census cost the government just $45,000!

Until 1960, the population survey was conducted mostly door to door. At that time, bureau officials realized that this method was inefficient because the population was growing rapidly and becoming more mobile and diverse. By 1980, 90% of the census was taken by mail.

RETEACH 4-1

Reteach
4-1 Samples and Surveys

A *survey* uses a *small sample* to represent a *large population*.

A state senator's office sends workers to ask constituents shopping at a local mall how they feel about floating a bond to acquire land around a reservoir to be preserved as open space.

The population in this survey are all the eligible voters of the state. The shoppers at the mall are the sample of that population.

Identify the population and the sample.

1. Kennedy HS seniors planning to attend the prom were asked if senior dues should include a photo taken at the prom.
   Population ___**all seniors at Kennedy HS**___
   Sample ___**Kennedy seniors planning to attend the prom**___

A *biased sample* is not a good representative of the population.

A school principal asks parents attending an art workshop if funding for a theater arts program should be included in the school budget. This is a biased sample since the parents attending an art workshop are likely to be in favor of additional art programs.

Identify the population and the sample. Give a reason why the sample could be biased.

2. NY homeowners within a 10-mile radius of a nuclear power plant were asked if they think the plant should be closed.
   Population ___**NY homeowners**___
   Sample ___**homeowners in a 10-mi radius of the plant**___
   Possible Bias: **Homeowners close to the plant are more likely to want it closed.**

Sampling Method	How Members Are Chosen
Random	By chance; members have an equal chance of selection
Systematic	According to a rule or pattern
Stratified	At random from randomly chosen subgroups

Identify the sampling method used.

3. A survey calls every 10th name listed in a local phone book.
   systematic

PRACTICE 4-1

Practice B
4-1 Samples and Surveys

Identify the population and sample. Give a reason why the sample could be biased.

1. At a convention of science teachers, various attendees are asked to name their favorite subject in high school.
   population ___**teachers at the convention**___
   sample ___**teachers surveyed**___
   possible bias ___**most will say science**___

2. Donors participating in a blood drive are given a small amount of money for their blood donation. Before they can give blood, each person is surveyed to find out if they are eligible to give blood.
   population ___**blood donors**___
   sample ___**blood donors (entire population)**___
   possible bias ___**people may lie to get money**___

3. Interviewers at the mall are surveying girls with red hair to find out if a correlation exists between personality and red hair.
   population ___**girls with red hair**___
   sample ___**girls surveyed**___
   possible bias ___**some girls color their hair**___

Identify the sampling method used.

4. People in the security line at the airport are asked to step out of the line for a more detailed search. The people pulled out of the line have not necessarily done anything wrong, and they are not chosen according to any particular rule.
   random

5. At the 1-mile marker of a marathon, a timekeeper shouts out the time elapsed to every 10th runner that passes by. A statistician records the times shouted.
   systematic

6. A geologist visits 10 randomly selected lakes in the region and collects soil samples in randomly selected areas along each shoreline.
   stratified

Identify the sampling method used.

14. A manufacturer tests every sixtieth item from an assembly line. **systematic**

15. Every third student signing up for an astronomy class is asked about telescope preferences. **systematic**

16. Fifteen classes are randomly chosen. Ten students are randomly chosen from each class. **stratified**

17. **BUSINESS** For an advertising campaign, you need to survey people to find out why they like to visit the San Diego Zoo.
    a. How can you select an unbiased sample for this survey?
    b. How can you make your sampling method systematic?
    c. Why would surveying only families with children be biased?

18. **MONEY** Martin sorted coins he had been collecting in a jar for 15 years, and decided that most coins in circulation are dated 1980.
    a. What is the population of this survey?
    b. Give a reason why the sample could be biased.

19. Possible answer: I would choose a systematic sampling method. I would ask every third person on the roster to fill out a form. This way I would get a good idea of my class's overall preference for location.

19. **WRITE ABOUT IT** To help plan your annual class picnic, you survey students in your class about where they want to have the picnic. Choose a sampling method and explain your choice.

20. **WHAT'S THE ERROR?** A distributor planned to take a stratified sample of restaurants to find out what the most commonly ordered food product was. Five restaurants were chosen at random, and at each restaurant every tenth customer was surveyed. Why isn't this a stratified sample?

20. Possible answer: To be a stratified sample, the customers at each restaurant should have been chosen at random, but instead they were chosen systematically.

21. **CHALLENGE** The diagrams show the locations that have been chosen where soil samples will be taken to test for pollution. Identify the type of sample each diagram represents.

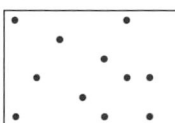 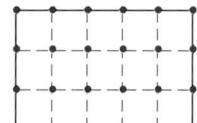 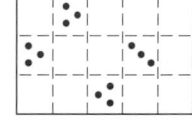

Spiral Review

Solve. (Lesson 3-6)

22. $x + \frac{1}{6} = -\frac{5}{6}$
$x = -1$

23. $\frac{y}{2.4} = -3$
$y = -7.2$

24. $y - 11.6 = -21$
$y = -9.4$

25. $23\frac{5}{7} - 24 = c$
$c = -\frac{2}{7}$

Solve. (Lesson 3-7)

26. $w + (-5.7) > -18.9$ $w > -13.2$

27. $-14.9x < -381.44$ $x > 25.6$

28. **TEST PREP** What is the next term in the sequence 5, 12, 26, 54, …? (Previous course) **D**

A   82   B   159   C   120   D   110

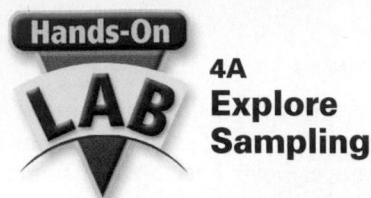

Explore Sampling

Use with Lesson 4-1

Pacing: Traditional $\frac{1}{2}$ day
Block $\frac{1}{3}$ day

Objective: To choose a sampling method, collect data, and summarize the results

Lab Resources

Hands-On Lab Activities . . . . . pp. 25–26

Using the Page

Discuss with students how the population choices, sampling-method choices, and uniform choices might affect the survey results differently.

Answers

Activity

1a. Possible answer: every student in the school

b. Possible answer: A random sample takes the least planning but could be unintentionally biased. A systematic sample is a little more difficult to execute but could give more reliable results. A stratified sample requires a great deal of planning but will ensure that all groups (e.g., grade levels) are represented fairly in the sample.

c. Possible answer: school colors (maroon and gray) and navy blue; pants, skirts, and vests

Think and Discuss

1. Possible answer: The teachers will not be wearing the uniforms and their clothing preferences may be different from the students'.

2. Possible answer: by selecting colors that seem the most popular when students choose their own school clothes

Try This

1. Check students' forms and tables.

REMEMBER
- Be organized before starting.
- Be sure that your sample reflects your population.

You can predict data about a population by collecting data from a representative sample.

Activity

Your school district has been discussing the possibility of school uniforms. Each school will get to choose its uniform and colors. Your class has been chosen to make the selection for your school. To be fair, you want to be sure that the other students in the school have input. Therefore, you take a survey to see what the majority of the students in your school want.

1 Follow the steps below to model conducting the survey.

  a. Choose your population.

    - every student in the school   • all boys
    - all girls   • all 8th grade students
    - only your class   • teachers

  b. Decide what kind of sample you will use. Discuss pros and cons of each.

    - random   • systematic   • stratified

  c. Decide what colors and what uniform choices to present to your sample.

    - pants   • sweaters   • school colors
    - shorts   • jackets   • navy blue
    - skirts   • vests   • forest green

Think and Discuss

1. Explain why choosing the teachers as your population might not be the best choice.

2. How did you decide which colors to present to your sample?

Try This

1. Create forms for your survey showing the options from which you want your sample to choose. Then survey your sample. Make a table of your results. Explain what your table tells you about the population.

Learn to organize data in tables and stem-and-leaf plots.

Vocabulary

stem-and-leaf plot

back-to-back stem-and-leaf plot

When you graduate and start looking for a job, you may have to keep track of a lot of information.

A table is one way to organize and display data so that you can understand the meaning and recognize any relationships.

Mathematics, physics, computer science, chemistry, and English are good courses to take if you want to be an airline mechanic.

EXAMPLE 1 Organizing Data in Tables

Use the given data to make a table.

Greg has received job offers as a mechanic at three airlines. The first has a salary range of $20,000–$34,000, benefits worth $12,000, and 10 days' vacation. The second has 15 days' vacation, benefits worth $10,500, and a salary range of $18,000–$50,000. The third has benefits worth $11,400, a salary range of $14,000–$40,000, and 12 days' vacation.

	Job 1	Job 2	Job 3
Salary Range	$20,000–$34,000	$18,000–$50,000	$14,000–$40,000
Benefits	$12,000	$10,500	$11,400
Vacation Days	10	15	12

A **stem-and-leaf plot** is another way to display data. The values are grouped so that all but the last digit is the same in each category.

$Stem = \text{first digit(s)}$

$$2 \,|\, 5 = 25$$

$Leaf = \text{last digit}$

EXAMPLE 2 Reading Stem-and-Leaf Plots

List the data values in the stem-and-leaf plot.

```
0 | 2 5
1 | 3 3 7 8
2 | 0 2 6
3 | 1 7        Key: 3|1 means 31
```

The data values are 2, 5, 13, 13, 17, 18, 20, 22, 26, 31, and 37.

4-2 Organizer

Pacing: Traditional 1 day
Block $\frac{1}{2}$ day

Objective: Students organize data in tables and stem-and-leaf plots.

Warm Up

Compare. Write $<$ or $>$.

1. $3(6)$ ▪ 15 $>$ **2.** $53 - 37$ ▪ 19 $<$

3. 27 ▪ $2(14)$ $<$ **4.** $49 - (-4)$ ▪ 51 $>$

Problem of the Day

If today is Wednesday, what day of the week will it be in 100 days?
Friday

Available on Daily Transparency in CRB

Math Humor

First the stem and the leaf conspired to steal water from the tree. Then they decided to take over the whole garden. It was a case of back-to-back stem-and-leaf plots.

---

1 Introduce

Alternate Opener

EXPLORATION

4-2 Organizing Data

1. Students in a class completed a survey in which they were asked how many hours they watched TV during one month. Their responses are shown below.

15	33	10	25	8
72	21	2	30	21
10	20	36	6	16
27	14	47	25	14
33	14	30	15	25

a. Write the numbers in order from least to greatest.

b. Describe how the numbers are spread out.

c. One student from the class was absent on the day of the survey. If the student were asked the same question, what number of hours would you guess might be close to the student's answer?

Think and Discuss

2. **Explain** how writing the numbers in order from least to greatest helps to organize data.

3. **Describe** another way to organize the data using a table.

Motivate

To introduce students to the opening scenario of the lesson, ask some of them to share with the class their career goals for the future. Ask if there are factors other than salary to consider when choosing among several job offers. Point out to students that a table makes it easy to compare data, especially when the data represents more than one category (such as salary, benefits, and vacation days).

Exploration worksheet and answers on Chapter 4 Resource Book pp. 16 and 82

---

2 Teach

Lesson Presentation

Guided Instruction

In this lesson, students learn to organize data in tables and stem-and-leaf plots. Discuss the roles of the column and row headings in tables.

Stem-and-leaf plots are effective for organizing a data set in which all the values represent the same characteristic. Show students how to order the data first. Then they can determine the stems and the leaves, create a key, and draw the plot. You may want to have students create a stem-and-leaf plot from data that include 3-digit values (e.g., 104, 95, 87, 98, 108, and 100).

Example 1

Use the given data to make a table.

Jack timed his bus rides to and from school. On Monday, it took 7 minutes to get to school and 9 minutes to get home. On Tuesday, it took 5 minutes and 9 minutes, respectively, and on Wednesday, it took 8 minutes and 7 minutes.

Day	To school	To home
Monday	7 min	9 min
Tuesday	5 min	9 min
Wednesday	8 min	7 min

Example 2

List the data values in the stem-and-leaf plot.

```
1 | 2 5
4 | 0 1 1
5 | 2 7 9    Key: 1|2 means 12
```

12, 15, 40, 41, 41, 52, 57, 59

Example 3

Use the given data to make a stem-and-leaf plot.

Top Speeds of Animals (mi/h)			
Cheetah	64	Elk	45
Wildebeest	61	Coyote	43
Lion	50	Gray Fox	42

```
4 | 2 3 5
5 | 0
6 | 1 4    Key: 4|2 means 42 mi/h
```

Example 4

Use the given data to make a back-to-back stem-and-leaf plot.

U.S. Representatives for Selected States, 1950 and 2000					
	IL	MA	MI	NY	PA
1950	25	14	18	43	31
2000	19	10	15	29	19

```
1950 |   | 2000
4 8  | 1 | 0 5 9 9
   5 | 2 | 9
   1 | 3 |       Key: |2|9 means 29
   3 | 4 |            8|1| means 18
```

EXAMPLE **3** Organizing Data in Stem-and-Leaf Plots

Use the given data to make a stem-and-leaf plot.

Heights of Tallest Trees in U.S. (m)					
Ash	47	Elm	38	Red maple	55
Beech	40	Grand fir	77	Sequoia	84
Black maple	40	Hemlock	74	Spruce	63
Cedar	67	Hickory	58	Sycamore	40
Cherry	42	Oak	61	Western pine	48
Douglas fir	91	Pecan	44	Willow	35

Heights range from 35 to 91, so stems are 3 to 9.

```
3 | 5 8
4 | 0 0 0 2 4 7 8
5 | 5 8
6 | 1 3 7
7 | 4 7
8 | 4
9 | 1          Key: 9|1 means 91 m
```

A **back-to-back stem-and-leaf plot** is used to compare two sets of data. The stems are in the center, and the left leaves are read in reverse.

EXAMPLE **4** Organizing Data in Back-to-Back Stem-and-Leaf Plots

Use the given data to make a back-to-back stem-and-leaf plot.

Super Bowl Scores, 1990–2000											
	1990	1991	1992	1993	1994	1995	1996	1997	1998	1999	2000
Winning	55	20	37	52	30	49	27	35	31	34	23
Losing	10	19	24	17	13	26	17	21	24	19	16

```
        Losing   |   | Winning
9 9 7 7 6 3 0    | 1 |
       6 4 4 1   | 2 | 0 3 7
                 | 3 | 0 1 4 5 7
                 | 4 | 9        Key:  |5|2 means 52 points
                 | 5 | 2 5           1|2| means 21 points
```

Think and Discuss

1. Tell which is always the same as the number of data values in a stem-and-leaf plot: the number of stems or the number of leaves.

Teach

Reaching All Learners
Through Grouping Strategies

Create groups of four or five students. Have them use the tape measures provided in the Manipulatives Kit to measure one another's height in inches and record the data. Then have each group create a stem-and-leaf plot to display the data. You may also want to combine the data and create a stem-and-leaf plot for the whole class.

3 Close

Summarize

Remind students that tables and stem-and-leaf plots are good ways to organize data and make it easy to understand. Review the three types of data displays in the lesson. Ask students to give an example of a situation that could be represented by each type of display.

Possible answer: A table could represent the nutritional content of different foods. A stem-and-leaf plot could show the number of people attending school functions. A back-to-back stem-and-leaf plot could be used to compare audience age at two different movies.

FOR EXTRA PRACTICE
see page 738

internet connect
Homework Help Online
go.hrw.com   Keyword: MP4 4-2

go.hrw.com

4-2 **PRACTICE & ASSESS**

GUIDED PRACTICE

See Example 1   **1.** Use the given data to make a table.

A 100 g serving of baked potato has 2.4 g fiber, 10 mg calcium (Ca), and 27 mg magnesium (Mg).

A 100 g serving of french fries has 3.2 g fiber, 10 mg Ca, and 22 mg Mg.

A 100 g serving of potato chips has 4.5 g fiber, 24 mg Ca, and 67 mg Mg. (*Source:* USDA)

See Example 2   List the data values in the stem-and-leaf plot.

2.
```
0 | 2 3 3 7        2, 3, 3, 7, 11, 13,
1 | 1 3 7 7 8      17, 17, 18, 20, 20,
2 | 0 0 7          27, 34, 34, 35, 35
3 | 4 4 5 5     Key: 3|5 means 35
```

3.
```
6 | 3 6 8        63, 66, 68, 73, 73, 75,
7 | 3 3 5 7      77, 80, 80, 81, 81, 90,
8 | 0 0 1 1      94, 95, 99
9 | 0 4 5 9     Key: 9|9 means 99
```

See Example 3   **4.** Use the given data to make a stem-and-leaf plot.

Atomic Numbers of Some Elements							
Hydrogen	1	Silver	47	Carbon	6	Titanium	22
Nitrogen	7	Barium	56	Argon	18	Bromine	35
Calcium	20	Iron	26	Krypton	36	Iodine	53

See Example 4   **5.** Use the given data to make a back-to-back stem-and-leaf plot.

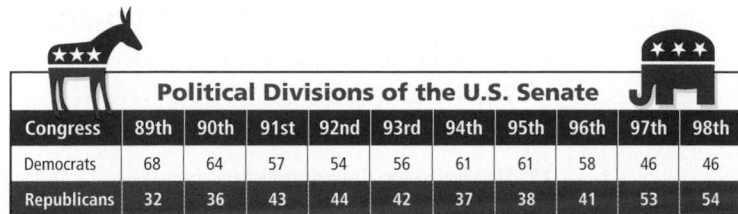

Political Divisions of the U.S. Senate									
Congress	89th	90th	91st	92nd	93rd	94th	95th	96th	97th	98th
Democrats	68	64	57	54	56	61	61	58	46	46
Republicans	32	36	43	44	42	37	38	41	53	54

INDEPENDENT PRACTICE

See Example 1   **6.** Use the given data to make a table.

New passenger car sales in 1970: 7,110,000 domestic, 313,000 Japanese imports, 750,000 German imports

New passenger car sales in 1980: 6,581,000 domestic, 1,906,000 Japanese imports, 305,000 German imports

New passenger car sales in 1990: 6,897,000 domestic, 1,719,000 Japanese imports, 265,000 German imports

Answers to Think and Discuss

1. Possible answer: the number of leaves; the number of stems is always less than or equal to the number of data values.

Students may want to refer back to the lesson examples.

Assignment Guide

If you finished Example 1 assign:
Core 1, 6, 13, 18–24
Enriched 1, 13, 15, 18–24

If you finished Example 2 assign:
Core 1–3, 6–8, 18–24
Enriched 1, 3, 6, 7, 13, 15, 18–24

If you finished Example 3 assign:
Core 1–4, 6–9, 11, 13, 18–24
Enriched 6–9, 11–13, 15, 17–24

If you finished Example 4 assign:
Core 1–10, 11, 13, 16, 18–24
Enriched 6–24

Answers

1.

Nutrition in Potatoes			
	Baked Potato (100 g)	French Fries (100 g)	Potato Chips (100 g)
Fiber	2.4 g	3.2 g	4.5 g
Ca	10 mg	10 mg	24 mg
Mg	27 mg	22 mg	67 mg

4.
```
0 | 1 6 7
1 | 8
2 | 0 2 6
3 | 5 6
4 | 7        Key: 1|8 means 18
5 | 3 6
```

5.
```
Democrats  |   | Republicans
           | 3 | 2 6 7 8
      6 6  | 4 | 1 2 3 4
  8 7 6 4  | 5 | 3 4            Key: |4|1 means 41
  8 4 1 1  | 6 |                     6|4| means 46
```

6.

New Passenger Car Sales			
	1970	1980	1990
Domestic	7,110,000	6,581,000	6,897,000
Japanese Imports	313,000	1,906,000	1,719,000
German Imports	750,000	305,000	265,000

9.

Tens of cents	Cents
9	3 5 5
10	2 6
11	1 1 3 4 7
12	1 3 3 4
13	0 8

Key: 11|1 means $1.11

10.

Coldest		Warmest
7 0	1	
7 5	2	
6 6 2	3	
8 6	4	0 3 4 7
	5	0 1 2 9
	6	2

Key: 7|1| means 17°
|4|0 means 40°

11.

4	3
5	7
6	5 8
7	2 2 3 5 6
8	1 2 4 8
9	1

Key: 5|7 means 57

12.

2	7
3	2 8
4	1 1 5 9
5	3 5 9
6	0 3 4 8

Key: 3|2 means 3.2

13.

Energy Use in U.S.			
	1980	1990	2000
Fossil Fuels	89%	86%	85%
Nuclear Power	3%	7%	8%
Renewable Resources	7%	7%	7%

14.

City		Highway
9 8 8 7 5 4 1	1	5 7
8 2 0	2	0 3 4 5 6 8 9
	3	6

Key: 7|1| means 17
|2|3 means 23

See Example 2 List the data values in the stem-and-leaf plot.

7.

5	0 1 4 8
6	2 6 7
7	1 4 5 6 6
8	2

50, 51, 54, 58, 62, 66, 67, 71, 74, 75, 76, 76, 82

Key: 6|2 means 62

8.

0	1 5 7
1	2 4 6 8
2	0 1 7 9
3	3 3 4 6

1, 5, 7, 12, 14, 16, 18, 20, 21, 27, 29, 33, 33, 34, 36

Key: 2|1 means 21

See Example 3 **9.** Use the given data to make a stem-and-leaf plot.

Average Price per Gallon of Unleaded Regular Gasoline by Year							
1981	$1.38	1986	$0.93	1991	$1.14	1996	$1.23
1982	$1.30	1987	$0.95	1992	$1.13	1997	$1.23
1983	$1.24	1988	$0.95	1993	$1.11	1998	$1.06
1984	$1.21	1989	$1.02	1994	$1.11	1999	$1.17

See Example 4 **10.** Use the data given in the map to make a back-to-back stem-and-leaf plot.

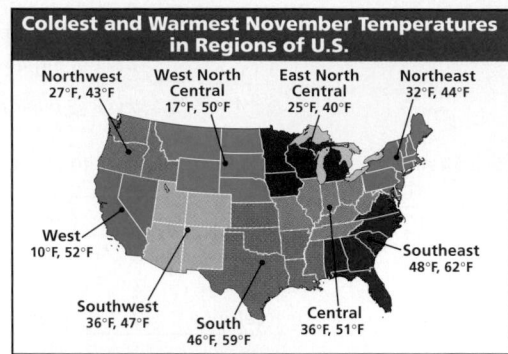

Coldest and Warmest November Temperatures in Regions of U.S.

Northwest 27°F, 43°F
West North Central 17°F, 50°F
East North Central 25°F, 40°F
Northeast 32°F, 44°F
West 10°F, 52°F
Southwest 36°F, 47°F
South 46°F, 59°F
Central 36°F, 51°F
Southeast 48°F, 62°F

PRACTICE AND PROBLEM SOLVING

Create a stem-and-leaf plot of the data values.

11. 72, 43, 75, 57, 81, 65, 68, 72, 73, 84, 91, 76, 82, 88

12. 5.3, 6.8, 3.2, 6.4, 2.7, 4.9, 6.3, 5.5, 4.1, 3.8, 6.0, 4.1, 4.5, 5.9

13. Use the given data to make a table.

In 1980, 89% of energy used in the United States was from fossil fuels, 3% from nuclear power, and 7% from renewable sources. In 1990, 86% was from fossil fuels, 7% from nuclear power, and 7% from renewable sources. In 2000, 85% was from fossil fuels, 8% from nuclear power, and 7% from renewable sources.

14. Use the given data to make a back-to-back stem-and-leaf plot.

Miles per Gallon Ratings of a Car Company's Models										
Model	A	B	C	D	E	F	G	H	I	J
City Miles	11	17	28	19	18	15	18	22	14	20
Highway Miles	15	24	36	28	26	20	23	25	17	29

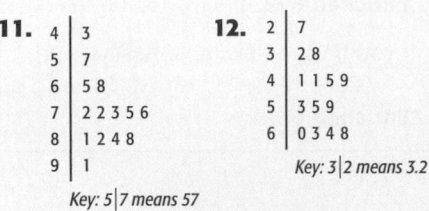

Math Background

Data is easier to analyze if it is organized in some way. Graphs, charts, and tables are three of the most popular ways to organize data. Many people may not be as familiar with stem-and-leaf plots.

Using a stem-and-leaf plot helps to see how the data is distributed. The median and the mode, two of the measures of central tendency studied in the next section, are easily determined from a stem-and-leaf plot.

RETEACH 4-2

LESSON **Reteach**
4-2 *Organizing Data*

Horizontal displays of numbers can be written in a compact form by eliminating repetition.

60 61 63 66 66 67 can be written 6 | 0 1 3 6 6 7

All the numbers start with 6, the **stem**.

Just write the second digit for each number, the **leaves**.

Display each set of numbers in compact form, using a stem and leaves.

1. 72 75 75 76 76 76 79
7 | 2 5 5 6 6 6 9

2. 120 123 124 125 125 127
12 | 0 3 4 5 5 7

Here are the scores on the last test in Ms. Kahn's math class.
76 84 88 93 97 65 100 86 91 97
93 79 81 99 92 78 78 79 87 100

To display these scores in a **stem-and-leaf plot**:
Use the given order to record scores.

6	5
7	6 9 8 8 9
8	4 8 6 1 7
9	3 7 1 7 3 9 2
10	0 0

Key: 7 | 6 represents 76.

Order each set of leaves.

6	5
7	6 8 8 9 9
8	1 4 6 7 8
9	1 2 3 3 7 7 9
10	0 0

Key: 7 | 6 represents 76.

Complete a stem-and-leaf plot for the data.

3. Daily High Temperatures
46 52 48 47 56 59 61 50 37 35 34 47 44 49 43
43 44 50 50 52 53 50 48 44 46 39 37 32 32 32

3	7 5 4 7 9 7 2 2 2
4	6 8 7 4 9 3 4 4 8 6 4
5	2 6 9 0 0 0 2 3 0
6	1

Key: 5 | 2 represents 52.

3	2 2 2 4 5 7 7 7 9
4	3 3 4 4 4 6 6 7 8 8 9
5	0 0 0 0 2 2 3 6 9
6	1

Key: 5 | 2 represents 52.

4. Create a stem-and-leaf plot for heights in inches of students in Mrs. Gray's class.
40 48 49 53 60 42 48 62 55 53 54 60 65 63 49 55

4	0 8 8 9 9
5	3 3 4 5 5
6	0 0 2 2 3 5

Key: 4 | 0 represents 40.

PRACTICE 4-2

LESSON **Practice B**
4-2 *Organizing Data*

1. Use the given data to make a table.

In the 2000 basketball season Nick scored 57 free throws, 137 two-pointers, and 18 three-pointers. In the 2001 season Nick scored 61 free throws, 176 two-pointers, and 35 three-pointers.

Season	Free-throws	Two-pointers	Three-pointers
2000	57	137	18
2001	61	176	35

List the data values in the stem-and-leaf plot.

2.

2	0 1 5 7
3	2 2 9
4	5 6 7 9
5	1 1 3

Key: 5 | 1 = 51

20, 21, 25, 27, 32, 32, 39, 45, 46, 47, 49, 51, 53

3.

6	1 5 5
7	0 3 4 7
8	2
9	0 3 8 9

Key: 9 | 0 = 90

61, 65, 65, 70, 73, 74, 77, 82, 90, 93, 98, 99

4. Use the given data to make a stem-and-leaf plot.

Raquel's Quiz Scores			
84	83	94	71
76	90	88	83
75	87	68	96

6	8
7	1 5 6
8	3 3 4 7 8
9	0 4 6

key: 9 | 0 = 90

5. Use the given data to make a back-to-back stem-and-leaf plot.

NBA Midwest Division 2000–2001 Final Standings					
NBA Team	Wins	Losses	NBA Team	Wins	Losses
San Antonio Spurs	58	24	Houston Rockets	45	37
Utah Jazz	53	29	Denver Nuggets	40	42
Dallas Mavericks	53	29	Vancouver Grizzlies	23	59
Minnesota Timberwolves	47	35			

Wins		Losses
3	2	4 9 9
	3	5 7
7 5 0	4	2
8 3 3	5	9

Key: |5|9 represents 59

3|5| represents 53

An author's writing style is as unique as a fingerprint. Punctuation, spelling, and word usage can be used to determine authorship.

Don Foster used this fact to analyze the 350-year-old poem "A Funeral Elegy." The analysis confirmed that the poem of previously unknown authorship was actually written by William Shakespeare.

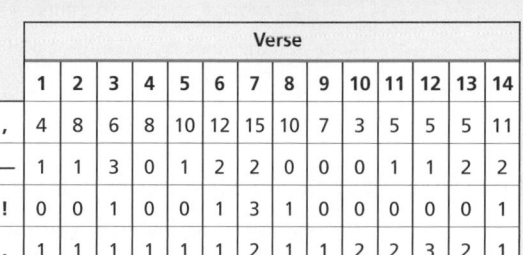

Don Foster's methods have also been used to analyze ransom notes and evidence in court cases.

15. Act 5 of Shakespeare's *A Midsummer Night's Dream* has the following references to numbers: 1 nine times, 2 three times, 3 six times, 10 two times, 12 one time, and 14 one time. There are also references to time: night 12 times, day 4 times, supper-time 1 time, bed-time 1 time, and evening 1 time. Use the data to make a table.

16. The table shows the punctuation in Henry Wadsworth Longfellow's poem "Paul Revere's Ride." Make a back-to-back stem-and-leaf plot of the number of commas and periods in each verse.

	Verse													
	1	2	3	4	5	6	7	8	9	10	11	12	13	14
,	4	8	6	8	10	12	15	10	7	3	5	5	5	11
—	1	1	3	0	1	2	2	0	0	0	1	1	2	2
!	0	0	1	0	0	1	3	1	0	0	0	0	0	1
.	1	1	1	1	1	1	2	1	1	2	2	3	2	1

17. ⭐ **CHALLENGE** Select two paragraphs from a work by your favorite author and a third paragraph by a different author. Compare word choices or punctuation use in the three paragraphs, and explain the similarities and differences. Use a table or stem-and-leaf plot to support your argument. **Check students' work.**

Spiral Review

Multiply or divide. Write the product or quotient as one power. (Lesson 2-7)

18. $\frac{7^4}{7^2}$ **7^2**

19. $5^3 \cdot 5^8$ **5^{11}**

20. $\frac{t^8}{t^5}$ **t^3**

21. $\frac{10^9}{9^3}$ **cannot combine**

Identify the population and sample. (Lesson 4-1)

22. A cable company surveys customers whose last names begin with an *S*.

23. The principal asks every other busload of students if their ride was comfortable.

24. **TEST PREP** Which number is less than 10^3? (Lesson 2-6) **B**

A 2^{10} **B** 25^2 **C** 8^4 **D** 7^5

Answers

15.

Numbers		Time	
One	9	Night	12
Two	3	Day	4
Three	6	Supper-time	1
Ten	2	Bed-time	1
Twelve	1	Evening	1
Fourteen	1		

16.

Commas		Periods
8 8 7 6 5 5 5 4 3	0	1 1 1 1 1 1 1 1 1 2 2 2 2 3
5 2 1 0 0	1	

Key: |0|2 means 2
5|1| means 15

22. population: cable customers; sample: customers whose last names begin with an *S*

23. population: students; sample: students on every other bus

Journal

Ask students to write why a back-to-back stem-and-leaf plot is a good way to show the data in Example 4 of the lesson.

Test Prep Doctor

For Exercise 24, encourage students to evaluate 10^3, and then evaluate each of the answer choices. For 2^{10}, suggest grouping as follows: $2^5 \cdot 2^5 = 32 \cdot 32$.

Lesson Quiz

1. Use the given data to make a table. There are three houses. The first house has 3 bedrooms, 2 bathrooms, and a full basement. The second house has 4 bedrooms, 3 bathrooms, and a partial basement. The third house has 5 bedrooms, 2.5 bathrooms and no basement.

	House 1	House 2	House 3
Bedrooms	3	4	5
Bathrooms	2	3	2.5
Basement	Full	Partial	None

2. List the data values in the stem-and-leaf plot. **6, 9, 14, 17, 22, 25**

```
0 | 6 9
1 | 4 7
2 | 2 5   Key: 1 | 5 means 15
```

Available on Daily Transparency in CRB

CHALLENGE 4-2

LESSON 4-2 Challenge
Think Out of the Box

To assist in drawing a conclusion, a table can be used to store facts. Bill Doctor, Jill Mason, and Phil Tailor are a doctor, a mason, and a tailor. Their occupations do not match their last names. Bill Doctor is the tailor's nephew. Who is the doctor?

Construct a table. List the facts, insert them in the table, and draw conclusions until you answer the question.

	doctor	mason	tailor
Bill Doctor	a. no	c. yes	b. no
Jill Mason		a. no	
Phil Tailor	e. yes	d. no	a. no

a. The occupation does not match the name.
b. Bill is the tailor's nephew. So, Bill is not the tailor.
c. Bill is not the doctor or the tailor. So, Bill is the mason.
d. Since Bill is the mason, Phil is not the mason.
e. Since Phil is not the mason or the tailor, **Phil is the doctor.**

Make a table to reason out each situation.

1. Jessie, Rachi, and Sara are three sisters who enjoy different sports. One likes swimming, another plays soccer, and the third plays softball. Find the favorite sport of each sister if: **a.** Sara does not like ball games. **b.** Rachi is the youngest. **c.** The oldest likes soccer.

	swimming	soccer	softball
Jessie	a. no	c, b. yes	no
Rachi	a. no	c, b. no	yes
Sara	a. yes	a. no	a. no

Jessie likes: __soccer__   Rachi likes: __softball__   Sara likes: __swimming__

2. Al, Cal, and Sal are married to Fran, Jan, and Nan, but not necessarily in that order. **a.** Cal is not married to Nan, and Jan is not married to Sal. **b.** Cal's wife is Fran's best friend. **c.** Nan is Al's sister. Name the married couples.

Al/Fran; Cal/Jan; Sal/Nan

	Fran	Jan	Nan
Al	yes	b. no	c. no
Cal	b. no	b. yes	a. no
Sal	no	a. no	yes

PROBLEM SOLVING 4-2

LESSON 4-2 Problem Solving
Organizing Data

A consumer survey in July of 2000 gathered the following data about what teens do while on online. 95% use e-mail, 86% use search engines, 82% use instant messaging, 73% visit music sites, 73% enter contests.

1. Make a table of the data.

Teens' Activities Online

Activity	Percent
E-mail	95
Search Engines	86
Instant Messaging	82
Music Sites	73
Enter Contests	73

2. Make a stem-and-leaf plot of the data.

Teens' Activities Online
```
7 | 3 3
8 | 2 6
9 | 5
```

Use the stem-and-leaf plot that shows the total medals won by different countries in the 2000 Summer Olympics. Choose the letter for the best answer.

3. List all the data values in the stem-and-leaf plot.

A 2, 4, 5, 6, 7, 8, 9
B 23, 25, 26, 28, 29, 34, 38, 40, 57, 58, 59, 60, 70, 88, 97
C 23, 25, 26, 28, 29, 34, 38, 57, 58, 59, 88, 97
D 23, 25, 26, 28, 29, 34, 38, 57, 58, 59, 88, 97

2000 Olympic Medals
```
2 | 3 5 6 8 8 9
3 | 4 8
4 |
5 | 7 8 9
6 |
7 |
8 | 8
9 | 5
```

4. What is the smallest number of medals won by a country represented in the stem-and-leaf plot?

F 3   **H** 23
G 4   **J** 97

5. What is the greatest number of medals won by a country represented in the stem-and-leaf plot?

A 9   **C** 79
B 70   **D** 97

Pacing: Traditional 1 day
Block $\frac{1}{2}$ day

Objective: Students find appropriate measures of central tendency.

Warm Up

Order the values from least to greatest.

1. 9, 4, 8, 7, 6, 8, 5, 3, 7
3, 4, 5, 6, 7, 7, 8, 8, 9

2. 36, 22, 35, 46, 37, 47, 30
22, 30, 35, 36, 37, 46, 47

Divide.

3. $\frac{198}{3}$ **66**   **4.** $\frac{576}{4}$ **144**

Problem of the Day

A mom buys a white, a green, a blue, and a yellow sweater for her 4 children. Bill and Bob refuse to wear yellow. Barb doesn't like green. Beth hates green and white. Mom will not put the boys in white, and Bob won't wear blue. Which sweater will each child wear? Barb: white, Beth: yellow, Bob: green, Bill: blue

Available on Daily Transparency in CRB

4-3 Measures of Central Tendency

Vocabulary
mean
median
mode
outlier

A measure of central tendency is an attempt to describe a data set using only one number. This number represents the "middle" of the set.

	Measures of Central Tendency	
	Definition	**Use to Answer**
Mean	The sum of the values, divided by the number of values	"What is the average?" "What single number best represents the data?"
Median	If an odd number of values: the middle value If an even number of values: the average of the two middle values	"What is the halfway point of the data?"
Mode	The value or values that occur most often	"What is the most common value?"

EXAMPLE 1 Finding Measures of Central Tendency

Find the mean, median, and mode of each data set.

A 4, 8, 8, 3, 6, 8, 3

mean: $4 + 8 + 8 + 3 + 6 + 8 + 3 = 40$ *Add the values.*

   $\frac{40}{7} \approx 5.7$ *Divide by 7, the number of values.*

median: 3 3 4 ⑥ 8 8 8   *Order the values.*
   3 values 3 values
   The median is 6.

mode: 8   *The value 8 occurs three times.*

B 9, 6, 91, 5, 7, 6, 8, 8, 7, 9

mean: $9 + 6 + 91 + 5 + 7 + 6 + 8 + 8 + 7 + 9 = 156$

   $\frac{156}{10} = 15.6$ *Divide by 10.*

median: 5 6 6 7 ⑦⑧ 8 9 9 91   *Order the values.*
   5 values 5 values

   $\frac{7 + 8}{2} = 7.5$ *Average the two middle values.*

mode: 6, 7, 8, 9   *Four values occur twice each.*

---

1 Introduce

Alternate Opener

EXPLORATION

4-3 Measures of Central Tendency

1. Students in a class completed a survey in which they were asked what their heights in inches are. Their responses are shown below.

49	53	60	55	48
72	65	66	58	68
75	65	64	57	59
61	67	64	58	62
63	59	61	55	65

a. Write the numbers in order from least to greatest.
b. What number appears most often?
c. What number is in the middle of the data set?
d. Add the numbers and divide the total by 25 to find the mean.

Think and Discuss
2. Discuss which number from 1b, 1c, and 1d best represents the entire set of numbers. Why?
3. Describe a set of numbers arranged from least to greatest in which the middle number is not close to the value of the mean.

Motivate

Ask students how teachers decide what grade to give each student at the end of a course. Explain that final grades are often based on the average of the grades earned during a grading period. Discuss the meaning of *average* and ask students for some other situations in which averages might be used.

Exploration worksheet and answers on Chapter 4 Resource Book pp. 26 and 84

---

2 Teach

Lesson Presentation

Guided Instruction

In this lesson, students learn to find appropriate measures of central tendency. Explain that the purpose of a measure of central tendency is to represent a data set with a single value. Point out that there are three different values that can generally be used to represent a set: *mean* (also called *average*), *median,* and *mode.* Discuss the definitions of these three terms (Teaching Transparency T10, CRB).

Emphasize that students must order the data values before finding the median. This will also help them to identify the mode. Remind students that some data sets will have more than one mode and others will have none.

Find the mean, median, and mode.

C 28, 12, 101, 53

mean: $28 + 12 + 101 + 53 = 194$

$$\frac{194}{4} = 48.5$$

median: $12 \; (28 \; | \; 53) \; 101$

$$\frac{28 + 53}{2} = 40.5$$

mode: No mode *No value occurs more than any other.*

Notice that the mean in Example 1B is much greater than most of the data values. This is because 91 is so far from the other data values. An extreme value such as this is called an **outlier**. An outlier can have a strong effect on the mean of a data set.

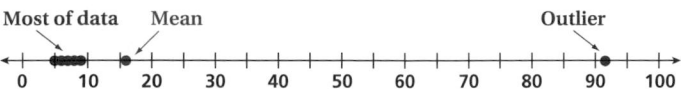

Most of data   Mean   Outlier

0   10   20   30   40   50   60   70   80   90   100

EXAMPLE 2 *Astronomy Application*

Astronomy LINK

Terrestrial planets are small, rocky planets that are close to the Sun. Gas giants are much larger and do not have a solid surface.

go.hrw.com
KEYWORD:
MP4 Moons

CNN student News.

Use the data to find each answer.

A Find the average number of moons for the *terrestrial planets*: Mercury, Venus, Earth, and Mars.

Use the mean to answer, "What's the average?"

$$\frac{0 + 0 + 1 + 2}{4} = \frac{3}{4} = 0.75$$

B Find the average number of moons for the *gas giants*: Jupiter, Saturn, Uranus, and Neptune.

$$\frac{39 + 30 + 21 + 8}{4} = \frac{98}{4} = 24.5$$

C Find the average number of moons per planet.

$$\frac{0 + 0 + 1 + 2 + 39 + 30 + 21 + 8 + 1}{9} = \frac{102}{9} \approx 11.33$$

Planet	Known Moons
Mercury	0
Venus	0
Earth	1
Mars	2
Jupiter	39
Saturn	30
Uranus	21
Neptune	8
Pluto	1

Source: NASA, 2002

Think and Discuss

1. Compare the mean and median of the set 1, 2, 3, and 4 to the mean and median of the set 1, 2, 3, and 40. Explain the difference.

2. Give a data set with the same mean, median, and mode.

---

3 Close

Summarize

Ask students to define *mean, median,* and *mode* in their own words. Discuss which measure of central tendency would be the best indicator of a student's grade in a class.

Possible answers: The mean is the sum of the values divided by the number of values. The median is the number in the middle or the average of the two numbers in the middle. The mode is the number or numbers that appear most often. The mean is most often used for a grade because it is based on all grades earned. The median is based on the number of grades but does not reflect the range, and there might not be a mode for the grades.

Answers to Think and Discuss

Possible answers:

1. The medians are the same: 2.5. Because of the outlier, the mean of the second set is much larger than the mean of the first set: 11.5 and 2.5.

2. {1, 1, 2, 2, 2, 3, 3}

Additional Examples

Example 1
Find the mean, median, and mode of each data set.

A. 16, 25, 31, 14, 14, 18
mean $\approx$ 19.67; median = 17; mode = 14

B. 83, 45, 19, 33
mean = 45; median = 39; mode = none

C. 21, 21, 28, 29, 30, 28, 32
mean = 27; median = 28; mode = 21 and 28

Example 2
Use the data to find each answer.

Building	Stories
Chongqing Tower, China	114
Sears Tower, U.S.	110
Empire State Building, U.S.	102
Bank of China, China	72
Amoco Building, U.S.	80
Chrysler Building, U.S.	77

A. Find the average number of stories for the buildings that have more than 100 stories. $\approx$**108.67**

B. Find the average number of stories for the buildings in the United States. **92.25**

C. Find the average number of stories for all of the buildings in the table. **92.5**

Answers

24. range: 14,100, 000; first quartile: 3,800,000; third quartile: 11,700,000

26. Possible answer: The student subtracted the first and last data values without ordering them first. The correct range is 54 − 14 = 40.

27. Possible answer: Box-and-whisker plots show how the data are spread out from the median. They also show the range and quartiles.

Journal

Refer students to Example 3 in the lesson. Ask students to write about whether they think Babe Ruth or Mark McGwire was the better home run hitter and why.

Test Prep Doctor ✚

For Exercise 36, students might simply find the product of the three numbers given and choose answer choice **D**. Encourage students to draw a chart to solve the problem. The correct answer is the least common multiple of the numbers, 12, which is choice **C**.

---

24. GEOGRAPHY Find the range and quartiles of the areas of Earth's continents, in square miles: Africa, 11,700,000; Antarctica, 5,400,000; Asia, 17,400,000; Europe, 3,800,000; North America, 9,400,000; Oceania, 3,300,000; South America, 6,900,000.

25. Match each set of data with a box-and-whisker graph.

a. range: 16
first quartile: 22
third quartile: 34   **Data set C**

b. range: 48
first quartile: 5
third quartile: 40   **Data set A**

c. range: 35
first quartile: 10
third quartile: 35
Data set B

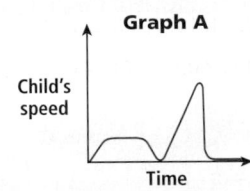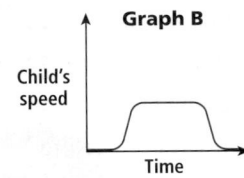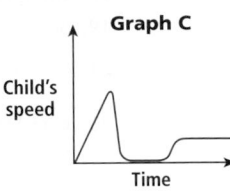

26. WHAT'S THE ERROR? A student wrote that the data set 22, 16, 45, 17, 18, 29, 22, 14, 32, 54 has a range of 32. What's the error?

27. WRITE ABOUT IT What do box-and-whisker plots tell you about data that measures of central tendency do not?

28. CHALLENGE What would an exceptionally short box with extremely long whiskers tell you about a data set?
Possible answer: The minimum and maximum values are outliers.

Spiral Review

Give the missing *y*-coordinates that are solutions to *y* = 4*x* − 2. (Lesson 1-8)
29. (0, *y*) **−2**   **30.** (1, *y*) **2**   **31.** (3, *y*) **10**   **32.** (7, *y*) **26**

Match each graph to one of the given situations. (Lesson 1-9)

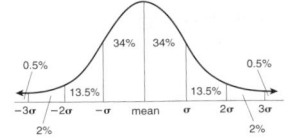

33. Emily sits on bench. Emily runs to buy treat. Emily sits to eat the treat. **Graph B**

34. Zen climbs ladder. Zen slides down slide. Zen sits and laughs. **Graph A**

35. Josh runs to catch bus. Josh sits on bus. Josh walks into school. **Graph C**

36. TEST PREP Claire visits her grandmother every 4 weeks, washes her car every 3 weeks, and gets paid every 2 weeks. How often will all three things happen in the same week? (Previous course) **C**

A every 8 weeks   **B** every 9 weeks   **C** every 12 weeks   **D** every 24 weeks

---

Lesson Quiz

Find the range and the first and third quartiles for each data set.

1. 48, 52, 68, 32, 53, 47, 51
range = 36; Q1 = 47; Q3 = 53

2. 3, 18, 11, 2, 7, 5, 9, 6, 13, 1, 17, 8, 0
range = 18; Q1 = 2.5; Q3 = 12

Use the following data for problems 3 and 4.

91, 87, 98, 93, 89, 78, 94

3. Make a box-and-whisker plot.

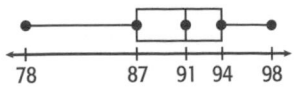

4. What is the mean? **90**

Available on Daily Transparency in CRB

CHALLENGE 4-4

LESSON 4-4 Challenge
What's Normal?

Standard deviation (symbol σ, sigma) is a measure of variability that tells how far average data is from the mean of a data set.

In many situations, such as scores on the SAT or other standardized tests, the data cluster around the mean in such a way that if they are graphed to show the frequency of measures, the graph appears as a **bell-shaped curve**, also called the **normal curve**.

If the mean math score for males on the 2001 SAT I was 533 and the standard deviation was 115, determine the scores achieved by about 68% of the male participants.

According to the normal curve, 68% of scores fall between − σ and σ.
mean − σ = 533 − 115 = 418   mean + σ = 533 + 115 = 648
So, the scores for about 68% of the males fell between 418 and 648.

Assume a normal distribution for each situation.

1. A survey of 16-year-olds showed that they watched an average (mean) of 9.4 hours of TV per week, with a standard deviation of 1.2 hours. Determine how many TV hours were watched by about:
   a. 68% of the participants ___between 8.2 and 10.6 hours___
   b. 95% of the participants ___between 7 and 11.8 hours___

2. On a certain standardized test, the mean score was 50 and the standard deviation 3. About what percent of the participants scored:
   a. between 50 and 56? ___47.5%___
   b. 44 and 47? ___13.5%___

PROBLEM SOLVING 4-4

LESSON 4-4 Problem Solving
Variability

Write the correct answer.

1. Find the range of the data.
   ___28 points___

2. Find the first and third quartiles of the data.
   Q₁ = 10, Q₃ = 23

3. Make a box-and-whisker plot of the data.

Super Bowl Point Differences	
Year	Point Difference
2001	27
2000	7
1999	15
1998	7
1997	14
1996	10
1995	23
1994	17
1993	35
1992	13

The box-and-whisker plots compare the highest recorded Fahrenheit temperatures on the seven continents with the lowest recorded temperatures. Choose the letter for the best answer.

4. Which statement is true?
   A The median of the high temperatures is smaller than the median of the low temperatures.
   B The range of low temperatures is greater than the range of high temperatures.
   C The range of the middle half of the data is greater for the high temperatures.
   D The median of the high temperatures is 49°F.

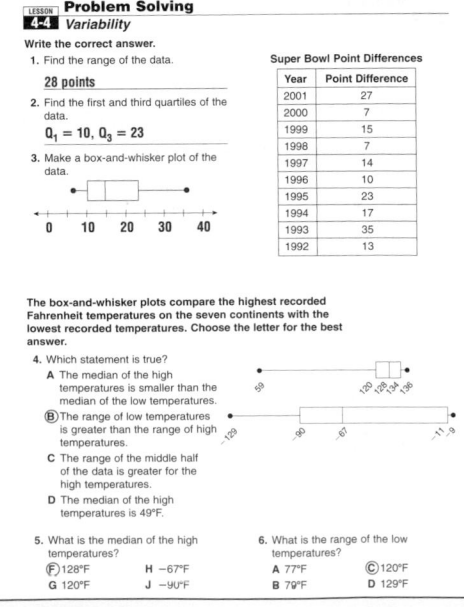

5. What is the median of the high temperatures?
   F 128°F   H −67°F
   G 120°F   J −90°F

6. What is the range of the low temperatures?
   A 77°F   **C** 120°F
   B 79°F   D 129°F

Technology LAB 4B

Create Box-and-Whisker Plots

Use with Lesson 4-4

internet connect
Lab Resources Online
go.hrw.com
KEYWORD: MP4 Lab4B

The data below are the heights in inches of the 15 girls in Mrs. Lopez's 8th-grade class.

57, 62, 68, 52, 53, 56, 58, 56, 57, 50, 56, 59, 50, 63, 52

Activity

1 Graph the heights of the 15 girls in Mrs. Lopez's class on a box-and-whisker plot.

Press **STAT** **Edit** to enter the values into List 1 (**L1**). If necessary, press the up arrow and then **CLEAR** **ENTER** to clear old data. Enter the data from the class into **L1**. Press **ENTER** after each value.

Use the **STAT PLOT** editor to obtain the plot setup menu.

Press **2nd** **Y=** (STAT PLOT) **ENTER**. Use the arrow keys and **ENTER** to select **On** and then the fifth type. **Xlist** should be **L1** and **Freq** should be 1, as shown. Press **ZOOM** **9:ZoomStat**.

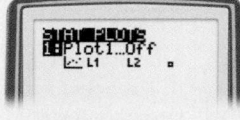

Use the **TRACE** key and the ◀ and ▶ keys to see all five summary statistical values (minimum: **MinX**, first quartile: **Q1**, median: **MED**, third quartile: **Q3**, and maximum: **MaxX**). The minimum value in the data set is 50 in., the first quartile is 52 in., the median is 56 in., the third quartile is 59 in., and the maximum is 68 in.

Think and Discuss

1. Explain how the box-and-whisker plot gives information that is hard to see by just looking at the numbers.

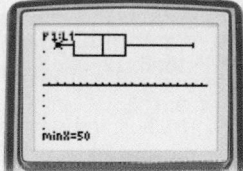

Try This

1. The shoe sizes of the 15 girls from Mrs. Lopez's 8th grade class are the following:
5.5, 6, 7, 5, 5, 5.5, 6, 6, 6.5, 4, 6, 7, 5, 8, and 5

Make a box-and-whisker plot of this data. What are the minimum, first quartile, median, third quartile, and maximum values of the data set?

Answers

Think and Discuss

1. Possible answer: The box-and-whisker plot makes it easy to see the minimum and maximum values, the position of the quartiles, and the median.

Try This

1.

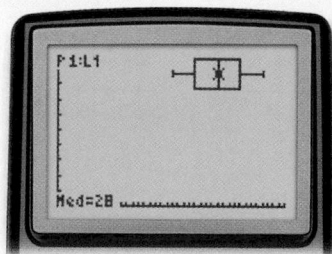

minimum: 4
first quartile: 5
median: 6
third quartile: 6.5
maximum: 8

Technology LAB

4B Box-and-Whisker Plots

Pacing: Traditional 1 day
Block $\frac{1}{2}$ day

Objective: To use a graphing calculator to produce box-and-whisker plots

Materials: Graphing calculator

Lab Resources

Technology Lab Activities . . . . p. 22

Using the Page

This technology activity shows students how to enter data and produce a box-and-whisker plot on a graphing calculator. Specific keystrokes may vary, depending on the make and model of the graphing calculator used. The keystrokes given are for a TI-83 model. For keystrokes to other models, visit go.hrw.com.

The Think and Discuss problem is meant to point out the usefulness of a box-and-whisker plot. While the Try This problem can be done without a graphing calculator, it is meant to help students become familiar with using a graphing calculator to enter data and produce box-and-whisker plots.

Assessment

Create a box-and-whisker plot from the following data:

30, 27, 29, 24, 20, 36, 32, 24, 21, 35

minimum: 20
first quartile: 24
median: 28
third quartile: 32
maximum: 36

Chapter
4
Mid-Chapter
Quiz

Purpose: *To assess students' mastery of concepts and skills in Lessons 4-1 through 4-4*

Assessment Resources

Section 4A Quiz
Assessment Resources .... p. 11

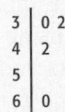 **Test and Practice Generator
CD-ROM**

Additional mid-chapter assessment items in both multiple-choice and free-response format may be generated for any objective in Lessons 4-1 through 4-4.

Answers

1. population: all the VCRs in an assembly line; sample: every 30th VCR in the assembly line; sampling method: systematic

2. population: all registered voters; sample: those whose names are chosen; sampling method: random

6. **Tall Buildings in Charlotte, NC**

```
3 | 0 2 2 2
4 | 2
5 |
6 | 0
```
Key: 3|2 means 32 stories

14.
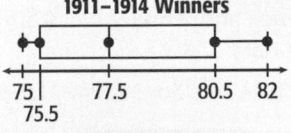
1911–1914 Winners
75 75.5 77.5 80.5 82

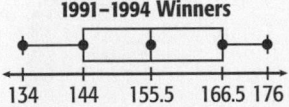
1991–1994 Winners
134 144 155.5 166.5 176

Mid-Chapter Quiz (side tab)

LESSON **4-1** (pp. 174–177)

Identify the population, sample, and sampling method.

1. Every thirtieth VCR out of 500 in an assembly line is tested.

2. Names are chosen randomly from a voter registration list.

Identify the sampling method used.

3. Postcards of contest entrants are put in a revolving drum. A celebrity draws a postcard. **random**

4. Ten schools are randomly chosen and ten students are randomly chosen from each school. **stratified**

5. Every thirteenth person who enters a local video store is polled. **systematic**

LESSON **4-2** (pp. 179–183)

6. Use the given data to make a stem-and-leaf plot.

Tall Buildings in Charlotte, NC			
Building Name	**Stories**	**Building Name**	**Stories**
Bank of America Center	60	One Wachovia Center	42
IJL Financial Center	30	Two Wachovia Center	32
Interstate Tower	32	Wachovia Center	32

LESSON **4-3** (pp. 184–187)

Find the mean, median, and mode of each data set.

7. 60, 70, 70, 80, 75
   71; 70; 70

8. 5, 2, 1, 7, 4, 6, 9
   ≈ 4.86; 5; no mode

9. 9.1, 8.7, 9.2, 9.0, 8.7, 8.9
   ≈ 8.93; 8.95; 8.7

LESSON **4-4** (pp. 188–192)

Find the range and the first and third quartiles for each data set.

10. 8, 5, 12, 9, 6, 2, 14, 7, 10, 17, 11
    15; 6; 12

11. 67, 70, 72, 77, 78, 78, 80, 84, 86
    19; 71; 82

12. 0, 0, 3, 3, 3, 1, 3, 1, 3, 7, 9, 9
    9; 1; 5

13. 3.6, 5.0, 4.0, 4.9, 4.2, 4.5, 4.3, 4.8
    1.4; 4.1; 4.85

14. Use box-and-whisker plots to compare the speeds of 1911–1914 with those of 1991–1994.

Indianapolis 500 Winners					
Year	Winner	Speed (mi/h)	Year	Winner	Speed (mi/h)
1911	Ray Harroun	75	1991	Rick Mears	176
1912	Joe Dawson	79	1992	Al Unser, Jr.	134
1913	Jules Goux	76	1993	Emerson Fittipaldi	157
1914	Rene Thomas	82	1994	Jacques Villeneuve	154

Focus on Problem Solving

Make a Plan
• **Identify too much/too little information**

When you read a problem, you must decide if the problem has too much or too little information. If the problem has too much information, you must decide what information to use to solve the problem. If the problem has too little information, then you should determine what additional information you need to solve the problem.

Read the problems below and decide if there is too much or too little information in each problem. If there is too much information, tell what information you would use to solve the problem. If there is too little information, tell what additional information you would need to solve the problem.

1 Mrs. Robinson has 35 students in her class. On the last test, there were 7 A's, 15 B's, 10 C's, and 2 D's. What was the average test score?

2 The average elevation in the United States is about 2500 ft above sea level. The highest point, Mt. McKinley, Alaska, has an elevation of 20,320 ft above sea level. The lowest point, in Death Valley, California, has an elevation of 282 ft below sea level. What is the range of elevations in the United States?

3 Use the table to find the median number of marriages per year in the United States for the years between 1940 and 1990.

4 George spent 1.5 hours doing homework on Tuesday, 1 hour doing homework on Wednesday, and 2.7 hours doing homework over the weekend. On Monday, Thursday, and Friday, he did not have homework and spent 1 hour each day reading or watching TV. What was the average amount of time per day George spent on homework last week?

Number of Marriages in the United States						
Year	1940	1950	1960	1970	1980	1990
Number (thousands)	1596	1667	1523	2159	2390	2443

Source: National Center for Health Statistics

Answers

1. too little information to solve

2. 20,602 ft

3. too little information to solve

4. about 0.74 hours per day

Focus on Problem Solving

Purpose: *To focus on identifying too much or too little information*

Problem Solving Resources

Interactive Problem Solving. . pp. 29–35

Math: Reading and Writing in the Content Area. . . . . pp. 29–35

Problem Solving Process

This page focuses on the second step of the problem-solving process:
Make a Plan

Discuss

Have students identify whether the problem gives too little, too much, or just the right amount of information. Then ask them to explain what information is needed.

Possible answers:

1. too little information; need the individual test scores and the total number of students

2. too much information; use the lowest elevation and the highest elevation

3. too little information; need the number of marriages each year from 1941 to 1989

4. too much information; use the number of hours spent doing homework each day

Section 4B

Displaying Data

One-Minute Section Planner

Lesson	Materials	Resources
Lesson 4-5 Displaying Data **NCTM:** Algebra, Data Analysis and Probability, Communication, Representation **NAEP:** Data Analysis and Probability 1b ☑ SAT-9 ☑ SAT-10 ☑ ITBS ☑ CTBS ☑ MAT ☑ CAT	**Optional** Social studies books	• *Chapter 4 Resource Book,* pp. 44–53 • Daily Transparency T19, CRB • Additional Examples Transparencies T20–T22, CRB • *Alternate Openers: Explorations,* p. 33
Lesson 4-6 Misleading Graphs and Statistics **NCTM:** Data Analysis and Probability, Reasoning and Proof, Communication, Representation **NAEP:** Data Analysis and Probability 1d ☐ SAT-9 ☐ SAT-10 ☐ ITBS ☐ CTBS ☐ MAT ☐ CAT	**Optional** Newspapers or magazines	• *Chapter 4 Resource Book,* pp. 54–62 • Daily Transparency T23, CRB • Additional Examples Transparencies T24–T27, CRB • *Alternate Openers: Explorations,* p. 34
Lesson 4-7 Scatter Plots **NCTM:** Data Analysis and Probability, Reasoning and Proof, Communication, Representation **NAEP:** Data Analysis and Probability 1b ☑ SAT-9 ☑ SAT-10 ☑ ITBS ☑ CTBS ☑ MAT ☑ CAT	**Optional** Teaching Transparency T29 (CRB) Science books	• *Chapter 4 Resource Book,* pp. 63–71 • Daily Transparency T28, CRB • Additional Examples Transparencies T30–T32, CRB • *Alternate Openers: Explorations,* p. 35
Extension Average Deviation **NCTM:** Data Analysis and Probability **NAEP:** Data Analysis and Probability 2d ☐ SAT-9 ☐ SAT-10 ☐ ITBS ☐ CTBS ☐ MAT ☐ CAT	**Optional** Teaching Transparency T33 (CRB)	• Additional Examples Transparency T34, CRB
Section 4B Assessment		• Section 4B Quiz, AR p. 12 • *Test and Practice Generator* CD-ROM

SAT = *Stanford Achievement Tests* **ITBS** = *Iowa Test of Basic Skills* **CTBS** = *Comprehensive Test of Basic Skills/Terra Nova*
MAT = *Metropolitan Achievement Test* **CAT** = *California Achievement Test*

NCTM = Complete standards can be found on pages T29–T35. **NAEP** = Complete standards can be found on pages A54–A58.

SE = *Student Edition* **TE** = *Teacher's Edition* **AR** = *Assessment Resources* **CRB** = *Chapter Resource Book* **MK** = *Manipulatives Kit*

Section Overview

Types of Graphs, Misleading Graphs and Statistics

Lessons 4-5, 4-6

Why? To analyze and interpret data, it is useful to have various types of graphs for displaying the data. Some graphs and some statements about data are misleading. Students should learn to recognize misleading graphs and statements.

Types of Graphs

Bar Graph

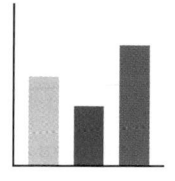

For data that can be grouped in categories

Histogram

For data that can be grouped in intervals

Line Graph

To show trends or to make estimates

Misleading Graphs and Statistics

Example

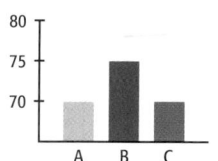

The vertical scale does not begin at zero.

Example

"According to a recent survey, two out of three people preferred our brand over brand *x*."

The survey may have included only a few people.

Scatter Plots, Correlations, Lines of Best Fit

Lesson 4-7

Why? A scatter plot is one of the best ways to show a correlation between two sets of data. A correlation is a description of a relationship. A line of best fit through a scatter plot is useful for making predictions.

Positive correlation; both data sets increase together.

Negative correlation; as one data set increases, the other decreases.

No correlation; changes in one data set do not affect the other data set.

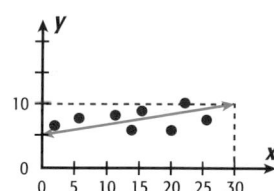

Predict the value of *y* when *x* = 30. (Using a line of best fit, you can predict that when *x* = 30, *y* will be approximately 10.)

Average Deviation

Extension

Why? Average deviation is another measure of variation. It tells how far the individual data values are from the mean.

Data set: 21 23 26 30 30

Step 1. Find the mean. $\dfrac{21 + 23 + 26 + 30 + 30}{5} = \dfrac{130}{5} = 26$

Step 2. Subtract each data value from the mean, and use the absolute values of these differences.

$|26 - 21| = 5$ $|26 - 23| = 3$ $|26 - 26| = 0$ $|26 - 30| = 4$ $|26 - 30| = 4$

Step 3. Find the average of all the differences.

average deviation $= \dfrac{5 + 3 + 0 + 4 + 4}{5} = \dfrac{16}{5} = 3.2$

The individual data items are an average of 3.2 units away from the mean.

Game Resources

Puzzles, Twisters & Teasers

Chapter 4 Resource Book

Distribution of Primes

Purpose: *To apply the skill of creating and interpreting scatter plots to the study of prime numbers*

Discuss: Ask students to explain how the sieve works. Possible answer: Start at 1. Circle 2, and then cross out all its multiples. Circle 3, and then cross out all its multiples. Continue until all the numbers are either circled or crossed out. Ask students why the entire second column is able to be crossed out. All of the numbers in the second column are multiples of 2.

Extend: Have students use their scatter plots and lines of best fit to guess the number of primes under 200. Have them use the Internet to check their guesses. There are 46 prime numbers under 200.

Math in the Middle

Purpose: *To practice finding the mean, median, and mode in a game format*

Discuss: Ask students how they will determine which measure (mean, median, or mode) to use. Possible answer: Choose the measure that will allow your game piece to land in the most favorable position.

Extend: Have students play the game again, using 8 number cubes instead of 5. Have students round the mean and the median to the nearest whole number.

Distribution of Primes

Remember that a prime number is only divisible by 1 and itself. There are infinitely many prime numbers, but there is no algebraic formula to find them. The largest known prime number, discovered on November 14, 2001, is $2^{13,466,917} - 1$. In standard form, this number would have 4,053,946 digits.

Sieve of Eratosthenes

One way to find prime numbers is called the sieve of Eratosthenes. Use a list of whole numbers in order. Cross off 1. The next number, 2, is prime. Circle it, and then cross off all multiples of 2, because they are not prime. Circle the next number on the list, and cross off all of its multiples. Repeat this step until all of the numbers are circled or crossed off. The circled numbers will all be primes.

1̸	②	3	4̸	5	6̸	7	8̸	9	1̸0̸
11	1̸2̸	13	1̸4̸	15	1̸6̸	17	1̸8̸	19	2̸0̸
21	2̸2̸	23	2̸4̸	25	2̸6̸	27	2̸8̸	29	3̸0̸
31	3̸2̸	33	3̸4̸	35	3̸6̸	37	3̸8̸	39	4̸0̸
41	4̸2̸	43	4̸4̸	45	4̸6̸	47	4̸8̸	49	5̸0̸

1. Use the sieve of Eratosthenes to find all prime numbers less than 50.

2. Create a scatter plot of the first 15 prime numbers. Use the prime numbers as the *x*-coordinates and their positions in the sequence as the *y*-coordinates; 2 is the 1st prime, 3 is the 2nd prime, and so on.

Prime Number	2	3	5	7	11	13	17	19	23	29	31	37	41	43	47
Position in Sequence	1	2	3	4	5	6	7	8	9	10	11	12	13	14	15

3. Estimate the line of best fit and use it to guess the number of primes under 100. Use the sieve of Eratosthenes to check your guess.

Math in the Middle

This game can be played by two or more players. On your turn, roll 5 number cubes. The number of spaces you move is your choice of the mean, rounded to the nearest whole number; the median; or the mode, if it exists. The winner is the first player to land on the *Finish* square by exact count.

🔲 **internet** connect

Go to **go.hrw.com** for a complete set of rules and the game board.

KEYWORD: MP4 Game4

Answers

1. 2, 3, 5, 7, 11, 13, 17, 19, 23, 29, 31, 37, 41, 43, and 47

2.

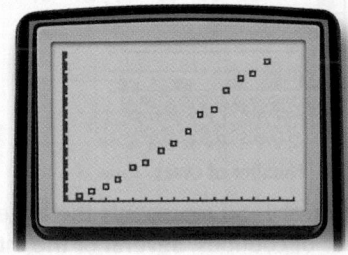

3. There are 25 prime numbers under 100.

Technology LAB

Mean, Median, and Mode

☑ internet connect
Lab Resources Online
go.hrw.com
KEYWORD: MP4 TechLab4

The National Collegiate Athletic Association (NCAA) tournaments determine the champions of women's and men's college basketball. The victory margins for the championship games from 1995 through 2001 are shown below.

Margin of Victory, NCAA Championship Games							
Year	1995	1996	1997	1998	1999	2000	2001
Men's Game (points)	11	9	5	9	3	13	10
Women's Game (points)	6	18	9	18	17	19	2

Activity

1 Use a spreadsheet to find the mean, median, and mode of the men's championship-game victory margins from the table. Fill in rows 1 and 2 with the data and labels shown in the spreadsheet below.

The **AVERAGE, MEDIAN,** and **MODE** functions find the mean, median, and mode of the data in a given range of spreadsheet cells.

- Enter **=AVERAGE(B2:H2)** into cell H3 to find the mean of the data in cells B2 through H2.

- Enter **=MEDIAN(B2:H2)** into cell H4 to find the median of the data.

- Enter **=MODE(B2:H2)** into cell H5 to find the mode of the data.

	A	B	C	D	E	F	G	H
1	Year	1995	1996	1997	1998	1999	2000	2001
2	Margin (points)	11	9	5	9	3	13	10
3							Mean	8.571429
4							Median	9
5							Mode	9

Think and Discuss

1. If an eighth game with a victory margin of 30 points were added, what would happen to these three calculated values?

Try This

1. Use a spreadsheet to find the mean, median, and mode for the women's championship games (shown in the table above).

Answers

Think and Discuss

1. The mean would increase from 8.6 to 11.25. The median would increase from 9 to 9.5. The mode would not be affected.

Try This

1. mean ≈ 12.71
   median = 17
   mode = 18

Technology LAB Mean, Median, and Mode

Objective: To use a spreadsheet to find the mean, median, and mode of a set of data

Materials: Computer with spreadsheet software

Lab Resources

Technology Lab Activities . . . . p. 23

Using the Page

This technology activity shows students how to use a spreadsheet to find the mean, median, and mode of a set of data. The instructions given are for Microsoft Excel. For instructions for other software, visit go.hrw.com.

The Think and Discuss problem can be used to assess students' understanding of the technology activity. While the Try This problem can be done without a spreadsheet, it is meant to help students become familiar with using a spreadsheet to calculate mean, median, and mode.

Assessment

1. What number appears in cell D2? 5 What does the number represent? the margin of victory in the 1997 men's game

2. How could you use the spreadsheet to find the average for the games from 1995 to 2000 only? Enter = AVERAGE(B2:G2) into an empty cell.

Purpose: *To help students review and practice concepts and skills presented in Chapter 4*

Assessment Resources

Chapter Review
Chapter 4 Resource Book.... pp 72–74

 Test and Practice Generator CD-ROM

Additional review assessment items in both multiple-choice and free-response format may be generated for any objective in Chapter 4.

Answers

1. median; mode

2. variability; variability; range

3. line of best fit; scatter plot; correlation

4. Population: moviegoers; sample: 25 people in line for *Star Wars;* possible bias: people in line for *Star Wars* might have a preference for science fiction movies.

5. Population: community members; sample: 50 parents of middle-school-aged children; possible bias: parents of middle-school-aged children may support the field more than other community members.

6. Population: constituents; sample: 75 constituents who visited the office; possible bias: constituents who visit the senator probably are strong supporters of the senator.

Vocabulary

back-to-back stem-and-leaf plot 179	line graph 197	random sample 175
bar graph 196	line of best fit 204	range 188
biased sample 174	mean 184	sample 174
box-and-whisker plot ..189	median 184	scatter plot 204
correlation 204	mode 184	stem-and-leaf plot 179
frequency table 196	outlier 185	stratified sample 175
histogram 196	population 174	systematic sample 175
	quartile 188	variability 188

Complete the sentences below with vocabulary words from the list above. Words may be used more than once.

1. The ___?___ of a data set is the middle value, while the ___?___ is the value that occurs most often.

2. ___?___ describes how spread out a data set is. One measure of ___?___ is the ___?___.

3. The ___?___ is the line that comes closest to all the points on a(n) ___?___. ___?___ describes the type of relationship between two data sets.

4-1 Samples and Surveys (pp. 174–177)

EXAMPLE

■ Identify the population and sample. Give a reason why the sample could be biased.

In a community of 1250 people, a pollster asks 250 people living near a railroad track if they want the tracks moved.

Population	Sample	Possible bias
1250 people who live in a community	250 residents living near tracks	People living near tracks are annoyed by the noise and want tracks moved.

EXERCISES

Identify the population and sample. Give a reason why the sample could be biased.

4. Of the 125 people in line for a *Star Wars* movie, 25 are asked to name their favorite type of movie.

5. Fifty parents of children attending Park Middle School are asked if the community should build a new Little League field.

6. This week, a U.S. senator asked 75 of the constituents who visited her office if she should run for reelection.

4-2 Organizing Data (pp. 179–183)

EXAMPLE

■ Make a back-to-back stem-and-leaf plot.

American League East Final Standings 2000

Team	Wins	Losses
New York	87	74
Boston	85	77
Toronto	83	79
Baltimore	74	88
Tampa Bay	69	92

Wins		Losses
9	6	
4	7	4 7 9
7 5 3	8	8
	9	2

Key:
$9|2$ *means 92*
$9|6|$ *means 69*

EXERCISES

Make a back-to-back stem-and-leaf plot.

7.

President	Inaugural Age	Age at Death
George Washington	57	67
Thomas Jefferson	57	83
Abraham Lincoln	52	56
Franklin D. Roosevelt	51	63
John F. Kennedy	43	46

4-3 Central Tendency (pp. 184–187)

EXAMPLE

■ Find the mean, median, and mode.

30, 41, 46, 39, 46

mean: $\dfrac{30 + 41 + 46 + 39 + 46}{5} = \dfrac{202}{5} = 40.4$

median: 30 39 ⑪ 46 46

mode: 46

EXERCISES

Find the mean, median, and mode.

8. 450, 500, 500, 570, 650, 700, 1950

9. 8, 8, 8.5, 10, 10, 9, 9, 11.5

10. 2, 6, 6, 10, 2, 6, 6, 10

11. 1.1, 3.1, 3.1, 3.1, 7.1, 1.1, 3.1, 3.1

4-4 Variability (pp. 188–192)

EXAMPLE

■ Find the range and quartiles.

7, 10, 14, 16, 17, 17, 18, 20, 20

range = 20 − 7 = 13   *largest − smallest*

lower half   *upper half*

1st quartile   *3rd quartile*

$\dfrac{10 + 14}{2} = 12$ $\dfrac{18 + 20}{2} = 19$

EXERCISES

Find the range and quartiles.

12. 80, 80, 80, 82, 85, 87, 87, 90, 90, 90

13. 67, 68, 68, 80, 92, 99, 80, 99, 99, 99

Answers

7.

Inaugural Age		Age at Death
3	4	6
7 7 2 1	5	6
	6	3 7
	7	
	8	3

Key: $3|4|$ means 43
$|4|6$ means 46

8. 760; 570; 500

9. 9.25; 9; 8, 9, and 10

10. 6; 6; 6

11. 3.1; 3.1; 3.1

12. 10; 80; 90

13. 32; 68; 99

Answers

14.

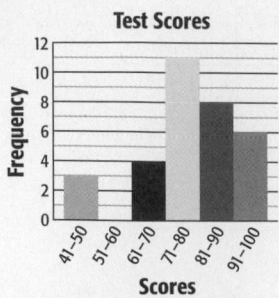

Test Scores

15.

TV Viewing

16. Possible answer: The symbols are different sizes even though they represent the same number of sightings.

17. positive

18. no correlation

4-5 **Displaying Data** (pp. 196–199)

EXAMPLE

■ Make a histogram of the data set.

Heights of 20 people, in inches:

72, 64, 56, 60, 66, 72, 48, 66, 58, 60,
60, 50, 68, 72, 68, 62, 72, 58, 60, 68

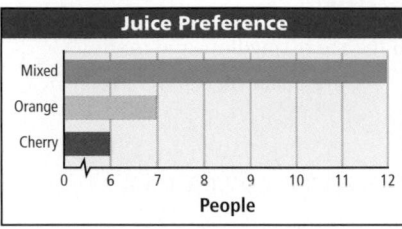

Heights (in.)

EXERCISES

Make a histogram of each data set.

14.

Test Scores	Frequency
91–100	6
81–90	8
71–80	11
61–70	4
51–60	0
41–50	3

15. TV viewing (hr/week): 19, 17, 11, 17, 3, 12, 27, 12, 20, 17, 25, 18, 23, 15, 16, 25, 23, 1, 14, 23, 17, 13, 19, 10, 21

4-6 **Misleading Graphs and Statistics** (pp. 200–203)

EXAMPLE

■ Explain why the graph is misleading.

Juice Preference

The bar for mixed juice is 7 times longer than the bar for cherry juice, but it is only preferred by 2 times as many people.

EXERCISES

Explain why the graph is misleading.

16.

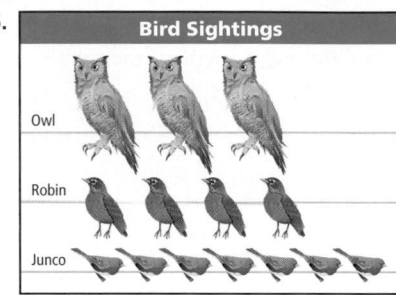

Bird Sightings

Each bird = 100 sightings

4-7 **Scatter Plots** (pp. 204–207)

EXAMPLE

■ Does the data set have a positive, a negative, or no correlation? Explain.

The age of a battery in a flashlight and the intensity of the flashlight beam.

Negative: The older the battery is, the less intense the flashlight beam will be.

EXERCISES

Do the data sets have a positive, a negative, or no correlation? Explain.

17. The price of an item and the dollar amount paid in sales tax.

18. Your height and the last digit of your phone number.

Identify the sampling method used.

1. Twenty U.S. cities are randomly chosen and 100 people are randomly chosen from each city.   **stratified**

Use the data: 59, 21, 32, 33, 40, 51, 23, 23, 28, 26, 35, 49, 48, 41, 37, 39, 44, 54, 53, 29, 28, 29, 57, 58, 46

2. Find the mean.   **39.32**
3. Find the median.   **39**
4. Find the mode.   **23, 28 and 29**
5. Make a stem-and-leaf plot.
6. Find the range.   **38**
7. Find the first quartile.   **28.5**
8. Find the third quartile.   **50**
9. Make a box-and-whisker plot.

Use the data: 7, 7, 7, 7, 8, 8, 8, 5, 5, 8, 6, 6, 7, 7, 8, 8, 8, 5, 7, 5, 6, 7, 7, 6, 6, 6, 7, 7, 7, 7, 8

10. Make a frequency table.
11. Make a bar graph.

Use the data: 155, 162, 168, 147, 152, 153, 178, 151, 180, 158, 163, 177, 171, 168, 183, 154, 180, 158, 157, 160, 171, 164, 171

12. Make a frequency table.
13. Make a histogram.

Use the data in the table.

14. Make a line graph.
15. Use the line graph to estimate the population of Africa in the year 1800. **about 95,000,000**
16. Use the line graph to estimate the population of Africa in the year 1900. **about 160,000,000**

Year	Population of Africa
1650	100,000,000
1750	95,000,000
1850	95,000,000
1950	229,000,000
2000	805,000,000

17. **Give a reason why the statistic could be misleading.**

A sign reads "Work at home—earn up to $1000 per week!"

Use the data in the table.

18. Make a scatter plot.
19. Draw the line of best fit.
20. Do the data sets have a positive, a negative, or no correlation? Explain.

Animal	Gestation Period (d)	Average Life (yr)
Baboon	187	20
Chipmunk	31	6
Elephant	645	40
Fox	52	7
Horse	330	20
Lion	100	15
Mouse	19	3

14.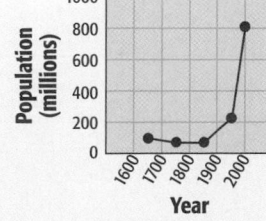
Population of Africa

18–19.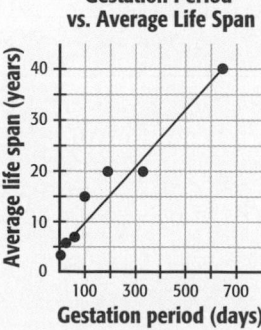
Gestation Period vs. Average Life Span

17. Possible answer: The income may depend on the hours worked; $1000 is probably an outlier.

20. Positive; animals with longer gestation periods have longer average life spans.

Purpose: *To assess students' mastery of concepts and skills in Chapter 4*

Assessment Resources ✓

Chapter 4 Tests (Levels A, B, C)
Assessment Resources . . . . p 51–56

***Test and Practice Generator* CD-ROM**

Additional assessment items in both multiple-choice and free-response format may be generated for any objective in Chapter 4.

Answers

5.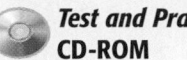

2	1 3 3 6 8 8 9 9
3	2 3 5 7 9
4	0 1 4 6 8 9
5	1 3 4 7 8 9

Key: 4|4 means 44

9.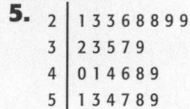
21   28.5   39   50   59

10.

Data	5	6	7	8
Frequency	4	6	13	8

11.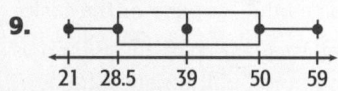

12.

	Tally	Frequency
180–189	|||	3
170–179	⊬⊬	5
160–169	⊬⊬	6
150–159	⊬⊬ |||	8
140–149			1

13.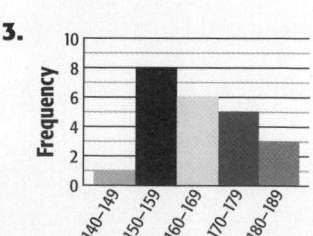

Chapter 4

Performance Assessment

Purpose: *To assess students' understanding of concepts in Chapter 4 and combined problem-solving skills*

Assessment Resources

Performance Assessment
Assessment Resources . . . . p. 124

Performance Assessment Teacher Support
Assessment Resources . . . . p. 123

Answers

1–3. See p. A4.

4. See Level 3 work sample below.

Scoring Rubric for Problem Solving Item 4

Level 3

Accomplishes the purpose of the task.

Student gives clear explanations, shows understanding of mathematical ideas and processes, and computes accurately.

Level 2

Purposes of the task not fully achieved.

Student demonstrates satisfactory but limited understanding of the mathematical ideas and processes.

Level 1

Purposes of the task not accomplished.

Student shows little evidence of understanding the mathematical ideas and processes and makes computational and/or procedural errors.

Chapter 4

Performance Assessment

Performance Assessment

Show What You Know

Create a portfolio of your work from this chapter. Complete this page and include it with your four best pieces of work from Chapter 4. Choose from your homework or lab assignments, mid-chapter quiz, or any journal entries you have done. Put them together using any design you want. Make your portfolio represent what you consider your best work.

Short Response

1. Determine the mean, median, and mode for the data set 2, 1, 8, 3, 500, 3, 1. Show your work.

2. Write a numeric expression that could be used to find the mean of the data in the frequency table. What is the mean of the data?

Number	1	2	3	4	5
Frequency	4	7	1	6	2

3. Name two ordered pairs (x, y) that satisfy these conditions: The mean of 0, x, and y is twice the median; $0 < x < y$; and $y = nx$ (y is a multiple of x). What is the value of n? Show your work or explain in words how you determined your answer.

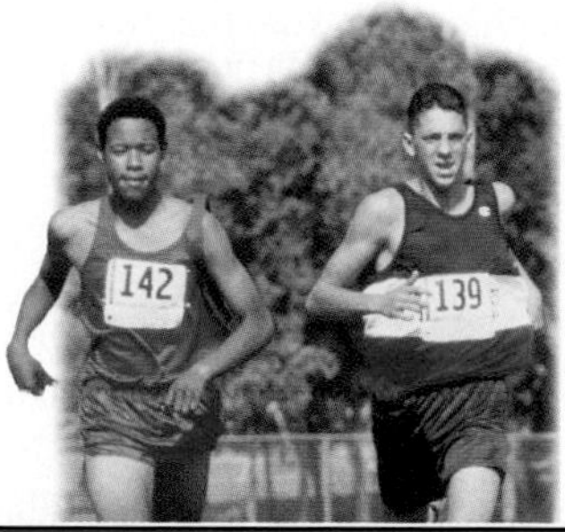

142
139

Extended Problem Solving

4. Twenty students in a gym class kept a record of their jogging. The results are shown in the scatter plot.
   a. Describe the correlation of the data in the scatter plot.
   b. Find the average speeds of joggers who run 1, 2, 3, 4, 5, and 6 miles.
   c. Explain the relationship between your answer from part **a** and your answers from part **b**.

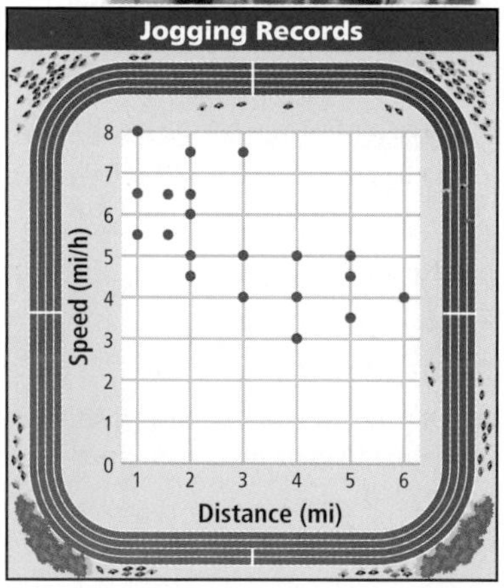

Jogging Records
Speed (mi/h)
0
1
2
3
4
5
6
7
8
Distance (mi)
1
2
3
4
5
6

Student Work Samples for Item 4

Level 3

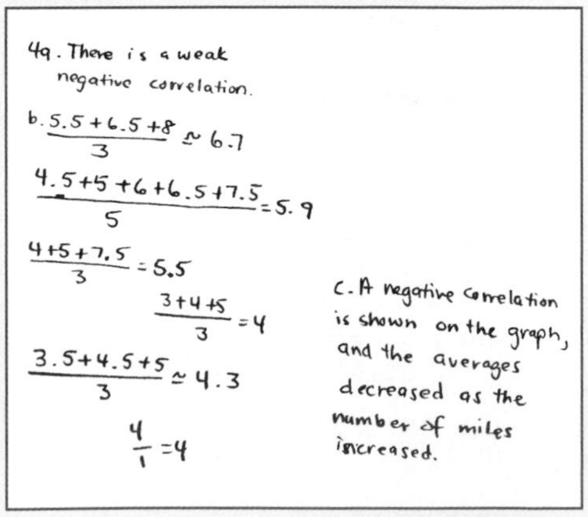

4a. There is a weak negative correlation.
b. 5.5+6.5+8 / 3 ≈ 6.7
4.5+5+6+6.5+7.5 / 5 = 5.9
4+5+7.5 / 3 = 5.5
3+4+5 / 3 = 4
3.5+4.5+5 / 3 ≈ 4.3
4 / 1 = 4
c. A negative correlation is shown on the graph, and the averages decreased as the number of miles increased.

The student understood differences in types of correlations, correctly identified the weak negative correlation, found average speeds, and understood relationships.

Level 2

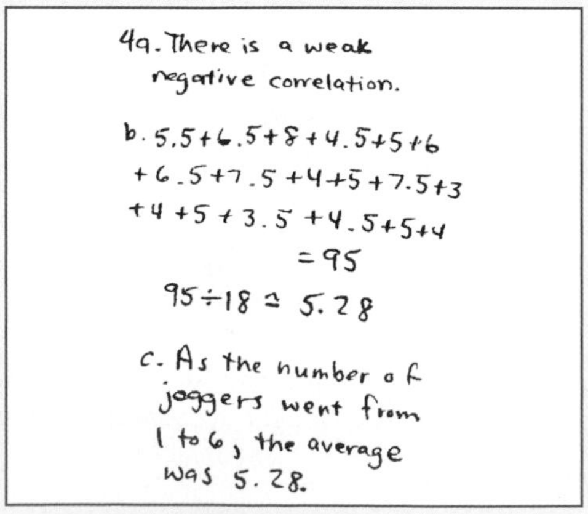

4a. There is a weak negative correlation.
b. 5.5+6.5+8+4.5+5+6
+6.5+7.5+4+5+7.5+3
+4+5+3.5+4.5+5+4
= 95
95÷18 ≈ 5.28
c. As the number of joggers went from 1 to 6, the average was 5.28.

The student correctly identified the weak negative correlation but used wrong data and showed limited understanding of relationships.

Level 1

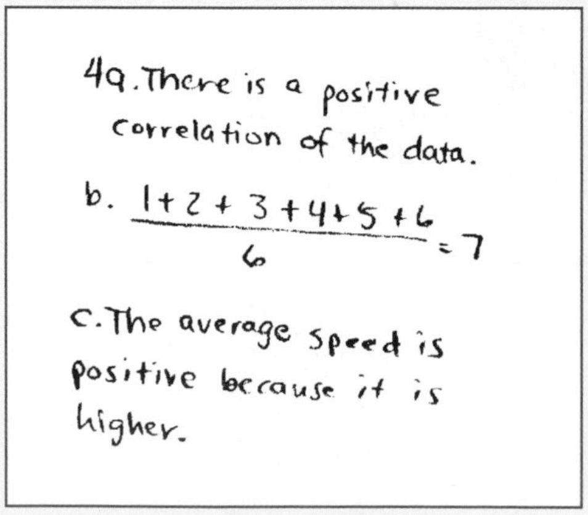

4a. There is a positive correlation of the data.
b. 1+2+3+4+5+6 / 6 = 7
c. The average speed is positive because it is higher.

No evidence of understanding the mathematical idea of correlation is shown.

Cumulative Assessment, Chapters 1–4

1. Dana bought 9 comic books for a total of $30.50. Which equation is equivalent to the equation $9c = 30.5$? **B**

   (A) $c = 30.5 - 9$ (C) $c = 9 - 30.5$

   (B) $c = \frac{30.5}{9}$ (D) $c = \frac{9}{30.5}$

2. On the number line, what number is the coordinate of point R? **H**

   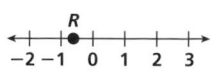

   (F) $-1\frac{3}{4}$ (H) $-\frac{3}{4}$

   (G) $-1\frac{1}{4}$ (J) $-\frac{1}{4}$

3. If the product of five integers is negative, then, at most, how many of the five integers could be negative? **A**

   (A) five   (C) three

   (B) four   (D) two

4. Which is equivalent to $3^8 \cdot 3^4$? **J**

   (F) 9^{32} (H) 3^{32}

   (G) 9^{12} (J) 3^{12}

5. What is the value of $32 - 2 \cdot 4^2$? **C**

   (A) 14,400   (C) 0

   (B) 480   (D) -32

   TEST TAKING TIP!

   To calculate the median, the data must be in order.

6. For which set of data are the mean, median, and mode all the same? **F**

   (F) 3, 1, 3, 3, 5   (H) 2, 1, 1, 1, 5

   (G) 1, 1, 2, 5, 6   (J) 10, 1, 3, 5, 1

7. Which is true for the data 6, 6, 6.5, 8, 8.5? **C**

   (A) median < mode

   (B) median = mean

   (C) median < mean

   (D) median = mode

8. The stem-and-leaf plot shows test scores for a teacher's first and second periods. What can you conclude? **J**

1st period		2nd period
7	6	5 8
6 4 2	7	5 6 9
9 8 6 4 2 0	8	1 3 5 7 7 8 8
9 7 7 2 1	9	0 6 7 8 9

   Key: | 9 | 0 means 90
   7 | 6 | means 67

   (F) More first period students scored in the 90's.

   (G) Fewer first period students scored 80 or below.

   (H) More second period students scored in the 70's.

   (J) More second period students scored in the 80's.

9. **SHORT RESPONSE** Julie wants to make homemade bows for her presents. She buys $\frac{1}{2}$ yard of red ribbon and $\frac{3}{4}$ yard of green. If each bow takes $\frac{1}{8}$ yard to make, how many total bows can she create? Justify your answer.

10. **SHORT RESPONSE** Max scored 75, 73, 71, 70, and 71 on his last 5 tests. Max wants to bring up his test average to a 75. What would Max need to make on his next test to bring his average up to a 75? Show your work.

Purpose: *To provide review and practice for Chapters 1–4 and standardized tests*

Assessment Resources

Cumulative Tests (Levels A, B, C)
Assessment Resources . . . . pp. 181–192

State-Specific Test Practice Online
KEYWORD: MP4 TestPrep

Test Prep Doctor

Point out to students that for item 1, the introductory sentence is not needed. If a question is in the form "Which equation is equivalent to the equation _____ = _____?" then the correct answer can usually be identified by performing the expected operation on each side of the given equation. In this case, divide both sides of the equation by 9 to get choice **B.**

Expand on the test-taking tip given for item 8 by reminding students of the following ways that may help them remember the definitions of *median* and *mode*: *median* strip (in the *middle* of a road), and *most* (same "o" sound as in *mode*).

Answers

9. 10 bows;
   She can make
   $\frac{1}{2} \div \frac{1}{8} = \frac{1}{2} \cdot \frac{8}{1} = 4$ red bows, and
   $\frac{3}{4} \div \frac{1}{8} = \frac{3}{4} \cdot \frac{8}{1} = 6$ green bows,
   for a total of 10 bows.

10. 90 points;
    To have an average of 75 on six tests, his total score must be $75 \cdot 6 = 450$. Let x be his score on the sixth test.

    $75 + 73 + 71 + 70 + 71 + x = 450$

    $360 + x = 450$

    $-360 \qquad = -360$

    $x = 90$

Standardized Test Prep (side tab)

Resource Options

Chapter 5 Resource Book

Student Resources

Teacher and Parent Resources

Reaching All Learners

English Language Learners

Individual Needs

Hands-On

Applications and Connections

Transparencies

Technology

Teacher Resources

Student Resources

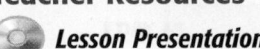

⤤ internet connect

Homework Help Online	KEYWORD: MP4 HWHelp5
Math Tools Online	KEYWORD: MP4 Tools
Glossary Online	KEYWORD: MP4 Glossary
Chapter Project Online	KEYWORD: MP4 PSProject5
Chapter Opener Online	KEYWORD: MP4 Ch5

KEYWORD: MP4 CNN5

SE = *Student Edition* **TE** = *Teacher's Edition* **AR** = *Assessment Resources* **CRB** = *Chapter Resource Book* **MK** = *Manipulatives Kit*

Assessment Options

Assessing Prior Knowledge

Determine whether students have the required prerequisite concepts and skills.

Are You Ready?...............................SE p. 221
Inventory Test............................AR pp. 1–4

Test Preparation

Provide review and practice for chapter and standardized tests.

Standardized Test Prep..........................SE p. 277
Spiral Review with Test Prep.....SE, last page of each lesson
Study Guide and Review...................SE pp. 272–274
Test Prep Tool Kit

Technology

 Test and Practice Generator CD-ROM

internet connect

State-Specific Test Practice Online KEYWORD: MP4 TestPrep

Performance Assessment

Assess students' understanding of chapter concepts and combined problem-solving skills.

Performance Assessment.......................SE p. 276
    Includes scoring rubric in TE
Performance Assessment.......................AR p. 126
Performance Assessment Teacher Support.........AR p. 125

Portfolio

Portfolio opportunities appear throughout the Student and Teacher's Editions.

Suggested work samples:

Problem Solving Project.......................TE p. 220
Performance Assessment.......................SE p. 276
Portfolio Guide...........................AR p. xxxvii
Journal....................TE, last page of each lesson
Write About It . . SE pp. 226, 231, 238, 243, 247, 253, 257, 262

Daily Assessment

Obtain daily feedback on students' understanding of concepts.

Spiral Review and Test Prep......SE, last page of each lesson

Also Available on Transparency In Chapter 5 Resource Book

Warm Up.....................TE, first page of each lesson
Problem of the Day............TE, first page of each lesson
Lesson Quiz...................TE, last page of each lesson

Student Self-Assessment

Have students evaluate their own work.

Group Project Evaluation ....AR p. xxxiv
Individual Group Member Evaluation ....AR p. xxxv
Portfolio Guide..............................AR p. xxxvii
Journal........................TE, last page of each lesson

Formal Assessment

Assess students' mastery of concepts and skills.

Section Quizzes ....AR pp. 13, 14
Mid-Chapter Quiz..............................SE p. 248
Chapter Test ....SE p. 275
Chapter Tests (Levels A, B, C) ....AR pp. 57–62
Cumulative Tests (Levels A, B, C) ....AR pp. 193–204
Standardized Test Prep
    Cumulative Assessment ....SE p. 277
End-of-Year Test.........................AR pp. 313–316

Technology

 Test and Practice Generator CD-ROM

Make tests electronically. This software includes:

• Dynamic practice for Chapter 5
• Customizable tests
• Multiple-choice items for each objective
• Free-response items for each objective
• Teacher management system

SE = *Student Edition*   **TE** = *Teacher's Edition*   **AR** = *Assessment Resources*   **CRB** = *Chapter Resource Book*   **MK** = *Manipulatives Kit*

Chapter 5 Tests

Three levels (A,B,C) of tests are available for each chapter in the *Assessment Resources.*

LEVEL A

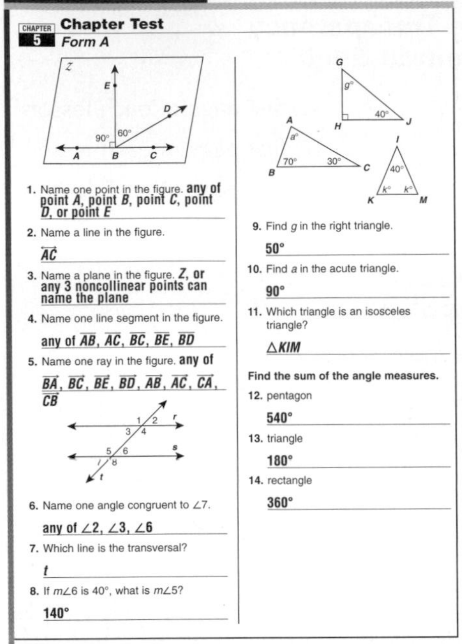

Chapter Test
CHAPTER 5 Form A

1. Name one point in the figure. **any of point A, point B, point C, point D, or point E**
2. Name a line in the figure.
   AC
3. Name a plane in the figure. **Z, or any 3 noncollinear points can name the plane**
4. Name one line segment in the figure.
   any of AB, AC, BC, BE, BD
5. Name one ray in the figure. **any of BA, BC, BE, BD, AB, AC, CA, CB**
6. Name one angle congruent to ∠7.
   any of ∠2, ∠3, ∠6
7. Which line is the transversal?
   t
8. If m∠6 is 40°, what is m∠5?
   140°

9. Find g in the right triangle.
   50°
10. Find a in the acute triangle.
    90°
11. Which triangle is an isosceles triangle?
    △KIM

Find the sum of the angle measures.
12. pentagon
    540°
13. triangle
    180°
14. rectangle
    360°

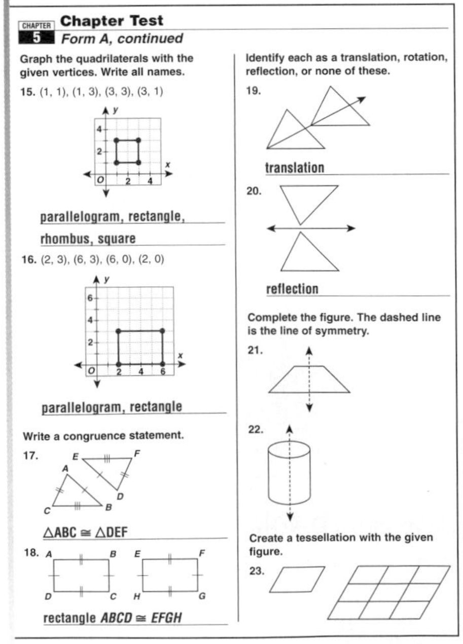

Chapter Test
CHAPTER 5 Form A, continued

Graph the quadrilaterals with the given vertices. Write all names.
15. (1, 1), (1, 3), (3, 3), (3, 1)

parallelogram, rectangle, rhombus, square
16. (2, 3), (6, 3), (6, 0), (2, 0)

parallelogram, rectangle

Write a congruence statement.
17.

△ABC ≅ △DEF
18.

rectangle ABCD ≅ EFGH

Identify each as a translation, rotation, reflection, or none of these.
19.

translation
20.

reflection

Complete the figure. The dashed line is the line of symmetry.
21.

22.

Create a tessellation with the given figure.
23.

LEVEL B

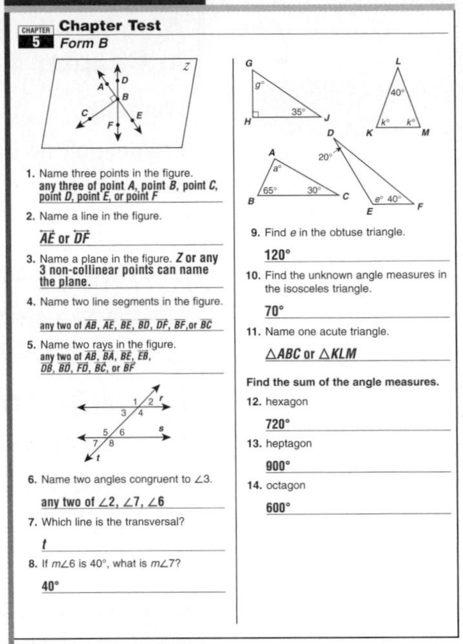

Chapter Test
CHAPTER 5 Form B

1. Name three points in the figure.
   any three of point A, point B, point C, point D, point E, or point F
2. Name a line in the figure.
   AE or DF
3. Name a plane in the figure. **Z or any 3 non-collinear points can name the plane.**
4. Name two line segments in the figure.
   any two of AB, AE, BE, BD, DF, BF, or BC
5. Name two rays in the figure. **any two of AB, BA, BE, EB, DB, BD, FD, BC, or BF**
6. Name two angles congruent to ∠3.
   any two of ∠2, ∠7, ∠6
7. Which line is the transversal?
   t
8. If m∠6 is 40°, what is m∠7?
   40°

9. Find e in the obtuse triangle.
   120°
10. Find the unknown angle measures in the isosceles triangle.
    70°
11. Name one acute triangle.
    △ABC or △KLM

Find the sum of the angle measures.
12. hexagon
    720°
13. heptagon
    900°
14. octagon
    600°

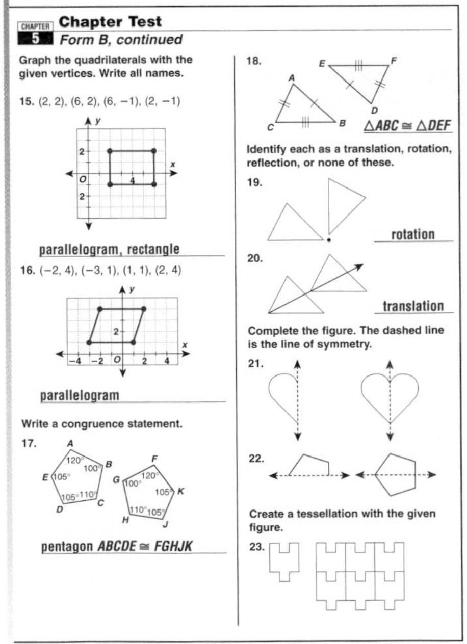

Chapter Test
CHAPTER 5 Form B, continued

Graph the quadrilaterals with the given vertices. Write all names.
15. (2, 2), (6, 2), (6, −1), (2, −1)

parallelogram, rectangle
16. (−2, 4), (−3, 1), (1, 1), (2, 4)

parallelogram

Write a congruence statement.
17.

pentagon ABCDE ≅ FGHJK

18.

△ABC ≅ △DEF

Identify each as a translation, rotation, reflection, or none of these.
19.

rotation
20.

translation

Complete the figure. The dashed line is the line of symmetry.
21.

22.

Create a tessellation with the given figure.
23.

LEVEL C

Chapter Test
CHAPTER 5 Form C

1. Name all points in the figure. **point A, point B, point C, point D, point E, point F, point G, and point H**
2. Name all lines in the figure. **AH, AG, AD, DA, GA, HA, FE, FG, FC, CF, GF, EF**
3. Name a plane in the figure. **Z, or any 3 noncollinear points can name the plane.**
4. Name eight line segments in the figure.
   any eight of AH, AG, AD, BH, BE, CG, CE, CF, DG, DH, EG, EF, FH, FG, or GH
5. Name all acute angles in the figure.
   ∠AHB, ∠EHG, and ∠HEG
6. Name all angles congruent to ∠8.
   ∠5, ∠1, ∠4
7. Which line is the transversal? **t**
8. If m∠7 is 35°, what is m∠4? **145°**

9. Find a in the acute triangle.
   85°
10. Find e in the obtuse triangle.
    120°
11. Which triangles are scalene triangles?
    △DEF, △ABC, and △GHJ

Find the sum of the angle measures.
12. hexagon
    720°
13. heptagon
    900°
14. octagon
    1080°

Chapter Test
CHAPTER 5 Form C, continued

Graph the quadrilaterals with the given vertices. Write all names.
15. (−3, 0), (−2, −2), (−3, −4), (−4, −2)

parallelogram, rhombus
16. (−4, 3), (0, 3), (1, 1), (−3, 1)

parallelogram

Write a congruence statement.
17.

quadrilateral ABDC ≅ KLMN

18.

pentagon ABCDE ≅ LMNOP

Identify each as a translation, rotation, reflection, or none of these.
19.

rotation
20.

reflection

Complete the figure. The dashed line is the line of symmetry.
21.

22.

Create a tessellation with the given figure.
23.

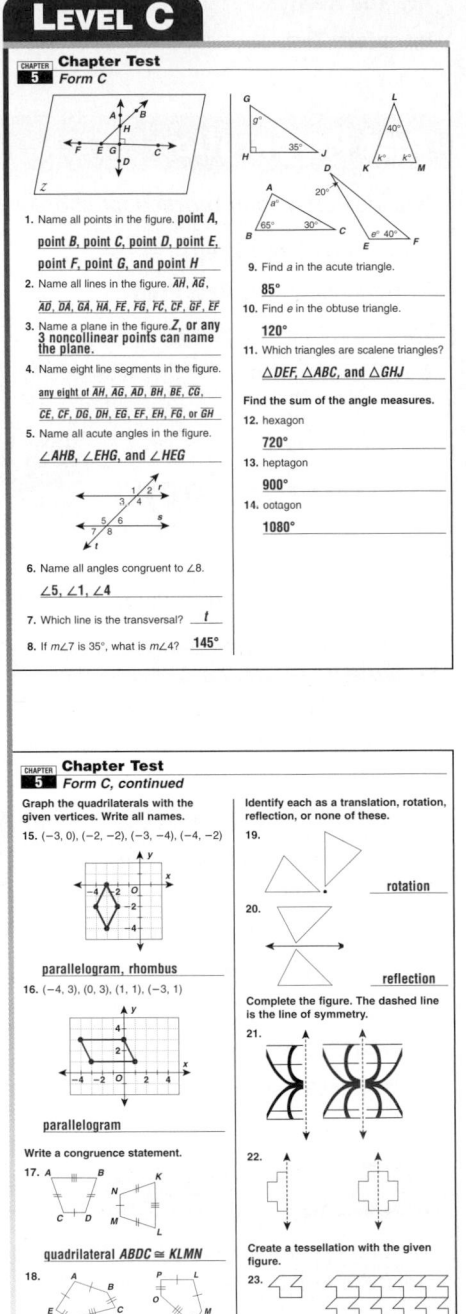

Test and Practice Generator
CD-ROM

Create and customize multiple versions of the same tests with corresponding answers for any chosen chapter objectives.

Chapter 5 State and Standardized Test Preparation

Test Taking Skill Builder and Standardized Test Practice
are provided for each chapter in the *Test Prep Tool Kit.*

TEST TAKING SKILL BUILDER

Test Taking Strategy   **Short Response**
Chapter 5

Short Response questions require you to find the solution to a
problem but do not provide answer choices or a grid. You need to
show each step of your calculations, when appropriate.

Example Short Response Sketch a quadrilateral with exactly two
lines of symmetry. Show the lines of symmetry in your sketch.
Explain your reasoning.

Scoring Rubric

- 2 points: Student sketches a rectangle or rhombus with two lines
  of symmetry and explains his or her reasoning.
- 1 point: Student sketches a rectangle or rhombus and only
  includes one of its lines of symmetry, or sketches a parallelogram
  and incorrectly identifies the diagonals as symmetry lines.
- 0 points: There is no response or it is completely incorrect.

Solution

2-point response: The sketch can be either a rectangle or a
rhombus, but not a square, parallelogram (unless it is a
rectangle or rhombus), trapezoid, or kite. A rectangle and a
rhombus are the only two quadrilaterals with only two lines of
symmetry.

1-point response: The sketch can be either a rectangle or a
rhombus, since they are the only two quadrilaterals with only
two lines of symmetry.

The student only received 1-point because only one line of
symmetry is shown.

0-point response: The student received 0-points because
there are no lines of symmetry and there are no other
markings on the drawing.

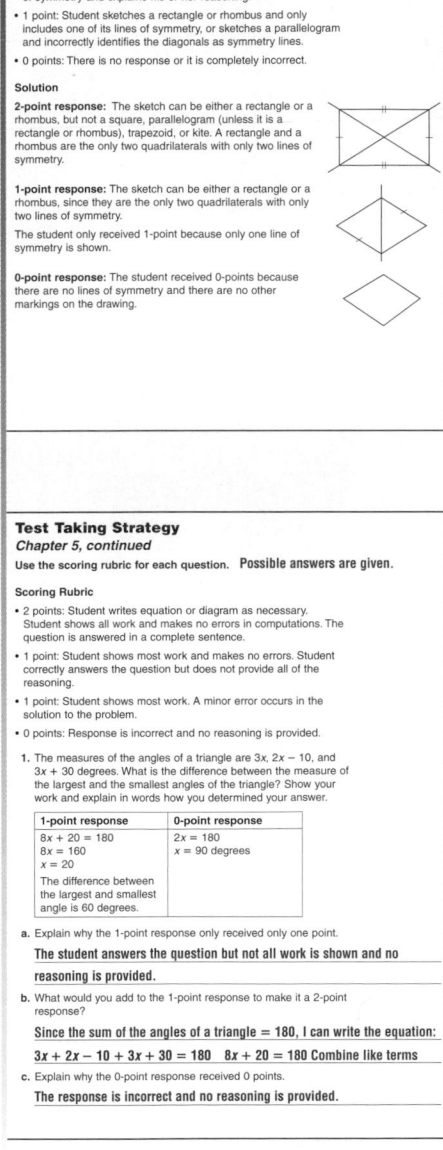

Test Taking Strategy
Chapter 5, continued

Use the scoring rubric for each question. **Possible answers are given.**

Scoring Rubric

- 2 points: Student writes equation or diagram as necessary.
  Student shows all work and makes no errors in computations. The
  question is answered in a complete sentence.
- 1 point: Student shows most work and makes no errors. Student
  correctly answers the question but does not provide all of the
  reasoning.
- 1 point: Student shows most work. A minor error occurs in the
  solution to the problem.
- 0 points: Response is incorrect and no reasoning is provided.

1. The measures of the angles of a triangle are $3x$, $2x - 10$, and
   $3x + 30$ degrees. What is the difference between the measure of
   the largest and the smallest angles of the triangle? Show your
   work and explain in words how you determined your answer.

1-point response	0-point response
$8x + 20 = 180$ $8x = 160$ $x = 20$ The difference between the largest and smallest angle is 60 degrees.	$2x = 180$ $x = 90$ degrees

a. Explain why the 1-point response only received only one point.

   The student answers the question but not all work is shown and no
   reasoning is provided.

b. What would you add to the 1-point response to make it a 2-point
   response?

   Since the sum of the angles of a triangle = 180, I can write the equation:

   $3x + 2x - 10 + 3x + 30 = 180$ $8x + 20 = 180$ Combine like terms

c. Explain why the 0-point response received 0 points.

   The response is incorrect and no reasoning is provided.

STANDARDIZED TEST PRACTICE

Standardized Test Practice
Chapter 5

Select the best answer for Questions 1–7.

1. Trina is a carpet designer and is
   designing a new carpet. She wants to
   use several figures. Which of the
   following shapes would NOT cover
   the design paper with a tessellation?

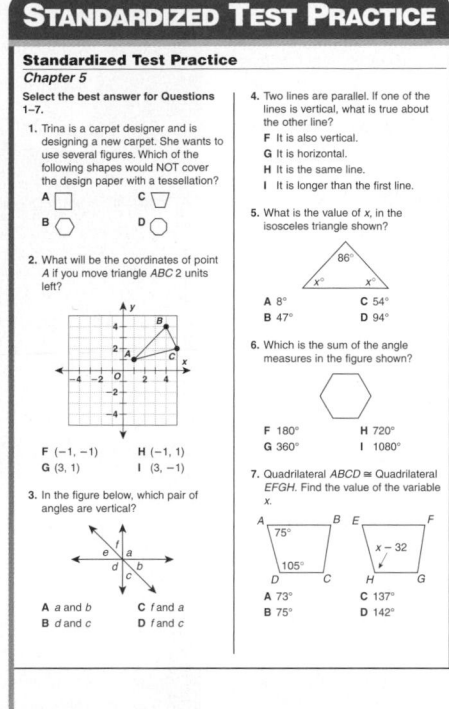

2. What will be the coordinates of point
   A if you move triangle ABC 2 units
   left?

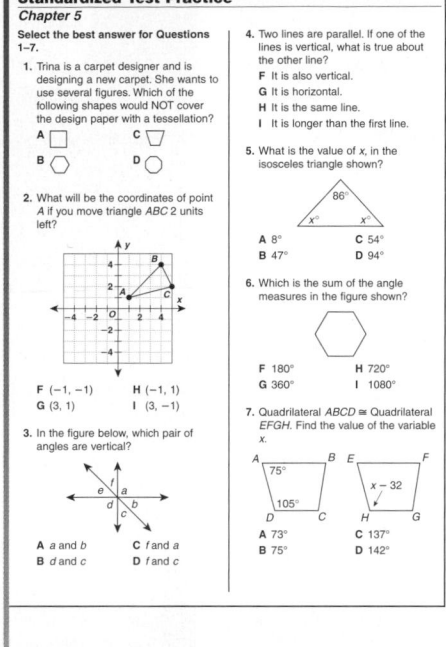

   F $(-1, -1)$ H $(-1, 1)$
   G $(3, 1)$ I $(3, -1)$

3. In the figure below, which pair of
   angles are vertical?

   A a and b C f and a
   B d and c D f and c

4. Two lines are parallel. If one of the
   lines is vertical, what is true about
   the other line?

   F It is also vertical.
   G It is horizontal.
   H It is the same line.
   I It is longer than the first line.

5. What is the value of x, in the
   isosceles triangle shown?

   A 8°   C 54°
   B 47°   D 94°

6. Which is the sum of the angle
   measures in the figure shown?

   F 180°   H 720°
   G 360°   I 1080°

7. Quadrilateral $ABCD \cong$ Quadrilateral
   $EFGH$. Find the value of the variable
   x.

   A 73°   C 137°
   B 75°   D 142°

Standardized Test Practice
Chapter 5, continued

Gridded Response
Solve the problems. Use the answer
sheet to write and grid-in your answer.

8. How many lines of symmetry does
   a regular octagon have?

9. What is the measure of each angle
   in a regular hexagon?

Short Response
Solve the problems. Use the answer
sheet to write your answers.

10. Determine if the slope of the line is
    positive, negative, 0, or undefined.
    Explain in words how you determined
    your answer. Then determine the
    actual slope.

11. One type of transformation is a
    reflection. Explain in words what a
    reflection is.

12. Explain in words the difference
    between supplementary angles and
    complementary angles.

Extended Response

13. The Spirit Club is making a banner
    for the pep rally. They finished
    one-half of the banner and plan to
    finish the other half tomorrow. The
    other half of the banner will be a
    mirror image of the first half.

    a. Complete the figure. The dashed line
       is the line of symmetry.

    b. Which of the following words could
       the Spirit Club display on each half of
       the banner so that the banner stays
       symmetrical? Explain in words how
       you determined your answer.

       GO WOW WIN RAH

    c. Explain what it means for a figure to
       have rotational symmetry.

Customized answer sheets give
students realistic practice for
actual standardized tests.

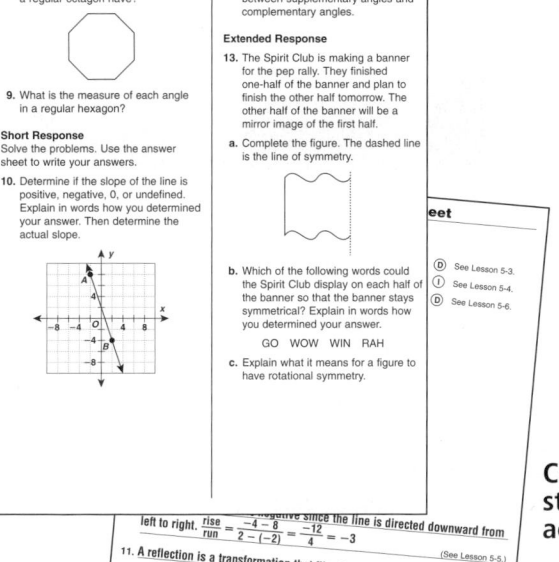

...eet

ⓓ See Lesson 5-3.
ⓘ See Lesson 5-4.
ⓓ See Lesson 5-6.

left to right. $\dfrac{rise}{run} = \dfrac{-4-8}{2-(-2)} = \dfrac{-12}{4} = -3$

                                    (See Lesson 5-5.)

11. A reflection is a transformation that flips the figure across a line to create
    a mirror image.

                                    (See Lesson 5-7.)

12. The measures of supplementary angles add up to 180°, while the
    measures of complementary angles add up to 90°.

                                    (See Lesson 5-2.)

Extended Response
Write your answers for Problem 13 on the back of this paper.
See Lesson 5-8.

Plane Geometry

Plane Geometry

Why Learn This?

Tell students that the word *geometry* comes from a Greek word meaning "to measure the earth." Learning geometry vocabulary terms can help you describe many shapes and designs that occur in nature as well as in man-made objects. For example, most playground equipment can be described and designed using lines, polygons, and circles.

Using Data

To begin the study of this chapter, have students:

• Sketch the ground shape for each piece of equipment listed. Possible answers:

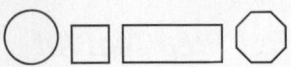

• Make a sketch of a swing set using only lines. Have them identify any pairs of lines that have the same length. Possible answers:

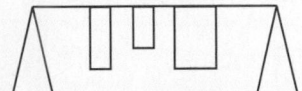

each side of a swing, each end support, two of the swing seats

☑ **internet** connect

Chapter Opener Online
go.hrw.com
KEYWORD: MP4 CH5

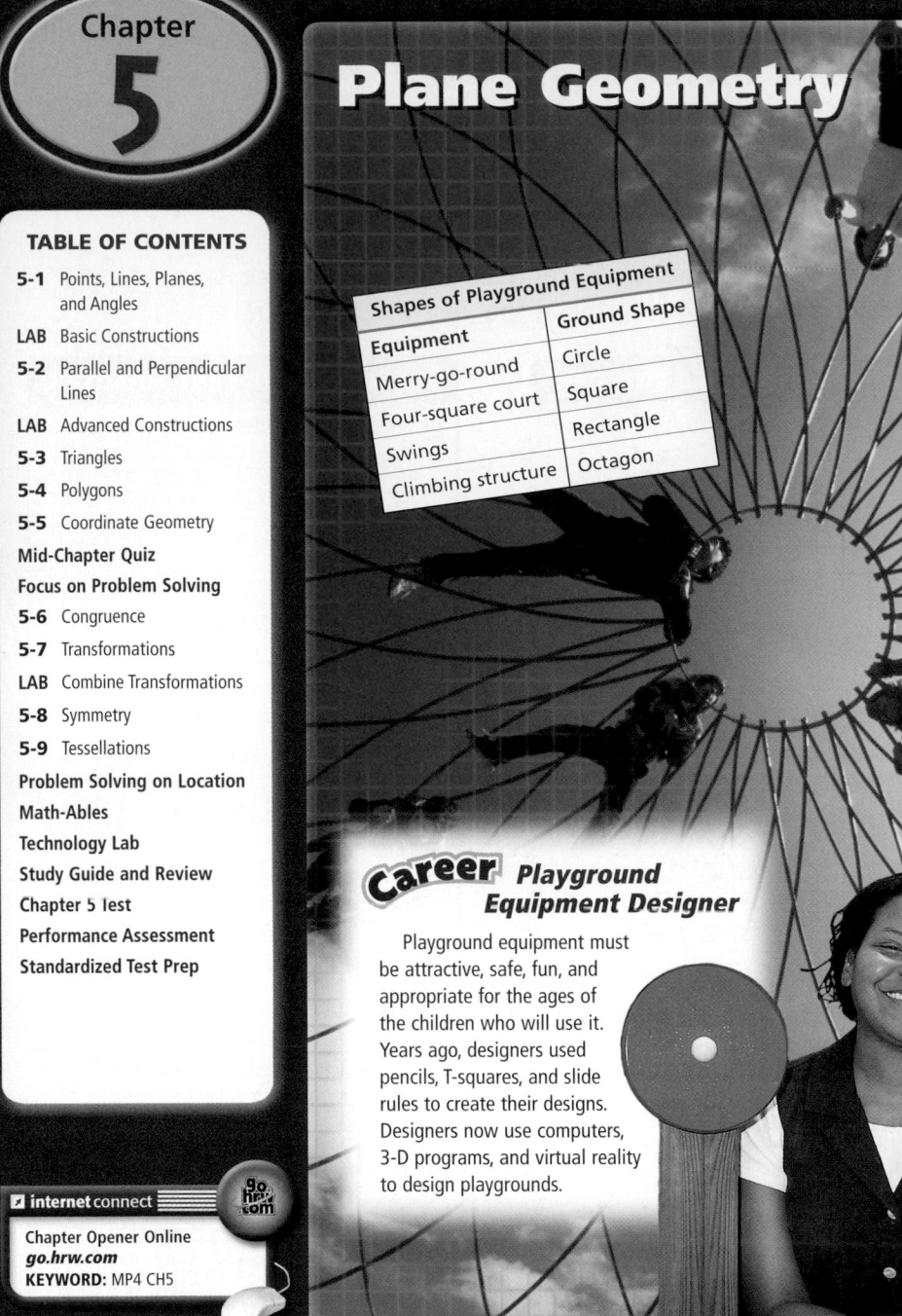

Shapes of Playground Equipment	
Equipment	**Ground Shape**
Merry-go-round	Circle
Four-square court	Square
Swings	Rectangle
Climbing structure	Octagon

Career *Playground Equipment Designer*

Playground equipment must be attractive, safe, fun, and appropriate for the ages of the children who will use it. Years ago, designers used pencils, T-squares, and slide rules to create their designs. Designers now use computers, 3-D programs, and virtual reality to design playgrounds.

PROBLEM SOLVING

Problem Solving Project

Physical Science and Social Studies Connection

Purpose: To solve problems by combining plane figures

Materials: The Ultimate Playground worksheet, construction materials (straws, toothpicks, pipe cleaners, tape, glue)

Understand, Plan, Solve, and Look Back

Have students:

✔ Complete The Ultimate Playground worksheet to analyze and create playground equipment using points, lines, planes, and angles.

✔ Make a chart listing playground equipment and the lines and angles that they see.

✔ Check students' work.

☑ **internet** connect

Chapter Project Online: *go.hrw.com*
KEYWORD: MP4 PSProject5

ARE YOU READY?

Choose the best term from the list to complete each sentence.

1. In the __?__ (4, −3), 4 is the __?__, and −3 is the __?__. **ordered pair; *x*-coordinate; *y*-coordinate**

2. The __?__ divide the __?__ into four sections. **coordinate axes; coordinate plane**

3. The point (0, 0) is called the __?__. **origin**

4. The point (0, −3) lies on the __?__, while the point (−2, 0) lies on the __?__. ***y*-axis; *x*-axis**

coordinate axes
coordinate plane
origin
ordered pair
x-axis
y-axis
x-coordinate
y-coordinate

Complete these exercises to review skills you will need for this chapter.

✔ Ordered Pairs

Write the coordinates of the indicated points.

5. point *A* **(2, 3)**   6. point *B* **(−1, 0)**

7. point *C* **(−3, −2)**   8. point *D* **(0, −3)**

9. point *E* **(5, −4)**   10. point *F* **(0, 5)**

11. point *G* **(−4, 0)**   12. point *H* **(5, 0)**

✔ Combine Like Terms

Simplify each expression by combining the like terms.

13. $5m + 7 - 2m - 1$ **$3m + 6$**

14. $2x - 4 - 6x + 1$ **$-4x - 3$**

15. $6w + z - 5w - z$ **w**

16. $3r + 11s$ **cannot be simplified**

17. $12h - 9 + 2 - 3h$ **$9h - 7$**

18. $4y + 1 - 2y - x$ **$2y - x + 1$**

✔ Equations

Solve each equation.

19. $2p = 18$ **$p = 9$**

20. $7 + h = 21$ **$h = 14$**

21. $\frac{x}{3} = 9$ **$x = 27$**

22. $y - 6 = 16$ **$y = 22$**

23. $4d + 1 = 13$ **$d = 3$**

24. $-2q - 3 = 3$ **$q = -3$**

25. $4(z - 1) = 16$ **$z = 5$**

26. $x + 3 + 4x = 23$ **$x = 4$**

Determine whether the given values are solutions of the given equations.

27. $\frac{2}{3}x + 1 = 7$ $x = 9$ **yes**

28. $2x - 4 = 6$ $x = -1$ **no**

29. $8 - 2x = -4$ $x = 5$ **no**

30. $\frac{1}{2}x + 5 = -2$ $x = -14$ **yes**

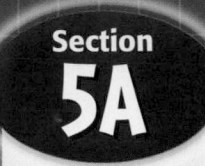

Plane Figures

One-Minute Section Planner

Lesson	Materials	Resources
Lesson 5-1 Points, Lines, Planes, and Angles **NCTM:** Geometry, Reasoning and Proof, Communication **NAEP:** Geometry 1c ☑ SAT-9 ☑ SAT-10 ☑ ITBS ☑ CTBS ☑ MAT ☑ CAT	**Optional** Teaching Transparency T2 (CRB) Straightedges (MK) Recording Sheet for Reaching All Learners (CRB, p. 103)	• *Chapter 5 Resource Book*, pp. 7–16 • *Daily Transparency* T1, CRB • *Additional Examples Transparencies* T3–T5, CRB • *Alternate Openers: Explorations*, p. 36
Hands-On Lab 5A Basic Constructions **NCTM:** Geometry **NAEP:** Geometry 1f ☐ SAT-9 ☐ SAT-10 ☐ ITBS ☐ CTBS ☐ MAT ☐ CAT	**Required** Compasses (MK) Straightedges (MK) Protractors (MK)	• *Hands-On Lab Activities*, pp. 31–32
Lesson 5-2 Parallel and Perpendicular Lines **NCTM:** Geometry, Reasoning and Proof, Communication **NAEP:** Geometry 3g ☑ SAT-9 ☑ SAT-10 ☑ ITBS ☑ CTBS ☑ MAT ☑ CAT	**Required** Protractors (MK or CRB, p. 105) Recording Sheet for Exercises 1 and 6 (CRB, p. 104) **Optional** Teaching Transparency T7 (CRB) Straightedges (MK)	• *Chapter 5 Resource Book*, pp. 17–25 • *Daily Transparency* T6, CRB • *Additional Examples Transparencies* T8–T9, CRB • *Alternate Openers: Explorations*, p. 37
Hands-On Lab 5B Advanced Constructions **NCTM:** Geometry **NAEP:** Geometry 1f ☐ SAT-9 ☐ SAT-10 ☐ ITBS ☐ CTBS ☐ MAT ☐ CAT	**Required** Compasses (MK) Straightedges (MK) Protractors (MK)	• *Hands-On Lab Activities*, pp. 33–35
Lesson 5-3 Triangles **NCTM:** Geometry, Reasoning and Proof, Communication **NAEP:** Geometry 3b ☑ SAT-9 ☑ SAT-10 ☑ ITBS ☑ CTBS ☑ MAT ☑ CAT	**Optional** Push pin Rubber band Paper triangle Teaching Transparencies T11–T12 (CRB)	• *Chapter 5 Resource Book*, pp. 26–34 • *Daily Transparency* T10, CRB • *Additional Examples Transparencies* T13–T16, CRB • *Alternate Openers: Explorations*, p. 38
Lesson 5-4 Polygons **NCTM:** Geometry, Reasoning and Proof, Communication **NAEP:** Geometry 3f ☑ SAT-9 ☑ SAT-10 ☑ ITBS ☑ CTBS ☑ MAT ☑ CAT	**Optional** Teaching Transparencies T18–T19 (CRB) Protractors (MK) Straightedges (MK) Coordinate planes (CRB, p. 106)	• *Chapter 5 Resource Book*, pp. 35–45 • *Daily Transparency* T17, CRB • *Additional Examples Transparencies* T20–T22, CRB • *Alternate Openers: Explorations*, p. 39
Lesson 5-5 Coordinate Geometry **NCTM:** Algebra, Geometry, Reasoning and Proof, Communication, Connections, Representation **NAEP:** Geometry 4d ☐ SAT-9 ☐ SAT-10 ☐ ITBS ☐ CTBS ☐ MAT ☑ CAT	**Optional** Teaching Transparencies T24–T25 (CRB) Coordinate planes (CRB, p. 106)	• *Chapter 5 Resource Book*, pp. 46–55 • *Daily Transparency* T23, CRB • *Additional Examples Transparencies* T26–T28, CRB • *Alternate Openers: Explorations*, p. 40
Section 5A Assessment	**Optional** Coordinate planes (CRB, p. 106)	• Mid-Chapter Quiz, SE p. 248 • Section 5A Quiz, AR p. 13 • *Test and Practice Generator* CD-ROM

SAT = *Stanford Achievement Tests*　**ITBS** = *Iowa Test of Basic Skills*　**CTBS** = *Comprehensive Test of Basic Skills/Terra Nova*
MAT = *Metropolitan Achievement Test*　**CAT** = *California Achievement Test*
NCTM = Complete standards can be found on pages T29–T35.　**NAEP** = Complete standards can be found on pages A54–A58.

Section Overview

Basic Geometric Figures *Hands-on Labs 5A and 5B, Lessons 5-1, 5-2*

Why? Points, lines, and planes are the building blocks of geometry.

Point *A*   Line *BC*   Segment *DE*

Perpendicular lines form 90° angles.

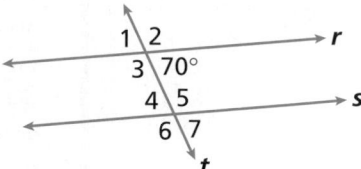

 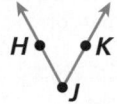 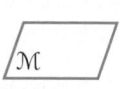

Ray *FG*   Angle *HJK*   Plane *M*

Parallel lines *r* and *s* are intersected by tranversal *t*.
m∠1 = m∠4 = m∠7 = 70°
m∠2 = m∠3 = m∠5 = m∠6 = 110°

Polygons *Lessons 5-3, 5-4*

Why? Polygons are all around us.

> A **polygon** is a closed plane figure formed by three or more segments.

Polygon	Number of Sides	Sum of Angle Measures
	n	180° (*n* − 2)
Triangle	3	180° (3 − 2) = 180°
Quadrilateral	4	180° (4 − 2) = 360°
Pentagon	5	180° (5 − 2) = 540°
Hexagon	6	180° (6 − 2) = 720°
Heptagon	7	180° (7 − 2) = 900°
Octagon	8	180° (8 − 2) = 1080°

Triangles

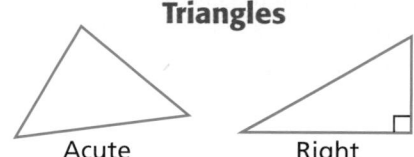 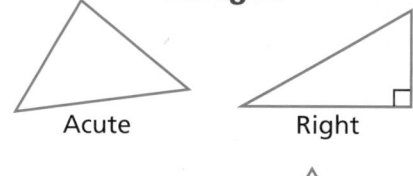

Acute   Right

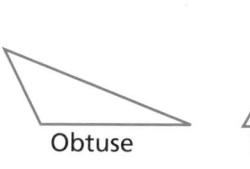 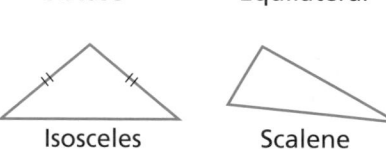

Obtuse   Equilateral

Isosceles   Scalene

Special Quadrilaterals

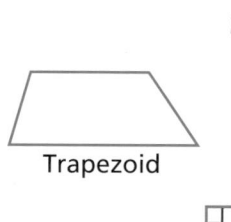 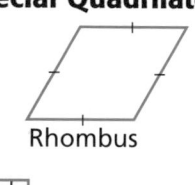 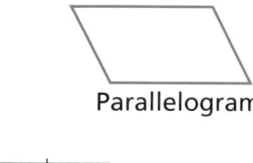

Trapezoid   Rhombus   Parallelogram

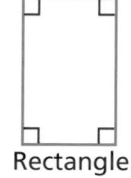 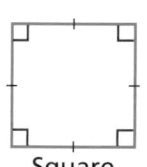

Rectangle   Square

Coordinate Geometry *Lesson 5-5*

Why? Coordinate geometry is the application of geometry and algebra together on a coordinate plane.

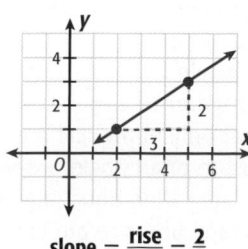

slope = $\frac{rise}{run}$ = $\frac{2}{3}$

Parallel Lines
equal slopes

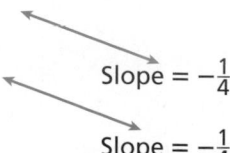

Slope = $-\frac{1}{4}$

Slope = $-\frac{1}{4}$

Perpendicular Lines
product of slopes = −1

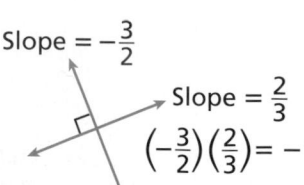

Slope = $-\frac{3}{2}$

Slope = $\frac{2}{3}$

$\left(-\frac{3}{2}\right)\left(\frac{2}{3}\right) = -1$

222B

5-1 Organizer

Pacing: Traditional 1 day
Block $\frac{1}{2}$ day

Objective: Students classify and name figures.

Warm Up

Solve.

1. $x + 30 = 90$ $x = 60$
2. $103 + x = 180$ $x = 77$
3. $32 + x = 180$ $x = 148$
4. $90 = 61 + x$ $x = 29$
5. $x + 20 = 90$ $x = 70$

Problem of the Day

Mrs. Meyer's class is having a pizza party. Half the class wants pepperoni on the pizza, $\frac{1}{3}$ of the class wants sausage on the pizza, and the rest want only cheese on the pizza. What fraction of Mrs. Meyer's class wants just cheese on the pizza? $\frac{1}{6}$

Available on Daily Transparency in CRB

Math Fact !

Plane geometry is known among mathematicians as Euclidean geometry. It is named for the Greek mathematician Euclid, who proposed five postulates that laid the foundations for geometry.

5-1 Points, Lines, Planes, and Angles

Learn to classify and name figures.

Vocabulary
point
line
plane
segment
ray
angle
right angle
acute angle
obtuse angle
complementary angles
supplementary angles
congruent
vertical angles

Points, lines, and planes are the building blocks of geometry. Segments, rays, and angles are defined in terms of these basic figures.

A **point** names a location.	• A point A
A **line** is perfectly straight and extends forever in both directions.	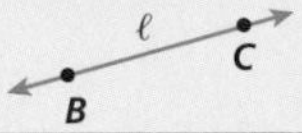 line ℓ, or $\overleftrightarrow{BC}$
A **plane** is a perfectly flat surface that extends forever in all directions.	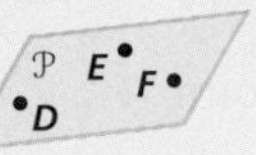 plane $\mathcal{P}$, or plane DEF
A **segment**, or line segment, is the part of a line between two points.	 $\overline{GH}$
A **ray** is part of a line that starts at one point and extends forever in one direction.	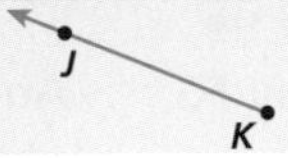 $\overrightarrow{KJ}$

$\overleftrightarrow{BC}$ is read "line BC." $\overline{GH}$ is read "segment GH." $\overrightarrow{KJ}$ is read "ray KJ." To name a ray, always write the endpoint first.

EXAMPLE 1   **Naming Points, Lines, Planes, Segments, and Rays**

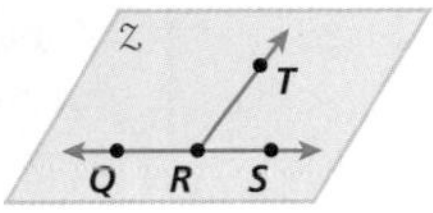

A Name four points in the figure.
point Q, point R, point S, point T

B Name a line in the figure.
$\overleftrightarrow{QS}$ or $\overleftrightarrow{QR}$ or $\overleftrightarrow{RS}$
Any 2 points on the line can be used.

C Name a plane in the figure.
plane $\mathcal{Z}$ or plane QRT   *Any 3 points in the plane that form a triangle can be used.*

D Name four segments in the figure.
$\overline{QR}, \overline{RS}, \overline{RT}, \overline{QS}$

E Name five rays in the figure.
$\overrightarrow{RQ}, \overrightarrow{RS}, \overrightarrow{RT}, \overrightarrow{SQ}, \overrightarrow{QS}$

An **angle** (∠) is formed by two rays with a common endpoint called the *vertex* (plural, *vertices*). Angles can be measured in degrees. One degree, or 1°, is $\frac{1}{360}$ of a circle. m∠1 means the measure of ∠1. The angle can be named ∠XYZ, ∠ZYX, ∠1, or ∠Y. The vertex must be the middle letter.

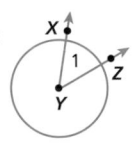

m∠1 = 50°

---

1 Introduce

Alternate Opener

5-1 Points, Lines, Planes, and Angles

In each group, one picture is different from the others. Identify the picture that is different and explain why it is different.

1.
2.

3. Look at the picture for each geometry term and write a real-world example for each term.

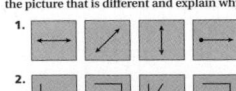

Term	Picture	Example
Point	•	
Segment	•—•	
Ray	•—→	
Line	←—→	
Angle	∧	

Think and Discuss
4. **Explain** the difference between a *segment* and a *ray*.
5. **Explain** the difference between a *ray* and a *line*.

Motivate

On the board, draw a square, a triangle, and a rectangle. Ask the students to identify each figure. Explain to them that they are geometric figures comprised of points, line segments, and angles.

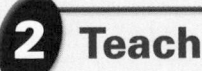

Exploration worksheet and answers on Chapter 5 Resource Book pp. 8 and 118

---

2 Teach

Lesson Presentation

Guided Instruction

In this lesson, students learn to classify and name figures. Discuss the concepts of points, lines, and planes as the building blocks of geometry (Teaching Transparency T2, CRB). Segments and rays are parts of lines, and angles consist of rays. Show students how to classify an angle by using the corner of a piece of paper to represent a right angle. Emphasize the difference between congruence and equality. Remind students that only numerical measures (such as the measure of an angle or the length of a segment) are called equal; figures that are the same shape and size are congruent.

The measures of angles that fit together to form a straight line, such as ∠FKG, ∠GKH, and ∠HKJ, add to 180°.

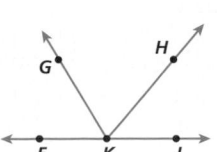

The measures of angles that fit together to form a complete circle, such as ∠MRN, ∠NRP, ∠PRQ, and ∠QRM, add to 360°.

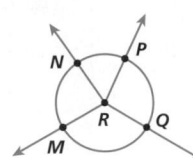

A **right angle** measures 90°. An **acute angle** measures less than 90°. An **obtuse angle** measures greater than 90° and less than 180°. **Complementary angles** have measures that add to 90°. **Supplementary angles** have measures that add to 180°.

EXAMPLE 2 **Classifying Angles**

Reading Math

A right angle can be labeled with a small box at the vertex.

A Name a right angle in the figure.
∠DEC

B Name two acute angles in the figure.
∠AED, ∠CEB

C Name two obtuse angles in the figure.
∠AEC, ∠DEB

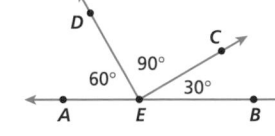

D Name a pair of complementary angles in the figure.
∠AED, ∠CEB $m∠AED + m∠CEB = 60° + 30° = 90°$

E Name two pairs of supplementary angles in the figure.
∠AED, ∠DEB $m∠AED + m∠DEB = 60° + 120° = 180°$
∠AEC, ∠CEB $m∠AEC + m∠CEB = 150° + 30° = 180°$

Congruent figures have the same size and shape.
• Segments that have the same length are congruent.
• Angles that have the same measure are congruent.
• The symbol for congruence is ≅, which is read "is congruent to."

Intersecting lines form two pairs of **vertical angles**. Vertical angles are always congruent, as shown in the next example.

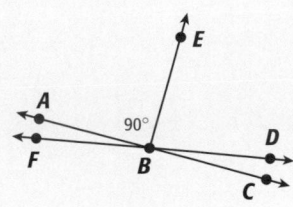

Reaching All Learners
Through Hands-On Experience

Give each student a recording sheet (Chapter 5 Resource Book p. 103) and a straightedge. Have students follow the instructions to draw the figure below, one step at a time. Then have them classify each angle and identify any special angle pairs.

Additional Examples

Example 1

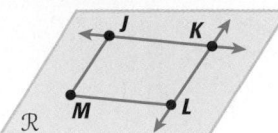

A. Name 4 points in the figure.
point *J*, point *K*, point *L*, point *M*

B. Name a line in the figure.
$\overleftrightarrow{KL}$ or $\overleftrightarrow{JK}$

C. Name a plane in the figure.
plane ℛ or plane *JKL*

D. Name four segments in the figure. $\overline{JK}$, $\overline{KL}$, $\overline{LM}$, $\overline{JM}$

E. Name four rays in the figure.
$\overrightarrow{KJ}$, $\overrightarrow{KL}$, $\overrightarrow{JK}$, $\overrightarrow{LK}$

Example 2

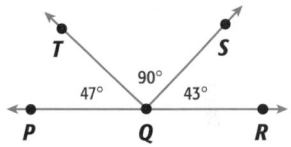

A. Name a right angle in the figure.
∠*TQS*

B. Name two acute angles in the figure. ∠*TQP*, ∠*RQS*

C. Name two obtuse angles in the figure. ∠*SQP*, ∠*RQT*

D. Name a pair of complementary angles in the figure.
∠*TQP*, ∠*RQS*

E. Name two pairs of supplementary angles in the figure. ∠*TQP*, ∠*TQR* and ∠*SQP*, ∠*SQR*

Example 3

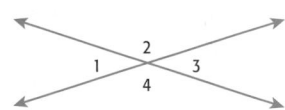

In the figure, ∠1 and ∠3 are vertical angles, and ∠2 and ∠4 are vertical angles.

A. If m∠1 = 37°, find m∠3.   37°

B. If m∠4 = *y*°, find m∠2.   *y*°

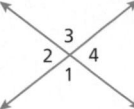

EXAMPLE 3 Finding the Measures of Vertical Angles

In the figure, ∠1 and ∠3 are vertical angles, and ∠2 and ∠4 are vertical angles.

A If m∠2 = 75°, find m∠4.

The measures of ∠2 and ∠3 add to 180° because they are supplementary, so m∠3 = 180° − 75° = 105°.

The measures of ∠3 and ∠4 add to 180° because they are supplementary, so m∠4 = 180° − 105° = 75°.

B If m∠3 = x°, find m∠1.

m∠4 = 180° − x°
m∠1 = 180° − (180° − x°)
    = 180° − 180° + x°   *Distributive Property*
    = x°   *m∠1 = m∠3*

Think and Discuss

1. Tell which statements are correct if ∠X and ∠Y are congruent.

   a. ∠X = ∠Y   **b.** m∠X = m∠Y   **c.** ∠X ≅ ∠Y   **d.** m∠X ≅ m∠Y

2. Explain why vertical angles must always be congruent.

5-1 PRACTICE & ASSESS

Assignment Guide

If you finished Example **1** assign:
  Core 1–5, 13–17, 25, 39–43
  Enriched 1–5, 13–17, 26, 39–43

If you finished Example **2** assign:
  Core 1–10, 13–22, 25–31 odd, 39–43
  Enriched 1–9 odd, 13–22, 25–34, 39–43

If you finished Example **3** assign:
  Core 1–24, 25–31 odd, 39–43
  Enriched 13–43

5-1 Exercises

FOR EXTRA PRACTICE
see page 740

internet connect
Homework Help Online
go.hrw.com Keyword: MP4 5-1

GUIDED PRACTICE

See Example **1**
1. Name three points in the figure. **points A, B, C**
2. Name a line in the figure. **$\overleftrightarrow{BC}$**
3. Name a plane in the figure. **plane z or plane ABC**

4. $\overline{AB}, \overline{BC}, \overline{AC}$
4. Name three segments in the figure.
5. Name three rays in the figure. **$\overrightarrow{BA}, \overrightarrow{BC}, \overrightarrow{CB}$**

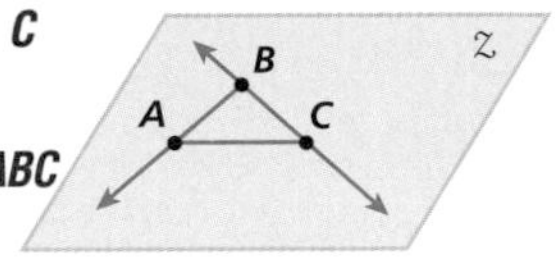

See Example **2**
6. Name a right angle in the figure. **∠HJL or ∠LJK**
7. Name two acute angles in the figure. **∠LJM, ∠MJK**
8. Name an obtuse angle in the figure. **∠HJM**
9. Name a pair of complementary angles in the figure. **∠LJM and ∠MJK**

10. ∠HJL and ∠LJK, ∠HJM and ∠MJK
10. Name two pairs of supplementary angles in the figure.

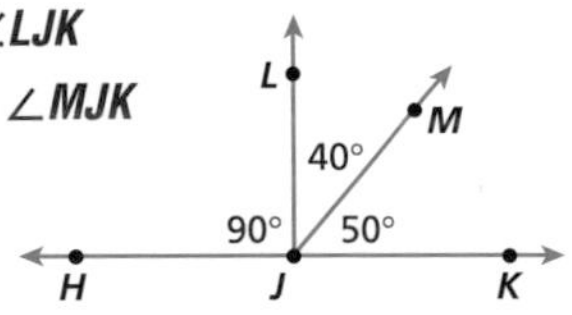

Math Background

Points, lines, and planes are mathematical ideas rather than real objects. A point has no size, a line has no width, and a plane has no thickness. The drawings we use to represent points, lines, and planes are real objects; for example, they are composed of chalk dust or ink on paper. Such idealizations of actual objects are called *mathematical abstractions*.

3 Close

Summarize

Review the vocabulary terms in the lesson. Ask volunteers to define each term and point out an example in the book. This will reinforce the vocabulary and encourage students to correctly name geometric figures. Discuss ways to remember the names of the different types of angles (e.g., an angle with a small measure is "a cute" little angle, and supplementary begins with s for *straight line*).

Answers to Think and Discuss

1. **b** and **c**; the = sign is used for numerical values, and the ≅ symbol is used for geometric figures.

2. Possible answer: They are always supplementary to the same angle, so they must have the same measure.

See Example 3 **In the figure, $\angle 1$ and $\angle 3$ are vertical angles, and $\angle 2$ and $\angle 4$ are vertical angles.**

11. If $m\angle 3 = 115°$, find $m\angle 1$. **115°**
12. If $m\angle 2 = a°$, find $m\angle 4$. **$a°$**

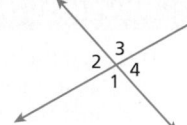

INDEPENDENT PRACTICE

See Example 1

13. Name four points in the figure. **points V, W, X, Y**
14. Name two lines in the figure. **$\overleftrightarrow{VW}$, $\overleftrightarrow{WY}$**
15. Name a plane in the figure. **plane $\mathcal{N}$ or plane VWX**
16. Name three segments in the figure. **$\overline{WV}$, $\overline{WX}$, $\overline{WY}$**
17. Name five rays in the figure. **$\overrightarrow{WV}$, $\overrightarrow{VW}$, $\overrightarrow{WY}$, $\overrightarrow{YW}$, $\overrightarrow{WX}$**

See Example 2

18. Name a right angle in the figure. **$\angle GEH$**
19. Name two acute angles in the figure. **$\angle DEH$, $\angle GEF$**
20. Name two obtuse angles in the figure. **$\angle DEG$, $\angle HEF$**
21. Name a pair of complementary angles in the figure. **$\angle FEG$, $\angle HED$**
22. Name two pairs of supplementary angles in the figure. **$\angle DEH$ and $\angle HEF$, $\angle DEG$ and $\angle GEF$**

See Example 3 **In the figure, $\angle 1$ and $\angle 3$ are vertical angles, and $\angle 2$ and $\angle 4$ are vertical angles.**

23. If $m\angle 2 = 117°$, find $m\angle 4$. **117°**
24. If $m\angle 1 = n°$, find $m\angle 3$. **$n°$**

PRACTICE AND PROBLEM SOLVING

Use the figure for Exercises 25–34. Write *true* or *false*. If a statement is false, rewrite it so it is true.

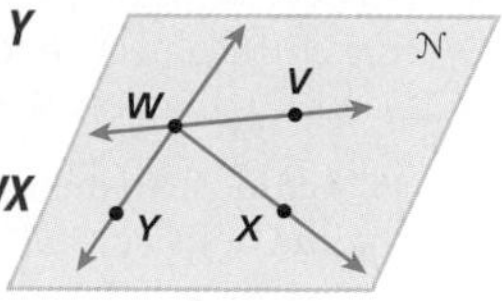

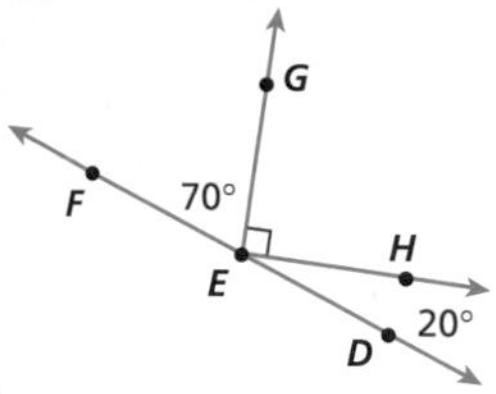

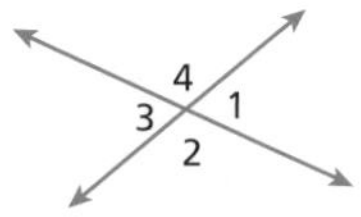

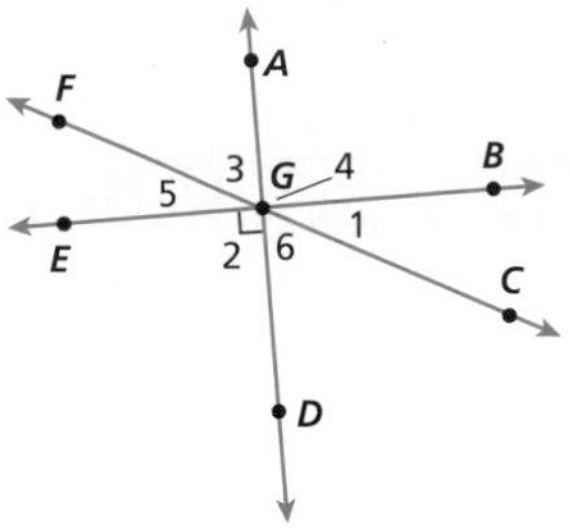

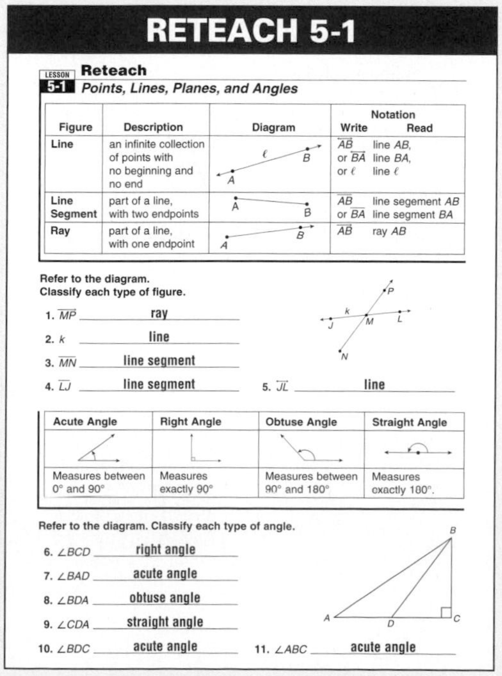

25. $\overleftrightarrow{AE}$ is a line in the figure.
26. Rays $\overrightarrow{GB}$ and $\overrightarrow{GE}$ make up line $\overleftrightarrow{EB}$. **true**
27. $\angle EGD$ is an obtuse angle.
28. $\angle 4$ and $\angle 2$ are supplementary. **true**
29. $\angle 3$ and $\angle 5$ are supplementary.
30. $\angle 6$ and $\angle 5$ are complementary. **true**
31. If $m\angle 1 = 30°$, then $m\angle 6 = 45°$.
32. If $m\angle FGD = 130°$, then $m\angle DGC = 130°$.
33. If $m\angle 3 = x°$, then $m\angle FGE = 180° - x°$.
34. $m\angle 1 + m\angle 3 + m\angle 5 + m\angle 6 = 180°$. **true**

Answers

25. False; $\overleftrightarrow{AD}$ is a line in the figure.
27. False; $\angle EGD$ is a right angle.
29. False; $\angle 3$ and $\angle 5$ are complementary.
31. False; If $m\angle 1 = 30°$, then $m\angle 6 = 60°$.
32. False; If $m\angle FGD = 130°$, then $m\angle DGC = 50°$.
33. False; If $m\angle 3 = x°$, then $m\angle FGE = 90° - x°$.
37. Possible answer: When light approaches in a direction perpendicular to the surface, there will be no refraction.
38. Possible answer: The person on the shore sees the fish in a position that is above where the fish actually is; the fish sees the person above where the person actually is.

Notes

RETEACH 5-1

LESSON 5-1 **Reteach**
Points, Lines, Planes, and Angles

Figure	Description	Diagram	Notation Write	Read
Line	an infinite collection of points with no beginning and no end		$\overleftrightarrow{AB}$ or $\overleftrightarrow{BA}$ or ℓ	line AB, line BA, line ℓ
Line Segment	part of a line, with two endpoints		$\overline{AB}$ or $\overline{BA}$	line segement AB line segment BA
Ray	part of a line, with one endpoint		$\overrightarrow{AB}$	ray AB

Refer to the diagram.
Classify each type of figure.

1. $\overrightarrow{MP}$ **ray**
2. k **line**
3. $\overline{MN}$ **line segment**
4. $\overline{LJ}$ **line segment**
5. $\overleftrightarrow{JL}$ **line**

Acute Angle	Right Angle	Obtuse Angle	Straight Angle
Measures between 0° and 90°	Measures exactly 90°	Measures between 90° and 180°	Measures exactly 180°.

Refer to the diagram. Classify each type of angle.

6. $\angle BCD$ **right angle**
7. $\angle BAD$ **acute angle**
8. $\angle BDA$ **obtuse angle**
9. $\angle CDA$ **straight angle**
10. $\angle BDC$ **acute angle**
11. $\angle ABC$ **acute angle**

PRACTICE 5-1

LESSON 5-1 **Practice B**
Points, Lines, Planes, and Angles

1. Name four points in the figure.
R, S, T, W
2. Name a line in the figure.
$\overleftrightarrow{WT}$
3. Name a plane in the figure.
Possible answer: RST
4. Name three segments in the figure.
$\overline{RS}$, $\overline{ST}$, $\overline{TW}$, or $\overline{RW}$
5. Name four rays in the figure.
$\overrightarrow{RS}$, $\overrightarrow{WT}$, $\overrightarrow{WR}$, $\overrightarrow{TW}$
6. Name a right angle in the figure.
$\angle NXP$, $\angle MXR$, $\angle MXN$, or $\angle RXP$
7. Name two acute angles in the figure.
$\angle MXS$ and $\angle RXS$
8. Name two obtuse angles in the figure.
$\angle NXS$ and $\angle PXS$
9. Name a pair of complementary angles in the figure.
$\angle RXS$ and $\angle SXM$
10. Name three pairs of supplementary angles in the figure. **Possible answer:**
$\angle NXP$ and $\angle MXN$, $\angle RXM$ and $\angle MXN$, $\angle RXP$ and $\angle PXN$

In the figure, $\angle 1$ and $\angle 2$ are vertical angles.

11. If $m\angle 1 = 110°$, find $m\angle 2$.
$m\angle 2 = 110°$
12. If $m\angle 2 = 80°$, find $m\angle 1$.
$m\angle 1 = 80°$

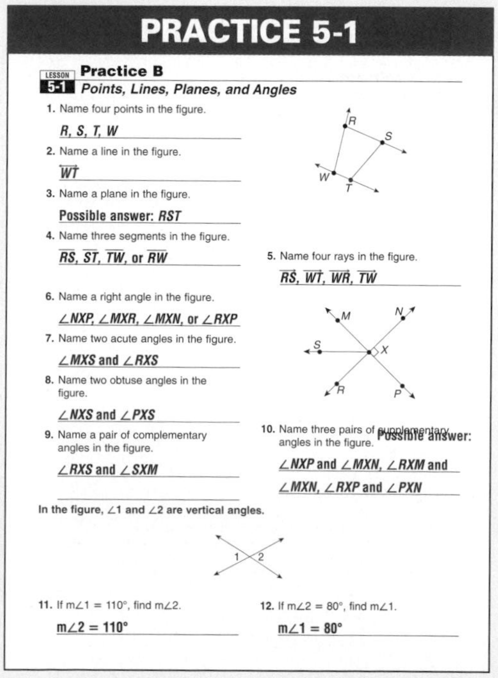

Physical Science

Exercises 35–38 focus on the refraction of light caused by water. The many interactions of light waves are studied in middle school physical science programs, such as *Holt Science & Technology*.

Answers

37–38. See previous page.

Journal

Have students look around the classroom and find examples of angles. Have them describe the location of the angles and tell whether the angles are acute, right, or obtuse. Examples might include window frames, the angle of the flagpole to the wall, and the angles of the floor tiles.

Test Prep Doctor

For Exercise 43, remind students that if two positive fractions have the same numerator, then the fraction with the lesser denominator has the greater value. The given fraction can be rewritten and compared with all of the answer choices.

$$\frac{1}{4} = \frac{1 \cdot 12}{4 \cdot 12} = \frac{12}{48}, \frac{12}{49} < \frac{12}{48}$$

$$\frac{1}{4} = \frac{1 \cdot 6}{4 \cdot 6} = \frac{6}{24}, \frac{6}{23} > \frac{6}{24}$$

$$\frac{1}{4} = \frac{1 \cdot 15}{4 \cdot 15} = \frac{15}{60}, \frac{15}{68} < \frac{15}{60}$$

$$\frac{1}{4} = \frac{1 \cdot 17}{4 \cdot 17} = \frac{17}{68}, \frac{17}{99} < \frac{17}{68}$$

Lesson Quiz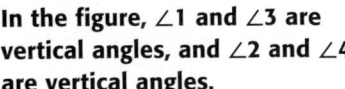

In the figure, ∠1 and ∠3 are vertical angles, and ∠2 and ∠4 are vertical angles.

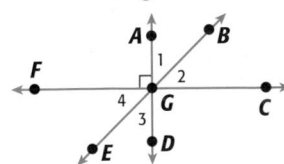

1. Name three points in the figure. Possible answer: *A, B,* and *C*

2. Name two lines in the figure. Possible answer: $\overrightarrow{AD}$ and $\overrightarrow{BE}$

3. Name a right angle in the figure. Possible answer: ∠AGF

4. Name a pair of complementary angles. Possible answer: ∠1 and ∠2

5. If m∠1 = 47°, then find m∠3. 47°

Available on Daily Transparency in CRB

The archerfish can spit a stream of water up to 3 meters in the air to knock its prey into the water. This job is made more difficult by *refraction,* the bending of light waves as they pass from one substance to another. When you look at an object through water, the light between you and the object is refracted. Refraction makes the object appear to be in a different location. Despite refraction, the archerfish still catches its prey.

35. Suppose that the measure of the angle between the bug's actual location and the bug's apparent location is 35°.

 a. Refer to the diagram. Along the fish's line of vision, what is the measure of the angle between the fish and the bug's apparent location? **145°**

 b. What is the relationship of the angles in the diagram? **They are supplementary angles.**

36. In the photograph, the underwater part of the net appears to be 40° to the right of where it actually is. What is the measure of the angle formed by the image of the underwater part of the net and the part of the net above the water? **140°**

37. ✎ **WRITE ABOUT IT** Suppose an archerfish is directly below its prey. Explain why there would be little or no distortion.

38. ✎ **CHALLENGE** A person on the shore is looking at a fish in the water. At the same time, the fish is looking at the person from below the surface. Describe what each observer sees, and where the person and the fish actually are in relation to where they appear to be.

Spiral Review

Find the mean, median, and mode of each data set. Round to the nearest tenth. (Lesson 4-3)

39. 16, 16, 14, 13, 20, 29, 14, 13, 16 **16.8; 16; 16** 40. 2.1, 2.3, 3.2, 2.2, 1.9, 2.3, 2.2 **2.3; 2.2; 2.3 and 2.2**

Find the range and the first and third quartiles of each data set. (Lesson 4-4)

41. 32, 26, 24, 14, 20, 32, 16, 25, 26 **18; 18; 29** 42. 221, 223, 352, 202, 139, 243, 232 **213; 202; 243**

43. **TEST PREP** Which fraction is greater than $\frac{1}{4}$? (Previous course) **B**

 A $\frac{12}{49}$ B $\frac{6}{23}$ C $\frac{15}{68}$ D $\frac{17}{99}$

CHALLENGE 5-1

LESSON 5-1 Challenge
Let's Meet!

Materials needed: paper strips, index cards, and scissors

1. Use a flat surface such as the top of your desk to represent a plane. Use strips of paper to represent lines. Move the lines around in the plane (**coplanar lines**) to determine the number of intersections that are possible. Summarize your results in a table.

Number of Coplanar Lines	Possible Number of Points of Intersection
2	0 or 1
3	0, 1, 2, or 3
4	0, 1, 3, 4, 5, or 6
5	0, 1, 3, 4, 5, 6, 7, 8, 9, or 10

2. Slit one index card and connect two cards to model two intersecting planes.

 a. What is the intersection of two planes?

 _____ **a line** _____

 b. Mark the diagram to illustrate the intersection of the two planes.

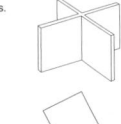

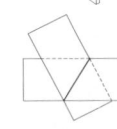

3. Using index cards to represent planes, determine the number of intersections that are possible. Summarize your results in a table.

Number of Planes	Possible Number of Lines of Intersection
2	0 or 1
3	0, 1, 2, or 3
4	0, 1, 3, 4, 5, or 6
5	0, 1, 4, 5, 6, 7, 8, 9, or 10

PROBLEM SOLVING 5-1

LESSON 5-1 Problem Solving
Points, Lines, Planes, and Angles

Use the flag of the Bahamas to solve the problems.

1. Name four points in the flag.

 Possible answers: A, B, C, D

2. Name four segments in the flag.

 Possible answers: $\overline{AB}$, $\overline{BH}$, $\overline{HI}$, $\overline{IC}$

3. Name a right angle in the flag.

 Possible answer: ∠DAB

4. Name two acute angles in the flag.

 Possible answers: ∠AED, ∠DAE

5. Name a pair of complementary angles in the flag.

 Possible answer: ∠DAE, ∠EAB

6. Name a pair of supplementary angles in the flag.

 Possible answer: ∠DGI, ∠IGE

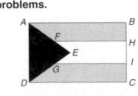

The diagram illustrates a ray of light being reflected off a mirror. The angle of incidence is congruent to the angle of reflection.

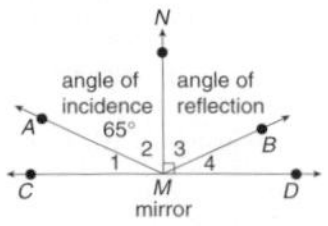

7. Name two rays in the diagram.
 A $\overrightarrow{AM}$, $\overrightarrow{MB}$ C $\overrightarrow{MA}$, $\overrightarrow{MB}$
 B $\overrightarrow{MA}$, $\overrightarrow{BM}$ D $\overrightarrow{MA}$, $\overrightarrow{MB}$

8. Name a pair of complementary angles.
 F ∠NMB, ∠BMD H ∠CMA, ∠AMD
 G ∠AMN, ∠NMB J ∠CMA, ∠DMB

9. Which angle is congruent to ∠2?
 A ∠1 C ∠3
 B ∠4 D none

10. Find the measure of ∠4.
 F 65° H 25°
 G 35° J 90°

11. Find the measure of ∠1.
 A 65° C 25°
 B 35° D 90°

12. Find the measure of ∠3.
 F 90° H 35°
 G 45° J 65°

Hands-On LAB 5A

Basic Constructions

Use with Lesson 5-1

internet connect
Lab Resources Online
go.hrw.com
KEYWORD: MP4 Lab5A

When you *bisect* a figure, you divide it into two congruent parts.

Activity

1 Follow the steps below to bisect a segment.

a. Draw $\overline{JK}$ on your paper. Place your compass point on J and draw an arc. Without changing your compass opening, place your compass point on K and draw an arc.

b. Connect the intersections of the arcs with a line. Measure $\overline{JM}$ and $\overline{KM}$. What do you notice? **$\overline{JM} \cong \overline{KM}$**

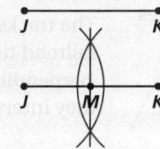

2 Follow the steps below to bisect an angle.

a. Draw acute ∠H on your paper.

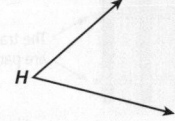

b. Place your compass point on H and draw an arc through both sides of the angle.

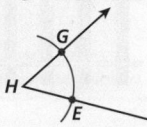

c. Without changing your compass opening, draw intersecting arcs from G and E. Label the intersection D.

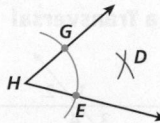

d. Draw $\overline{HD}$. Use a protractor to measure ∠GHD and ∠DHE. What do you notice?

∠$GHD \cong$ ∠DHE

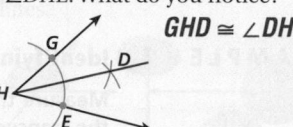

Think and Discuss

1. Explain how to use a compass and a straightedge to divide a segment into four congruent segments. Prove that the segments are congruent.

Try This

Draw each figure, and then use a compass and a straightedge to bisect it. Verify by measuring.

1. a 2-inch segment
2. a 1-inch segment
3. a 4-inch segment
4. a 64° angle
5. a 90° angle
6. a 120° angle

Hands-On LAB 5A Basic Constructions

Pacing: Traditional 1 day
Block $\frac{1}{2}$ day

Objective: To use a compass and straightedge to bisect line segments and angles

Materials: Compass, ruler with straightedge, protractor

Lab Resources

Hands-On Lab Activities . . . . . pp. 31–32

Using the Page

Allow students several minutes to get comfortable with the compass and straightedge.

Sketch each figure.

1. circle
2. 4-inch line segment
3. circle with radius 2 inches
4. line segment $\overline{AB}$ with length 3 inches; circle A containing point B

Check students' constructions.

Answers

Think and Discuss

Possible answers:

1. Possible answer: Begin by constructing a segment. Use the procedure to bisect the segment. Then bisect each half. The four segments are congruent because each one represents an equal part of the length of the original segment.

Try This

1–6. Check students' work.

Hands-On LAB 5B
Advanced Constructions

Pacing: Traditional 1 day
Block $\frac{1}{2}$ day

Objective: To use a compass and straightedge to bisect line segments and angles

Materials: Compass, ruler with straightedge, protractor

Lab Resources

Hands-On Lab Activities . . . . . pp. 33–35

Using the Page

Allow students several minutes to get comfortable with the compass and straightedge.

Use a compass and straightedge to perform each construction.

1. Construct circle *B* with diameter 4 inches.
2. Bisect obtuse ∠*ABC*.
3. Construct segment $\overline{RT}$ and bisect it.
   Check students' constructions.

Hands-On LAB 5B
Advanced Constructions

Use with Lesson 5-2

Copying an angle is an important step in the construction of parallel lines.

Activity

❶ **Follow the steps below to copy an angle.**

 a. Draw acute ∠*ABC* on your paper. Draw $\overrightarrow{DE}$.

 b. With your compass point on *B*, draw an arc through ∠*ABC*. With the same compass opening, place your compass point on *D* and draw an arc through $\overrightarrow{DE}$.

 c. Adjust your compass to the width of the arc intersecting ∠*ABC*. Place your compass point on *F* and draw an arc that intersects the arc through $\overrightarrow{DE}$ at *G*. Draw $\overrightarrow{DG}$. Use your protractor to measure ∠*ABC* and ∠*GDF*.

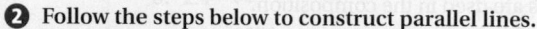

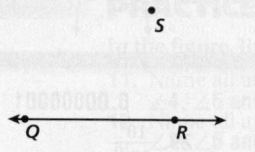
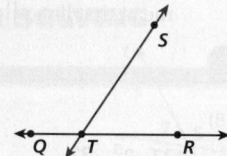

❷ **Follow the steps below to construct parallel lines.**

1. Draw $\overrightarrow{QR}$ on your paper. Draw point *S* above or below $\overrightarrow{QR}$. Draw a line through point *S* that intersects $\overrightarrow{QR}$. Label the intersection *T*.

2. Make a copy of ∠*STR* with its vertex at *S* using the method described in the first Activity. How do you know the lines are parallel? **Corresponding angles are congruent.**

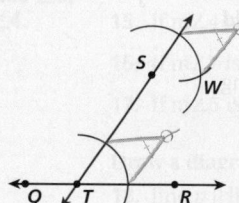
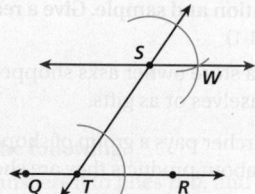

❸ Follow the steps below to construct perpendicular lines.

a. Draw $\overleftrightarrow{MN}$ on your paper. Draw point P above or below $\overleftrightarrow{MN}$.

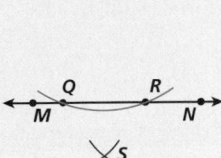

b. With your compass point at P, draw an arc intersecting $\overleftrightarrow{MN}$ at points Q and R.

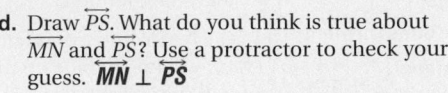

c. Draw arcs from points Q and R, using the same compass opening, that intersect at point S.

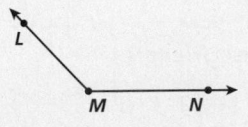

d. Draw $\overleftrightarrow{PS}$. What do you think is true about $\overleftrightarrow{MN}$ and $\overleftrightarrow{PS}$? Use a protractor to check your guess. $\overleftrightarrow{MN} \perp \overleftrightarrow{PS}$

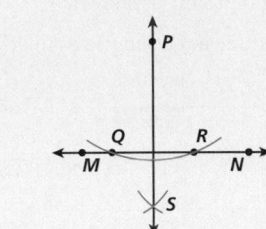

Think and Discuss

1. How many lines can be drawn that are perpendicular to a given line? Explain your answer.

2. Name three ways that you can determine if two lines are parallel.

Try This

Use a compass and a straightedge to construct each figure.

1. an angle congruent to $\angle LMN$

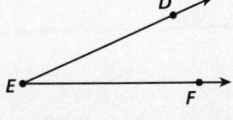

2. a line parallel to $\overleftrightarrow{ST}$

3. a line perpendicular to $\overleftrightarrow{GH}$

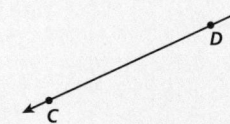

4. an angle congruent to $\angle DEF$

5. a line parallel to $\overleftrightarrow{AB}$

6. a line perpendicular to $\overleftrightarrow{CD}$

Answers

Warm Up

Solve each equation.

1. $62 + x + 37 = 180$ $x = 81$

2. $x + 90 + 11 = 180$ $x = 79$

3. $x + x + 18 = 180$ $x = 81$

4. $180 = 2x + 72 + x$ $x = 36$

Problem of the Day

What is the one hundred fiftieth day of a non-leap year?   May 30
Available on Daily Transparency in CRB

Math Humor

The obtuse angle was always complaining about the heat, "I can't take this heat anymore. It must be over 90° in here!"

5-3 Triangles

Learn to find unknown angles in triangles.

Vocabulary

Triangle Sum Theorem
acute triangle
right triangle
obtuse triangle
equilateral triangle
isosceles triangle
scalene triangle

If you tear off two corners of a triangle and place them next to the third corner, the three angles seem to form a straight line.

Draw a triangle and extend one side. Then draw a line parallel to the extended side, as shown.

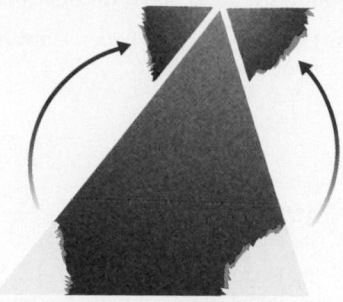

This torn triangle demonstrates an important geometry theorem called the Triangle Sum Theorem.

The three angles in the triangle can be arranged to form a straight line, or 180°.

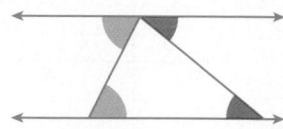

The sides of the triangle are transversals to the parallel lines.

TRIANGLE SUM THEOREM		
Words	**Numbers**	**Algebra**
The angle measures of a triangle in a plane add to 180°.	58° 43° 79° $43° + 58° + 79° = 180°$	$r°$ $t°$ $s°$ $r° + s° + t° = 180°$

An **acute triangle** has 3 acute angles. A **right triangle** has 1 right angle. An **obtuse triangle** has 1 obtuse angle.

EXAMPLE 1   Finding Angles in Acute, Right, and Obtuse Triangles

A Find x in the acute triangle.

$$62° + 33° + x° = 180°$$
$$95° + x° = 180°$$
$$\underline{-95° \qquad -95°}$$
$$x° = 85°$$

B Find y in the right triangle.

$$28° + 90° + y° = 180°$$
$$118° + y° = 180°$$
$$\underline{-118° \qquad -118°}$$
$$y° = 62°$$

---

1 Introduce

Alternate Opener

EXPLORATION

5-3 Triangles

The figure below shows two parallel lines *m* and *n*, and a triangle.

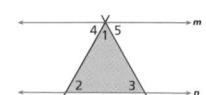

1. What is the sum of the measures of ∠1, ∠2, and ∠3?
2. What is the sum of the measures of ∠4, ∠1, and ∠5?
3. Name an angle with the same measure as ∠4.
4. Name an angle with the same measure as ∠5.

Think and Discuss

5. **Discuss** what would happen to the sum of the measures of ∠1, ∠2, and ∠3 if the triangle above were bigger.
6. **Explain** why each angle measure in an *equilateral triangle* is 60°. (In an equilateral triangle, the three sides are equal.)

Motivate

Use a rubber band to form triangles with different shapes. (Use a push pin and two pencils at a bulletin board, or use three pencils at the chalkboard with a student volunteer.) Help students to see that if one of the angles gets larger, then another angle gets smaller. This fact supports the Triangle Sum Theorem, which states that the sum of the angle measures in any triangle is 180°.

Exploration worksheet and answers on Chapter 5 Resource Book pp. 27 and 122

---

2 Teach

Lesson Presentation

Guided Instruction

In this lesson, students learn to find unknown angles in triangles. The lesson is based on the Triangle Sum Theorem, one of the most important theorems in geometry (Teaching Transparency T11, CRB). Illustrate the Triangle Sum Theorem by tearing off two corners of a paper triangle and placing them next to the third corner to form a straight line, which measures 180°. Explain that a triangle can have at most one right angle or one obtuse angle because of the Triangle Sum Theorem.

Teaching Tip   For Example 2, explain that in a triangle, angles opposite congruent sides are congruent.

C Find z in the obtuse triangle.

$$14° + 51° + z° = 180°$$
$$65° + z° = 180°$$
$$\underline{-65° \qquad -65°}$$
$$z° = 115°$$

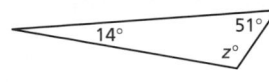

An **equilateral triangle** has 3 congruent sides and 3 congruent angles. An **isosceles triangle** has at least 2 congruent sides and 2 congruent angles. A **scalene triangle** has no congruent sides and no congruent angles.

EXAMPLE 2 Finding Angles in Equilateral, Isosceles, and Scalene Triangles

A Find the angle measures in the equilateral triangle.

$$3m° = 180° \qquad \textit{Triangle Sum Theorem}$$
$$\frac{3m°}{3} = \frac{180°}{3}$$
$$m° = 60°$$

All three angles measure 60°.

B Find the angle measures in the isosceles triangle.

$$77° + n° + n° = 180° \qquad \textit{Triangle Sum Theorem}$$
$$77° + 2n° = 180° \qquad \textit{Combine like terms.}$$
$$\underline{-77° \qquad\qquad -77°} \qquad \textit{Subtract 77° from both sides.}$$
$$2n° = 103°$$
$$\frac{2n°}{2} = \frac{103°}{2} \qquad \textit{Divide both sides by 2.}$$
$$n° = 51.5°$$

The angles labeled $n°$ measure 51.5°.

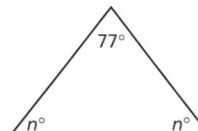

C Find the angle measures in the scalene triangle.

$$p° + 2p° + 3p° = 180° \qquad \textit{Triangle Sum Theorem}$$
$$\frac{6p°}{6} = \frac{180°}{6} \qquad \textit{Combine like terms.}$$
$$p° = 30°$$

The angle labeled $p°$ measures 30°, the angle labeled $2p°$ measures $2(30°) = 60°$, and the angle labeled $3p°$ measures $3(30°) = 90°$.

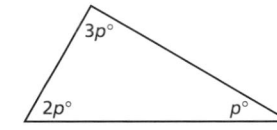

Reaching All Learners
Through Critical Thinking

Show drawings of a square, a rectangle, and a parallelogram (Teaching Transparency T12, CRB).

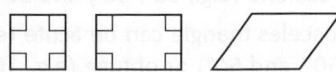

Ask the students, "If a diagonal is drawn in each figure, what types of triangles will be created?"

square: isosceles right triangles

rectangle: scalene right triangles

parallelogram: scalene obtuse triangles

Example 1

A. Find p in the acute triangle.   **63°**

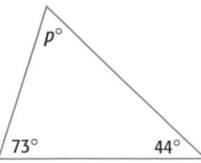

B. Find c in the right triangle.   **48°**

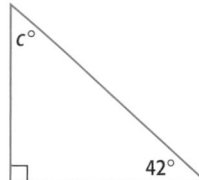

C. Find m in the obtuse triangle. **95°**

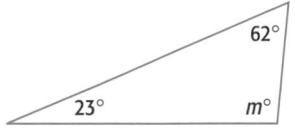

Example 2

A. Find the angle measures in the equilateral triangle.   **60°**

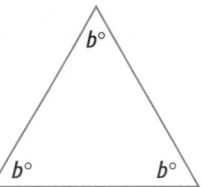

B. Find the angle measures in the isosceles triangle.   **59°**

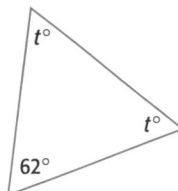

C. Find the angle measures in the scalene triangle.   **36°, 54°, 90°**

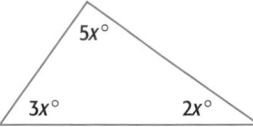

Example 3

The second angle in a triangle is six times as large as the first. The third angle is half as large as the second. Find the angle measures and draw a possible picture.   **18°, 108°, 54°**

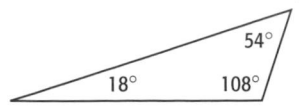

Answers

35a. $w° = 75°$; $y° = 75°$; two right angles

b. $x° = 30°$; $z° = 75°$; $m° = 75°$

c. The two blue triangles are right scalene triangles, and the white triangle is an acute isosceles triangle.

36. Possible answer: Two angles of an isosceles triangle must be the same, so the triangle would have angles measuring 50°, 50°, and 70° or 50°, 70°, and 70°. Neither combination adds up to 180°, so the triangle is impossible.

37. Possible answer: Cut the square in half diagonally. The angle measures are 90°, 45°, and 45°. Cut the triangle from one vertex to the midpoint of the opposite side. The angles are 30°, 60°, and 90°.

Journal

Ask students to sketch and label an acute triangle, a right triangle, and an obtuse triangle.

Note: You may want to suggest that students use the vertical margin line and a horizontal line on notebook paper to draw a right angle.

Test Prep Doctor

For Exercise 41, remind students to consider what they know about a rectangle. They should recall that a rectangle has four right angles. When a diagonal cuts two angles, there is one right angle remaining in each triangle.

Lesson Quiz

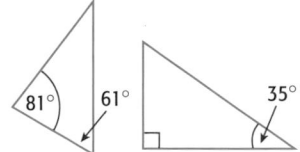

1. Find the missing angle measure in the acute triangle shown. **38°**

2. Find the missing angle measure in the right triangle shown. **55°**

3. Find the missing angle measure in an acute triangle with angle measures of 67° and 63°. **50°**

4. Find the missing angle measure in an obtuse triangle with angle measures of 10° and 15°. **155°**

Available on Daily Transparency in CRB

Describe each statement as always, sometimes, or never true.

27. An equilateral triangle is an acute triangle. **always**

28. An equilateral triangle is an isosceles triangle. **always**

29. An acute triangle is an equilateral triangle. **sometimes**

30. An isosceles triangle is an equilateral triangle. **sometimes**

31. A scalene triangle is an equilateral triangle. **never**

32. An obtuse triangle is an isosceles triangle. **sometimes**

33. A right triangle is an obtuse triangle. **never**

34. An obtuse triangle has two acute angles. **always**

35. **SOCIAL STUDIES** American Samoa is a territory of the United States made up of a group of islands in the South Pacific Ocean, about halfway between Hawaii and New Zealand. The flag of American Samoa is shown.

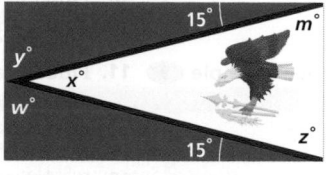

   a. Find the measure of each angle in the blue triangles.

   b. Use your answers to part **a** to find the angle measures in the white triangle.

   c. Classify the triangles in the flag by their sides and angles.

 36. **WHAT'S THE ERROR?** An isosceles triangle has one angle that measures 50° and another that measures 70°. Why can't this triangle be drawn?

 37. **WRITE ABOUT IT** Explain how to cut a square or an equilateral triangle in half to form two identical triangles. What are the angle measures in the resulting triangles in each case?

 38. **CHALLENGE** Find x, y, and z. $x = 30$, $y = 70$, $z = 80$

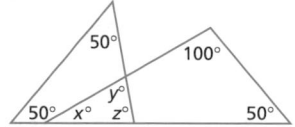

Spiral Review

Evaluate each expression for the given values of the variables. (Lesson 1-1)

39. $7x - 4y$ for $x = 5$ and $y = 6$ **11**

40. $6.5p - 9.1q$ for $p = 2.5$ and $q = 0$ **16.25**

41. **TEST PREP** The rectangle shown is cut by a diagonal. What two figures are formed? (Lesson 5-3) **C**

    A Two acute triangles   **C** Two right triangles

    B Two equilateral triangles   **D** Two isosceles triangles

CHALLENGE 5-3

LESSON 5-3 Challenge
Change a This into a That

A geometric dissection involves cutting a figure into pieces that can then be rearranged to form another figure.

Trace each figure. Cut up the figure you have traced and rearrange the numbered pieces to form the indicated figure. Sketch your solution.

1. Rearrange the pieces of the equilateral triangle to form a square.

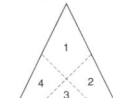 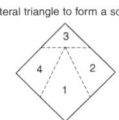

2. Rearrange the pieces of the star to form an equilateral triangle.

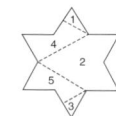 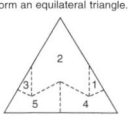

3. Rearrange the pieces of the cross to form an equilateral triangle.

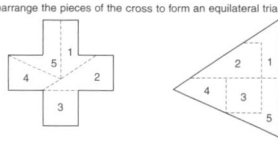

PROBLEM SOLVING 5-3

LESSON 5-3 Problem Solving
Triangles

The American flag must be folded according to certain rules that result in the flag being folded into the shape of a triangle. The figure shows a frame designed to hold an American flag.

1. Is the triangle acute, right, or obtuse?

    right

2. Is the triangle equilateral, isosceles, or scalene?

    isosceles

3. Find x.

    $x = 45°$

4. Find y.

    $y = 45°$

The figure shows a map of three streets.

5. Find x.
  A 22°   C 30°
  Ⓑ 128°   D 68°

6. Find w.
  F 22°   Ⓗ 30°
  G 128°   J 52°

7. Find y.
  A 22°
  B 30°
  Ⓒ 128°
  D 143°

8. Find z.
  Ⓕ 22°
  G 30°
  H 128°
  J 143°

  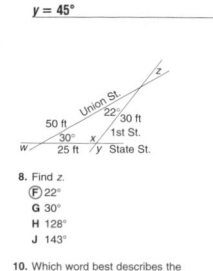

9. Which word best describes the triangle formed by the streets?
  A acute
  B right
  Ⓒ obtuse
  D equilateral

10. Which word best describes the triangle formed by the streets?
  F equilateral
  G isosceles
  Ⓗ scalene
  J acute

5-4 Polygons

Learn to classify and find angles in polygons.

Vocabulary

polygon

regular polygon

trapezoid

parallelogram

rectangle

rhombus

square

The cross section of a brilliant-cut diamond is a *pentagon*. The most beautiful and valuable diamonds have precisely cut angles that maximize the amount of light they reflect.

A **polygon** is a closed plane figure formed by three or more segments. A polygon is named by the number of its sides.

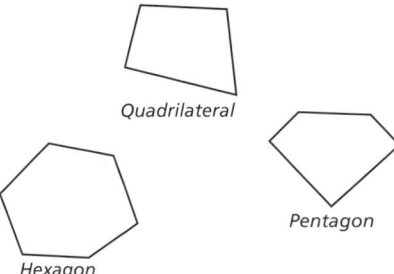

Too shallow   Ideal   Too deep

Polygon	Number of Sides
Triangle	3
Quadrilateral	4
Pentagon	5
Hexagon	6
Heptagon	7
Octagon	8
n-gon	*n*

Quadrilateral

Pentagon

Hexagon

EXAMPLE **1** **Finding Sums of the Angle Measures in Polygons**

Find the sum of the angle measures in each figure.

A Find the sum of the angle measures in a quadrilateral.
Divide the figure into triangles.
$2 \cdot 180° = 360°$ *2 triangles*

B Find the sum of the angle measures in a pentagon.
Divide the figure into triangles.
$3 \cdot 180° = 540°$ *3 triangles*

Look for a pattern between the number of sides and the number of triangles.

Hexagon:
6 sides
4 triangles

Heptagon:
7 sides
5 triangles

1 Introduce

Alternate Opener

5-4 Polygons

Use a protractor to measure each angle in each figure. Then find the sum of the angle measures.

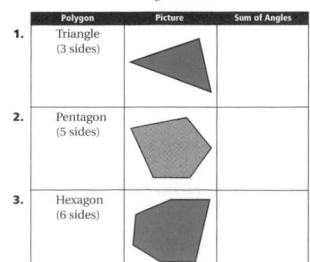

	Polygon	Picture	Sum of Angles
1.	Triangle (3 sides)		
2.	Pentagon (5 sides)		
3.	Hexagon (6 sides)		

Think and Discuss

4. **Explain** how you know the sum of the angle measures in a square.

5. **Discuss** how you can use the two triangles to show that the sum of the angle measures in a four-sided figure is 360°.

Motivate

Show students the following prefixes: *tri-, quad-, penta-, hexa-, hepta-,* and *octa-*. Ask students to think of words that begin with these prefixes, such as tricycle, quadruple, and octopus. Discuss with students what each prefix means. Tell students that they will be using these prefixes to classify polygons.

Exploration worksheet and answers on Chapter 5 Resource Book pp. 36–37 and 124–125

5-4 **Organizer**

Pacing: Traditional 1 day
Block $\frac{1}{2}$ day

Objective: Students classify and find angles in polygons.

Warm Up

1. How many sides does a hexagon have? **6**

2. How many sides does a pentagon have? **5**

3. How many angles does an octagon have? **8**

4. Evaluate $(n - 2)180$ for $n = 7$. **900**

Problem of the Day

Jeffrey planted four carnations, three dahlias, seven marigolds, five cornflowers, one geranium, and four mums. He forgot to water them and on each of the two following days, half the remaining flowers died. How many flowers were still living at the end of the second day? **6**

Available on Daily Transparency in CRB

Math Humor

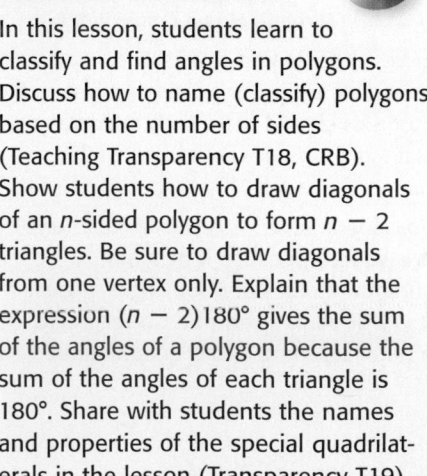

Teacher: What is the definition of a rectangle?

Student: An angle that's been in an accident.

2 Teach

Lesson Presentation

Guided Instruction

In this lesson, students learn to classify and find angles in polygons. Discuss how to name (classify) polygons based on the number of sides (Teaching Transparency T18, CRB). Show students how to draw diagonals of an *n*-sided polygon to form $n - 2$ triangles. Be sure to draw diagonals from one vertex only. Explain that the expression $(n - 2)180°$ gives the sum of the angles of a polygon because the sum of the angles of each triangle is 180°. Share with students the names and properties of the special quadrilaterals in the lesson (Transparency T19).

Additional Examples

Example 1

Find the sum of the angle measures in each figure.

A. Find the sum of the angle measures in a hexagon. **720°**

B. Find the sum of the angle measures in an octagon. **1080°**

Example 2

Find the angle measures in each regular polygon.

A. 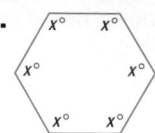 $x° = 120°$

B. 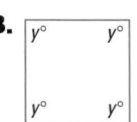 $y° = 90°$

Example 3

Give all the names that apply to each figure.

A. 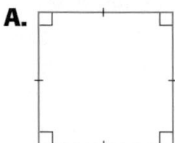 quadrilateral, parallelogram, rectangle, rhombus, square

B. 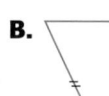 quadrilateral, parallelogram, rhombus

---

The pattern is that the number of triangles is always 2 less than the number of sides. So an n-gon can be divided into $n - 2$ triangles. The sum of the angle measures of any n-gon is $180°(n - 2)$.

All the sides and angles of a **regular polygon** have equal measures.

EXAMPLE 2 Finding the Measure of Each Angle in a Regular Polygon

Find the angle measures in each regular polygon.

A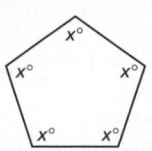

5 congruent angles
$5x° = 180°(5 - 2)$
$5x° = 180°(3)$
$5x° = 540°$
$\dfrac{5x°}{5} = \dfrac{540°}{5}$
$x° = 108°$

B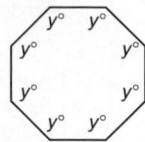

8 congruent angles
$8y° = 180°(8 - 2)$
$8y° = 180°(6)$
$8y° = 1080°$
$\dfrac{8y°}{8} = \dfrac{1080°}{8}$
$y° = 135°$

Quadrilaterals with certain properties are given additional names. A **trapezoid** has exactly 1 pair of parallel sides. A **parallelogram** has 2 pairs of parallel sides. A **rectangle** has 4 right angles. A **rhombus** has 4 congruent sides. A **square** has 4 congruent sides and 4 right angles.

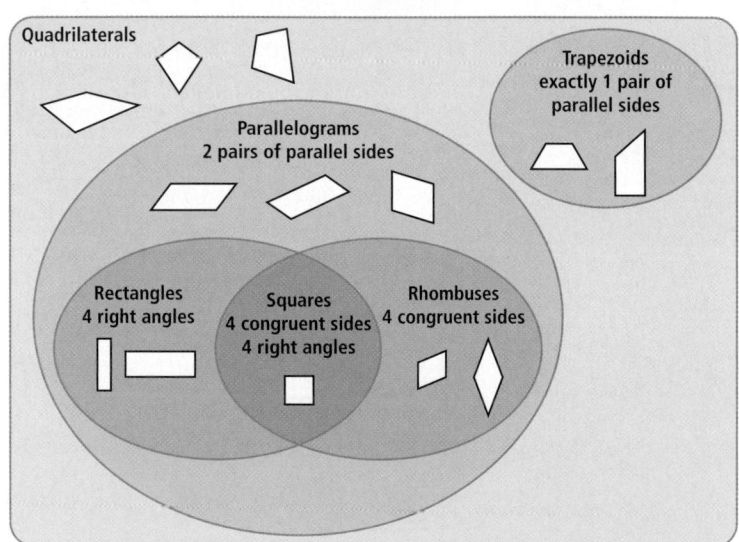

Teach

 **Reaching All Learners**
Through Home Connection

Have students find at least 5 quadrilaterals in their everyday lives. Students should draw each quadrilateral, state its properties, and write all the names that apply.

Possible answer: A baseball "diamond" is a square because it has four congruent sides and four right angles. It is a square, a rhombus, a rectangle, a parallelogram, and a quadrilateral.

EXAMPLE 3 Classifying Quadrilaterals

Give all of the names that apply to each figure.

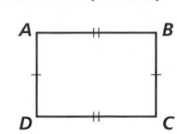

Helpful Hint

Marks on the sides of a figure can be used to show congruence.

$\overline{AB} \cong \overline{CD}$ (2 marks)
$\overline{AD} \cong \overline{BC}$ (1 mark)

A

$\overline{EF} \parallel \overline{GH}$

quadrilateral	*Four-sided polygon*
trapezoid	*1 pair of parallel sides*

B

quadrilateral	*Four-sided polygon*
parallelogram	*2 pairs of parallel sides*
rectangle	*4 right angles*

Think and Discuss

1. **Choose** which is larger, an angle in a regular heptagon or an angle in a regular octagon.

2. **Explain** why all rectangles are parallelograms and why all squares are rectangles.

3. **Give** another name for a regular triangle and for a regular quadrilateral.

5-4 Exercises

FOR EXTRA PRACTICE

see page 740

✔ **internet** connect

Homework Help Online
go.hrw.com   Keyword: MP4 5-4

5-4 PRACTICE & ASSESS

Assignment Guide

If you finished Example **1** assign:
  Core 1, 2, 7, 8, 25, 26, 39–43
Enriched 1, 2, 7, 8, 25, 37, 39–43

If you finished Example **2** assign:
  Core 1–4, 7–10, 13–24, 25, 27, 39–43
Enriched 7, 9, 13–30, 35, 37, 39–43

If you finished Example **3** assign:
  Core 1–24, 25–33 odd, 39–43
Enriched 7, 9, 11, 13–43

GUIDED PRACTICE

See Example **1** Find the sum of the angle measures in each figure.

1. **360°**

2. **720°**

See Example **2** Find the angle measures in each regular polygon.

3.  $t° = 90°$

4. 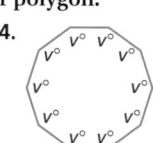 $v° = 144°$

3 Close

Summarize

Review the prefixes for classifying polygons and the expression $(n - 2)180°$ for the angle sum of an n-sided polygon. Ask for volunteers to find the angle sum and the measure of an individual angle in a regular polygon with ten sides (decagon).

Possible answers: The angle sum is $(10 - 2)180° = 1440°$. Each angle measures 144°.

Answers to Think and Discuss

1. An angle in a regular octagon (135°) is larger than an angle in a regular heptagon $(128\frac{4}{7}°)$.

2. Possible answer: All rectangles have two pairs of parallel sides because they have four right angles. This makes them parallelograms. All squares have four right angles, which makes them rectangles.

3. equilateral triangle and square

Answers

28.

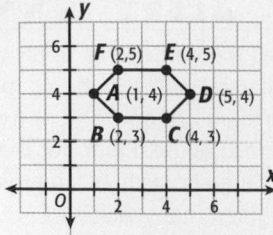

hexagon

29.

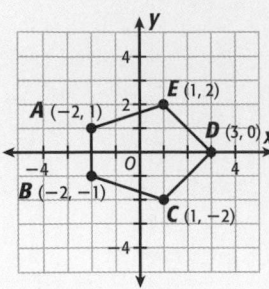

pentagon

30.

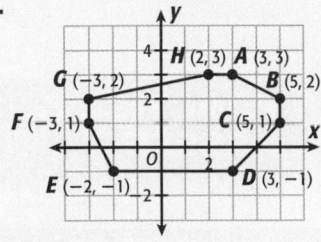

octagon

31. Possible answer:

33. Possible answer:

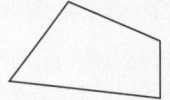

See Example **3** Give all of the names that apply to each figure.

5.

quadrilateral, trapezoid

6.

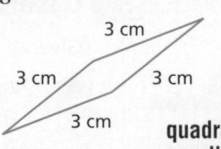

3 cm

3 cm   3 cm

3 cm

quadrilateral, parallelogram, rhombus

INDEPENDENT PRACTICE

See Example **1** Find the sum of the angle measures in each figure.

7. 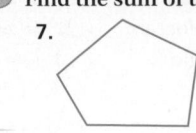 540°

8. 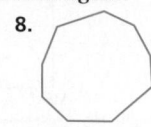 1260°

See Example **2** Find the angle measures in each regular polygon.

9. 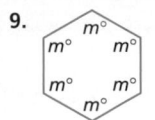 $m° = 120°$

10. 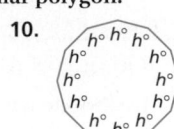 $h° = 150°$

See Example **3** Give all of the names that apply to each figure.

11.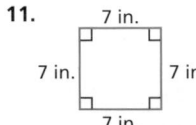
7 in.

7 in.   7 in.

7 in.

quadrilateral, parallelogram, rhombus, rectangle, square

12.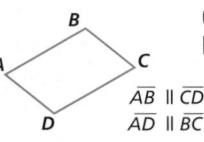
B

A   C

D

$\overline{AB} \parallel \overline{CD}$
$\overline{AD} \parallel \overline{BC}$

quadrilateral, parallelogram

PRACTICE AND PROBLEM SOLVING

Find the sum of the angle measures in each polygon. Then, if the polygon is regular, find the measure of each angle.

13. 20-gon **3240°; 162°** **14.** 11-gon **1620°; 147.3°** **15.** 72-gon **12,600°; 175°**

16. pentagon **540°; 108°** **17.** 18-gon **2880°; 160°** **18.** n-gon **$180°(n-2)$;** $\dfrac{180°(n-2)}{n}$

Find the value of each variable.

19. 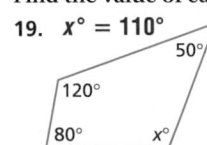 $x° = 110°$
50°
120°
80°   x°

20. 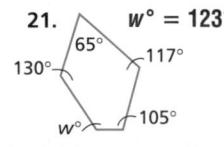 $y° = 100°$
45°
35°   y°

21. $w° = 123°$
65°
130°   117°
w°   105°

22. $z° = 102°$
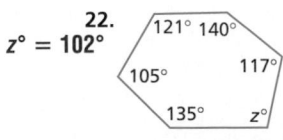
121° 140°
105°   117°
135°   z°

23. $x° = 130°$
x° x°
50°   50°

24. $m° = 50°$
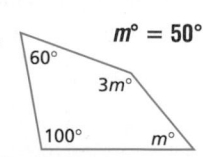
60°
3m°
100°   m°

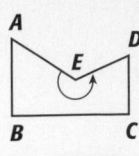 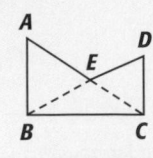

Math Background

The lesson focuses on the expression $(n - 2)180°$ for the sum of the angles of a polygon. The implicit assumption is that we are discussing *interior* angles of a *convex* polygon.

The same expression gives the sum of the interior angles of a *concave* polygon, with the understanding that at least one of the interior angles has a measure greater than 180° (such an angle is called a *reflex* angle).

A   A
     D   D
  E   E
B   C   B   C

$m\angle E > 180°$ $(n - 2)$ **triangles**

RETEACH 5-4

LESSON 5-4 Reteach
Polygons

A polygon of *n* sides (an *n*-gon) can be divided into $(n - 2)$ triangles.
The sum of the angle measures of an *n*-gon = $(n - 2)180°$.

A polygon of 5 sides (pentagon) can be divided into 3 triangles.

Sum of angle measures of pentagon
$= (n - 2)\,180°$
$= (5 - 2)\,180° = (3)180° = 540°$

Find the sum of the measures of the angles.

	1. quadrilateral	2. hexagon
How many sides in the polygon?	**4**	**6**
How many triangles can be formed?	**4** − 2 = **2**	**6** − 2 = **4**
Multiply the number of triangles by 180°.	180° × **2**	180° × **4**
sum of the measures of the angles	**360°**	**720°**

In a **regular polygon**, all sides and all angles are congruent.
The measure of each angle of a regular polygon = $\dfrac{\text{sum of the angles}}{\text{number of sides}}$

The measure of each angle of a regular pentagon $= \dfrac{(5-2)180°}{5} = \dfrac{(3)180°}{5} = \dfrac{540°}{5} = 108°$

Find the measure of each angle.

	3. regular octagon	4. regular decagon
How many sides (angles) in the polygon?	**8**	**10**
How many triangles can be formed?	**8** − 2 = **6**	**10** − 2 = **8**
Multiply the number of triangles by 180°.	180° × **6**	180° × **8**
Sum of the measures of the angles	**1080°** **1080°**	**1440°** **1440°**
Divide the sum by the number of angles.	**8**	**10**
Measure of each angle of the polygon	**135°**	**144°**

PRACTICE 5-4

LESSON 5-4 Practice B
Polygons

Find the sum of the angle measures in each figure.

1.   **360°**
2.   **720°**
3.   **900°**

4.   **180°**
5.   **1080°**
6.   **540°**

Find the angle measure in each regular polygon.

7.   **120°**
8.   **90°**
9.   **135°**

10.   **108°**
11.   **60°**
12.   **144°**

Write all the names that apply to each figure.

13. quadrilateral, parallelogram, rhombus
14. quadrilateral
15. quadrilateral, trapezoid

The sum of the angle measures of a polygon is given. Name the polygon.

25. 720° **hexagon**   **26.** 360° **quadrilateral**   **27.** 1980° **13-gon**

Graph the given vertices on a coordinate plane. Connect the points to draw a polygon and classify it by the number of its sides.

28. $A(1, 4)$, $B(2, 3)$, $C(4, 3)$, $D(5, 4)$, $E(4, 5)$, $F(2, 5)$

29. $A(-2, 1)$, $B(-2, -1)$, $C(1, -2)$, $D(3, 0)$, $E(1, 2)$

30. $A(3, 3)$, $B(5, 2)$, $C(5, 1)$, $D(3, -1)$, $E(-2, -1)$, $F(-3, 1)$, $G(-3, 2)$, $H(2, 3)$

Sketch a quadrilateral to fit each description. If no quadrilateral can be drawn, write *not possible*.

31. a parallelogram that is not a rectangle

32. a square that is not a rhombus   **not possible**

33. a quadrilateral that is not a trapezoid or a parallelogram

34. a rectangle that is not a square

The master jeweler of Great Britain's crown jewels has thousands of diamonds in his care, including the world's two largest cut diamonds. The Imperial State Crown contains over 3000 precious stones, including 2800 diamonds.

35. EARTH SCIENCE Precious stones are often cut in a *brilliant cut* to maximize the light they reflect. The best angles for a cut depend on the type of stone. The best angles for a diamond are shown in the figure.

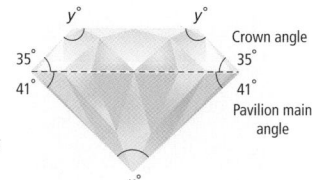

Crown angle
Pavilion main angle

a. Use the fact that the pavilion main angle is 41° to find x. $x° = 98°$

b. Use the fact that the crown angle is 35° to find y. $y° = 145°$

36. WHAT'S THE ERROR? A student said that all squares are rectangles, but not all squares are rhombuses. What was the error?

37. WRITE ABOUT IT Why is it possible to find the sum of the angle measures of an *n*-gon using the formula $(180n - 360)°$?

38. CHALLENGE Use properties of parallel lines to explain which angles in a parallelogram must be congruent.

Spiral Review

Write each number in scientific notation. (Lesson 2-9)

39. 0.00000064
6.4×10^{-7}

40. 7,390,000,000
7.39×10^9

41. −0.0000016
-1.6×10^{-6}

42. −4,100,000
-4.1×10^6

43. TEST PREP If the measure of one acute angle of a right triangle is 32°, then the measure of the other acute angle is ▇. (Lesson 5-3)   **C**

A 32°   B 148°   C 58°   D 48°

Answers

34. Possible answer:

36. Possible answer: A rectangle has four right angles; a rhombus has four congruent sides. A square has four right angles and four congruent sides, so it is a rectangle and a rhombus.

37. Possible answer: The formula for the sum of the angle measures of a polygon, $180(n - 2)$, can be written as $180n - 360$ using the Distributive Property.

38. Possible answer: If you extend the sides of a parallelogram, each side becomes a transversal to a pair of parallel lines, so all of the acute angles are congruent and all of the obtuse angles are congruent. That means the angles in the opposite corners of a parallelogram are congruent.

Journal

Have students write about how the expression $(n - 2)180°$ is related to the number of triangles formed by the diagonals of a polygon.

Test Prep Doctor

For Exercise 43, students might subtract 32 from 180 and choose **B**. Encourage students to draw a picture so they can see the three angles.

Lesson Quiz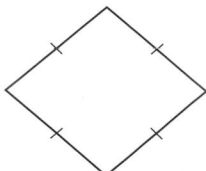

1. Find the sum of the angle measures in a quadrilateral.   **360°**

2. Find the sum of the angle measures in a hexagon.   **720°**

3. Find the measure of each angle in a regular octagon.   **135°**

4. Write all of the names that apply to the figure below.   **quadrilateral, rhombus, parallelogram**

Available on Daily Transparency in CRB

Pacing: Traditional 1 day
Block $\frac{1}{2}$ day

Objective: Students identify polygons on the coordinate plane.

Warm Up

Complete each statement.

1. Two lines in a plane that never meet are called ____?____ lines.
parallel

2. ____?____ lines intersect at right angles.
Perpendicular

3. The symbol ∥ means that lines are ____?____.
parallel

4. When a transversal intersects two ____?____ lines, all of the acute angles are congruent.
parallel

Problem of the Day

What type of polygon am I? My opposite angles have equal measure. I do not have a right angle. All my sides are congruent. **rhombus**

Available on Daily Transparency in CRB

What kind of geometric figure is like a runaway parrot? A polygon

---

5-5 Coordinate Geometry

Learn to identify polygons in the coordinate plane.

Vocabulary
slope
rise
run

In computer graphics, a coordinate system is used to create images, from simple geometric figures to realistic figures used in movies.

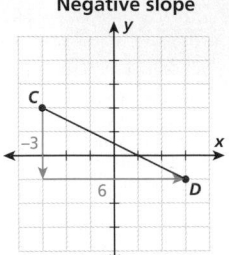

Properties of the coordinate plane can be used to find information about figures in the plane, such as whether lines in the plane are parallel.

Slope is a number that describes how steep a line is.

$$\text{slope} = \frac{\text{vertical change}}{\text{horizontal change}} = \frac{\text{rise}}{\text{run}}$$

Positive slope

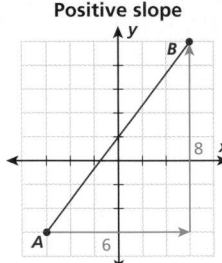

slope of $\overline{AB} = \frac{8}{6} = \frac{4}{3}$

slope of $\overline{CD} = \frac{-3}{6} = \frac{-1}{2}$

Negative slope

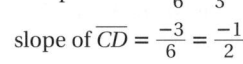

The slope of a horizontal line is 0. The slope of a vertical line is undefined.

EXAMPLE 1 Finding the Slope of a Line

Determine if the slope of each line is positive, negative, 0, or undefined. Then find the slope of each line.

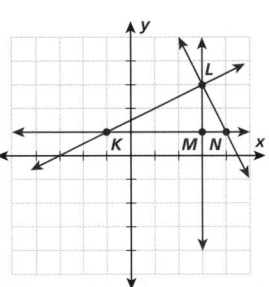

A $\overleftrightarrow{KL}$
positive slope; slope of $\overleftrightarrow{KL} = \frac{2}{4} = \frac{1}{2}$

B $\overleftrightarrow{LM}$
slope of $\overleftrightarrow{LM}$ is undefined

C $\overleftrightarrow{LN}$
negative slope; slope of $\overleftrightarrow{LN} = \frac{-2}{1} = -2$

D $\overleftrightarrow{KM}$
slope of $\overleftrightarrow{KM} = 0$

---

1 Introduce
Alternate Opener

EXPLORATION

5-5 Coordinate Geometry

Quadrilaterals are figures with four sides. Use the coordinate grid to draw each of the following quadrilaterals. Label vertices with A, B, C, and D.

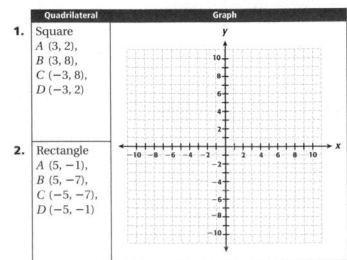

	Quadrilateral	Graph
1.	Square A (3, 2), B (3, 8), C (−3, 8), D (−3, 2)	
2.	Rectangle A (5, −1), B (5, −7), C (−5, −7), D (−5, −1)	

Think and Discuss

3. Describe the relationship between the number of sides and the number of vertices of quadrilaterals.
4. Discuss whether the relationship you described in number 3 is also true for other polygons.

Motivate

Show students two lines that appear to be parallel, and ask them if the lines are parallel. Tell students that although they appear to be parallel, the appearance does not prove that they are. Explain that by plotting the lines on a coordinate plane, you can find a measure of the "steepness" of each line, called *slope*. Slope can be used to determine if the two lines are truly parallel.

Exploration worksheet and answers on Chapter 5 Resource Book pp. 47 and 127

2 Teach

Lesson Presentation

Guided Instruction

In this lesson, students learn to identify polygons on the coordinate plane. To begin, you may want to review plotting points and lines on a coordinate plane. Show students how to find the slope of a line by using rise and run (Teaching Transparency T24, CRB). Explain that slope indicates how steep a line is and whether it slants up to the right (positive slope) or down to the right (negative slope). You may want to use the Slope Transparency in the Teacher's MK. Discuss how to use slope to determine whether a pair of lines are parallel or perpendicular. Then show students how to apply these concepts to classify quadrilaterals.

Slopes of Parallel and Perpendicular Lines
Two lines with equal slopes are parallel.
Two lines whose slopes have a product of -1 are perpendicular.

EXAMPLE 2 Finding Perpendicular and Parallel Lines

Which lines are parallel?
Which lines are perpendicular?

slope of $\overleftrightarrow{PQ} = \frac{4}{3}$

slope of $\overleftrightarrow{RS} = \frac{5}{4}$

Helpful Hint

If a line has slope $\frac{a}{b}$, then a line perpendicular to it has slope $-\frac{b}{a}$.

slope of $\overleftrightarrow{AB} = \frac{4}{3}$

slope of $\overleftrightarrow{PA} = \frac{-3}{3}$ or -1

slope of $\overleftrightarrow{GH} = \frac{-4}{5}$

slope of $\overleftrightarrow{XY} = \frac{-7}{9}$

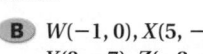

$\overleftrightarrow{PQ} \parallel \overleftrightarrow{AB}$ *The slopes are equal:* $\frac{4}{3} = \frac{4}{3}$

$\overleftrightarrow{RS} \perp \overleftrightarrow{GH}$ *The slopes have a product of -1:* $\frac{5}{4} \cdot \frac{-4}{5} = -1$

EXAMPLE 3 Using Coordinates to Classify Quadrilaterals

Graph the quadrilaterals with the given vertices. Give all of the names that apply to each quadrilateral.

A $J(-6, 3), K(-2, 3),$
$L(-2, -1), M(-6, -1)$

B $W(-1, 0), X(5, -4),$
$Y(3, -7), Z(-3, -3)$

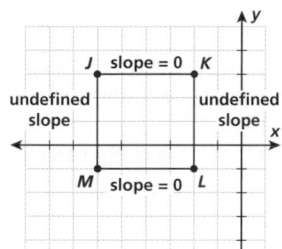

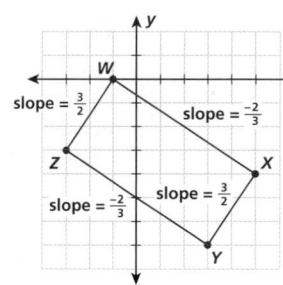

$\overleftrightarrow{JK} \parallel \overleftrightarrow{ML}$ and $\overleftrightarrow{MJ} \parallel \overleftrightarrow{LK}$
$\overleftrightarrow{JK} \perp \overleftrightarrow{LK}, \overleftrightarrow{JK} \perp \overleftrightarrow{MJ},$
$\overleftrightarrow{ML} \perp \overleftrightarrow{LK}$ and $\overleftrightarrow{ML} \perp \overleftrightarrow{MJ}$
parallelogram, rectangle,
square, rhombus

$\overleftrightarrow{WX} \parallel \overleftrightarrow{ZY}$ and $\overleftrightarrow{ZW} \parallel \overleftrightarrow{YX}$
$\overleftrightarrow{ZW} \perp \overleftrightarrow{WX}, \overleftrightarrow{ZW} \perp \overleftrightarrow{ZY},$
$\overleftrightarrow{YX} \perp \overleftrightarrow{WX}$ and $\overleftrightarrow{YX} \perp \overleftrightarrow{ZY}$
parallelogram, rectangle

3 Close

Reaching All Learners
Through Grouping Strategies

Have students work in pairs. Ask each student to create a quadrilateral *ABCD* on a coordinate plane (CRB p. 105) that satisfies all of the following conditions:

$\overline{AB} \parallel \overline{CD}, \overline{AB} \perp \overline{CB},$ and $\overline{AD}$ not $\parallel \overline{CB}$.

Have partners exchange papers and write the slope of each segment on the diagram. Then have them classify the quadrilateral and decide whether it satisfies the conditions. Finally, have the partners return papers to each other for comparison and discussion.

Possible answers: slopes of zero, undefined, zero, and $\frac{2}{3}$; trapezoid; yes

Summarize

Discuss with students how slope can be used to determine whether lines in a coordinate plane are parallel or perpendicular. Discuss the slopes of horizontal and vertical lines. Remind students that they can use this information to classify quadrilaterals.

Possible answers: If two lines have the same slope, they are parallel. If the product of the slopes of two lines equals -1, the lines are perpendicular. Also, if one line is horizontal and another is vertical, the lines are perpendicular. Horizontal lines have a slope of zero. Vertical lines have undefined slope.

Additional Examples

Example 1

Determine if the slope of each line is positive, negative, 0, or undefined. Then find the slope of each line.

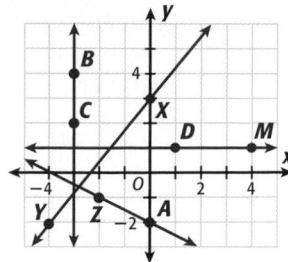

A. $\overleftrightarrow{XY}$ positive; $\frac{5}{4}$ **B.** $\overleftrightarrow{ZA}$ negative; $-\frac{1}{2}$

C. $\overleftrightarrow{BC}$ undefined   **D.** $\overleftrightarrow{DM}$ 0

Example 2

Which lines are parallel?
Which lines are perpendicular?

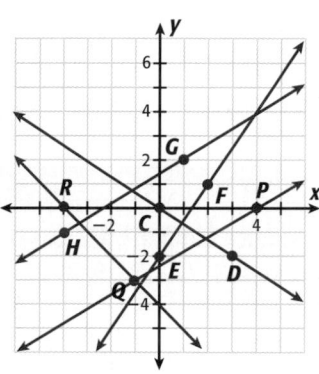

$\overleftrightarrow{GH} \parallel \overleftrightarrow{PQ}, \overleftrightarrow{EF} \perp \overleftrightarrow{CD}$

Example 3

Graph the quadrilateral with the given vertices. Give all of the names that apply to the quadrilateral.

$A(3, -2), B(2, -1), C(4, 3), D(5, 2)$

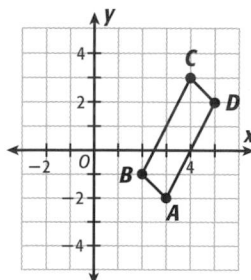

parallelogram

Answers to Think and Discuss

Possible answers:

1. A horizontal line has a rise of 0, so the numerator of the slope is 0. A fraction with a zero in the numerator and a nonzero denominator equals 0.

2. A vertical line has a run of 0, so the denominator of the slope is 0. A fraction with a zero in the denominator is undefined.

Graph the quadrilaterals with the given vertices. Give all of the names that apply to each quadrilateral.

C $E(-1, 6), F(5, 6),$ $G(3, 4), H(-3, 4)$

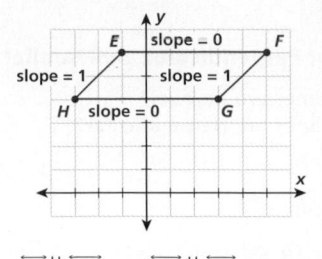

$\overleftrightarrow{EF} \parallel \overleftrightarrow{HG}$ and $\overleftrightarrow{HE} \parallel \overleftrightarrow{GF}$
parallelogram

D $P(4, 3), Q(9, 2),$ $R(4, -3), S(1, 0)$

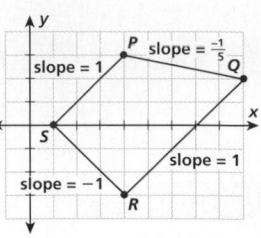

$\overleftrightarrow{SP} \parallel \overleftrightarrow{RQ}$
trapezoid

Think and Discuss

1. **Explain** why the slope of a horizontal line is 0.

2. **Explain** why the slope of a vertical line is undefined.

5-5 PRACTICE & ASSESS

Assignment Guide

If you finished Example **1** assign:
Core 1–4, 9–12, 17, 19, 26–30
Enriched 9–12, 17–22, 24, 26–30

If you finished Example **2** assign:
Core 6, 9–14, 17, 19, 26–30
Enriched 9–14, 17–22, 24, 26–30

If you finished Example **3** assign:
Core 1–16, 17, 19, 26–30
Enriched 9–30

5-5 Exercises

FOR EXTRA PRACTICE
see page 740

internet connect
Homework Help Online
go.hrw.com   Keyword: MP4 5-5

GUIDED PRACTICE

See Example **1** Determine if the slope of each line is positive, negative, 0, or undefined. Then find the slope of each line.

1. $\overleftrightarrow{AD}$ **0**
2. $\overleftrightarrow{BE}$ **Slope is undefined.**
3. $\overleftrightarrow{MN}$ **positive slope; 1**
4. $\overleftrightarrow{EF}$ **negative slope; $-\frac{1}{2}$**

See Example **2** 5. Which lines are parallel? $\overleftrightarrow{AB} \parallel \overleftrightarrow{CD}$

6. Which lines are perpendicular?
$\overleftrightarrow{MN} \perp \overleftrightarrow{AB}, \overleftrightarrow{MN} \perp \overleftrightarrow{CD},$ and $\overleftrightarrow{AD} \perp \overleftrightarrow{BE}$

See Example **3** Graph the quadrilaterals with the given vertices. Give all of the names that apply to each quadrilateral.

7. $D(-3, -2), E(-3, 3), F(2, 3), G(2, -2)$

8. $R(3, -2), S(3, 1), T(-3, 5), V(-3, -2)$

Math Background

The algebraic formula for the slope m of a line through points (x_1, y_1) and (x_2, y_2) is given by the following equation:

$$m = \frac{(y_2 - y_1)}{(x_2 - x_1)}.$$

The equation of any line can be written by using the slope and one point on the line. These concepts will be further explored in Chapter 11.

The concept of slope has many applications outside of coordinate geometry. Real-world examples include the pitch of a roof or the grade of a highway. In many other graphs, the slopes of lines or segments represent rates of change.

RETEACH 5-5

Reteach
5-5 Coordinate Geometry

Possible Values for Slope

Slope is Positive	Slope is Negative	Slope = 0	Slope is Undefined
Line slants up. Forms acute angle with the positive direction of x-axis.	Line slants down. Forms obtuse angle with positive direction of x-axis.	Horizonal Line Parallel to x-axis.	Vertical Line Perpendicular to x-axis

Plot the given points. Describe the slope of the line that joins them.

1. (−2, 2) and (2, 5)

slope is: **positive**

2. (−2, −5) and (−2, 2)

slope is: **undefined**

3. (1, 2) and (5, −2)

slope is: **negative**

4. (−2, −2) and (4, −2)

slope is: **0**

PRACTICE 5-5

Practice B
5-5 Coordinate Geometry

Determine if the slope of each line is positive, negative, or undefined. Then find the slope of each line.

1. $\overleftrightarrow{AB}$ **positive; 2**

2. $\overleftrightarrow{CD}$ **positive; 2**

3. $\overleftrightarrow{RS}$ **negative; −2**

4. $\overleftrightarrow{TC}$ **undefined**

5. $\overleftrightarrow{DR}$ **0**

6. $\overleftrightarrow{TX}$ **negative; $-\frac{1}{2}$**

7. Which lines are parallel?
$\overleftrightarrow{AB} \parallel \overleftrightarrow{CD}; \overleftrightarrow{TC} \parallel \overleftrightarrow{AR}; \overleftrightarrow{XB} \parallel \overleftrightarrow{DR}$

8. Which lines are perpendicular?
$\overleftrightarrow{AB} \perp \overleftrightarrow{TX}; \overleftrightarrow{CD} \perp \overleftrightarrow{TX}; \overleftrightarrow{DR} \perp \overleftrightarrow{TC}; \overleftrightarrow{DR} \perp \overleftrightarrow{AR}; \overleftrightarrow{TC} \perp \overleftrightarrow{XB}; \overleftrightarrow{AR} \perp \overleftrightarrow{XB}$

Graph the quadrilaterals with the given vertices. Write all the names that apply for each quadrilateral.

9. (−1, 1), (4, 1), (1, −3), (−4, −3)

quadrilateral, parallelogram, rhombus

10. (0, 4), (4, 1), (0, −2), (−4, 1)

quadrilateral, parallelogram, rhombus

INDEPENDENT PRACTICE

See Example 1 Determine if the slope of each line is positive, negative, 0, or undefined. Then find the slope of each line.

9. $\overleftrightarrow{AB}$ **positive slope, 1**   10. $\overleftrightarrow{EG}$ **negative slope, −2**

11. $\overleftrightarrow{HG}$ **0**   12. $\overleftrightarrow{CH}$ **Slope is undefined.**

See Example 2 13. Which lines are parallel? $\overleftrightarrow{CD} \parallel \overleftrightarrow{AB}$

14. Which lines are perpendicular?
$\overleftrightarrow{HG} \perp \overleftrightarrow{CH}$, $\overleftrightarrow{FE} \perp \overleftrightarrow{CD}$, and $\overleftrightarrow{FE} \perp \overleftrightarrow{AB}$

See Example 3 Graph the quadrilaterals with the given vertices. Give all of the names that apply to each quadrilateral.

15. $D(-3, 5)$, $E(3, 5)$, $F(3, -1)$, $G(-3, -1)$

16. $W(-2, 1)$, $X(-2, -2)$, $Y(4, 1)$, $Z(0, 2)$

PRACTICE AND PROBLEM SOLVING

Draw the line through the given points and find its slope.

17. $A(2, 1)$, $B(4, 7)$ **3**

18. $C(-2, 0)$, $D(-2, -5)$ **undefined**

19. $G(5, -4)$, $H(-2, -4)$ **0**

20. $E(-3, 1)$, $F(4, -2)$ $-\dfrac{3}{7}$

21. On a coordinate grid draw a line s with slope 0 and a line t with slope 1. Then draw three lines through the intersection of lines s and t that have slopes between 0 and 1.

22. On a coordinate grid draw a line m with slope 0 and a line n with slope −1. Then draw three lines through the intersection of lines m and n that have slopes between 0 and −1.

23. **WHAT'S THE ERROR?** Points $P(3, 7)$, $Q(5, 2)$, $R(3, -3)$, and $S(1, 2)$ are vertices of a square. What is the error?

24. **WRITE ABOUT IT** Explain how using different points on a line to find the slope affects the answer.

25. **CHALLENGE** Use a square in a coordinate plane to explain why a line with slope 1 makes a 45° angle with the x-axis.

Spiral Review

The measures of two angles of a triangle are given. Find the measure of the third angle. (Lesson 5-3)

26. 45°, 45° **90°**   27. 30°, 60° **90°**   28. 21°, 82° **77°**   29. 105°, 42° **33°**

30. **TEST PREP** Evaluate $[(4 \cdot 5) - 5] \div 2$. (Previous course) **C**

A 2   B 5   C 7.5   D 0

Answers

7–8, 15–16. Complete answers on p. A4

7. parallelogram, rhombus, rectangle, square

8. trapezoid

15. parallelogram, rhombus, rectangle, square

16. trapezoid

17–25. See pp. A4–A5.

Journal

Have students give real-world examples of things with very steep slopes, gentle slopes, zero slopes, and undefined slopes. Have them estimate the mathematical value for the slope of each example.

Test Prep Doctor

For Exercise 30, remind students that when there are nested grouping symbols, it is necessary to evaluate the expression in the innermost group first.

Lesson Quiz

Determine the slope of each line.

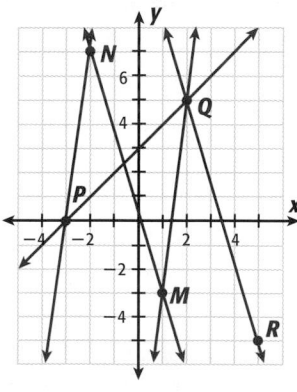

1. $\overleftrightarrow{PQ}$ **1**   2. $\overleftrightarrow{MN}$ $-\dfrac{10}{3}$

3. $\overleftrightarrow{MQ}$ **8**   4. $\overleftrightarrow{NP}$ **7**

5. Which pairs of lines are parallel?
$\overleftrightarrow{MN}$, $\overleftrightarrow{RQ}$

Available on Daily Transparency in CRB

CHALLENGE 5-5

Challenge
5-5 *Are They Lined Up?*

You can find the slope of a line by using the coordinates of two points on the line.

$\text{slope} = \dfrac{\text{difference of } y\text{-values}}{\text{difference of } x\text{-values}}$

Be sure to take the differences in the same order.

To find the slope of $\overleftrightarrow{AB}$ with $A(-2, 5)$ and $B(6, 7)$:

slope of $\overleftrightarrow{AB} = \dfrac{7 - 5}{6 - (-2)} = \dfrac{2}{8} = \dfrac{1}{4}$, or

slope of $\overleftrightarrow{AB} = \dfrac{5 - 7}{-2 - 6} = \dfrac{-2}{-8} = \dfrac{1}{4}$

Find the slope of the line joining each pair of points. Verify your result on a graph.

1. (5, 5) and (2, 1) $\dfrac{4}{3}$

2. (4, 1) and (6, −2) $-\dfrac{3}{2}$

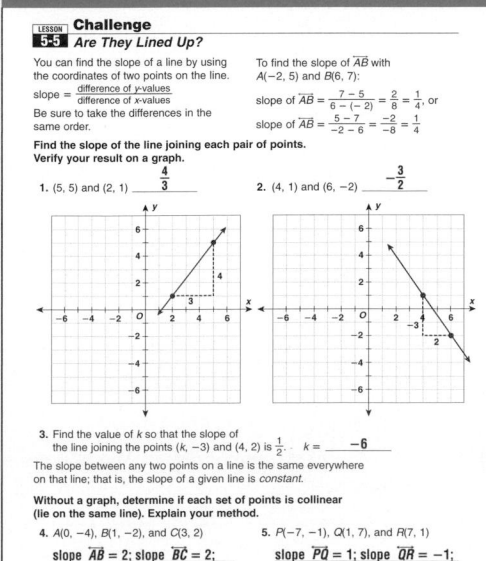

3. Find the value of k so that the slope of the line joining the points $(k, -3)$ and $(4, 2)$ is $\dfrac{1}{2}$. $k = $ **−6**

The slope between any two points on a line is the same everywhere on that line; that is, the slope of a given line is *constant*.

Without a graph, determine if each set of points is collinear (lie on the same line). Explain your method.

4. $A(0, -4)$, $B(1, -2)$, and $C(3, 2)$
slope $\overline{AB}$ = 2; slope $\overline{BC}$ = 2; **yes**

5. $P(-7, -1)$, $Q(1, 7)$, and $R(7, 1)$
slope $\overline{PQ}$ = 1; slope $\overline{QR}$ = −1; **no**

6. Find the value of k so that the points $L(-1, 5)$, $M(0, k)$ and $N(1, -1)$ are collinear. $k = $ **2**

PROBLEM SOLVING 5-5

Problem Solving
5-5 *Coordinate Geometry*

The Uniform Federal Accessibility Standards describes the standards for making buildings accessible for the handicapped. The standards say that the least possible slope should be used for a ramp and that the maximum slope of a ramp should be $\dfrac{1}{12}$.

1. What is the slope of the pictured ramp? Does the ramp meet the standard?

ramp 12 in. 12 ft

$\dfrac{1}{12}$; yes

2. What is the slope of the pictured ramp? Does the ramp meet the standard?

ramp 12 in. 10 ft

$\dfrac{1}{10}$; no

Write the correct answer.

3. Find the slope of the roof.

8 ft 24 ft

Slope = $\dfrac{2}{3}$ or $-\dfrac{2}{3}$

Choose the letter that represents the slope.

4. Many building codes require that a staircase be built with a maximum rise of 8.25 inches for a minimum tread width (run) of 9 inches.

A $\dfrac{8}{9}$ C $\dfrac{9}{8.25}$
(B) $\dfrac{11}{12}$ D $\dfrac{12}{11}$

5. Hills that have a rise of about 10 feet for every 17 feet horizontally are too steep for most cars.

(F) $\dfrac{10}{17}$ H $\dfrac{17}{10}$
G $\dfrac{2}{5}$ J $\dfrac{5}{3}$

6. At its steepest part, an intermediate ski run has a rise of about 4 feet for 10 feet horizontally.

(A) $\dfrac{2}{5}$ C $\dfrac{5}{2}$
B $\dfrac{4}{5}$ D $\dfrac{5}{4}$

7. Black diamond, or expert, ski slopes often have a rise of 10 feet for every 14 feet horizontally.

F $\dfrac{7}{5}$ (H) $\dfrac{5}{7}$
G $\dfrac{2}{7}$ J $\dfrac{7}{2}$

Chapter 5
Mid-Chapter Quiz

Purpose: *To assess students' mastery of concepts and skills in Lessons 5-1 through 5-5*

Assessment Resources

Section 5A Quiz
Assessment Resources .... p. 13

Test and Practice Generator CD-ROM

Additional mid-chapter assessment items in both multiple-choice and free-response format may be generated for any objective in Lessons 5-1 through 5-5.

Answers

1. ∠ABD and ∠DBE, ∠EBF and ∠FBC

2. ∠ABD and ∠DBC, ∠ABE and ∠EBC, ∠ABF and ∠FBC

12.
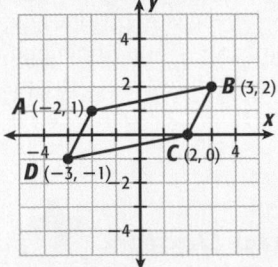
parallelogram

13.
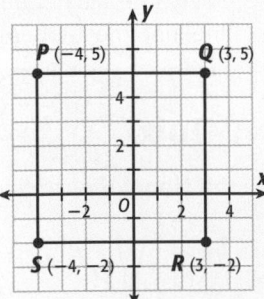
parallelogram, rhombus, rectangle, square

14.
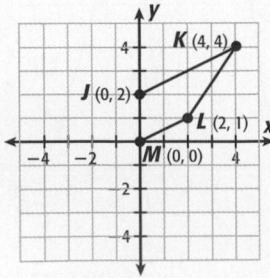
trapezoid

Mid-Chapter Quiz

LESSON 5-1 (pp. 222–226)

Refer to the figure.

1. Name two pairs of complementary angles.

2. Name three pairs of supplementary angles.

3. Name two right angles. **∠ABE, ∠EBC**

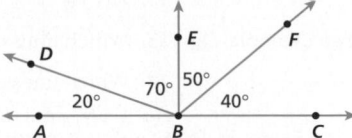

LESSON 5-2 (pp. 228–231)

In the figure, line *m* ∥ line *n*. Find the measure of each angle.

4. ∠1   **135°**

5. ∠2   **45°**

6. ∠3   **45°**

7. ∠4   **135°**

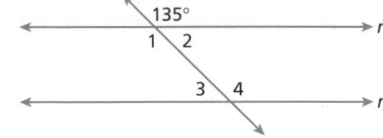

LESSON 5-3 (pp. 234–238)

Find *x* in each triangle.

8. $x° = 23°$

9. $x° = 57°$

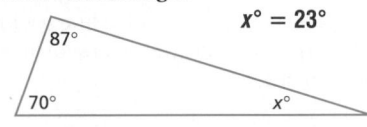
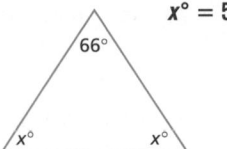

LESSON 5-4 (pp. 239–243)

Give all of the names that apply to each figure.

10. $\overline{AB} \parallel \overline{CD}$ **quadrilateral, trapezoid**

11.   **quadrilateral, parallelogram, rhombus**

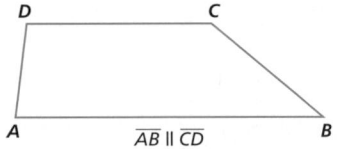
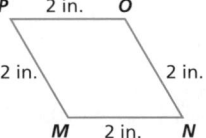

LESSON 5-5 (pp. 244–247)

Graph the quadrilaterals with the given vertices. Give all of the names that apply to each quadrilateral.

12. A(−2, 1), B(3, 2), C(2, 0), D(−3, −1)

13. P(−4, 5), Q(3, 5), R(3, −2), S(−4, −2)

14. J(0, 2), K(4, 4), L(2, 1), M(0, 0)

15. U(4, 2), V(−2, 4), W(−3, 1), X(3, −1)

15.
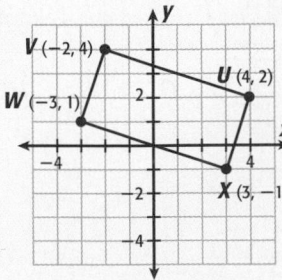
parallelogram, rectangle

Focus on Problem Solving

 Understand the Problem

• Restate the problem in your own words

If you write a problem in your own words, you may understand it better. Before writing a problem in your own words, you may need to read it over several times—perhaps aloud, so you can hear yourself say the words.

Once you have written the problem in your own words, you may want to make sure you included all of the necessary information to solve the problem.

Focus on Problem Solving

Purpose: To focus on understanding the problem

Problem Solving Resources

Interactive Problem Solving . . pp. 36–44

Math: Reading and Writing in the Content Area . . . . pp. 36–44

Problem Solving Process

This page focuses on the first step of the problem-solving process: **Understand the Problem**

Discuss

Have students discuss which facts are necessary to solve each problem, and then write each problem in their own words.

Possible answers:

1. m∠1 and m∠2 add up to 90°. m∠2 and m∠3 add up to 180°. If m∠2 = 50°, find m∠4 + m∠5.

2. Find out if m∠C = 90° when m∠A = 25° and m∠B = 65°

3. First angle = x; second angle = 6x; third angle = 3x; fourth angle = 4x; the angles add up to 360°; find the measure of each angle.

4. Find the supplement of 45°.

Write each problem in your own words. Check to make sure you have included all of the information needed to solve the problem.

1 In the figure, ∠1 and ∠2 are complementary, and ∠2 and ∠3 are supplementary. If m∠2 = 50°, find m∠4 + m∠5.

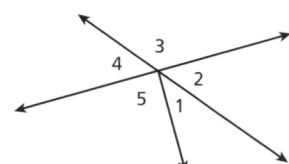

2 In triangle *ABC*, m∠A = 25° and m∠B = 65°. Use the Triangle Sum Theorem to determine whether triangle *ABC* is a right triangle.

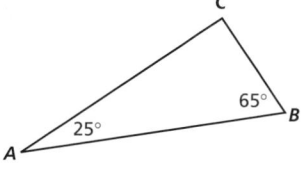

3 The second angle in a quadrilateral is six times as large as the first angle. The third angle is half as large as the second. The fourth angle is as large as the first angle and the third angle combined. Find the angle measures in the quadrilateral.

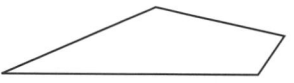

4 Parallel lines *m* and *n* are intersected by a transversal, line *p*. The acute angles formed by line *m* and line *p* measure 45°. Find the measure of the obtuse angles formed by the intersection of line *n* and line *p*.

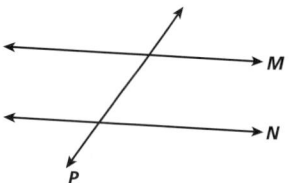

Answers

1. 140°

2. Triangle *ABC* is a right triangle.

3. ≈ 25.7°, ≈ 154.3°, ≈ 77.1°, and ≈ 102.9°

4. 135°

Patterns in Geometry

One-Minute Section Planner

Lesson	Materials	Resources
Lesson 5-6 Congruence **NCTM:** Geometry, Reasoning and Proof, Communication **NAEP:** Geometry 2e ☑SAT-9 ☑SAT-10 ☑ITBS ☑CTBS ☑MAT ☑CAT	**Optional** Congruent paper triangles *(CRB, p. 107)* Recording Sheet for Reaching All Learners *(CRB, p. 108)* Protractors *(MK)* Rulers *(MK)* Scissors *(MK)*	● *Chapter 5 Resource Book,* pp. 57–66 ● Daily Transparency T29, CRB ● Additional Examples Transparencies T30–T32, CRB ● *Alternate Openers: Explorations,* p. 41
Lesson 5-7 Transformations **NCTM:** Algebra, Geometry, Communication **NAEP:** Geometry 2c ☑SAT-9 ☑SAT-10 ☐ITBS ☑CTBS ☑MAT ☑CAT	**Required** Protractors *(MK)* Straightedges *(MK)* Coordinate planes *(CRB, p. 106)* Recording Sheet for Exercises 3–4, 10–11, and 15–17 *(CRB, p. 109)* **Optional** Teaching Transparency T34 *(CRB)* Pattern blocks *(MK or CRB, p. 110)*	● *Chapter 5 Resource Book,* pp. 67–76 ● Daily Transparency T33, CRB ● Additional Examples Transparencies T35–T37, CRB ● *Alternate Openers: Explorations,* p. 42
Hands-On Lab 5C Combine Transformations **NCTM:** Algebra, Geometry, Reasoning and Proof, Representation **NAEP:** Geometry 2c ☐SAT-9 ☐SAT-10 ☐ITBS ☐CTBS ☐MAT ☐CAT	**Required** Pattern blocks *(MK)*	● *Hands-On Lab Activities,* pp. 38, 96
Lesson 5-8 Symmetry **NCTM:** Geometry, Communication **NAEP:** Geometry 2a ☑SAT-9 ☑SAT-10 ☐ITBS ☐CTBS ☑MAT ☑CAT	**Required** Recording Sheet for Exercises 1–12 and 17 *(CRB, pp. 111–112)* **Optional** Rotational Symmetry Transparency *(MK)* Cut-out figures *(CRB, p. 113)* Construction paper Scissors	● *Chapter 5 Resource Book,* pp. 77–85 ● Daily Transparency T38, CRB ● Additional Examples Transparencies T39–T40, CRB ● *Alternate Openers: Explorations,* p. 43
Lesson 5-9 Tessellations **NCTM:** Geometry, Problem Solving, Communication **NAEP:** Geometry 1f ☐SAT-9 ☐SAT-10 ☐ITBS ☐CTBS ☐MAT ☐CAT	**Required** Cut-out shapes for Exercises 2, 3, 5, 6–12, and 14 *(CRB, pp. 114–115)* Scissors **Optional** Teaching Transparencies T42–T43 *(CRB)* Pattern blocks *(MK)* Protractors *(MK)*	● *Chapter 5 Resource Book,* pp. 86–95 ● Daily Transparency T41, CRB ● Additional Examples Transparencies T44–T47, CRB ● *Alternate Openers: Explorations,* p. 44
Section 5B Assessment		● Section 5B Quiz, AR p. 14 ● *Test and Practice Generator* CD-ROM

SAT = *Stanford Achievement Tests*   **ITBS** = *Iowa Test of Basic Skills*   **CTBS** = *Comprehensive Test of Basic Skills/Terra Nova*
MAT = *Metropolitan Achievement Test*   **CAT** = *California Achievement Test*
NCTM = Complete standards can be found on pages T29–T35.   **NAEP** = Complete standards can be found on pages A54–A58.

Section Overview

Congruence

Lesson 5-6

Why? Replacement parts in machines must be congruent.

> **Congruent polygons** are the same shape and size. Their corresponding sides and angles have equal measures.

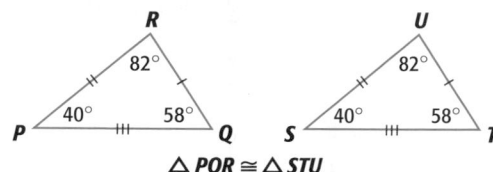

$$\triangle PQR \cong \triangle STU$$

You can find unknown values in congruent polygons.

If $RQ = x + 8$ and $UT = 20$, then $RQ = UT$.

$$RQ = UT$$
$$x + 8 = 20$$
$$x = 12$$

Transformations

Lesson 5-7

Why? Transformations are useful for creating patterns and designs.

Translation	**Rotation**	**Reflection**
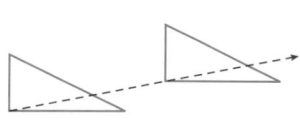	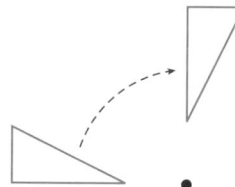	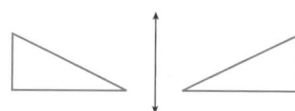
A **translation** slides a figure along a line without turning.	A **rotation** turns the figure around a point, called the **center of rotation**.	A **reflection** flips the figure across a line to create a mirror image.

Symmetry and Tessellations

Lessons 5-8 and 5-9

Why? Symmetry occurs in our natural environment. Tessellations are often used to create patterns in art, architecture, and home furnishings.

> The repeating pattern of red and white stripes forms a tessellation.

Line Symmetry	**Rotational Symmetry**	**Tessellation**
	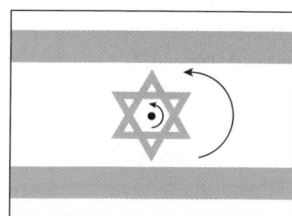	
A figure has **line symmetry** if you can draw a line through it so that the two sides are mirror images of each other.	A figure has **rotational symmetry** if you can rotate the figure less than one full turn around some point so that it coincides with itself.	A **tessellation** is a repeating pattern of plane figures that completely covers a plane with no gaps or overlaps.

FOR EXTRA PRACTICE
see page 741

✓ internet connect
Homework Help Online
go.hrw.com   Keyword: MP4 5-6

go.hrw.com

GUIDED PRACTICE

> Students may want to refer back to the lesson examples.

See Example 1   Write a congruence statement for each pair of polygons.

1.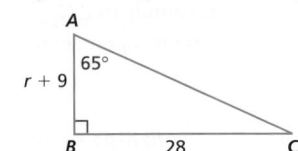

triangle *ABC* ≅ triangle *FED*

2.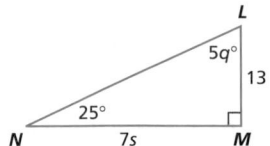

quadrilateral *LMNO* ≅ quadrilateral *STQR*

See Example 2   In the figure, triangle *ABC* ≅ triangle *LMN*.

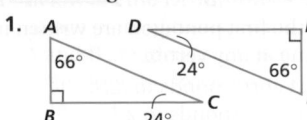

3. Find *q*.   *q* = 13   4. Find *r*.   *r* = 4   5. Find *s*.   *s* = 4

Assignment Guide

If you finished Example **1** assign:
   Core 1, 2, 6, 7, 18–26
   Enriched 6, 7, 15, 16, 18–26

If you finished Example **2** assign:
   Core 1–12, 18–26
   Enriched 6–26

INDEPENDENT PRACTICE

See Example 1   Write a congruence statement for each pair of poygons.

6.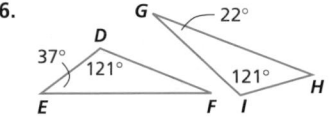

triangle *DEF* ≅ triangle *IHG*

7.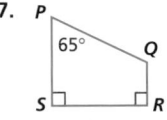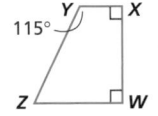

quadrilateral *PQRS* ≅ quadrilateral *ZYXW*

See Example 2   In the figure, quadrilateral *ABCD* ≅ quadrilateral *LMNO*.

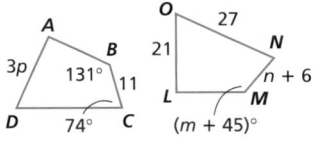

8. Find *m*.   *m* = 86   9. Find *n*.   *n* = 5   10. Find *p*.   *p* = 7

PRACTICE AND PROBLEM SOLVING

Find the value of each variable.

11. pentagon *ABCDE* ≅ pentagon *PQRST*   *x* = 16; *y* = 25; *z* = 14.2

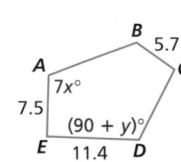 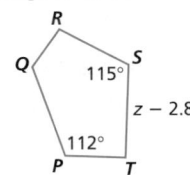

Math Background

The use of congruent figures is a dominant aspect in geometric patterns and designs. Fabric and wallpaper designs in particular make use of congruent figures by repeating a given figure throughout the material. A repeating pattern of congruent figures that completely covers a surface is a *tessellation* (Lesson 5-9). Sometimes the figures are rotated or reflected throughout the design. Rotations and reflections are types of *transformations* (Lesson 5-7).

RETEACH 5-6

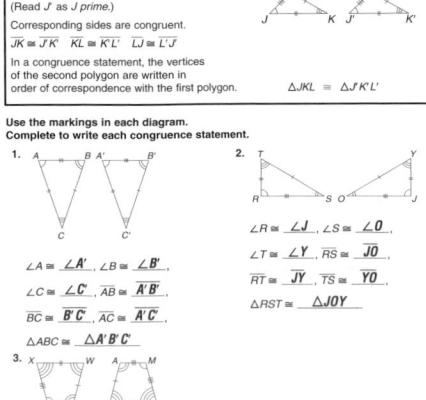

PRACTICE 5-6

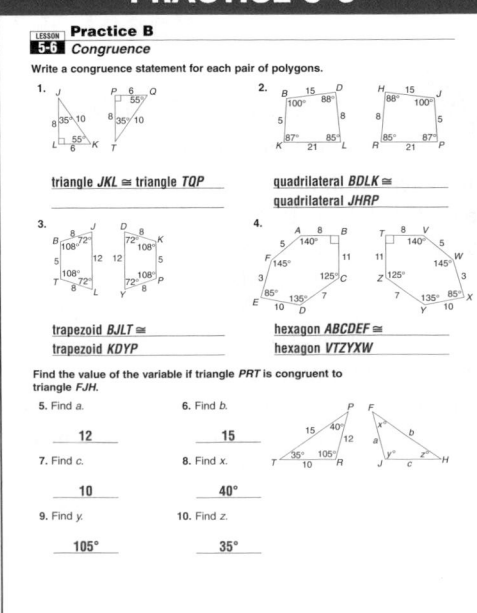

12. hexagon *ABCDEF* ≅ hexagon *LMNOPQ*   **m = 14; n = 32; p = 64**

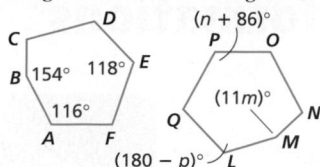

13. quadrilateral *ABCD* ≅ quadrilateral *EFGH*   **s = 120, t = 33, r = 33**

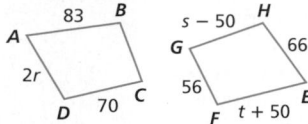

14. heptagon *ABCDEFG* ≅ heptagon *JKLMNOP*   **x = 309, y = 292, w = 218**

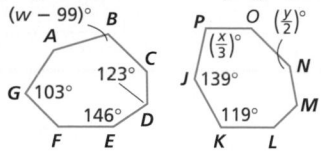

15. **WHAT'S THE ERROR?** Explain the error in this congruence statement and write a correct congruence statement.

triangle *ABC* ≅ triangle *DEF*

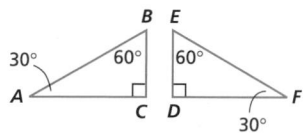

16. **WRITE ABOUT IT** How can knowing two polygons are congruent help you find angle measures of the polygons?

17. **CHALLENGE** Triangle *ABC* ≅ triangle *LMN* and $\overline{AE} \parallel \overline{BD}$. Find m∠*ACD*.   **∠ACD = 110°**

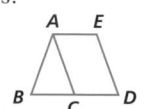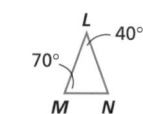

Spiral Review

Solve. (Lesson 2-4)

18. $\frac{m}{-3} = 4$ **m = −12**   19. $64 = 4x$ **16 = x**   20. $\frac{x}{-6} = -2$ **x = 12**   21. $-60 = 4m$ **−15 = m**

22. $21 = 6p$ **3.5 = p**   23. $\frac{b}{3} = -2$ **b = −6**   24. $-95 = 19y$ **−5 = y**   25. $\frac{a}{4} = -8$ **a = −32**

26. **TEST PREP** Determine the angle measures of the following triangle: The first angle is less than 90°. The second angle is $\frac{3}{4}$ as large as the first angle. The third angle is $\frac{2}{3}$ as large as the second angle. (Lesson 5-3)   **D**

  A 60°, 45°, 75°   B 75°, 60°, 45°   C 75°, 50°, 35°   D 80°, 60°, 40°

Journal

Ask students to find sets of congruent figures in the classroom, around the school, or at home. Suggest that they look at floor patterns, wall decorations, clothing, art, and books. Ask students to record several examples in their journals.

Test Prep Doctor

For Exercise 26, encourage students to check the angles one at a time. The second angle must be $\frac{3}{4}$ of the first. Only choices **A** and **D** satisfy this requirement. The third angle must be $\frac{2}{3}$ the second. Only choice **D** satisfies this requirement.

Lesson Quiz

In the figure, *WXYZ* ≅ *ABCD*.

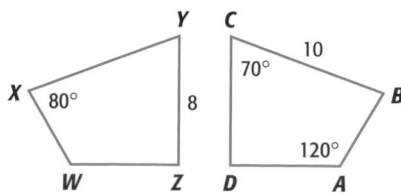

1. Find *XY*.   **10**   2. Find m∠*B*. **80°**
3. Find *CD*.   **8**   4. Find m∠*Z*. **90°**

Available on Daily Transparency in CRB

Pacing: Traditional 1 day
Block $\frac{1}{2}$ day

Objective: Students transform plane figures using translations, rotations, and reflections.

Warm Up

Determine if the following sets of points form a parallelogram.

1. $(-3, 0), (1, 4), (6, 0), (2, -4)$
yes

2. $(1, 2), (-2, 2), (-2, 1), (1, -2)$
no

3. $(2, 3), (-3, 1), (1, -4), (6, -2)$
yes

Problem of the Day

How can you move just one number to a different triangle to make the sum of the numbers in each triangle equal? (*Hint:* There do not have to be exactly 3 numbers in each triangle.)

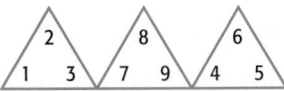

Move 9 into the first triangle.
Available on Daily Transparency in CRB

5-7 Transformations

Learn to transform plane figures using translations, rotations, and reflections.

When you are on an amusement park ride, you are undergoing a *transformation*. Ferris wheels and merry-go-rounds are *rotations*. Free-fall rides and water slides are *translations*. Translations, rotations, and reflections are types of **transformations**.

Vocabulary
transformation
translation
rotation
center of rotation
reflection
image

Translation	Rotation	Reflection
A **translation** slides a figure along a line without turning.	A **rotation** turns the figure around a point, called the **center of rotation**.	A **reflection** flips the figure across a line to create a mirror image.

The resulting figure, or **image**, of a translation, rotation, or reflection is congruent to the original figure.

EXAMPLE 1 **Identifying Transformations**

Identify each as a translation, rotation, reflection, or none of these.

Reading Math
A' is read "*A* prime." The point *A'* is the image of point *A*.

A
translation

B
none of these

C
rotation

D
reflection

---

1 Introduce

Alternate Opener

EXPLORATION

5-7 Transformations

1. Draw arrows from all the vertices (corners) of each original figure (blue) to the corresponding vertices of its *image* (red).

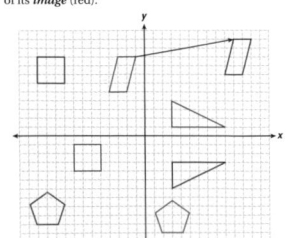

Think and Discuss

2. **Explain** whether the transformations in number **1** are congruent, which means that the image is the same shape and size as the original figure.

3. **Compare** the transformation of the triangle with the transformations of the other figures. How is it different? Find a name for this transformation.

Motivate

Discuss with students some real-world examples of transformations. You may mention that a pinwheel illustrates the concept of a rotation. You may also point out that the image of trees on a lake represents a reflection. Translations can often be found in tile or clothing patterns. Ask students if they can translate (yes), rotate (yes), or reflect (no) themselves.

Exploration worksheet and answers on Chapter 5 Resource Book pp. 68 and 131

2 Teach

Lesson Presentation

Guided Instruction

In this lesson, students learn to transform plane figures using translations, rotations, and reflections. Illustrate each of the three types of transformations with simple polygons (Teaching Transparency T34, CRB). Explain that when each of these transformations is performed, the resulting polygon (the image) is congruent to the original polygon. Each vertex in the original polygon corresponds to a vertex in the image. The correspondence is indicated by pairing a given point *A* with its image point *A'*, a given point *B* with its image point *B'*, and so on. Show how to draw transformation images on a coordinate system.

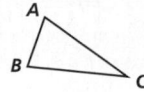

EXAMPLE 2 **Drawing Transformations**

Draw the image of the triangle after each transformation.

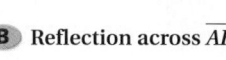

A Translation along $\overline{BC}$ so that B' coincides with C

B Reflection across $\overline{AB}$

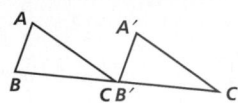
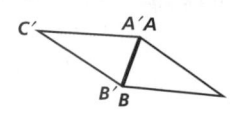

C 90° counterclockwise rotation around point C

Trace the figure. Place your pencil at point C and rotate the tracing 90° counterclockwise.

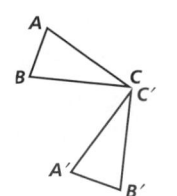

EXAMPLE 3 **Graphing Transformations**

Draw the image of a triangle with vertices (2, 1), (3, 3), and (1, 2) after each transformation.

A Translation 3 units down

B Reflection across the y-axis

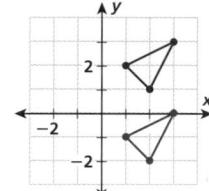

Helpful Hint

The image of the point (x, y) after a rotation of 180° around (0, 0) is $(-x, -y)$.

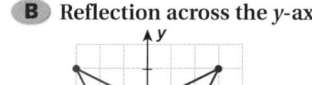

C 180° rotation around (0, 0)

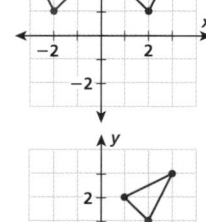

Think and Discuss

1. Tell whether the image of a vertical line is sometimes, always, or never vertical after a translation, a reflection, or a rotation.

2. Give the image of point $A(a, b)$ after a reflection across the x-axis.

Example 1

Identify as a translation, rotation, reflection, or none of these.

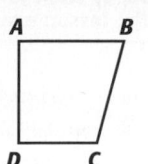
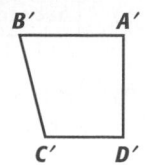

reflection

Example 2

Draw the image of the triangle after the transformation.

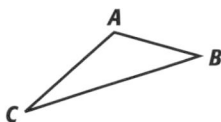

translation along $\overline{AB}$ so that A' coincides with B

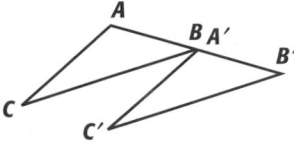

Example 3

Draw the image of a triangle with vertices (1, 1), (2, −2), and (5, 0) after a 180° counterclockwise rotation around (0, 0).

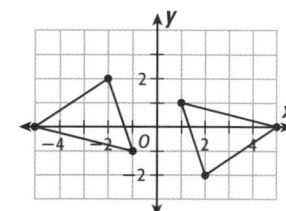

---

3 Close

Reaching All Learners

Through Hands-on Experience

Give each student three copies of a coordinate plane (CRB p. 106) and a pattern block of a geometric figure (Manipulatives Kit or CRB p. 110). On the first plane, ask students to trace the figure and then to translate the figure according to a given rule (e.g., two inches left and one inch down). On the second plane, ask students to trace each figure and reflect it across a line. On the third plane, ask students to rotate each figure according to a given rule (e.g., rotation of 180°). Students should perform the transformations by tracing their pattern blocks in the appropriate places on the coordinate planes.

Summarize

Ask students to describe the steps they would take to perform each of the following transformations on a simple figure on a coordinate plane:

1. translation two units to the left

2. reflection across the y-axis

3. rotation of 180° around the origin

Possible answers: **1.** Subtract 2 from the x-coordinate of each point, plot the points, and connect them. **2.** Change each x-coordinate to its opposite, plot the points, and connect them. **3.** Change both coordinates to their opposites, plot the points, and connect them.

Answers to Think and Discuss

1. always; sometimes (if the line of reflection is vertical or horizontal); sometimes (if the angle of rotation is a multiple of 180°)

2. $A'(a, -b)$

FOR EXTRA PRACTICE
see page 741

internet connect
Homework Help Online
go.hrw.com   Keyword: MP4 5-7

GUIDED PRACTICE

> Students may want to refer back to the lesson examples.

See Example 1 Identify each as a translation, rotation, reflection, or none of these.

1.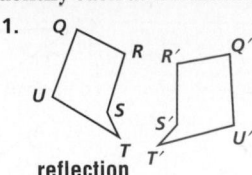

reflection

2.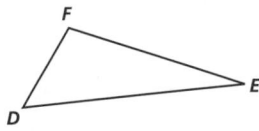

rotation

See Example 2 Draw the image of the triangle after each transformation.

3. translation along $\overline{AC}$ so that C' coincides with A

4. reflection across $\overline{ED}$

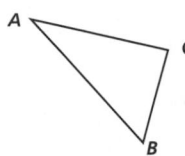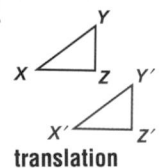

Assignment Guide

If you finished Example **1** assign:
Core 1, 2, 8, 9, 31–39
Enriched 1, 2, 8, 9, 31–39

If you finished Example **2** assign:
Core 1–4, 8–11, 15, 17, 31–39
Enriched 8–11, 15–17, 27, 29, 31–39

If you finished Example **3** assign:
Core 1–14, 15–27 odd, 31–39
Enriched 8–39

See Example 3 Draw the image of the parallelogram with vertices $(-3, 6)$, $(-4, 2)$, $(4, 4)$, and $(3, 0)$ after each transformation.

5. translation 2 units up

6. reflection across the x-axis

7. $180°$ rotation around $(0, 0)$

You may use the Recording Sheet on Chapter 5 Resource Book p. 109 for Exercises 3–4, 10–11, and 15–17.

INDEPENDENT PRACTICE

See Example 1 Identify each as a translation, rotation, reflection, or none of these.

Answers

3.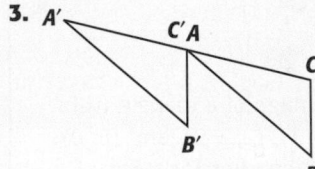

4.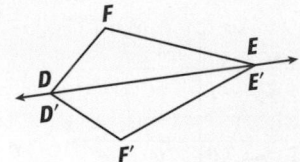

5–7, 10–11. See p. A5.

8.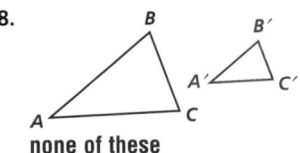

none of these

9.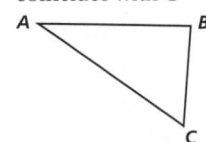

translation

See Example 2 Draw the image of the triangle after each transformation.

10. translation along $\overline{BC}$ so that B' coincides with C

11. reflection across $\overline{AB}$

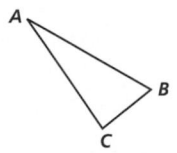

Math Background

The three types of transformations discussed in this section are *rigid transformations;* that is, all result in images that maintain the shape and size of the original figures.

A *dilation* (Lesson 7-5) is a nonrigid transformation that preserves the shape of the figure but not the size. In a dilation, the figure is contracted or expanded by a constant called the *scale factor.* An example of this type of transformation is an enlargement or reduction of a photograph.

RETEACH 5-7

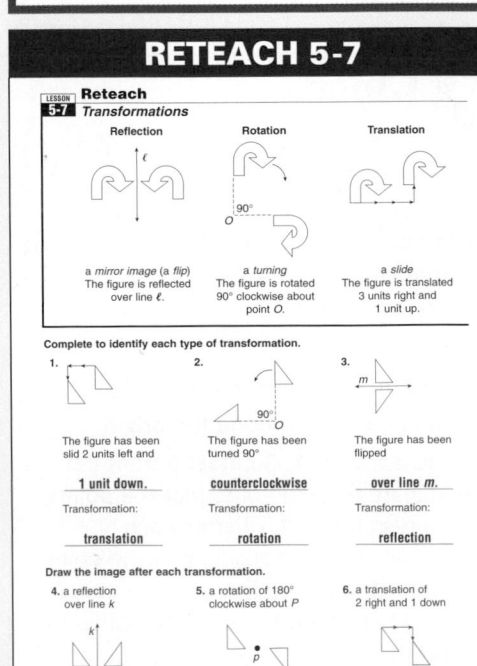

PRACTICE 5-7

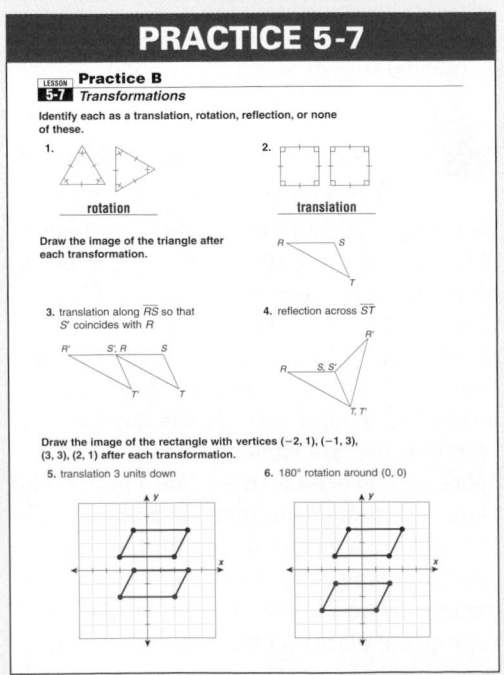

See Example **3** — Draw the image of the quadrilateral with vertices $(1, 2)$, $(5, 4)$, $(5, 1)$, and $(3, 5)$ after each transformation.

12. translation 3 units down

13. reflection across the y-axis

14. $180°$ rotation around $(0, 0)$

PRACTICE AND PROBLEM SOLVING

Copy each figure and perform the given transformations.

15. Reflect across line m. **16.** Reflect across line n. **17.** Rotate clockwise $90°$.

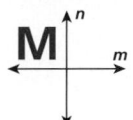 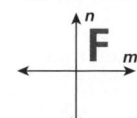 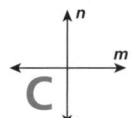

Give the coordinates of each point after a reflection across the given axis.

18. $(3, 5)$; x-axis
(3, −5)

19. $(−2, 1)$; x-axis
(−2, −1)

20. (m, n); x-axis
(m, −n)

21. $(4, −3)$; y-axis
(−4, −3)

22. $(−5, 2)$; y-axis
(5, 2)

23. (m, n); y-axis
(−m, n)

Give the coordinates of each point after a $180°$ rotation around $(0, 0)$.

24. $(2, 3)$ **(−2, −3)**   **25.** $(−6, 1)$ **(6, −1)**   **26.** (m, n) **(−m, −n)**

27. *ART* A rubber stamp is a reflection of the image the ink makes on the page. Draw a rubber stamp that would print the name **EMILY**. Is the image a reflection across a vertical line or a horizontal line?

28. *WRITE A PROBLEM* Write a problem involving transformations on a coordinate grid that result in a pattern.

29. *WRITE ABOUT IT* Explain how each type of transformation performed on the arrow would affect the direction the arrow is pointing.

30. *CHALLENGE* A triangle has vertices at $(−1, 1)$, $(1, 3)$, and $(4, −2)$. After a reflection and a translation, the coordinates of the image are $(5, 3)$, $(3, 5)$, and $(0, 0)$. Describe the transformations.

Spiral Review

Evaluate. (Lesson 2-6)

31. 2^5 **32**   **32.** $(−3)^2$ **9**   **33.** $(−7)^3$ **−343**   **34.** 4^0 **1**

35. $(−2)^7$ **−128**   **36.** 5^3 **125**   **37.** $(−4)^2$ **16**   **38.** 8^1 **8**

39. TEST PREP Each angle of a regular polygon with 15 sides measures ▇. (Lesson 5-4) **A**

  A $156°$ **B** $146°$ **C** $150°$ **D** $148°$

Journal

Have students think about a career that might involve transformations. Have them write how they might use transformations if they had a job in that field. Examples might include art, architecture, sewing, computer graphics, construction, and design.

Test Prep Doctor

For Exercise 39, remind students how to find the sum of the measures of the interior angles in a 15-sided polygon. Then remind the students that when the polygon is a regular polygon, each of the interior angles will have the same measure.

$$\frac{(15 - 2) \cdot 180}{15} = \frac{13 \cdot 180}{15} = 156$$

CHALLENGE 5-7

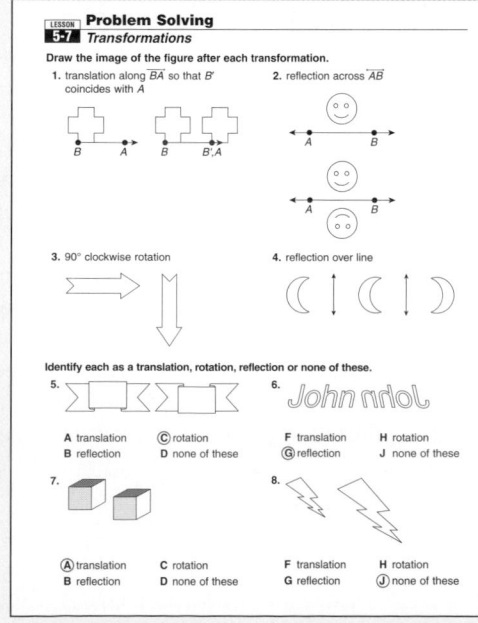

PROBLEM SOLVING 5-7

Lesson Quiz

Given the coordinates for the vertices of each pair of quadrilaterals, determine whether each pair represents a translation, rotation, reflection, or none of these.

1. $(2, 2)$, $(4, 0)$, $(3, 5)$, $(6, 4)$ and $(3, −1)$, $(5, −3)$, $(4, 2)$, $(7, 1)$
  translation

2. $(2, 3)$, $(5, 5)$, $(1, −2)$, $(5, −4)$ and $(−2, 3)$, $(−5, 5)$, $(−1, −2)$, $(−5, −4)$ **reflection**

3. $(1, 3)$, $(−1, 2)$, $(2, −3)$, $(4, 0)$ and $(1, −3)$, $(−1, 2)$, $(−2, 3)$, $(−4, 0)$ **none**

4. $(4, 1)$, $(1, 2)$, $(4, 5)$, $(1, 5)$ and $(−4, −1)$, $(−1, −2)$, $(−4, −5)$, $(−1, −5)$ **rotation**

Available on Daily Transparency in CRB

Pacing: Traditional 1 day
Block $\frac{1}{2}$ day

Objective: To use pattern blocks and a coordinate plane to explore compound transformations

Materials: Pattern blocks, coordinate planes

Lab Resources

Hands-On Lab Activities . . . . . pp. 38, 88

Using the Page

Let students practice placing the pattern blocks on the coordinate plane.

1. Place a red trapezoid so that its longer side rests on the x-axis.

2. Place a green triangle so that one of its vertices is at the origin and the opposite side is parallel to the y-axis. What are the coordinates of the other two vertices?

3. Place a rhombus so that one of its vertices is at (−3, −1). What are the coordinates of the other two vertices?

Check students' work.

Answers

Think and Discuss

Possible answer:

1. Yes; the result of doing multiple transformations will not be the same if the order is switched. For example, in the Activity, the vertex at (2, 5) would end up at (7, −8) instead of at (7, −2).

Try This

1. Check students' work.

 **Hands-On** **LAB 5C**

Combine Transformations

Use with Lesson 5-7

KEY
Pattern blocks =

triangle   rhombus   trapezoid

 **internet connect**
Lab Resources Online
go.hrw.com
KEYWORD: MT4 Lab5C

You can use a coordinate plane when transforming a geometric figure.

Activity

❶ Follow the steps below to transform a figure.

 a. Place a red pattern block on a coordinate plane. Trace the block, and label the vertices.

 b. Translate the figure 3 units down and 5 units right, and then reflect the resulting figure across the x-axis. Draw the image and label the vertices.

 c. Now place a green pattern block on the same coordinate plane. Trace the block and label the vertices. Rotate the figure 180° around the point (0, 0), and then translate it 4 units up and 3 units right. Draw the image and label the vertices.

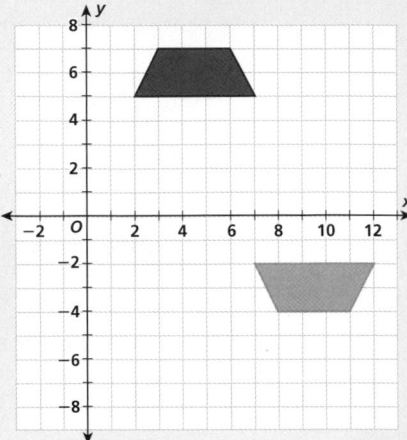

Think and Discuss

1. When you perform two or more transformations on a figure, does it matter in which order the transformations are performed? Explain.

Try This

1. Place a blue pattern block on a coordinate plane. Trace the block, and label the vertices. Perform two different transformations on the figure. Draw the image and label the vertices. Trade with a classmate. Describe the transformations your classmate used.

Teacher to Teacher

Lynn Bodet
San Antonio, TX

I teach coordinate plane concepts using a set of "integer ropes." We take the ropes outside and lay out the coordinate plane using one rope as the x-axis and one as the y-axis. For this activity, the students were given pieces of red yarn so that they could form the red trapezoids. One group of students was the original, a second group was the translated image, and a third group was the reflected image. The second group walked to the right and down from the original group. The third group helped the class see that the reflected image really is a mirror image of the original.

5-8 Symmetry

Learn to identify symmetry in figures.

Vocabulary
line symmetry
line of symmetry
rotational symmetry

Nature provides many beautiful examples of *symmetry*, such as the wings of a butterfly or the petals of a flower. Symmetric objects have parts that are congruent.

A figure has **line symmetry** if you can draw a line through it so that the two sides are mirror images of each other. The line is called the **line of symmetry** .

EXAMPLE 1 Drawing Figures with Line Symmetry

Complete each figure. The dashed line is the line of symmetry.

Helpful Hint
If you fold a figure on the line of symmetry, the halves match exactly.

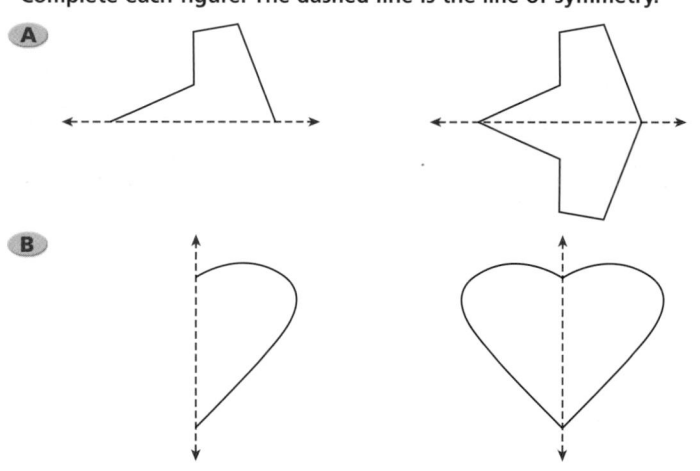

5-8 Organizer

Pacing: Traditional 1 day
Block $\frac{1}{2}$ day

Objective: Students identify symmetry in figures.

Warm Up

Multiply.

1. $\frac{1}{4} \times 360$ **90** 2. $\frac{1}{2} \times 360$ **180**

3. $\frac{1}{8} \times 360$ **45** 4. $\frac{3}{5} \times 360$ **216**

5. $\frac{2}{3} \times 360$ **240**

Problem of the Day

Name four plane figures that can be rotated 180° around a center point and look the same after the rotation as they did before.

Possible answer: rectangle, square, parallelogram, regular hexagon

Available on Daily Transparency in CRB

Math Fact !!!

A circle has both line symmetry and rotational symmetry. It has an infinite number of lines of symmetry and any angle of rotation can be used for its rotational symmetry.

---

1 Introduce

Alternate Opener

EXPLORATION

5-8 Symmetry

1. Draw one line through each hexagon to form two congruent figures. Use a different line for each hexagon.

2. Draw one line through each figure to form two congruent figures.

Think and Discuss

3. **Draw** a figure that cannot be cut into two congruent figures with one line.

4. **Describe** the characteristics of a figure that can be cut into two congruent figures.

Motivate

Fold a sheet of red construction paper in half. Cut half of a heart shape along the folded edge. Open the folded piece of paper to reveal a complete heart shape. Ask students whether they think there is a benefit to folding the paper before cutting it. The benefit is that both halves of the heart are symmetrical.

Exploration worksheet and answers on Chapter 5 Resource Book pp. 78 and 134

2 Teach

Lesson Presentation

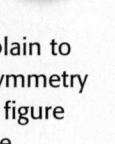

Guided Instruction

In this lesson, students learn to identify symmetry in figures. Explain to students that a figure has line symmetry if there is a line that divides the figure into two congruent parts that are reflected images of each other. Point out to students that some figures have more than one line of symmetry. Show an example of rotational symmetry. You may want to use the Rotational Symmetry Transparency in the Teacher's Manipulative Kit.

Example 1

Complete each figure. The dashed line is the line of symmetry.

A.

B.

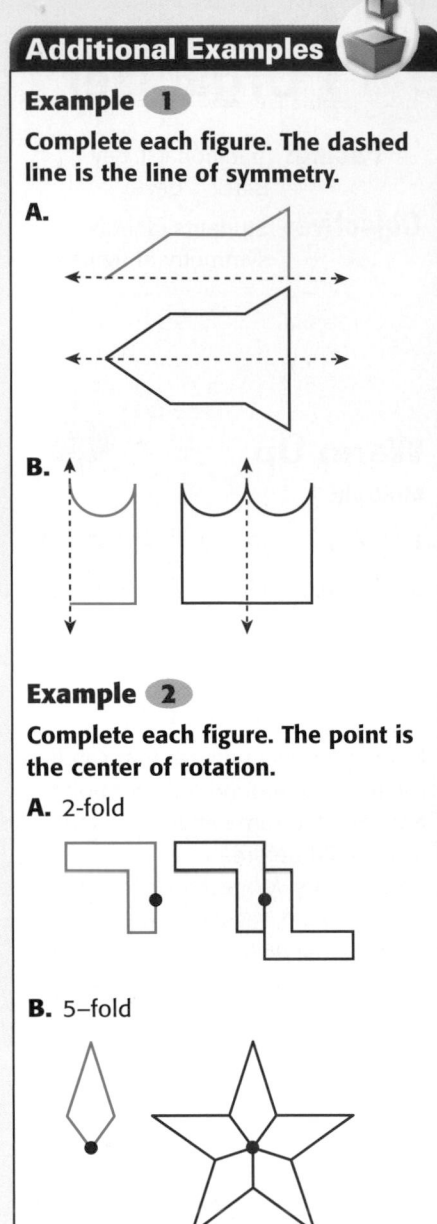

Example 2

Complete each figure. The point is the center of rotation.

A. 2-fold

B. 5–fold

A figure has **rotational symmetry** if you can rotate the figure around some point so that it coincides with itself. The point is the center of rotation, and the amount of rotation must be less than one full turn, or 360°.

7-fold rotational symmetry

6-fold rotational symmetry

EXAMPLE 2 Drawing Figures with Rotational Symmetry

Complete each figure. The point is the center of rotation.

A 2-fold

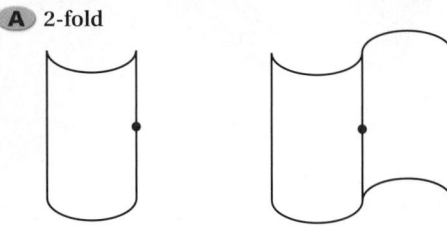

Figure coincides with itself every 180°.

B 8-fold

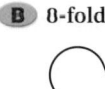

Figure coincides with itself every 45°.

Think and Discuss

1. **Explain** what it means for a figure to be symmetric.
2. **Tell** which letters of the alphabet have line symmetry.
3. **Tell** which letters of the alphabet have rotational symmetry.

Teach

![Reaching All Learners icon]

Reaching All Learners
Through Hands-On Experience

Provide students with some construction paper, scissors, and two cut-out plane figures (Chapter 5 Resource Book p. 113). One of the cutouts should be half of an image, such as half of a heart or half of a butterfly. Have students fold the paper in half, trace the figure, and cut it out to see what shape it reveals. The second cutout should be a piece of a figure with rotational symmetry, such as a petal from a flower or the point of a star. Instruct students to trace the figure, rotate it, and trace it again until they have completed a circle. Then have them cut out the figure.

3 Close

Summarize

Remind the students that they have been examining two different kinds of symmetry. Both types of symmetry are related to the transformations studied in Lesson 5-6. Line symmetry is related to reflection, and rotational symmetry is related to rotation. Ask students to point out examples of symmetry in the classroom.

Answers to Think and Discuss

Possible answers:

1. The figure coincides with itself after certain transformations, such as reflections or rotations.
2. capitals: A, B, C, D, E, H, I, K, M, O, T, U, V, W, X, and Y; lower case: c, i, l, o, t, u, v, w, and x (Answers may vary due to font or handwriting styles.)
3. capitals: H, I, N, O, S, X, Z; lower case: l, o, s, x, and z (Answers may vary due to font or handwriting styles.)

FOR EXTRA PRACTICE
see page 741

internet connect
Homework Help Online
go.hrw.com   Keyword: MP4 5-8

go.hrw.com

GUIDED PRACTICE

See Example ① Complete each figure. The dashed line is the line of symmetry.

1.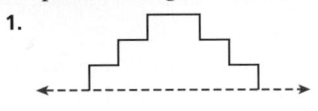

2.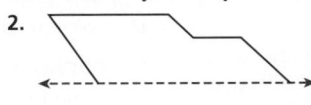

3.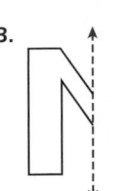

4.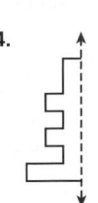

See Example ② Complete each figure. The point is the center of rotation.

5. 4-fold

6. 6-fold

INDEPENDENT PRACTICE

See Example ① Complete each figure. The dashed line is the line of symmetry.

7.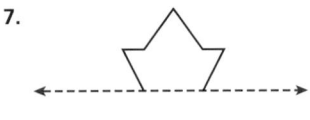

8.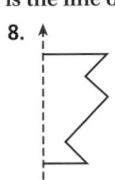

9.

10.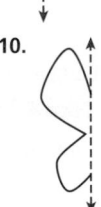

See Example ② Complete each figure. The point is the center of rotation.

11. 3-fold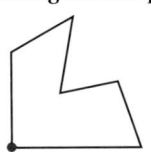

12. 5-fold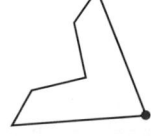

Students may want to refer back to the lesson examples.

Assignment Guide

If you finished Example ① assign:
Core 1–4, 7–10, 21–27
Enriched 1–4, 7–10, 21–27

If you finished Example ② assign:
Core 1–15, 21–27
Enriched 7–27

You may use the Recording Sheet on Chapter 5 Resource Book pp. 111–112 for Exercises 1–12 and 17.

Answers

1.

2.

3.

4.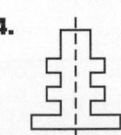

5–12. See pp. A5–A6.

Math Background

A special type of symmetry, called *point symmetry*, occurs when a figure has two-fold rotational symmetry. From any point on a figure with point symmetry, a line segment can be drawn through the center of rotation to a corresponding point on the figure. The center of rotation is the midpoint of that line segment.

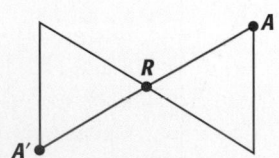

The letter *N* has point symmetry with center of rotation *R*. *R* is the midpoint of segment *AA'*.

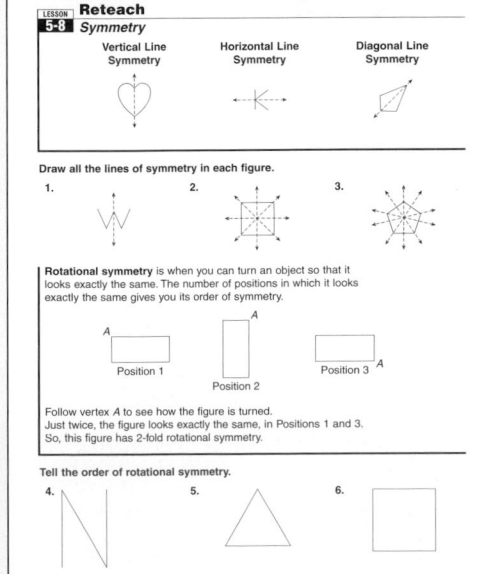
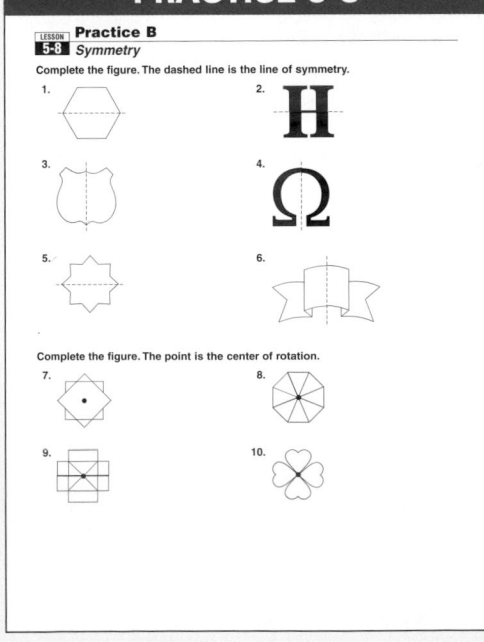

Answers

13–17. See p. A6.

19. Possible answer: Each part should be rotated $\left(\frac{360}{n}\right)$ degrees.

Journal

Have students find three words that display symmetry. Have them write the words and draw the line of symmetry or identify the rotational symmetry. Examples might include MOM (line symmetry) or pod (2-fold rotational symmetry).

Test Prep Doctor

For Exercise 27, remind students to check each ordered pair in the equation. Even after determining that **C** is the correct choice, they should verify that each of the other choices is incorrect.

Lesson Quiz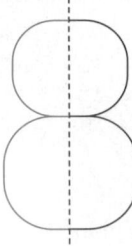

Does the figure described have symmetry?

1. a regular hexagon   yes

2. the letter F   no

3. the letter M   yes

4. Complete the figure. The dashed line is the line of symmetry.

Available on Daily Transparency in CRB

PRACTICE AND PROBLEM SOLVING

Draw an example of a figure with each type of symmetry.

13. line symmetry and rotational symmetry

14. line symmetry but not rotational symmetry

15. rotational symmetry but not line symmetry

16. no symmetry

17. SOCIAL STUDIES Family crests called *ka-mon* have been in use in Japan for many centuries. Copy each crest below. Describe the symmetry, and draw any lines of symmetry or the center of rotation.

In Japan, a kimono that displays the wearer's family crest (*ka-mon*) is often worn for ceremonial occasions.

a.
Kage Asa no ha

b.
Maru ni shichiyo

c.
Nito Nami

d.
Chukage itsutsu nenji Aoi

e.
Tsuki ni sansei

f.
Teuno ke

18. WRITE A PROBLEM Signal flags are hung from lines of rigging on ships. Research the full alphabet of signal flags, and write a problem about the types of symmetry in the flags.   **Check students' work.**

19. WRITE ABOUT IT To complete a figure with *n*-fold rotational symmetry, explain how much you rotate each part.

20. CHALLENGE Many countries' flags have symmetry. The flag of Japan has a rotational symmetry of 180°. Identify at least three other countries that have flags with rotational symmetry of 180°.
Possible answers: Israel, Macedonia, Micronesia, and United Kingdom.

Spiral Review

Write each number in standard notation. (Lesson 2-9)

21. 8.21×10^5 **821,000**   **22.** 2.07×10^{-7} **23.** -1.4×10^3 **−1400**
    0.000000207

Write each number in scientific notation. (Lesson 2-9)

24. 4,080,000   **4.08×10^6**   **25.** -0.000035 **-3.5×10^{-5}**   **26.** 5,910,000,000   **5.91×10^9**

27. TEST PREP Which ordered pair lies on the line with the equation $y = 2x + 1$? (Lesson 1-8)   **C**

   A $(0, 0)$ **B** $(2, 6)$ **C** $(0, 1)$ **D** $(5, 13)$

CHALLENGE 5-8

LESSON 5-8 Challenge
Inside, Outside

Materials: sheet of paper (8.5 in. by 11 in.), scissors, tape

1. a. Cut a strip of paper 2 in. by 11 in. Tape the ends together to form a loop.

b. Use a pencil to create a center line around the outside of the loop. Cut the loop along its center line. How many loops? **2**

2. a. Cut a strip of paper 2 in. by 11 in. Before taping the ends together, give the strip a half-twist by turning one end over 180°.

b. Use a pencil to create a center line around the loop. Continue drawing until you reach your starting point. Describe the result. How many surfaces (sides) does this strip have?
The line appears on both sides of the strip; one surface.

c. Cut the loop along its center line. How many loops do you have?
1 loop

This unusual surface is called a **Mobius Strip**, named after A. F. Mobius, a 19th century German mathematician and astronomer, who pioneered *topology* (how geometric figures act when distorted by such things as twists and stretches).

3. Prepare another Mobius Strip, a loop with a half-twist. Cut this strip parallel to one edge about one-third of the way from the edge. Continue cutting until you get back to your starting point. Describe the result.
2 loops of unequal lengths linked together

4. Prepare a loop with a full twist. Cut this strip parallel to one edge about one-third of the way from the edge. Continue cutting until you get back to your starting point. Describe the result.
2 loops of same length linked together; one loop is twice as wide as the other

5. Prepare a loop with three half-twists. Cut this strip parallel to one edge about one-half of the way from the edge. Continue cutting until you get back to your starting point. Describe the result.
1 loop with a knot

PROBLEM SOLVING 5-8

LESSON 5-8 Problem Solving
Symmetry

Complete the figure. A dashed line is a line of symmetry and a point is a center of rotation.

1. 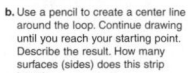   **2.**

3. 2 fold symmetry   **4.** 4 fold symmetry

Use the flag of Switzerland to answer the questions.

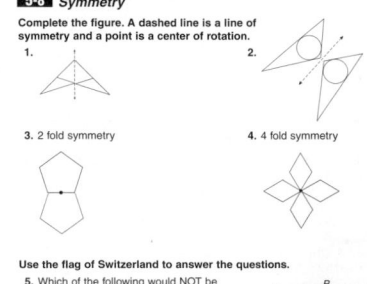

5. Which of the following would NOT be a line of symmetry?
   A $\overline{HD}$ **C** $\overline{AE}$
   B $\overline{BF}$ **(D)** $\overline{HB}$

6. How many lines of symmetry does the flag have?
   F 2   **(H)** 4
   G 6   **J** 8

7. How many folds of rotational symmetry does the flag have?
   A 0   **C** 2
   (B) 4   **D** 8

8. Which lists all lines of symmetry of the flag?
   F $\overline{AE}$, $\overline{GC}$
   G $\overline{HD}$, $\overline{BF}$
   (H) $\overline{HD}$, $\overline{BF}$, $\overline{AE}$, $\overline{GC}$
   J $\overline{HB}$, $\overline{DF}$, $\overline{AE}$, $\overline{GC}$

9. Which describes the center of rotation?
   (A) intersection of $\overline{BF}$ and $\overline{HD}$
   B intersection of $\overline{AE}$ and $\overline{HB}$
   C A
   D There is no center of rotation

5-9 Tessellations

Learn to predict and verify patterns involving tessellations.

Vocabulary
tessellation
regular tessellation
semiregular tessellation

Fascinating designs can be made by repeating a figure or group of figures. These designs are often used in art and architecture.

A repeating pattern of plane figures that completely covers a plane with no gaps or overlaps is a **tessellation**.

In a **regular tessellation**, a regular polygon is repeated to fill a plane. The angles at each vertex add to 360°, so exactly three regular tessellations exist.

The Alhambra Palace in Granada, Spain, is decorated by many tessellations and other geometric patterns.

Equilateral triangles	Squares	Regular hexagons
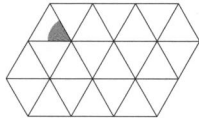	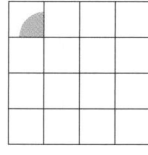	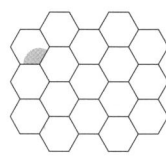
$6 \cdot 60° = 360°$	$4 \cdot 90° = 360°$	$3 \cdot 120° = 360°$

In a **semiregular tessellation**, two or more regular polygons are repeated to fill the plane and the vertices are all identical.

EXAMPLE 1   PROBLEM SOLVING

Find all the possible semiregular tessellations that use triangles and hexagons.

 Understand the Problem

List the **important information**:
- The angles at each vertex add to 360°.
- All of the angles in a regular hexagon measure 120°.
- All of the angles in an equilateral triangle measure 60°.

---

1 Introduce

Alternate Opener

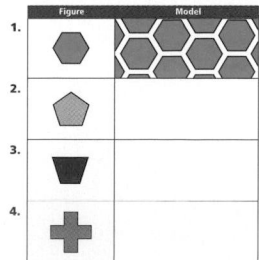

5-9 Tessellations

Tile floors come in many different designs. For each figure, determine whether or not it could be used as a floor tile. Draw a model to show a possible tiling pattern.

	Figure	Model
1.		
2.		
3.		
4.		

Think and Discuss

5. **Draw** a figure that cannot be used as a tile for a floor.
6. **Describe** the common characteristics that make some of the figures above suitable for floor tiles.

Motivate

Show the students an image of the inside of a beehive (Teaching Transparency T43, CRB). Explain that when the bees construct their homes, they don't want the "rooms" to overlap, and they don't want any space between the "rooms." Bees are able to do this by building "rooms" out of hexagons. The inside of a beehive is an example of a tessellation.

Exploration worksheet and answers on Chapter 5 Resource Book pp. 87 and 136

5-9 Organizer

Pacing: Traditional 1 day
Block $\frac{1}{2}$ day
Objective: Students predict and verify patterns involving tessellations.

Warm Up

Identify each polygon.

1. polygon with 10 sides   decagon
2. polygon with 3 congruent sides   equilateral triangle
3. polygon with 4 congruent sides and no right angles   rhombus

Problem of the Day

If each of the capital letters of the alphabet is rotated 180° around its center, which of them will look the same?   H, I, N, O, S, X, Z

Available on Daily Transparency in CRB

Math Fact

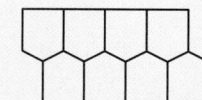

Even though regular pentagons cannot tessellate a plane (see Think and Discuss 2), there are irregular pentagons that can tessellate a plane.

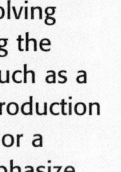

---

2 Teach

Lesson Presentation

Guided Instruction

In this lesson, students learn to predict and verify patterns involving tessellations. Begin by showing the students tessellating designs such as a picture of a honeycomb, a reproduction of an artwork by M. C. Escher, or a piece of clothing or fabric. Emphasize that in any tessellation, there are no gaps or overlaps. Show students the three regular tessellations (Teaching Transparency T42, CRB). Explain that for a regular polygon to tessellate, the measure of an interior angle must be a factor of 360°.

Example 1

Find all the possible semiregular tessellations that use triangles and squares.

There are two arrangements of three triangles and two squares around a vertex.

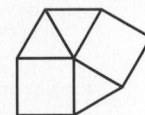

Repeat each arrangement to create a tessellation.

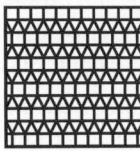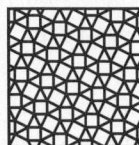

Example 1 note: As you work through Example 1, explain to students that each vertex in a semiregular tessellation must have the same number and arrangement of figures.

---

2 Make a Plan

Account for all possibilities: List all possible combinations of triangles and hexagons around a vertex that add to 360°. Then see which combinations can be used to create a semiregular tessellation.

6 triangles, 0 hexagons	$6(60°) = 360°$	*regular*
4 triangles, 1 hexagon	$4(60°) + 120° = 360°$	
2 triangles, 2 hexagons	$2(60°) + 2(120°) = 360°$	
0 triangles, 3 hexagons	$3(120°) = 360°$	*regular*

3 Solve

There is one arrangement of 4 triangles and 1 hexagon around a vertex. There are two arrangements of 2 triangles and 2 hexagons around a vertex.

4 triangles, 1 hexagon   *2 triangles, 2 hexagons*

Repeat each arrangement around every vertex, if possible, to create a tessellation.

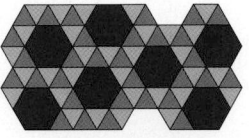

 *If you try to repeat the third arrangement around the blue vertex, the green vertex has 3 triangles. So this arrangement does not produce a semiregular tessellation.*

There are exactly two semiregular tessellations that use triangles and hexagons.

4 Look Back

When the third arrangement is repeated, a vertex is created that is not identical to the other vertices, so this arrangement cannot be used to produce a semiregular tessellation.

Teach

 **Reaching All Learners**
Through Hands-On Experience

Have students work in small groups. Give each group a set of pattern blocks (in Manipulatives Kit) containing equilateral triangles, regular hexagons, squares, rhombuses, trapezoids, and parallelograms. Ask each group to create as many different tessellations as they can with the pattern blocks and to draw each tessellation they find on a sheet of paper. If you want to provide an additional challenge, ask students to use a protractor to show that the sum of the measures of the angles around each vertex is 360°.

It is also possible to tessellate with polygons that are not regular. Any triangle or quadrilateral can be used to create a tessellation.

EXAMPLE **2** **Creating a Tessellation**

Create a tessellation with quadrilateral *ABCD*.

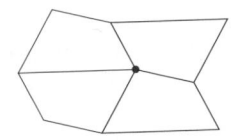

There must be a copy of each angle of quadrilateral *ABCD* at every vertex.

EXAMPLE **3** **Creating a Tessellation by Transforming a Polygon**

Use rotations to create a variation of the tessellation in Example 2.

Step 1: Find the midpoint of a side.

Step 2: Make a new edge for half of the side.

Step 3: Rotate the new edge around the midpoint to form the edge of the other half of the side.

Step 4: Repeat with the other sides.

Step 5: Use the figure to make a tessellation.

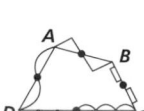

Think and Discuss

1. Compare regular tessellations with semiregular tessellations.

2. Explain why a regular pentagon cannot be used to create a regular tessellation.

Additional Examples

Example **2**

Create a tessellation with quadrilateral *EFGH*.

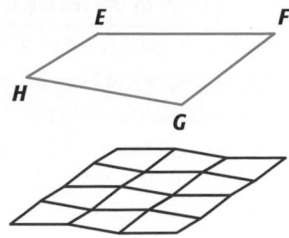

Example **3**

Use rotations to create a variation of the tessellation in Additional Example 2.

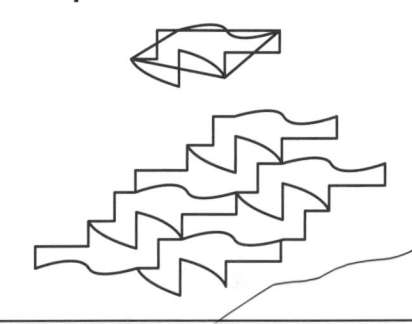

---

3 **Close**

Summarize

Review the requirements for a design to be a tessellation. The design must be a repeating pattern of plane figures that completely covers a plane with no gaps or overlaps.

Remind students that tessellations are found in nature, art, and manufactured structures. The lesson focused mainly on tessellations formed by polygons, but there are many tessellations formed by figures other than polygons.

Answers to Think and Discuss

Possible answers:

1. Both are repeated patterns that completely fill a plane. However, regular tessellations involve only one type of regular polygon while semiregular tessellations involve two or more types of regular polygons.

2. The angles in a regular pentagon all measure 108°, and 108 does not divide evenly into 360.

FOR EXTRA PRACTICE
see page 741

internet connect
Homework Help Online
go.hrw.com   Keyword: MP4 5-9

> Students may want to refer back to the lesson examples.

Assignment Guide

If you finished Example **1** assign:
  Core 1, 4, 7, 18–26
  Enriched 7–9, 18–26

If you finished Example **2** assign:
  Core 1, 2, 4, 5, 7, 11, 13, 18–26
  Enriched 7–13, 18–26

If you finished Example **3** assign:
  Core 1–6, 7–13 odd, 18–26
  Enriched 7–26

You may use the cut-out shapes on Chapter 5 Resource Book pp. 114–115 for Exercises 2–3, 5–12, and 14.

Answers

1. The angles in a regular octagon all measure 135°; the angles in a square all measure 90°. There is only one possibility: 1 square and 2 octagons; 1(90°) + 2(135°) = 360°.

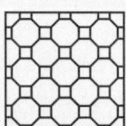

2. Possible answer:

GUIDED PRACTICE

See Example **1** 1. Find all the possible semiregular tessellations that use squares and octagons.

See Example **2** 2. Create a tessellation with quadrilateral *QRST*.

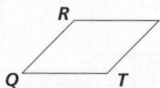

See Example **3** 3. Use rotations to create a variation of the tessellation in Exercise 2.

INDEPENDENT PRACTICE

See Example **1** 4. Find all the possible semiregular tessellations that use triangles and squares.

See Example **2** 5. Create a tessellation with triangle *PQR*.

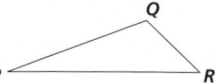

See Example **3** 6. Use rotations to create a variation of the tessellation in Exercise 5.

PRACTICE AND PROBLEM SOLVING

Use each arrangement of regular polygons to create a semiregular tessellation.

7.
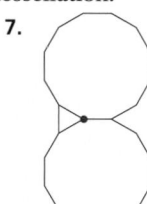

8.
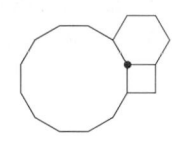

9.
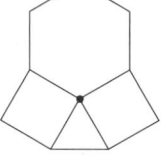

Use each shape to create a tessellation.

10.
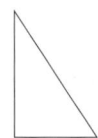

11.

12.

13. A piece is removed from one side of a rectangle and translated to the opposite side. Will this shape tessellate?

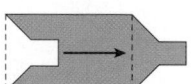

Math Background

Allowing students to learn mathematics through its application in art is a powerful motivator for some students. Other students, however, find the creative side of this work quite challenging. Computer programs can help these students create designs that tessellate. These programs allow a student to experiment with various designs and to see many more possibilities than would be practical with paper and pencil.

It might be useful to investigate the work of M. C. Escher, the artist most often associated with tessellations.

RETEACH 5-9

LESSON 5-9 Reteach
Tessellations

Tessellation: repeating pattern of plane figures that completely covers a plane with no gaps or overlaps.

Regular Tessellation
A single regular polygon is used.
The measure of the interior angle of the regular polygon is a divisor of 360°.
This regular tessellation uses an equilateral triangle (measure of interior angle = 60°) as the basic unit.

Tell why each tessellation is regular.

1.   2.

What is the regular polygon used?   **square**   **hexagon**
What is the measure of an interior angle?   **90°**   **120°**
Is the angle measure a divisor of 360°?   **yes**   **yes**

Semiregular Tessellation
Two or more regular polygons are used.
The sum of the angle measures at each vertex is 360°.
This semiregular tessellation uses a square (each angle = 90°) and a regular octagon (each angle = 135°) as the basic units. The sum of the angle measures at each vertex is 360° (one 90°-angle and two 135°-angles).

Tell why the tessellation is semiregular.

3.

What are the regular polygons used?   **hexagon, triangle**
What are their interior-angle measures?   **120°, 60°**
Explain a sum of 360° at each vertex.   two **120°** angles + two **60°** angles

PRACTICE 5-9

LESSON 5-9 Practice B
Tessellations

1. Find all the possible semiregular tessellations that use hexagons and triangles.
   3 hexagons, 0 triangles; 6 triangles, 0 hexagons; 1 hexagon, 4 triangles; 2 hexagons, 2 triangles

2. Create a semiregular tessellation that uses hexagons and triangles.
   sample answer:
   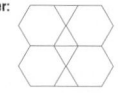

3. Create another semiregular tessellation that uses hexagons and triangles.
   sample answer:

4. Create a tessellation with quadrilateral *ABCD*.
   sample answer:
   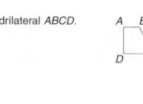

5. Use rotations to create a variation of the tessellation in Exercise 4.
   sample answer:
   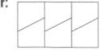

 **Art LINK**

M. C. Escher created works of art by repeating interlocking shapes. He used both regular and nonregular tessellations. He often used what he called *metamorphoses*, in which shapes change into other shapes. Escher used his reptile pattern in many hexagonal tessellations. One of the most famous is entitled simply *Reptiles*.

14. The steps below show the method Escher used to make a bird out of a triangle. Use the bird to create a tessellation.

Step 1   Step 2

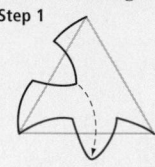 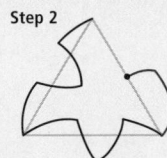

Step 3   Step 4

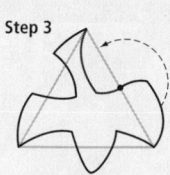 

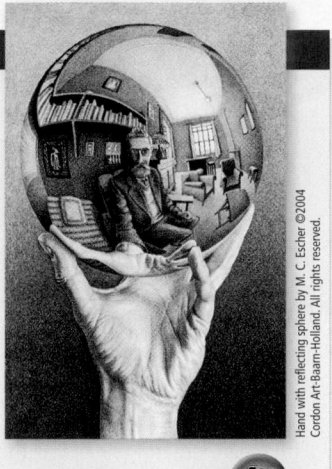

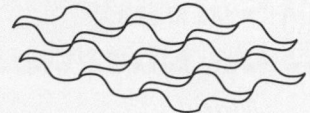
go.hrw.com
KEYWORD: MP4 ESCHER
CNN student News.

Refer to the sketch for *Reptiles* for Exercises 15–16.

15. What regular polygon do you think Escher used to begin the sketch? **hexagon**

16. Describe the process he used to create each figure from the basic shape.

17. **CHALLENGE** Create an Escher-like tessellation of your own design. **Check students' work.**

Spiral Review

Solve and graph each inequality. (Lesson 1-5)

18. $y + 4 > 1$ $y > -3$ 19. $4p \le 12$ $p \le 3$ 20. $f - 3 \ge 2$ $f \ge 5$ 21. $4 < \frac{w}{3}$ $12 < w$

22. $p - 1 \ge 4$ $p \ge 5$ 23. $m + 3 \le 3$ $m \le 0$ 24. $3 > \frac{n}{2}$ $6 > n$ 25. $3z < 6$ $z < 2$

26. **TEST PREP** Which word phrase represents the expression $8 - 6p$? (Lesson 1-2) **D**

   A   Eight less than six times a number   C   Six times a number minus eight

   B   Eight minus six, times a number   D   Six times a number, subtracted from eight

---

CHALLENGE 5-9

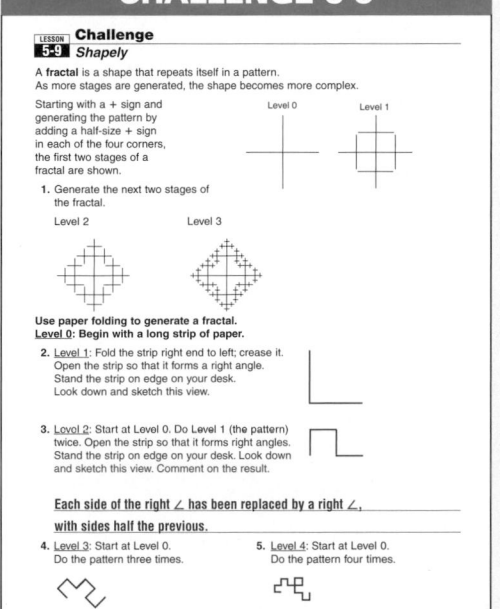

PROBLEM SOLVING 5-9

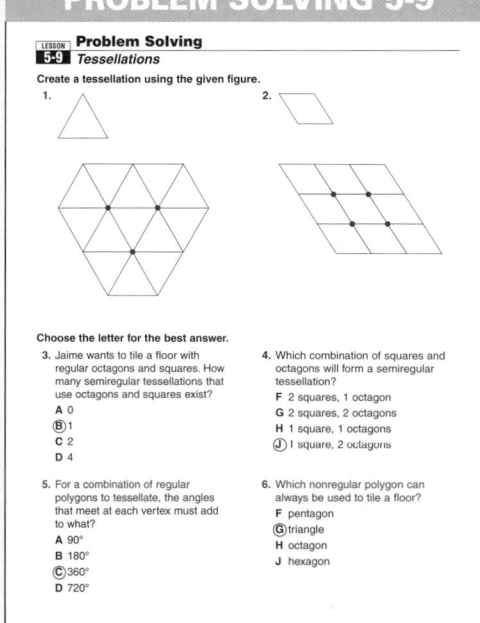

Answers

3. Possible answer:

4–14. See pp. A6–A7.

16. Possible answer: He probably made the hexagon look like a lizard by creating partial sides and rotating them as in Example 3.

18–25. Complete answers on p. A7

Journal

Have each student choose a favorite tessellation from the examples shown in class today. Ask students to write about their choices.

Test Prep Doctor

For Exercise 26, suggest to students that for each choice they should decide which term is being subtracted. In both choice **A** and choice **C**, the eight is being subtracted. The expression indicates that $6p$ is being subtracted, so choice **D** is correct.

Lesson Quiz

1. Find all possible semiregular tessellations that use squares and regular hexagons.   **none**

2. Explain why a regular tessellation with regular octagons is impossible. **Each angle measure in a regular octagon is 135° and 135 is not a factor of 360.**

3. Can a semiregular tessellation be formed using a regular 12-sided polygon and a regular hexagon? Explain. **No; a regular 12-sided polygon has angles that measure 150° and a regular hexagon has angles that measure 120°. No combinations of 120° and 150° add to 360°.**

Available on Daily Transparency in CRB

Problem Solving on Location

Maryland

Purpose: *To provide additional practice for problem-solving skills in Chapters 1–5*

Maryland State Flag

- After problem 2, ask students if there are any other plane figures they can identify in the flag.

  Possible answer: Some of the Calvert crest pattern is made up of rhombuses. There are also small triangles.

- After problem 5, have students consider whether they could use transformations to describe any part of the flag.

  Possible answers: The Calvert sections are identical, so one can be described as a translation of the other, and likewise for the Crossland sections. The cross-like part of the Crossland crest uses different shades to distinguish images from their reflections.

Extension Have students create a flag that uses plane figures and transformations. Have them color their flag and write a description of it using geometric vocabulary. Check students' work.

Problem Solving on Location

MARYLAND

Maryland State Flag

The colony of Maryland was founded in 1634 by Cecil Calvert, the second Lord Baltimore. The Maryland flag is made up of two family crests.

- The black and gold design is the crest of the Calvert family.

- The red and white design is the crest of the Crossland family, the family of Cecil Calvert's mother.

Copy the flag onto your paper.

1. Label all parallel lines, perpendicular lines, and transversals.

2. Identify at least one trapezoid, one parallelogram, and one rectangle in the flag.

3. Describe the symmetry in the Calvert family crest.

4. Describe the symmetry in the Crossland family crest.

5. Describe the symmetry in the entire Maryland flag.

6. Describe any other interesting geometric features of the Maryland flag.

Answers

1. Possible answer: Opposite sides of the flag are parallel. Adjacent sides of the flag are perpendicular. In the Calvert flag, the pattern is created when parallel lines are crossed by a transversal.

2. Possible answer: The lower-right figure on the Calvert flag is a trapezoid. The flag is a parallelogram. Each of the four portions of the flag is a rectangle.

3. Disregarding color, the Calvert crest has 2-fold rotational symmetry; otherwise there is no symmetry.

4. Disregarding color, the Crossland crest has a horizontal and vertical line of symmetry and 2-fold rotational symmetry. If color is considered, the Crossland crest has only 2-fold rotational symmetry.

5. Disregarding color, the flag as a whole has 2-fold rotational symmetry. If color is considered, the flag has no symmetry.

6. The two Calvert crests are translations of each other, and so are the two Crossland crests.

Pride of Baltimore II

The *Pride of Baltimore II* is a replica of a kind of 1812-era sailing ship called a Baltimore Clipper. It is a square topsail schooner. Three basic maneuvers are used in sailing: sailing into the wind, sailing across the wind, and sailing with the wind. The result is a zigzagging course of alternating directions that the vessel follows to move in the desired direction. A ship's sails must be adjusted in relation to the wind to achieve the desired heading, or direction of travel.

1. A sailing ship starts at *A* and sails to *E* using the indicated zigzag course. Find all the values of *p, q, r, s,* and *t.*
   p = 45°; q = 45°; r = 55°; s = 55°; t = 50°

2. Suppose that ∠*C* is changed to 100°. Find all the values of *p, q, r, s,* and *t.* **p = 45°; q = 45°; r = 35°; s = 35°; t = 70°**

3. Suppose that ∠*B* is changed to 85°. Find all the values of *p, q, r, s,* and *t.* **p = 45°; q = 45°; r = 55°; s = 55°; t = 40°**

4. Draw a similar zigzag course from point *A* to point *E*, using 45° for ∠*B*, 70° for ∠*C*, 65° for ∠*D*, and 50° for ∠*E*. Find the values of *p, q, r, s,* and *t.* **p = 65°, q = 65°, r = 45°, s = 45°, t = 90°**

5. Measure the distance from point *A* to point *F*, and draw it on your paper. Now make up your own angles of adjustment to travel from point *A* to point *F* exactly, and sketch them on your paper.
   Check students' work.

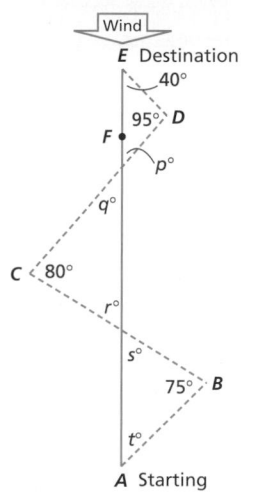

Diagram labels: Wind | E Destination | 40° | 95° D | F | p° | q° | C 80° | r° | s° | 75° B | t° | A Starting

Pride of Baltimore II

- After problem 1, have students identify how they determined what to do first. Possible answer: I found *p* first because I knew the measurements of the other two angles in that triangle. Ask: How did knowing *p* help you find *q*? *p* and *q* are congruent because they are vertical angles.

- After problem 3, have students consider the following problem: What kind of triangles are formed by the ship's path? How do you know?

  Possible answer: They are all scalene triangles. The bottom two are acute, and the top one is obtuse. The measures of the angles determine the classification of the triangles.

Extension Have students use a protractor to accurately draw the diagram. Have them check their angle measurements using the protractor.

Check students' work.

Game Resources

Puzzles, Twisters & Teasers
Chapter 5 Resource Book

Coloring Tessellations

Purpose: *To extend the study of tessellations to a coloring problem*

Discuss: Ask students to explain the rule for coloring the tessellations. No two figures that share an edge can be colored with the same color. Explain why the tessellation of hexagons requires 3 colors. Choose any hexagon and color it with color 1. Color 1 can not be used to color any of the surrounding hexagons because the surrounding hexagons all share an edge with the center hexagon. So you must color the 6 surrounding hexagons in order with colors 2, 3, 2, 3, 2, and 3.

Extend: Give students a map of the United States. Ask them to color the map with as few colors as possible, using a different color for each state so that no states that share a border are the same color. The map can be colored with 4 colors.

Polygon Rummy

Purpose: *To practice using the properties of polygons in a card game*

Discuss: Ask: Is there any set of 3 cards for which there is no polygon that has the properties on all 3 cards? Give an example. yes; triangle, quadrilateral, all angles obtuse What figure satisfies all the properties "pentagon," "all sides congruent," and "all angles congruent"? regular pentagon

Extend: Ask students to name the cards that describe a square and those that describe a regular hexagon. Which figure meets the requirements of more cards? 8 cards describe a square; 6 cards describe a regular hexagon; square.

MATH-ABLES

Coloring Tessellations

Two of the three regular tessellations—triangles and squares—can be colored with two colors so that no two polygons that share an edge are the same color. The third—hexagons—requires three colors.

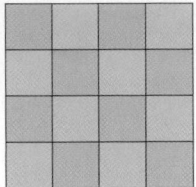 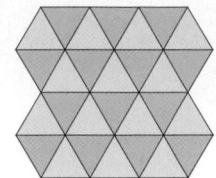 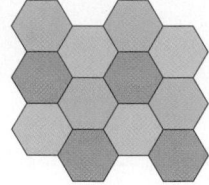

1. Determine if each semiregular tessellation can be colored with two colors. If not, tell the minimum number of colors needed.

 yes   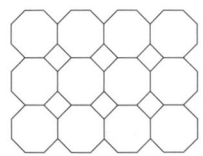 no (3)   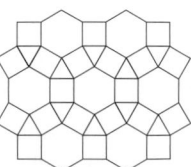 yes

2. Try to write a rule about which tessellations can be colored with two colors. **A tessellation can be colored with two colors if each figure borders other figures that do not border each other.**

Polygon Rummy

The object of this game is to create geometric figures. Each card in the deck shows a property of a geometric figure. To create a figure, you must draw a polygon that matches at least three cards in your hand. For example, if you have the cards "quadrilateral," "a pair of parallel sides," and "a right angle," you could draw a rectangle.

internet connect
Go to *go.hrw.com* for a complete set of rules and game cards.
KEYWORD: MP4 Game5

Technology LAB

Exterior Angles of a Polygon

internet connect
Lab Resources Online
go.hrw.com
KEYWORD: MP4 TechLab5

The **exterior angles** of a polygon are formed by extending the polygon's sides. Every exterior angle is supplementary to the angle next to it inside the polygon.

Exterior angle

Activity

❶ Follow the steps to find the sum of the exterior angle measures for a polygon.

  a. Use geometry software to make a pentagon. Label the vertices *A* through *E*.

  b. Use the **LINE-RAY** tool to extend the sides of the pentagon. Add points *F* through *J* as shown.

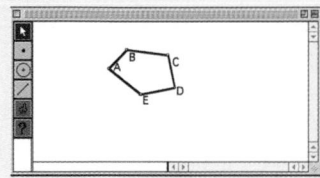

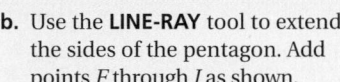

  c. Use the **ANGLE MEASURE** tool to measure each exterior angle and the **CALCULATOR** tool to add the measures. Notice the sum.

  d. Drag vertices *A* through *E* and watch the sum. Notice that the sum of the angle measures is *always* 360°.

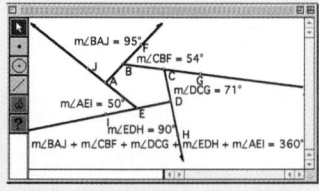

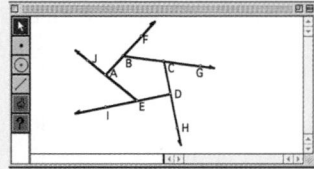

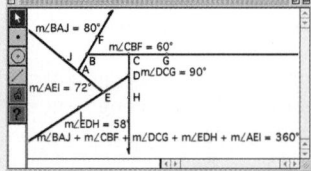

Think and Discuss

1. Suppose you were to drag the vertices of a polygon so that the polygon almost vanishes. How would this show that the sum of the exterior angle measures is 360°?

Try This

1. Use geometry software to draw a quadrilateral. Find the sum of its exterior angle measures. Drag its vertices to check that the sum is always the same.

Answers

Think and Discuss

1. Possible answer: The resulting figure would look like a point with 5 line segments radiating from it. The sum of the measures of the angles formed would be 360° since they form a complete circle.

Try This

1. 360°; the sum is always 360°.

Technology LAB
Exterior Angles of a Polygon

Objective: To use a geometry program to study the exterior angles of a polygon

Materials: Computer with geometry software

Lab Resources
Technology Lab Activities . . . . p. 34

Using the Page

This technology activity shows students how to use geometry software to explore the properties of polygons. Specific instructions may vary, depending on the geometry software used. The instructions given are for Geometer's Sketch Pad. For instructions to other software, visit go.hrw.com.

The Think and Discuss problem can be used to assess students' understanding of the technology activity. While the Try This problem can be done without geometry software, it is meant to help students become familiar with using geometry software to explore the properties of polygons.

Assessment

1. Use the software to draw a triangle and to measure its exterior angles. What is the sum of the measures of the exterior angles of a triangle? **360°**

2. What general statement can you make about the sum of the measures of the exterior angles of any polygon? The sum of the measures of the exterior angles of any polygon is 360°.

Technology Lab   **271**

Chapter 5

Study Guide and Review

Purpose: *To help students review and practice concepts and skills presented in Chapter 5*

Assessment Resources

Chapter Review

Chapter 5 Resource Book . . . pp. 96–98

***Test and Practice Generator* CD-ROM**

Additional review assessment items in both multiple-choice and free-response format may be generated for any objective in Chapter 5.

You may use the Recording Sheet on Chapter 5 Resource Book p. 116 for problems 26–27.

Answers

1. parallel lines; perpendicular lines
2. rectangle; rhombus
3. 108°
4. 72°
5. 108°

Chapter 5

Study Guide and Review

Study Guide and Review

Vocabulary

acute angle 223
acute triangle 234
angle 222
center of rotation 254
complementary angles . . 223
congruent 223
correspondence 250
equilateral triangle 235
image 254
isosceles triangle 235
line 222
line of symmetry 259
line symmetry 259
obtuse angle 223
obtuse triangle 234
parallel lines 228
parallelogram 240
perpendicular lines 228
plane 222
point 222
polygon 239
ray 222
rectangle 240
reflection 254
regular polygon 240
regular tessellation 263
rhombus 240
right angle 223
right triangle 234
rise 244
rotation 254
rotational symmetry . . . 260
run 244
scalene triangle 235
segment 222
semiregular tessellation 263
slope 244
square 240
supplementary angles . . 223
tessellation 263
transformation 254
translation 254
transversal 228
trapezoid 240
Triangle Sum Theorem . 234
vertical angles 223

Complete the sentences below with vocabulary words from the list above. Words may be used more than once.

1. Lines in the same plane that never meet are called __?__. Lines that intersect at 90° angles are called __?__.
2. A quadrilateral with 4 congruent angles is called a __?__. A quadrilateral with 4 congruent sides is called a __?__.

5-1 Points, Lines, Planes, and Angles (pp. 222–226)

EXAMPLE

■ **Find the angle measure.**

$m\angle 1$

$$m\angle 1 + 122° = 180°$$
$$-122° \quad -122°$$
$$m\angle 1 = 58°$$

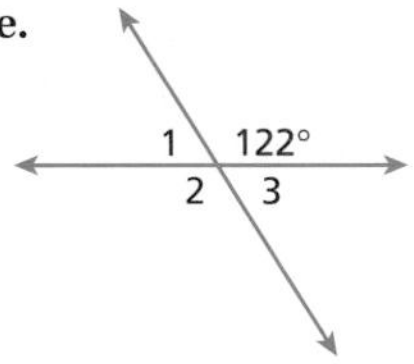

EXERCISES

Find each angle measure.

3. $m\angle 1$
4. $m\angle 2$
5. $m\angle 3$

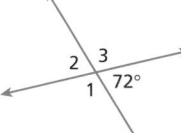

5-2 Parallel and Perpendicular Lines (pp. 228–231)

EXAMPLE

Line $j \parallel$ line k. Find each angle measure.

- $m\angle 1$
  $m\angle 1 = 143°$
- $m\angle 2$
  $\begin{aligned} m\angle 2 + 143° &= 180° \\ -143° \quad -143° & \\ \hline m\angle 2 \quad &= 37° \end{aligned}$

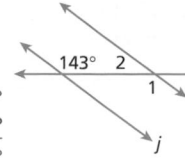

EXERCISES

Line $p \parallel$ line q. Find each angle measure.

6. $m\angle 1$
7. $m\angle 2$
8. $m\angle 3$
9. $m\angle 4$
10. $m\angle 5$

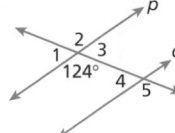

5-3 Triangles (pp. 234–238)

EXAMPLE

- Find n.

  $\begin{aligned} n° + 50° + 90° &= 180° \\ n° + 140° &= 180° \\ -140° \quad -140° & \\ \hline n° \quad &= 40° \end{aligned}$

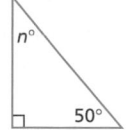

EXERCISES

11. Find $m°$.

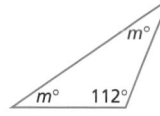

5-4 Polygons (pp. 239–243)

EXAMPLE

- Find the sum of the angle measures in a regular 12-gon.

  $\begin{aligned} \text{sum of angle measures} &= 180°(n - 2) \\ &= 180°(12 - 2) \\ &= 180°(10) = 1800° \end{aligned}$

EXERCISES

Find the angle measures in each regular polygon.

12. a regular hexagon
13. a regular 10-gon

5-5 Coordinate Geometry (pp. 244–247)

EXAMPLE

- Graph the quadrilateral with the given vertices. Give all the names that apply.
  $D(-2, 1), E(2, 3), F(3, 1), G(-1, -1)$

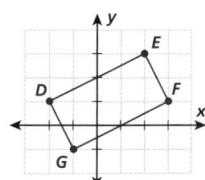

$\overline{DE} \parallel \overline{FG}$
$\overline{EF} \parallel \overline{GD}$
$\overline{DE} \perp \overline{EF}$

quadrilateral, parallelogram, rectangle

EXERCISES

Graph the quadrilaterals with the given vertices. Give all the names that apply.

14. $Q(2, 0), R(-1, 1), S(3, 3), T(8, 3)$
15. $K(0, 3), L(1, 0), M(0, -3), N(-1, 0)$
16. $W(2, 3), X(2, -2), Y(-1, -3), Z(-1, 2)$

Answers

6. $56°$
7. $124°$
8. $56°$
9. $56°$
10. $124°$
11. $m° = 34°$
12. $120°$
13. $144°$
14.

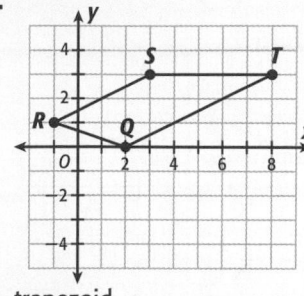

trapezoid

15.

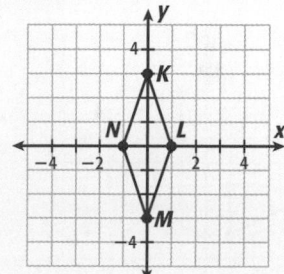

parallelogram, rhombus

16.

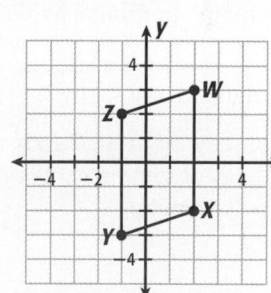

parallelogram

Answers

17. $x = 23$

18. $t = 3.2$

19. $q = 5$

20.

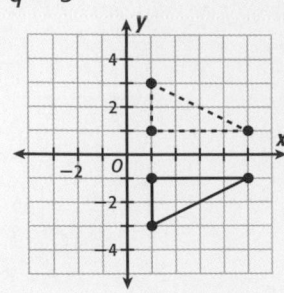

21.

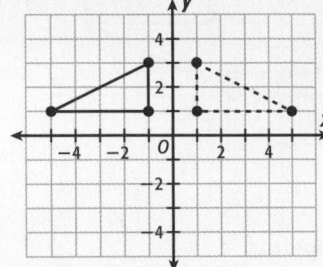

22.

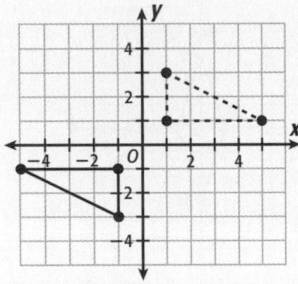

23. line symmetry: horizontal line of symmetry

24. 2-fold rotational symmetry

25. line symmetry: horizontal and vertical lines of symmetry; 2-fold rotational symmetry

5-6 Congruence (pp. 250–253)

EXAMPLE

■ Triangle $ABC \cong$ triangle FDE. Find x.

$\overline{AC} \cong \overline{FE}$

$$\begin{array}{rcl} x - 4 & = & 4 \\ + 4 & & + 4 \\ \hline x & = & 8 \end{array}$$

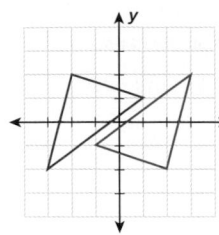

EXERCISES

Triangle $JQZ \cong$ triangle VTZ.

17. Find x.

18. Find t.

19. Find q.

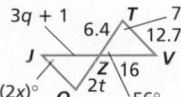

5-7 Transformations (pp. 254–257)

EXAMPLE

■ Draw the image of a triangle with vertices $(-2, 2)$, $(1, 1)$, $(-3, -2)$ after a 180° rotation around $(0, 0)$.

EXERCISES

Draw the image of a triangle with vertices $(1, 3)$, $(5, 1)$, $(1, 1)$ after each transformation.

20. a reflection across the x-axis

21. a reflection across the y-axis

22. a 180° rotation around $(0, 0)$

5-8 Symmetry (pp. 259–262)

EXAMPLE

Describe the symmetry in each letter.

■ M
line symmetry; vertical line of symmetry

■ N
2-fold rotational symmetry

EXERCISES

Describe the symmetry in each letter.

23. D

24. S

25. H

5-9 Tessellations (pp. 263–267)

EXAMPLE

■ Create a tessellation with the figure.

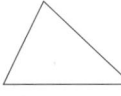

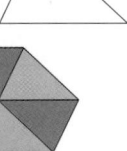

EXERCISES

Create a tessellation with each figure.

26.

27.

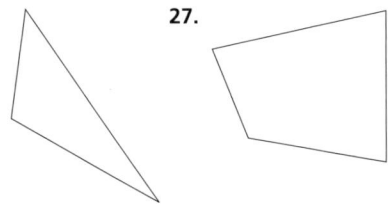

26. Possible answer:

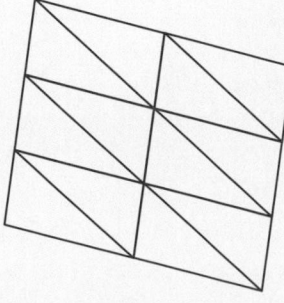

27. Possible answer:

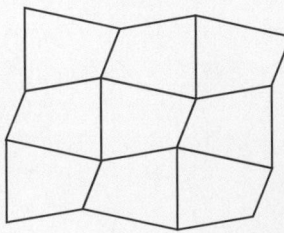

Refer to the figure. ∠*AEB* and ∠*BEC*

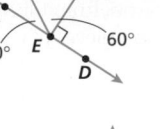

1. Name a pair of complementary angles.

2. Name a pair of supplementary angles.
   ∠*AEB* and ∠*BED* or ∠*AEC* and ∠*CED*

Line *w* ∥ line *v*. Find each angle measure.

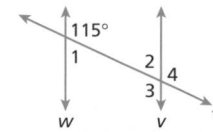

3. ∠1 **65°**   4. ∠2 **65°**   5. ∠3 **115°**

6. The second angle in a triangle is three times as large as the first. The measure of the third angle is 60° less than twice the measure of the first. Find the angle measures.   **40°; 120°; 20°**

Find the angle measures in each regular polygon.

7. 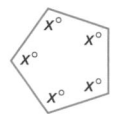   *x*° = 108°

8.   *y*° ≈ 128.6°

Graph the quadrilaterals with the given vertices. Give all of the names that apply to each.

9. (0, 1), (−2, 2), (−1, 0), (3, −2)

10. (4, 0), (0, 4), (−4, 0), (0, −4)

Write a congruence statement for each pair of polygons.

11.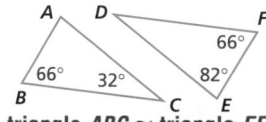
    triangle *ABC* ≅ triangle *EFD*

12.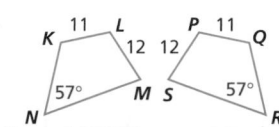
    quadrilateral *KLMN* ≅ quadrilateral *QPSR*

Draw the image of a triangle with vertices (0, 0), (3, 0), and (3, 4) after each transformation.

13. translation 3 units left

14. reflection across the *y*-axis

15. 180° rotation around (3, 0)

16. translation 2 units down

17. Complete the figure. The dashed line is the line of symmetry.
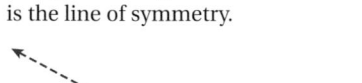

18. Create a tessellation with the given figure.
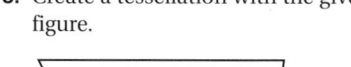

15.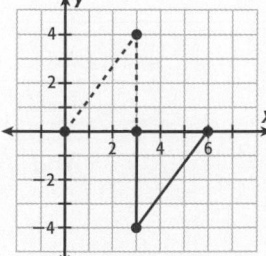

16.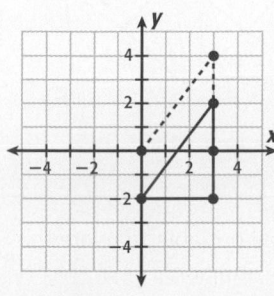

17.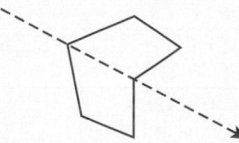

18. Possible answer:

Purpose: *To assess students' mastery of concepts and skills in Chapter 5*

Assessment Resources

Chapter 5 Tests (Levels A, B, C)
Assessment Resources...... pp. 57–62

 Test and Practice Generator CD-ROM

Additional assessment items in both multiple-choice and free-response format may be generated for any objective in Chapter 5.

You may use the Recording Sheet on Chapter 5 Resource Book p. 117 for problems 17–18.

Answers

9.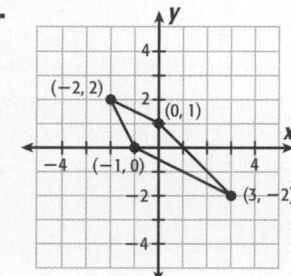

trapezoid

10.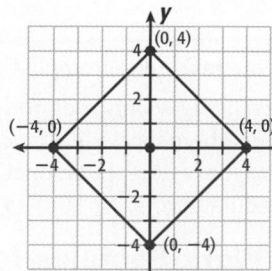

parallelogram, rhombus, rectangle, square

13.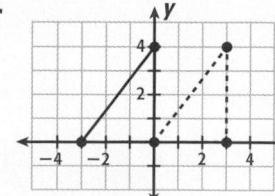

14.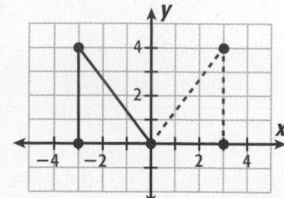

Purpose: *To assess students' under-standing of concepts in Chapter 5 and combined problem-solving skills*

Assessment Resources

Performance Assessment
Assessment Resources .... p. 126

Performance Assessment Teacher Support
Assessment Resources .... p. 125

Answers

1–3. See p. A7.

4. See Level 3 work sample below.

Scoring Rubric for Problem Solving Item 4

Level 3
Accomplishes the purpose of the task.

Student gives clear explanations, shows understanding of mathematical ideas and processes, and computes accurately.

Level 2
Purposes of the task not fully achieved.

Student demonstrates satisfactory but limited understanding of the mathematical ideas and processes.

Level 1
Purposes of the task not accomplished.

Student shows little evidence of under-standing the mathematical ideas and processes and makes computational and/or procedural errors.

Performance Assessment

Show What You Know

Create a portfolio of your work from this chapter. Complete this page and include it with your four best pieces of work from Chapter 5. Choose from your homework or lab assignments, mid-chapter quiz, or any journal entries you have done. Put them together using any design you want. Make your portfolio represent what you consider your best work.

Short Response

For 1–2, refer to the figure.

1. What is the measure of ∠*BAH*? Explain in words how you determined your answer.

2. Name all the pairs of supplementary angles. Explain how you know that you have named all the pairs.

3. Complete the table to show the number of diagonals for the polygons with the numbers of sides listed.

Number of Sides	3	4	5	6	7	*n*
Number of Diagonals	0					

Extended Problem Solving

Choose any strategy to solve each problem.

4. Four people are introduced to each other at a party, and they all shake hands.

   a. Explain in words how the diagram can be used to determine the number of handshakes exchanged at the party.

   b. How many handshakes are exchanged?

   c. Suppose that 6 people were introduced to each other at a party. Draw a diagram similar to the one shown that could be used to determine the number of handshakes exchanged.

Student Work Samples for Item 4

Level 3	Level 2	Level 1

Level 3

4a. The vertices represent the people, so the sum of the sides and diagonals is the number of handshakes.

b. There are 6 handshakes.

c.
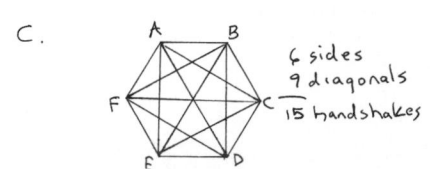
6 sides
9 diagonals
15 handshakes

The student related the drawing correctly to the problem, found 6 handshakes in part **b**, and used a hexagon with all diago-nals for handshakes between 6 people.

Level 2

4a. The letters are the people. The handshakes are AB, AD, AC, BC, BD, CD.

b. There are 6 altogether.

c.

The student related the drawing correctly to the problem and found 6 handshakes in part **b** but was not able to expand the model correctly.

Level 1

4a. The sides are used.

b. 4 of them

c. A

The student was not able to relate the drawing to the situation.

Cumulative Assessment, Chapters 1–5

1. Which of the following is $3.1415 \cdot 10^3$ written in standard notation? **C**

   Ⓐ 31,415,000
   Ⓒ 3141.5
   Ⓑ 31,415
   Ⓓ 314.5

2. Which number is equivalent to 5^{-2}? **G**

   Ⓕ $\frac{1}{10}$
   Ⓗ $\frac{1}{-10}$
   Ⓖ $\frac{1}{25}$
   Ⓙ $\frac{1}{-25}$

3. The cost of 3 sweatshirts is d dollars. At this rate, what is the cost in dollars of 30 sweatshirts? **C**

   Ⓐ $30d$
   Ⓒ $10d$
   Ⓑ $\frac{10d}{3}$
   Ⓓ $\frac{30}{d}$

TEST TAKING TIP!
When a letter is used more than once in a statement, it always has the same value.

4. If $a \cdot k = a$ for all values of a, what is the value of k? **J**

   Ⓕ $-a$
   Ⓗ 0
   Ⓖ -1
   Ⓙ 1

5. If $m^x \cdot m^7 = m^{28}$ and $\frac{m^y}{m^5} = m^3$ for all values of m, what is the value of $x + y$? **B**

   Ⓐ 19
   Ⓒ 12
   Ⓑ 29
   Ⓓ 31

6. Laura wants to tile her kitchen floor. Which of the following shapes would **not** cover her floor with a tessellation? **G**

   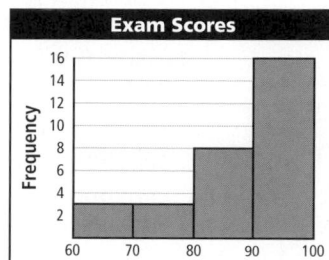

7. The solution of $9x = -72$ is __?__. **D**

   Ⓐ $x = 8$
   Ⓒ $x = -648$
   Ⓑ $x = 648$
   Ⓓ $x = -8$

8. In the histogram below, which interval contains the median score? **J**

   Exam Scores

   (histogram with Frequency on y-axis from 2 to 16, and x-axis labeled 60, 70, 80, 90, 100)

   Ⓕ 60–69
   Ⓗ 80–89
   Ⓖ 70–79
   Ⓙ 90–99

9. **SHORT RESPONSE** Triangle ABC, with vertices $A(2, 3)$, $B(4, -5)$, $C(6, 8)$, is reflected across the x-axis to trangle $A'B'C'$. On a coordinate grid, draw and label triangle ABC and triangle $A'B'C'$. Give the new coordinates for triangle $A'B'C'$.

10. **SHORT RESPONSE** Stephen bought 3 fish for his pond at a total cost of d dollars. At this rate what is the cost in dollars if he purchased 12 more fish? Show your work.

Purpose: *To provide review and practice for Chapters 1–5 and standardized tests*

Assessment Resources

Cumulative Tests (Levels A, B, C)
Assessment Resources . . . pp. 193–204

State-Specific Test Practice Online
KEYWORD: MP4 TestPrep

Test Prep Doctor

For item 9, remind students of the properties of exponents. Because $m^x \cdot m^7 = m^{28}$, $x + 7 = 28$. Therefore, $x = 21$. Because $\frac{m^y}{m^5} = m^3$, $y - 5 = 3$. Therefore, $y = 8$. So $x + y = 21 + 8 = 29$.

Answers

9. $A'(2, -3)$, $B'(4, 5)$, $C'(6, -8)$

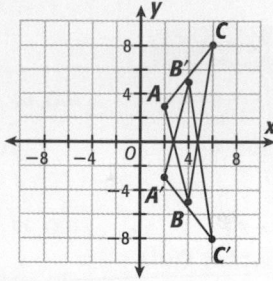

10. $4d$;
    If 3 fish cost d dollars, then 1 fish costs $\frac{d}{3}$ dollars. So, 12 fish cost $12 \cdot \frac{d}{3} = 4d$ dollars.

Perimeter, Area, and Volume

Pacing Guide for 45-Minute Classes

Chapter 6

DAY 64	DAY 65	DAY 66	DAY 67	DAY 68
Lesson 6-1	Lesson 6-2	Hands-On Lab 6A	Lesson 6-3	Lesson 6-4
DAY 69	**DAY 70**	**DAY 71**	**DAY 72**	**DAY 73**
Mid-Chapter Quiz Hands-On Lab 6B	Lesson 6-5	Lesson 6-6	Lesson 6-7	Lesson 6-8
DAY 74	**DAY 75**	**DAY 76**	**DAY 77**	**DAY 78**
Lesson 6-9	Lesson 6-10	Extension	Chapter 6 Review	Chapter 6 Assessment

Pacing Guide for 90-Minute Classes

Chapter 6

DAY 32	DAY 33	DAY 34	DAY 35	DAY 36
Chapter 5 Assessment Lesson 6-1	Lesson 6-2 Hands-On Lab 6A	Lesson 6-3 Lesson 6-4	Mid-Chapter Quiz Hands-On Lab 6B Lesson 6-5	Lesson 6-6 Lesson 6-7
DAY 37	**DAY 38**	**DAY 39**	**DAY 40**	
Lesson 6-8 Lesson 6-9	Lesson 6-10 Extension	Chapter 6 Review Lesson 7-1 Lesson 7-2	Chapter 6 Assessment Lesson 7-3	

COURSE 1

- Describe figures by using the terms of geometry.
- Name, measure, classify, estimate, and draw angles, and understand line and angle relationships.
- Identify, classify, and solve problems involving triangles, quadrilaterals, and regular and irregular polygons.

COURSE 2

- Find perimeter or circumference and area of rectangles, parallelograms, triangles, trapezoids, and circles.
- Use the Pythagorean Theorem to find the measure of a side of a right triangle.
- Find the surface area and volume of prisms, cylinders, pyramids, cones, and spheres.
- Find the surface area, volume, and weight of proportional three-dimensional figures.

PRE-ALGEBRA

- **Find perimeter or circumference and area of rectangles, parallelograms, triangles, trapezoids, and circles.**
- **Use the Pythagorean Theorem and its converse to solve problems.**
- **Draw and identify parts of three-dimensional figures.**
- **Find the volume and surface area of prisms, cylinders, pyramids, cones, and spheres.**

Across the Curriculum

LANGUAGE ARTS LINK

Math: Reading and Writing in the Content Area . . . . pp. 45–54

Focus on Problem Solving
Look Back . . . . SE p. 299
Journal . . . . TE, last page of each lesson
Write About It . . . . SE pp. 284, 293, 297, 306, 311, 315, 319, 323

SOCIAL STUDIES LINK

Architecture . . . . SE p. 315
Social Studies . . . . SE pp. 284, 293, 311, 313, 323

SCIENCE LINK

Life Science . . . . SE pp. 311, 321, 327
Earth Science . . . . SE p. 323
Physical Science . . . . SE p. 288

TE = *Teacher's Edition*   **SE** = *Student Edition*

Interdisciplinary

Bulletin Board

Physical Science

Boyle's law states that when the pressure applied to a gas increases, the volume of the gas decreases, and when the pressure decreases, the volume increases. If the volume of a gas is 2.0 liters at 1.1 atmospheres, would the pressure be more or less than 1.1 atmospheres if the volume changed to 2.5 liters? less than 1.1 atmospheres

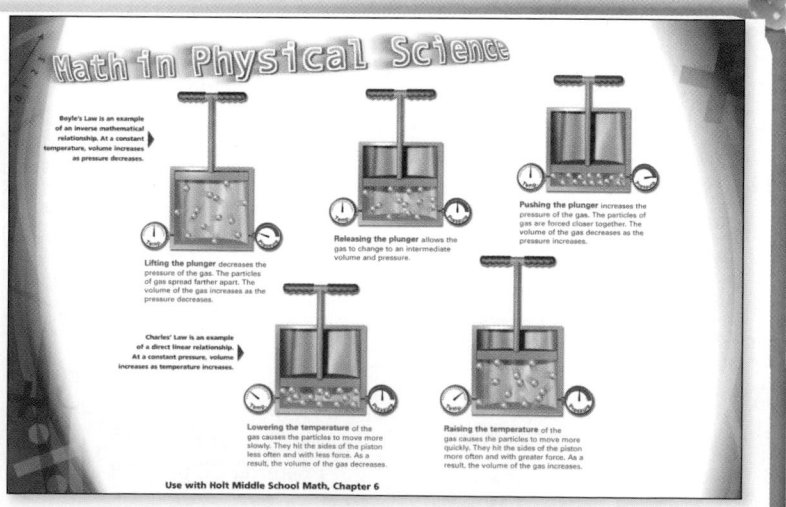

Interdisciplinary posters and worksheets are provided in your resource material.

Resource Options

Chapter 6 Resource Book

Student Resources

Practice (Levels A , B, C).... pp. 8–10, 18–20, 28–30, 38–40, 48–50, 58–60, 68–70, 78–80, 88–90, 98–100

Reteach....... pp. 11–12, 21–22, 31–32, 41, 51–52, 61–62 71–72, 81–82, 91–92, 101

Challenge........ pp. 13, 23, 33, 42, 53, 63, 73, 83, 93, 102

Problem Solving... pp. 14, 24, 34, 43, 54, 64, 74, 84, 94, 103

Puzzles, Twisters & Teasers....... pp. 15, 25, 35, 44, 55, 65, 75, 85, 95, 104

Recording Sheets....... pp. 3, 7, 17, 27, 37, 47, 57, 67, 77, 87, 97, 108, 112, 114–115, 118

Chapter Review........................... pp. 105–107

Reaching All Learners

English Language Learners

Success for English Language Learners .... pp. 89–108

Math: Reading and Writing in the Content Area .... pp. 45–54

Spanish Homework and Practice .... pp. 45–54

Spanish Interactive Study Guide .... pp. 45–54

Spanish Family Involvement Activities........ pp. 41–48

Multilingual Glossary

Individual Needs

Are You Ready? Intervention and Enrichment .. pp. 53–56, 169–172, 185–192, 415–416

Alternate Openers: Explorations .... pp. 45–54

Family Involvement Activities .... pp. 41–48

Interactive Problem Solving................. pp. 45–54

Interactive Study Guide .... pp. 45–54

Readiness Activities .... pp. 11–12

Math: Reading and Writing in the Content Area .... pp. 45–54

Challenge. CRB pp. 13, 23, 33, 42, 53, 63, 73, 83, 93, 102

Hands-On

Hands-On Lab Activities.................... pp. 39–50

Technology Lab Activities.................. pp. 35–43

Alternate Openers: Explorations .... pp. 45–54

Family Involvement Activities .... pp. 41–48

Applications and Connections

Consumer and Career Math.................. pp. 21–24

Interdisciplinary Posters .... Poster 6, TE p. 278B

Interdisciplinary Poster Worksheets........... pp. 16–18

Teacher and Parent Resources

Chapter Planning and Pacing Guide.................. p. 4

Section Planning Guides .... pp. 5, 45

Parent Letter .... pp. 1–2

Teaching Tools .... pp. 111–122

Teacher Support for Chapter Project .... p. 109

Transparencies .... pp. T1–T52

• Daily Transparencies

• Additional Examples Transparencies

• Teaching Transparencies

Transparencies

Alternate Openers: Explorations............... pp. 45–54

Exercise Answers Transparencies

Chapter 6 Resource Book .... pp. T1–T52

• Daily Transparencies

• Additional Examples Transparencies

• Teaching Transparencies

Technology

Teacher Resources

Lesson Presentations CD-ROM............. Chapter 6

Test and Practice Generator CD-ROM .... Chapter 6

One-Stop Planner CD-ROM .... Chapter 6

Student Resources

Are You Ready? Intervention CD-ROM
Skills 11, 40, 44, 45

⏎ internet connect

Homework Help Online	KEYWORD: MP4 HWHelp6
Math Tools Online	KEYWORD: MP4 Tools
Glossary Online	KEYWORD: MP4 Glossary
Chapter Project Online	KEYWORD: MP4 PSProject6
Chapter Opener Online	KEYWORD: MP4 Ch6

 KEYWORD: MP4 CNN6

SE = *Student Edition*   **TE** = *Teacher's Edition*   **AR** = *Assessment Resources*   **CRB** = *Chapter Resource Book*   **MK** = *Manipulatives Kit*

Assessment Options

Assessing Prior Knowledge

Determine whether students have the required prerequisite concepts and skills.

Are You Ready?. . . . . SE p. 279
Inventory Test. . . . . AR pp. 1–4

Test Preparation

Provide review and practice for chapter and standardized tests.

Standardized Test Prep. . . . . SE p. 339
Spiral Review with Test Prep. . . . . SE, last page of each lesson
Study Guide and Review. . . . . SE pp. 334–336
Test Prep Tool Kit

Technology

 ***Test and Practice Generator* CD-ROM**

⚡ internet connect

State-Specific Test Practice Online KEYWORD: MP4 TestPrep

Performance Assessment

Assess students' understanding of chapter concepts and combined problem-solving skills.

Performance Assessment. . . . . SE p. 338
   Includes scoring rubric in TE
Performance Assessment. . . . . AR p. 128
Performance Assessment Teacher Support. . . . . AR p. 127

Portfolio

Portfolio opportunities appear throughout the Student and Teacher's Editions.

Suggested work samples:

Problem Solving Project. . . . . TE p. 278
Performance Assessment. . . . . SE p. 338
Portfolio Guide. . . . . AR p. xxxvii
Journal. . . . . TE, last page of each lesson
Write About It. . SE pp. 284, 293, 297, 306, 311, 315, 319, 323

Daily Assessment

Obtain daily feedback on students' understanding of concepts.

Spiral Review and Test Prep. . . . . . SE, last page of each lesson

Also Available on Transparency In Chapter 6 Resource Book

Warm Up. . . . . TE, first page of each lesson
Problem of the Day. . . . . TE, first page of each lesson
Lesson Quiz. . . . . TE, last page of each lesson

Student Self-Assessment

Have students evaluate their own work.

Group Project Evaluation. AR p. xxxiv
Individual Group Member Evaluation. . . . . AR p. xxxv
Portfolio Guide. . . . . AR p. xxxvii
Journal. TE, last page of each lesson

Formal Assessment

Assess students' mastery of concepts and skills.

Section Quizzes. . . . . AR pp. 15, 16
Mid-Chapter Quiz. . . . . SE p. 298
Chapter Test. . . . . SE p. 337
Chapter Tests (Levels A, B, C). . . . . AR pp. 63–68
Cumulative Tests (Levels A, B, C). . . . . AR pp. 205–216
Standardized Test Prep
   Cumulative Assessment. . . . . SE p. 339
End-of-Year Test. . . . . AR pp. 313–316

Technology

 ***Test and Practice Generator* CD-ROM**

Make tests electronically. This software includes:

- Dynamic practice for Chapter 6
- Customizable tests
- Multiple-choice items for each objective
- Free-response items for each objective
- Teacher management system

SE = *Student Edition*   **TE** = *Teacher's Edition*   **AR** = *Assessment Resources*   **CRB** = *Chapter Resource Book*   **MK** = *Manipulatives Kit*

Chapter 6 Tests

Three levels (A,B,C) of tests are available for each chapter in the *Assessment Resources.*

LEVEL A

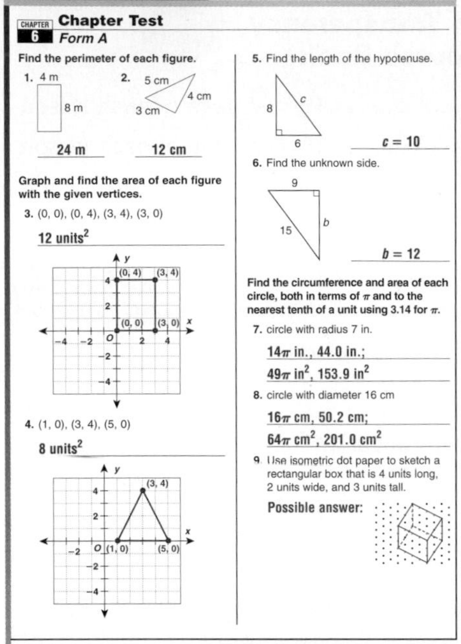

CHAPTER 6 Chapter Test
Form A

Find the perimeter of each figure.

1. 4 m
   8 m
   24 m

2. 5 cm
   3 cm   4 cm
   12 cm

5. Find the length of the hypotenuse.
   8   c
   6
   c = 10

6. Find the unknown side.
   9
   15   b
   b = 12

Graph and find the area of each figure with the given vertices.

3. (0, 0), (0, 4), (3, 4), (3, 0)
   12 units²

4. (1, 0), (3, 4), (5, 0)
   8 units²

Find the circumference and area of each circle, both in terms of π and to the nearest tenth of a unit using 3.14 for π.

7. circle with radius 7 in.
   14π in., 44.0 in.;
   49π in², 153.9 in²

8. circle with diameter 16 cm
   16π cm, 50.2 cm;
   64π cm², 201.0 cm²

9. Use isometric dot paper to sketch a rectangular box that is 4 units long, 2 units wide, and 3 units tall.
   Possible answer:

CHAPTER 6 Chapter Test
Form A, continued

10. Sketch a one-point perspective drawing of a cube.
    Possible answer:

Find the volume of each figure to the nearest tenth. Use 3.14 for π.

11. 8 cm, 3 cm
    226.1 cm³

12. 4 in., 2 in., 6 in.
    48.0 in³

13. 5 cm, 3 cm, 4 cm
    20.0 cm³

14. 6 in., 4 in.
    100.5 in³

Find the surface area of each figure to the nearest tenth. Use 3.14 for π.

15. the figure in Exercise 11
    207.2 cm²

16. the figure in Exercise 12
    88.0 in²

17. 7 ft, 9 ft, 9 ft
    207.0 ft²

18. 8 m, 3 m
    103.6 m²

19. Find the surface area of the sphere, both in terms of π and to the nearest tenth of a unit using 3.14 for π. 6 cm
    144π cm², 452.2 cm²

20. Find the volume of the sphere, both in terms of π and to the nearest tenth of a unit using 3.14 for π.
    288π cm³; 904.3 cm³

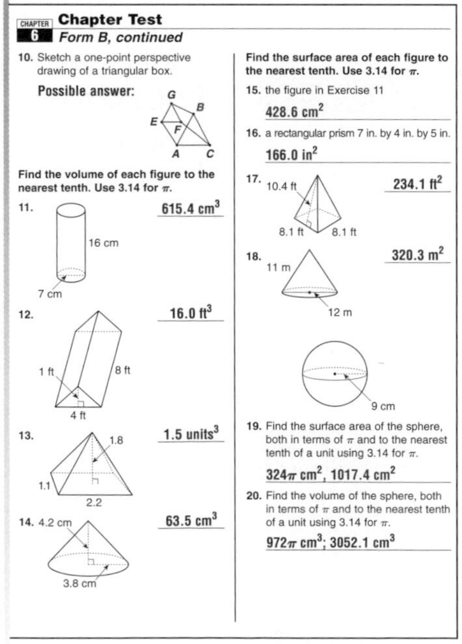

LEVEL B

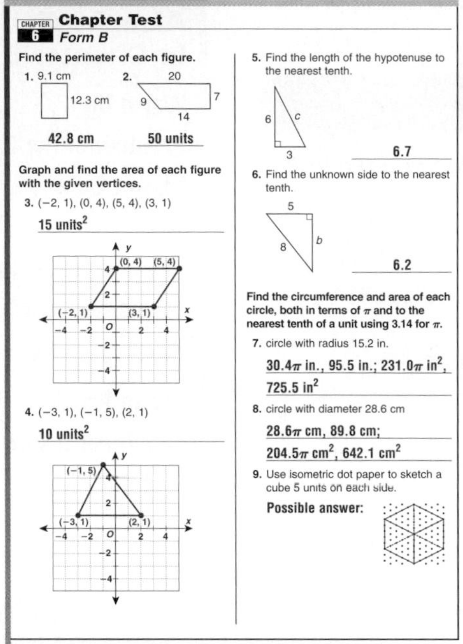

CHAPTER 6 Chapter Test
Form B

Find the perimeter of each figure.

1. 9.1 cm
   12.3 cm
   42.8 cm

2. 20
   9   7
   14
   50 units

5. Find the length of the hypotenuse to the nearest tenth.
   6   c
   3
   6.7

6. Find the unknown side to the nearest tenth.
   5
   8   b
   6.2

Graph and find the area of each figure with the given vertices.

3. (−2, 1), (0, 4), (5, 4), (3, 1)
   15 units²

4. (−3, 1), (−1, 5), (2, 1)
   10 units²

Find the circumference and area of each circle, both in terms of π and to the nearest tenth of a unit using 3.14 for π.

7. circle with radius 15.2 in.
   30.4π in., 95.5 in.; 231.0π in²,
   725.5 in²

8. circle with diameter 28.6 cm
   28.6π cm, 89.8 cm;
   204.5π cm², 642.1 cm²

9. Use isometric dot paper to sketch a cube 5 units on each side.
   Possible answer:

CHAPTER 6 Chapter Test
Form B, continued

10. Sketch a one-point perspective drawing of a triangular box.
    Possible answer:
    G   B
    E   F
    A   C

Find the volume of each figure to the nearest tenth. Use 3.14 for π.

11. 16 cm, 7 cm
    615.4 cm³

12. 1 ft, 8 ft, 4 ft
    16.0 ft³

13. 1.8, 1.1, 2.2
    1.5 units³

14. 4.2 cm, 3.8 cm
    63.5 cm³

Find the surface area of each figure to the nearest tenth. Use 3.14 for π.

15. the figure in Exercise 11
    428.6 cm²

16. a rectangular prism 7 in. by 4 in. by 5 in.
    166.0 in²

17. 10.4 ft, 8.1 ft, 8.1 ft
    234.1 ft²

18. 11 m, 12 m
    320.3 m²

19. Find the surface area of the sphere, both in terms of π and to the nearest tenth of a unit using 3.14 for π. 9 cm
    324π cm², 1017.4 cm²

20. Find the volume of the sphere, both in terms of π and to the nearest tenth of a unit using 3.14 for π.
    972π cm³; 3052.1 cm³

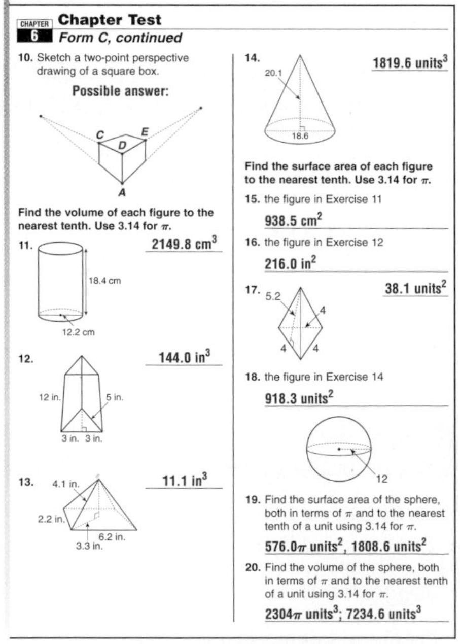

LEVEL C

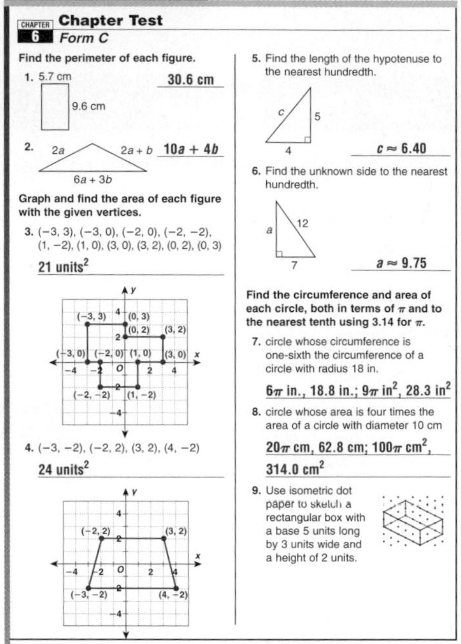

CHAPTER 6 Chapter Test
Form C

Find the perimeter of each figure.

1. 5.7 cm   30.6 cm
   9.6 cm

2. 2a   2a + b   **10a + 4b**
   6a + 3b

5. Find the length of the hypotenuse to the nearest hundredth.
   c   5
   4
   c ≈ 6.40

6. Find the unknown side to the nearest hundredth.
   a   12
   7
   a ≈ 9.75

Graph and find the area of each figure with the given vertices.

3. (−3, 3), (−3, 0), (−2, 0), (−2, −2), (1, −2), (1, 0), (3, 0), (3, 2), (0, 2), (0, 3)
   21 units²

4. (−3, −2), (−2, 2), (3, 2), (4, −2)
   24 units²

Find the circumference and area of each circle, both in terms of π and to the nearest tenth using 3.14 for π.

7. circle whose circumference is one-sixth the circumference of a circle with radius 18 in.
   6π in., 18.8 in.; 9π in², 28.3 in²

8. circle whose area is four times the area of a circle with diameter 10 cm
   20π cm, 62.8 cm; 100π cm²,
   314.0 cm²

9. Use isometric dot paper to sketch a rectangular box with a base 5 units long by 3 units wide and a height of 2 units.

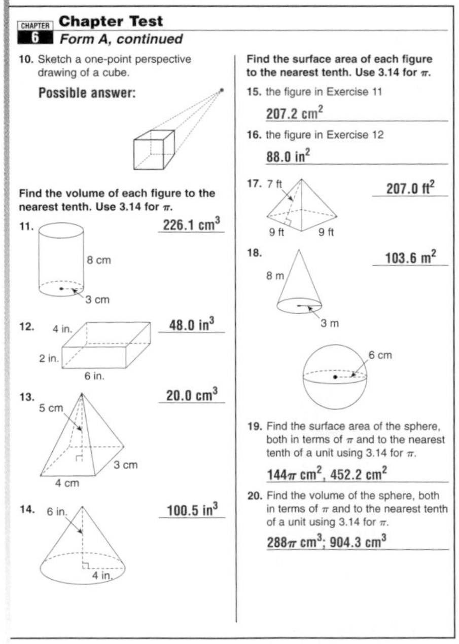

CHAPTER 6 Chapter Test
Form C, continued

10. Sketch a two-point perspective drawing of a square box.
    Possible answer:
    C   E
    D
    A

Find the volume of each figure to the nearest tenth. Use 3.14 for π.

11. 18.4 cm, 12.2 cm
    2149.8 cm³

12. 12 in., 5 in., 3 in., 3 in.
    144.0 in³

13. 4.1 in., 2.2 in., 6.2 in., 3.3 in.
    11.1 in³

14. 20.1, 18.6
    1819.6 units³

Find the surface area of each figure to the nearest tenth. Use 3.14 for π.

15. the figure in Exercise 11
    938.5 cm²

16. the figure in Exercise 12
    216.0 in²

17. 5.2, 4, 4
    38.1 units²

18. the figure in Exercise 14
    918.3 units²

19. Find the surface area of the sphere, both in terms of π and to the nearest tenth of a unit using 3.14 for π. 12
    576.0π units², 1808.6 units²

20. Find the volume of the sphere, both in terms of π and to the nearest tenth of a unit using 3.14 for π.
    2304π units³; 7234.6 units³

Test and Practice Generator
CD-ROM

Create and customize multiple versions of the same tests with corresponding answers for any chosen chapter objectives.

Chapter 6 State and Standardized Test Preparation

Test Taking Skill Builder and Standardized Test Practice
are provided for each chapter in the *Test Prep Tool Kit.*

TEST TAKING SKILL BUILDER

Test Taking Strategy   **Sketch a Picture or Diagram**
Chapter 6

Sketching a picture or diagram can help you find the solution to some problems.

Example 1 Multiple Choice What are the dimensions of the bottom of a box in which you are to pack 12 soup cans? The cans are to be packed so that the cans touch each other. There are to be 3 rows and 4 columns. Each can has a radius of 1.75 in.

A   14 in. × 10.5 in.   B   21 in. × 1.75 in.   C   7 in. × 5.25 in.   D   10 in. × 8 in.

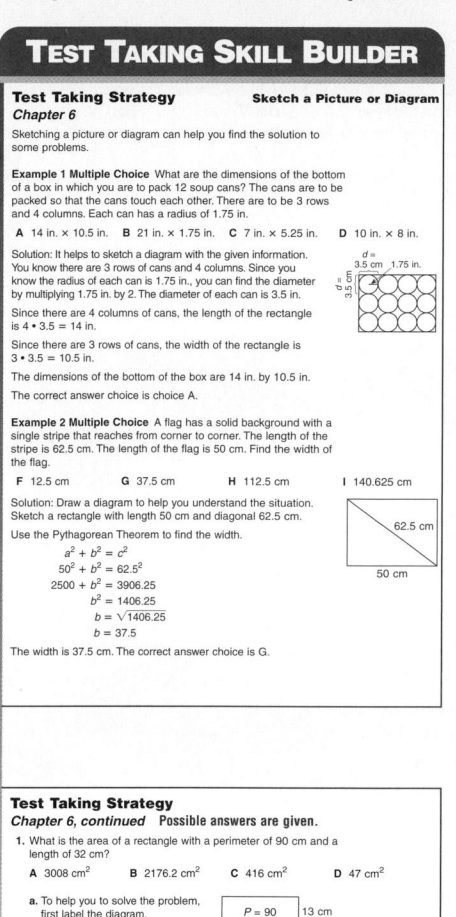

Solution: It helps to sketch a diagram with the given information. You know there are 3 rows of cans and 4 columns. Since you know the radius of each can is 1.75 in., you can find the diameter by multiplying 1.75 in. by 2. The diameter of each can is 3.5 in.

Since there are 4 columns of cans, the length of the rectangle is 4 • 3.5 = 14 in.

Since there are 3 rows of cans, the width of the rectangle is 3 • 3.5 = 10.5 in.

The dimensions of the bottom of the box are 14 in. by 10.5 in.

The correct answer choice is choice A.

Example 2 Multiple Choice A flag has a solid background with a single stripe that reaches from corner to corner. The length of the stripe is 62.5 cm. The length of the flag is 50 cm. Find the width of the flag.

F   12.5 cm   G   37.5 cm   H   112.5 cm   I   140.625 cm

Solution: Draw a diagram to help you understand the situation. Sketch a rectangle with length 50 and diagonal 62.5 cm.

Use the Pythagorean Theorem to find the width.

$$a^2 + b^2 = c^2$$
$$50^2 + b^2 = 62.5^2$$
$$2500 + b^2 = 3906.25$$
$$b^2 = 1406.25$$
$$b = \sqrt{1406.25}$$
$$b = 37.5$$

The width is 37.5 cm. The correct answer choice is G.

Test Taking Strategy
Chapter 6, continued   Possible answers are given.

1. What is the area of a rectangle with a perimeter of 90 cm and a length of 32 cm?

   A   3008 cm²   B   2176.2 cm²   C   416 cm²   D   47 cm²

   a. To help you solve the problem, first label the diagram.

     P = 90   13 cm   32 cm

   b. Explain how you determined the width.

     The length of the rectangle is 32 cm, so the opposite side is also 32 cm.

     90 − 64 = 26. The width must be 13 cm on both sides of the rectangle.

   c. How does using the diagram help you to solve the problem? What is the correct answer?

     Once I labeled the diagram with the length and the width, I then used

     the area formula, length times width, to determine the area of the

     rectangle. The correct answer is choice C.

2. What is the area of a figure with vertices (−6, 2), (−2, 2), and (−6, 5)?

   F   24 units²   G   12 units²   H   8 units²   I   6 units²

   a. Sketch the figure.

   b. What figure did you draw? **a right triangle**

   c. Explain how to use the diagram to find the height of the figure.

     Subtract the *y*-values of the vertical

     leg of the triangle; *h* = 3 units.

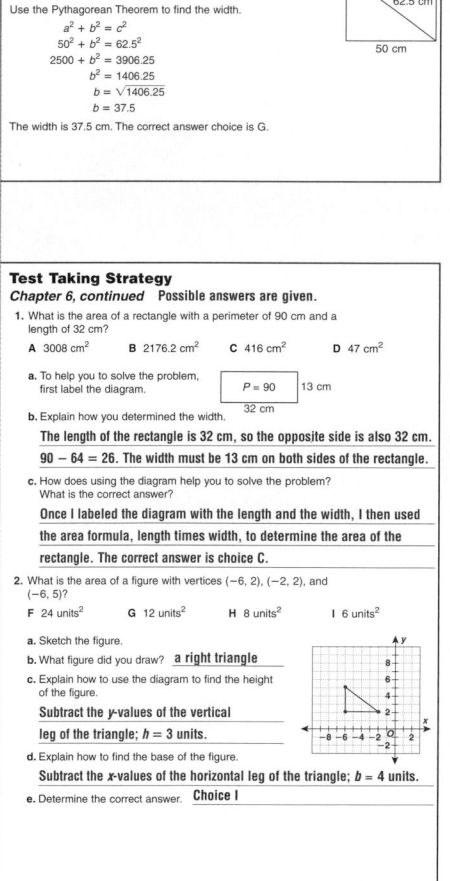

   d. Explain how to find the base of the figure.

     Subtract the *x*-values of the horizontal leg of the triangle; *b* = 4 units.

   e. Determine the correct answer. **Choice I**

STANDARDIZED TEST PRACTICE

Standardized Test Practice
Chapter 6

Select the best answer for Questions 1–6.

1. What is the area of the *unshaded* portion of the figure below?

   A   52 units   C   75 units
   B   62 units   D   92 units

2. A decorative candle is made in a pyramid shape with a square base. The height of the candle is 12.5 cm, and the sides of the base are 6 cm long. How much wax would be needed to make one candle?

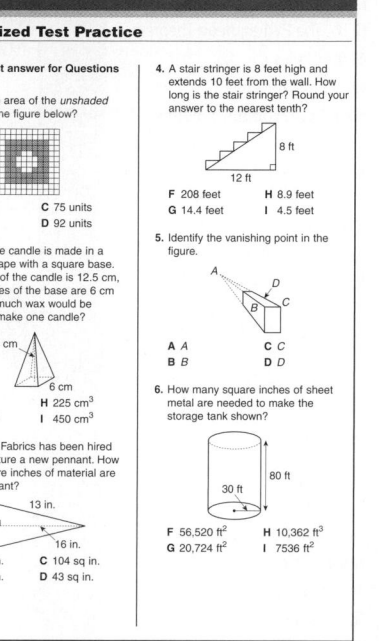

   F   75 cm³   H   225 cm³
   G   150 cm³   I   450 cm³

3. Westerville Fabrics has been hired to manufacture a new pennant. How many square inches of material are in the pennant?

   A   224 sq in.   C   104 sq in.
   B   112 sq in.   D   43 sq in.

4. A stair stringer is 8 feet high and extends 10 feet from the wall. How long is the stair stringer? Round your answer to the nearest tenth?

   F   208 feet   H   8.9 feet
   G   14.4 feet   I   4.5 feet

5. Identify the vanishing point in the figure.

   A   A   C   C
   B   B   D   D

6. How many square inches of sheet metal are needed to make the storage tank shown?

   F   56,520 ft²   H   10,362 ft³
   G   20,724 ft²   I   7536 ft²

Standardized Test Practice
Chapter 6, continued

Gridded Response
Solve the problems. Use the answer sheet to write and grid-in your answer.

7. A can of soup had a diameter of 3 inches and a height of 5 inches. What is the volume of the can in cubic inches?

8. The following pyramid has a square base. What is the surface area of the figure in square centimeters?

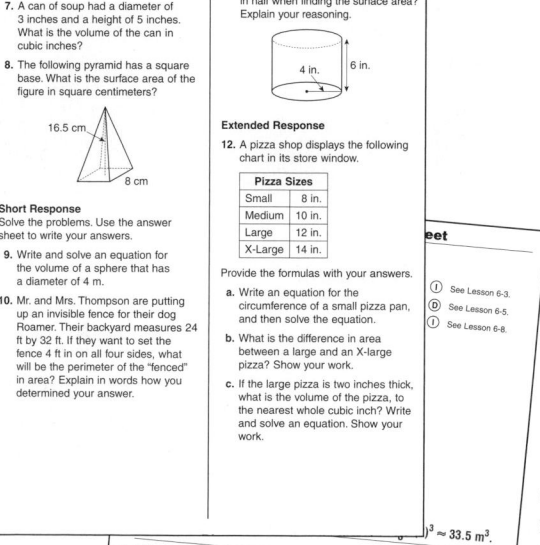

Short Response
Solve the problems. Use the answer sheet to write your answers.

9. Write and solve an equation for the volume of a sphere that has a diameter of 4 m.

10. Mr. and Mrs. Thompson are putting up an invisible fence for their dog Roamer. Their backyard measures 24 ft by 32 ft. If they want to set the fence 4 ft in on all four sides, what will be the perimeter of the "fenced" in area? Explain in words how you determined your answer.

11. Look at the following cylinder. Would decreasing the height by half have the same effect as decreasing the radius in half when finding the surface area? Explain your reasoning.

Extended Response

12. A pizza shop displays the following chart in its store window.

Pizza Sizes	
Small	8 in.
Medium	10 in.
Large	12 in.
X-Large	14 in.

Provide the formulas with your answers.

   a. Write an equation for the circumference of a small pizza pan, and then solve the equation.

   b. What is the difference in area between a large and an X-large pizza? Show your work.

   c. If the large pizza is two inches thick, what is the volume of the pizza, to the nearest whole cubic inch? Write and solve an equation. Show your work.

     ⓘ See Lesson 6-3.
     Ⓓ See Lesson 6-5.
     ⓘ See Lesson 6-8.

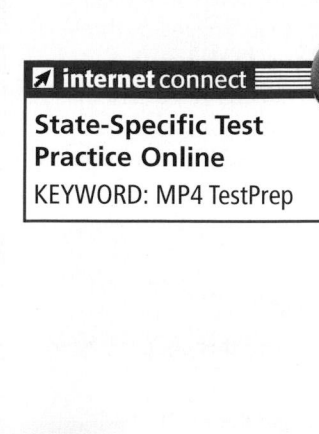

Test Prep Tool Kit

- Standardized Test Prep Workbook
- Countdown to Testing transparencies
- State Test Prep CD-ROM
- Standardized Test Prep Video

Customized answer sheets give students realistic practice for actual standardized tests.

---

³ ≈ 33.5 m³.

10. **If they set the fence 4 ft in on all sides, the new dimensions would be**   (See Lesson 6-10.)
20 ft by 28 ft. The perimeter would be 20(2) + 28(2) = 96 ft. Since
the yard is in the shape of the rectangle, I used the perimeter formula
2*l* + 2*w* to find the perimeter.

11. **They would not have the same effect. Decreasing the radius would**   (See Lesson 6-1.)
decrease the surface area by more than half.   (See Lesson 6-8.)

Extended Response
Write your answers for Problem 12 on the back of this paper.
See Lesson 6-4.

Perimeter, Area, and Volume

Why Learn This?

Tell students that in many careers it is essential to have the ability to recognize two-dimensional (flat) images as three-dimensional objects. Surgeons use flat on-screen images to determine the shapes of masses before operating on them. Also, builders use two-dimensional plans consisting of front and side views of buildings.

Using Data

To begin the study of this chapter, have students:

- Identify each shape in the table.
  A: cone; B: cylinder; C: rectangular prism (box)

- Draw the front, top, and side views of their pencil. Possible answers:

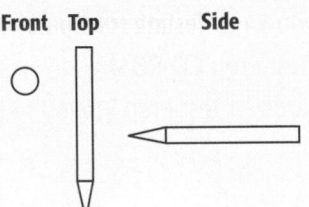

Front   Top   Side

Perimeter, Area, and Volume

internet connect

go.hrw.com

Chapter Opener Online
go.hrw.com
KEYWORD: MP4 Ch6

Mystery Solid	Front View	Side View	Top View
A	△	△	○
B	□	□	○
C	□	□	□

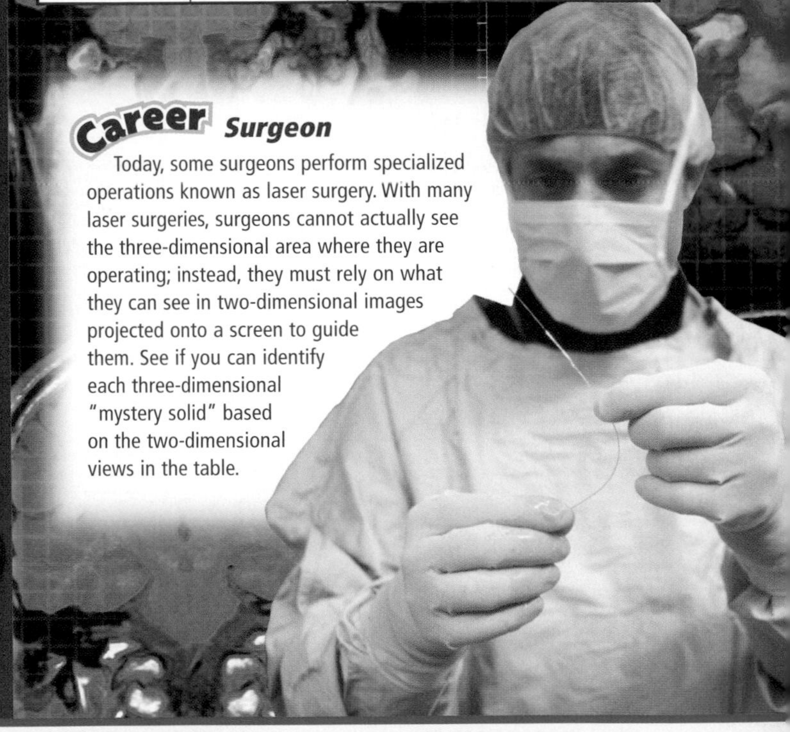

Career *Surgeon*

Today, some surgeons perform specialized operations known as laser surgery. With many laser surgeries, surgeons cannot actually see the three-dimensional area where they are operating; instead, they must rely on what they can see in two-dimensional images projected onto a screen to guide them. See if you can identify each three-dimensional "mystery solid" based on the two-dimensional views in the table.

PROBLEM SOLVING

Problem Solving Project

Physical Science and Life Science Connection

Purpose: To solve problems by combining plane figures

Materials: The Mystery Solids worksheet, geometrical and common-item mystery solids, overhead projector

internet connect

go.hrw.com

Chapter Project Online: *go.hrw.com*
KEYWORD: MP4 PSProject6

Understand, Plan, Solve, and Look Back

Have students:

✔ Complete the Mystery Solids worksheet to become familiar with the characteristics of geometric solids.

✔ Play a game by taking turns showing parts of a three-dimensional object on the overhead and having classmates see how much of the object must be revealed before they can identify the object.

✔ Do research to discover information on laparoscopic surgery. Interview people who have had laparoscopic surgery or a surgeon who performs the surgery.

✔ Check students' work.

ARE YOU READY?

Choose the best term from the list to complete each sentence.

1. A(n) __fraction__ is a number that represents a part of a whole.

2. A(n) __decimal__ is another way of writing a fraction.

3. To multiply 7 by the fraction $\frac{2}{3}$, multiply 7 by the __?__ of the fraction and then divide the result by the __?__ of the fraction. __numerator; denominator__

4. To round 7.836 to the nearest tenth, look at the digit in the __?__ place. __hundredths__

decimal
denominator
fraction
numerator
tenths
hundredths

Complete these exercises to review skills you will need for this chapter.

✔ Square and Cube Numbers

Evaluate.

5. 16^2 **256** 6. 9^3 **729** 7. $(4.1)^2$ **16.81** 8. $(0.5)^3$ **0.125**

9. $\left(\frac{1}{4}\right)^2$ **$\frac{1}{16}$** 10. $\left(\frac{2}{5}\right)^2$ **$\frac{4}{25}$** 11. $\left(\frac{1}{2}\right)^3$ **$\frac{1}{8}$** 12. $\left(\frac{2}{3}\right)^3$ **$\frac{8}{27}$**

✔ Multiply with Fractions

Multiply.

13. $\frac{1}{2}(8)(10)$ **40** 14. $\frac{1}{2}(3)(5)$ **$7\frac{1}{2}$** 15. $\frac{1}{3}(9)(12)$ **36** 16. $\frac{1}{3}(4)(11)$ **$14\frac{2}{3}$**

17. $\frac{1}{2}(8^2)16$ **512** 18. $\frac{1}{2}(5^2)24$ **300** 19. $\frac{1}{2}(6)(3+9)$ **36** 20. $\frac{1}{2}(5)(7+4)$ **$27\frac{1}{2}$**

✔ Multiply with Decimals

Multiply. Write each answer to the nearest tenth.

21. $2(3.14)(12)$ **75.4** 22. $3.14(5^2)$ **78.5** 23. $3.14(4^2)(7)$ **351.7** 24. $3.14(2.3)^2(5)$ **83.1**

✔ Multiply with Fractions and Decimals

Multiply. Write each answer to the nearest tenth.

25. $\frac{1}{3}(3.14)(5^2)(7)$ **183.2** 26. $\frac{1}{3}(3.14)(5^3)$ **130.8**

27. $\frac{1}{3}(3.14)(3.2)^2(2)$ **21.4** 28. $\frac{4}{3}(3.14)(2.7)^3$ **82.4**

29. $\frac{1}{5}\left(\frac{22}{7}\right)(4^2)(5)$ **50.3** 30. $\frac{4}{11}\left(\frac{22}{7}\right)(3.2^3)$ **37.4**

31. $\frac{1}{2}\left(\frac{22}{7}\right)(1.7)^2(4)$ **18.2** 32. $\frac{7}{11}\left(\frac{22}{7}\right)(9.5)^3$ **1714.8**

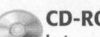

Perimeter and Area

One-Minute Section Planner

Lesson	Materials	Resources
Lesson 6-1 Perimeter and Area of Rectangles and Parallelograms **NCTM:** Geometry, Measurement, Reasoning and Proof, Communication, Connections, Representation **NAEP:** Measurement 1h ☑ SAT-9 ☑ SAT-10 ☑ ITBS ☑ CTBS ☑ MAT ☑ CAT	**Required** Graph paper *(CRB, p. 111)* **Optional** Teaching Transparency T2 *(CRB)* Recording Sheet for Reaching All Learners *(CRB, p. 112)*	• *Chapter 6 Resource Book,* pp. 6–15 • Daily Transparency T1, CRB • Additional Examples Transparencies T3–T6, CRB • *Alternate Openers: Explorations,* p. 45
Lesson 6-2 Perimeter and Area of Triangles and Trapezoids **NCTM:** Geometry, Measurement, Reasoning and Proof, Communication, Connections, Representation **NAEP:** Measurement 1h ☑ SAT-9 ☑ SAT-10 ☑ ITBS ☑ CTBS ☑ MAT ☑ CAT	**Required** Graph paper *(CRB, p. 111)* **Optional** Paper shapes *(CRB, p. 113)* Teaching Transparencies T8–T9 *(CRB)* Recording Sheet for Reaching All Learners *(CRB, p. 114)*	• *Chapter 6 Resource Book,* pp. 16–25 • Daily Transparency T7, CRB • Additional Examples Transparencies T10–T12, CRB • *Alternate Openers: Explorations,* p. 46
Hands-On Lab 6A Explore Right Triangles **NCTM:** Algebra, Geometry, Reasoning and Proof, Representation **NAEP:** Geometry 2d ☐ SAT-9 ☐ SAT-10 ☐ ITBS ☐ CTBS ☐ MAT ☐ CAT	**Required** Scissors Rulers *(MK, CRB p. 117)*	• *Hands-On Lab Activities,* pp. 42, 95
Lesson 6-3 The Pythagorean Theorem **NCTM:** Geometry, Reasoning and Proof, Communication, Connections, Representation **NAEP:** Geometry 3d ☐ SAT-9 ☐ SAT-10 ☐ ITBS ☑ CTBS ☑ MAT ☐ CAT	**Required** Graph paper *(CRB, p. 111)* **Optional** 2 ft long piece of yarn Teaching Transparency T14 *(CRB)* Scissors Recording Sheet for Reaching All Learners *(CRB, p. 115)*	• *Chapter 6 Resource Book,* pp. 26–35 • Daily Transparency T13, CRB • Additional Examples Transparencies T15–T16, CRB • *Alternate Openers: Explorations,* p. 47
Lesson 6-4 Circles **NCTM:** Geometry, Measurement, Reasoning and Proof, Communication, Connections, Representation **NAEP:** Measurement 1h ☑ SAT-9 ☑ SAT-10 ☐ ITBS ☑ CTBS ☑ MAT ☑ CAT	**Required** Graph paper *(CRB, p. 111)* **Optional** Teaching Transparency T18 *(CRB)* Calculators Cans Tape measures *(MK)*	• *Chapter 6 Resource Book,* pp. 36–44 • Daily Transparency T17, CRB • Additional Examples Transparencies T19–T21, CRB • *Alternate Openers: Explorations,* p. 48
Section 6A Assessment		• Mid-Chapter Quiz, SE p. 298 • Section 6A Quiz, AR p. 15 • *Test and Practice Generator* CD-ROM

SAT = *Stanford Achievement Tests* **ITBS** = *Iowa Test of Basic Skills* **CTBS** = *Comprehensive Test of Basic Skills/Terra Nova*
MAT = *Metropolitan Achievement Test* **CAT** = *California Achievement Test*

NCTM = Complete standards can be found on pages T29–T35. **NAEP** = Complete standards can be found on pages A54–A58.

SE = *Student Edition* **TE** = *Teacher's Edition* **AR** = *Assessment Resources* **CRB** = *Chapter Resource Book* **MK** = *Manipulatives Kit*

Section Overview

Perimeter and Area of Polygons
Lessons 6-1, 6-2

Why? It is necessary to understand perimeter and area for many real-world applications.

Perimeter is the distance around a figure.	Area is the number of square units inside a figure.

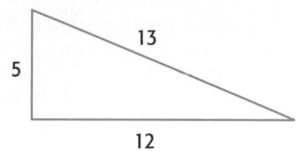

$P = 5 + 12 + 13 = 30$ units

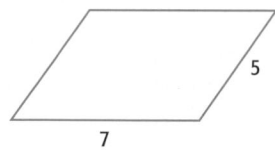

$P = 5 + 5 + 7 + 7 = 24$ units

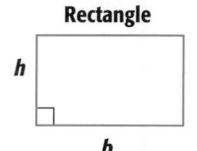

Rectangle   **Parallelogram**

$A = bh$ $A = bh$

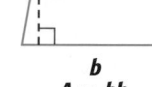
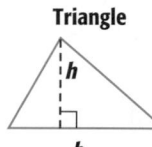

Triangle   **Trapezoid**

$A = \frac{1}{2}bh$ $A = \frac{1}{2}h(b_1 + b_2)$

The Pythagorean Theorem
Hands-On Lab 6A, Lesson 6-3

Why? You can use the Pythagorean Theorem to find information about triangles, such as the area of a triangle whose height is unknown.

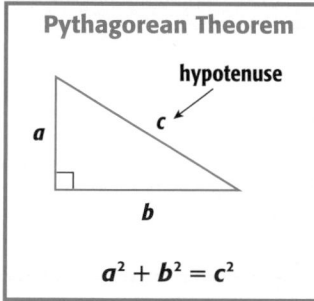

Pythagorean Theorem

hypotenuse

$a^2 + b^2 = c^2$

Find the area of the triangle.

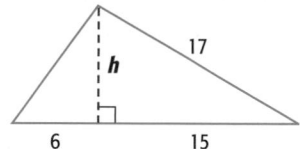

First, use the right triangle to find h.

$$h^2 + 15^2 = 17^2$$
$$h^2 + 225 = 289$$
$$h^2 = 64$$
$$h = 8$$

Then, use the area formula to find the area of the large triangle.

$$A = \frac{1}{2}bh$$
$$A = \frac{1}{2}(21)(8)$$
$$A = 84 \text{ units}^2$$

Area and Circumference of Circles
Lesson 6-4

Why? You can find the area and circumference of a circle if you know its radius or diameter.

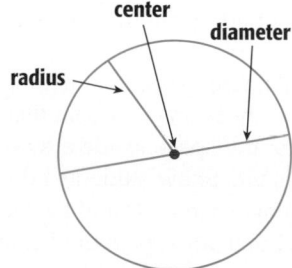

center

diameter

radius

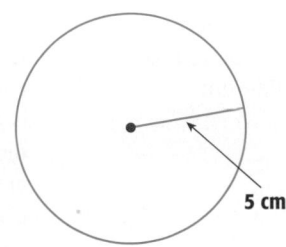

5 cm

Circumference	Area
$C = \pi d$, or $2\pi r$	$A = \pi r^2$
$= \pi(10)$	$= \pi(5)^2$
$\approx 3.14(10)$	$\approx 3.14(25)$
≈ 31.4 cm	≈ 78.5 cm^2

Pacing: Traditional 1 day
Block $\frac{1}{2}$ day

Objective: Students find perimeter and area of rectangles and parallelograms.

Learn to find the perimeter and area of rectangles and parallelograms.

In inlaid woodworking, artists use geometry to create a variety of beautiful patterns. One design can have thousands of pieces made from many different kinds of wood. In a design made entirely of parallelograms, the total area of the design is the sum of the areas of the parallelograms in the design.

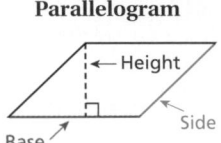

Vocabulary
perimeter
area

Any side of a rectangle or parallelogram can be chosen as the base. The height is measured along a line perpendicular to the base.

Rectangle   **Parallelogram**

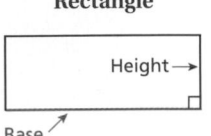

Perimeter is the distance around the outside of a figure. To find the perimeter of a figure, add the lengths of all its sides.

Warm Up

Graph the line segment for each set of ordered pairs. Then find the length of the line segment.

1. $(-7, 0), (0, 0)$ 7 units

2. $(0, 3), (0, 6)$ 3 units

3. $(-4, -2), (1, -2)$ 5 units

4. $(-5, 4), (-5, -2)$ 6 units

Problem of the Day

Six pennies are placed around a seventh so that there are no gaps. What figure is formed by connecting the centers of the six outer pennies?
a regular hexagon
Available on Daily Transparency in CRB

EXAMPLE **1** **Finding the Perimeter of Rectangles and Parallelograms**

Find the perimeter of each figure.

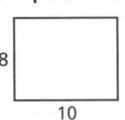

A   8   10

$P = 10 + 10 + 8 + 8$ *Add all side lengths.*
$= 36$ units

or $P = 2b + 2h$ *Perimeter of rectangle*
    $= 2(10) + 2(8)$ *Substitute 10 for b and 8 for h.*
    $= 20 + 16 = 36$ units

Helpful Hint

The formula for the perimeter of a rectangle can be written as $P = 2b + 2h$, where b is the length of the base and h is the height.

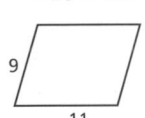

B   9   11

$P = 9 + 9 + 11 + 11$ *Add all side lengths.*
$= 40$ units

Math Fact !!!

An *altitude* is a segment that is perpendicular to one or both bases of a figure and shows the height of the figure. Often the term *altitude* is used interchangeably with height.

1 Introduce

Alternate Opener

EXPLORATION

6-1 Perimeter and Area of Rectangles and Parallelograms

Rectangle   Parallelogram

1. Compare the two figures. How are they alike and how are they different?

2. Find the perimeter of each figure. Use a graph paper ruler to measure the two slanted sides of the parallelogram.

3. Count the number of squares inside each figure to determine its area in square units. How can you count the incomplete squares on each slanted side of the parallelogram?

4. Draw two different parallelograms with the same area on graph paper. Are the perimeters the same?

Think and Discuss

5. **Explain** how you can cut a parallelogram into two pieces and reassemble the pieces to form a rectangle.

6. **Explain** how to construct rectangles and parallelograms that have the same area.

Motivate

Ask students how many measurements it would take to find the length of a wall in the classroom. one Explain that length is a *one-dimensional* measure because it requires only one measurement. Then ask them how many measurements it would take to find the size (area) of the floor. two Explain that area is *two-dimensional* because it requires two different measurements.

Exploration worksheet and answers on Chapter 6 Resource Book pp. 7 and 123

2 Teach

Lesson Presentation

Guided Instruction

In this lesson, students learn to find perimeter and area of rectangles and parallelograms. Explain to students that *perimeter* is the distance around the outside of a figure. Explain that the perimeter of a figure can be found by adding up all of its side lengths, or for some special figures, by using a formula. Emphasize that perimeter is measured in linear units. Explain that area is a measure of the space inside a two-dimensional figure. Show students how the area of a rectangle can be found by counting the number of unit squares in the figure (Teaching Transparency T2, CRB). For this reason, area is measured in square units.

Area is the number of square units in a figure. A parallelogram can be cut and the cut piece shifted to form a rectangle with the same base length and height as the original parallelogram. So a parallelogram has the same area as a rectangle with the same base length and height.

AREA OF RECTANGLES AND PARALLELOGRAMS

Words	Numbers		Formula
The area A of a rectangle or parallelogram is the base length b times the height h.	5 3 $5 \cdot 3 = 15$ units^2 **Rectangle**	5 3 $5 \cdot 3 = 15$ units^2 **Parallelogram**	$A = bh$

EXAMPLE 2 Using a Graph to Find Area

Graph each figure with the given vertices. Then find the area of each figure.

A $(-2, -1), (2, -1), (2, 2), (-2, 2)$

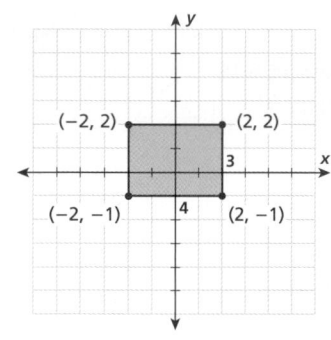

$A = bh$ *Area of rectangle*
$= 4 \cdot 3$ *Substitute 4 for b and 3 for h.*
$= 12$ units^2

Reaching All Learners
Through Concrete Manipulatives

Give each student or group of students a sheet of graph paper and a recording sheet (Chapter 6 Resource Book pp. 111–112). Have students use graph paper to find as many different rectangles as possible that have a perimeter of 16 units and then calculate the area of each. Then have them find as many different rectangles as possible that have an area of 24 units^2 and calculate the perimeter of each. Finally, have them describe any patterns they notice in the tables.

Additional Examples

Example 1

Find the perimeter of each figure.

A.

5

14

$P = 38$ units

B.

16

20

$P = 72$ units

Example 2

Graph each figure with the given vertices. Then find the area of each figure.

A. $(-1, -2), (2, -2), (2, 3), (-1, 3)$

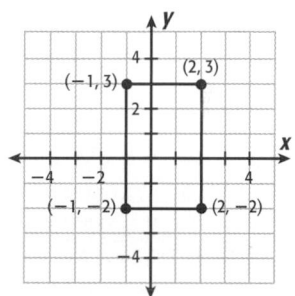

$A = 15$ units^2

B. $(0, 0), (5, 0), (6, 4), (1, 4)$

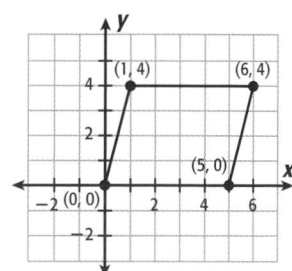

$A = 20$ units^2

Example 3

Find the perimeter and area of the figure.

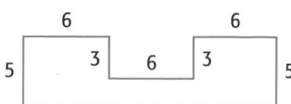

$P = 52$ units; $A = 72$ units^2

Graph each figure with the given vertices. Then find the area of the figure.

B $(-4, 0), (2, 0), (4, 3), (-2, 3)$

$A = bh$ *Area of parallelogram*

$= 6 \cdot 3$ *Substitute 6 for b and 3 for h.*

$= 18 \text{ units}^2$

EXAMPLE 3 **Finding Area and Perimeter of a Composite Figure**

Find the perimeter and area of the figure.

The length of the side that is not labeled is the same as the length of the opposite side, 3 units.

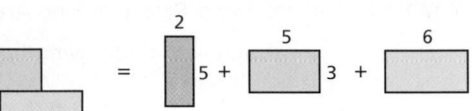

$P = 5 + 2 + 2 + 5 + 3 + 3 + 3 + 6 + 3 + 4$

$= 36 \text{ units}$

$A = 5 \cdot 2 + 5 \cdot 3 + 6 \cdot 3$ *Add the areas together.*

$= 10 + 15 + 18$

$= 43 \text{ units}^2$

Think and Discuss

1. Compare the area of a rectangle with base b and height h with the area of a rectangle with base $2b$ and height $2h$.

2. Express the formulas for the area and perimeter of a square using s for the length of a side.

③ Close

Summarize

Review the important differences between perimeter and area. Remind students that perimeter is the distance around a figure and is measured in linear units. Remind them that area is the space inside a figure and is measured in square units. Show students some concrete examples, such as a photograph. Ask students which measure would help you find how much wood you would need to frame the photograph. perimeter Then ask which would tell you how much glass you would need. area As a challenge, you may want to ask them to estimate both measures.

Answers to Think and Discuss

1. Possible answer: The area of the rectangle with base $2b$ and height $2h$ is four times the area of the rectangle with base b and height h.

2. Because the side lengths of a square are equal, both the base and the height can be represented by the same variable, s. Substituting s into both formulas for b and h gives $A = s^2$ and $P = 4s$.

FOR EXTRA PRACTICE
see page 742

⚡ internet connect
Homework Help Online
go.hrw.com   Keyword: MP4 6-1

6-1 PRACTICE & ASSESS

GUIDED PRACTICE

See Example **1** Find the perimeter of each figure.

1.
3
7
20 units

2.
8
10
36 units

3.
3.2x
6.5x
19.4x units

See Example **2** Graph each figure with the given vertices. Then find the area of each figure.

4. $(-3, 2), (0, 2), (3, -3), (0, -3)$
15 units²

5. $(-4, 0), (-4, 4), (3, 4), (3, 0)$
28 units²

6. $(-4, 1), (4, 1), (3, -3), (-5, -3)$
32 units²

7. $(-2, 3), (0, 3), (0, -4), (-2, -4)$
14 units²

See Example **3** **8.** Find the perimeter
and area of the figure.
44 units; 53 units²

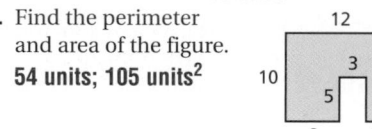
10
4   4
4   2   4
3   5   2
7

INDEPENDENT PRACTICE

See Example **1** Find the perimeter of each figure.

9.
11
6
34 units

10.
1.0
0.7
3.4 units

11.
5x
8x
26x units

See Example **2** Graph each figure with the given vertices. Then find the area of each figure.

12. $(-5, -1), (2, -1), (2, -5), (-5, -5)$
28 units²

13. $(0, 3), (6, 3), (3, -1), (-3, -1)$
24 units²

14. $(3, 5), (5, 3), (-3, 3), (-5, 5)$
16 units²

15. $(2, 5), (5, 5), (5, -1), (2, -1)$
18 units²

See Example **3** **16.** Find the perimeter
and area of the figure.
54 units; 105 units²

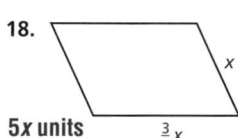
12
3
10   10
5   5
6   3

PRACTICE AND PROBLEM SOLVING

Find the perimeter of each figure.

17.
9
23
64 units

18.
x
5x units
$\frac{3}{2}x$

Assignment Guide

If you finished Example **1** assign:
  Core 1–3, 9–11, 30–34
  Enriched 1–3, 9–11, 30–34

If you finished Example **2** assign:
  Core 1–7, 9–15, 24–26, 30–34
  Enriched 1–7 odd, 9–15, 22–27,
    30–34

If you finished Example **3** assign:
  Core 1–16, 17, 20, 24–26,
    30–34
  Enriched 9–34

Graph paper for Exercises 4–7 and
12–15 is provided on Chapter 6
Resource Book p. 111.

Answers

4.
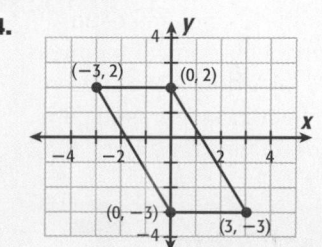
(-3, 2)   (0, 2)
(0, -3)   (3, -3)

5.
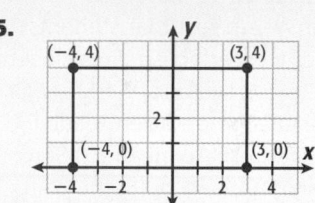
(-4, 4)   (3, 4)
(-4, 0)   (3, 0)

6–7, 12–15. Complete answers on
p. A7

RETEACH 6-1

Perimeter = distance around a figure.
To find the perimeter of a figure, add the lengths of all its sides.

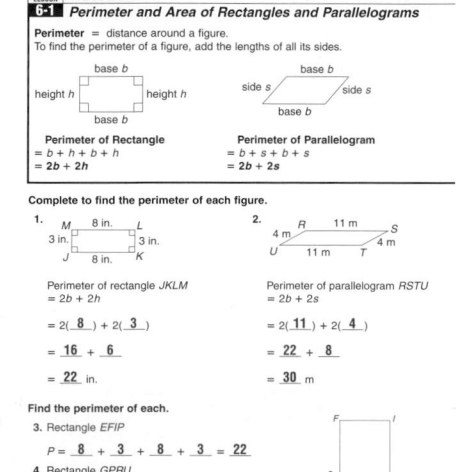

base b
height h   height h
base b

base b
side s   side s
base b

Perimeter of Rectangle
= b + h + b + h
= **2b + 2h**

Perimeter of Parallelogram
= b + s + b + s
= **2b + 2s**

Complete to find the perimeter of each figure.

1.
M   8 in.   L
3 in.   3 in.
J   8 in.   K

Perimeter of rectangle JKLM
= 2b + 2h
= 2(**8**) + 2(**3**)
= **16** + **6**
= **22** in.

2.
4 m   R   11 m   S
U   11 m   T   4 m

Perimeter of parallelogram RSTU
= 2b + 2s
= 2(**11**) + 2(**4**)
= **22** + **8**
= **30** m

Find the perimeter of each.

3. Rectangle EFIP
P = **8** + **3** + **8** + **3** = **22**

4. Rectangle GPRU
P = **4** + **3** + **4** + **3** = **14**

5. The combined rectangles EFIP and GPRU as
shown in the figure.
P = **8** + **7** + **3** + **4** + **5** + **3** = **30**

F   I
8   G   U
3
E   3   P   3   R
7

PRACTICE 6-1

Find the perimeter of each figure.

1.
M   27   A
16
H   T
86 units

2.
R   S
11
W   20   T
62 units

3. S 13.9 T
26.7
Y   X
81.2 units

Graph each figure with the given vertices. Then find the area of
each figure.

4. $(-3, 4), (3, 4), (3, -4), (-3, -4)$
48 units²

5. $(-1, 3), (2, 3), (-1, -4), (-4, -4)$
21 units²

6. Sloppi and Sons Painting Co. charges its
customers $1.50 per square foot. How much
would Sloppi and Sons charge to paint the
rooms of this house if the walls in each room
are 9 ft high?
$3024

14 ft   12 ft   15 ft
10 ft   16 ft
9 ft

Math Background

The concepts in this lesson are
important for more advanced studies
in geometry. Students should begin to
develop an understanding of dimensions and appropriate units. An object
that has one directional measure
(e.g., length) is one-dimensional, and
its length is measured in linear units.
Similarly, an object with two different
directional measures (e.g., length and
width) is two-dimensional, and its area
is measured in square units. It is also
important that students be able to
apply basic formulas, such as those
presented in this lesson.

Answers

28. Possible answer: The figures have the same area, but the figure with the cutout has a larger perimeter because it has additional sides.

30.

31. $-6\ -4\ -2\ \ 0\ \ 2\ \ 4$

32. $-8\ -6\ -4\ -2\ \ 0\ \ 2$

33. $-5\ \ \ \ \ 0\ \ \ \ 3\ \ 5$

Journal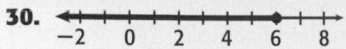

Ask students to write about a real-world situation in which they would need to find the perimeter and area of an object. Have them estimate the measures of the object they have chosen. Examples might include framing a picture or decorating a wall or floor.

Test Prep Doctor

For Exercise 34, suggest that students evaluate the two perfect squares closest to 46. Since $\sqrt{36} = 6$ and $\sqrt{49} = 7$, the answer must be between 6 and 7, so choices **A** and **D** are incorrect. Since 46 is closer to 49 than to 36, the correct answer should be closer to 7 than to 6. Thus, choice **B** (6.78) is correct.

Lesson Quiz

1. Find the perimeter of the figure.
44 ft

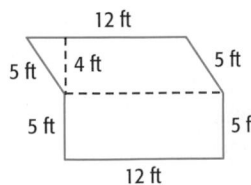

2. Find the area of the figure.
108 ft²

Graph the figure with the given vertices and find its area.

Complete answers on p. A24

3. $(-4, 2), (6, 2), (6, -3), (-4, -3)$
50 units²

4. $(4, -2), (-2, -2), (-3, 5), (3, 5)$
42 units²

Available on Daily Transparency in CRB

Find the perimeter and area of each figure.

19. 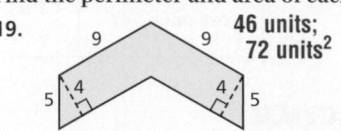 **46 units; 72 units²**

20. 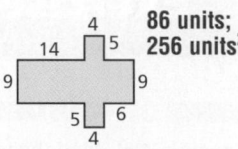 **86 units; 256 units²**

21. Find the perimeter and area of the figure with vertices $A(-8, 5)$, $B(-4, 5)$, $C(-4, 2)$, $D(3, 2)$, $E(3, -2)$, $F(6, -2)$, $G(6, -4)$, $H(-8, -4)$. **46 units; 84 units²**

22. If the area of a parallelogram is 52.7 cm² and the height is 6.2 cm, what is the length of the base? **8.5 cm**

23. Find the height of a rectangle with perimeter 114 in. and base length 24 in. What is the area? **33 in.; 792 in²**

24. Find the height of a rectangle with area 143 cm² and base length 11 cm. What is the perimeter? **13 cm; 48 cm**

25. A rectangular ice-skating rink measures 50 ft by 75 ft.
   a. If it costs $4.50 per foot to build a railing, how much would it cost to completely enclose the rink with a railing? **$1125**
   b. If the skating rink allows one person for every 10 ft² of ice, how many people are allowed in the rink at one time? **375 people**

26. **SOCIAL STUDIES** The state of Tennessee is shaped approximately like a parallelogram. Estimate the area of the state. **42,000 mi²**

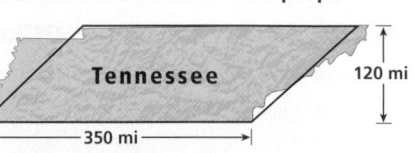

27. 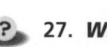 **WHAT'S THE QUESTION?** A rectangle has base 4 mm and height 3.7 mm. If the answer is 14.8 mm², what is the question? **Possible answer: What is the area?**

28. **WRITE ABOUT IT** A rectangle and an identical rectangle with a smaller rectangle cut from the bottom and placed on top are shown. Do the two figures have the same area? Do they have the same perimeter? Explain.

29. **CHALLENGE** A ruler is 12 in. long by 1 in. wide. How many rulers this size can be cut from a 72 in² rectangular piece of wood with base length 15 in.? **4 rulers**

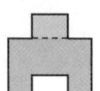

Spiral Review

Solve and graph. (Lesson 2-5)

30. $\frac{2}{3}n \le 4$ **$n \le 6$**   **31.** $y + 4 < 2$ **$y < -2$**   **32.** $-4x \ge 16$ **$x \le -4$**   **33.** $w - 5 > -2$ **$w > 3$**

34. **TEST PREP** Estimate $\sqrt{46}$ to two decimal places. (Lesson 3-9) **B**

   A 7.12   **B** 6.78   **C** 6.05   **D** 5.98

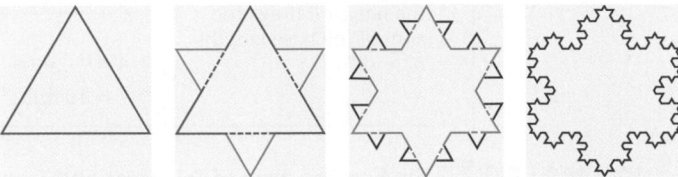

6-2 Perimeter and Area of Triangles and Trapezoids

Learn to find the area of triangles and trapezoids.

The figures show a *fractal* called the Koch snowflake. It is constructed by first drawing an equilateral triangle. Then triangles with sides one-third the length of the original sides are added to the middle of each side. The second step is then repeated over and over again.

The area and perimeter of each figure is larger than that of the one before it. However, the area of any figure is never greater than the area of the shaded box, while the perimeters increase without bound. To find the area and perimeter of each figure, you must be able to find the area of a triangle.

EXAMPLE 1 **Finding the Perimeter of Triangles and Trapezoids**

Find the perimeter of each figure.

A
$P = 14 + 10 + 8$ *Add all sides.*
$= 32$ units

B
$P = 7 + 11 + 2 + 4$ *Add all sides.*
$= 24$ units

A triangle or a trapezoid can be thought of as half of a parallelogram.

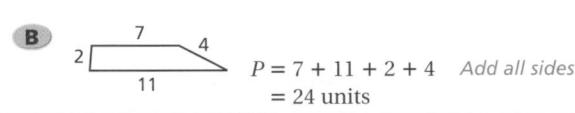

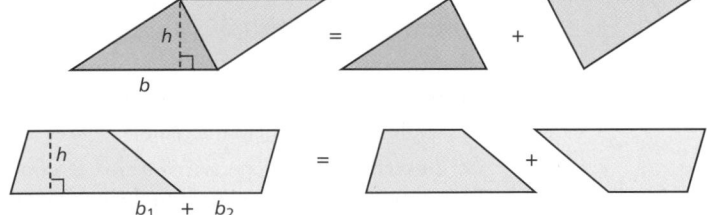

6-2 Organizer

Pacing: Traditional 1 day
Block $\frac{1}{2}$ day
Objective: Students find the area of triangles and trapezoids.

Warm Up

1. Find the perimeter of a rectangle with side lengths 12 ft and 20 ft. **64 ft**

2. Find the area of a rectangle with side lengths 24 in. and 32 in. **768 in²**

3. Find the area of a parallelogram with height 9 in. and base length 15 in. **135 in²**

Problem of the Day

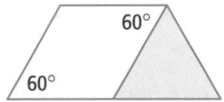

The area of a rhombus with two 60° angles is 24 in². An equilateral triangle is drawn so that one side is a side of the rhombus. What is the area of the triangle? (*Hint:* You don't need to use a formula.) **12 in²**

Available on Daily Transparency in CRB

1 Introduce

Alternate Opener

EXPLORATION

6-2 Perimeter and Area of Triangles and Trapezoids

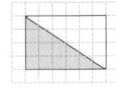   

Right triangle   Triangle

1. Compare the two triangles. How are they alike and how are they different?

2. Compare each triangle to the red rectangle around it. Use this comparison to help determine the area of each triangle in square units.

3. Draw two different triangles with the same area on graph paper. Are the perimeters the same? (Use a graph paper ruler to measure the perimeters.)

Think and Discuss

4. **Discuss** how to find the area of a triangle.

5. **Explain** why the areas of the two triangles from number 3 are the same.

Motivate

Cut a parallelogram along a diagonal to form a pair of congruent triangles. (Paper shapes are available on Chapter 6 Resource Book p. 113.) Ask students how the area of each triangle compares to the area of the parallelogram. Help them see that each triangle is half of the original figure. Then place two congruent trapezoids side by side to form a parallelogram. Explain that the area of one of the trapezoids is half the area of the parallelogram.

Exploration worksheet and answers on Chapter 6 Resource Book pp. 17 and 125

2 Teach

Lesson Presentation

Guided Instruction

In this lesson, students learn to find the area of triangles and trapezoids. Remind students that the perimeter of any polygon can be found by adding up all its side lengths. Review the formula for the area of a parallelogram. Use the diagrams in the lesson to show how a triangle or trapezoid can be thought of as half of a parallelogram (Teaching Transparency T8, CRB). Finally, introduce the formulas for area of triangles and trapezoids (Teaching Transparency T9, CRB). Point out that to find the area of a trapezoid, students must find the sum of the lengths of its bases.

 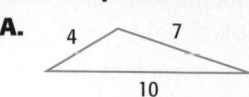
Example 1

Find the perimeter of each figure.

A.
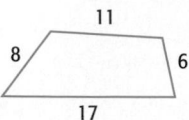

4   7
10

$P = 21$ units

B.

11
8   6
17

$P = 42$ units

Example 2

Graph and find the area of each figure with the given vertices.

A. $(-2, 2)$, $(4, 2)$, $(0, 5)$

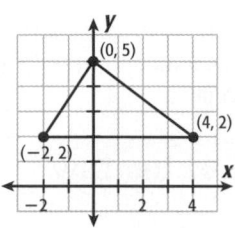

(0, 5)
(4, 2)
(−2, 2)

$A = 9$ units^2

B. $(-1, -2)$, $(5, -2)$, $(5, 2)$, $(-1, 6)$

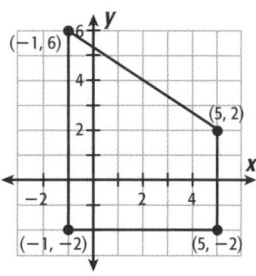

(−1, 6)
(5, 2)
(−1, −2)   (5, −2)

$A = 36$ units^2

AREA OF TRIANGLES AND TRAPEZOIDS

Words	Numbers	Formula
Triangle: The area A of a triangle is one-half of the base length b times the height h.	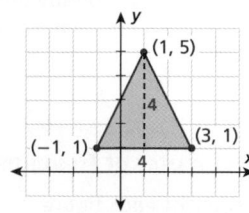 4 / 8 $A = \frac{1}{2}(8)(4)$ $= 16$ units^2	$A = \frac{1}{2}bh$
Trapezoid: The area of a trapezoid is one-half the height h times the sum of the base lengths b_1 and b_2.	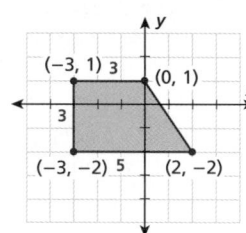 3 / 2 / 7 $A = \frac{1}{2}(2)(3 + 7)$ $= 10$ units^2	$A = \frac{1}{2}h(b_1 + b_2)$

EXAMPLE 2 **Finding the Area of Triangles and Trapezoids**

Graph and find the area of each figure with the given vertices.

A $(-1, 1)$, $(3, 1)$, $(1, 5)$

(1, 5)
4
(−1, 1)   (3, 1)
4

$A = \frac{1}{2}bh$ *Area of a triangle*

$= \frac{1}{2} \cdot 4 \cdot 4$ *Substitute for b and h.*

$= 8$ units^2

B $(-3, -2)$, $(-3, 1)$, $(0, 1)$, $(2, -2)$

(−3, 1) 3 (0, 1)
3
(−3, −2) 5 (2, −2)

$A = \frac{1}{2}h(b_1 + b_2)$ *Area of a trapezoid*

$= \frac{1}{2} \cdot 3(3 + 5)$ *Substitute for h, b_1, and b_2.*

$= 12$ units^2

Think and Discuss

1. **Describe** what happens to the area of a triangle when the base is doubled and the height remains the same.

2. **Describe** what happens to the area of a trapezoid when the length of both bases are doubled but the height remains the same.

Teach

 **Reaching All Learners**
Through Critical Thinking

Give students a recording sheet (Chapter 6 Resource Book p. 114) that contains some ordered pairs and a coordinate grid. Have students plot the points and connect them in order to create a geometric figure. Then have them divide the shape into any combination of rectangles, trapezoids, and triangles to find the area of the figure.

3 Close

Summarize

Show students a trapezoid that is divided into a rectangle and two triangles, such as the one below.

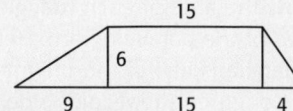
15
6
9   15   4

Have volunteers find the area of the trapezoid and of each individual section. Show them that the area of the trapezoid is equal to the sum of the areas of the sections.

trapezoid: 129 units^2; large triangle: 27 units^2; rectangle: 90 units^2; small triangle: 12 units^2; $27 + 90 + 12 = 129$ units^2

Answers to Think and Discuss

1. The area doubles.

2. The area doubles.

FOR EXTRA PRACTICE
see page 742

internet connect
Homework Help Online
go.hrw.com   Keyword: MP4 6-2

6-2 PRACTICE & ASSESS

GUIDED PRACTICE

See Example **1**   Find the perimeter of each figure.

1. 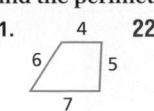 **22 units**

2. 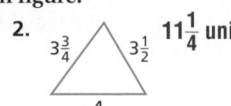 $11\frac{1}{4}$ units

3. 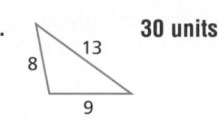 **30 units**

> Students may want to refer back to the lesson examples.

4. 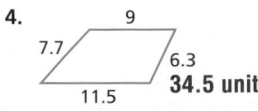 **34.5 units**

5. 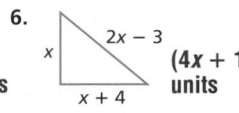 **84 units**

6. x $2x - 3$ $x + 4$ $(4x + 1)$ units

See Example **2**   Graph and find the area of each figure with the given vertices.

7. $(-2, 3), (2, -3), (-3, -3)$ **15 units^2**

8. $(5, 2), (2, -2), (-3, -2), (-4, 2)$ **28 units^2**

9. $(4, 2), (5, -6), (2, -6)$ **12 units^2**

10. $(0, -1), (-7, -1), (-5, 4), (-2, 4)$ **25 units^2**

INDEPENDENT PRACTICE

See Example **1**   Find the perimeter of each figure.

11. 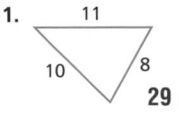 **29 units**

12. 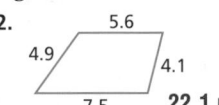 **22.1 units**

13. 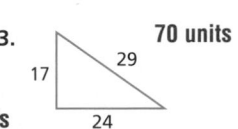 **70 units**

14. 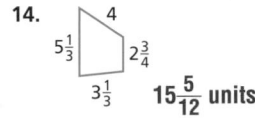 $15\frac{5}{12}$ units

15. 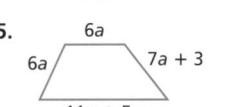 $(30a + 8)$ units

16. 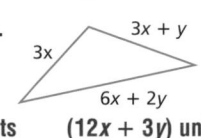 $(12x + 3y)$ units

See Example **2**   Graph and find the area of each figure with the given vertices.

17. $(1, 5), (1, 1), (-3, 1), (-5, 5)$

18. $(-5, 2), (1, -3), (-3, -3)$

19. $(2, -3), (-1, -6), (-6, -3)$

20. $(1, 4), (4, -5), (-5, -5), (-3, 4)$

PRACTICE AND PROBLEM SOLVING

Find the area of each figure with the given dimensions.

21. triangle: $b = 9, h = 11$

22. trapezoid: $b_1 = 6, b_2 = 10, h = 5$

23. triangle: $b = 7x, h = 6$

24. trapezoid: $b_1 = 4.5, b_2 = 8, h = 6.7$

25. The perimeter of a triangle is 37.4 ft. Two of its sides measure 16.4 ft and 11.9 ft, respectively. What is the length of its third side? **9.1 ft**

26. The area of a triangle is 63 mm^2. If its height is 14 mm, what is the length of its base? **9 mm**

Assignment Guide

If you finished Example **1** assign:
 Core 1–6, 11–16, 25, 33–39
 Enriched 1–6, 11–16, 25, 33–39

If you finished Example **2** assign:
 Core 1–20, 21–25 odd, 33–39
 Enriched 11–39

Graph paper for Exercises 7–10 and 17–20 is provided on Chapter 6 Resource Book p. 111.

Answers

7–10, 17–20. Complete answers on pp. A7–A8

17. 20 units^2

18. 10 units^2

19. 12 units^2

20. 58.5 units^2

21. 49.5 units^2

22. 40 units^2

23. 21x units^2

24. 41.875 units^2

RETEACH 6-2

LESSON 6-2 *Reteach*
Perimeter and Area of Triangles and Trapezoids

To find the perimeter of a figure, add the lengths of all its sides.

Complete to find the perimeter of each figure.

1.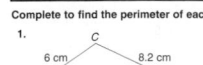

Perimeter of triangle *ABC*
= __12.4__ + __8.2__ + __6__
= __26.6__ cm

2.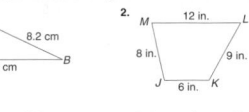

Perimeter of trapezoid *JKLM*
= __6__ + __9__ + __12__ + __8__
= __35__ in.

Area of Triangle = $\frac{1}{2}bh$
The area of a triangle is one-half the product of a base length *b* and the height *h* drawn to that base.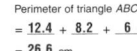

Complete to find the area of each triangle.

3. Area of triangle *PQR*
= $\frac{1}{2}bh$
= $\frac{1}{2}$ × __20__ × __4__
= $\frac{1}{2}$ × __80__ = __40__ in^2

4. In triangle *ABC*:
base $AC = 10 - 3 =$ __7__ units
height $BH = 4 - (-2) =$ __6__ units.
Area of triangle *ABC*
= $\frac{1}{2}$ × base *AC* × height *BH*
= $\frac{1}{2}$ × __7__ × __6__
= $\frac{1}{2}$ × __42__ = __21__ units^2

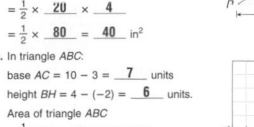
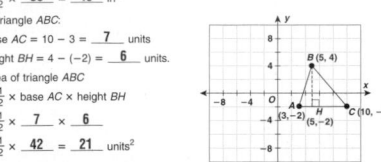

PRACTICE 6-2

LESSON 6-2 *Practice B*
Perimeter and Area of Triangles and Trapezoids

Find the perimeter of each figure.

1. 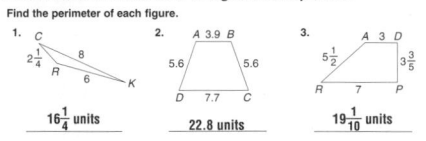 $16\frac{1}{4}$ units

2. **22.8 units**

3. $19\frac{1}{10}$ units

Graph and find the area of each figure with the given vertices.

4. $(-1, 3), (4, 3), (4, -4), (-4, -4)$ **45.5 units^2**

5. $(-1, 2), (-4, -2), (4, -2)$ **16 units^2**

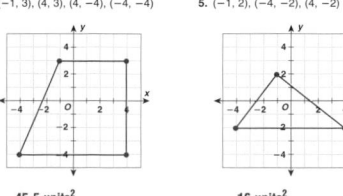

6. Find the area of a triangle with height 16 cm and base $12\frac{1}{3}$ cm. $98\frac{2}{3}$ cm^2

7. Find the area of a trapezoid with height 20.4 in. and bases 14.4 in. and 9.7 in. **245.82 in^2**

8. Find the height of a trapezoid with area 96 m^2 and bases 13 m and 19 m. **6 m**

Math Background

The formula for the area of a trapezoid is explained in the lesson by forming a parallelogram with two congruent trapezoids. An alternate method is to use a diagonal of a trapezoid to form two triangles.

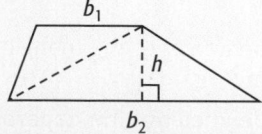

The area of the trapezoid is the sum of the areas of two triangles that share the same height:

$$A = \frac{1}{2}b_1 h + \frac{1}{2}b_2 h = \frac{1}{2}h(b_1 + b_2)$$

Physical Science

Exercises 27–32 involve using diagrams to solve problems about the perimeters and areas of wings. Wing shape and lift are studied in middle school science programs, such as *Holt Science and Technology*.

Answers

32. Possible answer: No; the area of the Boeing 747 wing is more than 11 times the area of the Wright brothers' wing.

Journal

In the lesson, students used formulas to find areas of triangles and trapezoids that they drew on graph paper. Have students write about why it is not always possible to find areas by counting the square units inside a figure.

Test Prep Doctor

For Exercise 39, students should draw the quadrilateral. The figure does not contain right angles, so choice **D** is incorrect. Both pairs of sides are parallel, so choice **A** is incorrect. All sides are not equal in length, so choice **C** is incorrect. Therefore, the correct choice is **B**.

Physical Science

To fly, a plane must overcome gravity and achieve *lift*, the force that allows a flying object to have upward motion. The shape and size of a plane's wings affect the amount of lift that is created. The wings of high-speed airplanes are thin and usually angled back to give the plane more lift.

27. a. Find the area of a Concorde wing to the nearest tenth of a square foot. **1929.5 ft²**

b. Find the total perimeter of the two wings of a Concorde to the nearest tenth of a foot. **466.6 ft**

28. What is the area of a Boeing 747 wing to the nearest tenth of a square foot? **2747.9 ft²**

29. What is the perimeter of an F-18 wing to the nearest tenth of a foot? **49.8 ft**

30. What is the total area of the two wings of an F-18? **273 ft²**

31. Find the area and perimeter of the wing of a space shuttle rounded to the nearest tenth. **874.6 ft²; 160.4 ft**

32. ⭐ **CHALLENGE** The wing of the Wright brothers' plane is about half the length of a Boeing 747 wing. Compare the area of the Wright brothers' wing with the area of a Boeing 747 wing. Is the area of the Wright brothers' wing half the area of the 747 wing? Explain.

go.hrw.com
KEYWORD: MP4 Lift
CNN Student News.

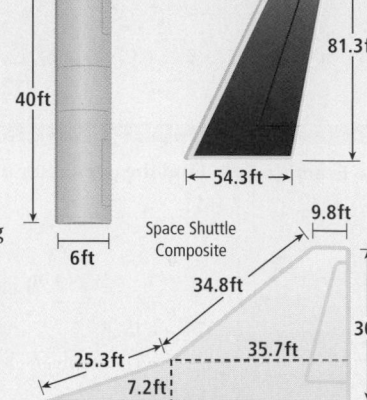

F-18 trapezoid 6ft
15.8ft 13ft
←15ft→

Concorde triangle 100ft
42.5ft
←─ 90.8ft ─→

Wright brothers rectangle
40ft
6ft

Boeing 747 trapezoid 13.3ft
81.3ft
←─ 54.3ft ─→

Space Shuttle Composite
9.8ft
34.8ft
30.5ft
25.3ft 35.7ft
7.2ft
←──── 60ft ────→

Spiral Review

Write each fraction as a decimal. (Lesson 3-1)

33. $\frac{3}{4}$ **0.75**   **34.** $\frac{1}{8}$ **0.125**   **35.** $\frac{10}{4}$ **2.5**   **36.** $\frac{9}{15}$ **0.6**

Do the data sets have a positive, a negative, or no correlation? (Lesson 4-7)

37. the number of shoes purchased and the amount of money left over **negative**

38. the length of a sub sandwich and the price of the sandwich **positive**

39. TEST PREP What name best describes a quadrilateral with vertices at (2, 4), (4, 1), (−3, 1), and (−5, 4)? (Lesson 5-5) **B**

   A Trapezoid   **B** Parallelogram   **C** Rhombus   **D** Rectangle

---

Lesson Quiz

Use the figure to find the following measurements.

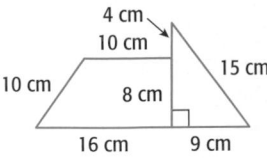

4 cm
10 cm
15 cm
10 cm
8 cm
16 cm   9 cm

1. the perimeter of the triangle **36 cm**

2. the perimeter of the trapezoid **44 cm**

3. the perimeter of the combined figure **64 cm**

4. the area of the triangle **54 cm²**

5. the area of the trapezoid **104 cm²**

Available on Daily Transparency in CRB

CHALLENGE 6-2

Challenge
6-2 *Fence Me In!*

You can find the area of a triangle in the coordinate plane that has no horizontal or vertical side.

Consider △ABC with vertices A(−3, 2), B(8, −3), and C(5, 6).

By drawing horizontal and vertical lines, △ABC is enclosed in rectangle PQBR.

Write the coordinates of the remaining vertices of the rectangle.

1. P **(−3, 6)**

   Q **(8, 6)**

   R **(−3, −3)**

Count boxes or subtract coordinates to find the indicated dimensions. Then find the indicated areas.

2. For rectangle PQBR:

   base RB = **11** units   height RP = **9** units   area = **99** units²

3. For right triangle APC:

   base PC = **8** units   height AP = **4** units   area = **16** units²

4. For right triangle CQB:

   base CQ = **3** units   height BQ = **9** units   area = **13.5** units²

5. For right triangle ARB:

   base RB = **11** units   height AR = **5** units   area = **27.5** units²

6. Explain how to combine the areas of the rectangle and the three right triangles to find the area of △ABC. Then find the area of △ABC.

△ABC = rectangle PQBR − (△APC + △CQB + △ARB)

△ABC = 99 − (16 + 13.5 + 27.5) = 42 units²

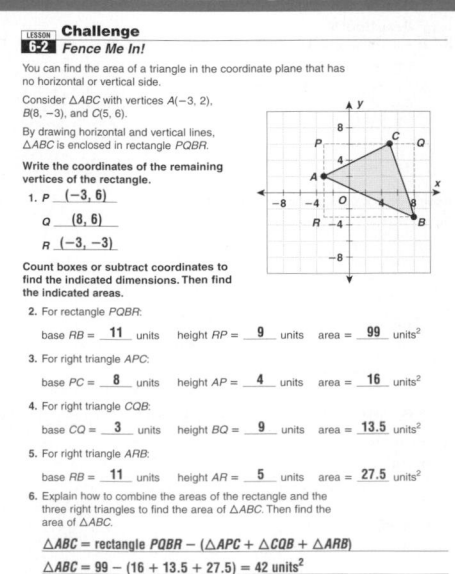

PROBLEM SOLVING 6-2

Problem Solving
6-2 *Perimeter and Area of Triangles and Trapezoids*

Write the correct answer.

1. Find the area of the material required to cover the kite pictured below.

2 ft
1 ft
3 ft

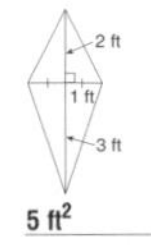

5 ft²

2. Find the area of the material required to cover the kite pictured below.

2 ft
1 ft
2 ft   2 ft
3 ft

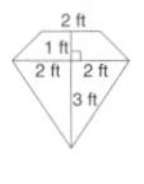

9 ft²

3. Find the approximate area of the state of Nevada.

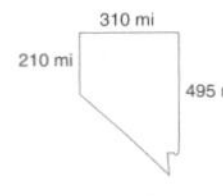

310 mi
210 mi
495 mi

109,275 mi²

4. Find the area of the hexagonal gazebo floor.

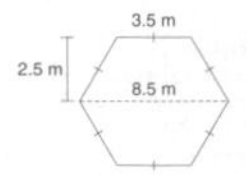

3.5 m
2.5 m   8.5 m

30 m²

Choose the letter for the best answer.

5. Find the amount of flooring needed to cover the stage pictured below.

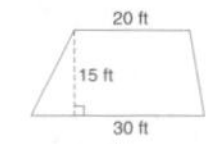

20 ft
15 ft
30 ft

A 4500 ft²
B 750 ft²
C 525 ft²
Ⓓ 375 ft²

6. Find the combined area of the congruent triangular gables.

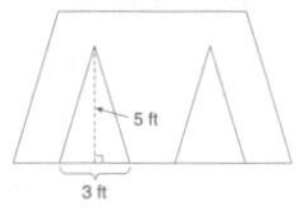

5 ft
3 ft

F 7.5 ft²
Ⓖ 15 ft²
J 60 ft²
H 30 ft²

Hands-On LAB 6A

Explore Right Triangles

Use with Lesson 6-3

WHAT YOU NEED
- scissors
- paper

REMEMBER
Right triangles have 1 right angle and 2 acute angles.

internet connect
Lab Resources Online
go.hrw.com
KEYWORD: MP4 Lab6A

Activity

1 The Pythagorean Theorem states that if a and b are the lengths of the legs of a right triangle, then c is the length of the hypotenuse, where $a^2 + b^2 = c^2$. Prove the Pythagorean Theorem using the following steps.

a. Draw two squares side by side. Label one with side a and one with side b.

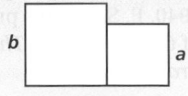

Notice that the area of this composite figure is $a^2 + b^2$.

b. Draw hypotenuses of length c, so that we have right triangles with sides a, b, and c.

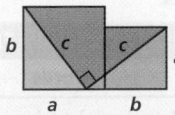

c. Cut out the triangles and the remaining piece.

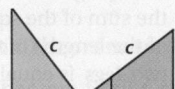 

d. Fit the pieces together to make a square with sides c and area c^2. You have shown that the area $a^2 + b^2$ can be cut up and rearranged to form the area c^2, so $a^2 + b^2 = c^2$.

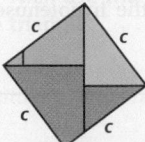

Think and Discuss

1. Does the Pythagorean Theorem work for triangles that are not right triangles?
   No; for example, an equilateral triangle with side lengths of 1 does not satisfy the Pythagorean Theorem. $1^2 + 1^2 \neq 1^2$

Try This

1. If you know that the lengths of two legs of a right triangle are 9 and 12, can you find the length of the hypotenuse? Show your work.
   Yes; $9^2 + 12^2 = 81 + 144 = 225 = c^2$; $c = \sqrt{225} = 15$
2. Take a piece of paper and fold the right corner down so that the top edge of the paper matches the side edge. Crease the paper. Without measuring, find the diagonal's length.

Teacher to Teacher

I find that students have trouble drawing the hypotenuses of length c on regular paper, so I pass out large sheets of graph paper to each student. The students are then able to draw the squares and hypotenuses with ease. I have them find the areas of the squares with side lengths a and b as well as the square with side length c. Doing this helps them to see the Pythagorean Theorem more clearly. After the activity is finished, I display the students' work and use it as a teaching tool for later lessons.

Miguel Carrizales
San Antonio, Texas

Hands-On LAB

6A Explore Right Triangles

Pacing: Traditional 1 day
Block $\frac{1}{2}$ day

Objective: To use scissors and paper to explore right triangles

Materials: Scissors, paper

Lab Resources

Hands-On Lab Activities . . . . pp. 42, 95

Using the Page

Discuss with students the importance of accuracy when they are making their drawings. Any errors in their initial drawings may affect the results of the lab.

1. What properties do squares have?
   Squares have four right angles and four congruent sides.

2. How can you be sure your squares have four right angles and four congruent sides?

   Possible answer: Use the corner of a separate piece of paper to test the four angles of the square; use a compass or ruler to make sure all four sides are congruent.

Answers

Try This

2. Possible answer:
   using a sheet of $8\frac{1}{2}$ in. by 11 in. paper:

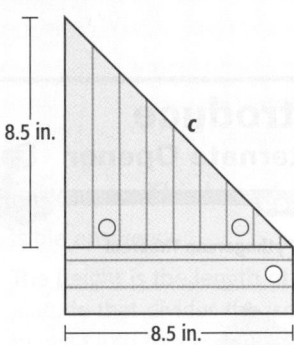

8.5 in.

c

8.5 in.

$8.5^2 + 8.5^2 = c^2$

$144.5 = c^2$

$12.02 = c$

The length of the folded edge is about 12.02 inches.

6-3 Exercises

FOR EXTRA PRACTICE
see page 742

internet connect
Homework Help Online
go.hrw.com   Keyword: MP4 6-3

Students may want to refer back to the lesson examples.

GUIDED PRACTICE

See Example **1** Find the length of the hypotenuse in each triangle to the nearest tenth.

1. 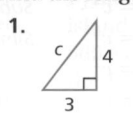 **5**

2. 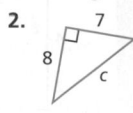 **10.6**

3. triangle with coordinates $(-5, 0)$, $(-5, 6)$, and $(0, 6)$ **7.8**

See Example **2** Solve for the unknown side in each right triangle to the nearest tenth.

4. 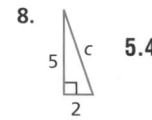 **5**

5. **6** **5.3**

6. b **20**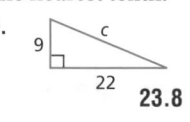

See Example **3** 7. Use the Pythagorean Theorem to find the height of the triangle. Then use the height to find the area of the triangle. $\sqrt{24} \approx$ **4.9 units; 19.6 units²**

INDEPENDENT PRACTICE

See Example **1** Find the length of the hypotenuse in each triangle to the nearest tenth.

8. 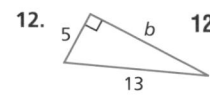 **5.4**

9. **15   8** **17**

10. c **9   22** **23.8**

11. triangle with coordinates $(-4, 2)$, $(4, -2)$, and $(-4, -2)$ **8.9**

See Example **2** Solve for the unknown side in each right triangle to the nearest tenth.

12. b **12** **5   13**

13. b **9.2** **11   6**

14. a **10   9.8** **14**

See Example **3** 15. Use the Pythagorean Theorem to find the height of the triangle. Then use the height to find the area of the triangle.
$\sqrt{80} \approx$ **8.9 units; 71.2 units²**

PRACTICE AND PROBLEM SOLVING

Find the missing length for each right triangle.

16. $a = 3$, $b = 6$, $c = \blacksquare$ $\sqrt{45} \approx$ **6.7**   17. $a = \blacksquare$, $b = 24$, $c = 25$ **7**

18. $a = 30$, $b = 72$, $c = \blacksquare$ **78**   19. $a = 20$, $b = \blacksquare$, $c = 46$ $\sqrt{1716} \approx$ **41.4**

$\sqrt{2091} \approx$ **45.7**   20. $a = \blacksquare$, $b = 53$, $c = 70$ 21. $a = 65$, $b = \blacksquare$, $c = 97$ **72**

Assignment Guide

If you finished Example **1** assign:
Core 1–3, 8–11, 16, 18, 23–29 odd, 37–41
Enriched 8–11, 16, 18, 22–29, 37–41

If you finished Example **2** assign:
Core 1–6, 8–14, 17–29 odd, 37–41
Enriched 8–14, 16–29, 37–41

If you finished Example **3** assign:
Core 1–21, 23–29 odd, 31–33, 37–41
Enriched 8–41

Graph paper for Exercises 3 and 11 is available on Chapter 6 Resource Book p. 111.

Math Background

The distinction between the Pythagorean Theorem and its converse is sometimes overlooked. The Pythagorean Theorem states that if a triangle is a right triangle, then the lengths of its sides satisfy the equation $a^2 + b^2 = c^2$. The converse says that if you have three numbers that satisfy the equation $a^2 + b^2 = c^2$, then those three numbers are side lengths of a right triangle. Because the Pythagorean Theorem and its converse are both true, they can be combined into the following *biconditional* statement: $a^2 + b^2 = c^2$ if and only if a and b are lengths of the legs of a right triangle and c is the length of the hypotenuse.

RETEACH 6-3

Reteach
6-3 *The Pythagorean Theorem*

In a right triangle,
the sum of the areas of the squares on the legs is equal to the area of the square on the hypotenuse.

$3^2 + 4^2 = 5^2$
$9 + 16 = 25$

Given the squares that are on the legs of a right triangle, draw the square for the hypotenuse.

1. leg   leg   hypotenuse

Without drawing the squares, you can find the length of a side.

$a^2 + b^2 = c^2$
$3^2 + 4^2 = c^2$
$9 + 16 = c^2$
$25 = c^2$
$c = 5$ in.

Complete to find the length of each hypotenuse.

2. $\dfrac{5^2} + \dfrac{12^2} = c^2$
$\underline{25} + \underline{144} = c^2$
$\underline{169} = c^2$
$c = \underline{13c}$ ft

3. $\dfrac{8^2} + \dfrac{15^2} = c^2$
$\underline{64} + \underline{225} = c^2$
$\underline{289} = c^2$
$c = \underline{17}$ in.

PRACTICE 6-3

Practice B
6-3 *The Pythagorean Theorem*

Find the length of the hypotenuse in each triangle.

1. c 12   **13**
2. c 8.4   **10.5**
3. c 45   **51**

Solve for the unknown side in each right triangle. Round the answers to the nearest hundredth.

4. c 10   **14.14**
5. a 18   **13.42**
6. b 29   **20**

7. b **18.25**
8. a **17.66**
9. b **72**

10. Use the Pythagorean Theorem to find the height of this triangle. Then use the height to find the area of the triangle.
$h = 42$; area = **661.5 units²**

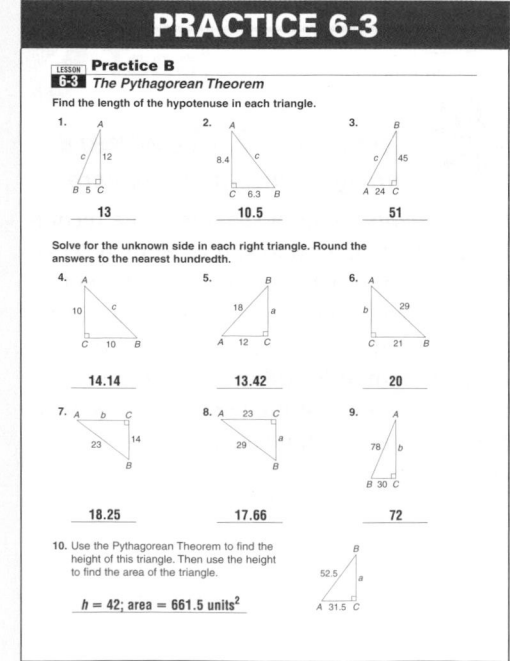

The *converse* of the Pythagorean Theorem states that any three positive numbers that make the equation $a^2 + b^2 = c^2$ true are the side lengths of a right triangle. If the side lengths are all whole numbers, they are called *Pythagorean triples.* Determine whether each set is a Pythagorean triple.

22. 2, 6, 8   **no**   23. 3, 4, 5   **yes**   24. 8, 15, 17   **yes** 25. 12, 16, 20   **yes**

26. 10, 24, 26   **yes** 27. 9, 13, 16   **no**   28. 11, 17, 23   **no** 29. 24, 32, 40   **yes**

30. Use the Pythagorean Theorem to find the height of the figure. Then find the area, to the nearest whole number.
    15 cm; 341 cm²

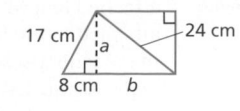

31. How far is the sailboat from the lighthouse, to the nearest kilometer?   **139 km**

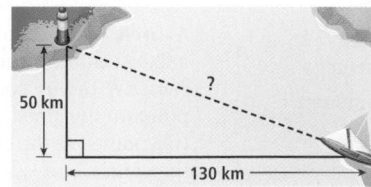

32. **CONSTRUCTION** A construction company is pouring a concrete foundation. The measures of two sides that meet in a corner are 33 ft and 56 ft. For the corner to be square (a right angle), what would the length of the diagonal have to be? (*Hint:* Draw a diagram.)   **65 ft**

33. **SOCIAL STUDIES** The state of Colorado is shaped approximately like a rectangle. To the nearest mile, what is the distance between opposite corners of the state?   **475 mi**

34. **WRITE A PROBLEM** Use a street map to write and solve a problem that requires the use of the Pythagorean Theorem.

35. **WRITE ABOUT IT** Explain how to use the converse of the Pythagorean Theorem to show that a triangle is a right triangle. (See Exercises 22–29.)

36. **CHALLENGE** A right triangle has legs of length $6x$ m and $8x$ m and hypotenuse of length 90 m. Find the lengths of the legs of the triangle.   **54 m and 72 m**

Spiral Review

Solve. (Lesson 2-4)

37. $x + 13 = 22$ **$x = 9$**   38. $b + 5 = -2$ **$b = -7$**   39. $2y + 9 = 19$ **$y = 5$**   40. $4a + 2 = -18$
    $a = -5$

41. **TEST PREP** Which real number lies between $3\frac{1}{5}$ and $3\frac{4}{7}$? (Lesson 3-10)   **A**

   A 3.216   B 3.59   C 3.701   D 3.9

CHALLENGE 6-3

Challenge
6-3 *Triple Play*

Three numbers connected by the Pythagorean relation are called Pythagorean triples.

Since $3^2 + 4^2 = 5^2$, the numbers 3-4-5 are a Pythagorean Triple.

Consider the Pythagorean triples shown in the table.

	Column A	Column B	Column C
row 1	3	4	5
row 2	5	12	13
row 3	7	24	25
row 4	9	40	41
row 5	11	60	61

1. Make an observation about the numbers in Column A.
   consecutive odd numbers

2. How are the numbers in Column C related to those in Column B?
   $C = B + 1$

3. Complete this table by carrying out the indicated calculation. Two calculations are done.

	Column A	row × A + row
row 1	3	1 × 3 + 1 = 4
row 2	5	2 × 5 + 2 = 12
row 3	7	3 × 7 + 3 = 24
row 4	9	4 × 9 + 4 = 40
row 5	11	5 × 11 + 5 = 60

Compare the results to the Pythagorean triples in Columns A, B, and C of the original table.
   results = Column B

4. In the original table, how do the squares of the numbers in Column A relate to the numbers in Columns B and C?
   $A^2 = B + C$

5. Using the relationships you have observed, calculate rows 6 and 10 of the table of Pythagorean triples. Verify your results by applying the Pythagorean Theorem.

	Column A	Column B	Column C	Verify $A^2 + B^2 = C^2$
row 6	13	84	85	$13^2 + 84^2 \overset{?}{=} 85^2$; 7225 = 7225 ✓
row 10	21	220	221	$21^2 + 220^2 \overset{?}{=} 221^2$; 48,841 = 48,841 ✓

PROBLEM SOLVING 6-3

Problem Solving
6-3 *The Pythagorean Theorem*

Write the correct answer. Round to the nearest tenth.

1. A utility pole 10 m high is supported by two guy wires. Each guy wire is anchored 3 m from the base of the pole. How many meters of wire are needed for the guy wires?
   20.9 m

2. A 12 foot-ladder is resting against a wall. The base of the ladder is 2.5 feet from the base of the wall. How high up the wall will the ladder reach?
   11.7 ft

3. The base-path of a baseball diamond form a square. If it is 90 ft from home to first, how far does the catcher have to throw to catch someone stealing second base?
   127.3 ft

4. A football field is 100 yards with 10 yards at each end for the end zones. The field is 45 yards wide. Find the length of the diagonal of the entire field, including the end zones.
   128.2 yd

Choose the letter for the best answer.

5. The frame of a kite is made from two strips of wood, one 27 inches long, and one 18 inches long. What is the perimeter of the kite? Round to the nearest tenth.

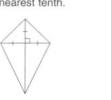

   A 18.8 in.   C 65.7 in.
   B 32.8 in.   D 131.2 in.

6. The glass for a picture window is 8 feet wide. The door it must pass through is 3 feet wide. How tall must the door be for the glass to pass through the door? Round to the nearest tenth.
   F 3.3 ft   H 7.4 ft
   G 6.7ft   J 8.5 ft

7. A television screen measures approximately 15.5 in. high and 19.5 in. wide. A television is advertised by giving the approximate length of the diagonal of its screen. How should this television be advertised?
   A 25 in.   C 12 in.
   B 21 in.   D 6 in.

8. To meet federal guidelines, a wheelchair ramp that is constructed to rise 1 foot off the ground must extend 12 feet along the ground. How long will the ramp be? Round to the nearest hundredth.
   F 11.96 ft   H 13.21 ft
   G 12.04 ft   J 15.00 ft

Lesson Quiz

Use the figure For Problems 1–3.

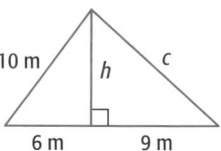

1. Find the height h of the triangle.
   8 m

2. Find the length of side c to the nearest meter.
   12 m

3. Find the area of the largest triangle.
   60 m²

4. One leg of a right triangle is 48 units long, and the hypotenuse is 50 units long. How long is the other leg?
   14 units

Available on Daily Transparency in CRB

Objective: Students find the area and circumference of circles.

Warm Up

1. Find the length of the hypotenuse of a right triangle that has legs 3 in. and 4 in. long.   **5 in.**

2. The hypotenuse of a right triangle measures 17 in., and one leg measures 8 in. How long is the other leg?   **15 in.**

3. To the nearest centimeter, what is the height of an equilateral triangle with sides 9 cm long? **8 cm**

Problem of the Day

A rectangular box is 3 ft by 4 ft by 12 ft. What is the distance from a top corner to the opposite bottom corner?

13 ft

Available on Daily Transparency in CRB

Math Humor

The circle had a radius of 4. So what did the hungry circumference do?
Ate pi.

 **Circles**

Learn to find the area and circumference of circles.

Vocabulary
circle
radius
diameter
circumference

A bicycle odometer uses a magnet attached to a wheel and a sensor attached to the bicycle frame. Each time the magnet passes the sensor, the odometer registers the distance traveled. This distance is the *circumference* of the wheel.

A **circle** is the set of points in a plane that are a fixed distance from a given point, called the *center*. A **radius** connects the center to any point on the circle, and a **diameter** connects two points on the circle and passes through the center.

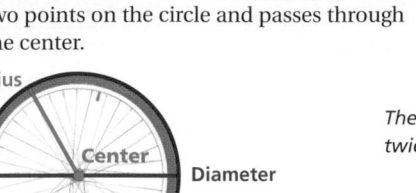

Radius

Center

Diameter

Circumference

The diameter d is twice the radius r.

$d = 2r$

The **circumference** of a circle is the distance around the circle.

Remember!

Pi (π) is an irrational number that is often approximated by the rational numbers 3.14 and $\frac{22}{7}$.

CIRCUMFERENCE OF A CIRCLE		
Words	**Numbers**	**Formula**
The circumference C of a circle is π times the diameter d, or 2π times the radius r.	$C = \pi(6)$ $= 2\pi(3)$ ≈ 18.8 units	$C = \pi d$ or $C = 2\pi r$

EXAMPLE **1** **Finding the Circumference of a Circle**

Find the circumference of each circle, both in terms of π and to the nearest tenth. Use 3.14 for π.

A circle with radius 5 cm
$$C = 2\pi r$$
$$= 2\pi(5)$$
$$= 10\pi \text{ cm} \approx 31.4 \text{ cm}$$

B circle with diameter 1.5 in.
$$C = \pi d$$
$$= \pi(1.5)$$
$$= 1.5\pi \text{ in.} \approx 4.7 \text{ in.}$$

---

1 Introduce

Alternate Opener

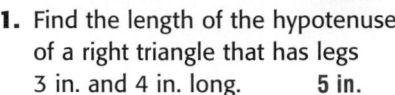

EXPLORATION

6-4 Circles

First estimate the area of each circle by counting squares. Then square each circle's radius. Then compare each estimated area with the square of the radius by computing $\frac{A}{r^2}$.

		Radius	Estimated Area	r^2	$\frac{A}{r^2}$
1.		$r = 1$			
2.		$r = 2$			
3.		$r = 3$			
4.		$r = 4$			

5. Is the estimated area more accurate for smaller circles than for larger circles? How do you know?

Think and Discuss
6. **Discuss** how you can make a generalization about how to find the area of a circle if you know the radius.

Motivate

Ask students if they like pie. Ask them to describe the shape of a pie. circle Then explain that this will help them remember an important number called *pi*. Tell them pi is a special number that they will use to solve problems involving circles.

Exploration worksheet and answers on Chapter 6 Resource Book pp. 37 and 130

2 Teach

 *Lesson Presentation*

Guided Instruction

In this lesson, students learn to find the area and circumference of circles. Review the diagram of the circular wheel and the vocabulary terms. Point out the important relationship between radius and diameter ($d = 2r$). Explain that circumference, a measure of the distance around the outside of a circle, is like perimeter. Show them how to find circumference using the formula (Teaching Transparency T18, CRB). Explain that a formula is used to find circumference because unlike the perimeter of a polygon, it is not easy to measure. Then show students how to find the area of a circle. Point out that the radius must be squared when finding area.

<table>
<tr><td colspan="3" align="center">**AREA OF A CIRCLE**</td></tr>
<tr><td align="center">**Words**</td><td align="center">**Numbers**</td><td align="center">**Formula**</td></tr>
<tr>
<td>The area A of a circle is π times the square of the radius r.</td>
<td> $A = \pi(3^2)$
$= 9\pi$
$\approx 28.3 \text{ units}^2$</td>
<td>$A = \pi r^2$</td>
</tr>
</table>

E X A M P L E 2 Finding the Area of a Circle

Find the area of each circle, both in terms of π and to the nearest tenth. Use 3.14 for π.

A circle with radius 5 cm

$A = \pi r^2 = \pi(5^2)$
$= 25\pi \text{ cm}^2 \approx 78.5 \text{ cm}^2$

B circle with diameter 1.5 in.

$A = \pi r^2 = \pi(0.75^2)$ $\frac{d}{2} = 0.75$
$= 0.5625\pi \text{ in}^2 \approx 1.8 \text{ in}^2$

E X A M P L E 3 Finding Area and Circumference on a Coordinate Plane

Graph the circle with center $(-1, 1)$ that passes through $(-1, 3)$. Find the area and circumference, both in terms of π and to the nearest tenth. Use 3.14 for π.

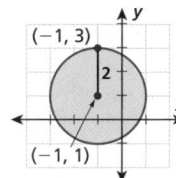

$A = \pi r^2$
$= \pi(2^2)$
$= 4\pi \text{ units}^2$
$\approx 12.6 \text{ units}^2$

$C = \pi d$
$= \pi(4)$
$= 4\pi \text{ units}$
$\approx 12.6 \text{ units}$

E X A M P L E 4 *Transportation Application*

A bicycle odometer recorded 147 revolutions of a wheel with diameter $\frac{4}{3}$ ft. How far did the bicycle travel? Use $\frac{22}{7}$ for π.

$C = \pi d = \pi\left(\frac{4}{3}\right) \approx \frac{22}{7}\left(\frac{4}{3}\right) = \frac{88}{21}$ *Find the circumference.*

The distance traveled is the circumference of the wheel times the number of revolutions, or about $\frac{88}{21} \cdot 147 = 616$ ft.

Think and Discuss

1. **Compare** the circumference of a circle with diameter x to the circumference of a circle with diameter $2x$.

2. **Give** the formula for area of a circle in terms of the diameter d.

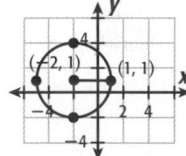
3 Close

Reaching All Learners
Through Hands-On Experience

Provide each pair or group of students with cans of various sizes and a tape measure (Manipulatives Kit). Have students measure the diameter and circumference of the top of each can and enter the data in a table like the one shown.

Can	C	d	$\frac{C}{d}$

After the measurements are complete, have students use calculators to divide the circumference by the diameter for each circle and enter these results in the last column. Results should be close to pi.

Summarize

Review the vocabulary terms and the formulas from the lesson. Stress the importance of the difference between radius and diameter. Remind students that pi is the value of the circumference of a circle divided by its diameter. Show them that they can see that the equations $\pi = \frac{C}{d}$ and $C = \pi d$ are equivalent based on their knowledge of solving algebraic equations.

Answers to Think and Discuss

1. Possible answer: The circumference of the circle with diameter $2x$ is twice the circumference of the circle with diameter x.

2. $A = \pi\left(\frac{d}{2}\right)^2$ or $A = \frac{\pi d^2}{4}$

FOR EXTRA PRACTICE
see page 742

internet connect
Homework Help Online
go.hrw.com   Keyword: MP4 6-4

Students may want to refer back to the lesson examples.

Assignment Guide

If you finished Example **1** assign:
Core 1, 2, 7–10, 21, 34–38
Enriched 7–10, 20–22, 34–38

If you finished Example **2** assign:
Core 1–4, 7–14, 19–25 odd, 28, 29, 34–38
Enriched 7–13 odd, 17–29, 31, 34–38

If you finished Example **3** assign:
Core 1–5, 7–15, 17–25 odd, 28, 29, 34–38
Enriched 7–15 odd, 17–29, 31–38

If you finished Example **4** assign:
Core 1–16, 17–25 odd, 28–30, 34–38
Enriched 7–15 odd, 16–38

Graph paper for Exercises 5 and 15 is provided on Chapter 6 Resource Book p. 111.

Answers

5, 15. Complete answers on p. A8

5. $A = 9\pi$ units^2 ≈ 28.3 units^2;
$C = 6\pi$ units ≈ 18.8 units

13. 324π in^2; 1017.4 in^2

14. 72.25π ft^2; 226.9 ft^2

15. $A = 36\pi$ units^2 ≈ 113.0 units^2;
$C = 12\pi$ units ≈ 37.7 units

GUIDED PRACTICE

See Example **1**   Find the circumference of each circle, both in terms of π and to the nearest tenth. Use 3.14 for π.

1. circle with diameter 8 cm
   8π cm; 25.1 cm

2. circle with radius 3.2 in.
   6.4π in.; 20.1 in.

See Example **2**   Find the area of each circle, both in terms of π and to the nearest tenth. Use 3.14 for π.   **2.25π ft^2; 7.1 ft^2**   **56.25π cm^2; 176.6 cm^2**

3. circle with radius 1.5 ft

4. circle with diameter 15 cm

See Example **3**   **5.** Graph a circle with center (3, −1) that passes through (0, −1). Find the area and circumference, both in terms of π and to the nearest tenth. Use 3.14 for π.

See Example **4**   **6.** Estimate the diameter of a wheel that makes 9 revolutions and travels 50 feet. Use $\frac{22}{7}$ for π. $\frac{175}{99}$ **≈ 1.8 ft**

INDEPENDENT PRACTICE

See Example **1**   Find the circumference of each circle, both in terms of π and to the nearest tenth. Use 3.14 for π.   **11.5π m; 36.1 m**

7. circle with radius 7 in.
   14π in.; 44.0 in.

8. circle with diameter 11.5 m

9. circle with radius 20.2 cm
   40.4π cm; 126.9 cm

10. circle with diameter 2 ft
   2π ft; 6.3 ft

See Example **2**   Find the area of each circle, both in terms of π and to the nearest tenth. Use 3.14 for π.   **144π cm^2; 452.2 cm^2**   **1.96π yd^2; 6.2 yd^2**

11. circle with diameter 24 cm

12. circle with radius 1.4 yd

13. circle with radius 18 in.

14. circle with diameter 17 ft

See Example **3**   **15.** Graph a circle with center (−4, 2) that passes through (−4, −4). Find the area and circumference, both in terms of π and to the nearest tenth. Use 3.14 for π.

See Example **4**   **16.** If the diameter of a wheel is 2 ft, about how many revolutions does the wheel make for every mile driven? Use $\frac{22}{7}$ for π. (*Hint:* 1 mi = 5280 ft.)   **840**

PRACTICE AND PROBLEM SOLVING

Find the circumference and area of each circle to the nearest tenth. Use 3.14 for π.   **$C ≈ 25.1$ in.; $A ≈ 50.2$ in^2**

17.   1.2 m $C ≈ 7.5$ m; $A ≈ 4.5$ m^2

18.   14 ft $C ≈ 44.0$ ft; $A ≈ 153.9$ ft^2

19.    4 in.

Math Background

Although pi is often approximated as 3.14, it is actually an irrational number with an infinite number of nonrepeating decimal digits. Mathematicians have been interested in the value of π for thousands of years.

From 2000 to 1600 B.C., the Babylonians believed that a circle's circumference was three times the diameter and that its area was one-twelfth of the square of the circumference. These formulas reveal the Babylonian approximation for pi as 3. Since then, the approximations for pi have become more and more accurate. Modern computers have calculated pi to more than 68 billion decimal places.

RETEACH 6-4

CHAPTER 6-4 Reteach
Circles

A **radius** connects the **center** of a **circle** to any point on the circle.
A **diameter** passes through the center and connects two points on the circle.
diameter d = twice radius r
$d = 2r$
Circumference is the distance around a circle.
(The symbol ≈ means *is approximately equal to.*)
Circumference $C ≈ 3$(diameter d)
$C ≈ \pi d$
Circumference $C ≈ 6$(radius r)
$C ≈ 2\pi r$

For a circle with diameter = 8 in.
$C ≈ \pi d$
$C ≈ \pi(8)$
$C ≈ 8\pi$ in.
$\pi ≈ 3.14$ $C ≈ 8(3.14) ≈ 25.12$ in.

For a circle with radius = 8 in.
$C ≈ 2\pi r$
$C ≈ 2\pi(8)$
$C ≈ 16\pi$ in.
$\pi ≈ 3.14$ $C ≈ 16(3.14) ≈ 50.24$ in.

Find the circumference of each circle, exactly in terms of π and approximately when $\pi = 3.14$.

1. diameter = 15 ft
$C ≈ \pi d$
$C = \pi(\underline{15}) = \underline{15\pi}$ ft
$C ≈ 3.14(\underline{15}) = \underline{47.1}$ ft

2. radius = 4 m
$C = 2\pi r$
$C = 2\pi(\underline{4}) = \underline{8\pi}$ m
$C ≈ \underline{8}(3.14) = \underline{25.12}$ m

Area $A ≈ 3$(the square of radius r)
$A ≈ \pi r^2$
For a circle with radius = 5 in.: $A = \pi r^2 = \pi(5^2) = 25\pi$ in^2 $A ≈ 25(3.14) ≈ 78.5$ in^2

Find the area of each circle, exactly in terms of π and approximately when $\pi = 3.14$.

3. radius = 9 ft
$A ≈ \pi r^2$
$A = \pi(\underline{9^2}) = \underline{81}\pi$ ft^2
$A ≈ \underline{81}(3.14) = \underline{254.34}$ ft^2

4. diameter = 10 m, radius = $\underline{5}$ m
$A = \pi r^2$
$A = \pi(\underline{5^2}) = \underline{25\pi}$ m^2
$A ≈ \underline{25}(3.14) = \underline{78.5}$ m^2

PRACTICE 6-4

LESSON 6-4 Practice B
Circles

Find the circumference of each circle, both in terms of π and to the nearest tenth of a unit. Use 3.14 for π.

1. circle with radius 10 in.
   20π in. or 62.8 in.

2. circle with diameter 13 cm
   13π cm or 40.8 cm

3. circle with diameter 18 m
   18π m or 56.5 m

4. circle with radius 15 ft
   30π ft or 94.2 ft

5. circle with radius 11.5 in.
   23π in. or 72.2 in.

6. circle with diameter 16.4 cm
   16.4π cm or 51.5 cm

Find the area of each circle, both in terms of π and to the nearest tenth of a unit. Use 3.14 for π.

7. circle with radius 9 in.
   81π in^2 or 254.3 in^2

8. circle with diameter 14 cm
   49π cm^2 or 153.9 cm^2

9. circle with radius 20 ft
   400π ft^2 or 1256 ft^2

10. circle with diameter 17 m
   72.3π m^2 or 226.9 m^2

11. circle with diameter 15.4 m
   59.3π m^2 or 186.2 m^2

12. circle with radius 22 yd
   484π yd^2 or 1519.8 yd^2

13. Graph a circle with center (0, 0) that passes through (0, −3). Find the area and circumference, both in terms of π and to the nearest tenth of a unit. Use 3.14 for π.
   $A = 9\pi$ units^2 or 28.3 units^2;
   $C = 6\pi$ units or 18.8 units

14. Kilauea, a crater in Hawaii Volcanoes National Park, is one of the largest active craters in the world. It is 1111 m deep and has a circumference of 13 km. The crater is surrounded by a wall of volcanic rock 61–152 m high. Find the diameter of the crater to the nearest hundredth of a unit. Use 3.14 for π.
   4.14 km

Find the radius of each circle with the given measurement.

20. $C = 18\pi$ in. **9 in.**   **21.** $C = 12.8\pi$ cm **6.4 cm** **22.** $C = 25\pi$ ft   **12.5 ft**

23. $A = 16\pi$ cm^2 **4 cm**   **24.** $A = 169\pi$ in^2 **13 in.**   **25.** $A = 136.89\pi$ m^2
11.7 m

Find the shaded area to the nearest tenth. Use 3.14 for π.

26.   12 cm   8 cm   12 cm   12 cm   12 cm   **49.0 cm^2**

27.   4 m   8 m   7 m   4 m   **297.7 m^2**

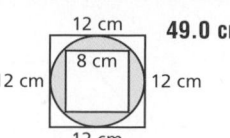

28. **ENTERTAINMENT** The London Eye is an observation wheel with a diameter greater than 135 meters and less than 140 meters. Describe the range of the possible circumferences of the wheel to the nearest meter.   **424 m $< C <$ 440 m**

29. **SPORTS** The radius of the free-throw circle on an NBA basketball court is 6 ft. What is its circumference and area to the nearest tenth?
$C = 12\pi \approx 37.7$ ft; $A = 36\pi \approx 113.1$ ft^2

30. **FOOD** A pancake restaurant serves small silver dollar pancakes and regular-size pancakes.

   a. What is the area of a silver dollar pancake to the nearest tenth?   **9.6 in^2**

   b. What is the area of a regular pancake to the nearest tenth?   **28.3 in^2**

3.5 in.   6 in.

   c. If 6 silver dollar pancakes are the same price as 3 regular pancakes, which is a better deal?
   three regular pancakes

31. **WHAT'S THE ERROR?** The area of a circle is 169π in^2. A student says that this means the diameter is 13 in. What is the error?

32. **WRITE ABOUT IT** Explain how you would find the area of the composite figure shown. Then find the area.
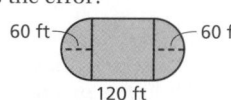
60 ft   60 ft   120 ft

33. **CHALLENGE** Graph the circle with center (1, 2) that passes through the point (4, 6). Find its area and circumference, both in terms of π and to the nearest tenth.

Spiral Review

Multiply. Write each answer in simplest form. (Lesson 3-3)

34. $-8\left(3\frac{3}{4}\right)$ **-30**   **35.** $\frac{6}{7}\left(\frac{7}{19}\right)$ **$\frac{6}{19}$**   **36.** $-\frac{5}{8}\left(-\frac{6}{15}\right)$ **$\frac{1}{4}$**   **37.** $-\frac{9}{10}\left(\frac{7}{12}\right)$ **$-\frac{21}{40}$**

38. **TEST PREP** $\angle 1$ and $\angle 3$ are supplementary angles. If $m\angle 1 = 63°$, find $m\angle 3$. (Lesson 5-1)   **D**

   A 27°   **B** 63°   **C** 87°   **D** 117°

Answers

31. Possible answer: The student confused diameter with radius. The radius of the circle is 13 in.

32. Possible answer: First find the area of the square with a side length of 120 ft. Then find the area of the semicircles. Since the two ends are half circles with the same radius, they have the same area as a circle with a radius of 60 ft. Finally, find the sum of the two areas. $A = 14,400 + 3600\pi \approx 25,710$ ft^2

33.

(4, 6)   (1, 2)

$A = 25\pi$ units$^2 \approx 78.5$ units^2
$C = 10\pi$ units ≈ 31.4 units

Journal

Ask students to write about any ideas they have for remembering the vocabulary terms and formulas from the lesson.

Test Prep Doctor

For Exercise 38, remind students that supplementary angles are two angles whose measures have a sum of 180°. If students chose **A,** they may have confused complementary with supplementary. If students chose **B,** they may have been thinking of vertical angles, or some other congruent angle pair. If students chose **C,** they may have thought that supplementary angles have a sum of 150° instead of 180°.

CHALLENGE 6-4

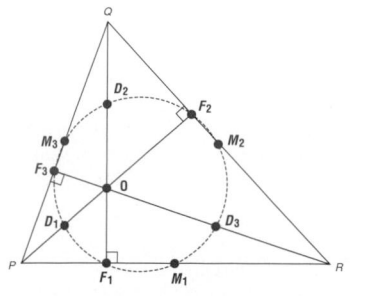
PROBLEM SOLVING 6-4

Lesson Quiz

Find the circumference of each circle, both in terms of π and to the nearest tenth. Use 3.14 for π.

1. radius 5.6 m   **11.2π m; 35.2 m**

2. diameter 113 mm
113π mm; 354.8 mm

Find the area of each circle, both in terms of π and to the nearest tenth. Use 3.14 for π.

3. radius 3 in.   **9π in^2; 28.3 in^2**

4. diameter 1 ft   **0.25π ft^2; 0.8 ft^2**

Available on Daily Transparency in CRB

Chapter 6 Mid-Chapter Quiz

Purpose: *To assess students' mastery of concepts and skills in Lessons 6-1 through 6-4*

Assessment Resources

Section 6A Quiz
Assessment Resources . . . . p. 15

 Test and Practice Generator
CD-ROM

Additional mid-chapter assessment items in both multiple-choice and free-response format may be generated for any objective in Lessons 6-1 through 6-4.

Answers

3.
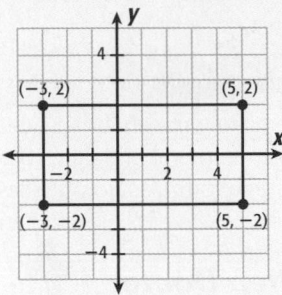

32 units^2

4.
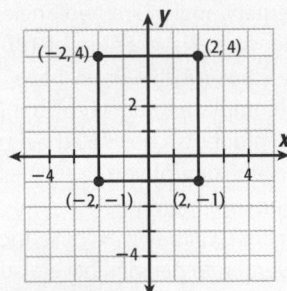

20 units^2

5.
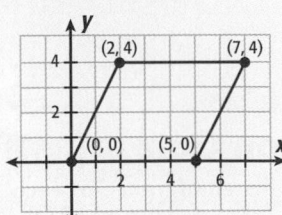

20 units^2

6.
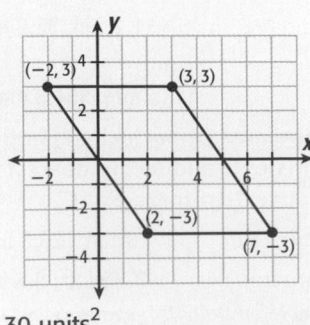

30 units^2

Mid-Chapter Quiz

LESSON 6-1 (pp. 280–284)

Find the perimeter of each figure.

1.
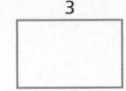
3
2
10 units

2.
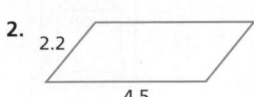
2.2
4.5
13.4 units

Graph and find the area of each figure with the given vertices.

3. (−3, 2), (−3, −2), (5, −2), (5, 2)

4. (−2, 4), (−2, −1), (2, −1), (2, 4)

5. (2, 4), (7, 4), (5, 0), (0, 0)

6. (7, −3), (2, −3), (−2, 3), (3, 3)

LESSON 6-2 (pp. 285–288)

Find the perimeter of each figure.

7.
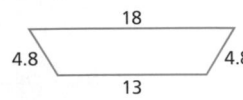
18
4.8   4.8
13
40.6 units

8.
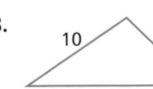
10   8
14
32 units

Graph and find the area of each figure with the given vertices.

9. (−6, −2), (4, −2), (−3, 3)

10. (−5, 0), (0, 0), (4, 4)

11. (2, −2), (3, 3), (−4, 3), (−3, −2)

12. (0, 4), (3, 6), (3, −3), (0, −3)

LESSON 6-3 (pp. 290–293)

Use the Pythagorean Theorem to find the height of each figure. Then find the area of each figure. If necessary, round the area to the nearest tenth of a square unit.

13.
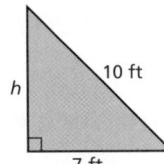
h
10 ft
7 ft
$h \approx 7.1$ ft; $A \approx 24.9$ ft^2

14.
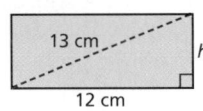
13 cm
h
12 cm
$h = 5$ cm; $A = 60$ cm^2

15.
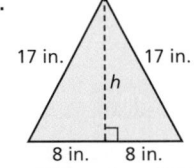
17 in.   17 in.
h
8 in.   8 in.
$h = 15$ in.; $A = 120$ in^2

LESSON 6-4 (pp. 294–297)

Find the area and circumference of each circle, both in terms of π and to the nearest tenth. Use 3.14 for π.

16. radius = 15 cm

17. diameter = 6.5 ft

18. radius = $7\frac{1}{2}$ ft

9.
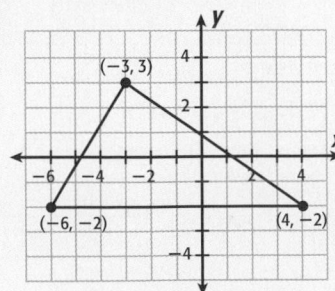

25 units^2

10.
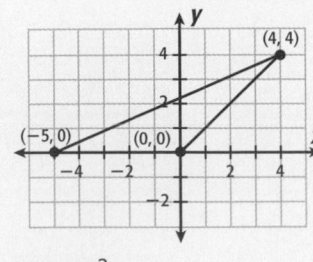

10 units^2

11–12. Complete answers on p. A8

11. 30 units^2

12. 24 units^2

16. $A = 225\pi$ cm$^2 \approx 706.5$ cm^2;
$C = 30\pi$ cm ≈ 94.2 cm

17. $A = 10.6\pi$ ft$^2 \approx 33.3$ ft^2;
$C = 6.5\pi$ ft ≈ 20.4 ft

18. $A = 56.3\pi$ ft$^2 \approx 176.8$ ft^2;
$C = 15\pi$ ft ≈ 47.1 ft

Focus on Problem Solving

Look Back

• **Does your solution answer the question?**

When you think you have solved a problem, think again. Your answer may not really be the solution to the problem. For example, you may solve an equation to find the value of a variable, but to find the answer the problem is asking for, the value of the variable may need to be substituted into an expression.

Purpose: To focus on looking back to check that a solution answers the question

Problem Solving Resources

Interactive Problem Solving . . pp. 45–54

Math: Reading and Writing in the Content Area . . . . pp. 45–54

Problem Solving Process

This page focuses on the fourth step of the problem-solving process:
Look Back

Write and solve an equation for each problem. Check to see whether the value of the variable is the answer to the question. If not, give the answer to the question.

❶ Triangle *ABC* is an isosceles triangle. Find its perimeter.

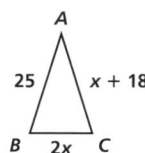

❷ Find the measure of the smallest angle in triangle *DEF*.

❸ Find the measure of the largest angle in triangle *DEF*.

❹ Find the area of right triangle *GHI*.

❺ A *pediment* is a triangular space filled with statues on the front of a building. The approximate measurements of an isosceles triangular pediment are shown below. Find the area of the pediment.

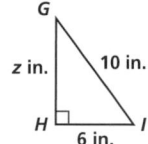

Discuss

Each problem requires that the student solve for a variable. Have students discuss what they need to know in order to find the given measurement and tell whether the value of the variable is the solution to the problem.

1. perimeter Lengths of all three sides of the triangle (25, 2*x*, and *x* + 18); triangle perimeter formula; no, but the value of the variable is needed to find the answer.

2. measure of smallest angle Triangle angle sum property; value of *y*; yes, the value of the variable is the answer since the smallest angle measures *y*°.

3. measure of largest angle Value of 7*y*; no, but the value of the variable is needed to find the answer.

4. area Pythagorean Theorem; length of the base (6 in.) and height (*z* in.); triangle area formula; no, but the value of the variable is needed to find the answer.

5. area Length of half the base; value of *h*; length of base; triangle area formula; no, but the value of the variable is needed to find the answer.

Answers

1. $x + 18 = 25$; $x = 7$; $25 + 25 + 2(7) = 64$

2. $180 = 12y$; $y = 15°$

3. $7y = 7(15) = 105°$

4. $6^2 + z^2 = 10^2$; $z = 8$; $\frac{1}{2}(6)(8) = 24$ in^2

5. $48^2 + h^2 = 50^2$; $h = 14$; $\frac{1}{2}(96)(14) = 672$ ft^2

Three-Dimensional Geometry

One-Minute Section Planner

Lesson	Materials	Resources
Hands-On Lab 6B Patterns of Solid Figures **NCTM:** Geometry, Representation **NAEP:** Geometry 1f ☐ SAT-9 ☐ SAT-10 ☐ ITBS ☐ CTBS ☐ MAT ☐ CAT	**Required** Rulers *(MK, CRB p. 117)* Protractors *(MK)* Tape	• *Hands-On Lab Activities*, pp. 45–46, 95
Lesson 6-5 Drawing Three-Dimensional Figures **NCTM:** Geometry, Measurement, Communication, Representation **NAEP:** Geometry 1e ☐ SAT-9 ☐ SAT-10 ☐ ITBS ☐ CTBS ☐ MAT ☐ CAT	**Required** Isometric dot paper *(CRB, p. 116)* **Optional** Teaching Transparency T23 *(CRB)* Boxes and rulers *(MK)*	• *Chapter 6 Resource Book*, pp. 46–55 • Daily Transparency T22, CRB • Additional Examples Transparencies T24–T26, CRB • *Alternate Openers: Explorations*, p. 49
Lesson 6-6 Volume of Prisms and Cylinders **NCTM:** Algebra, Geometry, Measurement, Reasoning and Proof, Communication, Connections **NAEP:** Measurement 1j ☑ SAT-9 ☑ SAT-10 ☐ ITBS ☑ CTBS ☑ MAT ☑ CAT	**Optional** Prisms and cylinders Teaching Transparency T28 *(CRB)* Centimeter cubes *(MK)*	• *Chapter 6 Resource Book*, pp. 56–65 • Daily Transparency T27, CRB • Additional Examples Transparencies T29–T33, CRB • *Alternate Openers: Explorations*, p. 50
Lesson 6-7 Volume of Pyramids and Cones **NCTM:** Algebra, Geometry, Measurement, Reasoning and Proof, Communication **NAEP:** Measurement 1j ☐ SAT-9 ☐ SAT-10 ☐ ITBS ☐ CTBS ☐ MAT ☑ CAT	**Optional** Cylinders and cones Teaching Transparency T35 *(CRB)* Recording Sheet for Reaching All Learners *(CRB, p. 118)* Scissors, tape, rulers *(MK)*	• *Chapter 6 Resource Book*, pp. 66–75 • Daily Transparency T34, CRB • Additional Examples Transparencies T36–T37, CRB • *Alternate Openers: Explorations*, p. 51
Lesson 6-8 Surface Area of Prisms and Cylinders **NCTM:** Algebra, Geometry, Measurement, Reasoning and Proof, Communication **NAEP:** Measurement 1j ☐ SAT-9 ☐ SAT-10 ☐ ITBS ☐ CTBS ☐ MAT ☑ CAT	**Optional** Boxes Teaching Transparency T39 *(CRB)*	• *Chapter 6 Resource Book*, pp. 76–85 • Daily Transparency T38, CRB • Additional Examples Transparencies T40–T41, CRB • *Alternate Openers: Explorations*, p. 52
Lesson 6-9 Surface Area of Pyramids and Cones **NCTM:** Algebra, Geometry, Measurement, Reasoning and Proof, Communication, Connections **NAEP:** Measurement 1j ☐ SAT-9 ☐ SAT-10 ☐ ITBS ☐ CTBS ☐ MAT ☐ CAT	**Optional** Paper cone Teaching Transparency T43 *(CRB)* Nets of pyramids and cones *(CRB, pp. 119–122)* Rulers *(MK)*	• *Chapter 6 Resource Book*, pp. 86–95 • Daily Transparency T42, CRB • Additional Examples Transparencies T44–T45, CRB • *Alternate Openers: Explorations*, p. 53
Lesson 6-10 Spheres **NCTM:** Geometry, Measurement, Reasoning and Proof, Communication **NAEP:** Measurement 1j ☐ SAT-9 ☐ SAT-10 ☐ ITBS ☐ CTBS ☐ MAT ☐ CAT	**Optional** Teaching Transparency T47 *(CRB)* Models of spheres	• *Chapter 6 Resource Book*, pp. 96–104 • Daily Transparency T46, CRB • Additional Examples Transparencies T48–T49, CRB • *Alternate Openers: Explorations*, p. 54
Extension Symmetry in Three Dimensions **NCTM:** Geometry, Connections **NAEP:** Geometry 4c ☐ SAT-9 ☐ SAT-10 ☐ ITBS ☐ CTBS ☐ MAT ☐ CAT	**Optional** Cheese and bread Teaching Transparency T50 *(CRB)*	• Additional Examples Transparencies T51–T52, CRB
Section 6B Assessment		• Section 6B Quiz, AR p. 16 • *Test and Practice Generator* CD-ROM

SAT = *Stanford Achievement Tests*   **ITBS** = *Iowa Test of Basic Skills*   **CTBS** = *Comprehensive Test of Basic Skills/Terra Nova*
MAT = *Metropolitan Achievement Test*   **CAT** = *California Achievement Test*
NCTM = Complete standards can be found on pages T29–T35.   **NAEP** = Complete standards can be found on pages A54–A58.

Section Overview

Three-Dimensional Figures   *Hands-On Lab 6B, Lesson 6-5, Extension*

Why? Visualizing and drawing three-dimensional figures helps you understand the mathematical concepts and relationships in three-dimensional space.

Nets of Three-Dimensional Figures

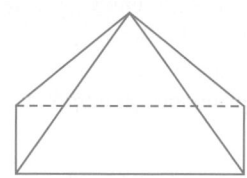

Rectangular pyramid

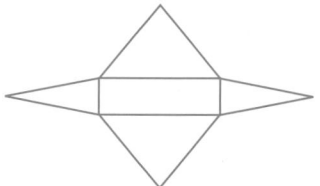

Net of a rectangular pyramid

Drawing Three-Dimensional Figures

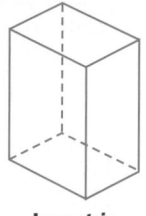

Isometric

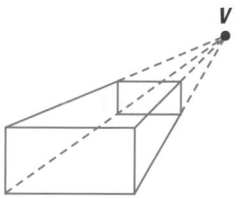

One-point perspective

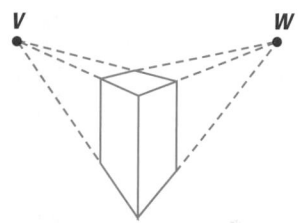

Two-point perspective

Volume and Surface Area   *Lessons 6-6 through 6-10*

Why? Prisms, cylinders, pyramids, cones, and spheres are used in art, architecture, and manufacturing.

| **Volume** is the number of cubic units needed to fill a three-dimensional figure. | **Surface area** is the number of **square units** needed to **cover** all surfaces of a three-dimensional figure. |

Volume and Surface Area Formulas

Solid	Prism	Cylinder	Pyramid	Cone	Sphere
Volume	$V = Bh$	$V = Bh = (\pi r^2)h$	$V = \frac{1}{3}Bh$	$V = \frac{1}{3}Bh$ $= \frac{1}{3}(\pi r^2)h$	$V = \frac{4}{3}\pi r^3$
Surface area	$S = 2B + F$ $= 2B + Ph$	$S = 2B + L$ $= 2\pi r^2 + 2\pi rh$	$S = B + F$ $= B + \frac{1}{2}p\ell$	$S = B + L$ $= \pi r^2 + \pi r\ell$	$S = 4\pi r^2$

Key for Variables:

B area of base (circle or polygon)
F lateral area of polygon surfaces
L lateral area of curved surface

h height (always perpendicular to base)
r radius of circular base

P perimeter of base
ℓ slant height

Hands-On LAB
6B Patterns of Solid Figures

Pacing: Traditional 1 day
Block $\frac{1}{2}$ day

Objective: To use paper models to explore patterns of solid figures

Materials: Ruler, protractor, tape, paper

Lab Resources

Hands-On Lab Activities . pp. 45–46, 95

Using the Pages

Discuss with students how to use a ruler and protractor to precisely create each figure. Discuss the importance of drawing figures in making nets.

Use a ruler and protractor to accurately draw each figure. Test angle measurements and side lengths when you are finished.
Check students' drawings.

1. equilateral triangle

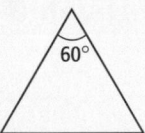

2. square

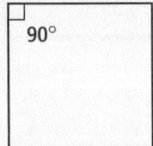

3. regular pentagon

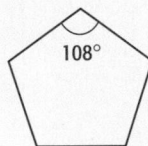

Patterns of Solid Figures

Use with Lesson 6-5

WHAT YOU NEED
• Ruler
• Protractor
• Tape
• Paper

REMEMBER
• A polygon is a closed plane figure formed by three or more line segments.
• The faces of a regular polyhedron are congruent polygons.

internet connect
Lab Resources Online
go.hrw.com
KEYWORD: MP4 Lab6B

A **polyhedron** is a solid figure in which every surface is a polygon. A net is a pattern of polygons used to model a regular polyhedron.

There are 5 regular polyhedra: tetrahedron, cube, octahedron, dodecahedron, and icosahedron.

Activity

1 Follow the directions to make each net. Then fold your nets into three-dimensional figures.

a. Draw a 2 in. equilateral triangle.
The measurement of each angle will be 60°.
Draw three more of them to look like Figure 1.
There will be 4 triangles.
Join the common edges and tape them together.
This is the net of a **tetrahedron**.

Figure 1

b. Draw a 2 in. square.
Draw 5 more of them to look like Figure 2.
There will be **6** squares.
Join the common edges and tape them together.
This is the net of a **cube.**

Figure 2

c. Draw a 2 in. equilateral triangle.
Draw 7 more of them to look like Figure 3.
There will be **8** triangles.
Join the common edges and tape them together.
This is the net of an **octahedron.**

Figure 3

d. Draw a regular pentagon with each side measuring 2 inches. The measurement of each angle will be 108°.
Draw 11 more of them to look like Figure 4.
There will be **12** pentagons.
Join the common edges and tape them together.
This is the net of a **dodecahedron.**

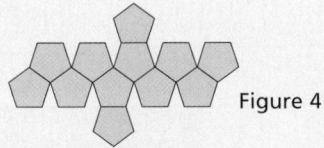
Figure 4

e. Draw a 2 in. equilateral triangle.
Draw 19 more of them to look like Figure 5.
There will be **20** triangles.
Join the common edges and tape them together.
This is the net of an **icosahedron.**

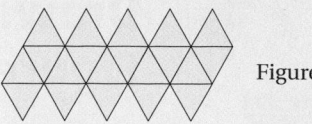 Figure 5

Copy the following table. Compare the number of vertices, faces, and edges in your polyhedra to the numbers listed in the table.

Polyhedron	Number of Vertices (V)	Number of Faces (F)	Number of Edges (E)
Tetrahedron	4	4	6
Cube	8	6	12
Octahedron	6	8	12
Dodecahedron	20	12	30
Icosahedron	12	20	30

Think and Discuss

1. Look for patterns in the table. What relationship can you find between the number of vertices, the number of faces, and the number of edges of regular polyhedra?   **V + F − 2 = E**

2. Can you make a net for an octahedron that is different from the net in Figure 3? Show the new net.   **Possible answer:**

Try This

Copy and fold each net to determine whether it is the net of a polyhedron. If so, name the regular polyhedron that the net forms.

1.   **no**   **2.**   **yes; tetrahedron**   **3.**   **yes; octahedron**

4.   **yes; cube**   **5.**   **no**   **6.**   **yes; dodecahedron**

Give the missing number for each regular polygon.

7. 12 edges **6**
   ▮ vertices
   8 faces

8. ▮ edges **30**
   12 vertices
   20 faces

9. 30 edges **12**
   20 vertices
   ▮ faces

10. 6 edges **4**
   ▮ vertices
   4 faces

Pacing: Traditional 1 day
Block $\frac{1}{2}$ day

Objective: Students draw and identify parts of three-dimensional figures.

Warm Up

Find the circumference of each circle, both in terms of π and to the nearest tenth. Use 3.14 for π.

1. radius 2.5 m 5π m; 15.7 m

2. diameter 8.8 cm 8.8π cm; 27.6 cm

Find the area of each circle, both in terms of π and to the nearest tenth. Use 3.14 for π.

3. radius 14 ft 196π ft^2; 615.4 ft^2

4. diameter 14 ft 49π ft^2; 153.9 ft^2

Problem of the Day

What is the least number of lines needed to draw 5 squares?   **6**

Available on Daily Transparency in CRB

Teacher: Why is your perspective drawing a blank sheet of paper?

Student: You told me to draw it with a vanishing point!

6-5 Drawing Three-Dimensional Figures

Learn to draw and identify parts of three-dimensional figures.

Vocabulary

face
edge
vertex
perspective
vanishing point
horizon line

Architects use drawings to show what the exteriors of buildings will look like. Since they are drawing three-dimensional objects on two-dimensional surfaces, they must use special techniques to give the appearance of three dimensions.

Three-dimensional figures have *faces*, *edges*, and *vertices*. A **face** is a flat surface, an **edge** is where two faces meet, and a **vertex** is where three or more edges meet.

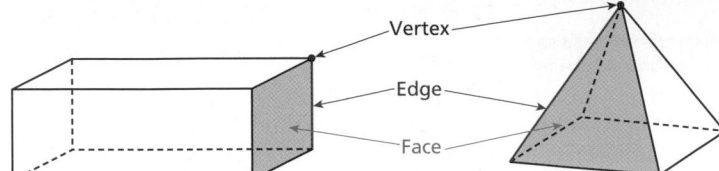

Isometric dot paper can be used to draw three-dimensional figures.

EXAMPLE **Drawing a Rectangular Box**

Use isometric dot paper to sketch a rectangular box that is 4 units long, 2 units wide, and 3 units high.

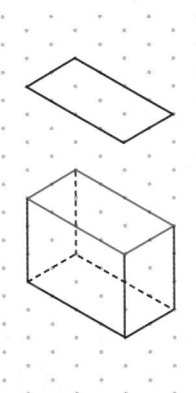

Step 1: Lightly draw the edges of the bottom face. It will look like a parallelogram.
2 units by 4 units

Step 2: Lightly draw the vertical line segments from the vertices of the base.
3 units high

Step 3: Lightly draw the top face by connecting the vertical lines to form a parallelogram.
2 units by 4 units

Step 4: Darken the lines.
Use solid lines for the edges that are visible and dashed lines for the edges that are hidden.

---

1 Introduce

Alternate Opener

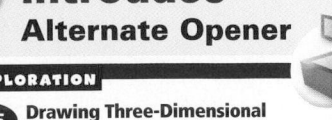
 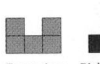

Motivate

Show students a simple drawing of a square. Draw another square of the same size with a corner at the center of the first square. Then connect the corners of each square to create a cube. Erase the lines that would be hidden if the cube were solid. Explain to students that the technique of making a flat drawing look three-dimensional is called *perspective*.

Exploration worksheet and answers on Chapter 6 Resource Book pp. 47 and 132

---

2 Teach

Guided Instruction

In this lesson, students learn to draw and identify parts of three-dimensional figures. Show students a rectangular box and define the terms *face*, *edge*, and *vertex*. Lead students through the steps in Example 1, using isometric dot paper to create sketches of three-dimensional figures. Then guide them through the examples for drawing sketches with one-point and two-point perspective (Teaching Transparency T23, CRB).

 Show students pieces of art in which perspective is used (and if possible, some in which it is not used).

Perspective is a technique used to make drawings of three-dimensional objects appear to have depth and distance. In one-point perspective drawings, there is one **vanishing point**.

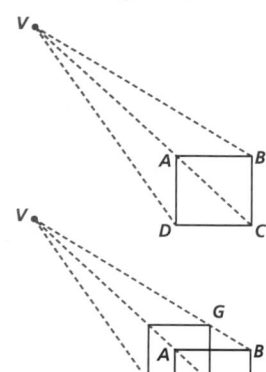

Vanishing point

Visible edge

Hidden edge

EXAMPLE 2 Sketching a One-Point Perspective Drawing

Sketch a one-point perspective drawing of a rectangular box.

Step 1: Draw a rectangle. This will be the front face.
Label the vertices A through D.

Step 2: Mark a vanishing point *V* somewhere above your rectangle, and draw a dashed line from each vertex to *V*.

Step 3: Choose a point *G* on $\overline{BV}$. Lightly draw a smaller rectangle that has *G* as one of its vertices.

Step 4: Connect the vertices of the two rectangles along the dashed lines.

Step 5: Darken the visible edges, and draw dashed segments for the hidden edges. Erase the vanishing point and all the lines connecting it to the vertices.

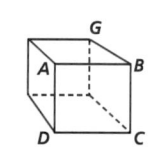

You can also draw a figure in two-point perspective by using two vanishing points and a **horizon line**.

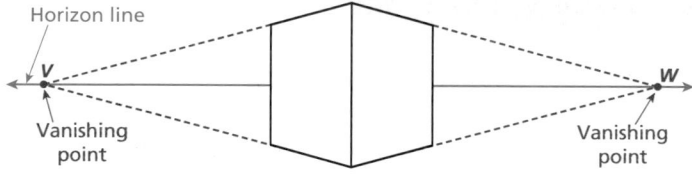

Horizon line

Vanishing point

Vanishing point

Moving the horizon line up and down gives you different views of the figure.

Additional Examples

Example 1

Use isometric dot paper to sketch a rectangular box that is 5 units long, 3 units wide, and 2 units high.

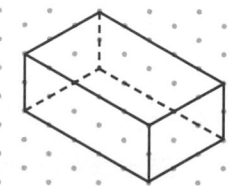

Example 2

Sketch a one-point perspective drawing of a cube.

Example 3

Sketch a two-point perspective drawing of a cube.

Reaching All Learners
Through Hands-On Experience

Have boxes of different sizes and shapes available for students to measure (e.g., cereal boxes, paper clip boxes, etc.). Have students work in pairs. Have each student choose a box and measure its dimensions. Ask the students to sketch the box using isometric dot paper (Chapter 6 Resource Book p. 116), letting each space on the paper represent one inch (or one centimeter). When sketches are done, have each student's partner try to determine which box he or she sketched.

Answers

27. Possible answer:

28. Possible answer: Only three faces and nine edges can be seen in an isometric drawing of a cube.

29. Possible answer: All parallel edges of a cube are parallel on dot paper. In a perspective drawing, the edges drawn on lines that intersect at the vanishing point are not parallel. The lengths of the edges are equal on dot paper but not in a perspective drawing.

30. Possible answer:

Journal

Have students look at some photographs or paintings and describe what they see that pertains to this lesson. They may describe the type of perspective, identify some vertices, faces, and edges, or identify an area of the picture that serves as a vanishing point.

Test Prep Doctor

For Exercise 35, because the triangle has two congruent sides and a 50° angle between them, it also has two congruent angles that each measure 65°. Choice **A** is not correct because an equilateral triangle has three 60° angles. Choice **C** is not correct because a scalene triangle has no congruent sides. Choice **D** is not correct because none of the angles are obtuse. The correct choice is **B**.

Lesson Quiz

1. Use isometric dot paper to sketch a rectangular box 3 units tall with a base 2 units by 5 units.

2. Sketch a cube in one-point perspective.

3. Sketch a brick in two-point perspective.

Answers on p. A25

Available on Daily Transparency in CRB

Transportation LINK

EURO TUNNEL

The Chunnel was built from both ends at the same time. Specially designed tunneling machines completed an average of 125 m per week.

25. *TRANSPORTATION* Engineers long dreamed of linking England with the European mainland. In 1994, the dream became a reality with the opening of the Channel Tunnel, or Chunnel, which links Britain and France. The drawing shows the train *Eurostar* in the Chunnel. Is this an example of one-point or two-point perspective? **one-point**

26. *TECHNOLOGY* Architects often use CADD (Computer Aided Design/Drafting) programs to create 3-D images of their ideas. Is the image an example of one-point or two-point perspective? **two-point**

27. Copy the drawing below, and add another building like the one shown, with its lower front edge at $\overline{AB}$.

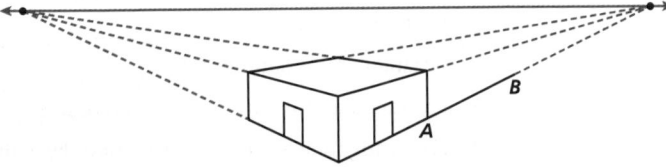

28. *WHAT'S THE ERROR?* A student sketched a 3-unit cube on dot paper. The student said that four faces and eight edges were visible in the sketch. What was the student's error?

29. *WRITE ABOUT IT* Describe the differences between a dot-paper drawing of a cube and a perspective drawing of a cube.

30. *CHALLENGE* Use one-point perspective to create a block-letter sign of your name.

Spiral Review

Write each number in standard notation. (Lesson 2-9)

31. 2.75×10^3 **2750**   **32.** -4.2×10^2 **−420**   **33.** 6.3×10^{-7} **0.00000063**   **34.** -1.9×10^{-4} **−0.00019**

35. TEST PREP Which type of triangle can be constructed with a 50° angle between two 8-inch sides? (Lesson 5-3) **B**

  A Equilateral   **B** Isosceles   **C** Scalene   **D** Obtuse

---

CHALLENGE 6-5

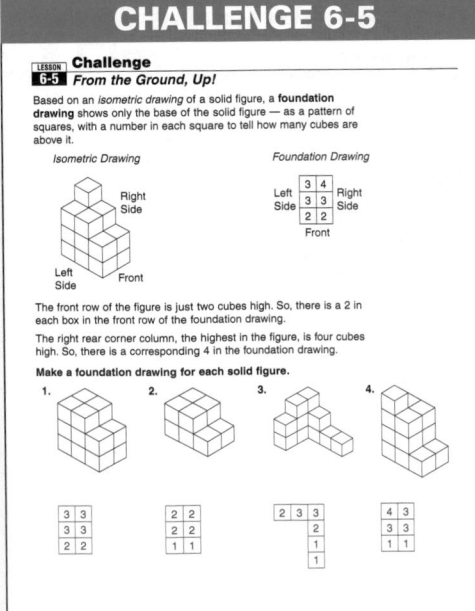

LESSON 6-5 Challenge

From the Ground, Up!

Based on an *isometric* drawing of a solid figure, a **foundation drawing** shows only the base of the solid figure — as a pattern of squares, with a number in each square to tell how many cubes are above it.

Isometric Drawing

Foundation Drawing

Left Side	3	4	Right Side
	3	3	
	2	2	

Front

The front row of the figure is just two cubes high. So, there is a 2 in each box in the front row of the foundation drawing.

The right rear corner column, the highest in the figure, is four cubes high. So, there is a corresponding 4 in the foundation drawing.

Make a foundation drawing for each solid figure.

1.   2.   3.   4.

3	3
3	3
2	2

2	2
3	
1	1

2	3	3
2		
1		
1		

4	3
3	3
1	1

PROBLEM SOLVING 6-5

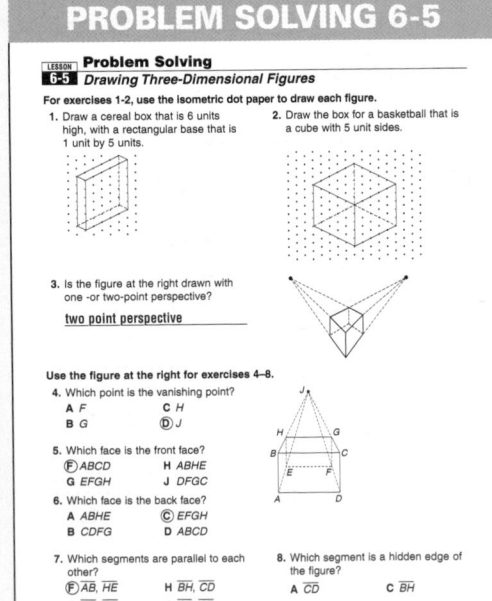

LESSON 6-5 Problem Solving

Drawing Three-Dimensional Figures

For exercises 1-2, use the isometric dot paper to draw each figure.

1. Draw a cereal box that is 6 units high, with a rectangular base that is 1 unit by 5 units.

2. Draw the box for a basketball that is a cube with 5 unit sides.

3. Is the figure at the right drawn with one -or two-point perspective?

two point perspective

Use the figure at the right for exercises 4–8.

4. Which point is the vanishing point?
  A F   C H
  B G   **D** J

5. Which face is the front face?
  F ABCD   H ABHE
  G EFGH   J DFGC

6. Which face is the back face?
  A ABHE   **C** EFGH
  B CDFG   D ABCD

7. Which segments are parallel to each other?
  F $\overline{AB}$, $\overline{HE}$ H $\overline{BH}$, $\overline{CD}$
  G $\overline{HE}$, $\overline{AD}$ J $\overline{AE}$, $\overline{CD}$

8. Which segment is a hidden edge of the figure?
  A $\overline{CD}$ C $\overline{BH}$
  B $\overline{AE}$ D $\overline{AB}$

6-6 Volume of Prisms and Cylinders

Learn to find the volume of prisms and cylinders.

Vocabulary
prism
cylinder

Kansai International Airport, in Japan, is built on the world's largest man-made island. To find the amount of rock, gravel, and concrete needed to build the island, you need to know how to find the volume of a *rectangular prism*.

A **prism** is a three-dimensional figure named for the shape of its bases. The two bases are congruent polygons. All of the other faces are parallelograms. A **cylinder** has two circular bases.

Remember!

If all six faces of a rectangular prism are squares, it is a cube.

Triangular prism   Rectangular prism   Cylinder

Height →   Height →   Height →

Base   Base   Base

VOLUME OF PRISMS AND CYLINDERS		
Words	**Numbers**	**Formula**
Prism: The volume V of a prism is the area of the base B times the height h.	$B = 2(5)$ $= 10 \text{ units}^2$ $V = (10)(3)$ $= 30 \text{ units}^3$	$V = Bh$
Cylinder: The volume of a cylinder is the area of the base B times the height h.	$B = \pi(2^2)$ $= 4\pi \text{ units}^2$ $V = (4\pi)(6) = 24\pi$ $\approx 75.4 \text{ units}^3$	$V = Bh$ $= (\pi r^2)h$

E X A M P L E **1**   **Finding the Volume of Prisms and Cylinders**

Find the volume of each figure to the nearest tenth.

Helpful Hint

Area is measured in *square units.* Volume is measured in *cubic units.*

A A rectangular prism with base 1 m by 3 m and height 6 m.

$B = 1 \cdot 3 = 3 \text{ m}^2$ *Area of base*
$V = Bh$ *Volume of prism*
$\quad = 3 \cdot 6 = 18 \text{ m}^3$

---

1 Introduce

Alternate Opener

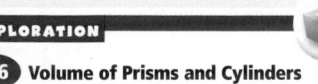

EXPLORATION

6-6 Volume of Prisms and Cylinders

The drawing below represents a box that has been unfolded and laid flat. The blue rectangles represent the bottom and the lid of the box. The green rectangles represent the sides of the box.

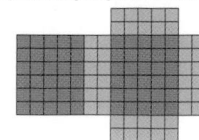

1. What are the dimensions of the bottom and the lid of the box?

2. What is the height of the box when it is assembled?

3. What are the dimensions of each side of the box?

4. How many 1-by-1-by-1 cubes will it take to fill the box when it is assembled? (*Hint:* This is the *volume* of the box.)

Think and Discuss

5. **Explain** how you can determine the volume of a box if you know the dimensions.

6. **Discuss** whether you could figure out the height of a box if you know the volume and the dimensions of the base.

Motivate

Show students prisms and cylinders of different shapes and sizes. Include prisms other than rectangular prisms. Point out that all of the solids you show have at least two congruent surfaces, which are called *bases.* Explain that to find the volume of one of these solids, you would multiply the area of the base by the height.

Exploration worksheet and answers on Chapter 6 Resource Book pp. 57 and 134

6-6 Organizer

Pacing: Traditional 1 day
         Block $\frac{1}{2}$ day

Objective: Students find the volume of prisms and cylinders.

Warm Up

Make a sketch of a closed book using two-point perspective.

Possible answer:

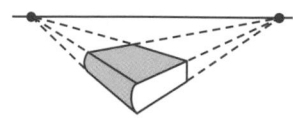

Problem of the Day

You are painting identical wooden cubes red and blue. Each cube must have 3 red faces and 3 blue faces. How many cubes can you paint that can be distinguished from one another?   only 2

Available on Daily Transparency in CRB

Math Humor

What happens to a mathematician who commits a crime? She goes to prism!

---

2 Teach

Lesson Presentation

Guided Instruction

In this lesson, students learn to find the volume of prisms and cylinders. Point out that the congruent bases of a prism are polygons (such as rectangles and triangles) and that the bases of a cylinder are circles (Teaching Transparency T28, CRB). Review the area formulas for these figures. Explain that the volume of any prism or cylinder is found by multiplying the area of its base by its height. Explain that because these figures are three-dimensional, volume is measured in cubic units.

Example 1

Find the volume of each figure to the nearest tenth.

A. a rectangular prism with base 2 cm by 5 cm and a height 3 cm

$V = 30$ cm^3

B.

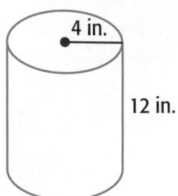
4 in.
12 in.

$V = 192\pi \approx 602.9$ in^3

C.

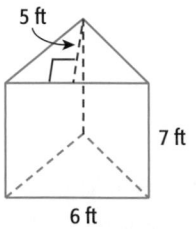
5 ft
7 ft
6 ft

$V = 105$ ft^3

Example 2

A. A juice box measures 3 in. by 2 in. by 4 in. Explain whether tripling the length, width, or height of the box would triple the amount of juice the box holds.
The original box has a volume of 24 in^3. You could triple the volume to 72 in^3 by tripling any one of the dimensions. So tripling the length, width, or height would triple the amount of juice the box would hold.

Find the volume of each figure to the nearest tenth.

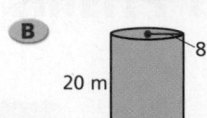

B

8 m
20 m

$B = \pi(8^2) = 64\pi$ m^2 *Area of base*
$V = Bh$ *Volume of a cylinder*
$= 64\pi \cdot 20$
$= 1280\pi \approx 4021.2$ m^3

C

4 ft
7 ft
11 ft

$B = \frac{1}{2} \cdot 4 \cdot 7 = 14$ ft^2 *Area of base*
$V = Bh$ *Volume of a prism*
$= 14 \cdot 11$
$= 154$ ft^3

The volume of a rectangular prism can be written as $V = \ell wh$, where ℓ is the length, w is the width, and h is the height.

EXAMPLE 2 Exploring the Effects of Changing Dimensions

A A juice box measures 3 in. by 2 in. by 4 in. Explain whether doubling the length, width, or height of the box would double the amount of juice the box holds.

Original Dimensions	Double the Length	Double the Width	Double the Height
$V = \ell wh$	$V = (2\ell)wh$	$V = \ell(2w)h$	$V = \ell w(2h)$
$= 3 \cdot 2 \cdot 4$	$= 6 \cdot 2 \cdot 4$	$= 3 \cdot 4 \cdot 4$	$= 3 \cdot 2 \cdot 8$
$= 24$ in^3	$= 48$ in^3	$= 48$ in^3	$= 48$ in^3

The original box has a volume of 24 in^3. You could double the volume to 48 in^3 by doubling any one of the dimensions. So doubling the length, width, or height would double the amount of juice the box holds.

B A juice can has a radius of 1.5 in. and a height of 5 in. Explain whether doubling the height of the can would have the same effect on the volume as doubling the radius.

Original Dimensions	Double the Radius	Double the Height
$V = \pi r^2 h$	$V = \pi(2r)^2 h$	$V = \pi r^2(2h)$
$= 1.5^2\pi \cdot 5$	$= 3^2\pi \cdot 5$	$= 1.5^2\pi \cdot (2 \cdot 5)$
$= 11.25\pi$ in^3	$= 45\pi$ in^3	$= 22.5\pi$ in^3

By doubling the height, you would double the volume. By doubling the radius, you would increase the volume to four times the original.

Teach

Reaching All Learners
Through Number Sense

Have students build a variety of prisms using centimeter cubes. Ask students to count the cubes to find the height, area of the base, and volume of each prism and to record the data in a table as shown.

Area of base	Height	Volume

Ask students to find a relationship between the area of the base, the height, and the volume. Possible answer: The volume is the product of the area of the base and the height, or $V = Bh$.

EXAMPLE **3** *Construction Application*

Kansai International Airport is on a man-made island that is a rectangular prism measuring 60 ft deep, 4000 ft wide, and 2.5 miles long. What is the volume of rock, gravel, and concrete that was needed to build the island?

$$\text{length} = 2.5 \text{ mi} = 2.5(5280) \text{ ft}$$
$$= 13,200 \text{ ft}$$ *1 mi = 5280 ft*

$$\text{width} = 4000 \text{ ft}$$

$$\text{height} = 60 \text{ ft}$$

$$V = 13,200 \cdot 4000 \cdot 60 \text{ ft}^3$$ *V = lwh*
$$= 3,168,000,000 \text{ ft}^3$$

The volume of rock, gravel, and concrete needed was 3,168,000,000 ft³, which is equivalent to nearly 24 billion gallons of water.

To find the volume of a composite three-dimensional figure, find the volume of each part and add the volumes together.

EXAMPLE **4** Finding the Volume of Composite Figures

Find the volume of the milk carton.

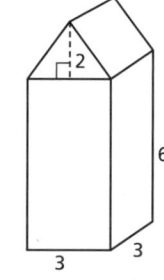

Volume of milk carton	=	Volume of rectangular prism	+	Volume of triangular prism
V	=	$(3)(3)(6)$	+	$\frac{1}{2}(3)(2)(3)$
	=	54	+	9
	=	63 in³		

The volume is 63 in³, or about 0.27 gallons.

Think and Discuss

1. **Give an example** that shows that two rectangular prisms can have different heights but the same volume.

2. **Apply** your results from Example 2 to make a conclusion about changing dimensions in a triangular prism.

3. **Describe** what happens to the volume of a cylinder when the diameter of the base is tripled.

Additional Examples

B. A juice can has a radius of 2 in. and a height of 5 in. Explain whether tripling the height of the can would have the same effect on the volume as tripling the radius.

By tripling the height, you would triple the volume. By tripling the radius, you would increase the volume to nine times the original volume.

Example **3**

A section of an airport runway is a rectangular prism measuring 2 feet thick, 100 feet wide, and 1.5 miles long. What is the volume of material that was needed to build the runway?

$V = 1,584,000 \text{ ft}^3$

Example **4**

Find the volume of the barn.

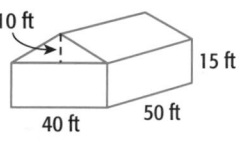

$V = 40,000 \text{ ft}^3$

---

3 **Close**

Summarize

Review the definitions of *prism* and *cylinder*. Review the formula for finding the volume of each. Discuss the differences between area and volume. You may want to show some examples of flat and solid figures and ask students which measure applies to each one. As a challenge, you may want them to estimate the value of each measure.

Possible answer: Area applies to flat (two-dimensional) figures and is measured in square units. Volume applies to solid (three-dimensional) figures and is measured in cubic units.

Answers to Think and Discuss

Possible answers:

1. A prism with a 2 × 3 unit base and a height of 2 units has the same volume (12 units³) as a prism with a 6 × 2 unit base and a height of 1 unit.

2. If you double one dimension of a triangular prism, the prism's volume doubles. If you double two dimensions, the volume becomes four times the original volume. If you double three dimensions, the volume becomes eight times the original volume.

3. The volume will increase by a factor of 3², or 9.

FOR EXTRA PRACTICE
see page 743

internet connect
Homework Help Online
go.hrw.com   Keyword: MP4 6-6

GUIDED PRACTICE

See Example **1**   Find the volume of each figure to the nearest tenth. Use 3.14 for π.

1.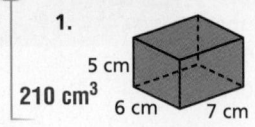
5 cm, 6 cm, 7 cm
210 cm³

2.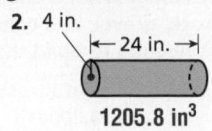
4 in., ← 24 in. →
1205.8 in³

3.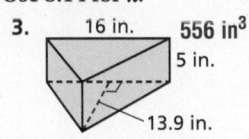
16 in., 556 in³, 5 in., 13.9 in.

See Example **2**   4. A box measures 4 in. by 3 in. by 5 in. Explain whether tripling a side from 4 in. to 12 in. would triple the volume of the box.

5. A can of vegetables has radius 2 in. and height 4 in. Explain whether tripling the radius would triple the volume of the can.

See Example **3**   6. Grain is stored in cylindrical structures called *silos*. What is the volume of a silo with diameter 15 feet and height 25 feet?
1406.25π ft³ ≈ 4417.9 ft³

See Example **4**   7. Find the volume of the barn. **4725 ft³**

25 ft, 20 ft, 10 ft, 18 ft, 15 ft

INDEPENDENT PRACTICE

See Example **1**   Find the volume of each figure to the nearest tenth. Use 3.14 for π.

8.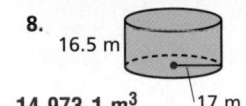
16.5 m, 17 m
14,973.1 m³

9.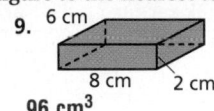
6 cm, 8 cm, 2 cm
96 cm³

10.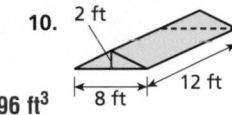
2 ft, 8 ft, 12 ft
96 ft³

See Example **2**   11. A toy box measures 4 ft by 3 ft by 2 ft. Explain whether increasing the height by four times, from 2 ft to 8 ft, would increase the volume by four times.

12. A cylindrical oatmeal box has diameter 4 in. and height 7 in. Explain whether increasing the diameter by 1.5 times would increase the volume by 1.5 times.

See Example **3**   13. An ink cartridge for a printer is 5 cm by 3 cm by 4 cm. What is the volume of the ink cartridge? **60 cm³**

See Example **4**   14. Find the volume of the box containing the ink cartridge. **94.5 cm³**

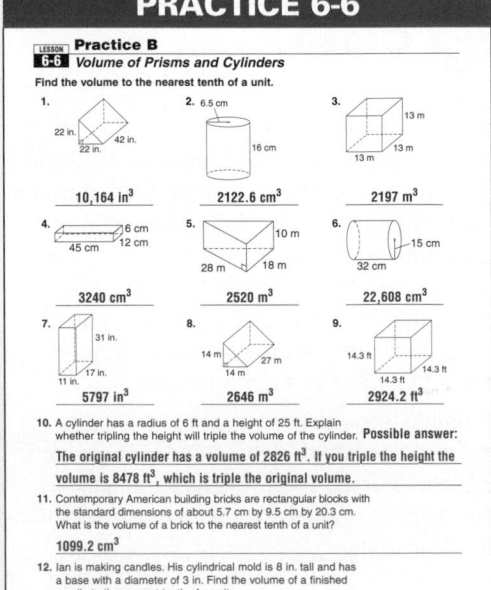
3.5 cm, 4.5 cm, 6 cm

Assignment Guide

If you finished Example **1** assign:
Core 1–3, 8–10, 22–25
Enriched 1–3, 8–10, 22–25

If you finished Example **2** assign:
Core 1–5, 8–12, 22–25
Enriched 1–5 odd, 8–12, 19, 20, 22–25

If you finished Example **3** assign:
Core 1–6, 8–13, 15, 16, 22–25
Enriched 1–5 odd, 8–13, 15–20, 22–25

If you finished Example **4** assign:
Core 1–16, 22–25
Enriched 1–7 odd, 8–25

Answers

4. Yes; the volume of the box is 60 in³. Tripling the length gives a volume of 180 in³. The volume is tripled.

5. No; the volume of the can is 16π in³. Tripling the radius gives a volume of 144π in³, which is nine times the original.

11. Yes; the volume of the box is 24 ft³. Increasing the height by four times gives a volume of 96 ft³, which is four times the original.

12. No; the volume of the can is 28π in³. Increasing the diameter by 1.5 times gives a volume of 63π in³, which is 2.25 times the original.

Math Background

The volume formula, $V = Bh$, applies to *right* cylinders and prisms, as studied in the lesson, and also to *oblique* cylinders and prisms, in which the axis is not perpendicular to the base.

Oblique cylinder

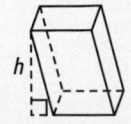
Oblique prism

This fact is formally stated by Cavalieri's principle: If two solids have equal heights and the cross sections formed by every plane parallel to the bases of both solids have equal areas, then the two solids have equal volumes.

RETEACH 6-6

LESSON 6-6 Reteach
Volume of Prisms and Cylinders

Volume = number of cubic units inside a solid figure

To find the volume of this solid figure:
3 cm, 6 cm, 4 cm

Count the number of cubic centimeters in one "slice" of the figure.
4 × 3 = 12
Multiply by the number of "slices." 12 × 6 = 72 cm³

Complete to find the volume of each solid figure.

1. 4 in., 2 in., 2 in.
number in³ in a slice
= 4 × 2 = 8
number of slices = 2
volume = 16 in³

2. 3 cm, 5 cm, 4 cm
number cm³ in a slice
= 3 × 5 = 15
number of slices = 4
volume = 60 cm³

3. 5 mm, 6 mm, 4 mm
number mm³ in a slice
= 5 × 6 = 30
number of slices = 4
volume = 120 mm³

PRACTICE 6-6

LESSON 6-6 Practice B
Volume of Prisms and Cylinders

Find the volume to the nearest tenth of a unit.

1. 22 in., 22 in., 42 in.
10,164 in³

2. 6.5 cm, 16 cm
2122.6 cm³

3. 13 m, 13 m, 13 m
2197 m³

4. 6 cm, 45 cm, 12 cm
3240 cm³

5. 10 m, 28 m, 18 m
2520 m³

6. 32 cm, 15 cm
22,608 cm³

7. 31 in., 11 in., 17 in.
5797 in³

8. 14 m, 27 m, 14 m
2646 m³

9. 14.3 ft, 14.3 ft, 14.3 ft
2924.2 ft³

10. A cylinder has a radius of 6 ft and a height of 25 ft. Explain whether tripling the height will triple the volume of the cylinder. **Possible answer: The original cylinder has a volume of 2826 ft³. If you triple the height the volume is 8478 ft³, which is triple the original volume.**

11. Contemporary American building bricks are rectangular blocks with the standard dimensions of about 5.7 cm by 9.5 cm by 20.3 cm. What is the volume of a brick to the nearest tenth of a unit?
1099.2 cm³

12. Ian is making candles. His cylindrical mold is 8 in. tall and has a base with a diameter of 3 in. Find the volume of a finished candle to the nearest tenth of a unit.
56.5 in³

PRACTICE AND PROBLEM SOLVING

15. *LIFE SCIENCE* The cylindrical Giant Ocean Tank at the New England Aquarium in Boston has a volume of 200,000 gallons.

a. One gallon of water equals 231 cubic inches. How many cubic inches of water are in the Giant Ocean Tank? **46,200,000 in³**

b. Use your answer from part **a** as the volume. The tank is 24 ft deep. Find the radius in feet of the Giant Ocean Tank. **about 18.8 ft**

16. *ENTERTAINMENT* An outdoor theater group sets up a portable stage. The stage comes in sections that are 48 in. by 96 in. by 36 in.

a. What are the dimensions in feet of one stage section? **4 ft by 8 ft by 3 ft**

b. What is the volume in cubic feet of one section? **96 ft³**

c. If the stage has a total volume of 864 ft³, how many sections make up the stage? **9**

17. *SOCIAL STUDIES* The tablet held by the Statue of Liberty is approximately a rectangular prism with volume 1,107,096 in³. Estimate the thickness of the tablet. **about 20.5 in.**

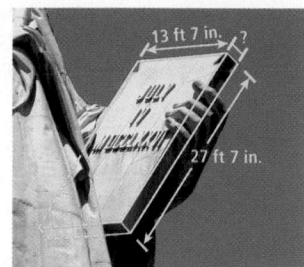
13 ft 7 in.
27 ft 7 in.

18. *LIFE SCIENCE* Air has about 4000 bacteria per cubic meter. There are about 120,000 bacteria in a room that is 3 m long by 4 m wide. What is the height of the room? **2.5 m**

19. *WHAT'S THE ERROR?* A student read this statement in a book: "The volume of a triangular prism with height 10 cm and base area 25 cm is 250 cm³." Correct the error in the statement.

20. *WRITE ABOUT IT* Explain why one cubic foot equals 1728 cubic inches.

21. *CHALLENGE* A 6 cm section of plastic water pipe has inner diameter 12 cm and outer diameter 15 cm. Find the volume of the plastic pipe, not the hollow interior, to the nearest tenth. **121.5π ≈ 381.7 cm³**

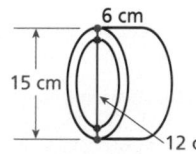
6 cm
15 cm
12 cm

Spiral Review

Find the mean, median, and mode of each data set to the nearest tenth. (Lesson 4-3)

22. 3, 5, 5, 6, 9, 3, 5, 2, 5
4.8; 5; 5

23. 17, 15, 14, 16, 18, 13
15.5; 15.5; no mode

24. 100, 75, 48, 75, 48, 63, 45
64.9; 63; 48 and 75

25. TEST PREP Find the sum of the angle measures of an octagon. (Lesson 5-4) **C**

A 8°   B 135°   C 1080°   D 1440°

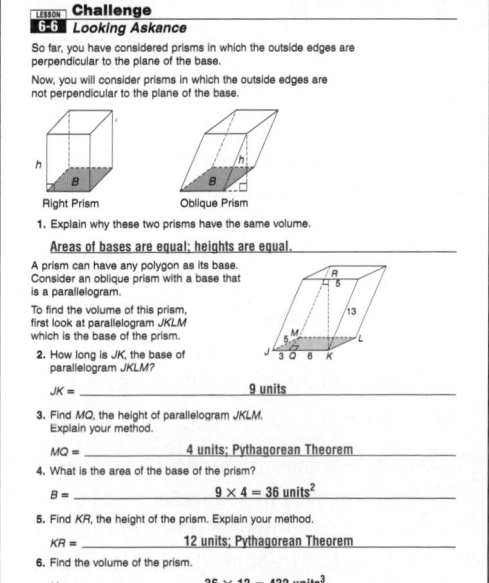

CHALLENGE 6-6

Challenge
6-6 Looking Askance

So far, you have considered prisms in which the outside edges are perpendicular to the plane of the base.

Now, you will consider prisms in which the outside edges are not perpendicular to the plane of the base.

Right Prism   Oblique Prism

1. Explain why these two prisms have the same volume.
   Areas of bases are equal; heights are equal.

A prism can have any polygon as its base. Consider an oblique prism with a base that is a parallelogram.

To find the volume of this prism, first look at parallelogram *JKLM* which is the base of the prism.

2. How long is *JK*, the base of parallelogram *JKLM*?
   JK = _____ **9 units**

3. Find *MQ*, the height of parallelogram *JKLM*. Explain your method.
   MQ = _____ **4 units; Pythagorean Theorem**

4. What is the area of the base of the prism?
   B = _____ **9 × 4 = 36 units²**

5. Find *KR*, the height of the prism. Explain your method.
   KR = _____ **12 units; Pythagorean Theorem**

6. Find the volume of the prism.
   V = _____ **36 × 12 = 432 units³**

PROBLEM SOLVING 6-6

Problem Solving
6-6 Volume of Prisms and Cylinders

Round to the nearest tenth. Write the correct answer.

1. A contractor pours a sidewalk that is 4 inches deep, 1 yard wide, and 20 yards long. How many cubic yards of concrete will be needed? (Hint: 36 inches = 1 yard.)
   2.2 yd³

2. A refrigerator has inside measurements of 50 cm by 118 cm by 44 cm. What is the capacity of the refrigerator?
   259,600 cm³

A rectangular box is 2 inches high, 3.5 inches wide and 4 inches long. A cylindrical box is 3.5 inches high and has a diameter of 3.2 inches. Use 3.14 for π. Round to the nearest tenth.

3. Which box has a larger volume?
   Cylinder

4. How much bigger is the larger box?
   0.1 in³

Use 3.14 for π. Choose the letter for the best answer.

5. A child's wading pool has a diameter of 5 feet and a height of 1 foot. How much water would it take to fill the pool? Round to the nearest gallon. (Hint: 1 cubic foot of water is approximately 7.5 gallons.)
   A 79 gallons
   B 589 gallons
   C 59 gallons
   D 147 gallons

6. How many cubic feet of air are in a room that is 15 feet long, 10 feet wide and 8 feet high?
   F 33 ft³
   G 1200 ft³
   H 1500 ft³
   J 3768 ft³

7. How many gallons of water will the water trough hold? Round to the nearest gallon. (Hint: 1 cubic foot of water is approximately 7.5 gallons.)
   A 19 gallons   C 141 gallons
   B 71 gallons   D 565 gallons

8. A can has diameter of 9.8 cm and is 13.2 cm tall. What is the capacity of the can? Round to the nearest tenth.
   F 203.1 cm³
   G 995.2 cm³
   H 3980.7 cm³
   J 959.2 cm³

Answers

19. Possible answer: The base area should be in cm².

20. Possible answer: Since there are 12 inches in every foot, a one-foot cube would be the same as a 12-inch cube. The volume of a 12-inch cube is 1728 cubic inches.

Journal

Have students describe the steps they would take to measure and calculate the volume of the classroom in cubic feet.

Test Prep Doctor

For Exercise 25, review the expression for the sum of the angle measures in a polygon, (*n* − 2)180°. For an octagon, the angle sum is (8 − 2)180° = (6)180° = 1080°, choice **C**. Students who chose **D** may have forgotten to subtract 2 before multiplying by 180°.

Lesson Quiz

Find the volume of each figure to the nearest tenth. Use 3.14 for π.

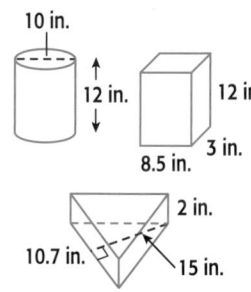
10 in.
12 in.
8.5 in.
12 in.
3 in.
10.7 in.
2 in.
15 in.

1. the cylinder **942 in³**

2. the rectangular prism **306 in³**

3. the triangular prism **160.5 in³**

4. Explain whether doubling the radius of the cylinder above will double the volume. **No; the volume would be quadrupled because you have to use the square of the radius to find the volume.**

Available on Daily Transparency in CRB

6-6 Volume of Prisms and Cylinders **311**

Warm Up

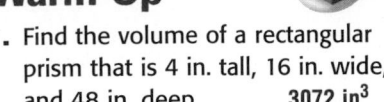

1. Find the volume of a rectangular prism that is 4 in. tall, 16 in. wide, and 48 in. deep. **3072 in³**

2. A cylinder has a height of 4.2 m and a diameter of 0.6 m. To the nearest tenth of a cubic meter, what is the volume of the cylinder? Use 3.14 for π. **1.2 m³**

3. A triangular prism's base is an equilateral triangle. The sides of the equilateral triangle are 4 ft, and the height of the prism is 8 ft. To the nearest cubic foot, what is the volume of the prism? **55.4 ft³**

Problem of the Day

A ream of paper (500 sheets) forms a rectangular prism 11 in. by 8.5 in. by 2 in. What is the volume of one sheet of paper? **0.374 in³**

Available on Daily Transparency in CRB

6-7 Volume of Pyramids and Cones

Learn to find the volume of pyramids and cones.

Vocabulary
pyramid
cone

The Great Pyramid of Giza was built using about 2.5 million blocks of stone, each weighing at least two tons. It is believed that 20,000 to 30,000 workers took about 20 years to complete the pyramid.

A **pyramid** is named for the shape of its base. The base is a polygon, and all of the other faces are triangles. A **cone** has a circular base. The height of a pyramid or cone is measured from the highest point to the base along a perpendicular line.

The Great Pyramid's height is equivalent to that of a forty-story skyscraper. The pyramid covers an area of thirteen acres.

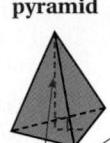 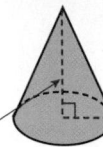

Rectangular pyramid   Triangular pyramid   Cone

Height

VOLUME OF PYRAMIDS AND CONES

Words	Numbers	Formula
Pyramid: The volume V of a pyramid is one-third of the area of the base B times the height h.	$B = 3(3)$ $= 9$ units² $V = \frac{1}{3}(9)(4)$ $= 12$ units³	$V = \frac{1}{3}Bh$
Cone: The volume of a cone is one-third of the area of the circular base B times the height h.	$B = \pi(2^2)$ $= 4\pi$ units² $V = \frac{1}{3}(4\pi)(3)$ $= 4\pi$ ≈ 12.6 units³	$V = \frac{1}{3}Bh$ or $V = \frac{1}{3}\pi r^2 h$

EXAMPLE 1 Finding the Volume of Pyramids and Cones

Find the volume of each figure.

A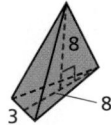

$B = \frac{1}{2}(3 \cdot 8) = 12$ units²

$V = \frac{1}{3} \cdot 12 \cdot 8$ $V = \frac{1}{3}Bh$

$V = 32$ units³

1 Introduce

Alternate Opener

EXPLORATION

6-7 Volume of Pyramids and Cones

The tip of a sharpened pencil is shaped like a cone. How much of the pencil is lost after the tip is formed? To answer this question, you should know that the volume of a cone is $\frac{1}{3}$ the volume of the cylinder from which it was formed.

The tip of this pencil was formed out of a cylinder with a height of 0.8 cm and a diameter of 1cm. The cylinder had a volume of approximately 0.63 cm³.

volume of cone = $\frac{1}{3}$ · volume of cylinder

volume of cone = $\frac{1}{3}$ · 0.63 = 0.21

Since the tip of the pencil has a volume of 0.21 cm³, 0.42 cm³ was lost when the tip of the pencil was formed.

Find the volume of each cone.

	Volume of Cylinder	Volume of Cone = $\frac{1}{3}$ Volume of Cylinder
1.	66.9 in³	
2.	99 cm³	
3.	108 in³	

Think and Discuss

4. **Name** real objects that are shaped like a cone.
5. **Explain** how to estimate the volume of one of the cone-shaped objects you named in number 4.

Motivate

Show students an empty cylinder and an empty cone that have congruent circular bases and equal heights. Have students guess how many cones it will take to fill the cylinder. Fill the cone and pour it into the cylinder. (You can use water or uncooked rice or beans.) Repeat until the cylinder is full. Students will see that the volume of the cylinder is three times the volume of the cone.

Exploration worksheet and answers on Chapter 6 Resource Book pp. 67 and 136

2 Teach

Lesson Presentation

Guided Instruction

In this lesson, students learn to find the volume of pyramids and cones. Remind students how they found the volumes of prisms and cylinders. Illustrate that the volume of a pyramid is $\frac{1}{3}$ the volume of a prism with an equal height and a congruent base (Teaching Transparency T35, CRB). Show that the volume of a cone is $\frac{1}{3}$ the volume of a cylinder with an equal height and a congruent base.

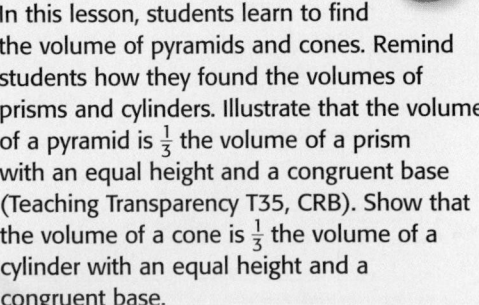

Teaching Tip Explain to students that the activity in Motivate is not a mathematical proof of the volume formula. The formula will, however, be proven in future math classes.

Find the volume of each figure.

 B

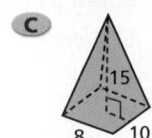

$B = \pi(2^2) = 4\pi$ units^2

$V = \frac{1}{3} \cdot 4\pi \cdot 12$ $V = \frac{1}{3}Bh$

$V = 16\pi \approx 50.3$ units^3 *Use 3.14 for π.*

 C

$B = 10 \cdot 8 = 80$ units^2

$V = \frac{1}{3} \cdot 80 \cdot 15$ $V = \frac{1}{3}Bh$

$V = 400$ units^3

EXAMPLE 2   **Exploring the Effects of Changing Dimensions**

A cone has radius 7 ft and height 14 ft. Explain whether doubling the height would have the same effect on the volume of the cone as doubling the radius.

Original Dimensions	Double the Height	Double the Radius
$V = \frac{1}{3}\pi r^2 h$	$V = \frac{1}{3}\pi r^2(2h)$	$V = \frac{1}{3}\pi(2r)^2 h$
$= \frac{1}{3}\pi(7^2)(14)$	$= \frac{1}{3}\pi(7^2)(2 \cdot 14)$	$= \frac{1}{3}\pi(2 \cdot 7)^2(14)$
≈ 718.01 ft^3	≈ 1436.03 ft^3	≈ 2872.05 ft^3

When the height of the cone is doubled, the volume is doubled. When the radius is doubled, the volume becomes 4 times the original volume.

EXAMPLE 3   *Social Studies Application*

The Great Pyramid of Giza is a square pyramid. Its height is 481 ft, and its base has 756 ft sides. Find the volume of the pyramid.

$B = 756^2 = 571,536$ ft^2 $A = bh$

$V = \frac{1}{3}(571,536)(481)$ $V = \frac{1}{3}Bh$

$V = 91,636,272$ ft^3

Think and Discuss

1. Describe two or more ways that you can change the dimensions of a rectangular pyramid to double its volume.

2. Compare the volume of a cube with 1 in. sides with a pyramid that is 1 in. high and has a square base with 1 in. sides.

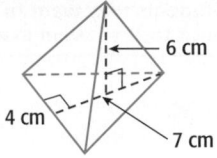
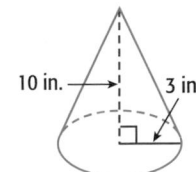
---

3 Close

Reaching All Learners

Through Hands-On Experience

Equip each student or group of students with a pair of scissors, tape, and a ruler. Give them a recording sheet (Chapter 6 Resource Book p. 118) that contains the net for a pyramid. Have them cut out the net and construct the pyramid. Then have them find the pyramid's approximate volume by taking the appropriate measurements.

Summarize

Remind students that the volumes of pyramids and cones are related to the volumes of prisms and cylinders. Show them the table below to emphasize the connection.

Prisms	Pyramids
$V = Bh$	$V = \frac{1}{3}Bh$
Cylinders	**Cones**
$V = Bh$ or $V = \pi r^2 h$	$V = \frac{1}{3}Bh$ or $V = \frac{1}{3}\pi r^2 h$

Answers to Think and Discuss

Possible answers:

1. You can double the height of the pyramid. You can double either dimension of the rectangular base. (In fact, you can multiply one dimension by a, the second by b, and the third by c, as long as $abc = 2$. For example, $a = 3$, $b = 2$, and $c = \frac{1}{3}$.)

2. The volume of the pyramid is one-third the volume of the cube. The volume of the cube is $1 \times 1 \times 1 = 1$ in^3. The volume of the square pyramid is $\frac{1}{3} \times (1 \times 1)(1) = \frac{1}{3}$ in^3.

FOR EXTRA PRACTICE
see page 743

📶 **internet** connect

Homework Help Online
go.hrw.com   Keyword: MP4 6-7

Students may want to refer back to the lesson examples.

Assignment Guide

If you finished Example **1** assign:
Core 1–6, 9–14, 17, 18, 27–31
Enriched 1–5 odd, 9–14, 17–20, 24, 27–31

If you finished Example **2** assign:
Core 1–7, 9–15, 17, 18, 27–31,
Enriched 7, 9–15, 17–20, 24–31

If you finished Example **3** assign:
Core 1–18, 21, 27–31
Enriched 9–31

Answers

7. Yes, the volume is doubled. The volume of the pyramid is 12 ft³. Doubling the height gives a volume of 24 ft³.

15. Yes, the volume is doubled. The volume of the pyramid is 36 ft³. Doubling the height of the base gives a volume of 72 ft³.

GUIDED PRACTICE

See Example **1** Find the volume of each figure to the nearest tenth. Use 3.14 for π.

1. 7 **70 units³** 6 5

2. **52.5 units³** 5 7 9

3. **14.8 units³** 4.9 1.7

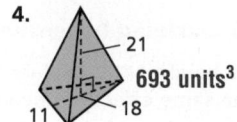

4. 21 **693 units³** 11 18

5. **213.4 units³** 5.6 6.5

6. 13 27 27 **3159 units³**

See Example **2** **7.** A square pyramid has height 4 ft and a base that measures 3 ft on each side. Explain whether doubling the height would double the volume of the pyramid.

See Example **3** **8.** The Transamerica Pyramid in San Francisco has a base area of 22,000 ft² and a height of 853 ft. What is the volume of the building? **6,255,333⅓ ft³**

INDEPENDENT PRACTICE

See Example **1** Find the volume of each figure to the nearest tenth. Use 3.14 for π.

9. **0.2 units³** 1.6 0.4 0.8

10. **5.1 units³** 4.6 1.7 3.9

11. **359.0 units³** 7 7

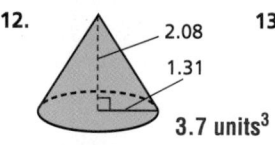

12. 2.08 1.31 **3.7 units³**

13. 14 **168 units³** 6 12

14. **5494.5 units³** 13.5 37 33

See Example **2** **15.** A triangular pyramid has a height of 6 ft. The triangular base has a height of 6 ft and a width of 6 ft. Explain whether doubling the height of the base would double the volume of the pyramid.

See Example **3** **16.** A cone-shaped building is commonly used to store rock salt. What would be the volume of a cone-shaped building with diameter 70 ft and height 50 ft, to the nearest hundredth? **64,140.85 ft³**

Math Background

A polyhedron is any three-dimensional figure whose surfaces are all polygons. A regular polyhedron is one whose faces are all congruent regular polygons and whose polyhedral angles are all congruent. There are only five different regular polyhedrons: A tetrahedron has 4 triangular faces, a hexahedron has 6 square faces, an octahedron has 8 triangular faces, a dodecahedron has 12 pentagonal faces, and an icosahedron has 20 triangular faces. A tetrahedron is a triangular pyramid. A hexahedron is a cube. An octahedron can be formed by joining two square pyramids at their bases.

RETEACH 6-7

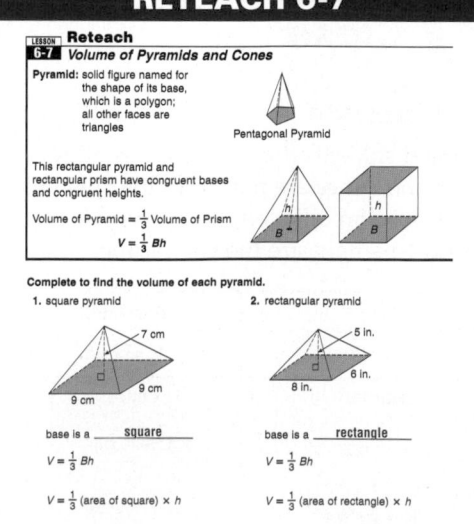

PRACTICE 6-7

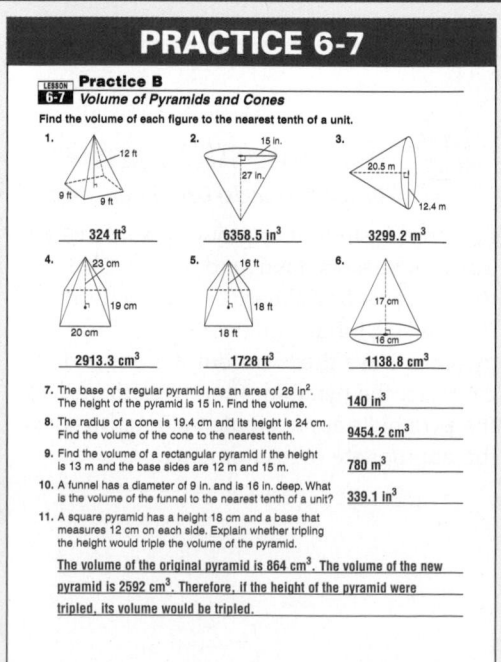

PRACTICE AND PROBLEM SOLVING

Find the missing measure to the nearest tenth. Use 3.14 for π.

17. cone:
    radius = 3 cm
    height = ■ **4.0 cm**
    volume = 37.7 cm³

18. cylinder:
    radius = ■ **3.5 cm**
    height = 2 cm
    volume = 75.36 cm³

19. triangular pyramid:
    base height = ■ **9 ft**
    base width = 10 ft
    height = 7 ft
    volume = 105 ft³

20. rectangular pyramid:
    base length = 3 ft
    base width = ■ **6 ft**
    height = 7 ft
    volume = 42 ft³

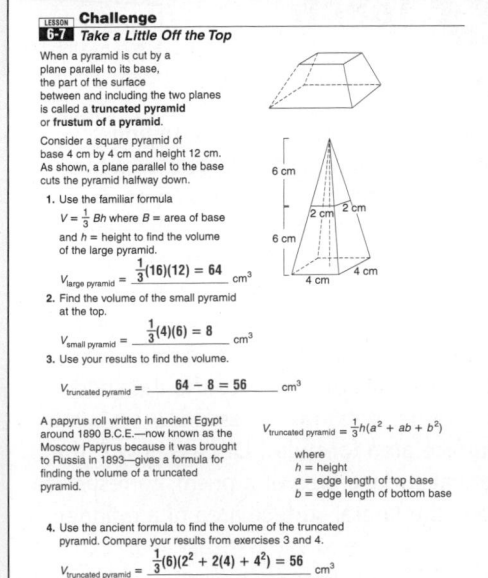
21. **ARCHITECTURE** The pyramid at the entrance to the Louvre in Paris has a height of 72 feet and a square base that is 112 feet long on each side. What is the volume of this pyramid? **301,056 ft³**

22. **TRANSPORTATION** Orange traffic cones, or pylons, come in a variety of sizes. What is the volume in cubic inches of a pylon with height 3 feet and diameter 9 inches? **243π ≈ 763.4 in³**

23. **ARCHITECTURE** The Pyramid Arena in Memphis, Tennessee, is 321 feet tall and has a square base that is 200 yards on each side.

    a. What is the volume in cubic feet of the arena? **38,520,000 ft³**

    b. How many cubic feet are in one cubic yard? **27**

    c. What is the volume in cubic yards of the arena to the nearest hundredth? **1,426,666.67 yd³**

24. **WHAT'S THE ERROR?** A student says that the formula for the volume of a cylinder is the same as the formula for the volume of a pyramid, $\frac{1}{3}Bh$. What error did this student make?

25. **WRITE ABOUT IT** How would a cone's volume be affected if you doubled the height? the radius?

26. **CHALLENGE** The diameter of a cone is x in., the height is 12 in., and the volume is 36π in³. What is x? **$x = 6$**

Spiral Review

Use guess and check to estimate each square root to two decimal places. (Lesson 3-9)

27. $\sqrt{35}$ **5.92**   28. $\sqrt{45}$ **6.71**   29. $\sqrt{55}$ **7.42**   30. $\sqrt{65}$ **8.06**

31. **TEST PREP** Write $\frac{15^3 \cdot 15^{11}}{15^{-13}}$ as one power. (Lesson 2-7) **C**

    A 1   B 15^1 C 15^{27} D 15^{46}

Answers

24. Possible answer: The student probably confused a cylinder with a cone. The volume formulas for a cylinder ($V = Bh$) and a pyramid ($V = \frac{1}{3}Bh$) are not the same. However, the volume formulas for a cone and a pyramid are the same ($V = \frac{1}{3}Bh$).

25. Possible answer: When the height of the cone is doubled, the volume is doubled. When the radius is doubled, the volume is increased by a factor of 4.

Journal

Ask students to write about the relationships among prisms, cylinders, pyramids, and cones. They may write about the shapes themselves or the formulas for their volumes.

Test Prep Doctor

For Exercise 31, remind students of the rules for multiplying and dividing powers with the same base. To multiply the factors in the numerator, add the exponents: $15^3 \cdot 15^{11} = 15^{14}$.

To divide, subtract the exponents: $15^{14} \div 15^{-13} = 15^{27}$. The correct answer is **C**. If students selected **A** or **B**, they probably do not understand negative exponents. If students selected **D**, they may have forgotten how to multiply exponents.

Lesson Quiz

Find the volume of each figure to the nearest tenth. Use 3.14 for π.

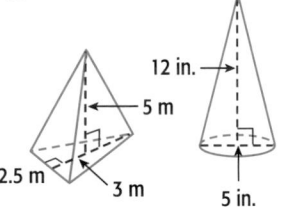

1. the triangular pyramid **6.3 m³**

2. the cone **78.5 in³**

3. Explain whether tripling the height of the square pyramid above would triple the volume.
   Yes; the volume is one-third the product of the base area and the height. So if you triple the height, the product would be tripled.

Available on Daily Transparency in CRB

Pacing: Traditional 1 day
Block $\frac{1}{2}$ day

Objective: Students find the surface area of prisms and cylinders.

Warm Up

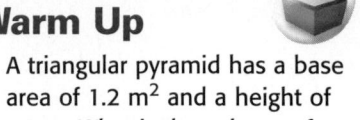

1. A triangular pyramid has a base area of 1.2 m² and a height of 7.5 m. What is the volume of the pyramid?   **3 m³**

2. A cone has a radius of 4 cm and a height of 10 cm. What is the volume of the cone to the nearest cubic centimeter? Use 3.14 for π.   **167 cm³**

Problem of the Day

An ice cream cone is filled halfway to the top. The radius of the filled part is half the radius at the top. What fraction of the cone's volume is filled? $\frac{1}{8}$

Available on Daily Transparency in CRB

Math Humor

Teacher: How much foil would you need to cover a cylindrical cake with a radius of 6 inches and a height of 4 inches?

Student: None; I'd eat the whole thing!

6-8 Surface Area of Prisms and Cylinders

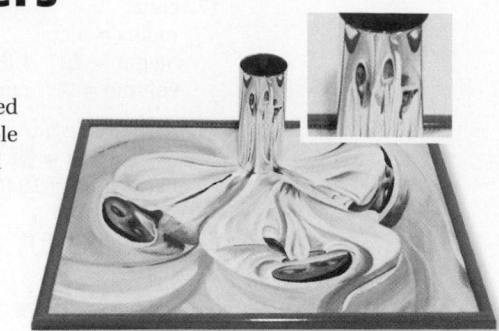

Learn to find the surface area of prisms and cylinders.

Vocabulary
surface area
lateral face
lateral surface

An *anamorphic image* is a distorted picture that becomes recognizable when reflected onto a cylindrical mirror.

Surface area is the sum of the areas of all surfaces of a figure. The **lateral faces** of a prism are parallelograms that connect the bases. The **lateral surface** of a cylinder is the curved surface.

SURFACE AREA OF PRISMS AND CYLINDERS

Words	Numbers	Formula
Prism: The surface area *S* of a prism is twice the base area *B* plus the lateral area *F*. The lateral area is the base perimeter *P* times the height *h*.	$S = 2(3 \cdot 2) + (10)(5) = 62$ units²	$S = 2B + F$ or $S = 2B + Ph$
Cylinder: The surface area *S* of a cylinder is twice the base area *B* plus the lateral area *L*. The lateral area is the base circumference $2\pi r$ times the height *h*.	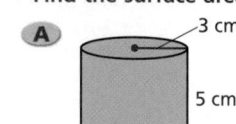 $S = 2\pi(5^2) + 2\pi(5)(6) \approx 345.4$ units²	$S = 2B + L$ or $S = 2\pi r^2 + 2\pi rh$

EXAMPLE 1 Finding Surface Area

Find the surface area of each figure.

Ⓐ   3 cm / 5 cm

$S = 2\pi r^2 + 2\pi rh$
$= 2\pi(3^2) + 2\pi(3)(5)$
$= 48\pi$ cm² ≈ 150.8 cm²

---

1 Introduce

Alternate Opener

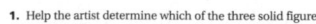
Motivate

Show students an empty cereal box. Point out that if they find the volume, they will find the amount of cereal the box can hold. Rip the seams so the box will lie flat. Tell the students that now they will find the surface area of shapes like the box. The surface area determines how much cardboard is needed to make the box.

Exploration worksheet and answers on Chapter 6 Resource Book pp. 77 and 138

2 Teach

Lesson Presentation

Guided Instruction

In this lesson, students learn to find the surface area of prisms and cylinders. Have students open samples of prisms and cylinders so they can see the two-dimensional surfaces that form the three-dimensional figures (Teaching Transparency T39, CRB). Explain that the surface area is the sum of all the areas of these two-dimensional surfaces. Use the flattened-out diagrams (nets) in the lesson to explain the surface area formulas. Discuss how *F*, the lateral surface area of a prism, corresponds to *L*, the lateral surface area of a cylinder.

Find the surface area of each figure.

B

6 in.
2.4 in.
4 in.   3 in.
5 in.

$S = 2B + Ph$
$= 2(\frac{1}{2} \cdot 5 \cdot 2.4) + (12)(6)$
$= 84 \text{ in}^2$

EXAMPLE 2 **Exploring the Effects of Changing Dimensions**

A cylinder has diameter 8 in. and height 3 in. Explain whether doubling the height would have the same effect on the surface area as doubling the radius.

Original Dimensions	Double the Height	Double the Radius
$S = 2\pi r^2 + 2\pi rh$	$S = 2\pi r^2 + 2\pi rh$	$S = 2\pi r^2 + 2\pi rh$
$= 2\pi(4)^2 + 2\pi(4)(3)$	$= 2\pi(4)^2 + 2\pi(4)(6)$	$= 2\pi(8)^2 + 2\pi(8)(3)$
$= 56\pi \text{ in}^2 \approx 175.8 \text{ in}^2$	$= 80\pi \text{ in}^2 \approx 251.2 \text{ in}^2$	$= 176\pi \text{ in}^2 \approx 552.6 \text{ in}^2$

They would not have the same effect. Doubling the radius would increase the surface area more than doubling the height.

EXAMPLE 3 **Art Application**

A Web site advertises that it can turn your photo into an anamorphic image. To reflect the picture, you need to cover a cylinder that is 32 mm in diameter and 100 mm tall with reflective material. How much reflective material do you need?

$L = 2\pi rh$ *Only the lateral surface needs to be covered.*
$= 2\pi(16)(100)$ *The diameter is 32 mm, so r = 16mm.*
$\approx 10,048 \text{ mm}^2$

Think and Discuss

1. **Compare** the formula for the surface area of a cylinder to the formula for the surface area of a prism.

2. **Explain** how finding the surface area of a cylindrical drinking glass would be different from finding the surface area of a cylinder.

3. **Compare** the amount of paint needed to cover a cube with 1 ft sides to the amount needed to cover a cube with 2 ft sides.

 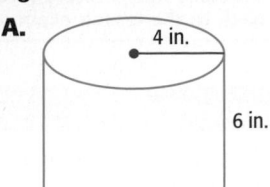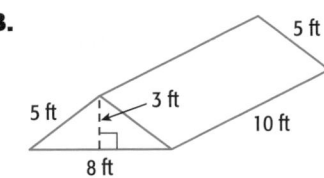
3 Close

**Reaching All Learners**
Through Critical Thinking

Discuss with students descriptions of real-world examples of area and volume, such as the amount of soup in a can, the amount of paper needed to make a label around a can, or the amount of wallpaper needed to cover the walls of a room. Have students identify the measure that applies to each situation and the appropriate formula.

Possible answers: volume of a cylinder $V = Bh$; lateral surface area of a cylinder $L = 2\pi rh$; area of lateral faces of a prism $F = Ph$

Summarize

Point out that the formulas for surface area of prisms and cylinders are based on area formulas for rectangles and circles. Discuss the difference between surface area and volume.

Possible answer: Surface area is the sum of the areas of all surfaces of a figure and is measured in square units. Volume is the space that a figure occupies and is measured in cubic units.

Answers to Think and Discuss

Possible answers:

1. Both formulas contain the sum of a lateral surface area and two base areas.

2. The drinking glass only has one base, but a cylinder has two bases. So the surface area of the glass would be $\pi r^2 + 2\pi rh$ instead of $2\pi r^2 + 2\pi rh$.

3. The smaller cube has a surface area of 6 ft^2 and the larger cube has a surface area of 24 ft^2, so the larger cube requires 4 times as much paint.

FOR EXTRA PRACTICE
see page 743

internet connect
Homework Help Online
go.hrw.com   Keyword: MP4 6-8

go.hrw.com

GUIDED PRACTICE

See Example ① Find the surface area of each figure to the nearest tenth. Use 3.14 for π.

1. 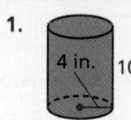 **351.7 in²**
   4 in. 10 in.

2. **356 cm²**
   3 cm 14 cm 8 cm

See Example ② 3. A rectangular prism is 6 cm by 8 cm by 9 cm. Explain whether doubling all of the dimensions would double the surface area.

See Example ③ 4. To the nearest tenth of a square inch, how much paper is needed for a soup can label if the can is 6.4 in. tall and has a diameter of 4 in.? **80.4 in²**

INDEPENDENT PRACTICE

See Example ① Find the surface area of each figure to the nearest tenth. Use 3.14 for π.

5. 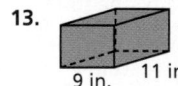 12 in. 16 in. 12 in. 20 in. **768 in²**

6. 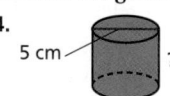 7 ft 9 ft **325.0 ft²**

See Example ② 7. A cylinder has diameter 4 ft and height 9 ft. Explain whether halving the diameter has the same effect on the surface area as halving the height.

See Example ③ 8. How much aluminum foil, to the nearest tenth of a square inch, would it take to cover a loaf of banana-nut bread that is a rectangular prism measuring 8.5 in. by 4 in. by 3.5 in.? **155.5 in²**

PRACTICE AND PROBLEM SOLVING

Find the surface area of each figure with the given dimensions. Use 3.14 for π.

9. rectangular prism: 9 in. by 12 in. by 15 in. **846 in²**

10. cylinder: $d = 20$ mm, $h = 37$ mm **940π ≈ 2951.6 mm²**

11. cylinder: $r = 7.8$ cm, $h = 8.2$ cm **249.6π ≈ 783.7 cm²**

12. rectangular prism: $4\frac{1}{2}$ ft by 6 ft by 11 ft **285 ft²**

Find the missing dimension in each figure with the given surface area.

13. ? $S = 438$ in²
    9 in. 11 in. **6 in.**

14. 21.5 cm
    5 cm ? $S = 120π$ cm²

Assignment Guide

If you finished Example ① assign:
Core 1, 2, 5, 6, 9–13 odd, 24–29
Enriched 5, 6, 9–14, 24–29

If you finished Example ② assign:
Core 1–3, 5–7, 9–13 odd, 24–29
Enriched 5–7, 9–14, 23, 24–29

If you finished Example ③ assign:
Core 1–19, 20, 24–29
Enriched 5–29

Answers

3. No; doubling the dimensions results in a surface area that is four times the original. The prism has a surface area of 348 cm². Doubling the dimensions results in a surface area of 1392 cm².

7. No; they do not have the same effect. Halving the diameter decreases the surface area more than halving the height.

Math Background

Research has found that children who play with blocks and other building toys have a better-developed spatial sense than those who do not. Even older students can benefit from hands-on activities in their study of three-dimensional figures. Many students in high school geometry classes are at a disadvantage that can be partially attributed to a lack of hands-on experience in prior grades. Students should, however, be familiar with basic concepts and have a clear understanding of the objectives of hands-on activities.

RETEACH 6-8

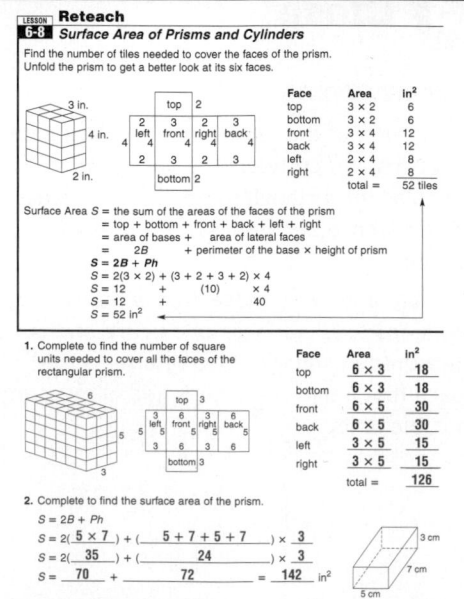

PRACTICE 6-8

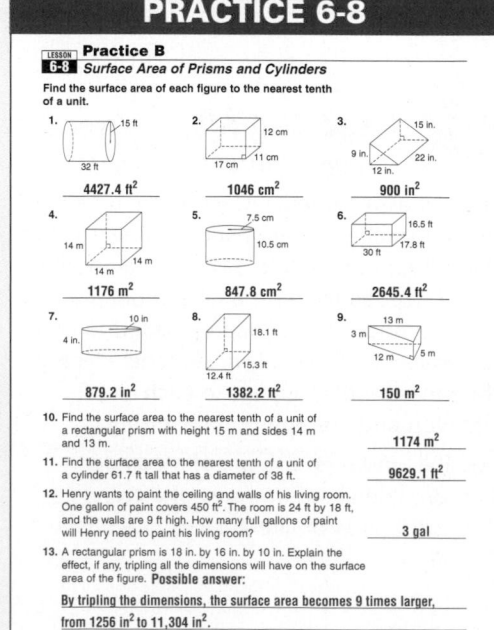

Find the surface area of each battery to the nearest tenth of a square centimeter.

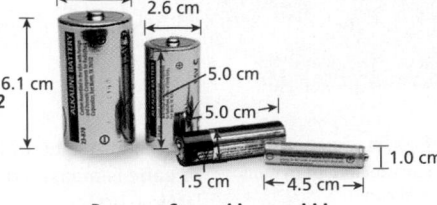

3.4 cm 2.6 cm
6.1 cm
5.0 cm
5.0 cm
1.0 cm
1.5 cm ⊢4.5 cm⊣
D   C   AA   AAA

15. D **83.3 cm²**   **16.** C **51.5 cm²**

17. AA **27.1 cm²**   **18.** AAA **15.7 cm²**

19. Jesse makes rectangular tin boxes measuring 4 in. by 6 in. by 6 in. If tin costs $0.09 per in², how much will the tin for one box cost? **$15.12**

20. *SPORTS* In the snowboard half-pipe, competitors ride back and forth on a course shaped like a cylinder cut in half lengthwise. What is the surface area of this half-pipe course?

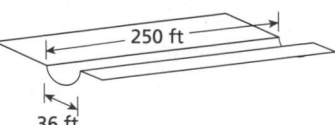

250 ft

36 ft

21. *CHOOSE A STRATEGY* Which of the following unfolded figures can be folded into the given three-dimensional figure? **B**

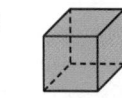

A 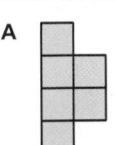   B 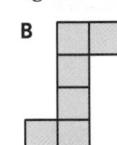   C 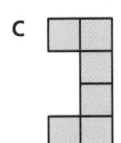   D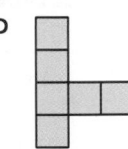

22. *WRITE ABOUT IT* Compare the formulas for surface area of a prism and surface area of a cylinder.

23. *CHALLENGE* A rectangular wood block that is 12 cm by 9 cm by 5 cm has a hole drilled through the center with diameter 4 cm. What is the total surface area of the wood block? **426 + 12π ≈ 463.7 cm²**

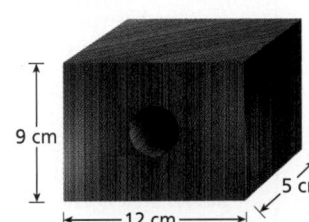

9 cm
5 cm
⊢12 cm⊣

Spiral Review

Divide. Write each answer in simplest form. (Lesson 3-4)

24. $-\frac{4}{11} \div \frac{2}{7}$ $-1\frac{3}{11}$ **25.** $\frac{4}{9} \div 8$ $\frac{1}{18}$ **26.** $-\frac{7}{15} \div \frac{14}{45}$ $-1\frac{1}{2}$ **27.** $3\frac{1}{3} \div \frac{7}{9}$ $4\frac{2}{7}$

28. In a right triangle, if $a = 9$ and $b = 12$, what is the value of the hypotenuse c? (Lesson 6-3) **15**

29. TEST PREP Determine which ordered pair is a solution of $y = 5x - 3$.
(Lesson 1-7) **D**

A $(-8, -1)$ B $(2, 0)$ C $(-3, -16)$ D $(2, 7)$

Answers

20. $4500\pi \approx 14{,}137$ ft²

22. Possible answer: The formulas are similar in that you find the area of the two bases and then add that to the product of the height and the perimeter or circumference. However, the formulas you use to find each of these parts are different.

Journal

Have students consider the question, "Is it possible for two rectangular prisms to have the same volume and different surface areas?" Have them answer the question and include examples to support their answers.

Test Prep Doctor

For Exercise 29, suggest that students substitute the ordered pairs into the given equation. Remind students that the first coordinate in each pair should be substituted for x in the equation. The only ordered pair that satisfies the equation is (2, 7), choice **D.** If students chose choice **A,** they may have substituted the values in the wrong order.

CHALLENGE 6-8

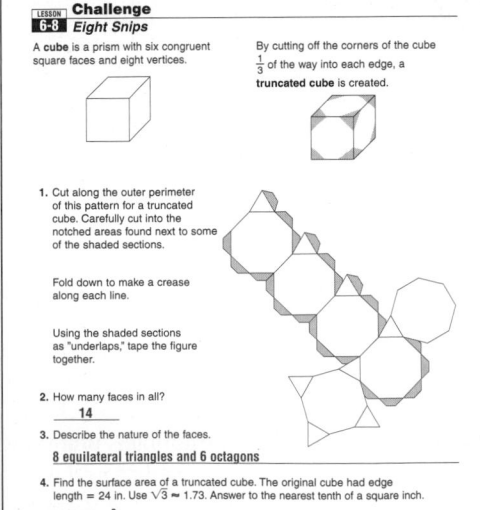

LESSON 6-8 Challenge
Eight Snips

A **cube** is a prism with six congruent square faces and eight vertices.

By cutting off the corners of the cube $\frac{1}{3}$ of the way into each edge, a **truncated cube** is created.

1. Cut along the outer perimeter of this pattern for a truncated cube. Carefully cut into the notched areas found next to some of the shaded sections.

Fold down to make a crease along each line.

Using the shaded sections as "underlaps," tape the figure together.

2. How many faces in all?
14

3. Describe the nature of the faces.
8 equilateral triangles and 6 octagons

4. Find the surface area of a truncated cube. The original cube had edge length = 24 in. Use $\sqrt{3} \approx 1.73$. Answer to the nearest tenth of a square inch.
3130.88 in²

PROBLEM SOLVING 6-8

LESSON 6-8 Problem Solving
Surface Area of Prisms and Cylinders

An important factor in designing packaging for a product is the amount of material required to make the package. Consider the three figures described in the table below. Use 3.14 for π. Round to the nearest tenth. Write the correct answer.

1. Find the surface area of each package given in the table.

2. Which package has the lowest materials cost? Assume all of the packages are made from the same material.

cylinder

Package	Dimensions	Volume	Surface Area
Prism	Base: 2" × 16" Height = 2"	64 in³	136 in²
Prism	Base: 4" × 4" Height = 4"	64 in³	96 in²
Cylinder	Radius = 2" Height = 5.1"	64.06 in³	89.2 in²

Use 3.14 for π. Round to the nearest hundredth.

3. How much cardboard material is required to make a cylindrical oatmeal container that has a diameter of 12.5 cm and a height of 24 cm, assuming there is no overlap? The container will have a plastic lid.
1064.66 cm²

4. What is the surface area of a rectangular prism that is 5 feet by 6 feet by 10 feet?
280 ft²

Use 3.14 for π. Round to the nearest tenth. Choose the letter for the best answer.

5. How much metal is required to make the trough pictured below?

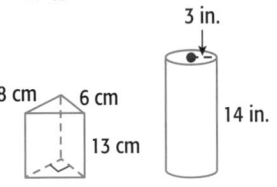

2 ft
6 ft

Ⓐ 22.0 ft²   C 44.0 ft²
B 34.0 ft²   D 56.7 ft²

6. A can of vegetables has a diameter of 9.8 cm and is 13.2 cm tall. How much paper is required to make the label, assuming there is no overlap? Round to the nearest tenth.
F 203.1 cm²
Ⓖ 406.2 cm²
H 557.0 cm²
J 812.4 cm²

Lesson Quiz

Find the surface area of each figure to the nearest tenth. Use 3.14 for π.

3 in.
8 cm   6 cm
13 cm
14 in.

1. the triangular prism   **360 cm²**

2. the cylinder   **320.3 in²**

3. All outer surfaces of a box are covered with gold foil, except the bottom. The box measures 6 in. long, 4 in. wide, and 3 in. high. How much gold foil was used? **84 in²**

Available on Daily Transparency in CRB

Pacing: Traditional 1 day
Block $\frac{1}{2}$ day

Objective: Students find the surface area of pyramids and cones.

6-9 Surface Area of Pyramids and Cones

Learn to find the surface area of pyramids and cones.

Vocabulary
slant height
regular pyramid
right cone

The **slant height** of a pyramid or cone is measured along its lateral surface.

The base of a **regular pyramid** is a regular polygon, and the lateral faces are all congruent.

In a **right cone**, a line perpendicular to the base through the tip of the cone passes through the center of the base.

Right cone

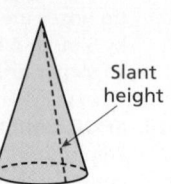

Slant height

Regular pyramid

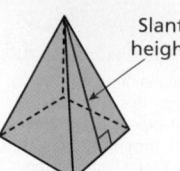

Slant height

Warm Up

1. A rectangular prism is 0.6 m by 0.4 m by 1.0 m. What is the surface area? **2.48 m²**

2. A cylindrical can has a diameter of 14 cm and a height of 20 cm. What is the surface area to the nearest tenth? Use 3.14 for π. **1186.9 cm²**

Problem of the Day

Sandy is building a model of a pyramid with a hexagonal base. If she uses a toothpick for each edge, how many toothpicks will she need? **12**

Available on Daily Transparency in CRB

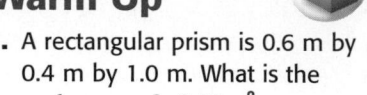

Math Humor

Pharaoh (to Architect): I want the burial chamber to be built in the center of my tomb.

Architect: You mean in the *pyramiddle*?

SURFACE AREA OF PYRAMIDS AND CONES		
Words	**Numbers**	**Formula**
Pyramid: The surface area S of a regular pyramid is the base area B plus the lateral area F. The lateral area is one-half the base perimeter P times the slant height ℓ.	$S = (12 \cdot 12) + \frac{1}{2}(48)(8) = 336$ units²	$S = B + F$ or $S = B + \frac{1}{2}P\ell$
Cone: The surface area S of a right cone is the base area B plus the lateral area L. The lateral area is one-half the base circumference $2\pi r$ times the slant height ℓ.	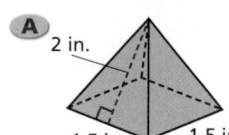 $S = \pi(3^2) + \pi(3)(4) = 21\pi \approx 65.94$ units²	$S = B + L$ or $S = \pi r^2 + \pi r\ell$

EXAMPLE 1 Finding Surface Area

Find the surface area of each figure.

A

2 in.

1.5 in. 1.5 in.

$S = B + \frac{1}{2}P\ell$

$= (1.5 \cdot 1.5) + \frac{1}{2}(6)(2)$

$= 8.25$ in²

1 Introduce

Alternate Opener

 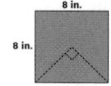
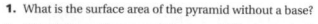
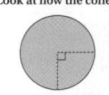

Exploration worksheet and answers on Chapter 6 Resource Book pp. 87 and 141

2 Teach

Lesson Presentation

Motivate

Show students an ice cream cone wrapped in paper, or a cone-shaped paper cup. Ask students for the word that describes how much the cone holds. volume Ask students for the term that describes the amount of paper. lateral surface area Tell students that they will find surface area of cones and pyramids.

Guided Instruction

In this lesson, students learn to find the surface area of pyramids and cones. Show students nets for a pyramid and a cone (Chapter 6 Resource Book pp. 119–120). Remind students that they found the total surface area of prisms and cylinders by adding the lateral area and *two* base areas. Note that for pyramids and cones, students will add the lateral area and *one* base area. You may want to display Teaching Transparency T43 (CRB). As you discuss the lesson examples, point out that the same variables are used as in the previous lesson: B for area of a base, F for lateral area of polygon surfaces, and L for lateral area of a curved surface.

Find the surface area of each figure.

B   5 m   2 m

$S = \pi r^2 + \pi r \ell$
$= \pi(2)^2 + \pi(2)(5)$
$= 14\pi \approx 44.0 \text{ m}^2$

EXAMPLE 2 **Exploring the Effects of Changing Dimensions**

A cone has diameter 8 in. and slant height 5 in. Explain whether doubling the slant height would have the same effect on the surface area as doubling the radius.

Original Dimensions	Double the Slant Height	Double the Radius
$S = \pi r^2 + \pi r \ell$	$S = \pi r^2 + \pi r(2\ell)$	$S = \pi(2r)^2 + \pi(2r)\ell$
$= \pi(4)^2 + \pi(4)(5)$	$= \pi(4)^2 + \pi(4)(10)$	$= \pi(8)^2 + \pi(8)(5)$
$= 36\pi \text{ in}^2 \approx 113.1 \text{ in}^2$	$= 56\pi \text{ in}^2 \approx 175.9 \text{ in}^2$	$= 104\pi \text{ in}^2 \approx 326.7 \text{ in}^2$

They would not have the same effect. Doubling the radius would increase the surface area more than doubling the slant height.

EXAMPLE 3 **Life Science Application**

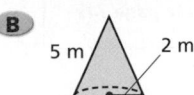

An ant lion pit is an inverted cone with the dimensions shown. What is the lateral surface area of the pit?

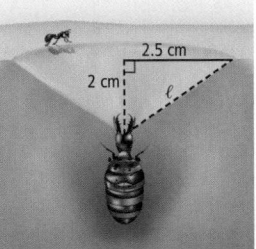
2.5 cm
2 cm
ℓ

The slant height, radius, and depth of the pit form a right triangle.

$a^2 + b^2 = \ell^2$ *Pythagorean Theorem*

$(2.5)^2 + 2^2 = \ell^2$

$10.25 = \ell^2$

$\ell \approx 3.2$

$L = \pi r \ell$ *Lateral surface area*

$= \pi(2.5)(3.2) \approx 25.1 \text{ cm}^2$

Ant lions are the larvae of an insect similar to a dragonfly. They dig cone-shaped pits in the sand to trap ants and other crawling insects.

Think and Discuss

1. **Compare** the formula for surface area of a pyramid to the formula for surface area of a cone.

2. **Explain** how you would find the slant height of a square pyramid with base edge length 6 cm and height 4 cm.

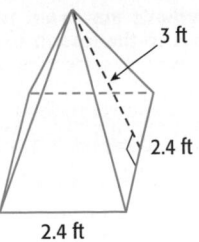
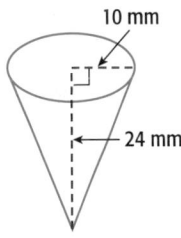
---

3 Close

Reaching All Learners
Through Number Sense

Create a few models of pyramids and cones. (Nets are provided on Chapter 6 Resource Book pp. 119–122.) Give students a list of the surface areas. Ask students to use estimation to match the surface areas and the models. Then allow the students to use rulers (MK) to measure one of the models and calculate its surface area. Based on this result, allow students to revise their estimates. Then ask students to measure the remaining models, calculate the surface areas, and compare the actual surface areas with their estimates.

Summarize

Suggest to students that being able to visualize how a three-dimensional figure can be formed from a two-dimensional figure will help them understand surface area. Draw (or have drawn ahead of time) the two-dimensional patterns for a prism, a cylinder, a pyramid, and a cone, but do not label them. Ask students to identify each.

Answers to Think and Discuss

Possible answers:

1. Both formulas require adding the base area to the product of one half the distance around the base and the slant height.

2. Make a right triangle with its hypotenuse as the slant height of the pyramid, one leg as the height (4), and the other leg as half the base (3). Use the Pythagorean Theorem to find that the slant height is 5.

FOR EXTRA PRACTICE
see page 743

☐ **internet** connect ══
Homework Help Online
go.hrw.com   Keyword: MP4 6-9

GUIDED PRACTICE

Students may want to refer back to the lesson examples.

See Example **1** Find the surface area of each figure to the nearest tenth. Use 3.14 for π.

1. 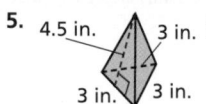 9 m **144 m²** 6 m 6 m

2. 7 ft 2.5 ft **74.6 ft²**

See Example **2** **3.** A cone has diameter 10 in. and slant height 8 in. Tell whether doubling both dimensions would double the surface area.

See Example **3** **4.** The cone-shaped wigwams at the Wigwam Village Motel in Cave City, Kentucky, are about 20 ft high and have a diameter of about 20 ft. Estimate the lateral surface area of a wigwam. ≈ **702.5 ft²**

INDEPENDENT PRACTICE

See Example **1** Find the surface area of each figure to the nearest tenth. Use 3.14 for π.

5. 4.5 in. 3 in. **24.1 in²** 3 in. 3 in.

6. 5 mm 8 mm **204.1 mm²**

See Example **2** **7.** A regular square pyramid has a base with 10 yd sides and has slant height 6 yd. Tell whether doubling both dimensions would double the surface area.

See Example **3** **8.** In the late 1400s, Leonardo daVinci designed a parachute shaped like a pyramid. His design called for a tent-like structure made of linen, measuring 21 feet on each side and 12 feet high. How much material would be needed to make the parachute? ≈ **669.7 ft²**

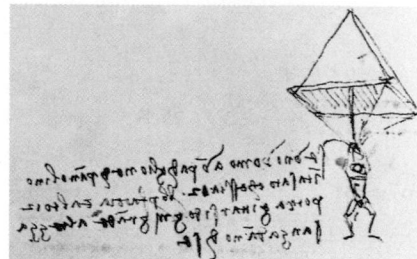

PRACTICE AND PROBLEM SOLVING

Find the surface area of each figure with the given dimensions. Use 3.14 for π.

9. regular square pyramid:
base perimeter = 60 cm
slant height = 18 cm **765 cm²**

10. regular triangular pyramid:
base area: 0.04 km²
base perimeter 0.9 km
slant height = 0.2 km **0.13 km²**

11. cone: d = 38 ft
slant height = 53 ft

12. cone: r = $12\frac{1}{2}$ mi
slant height = $44\frac{1}{4}$ mi

Assignment Guide

If you finished Example **1** assign:
Core 1, 2, 5, 6, 9, 11, 19–24
Enriched 5, 6, 9–12, 19–24

If you finished Example **2** assign:
Core 1–3, 5–7, 9, 11, 19–24
Enriched 5–7, 9–12, 16, 19–24

If you finished Example **3** assign:
Core 1–8, 9–15 odd, 19–24
Enriched 5–24

Answers

3. No; doubling the dimensions results in a surface area that is four times the original. The cone has a surface area of 204.2 in². Doubling the dimensions results in a surface area of 816.8 in².

7. No; doubling the dimensions results in a surface area that is four times the original. The pyramid has a surface area of 220 yd². Doubling the dimensions results in a surface area of 880 yd².

11. $1368\pi \approx 4295.5$ ft²

12. $709\frac{3}{8}\pi \approx 2227.4$ mi²

Math Background

When you are working with problems involving irrational numbers, such as *pi* and some square roots, different methods of rounding can result in different answers. This is especially true for problems that contain large numbers. For example, the answer to Exercise 14 is ≈ 58,468 m² if calculated on a calculator and rounded at the last step. However, the answer is 58,455 m² if the values are rounded to the nearest tenth at each step. Unless otherwise indicated, given answers have been calculated on a calculator and rounded at the last step.

RETEACH 6-9

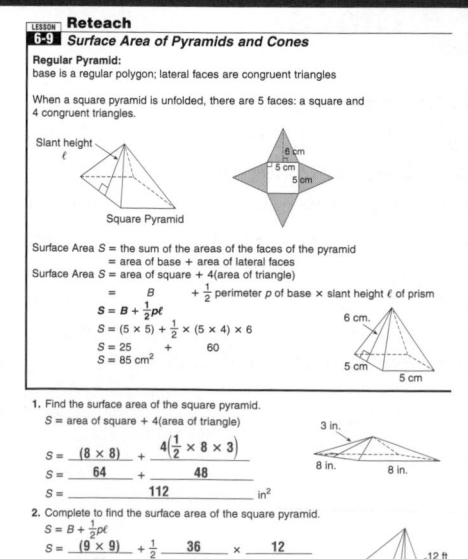

PRACTICE 6-9

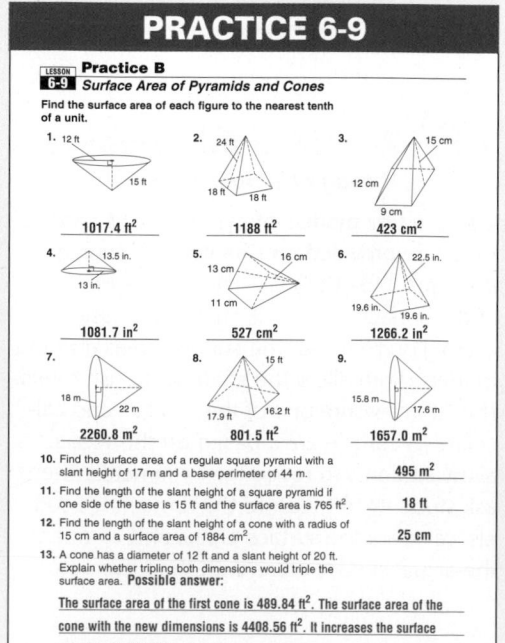

---

13. **EARTH SCIENCE** When the Moon is between the Sun and Earth, it casts a conical shadow called the *umbra*. If the shadow is 2140 mi in diameter and 260,955 mi along the edge, what is the lateral surface area of the shadow? ≈ 877,201,312 mi²

14. **SOCIAL STUDIES** The Pyramid of the Sun, in Teotihuacán, Mexico, is about 65 m tall and has a square base with side length 225 m. What is the lateral surface area of the pyramid? ≈ **58,468 m²**

15. The table shows the dimensions of three square pyramids.
    a. Complete the table. ≈ **588; ≈ 216**
    b. Which pyramid has the greatest lateral surface area? What is its lateral surface area? **Khufu; 925,344 ft²**
    c. Which pyramid has the least volume? What is its volume? **Menkaure; ≈ 8,619,552 ft³**

Dimensions of Giza Pyramids (ft)			
Pyramid	Height	Slant Height	Side of Base
Khufu	481	612	756
Khafre	471		704
Menkaure		277	346

16. **WHAT'S THE ERROR?** Correct the error in this statement: The lateral surface area of a cone is π times the radius of the base times the height of the cone.

17. **WRITE ABOUT IT** The dimensions of a square pyramid give its height and base dimensions. Explain how to find the slant height.

18. **CHALLENGE** The oldest pyramid is said to be the Step Pyramid of King Zoser, built around 2650 B.C. in Saqqara, Egypt. The base is a rectangle that measures 358 ft by 411 ft, and the height of the pyramid is 204 ft. Find the lateral surface area of the pyramid. ≈ **215,208 ft²; Answers may vary due to rounding.**

Spiral Review

Find each sum. (Lesson 3-2)

19. $-1.7 + 2.3$ **0.6**
20. $-\frac{2}{3} + \left(-\frac{1}{6}\right)$ $-\frac{5}{6}$
21. $23.75 + (-25.15)$ **−1.4**
22. $-\frac{4}{9} + \frac{2}{9}$ $-\frac{2}{9}$

23. Find the length of the hypotenuse of a right triangle with legs measuring 15 m and 22 m. (Lesson 6-3) $\sqrt{709} \approx 26.63$ m

24. **TEST PREP** Triangle $EFG \cong$ triangle JIH. Find the value of x. (Lesson 5-6) **C**

    A 5.67
    B 30
    C 63
    D 71

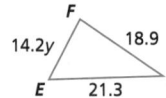 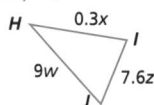

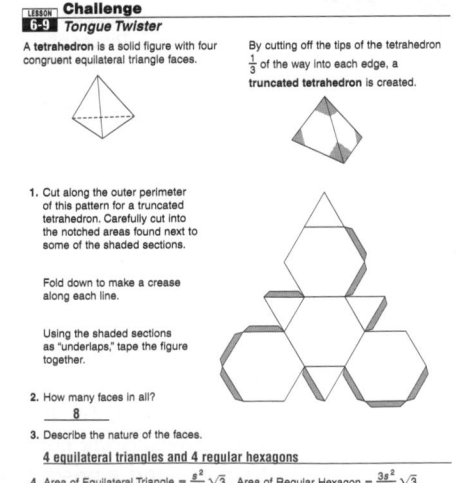
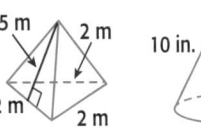 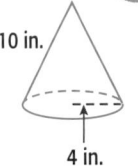

6-10 PRACTICE & ASSESS

6-10 Exercises

FOR EXTRA PRACTICE
see page 743

internet connect
Homework Help Online
go.hrw.com   Keyword: MP4 6-10

Students may want to refer back to the lesson examples.

Assignment Guide

If you finished Example **1** assign:
  Core 1–4, 10–13, 30–35
  Enriched 1–4, 10–13, 30–35

If you finished Example **2** assign:
  Core 1–8, 10–17, 19–23 odd,
    30–35
  Enriched 10–17, 19–35

If you finished Example **3** assign:
  Core 1–18, 19–27 odd, 30–35
  Enriched 10–35

Answers

9. The volume of the sphere and the cube are about equal (≈ 268 in^3). The surface area of the sphere is about 201 in^2, and the surface area of the cube is about 250 in^2.

18. The volume of the sphere is 4.5π ft^3, and its surface area is 9π ft^2. The volume of the cylinder is 4π ft^3, and its surface area is 12π ft^2.

21. $V = 39.72\pi \approx 124.72$ yd^3; $S = 38.44\pi \approx 120.70$ yd^2

22. 9 in.; $324\pi \approx 1017.36$ in^2

23. 30 km; $36,000\pi \approx 113,040$ km^3

24. 20.6 mi; 41.2 mi

GUIDED PRACTICE

See Example **1** Find the volume of each sphere, both in terms of π and to the nearest tenth. Use 3.14 for π.
10.7π cm^3; 33.6 cm^3
85.3π mi^3; 267.8 mi^3
1. $r = 2$ cm   2. $r = 10$ ft   3. $d = 3.4$ m   4. $d = 8$ mi
1333.3π ft^3; 4186.6 ft^3 6.6π m^3; 20.7 m^3

See Example **2** Find the surface area of each sphere, both in terms of π and to the nearest tenth. Use 3.14 for π. 174.2π mm^2; 547.0 mm^2 225π yd^2; 706.5 yd^2
5. 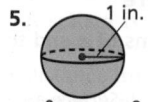 1 in. 6. 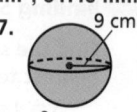 6.6 mm 7. 9 cm 8.  15 yd
4π in^2; 12.6 in^2 324π cm^2; 1017.4 cm^2

See Example **3** 9. Compare the volume and surface area of a sphere with radius 4 in. with that of a cube with sides measuring 6.45 in.

INDEPENDENT PRACTICE

See Example **1** Find the volume of each sphere, both in terms of π and to the nearest tenth. Use 3.14 for π.
1774.7π mm^3; 5572.6 mm^3
10. $r = 12$ ft   11. $r = 4.8$ cm   12. $d = 22$ mm   13. $d = 1$ in.
2304π ft^3; 7234.6 ft^3 147.5π cm^3; 463.2 cm^3 0.17π in^3; 0.5 in^3

See Example **2** Find the surface area of each sphere, both in terms of π and to the nearest tenth. Use 3.14 for π.
2500π cm^2; 7850 cm^2
14. 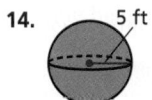 5 ft 15. 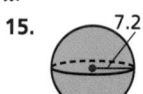 7.2 m 16. 9 km 17. 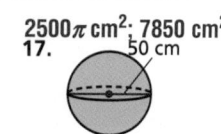 50 cm
100π ft^2; 314 ft^2 207.4π m^2; 651.2 m^2 81π km^2; 254.3 km^2

See Example **3** 18. Compare the volume and surface area of a sphere with diameter 3 ft with that of a cylinder with height 1 ft and a base with radius 2 ft.

PRACTICE AND PROBLEM SOLVING

Find the missing measurements of each sphere, both in terms of π and to the nearest hundredth. Use 3.14 for π.
466.56π m^2; 1465.00 m^2
19. radius = 5.5 in. 221.83π in^3;   20. radius = 10.8 m
   volume = ■   696.55 in^3 volume = 1679.62π m^2
   surface area = 121π in.^{2} surface area = ■

21. diameter = 6.2 yd   22. radius = ■
   volume = ■   diameter = 18 in.
   surface area = ■   surface area = ■

23. radius = ■   24. radius = ■
   volume = ■   diameter = ■
   surface area = 3600π km^2 surface area = 1697.44π mi^2

Math Background

Traditional (Euclidean) geometry is based on points, straight lines, and flat planes. In spherical geometry, a point has the traditional meaning, but flat planes are replaced by spherical surfaces, and straight lines are replaced by great circles of the sphere.

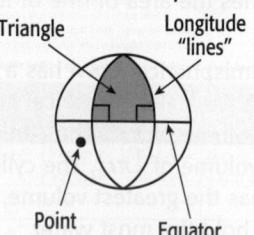

Triangle   Longitude "lines"

Point   Equator

In spherical geometry, there are no parallel lines, and there are "triangles" with two right angles.

Eggs come in many different shapes. The eggs of birds that live on cliffs are often extremely pointed to keep the eggs from rolling. Other birds, such as great horned owls, have eggs that are nearly spherical. Turtles and crocodiles also have nearly spherical eggs, and the eggs of many dinosaurs were spherical.

25. To lay their eggs, green turtles travel hundreds of miles to the beach where they were born. The eggs are buried on the beach in a hole about 40 cm deep. The eggs are approximately spherical, with an average diameter of 4.5 cm, and each turtle lays an average of 113 eggs at a time. Estimate the total volume of eggs laid by a green turtle at one time. **≈ 5392 cm³**

26. Fossilized embryos of dinosaurs called titanosaurid sauropods have recently been found in spherical eggs in Patagonia. The eggs were 15 cm in diameter, and the adult dinosaurs were more than 12 m in length. Find the volume of an egg. **≈ 1767.15 cm³**

27. The glasshouse spider mite lays spherical eggs that are translucent and about 0.1 mm in diameter. Find the surface area of an egg. **≈ 0.0314 mm²**

28. Hummingbirds lay eggs that are nearly spherical and about 1 cm in diameter. Find the surface area of an egg. **≈ 3.14 cm²**

29. ⬛ *CHALLENGE* An ostrich egg has about the same volume as a sphere with a diameter of 5 inches. If the shell is about $\frac{1}{12}$ inch thick, estimate the volume of just the shell, not including the interior of the egg. **≈ 6.33 in³**

Spiral Review

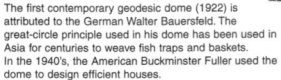

Multiply or divide. Write each answer in simplest form. (Lessons 3-3 and 3-4)

30. $\frac{2}{3} \cdot \frac{9}{10}$ **$\frac{3}{5}$**   31. $\frac{4}{5} \cdot \frac{3}{8}$ **$\frac{3}{10}$**   32. $\frac{1}{3} \div \frac{2}{3}$ **$\frac{1}{2}$**   33. $\frac{11}{15} \div \frac{5}{22}$ **$3\frac{17}{75}$**

34. **TEST PREP** Two angles are complementary if the sum of their measures equals ___?___. (Lesson 5-1) **A**

   A 90°   B 180°   C 270°   D 360°

35. **TEST PREP** Two angles are supplementary if the sum of their measures equals ___?___. (Lesson 5-1) **G**

   F 90°   G 180°   H 270°   J 360°

CHALLENGE 6-10

LESSON 6-10 Challenge
Useful and Intriguing

A **geodesic dome** is a structure made of a complex network of triangles that form a roughly spherical shape. The dome gets its efficiency from the characteristics of a sphere.

The first contemporary geodesic dome (1922) is attributed to the German Walter Bauersfeld. The great-circle principle used in his dome has been used in Asia for centuries to weave fish traps and baskets. In the 1940's, the American Buckminster Fuller used the dome to design efficient houses.

The classic geodesic dome takes its form from the **icosahedron**, a regular solid with 20 equilateral triangles as faces, 30 congruent edges, and 12 vertices.

Consider an icosahedron with edge $s = 12$ ft.

1. Find the surface area with the formula Area of Equilateral Triangle $= \frac{s^2}{4}\sqrt{3}$. Use $\sqrt{3} \approx 1.73$ to answer to the nearest tenth of a square foot.
$20\left(\frac{12^2}{4}\sqrt{3}\right) = 720\sqrt{3} \approx 720(1.73) \approx 1245.6$ ft²

2. Find the volume with the formula Volume of Icosahedron $= \frac{5}{12}(3 + \sqrt{5})s^2$. Use $\sqrt{5} \approx 2.24$ to answer to the nearest tenth of a cubic foot.
$\frac{5}{12}(3 + \sqrt{5})12^2 = 60(3 + \sqrt{5}) \approx 60(3 + 2.24) \approx 314.4$ ft³

3. Find an approximate value for the radius r of the sphere that has approximately the same volume as the icosahedron.
$\frac{4}{3}\pi r^3 \approx 314.4 \rightarrow \frac{4}{3}(3.14)r^3 \approx 314.4 \rightarrow 4.19r^3 \approx 314.4 \rightarrow r^3 \approx$
$75 \rightarrow r \approx 4.2$

4. Using your value for r, find the surface area of that sphere.
$4(3.14)(4.2^2) \approx 221.6$ ft³

5. Use your results to make an observation about why a sphere is more efficient than an icosahedron. **Possible answer:**
For a given volume, a sphere exposes less surface area than an icosahedron.

PROBLEM SOLVING 6-10

LESSON 6-10 Problem Solving
Spheres

Early golf balls were smooth spheres. Later it was discovered that golf balls flew better when they were dimpled. On January 1, 1932, the United States Golf Association set standards for the weight and size of a golf ball. The minimum diameter of a regulation golf ball is 1.680 inches. Use 3.14 for π. Round to the nearest hundredth.

1. Find the volume of a smooth golf ball with the minimum diameter allowed by the United States Golf Association.
2.48 in³

2. Find the surface area of a smooth golf ball with the minimum diameter allowed by the United States Golf Association.
8.86 in²

3. Would the dimples on a golf ball increase or decrease the volume of the ball?
decrease

4. Would the dimples on a golf ball increase or decrease the surface area of the ball?
increase

Use 3.14 for π. Use the following information for Exercises 5–6. A track and field expert recommends changes to the size of a shot put. One recommendation is that a shot put should have a diameter between 90 and 110 mm. Choose the letter for the best answer.

5. Find the surface area of a shot put with a diameter of 90 mm.
Ⓐ 25,434 mm²
B 101,736 mm²
C 381,520 mm²
D 3,052,080 mm²

6. Find the surface area of a shot put with diameter 110 mm.
F 9,499 mm²
G 22,834 mm²
Ⓗ 37,994 mm²
J 151,976 mm²

7. Find the volume of the earth if the average diameter of the earth is 7926 miles.
A 2.0 × 10⁸ mi³
Ⓑ 2.6 × 10¹¹ mi³
C 7.9 × 10⁸ mi³
D 2.1 × 10¹² mi³

8. An ice cream cone has a diameter of 4.2 cm and a height of 11.5 cm. One spherical scoop of ice cream is put on the cone that has a diameter of 5.6 cm. If the ice cream were to melt in the cone, by how much it would overflow the cone? Round to the nearest tenth.
F 0 cm³
Ⓗ 38.8 cm³
G 12.3 cm³
J 54.3 cm³

Lesson Quiz

Find the volume of each sphere, both in terms of π and to the nearest tenth. Use 3.14 for π.

1. $r = 4$ ft   85.3π ft³, 267.8 ft³

2. $d = 6$ m   36π m³, 113.0 m³

Find the surface area of each sphere, both in terms of π and to the nearest tenth. Use 3.14 for π.

3. $r = 22$ in.   1936π in², 6079.0 in²

4. $d = 1.5$ mi   2.25π mi², 7.1 mi²

5. A basketball has a circumference of 29 in. To the nearest cubic inch, what is its volume?   412 in³

Available on Daily Transparency in CRB

Problem Solving on Location

Minnesota

Purpose: To provide additional practice for problem-solving skills in Chapters 1–6

Mall of America

- After problem 1a, have students write an equation to find the number of feet in a mile. Have them solve the equation. $4.3x = 22,704$; $x = 5280$ ft Ask: How many square feet are in a square mile? 1 square mile = 27,878,400 square feet How many square miles is the Mall of America? 4,200,000 square feet = about 0.15 square miles

- After problem 2b, have students discuss whether they think it is likely that all the LEGO® bricks in the blimp are congruent.

  Possible answer: Probably not; it is more likely that the figure is made up of bricks of many different sizes and shapes.

Extension Have students use LEGO® or other building blocks to build a simple structure. Have them find the volume and surface area of the structure. Check students' work.

Problem Solving on Location

MINNESOTA

Mall of America

The 4.2-million-square-foot Mall of America is located in Bloomington, Minnesota. The mall is so large that 24,336 school buses or 32 Boeing 747's could fit inside it. There are plans to make the mall even bigger with the addition of 5 million square feet. The mall contains 520 stores. Each year, over 42.5 million people visit the mall. The mall houses Camp Snoopy, a 7-acre indoor amusement park with 30,000 live plants and 400 live trees, and Underwater Adventures, an aquarium with over 3000 sea creatures.

1. The Mall of America has 2.5 million square feet of retail space.
   a. If you lined up all of the stores, they would form an approximate rectangle with a base equal to 4.3 miles (22,704 ft). Find the height of this rectangle.
   b. What is the average area of the 520 stores, based on the total area of all the stores?

2. Within the LEGO® imagination center in the Mall of America is a blimp that was built with 138,240 Lego bricks.
   a. Find the volume of a standard rectangular brick with length $1\frac{1}{4}$ in., width $\frac{5}{8}$ in., and height $\frac{3}{8}$ in.
   b. Assume that the blimp is made of all standard rectangular bricks with the dimensions from part **a**. Find the approximate volume of the Legos that make up the blimp.

3. Camp Snoopy has a 74-foot-diameter Ferris wheel. What is the area and circumference of the Ferris wheel?

Answers

1. **a.** about 110.11 ft
   b. about 4808 ft^2

2. **a.** $\frac{75}{256}$, or about 0.29 in^3
   b. 40,500 in^3

3. area: about 4298.66 ft^2
   circumference: about 232.36 ft

Great Lakes Aquarium

The Great Lakes Aquarium in Duluth, Minnesota, is the only all-freshwater aquarium in the United States. It is located on the shore of Lake Superior and contains 170,000 gallons of water.

1. The aquarium has a glass water wall with etched panels. The water wall is 5 panels across and 7 panels tall.

   a. Find the number of panels that make up the water wall.   **35**

   b. If each panel is 10 feet across and 4 feet high, what is the area of one panel? What is the total area of the water wall?   **40 ft²; 1400 ft²**

2. You can compare Lake Superior to 40 different lakes, icecaps, and rivers at the 5 ft diameter globe and computer station. Find the surface area and volume of the aquarium's globe.
   surface area: 25π ft² ≈ 78.5 ft²; volume: $\frac{125}{6}\pi$ ft³ ≈ 65.4 ft³

3. The St. Louis River tank is a trapezoidal prism. The height of the trapezoidal base is 24 ft, and the two bases are 3 ft and 7 ft. If the height of the tank is 3 ft 6 in., what is the volume of the tank?   **420 ft³**

The 85,000-gallon Isle Royale exhibit is made up of three back-to-back tanks. Each tank is a prism with a different-shaped base.

4. Isle Royale of the Present is the largest of the three tanks and has samples of every kind of fish living in Lake Superior today. The base of this tank is approximately 447.6 ft², and the tank has a water level of 23.33 ft. What is the volume of water in the tank?   **10,442.5 ft³**

5. The Isle Royale of the Past houses the fish that are native to the lake. The water level of the tank is 23.33 ft. Find the area of the base using the dimensions on the diagram, and then find the volume of water in the tank.   **base area: 124.6 ft²; volume: 2907.5 ft³**

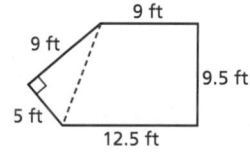

6. The Lake Herring tank has a height of about 17 ft and a trapezoidal base with a perimeter of 49 ft and an area of 294 ft². Find the area of the glass needed to construct the sides of this tank.   **833 ft²**

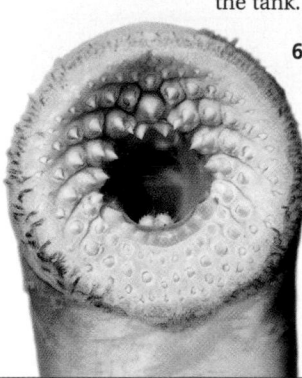

The sea lamprey is known as the "vampire of the Great Lakes," because of its blood-sucking method of eating fish.

Great Lakes Aquarium

- After problem 4, ask students why it is not necessary to know the shape of the tank's base. The volume of a prism is $V = Bh$, where B is the area of the prism's base. Since B is given, knowing its shape is not necessary because we don't have to find its area.

- After problem 5, have students consider different ways to break up the polygon in order to find its area.

  Possible answer: trapezoid + triangle; triangle + triangle + rectangle

Extension Have students make a model or drawing of one of the tanks described in problems 4–6. Check students' work.

Game Resources

Puzzles, Twisters & Teasers
Chapter 6 Resource Book

Planes in Space

Purpose: *To apply knowledge of three-dimensional figures to visualizing solids of revolution*

Discuss: Ask students to describe the technique used to create the figures in the examples.

Possible answer: Begin with a two-dimensional figure (circle, polygon, etc.). Rotate the figure around a line, or translate it along a line to form a three-dimensional figure. What kind of figure is formed when the figure below is rotated around the line shown? cup shape shown below

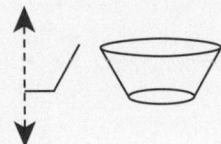

Extend: Challenge students to identify objects in the real world that can be described as having been generated by rotating a two-dimensional figure around a line. Have them sketch the figures and lines.

Possible answer:

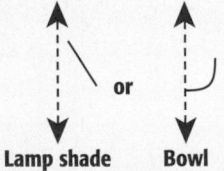

Lamp shade   **Bowl**

Triple Concentration

Purpose: *To use the Pythagorean Theorem to test sets of numbers*

Discuss: When a student draws three cards, ask: How do you know which numbers to substitute for the variables *a, b,* and *c* in the theorem?

Possible answer: The greatest length is always *c*, the hypotenuse. The lengths of *a* and *b*, the legs, are interchangeable.

Extend: Have students create and post a table of Pythagorean Triples. Have them do research to find as many triples as they can. Check students' tables.

Planes in Space

Some three-dimensional figures can be generated by plane figures.

Experiment with a circle first. Move the circle around. See if you recognize any three-dimensional shapes.

If you rotate a circle around a diameter, you get a sphere.

If you translate a circle up along a line perpendicular to the plane that the circle is in, you get a cylinder.

If you rotate a circle around a line outside the circle but in the same plane as the circle, you get a donut shape called a *torus*.

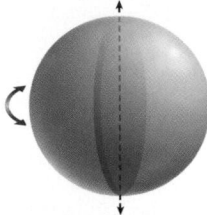
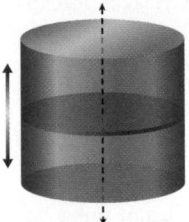
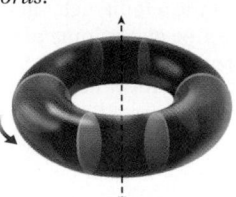

Draw or describe the three-dimensional figure generated by each plane figure.

1. a square translated along a line perpendicular to the plane it is in   **rectangular prism**

2. a rectangle rotated around one of its edges   **cylinder**

3. a right triangle rotated around one of its legs   **cone**

Triple Concentration

The goal of this game is to form *Pythagorean triples*, which are sets of three whole numbers *a*, *b*, and *c* such that $a^2 + b^2 = c^2$. A set of cards with numbers on them are arranged face down. A turn consists of drawing 3 cards to try to form a Pythagorean triple. If the cards do not form a Pythagorean triple, they are replaced in their original positions.

☑ internet connect

Go to **go.hrw.com** for a complete set of rules and cards.
KEYWORD: MP4 Game6

Pythagorean Triples

Use with Lesson 6-3

internet connect

Lab Resources Online
go.hrw.com
KEYWORD: MP4 TechLab6

Three positive integers a, b, and c that satisfy the equation $a^2 + b^2 = c^2$ are called **Pythagorean triples.** You know that $3^2 + 4^2 = 5^2$. So 3, 4, and 5 are Pythagorean triples.

Activity

You can generate Pythagorean triples a, b, and c by starting with two different whole numbers m and n, where m is the larger number. The Pythagorean triple will be as follows:

$a = m^2 - n^2$
$b = 2mn$
$c = m^2 + n^2$

> **Example:** Using $m = 2$ and $n = 1$
> $a = 2^2 - 1^2 = 4 - 1 = 3$
> $b = 2(2)(1) = 4$
> $c = 2^2 + 1^2 = 5$

Using a spreadsheet, enter the letters m, n, a, b, and c, respectively, in cells A1 to E1.

Then enter the formula for a as **=A2^2–B2^2** in cell C2.

For b, enter **=2*A2*B2** in cell D2, and for c, enter **=A2^2+B2^2** in cell E2.

	A	B	C	D	E
1	m	n	a	b	c
2			=A2^2-B2^2		0

Highlight cells C2, D2, and E2, and click the **Copy** button on the toolbar. Then select cells C2, D2, and E2, and drag down to highlight 7 rows. Click the **Paste** button on the toolbar.

	A	B	C	D	E
1	m	n	a	b	c
2			0	0	0
3			0	0	0
4			0	0	0
5			0	0	0
6			0	0	0
7			0	0	0
8			0	0	0

Next, enter 5 and 4 in cells A2 and B2, 5 and 3 in cells A3 and B3, and so forth until you complete the seventh row for the patterns given.

All integers a, b, and c shown in the last three columns are Pythagorean triples.

	A	B	C	D	E
1	m	n	a	b	c
2	5	4	9	40	41
3	5	3	16	30	34
4	5	2	21	20	29
5	5	1	24	10	26
6	4	3	7	24	25
7	4	2	12	16	20
8	4	1	15	8	17

Think and Discuss

1. Is order important in a Pythagorean triple? Why? **Yes and no; the lengths of the legs a and b are interchangeable, but c must always represent the length of the hypotenuse.**

Try This

1. Using a spreadsheet, generate 30 Pythagorean triples. **Check students' work.**

Objective: To use a spreadsheet to find Pythagorean triples

Materials: Spreadsheet software

Lab Resources

Technology Lab Activities . . . . p. 43

Using the Page

This technology activity shows students how to use a spreadsheet to find Pythagorean triples. Specific instructions may vary, depending on the spreadsheet software used. The instructions given are for Microsoft Excel. For instructions for other software, visit go.hrw.com.

The Think and Discuss problem is meant to point out that the order of the variables a and b (the length of the legs) is not important. The Try This problem shows students how quickly the computer can find Pythagorean triples once the correct formula is entered.

Assessment

1. What is meant by the input **A2^2–B2^2**? This input represents the value of a, which is $m^2 - n^2$.

2. Why is 0 not a valid input for m or n? Length must be greater than 0, and m and n represent the lengths of the triangle's legs.

Purpose: To help students review and practice concepts and skills presented in Chapter 6

Assessment Resources

Chapter Review
Chapter 6 Resource Book . pp. 105–107

Test and Practice Generator
CD-ROM

Additional review assessment items in both multiple-choice and free-response format may be generated for any objective in Chapter 6.

Answers

1. perimeter; area

2. edge; vertex

3. great circle; hemispheres

4. $13\frac{2}{9}$ in^2; 16 in.

5. 208 m^2; 80m

Vocabulary

area 281	horizon line 303	radius 294
circle 294	hypotenuse 290	regular pyramid 320
circumference 294	lateral face 316	right cone 320
cone 312	lateral surface 316	slant height 320
cylinder 307	leg 290	sphere 324
diameter 294	perimeter 280	surface area 316
edge 302	perspective 303	vanishing point 303
face 302	prism 307	vertex 302
great circle 324	pyramid 312	
hemisphere 324	Pythagorean Theorem .. 290	

Complete the sentences below with vocabulary words from the list above. Words may be used more than once.

1. In a two-dimensional figure, ___?___ is the distance around the outside of the figure, while ___?___ is the number of square units in the figure.

2. In a three-dimensional figure, a(n) ___?___ is where two faces meet, and a(n) ___?___ is where three or more edges meet.

3. A(n) ___?___ divides a sphere into two halves, or ___?___.

 6-1 Perimeter and Area of Rectangles and Parallelograms (pp. 280–284)

EXAMPLE

■ Find the area and perimeter of a rectangle with base 2 ft and height 5 ft.

$A = bh$ $P = 2l + 2w$
$= 5(2)$ $= 2(5) + 2(2)$
$= 10$ ft^2 $= 10 + 4$
                $= 14$ ft

EXERCISES

Find the area and perimeter of each figure.

4. a rectangle with base $2\frac{1}{3}$ in. and height $5\frac{2}{3}$ in.

5. a parallelogram with base 16 m, side length 24 m, and height 13 m.

6-2 Perimeter and Area of Triangles and Trapezoids (pp. 285–288)

EXAMPLE

■ Find the area of a right triangle with base 6 cm and height 3 cm.
$A = \frac{1}{2}bh = \frac{1}{2}(6)(3) = 9$ cm^2

EXERCISES

Find the area and perimeter of each figure.

6. a triangle with base 8 cm, sides 4.1 cm and 8.1 cm, and height 4 cm

7. trapezoid $DEFG$ with $DE = 4.5$ in., $EF = 10.1$ in., $FG = 16.5$ in., and $DG = 2.9$ in., where $\overline{DE} \parallel \overline{FG}$ and $h = 2.0$ in.

6-3 The Pythagorean Theorem (pp. 290–293)

EXAMPLE

■ Find the length of side b in the right triangle where $a = 8$ and $c = 17$.
$a^2 + b^2 = c^2$
$8^2 + b^2 = 17^2$
$64 + b^2 = 289$
$b^2 = 225$
$b = \sqrt{225} = 15$

EXERCISES

Solve for the unknown side in each right triangle.

8. If $a = 6$ and $b = 8$, find c.

9. If $b = 24$ and $c = 26$, find a.

6-4 Circles (pp. 294–297)

EXAMPLE

■ Find the area and circumference of a circle with radius 3.1 cm.
$A = \pi r^2$ $C = 2\pi r$
$= \pi(3.1)^2$ $= 2\pi(3.1)$
$= 9.61\pi \approx 30.2$ cm^2 $= 6.2\pi \approx 19.5$ cm

EXERCISES

Find the area and circumference of each circle, both in terms of π and to the nearest tenth. Use 3.14 for π.

10. $r = 15$ in.   11. $r = 2.4$ cm

12. $d = 8$ m   13. $d = 1.2$ ft

6-5 Drawing Three-Dimensional Figures (pp. 302–306)

EXAMPLE

■ Use isometric dot paper to sketch a rectangular prism that is 3 units long, 1 unit wide, and 2 units high.

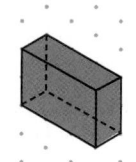

EXERCISES

Use isometric dot paper to sketch each figure.

14. a rectangular box that is 4 units long, 3 units deep, and 1 unit high

15. a cube with 3-unit sides

16. a box with a 2-unit square base and a height of 4 units

Answers

6. 16 cm^2; 20.2 cm

7. 21 in^2; 34 in.

8. $c = 10$

9. $a = 10$

10. $A = 225\pi \approx 706.5$ in^2; $C = 30\pi \approx 94.2$ in.

11. $A = 5.8\pi \approx 18.2$ cm^2; $C = 4.8\pi \approx 15.1$ cm

12. $A = 16\pi \approx 50.2$ m^2; $C = 8\pi \approx 25.1$ m

13. $A = 0.4\pi \approx 1.3$ ft^2; $C = 1.2\pi \approx 3.8$ ft

14.

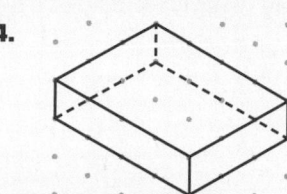

15.

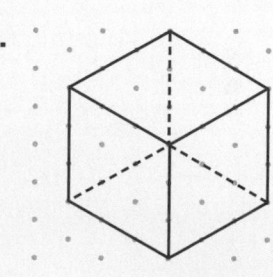

16.

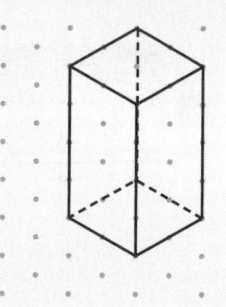

Study Guide and Review

6-6 Volume of Prisms and Cylinders (pp. 307–311)

EXAMPLE

■ Find the volume.
$$V = Bh = (\pi r^2)h$$
$$= \pi(4^2)(6)$$
$$= (16\pi)(6) = 96\pi \text{ cm}^3$$
$$\approx 301.6 \text{ cm}^3$$

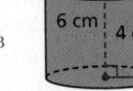

EXERCISES

Find the volume of each figure.

17.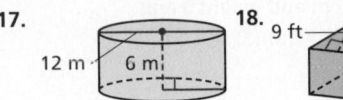

18.

6-7 Volume of Pyramids and Cones (pp. 312–315)

EXAMPLE

■ Find the volume.
$$V = \frac{1}{3}Bh = \frac{1}{3}(6)(4)(8)$$
$$= \frac{1}{3}(24)(8) = 64 \text{ in}^3$$

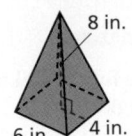

EXERCISES

Find the volume of each figure.

19. **20.**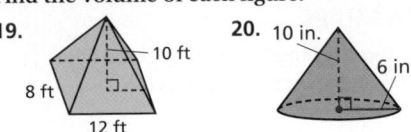

6-8 Surface Area of Prisms and Cylinders (pp. 316–319)

EXAMPLE

■ Find the surface area.
$$S = 2B + Ph$$
$$= 2(6) + (10)(4)$$
$$= 52 \text{ in}^2$$

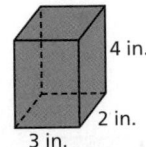

EXERCISES

Find the surface area of the figure.

21.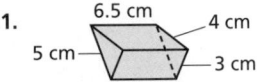

6-9 Surface Area of Pyramids and Cones (pp. 320–323)

EXAMPLE

■ Find the surface area.
$$S = B + \frac{1}{2}P\ell$$
$$= 16 + \frac{1}{2}(16)(5)$$
$$= 56 \text{ in}^2$$

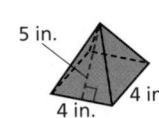

EXERCISES

Find the surface area.

22. 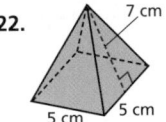 **23.**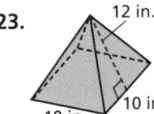

6-10 Spheres (pp. 324–327)

EXAMPLE

■ Find the volume of a sphere of radius 12 cm.
$$V = \frac{4}{3}\pi r^3 = \frac{4}{3}\pi(12^3)$$
$$= 2304\pi \text{ cm}^3 \approx 7234.6 \text{ cm}^3$$

EXERCISES

Find the volume of each sphere, both in terms of π and to the nearest tenth. Use 3.14 for π.

24. $r = 9$ in. **25.** $d = 30$ m

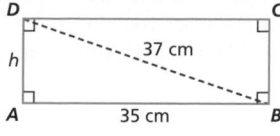
Graph and find the area of each figure with the given vertices.

1. $(4, 1), (-3, 1), (-3, -4), (4, -4)$

2. $(0, 4), (2, 3), (2, -3), (0, -2)$

3. $(-3, 0), (2, 0), (4, -2)$

4. $(2, 3), (6, -2), (-5, -2), (-2, 3)$

5. Use the Pythagorean Theorem to find the height of rectangle $ABCD$. **12 cm**

6. Find the area of rectangle $ABCD$. **420 cm²**

D   C
h   37 cm
A   35 cm   B

7. Use the Pythagorean Theorem to find the height of equilateral triangle PQR to the nearest hundredth. **6.93 in.**

8. Find the area of equilateral triangle PQR to the nearest tenth. **27.7 in²**

Q
8 in. h 8 in.
P 4 in. 4 in. R

Find the area of the circle to the nearest tenth. Use 3.14 for π.

9. radius = 11 in. **379.9 in²**

10. diameter = 26 cm **530.7 cm²**

Find the volume of each figure.

11. a sphere of radius 8 cm **682.7π ≈ 2143.7 cm³**

12. a cylinder of height 10 in. and radius 6 in. **360π ≈ 1130.4 in³**

13. a pyramid with a 3 ft by 3 ft square base and height 5 ft **15 ft³**

14. a cone of diameter 12 in. and height 18 in. **216π ≈ 678.2 in³**

Find the surface area of each figure.

15. 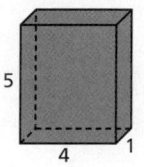 **58 units²**

16. 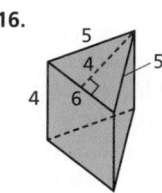 **88 units²**

17. 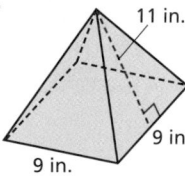 **279 in²**

18. 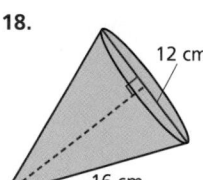 **132π ≈ 414.5 cm²**

Chapter Test

Purpose: *To assess students' mastery of concepts and skills in Chapter 6*

Assessment Resources ✔

Chapter Tests (Levels A, B, C)
Assessment Resources . . . . . pp. 63–68

Test and Practice Generator CD-ROM

Additional assessment items in both multiple-choice and free-response format may be generated for any objective in Chapter 6.

Answers

1.
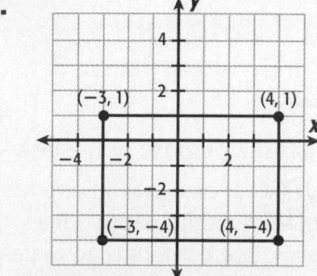
35 units²

2.
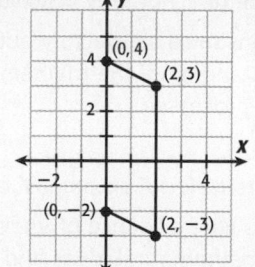
12 units²

3.
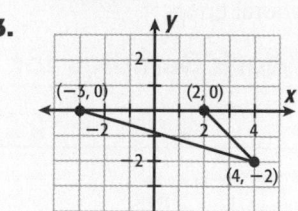
5 units²

4.
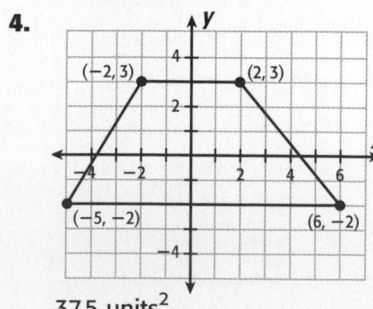
37.5 units²

Purpose: To assess students' understanding of concepts in Chapter 6 and combined problem-solving skills

Assessment Resources

Performance Assessment
Assessment Resources .... p. 128

Performance Assessment Teacher Support
Assessment Resources .... p. 127

Answers

1–2. See p. A8.

3. See Level 3 work sample below.

Scoring Rubric for Problem Solving Item 3

Level 3
Accomplishes the purpose of the task.

Student gives clear explanations, shows understanding of mathematical ideas and processes, and computes accurately.

Level 2
Purposes of the task not fully achieved.

Student demonstrates satisfactory but limited understanding of the mathematical ideas and processes.

Level 1
Purposes of the task not accomplished.

Student shows little evidence of understanding the mathematical ideas and processes and makes computational and/or procedural errors.

🥄 Show What You Know

Create a portfolio of your work from this chapter. Complete this page and include it with your four best pieces of work from Chapter 6. Choose from your homework or lab assignments, mid-chapter quiz, or any journal entries you have done. Put them together using any design you want. Make your portfolio represent what you consider your best work.

⭐ Short Response

Trace each figure, and then locate the vanishing point or horizon line.

1. Draw a rectangle with base length 7 cm and height 4 cm. Then draw a rectangle with base length 14 cm and height 1 cm. Which rectangle has the larger area? Which rectangle has the larger perimeter? Show your work or explain in words how you determined your answers.

2. A cylinder with a height of 6 in. and a diameter of 4 in. is filled with water. A cone with a height of 6 in. and a diameter of 2 in. is placed in the cylinder, point down, with its base even with the top of the cylinder. Draw a diagram to illustrate the situation described, and then determine how much water is left in the cylinder. Show your work.

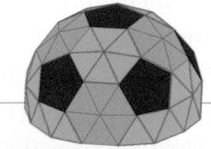

🧩 Extended Problem Solving

3. A *geodesic dome* is constructed of triangles. The surface is approximately spherical.
   a. A pattern for a geodesic dome that approximates a hemisphere uses 30 triangles with base 8 ft and height 5.63 ft and 75 triangles with base 8 ft and height 7.13 ft. Find the surface area of the dome.
   b. The base of the dome is approximately a circle with diameter 41 ft. Use a hemisphere with this diameter to estimate the surface area of the dome.
   c. Compare your answer from part **a** to your estimate from part **b**. Explain the difference.

Richard Buckminster Fuller created the *geodesic dome* and designed the Dymaxion™ house, car, and map.

Performance Assessment

Student Work Samples for Item 3

Level 3

> 3a. Area of Triangles = Surface Area
> 30•½(8)•5.63 = 675.6
> +75•½(8)•7.13 = 2139
> _____
> 2814.6 ft²
>
> b. Hemisphere Surface Area = 2πr²
>
> r = ½(41) = 20.5
>
> 2πr² = 840.5π ≈ 2640.5 ft²
>
> c. 2814.6
>   −2640.5
>   _____
>   174.1   The surface area of the dome is 174.1 ft² greater. A sphere is smooth and the dome is not.

The student accurately found the surface areas and gave a plausible reason for their differences.

Level 2

> 3a. 30•8•5.63 = 1351.2
>   75•8•7.13 = 4278
>   _____
>   Surface Area 5629.2
>
> b. 4πr² Surface Area
>   = 4π(d/2)²
>   = 4π(20.5)² = 1681π = 5281.0
>
> c. The hemisphere is slightly smaller.

The student left key information out of a formula and calculated correctly but did not give a plausible reason for differences.

Level 1

> 3a. ½ 30×8×5.63×75×7.13
>   = 361,227.1
>
> b. πr² = 41²×π
>   = 5281
>
> c. They are a lot different!

The student did not understand how to find surface area and was not able to give a plausible reason for differences.

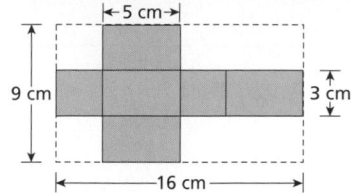
Cumulative Assessment, Chapters 1–6

1. The shaded figure below is a net that can be used to form a rectangular prism. What is the surface area of the prism? **C**

- Ⓐ 15 cm²
- Ⓒ 78 cm²
- Ⓑ 144 cm²
- Ⓓ 180 cm²

2. What is the value of x in the table below? **H**

Number of Inches	5	10	x
Number of Centimeters	12.7	25.4	50.8

- Ⓕ 15
- Ⓗ 20
- Ⓖ 18
- Ⓙ 22

3. The quantity (3×8^{12}) is how many times the quantity (3×8^5)? **D**
- Ⓐ 7
- Ⓒ 21
- Ⓑ 8
- Ⓓ 8^7

4. The arithmetic mean of 3 numbers is 60. If two of the numbers are 50 and 60, what is the third number? **J**
- Ⓕ 55
- Ⓗ 65
- Ⓖ 60
- Ⓙ 70

5. What is the value of $26 - 24 \cdot 2^3$? **D**
- Ⓐ 18
- Ⓒ −118
- Ⓑ 16
- Ⓓ −166

6. If $p = 3$, what is $4r(3 - 2p)$ in terms of r? **F**
- Ⓕ −12r
- Ⓗ −7r
- Ⓖ −8r
- Ⓙ 12r − 6

7. Point A' is formed by reflecting $A(-9, -8)$ across the y-axis. Find the coordinates of A'. **B**
- Ⓐ (9, 8)
- Ⓒ (−9, 8)
- Ⓑ (9, −8)
- Ⓓ (−8, −9)

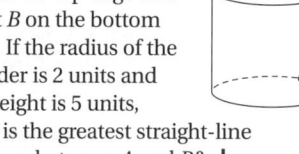

TEST TAKING TIP!
Look for a pattern in the data set to help you find the answer.

8. In the cylinder, point A lies on the top edge and point B on the bottom edge. If the radius of the cylinder is 2 units and the height is 5 units, what is the greatest straight-line distance between A and B? **J**

- Ⓕ 5
- Ⓗ $\sqrt{29}$
- Ⓖ 7
- Ⓙ $\sqrt{41}$

9. ***SHORT RESPONSE*** On a number line, point A has the coordinate −3 and point B has the coordinate 12. Point P is $\frac{2}{3}$ of the way from A to B. Draw and label the three points on a number line.

10. ***SHORT RESPONSE*** The tip of a blade on an electric fan is 1.5 feet from the axis of rotation. If the fan spins at a full rate of 1760 revolutions per minute, how many miles will a point at the tip of a blade travel in one hour? (1 mile = 5280 feet) Show your work.

Standardized Test Prep

Purpose: *To provide review and practice for Chapters 1–6 and standardized tests*

Assessment Resources ✓

Cumulative Tests (Levels A, B, C)
Assessment Resources.... pp. 205–216

State-Specific Test Practice Online
KEYWORD: MP4 TestPrep

Test Prep Doctor ✚

For item 2, encourage students to sketch the point and its image on a grid. With the visual aid, they should see that the y-coordinate remains the same while the x-coordinate is changed to its opposite. The correct answer is choice **G** (9, −8). If students chose choice **H,** they probably reflected the point over the wrong axis.

For item 8, encourage students to preview the answer choices before they work the problem. Students should notice that the answer choices are given in terms of *pi,* so they will not have to substitute a value for *pi* to solve the problem.

Answers

9.

10. 188.5 miles;

The tip of the blade travels $2\pi r = 2\pi(1.5) = 3\pi$ feet in one revolution. In one minute, the blade's tip travels $1760 \cdot 3\pi = 5280\pi$ feet, or π miles. In one hour, the blade's tip travels $60\pi \approx 188.5$ miles.

Chapter 7

Ratios and Similarity

Section 7A: Ratios, Rates, and Proportions	Section 7B: Similarity and Scale
Lesson 7-1 Ratios and Proportions	**Lesson 7-5** Dilations
Lesson 7-2 Ratios, Rates, and Unit Rates	**Hands-On Lab 7B** Explore Similarity
Lesson 7-3 Analyze Units	**Lesson 7-6** Similar Figures
Hands-On Lab 7A Model Proportions	**Lesson 7-7** Scale Drawings
Lesson 7-4 Solving Proportions	**Lesson 7-8** Scale Models
	Hands-On Lab 7C Make a Scale Model
	Lesson 7-9 Scaling Three-Dimensional Figures
	Extension Trigonometric Ratios

Pacing Guide for 45-Minute Classes

Chapter 7

DAY 79 Lesson 7-1, Lesson 7-2	**DAY 80** Lesson 7-3	**DAY 81** Hands-On Lab 7A	**DAY 82** Lesson 7-4	**DAY 83** Mid-Chapter Quiz, Lesson 7-5
DAY 84 Hands-On Lab 7B	**DAY 85** Lesson 7-6	**DAY 86** Lesson 7-7	**DAY 87** Lesson 7-8	**DAY 88** Hands-On Lab 7C
DAY 89 Lesson 7-9	**DAY 90** Extension	**DAY 91** Chapter 7 Review	**DAY 92** Chapter 7 Assessment	

Pacing Guide for 90-Minute Classes

Chapter 7

DAY 39 Chapter 6 Review, Lesson 7-1, Lesson 7-2	**DAY 40** Chapter 6 Assessment, Lesson 7-3	**DAY 41** Hands-On Lab 7A, Lesson 7-4	**DAY 42** Mid-Chapter Quiz, Lesson 7-5, Hands-On Lab 7B	**DAY 43** Lesson 7-6, Lesson 7-7
DAY 44 Lesson 7-8, Hands-On Lab 7C	**DAY 45** Lesson 7-9, Extension	**DAY 46** Chapter 7 Review, Lesson 8-1	**DAY 47** Chapter 7 Assessment, Hands-On Lab 8A	

COURSE 1

- Find unit rates.
- Identify similar figures.
- Write, solve, and use proportions.
- Make measurement conversions.

COURSE 2

- Find and compare rates and ratios.
- Identify similar figures.
- Write, solve, and use proportions.
- Understand and use scale drawings, scale factors, and scale maps.
- Use dimensional analysis to solve problems.

PRE-ALGEBRA

- **Find equivalent rates and ratios, and write and solve proportions.**
- **Identify similar figures and create dilations of plane figures.**
- **Compare and use scale drawings and scale factors, and make scale models.**
- **Use dimensional analysis to solve problems.**
- **Use trigonometric ratios.**

Across the Curriculum

LANGUAGE ARTS LINK

Math: Reading and Writing in the Content Area . . . . pp. 55–63

Focus on Problem Solving
  Solve . . . . SE p. 361
Journal . . . . TE, last page of each lesson
Write About It . . . . SE pp. 345, 349, 354, 365, 371, 379, 385

SCIENCE LINK

Life Science . . . . SE pp. 354, 373, 377, 378
Earth Science . . . . SE p. 343
Physical Science . . . . SE pp. 352, 357, 358, 371, 385
Health . . . . SE p. 359

TE = *Teacher's Edition*   **SE** = *Student Edition*

Interdisciplinary

Bulletin Board

Life Science

According to the results of Gregor Mendel's experiment, the ratio of plants with purple flowers to plants with white flowers can be written as 705 to 224 or 705:224, or in the simplified form of approximately 3.15 to 1. Find the simplified ratio for the seed color. 3 yellow:1 green or 3:1

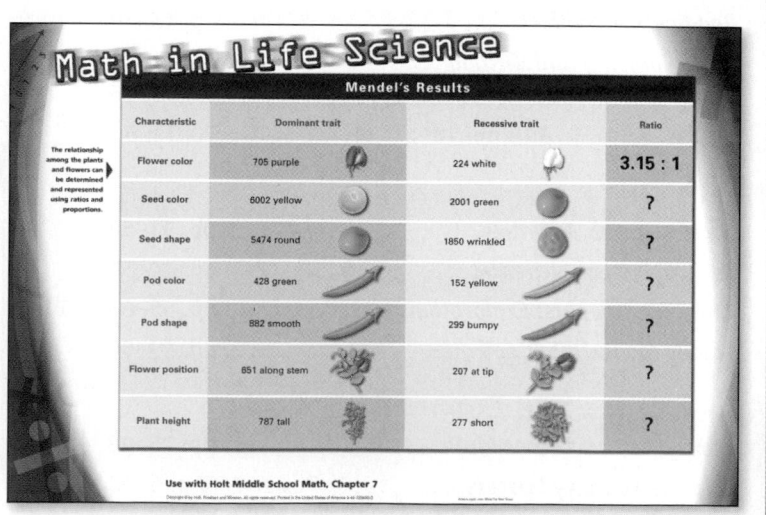

Interdisciplinary posters and worksheets are provided in your resource material.

Resource Options

Chapter 7 Resource Book

Student Resources

Teacher and Parent Resources

- Daily Transparencies
- Additional Examples Transparencies
- Teaching Transparencies

Reaching All Learners

English Language Learners

Individual Needs

Hands-On

Applications and Connections

Transparencies

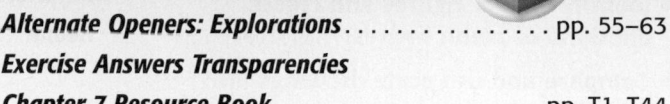

- Daily Transparencies
- Additional Examples Transparencies
- Teaching Transparencies

Technology

Teacher Resources

Student Resources

internet connect

Homework Help Online	**KEYWORD:** MP4 HWHelp7
Math Tools Online	**KEYWORD:** MP4 Tools
Glossary Online	**KEYWORD:** MP4 Glossary
Chapter Project Online	**KEYWORD:** MP4 PSProject7
Chapter Opener Online	**KEYWORD:** MP4 Ch7

KEYWORD: MP4 CNN7

SE = *Student Edition*　　**TE** = *Teacher's Edition*　　**AR** = *Assessment Resources*　　**CRB** = *Chapter Resource Book*　　**MK** = *Manipulatives Kit*

Assessing Prior Knowledge

Determine whether students have the required prerequisite concepts and skills.

Are You Ready?.................................. SE p. 341
Inventory Test................................ AR pp. 1–4

Test Preparation

Provide review and practice for chapter and standardized tests.

Standardized Test Prep......................... SE p. 397
Spiral Review with Test Prep..... SE, last page of each lesson
Study Guide and Review................. SE pp. 392–394
Test Prep Tool Kit

Technology

🔘 **Test and Practice Generator CD-ROM**

📶 **internet** connect
State-Specific Test Practice Online KEYWORD: MP4 TestPrep

Performance Assessment

Assess students' understanding of chapter concepts and combined problem-solving skills.

Performance Assessment...................... SE p. 396
   Includes scoring rubric in TE
Performance Assessment...................... AR p. 130
Performance Assessment Teacher Support......... AR p. 129

Portfolio

Portfolio opportunities appear throughout the Student and Teacher's Editions.

Suggested work samples:

Problem Solving Project.................... TE p. 340
Performance Assessment.................... SE p. 396
Portfolio Guide......................... AR p. xxxvii
Journal................... TE, last page of each lesson
Write About It...... SE pp. 345, 349, 354, 365, 371, 379, 385

Daily Assessment

Obtain daily feedback on students' understanding of concepts.

Spiral Review and Test Prep...... SE, last page of each lesson

**Also Available on Transparency
In Chapter 7 Resource Book**

Warm Up..................... TE, first page of each lesson
Problem of the Day............ TE, first page of each lesson
Lesson Quiz................... TE, last page of each lesson

Student Self-Assessment

Have students evaluate their own work.

Group Project Evaluation .... AR p. xxxiv
Individual Group Member Evaluation .... AR p. xxxv
Portfolio Guide............................ AR p. xxxvii
Journal..................... TE, last page of each lesson

Formal Assessment

Assess students' mastery of concepts and skills.

Section Quizzes .... AR pp. 17–18
Mid-Chapter Quiz............................ SE p. 360
Chapter Test .... SE p. 395
Chapter Tests (Levels A, B, C) .... AR pp. 69–74
Cumulative Tests (Levels A, B, C)........... AR pp. 217–228
Standardized Test Prep
   Cumulative Assessment .... SE p. 397
End-of-Year Test.......................... AR pp. 313–316

Technology

🔘 **Test and Practice Generator CD-ROM**

Make tests electronically. This software includes:

- Dynamic practice for Chapter 7
- Customizable tests
- Multiple-choice items for each objective
- Free-response items for each objective
- Teacher management system

SE = *Student Edition*   **TE** = *Teacher's Edition*   **AR** = *Assessment Resources*   **CRB** = *Chapter Resource Book*   **MK** = *Manipulatives Kit*

340D

Chapter 7 Tests

Three levels (A,B,C) of tests are available for each chapter in the *Assessment Resources.*

LEVEL A

CHAPTER 7 Chapter Test
Form A

Find two ratios that are equivalent to each given ratio.

1. $\frac{5}{10}$ Possible answer: $\frac{1}{2}, \frac{10}{20}$

2. $\frac{4}{6}$ Possible answer: $\frac{2}{3}, \frac{8}{12}$

3. $\frac{16}{4}$ Possible answer: $\frac{4}{1}, \frac{8}{2}$

4. $\frac{21}{27}$ Possible answer: $\frac{7}{9}, \frac{14}{18}$

Simplify to tell whether the ratios form a proportion.

5. $\frac{4}{12}$ and $\frac{2}{8}$ **no**

6. $\frac{1}{2}$ and $\frac{4}{8}$ **yes**

7. $\frac{1}{3}$ and $\frac{2}{6}$ **yes**

Find the unit price for each offer and tell which is the better buy.

8. 20-oz box of cereal for $3.80; 15-oz box of cereal for $3.15

The 20-oz box costs $0.19/oz and the 15-oz box costs $0.21/oz. The 20-oz box is a better buy.

9. 10 blank CDs for $2.50; 15 blank CDs for $3.00

10 CDs cost $0.25 per CD and 15 CDs cost $0.20 per CD. 15 CDs is a better buy.

Find the appropriate factor for each conversion.

10. kilometers to meters

$$\frac{1000 \text{ m}}{1 \text{ kilometer}}$$

11. gallons to quarts

$$\frac{4 \text{ quarts}}{1 \text{ gallon}}$$

12. A car travels 5 miles in 6 minutes. What is its speed in miles per hour?

50 mi/h

13. A home improvement store sells 2355 feet of wire per day. How many yards of wire does the store sell per day?

785 yards

Solve each proportion.

14. $\frac{3}{9} = \frac{x}{3}$ $x = 1$

15. $\frac{32}{t} = \frac{4}{1}$ $t = 8$

16. $\frac{12}{22} = \frac{18}{p}$ $p = 33$

CHAPTER 7 Chapter Test
Form A, continued

Dilate each figure by the given scale factor with the origin as the center of dilation.

17. Triangle with vertices $A(1, 2)$, $B(4, 1)$, $C(4, 4)$, scale factor = 2

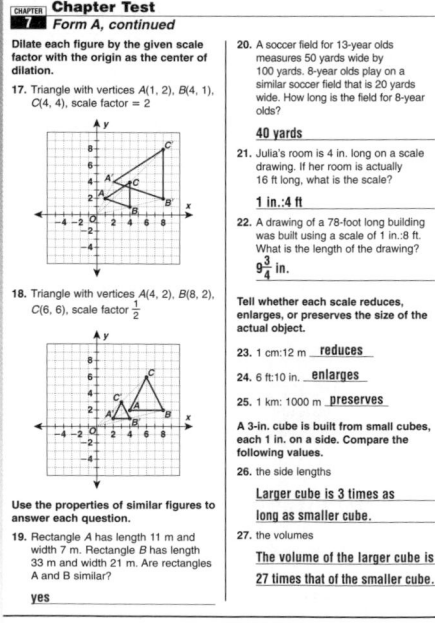

18. Triangle with vertices $A(4, 2)$, $B(8, 2)$, $C(6, 6)$, scale factor $\frac{1}{2}$

Use the properties of similar figures to answer each question.

19. Rectangle *A* has length 11 m and width 7 m. Rectangle *B* has length 33 m and width 21 m. Are rectangles A and B similar?

yes

20. A soccer field for 13-year olds measures 50 yards wide by 100 yards. 8-year olds play on a similar soccer field that is 20 yards wide. How long is the field for 8-year olds?

40 yards

21. Julia's room is 4 in. long on a scale drawing. If her room is actually 16 ft long, what is the scale?

1 in.:4 ft

22. A drawing of a 78-foot long building was built using a scale of 1 in.:8 ft. What is the length of the drawing?

$9\frac{3}{4}$ in.

Tell whether each scale reduces, enlarges, or preserves the size of the actual object.

23. 1 cm:12 m   **reduces**

24. 6 ft:10 in.   **enlarges**

25. 1 km: 1000 m   **preserves**

A 3-in. cube is built from small cubes, each 1 in. on a side. Compare the following values.

26. the side lengths

Larger cube is 3 times as long as smaller cube.

27. the volumes

The volume of the larger cube is 27 times that of the smaller cube.

LEVEL B

CHAPTER 7 Chapter Test
Form B

Find two ratios that are equivalent to each given ratio.

1. $\frac{2}{9}$ Possible answer: $\frac{4}{18}, \frac{6}{27}$

2. $\frac{50}{15}$ Possible answer: $\frac{10}{3}, \frac{100}{30}$

3. $\frac{11}{17}$ Possible answer: $\frac{22}{34}, \frac{33}{51}$

4. $\frac{18}{16}$ Possible answer: $\frac{9}{8}, \frac{36}{32}$

Simplify to tell whether the ratios form a proportion.

5. $\frac{4}{26}$ and $\frac{2}{13}$ **yes**

6. $\frac{18}{60}$ and $\frac{3}{10}$ **yes**

7. $\frac{5}{25}$ and $\frac{15}{50}$ **no**

Find the unit price for each offer and tell which is the better buy.

8. 20 blank CDs for $2.79; 12 blank CDs for $1.20

20 CDs cost $0.14 per CD and 12 CDs cost $0.10 per CD. 12 CDs is the better buy.

9. 6 paperback books for $19.00; 8 paperback books for $26.00

6 books cost $3.17 per book and 8 books cost $3.25 per book. 6 books is the better buy.

Find the appropriate factor for each conversion.

10. months to years

$$\frac{1 \text{ yr}}{12 \text{ months}}$$

11. pounds to ounces

$$\frac{16 \text{ oz}}{1 \text{ lb}}$$

12. A woodworker can put together 2 wood toy trains per day. How many trains could the woodworker make in 8 weeks?

112

13. A store sells 8-ounce packages of mushrooms for $1.29. What is the cost of 3 pounds of mushrooms?

$7.74

Solve each proportion.

14. $\frac{4}{10} = \frac{y}{20}$ $y = 8$

15. $\frac{r}{0.32} = \frac{3}{2}$ $r = 0.48$

16. $\frac{11}{q} = \frac{5}{2}$ $q = 4.4$

CHAPTER 7 Chapter Test
Form B, continued

Dilate each figure by the given scale factor with the origin as the center of dilation.

17. Triangle with vertices $A(2, 2)$, $B(8, 4)$, $C(4, 8)$, scale factor $\frac{1}{4}$

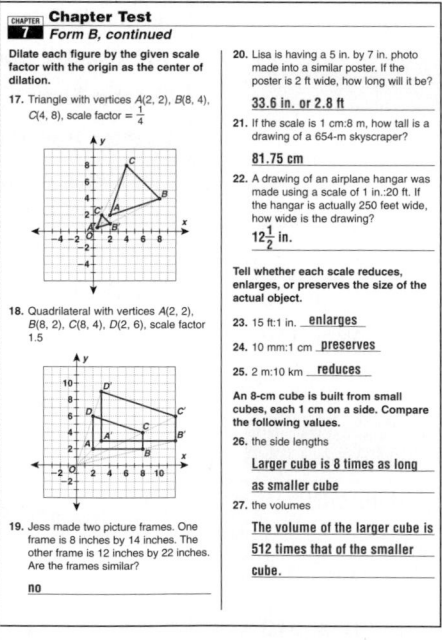

18. Quadrilateral with vertices $A(2, 2)$, $B(8, 2)$, $C(8, 4)$, $D(2, 6)$, scale factor 1.5

19. Jess made two picture frames. One frame is 8 inches by 14 inches. The other frame is 12 inches by 22 inches. Are the frames similar?

no

20. Lisa is having a 5 in. by 7 in. photo made into a similar poster. If the poster is 2 ft wide, how long will it be?

33.6 in. or 2.8 ft

21. If the scale is 1 cm:8 m, how tall is a drawing of a 654-m skyscraper?

81.75 cm

22. A drawing of an airplane hangar was made using a scale of 1 in.:20 ft. If the hangar is actually 250 feet wide, how wide is the drawing?

$12\frac{1}{2}$ in.

Tell whether each scale reduces, enlarges, or preserves the size of the actual object.

23. 15 ft:1 in.   **enlarges**

24. 10 mm:1 cm   **preserves**

25. 2 m:10 km   **reduces**

An 8-cm cube is built from small cubes, each 1 cm on a side. Compare the following values.

26. the side lengths

Larger cube is 8 times as long as smaller cube.

27. the volumes

The volume of the larger cube is 512 times that of the smaller cube.

LEVEL C

CHAPTER 7 Chapter Test
Form C

Find two ratios that are equivalent to each given ratio.

1. $\frac{5}{9}$ Possible answer: $\frac{10}{18}, \frac{15}{27}$

2. $\frac{2}{12}$ Possible answer: $\frac{1}{6}, \frac{3}{18}$

3. $\frac{15}{6}$ Possible answer: $\frac{5}{2}, \frac{10}{4}$

4. $\frac{14}{8}$ Possible answer: $\frac{7}{4}, \frac{21}{12}$

Simplify to tell whether the ratios form a proportion.

5. $\frac{6}{16}$ and $\frac{9}{24}$ **yes**

6. $\frac{36}{28}$ and $\frac{10}{7}$ **no**

7. $\frac{21}{27}$ and $\frac{7}{9}$ **no**

Find the unit price for each offer and tell which is the better buy.

8. $7.98 for a 3-pound ham; $10.84 for a 5-pound ham

The 3-lb ham costs $2.66 per lb and the 5-lb ham costs $2.17 per pound. The 5-pound ham is the better buy.

9. A 12-ounce drink for $1.40; a 20-ounce drink for $2.70.

The 12-oz drink costs $0.12 per ounce and the 20-oz drink costs $0.14 per ounce. The 12-ounce drink is the better buy.

Find the appropriate factor for each conversion.

10. weeks to hours

$$\frac{168 \text{ hours}}{1 \text{ week}}$$

11. miles to yards

$$\frac{1760 \text{ yd}}{1 \text{ mi}}$$

12. A car wash cleans automobiles at a rate of 35 per hour. How many cars do they clean in an 8-hour day?

280 automobiles

13. Zara biked for 2 hours at an average rate of 15 meters per second. How many kilometers did she bike?

108 km

Solve each proportion.

14. $\frac{6}{4} = \frac{x}{5}$ $x = 7.5$

15. $\frac{33}{t} = \frac{4}{1}$ $t = 8.25$

16. $\frac{12}{1.5} = \frac{40}{p}$ $p = 5$

CHAPTER 7 Chapter Test
Form C, continued

Identify the scale factor used in each dilation.

17.

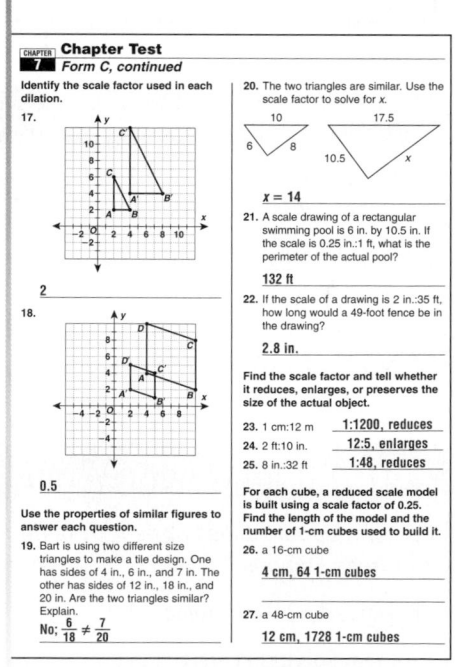

2

18.

0.5

Use the properties of similar figures to answer each question.

19. Bart is using two different size triangles to make a tile design. One has sides of 4 in., 6 in., and 7 in. The other has sides of 12 in., 18 in., and 20 in. Are the two triangles similar? Explain.

No; $\frac{6}{18} \neq \frac{7}{20}$

20. The two triangles are similar. Use the scale factor to solve for *x*.

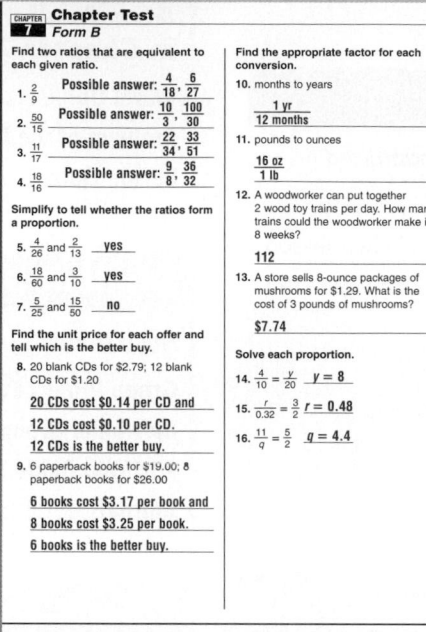

$x = 14$

21. A scale drawing of a rectangular swimming pool is 6 in. by 10.5 in. If the scale is 0.25 in.:1 ft, what is the perimeter of the actual pool?

132 ft

22. If the scale of a drawing is 2 in.:35 ft, how long would a 49-foot fence be in the drawing?

2.8 in.

Find the scale factor and tell whether it reduces, enlarges, or preserves the size of the actual object.

23. 1 cm:12 m   **1:1200, reduces**

24. 2 ft:10 in.   **12:5, enlarges**

25. 8 in.:32 ft   **1:48, reduces**

For each cube, a reduced scale model is built using a scale factor of 0.25. Find the length of the model and the number of 1-cm cubes used to build it.

26. a 16-cm cube

4 cm, 64 1-cm cubes

27. a 48-cm cube

12 cm, 1728 1-cm cubes

Test and Practice Generator
CD-ROM

Create and customize multiple versions of the same tests with corresponding answers for any chosen chapter objectives.

Chapter 7 State and Standardized Test Preparation

Test Taking Skill Builder and Standardized Test Practice are provided for each chapter in the *Test Prep Tool Kit.*

TEST TAKING SKILL BUILDER

Test Taking Strategy
Short Response Questions
Chapter 7

To receive full credit to a short response question, you need to show each step of your calculations. When appropriate, answer the question in a complete sentence and provide your reasoning.

The scoring rubric shown is used to score the following example.

Scoring Rubric
- 2 points: Student writes appropriate equation and defines the variable. Student shows all work and makes no errors in computations. Student explains his or her reasoning. The question is answered in a complete sentence.
- 1 point: Student shows most work and makes no errors or has a minor error. The student answers the question but does not explain how the answer was determined.
- 0 points: The answer is incorrect and no work is provided.

Example Short Response Dean has a recipe for lemonade. To make 14 servings, Dean will need 4 cups of lemon juice. How many cups of lemon juice will Dean need if he wants to make 35 servings? Show your work and explain how you determined your answer.

A possible 2-point response is shown below.
Write a proportion to determine the number of cups of lemon juice needed for 35 servings. Let x = the number of cups needed.

$\frac{4}{14} = \frac{x}{35}$

$4 \cdot 35 = 14 \cdot x$

$140 = 14x$

$\frac{140}{14} = \frac{14x}{14}$

$10 = x$

Dean will need 10 cups of lemon juice to make 35 servings.

Below is a 1-point response.

$\frac{4}{14} = \frac{x}{35}$

$140 = 14x$

$10 = x$

The student receives only 1 point because all of the work is not shown, no explanation is provided, and the question isn't answered in a complete sentence.

Test Taking Strategy
Chapter 7, continued
Exercises
Use the scoring rubric for each question.

Scoring Rubric
- 2 points: Student writes appropriate equation and defines the variable. Student shows all work and makes no errors in computations. Student explains his or her reasoning. The question is answered in a complete sentence.
- 1 point: The student answers the question correctly, shows his or her work, but does not explain how the answer was determined or has a minor computational error.
- 0 points: The answer is incorrect and no work is provided.

1. Emma is standing on a sidewalk next to a lamppost. The lamppost is 7.5 ft tall and casts a 6-ft shadow. Emma's is 5.5 ft tall. How long is her shadow? Show your work and explain in words how you determined your answer.

2-point response	1-point response
Use a proportion to determine the length of Emma's shadow.	$\frac{6 \text{ ft}}{7.5 \text{ ft}} = \frac{5.5 \text{ ft}}{x \text{ ft}}$
Let x = the length of Emma's shadow.	$6 \cdot x = 7.5 \cdot 5.5$
$\frac{7.5 \text{ ft}}{6 \text{ ft}} = \frac{5.5 \text{ ft}}{x \text{ ft}}$	$6x = 41.25$
$7.5 \cdot x = 6 \cdot 5.5$	$x = 6.875$
$7.5x = 33$	Emma's shadow is 6.875 feet long.
$\frac{7.5x}{7.5} = \frac{33}{7.5}$	
$x = 4.4$	
Emma's shadow is 4.4 ft long.	

a. Explain why the 1-point response only received one point.

Possible answer: The student answers the question but not all work is shown. The proportion is set up incorrectly but all work is done correctly. There is no explanation given.

b. In the 1-point response, what did the student fail to define?

the variable

STANDARDIZED TEST PRACTICE

Standardized Test Practice
Chapter 7
Select the best answer for Questions 1–8.

1. If a glass holds 12-oz and there is 4-oz of water in it, how full is the glass?

A $\frac{1}{4}$ C $\frac{1}{2}$

B $\frac{1}{3}$ D $\frac{2}{3}$

2. A teacher has a transparency with a rectangle drawn on it. The rectangle has dimensions 7 in. long by 4 in. wide. If when projected on the wall the width is 2 ft, what is the length of the rectangle?

F 2.6 ft   H 3.2 ft

G 2.8 ft   I 3.5 ft

3. What is the scale factor of the dilation?

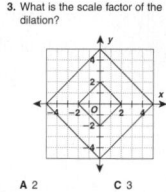

A 2   C 3

B 2.5   D 5

4. A farmer has a tractor with a wheelbase of 25 ft. If his grandson has a 1:8 scale replica pedal tractor, what is the wheelbase of the toy tractor?

F 0.32 ft   H 7.625 ft

G 3.125 ft   I 200 ft

5. If a heart beats 8 times in 6 seconds, how many times does it beat in one minute?

A 60   C 80

B 72   D 86

6. Solve for x.

$\frac{x}{12} = \frac{45}{18}$

F 4.8   H 30

G 22   I 67.5

7. To make a batch of purple frosting, you need 16 drops of blue food coloring and 12 drops of red food coloring. What is the ratio of red to blue in simplest form?

A 3:4   C 4:3

B 8:6   D 12:16

8. If there are about 18 ounces of blueberries in a pint, approximately how many pounds of blueberries are there in a gallon?

F 4.5 lb   H 9 lb

G 7 lb   I 13.5 lb

Standardized Test Practice
Chapter 7, continued
Gridded Response
Solve the problems. Use the answer sheet to write and grid-in your answer.

9. A 5 cm cube is built from small cubes, each 1 cm on a side. Complete the surface area ratio.

$\frac{5 \text{ cm cube}}{1 \text{ cm cube}} \rightarrow \frac{? \text{ cm}^2}{6 \text{ cm}^2}$

10. The two triangles shown are similar. Find x.

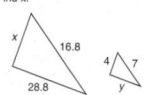

Short Response
Solve the problems. Use the answer sheet to write your answers.

The students in an art class are making a scale drawing of their school. The scale they have chosen is 1 inch = 4 feet. When they drew the art room, their drawing was 22 inches long and 18 inches wide.

11. What were the actual dimensions of the art room? Explain in words how you determined your answer.

12. If the actual dimensions of the cafeteria are 100 feet by 75 feet how large will the scale drawing be? Draw a diagram to help you determine the answer. Show your work.

Extended Response

13. A machine produces steel bars for a construction company. The bars need to be produced to within a tolerance of 0.02 inch to fit correctly. A recent inspection found that 6 out of 150 parts were defective. In the next run 250 steel bars are to be made.

a. How many steel bars are expected to be defective? Write a proportion that can be used to solve this problem and then solve the proportion.

b. If a part costs $1.59 to make and each defective part costs an additional $2.99 to correct, find the total cost of the 250 steel bars produced. Show your work.

c. Suppose that the next run is for 1275 steel bars. Write a proportion that can be used to determine how many will be defective and determine the total cost of producing all 1275 steel bars. Show your work.

State-Specific Test Practice Online
KEYWORD: MP4 TestPrep

Test Prep Tool Kit

- Standardized Test Prep Workbook
- Countdown to Testing transparencies
- State Test Prep CD-ROM
- Standardized Test Prep Video

Customized answer sheets give students realistic practice for actual standardized tests.

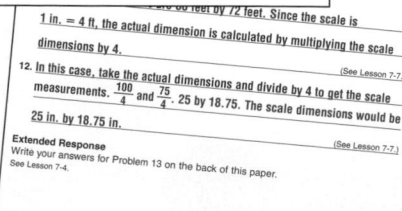

...o 60 feet by 72 feet. Since the scale is

1 in. = 4 ft, the actual dimension is calculated by multiplying the scale dimensions by 4.

12. In this case, take the actual dimensions and divide by 4 to get the scale measurements. $\frac{100}{4}$ and $\frac{75}{4}$ · 25 by 18.75. The scale dimensions would be 25 in. by 18.75 in. (See Lesson 7-7.)

Extended Response (See Lesson 7-4.)
Write your answers for Problem 13 on the back of this paper.
See Lesson 7-4.

Ⓓ See Lesson 7-3.
ⓘ See Lesson 7-4.
Ⓓ See Lesson 7-2.
ⓘ See Lesson 7-3.

Chapter 7

Ratios and Similarity

Why Learn This?

Tell students that a horticulturist creating a bonsai plant may use ratios and proportions to determine the ideal height of the bonsai. For example, the height of the bonsai juniper tree is $\frac{6}{120}$, or $\frac{1}{20}$, of the height of the natural juniper tree. The horticulturist might use this ratio in a proportion to determine the seedling height, adult height, pruning guidelines, etc.

Using Data

To begin the study of this chapter, have students:

- Write the height of the bonsai pitch pine as a fraction of the height of the natural pitch pine. $\frac{14}{2400}$ Write the fraction in simplest form. $\frac{7}{1200}$

- Write an equation to determine how many times taller the natural brush cherry is than the bonsai brush cherry. $8t = 50(12); 8t = 600$ Have them solve the equation. $t = 75$

Chapter 7

Ratios and Similarity

Tree	Natural Height (ft)	Bonsai Height (in.)
Chinese elm	60	10
Brush cherry	50	8
Juniper	10	6
Pitch pine	200	14
Eastern hemlock	80	18

Career Horticulturist

Chances are that a horticulturist helped create many of the varieties of plants at your local nursery. Horticulturists work in vegetable development, fruit growing, flower growing, and landscape design. Horticulturists who are also scientists work to develop new types of plants or ways to control plant diseases.

The art of *bonsai*, or making miniature plants, began in China and became popular in Japan. Now bonsai is practiced all over the world.

🎵 internet connect
Chapter Opener Online
go.hrw.com
KEYWORD: MP4 Ch7

Problem Solving Project

Life Science Connection

Purpose: To use ratios, proportions, and scale drawings and models to solve problems

Materials: Growing Tiny Trees worksheet, drawing and modeling materials

🎵 internet connect
Chapter Project Online: *go.hrw.com*
KEYWORD: MP4 PSProject7

Understand, Plan, Solve, and Look Back

Have students:

✔ Complete the Growing Tiny Trees worksheet to learn more about ratios, proportions, and scale.

✔ Create a full-size drawing of a tree that they selected and researched. Then have them divide into groups and create scale drawings of the tree using different scales that they select. Post all the scale drawings, and note their similarities and differences, as well as their similarities to the original full-size tree drawing.

✔ Do research on the invention of bonsai. Invite some bonsai growers to class and to bring some of their trees.

✔ Check students' work.

ARE YOU READY?

Choose the best term from the list to complete each sentence.

1. To solve an equation, you use __?__ to isolate the variable. So to solve the __?__ $3x = 18$, divide both sides by 3. **inverse operations; multiplication equation**

2. In the fractions $\frac{2}{3}$ and $\frac{1}{6}$, 18 is a __?__, but 6 is the __?__. **common denominator; least common denominator**

3. If two polygons are congruent, all of their __?__ sides and angles are congruent. **corresponding**

common denominator

corresponding

inverse operations

least common denominator

multiplication equation

Complete these exercises to review skills you will need for this chapter.

✔ Simplify Fractions

Write each fraction in simplest form.

4. $\frac{8}{24}$ **$\frac{1}{3}$**
5. $\frac{15}{50}$ **$\frac{3}{10}$**
6. $\frac{18}{72}$ **$\frac{1}{4}$**
7. $\frac{25}{125}$ **$\frac{1}{5}$**

✔ Use a Least Common Denominator

Find the least common denominator for each set of fractions.

8. $\frac{2}{3}$ and $\frac{1}{5}$ **15**
9. $\frac{3}{4}$ and $\frac{1}{8}$ **8**
10. $\frac{5}{7}, \frac{3}{7}$, and $\frac{1}{14}$ **14**
11. $\frac{1}{2}, \frac{2}{3}$, and $\frac{3}{5}$ **30**

✔ Order Decimals

Write each set of decimals in order from least to greatest.

12. $4.2, 2.24, 2.4, 0.242$ **0.242, 2.24, 2.4, 4.2**
13. $1.1, 0.1, 0.01, 1.11$ **0.01, 0.1, 1.1, 1.11**
14. $1.4, 2.53, 1.\overline{3}, 0.\overline{9}$ **0.9, 1.3, 1.4, 2.53**

✔ Solve Multiplication Equations

Solve.

15. $5x = 60$ **$x = 12$**
16. $0.2y = 14$ **$y = 70$**
17. $\frac{1}{2}t = 10$ **$t = 20$**
18. $\frac{2}{3}z = 9$ **$z = 13.5$**

✔ Identify Corresponding Parts of Congruent Figures

If $\triangle ABC \cong \triangle JRW$, complete each congruence statement.

19. $\overline{AB} \cong$ __?__ **$\overline{JR}$**
20. $\angle R \cong$ __?__ **$\angle B$**
21. $\overline{AC} \cong$ __?__ **$\overline{JW}$**
22. $\angle C \cong$ __?__ **$\angle W$**

ARE YOU READY?
Were students successful with Are You Ready?

 **NO** INTERVENE   **YES** ENRICH

 **Simplify Fractions** 4–7
Intervention Practice, Skill 19
CD-ROM
Intervention Activities, Skill 19

 **Order Decimals** 12–14
Intervention Practice, Skill 17
CD-ROM
Intervention Activities, Skill 17

✔ **Identify Corresponding Parts of Congruent Figures** 19–22
Intervention Practice, Skill 81
CD-ROM
Intervention Activities, Skill 81

✔ **Use a Least Common Denominator** 8–11
Intervention Practice, Skill 23
 CD-ROM
Intervention Activities, Skill 23

✔ **Solve Multiplication Equations** 15–18
Intervention Practice, Skill 59
 CD-ROM
Intervention Activities, Skill 59

Are You Ready? Enrichment, pp. 417–418

Ratios, Rates, and Proportions

One-Minute Section Planner

Lesson	Materials	Resources
Lesson 7-1 Ratios and Proportions **NCTM:** Number and Operations, Communication, Representation **NAEP:** Number Properties 4a ☑ SAT-9 ☑ SAT-10 ☑ ITBS ☑ CTBS ☑ MAT ☑ CAT	**Optional** Teaching Transparency T2 *(CRB)* Recording Sheet for Reaching All Learners *(CRB, p. 98)*	• *Chapter 7 Resource Book,* pp. 6–14 • *Daily Transparency T1, CRB* • *Additional Examples Transparencies T3–T5, CRB* • *Alternate Openers: Explorations,* p. 55
Lesson 7-2 Ratios, Rates, and Unit Rates **NCTM:** Number and Operations, Measurement, Communication, Representation **NAEP:** Number Properties 4c ☑ SAT-9 ☑ SAT-10 ☑ ITBS ☑ CTBS ☑ MAT ☑ CAT		• *Chapter 7 Resource Book,* pp. 15–23 • *Daily Transparency T6, CRB* • *Additional Examples Transparencies T7–T9, CRB* • *Alternate Openers: Explorations,* p. 56
Lesson 7-3 Analyze Units **NCTM:** Number and Operations, Measurement, Problem Solving, Communication, Representation **NAEP:** Measurement 2c ☑ SAT-9 ☑ SAT-10 ☑ ITBS ☐ CTBS ☑ MAT ☑ CAT		• *Chapter 7 Resource Book,* pp. 24–33 • *Daily Transparency T10, CRB* • *Additional Examples Transparencies T11–T15, CRB* • *Alternate Openers: Explorations,* p. 57
Hands-On Lab 7A Model Proportions **NCTM:** Number and Operations, Algebra, Geometry, Connections, Representation **NAEP:** Number Properties 4c ☐ SAT-9 ☐ SAT-10 ☐ ITBS ☐ CTBS ☐ MAT ☐ CAT	**Required** Pattern blocks *(MK)* Rulers *(MK)*	• *Hands-On Lab Activities,* pp. 53, 94–95
Lesson 7-4 Solving Proportions **NCTM:** Algebra, Communication **NAEP:** Number Properties 4c ☑ SAT-9 ☑ SAT-10 ☑ ITBS ☑ CTBS ☑ MAT ☑ CAT	**Optional** Teaching Transparency T17 *(CRB)*	• *Chapter 7 Resource Book,* pp. 34–42 • *Daily Transparency T16, CRB* • *Additional Examples Transparencies T18–T19, CRB* • *Alternate Openers: Explorations,* p. 58
Section 7A Assessment		• Mid-Chapter Quiz, SE p. 360 • Section 7A Quiz, AR p. 17 • *Test and Practice Generator* CD-ROM

SAT = *Stanford Achievement Tests*   **ITBS** = *Iowa Test of Basic Skills*   **CTBS** = *Comprehensive Test of Basic Skills/Terra Nova*
MAT = *Metropolitan Achievement Test*   **CAT** = *California Achievement Test*

NCTM = Complete standards can be found on pages T29–T35.   **NAEP** = Complete standards can be found on pages A54–A58.

SE = *Student Edition*   **TE** = *Teacher's Edition*   **AR** = *Assessment Resources*   **CRB** = *Chapter Resource Book*   **MK** = *Manipulatives Kit*

Section Overview

Ratios, Rates, and Unit Rates *Lessons 7-1, 7-2, 7-3*

Why? Unit rates can be used to compare ratios, such as unit prices for a grocery item.

A **rate** is a ratio of two quantities with different units. $$\frac{66 \text{ mi}}{3 \text{ gal}}$$	A **unit rate** is a rate with a denominator of 1. $$\frac{22 \text{ mi}}{1 \text{ gal}}, \text{ or 22 miles per gallon}$$	A **conversion factor** is a ratio of equal quantities. $$\frac{1 \text{ ft}}{12 \text{ in.}}$$

Equivalent Ratios

$$\frac{10}{20} = \frac{10 \div 10}{20 \div 10} = \frac{1}{2} \qquad \frac{10}{20} = \frac{10 \cdot 5}{20 \cdot 5} = \frac{50}{100}$$

The ratios $\frac{10}{20}$, $\frac{1}{2}$, and $\frac{50}{100}$ are equivalent.

Ordering Ratios

To order ratios, write them as fractions or decimals.

$$3{:}2 = \frac{3}{2} = 1.5 \qquad 20{:}13 = \frac{20}{13} \approx 1.54 \qquad 13{:}9 = \frac{13}{9} = 1.\overline{4}$$

Order from least to greatest: $\quad 1.\overline{4} \qquad 1.5 \qquad 1.54$

$$\downarrow \qquad\qquad \downarrow \qquad\qquad \downarrow$$

$$13{:}9 \qquad\quad 3{:}2 \qquad\quad 20{:}13$$

Using Conversion Factors

If a car travels 1800 feet in one minute on a residential street, is it going too fast?

$$\frac{1800 \text{ ft}}{1 \text{ min.}} = \frac{1800 \text{ ft}}{1 \text{ min.}} \cdot \frac{1 \text{ mi}}{5280 \text{ ft}} \cdot \frac{60 \text{ min}}{1 \text{ h}} = \frac{1800 \cdot 60 \text{ mi}}{5280 \text{ h}} \approx 20 \text{ mi/h}$$

The car is traveling about 20 miles per hour, which is probably not too fast.

Proportions *Hands-On Lab 7A, Lessons 7-1, 7-4*

 Why? Proportions can be used to solve problems involving mixtures, such as mixing cement and sand in a given ratio.

A **proportion** is a statement that two ratios are equivalent.

$\dfrac{2}{8} = \dfrac{5}{20}$ is a **proportion**.

$\dfrac{2}{8}$ and $\dfrac{5}{20}$ are **proportional** because $\dfrac{2}{8} = \dfrac{1}{4}$ and $\dfrac{5}{20} = \dfrac{1}{4}$.

In a proportion, **cross products are equal.**

$$2 \times 20 = 8 \times 5$$
$$40 = 40$$

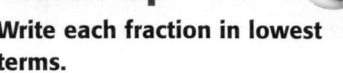

Pacing: Traditional $\frac{1}{2}$ day
Block $\frac{1}{3}$ day

Objective: Students find equivalent ratios to create proportions.

Warm Up

Write each fraction in lowest terms.

1. $\frac{14}{16}$ $\frac{7}{8}$ 2. $\frac{24}{64}$ $\frac{3}{8}$

3. $\frac{9}{72}$ $\frac{1}{8}$ 4. $\frac{45}{120}$ $\frac{3}{8}$

Problem of the Day

A magazine has page numbers from 1 to 80. What fraction of those page numbers include the digit 5? $\frac{17}{80}$

Available on Daily Transparency in CRB

Math Humor

Mathematics is made up of 50 percent formulas, 50 percent proofs, and 50 percent imagination.

Learn to find equivalent ratios to create proportions.

Vocabulary

ratio

equivalent ratio

proportion

Relative density is the ratio of the density of a substance to the density of water at 4°C. The relative density of silver is 10.5. This means that silver is 10.5 times as heavy as an equal volume of water.

The comparisons of water to silver in the table are *ratios* that are all equivalent.

Mexico and Peru are the world's largest silver producers.

Comparisons of Mass of Equal Volumes of Water and Silver				
Water	1 g	2 g	3 g	4 g
Silver	10.5 g	21 g	31.5 g	42 g

Reading Math

Ratios can be written in several ways. A colon is often used. 90:3 and $\frac{90}{3}$ name the same ratio.

A **ratio** is a comparison of two quantities by division. In one rectangle, the ratio of shaded squares to unshaded squares is 7:5. In the other rectangle, the ratio is 28:20. Both rectangles have equivalent shaded areas. Ratios that make the same comparison are **equivalent ratios**.

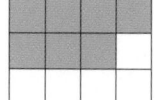 

 EXAMPLE 1 **Finding Equivalent Ratios**

Find two ratios that are equivalent to each given ratio.

A $\frac{6}{8}$

$\frac{6}{8} = \frac{6 \cdot 2}{8 \cdot 2} = \frac{12}{16}$ *Multiply or divide the numerator*
$\frac{6}{8} = \frac{6 \div 2}{8 \div 2} = \frac{3}{4}$ *and denominator by the same nonzero number.*

Two ratios equivalent to $\frac{6}{8}$ are $\frac{12}{16}$ and $\frac{3}{4}$.

B $\frac{48}{27}$

$\frac{48}{27} = \frac{48 \cdot 2}{27 \cdot 2} = \frac{96}{54}$
$\frac{48}{27} = \frac{48 \div 3}{27 \div 3} = \frac{16}{9}$

Two ratios equivalent to $\frac{48}{27}$ are $\frac{96}{54}$ and $\frac{16}{9}$.

---

1 Introduce

Alternate Opener

EXPLORATION

7-1 Ratios and Proportions

Sam and Jill counted the number of foreign cars and the number of domestic cars that entered the parking lot of a movie theater between 5:30 P.M. and 5:45 P.M. for a week. Their results are listed in the table.

	Mon	Tue	Wed	Thu	Fri	Sat	Sun
Foreign	16	25	15	20	40	40	45
Domestic	8	25	10	25	20	50	45

They compared the number of foreign cars to the number of domestic cars by using ratios. For example, the ratio for Monday was $\frac{foreign}{domestic} = \frac{16}{8} = \frac{2}{1}$.

1. Write a ratio for each day.
2. Which ratios are greater than the Monday ratio?
3. Which ratios are less than the Monday ratio?
4. Which ratio is the same as the Monday ratio?

Think and Discuss

5. **Explain** what it means if two ratios are equal.
6. **Describe** how the ratios are different if you write them as $\frac{domestic}{foreign}$.

Motivate

Pose the following question to the students: "If there are 16 cookies for 8 children, how many cookies will each child receive?" 2 cookies Show students that the number of cookies can be compared with the number of students by using the *ratio* $\frac{16 \text{ cookies}}{8 \text{ children}}$. Show them that the ratio can be simplified to $\frac{2 \text{ cookies}}{1 \text{ child}}$. Explain that the ratios are equivalent because they express the same relationship between the two quantities.

Exploration worksheet and answers on Chapter 7 Resource Book pp. 7 and 102

2 Teach

Lesson Presentation

Guided Instruction

In this lesson, students learn to find equivalent ratios to create proportions. Explain that ratios are comparisons between two numbers. Point out that ratios are similar to fractions but that ratios can have numbers other than integers in both the numerator and denominator. Teach students to find equivalent ratios by either multiplying or dividing the numerator and denominator by the same number. You may want to use the shaded square diagram as an example of equivalent ratios (Teaching Transparency T2, CRB). Explain that if two ratios can be simplified to equivalent fractions, then they are *in proportion*.

Ratios that are equivalent are said to be *proportional*, or in **proportion** . Equivalent ratios are identical when they are written in simplest form.

EXAMPLE 2 **Determining Whether Two Ratios are in Proportion**

Simplify to tell whether the ratios form a proportion.

A $\frac{7}{21}$ and $\frac{2}{6}$

$$\frac{7}{21} = \frac{7 \div 7}{21 \div 7} = \frac{1}{3}$$

$$\frac{2}{6} = \frac{2 \div 2}{6 \div 2} = \frac{1}{3}$$

Since $\frac{1}{3} = \frac{1}{3}$, the ratios are in proportion.

B $\frac{9}{12}$ and $\frac{16}{24}$

$$\frac{9}{12} = \frac{9 \div 3}{12 \div 3} = \frac{3}{4}$$

$$\frac{16}{24} = \frac{16 \div 8}{24 \div 8} = \frac{2}{3}$$

Since $\frac{3}{4} \neq \frac{2}{3}$, the ratios are *not* in proportion.

EXAMPLE 3 *Earth Science Application*

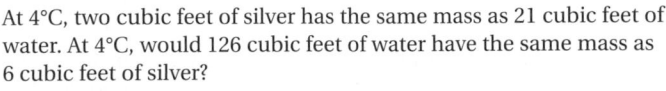

Earth Science LINK

Silver is a rare mineral usually mined along with lead, copper, and zinc.

At 4°C, two cubic feet of silver has the same mass as 21 cubic feet of water. At 4°C, would 126 cubic feet of water have the same mass as 6 cubic feet of silver?

$$\frac{2}{21} \overset{?}{=} \frac{6}{126}$$

$$\frac{2}{21} \overset{?}{=} \frac{6 \div 6}{126 \div 6} \qquad \textit{Simplify.}$$

$$\frac{2}{21} \neq \frac{1}{21}$$

Since $\frac{2}{21}$ is not equal to $\frac{1}{21}$, 126 cubic feet of water would not have the same mass at 4°C as 6 cubic feet of silver.

Think and Discuss

1. **Describe** how two ratios can form a proportion.

2. **Give** three ratios equivalent to 12:24.

3. **Explain** why the ratios 2:4 and 6:10 do not form a proportion.

4. **Give an example** of two ratios that are proportional and have numerators with different signs.

Additional Examples

Example 1

Find two ratios that are equivalent to each given ratio.

A. $\frac{9}{27}$ $\frac{18}{54}, \frac{1}{3}$

B. $\frac{64}{24}$ $\frac{128}{48}, \frac{8}{3}$

Example 2

Simplify to tell whether the ratios form a proportion.

A. $\frac{3}{27}$ and $\frac{2}{18}$ $\frac{1}{9} = \frac{1}{9}$; yes

B. $\frac{12}{15}$ and $\frac{27}{36}$ $\frac{4}{5} \neq \frac{3}{4}$; no

Example 3

At 4°C, four cubic feet of silver has the same mass as 42 cubic feet of water. At 4°C, would 210 cubic feet of water have the same mass as 20 cubic feet of silver? yes

3 Close

Reaching All Learners
Through Critical Thinking

Have pairs of students find the missing number to make each pair of ratios form a proportion. You may want to use the recording sheet on Chapter 7 Resource Book p. 98.

1. $\frac{5}{8} = \frac{?}{40}$ 25   2. $\frac{?}{21} = \frac{3}{7}$ 9

3. $\frac{10}{?} = \frac{20}{40}$ 20   4. $\frac{6}{27} = \frac{18}{?}$ 81

Then have students explain the strategies they used to find their answers.

Summarize

Remind students that a ratio can be expressed with a colon or as a fraction. Ask students to name two ways to find equivalent ratios. Ask students how many equivalent ratios can be found for the ratio 2:5. Recall that two equivalent ratios can be used to form a proportion.

Possible answers: Multiply the numerator and denominator by the same number or divide the numerator and denominator by the same number. Because you can multiply or divide the quantities in a ratio by any number, there are an infinite number of equivalent ratios for 2:5.

Answers to Think and Discuss

Possible answers:

1. Two ratios can be written to form a proportion if they are equivalent ratios.

2. 1:2, 4:8, 24:48

3. 2:4 simplifies to 1:2, and 6:10 simplifies to 3:5. In their simplest forms, they are not equivalent, so they do not form a proportion.

4. $\frac{-2}{4} = \frac{5}{-10}$

FOR EXTRA PRACTICE
see page 744

internet connect
Homework Help Online
go.hrw.com   Keyword: MP4 7-1

> Students may want to refer back to the lesson examples.

GUIDED PRACTICE

See Example 1 — Find two ratios that are equivalent to each given ratio.   **Possible answers:**

1. $\frac{4}{10}$ $\frac{2}{5}, \frac{8}{20}$ 2. $\frac{3}{9}$ $\frac{1}{3}, \frac{6}{18}$ 3. $\frac{21}{7}$ $\frac{3}{1}, \frac{42}{14}$ 4. $\frac{40}{32}$ $\frac{20}{16}, \frac{10}{8}$

See Example 2 — Simplify to tell whether the ratios form a proportion.

5. $\frac{6}{30}$ and $\frac{3}{15}$ 6. $\frac{6}{9}$ and $\frac{10}{18}$ 7. $\frac{35}{21}$ and $\frac{20}{12}$

See Example 3 — 8. A recipe calls for 1.5 cups of mix to make 8 pancakes. Mike wants to make 12 pancakes and uses 2 cups of mix. Does Mike have the correct ratio for the recipe? Explain.   **No; $2\frac{1}{4}$ cups are needed.**

Assignment Guide

If you finished Example **1** assign:
  Core 1–4, 9–12, 34–42
  Enriched 1–4, 9–12, 34–42

If you finished Example **2** assign:
  Core 1–7, 9–15, 17–25 odd, 34–42
  Enriched 9–15, 17–25, 32–42

If you finished Example **3** assign:
  Core 1–16, 17–25 odd, 26–28, 34–42
  Enriched 9–42

INDEPENDENT PRACTICE

See Example 1 — Find two ratios that are equivalent to each given ratio.   **Possible answers:**

9. $\frac{1}{7}$ $\frac{2}{14}, \frac{3}{21}$ 10. $\frac{5}{11}$ $\frac{10}{22}, \frac{15}{33}$ 11. $\frac{16}{14}$ $\frac{8}{7}, \frac{32}{28}$ 12. $\frac{65}{15}$ $\frac{13}{3}, \frac{130}{30}$

See Example 2 — Simplify to tell whether the ratios form a proportion.

13. $\frac{7}{14}$ and $\frac{13}{28}$ 14. $\frac{80}{100}$ and $\frac{4}{5}$ 15. $\frac{1}{3}$ and $\frac{15}{45}$

See Example 3 — 16. A molecule of carbonic acid contains 3 atoms of oxygen for every 2 atoms of hydrogen. Could a compound containing 81 hydrogen atoms and 54 oxygen atoms be carbonic acid? Explain.
    No; the ratio is reversed.

Answers

5. $\frac{1}{5} = \frac{1}{5}$; yes   6. $\frac{2}{3} \neq \frac{5}{9}$; no

7. $\frac{5}{3} = \frac{5}{3}$; yes   13. $\frac{1}{2} \neq \frac{13}{28}$; no

14. $\frac{4}{5} = \frac{4}{5}$; yes   15. $\frac{1}{3} = \frac{1}{3}$; yes

26. No; February is the only month that is equivalent to 4 weeks (28 days). Other months have 30 or 31 days.

27. no; 4 gallons

PRACTICE AND PROBLEM SOLVING

Tell whether the ratios form a proportion. If not, find a ratio that would form a proportion with the first ratio.   **Possible answers:**

17. $\frac{8}{14}$ and $\frac{6}{21}$ no; $\frac{4}{7}$ 18. $\frac{7}{9}$ and $\frac{140}{180}$ yes   19. $\frac{4}{7}$ and $\frac{12}{49}$ no; $\frac{8}{14}$

20. $\frac{30}{36}$ and $\frac{15}{16}$ no; $\frac{5}{6}$ 21. $\frac{13}{12}$ and $\frac{39}{36}$ yes   22. $\frac{11}{20}$ and $\frac{22}{40}$ yes

23. $\frac{16}{84}$ and $\frac{6}{62}$ no; $\frac{8}{42}$ 24. $\frac{24}{10}$ and $\frac{44}{18}$ no; $\frac{12}{5}$ 25. $\frac{11}{121}$ and $\frac{33}{363}$ yes

26. **BUSINESS** Cal pays his employees weekly. He would like to start paying them four times the weekly amount on a monthly basis. Is a month equivalent to four weeks? Explain.

27. **TRANSPORTATION** Aaron's truck has a 12-gallon gas tank. He just put 3 gallons of gas into the tank. Is this equivalent to a third of a tank? If not, what amount of gas is equivalent to a third of a tank?

Math Background

The proportion $\frac{3}{4} = \frac{6}{8}$ can also be written 3:4 = 6:8. It is read "3 is to 4 as 6 is to 8." The four terms of the proportion are 3, 4, 6, and 8. The 3 and 8 are called the *extremes*, and the 4 and 6 are called the *means*. Notice that the extremes are the "outside" numbers and the means are the "inside" numbers in the statement 3:4 = 6:8. An important property of proportions is that the product of the means equals the product of the extremes (the cross-products are equal). For the proportion $\frac{a}{b} = \frac{c}{d}$, $bc = ad$.

RETEACH 7-1

LESSON 7-1 Reteach
Ratios and Proportions

A **ratio** compares two quantities by division. **Equivalent ratios** make the same comparison.

$$\frac{8}{32} = \frac{2}{8} = \frac{1}{4}$$

To find equivalent ratios:

Divide by a common factor.   Multiply by a common factor.

$$\frac{10 \div 5}{15 \div 5} = \frac{2}{3} \qquad \frac{10 \times 2}{15 \times 2} = \frac{20}{30}$$

So, two ratios equivalent to $\frac{10}{15}$ are $\frac{2}{3}$ and $\frac{20}{30}$.

Complete to find two ratios equivalent to each given ratio.

1. $\frac{16 \div 4}{40 \div 4} = \frac{4}{10}$ $\frac{16 \times 3}{40 \times 3} = \frac{48}{120}$ 2. $\frac{75 \div 25}{100 \div 25} = \frac{3}{4}$ $\frac{75 \times 2}{100 \times 2} = \frac{150}{200}$

Name two ratios equivalent to the given ratio. Sample answers given.

3. $\frac{6}{9}$ $\frac{2}{3}, \frac{18}{27}$ 4. $\frac{1}{3}$ $\frac{3}{9}, \frac{9}{27}$ 5. $\frac{2}{5}$ $\frac{12}{30}, \frac{18}{20}$ 6. $\frac{16}{20}$ $\frac{8}{10}, \frac{4}{5}$

When two equivalent ratios are set equal to each other, they form a **proportion**.

To tell whether two ratios form a proportion, write each in simplest form.

Does $\frac{16}{32} = \frac{50}{100}$?   Does $\frac{8}{18} = \frac{10}{15}$?

$$\frac{16 \div 16}{32 \div 16} \qquad \frac{50 \div 50}{100 \div 50} \qquad\qquad \frac{8 \div 2}{18 \div 2} \qquad \frac{10 \div 5}{15 \div 5}$$

$$\frac{1}{2} \qquad \frac{1}{2} \qquad\qquad\qquad \frac{4}{9} \qquad \frac{2}{3}$$

The ratios form a proportion.   The ratios do not form a proportion.

Tell whether the ratios form a proportion.

7. $\frac{14}{21} \overset{?}{=} \frac{50}{75}$ 8. $\frac{27}{48} \overset{?}{=} \frac{54}{72}$
    yes   no

PRACTICE 7-1

LESSON 7-1 Practice B
Ratios and Proportions

Find two ratios that are equivalent to each given ratio. **Sample answers given.**

1. $\frac{9}{12}$ $\frac{3}{4}, \frac{18}{24}$ 2. $\frac{4}{20}$ $\frac{1}{5}, \frac{6}{30}$ 3. $\frac{15}{25}$ $\frac{3}{5}, \frac{6}{10}$

4. $\frac{7}{12}$ $\frac{14}{24}, \frac{21}{36}$ 5. $\frac{14}{7}$ $\frac{2}{1}, \frac{8}{4}$ 6. $\frac{11}{22}$ $\frac{1}{2}, \frac{5}{10}$

7. $\frac{10}{3}$ $\frac{20}{6}, \frac{30}{9}$ 8. $\frac{18}{28}$ $\frac{9}{14}, \frac{27}{42}$ 9. $\frac{12}{27}$ $\frac{4}{9}, \frac{8}{18}$

Simplify to tell whether the ratios form a proportion.

10. $\frac{13}{39}$ and $\frac{16}{48}$ 11. $\frac{21}{49}$ and $\frac{28}{56}$ 12. $\frac{12}{42}$ and $\frac{18}{56}$ 13. $\frac{18}{27}$ and $\frac{10}{15}$
    yes, $\frac{1}{3} = \frac{1}{3}$ no, $\frac{3}{7} \neq \frac{1}{2}$ yes, $\frac{2}{7} = \frac{2}{7}$ yes, $\frac{2}{3} = \frac{2}{3}$

14. $\frac{24}{32}$ and $\frac{27}{30}$ 15. $\frac{14}{10}$ and $\frac{35}{25}$ 16. $\frac{10}{32}$ and $\frac{26}{80}$ 17. $\frac{16}{45}$ and $\frac{15}{45}$
    no, $\frac{8}{9} \neq \frac{9}{10}$ yes, $\frac{7}{5} = \frac{7}{5}$ yes, $\frac{5}{16} = \frac{5}{16}$ yes, $\frac{1}{3} = \frac{1}{3}$

18. Mrs. Walters wanted one daffodil plant for every 2 tulip plants in her garden. If she planted 20 daffodil bulbs, how many tulip bulbs did she plant?
    40 tulip bulbs

19. In a survey, 9 out of 10 doctors recommended a certain medicine. If 80 doctors were surveyed, how many doctors recommended the medicine?
    72 doctors

---

28. **ENTERTAINMENT** The table lists prices for movie tickets.

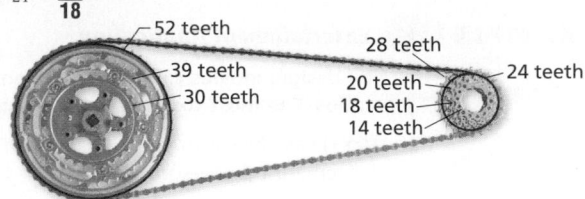

a. Are the ticket prices proportional? **yes**

b. How much do 6 movie tickets cost? **$49.50**

c. If Suzie paid $57.75 for movie tickets, how many did she buy? **7**

Movie Ticket Prices			
Number of Tickets	1	2	3
Price	$8.25	$16.50	$24.75

29. **HOBBIES** A bicycle chain moves between two sprockets when you shift gears. The number of teeth on the front sprocket and the number of teeth on the rear sprocket form a ratio. Equivalent ratios provide equal pedaling power. Find a ratio equivalent to the ratio shown, $\frac{52}{24}$. **$\frac{39}{18}$**

- 52 teeth
- 39 teeth
- 30 teeth
- 28 teeth
- 20 teeth
- 18 teeth
- 14 teeth
- 24 teeth

30. **COMPUTERS** While a file downloads, a computer displays the total number of kilobytes downloaded and the number of seconds that have passed. If the display shows 42 kilobytes after 7 seconds, is the file downloading at about 6 kilobytes per second? Explain.

 31. **WRITE A PROBLEM** The ratio of the number of bones in the human skull to the number of bones in the ears is 11:3. There are 22 bones in the skull and 6 in the ears. Use this information to write a problem using equivalent ratios. Explain your solution.

 32. **WRITE ABOUT IT** Describe at least two ways, given a ratio, to create an equivalent ratio.

 33. **CHALLENGE** Write all possible proportions using each of the numbers 2, 4, 8, and 16 once.

Spiral Review

Add or subtract. (Lesson 3-5)

34. $\frac{5}{7} + \frac{2}{3}$ $1\frac{8}{21}$

35. $\frac{4}{9} + \left(-1\frac{3}{4}\right)$ $-1\frac{11}{36}$

36. $\frac{3}{5} - \frac{7}{10}$ $-\frac{1}{10}$

37. $2\frac{7}{9} - 1\frac{8}{11}$ $1\frac{5}{99}$

Find the two square roots of each number. (Lesson 3-8)

38. 49 **±7**

39. 9 **±3**

40. 81 **±9**

41. 169 **±13**

42. **TEST PREP** Name the two integers that $-\sqrt{74}$ lies between. (Lesson 3-9) **B**

A -7 and -6 B -9 and -8 C -10 and -11 D -8 and -7

CHALLENGE 7-1

LESSON
7-1 Challenge
Mixing It Up

In a proportion, there are 4 terms.

The 1st and 4th are called **extremes**.
The 2nd and 3rd are called **means**.

$\frac{\text{1st (extreme)}}{\text{2nd (mean)}} = \frac{\text{3rd (mean)}}{\text{4th (extreme)}}$

In the following exercises, you will explore some properties of proportions.

1. Explain why the ratios $\frac{6}{9}$ and $\frac{8}{12}$ are in proportion.

   Both ratios are equivalent to $\frac{2}{3}$.

2. a. Rewrite the proportion $\frac{6}{9} = \frac{8}{12}$ by interchanging the means and extremes. Determine if the resulting statement is a proportion. Explain.

   $\frac{9}{6} = \frac{12}{8}$; **yes, since each ratio is equivalent to $\frac{3}{2}$**

   b. Generalize these results by completing this statement:
   If $\frac{a}{b} = \frac{c}{d}$, then it is also true that:

   $\frac{b}{a} = \frac{d}{c}$

3. a. Consider changing each ratio of the proportion $\frac{6}{9} = \frac{8}{12}$ by addition. Determine if the ratios $\frac{6+9}{9}$ and $\frac{8+12}{12}$ are in proportion. Explain.

   yes; each ratio is equivalent to $\frac{5}{3}$

   b. Generalize these results by completing this statement:
   If $\frac{a}{b} = \frac{c}{d}$, then it is also true that:

   $\frac{a+b}{b} = \frac{c+d}{d}$

   c. Determine if the ratios $\frac{6-9}{9}$ and $\frac{8-12}{12}$ are in proportion. Explain.

   yes; each ratio is equivalent to $\frac{-1}{3}$

   d. Generalize these results. If $\frac{a}{b} = \frac{c}{d}$, then it is also true that:

   $\frac{a-b}{b} = \frac{c-d}{d}$

PROBLEM SOLVING 7-1

LESSON
7-1 Problem Solving
Ratios and Proportions

A medicine for dogs indicates that the medicine should be administered in the ratio 0.5 tsp per 5 lb, based on the weight of the dog. Write the correct answer.

1. Jaime has a 60 lb dog. She plans to give the dog 12 teaspoons of medicine. Is she administering the medicine correctly?

   no

2. Jaime also has a 15 lb puppy. She plans to give the puppy 1.5 teaspoons of medicine. Is she administering the medicine correctly?

   yes

Sports statistics can be given as ratios. Find the ratios for the given statistics. Reduce each ratio.

3. In 69 games Darrel Armstrong of the Orlando Magic had 136 steals and 144 turnovers. What is his steals per turnover ratio?

   $\frac{17}{18}$

4. In 69 games, Ben Wallace of the Detroit Pistons blocked 234 shots. What is his blocks per game ratio?

   $\frac{78}{23}$

Choose the letter for the best answer.

5. There are 675 students and 30 teachers in the middle school. What is the ratio of teachers to students?

   A $\frac{45}{2}$ C $\frac{1}{27}$
   B $\frac{2}{45}$ D $\frac{27}{1}$

6. In a science experiment, out of a sample of seeds, 13 sprouted and 7 didn't. What is the ratio of seeds that sprouted to the number of seeds planted?

   F $\frac{13}{7}$ H $\frac{13}{20}$
   G $\frac{7}{13}$ J $\frac{7}{20}$

7. Many Internet services advertise their customer to modem ratio. One company advertises a 10 to 1 customer to modem ratio. Find a ratio that is equivalent to the $\frac{10}{1}$.

   A $\frac{40}{4}$ C $\frac{400}{4}$
   B $\frac{2}{20}$ D $\frac{50}{10}$

8. In one season, the Denver Broncos won 10 games and lost 4. What was their win/loss ratio?

   F $\frac{4}{10}$ H $\frac{2}{5}$
   G $\frac{5}{2}$ J $\frac{10}{14}$

Answers

30. Possible answer: Yes; the ratio $\frac{42 \text{ Kb}}{7 \text{ s}}$ is equivalent to $\frac{6 \text{ Kb}}{1 \text{ s}}$.

31. Possible answer: The ratio of the number of bones in the skull to the number of bones in the ears is 11:3. If there are 22 bones in the skull, are there 6 bones in the ears? Solution: yes; $\frac{11}{3} = \frac{22}{6}$

32. Possible answer: Multiply the numerator and denominator by the same number, or divide the numerator and denominator by the same number.

33. $\frac{2}{4} = \frac{8}{16}, \frac{4}{2} = \frac{16}{8}, \frac{16}{8} = \frac{4}{2}, \frac{8}{16} = \frac{2}{4}$,
$\frac{2}{8} = \frac{4}{16}, \frac{8}{2} = \frac{16}{4}, \frac{16}{4} = \frac{8}{2}, \frac{4}{16} = \frac{2}{8}$

Journal

Have students write about some real-world situations in which ratios would be useful. Examples might include figuring the cost of a number of items or changing the quantities of ingredients for a recipe.

Test Prep Doctor

For Exercise 42, students can find the two integers that $\sqrt{74}$ lies between and then use opposites. Suggest that students extend the list of perfect squares until they get to 64 and 81, which have the square roots 8 and 9. Taking opposites, the two integers are -8 and -9, choice **B**.

Lesson Quiz

Find two ratios that are equivalent to each given ratio.

1. $\frac{4}{15}$ Possible answer: $\frac{8}{30}, \frac{12}{45}$

2. $\frac{8}{21}$ Possible answer: $\frac{16}{42}, \frac{24}{63}$

Simplify to tell whether the ratios form a proportion.

3. $\frac{16}{10}$ and $\frac{32}{20}$ $\frac{8}{5} = \frac{8}{5}$; yes

4. $\frac{36}{24}$ and $\frac{28}{18}$ $\frac{3}{2} \neq \frac{14}{9}$; no

5. Kate poured 8 oz of juice from a 64 oz bottle. Brian poured 16 oz of juice from a 128 oz bottle. What ratio of juice is missing from each bottle? Are the ratios proportional?

   $\frac{8}{64}$ and $\frac{16}{128}$; yes, both equal $\frac{1}{8}$.

Available on Daily Transparency in CRB

Warm Up

Divide. Round answers to the nearest tenth.

1. $\frac{420}{18}$ 23.3   2. $\frac{73}{21}$ 3.5

3. $\frac{380}{16}$ 23.8   4. $\frac{430}{18}$ 23.9

Problem of the Day

There are 3 bags of flour for every 2 bags of sugar in a freight truck. A bag of flour weighs 60 pounds, and a bag of sugar weighs 80 pounds. Which part of the truck's cargo is heavier, the flour or the sugar? flour

Available on Daily Transparency in CRB

Math Humor

Wow! 9 out of 10 cars this company has built in the past 20 years are still on the road.

Amazing. Except that the company just started manufacturing cars last year!

7-2 Ratios, Rates, and Unit Rates

Learn to work with rates and ratios.

Vocabulary
rate
unit rate
unit price

Movie and television screens range in shape from almost perfect squares to wide rectangles. An *aspect ratio* describes a screen by comparing its width to its height. Common aspect ratios are 4:3, 37:20, 16:9, and 47:20.

Most high-definition TV screens have an aspect ratio of 16:9.

EXAMPLE 1 *Entertainment Application*

By design, movies can be viewed on screens with varying aspect ratios. The most common ones are 4:3, 37:20, 16:9, and 47:20.

A Order the width-to-height ratios from least (standard TV) to greatest (wide-screen).

$4:3 = \frac{4}{3} = 1.\overline{3}$ *Divide.* $\frac{4}{3} = \frac{1.\overline{3}}{1}$

$37:20 = \frac{37}{20} = 1.85$

$16:9 = \frac{16}{9} = 1.\overline{7}$

$47:20 = \frac{47}{20} = 2.35$

The decimals in order are $1.\overline{3}$, $1.\overline{7}$, 1.85, and 2.35.
The width-to-height ratios in order from least to greatest are 4:3, 16:9, 37:20, and 47:20.

B A wide-screen television has screen width 32 in. and height 18 in. What is the aspect ratio of this screen?

The ratio of the width to the height is 32:18.

The ratio $\frac{32}{18}$ can be simplified: $\frac{32}{18} = \frac{2(16)}{2(9)} = \frac{16}{9}$.

The screen has the aspect ratio 16:9.

A ratio is a comparison of two quantities. A **rate** is a comparison of two quantities that have different units.

$$\text{ratio: } \frac{90}{3} \qquad \text{rate: } \frac{90 \text{ miles}}{3 \text{ hours}} \longleftarrow \text{Read as "90 miles per 3 hours."}$$

Unit rates are rates in which the second quantity is 1. The ratio $\frac{90}{3}$ can be simplified by dividing: $\frac{90}{3} = \frac{30}{1}$.

$$\text{unit rate: } \frac{30 \text{ miles}}{1 \text{ hour}}, \text{ or } 30 \text{ mi/h}$$

---

1 Introduce

Alternate Opener

7-2 Ratios, Rates, and Unit Rates

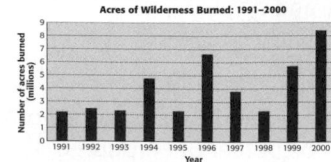

The bar graph shows the number of acres of wilderness burned each year from 1991 to 2000. Each bar represents a *unit rate*, because it shows the number of acres burned in one year.

Acres of Wilderness Burned: 1991–2000

1. The National Interagency Fire Center (http://www.nifc.gov) reported that an average of 3,647,883 acres were burned per year from 1991 to 2000.

   a. Is the average also a unit rate?

   b. In what years were the number of acres burned above the average?

   c. In what years were the number of acres burned below the average?

Think and Discuss

2. **Explain** what a unit rate is.
3. **Describe** real-world situations that involve unit rates.

Motivate

Present the following problem: A package of 8 rolls costs $2.00. A package of 10 rolls costs $2.79. Which is the better buy?

The 8-roll package; unit prices are 25 cents and 27.9 cents, respectively.

Encourage students to give answers and explain how they got their answers. Tell students that one way to compare prices is to use *unit prices*, such as the price per roll. A unit price is a type of *unit rate*, one of the lesson topics.

Exploration worksheet and answers on Chapter 7 Resource Book pp. 16 and 104

---

2 Teach

Lesson Presentation

Guided Instruction

In this lesson, students learn to work with rates and ratios. Review simplifying fractions and writing fractions as decimals. Remind students how to order decimals. Discuss the concepts of *rate* and *unit rate*, and use the data in the Example 2 bar graph to explain how to convert from a rate to a unit rate. Point out that a *unit price* is a type of unit rate, and use Example 3 to find and compare unit prices.

Teaching Tip
To help students remember the distinction between rate and ratio, point out that a heart *rate* compares *beats* and *minutes*, two quantities that have different units.

EXAMPLE 2 Using a Bar Graph to Determine Rates

The number of acres destroyed by wildfires in 2000 is shown for the states with the highest totals. Use the bar graph to find the number of acres, to the nearest acre, destroyed in each state per day.

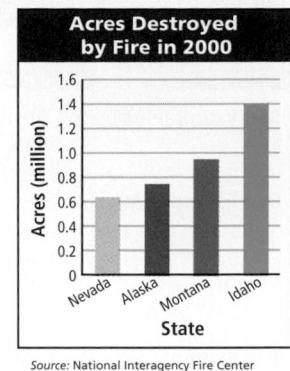

Acres Destroyed by Fire in 2000

Source: National Interagency Fire Center

Nevada $= \dfrac{640,000 \text{ acres}}{366 \text{ days}} \approx \dfrac{1749 \text{ acres}}{1 \text{ day}}$

Alaska $= \dfrac{750,000 \text{ acres}}{366 \text{ days}} \approx \dfrac{2049 \text{ acres}}{1 \text{ day}}$

Montana $= \dfrac{950,000 \text{ acres}}{366 \text{ days}} \approx \dfrac{2596 \text{ acres}}{1 \text{ day}}$

Idaho $= \dfrac{1,400,000 \text{ acres}}{366 \text{ days}} \approx \dfrac{3825 \text{ acres}}{1 \text{ day}}$

Nevada: 1749 acres/day; Alaska: 2049 acres/day; Montana: 2596 acres/day; Idaho: 3825 acres/day

Unit price is a unit rate used to compare costs per item.

EXAMPLE 3 Finding Unit Prices to Compare Costs

A Blank videotapes can be purchased in packages of 3 for $4.99, or 10 for $15.49. Which is the better buy?

$\dfrac{\text{price for package}}{\text{number of videotapes}} = \dfrac{\$4.99}{3} \approx \$1.66$ *Divide the price by the number of tapes.*

$\dfrac{\text{price for package}}{\text{number of videotapes}} = \dfrac{\$15.49}{10} \approx \$1.55$

The better buy is the package of 10 for $15.49.

B Leron can buy a 64 oz carton of orange juice for $2.49 or a 96 oz carton for $3.99. Which is the better buy?

$\dfrac{\text{price for carton}}{\text{number of ounces}} = \dfrac{\$2.49}{64} \approx \$0.0389$ *Divide the price by the number of ounces.*

$\dfrac{\text{price for carton}}{\text{number of ounces}} = \dfrac{\$3.99}{96} \approx \$0.0416$

The better buy is the 64 oz carton for $2.49.

Think and Discuss

1. Choose the quantity that has a lower unit price: 6 oz for $1.29 or 15 oz for $3.00. Explain your answer.

2. Explain why an aspect ratio is not considered a rate.

3. Determine two different units of measurement for speed.

Additional Examples

Example 1

A. Order the ratios 4:3, 23:10, 13:9, and 47:20 from least to greatest. **4:3, 13:9, 23:10, 47:20**

B. A television has screen width 20 in. and height 15 in. What is the aspect ratio of this screen? **4:3**

Example 2

Use the bar graph on page 347 to find the number of acres, to the nearest acre, destroyed in Nevada and Alaska per week.

Nevada: $\approx \dfrac{12,308 \text{ acres}}{1 \text{ week}}$

Alaska: $\approx \dfrac{14,423 \text{ acres}}{1 \text{ week}}$

Example 3

A. Pens can be purchased in a 5-pack for $1.95 or a 15-pack for $6.20. Which is the better buy? **5-pack**

B. Jamie can buy a 15 oz jar of peanut butter for $2.19 or a 20 oz jar for $2.78. Which is the better buy? **20 oz jar**

Example 2 note: You may want to remind students that the year 2000 had 366 days because it was a leap year.

---

3 Close

Reaching All Learners
Through Home Connection

Ask students to go to the grocery store with a family member and make a list of five products. For each product, students should record the brand and quantity they purchase (or would purchase) and its unit price. For each product, students should also record one other brand or quantity they decided not to purchase (or would not purchase) and its unit price. Ask students to give a reason for each decision. Suggest that not all decisions must be based only on cost, but that cost is one consideration along with preference, quality, and other considerations.

Summarize

Have the students match each term below with its corresponding example. Ask them to be as specific as possible and to use each term only once. Ask them to explain their choices.

Ratio — 120 miles per 3 hours
Rate — 15 ft/s
Unit rate — $1.15 per pound
Unit price — 4:2

A ratio is a comparison of two quantities. A rate is a ratio that has two different units. A unit rate is a rate in which the second quantity is 1. A unit price is a unit rate involving the cost of an item.

Answers to Think and Discuss

Possible answers:

1. 15 oz has the lower unit price at $0.20 per ounce. The unit price for 6 oz is $0.22 per ounce.

2. An aspect ratio is not a rate because the quantities have the same units.

3. miles per hour; feet per second

FOR EXTRA PRACTICE
see page 744

▱ internet connect
Homework Help Online
go.hrw.com   Keyword: MP4 7-2

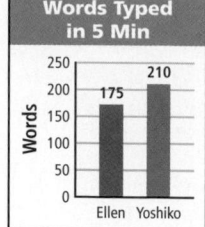

Students may want to refer back to the lesson examples.

GUIDED PRACTICE

See Example **1**   **1.** The height of a bridge is 68 ft, and its length is 340 ft. Find the ratio of its height to its length in simplest form.   **1:5**

See Example **2**   For Exercises 2 and 3, use the bar graph to find each unit rate.

2. Ellen's words per minute   **35 wpm**

3. Yoshiko's words per minute   **42 wpm**

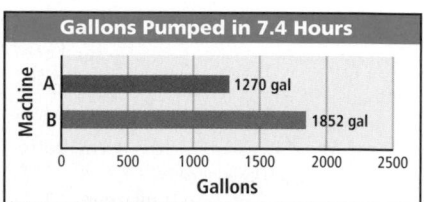

Words Typed in 5 Min

See Example **3**   Determine the better buy.

4. a 15 oz can of corn for $1.39 or a 22 oz can for $1.85   **22 oz can**

5. a dozen golf balls for $22.99 or 20 golf balls for $39.50   **dozen golf balls**

INDEPENDENT PRACTICE

See Example **1**   **6.** A child's basketball hoop is 6 ft tall. Find the ratio of its height to the height of a regulation basketball hoop, which is 10 ft tall. Express the ratio in simplest form.   **3:5**

See Example **2**   For Exercises 7 and 8, use the bar graph to find each unit rate.

7. gallons per hour for machine A   **≈ 171.6 gal/h**

8. gallons per hour for machine B   **≈ 250.3 gal/h**

Gallons Pumped in 7.4 Hours

See Example **3**   Determine the better buy.

9. 4 boxes of cereal for $9.56; 2 boxes of cereal for $4.98   **4 boxes**

10. 8 oz jar of soup for $2.39; 10 oz jar of soup for $2.69   **10 oz jar**

PRACTICE AND PROBLEM SOLVING

Find each unit rate.

11. $525 for 20 hours of work   **$26.25 per hour**

12. 96 chairs in 8 rows   **12 chairs per row**

13. 12 slices of pizza for $9.25   **$0.77 per slice**

14. 64 beats in 4 measures of music   **16 beats per measure**

Find each unit price and tell which is the better buy.

15. $7.47 for 3 yards of fabric; $11.29 for 5 yards of fabric

16. A $\frac{1}{2}$-pound hamburger for $3.50; a $\frac{1}{3}$-pound hamburger for $3.25

17. 10 gallons of gasoline for $13.70; 12.5 gallons of gasoline for $17.75

18. $1.65 for 5 pounds of bananas; $3.15 for 10 pounds of bananas

Assignment Guide

If you finished Example **1** assign:
Core 1, 6, 25–31
Enriched 1, 6, 25–31

If you finished Example **2** assign:
Core 1–3, 6–8, 11, 13, 20, 25–31
Enriched 6–8, 11–14, 20, 21, 25–31

If you finished Example **3** assign:
Core 1–10, 11–21 odd, 25–31
Enriched 6–31

Answers

15. $2.49/yd; $2.26/yd; 5 yd

16. $7/lb; $9.75/lb; $\frac{1}{2}$ lb hamburger

17. $1.37/gal; $1.42/gal; 10 gal

18. $0.33/lb; $0.32/lb; 10 lb

Math Background

The golden ratio has fascinated artists, architects, and mathematicians for centuries. If a line segment is divided into sections of lengths a and b so that the proportion $\frac{a}{b} = \frac{b}{a+b}$ is satisfied, then those lengths form the golden ratio $\frac{b}{a}$.
The value of the golden ratio is
$\frac{b}{a} = \frac{1+\sqrt{5}}{2} \approx 1.62$.

$$\underline{\qquad a \qquad | \qquad b \qquad}$$

A rectangle whose length and width are in the golden ratio is called a golden rectangle. The ancient Greeks believed that the golden rectangle was the most visually appealing rectangle and incorporated it into their art and architecture.

RETEACH 7-2

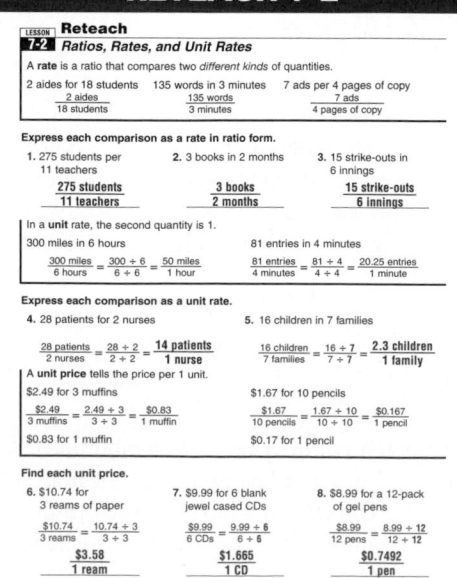

PRACTICE 7-2

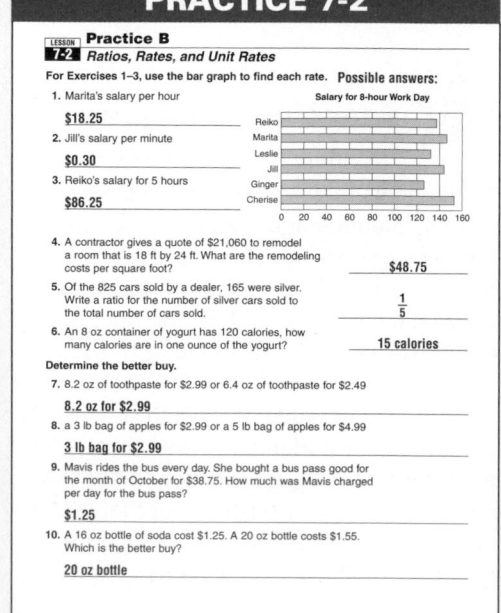

19. COMMUNICATIONS Super-Cell offers a wireless phone plan that includes 250 base minutes for $24.99 a month. Easy-Phone has a plan that includes 325 base minutes for $34.99.

   a. Find the unit rate for the base minutes for each plan.

   b. Which company offers a lower rate for base minutes?

20. BUSINESS A cereal company pays $59,969 to have its new cereal placed in a grocery store display for one week. Find the daily rate for this display. **$8567 per day**

21a. Tom: $25\frac{3}{8}$ frames per hour; Cherise: 27 frames per hour; Tina $28\frac{3}{8}$ frames per hour

21. ENTERTAINMENT Tom, Cherise, and Tina work as film animators. The circle graph shows the number of frames they each rendered in an 8-hour day.

   a. Find the hourly unit rendering rate for each employee.

   b. Who was the most efficient employee? **Tina**

   c. How many more frames per hour did Cherise render than Tom? $1\frac{5}{8}$

   d. How many more frames per hour did Tom and Cherise together render than Tina? **24**

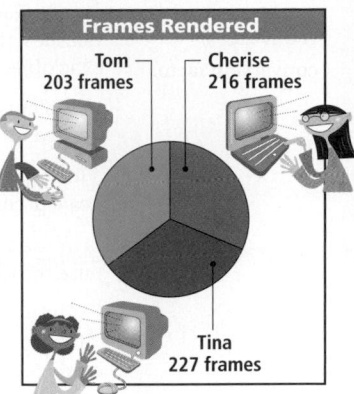

Frames Rendered

Tom 203 frames   Cherise 216 frames   Tina 227 frames

22. WHAT'S THE ERROR? A clothing store charges $30 for 12 pairs of socks. A student says that the unit price is $0.40 per pair. What is the error? What is the correct unit price?

23. WRITE ABOUT IT Explain how to find unit rates. Give an example and explain how consumers can use unit rates to save money.

24. CHALLENGE The size of a television (13 in., 25 in., 32 in., and so on) represents the length of the diagonal of the television screen. A 25 in. television has an aspect ratio of 4:3. What is the width and height of the screen? **width: 20 in.; height: 15 in.**

Spiral Review

Evaluate each expression for the given value of the variable. (Lesson 2-1)

25. $c + 4$ for $c = -8$ **−4**   **26.** $m - 2$ for $m = 13$ **11**   **27.** $5 + d$ for $d = -10$ **−5**

Evaluate each expression for the given value of the variable. (Lesson 3-2)

28. $45.6 + x$ for $x = -11.1$ **34.5** **29.** $17.9 - b$ for $b = 22.3$ **−4.4** **30.** $r + (-4.9)$ for $r = 31.8$ **26.9**

31. TEST PREP How much fencing, to the nearest foot, is needed to enclose a square lot with an area of 350 ft²? (Lesson 3-9) **D**

  A 74 ft   **B** 65 ft   **C** 68 ft   **D** 75 ft

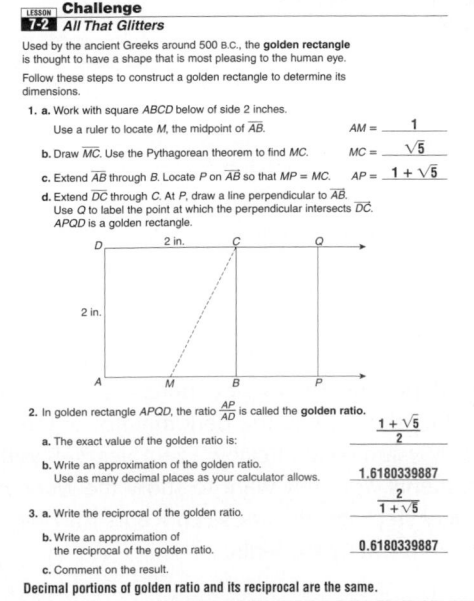
Answers

19. a. Super-Cell: ≈ $0.10/min; Easy-Phone: ≈ $0.11/min

  b. Super-Cell offers a better rate.

22. Possible answer: The student may have divided 12 by 30 instead of dividing 30 by 12 to find the actual unit cost, $2.50 per pair.

23. Possible answer: Simplify the ratio so that the second quantity is 1. An example is the cost per pound of fruit. Consumers can use the unit rate, or unit price, to find the better buy.

Journal

Ask students to describe why they think it is or is not helpful for supermarkets to include unit prices on price labels.

Test Prep Doctor

For Exercise 31, remind students that it takes two steps to find the correct solution. Students must first find the approximate square root of 350 and then multiply that result by 4 ($\sqrt{350} \approx 18.7$, and $4(18.7) = 74.8$). Rounded to the nearest whole number, 74.84 is 75, which is choice **D**. Students who chose **A** may have rounded incorrectly.

Lesson Quiz

1. At a family golf outing, a father drove the ball 285 ft. His daughter drove the ball 95 ft. Express the ratio of the father's distance to his daughter's in simplest terms. **3:1**

2. Find the unit price of 6 stamps for $2.22. **$0.37 per stamp**

3. Find the unit rate of 8 heartbeats in 6 seconds. **≈ 1.3 beats/s**

4. What is the better buy, a half dozen carnations for $4.75 or a dozen for $9.24? **a dozen**

5. Which is the better buy, four pens for $5.16 or a ten-pack for $12.90? **They cost the same.**

Available on Daily Transparency in CRB

7-3 Organizer

Pacing: Traditional 1 day
Block $\frac{1}{2}$ day

Objective: Students use one or more conversion factors to solve rate problems.

Warm Up

Find each unit rate.

1. jump rope 192 times in 6 minutes
32 jumps/min

2. four pounds of bananas for $2.36
$0.59/lb

3. 16 anchor bolts for $18.56
$1.16/bolt

4. 288 movies on 9 shelves
32 movies/shelf

Problem of the Day

Replace each ● with a digit from 0 to 6 to make equivalent ratios. Use each digit only once.

$\frac{••}{••} = \frac{•}{••}$ Possible answer: $\frac{13}{65} = \frac{4}{20}$

Available on Daily Transparency in CRB

A *statute mile* is another term for *mile*. A *nautical mile* is 1.15 statute miles. A *knot* is one nautical mile per hour, a rate of speed.

7-3 Analyze Units
🧩 Problem Solving Skill

Learn to use one or more conversion factors to solve rate problems.

Vocabulary
conversion factor

You can measure the speed of an object using a strobe lamp and a camera in a dark room. Each time the lamp flashes, the camera records the object's position.

Problems often require *dimensional analysis*, also called *unit analysis*, to convert from one unit to another unit.

To convert units, multiply by one or more ratios of equal quantities called **conversion factors**.

For example, to convert inches to feet you would use the ratio at right as a conversion factor.

$\frac{1 \text{ ft}}{12 \text{ in.}}$

Multiplying by a conversion factor is like multiplying by a fraction that reduces to 1, such as $\frac{5}{5}$.

$\frac{1 \text{ ft}}{12 \text{ in.}} = \frac{12 \text{ in.}}{12 \text{ in.}}$, or $\frac{1 \text{ ft}}{1 \text{ ft}} = 1$

EXAMPLE 1 Finding Conversion Factors

Find the appropriate factor for each conversion.

Helpful Hint

The conversion factor
• must introduce the unit desired in the answer and
• must cancel the original unit so that the unit desired is all that remains.

A quarts to gallons

There are 4 quarts in 1 gallon. To convert quarts to gallons, multiply the number of **quarts** by $\frac{1 \text{ gal}}{4 \text{ qt}}$.

B meters to centimeters

There are 100 centimeters in 1 meter. To convert meters to centimeters, multiply the number of **meters** by $\frac{100 \text{ cm}}{1 \text{ m}}$.

EXAMPLE 2 Using Conversion Factors to Solve Problems

The average American eats 23 pounds of pizza per year. Find the number of ounces of pizza the average American eats per year.

The problem gives the ratio 23 *pounds* to 1 year and asks for an answer in *ounces* per year.

$\frac{23 \text{ lb}}{1 \text{ yr}} \cdot \frac{16 \text{ oz}}{1 \text{ lb}}$ *Multiply the ratio by the conversion factor.*

$= \frac{23 \cdot 16 \text{ oz}}{1 \text{ yr}}$ *Cancel lb units.* $\frac{lb}{yr} \cdot \frac{oz}{lb} = \frac{oz}{yr}$

$= 368$ oz per year   *Multiply 23 by 16 oz.*

The average American eats 368 ounces of pizza per year.

---

1 Introduce
Alternate Opener

EXPLORATION

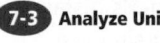

7-3 Analyze Units

The radius of a planet is called the *equatorial radius* and is the imaginary line between the center of the planet and a point on its equator.

The table shows the radius measured in two different units of each planet in our solar system. In the second column, each radius is measured using the radius of Earth. In the third column, each radius is measured in kilometers.

Calculate the equatorial radius of each planet.

	Planet	Equatorial Radius (number of Earth radii)	Equatorial Radius (km)
1.	Mercury	0.38	
2.	Venus	0.95	
3.	Earth	1	6378.14
4.	Mars	0.53	
5.	Jupiter	11	
6.	Saturn	9	
7.	Uranus	4	
8.	Neptune	4	
9.	Pluto	0.19	

Think and Discuss

10. **Explain** how you found the equatorial radius of each planet.

11. **Discuss** whether using Earth's radius or the kilometer makes comparing the radii of the planets easier.

Motivate

Have students determine the missing values.

_?_	inches = 1 foot	**12**
_?_	centimeters = 1 meter	**100**
_?_	ounces = 1 pound	**16**
_?_	seconds = 1 minute	**60**
_?_	minutes = 1 hour	**60**

Tell students that these facts allow people to convert from one measurement to another.

Exploration worksheet and answers on Chapter 7 Resource Book pp. 25 and 106

2 Teach

Lesson Presentation

Guided Instruction

In this lesson, students learn to use one or more conversion factors to solve rate problems. Show students how to write common rates in fractional form (e.g., 40 miles per hour = $\frac{40 \text{ mi}}{1 \text{ h}}$). Then explain how to choose the correct conversion factors by setting up rates that cancel the appropriate units. In Examples 1 and 2, show students how the original units are canceled out by the units in the denominator of the conversion factor. Review Examples 3–5 with students. You may want to show the work at every step because these concepts may be new for many students.

EXAMPLE 3

PROBLEM SOLVING APPLICATION

A car traveled 990 feet down a road in 15 seconds. How many miles per hour was the car traveling?

1 Understand the Problem

The problem is stated in units of **feet** and **seconds**. The question asks for the **answer** in units of **miles** and **hours**. You will need to use several conversion factors.

List the important information:

- Feet to miles $\longrightarrow \dfrac{1\text{ mi}}{5280\text{ ft}}$
- Seconds to minutes $\longrightarrow \dfrac{60\text{ s}}{1\text{ min}}$; minutes to hours $\longrightarrow \dfrac{60\text{ min}}{1\text{ h}}$

2 Make a Plan

Multiply by each conversion factor separately, or **simplify the problem** and multiply by several conversion factors at once.

3 Solve

First, convert 990 feet in 15 seconds into a unit rate.

$$\frac{990\text{ ft}}{15\text{ s}} = \frac{(990 \div 15)\text{ ft}}{(15 \div 15)\text{ s}} = \frac{66\text{ ft}}{1\text{ s}}$$

Create a single conversion factor to convert seconds directly to hours:

seconds to minutes $\longrightarrow \dfrac{60\text{ s}}{1\text{ min}}$; minutes to hours $\longrightarrow \dfrac{60\text{ min}}{1\text{ h}}$

seconds to hours $= \dfrac{60\text{ s}}{1\text{ min}} \cdot \dfrac{60\text{ min}}{1\text{ h}} = \dfrac{3600\text{ s}}{1\text{ h}}$

$\dfrac{66\text{ ft}}{1\text{ s}} \cdot \dfrac{1\text{ mi}}{5280\text{ ft}} \cdot \dfrac{3600\text{ s}}{1\text{ h}}$ *Set up the conversion factors.*

Do not include the numbers yet. Notice what happens to the units.

$\dfrac{\cancel{ft}}{\cancel{s}} \cdot \dfrac{mi}{\cancel{ft}} \cdot \dfrac{\cancel{s}}{h}$ *Simplify. Only $\frac{mi}{h}$ remain.*

$\dfrac{66\text{ ft}}{1\text{ s}} \cdot \dfrac{1\text{ mi}}{5280\text{ ft}} \cdot \dfrac{3600\text{ s}}{1\text{ h}}$ *Multiply.*

$\dfrac{66 \cdot 1\text{ mi} \cdot 3600}{1 \cdot 5280 \cdot 1\text{ h}} = \dfrac{237{,}600\text{ mi}}{5280\text{ h}} = \dfrac{45\text{ mi}}{1\text{ h}}$

The car was traveling 45 miles per hour.

4 Look Back

A rate of 45 mi/h is less than 1 mi/min. 15 seconds is $\frac{1}{4}$ min. A car traveling 45 mi/h would go less than $\frac{1}{4}$ of 5280 ft in 15 seconds. It goes 990 ft, so 45 mi/h is a reasonable speed.

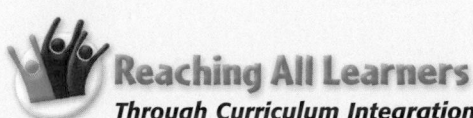

Reaching All Learners
Through Curriculum Integration

Physical Science The speed of sound in air is approximately 770 mi/h. Have students use the conversion factors given below to express the speed of sound in ft/s and m/s to the nearest whole unit. The speed of light is approximately 300,000 km/s. Have students express the speed of light in cm/s, mi/s, and mi/h in scientific notation.

conversion factors: 1 mi = 5280 ft, 1 m = 3.28 ft, 1 km = 10^5 cm, 1 km = 0.62 mi

speed of sound: 1129 ft/s; 344 m/s
speed of light: 3.0×10^{10} cm/s; 1.86×10^5 mi/s; 6.696×10^8 mi/h

---

Additional Examples

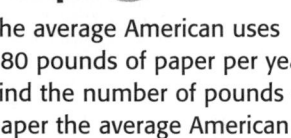

Example 1

Find the appropriate factor for each conversion.
A. feet to yards $\dfrac{1\text{ yd}}{3\text{ ft}}$
B. pounds to ounces $\dfrac{16\text{ oz}}{1\text{ lb}}$

Example 2

The average American uses 580 pounds of paper per year. Find the number of pounds of paper the average American uses per month, to the nearest tenth. **48.3 lb/mo**

Example 3

A car traveled 60 miles on a road in 2 hours. How many feet per second was the car traveling? **44 ft/s**

Example 4

A strobe lamp can be used to measure the speed of an object. The lamp flashes every $\frac{1}{100}$ of a second. A camera records the object moving 52 cm between flashes. How fast is the object moving in m/s? **52 m/s**

Example 5

The rate 1 knot equals 1 nautical mile per hour. One nautical mile is 1852 meters. What is the speed in kilometers per hour of a ship traveling at 5 knots? **9.26 km/h**

EXAMPLE **4** *Physical Science Application*

Physical Science LINK

A strobe lamp can be used to measure the speed of an object. The lamp flashes every $\frac{1}{1000}$ s. A camera records the object moving 7.5 cm between flashes. How fast is the object moving in m/s?

$$\frac{7.5 \text{ cm}}{\frac{1}{1000} \text{ s}}$$ *Use rate = $\frac{distance}{time}$.*

It may help to eliminate the fraction $\frac{1}{1000}$ first.

$$\frac{7.5 \text{ cm}}{\frac{1}{1000} \text{ s}} = \frac{1000 \cdot 7.5 \text{ cm}}{1000 \cdot \frac{1}{1000} \text{ s}}$$ *Multiply top and bottom by 1000.*

$$= \frac{7500 \text{ cm}}{1 \text{ s}}$$

Now convert centimeters to meters.

$$\frac{7500 \text{ cm}}{1 \text{ s}}$$

$$= \frac{7500 \text{ cm}}{1 \text{ s}} \cdot \frac{1 \text{ m}}{100 \text{ cm}}$$ *Multiply by the conversion factor.*

$$= \frac{7500 \text{ m}}{100 \text{ s}} = \frac{75 \text{ m}}{1 \text{ s}}$$

The object is traveling 75 m/s.

A strobe light flashing on dripping liquid can make droplets appear to stand still or even move upward.

EXAMPLE **5** *Transportation Application*

The rate of one knot equals one nautical mile per hour. One nautical mile is 1852 meters. What is the speed in meters per second of a ship traveling at 20 knots?

20 knots = 20 nautical mi/h

Set up the units to obtain m/s in your answer.

$$\frac{\cancel{\text{nautical mi}}}{\cancel{\text{h}}} \cdot \frac{\text{m}}{\cancel{\text{nautical mi}}} \cdot \frac{\cancel{\text{h}}}{\text{s}}$$ *Examine the units.*

$$\frac{20 \text{ nautical mi}}{\text{h}} \cdot \frac{1852 \text{ m}}{\text{nautical mi}} \cdot \frac{1 \text{ h}}{3600 \text{ s}}$$

$$\frac{20 \cdot 1852}{3600} \approx 10.3$$

The ship is traveling about 10.3 m/s.

Think and Discuss

1. **Give** the conversion factor for converting $\frac{\text{lb}}{\text{yr}}$ to $\frac{\text{lb}}{\text{mo}}$.

2. **Explain** how to find whether 10 miles per hour is faster than 15 feet per second.

3. **Give an example** of a conversion between units that includes ounces as a unit in the conversion.

3 Close

Summarize

Review the process for using conversion factors to convert rates from one set of units to another. Have students choose the correct conversion factor for each conversion:

1. $\frac{\text{mi}}{\text{h}}$ to $\frac{\text{ft}}{\text{h}}$ A. $\frac{1 \text{ yr}}{12 \text{ mo}}$ E. $\frac{1 \text{ kg}}{1000 \text{ g}}$

2. $\frac{\text{g}}{\text{yr}}$ to $\frac{\text{g}}{\text{mo}}$ B. $\frac{1 \text{ mi}}{5280 \text{ ft}}$ F. $\frac{1 \text{ h}}{3600 \text{ s}}$

3. $\frac{\text{ft}}{\text{h}}$ to $\frac{\text{ft}}{\text{s}}$ C. $\frac{1 \text{ h}}{3600 \text{ ft}}$ G. $\frac{5280 \text{ ft}}{1 \text{ mi}}$

4. $\frac{\text{g}}{\text{yr}}$ to $\frac{\text{kg}}{\text{yr}}$ D. $\frac{12 \text{ mo}}{1 \text{ yr}}$ H. $\frac{1000 \text{ g}}{1 \text{ kg}}$

1. G   2. A   3. F   4. E

Answers to Think and Discuss

1. $\frac{1 \text{ yr}}{12 \text{ mo}}$

2. Possible answer: Convert 10 miles per hour to feet per second by multiplying by the conversion factors $\frac{1 \text{ h}}{3600 \text{ s}}$ and $\frac{5280 \text{ ft}}{1 \text{ mi}}$. (10 miles per hour $\approx$ 14.67 feet per second, so 10 miles per hour is slower than 15 feet per second.)

3. Possible answer:
$48 \text{ oz} = 48 \text{ oz} \cdot \frac{1 \text{ lb}}{16 \text{ oz}} = 3 \text{ lb}$

FOR EXTRA PRACTICE
see page 744

⊿ internet connect
Homework Help Online
go.hrw.com   Keyword: MP4 7-3

7-3 PRACTICE & ASSESS

GUIDED PRACTICE

See Example ① Find the appropriate factor for each conversion.

1. feet to inches $\dfrac{12 \text{ in.}}{1 \text{ ft}}$ 2. gallons to pints $\dfrac{8 \text{ pt}}{1 \text{ gal}}$ 3. centimeters to meters $\dfrac{1 \text{ m}}{100 \text{ cm}}$

See Example ② 4. Aihua drinks 4 cups of water a day. Find the total number of gallons of water she drinks in a year. **91.25 gal**

See Example ③ 5. A model airplane flies 22 feet in 2 seconds. What is the airplane's speed in miles per hour? **7.5 mi/h**

See Example ④ 6. If a fish swims 0.09 centimeter every hundredth of a second, how fast in meters per second is it swimming? **0.09 m/s**

See Example ⑤ 7. There are about 400 cocoa beans in a pound. There are 2.2 pounds in a kilogram. About how many grams does a cocoa bean weigh? $\approx$ **1.14 g**

INDEPENDENT PRACTICE

See Example ① Find the appropriate factor for each conversion.

8. kilometers to meters   9. inches to yards   10. days to weeks

See Example ② 11. A theme park sells 71,175 yards of licorice each year. How many feet per day does the park sell? **585 ft**

See Example ③ 12. A yellow jacket can fly 4.5 meters in 9 seconds. How fast in kilometers per hour can a yellow jacket fly? **1.8 km/h**

See Example ④ 13. Brilco Manufacturing produces 0.2 of a brick every tenth of a second. How many bricks can be produced in an 8-hour day? **57,600 bricks**

See Example ⑤ 14. Assume that one dollar is equal to 1.14 euros. If 500 g of an item is selling for 25 euros, what is its price in dollars per kg? **$43.86/kg**

PRACTICE AND PROBLEM SOLVING

Use conversion factors to find each specified amount.

15. radios produced in 5 hours at a rate of 3 radios per minute **900 radios**

16. distance traveled (in feet) after 12 seconds at 87 miles per hour **1531.2 ft**

17. hot dogs eaten in a month at a rate of 48 hot dogs eaten each year **4 hot dogs**

18. umbrellas sold in a year at a rate of 5 umbrellas sold per day **1825 umbrellas**

19. miles jogged in 1 hour at an average rate of 7.3 feet per second $\approx$ **4.98 mi**

20. states visited in a two-week political campaign at a rate of 2 states per day **28 states**

Assignment Guide

If you finished Example ① assign:
  Core 1–3, 8–10, 28–34
  Enriched 1–3, 8–10, 28–34

If you finished Example ② assign:
  Core 1–4, 8–11, 15–19 odd, 28–34
  Enriched 8–11, 15–20, 25, 28–34

If you finished Example ③ assign:
  Core 1–5, 8–12, 15–21 odd, 28–34
  Enriched 8–12, 15–21, 25, 26, 28–34

If you finished Example ④ assign:
  Core 1–6, 8–13, 15–21 odd, 22, 28–34
  Enriched 8–13, 15–34

If you finished Example ⑤ assign:
  Core 1–14, 15–21 odd, 22, 28–34
  Enriched 8–34

Answers

8. $\dfrac{1000 \text{ m}}{1 \text{ km}}$

9. $\dfrac{1 \text{ yd}}{36 \text{ in.}}$

10. $\dfrac{1 \text{ wk}}{7 \text{ days}}$

RETEACH 7-3

LESSON 7-3 *Reteach*
Analyze Units

A **conversion relation** gives equivalent units.

1 foot = 12 inches   1 week = 7 days   1 pound = 16 ounces

Complete each conversion relation.

1. 1 gallon = __4__ quarts

2. 1 meter = __100__ centimeters

3. 1 yard = __3__ feet

To change from one unit to another, multiply by a fraction of value 1.

Convert 7 ft to in.

Use 1 foot = 12 inches to write a fraction of value 1.

Put the unit to be changed in the denominator. $\dfrac{12 \text{ in.}}{1 \text{ ft.}}$

7 ft $\times \dfrac{12 \text{ in.}}{1 \text{ ft}} = (7 \times 12)$ in. = 84 in.

When converting from a larger unit to a smaller, the result is a larger number.

7 ft = 84 in.

When converting from a smaller unit to a larger, the result is a smaller number.

84 in. = 7 ft

Complete to convert each unit.

4. 72 inches to feet
  1 ft = __12 in.__
  unit for denominator of fraction is
  __inches__
  72 in. $\times \dfrac{1 \text{ ft}}{12 \text{ in.}} =$ __6__ ft

5. 40 ounces to pounds
  1 lb = __16 oz__
  unit for denominator of fraction is
  __ounces__
  40 oz $\times \dfrac{1 \text{ lb}}{16 \text{ oz}} =$ __2.5__ lb

6. 735 meters to centimeters
  1 m = __100 cm__
  unit for denominator of fraction is
  __meters__
  735 m $\times \dfrac{100 \text{ cm}}{1 \text{ m}} =$ __73,500__ cm

7. 150 pounds to kilograms
  1 kg = 2.2 lb
  unit for denominator of fraction is
  __pounds__
  150 lb $\times \dfrac{1 \text{ kg}}{2.2 \text{ lb}} \approx$ __68.2__ kg

---

PRACTICE 7-3

LESSON 7-3 *Practice B*
Analyze Units

Find the appropriate factor for each conversion.

1. grams to kilograms   2. quarts to gallons   3. minutes to seconds
  $\dfrac{1 \text{ kg}}{1000 \text{ g}}$ $\dfrac{1 \text{ gal}}{4 \text{ qt}}$ $\dfrac{60 \text{ sec}}{1 \text{ min}}$

4. David takes 300 milligrams of medicine every day. How many grams is this?
  0.3 g

5. Jody runs the 500-yard dash for his school's track team. How many feet does he run in each 500-yard dash?
  1500 ft

6. Sean drinks six 12-ounce cans of soda a week. How many pints of soda does he drink in a week?
  4.5 pt

7. A recipe for punch requires diluting the punch concentrate with 7 quarts of water. How many gallons of water are required to dilute the concentrate according to the directions?
  1.75 gal

8. Jesse's dog Angel weighs $18\frac{1}{2}$ pounds. How many ounces does Angel weigh?
  296 oz

9. A roll of tape contains 32.9 meters of tape. How many millimeters of tape does the roll contain?
  32,900 mm

10. There are two types of lifts in the sport of weightlifting, the *snatch* and the *clean and jerk*. Winners are determined by the combined weights of the two type of lifts. In the 2002 Collegiate Weightlifting Competition, Timothy Leancu from the U.S. Naval Academy competed in the 94-kilogram weight class. He lifted 100 kg in the *snatch* and 132.5 kg in the *clean and jerk*. What was the combined weight of his lifts in grams?
  232,500 g

Math Background

The key to solving the problems in this lesson is finding the correct conversion factors. Emphasize that a conversion factor is a rate in which the numerator and the denominator represent the same quantity with different units. Any conversion factor can be simplified to a 1:1 ratio. For example, the conversion factor $\frac{1 \text{ gal}}{4 \text{ qt}}$ is equal to $\frac{1 \text{ gal}}{1 \text{ gal}}$ or $\frac{4 \text{ qt}}{4 \text{ qt}}$, both of which simplify to 1:1. Conversion factors can be applied to introduce desired units and cancel unwanted units. For example, to convert 60 mi/h to ft/s, apply three conversion factors.

$$\dfrac{60 \text{ mi}}{1 \text{ h}} \cdot \dfrac{5280 \text{ ft}}{1 \text{ mi}} \cdot \dfrac{1 \text{ h}}{60 \text{ min}} \cdot \dfrac{1 \text{ min}}{60 \text{ s}} = \dfrac{88 \text{ ft}}{1 \text{ s}}$$

Answers

25. Possible answer: In the last ratio, the student used 60 seconds per hour, but there are 3600 seconds in one hour. The correct answer is about 17.05 mi/h.

26. Possible answer: Conversion factors allow you to change from one rate to another. For example, to change mi/h into ft/s, multiply the speed by $\frac{5280 \text{ ft}}{1 \text{ mi}} \cdot \frac{1 \text{ h}}{3600 \text{ s}}$.

Journal

Have students write about a situation in which they might need to convert a rate to a different rate.

Test Prep Doctor ✚

For Exercise 34, students who chose answer choice **B** multiplied the volume of the original cylinder by $\frac{1}{2}$. For those students, point out that cutting the radius in half does not cut the volume of the cylinder in half. To find the correct answer they must use a radius of 3 (half of the given radius) in the volume formula. The correct choice is **A.**

Lesson Quiz

Find the appropriate factor for each conversion.

1. kilograms to grams $\frac{1000 \text{ g}}{\text{kg}}$

2. pints to gallons $\frac{1 \text{ gal}}{8 \text{ pt}}$

3. You drive 136 miles from your house to your aunt's house at the lake. You use 8 gallons of gas. How many yards does your car get to the gallon? $\frac{29,920 \text{ yd}}{\text{gal}}$

4. A cheetah was timed running 200 yards in 6 seconds. What was its average speed in miles per hour? ≈ 68 mi/h

Available on Daily Transparency in CRB

---

21. A. ≈ 22.88 mi/h;
B. ≈ 23.16 mi/h;
C. ≈ 21.76 mi/h

21. SPORTS Use the graph to find each world-record speed in miles per hour. (*Hint:* 1 mile ≈ 1609 m.)

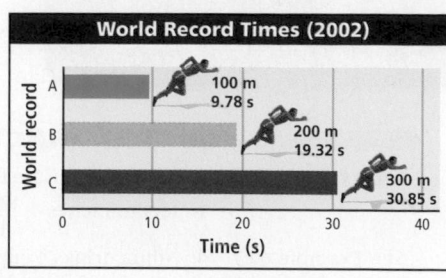

World Record Times (2002)

22. LIFE SCIENCE The Kelp Forest exhibit at the Monterey Bay Aquarium holds 335,000 gallons. How many days would it take to fill it at a rate of 1 gallon per second? ≈ **3.88 days**

23. TRANSPORTATION An automobile engine is turning at 3000 revolutions per minute. During each revolution, each of the four spark plugs fires. How many times do the spark plugs fire in one second? **200 times**

24. CHOOSE A STRATEGY The label on John's bottle of cough syrup says a person should take 3 teaspoons. Which spoon could John use to take the cough medicine? (*Hint:* 1 teaspoon = $\frac{1}{6}$ oz.) **B**

A   A 1.5 oz spoon   C   A 1 oz spoon

B   A 0.5 oz spoon   D   None of these

25. WHAT'S THE ERROR? To convert 25 feet per second to miles per hour, a student wrote $\frac{25 \text{ ft}}{1 \text{ s}} \cdot \frac{1 \text{ mile}}{5280 \text{ ft}} \cdot \frac{60 \text{ s}}{1 \text{ h}} \approx 0.28$ mi/h. What error did the student make? What should the correct answer be?

26. WRITE ABOUT IT Describe the important role that conversion factors play in solving rate problems. Give an example.

27. CHALLENGE Anthony the anteater requires 1800 calories each day. He gets 1 calorie from every 50 ants that he eats. If he sticks his tongue out 150 times per minute and averages 2 ants per lick, how many hours will it take for him to get 1800 calories? **5 hr**

Spiral Review

Find the area of the quadrilateral with the given vertices. (Lesson 6-1)

28. (0, 0), (0, 9), (5, 9), (5, 0) **45 units²**   **29.** (−3, 1), (4, 1), (6, 3), (−1, 3) **14 units²**

Find the area of each circle to the nearest tenth. Use 3.14 for π. (Lesson 6-4)

30. circle with radius 7 ft **153.9 ft²**   **31.** circle with diameter 17 in. **226.9 in²**

32. circle with radius 3.5 cm **38.5 cm²**   **33.** circle with diameter 2.2 mi **3.8 mi²**

34. TEST PREP A cylinder has radius 6 cm and height 14 cm. If the radius were cut in half, what would the volume of the cylinder be? Use 3.14 for π and round to the nearest tenth. (Lesson 6-6) **A**

A   395.6 cm³   B   791.3 cm³   C   422.3 cm³   D   393.5 cm³

CHALLENGE 7-3

Challenge
7-3 *Water, Water, Everywhere*

Next time you are in a supermarket, pick up a gallon of water to verify what you will now determine.

How many pounds does a gallon of water weigh?

The amount of space between particles of a substance is what determines its **density**.

Lead is more dense than wood, which is more dense than foam rubber.

Density *d* is equal to the mass (weight) *m* of a substance divided by its volume *V*. $d = \frac{m}{V}$

Water was used as the basis for establishing the metric unit of mass.

The density of water is 1 gram per milliliter: $\frac{1 \text{ g}}{1 \text{ mL}}$.

1. Use the conversion relation 1 mL = 1 cm³ to write the density of water in terms of grams per cubic centimeter.

density of water = $\frac{1 \text{ g}}{1 \text{ mL}} \times \frac{1 \text{ mL}}{1 \text{ cm}^3} = \frac{1 \text{ g}}{1 \text{ cm}^3}$

2. Since 1 m = 100 cm, how many cubic centimeters are there in 1 cubic meter?

1 m³ = **1,000,000** cm³

3. Use your results from Exercises 1 and 2, and the conversion relation 1 kg = 1000 g to write the density of water in terms of kilograms per cubic meters. Round your answer to the nearest hundredth.

density of water = $\frac{1 \text{ g}}{1 \text{ cm}^3} \times \frac{1 \text{ kg}}{1000 \text{ g}} \times \frac{1,000,000 \text{ cm}^3}{1 \text{ m}^3} = \frac{1000 \text{ kg}}{1 \text{ m}^3}$

4. Use your result from Exercise 3 and these conversion relations to write the density of water as a unit rate in terms of pounds per gallon:

1 kg = 2.205 lb and 1 m³ = 264.2 gal

density of water = $\frac{1000 \text{ kg}}{1 \text{ m}^3} \times \frac{2.205 \text{ lb}}{1 \text{ kg}} \times \frac{1 \text{ m}^3}{264.2 \text{ gal}} \approx \frac{8.35 \text{ lb}}{1 \text{ gal}}$

5. So, a gallon of water weighs about **8.35** pounds.

PROBLEM SOLVING 7-3

Problem Solving
7-3 *Analyze Units*

Use the following: 1 mile = 1.609 km; 1 kg = 2.2046 lb. Round to the nearest tenth.

1. Worker bees travel up to 14 km to find pollen and nectar. How far will a worker bee travel in miles?

8.7 mi

2. Worker bees can travel at 24 km/h. How fast can the worker bee travel in miles per hour?

14.9 miles an hour

3. The average hippopotamus weighs 1800 kg. How many pounds does the average hippopotamus weigh?

3,968.3 lb

4. At the age of 45, an elephant grows teeth, each weighing 4 kg. How many pounds do these teeth weigh?

8.8 lb

Paraceratherium was the biggest land mammal there has ever been. It lived about 35 million years ago and was 8 m tall and 11 m long. It looked like a gigantic rhinoceros but had a long neck like a giraffe. 1 foot = 0.3048 meters. Round to the nearest tenth.

5. How tall was the paraceratherium in feet?

26.2 ft

6. How long was the paraceratherium in feet?

36.1 ft

Round to the nearest tenth. Choose the letter for the best answer.

7. The fastest sporting animal is the racing pigeon that flies up to 110 mi an hour. How fast is the racing pigeon in feet each second?

A 75.0 ft/s   C 543.2 ft/s
Ⓑ 161.3 ft/s   D 9,680 ft/s

8. The longest gloved fight between two Americans lasted for more than seven hours before being declared a draw. How many seconds did the fight last?

F 127 s   H 420 s
G 385 s   Ⓙ 25,200 s

9. The average person falls asleep in seven minutes. How many seconds does it take the average person to fall asleep?

A 127 s   Ⓒ 420 s
B 385 s   D 25,200 s

10. The brain of an average adult male weighs 55 oz. How many pounds does the average male brain weigh?

Ⓕ 3.4 lb   H 13.8 lb
G 5.8 lb   J 880 lb

Hands-On LAB 7A

Model Proportions

Use with Lesson 7-4

WHAT YOU NEED:
- Ruler
- Pattern blocks

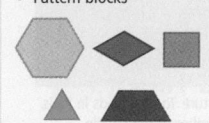

REMEMBER
- Use the area formulas to find the area of each pattern block except the hexagon.

To find the area of the hexagon, think of the pieces that can fit together to make the hexagon.

internet connect

Lab Resources Online
go.hrw.com
KEYWORD: MP4 Lab7A

Activity

❶ Measure each type of pattern block to the nearest eighth of an inch to determine its area. Use pattern blocks to find several area relationships that represent fractions equivalent to one-half. For example,

 $= \frac{1}{2}$.

❷ Above, you related area of pattern blocks to a ratio. Now make a proportion based upon area that uses only pattern blocks on both sides of the equal sign. Then write these proportions using numbers based on your measurements for area. Use cross products to check your work.

Think and Discuss

Possible answer:

1. Which pattern-block area relationships equal $\frac{5}{6}$?

2. What area relationships can you make with a triangle and a trapezoid? $\frac{1}{3}; \frac{3}{1}$

3. What area relationships can you make with only a triangle? $\frac{1}{1}$

Try This

Use pattern blocks to complete each proportion based on area. Then write these proportions using numbers based on your measurements for area.

1. 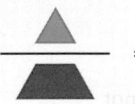 $\frac{1}{3}$

2. 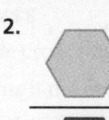 $\frac{3}{2}$

Hands-On LAB 7A

Model Proportions

Pacing: Traditional 1 day
Block $\frac{1}{2}$ day

Objective: To use pattern blocks to model proportions

Materials: Pattern blocks, ruler

Lab Resources
Hands-On Lab Activities . pp. 53, 94–95

Using the Page
Let students become familiar with the pattern blocks before beginning the activity.

1. How many green triangles are needed to completely cover a yellow hexagon? **6**

2. How many green triangles are needed to completely cover a blue rhombus? a red trapezoid? **2; 3**

3. If a yellow hexagon represents 1 whole, how can you model the fraction $\frac{1}{3}$? Use two green triangles or a blue rhombus to model $\frac{2}{6}$, or $\frac{1}{3}$.

Answers
Activity

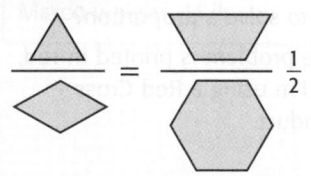 $= \frac{1}{2}$

Sara Fox
New City, New York

Teacher to Teacher

Discuss with the class the usefulness of proportions in the real world. How are proportions useful for problem solving in real-life situations? Students' responses could include the following: enlarging or decreasing photographs, watching a news report on television requiring transmitting images by scaling proportions from the video camera to a satellite to the television screen, creating a scale model/plan of a house, converting cooking recipes from home-use quantities to class-size quantities, and using computer technology to enlarge and decrease font size. This discussion may also lead to individual student projects.

FOR EXTRA PRACTICE
see page 744

internet connect
Homework Help Online
go.hrw.com   Keyword: MP4 7-4

Students may want to refer back to the lesson examples.

Assignment Guide

If you finished Example **1** assign:
Core 1–5, 15–19, 29, 40–44
Enriched 5, 15–19, 29–34, 40–44

If you finished Example **2** assign:
Core 1–13, 15–27, 29, 40–44
Enriched 5, 14, 15–27, 29–44

If you finished Example **3** assign:
Core 1–29, 35, 40–44
Enriched 5, 14, 15–44

Notes

GUIDED PRACTICE

See Example **1**   Tell whether the ratios in each pair are proportional.

1. $\frac{7}{14} \overset{?}{=} \frac{14}{28}$ **yes**   2. $\frac{2}{9} \overset{?}{=} \frac{6}{27}$ **yes**   3. $\frac{3}{7} \overset{?}{=} \frac{6}{15}$ **no**   4. $\frac{15}{25} \overset{?}{=} \frac{9}{15}$ **yes**

5. A bubble solution can be made with a ratio of one part detergent to eight parts water. Would a mixture of 56 oz water and 8 oz detergent be proportional to this ratio? Explain. **no; $\frac{1}{8} \neq \frac{8}{56}$**

See Example **2**   Solve each proportion.

6. $\frac{x}{5} = \frac{2}{10}$ **x = 1**   7. $\frac{4}{9} = \frac{n}{18}$ **n = 8**   8. $\frac{11}{d} = \frac{66}{12}$ **d = 2**   9. $\frac{21}{7} = \frac{h}{2}$ **h = 6**

10. $\frac{12}{f} = \frac{16}{13}$ **f = 9.75**   11. $\frac{t}{7} = \frac{8}{28}$ **t = 2**   12. $\frac{1}{2} = \frac{s}{18}$ **s = 9**   13. $\frac{28}{7} = \frac{50}{q}$ **q = 12.5**

See Example **3**   14. A 10 kg weight is positioned 5 cm from a fulcrum. At what distance from the fulcrum must a 15 kg weight be positioned to keep the scale balanced? **≈ 3.3 cm**

INDEPENDENT PRACTICE

See Example **1**   Tell whether the ratios in each pair are proportional.

15. $\frac{12}{49} \overset{?}{=} \frac{4}{7}$ **no**   16. $\frac{17}{51} \overset{?}{=} \frac{2}{6}$ **yes**   17. $\frac{30}{36} \overset{?}{=} \frac{15}{16}$ **no**   18. $\frac{7}{8} \overset{?}{=} \frac{35}{40}$ **yes**

19. A class had 18 girls and 12 boys. Then 2 boys and 3 girls transferred out of the class. Did the ratio of girls to boys stay the same? Explain.
yes; $\frac{18}{12} = \frac{15}{10}$

See Example **2**   Solve each proportion.

20. $\frac{3}{9} = \frac{b}{21}$ **b = 7**   21. $\frac{27}{90} = \frac{b}{10}$ **b = 3**   22. $\frac{4}{1} = \frac{0.56}{m}$ **m = 0.14**   23. $\frac{y}{5} = \frac{42}{35}$ **y = 6**

24. $\frac{r}{7} = \frac{3}{2}$ **r = 10.5**   25. $\frac{48}{16} = \frac{12}{n}$ **n = 4**   26. $\frac{p}{9} = \frac{2}{12}$ **p = 1.5**   27. $\frac{2}{d} = \frac{6}{1.5}$ **d = 0.5**

See Example **3**   28. Jo weighs 65 lb and Tim weighs 78 lb. If Tim is seated 6 ft from the center of a balanced seesaw, how far is Jo seated from the center? **7.2 ft**

PRACTICE AND PROBLEM SOLVING

For each set of ratios, find the two that are proportional.

29. $\frac{6}{3}, \frac{18}{9}, \frac{51}{25}$ **$\frac{6}{3}, \frac{18}{9}$**   30. $\frac{1}{4}, \frac{11}{44}, \frac{111}{440}$ **$\frac{1}{4}, \frac{11}{44}$**   31. $\frac{30}{14}, \frac{66}{21}, \frac{22}{7}$ **$\frac{66}{21}, \frac{22}{7}$**

32. $\frac{54}{168}, \frac{9}{28}, \frac{52}{142}$ **$\frac{54}{168}, \frac{9}{28}$**   33. $\frac{0.25}{4}, \frac{0.125}{6}, \frac{1}{16}$ **$\frac{0.25}{4}, \frac{1}{16}$**   34. $\frac{a}{c}, \frac{a}{b}, \frac{4a}{4b}$ **$\frac{a}{b}, \frac{4a}{4b}$**

35. **PHYSICAL SCIENCE** Each molecule of sulfuric acid reacts with 2 molecules of ammonia. How many molecules of sulfuric acid react with 24 molecules of ammonia? **12 molecules**

Math Background

Proportions are used to represent *direct* relationships and *inverse* relationships.

The dimensions of similar polygons (to be studied in Lesson 7-6) are directly proportional. The dimensions of two different rectangles that have the same area are *inversely proportional*, as shown below.

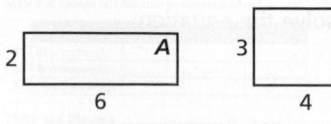

$$\frac{\text{length}_A}{\text{length}_B} = \frac{\text{width}_B}{\text{width}_A}, \text{ or } \frac{6}{4} = \frac{3}{2}$$

RETEACH 7-4

LESSON 7-4 **Reteach**
Solving Proportions

In a proportion, the **cross products** are equal.

These ratios are proportional, since their cross products are equal.
$\frac{7}{14} \overset{?}{=} \frac{8}{16}$
$7 \times 16 \overset{?}{=} 14 \times 8$
$112 = 112$

These ratios are not proportional, since their cross products are not equal.
$\frac{4}{9} \overset{?}{=} \frac{2}{3}$
$4 \times 3 \overset{?}{=} 9 \times 2$
$12 \neq 18$

Complete to tell if the ratios are proportional.

1. $\frac{15}{45} \overset{?}{=} \frac{6}{18}$ $15 \times \underline{18} \overset{?}{=} 45 \times \underline{6}$ $\underline{270} = \underline{270}$
   The ratios **are** in proportion.

2. $\frac{75}{100} \overset{?}{=} \frac{4}{8}$ $75 \times 8 \overset{?}{=} 100 \times \underline{4}$ $\underline{600} \neq \underline{400}$
   The ratios **are not** in proportion.

3. $\frac{7}{2} \overset{?}{=} \frac{21}{6}$ $7 \times 6 \overset{?}{=} 2 \times 21$ $\underline{42} = \underline{42}$
   The ratios **are** in proportion.

To solve for one member of a proportion, set the cross products equal.
$\frac{n}{32} = \frac{9}{16}$
$16n = 32 \times 9$
$\frac{16n}{16} = \frac{288}{16}$
$n = 18$

To check the result, substitute and see if the ratios are equivalent.
$\frac{18}{32} \overset{?}{=} \frac{9}{16}$
$\frac{18 \div 2}{32 \div 2} \overset{?}{=} \frac{9}{16}$
$\frac{9}{16} = \frac{9}{16}$ ✓

Solve and check.

4. $\frac{8}{24} = \frac{2}{n}$
   $8n = \underline{24 \times 2}$
   $\frac{8n}{8} = \frac{48}{8}$
   $n = \underline{6}$

   Check: $\frac{8}{24} \overset{?}{=} \frac{2}{n}$
   $\frac{8}{24} \overset{?}{=} \frac{2}{6}$
   $\frac{8 \div 8}{24 \div 8} \overset{?}{=} \frac{2 \div 2}{6 \div 2}$
   $\frac{1}{3} = \frac{1}{3}$ ✓

PRACTICE 7-4

LESSON 7-4 **Practice B**
Solving Proportions

Tell whether each pair of ratios is proportional.

1. $\frac{3}{4} \overset{?}{=} \frac{9}{12}$ **yes**
2. $\frac{9}{24} \overset{?}{=} \frac{18}{48}$ **yes**
3. $\frac{16}{24} \overset{?}{=} \frac{10}{18}$ **no**
4. $\frac{13}{25} \overset{?}{=} \frac{26}{50}$ **yes**
5. $\frac{10}{32} \overset{?}{=} \frac{16}{38}$ **no**
6. $\frac{20}{36} \overset{?}{=} \frac{50}{90}$ **yes**
7. $\frac{20}{28} \overset{?}{=} \frac{28}{36}$ **no**
8. $\frac{14}{42} \overset{?}{=} \frac{16}{48}$ **no**
9. $\frac{28}{20} \overset{?}{=} \frac{42}{30}$ **yes**
10. $\frac{24}{33} \overset{?}{=} \frac{14}{11}$ **no**
11. $\frac{9}{24} \overset{?}{=} \frac{8}{21}$ **no**
12. $\frac{18}{23} \overset{?}{=} \frac{28}{33}$ **no**

Solve each proportion.

13. $\frac{3}{19} = \frac{x}{38}$ **x = 6**
14. $\frac{18}{45} = \frac{8}{y}$ **y = 20**
15. $\frac{18}{19} = \frac{36}{w}$ **w = 38**
16. $\frac{13}{a} = \frac{39}{45}$ **a = 15**
17. $\frac{d}{19} = \frac{12}{57}$ **d = 4**
18. $\frac{16}{y} = \frac{4}{25}$ **y = 100**
19. $\frac{16}{42} = \frac{x}{36}$ **x = 12**
20. $\frac{15}{r} = \frac{30}{42}$ **r = 21**
21. $\frac{19}{20} = \frac{57}{x}$ **x = 60**
22. $\frac{16}{25} = \frac{48}{n}$ **n = 75**
23. $\frac{30}{19} = \frac{15}{32}$ **a = 64**
24. $\frac{16}{40} = \frac{x}{65}$ **x = 26**

25. The Supreme Court of the United States is composed of a chief justice and eight associate justices. The President of the United States appoints these positions, which are life terms. In March 2002, the Court had two women serving as justices. What is the ratio of women to men on this Supreme Court? **$\frac{2}{7}$**

26. If 1 mi ≈ 1.6 km, how many miles is 112.6 km? **70.375 mi**

Health LINK

A doctor reports blood pressure in millimeters of mercury (mm Hg) as a ratio of *systolic* blood pressure to *diastolic* blood pressure (such as 140 over 80). Systolic pressure is measured when the heart beats, and diastolic pressure is measured when it rests. Refer to the table of blood pressure ranges for adults for Exercises 36–39.

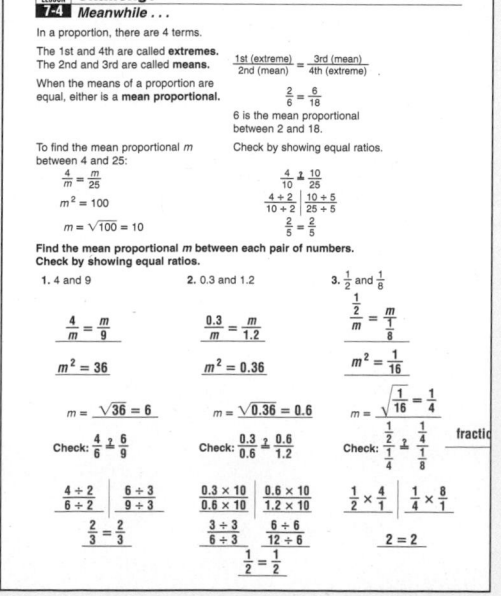

Blood Pressure Ranges

	Optimal	Normal–High	Hypertension (very high)
Systolic	under 120 mm Hg	120–140 mm Hg	over 140 mm Hg
Diastolic	under 80 mm Hg	80–90 mm Hg	over 90 mm Hg

The disc-like shape of red blood cells allows them to pass through tiny capillaries.

36. Eduardo is a healthy 37-year-old man whose blood pressure is in the optimal category.

   a. Calculate an approximate ratio of systolic to diastolic blood pressure in the optimal range.   **about 3:2**

   b. If Eduardo's systolic blood pressure is 102 mm Hg, use the ratio from part **a** to predict his diastolic blood pressure.   **about 68 mm Hg**

37. The midpoint of a range of values can be found by adding the highest and lowest numbers together and dividing by 2.

   a. Calculate an approximate ratio of systolic to diastolic blood pressure for the normal–high category.   **about 1.53:1**

   b. Tyra's diastolic blood pressure is 88 mm Hg. Use the ratio from part **a** to predict her systolic blood pressure.   **about 134.6 mm Hg**

38. Another ratio related to heart health is the ratio of LDL cholesterol to HDL cholesterol. The optimal ratio of LDL to HDL is below 3. If a patient's total cholesterol is 168 and HDL is 44, is the ratio optimal? Explain.   **Yes; the ratio is less than 2.82:1.**

39. ⭐ *CHALLENGE* The sum of Ken's LDL and HDL cholesterol is 210, and his LDL to HDL ratio is 2.75. What are his LDL and HDL?   **154 and 56**

About $\frac{9}{20}$ of your blood is made of cells; the rest is plasma.

go.hrw.com
KEYWORD: MP4 Health
CNN Student News.

Spiral Review

Write each decimal as a fraction in simplest form. (Lesson 3-1)

40. 0.65 $\frac{13}{20}$

41. −1.25 $-1\frac{1}{4}$

42. 0.723 $\frac{723}{1000}$

43. 11.17 $11\frac{17}{100}$

44. **TEST PREP** A $4\frac{5}{8}$ ft section of wood is cut from a $7\frac{1}{2}$ ft board. How much of the original board remains? (Section 3-5)   **C**

   A $3\frac{5}{8}$ ft   B $2\frac{3}{8}$ ft   C $2\frac{7}{8}$ ft   D $3\frac{9}{16}$ ft

Interdisciplinary LINK

Health

Exercises 36–39 involve using information about heart health. Students learn about blood pressure and cholesterol in middle school health programs such as Holt, Rinehart & Winston's *Decisions for Health*.

Journal

An *analogy* is a comparison of pairs of words or ideas that have a similar relationship. For example, in the analogy "*good* is to *bad* as *big* is to *small*," both sets of words have opposite meanings. Have students write about how proportions are like analogies.

Test Prep Doctor ✚

For Exercise 44, tell students that they can estimate $7\frac{1}{2} - 4\frac{5}{8}$ by evaluating $7\frac{1}{2} - 4\frac{1}{2}$, which is 3. The closest answer choice is $2\frac{7}{8}$, choice **C**. Students can check choice **C** by adding $2\frac{7}{8}$ and $4\frac{5}{8}$. The sum, $7\frac{1}{2}$, is the original length, so this choice is correct.

CHALLENGE 7-4

LESSON 7-4 **Challenge**
Meanwhile . . .

In a proportion, there are 4 terms.

The 1st and 4th are called *extremes*. The 2nd and 3rd are called *means*.

$$\frac{\text{1st (extreme)}}{\text{2nd (mean)}} = \frac{\text{3rd (mean)}}{\text{4th (extreme)}}$$

When the means of a proportion are equal, either is a *mean proportional*.

$$\frac{2}{6} = \frac{6}{18}$$

6 is the mean proportional between 2 and 18.

To find the mean proportional m between 4 and 25:

$$\frac{4}{m} = \frac{m}{25}$$

$$m^2 = 100$$

$$m = \sqrt{100} = 10$$

Check by showing equal ratios.

$$\frac{4}{10} \stackrel{?}{=} \frac{10}{25}$$

$$\frac{4 \div 2}{10 \div 2} \mid \frac{10 \div 5}{25 \div 5}$$

$$\frac{2}{5} = \frac{2}{5}$$

Find the mean proportional m between each pair of numbers. Check by showing equal ratios.

1. 4 and 9

$$\frac{4}{m} = \frac{m}{9}$$

$$m^2 = 36$$

$$m = \sqrt{36} = 6$$

Check: $\frac{4}{6} \stackrel{?}{=} \frac{6}{9}$

$$\frac{4 \div 2}{6 \div 2} \mid \frac{6 \div 3}{9 \div 3}$$

$$\frac{2}{3} = \frac{2}{3}$$

2. 0.3 and 1.2

$$\frac{0.3}{m} = \frac{m}{1.2}$$

$$m^2 = 0.36$$

$$m = \sqrt{0.36} = 0.6$$

Check: $\frac{0.3}{0.6} \stackrel{?}{=} \frac{0.6}{1.2}$

$$\frac{0.3 \times 10}{0.6 \times 10} \mid \frac{0.6 \times 10}{1.2 \times 10}$$

$$\frac{3 \div 3}{6 \div 3} \mid \frac{6 \div 6}{12 \div 6}$$

$$2 = 2$$

3. $\frac{1}{2}$ and $\frac{1}{8}$

$$\frac{\frac{1}{2}}{m} = \frac{m}{\frac{1}{8}}$$

$$m^2 = \frac{1}{16}$$

$$m = \sqrt{\frac{1}{16}} = \frac{1}{4}$$

Check: $\frac{\frac{1}{2}}{\frac{1}{4}} \stackrel{?}{=} \frac{\frac{1}{4}}{\frac{1}{8}}$ fracti

$$\frac{1}{2} \times \frac{4}{1} \mid \frac{1}{4} \times \frac{8}{1}$$

$$\frac{1}{2} = \frac{1}{2}$$

PROBLEM SOLVING 7-4

LESSON 7-4 **Problem Solving**
Solving Proportions

Scientists have researched the ratio of human body parts and height. Use the ratios in the table to answer each question. Round to the nearest tenth.

Body Part	Body Part Height
Femur	$\frac{1}{4}$
Tibia	$\frac{1}{5}$
Hand span	$\frac{2}{17}$
Arm span	$\frac{1}{1}$
Head circumference	$\frac{1}{3}$

1. Which body part is the same length as the person's height?

arm span

2. If a person's tibia is 13 inches, how tall would you expect the person to be?

65 inches

3. If a person's hand span is 8.5 inches, about how tall would you expect the person to be?

72.3 inches

4. If a femur is 18 inches long, how many feet tall would you expect the person to be?

6 feet

5. What would you expect the head circumference to be of a person who is 5.5 feet tall?

1.8 feet

6. What would you expect the hand span to be of a person who is 5 feet tall?

0.6 feet

Choose the letter for the best answer.

7. Five milliliters of a children's medicine contains 400 mg of the drug amoxicillin. How many mg of amoxicillin does 25 mL contain?

A 0.3 mg   Ⓒ 2000 mg
B 80 mg   D 2500 mg

8. Vladimir Radmanovic of the Seattle Supersonics makes, on average, about 2 three-pointers for every 5 he shoots. If he attempts 10 three-pointers in a game, how many would you expect him to make?

Ⓕ 4   H 8
G 5   J 25

9. In 2002, a 30-second commercial during the Super Bowl cost an average of $1,900,000. At this rate, how much would a 45-second commercial cost?

A $1,266,666   C $3,500,000
Ⓑ $2,850,000   D $4,000,000

10. A medicine for dogs indicates that the medicine should be administered in the ratio 2 teaspoons per 5 lb, based on the weight of the dog. How much should be given to a 70 lb dog?

F 5 teaspoons   H 14 teaspoons
G 12 teaspoons   Ⓙ 28 teaspoons

Lesson Quiz

Tell whether each pair of ratios is proportional.

1. $\frac{48}{42} \stackrel{?}{=} \frac{16}{14}$ **yes**

2. $\frac{20}{15} \stackrel{?}{=} \frac{3}{4}$ **no**

Solve each proportion.

3. $\frac{n}{12} = \frac{45}{18}$ **$n = 30$**

4. $\frac{6}{9} = \frac{n}{24}$ **$n = 16$**

5. Two weights are balanced on a fulcrum. If a 6 lb weight is positioned 1.5 ft from the fulcrum, at what distance from the fulcrum must an 18 lb weight be placed to keep the weights balanced?   **0.5 ft**

Available on Daily Transparency in CRB

Chapter
7 Mid-Chapter Quiz

Purpose: *To assess students' mastery of concepts and skills in Lessons 7-1 through 7-4*

Assessment Resources

Section 7A Quiz
Assessment Resources .... p. 17

Test and Practice Generator CD-ROM

Additional mid-chapter assessment items in both multiple-choice and free-response format may be generated for any objective in Lessons 7-1 through 7-4.

Answers

9. $\frac{4 \text{ qt}}{1 \text{ gal}}$

10. $\frac{1 \text{ cm}}{10 \text{ mm}}$

11. $\frac{1 \text{ day}}{1440 \text{ min}}$

LESSON 7-1 (pp. 342–345)

Simplify to tell whether the ratios form a proportion.

1. $\frac{4}{5}$ and $\frac{16}{20}$ **yes**
2. $\frac{33}{60}$ and $\frac{11}{21}$ **no**
3. $\frac{12}{42}$ and $\frac{6}{21}$ **yes**
4. $\frac{8}{20}$ and $\frac{4}{25}$ **no**

5. Josh is following a recipe that calls for 2.5 cups of sugar to make 2 dozen cookies. He uses 3.5 cups of sugar to make 3 dozen cookies. Has he followed the recipe? Explain. **no;** $\frac{2.5}{2} \neq \frac{3.5}{3}$

LESSON 7-2 (pp. 346–349)

Find the unit price for each offer and tell which is the better buy.

6. a long distance phone charge of $1.40 for 10 min or $4.50 for 45 min **$0.14 per min; $0.10 per min; 45 min**

7. Buy one 10 pack of AAA batteries for $5.49 and get one free, or buy two 4 packs for $2.98. **$0.27 per battery; $0.37 per battery; 10-pack**

8. A 64 oz bottle of juice costs $2.39, and a 20 oz bottle costs $0.79. You can use a 20-cents-off coupon if you buy four 20 oz bottles or a 15-cents-off coupon if you buy a 64 oz bottle. Which is the better buy? **$0.035 per oz; $0.037 per oz; one 64 oz bottle**

LESSON 7-3 (pp. 350–354)

Find the appropriate factor for each conversion.

9. gallons to quarts
10. millimeters to centimeters
11. minutes to days

Convert to the indicated unit to the nearest hundredth.

12. Change 60 ounces to pounds. **3.75 lb**

13. Change 25 pounds to ounces. **400 oz**

14. Change 5 feet per minute to feet per second. **0.08 ft/s**

15. Change 40 miles per hour to miles per second. **0.01 mi/s**

16. Driving at a constant rate, Noah covered 140 miles in 3.5 hours. Express his driving rate in feet per minute. **3520 ft/min**

LESSON 7-4 (pp. 356–359)

Solve.

17. $\frac{6}{9} = \frac{n}{72}$ $n = 48$
18. $\frac{18}{12} = \frac{3}{x}$ $x = 2$
19. $\frac{0.7}{1.4} = \frac{z}{28}$ $z = 14$
20. $\frac{12}{y} = \frac{32}{16}$ $y = 6$
21. $\frac{c}{5} = \frac{9}{24}$ $c = 1.875$
22. $\frac{5}{3} = \frac{g}{27}$ $g = 45$
23. $\frac{0.5}{h} = \frac{2}{3}$ $h = 0.75$
24. $\frac{9}{0.9} = \frac{72}{b}$ $b = 7.2$

25. Tim can input 110 data items in 2.5 minutes. Typing at the same rate, how many data items can he input in 7 minutes? **308 data items**

Focus on Problem Solving

Solve

Solve the Problem

• Choose an operation (· or ÷)

When you are converting units, think about whether the number in the answer will be larger or smaller than the number given in the question. This will help you to decide whether to multiply or divide to convert the units.

For example, if you are converting feet to inches, you know that the number of inches will be larger than the number of feet because each foot is 12 inches. So you know that you should multiply by 12 to get a larger number.

In general, if you are converting to smaller units, the number of units will have to be larger to represent the same quantity.

Purpose: To focus on choosing an operation to solve the problem

Problem Solving Resources

Interactive Problem Solving . . . pp. 55–63

Math: Reading and Writing in the Content Area . . . . pp. 55–63

Problem Solving Process

This page focuses on the third step of the problem-solving process: **Solve**

Discuss

Discuss whether the number in the answer will be greater or less than the number given in the problem, and discuss how knowing which measurement is less can help determine whether to multiply or divide.

Possible answers:

1. Greater; one mile per hour is smaller than a knot, so you should multiply.

2. Less; a meter is greater than a foot, so you should divide.

3. Greater; a liter is less than a gallon, so you should multiply.

4. Less; a mile is greater than a kilometer, so you should divide.

For each problem, determine whether the number in the answer will be larger or smaller than the number given in the question. Use your answer to decide whether to multiply or divide by the conversion factor, and solve the problem.

1 The speed a boat travels is usually measured in nautical miles, or knots. The Golden Gate Sausalito Ferry in California, which provides service between Sausalito and San Francisco, can travel at 20.5 knots. Find the speed in miles per hour. (1 knot 5 1.15 miles per hour)

2 When it is finished, the Crazy Horse Memorial in the Black Hills of South Dakota will be the world's largest sculpture. The sculpture's finished height will be 563 feet. Find the height in meters. (1 meter = 3.28 feet)

3 The amounts of water typically used for common household tasks are given in the table below. Find the number of liters needed for each task. (1 gallon = 3.79 liters)

Task	Water Used (gal)
Laundry (1 load)	40
5-minute shower	12.5
Washing hands	0.5
Flushing toilet	3.5

4 Lake Baikal, in Siberia, is so large that it would take all of the rivers on earth combined an entire year to fill it. At 1.62 kilometers deep, it is the deepest lake in the world. Find the depth of Lake Baikal in miles. (1 mile = 1.61 kilometers)

Answers

1. 23.575 mi/h

2. 171.65 m

3.

Task	Water Used (L)
Laundry (1 load)	151.60
5-minute shower	47.375
Washing hands	1.895
Flushing toilet	13.265

4. 1.01 mi

Similarity and Scale

One-Minute Section Planner

Lesson	Materials	Resources
Lesson 7-5 Dilations **NCTM:** Geometry, Communication **NAEP:** Geometry 2c ☐ SAT-9 ☐ SAT-10 ☐ ITBS ☐ CTBS ☑ MAT ☑ CAT	**Optional** Coordinate plane (CRB, p. 99) Teaching Transparency T21 (CRB)	• *Chapter 7 Resource Book,* pp. 44–53 • *Daily Transparency T20, CRB* • *Additional Examples Transparencies* T22–T24, CRB • *Alternate Openers: Explorations,* p. 59
Hands-On Lab 7B Explore Similarity **NCTM:** Algebra, Geometry, Measurement, Reasoning and Proof, Connections, Representation **NAEP:** Geometry 2e ☑ SAT-9 ☑ SAT-10 ☑ ITBS ☑ CTBS ☑ MAT ☑ CAT	**Required** Two sizes of graph paper Number cubes (MK) Rulers (MK, CRB p. 100) Protractors (MK)	• *Hands-On Lab Activities,* pp. 54–56, 95
Lesson 7-6 Similar Figures **NCTM:** Geometry, Measurement, Reasoning and Proof, Communication, Connections, Representation **NAEP:** Geometry 2f ☑ SAT-9 ☑ SAT-10 ☑ ITBS ☑ CTBS ☑ MAT ☑ CAT	**Optional** Teaching Transparency T26 (CRB)	• *Chapter 7 Resource Book,* pp. 54–63 • *Daily Transparency T25, CRB* • *Additional Examples Transparencies* T27–T29, CRB • *Alternate Openers: Explorations,* p. 60
Lesson 7-7 Scale Drawings **NCTM:** Geometry, Measurement, Reasoning and Proof, Communication, Connections, Representation **NAEP:** Geometry 2f ☐ SAT-9 ☐ SAT-10 ☐ ITBS ☐ CTBS ☑ MAT ☑ CAT	**Required** Rulers (MK, CRB p. 100) **Optional** Tape measures (MK) Graph paper (CRB p. 101)	• *Chapter 7 Resource Book,* pp. 64–72 • *Daily Transparency T30, CRB* • *Additional Examples Transparencies* T31–T33, CRB • *Alternate Openers: Explorations,* p. 61
Lesson 7-8 Scale Models **NCTM:** Geometry, Measurement, Communication, Connections, Representation **NAEP:** Number Properties 4c ☐ SAT-9 ☐ SAT-10 ☐ ITBS ☐ CTBS ☑ MAT ☑ CAT	**Optional** Scale drawings Scale models	• *Chapter 7 Resource Book,* pp. 73–81 • *Daily Transparency T34, CRB* • *Additional Examples Transparencies* T35–T37, CRB • *Alternate Openers: Explorations,* p. 62
Hands-On Lab 7C Make a Scale Model **NCTM:** Algebra, Geometry, Measurement, Connections, Representation **NAEP:** Geometry 1f ☐ SAT-9 ☐ SAT-10 ☐ ITBS ☐ CTBS ☐ MAT ☐ CAT	**Required** Cardstock Scissors Tape Rulers (MK, CRB p. 100)	• *Hands-On Lab Activities,* pp. 57–58, 95
Lesson 7-9 Scaling Three-Dimensional Figures **NCTM:** Algebra, Measurement, Communication, Connections, Representation **NAEP:** Number Properties 4c ☐ SAT-9 ☐ SAT-10 ☐ ITBS ☐ CTBS ☑ MAT ☑ CAT	**Optional** Cube Rectangular prism Triangular prism Teaching Transparency T39 (CRB) Centimeter cubes (MK)	• *Chapter 7 Resource Book,* pp. 82–91 • *Daily Transparency T38, CRB* • *Additional Examples Transparencies* T40–T42, CRB • *Alternate Openers: Explorations,* p. 63
Extension Trigonometric Ratios **NCTM:** Geometry, Reasoning and Proof, Connections **NAEP:** Number Properties 4a ☐ SAT-9 ☐ SAT-10 ☐ ITBS ☐ CTBS ☐ MAT ☐ CAT	**Required** Calculators **Optional** Teaching Transparency T43 (CRB)	• *Additional Examples Transparency* T44, CRB
Section 7B Assessment		• *Section 7B Quiz, AR p. 18* • *Test and Practice Generator CD-ROM*

SAT = *Stanford Achievement Tests*   **ITBS** = *Iowa Test of Basic Skills*   **CTBS** = *Comprehensive Test of Basic Skills/Terra Nova*
MAT = *Metropolitan Achievement Test*   **CAT** = *California Achievement Test*

NCTM = Complete standards can be found on pages T29–T35.   **NAEP** = Complete standards can be found on pages A54–A58.

SE = *Student Edition*   **TE** = *Teacher's Edition*   **AR** = *Assessment Resources*   **CRB** = *Chapter Resource Book*   **MK** = *Manipulatives Kit*

Section Overview

Similar Figures and Scale Drawings   *Hands-On Lab 7B, Lessons 7-5, 7-6, 7-7*

Why? Similar figures, dilations, and scale drawings are used in blueprints for construction, photography, medical research, and many other applications.

> **Similar figures** have congruent angles and proportional sides.

> A **dilation** produces an image similar to the original figure.
> Scale factor > 1 ⟶ **Enlargement**
> Scale factor < 1 ⟶ **Reduction**

Similar Figures

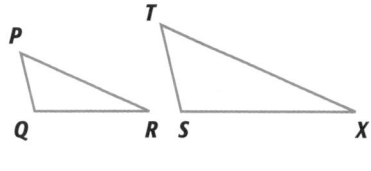

$\angle Q \cong \angle S$

$\angle P \cong \angle T$

$\angle R \cong \angle X$

$$\frac{PQ}{TS} = \frac{RP}{XT} = \frac{QR}{SX}$$

Dilations

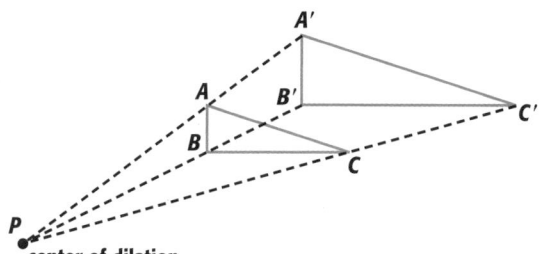

center of dilation

Scale factor = 1.5

$$\frac{PA'}{PA} = \frac{PB'}{PB} = \frac{PC'}{PC} = 1.5$$

$$\frac{A'B'}{AB} = \frac{A'C'}{AC} = \frac{B'C'}{BC} = 1.5$$

Scale Drawings

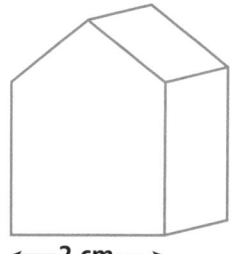

←2 cm→

Scale: 1 cm:5 m

$$\frac{\text{model}}{\text{building}} \quad \frac{1 \text{ cm}}{5 \text{ m}} = \frac{2 \text{ cm}}{x \text{ m}}$$

Width of building = 10 m

Scale Models   *Hands-On Lab 7C, Lessons 7-8, 7-9*

Why? A scale model is an accurate and similar three-dimensional representation of an object.

The building is shown with its dimensions. What should be the height of the model with a scale of 2 in. = 5 ft?

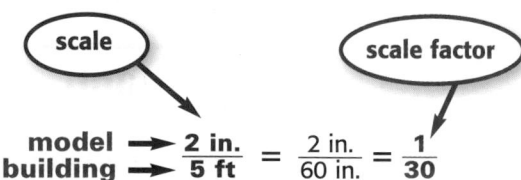

scale ⟶

scale factor ⟶

$$\begin{array}{l}\text{model} \rightarrow \\ \text{building} \rightarrow\end{array} \frac{2 \text{ in.}}{5 \text{ ft}} = \frac{2 \text{ in.}}{60 \text{ in.}} = \frac{1}{30}$$

The height of the building is 40 ft, or 480 inches. To find the height of the model, solve a proportion.

$$\frac{1}{30} = \frac{h \text{ in.}}{480 \text{ in.}} \quad \longleftarrow \text{ height of the model}$$
$$\longleftarrow \text{ height of the building}$$

$h = 16$ in.

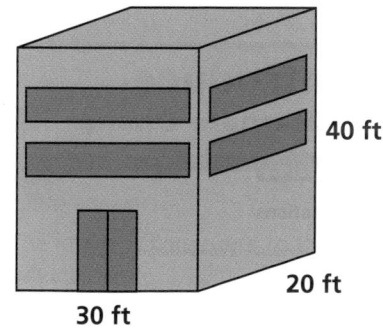

40 ft

20 ft

30 ft

Scale: 2 in. = 5 ft

Pacing: Traditional 1 day
Block $\frac{1}{2}$ day

Objective: To explore similarity using a number cube and graph paper

Materials: Two pieces of graph paper with different-sized boxes (such as 1 cm graph paper and $\frac{1}{4}$ in. graph paper), number cube, metric ruler, protractor

Lab Resources

Hands-On Lab Activities . pp. 54–56, 95

Using the Pages

Let students become familiar with translating points horizontally and vertically. Have students begin by plotting the point $(1, -1)$.

1. What are the coordinates of the point 5 units down and 3 units right from $(1, -1)$? **(4, −6)**

2. What are the coordinates of the point 2 units up and 4 units left from $(1, -1)$? **(−3, 1)**

3. What are the coordinates of the point 1 unit up and 6 units right from $(1, -1)$? **(7, 0)**

Explore Similarity

Use with Lesson 7-6

Triangles that have the same shape have some interesting relationships.

Activity

❶ Follow the steps below to draw two triangles.

a. On a sheet of graph paper, plot a point below and to the left of the center of the paper. Label the point *A*. On the other sheet of paper, plot a point below and to the left of the center and label this point *D*.

b. Roll a number cube twice. On each sheet of graph paper, move up the number on the first roll, move right the number on the second roll, and plot this location as point *B* on the first sheet and point *E* on the second sheet.

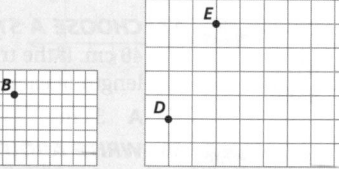

c. Roll the number cube twice again. On each sheet of graph paper, move down the number on the first roll, move right the number on the second roll, and plot point *C* on the first sheet and point *F* on the second sheet.

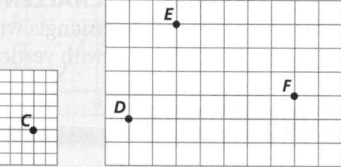

d. Connect the three points on each sheet of graph paper to form triangles *ABC* and *DEF*.

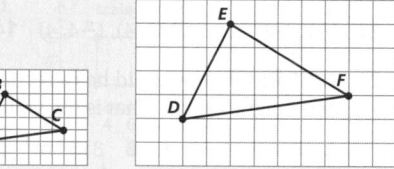

e. Measure the angles of each triangle. Measure the side lengths of each triangle to the nearest millimeter. Find the following:

m∠A	m∠D	m∠B	m∠E	m∠C	m∠F
AB	DE	$\frac{AB}{DE}$	BC	EF	$\frac{BC}{EF}$
AC	DF	$\frac{AC}{DF}$			

❷ Follow the steps below to draw two triangles.

a. On one sheet of graph paper, plot a point below and to the left of the center of the paper. Label the point A.

b. Roll a number cube twice. Move up the number on the first roll, move right the number on the second roll, and plot this location as point B. From B, move up the number on the first roll, move right the number on the second roll, and label this point D.

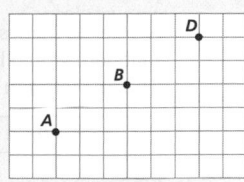

c. Roll a number cube twice. From B, move down the number on the first roll, move right the number on the second roll, and plot this location as point C.

d. From D, move down twice the number on the first roll, move right twice the number on the second roll, and label this point E.

e. Connect points to form triangles ABC and ADE.

f. Measure the angles of each triangle. Measure the side lengths of each triangle to the nearest millimeter.

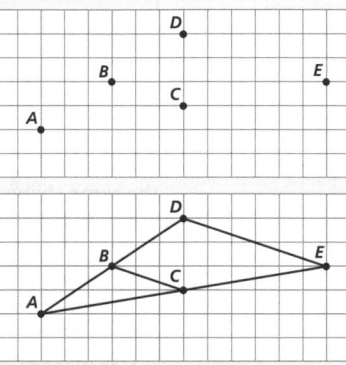

Think and Discuss

1. How do corresponding angles of triangles with the same shape compare?

2. How do corresponding side lengths of triangles with the same shape compare?

3. Suppose you enlarge a triangle on a copier machine. What measurements or values would be the same on the enlargement?

Try This

1. Make a small trapezoid on graph paper and triple the length of each side. Compare the angle measures and side lengths of the trapezoids.

2. Make a large polygon on graph paper. Use a copier to reduce the size of the polygon. Compare the angle measures and side lengths of the polygons.

Answers

Think and Discuss

1. They are equal.

2. They are proportional.

3. The angles will be the same.

Try This

1. Check students' work. The angles should be the same, and side lengths should be proportional.

2. Check students' work. The angles should be the same, and side lengths should be proportional.

Objective: Students make comparisons between and find dimensions of scale drawings and actual objects.

Warm Up

Evaluate the following for $x = 16$.

1. $3x$ 48   **2.** $\frac{3}{4}x$ 12

Evaluate the following for $x = \frac{2}{5}$.

3. $10x$ 4   **4.** $\frac{1}{4}x$ $\frac{1}{10}$

Problem of the Day

An isosceles triangle with a base length of 6 cm and side lengths of 5 cm is dilated by a scale factor of 3. What is the area of the image?
108 cm²

Available on Daily Transparency in CRB

Math Humor

Teacher: Why did you turn in a picture of a snake?

Student: You said to make a *scale drawing.*

7-7 Scale Drawings

Learn to make comparisons between and find dimensions of scale drawings and actual objects.

Vocabulary

scale drawing

scale

reduction

enlargement

Stan Herd is a crop artist and farmer who has created works of art that are as large as 160 square acres. Herd first makes a *scale drawing* of each piece, and then he determines the actual lengths of the parts that make up the art piece.

A **scale drawing** is a two-dimensional drawing that accurately represents an object. The scale drawing is mathematically similar to the object.

To get an idea of scale, notice the red tractor at the lower right.

A **scale** gives the ratio of the dimensions in the drawing to the dimensions of the object. All dimensions are reduced or enlarged using the same scale. Scales can use the same units or different units.

Reading Math

The scale *a:b* is read "a to b." For example, the scale 1 cm:3 ft is read "one centimeter to three feet."

Scale	Interpretation
1:20	1 unit on the drawing is 20 units.
1 cm:1 m	1 cm on the drawing is 1 m.
$\frac{1}{4}$ in. = 1 ft	$\frac{1}{4}$ in. on the drawing is 1 ft.

EXAMPLE 1   **Using Proportions to Find Unknown Scales or Lengths**

A The length of an object on a scale drawing is 5 cm, and its actual length is 15 m. The scale is 1 cm:▓ m. What is the scale?

$\dfrac{1 \text{ cm}}{x \text{ m}} = \dfrac{5 \text{ cm}}{15 \text{ m}}$ Set up proportion using $\frac{\text{scale length}}{\text{actual length}}$.

$1 \cdot 15 = x \cdot 5$ Find the cross products.

$x = 3$ Solve the proportion.

The scale is 1 cm:3 m.

B The length of an object on a scale drawing is 3.5 in. The scale is 1 in:12 ft. What is the actual length of the object?

$\dfrac{1 \text{ in.}}{12 \text{ ft}} = \dfrac{3.5 \text{ in.}}{x \text{ ft}}$ Set up proportion using $\frac{\text{scale length}}{\text{actual length}}$.

$1 \cdot x = 3.5 \cdot 12$ Find the cross products.

$x = 42$ Solve the proportion.

The actual length is 42 ft.

---

1 Introduce

Alternate Opener

EXPLORATION

7-7 Scale Drawings

A coffeehouse rents the floor space modeled by the figure below.

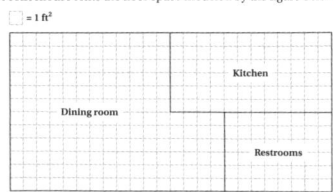

Count the number of squares to answer each question.

1. What is the actual area of the dining room?

2. What is the actual area of the kitchen?

3. What is the actual area of the restrooms?

Think and Discuss

4. Show a possible arrangement of square tables in the dining area if the tabletops each measured 2 ft by 2 ft.

5. Discuss real-world situations in which scale drawings are used.

Motivate

On the chalkboard, sketch a rough scale drawing of the classroom floor. (You will need to measure the room ahead of time and decide on a scale; you may want to use a scale of 1:10 or 1:20.) Ask students how much larger they think the actual floor is than the drawing. Have students measure the drawing and the actual floor and determine the scale.

Exploration worksheet and answers on Chapter 7 Resource Book pp. 65 and 115

2 Teach

Lesson Presentation

Guided Instruction

In this lesson, students learn to make comparisons between and find dimensions of scale drawings and actual objects. Explain that a *scale drawing* represents an actual object that is either larger or smaller than the drawing. Define *scale*, and show that a scale may contain different units or no units at all. In Example 1A, use a proportion to find a missing scale. In Example 1B, use a proportion to find an actual length. Discuss *reductions* and *enlargements*, and point out that Example 2 is about an enlargement. Explain the meaning of $\frac{1}{4}$ in. scale ($\frac{1}{4}$ in. = 1 ft) and $\frac{1}{2}$ in. scale ($\frac{1}{2}$ in. = 1 ft) before discussing Example 3.

A scale drawing that is smaller than the actual object is called a **reduction**. A scale drawing can also be larger than the object. In this case, the drawing is referred to as an **enlargement**.

Example 1

A. The length of an object on a scale drawing is 2 cm, and its actual length is 8 m. The scale is 1 cm: ■ m. What is the scale?

1 cm:4 m

B. The length of an object on a scale drawing is 1.5 inches. The scale is 1 in:6 ft. What is the actual length of the object? **9 ft**

Example 2

Under a 1000:1 microscope view, an amoeba appears to have a length of 8 mm. What is its actual length?

0.008 mm

Example 3

A. If a wall in a $\frac{1}{4}$ in. scale drawing is 4 in. tall, how tall is the actual wall? **16 ft**

B. How tall is the wall if a $\frac{1}{2}$ in. scale is used? **8 ft**

EXAMPLE 2 **Life Science Application**

Under a 1000:1 microscope view, a paramecium appears to have length 39 mm. What is its actual length?

$$\frac{1000}{1} = \frac{39 \text{ mm}}{x \text{ mm}} \begin{matrix}\leftarrow \text{ scale length} \\ \leftarrow \text{ actual length}\end{matrix}$$

$1000 \cdot x = 1 \cdot 39$ *Find the cross products.*

$x = 0.039$ *Solve the proportion.*

The actual length of the paramecium is 0.039 mm.

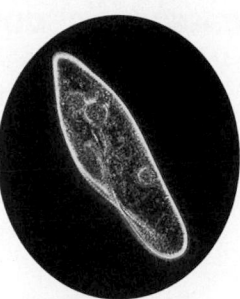
A paramecium is a cylindrical or foot-shaped microorganism.

A drawing that uses the scale $\frac{1}{4}$ in. = 1 ft is said to be in $\frac{1}{4}$ in. scale. Similarly, a drawing that uses the scale $\frac{1}{2}$ in. = 1 ft is in $\frac{1}{2}$ in. scale.

EXAMPLE 3 **Using Scales and Scale Drawings to Find Heights**

A If a wall in a $\frac{1}{4}$ in. scale drawing is 3 in. tall, how tall is the actual wall?

$$\frac{0.25 \text{ in.}}{1 \text{ ft}} = \frac{3 \text{ in.}}{x \text{ ft}} \begin{matrix}\leftarrow \text{ scale length} \\ \leftarrow \text{ actual length}\end{matrix}$$ *Length ratios are equal.*

$0.25 \cdot x = 1 \cdot 3$ *Find the cross products.*

$x = 12$ *Solve the proportion.*

The wall is 12 ft tall.

B How tall is the wall if a $\frac{1}{2}$ in. scale is used?

$$\frac{0.5 \text{ in.}}{1 \text{ ft}} = \frac{3 \text{ in.}}{x \text{ ft}} \begin{matrix}\leftarrow \text{ scale length} \\ \leftarrow \text{ actual length}\end{matrix}$$ *Length ratios are equal.*

$0.5 \cdot x = 1 \cdot 3$ *Cross multiply.*

$x = 6$ *Solve the proportion.*

The wall is 6 ft tall.

Think and Discuss

1. **Describe** which scale would produce the largest drawing of an object: 1:20, 1 in. = 1 ft, or $\frac{1}{4}$ in. = 1 ft.

2. **Describe** which scale would produce the smallest drawing of an object: 1:10, 1 cm = 10 cm, or 1 mm:1 m.

3 Close

Reaching All Learners
Through Home Connection

Have students use tape measures (Manipulatives Kit) and graph paper (Chapter 7 Resource Book p. 101) to make a scale drawing of an athletic field (or court) or one room or floor of their home. Students will need to measure important lines, walls, or objects, and decide on an appropriate scale. For example, on a basketball court they should include free-throw lines and the center court line. For a room, they might include a bed and other large furniture. Ask students to include the scale with the scale drawing. Discuss the importance of scale drawings with the students.

Summarize

Draw the sketch below so that the arrow measures 22 in. and each side of the square measures 6 in.

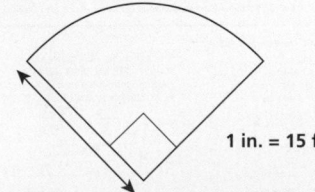
1 in. = 15 ft

Show students the sketch and tell them it is a scale drawing of a baseball field with a scale of 1 in. = 15 ft. Ask students to find the actual distance indicated by the arrow and the distance from home plate to first base. **330 ft; 90 ft**

Answers to Think and Discuss

1. The scale 1 in. = 1 ft would produce the largest drawing.

2. The scale 1 mm:1 m would produce the smallest drawing.

7-7 Exercises

FOR EXTRA PRACTICE
see page 745

internet connect
Homework Help Online
go.hrw.com   Keyword: MP4 7-7

Students may want to refer back to the lesson examples.

Assignment Guide

If you finished Example **1** assign:
Core 1, 2, 7, 8, 13–19 odd,
    30–34
Enriched 7, 8, 13–20, 30–34

If you finished Example **2** assign:
Core 1–4, 7–10, 13–21 odd,
    30–34
Enriched 7–10, 13–21, 30–34

If you finished Example **3** assign:
Core 1–12, 13–21 odd, 22–25,
    30–34
Enriched 7–34

You may use the rulers provided in the Manipulatives Kit or on Chapter 7 Resource Book p. 100 for Exercises 21–29.

Notes

GUIDED PRACTICE

See Example **1** 1. A 10 ft fence is 8 in. long on a scale drawing. What is the scale? **1 in:1.25 ft**

2. Using a scale of 2 cm:9 m, how long is an object that is 4.5 cm long in a drawing? **20.25 m**

See Example **2** 3. Under a 100:1 microscope view, a microorganism appears to have a length of 0.85 in. How long is the microorganism? **0.0085 in.**

4. Using the microscope from Exercise 3, how long would a 0.075 mm microorganism appear to be under the microscope? **7.5 mm**

See Example **3** 5. On a $\frac{1}{4}$ in. scale, a tree is 13 in. tall. How tall is the actual tree? **52 ft**

6. How high is a 54 ft bridge on a $\frac{1}{2}$ in. scale drawing? **27 in.**

INDEPENDENT PRACTICE

See Example **1** 7. What is the scale of a drawing where a 6 m wall is 4 cm long? **1 cm = 1.5 m**

8. If a scale of 2 in:10 ft is used, how long is an object that is 14 in. long in a drawing? **70 ft**

See Example **2** 9. Using a 1000:1 magnification microscope, a paramecium has length 23 mm. What is the actual length of the paramecium? **0.023 mm**

10. If a 0.27 cm long crystal appears to be 13.5 cm long under a microscope, what is the power of the microscope? **50:1**

See Example **3** 11. Using a $\frac{1}{2}$ in. scale, how tall would a 40 ft statue be in a drawing? **20 in.**

12. How wide is a 3 ft doorway in a $\frac{1}{4}$ in. scale drawing? **0.75 in.**

PRACTICE AND PROBLEM SOLVING

The scale of a map is 1 in. = 15 mi. Find each length on the map.
13. 30 mi **2 in.** 14. 45 mi **3 in.** 15. 7.5 mi **0.5 in.** 16. 153.75 mi **10.25 in.**

The scale of a drawing is 3 in. = 27 ft. Find each actual measurement.
17. 2 in. **18 ft** 18. 5 in. **45 ft** 19. 6.5 in. **58.5 ft** 20. 11.25 in. **101.25 ft**

21. Use the scale of the map and a ruler to find the distance in miles between Two Egg, Florida, and Gnaw Bone, Indiana. **about 550 mi**

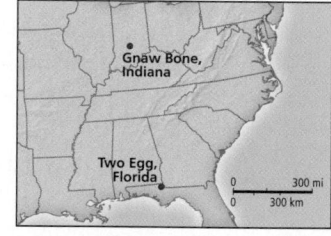

Math Background

One of the customary ways to write a scale is with an = symbol. Explain to students that the = symbol in this context has exactly the same meaning as a colon. (So the scale $\frac{1}{4}$ in. = 1 ft could be written as $\frac{1}{4}$ in:1 ft). The use of the = symbol in this way is unfortunate from a strictly mathematical perspective, but it is traditional. It is analogous to the use of the word *yard* in construction or landscaping instead of the proper term *cubic yard*. A quantity such as 2 cubic yards of concrete or mulch is often called "2 yards."

RETEACH 7-7

LESSON
7-1 Reteach
Scale Drawings

In a **scale drawing**, all the dimensions of the actual object are reduced or enlarged proportionally.

A map is a scale drawing in which actual distance is reduced.

The towns of Ardon and Bacton are on a map with scale 1 cm = 15 km.

If the map distance between Ardon and Bacton is 4.5 cm, what is the actual distance?

$\frac{\text{actual distance}}{\text{map distance}} = \frac{\text{actual distance}}{\text{map distance}}$

$\frac{15 \text{ km}}{1 \text{ cm}} = \frac{x \text{ km}}{4.5 \text{ cm}}$

$1(x) = 15(4.5)$

$x = 67.5 \text{ km}$ ← actual distance between Ardon and Bacton

Complete to find each unknown measure.

1. A map scale is 1 in. = 75 mi. The map distance between two towns is 3.5 in. Find the actual distance x between the towns.

$\frac{\text{actual distance}}{\text{map distance}} = \frac{\text{actual distance}}{\text{map distance}}$

$\frac{75 \text{ mi}}{1 \text{ in.}} = \frac{x \text{ mi}}{3.5 \text{ in.}}$

$x = \underline{262.5}$

actual distance = $\underline{262.5 \text{ mi}}$

2. The actual distance between two towns is 175 km. If the distance between them on a map is 7 cm, what is the map scale?

$\frac{x \text{ km}}{1 \text{ cm}} = \frac{175 \text{ km}}{7 \text{ cm}}$

$7x = 175$

$x = \underline{25}$

map scale: 1 cm = $\underline{25 \text{ km}}$

3. An archway in a $\frac{1}{2}$ in. scale drawing is 4.5 in. tall. Find the actual height x.

$\frac{\text{actual height}}{\text{scale height}} = \frac{\text{actual height}}{\text{scale height}}$

$\frac{1 \text{ ft}}{0.5 \text{ in.}} = \frac{x \text{ ft}}{4.5 \text{ in.}}$

$0.5x = 4.5$

$x = 9$

actual height = $\underline{9 \text{ ft}}$

4. Under a 7:1 magnification, this letter F appears to be 84 points high. Find the actual height x.

$\frac{\text{actual height}}{\text{scale height}} = \frac{\text{actual height}}{\text{scale height}}$

$\frac{1}{7} = \frac{x \text{ points}}{84 \text{ points}}$

$7x = 84$

$x = 12$

actual height = $\underline{12 \text{ points}}$

PRACTICE 7-7

LESSON
7-1 Practice B
Scale Drawings

The scale of a drawing is $\frac{1}{4}$ in. = 15 ft. Find the actual measurement.

1. 9 in. **540 ft** 2. 12 in. **720 ft** 3. 14 in. **840 ft** 4. 15 in. **900 ft**

5. 18 in. **1080 ft** 6. 20 in. **1200 ft** 7. 16.5 in. **990 ft** 8. 10.8 in. **648 ft**

The scale is 2 cm = 25 m. Find the length each measurement would be on a scale drawing.

9. 150 m **12 cm** 10. 475 m **38 cm** 11. 350 m **28 cm** 12. 500 m **40 cm**

13. 625 m **50 cm** 14. 262.5 m **21 cm** 15. 387.5 m **31 cm** 16. 437.5 m **35 cm**

17. On a map the distance between Atlanta, Georgia, and Nashville, Tennessee, is 12.5 in. The scale is 1 in. = 20 mi. What is the actual distance between these two cities?

250 mi

18. Blueprints of a house are drawn to the scale of $\frac{1}{4}$ in. = 1 ft. A kitchen measures 3.5 in. by 5 in. on the blueprints. What is the actual size of the kitchen?

14 ft × 20 ft

19. A scale drawing has a scale of 1 in. = 10 ft. How long is a line on the drawing that represents an actual length of 47.5 ft?

4.75 in.

20. A scale drawing of a square with area 64 m² is made using a scale of 1 cm = 4 m. Find the dimensions of the scale drawing.

2 cm × 2 cm

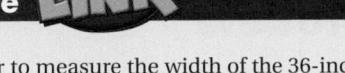
Use a metric ruler to measure the width of the 36-inch-wide door on the blueprint of the family room below.

The scale is about 1.2 cm:36 in.

For Exercises 22–28, indicate the scale that you used.

22. How wide are the pocket doors (shown by the red line)?
≈ **63 in.**

23. What is the distance *s* between two interior studs?
≈ **18 in.**

24. How long is the oak mantle? (The right side ends just above the *B* in the word *BRICK*.) ≈ **81 in.**

25. Could a 4 ft wide bookcase fit along the right-hand wall without blocking the pocket doors? Explain.

26. What is the area of the tiled hearth in in²? in ft²?
≈ **945 in²; ≈ 6.6 ft²**

27. What is the area of the entire family room in ft²? ≈ **298 ft²**

28. Blueprint paper has a maximum width of 36 in., or about 91.4 cm. What does this width represent in the real world corresponding to the scale that you used?
≈ **229 ft**

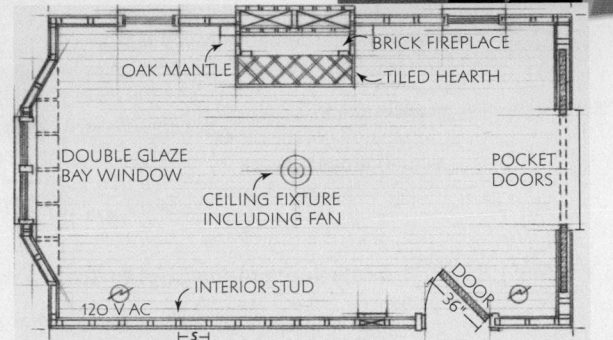

BRICK FIREPLACE
OAK MANTLE — TILED HEARTH
DOUBLE GLAZE BAY WINDOW
CEILING FIXTURE INCLUDING FAN
POCKET DOORS
INTERIOR STUD
120 V AC
DOOR 36"
⊢S⊣

29. **CHALLENGE** Suppose the architect used a $\frac{1}{8}$ in. = 1 ft scale.

   a. What would the dimensions of the family room be? **28 ft by 16 ft**

   b. Use the result from part **a** to find the area of the family room. **448 ft²**

   c. If the carpet the Andersons want costs $4.99 per square foot, how much would it cost to carpet the family room? **$2235.52**

go.hrw.com
KEYWORD: MP4 Scale
CNN Student News.

Architecture

Exercises 22–29 involve using a scale drawing, or an architectural blueprint, to find actual dimensions.

Answers

25. No; the bookcase is 48 in. wide, but each wall next to the doors is only about 45 in. wide.

Journal

Explain to students that one kind of common scale drawing is a highway map. Ask students to write about how the scale on a map could be helpful to them when they are taking a trip.

Test Prep Doctor

For Exercise 34, suggest to students that they use a proportion to solve the problem. Because the growth rate is constant, either of the given ratios can be used.

$$\frac{3.5 \text{ ft}}{2 \text{ yr}} = \frac{x \text{ ft}}{3 \text{ yr}} \quad \text{or} \quad \frac{8.75 \text{ ft}}{5 \text{ yr}} = \frac{x \text{ ft}}{3 \text{ yr}}$$
$$x = 5.25 \text{ ft} \qquad x = 5.25 \text{ ft}$$

The tree is 5.25 ft tall after 3 years. The correct choice is **B**.

Spiral Review

State whether the ratios in each pair are in proportion. (Lesson 7-1)

30. $\frac{3}{7}$ and $\frac{6}{14}$ **yes** 31. $\frac{5}{8}$ and $\frac{10}{4}$ **no** 32. $\frac{13}{4}$ and $\frac{52}{16}$ **yes** 33. $\frac{22}{7}$ and $\frac{11}{3}$ **no**

34. **TEST PREP** A tree was 3.5 ft tall after 2 years and 8.75 ft tall after 5 years. If the tree grew at a constant rate, how tall was it after 3 years? (Lesson 7-4) **B**

   A 5 ft   **B** 5.25 ft   **C** 6.5 ft   **D** 5.75 ft

CHALLENGE 7-7

Challenge
7-7 Too Small, Too Big

A scientist invented a machine that shrinks an object to half its original size. The pictures below show Spot before he was shrunk and after he was shrunk 1, 2, and 3 times.

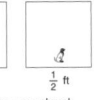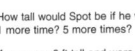

2 ft   1 ft $\frac{1}{2}$ ft $\frac{1}{4}$ ft

1. How tall would Spot be if he were shrunk 1 more time? 5 more times?
$\frac{1}{8}$ ft; $\frac{1}{128}$ ft

2. If you were 6 ft tall and were shrunk 10 times, how tall would you be?
$\frac{3}{512}$ ft

3. Write an expression for Spot's size after he has been shrunk *n* times.
$2 \times \left(\frac{1}{2}\right)^n$ ft

The scientist then invented a machine to triple the size of an object. The pictures below show Spot before his size was tripled and after being tripled 1, 2, and 3 times.

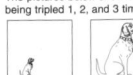

2 ft   6 ft   18 ft   54 ft

4. How tall would Spot be if he were tripled 1 more time? 5 more times?
162 ft; 13,122 ft

5. If you were 6 ft tall and the machine tripled your size 10 times, how tall would you be?
354,294 ft

6. Write an expression for Spot's size after he has been tripled *n* times.
2×3^n ft

7. Could Spot be tripled in size in one machine and then returned to normal size in the shrinking machine? Explain. **Possible answer:**

2 and 3 are relatively prime so there are no positive integers *n* and *m* such that $\frac{3^m}{2^n} = 1$.

PROBLEM SOLVING 7-7

Problem Solving
7-7 Scale Drawings

1. Make a scale drawing of the deck pictured below. Use a $\frac{1}{4}$ in. scale. Each block on the grid is $\frac{1}{4}$ in. by $\frac{1}{4}$ in.

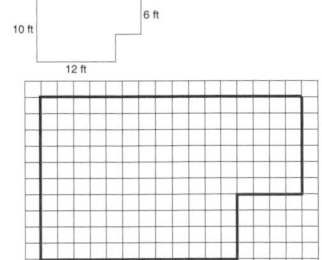
16 ft
6 ft
10 ft
12 ft

Choose the letter for the best answer.

2. On a scale drawing, a 14 ft room is pictured as 3.5 inches. What is the scale of the drawing?
Ⓐ $\frac{1}{4}$ in.   C $\frac{1}{4}$:14
B $\frac{1}{2}$ in.   D 1:56

3. On a scale drawing, $\frac{1}{2}$ inch = 1 foot. A room is pictured as 7.5 inches by 6 inches. How many square yards of carpet are needed for the room?
F 5 yd²   H 45 yd²
Ⓖ 20 yd²   J 90 yd²

4. On a scale drawing of a computer component, $\frac{1}{4}$ in. = 4 in. On the drawing, a piece is $\frac{3}{8}$ in. long. How long is the actual piece?
A 1.5 in.   Ⓒ 6 in.
B 3 in.   D 7.5 in.

5. A scale drawing has a $\frac{1}{4}$ inch scale. The width of a 12 foot room is going to be increased by 4 feet. How much wider will the room be on the drawing?
F $\frac{1}{4}$ in.   Ⓗ 1 in.
G $\frac{1}{2}$ in.   J 4 in.

Lesson Quiz

1. What is the scale of a drawing in which a 9 ft wall is 6 cm long? **1 cm = 1.5 ft**

2. Using a $\frac{1}{4}$ in. = 1 ft scale, how long would a drawing of a 22 ft car be? **5.5 in.**

3. The height of a person on a scale drawing is 4.5 in. The scale is 1:16. What is the actual height of the person? **72 in.**

The scale of a map is 1 in. = 21 mi. Find each length on the map.

4. 147 mi   **7 in.**

5. 5.25 mi   **0.25 in.**

Available on Daily Transparency in CRB

Pacing: Traditional 1 day
Block $\frac{1}{2}$ day

Objective: Students make comparisons between and find dimensions of scale models and actual objects.

Warm Up

The scale of a drawing is 4 in. = 12 ft. Find each actual measurement.

1. 6 in. **18 ft**   **2.** 2.5 in. **7.5 ft**

The scale of a map is 1 in. = 3.5 mi. Find each length on the map.

3. 21 mi **6 in.**   **4.** 1.75 mi **0.5 in.**

Problem of the Day

There is a 2.5-mile racetrack at the Indianapolis Motor Speedway. If a drawing of the racetrack has a scale of 1:1760, what is the length of the track on the drawing? **7.5 ft**

Available on Daily Transparency in CRB

For the movie *Titanic*, the filmmakers built a 0.9 scale model of the actual ship. It was 775 feet long and weighed about 2 million pounds.

Scale Models

Learn to make comparisons between and find dimensions of scale models and actual objects.

Vocabulary
scale model

Mammoths weighing 4 to 6 tons roamed the earth from 3.75 million to 4000 years ago.

Very large and very small objects are often modeled. A **scale model** is a three-dimensional model that accurately represents a solid object. The scale model is mathematically similar to the solid object.

A scale gives the ratio of the dimensions of the model to the actual dimensions.

One species of mammoth survived until the time of the Egyptian pharaohs.

EXAMPLE 1 **Analyzing and Classifying Scale Factors**

Tell whether each scale reduces, enlarges, or preserves the size of the actual object.

A 1 yd:1 ft

$$\frac{1 \text{ yd}}{1 \text{ ft}} = \frac{3 \text{ ft}}{1 \text{ ft}} = 3 \qquad \textit{Convert: 1 yd = 3 ft. Simplify.}$$

The scale enlarges the size of the actual object 3 times.

B 100 cm:1 m

$$\frac{100 \text{ cm}}{1 \text{ m}} = \frac{1\text{m}}{1\text{m}} = 1 \qquad \textit{Convert: 100 cm = 1 m. Simplify.}$$

The scale preserves the size of the object since the scale factor is 1.

EXAMPLE 2 **Finding Scale Factors**

What scale factor relates a 20 in. scale model to an 80 ft apatosaurus?

20 in:80 ft   *State the scale.*

$$\frac{20 \text{ in.}}{80 \text{ ft}} = \frac{1 \text{ in.}}{4 \text{ ft}} = \frac{1 \text{ in.}}{48 \text{ in.}} \qquad \textit{Write the scale as a ratio and simplify.}$$

The scale factor is $\frac{1}{48}$, or 1:48.

---

① Introduce

Alternate Opener

EXPLORATION

7-8 Scale Models

The table shows the approximate measurement of the diameter of a scale model of each planet of our solar system.

Planet	Model Diameter (mm)	Actual Diameter (mi)
Mercury	0.4	
Venus	1.0	
Earth	1.0	8000
Mars	0.5	
Jupiter	11.3	
Saturn	9.4	
Uranus	4.1	
Neptune	3.9	
Pluto	0.2	

1. Use the model Earth's scale diameter of 1.0 mm and Earth's actual diameter of about 8,000 mi to estimate the diameter of each planet.

2. Use the actual diameters to complete each ratio.

a. $\frac{\text{Mercury}}{\text{Earth}} =$ ___ b. $\frac{\text{Venus}}{\text{Earth}} =$ ___ c. $\frac{\text{Mars}}{\text{Earth}} =$ ___

d. $\frac{\text{Jupiter}}{\text{Earth}} =$ ___ e. $\frac{\text{Saturn}}{\text{Earth}} =$ ___ f. $\frac{\text{Uranus}}{\text{Earth}} =$ ___

g. $\frac{\text{Neptune}}{\text{Earth}} =$ ___ h. $\frac{\text{Pluto}}{\text{Earth}} =$ ___

Think and Discuss

3. Explain what each ratio tells you about each pair of planets.

Motivate

Show students a two-dimensional scale drawing and a three-dimensional scale model. Ask students to describe possible benefits of each type of representation.

Possible responses: A scale drawing is more portable; you can fold it or roll it. A scale model can be viewed from different angles.

Exploration worksheet and answers on Chapter 7 Resource Book pp. 74 and 117

---

② Teach

Lesson Presentation

Guided Instruction

In this lesson, students learn to make comparisons between and find dimensions of scale models and actual objects. Explain that a scale model is similar to a scale drawing except that it shows all three dimensions. Show students how to simplify scale factors by writing the ratios with the same units. You may want to review the use of conversion factors (Lesson 7-3).

EXAMPLE **3** **Finding Unknown Dimensions Given Scale Factors**

A model of a 27 ft tall house was made using the scale 2 in:3 ft. What is the height of the model?

$\frac{2\text{ in.}}{3\text{ ft}} = \frac{2\text{ in.}}{36\text{ in.}} = \frac{1\text{ in.}}{18\text{ in.}}$ *First find the scale factor.*

The scale factor for the model is $\frac{1}{18}$. Now set up a proportion.

$\frac{1}{18} = \frac{h\text{ in.}}{324\text{ in.}}$ *Convert: 27 ft = 324 in.*

$324 = 18h$ *Cross multiply.*

$h = 18$ *Solve for the height.*

The height of the model is 18 in.

EXAMPLE **4** *Life Science Application*

A DNA model was built using the scale 2 cm:0.0000001 mm. If the model of the DNA chain is 17 cm long, what is the length of the actual chain? Find the scale factor.

$\frac{2\text{ cm}}{0.0000001\text{ mm}} = \frac{20\text{ mm}}{0.0000001\text{ mm}} = 200{,}000{,}000$

The scale factor for the model is 200,000,000. This means the model is 200 million times larger than the actual chain.

$\frac{200{,}000{,}000}{1} = \frac{17\text{ cm}}{x\text{ cm}}$ *Set up a proportion.*

$200{,}000{,}000x = 17(1)$ *Cross multiply.*

$x = 0.000000085$ *Solve for the length.*

The length of the DNA chain is 8.5×10^{-8} cm.

Think and Discuss

1. **Explain** how you would find the width of the model house in Example 3.

2. **Describe** how you would find the scale factor for a model of the Statue of Liberty. What information would you need to have?

3. **Explain** why comparing models with different scale factors, such as the apatosaurus in Example 2 and the house in Example 3, can be misleading.

3 Close

Reaching All Learners
Through Home Connection

Ask each student to work with an adult or family member to create a scale model of a room in his or her home. Students should measure the room and furniture, choose scale factors, and then design the scale models using materials of their choice (sugar cubes, balsa wood, cardstock, etc.). Each student should submit the scale factor and all calculations, along with the model.

Summarize

Review the process of simplifying scale factors. For each scale below, ask students to write the scale factor. Ask students to also write whether the actual object would be larger or smaller than the scale model.

1. 1 in:4 ft
2. 2 ft:6 in.
3. 2 cm:1 km

1. 1:48; actual object is larger.
2. 4:1; actual object is smaller.
3. 1:50,000; actual object is larger.

Answers to Think and Discuss

1. Convert the width of the house to inches, and then set up a proportion using the scale factor $\frac{1}{18}$.

2. You would need to know at least one dimension of the Statue of Liberty. Then you could measure the corresponding dimension on the model and set up a ratio to determine the scale factor.

3. The different units make the model and the object seem a lot closer in size than they actually are.

7-8 **Exercises**

FOR EXTRA PRACTICE
see page 745

🔲 internet connect
Homework Help Online
go.hrw.com Keyword: MP4 7-8

Assignment Guide

If you finished Example **1** assign:
Core 1–6, 10–15, 32–39
Enriched 1–6, 10–15, 32–39

If you finished Example **2** assign:
Core 1–7, 10–16, 19–25 odd,
32–39
Enriched 10–16, 19–25, 29–39

If you finished Example **3** assign:
Core 1–8, 10–17, 19–27 odd,
32–39
Enriched 10–17, 19–39

If you finished Example **4** assign:
Core 1–18, 19–27, 32–39
Enriched 10–39

Students may want to refer back to the lesson examples.

GUIDED PRACTICE

See Example **1** Tell whether each scale reduces, enlarges, or preserves the size of the actual object.

1. 1 in:18 in. **reduces** **2.** 4 ft:15 in. **enlarges** **3.** 1 m:1000 mm **preserves**

4. 1 cm:10 mm **preserves** **5.** 6 in:100 ft **reduces** **6.** 80 ft:20 in. **enlarges**

See Example **2** **7.** What scale factor relates a 15 in. tall model boat to a 30 ft tall yacht? $\frac{1}{24}$

See Example **3** **8.** A model of a 42 ft tall shopping mall was built using the scale 1 in:3 ft. What is the height of the model? **14 in.**

See Example **4** **9.** A molecular model uses the scale 2.5 cm:0.00001 mm. If the model is 7 cm long, how long is the molecule? **0.000028 mm**

INDEPENDENT PRACTICE

See Example **1** Tell whether each scale reduces, enlarges, or preserves the size of the actual object.

10. 10 ft:24 in. **enlarges** **11.** 1 mi:5280 ft **preserves** **12.** 6 in:100 ft **reduces**

13. 0.25 in:1 ft **reduces** **14.** 50 ft:1 in. **enlarges** **15.** 250 cm:1 km **reduces**

See Example **2** **16.** What scale factor was used to build a 55 ft wide billboard from a 25 in. wide model? $\frac{1}{26.4}$

See Example **3** **17.** A model of a house was built using the scale 5 in:25 ft. If a window in the model is 1.5 in. wide, how wide is the actual window? **7.5 ft**

See Example **4** **18.** To create a model of an artery, a health teacher uses the scale 2.5 cm:0.75 mm. If the diameter of the artery is 2.7 mm, what is the diameter on the model? **9 cm**

PRACTICE AND PROBLEM SOLVING

Change both measurements to the same unit of measure, and find the scale factor.

19. 1 ft model of a 1 in. fossil $\frac{12}{1}$ **20.** 8 cm model of a 24 m rocket $\frac{1}{300}$

21. 2 ft model of a 30 yd sports field $\frac{1}{45}$ **22.** 4 ft model of a 6 yd whale $\frac{1}{4.5}$

23. 40 cm model of a 5 m tree $\frac{1}{12.5}$ **24.** 6 in. model of a 6 ft sofa $\frac{1}{12}$

25. *LIFE SCIENCE* Wally has an 18 in. model of a 42 ft dinosaur, the *Tyrannosaurus rex*. What scale factor does this represent? $\frac{1}{28}$

Math Background

The Museum of Science in Boston, Massachusetts, recently installed a full-scale version of *Tyrannosaurus rex*. The process used to create the full-scale version involved a reduction followed by an enlargement. First, a scale-model sculpture was created using a 1:12 scale. Then the model was sliced from snout to tail, like a banana. Next, images of each slice were projected using a 12:1 scale. The enlarged images were then traced and cut from foam. The foam pieces were glued together to create a dinosaur display 14 feet tall and 39 feet long.

RETEACH 7-8

LESSON **Reteach**
7-8 *Scale Models*

In a **scale model**, all the dimensions of the actual object are reduced or enlarged proportionally.

The **scale** used in a model train, which is a reduction of an actual train, gives the ratio of the dimensions of the model to the actual dimensions.

The average tie height of actual railroad tracks is 7 in. Using the HO scale for model trains, the average tie height is 0.0804 in. What is the HO scale factor for model trains?

$\frac{7}{0.0804} = 87.06$ So, the HO scale factor is 1:87.1.

Find each standard scale factor for model trains.

1. Diameter of actual freight car wheel is 33 in. On O scale, diameter is 0.69 in.

$\frac{33}{0.69} = 47.8$

O scale = 1: **47.8**

2. Average tie width of actual tracks is 9 in. Average tie width on Z scale is 0.041 in.

$\frac{9}{0.041} = 219.5$

Z scale = 1: **219.5**

3. Centers of actual mainline tracks are at 15 in. On G scale, centers are at 8 in.

15 ft = **180** in.

$\frac{180}{8} = 22.5$

G scale = 1: **22.5**

A model railroad track was built using the N scale of 1 in. : 160 in.
The average length of the actual track is 8.5 ft. Find the tie length of the model track.

8.5 ft = 102 in. $\frac{102}{160} = 0.6375 \approx 0.64$ in. ← tie length of model track

Find the unknown measures.

4. Actual coupler height is 34.5 in. Find the coupler height of the model using S scale of 1 in. : 64 in.

$\frac{34.5}{64} = 0.539$

5. Actual branch pipe length is 1.315 in. Find the length using O scale of 1 in : 4 ft

4 ft = **48** in.

$\frac{1.315}{48} = 0.027$

6. The actual diameter of a wheel is 3 ft. Find the diameter of a model wheel using a scale of 1 in. : 32 in.

3 ft = **36** in.

$\frac{36}{32} = 1.125$

PRACTICE 7-8

LESSON **Practice B**
7-8 *Scale Models*

Tell whether each scale reduces, enlarges, or preserves the size of the actual object.

1. 1 m : 25 cm
enlarges
2. 8 in. : 1 ft
reduces
3. 12 in. : 1 ft
preserves

4. 9 ft : 1 yd
enlarges
5. 2.54 cm : 1 in.
preserves
6. 2 yd : 1 mi
reduces

Change both measurements to the same unit of measure, and find the scale factor

7. 4 in. model of a 6 ft cabinet
1:18
8. 12 ft model of a 100 yd playground
1:25

9. 4 cm model of a 2 m door
1:50
10. 20 m model of a 3 km street
1:150

11. A scale model of a house is 1 ft long. The actual house is 50 ft long. In the model, the window is $1\frac{1}{5}$ in. high. How many feet high is the actual window?
5 ft

12. A model of a skyscraper is 1.6 in. long, 2.8 in. wide, and 11.2 in. high. The scale factor is 8 in. : 250 ft. What are the actual dimensions of the skyscraper?
50 ft long, 87.5 ft wide, 350 ft high

13. A 10 in. model of a Chrysler PT Cruiser is made with a scale factor of $\frac{1}{18}$. Estimate the actual length of a Chrysler PT Cruiser.
180 in.

14. A replica of a 30 in. wide stove is made to a scale of 1 in. : 18 in. for a dollhouse. How wide is the stove for the dollhouse?
$1\frac{2}{3}$ **in.**

15. Mr. Knott has a rectangular garden with length 12 ft and a perimeter of 44 ft. A model of the garden is made with a scale of 1 in. : 2 ft. What is the area of the model of the garden?
30 in²

The Gateway Arch in St. Louis, Missouri, consists of 143 triangular sections, each about 12 feet tall. The sections decrease in cross section from 54 feet at the base to 17 feet at the top.

go.hrw.com
KEYWORD:
MP4 Arch

Architecture LINK

26. BUSINESS Engineers designed a theme park by creating a model using the scale 0.5 in:32 ft.

   a. If the dimensions of the model are 41.25 in. by 82.5 in., what are the dimensions of the park? **2640 ft by 5280 ft**

   b. What is the area of the park in square feet? **13,939,200 ft²**

   c. If the builders estimate that it will cost $250 million to build the park, how much will it cost per square foot? **$17.94 per ft²**

27. ARCHITECTURE Maurice is building a 2 ft high model of the Gateway Arch in St. Louis, Missouri. If he is using a 3 in:78.75 ft scale, how high is the actual arch? **630 ft**

28. ENTERTAINMENT At Tobu World Square, a theme park in Japan, there are more than 100 scale models of world-famous landmarks, $\frac{1}{25}$ the size of the originals. Using this scale factor,

   a. how tall in inches would a scale model of Big Ben's 320 ft clock tower be?

   b. how tall would a 5 ft tall person be in the model?

The models in Tobu World Square are often seen in movies and television.

29. WHAT'S THE ERROR? A student is asked to find the scale factor that relates a 10 in. scale model to a 45 ft building. She solves the problem by writing $\frac{10 \text{ in.}}{45 \text{ ft}} = \frac{2}{9} = \frac{1}{4.5}$. What error did the student make? What is the correct scale factor?

30. WRITE ABOUT IT Explain how you can tell whether a scale factor will make an enlarged scale model or a reduced scale model.

31. CHALLENGE A scientist wants to build a model, reduced 11,000,000 times, of the Moon revolving around Earth. Will the scale 48 ft:100,000 mi give the desired reduction? **yes**

Spiral Review

Find the surface area of each sphere. Use 3.14 for π. (Lesson 6-10)

32. radius 5 mm **314 mm²**
33. radius 12.2 ft **≈ 1869.4 ft²**
34. diameter 4 in. **≈ 50.2 in²**
35. diameter 20 cm **1256 cm²**

Find each unit rate. (Lesson 7-2)

36. $90 for 8 hours of work **$11.25 per hour**
37. 5 apples for $0.85 **$0.17 per apple**
38. 24 players on 2 teams **12 players per team**

39. TEST PREP How long would it take to drain a 750-gallon hot tub at a rate of 12.5 gallons per minute? (Lesson 7-3) **A**

   A 1 hour   **B** 45 minutes   **C** 80 minutes   **D** 55 minutes

---

Answers

28. a. 153.6 in.
   b. 2.4 in.

29. Possible answer: The student did not convert 45 ft to 540 in. The correct scale factor is $\frac{10 \text{ in.}}{540 \text{ in.}} = \frac{1}{54}$.

30. Possible answer: Convert both measures into the same unit. If the first is smaller, then it is a reduced model. If the first is larger, then it is an enlarged model. If they are equal, the model preserves the size.

Journal

Explain to students that scale models are useful in many career fields, including biology, chemistry, architecture, and landscaping. Have students write about how scale models are useful in one of these fields.

Test Prep Doctor

For Exercise 39, remind students how ratios can be used to cancel out unwanted units. Because the question asks, "how long," the answer should be in units of time. Students should set up ratios to cancel out the gallons and leave minutes remaining.

$$\frac{750 \text{ gal}}{1} \cdot \frac{1 \text{ min}}{12.5 \text{ gal}} = \frac{750 \text{ min}}{12.5} = 60 \text{ min}$$

Because 60 min = 1 hr, choice **A** is correct.

CHALLENGE 7-8

Challenge
7-8 Fast Facts

In the following exercises, you will determine actual measures of some notable structures from around the world.

The Sears Tower in Chicago, Illinois, completed in 1974, is one of the tallest buildings in the world.

  1. The height of a model of the Sears Tower, built with a scale of 1 in. = 50 ft., is 29 in. What is the actual height of the Sears Tower?

  1450 ft

The Eiffel Tower was built in 1889 to commemorate the 100th anniversary of the French Revolution. It was the world's tallest building until 1930, when the Empire State Building was erected in New York City.

  2. The height of a model of the Eiffel Tower including antenna, built with a scale of 1 cm = 35 m, is 16.04 cm. Find the actual height of the Eiffel Tower.

  561.4 m

The Great Pyramid of Khufu in Giza, Egypt, is the largest of the Seven Wonders of the ancient world that is basically still intact. The Greek historian Herodotus thought it took 100,000 men 20 years to build it.

  3. The length of a side of the square base of a model of the Great Pyramid, built with a scale of 1 in. = 25 ft, is 30.22 in.

    a. Find the length of a side of the square base of the actual pyramid. **755.5 ft**

    b. Find the area of the square base of the actual pyramid. **570,780.25 ft²**

PROBLEM SOLVING 7-8

Problem Solving
7-8 Scale Models

Round to the nearest tenth. Write the correct answer.

1. The Statue of Liberty is approximately 305 feet tall. A scale model of the Statue of Liberty is 5 inches tall. The scale of the model is 1 in. : ___ .

  61

2. The right arm of the Statue of Liberty is 42 feet long. How long is the right arm of the Statue of Liberty model given in Exercise 1?

  0.7 inches

3. The Sears Tower is 1454 feet tall. A scale model of the Sears Tower is 6 inches tall. The scale of the model is 1 in. : ___ ft?

  2908

4. The Empire State Building is 1250 feet tall. A model of the Empire State Building that uses the same scale as the Sears Tower in Exercise 3 would be how much shorter than the model of the Sears Tower?

  0.8 inches

5. The diameter of an atom is 10^{-9} cm. If a scale model of an atom has a diameter of 10 cm, the scale of the model is 1 cm : ___ cm?

  10^{-10}

6. The diameter of the nucleus of an atom is 10^{-13} cm. If a scale model of the nucleus of an atom has a diameter of 1 cm, the scale of the model is 1 cm : ___ cm?

  10^{-13}

A toy car has a scale of $\frac{1}{40}$. Choose the letter for the best answer.

7. The diameter of the steering wheel of the actual car is 15 inches. What is the diameter of the toy car's steering wheel?
  (A) $\frac{3}{8}$ in.   **C** $1\frac{1}{2}$ in.
  B $\frac{1}{2}$ in.   **D** $2\frac{2}{3}$ in.

8. The diameter of the toy car's tire is $\frac{5}{8}$ in. What is the diameter of the tire of the actual car?
  F $12\frac{1}{2}$ in.   (H) 25 in.
  G 16 in.   **J** 64 in.

9. The length of the actual car is 15 feet. What is the length of the toy car?
  (A) $\frac{3}{8}$ in.   **C** 12 in.
  B $4\frac{1}{2}$ in.   **D** 50 in.

10. The width of the toy car is $1\frac{1}{2}$ inches. What is the width of the actual car?
  F $3\frac{1}{2}$ in.   **H** 7 ft
  (G) 5 ft   **J** 60 ft

Lesson Quiz

Tell whether each scale reduces, enlarges, or preserves the size of the actual object.

1. 75 ft:40 in. **enlarges**

2. 1 mi:1760 yd **preserves**

3. 400 m:1 km **reduces**

4. What scale factor was used to build a 5 in. model of a 60 ft statue? **1:144**

5. To create a model of the Eustachian tube of the human ear, an audiologist used the scale 1.5 cm = 0.6 mm. If the diameter of the Eustachian tube is 1.8 mm, what is the diameter of the model? **4.5 cm**

Available on Daily Transparency in CRB

7-8 Scale Models   **379**

7C
Make a Scale Model

Pacing: Traditional 1 day
Block $\frac{1}{2}$ day

Objective: To become familiar with the concept of scale by making scale models

Materials: Card stock, scissors, tape, ruler

Lab Resources

Hands-On Lab Activities. . pp. 57–58, 95

Using the Pages

Discuss with students how to choose the best scale, depending upon what size paper the model is being made from.

1. What scale would you use to make a model of a building 850 ft tall with a rectangular base 300 ft by 150 ft out of the following:

   a. 8.5 in. by 11 in. card stock
   Possible answer: 1 in. = 100 ft

   b. 11 in. by 17 in. card stock
   Possible answer: 1 in. = 75 ft

   c. 30 in. by 40 in. poster board
   Possible answer: 1 in. = 28 ft

Make a Scale Model

Use with Lesson 7-9

WHAT YOU NEED	REMEMBER
• Card stock • Scissors • Ruler • Tape	A scale such as 1 in. = 200 ft results in a smaller-scale model than a scale of 1 in. = 20 feet.

internet connect
Lab Resources Online
go.hrw.com
KEYWORD: MP4 Lab7C

You can make a scale model of a solid object, such as a rectangular prism, in many ways; you can make a net and fold it, or you can cut card stock and tape the pieces together. The most important thing is to find a good scale.

Activity 1

The Trump Tower in New York City is a rectangular prism with these approximate dimensions: height, 880 feet; base length, 160 feet; base width, 80 feet.

❶ Make a scale model of the Trump Tower.

First determine the appropriate height for your model and find a good scale.

To use $8\frac{1}{2}$ in. by 11 in. card stock, divide the longest dimension by 11 to find a scale.

$$\frac{880 \text{ ft}}{11 \text{ in.}} = \frac{80 \text{ ft}}{1 \text{ in.}}$$

Let 1 in. = 80 ft.

The dimensions of the model using this scale are

$\frac{880}{80} = 11$ in., $\frac{160}{80} = 2$ in., and $\frac{80}{80} = 1$ in.

So you will need to cut the following:

Two 11 in. × 2 in. rectangles

Two 11 in. × 1 in. rectangles

Two 2 in. × 1 in. rectangles

Tape the pieces together to form the model.

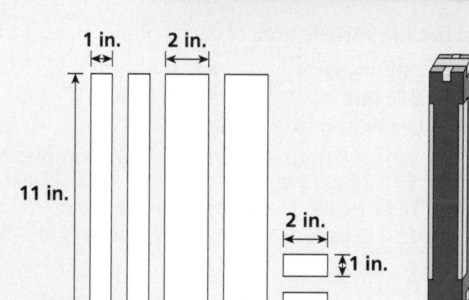

Think and Discuss

1. How tall would a model of a 500 ft tall building be if the same scale were used?

2. Why would a building stand more solidly than your model?

3. What could be another scale of the model if the numbers were without units?

Try This

1. Build a scale model of a four-wall handball court. The court is an open-topped rectangular prism 20 feet wide and 40 feet long. Three of the walls are 20 feet tall, and the back wall is 14 feet tall.

A scale model can also be used to make a model that is larger than the original object.

Activity 2

❶ A size-AA battery has a diameter of about 0.57 inches and a height of about 2 inches. Make a scale model of a AA battery.

You can roll up paper or card stock to create a cylinder. Find the circumference of the battery: $0.57\pi \approx 1.8$ in.

Note that the height is greater than the circumference, so use the height to find a scale.

$$\frac{11 \text{ in.}}{2 \text{ in.}} = 5.5$$

To use $8\frac{1}{2}$ in. by 11 in. paper or card stock, try multiplying the dimensions of the battery by 5.5.

$2(5.5) = 11$ in. $1.8(5.5) = 9.9$ in.

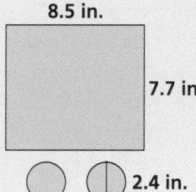

Note that 9.9 in. by 11 in. is larger than an 8.5 in. by 11 in. piece of paper. Divide the width of the paper by the height of the battery to find a smaller scale. $8.5 \div 2 = 4.25$. Use the scale to find the new dimensions: diameter ≈ 2.4 in., circumference ≈ 7.7 in., and height $= 8.5$ in. The pieces for the scale model are shown.

8.5 in.

7.7 in.

2.4 in.

Think and Discuss

1. A salt crystal is one-sixteenth inch long on each side. What would a good scale be for a model of the crystal?

Try This

1. Measure the diameter of the terminal at the top of the battery. Make a scale model of the terminal using the same scale used to make a model of the battery.

Answers

Activity 1

Think and Discuss

1. 6.25 in. tall

2. Possible answer: The building is made of concrete and has a solid base.

3. 1:80

Try This

1. Check students' work. Possible scale is 1 in. = 5 ft. Then you will need the following pieces:

   Three 8 in. × 4 in. rectangles
   One 4 in. × 4 in. square
   One 4 in. × 2.8 in. rectangle

Activity 2

Think and Discuss

1. Possible answer: A good scale for 8.5 x 11 in. card stock would be $4\frac{1}{4}$ in. = $\frac{1}{16}$ in.

Try This

1. Check students' work.

Hands-On Lab 7C   **381**

Warm Up

Find the surface area of each rectangular prism.

1. length 14 cm, width 7 cm, height 7 cm **490 cm²**

2. length 30 in., width 6 in., height 21 in. **1872 in²**

3. length 3 mm, width 6 mm, height 4 mm **108 mm²**

4. length 37 in., width 9 in., height 18 in. **2322 in²**

Problem of the Day

A model of a solid-steel machine tool is built to a scale of 1 cm = 10 cm. The real object will weigh 2500 grams. How much does the model, also made of solid steel, weigh? **2.5 g**

Available on Daily Transparency in CRB

Math Humor

When the student was caught climbing the statues in the city park, she explained that she was just doing her math homework—she was *scaling three-dimensional figures.*

7-9 Scaling Three-Dimensional Figures

Learn to make scale models of solid figures.

Vocabulary
capacity

A popcorn company sells a small box of popcorn that measures 1 ft × 1 ft × 1 ft. They also sell a large box that measures 3 ft × 3 ft × 3 ft. It takes 5 seconds for a machine to fill the smaller box with popcorn. It takes quite a bit longer to fill the larger box.

  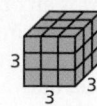

Edge Length	1 ft	2 ft	3 ft
Volume	1 × 1 × 1 = 1 ft³	2 × 2 × 2 = 8 ft³	3 × 3 × 3 = 27 ft³
Surface Area	6 · 1 × 1 = 6 ft²	6 · 2 × 2 = 24 ft²	6 · 3 × 3 = 54 ft²

Helpful Hint

Multiplying the linear dimensions of a solid by n creates n^2 as much surface area and n^3 as much volume.

Corresponding edge lengths of any two cubes are in proportion to each other because the cubes are similar. However, volumes and surface areas do not have the same scale factor as edge lengths.

Each edge of the 2 ft cube is 2 times as long as each edge of the 1 ft cube. However, the cube's volume, or **capacity**, is 8 times as large, and its surface area is 4 times as large as the 1 ft cube's.

EXAMPLE 1 Scaling Models That Are Cubes

A 5 cm cube is built from small cubes, each 1 cm on an edge. Compare the following values.

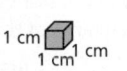 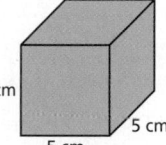

A the edge lengths of the large and small cubes

$$\frac{5 \text{ cm cube}}{1 \text{ cm cube}} \longrightarrow \frac{5 \text{ cm}}{1 \text{ cm}} = 5 \qquad \textit{Ratio of corresponding edges}$$

The edges of the large cube are 5 times as long as those of the small cube.

B the surface areas of the two cubes

$$\frac{5 \text{ cm cube}}{1 \text{ cm cube}} \longrightarrow \frac{150 \text{ cm}^2}{6 \text{ cm}^2} = 25 \qquad \textit{Ratio of corresponding areas}$$

The surface area of the large cube is 25 times that of the small cube.

C the volumes of the two cubes

$$\frac{5 \text{ cm cube}}{1 \text{ cm cube}} \longrightarrow \frac{125 \text{ cm}^3}{1 \text{ cm}^3} = 125 \qquad \textit{Ratio of corresponding volumes}$$

The volume of the large cube is 125 times that of the small cube.

---

1 Introduce

Alternate Opener

EXPLORATION

7-9 Scaling Three-Dimensional Figures

A rectangular box has a volume of 2 cubic units. If one of its dimensions is doubled, the volume of the box is doubled. If all three dimensions are doubled, the volume is 8 times the volume of the original box.

	Length: 2	Length: 2 · 2 = 4	Length: 4	Length: 4
	Width: 1	Width: 1	Width: 1 · 2 = 2	Width: 2
	Height: 1	Height: 1	Height: 1	Height: 1 · 2 = 2
	Volume:	Volume:	Volume:	Volume:
	2 · 1 · 1 = 2	4 · 1 · 1 = 4	4 · 2 · 1 = 8	4 · 2 · 2 = 16

A box has a length of 3 in., a width of 2 in., and a height of 1 in. Multiply the dimensions of the box by each scale factor. Then find the volume.

	Scale Factor	Length (ℓ)	Width (w)	Height (h)	Volume ℓ · w · h
	2	3 · 2 = 6 in.	2 · 2 = 4 in.	1 · 2 = 2 in.	6 · 4 · 2 = 48 in³
1.	3				
2.	4				
3.	5				

Think and Discuss

4. **Discuss** how each scale factor affects the volume of the rectangular box.

5. **Predict** what would happen to the volume if you were to multiply each dimension of the box by a scale factor of 0.5.

Motivate

Show students a cube, a rectangular prism that is not a cube, and a triangular prism. Remind students that a cube is a rectangular prism with six congruent square faces. Remind students that a triangular prism is a prism with two parallel congruent triangular bases. Review the formulas for surface area and volume.

Exploration worksheet and answers on Chapter 7 Resource Book pp. 83 and 119

---

2 Teach

Lesson Presentation

Guided Instruction

In this lesson, students learn to make scale models of solid figures. Review with students how to find the volume and surface area of a 1 ft cube, a 2 ft cube, and a 3 ft cube. Share with students that capacity is the same as volume. In Example 1, compare the ratios of corresponding edges, surface areas, and volumes for a 1 ft cube and a 5 ft cube. Point out that the ratio of surface areas is the *square* of the ratio of corresponding edges, and the ratio of volumes is the *cube* of the ratio of corresponding edges. In Example 2, point out that the length of the paper needed to model the prism is determined by the perimeter of the triangular base.

EXAMPLE 2 · Scaling Models That Are Other Solid Figures

The Fuller Building in New York, also known as the Flatiron Building, can be modeled as a trapezoidal prism with the approximate dimensions shown. For a 10 cm tall model of the Fuller Building, find the following.

A What is the scale factor of the model?

$$\frac{10 \text{ cm}}{93 \text{ m}} = \frac{10 \text{ cm}}{9300 \text{ cm}} = \frac{1}{930}$$ *Convert and simplify.*

The scale factor of the model is 1:930.

B What are the other dimensions of the model?

left side: $\frac{1}{930} \cdot 65 \text{ m} = \frac{6500}{930} \text{ cm} \approx 6.99 \text{ cm}$

back: $\frac{1}{930} \cdot 30 \text{ m} = \frac{3000}{930} \text{ cm} \approx 3.23 \text{ cm}$

right side: $\frac{1}{930} \cdot 60 \text{ m} = \frac{6000}{930} \text{ cm} \approx 6.45 \text{ cm}$

front: $\frac{1}{930} \cdot 2 \text{ m} = \frac{200}{930} \text{ cm} \approx 0.22 \text{ cm}$

The trapezoidal base has side lengths 6.99 cm, 3.23 cm, 6.45 cm, and 0.22 cm.

EXAMPLE 3 · *Business Application*

A machine fills a cubic box that has edge lengths of 1 ft with popcorn in 5 seconds. How long does it take the machine to fill a cubic box that has edge lengths of 3 ft?

$V = 3 \text{ ft} \cdot 3 \text{ ft} \cdot 3 \text{ ft} = 27 \text{ ft}^3$ *Find the volume of the larger box.*

Set up a proportion and solve.

$\frac{5}{1 \text{ ft}^3} = \frac{x}{27 \text{ ft}^3}$ *Cancel units.*

$5 \cdot 27 = x$ *Multiply.*

$135 = x$ *Calculate the fill time.*

It takes 135 seconds to fill the larger box.

Think and Discuss

1. **Describe** how the volume of a model compares to the original object if the linear scale factor of the model is 1:2.

2. **Explain** one possible way to double the surface area of a rectangular prism.

Close

Reaching All Learners
Through Concrete Manipulatives

Have students work in pairs. Provide each pair with a set of 8 to 125 centimeter cubes (provided in the Manipulatives Kit). Ask students to build larger cubes from the centimeter cubes. For each larger cube, students should find the length of the edge, the surface area (by counting the visible square centimeters on the faces of the cube), and the volume (by counting the number of centimeter cubes needed to build the larger cube). Have them record their data and compare with another group's data.

Summarize

Show students models or drawings of a 1 in. cube and a 1 ft cube. Ask students to find the surface area of each figure in square inches and the volume of each figure in cubic inches. Have them also identify the scale factor if the small cube is a model of the larger cube.

surface areas: 6 in² and 864 in²; volumes: 1 in³ and 1728 in³; scale factor: $\frac{1}{12}$

Answers to Think and Discuss

1. The volume of the model is $\frac{1}{8}$ the volume of the original object.

2. Possible answer: Multiply each of the dimensions of the prism by the square root of 2.

Pacing: Traditional 1 day
Block $\frac{1}{2}$ day

Objective: Students find the three basic trigonometric ratios for a right triangle and use them to find missing lengths.

Using the Pages

In Lesson 7-6, students found missing dimensions in similar figures. In this extension, students will find missing lengths in right triangles by using the three basic trigonometric ratios.

EXTENSION # Trigonometric Ratios

Learn to find the three basic trigonometric ratios for a right triangle and to use them to find missing lengths.

Vocabulary
trigonometric ratios
sine
cosine
tangent

Look at the ratios of the side lengths in the two similar right triangles, *ABC* and *DEF.*

The ratios of corresponding sides are equal.

Special ratios called **trigonometric ratios** compare the lengths of the side *opposite* an acute angle in a right triangle, the side *adjacent* (next to) the acute angle, and the length of the hypotenuse. The hypotenuse is never the adjacent side.

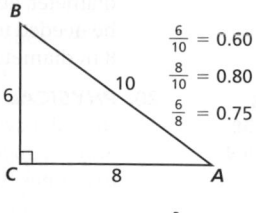

$\frac{6}{10} = 0.60$
$\frac{8}{10} = 0.80$
$\frac{6}{8} = 0.75$

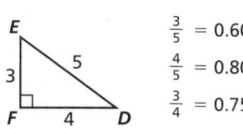

$\frac{3}{5} = 0.60$
$\frac{4}{5} = 0.80$
$\frac{3}{4} = 0.75$

sine of $\angle A$ = $\sin A = \dfrac{\text{length of side opposite } \angle A}{\text{hypotenuse}}$

cosine of $\angle A$ = $\cos A = \dfrac{\text{length of side adjacent to } \angle A}{\text{hypotenuse}}$

tangent of $\angle A$ = $\tan A = \dfrac{\text{length of side opposite } \angle A}{\text{length of side adjacent to } \angle A}$

Trigonometric ratios are constant for a given angle measure.

EXAMPLE **1** **Finding the Value of a Trigonometric Ratio**

Find the cosine of 50°.

In triangle *ABC*: $\cos A = \frac{AC}{AB} = \frac{54}{84} \approx 0.64$

On a calculator: [cos] 50 = 0.64278761

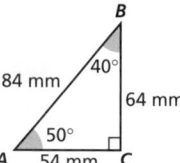

EXAMPLE **2** **Using Trigonometric Ratios to Find Missing Lengths**

Find the height of the Washington Monument to the nearest foot.

$\tan 70° = \frac{x}{202}$ *Write the tangent ratio for a 70° angle.*

$2.75 \approx \frac{x}{202}$ *Use a calculator to find the value of tan 70°.*

$x \approx 2.75(202) \approx 555.5$ *Solve the equation.*

The height of the Washington Monument is about 556 ft.

---

1 Introduce

Motivate

Ask if any students know how mapmakers find distances and mountain heights. Share with all students that surveying is the method of finding distances and angles that cannot be measured directly. Point out that surveying techniques are based on trigonometry, the study of right triangle relationships.

2 Teach

Lesson Presentation

Guided Instruction

In this lesson, students learn to find the three basic trigonometric ratios for a right triangle and to use them to find missing lengths. Review the fact that every right triangle has two legs and one hypotenuse. Point out that for each acute angle, there is an opposite leg and an adjacent leg. Define the three basic trigonometric ratios that apply to any acute angle of any right triangle (Teaching Transparency T43, CRB). Emphasize that these ratios depend only on the angles and not on the size of the triangles. The students will need calculators for the exercises.

Find the value of each trigonometric ratio to the nearest thousandth.

1. sin 51° **0.777**   2. tan 72° **3.078**   3. cos 89° **0.017**

Find each indicated height to the nearest foot.

4.
282 ft
65°
605 ft

5.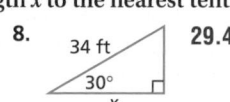
x ft
75°
12 ft
45 ft

Use trigonometric ratios to find each unknown length x to the nearest tenth.

6. 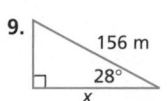 **25.0 cm**
45°
x
25 cm

7. **16.7 m**
x
20°
46 m

8. **29.4 ft**
34 ft
30°
x

9. 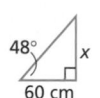 **137.7 m**
156 m
28°
x

10. **66.6 cm**
48°
x
60 cm

11. **11.7 yd**
x
38°
19 yd

12. Joaquim puts a flagpole on his front porch. He attaches a support wire from the house to hold the flagpole in place. The wire attaches to the house at a right angle. Find, to the nearest tenth of a foot, the length of the support wire. **1.6 ft**

28°
3.5 ft

13. Samantha is building a shed. She wants the pitch of the roof to be 36°. Find, to the nearest foot, how high from the ground the peak of the roof will be. **10 ft**

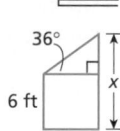
36°
x
6 ft
6 ft

14. Since the hypotenuse is always the longest side of a right triangle, which trigonometric ratio(s) cannot be greater than 1? **sine and cosine**

15. What angle has a tangent of 1? Explain why this is true.

16. A right triangle has acute angles A and B, where m∠A = 36° and m∠B = 54°. Compare sin A to cos B. Compare cos A to sin B. Explain your findings.

Additional Examples

Example 1

Find the sine of 39°.

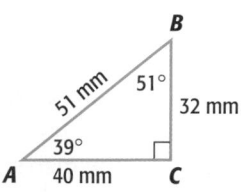
B
51°
51 mm
32 mm
39°
A   40 mm   C

sin A ≈ 0.63

Example 2

Find the height of the tree to the nearest meter.

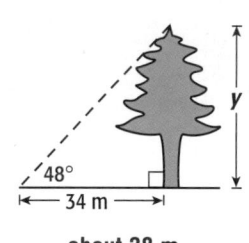
y
48°
34 m

about 38 m

---

3 Close

Summarize

Review the vocabulary from the lesson. Ask students to state each of the basic trigonometric ratios. Have students use the diagram to complete each ratio.

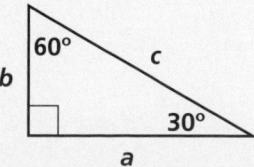
60°
b   c
30°
a

1. sin 30° = $\frac{?}{?}$ $\frac{b}{c}$ 2. cos 30° = $\frac{?}{?}$ $\frac{a}{c}$

3. tan 60° = $\frac{?}{?}$ $\frac{a}{b}$ 4. tan 30° = $\frac{?}{?}$ $\frac{b}{a}$

Answers

15. 45°; If one angle in a right triangle measures 45°, then the remaining angle also measures 45°. Thus, the triangle is isosceles, so the opposite and adjacent sides must be congruent.

16. Sin A is equivalent to cos B, and cos A is equivalent to sin B. In a right triangle, if you take the sine of one acute angle and the cosine of the other acute angle, you use the same sides to form the ratios. So these ratios are congruent.

Problem Solving on Location

Virginia

Purpose: *To provide additional practice for problem-solving skills in Chapters 1–7*

The People's Marathon®

- After problem 1, have students consider the following problem: If each step a runner takes is about 4 feet long, how many steps does the runner take in the marathon? about 34,609 steps

- After problem 4, have students compare the average time it took the women's and men's winners to run 1 mile. The women's winner ran a 6-minute mile; the men's winner ran a 5-minute mile.

Extension Have students design a marathon course. Have them use scale to position the 10-, 15-, and 20-mile marks on the course.
Check students' work.

Problem Solving on Location
VIRGINIA

The People's Marathon®

The People's Marathon, or the "Marathon of the Monuments," is the fourth-largest marathon in the United States. The 26-mile, 385-yard race starts and ends at the Marine Corps War Memorial in Arlington, Virginia. About 16,000 runners participate in the marathon, which is usually held on the fourth Sunday of October each year. In 2000, the 225th anniversary of the Marine Corps, more than 25,000 runners participated.

The race course winds past many historical national monuments and buildings, such as the Washington Monument, the Pentagon, the Lincoln Memorial, Kennedy Center, Union Station, the U.S. Capitol, the Smithsonian Institution buildings and the Jefferson Memorial.

1. How many feet long is the 26-mile, 385-yard race?
   (*Hint:* 1 mi = 5280 ft, and 1 yd = 3 ft.)
   138,435 ft
2. How many yards long is the 26-mile, 385-yard race? **46,145 yd**

3. In 2001, Olga Markova set the women's record time by running the People's Marathon in 2 hours 37 minutes. To the nearest whole number, how many miles per hour did she run? **10 mi/h**

4. In 2001, Jeff Scuffins set the men's record time by running the marathon in 2 hours 14 minutes 1 second. To the nearest whole number, how many miles per hour did he run? **12 mi/h**

Kings Dominion

The 400-acre Kings Dominion theme park is located in Doswell and includes a 33-story replica of the Eiffel Tower, along with 50 rides located in eight different themed areas.

1. The 331 ft 6 in. Kings Dominion Eiffel Tower is built at about a 1:3 scale to the Eiffel Tower in Paris. Approximately how tall is the Eiffel Tower in Paris? **994.5 ft**

For 2–6, use the table.

Roller Coasters at Kings Dominion			
Roller Coaster	Length (ft)	Height (ft)	Duration
Anaconda	▓	128	1 min 50 s
HyperSonic XLC	1560	▓	20 s
Rebel Yell	3368.5	85	2 min 15 s
Scooby-Doo's Ghoster Coaster	1385	35	▓

2. The height-to-length ratio of the Anaconda roller coaster at Kings Dominion is $\frac{32}{675}$. Approximately how long is the Anaconda? **2700 ft**

3. The height-to-length ratio of the HyperSonic XLC roller coaster at Kings Dominion is $\frac{11}{104}$. Approximately how tall is the HyperSonic XLC? **165 ft**

4. The duration of the Hypersonic XLC has a ratio of 1:5 with the duration of Scooby-Doo's Ghoster Coaster. What is the duration of Scooby-Doo's Ghoster Coaster in minutes and seconds? **1 min 40 s**

5. Convert the length of the Rebel Yell roller coaster to miles, and find the maximum number of rides the Rebel Yell could give in an hour. **≈ 0.64 mi; 26**

6. A scale model of Scooby-Doo's Ghoster Coaster had a length of 277 feet and a height of 7 feet. What was the scale factor? **1:5**

Kings Dominion

- After problem 4, have students determine the time it takes to ride the Anaconda ride 4 times. **7 min 20 s**

- After problem 5, have students calculate the speed in feet per second of the HyperSonic XLC. **78 ft/s** Convert this speed to miles per hour. **53.2 mi/h**

Extension Encourage students to write a problem about the Kings Dominion roller coasters, using the data in the table. Then have them exchange problems and solve. **Check students' work.**

Game Resources

Puzzles, Twisters & Teasers

Chapter 7 Resource Book

Copy-Cat

Purpose: *To apply drawing skills to creating similar figures*

Discuss: What geometry word describes the relationship between the original and the copy? similar If 1-inch square gridlines are drawn on a 10 in. × 13 in. photo and the photo is copied so that its dimensions are 5 in. × $6\frac{1}{2}$ in., what size grid was used on the copy? $\frac{1}{2}$ in. × $\frac{1}{2}$ in. squares

Extend: Let students explore how an overhead projector can be used to create similar figures. Have them use an overhead projector to enlarge a drawing or picture for display in the classroom. Check students' work.

Tic-Frac-Toe

Purpose: *To practice forming proportions in a game format*

Discuss: What would a player have to spin in order to win a square containing the equation $\frac{2}{3} = \frac{8}{\square}$? 12 What would a player have to spin in order to block a square containing the equation $\frac{2}{3} = \frac{\square}{\square}$? Possible answer: The player could spin a 10 and place it in the missing denominator.

Extend: Have students model each proportion formed in the game using fraction strips or other manipulatives.

MATH-ABLES

Copy-Cat

You can use this method to copy a well-known work of art or any drawing. First, draw a grid over the work you want to copy, or draw a grid on tracing paper and tape it over the picture.

Next, on a separate sheet of paper draw a blank grid with the same number of squares. The squares do not have to be the same size. Copy each square from the original exactly onto the blank grid. Do not look at the overall picture as you copy. When you have copied all of the squares, the drawing on your finished grid should look just like the original work.

Suppose you are copying an image from a 12 in. by 18 in. print, and that you use 1-inch squares on the first grid.

1. If you use 3-inch squares on the blank grid, what size will your finished copy be? **36 in. by 54 in.**
2. If you want to make a copy that is 10 inches tall, what size should you make the squares on your blank grid? How wide will the copy be? $\frac{5}{6}$ **in.; 15 in.**

3. Choose a painting, drawing, or cartoon, and copy it using the method above. **Check students' work.**

Tic-Frac-Toe

Draw a large tic-tac-toe board. In each square, draw a blank proportion, $\frac{\square}{\square} = \frac{\square}{\square}$. Players take turns using a spinner with 12 sections or a 12-sided die. A player's turn consists of placing a number anywhere in one of the proportions. The player who correctly completes the proportion can claim that square. A square may also be blocked by filling in three parts of a proportion that cannot be completed with a number from 1 to 12. The first player to claim three squares in a row wins.

internet connect
Go to **go.hrw.com** for a copy of the game board.
KEYWORD: MP4 Game7

Technology LAB
Dilations of Geometric Figures

Use with Lesson 7-5

A **dilation** is a geometric transformation that changes the size but not the shape of a figure.

internet connect
Lab Resources Online
go.hrw.com
KEYWORD: MP4 TechLab7

Activity

❶ Construct a triangle similar to the one shown below. Label the vertices *A*, *B*, and *C*.

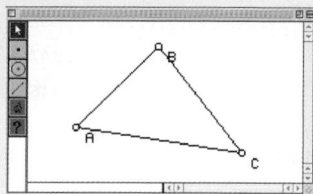

❷ Next pick a center of dilation inside triangle *ABC* and label it point *D*.

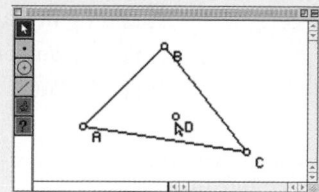

❸ Use the dilation tool on your software to shrink the triangle by a ratio of 1 to 2.

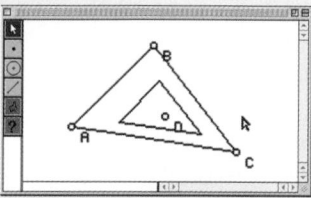

❹ Use the dilation tool again to stretch the original triangle by a ratio of 4 to 3.

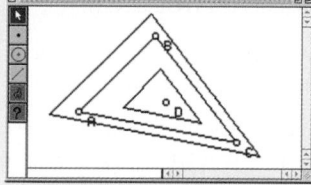

Notice that the dilations of triangle *ABC* are exactly the same *shape* as the original triangle, but they are different *sizes*.

Think and Discuss

1. Are all of the triangles shown in the last figure similar? **yes**

2. If the center of dilation is inside the triangle, and the dilated triangle is shrunk, is the smaller triangle always completely inside the original triangle? **yes**

Try This

1. Use geometry software to construct a quadrilateral *ABCD*.

   a. Choose a center of dilation inside *ABCD*. Shrink *ABCD* by a factor of 1 to 3.

   b. Choose a center of dilation outside *ABCD*. Stretch *ABCD* by a factor of 3 to 2.

Answers

Try This

Check students' work.

1. a.

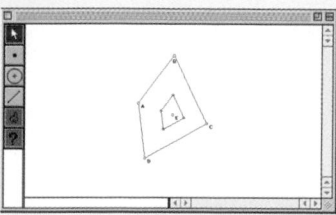

b.

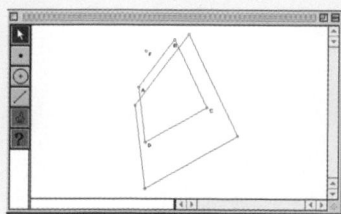

Technology LAB
Dilations of Geometric Figures

Objective: To use geometry software to create dilations of geometric figures

Materials: Computer with geometry software

Lab Resources

Technology Lab Activities . . . . p. 51

Using the Page

This technology activity shows students how to use geometry software to create dilations of geometric figures. Specific instructions may vary, depending on the geometry software used. The instructions given are for Geometer's Sketchpad. For instructions for other software, visit go.hrw.com.

The Think and Discuss problems can be used to assess students' understanding of the technology activity. While the Try This problem can be done without a computer, it is meant to help students become familiar with using a computer to create dilations.

Assessment

1. After the Activity, use the software to draw a line through each set of 3 corresponding vertices. Where do the three lines intersect? **at point *D***

2. Draw a square and shrink it by a ratio of 2:1. Use a point outside the square as the center of dilation. Check students' work.

Purpose: *To help students review and practice concepts and skills presented in Chapter 7*

Assessment Resources

Chapter Review
Chapter 7 Resource Book ... pp. 92–94

 Test and Practice Generator CD-ROM

Additional review assessment items in both multiple-choice and free-response format may be generated for any objective in Chapter 7.

Answers

1. ratio; proportion

2. rate; unit rate

3. similar; scale factor

4. dilation; enlargement; reduction

5. Possible answers: $\frac{1}{2}$, $\frac{2}{4}$

6. Possible answers: $\frac{3}{6}$, $\frac{4}{8}$

7. Possible answers: $\frac{7}{12}$, $\frac{14}{24}$

8. yes

9. no

10. yes

11. no

Study Guide and Review

Vocabulary

Complete the sentences below with vocabulary words from the list above. Words may be used more than once.

1. A __?__ is a comparison of two quantities by division. Two ratios that are equivalent are said to be in __?__.

2. A __?__ is a comparison of two quantities that have different units. A rate in which the second quantity is 1 is called a(n) __?__.

3. A scale drawing is mathematically __?__ to the actual object. All dimensions are reduced or enlarged using the same __?__.

4. A transformation that changes the size but not the shape of a figure is called a __?__. A scale factor greater than 1 results in a(n) __?__ of the figure, while a scale factor between 0 and 1 results in a(n) __?__ of the figure.

7-1 Ratios and Proportions (pp. 342–345)

EXAMPLE

- Find two ratios that are equivalent to $\frac{4}{12}$.

$$\frac{4 \cdot 2}{12 \cdot 2} = \frac{8}{24} \qquad \frac{4 \div 2}{12 \div 2} = \frac{2}{6}$$

8:24 and 2:6 are equivalent to 4:12.

- Simplify to tell whether $\frac{5}{15}$ and $\frac{6}{24}$ form a proportion.

$$\frac{5 \div 5}{15 \div 5} = \frac{1}{3} \qquad \frac{6 \div 6}{24 \div 6} = \frac{1}{4}$$

Since $\frac{1}{3} \neq \frac{1}{4}$, the ratios are not in proportion.

EXERCISES

Find two ratios that are equivalent to each given ratio.

5. $\frac{8}{16}$ **6.** $\frac{9}{18}$ **7.** $\frac{35}{60}$

Simplify to tell whether the ratios in each pair form a proportion.

8. $\frac{8}{24}$ and $\frac{2}{6}$ **9.** $\frac{3}{12}$ and $\frac{6}{18}$

10. $\frac{25}{125}$ and $\frac{5}{25}$ **11.** $\frac{6}{8}$ and $\frac{9}{16}$

7-2 Ratios, Rates, and Unit Rates (pp. 346–349)

EXAMPLE

■ Alex can buy a 4 pack of AA batteries for $2.99 or an 8 pack for $4.98. Which is the better buy?

$$\frac{\text{price per package}}{\text{number of batteries}} = \frac{\$2.99}{4} \approx \$0.75 \text{ per battery}$$

$$\frac{\text{price per package}}{\text{number of batteries}} = \frac{\$4.98}{8} \approx \$0.62 \text{ per battery}$$

The better buy is the 8 pack for $4.98.

EXERCISES

Find the unit price for each offer and tell which is the better buy.

12. 50 formatted computer disks for $14.99 or 75 disks for $21.50

13. 6 boxes of 3-inch incense sticks for $22.50 or 8 boxes for $30

14. a package of 8 multicolored binder dividers for $23.09 or a 25 pack for $99.99

7-3 Analyze Units (pp. 350–354)

EXAMPLE

■ At a rate of 75 kilometers per hour, how many meters does a car travel in 1 minute?

km to m

$$\longrightarrow \frac{1000 \text{ m}}{1 \text{ km}}$$

h to min

$$\longrightarrow \frac{1 \text{ h}}{60 \text{ min}}$$

$$\frac{75 \text{ km}}{1 \text{ h}} \cdot \frac{1000 \text{ m}}{1 \text{ km}} \cdot \frac{1 \text{ h}}{60 \text{ min}} = \frac{75 \cdot 1000 \text{ m}}{60 \text{ min}}$$

$$= \frac{1250 \text{ m}}{1 \text{ min}}$$

The car travels 1250 meters in 1 minute.

EXERCISES

Convert each rate.

15. 90 km/h to m/h

16. 75 feet per second to feet per minute

17. 35 kilometers per hour to meters per minute

18. 55 miles per hour to feet per second

19. 60 cm/s to m/h

7-4 Solving Proportions (pp. 356–359)

EXAMPLE

■ Solve the proportion $\frac{18}{12} = \frac{x}{2}$.

$12x = 18 \cdot 2$ *Find the cross products.*

$\frac{12x}{12} = \frac{36}{12}$ *Solve for x.*

$x = 3$

EXERCISES

Solve each proportion.

20. $\frac{3}{5} = \frac{9}{x}$

21. $\frac{24}{h} = \frac{16}{4}$

22. $\frac{w}{6} = \frac{7}{2}$

23. $\frac{3}{8} = \frac{11}{y}$

7-5 Dilations (pp. 362–365)

EXAMPLE

■ Dilate triangle *ABC* by a scale factor of 2 with *O*(0, 0) as the center of dilation.

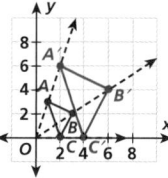

EXERCISES

Dilate each triangle *ABC* by the given scale factor with *O*(0, 0) as the center of dilation.

24. *A*(1, 0), *B*(1, 2), *C*(3, 1); scale factor = 3

25. *A*(4, 6), *B*(8, 4), *C*(6, 2); scale factor = 0.5

26. *A*(2, 2), *B*(6, 2), *C*(4, 4); scale factor = 1.5

Study Guide and Review

Answers

12. $0.30 per disk; $0.29 per disk; 75 disks

13. $3.75 per box; $3.75 per box; unit prices are the same.

14. $2.89 per divider; $4.00 per divider; 8-pack

15. 90,000 m/h

16. 4500 ft/min

17. $583\frac{1}{3}$ m/min

18. $80\frac{2}{3}$ ft/s

19. 2160 m/h

20. $x = 15$

21. $h = 6$

22. $w = 21$

23. $y = 29\frac{1}{3}$

24.

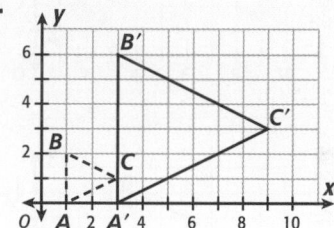

25.

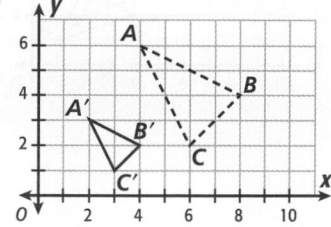

26.

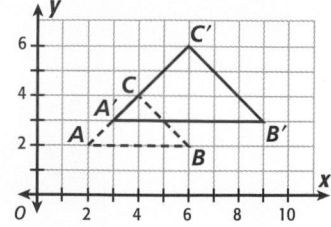

Chapter 8 Resource Book

Student Resources

Practice (Levels A, B, C) .... pp. 8–10, 17–19, 27–29, 37–39, 47–49, 57–59, 67–69
Reteach .... pp. 11, 20–21, 30, 40–41, 50–51, 60–61, 70–71
Challenge .... pp. 12, 22, 31, 42, 52, 62, 72
Problem Solving .... pp. 13, 23, 32, 43, 53, 63, 73
Puzzles, Twisters & Teasers..... pp. 14, 24, 33, 44, 54, 64, 74
Recording Sheets .... pp. 3, 7, 16, 26, 36, 46, 56, 66, 77, 81–83
Chapter Review .... pp. 75–76

Teacher and Parent Resources

Chapter Planning and Pacing Guide................... p. 4
Section Planning Guides .... pp. 5, 34
Parent Letter .... pp. 1–2
Teaching Tools .... pp. 80–84
Teacher Support for Chapter Project .... p. 78
Transparencies .... pp. T1–T35
• Daily Transparencies
• Additional Examples Transparencies
• Teaching Transparencies

Reaching All Learners

English Language Learners
Success for English Language Learners ...... pp. 127–140
Math: Reading and Writing in the Content Area .... pp. 64–70
Spanish Homework and Practice .... pp. 64–70
Spanish Interactive Study Guide .... pp. 64–70
Spanish Family Involvement Activities .... pp. 57–64
Multilingual Glossary

Individual Needs
Are You Ready? Intervention and Enrichment .. pp. 65–68, 113–116, 269–272, 389–392, 419–420
Alternate Openers: Explorations .... pp. 64–70
Family Involvement Activities .... pp. 57–64
Interactive Problem Solving................. pp. 64–70
Interactive Study Guide .... pp. 64–70
Readiness Activities .... pp. 15–16
Math: Reading and Writing in the Content Area .... pp. 64–70
Challenge .... CRB pp. 12, 22, 31, 42, 52, 62, 72

Hands-On
Hands-On Lab Activities................... pp. 59–63
Technology Lab Activities................. pp. 52–58
Alternate Openers: Explorations .... pp. 64–70
Family Involvement Activities .... pp. 57–64

Applications and Connections
Consumer and Career Math................. pp. 29–32
Interdisciplinary Posters .... Poster 8, TE p. 398B
Interdisciplinary Poster Worksheets .... pp. 22–24

Transparencies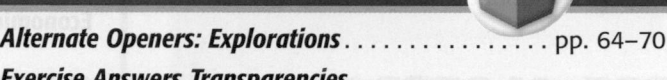

Alternate Openers: Explorations............... pp. 64–70
Exercise Answers Transparencies
Chapter 8 Resource Book .... pp. T1–T35
• Daily Transparencies
• Additional Examples Transparencies
• Teaching Transparencies

Technology

Teacher Resources
Lesson Presentations CD-ROM.............. Chapter 8
Test and Practice Generator CD-ROM .... Chapter 8
One-Stop Planner CD-ROM .... Chapter 8

Student Resources
Are You Ready? Intervention CD-ROM
Skills 14, 26, 65, 95

 internet connect

Homework Help Online	**KEYWORD:** MP4 HWHelp8
Math Tools Online	**KEYWORD:** MP4 Tools
Glossary Online	**KEYWORD:** MP4 Glossary
Chapter Project Online	**KEYWORD:** MP4 PSProject8
Chapter Opener Online	**KEYWORD:** MP4 Ch8

 **KEYWORD:** MP4 CNN8

SE = *Student Edition*   **TE** = *Teacher's Edition*   **AR** = *Assessment Resources*   **CRB** = *Chapter Resource Book*   **MK** = *Manipulatives Kit*

Assessment Options

Assessing Prior Knowledge

Determine whether students have the required prerequisite concepts and skills.

Are You Ready?. . . . . SE p. 399

Inventory Test. . . . . AR pp. 1–4

Test Preparation

Provide review and practice for chapter and standardized tests.

Standardized Test Prep. . . . . SE p. 443

Spiral Review with Test Prep . . . . . SE, last page of each lesson

Study Guide and Review . . . . SE pp. 438–440

Test Prep Tool Kit

Technology

 Test and Practice Generator CD-ROM

 **internet** connect

State-Specific Test Practice Online KEYWORD: MP4 TestPrep

Performance Assessment

Assess students' understanding of chapter concepts and combined problem-solving skills.

Performance Assessment . . . . SE p. 442
   Includes scoring rubric in TE

Performance Assessment . . . . AR p. 132

Performance Assessment Teacher Support. . . . . AR p. 131

Portfolio

Portfolio opportunities appear throughout the Student and Teacher's Editions.

Suggested work samples:

Problem Solving Project . . . . TE p. 398

Performance Assessment . . . . SE p. 442

Portfolio Guide. . . . . AR p. xxxvii

Journal. . . . . TE, last page of each lesson

Write About It . . . . SE pp. 403, 408, 419, 423, 431

Daily Assessment

Obtain daily feedback on students' understanding of concepts.

Spiral Review and Test Prep . . . . . . SE, last page of each lesson

Also Available on Transparency In Chapter 8 Resource Book

Warm Up. TE, first page of each lesson

Problem of the Day. . . . . TE, first page of each lesson

Lesson Quiz . . . . TE, last page of each lesson

Student Self-Assessment

Have students evaluate their own work.

Group Project Evaluation . . . . AR p. xxxiv

Individual Group Member Evaluation . . . . AR p. xxxv

Portfolio Guide. . . . . AR p. xxxvii

Journal. . . . . TE, last page of each lesson

Formal Assessment

Assess students' mastery of concepts and skills.

Section Quizzes . . . . AR pp. 19–20

Mid-Chapter Quiz . . . . SE p. 414

Chapter Test . . . . SE p. 441

Chapter Tests (Levels A, B, C) . . . . AR pp. 75–80

Cumulative Tests (Levels A, B, C) . . . . AR pp. 229–240

Standardized Test Prep
   Cumulative Assessment . . . . SE p. 443

End-of-Year Test. . . . . AR pp. 313–316

Technology

 Test and Practice Generator CD-ROM

Make tests electronically. This software includes:

• Dynamic practice for Chapter 8

• Customizable tests

• Multiple-choice items for each objective

• Free-response items for each objective

• Teacher management system

SE = *Student Edition*   **TE** = *Teacher's Edition*   **AR** = *Assessment Resources*   **CRB** = *Chapter Resource Book*   **MK** = *Manipulatives Kit*

Chapter 8 Tests

Three levels (A,B,C) of tests are available for each chapter in the *Assessment Resources.*

LEVEL A

CHAPTER 8 Chapter Test
Form A

Find the missing ratio or percent equivalent for each letter on the number line.

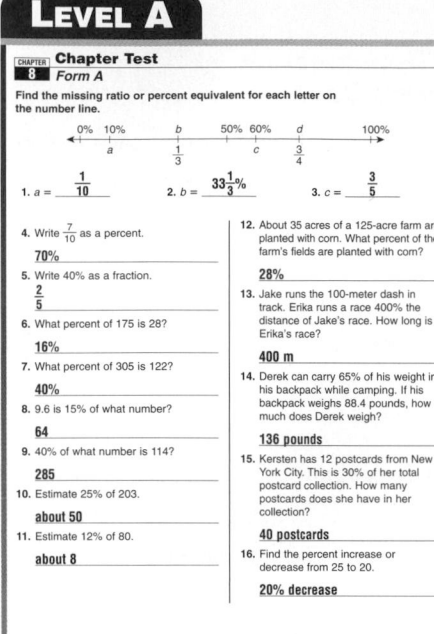

1. $a = \dfrac{1}{10}$ 2. $b = 33\dfrac{1}{3}\%$ 3. $c = \dfrac{3}{5}$

4. Write $\dfrac{7}{10}$ as a percent.
70%

5. Write 40% as a fraction.
$\dfrac{2}{5}$

6. What percent of 175 is 28?
16%

7. What percent of 305 is 122?
40%

8. 9.6 is 15% of what number?
64

9. 40% of what number is 114?
285

10. Estimate 25% of 203.
about 50

11. Estimate 12% of 80.
about 8

12. About 35 acres of a 125-acre farm are planted with corn. What percent of the farm's fields are planted with corn?
28%

13. Jake runs the 100-meter dash in track. Erika runs a race 400% the distance of Jake's race. How long is Erika's race?
400 m

14. Derek can carry 65% of his weight in his backpack while camping. If his backpack weighs 88.4 pounds, how much does Derek weigh?
136 pounds

15. Kersten has 12 postcards from New York City. This is 30% of her total postcard collection. How many postcards does she have in her collection?
40 postcards

16. Find the percent increase or decrease from 25 to 20.
20% decrease

CHAPTER 8 Chapter Test
Form A, continued

Give answers to the nearest percent.

17. A toy store sold a toy train set for $49.95 last year. This year, the same train set costs $52.50. What was the percent increase of the cost of the train set?
5%

18. The president of a small company made $72,000 last year. She cut her salary this year to $60,000 because the company was not doing as well. What was the percent decrease in her salary to the nearest percent?
17%

19. A rain jacket costs $52. It is on sale for 20% off. Estimate the discount on the jacket.
about $10

20. In a state with a sales tax rate of 6%, Alex bought paper for his printer for $17.99. How much was his sales tax to the nearest penny?
$1.08

21. Theo earned $3045 over the summer as a lifeguard. Of this, $669.90 was withheld for taxes. What percent of his income was withheld?
22%

22. Clarence earns a 10% commission on sales plus a $200 weekly salary. In one week, his sales totaled $2100. What was his total pay that week?
$410

23. Shannon invested $5000 in a bond. Her total simple interest on her investment after 3 years was $1200. What was the yearly interest rate on her investment?
8%

24. Tess borrowed $10,500 from the bank to buy a car. The length of her loan is 4 years with a simple interest rate of 7.5%. How much will she pay in interest if she pays off the loan at the end of the 4 years?
$3150

25. Stephen deposited $2500 in a savings account that earned an annual simple interest rate of 4%. When he closed his account, he had $3000. How long did he have his account?
5 years

LEVEL B

CHAPTER 8 Chapter Test
Form B

Find the missing ratio or percent equivalent for each letter on the number line.

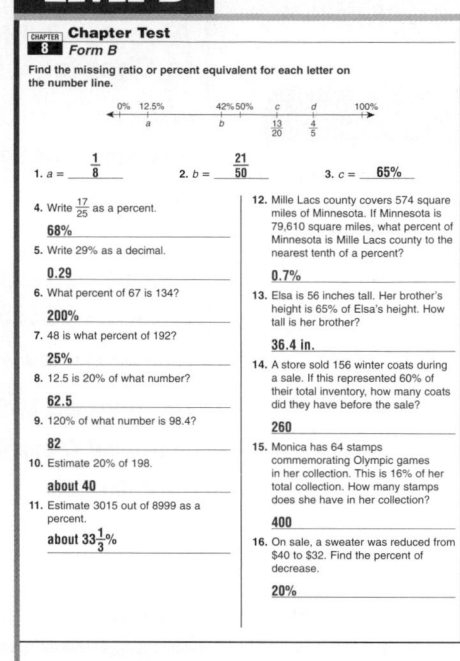

1. $a = \dfrac{1}{8}$ 2. $b = \dfrac{21}{50}$ 3. $c = 65\%$

4. Write $\dfrac{17}{25}$ as a percent.
68%

5. Write 29% as a decimal.
0.29

6. What percent of 67 is 134?
200%

7. 48 is what percent of 192?
25%

8. 12.5 is 20% of what number?
62.5

9. 120% of what number is 98.4?
82

10. Estimate 20% of 198.
about 40

11. Estimate 3015 out of 8999 as a percent.
about $33\dfrac{1}{3}\%$

12. Mille Lacs county covers 574 square miles of Minnesota. If Minnesota is 79,610 square miles, what percent of Minnesota is Mille Lacs county to the nearest tenth of a percent?
0.7%

13. Elsa is 56 inches tall. Her brother's height is 65% of Elsa's height. How tall is her brother?
36.4 in.

14. A store sold 156 winter coats during a sale. If this represented 60% of their total inventory, how many coats did they have before the sale?
260

15. Monica has 64 stamps commemorating Olympic games in her collection. This is 16% of her total collection. How many stamps does she have in her collection?
400

16. On sale, a sweater was reduced from $40 to $32. Find the percent of decrease.
20%

CHAPTER 8 Chapter Test
Form B, continued

Give answers to the nearest percent.

17. A toy store sold a toy train set for $49.79 last year. This year, the same train set costs $51.29. What was the percent increase of the cost of the train set?
3%

18. A small company had profits of $550,000 last year. This year, their profits were only $484,000. What was the percent decrease in their profits?
12%

19. A pair of shoes are on sale for 25% off. They normally cost $42.95. Estimate the discount on the shoes.
about $10

20. In a state with a sales tax rate of 6.5%, Thomas bought a new DVD player for $256.99 and a DVD movie for $24.99. How much is the sales tax on his purchase to the nearest penny?
$18.33

21. Kirk earned $4147 over the summer working as a waiter. $829.40 was taken out for taxes. What percent of his income was withheld for taxes?
20%

22. Justin works as a car salesman where he earns 8% commission on his sales and no weekly salary. What will his weekly sales have to be to earn $3280 for the week?
$41,000

23. Karen deposited $8500 in a college savings account for her grandson that earns an annual simple interest rate of 6.5%. What will be the total amount in the account in 10 years?
$14,025

24. Sadie borrowed $3500 from a bank at an annual simple interest rate for a home remodeling project. After 4 years, she repaid the bank $4200. What was the interest rate of the loan?
5%

25. If Aisha deposits $2250 in a savings account that earns 3.5% annual simple interest, how long must she keep the money in the account for its total to reach $2722.50?
6 yr

LEVEL C

CHAPTER 8 Chapter Test
Form C

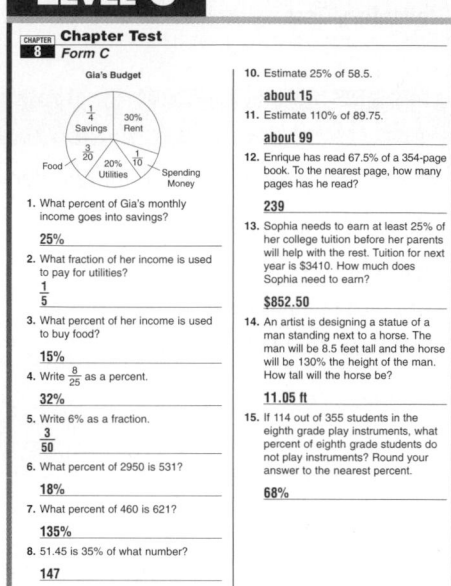

Gia's Budget

1. What percent of Gia's monthly income goes into savings?
25%

2. What fraction of her income is used to pay for utilities?
$\dfrac{1}{5}$

3. What percent of her income is used to buy food?
15%

4. Write $\dfrac{8}{25}$ as a percent.
32%

5. Write 6% as a fraction.
$\dfrac{3}{50}$

6. What percent of 2950 is 531?
18%

7. What percent of 460 is 621?
135%

8. 51.45 is 35% of what number?
147

9. 23.5% of what number is 42.3?
180

10. Estimate 25% of 58.5.
about 15

11. Estimate 110% of 89.75.
about 99

12. Enrique has read 67.5% of a 354-page book. To the nearest page, how many pages has he read?
239

13. Sophia needs to earn at least 25% of her college tuition before her parents will help with the rest. Tuition for next year is $3410. How much does Sophia need to earn?
$852.50

14. An artist is designing a statue of a man standing next to a horse. The man will be 8.5 feet tall and the horse will be 130% the height of the man. How tall will the horse be?
11.05 ft

15. If 114 out of 355 students in the eighth grade play instruments, what percent of eighth grade students do not play instruments? Round your answer to the nearest percent.
68%

CHAPTER 8 Chapter Test
Form C, continued

Give answers to the nearest hundredth of a percent.

16. The number of children taking swimming lessons at the community pool last summer was 274. The number registered for lessons this summer is 292. What is the percent increase?
6.57%

17. A pair of Kelly's new pants shrunk in the wash. The inseam was 37.25 inches before they were washed and 36.375 inches after they were washed. What is the percent decrease in the length of the pants?
2.35%

18. During a sale, the price of a DVD was decreased by 50%. By what percent must the sale price be increased to restore the original price?
200%

19. The total area of the state of Florida is 58,560 square miles. The total land area of Florida is 54,252 square miles. Estimate what percent of Florida's total area is water.
about 8%

20. Janine is paid $200 a week plus 12.5% commission on sales. What were her weekly sales if she earned $950?
$6000

21. Shaneece made $42,500 last year. Isak made $35,450 last year. If they are both taxed at 27.5% of the amount over $27,050, how much more tax did Shaneece pay than Isak?
$1938.75

22. 19.5% of Alisha's paycheck is withheld each week for taxes. If she earns $568 per week, how much money is her check written for?
$457.24

23. A credit union advertises that if you put $3000 in a certificate of deposit, you can earn $1800 in ten years. What yearly simple interest rate does the certificate of deposit offer?
6%

24. Salim borrowed $12,375 for 4 years at an annual simple interest rate of 7.5%. If he makes one payment at the end of the loan, how much interest will he have to repay?
$3712.50

25. Todd had to pay $1147.50 of interest on a loan with an annual simple interest rate of 8.5% that he had for 5 years. Assuming he makes one payment at the end of the loan, what was the principal of the loan?
$2700

Test and Practice Generator
CD-ROM

Create and customize multiple versions of the same tests with corresponding answers for any chosen chapter objectives.

Chapter 8 State and Standardized Test Preparation

Test Taking Skill Builder and Standardized Test Practice
are provided for each chapter in the *Test Prep Tool Kit*.

TEST TAKING SKILL BUILDER

Test Taking Strategy
Chapter 8

Multiple Choice Questions—Elimination Method

Eliminating answer choices that you know are incorrect is an excellent test-taking strategy. Use mental math and estimation techniques to help you decide which answer choices to eliminate.

Example 1 Multiple Choice A portable CD player regularly costs $63.99. The tax in your area is 8.25%. What is the total cost of the CD player?

A $55.74   B $58.71   C $69.27   D $72.24

Solution:
Consider choices A and B: Eliminate these choices because they are too small. They are less than the cost of the CD player.

Use mental math to estimate the amount of tax.
8.25% is approximately $\frac{1}{10}$
$\frac{1}{10}$ (64) = 6.4
The amount of tax will be less than $6.40, so the total price will be less than $70.40.
You can now eliminate choice D.
The correct answer choice is C.

Example 2 Multiple Choice A graph of the Tyler family's monthly budget is shown. Their monthly income is $2,500. How much do the Tyler's spend for groceries every month?

F $180   H $480
G $450   I $500

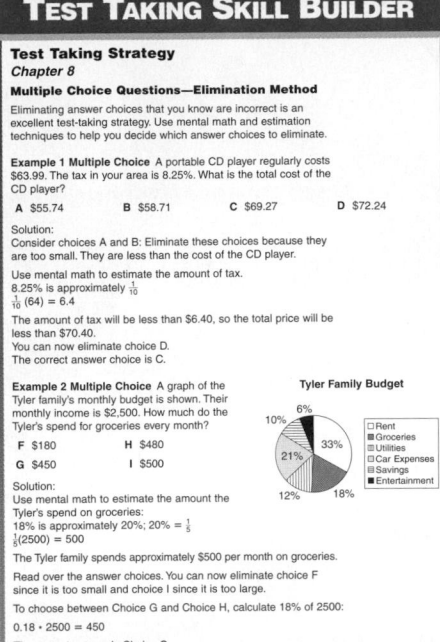

Tyler Family Budget
- Rent 33%
- Groceries 18%
- Utilities 12%
- Car Expenses 21%
- Savings 10%
- Entertainment 6%

Solution:
Use mental math to estimate the amount the Tyler's spend on groceries:
18% is approximately 20%; 20% = $\frac{1}{5}$
$\frac{1}{5}$(2500) = 500
The Tyler family spends approximately $500 per month on groceries.

Read over the answer choices. You can now eliminate choice F since it is too small and choice I since it is too large.
To choose between Choice G and Choice H, calculate 18% of 2500:
0.18 • 2500 = 450
The correct answer is Choice G.

Test Taking Strategy
Chapter 8, continued

Exercises Possible answers are given.
Identify two answer choices you can eliminate and explain why you eliminated them. Then solve the problem.

1. To finance her education, Sara takes out a loan for $3500. After a year, Sara decides to pay off the interest, which is 8%. How much will she pay?

A $280   B $320   C $2800   D $3780

Choice 1: Choice C is too large, 10% of 3500 is $350.

Choice 2: Choice D is also too large, 10% of 3500 is $350.

Answer: A

2. The circle graph represents the sales of 5 different vehicles. Which vehicle accounted for about 25% of the sales?

F V   G X   H Y   I Z

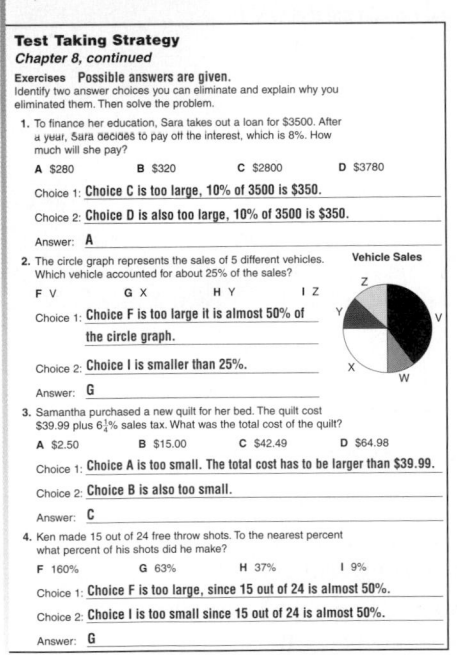

Vehicle Sales

Choice 1: Choice F is too large it is almost 50% of the circle graph.

Choice 2: Choice I is smaller than 25%.

Answer: G

3. Samantha purchased a new quilt for her bed. The quilt cost $39.99 plus 6$\frac{1}{4}$% sales tax. What was the total cost of the quilt?

A $2.50   B $15.00   C $42.49   D $64.98

Choice 1: Choice A is too small. The total cost has to be larger than $39.99.

Choice 2: Choice B is also too small.

Answer: C

4. Ken made 15 out of 24 free throw shots. To the nearest percent what percent of his shots did he make?

F 160%   G 63%   H 37%   I 9%

Choice 1: Choice F is too large, since 15 out of 24 is almost 50%.

Choice 2: Choice I is too small since 15 out of 24 is almost 50%.

Answer: G

STANDARDIZED TEST PRACTICE

Standardized Test Practice
Chapter 8

Select the best answer for Questions 1–7.

1. A dime is 10% of a dollar. How is this expressed as both a decimal and fraction?

A 0.25 and $\frac{1}{4}$

B 0.1 and $\frac{1}{10}$

C 0.10 and $\frac{1}{25}$

D 0.1 and $\frac{1}{100}$

2. Mr. Roberto is paying for a business luncheon. He normally leaves a tip of 15%. If the bill came to $45, how much should he leave for a tip?

F $4.50
G $6.75
H $9.00
I $9.50

3. A box of ceramic tile weighs 58 lb. This is 45% more than a box of plastic tile. How much does a box of plastic tile weigh?

A 26 lb
B 40 lb
C 50 lb
D 84 lb

4. Animal Care borrows $6500 for 3 months at 7% interest from National Money Loans. What is the total amount Animal Care will have to pay back to National Money Loans?

F $113.75   H $6613.75
G $1365   I $11,375

5. Lori has a lemonade stand with a 64-ounce pitcher of lemonade. If she sells five 8-ounce glasses of lemonade, what percentage of the lemonade remains?

A 50%
B 37.5%
C 62.5%
D 12.8%

6. Keegan earned a commission of $32,000 on the sale of $620,000 worth of farm machinery. What is his rate of commission?

F 1%
G 5.2%
H 7.4%
I 19.375%

7. Alex recorded his expenses for the month of July. What percent of his total expense is spent on Groceries?

Expenses for July	
Housing	$450.00
Groceries	120.00
Auto	85.00
Clothing	60.00
Entertainment	50.00
Misc.	35.00
Total	**$800.00**

A 12%
B 66.6%
C 15%
D 25%

Standardized Test Practice
Chapter 8, continued

Gridded Response
Solve the problems. Use the answer sheet to write and grid-in your answer.

8. Miguel earns $320 per week plus an 8% sales commission. What is Miguel's pay for a week in which he has sales totaling $4209?

9. Last Friday, 30% of the 20-member cheerleading squad missed practice. How many cheerleaders missed practice on Friday?

Short Response
Solve the problems. Use the answer sheet to write your answers.

10. In a recent state election, only 48.2% of all registered voters actually voted. If there were 2,987,650 registered voters in the state, estimate the number of people who voted. Explain in words how you determined your answer.

11. Mrs. Wong's garden is 10 feet by 15 feet. She plans to increase it to 20 feet by 25 feet. Find the percent of increase of the area of the garden. Show your work.

Extended Response

12. D & D's Hardware borrows $8200 for 4 months at a 6% annual simple interest from Farmer's Bank.

a. What is the amount of interest D & D's will have to pay? Show your work.

b. What is the total amount D & D's will have to pay back to Farmer's Bank if they take a one-year loan instead? Show your work.

c. National Loans will loan D & D's the $8200 for 5 months at an interest rate of 5.0%. Which loan is the best deal for D & D's, the loan from National or the loan from Farmer's Bank? Explain in words how you determined your answer.

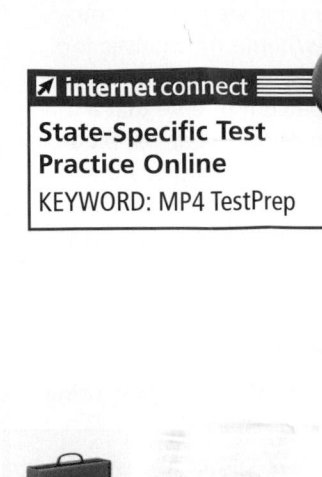

State-Specific Test Practice Online
KEYWORD: MP4 TestPrep

Test Prep Tool Kit

- Standardized Test Prep Workbook
- Countdown to Testing transparencies
- State Test Prep CD-ROM
- Standardized Test Prep Video

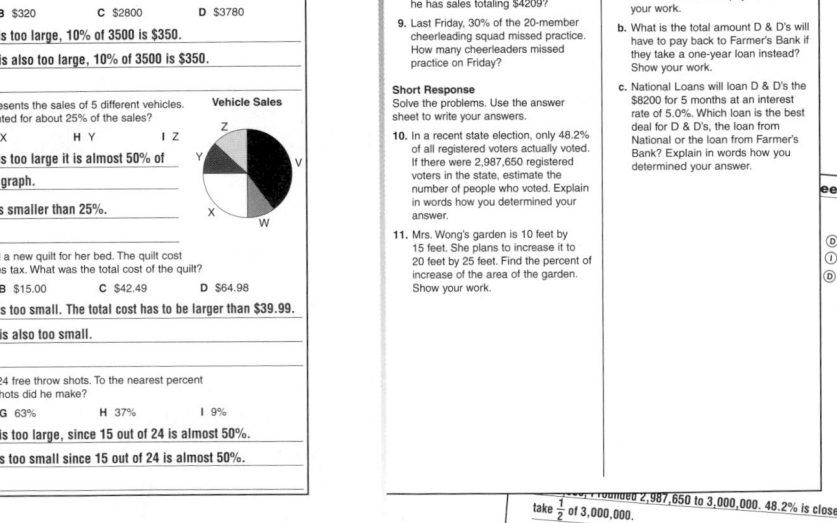

...rounded 2,987,650 to 3,000,000. 48.2% is close to 50% so take $\frac{1}{2}$ of 3,000,000.

(See Lesson 8-5.)

11. The square footage of the original garden is 15 × 10 = 150 square feet. The proposed new garden would be 20 × 25 = 500 square feet. 500 − 150 = 350. $\frac{350}{150}$ = 2.333. This means that increasing the garden 350 square feet is a 233.33% increase.

(See Lesson 8-4.)

Extended Response Write your answers for Problem 12 on the back of this paper. See Lesson 8-7.

(D) See Lesson 8-2.
(I) See Lesson 8-6.
(D) See Lesson 8-1.

Customized answer sheets give students realistic practice for actual standardized tests.

Percents

Why Learn This?

Tell students that statistics are used to compare players. For example, in order to determine the hitter with the most frequent home runs, statisticians compare the at bats/home run statistic for each player. These statistics help statisticians find the percent of time that a player who comes up to bat will hit a home run. Point out to students that the player with the best at bats/home run statistic may not be the player with the most home runs.

Using Data

To begin the study of this chapter, using the table, have students:

- Order the players from fewest to most home runs. Rodriguez, Sosa, Griffey, Bonds

- Predict the number of At Bats each player would need to hit 73 home runs.

  Bonds: 73(14.0) = 1022

  Sosa: 73(14.4) ≈ 1051

  Griffey: 73(14.6) ≈ 1066

  Rodriguez: 73(15.6) ≈ 1139

Percents

internet connect

Chapter Opener Online
go.hrw.com
KEYWORD: MP4 Ch8

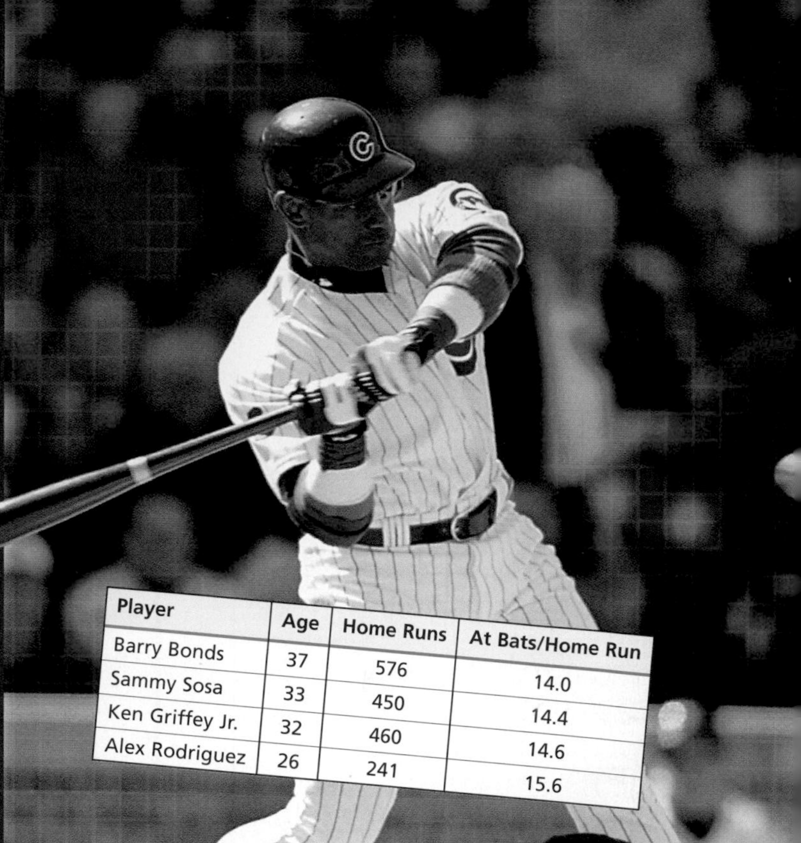

Player	Age	Home Runs	At Bats/Home Run
Barry Bonds	37	576	14.0
Sammy Sosa	33	450	14.4
Ken Griffey Jr.	32	460	14.6
Alex Rodriguez	26	241	15.6

Career *Sports Statistician*

Statisticians are mathematicians who work with data, creating statistics, graphs, and tables that describe and explain the real world. Sports statisticians combine their love of sports with their ability to use mathematics.

Statistics not only explain what has happened, but can help you predict what may happen in the future. The table describes the home run hitting of some active Major League baseball players.

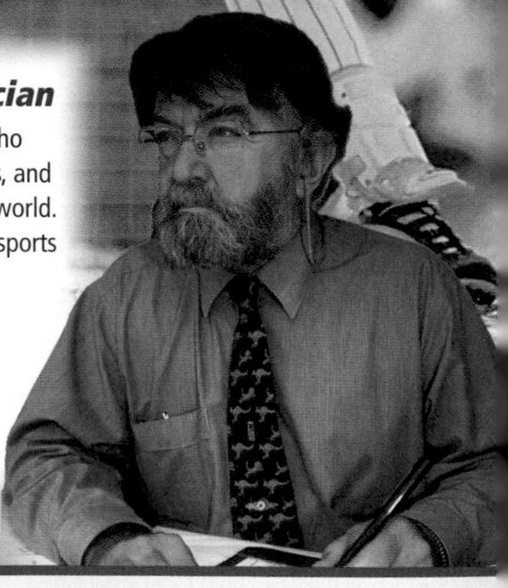

Problem Solving Project

Social Studies and Sports Connection

Purpose: To solve problems using percents and the various methods of displaying them

Materials: Home Run Derby worksheet

internet connect

Chapter Project Online: go.hrw.com
KEYWORD: MP4 PSProject8

Understand, Plan, Solve, and Look Back

Have students:

✔ Complete the Home Run Derby worksheet to learn more about working with percents.

✔ Estimate the number of home runs each player would hit if they each had 500 at bats in a given season.

✔ Research the number of home runs hit by their favorite players. What percent of their favorite players' total hits were home runs? What other kinds of hits did they have? Create a circle graph to show the results.

✔ Check students' work.

ARE YOU READY?

Choose the best term from the list to complete each sentence.

1. A _?_ is a comparison of two quantities by division. **ratio**

2. Ratios that make the same comparison are _?_.
   equivalent ratios

3. Two ratios that are equivalent are in _?_. **proportion**

4. To solve a proportion, you _?_. **cross multiply**

cross multiply

equivalent ratios

proportion

ratio

Complete these exercises to review skills you will need for this chapter.

✔ Write Fractions as Decimals
Write each fraction as a decimal.

5. $\frac{3}{4}$ **0.75**   6. $\frac{5}{8}$ **0.625**   7. $\frac{2}{5}$ **0.4**   8. $\frac{2}{3}$ **0.$\overline{6}$**

✔ Write Decimals as Fractions
Write each decimal as a fraction in simplest form.

9. 0.7 $\frac{7}{10}$ 10. 0.6 $\frac{3}{5}$ 11. 0.25 $\frac{1}{4}$ 12. 0.375 $\frac{3}{8}$

13. 0.2 $\frac{1}{5}$ 14. 0.9 $\frac{9}{10}$ 15. 0.86 $\frac{43}{50}$ 16. 0.99 $\frac{99}{100}$

✔ Solve Proportions
Solve each proportion.

17. $\frac{x}{3} = \frac{9}{27}$ **$x = 1$**   18. $\frac{7}{8} = \frac{h}{4}$ **$h = 3.5$**   19. $\frac{9}{n} = \frac{2}{3}$ **$n = 13.5$**

20. $\frac{3}{8} = \frac{12}{t}$ **$t = 32$**   21. $\frac{4}{5} = \frac{28}{z}$ **$z = 35$**   22. $\frac{100}{p} = \frac{90}{45}$ **$p = 50$**

✔ Read Circle Graphs
Refer to the graph to answer each question.

23. Which item accounts for nearly half the budget?
    programs

24. What dollar amount is spent on computer equipment? **$3750**

25. What dollar amount is spent on new books and programs? **$17,250**

26. What dollar amount is spent on other expenses?
    $4000

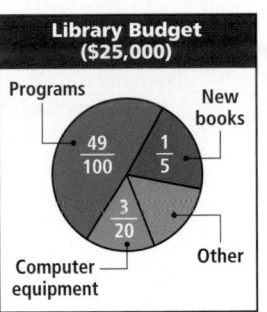

Library Budget ($25,000)

Programs — $\frac{49}{100}$

New books — $\frac{1}{5}$

Computer equipment — $\frac{3}{20}$

Other

Assessing Prior Knowledge

INTERVENTION

Diagnose and Prescribe

Evaluate your students' performance on this page to determine whether intervention is necessary or whether enrichment is appropriate. Options that provide instruction, practice, and a check are listed below.

Resources for Are You Ready?

• *Are You Ready? Intervention and Enrichment*

• *Recording Sheet for Are You Ready?*
  Chapter 8 Resource Book...... p. 3

  Are You Ready? Intervention **CD-ROM**

ARE YOU READY?
Were students successful with Are You Ready?

NO INTERVENE      **YES ENRICH**

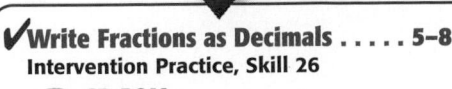
✔ **Write Fractions as Decimals** 5–8
Intervention Practice, Skill 26
 CD-ROM
Intervention Activities, Skill 26

✔ **Write Decimals as Fractions** 9–16
Intervention Practice, Skill 14
 CD-ROM
Intervention Activities, Skill 14

✔ **Solve Proportions** 17–22
Intervention Practice, Skill 65
 CD-ROM
Intervention Activities, Skill 65

✔ **Read Circle Graphs** 23–26
Intervention Practice, Skill 95
 CD-ROM
Intervention Activities, Skill 95

Are You Ready? Enrichment, pp. 419–420

Section 8A

Numbers and Percents

 **One-Minute Section Planner**

Lesson	Materials	Resources
Lesson 8-1 Relating Decimals, Fractions, and Percents **NCTM:** Number and Operations, Communication, Representation **NAEP:** Number Properties 1e ☑ SAT-9 ☑ SAT-10 ☑ ITBS ☑ CTBS ☑ MAT ☑ CAT	**Optional** Magazine or newspaper ads containing percents Number cards for Reaching All Learners (CRB, p. 80)	• *Chapter 8 Resource Book,* pp. 6–14 • Daily Transparency T1, CRB • Additional Examples Transparencies T2–T5, CRB • *Alternate Openers: Explorations,* p. 64
Hands-On Lab 8A Make a Circle Graph **NCTM:** Algebra, Geometry, Connections, Representation **NAEP:** Data Analysis and Probability 1b ☐ SAT-9 ☐ SAT-10 ☐ ITBS ☐ CTBS ☐ MAT ☐ CAT	**Required** Compasses *(MK)* Rulers *(MK)* Protractors *(MK)*	• *Hands-On Lab Activities,* pp. 61, 95
Lesson 8-2 Finding Percents **NCTM:** Number and Operations, Communication, Connections, Representation **NAEP:** Number Properties 4d ☑ SAT-9 ☑ SAT-10 ☑ ITBS ☑ CTBS ☑ MAT ☑ CAT	**Optional** Recording Sheet for Reaching All Learners (CRB, p. 81)	• *Chapter 8 Resource Book,* pp. 15–24 • Daily Transparency T6, CRB • Additional Examples Transparencies T7–T9, CRB • *Alternate Openers: Explorations,* p. 65
Technology Lab 8B Find Percent Error **NCTM:** Number and Operations, Measurement **NAEP:** Number Properties 4d ☐ SAT-9 ☐ SAT-10 ☐ ITBS ☐ CTBS ☐ MAT ☐ CAT	**Required** Graphing calculators	• *Technology Lab Activities,* p. 52
Lesson 8-3 Finding a Number When the Percent Is Known **NCTM:** Number and Operations, Communication, Representation **NAEP:** Number Properties 4d ☑ SAT-9 ☑ SAT-10 ☑ ITBS ☑ CTBS ☑ MAT ☑ CAT	**Optional** Recording Sheet for Reaching All Learners (CRB, p. 82)	• *Chapter 8 Resource Book,* pp. 25–33 • Daily Transparency T10, CRB • Additional Examples Transparencies T11–T14, CRB • *Alternate Openers: Explorations,* p. 66
Section 8A Assessment		• Mid-Chapter Quiz, SE p. 414 • Section 8A Quiz, AR p. 19 • *Test and Practice Generator* CD-ROM

SAT = *Stanford Achievement Tests*   **ITBS** = *Iowa Test of Basic Skills*   **CTBS** = *Comprehensive Test of Basic Skills/Terra Nova*
MAT = *Metropolitan Achievement Test*   **CAT** = *California Achievement Test*
NCTM = Complete standards can be found on pages T29–T35.   **NAEP** = Complete standards can be found on pages A54–A58.
SE = *Student Edition*   **TE** = *Teacher's Edition*   **AR** = *Assessment Resources*   **CRB** = *Chapter Resource Book*   **MK** = *Manipulatives Kit*

Section Overview

Decimals, Fractions, and Percents
Lesson 8-1, Hands-On Lab 8A

Why? Any rational number can be represented by a decimal, a fraction, or a percent.

A **percent** is a ratio that compares a number to 100.

> To write a **fraction as a decimal,** divide the numerator by the denominator.
> $$\frac{1}{8} = 1 \div 8 = 0.125$$

> To write a **decimal as a percent,** multiply by 100 and insert the percent symbol, which means "divided by 100."
> $$0.125 \cdot \mathbf{100} = 12.5\mathbf{\%}$$

> **Common percents and their equivalent fractions:**
> $10\% = \frac{1}{10}$
> $12.5\% = \frac{1}{8}$
> $16\frac{2}{3}\% = \frac{1}{6}$
> $20\% = \frac{1}{5}$
> $25\% = \frac{1}{4}$
> $33\frac{1}{3}\% = \frac{1}{3}$
> $50\% = \frac{1}{2}$
> $66\frac{2}{3}\% = \frac{2}{3}$
> $75\% = \frac{3}{4}$

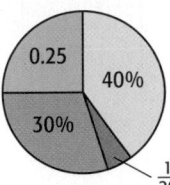

Fraction	Decimal	Percent
$\frac{25}{100} = \frac{1}{4}$	$\frac{1}{4} = 0.25$	25%
$\frac{40}{100} = \frac{2}{5}$	$\frac{2}{5} = 0.4$	40%
$\frac{5}{100} = \frac{1}{20}$	$\frac{1}{20} = 0.05$	5%
$\frac{30}{100} = \frac{3}{10}$	$\frac{3}{10} = 0.3$	30%

Percent Problems
Technology Lab 8B, Lessons 8-2, 8-3

 Why? Using percents is a way of comparing numbers.

Percent problems generally involve two numbers and a percent comparing the two numbers. To solve the problem, you may need to find one of the numbers, or you may need to find the percent.

Finding a Percent of a Number	Finding the Percent One Number Is of Another	Finding a Number When the Percent Is Known
What number is 25% of 32?	What percent of 120 is 90?	36 is 4% of what number?

Finding a Percent of a Number

What number is 25% of 32?

Set up an equation.

What number is 25% of 32?

$n = 25\% \cdot 32$
$n = 0.25 \cdot 32$
$n = 8$

So **8** is 25% of 32.

Set up a proportion.

25 is to 100 as **what number** is to 32?

$$\frac{25}{100} = \frac{n}{32}$$
$100n = 25 \cdot 32$
$100n = 800$
$n = 8$

So **8** is 25% of 32.

Finding the Percent One Number Is of Another

What percent of 120 is 90?

Set up an equation.

What percent of 120 is 90?

$p \cdot 120 = 90$
$p = \frac{90}{120}$
$p = 0.75$

So 90 is **75%** of 120.

Set up a proportion.

What number is to 100 as 90 is to 120?

$$\frac{n}{100} = \frac{90}{120}$$
$120n = 100 \cdot 90$
$120n = 9000$
$n = 75$

So 90 is **75%** of 120.

Finding a Number When the Percent Is Known

36 is 4% of what number?

Set up an equation.

36 is 4% of **what number**?

$36 = 4\% \cdot n$
$36 = 0.04n$
$\frac{36}{0.04} = \frac{0.04}{0.04}n$
$900 = n$

So 36 is 4% of **900**.

Set up a proportion.

4 is to 100 as 36 is to **what number**?

$$\frac{4}{100} = \frac{36}{n}$$
$4n = 100 \cdot 36$
$4n = 3600$
$n = 900$

So 36 is 4% of **900**.

Objective: Students relate decimals, fractions, and percents.

Warm Up

Evaluate.

1. $\frac{2}{15} + \frac{3}{15}$ $\frac{1}{3}$ 2. $\frac{7}{12} - \frac{3}{12}$ $\frac{1}{3}$

3. $\frac{4}{5} \cdot \frac{7}{2}$ $\frac{14}{5}$ or $2\frac{4}{5}$ 4. $3\frac{1}{2} \div \frac{1}{4}$ 14

Problem of the Day

A fast-growing flower grows to a height of 12 inches in 12 weeks by doubling its height every week. If you want your flower to be only 6 inches tall, after how many weeks should you pick it? 11 weeks

Available on Daily Transparency in CRB

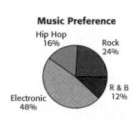

Math Humor

Which of these numbers is under the most stress, 30%, $\frac{1}{3}$, or 0.2? The last, because it's *too tense*.

8-1 Relating Decimals, Fractions, and Percents

Learn to relate decimals, fractions, and percents.

Vocabulary

percent

Reading Math

Think of the % symbol as meaning /100.
0.75 = 75% = 75/100

In an average day, a typical koala sleeps 20 out of 24 hours. The part of a day the koala sleeps can be shown in several ways:

$$\frac{20}{24} = 0.83\overline{3} = 83.\overline{3}\%$$

So koalas sleep over 80% of the time.

Percents are ratios that compare a number to 100.

Koalas usually sleep in the fork of a tree. They are most active after sunset.

Ratio	Equivalent Ratio with Denominator of 100	Percent
$\frac{3}{10}$	$\frac{30}{100}$	30%
$\frac{1}{2}$	$\frac{50}{100}$	50%
$\frac{3}{4}$	$\frac{75}{100}$	75%

To convert a fraction to a decimal, divide the numerator by the denominator.

$$\frac{1}{8} = 1 \div 8 = 0.125$$

To convert a decimal to a percent, multiply by 100 and insert the percent symbol.

$$0.125 \cdot 100 \rightarrow 12.5\%$$

$$\begin{array}{r} 0.125 \\ 8\overline{)1.000} \\ \underline{8} \\ 20 \\ \underline{16} \\ 40 \\ \underline{40} \\ 0 \end{array}$$

EXAMPLE 1 Finding Equivalent Ratios and Percents

Remember!

Here are some percents and their equivalent ratios:

$10\% = \frac{1}{10}$ $33\frac{1}{3}\% = \frac{1}{3}$

$12\frac{1}{2}\% = \frac{1}{8}$ $40\% = \frac{2}{5}$

$16\frac{2}{3}\% = \frac{1}{6}$ $50\% = \frac{1}{2}$

$20\% = \frac{1}{5}$ $66\frac{2}{3}\% = \frac{2}{3}$

$25\% = \frac{1}{4}$ $75\% = \frac{3}{4}$

Find the missing ratio or percent equivalent for each letter a–g on the number line.

$$0\% \quad b \quad 20\% \quad 33\frac{1}{3}\% \quad e \quad 62\frac{1}{2}\% \quad 100\%$$
$$a \quad \frac{1}{8} \quad c \quad d \quad \frac{1}{2} \quad f \quad g$$

a: $0\% = \frac{0}{100} = 0$

b: $\frac{1}{8} = 0.125 = 12.5\% = 12\frac{1}{2}\%$

c: $20\% = \frac{20}{100} = \frac{2}{10} = \frac{1}{5}$

d: $33\frac{1}{3}\% = 0.33\overline{3} = \frac{1}{3}$

e: $\frac{1}{2} = 0.5 = 50\%$

f: $62\frac{1}{2}\% = 0.625 = \frac{625}{1000} = \frac{5}{8}$

g: $100\% = \frac{100}{100} = 1$

---

1 Introduce

Alternate Opener

EXPLORATION

8-1 Relating Decimals, Fractions, and Percents

A *percent* is a ratio that compares a number to 100. Percents can be modeled on circle graphs. The circle graph below shows the results of a survey in which 25 students were asked which type of music they preferred.

- 16% means 16 per 100.
- 16% as a decimal is 0.16.
- 16% as a fraction is $\frac{16}{100} = \frac{4}{25}$.

Music Preference

Hip Hop 16%
Rock 24%
R & B 12%
Electronic 48%

Use the percents in the circle graph to complete the table.

	Music Type	Percent	Decimal	Fraction
1.	Electronic	48%		
2.	Rock	24%		
3.	Hip Hop	16%		
4.	R & B	12%		

Think and Discuss

5. **Explain** how you wrote each percent as a decimal.
6. **Explain** how you wrote each decimal as a fraction.

Motivate

Show students magazines or newspapers that advertise sales using percents. Ask the students whether they can explain how much money would be saved and what the sale price would be as a result of the percent savings.

Exploration worksheet and answers on Chapter 8 Resource Book pp. 7 and 85

2 Teach

Lesson Presentation

Guided Instruction

In this lesson, students learn to relate decimals, fractions, and percents. Demonstrate to students that fractions, decimals, and percents are all ratios. Remind students how to convert between fractions and decimals (Lesson 3-1). Explain that percents are ratios in the form of *parts per hundred*.

Teaching Tip

Ask students how many cents there are in a dollar. 100 Then ask what the word *per* means. "for each" or "divided by" Then explain that the word *percent* means "divided by one hundred" (e.g., 20% literally means "20 divided by 100" or "20 for each 100").

EXAMPLE 2 Finding Equivalent Fractions, Decimals, and Percents

Find the equivalent value missing from the table for each value given on the circle graph.

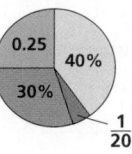

Fraction	Decimal	Percent
$\frac{25}{100} = \frac{1}{4}$	**0.25**	$0.25(100) = 25\%$
$\frac{40}{100} = \frac{2}{5}$	$\frac{2}{5} = 0.4$	**40%**
$\frac{1}{20}$	$\frac{1}{20} = 0.05$	$0.05(100) = 5\%$
$\frac{30}{100} = \frac{3}{10}$	$\frac{3}{10} = 0.3$	**30%**

You can use the information in each column of Example 2 to make three equivalent circle graphs. One shows the breakdown by fractions, one shows the breakdown by decimals, and one shows the breakdown by percents.

Fraction
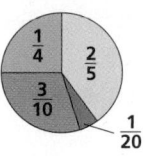

Decimal
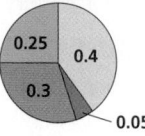

Percent
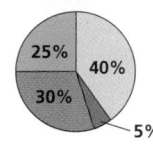

The sum of the fractions should be 1.

The sum of the decimals should be 1.

The sum of the percents should be 100%.

EXAMPLE 3 *Physical Science Application*

Gold that is 24 karat is 100% pure gold. Gold that is 18 karat is 18 parts pure gold and 6 parts another metal, such as copper, zinc, silver, or nickel.

What percent of 18-karat gold is pure gold?

$\dfrac{\text{parts pure gold}}{\text{total parts}} \rightarrow \dfrac{18}{24} = \dfrac{3}{4}$ *Set up a ratio and reduce.*

$\dfrac{3}{4} = 3 \div 4 = 0.75 = 75\%$ *Find the percent.*

So 18-karat gold is 75% pure gold.

Think and Discuss

1. Give an example of a real-world situation in which you would use (1) decimals (2) fractions, and (3) percents.

2. Show 25 cents as a part of a dollar in terms of (1) a reduced fraction (2) a percent, and (3) a decimal. Which is most common?

Additional Examples

Example 1

Find the missing ratio or percent equivalent for each letter *a–g* on the number line.

a: $\frac{1}{10}$ b: 25%   c: $\frac{2}{5}$

d: 60%   e: $\frac{2}{3}$ f: $\frac{7}{8}$

g: $\frac{5}{4}$ or $1\frac{1}{4}$

Example 2

Find the equivalent fraction, decimal, or percent for each value given on the circle graph.

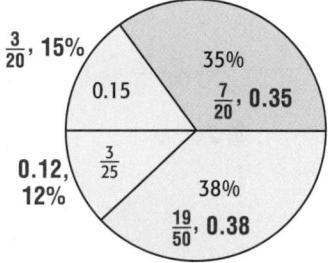

Example 3

Gold that is 24 karat is 100% pure gold. Gold that is 14 karat is 14 parts pure gold and 10 parts another metal, such as copper, zinc, silver, or nickel. What percent of 14 karat gold is pure gold? $58\frac{1}{3}\%$

---

3 Close

Reaching All Learners
Through Number Sense

Prepare sets of number cards that contain three equivalent numbers, such as $\frac{1}{4}$, 0.25, and 25% (Chapter 8 Resource Book p. 80). Distribute the numbers randomly so that each student gets a decimal, a fraction, or a percent. Have each student find the two students in the class with the numbers equivalent to his or her number. Once students have formed their groups of three, have the class form a human number line (three deep) so that the numbers are in increasing order.

Summarize

Remind students that fractions, terminating or repeating decimals, and percents are all ratios. In a fraction, the denominator can be any nonzero integer. A percent is equivalent to a ratio with a denominator of 100. A terminating decimal is equivalent to a ratio with a denominator of 10, 100, 1000, or some other power of 10. Any of the three forms can be changed to the other two forms.

Answers to Think and Discuss

Possible answers:

1. (1) money and prices; (2) shoe sizes or measurements in recipes; (3) sales tax or discounts.

2. (1) $\frac{1}{4}$; (2) 25%; (3) 0.25; $\frac{1}{4}$ (one-quarter) is most common.

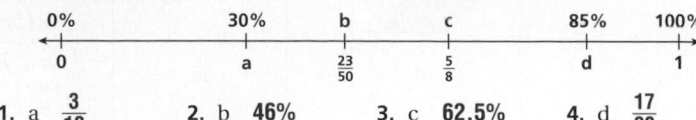

8-1 Exercises

FOR EXTRA PRACTICE
see page 746

internet connect
Homework Help Online
go.hrw.com   Keyword: MP4 8-1

Students may want to refer back to the lesson examples.

GUIDED PRACTICE

See Example **1** Find the missing ratio or percent equivalent for each letter on the number line.

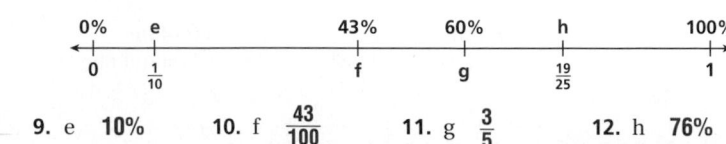

1. a $\frac{3}{10}$ 2. b **46%**   3. c **62.5%**   4. d $\frac{17}{20}$

See Example **2** Find each equivalent value.

5. $\frac{2}{5}$ as a percent **40%** 6. 32% as a fraction $\frac{8}{25}$ 7. $\frac{7}{8}$ as a decimal **0.875**

See Example **3** 8. A molecule of water is made up of 2 atoms of hydrogen and 1 atom of oxygen. What percent of the atoms of a water molecule is oxygen? $33\frac{1}{3}\%$

INDEPENDENT PRACTICE

See Example **1** Find the missing ratio or percent equivalent for each letter on the number line.

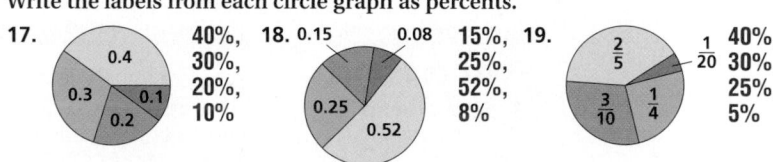

9. e **10%**   10. f $\frac{43}{100}$ 11. g $\frac{3}{5}$ 12. h **76%**

See Example **2** Find each equivalent value as indicated.

13. 32% as a decimal **0.32** 14. $\frac{23}{25}$ as a percent **92%** 15. 0.545 as a fraction $\frac{109}{200}$

See Example **3** 16. Sterling silver is an alloy combining 925 parts pure silver and 75 parts of another metal, such as copper. What percent of sterling silver is not pure silver? **7.5%**

PRACTICE AND PROBLEM SOLVING

Write the labels from each circle graph as percents.

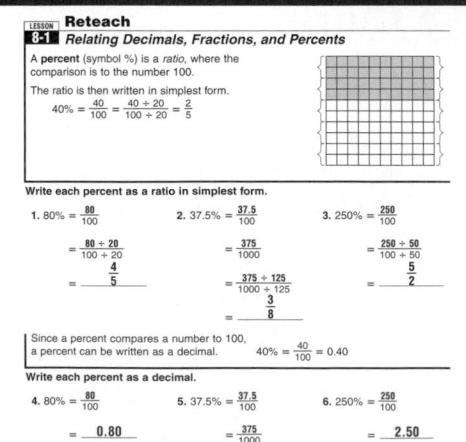

17. **40%, 30%, 20%, 10%**

18. **15%, 25%, 52%, 8%**

19. **40%, 30%, 25%, 5%**

20. **$0.05,** $\frac{1}{20}$ **dollar** 20. A nickel is 5% of a dollar. Write the value of a nickel as a decimal and as a fraction.

Assignment Guide

If you finished Example **1** assign:
Core 1–4, 9–12, 27–32
Enriched 1–4, 9–12, 27–32

If you finished Example **2** assign:
Core 1–7, 9–15, 17–19, 27–32
Enriched 1–7 odd, 9–15, 17–19, 24–32

If you finished Example **3** assign:
Core 1–21, 27–32
Enriched 1–7 odd, 9–32

Notes

Math Background

Using 10-by-10 grids is an excellent way to reinforce the meaning of *percent*. On the grid, each square represents 1%, or $\frac{1}{100}$, of the grid's area. You can shade different percents on the grid and then divide the grid into equal areas to show fractional equivalents. The example below shows that 20% is equal to $\frac{1}{5}$.

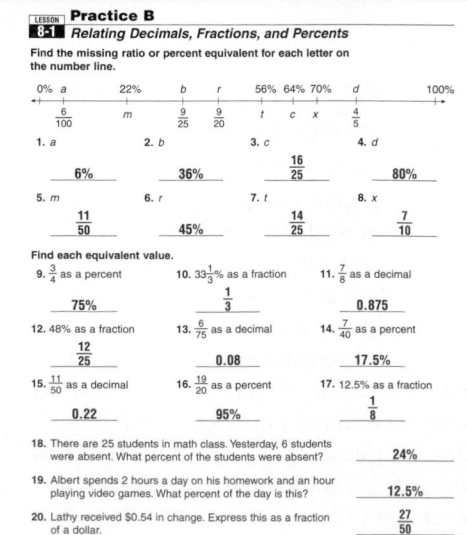

RETEACH 8-1

Reteach
8-1 *Relating Decimals, Fractions, and Percents*

A **percent** (symbol %) is a *ratio*, where the comparison is to the number 100.

The ratio is then written in simplest form.
$40\% = \frac{40}{100} = \frac{40 \div 20}{100 \div 20} = \frac{2}{5}$

Write each percent as a ratio in simplest form.

1. $80\% = \frac{80}{100}$ 2. $37.5\% = \frac{37.5}{100}$ 3. $250\% = \frac{250}{100}$
$= \frac{80 \div 20}{100 \div 20}$ $= \frac{375}{1000}$ $= \frac{250 \div 50}{100 \div 50}$
$= \frac{4}{5}$ $= \frac{375 \div 125}{1000 \div 125}$ $= \frac{5}{2}$
$= \frac{3}{8}$

Since a percent compares a number to 100, a percent can be written as a decimal. $40\% = \frac{40}{100} = 0.40$

Write each percent as a decimal.

4. $80\% = \frac{80}{100}$ 5. $37.5\% = \frac{37.5}{100}$ 6. $250\% = \frac{250}{100}$
$= 0.80$ $= \frac{375}{1000}$ $= 2.50$
$= 0.375$

Use the results of Exercises 4–6 to complete.

	Percent	Fraction	Decimal
7.	80%	$\frac{4}{5}$	0.80
8.	37.5%	$\frac{3}{8}$	0.375
9.	250%	$\frac{5}{2}$	2.50

PRACTICE 8-1

Practice B
8-1 *Relating Decimals, Fractions, and Percents*

Find the missing ratio or percent equivalent for each letter on the number line.

1. a **6%**   2. b **36%**   3. c $\frac{16}{25}$ 4. d **80%**

5. m $\frac{11}{50}$ 6. r **45%**   7. t $\frac{14}{25}$ 8. x $\frac{7}{10}$

Find each equivalent value.

9. $\frac{3}{4}$ as a percent **75%** 10. $33\frac{1}{3}\%$ as a fraction $\frac{1}{3}$ 11. $\frac{7}{8}$ as a decimal **0.875**

12. 48% as a fraction $\frac{12}{25}$ 13. $\frac{6}{75}$ as a decimal **0.08** 14. $\frac{7}{40}$ as a percent **17.5%**

15. $\frac{11}{50}$ as a decimal **0.22** 16. $\frac{19}{20}$ as a percent **95%** 17. 12.5% as a fraction $\frac{1}{8}$

18. There are 25 students in math class. Yesterday, 6 students were absent. What percent of the students were absent? **24%**

19. Albert spends 2 hours a day on his homework and an hour playing video games. What percent of the day is this? **12.5%**

20. Lathy received $0.54 in change. Express this as a fraction of a dollar. $\frac{27}{50}$

21. Ragu ran the first 3 miles of a 5 mile race in 24 minutes. What percent of the race has he run? **60%**

21. **PHYSICAL SCIENCE** Of the 20 highest mountains in the United States, 17 are located in Alaska. What percent of the highest mountains in the United States are in Alaska? **85%**

22. **LIFE SCIENCE** When collecting plant specimens, it is a good idea to remove no more than 5% of a population of plants. A botanist wants to collect plants from an area with 60 plants. What is the greatest number of plants she should remove? **3**

23. The graph shows the percents of the total U.S. land area taken up by the five largest states. The sixth section of the graph represents the area of the remaining 45 states.

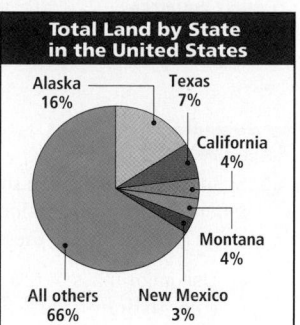

Total Land by State in the United States

Alaska 16% · Texas 7% · California 4% · Montana 4% · New Mexico 3% · All others 66%

   a. Alaska is the largest state in total land area. Write Alaska's portion of the total U.S. land area as a fraction and as a decimal. $\frac{4}{25}$; **0.16**

23b. 23%; Possible answer: Alaska and Texas make up nearly $\frac{1}{4}$ of the total land area in the United States.

   b. What percent of the total U.S. land area is Alaska and Texas combined? How might you describe this percent?

24. **WHAT'S THE ERROR?** An analysis showed that 0.03% of the video games produced by one company were defective. Wynn says this is 3 out of every 100. What is Wynn's error?

25. **WRITE ABOUT IT** How can you find a fraction, decimal, or percent when you have only one form of a number?

26. **CHALLENGE** Luke and Lissa were asked to solve a percent problem using the numbers 17 and 45. Luke found 17% of 45, but Lissa found 45% of 17. Explain why they both got the same answer. Would this work for other numbers as well? Why or why not?

Spiral Review

Tell whether the two lines described in each exercise are parallel, perpendicular, or neither. (Lesson 5-5)

27. $\overleftrightarrow{PQ}$ has slope $\frac{3}{2}$. $\overleftrightarrow{EF}$ has slope $-\frac{2}{3}$. **perpendicular** 28. $\overleftrightarrow{AB}$ has slope $\frac{9}{11}$. $\overleftrightarrow{CD}$ has slope $-\frac{3}{4}$. **neither**

29. $\overleftrightarrow{XY}$ has slope $\frac{13}{25}$. $\overleftrightarrow{QR}$ has slope $\frac{13}{25}$. **parallel** 30. $\overleftrightarrow{MN}$ has slope $-\frac{1}{8}$. $\overleftrightarrow{OP}$ has slope 8. **perpendicular**

31. **TEST PREP** A cone has diameter 12 cm and height 9 cm. Using 3.14 for π, find the volume of the cone to the nearest tenth. (Lesson 6-7) **D**

   A 56.5 cm³   B 118.3 cm³   C 1356.5 cm³   D 339.1 cm³

32. **TEST PREP** Evaluate $Q - 1\frac{2}{3}$ for $Q = 4\frac{3}{4}$. (Lesson 3-5) **F**

   F $3\frac{1}{12}$ G $5\frac{1}{12}$ H $1\frac{1}{6}$ J $3\frac{5}{12}$

CHALLENGE 8-1

LESSON 8-1 Challenge

100% Filled

Materials needed: colored pencils or pens

For each exercise, select from the box a different combination of numbers whose sum is 100%. An item may be used only once in a combination, but may be used again in a different combination. Write your selection on the line below the grid. Shade the squares in the grid with a different color for each number you selected.

68%	$\frac{1}{2}$	60%	$\frac{1}{20}$	1%	0.9	0.04	$\frac{13}{20}$	$\frac{1}{7}$
$\frac{3}{25}$	16%	55%	0.02	$\frac{1}{9}$	0.15	$\frac{1}{8}$	24%	0.44
0.06	$\frac{1}{6}$	$\frac{1}{4}$	29%	0.037	$\frac{1}{3}$	19%	$\frac{17}{50}$	

1.

2.

3.

Possible Combinations

0.15, 60%, $\frac{1}{4}$

55%, 0.44, 1%

19%, 1%, $\frac{1}{2}$, 24%, 0.06

PROBLEM SOLVING 8-1

LESSON 8-1 Problem Solving

Relating Decimals, Fractions, and Percents

Write the correct answer. Round to the nearest hundredth.

1. A survey showed that 42% of people responded that gathering at a campfire was their favorite camping activity. Write this percent as a decimal and as a fraction. **0.42; $\frac{21}{50}$**

2. Scientists have researched the ratio of brain weight to body size in different animals. The results are in the table below. What percent of each animal's body weight is brain weight?

Animal	Brain Weight / Body Weight	Percent
Cat	$\frac{1}{100}$	1%
Dog	$\frac{1}{125}$	0.8%
Elephant	$\frac{1}{560}$	0.18%
Horse	$\frac{1}{600}$	0.17%
Human	$\frac{1}{40}$	2.5%
Mouse	$\frac{1}{40}$	2.5%

3. A survey showed that 24% of people responded that enjoying the scenery was their favorite camping activity. Write this percent as a decimal and as a fraction. **0.24; $\frac{6}{25}$**

The top rated passers in the NFL for the 2001 regular season are given in the table. Choose the letter for the best answer. Round to the nearest hundredth.

Player	Attempts	Completions
Kurt Warner	546	375
Rich Gannon	549	361
Jeff Garcia	504	316
Brett Favre	510	314

Top Rated Passers: 2001 Regular Season

4. Which player listed completed the highest percentage of passes in the 2001 season?

   A Brett Favre   C Jeff Garcia
   B Rich Gannon   D Kurt Warner

5. Which decimal is equivalent to the pass completion percentage of Kurt Warner?

   F 0.63   H 0.69
   G 0.66   J 0.62

6. What fraction of passes did Jeff Garcia complete?

   A $\frac{31}{50}$ C $\frac{504}{316}$
   B $\frac{79}{126}$ D $\frac{1}{2}$

7. What percent of passes did Rich Gannon complete?

   F 62.70%   H 75%
   G 65.76%   J 162.08%

Lesson Quiz

Find each equivalent value.

1. $\frac{3}{8}$ as a percent **37.5%**

2. 20% as a fraction $\frac{1}{5}$

3. $\frac{5}{8}$ as a decimal **0.625**

4. $\frac{14}{25}$ as a percent **56%**

5. About 342,000 km² of Greenland's total area (2,175,000 km²) is not covered with ice. To the nearest percent, what percent of Greenland's total area is not covered with ice? **16%**

Available on Daily Transparency in CRB

Pacing: Traditional 1 day
Block $\frac{1}{2}$ day

Objective: To use a compass, ruler, protractor, and paper to make circle graphs

Materials: Compass, ruler, protractor

Lab Resources

Hands-On Lab Activities . . . . . pp. 61, 95

Using the Page

Discuss with students how to determine the angle measurement for each sector of the circle.

1. How many states are there? 50 How many degrees are in a circle? 360°

2. Write a proportion for finding the angle measurement for the sector representing states where skunks are legal. $\frac{6}{50} = \frac{x}{360}$

Answers

Try This

1. **Skunks as Pets by State (Where Legal)**

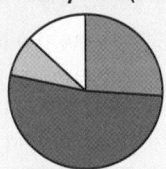

■ Legal (no restrictions)
■ Legal with permit
□ Legal in some areas
□ Other conditions for legality

Legal (no restrictions): 94°

Legal with permit: 188°

Legal in some areas: 31°

Other conditions for legality: 47°

Hands-On
LAB
8A

Make a Circle Graph

Use with Lesson 8-1

WHAT YOU NEED:
• Compass • Protractor
• Ruler • Paper

REMEMBER
• A circle measures 360°.
• Percent compares a number to 100.

▢ **internet** connect
Lab Resources Online
go.hrw.com
KEYWORD: MP4 Lab8A

Activity

1 Skunks are legal pets in some states but not in most. Use the information from the table to make a circle graph showing the percents for each category.

   a. Use a compass to draw a large circle. Use a ruler to draw a vertical radius.

   b. Extend the table to show the percent of states with each category of legality.

   c. Use the percents to determine the angle measure of each sector of the graph.

   d. Use a protractor to draw each angle clockwise from the radius.

   e. Label the graph and each sector. Color the sectors.

Skunks as Pets by State	
Legality	**Number of States**
Legal (no restrictions)	6
Legal with permit	12
Legal in some areas	2
Illegal	27
Other conditions	3

Legality	Number of States	Percent of States	Angle of Section
Legal (no restrictions)	6	$\frac{6}{50} = 12\%$	$\frac{12}{100} \cdot 360 = 43.2°$
Legal with permit	12	$\frac{12}{50} = 24\%$	$\frac{24}{100} \cdot 360 = 86.4°$
Legal in some areas	2	$\frac{2}{50} = 4\%$	$\frac{4}{100} \cdot 360 = 14.4°$
Illegal	27	$\frac{27}{50} = 54\%$	$\frac{54}{100} \cdot 360 = 194.4°$
Other conditions	3	$\frac{3}{50} = 6\%$	$\frac{6}{100} \cdot 360 = 21.6°$

Think and Discuss

1. How many states would need to legalize skunks for the largest sector to be 180°? **2**

Try This

1. Make a circle graph to show only the states where skunks are not illegal.

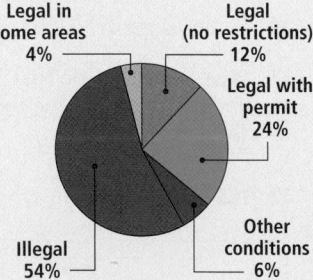

Legal in some areas 4%
Legal (no restrictions) 12%
Legal with permit 24%
Other conditions 6%
Illegal 54%

Kimberly Johnson-Green
Baltimore, Maryland

Teacher to Teacher

I use a budgeting project to help students learn about fractions, decimals, and percents. The students are divided into pairs and I distribute job descriptions with monthly incomes, family sizes, and monthly expense lists. The students have to calculate how much (what fraction, decimal, and percent) of their income is used for each expense (mortgage, food, gas & electric, telephone, car payment, car insurance, clothes, savings, entertainment, etc.). Then I have students create a circle graph to display their findings. In some cases, the students find that they do not make enough to cover all expenses, so they must decide which expenses in their budget can be reduced.

FOR EXTRA PRACTICE
see page 746

internet connect
Homework Help Online
go.hrw.com Keyword: MP4 8-2

8-2 PRACTICE & ASSESS

GUIDED PRACTICE

See Example 1 — **Find each percent to the nearest tenth.**

1. What percent of 71 is 35? **49.3%** 2. What percent of 1130 is 225? **19.9%**

3. Of Earth's 197 million mi² of surface area, about 139 million mi² is water. Find the percent of Earth's surface that is covered by water. **70.6%**

See Example 2 — 4. Jay's term paper is 18 pages long. If Madison's paper is 175% of the length of Jay's paper, find the length of Madison's paper. **31.5 pages**

INDEPENDENT PRACTICE

See Example 1 — **Find each percent to the nearest tenth.**

5. What percent of 74 is 222? **300%** 6. What percent of 150 is 25? **16.7%**

7. 12.5 is what percent of 1250? **1%** 8. 150 is what percent of 80? **187.5%**

9. About 600 mi² of the 700 mi² of the Okefenokee Swamp is located in Georgia. If Georgia is 57,906 mi², find the percent of that area that is part of the Okefenokee Swamp. **1.0%**

See Example 2 — 10. In Arkansas, the highest elevation is Mount Magazine, in west Arkansas, and the lowest is the Ouachita River, in the southeast corner of the state. Mount Magazine is 2753 ft above sea level, which is about 5098% of the elevation of the lowest portion of the state. Find the elevation of the Ouachita River area.
≈ 54.0 ft above sea level

PRACTICE AND PROBLEM SOLVING

Find each number to the nearest tenth.

11. What number is $66\frac{2}{3}\%$ of 45? **30** 12. What number is $22\frac{2}{3}\%$ of 320? **72.5**

13. What number is 44% of 6? **2.6** 14. What number is $2\frac{1}{2}\%$ of 11,960? **299**

15. What number is 133% of 200? **266** 16. What number is $66\frac{2}{3}\%$ of 750? **500**

Complete each statement.

17. Since 9 is 15% of 60,

   a. 18 is ▓% of 60. **30**

   b. 27 is ▓% of 60. **45**

   c. 90 is ▓% of 60. **150**

18. Since 8 is 5% of 160,

   a. 8 is ▓% of 80. **10**

   b. 8 is ▓% of 40. **20**

   c. 8 is ▓% of 20. **40**

19. Since 20 is 200% of 10,

   a. 20 is ▓% of 20. **100**

   b. 20 is ▓% of 40. **50**

   c. 20 is ▓% of 80. **25**

Students may want to refer back to the lesson examples.

Assignment Guide

If you finished Example 1 assign:
Core 1–3, 5–9, 24, 28–32
Enriched 5–9, 21, 24, 25, 27–32

If you finished Example 2 assign:
Core 1–10, 11–19 odd, 20, 22–24, 28–32
Enriched 5–9 odd, 10–32

Notes

Math Background

Percents are widely used in many disciplines. They are used extensively in business economics to describe many things, including taxes, economic growth, inflation, interest rates, and changes in the stock market. Percents are also useful in consumer economics, including the calculation of sales tax, discounts, withholding tax, and tips. Other applications include sports (e.g., winning percentage, field-goal percentage, and save percentage) and even grading (e.g., 90% or above is an A, etc.). Understanding percents will help students in many areas, including the study of probability (Chapter 9).

RETEACH 8-2

LESSON 8-2 Reteach
Finding Percents

Since a percent is a ratio, problems involving percent can be solved by using a proportion.

There are different possibilities for an unknown quantity in this proportion.

$$\frac{symbol\ number}{100} = \frac{is\ number}{of\ number}$$

Possibility 1: Find the symbol number.

$$\frac{symbol\ number}{100} = \frac{is\ number}{of\ number}$$
$$\frac{x}{100} = \frac{16}{80}$$
$$\frac{80x}{80} = \frac{1600}{80}$$
$$x = 20 \qquad \text{So, 16 is 20\% of 80.}$$

Find what percent one number is of another.

1. What percent of 64 is 16?
$$\frac{x}{100} = \frac{16}{64}$$
$$\frac{64 \cdot x}{64} = \frac{16 \cdot 100}{64}$$
$$\frac{64x}{64} = \frac{1600}{64}$$
$$x = \frac{25}{\ }$$
So, 16 is **25%** of 64.

2. What percent of 200 is 150?
$$\frac{x}{100} = \frac{150}{200}$$
$$\frac{200 \cdot x}{200} = \frac{150 \cdot 100}{200}$$
$$\frac{200x}{200} = \frac{15,000}{200}$$
$$x = \frac{75}{\ }$$
So, 150 is **75%** of 200.

3. What percent of 4 is 6?
$$\frac{x}{100} = \frac{6}{4}$$
$$\frac{4 \cdot x}{4} = \frac{6 \cdot 100}{4}$$
$$\frac{4x}{4} = \frac{600}{4}$$
$$x = \frac{150}{\ }$$
So, 6 is **150%** of 4.

4. About what percent of 115 is 40?
$$\frac{x}{100} = \frac{40}{115}$$
$$\frac{115 \cdot x}{115} = \frac{40 \cdot 100}{115}$$
$$\frac{115x}{115} = \frac{4000}{115}$$
$$x = \frac{34.7}{\ }$$
So, 40 is about **35%** of 115.

PRACTICE 8-2

LESSON 8-2 Practice B
Finding Percents

Find each percent.

1. What percent of 64 is 21? **25%**
2. 24 is what percent of 60? **40%**
3. What percent of 150 is 75? **50%**
4. What percent of 80 is 68? **85%**
5. 36 is what percent of 80? **45%**
6. 13 is what percent of 52? **25%**
7. What percent of 85 is 51? **60%**
8. What percent of 88 is 33? **37.5%**
9. 19 is what percent of 95? **20%**
10. 28.8 is what percent of 120? **24%**
11. What percent of 56 is 49? **87.5%**
12. What percent of 102 is 17? **$16\frac{2}{3}\%$**
13. What percent of 94 is 42.3? **45%**
14. 90 is what percent of 75? **120%**
15. Daphne bought a used car for $9200. She made a down payment of $1840. What is the percent of the amount of the down payment? **20%**
16. An airplane traveled from Boston to Las Vegas making a stop in St. Louis. The distance from Boston to St. Louis is 1040 miles. The distance from St. Louis to Las Vegas is 1370 miles. Once the plane landed in St. Louis, what percent of the trip was completed? Round the answer to the nearest tenth of a percent. **43.2%**
17. An item priced at $10.20 has a sales tax of $0.51. Find the sales tax rate expressed as a percent. **5%**
18. On average, Thad gets a hit 24 out of every 75 times at bat. What percent of the time does Thad get a hit? **32%**

8-2 Finding Percents

Learn to find percents.

Relative humidity is a measure of the amount of water vapor in the air. When the relative humidity is 100%, the air has the maximum amount of water vapor. At this point, any additional water vapor would cause precipitation. To find the relative humidity on a given day, you would need to find a percent.

The rainy season in some parts of Indochina extends from March to November, and the average humidity is close to 90%.

EXAMPLE 1 — **Finding the Percent One Number Is of Another**

A What percent of 162 is 90?

Method 1: Set up an equation to find the percent.

$$p \cdot 162 = 90 \qquad \text{Set up an equation.}$$
$$p = \frac{90}{162} \qquad \text{Solve for } p.$$
$$p = 0.\overline{5}, \text{ or approximately } 0.56. \qquad 0.56 \text{ is } 56\%$$

So 90 is approximately 56% of 162.

B Earth has a surface area of approximately 197 million square miles. About 58 million square miles of that surface area is land. Find the percent of Earth's surface area that is land.

Method 2: Set up a proportion to find the percent.

Think: What number is to 100 as 58 is to 197?

$$\frac{number}{100} = \frac{part}{whole} \qquad \text{Set up a proportion.}$$
$$\frac{n}{100} = \frac{58}{197} \qquad \text{Substitute.}$$
$$n \cdot 197 = 100 \cdot 58 \qquad \text{Find the cross products.}$$
$$197n = 5800$$
$$n = \frac{5800}{197} \qquad \text{Solve for } n.$$
$$n \approx 29.44, \text{ or approximately } 29.$$
$$\frac{29}{100} \approx \frac{58}{197} \qquad \text{The proportion is reasonable.}$$

So approximately 29% of Earth's surface area is land.

1 Introduce

Alternate Opener

EXPLORATION

8-2 Finding Percents

The circle graph shows results from a mock election in which 40 students were polled.

Percent of Votes Received by Each Candidate

Johnson 15%
Mosely 20%
Phillips 25%
Herrera 40%

Complete the table to find the number of votes each candidate received.

	Candidate	Percent of Votes	Number of Votes
	Mosely	20%	0.20 · 40 = 8
1.	Herrera	40%	
2.	Phillips	25%	
3.	Johnson	15%	

Think and Discuss

4. **Discuss** whether there is a clear winner in the election.
5. **Tell** which two candidates combined received the same total votes as another candidate. Did they also get the same percent of the votes as the other candidate?

Exploration worksheet and answers on Chapter 8 Resource Book pp. 16 and 87

Motivate

Ask students percent questions about themselves, such as: "What percent of the students in this class have brown hair?" or "What percent of the students are wearing blue jeans?" Tell students that before class is over, they will be able to answer questions like these.

8-2 Organizer

Pacing: Traditional 1 day
Block $\frac{1}{2}$ day

Objective: Students find percents.

Warm Up

Rewrite each value as indicated.

1. $\frac{24}{50}$ as a percent **48%**
2. 25% as a fraction **$\frac{1}{4}$**
3. $\frac{3}{8}$ as a decimal **0.375**
4. 0.16 as a fraction **$\frac{4}{25}$**

Problem of the Day

A number between 1 and 10 is halved, and the result is squared. This gives an answer that is double the original number. What is the starting number? **8**

Available on Daily Transparency in CRB

Math Humor

Teacher: What is 5% of the power of a 20-watt lightbulb?

Student: A what?

Teacher: Wow, you got that one fast.

2 Teach

Lesson Presentation

Guided Instruction

In this lesson, students learn to find percents. Explain the relationship among a part, a whole, and a percent (percent × whole = part). Show students that percent problems can be solved by setting up and solving an equation (Examples 1A and 2A) or by using a proportion (Examples 1B and 2B). Encourage students to become familiar with both methods. In the application problems, you may want to have students identify the part, whole, and percent before solving the problems. Point out that the word *of* can help students identify the whole.

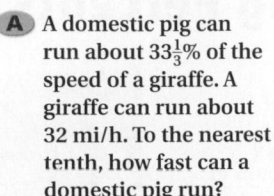
Example 1

A. What percent of 220 is 88? **40%**

B. Eddie weighs 160 lb, and his bones weigh 24 lb. Find the percent of his weight that his bones are. **15%**

Example 2

A. After a drought, a reservoir had only $66\frac{2}{3}$% of the average amount of water. If the average amount of water is 57,000,000 gallons, how much water was in the reservoir after the drought? **38,000,000 gal**

B. Ms. Chang deposited $550 in the bank. Four years later her account held 110% of the original amount. How much money did Ms. Chang have in the bank at the end of the four years? **$605**

EXAMPLE 2 Finding a Percent of a Number

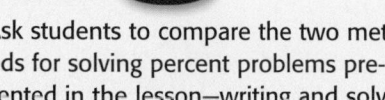

A A domestic pig can run about $33\frac{1}{3}$% of the speed of a giraffe. A giraffe can run about 32 mi/h. To the nearest tenth, how fast can a domestic pig run?

Choose a method: Set up an equation.

Think: What number is $33\frac{1}{3}$% of 32?

$n = 33\frac{1}{3}\% \cdot 32$ *Set up an equation.*

$n = \frac{1}{3} \cdot 32$ *$33\frac{1}{3}$% is equivalent to $\frac{1}{3}$.*

$n = \frac{32}{3} = 10\frac{2}{3} = 10.\overline{6}$

$n \approx 10.7$ *Round to the nearest tenth.*

A domestic pig can run about 10.7 miles per hour.

B The Chrysler Building in New York City is about 1046 feet tall. The height of the Empire State Building is approximately 120% of the height of the Chrysler Building. To the nearest foot, find the height of the Empire State Building.

Choose a method: Set up a proportion.

Think: 120 is to 100 as **what number** is to 1046?

$\frac{120}{100} = \frac{n}{1046}$ *Set up a proportion.*

$120 \cdot 1046 = 100 \cdot n$ *Find the cross products.*

$125,520 = 100n$

$1255.2 = n$ *Solve for n.*

$n \approx 1255$ *Round to the nearest whole number.*

The Empire State Building is about 1255 feet tall.

Helpful Hint

When solving a problem like this one, the number you are looking for will be greater than the number given, in this case, 1046.

Think and Discuss

1. **Show** why 5% of a number is less than $\frac{1}{10}$ of the number.

2. **Demonstrate** two ways to find 70% of a number.

3. **Give an example** of a situation in which one quantity is 300% of another quantity.

4. **Name** fractions in simplest form that are the same as 40% and as 250%.

Teach

Reaching All Learners
Through Curriculum Integration

Social Studies Give students a recording sheet (Chapter 8 Resource Book p. 81) that contains a circle graph showing election results. Have students use the percents given in the graph and the total number of voters to solve percent problems.

3 Close

Summarize

Review the two types of problems in the lesson: Find what percent a number is of another number, and find a percent of a number. Remind students that there are two methods that can be used to solve each type of problem. Ask students for some real-world situations in which they might solve these types of problems.

Possible answers: finding sales tax or discounts, calculating nutritional information, calculating sports statistics

Answers to Think and Discuss

Possible answers:

1. 5% is less than $\frac{1}{10}$ because 5% = 0.05 and $\frac{1}{10}$ = 0.10.

2. $0.70 \cdot 50 = 35$; $\frac{70}{100} = \frac{n}{50}$; $n = 35$.

3. Possible answer: A building is 3 times taller than another building.

4. $\frac{2}{5}, \frac{5}{2}$

Answers

26. Possible answer: This answer is not reasonable because 150% of 88 must be greater than 88. The correct answer should be 132.

Journal

Ask students to compare the two methods for solving percent problems presented in the lesson—writing and solving equations and setting up proportions.

Test Prep Doctor

For Exercise 32, suggest that students sketch the box. Each edge is 4 in. long, so the area of each face is 16 in². There are 6 faces, so the total surface area is 96 in². This is choice **A**. Students who selected **B** may have found the volume. Students who selected **C** may have found the area of a face and then doubled that number.

Lesson Quiz

Find each percent.

1. What percent of 33 is 22? **$66\frac{2}{3}$%**

2. What percent of 300 is 120? **40%**

3. 18 is what percent of 25? **72%**

4. The volume of Lake Superior is 2900 mi³ and the volume of Lake Erie is 116 mi³. What percent of the volume of Lake Superior is the volume of Lake Erie? **4%**

Available on Daily Transparency in CRB

---

20. **LANGUAGE ARTS** The Hawaiian words shown contain all of the letters of the Hawaiian alphabet. The ` is actually a consonant!

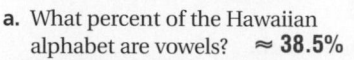
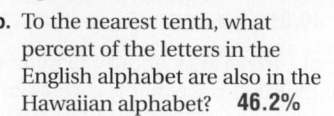

   alakahiki: pineapple

   Wai: water

   kahi: one

   Pohaku: rock, stone

   Mauna: mountain

   a. What percent of the Hawaiian alphabet are vowels? $\approx$ **38.5%**

   b. To the nearest tenth, what percent of the letters in the English alphabet are also in the Hawaiian alphabet? **46.2%**

21. **EARTH SCIENCE** If there are 3.87 cm³ of oxygen in an 18 cm³ sample of air, what percent of the sample is oxygen? **21.5%**

22. **SOCIAL STUDIES** According to the 2000 U.S. Census, approximately 2.5 million Americans spend $12\frac{1}{2}$% of the 24-hour day commuting. How many hours a day does a person in this group spend commuting? **3 hr**

23. **SOCIAL STUDIES** Of the 50 states in the Union, 32% have names that begin with either *M* or *N*. How many states have names beginning with either *M* or *N*? **16**

24. **LIFE SCIENCE** The General Sherman sequoia tree, in California, is thought to be the largest living thing on Earth by volume. It has a height of 275 ft. Its lowest large branch is at a height of 130 ft. What percent of the height of the tree would you need to climb to reach that branch? $\approx$ **47.3%**

25. **CHOOSE A STRATEGY** Demco Industries has total annual operating expenses of $12,585,000. Employee salaries cost Demco $5,034,000 each year. What percent of the company's operating expenses is employee salaries? **B**

   A 4%   **B** 40%   **C** 25%   **D** 250%

26. **WRITE ABOUT IT** A question on a math quiz asks, "What is 150% of 88?" Mark calculates 13.2 as the answer. Is this a reasonable answer? Explain why or why not.

27. **CHALLENGE** Tani cut 2 ft 6 in. from a board measuring 3 yd 1 ft. What percent of the board's original length did Tani remove, and what is the length of the board that remains? **25%; 7 ft 6 in.**

Spiral Review

State if each number is rational, irrational, or not a real number. (Lesson 3-10)

28. -14 **rational**

29. $\sqrt{13}$ **irrational**

30. $\frac{127}{46,191}$ **rational**

31. $\sqrt{-\frac{5}{6}}$ **not real**

32. **TEST PREP** Each edge of a gift box is 4 in. long. How much wrapping paper would it take to cover the surface of the gift box? (Lesson 6-8) **A**

   A 96 in²   **B** 64 in²   **C** 32 in²   **D** 128 in²

CHALLENGE 8-2

Challenge
8-2 Particular Percentages

The shaded portion of this rectangle represents $\frac{1}{4}$, or 25%, of the whole rectangle.

Find the percent of each figure that is shaded. Lines intersecting sides or circumferences divide them equally.

1. 12.5%	2. 37.5%	3. 33.33%
4. 60%	5. 6.25%	6. 16.67%
7. 50%	8. 50%	9. 25%
10. 6.25%	11. 25%	12. 25%

PROBLEM SOLVING 8-2

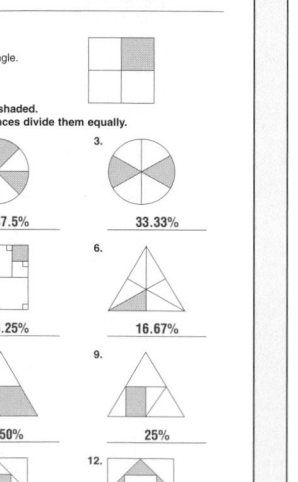

Problem Solving
8-2 Finding Percents

Write the correct answer.

1. Florida State University in Tallahassee, Florida has 27,014 students. Approximately 56% of the students are women. How many of the students are women?
   15,128 students

2. The yearly cost of tuition, room and board at Florida State University for a Florida resident is $7,164. If tuition is $1,554 a year, what percent of the yearly cost is tuition? Round to the nearest tenth of a percent.
   21.7%

3. The yearly cost of tuition, room and board at Florida State University for a non-Florida resident is $14,152. If tuition at $8,542 a year, what percent of the yearly cost is tuition for a non-resident? Round to the nearest tenth of a percent.
   60.4%

4. Approximately 54% of the students who apply to Florida State University are accepted. If 15,000 students apply to Florida State University, how many would you expect to be accepted?
   8100 students

The top four NBA field goal scorers for the 2001–2002 regular season are given in the table below. Choose the letter for the best answer.

NBA Field Goal Leaders 2001–2002 Season			
Player	Attempts	Made	Percent
Shaquille O'Neal	1109	639	
Donyell Marshall	603		52.7
Elton Brand	933	489	
Dale Davis	552		51.6

5. What percent of field goals did Shaquille O'Neal make? Round to the nearest tenth of a percent.
   A 0.6%   C 57.6%
   B 1.74%   D 59.2%

6. How many field goals did Donyell Marshall make in the 2001–2002 regular season?
   F 11   H 295
   G 114   J 318

7. What percent of field goals did Elton Brand make? Round to the nearest tenth of a percent.
   A 0.5%   C 51.9%
   B 50%   D 52.4%

8. How many field goals did Dale Davis make in the 2001–2002 regular season?
   F 107   H 299
   G 285   J 359

Technology LAB 8B

Find Percent Error

Use with Lesson 8-2

A measurement is only as precise as the device that is used to measure. There is often a difference between a measured value and an accepted or actual value. When the difference is given as a percent of the accepted value, this is called the *percent error*.

Percent error is always nonnegative, so use absolute value.

$$\text{percent error} = \frac{|\text{measured value} - \text{accepted value}|}{\text{accepted value}} \cdot 100$$

internet connect
Lab Resources Online
go.hrw.com
KEYWORD: MP4 Lab8B

Activity

1 A student uses an 8 oz cup and finds the volume of a container to the nearest 8 oz as 64 oz. The actual volume of the container is 67.6 oz. Find the percent error of the measurement to the nearest tenth of a percent.

   a. Store the measured volume on your calculator as *M* and the actual volume as *A*. Type 64 [STO▶] [ALPHA] M [ENTER] and 67.6 [STO▶] [ALPHA] A [ENTER].

   b. Find the percent error by using the following keystrokes:
     [(] [MATH] **NUM 1: ABS (** [ALPHA] M [−] [ALPHA] A [)]
     [÷] [ALPHA] A [)] [×] 100

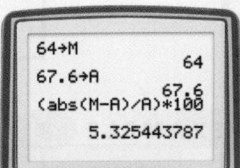

To the nearest tenth of a percent, the percent error is 5.3%.

Think and Discuss

1. Can percent error exceed 100%? Explain.

2. Tell why one measurement that is 0.1 cm from an actual length may have a larger percent error than another measurement that is 25 cm from a different actual length.

3. Describe why a ruler with centimeter markings can only measure accurately to within $\frac{1}{2}$ cm of an actual length.

Try This

Find the percent error to the nearest tenth of a percent.

1. measured length 3 cm; actual length 3.4 cm

2. measured length 250 ft; actual length 246.9 ft

Answers

Think and Discuss

1. Yes, if the accepted value is 0.4 mm and you measure 1 mm, the percent error would exceed 100%. (150%)

2. A measurement that is 0.1 cm from an actual measurement less than 0.1 cm would exceed 100%, while a measurement 25 cm from an actual measurement larger than 25 cm would be less than 100%.

3. Any measurement that falls between two centimeter marks would be a guess. So you must round to the nearest centimeter, which gives you a possible error of 0.5 cm.

Try This

1. 11.8%

2. 1.3%

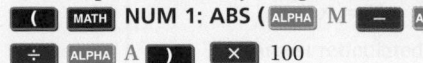

Technology LAB 8B

Find Percent Error

Pacing: Traditional 1 day
         Block $\frac{1}{2}$ day
Objective: To use a graphing calculator to compute percent error
Materials: Graphing calculator

Lab Resources
Technology Lab Activities . . . . p. 52

Using the Page

This technology activity shows students how to find percent error, which can be done on any graphing calculator. Specific keystrokes may vary, depending on the make and model of the graphing calculator used. The keystrokes given are for a TI-83 model. For keystrokes to other models, visit go.hrw.com.

The Think and Discuss problems can be used to assess students' understanding of percent error and the technology activity. While the Try This problems can be done without a graphing calculator, they are meant to help students become familiar with using a graphing calculator to compute compound interest.

Assessment

Suppose you measured a wall to be 15 ft long using a tape measure marked in feet. The floor plan of the room showed the wall as being 14.8 ft.

1. What is the percent error of your measurement compared with the floor plan? **1.4%**

2. What is the percent error of the floor plan compared with your measurement? **1.3%**

3. Why aren't they the same? Although the absolute value of the difference between 14.8 and 15 remains the same for both calculations, the accepted value in the denominator changes.

8-3 Exercises

FOR EXTRA PRACTICE
see page 746

internet connect
Homework Help Online
go.hrw.com   Keyword: MP4 8-3

go.hrw.com

> Students may want to refer back to the lesson examples.

Assignment Guide

If you finished Example **1** assign:
Core 1–4, 7–14, 17, 19, 26–30
Enriched 1–4, 7–14, 17–19, 26–30

If you finished Example **2** assign:
Core 1–5, 7–15, 17, 19, 26–30
Enriched 7–15, 17–20, 25–30

If you finished Example **3** assign:
Core 1–16, 17–23 odd, 26–30
Enriched 7–30

Notes

_____
_____
_____
_____
_____
_____
_____
_____
_____
_____
_____
_____
_____

GUIDED PRACTICE

See Example **1** Find each number to the nearest tenth.

1. 4.3 is $12\frac{1}{2}$% of what number? **34.4** **2.** 56 is $33\frac{1}{3}$% of what number? **168**

3. 18% of what number is 30? **166.7** **4.** 30% of what number is 96? **320**

See Example **2** **5.** The only kind of rock that floats in water is pumice. Chalk, although denser, absorbs more water than pumice does. How much water can a 5.2 oz piece of chalk absorb if it can absorb 32% of its weight? **≈ 1.7 oz**

See Example **3** **6.** At 3 P.M., a chimney casts a shadow that is 135% of its actual height. If the shadow is 37.8 ft, what is the actual height of the chimney? **28 ft**

INDEPENDENT PRACTICE

See Example **1** Find each number to the nearest tenth.

7. 105 is $33\frac{1}{3}$% of what number? **315** **8.** 77 is 25% of what number? **308**

9. 51 is 6% of what number? **850** **10.** 24 is 15% of what number? **160**

11. 84% of what number is 14? **16.7** **12.** 56% of what number is 39.2? **70**

13. 10% of what number is 57? **570** **14.** 180% of what number is 6? **3.3**

See Example **2** **15.** Manuel sold 42 of his baseball cards at a collectors show. If this represented $12\frac{1}{2}$% of his total collection, how many baseball cards did Manuel have before the show? **336**

See Example **3** **16.** When a tire is labeled "185/70/14," that means it is 185 mm wide, the sidewall height (from the rim to the road) is 70% of its width, and the wheel has a diameter of 14 in. What is the tire's sidewall height? **129.5 mm**

PRACTICE AND PROBLEM SOLVING

Complete each statement.

17. Since 1% of 600 is 6,   **18.** Since 100% of 8 is 8,   **19.** Since 5% of 80 is 4,

a. 2% of ▯ is 6. **300**   **a.** 50% of ▯ is 8. **16**   **a.** 10% of ▯ is 4. **40**

b. 4% of ▯ is 6. **150**   **b.** 25% of ▯ is 8. **32**   **b.** 20% of ▯ is 4. **20**

c. 8% of ▯ is 6. **75**   **c.** 10% of ▯ is 8. **80**   **c.** 40% of ▯ is 4. **10**

20. In a poll of 225 students, 36 said that their favorite Thanksgiving food was turkey, and 56 said that their favorite was stuffing. Give the percent of students who said that each food was their favorite. **16% turkey, ≈ 24.9% stuffing**

Math Background

As indicated at the end of the lesson, there are three basic types of percent problems. Equations for all three types can be written using a direct translation approach. If a question is stated using the word "what," then an equation can be written with "what" replaced by a variable, "of" by a multiplication symbol, and "is" by an equal sign. Some examples are shown below:

36 is 4% of what number?
→ $36 = 0.04 \cdot x$

What is 82.5% of 250?
→ $x = 0.825 \cdot 250$

What percent of 162 is 90?
→ $x \cdot 162 = 90$

RETEACH 8-3

LESSON 8-3 Reteach
Finding a Number When the Percent Is Known

Since a percent is a ratio, problems involving percent can be solved by using a proportion.

$$\frac{symbol\ number}{100} = \frac{is\ number}{of\ number}$$

To find a number when the percent is known, the variable appears in the *of* position in the proportion

16 is 20% of what number?

$$\frac{20}{100} = \frac{16}{x}$$
$$20 \cdot x = 16 \cdot 100$$
$$\frac{20x}{20} = \frac{1600}{20}$$
$$x = 80$$

So, 16 is 20% of 80.

Find each number whose percentage is given.

1. 18 is 75% of what number?

$$\frac{75}{100} = \frac{18}{x}$$
$$75 \cdot x = 18 \cdot 100$$
$$\frac{75x}{75} = \frac{1800}{75}$$
$$x = \underline{24}$$

So, 18 is 75% of __24__.

2. 96 is 40% of what number?

$$\frac{40}{100} = \frac{96}{x}$$
$$40 \cdot x = 96 \cdot 100$$
$$\frac{40x}{40} = \frac{9600}{40}$$
$$x = \underline{240}$$

So, 96 is 40% of __240__.

3. 7 is 125% of what number

$$\frac{125}{100} = \frac{7}{x}$$
$$125 \cdot x = 7 \cdot 100$$
$$\frac{125x}{125} = \frac{700}{125}$$
$$x = \underline{5.6}$$

So, 7 is 125% of __5.6__.

4. 40 is about 30% of what number?

$$\frac{30}{100} = \frac{40}{x}$$
$$30 \cdot x = 40 \cdot 100$$
$$\frac{30x}{30} = \frac{4000}{30}$$
$$x = \underline{133.3}$$

So, 40 is about 30% of __133.3__.

PRACTICE 8-3

LESSON 8-3 Practice B
Finding a Number When the Percent Is Known

Find each number.

1. 40% of what number is 18? __45__
2. 28 is 35% of what number? __80__
3. 21 is 60% of what number? __35__
4. 25% of what number is 19? __76__
5. 40% of what number is 22? __55__
6. 41 is 50% of what number? __82__
7. 48 is 15% of what number? __320__
8. 30% of what number is 24? __80__
9. 36 is 30% of what number? __120__
10. 26 is 65% of what number? __40__
11. 12.5% of what number is 14? __112__
12. 25% of what number is 28.25? __113__
13. 27 is $33\frac{1}{3}$% of what number? __81__
14. 54 is 150% of what number? __36__

15. There were 546 students at a school assembly. This was 65% of all students who attend Content Middle School. How many students attend Content Middle School?
__840 students__

16. On his last test Greg answered 64 questions correctly. This was 80% of the questions. How many questions were on the test?
__80 questions__

17. The price of a jacket is $48. If the sales tax rate is 5.5%, what is the amount of tax? What is the total cost of the jacket?
__tax is $2.64; total cost is $50.64__

18. Carla has finished swimming 14 laps in swim practice. This is 70% of the total number of laps she must swim. How many more laps must Carla swim to complete her practice?
__6 laps__

The U.S. census collects information about state populations, economics, income and poverty levels, births and deaths, and so on. This information can be used to study trends and patterns. For Exercises 21–23, round answers to the nearest tenth.

The New York counties with the greatest populations are Kings (Brooklyn) and Queens.

2000 U.S. Census Data			
	Population	Male	Female
Alaska	626,932	324,112	302,820
New York	18,976,457	9,146,748	9,829,709
Age 34 and Under	139,328,990	71,053,554	68,275,436
Age 35 and Over	142,092,916	67,000,009	75,092,907
Total U.S.	281,421,906	138,053,563	143,368,343

21. What percent of New York's population is male? **48.2%**

22. What percent of the entire country's population, to the nearest tenth of a percent, is made up of people in New York? **6.7%**

23. Tell what percent of the U.S. population each represents.
   a. people 34 and under **49.5%** b. people 35 and over **50.5%**
   c. male **49.1%**   d. female **50.9%**

24. American Indians and Native Alaskans make up about 15.6% of Alaska's population. What is their population, to the nearest thousand? **98,000**

25. ★ *CHALLENGE* About 71% of the U.S. population age 85 and over is female. Of the fractions that round to 71% when rounded to the nearest percent, which has the least denominator? $\frac{5}{7}$

go.hrw.com
KEYWORD: MP4 Census
CNN Student News.

Social Studies

Exercises 21–25 involve using U.S. Census data in percent problems. Students study census data from countries around the world in middle school social studies programs, such as Holt, Rinehart & Winston's *People, Places, and Change*.

Journal

Ask students to describe the process they would use to solve each of the three types of percent problems.

Test Prep Doctor

For Exercise 30, remind students how to dilate a figure with the origin as the center of dilation. Students who selected **A** may have used a scale factor of $\frac{1}{2}$. Students who selected **B** may have used a scale factor of -2. Students who selected **D** probably squared the original coordinates. The correct choice is **C**.

Spiral Review

Find the range of each set of data. (Lesson 4-4)

26. 16, 32, 1, 54, 30, 28 **53**   27. 105, 969, 350, 87, 410 **882** 28. 0.2, 0.8, 0.65, 0.7, 1.6, 1.1 **1.4**

Find the first and third quartiles of the data set. (Lesson 4-4)

29. 55, 60, 40, 45, 70, 65, 35, 40, 75, 50, 60, 80, 45, 55 **45, 65**

30. **TEST PREP** A triangle has vertices $A(4, 4)$, $B(6, -2)$, and $C(-4, -12)$. What are the vertices after dilating by a scale factor of 2 with the origin as the center of dilation? (Lesson 7-5) **C**

   A $A'(2, 2)$, $B'(3, -1)$, $C'(-2, -6)$ C $A'(8, 8)$, $B'(12, -4)$, $C'(-8, -24)$
   B $A'(-8, -8)$, $B'(-12, 4)$, $C'(8, 24)$ D $A'(16, 16)$, $B'(36, 4)$, $C'(16, 144)$

CHALLENGE 8-3

LESSON 8-3 Challenge
In the Chemistry Laboratory

When a chemist dilutes pure acid with another substance, the resulting mixture is no longer pure acid.

Consistent with the words, *pure acid* is 100% acid. So, there are 20 grams of pure acid in 20 grams of a pure-acid solution.

Laura, a chemist, has 20 grams of a solution that is only 40% acid.

1. How many grams of pure acid are there in Laura's acid solution?

   40% of 20 grams = 8 grams

Suppose, now, Laura wants to increase the acid content of the 40% acid solution to make it a 50%-acid solution.

2. What do you think Laura has to do to increase the acid content of the solution? **Possible answer:**

   Add some pure acid; also possible to evaporate.

Laura decides to add *n* ounces of pure acid to increase the acid content of the original 20 grams of 40%-acid solution to make it a 50%-acid solution.

3. Represent in terms of *n* the total number of grams in the new solution.   **20 + n**

4. Represent in terms of *n* the number of grams of pure acid in the new solution.   **0.50 (20 + n)**

Then, the amount of pure acid in the original solution plus the amount of pure acid added equals the amount of pure acid in the new solution.

5. Use the results of Exercises 1 and 4 to write an equation that will find the number *n* of grams of pure acid that will be added to the original solution to increase its acid content from 40% to 50%. Solve the equation.

   $8 + n = 0.50(20 + n)$
   $8 + n = 10 + 0.50n$
   $n - 0.50n = 10 - 8$
   $0.50n = 2$
   $n = 4$

6. Explain how to check your result. **Possible answer:**

   There are 24 g in all in the new solution; 50%, or 12 g, are pure acid.
   This is consistent with adding 4 g of pure acid to the original solution
   that had 8 g of pure acid.

PROBLEM SOLVING 8-3

LESSON 8-3 Problem Solving
Finding a Number When the Percent is Known

Write the correct answer.

1. The two longest running Broadway shows are *Cats* and *A Chorus Line*. *A Chorus Line* had 6137, or about 82% of the number of performances that *Cats* had. How many performances of *Cats* were there?

   7484

2. *Titanic* and *Star Wars* have made the most money at the box office. *Star Wars* made about 76.7% of the money that *Titanic* made at the box office. If *Star Wars* made about $461 million, how much did *Titanic* make? Round to the nearest million dollars.

   $601 million

Use the table below. Round to the nearest tenth of a percent.

3. What percent of students are in Pre-K through 8th grade?

   71.4%

4. What percent of students are in grades 9–12?

   28.6%

Public Elementary and Secondary School Enrollment, 1999	
Grades	Population (in thousands)
Pre-K through grade 8	33,437
Grades 9–12	13,375
Total	46,812

Choose the letter for the best answer.

5. In 2000, women earned about 72.2% of what men did. If the average woman's weekly earnings was $491 in 2000, what was the average man's weekly earnings? Round to the nearest dollar.
   A $355   C $680
   B $542   D $725

6. The highest elevation in North America is Mt. McKinley at 20,320 ft. The highest elevation in Australia is Mt. Kosciusko, which is about 36% of the height of Mt. McKinley. What is the highest elevation in Australia? Round to the nearest foot.
   F 5480 ft   H 12,825 ft
   G 7315 ft   J 56,444 ft

7. The Gulf of Mexico has an average depth of 4,874 ft. This is about 36.2% of the average depth of the Pacific Ocean. What is the average depth of the Pacific Ocean? Round to the nearest foot.
   A 1764 ft   C 10,280 ft
   B 5843 ft   D 13,464 ft

8. Karl Malone is the NBA lifetime leader in free throws. He attempted 11,703 and made 8,636. What percent did he make? Round to the nearest tenth of a percent.
   F 1.4%   H 73.8%
   G 58.6%   J 135.6%

Lesson Quiz

1. 10 is $12\frac{1}{2}\%$ of what number? **80**

2. 326 is 25% of what number? **1304**

3. 44% of what number is 11? **25**

4. 290% of what number is 145? **50**

5. Larry has 9 novels about the American Revolutionary War. This represents 15% of his total book collection. How many books does Larry have in all? **60**

Available on Daily Transparency in CRB

Purpose: *To assess students' mastery of concepts and skills in Lessons 8-1 through 8-3*

Assessment Resources

Section 8A Quiz
Assessment Resources . . . . p. 19

 Test and Practice Generator
CD-ROM

Additional mid-chapter assessment items in both multiple-choice and free-response format may be generated for any objective in Lessons 8-1 through 8-3.

Mid-Chapter Quiz

Chapter 8 **Mid-Chapter Quiz**

LESSON 8-1 (pp. 400–403)

Find the equivalent value missing from the table for each value given on the circle graph.

Fraction	Decimal	Percent
$\frac{1}{8}$	1. ▦	2. ▦
3. ▦	0.25	4. ▦
5. ▦	6. ▦	$37\frac{1}{2}\%$
$\frac{1}{4}$	7. ▦	8. ▦

0.125, 12.5%

$\frac{1}{4}$, 25%

$\frac{3}{8}$, 0.375

0.25, 25%

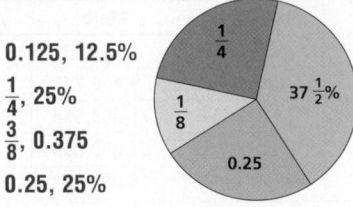

LESSON 8-2 (pp. 405–408)

9. What is 27% of 16? **4.32**

10. 48 is what percent of 384? **12.5%**

11. In the November 2001 election, only 191,411 of the 509,719 voters registered in Westchester County, New York, cast a ballot. This was the lowest turnout in at least a century. To the nearest tenth, what percent of registered voters actually voted in the election? **37.6%**

12. Use the height of the 88-story Jin Mao Tower in Shanghai and the information shown at right to find the heights of the Eiffel Tower and Russia's Motherland Statue. **≈ 985 ft; ≈ 270 ft**

13. Of Canada's total area of 9,976,140 km², 755,170 km² is water. To the nearest tenth of a percent, what part of Canada is water? **≈ 7.6%**

$x = 1378$ ft

71.5% of x

19.6% of x

Jin Mao Tower   Eiffel Tower   Motherland Statue

LESSON 8-3 (pp. 410–413)

14. 30 is 12.5% of what number? **240**
15. 244 is 250% of what number? **97.6**

16. The speed of sound in air at sea level at 32°F is 1088 ft/s. If that represents only 22.04% of the speed of sound in ice-cold water, what is the speed of sound in ice-cold water, to the nearest whole number? **4936 ft/s**

17. In 2000, U.S. imports from Canada totaled $230,838.3 million. This was about 129% of the total dollar value of the U.S. exports to Canada. To the nearest ten million dollars, what was the value of U.S. exports to Canada? **$178,940 million**

Focus on Problem Solving

Make a Plan

• **Do you need an estimate or an exact answer?**

When you are solving a word problem, ask yourself whether you need an exact answer or whether an estimate is sufficient. For example, if the amounts given in the problem are approximate, only an approximate answer can be given. If an estimate is sufficient, you may wish to use estimation techniques to save time in your calculations.

For each problem below, explain whether an exact answer is needed or whether an estimate is sufficient. Then find the answer.

1 In a poll of 3000 registered voters in a certain district, 1800 favored a proposed school bond package. What percent favored the bond package?

2 George needs to score 76% on his final exam to get a B in his math class. If the final is worth 200 points, how many points does he need?

3 Karou is trying to save about $3500 for a trip to Japan. If she has $1000 in an account that earns 8% interest and puts $100 per month in the account, will she have enough in 2 years?

4 Erik makes $7.60 per hour at his job. If he receives a 5% raise, how much will he be making per hour?

5 Jamie is planning to tile her kitchen floor. The room is 330 square feet. It is recommended that she buy enough tile for an area 15% greater than the actual kitchen floor in case of breakage. How many square feet of tile should she buy?

6 There are about 1,032,000 known species of animals on Earth. Of these, about 751,000 are insects. What percent of known species are insects?

Answers

1. 60%
2. 152 points
3. yes
4. $7.98
5. about 375 ft² of tile
6. about 75%

Purpose: *To focus on making a plan to solve a problem by deciding whether an exact answer or an estimate is needed*

Problem Solving Resources

Interactive Problem Solving . . . pp. 64–70

Math: Reading and Writing in the Content Area . . . . pp. 64–70

Problem Solving Process

This page focuses on the second step of the problem-solving process: **Make a Plan**

Discuss

For each problem, have students tell whether an exact answer is needed and then use estimation to check the reasonableness of their answers. If estimation was all that was required, have them explain how they estimated.

Possible answers:

1. An exact answer is needed. 1800 out of 3000 is a little more than half. The answer should be a little more than 50%.

2. An exact answer is needed. 70% of 200 is 140. 80% of 200 is 160. George's score should be between 140 and 160.

3. An estimation is sufficient because of the word *about.* Even without interest, Karou will have $1000 + 24($100) = $3400. It is likely that interest will be enough to make her balance over $3500.

4. An exact answer is needed. 5% of $7.00 is $0.35, so Erik's new pay rate should be around $8.00.

5. An estimate is sufficient because the 15% extra is an approximation of how many tiles may be broken. Round 330 to 300; 15% of 300 is 45, so she should buy about 330 + 45 = 375 tiles.

6. An estimation is sufficient because of the word *about.* Round each number: $\frac{750,000}{1,000,000} = \frac{3}{4}$. About 75% of the known species are insects.

Applying Percents

One-Minute Section Planner

Lesson	Materials	Resources
Lesson 8-4 Percent Increase and Decrease **NCTM:** Number and Operations, Algebra, Communication, Representation **NAEP:** Number Properties 4d ☐ SAT-9 ☑ SAT-10 ☑ ITBS ☑ CTBS ☑ MAT ☑ CAT		• *Chapter 8 Resource Book,* pp. 35–44 • *Daily Transparency* T15, CRB • Additional Examples Transparencies T16–T19, CRB • *Alternate Openers: Explorations,* p. 67
Lesson 8-5 Estimating with Percents **NCTM:** Number and Operations, Problem Solving, Communication, Representation **NAEP:** Number Properties 2b ☑ SAT-9 ☑ SAT-10 ☐ ITBS ☐ CTBS ☑ MAT ☑ CAT	**Optional** Recording Sheet for Reaching All Learners *(CRB, p. 83)* Calculators	• *Chapter 8 Resource Book,* pp. 45–54 • *Daily Transparency* T20, CRB • Additional Examples Transparencies T21–T23, CRB • *Alternate Openers: Explorations,* p. 68
Lesson 8-6 Applications of Percents **NCTM:** Number and Operations, Communication, Representation **NAEP:** Number Properties 4d ☐ SAT-9 ☑ SAT-10 ☐ ITBS ☐ CTBS ☑ MAT ☑ CAT	**Optional** Calculators	• *Chapter 8 Resource Book,* pp. 55–64 • *Daily Transparency* T24, CRB • Additional Examples Transparencies T25–T28, CRB • *Alternate Openers: Explorations,* p. 69
Lesson 8-7 More Applications of Percents **NCTM:** Number and Operations, Communication, Representation **NAEP:** Number Properties 4d ☐ SAT-9 ☑ SAT-10 ☐ ITBS ☐ CTBS ☑ MAT ☑ CAT	**Optional** Teaching Transparency T30 *(CRB)* Sample store circulars *(CRB, p. 84)* Number cubes *(MK)*	• *Chapter 8 Resource Book,* pp. 65–74 • *Daily Transparency* T29, CRB • Additional Examples Transparencies T31–T33, CRB • *Alternate Openers: Explorations,* p. 70
Extension Compound Interest **NCTM:** Number and Operations **NAEP:** Number Properties 4d ☐ SAT-9 ☐ SAT-10 ☐ ITBS ☐ CTBS ☐ MAT ☑ CAT	**Optional** Computers with spread-sheet software Calculators	• Additional Examples Transparencies T34–T35, CRB
Section 8B Assessment		• Section 8B Quiz, AR p. 20 • *Test and Practice Generator* CD-ROM

SAT = *Stanford Achievement Tests*   **ITBS** = *Iowa Test of Basic Skills*   **CTBS** = *Comprehensive Test of Basic Skills/Terra Nova*
MAT = *Metropolitan Achievement Test*   **CAT** = *California Achievement Test*

NCTM = Complete standards can be found on pages T29–T35.   **NAEP** = Complete standards can be found on pages A54–A59.

SE = *Student Edition*   **TE** = *Teacher's Edition*   **AR** = *Assessment Resources*   **CRB** = *Chapter Resource Book*   **MK** = *Manipulatives Kit*

Section Overview

Percent Increase and Decrease

Lesson 8-4

 Why? Percents are often used to describe changes.

> **Percent change** is the ratio of the *amount of change* to the *original amount*.
>
> **percent change** = $\frac{\text{amount of change}}{\text{original amount}}$

Find the percent change on a $24 shirt that is on sale for $20.

First, find the **amount of change.**

$24 − $20 = **$4**

Next, find the **percent change.**

$$\frac{\text{amount of change}}{\text{original amount}} = \frac{\$4}{\$24} = \frac{1}{6} = 16\frac{2}{3}\%$$

There was a $16\frac{2}{3}\%$ decrease in the price.

Estimating Percents

Lesson 8-5

 Why? Tips do not have to be calculated exactly, so estimation is often used.

> **Methods of Estimating**
> 1. Use compatible numbers.
> 2. Round to common percents (e.g., 10%, 25%, $33\frac{1}{3}\%$).
> 3. Break percents into smaller parts (e.g., 1%, 5%, 10%).

Your dinner bill is $35.74. Estimate the 15% tip.

Estimate the total:	$35.74 is close to $36.
Break the percent into smaller parts:	15% = 10% + 5%
Estimate the amounts:	10% of $36 is $3.60.
	5% is half of 10%,
	so 5% of $36 is $1.80.
Combine the amounts:	$3.60 + $1.80 = $5.40

A 15% tip on a $35.74 bill is about $5.40.

Applications of Percents

Lesson 8-6, 8-7, Extension

Why? Percents are used in banking and in sales.

Commission

Last month, a real estate agent sold one house for $90,000, earning a 3% commission on the sale. The agent is paid a monthly salary of $1200 plus commission. What was her total pay last month?

commission = commission rate · sales
= 0.03 · 90,000
= **$2700**

total pay = commission + salary
= **$2700** + $1200
= $3900

The agent's total pay for last month was $3900.

Simple Interest

Find the amount repaid on a 4-year $13,500 loan at an annual simple interest rate of 6%.

Find the simple interest:

simple interest = principal · rate · time
$I = Prt$
= 13,500 · 0.06 · 4
= **$3240**

Find the amount repaid:

amount = principal + **simple interest**
$A = P + I$
= $13,500 + **$3240**
= $16,740

The amount repaid is $16,740.

Pacing: Traditional $\frac{1}{2}$ day
Block $\frac{1}{3}$ day

Objective: Students find percent increase and decrease.

Warm Up

1. 14,000 is $2\frac{1}{2}$% of what number? **560,000**

2. 39 is 13% of what number? **300**

3. $37\frac{1}{2}$% of what number is 12? **32**

4. 150% of what number is 189? **126**

Problem of the Day

In a school survey, 45% of the students said orange juice was their favorite juice, 25% preferred apple, and 10% preferred grapefruit. The remaining 32 students preferred grape juice. How many students participated in the survey?
160 students

Available on Daily Transparency in CRB

Math Humor

Salesperson: I had zero sales this week, but next week I'm going after a 100% increase.

Manager: You should be going after a new job instead.

8-4 # Percent Increase and Decrease

Learn to find percent increase and decrease.

Vocabulary
percent change
percent increase
percent decrease

Many animals hibernate during the winter to survive harsh conditions and food shortages. While they sleep, their body temperatures drop, their breathing rates decrease, and their heart rates slow. They may even appear to be dead.

"He hums in his sleep."

Percents can be used to describe a change. **Percent change** is the ratio of the *amount of change* to the *original amount*.

$$\text{percent change} = \frac{\text{amount of change}}{\text{original amount}}$$

Percent increase describes how much the original amount increases.
Percent decrease describes how much the original amount decreases.

EXAMPLE 1 **Finding Percent Increase or Decrease**

Find the percent increase or decrease from 20 to 24.

This is percent increase.

$24 - 20 = 4$ *First find the amount of change.*

Think: What percent is 4 of 20?

$\dfrac{\text{amount of increase}}{\text{original amount}} \rightarrow \dfrac{4}{20}$ *Set up the ratio.*

$\dfrac{4}{20} = 0.2$ *Find the decimal form.*

$\quad\; = 20\%$ *Write as a percent.*

From 20 to 24 is a 20% increase.

EXAMPLE 2 **Life Science Application**

A The heart rate of a hibernating woodchuck slows from 80 to 4 beats per minute. What is the percent decrease?

$80 - 4 = 76$ *First find the amount of change.*

Think: What percent is 76 of 80?

$\dfrac{\text{amount of decrease}}{\text{original amount}} \rightarrow \dfrac{76}{80}$ *Set up the ratio.*

$\dfrac{76}{80} = 0.95$ *Find the decimal form.*

$\quad\; = 95\%$ *76 is 95% of 80.*

The woodchuck's heart rate decreases by 95% during hibernation.

---

1 ## Introduce
Alternate Opener

EXPLORATION

8-4 **Percent Increase and Decrease**

1. After his first year at a job, Andrew's original hourly wage of $7.95 increased to $9.54.

   a. Subtract the old hourly wage from the new hourly wage to find the amount by which Andrew's hourly wage increased.

   b. Divide the amount in 1a by Andrew's original hourly wage to compare the amount of increase to the original hourly wage.

   c. Write the decimal in 1b as a percent to find the percent increase in Andrew's hourly wage.

2. An electronics store offers a $100 discount on everything in the store that is priced between $500 and $900. Complete the table to determine the percent of decrease for each price.

Price	$500	$600	$700	$800	$900
Percent of Decrease	$\frac{100}{500} =$ ___%				

Think and Discuss

3. **Explain** how you wrote the decimal in 1c as a percent.
4. **Compare** the percents of decrease in the table in number 2.

Motivate

Ask students to name a salary they would like to earn if they had a job (perhaps a student has a part-time job and can name an actual amount earned). Tell students to imagine that they received a 12% increase. Ask if anyone can find the new salary. Point out to students that this lesson is about percent increase and percent decrease. You may want to discuss other situations involving various increases or decreases.

Exploration worksheet and answers on Chapter 8 Resource Book pp. 36 and 91

2 ## Teach

Lesson Presentation

Guided Instruction

In this lesson, students learn to find percent increase and decrease. Explain to students that the actual amount of increase or decrease is not the same as the percent of increase or decrease. Emphasize to students that when calculating the percent, they must compare the amount of increase or decrease with the original amount. Explain that percent increase or decrease problems often involve two steps: solving a percent problem and addition or subtraction. In Example 2, the addition or subtraction is the first step. In Example 3, solving the percent problem is the first step.

B According to the U.S. Census Bureau, 69.9 million children lived in the United States in 1998. It is estimated that there will be 77.6 million children in 2020. What is the percent increase, to the nearest percent?

$77.6 - 69.9 = 7.7$ *First find the amount of change.*

Think: What percent is 7.7 of 69.9?

$\dfrac{\text{amount of increase}}{\text{original amount}} = \dfrac{7.7}{69.9}$ *Set up the ratio.*

$\dfrac{7.7}{69.9} \approx 0.1102$ *Find the decimal form.*

$\approx 11.02\%$ *Write as a percent.*

The number of children in the United States is estimated to increase 11%.

EXAMPLE 3  **Using Percent Increase or Decrease to Find Prices**

A Anthony bought an LCD monitor originally priced at $750 that was reduced in price by 35%. What was the reduced price?

$750 \cdot 35\%$ *First find 35% of $750.*

$\$750 \cdot 0.35 = \$262.50$ *35% = 0.35*

The amount of decrease is $262.50.

Think: The reduced price is $262.50 *less than* $750.

$\$750 - \$262.50$ *Subtract the amount of decrease.*

$= \$487.50$

The reduced price of the monitor was $487.50.

B Mr. Salazar received a shipment of sofas that cost him $366 each. He marks the price of each sofa up $33\frac{1}{3}\%$ to find the *retail price*. What is the retail price of each sofa?

$\$366 \cdot 33\frac{1}{3}\%$ *First find $33\frac{1}{3}\%$ of $366.*

$\$366 \cdot \frac{1}{3} = \$122$ $33\frac{1}{3}\% = \frac{1}{3}$

The amount of increase is $122.

Think: The retail price is $122 *more than* $366.

$\$366 + \$122 = \$488$ *Add the amount of increase.*

The retail price of each sofa is $488.

Think and Discuss

1. Explain whether a 150% increase or a 150% decrease is possible.

2. Compare finding a 20% increase to finding 120% of a number.

3. Explain how you could find the percent of change if you knew the U.S. populations in 1990 and 2000.

---

3 Close

Reaching All Learners
Through Curriculum Integration

Ask students to find a fact in another subject area that involves a percent increase or decrease (e.g., social studies: population increase, economics: cost of living increase, health: increase in life expectancy, physical education: decrease in winning times for track events).

Have students prepare a brief report on the topic and include the percent of increase or decrease.

Summarize

Remind students that when solving percent increase or decrease problems, the original amount is always used as the denominator. Ask students to name as many words as possible that indicate an increase or a decrease. Ask students what a 100% increase means and a 100% decrease means.

Possible answers: Increase: *go up, raise, rise, was higher, inflation, marked up, growth;* decrease: *went down, lowered, discount, savings, reduced, decline;* a 100% increase doubles the original amount; a 100% decrease reduces the original amount to zero.

Answers to Think and Discuss

Possible answers:

1. It is possible to have a 150% increase (e.g., from 10 to 25). It is usually not possible to have a 150% decrease, because a 100% decrease would result in zero, and any decrease beyond that would require negative numbers.

2. The amount after a 20% increase is the same as 120% of the original number.

3. Find the difference between the 2000 population and the 1990 population, and divide that difference by the 1990 population.

8-4 **Exercises**

FOR EXTRA PRACTICE
see page 747

internet connect
Homework Help Online
go.hrw.com   Keyword: MP4 8-4

go.hrw.com

Students may want to refer back to the lesson examples.

GUIDED PRACTICE

See Example **1** Find each percent increase or decrease to the nearest percent.

1. from 40 to 55
38% increase
2. from 85 to 30
65% decrease
3. from 75 to 150
100% increase
4. from 55 to 90
64% increase
5. from 110 to 82
25% decrease
6. from 82 to 110
34% increase

See Example **2** **7.** A population of geese rose from 234 to 460 over a period of two years. What is the percent increase, to the nearest tenth of a percent? **96.6% increase**

See Example **3** **8.** An automobile dealer agrees to cut 5% off the $10,288 sticker price of a new car for a customer. What is the price of the car for the customer? **$9773.60**

INDEPENDENT PRACTICE

See Example **1** Find each percent increase or decrease to the nearest percent.

9. from 55 to 60
9% increase
10. from 111 to 200
80% increase
11. from 9 to 5
44% decrease
12. from 800 to 1500
88% increase
13. from 0.84 to 0.67
20% decrease
14. from 45 to 20
56% decrease

See Example **2** **15.** The boiling point of water is lower at higher altitudes. Water boils at 212°F at sea level and 193.7°F at 10,000 ft. What is the percent decrease in the temperatures, to the nearest tenth of a percent? **≈ 8.6%**

See Example **3** **16.** Mr. Simmons owns a hardware store and typically marks up merchandise by 28% over warehouse cost. How much would he charge for a hammer that costs him $13.50? **$17.28**

PRACTICE AND PROBLEM SOLVING

Find each percent increase or decrease to the nearest percent.

17. from $49.60 to $38.10
23% decrease
18. from $67 to $104
55% increase
19. from $575 to $405
30% decrease
20. from $822 to $766
7% decrease
21. from $0.23 to $0.19
17% decrease
22. from $12.50 to $14.75
18% increase

Find each missing number.

23. originally: $500
new price: ■ **$600**
20% increase
24. originally: 140
new amount: ■ **210**
50% increase
25. originally: ■ **200**
new amount: 230
15% increase

26. originally: ■ **$4.42**
new price: $4.20
5% decrease
27. originally: 32
new amount: 48
■ % increase **50**
28. originally: $65
new price: $52
■ % decrease **20**

29. Maria purchased a CD burner for $199. Six months later, the same burner was selling for $119. By what percent had the price decreased, to the nearest percent? **40%**

Math Background

If an amount of increase is followed by an equal amount of decrease, the percent increase and the percent decrease are not equal. Consider the example:

Find the percent of increase from 40 to 50: $\frac{increase}{original\ amount} = \frac{10}{40} = 25\%$.

Find the percent of decrease from 50 to 40: $\frac{decrease}{original\ amount} = \frac{10}{50} = 20\%$.

The explanation for this is that a percent change is based on a comparison to an original amount. When the same amount of change (10) is compared with two different original amounts (40 and 50), you get different percents of change.

RETEACH 8-4

LESSON **8-4** **Reteach**
Percent Increase and Decrease

To find the percent increase:
- Find the amount of increase by subtracting the lesser number from the greater.
- Write a fraction: percent increase = $\frac{amount\ of\ increase}{original\ amount}$
- If possible, simplify the fraction.
- Rewrite the fraction as a percent.

The temperature increased from 60°F to 75°F. Find the percent of increase.

percent of increase = $\frac{75° - 60°}{60°} = \frac{15°}{60°} = \frac{1}{4} = 25\%$

Complete to find each percent increase.

1. Membership increased from 80 to 100.

$\underline{100} - \underline{80}$

$= \underline{20}$

$\frac{20}{80} = \frac{1}{4}$

$= \underline{25}$ %

2. Savings increased from $500 to $750.

$\underline{750} - \underline{500}$

$= \underline{250}$

$\frac{250}{500} = \frac{1}{2}$

$= \underline{50}$ %

percent increase = $\frac{amount\ of\ increase}{original\ amount}$
Change the fraction to a percent.

3. Price increased from $20 to $23.

$\underline{23} - \underline{20} = \underline{3}$

$\frac{3}{20}$

$\frac{3}{20} = 20\overline{)3.00}$ $\underline{15}$ %

percent increase = $\frac{amount\ of\ increase}{original\ amount}$
Change the fraction to a percent.

Find the amount of increase.

Find the amount of increase.

PRACTICE 8-4

LESSON **8-4** **Practice B**
Percent Increase and Decrease

Find each percent increase or decrease to the nearest percent.

1. from 16 to 20
increase 25%
2. from 30 to 24
decrease 20%
3. from 15 to 30
increase 100%

4. from 35 to 21
decrease 40%
5. from 40 to 46
increase 15%
6. from 45 to 63
increase 40%

7. from 18 to 26.1
increase 45%
8. from 24.5 to 21.56
decrease 12%
9. from 90 to 72
decrease 20%

10. from 29 to 54
increase 86%
11. from 42 to 92.4
increase 120%
12. from 38 to 33
decrease 13%

13. from 64 to 36.4
decrease 43%
14. from 78 to 136.5
increase 75%
15. from 89 to 32.9
decrease 63%

16. Mr. Havel bought a car for $2400 and sold it for $2700. What was the percent of profit for Mr. Havel in selling the car? **12.5%**

17. A computer store buys a computer program for $24 and sells it for $91.20. What is the percent of increase in the price? **280%**

18. A manufacturing company with 450 employees begins a new product line and must add 81 more employees. What is the percent of increase in the number of employees? **18%**

19. Richard earns $2700 a month. He received a 3% raise. What is Richard's new annual salary? **$33,372**

20. Marlis has 765 cards in her baseball card collection. She sells 153 of the cards. What is the percent of decrease in the number of cards in the collection? **20%**

30. LIFE SCIENCE The *Carcharodon megaladon* shark of the Miocene era is believed to have been about 12 m long. The modern great white shark is about 6 m long. Write the change in length of these longest sharks over time as a percent increase or decrease. **50% decrease**

31. A sale ad shows a $240 winter coat discounted 35%.

    a. How much is the price decrease? **$84**

    b. What is the sale price of the coat? **$156**

    c. If the coat is reduced in price by an additional $33\frac{1}{3}$%, what will be the new sale price? **$104**

    d. What percent decrease does this final sale price represent? **$56\frac{2}{3}$%**

32. Is the percent change the same when a blouse is marked up from $15 to $20 as when it is marked down from $20 to $15? Explain.

33. EARTH SCIENCE After the Mount St. Helens volcano erupted in 1980, the elevation of the mountain decreased by about 13.6%. Its elevation had been 9677 ft. What was its elevation after the eruption? **about 8361 ft**

 34. CHOOSE A STRATEGY A printer originally sold for $199. Six months later, the price was reduced 45%. During a sale, the printer was discounted an additional 20% off the reduced price. What was the final price of the printer? **B**

    A $17.91   **B** $87.56   **C** $101.89   **D** $98.97

 35. WRITE ABOUT IT Describe how you can use mental math to find the percent increase from 80 to 100 and the percent decrease from 100 to 80.

 36. CHALLENGE During a sale, the price of a computer game was decreased by 40%. By what percent must the sale price be increased to restore the original price? **$66\frac{2}{3}$%**

Spiral Review

Find the surface area of each figure to the nearest tenth. Use 3.14 for π.
(Lesson 6-9)

37. a square pyramid with base 13 m by 13 m and slant height 7.5 m   **364 m²**

38. a cone with a diameter 90 cm and slant height 125 cm   **24,021 cm²**

39. a square pyramid with base length 6 yd and slant height 4 yd   **84 yd²**

40. TEST PREP A 1 lb 8 oz package of corn sells for $5.76. What is the unit price? (Lesson 7-2) **C**

    A $0.34 per oz   **B** $0.32 per oz   **C** $0.24 per oz   **D** $0.64 per oz

---

CHALLENGE 8-4

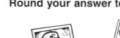 **Challenge**
8-4 *The Ups and Downs of the Marketplace*

Prices change. The price of a stock can change every few minutes. The price of a house changes over a longer period of time.

The *selling price* of an item is what someone is willing to pay for it. It is a good measure of market value.

Find the current value of each item. Round your answer to the nearest cent

1. a. Amy bought a baseball card for $12. To date, the value of the card increased by 30%, then decreased by 15%, and finally increased by 40%.

Joe bought a baseball card for $12. To date, the value of the card decreased by 10%, then increased by 70%, and finally decreased by 5%.

Whose card is currently worth more? by how much? Explain.

Amy's, by $1.12

$18.56 − $17.44 = $1.12

b. By about what percent must the currently lesser-valued card increase to be of equal value with the greater-valued card? Round your answer to the nearest tenth of a percent.

6.4%

2. a. Jorge's family bought a house for $125,000. To date, the value of the house increased by 5%, then decreased by 25%, and finally increased by 10%.

Gene's family bought a house for $125,000. To date, the value of the house decreased by 5%, then increased by 15%, and finally decreased by 20%.

Whose house is currently worth more? by how much? Explain.

Gene's, by $968.75

$109,250 − $108,281.25
= $968.75

b. By about what percent must the currently lesser-valued house increase to be of equal value with the greater-valued house? Round your answer to the nearest tenth of a percent.

0.9%

PROBLEM SOLVING 8-4

 Problem Solving
8-4 *Percent Increase and Decrease*

Use the table below. Write the correct answer.

1. What is the percent increase in the population of Las Vegas, NV from 1990 to 2000? Round to the nearest tenth of a percent.

83.3%

2. What is the percent increase in the population of Naples, FL from 1990 to 2000? Round to the nearest tenth of a percent.

65.3%

Fastest Growing Metropolitan Areas, 1990–2000

Metropolitan Area	Population 1990	Population 2000	Percent Increase
Las Vegas, NV	852,737	1,563,282	
Naples, FL	152,099	251,377	
Yuma, AZ	106,895		49.7%
McAllen-Edinburg-Mission, TX	383,545		48.5%

3. What was the 2000 population of Yuma, AZ to the nearest whole number?

160,022

4. What was the 2000 population of McAllen-Edinburg-Mission, TX metropolitan area to the nearest whole number?

569,564

For exercises 5–7, round to the nearest tenth. Choose the letter for the best answer.

5. The amount of money spent on automotive advertising in 2000 was 4.4% lower than in 1999. If the 1999 spending was $1812.3 million, what was the 2000 spending?
  A $79.7 million **C** $1892 million
  (B) $1732.6 million **D** $1923.5 million

6. In 1967, a 30-second Super Bowl commercial cost $42,000. In 2000, a 30-second commercial cost $1,900,000. What was the percent increase in the cost?
  F 1.7% **H** 442.4%
  G 44.2% **(J)** 4423.8%

7. In 1896 Thomas Burke of the U.S. won the 100-meter dash at the Summer Olympics in a time of 12.00 seconds. In 2000, Maurice Green of the U.S. won with a time of 9.87 seconds. What was the percent decrease in the winning time?
  A 5.5% **C** 21.6%
  (B) 17.8% **D** 32.1%

8. In 1928 Elizabeth Robinson won the 100-meter dash with a time of 12.20 seconds. In 2000, Marion Jones won with a time that was about 11.9% less than Robinson's winning time. What was Jones' time, rounded to the nearest hundredth?
  F 9.75 seconds **H** 12.08 seconds
  (G) 10.75 seconds **J** 13.65 seconds

Lesson Quiz

Find each percent increase or decrease to the nearest percent.

1. from 12 to 15   25% increase

2. from 1625 to 1400 14% decrease

3. from 37 to 125   238% increase

4. from 1.25 to 0.85 32% decrease

5. A computer game originally sold for $40 but is now on sale for 30% off. What is the sale price of the computer game? **$28**

Available on Daily Transparency in CRB

Pacing: Traditional $\frac{1}{2}$ day
Block $\frac{1}{3}$ day

Objective: Students estimate with percents.

Warm Up

Find each percent increase or decrease to the nearest percent.

1. from 1 to 0.62   38% decrease

2. from 162 to 177   9% increase

3. from 16 to 22   38% increase

4. from $\frac{3}{4}$ to $\frac{5}{8}$ 17% decrease

Problem of the Day

If you enlarge a picture by 25%, by what percent do you need to reduce it to return it to its original size? (*Hint:* Try using a simple number for the original area of the picture.) **20%**

Available on Daily Transparency in CRB

Math Fact

One thousand seconds pass in about 17 minutes. One million seconds pass in about 12 days. One billion seconds pass in about 32 years.

8-5 Estimating with Percents

Learn to estimate with percents.

Vocabulary
estimate
compatible numbers

Waiters, waitresses, and other restaurant employees depend upon tips for much of their income. Typically, a tip is 15% to 20% of the bill. Tips do not have to be calculated exactly, so estimation is often used. When the sales tax is about 8%, doubling the tax gives a good estimate for a tip.

Some problems require only an **estimate**. Estimates involving percents and fractions can be found by using **compatible numbers**, numbers that go well together because they have common factors.

$\frac{13}{24}$ The numbers 13 and 24 are not compatible numbers.

    Change 13 to 12. $\frac{13}{24}$ is nearly equivalent to $\frac{12}{24}$.

$\approx \frac{12}{24}$ 12 and 24 are compatible numbers. 12 is a common factor.

    The fraction $\frac{12}{24}$ simplifies to $\frac{1}{2}$. $\frac{13}{24} \approx \frac{1}{2}$

EXAMPLE 1   **Estimating with Percents**

Helpful Hint

Methods of estimating:
1. Use compatible numbers.
2. Round to common percents. (10%, 25%, $33\frac{1}{3}$%)
3. Break percents into smaller parts. (1%, 5%, 10%)

Estimate.

A 26% of 48

Instead of computing the exact answer of 26% · 48, estimate.

$26\% = \frac{26}{100} \approx \frac{25}{100}$ *Use compatible numbers, 25 and 100.*

$\approx \frac{1}{4}$ *Simplify.*

$\frac{1}{4} \cdot 48 = 12$ *Use mental math: 48 ÷ 4.*

So 26% of 48 is about 12.

B 14% of 20

Instead of computing the exact answer of 14% · 20, estimate.

$14\% \approx 15\%$ *Round.*

$\approx 10\% + 5\%$ *Break down the percent into smaller parts.*

$15\% \cdot 20 = (10\% + 5\%) \cdot 20$ *Set up an equation.*

$= 10\% \cdot 20 + 5\% \cdot 20$ *Use Distributive Property.*

$= 2 + 1$ *10% of 20 is 2, so 5% of 20 is 1.*

So 14% of 20 is about 3.

---

1 Introduce

Alternate Opener

EXPLORATION

8-5 Estimating with Percents

You can use compatible numbers when estimating with percents. For example, to find 24% of 82, use 25% of 80.

$0.25 \cdot 80 = 20$
So 24% of 82 is approximately 20.

For each problem, estimate using compatible numbers. Then use a calculator to find the actual answer.

		Estimate	Actual
1.	Of 610 students, 34% prefer domestic cars over foreign cars.		
2.	In a survey, 19% of 152 people selected juice as their favorite drink.		
3.	A family wishes to leave a 15% tip on a $32.15 bill at a restaurant.		
4.	In a country with a population of 10,036,724, 48% are males.		

Think and Discuss

5. Discuss your strategies for estimating.

6. Explain how to avoid overestimating or underestimating.

Motivate

Tell students to imagine that a store is celebrating its 19th anniversary, and it is having a sale in which every item is 19% off. Ask students if they know how to estimate the savings on an item priced at $19.99. **About $4**

Exploration worksheet and answers on Chapter 8 Resource Book pp. 46 and 93

2 Teach

Lesson Presentation

Guided Instruction

In this lesson, students learn to estimate with percents. The lesson presents three methods of estimating. One method involves compatible numbers, another involves rounding to common percents, and the other involves breaking a percent into smaller parts. Explain how to determine whether numbers are compatible, and if they are not, how to find compatible numbers that are appropriate. After each estimation, have students decide whether their estimate makes sense. Before students work the exercises, remind them that their estimates may vary.

EXAMPLE 2

 PROBLEM SOLVING

PROBLEM SOLVING APPLICATION

Angel Falls, in Venezuela, is the tallest waterfall in the world. Horseshoe Falls, which makes up the large portion of Niagara Falls, has a height of only 173 ft. This is about 5.3% of the height of Angel Falls. Approximately how tall is Angel Falls?

Angel Falls has one section that drops uninterrupted for one-half mile.

 Understand the Problem

The **answer** is the approximate height of Angel Falls.

List the **important information**:
- Horseshoe Falls is 173 ft tall.
- Horseshoe Falls is about 5.3% of the height of Angel Falls.

Let *a* represent the height of Angel Falls.

Height of Horseshoe Falls	≈	5.3%	•	Height of Angel Falls
173	≈	5.3%	•	*a*

 Make a Plan

Think: The numbers 173 and 5.3% are difficult to work with.
Use compatible numbers: 173 is close to 170; 5.3% is close to 5%.

$5\% = \frac{5}{100} = \frac{1}{20}$ *Find an equivalent ratio for 5%.*

3 **Solve**

Think: 170 is $\frac{1}{20}$ of what number?

$20 \cdot 170 = a$
Angel Falls is approximately 3400 ft tall.

4 **Look Back**

5% of 3400 ft is $\frac{3400}{20}$, or 170 ft. This is the approximate height of Horseshoe Falls.

Think and Discuss

1. **Determine** the ratios that are nearly equivalent to each of the following percents: 23%, 53%, 65%, 12%, and 76%.

2. **Describe** how to find 35% of a number when you know 10% of the number.

3. **Explain** a method for estimating a 15%–20% tip on a $24.89 bill.

COMMON ERROR ALERT

When estimating, students may use numbers that are compatible but are too far away from the original numbers to give an estimate that is reasonable.

Additional Examples

Example **1**

Estimate.

A. 21% of 66   about 13

B. 36% of 120   about 42

Example **2**

The diameter of the Moon is about 2160 miles. If the diameter of the Moon is about 27% of the diameter of Earth, what is the approximate diameter of Earth? about 8000 miles

---

3 **Close**

Reaching All Learners
Through Number Sense

Give students a recording sheet with a table of prices and sale information (Chapter 8 Resource Book p. 83). Have students estimate the answer to each question and then find the exact answers. Have students compare their answers with their estimates. This activity will give students the opportunity to evaluate their estimates by comparing them with the exact answers.

Summarize

Discuss situations in which finding an estimate may be more appropriate than finding exact calculations. Some examples might include estimating grocery purchases so that the total does not exceed $20 or tipping a waiter. Explain that estimates are not exact answers and that estimates may vary. Ask students to estimate a 15% tip on a check for $39.70.

Possible answer: $6.00

Answers to Think and Discuss

Possible answers:

1. $\frac{1}{4}, \frac{1}{2}, \frac{2}{3}, \frac{1}{8}, \frac{3}{4}$

2. 35% = 10% + 10% + 10% + 5%
   $= (3 \cdot 10\%) + \frac{1}{2}(10\%)$

   So you can multiply the known 10% value by 3 and then add $\frac{1}{2}$ of the 10% value to that product.

3. To estimate 15% of $24.89, find 10% of $25, and add half of that amount. To estimate 20%, double the 10% amount.

FOR EXTRA PRACTICE
see page 747

internet connect
Homework Help Online
go.hrw.com   Keyword: MP4 8-5

Students may want to refer back to the lesson examples.

GUIDED PRACTICE

See Example **1** Estimate.

1. 20% of 493 **100**
2. 15% of 162 **24**
3. 20 out of 81 **25%**
4. 35% of 61 **21**
5. 5 out of 11 **50%**
6. 60% of 1475 **900**

See Example **2** 7. A restaurant bill is for a total of $29.84. Estimate the amount to leave as a 15% tip. **$4.50**

INDEPENDENT PRACTICE

See Example **1** Estimate.

8. 25% of 494 **125**
9. 5021 out of 10,107 **50%**
10. 63 out of 82 **75%**
11. 55% of 810 **440**
12. 50% of 989 **500**
13. 103 out of 989 **10%**

See Example **2** 14. A low-flush toilet uses approximately 6 L water per flush while a standard toilet uses about 19 L water per flush. Estimate the percent of water that can be saved per flush with the more efficient toilet. $66\frac{2}{3}\%$

PRACTICE AND PROBLEM SOLVING

Choose the best estimate. Write A, B, or C.

15. 10% of 61.4 **B**
  A 0.6
  B 6
  C 60

16. 50% of 29.85 **C**
  A 3
  B 12
  C 15

17. 35.5% of 92 **A**
  A 30
  B 3
  C 45

18. 75% of $238.99 **B**
  A $150
  B $180
  C $230

19. 65% of $298.99 **C**
  A $20
  B $100
  C $200

20. 105% of $776.50 **C**
  A $80
  B $900
  C $800

Estimate each number or percent.

21. 50% of 297 is about what number? **150**
22. About what percent of 42 is 31? **75%**
23. 48 is 20% of about what number? **250**
24. 25% of 925 is about what number? **250**
25. 795 is 50% of about what number? **1600**
26. 9.1 is about what percent of 21? **50%**
27. About what percent of 73 is 24? **33%**
28. 9.5% of 88 is about what number? **9**
29. 98 is 26% of about what number? **400**
30. 88 is about what percent of 180? **50%**

Assignment Guide

If you finished Example **1** assign:
  Core 1–6, 8–13, 15–29 odd, 39–45
  Enriched 9–13 odd, 15–30, 39–45

If you finished Example **2** assign:
  Core 1–14, 15–35 odd, 39–45
  Enriched 14–45

Notes

Math Background

Estimation is an important skill that students should be encouraged to work on continuously. Students should use estimation to check the reasonableness of their answers to problems involving calculations in mathematics, science, and other subjects. Encourage students to estimate problems before they work them and to check answers after they have worked them. Estimation can also be an effective tool for students to use on standardized tests.

RETEACH 8-5

LESSON **Reteach**
8-5 *Estimating with Percents*

You can estimate the solutions to different types of problems involving percents by rounding numbers.

Type 1: Finding a percent of a number.

Estimate 38% of 470.

First, round the percent to a common percent with an easy fractional equivalent.

$38\% \approx 40\% = \frac{40}{100} = \frac{2}{5}$

Then, round the number to a number divisible by the denominator of the fraction.

500 is divisible by 5.

$470 \approx 500$

Use mental math to multiply.

$\frac{2}{5} \cdot 500$ Think: If $500 \div 5 = 100$, then $100 \cdot 2 = 200$

So, 38% of 470 is about 200.

Complete to estimate each percent. **Estimates may vary.**

1. 32% of 872

Round to a percent with an easy fractional equivalent: $32\% \approx \underline{33\frac{1}{3}}\% = \frac{1}{3}$

Round to hundreds place, divisible by 3: 872 ≈ __900__

Do mental math to multiply: $\frac{1}{3} \times$ __900__ = __300__

So, 32% of 872 is about __300__.

2. 78% of 495

% to easy fraction:

$78\% \approx$ __80__ $\% = \frac{4}{5}$

divisible by denominator:

495 ≈ __500__

Do mental math to multiply.

$\frac{4}{5} \times$ __500__ = __400__

So, 78% of 495 is about __400__.

3. 73% of 1175

% to easy fraction:

$73\% \approx$ __75__ $\% = \frac{3}{4}$

divisible by denominator:

1175 ≈ __1200__

Do mental math to multiply.

$\frac{3}{4} \times$ __1200__ = __900__

So, 73% of 1175 is about __900__.

PRACTICE 8-5

LESSON **Practice B**
8-5 *Estimating with Percents*

Estimate. **Estimates may vary.**

1. 74 out of 99 __about 75%__
2. 25% of 39 __about 10__
3. 52% of 10 __about 5__
4. 21% of 50 __about 10__
5. 30% of 61 __about 20__
6. 10 out of 19 __about 50%__
7. 5 out of 26 __about 20%__
8. 50% of 17.8 __about 9__
9. 16 out of 62 __about 25%__

Estimate each number or percent.

10. 48% of 30 is about what number? __about 15__
11. 9 is 26% of about what number? __about 36__
12. About what number is 30% of 22? __about 7__
13. 15 is 21% of about what number? __about 75__

14. Rodney's weekly gross pay is $91. He must pay about 32% in taxes and deductions. Estimate Rodney's weekly take-home pay after deductions. __about $60__

15. In the last school election, 452 students voted. Mary received 219 votes. About what percent of the votes did she receive? __about 50%__

16. The gas tank of a car holds 20 gallons of gas. A gallon of gas cost $1.22. When the tank was filled it costs $20. About what percent of the tank was filled. __about 80%__

17. The Boston Marathon, America's oldest regularly contested foot race, is held on Patriots' Day every April. The present race is a distance of a little more than 26 miles. If a runner completes only 21 miles of the race, about what percent of the race was run? __about 80%__

Physical Science LINK

Freezing a light stick may make it glow longer, but not as brightly.

31. Yesterday, 294 books were checked out of the library. This is only 42% of the number usually checked out in a day. About how many books are usually checked out in a day? **750**

32. A jury wants to give an award of about 5% of $788,116. What is a good estimate of the award? **$40,000**

33. *PHYSICAL SCIENCE* When you snap a light stick, you break a barrier between two chemical compounds. This causes a reaction that releases energy as light. If an improvement allows a 9 hr light stick to glow for 13 hr 4 min, about what percent increase is this? **50%**

34. *EARTH SCIENCE* Alaska is the largest state in the United States in total land area, and Rhode Island is the smallest.

Area and Population: 2000		
	Total Land (mi²)	Population
Alaska	570,374	626,932
Rhode Island	1045	1,048,319

  a. Rhode Island is about what percent of the size of Alaska? **≈ 0.2%**

  b. Although much smaller than Alaska, Rhode Island has a larger population. About what percent of the population of Rhode Island is the population of Alaska? **≈ 60%**

  c. Estimate the number of people per square mile in Alaska and in Rhode Island. **≈ 1 per mi²; ≈ 1000 per mi²**

35. *SPORTS* In 2001, Barry Bonds reached base on 342 of 664 plate appearances. About what percent of the time did he reach base? **50%**

 36. *WRITE A PROBLEM* Write a percent estimation problem using the following data: The diameter of Earth is about 12,756 km, and the diameter of the Moon is about 3475 km.

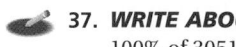 **37.** *WRITE ABOUT IT* Explain how you can estimate 1%, 10%, and 100% of 3051.

38. *CHALLENGE* How could you estimate the percent of words in the English language that begin with *Q*?

Answers

36. Possible answer: Estimate what percent the diameter of the Moon is of the diameter of Earth. Answer: about 25%

37. Possible answer: To find 1%, move the decimal point in the number two places left (about 30). To find 10%, move the decimal 1 place (about 300). 100% is the same as the original number (3051).

38. Possible answer: Count the pages in the *Q* section of the dictionary, and then divide by the total number of pages in the dictionary.

Journal

When selling products or services, many businesses give the customer an estimate of the cost of the job before the work is done. Have students write about why businesses provide estimates and about possible problems that these estimates could cause.

Test Prep Doctor

For Exercise 45, review how to use the necessary conversion factors (Lesson 7-3).

$$\frac{110 \text{ ft}}{5 \text{ s}} = \frac{22 \text{ ft}}{1 \text{ s}}$$

$$\frac{22 \text{ ft}}{1 \text{ s}} = \frac{22 \text{ ft}}{1 \text{ s}} \cdot \frac{3600 \text{ s}}{1 \text{ h}} \cdot \frac{1 \text{ mi}}{5280 \text{ ft}} = \frac{15 \text{ mi}}{1 \text{ h}}$$

The correct answer is **D.**

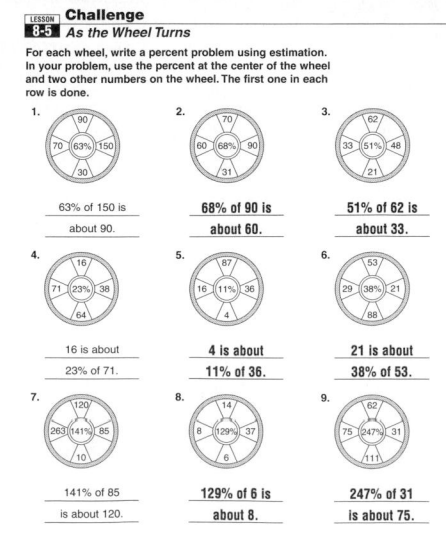
Lesson Quiz

Estimate. Possible answers:

1. 34% of 12 **4**
2. 47 out of 236 **20%**
3. 113% of 80 **90**
4. 22 out of 51 **40%**
5. According to the last census, 33,871,648 people live in California, and 3,694,820 of those people live in Los Angeles. Estimate the percent of Californians who live in Los Angeles. **10%**

Available on Daily Transparency in CRB

Warm Up

Estimate. Possible answers:

1. 20% of 602   **120**

2. 133 out of 264   **50%**

3. 151% of 78   **120**

4. 0.28 out of 0.95   **30%**

Problem of the Day

What is the percent discount on a purchase of three shirts if you take advantage of the shirt sale? $16\frac{2}{3}$%

> **All Shirts on Sale!**
> Buy 2—Get the Third for Half Price!

Available on Daily Transparency in CRB

Math Humor

Student: Would you ever punish me for something that I didn't do?

Teacher: Of course not.

Student: Good…I didn't do my homework!

8-6 Applications of Percents

Learn to find commission, sales tax, and withholding tax.

Vocabulary
commission
commission rate
sales tax
withholding tax

Real estate agents often work for *commission*. A **commission** is a fee paid to a person who makes a sale. It is usually a percent of the selling price. This percent is called the **commission rate**.

Often agents are paid a commission plus a regular salary. The total pay is a percent of the sales they make plus a salary.

commission rate · sales = commission

EXAMPLE 1   **Multiplying by Percents to Find Commission Amounts**

A real-estate agent is paid a monthly salary of $1200 plus commissions. Last month she sold one house for $97,500, earning a 3% commission on the sale. How much was her commission? What was her total pay for last month?

First find her commission.

$3\% \cdot \$97{,}500 = c$ *commission rate · sales = commission.*

$0.03 \cdot 97{,}500 = c$ *Change the percent to a decimal.*

$2925 = c$ *Solve for c.*

She earned a commission of $2925 on the sale.

Now find her total pay for last month.

$\$2925 + \$1200 = \$4125$ *commission + salary = total pay.*

Her total pay for last month was $4125.

Sales tax is the tax on the sale of an item or service. It is a percent of the purchase price and is collected by the seller.

EXAMPLE 2   **Multiplying by Percents to Find Sales Tax Amounts**

If the sales tax rate is 8.25%, how much tax would Alexis pay if she bought one twin pack of black refill cartridges for her printer for $52.88 and two color refill cartridges for $34.79 each?

black refills: 1 at $52.88 → $52.88

color refills: 2 at $34.79 → $69.58

$122.46   *Total price*

$0.0825 \cdot 122.46 = 10.10295$ *Convert tax rate to a decimal and multiply by the total price.*

Alexis would pay $10.10 in sales tax.

---

1 Introduce

Alternate Opener

You often need to calculate percents when making a purchase.

Use a calculator to find the tax on each item and the total cost of the item including the tax.

	Item	Cost	Tax Rate	Tax	Total Cost = Cost + Tax
1.	CD	$13.95	8%		
2.	DVD	$24.99	8%		
3.	Headphones	$29.95	8%		

Use a calculator to find the total cost of each item.

	Item	Cost	Tax Rate	Total Cost = 1.08 · Cost
4.	CD	$13.95	8%	
5.	DVD	$24.99	8%	
6.	Headphones	$29.95	8%	

Think and Discuss

7. Explain how you calculated the tax on each item in numbers 1–3.

8. Explain why the total cost is the same whether you use the formula *total cost = cost + tax* or the formula *total cost = 1.08 · cost*.

Motivate

Ask students if they've ever shopped in a store where the salesperson seemed especially helpful. Explain that in some stores, the amount an employee earns is based on how much merchandise he or she sells. Tell students that an amount earned based on sales is called a *commission*.

Exploration worksheet and answers on Chapter 8 Resource Book pp. 56 and 96

2 Teach

> **Lesson Presentation**

Guided Instruction

In this lesson, students learn to find commission, sales tax, and withholding tax. Explain the meanings of *commission, sales tax,* and *withholding tax*. Explain that each is a percent of some amount of money. Students use given percents to find these amounts in Examples 1 and 2. In Example 3, students find what percent of earnings a given amount of withholding tax is. In Example 4, students use a given percent and a given amount of earnings to find an amount sold.

A tax deducted from a person's earnings as an advance payment of income tax is called **withholding tax**.

EXAMPLE **3** **Using Proportions to Find the Percent of Tax Withheld**

Joseph earns $1070 monthly. Of that, $160.50 is withheld for taxes. What percent of Joseph's earnings is withheld?

Think: What percent of $1070 is $160.50?

Solve by proportion:

$$\frac{n}{100} = \frac{160.50}{1070}$$

$n \cdot 1070 = 100 \cdot 160.50$ *Find the cross products.*

$1070n = 16{,}050$

$n = \frac{16{,}050}{1070}$ *Divide both sides by 1070.*

$n = 15$

So 15% of Joseph's earnings is withheld.

EXAMPLE **4** **Dividing by Percents to Find Total Sales**

Students in Sele's class sell gift wrap to raise funds for class trips. The class earns 14% on all sales. If the class made $791.70 on sales of wrapping paper, how much were the total sales?

Think: 791.70 is 14% of what number?

Solve by equation:

$791.70 = 0.14 \cdot s$ *Let s = total sales.*

$\frac{791.70}{0.14} = s$ *Divide each side by 0.14.*

$5655 = s$

The total sales of gift wrap for Sele's class were $5655.

Think and Discuss

1. Tell how finding commission is similar to finding sales tax.

2. Explain whether adding 6% sales tax to a total gives the same result as finding 106% of the total.

3. Explain how to find the price of an item if you know the total cost after 5% sales tax.

Additional Examples

Example **1**

A real-estate agent is paid a monthly salary of $900 plus commission. Last month he sold one condominium for $65,000, earning a 4% commission on the sale. How much was his commission? What was his total pay last month? **$2600; $3500**

Example **2**

If the sales tax rate is 6.75%, how much tax would Adrian pay if he bought two CDs at $16.99 each and one DVD for $36.29? **$4.74**

Example **3**

Anna earns $1500 monthly. Of that, $114.75 is withheld for Social Security and Medicare. What percent of Anna's earnings are withheld for Social Security and Medicare? **7.65%**

Example **4**

A furniture sales associate earned $960 in commission in May. If his commission is 12% of sales, how much were his sales in May? **$8000**

3 Close

Reaching All Learners
Through Critical Thinking

Have students work in pairs. Tell students that they have been hired to sell cars. They can choose between three salary packages: A) $2500 per month with no commission, B) $1000 per month with 2% commission, or C) no monthly salary and 4% commission. If an average car sells for $20,000, have students calculate the salary they would receive if they sold 1, 3, or 5 cars in a month. Then have them discuss which salary package they would choose.

A) $2500; B) $1400, $2200, or $3000;
C) $800, $2400, or $4000

Summarize

Ask for volunteers to define each of the vocabulary terms. Remind students that commissions and sales taxes are based on the price of an item. Explain that withholding taxes are also called income taxes and that generally this tax is taken before you get your paycheck.

Answers to Think and Discuss

Possible answers:

1. Both commission and sales tax are based on percents of the price.

2. Yes, 6% of a prices plus the price is equal to 106% of the price.

3. Solve the equation $x + 0.05x = $ total cost.

FOR EXTRA PRACTICE
see page 747

internet connect
Homework Help Online
go.hrw.com   Keyword: MP4 8-6

Students may want to refer back to the lesson examples.

Assignment Guide

If you finished Example **1** assign:
 Core 1, 5, 9, 11, 20–24
 Enriched 5, 9–12, 20–24

If you finished Example **2** assign:
 Core 1, 2, 5, 6, 9, 11, 20–24
 Enriched 5, 6, 9–12, 20–24

If you finished Example **3** assign:
 Core 1–3, 5–7, 9, 11, 20–24
 Enriched 5, 7, 9–12, 15, 20–24

If you finished Example **4** assign:
 Core 1–8, 9–15 odd, 20–24
 Enriched 5–24

Notes

GUIDED PRACTICE

See Example **1**
1. Josh earns a weekly salary of $300 plus a 6% commission on sales. Last week, his sales totaled $3500. What was his total pay? **$510**

See Example **2**
2. In a state with a sales tax rate of 7%, Hernando buys a radio for $59.99 and a CD for $13.99. How much is the sales tax? **$5.18**

See Example **3**
3. Last year, Janell earned $33,095. From this amount, $7,446.38 was withheld for taxes. What percent of her income was withheld, to the nearest tenth of a percent? **22.5%**

See Example **4**
4. Chuck works as a salesperson at an electronics store. If he earns $29.94 from a 6% commission on the sale of a video camera, what is the price of the camera? **$499**

INDEPENDENT PRACTICE

See Example **1**
5. Marta earns a weekly salary of $110 plus a 6.5% commission on sales at a hobby store. How much would she make in a week if she sold $4300 worth of merchandise? **$389.50**

See Example **2**
6. The sales tax rate in Lisa's town is 5.75%. If she purchases 4 chairs for $124.99 each and an area rug for $659.99, how much sales tax does she owe? **$66.70**

See Example **3**
7. Jan typically earns $435 each week, of which $78.30 is withheld for taxes. What percent of Jan's earnings are withheld each week? **18%**

See Example **4**
8. Heather works in a clothes shop where she earns a commission of 5% and no weekly salary. What will Heather's weekly sales have to be for her to earn $375? **$7500**

PRACTICE AND PROBLEM SOLVING

Find each commission or sales tax to the nearest cent.

9. total sales: $12,000 **$330**
 commission rate: 2.75%

10. total sales: $125.50 **$7.84**
 sales tax rate: 6.25%

11. total sales: $26.98 **$2.16**
 sales tax rate: 8%

12. total sales: $895.75 **$38.07**
 commission rate: 4.25%

Find the total sales to the nearest cent.

13. commission: $78.55 **$1963.75**
 commission rate: 4%

14. commission: $2842 **$81,200**
 commission rate: 3.5%

15. $2800 plus 3% of sales: $3100 to $3400 a month

15. Elena can choose between a monthly salary of $2200 plus 5.5% of sales or $2800 plus 3% of sales. She expects sales between $10,000 and $20,000 a month. Which salary option should she choose?

Math Background

Although students may not have any first-hand experience with commission or withholding tax, they most likely will be aware of other applications of percents, such as sales tax, percent off sales, and percents used in tipping. One of the most important steps in solving problems is checking to see whether an answer is reasonable. Encourage students to develop the habit of using estimation to check their answers. For example, if an answer is supposed to be 19% of $582, that answer should be close to $120 because 20% of $600 = $120. The estimate should help them see that the actual answer, $110.58, is reasonable.

RETEACH 8-6

LESSON 8-6 Reteach
Applications of Percents

Salespeople often earn a **commission**, a percent of their total sales.

Find the commission on a real-estate sale of $125,000 if the commission rate is 4%.

Write the percent as a decimal and multiply.

commission rate × amount of sale = amount of commission
0.04 × $125,000   = $5000

If, in addition to the commission, the salesperson earns a salary of $1000, what is the total pay?

commission+ salary = total pay
$5000 + $1000   = $6000

Complete to find each total monthly pay.

1. total monthly sales = $170,000; commission rate = 3%; salary = $1500

 amount of commission = 0.03 × $ __170,000__ = $ __5100__

 total pay = $ __5100__ + $1500 = $ __6600__

2. total monthly sales = $16,000; commission rate = 5.5%; salary = $1750

 amount of commission = __0.055__ × $ __16,000__ = $ __880__

 total pay = $ __880__ + __1750__ = $ __2630__

A **tax** is a charge, usually a percentage, generally imposed by a government.
Sales tax is the tax on the sale of an item or service.

If the sales tax rate is 7%, find the tax on a sale of $9.49.

Write the tax rate as a decimal and multiply.

tax rate × amount of sale = amount of tax
0.07   × $9.49   = $0.6643 ≈ $0.66

Complete to find each amount of sales tax.

3. item price = $5.19; sales tax rate = 6%

 amount of sales tax = 0.06 × $ __5.19__ = $ __0.3114__ ≈ $ __0.31__

4. item price = $250; sales tax rate = 6.75%

 amount of sales tax = __0.0675__ × $ __250__ = $ __16.875__ ≈ $ __16.88__

PRACTICE 8-6

LESSON 8-6 Practice B
Applications of Percents

Complete the table to find the amount of sales tax for each sale amount to the nearest cent.

1.	Sale amount	5% sales tax	8% sales tax	6.5% sales tax
	$67.50	$3.38	$5.40	$4.39
	$98.75	$4.94	$7.90	$6.42
	$399.79	$19.99	$31.98	$25.99
	$1250.00	$62.50	$100.00	$81.25

Complete the table to find the commission for each sale amount to the nearest cent.

2.	Sale amount	6% commision	9% commision	8.5% commission
	$475.00	$28.50	$42.75	$40.38
	$2450.00	$147.00	$220.50	$208.25
	$12,500.00	$750.00	$1125.00	$1062.50
	$98,900.00	$5934.00	$8901.00	$8406.50

3. Alice earns a monthly salary of $315 plus a commission on her total sales. Last month her total sales were $9640, and she earned a total of $1182.60. What is her commission rate? **9%**

4. Phillipe works for a computer store that pays a 12% commission and no salary. What will Phillipe's weekly sales have to be for him to earn $360? **$3000**

5. The purchase price of a book is $35.85. The sales tax rate is 6.5%. How much is the sales tax to the nearest cent? What is the total cost of the book?

 sales tax is $2.33; total cost is $38.18

6. Who made more commission this month? How much did she make? Salesperson A made 11% of $67,530. Salesperson B made 8% of $85,740.

 Salesperson A $7428.30

7. Jon bought groceries for $105.30 and paper products for $35.50. The total bill including tax was $142.93. There was no tax charged on the groceries. What was the tax rate on the paper products? **6%**

8. The Cougars won 62% of their games. They won 93 games. How many games did they lose? **57 games**

Tax brackets are used to determine how much income tax you pay. Depending upon your taxable income, your tax is given by the formula base tax + tax rate(amount over). "Amount over" refers only to the income above the amount listed. Refer to the table for Exercises 16–19.

2001 IRS Income Tax Brackets (Single)			
Taxable Income Range	Base Tax	Tax Rate	Amount Over
$0–$27,050	$0	15%	$0
$27,050–$65,550	$4057.50	27.5%	$27,050
$65,550–$136,750	$14,645	30.5%	$65,550
$136,750–$297,350	$36,361	35.5%	$136,750
$297,350 and up	93,374	39.1%	$297,350

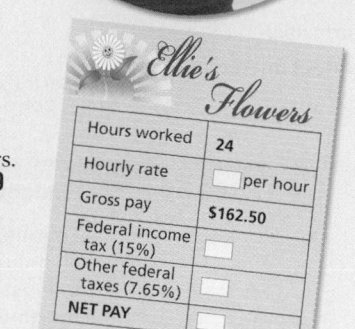

16. Tina's pay stub is shown at right. Find the missing numbers.
$6.77; $24.38; $12.43; $125.69

Ellie's Flowers	
Hours worked	24
Hourly rate	☐ per hour
Gross pay	$162.50
Federal income tax (15%)	☐
Other federal taxes (7.65%)	☐
NET PAY	☐

17. Anna earned $71,458 total in 2001. However, she was able to deduct $7250 for job-related expenses. This amount is subtracted from her total income to determine her taxable income.

a. What was Anna's taxable income in 2001? **$64,208**

b. How much income tax did she owe? **$14,275.95**

c. What percent of Anna's total income did the tax represent? **≈ 20.0%**

d. What percent of her taxable income did the tax represent? **≈ 22.2%**

18. How much more tax would someone who made $27,100 pay than someone who made $27,000? What percent would they pay on the additional $100 of income? **$21.25; 21.25%**

19. ◆ CHALLENGE Charlena paid $10,050 in taxes in 2001. How much taxable income did she earn that year? (*Hint:* Which tax bracket must she have been in to have paid $10,050 in taxes?) **$48,840.91**

Spiral Review

Find the scale factor that relates each model to the actual object. (Lesson 7-8)

20. 14 in. model, 70 in. object **1:5**

21. 8 cm model, 6 mm object **40:3**

22. 4 in. model, 6 ft 8 in. object **1:20**

23. 0.25 m model, 0.0025 cm object **10,000:1**

24. **TEST PREP** Of the 32 students in Mr. Smith's class, 14 have jobs during the summer. What percent of the students have a summer job? (Lesson 8-1) **A**

A 43.75%   B 68.56%   C 56.25%   D 35.65%

CHALLENGE 8-6

LESSON 8-6 Challenge
Shoppers' Delight

Shoppers save money by buying items on sale.
The amount by which the regular price is reduced is called a **discount**.

amount of discount = discount rate × regular price
sale price = regular price − amount of discount

Find the sale price after each discount.

1. regular price = $899;
   discount rate = 20%
   amount of discount = **$179.80**
   sale price = **$719.20**

2. regular price = $14.99;
   discount rate = 15%
   amount of discount = **$2.25**
   sale price = **$12.74**

Stores may offer discounts in a variety of ways.

Use $100 as the regular price for the item to write your explanations. **Possible answers are given:**

3. Buy one at regular price. Get a second one for half price. Explain how this is different from getting a 50% discount.
   50% discount: $50 for 1 item and $100 for 2 items
   2nd item half price: $100 for first item and $150 for 2 items

4. Buy two. Get one free. Explain how this is different from getting a 33⅓% discount.
   Buy 2, get 1 free: $200 for 3 items
   33⅓% discount: $200 for 3 items

5. This item is marked down by 10%. Use a coupon and get an additional 10% off. Explain how this is different from getting a 20% discount.
   20% discount on $100 item: pay $80
   After first 10% discount, price is $90.
   Now, take 10% off $90 and final price is $81.

6. An item is marked "50% off — Today Only Get Another 50% off". Explain why the item is not free.
   The additional 50% is off 50% of the reduced price.
   The item is 25% of the original price.

PROBLEM SOLVING 8-6

LESSON 8-6 Problem Solving
Applications of Percents

Write the correct answer.

1. The sales tax rate for a community is 6.75%. If you purchase an item for $500, how much will you pay in sales tax?
   $33.75

2. A community is considering increasing the sales tax rate 0.5% to fund a new sports arena. If the tax rate is raised, how much more will you pay in sales tax on $500?
   $2.50

3. If you earn 5% commission on your sales at a computer software store, how much will you earn in commission on sales of $2,500?
   $125

4. Julie has been offered two jobs. The first pays $400 per week. The second job pays $175 per week plus 15% commission on her sales. How much will she have to sell in order to pay as much as the first?
   $1500

Choose the letter for the best answer. Round to the nearest cent.

5. John's dad earns $2250 per month. His employer deducts 6% of his pay for his retirement program. How much is deducted from John's dad's paycheck for his retirement?
   A $125.00
   Ⓑ $135.00
   C $375.00
   D $2115.00

6. Susan's parents have offered to help her pay for a new computer. They will pay 30% and Susan will pay 70% of the cost of a new computer. Susan has saved $550 for a new computer. With her parents help, how expensive of a computer can she afford?
   F $165.00   H $1650.00
   Ⓖ $785.71   J $1833.33

7. Kellen's bill at a restaurant before tax and tip is $22.00. If tax is 5.25% and he wants to leave 15% of the bill including the tax for a tip, how much will he spend in total?
   A $22.17   Ⓒ $26.63
   B $26.46   D $27.82

8. The 8th grade class is trying to raise money for a field trip. They need to raise $600 and the fundraiser they have chosen will give them 20% of the amount that they sell. How much do they need to sell to raise the money for the field trip?
   F $120.00   Ⓗ $3000.00
   G $857.14   J $3200.00

Interdisciplinary LINK

Economics

Exercises 16–19 involve using data from pay stubs and tax tables to solve percent problems. Understanding these types of problems is important in the study of economics.

Journal

Ask students to write whether they would prefer to have a job that pays commission or one that pays a straight salary. Have them state the reasons for their preferences.

Test Prep Doctor ➕

For Exercise 24, students should write as the fraction $\frac{14}{32}$ as a percent. This will result in 43.75%, choice **A.** If they selected choice **C,** they may have found the percent of students who do *not* have a summer job.

Lesson Quiz

1. The lunch bill was $8, and you want to leave a 15% tip. How much should you tip? **$1.20**

2. The sales tax is 5.75%, and the shirt costs $20. What is the total cost of the shirt? **$21.15**

3. As of 2001, the minimum hourly wage was $5.15. Congress proposed to increase it to $6.15 per hour. To the nearest percent, what is the proposed percent increase in the minimum wage? **19%**

4. It costs a business $13.30 to make its product. To satisfy investors, the company needs to make $4 profit per unit. To the nearest percent, what should be the company's markup? **30%**

Available on Daily Transparency in CRB

Pacing: Traditional 1 day
Block $\frac{1}{2}$ day

Objective: Students compute simple interest.

Warm Up

1. What is 35 increased by 8%? **37.8**

2. What is the percent of decrease from 144 to 120? **$16\frac{2}{3}\%$**

3. What is 1500 decreased by 75%? **375**

4. What is the percent of increase from 0.32 to 0.64? **100%**

Problem of the Day

Maggie is running for class president. A poll revealed that 40% of her classmates have decided to vote for her, 32% have decided to vote for her opponent, and 7 voters are undecided. If she needs 50% of the vote to win, how many of the undecided voters must vote for Maggie for her to win the election? **3**

Available on Daily Transparency in CRB

Math Humor

Old bankers never die. They just lose their interest.

Learn to compute simple interest.

Vocabulary
interest
simple interest
principal
rate of interest

When you borrow money from a bank, you pay **interest** for the use of the bank's money. When you deposit money into a savings account, you are paid interest. **Simple interest** is one type of fee paid for the use of money.

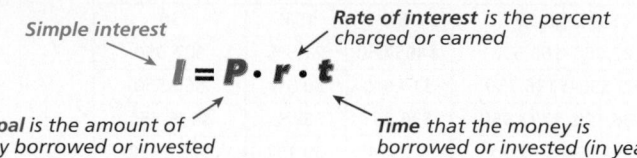

Simple interest

Rate of interest is the percent charged or earned

$$I = P \cdot r \cdot t$$

Principal is the amount of money borrowed or invested

Time that the money is borrowed or invested (in years)

EXAMPLE 1 Finding Interest and Total Payment on a Loan

Thurman borrowed $13,500 from his brother-in-law for 4 years at an annual simple interest rate of 6% to buy a car. How much interest will he pay if he pays the entire loan off at the end of the fourth year? What is the total amount he will repay?

First, find the interest he will pay.

$I = P \cdot r \cdot t$	Use the formula.
$I = 13{,}500 \cdot 0.06 \cdot 4$	Substitute. Use 0.06 for 6%.
$I = 3240$	Solve for I.

Thurman will pay $3240 in interest.

You can find the total amount A to be repaid on a loan by adding the principal P to the interest I.

$P + I = A$	principal + interest = amount
$13{,}500 + 3240 = A$	Substitute.
$16{,}740 = A$	Solve for A.

Thurman will repay a total of $16,740 on his loan.

EXAMPLE 2 Determining the Amount of Investment Time

Tony invested $3000 in a mutual fund at a yearly rate of 5%. He earned $525 in interest. How long was the money invested?

$I = P \cdot r \cdot t$	Use the formula.
$525 = 3000 \cdot 0.05 \cdot t$	Substitute values into equation.
$525 = 150t$	Solve for t.
$3.5 = t$	

The money was invested for 3.5 years, or 3 years and 6 months.

---

1 Introduce

Alternate Opener

EXPLORATION

8-7 More Applications of Percents

Simple interest is the amount earned on money deposited in some savings accounts.

The interest your account earns is calculated using the formula $I = Prt$. For example, if you start a savings account with $100.00 (**principal P**) and your savings account pays 5% (**interest rate r**), the interest (**I**) you will have earned at the end of 1 year (**time t**) will be $0.05 \cdot 100 = \$5$. Your total balance at the end of 1 year will be $100.00 + $5.00 = $105.00.

Complete the table.

	Savings	Interest Rate	Interest	Total Balance
	$100	5%	$0.05 \cdot 100 = \$5$	$100 + 5 = \$105$
1.	$200	6%		
2.	$300		$24	
3.	$500			$550

Think and Discuss

4. **Describe** your strategies for completing the table.
5. **Explain** how to use the formula $I = Prt$ when you need to find an interest rate.

Motivate

Show students some bank advertisements or credit card applications that offer different interest rates. Explain that it is important to understand interest if you want to avoid paying more than necessary for something.

Exploration worksheet and answers on Chapter 8 Resource Book pp. 66 and 98

2 Teach

Guided Instruction

[Lesson Presentation]

In this lesson, students learn to compute simple interest. Begin by discussing *interest*. Explain to students that there are different kinds of interest but that the principle of simple interest provides the basis for all types of interest. (Compound interest is addressed in the Extension.) Show them the formula $I = Prt$, and discuss what each of the four variables represents (Teaching Transparency T30, CRB). Emphasize the fact that the time is generally in years and that the rate should be changed to a decimal before using it in the formula. Point out that the formula can be used to solve for any of the four variables.

EXAMPLE 3 Computing Total Savings

Rebecca's grandmother deposited $2000 into a savings account as a college fund. How much will Rebecca have in this account after 3 years at a yearly simple interest rate of 2.5%?

$I = P \cdot r \cdot t$	*Use the formula.*
$I = 2000 \cdot 0.025 \cdot 3$	*Substitute. Use 0.025 for 2.5%.*
$I = 150$	*Solve for I.*

Now you can find the total.

$P + I = A$	*Use the formula.*
$2000 + 150 = A$	
$2150 = A$	

Rebecca will have $2150 in her savings account after three years.

EXAMPLE 4 Finding the Rate of Interest

Suzanne borrowed $5000 for 5 years at simple interest to pay for her college classes. If Suzanne repaid a total of $6187.50, at what interest rate did she borrow the money?

$P + I = A$
$5000 + I = 6187.5$ *Find the amount of interest.*
$I = 6187.5 - 5000 = 1187.5$

She paid $1187.50 in interest. Use the amount of interest to find the interest rate.

$I = P \cdot r \cdot t$	*Use the formula.*
$1187.5 = 5000 \cdot r \cdot 5$	*Substitute.*
$1187.5 = 25,000r$	*Multiply.*
$\frac{1187.5}{25,000} = r$	
$0.0475 = r$	

Suzanne borrowed the money at an annual rate of 4.75%, or $4\frac{3}{4}$%.

Think and Discuss

1. **Explain** the meaning of each variable in the interest formula.

2. **Tell** what value should be used for *t* when referring to 6 months.

3. **Name** the variables in the simple interest formula that represent dollar amounts.

4. **Demonstrate** that doubling the time while halving the interest rate results in the same amount of simple interest.

COMMON ERROR ALERT

Students may forget to write a percent as a decimal before using it in the interest formula. For example, they may use 7.25 instead of 0.0725 when the interest rate is 7.25%.

Additional Examples

Example 1

To buy a car, Jessica borrowed $15,000 for 3 years at an annual simple interest rate of 9%. How much interest will she pay if she pays the entire loan off at the end of the third year? What is the total amount that she will repay? **$4050; $19,050**

Example 2

Nancy invested $6000 in a bond at a yearly rate of 3%. She earned $450 in interest. How long was the money invested? **2.5 yr, or 2 yr 6 mo**

Example 3

John's parents deposited $1000 into a savings account as a college fund when he was born. How much will John have in this account after 18 years at a yearly simple interest rate of 3.25%? **$1585**

Example 4

Mr. Johnson borrowed $8000 for 4 years to make home improvements. If he repaid a total of $10,320, at what interest rate did he borrow the money? **7.25%, or $7\frac{1}{4}$%**

---

3 Close

Reaching All Learners
Through Grouping Strategies

Have the students work in pairs. Give each pair a sample store circular with an advertised interest rate (Chapter 8 Resource Book p. 84) and a number cube (Manipulative Kit). Have students "buy" items on the store circular. Tell students to roll the die to determine the number of years the item will remain charged on their credit cards. Using the interest rate on the circular and the number of years rolled on the number cube, students are to calculate the interest they will owe on that item.

Summarize

Remind students that they can earn interest on money that they deposit in the bank or invest in some other way. If they borrow money, they must pay interest as a fee to the lender. As the amount of time that the money is deposited or borrowed increases, so does the amount of interest that is paid.

Answers to Think and Discuss

Possible answers:

1. *I* (interest) is simple interest earned, *P* (principal) is the amount of money invested or borrowed, *r* (rate) is the percent earned or charged, and *t* (time) is the number of years the money is borrowed or invested.

2. Since *t* is always written in years, $t = 0.5$, or $\frac{1}{2}$, when the time is 6 months, or half a year.

3. *I* and *P*

4. The interest on $500 at 8% for 1 year is $500 \cdot 0.08 \cdot 1 = \$40$. The interest on the same amount at 4% for 2 years is $500 \cdot 0.04 \cdot 2 = \$40$.

8-7 Exercises

FOR EXTRA PRACTICE
see page 747

✓ internet connect

Homework Help Online
go.hrw.com   Keyword: MP4 8-7

go.
hrw.
.com

Students may want to refer back to the lesson examples.

Assignment Guide

If you finished Example ❶ assign:
Core 1, 5, 9–16, 24–30
Enriched 5, 9–17, 24–30

If you finished Example ❷ assign:
Core 1, 2, 5, 6, 9–18, 24–30
Enriched 5, 6, 9–18, 21, 22, 24–30

If you finished Example ❸ assign:
Core 1–3, 5–7, 9–19, 24–30
Enriched 5–7, 9–19, 21–30

If you finished Example ❹ assign:
Core 1–19, 24–30
Enriched 5–30

Notes

GUIDED PRACTICE

See Example ❶ 1. Leroy borrowed $8250 to be repaid after 3 years at an annual simple interest rate of 7.25%. How much interest will be due after 3 years? How much will Leroy have to repay? **$1794.38; $10,044.38**

See Example ❷ 2. Mr. Williams invested $4000 in a bond with a yearly interest rate of 4%. His total interest on the investment was $800. What was the length of the investment? **5 years**

See Example ❸ 3. Kim deposited $1422 in a savings account. How much would she have in the account after 5 years at an annual simple interest rate of 3%? **$1635.30**

See Example ❹ 4. Hank borrowed $25,000 for 3 years to remodel his house. At the end of the loan, he had repaid a total of $29,125. At what simple interest rate did he borrow the money? **5.5%**

INDEPENDENT PRACTICE

See Example ❶ 5. A bank offers an annual simple interest rate of 7% on home improvement loans. How much would Nick owe if he borrows $18,500 over a period of 3.5 years? **$23,032.50**

See Example ❷ 6. Anne deposits $7500 in a college fund for her niece. If the fund earns an annual simple interest rate of 5.5%, how much will be in the fund after 15 years? **$13,687.50**

See Example ❸ 7. Olivia gave a security deposit of $1500 to her landlord, Mr. Rey, 6 years ago. Mr. Rey will give her the deposit back with simple interest of 3.85%. How much will he return to her? **$1846.50**

See Example ❹ 8. First Bank loaned a construction company $125,000 at an annual simple interest rate. After 3 years, the company repaid the bank $149,375. What was the loan's interest rate? **6.5%**

PRACTICE AND PROBLEM SOLVING

Find the interest and the total amount to the nearest cent.

9. $225 at 5% per year for 3 years **$33.75, $258.75**
10. $775 at 8% per year for 1 year **$62, $837**
11. $4250 at 7% per year for 1.5 years **$446.25, $4696.25**
12. $650 at 4.5% per year for 2 years **$58.50, $708.50**
13. $397 at 5% per year for 9 months **$14.89, $411.89**
14. $2975 at 6% per year for 5 years **$892.50, $3867.50**
15. $700 at 6.25% per year for 2 years **$87.50; $787.50**
16. $500 at 9% per year for 3 months **$11.25; $511.25**
17. Akule borrowed $1500 for 18 months at a 12% annual simple interest rate. How much interest will he have to repay? What is the total amount he will repay? **$270, $1770**

Math Background

While the principle of simple interest is mathematically important, compound interest is used for most applications in the real world. Students may benefit from studying compound interest (Chapter 8 Extension) as it will give them a better understanding of how interest works in the real world.

The contrast between simple and compound interest becomes increasingly evident as the length of time increases. For example, an investment of $1000 at 8% for 30 years would be worth $3400 using simple interest and $10,765.16 using interest compounded quarterly.

RETEACH 8-7

Reteach
8-7 *More Applications of Percents*

Interest is money paid on an investment. A borrower pays the interest. An investor earns the interest.

Simple interest, *I*, is earned when an amount of money, the *principal P*, is borrowed or invested at a *rate of interest r* for a *period of time t*.

| Interest = Principal • Rate • Time |
| $I = P \cdot r \cdot t$ |

Situation 1: Find *I* given *P, r,* and *t.*

Calculate the simple interest on a loan of $3500 for a period of 6 months at a yearly rate of 5%.

Write the interest rate as a decimal. 5% = 0.05
Write the time period in terms of years. 6 months = 0.5 year
$I = P \cdot r \cdot t$
$I = 3500 \cdot 0.05 \cdot 0.5 = \$87.50$ ◄— interest earned

Find the interest in each case.

1. principal $P = \$5000$; time $t = 2$ years; interest rate $r = 6\%$
$I = P \cdot r \cdot t = \underline{5000} \cdot 0.06 \cdot \underline{2} = \$ \underline{600}$

2. principal $P = \$2500$; time $t = 3$ months; interest rate $r = 8\%$
$I = P \cdot r \cdot t = \underline{2500} \cdot \underline{0.08} \cdot \underline{0.25} = \$ \underline{50}$

Situation 2: Find *t* given *I, P,* and *r.*

An investment of $3000 at a yearly rate of 6.5% earned $390 in interest. Find the period of time for which the money was invested.
$I = P \cdot r \cdot t$
$390 = 3000 \cdot 0.065 \cdot t$
$390 = 195t$
$\frac{390}{195} = \frac{195t}{195}$
$2 = t$

The investment was for 2 years.

Find the time in each case.

3. $I = \$1120$; $P = \$4000$; $r = 7\%$
$I = P \cdot r \cdot t$
$1120 = \underline{4000} \cdot 0.07 \cdot t$
$1120 = \underline{280} \cdot t$
$\frac{1120}{280} = \frac{280t}{280}$
$\underline{4}$ years = t

4. $I = \$812.50$; $P = \$5000$; $r = 6.5\%$
$I = P \cdot r \cdot t$
$812.50 = \underline{5000} \cdot \underline{0.065} \cdot t$
$812.50 = \underline{325} \cdot t$
$\frac{812.50}{325} = \frac{325t}{325}$
$\underline{2.5}$ years = t

PRACTICE 8-7

Practice B
8-7 *More Applications of Percents*

Find the missing value.

1. principal = $125
rate = 4%
time = 2 years
interest = ?
$10

2. principal = ?
rate = 5%
time = 4 years
interest = $90
$450

3. principal = $150
rate = 6%
time = ? years
interest = $54
6 years

4. principal = $200
rate = 7%
time = 3 years
interest = $30
5%

5. principal = $550
rate = ?%
time = 3 years
interest = $57.75
3.5%

6. principal = ?
rate = $3\frac{1}{4}$%
time = 2 years
interest = $63.05
$970

7. Kwang deposits money in an account that earns 5% simple interest. He earned $546 in interest 2 years later. How much did he deposit? **$5460**

8. Simon opened a certificate of deposit with the money from his bonus check. The bank offered 4.5% interest for 3 years of deposit. Simon calculated that he would earn $87.75 interest in that time. How much did Simon deposit to open the account? **$650**

9. Douglas borrowed $1000 from Patricia. He agreed to repay her $1150 after 3 years. What was the interest rate of the loan? **5%**

10. What is the interest paid for a loan of $800 at 5% annual interest for 9 months? **$30**

Money **LINK**

Many bank ATMs in Bangkok, Thailand, are located in sculptures to attract customers.

go.hrw.com
KEYWORD: MP4 Money

CNN Student News.

20b. A: 8.25%, B: 7.75%

18. Dena borrowed $7500 to buy a used car. The credit union charged 9% simple interest per year. She paid $2025 in interest. For what period of time did she borrow the money? **3 years**

19. At Thrift Bank, if you keep $675 in a savings account for 12 years, your money will earn $486 in interest. What yearly interest rate does the account offer? **6%**

20. The Smiths will borrow $35,500 from a bank to start a business. They have two loan options. Option A is a 5-year loan; option B is a 4-year loan. Use the graph to answer the following questions.

a. What is the total amount the Smiths would pay under each loan option? **A: $50,143.75, B: $46,505**

b. What would be the interest rate under each loan option?

c. What would be the monthly payment under each loan option? **A: $835.73, B: $968.85**

d. How much interest will the Smiths save by choosing loan option B? **$3638.75**

Loan Options

A: $35,500 — $14,643.75
B: $35,500 — $11,005.00

0 10,000 20,000 30,000 40,000 50,000
Amount of loan ($)

■ Initial loan ■ Interest

21. *WHAT'S THE ERROR?* On a quiz, a student is asked to calculate the total interest owed on a $4360 loan at a yearly rate of 4.5% over 3 years. The student's answer is $4,948.60. What error has the student made, and what is the correct answer?

22. *WRITE ABOUT IT* Which loan would cost the borrower less: $2000 at 8% for 3 years or $2000 at 9.5% for 2 years? How much interest would the borrower save by taking the cheaper loan?

23. *CHALLENGE* How would the total payment on a 3-year loan at 6% annual simple interest compare with the total payment on a 3-year loan where one-twelfth of that simple interest, 0.5%, is calculated monthly? Give an example.

Spiral Review

Find each number or percent. (Lesson 8-2)

24. What percent of 82 is 20.5? **25%**

25. What is 15% of 96? **14.4**

26. What is 146% of 12,500? **18,250**

27. What percent of 750 is 125? **$16\frac{2}{3}$%**

28. What percent of 0.26 is 0.0338? **13%**

29. What is 0.5% of 1000? **5**

30. **TEST PREP** A washing machine that usually sells for $459 goes on sale for $379. What is the percent decrease, to the nearest tenth of a percent? (Lesson 8-4) **D**

A 20.3%   B 82.6%   C 32.8%   D 17.4%

Answers

21. Possible answer: The student calculated the interest correctly but then answered with the total amount owed. The interest owed is $588.60.

22. Possible answer: The 2-year loan would cost the borrower less ($380 compared to $480). The interest saved would be $100.

23. Possible answer: The payments are equal. For example, the interest on a $1000 loan for 3 years at 6% is $1000 \cdot 0.06 \cdot 3 = \$180$. The interest on the same loan with a monthly rate of 0.5% is $1000 \cdot 0.005 \cdot 36 = \$180$.

Journal

Ask students to write about why they think that interest rates on savings accounts are low and interest rates on credit cards are high.

Test Prep Doctor

For Exercise 30, remind students that they first need to find the amount of decrease by subtracting, and then they need to divide by the original amount.

$$\frac{459 - 379}{459} = \frac{80}{459} \approx 0.174 = 17.4\%$$

The correct answer is **D.** Students who chose **B** probably did not subtract first.

Lesson Quiz

1. A bank is offering 2.5% simple interest on a savings account. If you deposit $5000, how much interest will you earn in one year? **$125**

2. Joshua borrowed $1000 from his friend and paid him back $1050 in six months. What simple annual interest did Joshua pay his friend? **10%**

3. The Hemmings borrowed $3000 for home improvements. They repaid the loan and $600 in simple interest four years later. What simple annual interest rate did they pay? **5%**

4. Mr. Berry had $120,000 in a retirement account. The account paid 4.25% simple interest. How much money was in the account at the end of 10 years? **$171,000**

Available on Daily Transparency in CRB

Pacing: Traditional 1 days
Block $\frac{1}{2}$ day
Objective: Students compute compound interest.

Using the Pages

In Lesson 8-7, students computed simple interest. In this extension, students will compute compound interest. Compound interest is used in almost all real-world applications, but understanding simple interest is a prerequisite to understanding compound interest.

EXTENSION **Compound Interest**

Learn to compute compound interest.

Vocabulary
compound interest

After you deposit money in a savings account, the bank pays you interest. You will probably be paid *compound interest*. When you borrow money or use a credit card, the interest you pay is also *compounded*.

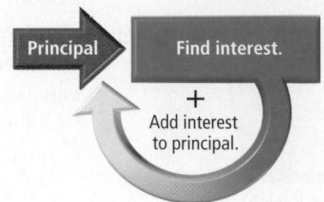

Compound interest is computed on the principal plus any interest already earned in a previous period.

Interest may compound *annually* (once a year), *semiannually* (twice a year), *quarterly* (four times a year), or *daily*.

EXAMPLE 1 Calculating Compound Interest Using a Spreadsheet

You deposit $1000 in a saving account paying 5% interest, compounded annually. Use a spreadsheet or calculator to find how much money you would have after 3 years.

You can find the total after each year several different ways:

Method 1: Find the compound interest each year and add it to the total.

Year	Principal ($)	Compound Interest ($)	Total at End of Year ($)
1	1000	1000 × 0.05 = 50	1000 + 50 = **1050**
2	**1050**	1050 × 0.05 = 52.50	1050 + 52.50 = **1102.50**
3	**1102.50**	1102.50 × 0.05 = 55.125	1102.50 + 55.125 = **1157.625**

You would have a total of $1157.63 at the end of 3 years.
You can also use the Distributive Property to multiply quickly.
$1000 + 1000(0.05) = 1000(1) + 1000(0.05) = 1000(1.05)$

Method 2: Find the total for each year and add it to the previous total.

Year	Principal ($)	Total at End of Year ($)
1	1000	1000(1.05) = **1050**
2	**1050**	1050(1.05) = **1102.50**
3	**1102.50**	1102.50(1.05) = **1157.625**

You would have a total of $1157.63 at the end of 3 years.

1 Introduce

Motivate

Tell students that simple interest is rarely used in real-world situations. *Compound interest* is used for loans, investments, bank accounts, and in almost all other real-world applications. Emphasize, however, that it was necessary to learn how to compute simple interest before trying to learn how to compute compound interest. Tell students that now they are prepared to learn how to compute compound interest.

2 Teach

Lesson Presentation

Guided Instruction

In this lesson, students learn how to compute compound interest. Before beginning Example 1, you may want to ask students to find how much $1000 would be worth if it earns 5% simple interest for 3 years. $1000 + $150 = $1150 Work through Example 1, and compare the $1157.63 total with the $1150 total using simple interest. Encourage students to see that they perform the same operations repeatedly when they compute compound interest. Explain to students that when the same operations are repeated in mathematics, there is often a formula that incorporates those operations.

You can calculate compound interest using a formula.

$A = P\left(1 + \frac{r}{k}\right)^{n \cdot k}$, where A = amount (new balance),

$\quad\quad P$ = principal (original amount of account),
$\quad\quad r$ = rate of annual interest,
$\quad\quad n$ = number of years, and
$\quad\quad k$ = number of compounding periods per year.

EXAMPLE 2 Calculating Compound Interest Using a Formula

Use the formula to find the amount after 3 years if $5000 is invested at 3% annual interest that is compounded semiannually.

$A = 5000\left(1 + \frac{0.03}{2}\right)^{3 \cdot 2}$ *Substitute P = 5000, r = 0.03, k = 2, n = 3.*

$A = 5000(1.015)^6$ *Evaluate in the parentheses and the exponent.*

$A = 5000(1.093443264)$ *Evaluate the power. Use a calculator.*

$A = \$5467.22$ *Evaluate the product, and round.*

There would be a total of $5467.22 at the end of 3 years.

EXTENSION

Exercises

Use a spreadsheet or calculator to find the value of each investment after 3 years, compounded annually.

1. $10,000 at 8% annual interest
$12,597.12

2. $1000 at 6% annual interest
$1191.02

Use the compound interest formula to find the value of each investment after 5 years, compounded semiannually.

3. $10,000 at 8% annual interest
$14,802.44

4. $1000 at 6% annual interest
$1343.92

Use the compound interest formula to find the value of the investment.

5. $12,500 at 4% annual interest, compounded annually, for 5 years
$15,208.16

6. $800 at $5\frac{1}{2}$% annual interest, compounded semiannually, for 7 years
$1169.60

7. $2000 at 7% annual interest, compounded quarterly, for 3 years
$2462.88

8. Determine the value of a $20,000 inheritance after 20 years if it is invested at a 4% annual rate of interest that is compounded annually, semiannually, and quarterly. **$43,822.46, $44,160.79, $44,334.30**

9. Determine the value of a $5000 savings account paying 6% interest, compounded monthly, over a 5-year period, assuming that no additional deposits or withdrawals are made during that time. **$6744.25**

10. Explain whether money earns more compounded annually or quarterly.

3 Close

Summarize

Review the compound interest formula. Discuss compounding periods, and show students how the number of periods fits in the formula in two places. Use the formula to work Example 2.

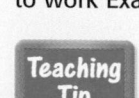 **Teaching Tip** You may want to show students a comparison between simple and compound interest over a long period of time. For example, an investment of $1000 at 8% for 30 years would be worth $3400 with simple interest and $10,765.16 with interest compounded quarterly.

Remind students that compound interest is computed on the principal plus all interest earned in previous periods. Point out that the compound interest formula gives the total amount of investment rather than just the interest, as the simple interest formula does.

Problem Solving on Location

Pennsylvania

Purpose: To provide additional practice for problem-solving skills in Chapters 1–8

Punxsutawney Phil

- After problem 1, have students consider the following problem: About what percent of the time does Punxsutawney Phil see his shadow? About 90%; an exact answer would be $100 - 13.2 = 86.8\%$. Why is it easy to estimate the percent? because the total number of trials is around 100

- After problem 2, have students consider the following problem: How can you use your answer in problem 2 to determine the percent of the time between 1980 and 1988 that Phil did *not* see his shadow? Explain. Subtract the answer from problem 2 from 100. There are only two outcomes (saw and didn't see), so the percent of time he saw his shadow and the percent of time he did not see his shadow must add up to 100%.

Extension Have students take a trip outside to record whether they see their shadows each day at the same time for a number of school days. Have them write percent problems based on their results. Check students' work.

Answers

3. Since Phil saw his shadow about 68% of the time from 1980 to 1998, he did not see his shadow 32% of the time during this period. So the percent of times he did not see his shadow from 1980 to 1998 is about twice the percent of times he did not see his shadow over the entire period from 1887 to 2002.

Problem Solving on Location

PENNSYLVANIA

Punxsutawney Phil

Punxsutawney Phil is America's most famous groundhog. According to tradition, if he sees his shadow on Groundhog Day, there will be six more weeks of winter. If he doesn't see his shadow, there will be an early spring.

Groundhog Day began as Candlemas Day, which was around February 2. If the day was clear and sunny, people said it meant a longer winter. Because early German settlers in Pennsylvania found groundhogs in many parts of the state, the tradition gradually changed to include the groundhog. Today, tens of thousands of visitors trek to Punxsutawney, Pennsylvania, each year to await the famous groundhog's appearance.

The first official record of Groundhog Day in Punxsatawney was made in 1887. From 1887 to 2002, Phil saw his shadow 92 times and didn't see it 14 times. There are 10 years with no record.

1. Ignoring the years when there was no record, what percent of the time did Phil not see his shadow? $\frac{7}{53} \approx$ **13.2%**

2. The table shows Punxsatawney Phil's shadow sightings for the years 1980–1998. What percent of the time did Phil see his shadow? $\frac{13}{19} \approx$ **68.4%**

3. Compare the results from 1980 to 1998 with the results from 1887 to 2002.

4. According to records from the National Climatic Data Center in Asheville, North Carolina, Phil's accuracy rate from 1980 to 1998 was about 59 percent. How many times did Phil correctly predict the length of winter during this time period? **about 11 times**

Phil's Shadow sightings

Year	Saw	Year	Saw
1980	yes	1990	no
1981	yes	1991	yes
1982	yes	1992	yes
1983	no	1993	yes
1984	yes	1994	yes
1985	yes	1995	no
1986	no	1996	yes
1987	yes	1997	no
1988	no	1998	yes
1989	yes		

Mural Arts Program

The Mural Arts Program in Philadelphia was founded in 1984. Since then, more than 2000 murals have been painted in the city.

1. The average mural is 45 ft tall (3 stories) by 30 ft wide. What percent of the height is the width? **$66\frac{2}{3}\%$**

A common method of painting a mural is using a grid system. A grid is drawn onto the scale drawing of the image, and a larger, similar grid is drawn onto the wall. The squares of the smaller grid are then copied exactly onto the wall.

2. A 30-ft-by-45-ft mural is divided into 1 ft squares, and the 10-in.-by-15-in. scale drawing is divided into $\frac{1}{3}$ in. squares. What is the percent increase from the area of one grid square of the scale drawing to the area of one grid square on the wall? What is the percent increase in area of the entire drawing? Are the percent increases equal? Explain.

3. If a 44-ft-by-24-ft mural is transferred from an 11-in.-by-6-in. scale drawing, what is the percent increase from the height of the drawing to the height of the mural? **4700%**

4. The 7500 ft² mural *Common Threads* is the Mural Arts Program's largest mural, at 120 ft (8 stories) tall.
   a. Find the width of the mural. **62.5 ft**
   b. If a scale drawing of the mural is 18 in. tall, how wide must it be? **9.375 in.**
   c. What percent of the area of the mural is the area of the scale drawing? **0.0225%**

MATH-ABLES

Game Resources

Puzzles, Twisters & Teasers
Chapter 8 Resource Book

Percent Puzzlers

Purpose: *To apply the skill of solving percent problems to perplexing puzzles*

Discuss: Instruct students that it is often helpful to express percents as decimals or fractions in order to solve a problem. In problem 1, what fraction of his sheep did the farmer put in each pen? $\frac{1}{5}$ in the first pen, $\frac{3}{10}$ in the second pen, $\frac{3}{8}$ in the third pen, and $1 - (\frac{1}{5} + \frac{3}{10} + \frac{3}{8}) = \frac{5}{40} = \frac{1}{8}$ in the fourth pen

Extend: Have students search the Internet for more tricky percent problems. Have them prepare their own percent puzzlers to test on a parent or classmate. Check students' work.

Percent Tiles

Purpose: *To practice finding percents in a game format*

Discuss: When a student collects a card, have him or her write an equation to show that the card has been correctly completed.

Extend: Have students play again using tiles containing numbers from 1 to 25.

MATH-ABLES

Percent Puzzlers

Prove your precision with these perplexing percent puzzlers!

1. A farmer is dividing his sheep among four pens. He puts 20% of the sheep in the first pen, 30% in the second pen, 37.5% in the third pen, and the rest in the fourth pen. What is the smallest number of sheep he could have? **40**

2. Karen and Tina are on the same baseball team. Karen has hit in 35% of her 200 times at bat. Tina has hit in 30% of her 20 times at bat. If Karen hits in 100% of her next five times at bat and Tina hits in 80% of her next five times at bat, who will have the higher percentage of hits? **Tina**

3. Joe was doing such a great job at work that his boss gave him a 10% raise! Then he made such a huge mistake that his boss gave him a 10% pay cut. What percent of his original salary does Joe make now? **99%**

4. Suppose you have 100 pounds of saltwater that is 99% water (by weight) and 1% salt. Some of the water evaporates so that the remaining liquid is 98% water and 2% salt. How much does the remaining liquid weigh? **50 lb**

Percent Tiles

Use cardboard or heavy paper to make 100 tiles with a digit from 0 through 9 (10 of each) on each tile, and print out a set of cards. Each player draws seven tiles. Lay four cards out on the table as shown. The object of the game is to collect as many cards as possible.

internet connect

Go to *go.hrw.com* for cards and a complete set of rules.
KEYWORD: MP4 Game8

To collect a card, use numbered tiles to correctly complete the statement on the card.

Technology

Compute Compound Interest

Use with Chapter 8 Extension

The formula for compound interest is $A = P\left(1 + \frac{r}{k}\right)^{nk}$, where A is the final dollar value, P is the initial dollar investment, r is the rate for each interest period, n is the number of interest periods, and k is the number of compounding periods per year.

internet connect
Lab Resources Online
go.hrw.com
KEYWORD: MP4 TechLab8

Activity

❶ Use a calculator to find the value after 9 years of $1500 invested in a savings bank that pays 3% interest compounded annually.

The initial investment P is $1500. The rate r is 3% = 0.03. The interest period is one year. The number of interest periods n is 9, and $k = 1$.

$$A = 1500\left(1 + \frac{0.03}{1}\right)^{9\cdot 1} = 1500(1.03)^9$$

On your graphing calculator, press

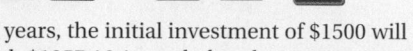 1500 [×] [(] 1.03 [)] [^] 9 [ENTER].

After 9 years, the initial investment of $1500 will be worth $1957.16 (rounded to the nearest cent).

```
1500*(1.03)^9
        1957.159776
```

❷ Use a calculator to find the value after 9 years of $1500 invested in a savings bank that pays 6% interest compounded semi-annually (twice a year).

The initial investment P is $1500. Since interest is compounded twice a year, there are 18 interest periods in 9 years, and $n = 9$. The interest rate for each period r is 6% divided by 2, or 3% = 0.03.

$$A = 1500 \times \left(1 + \frac{0.06}{2}\right)^{9\cdot 2} = 1500 \times (1.03)^{18}$$

On your calculator, press 1500 [×] [(] 1.03 [)] [^] 18 [ENTER]. You should find that A = $2553.65.

Think and Discuss

1. Compare the value of an initial deposit of $1000 at 6% simple interest for 10 years with the same initial deposit at 6% annual compound interest for 10 years. Which is greater? Why? **Simple interest: $1600; compound interest: $1790.85; the compound interest value is greater. With simple interest you earn interest only on the initial deposit. With compound interest you also earn interest on any interest you've already earned.**

Try This

1. Find the value of an initial investment of $2500 for the specified term and interest rate.

$3693.64   **$6781.60**

a. 8 years, 5% compounded annually   **b.** 20 years, 5% compounded monthly

---

Technology

Compute Compound Interest

Objective: To use a graphing calculator to compute compound interest

Materials: Graphing calculator

Lab Resources

Technology Lab Activities . . . . p. 58

Using the Page

This technology activity shows students how to compute compound interest, which can be done on any graphing calculator. Specific keystrokes may vary, depending on the make and model of the graphing calculator used. The keystrokes given are for a TI-83 model. For keystrokes to other models, visit go.hrw.com.

The Think and Discuss problem can be used to assess students' understanding of the technology activity. While the Try This problem can be done without a graphing calculator, it is meant to help students become familiar with using a graphing calculator to compute compound interest.

Assessment

Suppose you made the following investment: $525 for $1\frac{1}{2}$ years at 3% interest compounded monthly.

1. Give the value of each variable in the formula.

A: unknown

P: $525

r: 0.03

n: 18

Purpose: *To help students review and practice concepts and skills presented in Chapter 8*

Assessment Resources

Chapter Review
Chapter 8 Resource Book . . . pp. 75–76

 Test and Practice Generator
CD-ROM

Additional review assessment items in both multiple-choice and free-response format may be generated for any objective in Chapter 8.

Answers

1. percent

2. percent change

3. commission

4. simple interest; principal; rate of interest

5. 0.4375

6. 43.75%

7. $1\frac{1}{8}$

8. 112.5%

9. $\frac{7}{10}$

10. 0.7

11. $\frac{1}{250}$

12. 0.4%

Study Guide and Review

Vocabulary

Complete the sentences below with vocabulary words from the list above. Words may be used more than once.

1. A ratio that compares a number to 100 is called a(n) ___?___.

2. The ratio $\frac{\text{amount of change}}{\text{original amount}}$ is called the ___?___.

3. Percent is used to calculate ___?___, a fee paid to a person who makes a sale.

4. The formula $I = Prt$ is used to calculate ___?___. In the formula, P represents the amount borrowed or invested, which is called the ___?___, r is the ___?___, and t is the period of time that the money is borrowed or invested.

8-1 Relating Decimals, Fractions, and Percents (pp. 400–403)

EXAMPLE

■ Complete the table.

Fraction	Decimal	Percent
$\frac{3}{4}$	0.75	0.75(100) = 75%
$\frac{625}{1000} = \frac{5}{8}$	0.625	0.625(100) = 62.5%
$\frac{80}{100} = \frac{4}{5}$	$\frac{80}{100} = 0.80$	80%

EXERCISES

Complete the table.

Fraction	Decimal	Percent
$\frac{7}{16}$	**5.**	**6.**
7.	1.125	**8.**
9.	**10.**	70%
11.	0.004	**12.**

8-2 Finding Percents (pp. 405–408)

EXAMPLE

- A raw apple weighing 5.3 oz contains about 4.45 oz of water. What percent of an apple is water?

$$\frac{number}{100} = \frac{part}{whole} \quad \textit{Set up a proportion.}$$
$$\frac{n}{100} = \frac{4.45}{5.3} \quad \textit{Substitute.}$$
$$5.3n = 445 \quad \textit{Cross multiply.}$$
$$n = \frac{445}{5.3} \approx 83.96 \approx 84\%$$

An apple is about 84% water.

EXERCISES

13. The length of a year on Mercury is about 88 Earth days. The length of a year on Venus is about 225 Earth days. About what percent of the length of Venus's year is Mercury's year?

14. The main span of the Brooklyn Bridge is 1595 feet long. The Golden Gate Bridge is about 263% the length of the Brooklyn Bridge. To the nearest hundred feet, how long is the Golden Gate Bridge?

8-3 Finding a Number When the Percent Is Known (pp. 410–413)

EXAMPLE

- The population of Fairbanks, Alaska, is 30,224. This is about 477% of the population of Kodiak, Alaska. To the nearest ten people, find the population of Kodiak.

$$\frac{477}{100} = \frac{30,224}{n} \quad \textit{Set up a proportion.}$$
$$477n = 3,022,400 \quad \textit{Cross multiply.}$$
$$n = \frac{3,022,400}{477} \approx 6336.2683 \approx 6340$$

The population of Kodiak is about 6340.

EXERCISES

15. The diameter at the equator of the planet Jupiter is 88,846 miles. This is about 2930% of the diameter of Mercury at its equator. To the nearest ten miles, find the diameter of Mercury at its equator.

16. At the age of 12 weeks, Rachel weighed 8 lb 2 oz. Her birth weight was about $66\frac{2}{3}\%$ of her 12-week weight. To the nearest ounce, what was her birth weight?

8-4 Percent Increase and Decrease (pp. 416–419)

EXAMPLE

- In 1990, there were 639,270 robberies reported in the United States. This number decreased to 409,670 in 1999. What was the percent decrease?

$$639,270 - 409,670 = 229,600 \quad \textit{Amount of decrease}$$
$$\frac{amount\ of\ decrease}{original\ amount} = \frac{229,600}{639,270}$$
$$\approx 0.3592 \approx 35.92\%$$

From 1990 to 1999, the number of reported robberies in the United States decreased by 35.92%.

EXERCISES

17. On sale, a shirt was reduced from $20 to $16. Find the percent decrease.

18. In 1900, the U.S. public debt was $1.2 billion dollars. This number increased to $5674.2 billion dollars in the year 2000. Find the percent increase.

19. At the beginning of a 10-week medically supervised diet, Ken weighed 202 lb. After the diet, Ken weighed 177 lb. Find the percent decrease.

Answers

13. 39%

14. 4200 ft

15. 3030 mi

16. 5 lb 7 oz

17. 20%

18. 472,750%

19. ≈ 12.38%

Answers

Answers

20. $\approx 25\%$

21. $\approx 25\%$

22. ≈ 13

23. ≈ 16

24. ≈ 6

25. ≈ 4.5

26. $10,990

27. $3.04

28. $1796.88

29. $500

30. 7%

31. $\frac{1}{2}$ yr, or 6 mo

32. 2-year loan; $50

8-5 Estimating with Percents (pp. 420–423)

EXAMPLE

■ Estimate the percent that 5 is of 17.

$\frac{5}{17} \approx \frac{5}{15}$ *Use compatible numbers.*

$\frac{1}{3} = 33\frac{1}{3}\%$ *Simplify; change to %.*

So 5 is about $33\frac{1}{3}\%$ of 17.

EXERCISES

Use compatible numbers to estimate.

20. the percent that 6 is of 25

21. the percent that 7 is of 33

22. 23% of 64   23. 78% of 19

24. 14% of 40   25. 16% of 30

8-6 Applications of Percents (pp. 424–427)

EXAMPLE

■ As an appliance salesman, Jim earns a base pay of $450 per week plus an 8% commission on his weekly sales. Last week, his sales totaled $2750. How much did he earn for the week?

Find the amount of commission.

$8\% \cdot \$2750 = 0.08 \cdot \$2750 = \$220$

Add the commission amount to his base pay.

$\$220 + \$450 = \$670$

Last week Jim earned $670.

EXERCISES

26. As a real-estate agent, Hal earns $3\frac{1}{2}\%$ commission on the houses he sells. In the first quarter of this year, he sold two houses, one for $125,000 and the other for $189,000. How much was Hal's commission for this quarter?

27. If the sales tax is $6\frac{3}{4}\%$, how much tax would Raymond pay if he bought a radio for $19.99 and a camera for $24.99?

8-7 More Applications of Percents (pp. 428–431)

EXAMPLE

■ For home improvements, the Walters borrowed $10,000 for 3 years at simple interest. They repaid a total of $11,050. What was the interest rate of the loan?

Find the amount of interest.

$P + I = A$

$10,000 + I = 11,050$

$I = 11,050 - 10,000 = 1050$

Substitute into the simple interest formula.

$I = \quad P \cdot r \cdot t$

$1050 = 10,000 \cdot r \cdot 3$

$1050 = 30,000r$

$\frac{1050}{30,000} = r$

$0.035 = r$

The interest rate of the loan was 3.5%.

EXERCISES

Using the simple interest formula, find the missing number.

28. interest = ▨; principal = $12,500; rate = $5\frac{3}{4}\%$ per year; time = $2\frac{1}{2}$ years

29. interest = $90; principal = ▨; rate = 3% per year; time = 6 years

30. interest = $367.50; principal = $1500; rate per year = ▨; time = $3\frac{1}{2}$ years

31. interest = $1237.50; principal = $45,000; rate = $5\frac{1}{2}\%$ per year; time = ▨

Which simple-interest loan would cost the borrower less? How much less?

32. $2000 at 4% for 3 years or $2000 at 4.75% for 2 years

1. Write the percent 125% as a decimal. **1.25**

2. Write the fraction $\frac{7}{20}$ as a percent. **35%**

3. Write the decimal 0.0375 as a percent. **3.75%**

4. Write the percent $87\frac{1}{2}$% as a fraction in simplest form. $\frac{7}{8}$

Calculate.

5. What percent of 72 is 9? **12.5%**

6. What is 25% of 48? **12**

7. 15.9 is $33\frac{1}{3}$% of what number? **47.7**

8. What percent of 19 is 61.75? **325%**

Use compatible numbers to estimate.

9. the percent that 7 is of 23
   ≈ **33%**

10. the percent that 110 is of 48
    ≈ **200%**

11. 83% of 197
    ≈ **160**

Using the simple interest formula, find the missing number.

12. interest = ; principal = $15,500;
    rate = $4\frac{1}{2}$% per year; time = 3 years **$2092.50**

13. interest = $87.50; principal = ▨; **$5000**
    rate = $3\frac{1}{2}$% per year; time = 6 months

14. interest = $401.63; principal = $2550;
    rate per year = ▨; time = $3\frac{1}{2}$ years **4.5%**

15. interest = $562.50; principal = $20,000;
    rate = $3\frac{3}{4}$% per year, time = ▨
    $\frac{3}{4}$ **yr, or 9 mo**

Solve each problem. Give percents to the nearest hundredth.

16. The mean distance of Earth from the Sun is 92,960,000 miles. This is about 258% of the mean distance of Mercury from the Sun. To the nearest ten million miles, what is Mercury's mean distance from the Sun? **40,000,000 mi**

17. In 2000, U.S. trade with Saudi Arabia totaled $20.6 billion. Of this total, $6.2 billion were U.S. exports to Saudi Arabia. What percent of its total trade with Saudi Arabia were U.S. exports? **30.10%**

18. In the third quarter of 2001, the median sale price for a single-family home in Putnam County, New York, was $259,970. A year earlier, the price was $242,555. What was the percent increase? **7.18%**

19. As a real-estate agent, Walter Jordan earns $3\frac{3}{4}$% commission on the houses he sells. In the last quarter of this year, he sold two houses, one for $225,000 and the other for $199,000. How much was Walter Jordan's commission for this quarter? **$15,900**

20. If the sales tax rate is $7\frac{1}{4}$% and Jessica paid $1.45 in sales tax for a sweater, what was the price of the sweater? **$20**

21. Determine the amount of simple interest on a $1250 loan at $6\frac{1}{2}$% simple interest for 3 years. **$243.75**

Chapter Test

Purpose: *To assess students' mastery of concepts and skills in Chapter 8*

Assessment Resources ✓

Chapter 8 Tests (Levels A, B, C)
Assessment Resources . . . . . pp. 75–80

Test and Practice Generator CD-ROM

Additional assessment items in both multiple-choice and free-response format may be generated for any objective in Chapter 8.

Purpose: *To assess students' under-standing of concepts in Chapter 8 and combined problem-solving skills*

Assessment Resources

Performance Assessment
Assessment Resources . . . . p. 132

Performance Assessment Teacher Support
Assessment Resources . . . . p. 131

Answers

1–2. See p. A9.

3. See Level 3 work sample below.

Scoring Rubric for Problem Solving Item 3

Level 3
Accomplishes the purposes of the task.

Student gives clear explanations, shows understanding of mathematical ideas and processes, and computes accurately.

Level 2
Purposes of the task not fully achieved.

Sudent demonstrates satisfactory but limited understanding of the mathematical ideas and processes.

Level 1
Purposes of the task not accomplished.

Student shows little evidence of understanding the mathematical ideas and processes and makes computational and/or procedural errors.

Performance Assessment

Show What You Know

Create a portfolio of your work from this chapter. Complete this page and include it with your four best pieces of work from Chapter 8. Choose from your homework or lab assignments, mid-chapter quiz, or any journal entries you have done. Put them together using any design you want. Make your portfolio represent what you consider your best work.

Short Response

1. If 10 kg of pure acid are added to 15 kg of pure water, what percent of the resulting solution is acid? Show your work.

2. In the chemistry laboratory, Jim is working with six large jars of capacities 5 L, 4 L, 3 L, 2 L, 1 L, and 10 L. The 5 L jar is filled with an acid mix, and the rest of the jars are empty. Jim uses the 5 L jar to fill the 4 L jar and pours the excess into the 10 L jar. Then he uses the 4 L jar to fill the 3 L jar and pours the excess into the 10 L jar. He repeats the process until all but the 1 L and 10 L jars are empty. What percent of the 10 L jar is filled? Show your work.

Extended Problem Solving

3. The 60 students in a physical education class were asked to choose an elective. The results of the selection are shown in the table. Make a circle graph to display the results as percents. When making your graph, use a protractor to draw the angles at the center of the circle for each sector of the circle.

   a. Which two groups make up 50% of the graph?

   b. Could you make all the activities have an equal percent of participation? Explain.

Activity	Number of Students
Badminton	15
Basketball	12
Gymnastics	6
Volleyball	15
Wrestling	12

Student Work Samples for Item 3

Level 3

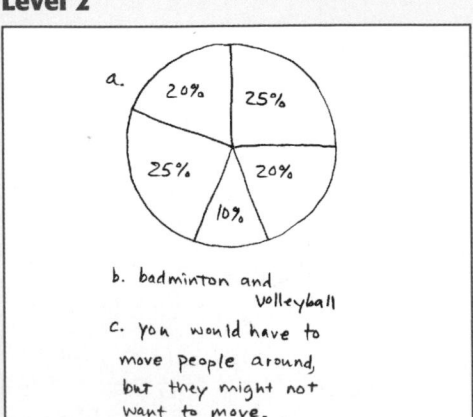

The student created an accurate graph and labeled it correctly. The student also provided correct answers to the problems.

Level 2

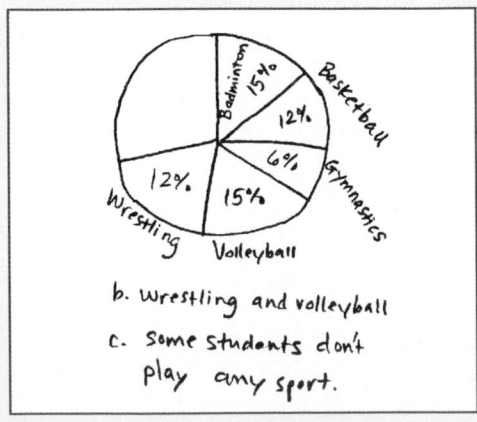

The student's graph is accurate but does not contain complete labels. The student answered part **b** correctly, but did not effectively answer part **c**.

Level 1

The student's graph is inaccurate, and the student did not find the correct percents. The answers to parts **b** and **c** are also incorrect.

Cumulative Assessment, Chapters 1–8

1. A club with 30 girls and 40 boys sponsored a boat ride. If 60% of the girls and 25% of the boys went on the boat ride, what percent of the club went on the ride? **C**
   - (A) 30%
   - (B) 35%
   - (C) 40%
   - (D) 60%

2. For every 1000 m^3 of air that goes through the filtering system in Ken's bedroom, 0.05 g of dust is removed. How many grams of dust are removed when $10^7 \, m^3$ of air are filtered? **H**
   - (F) 50,000 g
   - (G) 5,000 g
   - (H) 500 g
   - (J) 50 g

3. Let operation ◆ be defined as $x \blacklozenge y = x^y$. What is the value of $3 \blacklozenge (-1)$? **D**
   - (A) -3
   - (B) $-\frac{1}{3}$
   - (C) -1
   - (D) $\frac{1}{3}$

4. If $\frac{1}{2}$ of a number is 2 more than $\frac{1}{3}$ of the number, what is the number? **G**
   - (F) 6
   - (G) 12
   - (H) 20
   - (J) 24

5. When a certain rectangle is divided in half, two squares are formed, each with perimeter 48 inches. What is the perimeter of the original rectangle? **D**
   - (A) 24 inches
   - (B) 36 inches
   - (C) 48 inches
   - (D) 72 inches

6. Consider this equation: $\frac{20}{x} = \frac{4}{x-5}$. Which of the following is equivalent to the given equation? **H**
   - (F) $x(x - 5) = 80$
   - (G) $20x = 4(x - 5)$
   - (H) $20(x - 5) = 4x$
   - (J) $24 = x + (x - 5)$

7. A 1000-ton load is increased by 1%. What is the weight of the adjusted load? **B**
   - (A) 1001 tons
   - (B) 1010 tons
   - (C) 1100 tons
   - (D) 1110 tons

TEST TAKING TIP!
Using diagrams: Remember that no inferences are to be taken from a diagram that are not stated as given.

8. In the figure, what is y in terms of x? **F**
   - (F) $90 + x$
   - (G) $90 + 2x$
   - (H) $180 - x$
   - (J) $180 - 2x$

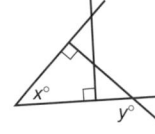

9. **SHORT RESPONSE** If $(x + 3)(9 - 5) = 16$, what is the value of x? Show your work or explain in words how you determined your answer.

10. **SHORT RESPONSE** The table shows the test scores of four students.

	Test 1	Test 2	Test 3	Test 4
Ann	80	100	100	90
Dan	60	90	90	100
Juan	100	80	100	60
Leon	100	100	100	65

Find the students' mean scores. What is the mode of the mean scores?

Purpose: *To provide review and practice for Chapters 1–8 and standardized tests*

Assessment Resources

Cumulative Tests (Levels A, B, C)
Assessment Resources . . . . . pp. 229–240

State-Specific Test Practice Online
KEYWORD: MP4 TestPrep

Test Prep Doctor ✚

For item 3, encourage students not to be distracted by the unfamiliar sign. Point out that the problem can be set up by using substitution.

$$x \blacklozenge y = x^y$$
$$3 \blacklozenge (-1) = 3^{(-1)}$$

Remind students that a negative exponent indicates a reciprocal relationship, so $3^{(-1)} = \frac{1}{3^1} = \frac{1}{3}$.

Answers

9. $(x + 3)(9 - 5) = 16$
   $(x + 3)4 = 16$
   $4x + 12 = 16$
   $4x = 4$
   $x = 1$

10. Dan: 85, Juan: 85, Leon: 91.25, Ann: 92.5; the mode of the scores is 85.

Chapter 9

Probability

Section 9A	Section 9B
Experimental Probability	**Theoretical Probability and Counting**
Lesson 9-1 Probability	**Lesson 9-4** Theoretical Probability
Lesson 9-2 Experimental Probability	**Lesson 9-5** The Fundamental Counting Principle
Technology Lab 9A Generate Random Numbers	**Lesson 9-6** Permutations and Combinations
Lesson 9-3 Use a Simulation	**Hands-On Lab 9B** Pascal's Triangle
	Lesson 9-7 Independent and Dependent Events
	Lesson 9-8 Odds

Pacing Guide for 45-Minute Classes

Chapter 9

DAY 104	DAY 105	DAY 106	DAY 107	DAY 108
Lesson 9-1	Lesson 9-2	Technology Lab 9A	Lesson 9-3	Mid-Chapter Quiz Lesson 9-4

DAY 109	DAY 110	DAY 111	DAY 112	DAY 113
Lesson 9-5	Lesson 9-6	Hands-On Lab 9B	Lesson 9-7	Lesson 9-8

DAY 114	DAY 115
Chapter 9 Review	Chapter 9 Assessment

Pacing Guide for 90-Minute Classes

Chapter 9

DAY 52	DAY 53	DAY 54	DAY 55	DAY 56
Chapter 8 Assessment Lesson 9-1	Lesson 9-2 Technology Lab 9A	Lesson 9-3 Lesson 9-4	Mid-Chapter Quiz Lesson 9-5 Lesson 9-6	Hands-On Lab 9B Lesson 9-7

DAY 57	DAY 58
Lesson 9-8 Chapter 9 Review	Chapter 9 Assessment Lesson 10-1

COURSE 1

- Find experimental and theoretical probabilities, including compound events.
- Model experiments using simulation.
- Use an organized list, permutations, and combinations to find all possible outcomes of an experiment.
- Find the odds of a specified outcome.

COURSE 2

- Find theoretical probabilities, including dependent and independent events.
- Use an organized list, permutations, and combinations to find all possible outcomes of an experiment.
- Use Pascal's Triangle, permutations, and combinations to find probabilities.
- Find the odds of a specified outcome.

PRE-ALGEBRA

- **Find theoretical probabilities, including dependent and independent events.**
- **Estimate probabilities using experiments and simulations.**
- **Use The Fundamental Counting Principle, permutations, and combinations to find probabilities.**
- **Convert between probability and odds of a specified outcome.**

Across the Curriculum

LANGUAGE ARTS

Math: Reading and Writing in the Content Area . . . . pp. 71–78

Focus on Problem Solving
  Understand the Problem. . . . . SE p. 461

Journal . . . . TE, last page of each lesson

Write About It . . . . SE pp. 450, 454, 459, 470, 475, 481, 485

SCIENCE LINK

Life Science. . . . . SE pp. 459, 466, 475

Earth Science. . . . . SE p. 454

TE = *Teacher's Edition*   **SE** = *Student Edition*

Interdisciplinary

Bulletin Board

Life Science

A typical human cell has 46 chromosomes in 23 pairs. If the cell undergoes meiosis, the process by which eggs and sperm are formed, each of the resulting reproductive cells receives one chromosome from each pair. How many chromosomal combinations are possible in the resulting cells?

$2^{23} = 8,388,608$

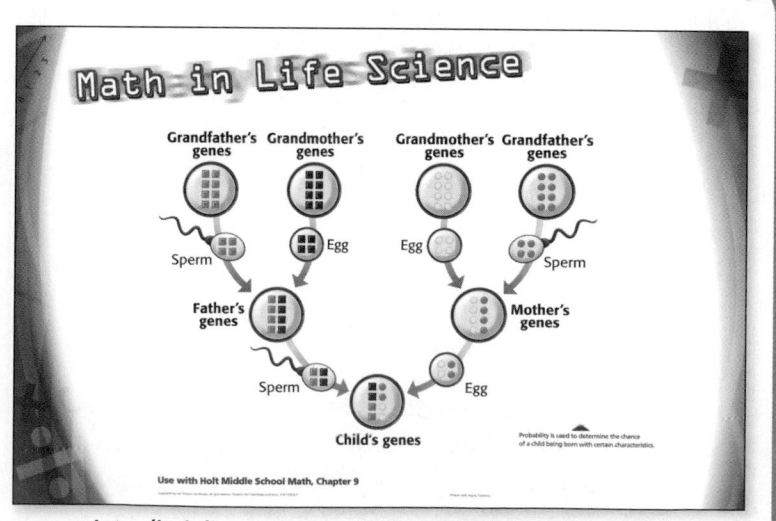

Interdisciplinary posters and worksheets are provided in your resource material.

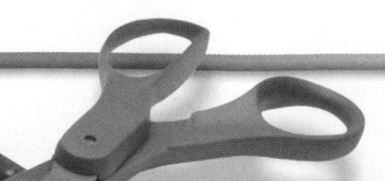

Chapter 9 Resource Book

Student Resources

Reaching All Learners

English Language Learners

Individual Needs

Hands-On

Applications and Connections

Teacher and Parent Resources

- Daily Transparencies
- Additional Examples Transparencies
- Teaching Transparencies

Transparencies

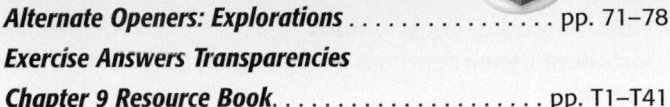

- Daily Transparencies
- Additional Examples Transparencies
- Teaching Transparencies

Technology

Teacher Resources

Student Resources

▲ internet connect

Homework Help Online	**KEYWORD:** MP4 HWHelp9
Math Tools Online	**KEYWORD:** MP4 Tools
Glossary Online	**KEYWORD:** MP4 Glossary
Chapter Project Online	**KEYWORD:** MP4 PSProject9
Chapter Opener Online	**KEYWORD:** MP4 Ch9

KEYWORD: MP4 CNN9

SE = *Student Edition*   **TE** = *Teacher's Edition*   **AR** = *Assessment Resources*   **CRB** = *Chapter Resource Book*   **MK** = *Manipulatives Kit*

444C Chapter 9 Probability

Assessment Options

Assessing Prior Knowledge

Determine whether students have the required prerequisite concepts and skills.

Are You Ready?. . . . . SE p. 445
Inventory Test. . . . . AR pp. 1–4

Test Preparation

Provide review and practice for chapter and standardized tests.

Standardized Test Prep. . . . . SE p. 495
Spiral Review with Test Prep . . . . . SE, last page of each lesson
Study Guide and Review . . . . SE pp. 490–492
Test Prep Tool Kit

Technology

 Test and Practice Generator **CD-ROM**

▸ internet connect
State-Specific Test Practice Online **KEYWORD:** MP4 TestPrep

Performance Assessment

Assess students' understanding of chapter concepts and combined problem-solving skills.

Performance Assessment . . . . SE p. 494
   Includes scoring rubric in TE
Performance Assessment . . . . AR p. 134
Performance Assessment Teacher Support. . . . . AR p. 133

Portfolio

Portfolio opportunities appear throughout the Student and Teacher's Editions.

Suggested work samples:

Problem Solving Project . . . . TE p. 444
Performance Assessment . . . . SE p. 494
Portfolio Guide. . . . . AR p. xxxvii
Journal. TE, last page of each lesson
Write About It . . . . . . SE pp. 450, 454, 459, 470, 475, 481, 485

Daily Assessment

Obtain daily feedback on students' understanding of concepts.

Spiral Review and Test Prep . . . . . . SE, last page of each lesson

**Also Available on Transparency
In Chapter 9 Resource Book**

Warm Up. TE, first page of each lesson
Problem of the Day. . . . . TE, first page of each lesson
Lesson Quiz . . . . TE, last page of each lesson

Student Self-Assessment

Have students evaluate their own work.

Group Project Evaluation . . . . AR p. xxxiv
Individual Group Member Evaluation . . . . AR p. xxxv
Portfolio Guide. . . . . AR p. xxxvii
Journal. . . . . TE, last page of each lesson

Formal Assessment

Assess students' mastery of concepts and skills.

Section Quizzes . . . . AR pp. 21–22
Mid-Chapter Quiz. . . . . SE p. 460
Chapter Test . . . . SE p. 493
Chapter Tests (Levels A, B, C) . . . . AR pp. 81–86
Cumulative Tests (Levels A, B, C). . . . . AR pp. 241–252
Standardized Test Prep
   Cumulative Assessment . . . . SE p. 495
End-of-Year Test. . . . . AR pp. 313–316

Technology

 Test and Practice Generator **CD-ROM**

Make tests electronically. This software includes:

- Dynamic practice for Chapter 9
- Customizable tests
- Multiple-choice items for each objective
- Free-response items for each objective
- Teacher management system

SE = *Student Edition*   **TE** = *Teacher's Edition*   **AR** = *Assessment Resources*   **CRB** = *Chapter Resource Book*   **MK** = *Manipulatives Kit*

Chapter 9

Chapter 9 Tests

Three levels (A,B,C) of tests are available for each chapter in the *Assessment Resources.*

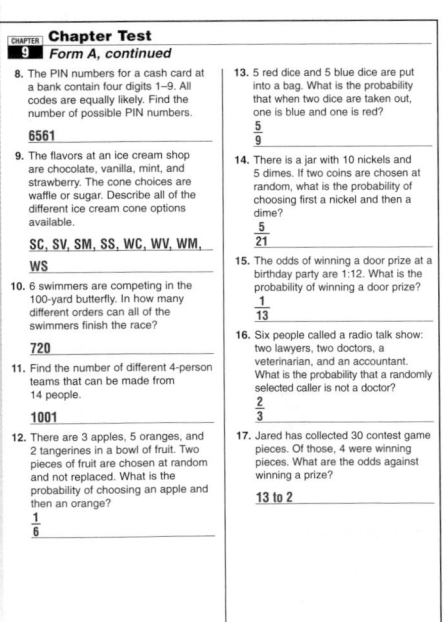

Level A

CHAPTER 9 Chapter Test
Form A

An experiment consists of drawing 4 balls from a bag and counting the number of red balls. The table gives the probability of each outcome.

Number of Red Balls	0	1	2	3	4
Probability	0.025	0.36	0.27	0.282	0.063

1. What is the probability of drawing fewer than 2 red balls?
   0.385
2. What is the probability of drawing more than 1 red ball?
   0.615
3. Erin's soccer team has won 17 of their 20 games this season. What is the probability that they will win their next game?
   0.85 or 85%
4. A researcher conducted a survey of 324 high school students and found that 54 of them were enrolled in advanced chemistry. What is the probability that a randomly selected student is enrolled in advanced chemistry?
   $\frac{1}{6}$

Use the table of random numbers to simulate each situation. Use at least 10 trials for each simulation.

33	35	71	65	22	33	04	35	56	99
63	41	51	27	76	48	30	84	63	20
57	62	81	29	54	61	35	22	35	44
62	61	22	24	35	12	73	42	64	46
33	20	52	77	88	15	73	82	19	97
31	96	04	29	74	36	44	42	38	26
53	14	76	41	98	83	53	64	15	91
24	83	42	19	61	12	52	62	28	32

5. Katy hits the ball 58% of the time she bats. Estimate the probability that she will hit the ball at least 3 times in her next 4 at bats.
   possible answer: 40%
6. Sandra hits a golf ball over 120 yards on her first drive 67% of the time. Estimate the probability that she will hit the ball over 120 yards at least 4 times in the next 7 times she drives.
   possible answer: 90%

An experiment consists of spinning a spinner with equal chances of landing on one of four colors – red, blue, green, or yellow.

7. What is the probability of landing on blue or yellow?
   $\frac{1}{2}$

CHAPTER 9 Chapter Test
Form A, continued

8. The PIN numbers for a cash card at a bank contain four digits 1–9. All codes are equally likely. Find the number of possible PIN numbers.
   6561
9. The flavors at an ice cream shop are chocolate, vanilla, mint, and strawberry. The cone choices are waffle or sugar. Describe all of the different ice cream cone options available.
   SC, SV, SM, SS, WC, WV, WM, WS
10. 6 swimmers are competing in the 100-yard butterfly. In how many different orders can all of the swimmers finish the race?
    720
11. Find the number of different 4-person teams that can be made from 14 people.
    1001
12. There are 3 apples, 5 oranges, and 2 tangerines in a bowl of fruit. Two pieces of fruit are chosen at random and not replaced. What is the probability of choosing an apple and then an orange?
    $\frac{1}{6}$
13. 5 red dice and 5 blue dice are put into a bag. What is the probability that when two dice are taken out, one is blue and one is red?
    $\frac{5}{9}$
14. There is a jar with 10 nickels and 5 dimes. If two coins are chosen at random, what is the probability of choosing first a nickel and then a dime?
    $\frac{5}{21}$
15. The odds of winning a door prize at a birthday party are 1:12. What is the probability of winning a door prize?
    $\frac{1}{13}$
16. Six people called a radio talk show: two lawyers, two doctors, a veterinarian, and an accountant. What is the probability that a randomly selected caller is not a doctor?
    $\frac{2}{3}$
17. Jared has collected 30 contest game pieces. Of those, 4 were winning pieces. What are the odds against winning a prize?
    13 to 2

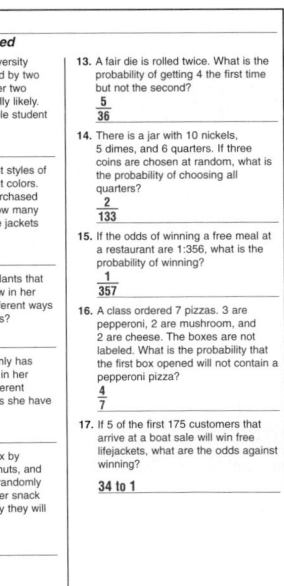

Level B

CHAPTER 9 Chapter Test
Form B

Trisha made an educated guess on four of the multiple choice questions on her drivers license examination. The table gives the probability of each possible result.

Number of Correct Answers	Probability
0	0.025
1	0.271
2	0.359
3	0.282
4	0.063

1. What is the probability of getting 3 or more correct?
   0.345
2. What is Trisha's probability of failing this part of the exam (getting fewer than 2 correct)?
   0.296
3. A ball was randomly drawn from a bag and then replaced. In 300 experiments, a green ball was chosen 58 times, a red ball was chosen 118 times, a yellow ball was chosen 99 times, and a blue ball was chosen 25 times. What is the probability of choosing a yellow ball?
   0.33 or 33%
4. A researcher conducted a survey of 425 high school students and found that 356 of them planned to attend college. Estimate to the nearest percent the probability that a randomly selected student plans to attend college.
   0.84 or 84%

Use the table of random numbers to simulate each situation. Use at least 10 trials for each simulation.

33	35	71	65	22	33	04	35	56	99
65	41	51	27	76	48	30	84	63	20
57	62	81	29	74	61	35	22	35	44
42	61	22	24	35	12	73	42	69	46
33	20	52	57	88	15	73	82	19	97
31	96	04	29	74	36	48	42	38	26
53	14	76	41	18	43	53	68	15	91
24	83	42	19	61	12	52	62	28	32

5. In a city in Alaska, snow falls on 64% of winter days. Estimate the probability that it will snow at least 6 out of 7 days during a week in January.
   possible answer: 20%
6. At a pizza place, about 45% of the customers order pepperoni on their pizza. Estimate the probability that at least 5 out of the next 6 customers will order pepperoni.
   possible answer: 10%

An experiment consists of rolling a fair eight-sided die.

7. What is the probability of rolling a 3, 4, or 5?
   $\frac{3}{8}$

CHAPTER 9 Chapter Test
Form B, continued

8. Student ID codes at a university contain two letters followed by two digits 0–9 and then another two letters. All codes are equally likely. Find the number of possible student ID codes.
   45,697,600
9. A store sells three different styles of fleece jackets in 8 different colors. Each style can also be purchased with or without a hood. How many different versions of fleece jackets does the store sell?
   48
10. Maureen has 7 different plants that she wants to plant in a row in her flower bed. How many different ways can she arrange the plants?
    5040
11. If Maureen decides she only has room for 3 of the 7 plants in her flower bed, how many different selections of 3 plants does she have to choose from?
    35
12. Karleen made a snack mix by mixing 30 raisins, 45 peanuts, and 15 marshmallows. If she randomly selects two pieces from her snack mix, what is the probability they will both be peanuts?
    $\frac{22}{89}$
13. A fair die is rolled twice. What is the probability of getting 4 the first time but not the second?
    $\frac{5}{36}$
14. There is a jar with 10 nickels, 5 dimes, and 6 quarters. If three coins are chosen at random, what is the probability of choosing all quarters?
    $\frac{2}{133}$
15. If the odds of winning a free meal at a restaurant are 1:356, what is the probability of winning?
    $\frac{1}{357}$
16. A class ordered 7 pizzas. 3 are pepperoni, 2 are mushroom, and 2 are cheese. The boxes are not labeled. What is the probability that the first box opened will not contain a pepperoni pizza?
    $\frac{4}{7}$
17. If 5 of the first 175 customers that arrive at a boat sale will win free lifejackets, what are the odds against winning?
    34 to 1

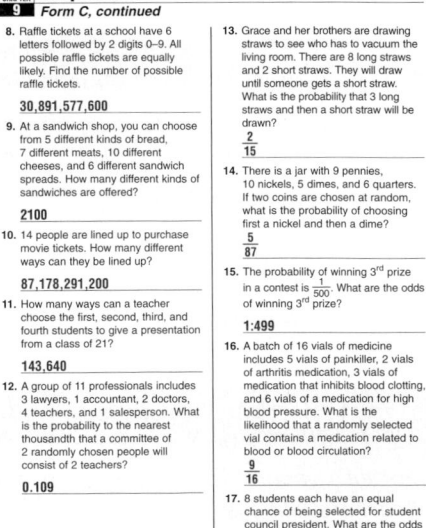

Level C

CHAPTER 9 Chapter Test
Form C

Outcome	Probability
1	0.071
2	0.271
3	0.314
4	0.282
5	0.062

1. What is the probability of outcome 1 or 2 occurring?
   0.342
2. What is the probability of neither outcome 1 nor outcome 4?
   0.647
3. Among 692 patients who were tested for a certain disease, 295 tested negative, 104 tested positive, and 293 had inconclusive results. Estimate to the nearest thousandth the probability of a patient's testing negative.
   0.426 or 42.6%

Sport	Number of Students
Swimming	28
Basketball	31
Football	45
Tennis	24

4. 165 students were polled about the sports they participate in. Estimate to the nearest thousandth the probability that a randomly polled student does not participate in any sports.
   0.224 or 22.4%

Use the table of random numbers to simulate each situation. Use at least 10 trials for each simulation.

12	77	64	83	08	46	32	03	19	60
42	97	43	64	44	23	82	66	34	15
16	52	23	30	43	55	69	93	32	05
25	78	81	57	38	21	99	73	33	57
22	43	51	27	34	26	44	62	48	34
21	96	11	25	26	22	83	24	26	72
12	95	46	13	52	23	54	45	01	47
64	11	84	35	36	41	59	15	76	28

5. Mr. Guanara gives a pop quiz about 25% of the days his class meets. What is the probability that there will be a pop quiz at least 3 out of the next 6 times his class meets.
   possible answer: 20%
6. An inspector finds that about 3% of computer motherboards assembled are defective. What is the probability that 1 of the next 8 motherboards will be defective?
   possible answer: 20%

Three fair coins with sides marked A and B are tossed.

7. Find each probability. $P(ABA)$
   $\frac{1}{8}$

CHAPTER 9 Chapter Test
Form C, continued

8. Raffle tickets at a school have 6 letters followed by 2 digits 0–9. All possible raffle tickets are equally likely. Find the number of possible raffle tickets.
   30,891,577,600
9. At a sandwich shop, you can choose from 5 different kinds of bread, 7 different meats, 10 different cheeses, and 6 different sandwich spreads. How many different kinds of sandwiches are offered?
   2100
10. 14 people are lined up to purchase movie tickets. How many different ways can they be lined up?
    87,178,291,200
11. How many ways can a teacher choose the first, second, third, and fourth students to give a presentation from a class of 21?
    143,640
12. A group of 11 professionals includes 3 lawyers, 1 accountant, 2 doctors, 4 teachers, and 1 salesperson. What is the probability to the nearest thousandth that a committee of 2 randomly chosen people will consist of 2 teachers?
    0.109
13. Grace and her brothers are drawing straws to see who has to vacuum the living room. There are 8 long straws and 2 short straws. They will draw until someone gets a short straw. What is the probability that 3 long straws and then a short straw will be drawn?
    $\frac{2}{15}$
14. There is a jar with 9 pennies, 10 nickels, 5 dimes, and 6 quarters. If two coins are chosen at random, what is the probability of choosing first a nickel and then a dime?
    $\frac{5}{87}$
15. The probability of winning 3rd prize in a contest is $\frac{1}{500}$. What are the odds of winning 3rd prize?
    1:499
16. A batch of 16 vials of medicine includes 5 vials of painkiller, 2 vials of arthritis medication, 3 vials of medication that inhibits blood clotting, and 6 vials of a medication for high blood pressure. What is the likelihood that a randomly selected vial contains a medication related to blood or blood circulation?
    $\frac{9}{16}$
17. 8 students each have an equal chance of being selected for student council president. What are the odds against being selected?
    7:1

Test and Practice Generator CD-ROM

Create and customize multiple versions of the same tests with corresponding answers for any chosen chapter objectives.

Chapter 9 State and Standardized Test Preparation

Test Taking Skill Builder and Standardized Test Practice are provided for each chapter in the *Test Prep Tool Kit.*

TEST TAKING SKILL BUILDER

Test Taking Strategy
Chapter 9   **Gridded Response**

Gridded Reponse questions require that you fill in the grid on your answer sheet.

To show a gridded response to a problem with a decimal or fraction answer follow these steps.

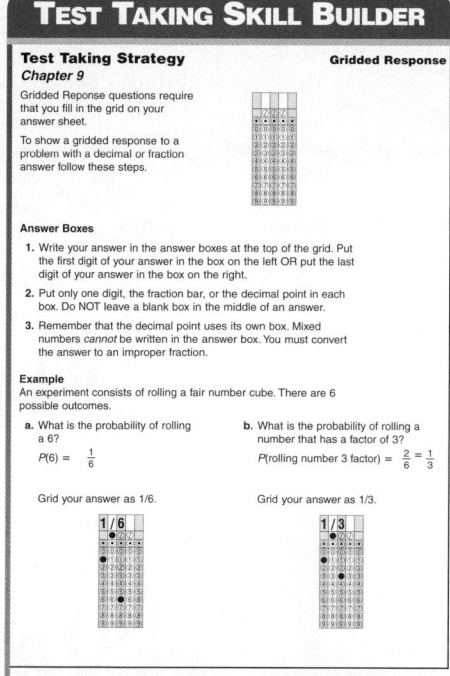

Answer Boxes

1. Write your answer in the answer boxes at the top of the grid. Put the first digit of your answer in the box on the left OR put the last digit of your answer in the box on the right.

2. Put only one digit, the fraction bar, or the decimal point in each box. Do NOT leave a blank box in the middle of an answer.

3. Remember that the decimal point uses its own box. Mixed numbers *cannot* be written in the answer box. You must convert the answer to an improper fraction.

Example
An experiment consists of rolling a fair number cube. There are 6 possible outcomes.

a. What is the probability of rolling a 6?

$P(6) = \frac{1}{6}$

Grid your answer as 1/6.

b. What is the probability of rolling a number that has a factor of 3?

$P(\text{rolling number 3 factor}) = \frac{2}{6} = \frac{1}{3}$

Grid your answer as 1/3.

Test Taking Strategy
Chapter 9, continued

Exercises
What should go in the first box on the left for each gridded response answer?

1. 0.35 ___0___   2. 1.8 ___1___   3. 234 ___2___   4. 73.5 ___7___

If you grid your answer starting in the far left column, which column should the decimal point or fraction bar go in for each answer below?

5. 20.05 ___third___   6. 0.125 ___second___   7. 246 ___none___   8. $\frac{1}{24}$ ___second___

Tell what error was made in each gridded response below.

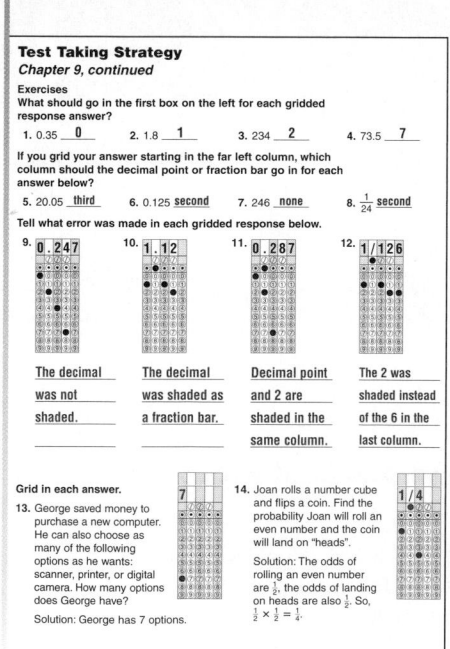

9. The decimal was not shaded.

10. The decimal was shaded as a fraction bar.

11. Decimal point and 2 are shaded in the same column.

12. The 2 was shaded instead of the 6 in the last column.

Grid in each answer.

13. George saved money to purchase a new computer. He can also choose as many of the following options as he wants: scanner, printer, or digital camera. How many options does George have?

Solution: George has 7 options.

14. Joan rolls a number cube and flips a coin. Find the probability Joan will roll an even number and the coin will land on "heads".

Solution: The odds of rolling an even number are $\frac{1}{2}$, the odds of landing on heads are also $\frac{1}{2}$. So, $\frac{1}{2} \times \frac{1}{2} = \frac{1}{4}$.

STANDARDIZED TEST PRACTICE

Standardized Test Practice
Chapter 9

Select the best answer for Questions 1–7.

1. Jack has thrown 24 strikes out of his last 52 pitches. Estimate the probability that his next throw will be a strike.
   A 36%
   B 46%
   C 54%
   D 216%

2. If you have 3 kinds of bread, 2 kinds of meat, and 4 kinds of cheese, how many different kinds of sandwiches could you make containing one kind of each?
   F 16   H 22
   G 18   I 24

3. Give the probability of the spinner landing on the star.

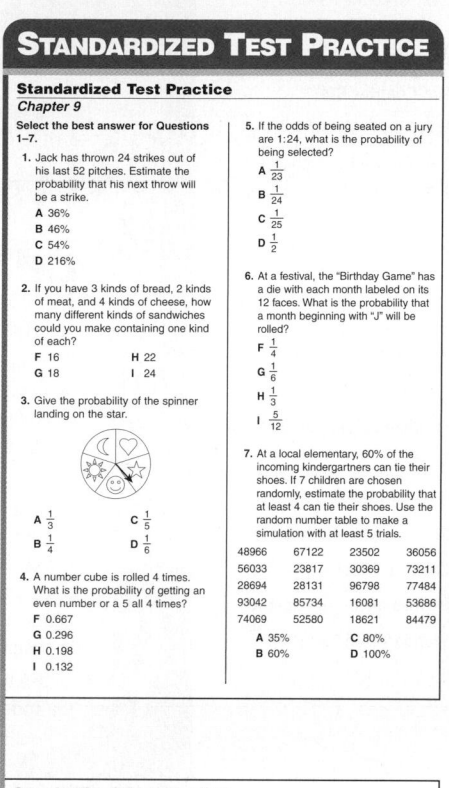

   A $\frac{1}{3}$ C $\frac{1}{5}$
   B $\frac{1}{4}$ D $\frac{1}{6}$

4. A number cube is rolled 4 times. What is the probability of getting an even number or a 5 all 4 times?
   F 0.667
   G 0.296
   H 0.198
   I 0.132

5. If the odds of being seated on a jury are 1:24, what is the probability of being selected?
   A $\frac{1}{23}$
   B $\frac{1}{24}$
   C $\frac{1}{25}$
   D $\frac{1}{2}$

6. At a festival, the "Birthday Game" has a die with each month labeled on its 12 faces. What is the probability that a month beginning with "J" will be rolled?
   F $\frac{1}{4}$
   G $\frac{1}{6}$
   H $\frac{1}{3}$
   I $\frac{5}{12}$

7. At a local elementary, 60% of the incoming kindergartners can tie their shoes. If 7 children are chosen randomly, estimate the probability that at least 4 can tie their shoes. Use the random number table to make a simulation with at least 5 trials.

48966	67122	23502	36056
56033	23817	30369	73211
28694	28131	96798	77484
93042	85734	16081	53686
74069	52580	18621	84479

   A 35%   C 80%
   B 60%   D 100%

Standardized Test Practice
Chapter 9, continued

Gridded Response
Solve the problems. Use the answer sheet to write and grid-in your answer.

8. How many possible outcomes are there?
   Bagel: plain, wheat, salt, blueberry
   Cream cheese: plain, strawberry, light

9. A store is promoting a new line of juice. There are 4 kinds of juice and they want to display them in a row on a shelf in the front of the store. How many different ways can the juice be displayed?

Short Response
Solve the problems. Use the answer sheet to write your answers.

10. At the Fall Sports Banquet, three different pasta dishes are offered. Of the first 30 people through the buffet line, 12 choose lasagna, 8 choose manicotti, and the remainder chose ravioli. Estimate the probability that the next person will choose ravioli. Explain in words how you determined your answer.

11. Alyce says the probability of spinning an *A* is 3 times the probability of spinning a *B*. Is she correct? Explain why or why not. Show your work.

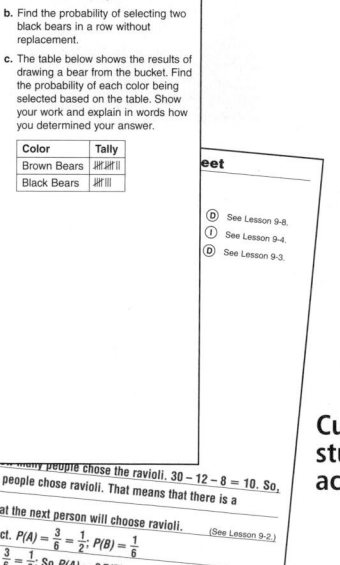

Extended Response

12. One of the children's games at the school carnival has a bucket that contains 30 brown bears and 22 black bears. You close your eyes and select a bear from the bucket.

   a. Find the probability of selecting a brown bear. Show your work.

   b. Find the probability of selecting two black bears in a row without replacement.

   c. The table below shows the results of drawing a bear from the bucket. Find the probability of each color being selected based on the table. Show your work and explain in words how you determined your answer.

Color	Tally
Brown Bears	
Black Bears	

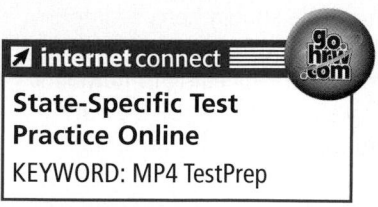

State-Specific Test Practice Online
KEYWORD: MP4 TestPrep

Test Prep Tool Kit

- Standardized Test Prep Workbook
- Countdown to Testing transparencies
- State Test Prep CD-ROM
- Standardized Test Prep Video

...eet

   D See Lesson 9-8.
   I See Lesson 9-4.
   D See Lesson 9-3.

Customized answer sheets give students realistic practice for actual standardized tests.

$\frac{10}{30} = \frac{1}{3}$; ... $\frac{1}{3}$ of the people chose ravioli. That means that there is a

33.33% chance that the next person will choose ravioli.

11. **Yes, Alyce is correct.** $P(A) = \frac{3}{6} = \frac{1}{2}$; $P(B) = \frac{1}{6}$ (See Lesson 9-2.)

$3 \cdot P(B) = 3 \cdot \frac{1}{6} = \frac{3}{6} = \frac{1}{2}$; So $P(A) = 3P(B)$.

Extended Response   (See Lesson 9-1.)
Write your answers for Problem 12 on the back of this paper.
See Lesson 9-7.

Probability

Probability

Why Learn This?

Tell students that cryptography is an important field, especially for military intelligence and information technology. Cryptographers create and crack codes. Using the code shown here, a simple message can be encoded as a complicated-looking string of numbers. A cryptographer's job is to study the message to try to identify recurring patterns as specific letters.

Using Data

To begin the study of this chapter, have students:

- Use the codes from the table to encrypt the word *movie*. **1001101 1001111101011010010011000101**

- Determine another word that can be spelled with the given letters, encrypt it, and pass it to another student to decode.

  Possible answer:
  heal: **10010001000101110000001 1001100**

  Puzzle answer:
  I LOVE MATH

internet connect
go.hrw.com
Chapter Opener Online
go.hrw.com
KEYWORD: MP4 Ch9

Letter	Code
A	1000001
E	1000101
H	1001000
I	1001001
L	1001100
M	1001101
O	1001111
T	1010100
V	1010110

Career Cryptographer

1001001100110010011111010110
1000101100110110000011010100100100 1000

Is this pattern of zeros and ones some kind of message or secret code? A cryptographer could find out. Cryptographers create and break codes by assigning number values to letters of the alphabet.

Almost all text sent over the Internet is encrypted to ensure security for the sender. Codes made up of zeros and ones, or *binary codes*, are frequently used in computer applications.

Use the table to break the code above.

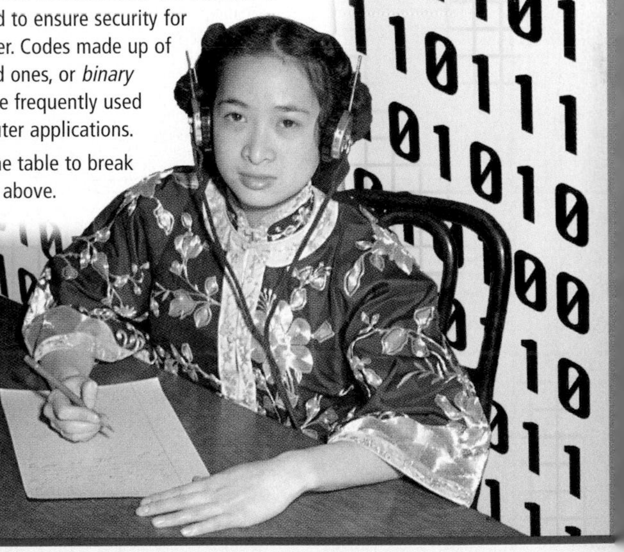

Problem Solving Project

Social Studies and Technology Connection

Purpose: To solve problems involving permutations and/or combinations

Materials: Code Breaking worksheet

internet connect
go.hrw.com
Chapter Project Online: *go.hrw.com*
KEYWORD: MP4 PSProject9

Understand, Plan, Solve, and Look Back

Have students:

- ✔ Examine the string of 1's and 0's. It's a code containing a message. Tell what patterns, if any, they see. Can they break the code?

- ✔ Complete the Code Breaking worksheet.

- ✔ Create a possible code for the rest of the alphabet, or a new code for the entire alphabet. Use the codes to create messages for their classmates to break.

- ✔ Research the history of cryptography and historical events in which cryptographers played an important role.

- ✔ Check students' work.

ARE YOU READY?

Choose the best term from the list to complete each sentence.

1. The term __?__ means "per hundred." **percent**
2. A __?__ is a comparison of two numbers. **ratio**
3. In a set of data, the __?__ is the largest number minus the smallest number. **range**
4. A __?__ is in simplest form when its numerator and denominator have no common factors other than 1. **fraction**

fraction
percent
range
ratio

Complete these exercises to review skills you will need for this chapter.

✔ Simplify Ratios

Write each ratio in simplest form.

5. 5:50 **1:10**   6. 95 to 19 **5:1**   7. $\frac{20}{100}$ **$\frac{1}{5}$**   8. $\frac{192}{80}$ **$\frac{12}{5}$**

✔ Write Fractions as Decimals

Express each fraction as a decimal.

9. $\frac{52}{100}$ **0.52**   10. $\frac{7}{1000}$ **0.007**   11. $\frac{3}{5}$ **0.6**   12. $\frac{2}{9}$ **$0.\overline{2}$**

✔ Write Fractions as Percents

Express each fraction as a percent.

13. $\frac{19}{100}$ **19%**   14. $\frac{1}{8}$ **12.5%**   15. $\frac{5}{2}$ **250%**   16. $\frac{2}{3}$ **$66\frac{2}{3}$% or $66.\overline{6}$%**

17. $\frac{3}{4}$ **75%**   18. $\frac{9}{20}$ **45%**   19. $\frac{7}{10}$ **70%**   20. $\frac{2}{5}$ **40%**

✔ Operations with Fractions

Add. Write each answer in simplest form.

21. $\frac{3}{8} + \frac{1}{4} + \frac{1}{6}$ **$\frac{19}{24}$**   22. $\frac{1}{6} + \frac{2}{3} + \frac{1}{9}$ **$\frac{17}{18}$**   23. $\frac{1}{8} + \frac{1}{4} + \frac{1}{8} + \frac{1}{2}$ **1**   24. $\frac{1}{3} + \frac{1}{4} + \frac{2}{5}$ **$\frac{59}{60}$**

Multiply. Write each answer in simplest form.

25. $\frac{3}{8} \cdot \frac{1}{5}$ **$\frac{3}{40}$**   26. $\frac{2}{3} \cdot \frac{6}{7}$ **$\frac{4}{7}$**   27. $\frac{3}{7} \cdot \frac{14}{27}$ **$\frac{2}{9}$**   28. $\frac{13}{52} \cdot \frac{3}{51}$ **$\frac{1}{68}$**

29. $\frac{4}{5} \cdot \frac{11}{4}$ **$\frac{11}{5}$**   30. $\frac{5}{2} \cdot \frac{3}{4}$ **$\frac{15}{8}$**   31. $\frac{27}{8} \cdot \frac{4}{9}$ **$\frac{3}{2}$**   32. $\frac{1}{15} \cdot \frac{30}{9}$ **$\frac{2}{9}$**

ARE YOU READY?
Were students successful with Are You Ready?

NO INTERVENE ← → **YES** ENRICH

✔ Simplify Ratios 5–8
Intervention Practice, Skill 28
 CD-ROM
Intervention Activities, Skill 28

✔ Write Fractions as Percents 13–20
Intervention Practice, Skill 31
CD-ROM
Intervention Activities, Skill 31

✔ Write Fractions as Decimals 9–12
Intervention Practice, Skill 26
 CD-ROM
Intervention Activities, Skill 26

✔ Operations with Fractions 21–32
Intervention Practice, Skill 42
CD-ROM
Intervention Activities, Skill 42

Are You Ready? Enrichment, pp. 421–422

Pacing: Traditional 1 day
Block $\frac{1}{2}$ day

Objective: Students find the probability of an event by using the definition of probability.

Warm Up

Write each fraction in simplest form.

1. $\frac{16}{20}$ $\frac{4}{5}$
2. $\frac{12}{36}$ $\frac{1}{3}$
3. $\frac{8}{64}$ $\frac{1}{8}$
4. $\frac{39}{195}$ $\frac{1}{5}$

Problem of the Day

A careless reader mixed up some encyclopedia volumes on a library shelf. The *Q* volume is to the right of the *X* volume, and the *C* is between the *X* and *D* volumes. The *Q* is to the left of the *G*. *X* is to the right of *C*. From left to right, in what order are the five volumes?

D, C, X, Q, G

Available on Daily Transparency in CRB

Math Humor

Did you hear about the man who wore half a raincoat to work? The weather report said there was a 50% chance of rain.

9-1 Probability

Learn to find the probability of an event by using the definition of probability.

Vocabulary
experiment
trial
outcome
sample space
event
probability
impossible
certain

An **experiment** is an activity in which results are observed. Each observation is called a **trial** , and each result is called an **outcome**. The **sample space** is the set of all possible outcomes of an experiment.

Experiment	Sample space
• flipping a coin	• heads, tails
• rolling a number cube	• 1, 2, 3, 4, 5, 6
• guessing the number of jelly beans in a jar	• whole numbers

An **event** is any set of one or more outcomes. The **probability** of an event, written *P*(event), is a number from 0 (or 0%) to 1 (or 100%) that tells you how likely the event is to happen.

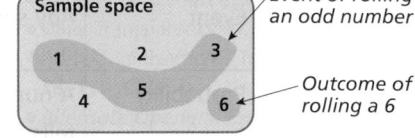

Sample space — Event of rolling an odd number — Outcome of rolling a 6

• A probability of 0 means the event is **impossible** , or can never happen.

• A probability of 1 means the event is **certain** , or has to happen.

• The probabilities of all the outcomes in the sample space add up to 1.

Never happens		Happens about half the time		Always happens
0	$\frac{1}{4}$	$\frac{1}{2}$	$\frac{3}{4}$	1
0	0.25	0.5	0.75	1
0%	25%	50%	75%	100%

EXAMPLE **1** Finding Probabilities of Outcomes in a Sample Space

Give the probability for each outcome.

A The weather forecast shows a 40% chance of rain.

Outcome	Rain	No rain
Probability		

The probability of rain is *P*(rain) = 40% = 0.4. The probabilities must add to 1, so the probability of no rain is *P*(no rain) = 1 − 0.4 = 0.6, or 60%.

---

1 Introduce

Alternate Opener

Exploration worksheet and answers on Chapter 9 Resource Book pp. 8 and 92

Motivate

Ask students to count the number of students in the room. Ask them if you put all of their names in a hat and pulled one out, how likely it would be for each one to be chosen. Ask how likely it would be for the name of someone from the first row to be chosen. Explain that *probability* is a branch of mathematics that predicts the likelihood of events like these.

2 Teach

Lesson Presentation

Guided Instruction

In this lesson, students learn to find the probability of an event by using the definition of probability. Discuss the new vocabulary. Explain that if an event has a probability of 0, that event *can never* happen (e.g., rolling a 7 on a number cube), and if an event has a probability of 1, that event *will certainly* happen (e.g., rolling a number less than 10 on a number cube). All other possible events have a probability greater than 0 and less than 1 (Teaching Transparency T2, CRB). Emphasize that the sum of the probabilities of all the possible outcomes in the sample space equals 1.

Give the probability for each outcome.

B

Outcome	Red	Yellow	Blue
Probability	■	■	■

Half of the spinner is red, so a reasonable estimate of the probability that the spinner lands on red is $P(\text{red}) = \frac{1}{2}$.

One-fourth of the spinner is yellow, so a reasonable estimate of the probability that the spinner lands on yellow is $P(\text{yellow}) = \frac{1}{4}$.

One-fourth of the spinner is blue, so a reasonable estimate of the probability that the spinner lands on blue is $P(\text{blue}) = \frac{1}{4}$.

Check The probabilities of all the outcomes must add to 1.

$\frac{1}{2} + \frac{1}{4} + \frac{1}{4} = 1$ ✔

To find the probability of an event, add the probabilities of all the outcomes included in the event.

EXAMPLE 2 **Finding Probabilities of Events**

A quiz contains 5 multiple-choice questions. Suppose you guess randomly on every question. The table below gives the probability of each score.

Score	0	1	2	3	4	5
Probability	0.237	0.396	0.264	0.088	0.014	0.001

A What is the probability of guessing one or more correct?

The event "one or more correct" consists of the outcomes 1, 2, 3, 4, 5.

$P(\text{one or more correct}) = 0.396 + 0.264 + 0.088 + 0.014 + 0.001$
$= 0.763, \text{ or } 76.3\%$

B What is the probability of guessing fewer than 2 correct?

The event "fewer than 2 correct" consists of the outcomes 0 and 1.

$P(\text{fewer than 2 correct}) = 0.237 + 0.396$
$= 0.633, \text{ or } 63.3\%$

C What is the probability of passing the quiz (getting 4 or 5 correct) by guessing?

The event "passing the quiz" consists of the outcomes 4 and 5.

$P(\text{passing the quiz}) = 0.014 + 0.001$
$= 0.015, \text{ or } 1.5\%$

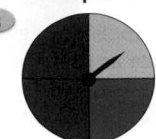
Reaching All Learners
Through Grouping Strategies

Have students work in pairs or small groups. Give each student a recording sheet (Chapter 9 Resource Book p. 90) that contains the descriptions for various experiments. Have each group write the complete sample space for each experiment. An example is given below.

1. Choose two coins from a jar that contains a penny, a nickel, and a dime. penny and nickel, penny and dime, nickel and dime

Example 1

Give the probability for each outcome.

A. The basketball team has a 70% chance of winning.

Outcome	Win	Lose
Probability	■	■
	0.7	0.3

B.

Outcome	1	2	3
Probability	■	■	■
	$\frac{3}{8}$	$\frac{3}{8}$	$\frac{1}{4}$

Example 2

A quiz contains 5 true or false questions. Suppose you guess randomly on every question. The table below gives the probability of each score.

Score	Probability
0	0.031
1	0.156
2	0.313
3	0.313
4	0.156
5	0.031

A. What is the probability of guessing 3 or more correct? **0.5**

B. What is the probability of guessing fewer than 2 correct? **0.187**

C. What is the probability of passing the quiz (getting 4 or 5 correct) by guessing? **0.187**

Example 3

Six students are in a race. Ken's probability of winning is 0.2. Lee is twice as likely to win as Ken. Roy is $\frac{1}{4}$ as likely to win as Lee. Tracy, James, and Kadeem all have the same chance of winning. Create a table of probabilities for the sample space.

Outcome	Probability
Ken	0.2
Lee	0.4
Roy	0.1
Tracy	0.1
James	0.1
Kadeem	0.1

Earth Science

Exercises 12–16 involve calculating probabilities using data about earthquakes. Students study earthquakes in middle school Earth science programs, such as *Holt Science & Technology*.

Answers

15. Possible answer: I would need to know the number of earthquakes in that country for at least the past 20 years. I would estimate the probability by dividing the number of years in which there were more than 5 earthquakes by the number of years in the sample.

16. Possible answer: 0.654

Journal

Ask students to write how the data on auto accidents in Example 2 helps explain why insurance rates are different for different age groups.

Test Prep Doctor

For Exercise 21, remind students that in similar triangles, the ratios of lengths of corresponding sides are constant. Because the base of the second triangle is half as long as the first, the other sides must be in the same ratio. So, the sides of the second triangle are $\frac{1}{2} \cdot 4.5 = 2.25$ cm long. The correct choice is **D**.

Lesson Quiz

1. Of 425, 234 seniors were enrolled in a math course. Estimate the probability that a randomly selected senior is enrolled in a math course. **0.55, or 55%**

2. Mason made a hit 34 out of his last 125 times at bat. Estimate the probability that he will make a hit his next time at bat. **0.27, or 27%**

3. Christina polled 176 students about their favorite ice cream flavor. 63 students' favorite flavor is vanilla and 40 students' favorite flavor is strawberry. Compare the probability of a student's liking vanilla to a student's liking strawberry. **about 36% to about 23%**

Available on Daily Transparency in CRB

Earth Science

The strength of an earthquake is measured on the Richter scale. A *major* earthquake measures between 7 and 7.9 on the Richter scale, and a *great* earthquake measures 8 or higher. The table shows the number of major and great earthquakes per year worldwide from 1970 to 1995.

12. Estimate the probability that there will be more than 15 major earthquakes next year. **0.192**

13. Estimate the probability that there will be fewer than 12 major earthquakes next year. **0.308**

14. Estimate the probability that there will be no great earthquakes next year. **0.5**

15. WRITE ABOUT IT Suppose you want to know the probability that there will be more than five earthquakes next year in a certain country. What would you need to know, and how would you estimate the probability?

16. CHALLENGE Estimate the probability that there will be more than one major earthquake in the next month.

go.hrw.com
KEYWORD: MP4 Quake
CNN Student News.

Number of Earthquakes Worldwide

Year	Major	Great	Year	Major	Great
1970	20	0	1983	14	0
1971	19	1	1984	8	0
1972	15	0	1985	13	1
1973	13	0	1986	5	1
1974	14	0	1987	11	0
1975	14	1	1988	8	0
1976	15	2	1989	6	1
1977	11	2	1990	12	0
1978	16	1	1991	11	0
1979	13	0	1992	23	0
1980	13	1	1993	15	1
1981	13	0	1994	13	2
1982	10	1	1995	22	3

Spiral Review

Solve each proportion. (Lesson 7-4)

17. $\frac{x}{3} = \frac{8}{12}$ $x = 2$ **18.** $\frac{7}{y} = \frac{49}{98}$ $y = 14$ **19.** $\frac{10}{12} = \frac{b}{6}$ $b = 5$ **20.** $\frac{12}{36} = \frac{4}{c}$ $c = 12$

21. **TEST PREP** An isosceles triangle has two sides that are 4.5 cm long and a base that is 3 cm long. A similar triangle has a base that is 1.5 cm long. How long are the other two legs of the similar triangle? (Lesson 7-6) **D**

A 150 cm   B 3.75 cm   C 4.5 cm   D 2.25 cm

CHALLENGE 9-2

LESSON 9-2 **Challenge**
Tossing and Spinning

The more times you repeat an experiment, the closer the experimental probability and the theoretical probability become.

Toss a penny 200 times.

	Heads	Tails

1. Record your results in the table. **Results will vary.**

2. What is the theoretical probability of:
   getting heads? $\frac{1}{2}$ getting tails? $\frac{1}{2}$

3. What is your experimental probability of:
   getting heads? **near $\frac{1}{2}$** getting tails? **near $\frac{1}{2}$**

4. How close are your experimental probabilities to the theoretical probabilities?
   Possible answer: very close

Spin a penny 200 times.

	Heads	Tails

5. Record your results in the table. **Results will vary.**

6. What is your experimental probability of:
   getting heads? $\frac{1}{2}$ getting tails? $\frac{1}{2}$

7. Compare your experimental probabilities for tossing the penny and spinning the penny. Are they close? Explain.
   Answers will vary.

Roll a number cube 200 times. **Results will vary.**

	1	2	3	4	5	6

8. Record your results in the table.

9. What is the theoretical probability of:
   getting a 1? $\frac{1}{6}$ a 2? $\frac{1}{6}$ a 3? $\frac{1}{6}$ a 4? $\frac{1}{6}$ a 5? $\frac{1}{6}$ a 6? $\frac{1}{6}$

10. What is your experimental probability of getting:
   a 1? **vary** a 2? **vary** a 3? **vary** a 4? **vary** a 5? **vary** a 6? **vary**

11. How close are your experimental probabilities to the theoretical probabilities?
   Possible answer: very close

PROBLEM SOLVING 9-2

LESSON 9-2 **Problem Solving**
Experimental Probability

Use the table below. Round to the nearest percent. Write the correct answer.

1. Estimate the probability of sunshine in Buffalo, NY.
   48%

2. Estimate the probability of sunshine in Fort Wayne, IN.
   59%

Average Number of Days of Sunshine Per Year for Selected Cities

City	Number of Days
Buffalo, NY	175
Fort Wayne, IN	215
Miami, FL	256
Raleigh, NC	212
Richmond, VA	230

3. Estimate the probability of sunshine in Miami, FL.
   70%

4. Estimate the probability that it will not be sunny in Raleigh, NC.
   42%

5. Estimate the probability that it will not be sunny in Miami, FL.
   30%

6. Estimate the probability of sunshine in Richmond, VA.
   63%

Use the table below that shows the number of deaths and injuries caused by lightning strikes. Choose the letter for the best answer.

7. Estimate the probability of being injured by a lightning strike in New York.
   A 0.0000007% C 0.00007%
   B 0.0000002% D 0.000002%

States with Most Lightning Deaths

State	Average deaths per year	Average injuries per year	Population
Florida	9.6	32.7	15,982,378
North Carolina	4.6	12.9	8,049,313
Texas	4.6	9.3	20,851,820
New York	3.6	12.5	18,976,457
Tennessee	3.4	9.7	5,689,283

8. Estimate the probability of being killed by lightning in North Carolina.
   F 0.0000006% H 0.00002%
   G 0.00006% J 0.000002%

9. Estimate the probability of being struck by lightning in Florida.
   A 0.000006%
   B 0.00026%
   C 0.0000026%
   D 0.0006%

10. In which two states do you have the highest probability of being struck by lightning?
   F Florida, North Carolina
   G Florida, Tennessee
   H Texas, New York
   J North Carolina, Tennessee

Technology LAB 9A

Generate Random Numbers

Use with Lesson 9-3

A spreadsheet can be used to generate random decimal numbers that are greater than or equal to 0 but less than 1. By using formulas, you can shift these numbers into a useful range.

internet connect

Lab Resources Online
go.hrw.com
KEYWORD: MP4 Lab9A

Activity

1 Use a spreadsheet to generate five random decimal numbers that are between 0 and 1. Then convert these numbers to integers from 1 to 10.

a. Type **=RAND()** into cell A1 and press **ENTER**. A random decimal number appears.

	A
1	0.063515
2	

b. Click to highlight cell A1. Go to the **Edit** menu and **Copy** the contents of A1. Then click and drag to highlight cells A2 through A5. Go to the **Edit** menu and use **Paste** to fill cells A2 through A5.

	A
1	0.20589
2	0.837083
3	0.445334
4	0.939134
5	0.993354
6	

Notice that the random number in cell A1 changed when you filled the other cells.

RAND() gives a decimal number greater than or equal to 0, but less than 1. To generate random integers from 1 to 10, you need to do the following:

- Multiply **RAND()** by 10 (to give a number greater than or equal to 0 but less than 10).

- Use the **INT** function to drop the decimal part of the result (to give an integer from 0 to 9).

- Add 1 (to give an integer from 1 to 10).

c. Change the formula in A1 to **=INT(10*RAND()) + 1** and press **ENTER**. Repeat the process in part **b** to fill cells A2 through A5.

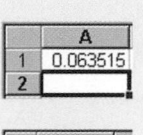

A2		▼		=	=INT(10*RAND()) + 1
	A	B	C	D	
1	9				
2	1				
3	7				
4	7				
5	6				
6					

The formula **=INT(10*RAND()) + 1** generates random integers from 1 to 10.

Think and Discuss

1. Explain how **INT(10*RAND()) + 1** generates random integers from 1 to 10.

Try This

1. Use a spreadsheet to simulate three rolls of a number cube.

Technology LAB 9A

Generate Random Numbers

Pacing: Traditional 1 day
Block ½ day

Objective: To use spreadsheet software to generate random numbers

Materials: Spreadsheet software

Lab Resources

Technology Lab Activities . . . . p. 59

Using the Page

This technology activity shows students how to use a spreadsheet to generate random numbers. Specific instructions may vary, depending on the spreadsheet software used. The instructions given are for Microsoft Excel. For instructions for other software, visit go.hrw.com.

The Think and Discuss problem can be used to assess students' understanding of spreadsheet functions. The Try This problem is meant to assess students' ability to write appropriate commands to generate random numbers.

Assessment

Assess students' understanding of spreadsheet commands with the following exercises.

1. What command creates a random number? **RAND()**

2. How can you describe the random numbers generated by the command RAND()? They are decimal values between 0 and 1.

3. What command creates a random whole number between 1 and 7? **INT(7*RAND()) + 1**

Answers

Think and Discuss

1. **RAND()** returns a random decimal number greater than or equal to zero but less than 1.
(10*RAND()) multiplies 10 by the decimal generated, resulting in a random decimal greater than or equal to zero but less than 10.
INT(10*RAND()) tells the spreadsheet to give only integers from zero to 9, meaning only 0, 1, 2, 3, 4, 5, 6, 7, 8, or 9.
INT(10*RAND()) + 1 adds 1 to the random integer from 0 to 9, resulting in a random integer from 1 to 10.

Try This

1. The formula is **INT(6*RAND()) + 1.**
The numbers generated will vary.

Warm Up

1. There are 25 out of 216 sopho-mores enrolled in a physical-education course. Estimate the probability that a randomly selected sophomore is enrolled in a physical-education course. **0.12**

2. A spinner was spun 230 times. It landed on red 120 times, green 65 times, and yellow 45 times. Estimate the probability of its landing on red. **0.52**

Problem of the Day

If a triangle is worth 7 and a rec-tangle is worth 8, how much is a hexagon worth? **10**

Available on Daily Transparency in CRB

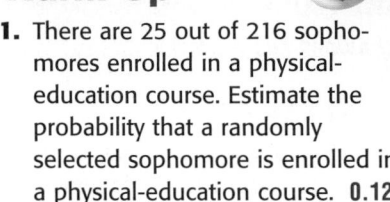
Math Humor

Teacher: Why are you tickling your math homework?

Student: Aren't we supposed to *stimu-late* the problem?

9-3 Use a Simulation
 Problem Solving Strategy

Learn to use a simulation to estimate probability.

Vocabulary
simulation
random numbers

In basketball, free throws are worth only one point, but they can make a big difference. In a close game, the coach may put in players with good free-throw shooting records.

If a player shoots 78% from the free-throw line, he makes about 78 out of every 100 free throws. What is the probability that he will make at least 7 out of 10 free throws? A *simulation* can help you estimate this probability.

A **simulation** is a model of a real situation. In a set of **random numbers**, each number has the same probability of occurring as every other number, and no pattern can be used to predict the next number. Random numbers can be used to simulate random events in real situations. The table is a set of 280 random digits.

In the 2001–2002 season, Utah's Karl Malone had a free-throw percentage of 79.7%.

87244	11632	85815	61766	19579	28186	18533	42633
74681	65633	54238	32848	87649	85976	13355	46498
53736	21616	86318	77291	24794	31119	48193	44869
86585	27919	65264	93557	94425	13325	16635	28584
18394	73266	67899	38783	94228	23426	76679	41256
39917	16373	59733	18588	22545	61378	33563	65161
96916	46278	78210	13906	82794	01136	60848	98713

EXAMPLE 1 **PROBLEM SOLVING APPLICATION**

A player has a free-throw rate of 78%. Estimate the probability that he will make at least 7 out of his next 10 shots.

1 Understand the Problem

The **answer** will be the probability that he will make at least 7 out of his next 10 free throws. It must be a number between 0 and 1.

List the **important information:**
• The probability that the player will make a free throw is 0.78.

---

1 Introduce
Alternate Opener

Motivate

Ask every student to write any single digit on a piece of paper. Make a list on the chalk-board of all the students' digits. Tell the class that they have just generated a list of ran-dom digits. Explain that there are computer programs that will generate lists of random numbers and that such lists can be used to simulate random events.

Exploration worksheet and answers on Chapter 9 Resource Book pp. 27 and 96

---

2 Teach

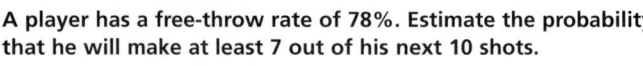 **Lesson Presentation**

Guided Instruction

In this lesson, students learn to use a simulation to estimate probability. Discuss the meaning of *simulation*. Explain that the number of random digits in the list in the lesson (280) is arbitrary and that the way they are grouped on the page is only for ease of reading. Also explain that they could choose any number on the list as a starting point. Work carefully through the example as this subject may be new for many students. You may want to use the table of random numbers on Teaching Transparency T12 (CRB).

2 Make a Plan

Use a simulation to model the situation. Use digits from the table, grouped in pairs. The numbers 01–78 represent a successful free throw, and the numbers 79–00 represent a missed free throw. Each group of 20 digits represents one trial. You can start anywhere on the table.

3 Solve

The first 20 digits in the table are shown below.

87244 11632 85815 61766

The digits can be grouped in ten pairs, as shown below.

87 24 41 16 32 85 81 56 17 66

This represents 7 successful free throws out of 10.

If you continue using the table, the next nine trials are as follows.

19	57	92	81	86	18	53	34	26	33	*7 successful free throws*
74	68	16	56	33	54	23	83	28	48	*9 successful free throws*
87	64	98	59	76	13	35	54	64	98	*7 successful free throws*
53	73	62	16	16	86	31	87	72	91	*7 successful free throws*
24	79	43	11	19	48	19	34	48	69	*9 successful free throws*
86	58	52	79	19	65	26	49	35	57	*8 successful free throws*
94	42	51	33	25	16	63	52	85	84	*7 successful free throws*
18	39	47	32	66	67	89	93	87	83	*6 successful free throws*
94	22	82	34	26	76	67	94	12	56	*7 successful free throws*

Out of the 10 trials, 9 represented 7 or more successful free throws. Based on this simulation, the probability of making at least 7 out of 10 free throws is about $\frac{9}{10}$, or 90%.

4 Look Back

A free-throw shooting rate of 78% means the player makes about 78 out of every 100 free throws. This ratio is equivalent to 7.8 out of 10 free throws, so he should make at least 7 free throws most of the time. The answer is reasonable.

Helpful Hint

Calculators and computers can generate sets of approximately random numbers. A formula is used to generate the numbers, so they are not truly random, but they work for most simulations.

Additional Examples

Example 1

A dart player hits the bull's-eye 25% of the times that he throws a dart. Estimate the probability that he will make at least 2 bull's-eyes out of his next 5 throws. about 20%

Think and Discuss

1. **Explain** why a random number generator on a computer or calculator is useful for estimating probability by simulation.

2. **Tell** how you could use a simulation to estimate the probability that a player who shoots 50% from the line will make at least 6 out of 10 free throws.

---

3 Close

Reaching All Learners
Through Critical Thinking

Have students use the data from Example 1 to perform another simulation. Instruct them to start at a different place in the table of random digits, or if possible, have them use a calculator or computer to generate different random digits. When they have completed their simulations, discuss the different results with the class. If some results are different, ask students to explain why.

Possible answer: Probability is only an estimate. It never guarantees that a specific event will happen.

Summarize

Review the steps for creating and using a simulation. Remind students that to use a simulation to estimate a probability, they need a given probability, such as a player's having already made 78% of his free throws. Emphasize that simulations are only estimates and that they cannot be used to predict exactly what will happen.

Answers to Think and Discuss

Possible answers:

1. Computers or calculators can generate random numbers quickly, and they can generate multiple-digit numbers. In addition, they can generate as many numbers as you need.

2. Use a coin. You could let heads represent a successful free throw. Flip a coin 10 times (or 10 coins at once) for each trial. Then find in how many trials at least 6 of the outcomes were heads.

FOR EXTRA PRACTICE
see page 748

internet connect
Homework Help Online
go.hrw.com   Keyword: MP4 9-3

Students may want to refer back to the lesson examples.

Assignment Guide

If you finished Example **1** assign:
Core 1–9, 13–17
Enriched 4–17

GUIDED PRACTICE

See Example **1** Use the table of random numbers to simulate each situation. Use at least 10 trials for each simulation. **Possible answers are given.**

49064	12830	66783	14965	81537	24935	69675	32681
42893	42668	70963	58827	17354	42190	36165	29827
21705	89446	38703	21274	90049	19036	37971	05322
52737	40117	54132	11152	02985	82873	28197	89796

1. Carlos completes a sale with approximately 42% of the customers he meets. If he has 8 customer appointments tomorrow, estimate the probability that he will complete at least 3 sales. **90%**

2. A city typically has rain on 30% of summer days. Estimate the probability that it will rain in the city on at least 2 days during the first full week in July. **50%**

3. Customers at a carnival game win about 25% of the time. Estimate the probability that no more than 1 of the next 6 customers will win the game. **60%**

INDEPENDENT PRACTICE

See Example **1** Use the table of random numbers to simulate each situation. Use at least 10 trials for each simulation. **Possible answers are given.**

63415	12776	31960	42974	36444	23826	46320	48308
41591	43536	64118	53147	23544	61352	12954	57628
26446	12734	22435	42612	24834	21961	12526	22832
16522	33043	21997	15738	25788	33205	55699	33357
53040	39923	29591	64384	58166	39164	54474	38970

4. Michelle gets a hit 28% of the time she bats. Estimate the probability that she will get at least 4 hits in her next 9 at bats. **30%**

5. At a local fast-food restaurant, about 67% of the customers order french fries. Estimate the probability that 4 out of the next 5 customers will order french fries. **30%**

6. A local radio station is having a contest. Each time you call in, your chances of winning are 6%. If you call in 10 times, estimate the probability that you will win more than once. **10%**

7. Liam works at a video store. He knows about 55% of the customers by name. Estimate the probability that he will know the names of at least 8 of the next 10 customers. **30%**

Math Background

Example 1 in the lesson refers to a 78% free-throw rate. This percent is similar to an experimental probability because it is based on the ratio

$$\frac{\text{free throws made}}{\text{free throws attempted}}.$$

A batting average in baseball is also similar to an experimental probability. If a player gets 24 hits in 50 at bats, his batting average is $\frac{\text{number of hits}}{\text{number of at bats}} = \frac{24}{50} = .480$. As with experimental probability, sports statistics such as these can be used to estimate the probability of an event, whether that event is making a free throw or getting a hit.

RETEACH 9-3

LESSON 9-3 Reteach
Use a Simulation

Situation: Strout's Market is having a contest. They give a puzzle piece to each customer at the checkout. A customer who collects all 10 different puzzle pieces gets $100 in store credit.

Using a table of random numbers, you can model the situation to estimate how many times a customer would have to shop to collect all 10 different puzzle pieces.

3	1	9	4	1	1	8	8
5	7	4	5	7	7	9	0
7	0	3	0	1	3	5	0
0	4	3	8	9	5	3	8
2	6	1	7	6	7	6	9
0	8	2	6	5	5	9	2

• Start anywhere in the table. Count the numbers you pass as you "collect" the digits 0-9.

Suppose you start at the top of Column 3 and move to the right. List each number until you have collected all the numbers 0–9.

9 4 1 1 8 8 5 7 4 5 7 7 9 0 7 0 3 0 1 3 5 0 0 4 3 8 9 5 3 8 2 6

You had to go through 32 numbers to get each number at least once (underscored).

• Do the experiment again.

Suppose you start at the bottom of Column 4 and move to the right. When you reach the end of the row, go to the beginning of the table.

6 5 5 9 2 3 1 9 4 1 1 8 8 5 7 4 5 7 7 9 0

You had to go through 21 numbers to get each number at least once (underscored).

• Find the average of your results. $\frac{32 + 21}{2} = \frac{53}{2} = 26.5$

So, on average, you need to shop 27 times to get all 10 pieces to win $100 credit.

Model each situation. Use the list of random numbers shown above. Do two trials. Tell where you start for each trial. **Possible answers shown.**

1. A box of Whammos contains a toy dinosaur. If there are 10 different model dinosaurs in the collection, estimate how many boxes of Whammos you would have to buy to get all 10 different dinosaurs.

top Col. 2, go right, 33; bottom Col. 5, go right, 38; on average, about 36 boxes

2. For this spinner, estimate how many times you would have to spin the pointer to get the numbers 1–10.

top Col.8, next row, 27; bottom Col. 1, right, 18; about 23 times

PRACTICE 9-3

LESSON 9-3 Practice B
Use a Simulation

Use the table of random numbers for the problems below.

8125	4764	7693	3675	1642	7988	7048	9135	3138	3256
9566	4413	7215	7992	4320	7438	3805	5473	8847	2397
7336	5393	8623	8570	5095	5685	6695	3570	3605	4656
6470	6065	8239	2953	5942	6496	8899	0701	5368	2106
5210	2570	8137	3587	3578	6657	6636	7188	5717	1770
4329	4110	2655	8258	9928	3873	5609	3695	7091	0368
5315	2654	0484	4601	4336	6624	5403	5870	8545	3905
2361	9097	3753	2498	0544	0923	6099	1737	4025	1221
2677	7741	5342	9844	3722	5120	8742	1382	2842	7386
3292	5084	1130	2747	0664	9718	6072	9432	7008	2024

Mr. Domino gave the same math test to all three of his math classes. In the first two classes, 80% of the students passed the test. If the third class has 20 students, estimate the number of students who will pass the test.

1. Using the first row as the first trial, count the successful outcomes and name the unsuccessful outcomes.

16 out of 20 successful; 81, 93, 88, 91

2. Count and name the successful outcomes in the second row as the second trial.

16 out of 20 successful; 95, 92, 88, 97

Determine the successful outcomes in the remaining rows of the random number table.

3. third row   **4.** fourth row   **5.** fifth row   **6.** sixth row

  14   16   17   16

7. seventh row   **8.** eighth row   **9.** ninth row   **10.** tenth row

  18   16   16   16

11. Based on the simulation, estimate the probability that 80% of the class will pass the math test.   90%

PRACTICE AND PROBLEM SOLVING

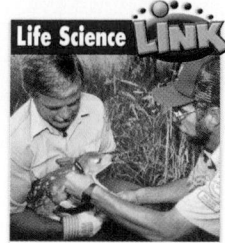

Life Science LINK

The capture-release-recapture method uses ratios to estimate the size of wild populations.

Use the table of random numbers for Exercises 8 and 9. Use at least 10 trials to simulate each situation.

19067	26149	88557	80696	88246	56652	73023	56838
98048	26387	65953	94163	66233	57325	65618	76782
32958	47253	24960	32052	16921	54925	44766	33115
89164	06342	98577	44523	72304	38221	33506	63923
48117	18686	54621	65793	70299	20622	81309	76106

8. Suppose your math teacher assigns homework about 75% of the time. What is the probability that you will have math homework at least 4 days next week? **Possible answer: 60%**

9. **LIFE SCIENCE** About 7% of the deer population in a particular region were captured, tagged for research, and released. If a group of 10 deer were encountered randomly at a later date, estimate the probability that at least 1 of them would have a tag. **Possible answer: 50%**

 10. **WHAT'S THE ERROR?** A student is doing a simulation that involves an outcome with a probability of 0.12, using a random number table. He lets the numbers 00–12 represent the outcome occurring, and 13–99 represent the outcome not occurring. Why won't this be an accurate simulation?

 11. **WRITE ABOUT IT** A manufacturer tests items from the assembly line for quality control. If 2% of the items were defective, how would you estimate the probability that no more than 1 item in each case of 144 would be defective?

 12. **CHALLENGE** A box of chocolates contains 12 chocolate creams in 5 flavors. The probabilities for each flavor are given below. Lindsay likes chocolate and vanilla best. April likes orange and vanilla best. If they share the box by each choosing 6 creams randomly, estimate the probability that they will each get at least one of their favorites. **Possible answer: 40%**

Flavor	Chocolate	Vanilla	Orange	Cherry	Raspberry
Probability	0.4	0.3	0.1	0.1	0.1

Spiral Review

Using the scale 1 in. = 6 ft, find the height or length of each object. (Lesson 7-7)

13. a 14 in. tall model of an office building **84 ft tall**

14. a 2.5 ft long model of a train **180 ft long**

15. a 7 in. tall model of a billboard **42 ft tall**

16. a 14 in. long model of an airplane **84 ft long**

17. **TEST PREP** A machine can fill a box measuring 2 ft by 3 ft by 5 ft in 42 seconds. How long would it take for the machine to fill a box measuring 4 ft by 6 ft by 10 ft? (Lesson 7-9) **B**

    A 168 seconds   **B** 336 seconds   **C** 442 seconds   **D** 84 seconds

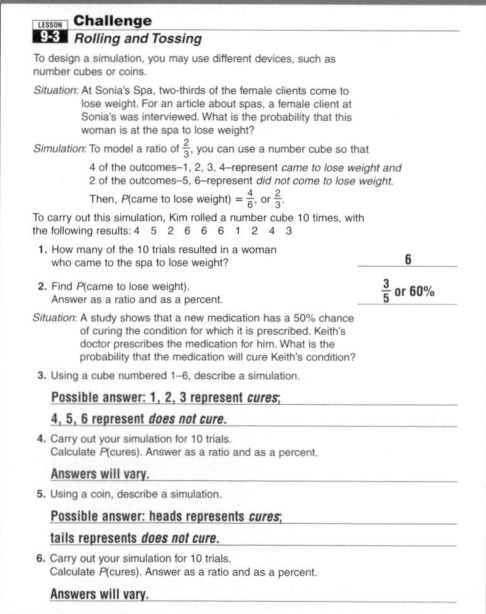

CHALLENGE 9-3

LESSON 9-3 Challenge
Rolling and Tossing

To design a simulation, you may use different devices, such as number cubes or coins.

Situation: At Sonia's Spa, two-thirds of the female clients come to lose weight. For an article about spas, a female client at Sonia's was interviewed. What is the probability that this woman is at the spa to lose weight?

Simulation: To model a ratio of $\frac{2}{3}$, you can use a number cube so that

  4 of the outcomes—1, 2, 3, 4—represent *came to lose weight* and
  2 of the outcomes—5, 6—represent *did not come to lose weight.*

Then, P(came to lose weight) = $\frac{4}{6}$, or $\frac{2}{3}$.

To carry out this simulation, Kim rolled a number cube 10 times, with the following results: 4 5 2 6 6 6 1 2 4 3

1. How many of the 10 trials resulted in a woman who came to the spa to lose weight?   **6**

2. Find P(came to lose weight). Answer as a ratio and as a percent.   **$\frac{3}{5}$ or 60%**

Situation: A study shows that a new medication has a 50% chance of curing the condition for which it is prescribed. Keith's doctor prescribes the medication for him. What is the probability that the medication will cure Keith's condition?

3. Using a cube numbered 1–6, describe a simulation.
  Possible answer: 1, 2, 3 represent *cures*;
  4, 5, 6 represent *does not cure.*

4. Carry out your simulation for 10 trials. Calculate P(cures). Answer as a ratio and as a percent.
  Answers will vary.

5. Using a coin, describe a simulation.
  Possible answer: heads represents *cures*;
  tails represents *does not cure.*

6. Carry out your simulation for 10 trials. Calculate P(cures). Answer as a ratio and as a percent.
  Answers will vary.

PROBLEM SOLVING 9-3

LESSON 9-3 Problem Solving
Use a Simulation

Use the table of random numbers below. Use at least 10 trials to simulate each situation. Write the correct answer.

1. Of people 18–24 years of age, 49% do volunteer work. If 10 people ages 18–24 were chosen at random, estimate the probability that at least 4 of them do volunteer work.

87244	11632	85815	61766
19579	28186	18533	24633
74581	65633	54238	32848
87549	85976	13355	46498
53736	21616	86318	77291
24794	31119	48193	44869
86585	27919	65264	93557
94425	13325	16635	25840
18394	73266	67899	38783
94228	23426	76679	41256

Possible answer: 80%

2. In the 2000 Presidential election, 56% of the population of North Carolina voted for George W. Bush. If 10 people were chosen at random from North Carolina, estimate the probability that at least 8 of them voted for Bush.
Possible answer: 10%

3. Forty percent of households with televisions watched the 2001 Super Bowl game. If 10 households with televisions are chosen at random, estimate the probability that at least 3 watched the 2001 Super Bowl.
Possible answer: 90%

Use the table above and at least 10 trials to simulate each situation. Choose the letter for the best estimate.

4. As of August 2000, 42% of U.S. households had Internet access. If 10 households are chosen at random, estimate the probability that at least 5 of them will have Internet access.
  A 60%   C 60%
  (B)30%   D 90%

5. On average, there is rain 20% of the days in April in Orlando, FL. Estimate the probability that it will rain at least once during your 7-day vacation in Orlando in April.
  F 20%   (H)70%
  G 50%   J 40%

6. Kareem Abdul-Jabaar is the NBA lifetime leader in field goals. During his career, he made 56% of the field goals he attempted. In a given game, estimate the probability that he would make at least 6 of 10 field goals.
  (A)40%   C 80%
  B 60%   D 100%

7. At the University of Virginia 39% of the applicants are accepted. If 10 applicants to the University of Virginia are chosen at random, estimate the probability that at least 4 of them are accepted to the University of Virginia.
  F 10%   H 80%
  (G)40%   J 70%

Answers

10. Possible answer: There are 13 numbers from 00 to 12, so this would make the outcome have a probability of 13%. The student should use 01–12 to represent the outcome occurring.

11. Possible answer: Use a simulation with each trial using 144 two-digit numbers from a random number table. Let the numbers 01 and 02 represent defective items and the numbers 03–00 represent items that are not defective.

Journal

Ask students to describe a way to generate random numbers without using a computer or a calculator.

Test Prep Doctor

For Exercise 17, point out that because the ratio of each pair of corresponding dimensions is $\frac{2}{1}$, the ratio of volumes is $\frac{2^3}{1^3}$, or $\frac{8}{1}$. Therefore, it takes 8 times longer to fill the larger box.

$8 \times 42 = 336$, so the correct answer is **B**. Students who selected **A** may have multiplied the time by 4, and students who selected **D** may have doubled the time.

Lesson Quiz

1. Use the table of random numbers to simulate the situation.

38094	76211	43659	29272
76005	93391	19587	47380
33442	40809	27904	95412
69632	48461	25654	55889
42231	39983	13802	24483
52730	15604	80949	46351
10580	59765	76431	38586
62987	40440	93594	30198
64926	17672	68735	35168
19085	35497	30798	21966

Lydia gets a hit 34% of the time she bats. Estimate the probability that she will get **at least** 4 hits in her next 10 at bats.
Possible answer: 30%

Available on Daily Transparency in CRB

Purpose: *To assess students' mastery of concepts and skills in Lessons 9-1 through 9-3*

Assessment Resources

Section 9A Quiz
Assessment Resources . . . . p. 21

 Test and Practice Generator CD-ROM

Additional mid-chapter assessment items in both multiple-choice and free-response format may be generated for any objective in Lessons 9-1 through 9-3.

Answers

4.

Racer	Probability
Jennifer	0.3
Anjelica	0.3
Debra	0.2
Yolanda	0.2

5.

Marble	Probability
red	23%
green	18%
blue	47%
yellow	12%

Mid-Chapter Quiz

LESSON 9-1 (pp. 446–450)

Use the table of probabilities for the sample space to find the probability of each event.

Outcome	A	B	C	D
Probability	0.4	0.3	0.2	0.1

1. $P(D)$ **0.1**
2. $P(\text{not } C)$ **0.8**
3. $P(A \text{ or } B)$ **0.7**

4. There are 4 students in a race. Jennifer has a 30% chance of winning. Anjelica has the same chance as Jennifer. Debra and Yolanda have equal chances. Create a table of probabilities for the sample space.

LESSON 9-2 (pp. 451–454)

An experiment consists of drawing a marble from a bag and putting it back. The experiment is repeated 100 times, with the following results.

Outcome	Red	Green	Blue	Yellow
Draws	23	18	47	12

5. Estimate the probability of each outcome. Create a table of probabilities for the sample space.

6. Estimate $P(\text{red or blue})$. **70%**

7. Estimate $P(\text{not green})$. **82%**

LESSON 9-3 (pp. 456–459)

Use the table of random numbers to simulate each situation. Use at least 10 trials for each simulation.

93840	03363	31168	57602	19464	52245	98744	61040
68395	76832	56386	45060	57512	38816	51623	23252
16805	92120	74443	49176	49898	62042	65847	15380
85178	78842	16598	28335	84837	76406	53436	45043

8. At a local school, 68% of the tenth grade students are studying geometry. Estimate the probability that at least 6 out of 8 randomly selected tenth grade students are studying geometry. **Possible Answer: 60%**

9. Kayla has a package of 100 multicolored beads that contains 15 purple beads. If she randomly selects 8 beads to make a friendship bracelet, estimate the probability that she will get more than 1 purple bead.
Possible Answer: 0%

Focus on Problem Solving

Understand the Problem

• **Understand the words in the problem**

Words that you don't understand can make a simple problem seem difficult. Before you try to solve a problem, you will need to know the meaning of the words in it.

If a problem gives a name of a person, place, or thing that is difficult to understand, such as *Eulalia*, you can use another name or a pronoun in its place. You could replace *Eulalia* with *she*.

Read the problems so that you can hear yourself saying the words.

Copy each problem, and circle any words that you do not understand. Look up each word and write its definition, or use context clues to replace the word with a similar word that is easier to understand.

1 A point in the circle is chosen randomly. What is the probability that the point is in the inscribed triangle?

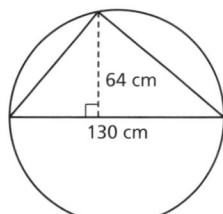
64 cm
130 cm

2 A chef has observed the number of people ordering each entrée from the evening's specials. Estimate the probability that the next customer will order Boeuf Bourguignon.

Entrée	Boeuf Bourguignon	Chateaubriand	Rabbit Provençal
Number Ordered	23	15	12

3 Eulalia and Nunzio play cribbage 5 times a week. Eulalia skunked Nunzio 3 times in the last 12 weeks. Estimate the probability that Eulalia will skunk Nunzio the next time they play cribbage.

4 A pula has a coat of arms on the obverse and a running zebra on the reverse. If a pula is tossed 150 times and lands with the coat of arms facing up 70 times, estimate the probability of its landing with the zebra facing up.

Answers

1. about 31%

2. 46%

3. 5%

4. $53\frac{1}{3}\%$

Focus on Problem Solving

Purpose: *To focus on understanding the words in a problem*

Problem Solving Resources

Interactive Problem Solving . . pp. 71–78

Math: Reading and Writing in the Content Area . . . . pp. 71–78

Problem Solving Process

This page focuses on the first step of the problem-solving process: **Understand the Problem**

Discuss

Have students rewrite each problem, replacing unknown words with alternative words or phrases or with the definition of the word.

Possible answers:

1. *Inscribed triangle:* A triangle is inscribed in a circle if each of its vertices lies on the circle.
A point in the circle is chosen randomly. What is the probability that the point is in the triangle shown?

2. Replace *entrée* with *item*. Replace *Boeuf Bourguignon* with *chili*.
A chef has observed the number of people ordering each item from the evening's specials. Estimate the probability that the next customer will order chili.

3. Replace *Eulalia* with *Ellen*. Replace *Nunzio* with *Neal*. Replace *cribbage* with *chess*. Replace *skunked* with *beat*.
Ellen and Neal play chess 5 times a week. Ellen beat Neal 3 times in the last 12 weeks. Estimate the probability that Ellen will beat Neal the next time they play chess.

4. Replace *pula* with *coin*. Replace *obverse* with *one side*. Replace *reverse* with *other side*.
A coin has a coat of arms on one side and a running zebra on the other side. If the coin is tossed 150 times and lands with the coat of arms facing up 70 times, estimate the probability of its landing with the zebra facing up.

Theoretical Probability and Counting

One-Minute Section Planner

Lesson	Materials	Resources
Lesson 9-4 Theoretical Probability **NCTM:** Data Analysis and Probability, Communication **NAEP:** Data Analysis and Probability 4b ☑ SAT-9 ☑ SAT-10 ☑ ITBS ☑ CTBS ☑ MAT ☑ CAT	**Optional** Teaching Transparency T16 (CRB) Index cards Playing cards	• *Chapter 9 Resource Book,* pp. 36–45 • Daily Transparency T15, CRB • Additional Examples Transparencies T17–T20, CRB • Alternate Openers: Explorations, p. 74
Lesson 9-5 The Fundamental Counting Principle **NCTM:** Data Analysis and Probability, Communication **NAEP:** Data Analysis and Probability 4b ☑ SAT-9 ☑ SAT-10 ☐ ITBS ☐ CTBS ☐ MAT ☑ CAT	**Optional** Teaching Transparency T22 (CRB)	• *Chapter 9 Resource Book,* pp. 46–55 • Daily Transparency T21, CRB • Additional Examples Transparencies T23–T25, CRB • Alternate Openers: Explorations, p. 75
Lesson 9-6 Permutations and Combinations **NCTM:** Number and Operations, Communication **NAEP:** Data Analysis and Probability 4e ☑ SAT-9 ☑ SAT-10 ☐ ITBS ☐ CTBS ☐ MAT ☑ CAT	**Optional** Teaching Transparency T27 (CRB) Recording sheet for Reaching All Learners (CRB p. 91)	• *Chapter 9 Resource Book,* pp. 56–65 • Daily Transparency T26, CRB • Additional Examples Transparencies T28–T30, CRB • Alternate Openers: Explorations, p. 76
Hands-On Lab 9B Pascal's Triangle **NCTM:** Algebra **NAEP:** Algebra 1a ☐ SAT-9 ☐ SAT-10 ☐ ITBS ☐ CTBS ☐ MAT ☐ CAT		• *Hands-On Lab Activities,* pp. 66–67
Lesson 9-7 Independent and Dependent Events **NCTM:** Data Analysis and Probability, Reasoning and Proof, Communication **NAEP:** Data Analysis and Probability 4h ☐ SAT-9 ☑ SAT-10 ☐ ITBS ☐ CTBS ☐ MAT ☑ CAT	**Optional** Teaching Transparency T32 (CRB) Red and yellow counters (MK)	• *Chapter 9 Resource Book,* pp. 66–74 • Daily Transparency T31, CRB • Additional Examples Transparencies T33–T37, CRB • Alternate Openers: Explorations, p. 77
Lesson 9-8 Odds **NCTM:** Data Analysis and Probability, Reasoning and Proof, Communication **NAEP:** Data Analysis and Probability 4j ☐ SAT-9 ☐ SAT-10 ☐ ITBS ☐ CTBS ☐ MAT ☑ CAT	**Optional** Teaching Transparency T39 (CRB)	• *Chapter 9 Resource Book,* pp. 75–84 • Daily Transparency T38, CRB • Additional Examples Transparencies T40–T41, CRB • Alternate Openers: Explorations, p. 78
Section 9B Assessment		• Section 9B Quiz, AR p. 22 • *Test and Practice Generator* CD-ROM

SAT = *Stanford Achievement Tests*   **ITBS** = *Iowa Test of Basic Skills*   **CTBS** = *Comprehensive Test of Basic Skills/Terra Nova*
MAT = *Metropolitan Achievement Test*   **CAT** = *California Achievement Test*
NCTM = Complete standards can be found on pages T29–T35.   **NAEP** = Complete standards can be found on pages A54–A58.
SE = *Student Edition*   **TE** = *Teacher's Edition*   **AR** = *Assessment Resources*   **CRB** = *Chapter Resource Book*   **MK** = *Manipulatives Kit*

Section Overview

Probability and Odds
Lessons 9-4, 9-8

Why? Understanding probability and odds will help you to make informed choices.

Theoretical probability	Odds in Favor	Odds Against
$\dfrac{\text{number of favorable outcomes}}{\text{number of possible outcomes}}$ Because a number cube has six equally likely outcomes, the probability of rolling a 6 is $\frac{1}{6}$.	**a:b** a = number of favorable outcomes b = number of unfavorable outcomes If the probability of an event is $\frac{1}{6}$, then the odds in favor of the event are 1:5.	**b:a** a = number of favorable outcomes b = number of unfavorable outcomes If the probability of an event is $\frac{1}{6}$, then the the odds against the event are 5:1.

Counting Methods
Hands-On Lab 9B, Lessons 9-5, 9-6

Why? Understanding counting methods will help you to handle large numbers.

> The **Fundamental Counting Principle** If there are m ways to choose a first item and n ways to choose a second, then there are $m \cdot n$ ways to choose both items.

Combinations are ways to choose things from a group if **order does not matter.**

> The number of **combinations** of n things taken r at a time is given by the formula
> $$_nC_r = \frac{n!}{r!(n-r)!}.$$

The number of combinations of 5 things taken 3 at a time

$$_5C_3 = \frac{5!}{3!2!} = \frac{5 \cdot 4 \cdot 3 \cdot 2 \cdot 1}{(3 \cdot 2 \cdot 1)(2 \cdot 1)} = 10$$

Permutations are ways to choose things from a group if **order does matter.**

> The number of **permutations** of n things taken r at a time is given by the formula
> $$_nP_r = \frac{n!}{(n-r)!}.$$

The number of permutations of 5 things taken 3 at a time

$$_5P_3 = \frac{5!}{2!} = \frac{5 \cdot 4 \cdot 3 \cdot 2 \cdot 1}{2 \cdot 1} = 60$$

Independent and Dependent Events
Lesson 9-7

Why? Understanding how one event affects another will help you plan.

Independent Events

The occurrence of one event **does not** affect the probability of the other.

Example:

Roll a number cube and toss a coin. Find the probability of rolling a number less than 3 and getting heads.

$$P(3, \text{heads}) = \frac{2}{6} \cdot \frac{1}{2} = \frac{1}{6} = 16\frac{2}{3}\%$$

Dependent Events

The occurrence of one event **does** affect the probability of the other.

Example:

Pick two marbles from a bag containing 4 red marbles and 1 blue marble without replacing the first. Find the probability of picking two red marbles.

$$P(\text{red, red}) = \frac{4}{5} \cdot \frac{3}{4} = \frac{12}{20} = 60\%$$

Pacing: Traditional 1 day
Block $\frac{1}{2}$ day

Objective: Students estimate probability using theoretical methods.

Warm Up

1. If you roll a number cube, what are the possible outcomes?
   1, 2, 3, 4, 5, or 6

2. Add $\frac{3}{36} + \frac{1}{6}$. $\frac{1}{4}$ 3. Add $\frac{1}{2} + \frac{2}{36}$. $\frac{5}{9}$

Problem of the Day

A spinner is divided into 4 different-colored sections. It is designed so that the probability of spinning red is twice the probability of spinning green, the probability of spinning blue is 3 times the probability of spinning green, and the probability of spinning yellow is 4 times the probability of spinning green. What is the probability of spinning yellow?
0.4

Available on Daily Transparency in CRB

Math Humor

If you want to win the flip of a coin, call, "Heads, I win. Tails, you lose."

9-4 Theoretical Probability

Learn to estimate probability using theoretical methods.

Vocabulary
theoretical probability
equally likely
fair
mutually exclusive

In the game of Monopoly®, you can get out of jail if you roll doubles, but if you roll doubles three times in a row, you have to go to jail. Your turn is decided by the probability that both dice will be the same number.

Theoretical probability is used to estimate probabilities by making certain assumptions about an experiment. Suppose a sample space has 5 outcomes that are **equally likely**, that is, they all have the same probability, x. The probabilities must add to 1.

$$x + x + x + x + x = 1$$
$$5x = 1$$
$$x = \frac{1}{5}$$

THEORETICAL PROBABILITY FOR EQUALLY LIKELY OUTCOMES

Suppose there are n equally likely outcomes in the sample space of an experiment.

- The probability of each outcome is $\frac{1}{n}$.
- The probability of an event is $\dfrac{\text{number of outcomes in the event}}{n}$.

A coin, die, or other object is called **fair** if all outcomes are equally likely.

EXAMPLE 1 Calculating Theoretical Probability

An experiment consists of rolling a fair die. There are 6 possible outcomes: 1, 2, 3, 4, 5, and 6.

A **What is the probability of rolling a 3?**

The die is fair, so all 6 outcomes are equally likely. The probability of the outcome of rolling a 3 is $P(3) = \frac{1}{6}$.

B **What is the probability of rolling an odd number?**

There are 3 outcomes in the event of rolling an odd number: 1, 3, and 5.
$$P(\text{rolling an odd number}) = \frac{\text{number of possible odd numbers}}{6} = \frac{3}{6} = \frac{1}{2}$$

1 Introduce
Alternate Opener

EXPLORATION

9-4 Theoretical Probability

The *theoretical probability* of an event tells you the probability of the event without your having to conduct an experiment.

For example, the experiment of rolling two dice and adding the two numbers that each die shows does not have to be conducted to know the possible sums of numbers.

1. Use the number of times each sum occurs to complete the table.

Sum	2	3	4	5	6	7	8	9	10	11	12
Outcomes	1	2									
Theoretical Probability	$\frac{1}{36}$	$\frac{2}{36}$									

Think and Discuss

2. **Explain** which sum is most likely to occur.

Exploration worksheet and answers on Chapter 9 Resource Book pp. 37 and 98

Motivate

Show the students five index cards, numbered 1, 2, 3, 3, and 3. Put the five cards in a bag and ask the students which number they think you will most likely pull out. Now show three cards numbered 1, 2, and 3. Put the three cards in a bag and ask the students which number they think you will most likely pull out.

2 Teach

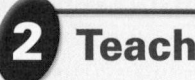

Guided Instruction

In this lesson, students learn to estimate probability using theoretical methods. Remind the students that an experimental probability ratio compares the number of times an event *actually* occurs to the total number of *actual trials*. In theoretical probability, there are no actual trials or actual results. If you can identify all possible outcomes and assume that all those possible outcomes are *equally likely,* you can create a theoretical probability ratio (Teaching Transparency T16, CRB).

An experiment consists of rolling a fair die. There are 6 possible outcomes: 1, 2, 3, 4, 5, and 6.

C **What is the probability of rolling a number less than 5?**

There are 4 outcomes in the event of rolling a number less than 5: 1, 2, 3, and 4.

$P(\text{rolling a number less than 5}) = \frac{4}{6} = \frac{2}{3}$

Suppose you roll two fair dice. Are all outcomes equally likely? It depends on how you consider the outcomes. You could look at the number on each die or at the total shown on the dice.

If you look at the total, all outcomes are not equally likely. For example, there is only one way to get a total of 2, 1 + 1, but a total of 5 can be 1 + 4, 2 + 3, 3 + 2, or 4 + 1.

E X A M P L E 2 **Calculating Theoretical Probability for Two Fair Dice**

An experiment consists of rolling two fair dice.

A **Show a sample space that has all outcomes equally likely.**

Suppose the dice are two different colors, red and blue.

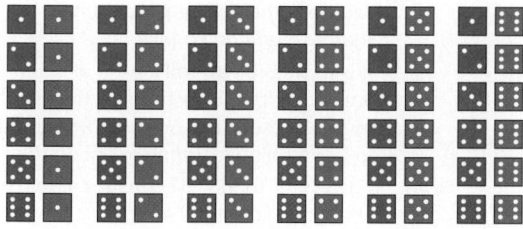

The outcome of a red 3 and a blue 6 can be written as the ordered pair (3, 6). There are 36 possible outcomes in the sample space.

B **What is the probability of rolling doubles?**

There are 6 outcomes in the event "rolling doubles": (1, 1), (2, 2), (3, 3), (4, 4), (5, 5) and (6, 6).

$P(\text{doubles}) = \frac{6}{36} = \frac{1}{6}$

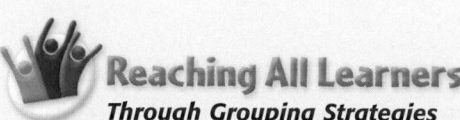

C **What is the probability that the total shown on both dice is 10?**

There are 3 outcomes in the event "a total of 10": (4, 6), (5, 5), and (6, 4).

$P(\text{total} = 10) = \frac{3}{36} = \frac{1}{12}$

D **What is the probability that the total shown is less than 5?**

There are 6 outcomes in the event "a total less than 5": (1, 1), (1, 2), (1, 3), (2, 1), (2, 2), and (3, 1).

$P(\text{total} < 5) = \frac{6}{36} = \frac{1}{6}$

Reaching All Learners
Through Grouping Strategies

Provide each group of students with a deck of playing cards. Have each group write five probability problems for another group to solve. Examples include finding the probability of drawing an eight, a diamond, or a red jack. After each group has written five questions, have them exchange lists with another group. Students can then solve the problems by identifying the outcomes that make each event occur. You may want each group to return the solved problems to the original group for review.

Additional Examples

Example **1**

An experiment consists of spinning this spinner once.

A. What is the probability of spinning a 4? $\frac{1}{5}$

B. What is the probability of spinning an even number? $\frac{2}{5}$

C. What is the probability of spinning a number less than 4? $\frac{3}{5}$

Example **2**

An experiment consists of rolling one fair die and flipping a coin.

A. Show a sample space that has all outcomes equally likely.

1H	2H	3H	4H	5H	6H
1T	2T	3T	4T	5T	6T

B. What is the probability of getting tails? $\frac{1}{2}$

C. What is the probability of getting an even number and heads? $\frac{1}{4}$

D. What is the probability of getting a prime number? $\frac{1}{2}$

Example **3**

Suppose you are playing a game in which you roll two fair dice. If you roll a total of five you will win. If you roll a total of two, you will lose. If you roll anything else, the game continues. What is the probability that the game will end on your next roll? $\frac{5}{36}$

Two events are **mutually exclusive** if they cannot both occur in the same trial of an experiment. Suppose A and B are two mutually exclusive events.

- P(both A *and* B will occur) $= 0$
- P(either A *or* B will occur) $= P(A) + P(B)$

Examples 2C and 2D are mutually exclusive, because the total cannot be less than 5 and equal to 10 at the same time. Examples 2B and 2C are *not* mutually exclusive, because the outcome (5, 5) is a double *and* has a sum of 10.

EXAMPLE 3 Finding the Probability of Mutually Exclusive Events

Suppose you are playing a game of Monopoly and have just rolled doubles two times in a row. If you roll doubles again, you will go to jail. You will also go to jail if you roll a total of 3, because you are 3 spaces away from the "Go to Jail" square. What is the probability that you will go to jail?

It is impossible to roll a total of 3 and doubles at the same time, so the events are mutually exclusive. Add the probabilities to find the probability of going to jail on the next roll.

The event "total = 3" consists of two outcomes, (1, 2) and (2, 1), so P(total of 3) $= \frac{2}{36}$. From Example 2B, P(doubles) $= \frac{6}{36}$.

P(going to jail) $= P$(doubles) $+ P$(total = 3)

$$= \frac{6}{36} + \frac{2}{36}$$

$$= \frac{8}{36}$$

The probability of going to jail is $\frac{8}{36} = \frac{2}{9}$, or about 22.2%.

Think and Discuss

1. **Describe** a sample space for tossing two coins that has all outcomes equally likely.

2. **Give an example** of an experiment in which it would not be reasonable to assume that all outcomes are equally likely.

❸ Close

Summarize

Review the concepts of theoretical probability. Ask students to define *mutually exclusive*. Ask them to provide examples of events that are mutually exclusive and events that are not mutually exclusive.

Possible answers: Mutually exclusive events cannot occur in the same trial of an experiment (e.g., rolling a 5 and a 6 on one roll of a number cube). Events that are not mutually exclusive include rolling an odd number or a number greater than 4 on a number cube and drawing a queen or a heart from a deck of cards.

Answers to Think and Discuss

1. There are four outcomes in the sample space: (H, H), (H, T), (T, H), and (T, T).

2. Possible answer: choosing letters in Scrabble®, which has different numbers of tiles for different letters

FOR EXTRA PRACTICE
see page 749

⌐ internet connect
Homework Help Online
go.hrw.com Keyword: MP4 9-4

9-4 **PRACTICE & ASSESS**

GUIDED PRACTICE

See Example ① An experiment consists of rolling a fair die.

1. What is the probability of rolling an even number? $\frac{1}{2}$

2. What is the probability of rolling a 3 or a 5? $\frac{1}{3}$

See Example ② An experiment consists of rolling two fair dice. Find each probability.

3. $P(\text{total shown} = 7)$ $\frac{1}{6}$ 4. $P(\text{rolling two 5's})$ $\frac{1}{36}$

5. $P(\text{rolling two even numbers})$ $\frac{1}{4}$ 6. $P(\text{total shown} > 8)$ $\frac{5}{18}$

See Example ③ 7. Suppose you are playing a game in which two fair dice are rolled. To make the first move, you need to roll doubles or a sum of 3 or 11. What is the probability that you will be able to make the first move? $\frac{5}{18}$

INDEPENDENT PRACTICE

See Example ① An experiment consists of rolling a fair die.

8. What is the probability of rolling a 7? **0**

9. What is the probability of not rolling a 6? $\frac{5}{6}$

10. What is the probability of rolling a number greater than 2? $\frac{2}{3}$

See Example ② An experiment consists of rolling two fair dice. Find each probability.

11. $P(\text{total shown} = 12)$ $\frac{1}{36}$ 12. $P(\text{not rolling doubles})$ $\frac{5}{6}$

13. $P(\text{total shown} > 0)$ **1**   14. $P(\text{total shown} < 4)$ $\frac{1}{12}$

See Example ③ 15. Suppose you are playing a game in which two fair dice are rolled. You need 7 to land on the finish by an exact count, or 4 to land on a "roll again" space. What is the probability of landing on the finish or rolling again? $\frac{1}{4}$

PRACTICE AND PROBLEM SOLVING

Three fair coins are tossed: a penny, a dime, and a quarter. The table shows a sample space with all outcomes equally likely. Find each probability.

16. $P(\text{HTT})$ $\frac{1}{8}$ 17. $P(\text{THT})$ $\frac{1}{8}$

18. $P(\text{TTT})$ $\frac{1}{8}$ 19. $P(2\text{ heads})$ $\frac{3}{8}$

20. $P(0\text{ tails})$ $\frac{1}{8}$ 21. $P(\text{at least 1 head})$ $\frac{7}{8}$

22. $P(1\text{ tail})$ $\frac{3}{8}$ 23. $P(\text{all the same})$ $\frac{1}{4}$

Penny	Dime	Quarter	Outcome
H	H	H	HHH
H	H	T	HHT
H	T	H	HTH
H	T	T	HTT
T	H	H	THH
T	H	T	THT
T	T	H	TTH
T	T	T	TTT

Students may want to refer back to the lesson examples.

Assignment Guide

If you finished Example ① assign:
Core 1, 2, 8–10, 27–34
Enriched 1, 2, 8–10, 27–34

If you finished Example ② assign:
Core 1–6, 8–14, 17–23 odd, 27–34
Enriched 8–14, 16–34

If you finished Example ③ assign:
Core 1–15, 17–25 odd, 27–34
Enriched 8–34

Notes

RETEACH 9-4

LESSON Reteach
9-4 *Theoretical Probability*

The sample space for a fair coin has 2 possible outcomes: heads or tails. Both possibilities have the same chance of occurring; they are **equally likely**.

The probability of each outcome is $\frac{1}{2}$.

$$P(\text{heads}) = P(\text{tails}) = \frac{1}{2}$$

Complete to find each probability.

1.   2.

How many outcomes in the sample space? **3**   **6**

Are the outcomes equally likely? **yes**   **yes**

What is the probability for each outcome? $\frac{1}{3}$ $\frac{1}{6}$

For this spinner, there are 10 possible outcomes in the sample space. The outcomes are equally likely.

$P(7) = \frac{1}{10}$ $P(\text{even number}) = \frac{5}{10}$, or $\frac{1}{2}$

$P(\text{a number greater than 4}) = \frac{6}{10}$, or $\frac{3}{5}$

When the possible outcomes are equally likely, you calculate the probability that an event E will occur by using a ratio.

$$P(E) = \frac{\text{number of favorable outcomes}}{\text{total number of possible outcomes}}$$

Find each probability.

3.   4.   5.

$P(C) = \frac{5}{8}$ $P(1) = \frac{2}{4}$, or $\frac{1}{2}$ $P(2) = \frac{6}{6}$, or $\frac{1}{3}$

$P(A) = \frac{2}{8}$ $P(\text{even}) = \frac{1}{4}$ $P(\text{odd}) = \frac{6}{6}$, or $\frac{2}{3}$

PRACTICE 9-4

LESSON Practice B
9-4 *Theoretical Probability*

An experiment consists of rolling two fair number cubes. Find each probability.

1. $P(\text{total shown} = 3)$ 2. $P(\text{total shown} = 7)$ 3. $P(\text{total shown} = 9)$
$\frac{1}{18}$ $\frac{1}{6}$ $\frac{1}{9}$

4. $P(\text{total shown} = 2)$ 5. $P(\text{total shown} = 4)$ 6. $P(\text{total shown} = 13)$
$\frac{1}{36}$ $\frac{1}{12}$ 0

7. $P(\text{total shown} > 8)$ 8. $P(\text{total shown} \le 12)$ 9. $P(\text{total shown} < 7)$
$\frac{5}{18}$ 1 $\frac{5}{12}$

Find the theoretical probability of having a thrown dart land in the indicated area.

10. $P(\text{landing in area 1})$ 11. $P(\text{landing in area 3})$
$\frac{1}{2}$ $\frac{1}{8}$

12. $P(\text{landing in area 5})$ 13. $P(\text{landing in area 2})$ 14. $P(\text{landing in area 4})$
$\frac{1}{8}$ $\frac{1}{4}$ $\frac{1}{16}$

15. What is the theoretical probability of winning a split-the-pot prize if you bought 5 tickets and 150 were sold? $\frac{1}{30}$

16. Darnell won the class spelling bee. His teacher put each prize in separate but equal size boxes. The prizes were 10 candy bars, 1 homework pass, and 3 pencils. Darnell could have his choice of the boxes. What is the theoretical probability that he will choose a homework pass as his prize? $\frac{1}{14}$

Math Background

It is important to distinguish between common meanings of words and strict mathematical meanings. In probability study, the words *outcome* and *event* have very precise meanings. An outcome is a particular result of one trial of an experiment. An event is either a single outcome or a set of outcomes. Suppose an experiment is to roll two fair dice, and we are interested in the probabilities of the sums that could occur. We call rolling a sum of 5 an event, and it consists of the following possible outcomes: (1, 4), (2, 3), (3, 2), and (4, 1). We also call rolling a sum of 2 an event, and it consists of the single outcome (1, 1).

9B Pascal's Triangle

Pacing: Traditional 1 day
Block $\frac{1}{2}$ day

Objective: To use Pascal's Triangle to find combinations

Materials: Pencil and paper

Lab Resources

Hands-On Lab Activities . . . . . pp. 66–67

Using the Page

Discuss with students how to form each row of Pascal's Triangle. Discuss how the numbers in each row can be used in place of the combination formula to find combinations.

Use rows 0–6 of Pascal's Triangle to answer the following questions.

1. What pattern do you see within each row of the triangle? Each row is symmetrical. What pattern do you see in column 2? As you go down the rows in that column, the difference between the number on that row and the previous row's entry increases by 1.

2. How can you use Pascal's Triangle to determine the number of different ways you can choose 2 items from 6 items? Find the number in row 6 column 2; $_6C_2 = 15$.

Answers

Think and Discuss

Possible answer:

1. You could use Pascal's Triangle to find $_{31}C_8$, but it would take a lot of time to write out 31 rows of Pascal's Triangle. It would be easier to use the combination formula.

Try This

1. Each row's entry is one more than the previous row's entry; 7.

2. Row 0: $1 = 2^0$
   Row 1: $2 = 2^1$
   Row 2: $4 = 2^2$
   Row 3: $8 = 2^3$
   Row n: 2^n

3. Look at the entry in row 7 column 2 to find $_7C_2$; $_7C_2 = 21$.

Pascal's Triangle

Use with Lesson 9-6

Pascal's Triangle is a triangular array of counting numbers. The first row of the triangle is a 1, and every other number in the triangle is the sum of the two numbers diagonal to it in the row above it. On the outer edge of the triangle, each number has only one number diagonal to it, so each row of the triangle begins and ends with the number 1.

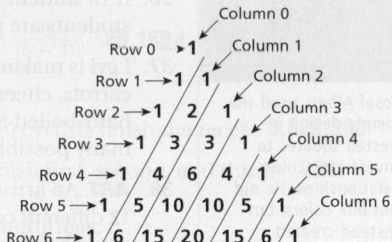

Activity

1. Copy Pascal's Triangle onto your paper.
2. Add two more rows to Pascal's Triangle.
3. Look at row 5 of Pascal's Triangle.

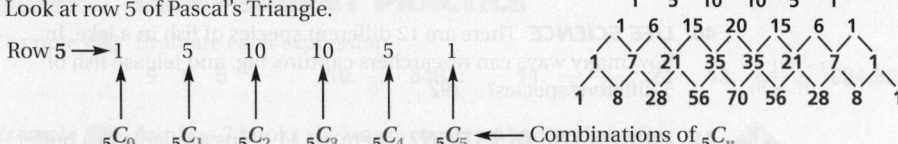

This row shows all possible combinations of 5 items taken 1, 2, 3, 4, or 5 at a time ($_5C_n$).

If there are **5** people in a club, how many different combinations of 2 club members can bring refreshments to a meeting?

$$_5C_2 = 10$$

Think and Discuss

1. Can you use Pascal's Triangle to find $_{31}C_8$? Would this be easier than using the combination formula? Explain your answer.

Try This

1. Find a pattern for the numbers in column 1 of Pascal's Triangle. What will the number in row 7 column 1 be?

2. Find the sum for each row of numbers. Write each sum as a power of 2.

3. Misha has 7 pens that are each a different color. He brings 2 to school each day. How many days can he have a different combination of pen colors before he must start repeating combinations? Explain how Pascal's triangle can help you answer this question.

9-7 Independent and Dependent Events

Learn to find the probabilities of independent and dependent events.

Vocabulary
independent events
dependent events

It is critical that the engine of a single-engine airplane not fail during flight. These planes often have two *independent* electrical systems. In the event that one electrical system fails, for example, due to a faulty spark plug, the second system will still be able to keep the plane in flight.

Events are **independent events** if the occurrence of one event does not affect the probability of the other. Events are **dependent events** if the occurrence of one does affect the probability of the other.

EXAMPLE 1 Classifying Events as Independent or Dependent

Determine if the events are dependent or independent.

A a coin landing heads on one toss and tails on another toss
The result of one toss does not affect the result of the other, so the events are independent.

B drawing a heart and a spade from a deck at the same time
The cards drawn cannot be the same card, so the events are dependent.

FINDING THE PROBABILITY OF INDEPENDENT EVENTS

If A and B are independent events, then $P(A \text{ and } B) = P(A) \cdot P(B)$.

EXAMPLE 2 Finding the Probability of Independent Events

An experiment consists of spinning the spinner 3 times. For each spin, all outcomes are equally likely.

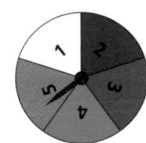

A What is the probability of spinning a 5 all 3 times?
The result of each spin does not affect the results of the other spins, so the spin results are independent.
For each spin, $P(5) = \frac{1}{5}$.
$P(5, 5, 5) = \frac{1}{5} \cdot \frac{1}{5} \cdot \frac{1}{5} = \frac{1}{125} = 0.008$ *Multiply.*

9-7 Organizer

Pacing: Traditional 1 day
Block $\frac{1}{2}$ day

Objective: Students find the probabilities of independent and dependent events.

Warm Up
Evaluate each expression.
1. 8! 40,320
2. $\frac{10!}{7!}$ 720
3. Find the number of permutations of the letters in the word *quiet* if no letters are used more than once. 120

Problem of the Day
The area of a spinner is 75% red and 25% blue. However, the probability of its landing on red is only 50%. Sketch a spinner to show how this can be.
Possible answer:

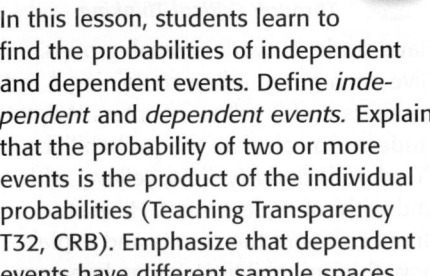

Available on Daily Transparency in CRB

Math Humor

Teacher: Give an example of an independent event.
Student: A Fourth-of-July fireworks display.

---

1 Introduce

Alternate Opener

EXPLORATION

9-7 Independent and Dependent Events

When the occurrence of one event affects the probability of a second event, the events are **dependent**. Otherwise, they are **independent**.

Classify each pair of events as dependent or independent.

	Events	Independent	Dependent
1.	Event A: Drawing a 3 from a deck without replacing it Event B: Drawing another 3 from the same deck		
2.	Event A: Tossing heads on a flip of a coin Event B: Tossing heads on a second flip of the same coin		
3.	Event A: Running between 5:00 P.M. and 5:30 P.M. Event B: Drinking water between 5:30 P.M. and 6:00 P.M.		
4.	Event A: Drawing a blue marble and putting it back in a bag that contains 6 blue marbles and 4 red marbles Event B: Drawing another blue marble from the same bag		

Think and Discuss
5. **Discuss** real-world examples of dependent events and independent events.

Motivate

Put two pairs of identical items in a container (e.g., two pennies and two nickels). Ask students to find the probability of choosing one of the items. 50% Choose one item and without replacing it, repeat the same question. 33% or 67% depending on the first event Help students understand that the probability has changed because the sample space has changed.

Exploration worksheet and answers on
Chapter 9 Resource Book pp. 67 and 105

---

2 Teach   Lesson Presentation

Guided Instruction

In this lesson, students learn to find the probabilities of independent and dependent events. Define *independent* and *dependent events*. Explain that the probability of two or more events is the product of the individual probabilities (Teaching Transparency T32, CRB). Emphasize that dependent events have different sample spaces.

Teaching Tip You may want to review multiplication of fractions with the students.

Game Resources

Puzzles, Twisters & Teasers
Chapter 9 Resource Book

The Paper Chase

Purpose: *To apply the skill of finding probability to a brainteaser*

Discuss: Ask students to explain how to determine the answer to problem 1. (Hint: You should treat the $\frac{2}{10}$ chance that the paper gets lost as if there were 2 more drawers that you will never be able to check, for a total of 10 drawers.)

The probability that the paper is in any searchable drawer is $\frac{8}{10}$. If it is not in the first drawer, then there are only 9 possibilities left for where it might be, and 7 of those are the remaining searchable drawers. So the probability that the paper is in one of the remaining searchable drawers is $\frac{7}{9}$.

Extend: Challenge students to write a similar problem, changing the probability that a paper gets lost and the number of drawers. Have them use their information to determine a formula for the probability of finding a paper.

Check students' work.

Answers

Let n be the number of drawers that have already been checked. The probability of finding the paper in a specific unchecked drawer is $\frac{(8 - n)}{(10 - n)}$, a decreasing function, and the probability of finding the paper in the next drawer is $\frac{1}{(10 - n)}$, an increasing function.

Permutations

Purpose: *To practice finding permutations in a game format*

Discuss: Have students practice making English words using permutations of the letters *A, I, M, R,* and *N*.

Possible answers: *MAN, RAIN, RAN, RAM, RIM,* and *AIR.*

Extend: Have students repeat the game, using 2 vowels and 5 consonants. The first player to reach 150 points wins.

The Paper Chase

Stephen's desk has 8 drawers. When he receives a paper, he usually chooses a drawer at random to put it in. However, 2 out of 10 times he forgets to put the paper away, and it gets lost.

The probability that a paper will get lost is $\frac{2}{10}$, or $\frac{1}{5}$.

- What is the probability that a paper will get put into a drawer? $\frac{4}{5}$

- If all drawers are equally likely to be chosen, what is the probability that a paper will get put in drawer 3? $\frac{1}{10}$

When Stephen needs a document, he looks first in drawer 1 and then checks each drawer in order until the paper is found or until he has looked in all the drawers.

1. If Stephen checked drawer 1 and didn't find the paper he was looking for, what is the probability that the paper will be found in one of the remaining 7 drawers? $\frac{7}{9}$

2. If Stephen checked drawers 1, 2, and 3, and didn't find the paper he was looking for, what is the probability that the paper will be found in one of the remaining 5 drawers? $\frac{5}{7}$

3. If Stephen checked drawers 1–7 and didn't find the paper he was looking for, what is the probability that the paper will be found in the last drawer? $\frac{1}{3}$

Try to write a formula for the probability of finding a paper.

Permutations

Use a set of Scrabble™ tiles, or make a similar set of lettered cards. Draw 2 vowels and 3 consonants, and place them face up in the center of the table. Each player tries to write as many permutations as possible in 60 seconds. Score 1 point per permutation, with a bonus point for each permutation that forms an English word.

📶 internet connect

Go to *go.hrw.com* for a complete set of rules and game pieces.
KEYWORD: MP4 Game9

Technology LAB

Permutations and Combinations

Use with Lesson 9-6

Graphing calculators have features to help with computing factorials, permutations, and combinations.

internet connect
Lab Resources Online
go.hrw.com
KEYWORD: MP4 TechLab9

Activity

1 In a stock-car race, 11 cars finish the race. The number of different orders in which they can finish is 11! A calculator can help you do the computation. Both ways are shown—the direct way, using the definition of *factorial,* and the calculator factorial command.

To compute 11! on a graphing calculator, enter 11 [MATH], press ▶ to go to the **PRB** menu, and select **4:!** [ENTER].

The number of ways the 11 cars can finish first, second, third, and fourth is given by $11 \cdot 10 \cdot 9 \cdot 8$, or in *permutation* notation, $_{11}P_4$, 11 things taken 4 at a time. Both the direct and calculator *nPr* command methods are shown. The *nPr* command is also found in the **PRB** menu.

To compute $_{11}P_4$, enter 11 [MATH], press ▶ to go to the **PRB** menu, select **2:nPr**, type 4, and press [ENTER].

2 Twenty girls try out for 5 open places on a hockey team. Since order is not considered, the number of different *combinations* of these girls that can be chosen is given by $_{20}C_5$, the number of combinations of 20 things taken 5 at a time. Both the direct and calculator *nCr* command computations are shown.

To compute $_{20}C_5$, press 20 [MATH], press ▶ to go to the **PRB** menu, select **3:nCr**, and press 5 [ENTER].

Think and Discuss
nPr is usually greater than nCr (for the same n and r) because AB and BA represent different permutations but the same combination.

1. Explain why *nPr* is usually greater than *nCr* for the same values of *n* and *r*.

2. Can *nPr* ever equal *nCr*? *nPr can equal nCr when r = 1 and n is any positive integer.*

Try This

Compute each value by direct calculator multiplication and division and by using the calculator permutation and combination commands.

1. $_{14}P_6$ 2. $_{25}P_{17}$ 3. $_8P_3$ 4. $_8C_3$ 5. $_{16}C_4$ 6. $_{40}C_6$

Answers

Try This

1. 2,162,160
2. 3.8470263E20
3. 336
4. 56
5. 1820
6. 3,838,380

Technology LAB

Permutations and Combinations

Objective: To use a graphing calculator to find permutations and combinations

Materials: Graphing calculator

Lab Resources
Technology Lab Activities . . . . p. 64

Using the Page

This technology activity shows students how to find permutations and combinations on a graphing calculator. Specific keystrokes may vary, depending on the make and model of the graphing calculator used. The keystrokes given are for a TI-83 model. For keystrokes to other models, visit go.hrw.com.

The Think and Discuss problems can be used to assess students' understanding of the ideas of *permutation* and *combination.* Try This problems 1–6 are meant to show students the benefit of using the *nPr* and *nCr* functions on the graphing calculator.

Assessment

Have students practice using the probability commands with the following exercises.

Give the keystrokes needed to solve each problem. Find the answer.

1. A teacher wants to select 4 of her 20 students to run an errand. How many different groups of 4 students can she select?

   To compute $_{20}C_4$, type 20, press 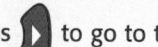, press ▶ to go to the **PRB** menu, select option 3: **nCr**, type 4, and press [ENTER]. The teacher can select 4,845 different groups.

2. Find $_{10}P_3$.
   To compute $_{10}P_3$, type 10, press , press ▶ to go to the **PRB** menu, select option 2: **nPr**, type 3, and press [ENTER]. $_{10}P_3 = 720$

Chapter
9
Study Guide and Review

Purpose: *To help students review and practice concepts and skills presented in Chapter 9*

Assessment Resources

Chapter Review
Chapter 9 Resource Book . . . pp. 85–86

Test and Practice Generator
CD-ROM

Additional review assessment items in both multiple-choice and free-response format may be generated for any objective in Chapter 9.

Answers

1. probability; impossible; certain

2. sample space

3. permutation; combination

4. 0.75; 0.25

Vocabulary

Complete the sentences below with vocabulary words from the list above. Words may be used more than once.

1. The ___?___ of an event tells you how likely the event is to happen.
   • A probability of 0 means it is ___?___ for the event to occur.
   • A probability of 1 means it is ___?___ that the event will occur.

2. The set of all possible outcomes of an experiment is called the ___?___.

3. A(n) ___?___ is an arrangement where order is important.
   A(n) ___?___ is an arrangement where order is not important.

9-1 Probability (pp. 446–450)

EXAMPLE

■ Of the raw diamonds received by a diamond cutter, it is expected that about $\frac{1}{8}$ of them will be acceptable.

Outcome	Acceptable	Unacceptable
Probability		

$P(\text{acceptable}) = \frac{1}{8} = 0.125 = 12.5\%$

$P(\text{unacceptable}) = 1 - \frac{1}{8} = \frac{7}{8} = 0.875 = 87.5\%$

EXERCISES

Give the probability for each outcome.

4. About 75% of the people attending a book signing have already read the book.

Outcome	Read	Not read
Probability		

Study Guide and Review

9-2 Experimental Probability (pp. 451–454)

EXAMPLE

■ The table shows the results of spinning a spinner 80 times. Estimate the probability of the spinner landing on blue.

Outcome	White	Red	Blue	Black
Spins	32	17	24	7

probability $\approx \dfrac{\text{spins that landed on blue}}{\text{total number of spins}}$

$= \dfrac{24}{80} = \dfrac{3}{10} = 0.3$

The probability of the spinner landing on blue is about 0.3, or 30%.

EXERCISES

5. The table shows the results of spinning a spinner 100 times. Estimate the probability of the spinner landing on 5.

Outcome	1	2	3	4	5	6
Spins	17	22	11	18	17	15

6. The table shows the results of a survey of 500 students. Estimate the probability that a randomly selected student's favorite subject is math.

Favorite Subject	Math	Science	Art	Other
Number of Students	140	105	75	180

9-3 Use a Simulation (pp. 456–459)

EXAMPLE

■ At a local school, 75% of the students study a foreign language. If 5 students are chosen randomly, estimate the probability that at least 4 study a foreign language. Use the random number table to make a simulation with at least 10 trials.

08	57	09	92	75		27	37	87	52	36
16	73	29	39	73		78	65	88	02	42
53	19	18	65	79		64	46	47	60	51
73	16	79	89	12		63	84	60	59	57
13	89	68	35	51		22	56	51	23	81

The probability is about $\frac{8}{10}$, or 80%.

EXERCISES

08570	99275	27378	75236	16732
93973	78658	80242	53191	86579
64464	76051	73167	98912	63846
05957	13896	83551	22565	12381
93861	72073	87891	19845	71302

7. On an assembly line, 25% of the items are rejected. Estimate the probability that at least 2 of the next 6 items will be rejected. Use the random number table to make a simulation with at least 10 trials.

9-4 Theoretical Probability (pp. 462–466)

EXAMPLE

■ A fair die is rolled once. Find the probability of getting an odd number or a 4.

$P(\text{odd or } 4) = P(\text{odd}) + P(4)$

$= \dfrac{3}{6} + \dfrac{1}{6} = \dfrac{4}{6} = \dfrac{2}{3}$

EXERCISES

8. A marble is drawn at random from a box that contains 7 red, 12 blue, and 5 white marbles. What is the probability of getting a red or a white marble?

Answers

5. 0.17, or 17%

6. 0.28, or 28%

7. Possible answer: 40%

8. $\dfrac{1}{2}$

Study Guide and Review

Answers

9. 36

10. 30

11. 210

12. 126

13. $\frac{1}{216}$

14. $\frac{13}{51}$

15. 5:21

9-5 The Fundamental Counting Principle (pp. 467–470)

EXAMPLE

- A code contains 4 letters. How many possible codes are there?

  $26 \cdot 26 \cdot 26 \cdot 26 = 456,976$ codes

EXERCISES

A building has 6 doors to the outside.

9. How many ways can you enter and leave the building?

10. How many ways can you enter by one door and leave by a different door?

9-6 Permutations and Combinations (pp. 471–475)

EXAMPLE

- Blaire has 5 plants to arrange on a shelf that will hold 3 plants. How many ways are there to arrange the plants, if the order is important? if the order is not important?

  order important: $_5P_3 = \frac{5!}{(5-3)!} = \frac{5!}{2!} = 60$ ways

  order not important: $_5C_3 = \frac{5!}{3!\,(5-3)!} = 10$ ways

EXERCISES

11. Seven people are arranged in a row of 3 seats. How many different arrangements are possible?

12. A school's debate club has 9 members. A team of 4 students will be chosen to represent the school at a competition. How many different teams are possible?

9-7 Independent and Dependent Events (pp. 477–481)

EXAMPLE

- Two marbles are drawn from a jar containing 3 red marbles and 4 black marbles. What is P(red, black) if the first marble is replaced? if the first marble is not replaced?

	P(red)	P(black)	P(red, black)
Replaced	$\frac{3}{7}$	$\frac{4}{7}$	$\frac{12}{49} \approx 0.24$
Not replaced	$\frac{3}{7}$	$\frac{4}{6}$	$\frac{12}{42} \approx 0.29$

EXERCISES

13. A number cube is rolled three times. What is the probability of getting a 4 all three times?

14. Two cards are drawn at random from a deck that has 26 red and 26 black cards. What is the probability that the first card is red and the second is black?

9-8 Odds (pp. 482–485)

EXAMPLE

- A digit from 1 to 9 is selected at random. What are the odds in favor of selecting an even number?

  favorable → 4:5 ←unfavorable

EXERCISES

15. A letter is selected at random from the alphabet. What are the odds in favor of getting a vowel (A, E, I, O, U)?

1. Outcomes A, C, D, and F have the same probability. Complete the probability table.

Outcome	A	B	C	D	E	F
Probability	$\blacksquare\frac{1}{8}$	$\frac{1}{6}$	$\blacksquare\frac{1}{8}$	$\blacksquare\frac{1}{8}$	$\frac{1}{3}$	$\blacksquare\frac{1}{8}$

2. Madeline is choosing 3 of her 10 best flower displays to be entered in a competition. How many different selections are possible? **120**

3. Jim wants to hang 4 pictures in a row on his wall. If he has 6 pictures to choose from, how many different arrangements are possible? **360**

4. A fair coin is tossed three times. What is the probability of getting heads all three times? $\frac{1}{8}$

5. In the Westcreek neighborhood, 37% of the families have a dog. Each block has 16 families, 8 on each side. Estimate the probability that 3 or more families on one side of a given block have a dog. Use the random number table to make a simulation with at least 10 trials. **Possible answer: 70%**

   97120 08320 17871 21826 74838 37240 36810 20423

   12562 45677 88983 94930 31599 76585 61429 05379

   34628 46304 66531 96270 21309 31567 30762 47240

   30883 71946 25948 97988 26267 21350 59356 43952

6. Jill has 6 cans of food without labels. She knows there are 2 cans of fruit, 3 of corn, and 1 of beans. If she chooses a can at random, what is the probability that it will not be fruit? $\frac{2}{3}$

7. Julio's parents write down 10 different chores on slips of paper and put them in a box. If Julio has to draw 2 different chores from the box, what is the probability that he will draw his 2 least favorite chores, vacuuming and pulling weeds? $\frac{1}{45}$

8. The table shows the results of an interview in which 1000 college students were asked whether they went home during spring and winter breaks. Estimate the probability a student will go home during winter break. **69%**

	Spring (yes)	Spring (no)
Winter (yes)	170	520
Winter (no)	233	77

9. A frame shop has a special offer. Pictures can be framed in gold, silver, or brass, the mat can be any one of 16 colors, and the glass can be regular or nonglare. How many ways can a picture be framed with this offer? **96**

10. At a bazaar, the odds in favor of winning a door prize is 1:15. What is the probability of winning a door prize? $\frac{1}{16}$

Chapter Test

Purpose: *To assess students' mastery of concepts and skills in Chapter 9*

Assessment Resources

Chapter 9 Tests (Levels A, B, C)
Assessment Resources . . . . . . pp. 81–86

 ***Test and Practice Generator* CD-ROM**

Additional assessment items in both multiple-choice and free-response format may be generated for any objective in Chapter 9.

Purpose: To assess students' mastery of concepts and skills in Chapter 9 and combined problem-solving skills

Assessment Resources

Performance Assessment
Assessment Resources . . . . p. 134

Performance Assessment Teacher Support
Assessment Resources . . . . p. 133

Answers

1–2. See p. A10.

3. See Level 3 work sample below.

Scoring Rubric for Problem Solving Item 3

Level 3
Accomplishes the purposes of the task.

Student gives clear explanations, shows understanding of mathematical ideas and processes, and computes accurately.

Level 2
Purposes of the task not fully achieved.

Student demonstrates satisfactory but limited understanding of the mathematical ideas and processes.

Level 1
Purposes of the task not accomplished.

Student shows little evidence of understanding the mathematical ideas and processes and makes computational and/or procedural errors.

Performance Assessment

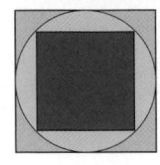 **Show What You Know**

Create a portfolio of your work from this chapter. Complete this page and include it with your four best pieces of work from Chapter 9. Choose from your homework or lab assignments, mid-chapter quiz, or any journal entries you have done. Put them together using any design you want. Make your portfolio represent what you consider your best work.

 Short Response

1. A dart thrown at the square board shown lands in a random spot on the board. What is the probability it lands in the blue square? Show your work.

2. The pilot of a hot air balloon is trying to land in a 2 km square field. The field has a large tree in each corner. The balloon's ropes will tangle in a tree if it lands within $\frac{1}{7}$ km of its trunk. What is the probability the balloon will land in the field without getting caught in a tree? Express your answer to the nearest tenth of a percent. Show your work.

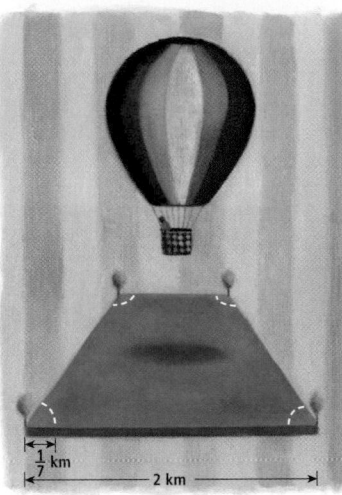

Extended Problem Solving

3. The students at a new high school are choosing a school mascot and school color. The mascot choices are a bear, a lion, a jaguar, or a tiger. The color choices are red, orange, or blue.

   a. How many different combinations do the students have to choose from? Show your work.

   b. If a second school color is added, either gold or silver, how many different combinations do the students have to choose from? Show your work.

   c. How would adding a choice from among n names change the number of combinations to choose from?

Student Work Samples for Item 3

Level 3

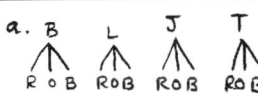

a.
```
  B     L     J     T
 /\    /\    /\    /\
ROB   ROB   ROB   ROB
```
12 combinations

b. There will be 2 more choices for each branch of the tree.
   12 × 2 = 24 combinations.

c. There will be n more choices for each branch of the tree. 24 × n = 24n combinations.

The student correctly used a tree diagram in part **a**. The student showed understanding of the Fundamental Counting Principle in parts **b** and **c**.

Level 2

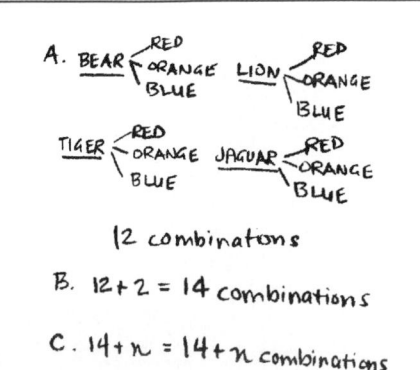

A. BEAR — RED, ORANGE, BLUE   LION — RED, ORANGE, BLUE
TIGER — RED, ORANGE, BLUE   JAGUAR — RED, ORANGE, BLUE

12 combinations

B. 12 + 2 = 14 combinations

C. 14 + n = 14 + n combinations

The student correctly used a tree diagram in part **a**, but added instead of multiplying in parts **b** and **c**.

Level 1

a. 4 + 3 = 7

b. 7 + 2 = 9

c. n + 7 + 9 = 16 + n

The student incorrectly answered all parts, showing no understanding of combinations or the Fundamental Counting Principle.

Cumulative Assessment, Chapters 1–9

1. In a box containing gumdrops, 78 are red, 24 are green, and the rest are yellow. If the probability of selecting a yellow gumdrop is $\frac{1}{3}$, how many yellow gumdrops are in the box? **B**
   (A) 34 (C) 54
   (B) 51 (D) 102

2. $P(-2, -3)$ is reflected across the y-axis to P'. What are the coordinates of P'? **J**
   (F) $(-3, -2)$ (H) $(-2, 3)$
   (G) $(-3, 2)$ (J) $(2, -3)$

3. If 125% of x is equal to 80% of y, and $y \neq 0$, what is the value of $\frac{x}{y}$? **A**
   (A) $\frac{16}{25}$ (C) $\frac{25}{16}$
   (B) $\frac{4}{5}$ (D) $\frac{5}{4}$

4. Mia made 5 payments on a loan, with each payment being twice the amount of the one before it. If the total of all 5 payments was $465, how much was the first payment? **G**
   (F) $5 (H) $31
   (G) $15 (J) $93

5. An electric pump can fill a 45-gallon tub in half an hour. At this rate, how long would it take to fill a 60-gallon tub? **C**
   (A) 35.0 minutes (C) 40.0 minutes
   (B) 37.5 minutes (D) 42.5 minutes

6. Which of the following ratios is equivalent to the ratio 1.2:1? **J**
   (F) 1:2 (H) 5:6
   (G) 12:1 (J) 6:5

7. If $20 \cdot 3000 = 6 \cdot 100^x$, what is the value of x? **A**
   (A) 2 (C) 4
   (B) 3 (D) 5

8. In the chart below, the amount represented by each shaded square is twice that represented by each unshaded square.

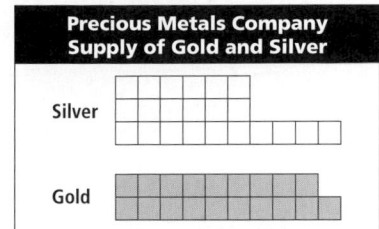

Precious Metals Company Supply of Gold and Silver

Silver

Gold

What is the ratio of the amount of gold to the amount of silver? **J**
   (F) $\frac{19}{22}$ (H) $\frac{22}{19}$
   (G) $\frac{13}{19}$ (J) $\frac{19}{11}$

 TEST TAKING TIP!
To check that answers are reasonable, you can sometimes draw a graph or a diagram on graph paper.

9. **SHORT RESPONSE** Which type of quadrilateral is a figure with vertices $(-3, 4)$, $(3, 4)$, $(3, -2)$, and $(-3, -2)$? What is its area? Explain.

10. **SHORT RESPONSE** The heights of two similar triangles are 4 in. and 5 in. What percent of the height of the smaller triangle is the height of the larger triangle? Show your work.

Standardized Test Prep

Standardized Test Prep

Chapter **9**

Purpose: *To provide review and practice for Chapters 1–9 and standardized tests*

Assessment Resources

Cumulative Tests (Levels A, B, C)
Assessment Resources . . . . pp. 241–252

State-Specific Test Practice Online
KEYWORD: MP4 TestPrep

Test Prep Doctor

For item 6, point out that the first number in the ratio is greater than the second, so any equivalent ratio must follow the same pattern. Thus, choices **F** and **H** can be eliminated. In choice **G**, one of the numbers is the same but the other is not, so this cannot be an equivalent ratio. The correct answer is choice **J**.

Supplement the test-taking tip given for item 8 by encouraging students to plot the point, connect the lines, and determine the base(s) and height of the figure.

Answers

9. square; 36 units^2; each side of the square is 6 units long, so the area is $6^2 = 36$ units.

10. $\frac{x}{100} = \frac{5}{4}$

    $4x = 500$

    $x = 125$

    5 is 125% of 4

More Equations and Inequalities

Section 10A	Section 10B
Solving Linear Equations	**Solving Equations and Inequalities**
Lesson 10-1 Solving Two-Step Equations	**Lesson 10-4** Solving Multistep Inequalities
Lesson 10-2 Solving Multistep Equations	**Lesson 10-5** Solving for a Variable
Hands-On Lab 10A Model Equations with Variables on Both Sides	**Lesson 10-6** Systems of Equations
Lesson 10-3 Solving Equations with Variables on Both Sides	

Pacing Guide for 45-Minute Classes

Chapter 10

DAY 116	DAY 117	DAY 118	DAY 119	DAY 120
Lesson 10-1	**Lesson 10-2**	**Hands-On Lab 10A**	**Lesson 10-3**	**Mid-Chapter Quiz** **Lesson 10-4**

DAY 121	DAY 122	DAY 123	DAY 124	
Lesson 10-5	**Lesson 10-6**	**Chapter 10 Review**	**Chapter 10 Assessment**	

Pacing Guide for 90-Minute Classes

Chapter 10

DAY 58	DAY 59	DAY 60	DAY 61	DAY 62
Chapter 9 Assessment **Lesson 10-1**	**Lesson 10-2** **Hands-On Lab 10A**	**Lesson 10-3** **Lesson 10-4**	**Mid-Chapter Quiz** **Lesson 10-5** **Lesson 10-6**	**Chapter 10 Review** **Lesson 11-1**

DAY 63				
Chapter 10 Assessment **Lesson 11-2**				

COURSE 1
- Solve one-step equations and inequalities.

COURSE 2
- Solve two-step and multistep equations.
- Solve equations with variables on both sides of the equation.
- Solve two-step inequalities and graph the solutions.
- Solve equations for a variable.

PRE-ALGEBRA
- Solve two-step and multistep equations.
- Solve equations with variables on both sides of the equation.
- Solve two-step inequalities and graph the solutions.
- Solve equations for a variable.
- Solve systems of equations.

Across the Curriculum

LANGUAGE ARTS

SOCIAL STUDIES

SCIENCE

TE = *Teacher's Edition*   **SE** = *Student Edition*

Interdisciplinary

Bulletin Board

Physical Science

Pressure is the force pushing on an object from all directions. At sea level, the pressure is approximately 15 pounds per square inch (psi). As the depth of the water increases, pressure increases. The pressure increases by about 15 psi for every 30 feet of seawater. Write an equation to express this relationship.

$$\text{pressure} = 15 \text{ psi} \left(\frac{d}{30 \text{ ft}} \right) + 15 \text{ psi, or } p = \tfrac{1}{2}d + 15,$$

where d represents depth in feet

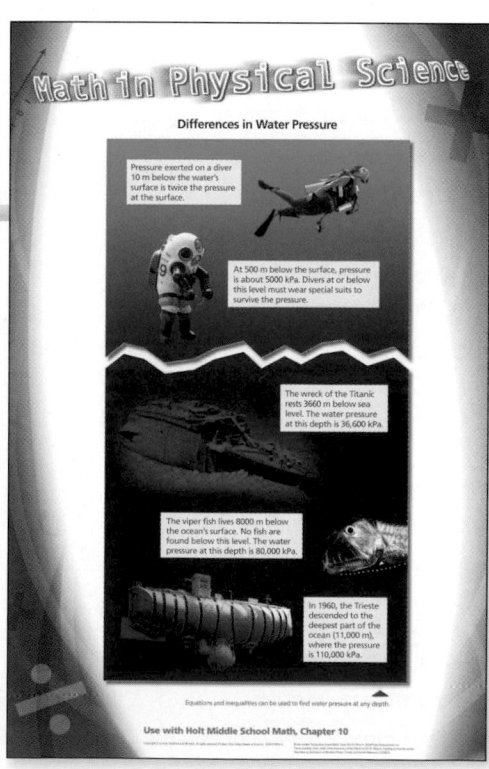

Interdisciplinary posters and worksheets are provided in your resource material.

Resource Options

Chapter 10 Resource Book

Student Resources

Practice (Levels A, B, C) . . . . . pp. 8–10, 18–20, 28–30, 39–41 49–51, 59–61

Reteach . . . . pp. 11–12, 21–22, 31–32, 42–43, 52–53, 62–63

Challenge . . . . pp. 13, 23, 33, 44, 54, 64

Problem Solving . . . . pp. 14, 24, 34, 45, 55, 65

Puzzles, Twisters & Teasers . . . . pp. 15, 25, 35, 46, 56, 66

Recording Sheets . . . . . . pp. 3, 7, 17, 27, 38, 48, 58, 69, 72–77

Chapter Review . . . . pp. 67–68

Teacher and Parent Resources

Chapter Planning and Pacing Guide . . . . p. 4

Section Planning Guides . . . . pp. 5, 36

Parent Letter . . . . pp. 1–2

Teaching Tools . . . . pp. 72–77

Teacher Support for Chapter Project . . . . p. 70

Transparencies . . . . pp. T1–T33

- Daily Transparencies
- Additional Examples Transparencies
- Teaching Transparencies

Reaching All Learners

English Language Learners

Success for English Language Learners . . . . . . pp. 157–168

Math: Reading and Writing in the Content Area . . . . pp. 79–84

Spanish Homework and Practice . . . . pp. 79–84

Spanish Interactive Study Guide . . . . pp. 79–84

Spanish Family Involvement Activities . . . . pp. 73–80

Multilingual Glossary

Individual Needs

Are You Ready? Intervention and Enrichment . pp. 205–208, 229–236, 423–424

Alternate Openers: Explorations . . . . pp. 79–84

Family Involvement Activities . . . . pp. 73–80

Interactive Problem Solving . . . . pp. 79–84

Interactive Study Guide . . . . pp. 79–84

Readiness Activities . . . . pp. 19–20

Math: Reading and Writing in the Content Area . . . . pp. 79–84

Challenge . . . . CRB pp. 13, 23, 33, 44, 54, 64

Hands-On

Hands-On Lab Activities . . . . pp. 71–76

Technology Lab Activities . . . . pp. 65–71

Alternate Openers: Explorations . . . . pp. 79–84

Family Involvement Activities . . . . pp. 73–80

Applications and Connections

Consumer and Career Math . . . . pp. 37–40

Interdisciplinary Posters . . . . Poster 10, TE p. 496B

Interdisciplinary Poster Worksheets . . . . pp. 28–30

Transparencies

Alternate Openers: Explorations . . . . pp. 79–84

Exercise Answers Transparencies

Chapter 10 Resource Book . . . . pp. T1–T33

- Daily Transparencies
- Additional Examples Transparencies
- Teaching Transparencies

Technology

Teacher Resources

Lesson Presentations **CD-ROM** . . . . Chapter 10

Test and Practice Generator **CD-ROM** . . . . . . Chapter 10

One-Stop Planner **CD-ROM** . . . . Chapter 10

Student Resources

Are You Ready? Intervention **CD-ROM**
Skills 49, 55, 56

 internet connect

Homework Help Online	**KEYWORD:** MP4 HWHelp10
Math Tools Online	**KEYWORD:** MP4 Tools
Glossary Online	**KEYWORD:** MP4 Glossary
Chapter Project Online	**KEYWORD:** MP4 PSProject10
Chapter Opener Online	**KEYWORD:** MP4 Ch10

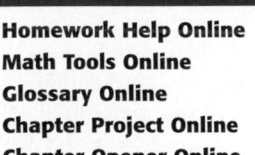

 **KEYWORD:** MP4 CNN10

SE = *Student Edition*   **TE** = *Teacher's Edition*   **AR** = *Assessment Resources*   **CRB** = *Chapter Resource Book*   **MK** = *Manipulatives Kit*

Assessment Options

Assessing Prior Knowledge

Determine whether students have the required prerequisite concepts and skills.

Are You Ready?...............................SE p. 497
Inventory Test...............................AR pp. 1–4

Test Preparation

Provide review and practice for chapter and standardized tests.

Standardized Test Prep...........................SE p. 537
Spiral Review with Test Prep ..... SE, last page of each lesson
Study Guide and Review ....SE pp. 532–534
Test Prep Tool Kit

Technology

 Test and Practice Generator CD-ROM

☑ internet connect
State-Specific Test Practice Online KEYWORD: MP4 TestPrep

Performance Assessment

Assess students' understanding of chapter concepts and combined problem-solving skills.

Performance Assessment ....SE p. 536
   Includes scoring rubric in TE
Performance Assessment ....AR p. 136
Performance Assessment Teacher Support ....AR p. 135

Portfolio

Portfolio opportunities appear throughout the Student and Teacher's Editions.

Suggested work samples:

Problem Solving Project...................TE p. 496
Performance Assessment ....SE p. 536
Portfolio Guide...........................AR p. xxxvii
Journal.......................TE, last page of each lesson
Write About It ....SE pp. 505, 511, 518, 527

Daily Assessment

Obtain daily feedback on students' understanding of concepts.

Spiral Review and Test Prep ...... SE, last page of each lesson

Also Available on Transparency In Chapter 10 Resource Book

Warm Up......................TE, first page of each lesson
Problem of the Day.............TE, first page of each lesson
Lesson Quiz...................TE, last page of each lesson

Student Self-Assessment

Have students evaluate their own work.

Group Project Evaluation ....AR p. xxxiv
Individual Group Member Evaluation ....AR p. xxxv
Portfolio Guide............................AR p. xxxvii
Journal.......................TE, last page of each lesson

Formal Assessment

Assess students' mastery of concepts and skills.

Section Quizzes ....AR pp. 23–24
Mid-Chapter Quiz...............................SE p. 512
Chapter Test ....SE p. 535
Chapter Tests (Levels A, B, C) ....AR pp. 87–92
Cumulative Tests (Levels A, B, C) ....AR pp. 253–264
Standardized Test Prep
   Cumulative Assessment ....SE p. 537
End-of-Year Test..........................AR pp. 313–316

Technology

 Test and Practice Generator CD-ROM

Make tests electronically. This software includes:

- Dynamic practice for Chapter 10
- Customizable tests
- Multiple-choice items for each objective
- Free-response items for each objective
- Teacher management system

SE = Student Edition   **TE** = Teacher's Edition   **AR** = Assessment Resources   **CRB** = Chapter Resource Book   **MK** = Manipulatives Kit

Chapter 10 Tests

Three levels (A,B,C) of tests are available for each chapter in the *Assessment Resources.*

LEVEL A

CHAPTER 10 Chapter Test
Form A

Solve.

1. $4t + 5 = 13$ $t = 2$

2. $3h - 2 = 1$ $h = 1$

3. $-2.1a + 1.3 = 5.5$ $a = -2$

4. $\frac{x}{2} + 1 = 4$ $x = 6$

5. $5p + 2p - 3 = 11$ $p = 2$

6. $-3b - 6 + 7b = 6$ $b = 3$

7. $\frac{s}{3} - \frac{2}{3} = \frac{1}{3}$ $s = 3$

8. $\frac{x}{2} + \frac{x}{6} = 2$ $x = 3$

9. $x - 1 = x + 7$ no solution

10. $5t + 6 = 2t - 3$ $t = -3$

11. $4(g + 1) = 2g - 6$ $g = -5$

12. $\frac{a}{3} + \frac{a}{3} - 1 = a$ $a = -3$

Solve and graph.

13. $3c - 2 \geq 4$ $c \geq 2$

14. $-1 < 5z + 4$ $-1 < z$

15. $t - 3t - 1 > 7$ $t < -4$

Solve each equation for the indicated variable.

16. Solve $a + b + 3 = 5$ for b. $b = 2 - a$

17. Solve $P = 2(l + w)$ for l. $l = \frac{P - 2w}{2}$

18. Solve $d = r \cdot t$ for t. $t = \frac{d}{r}$

19. Solve $y = x^2$ for x. $x = \sqrt{y}$

Solve each equation for y and graph.

20. $y + 2x = 4$ $y = -2x + 4$

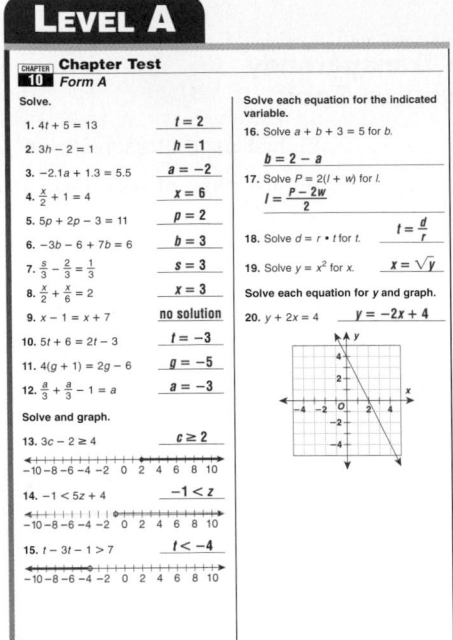

CHAPTER 10 Chapter Test
Form A, continued

21. $y - 3x = -2$ $y = 3x - 2$

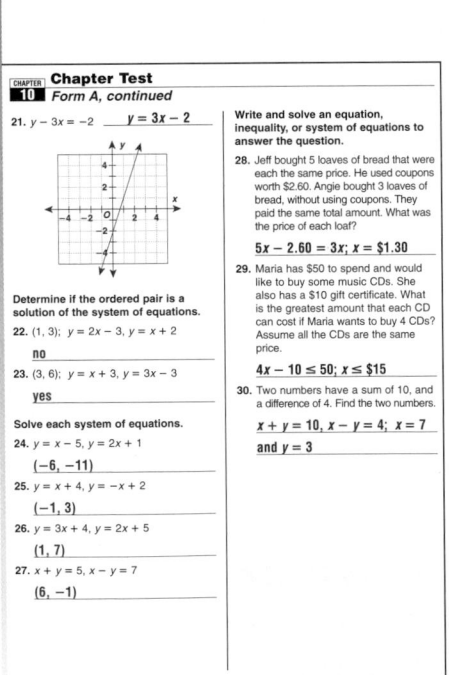

Determine if the ordered pair is a solution of the system of equations.

22. $(1, 3)$; $y = 2x - 3$, $y = x + 2$ no

23. $(3, 6)$; $y = x + 3$, $y = 3x - 3$ yes

Solve each system of equations.

24. $y = x - 5$, $y = 2x + 1$ $(-6, -11)$

25. $y = x + 4$, $y = -x + 2$ $(-1, 3)$

26. $y = 3x + 4$, $y = 2x + 5$ $(1, 7)$

27. $x + y = 5$, $x - y = 7$ $(6, -1)$

Write and solve an equation, inequality, or system of equations to answer the question.

28. Jeff bought 5 loaves of bread that were each the same price. He used coupons worth $2.60. Angie bought 3 loaves of bread, without using coupons. They paid the same total amount. What was the price of each loaf?

   $5x - 2.60 = 3x$; $x = \$1.30$

29. Maria has $50 to spend and would like to buy some music CDs. She also has a $10 gift certificate. What is the greatest amount that each CD can cost if Maria wants to buy 4 CDs? Assume all the CDs are the same price.

   $4x - 10 \leq 50$; $x \leq \$15$

30. Two numbers have a sum of 10, and a difference of 4. Find the two numbers.

   $x + y = 10$, $x - y = 4$; $x = 7$ and $y = 3$

LEVEL B

CHAPTER 10 Chapter Test
Form B

Solve.

1. $8t + 5 = 37$ $t = 4$

2. $\frac{-b}{2} - 10 = 5$ $b = -30$

3. $-5.9a - 5.5 = 12.2$ $a = -3$

4. $\frac{g + 3}{5} = 9$ $g = 42$

5. $-5p - 3p + 4 = 36$ $p = -4$

6. $\frac{4c}{5} - \frac{3c}{5} + \frac{1}{5} = \frac{-2}{5}$ $c = -3$

7. $\frac{x}{2} + \frac{5x}{6} = 4$ $x = 3$

8. $\frac{2y}{7} + \frac{4y}{7} - \frac{3}{14} = \frac{1}{14}$ $y = \frac{1}{3}$

9. $-5x - 9 = -2x + 6$ $x = -5$

10. $3t + 6 = 3t - 14$ no solution

11. $5(g - 3) = 2g + 3$ $g = 6$

12. $\frac{8a}{3} + \frac{4a}{3} - 1 = a + 5$ $a = 2$

Solve and graph.

13. $4d - 5 > 11$ $d > 4$

14. $-1 < \frac{k}{4} + \frac{1}{2} + \frac{k}{2}$ $-2 < k$

15. $-5s - 9 - 3s \geq 15$ $s \leq -3$

Solve each equation for the indicated variable.

16. Solve $P = a + b + c$ for b. $b = P - a - c$

17. Solve $A = \frac{1}{2} rp$ for p. $p = \frac{2A}{r}$

18. Solve $y = 2x + 3$ for x. $x = \frac{y - 3}{2}$

19. Solve $A = \pi r^2$ for r. $r = \sqrt{\frac{A}{\pi}}$

Solve each equation for y and graph.

20. $2y - 3x = 4$ $y = \frac{3x}{2} + 2$

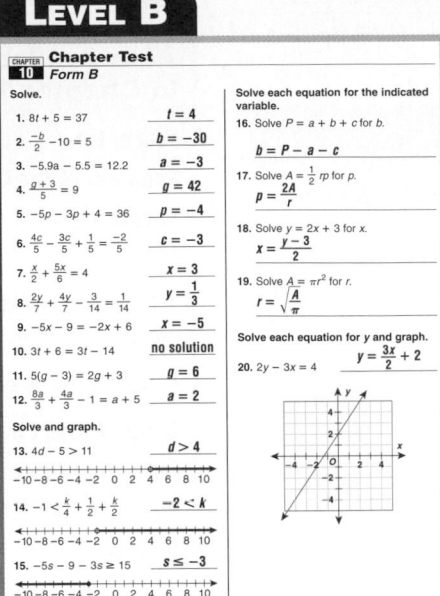

CHAPTER 10 Chapter Test
Form B, continued

21. $2y - 5x = 8$ $y = \frac{5x}{2} + 4$

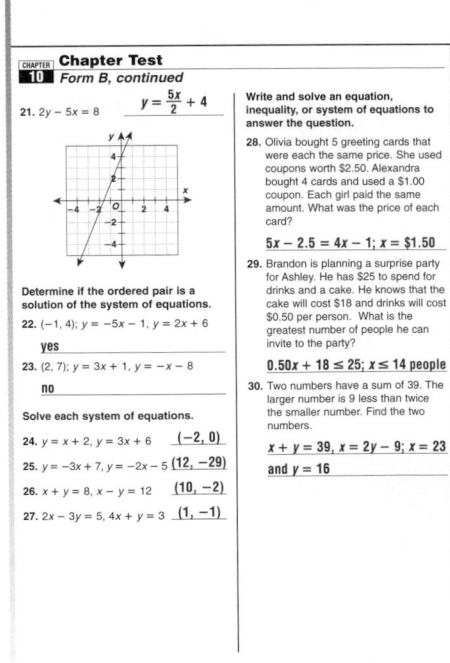

Determine if the ordered pair is a solution of the system of equations.

22. $(-1, 4)$; $y = -5x - 1$, $y = 2x + 6$ yes

23. $(2, 7)$; $y = 3x + 1$, $y = -x - 8$ no

Solve each system of equations.

24. $y = x + 2$, $y = 3x + 6$ $(-2, 0)$

25. $y = -3x + 7$, $y = -2x - 5$ $(12, -29)$

26. $x + y = 8$, $x - y = 12$ $(10, -2)$

27. $2x - 3y = 5$, $4x + y = 3$ $(1, -1)$

Write and solve an equation, inequality, or system of equations to answer the question.

28. Olivia bought 5 greeting cards that were each the same price. She used coupons worth $2.50. Alexandra bought 4 cards and used a $1.00 coupon. Each girl paid the same amount. What was the price of each card?

   $5x - 2.5 = 4x - 1$; $x = \$1.50$

29. Brandon is planning a surprise party for Ashley. He has $25 to spend for drinks and a cake. He knows that the cake will cost $18 and drinks will cost $0.50 per person. What is the greatest number of people he can invite to the party?

   $0.50x + 18 \leq 25$; $x \leq 14$ people

30. Two numbers have a sum of 39. The larger number is 9 less than twice the smaller number. Find the two numbers.

   $x + y = 39$, $x = 2y - 9$; $x = 23$ and $y = 16$

LEVEL C

CHAPTER 10 Chapter Test
Form C

Solve.

1. $\frac{3}{4} - \frac{5t}{6} = \frac{5}{12}$ $t = \frac{2}{5}$

2. $-6.8h + 15.3 = -39.1$ $h = 8$

3. $\frac{8 - a}{12} = -6$ $a = -64$

4. $\frac{12 + x}{5} + \frac{4}{5} = -20$ $x = -116$

5. $12p + 8 - 7p - 3 = -20$ $p = -5$

6. $\frac{3b}{4} - \frac{2}{3} + \frac{5b}{12} = -3$ $b = -2$

7. $\frac{2x}{27} - \frac{2}{9} + \frac{y}{3} = -\frac{5}{17}$ $y = \frac{3}{17}$

8. $\frac{x + 4}{3} - \frac{1}{5} + \frac{4x}{5} = \frac{1}{2}$ $x = \frac{1}{30}$

9. $2(x - 5) + \frac{3x}{4} = 4x$ $x = -8$

10. $-(t + 5) - 4 = -2(t - 6)$ $t = 21$

11. $3(4g + 1) + 5 = 12g - 7$ no solution

12. $-4\left(2a - \frac{1}{2}\right) + 5 = 2\left(4a - \frac{1}{2}\right)$ $a = \frac{1}{2}$

Solve and graph.

13. $\frac{3n}{11} - 2 > 4$ $n > 22$

14. $-\frac{x}{2} < \frac{x}{4} + \frac{3}{8}$ $x > -\frac{1}{2}$

15. $2\left(\frac{v}{8} + \frac{1}{4}\right) \leq \frac{v + 1}{6}$ $v \leq -4$

Solve each equation for the indicated variable.

16. Solve $C = \frac{5}{9}(F - 32)$ for F. $F = \frac{9}{5}C + 32$

17. Solve $A = \frac{1}{2}h(b_1 + b_2)$ for b_1. $b_1 = \frac{2A}{h} - b_2$

18. Solve $S = 6rs + 6sh$ for r. $r = \frac{S - 6sh}{6s}$

19. Solve $V = \frac{1}{3}\pi r^2 h$ for r. $r = \sqrt{\frac{3V}{\pi h}}$

CHAPTER 10 Chapter Test
Form C, continued

Solve each equation for y and graph.

20. $5y + 2x = 15$ $y = -\frac{2x}{5} + 3$

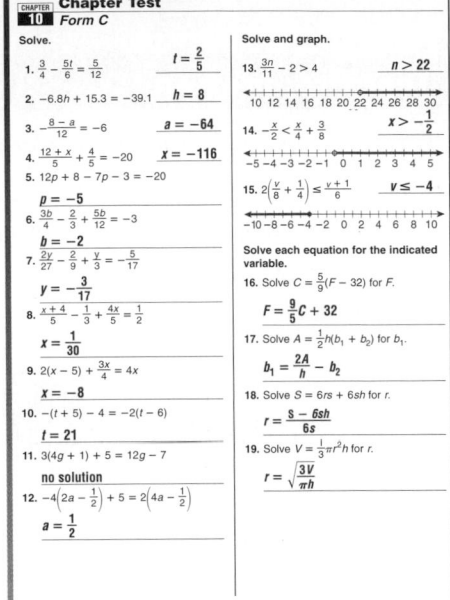

21. $y - 4 = -2(x + 3)$ $y = -2x - 2$

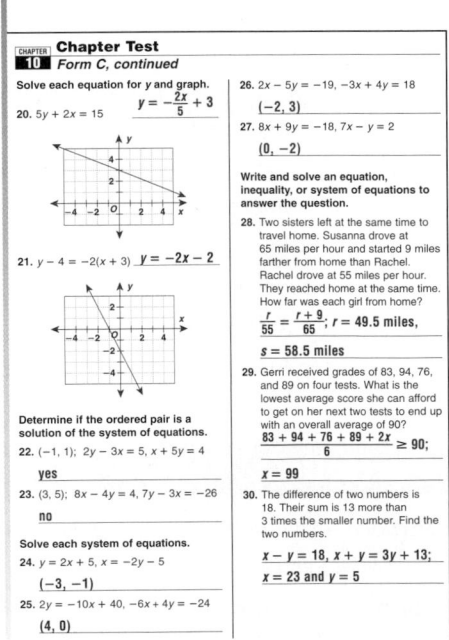

Determine if the ordered pair is a solution of the system of equations.

22. $(-1, 1)$; $2y - 3x = 5$, $x + 5y = 4$ yes

23. $(3, 5)$; $8x - 4y = 4$, $7y - 3x = -26$ no

Solve each system of equations.

24. $y = 2x + 5$, $x = -2y - 5$ $(-3, -1)$

25. $2y = -10x + 40$, $-6x + 4y = -24$ $(4, 0)$

26. $2x - 5y = -19$, $-3x + 4y = 18$ $(-2, 3)$

27. $8x + 9y = -18$, $7x - y = 2$ $(0, -2)$

Write and solve an equation, inequality, or system of equations to answer the question.

28. Two sisters left at the same time to travel home. Susanna drove at 65 miles per hour and started 9 miles farther from home than Rachel. Rachel drove at 55 miles per hour. They reached home at the same time. How far was each girl from home?

   $\frac{r}{55} = \frac{r + 9}{65}$; $r = 49.5$ miles, $s = 58.5$ miles

29. Gerri received grades of 83, 94, 76, and 89 on four tests. What is the lowest average score she can afford to get on her next two tests to end up with an overall average of 90?

   $\frac{83 + 94 + 76 + 89 + 2x}{6} \geq 90$; $x = 99$

30. The difference of two numbers is 18. Their sum is 13 more than 3 times the smaller number. Find the two numbers.

   $x - y = 18$, $x + y = 3y + 13$; $x = 23$ and $y = 5$

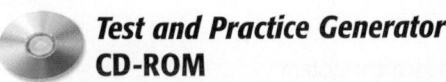

Test and Practice Generator
CD-ROM

Create and customize multiple versions of the same tests with corresponding answers for any chosen chapter objectives.

Chapter 10 State and Standardized Test Preparation

Test Taking Skill Builder and Standardized Test Practice
are provided for each chapter in the *Test Prep Tool Kit.*

TEST TAKING SKILL BUILDER

Test Taking Strategy
Chapter 10   Grid Accurately

Some questions on standardized tests require you to grid in your
solution on a grid. Be sure you know how to correctly grid your
solution.

Example

Grid-In Question Solve for *x*. $6x - 3 = 3x + 8$
Solution:
Solve for *x*.

$$6x - 3 = 3x + 8$$
$$6x - 3 - 3 = 3x - 3x + 8$$
$$3x - 3 = 8$$
$$3x - 3 + 3 = 8 + 3$$
$$3x = 11$$
$$x = 3\frac{2}{3}$$

The correct way to bubble $3\frac{2}{3}$ is to write the mixed number as
an improper fraction.

Multiply $3 \cdot 3$ then add 2 to find the numerator. The
denominator is 3.

$3\frac{2}{3} = \frac{11}{3}$

Two correct ways to bubble $\frac{11}{3}$ are shown at the right.

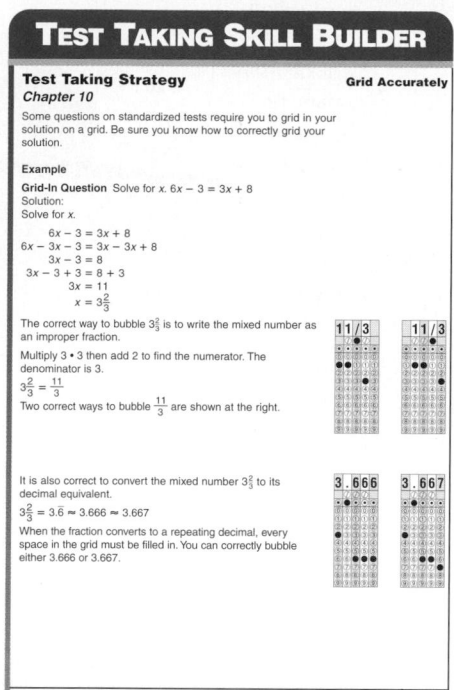

It is also correct to convert the mixed number $3\frac{2}{3}$ to its
decimal equivalent.

$3\frac{2}{3} = 3.6 \approx 3.666 \approx 3.667$

When the fraction converts to a repeating decimal, every
space in the grid must be filled in. You can correctly bubble
either 3.666 or 3.667.

Test Taking Strategy
Chapter 10, continued
Exercises

Grid each value. **Possible answers shown.**

1. $4\frac{4}{5}$ 2. $0.5\overline{5}$ 3. $9\frac{1}{2}$ 4. $2\frac{1}{3}$

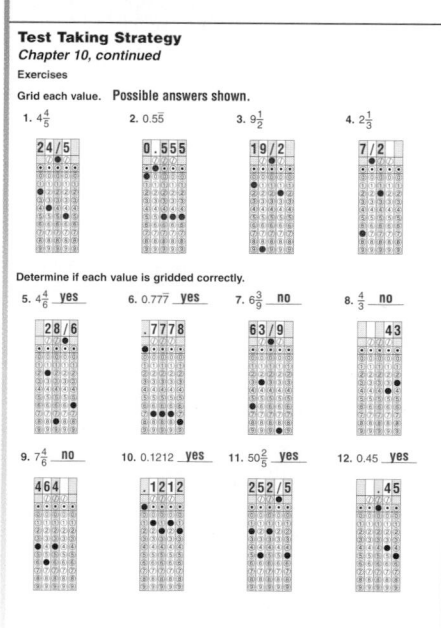

Determine if each value is gridded correctly.

5. $4\frac{4}{6}$ **yes**   6. $0.77\overline{7}$ **yes**   7. $6\frac{3}{9}$ **no**   8. $\frac{4}{3}$ **no**

9. $7\frac{4}{6}$ **no**   10. $0.12\overline{12}$ **yes**   11. $50\frac{2}{5}$ **yes**   12. 0.45 **yes**

STANDARDIZED TEST PRACTICE

Standardized Test Practice
Chapter 10

Select the best answer for Questions 1–8.

1. Solve. $37 = 9x + 1$
   A $x = \frac{38}{9}$
   B $x = 4$
   C $x = 5$
   D $x = 6$

2. There were 1678 people at a play
   before the curtain opened. At
   intermission there were 1945 people.
   How many people came after the
   curtain opened?
   F 3623 people
   G 1267 people
   H 267 people
   I 154 people

3. Solve for *b*. $a^2 + b^2 = c^2$
   A $b = c^2 - a^2$
   B $b = c^2 + a^2$
   C $b = \sqrt{c^2 - a^2}$
   D $b = \sqrt{a^2 - c^2}$

4. Which of the following would be the
   first step when solving the equation
   $\frac{4y}{7} + \frac{3}{7} = \frac{-1}{7}$?
   F Subtract $\frac{3}{4}$ from $\frac{4y}{7}$.
   G Multiply both sides by 7.
   H Add $\frac{1}{7}$ to both sides.
   I Add $\frac{3}{7}$ to $\frac{4y}{7}$.

5. Solve. $4x + 5 + 3x - 7 = 19$
   A $x = 2$
   B $x = 3$
   C $x = 4.43$
   D $x = 6$

6. Solve. $4x - 42 = -12 + x + 15$
   F $x = -13$
   G $x = 9$
   H $x = 13$
   I $x = 15$

7. The rectangle and isosceles triangle
   shown below have the dimensions
   shown and equal perimeters. How
   long is each side of the triangle?

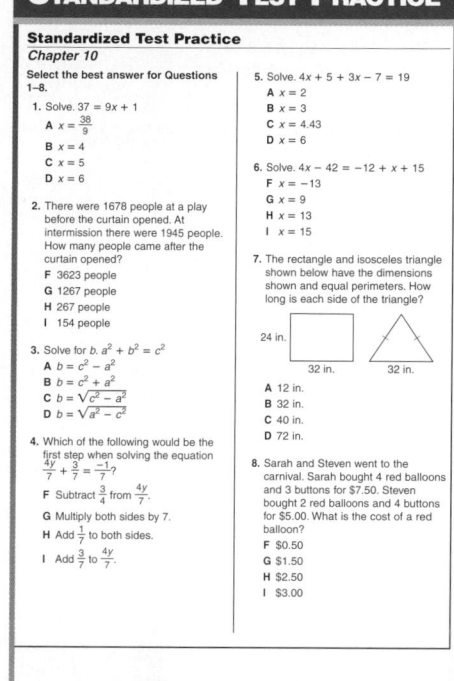

24 in.

32 in.   32 in.

   A 12 in.
   B 32 in.
   C 40 in.
   D 72 in.

8. Sarah and Steven went to the
   carnival. Sarah bought 4 red balloons
   and 3 buttons for $7.50. Steven
   bought 2 red balloons and 4 buttons
   for $5.00. What is the cost of a red
   balloon?
   F $0.50
   G $1.50
   H $2.50
   I $3.00

Standardized Test Practice
Chapter 10, continued

Gridded Response
Solve the problems. Use the answer
sheet to write and grid-in your answer.

9. Martin's automobile repair bill is
   $272.50. The parts cost $122 and the
   labor cost $43 per hour. How long did
   the auto technician work
   on Martin's car?

10. Larissa is paid 1.5 times her normal
    hourly rate for each hour of overtime
    she works. Her normal workweek is
    40 hours. Last week she worked
    46 hours and earned $845.25.
    What is her normal hourly rate?

Short Response
Solve the problems. Use the answer
sheet to write your answers.

11. Solve $12 - 8a < -12a - 16$ and
    graph your solution.

12. Determine if the ordered pair is a
    solution of the system of equations
    below. Explain in words how you
    determined your answer.

    $(4, -2)$
    $3x - 4y = 20$
    $x - 2y = 8$

Extended Response

13. Martin wants a 42-inch tree stump
    removed from their lawn. They
    called two different stump removal
    companies. Fresh Lawn charges
    $98 with no per inch charge. Spikes
    Stump Removal charges $65 and
    52 cents per inch of stump removal.

a. How many inches of stump could
   be removed before the Spikes
   price is less than Fresh Lawn?
   Show your work.

b. How much will Martin save by
   using the least expensive company?
   Show your work.

c. The total stump removal will take
   3.75 hours. What would a third
   company need to charge in per
   hour to be the least expensive
   company by $5? Explain in words
   how you determined your answer.

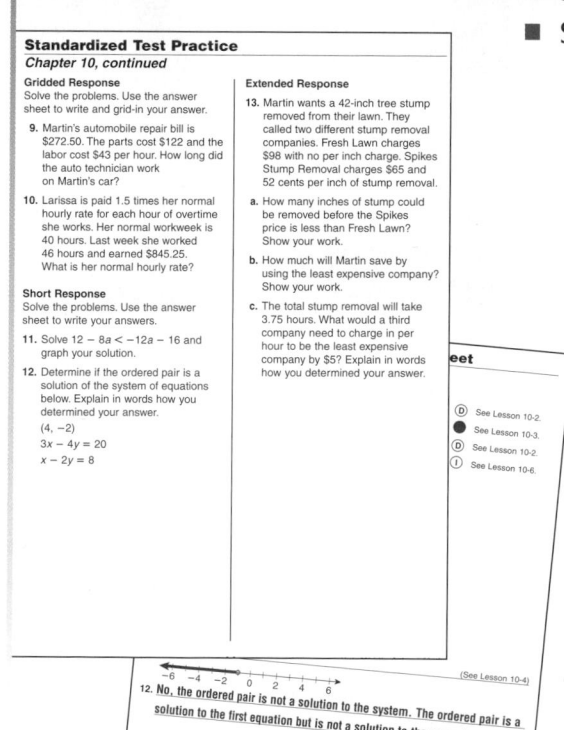

○Ⓓ See Lesson 10-2.
●   See Lesson 10-3.
○Ⓓ See Lesson 10-2.
○Ⓘ See Lesson 10-6.

-6 -4 -2   0   2   4   6   *(See Lesson 10-4)*

12. **No, the ordered pair is not a solution to the system. The ordered pair is a
solution to the first equation but is not a solution to the second equation.**

Extended Response   *(See Lesson 10-6)*
Write your answers for Problem 13 on the back of this paper.
See Lesson 10-4.

**State-Specific Test
Practice Online**
KEYWORD: MP4 TestPrep

Test Prep Tool Kit

- Standardized Test Prep Workbook
- Countdown to Testing transparencies
- State Test Prep CD-ROM
- Standardized Test Prep Video

Customized answer sheets give
students realistic practice for
actual standardized tests.

More Equations and Inequalities

Why Learn This?

Tell students that equations and inequalities can be used in many real-world applications. For example, a hydrologist might use an equation to determine the difference between two rivers' discharge rates. The discharge rate of each river is known. The difference is unknown and might be represented with the variable d. To find the difference between the discharge rates of the Colorado and Snake Rivers, the equation $726.04 = d + 314.6$ can be used.

Using Data

To begin the study of this chapter, have students:

- Write the rivers in order from greatest discharge rate to least discharge rate. Columbia, Missouri, Snake, Colorado

- Find the discharge rate in m³/h of the Snake River. 2,613,744 m³/h

- Determine which of the rivers' discharge rates is a solution to the inequality $d \geq 726$. Snake, Missouri, and Columbia Rivers

More Equations and Inequalities

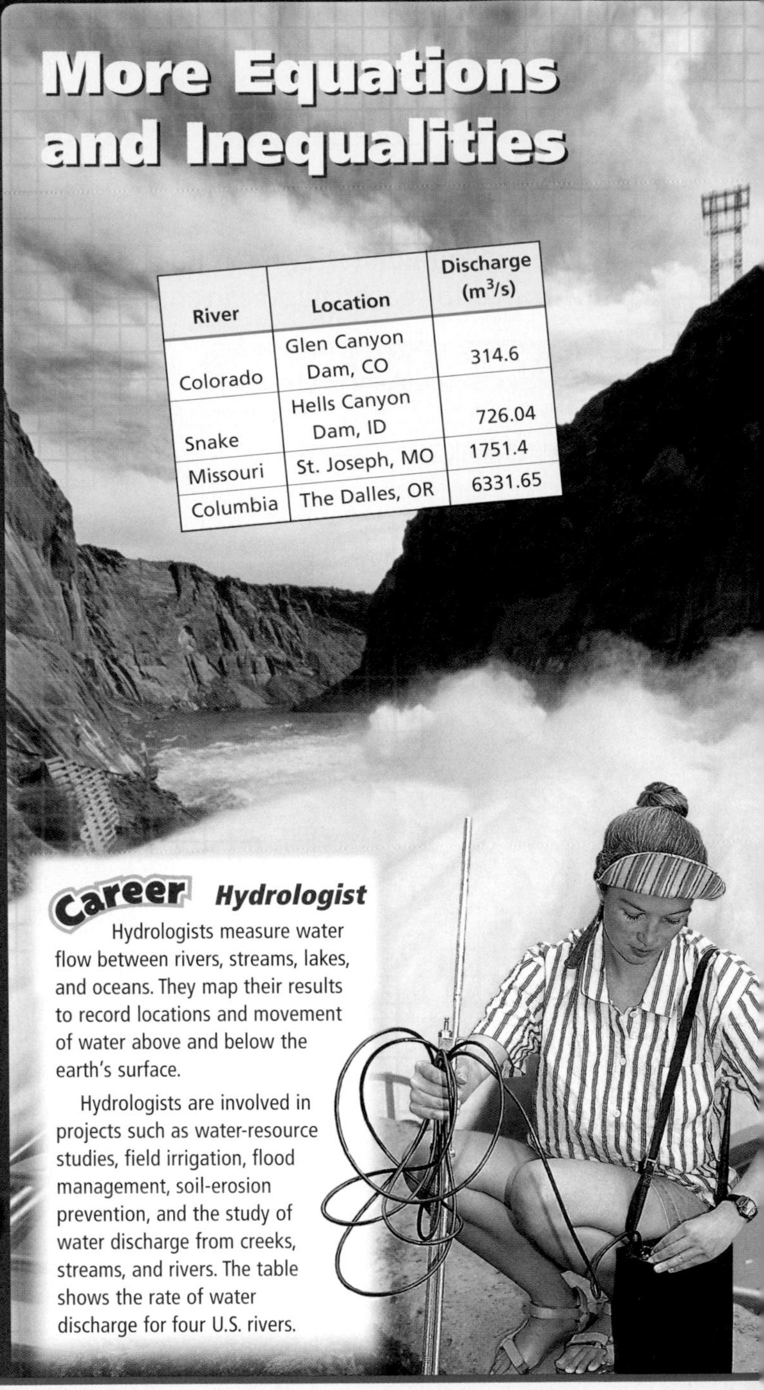

River	Location	Discharge (m³/s)
Colorado	Glen Canyon Dam, CO	314.6
Snake	Hells Canyon Dam, ID	726.04
Missouri	St. Joseph, MO	1751.4
Columbia	The Dalles, OR	6331.65

Career   *Hydrologist*

Hydrologists measure water flow between rivers, streams, lakes, and oceans. They map their results to record locations and movement of water above and below the earth's surface.

Hydrologists are involved in projects such as water-resource studies, field irrigation, flood management, soil-erosion prevention, and the study of water discharge from creeks, streams, and rivers. The table shows the rate of water discharge for four U.S. rivers.

internet connect
Chapter Opener Online
go.hrw.com
KEYWORD: MP4 Ch10

Problem Solving Project

Earth Science Connection

Purpose: To solve problems using equations
Materials: Water Flows Downhill worksheet

internet connect
Chapter Project Online: *go.hrw.com*
KEYWORD: MP4 PSProject10

Understand, Plan, Solve, and Look Back

Have students:

✔ Complete the Water Flows Downhill worksheet.

✔ Create an equation for calculating river discharge by relating water velocity (m/s) and stream cross section (m²).

✔ Draw a picture to visualize the equation, showing river discharge, water velocity, and stream cross section.

✔ Research to discover facts about a local river.

✔ Check students' work.

ARE YOU READY?

Choose the best term from the list to complete each sentence.

1. A letter that represents a value that can change is called a(n) __?__ . **variable**

2. A(n) __?__ has one or more variables. **algebraic expression**

3. The algebraic expression $5x^2 - 3y + 4x^2 + 7$ has four __?__ . Since they have the same variable raised to the same power, $5x^2$ and $4x^2$ are __?__ . **terms; like terms**

4. When you individually multiply the numbers inside parentheses by the factor outside the parentheses, you are applying the __?__ . **Distributive Property**

algebraic expression

Distributive Property

like terms

terms

variable

Complete these exercises to review skills you will need for this chapter.

✔ **Distribute Multiplication**

Replace each ▨ with a number so that each equation illustrates the Distributive Property.

5. $6 \cdot (11 + 8) = 6 \cdot 11 + 6 \cdot \blacksquare$ **8**
6. $7 \cdot (14 + 12) = \blacksquare \cdot 14 + \blacksquare \cdot 12$ **7**
7. $9 \cdot (6 - \blacksquare) = 9 \cdot 6 - 9 \cdot 2$ **2**
8. $14 \cdot (\blacksquare - 7) = 14 \cdot 20 - 14 \cdot 7$ **20**

✔ **Simplify Algebraic Expressions**

Simplify each expression by applying the Distributive Property and combining like terms.

9. $3(x + 2) + 7x$ **$10x + 6$**
10. $4(y - 3) + 8y$ **$12y - 12$**
11. $2(z - 1) - 3z$ **$-z - 2$**
12. $-4(t - 6) - t$ **$-5t + 24$**
13. $-(r - 3) - 8r$ **$-9r + 3$**
14. $-5(4 - 2m) + 7$ **$10m - 13$**

✔ **Connect Words and Equations**

Write an equation to represent each situation.

15. The perimeter P of a rectangle is the sum of twice the length ℓ and twice the width w.   **$P = 2\ell + 2w$**

16. The volume V of a rectangular prism is the product of its three dimensions: length ℓ, width w, and height h.   **$V = \ell wh$**

17. The surface area S of a sphere is the product of 4π and the square of the radius r.   **$S = 4\pi r^2$**

18. The cost c of a telegram of 18 words is the cost f of the first 10 words plus the cost a of each additional word.   **$c = f + 8a$**

Assessing Prior Knowledge

INTERVENTION

Diagnose and Prescribe

Evaluate your students' performance on this page to determine whether intervention is necessary or whether enrichment is appropriate. Options that provide instruction, practice, and a check are listed below.

Resources for Are You Ready?

- *Are You Ready? Intervention and Enrichment*
- **Recording Sheet for Are You Ready?**
  Chapter 10 Resource Book . . . . . p. 3

 Are You Ready? Intervention **CD-ROM**

ARE YOU READY?
Were students successful with Are You Ready?

 NO INTERVENE   **YES** ENRICH

✔**Distribute Multiplication** 5–8
Intervention Practice, Skill 49
 CD-ROM
Intervention Activities, Skill 49

✔**Simplify Algebraic Expressions** 9–14
Intervention Practice, Skill 55
CD-ROM
Intervention Activities, Skill 55

✔**Connect Words and Equations** 15–18
Intervention Practice, Skill 56
CD-ROM
Intervention Activities, Skill 56

Are You Ready? Enrichment, pp. 423–424

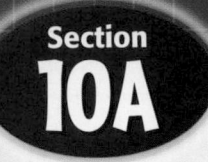

Section 10A

Solving Linear Equations

One-Minute Section Planner

Lesson	Materials	Resources

Lesson 10-1 Solving Two-Step Equations
NCTM: Number and Operations, Algebra, Problem Solving, Communication, Connections
NAEP: Algebra 4a
☐ SAT-9 ☐ SAT-10 ☐ ITBS ☐ CTBS ☐ MAT ☑ CAT
	Optional
	Wrapped gift
	Recording Sheet for
	  Reaching All Learners
	  (CRB, p. 72)
		• *Chapter 10 Resource Book*, pp. 6–15
		• Daily Transparency T1, CRB
		• Additional Examples Transparencies
		  T2–T5, CRB
		• *Alternate Openers: Explorations*, p. 79

Lesson 10-2 Solving Multistep Equations
NCTM: Number and Operations, Algebra, Communication, Connections
NAEP: Algebra 4a
☐ SAT-9 ☐ SAT-10 ☐ ITBS ☐ CTBS ☐ MAT ☑ CAT
	Optional
	Algebra tiles *(MK)*
	Equation tic-tac-toe
	  (CRB, p. 73)
		• *Chapter 10 Resource Book*, pp. 16–25
		• Daily Transparency T6, CRB
		• Additional Examples Transparencies
		  T7–T10, CRB
		• *Alternate Openers: Explorations*, p. 80

Hands-On Lab 10A Model Equations with Variables on Both Sides **NCTM:** Number and Operations, Algebra, Representation **NAEP:** Algebra 4a ☐ SAT-9 ☐ SAT-10 ☐ ITBS ☐ CTBS ☐ MAT ☐ CAT	**Required** Algebra tiles *(MK)*	• *Hands-On Lab Activities*, pp. 73–74, 93

Lesson 10-3 Solving Equations with Variables on Both Sides
NCTM: Algebra, Communication
NAEP: Algebra 4a
☐ SAT-9 ☐ SAT-10 ☐ ITBS ☐ CTBS ☐ MAT ☑ CAT
	Optional
	Algebra tiles *(MK)*
	Cards for Reaching All
	  Learners (CRB, p. 74)
		• *Chapter 10 Resource Book*, pp. 26–35
		• Daily Transparency T11, CRB
		• Additional Examples Transparencies
		  T12–T16, CRB
		• *Alternate Openers: Explorations*, p. 81

Section 10A Assessment		• Mid-Chapter Quiz, SE p. 512 • Section 10A Quiz, AR p. 23 • *Test and Practice Generator* CD-ROM

SAT = *Stanford Achievement Tests*   **ITBS** = *Iowa Test of Basic Skills*   **CTBS** = *Comprehensive Test of Basic Skills/Terra Nova*
MAT = *Metropolitan Achievement Test*   **CAT** = *California Achievement Test*

NCTM = Complete standards can be found on pages T29–T35.   **NAEP** = Complete standards can be found on pages A54–A58.

SE = *Student Edition*   **TE** = *Teacher's Edition*   **AR** = *Assessment Resources*   **CRB** = *Chapter Resource Book*   **MK** = *Manipulatives Kit*

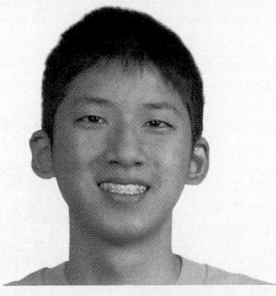

$$4(x + 3) - 2x = 16$$
$$4x + 12 - 2x = 16$$
$$2x + 12 = 16$$
$$2x = 4$$
$$x = 2$$

Section Overview

Two-Step Equations — Lesson 10-1

Why? You can solve some problems by using two-step equations.

Jill's auto repair bill was $225. The parts cost $95 and the labor cost $52 per hour. For how many hours of labor was Jill charged?

Solve

total bill = parts + labor

$$\begin{aligned} 225 &= 95 + 52h \\ -95 & \quad -95 \\ 130 &= 52h \\ \frac{130}{52} &= \frac{52h}{52} \\ 2.5 &= h \end{aligned}$$

Jill was charged for 2.5 hours of labor.

Check

$$\begin{aligned} 225 &= 95 + 52h \\ 225 &\stackrel{?}{=} 95 + 52(2.5) \\ 225 &\stackrel{?}{=} 95 + 130 \\ 225 &\stackrel{?}{=} 225 \ ✔ \end{aligned}$$

MultiStep Equations — Lesson 10-2

Why? Some problems require equations that have more than two steps.

Jack had a $5 gift certificate for a restaurant. After a 15% tip was added to the bill, the $5 was deducted. Jack actually paid $18. What was the original bill before the tip was added?

Solve

$$\begin{aligned} b + 0.15b - 5 &= 18 \\ 1.15b - 5 &= 18 \\ +5 & \quad +5 \\ 1.15b &= 23 \\ \frac{1.15b}{1.15b} &= \frac{23}{1.15} \\ b &= 20 \end{aligned}$$

The original bill was $20.

Check

$$\begin{aligned} b + 0.15b - 5 &= 18 \\ 20 + 0.15(20) - 5 &\stackrel{?}{=} 18 \\ 20 + 3 - 5 &\stackrel{?}{=} 18 \\ 18 &\stackrel{?}{=} 18 \ ✔ \end{aligned}$$

Equations with Variables on Both Sides — Hands-On Lab 10A, Lesson 10-3

Why? A problem may require an equation that has a variable on both sides of the equal sign.

The members of a book club spend the same amount for refreshments at each meeting. At one meeting they bought 6 bagels and spent $9.90 on beverages. At the next meeting they bought 8 bagels and spent $8.20 on beverages. What was the cost of each bagel?

Solve

$$\begin{aligned} 6x + 9.90 &= 8x + 8.20 \\ -6x & \quad -6x \\ 9.90 &= 2x + 8.20 \\ -8.20 & \quad -8.20 \\ 1.70 &= 2x \\ \frac{1.70}{2} &= \frac{2x}{2} \\ 0.85 &= x \end{aligned}$$

The cost of each bagel was $0.85.

Check

$$\begin{aligned} 6x + 9.90 &= 8x + 8.20 \\ 6(0.85) + 9.90 &\stackrel{?}{=} 8(0.85) + 8.20 \\ 5.10 + 9.90 &\stackrel{?}{=} 6.80 + 8.20 \\ 15 &\stackrel{?}{=} 15 \ ✔ \end{aligned}$$

Warm Up

Solve.

1. $x + 12 = 35$ $x = 23$
2. $8x = 120$ $x = 15$
3. $\frac{y}{9} = 7$ $y = 63$
4. $-34 = y + 56$ $y = -90$

Problem of the Day

x is an odd integer. If you triple x and
then subtract 7, you get a prime
number. What is x? (*Hint:* Think
about what the prime number must
be in order for x to be odd.) $x = 3$

Available on Daily Transparency in CRB

Math Humor

Several surgeries only made the
author's condition worse. So he wrote
his new novel from the end to the
beginning. He hoped to undo the
operations by working backward.

10-1 Solving Two-Step Equations

Learn to solve
two-step equations.

Sometimes more than one inverse operation
is needed to solve an equation. Before
solving, ask yourself, "What is being done
to the variable, and in what order?" Then
work backward to undo the operations.

Landscapers charge an hourly rate for labor,
plus the cost of the plants. The number of
hours a landscaper worked can be found by
solving a two-step equation.

EXAMPLE **1**  **PROBLEM SOLVING APPLICATION**

Chris's landscaping bill is $380. The plants cost $212, and the labor
cost $48 per hour. How many hours did the landscaper work?

1 **Understand the Problem**

The **answer** is the number of hours the landscaper worked on the yard.
List the **important information:** The plants cost $212, the labor cost
$48 per hour, and the total bill is $380.

Let h represent the hours the landscaper worked.

Total bill	=	Plants	+	Labor
380	=	212	+	$48h$

2 **Make a Plan**

Think: First the variable is multiplied by 48, and then 212 is added to
the result. Work backward to solve the equation. Undo the operations
in reverse order: First subtract 212 from both sides of the equation,
and then divide both sides of the new equation by 48.

3 **Solve**

$$380 = 212 + 48h$$
$$\underline{-212 \quad -212} \qquad \textit{Subtract to undo addition.}$$
$$168 = 48h$$

$$\frac{168}{48} = \frac{48h}{48} \qquad \textit{Divide to undo multiplication.}$$
$$3.5 = h$$

The landscaper worked 3.5 hours.

4 **Look Back**

If the landscaper worked 3.5 hours, the labor would be $48(3.5) = $168.
The sum of the plants and the labor would be $212 + $168 = $380.

---

1 Introduce

Alternate Opener

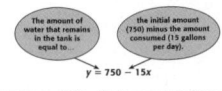
Motivate

Show the students a wrapped gift. You may
want to wrap a small box in front of the
class. Demonstrate that to open it, the ribbon is removed, then the wrapping paper is
removed, and then the box is opened. Ask
the students to think about the steps that
were taken when the gift was wrapped.
Point out that to unwrap the gift, the steps
must be reversed.

*Exploration worksheet and answers on
Chapter 10 Resource Book pp. 7 and 78*

2 Teach

Lesson Presentation

Guided Instruction

In this lesson, students learn to solve
two-step equations. Remind students of the
order of operations for evaluating expressions.
Tell students that to solve an equation, the
order of operations must be reversed. Point
out that they should usually undo addition
and subtraction first and then undo multiplication and division.

Teaching Tip

Remind students to check solutions by substituting the value
into the original equation and
evaluating to make sure both sides are equal.

EXAMPLE 2 **Solving Two-Step Equations**

Solve.

A $\frac{p}{4} + 5 = 13$

Think: First the variable is **divided by 4,** and then **5** is added. To isolate the variable, **subtract 5,** and then **multiply by 4.**

$$\begin{array}{rcl} \frac{p}{4} + 5 &=& 13 \\ \underline{-5} && \underline{-5} \\ \frac{p}{4} &=& 8 \end{array}$$ *Subtract to undo addition.*

$4 \cdot \frac{p}{4} = 4 \cdot 8$ *Multiply to undo division.*

$p = 32$

Check $\frac{p}{4} + 5 \stackrel{?}{=} 13$

$\frac{32}{4} + 5 \stackrel{?}{=} 13$ *Substitute 32 into the original equation.*

$8 + 5 \stackrel{?}{=} 13$ ✔

B $1.8 = -2.5m - 1.7$

Think: First the variable is **multiplied by −2.5,** and then **1.7 is subtracted.** To isolate the variable, **add 1.7,** and then **divide by −2.5.**

$$\begin{array}{rcl} 1.8 &=& -2.5m - 1.7 \\ \underline{+1.7} && \underline{+1.7} \\ 3.5 &=& -2.5m \end{array}$$ *Add to undo subtraction.*

$\frac{3.5}{-2.5} = \frac{-2.5m}{-2.5}$ *Divide to undo multiplication.*

$-1.4 = m$

C $\frac{k+4}{9} = 6$

Think: First **4 is added** to the variable, and then the result is **divided by 9.** To isolate the variable, **multiply by 9,** and then **subtract 4.**

$$\frac{k+4}{9} = 6$$

$9 \cdot \frac{k+4}{9} = 9 \cdot 6$ *Multiply to undo division.*

$$\begin{array}{rcl} k + 4 &=& 54 \\ \underline{-4} && \underline{-4} \\ k &=& 50 \end{array}$$ *Subtract to undo addition.*

Think and Discuss

1. Describe how you would solve $4(x - 2) = 16$.

COMMON ERROR ALERT

In two-step equations, some students may try to multiply or divide first and forget to apply the operation to every term. For example, a student may try to solve Example 1 by multiplying each side of the equation by 4.

$$4 \cdot \frac{p}{4} + 5 = 13 \cdot 4$$
$$p + 5 = 52$$
$$p = 47$$

Because the student did not multiply the middle term by 4, the answer is incorrect. Encourage students to perform addition or subtraction first to avoid this problem.

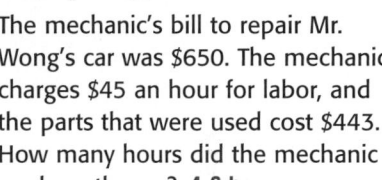

Additional Examples

Example 1

The mechanic's bill to repair Mr. Wong's car was $650. The mechanic charges $45 an hour for labor, and the parts that were used cost $443. How many hours did the mechanic work on the car? **4.6 hr**

Example 2

Solve.

A. $\frac{n}{3} + 7 = 22$ $n = 45$

B. $2.7 = -1.3m + 6.6$ $3 = m$

C. $\frac{y - 4}{3} = 9$ $y = 31$

---

3 Close

Reaching All Learners
Through Grouping Strategies

Group students in circles of three. Give each group a recording sheet that contains three real-world problems that can be solved by writing and solving two-step equations (Chapter 10 Resource Book p. 72). Each student takes one problem and writes an equation for it. Then the student passes the problem to the student on his or her left, who solves the equation. Then it is passed again to the next student, who checks the solution. Finally, the problems are returned to the original student, who reviews the work.

Summarize

Remind students to reverse the order of operations to solve equations. Point out that addition and subtraction are not always the operations that should be undone first. For an equation such as the one in Example 2C, division should be undone first.

Answers to Think and Discuss

1. Divide both sides by 4, and then add 2 to both sides.

10-1 Exercises

FOR EXTRA PRACTICE
see page 750

☑ internet connect
Homework Help Online
go.hrw.com Keyword: MP4 10-1

Students may want to refer back to the lesson examples.

Assignment Guide

If you finished Example **1** assign:
  Core 1, 8, 30–32, 37–41
  Enriched 1, 8, 30–32, 37–41

If you finished Example **2** assign:
  Core 1–20, 21–29 odd, 30–33, 37–41
  Enriched 8–41

Notes

GUIDED PRACTICE

See Example **1** **1.** Joe is paid a weekly salary of $520. He is paid an additional $21 for every hour of overtime he works. This week his total pay, including regular salary and overtime pay, was $604. How many hours of overtime did Joe work this week? **4 hr**

See Example **2** Solve.

2. $9t + 12 = 75$ $t = 7$ **3.** $-2.4 = -1.2x + 1.8$ $x = 3.5$ **4.** $\frac{r}{7} + 11 = 25$ $r = 98$

5. $\frac{b + 24}{2} = 13$ $b = 2$ **6.** $14q - 17 = 39$ $q = 4$ **7.** $\frac{a - 3}{28} = 3$ $a = 87$

INDEPENDENT PRACTICE

See Example **1** **8.** The cost of a family membership at a health club is $58 per month plus a one-time $129 start-up fee. If a family spent $651, how many months is their membership? **9 mo**

See Example **2** Solve.

9. $\frac{m}{-3} - 2 = 8$ $m = -30$ **10.** $\frac{c - 1}{2} = 12$ $c = 25$ **11.** $15g - 4 = 46$ $g = 3\frac{1}{3}$

12. $\frac{h + 19}{19} = 2$ $h = 19$ **13.** $6y + 3 = -27$ $y = -5$ **14.** $9.2 = 4.4z - 4$ $z = 3$

PRACTICE AND PROBLEM SOLVING

Solve.

15. $5w + 3.8 = 16.3$ $w = 2.5$ **16.** $15 - 3x = -6$ $x = 7$ **17.** $\frac{m}{5} + 6 = 9$ $m = 15$

18. $2.3a + 8.6 = -5.2$ $a = -6$ **19.** $\frac{q + 4}{7} = 1$ $q = 3$ **20.** $9 = -5g - 23$ $g = -6.4$

21. $6z - 2 = 0$ $z = \frac{1}{3}$ **22.** $\frac{5}{2}d - \frac{3}{2} = -\frac{1}{2}$ $d = \frac{2}{5}$ **23.** $47k + 83 = 318$ $k = 5$

24. $8 = 6 + \frac{p}{4}$ $p = 8$ **25.** $46 - 3n = -23$ $n = 23$ **26.** $\frac{7 + s}{5} = -4$ $s = -27$

27. $9y - 7.2 = 4.5$ $y = 1.3$ **28.** $\frac{2}{3} - 6h = -\frac{11}{6}$ $h = \frac{5}{12}$ **29.** $-1 = \frac{3}{5}b + \frac{1}{5}$ $b = -2$

Write an equation for each sentence, and then solve it.

30. The quotient of a number and 2, minus 9, is 14. $\frac{x}{2} - 9 = 14$; $x = 46$

31. A number increased by 5 and then divided by 7 is 12. $\frac{x + 5}{7} = 12$; $x = 79$

32. The sum of 10 and 5 times a number is 25. $10 + 5x = 25$; $x = 3$

Math Background

The method of reversing the order of operations suggested in the lesson is not the only possible method for solving equations. Some equations may be solved with different steps in different orders as long as algebraic properties are applied correctly. For example, the equation in Think and Discuss can be solved in at least two different ways:

$$4(x - 2) = 16$$
$$4x - 8 = 16$$
$$\underline{+ 8 \quad + 8}$$
$$4x = 24$$
$$\frac{4x}{4} = \frac{24}{4}$$
$$x = 6$$

$$4(x - 2) = 16$$
$$\frac{4(x - 2)}{4} = \frac{16}{4}$$
$$x - 2 = 4$$
$$\underline{+ 2 + 2}$$
$$x = 6$$

RETEACH 10-1

LESSON 10-1 Reteach
Solving Two-Step Equations

To solve an equation, it is important to first note how it is formed.
Then, work backward to undo each operation.

$4z + 3 = 15$	$\frac{z}{4} - 3 = 7$	$\frac{z + 3}{4} = 7$
The variable is multiplied by 4 and then 3 is added.	The variable is divided by 4 and then 3 is subtracted.	3 is added to the variable and then the result is divided by 4.
To solve, first subtract 3 and then divide by 4.	To solve, first add 3 and then multiply by 4.	To solve, multiply by 4 and then subtract 3.

Describe how each equation is formed.
Then, tell the steps needed to solve.

1. $3x - 5 = 7$
  The variable is __multiplied by 3__ and then __5 is subtracted__
  To solve, first __add 5__ and then __divide by 3__

2. $\frac{x}{3} + 5 = 7$
  The variable is __divided by 3__ and then __5 is added__
  To solve, first __subtract 5__ and then __multiply by 3__

3. $\frac{x + 5}{3} = 7$
  5 is __added to the variable__ and then the result is __divided by 3__
  To solve, first __multiply by 3__ and then __subtract 5__

4. $10 = -3x - 2$
  The variable is __multiplied by -3__ and then __2 is subtracted__
  To solve, first __add 2__ and then __divide by -3__

5. $10 = \frac{x - 2}{5}$
  2 is __subtracted from__ the variable and then the result is __divided by 5__
  To solve, first __multiply by 5__ and then __add 2__

PRACTICE 10-1

LESSON 10-1 Practice B
Solving Two-Step Equations

Write and solve a two-step equation to answer the following questions.

1. The school purchased baseball equipment and uniforms for a total cost of $1762. The equipment costs $598 and the uniforms were $24.25 each. How many uniforms did the school purchase?

  $x = $ # of uniforms
  $1762 = 598 + 24.25x$
  $1762 - 598$
  $= 598 - 598 + 24.25x$
  $1164 = 24.25x$
  $\frac{1164}{24.25} = \frac{24.25x}{24.25}$
  $48 = x$

2. Carla runs 4 miles every day. She jogs from home to the school track, which is $\frac{3}{4}$ mile away. She then runs laps around the $\frac{1}{4}$-mile track. Carla then jogs home. How many laps does she run at the school?

  $x = $ # of laps
  $4 = \frac{3}{4} + \frac{1}{4}x + \frac{3}{4}$
  $4 - \frac{6}{4} = \frac{6}{4} - \frac{6}{4} + \frac{1}{4}x$
  $\frac{5}{2} = \frac{1}{4}x$
  $4(\frac{5}{2}) = (\frac{1}{4}x)4$
  $10 = x$

Solve.

3. $-\frac{1}{3}x + 2 = -2$ $x = 12$ 4. $\frac{-t}{-2} - 1 = 4$ $t = -10$ 5. $9 = 2x + 1$ $x = 4$ 6. $3 - 4k = -1$ $k = 1$

7. $0.5x - 6 = -4$ $x = 4$ 8. $\frac{x}{2} + 3 = -4$ $x = -14$ 9. $\frac{1}{5}n + 3 = 6$ $n = 15$ 10. $2a - 7 = -9$ $a = -1$

11. $\frac{3x - 1}{4} = 2$ $x = 3$ 12. $-7.8 = 4.4 + 2r$ $r = -6.1$ 13. $\frac{-4w + 5}{-3} = -7$ $w = -4$ 14. $5 - \frac{r}{1.3} = 7.4$ $r = -3.12$

15. A phone call costs $0.58 for the first 3 minutes and $0.15 for each additional minute. If the total charge for the call was $4.78, how many minutes was the call? **31 minutes**

16. Seventeen less than four times a number is twenty-seven. Find the number. **11**

About 20% of the more than 2500 species of snakes are venomous. The United States has 20 domestic venomous snake species, including coral snakes, rattlesnakes, copperheads, and cottonmouths.

33. The inland taipan of central Australia is the world's most toxic venomous snake. Just 1 mg of its venom is enough to kill 1000 mice. One bite contains up to 110 mg of venom. About how many mice could be killed with the venom contained in just one inland taipan bite? **110,000**

34. A rattlesnake grows a new rattle segment each time it sheds its skin. Rattlesnakes shed their skin an average of three times per year. However, segments often break off. If a rattlesnake had 44 rattle segments break off in its lifetime and it had 10 rattles when it died, approximately how many years did the rattlesnake live? **18 yr**

35. All snakes shed their skin as they grow. The shed skin of a snake is an average of 10% longer than the actual snake. If the shed skin of a coral snake is 27.5 inches long, estimate the length of the coral snake. **25 in.**

36. ⭐ **CHALLENGE** Black mambas feed mainly on small rodents and birds. Suppose a black mamba is 100 feet away from an animal that is running at 8 mi/h. About how long will it take for the mamba to catch the animal? (*Hint*: 1 mile = 5280 feet) **17 s**

go.hrw.com
KEYWORD: MP4 Snakes
CNN student News.

Venom is collected from snakes and injected into horses, which develop antibodies. The horses' blood is sterilized to make antivenom.

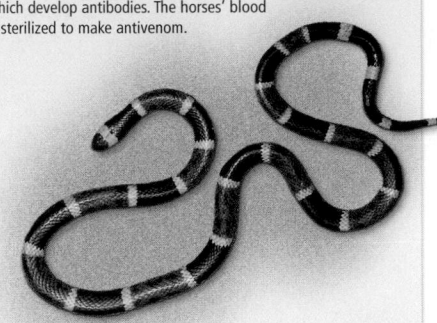

Records of World's Most Venomous Snakes		
Category	Record	Type of Snake
Fastest	12 mi/h	Black mamba
Longest	18 ft 9 in.	King cobra
Heaviest	34 lb	Eastern diamondback rattlesnake
Longest fangs	2 in.	Gaboon viper

Spiral Review

Simplify. (Lesson 1-6)

37. $x + 4x + 3 + 7x$
$12x + 3$

38. $-2m + 4 + 2m$ **4**

39. $w - 17 + 2$
$w - 15$

40. $5s + 3r + s - 5r$
$6s - 2r$

41. **TEST PREP** Find the area of the parallelogram. (Lesson 6-1) **C**

24 cm
14 cm | 12 cm

A   38 cm²
B   76 cm²
C   288 cm²
D   336 cm²

Life Science

Exercises 33–36 involve using data about venomous snakes to solve two-step equations. Snakes and other reptiles are studied in middle school life science programs, such as *Holt Science & Technology*.

Journal

Ask students to write about how they would explain the process of solving two-step equations to a friend who has not seen them yet.

Test Prep Doctor ➕

For Exercise 41, remind students that the formula for finding the area of a parallelogram is $A = bh$. Point out that the length 14 cm is not needed. The area is $12 \cdot 24 = 288$ cm². The correct answer is **C**. Students who chose **A** may have added 14 and 24. Students who chose **B** may have found the perimeter. Students who chose **D** may have multiplied 14 and 24.

CHALLENGE 10-1

LESSON 10-1 Challenge
Work It Algebraically!

An equation may be used to solve a problem involving probability.

A bag contains marbles of four colors: red, white, blue, and yellow. There are 3 more blue marbles than red, and 48 marbles in all. How many blue marbles are there if, in one draw, the probability of getting a blue marble is $\frac{5}{12}$?

Let x = the number of blue marbles.

$P(\text{blue}) = \frac{\text{number of successes}}{\text{total number}}$

$\frac{5}{12} = \frac{x}{48}$

$\frac{5}{12} \cdot 48 = 48 \cdot \frac{x}{48}$ Multiply by 48.

$20 = x$

So, there are 20 blue marbles in the bag.

Write and solve an equation for each problem.

1. A box has four kinds of candies: lime, orange, cherry, and mint. There are 9 more lime candies than orange, and 36 candies in all. How many lime candies are there if, in one draw, the probability of getting a lime candy is $\frac{4}{9}$?

Let x = number of lime candies.

$\frac{x}{36} = \frac{4}{9}$

$36 \cdot \frac{x}{36} = \frac{4}{9} \cdot 36$

$x = 16$

There are __16__ lime candies.

2. A carton has four kinds of cookies: lemon, mint, vanilla, and chocolate. There are 7 fewer mint cookies than lemon, and 64 cookies in all. How many mint cookies are there if, in one draw, the probability of getting a lemon cookie is $\frac{5}{8}$?

Let x = number of lemon cookies.

Then $x - 7$ = number of mint cookies.

$\frac{x}{64} = \frac{5}{8}$

$64 \cdot \frac{x}{64} = \frac{5}{8} \cdot 64$

$x = 24$ ⟵ lemon cookies

There are __17__ mint cookies.

PROBLEM SOLVING 10-1

LESSON 10-1 Problem Solving
Solving Two-Step Equations

The chart below describes three different long distance calling plans. Jamie has budgeted $20 per month for long distance calls. Write the correct answer.

1. How many minutes will Jamie be able to use per month with plan A? Round to the nearest minute.
201 min

Plan	Monthly Access Fee	Charge per minute
A	$3.95	$0.08
B	$8.95	$0.06
C	$0	$0.10

2. How many minutes will Jamie be able to use per month with plan B? Round to the nearest minute.
184 min

3. How many minutes will Jamie be able to use per month with plan C? Round to the nearest minute.
200 min

4. Which plan is the best deal for Jamie's budget?
Plan A

5. Nolan has budgeted $50 per month for long distance. Which plan is the best deal for Nolan's budget?
Plan B

The table describes four different car loans that Susana can get to finance her new car. The total column gives the amount she will end up paying for the car including the down payment and the payments with interest. Choose the letter for the best answer.

6. How much will Susana pay each month with loan A?
A $252.04   C $330.35
Ⓑ $297.02   D $353.68

Loan	Down Payment	Number of Months	Total
A	$2000	60	$19,821.20
B	$1000	48	$19,390.72
C	$0	60	$20,197.20

7. How much will Susana pay each month with loan B?
F $300.85   H $323.17
G $306.50   Ⓙ $383.14

8. How much will Susana pay each month with loan C?
Ⓐ $336.62   C $369.95
B $352.28   D $420.78

9. Which loan will give Susana the smallest monthly payment?
Ⓕ Loan A   H Loan C
G Loan B   J They are equal

Lesson Quiz

Solve.

1. $\frac{x}{-9} - 3 = 10$ $x = -117$

2. $7y + 25 = -24$ $y = -7$

3. $-8.3 = -3.5x + 13.4$ $x = 6.2$

4. $\frac{y + 5}{11} = 3$ $y = 28$

5. The cost for a new cell phone plan is $39 per month plus a one-time start-up fee of $78. If you are charged $1014, how many months will the contract last?
24 months

Available on Daily Transparency in CRB

Warm Up

Solve.

1. $3x = 102$ $x = 34$

2. $\frac{y}{15} = 15$ $y = 225$

3. $z - 100 = -1$ $z = 99$

4. $1.1 + 5w = 98.6$ $w = 19.5$

Problem of the Day

Ana has twice as much money as Ben, and Ben has three times as much as Clio. Together they have $160. How much does each person have? Ana, $96; Ben, $48; Clio, $16

Available on Daily Transparency in CRB

Math Humor

What did one math book say to the other? Leave me alone—I've got my own problems!

10-2 Solving Multistep Equations

Learn to solve multistep equations.

To solve a complicated equation, you may have to simplify the equation first by combining like terms.

EXAMPLE 1 Solving Equations That Contain Like Terms

Solve.

$$2x + 4 + 5x - 8 = 24$$
$$2x + 4 + 5x - 8 = 24$$
$$7x - 4 = 24 \qquad \text{Combine like terms.}$$
$$\underline{+4 \qquad +4} \qquad \text{Add to undo subtraction.}$$
$$7x = 28$$
$$\frac{7x}{7} = \frac{28}{7} \qquad \text{Divide to undo multiplication.}$$
$$x = 4$$

Check

$$2x + 4 + 5x - 8 = 24$$
$$2(4) + 4 + 5(4) - 8 \overset{?}{=} 24 \qquad \text{Substitute 4 for } x.$$
$$8 + 4 + 20 - 8 \overset{?}{=} 24$$
$$24 \overset{?}{=} 24 \ ✔$$

If an equation contains fractions, it may help to multiply both sides of the equation by the least common denominator (LCD) to clear the fractions before you isolate the variable.

EXAMPLE 2 Solving Equations That Contain Fractions

Solve.

 $\frac{3y}{7} + \frac{5}{7} = -\frac{1}{7}$

Multiply both sides by 7 to clear fractions, and then solve.

$$7\left(\frac{3y}{7} + \frac{5}{7}\right) = 7\left(-\frac{1}{7}\right)$$
$$7\left(\frac{3y}{7}\right) + 7\left(\frac{5}{7}\right) = 7\left(-\frac{1}{7}\right) \qquad \text{Distributive Property}$$
$$3y + 5 = -1$$
$$\underline{-5 \qquad -5} \qquad \text{Subtract to undo addition.}$$
$$3y = -6$$
$$\frac{3y}{3} = \frac{-6}{3} \qquad \text{Divide to undo multiplication.}$$
$$y = -2$$

---

1 Introduce

Alternate Opener

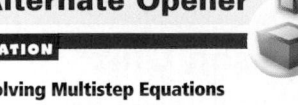
Motivate

Show students an equation, such as $2x + 1 = 7$, and represent the equation using algebra tiles. You may want to use the overhead algebra tiles provided in the Teacher's Manipulatives Kit.

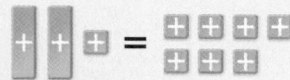

Ask students to solve the equation, always keeping it balanced. $x = 3$

Exploration worksheet and answers on Chapter 10 Resource Book pp. 17 and 80

2 Teach

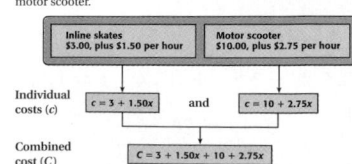 Lesson Presentation

Guided Instruction

In this lesson, students learn to solve multistep equations. Remind students that they can use the Commutative Property of Addition to rearrange terms; this will be helpful for combining like terms. Before discussing the equations that have fractions, you may want to review how to find the LCD of a set of fractions.

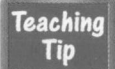 **Teaching Tip** Remind students to check each equation by substituting the solution into the *original equation*. This will help them catch mistakes they may have made early in the solution process.

Solve.

B $\frac{2p}{3} + \frac{p}{4} - \frac{1}{6} = \frac{7}{2}$

The LCD is 12.

$$12\left(\frac{2p}{3} + \frac{p}{4} - \frac{1}{6}\right) = 12\left(\frac{7}{2}\right) \qquad \textit{Multiply both sides by the LCD.}$$

$$12\left(\frac{2p}{3}\right) + 12\left(\frac{p}{4}\right) - 12\left(\frac{1}{6}\right) = 12\left(\frac{7}{2}\right) \qquad \textit{Distributive Property}$$

$$8p + 3p - 2 = 42$$

$$11p - 2 = 42 \qquad \textit{Combine like terms.}$$

$$\underline{\;+2\quad+2} \qquad \textit{Add to undo subtraction.}$$

$$11p = 44$$

$$\frac{11p}{11} = \frac{44}{11} \qquad \textit{Divide to undo multiplication.}$$

$$p = 4$$

Check

$$\frac{2p}{3} + \frac{p}{4} - \frac{1}{6} = \frac{7}{2}$$

$$\frac{2(4)}{3} + \frac{4}{4} - \frac{1}{6} \stackrel{?}{=} \frac{7}{2} \qquad \textit{Substitute 4 for p.}$$

$$\frac{8}{3} + 1 - \frac{1}{6} \stackrel{?}{=} \frac{7}{2}$$

$$\frac{16}{6} + \frac{6}{6} - \frac{1}{6} \stackrel{?}{=} \frac{21}{6} \qquad \textit{The LCD is 6.}$$

$$\frac{21}{6} \stackrel{?}{=} \frac{21}{6} \checkmark$$

EXAMPLE 3 *Money Application*

Carly had a $10 gift certificate for her favorite restaurant. After a 20% tip was added to the bill, the $10 was deducted. The amount she paid was $4.40. What was her original bill?

Let b represent the amount of the original bill.

$$b + 0.20b - 10 = 4.40 \qquad \textit{bill + tip − gift certificate = amount paid}$$

$$1.20b - 10 = 4.40 \qquad \textit{Combine like terms.}$$

$$\underline{\;+10\quad+10} \qquad \textit{Add 10 to both sides.}$$

$$1.20b = 14.40$$

$$\frac{1.20b}{1.20} = \frac{14.40}{1.20} \qquad \textit{Divide both sides by 1.20.}$$

$$b = 12 \qquad \textit{Her original bill was \$12.}$$

Think and Discuss

1. List the steps required to solve $3x - 4 + 2x = 7$.

2. Tell how you would clear the fractions in the equation $\frac{3x}{4} - \frac{2x}{3} + \frac{5}{8} = 1$.

Additional Examples

Example 1

Solve.

$8x + 6 + 3x - 2 = 37 \quad x = 3$

Example 2

Solve.

A. $\frac{5n}{4} + \frac{7}{4} = \frac{-3}{4} \qquad n = -2$

B. $\frac{7x}{9} + \frac{x}{2} - \frac{17}{9} = \frac{2}{3} \qquad x = 2$

Example 3

When Mr. and Mrs. Harris left for the mall, Mrs. Harris had twice as much money as Mr. Harris had. While shopping, Mrs. Harris spent $54 and Mr. Harris spent $26. When they arrived home, they had a total of $46. How much did Mr. Harris have when he left home? **$42**

---

3 Close

Reaching All Learners
Through Grouping Strategies

Have students work in pairs to play equation tic-tac-toe. Give each group a recording sheet containing nine equations and a tic-tac-toe board containing the solutions (Chapter 10 Resource Book p. 73). Have students cut out the equations and mix them up. Then have each student randomly draw one equation and solve it. Repeat until one student in each group gets three across, down, or diagonally.

Summarize

Ask students to write a reasonable first step to solve each of the following equations.

1. $\frac{2a}{5} - \frac{4a}{5} = 3$

2. $\frac{x}{6} - \frac{4x}{3} = \frac{x}{9}$

3. $3t - 1 + t - 5t = 1$

Possible answers:

1. Multiply both sides of the equation by 5.

2. Multiply both sides of the equation by 18.

3. Combine like terms.

Answers to Think and Discuss

Possible answers:

1. Combine like terms ($3x$ and $2x$). Add 4 to both sides. Divide both sides by 5. The answer is $x = \frac{11}{5}$.

2. Multiply both sides by 24.

10-2 PRACTICE & ASSESS

10-2 Exercises

FOR EXTRA PRACTICE
see page 750

internet connect
Homework Help Online
go.hrw.com Keyword: MP4 10-2

> Students may want to refer back to the lesson examples.

Assignment Guide

If you finished Example **1** assign:
Core 1–6, 12–17, 26, 44–50
Enriched 12–17, 26, 27, 29, 31, 32, 34, 36, 44–50

If you finished Example **2** assign:
Core 1–10, 12–23, 26, 30, 44–50
Enriched 12–23, 25–36, 44–50

If you finished Example **3** assign:
Core 1–24, 25–37 odd, 44–50
Enriched 12–50

Notes

GUIDED PRACTICE

See Example **1** Solve.

1. $8d - 11 + 3d + 2 = 13$ $d = 2$ 2. $2y + 5y + 4 = 25$ $y = 3$

3. $10e - 2e - 9 = 39$ $e = 6$ 4. $3c - 7 + 12c = 53$ $c = 4$

5. $4h + 8 + 7h - 2h = 89$ $h = 9$ 6. $8x - 3x + 2 = -33$ $x = -7$

See Example **2** 7. $\frac{5x}{11} + \frac{4}{11} = -\frac{1}{11}$ $x = -1$ 8. $\frac{y}{2} - \frac{3y}{8} + \frac{1}{4} = \frac{1}{2}$ $y = 2$

9. $\frac{4}{5} - \frac{2p}{5} = \frac{6}{5}$ $p = -1$ 10. $\frac{9}{4}z + \frac{1}{2} = 2$ $z = \frac{2}{3}$

See Example **3** 11. Joley used a $20 gift certificate to help pay for dinner for herself and a friend. After an 18% tip was added to the bill, the $20 was deducted. The amount she paid was $8.90. What was the original bill? **$24.49**

INDEPENDENT PRACTICE

See Example **1** Solve.

12. $6n + 4n - n + 5 = 23$ $n = 2$ 13. $-83 = 6k + 17 + 4k$ $k = -10$

14. $36 - 4c - 3c = 22$ $c = 2$ 15. $10 + 4w - 3w = 13$ $w = 3$

16. $28 = 10a - 5a - 2$ $a = 6$ 17. $30 = 7y - 35 + 6y$ $y = 5$

See Example **2** 18. $\frac{3}{8} + \frac{p}{8} = 3\frac{1}{8}$ $p = 22$ 19. $\frac{9h}{10} - \frac{3h}{10} = \frac{18}{10}$ $h = 3$

20. $\frac{4g}{14} - \frac{3}{7} - \frac{g}{14} = \frac{3}{14}$ $g = 3$ 21. $\frac{5}{18} = \frac{4m}{9} - \frac{m}{3} + \frac{1}{2}$ $m = -2$

22. $\frac{5}{11} = -\frac{3b}{11} + \frac{8b}{22}$ $b = 5$ 23. $\frac{3x}{4} - \frac{11x}{24} = -1\frac{1}{6}$ $x = -4$

See Example **3** 24. Pat bought 6 shirts that were all the same price. He used a traveler's check for $25, and then paid the difference of $86. What was the price of each shirt? **$18.50**

PRACTICE AND PROBLEM SOLVING

Solve and check.

25. $\frac{5n}{6} - \frac{1}{4} = \frac{3}{8}$ $n = \frac{3}{4}$ 26. $5n + 12 - 9n = -16$ $n = 7$

27. $6b - 1 - 10b = 51$ $b = -13$ 28. $\frac{x}{2} + \frac{2}{3} = \frac{5}{6}$ $x = \frac{1}{3}$

29. $-2x - 7 + 3x = 10$ $x = 17$ 30. $\frac{3r}{4} - \frac{2}{3} = \frac{5}{6}$ $r = 2$

31. $5y - 2 - 8y = 31$ $y = -11$ 32. $7n - 10 - 9n = -13$ $n = \frac{3}{2}$

33. $\frac{h}{6} + \frac{h}{6} = 1\frac{1}{3}$ $h = 4$ 34. $2a + 7 + 3a = 32$ $a = 5$

35. $\frac{b}{6} + \frac{3b}{8} = \frac{5}{12}$ $b = \frac{10}{13}$ 36. $-10 = 9m - 13 - 7m$ $m = \frac{3}{2}$

Math Background

Multistep equations can often be solved using different methods. Clearing fractions is one method of solving equations that contain fractions, but multiplying by the LCD is not the required first step for solving these types of equations. For example, the equation in Example 2A can be solved as follows:

$$\frac{3y}{7} + \frac{5}{7} = -\frac{1}{7}$$
$$\frac{3y}{7} + \frac{5}{7} - \frac{5}{7} = -\frac{1}{7} - \frac{5}{7}$$
$$(7)\frac{3y}{7} = -\frac{6}{7}(7)$$
$$3y = -6$$
$$y = -2$$

RETEACH 10-2

LESSON 10-2 Reteach
Solving Multistep Equations

To combine like terms, add (or subtract) coefficients.
$2m + 3m = (2 + 3)m = 5m$ $x - 3x = (1 - 3)x = -2x$
To solve an equation that contains like terms, first combine the like terms.

$2m + 3m = 35 - 25$ **Check:** Substitute into the original.
   $5m = 10$ Combine like terms. $2m + 3m = 35 - 25$
   $\frac{5m}{5} = \frac{10}{5}$ Divide by 5. $2(2) + 3(2) \stackrel{?}{=} 35 - 25$
     $m = 2$ $4 + 6 \stackrel{?}{=} 10$
                                $10 = 10$ ✔
$x + 6 - 3x + 5 = 13$ **Check:** $x + 6 - 3x + 5 = 13$
   $-2x + 11 = 13$ Combine like terms. $-1 + 6 - 3(-1) + 5 \stackrel{?}{=} 13$
   $-11 \quad -11$ Subtract 11. $-1 + 6 + 3 + 5 \stackrel{?}{=} 13$
   $-2x = 2$ $-1 + 14 \stackrel{?}{=} 13$
   $\frac{-2x}{-2} = \frac{2}{-2}$ Divide by −2. $13 = 13$ ✔
     $x = -1$

Complete to solve and check each equation.

1. $4z - 7z = -20 - 1$ **Check:** $4z - 7z = -20 - 1$
   $\underline{-3}\,z = \underline{-21}$ Combine like terms. $4(\underline{7}) - 7(\underline{7}) \stackrel{?}{=} -20 - 1$
   $\frac{-3z}{-3} = \underline{-3}$ Divide. $\underline{28} - \underline{49} \stackrel{?}{=} -21$
     $z = \underline{7}$ $\underline{-21} = -21$ ✔

2. $t + 1 - 4t + 8 = 21$ **Check:** $t + 1 - 4t + 8 = 21$
   $\underline{-3}\,t + \underline{9} = \underline{21}$ Combine like terms. $\underline{-4} + 1 - 4(\underline{-4}) + 8 \stackrel{?}{=} 21$
   $\underline{-9} \quad \underline{-9}$ Subtract. $\underline{-4} + 1 + \underline{16} + 8 \stackrel{?}{=} 21$
   $\underline{-3}\,t = \underline{12}$ $\underline{-4} + \underline{25} \stackrel{?}{=} 21$
   $\frac{-3t}{-3} = \underline{-3}$ Divide. $21 = 21$ ✔
     $t = \underline{-4}$

PRACTICE 10-2

LESSON 10-2 Practice B
Solving Multistep Equations

Solve.

1. $2x + 5x + 4 = 25$ 2. $9 + 3y - 2y = 14$ 3. $16 = 4w + 2w - 2$
   $\underline{x = 3}$ $\underline{y = 5}$ $\underline{w = 3}$

4. $26 = 3b - 2 - 7b$ 5. $31 + 4t - t = 40$ 6. $14 - 2x + 4x = 20$
   $\underline{b = -7}$ $\underline{t = 3}$ $\underline{x = 3}$

7. $\frac{5}{8}m - \frac{3}{5} + \frac{3}{8}m = \frac{2}{5}$ 8. -29.2 9. $7a + 16 - 3a - 7$
                                $= 1.7n + 5.8 + 5.3n$ $= -11$
   $\underline{m = 1}$ $\underline{n = -5}$ $\underline{a = -5}$

10. $\frac{1}{2}x + 1 + \frac{3}{4}x = -9$ 11. $7.5m + 3 - 3.5m = -15$ 12. $\frac{2}{5}x + 3 - \frac{4}{5}x = \frac{1}{5}$
    $\underline{x = -8}$ $\underline{m = -4.5}$ $\underline{x = 7}$

13. $7k - 12 - 4.5k = 18$ 14. $6y + 9 - 4y = -3$ 15. $5a - 1 + a = 11$
    $\underline{k = 12}$ $\underline{y = -6}$ $\underline{a = 2}$

16. The measure of an angle is 28° greater than its complement. Find the measure of each angle.
    $\underline{\text{angle} = 59°; \text{complement} = 31°}$

17. The measure of an angle is 21° more than twice its supplement. Find the measure of each angle.
    $\underline{\text{angle} = 127°; \text{supplement} = 53°}$

18. The perimeter of the triangle is 126. Find the measure of each side.
    $\underline{AC = 25; \ BC = 50; \ AB = 51}$

19. The base angles of an isosceles triangle are congruent. If the measure of each of the base angles is twice the measure of the third angle, find the measure of all three angles.
    $\underline{36°; 72°; 72°}$

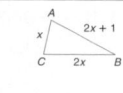

Sports **LINK**

You can estimate the weight in pounds of a fish that is L inches long and G inches around at the thickest part by using the formula $W \approx \frac{LG^2}{800}$.

37. Gina is paid 1.5 times her normal hourly rate for each hour she works over 40 hours in a week. Last week she worked 48 hours and earned $634.40. What is her normal hourly rate? **$12.20 per hr**

38. ***SPORTS*** The average weight of the top 5 fish at a fishing tournament was 12.3 pounds. The weights of the second-, third-, fourth-, and fifth-place fish are shown in the table. What was the weight of the heaviest fish? **14.6 lb**

Winning Entries	
Caught By	Weight (lb)
Wayne S.	
Carla P.	12.8
Deb N.	12.6
Virgil W.	11.8
Brian B.	9.7

39. ***PHYSICAL SCIENCE*** The formula $C = \frac{5}{9}(F - 32)$ is used to convert a temperature from degrees Fahrenheit to degrees Celsius. Water boils at 100°C. Use the formula to find the boiling point of water in degrees Fahrenheit. **212°F**

40. At a bulk food store, Kerry bought $\frac{2}{3}$ lb of coffee that cost $4.50/lb, $\frac{3}{4}$ lb of coffee that cost $5.20/lb, and $\frac{1}{5}$ lb of coffee that did not have a price marked. If her total cost was $8.18, what was the price per pound of the third type of coffee? **$6.40**

41. ***WHAT'S THE ERROR?*** A student's work in solving an equation is shown. What error has the student made, and what is the correct answer?

$$\frac{1}{3}x + 3x = 7$$
$$x + 3x = 21$$
$$4x = 21$$
$$x = \frac{21}{4}$$

42. ***WRITE ABOUT IT*** Compare the steps you would use to solve the following equations.

$$4x - 8 = 16 \qquad 4(x - 2) = 16$$

43. ***CHALLENGE*** List the steps you would use to solve the following equation.

$$\frac{5\left(\frac{1}{2}x - \frac{1}{3}\right) + \frac{7}{6}x}{2} + 2 = 3$$

Spiral Review

Evaluate each expression for the given value of the variable. (Lesson 3-2)

44. $19.4 - x$ for $x = -5.6$ **25**
45. $11 - r$ for $r = 13.5$ **−2.5**
46. $p + 65.1$ for $p = -42.3$ **22.8**
47. $-\frac{3}{7} - t$ for $t = 1\frac{5}{7}$ **−2$\frac{1}{7}$**
48. $3\frac{5}{11} + y$ for $y = -2\frac{4}{11}$ **1$\frac{1}{11}$**
49. $-\frac{1}{19} + g$ for $g = \frac{18}{19}$ **$\frac{17}{19}$**

50. **TEST PREP** $\overleftrightarrow{AB}$ has a slope of $\frac{2}{5}$. What is the slope of a line perpendicular to $\overleftrightarrow{AB}$? (Lesson 5-5) **C**

A $-\frac{2}{5}$　　B $\frac{5}{2}$　　C $-\frac{5}{2}$　　D $\frac{7}{5}$

Answers

41. Possible answer: The student forgot to multiply $3x$ times 3 to get $9x$. The correct answer is $x = 2.1$.

42. Possible answer: To solve the first equation, I would add 8 to both sides and then divide both sides by 4 to find that $x = 6$. To solve the second equation, I would divide both sides by 4 and then add 2 to both sides to find that $x = 6$. Alternately, I could distribute the 4 in the second equation and follow the steps for solving the first equation.

43. Possible answer: Subtract 2 from both sides. Multiply both sides by 2 to clear the large fraction. Distribute the 5. Multiply both sides by 6 to clear the fractions. Combine like terms to get $22x - 10 = 12$. Add 10 to both sides, and then divide both sides by 22 to find that $x = 1$.

Journal

Ask students to compare the process of solving a multistep equation with the process of checking the solution.

Test Prep Doctor ✚

For Exercise 50, remind students that perpendicular lines have slopes whose product is −1. If the slope of a line is $\frac{2}{5}$, then the slope of a perpendicular line is $-\frac{5}{2}$. The correct answer is **C**. If students chose **A**, they may have chosen the opposite of $\frac{2}{5}$. If students chose **B**, they may have used the reciprocal of $\frac{2}{5}$.

CHALLENGE 10-2

LESSON **Challenge**
10-2 *Use the Power of Algebra!*

An equation may be used to solve a problem involving angle measure in a triangle.

In isosceles triangle ABC, the measure of vertex angle C is 30° more than the measure of each base angle. Find the measure of each angle of the triangle.

Let x = the number of degrees in m∠A.
Then x = the number of degrees in m∠B.
And $x + 30$ = the number of degrees in m∠C.

The sum of the measures of the angles of a triangle is 180°.

$x + x + x + 30 = 180$
$3x + 30 = 180$　Combine like terms.
$\underline{\quad -30 \quad -30}$　Subtract 30.
$3x = 150$
$\frac{3x}{3} = \frac{150}{3}$　Divide by 3.
$x = 50$
$x + 30 = 80$

Check:
m base ∠ = 50°
m base ∠ = 50°
m vertex ∠ = 80°
180° ✔

So, the measure of each base angle is 50° and the measure of the vertex angle is 80°.

Write and solve an equation to find the measures of the angles of each triangle.

1. The measure of each of the base angles of an isosceles triangle is 9° less than 4 times the measure of the vertex angle.

$x + 4x - 9 + 4x - 9 = 180$
$9x - 18 = 180$
$\underline{+18 \quad +18}$
$9x = 198$
$\frac{9x}{9} = \frac{198}{9}$
$x = 22$
$4x - 9 = 79$

measure of each base angle = __79°__
measure of vertex angle = __22°__

2. The measure of the vertex angle of an isosceles triangle is one-fourth that of a base angle.

$x + x + \frac{x}{4} = 180$
$4 \cdot x + 4 \cdot x + 4 \cdot \frac{x}{4} = 4 \cdot 100$
$9x = 720$
$\frac{9x}{9} = \frac{720}{9}$
$x = 80$
$\frac{x}{4} = 20$

measure of each base angle = __80°__
measure of vertex angle = __20°__

PROBLEM SOLVING 10-2

LESSON **Problem Solving**
10-2 *Solving Multistep Equations*

A taxi company charges $2.25 for the first mile and then $0.20 per mile for each mile after the first, or $F = $2.25 + $0.20(m − 1)$ where F is the fare and m is the number of miles.

1. If Juan's taxi fare was $6.05, how many miles did he travel in the taxi? **20 miles**

2. If Juan's taxi fare was $7.65, how many miles did he travel in the taxi? **28 miles**

A new car loses 20% of its original value when you buy it and then 8% of its original value per year, or $D = 0.8V − 0.08Vy$ where D is the value after y years with an original value V.

3. If a vehicle that was valued at $20,000 new is now worth $9,600, how old is the car? **4 years**

4. A 6-year old vehicle is worth $12,000. What was the original value of the car? **$37,500**

The equation used to estimate typing speed is $S = \frac{1}{5}(w - 10e)$, where S is the accurate typing speed, w is the number of words typed in 5 minutes, and e is the number of errors. Choose the letter of the best answer.

5. Jane can type 55 words per minute (wpm). In 5 minutes, she types 285 words. How many errors would you expect her to make?
A 0
B 1
C 2
D 5

6. If Alex types 300 words in 5 minutes with 5 errors, what is his typing speed?
F 48 wpm
G 50 wpm
H 59 wpm
J 60 wpm

7. Johanna receives a report that says her typing speed is 65 words per minute. She knows that she made 4 errors in the 5-minute test. How many words did she type in 5 minutes?
A 285
B 329
C 365
D 1825

8. Cecil can type 35 words per minute. In 5 minutes, she types 255 words. How many errors would you expect her to make?
F 2
G 4
H 6
J 8

Lesson Quiz

Solve.

1. $6x + 3x - x + 9 = 33$ $x = 3$

2. $-9 = 5x + 21 + 3x$ $x = -3.75$

3. $\frac{5}{8} + \frac{x}{8} = \frac{33}{8}$ $x = 28$

4. $\frac{6x}{7} - \frac{2x}{21} = \frac{25}{21}$ $x = 1\frac{9}{16}$

5. Linda is paid double her normal hourly rate for each hour she works over 40 hours in a week. Last week she worked 52 hours and earned $544. What is her hourly rate? **$8.50**

Available on Daily Transparency in CRB

Hands-On

10A Model Equations with Variables on Both Sides

Pacing: Traditional 1 day
Block $\frac{1}{2}$ day

Objective: To use algebra tiles to model equations with variables on both sides

Materials: Algebra tiles

Lab Resources

Hands-On Lab Activities. . pp. 73–74, 93

Using the Page

Discuss with students what each algebra tile represents and how to represent equations with algebra tiles.

Represent each expression or equation with algebra tiles.

1. $-2x + 3$

2. $x - 5 = -x + 3$

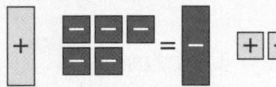

Answers

Think and Discuss

Possible answers:

1. Replace each yellow x bar with 2 yellow unit squares. Replace each red x bar with 2 red unit squares. See whether the equation balances.

2. to be able to see what number of unit tiles one variable tile is equal to

Try This

Check students' models.

1. $x = -1$

2. $x = 3$

3. $x = 1$

4. $x = 3$

Model Equations with Variables on Both Sides

Use with Lesson 10-3

KEY

Algebra tiles

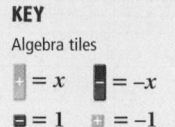

REMEMBER

It will not change the value of an expression if you add or remove zero.

internet connect
Lab Resources Online
go.hrw.com
KEYWORD: MP4 Lab10A

To solve an equation with the same variable on both sides of the equal sign, you must first add or subtract to eliminate the variable term from one side of the equation.

Activity

1 Model and solve the equation $-x + 2 = 2x - 4$.

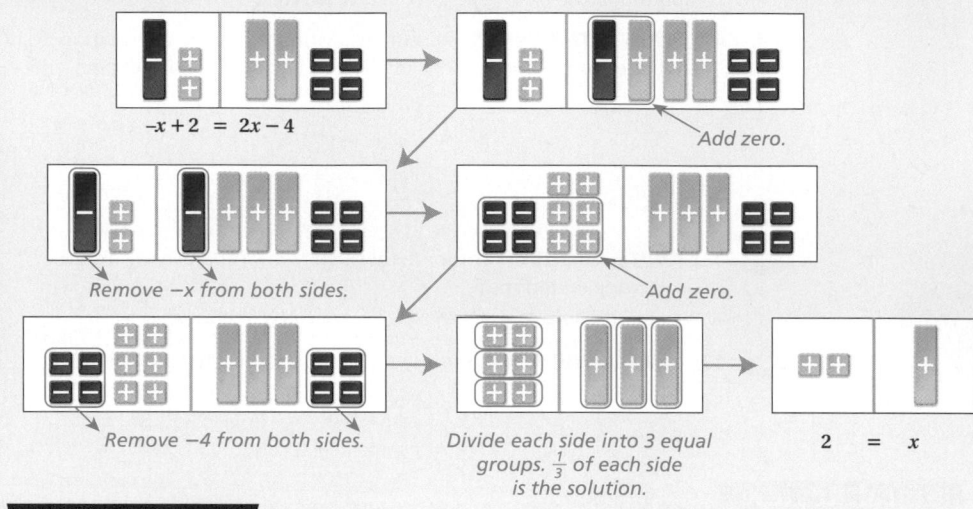

$-x + 2 = 2x - 4$

Add zero.

Remove $-x$ from both sides.

Add zero.

Remove -4 from both sides.

Divide each side into 3 equal groups. $\frac{1}{3}$ of each side is the solution.

$2 = x$

Think and Discuss

1. How would you check the solution to $-x + 2 = 2x - 4$ using algebra tiles?

2. Why must you isolate the variable terms by having them on only one side of the equation?

Try This

Model and solve each equation.

1. $x + 1 = -x - 1$ **2.** $3x = -3x + 18$ **3.** $4 - 2x = -5x + 7$ **4.** $2x + 2x + 1 = x + 10$

10-3 Solving Equations with Variables on Both Sides

Learn to solve equations with variables on both sides of the equal sign.

Some problems produce equations that have variables on both sides of the equal sign. For instance, Elaine runs the same distance each day. On Mondays, Fridays, and Saturdays, she runs 3 laps on the track and an additional 5 miles off the track. On Tuesdays and Thursdays, she runs 4 laps on the track and 2.5 miles off the track.

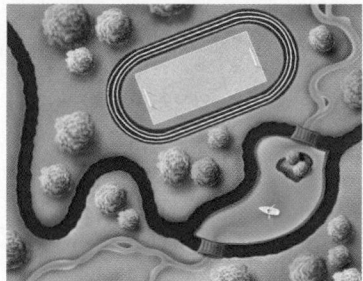

Expression for Mondays, Fridays, and Saturdays $3x+5$ $4x+2.5$ *Expression for Tuesdays and Thursdays*

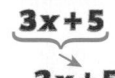

$$3x+5 = 4x+2.5$$

The variable x in these expressions is the length of one lap of the track. Since the total distance each day is the same, the two expressions are equal.

Solving an equation with variables on both sides is similar to solving an equation with a variable on only one side. You can add or subtract a term containing a variable on both sides of an equation.

EXAMPLE **1** **Solving Equations with Variables on Both Sides**

Solve.

A $2a + 3 = 3a$

$$\begin{aligned} 2a + 3 &= 3a \\ \underline{-2a \qquad -2a} & \qquad \text{Subtract } 2a \text{ from both sides.} \\ 3 &= a \end{aligned}$$

B $4v - 7 = 5 + 7v$

$$\begin{aligned} 4v - 7 &= 5 + 7v \\ \underline{-4v \qquad\qquad -4v} & \qquad \text{Subtract } 4v \text{ from both sides.} \\ -7 &= 5 + 3v \\ \underline{-5 \quad -5} & \qquad \text{Subtract 5 from both sides.} \\ -12 &= 3v \\ \frac{-12}{3} &= \frac{3v}{3} \qquad \text{Divide both sides by 3.} \\ -4 &= v \end{aligned}$$

---

1 Introduce

Alternate Opener

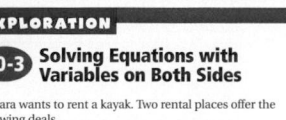

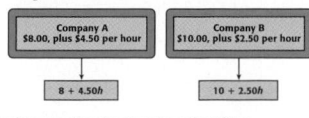

EXPLORATION

10-3 Solving Equations with Variables on Both Sides

Samara wants to rent a kayak. Two rental places offer the following deals.

Company A	Company B
$8.00, plus $4.50 per hour	$10.00, plus $2.50 per hour
$8 + 4.50h$	$10 + 2.50h$

Use the expressions above to solve each problem.

1. Find the cost of a 1-hour rental from company A ($h = 1$).
2. Find the cost of a 1-hour rental from company B ($h = 1$).
3. Find the cost of a 2-hour rental from company A ($h = 2$).
4. Find the cost of a 2-hour rental from company B ($h = 2$).
5. Find the number of hours that a kayak would need to be rented from both companies to make the costs equal by solving $8 + 4.50h = 10 + 2.50h$ for h. (*Hint:* Combine like terms first.)

Think and Discuss

6. **Explain** how you solved the equation in number 5.
7. **Discuss** what it means to set $8 + 4.50h$ equal to $10 + 2.50h$.

Motivate

Show students the following algebra tiles modeling an equation. You may want to use the overhead algebra tiles provided in the Teacher's Manipulatives Kit.

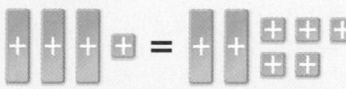

Ask them to write the equation that is represented by the tiles. $3x + 1 = 2x + 5$ Tell them that they will learn how to solve this kind of equation in the new lesson.

Exploration worksheet and answers on Chapter 10 Resource Book pp. 27 and 82

10-3 Organizer

Pacing: Traditional 1 day
Block $\frac{1}{2}$ day

Objective: Students solve equations with variables on both sides of the equal sign.

Warm Up

Solve.

1. $2x + 9x - 3x + 8 = 16$ $x = 1$
2. $-4 = 6x + 22 - 4x$ $x = -13$
3. $\frac{2}{7} + \frac{x}{7} = 5\frac{1}{7}$ $x = 34$
4. $\frac{9x}{16} - \frac{2x}{4} = 3\frac{1}{8}$ $x = 50$

Problem of the Day

An equilateral triangle and a regular pentagon have the same perimeter. Each side of the pentagon is 3 inches shorter than each side of the triangle. What is the perimeter of the triangle?   **22.5 in.**

Available on Daily Transparency in CRB

Math Humor

The chef who was taking an algebra class had a flair for creating veggie dishes with a symmetrical look. One day she had to solve the equation $4 - 2p = 3p + 9$. Naturally, she added p's to both sides.

---

2 Teach

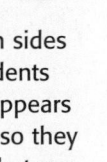

Lesson Presentation

Guided Instruction

In this lesson, students solve equations with variables on both sides of the equal sign. Explain to students that in Example 1, the variable appears on both sides of the equal sign, so they need to add or subtract a variable term so that the variable appears on one side only. The paragraph just before Example 2 is an excellent summary of the process of solving multistep equations. (See also Summarize.) Note that in Example 2A, there are like terms to combine, and in Example 2B, there are fractions to clear.

10-3 Solving Equations with Variables on Both Sides   **507**

Additional Examples

Example 1

Solve.

A. $4x + 6 = x$ $x = -2$

B. $9b - 6 = 5b + 18$ $b = 6$

C. $9w + 3 = 5w + 7 + 4w$
no solution

Example 2

Solve.

A. $10z - 15 - 4z = 8 - 2z - 15$

$z = 1$

B. $\frac{y}{5} + \frac{3y}{5} - \frac{3}{4} = y - \frac{7}{10}$

$y = -\frac{1}{4}$

Example 3

Jamie spends the same amount of money each morning. On Sunday, he bought a newspaper for $1.25 and also bought two doughnuts. On Monday, he bought a newspaper for fifty cents and bought five dough-nuts. On Tuesday, he spent the same amount of money and bought just doughnuts. How many doughnuts did he buy on Tuesday? 7 doughnuts

---

Helpful Hint

If the variables in an equation are eliminated and the resulting statement is false, the equation has no solution.

Solve.

C $g + 5 = g - 2$

$$\begin{array}{rcl} g + 5 &=& g - 2 \\ \underline{-g} & & \underline{-g} \\ 5 &\neq& -2 \end{array}$$ *Subtract g from both sides.*

No solution. There is no number that can be substituted for the variable g to make the equation true.

To solve multistep equations with variables on both sides, first combine like terms and clear fractions. Then add or subtract variable terms to both sides so that the variable occurs on only one side of the equation. Then use properties of equality to isolate the variable.

EXAMPLE 2 **Solving Multistep Equations with Variables on Both Sides**

Solve.

Ⓐ $2c + 4 - 3c = -9 + c + 5$

$$2c + 4 - 3c = -9 + c + 5$$

$$\begin{array}{rcl} -c + 4 &=& -4 + c \end{array}$$ *Combine like terms.*

$$\begin{array}{rcl} \underline{+c} & & \underline{+c} \\ 4 &=& -4 + 2c \end{array}$$ *Add c to both sides.*

$$\begin{array}{rcl} \underline{+4} & & \underline{+4} \\ 8 &=& 2c \end{array}$$ *Add to undo subtraction.*

$$\frac{8}{2} = \frac{2c}{2}$$ *Divide to undo multiplication.*

$$4 = c$$

Ⓑ $\frac{w}{2} - \frac{3w}{4} + \frac{1}{3} = w + \frac{7}{6}$

$$\frac{w}{2} - \frac{3w}{4} + \frac{1}{3} = w + \frac{7}{6}$$

$$12\left(\frac{w}{2} - \frac{3w}{4} + \frac{1}{3}\right) = 12\left(w + \frac{7}{6}\right)$$ *Multiply by LCD, 12.*

$$12\left(\frac{w}{2}\right) - 12\left(\frac{3w}{4}\right) + 12\left(\frac{1}{3}\right) = 12(w) + 12\left(\frac{7}{6}\right)$$

$$6w - 9w + 4 = 12w + 14$$

$$-3w + 4 = 12w + 14$$ *Combine like terms.*

$$\begin{array}{rcl} \underline{+3w} & & \underline{+3w} \\ 4 &=& 15w + 14 \end{array}$$ *Add 3w to both sides.*

$$\begin{array}{rcl} \underline{-14} & & \underline{-14} \\ -10 &=& 15w \end{array}$$ *Subtract 14 from both sides.*

$$\frac{-10}{15} = \frac{15w}{15}$$ *Divide both sides by 15.*

$$-\frac{2}{3} = w$$

Teach

 **Reaching All Learners**
Through Grouping Strategies

Give each student a card containing an expression like $2x + 4$ or $3x - 7$ (Chapter 10 Resource Book p. 74). Have the students form two concentric circles, each with an equal number of students, so that they are facing each other. Have each pair of students facing each other form an equation by setting their expressions equal to each other, and then have them solve that equa-tion. After each pair agrees on their solution, have the circles rotate two students to the right and continue the process.

EXAMPLE 3 *Sports Application*

Elaine runs the same distance every day. On Mondays, Fridays, and Saturdays, she runs 3 laps on the track, and then runs 5 more miles. On Tuesdays and Thursdays, she runs 4 laps on the track, and then runs 2.5 more miles. On Wednesdays, she just runs laps. How many laps does she run on Wednesdays?

First solve for the distance around the track.

$$3x + 5 = 4x + 2.5 \qquad \text{Let } x \text{ represent the distance around the track.}$$
$$\underline{-3x \qquad = -3x}\qquad \text{Subtract } 3x \text{ from both sides.}$$
$$5 = x + 2.5$$
$$\underline{-2.5 \qquad -2.5}\qquad \text{Subtract 2.5 from both sides.}$$
$$2.5 = x \qquad \text{The track is 2.5 miles around.}$$

Now find the total distance Elaine runs each day.

$$3x + 5 \qquad \text{Choose one of the original expressions.}$$
$$3(2.5) + 5 = 12.5 \qquad \text{Elaine runs 12.5 miles each day.}$$

Find the number of laps Elaine runs on Wednesdays.

$$2.5n = 12.5 \qquad \text{Let } n \text{ represent the number of 2.5-mile laps.}$$
$$\frac{2.5n}{2.5} = \frac{12.5}{2.5} \qquad \text{Divide both sides by 2.5.}$$
$$n = 5$$

Elaine runs 5 laps on Wednesdays.

Think and Discuss

1. Give an example of an equation that has no solution.

10-3 Exercises

FOR EXTRA PRACTICE
see page 750

internet connect
Homework Help Online
go.hrw.com Keyword: MP4 10-3

GUIDED PRACTICE

See Example **1** Solve.

1. $5x + 2 = x + 6$ $x = 1$
2. $6a - 6 = 8 + 4a$ $a = 7$
3. $3x + 9 = 10x - 5$ $x = 2$
4. $4y - 2 = 6y + 6$ $y = -4$

See Example **2**
5. $4x - 5 + 2x = 13 + 9x - 21$ $x = 1$
6. $\frac{2n}{5} + \frac{n}{10} - 4 = 6 + 3n - 15$ $n = 2$
7. $\frac{3}{10} + \frac{9d}{10} - 2 = 2d + 4 - 3d$ $d = 3$
8. $4(x - 5) + 2 = x + 3$ $x = 7$

3 Close

Summarize

Review the process for solving multistep equations to this point:

- Clear fractions.
- Combine like terms.
- Get the variable on only one side of the equation.
- Undo addition and subtraction.
- Undo multiplication and division.

You may want to create a poster of these steps to display in the classroom.

Answers to Think and Discuss

1. Possible answer: $3x + 4 = 3x - 2$; if the variables in an equation are eliminated and the resulting statement is false, then the equation has no solution. Another possible answer is $0x = 1$ because zero times any number must equal zero.

See Example ③ 9. June has a set of folding chairs for her flute students. If she arranges them in 5 rows for a recital, she has 2 chairs left over. If she arranges them in 3 rows of the same length, she has 14 left over. How many chairs does she have? **32 chairs**

INDEPENDENT PRACTICE

See Example ① Solve.

10. $2n + 12 = 5n$ **n = 4**   **11.** $9x - 2 = 10 - 3x$ **x = 1**

12. $5n + 3 = 14 - 6n$ **n = 1**   **13.** $9y - 6 = 7y + 8$ **y = 7**

14. $5x + 2 = x + 6$ **x = 1**   **15.** $2(4x + 15) = 8x + 3$ **no solution**

See Example ② **16.** $\frac{2p}{9} + \frac{5p}{18} - \frac{5}{6} = \frac{2}{3} + \frac{p}{12} + \frac{1}{6}$ **p = 4** **17.** $3(x - 4) - 4 = 5x + 6.9 - 3x$ **x = 22.9**

18. $\frac{1}{2}(2n + 6) = 5n - 12 - n$ **n = 5** **19.** $\frac{a}{22} - 4.5 + 2a = \frac{7}{11} + \frac{17a}{11} + \frac{4}{11}$ **a = 11**

See Example ③ **20.** Sean and Laura have the same number of action figures in their collections. Sean has 6 complete sets plus 2 individual figures, and Laura has 3 complete sets plus 20 individual figures. How many figures are in a complete set? **6 figures**

PRACTICE AND PROBLEM SOLVING

Solve and check.

21. $8y - 3 = 17 - 2y$ **y = 2**   **22.** $2n + 6 = 7n - 9$ **n = 3**

23. $2n + 12n = 2(n + 12)$ **n = 2**   **24.** $3(4x - 2) = 12x$ **no solution**

25. $100(x - 3) = 450 - 50x$ **x = 5**   **26.** $5p - 15 = 15 - 5p$ **p = 3**

27. $\frac{1}{2} - \frac{3m}{4} + 7 = 4m - 9 - \frac{m}{28}$ **m = 3.5**   **28.** $7(x - 1) = 3\left(x + \frac{1}{3}\right)$ **x = 2**

29. $4(x - 5) + 2 = \frac{1}{3}(x + 9) + \frac{2x}{3}$ **x = 7**   **30.** $12\left(4r - \frac{5r}{6}\right) + 20 = 19r - 15 + \frac{45r}{2}$ **r = 10**

Both figures have the same perimeter. Find each perimeter.

31. **360 units**   **32.** **28 units**

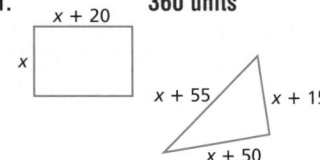

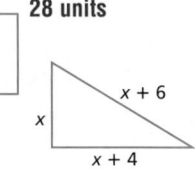

33. Find two consecutive whole numbers such that one-fourth of the first number is one more than one-fifth of the second number. (*Hint:* Let n represent the first number. Then $n + 1$ represents the next consecutive whole number.)   **24, 25**

34. Find three consecutive whole numbers such that the sum of the first two numbers equals the third number. (*Hint:* Let n represent the first number. Then $n + 1$ and $n + 2$ represent the next two whole numbers.)   **1, 2, 3**

Math Background

We speak of the steps taken to solve an equation but sometimes forget that the process of solving an equation is really the process of writing equivalent equations, each simpler than the one before, until the last one tells the solution.

With this understanding, we can see why an equation has no solution when we get a false statement.

$$x + 2 = x + 3$$
$$\underline{-x \qquad -x}$$
$$2 \neq 3$$

This final equation has no solution. Because it is equivalent to the original equation, the original equation has no solution.

RETEACH 10-3

LESSON **Reteach**
10-3 *Solving Equations with Variables on Both Sides*

If there are variable terms on both sides of an equation, first collect them on one side. Do this by adding or subtracting.

Go to the side where the resulting coefficient will be positive.

$5x = 2x + 12$	To go to left side,	**Check:** Substitute into the original and
$\underline{-2x \quad -2x}$	subtract 2x.	work to show equivalence.
$3x = 12$		$5x = 2x + 12$
$\frac{3x}{3} = \frac{12}{3}$	Divide by 3.	$5(4) \overset{?}{=} 2(4) + 12$
$x = 4$		$20 \overset{?}{=} 8 + 12$
		$20 = 20$ ✔

$-6z + 28 = 9z - 2$	To go to right side,	
$\underline{+6z \qquad +6z}$	add 6z.	
$28 = 15z - 2$		**Check:** $-6z + 28 = 9z - 2$
$\underline{+2 \qquad +2}$	Add 2.	$-6(2) + 28 \overset{?}{=} 9(2) - 2$
$30 = 15z$		$-12 + 28 \overset{?}{=} 18 - 2$
$\frac{30}{15} = \frac{15z}{15}$	Divide by 15.	$16 = 16$ ✔
$2 = z$		

Complete to solve and check each equation.

1. $9m = 4m - 25$ To go to left,
$\underline{-4m \quad -4m}$ subtract.   **Check:** $9m = 4m - 25$
$5m = -25$ $9(\underline{-5}) \overset{?}{=} 4(\underline{-5}) - 25$
$\frac{5m}{5} = \frac{-25}{5}$ Divide. $\underline{-45} \overset{?}{=} \underline{-20} - 25$
$m = \underline{-5}$ $-45 = -45$ ✔

2. $3h - 7 = 5h + 1$ To go to right,
$\underline{-3h \quad -3h}$ subtract.   **Check:** $3h - 7 = 5h + 1$
$-7 = \underline{2} h + 1$
$\underline{-1 \qquad -1}$ Subtract. $3(\underline{-4}) - 7 \overset{?}{=} 5(\underline{-4}) + 1$
$\underline{-8} = \underline{2} h$ $\underline{-12} - 7 \overset{?}{=} \underline{-20} + 1$
$\frac{-8}{2} = \frac{2h}{2}$ Divide. $-19 = -19$ ✔
$\underline{-4} = h$

PRACTICE 10-3

LESSON **Practice B**
10-3 *Solving Equations with Variables on Both Sides*

Solve.

1. $7x - 11 = -19 + 3x$ **2.** $11a + 9 = 4a + 30$ **3.** $4t + 14 = 1.2t + 7$

$x = -2$ $a = 3$ $t = -2.5$

4. $19c + 31 = 26c - 74$ **5.** $\frac{3}{8}y - 9 = 13 + \frac{1}{8}y$ **6.** $0.6k + 44 = 0.48k + 8$

$c = 15$ $y = 88$ $k = -300$

7. $10a - 37 = 6a + 51$ **8.** $5w + 9.9 = 4.8 + 8w$ **9.** $15 - x = 2(x + 3)$

$a = 22$ $w = 1.7$ $x = 3$

10. $15y + 14 = 2(5y + 6)$ **11.** $14 - \frac{1}{8}w = \frac{3}{4}w - 21$ **12.** $\frac{1}{2}(6x - 4) = 4x - 9$

$y = -0.4$ $w = 40$ $x = 7$

13. $4(3d - 2) = 8d - 5$ **14.** $\frac{1}{3}y + 11 = \frac{1}{2}y - 3$ **15.** $\frac{2x - 9}{3} = 8 - 3x$

$d = \frac{3}{4}$ $y = 84$ $x = 3$

16. Forty-eight decreased by a number is the same as the difference of four times a number and seven. Find the number.   **11**

17. The square and the equilateral triangle at the right have the same perimeter. Find the lengths of the sides of the triangle.   **12**

Physical Science LINK

Sodium and chlorine bond together to form sodium chloride, or salt. The atomic structure of sodium chloride causes it to form cubes.

35. PHYSICAL SCIENCE An atom of chlorine (Cl) has 6 more protons than an atom of sodium (Na). The atomic number of chlorine is 5 less than twice the atomic number of sodium. The atomic number of an element is equal to the number of protons per atom.

a. How many protons are in an atom of chlorine?   **17**

b. What is the atomic number of sodium?   **11**

36. EARTH SCIENCE *Specific gravity* compares the density of a mineral with the density of water. The following equation relates a mineral's specific gravity *s*, its weight in air *a*, and its weight in water *w*.

Mineral	Specific Gravity	Weight in Air	Weight in Water
Granite		152.3 g	97.2 g
Gold	19.3	10 g	
Quartz	2.65		6.5 g

$$s(a - w) = a$$

a. Find the specific gravity of a piece of granite.   ≈ **2.76**

b. Find the weight in water of a piece of gold that weighs 10 g in air.   ≈ **9.48 g**

c. Find the weight in air of a piece of quartz that weighs 6.5 g in water.   ≈ **10.44 g**

37. CHOOSE A STRATEGY Solve the following equation for *t*. How can you determine the solution once you have combined like terms?

$$2(t - 24) = 5t - 3(t + 16)$$

38. WRITE ABOUT IT Two cars are traveling in the same direction. The first car is going 45 mi/h, and the second car is going 60 mi/h. The first car left 2 hours before the second car. Explain how you could solve an equation with variables on both sides to find how long it will take the second car to catch up to the first car.

39. CHALLENGE Solve the equation $\frac{x+1}{7} = \frac{3}{4} + \frac{x-3}{5}$. $-\frac{1}{8}$

Spiral Review

Find both unit prices and tell which is the better buy. (Lesson 7-2)

40. $11.99 for 2 yd of fencing   ≈ **$2 per foot;**
$25 for 10 ft of fencing   **$2.50 per foot; 2 yd**

41. 20 oz of cereal for $3.49   ≈ **$0.175 per oz;**
16 oz of cereal for $2.99   ≈ **$0.187 per oz; 20 oz**

42. 4 tickets for $110   **$27.50 per ticket; $30**
6 tickets for $180   **per ticket; 4 tickets**

43. $2.39 for a 12 oz can of carrots   ≈ **$0.199 per**
$3.68 for a 20 oz can of carrots   **oz; $0.184 per oz; 20 oz**

44. $5.47 for a box of 100 nails   ≈ **$0.055 per nail;**
$13.12 for a box of 250 nails   ≈ **$0.052 per nail; 250 nails**

45. $747 for 3 computer monitors   **$249 per**
$550 for 2 computer monitors   **monitor; $275 per monitor; 3 monitors**

46. TEST PREP A square has a perimeter of 56 cm. If the square is dilated by a scale factor of 0.2, what is the length of each side of the new square? (Lesson 7-5)   **B**

A   11.2 cm   B   2.8 cm   C   5.6 cm   D   14 cm

Answers

37. Possible answer: Substitute different values for *t*. The equation is true for any value of *t* that is a real number, so the solution is all real numbers.

38. Possible answer: When the second car catches the first, the cars will have traveled an equal distance. Since distance can be calculated by multiplying rate and time, the situation can be represented by the equation $45(t + 2) = 60t$. Solving for *t* gives $t = 6$. The second car will catch the first car after 6 hours.

Journal

Have students describe the kind of equation that has no solution.

Test Prep Doctor

For Exercise 46, remind students that to find a dimension of a figure that is a dilation, they should multiply the corresponding dimension of the original figure by the scale factor. The original square has a perimeter of 56 cm, so each side is 14 cm. Each side of the new square is $(0.2)(14) = 2.8$ cm. The correct answer is **B**. Students who selected **A** may have multiplied 0.2 by the original perimeter. Students who selected **D** probably found the length of a side of the original square.

CHALLENGE 10-3

LESSON 10-3 Challenge
A Handy Tool!

A **lever** is a bar that can rotate about a fixed point called the **fulcrum**.

The ancient Greek mathematician Archimedes knew of the power of the *lever principle*. He has been quoted as saying "Give me a place to stand and I will move the Earth."

The Lever Principle

A weight w_1 is placed on one arm of a lever at a distance d_1 from the fulcrum. A second weight w_2 is placed on the other arm at a distance d_2 from the fulcrum.

$$w_1 \cdot d_1 = w_2 \cdot d_2$$

An equation may be used to solve a problem involving the lever principle.

A 14-foot plank is used as a seesaw by Jon, who weighs 120 lb, and Al, who weighs 90 lb. If the boys balance each other, how far from the fulcrum is each sitting?

Let x = Jon's distance from fulcrum.
Then $14 - x$ = Al's distance from fulcrum.

$$w_1 \cdot d_1 = w_2 \cdot d_2$$
$$120 \cdot x = 90 \cdot (14 - x)$$
$$120x = 1260 - 90x$$
$$\underline{+90x \qquad\qquad +90x}$$
$$210x = 1260$$
$$\frac{210x}{210} = \frac{1260}{210}$$
$$x = 6 \text{ ft} \leftarrow \text{Jon's distance from the fulcrum.}$$
$$14 - 6 = 8 \text{ ft} \leftarrow \text{Al's distance from the fulcrum.}$$

Write and solve an equation.

A 21-ft plank is used as a seesaw by Mia, who weighs 108 lb, and Ruth, who weighs 81 lb. If the girls balance each other, how far from the fulcrum is Mia sitting?

Mia is sitting __9 ft__ from the fulcrum.

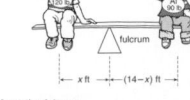

Let x = Mia's distance from fulcrum.
Then $21 - x$ = Ruth's distance.

$$108x = 81(21 - x)$$
$$108x = 1701 - 81x$$
$$189x = 1701$$
$$x = 9$$

PROBLEM SOLVING 10-3

LESSON 10-3 Problem Solving
Solving Equations with Variables on Both Sides

The chart below describes three different long distance calling plans. Round to the nearest minute. Write the correct answer.

1. For what number of minutes per month will plan A and plan B cost the same?

250 minutes

| | **Long Distance Plans** | |
Plan	Monthly Access Fee	Charge per minute
A	$3.95	$0.08
B	$8.95	$0.06
C	$0	$0.10

2. For what number of minutes per month will plan B and plan C cost the same?

224 minutes

3. For what number of minutes per month will plan A and plan C cost the same?

198 minutes

Choose the letter for the best answer.

4. Carpet Plus installs carpet for $100 plus $8 per square yard of carpet. Carpet World charges $75 for installation and $10 per square yard of carpet. Find the number of square yards of carpet for which the cost including carpet and installation is the same.

A 1.4 yd²   C 12.5 yd²
B 9.7 yd²   D 87.5 yd²

5. One shuttle service charges $10 for pickup and $0.10 per mile. The other shuttle service has no pickup fee but charges $0.35 per mile. Find the number of miles for which the cost of the two shuttle services is the same.

F 2.5 miles   H 10 miles
G 22 miles   J 48 miles

6. Joshua can purchase tile at one store for $0.99 per tile, but he will have to rent a tile saw for $25. At another store he can buy tile for $1.50 per tile and borrow a tile saw for free. Find the number of tiles for which the cost is the same. Round to the nearest tile.

A 10 tiles   C 25 tiles
B 13 tiles   D 49 tiles

7. One plumber charges a fee of $75 per service call plus $15 per hour. Another plumber has no flat fee, but charges $25 per hour. Find the number of hours for which the cost of the two plumbers is the same.

F 1.9 hours   H 10 hours
G 7.5 hours   J 25 hours

Lesson Quiz

Solve.

1. $4x + 16 = 2x$ $x = -8$

2. $8x - 3 = 15 + 5x$ $x = 6$

3. $2(3x + 11) = 6x + 4$ no solution

4. $\frac{1}{4}x = \frac{1}{2}x - 9$ $x = 36$

5. An apple has about 30 more calories than an orange. Five oranges have about as many calories as 3 apples. How many calories are in each? An orange has 45 calories. An apple has 75 calories.

Available on Daily Transparency in CRB

Chapter 10 Mid-Chapter Quiz

Chapter 10 Mid-Chapter Quiz

Purpose: To assess students' mastery of concepts and skills in Lessons 10-1 through 10-3

Assessment Resources

Section 10A Quiz
Assessment Resources . . . . p. 23

Test and Practice Generator CD-ROM

Additional mid-chapter assessment items in both multiple-choice and free-response format may be generated for any objective in Lessons 10-1 through 10-3.

Mid-Chapter Quiz

LESSON 10-1 (pp. 498–501)

Solve.

1. $5x + 17 = 47$ $x = 6$ 2. $4y + 1 = -15$ $y = -4$ 3. $16 - z = 12$ $z = 4$

4. $\frac{1}{2}t + 9 = 25$ $t = 32$ 5. $-32 = \frac{7}{3}w - 11$ $w = -9$ 6. $\frac{2}{3}q - 9 = -1$ $q = 12$

7. $\frac{x + 8}{4} = -10$ $x = -48$ 8. $5 = \frac{21 - z}{3}$ $z = 6$ 9. $\frac{a - 4}{3} = 5$ $a = 19$

10. A car rental company charges $39.99 per day plus $0.20 per mile. Jill rented a car for one day and the charges were $47.39, before tax. How many miles did Jill drive?   **37 mi**

LESSON 10-2 (pp. 502–505)

Solve.

11. $4c + 2c + 6 = 24$ $c = 3$ 12. $\frac{2x}{5} - \frac{3}{5} = \frac{11}{5}$ $x = 7$ 13. $\frac{t}{5} + \frac{t}{3} = \frac{8}{15}$ $t = 1$

14. $\frac{4m}{3} - \frac{m}{6} = \frac{7}{2}$ $m = 3$ 15. $8 - 6g + 15 = 19$ $g = \frac{2}{3}$ 16. $\frac{2}{5}b - \frac{1}{4}b = 3$ $b = 20$

17. $\frac{r}{3} + 7 - \frac{r}{5} = -3$ $r = -75$ 18. $5k + 9.3 = 21.8$ $k = 2.5$ 19. $\frac{x}{4} - \frac{x}{5} - \frac{1}{3} = \frac{16}{15}$ $x = 28$

20. On his last three math tests, Mark scored 85, 95, and 80. What grade must he get on his next test to have an average of 90 for all four tests?   **100**

LESSON 10-3 (pp. 507–511)

Solve.

21. $3x + 13 = x + 1$ $x = -6$ 22. $q + 7 = 2q + 5$ $q = 2$ 23. $8n + 24 = 3n + 59$ $n = 7$

24. $m + 5 = m - 3$ **no solution**   25. $9w - 2w + 8 = 4w + 38$ $w = 10$ 26. $-2a - a + 9 = 3a - 9$ $a = 3$

27. $\frac{5c}{4} = \frac{2c}{3} + 7$ $c = 12$ 28. $\frac{3z}{2} - \frac{17}{3} = \frac{2z}{3} - \frac{3}{2}$ $z = 5$ 29. $\frac{7}{12}y - \frac{1}{4} = 2y - \frac{5}{3}$ $y = 1$

30. The rectangle and the triangle have the same perimeter. Find the perimeter of each figure.   **58 units**

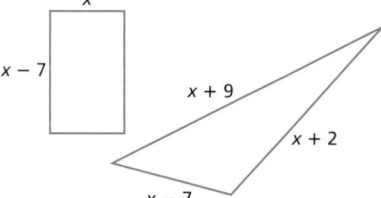

Focus on Problem Solving

Make a Plan

• **Write an equation**

Several steps may be needed to solve a problem. It often helps to write an equation that represents the steps.

Example:

Juan's first 3 exam scores are 85, 93, and 87. What does he need to score on his next exam to average 90 for the 4 exams?

Let x be the score on his next exam. The average of the exam scores is the sum of the 4 scores, divided by 4. This amount must equal 90.

$$\text{Average of exam scores} = 90$$

$$\frac{85 + 93 + 87 + x}{4} = 90$$

$$\frac{265 + x}{4} = 90$$

$$4\left(\frac{265 + x}{4}\right) = 4(90)$$

$$
\begin{aligned}
265 + x &= 360 \\
-265 &\quad -265 \\
x &= 95
\end{aligned}
$$

Juan needs a 95 on his next exam.

Read each problem and write an equation that could be used to solve it.

① The average of two numbers is 27. The first number is twice the second number. What are the two numbers?

② Nancy spends $\frac{1}{3}$ of her monthly salary on rent, $\frac{1}{10}$ on her car payment, $\frac{1}{12}$ on food, $\frac{1}{5}$ on other bills, and has $680 left for other expenses. What is Nancy's monthly salary?

③ A vendor at a concert sells caps and T-shirts. The T-shirts cost 1.5 times as much as the caps. If 5 caps and 7 T-shirts cost $248, what is the price of each item?

④ Amanda and Rick have the same amount to spend on school supplies. Amanda buys 4 notebooks and has $8.60 left. Rick buys 7 notebooks and has $7.55 left. How much does each notebook cost?

Answers

1. 18 and 36

2. $2400

3. Caps cost $16;
T-shirts cost $24.

4. $0.35

Focus on Problem Solving

Purpose: *To focus on making a plan, by writing an equation, to solve a problem*

Problem Solving Resources

Interactive Problem Solving . . pp. 79–84

Math: Reading and Writing in the Content Area . . . . pp. 79–84

Problem Solving Process

This page focuses on the second step of the problem-solving process: **Make a Plan**

Discuss

Have students identify what the variable in each equation represents and give the equation that can be used to solve the problem.

Possible answers:

1. Let x represent the second number; $\frac{2x + x}{2} = 27$.

2. Let x represent Nancy's monthly salary; $\frac{1}{3}x + \frac{1}{10}x + \frac{1}{12}x + \frac{1}{5}x + 680 = x$.

3. Let x represent the cost of one cap; $5x + 7(1.5x) = 248$.

4. Let x represent the cost of each notebook; $8.60 + 4x = 7.55 + 7x$.

Section 10B
Solving Equations and Inequalities

One-Minute Section Planner

Lesson	Materials	Resources
Lesson 10-4 Solving Multistep Inequalities **NCTM:** Number and Operations, Algebra, Communication **NAEP:** Algebra 4a ☐ SAT-9 ☐ SAT-10 ☐ ITBS ☐ CTBS ☐ MAT ☑ CAT	Optional Recording Sheet for Reaching All Learners *(CRB, p. 75)*	• *Chapter 10 Resource Book*, pp. 37–46 • Daily Transparency T17, CRB • Additional Examples Transparencies T18–T23, CRB • *Alternate Openers: Explorations*, p. 82
Lesson 10-5 Solving for a Variable **NCTM:** Algebra, Communication **NAEP:** Algebra 4a ☐ SAT-9 ☐ SAT-10 ☐ ITBS ☐ CTBS ☐ MAT ☑ CAT	Optional Recording Sheet for Reaching All Learners *(CRB, p. 76)*	• *Chapter 10 Resource Book*, pp. 47–56 • Daily Transparency T24, CRB • Additional Examples Transparencies T25–T27, CRB • *Alternate Openers: Explorations*, p. 83

Lesson 10-6 Systems of Equations
NCTM: Algebra, Communication
NAEP: Algebra 4a
☐ SAT-9 ☐ SAT-10 ☐ ITBS ☐ CTBS ☑ MAT ☐ CAT
	Optional
	Teaching Transparency T29
	  (CRB)
	Recording Sheet for
	Reaching All Learners
	  (CRB, p. 77)
	Calculators
		• *Chapter 10 Resource Book*, pp. 57–66
		• Daily Transparency T28, CRB
		• Additional Examples Transparencies
		  T30–T33, CRB
		• *Alternate Openers: Explorations*, p. 84

Section 10B Assessment		• Section 10B Quiz, AR p. 24 • *Test and Practice Generator* CD-ROM

SAT = Stanford Achievement Tests   **ITBS** = Iowa Test of Basic Skills   **CTBS** = Comprehensive Test of Basic Skills/Terra Nova
MAT = Metropolitan Achievement Test   **CAT** = California Achievement Test

NCTM = Complete standards can be found on pages T29–T35.   **NAEP** = Complete standards can be found on pages A54–A58.

SE = Student Edition   **TE** = Teacher's Edition   **AR** = Assessment Resources   **CRB** = Chapter Resource Book   **MK** = Manipulatives Kit

Section Overview

Solving and Graphing Two-Step Inequalities
Lesson 10-4

Why? To solve some problems, you need to write and solve inequalities.

A T-shirt retailer must pay $120 for a design and $4 per shirt. How many T-shirts would he have to sell at $9 per shirt to make a profit?

$$R > C$$

He will make a profit if his revenue R is greater than his cost C.

$$9x > 120 + 4x$$
$$\underline{-4x \qquad\qquad -4x}$$
$$5x > 120$$
$$\frac{5x}{5} > \frac{120}{5}$$
$$x > 24$$

The retailer would have to sell more than 24 T-shirts to make a profit.

```
<--+---+---⊕---+---+---+-->
  22  23  24  25  26  27
```

Solving a Variable, Graphing Solutions
Lesson 10-5

Why? Sometimes you need to solve for a variable in terms of other variables.

Solve for y and graph $x + 2y = 6$.

$$x + 2y = 6$$
$$\underline{-x \qquad\quad -x}$$
$$2y = -x + 6$$
$$\frac{2y}{2} = \frac{-x}{2} + \frac{6}{2}$$
$$y = -\frac{1}{2}x + 3$$

x	y
−4	5
−2	4
0	3
2	2
4	1

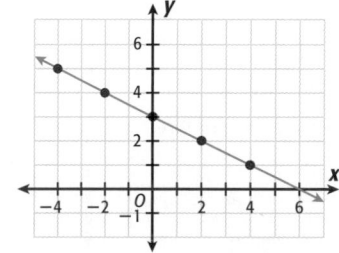

Systems of Equations
Lesson 10-6

Why? If a problem has more than one condition to be satisfied, it may require a system of equations.

Is $(3, -2)$ a solution of the system of equations?

$$2x + y = 4$$
$$x - y = 8$$

$$\begin{array}{ll}
2x + y = 4 & x - y = 8 \\
2(3) + (-2) \stackrel{?}{=} 4 & 3 - (-2) \stackrel{?}{=} 8 \\
6 + (-2) \stackrel{?}{=} 4 & 5 \neq 8 \text{✗} \\
4 \stackrel{?}{=} 4 \text{✔} &
\end{array}$$

$(3, -2)$ is *not* a solution because it does *not* satisfy *both* equations.

Is $(4, -4)$ a solution of the system of equations?

$$2x + y = 4$$
$$x - y = 8$$

$$\begin{array}{ll}
2x + y = 4 & x - y = 8 \\
2(4) + (-4) \stackrel{?}{=} 4 & 4 - (-4) \stackrel{?}{=} 8 \\
8 + (-4) \stackrel{?}{=} 4 & 8 = 8 \text{✔} \\
4 \stackrel{?}{=} 4 \text{✔} &
\end{array}$$

$(4, -4)$ *is* a solution because it satisfies *both* equations.

Solve the system.

$$y = 2x + 3$$
$$y = x - 5$$

$$2x + 3 = x - 5$$
$$\underline{-x \qquad\quad -x}$$
$$x + 3 = -5$$
$$\underline{-3 \quad\; -3}$$
$$x = -8$$

Now substitute -8 for x into either original equation, and solve for y.

$$y = x - 5$$
$$y = -8 - 5$$
$$y = -13$$

The solution is $(-8, -13)$.

Pacing: Traditional 1 day
Block $\frac{1}{2}$ day

Objective: Students solve two-step inequalities and graph the solutions of an inequality on a number line.

Warm Up

Solve.

1. $6x + 36 = 2x$ $x = -9$
2. $4x - 13 = 15 + 5x$ $x = -28$
3. $5(x - 3) = 2x + 3$ $x = 6$
4. $\frac{7}{8} + x = \frac{3}{16}$ $x = -\frac{11}{16}$

Problem of the Day

Find an integer x that makes the following two inequalities true:
$4 < x^2 < 16$ and $x < 2.5$ $x = -3$

Available on Daily Transparency in CRB

Math Humor

What did the marine biologist call an algebraic expression consisting of 8 terms added together? An *octo-plus*

10-4 Solving Multistep Inequalities

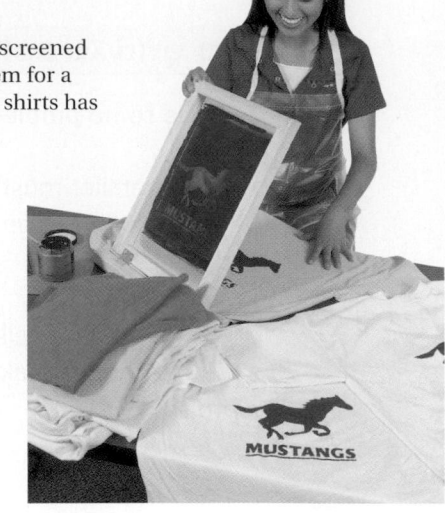

Learn to solve two-step inequalities and graph the solutions of an inequality on a number line.

The student council is making silk-screened T-shirts with the school logo on them for a fund-raiser. The cost of making the shirts has two parts.

1. fixed costs (silk screen equipment, etc.)
2. unit costs (shirts, ink, etc.)

Revenue is the price each unit is sold for multiplied by the number of units sold. The student council makes a profit when the revenue is greater than the cost. To find out how many units they need to sell for the revenue to be greater than the cost you can write and solve a multistep inequality.

Solving a multistep inequality uses the same inverse operations as solving a multistep equation. Multiplying or dividing the inequality by a negative number reverses the inequality symbol.

> **EXAMPLE** 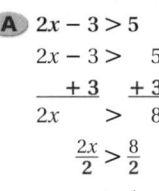 **Solving Two-Step Inequalities**
>
> Solve and graph.
>
> **A** $2x - 3 > 5$
>
> $\begin{array}{r} 2x - 3 > 5 \\ \underline{+3 +3} \\ 2x > 8 \end{array}$ *Add 3 to both sides.*
>
> $\frac{2x}{2} > \frac{8}{2}$ *Divide both sides by 2.*
>
> $x > 4$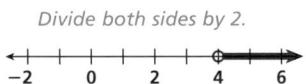
>
> **B** $-10 < 3x + 2$
>
> $\begin{array}{r} -10 < 3x + 2 \\ \underline{-2 -2} \\ -12 < 3x \end{array}$ *Subtract 2 from both sides.*
>
> $\frac{-12}{3} < \frac{3x}{3}$ *Divide both sides by 3.*
>
> $-4 < x$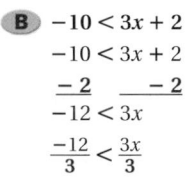

---

1 Introduce

Alternate Opener

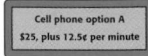
Motivate

Tell students that a goal in a fund-raiser is to take in more money than is spent. To accomplish that, you may need to determine the *least number* of items that must be sold to make a profit. One way to determine the least number of items is to write and solve an inequality.

Exploration worksheet and answers on Chapter 10 Resource Book pp. 38 and 85

2 Teach

 Lesson Presentation

Guided Instruction

In this lesson, students solve two-step inequalities and graph the solutions of an inequality on a number line. Remind students that the process of solving inequalities is the same as the process of solving equations, with one exception. The exception is that if you multiply or divide both sides of an inequality by a negative number, you must reverse the inequality symbol. (See Math Background.) Remind students how to graph inequalities on a number line.

Solve and graph.

C $-2x + 4 \le 3$

$$-2x + 4 \le \quad 3$$
$$\underline{\quad -4 \quad -4}$$ *Subtract 4 from both sides.*
$$-2x \quad \le -1$$

$$\frac{-2x}{-2} \ge \frac{-1}{-2}$$ *Divide each side by −2; change ≤ to ≥.*

$$x \ge \frac{1}{2}$$

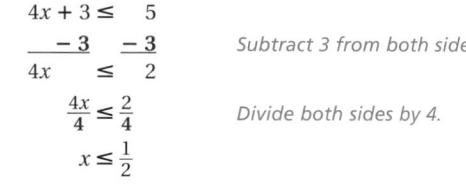

E X A M P L E **2** Solving Multistep Inequalities

Solve and graph.

A $3x - 2 - 4x > 5$

$$3x - 2 - 4x > \quad 5$$
$$-1x - 2 \quad > \quad 5$$ *Combine like terms.*
$$\underline{\quad +2 \quad\quad +2}$$ *Add 2 to both sides.*
$$-1x \quad > \quad 7$$

$$\frac{-1x}{-1} < \frac{7}{-1}$$ *Divide both sides by −1; change > to <.*

$$x < -7$$

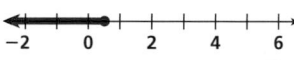

B $\frac{2x}{3} + \frac{1}{2} \le \frac{5}{6}$

$$\frac{2x}{3} + \frac{1}{2} \le \frac{5}{6}$$

$$6\left(\frac{2x}{3} + \frac{1}{2}\right) \le 6\left(\frac{5}{6}\right)$$ *Multiply by LCD, 6.*

$$6\left(\frac{2x}{3}\right) + 6\left(\frac{1}{2}\right) \le 6\left(\frac{5}{6}\right)$$

$$4x + 3 \le \quad 5$$
$$\underline{\quad -3 \quad -3}$$ *Subtract 3 from both sides.*
$$4x \quad \le \quad 2$$

$$\frac{4x}{4} \le \frac{2}{4}$$ *Divide both sides by 4.*

$$x \le \frac{1}{2}$$

Reaching All Learners

Through Critical Thinking

Give students a recording sheet (Chapter 10 Resource Book p. 75) with pairs of inequalities and graphs of possible solution sets. Have students check the graph of each solution set by choosing a number from the shaded portion and substituting that number into the inequality. For each incorrect graph, students should solve the inequality and graph the solution set correctly.

COMMON ERROR
///// **ALERT** \\\\\

Students will sometimes reverse the inequality symbol when they divide a negative number by a positive number.

Incorrect	Correct
$-12 < 3x$	$-12 < 3x$
$\frac{-12}{3} > \frac{3x}{3}$	$\frac{-12}{3} < \frac{3x}{3}$
$-4 > x$	$-4 < x$

Additional Examples

Example 1

Solve and graph.

A. $4x + 1 > 13$ $x > 3$

B. $-7 < 3x + 8$ $x > -5$

C. $-9x + 7 \ge 25$ $x \le -2$

Example 2

Solve and graph.

A. $10x + 21 - 4x < -15$ $x < -6$

B. $\frac{2x}{5} + \frac{3}{4} \ge \frac{9}{10}$ $x \ge \frac{3}{8}$

C. $8x + 8 > 11x - 1$ $x < 3$

Example 3

A school's Spanish club is selling bumper stickers. They bought 100 for $55, and have to give the company 15 cents for every sticker sold. If they plan to sell each bumper sticker for $1.25, how many do they have to sell to make a profit?

more than 50 bumper stickers

Solve and graph.

C $2x + 3 > 5x - 6$

$$2x + 3 > 5x - 6$$
$$\underline{-2x \qquad -2x}$$ *Subtract 2x from both sides.*
$$3 > 3x - 6$$
$$\underline{+6 \qquad +6}$$ *Add 6 to both sides.*
$$9 > 3x$$
$$\frac{9}{3} > \frac{3x}{3}$$ *Divide both sides by 3.*
$$3 > x$$

EXAMPLE 3 *Business Application*

The student council sells T-shirts with the school logo on them. The unit cost is $10.50 for the shirt and the ink. They have a fixed cost of $60 for silk screen equipment. If they sell the shirts for $12 each, how many must they sell to make a profit?

Let R represent the revenue and C represent the cost. In order for the student council to make a profit, the revenue must be greater than the cost.

$$R > C$$

The revenue from selling x shirts at $12 each is $12x$. The cost of producing x shirts is the fixed cost plus the unit cost times the number of shirts produced, or $60 + 10.50x$. Substitute the expressions for R and C.

$$12x > 60 + 10.50x$$ *Let x represent the number of shirts sold.*
$$\underline{-10.50x \qquad -10.50x}$$ *Subtract 10.50x from both sides.*
$$1.5x > 60$$
$$\frac{1.5x}{1.5} > \frac{60}{1.5}$$ *Divide both sides by 1.5.*
$$x > 40$$

The student council must sell more than 40 shirts to make a profit.

Think and Discuss

1. **Compare** solving a multistep equation with solving a multistep inequality.

2. **Describe** two situations in which you would have to reverse the inequality symbol when solving a multistep inequality.

3 Close

Summarize

Remind students that to solve a multistep inequality, they should do the following:

• Clear fractions.

• Combine like terms.

• Get the variable on one side of the inequality only.

• Undo addition and subtraction.

• Undo multiplication and division, remembering to reverse the inequality symbol if they multiply or divide both sides of an inequality by a negative number.

Answers to Think and Discuss

Possible answers:

1. Use the same steps for both, but if you multiply or divide an inequality by a negative number, you must reverse the inequality symbol.

2. To solve the inequality $-2x + 1 > 7$, subtract 1 from both sides, divide both sides by -2, and reverse the inequality symbol. To solve the inequality $2 - \frac{x}{5} < 3$, subtract 2 from both sides, multiply both sides by -5, and reverse the inequality symbol.

10-4 Exercises

FOR EXTRA PRACTICE
see page 750

☑ internet connect
Homework Help Online
go.hrw.com   Keyword: MP4 10-4

10-4 PRACTICE & ASSESS

GUIDED PRACTICE

See Example ① **Solve and graph.**

1. $2k + 4 > 10$ $k > 3$
2. $\frac{1}{2}z - 5.5 \le 4.5$ $z \le 20$
3. $5y + 10 < -25$ $y < -7$
4. $-4x + 6 \ge 14$ $x \le -2$
5. $4y + 1.5 \ge 13.5$ $y \ge 3$
6. $3k - 2 > 13$ $k > 5$

See Example ②
7. $4x - 3 + x < 12$ $x < 3$
8. $\frac{4b}{5} + \frac{7}{10} \ge \frac{1}{2}$ $b \ge -\frac{1}{4}$
9. $4 + 9h - 7 \le 3h + 3$

10. $14c + 2 - 3c > 8 + 8c$
11. $\frac{1}{9} + \frac{d}{3} < \frac{1}{2} - \frac{2d}{3}$
12. $\frac{5}{6} \ge \frac{4m}{9} - \frac{1}{3} + \frac{2m}{9}$

See Example ③ 13. A school's Spanish club is selling printed caps to raise money for a trip. The printer charges $150 in advance plus $3 for every cap ordered. If the club sells caps for $12.50 each, at least how many caps do they need to sell to make a profit? **at least 16 caps**

INDEPENDENT PRACTICE

See Example ① **Solve and graph.**

14. $6k - 8 > 22$ $k > 5$
15. $10x + 2 > 42$ $x > 4$
16. $5p - 5 \le 45$ $p \le 10$
17. $14 \ge 13q - 12$ $q \le 2$
18. $3.6 + 7.2n < 25.2$ $n < 3$
19. $-8x - 12 \ge 52$ $x \le -8$

See Example ②
20. $7p + 5 < 6p - 12$ $p < -17$
21. $11 + 17a \ge 13a - 1$ $a \ge -3$
22. $\frac{11}{13} + \frac{n}{2} > \frac{25}{26}$ $n > \frac{3}{13}$
23. $\frac{2}{3} \le \frac{1}{2}k - \frac{5}{6}$ $k \ge 3$
24. $\frac{n}{7} + \frac{11}{14} \le -\frac{17}{14}$ $n \le -14$
25. $3r - 16 + 7r < 14$ $r < 3$

See Example ③ 26. Josef is on the planning committee for the eighth-grade holiday party. The food, decoration, and entertainment costs total $350. The committee has $75 in the treasury. If the committee expects to sell the tickets for $5 each, at least how many tickets must be sold to cover the remaining cost of the party? **at least 55 tickets**

PRACTICE AND PROBLEM SOLVING

Solve and graph.

27. $3p - 3 \le 19$ $p \le \frac{22}{3}$
28. $12n + 26 > -10$ $n > -3$
29. $4 - 9w < 13$ $w > -1$
30. $-8x - 18 \ge 14$ $x \le -4$
31. $16a + 3 > 11$ $a > \frac{1}{2}$
32. $-2y + 1 \ge 8$ $y \le -\frac{7}{2}$
33. $3q - 5q > -12$ $q < 6$
34. $\frac{3m}{4} + \frac{2}{3} > \frac{m}{2} + \frac{7}{8}$ $m > \frac{5}{6}$
35. $7b - 4.6 < 3b + 6.2$ $b < 2.7$
36. $6k + 4 - 3k \ge 2$ $k \ge -\frac{2}{3}$
37. $26 - \frac{33}{4} \le -\frac{2}{3}f - \frac{1}{4}$ $f \le -27$
38. $\frac{7}{9}v + \frac{5}{12} - \frac{3}{18}v \ge \frac{3}{4}v + \frac{1}{3}$ $v \le \frac{3}{5}$

39. **ENTERTAINMENT** A concert is being held in a gymnasium that can hold no more than 550 people. A permanent bleacher will seat 30 people. The event organizers are setting up 20 rows of chairs. At most, how many chairs can be in each row? **at most 26 chairs**

Students may want to refer back to the lesson examples.

Assignment Guide

If you finished Example ① assign:
Core 1–6, 14–19, 46–50
Enriched 14–19, 27–32, 46–50

If you finished Example ② assign:
Core 1–12, 14–25, 46–50
Enriched 14–25, 27–38, 46–50

If you finished Example ③ assign:
Core 1–26, 27–41 odd, 46–50
Enriched 14–50

Answers

1–12, 14–25, 27–38. Complete answers on p. A9

9. $h \le 1$
10. $c > 2$
11. $d < \frac{7}{18}$
12. $m \le 1\frac{3}{4}$

RETEACH 10-4

LESSON Reteach
10-4 Solving Multistep Inequalities

To solve an inequality, undo operations as with an equation. But, when multiplying or dividing by a negative number, reverse the inequality symbol.

$3x + 2 > 11$	To undo addition,	$-3x + 2 > 11$	To undo addition,
$-2\ -2$	subtract 2.	$-2\ -2$	subtract 2.
$3x\ > 9$	To undo multiplication,	$-3x\ > 9$	To undo multiplication,
$\frac{3x}{3}\ > \frac{9}{3}$	divide by 3.	$\frac{-3x}{-3}\ < \frac{9}{-3}$	divide by -3.
$x\ > 3$	Same direction.	$x\ < -3$	Change > to <.

The solution set contains all real numbers greater than 3.

The solution set contains all real numbers less than -3.

Complete to solve and graph.

1. $2t + 1 \le 9$ To undo addition,
   $-1\ -1$ subtract.
   $2t\ \le 8$ To undo multiplication,
   $\frac{2t}{2} \le \frac{8}{2}$ divide.
   $t\ \le 4$ Same direction.

2. $-2t + 1 \le 9$ To undo addition,
   $-1\ -1$ subtract.
   $-2t\ \le 8$ To undo multiplication,
   $\frac{-2t}{-2} \ge \frac{8}{-2}$ divide.
   $t\ \ge -4$ Change $\le$ to $\ge$.

3. $\frac{z+2}{3} > 1$ To undo division,
   $3 \cdot \frac{z+2}{3} \ge 3 \cdot 1$ multiply.
   $z + 2 \ge 3$ To undo addition,
   $-2\ -2$ subtract.
   $z\ \ge 1$ Same direction.

4. $\frac{z}{-3} + 2 > 1$ To undo addition,
   $-2\ -2$ subtract.
   $\frac{z}{-3}\ > -1$ To undo division,
   $-3 \cdot \frac{z}{-3} < -3 \cdot (-1)$ multiply.
   $z\ < 3$ Change > to <.

PRACTICE 10-4

LESSON Practice B
10-4 Solving Multistep Inequalities

Solve and graph.

1. $4x - 2 < 26$ $x < 7$
2. $6 - \frac{1}{5}y \le 7$ $y \ge -5$
3. $3x + 27 - x \ge 15$ $x \ge -6$
4. $10x > 14x + 8$ $x < -2$
5. $7 - 2w \le 2w + 19$ $w \ge -3$
6. $-\frac{2}{3}k + 13 < 15$ $k > -3$
7. $3(2x - 5) \le 4x - 17$ $x \le -1$
8. $12 - \frac{1}{8}y < 19 + \frac{3}{4}y$ $y > -8$

9. Two-thirds of a number, decreased by thirty-six, is at most twenty-two. Find the number. **87**

10. Jack wants to weigh at most 176 pounds before the baseball season begins. He has already lost 15 pounds. He believes he can lose 1.5 pounds a week. If Jack weighed 197 pounds before his weight loss, about how many weeks will it take for Jack to get to his desired weight for the baseball season? **4 weeks**

Math Background

To better understand why an inequality symbol must be reversed when multiplying or dividing by a negative number, use an inequality such as $10 > 8$ to examine the following four possibilities.

- Multiply each side by 2:
  $20 > 16$ (true).
- Divide each side by 2:
  $5 > 4$ (true).
- Multiply each side by -2:
  $-20 < -16$. (The inequality sign must be reversed to make it true.)
- Divide each side by -2:
  $-5 < -4$. (The inequality sign must be reversed to make it true.)

43. Possible answer: Sergio paid $7.95 for the shipping on his last purchase, and his total was less than $49.45. Write and solve an inequality to describe the cost of the merchandise he bought. Answer: $x + \$7.95 < \$49.45$; $x < \$41.50$.

44. Possible answer: Method 1: Subtract x from both sides, add 3 to both sides, divide both sides by -3, and reverse the inequality symbol. Method 2: Add $2x$ to both sides, subtract 4 from both sides, and divide both sides by 3.

Journal

Ask students to describe some real-world situations in which an inequality might be more useful than an equation.

Test Prep Doctor

For Exercise 50, remind students that a fair number cube has six faces, numbered 1 through 6. Of these six numbers, three are odd, so the probability of rolling an odd number is $\frac{3}{6}$, or $\frac{1}{2}$. The correct answer is **A**.

Lesson Quiz

Solve and graph.

1. $4x - 6 > 10$ $x > 4$

(number line: 0, 4, 8 — open circle at 4, shaded right)

2. $7x + 9 < 3x - 15$ $x < -6$

(number line: -10, -6, -2 — open circle at -6, shaded left)

3. $w - 3w < 32$ $w > -16$

(number line: -20, -16, -12 — open circle at -16, shaded right)

4. $\frac{2}{3}w + \frac{1}{4} \leq \frac{1}{2}$ $w \leq \frac{3}{8}$

(number line: 0, $\frac{3}{8}$, 1 — closed circle at $\frac{3}{8}$, shaded left)

5. Antonio has budgeted an average of $45 a month for entertainment. For the first five months of the year he has spent $48, $39, $60, $48, and $33. How much can Antonio spend in the sixth month without exceeding his average budget? no more than $42

Available on Daily Transparency in CRB

---

40. Katie and April are making a string of pi beads for pi day (March 14). They use 10 colors of beads that represent the digits 0–9, and the beads are strung in the order of the digits of π. The string already has 70 beads. If they have 30 days to string the beads, and they want to string 1000 beads by π day, at least how many beads do they have to string each day?
at least 31 beads

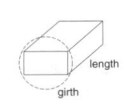

41. **SPORTS** The Cubs have won 44 baseball games and have lost 65 games. They have 53 games remaining. At least how many of the remaining 53 games must the Cubs win to have a winning season? (A winning season means they win more than 50% of their games.)
at least 38 games

42. **ECONOMICS** Satellite TV customers can either purchase a dish and receiver for $249 or pay a $50 fee and rent the equipment for $12 a month.
  a. How much would it cost to rent the equipment for 9 months? **$158**
  b. How many months would it take for the rental charges to exceed the purchase price? **17 mo**

43. **WRITE A PROBLEM** Write and solve an inequality using the following shipping rates for orders from a mail-order catalog.

Mail-Order Shipping Rates				
Merchandise Amount	$0.01–$20	$20.01–$30	$30.01–$45	$45.01–$60
Shipping Cost	$4.95	$5.95	$7.95	$8.95

44. **WRITE ABOUT IT** Describe two ways to solve the inequality below. In one way, you must reverse the inequality symbol, but in the other way, you do not need to reverse the symbol.

$$-2x - 3 < x + 4$$

45. **CHALLENGE** Solve the inequality $\frac{x-1}{5} - \frac{x+2}{6} \geq \frac{7}{15}$. $x \geq 30$

Spiral Review

Find each number. (Lesson 8-3)

46. 19 is 20% of what number?   **95**

47. 74% of what number is 481?   **650**

48. 32% of what number is 58.88?   **184**

49. 0.7488 is 52% of what number?   **1.44**

50. **TEST PREP** What is the probability of rolling an odd number on a fair number cube? (Lesson 9-4)   **A**

  A $\frac{1}{2}$ B $\frac{2}{3}$ C $\frac{1}{6}$ D $\frac{1}{3}$

CHALLENGE 10-4

LESSON 10-4 Challenge
Updated Pony Express

Pat wants to send some copies of her newly published book to friends.

According to the U.S. Postal Service:

Rates are based on the weight of the piece and the zone (distance from origin to destination ZIP code).

The combined length and girth (perimeter of an end) of a package may not exceed 108 inches.

(diagram of a box labeled *length* and *girth*)

1. Pat wants the box that contains books to be 6 inches high, and twice as long as it is wide.

Let x represent the width of a box that Pat might use.

$2(x) + 2(6) + 2x \leq 108$
$4x + 12 \leq 108$
$4x \leq 96$
$x \leq 24$

Write and solve an inequality to find all possible widths for a box that will satisfy the postal requirements and Pat's conditions.

possible width: **≤ 24 inches**

2. Pat's husband, Mike, suggests that the box be 8 inches high and that the length be 3 times the width.

Let z represent the length of a box that Mike suggests.

$2\left(\frac{z}{3}\right) + 2(8) + z \leq 108$
$2z + 48 + 3z \leq 324$
$5z + 48 \leq 324$
$5z \leq 276$
$z \leq 55.2$

Write and solve an inequality to find, to the nearest inch, the maximum length for a box that will satisfy.

maximum length: **55 inches**

3. On May 1, 2002, Pat shipped a box containing a book to a friend who lives in Zone 4. Pat paid $2.08 to ship this package.

According to the table below, write an inequality to show the weight of this package.   **2.5 < x ≤ 3**

Bound Printed Matter Rates

Weight Not Over (pounds)	Local, Zones 1&2	Zone 3	Zone 4	Zone 5	Zone 6	Zone 7	Zone 8
1.0	$1.80	$1.83	$1.87	$1.93	$1.99	$2.06	$2.21
1.5	1.80	1.83	1.87	1.93	1.99	2.06	2.21
2.0	1.84	1.88	1.94	2.02	2.10	2.19	2.38
2.5	1.90	1.95	2.00	2.11	2.21	2.33	2.57
3.0	1.94	2.00	2.08	2.20	2.32	2.46	2.75
3.5	1.99	2.06	2.15	2.29	2.43	2.60	2.93
4.0	2.03	2.11	2.21	2.37	2.55	2.72	3.11

PROBLEM SOLVING 10-4

LESSON 10-4 Problem Solving
Solving Multistep Inequalities

A school club wants to sell printed T-shirts for a fundraiser. The chart below describes the cost to purchase shirts and the price for which the students could sell the shirts. In addition, there is a $25 setup fee to print the shirts.

	Shirt	Cost	Price
	50/50	$4.33 ea	$9.50 ea
	100% cotton	$5.42 ea	$12 ea

1. How many 50/50 shirts would the club have to sell to make a profit?
5 shirts

2. How many 100% cotton shirts would the club have to sell to make a profit?
4 shirts

3. How many of each kind of shirt would the club have to sell to make at least $1000?
50/50: 199; 100% cotton: 156

For Exercises 4–5, use this equation to estimate typing speed, $S = \frac{1}{5}(w - 10e)$, where S is the accurate typing speed, w is the number of words typed in 5 minutes, and e is the number of errors.

4. One of the qualifications for a job is a typing speed of at least 65 words per minute. If Jordan knows that she will be able to type 350 words in five minutes, what is the maximum number of errors she can make?
  A 0   C 3
  B 2   D 4

5. Tanner usually makes 3 errors every 5 minutes when he is typing. If his goal is an accurate typing speed of at least 55 words per minute, how many words does he have to be able to type in 5 minutes?
  F 61 words   **H 305 words**
  G 325 words   J 325 words

6. A taxi charges $2.25 for the first mile and $0.20 for each mile after the first, or $F = \$2.25 + \$0.20(m - 1)$. How many miles can you travel in the cab and have the fare be less than $10?
  A 15   **C 39**
  B 25   D 43

7. Celia's long distance company charges $5.95 per month plus $0.06 per minute. If Celia has budgeted $30 for long distance, what is the maximum number of minutes she can call long distance per month?
  F 375 minutes   H 405 minutes
  G 400 minutes   J 420 minutes

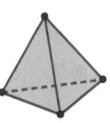

10-5 Solving for a Variable

Learn to solve an equation for a variable.

Euler's formula relates the number of vertices V, the number of edges E, and the number of faces F of a polyhedron.

$$V - E + F = 2$$

Tetrahedron:
4 faces   6 edges   4 vertices
$4 - 6 + 4 = 2$

$V - E + F = 2$
$e^{i\theta} = \cos\theta + i\sin\theta$

Leonard Euler (1707–1783) made major contributions to nearly every area of mathematics, including algebra, geometry, and calculus.

Suppose a polyhedron has 8 vertices and 12 edges. How many faces does it have? One way to find the answer is to substitute values into the formula and solve. Another way to find the answer is to solve for the variable first and then substitute the values.

Substitute, then solve:

$$V - E + F = \quad 2$$
$$8 - 12 + F = \quad 2$$
$$-4 + F = \quad 2$$
$$\underline{+4 \qquad +4}$$
$$F = \quad 6$$

Solve, then substitute:

$$V - E + F = 2$$
$$\underline{-V + E \qquad -V + E}$$
$$F = 2 - V + E$$
$$F = 2 - 8 + 12$$
$$F = 6$$

If an equation contains more than one variable, you can sometimes isolate one of the variables by using inverse operations. You can add and subtract any variable quantity on both sides of an equation.

EXAMPLE 1   **Solving for a Variable by Addition or Subtraction**

Solve for the indicated variable.

A Solve $V - E + F = 2$ for V.

$$V - E + F = 2$$
$$\underline{+ E - F \qquad + E - F}$$ *Add E and subtract F from both sides.*
$$V \qquad = 2 + E - F$$ *Isolate V.*

B Solve $V - E + F = 2$ for E.

$$V - E + F = \quad 2$$
$$\underline{-V \quad - F \qquad - V - F}$$ *Subtract V and F from both sides.*
$$-E \quad = \quad 2 - V - F$$
$$-1 \cdot (-E) \quad = -1 \cdot (2 - V - F)$$ *Multiply both sides by −1.*
$$E \quad = \quad -2 + V + F$$ *Isolate E.*

---

1 Introduce

Alternate Opener

10-5 Solving for a Variable

Use the formula $A = \ell \cdot w$ to find each missing dimension.

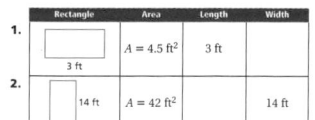

Rectangle	Area	Length	Width
1.	$A = 4.5 \text{ ft}^2$ 3 ft	3 ft	
2.	14 ft $A = 42 \text{ ft}^2$		14 ft

Use the formula $A = \frac{1}{2}b \cdot h$ to find each missing dimension.

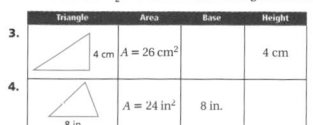

Triangle	Area	Base	Height
3.	4 cm $A = 26 \text{ cm}^2$		4 cm
4.	8 in.	$A = 24 \text{ in}^2$	8 in.

Think and Discuss
5. **Explain** how you used the formula $A = \ell \cdot w$ in numbers 1 and 2.
6. **Explain** how you used the formula $A = \frac{1}{2}b \cdot h$ in numbers 3 and 4.

Motivate

Show students these two equivalent equations that relate circumference and diameter: $C = \pi d$ and $d = \frac{C}{\pi}$.

Explain that the first equation is more convenient if you know the diameter and want to find the circumference, and the second is more convenient if you know the circumference and want to find the diameter.

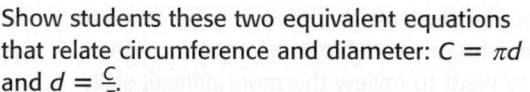
Exploration worksheet and answers on Chapter 10 Resource Book pp. 48 and 87

10-5 Organizer

Pacing: Traditional 1 day
Block $\frac{1}{2}$ day

Objective: Students solve an equation for a variable.

Warm Up

Solve.

1. $8x - 9 = 23$ $x = 4$
2. $9x + 12 = 4x + 37$ $x = 5$
3. $6x - 8 = 7x + 3$ $x = -11$
4. $\frac{1}{2}x + 3 = \frac{1}{8}$ $x = -\frac{23}{4}$, or $-5\frac{3}{4}$

Problem of the Day

The formula $A = 4\pi r^2$ gives the surface area of a geometric figure. Solve the formula for r. $r = \sqrt{\frac{A}{4\pi}}$

Can you identify what the geometric figure is?   sphere

Available on Daily Transparency in CRB

Math Humor

If a father gives his daughter 15 cents and his son 10 cents, what time is it?
A quarter to two

---

2 Teach   [Lesson Presentation]

Guided Instruction

In this lesson, students learn to solve an equation for a variable. Share with students that they will be solving equations by methods they already know, but that the solutions will be expressions with one or more variables. You may want to show how to solve the equation in Example 1A for V by adding E to both sides and then subtracting F from both sides. Point out that the method on the student page combines these two steps into one step. For Example 2A, you may point out that dividing both sides by b requires that $b \neq 0$.

Answers

29. $\ell = \dfrac{P - 2w}{2}$

30. $r = \dfrac{A - P}{Pt}$

31. $C = \dfrac{5}{9}(F - 32)$

32. $y = -\dfrac{2}{3}x + 3$

33. $y = \dfrac{7 - x}{3}$

34. $x = 3y + 2$

35. $(2, 9)$

36. $(8, 10)$

37. $(3, 5)$

38. $(3, -2)$

39. $(6, -2)$

40. $\left(-\dfrac{1}{2}, -2\right)$

10-5 Solving for a Variable (pp. 519–522)

EXAMPLE

Solve for the indicated variable.

■ Solve $A = 3b - 4c$ for b.

$$A = 3b - 4c$$

$$\underline{\quad + 4c \qquad + 4c\quad}$$ *Add 4c to both sides.*

$$A + 4c = 3b$$

$$\dfrac{A + 4c}{3} = \dfrac{3b}{3}$$ *Divide by 3.*

$$\dfrac{A}{3} + \dfrac{4c}{3} = b$$

■ Solve $m = \dfrac{100y}{x}$ for y.

$$m = \dfrac{100y}{x}$$

$$x \cdot m = x \cdot \left(\dfrac{100y}{x}\right)$$ *Multiply by x.*

$$xm = 100y$$

$$\dfrac{xm}{100} = \dfrac{100y}{100}$$ *Divide by 100.*

$$\dfrac{xm}{100} = y$$

EXERCISES

Solve for the indicated variable.

29. Solve $P = 2w + 2\ell$ for ℓ.

30. Solve $A = P + Prt$ for r.

31. Solve $F = \dfrac{9}{5}C + 32$ for C.

32. Solve $2x + 3y = 9$ for y.

33. Solve $x + 3y = 7$ for y.

34. Solve $4x - 12y = 8$ for x.

10-6 Systems of Equations (pp. 523–527)

EXAMPLE

■ Solve the system of equations.

$$4x + y = 3$$
$$x + y = 12$$

Solve both equations for y.

$$4x + y = 3 \qquad\qquad x + y = 12$$
$$\underline{-4x \qquad -4x} \qquad \underline{-x \qquad -x}$$
$$y = -4x + 3 \qquad\qquad y = -x + 12$$

$$-4x + 3 = -x + 12$$
$$\underline{+4x \qquad\qquad +4x}$$ *Add 4x.*
$$3 = 3x + 12$$
$$\underline{-12 \qquad\qquad -12}$$ *Subtract 12.*
$$-9 = 3x$$
$$\dfrac{-9}{3} = \dfrac{3x}{3}$$ *Divide by 3.*
$$-3 = x$$

$$y = -4x + 3$$
$$= -4(-3) + 3$$ *Substitute −3 for x.*
$$= 12 + 3$$
$$= 15$$

The solution is $(-3, 15)$.

EXERCISES

Solve each system of equations.

35. $y = x + 7$
$\quad\ y = 2x + 5$

36. $x - y = -2$
$\quad\ x + y = 18$

37. $4x + 3y = 27$
$\quad\ 2x - y = 1$

38. $4x + y = 10$
$\quad\ x - 2y = 7$

39. $3x - 4y = 26$
$\quad\ x + 2y = 2$

40. $4x - 3y = 4$
$\quad\ 2x - y = 1$

Solve each equation.

1. $3t - 1 = 92$ $t = 31$

2. $\frac{2}{5}y - 9 = 1$ $y = 25$

3. $\frac{z - 3}{5} = -4$ $z = -17$

4. $\frac{7x}{9} - \frac{2}{9} = \frac{19}{9}$ $x = 3$

5. $\frac{2v}{5} + \frac{v}{4} = \frac{9}{5} + \frac{3}{20}$ $v = 3$

6. $\frac{r}{3} - \frac{2r}{5} - \frac{1}{4} = \frac{5}{12}$ $r = -10$

7. $16z - 3z + 9 = 2z + 86$ $z = 7$

8. $\frac{1}{4}w = 2w + \frac{35}{2}$ $w = -10$

9. $15n = 29 + 2(3n - 1)$ $n = 3$

10. $3(s + 1) - (s - 2) = 22s$ $s = \frac{1}{4}$

11. $\frac{3}{5}(15k + 10) = 12k - 9$ $k = 5$

12. $\frac{3m}{4} - \frac{1}{9} = \frac{5m}{12} + \frac{14}{9}$ $m = 5$

13. A delivery service charges $2.50 for the first pound and $0.75 for each additional pound. If Carl paid $7.75 for his package, how many pounds did the package weigh?   **8 lb**

14. A wireless phone company offers two plans. In one plan, there is a monthly fee of $40 and a charge of $0.25 per minute. In the other plan, there is a monthly fee of $25 and a charge of $0.40 per minute. For what number of minutes are the costs equal?   **100 min**

15. The rectangle and the triangle shown have the same perimeter. Find the perimeter of each figure.   **48 units**

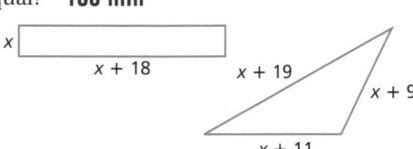

Solve and graph each inequality.

16. $6m + 4 > 2$

17. $z + 3z + 4 \geq -8$

18. $3x - 5x - 4 > 2$

19. $8 - 3p > 14$

Solve for the indicated variable.

20. Solve for w. $P = 2(\ell + w)$ $w = \frac{P}{2} - \ell$

21. Solve for r. $s = c + rc$ $r = \frac{s - c}{c}$

22. Solve for b. $A = \pi ab$ $b = \frac{A}{\pi a}$

23. Solve for d. $x^2 + d^2 = 1$ $d = \sqrt{1 - x^2}$

Solve each equation for y.

24. $x + 2y = 12$ $y = -\frac{1}{2}x + 6$

25. $10 - x + y = 0$ $y = x - 10$

26. $3y - x = -6$ $y = \frac{1}{3}x - 2$

27. $4x + 3y = 6$ $y = -\frac{4}{3}x + 2$

Solve each system of equations.

28. $x - 2y = 16$ $(2, -7)$
    $4x + y = 1$

29. $x + 2y = 6$ $(-2, 4)$
    $4x + 3y = 4$

30. $3x - 2y = -3$
    $3x + y = 3$ $\left(\frac{1}{3}, 2\right)$

31. $x + 5y = 11$ $(1, 2)$
    $4x - y = 2$

 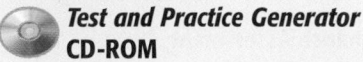

Purpose: *To assess students' mastery of concepts and skills in Chapter 10*

Assessment Resources ✔

Chapter 10 Tests (Levels A, B, C)
Assessment Resources . . . . . pp. 87–92

💿 *Test and Practice Generator* **CD-ROM**

Additional assessment items in both multiple-choice and free-response format may be generated for any objective in Chapter 10.

Answers

16. $m > -\frac{1}{3}$

17. $z \geq -3$

18. $x < -3$

19. $p < -2$

Chapter
10 **Performance Assessment**

Purpose: To assess students' under-standing of concepts in Chapter 10 and combined problem-solving skills

Assessment Resources ✓

Performance Assessment
Assessment Resources .... p. 136

Performance Assessment Teacher Support
Assessment Resources .... p. 135

Answers

1–3. See p. A10.

4. See Level 3 work sample below.

Scoring Rubric for Problem Solving Item 4

Level 3
Accomplishes the purpose of the task.

Student gives clear explanations, shows understanding of mathematical ideas and processes, and computes accurately.

Level 2
Purposes of the task not fully achieved.

Student demonstrates satisfactory but limited understanding of the mathematical ideas and processes.

Level 1
Purposes of the task not accomplished.

Student shows little evidence of under-standing the mathematical ideas and processes and makes computational and/or procedural errors.

Performance Assessment

✎ Show What You Know

Create a portfolio of your work from this chapter. Complete this page and include it with your four best pieces of work from Chapter 10. Choose from your homework or lab assignments, mid-chapter quiz, or any journal entries you have done. Put them together using any design you want. Make your portfolio represent what you consider your best work.

★ Short Response

1. Solve the inequality: $7x - 4 < 9x + 14$. Show your work or explain in words how you determined your answer.

2. Solve the system of equations. Show your work.
$$x - y = -3$$
$$2x - 4y = 22$$

3. Alfred and Eugene each spent $62 on campsite and gasoline expenses during their camping trip. Each campsite they stayed at had the same per-night charge. Alfred paid for 4 nights of campsites and $30 for gasoline. Eugene paid for 2 nights of campsites and $46 for gasoline. Write an equation that could be used to determine the cost of one night's stay at a campsite. What was the cost of one night's stay at a campsite?

🧩 Extended Problem Solving

4. You are designing a house to fit on a rectangular lot that has 90 feet of lake frontage and is 162 feet deep. The building codes require that the house not be built closer than 10 feet to the lot boundary lines.

   a. Write an inequality and solve it to find how long the front of the house facing the lake may be.

   b. If you want the house to cover no more than 20% of the lot, what would be the maximum square footage of the house?

   c. If you want to spend a maximum of $100,000 building the house, to the nearest whole dollar, what would be the maximum you could spend per square foot for a 1988-square-foot house?

90 ft
162 ft

Student Work Samples for Item 4

Level 3

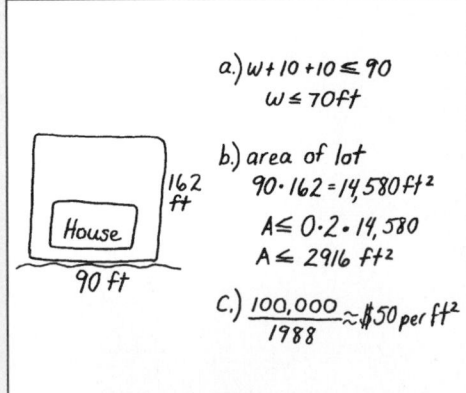

a.) $w + 10 + 10 \leq 90$
   $w \leq 70 ft$

b.) area of lot
   $90 \cdot 162 = 14,580 ft^2$
   $A \leq 0.2 \cdot 14,580$
   $A \leq 2916 ft^2$

c.) $\frac{100,000}{1988} \approx \$50 per ft^2$

House
162 ft
90 ft

The student drew a diagram, calculated each answer correctly, and used proper notation.

Level 2

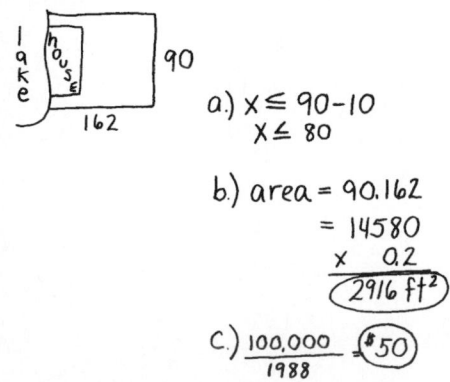

lake / house
90
162

a.) $x \leq 90 - 10$
   $x \leq 80$

b.) area = $90 \cdot 162$
   $= 14580$
   $\times \quad 0.2$
   $\boxed{2916 ft^2}$

c.) $\frac{100,000}{1988}$ $\boxed{\$50}$

The student drew a diagram and answered two of the three parts correctly but made a mistake in part **a.**

Level 1

a.) $90 - 10 = 80 feet$

b.) $20\% \cdot 80 = 1600$

c.) $\frac{100,000}{1600} = 62.5$

The student performed basic operations on the numbers in the problem without making any real progress.

Standardized Test Prep

Chapter 10

Cumulative Assessment, Chapters 1–10

1. Which of the following is the solution of the equation $5x = 4(x + 2)$?   **C**

Ⓐ $x = 2$ Ⓒ $x = 8$

Ⓑ $x = -2$ Ⓓ $x = -8$

2. If $x + y = 3 + k$ and $2x + 2y = 10$, what is the value of k?   **J**

Ⓕ 7 Ⓗ 3

Ⓖ 6 Ⓙ 2

3. If $3 = b^x$, what is the value of $3b$?   **A**

Ⓐ b^{x+1} Ⓒ b^{2x}

Ⓑ b^{x+2} Ⓓ b^{3x}

4. Tim is rolling a number cube with the numbers 1 through 6 on it. He rolls the cube twice. What is the probability that the two rolls will have a sum of 10?   **G**

Ⓕ $\frac{1}{36}$ Ⓗ $\frac{1}{10}$

Ⓖ $\frac{1}{12}$ Ⓙ $\frac{1}{9}$

5. The formula $M = \frac{P(rt + 1)}{12t}$ gives the monthly payment M on a loan with principal P, annual interest rate r, and length t years. What is the monthly payment on a 2-year loan for $3000 at an annual rate of 8%?   **C**

Ⓐ $605 Ⓒ $145

Ⓑ $480 Ⓓ $125

6. Mia earns a monthly base salary of $640 plus a 12% commission on her total monthly sales. At the end of last month, Mia earned $2380. What were her total monthly sales?   **J**

Ⓕ $145 Ⓗ $1740

Ⓖ $285.60 Ⓙ $14,500

7. Ten years from now, Cal will be x years old. How old was he 5 years ago?   **C**

Ⓐ $x - 5$ Ⓒ $x - 15$

Ⓑ $x - 10$ Ⓓ $x + 10$

TIP! **TEST TAKING TIP!**

One way to find an answer is to substitute the choices into the problem. Be sure to examine all the choices before deciding.

8. If n is the least positive integer for which $3n$ is both an even integer and the square of an integer, what is the value of n?   **J**

Ⓕ 3 Ⓗ 6

Ⓖ 4 Ⓙ 12

9. *SHORT RESPONSE* Between which two consecutive positive integers does $\sqrt{213}$ lie? Explain in words how you determined your answer.

10. *SHORT RESPONSE* The graph shows Richie's weekly budget.

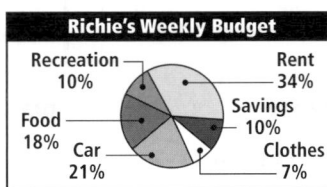

Richie's Weekly Budget

- Recreation 10%
- Rent 34%
- Savings 10%
- Clothes 7%
- Car 21%
- Food 18%

Write an equation that could be used to find how much more money m Richie allots for food than for recreation with a budget of d dollars. If Richie allots $6.40 more for food than for recreation, what is his total monthly budget?

Purpose: To provide review and practice for Chapters 1–10 and standardized tests

Assessment Resources

Cumulative Tests (Levels A, B, C)
Assessment Resources . . . . . pp. 253–264

State-Specific Test Practice Online
KEYWORD: MP4 TestPrep

Test Prep Doctor

For item 4, encourage students to draw a diagram showing the sample space for rolling a number cube twice. They can then find the number of outcomes with a sum of 10 and divide by the total number of outcomes in the sample space: $\frac{3}{36} = \frac{1}{12}$, choice **G**.

Answers

9. $\sqrt{213}$ lies between 14 and 15; 14^2 is 196, and 15^2 is 225; 213 is between 196 and 225.

10. $m = 0.18d - 0.1d$
$= 0.08d$
$6.40 = 0.08d$
$80 = d$
Richie's monthly budget is $80.

Resource Options

Chapter 11 Resource Book

Student Resources

Practice (Levels A, B, C) . . . . pp. 9–11, 19–21, 29–31, 39–41, 49–51, 58–60, 68–70

Reteach . . . . pp. 12–13, 22–23, 32–33, 42, 52, 61–62, 71–72

Challenge . . . . pp. 14, 24, 34, 43, 53, 63, 73

Problem Solving . . . . pp. 15, 25, 35, 44, 54, 64, 74

Puzzles, Twisters & Teasers . . . . . pp. 16, 26, 36, 45, 55, 65, 75

Recording Sheets . . . . pp. 3–4, 8, 18, 28, 38, 48, 57, 67, 78, 81, 83–87

Chapter Review . . . . pp. 76–77

Teacher and Parent Resources

Chapter Planning and Pacing Guide . . . . p. 5

Section Planning Guides . . . . pp. 6, 46

Parent Letter . . . . pp. 1–2

Teaching Tools . . . . pp. 81–87

Teacher Support for Chapter Project . . . . p. 79

Transparencies . . . . pp. T1–T45

• Daily Transparencies

• Additional Examples Transparencies

• Teaching Transparencies

Reaching All Learners

English Language Learners

Success for English Language Learners . . . . . . pp. 169–182

Math: Reading and Writing in the Content Area . . . . pp. 85–91

Spanish Homework and Practice . . . . pp. 85–91

Spanish Interactive Study Guide . . . . pp. 85–91

Spanish Family Involvement Activities . . . . pp. 81–88

Multilingual Glossary

Individual Needs

Are You Ready? Intervention and Enrichment . pp. 197–200, 241–244, 249–252, 257–260, 425–426

Alternate Openers: Explorations . . . . pp. 85–91

Family Involvement Activities . . . . pp. 81–88

Interactive Problem Solving . . . . pp. 85–91

Interactive Study Guide . . . . pp. 85–91

Readiness Activities . . . . pp. 21–22

Math: Reading and Writing in the Content Area . . . . pp. 85–91

Challenge . . . . CRB pp. 14, 24, 34, 43, 53, 63, 73

Hands-On

Hands-On Lab Activities . . . . pp. 77–81

Technology Lab Activities . . . . pp. 72–80

Alternate Openers: Explorations . . . . pp. 85–91

Family Involvement Activities . . . . pp. 81–88

Applications and Connections

Consumer and Career Math . . . . pp. 41–44

Interdisciplinary Posters . . . . Poster 11, TE p. 538B

Interdisciplinary Poster Worksheets . . . . pp. 31–33

Transparencies

Alternate Openers: Explorations . . . . pp. 85–91

Exercise Answers Transparencies

Chapter 11 Resource Book . . . . pp. T1–T45

• Daily Transparencies

• Additional Examples Transparencies

• Teaching Transparencies

Technology

Teacher Resources

Lesson Presentations CD-ROM . . . . Chapter 11

Test and Practice Generator CD-ROM . . . . . . Chapter 11

One-Stop Planner CD-ROM . . . . Chapter 11

Student Resources

Are You Ready? Intervention CD-ROM
Skills 47, 58, 60, 62

internet connect

Homework Help Online	KEYWORD: MP4 HWHelp11
Math Tools Online	KEYWORD: MP4 Tools
Glossary Online	KEYWORD: MP4 Glossary
Chapter Project Online	KEYWORD: MP4 PSProject11
Chapter Opener Online	KEYWORD: MP4 Ch11

 CNN student News™   KEYWORD: MP4 CNN11

SE = *Student Edition*   **TE** = *Teacher's Edition*   **AR** = *Assessment Resources*   **CRB** = *Chapter Resource Book*   **MK** = *Manipulatives Kit*

Assessment Options

Assessing Prior Knowledge

Determine whether students have the required prerequisite concepts and skills.

Test Preparation

Provide review and practice for chapter and standardized tests.

Test Prep Tool Kit

Technology

🔘 **Test and Practice Generator CD-ROM**

 internet connect ========================

State-Specific Test Practice Online KEYWORD: MP4 TestPrep

Performance Assessment

Assess students' understanding of chapter concepts and combined problem-solving skills.

Portfolio

Portfolio opportunities appear throughout the Student and Teacher's Editions.

Suggested work samples:

Daily Assessment

Obtain daily feedback on students' understanding of concepts.

Also Available on Transparency In Chapter 11 Resource Book

Student Self-Assessment

Have students evaluate their own work.

Formal Assessment

Assess students' mastery of concepts and skills.

Technology

🔘 **Test and Practice Generator CD-ROM**

Make tests electronically. This software includes:

- Dynamic practice for Chapter 11
- Customizable tests
- Multiple-choice items for each objective
- Free-response items for each objective
- Teacher management system

SE = *Student Edition*   **TE** = *Teacher's Edition*   **AR** = *Assessment Resources*   **CRB** = *Chapter Resource Book*   **MK** = *Manipulatives Kit*

538D

Chapter 11 Tests

Three levels (A,B,C) of tests are available for each chapter in the *Assessment Resources.*

LEVEL A

CHAPTER Chapter Test
11 Form A

Graph each equation and tell whether it is linear.

1. $y = x^2$ **Not linear**

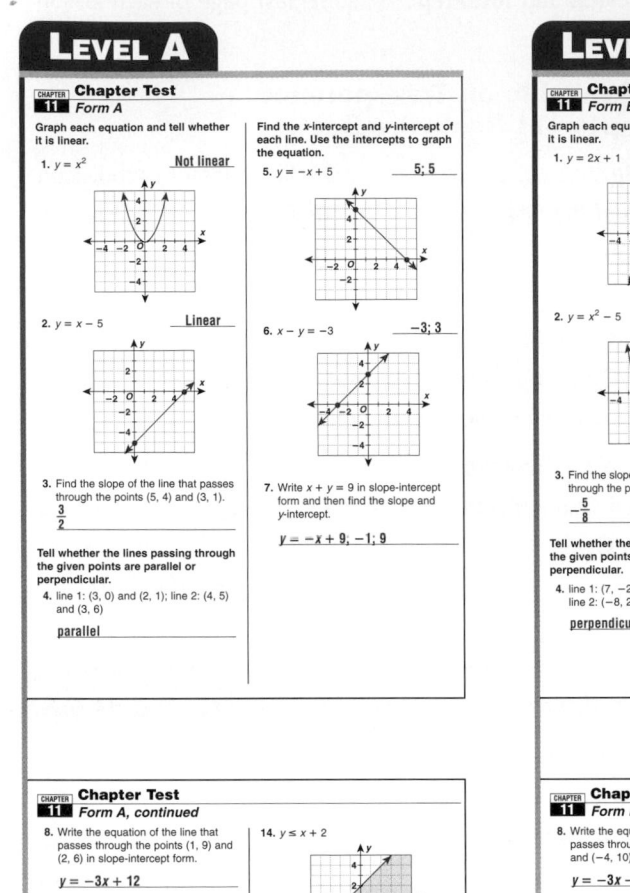

2. $y = x - 5$ **Linear**

3. Find the slope of the line that passes through the points (5, 4) and (3, 1).
$\frac{3}{2}$

Tell whether the lines passing through the given points are parallel or perpendicular.

4. line 1: (3, 0) and (2, 1); line 2: (4, 5) and (3, 6)

parallel

Find the *x*-intercept and *y*-intercept of each line. Use the intercepts to graph the equation.

5. $y = -x + 5$ **5; 5**

6. $x - y = -3$ **-3; 3**

7. Write $x + y = 9$ in slope-intercept form and then find the slope and *y*-intercept.
$y = -x + 9;\ -1;\ 9$

CHAPTER Chapter Test
11 Form A, continued

8. Write the equation of the line that passes through the points (1, 9) and (2, 6) in slope-intercept form.
$y = -3x + 12$

9. Identify a point the line for $y - 4 = 5(x - 3)$ passes through and identify the slope of the line.
Possible answer: (3, 4); 5

10. Write the point-slope form of the equation for a line with a slope -2 that passes through the point (-1, 1).
$y - 1 = -2(x + 1)$

Find each equation of direct variation, given that *y* varies directly with *x*.

11. *y* is 10 when *x* is 2
$y = 5x$

12. *y* is -3 when *x* is $\frac{1}{2}$
$y = -6x$

Graph each inequality.

13. $y > x - 1$

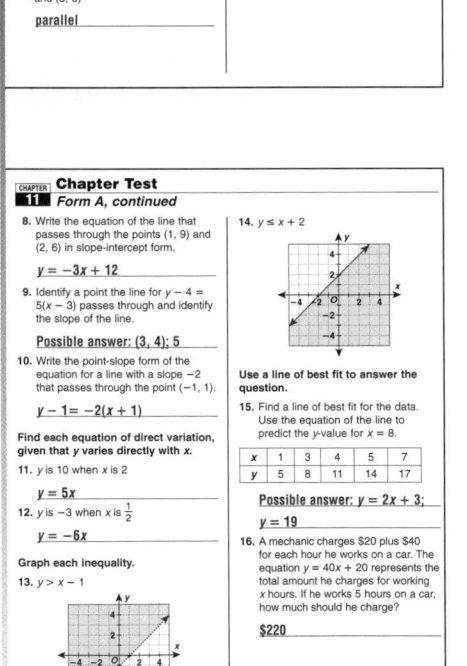

14. $y \le x + 2$

15. Find a line of best fit for the data. Use the equation of the line to predict the *y*-value for *x* = 8.

x	1	3	4	5	7
y	5	8	11	14	17

Possible answer: y = 2x + 3;
$y = 19$

16. A mechanic charges $20 plus $40 for each hour he works on a car. The equation $y = 40x + 20$ represents the total amount he charges for working *x* hours. If he works 5 hours on a car, how much should he charge?
$220

LEVEL B

CHAPTER Chapter Test
11 Form B

Graph each equation and tell whether it is linear.

1. $y = 2x + 1$ **Linear**

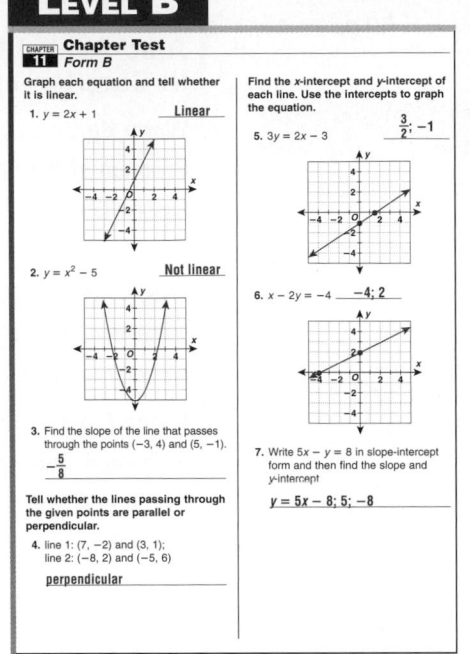

2. $y = x^2 - 5$ **Not linear**

3. Find the slope of the line that passes through the points (-3, 4) and (5, -1).
$-\frac{5}{8}$

Tell whether the lines passing through the given points are parallel or perpendicular.

4. line 1: (7, -2) and (3, 1); line 2: (-8, 2) and (-5, 6)

perpendicular

Find the *x*-intercept and *y*-intercept of each line. Use the intercepts to graph the equation.

5. $3y = 2x - 3$ $\frac{3}{2};\ -1$

6. $x - 2y = -4$ **-4; 2**

7. Write $5x - y = 8$ in slope-intercept form and then find the slope and *y*-intercept
$y = 5x - 8;\ 5;\ -8$

CHAPTER Chapter Test
11 Form B, continued

8. Write the equation of the line that passes through the points (-2, 4) and (-4, 10) in slope-intercept form.
$y = -3x - 2$

9. Identify a point the line for $y + 5 = -\frac{2}{3}(x - 3)$ passes through and identify the slope of the line.
Possible answer: (3, -5); $-\frac{2}{3}$

10. Write the point-slope form of the equation for a line with a slope of -5 that passes through the point (-4, 3).
$y - 3 = -5(x + 4)$

Find each equation of direct variation, given that *y* varies directly with *x*.

11. *y* is -3 when *x* is 12
$y = -\frac{1}{4}x$

12. *y* is 6 when *x* is 9
$y = \frac{2}{3}x$

Graph each inequality.

13. $y > 2x - 1$

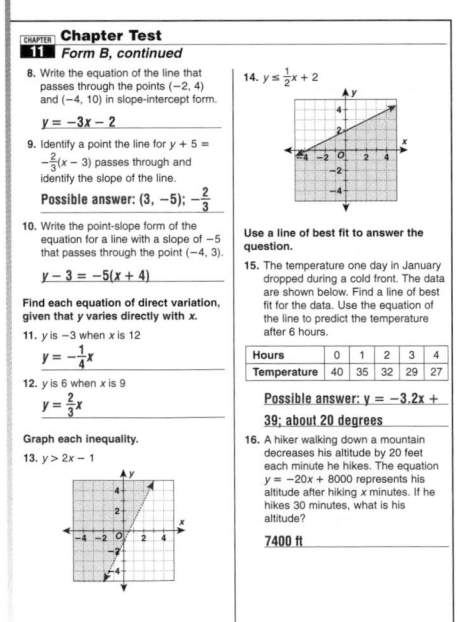

14. $y \le \frac{1}{2}x + 2$

15. The temperature one day in January dropped during a cold front. The data are shown below. Find a line of best fit for the data. Use the equation of the line to predict the temperature after 6 hours.

Hours	0	1	2	3	4
Temperature	40	35	32	29	27

Possible answer: y = -3.2x + 39; about 20 degrees

16. A hiker walking down a mountain decreases his altitude by 20 feet each minute he hikes. The equation $y = -20x + 8000$ represents his altitude after hiking *x* minutes. If he hikes 30 minutes, what is his altitude?
7400 ft

LEVEL C

CHAPTER Chapter Test
11 Form C

Graph each equation and tell whether it is linear.

1. $y = -\frac{1}{3}x - 2$ **Linear**

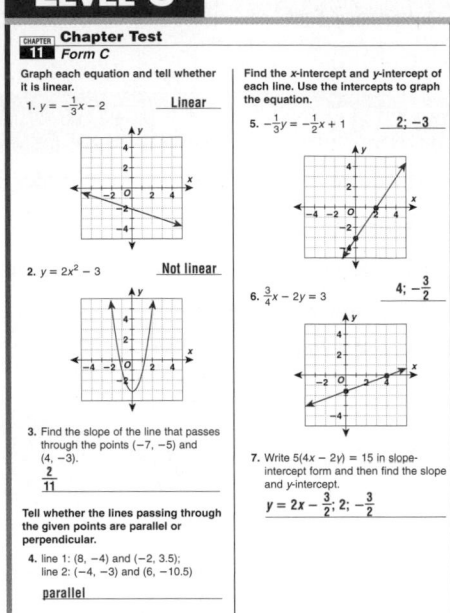

2. $y = 2x^2 - 3$ **Not linear**

3. Find the slope of the line that passes through the points (-7, -5) and (4, -3).
$\frac{2}{11}$

Tell whether the lines passing through the given points are parallel or perpendicular.

4. line 1: (8, -4) and (-2, 3.5); line 2: (-4, -3) and (6, -10.5)

parallel

Find the *x*-intercept and *y*-intercept of each line. Use the intercepts to graph the equation.

5. $-\frac{1}{3}y = -\frac{1}{2}x + 1$ **2; -3**

6. $\frac{3}{4}x - 2y = 3$ $4;\ -\frac{3}{2}$

7. Write $5(4x - 2y) = 15$ in slope-intercept form and then find the slope and *y*-intercept.
$y = 2x - \frac{3}{2};\ 2;\ -\frac{3}{2}$

CHAPTER Chapter Test
11 Form C, continued

8. Write the equation of the line that passes through the points (-10, 6) and (15, 1) in slope-intercept form.
$y = -\frac{1}{5}x + 4$

9. Identify a point the line for $y + 7 = -\frac{3}{5}(x - 4)$ passes through and identify the slope of the line.
Possible answer: (4, -7); $-\frac{3}{5}$

10. Write the point-slope form of the equation for a line with a slope of $-\frac{1}{7}$ that passes through the point $\left(-21, \frac{1}{2}\right)$
$y - \frac{1}{2} = -\frac{1}{7}(x + 21)$

Find each equation of direct variation, given that *y* varies directly with *x*.

11. *y* is -63 when *x* is 81
$y = -\frac{7}{9}x$

12. *y* is $\frac{3}{5}$ when *x* is $\frac{1}{2}$
$y = \frac{6}{5}x$

Graph each inequality.

13. $\frac{1}{4}y > -\frac{1}{6}x + \frac{1}{2}$

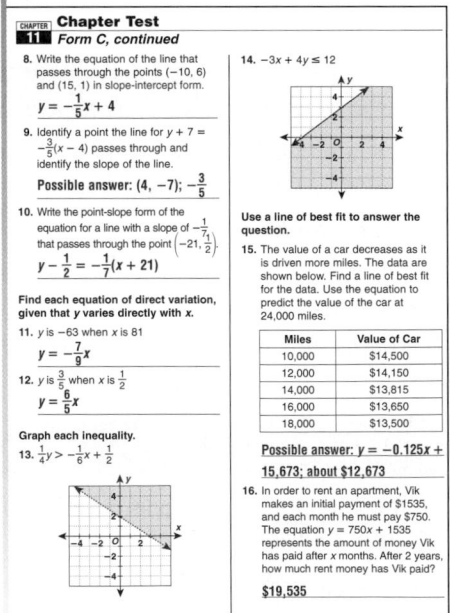

14. $-3x + 4y \le 12$

15. The value of a car decreases as it is driven more miles. The data are shown below. Find a line of best fit for the data. Use the equation to predict the value of the car at 24,000 miles.

Miles	Value of Car
10,000	$14,500
12,000	$14,150
14,000	$13,815
16,000	$13,650
18,000	$13,500

Possible answer: y = -0.125x + 15,673; about $12,673

16. In order to rent an apartment, Vik makes an initial payment of $1535, and each month he must pay $750. The equation $y = 750x + 1535$ represents the amount of money Vik has paid after *x* months. After 2 years, how much rent money has Vik paid?
$19,535

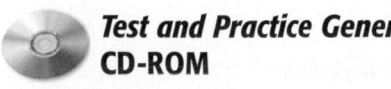
Test and Practice Generator
CD-ROM

Create and customize multiple versions of the same tests with corresponding answers for any chosen chapter objectives.

Chapter 11 State and Standardized Test Preparation

Test Taking Skill Builder and Standardized Test Practice are provided for each chapter in the *Test Prep Tool Kit.*

TEST TAKING SKILL BUILDER

Test Taking Strategy Context Based Multiple Choice Questions
Chapter 11

Use the information provided in a table, a diagram, or a graph to answer a context-based multiple-choice question.

Example
The length of a rectangle for a logo design has to be less than or equal to 3 times its width. Which rectangle dimensions are acceptable for the logo?

A 3 ft wide by 4 ft long
B 2 ft wide by 10 ft long
C 3 ft wide by 12 ft long
D 1 ft wide by 4 ft long

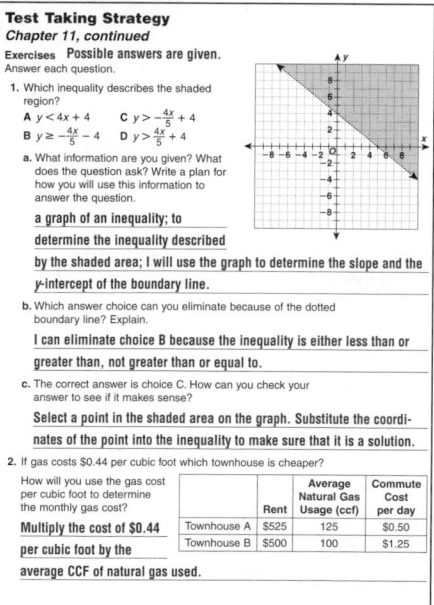

Solution:

First, read over the question again: What information does the question statement provide? Plan to use this information to solve the problem.

Given: The length of a rectangle is less than or equal to 3 times its width. The dimensions of this logo are modeled by the graph. You can use the graph to determine an inequality and then substitute the dimensions into the inequality to determine which are acceptable for the logo.

Determine the function: The slope of the graph is 3 and the y-intercept is 0. Since you know that the length is less than or equal to 3 times the width, the inequality can be written as $y \le 3x$, with y = to the length of the rectangle, and x = to the width.

Substitute the dimensions in the answer choices into the inequality $y \le 3x$ and see if the point is a solution.

Choice A (3, 4)   Choice B (2, 10)   Choice C (3, 12)   Choice D (1, 4)
$y \le 3x$ $y \le 3x$ $y \le 3x$ $y \le 3x$
$4 \le 3(3)$ $10 \le 3(2)$ $12 \le 3(3)$ $4 \le 3(1)$
$4 \le 9$ $10 \le 6$ $12 \le 9$ $4 \le 3$

Only Choice A is a solution to the inequality.

Check your answer. Does it make sense?
Yes, the point (3, 4) lies in the shaded area of the graph. All of the other points lie in the unshaded area. Also, the length of 4 ft is less than or equal to 3 times the width of 3 ft, since 4 is less than 9.

Test Taking Strategy
Chapter 11, continued
Exercises Possible answers are given. Answer each question.

1. Which inequality describes the shaded region?

A $y < 4x + 4$ C $y > -\frac{4x}{5} + 4$
B $y \ge -\frac{4x}{5} - 4$ D $y > \frac{4x}{5} + 4$

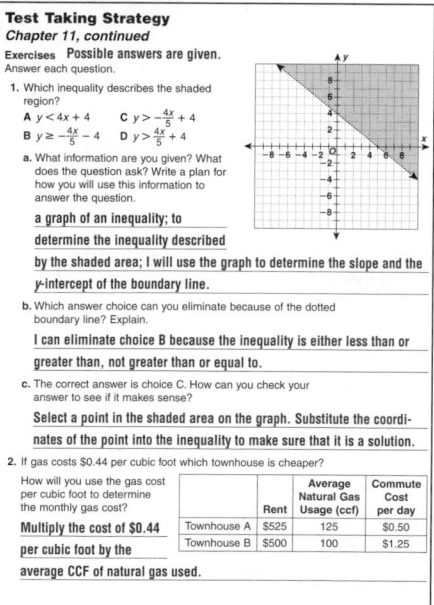

a. What information are you given? What does the question ask? Write a plan for how you will use this information to answer the question.

a graph of an inequality; to

determine the inequality described

by the shaded area; I will use the graph to determine the slope and the

y-intercept of the boundary line.

b. Which answer choice can you eliminate because of the dotted boundary line? Explain.

I can eliminate choice B because the inequality is either less than or

greater than, not greater than or equal to.

c. The correct answer is choice C. How can you check your answer to see if it makes sense?

Select a point in the shaded area on the graph. Substitute the coordi-

nates of the point into the inequality to make sure that it is a solution.

2. If gas costs $0.44 per cubic foot which townhouse is cheaper?

How will you use the gas cost per cubic foot to determine the monthly gas cost?

Multiply the cost of $0.44
per cubic foot by the
average CCF of natural gas used.

	Rent	Average Natural Gas Usage (ccf)	Commute Cost per day
Townhouse A	$525	125	$0.50
Townhouse B	$500	100	$1.25

STANDARDIZED TEST PRACTICE

Standardized Test Practice
Chapter 11

Select the best answer for Questions 1–6.

1. Which graph is a linear function?

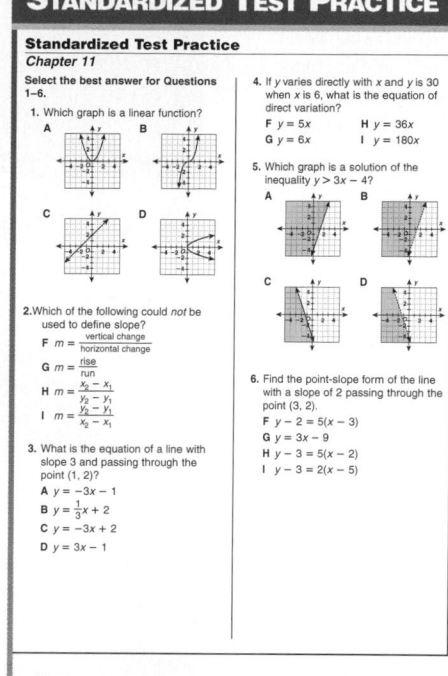

2. Which of the following could *not* be used to define slope?

F $m = \frac{\text{vertical change}}{\text{horizontal change}}$

G $m = \frac{\text{rise}}{\text{run}}$

H $m = \frac{x_2 - x_1}{y_2 - y_1}$

I $m = \frac{y_2 - y_1}{x_2 - x_1}$

3. What is the equation of a line with slope 3 and passing through the point (1, 2)?

A $y = -3x - 1$
B $y = \frac{1}{3}x + 2$
C $y = -3x + 2$
D $y = 3x - 1$

4. If y varies directly with x and y is 30 when x is 6, what is the equation of direct variation?

F $y = 5x$ H $y = 36x$
G $y = 6x$ I $y = 180x$

5. Which graph is a solution of the inequality $y > 3x - 4$?

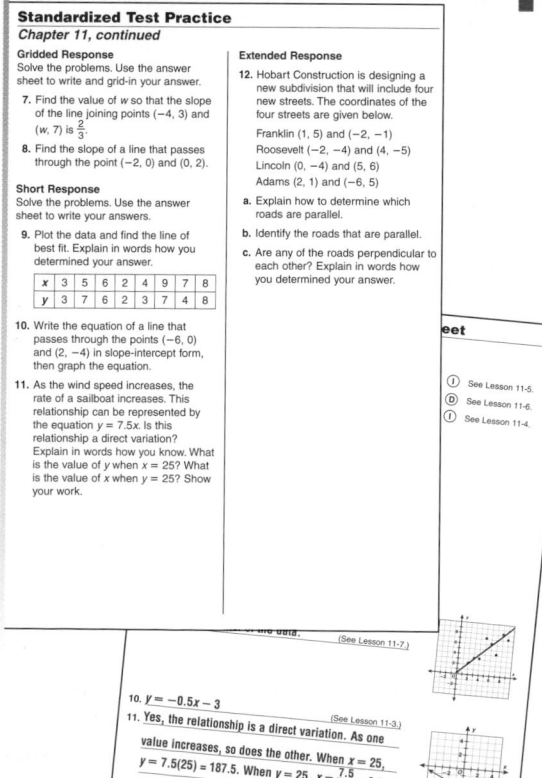

6. Find the point-slope form of the line with a slope of 2 passing through the point (3, 2).

F $y - 2 = 5(x - 3)$
G $y = 3x - 9$
H $y - 3 = 5(x - 2)$
I $y - 3 = 2(x - 5)$

Standardized Test Practice
Chapter 11, continued

Gridded Response
Solve the problems. Use the answer sheet to write and grid-in your answer.

7. Find the value of w so that the slope of the line joining points (−4, 3) and $(w, 7)$ is $\frac{2}{3}$.

8. Find the slope of a line that passes through the point (−2, 0) and (0, 2).

Short Response
Solve the problems. Use the answer sheet to write your answers.

9. Plot the data and find the line of best fit. Explain in words how you determined your answer.

x	3	5	6	2	4	9	7	8
y	3	7	6	2	3	7	4	8

10. Write the equation of a line that passes through the points (−6, 0) and (2, −4) in slope-intercept form, then graph the equation.

11. As the wind speed increases, the rate of a sailboat increases. This relationship can be represented by the equation $y = 7.5x$. Is this relationship a direct variation? Explain in words how you know. What is the value of y when $x = 25$? What is the value of x when $y = 25$? Show your work.

Extended Response

12. Hobart Construction is designing a new subdivision that will include four new streets. The coordinates of the four streets are given below.

Franklin (1, 5) and (−2, −1)
Roosevelt (−2, −4) and (4, −5)
Lincoln (0, −4) and (5, 6)
Adams (2, 1) and (−6, 5)

a. Explain how to determine which roads are parallel.

b. Identify the roads that are parallel.

c. Are any of the roads perpendicular to each other? Explain in words how you determined your answer.

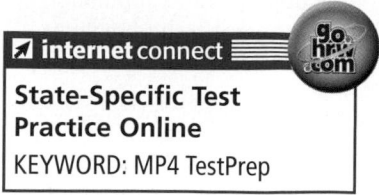
Test Prep Tool Kit

- Standardized Test Prep Workbook
- Countdown to Testing transparencies
- State Test Prep CD-ROM
- Standardized Test Prep Video

...eet

ⓘ See Lesson 11-5.
ⓓ See Lesson 11-6.
ⓘ See Lesson 11-4.

...he data. (See Lesson 11-7.)

10. $y = -0.5x - 3$
                              (See Lesson 11-3.)
11. Yes, the relationship is a direct variation. As one value increases, so does the other. When $x = 25$, $y = 7.5(25) = 187.5$. When $y = 25$, $x = \frac{7.5}{25} = 0.3$.

                              (See Lesson 11-5.)

Extended Response
Write your answers for Problem 12 on the back of this paper.
See Lesson 11-2.

Customized answer sheets give students realistic practice for actual standardized tests.

Graphing Lines

Why Learn This?

Remind students that line graphs are used to track the change in a quantity over time. A line graph would be useful to a wildlife ecologist studying data about the whooping crane population because a linear graph would make it easy to see what is happening to the population at a glance. The wildlife ecologist can use the line graph to show whether the population is increasing or decreasing and whether the rate of increase or decrease is fast or slow.

Using Data

To begin the study of this chapter, have students:

- Make a line graph to show the whooping crane population from 1940 to 2000.

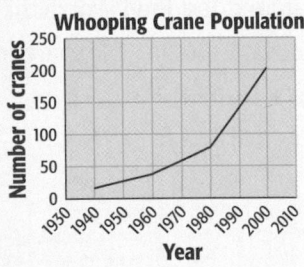

Whooping Crane Population

- Use their line graph to determine the time period in which the whooping crane population increased the most. 1980 to 2000

internet connect

Chapter Opener Online
go.hrw.com
KEYWORD: MP4 Ch11

Graphing Lines

Whooping Crane Population				
Year	1940	1960	1980	2000
Cranes	15	36	79	202

Career *Wildlife Ecologist*

Whatever happened to the Carolina parakeet and the passenger pigeon, two species of birds that once inhabited the United States? They are now as extinct as *Tyrannosaurus rex*. The primary focus of wildlife ecologists is to keep other animals from becoming extinct.

They have been successful with the whooping crane, the largest wild bird in North America. The table shows how the whooping crane has come back from the brink of extinction.

Problem Solving Project

Life Science Connection

Purpose: To use equations and graphs in order to solve problems

Materials: Whooping it Up worksheet

internet connect

Chapter Project Online: *go.hrw.com*
KEYWORD: MP4 PSProject11

Understand, Plan, Solve, and Look Back

Have students:

- ✔ Examine the table. Ask them what seems to have happened to the whooping crane over the last 60 years. Why did those changes occur?

- ✔ Complete the Whooping it Up worksheet.

- ✔ Estimate what the whooping crane population will be in 2050. Describe how they reached the conclusion.

- ✔ Do research on an endangered species. What environmental issues face the population? What is the prediction of population change in the future?

- ✔ Check students' work.

ARE YOU READY?

Choose the best term from the list to complete each sentence.

1. The expression $4 - 3$ is an example of a(n) __?__ expression. **subtraction**

2. When you divide both sides of the equation $2x = 20$ by 2, you are __?__. **solving for the variable**

3. An example of a(n) __?__ is $3x > 12$. **inequality in one variable**

4. The expression $7 - 6$ can be rewritten as the __?__ expression $7 + (-6)$. **addition**

addition
subtraction
inequality in one variable
solving for the variable
subtraction

Complete these exercises to review skills you will need for this chapter.

✔ Operations with Integers

Simplify.

5. $\frac{7-5}{-2}$ **−1**

6. $\frac{-3-5}{-2-3}$ **$\frac{8}{5}$**

7. $\frac{-8+2}{-2+8}$ **−1**

8. $\frac{-16}{-2}$ **8**

9. $\frac{-22}{2}$ **−11**

10. $-12 + 9$ **−3**

✔ Equations

Solve.

11. $3p - 4 = 8$ **$p = 4$**

12. $2(a + 3) = 4$ **$a = -1$**

13. $9 = -2k + 27$ **$k = 9$**

14. $3s - 4 = 1 - 3s$ **$s = \frac{5}{6}$**

15. $7x + 1 = x$ **$x = -\frac{1}{6}$**

16. $4m - 5(m + 2) = 1$ **$m = -11$**

Determine whether each ordered pair is a solution to $-\frac{1}{2}x + 3 = y$.

17. $(4, 1)$ **yes**

18. $\left(-\frac{8}{2}, 2\right)$ **no**

19. $(0, 5)$ **no**

20. $(-4, 5)$ **yes**

21. $(8, 1)$ **no**

22. $(2, 2)$ **yes**

23. $(-2, 4)$ **yes**

24. $(0, 1)$ **no**

✔ Solve for One Variable

Solve each equation for the indicated variable.

25. Solve for x: $5y - x = 4$. **$x = 5y - 4$**

26. Solve for y: $3y + 9 = 2x$. **$y = \frac{2}{3}x - 3$**

27. Solve for y: $2y + 3x = 6$. **$y = \frac{3}{-2}x + 3$**

28. Solve for x: $ax + by = c$. **$x = \frac{b}{-a}y + \frac{c}{a}$**

✔ Solve Inequalities in One Variable

Solve and graph each inequality.

29. $x + 4 > 2$ **$x > -2$**

30. $-3x < 9$ **$x > -3$**

31. $x - 1 \le -5$ **$x \le -4$**

Assessing Prior Knowledge

INTERVENTION

Diagnose and Prescribe

Evaluate your students' performance on this page to determine whether intervention is necessary or whether enrichment is appropriate. Options that provide instruction, practice, and a check are listed below.

Resources for Are You Ready?

• *Are You Ready? Intervention and Enrichment*

• **Recording Sheet for Are You Ready?**
  Chapter 11 Resource Book . . pp. 3–4

 Are You Ready? Intervention CD-ROM

Answers

29.
$-4 \quad -2 \quad 0 \quad 2 \quad 4$

30.
$-4 \quad -2 \quad 0 \quad 2 \quad 4 \quad 6$

31.
$-8 \quad -6 \quad -4 \quad -2 \quad 0$

ARE YOU READY?

Were students successful with Are You Ready?

NO INTERVENE ⟵ ⟶ **YES** ENRICH

✔ **Operations with Integers** 5–10
Intervention Practice, Skill 47
 CD-ROM Intervention Activities, Skill 47

✔ **Equations** 11–24
Intervention Practice, Skills 58, 60
 CD-ROM Intervention Activities, Skills 58, 60

✔ **Solve for One Variable** 25–28
Intervention Practice, Skill 60
 CD-ROM Intervention Activities, Skill 60

✔ **Solve Inequalities in One Variable** 29–31
Intervention Practice, Skill 62
 CD-ROM Intervention Activities, Skill 62

Are You Ready? Enrichment, pp. 425–426

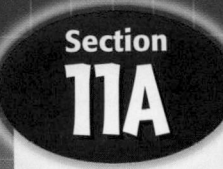

Lesson	Materials	Resources
Lesson 11-1 Graphing Linear Equations **NCTM:** Algebra, Communication, Connections, Representation **NAEP:** Algebra 4a ☐ SAT-9 ☑ SAT-10 ☐ ITBS ☐ CTBS ☐ MAT ☑ CAT	**Required** Graph paper *(CRB, p. 82)* **Optional** Teaching Transparency T2 *(CRB)* Graphing calculators Recording Sheet for Reaching All Learners *(CRB, p. 81)*	• *Chapter 11 Resource Book,* pp. 7–16 • Daily Transparency T1, CRB • Additional Examples Transparencies T3–T8, CRB • *Alternate Openers: Explorations,* p. 85
Lesson 11-2 Slope of a Line **NCTM:** Algebra, Geometry, Communication, Representation **NAEP:** Algebra 4a ☐ SAT-9 ☑ SAT-10 ☐ ITBS ☐ CTBS ☐ MAT ☑ CAT	**Required** Graph paper *(CRB, p. 82)* **Optional** Teaching Transparency T10 *(CRB)* Recording Sheet for Reaching All Learners *(CRB, p. 83)*	• *Chapter 11 Resource Book,* pp. 17–26 • Daily Transparency T9, CRB • Additional Examples Transparencies T11–T14, CRB • *Alternate Openers: Explorations,* p. 86
Lesson 11-3 Using Slopes and Intercepts **NCTM:** Algebra, Geometry, Communication, Representation **NAEP:** Algebra 4a ☐ SAT-9 ☑ SAT-10 ☐ ITBS ☐ CTBS ☐ MAT ☑ CAT	**Required** Graph paper, *(CRB, p. 82)* **Optional** Recording Sheet for Reaching All Learners *(CRB, p. 84)*	• *Chapter 11 Resource Book,* pp. 27–36 • Daily Transparency T15, CRB • Additional Examples Transparencies T16–T20, CRB • *Alternate Openers: Explorations,* p. 87
Technology Lab 11A Graph Equations in Slope-Intercept Form **NCTM:** Algebra, Representation **NAEP:** Algebra 4a ☐ SAT-9 ☑ SAT-10 ☐ ITBS ☐ CTBS ☐ MAT ☑ CAT	**Required** Graphing calculators	• *Technology Lab Activities,* pp. 76–77
Lesson 11-4 Point-Slope Form **NCTM:** Algebra, Communication, Representation **NAEP:** Algebra 4a ☐ SAT-9 ☐ SAT-10 ☐ ITBS ☐ CTBS ☐ MAT ☑ CAT	**Optional** Teaching Transparency T22 *(CRB)* Recording Sheet for Reaching All Learners *(CRB, p. 85)*	• *Chapter 11 Resource Book,* pp. 37–45 • Daily Transparency T21, CRB • Additional Examples Transparencies T23–T25, CRB • *Alternate Openers: Explorations,* p. 88
Section 11A Assessment		• Mid-Chapter Quiz, SE p. 560 • Section 11A Quiz, AR p. 25 • *Test and Practice Generator* CD-ROM

SAT = *Stanford Achievement Tests*   **ITBS** = *Iowa Test of Basic Skills*   **CTBS** = *Comprehensive Test of Basic Skills/Terra Nova*
MAT = *Metropolitan Achievement Test*   **CAT** = *California Achievement Test*

NCTM = Complete standards can be found on pages T29–T35.   **NAEP** = Complete standards can be found on pages A54–A58.

SE = *Student Edition*   **TE** = *Teacher's Edition*   **AR** = *Assessment Resources*   **CRB** = *Chapter Resource Book*   **MK** = *Manipulatives Kit*

Section Overview

Graphing Linear Equations and Using Slope
Lessons 11-1, 11-2

Why? Many relationships can be represented by linear equations. The slope
of a linear graph is a measure of the rate of change of one variable
with respect to the other variable.

Graph linear equation $y = 2x + 1$.

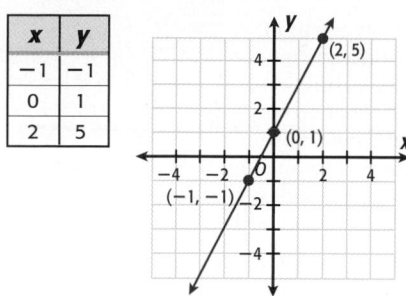

$$\text{Slope} = \frac{\text{rise}}{\text{run}} = \frac{y_2 - y_1}{x_2 - x_1}$$

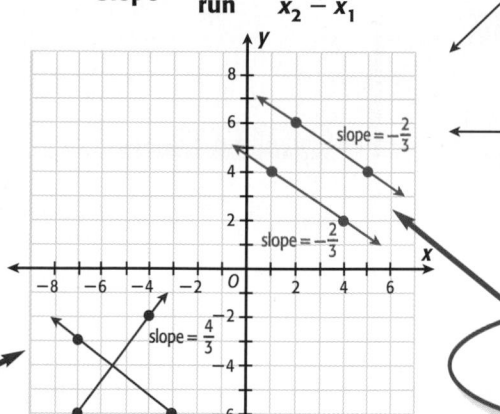

Positive slope   Negative slope

Zero slope   Undefined slope

Perpendicular lines have slopes that are **negative reciprocals**.

Parallel lines have **equal** slopes.

Slopes and Intercepts
Lesson 11-3, Technology Lab 11A

Why? The slope-intercept form of an equation of a line
is often used to express a relationship.

Intercepts of a Line

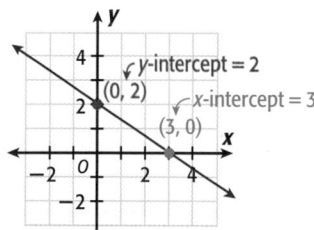

Slope-Intercept Form of a Linear Equation

Slope   y-intercept

$$y = mx + b$$

The graph of $y = -\frac{1}{4}x + 3$

has slope $-\frac{1}{4}$ and y-intercept **3**.

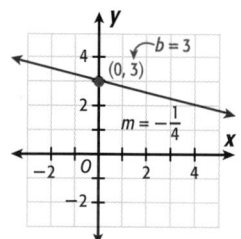

Point-Slope Form
Lesson 11-4

Why? The point-slope form of a linear equation
is convenient for some purposes.

Point-Slope Form of a Linear Equation

$$y - y_1 = m(x - x_1)$$
(x_1, y_1) is a point on
the line. m is the
slope of the line.

Write an equation of the line with
slope $\frac{1}{2}$ that passes through point $(-3, 5)$.

$$y - y_1 = m(x - x_1)$$
$$y - 5 = \frac{1}{2}(x - (-3))$$
$$y - 5 = \frac{1}{2}(x + 3)$$

Pacing: Traditional 1 day
Block $\frac{1}{2}$ day

Objective: Students identify and graph linear equations.

Warm Up

Solve each equation for y.

1. $6y - 12x = 24$ $y = 2x + 4$

2. $-2y - 4x = 20$ $y = -2x - 10$

3. $2y - 5x = 16$ $y = \frac{5}{2}x + 8$

4. $3y + 6x = 18$ $y = -2x + 6$

Problem of the Day

The same photo book of Niagara Falls costs $5.95 in the United States and $8.25 in Canada. If the exchange rate is $1.49 in Canadian dollars for each U.S. dollar, in which country is the book a better deal?   **Canada**

Available on Daily Transparency in CRB

Even though the graph maker didn't shave for a week, everyone still liked to hear him tell his tales. It was another example of the popularity of the *hairy plotter* stories.

11-1 Graphing Linear Equations

Learn to identify and graph linear equations.

Vocabulary
linear equation

In most bowling leagues, bowlers have a handicap added to their scores to make the game more competitive. For some leagues, the *linear equation* $h = 160 - 0.8s$ expresses the handicap h of a bowler who has an average score of s.

A **linear equation** is an equation whose solutions fall on a line on the coordinate plane. All solutions of a particular linear equation fall on the line, and all the points on the line are solutions of the equation. To find a solution that lies between two points (x_1, y_1) and (x_2, y_2), choose an x-value between x_1 and x_2 and find the corresponding y-value.

Reading Math

Read x_1 as "x sub one" or "x one."

If an equation is linear, a constant change in the x-value corresponds to a constant change in the y-value. The graph shows an example where each time the x-value increases by 3, the y-value increases by 2.

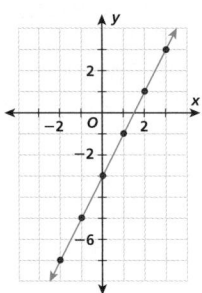

EXAMPLE **Graphing Equations**

Graph each equation and tell whether it is linear.

A $y = 2x - 3$

x	2x − 3	y	(x, y)
−2	2(−2) − 3	−7	(−2, −7)
−1	2(−1) − 3	−5	(−1, −5)
0	2(0) − 3	−3	(0, −3)
1	2(1) − 3	−1	(1, −1)
2	2(2) − 3	1	(2, 1)
3	2(3) − 3	3	(3, 3)

The equation $y = 2x - 3$ is a linear equation because it is the graph of a straight line and each time x increases by 1 unit, y increases by 2 units.

---

1 Introduce

Alternate Opener

EXPLORATION

11-1 Graphing Linear Equations

The calculator screens show the graphs of $y = x + 1$ and $y = x + 3$.

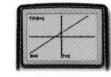 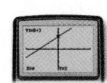

1. Why is the graph of $y = x + 3$ higher along the y-axis than the graph of $y = x + 1$?

The calculator screens show the graphs of $y = x + 2$ and $y = 2x + 2$.

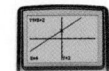 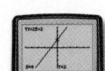

2. Why is the graph of $y = x + 2$ flatter than the graph of $y = 2x + 2$?

Think and Discuss

3. Discuss how the two graphs for number 1 are similar and different.

4. Discuss how the two graphs for number 2 are similar and different.

Motivate

Show students the three equations $y = x^2$, $y = 2x$, and $y = \frac{2}{x}$ (Teaching Transparency T2, CRB). Show them the three related graphs without telling them which graph belongs to which equation. Tell the students that one of the graphs is a line and one of the equations is a linear equation and that today they will learn how to graph an equation and determine whether it is linear.

Exploration worksheet and answers on Chapter 11 Resource Book pp. 8 and 88

2 Teach

Lesson Presentation

Guided Instruction

In this lesson, students identify and graph linear equations. Demonstrate how to find a solution to a linear equation by selecting any x-value and finding the corresponding y-value. Use the diagram to point out that a constant change in x-values corresponds to a constant change in y-values (3 right and 2 up). Show students some tips for choosing x-values. For example, if x has a fractional coefficient with a denominator of 3 (as in Example 1C), then convenient values for x would be multiples of 3.

Graph each equation and tell whether it is linear.

B $y = x^2$

x	x^2	y	(x, y)
−2	$(-2)^2$	4	(−2, 4)
−1	$(-1)^2$	1	(−1, 1)
0	$(0)^2$	0	(0, 0)
1	$(1)^2$	1	(1, 1)
2	$(2)^2$	4	(2, 4)

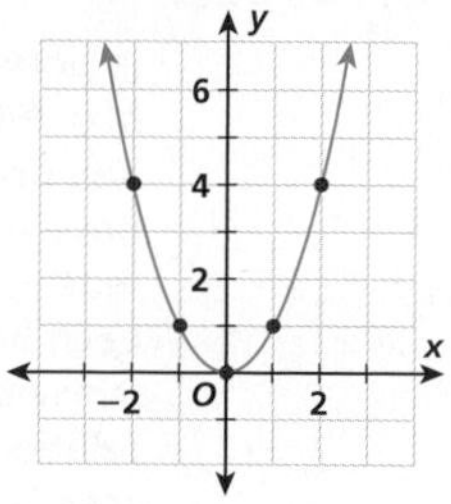

The equation $y = x^2$ is not a linear equation because its graph is not a straight line. Also notice that as x increases by a constant of 1, the change in y is not constant.

x	−2	−1	0	1	2
y	4	1	0	1	4

−3 −1 +1 +3

C $y = \frac{2x}{3}$

x	$\frac{2x}{3}$	y	(x, y)
−2	$\frac{2(-2)}{3}$	$-\frac{4}{3}$	$(-2, -\frac{4}{3})$
−1	$\frac{2(-1)}{3}$	$-\frac{2}{3}$	$(-1, -\frac{2}{3})$
0	$\frac{2(0)}{3}$	0	(0, 0)
1	$\frac{2(1)}{3}$	$\frac{2}{3}$	$(1, \frac{2}{3})$
2	$\frac{2(2)}{3}$	$\frac{4}{3}$	$(2, \frac{4}{3})$

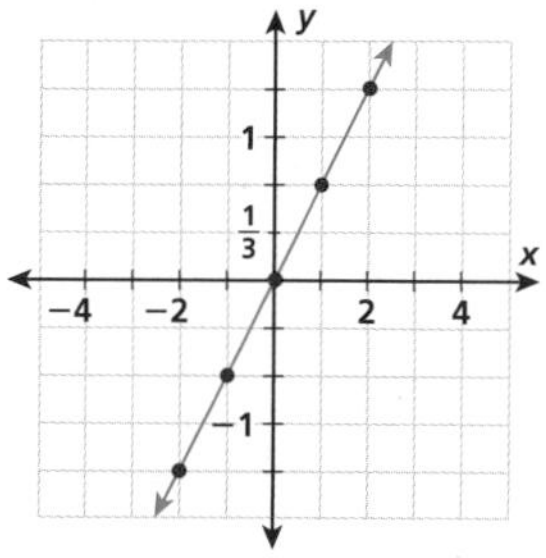

The equation $y = \frac{2x}{3}$ is a linear equation because the points form a straight line. Each time the value of x increases by 1, the value of y increases by $\frac{2}{3}$, or y increases by 2 each time x increases by 3.

D $y = -3$

x	−3	y	(x, y)
−2	−3	−3	(−2, −3)
−1	−3	−3	(−1, −3)
0	−3	−3	(0, −3)
1	−3	−3	(1, −3)
2	−3	−3	(2, −3)

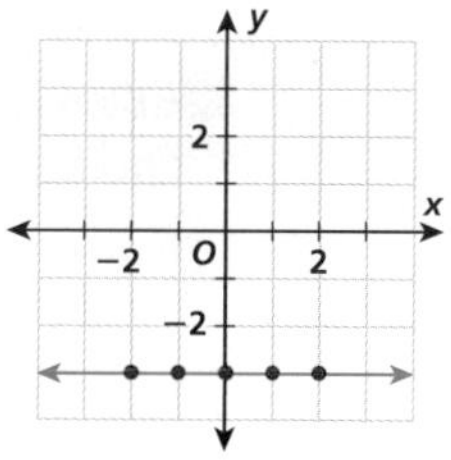

For any value of x, y = −3.

The equation $y = -3$ is a linear equation because the points form a straight line. As the value of x increases, the value of y has a constant change of 0.

Reaching All Learners
Through Critical Thinking

Give students a recording sheet with several tables of values (Chapter 11 Resource Book p. 81). Have students use the method of differences to determine which sets of values could show linear relationships. Then have students graph each set of points on a coordinate plane to confirm their answers.

COMMON ERROR ALERT

Students may incorrectly graph a point so that it is not collinear with the others. Encourage students to double check any points that are not collinear.

Additional Examples

Example 1

Graph each equation and tell whether it is linear.

A. $y = 3x - 1$

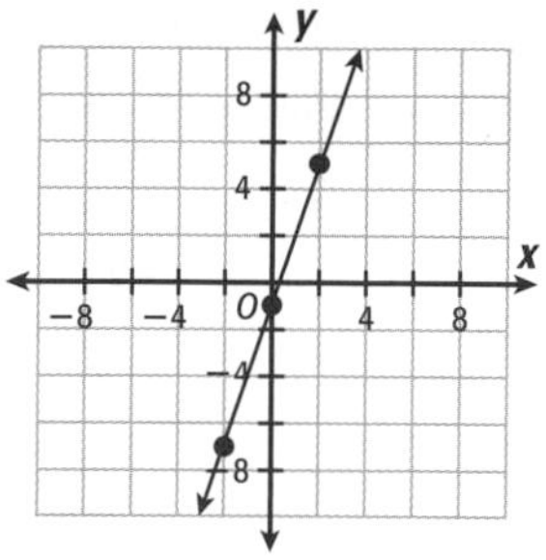

linear

B. $y = x^3$

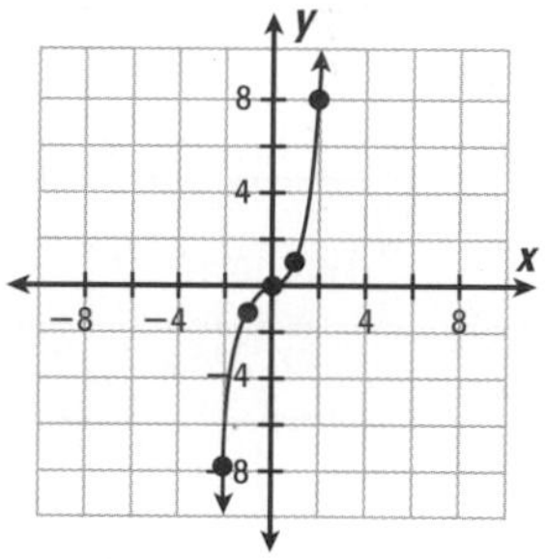

not linear

C. $y = -\frac{3x}{4}$

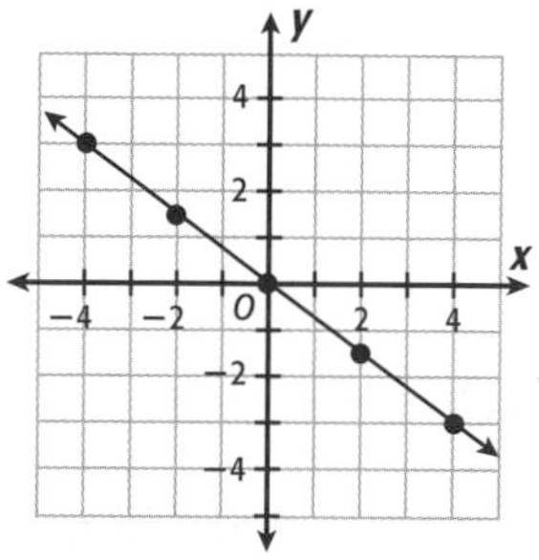

linear

D. $y = 2$

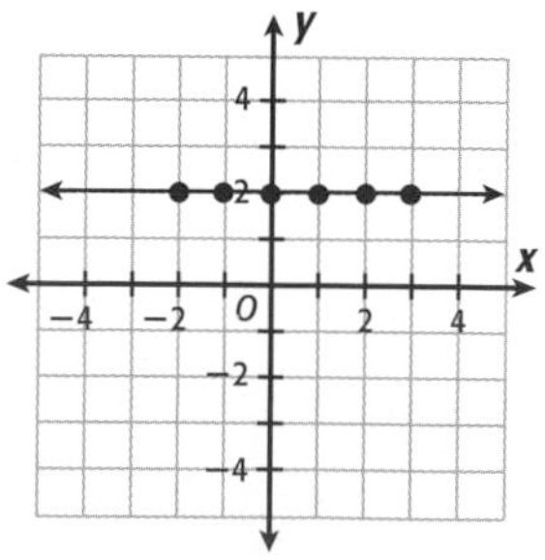

linear

Example 2

A lift on a ski slope rises according to the equation $a = 130t + 6250$, where a is the altitude in feet and t is the number of minutes that a skier has been on the lift. Five friends are on the lift. What is the altitude of each person if they have been on the ski lift for the times listed in the table? Draw a graph that represents the relationship between the time on the lift and the altitude.

Skier	Time on lift	
Anna	4 minutes	6770 ft
Tracy	3 minutes	6640 ft
Kwani	2 minutes	6510 ft
Tony	1.5 minutes	6445 ft
George	1 minute	6380 ft

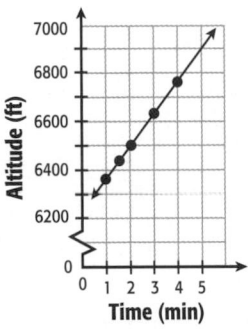

EXAMPLE **2** *Sports Application*

In bowling, the equation $h = 160 - 0.8s$ represents the handicap h calculated for a bowler with average score s. How much will the handicap be for each bowler listed in the table? Draw a graph that represents the relationship between the average score and the handicap.

Bowler	Average Score
Sandi	145
Dominic	125
Leo	160
Sheila	140
Tawana	175

s	$h = 160 - 0.8s$	h	(s, h)
145	$h = 160 - 0.8(145)$	44	(145, 44)
125	$h = 160 - 0.8(125)$	60	(125, 60)
160	$h = 160 - 0.8(160)$	32	(160, 32)
140	$h = 160 - 0.8(140)$	48	(140, 48)
175	$h = 160 - 0.8(175)$	20	(175, 20)

The handicaps are: Sandi, 44 pins; Dominic, 60 pins; Leo, 32 pins; Sheila, 48 pins; and Tawana, 20 pins. This is a linear equation because when s increases by 10 units, h decreases by 8 units. Note that a bowler with an average score of over 200 is given a handicap of 0.

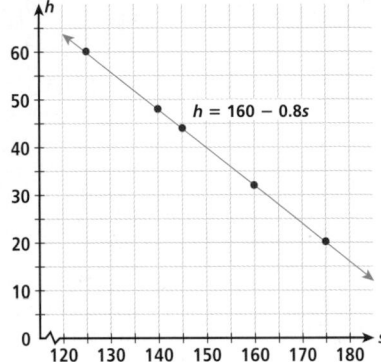

Think and Discuss

1. **Explain** whether an equation is linear if three ordered-pair solutions lie on a straight line but a fourth does not.

2. **Compare** the equations $y = 3x + 2$ and $y = 3x^2$. Without graphing, explain why one of the equations is not linear.

3. **Describe** why the ordered pair for a bowler with an average score of 210 would not fall on the line in Example 2.

---

3 Close

Summarize

Review the steps for graphing an equation.

- Choose a value for x.
- Substitute the x-value into the equation, and find the corresponding y-value.
- Form an ordered pair with the x-value and y-value.
- Graph the ordered pair.
- Repeat the process until you have at least 3 points to sketch the graph.

You may want to point out that this method is useful not only for linear equations but for many other equations as well.

Answers to Think and Discuss

Possible answers:

1. The equation is not linear because by the definition of a linear equation, *all* solutions must lie on a straight line when graphed on a coordinate plane.

2. The equation $y = 3x^2$ is not linear because constant changes in x do not produce constant changes in y.

3. The section of the graph for scores of 0 to 200 is linear, but the entire graph is not linear since any bowler with an average over 200 is given a handicap of 0. The point on the line for a bowler with a 210 average would be (210, −8), not (210, 0), which is the actual value on the graph.

FOR EXTRA PRACTICE
see page 752

internet connect
Homework Help Online
go.hrw.com Keyword: MP4 11-1

11-1 PRACTICE & ASSESS

GUIDED PRACTICE

See Example **1**   Graph each equation and tell whether it is linear.

1. $y = x + 2$ **linear**   2. $y = -2x$ **linear**   3. $y = x^3$ **not linear**

See Example **2**   4. Kelp is one of the fastest-growing plants in the world. It grows about 2 ft every day. If you found a kelp plant that was 124 ft long, the equation $\ell = 2d + 124$ would represent the length ℓ of the plant d days later. How long would the plant be after 3 days? after 4.5 days? after 6 days? Graph the equation. Is this a linear equation?
130 ft; 133 ft; 136 ft; linear

INDEPENDENT PRACTICE

See Example **1**   Graph each equation and tell whether it is linear.

5. $y = \frac{1}{3}x - 2$ **linear**   6. $y = -6$ **linear**   7. $y = \frac{1}{2}x^2$ **not linear**

8. $x = 3$ **linear**   9. $y = x^2 - 12$ **not linear** 10. $y = 2x + 1$ **linear**

See Example **2**   11. A catering service charges a $150 setup fee plus $7.50 for each guest at a reception. This is represented by the equation $C = 7.5g + 150$, where C is the total cost based on g guests. Find the total cost of catering for the following numbers of guests: 100, 150, 200, 250, 300. Is this a linear equation? Draw a graph that represents the relationship between the total cost and the number of guests.
$900, $1275, $1650, $2025, $2400; linear

PRACTICE AND PROBLEM SOLVING

Evaluate each equation for $x = -1, 0,$ and 1. Then graph the equation.

12. $y = 4x$ 13. $y = 2x + 5$ 14. $y = 6x - 3$

15. $y = x - 10$ 16. $y = 4x - 2$ 17. $y = 4x + 3$

18. $y = 2x - 4$ 19. $y = x + 7$ 20. $y = 3x + 2.5$

21. **PHYSICAL SCIENCE** The force exerted on an object by Earth's gravity is given by the formula $F = 9.8m$, where F is the force in newtons and m is the mass of the object in kilograms. How many newtons of gravitational force are exerted on a student with mass 52 kg? **509.6 N**

22. At a rate of $0.08 per kilowatt-hour, the equation $C = 0.08t$ gives the cost of a customer's electric bill for using t kilowatt-hours of energy. Complete the table of values and graph the energy cost equation for t ranging from 0 to 1000.

Kilowatt-hours (t)	540	580	620	660	700	740
Cost in Dollars (C)						

Students may want to refer back to the lesson examples.

Assignment Guide

If you finished Example **1** assign:
Core 1–3, 5–10, 12–20, 29–34
Enriched 5–10, 12–20, 27–34

If you finished Example **2** assign:
Core 1–25, 29–34
Enriched 5–34

Graph paper is provided on Chapter 11 Resource Book p. 82.

Answers

1–20, 22. Complete answers on pp. A10–A11

12. $(-1, -4), (0, 0), (1, 4)$

13. $(-1, 3), (0, 5), (1, 7)$

14. $(-1, -9), (0, -3), (1, 3)$

15. $(-1, -11), (0, -10), (1, -9)$

16. $(-1, -6), (0, -2), (1, 2)$

17. $(-1, -1), (0, 3), (1, 7)$

18. $(-1, -6), (0, -4), (1, -2)$

19. $(-1, 6), (0, 7), (1, 8)$

20. $(-1, -0.5), (0, 2.5), (1, 5.5)$

22. $43.20, $46.40, $49.60, $52.80, $56.00, $59.20

RETEACH 11-1

LESSON 11-1 *Graphing Linear Equations*

The graph of a **linear equation** is a straight line.
The line shown is the graph of $y = \frac{3}{2}x + 1$.

All the points on the line are solutions of the equation. There are infinitely many points.

For this line, each time the x-value increases by 2, the y-value increases by 3.

If a constant change in x-values corresponds to a constant change in y-values, then the graph must be a straight line and the equation must be linear.

The points in the table below are solutions of the equation $y = 3x - 4$.

The points in the table below are solutions of the equation $y = 3x^2$.

x	-2	-1	0	1	2
y	-10	-7	-4	-1	2

x	-2	-1	0	1	2
y	12	3	0	3	12

Since a constant change in x-values corresponds to a constant change in y-values, the equation $y = 3x - 4$ is a linear equation and its graph would be a straight line.

Since a constant change in x-values does not correspond to a constant change in y-values, the equation $y = 3x^2$ is not a linear equation and its graph would not be a straight line.

Each table has solutions to the equation below it. Indicate the changes in x-values and in y-values. Tell if the equation is linear.

1.
x	-2	-1	0	1	2
y	-9	-7	-5	-3	-1

$y = 2x - 5$ __is__ a linear equation.

2.
x	-2	-1	0	1	2
y	-16	-2	0	2	16

$y = 2x^3$ __is not__ a linear equation.

PRACTICE 11-1

LESSON 11-1 *Graphing Linear Equations*

Complete the table. Then graph each equation and tell whether it is linear.

1. $y = -2x - 5$

x	-2x - 5	y	(x, y)
-4	-2(-4) - 5	3	(-4, 3)
-3	-2(-3) - 5	1	(-3, 1)
-2	-2(-2) - 5	-1	(-2, -1)
-1	-2(-1) - 5	-3	(-1, -3)
0	-2(0) - 5	-5	(0, -5)
1	-2(1) - 5	-7	(1, -7)
2	-2(2) - 5	-9	(2, -9)

linear

2. $y = x^2 - 7$

x	x² - 7	y	(x, y)
-3	(-3)² - 7	2	(-3, 2)
-2	(-2)² - 7	-3	(-2, -3)
-1	(-1)² - 7	-6	(-1, -6)
0	(0)² - 7	-7	(0, -7)
1	(1)² - 7	-6	(1, -6)
2	(2)² - 7	-3	(2, -3)
3	(3)² - 7	2	(3, 2)

not linear

3. A real estate agent commission may be based on the equation $C = 0.06s + 450$, where s represents the total sales. If the agent sells a property for $125,000, what is the commission earned by the agent?

$7950

Math Background

Linear equations fit into the general category of *polynomial equations*. Linear equations (e.g., $y = x$ and $y = \frac{3}{2}x - 4$) are called *first* degree because the greatest power of x is 1.

Quadratic equations (e.g., $y = x^2$ and $y = 3x^2 + 2x - 7$) are second degree because the greatest power of x is 2.

Cubic equations (e.g., $y = x^3$ and $y = x^3 - 2x^2 + 7x + 3$) are third degree, and so on.

Answers

23–24. Complete answers on p. A11

25. a. Possible answer: $20w = 150$; $w = \$7.50$

  b. $E = 7.50h$

  c.

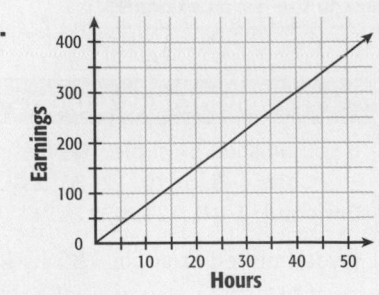

  d. yes

27. Possible answer: Choose several values for *x*, and solve the equation for *y*. Plot the points on a coordinate grid and see whether they form a straight line.

28. See p. A11.

28. See p. A11.

Journal

Ask students to write about how to tell whether a graph shows a linear equation.

Test Prep Doctor

For Exercise 33, remind students that if the probability of winning a raffle is $\frac{1}{1200}$, then there is one way to win and 1199 ways to lose. So the odds of winning are 1:1199, choice **B**. Students who chose **A** selected the probability of winning the raffle. Students who chose **C** selected the odds against winning the raffle.

Lesson Quiz

Graph each equation and tell whether it is linear.

1. $y = 3x - 1$ yes

2. $y = \frac{1}{4}x$ yes

3. $y = x^2 - 3$ no

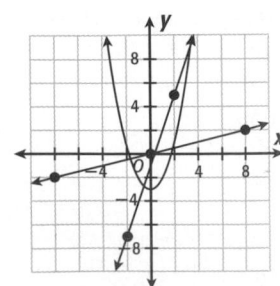

Available on Daily Transparency in CRB

---

23. The minute hand of a clock moves $\frac{1}{10}$ degree every second. If you look at the clock when the minute hand is 10 degrees past the 12, you can use the equation $y = \frac{1}{10}x + 10$ to find how many degrees past the 12 the minute hand is after *x* seconds. Graph the equation and tell whether it is linear.   **linear**

24. ENTERTAINMENT A bowling alley charges \$4 for shoe rental plus \$1.75 per game bowled. Write an equation that shows the total cost of bowling *g* games. Graph the equation. Is it linear?
  $C = 1.75g + 4$; **linear**

25. BUSINESS A car wash pays *d* dollars an hour. The table shows how much employees make based on the number of hours they work.

Car Wash Wages				
Hours Worked (*h*)	20	25	30	40
Earnings (*E*)	\$150.00	\$187.50	\$225.00	\$300.00

  a. Write and solve an equation to find the hourly wage.

  b. Write an equation that gives an employee's earnings *E* for *h* hours of work.

  c. Graph the equation for *h* between 0 and 50 hours.

  d. Is the equation linear?

 26. WHAT'S THE QUESTION? The equation $C = 9.5n + 1350$ gives the total cost of producing *n* trailer hitches. If the answer is \$10,850, what is the question? **Possible answer: How much does it cost to produce 1000 trailer hitches?**

 27. WRITE ABOUT IT Explain how you could show that $y = 5x + 1$ is a linear equation.

 28. CHALLENGE Three solutions of an equation are (1, 1), (3, 3), and (5, 5). Draw one possible graph that would show that the equation is not a linear equation.

Spiral Review

Two fair dice are rolled. Find each probability. (Lessons 9-4 and 9-7)

29. rolling two odd numbers $\frac{1}{4}$

30. rolling a two and a prime number $\frac{5}{36}$

31. rolling a pair of ones $\frac{1}{36}$

32. rolling a six and a seven   **0**

33. TEST PREP The probability of winning a raffle is $\frac{1}{1200}$. What are the odds in favor of winning the raffle? (Lesson 9-8)   **B**

  A   1:1200   C   1199:1

  B   1:1199   D   1200:1

34. TEST PREP A bag of 9 marbles has 3 red marbles and 6 blue marbles in it. What is the probability of drawing a red marble? (Lesson 9-4)   **H**

  F   1   H $\frac{1}{3}$

  G $\frac{2}{3}$ J $\frac{1}{2}$

CHALLENGE 11-1

LESSON 11-1 Challenge *A Recognition Factor*

Geometric figures have special properties that help tell them apart.

1. A triangle and a quadrilateral are both polygons. Name a characteristic that will distinguish between these two polygons.

Possible answer: 3 sides for a triangle, 4 sides for a quadrilateral

Equations for graphs have special properties that help tell them apart.

2. Make a table of values to graph each equation. Put all the graphs on the given grid. Write the equation near each graph.

  a. $y = 2x + 1$

x	y
−4	−7
−3	−5
−2	−3
−1	−1

  b. $y = x^2 + 1$

x	y
−1	2
0	1
1	2
2	5

  c. $y = \frac{6}{x}$

x	y
1	6
2	3
3	2
6	1

  d. $x + y = -1$

x	y
−5	4
−4	3
−3	2
−2	1

3. Use the graphs you have drawn here and in the lesson to make a conjecture about how to recognize a linear equation algebraically.

Possible answer: linear equation: both *x* and *y* to the first power variable terms separated by + or −, not × or ÷

PROBLEM SOLVING 11-1

LESSON 11-1 Problem Solving *Graphing Linear Equations*

Write the correct answer.

1. The distance in feet traveled by a falling object is found by the formula $d = 16t^2$ where *d* is the distance in feet and *t* is the time in seconds. Graph the equation at the right. Is the equation linear?

The equation is not linear.

2. The formula that relates Celsius to Fahrenheit is given by the formula $F = \frac{9}{5}C + 32$. Graph the equation on the grid. Is the equation linear?

The equation is linear.

Use the graph below that represents the wind chill *W* in degrees Fahrenheit for a 25 mi/h wind and temperature *T* in degrees Fahrenheit. Wind chill is the temperature that the air feels like with the effect of the wind.

3. If the temperature is 40° with a 25 mi/h wind, what is the wind chill?
  A 6°   C 29°
  B 20°   D 40°

4. If the temperature is 20° with a 25 mi/h wind, what is the wind chill?
  F 3°   H 13°
  G 10°   J 20°

5. If the temperature is 0° with a 25 mi/h wind, what is the wind chill?
  A −30°   C −15°
  B −24°   D 0°

6. If the wind chill is 10° and there is a 25 mi/h wind, what is the actual temperature?
  F −11°   H 15°
  G 0°   J 25°

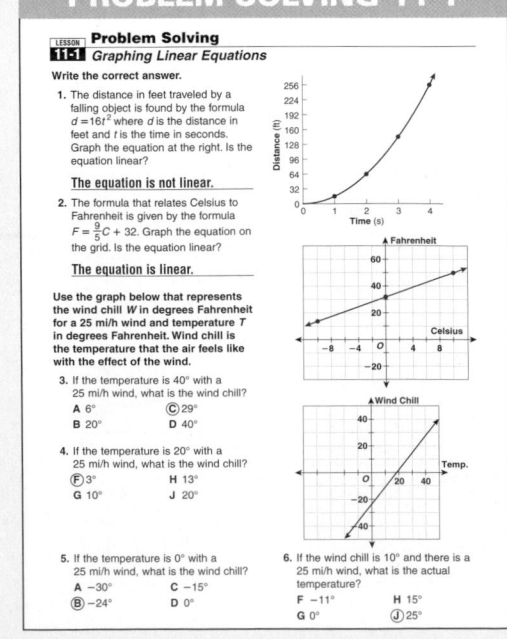

11-2 Slope of a Line

Learn to find the slope of a line and use slope to understand and draw graphs.

In skiing, the term *slope* refers to a slanted mountainside. The steeper a slope is, the higher its difficulty rating will be. In math, slope defines the "slant" of a line. The larger the absolute value of the slope of a line is, the "steeper," or more vertical, the line will be.

Remember!

You looked at slope on the coordinate plane in Lesson 5-5 (p. 244).

Linear equations have constant slope. For a line on the coordinate plane, slope is the following ratio:

$$\frac{\text{vertical change}}{\text{horizontal change}} = \frac{\text{change in } y}{\text{change in } x}$$

This ratio is often referred to as $\frac{\text{rise}}{\text{run}}$, or "rise over run,"

where *rise* indicates the number of units moved up or down and *run* indicates the number of units moved to the left or right. Slope can be positive, negative, zero, or undefined. A line with positive slope goes up from left to right. A line with negative slope goes down from left to right.

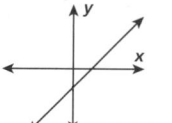 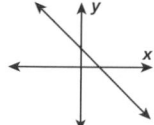 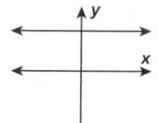 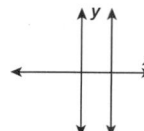

Positive slope   Negative slope   Zero slope   Undefined slope

If you know any two points on a line, or two solutions of a linear equation, you can find the slope of the line without graphing. The slope of a line through the points (x_1, y_1) and (x_2, y_2) is as follows:

$$\frac{y_2 - y_1}{x_2 - x_1}$$

EXAMPLE 1   **Finding Slope, Given Two Points**

Find the slope of the line that passes through **(2, 5)** and **(8, 1)**.

Let (x_1, y_1) be (2, 5) and (x_2, y_2) be (8, 1).

$$\frac{y_2 - y_1}{x_2 - x_1} = \frac{1 - 5}{8 - 2}$$ *Substitute 1 for y_2, 5 for y_1, 8 for x_2, and 2 for x_1.*

$$= \frac{-4}{6} = -\frac{2}{3}$$

The slope of the line that passes through (2, 5) and (8, 1) is $-\frac{2}{3}$.

---

1 Introduce

Alternate Opener

EXPLORATION

 Slope of a Line

Joseph plots a graph of his 300-mile trip to the coast.

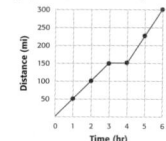

1. Calculate his average speed for the first 3 hours. (*Hint:* Divide the distance traveled during the first three hours by 3.)
2. What happens between hours 3 and 4?
3. Calculate his average speed for the last 2 hours. (*Hint:* Divide the distance traveled during the last two hours by 2.)

Think and Discuss

4. **Discuss** why numbers 1 and 3 refer to average speed and not actual speed. How is average speed represented on the graph?
5. **Discuss** what the graph of the actual speed would look like.

Exploration worksheet and answers on Chapter 11 Resource Book pp. 18 and 90

Motivate

Ask students if they have ever been skiing or seen it on television. Discuss the difference between the beginner's trail, often called the *bunny slope*, and the expert's trail, often called the *black diamond slope.* Generally the black diamond slope will be much steeper than the bunny slope.

11-2 Organizer

Pacing: Traditional 1 day
Block $\frac{1}{2}$ day

Objective: Students find the slope of a line and use slope to understand and draw graphs.

Warm Up

Evaluate each equation for $x = -1$, 0, and 1.

1. $y = 3x$ $-3, 0, 3$
2. $y = x - 7$ $-8, -7, -6$
3. $y = 2x + 5$ $3, 5, 7$
4. $y = 6x - 2$ $-8, -2, 4$

Problem of the Day

Write a linear equation that contains terms with x^2.

Possible answer:
$x^2 + y = x^2 + x + 4$

Available on Daily Transparency in CRB

Math Humor

Whenever the teacher mentioned *slope,* the student imagined a revolt in the bread bakery. He would picture all the commotion as the *ryes overrun* the bakery.

---

2 Teach

Lesson Presentation

Guided Instruction

In this lesson, students find the slope of a line and use it to understand and draw graphs. Remind students that in linear equations, a constant change in x-values corresponds to a constant change in y-values. Explain that this relationship is called *slope* (Teaching Transparency T10, CRB). Remind students that they have worked with slope before as $\frac{\text{rise}}{\text{run}}$ (Lesson 5-5). Emphasize that any two points on a line will yield the same slope and that the points can be used in the slope formula in either order, but the order must be the same for y's and x's.

Additional Examples

Example 1

Find the slope of the line that passes through $(-2, -3)$ and $(4, 6)$. $\frac{3}{2}$

Example 2

Use the graph of the line to determine its slope.

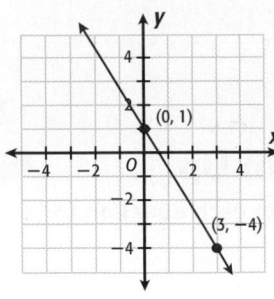

The slope is $-\frac{5}{3}$.

Example 3

Tell whether the lines passing through the given points are parallel or perpendicular.

A. line 1: $(-6, 4)$ and $(2, -5)$;

line 2: $(-1, -4)$ and $(8, 4)$

perpendicular

B. line 1: $(0, 5)$ and $(6, -2)$;

line 2: $(-1, 3)$ and $(5, -4)$

parallel

When choosing two points to evaluate the slope of a line, you can choose any two points on the line because slope is constant.

Below are two graphs of the same line.

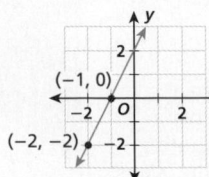   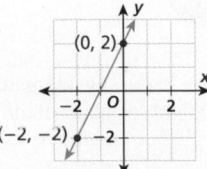

$$\frac{y_2 - y_1}{x_2 - x_1} = \frac{0 - (-2)}{-1 - (-2)} = \frac{2}{1} = 2 \qquad \frac{y_2 - y_1}{x_2 - x_1} = \frac{2 - (-2)}{0 - (-2)} = \frac{4}{2} = \frac{2}{1} = 2$$

The slope of the line is 2. Notice that although different points were chosen in each case, the slope formula still results in the same slope for the line.

EXAMPLE 2   **Finding Slope from a Graph**

Use the graph of the line to determine its slope.

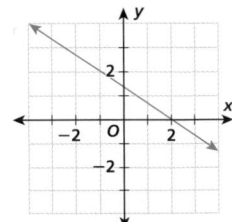

Helpful Hint

It does not matter which point is chosen as (x_1, y_1) and which point is chosen as (x_2, y_2).

Choose two points on the line: $(-1, 2)$ and $(2, 0)$.

Guess by looking at the graph:

$$\frac{\text{rise}}{\text{run}} = \frac{-2}{3} = -\frac{2}{3}$$

Use the slope formula.

Let $(2, 0)$ be (x_1, y_1) and $(-1, 2)$ be (x_2, y_2).

$$\frac{y_2 - y_1}{x_2 - x_1} = \frac{2 - 0}{-1 - 2} = \frac{2}{-3} = -\frac{2}{3}$$

Notice that if you switch (x_1, y_1) and (x_2, y_2), you get the same slope:

Let $(-1, 2)$ be (x_1, y_1) and $(2, 0)$ be (x_2, y_2).

$$\frac{y_2 - y_1}{x_2 - x_1} = \frac{0 - 2}{2 - (-1)} = \frac{-2}{3} = -\frac{2}{3}$$

The slope of the given line is $-\frac{2}{3}$.

Teach

 **Reaching All Learners**
Through Critical Thinking

Give students a recording sheet (Chapter 11 Resource Book p. 83) with the graphs of four lines (one with positive slope, one with negative slope, one with zero slope, and one with undefined slope) and five pairs of points (one for each line and one extraneous pair). Have students use the slope formula to determine the slope between each pair of points and to match the points with the appropriate graph.

Recall that two parallel lines have the same slope. The slopes of two perpendicular lines are negative reciprocals of each other.

EXAMPLE 3 **Identifying Parallel and Perpendicular Lines by Slope**

Tell whether the lines passing through the given points are parallel or perpendicular.

A line 1: $(1, 9)$ and $(-1, 5)$; line 2: $(-3, -5)$ and $(4, 9)$

slope of line 1: $\frac{y_2 - y_1}{x_2 - x_1} = \frac{5 - 9}{-1 - 1} = \frac{-4}{-2} = 2$

slope of line 2: $\frac{y_2 - y_1}{x_2 - x_1} = \frac{9 - (-5)}{4 - (-3)} = \frac{14}{7} = 2$

Both lines have a slope equal to 2, so the lines are parallel.

Remember!
The product of the slopes of perpendicular lines is −1.

B line 1: $(-10, 0)$ and $(20, 6)$; line 2: $(-1, 4)$ and $(2, -11)$

slope of line 1: $\frac{y_2 - y_1}{x_2 - x_1} = \frac{6 - 0}{20 - (-10)} = \frac{6}{30} = \frac{1}{5}$

slope of line 2: $\frac{y_2 - y_1}{x_2 - x_1} = \frac{-11 - 4}{2 - (-1)} = \frac{-15}{3} = -5$

Line 1 has a slope equal to $\frac{1}{5}$ and line 2 has a slope equal to -5. $\frac{1}{5}$ and -5 are negative reciprocals of each other, so the lines are perpendicular.

You can graph a line if you know one point on the line and the slope.

EXAMPLE 4 **Graphing a Line Using a Point and the Slope**

Graph the line passing through $(1, 1)$ with slope $-\frac{1}{3}$.

The slope is $-\frac{1}{3}$. So for every 1 unit down, you will move 3 units to the right, and for every 1 unit up, you will move 3 units to the left.

Plot the point $(1, 1)$. Then move 1 unit down, and right 3 units and plot the point $(4, 0)$. Use a straightedge to connect the two points.

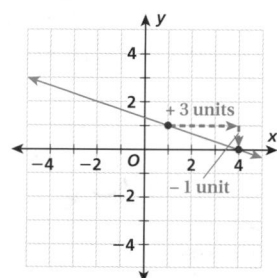

Think and Discuss

1. **Explain** why it does not matter which point you choose as (x_1, y_1) and which point you choose as (x_2, y_2) when finding slope.

2. **Give an example** of two pairs of points from each of two parallel lines.

COMMON ERROR ALERT

Students may reverse the order in the ratio and divide the difference of the x-values by the difference of the y-values. Remind students that slope is always rise over run (y-values over x-values).

Additional Examples

Example 4

Graph the line passing through $(3, 1)$ with slope 2.

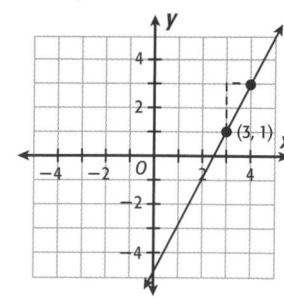

---

3 Close

Summarize

Remind students that slope describes the steepness of the line and is defined as the ratio of vertical change to horizontal change (rise over run). Show students a coordinate grid, and ask them to describe a line with each type of slope (positive, negative, zero, and undefined). As they describe each line, draw an example on the coordinate grid.

Answers to Think and Discuss

Possible answers:

1. Changing the order of the points will change both differences into their opposites. When these opposites are placed into a ratio, however, the ratio will be the same.

2. $(-1, 0)$ and $(0, 2)$ for one line and $(0, 0)$ and $(1, 2)$ for a parallel line

FOR EXTRA PRACTICE
see page 752

internet connect
Homework Help Online
go.hrw.com   Keyword: MP4 11-2

go.hrw.com

Students may want to refer back to the lesson examples.

GUIDED PRACTICE

See Example **1**   Find the slope of the line that passes through each pair of points.

1. (1, 3) and (2, 4)   **1**   **2.** (2, 6) and (0, 2)   **2**   **3.** (−1, 2) and (5, 5) $\frac{1}{2}$

See Example **2**   Use the graph of each line to determine its slope.

4.   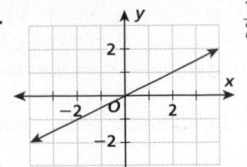 $\frac{1}{2}$ **5.**   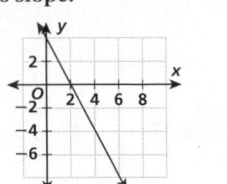   −2

See Example **3**   Tell whether the lines passing through the given points are parallel or perpendicular.

6. line 1: (2, 3) and (4, 7)   **perpendicular**   **7.** line 1: (−4, 1) and (0, 29)   **parallel**
line 2: (5, 2) and (9, 0)   line 2: (3, 3) and (5, 17)

See Example **4**   **8.** Graph the line passing through (0, 2) with slope $-\frac{1}{2}$.

9. Graph the line passing through (−2, 0) with slope $\frac{2}{3}$.

INDEPENDENT PRACTICE

See Example **1**   Find the slope of the line that passes through each pair of points.

10. (−1, −1) and (−3, 2) $-\frac{3}{2}$ **11.** (0, 0) and (6, −3) $-\frac{1}{2}$ **12.** (2, −5) and (1, −2)   **−3**

13. (3, 1) and (0, 3) $-\frac{2}{3}$ **14.** (−2, −3) and (2, 4) $\frac{7}{4}$ **15.** (0, −2) and (−6, 3) $-\frac{5}{6}$

See Example **2**   Use the graph of each line to determine its slope.

16.   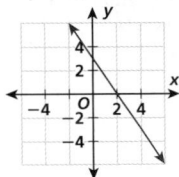 $-\frac{3}{2}$ **17.**   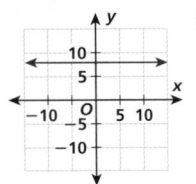   **0**

See Example **3**   Tell whether the lines passing through the given points are parallel or perpendicular.

18. line 1: (1, 4) and (6, 6)   **parallel**   **19.** line 1: (−1, −1) and (−3, 2)
line 2: (−1, −6) and (4, −4)   line 2: (7, −3) and (13, 1)
                                        perpendicular

See Example **4**   **20.** Graph the line passing through (−1, 3) with slope $\frac{1}{4}$.

21. Graph the line passing through (4, 2) with slope $-\frac{4}{5}$.

Assignment Guide

If you finished Example **1** assign:
Core 1–3, 10–15, 31–35
Enriched 10–15, 22, 28, 31–35

If you finished Example **2** assign:
Core 1–5, 10–17, 31–35
Enriched 10–17, 22, 28, 29, 31–35

If you finished Example **3** assign:
Core 1–7, 10–19, 23, 25, 31–35
Enriched 10–19, 22–26, 28, 29, 31–35

If you finished Example **4** assign:
Core 1–22, 23, 25, 31–35
Enriched 10–35

Answers

8.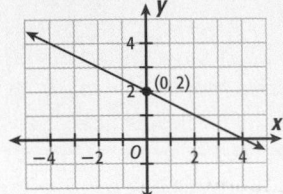

9.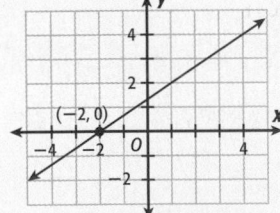

Math Background

Some important points about slope:

- The slope of a line is constant regardless of which points are used to calculate it.
- The order in which the points are used in the slope formula will not affect the slope.
- Vertical lines have an undefined slope. For any two points on the line, the run is zero, so
$\frac{rise}{run} = \frac{(any\ value)}{0} = $ undefined.
- Horizontal lines have a slope of zero. For any two points on the line, the rise is zero, so
$\frac{rise}{run} = \frac{0}{(any\ nonzero\ value)} = 0.$

RETEACH 11-2

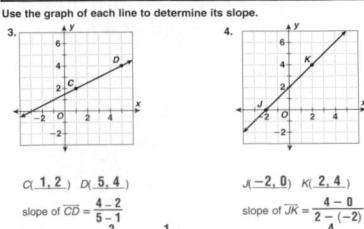

LESSON 11-2 Reteach
Slope of a Line

Slope is the *tilt*, or *steepness*, of a line.
The slope of a line is a constant ratio.

slope = $\frac{change\ in\ y}{change\ in\ x} = \frac{y_2 - y_1}{x_2 - x_1}$

slope of $\overline{AB} = \frac{5-1}{-1-3}$

$= \frac{4}{-4}$, or −1

Complete to find the slope of the line that passes through each set of points.

1. M(7, 8) and N(3, 2)

slope of $\overline{MN} = \frac{8-2}{7-3}$

$= \frac{6}{4}$, or $\frac{3}{2}$

2. R(−3, 5) and T(1, −6)

slope of $\overline{RT} = \frac{5-(-6)}{-3-(1)}$

$= \frac{11}{-4}$, or $-\frac{11}{4}$

Use the graph of each line to determine its slope.

3.   **4.**

C(_1_, _2_)   D(_5_, _4_)

slope of $\overline{CD} = \frac{4-2}{5-1}$

$= \frac{2}{4}$, or $\frac{1}{2}$

J(_−2_, _0_)   K(_2_, _4_)

slope of $\overline{JK} = \frac{4-0}{2-(-2)}$

$= \frac{4}{4}$, or _1_

PRACTICE 11-2

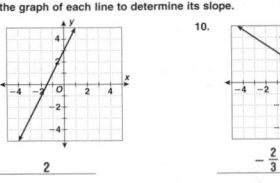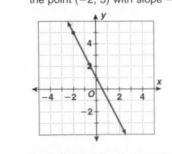

LESSON 11-2 Practice B
Slope of a Line

Find the slope of the line passing through each pair of points.

1. (−2, −8), (1, 4)   **2.** (−2, 0), (0, 4),   **3.** (0, 4), (4, 4)   **4.** (3, −6), (2, −4)
4   **2**   **0**   **−2**

5. (−3, 4), (3, −4)   **6.** (3, 0), (0, −6),   **7.** (3, 2), (3, −2)   **8.** (−4, 4), (3, −1)
$-\frac{4}{3}$ **2**   **undefined** $-\frac{5}{7}$

Use the graph of each line to determine its slope.

9.   **10.**

2 $-\frac{2}{3}$

Tell whether the lines passing through the given points are parallel or perpendicular.

11. line 1: (2, 2), (5, 5)   **12.** line 1: (2, 0), (4, 5)
line 2: (3, −6), (−5, 2)   line 2: (0, −2), (2, 3)

perpendicular   **parallel**

13. Graph the line passing through the point (−2, 5) with slope −2.

14. Graph the line passing through the point (1, −2) with slope $\frac{2}{3}$.

PRACTICE AND PROBLEM SOLVING

22. **SAFETY** To accommodate a 2.5 foot vertical rise, a wheelchair ramp extends horizontally for 30 feet. Find the slope of the ramp. $\frac{1}{12}$

For Exercises 23–26, find the slopes of each pair of lines. Use the slopes to determine whether the lines are perpendicular, parallel, or neither.

23.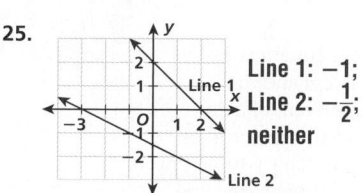

Line 1: 1;
Line 2: −2;
neither

24.

Line 1: 1;
Line 2: 1;
parallel

25.

Line 1: −1;
Line 2: $-\frac{1}{2}$;
neither

26.

Line 1: 2;
Line 2: $-\frac{1}{2}$;
perpendicular

27. The Luxor Hotel in Las Vegas, Nevada, has a 350 ft tall glass pyramid. The elevator of the pyramid moves at an incline, which has a slope of $-\frac{4}{5}$. Graph the line that describes the path it travels along. (*Hint:* The point (0, 350) is the top of the pyramid.)

28. **WHAT'S THE ERROR?** The slope of the line through the points (1, 4) and (−1, −4) is $\frac{1-(-1)}{4-(-4)} = \frac{1}{4}$. What is the error in this statement?

29. **WRITE ABOUT IT** The equation of a vertical line is $x = a$ where a is any number. Explain why the slope of a vertical line is undefined, using a specific vertical line.

30. **CHALLENGE** Graph the equations $y = 2x - 3$, $y = -\frac{1}{2}x$ and $y = 2x + 4$ on one coordinate plane. Find the slope of each line and determine whether each combination of two lines is parallel, perpendicular, or neither. Explain how to tell whether two lines are parallel, perpendicular, or neither by their equations.

Spiral Review

Find the area of each figure with the given dimensions. (Lesson 6-2)

31. triangle: $b = 4$, $h = 6$ **12 units²**

32. triangle: $b = 3$, $h = 14$ **21 units²**

33. trapezoid: $b_1 = 9$, $b_2 = 11$, $h = 12$ **120 units²**

34. trapezoid: $b_1 = 3.4$, $b_2 = 6.6$, $h = 1.8$ **9 units²**

35. **TEST PREP** A circular flower bed has radius 22 in. What is the circumference of the bed to the nearest tenth of an inch? Use 3.14 for π. (Lesson 6-4) **D**

A 1519.8 in.   **B** 69.1 in.   **C** 103.7 in.   **D** 138.2 in.

CHALLENGE 11-2

LESSON 11-2 Challenge
Aligned?

In these exercises, you will explore the results of the special property of the slope of a line: The slope of a line is *constant*.

1. Points A, B, and C are on the same line. Draw a conclusion about the slope between A and B and the slope between B and C.

slope between A and B = slope between B and C

2. Determine if the three points are collinear (lie on the same line).

a. $R(2, 5)$, $S(6, 15)$, $T(16, 18)$
slope between R and S =
$\frac{15-5}{6-2} = \frac{10}{4} = \frac{5}{2}$

slope between S and T =
$\frac{18-15}{16-6} = \frac{3}{10}$

R, S, T **are not** collinear.

b. $J(0, -4)$, $K(1, -2)$, $L(3, 2)$
slope between J and K =
$\frac{-2-(-4)}{1-0} = \frac{-2+4}{1} = 2$

slope between K and L =
$\frac{2-(-2)}{3-1} = \frac{2+2}{2} = \frac{4}{2} = 2$

J, K, L **are** collinear.

3. To find the value of k so that $U(-5, -1)$, $V(-1, -5)$, and $W(5, k)$ are collinear:

a. Find the slope between U and V. $\frac{-5-(-1)}{-1-(-5)} = \frac{-5+1}{-1+5} = \frac{-4}{4} = -1$

b. Find the slope between V and W. $\frac{k-(-5)}{5-(-1)} = \frac{k+5}{5+1} = \frac{k+5}{6}$

c. Set the results of parts a and b equal to each other and solve for k. Justify your result.

$\frac{-1}{1} = \frac{k+5}{6}$

$(k + 5)(1) = (-1)(6)$

$k + 5 = -6$

$k = -11$

Check: When $k = -11$, the slope between V and W should equal -1.

$\frac{k+5}{6} = \frac{-11+5}{6} = \frac{-6}{6} = -1$ ✓

4. The points $P(2, -3)$, $Q(2, 3)$ and $R(k, 0)$ are collinear. Find k. Justify your result.

Since P and Q have the same x-values, $\overline{PQR}$ is a vertical line. So, $k = 2$.

PROBLEM SOLVING 11-2

LESSON 11-2 Problem Solving
Slope of a Line

Write the correct answer.

1. The state of Kansas has a fairly steady slope from the east to west. At the eastern side, the elevation is 771 ft. At the western edge, 413 miles across the state, the elevation is 4039 ft. What is the approximate slope of Kansas?

−0.0015

2. The Feathered Serpent Pyramid in Teotihuacan, Mexico, has a square base. From the center of the base to the center of an edge of the pyramid is 32.5 m. The pyramid is 19.4 m high. What is the slope of each face of the pyramid?

$\frac{19.4}{32.5}$

3. On a highway, a 6% grade means a slope of 0.06. If a highway covers a horizontal distance of 0.5 miles and the elevation change is 184.8 feet, what is the grade of the road? (Hint: 5280 feet = 1 mile.)

7%

4. The roof of a house rises vertically 3 feet for every 12 feet of horizontal distance. What is the slope, or pitch of the roof?

$\frac{1}{4}$

Use the graph.

5. Find the slope of the line between 1990 and 1992.

A $\frac{2}{11}$ **C** $\frac{11}{2}$
B $\frac{35}{3982}$ **D** $\frac{11}{1992}$

6. Find the slope of the line between 1994 and 1996.

F $\frac{7}{2}$ **H** $\frac{2}{7}$
G $\frac{37}{3990}$ **J** $\frac{7}{1996}$

7. Find the slope of the line between 1998 and 2000.

A 1
B $\frac{1}{999}$
C $\frac{1}{1000}$
D 2

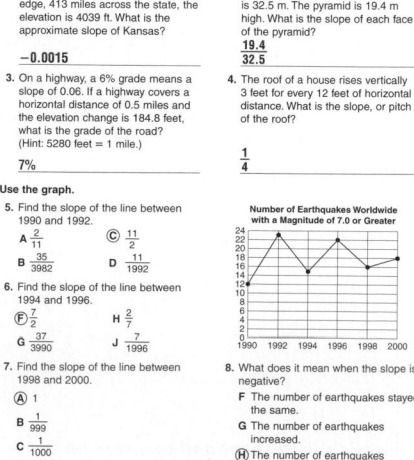

Number of Earthquakes Worldwide with a Magnitude of 7.0 or Greater

8. What does it mean when the slope is negative?

F The number of earthquakes stayed the same.
G The number of earthquakes increased.
H The number of earthquakes decreased.
J It means nothing.

20.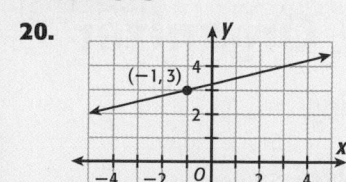

21.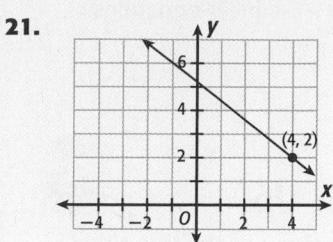

27–30. See p. A11.

Journal

Have students write about the difference between a line that has zero slope and a line that has an undefined slope.

Test Prep Doctor

For Exercise 35, remind students that the formula for the circumference of a circle is $C = 2\pi r \approx 2(3.14)(22) \approx 138.2$ in., choice **D**. Students who chose **A** may have found the area of the circle. Students who chose **B** may have confused radius with diameter.

Lesson Quiz

Find the slope of the line passing through each pair of points.

1. (4, 3) and (−1, 1) $\frac{2}{5}$

2. (−1, 5) and (4, 2) $-\frac{3}{5}$

3. Use the graph of the line to determine its slope. $-\frac{4}{3}$

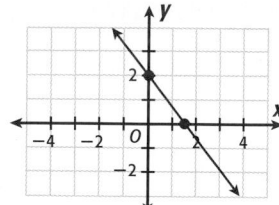

Tell whether the lines passing through the given points are parallel or perpendicular.

4. line 1: (−2, 1), (2, −1); line 2: (0, 0), (−1, −2)   **perpendicular**

5. line 1: (−3, 1), (−2, 3); line 2: (2, 1), (0, −3)   **parallel**

Available on Daily Transparency in CRB

Pacing: Traditional 1 day
Block $\frac{1}{2}$ day

Objective: Students use slopes and intercepts to graph linear equations.

Learn to use slopes and intercepts to graph linear equations.

Vocabulary
x-intercept
y-intercept
slope-intercept form

At an arcade, you buy a game card with 50 credit points on it. Each game of Skittle-ball reduces the number of points on your card by 3.5 points. The linear equation $y = -3.5x + 50$ relates the number of points *y* remaining on your card to the number of games *x* that you have played.

You can graph a linear equation easily by finding the *x-intercept* and the *y-intercept*. The **x-intercept** of a line is the value of *x* where the line crosses the *x*-axis (where $y = 0$). The **y-intercept** of a line is the value of *y* where the line crosses the *y*-axis (where $x = 0$).

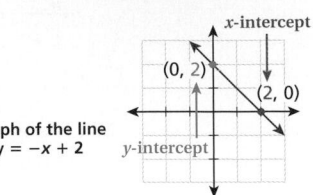

Graph of the line
y = −x + 2

Warm Up

Find the slope of the line that passes through each pair of points.

1. (3, 6) and (−1, 4) $\frac{1}{2}$

2. (1, 2) and (6, 1) $-\frac{1}{5}$

3. (4, 6) and (2, −1) $\frac{7}{2}$

4. (−3, 0) and (−1, 1) $\frac{1}{2}$

Problem of the Day

Write the equation of a straight line that passes through fewer than two quadrants on a coordinate plane.
x = 0 or y = 0

Available on Daily Transparency in CRB

EXAMPLE **1** **Finding x-intercepts and y-intercepts to Graph Linear Equations**

Find the *x*-intercept and *y*-intercept of the line $2x + 3y = 6$. Use the intercepts to graph the equation.

Find the *x*-intercept ($y = 0$).

$$2x + 3y = 6$$
$$2x + 3(0) = 6$$
$$2x = 6$$
$$\frac{2x}{2} = \frac{6}{2}$$
$$x = 3$$

The *x*-intercept is 3.

Find the *y*-intercept ($x = 0$).

$$2x + 3y = 6$$
$$2(0) + 3y = 6$$
$$3y = 6$$
$$\frac{3y}{3} = \frac{6}{3}$$
$$y = 2$$

The *y*-intercept is 2.

The graph of $2x + 3y = 6$ is the line that crosses the *x*-axis at the point (3, 0) and the *y*-axis at the point (0, 2).

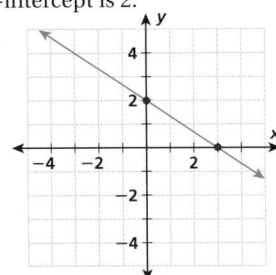

Math Humor

How do you clean the graph of a linear equation? With *slope suds*.

① Introduce
Alternate Opener

11-3 Using Slopes and Intercepts

On the graph below, 40 is called the *y-intercept* and 8 is called the *x-intercept*.

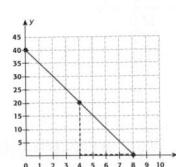

1. How do the rise and run compare to the intercepts?

2. Find the slope of the line by dividing the *y*-intercept by the *x*-intercept and then changing the sign of the quotient. Express the slope as a fraction in simplest form.

3. Consider the triangle drawn along the dashed lines below the line. Divide the vertical distance (20) by the horizontal distance (4), change the sign, and compare this fraction with the fraction you found in number 2.

Think and Discuss

4. **Explain** how to calculate the slope of a line using the *x*- and *y*-intercepts. Explain why this works.

5. **Explain** how to calculate the slope of a line using any two points on the line.

Motivate

Ask students, "What do all the points on the *y*-axis have in common?" The *x*-coordinate is zero. "What do all the points on the *x*-axis have in common?" The *y*-coordinate is zero. Tell students that those two facts will help them graph lines a new way.

Exploration worksheet and answers on Chapter 11 Resource Book pp. 28 and 93

② Teach
Lesson Presentation

Guided Instruction

In this lesson, students learn to use slopes and intercepts to graph linear equations. Review the definitions of *x*-intercept and *y*-intercept. Ask students to find the *x*-intercept and *y*-intercept of a line and to use the intercepts to graph the line. Have students use the slope-intercept form of a linear equation to find slopes and *y*-intercepts. Then show students how to write the equation of a line, given two points on the line.

In an equation written in **slope-intercept form**, $y = mx + b$, m is the slope and b is the y-intercept.

Slope y-intercept

$$y = mx + b$$

EXAMPLE 2 Using Slope-Intercept Form to Find Slopes and y-intercepts

Write each equation in slope-intercept form, and then find the slope and y-intercept.

A $y = x$

$y = x$

$y = 1x + 0$ *Rewrite the equation to show each part.*

$m = 1$ $b = 0$

The slope of the line $y = x$ is 1, and the y-intercept of the line is 0.

B $7x = 3y$

$7x = 3y$

$3y = 7x$ *Reverse the expressions.*

$\dfrac{3y}{3} = \dfrac{7x}{3}$ *Divide both sides by 3 to solve for y.*

$y = \dfrac{7}{3}x + 0$ *The equation is in slope-intercept form.*

$m = \dfrac{7}{3}$ $b = 0$

The slope of the line $7x = 3y$ is $\dfrac{7}{3}$, and the y-intercept is 0.

C $2x + 5y = 8$

$$2x + 5y = 8$$
$$\underline{-2x \qquad\quad -2x}$$ *Subtract 2x from both sides.*
$$5y = 8 - 2x$$

Rewrite to match slope-intercept form.

$5y = -2x + 8$

$\dfrac{5y}{5} = \dfrac{-2x}{5} + \dfrac{8}{5}$ *Divide both sides by 5.*

$y = -\dfrac{2}{5}x + \dfrac{8}{5}$ *The equation is in slope-intercept form.*

$m = -\dfrac{2}{5}$ $b = \dfrac{8}{5}$

The slope of the line $2x + 5y = 8$ is $-\dfrac{2}{5}$, and the y-intercept is $\dfrac{8}{5}$.

Helpful Hint

For an equation such as $y = x - 6$, write it as $y = x + (-6)$ to read the y-intercept, -6.

Reaching All Learners

Through Graphic Cues

Give each student a recording sheet (Chapter 11 Resource Book p. 84) with two related sets of linear equations, one with the same slopes and different y-intercepts (e.g., $y = 2x + 3$, $y = 2x - 2$, and $y = 2x$) and one with different slopes and the same y-intercepts (e.g., $y = 3x + 2$, $y = -\frac{1}{2}x + 2$, and $y = x + 2$). Have students graph each set of lines on the same coordinate plane. When students have completed the graphs, discuss the results with the class.

Example 1

Find the x-intercept and y-intercept of the line $4x - 3y = 12$. Use the intercepts to graph the equation. x-intercept: 3; y-intercept: -4

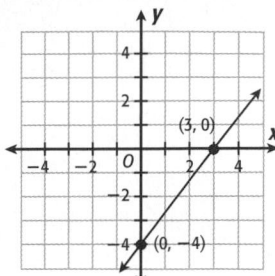

Example 2

Write each equation in slope-intercept form, and then find the slope and y-intercept.

A. $2x + y = 3$ $y = -2x + 3$; $m = -2$; $b = 3$

B. $5y = 3x$ $y = \dfrac{3}{5}x + 0$; $m = \dfrac{3}{5}$; $b = 0$

C. $4x + 3y = 9$ $y = -\dfrac{4}{3}x + 3$; $m = -\dfrac{4}{3}$; $b = 3$

Example 3

A video club charges $8 to join, and $1.25 for each DVD that is rented. The linear equation $y = 1.25x + 8$ represents the amount of money y spent after renting x DVDs. Graph the equation by first identifying the slope and y-intercept. $m = 1.25$; $b = 8$

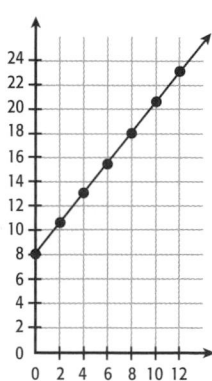

Example 4

Write the equation of the line that passes through $(3, -4)$ and $(-1, 4)$ in slope-intercept form. $y = -2x + 2$

EXAMPLE 3 Entertainment Application

An arcade deducts 3.5 points from your 50-point game card for each Skittle-ball game you play. The linear equation $y = -3.5x + 50$ represents the number of points y on your card after x games. Graph the equation using the slope and *y*-intercept.

$y = -3.5x + 50$ *The equation is in slope-intercept form.*

$$m = -3.5 \qquad b = 50$$

The slope of the line is −3.5, and the *y*-intercept is 50. The line crosses the *y*-axis at the point (0, 50) and moves down 3.5 units for every 1 unit it moves to the right.

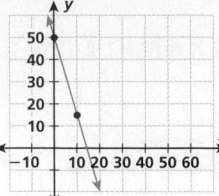

EXAMPLE 4 Writing Slope-Intercept Form

Write the equation of the line that passes through (−3, 1) and (2, −1) in slope-intercept form.

Find the slope.

$$\frac{y_2 - y_1}{x_2 - x_1} = \frac{-1-1}{2-(-3)} = \frac{-2}{5} = -\frac{2}{5} \qquad \textit{The slope is } -\frac{2}{5}.$$

Choose either point and substitute it along with the slope into the slope-intercept form.

$$y = mx + b$$

$$-1 = -\frac{2}{5}(2) + b \qquad \textit{Substitute 2 for x, −1 for y, and } -\frac{2}{5} \textit{ for m.}$$

$$-1 = -\frac{4}{5} + b \qquad \textit{Simplify.}$$

Solve for *b*.

$$-1 = -\frac{4}{5} + b$$

$$\underline{+\frac{4}{5} \quad +\frac{4}{5}} \qquad \textit{Add } \frac{4}{5} \textit{ to both sides.}$$

$$-\frac{1}{5} = b$$

Write the equation of the line, using $-\frac{2}{5}$ for *m* and $-\frac{1}{5}$ for *b*.

$$y = -\frac{2}{5}x + \left(-\frac{1}{5}\right), \text{ or } y = -\frac{2}{5}x - \frac{1}{5}$$

Think and Discuss

1. **Describe** the line represented by the equation $y = -5x + 3$.

2. **Give** a real-life example with a graph that has a slope of 5 and a *y*-intercept of 30.

 ## Close

Summarize

Review the different methods for graphing lines:

- using a table of values to plot points
- plotting a point and using $\frac{\text{rise}}{\text{run}}$ to plot other points
- plotting the *x*- and *y*-intercepts
- using the slope-intercept form of a line to determine the slope and to plot the *y*-intercept

Remind students that for a single equation, any of these methods will result in the same graph.

Answers to Think and Discuss

Possible answers:

1. The line has a *y*-intercept of 3 and a slope of −5. It crosses the *y*-axis at (0, 3) and slants down 5 units for every 1 unit right.

2. A student has $30 in a savings account. He plans to add $5 every week. The total amount *A* in the account can be represented by $A = 5w + 30$, where *w* is the number of weeks.

FOR EXTRA PRACTICE
see page 752

internet connect
Homework Help Online
go.hrw.com   Keyword: MP4 11-3

11-3 **PRACTICE & ASSESS**

GUIDED PRACTICE

See Example **1** Find the *x*-intercept and *y*-intercept of each line. Use the intercepts to graph the equation.

> **1.** $x - y = 5$ **2.** $2x + 3y = 12$ **3.** $3x + 5y = -15$ **4.** $-5x + 2y = -10$
> (5, 0), (0, −5)   (6, 0), (0, 4)   (−5, 0), (0, −3)   (2, 0), (0, −5)

See Example **2** Write each equation in slope-intercept form, and then find the slope and *y*-intercept.

> **5.** $2x = 4y$ **6.** $3x - y = 14$ **7.** $3x - 9y = 27$ **8.** $x + 2y = 8$

See Example **3** **9.** A freight company charges $22 plus $3.50 per pound to ship an item that weighs *n* pounds. The total shipping charges are given by the equation $C = 3.5n + 22$. Identify the slope and *y*-intercept, and use them to graph the equation for *n* between 0 and 100 pounds.
> $m = 3.5$; $b = 22$

See Example **4** Write the equation of the line that passes through each pair of points in slope-intercept form.

> **10.** (−1, −6) and (2, 6) **11.** (0, 5) and (3, −1) **12.** (3, 5) and (6, 6)
> $y = 4x - 2$ $y = -2x + 5$ $y = \frac{1}{3}x + 4$

INDEPENDENT PRACTICE

See Example **1** Find the *x*-intercept and *y*-intercept of each line. Use the intercepts to graph the equation.

> **13.** $2y = 20 - 4x$ **14.** $4x = 12 + 3y$ **15.** $-y = 18 - 6x$ **16.** $2x + y = 7$
> (5, 0), (0, 10)   (3, 0), (0, −4)   (3, 0), (0, −18)   (3.5, 0), (0, 7)

See Example **2** Write each equation in slope-intercept form, and then find the slope and *y*-intercept.

> **17.** $-y = 2x$ **18.** $5y + 2x = 15$ **19.** $-4y - 8x = 8$ **20.** $2y + 6x = -14$

See Example **3** **21.** A salesperson receives a weekly salary of $300 plus a commission of $15 for each TV sold. Total weekly pay is given by the equation $P = 15n + 300$. Identify the slope and *y*-intercept, and use them to graph the equation for *n* between 0 and 40 TVs. $m = 15$; $b = 300$

See Example **4** Write the equation of the line that passes through each pair of points in slope-intercept form.

> **22.** (0, −7) and (4, 25) **23.** (−1, 1) and (3, −3) **24.** (−6, −3) and (12, 0)
> $y = 8x - 7$ $y = -x$ $y = \frac{1}{6}x - 2$

PRACTICE AND PROBLEM SOLVING

Use the *x*-intercept and *y*-intercept of each line to graph the equation.

> **25.** $y = 2x - 10$ **26.** $y = \frac{1}{3}x + 2$ **27.** $y = 4x - 2.5$ **28.** $y = -\frac{4}{5}x + 15$

Students may want to refer back to the lesson examples.

Assignment Guide

If you finished Example **1** assign:
  Core 1–4, 13–16, 33–37
  Enriched 13–16, 25–28, 33–37

If you finished Example **2** assign:
  Core 1–8, 13–20, 33–37
  Enriched 1–7 odd, 13–20, 25–28, 33–37

If you finished Example **3** assign:
  Core 1–9, 13–21, 33–37
  Enriched 9, 13–21, 25–37

If you finished Example **4** assign:
  Core 1–24, 33–37
  Enriched 9–37

Graph paper is provided on Chapter 11 Resource Book p. 82.

Answers

1–4, 9, 13–16, 21. Complete answers on pp. A11–A12

5. $y = \frac{1}{2}x$; $m = \frac{1}{2}$; $b = 0$

6. $y = 3x - 14$; $m = 3$; $b = -14$

7. $y = \frac{1}{3}x - 3$; $m = \frac{1}{3}$; $b = -3$

8. $y = -\frac{1}{2}x + 4$; $m = -\frac{1}{2}$; $b = 4$

17. $y = -2x$; $m = -2$; $b = 0$

18. $y = -\frac{2}{5}x + 3$; $m = -\frac{2}{5}$; $b = 3$

19. $y = -2x - 2$; $m = -2$; $b = -2$

20. $y = -3x - 7$; $m = -3$; $b = -7$

25–28. See p. A12.

Math Background

The methods for graphing lines used in this section are mainly used for graphing lines that are not horizontal or vertical. For horizontal lines, there is a *y*-intercept, but no *x*-intercept (unless the horizontal line is the *x*-axis), and the slope is zero. For vertical lines, there is an *x*-intercept, but no *y*-intercept (unless the vertical line is the *y*-axis), and the slope is undefined.

RETEACH 11-3

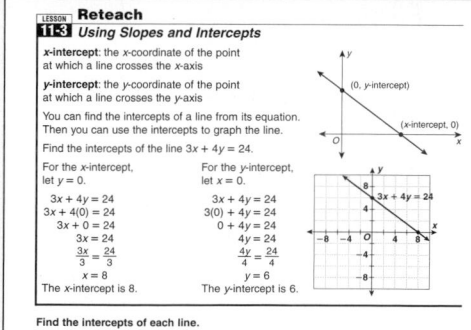

PRACTICE 11-3

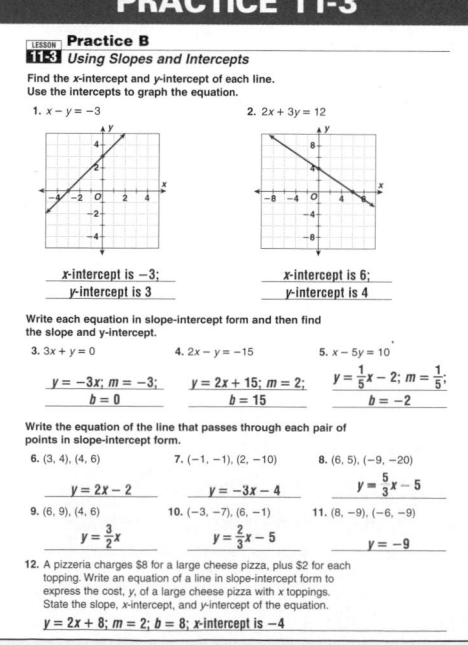

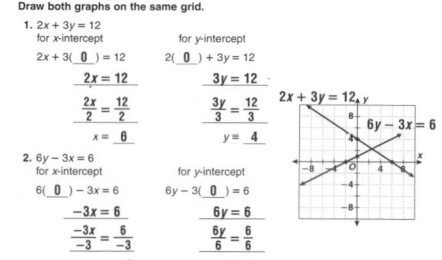

Life Science

Exercises 29–32 involve using linear equations to solve problems involving climbing rates and altitude sickness. The effects of altitude on respiration are studied in middle school life science programs, such as *Holt Science & Technology*.

Answers

29. a. See p. A12; $m = 2000$

   b. 8255 ft, the altitude of the base camp

   c. $y = 2000x + 8255$

   d. yes

30. $y = 544x + 9056$; yes; the 8th day

31. Slope: 955 ft per day; y-intercept: 16,500 ft; the slope is the number of feet climbed each day, and the y-intercept is the starting altitude.

32. See p. A12.

Journal

Ask students to describe some situations in which one method of graphing lines might be easier or more useful than another.

Test Prep Doctor

For Exercise 37, students can use logic to eliminate some choices and then test the remaining choices until they find the correct solution. If the number of dimes were odd, then the number of quarters would be even, in which case the total amount would be divisible by 10 cents. Since it is not, choices **B** and **D** can be eliminated. Testing choice **A** will reveal that it is correct.

Lesson Quiz

Write each equation in slope-intercept form, and then find the slope and y-intercept.

1. $2y - 6x = -10$ $y = 3x - 5$; $m = 3$; $b = -5$

2. $-5y - 15x = 30$ $y = -3x - 6$; $m = -3$; $b = -6$

Write the equation of the line that passes through each pair of points in slope-intercept form.

3. (0, 2) and (4, −1) $y = -\frac{3}{4}x + 2$

4. (−2, 2) and (4, −4) $y = -x$

Available on Daily Transparency in CRB

Acute Mountain Sickness (AMS) occurs if you ascend in altitude too quickly without giving your body time to adjust. It usually occurs at altitudes over 10,000 feet above sea level. To prevent AMS you should not ascend more than 1000 feet per day. And every time you climb a total of 3000 feet, your body needs two nights to adjust.

Often people will get sick at high altitudes because there is less oxygen and lower atmospheric pressure.

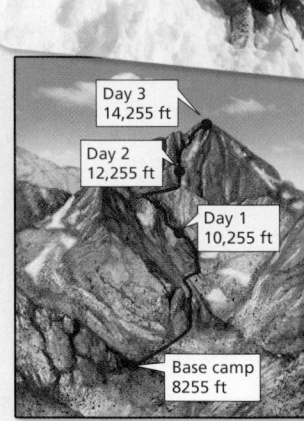

Day 3
14,255 ft

Day 2
12,255 ft

Day 1
10,255 ft

Base camp
8255 ft

29. The map shows a team's plan for climbing Long's Peak in Rocky Mountain National Park.

   a. Make a graph of the team's plan of ascent and find the slope of the line. (Day number should be your x-value, and altitude should be your y-value.)

   b. Find the y-intercept and explain what it means.

   c. Write the equation of the line in slope-intercept form.

   d. Does the team run a high risk of getting AMS?

30. An expedition starts at an altitude of 9056 ft and climbs at an average rate of 544 ft of elevation a day. Write an equation in slope-intercept form that describes the expedition's climb. Are the climbers likely to suffer from AMS at their present climbing rate? On what day of their climb will they be at risk?

31. The equation that describes a mountain climber's ascent up Mount McKinley in Alaska is $y = 955x + 16,500$, where x is the day number and y is the altitude at the end of the day. What are the slope and y-intercept? What do they mean in terms of the climb?

32. ⭐ **CHALLENGE** Make a graph of the ascent of a team that follows the rules to avoid AMS exactly and spends the minimum number of days climbing from base camp (17,600 ft) to the summit of Mount Everest (29,035 ft). Can you write a linear equation describing this trip? Explain your answer.

Spiral Review

Estimate the number or percent. (Lesson 8-5)

33. 25% of 398 is about what number?   **100**   **34.** 202 is about 50% of what number?   **400**

35. About what percent of 99 is 39?   **40%**   **36.** About what percent of 989 is 746?   **75%**

37. TEST PREP Carlos has $3.35 in dimes and quarters. If he has a total of 23 coins, how many dimes does he have? (Lesson 10-6)   **A**

   A 16   B 11   C 18   D 9

CHALLENGE 11-3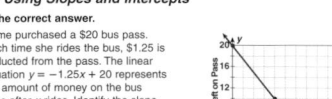

LESSON **Challenge**
11-3 *Another View*

The **intercepts** of a line are related to where the line crosses the coordinate axes.

You can find the intercepts of a line from its equation.

1. Find the intercepts of the line whose equation is $3x + 5y = 15$. Use the intercepts to graph the line.

For the x-intercept, let $y = 0$.
$$3x + 5(0) = 15$$
$$3x = 15$$
$$\frac{3x}{3} = \frac{15}{3}$$
$$x = 5$$

For the y-intercept, let $x = 0$.
$$3(0) + 5y = 15$$
$$5y = 15$$
$$\frac{5y}{5} = \frac{15}{5}$$
$$y = 3$$

The x-intercept is __5__. The y-intercept is __3__.

2. Divide each term of the equation $3x + 5y = 15$ by 15, and simplify. Compare the results to those obtained in Question 1.
$$\frac{3x}{15} + \frac{5y}{15} = \frac{15}{15}\qquad \frac{x}{5} + \frac{y}{3} = 1$$

The denominators of the variables are the intercepts.

3. Using b to represent the y-intercept and a to represent the x-intercept, write an equation that generalizes the observation you made in Question 2.
$$\frac{x}{a} + \frac{y}{b} = 1$$

4. a. Using the form of the equation you wrote in Question 3, find the intercepts of the linear equation $2x + 3y = 24$.
$$\frac{2x}{24} + \frac{3y}{24} = \frac{24}{24}\qquad \frac{x}{12} + \frac{y}{8} = 1$$

The x-intercept is __12__ and the y-intercept is __8__.

b. Check your result by using the intercept definitions.
$$2x + 3(0) = 24\qquad 2(0) + 3y = 24$$
$$2x = 24\qquad 3y = 24$$
$$x = 12\qquad y = 8$$

PROBLEM SOLVING 11-3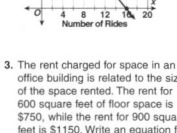

LESSON **Problem Solving**
11-3 *Using Slopes and Intercepts*

Write the correct answer.

1. Jaime purchased a $20 bus pass. Each time she rides the bus, $1.25 is deducted from the pass. The linear equation $y = -1.25x + 20$ represents the amount of money on the bus pass after x rides. Identify the slope and the x- and y-intercepts. Graph the equation at the right.

x-intercept = 16,
y-intercept = 20, slope = −1.25

2. A small business buys a piece of equipment for $2,500. After 10 years, the equipment will have a value of $500. Write the equation for the value of the equipment y after x years.

$y = -200x + 2500$

3. The rent charged for space in an office building is related to the size of the space rented. The rent for 600 square feet of floor space is $750, while the rent for 900 square feet is $1150. Write an equation for the rent y based on the square footage of the floor space x.

$y = \frac{4}{3}x - 50$

The linear equation that approximates the total prize money awarded at the Indianapolis 500 race is $y = 0.23x + 7$ where y is the prize money in millions of dollars and x is the year with $x = 1$ representing 1991. Choose the letter for the best answer.

4. What is the y-intercept?
   A 0.23   C 700
   Ⓑ 7   D 1997

5. What is the x-intercept?
   Ⓕ −30.4   H 7
   G 0.23   J 1997

6. What is the slope of the line?
   A −30.4   C 7
   Ⓑ 0.23   D 1997

7. What does the line predict for the total award money at the Indianapolis 500 in the year 2002?
   F $9.76   H $12,060,000
   Ⓖ $9,760,000   J $467,460,000

Technology LAB 11A

Graph Equations in Slope-Intercept Form

Use with Lesson 11-4

internet connect

Lab Resources Online
go.hrw.com
KEYWORD: MP4 Lab11A

To graph $y = x + 1$, a linear equation in slope-intercept form, in the standard graphing calculator window, press [Y=]; enter the right side of the equation, [X,T,θ,n] [+] 1; and press [ZOOM] **6:ZStandard.**

From the slope-intercept equation, you know that the slope of the line is 1. Notice that the standard window distorts the screen, and the line does not appear to have a great enough slope.

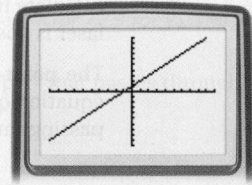

Press [ZOOM] **5:ZSquare.** This changes the scale for x from -10 to 10 to -15.16 to 15.16. The graph is shown at right. Or press [ZOOM] **8:ZInteger** [ENTER]. This changes the scale for x to -47 to 47 and the scale for y to -31 to 31.

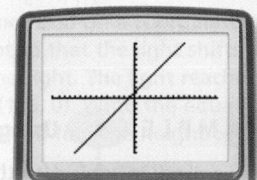

Activity

1 Graph $2x + 3y = 36$ in the integer window. Find the x- and y-intercepts of the graph.

First solve $3y = -2x + 36$ for y.

$y = \dfrac{-2x + 36}{3}$, so $y = \dfrac{-2}{3}x + 12$.

Press [Y=]; enter the right side of the equation, [(] [(−)] 2 [÷] 3 [)] [X,T,θ,n] [+] 12; and press [ZOOM] **8:ZInteger** [ENTER].

Press [TRACE] to see the equation of the line and the y-intercept. The graph in the **ZInteger** window is shown.

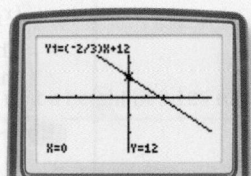

Think and Discuss

1. How do the ratios of the range of y to the range of x in the **ZSquare** and **ZInteger** windows compare?

Try This

Graph each equation in a square window.

1. $y = 2x$ 2. $2y = x$ 3. $2y - 4x = 12$ 4. $2x + 5y = 40$

Answers

Think and Discuss

1. The ratios 15.16 to 10 and 47 to 31 are roughly equal.

Try This

1.

2.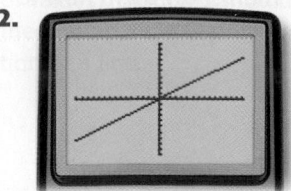

3.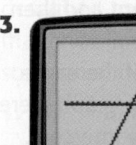

4.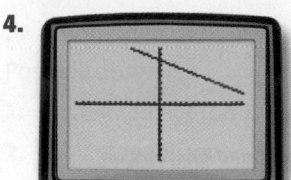

Technology LAB

11A Graph Equations in Slope-Intercept Form

Pacing: Traditional 1 day
Block $\frac{1}{2}$ day

Objective: To use a graphing calculator to graph equations in slope-intercept form

Materials: Graphing calculator

Lab Resources

Technology Lab Activities . . . . pp. 76–77

Using the Page

This technology activity shows students how to graph equations in slope-intercept form and observe the effects of changing the viewing window, which can be done on any graphing calculator. Specific keystrokes may vary, depending on the make and model of the graphing calculator used. The keystrokes given are for a TI-83 model. For keystrokes to other models, visit go.hrw.com.

The Think and Discuss problem can be used to assess students' understanding of the technology activity. Although the equations in Try This problems 1–4 can be graphed without a calculator, they are meant to help students become familiar with using a graphing calculator to graph equations in slope-intercept form.

Assessment

1. Use the graphing calculator to find the y-value of the equation $y = \frac{3}{4}x + 5$ when $x = 2$.
$y = 6\frac{1}{2}$, or $\frac{13}{2}$

2. Use the graphing calculator to find the x-intercept of the graph of the equation $y = -\frac{1}{3}x - 2$. **(−6, 0)**

11-4 PRACTICE & ASSESS

11-4 Exercises

FOR EXTRA PRACTICE
see page 752

☑ internet connect
Homework Help Online
go.hrw.com   Keyword: MP4 11-4

Students may want to refer back to the lesson examples.

Assignment Guide

If you finished Example **1** assign:
Core 1–6, 10–15, 29–33
Enriched 1–6, 10–15, 29–33

If you finished Example **2** assign:
Core 1–8, 10–17, 19, 21, 29–33
Enriched 1–7 odd, 10–17, 19–22, 29–33

If you finished Example **3** assign:
Core 1–18, 19–25 odd, 29–33
Enriched 10–33

Answers

10. $(-7, 1)$, $\frac{2}{3}$

11. $(-4, -7)$, 3

12. $(11, 2)$, $-\frac{1}{6}$

19. $y - 4 = 3(x + 1)$

20. $y + 3 = \frac{1}{2}(x - 7)$

21. $y + 8 = -1(x + 6)$

22. $y = -10(x + 3)$

GUIDED PRACTICE

See Example **1** Use the point-slope form of each equation to identify a point the line passes through and the slope of the line.   **Possible points are given.**

1. $y - 4 = -2(x + 7)$
$(-7, 4)$, -2

2. $y - 9 = 5(x - 12)$
$(12, 9)$, 5

3. $y + 2.4 = 2.1(x - 1.8)$
$(1.8, -2.4)$, 2.1

4. $y + 1 = 11(x - 1)$
$(1, -1)$, 11

5. $y + 8 = -6(x - 9)$
$(9, -8)$, -6

6. $y - 7 = 4(x + 3)$
$(-3, 7)$, 4

See Example **2** Write the point-slope form of the equation with the given slope that passes through the indicated point.

7. the line with slope 3 passing through $(0, 4)$ $y - 4 = 3x$

8. the line with slope -10 passing through $(-13, 8)$
$y - 8 = -10(x + 13)$

See Example **3** **9.** A pond is drained at a rate of 12.5 liters per minute. After 44 minutes, there are 2450 liters of water remaining. Write the equation of a line in point-slope form that models the situation. If the pond originally contained 3000 liters, how long does it take to drain the pond? $y - 2450 = -12.5(x - 44)$; $(240, 0)$ or 240 minutes

INDEPENDENT PRACTICE

See Example **1** Use the point-slope form of each equation to identify a point the line passes through and the slope of the line.   **Possible points are given.**

10. $y - 1 = \frac{2}{3}(x + 7)$

11. $y + 7 = 3(x + 4)$

12. $y - 2 = -\frac{1}{6}(x - 11)$

13. $y - 11 = 14(x - 8)$
$(8, 11)$, 14

14. $y - 3 = -1.8(x - 5.6)$
$(5.6, 3)$, -1.8

15. $y + 7 = 1(x - 5)$
$(5, -7)$, 1

See Example **2** Write the point-slope form of the equation with the given slope that passes through the indicated point.

16. the line with slope -5 passing through $(-3, -5)$
$y + 5 = -5(x + 3)$

17. the line with slope 4 passing through $(-1, 0)$
$y = 4(x + 1)$

See Example **3** **18.** A stretch of highway has a 5% grade, so the road rises 1 ft for each 20 ft of horizontal distance. The beginning of the highway ($x = 0$) has an elevation of 2344 ft. Write an equation in point-slope form, and find the highway's elevation 7500 ft from the beginning.
$y - 2344 = 0.05x$; 2719 ft above sea level

PRACTICE AND PROBLEM SOLVING

Write the point-slope form of each line described below.

19. the line parallel to $y = 3x - 4$ that passes through $(-1, 4)$

20. the line perpendicular to $y = -2x$ that passes through $(7, -3)$

21. the line perpendicular to $y = x + 1$ that passes through $(-6, -8)$

22. the line parallel to $y = -10x - 5$ that passes through $(-3, 0)$

Math Background

The many formulas associated with linear equations can lead to some confusion. It may help to see the relation of the equations to the slope formula,
$m = \frac{y_2 - y_1}{x_2 - x_1}$.

Multiplying both sides of this formula by $(x_2 - x_1)$ and replacing the point (x_2, y_2) with the variables x and y yields the point-slope form of a line,
$y - y_1 = m(x - x_1)$.

Substituting a point on the line with coordinates $(0, b)$ into this equation yields $y - b = mx$, which is easily changed to the slope-intercept form of a line, $y = mx + b$.

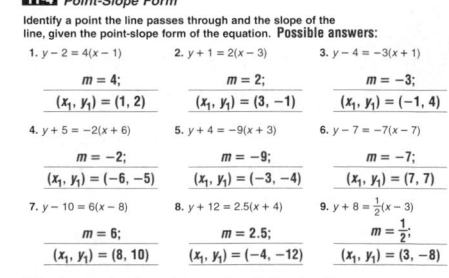

RETEACH 11-4

LESSON **11-4** Reteach
Point-Slope Form

$y - y_1 = m(x - x_1)$
$\quad\quad\quad$ slope

(x_1, y_1) are the coordinates of a known point on the line.

If there is a minus sign in front of a number that represents a coordinate, use that number.

If there is a plus sign, in front of a number that represents a coordinate, use the opposite of that number.

$y - 3 = 7(x - 1)$
$(1, 3)$ is on the line; slope $m = 7$

$y + 3 = 7(x - 1)$
$(1, -3)$ is on the line; slope $m = 7$

Identify the slope and a point on each line.

1. $y - 2 = 5(x - 3)$

2. $y + 4 = -3(x + 5)$

Slope is number in front of parentheses.
$m = \underline{\ 5\ }$ $m = \underline{\ -3\ }$
Which sign for each coordinate? (same or opposite) $\underline{\text{same}}$ $\underline{\text{opposite}}$
Coordinates of a point on the line? $\underline{(3, 2)}$ $\underline{(-5, -4)}$

To write an equation for the line with slope -4 that passes through $(6, -2)$, substitute $m = -4$, $x_1 = 6$, $y_1 = -2$.
$y - y_1 = m(x - x_1)$
$y - (-2) = -4(x - 6)$
$y + 2 = -4(x - 6)$

For each line, given its slope and a point on the line, write the point-slope form of its equation.

3. $m = 3$, $(x_1, y_1) = (7, 2)$
$y - y_1 = m(x - x_1)$
$y - \underline{\ 2\ } = \underline{\ 3\ }(x - \underline{\ 7\ })$

4. $m = -5$, $(x_1, y_1) = (2, 6)$
$y - y_1 = m(x - x_1)$
$y - \underline{\ 6\ } = \underline{\ -5\ }(x - \underline{\ 2\ })$

5. $m = \frac{1}{2}$, $(x_1, y_1) = (-8, 1)$
$y - y_1 = m(x - x_1)$
$y - \underline{\ 1\ } = \underline{\tfrac{1}{2}\ }(x - (\underline{\ -8\ }))$

6. $m = -\frac{3}{4}$, $(x_1, y_1) = (0, -1)$
$y - y_1 = m(x - x_1)$
$y - (\underline{\ -1\ }) = \underline{-\tfrac{3}{4}\ }(x - \underline{\ 0\ })$

PRACTICE 11-4

LESSON **11-4** Practice B
Point-Slope Form

Identify a point the line passes through and the slope of the line, given the point-slope form of the equation. **Possible answers:**

1. $y - 2 = 4(x - 1)$
$m = 4$;
$(x_1, y_1) = (1, 2)$

2. $y + 1 = 2(x - 3)$
$m = 2$;
$(x_1, y_1) = (3, -1)$

3. $y - 4 = -3(x + 1)$
$m = -3$;
$(x_1, y_1) = (-1, 4)$

4. $y + 5 = -2(x + 6)$
$m = -2$;
$(x_1, y_1) = (-6, -5)$

5. $y + 4 = -9(x + 3)$
$m = -9$;
$(x_1, y_1) = (-3, -4)$

6. $y - 7 = -7(x - 7)$
$m = -7$;
$(x_1, y_1) = (7, 7)$

7. $y - 10 = 6(x - 8)$
$m = 6$;
$(x_1, y_1) = (8, 10)$

8. $y + 12 = 2.5(x + 4)$
$m = 2.5$;
$(x_1, y_1) = (-4, -12)$

9. $y + 8 = \frac{1}{2}(x - 3)$
$m = \frac{1}{2}$;
$(x_1, y_1) = (3, -8)$

Write the point-slope form of the equation with the given slope that passes through the indicated point.

10. The line with slope -1 passing through $(2, 5)$
$y - 5 = -1(x - 2)$

11. The line with slope 2 passing through $(-1, 4)$
$y - 4 = 2(x + 1)$

12. The line with slope 4 passing through $(-3, -2)$
$y + 2 = 4(x + 3)$

13. The line with slope 3 passing through $(7, -6)$
$y + 6 = 3(x - 7)$

14. The line with slope -3 passing through $(-6, 4)$
$y - 4 = -3(x + 6)$

15. The line with slope -2 passing through $(5, 1)$
$y - 1 = -2(x - 5)$

16. Write an equation of a line in point-slope form that is parallel to $y = 2x - 8$ and passes through the $(-6, -2)$
$y + 2 = 2(x + 6)$

23. **EARTH SCIENCE** Jorullo is a cinder cone volcano in Mexico. Suppose Jorullo is 315 m tall, 50 m from the center of its base. Use the slope of a cinder cone to write a possible equation in point-slope form that approximately models the height of the volcano, x meters from the center of its base.

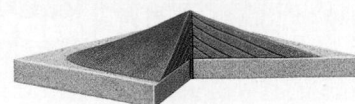
Shield volcano typical slope: 0.03–0.17

Composite volcano typical slope: 0.17–0.5

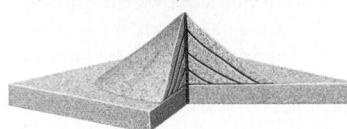
Cinder cone volcano typical slope: 0.5–0.65

24. **LIFE SCIENCE** Since a breed of finch was introduced to the United States, the population of the breed has increased by about 600 birds per year. After 4 years, there are roughly 2730 finches.

   a. Write an equation in point-slope form to model the finch population.

   b. What is the y-intercept of the equation in part **a**, and what does the y-intercept tell you about the finch population?

25. **LIFE SCIENCE** Moose antlers grow at the fastest rate of any animal bone. Each day, a moose antler grows about 1 in. Suppose you started observing a moose when its antlers were 15 in long. Write an equation in point-slope form that describes the length of the moose's antlers after d days of observation.

26. **WRITE A PROBLEM** Write a problem about the point-slope form of an equation using the data on a car's fuel economy.

Fuel Economy		
Gas Tank Capacity	City Efficiency	Highway Efficiency
16 gal	28 mi/gal	36 mi/gal

27. **WRITE ABOUT IT** Explain how you could convert an equation in point-slope form to slope-intercept form.

28. **CHALLENGE** The value of one line's x-intercept is the opposite of the value of its y-intercept. The line contains the point $(10, -5)$. Find the point-slope form of the equation.

Spiral Review

Solve each inequality. (Lesson 10-4)

29. $4x + 3 - x > 15$
    $x > 4$

30. $3 - 7x \leq 24$
    $x \geq -3$

31. $3x + 9 < 2x - 4$
    $x < -13$

32. $1 - x \geq 11 + x$
    $x \leq -5$

33. **TEST PREP** A landscaping company charges a $35 consultation fee, plus $50 per hour. How much would it cost to hire the company for 3 hours? (Lesson 11-1) **C**

   A $225
   B $150
   C $185
   D $135

CHALLENGE 11-4

PROBLEM SOLVING 11-4

Answers

23. Possible answer:
    $y - 315 = -0.6(x - 50)$

24. a. $y - 2730 = 600(x - 4)$
    b. 330; there was an initial population of 330 birds.

25. $\ell - 15 = d$

26. Possible answer: Write an equation in point-slope form for the number of highway miles y that a car can travel after using x gallons of gas.
    Answer: $y = -36(x - 16)$

27. Possible answer: Solve the equation for y, and simplify.

28. Possible answer: $y + 5 = x - 10$ or $y + 5 = -\frac{1}{2}(x - 10)$

Journal

Ask students to write about why they think there are so many different ways to write equations of lines. Ask students whether they think all of the forms are useful.

Test Prep Doctor

For Exercise 33, students can use estimation to eliminate choices. Students can mentally calculate that the cost for 3 hours at $50 per hour plus a fee will be more than $150, so choices **B** and **D** can be eliminated. Adding the $35 gives a total of $185, choice **C**.

Lesson Quiz

Use the point-slope form of each equation to identify a point the line passes through and the slope of the line.

1. $y + 6 = 2(x + 5)$
   $(-5, -6)$, 2

2. $y - 4 = -\frac{2}{5}(x - 6)$
   $(6, 4)$, $-\frac{2}{5}$

Write the point-slope form of the equation with the given slope that passes through the indicated point.

3. the line with slope 4 passing through $(3, 5)$
   $y - 5 = 4(x - 3)$

4. the line with slope -2 passing through $(-2, 4)$
   $y - 4 = -2(x + 2)$

Available on Daily Transparency in CRB

Chapter
11
Mid-Chapter
Quiz

Purpose: *To assess students' mastery of concepts and skills in Lessons 11-1 through 11-4*

Assessment Resources

Section 11A Quiz
Assessment Resources . . . . p. 25

Test and Practice Generator
CD-ROM

Additional mid-chapter assessment items in both multiple-choice and free-response format may be generated for any objective in Lessons 11-1 through 11-4.

Answers

1.
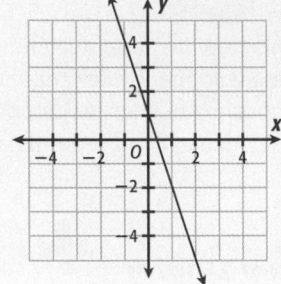
linear

2.
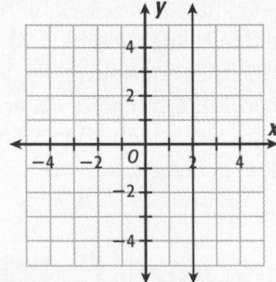
linear

3.
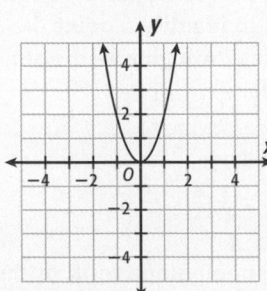
not linear

LESSON **11-1** (pp. 540–544)

Graph each equation and tell whether it is linear.

1. $y = 1 - 3x$ 2. $x = 2$ 3. $y = 2x^2$

Draw a graph that represents the relationship.

4. At Bob's Books, the equation $u = \frac{2}{3}n + 3$ represents the price for a used book u with a selling price n when the book was new. How much will a used copy cost for each of the listed new prices?

New Price	Used Price
$12	$11
$15	$13
$24	$19
$36	$27

LESSON **11-2** (pp. 545–549)

Find the slope of the line that passes through each pair of points.

5. (5, 2) and (1, 3) $-\frac{1}{4}$ 6. (1, 4) and (−1, −3) $\frac{7}{2}$ 7. (0, −2) and (−5, 0) $-\frac{2}{5}$

Tell whether the lines passing through the given points are parallel or perpendicular.

8. line 1: (−1, −3) and (3, −11)
   line 2: (−8, −3) and (6, 4) **perpendicular**

9. line 1: (0, −1) and (−2, −9)
   line 2: (2, 15) and (−1, 3) **parallel**

LESSON **11-3** (pp. 550–554)

Given two points through which a line passes, write the equation of each line in slope-intercept form.

10. (−4, 3) and (−2, 1)   11. (2, 7) and (5, 3)   12. (4, 0) and (2, −5)

Identify the slope and y-intercept, and use them to graph the equation.

13. An airline frequent-flyer plan offers a bonus of 5000 mi to new members plus 1.5 mi for every dollar charged on a credit card endorsed by the airline. The linear equation $y = 1.5x + 5000$ represents the number of miles earned after charging x dollars on the credit card.

LESSON **11-4** (pp. 556–559)

Use the point-slope form of each equation to identify a point the line passes through and the slope of the line.   **Possible points are given.**

14. $y + 4 = -2(x - 1)$ (1, −4); −2   15. $y = -(x + 4)$ (−4, 0); −1   16. $y - 7 = -3x$ (0, 7); −3

Write the point-slope form of each line with the given conditions.

17. slope −3, passing through (7, 2)   18. slope 4, passing through (−4, 1)

4.
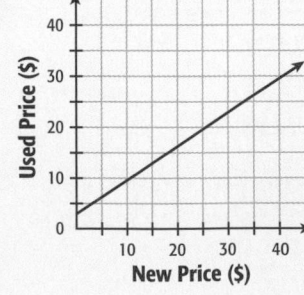

10. $y = -x - 1$

11. $y = -\frac{4}{3}x + \frac{29}{3}$

12. $y = \frac{5}{2}x - 10$

13. $m = 1.5; b = 5000$
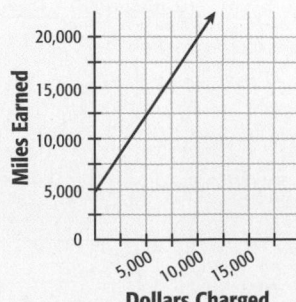

17. $y - 2 = -3(x - 7)$

18. $y - 1 = 4(x + 4)$

Mid-Chapter Quiz

Focus on Problem Solving

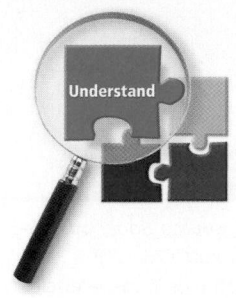

Understand the Problem
• Identify important details in the problem

When you are solving word problems, you need to find the information that is important to the problem.

You can write the equation of a line if you know the slope and one point on the line or if you know two points on the line.

Example:

A school bus carrying 40 students is traveling toward the school at **30 mi/hr**. After **15 minutes**, it has 20 miles to go. How far away from the school was the bus when it started?

You can write the equation of the line in point-slope form.

$$y - y_1 = m(x - x_1)$$
$$y - (-20) = 30(x - 0.25)$$ *The slope is the rate of change, or 30.*
$$y + 20 = 30x - 7.5$$ *15 minutes = 0.25 hours*
$$\underline{-20 \qquad\qquad -20}$$ *(0.25, −20) is a point on the line.*
$$y \qquad = 30x - 27.5$$

The *y*-intercept of the line is −27.5. At 0 minutes, the bus had 27.5 miles to go.

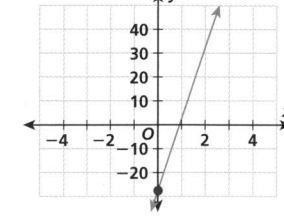

 Read each problem, and identify the information needed to write the equation of a line. Give the slope and one point on the line, or give two points on the line.

1 At sea level, water boils at 212°F. At an altitude of 2000 ft, water boils at 208°F. If the relationship is linear, estimate the temperature that water would boil at an altitude of 5000 ft.

2 Don earns a weekly salary of $480, plus a commission of 5% of his total sales. How many dollars in merchandise does he have to sell to make $500 in one week?

3 An environmental group has a goal of planting 10,000 trees. On Arbor Day, volunteers planted 4500 trees. If the group can plant 500 trees per week, how long will it take them to plant the remaining trees to reach their goal?

4 Kayla rents a booth at a craft fair. If she sells 50 bracelets, her profit is $25. If she sells 80 bracelets, her profit is $85. What would her profit be if she sold 100 bracelets?

Answers

1. 202°

2. $400

3. 11 weeks

4. $125

Focus on Problem Solving

Purpose: *To focus on better understanding the problem by identifying important details*

Problem Solving Resources
Interactive Problem Solving . . pp. 85–91
Math: Reading and Writing in the Content Area . . . . pp. 85–91

Problem Solving Process
This page focuses on the first step of the problem-solving process: **Understand the Problem**

Discuss
Have students discuss what information in the problem is necessary to write the equation and have them give the slope and one point on the line, or have them give two points on the line.

1. At sea level, water boils at 212°F. At an altitude of 2000 ft, water boils at 208°F.

(0, 212); (2000, 208)

2. He earns a weekly salary of $480, plus a commission of 5% of his total sales.

$m = 5\%$, or 0.05, or $\frac{5}{100}$; (0, 480)

3. The group planted 4500 trees and can plant 500 trees per week.

$m = 500$; (0, 4500)

4. If she sells 50 bracelets, her profit is $25.
If she sells 80 bracelets, her profit is $85.

(50, 25); (80, 85)

Linear Relationships

One-Minute Section Planner

Lesson	Materials	Resources
Lesson 11-5 Direct Variation **NCTM:** Algebra, Communication, Representation **NAEP:** Algebra 2g ☐ SAT-9 ☐ SAT-10 ☐ ITBS ☐ CTBS ☐ MAT ☐ CAT	**Required** Graph paper *(CRB, p. 82)* **Optional** Teaching Transparency T27 *(CRB)* Scales Paper cups Pennies	• *Chapter 11 Resource Book,* pp. 47–55 • Daily Transparency T26, CRB • Additional Examples Transparencies T28–T32, CRB • *Alternate Openers: Explorations,* p. 89
Lesson 11-6 Graphing Inequalities in Two Variables **NCTM:** Algebra, Communication, Connections, Representation **NAEP:** Algebra 4c ☐ SAT-9 ☑ SAT-10 ☐ ITBS ☐ CTBS ☐ MAT ☑ CAT	**Required** Graph paper *(CRB, p. 82)* **Optional** Teaching Transparency T34 *(CRB)* Recording Sheet for Reaching All Learners *(CRB, p. 86)*	• *Chapter 11 Resource Book,* pp. 56–65 • Daily Transparency T33, CRB • Additional Examples Transparencies T35–T39, CRB • *Alternate Openers: Explorations,* p. 90
Lesson 11-7 Lines of Best Fit **NCTM:** Algebra, Data Analysis and Probability, Reasoning and Proof, Communication, Representation **NAEP:** Data Analysis and Probability 2e ☐ SAT-9 ☐ SAT-10 ☐ ITBS ☐ CTBS ☑ MAT ☑ CAT	**Required** Graph paper *(CRB, p. 82)* **Optional** Recording Sheet for Reaching All Learners *(CRB, p. 87)* Rulers *(MK)*	• *Chapter 11 Resource Book,* pp. 66–75 • Daily Transparency T40, CRB • Additional Examples Transparencies T41–T43, CRB • *Alternate Openers: Explorations,* p. 91
Extension Systems of Equations **NCTM:** Algebra, Connections **NAEP:** Algebra 4c ☐ SAT-9 ☐ SAT-10 ☐ ITBS ☐ CTBS ☐ MAT ☐ CAT	**Required** Graph paper *(CRB, p. 82)*	• Additional Examples Transparencies T44–T45, CRB
Section 11B Assessment		• Section 11B Quiz, AR p. 26 • *Test and Practice Generator* CD-ROM

SAT = *Stanford Achievement Tests*   **ITBS** = *Iowa Test of Basic Skills*   **CTBS** = *Comprehensive Test of Basic Skills/Terra Nova*
MAT = *Metropolitan Achievement Test*   **CAT** = *California Achievement Test*
NCTM = Complete standards can be found on pages T29–T35.   **NAEP** = Complete standards can be found on pages A54–A58.
SE = *Student Edition*   **TE** = *Teacher's Edition*   **AR** = *Assessment Resources*   **CRB** = *Chapter Resource Book*   **MK** = *Manipulatives Kit*

Section Overview

Direct Variation
Lesson 11-5

Why? Data sets may be related by direct variation.

> *y* **varies directly** with *x* if there is some constant *k* such that **y = kx.**
> *k* is called the **constant of proportionality.**

Distance on map (in.)	x	3	5	8
Actual distance (mi)	y	36	60	96

There is direct variation. The constant of proportionality is 12.

$$y = k \cdot x$$
$$36 = 12 \cdot 3$$
$$60 = 12 \cdot 5$$
$$96 = 12 \cdot 8$$

What actual distance corresponds to a distance of 4 inches on the map?

$$y = k \cdot x$$
$$y = 12 \cdot 4$$
$$y = 48 \text{ miles}$$

Graphing Inequalities in Two Variables
Lesson 11-6

Why? An inequality in two variables has an infinite number of ordered pair solutions. The only way to indicate the solutions is to graph them.

> Draw a dashed boundary line for < and > symbols. Draw a solid boundary line for ≤ and ≥ symbols.

To graph $y < \frac{1}{3}x - 2$, graph the equation $y = \frac{1}{3}x - 2$ with a dashed line for the boundary line. Shade the side of the boundary line containing solutions.

Test point **A(3, 2).**
$$y < \frac{1}{3}x - 2$$
$$2 \overset{?}{<} \frac{1}{3}(3) - 2$$
$$2 \overset{?}{<} -1 \; ✗$$

(3, 2) *is not* a solution.

Test point **B(3, −4).**
$$y < \frac{1}{3}x - 2$$
$$-4 \overset{?}{<} \frac{1}{3}(3) - 2$$
$$-4 \overset{?}{<} -1 \; ✓$$

(3, −4) *is* a solution.

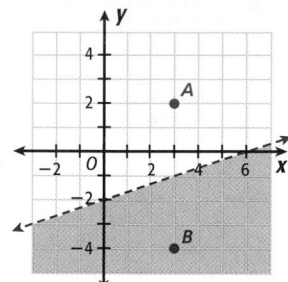

Lines of Best Fit
Lesson 11-7

Why? When there is a correlation in a data set, you can use a line of best fit to approximate the trend in the data and to make predictions.

The line of best fit is the line that comes closest to all the points on a scatter plot. There are about the same number of points on both sides of the line.

Data

x	1	3	3	5	6	6
y	1	1	4	2	4	6

Step 1 Find the coordinates of point **A** by finding the mean of all the x-coordinates and the mean of all the y-coordinates.

Step 2 Draw a line of best fit through **A.**

Step 3 Estimate the coordinates of some other point **B** on the line.

Step 4 Use points **A** and **B** to find the equation of the line of best fit.

$$m = \frac{5 - 3}{9 - 4} = \frac{2}{5} = 0.4$$

$$y - y_1 = m(x - x_1)$$
$$y - 3 = 0.4(x - 4)$$
$$y - 3 = 0.4x - 1.6$$
$$y = 0.4x + 1.4$$

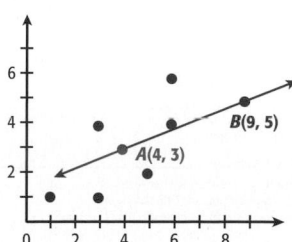

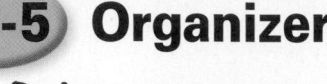

11-5 Organizer

Pacing: Traditional 1 day
Block $\frac{1}{2}$ day

Objective: Students recognize direct variation by graphing tables of data and checking for constant ratios.

Warm Up

Use the point-slope form of each equation to identify a point the line passes through and the slope of the line.

1. $y - 3 = -\frac{1}{7}(x - 9)$ $(9, 3), -\frac{1}{7}$
2. $y + 2 = \frac{2}{3}(x - 5)$ $(5, -2,), \frac{2}{3}$
3. $y - 9 = -2(x + 4)$ $(-4, 9), -2$
4. $y - 5 = -\frac{1}{4}(x + 7)$ $(-7, 5), -\frac{1}{4}$

Problem of the Day

Where do the lines defined by the equations $y = -5x + 20$ and $y = 5x - 20$ intersect? **(4, 0)**

Available on Daily Transparency in CRB

Teacher: Give an example of direct variation in business.
Student: Volume and spending: the *louder* I yell, the more time my mom makes me *spend* in my room.

11-5 Direct Variation

Learn to recognize direct variation by graphing tables of data and checking for constant ratios.

Vocabulary
direct variation
constant of proportionality

A satellite in orbit travels 8 miles in 1 second, 16 miles in 2 seconds, 24 miles in 3 seconds, and so on.

The ratio of distance to time is constant. The satellite travels 8 miles every 1 second.

$$\frac{\text{distance}}{\text{time}} = \frac{8 \text{ mi}}{1 \text{ s}} = \frac{16 \text{ mi}}{2 \text{ s}} = \frac{24 \text{ mi}}{3 \text{ s}}$$

DIRECT VARIATION

Words	Numbers	Algebra
For **direct variation**, two variable quantities are related proportionally by a constant positive ratio. The ratio is called the **constant of proportionality**.	$8 = k$ $16 = 2k$ $24 = 3k$	$y = kx$ $k = \frac{y}{x}$

The distance the satellite travels *varies directly* with time and is represented by the equation $y = kx$. The constant ratio k is 8.

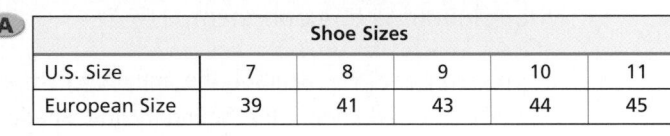

Determining Whether a Data Set Varies Directly

Determine whether the data set shows direct variation.

Helpful Hint

The graph of a direct-variation equation is always linear *and* always contains the point (0, 0). The variables x and y either increase together or decrease together.

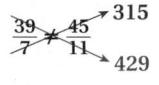

Shoe Sizes					
U.S. Size	7	8	9	10	11
European Size	39	41	43	44	45

Make a graph that shows the relationship between the U.S. sizes and the European sizes. The graph is not linear.

You can also compare ratios to see if a direct variation occurs.

$\dfrac{39}{7} \times \dfrac{45}{11}$ $\rightarrow 315$
  $\rightarrow 429$

315 ≠ 429
The ratios are not proportional.

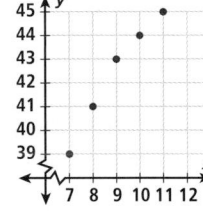

---

1 Introduce

Alternate Opener

EXPLORATION

11-5 Direct Variation

In a *direct variation*, when one quantity increases or decreases, the other quantity does the same. The table below shows the number of stamps x and the price y for each number.

Notice that when the number of stamps is doubled, the price is also doubled.

Stamps x	1	2	3	4	5	6
Price y	$0.37	$0.74	$1.11	$1.48	$1.85	$2.22

1. Use each pair of values (x, y) in the table to complete the graph.

2. To find the *constant of proportionality*, divide each price in the table by each number of stamps.

3. What feature of the graph tells you that the graph is of a direct variation?

Think and Discuss

4. **Explain** the relationship between a constant of proportionality and the slope of the line.

5. **Give** a real-world example of a direct variation.

Motivate

Give students a few examples of direct variation, for example, "The more groceries I buy, the more money I spend," or "The less sleep I get, the less energy I have." Tell students that if two variables are in direct variation, as one gets larger, the other gets proportionally larger, *or* as one gets smaller, the other gets proportionally smaller.

Exploration worksheet and answers on Chapter 11 Resource Book pp. 48 and 97

---

2 Teach

Lesson Presentation

Guided Instruction

In this lesson, students learn to recognize direct variation by graphing tables of data and checking for constant ratios. Explain that direct variation is very similar to linear equations. In direct variation, the constant of proportionality k replaces slope, and the y-intercept is always 0 (Teaching Transparency T27, CRB). Show students how to check for constant ratios to identify direct variation. Then show students how to find the constant of variation, given two values.

Determine whether the data set shows direct variation.

B
Distance Sound Travels at 20°C (m)					
Time (s)	0	1	2	3	4
Distance (m)	0	350	700	1050	1400

Make a graph that shows the relationship between the number of seconds and the distance sound travels.

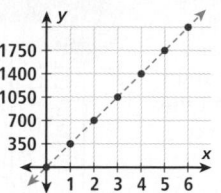

Plot the points.

The points lie in a straight line.

(0, 0) is included.

You can also compare ratios to see if a direct variation occurs.

$$\frac{350}{1} = \frac{700}{2} = \frac{1050}{3} = \frac{1400}{4}$$ *Compare ratios. The ratio is constant.*

The ratios are proportional. The relationship is a direct variation.

EXAMPLE 2 **Finding Equations of Direct Variation**

Find each equation of direct variation, given that y varies directly with x.

A y is 52 when x is 4

$y = kx$ *y varies directly with x.*

$52 = k \cdot 4$ *Substitute for x and y.*

$13 = k$ *Solve for k.*

$y = 13x$ *Substitute 13 for k in the original equation.*

B x is 10 when y is 15

$y = kx$ *y varies directly with x.*

$15 = k \cdot 10$ *Substitute for x and y.*

$\frac{3}{2} = k$ *Solve for k.*

$y = \frac{3}{2}x$ *Substitute $\frac{3}{2}$ for k in the original equation.*

C y is 5 when x is 2

$y = kx$ *y varies directly with x.*

$5 = k \cdot 2$ *Substitute for x and y.*

$\frac{5}{2} = k$ *Solve for k.*

$y = \frac{5}{2}x$ *Substitute $\frac{5}{2}$ for k in the original equation.*

Reaching All Learners
Through Hands-On Experience

Have students work in small groups. Give each group of students a scale, a paper cup, and 30 pennies. Have students adjust the scale so that the weight of the cup registers as zero. Have students weigh five pennies and record the number of pennies and the weight. Then have them repeat the procedure for 10, 15, 20, 25, and 30 pennies. Using the data recorded in this experiment, have students explain whether they believe the variables are in direct variation and explain why. You may want to allow for slight variations in the weights of individual pennies.

Additional Examples

Example 1

Determine whether the data set shows direct variation.

A.
Adam's Growth Chart				
Age (mo)	3	6	9	12
Length (in.)	22	24	25	27

no

B.
Distance Traveled by Train				
Time (min)	10	20	30	40
Distance (mi)	25	50	75	100

yes

Example 2

Find each equation of direct variation, given that y varies directly with x.

A. y is 54 when x is 6. $y = 9x$

B. x is 12 when y is 15. $y = \frac{5}{4}x$

C. y is 8 when x is 5. $y = \frac{8}{5}x$

Example 3

Mrs. Perez has $4000 in a CD and $4000 in a money market account. The amount of interest she has earned since the beginning of the year is organized in the following table. Determine whether there is a direct variation between either data set and time. If so, find the equation of direct variation.

Time (mo)	Interest from CD ($)	Interest from Money Market ($)
0	0	0
1	17	19
2	34	37
3	51	55
4	68	73

A. interest from CD and time

direct variation; $y = 17x$

B. interest from money market and time   no direct variation

EXAMPLE 3 *Physical Science Application*

When a driver applies the brakes, a car's total stopping distance is the sum of the reaction distance and the braking distance. The reaction distance is the distance the car travels before the driver presses the brake pedal. The braking distance is the distance the car travels after the brakes have been applied.

Determine whether there is a direct variation between either data set and speed. If so, find the equation of direct variation.

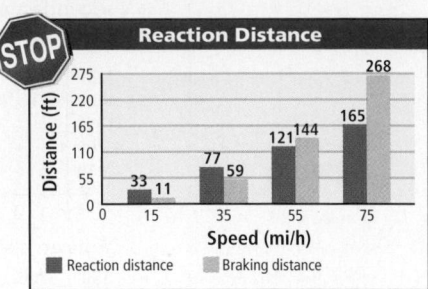

A reaction distance and speed

$$\frac{\text{reaction distance}}{\text{speed}} = \frac{33}{15} = 2.2 \qquad \frac{\text{reaction distance}}{\text{speed}} = \frac{77}{35} = 2.2$$

The first two pairs of data result in a common ratio. In fact, all of the reaction distance to speed ratios are equivalent to 2.2.

$$\frac{\text{reaction distance}}{\text{speed}} = \frac{33}{15} = \frac{77}{35} = \frac{121}{55} = \frac{165}{75} = 2.2$$

The variables are related by a constant ratio of 2.2 to 1, and (0, 0) is included. The equation of direct variation is $y = 2.2x$, where x is the speed, y is the reaction distance, and 2.2 is the constant of proportionality.

B braking distance and speed

$$\frac{\text{breaking distance}}{\text{speed}} = \frac{11}{15} = 0.7\overline{3} \qquad \frac{\text{breaking distance}}{\text{speed}} = \frac{59}{35} = 1.69$$

$$0.7\overline{3} \neq 1.69$$

If any of the ratios are not equal, then there is no direct variation. It is not necessary to compute additional ratios or to determine whether (0, 0) is included.

Think and Discuss

1. Describe the slope and the y-intercept of a direct variation equation.

2. Tell whether two variables that do not vary directly can result in a linear graph.

 Close

Summarize

Remind students that direct variation can be determined by checking for constant ratios or by graphing and seeing whether the graph is a straight line that passes through the origin. Point out that the phrase *direct variation* suggests that there is a direct relationship between the changes in the variables (i.e., as one variable increases, the other increases proportionally).

Answers to Think and Discuss

1. The slope is the constant of proportionality, k. The y-intercept is always 0.

2. Possible answer: Yes, the variables may have a linear relationship but have a y-intercept other than 0.

FOR EXTRA PRACTICE
see page 753

☑ internet connect
Homework Help Online
go.hrw.com Keyword: MP4 11-5

GUIDED PRACTICE

See Example **1** Make a graph to determine whether the data sets show direct variation.

1. The table shows an employee's pay per number of hours worked. **yes**

Hours Worked	0	1	2	3	4	5	6
Pay ($)	0	8.50	17.00	25.50	34.00	42.50	51.00

See Example **2** Find each equation of direct variation, given that y varies directly with x.

2. y is 10 when x is 2 **$y = 5x$**

3. y is 16 when x is 4 **$y = 4x$**

4. y is 12 when x is 15 **$y = \frac{4}{5}x$**

5. y is 3 when x is 6 **$y = \frac{1}{2}x$**

6. y is 220 when x is 2 **$y = 110x$**

7. y is 5 when x is 40 **$y = \frac{1}{8}x$**

See Example **3** 8. The following table shows how many hours it takes to travel 300 miles, depending on your speed in miles per hour. Determine whether there is direct variation between the two data sets. If so, find the equation of direct variation. **no direct variation**

Speed (mi/h)	5	6	7.5	10	15	30	60
Time (hr)	60	50	40	30	20	10	5

INDEPENDENT PRACTICE

See Example **1** Make a graph to determine whether the data sets show direct variation.

9. The table shows the amount of current flowing through a 12-volt circuit with various resistances. **no**

Resistance (ohms)	48	24	12	6	4	3	2
Current (amps)	0.25	0.5	1	2	3	4	6

See Example **2** Find each equation of direct variation, given that y varies directly with x.

10. y is 2.5 when x is 2.5 **$y = x$**

11. y is 2 when x is 8 **$y = \frac{1}{4}x$**

12. y is 93 when x is 3 **$y = 31x$**

13. y is 8 when x is 22 **$y = \frac{4}{11}x$**

14. y is 52 when x is 4 **$y = 13x$**

15. y is 10 when x is 100 **$y = \frac{1}{10}x$**

See Example **3** 16. The following table shows how many hours it takes to drive certain distances at a speed of 60 miles per hour. Determine whether there is direct variation between the two data sets. If so, find the equation of direct variation. **direct variation; $t = \frac{1}{60}d$**

Distance (mi)	15	30	60	90	120	150	180
Time (hr)	0.25	0.5	1	1.5	2	2.5	3

Students may want to refer back to the lesson examples.

Assignment Guide

If you finished Example **1** assign:
Core 1, 9, 23, 27–31
Enriched 1, 9, 23, 27–31

If you finished Example **2** assign:
Core 1–7, 9–15, 17, 19, 23, 27–31
Enriched 1–7 odd, 9–15, 17–20, 23, 27–31

If you finished Example **3** assign:
Core 1–16, 17–23 odd, 27–31
Enriched 9–31

Graph paper is provided on Chapter 11 Resource Book p. 82.

Answers

1.

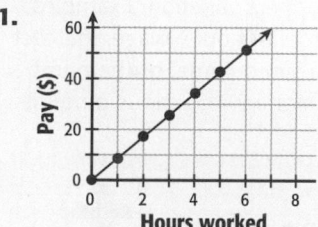

9.

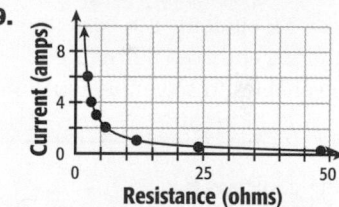

LESSON **11-5** Reteach
11-5 *Direct Variation*

Two quantities **vary directly** if their ratio is constant.
$y = kx$ where k is the constant of variation
y varies directly as x.
To determine if a data set varies directly, look for a constant ratio.

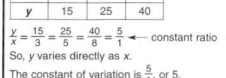

x	3	5	8
y	15	25	40

$\frac{y}{x} = \frac{15}{3} = \frac{25}{5} = \frac{40}{8} = \frac{5}{1}$ ← constant ratio

So, y varies directly as x.
The constant of variation is $\frac{5}{1}$, or 5.
The equation of direct variation is $y = 5x$.
On a graph, the points lie on a straight line.

Explore each data set for variation. If there is a constant ratio, identify it and write the equation of direct variation. Plot the points and tell if the graph is a line.

1.
x	1	2	4	8
y	8	4	2	1

constant ratio? **no**
If yes, equation.
Is the graph a line? **no**

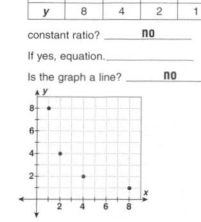

2.
x	0	2	3	5
y	0	20	30	50

constant ratio? **yes, 10**
If yes, equation. **$y = 10x$**
Is the graph a line? **yes**

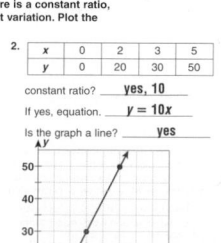

PRACTICE 11-5

LESSON **11-5** Practice B
11-5 *Direct Variation*

Make a graph to determine whether the data sets show direct variation.

1.
x	y
6	9
4	6
0	0
−2	−3
−8	−12

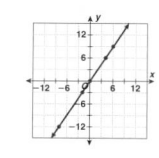

The data set shows a direct variation.

2. Write the equation of direct variation for Exercise 1.
$y = 1.5x$ or $y = \frac{3}{2}x$

Find each equation of direct variation, given that y varies directly with x.

3. y is 32 when x is 4
$y = 8x$

4. y is −10 when x is −20
$y = \frac{1}{2}x$

5. y is 63 when x is −7
$y = −9x$

6. y is 40 when x is 50
$y = \frac{4}{5}x$

7. y is 87.5 when x is 25
$y = 3.5x$

8. y is 90 when x is 270
$y = \frac{1}{3}x$

9. In 1942, Congress passed a law establishing specific rules for the flying of the U. S. flag for those individuals or groups not previously bound by government regulations for the display of the flag. The flying length of the flag, ℓ, must be in direct variation with the flying width, w, of the flag. If the width of the flag is 3 feet, then the length must be 5.7 feet. Write an equation of direct variation for the dimensions of a flying American flag.
$\ell = 1.9w$

Math Background

Variation is an important concept for real-world applications. Direct variation is one type of variation. Exercise 22 shows a physical science application of *inverse variation* (Lesson 12–8). Inverse variation has the form $y = \frac{k}{x}$, or $xy = k$. In inverse variation, as one variable increases, the other decreases.

Answers

22. No; if volume increases as pressure decreases, this cannot be direct variation.

23.

24. Check students' work.

25. Possible answer: The constant of proportionality represents the slope of the line graphed by a direct variation equation. The greater the constant, the steeper the slope of the line.

Journal

Ask students to describe a real-world example of direct variation. Examples might include an hourly wage, the cost of renting a video game, or the number of cookies made with various amounts of flour.

Test Prep Doctor ✚

For Exercise 31, students can solve the equation for b_1 and then substitute in the other variables to solve the equation.

$b_1 = \frac{2A}{h} - b_2 = \frac{2(60)}{6} - 5 = 15$

The correct choice is **B**.

Lesson Quiz

Find each equation of direct variation, given that y varies directly with x.

1. y is 78 when x is 3. $y = 26x$

2. x is 45 when y is 5. $y = \frac{1}{9}x$

3. y is 6 when x is 5. $y = \frac{6}{5}x$

4. The table shows the amount of money Bob makes for different amounts of time he works. Determine whether there is a direct variation between the two sets of data. If so, find the equation of direct variation.

Hours	4	5	6	7	8
Pay	$48	$60	$72	$84	$96

direct variation; $y = 12x$

Available on Daily Transparency in CRB

PRACTICE AND PROBLEM SOLVING

Tell whether each equation represents direct variation between x and y.

17. $y = 133x$ **yes** **18.** $y = -4x^2$ **no** **19.** $y = \frac{k}{x}$ **no** **20.** $y = 2\pi x$ **yes**

Life Science LINK

Most reptiles have a thick, scaly skin, which prevents them from drying out. As they grow, the outermost layer of this skin is shed. Although snakes shed their skins all in one piece, most reptiles shed their skins in much smaller pieces.

21. *LIFE SCIENCE* The weight of a person's skin is related to body weight by the equation $s = \frac{1}{16}w$, where s is skin weight and w is body weight.

a. Does this equation show direct variation between body weight and skin weight? **yes**

b. If a person calculates skin weight as $9\frac{3}{4}$ lb, what is the person's body weight? **156 lb**

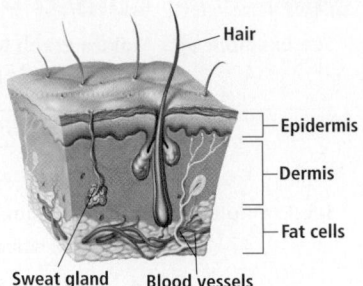

Hair
Epidermis
Dermis
Fat cells
Sweat gland   Blood vessels

22. *PHYSICAL SCIENCE* Boyle's law states that for a fixed amount at a constant temperature, the volume of a gas increases as its pressure decreases. Explain whether the relationship between volume and pressure described by Boyle's law is a direct variation.

23. *COOKING* A waffle recipe calls for different amounts of mix, depending on the number of servings. Graph the data set and determine whether it shows direct variation. **yes**

Number of Servings	2	4	6	8	10	12	14
Waffle Mix (c)	1.5	3	4.5	6	7.5	9	10.5

24. *WRITE A PROBLEM* In physical science, Charles's law states that for a fixed amount at a constant pressure, the volume of a gas increases as the temperature increases. Write a direct variation problem about Charles's law.

25. *WRITE ABOUT IT* Describe how the constant of proportionality k affects the appearance of the graph of a direct variation equation.

26. *CHALLENGE* Bananas are sold at 39¢ a pound. Determine what condition would need to be satisfied if the price paid and the number of bananas purchased represented a direct variation. **Each banana would need to be exactly the same weight.**

Spiral Review

Solve. (Lesson 10-1)

27. $5x + 2 = -18$ $x = -4$

28. $\frac{b}{-6} + 12 = 5$ $b = 42$

29. $\frac{a+4}{11} = -3$ $a = -37$

30. $\frac{1}{3}x - \frac{1}{4} = \frac{5}{12}$ $x = 2$

31. *TEST PREP* The area of a trapezoid is given by the formula $A = \frac{1}{2}(b_1 + b_2)h$. Find b_1 if $A = 60$ cm^2, $b_2 = 5$ m, and $h = 6$ m. (Lesson 10-5) **B**

  A 7 m   **B** 15 m   **C** 14.5 m   **D** 12 m

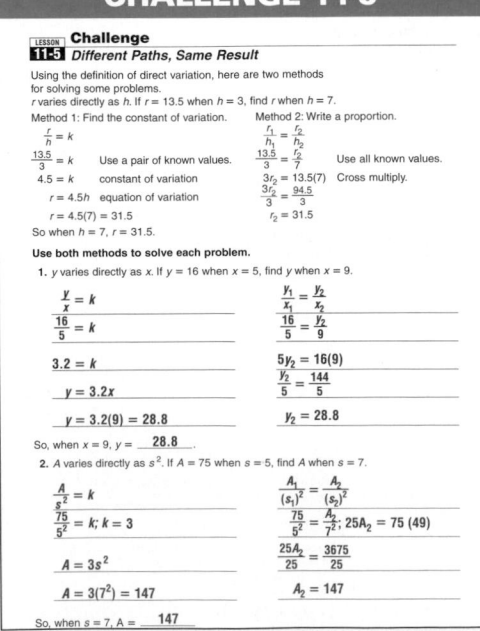

CHALLENGE 11-5

Challenge
11-5 *Different Paths, Same Result*

Using the definition of direct variation, here are two methods for solving some problems.
r varies directly as h. If $r = 13.5$ when $h = 3$, find r when $h = 7$.

Method 1: Find the constant of variation.

$\frac{r}{h} = k$

$\frac{13.5}{3} = k$ Use a pair of known values.

$4.5 = k$ constant of variation

$r = 4.5h$ equation of variation

$r = 4.5(7) = 31.5$

So when $h = 7$, $r = 31.5$.

Method 2: Write a proportion.

$\frac{r_1}{h_1} = \frac{r_2}{h_2}$

$\frac{13.5}{3} = \frac{r_2}{7}$ Use all known values.

$3r_2 = 13.5(7)$ Cross multiply.

$\frac{3r_2}{3} = \frac{94.5}{3}$

$r_2 = 31.5$

Use both methods to solve each problem.

1. y varies directly as x. If $y = 16$ when $x = 5$, find y when $x = 9$.

$\frac{y}{x} = k$

$\frac{16}{5} = k$

$3.2 = k$

$y = 3.2x$

$y = 3.2(9) = 28.8$

So, when $x = 9$, $y =$ ___28.8___

$\frac{y_1}{x_1} = \frac{y_2}{x_2}$

$\frac{16}{5} = \frac{y_2}{9}$

$5y_2 = 16(9)$

$\frac{5y_2}{5} = \frac{144}{5}$

$y_2 = 28.8$

2. A varies directly as s^2. If $A = 75$ when $s = 5$, find A when $s = 7$.

$\frac{A}{s^2} = k$

$\frac{75}{5^2} = k; k = 3$

$A = 3s^2$

$A = 3(7^2) = 147$

So, when $s = 7$, $A =$ ___147___

$\frac{A_1}{(s_1)^2} = \frac{A_2}{(s_2)^2}$

$\frac{75}{5^2} = \frac{A_2}{7^2}; 25A_2 = 75(49)$

$\frac{25A_2}{25} = \frac{3675}{25}$

$A_2 = 147$

PROBLEM SOLVING 11-5

Problem Solving
11-5 *Direct Variation*

Determine whether the data sets show a direct variation. If so, find the equation of direct variation.

1. The table shows the distance in feet traveled by a falling object.

Time (s)	0	0.5	1	1.5	2	2.5	3
Distance (ft)	0	4	16	36	64	100	144

The data does not show a direct variation.

2. The R-value of insulation gives the material's resistance to heat flow. The table shows the R-value for different thickness of fiberglass batting insulation.

Thickness (in)	1	2	3	4	5	6
R-value	3.14	6.28	9.42	12.56	15.7	18.84

Direct variation; $R = 3.14t$

3. The table shows the lifting power of hot air that is used to power hot air balloons.

Hot Air (ft³)	50	100	500	1000	2000	3000
Lift (lb)	1	2	10	20	40	60

Direct variation; $L = \left(\frac{1}{50}\right)H$

4. The table shows the relationship between Celsius and Fahrenheit.

Celsius	-10	-5	0	5	10	20	30
Fahrenheit	14	23	32	41	50	68	86

The data does not show a direct variation.

Weight depends on gravity. The relationship between your weight on earth and other planets is a direct variation. Use the table below that shows how much a person who weights 100 lb on Earth would weigh on different planets.

Planet	Weight (lb)
Moon	16.6
Jupiter	236.4
Pluto	6.7

5. Find the equation of direct variation for the weight on earth e and on the moon m.

Ⓐ $m = 0.166e$ C $m = 6.02e$
B $m = 16.6e$ D $m = 1660e$

6. How much would a 150 lb person weigh on Jupiter?

F 63.5 lb   Ⓗ 354.6 lb
G 286.4 lb   J 483.7 lb

7. How much would a 150 lb person weigh on Pluto?

A 5.8 lb   C 12.3 lb
Ⓑ 10.05 lb   D 2238.8 lb

Graphing Inequalities in Two Variables

Learn to graph inequalities on the coordinate plane.

Vocabulary
boundary line
linear inequality

Graphing can help you visualize the relationship between the maximum distance a Mars rover can travel and the number of Martian days.

A graph of a linear equation separates the coordinate plane into three parts: the points on one side of the line, the points on the **boundary line**, and the points on the other side of the line.

Solar-powered rovers landing on Mars in 2004 will have a range of up to 330 feet per Martian day.

Each point in the coordinate plane makes one of these three statements true:

Equality ⟶ $y = x + 2$

Inequality ⟶ $y > x + 2$
⟶ $y < x + 2$

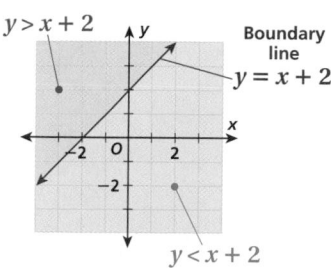

When the equality symbol is replaced in a linear equation by an inequality symbol, the statement is a **linear inequality**. Any ordered pair that makes the linear inequality true is a solution.

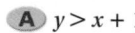 **EXAMPLE 1** **Graphing Inequalities**

Graph each inequality.

A $y > x + 1$

Helpful Hint
Any point on the line $y = x + 1$ is not a solution of $y > x + 1$ because the inequality symbol > means only "greater than" and does not include "equal to."

First graph the boundary line $y = x + 1$. Since no points that are on the line are solutions of $y > x + 1$, make the line *dashed*. Then determine on which side of the line the solutions lie.

$(0, 0)$ *Test a point not on the line.*

$y > x + 1$

$0 \overset{?}{>} 0 + 1$ *Substitute 0 for x and 0 for y.*

$0 \overset{?}{>} 1$

Since $0 > 1$ is not true, $(0, 0)$ is not a solution of $y > x + 1$. Shade the side of the line that does not include $(0, 0)$.

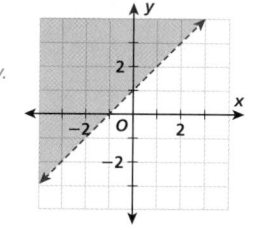

---

1 Introduce

Alternate Opener

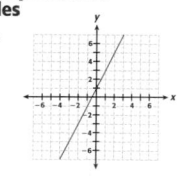
Motivate

Remind students of the inequalities they studied in earlier lessons (Lessons 1-5, 2-5, and 3-7). Remind them that the difference between a simple equation (e.g., $x + 4 = 6$) and a simple inequality (e.g., $x + 4 \geq 6$) is that the equation has a single solution but the inequality has many solutions.

Exploration worksheet and answers on Chapter 11 Resource Book pp. 57 and 99

11-6 Organizer

Pacing: Traditional 1 day
Block $\frac{1}{2}$ day
Objective: Students graph inequalities on the coordinate plane.

Warm Up

Find each equation of direct variation, given that y varies directly with x.

1. y is 18 when x is 3. $y = 6x$

2. x is 60 when y is 12. $y = \frac{1}{5}x$

3. y is 126 when x is 18. $y = 7x$

4. x is 4 when y is 20. $y = 5x$

Problem of the Day

The circumference of a pizza varies directly with its diameter. If you graph that direct variation, what will the slope be? π

Available on Daily Transparency in CRB

Math Humor

Teacher: Define *inequality*.

Student: When one motel is as good as another

---

2 Teach

Lesson Presentation

Guided Instruction

In this lesson, students learn to graph inequalities on the coordinate plane. Remind students how to graph linear equations. Explain that a graph of a linear equation separates the coordinate plane into three parts: the points on one side of the line, the points on the *boundary line,* and the points on the other side of the line (Teaching Transparency T34, CRB). Show students how to graph a linear inequality. Explain that if the inequality contains "equal to," then the boundary line is drawn solid. Otherwise, the boundary line is drawn dashed.

Example 1

Graph each inequality.

A. $y < x - 1$

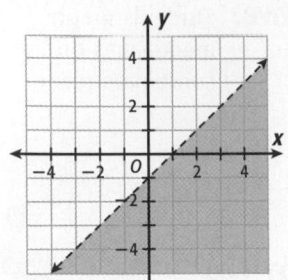

B. $y \geq 2x + 1$

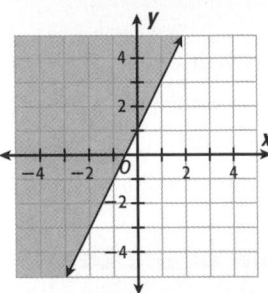

C. $2y + 5x < 6$

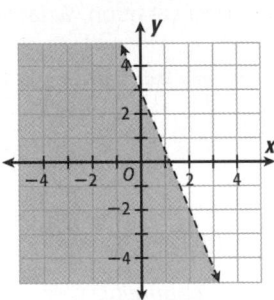

Graph each inequality.

B $y \leq x + 1$

First graph the boundary line $y = x + 1$. Since points that are on the line are solutions of $y \leq x + 1$, make the line **solid.** Then shade the part of the coordinate plane in which the rest of the solutions of $y \leq x + 1$ lie.

    $(2, 1)$ *Choose any point not on the line.*

    $y \leq x + 1$

    $1 \overset{?}{\leq} 2 + 1$ *Substitute 2 for x and 1 for y.*

    $1 \overset{?}{\leq} 3$

Since $1 \leq 3$ is true, $(2, 1)$ is a solution of $y \leq x + 1$. Shade the side of the line that includes the point $(2, 1)$.

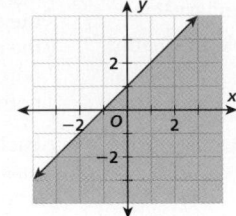

C $3y + 4x \leq 12$

First write the equation in slope-intercept form.

    $3y + 4x \leq 12$

        $3y \leq -4x + 12$ *Subtract 4x from both sides.*

        $y \leq -\frac{4}{3}x + 4$ *Divide both sides by 3.*

Then graph the line $y = -\frac{4}{3}x + 4$. Since points that are on the line are solutions of $y \leq -\frac{4}{3}x + 4$, make the line solid. Then shade the part of the coordinate plane in which the rest of the solutions of $y \leq -\frac{4}{3}x + 4$ lie.

    $(0, 0)$ *Choose any point not on the line.*

    $y \leq -\frac{4}{3}x + 4$

    $0 \overset{?}{\leq} 0 + 4$ *Substitute 0 for x and 0 for y.*

    $0 \overset{?}{\leq} 4$

Since $0 \leq 4$ is true, $(0, 0)$ is a solution of $y \leq -\frac{4}{3}x + 1$. Shade the side of the line that includes the point $(0, 0)$.

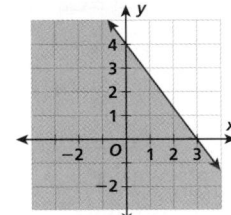

Reaching All Learners
Through Graphic Cues

Give students a recording sheet (Chapter 11 Resource Book p. 86) that has a coordinate grid and four linear inequalities. Have students graph the inequalities and identify the geometric shape that remains unshaded on the coordinate grid. trapezoid You may want to have students check each other's work after graphing each inequality.

EXAMPLE 2 *Science Application*

Helpful Hint

The phrase "up to 330 ft" can be translated as "less than or equal to 330 ft."

Solar-powered rovers landing on Mars in 2004 will have a range of up to 330 feet per Martian day. Graph the relationship between the distance a rover can travel and the number of Martian days. Can a rover travel 3000 feet in 8 days?

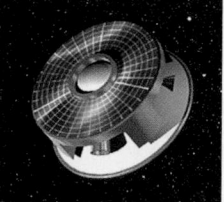

Mars rover in space.

First find the equation of the line that corresponds to the inequality.

In 0 days the rover travels 0 feet. ⟶ point (0, 0)

In 1 day the rover can travel up to 330 feet. ⟶ point (1, 330)

$m = \frac{330 - 0}{1 - 0} = \frac{330}{1} = 330$ *With two known points, find the slope.*

$y = 330x + 0$ *The y-intercept is 0.*

Graph the boundary line $y = 330x$. Since points on the line are solutions of $y \leq 330x$, make the line solid.

Shade the part of the coordinate plane in which the rest of the solutions of $y \leq 330x$ lie.

(5, 0)   *Choose any point not on the line.*

$y \leq 330x$

$0 \leq 330 \cdot 5$ *Substitute 5 for x and 0 for y.*

$0 \leq 1650$

Since $0 \leq 1650$ is true, (5, 0) is a solution of $y \leq 330x$. Shade the part on the side of the line that includes point (5, 0).

The point (8, 3000) is not included in the shaded area, so the rover cannot travel 3000 feet in 8 days.

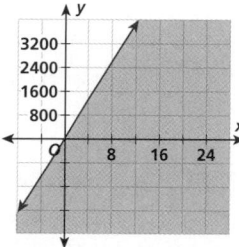

Think and Discuss

1. **Describe** the graph of $5x + y < 15$. Tell how it would change if < were changed to ≥.

2. **Compare and contrast** the use of an open circle, a closed circle, a dashed line, and a solid line when graphing inequalities.

3. **Explain** how you can tell if a point on the line is a solution of the inequality.

4. **Name** a linear inequality for which the graph is a horizontal dashed line and all points below it.

Additional Examples

Example 2

A successful screenwriter can write no more than seven and a half pages of dialogue each day. Graph the relationship between the number of pages the writer can write and the number of days. At this rate, would the writer be able to write a 200-page screenplay in 30 days?

$y \leq 7.5x$; yes

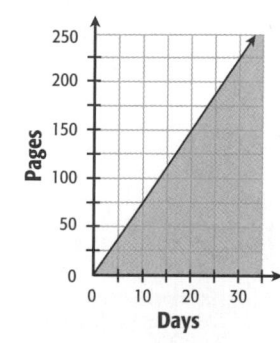

 Close

Summarize

Remind students that inequality graphs will have solid boundary lines if the inequality symbols are ≤ or ≥ and dashed boundary lines if the inequality symbols are < or >. They can determine which side of the boundary line to shade by testing points.

Answers to Think and Discuss

Possible answers:

1. The graph includes the dashed line $y = -5x + 15$ with shading below and left of the line. If the symbol were changed to ≥, the line would be solid and the shading would be on the opposite side of the line.

2. A closed circle and a solid line both mean that the points are included in the solution set. An open circle and a dashed line both mean that the points are not included in the solution set.

3. If the line is solid, the point is a solution.

4. $y < 2$

FOR EXTRA PRACTICE
see page 753

☑ internet connect
Homework Help Online
go.hrw.com   Keyword: MP4 11-6

Students may want to refer back to the lesson examples.

GUIDED PRACTICE

See Example **1**   Graph each inequality.

1. $y < x + 3$ **2.** $y \geq 2x - 1$ **3.** $y > -3x + 2$

4. $4x + y \leq 1$ **5.** $y \leq \frac{2}{3}x + 3$ **6.** $\frac{1}{2}x - \frac{1}{4}y < -1$

See Example **2**   **7. a.** The organizers of a golf outing have a prize budget of $150 to buy golf gloves and hats for the players. They can buy golf gloves for $10 each and hats for $12 each. Write and graph an inequality showing the different ways the organizers can spend their prize budget.

   b. Can the organizers of the golf outing purchase 7 hats and 6 golf gloves and still be within their prize budget?

INDEPENDENT PRACTICE

See Example **1**   Graph each inequality.

8. $y \leq -\frac{1}{2}x - 4$ **9.** $y < -1.5x + 2.5$ **10.** $-4(2x + y) \geq -8$

11. $3x - \frac{3}{4}y > -2$ **12.** $6x - 9y > 15$ **13.** $3\left(\frac{2}{3}x + \frac{1}{3}y\right) \leq -3$

See Example **2**   **14. a.** To avoid suffering from the bends, a diver should ascend no faster than 30 feet per minute. Write and graph an inequality showing the relationship between the depth of a diver and the time required to ascend to the surface. **d ≤ 30t**

   b. If a diver initially at a depth of 77 ft ascends to the surface in 2.6 minutes, is the diver in danger of developing the bends?   **no**

PRACTICE AND PROBLEM SOLVING

Tell whether the given ordered pair is a solution of each inequality shown.

15. $y \leq 2x + 4$, (2, 1)   **yes**   **16.** $y > -6x + 1$, (-3, 19)   **no**

17. $y \geq 3x - 3$, (5, 14)   **yes**   **18.** $y > -x + 12$, (0, 14)   **yes**

19. $y \geq 3.4x + 1.9$, (4, 22)   **yes**   **20.** $y \leq 7(x - 3)$, (3, 3)   **no**

21. a. Graph the inequality $y \geq x + 5$.

   b. Name an ordered pair that is a solution of the inequality.

   c. Is (3, 5) a solution of $y \geq x + 5$? Explain how to check your answer.

   d. Which side of the line $y = x + 5$ is shaded?

   e. Name an ordered pair that is a solution of $y < x + 5$.

22. *FOOD* The school cafeteria needs to buy no more than 30 pounds of potatoes. A supermarket sells 3-pound and 5-pound bags of potatoes. Write and graph an inequality showing the number of 3-pound and 5-pound bags of potatoes the cafeteria can buy. **3x + 5y ≤ 30**

Assignment Guide

If you finished Example **1** assign:
   Core 1–6, 8–13, 15–21 odd, 29–33
   Enriched 8–13, 15–21, 26–33

If you finished Example **2** assign:
   Core 1–14, 15–25 odd, 29–33
   Enriched 8–33

Answers

1–14. Complete answers on pp. A12–A13

7. a. $10g + 12h \leq 150$
   b. yes

21. a.

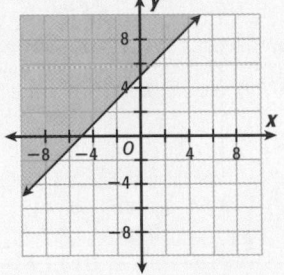

   b. Possible answer: (1,6)
   c. No; you can check by substituting (3, 5) into the inequality and evaluating.
   d. the upper side
   e. Possible answer: (0, 4)

22. Complete answers on p. A13

Math Background

The process of using test points to check solutions is important in graphing inequalities. Test points can often clarify which side of the graph should be shaded. If the coordinates of a test point make the original inequality true, then the region containing the test point should be shaded. If the coordinates make the original inequality false, the opposite region should be shaded. The process of checking solutions and solution regions by using substitution is an invaluable algebraic skill.

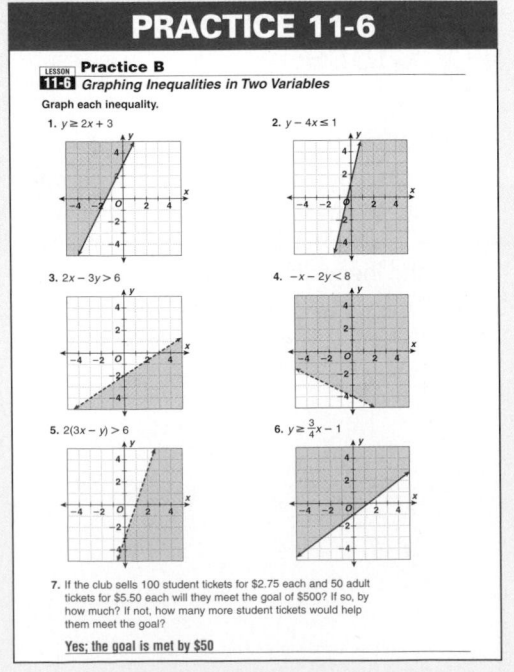

RETEACH 11-6

LESSON **11-6** Reteach
Graphing Inequalities in Two Variables

A **boundary line** divides the coordinate plane into two *half-planes*.

When $y = -x + 4$ is a boundary line:
All the points on the line satisfy the equation.
   (5, -1) is on the line since $-1 = -5 + 4$.
All the points in the half-plane <u>above the line</u> satisfy the linear inequality $y > -x + 4$.
   (5, 3) is in the half-plane above the line since $3 > -5 + 4$.
All the points in the half-plane <u>below the line</u> satisfy the linear inequality $y < -x + 4$.
   (-2, 1) lies in the half-plane below the line since $1 < -(-2) + 4$.

When a boundary line is in the form $y = mx + b$, points in the half-plane above the line satisfy the inequality $y > mx + b$ and points in the half-plane below the line satisfy the inequality $y < mx + b$.

Complete the linear inequality that each point satisfies.

1. (1, -2) is in the half-plane below the boundary line $y = 3x - 4$.

The boundary line is in the form $y = mx + b$; so, (1, -2) satisfies the linear inequality $y \_\_ 3x - 4$.

2. (-3, 7) is in the half-plane below the boundary line $y = -2x + 6$.

The boundary line is in the form $y = mx + b$; so, (-3, 7) satisfies the linear inequality $y \_\_ -2x + 6$.

Write the linear inequality whose solution set is shaded on each graph. The dashed boundary line is not included in the solution set.

3. $y < \frac{2}{5}x - 2$

4. $y > -\frac{2}{3}x + 4$

PRACTICE 11-6

LESSON **11-6** Practice B
Graphing Inequalities in Two Variables

Graph each inequality.

1. $y \geq 2x + 3$ **2.** $y - 4x \leq 1$

3. $2x - 3y > 6$ **4.** $-x - 2y < 8$

5. $2(3x - y) > 6$ **6.** $y \geq \frac{3}{4}x - 1$

7. If the club sells 100 student tickets for $2.75 each and 50 adult tickets for $5.50 each will they meet the goal of $500? If so, by how much? If not, how many more student tickets would help them meet the goal?

   Yes; the goal is met by $50

23. SPORTS A basketball player scored 18 points in a game. Some of her points may have been from free throws, so her points from 2-point and 3-point field goals could be at most 18. Write and graph an inequality showing the possible numbers of 2-point and 3-point field goals she scored. **$2x + 3y \le 18$**

24. BUSINESS It costs a manufacturing company $35 an hour to operate machine A and $25 an hour to operate machine B. The total cost of operating both machines can be no more than $250 each day.

   a. Write and graph an inequality showing the number of hours each machine can be used each day. **$35a + 25b \le 250$**

   b. If machine A is used for 4 hours, for how many hours can machine B be used without going over $250? **4.4 hr**

25. EARTH SCIENCE A weather balloon can ascend at a rate of up to 800 feet per minute.

   a. Write an inequality showing the relationship between the distance the balloon can ascend and the number of minutes. **$d \le 800t$**

   b. Graph the inequality for time between 0 and 30 minutes.

   c. Can the balloon ascend to a height of 2 miles within 15 minutes? (One mile is equal to 5280 feet.) **yes**

26. CHOOSE A STRATEGY Which of the following ordered pairs is NOT a solution of the inequality $4x + 9y \le 108$? **B**
   A $(0, 0)$ **B** $(-6, 15)$ **C** $(-4, -12)$ **D** $(7, 8)$

27. WRITE ABOUT IT When you graph a linear inequality that is solved for y, when do you shade above the boundary line and when do you shade below it? When do you use a dashed line?

28. CHALLENGE Graph the region that satisfies all three inequalities: $x \ge -2$, $y \ge 4$, and $y < -\frac{1}{2}x + 6$.

Answers

23–25. Complete answers on p. A13

27. Possible answer: If $y >$ or $\ge$ a quantity, shade above the line. If $y <$ or $\le$ a quantity, shade below the line. If the inequality is $\le$ or $\ge$, draw a solid line. If the inequality is $<$ or $>$, draw a dashed line.

28. See p. A13.

Journal

Ask students to explain which inequality symbol they would use to describe each of the following word phrases: *as low as, not more than, at least,* and *as much as.*

Test Prep Doctor

For Exercise 33, students can test the first point in the solution choices. This will eliminate choices **B** and **C**. Testing the second point will show that the correct answer is choice **A**.

Spiral Review

Solve for the indicated variable. (Lesson 10-5)

29. Solve $A = \frac{1}{2}bh$ for h. $h = \dfrac{2A}{b}$ **30.** Solve $2a + 2b + 2c = 2d$ for b. $b = d - a - c$

31. Solve $A = \frac{1}{2}(b_1 + b_2)h$ for b_2. $b_2 = \dfrac{2A}{h} - b_1$ **32.** Solve $W = X - 2Y + 4Z$ for Y. $Y = \dfrac{4Z + X - W}{2}$

33. TEST PREP What is the equation of the line that passes through points $(1, 6)$ and $(-1, -2)$ in slope-intercept form. (Lesson 11-3) **A**

   A $y = 4x + 2$ **B** $y = -3x + 6$ **C** $y = 4x - 2$ **D** $y = 2x + 4$

Lesson Quiz

Graph each inequality.

1. $y < -\frac{1}{3}x + 4$

2. $4y + 2x > 12$

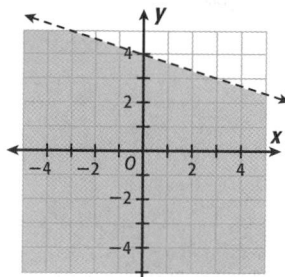

Tell whether the given ordered pair is a solution of each inequality.

3. $y < x + 15 \ (-2, 8)$ **yes**

4. $y \ge 3x - 1 \ (7, -1)$ **no**

Available on Daily Transparency in CRB

CHALLENGE 11-6

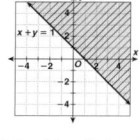

Challenge
11-6 *Two at a Time*

To solve a system of two linear inequalities graphically find the solution set for each linear inequality and mark off the part that overlaps.

Graph the solution set of the system: $x + y \ge 1$
                               $y < x - 3$

Work with the first inequality.
Rewrite $x + y = 1$ as $y = -x + 1$.
Since line is now in $y = mx + b$ form and given symbol is $\ge$, draw solid boundary line and shade half-plane above line.

Work with the second inequality.
Since boundary line $y = x - 3$ is in $y = mx + b$ form and given symbol is $<$, draw dashed boundary line and shade half-plane below the line.

Use a shading opposite to the first shading so that overlap is visible.

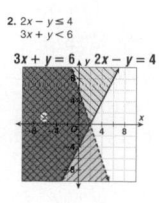

So, the solution set to the system is the cross-hatched region labeled S.

Graph the solution set S for each system.

1. $y \ge 3x + 1$
   $y < x + 1$

2. $2x - y \le 4$
   $3x + y < 6$

PROBLEM SOLVING 11-6

Problem Solving
11-6 *Graphing Inequalities in Two Variables*

The senior class is raising money by selling popcorn and soft drinks. They make $0.25 profit on each soft drink sold, and $0.50 on each bag of popcorn. Their goal is to make at least $500.

1. Write an inequality showing the relationship between the sales of x soft drinks and y bags of popcorn and the profit goal.

   $0.25x + 0.5y \ge 500$

2. Graph the inequality from exercise 1.

3. List three ordered pairs that represent a profit of exactly $500.

   Possible answers: (800, 600),
   (400, 800), (1600, 200).

4. List three ordered pairs that represent a profit of more than $500.

   Possible answers: (400, 900),
   (800, 700), (1600, 300).

5. List three ordered pairs that represent a profit of less than $500.

   Possible answers: (400, 200),
   (800, 400), (1200, 100).

The vehicle is rated to get 19 mpg in the city and 25 mpg on the highway. The vehicle has a 15-gallon gas tank. The graph below shows the number of miles you can drive using no more than 15 gallons.

6. Write the inequality represented by the graph.

   A $\frac{x}{19} + \frac{y}{25} < 15$
   B $\frac{x}{19} + \frac{y}{25} \le 15$
   C $\frac{x}{19} + \frac{y}{25} \ge 15$
   D $\frac{x}{19} + \frac{y}{25} > 15$

7. Which ordered pair represents city and highway miles that you can drive on one tank of gas?

   F (200, 150)   **H** (250, 75)
   G (50, 350)   **J** (100, 175)

8. Which ordered pair represents city and highway miles that you cannot drive on one tank of gas?

   A (100, 200)   **C** (50, 275)
   B (150, 200)   **D** (250, 25)

Pacing: Traditional 1 day
Block $\frac{1}{2}$ day

Objective: Students recognize relationships in data and find the equation of a line of best fit.

Warm Up

Answer the questions about the inequality $5x + 10y > 30$.

1. Would you use a solid or dashed boundary line? **dashed**

2. Would you shade above or below the boundary line? **above**

3. What are the intercepts of the graph? **(0, 3) and (6, 0)**

Problem of the Day

Write an inequality whose positive solutions form a triangular region with an area of 8 square units. (*Hint:* Sketch such a region on a coordinate plane.)
Possible answer: $y < -x + 4$

Available on Daily Transparency in CRB

The line wanted so much to be the line of best fit, but it didn't come close to any of the plotted data values. I guess you could say it was *dis-a-pointed.*

11-7 Lines of Best Fit

Learn to recognize relationships in data and find the equation of a line of best fit.

The graph shows the winning times for the women's 3000 meter Olympic speed skating event. As is the case with many Olympic sports, the athletes keep improving and setting new records, so there is a correlation between the year and the winning time.

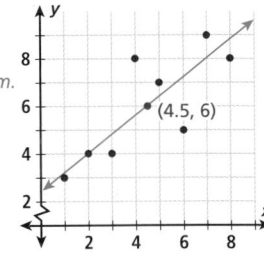

Winning Times for Women's 3000-Meter Olympic Speed Skating

When data show a correlation, you can estimate and draw a *line of best fit* that approximates a trend for a set of data and use it to make predictions.

To estimate the equation of a line of best fit:
- calculate the means of the x-coordinates and y-coordinates: (x_m, y_m).
- draw the line through (x_m, y_m) that appears to best fit the data.
- estimate the coordinates of another point on the line.
- find the equation of the line.

EXAMPLE 1 Finding a Line of Best Fit

Plot the data and find a line of best fit.

x	2	4	5	1	3	8	6	7
y	4	8	7	3	4	8	5	9

Remember!

The line of best fit is the line that comes closest to all the points on a scatter plot. Try to draw the line so that about the same number of points are above the line as below the line.

Plot the data points and find the mean of the x- and y-coordinates.

$$x_m = \frac{2+4+5+1+3+8+6+7}{8} = 4.5 \qquad y_m = \frac{4+8+7+3+4+8+5+9}{8} = 6$$

$$(x_m, y_m) = (4.5, 6)$$

Draw a line through (4.5, 6) that best represents the data.

Estimate and plot the coordinates of another point on that line, such as (7, 8). Find the equation of the line.

$$m = \frac{8-6}{7-4.5} = \frac{2}{2.5} = 0.8 \quad \text{Find the slope.}$$

$$y - y_1 = m(x - x_1) \quad \text{Use point-slope form.}$$

$$y - 6 = 0.8(x - 4.5) \quad \text{Substitute.}$$

$$y - 6 = 0.8x - 3.6$$

$$y = 0.8x + 2.4$$

The equation of a line of best fit is $y = 0.8x + 2.4$.

(4.5, 6)

---

1 Introduce

Alternate Opener

EXPLORATION

11-7 Lines of Best Fit

The table shows student enrollment at a college by year. The enrollment numbers are graphed below.

Year	1998	1999	2000	2001	2002	2003	2004	2005
Enrollment	950	995	1011	1020	1035			

1. Find x_m, the mean of the x-values of the points on the graph.

2. Find y_m, the mean of the y-values of the points on the graph.

3. Plot (x_m, y_m). Lay the edge of a ruler on the graph through (x_m, y_m). Pivot the ruler around (x_m, y_m) and draw the line that you think is closest to the line of best fit.

Think and Discuss

4. **Predict** the enrollment for 2003, 2004, and 2005 by using the line of best fit.

Motivate

Show students a scatter plot that approximates a line. Place a ruler or yardstick at different positions on the scatter plot, and ask students which position they think best approximates the data. When the class has agreed on a good position, explain that the line is a possible *line of best fit.* This is a line that has a minimal amount of distance between each of the data points and itself.

Exploration worksheet and answers on Chapter 11 Resource Book pp. 67 and 101

2 Teach

Lesson Presentation

Guided Instruction

In this lesson, students recognize relationships in data and find the equation of a line of best fit. Lead students through the steps for finding the equation of a line of best fit. Remind students that the equation is an approximation and that the line of best fit only approximates the data.

Teaching Tip Encourage students to understand that, depending on their choice of a second point, their line of best fit equation may be different from others' equations.

EXAMPLE 2 *Sports Application*

Find a line of best fit for the women's 3000-meter speed skating. Use the equation of the line to predict the winning time in 2006.

Year	1964	1968	1972	1976	1980	1984	1988	1992	1994	1998	2002
Winning Time (min)	5.25	4.94	4.87	4.75	4.54	4.41	4.20	4.33	4.29	4.12	3.96

Let 1960 represent year 0. The first point is then (4, 5.25), and the last point is (42, 3.96).

Plot the data points and find the mean of the *x*- and *y*-coordinates.

$$x_m = \frac{4 + 8 + 12 + 16 + 20 + 24 + 28 + 32 + 34 + 38 + 42}{11} \approx 23.5$$

$$y_m = \frac{5.25 + 4.94 + 4.87 + 4.75 + 4.54 + 4.41 + 4.20 + 4.33 + 4.29 + 4.12 + 3.96}{11} \approx 4.5$$

$$(x_m, y_m) = (\mathbf{23.5, 4.5})$$

Draw a line through (23.5, 4.5) that best represents the data.

Estimate and plot the coordinates of another point on that line, (8, 5).

Find the equation of that line.

$$m = \frac{5 - 4.5}{8 - 23.5} = \frac{0.5}{-15.5} \approx -0.03$$

$$y - y_1 = m(x - x_1)$$
$$y - 4.5 = -0.03(x - 23.5)$$
$$y - 4.5 = -0.03x + 0.7 \qquad \textit{Round 0.705 to 0.7.}$$
$$y = -0.03x + 5.2$$

Winning Times for Women's 3000-Meter Speed Skating

The equation of a line of best fit is $y = -0.03x + 5.2$.

Since 1960 represents year 0, 2006 represents year 46.

$$y = -0.03(46) + 5.2 \qquad \textit{Substitute.}$$
$$y = -1.38 + 5.2$$
$$y = 3.82$$

The equation predicts a winning time of 3.82 minutes for the year 2006.

Helpful Hint

If you substitute 2006 instead of 46 for the year, you get a negative value for *y*. The answer would not be reasonable.

Think and Discuss

1. Explain why selecting a different second point may result in a different equation.

2. Describe what a line of best fit can tell you.

3. Tell whether a line of best fit must include one or more points in the data.

Additional Examples

Example 1

Plot the data and find a line of best fit.

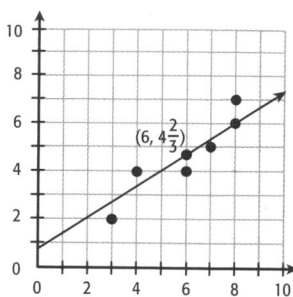

x	4	7	3	8	8	6
y	4	5	2	6	7	4

$$y = \frac{2}{3}x + \frac{2}{3}$$

Example 2

Find a line of best fit for the Main Street Elementary annual softball toss. Use the equation of the line to predict the winning distance in 2006. Let $x = 0$ represent the year 1990.

Year	Distance (ft)
1990	98
1992	101
1994	103
1997	106
2002	107

$y = 0.8x + 99$; about 112 ft

3 Close

Reaching All Learners
Through Hands-On Experience

Have students work in pairs. Give each pair of students a recording sheet that contains a set of data with which to draw a scatter plot (Chapter 11 Resource Book p. 87). Then have them find a line of best fit by placing a straightedge in different positions on the scatter plot until they believe they have found the best position for the line. Have students trace the line and write the equation of their line of best fit. To make the activity more interesting, you may want to give the groups a piece of dry spaghetti to use in place of a straightedge.

Summarize

Remind students that to find the line of best fit, they begin by finding the average values of the *x*- and *y*-coordinates. Then, using this point and an estimate of the coordinates of another point on the line, they can write an equation for the line.

Answers to Think and Discuss

Possible answers:

1. If a different second point is selected, the slope that is calculated might be different, and therefore the equation would also be different.

2. A line of best fit shows the trend of the data and can be used to make predictions.

3. A line of best fit does not have to include any of the points in the data.

MATH-ABLES

Game Resources

Puzzles, Twisters & Teasers
Chapter 11 Resource Book

Graphing in Space

Purpose: *To extend graphing skills into three dimensions*

Discuss: After graphing the point (3, 4, 1), have students explain how the point's actual location is different from where it appears to be on the graph. Discuss the difficulties in drawing three-dimensional graphs on two-dimensional paper.

Possible answer: The point seems to be in the plane formed by the *y*- and *z*-axes (the plane of the paper). Think of the *x*- and *y*-axes as being on the floor, and the *z*-axis as being a pole extending up from the floor. The point (3, 4, 1) is located above the floor, because the *z*-coordinate is 1.

Extend: Challenge students to create a model of the three coordinate axes in a three-dimensional coordinate system using dowels or straws held together with string or tape. Have students plot and show the locations of the points $(-2, 3, 1)$ and $(4, -1, -2)$ and the planes $y = 2$ and $x = -4$.

Check students' work.

Line Solitaire

Purpose: *To practice writing linear equations in a game format*

Discuss: Before playing, have students practice drawing 7 points on paper and drawing 3 lines so that each point lies in a different region.

Extend: Have students repeat the game, using only 4 plotted points and 2 dividing lines.

MATH-ABLES

Graphing in Space

You can graph a point in two dimensions using a coordinate plane with an *x*- and a *y*-axis. Each point is located using an ordered pair (x, y). In three dimensions, you need three coordinate axes, and each point is located using an ordered triple (x, y, z).

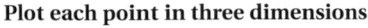

To graph a point, move along the *x*-axis the number of units of the *x*-coordinate. Then move left or right the number of units of the *y*-coordinate. Then move up or down the number of units of the *z*-coordinate.

Plot each point in three dimensions.

1. $(1, 2, 5)$ **2.** $(-2, 3, -2)$

3. $(4, 0, 2)$

The graph of the equation $y = 2$ in three dimensions is a plane that is perpendicular to the *y*-axis and is two units to the right of the origin.

Describe the graph of each plane in three dimensions.

4. $x = 3$ **5.** $z = 1$ **6.** $y = -1$

Line Solitaire

Use a red and a blue number cube and a coordinate plane. Roll the number cubes to generate the coordinates of points on the coordinate plane. The *x*-coordinate of each point is the number on the red cube, and the *y*-coordinate is the number on the blue cube. Generate seven ordered pairs and plot the points on the coordinate plane. Then try to write the equations of three lines that divide the plane into seven regions so that each point is in a different region.

Answers

1.

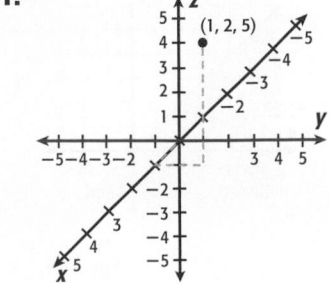

2.

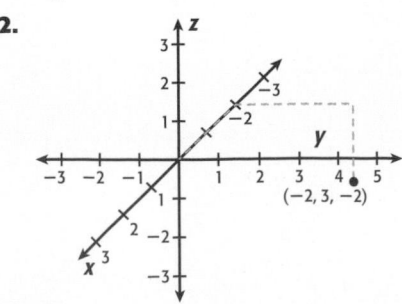

3.

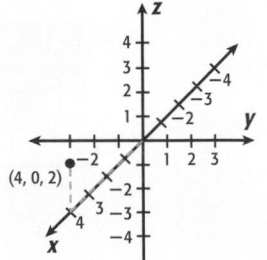

4. vertical plane parallel to the *y-z* plane that crosses through $x = 3$

5. horizontal plane parallel to the *x-y* plane that crosses through $z = 1$

6. vertical plane parallel to the *x-z* plane that crosses through $y = -1$

Technology LAB
Graph Inequalities in Two Variables

Use with Lesson 11-6

A graphing calculator can be used to graph the solution of an inequality in two variables.

internet connect
Lab Resources Online
go.hrw.com
KEYWORD: MP4 TechLab11

Activity

1 To graph the inequality $y > 2x - 4$ using a graphing calculator, use the **Y=** menu, and enter the equation $y = 2x - 4$.

Press **Y=** 2 **X,T,θ,n**   **−** 4 **GRAPH** .

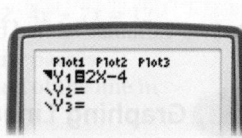

The line representing the graph of the equation represents the *boundary* of the solution region of the inequality. The graph of the inequality is either the region above the line or the region below the line. Use a test point to decide which region represents the graph of the inequality.

The point (0, 0) is a good test point if it is not on the line.

Substituting 0 for both x and y, $0 > 2 \cdot 0 - 4$, or $0 > -4$, which is *true*. The solution graph is the region above the line.

To graph this region, press **Y=** ◀ ◀ and notice that the edit cursor moves to the left of **Y1** onto an icon that looks like a small line segment, ＼.

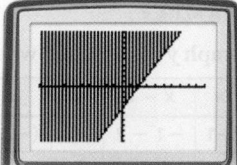

Now press the **ENTER** key several times and notice the different icons that are displayed. Choose the icon that looks like a shaded region above a line. Press **GRAPH** to display the shaded region. Any point (x, y) not on the line that is in the shaded region is a solution of $y > 2x - 4$.

Think and Discuss

1. What inequality would the graph with all points below the x-axis shaded represent? $y < 0$

2. How would you use your calculator to display a graph of the region that is the intersection of the solution graphs of **both** $y > x - 2$ and $y < x + 3$? Enter **X,T,θ,n** **−** 2 in **Y1** and **X,T,θ,n** **+** 3 in **Y2**. Test (0, 0) in each equation and determine whether to shade above or below. Select the appropriate icons, and graph. The area that has been shaded twice is the intersection (solution) to both inequalities.

Try This

Use a graphing calculator to graph each inequality.

1. $y < x - 4$

2. $y > 4 - x$

3. $y < 2x - 5$

4. $2x - 5y < 10$

5. $x + y < 4$

6. $3x + y > 6$

Answers

Try This

1.
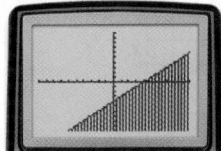

4.
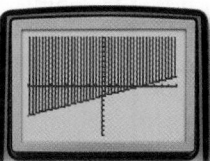

2.
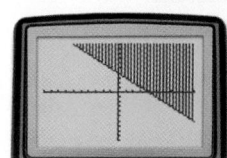

5.
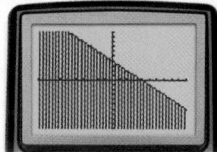

3.
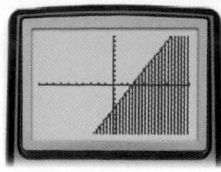

6.
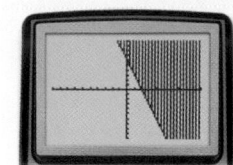

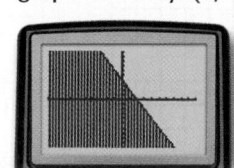

Chapter 12 Tests

Three levels (A,B,C) of tests are available for each chapter in the *Assessment Resources*.

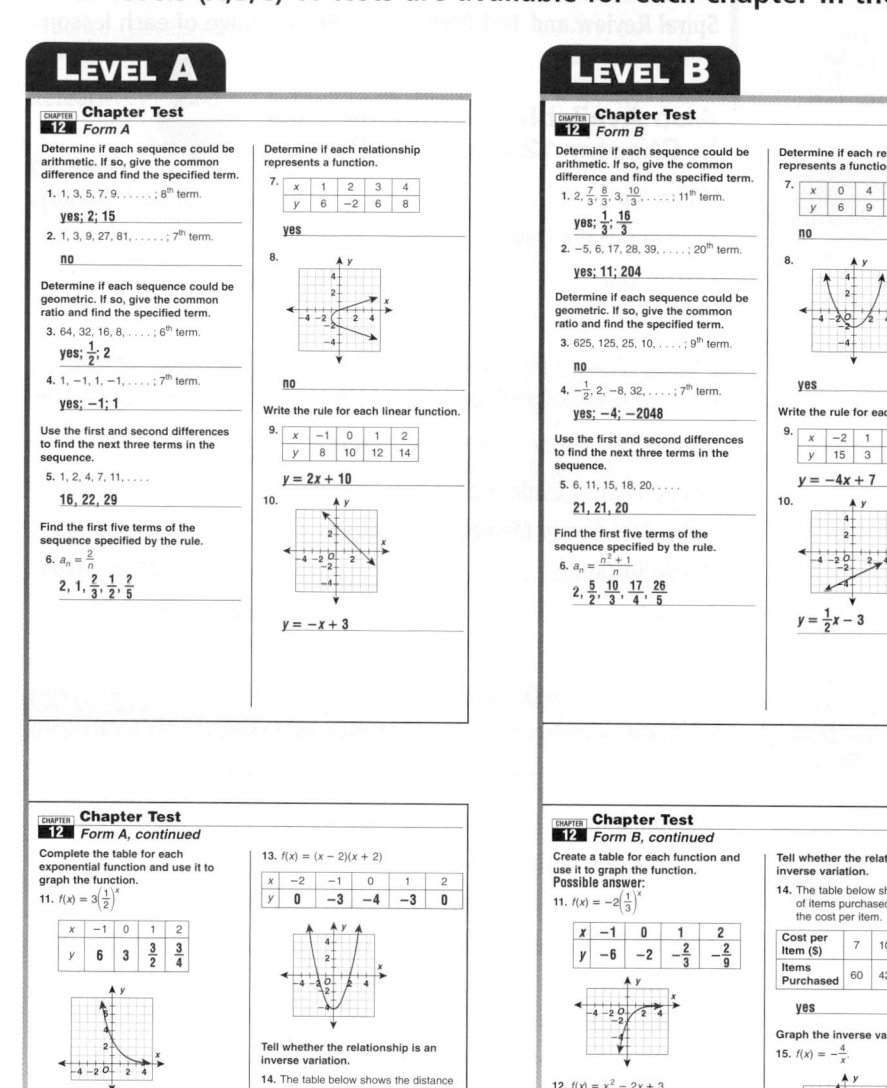

LEVEL A

CHAPTER 12 **Chapter Test**
Form A

Determine if each sequence could be arithmetic. If so, give the common difference and find the specified term.

1. 1, 3, 5, 7, 9, ... ; 8th term.

yes; 2; 15

2. 1, 3, 9, 27, 81, ... ; 7th term.

no

Determine if each sequence could be geometric. If so, give the common ratio and find the specified term.

3. 64, 32, 16, 8, ... ; 6th term.

yes; $\frac{1}{2}$; 2

4. 1, −1, 1, −1, ... ; 7th term.

yes; −1; 1

Use the first and second differences to find the next three terms in the sequence.

5. 1, 2, 4, 7, 11, ...

16, 22, 29

Find the first five terms of the sequence specified by the rule.

6. $a_n = \frac{2}{n}$

2, 1, $\frac{2}{3}$, $\frac{1}{2}$, $\frac{2}{5}$

Determine if each relationship represents a function.

7.
x	1	2	3	4
y	6	−2	6	8

yes

8.

Write the rule for each linear function.

9.
x	−1	0	1	2
y	8	10	12	14

y = 2x + 10

10.

y = −x + 3

CHAPTER 12 **Chapter Test**
Form A, continued

Complete the table for each exponential function and use it to graph the function.

11. $f(x) = 3\left(\frac{1}{2}\right)^x$

x	−1	0	1	2
y	6	3	$\frac{3}{2}$	$\frac{3}{4}$

Complete the table for each quadratic function and use it to make a graph.

12. $f(x) = -x^2 + 2x$

x	−2	−1	0	1	2	3
y	−8	−3	0	1	0	−3

13. $f(x) = (x − 2)(x + 2)$

x	−2	−1	0	1	2
y	0	−3	−4	−3	0

Tell whether the relationship is an inverse variation.

14. The table below shows the distance driven in a given time.

Time (hours)	2	4	5	7
Miles Driven	120	240	300	420

no

Graph the inverse variation.

15. $f(x) = \frac{9}{x}$

16. The height of a ball thrown horizontally from the top of a 75-meter tower is given by the function $f(t) = -5t^2 + 75$. What is the height after 2 seconds?

55 m

LEVEL B

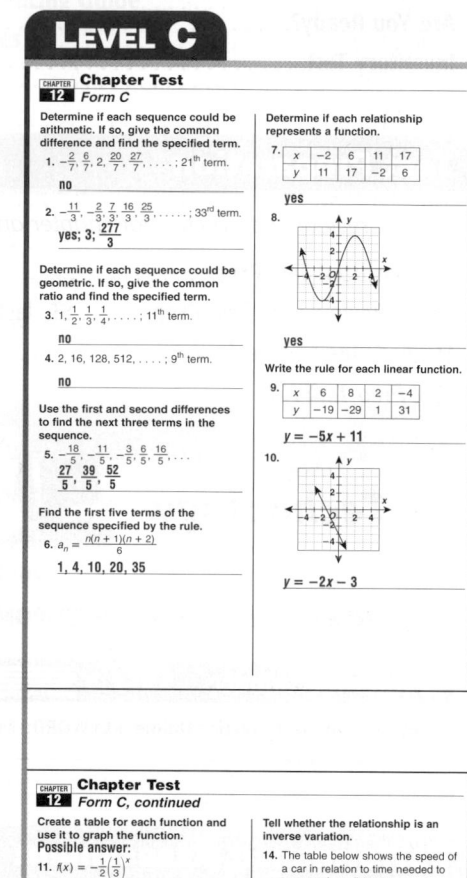

CHAPTER 12 **Chapter Test**
Form B

Determine if each sequence could be arithmetic. If so, give the common difference and find the specified term.

1. 2, $\frac{7}{3}$, $\frac{8}{3}$, 3, $\frac{10}{3}$, ... ; 11th term.

yes; $\frac{1}{3}$; $\frac{16}{3}$

2. −5, 6, 17, 28, 39, ... ; 20th term.

yes; 11; 204

Determine if each sequence could be geometric. If so, give the common ratio and find the specified term.

3. 625, 125, 25, 10, ... ; 9th term.

no

4. $-\frac{1}{2}$, 2, −8, 32, ... ; 7th term.

yes; −4; −2048

Use the first and second differences to find the next three terms in the sequence.

5. 6, 11, 15, 18, 20, ...

21, 21, 20

Find the first five terms of the sequence specified by the rule.

6. $a_n = \frac{n^2 + 1}{n}$

2, $\frac{5}{2}$, $\frac{10}{3}$, $\frac{17}{4}$, $\frac{26}{5}$

Determine if each relationship represents a function.

7.
x	0	4	0	9
y	6	9	5	−1

no

8.

Write the rule for each linear function.

9.
x	−2	1	4	6
y	15	3	−9	−17

y = −4x + 7

10.

y = $\frac{1}{2}$x − 3

CHAPTER 12 **Chapter Test**
Form B, continued

Create a table for each function and use it to graph the function.
Possible answer:

11. $f(x) = -2\left(\frac{1}{3}\right)^x$

x	−1	0	1	2
y	−6	−2	$-\frac{2}{3}$	$-\frac{2}{9}$

12. $f(x) = x^2 − 2x + 3$

x	−1	0	1	2	3
y	6	3	2	3	6

13. $f(x) = \frac{1}{2}(x − 2)(x + 1)$

x	−3	−2	−1	0	1	2	3
y	5	2	0	−1	−1	0	2

Tell whether the relationship is an inverse variation.

14. The table below shows the number of items purchased as a function of the cost per item.

Cost per Item ($)	7	10	12	15	20
Items Purchased	60	42	35	28	21

yes

Graph the inverse variation.

15. $f(x) = -\frac{4}{x}$

16. Ms. Suarez wants to earn $150 in interest over a 3-year period from a savings account. The principal she must deposit varies inversely with the interest rate of the account. If the interest rate is 0.08, she must deposit $625. If the interest rate is 0.064, how much must she deposit?

$781.25

LEVEL C

CHAPTER 12 **Chapter Test**
Form C

Determine if each sequence could be arithmetic. If so, give the common difference and find the specified term.

1. $-\frac{2}{7}$, $\frac{6}{7}$, 2, $\frac{20}{7}$, $\frac{27}{7}$, ... ; 21st term.

no

2. $-\frac{11}{3}$, $-\frac{2}{3}$, $\frac{7}{3}$, $\frac{16}{3}$, $\frac{25}{3}$, ... ; 33rd term.

yes; 3; $\frac{277}{3}$

Determine if each sequence could be geometric. If so, give the common ratio and find the specified term.

3. 1, $\frac{1}{2}$, $\frac{1}{3}$, $\frac{1}{4}$, ... ; 11th term.

no

4. 2, 16, 128, 512, ... ; 9th term.

no

Use the first and second differences to find the next three terms in the sequence.

5. $-\frac{18}{5}$, $-\frac{11}{5}$, $-\frac{3}{5}$, $\frac{6}{5}$, $\frac{16}{5}$, ...

$\frac{27}{5}$, $\frac{39}{5}$, $\frac{52}{5}$

Find the first five terms of the sequence specified by the rule.

6. $a_n = \frac{n(n + 1)(n + 2)}{6}$

1, 4, 10, 20, 35

Determine if each relationship represents a function.

7.
x	−2	6	11	17
y	11	17	−2	6

yes

8.

Write the rule for each linear function.

9.
x	6	8	2	−4
y	−19	−29	1	31

y = −5x + 11

10.

y = −2x − 3

CHAPTER 12 **Chapter Test**
Form C, continued

Create a table for each function and use it to graph the function.
Possible answer:

11. $f(x) = -\frac{1}{2}\left(\frac{1}{3}\right)^x$

x	−2	−1	0	1	2
y	$-\frac{9}{2}$	$-\frac{3}{2}$	$-\frac{1}{2}$	$-\frac{1}{6}$	$-\frac{1}{18}$

12. $f(x) = \frac{1}{2}x^2 + x − 2$

x	−3	−2	−1	0	1	2
y	$-\frac{1}{2}$	−2	$-\frac{5}{2}$	−2	$-\frac{1}{2}$	2

13. $f(x) = \frac{1}{2}(x − 3)(x + 2)$

x	−3	−2	−1	0	1	2	3
y	3	0	−2	−3	−3	−2	0

Tell whether the relationship is an inverse variation.

14. The table below shows the speed of a car in relation to time needed to cover a given distance.

Time (min)	10	15	24	30	50
Speed (kph)	60	40	25	20	12

yes

Graph the inverse variation.

15. $f(x) = \frac{-8}{(3x)}$

16. The resistance of a 30-m piece of wire varies inversely with the square of the diameter. If the diameter of the wire is 2 mm, it has a resistance of 0.2 ohms. What is the resistance of a wire with a diameter of 0.4 mm?

5 ohms

Test and Practice Generator
CD-ROM

Create and customize multiple versions of the same tests with corresponding answers for any chosen chapter objectives.

Chapter 12 State and Standardized Test Preparation

Test Taking Skill Builder and Standardized Test Practice
are provided for each chapter in the *Test Prep Tool Kit*.

TEST TAKING SKILL BUILDER

Test Taking Strategy   Patterns/Reasoning
Chapter 12

Extended Response questions are scored using a scoring rubric.
You are expected to show all your work in detail. Use complete
sentences to explain your thought process. The scoring rubric
shown is used to score the example on the page.

Scoring Rubric

- 3 points: Shows all work and included a through, comprehensible
  explanation.
- 2 points: Shows partial comprehension of mathematical concepts,
  contains the correct answer with incorrect or missing work.
  Explanation indicates some lack of understanding.
- 1 point: Does not answer all parts of the problem. Shows little work
  and contains errors.
- 0 points: The response is incorrect and no work is shown.

Example What should be the next number in the sequence?
Explain your reasoning.

    3   12   48   192

Solution:

When you explain your reasoning, show your calculations and give a
brief explanation of your thinking process.

A possible 3-point response is shown below:

> The next number should be 768.
> I found this by recognizing that $3 \cdot 4 = 12$.
> Next, I checked to see that $12 \cdot 4 = 48$ and that $48 \cdot 4 = 192$.
> Finally, I multiplied 192 by 4 to find that the next number in the
> sequence is 768.

The following is a 1-point response.

> The next number is 766.
> 192×4 is 766.

The student failed to calculate correctly. The student also failed to
explain the pattern that they recognized.

Test Taking Strategy
Chapter 12, continued

Exercises
Use the scoring rubric for each question.

Scoring Rubric

- 3 points: The answer is correct. Shows all work and includes an
  explanation.
- 2 points: Shows partial comprehension of mathematical concepts,
  contains the correct answer with incorrect or missing work.
  Explanation indicates some lack of understanding.
- 1 point: Does not answer all parts of the problem. Shows little work
  and contains errors.
- 0 points: The answer is incorrect and no work is shown.

1. Make a table and graph the function $y = 3(1 - x^2)$. Determine
   the domain and range of the function. Show your work.

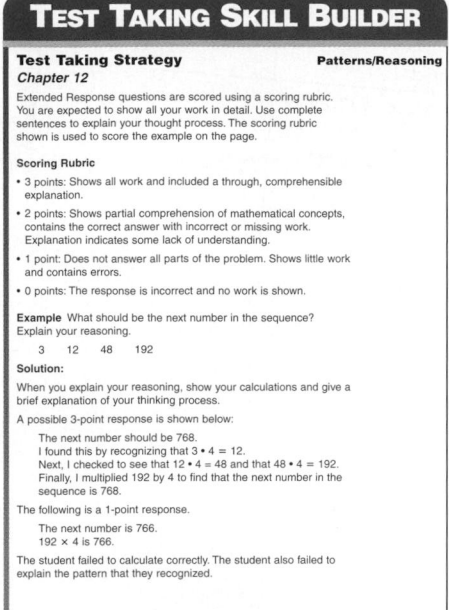

2-point response

x	1	0	-1
y	0	3	0

1-point response

x	1	0	-1
y	0	3	1

Domain = 0 and 3; Range = −1, 1 and 0

a. Read the 2-point response. Why did it only receive 2-points?

   Possible answer: Did not include enough points for the graph.

   The domain and range are incorrect and there is no reasoning.

b. What is wrong with the 1-point response?

   The domain and range are not given, and there is no explanation.

STANDARDIZED TEST PRACTICE

Standardized Test Practice
Chapter 12

Select the best answer for Questions 1–8.

1. Find the 8th term in the geometric
   sequence: $\frac{1}{3}, 1, 3, 9, \ldots$

   A 91.125   C 729
   B 216   D 5832

2. Complete the table for
   $f(x) = x^2 - x + 5$.

x	y
0	
2	
4	

   F 0, 5, 7   H 5, 7, 15
   G 5, 7, 12   I 5, 7, 17

3. The graph represents which inverse
   variation function?

   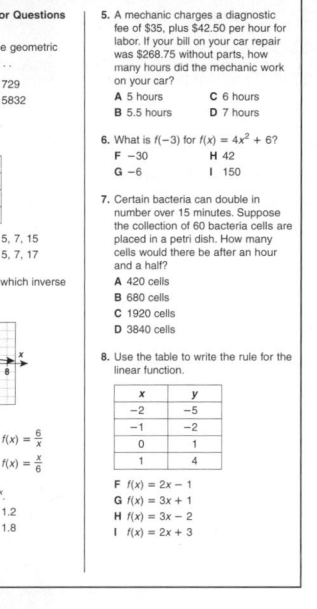

   A $f(x) = \frac{-6}{x}$ C $f(x) = \frac{6}{x}$
   B $f(x) = \frac{x}{-6}$ D $f(x) = \frac{x}{6}$

4. Find $f(3)$ for $f(x) = 0.4^x$.
   F 0.64   H 1.2
   G 0.064   I 1.8

5. A mechanic charges a diagnostic
   fee of $35, plus $42.50 per hour for
   labor. If your bill on your car repair
   was $268.75 without parts, how
   many hours did the mechanic work
   on your car?
   A 5 hours   C 6 hours
   B 5.5 hours   D 7 hours

6. What is $f(-3)$ for $f(x) = 4x^2 + 6$?
   F −30   H 42
   G −6   I 150

7. Certain bacteria can double in
   number over 15 minutes. Suppose
   the collection of 60 bacteria cells are
   placed in a petri dish. How many
   cells would there be after an hour
   and a half?
   A 420 cells
   B 680 cells
   C 1920 cells
   D 3840 cells

8. Use the table to write the rule for the
   linear function.

x	y
−2	−5
−1	−2
0	1
1	4

   F $f(x) = 2x - 1$
   G $f(x) = 3x + 1$
   H $f(x) = 3x - 2$
   I $f(x) = 2x + 3$

Standardized Test Practice
Chapter 12, continued

Gridded Response
Solve the problems. Use the answer
sheet to write and grid-in your answer.

9. The fifth term of an arithmetic
   sequence is 18. The common
   difference is −4. What is the first
   term in the sequence?

10. Find the second difference in the
    sequence.

    2, 5, 10, 17, 26, . . .

Short Response
Solve the problems. Use the answer
sheet to write your answers.

11. Determine whether the relationship
    shown in the table is an inverse
    variation. Explain in words how you
    determined your answer.

Miles Driven	72	120	168	216	264	312
Gallons of gas used	3	5	7	9	11	13

12. Graph $y = \frac{1}{2x}$ and determine the
    value of x when y equals 5. Show
    your work.

Extended Response

13. The rate of growth of a viral cell
    can be modeled by the function
    $f(x) = 2x^2 - 2$.

x	y
−1	
1	
2	
−2	
0	

    a. Complete the function table.

    b. Graph the function.

    c. How are the graphs of the function
       $f(x) = 2x^2$ and $f(x) = 2x^2 - 2$ similar
       to each other? How are they
       different?

(See Lesson 12-1.)
(See Lesson 12-4.)
(See Lesson 12-6.)
(See Lesson 12-5.)

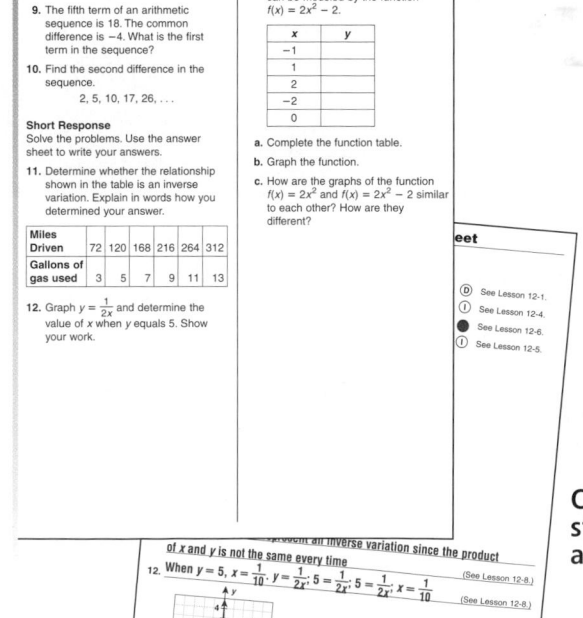

...problem an inverse variation since the product
of x and y is not the same every time

12. When $y = 5$, $x = \frac{1}{10}$. $y = \frac{1}{2x}$; $5 = \frac{1}{2x}$; $5 = \frac{1}{2x}$; $x = \frac{1}{10}$ (See Lesson 12-8.)

                                        (See Lesson 12-8.)

Extended Response
Write your answers for Problem 13 on the back of this paper.

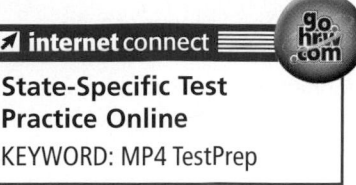

**State-Specific Test
Practice Online**
KEYWORD: MP4 TestPrep

Test Prep Tool Kit

- Standardized Test Prep Workbook
- Countdown to Testing transparencies
- State Test Prep CD-ROM
- Standardized Test Prep Video

Customized answer sheets give
students realistic practice for
actual standardized tests.

Sequences and Functions

Why Learn This?

Tell students that many professions require people to study change and record data. A bacteriologist studies doubling time (the amount of time necessary for a bacteria population to double) under different conditions by recording the population at fixed intervals. The increasing populations are written in a list called a *sequence*. The bacteriologist studies the sequence and determines how long it took the population to double in size. Each set of conditions produces a different sequence and a different doubling time.

Using Data

To begin the study of this chapter, have students:

• Identify the doubling time for *E. coli* bacteria under low-growth-medium conditions. 60 minutes

• Think about the relationship between doubling time and temperature. What is one way to control the population of E. coli bacteria? The lower the temperature, the longer the doubling time. To control the population growth, maintain a low temperature.

Sequences and Functions

internet connect

Chapter Opener Online
go.hrw.com
KEYWORD: MP4 Ch12

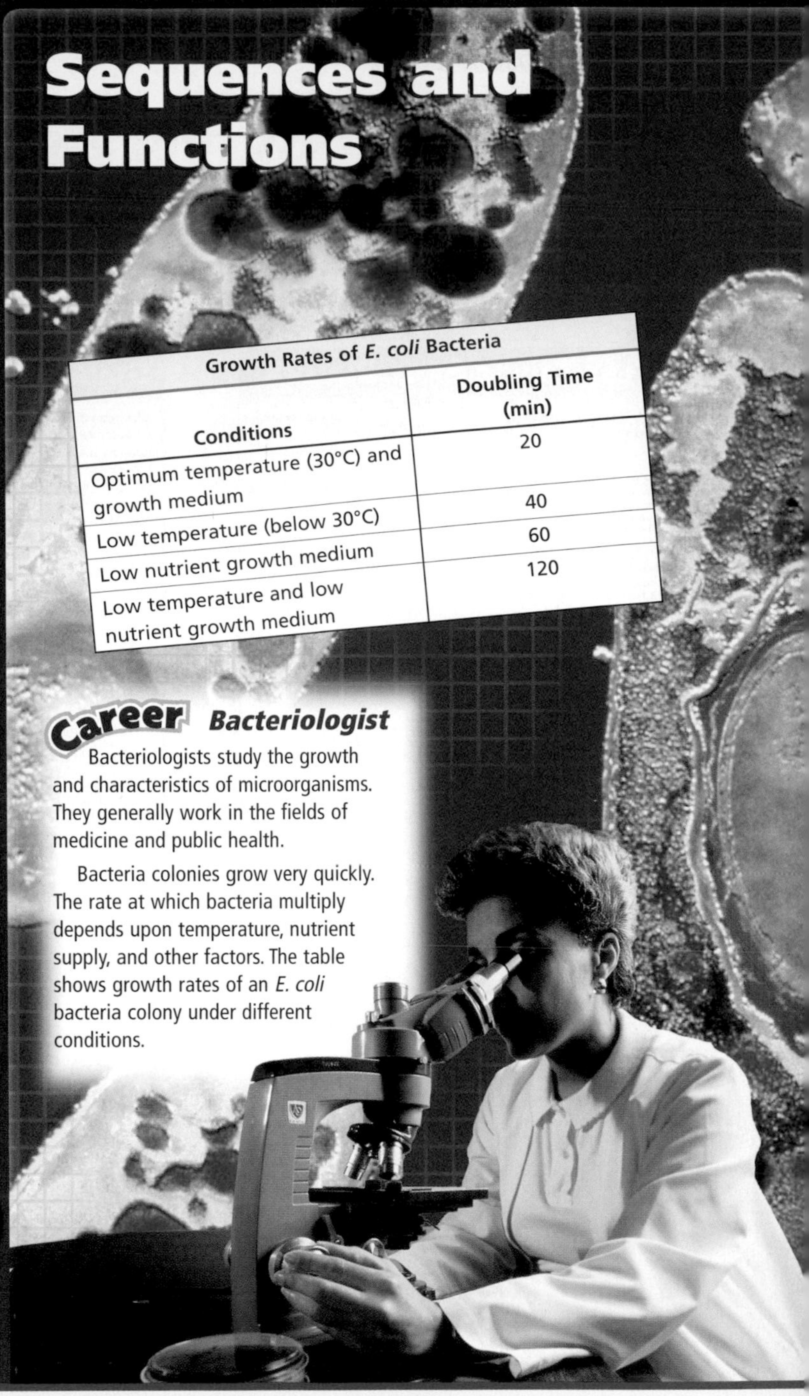

Growth Rates of *E. coli* Bacteria	
Conditions	Doubling Time (min)
Optimum temperature (30°C) and growth medium	20
Low temperature (below 30°C)	40
Low nutrient growth medium	60
Low temperature and low nutrient growth medium	120

Career   *Bacteriologist*

Bacteriologists study the growth and characteristics of microorganisms. They generally work in the fields of medicine and public health.

Bacteria colonies grow very quickly. The rate at which bacteria multiply depends upon temperature, nutrient supply, and other factors. The table shows growth rates of an *E. coli* bacteria colony under different conditions.

Problem Solving Project

Life Science Connection

Purpose: To graph, compare, and analyze arithmetic and geometric sequences to solve problems

Materials: The Addition by Division worksheet

internet connect

Chapter Project Online: *go.hrw.com*
KEYWORD: MP4 PSProject12

Understand, Plan, Solve, and Look Back

Have students:

✔ Complete the Addition by Division worksheet.

✔ Make a table showing the doubling of bacteria 20 times. Use the table to plot a growth curve.

✔ Plot curves to demonstrate the growth of an *E. coli* population under the conditions listed in the table. At what point would the population exceed that of people living in the United States? Estimate at what point the population would exceed the population of humans on Earth.

✔ Check students' work.

ARE YOU READY?

Choose the best term from the list to complete each sentence.

1. An equation whose solutions fall on a line on a coordinate plane is called a(n) __?__. **linear equation**

2. When the equation of a line is written in the form $y = mx + b$, m represents the __?__ and b represents the __?__. **slope; y-intercept**

3. To write an equation of the line that passes through $(1, 3)$ and has slope 2, you might use the __?__ of the equation of a line. **point-slope form**

- linear equation
- point-slope form
- slope
- x-intercept
- y-intercept

Complete these exercises to review skills you will need for this chapter.

✔ Number Patterns

Find the next three numbers in the pattern.

4. $\frac{1}{-3}, \frac{3}{-4}, \frac{5}{-5}, \ldots$

5. $2, 3, 6, 11, 18, \ldots$ **27, 38, 51; increase by next larger odd integer**

6. $-11, -8, -5, \ldots$ **−2, 1, 4; add 3**

7. $4, 2\frac{1}{2}, 1, \ldots$ $-\frac{1}{2}, -2, -3\frac{1}{2}$; **decrease by $1\frac{1}{2}$**

✔ Evaluate Expressions

Evaluate each expression for the given values of the variables.

8. $a + (b - 1)c$ **−2**   for $a = 6, b = 3, c = -4$

9. $a \cdot b^c$ **−32**   for $a = -2, b = 4, c = 2$

10. $(ab)^c$ **36**   for $a = 3, b = -2, c = 2$

11. $-(a + b) + c$ **−5**   for $a = -1, b = -4, c = -10$

✔ Graph Linear Equations

Use the slope and the y-intercept to graph each line.

12. $y = \frac{2}{3}x + 4$ 13. $y = -\frac{1}{2}x - 2$ 14. $y = 3x + 1$

15. $2y = 3x - 8$ 16. $3y + 2x = 6$ 17. $x - 5y = 5$

✔ Simplify Ratios

Write each ratio in simplest form.

18. $\frac{3}{9}$ $\frac{1}{3}$ 19. $\frac{21}{5}$ **already in simplest terms**   20. $\frac{-12}{4}$ **−3**

21. $\frac{27}{45}$ $\frac{3}{5}$ 22. $\frac{3}{-45}$ $-\frac{1}{15}$ 23. $\frac{20}{-8}$ $-\frac{5}{2}$

ARE YOU READY?
Were students successful with Are You Ready?

NO INTERVENE ← → **YES ENRICH**

✔ Number Patterns **4–7**
Intervention Practice, Skill 13
 **CD-ROM**
Intervention Activities, Skill 13

✔ Graph Linear Equations **12–17**
Intervention Practice, Skill 64
CD-ROM
Intervention Activities, Skill 64

✔ Evaluate Expressions **8–11**
Intervention Practice, Skill 54
 CD-ROM
Intervention Activities, Skill 54

✔ Simplify Ratios **18–23**
Intervention Practice, Skill 28
CD-ROM
Intervention Activities, Skill 28

Are You Ready? Enrichment, pp. 427–428

 **One-Minute Section Planner**

Lesson	Materials	Resources
Lesson 12-1 Arithmetic Sequences **NCTM:** Algebra, Reasoning and Proof, Communication, Representation **NAEP:** Algebra 1a ☑ SAT-9　☑ SAT-10　☑ ITBS　☑ CTBS　☑ MAT　☑ CAT	**Optional** Recording Sheet for Reaching All Learners *(CRB, p. 93)*	• *Chapter 12 Resource Book,* pp. 7–16 • Daily Transparency T1, CRB • Additional Examples Transparencies T2–T5, CRB • *Alternate Openers: Explorations,* p. 92
Lesson 12-2 Geometric Sequences **NCTM:** Algebra, Reasoning and Proof, Communication **NAEP:** Algebra 1a ☑ SAT-9　☑ SAT-10　☑ ITBS　☑ CTBS　☑ MAT　☑ CAT		• *Chapter 12 Resource Book,* pp. 17–26 • Daily Transparency T6, CRB • Additional Examples Transparencies T7–T10, CRB • *Alternate Openers: Explorations,* p. 93
Hands-On Lab 12A Fibonacci Sequence **NCTM:** Algebra, Reasoning and Proof, Representation **NAEP:** Algebra 1b ☐ SAT-9　☐ SAT-10　☐ ITBS　☐ CTBS　☐ MAT　☐ CAT	**Required** Square tiles *(MK)*	• *Hands-On Lab Activities,* p. 82
Lesson 12-3 Other Sequences **NCTM:** Algebra, Reasoning and Proof, Communication **NAEP:** Algebra 1b ☐ SAT-9　☐ SAT-10　☐ ITBS　☐ CTBS　☐ MAT　☐ CAT	**Optional** Recording Sheet for Reaching All Learners *(CRB, p. 94)* Toothpicks	• *Chapter 12 Resource Book,* pp. 27–35 • Daily Transparency T11, CRB • Additional Examples Transparencies T12–T15, CRB • *Alternate Openers: Explorations,* p. 94
Section 12A Assessment		• Mid-Chapter Quiz, SE p. 606 • Section 12A Quiz, AR p. 27 • *Test and Practice Generator* CD-ROM

SAT = *Stanford Achievement Tests*　　**ITBS** = *Iowa Test of Basic Skills*　　**CTBS** = *Comprehensive Test of Basic Skills/Terra Nova*
MAT = *Metropolitan Achievement Test*　　**CAT** = *California Achievement Test*
NCTM = Complete standards can be found on pages T29–T35.　　**NAEP** = Complete standards can be found on pages A54–A58.
SE = *Student Edition*　　**TE** = *Teacher's Edition*　　**AR** = *Assessment Resources*　　**CRB** = *Chapter Resource Book*　　**MK** = *Manipulatives Kit*

Section Overview

Arithmetic Sequences
Lesson 12-1

Why? A **sequence** is a list of numbers or objects, called **terms,** in a certain order. Many relationships can be represented by arithmetic sequences.

> In an **arithmetic sequence,** the difference between consecutive terms is constant. This difference is called the **common difference.**

> **nth Term of an Arithmetic Sequence**
> The nth term, a_n, of an arithmetic sequence with common difference d is
> $$a_n = a_1 + (n-1)d.$$

Example: 5, 7, 9, 11, ...
The **common difference** is $9 - 7 = 2$.

Find the 15th term of the sequence: 5, 7, 9, 11, ...
$a_n = a_1 + (n-1)d$
$a_{15} = 5 + (15-1)2$
$a_{15} = 33$

Geometric Sequences
Lesson 12-2

Why? Exponential growth and decay can be represented by geometric sequences.

> In a **geometric sequence,** the ratio of consecutive terms is constant. This ratio is called the **common ratio.**

> **nth Term of a Geometric Sequence**
> The nth term, a_n, of a geometric sequence with common ratio r is
> $$a_n = a_1 r^{n-1}.$$

Example: 6, 18, 54, 162, ...

The **common ratio** is $\frac{18}{6} = 3$.

Find the 12th term of the sequence: 6, 18, 54, 162, ...
$a_n = a_1 \cdot r^{n-1}$
$a_{12} = 6 \cdot 3^{12-1}$
$a_{12} = 6 \cdot 3^{11}$
$a_{12} = 1{,}062{,}882$

Other Sequences
Lesson 12-3

Why? Many patterns in nature, such as spirals in the centers of sunflowers, can be represented by the numbers of the Fibonacci sequence.

> If you subtract consecutive terms of a sequence, you get the **first differences.** If you subtract consecutive terms of the first differences, you get the **second differences.**

Term, n	1	2	3	4	5	6	7
Triangular Number	1	3	6	10	15	21	28

First differences   2   3   4   5   6   7
Second differences   1   1   1   1   1

$1 + 2$ $3 + 5$ $8 + 13$

1, 1, 2, 3, 5, 8, 13, 21, ...

> In the **Fibonacci sequence,** the first two terms are 1, 1. After this, every term is the sum of the two previous terms.

$1 + 1$ $2 + 3$ $5 + 8$

Pacing: Traditional 1 day
Block $\frac{1}{2}$ day

Objective: To use a graphing
calculator to explore
cubic functions

Materials: Graphing calculator

Lab Resources

Technology Lab Activities. . . . pp. 83–86

Using the Pages

This technology activity shows students
how to graph cubic functions, which
can be done on any graphing calculator.
Specific keystrokes may vary, depending
on the make and model of the graphing
calculator used. The keystrokes given
are for a TI-83 model. For keystrokes to
other models, visit go.hrw.com.

The Think and Discuss problems can be
used to assess students' understanding
of the technology activity. Although Try
This problems 1–4 can be done without
a graphing calculator, they are meant to
help students become familiar with
using a graphing calculator to graph
cubic functions.

Assessment

1. What sequence of keystrokes is
   needed to graph the cubic function
   $y = x^3 - 2x^2 + 3x - 4$?

2. How can you change the function so
   that the y-intercept is (0, 0) instead
   of (0, −4)? Add 4 to the function,
   making it $y = x^3 - 2x^2 + 3x$.

Answers

Activity 1

Think and Discuss

1. The sign determines whether the
   cubic curve rises or falls from left to
   right.

2. Use the **TRACE** feature to trace until
   $x = 7$. The y-value gives 7^3.

You can use your graphing calculator to explore cubic functions.
To graph the cubic equation $y = x^3$ in the standard graphing
calculator window, press `Y=` ; enter the right side of the
equation, `X,T,θ,n` `^` 3; and press `ZOOM` **6:ZStandard**.
Notice that the graph goes from the lower left to the upper right
and crosses the x-axis once, at $x = 0$.

Activity 1

❶ Graph $y = -x^3$. Describe the graph.

Press `Y=` , and enter the right side of the equation, `(−)` `X,T,θ,n`
`^` 3.

The graph goes from the upper left to the lower right
and crosses the x-axis once.

❷ Graph $y = x^3 + 3x^2 - 2$. Describe the graph.

Press `Y=` ; enter the right side of the equation, `X,T,θ,n` `^` 3
`+` 3 `X,T,θ,n` `x²` `−` 2; and press `ZOOM` **6:ZStandard**.
The graph goes from the lower left to the upper right and
crosses the x-axis three times.

Think and Discuss

1. How does the sign of the x^3 term affect the graph of a cubic
   function?

2. How could you find the value of 7^3 from the graph of $y = x^3$?

Try This

Graph each function and describe the graph.

1. $y = x^3 - 2$ **2.** $y = x^3 + 3x^2 - 2$ **3.** $y = (x - 2)^3$ **4.** $y = 5 - x^3$

Try This

1.

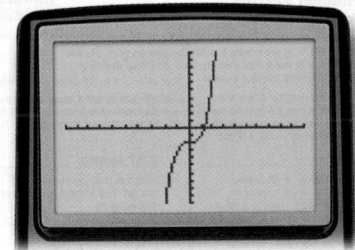

The graph goes from the lower left to the
upper right, crosses the y-axis at $y = -2$,
and crosses the x-axis once.

2.

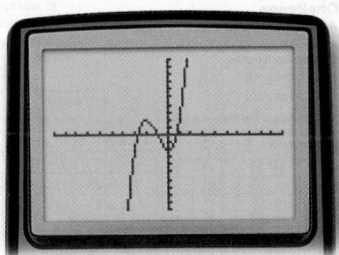

The graph goes from the lower left to the
upper right, crosses the y-axis at $y = -2$,
and crosses the x-axis three times.

Activity 2

1 Compare the graphs of $y = x^3$ and $y = x^3 + 3$.

Graph **Y₁=X^3** and **Y₂=X^3+3** on the same screen, as shown. Use the [TRACE] button and the [◄] and [►] buttons to trace to any integer value of x. Then use the [▲] and [▼] keys to move from one function to the other to compare the values of y for both functions for the value of x. You can also press [2nd] [GRAPH]^(TABLE) to see a table of values for both functions.

The graph of $y = x^3 + 3$ is translated up 3 units from the graph of $y = x^3$.

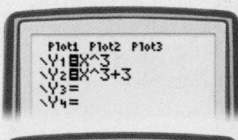

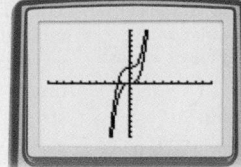

2 Compare the graphs of $y = x^3$ and $y = (x + 3)^3$.

Graph **Y₁=X^3** and **Y₂=(X+3)^3** on the same screen. Notice that the graph of $y = (x + 3)^3$ is the graph of $y = x^3$ moved left 3 units. Press [2nd] [GRAPH]^(TABLE) to see a table of values. The graph of $y = (x + 3)^3$ is translated left 3 units from the graph of $y = x^3$.

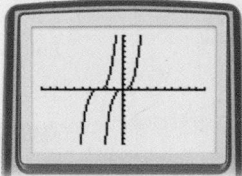

3 Compare the graphs of $y = x^3$ and $y = 2x^3$.

Graph **Y₁=X^3** and **Y₂=2X^3** on the same screen. Use the [TRACE] button and the arrow keys to see the values of y for any value of x. Press [2nd] [GRAPH]^(TABLE) to see a table of values.

The graph of $y = 2x^3$ is stretched upward from the graph of $y = x^3$. The y-value for $y = 2x^3$ increases twice as fast as it does for $y = x^3$. The table of values is shown.

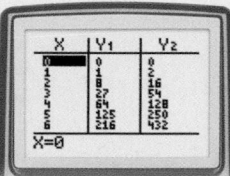

Think and Discuss

1. What function would translate $y = x^3$ right 6 units? $y = (x - 6)^3$

2. Do you think that the methods shown of translating a cubic function would have the same result on a quadratic function? Explain.
Yes; adding and subtracting numbers from a function will always have the same effect because the y-value will be changed in the same way.

Try This

Compare the graph of $y = x^3$ to the graph of each function.

1. $y = x^3 - 2$ **2.** $y = (x - 7)^3$ **3.** $y = \left(\frac{1}{2}\right)x^3$ **4.** $y = 5 - x^3$

Answers
Activity 2
Try This

1. This graph is translated down 2 units.

2. This graph is translated 7 units right.

3. This graph is flattened.

4. This graph is reflected over the x-axis and moved up 5 units.

Answers
Activity 1
Try This

3.

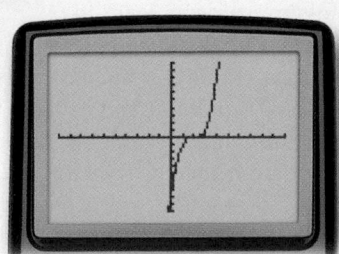

The graph goes from the lower left to the upper right, it is translated two units to the right, and crosses the x-axis once.

4.

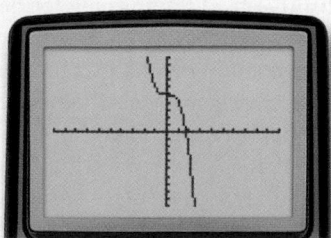

The graph goes from the upper left to the lower right, crosses the y-axis at $y = 5$, and crosses the x-axis once.

Pacing: Traditional 1 day
Block $\frac{1}{2}$ day

Objective: Students recognize inverse variation by graphing tables of data.

Warm Up

Find $f(-4)$, $f(0)$, and $f(3)$ for each quadratic function.

1. $f(x) = x^2 + 4$ 20, 4, 13
2. $f(x) = \frac{1}{4}x^2$ 4, 0, $\frac{9}{4}$
3. $f(x) = 2x^2 - x + 3$ 39, 3, 18

Problem of the Day

Use the digits 1–8 to fill in 3 pairs of values in the table of a direct variation function. Use each digit exactly once. The 2 and 3 have already been used.

x	8	32	56
y	1	4	7

Available on Daily Transparency in CRB

Math Humor

Teacher: Why did you turn in this poem?

Student: You said you wanted an example of *in verse* variation!

12-8 Inverse Variation

55Hz
110Hz
220Hz
440Hz

Learn to recognize inverse variation by graphing tables of data.

Vocabulary
inverse variation

The frequency of a piano string is related to its length. You can double a string's frequency by placing your finger at the halfway point of the string. The lowest note on the piano is A_1. As you place your finger at various fractions of the string's length, the frequency will *vary inversely*.

Full length: **55 Hz** $\frac{1}{2}$ the length: **110 Hz** $\frac{1}{4}$ the length: **220 Hz**

The fraction of the string length times the frequency is always 55.

INVERSE VARIATION

Words	Numbers	Algebra
An **inverse variation** is a relationship in which one variable quantity increases as another variable quantity decreases. The product of the variables is a constant.	$y = \frac{120}{x}$ $xy = 120$	$y = \frac{k}{x}$ $xy = k$

EXAMPLE 1   **Identifying Inverse Variation**

Tell whether each relationship is an inverse variation.

Helpful Hint

To determine if a relationship is an inverse variation, check if the product of x and y is always the same number.

A The table shows the number of days needed to construct a building based on the size of the work crew.

Crew Size	2	3	5	10	20
Days of Construction	90	60	36	18	9

$20(9) = 180; 10(18) = 180; 5(36) = 180; 3(60) = 180; 2(90) = 180$
$xy = 180$ *The product is always the same.*
The relationship is an inverse variation: $y = \frac{180}{x}$.

B The table shows the number of chips produced in a given time.

Chips Produced	36	60	84	108	120	144
Time (min)	3	5	7	9	10	12

$36(3) = 108; 60(5) = 300$ *The product is not always the same.*
The relationship is not an inverse variation.

1 Introduce

Alternate Opener

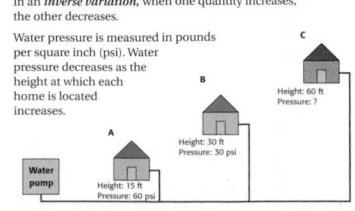

EXPLORATION

12-8 Inverse Variation

In an *inverse variation*, when one quantity increases, the other decreases.

Water pressure is measured in pounds per square inch (psi). Water pressure decreases as the height at which each home is located increases.

Height: 60 ft
Pressure: ?

Height: 30 ft
Pressure: 30 psi

Height: 15 ft
Pressure: 60 psi

Water pump

1. For house A and house B, multiply the water pressure by the height at which each house is located.
2. What do you notice about the products in number 1?
3. Predict the water pressure for house C.

Think and Discuss
4. **Explain** how you predicted the water pressure for house C.
5. **Give** another example of an inverse variation.

Motivate

Remind students of the formula for direct variation, $y = kx$ (Lesson 11-5). Point out that direct variation means that the quantities change *directly* with each other. That means if one increases, the other increases, and vice-versa. Ask students if they know the meaning of the word *inverse*. opposite or reverse Ask students to guess what happens in *inverse variation*. As one quantity increases, the other decreases.

Exploration worksheet and answers on Chapter 12 Resource Book pp. 78 and 115

2 Teach

Lesson Presentation

Guided Instruction

In this lesson, students learn to recognize inverse variation by graphing tables of data. Introduce students to the concept of *inverse variation* and its formula. As you work through the examples, point out that one quantity increases as the other decreases, but emphasize that this alone does not make inverse variation. In inverse variation, the product of the x- and y-values must be constant.

In the inverse variation relationship $y = \frac{k}{x}$, where $k \neq 0$, y is a function of x. The function is not defined for $x = 0$, so the domain is all real numbers except 0.

EXAMPLE 2 Graphing Inverse Variations

Graph each inverse variation function.

A $f(x) = \frac{1}{x}$

x	y
−3	$-\frac{1}{3}$
−2	$-\frac{1}{2}$
−1	−1
$-\frac{1}{2}$	−2
$\frac{1}{2}$	2
1	1
2	$\frac{1}{2}$
3	$\frac{1}{3}$

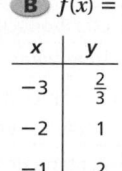

B $f(x) = \frac{-2}{x}$

x	y
−3	$\frac{2}{3}$
−2	1
−1	2
$-\frac{1}{2}$	4
$\frac{1}{2}$	−4
1	−2
2	−1
3	$-\frac{2}{3}$

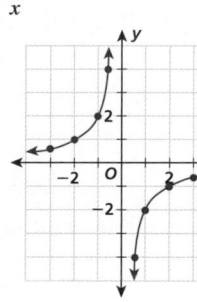

EXAMPLE 3 *Music Application*

The frequency of a piano string changes according to the fraction of its length that is allowed to vibrate. Find the inverse variation function, and use it to find the resulting frequency when $\frac{1}{16}$ of the string A_1 is allowed to vibrate.

Frequency of A_1 by Fraction of the Original String Length				
Frequency (Hz)	55	110	220	440
Fraction of the Length	1	$\frac{1}{2}$	$\frac{1}{4}$	$\frac{1}{8}$

You can see from the table that $xy = 55(1) = 55$, so $y = \frac{55}{x}$.
If the string is reduced to $\frac{1}{16}$ of its length, then its frequency will be
$y = 55 \div \left(\frac{1}{16}\right) = 16 \cdot 55 = 880$ Hz.

Think and Discuss

1. Identify k in the inverse variation $y = \frac{3}{x}$.

2. Describe how you know if a relationship is an inverse variation.

Additional Examples

Example 1

Tell whether the relationship is an inverse variation.

The table shows how 24 cookies can be divided equally among different numbers of students. yes

# of students	2	3	4	6	8
# of cookies	12	8	6	4	3

Example 2

Graph the inverse variation function $f(x) = \frac{4}{x}$.

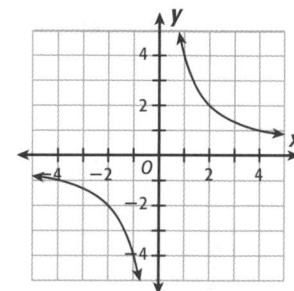

Example 3

As the pressure on the gas in a balloon changes, the volume of the gas changes.

Pressure (lb/in²)	5	10	15	20
Volume (in³)	300	150	100	75

Find the inverse variation function and use it to find the resulting volume when the pressure is 30 lb/in².
$y = \frac{1500}{x}$; 50 in³

---

3 Close

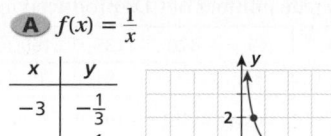

Reaching All Learners
Through Critical Thinking

Give students a recording sheet (Chapter 12 Resource Book p. 98) with examples of both direct and inverse variation. (For example, Gary is building a new patio with square tiles. The patio will be rectangular. He has enough tiles to build an 8-by-9-foot patio, a 6-by-12 foot patio, or an 18-by-4-foot patio.) Ask students to classify each example as direct or inverse variation and to write an equation to describe each situation.

Summarize

Remind students that in an inverse variation, the product of the x-value and the y-value in each ordered pair is a constant. When graphing an inverse variation, the graph is a pair of curves.

Answers to Think and Discuss

1. $k = 3$

2. Possible answer: As one quantity increases, the other will decrease. Then check to see that the product of the x-values and y-values is a constant number.

12-8 Exercises

FOR EXTRA PRACTICE

see page 755

☑ internet connect

Homework Help Online
go.hrw.com   Keyword: MP4 12-8

Students may want to refer back to the lesson examples.

Assignment Guide

If you finished Example **1** assign:
  Core 1–2, 7–8, 23–29
  Enriched 1–2, 7–8, 23–29

If you finished Example **2** assign:
  Core 1–5, 7–11, 13, 15, 23–29
  Enriched 7–11, 13–15, 20–21, 23–29

If you finished Example **3** assign:
  Core 1–12, 13–19 odd, 23–29
  Enriched 7–29

Answers

3–5, 9–11. See p. A18.

GUIDED PRACTICE

See Example **1** **Tell whether each relationship is an inverse variation.**

1. The table shows the number of CDs produced in a given time. **no**

CDs Produced	45	120	135	165	210
Time (min)	3	8	9	11	14

2. The table shows the construction time of a wall based on the number of workers. **yes**

Construction Time (hr)	5	9	15	22.5	45
Number of Workers	9	5	3	2	1

See Example **2** **Graph each inverse variation function.**

3. $f(x) = \dfrac{3}{x}$ 4. $f(x) = \dfrac{2}{x}$ 5. $f(x) = \dfrac{1}{2x}$

See Example **3** 6. Ohm's law relates the current in a circuit to the resistance. Find the inverse variation function, and use it to find the current in a 12-volt circuit with 9 ohms of resistance. $y = \dfrac{12}{x}$; $1\frac{1}{3}$ amps

Current (amps)	0.25	0.5	1	2	4
Resistance (ohms)	48	24	12	6	3

INDEPENDENT PRACTICE

See Example **1** **Tell whether each relationship is an inverse variation.**

7. The table shows the time it takes to throw a baseball from home plate to first base depending on the speed of the throw.   **yes**

Speed of Throw (ft/s)	30	36	45	60	90
Time (s)	3	2.5	2	1.5	1

8. The table shows the number of miles jogged in a given time.   **no**

Miles Jogged	1	1.5	3	4	5
Time (min)	8	12	24	32	40

See Example **2** **Graph each inverse variation function.**

9. $f(x) = -\dfrac{1}{x}$ 10. $f(x) = \dfrac{1}{3x}$ 11. $f(x) = -\dfrac{1}{2x}$

See Example **3** 12. According to Boyle's law, when the volume of a gas decreases, the pressure increases. Find the inverse variation function, and use it to find the pressure of the gas if the volume is decreased to 4 liters. $y = \dfrac{40}{x}$; 10 atm

Volume (L)	8	10	20	40	80
Pressure (atm)	5	4	2	1	0.5

Math Background

When a function of the form $y = \dfrac{k}{x}$ ($k \neq 0$) is graphed, it does not intersect the x-axis or the y-axis but continues to get closer and closer to each axis. When a curve gets and stays arbitrarily close to a line, the line is called an *asymptote*. The reason that the x-axis is an asymptote is that as the x-values increase, the y-values continue to decrease but never reach zero. The y-axis is an asymptote because as x-values decrease, the y-values continue to increase toward infinity but never reach infinity.

RETEACH 12-8

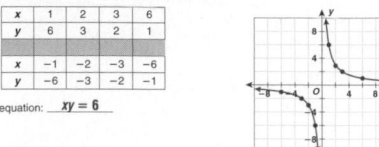

LESSON 12-8 Reteach
Inverse Variation

Two quantities **vary inversely** if their product is constant.
  y varies inversely as x. $xy = k$ ◄— constant of variation

To determine if a data set varies inversely, look for a constant product.

x	1	2	4	8	−1	−2	−4	−8
y	8	4	2	1	−8	−4	−2	−1

$xy = 1(8) = 2(4) = 4(2) = 8(1) = -1(-8)$
$= -2(-4) = -4(-2) = -8(-1) = 8$

So, y varies inversely as x.
The constant of variation is 8.
The equation of inverse variation is $xy = 8$.
On a graph, the points lie on a curve that has two branches and is not defined for $x = 0$.

Explore each data set for variation. If there is a constant product, identify it and write the equation of inverse variation.

1.
x	5	4	2	1	−1	−2
y	20	16	8	4	−4	−8

constant product? **no**

If yes, equation. _____

2.
x	1	2	3	−3	−2
y	12	6	4	−4	−6

constant product? **yes**

If yes, equation. __$xy = 12$__

Write the equation of inverse variation. Graph.

3.
x	1	2	3	6
y	6	3	2	1

x	−1	−2	−3	−6
y	−6	−3	−2	−1

equation: __$xy = 6$__

PRACTICE 12-8

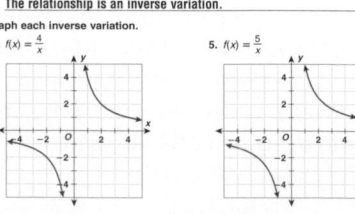

LESSON 12-8 Practice B
Inverse Variation

Tell whether each relationship is an inverse variation.

1. The table shows the length and width of a rectangle.

Length	6	8	12	16	24
Width	8	6	4	3	2

The relationship is an inverse variation.

2. The table shows the number of days needed to paint a house for the size of the work crew.

Crew Size	2	3	4	5	6
Days of Painting	21	14	10.5	8.5	7

The relationship is not an inverse variation.

3. The table shows the time traveled and speed traveled.

Hours	5	6	8	9	12
mi/h	72	60	45	40	30

The relationship is an inverse variation.

Graph each inverse variation.

4. $f(x) = \dfrac{4}{x}$ 5. $f(x) = \dfrac{5}{x}$

6. Amperes (abbreviated amp) measure the strength of electric current. An ohm is the unit of electrical resistance. In an electric circuit, the current varies inversely as the resistance. If the current is 24 amps when the resistance is 20 ohms, find the resistance in ohms when the current is 40 amps.   __12 ohms__

PRACTICE AND PROBLEM SOLVING

Find the inverse variation equation, given that x and y vary inversely.

13. $y = 2$ when $x = 2$ **14.** $y = 10$ when $x = 2$ **15.** $y = 8$ when $x = 4$

16. If y varies inversely with x and $y = 27$ when $x = 3$, find the constant of variation.   **81**

17. The height of a triangle with area 50 cm² varies inversely with the length of its base. If $b = 25$ cm when $h = 4$ cm, find b when $h = 10$ cm.   **10 cm**

18. *PHYSICAL SCIENCE* If a constant force of 30 N is applied to an object, the mass of the object varies inversely with its acceleration. The table contains data for several objects of different sizes.

Mass (kg)	3	6	30	10	5
Acceleration (m/s²)	10	5	1	3	6

   a. Use the table to write an inverse variation function. $y = \dfrac{30}{x}$

   b. What is the mass of an object if its acceleration is 15 m/s²?   **2 kg**

19. *FINANCE* Mr. Anderson wants to earn \$125 in interest over a 2-year period from a savings account. The principal he must deposit varies inversely with the interest rate of the account. If the interest rate is 6.25%, he must deposit \$1000. If the interest rate is 5%, how much must he deposit?   **\$1250**

 20. *WRITE ABOUT IT* Explain the difference between direct variation and inverse variation.

 21. *WRITE A PROBLEM* Write a problem that can be solved using inverse variation. Use facts and formulas from your science book.

 22. *CHALLENGE* The resistance of a 100 ft piece of wire varies inversely with the square of its diameter. If the diameter of the wire is 3 in., it has a resistance of 3 ohms. What is the resistance of a wire with a diameter of 1 in.?   **27 ohms**

Spiral Review

For each function, find $f(-1)$, $f(0)$, and $f(1)$. (Lesson 12-4)

23. $f(x) = 3x^2 - 5x + 1$ **9, 1, −1**   **24.** $f(x) = x^2 + 15x - 4$ **25.** $f(x) = 3(x - 9)^2$ **300, 243, 192**
                                           −18, −4, 12

26. $f(x) = 2x^3 - 6x - 2$ **27.** $f(x) = (x - 5)(x + 7)$ **28.** $f(x) = -144x^2 - 64x$
     2, −2, −6   **−36, −35, −32**   **−80, 0, −208**

29. **TEST PREP** The half-life of a particular radioactive isotope of thorium is 8 minutes. If 160 grams of the isotope are initially present, how many grams will remain after 40 minutes? (Lesson 12-6)   **C**

   A   10 grams   **B**   2.5 grams   **C**   5 grams   **D**   1.25 grams

---

CHALLENGE 12-8

PROBLEM SOLVING 12-8

Answers

13. $y = \dfrac{4}{x}$

14. $y = \dfrac{20}{x}$

15. $y = \dfrac{32}{x}$

20. Possible answer: In direct variation, the variables increase or decrease together. It is of the form $y = kx$. In inverse variation, one variable increases as the other decreases. It is of the form $y = \dfrac{k}{x}$.

21. Check students' work.

Journal

Ask students to explain the difference between direct variation and inverse variation and to give an example of each.

Test Prep Doctor

For Exercise 29, remind students that the formula for half-life is $f(x) = p \cdot \left(\dfrac{1}{2}\right)^x$, where p is the initial amount and x is the number of half-lives. If the half-life is 8 minutes, there are 5 half-lives in 40 minutes. So $f(x) = 160 \cdot \left(\dfrac{1}{2}\right)^5 = 5$ grams. This is choice **C**.

Lesson Quiz

Tell whether each relationship is an inverse variation.

1.

x	5	15	25	75
y	45	15	9	3

yes

2.

x	10	25	40	50
y	40	25	10	10

no

3. Graph the inverse variation function $f(x) = \dfrac{1}{4x}$.

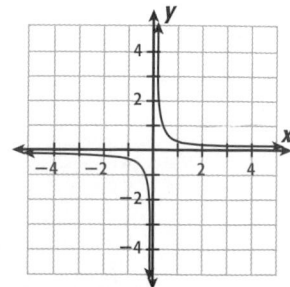

Available on Daily Transparency in CRB

Problem Solving on Location

Alabama

Purpose: *To provide additional practice for problem-solving skills in Chapters 1–12*

NASA Marshall Space Flight Center

- After problem 2, have students consider the following problem: With no air resistance, how long would it take an object to fall 1 kilometer on Earth? **about 14.29 seconds**

- After problem 3, have students find the gravitational constant necessary to cause an object to fall 100 meters in 4.2 seconds. **11.34 m/s²**

Extension Have students consider the following problem: The effect of gravity on Earth is known as $1g$. When we accelerate in a car or plane, we may feel a pull greater or less than $1g$ because of Earth's gravity, perhaps $2g$'s or $\frac{1}{2}g$. The acceleration is equal to the change in speed divided by the time it took to make this change in speed. The acceleration divided by Earth's gravitational constant gives the value in g's. What portion of Earth's gravitational pull is felt when a car accelerates from 0 m/s to 25 m/s (about 56 mi/h) in 10 seconds? in 5 seconds? **about 0.255 g's; about 0.510 g's**

Answers

2.

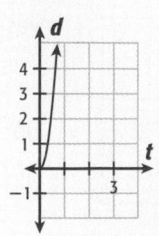

Problem Solving on Location

Alabama

NASA Marshall Space Flight Center

At NASA's Marshall Space Flight Center in Huntsville, Alabama, scientists work on the development of the International Space Station. One area of research that scientists at the Marshall Center specialize in is microgravity. Microgravity researchers try to minimize the effects of gravity in order to simulate the zero gravity of space.

To find the distance d in meters that a free-falling object travels in t seconds with no air resistance, you would use the function $d = \frac{1}{2}gt^2$. In this distance function, g is the gravitational constant. On Earth, this constant is $g = 9.8$ m/s².

1. What is the domain of the function $d = \frac{1}{2}gt^2$? What is the range?   **$t \geq 0$; $d \geq 0$**

2. Graph $d = \frac{1}{2}gt^2$.

3. In a microgravity experiment, NASA scientists recorded that it took 4.5 seconds for an object to fall 100 meters. Find the gravitational constant g in the experiment.   **9.88 m/s²**

NASA's KC-135 aircraft, referred to as the Weightless Wonder or the Vomit Comet, is used to create a microgravity environment.

4. While the KC-135 is climbing at a 45° angle, the equation of its path is $y = x$. While it is descending at a 45° angle, the equation of its path is $y = -x$. Are these linear or quadratic functions?   **linear**

5. While the KC-135 is in a microgravity environment, the equation of its path is $y = -x^2$. Is this function linear or quadratic?   **quadratic**

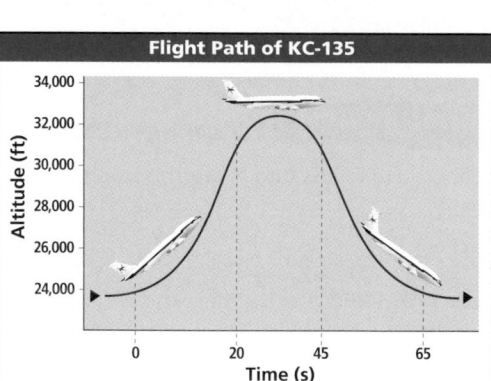

Muscle Shoals

Located on the Tennessee River, Muscle Shoals was once the site of the Muscle Shoals Canal. When the canal was built in the 1830s, its purpose was to connect Colbert and Lauderdale Counties with a passageway that was easy to travel. During later attempts to improve the canal, dams were built to control its water flow. Around 1924, two of these dams, Wilson Dam and Wheeler Dam, flooded the canal and created the lakes that are known today as Wilson Lake and Wheeler Lake.

When a lake's water level gets too low, a dam's floodgates can be opened to allow water to enter the lake. For 1–3, use the table.

1. What kind of sequence is formed by the total amount of water released after each second at 8 A.M. from Wheeler Dam? What is a possible rule for this sequence?

	Water Release at Wheeler Dam and Wilson Dam on March 15, 2002			
	Total Amount of Water Released (ft³)			
Number of Seconds	Wheeler Dam		Wilson Dam	
	8 A.M.	9 A.M.	8 A.M.	9 A.M.
1	683	9,520	7,310	20,300
2	1,366	19,040	14,620	40,600
3	2,049	28,560	21,930	60,900
4	2,732	38,080	29,240	81,200

2. Write a possible rule for the sequence formed by the total amount of water released after each second at 9 A.M. from Wilson Dam. If the pattern continues, what will the total amount of water released be after 6 seconds?

The generator hall at Wilson Dam in 2002 (above) and 1942 (below)

3. Suppose water was released from Wheeler Dam at 3 A.M. at a rate of 1000 cubic feet per second, at 4 A.M. water was released at a rate of 1200 cubic feet per second, at 5 A.M. water was released at a rate of 1440 cubic feet per second, and at 6 A.M. water was released at a rate of 1728 cubic feet per second.

   a. What kind of sequence do the rates of water release at each hour appear to form?

   b. Write a possible rule for the sequence.

   c. If the pattern continues, at what rate would you expect water to be released at 10 A.M.?

MATH-ABLES

Game Resources

Puzzles, Twisters & Teasers
Chapter 12 Resource Book

Squared Away

Purpose: *To apply the problem-solving skill of writing sequences to solve a puzzle*

Discuss: Ask students to explain how to determine the number of different squares in a square of side length *n*.

The number of $n \times n$ squares is 1^2, or 1. The number of $(n - 1) \times (n - 1)$ squares is 2^2, or 4. The number of $(n - 2) \times (n - 2)$ squares is 3^2, or 9. The number of $(n - a) \times (n - a)$ squares is $(a + 1)^2$. Add the number of squares of each size to find the total number.

Extend: Challenge students to notice other patterns in the tables used to record *Size of Square* and *Number of Squares*. Have them test hypotheses by testing their rule on the tables for 4×4, 5×5, 6×6, and 7×7 squares.

Possible answer: Find the area of each size square and multiply it by the number of squares of that size. The products are square numbers that form a palindrome pattern when listed in order. For example, for a 4×4 square, the products are as follows:

1   4×4 square = 16 square units

4   3×3 squares = 36 square units

9   2×2 squares = 36 square units

16   1×1 squares = 16 square units

What's Your Function?

Purpose: *To practice identifying functions in a game format*

Discuss: When a team guesses a function correctly, have them demonstrate that each input/output pair used is a solution to the function.

Extend: Have students create new function cards, including linear, quadratic, and cubic functions. Use the new equation cards to play again.

MATH-ABLES

Squared Away

How many squares can you find in the figure at right?

Did you find 30 squares?

There are four different-sized squares in the figure.

Size of Square	Number of Squares
4×4	1
3×3	4
2×2	9
1×1	16
Total	30

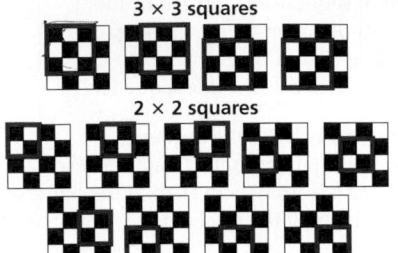

3 × 3 squares

2 × 2 squares

So the total number of squares is $1 + 4 + 9 + 16 = 1^2 + 2^2 + 3^2 + 4^2$.

Draw a 5×5 grid and count the number of squares of each size. Can you see a pattern?

What is the total number of squares on a 6×6 grid? a 7×7 grid? Can you come up with a general formula for the sum of squares on an $n \times n$ grid?

What's Your Function?

One member from the first of two teams draws a function card from the deck, and the other team tries to guess the rule of the function. The guessing team gives a function input, and the card holder must give the corresponding output. Points are awarded based on the type of function and number of inputs required. The first team to reach 20 points wins.

📶 **internet** connect

Go to ***go.hrw.com*** for a complete set of rules and game cards.
KEYWORD: MP4 Game12

Answers

$5 \times 5 \rightarrow 55$ squares
$6 \times 6 \rightarrow 91$ squares
$7 \times 7 \rightarrow 140$ squares
$8 \times 8 \rightarrow 204$ squares

Possible answer:

$$n \times n \rightarrow 1^2 + 2^2 + 3^2 + \cdots + n^2 = \frac{n(n + 1)(2n + 1)}{6}$$

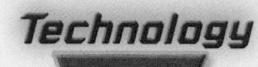 **Technology**

Generate Arithmetic and Geometric Sequences

Use with Lesson 12-2

Graphing calculators can be used to explore arithmetic and geometric sequences.

📶 **internet** connect
Lab Resources Online
go.hrw.com
KEYWORD: MP4 TechLab12

Activity

❶ The command **seq(** is used to generate a sequence.

a. Press `2nd` `STAT`(LIST) OPS 5:seq.

The **seq(** command is followed by the rule for generating the sequence, the variable used in the rule, and the positions of the first and last terms in the sequence. To find the first 20 terms of the arithmetic sequence generated by the rule $5 + (x - 1) \cdot 3$, enter **seq($5 + (x - 1) \cdot 3$, x, 1, 20)**:

5 `+` `(` `X,T,θ,n` `-` 1 `)` `×`

3 `,` `X,T,θ,n` `,` 1 `,` 20 `ENTER`

b. You can see all 20 terms by pressing the right arrow key ▶ repeatedly. From the calculator display, the first term is 5, the second is 8, the third is 11, the fourth is 14, and so on.

❷ Consider the *geometric* sequence whose nth term is $3\left(\frac{1}{4}\right)^{n-1}$. To use a graphing calculator to find the first 15 terms in fraction form, press

`2nd` `STAT`(LIST) ▶ 5:seq 3 `×` `(` 1 `÷` 4 `)` `^`

`(` `X,T,θ,n` `-` 1 `)` `,` `X,T,θ,n` `,` 1 `,` 15 `MATH` 1:Frac `ENTER`.

To see all 15 terms, press the right arrow key ▶ repeatedly.

Think and Discuss

1. Why is the seventh term of the sequence in ❷ *not* displayed as a fraction?

Try This

Find the first 15 terms of each sequence. Tell if the consecutive terms increase or decrease.

1. $-4 + (n - 1) \cdot 7$ **2.** $2\left(\frac{1}{5}\right)^{n-1}$ **3.** $9, 14, 19, 24, \ldots$ **4.** $2, \frac{2}{3}, \frac{2}{9}, \frac{2}{27}, \ldots$

Answers

Think and Discuss

1. The 7th term of the sequence in ❷ is not displayed as a fraction because **Frac** works only for a limited number of decimal places, usually 3 digits after the decimal point.

Try This

1. {−4, 3, 10, 17, 24, 31, 38, 45, 52, 59, 66, 73, 80, 87, 94}; increase

2. {2, $\frac{2}{5}$, $\frac{2}{25}$, $\frac{2}{125}$, $\frac{2}{625}$, 6.4 E −4, 1.28 E −4, 2.56 E −5, 5.12 E −6, 1.024 E −6, 2.048 E −7, 4.096 E −8, 8.192 E −9, 1.6384 E −9, 3.2768 E −10}; decrease

3. {9, 14, 19, 24, 29, 34, 39, 44, 49, 54, 59, 64, 69, 74, 79}; increase

4. {2, $\frac{2}{3}$, $\frac{2}{9}$, $\frac{2}{27}$, $\frac{2}{81}$, $\frac{2}{243}$, $\frac{2}{729}$, 9.144947417 E −4, 3.048315806 E −4, 1.016105269 E −4, 3.387017562 E −5, 1.129005854 E −5, 3.763352846 E −6, 1.254450949 E −6, 4.181503163 E −7}; decrease

Chapter
12 **Study Guide and Review**

Chapter
12 **Study Guide and Review**

Purpose: *To help students review and practice concepts and skills presented in Chapter 12*

Assessment Resources

Chapter 12 Review
Chapter 12 Resource Book .. pp 86–88

Test and Practice Generator CD-ROM

Additional review assessment items in both multiple-choice and free-response format may be generated for any objective in Chapter 12.

Answers

1. sequence

2. arithmetic sequence; geometric sequence

3. Fibonacci sequence

4. function; domain; range

5. 31

6. 0.65

7. $\frac{14}{3}$

Study Guide and Review

Vocabulary

Complete the sentences below with vocabulary words from the list above. Words may be used more than once.

1. A list of numbers or terms in a certain order is called a(n) __?__.

2. A sequence in which there is a common difference is a(n) __?__; a sequence in which there is a common ratio is a(n) __?__.

3. A famous sequence in which you add the two previous terms to find the next term is the __?__.

4. A rule that relates two quantities so that each input value corresponds to exactly one output value is a(n) __?__. The set of all input values is the __?__; the set of output values is the __?__.

12-1 Arithmetic Sequences (pp. 590–594)

EXAMPLE

■ Find the 10th term of the arithmetic sequence: 12, 10, 8, 6, ...

$d = 10 - 12 = -2$
$a_n = a_1 + (n - 1)d$
$a_{10} = 12 + (10 - 1)(-2)$
$a_{10} = 12 - 18$
$a_{10} = -6$

EXERCISES

Find the given term in each arithmetic sequence.

5. 8th term: 3, 7, 11, ...

6. 7th term: 0.05, 0.15, 0.25, ...

7. 9th term: $\frac{2}{3}, \frac{7}{6}, \frac{5}{3}, \ldots$

12-2 Geometric Sequences (pp. 595–599)

EXAMPLE

■ Find the 10th term of the geometric sequence: 6, 12, 24, 48, ...

$r = \frac{12}{6} = 2$

$a_n = a_1 r^{n-1}$

$a_{10} = 6(2)^{10-1} = 3072$

EXERCISES

Find the given term in each geometric sequence.

8. 8th term: 5, −10, 20, −40, ...

9. 7th term: $\frac{1}{2}, \frac{1}{3}, \frac{2}{9}, ...$

10. 50th term: 1, −1, 1, −1, ...

12-3 Other Sequences (pp. 601–605)

EXAMPLE

■ Find the first four terms of the sequence defined by $a_n = -2(-1)^{n-1} - 1$.

$a_1 = -2(-1)^{1-1} - 1 = -3$
$a_2 = -2(-1)^{2-1} - 1 = 1$
$a_3 = -2(-1)^{3-1} - 1 = -3$
$a_4 = -2(-1)^{4-1} - 1 = 1$
The first four terms are −3, 1, −3, 1.

EXERCISES

Find the first four terms of the sequence defined by each rule.

11. $a_n = 3n + 1$ **12.** $a_n = n^2 + 1$

13. $a_n = 8(-1)^n + 2n$ **14.** $a_n = n! + 2$

12-4 Functions (pp. 608–612)

EXAMPLE

■ For the function $f(x) = 3x^2 + 4$, find $f(0), f(3)$, and $f(-2)$.

$f(0) = 3(0)^2 + 4 = 4$
$f(3) = 3(3)^2 + 4 = 31$
$f(-2) = 3(-2)^2 + 4 = 16$

EXERCISES

For each function, find $f(0), f(2)$, and $f(-1)$.

15. $f(x) = 7x - 4$ **16.** $f(x) = 2x^3 + 1$

17. $f(x) = -x^2 + 3x$ **18.** $f(x) = -x^3 + 2x^2$

19. $f(x) = 3x^2 - x + 5$ **20.** $f(x) = -x^2 + x + 1$

12-5 Linear Functions (pp. 613–616)

EXAMPLE

■ Use the table to write the equation for the linear function.

x	y
−2	−10
−1	−3
0	4
1	11

The y-intercept is $f(0) = 4$.
$f(x) = mx + 4$ $f(x) = mx + b$
Substitute and solve for m.
$11 = m(1) + 4$ $(x, y) = (1, 11)$
$m = 7$
$f(x) = 7x + 4$

EXERCISES

Write the equation for each linear function.

21.

x	y
−2	−3
−1	−2
0	−1
1	0

22.

x	y
−4	2
−2	3
0	4
2	5

Answers

8. −640

9. $\frac{32}{729}$, or ≈ 0.0439

10. −1

11. 4, 7, 10, 13

12. 2, 5, 10, 17

13. −6, 12, −2, 16

14. 3, 4, 8, 26

15. −4, 10, −11

16. 1, 17, −1

17. 0, 2, −4

18. 0, 0, 3

19. 5, 15, 9

20. 1, −1, −1

21. $f(x) = x - 1$

22. $f(x) = \frac{1}{2}x + 4$

23.

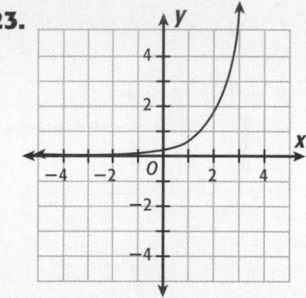

24.

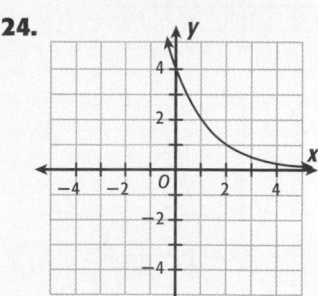

Answers

23–24. See previous page.

25.

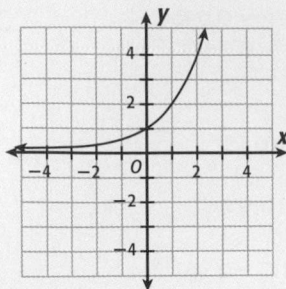

26.

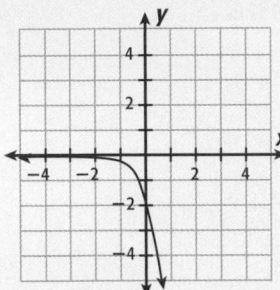

27.

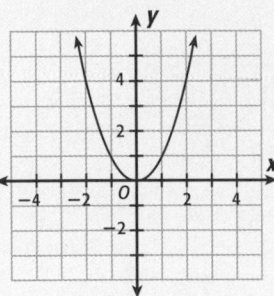

28.

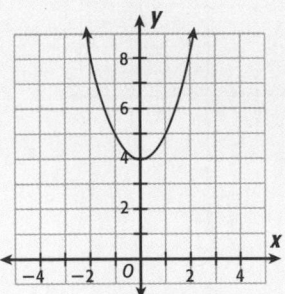

29.

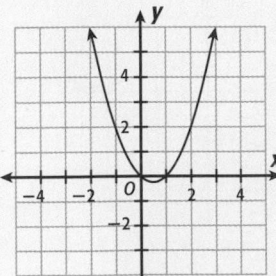

30.

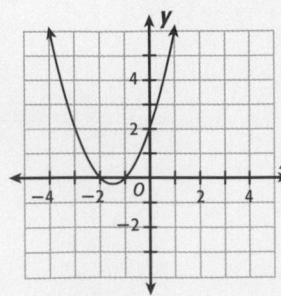

12-6 Exponential Functions (pp. 617–620)

EXAMPLE

- Graph the exponential function.
  $f(x) = 0.1 \cdot 4^x$

x	f(x)
−2	0.00625
−1	0.025
0	0.1
1	0.4
2	1.6

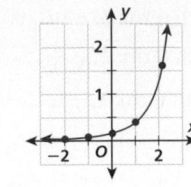

EXERCISES

Graph each exponential function.

23. $f(x) = 0.2 \cdot 3^x$

24. $f(x) = 4 \cdot \left(\frac{1}{2}\right)^x$

25. $f(x) = 2^x$

26. $f(x) = -2 \cdot 10^x$

12-7 Quadratic Functions (pp. 621–625)

EXAMPLE

- Graph the quadratic function.
  $f(x) = x^2 + 2x - 1$

x	f(x)
−3	2
−2	−1
−1	−2
0	−1
1	2
2	7
3	14

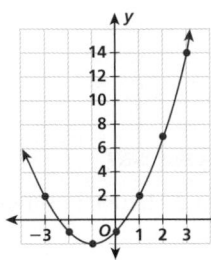

EXERCISES

Graph each quadratic function.

27. $f(x) = x^2$

28. $f(x) = x^2 + 4$

29. $f(x) = x^2 - x$

30. $f(x) = x^2 + 3x + 2$

12-8 Inverse Variation (pp. 628–631)

EXAMPLE

- Graph the inverse variation function.
  $f(x) = \frac{6}{x}$

x	y
−3	−2
−2	−3
−1	−6
1	6
2	3
3	2

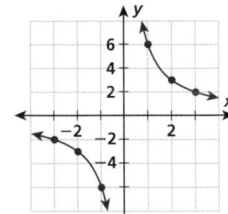

EXERCISES

Graph each inverse variation function.

31. $f(x) = \frac{12}{x}$

32. $f(x) = \frac{16}{x}$

33. $f(x) = -\frac{8}{x}$

34. $f(x) = -\frac{4}{x}$

31.

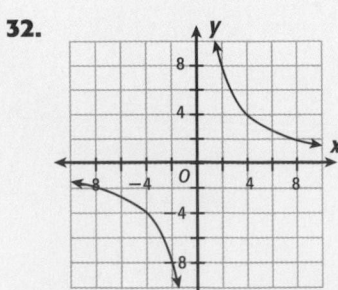

32.

33.

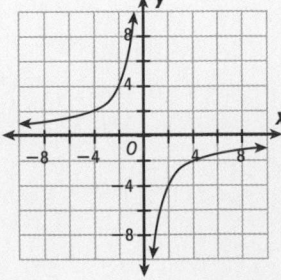

34.

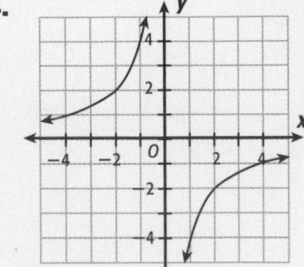

Find the given term in each arithmetic sequence.

1. 8th term: 7, 10, 13, … **28**

2. 13th term: $7, 7\frac{1}{5}, 7\frac{2}{5}, …$ **$9\frac{2}{5}$**

3. 11th term: 11, 10.9, 10.8, … **10**

4. 9th term: 75, 62, 49, 36, … **−29**

Find the given term in each geometric sequence.

5. 7th term: 8, 32, 128, … **32,768**

6. 8th term: 25, 5, 1 **$\frac{1}{3125}$, or 0.00032**

7. 6th term: 17, −0.34, 0.0068, … **−0.0000000544**

8. 10th term: 0.25, 1.25, 6.25, 31.25, … **488,281.25**

Find the first five terms of each sequence, given its rule.

9. $a_n = 6n − 2$ **4, 10, 16, 22, 28**

10. $a_n = 2 \cdot 3^n$ **6, 18, 54, 162, 486**

11. $a_n = (−1)^n \cdot 5 + 2n$ **−3, 9, 1, 13, 5**

Use first and second differences to find the next three terms in each sequence.

12. 7, 17, 32, 52, 77, … **107, 142, 182**

13. 10, 16, 20, 22, 22, … **20, 16, 10**

14. 1, 1, 1.05, 1.15, 1.30, … **1.50, 1.75, 2.05**

For each function, find $f(0)$, $f(4)$, and $f(−3)$.

15. $y = 5x − 3$ **−3; 17; −18**

16. $y = 3x^3 + 2x$ **0; 200; −87**

17. $y = −x^2 − 5$ **−5; −21; −14**

Write the equation for each linear function.

18. $y = −3x + 2$

x	y
−4	14
−1	5
0	2
3	−7

19. $y = 0.75x − 1$

x	y
−8	−7
−4	−4
0	−1
4	2

Graph each inverse variation function.

20. $f(x) = \frac{6}{x}$

21. $f(x) = \frac{10}{x}$

22. $f(x) = −\frac{12}{x}$

23. A microbiologist began a bacterial culture with 1000 *E. coli* bacteria. If the number of bacteria doubles every 20 minutes, find the number of bacteria in the culture after 2 hours. **64,000**

24. Carbon-14 (C14), a radioactive form of carbon, has a half-life of about 5730 years. C14 is used to date old objects made from plant material. If a wooden cup had 1000 grams of C14 when the tree it came from was cut, about how many grams of C14 would be present 1400 years later? **≈844 g**

Chapter Test

Purpose: *To assess students' mastery of concepts and skills in Chapter 12*

Assessment Resources

Chapter 12 Tests (Levels A, B, C)
Assessment Resources . . . . pp. 99–104

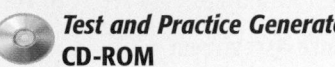
Test and Practice Generator CD-ROM

Additional assessment items in both multiple-choice and free-response format may be generated for any objective in Chapter 12.

Answers

20.

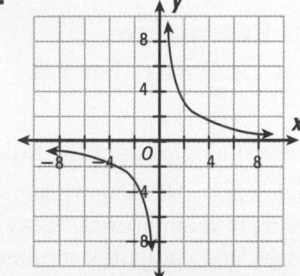

21.

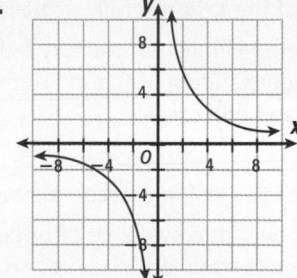

22.

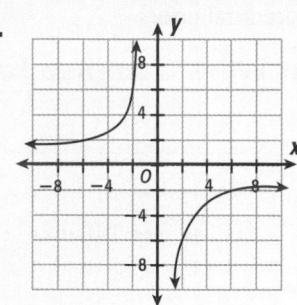

Purpose: *To assess students' under-standing of concepts in Chapter 12 and combined problem-solving skills*

Assessment Resources ✓

Performance Assessment
Assessment Resources . . . . p. 140

Performance Assessment Teacher Support
Assessment Resources . . . . p. 139

Answers

1–3. See p. A18.

4. See Level 3 work sample below.

Scoring Rubric for Problem Solving Item 4

Level 3
Accomplishes the purpose of the task.

Student gives clear explanations, shows understanding of mathematical ideas and processes, and computes accurately.

Level 2
Purposes of the task not fully achieved.

Student demonstrates satisfactory but limited understanding of the mathematical ideas and processes.

Level 1
Purposes of the task not accomplished.

Student shows little evidence of understanding the mathematical ideas and processes and makes computational and/or procedural errors.

Performance Assessment

✎ Show What You Know

Create a portfolio of your work from this chapter. Complete this page and include it with your four best pieces of work from Chapter 12. Choose from your homework or lab assignments, mid-chapter quiz, or any journal entries you have done. Put them together using any design you want. Make your portfolio represent what you consider your best work.

★ Short Response

1. Write out the next three terms of the sequence
$$\sqrt{2}, \sqrt{2 + \sqrt{2}}, \sqrt{2 + \sqrt{2 + \sqrt{2}}}, \sqrt{2 + \sqrt{2 + \sqrt{2 + \sqrt{2}}}}, \dots$$
Use your calculator to evaluate each term of the sequence. Describe what seems to be happening to the terms of the sequence.

2. A basketball player throws a basketball in a path defined by the function $f(x) = -16x^2 + 20x + 7$, where x is the time in seconds and $f(x)$ is the height in feet. Graph the function, and estimate how long it would take the basketball to reach its maximum height.

3. When playing the trombone, you produce different notes by changing the effective length of the tube by moving it in and out. This movement produces a sequence of lengths that form a geometric sequence. If the length is 119.3 inches in the 2nd position and 134.0 inches in the 4th position, what is the length in the 3rd position? Write a rule that would describe this relationship.

🧩 Extended Problem Solving

4. Consider the sequence 1, 2, 6, 24, 120, 720, . . .
   a. Determine whether the sequence is arithmetic, geometric, or neither.
   b. Find the ratio of each pair of consecutive terms. What pattern do you notice?
   c. Write a rule for the sequence. Use your rule to find the next two terms.

Student Work Samples for Item 4

Level 3

> 1, 2, 6, 24, 120, 720
> 1 4 18 96 600 Difference
> 2 3   4   5   6 Ratio
>
> a. neither
> b. 2,3,4,5,6 The ratio is increasing by 1.
> c. $a_n = a_{n-1} \cdot n$
> $a_7 = 720 \cdot 7 = 5040$
> $a_8 = 5040 \cdot 8 = 40,320$

The student correctly identified the sequence and the relationship of the ratios between terms. The student also wrote a correct rule and found the next two terms.

Level 2

> a. neither
> b. 2,3,4,5,6. The numbers are counting up by 1.
> c. $a_n = n(n+1)$
> $a_7 = 7(8) = 56$
> $a_8 = 8(9) = 72$

The student identified the sequence and described the ratios between terms; however, the rule and the terms in part **c** are incorrect.

Level 1

> a) Geometric
> b) 2,3,4,5 – It's a geometric sequence.
> c) 4320
>   25920

The student incorrectly identified the sequence, did not provide a rule, and found incorrect terms.

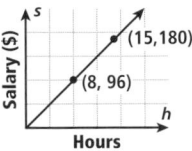
Cumulative Assessment, Chapters 1–12

1. What is the next term in the sequence?
   $1, 2, 4, 7, 11, \ldots$ **D**
   (A) 13   (C) 15
   (B) 14   (D) 16

2. A sequence is formed by doubling the preceding number: $2, 4, 8, 16, 32, \ldots$. What is the remainder when the 15th term of the sequence is divided by 6? **H**
   (F) 0   (H) 2
   (G) 1   (J) 4

3. Which equation describes the relationship shown in the graph? **B**

   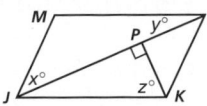

   (A) $h = 12s$ (C) $h = s + 88$
   (B) $s = 12h$ (D) $s = h + 88$

4. Which of the following is a solution of the system shown? $x > 3$
                                                               $x + y < 2$ **G**
   (F) $(4, -1)$ (H) $(5, 1)$
   (G) $(4, -3)$ (J) $(-5, 4)$

5. If $r = \frac{t}{5}$ and $10r = 32$, find the value of t. **C**
   (A) 64   (C) 16
   (B) 32   (D) 8

6. If $a = 3$ and $b = 4$, evaluate $b - ab^a$. **H**
   (F) 64   (H) -188
   (G) 8   (J) -1724

7. In parallelogram $JKLM$, $\overline{KP}$ is perpendicular to diagonal $\overline{JL}$. Which of the following is true? **C**

   (A) $x + y + z = 180$
   (B) $x + z = 90$
   (C) $y + z = 90$
   (D) $x + y = 90$

TEST TAKING TIP!
Reworking the given choices: It is sometimes useful to look at a choice in a form different from the given form.

8. If $2^{3x-1} = 8$, then what is the value of x? **H**
   (F) $\frac{2}{3}$ (H) $1\frac{1}{3}$
   (G) 2   (J) $2\frac{1}{3}$

9. **SHORT RESPONSE** The length of a rectangle is 8 ft less than twice its width w. Draw a diagram of the rectangle, and label each side length. What is the perimeter of the rectangle expressed in terms of w?

10. **SHORT RESPONSE** If two different numbers are selected at random from the set $\{1, 2, 3, 4, 5, 6\}$, what is the probability that their product will be 12? Show your work or explain in words how you determined your answer.

Purpose: To provide review and practice for Chapters 1–12 and standardized tests

Assessment Resources

Cumulative Tests (Levels A, B, C)
Assessment Resources . . . . pp. 277–288

State-Specific Test Practice Online
KEYWORD: MP4 TestPrep

Test Prep Doctor

For item 2, encourage students to divide each of the given terms by 6 to determine a pattern. They should notice that all of the terms in the pattern give a remainder of 2 or 4, eliminating choices **F** and **G**. All of the odd-numbered terms give a remainder of 2, so the 15th term will also give a remainder of 2, choice **H**.

Expand on the test-taking tip before item 8 by pointing out that in item 8, the value 8 is equal to 2^3. The equation can then be rewritten as $2^{3x-1} = 2^3$, which leads to the simpler equation $3x - 1 = 3$. The solution to this equation is $1\frac{1}{3}$, choice **H**.

Answers

9.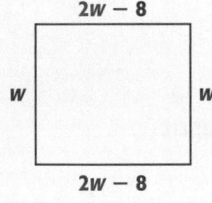

   Perimeter:
   $2(w) + 2(2w - 8)$
   $2w + 4w - 16$
   $6w - 16$

10. There are four ways two numbers will have a product of 12:
    $2 \times 6, 6 \times 2, 3 \times 4, 4 \times 3$

    There are 6 possible numbers to choose first and 5 possible numbers to choose second. There are $6 \cdot 5 = 30$ possible outcomes.
    $\frac{4}{30} = \frac{2}{15}$

Polynomials

Section 13A	Section 13B
Introduction to Polynomials	**Polynomial Operations**
Lesson 13-1 Polynomials	**Lesson 13-3** Adding Polynomials
Hands-On Lab 13A Model Polynomials	**Lesson 13-4** Subtracting Polynomials
Lesson 13-2 Simplifying Polynomials	**Lesson 13-5** Multiplying Polynomials by Monomials
	Hands-On Lab 13B Multiply Binomials
	Lesson 13-6 Multiplying Binomials
	Extension Dividing Polynomials by Monomials

Pacing Guide for 45-Minute Classes

Chapter 13

DAY 148	DAY 149	DAY 150	DAY 151	DAY 152
Lesson 13-1	Hands-On Lab 13A	Lesson 13-2	Mid-Chapter Quiz Lesson 13-3	Lesson 13-4

DAY 153	DAY 154	DAY 155	DAY 156	DAY 157
Lesson 13-5	Hands-On Lab 13B	Lesson 13-6	Extension	Chapter 13 Review

DAY 158
Chapter 13 Assessment

Pacing Guide for 90-Minute Classes

Chapter 13

DAY 74	DAY 75	DAY 76	DAY 77	DAY 78
Chapter 12 Assessment Lesson 13-1	Hands-On Lab 13A Lesson 13-2	Mid-Chapter Quiz Lesson 13-3 Lesson 13-4	Lesson 13-5 Hands-On Lab 13B	Lesson 13-6 Extension

DAY 79	DAY 80
Chapter 13 Review Lesson 14-1	Chapter 13 Assessment Lesson 14-2

Across the Series

COURSE 1

- Use the Distributive Property with whole-number exponents.
- Combine like terms to simplify first-degree algebraic expressions.
- Solve one-step first-degree equations and inequalities.

COURSE 2

- Use the Distributive Property with integer exponents.
- Combine like terms to simplify first-degree algebraic expressions.
- Solve multistep first-degree equations and inequalities.

PRE-ALGEBRA

- **Classify polynomials by degree and by the number of terms.**
- **Simplify polynomials.**
- **Add, subtract, multiply, and divide polynomials.**

Across the Curriculum

LANGUAGE ARTS LINK

Math: Reading and Writing in the Content Area . . . . pp. 100–105

Focus on Problem Solving
  Look Back . . . . SE p. 655
Journal . . . . TE, last page of each lesson
Write About It . . . . SE pp. 647, 659, 663, 667, 673

SOCIAL STUDIES LINK

Social Studies . . . . SE p. 642

SCIENCE LINK

Life Science . . . . SE pp. 652, 673
Physics . . . . SE p. 645
Health . . . . SE p. 667

TE = *Teacher's Edition*   **SE** = *Student Edition*

Interdisciplinary

Bulletin Board

Astronomy

Scientists are preparing to send a space probe to Mars. They estimate that the trip to Mars will require $0.2x^3 + 0.5x^2 + 0.4x + 64$ kilograms of fuel, where x represents the distance in millions of miles the probe will travel. The return trip is estimated to require $0.1x^3 + 0.7x^2 - 0.2x + 18$ kilograms of fuel. Write an expression for the amount of fuel in kilograms that the round-trip will require.

$0.3x^3 + 1.2x^2 + 0.2x + 82$ kg

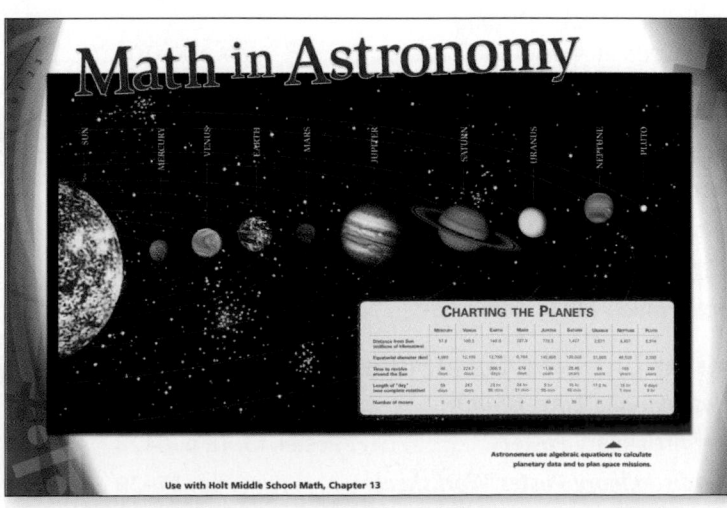

Interdisciplinary posters and worksheets are provided in your resource material.

Resource Options

Chapter 13 Resource Book

Student Resources

Teacher and Parent Resources

Reaching All Learners

English Language Learners

Individual Needs

Hands-On

Applications and Connections

Transparencies

Technology

Teacher Resources

Student Resources

 internet connect

Homework Help Online	**KEYWORD:** MP4 HWHelp13
Math Tools Online	**KEYWORD:** MP4 Tools
Glossary Online	**KEYWORD:** MP4 Glossary
Chapter Project Online	**KEYWORD:** MP4 PSProject13
Chapter Opener Online	**KEYWORD:** MP4 Ch13

KEYWORD: MP4 CNN13

SE = *Student Edition*   **TE** = *Teacher's Edition*   **AR** = *Assessment Resources*   **CRB** = *Chapter Resource Book*   **MK** = *Manipulatives Kit*

Assessment Options

Assessing Prior Knowledge

Determine whether students have the required prerequisite concepts and skills.

Are You Ready?. . . . . SE p. 643

Inventory Test. . . . . AR pp. 1–4

Test Preparation

Provide review and practice for chapter and standardized tests.

Standardized Test Prep. . . . . SE p. 685

Spiral Review with Test Prep . . . . . SE, last page of each lesson

Study Guide and Review . . . . SE pp. 680–682

Test Prep Tool Kit

Technology

 Test and Practice Generator CD-ROM

🔲 **internet** connect

State-Specific Test Practice Online KEYWORD: MP4 TestPrep

Performance Assessment

Assess students' understanding of chapter concepts and combined problem-solving skills.

Performance Assessment . . . . SE p. 684
    Includes scoring rubric in TE

Performance Assessment . . . . AR p. 142

Performance Assessment Teacher Support. . . . . AR p. 141

Portfolio

Portfolio opportunities appear throughout the Student and Teacher's Editions.

Suggested work samples:

Problem Solving Project . . . . TE p. 642

Performance Assessment . . . . SE p. 684

Portfolio Guide. . . . . AR p. xxxvii

Journal. TE, last page of each lesson

Write About It . . . . SE pp. 647, 659, 663, 667, 673

Daily Assessment

Obtain daily feedback on students' understanding of concepts.

Spiral Review and Test Prep . . . . . . SE, last page of each lesson

Also Available on Transparency In Chapter 13 Resource Book

Warm Up. . . . . TE, first page of each lesson

Problem of the Day. . . . . TE, first page of each lesson

Lesson Quiz. . . . . TE, last page of each lesson

Student Self-Assessment

Have students evaluate their own work.

Group Project Evaluation . . . . AR p. xxxiv

Individual Group Member Evaluation . . . . AR p. xxxv

Portfolio Guide. . . . . AR p. xxxvii

Journal. . . . . TE, last page of each lesson

Formal Assessment

Assess students' mastery of concepts and skills.

Section Quizzes . . . . AR pp. 29–30

Mid-Chapter Quiz. . . . . SE p. 654

Chapter Test . . . . SE p. 683

Chapter Tests (Levels A, B, C) . . . . AR pp. 105–110

Cumulative Tests (Levels A, B, C) . . . . AR pp. 289–300

Standardized Test Prep
    Cumulative Assessment . . . . SE p. 685

End-of-Year Test. . . . . AR pp. 313–316

Technology

 Test and Practice Generator CD-ROM

Make tests electronically. This software includes:

- Dynamic practice for Chapter 13
- Customizable tests
- Multiple-choice items for each objective
- Free-response items for each objective
- Teacher management system

SE = *Student Edition*   **TE** = *Teacher's Edition*   **AR** = *Assessment Resources*   **CRB** = *Chapter Resource Book*   **MK** = *Manipulatives Kit*

Chapter 13 Tests

Three levels (A,B,C) of tests are available for each chapter in the *Assessment Resources.*

LEVEL A

CHAPTER Chapter Test
13 *Form A*

Classify each expression as monomial, binomial, trinomial, or not a polynomial.

1. $5x^2$

    monomial

2. $9y^2$

    monomial

3. $3a^2 + b^2$

    binomial

4. $7x^2 + x^3 - 8x$

    trinomial

Find the degree of each polynomial.

5. $3x^2 + 4x$

    2

6. $4n^3 + 2n^8$

    8

7. $9a^5 - 3a^3 - 5a^2$

    5

8. $k^3 + k^4 + k^7 + k$

    7

Identify the like terms in each polynomial.

9. $4n^2 + 3 + 5n^2$

    $4n^2$ and $5n^2$

10. $6x^3 - 5x^2 + 5x^3$

    $6x^3$ and $5x^3$

Simplify.

11. $7y^2 + 3y - 4y^2$

    $3y^2 + 3y$

12. $4n^2 + 8 + 8n^2$

    $12n^2 + 8$

13. $5t^2 + t^2 - 3s^2$

    $6t^2 - 3s^2$

Add.

14. $(6x + 2) + (5x + 2)$

    $11x + 4$

15. $(8n + 8) + (6n - 5)$

    $14n + 3$

16. $(4x + 2x^2) + (4x^2 + 2x)$

    $6x + 6x^2$

17. $(5n^2 + 3n) + (2n^2 - 2n)$

    $7n^2 + n$

18. $(8a^3 - 6b) + (9b - 4a^3)$

    $4a^3 + 3b$

19. $(2x - 2x^3) + (8x^3 + 4x)$

    $6x + 6x^3$

20. $(7xy + 4x) + (3xy - 2x)$

    $10xy + 2x$

CHAPTER Chapter Test
13 *Form A, continued*

Find the opposite of each polynomial.

23. $5n^3$

    $-5n^3$

24. $6x^4y^4$

    $-6x^4y^4$

Subtract.

27. $5x - (2x - 4x^2)$

    $3x + 4x^2$

28. $(6n + 4n^2) - (2n^2 + 4n)$

    $2n + 2n^2$

29. $(4x^2 + 2y^2) - (3x^2 - 9y^2)$

    $x^2 + 11y^2$

30. $(8y^2 - 4) - (-5y^2 - 8)$

    $13y^2 + 4$

31. $(7n^2 - 5n) - (-7n + 5n^2)$

    $2n^2 + 2n$

32. $(9x + 7 + 10x^4) - (2x + 3 - 5x^4)$

    $7x + 4 + 15x^4$

33. $(4n - 2n^2 + 6a^3) - (-8n^2 - 6a^2)$

    $4n + 6n^2 + 12a^2$

Multiply.

34. $(3x^2)(5x^2)$

    $15x^4$

35. $(2n^2m^3)(3n^3m^2)$

    $6n^5m^5$

36. $5x(4x^2 + 8)$

    $20x^3 + 40x$

37. $9x^2(2x - 6x^3)$

    $18x^3 - 54x^5$

38. $3ab^2(2a^3b^3 + 2ab + 5b^2)$

    $6a^4b^5 + 6a^2b^3 + 15ab^4$

39. $(n + 3)(n + 4)$

    $n^2 + 7n + 12$

40. $(x + 2)(y + 7)$

    $xy + 7x + 2y + 14$

41. $(n + 6)(n - 6)$

    $n^2 - 36$

42. $(y + 5)(y - 4)$

    $y^2 + y - 20$

43. $(2a + b)(a - 4b)$

    $2a^2 - 7ab - 4b^2$

44. A rectangle has a length of $2w + 3$ and a width of $w - 5$. Write and simplify an expression for the area of the rectangle.

    $2w^2 - 7w - 15$

LEVEL B

CHAPTER Chapter Test
13 *Form B*

Classify each expression as monomial, binomial, trinomial, or not a polynomial.

1. $6a^2 - 4a$

    binomial

2. $\left(\frac{1}{7}\right)x^5y^3$

    monomial

3. $8p^2 + p^{22}$

    not a polynomial

4. $4m^2n^3 + 5n^3 - 4n$

    trinomial

Find the degree of each polynomial.

5. $8x^5 + 6x^4$

    5

6. $-6a^3 + 8a - 2a^5$

    5

7. $5 + 3m^4 - 2m^7 + m^8$

    8

8. $b^3 - b^9 - 3 - 3b^2$

    9

Identify the like terms in each polynomial.

9. $t^2 + 9s^3 - 5t^4 + 8 + 4s^3$

    $9s^3$ and $4s^3$

10. $4a^3b^2 - 5a^2b^2 + a^3b^2$

    $4a^3b^2$ and a^3b^2

Simplify.

11. $v^2 + 6v - 8v^2 + 4v^2$

    $-3v^2 + 6v$

12. $-4x^2 + 9 + 5x^2$

    $x^2 + 9$

13. $5m^2n^2 + n^3 - 3m^2n^2 + 7n^3$

    $2m^2n^2 + 8n^3$

Add.

14. $(13x - 4) + (5x + 5)$

    $18x + 1$

15. $(6a + 9) + (6a^2 - 5a - 3)$

    $a + 6a^2 + 6$

16. $(3x + 2x^3 - 7) + (4x^3 - 2x + 7)$

    $x + 6x^3$

17. $(6z^4 + 7y^3 + 4) + (2y^3 - z^4 - 3)$

    $5z^4 + 9y^3 + 1$

18. $(7r^3 + 7s + 7r^2) + (4s - 4r^3)$

    $3r^3 + 11s + 7r^2$

19. $(5x - 2x^3 - 9) + (8x^3 - 3x - 5)$

    $2x + 6x^3 - 14$

20. $(7ab + 6b^2 - 5a) + (ab - 3a)$

    $8ab + 6b^2 - 8a$

CHAPTER Chapter Test
13 *Form B, continued*

Find the opposite of each polynomial.

23. $9a^5b^5c^3$

    $-9a^5b^5c^3$

24. $9y^3 - 6x$

    $-9y^3 + 6x$

Subtract.

27. $8b - (8a^2 + 7b - 5)$

    $-8a^2 + b + 5$

28. $(4x + 5x^2) - (3x^2 + x)$

    $3x + 2x^2$

29. $(x^3 - 5x + 2x^2) - (3x - 3x^3 + 9)$

    $4x^3 + 3x + 2x^2 - 9$

30. $(6y^2z^6 + 3yz - 4z^2) - (-5yz - 9)$

    $6y^2z^6 + 8yz - 4z^2 + 9$

31. $(n^4 + 9n^3m^7 - 5n) - (-6n - n^3m^7)$

    $n^4 + 10n^3m^7 + n$

32. $(6x - 6 + 4) - (4 - 2x - 5x^3)$

    $8x - 10 + 6x^3$

33. $(-5s + 7s^2 + 4r^2 - 4) - (-3s^2 - 6s)$

    $s + 10s^2 + 4r^2 - 4$

Multiply.

34. $(5x^2)(2y^7)(4x^3)(2y^3)$

    $80x^5y^{10}$

35. $(9a^2b^5)(6ab^4)$

    $54a^3b^9$

36. $6n(3n^2 - 7)$

    $18n^3 - 42n$

37. $-4ab(5a^2 - 7b^3)$

    $-20a^3b + 28ab^4$

38. $6st^2(7s^3t^3 + st - 5t^2)$

    $42s^4t^5 + 6s^2t^3 - 30st^4$

39. $(a - 3)(y + 5)$

    $ay + 5a - 3y - 15$

40. $(x + 4)(x + 5)$

    $x^2 + 9x + 20$

41. $(b + 6)^2$

    $b^2 + 12b + 36$

42. $(y + 5)(y - 4)$

    $y^2 + y - 20$

43. $(5y - 7)^2$

    $25y^2 - 70y + 49$

44. A square has a side length of $3e + 4$. Write and simplify an expression for the area of the square.

    $9e^2 + 24e + 16$

LEVEL C

CHAPTER Chapter Test
13 *Form C*

Classify each expression as monomial, binomial, trinomial, or not a polynomial.

1. $9x^2 - \frac{7}{x^2}$

    not a polynomial

2. 88

    monomial

3. $5n^{11} + m^2 - nm$

    trinomial

4. $6a^2b^2 + \left(\frac{1}{2}b\right)n^{3.1} + 8n$

    not a polynomial

Find the degree of each polynomial.

5. $-2x + 8x$

    1

6. $-10n^8 + 10n^8 + n^8$

    8

7. $4 + 5y^3 - 8y^2 - 4y^9$

    9

8. $35j^4 + 28j^{12} - 15j^7$

    12

Identify the like terms in each polynomial.

9. $11x^2 + 11y^2 - t^4 + 9y^2 + 4x^{22}$

    $11y^2$ and $9y^2$

10. $g^2h^2 - 8gh^2 + g^5h + 8g^2h^2$

    g^2h^2 and $8g^2h^2$

Simplify.

11. $3(5x^2 + 3y) - 4x^2 + 4y^2$

    $11x^2 + 9y + 4y^2$

12. $5n^2 - 10m^3 - 4n^2 + m^3 + n^2$

    $2n^2 - 9m^3$

13. $3(3a^2b^2 + b^3) - 3a^2b^2 + 2(7a^3 + 4b^3)$

    $6a^2b^2 + 11b^3 + 14a^3$

Add.

14. $(3x^3 - 4x) + (-2x + 5 - x^3)$

    $2x^3 - 6x + 5$

15. $(9a + 5b^2) + (3a^2 - 2a) + (6b^2 - 5a^2)$

    $7a + 11b^2 - 2a^2$

16. $(4x + x^3) + (4x^3 - x + 7) + (3x + 9x^3)$

    $6x + 14x^3 + 7$

17. $(2a^3 + 6b^3 + 7) + (5b^3 - a^3b^4 - 3a^3)$

    $-a^3 + 11b^3 - a^3b^4 + 7$

18. $(6m^3n^3 + 2m + 12m^3n^3) + (-8m - n^3)$

    $18m^3n^3 - 6m - n^3$

19. $(xy^3 - 2x^3 - 9xy^3) + (8xy^3 - 2x^3 - 2)$

    $-4x^3 - 2$

20. $(7cd + 6d^6) + (cd - 3c) + (-8d - d^6)$

    $8cd + 5d^6 - 3c - 8d$

CHAPTER Chapter Test
13 *Form C, continued*

Find the opposite of each polynomial.

23. $12x^8yz^3 - 7xy + 23$

    $-12x^8yz^3 + 7xy - 23$

24. $-11a^3 - 6ab^3 + 12a^4b^3$

    $11a^3 + 6ab^3 - 12a^4b^3$

Subtract.

27. $(5b + 2a^2) - (9a^2 + 12b - 8)$

    $-7b - 7a^2 + 8$

28. $(4n^2 + 10n^5m^5 - 14n) - (3n - 8n^5m^5)$

    $4n^2 + 18n^5m^5 - 17n$

29. $(6r^4s^4 + 11r^2 + 9r^2s^2) - (14r^2s^2 - 5r^2)$

    $6r^4s^4 + 16r^2 - 5r^2s^2$

30. $(12y^2z - 9yz - 4z^5) - (-5yz - 9y^2z)$

    $21y^2z - 4yz - 4z^5$

31. $(9s - 13s^2 - r^2 + 4) - (5r^2 + s - 18s^2)$

    $8s + 5s^2 - 6r^2 + 4$

32. $(13rs^8 + 9s^4 - 3r^5s^2) - (-4r^5s^2 + 9s^4)$

    $13rs^8 + r^5s^2$

33. $(6yz^5 + 16yz - 4z^5) - (5yz^5 - 9z^5 + z)$

    $yz^5 + 16yz + 5z^5 - z$

Multiply.

34. $8x^2(5x^2y^2 + 9x^3y^3)$

    $40x^4y^2 + 72x^5y^3$

35. $(-rs^5)(7r^5s)$

    $-7r^6s^6$

36. $6x^8y(6x^2y^3 - 7xy^4 + 12x^3)$

    $36x^8y^4 - 42x^7y^5 + 72x^9y$

37. $13fg(-9g^2 - 5f^3 + f^5g^2)$

    $-117fg^3 - 65f^4g + 13f^5g^3$

38. $10x^2y^3(8xy - 5xy^3 + 4x^3y)$

    $80x^3y^3 - 50x^3y^5 + 40x^5y^3$

39. $(9x + 2)(x + 5)$

    $9x^2 + 47x + 10$

40. $(2x + 4)(4x - 6)$

    $8x^2 + 4x - 24$

41. $(5a - 4)^2$

    $25a^2 - 40a + 16$

42. $(6y + x)(y - 8x)$

    $6y^2 - 47xy - 8x^2$

43. $(8r - 6s)(4r - 10s)$

    $32r^2 - 104rs + 60s^2$

44. A parallelogram has a base length of $5p - 1$ and a height of $2p + 3$. Write and simplify an expression for the area of the parallelogram.

    $10p^2 + 13p - 3$

Test and Practice Generator
CD-ROM

Create and customize multiple versions of the same tests with corresponding answers for any chosen chapter objectives.

Chapter 13 State and Standardized Test Preparation

Test Taking Skill Builder and Standardized Test Practice
are provided for each chapter in the *Test Prep Tool Kit.*

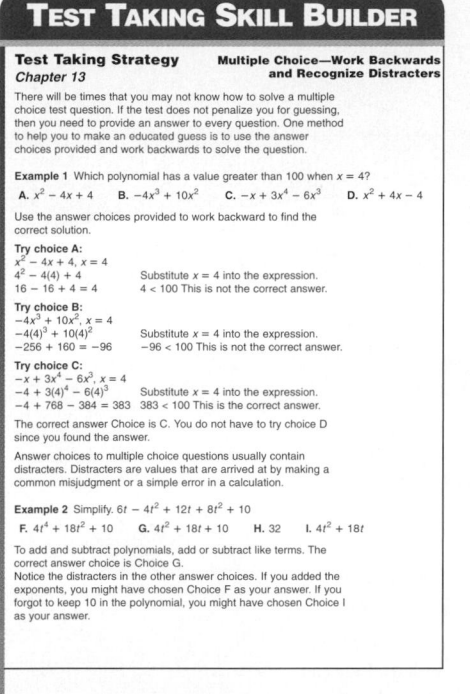

TEST TAKING SKILL BUILDER

Test Taking Strategy
Chapter 13

Multiple Choice—Work Backwards and Recognize Distracters

There will be times that you may not know how to solve a multiple choice test question. If the test does not penalize you for guessing, then you need to provide an answer to every question. One method to help you to make an educated guess is to use the answer choices provided and work backwards to solve the question.

Example 1 Which polynomial has a value greater than 100 when $x = 4$?

A. $x^2 - 4x + 4$ B. $-4x^3 + 10x^2$ C. $-x + 3x^4 - 6x^3$ D. $x^2 + 4x - 4$

Use the answer choices provided to work backward to find the correct solution.

Try choice A:
$x^2 - 4x + 4, x = 4$
$4^2 - 4(4) + 4$ Substitute $x = 4$ into the expression.
$16 - 16 + 4 = 4$ $4 < 100$ This is not the correct answer.

Try choice B:
$-4x^3 + 10x^2, x = 4$
$-4(4)^3 + 10(4)^2$ Substitute $x = 4$ into the expression.
$-256 + 160 = -96$ $-96 < 100$ This is not the correct answer.

Try choice C:
$-x + 3x^4 - 6x^3, x = 4$
$-4 + 3(4)^4 - 6(4)^3$ Substitute $x = 4$ into the expression.
$-4 + 768 - 384 = 383$ $383 < 100$ This is the correct answer.

The correct answer Choice is C. You do not have to try choice D since you found the answer.

Answer choices to multiple choice questions usually contain distracters. Distracters are values that are arrived at by making a common misjudgment or a simple error in a calculation.

Example 2 Simplify. $6t - 4t^2 + 12t + 8t^2 + 10$

F. $4t^4 + 18t^2 + 10$ G. $4t^2 + 18t + 10$ H. 32   I. $4t^2 + 18t$

To add and subtract polynomials, add or subtract like terms. The correct answer choice is Choice G.
Notice the distracters in the other answer choices. If you added the exponents, you might have chosen Choice F as your answer. If you forgot to keep 10 in the polynomial, you might have chosen Choice I as your answer.

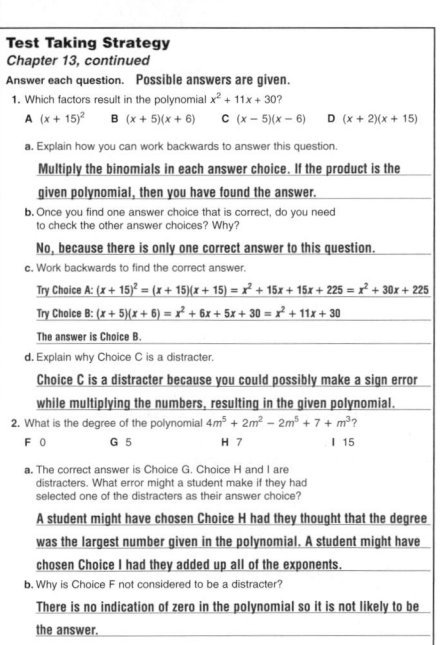

Test Taking Strategy
Chapter 13, continued

Answer each question. **Possible answers are given.**

1. Which factors result in the polynomial $x^2 + 11x + 30$?

   A $(x + 15)^2$ B $(x + 5)(x + 6)$ C $(x - 5)(x - 6)$ D $(x + 2)(x + 15)$

   a. Explain how you can work backwards to answer this question.

   Multiply the binomials in each answer choice. If the product is the given polynomial, then you have found the answer.

   b. Once you find one answer choice that is correct, do you need to check the other answer choices? Why?

   No, because there is only one correct answer to this question.

   c. Work backwards to find the correct answer.

   Try Choice A: $(x + 15)^2 = (x + 15)(x + 15) = x^2 + 15x + 15x + 225 = x^2 + 30x + 225$
   Try Choice B: $(x + 5)(x + 6) = x^2 + 6x + 5x + 30 = x^2 + 11x + 30$
   The answer is Choice B.

   d. Explain why Choice C is a distracter.

   Choice C is a distracter because you could possibly make a sign error while multiplying the numbers, resulting in the given polynomial.

2. What is the degree of the polynomial $4m^5 + 2m^2 - 2m^5 + 7 + m^3$?

   F 0   G 5   H 7   I 15

   a. The correct answer is Choice G. Choice H and I are distracters. What error might a student make if they had selected one of the distracters as their answer choice?

   A student might have chosen Choice H had they thought that the degree was the largest number given in the polynomial. A student might have chosen Choice I had they added up all of the exponents.

   b. Why is Choice F not considered to be a distracter?

   There is no indication of zero in the polynomial so it is not likely to be the answer.

STANDARDIZED TEST PRACTICE

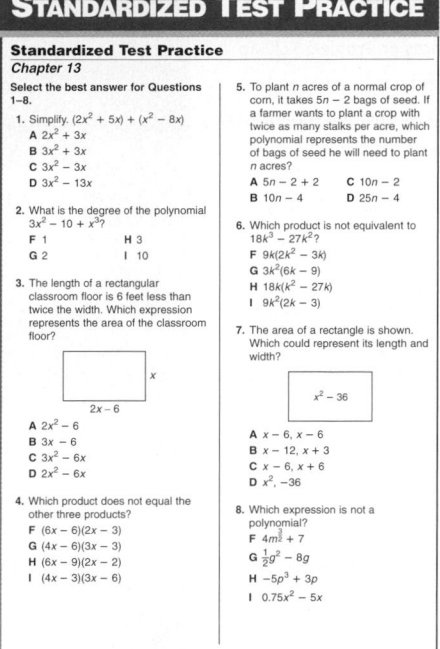

Standardized Test Practice
Chapter 13

Select the best answer for Questions 1–8.

1. Simplify. $(2x^2 + 5x) + (x^2 - 8x)$

   A $2x^2 + 3x$
   B $3x^2 + 3x$
   C $3x^2 - 3x$
   D $3x^2 - 13x$

2. What is the degree of the polynomial $3x^2 - 10 + x^3$?

   F 1   H 3
   G 2   I 10

3. The length of a rectangular classroom floor is 6 feet less than twice the width. Which expression represents the area of the classroom floor?

   A $2x^2 - 6$
   B $3x - 6$
   C $3x^2 - 6x$
   D $2x^2 - 6x$

4. Which product does not equal the other three products?

   F $(6x - 6)(2x - 3)$
   G $(4x - 6)(3x - 3)$
   H $(6x - 9)(2x - 2)$
   I $(4x - 3)(3x - 6)$

5. To plant n acres of a normal crop of corn, it takes $5n - 2$ bags of seed. If a farmer wants to plant a crop with twice as many stalks per acre, which polynomial represents the number of bags of seed he will need to plant n acres?

   A $5n - 2 + 2$ C $10n - 2$
   B $10n - 4$ D $25n - 4$

6. Which product is not equivalent to $18k^3 - 27k^2$?

   F $9k(2k^2 - 3k)$
   G $3k^2(6k - 9)$
   H $18k(k^2 - 27k)$
   I $9k^2(2k - 3)$

7. The area of a rectangle is shown. Which could represent its length and width?

   $x^2 - 36$

   A $x - 6, x - 6$
   B $x - 12, x + 3$
   C $x - 6, x + 6$
   D $x^2, -36$

8. Which expression is not a polynomial?

   F $4m^{\frac{3}{2}} + 7$
   G $\frac{1}{2}g^2 - 8g$
   H $-5p^3 + 3p$
   I $0.75x^2 - 5x$

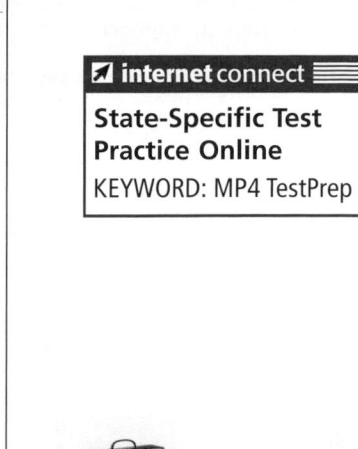

internet connect

go hrw .com

State-Specific Test Practice Online
KEYWORD: MP4 TestPrep

Standardized Test Practice
Chapter 13, continued

Gridded Response
Solve the problems. Use the answer sheet to write and grid-in your answer.

9. A group of tourists is taking a sightseeing trip around the city. The tour bus leaves the station at noon. The binomial $40t - 10t^2$ describes the bus's distance in miles from the station after it has traveled t hours. How far is the bus from the station at 3:00 P.M.?

10. If the cost in dollars of producing k kites is $18,000 + 2k$ and the revenue generated from sales is $400k - 2k^2$, what is the profit from making and selling 100 kites?

Short Response

11. Determine whether $12(a^2 + 3a) = 4(3a^2 + 9a)$. Explain in words how you used the Distributive Property to arrive at your conclusion.

12. Multiply: $(x + 6)(3x - 2)$. Explain in words how FOIL works when multiplying two binomials.

13. Joe wants to pour the contents of boxes 1 and 2 into box 3. If V represents volume, will they both fit? Describe in words how you arrived at your conclusion.

   Box 1
   $V = x^3 + 4x$
   $V = 2x^3$

   Box 2

   Box 3
   $V = 3x^3 + 4x + 6$

Extended Response

14. Triangle ABC is an isosceles triangle with side lengths of $4x + 2$, $4x + 2$, and $x - 1$.

   a. Draw and label a diagram to represent the information.

   b. Use your diagram to find a polynomial expression for the perimeter of triangle ABC. Calculate the perimeter of triangle ABC when $x = 8$.

   c. If the area of triangle ABC is 27 square units when $x = 7$, what is the height? Use an equation and show your work.

eet
D See Lesson 13-2.
I See Lesson 13-5.
D See Lesson 13-6.
I See Lesson 13-1.

Test Prep Tool Kit

- Standardized Test Prep Workbook
- Countdown to Testing transparencies
- State Test Prep CD-ROM
- Standardized Test Prep Video

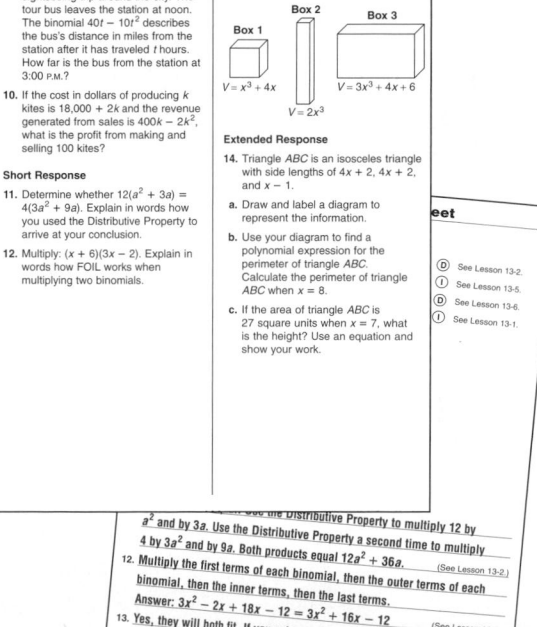

... use the Distributive Property to multiply 12 by a^2 and by $3a$. Use the Distributive Property a second time to multiply 4 by $3a^2$ and by $9a$. Both products equal $12a^2 + 36a$.
(See Lesson 13-2.)
12. Multiply the first terms of each binomial, then the outer terms of each binomial, then the inner terms, then the last terms.
Answer: $3x^2 - 2x + 18x - 12 = 3x^2 + 16x - 12$
(See Lesson 13-6.)
13. Yes, they will both fit. If you subtract the volumes of boxes 1 and 2 from the volume of box 3, 6 is left over, so there is room to spare.
(See Lesson 13-4.)

Extended Response
Write your answers for Problem 14 on the back of this paper.
See Lesson 13-3.

Customized answer sheets give students realistic practice for actual standardized tests.

Polynomials

Polynomials

Why Learn This?

Tell students that polynomial expressions are used in many real-world applications. The total production cost of a CD depends on the cost of the materials needed to produce it, the cost of labor, the payment made to the artist, and many other factors. These fixed and variable costs might be written into an expression that a record company could use to determine the selling price of the CD. The expression might take the form of a polynomial.

Using Data

To begin the study of this chapter, have students:

- Discuss the difference between fixed and variable costs. Fixed costs do not change with the number of CDs produced; variable costs do.

- Write an expression to show the cost of producing *n* CDs.
  $197 + 0.88n$

- Use the expression to find the cost of producing 1000 CDs.
  $197 + 0.88(1000) = \$1077$

internet connect

Chapter Opener Online
go.hrw.com
KEYWORD: MP4 Ch13

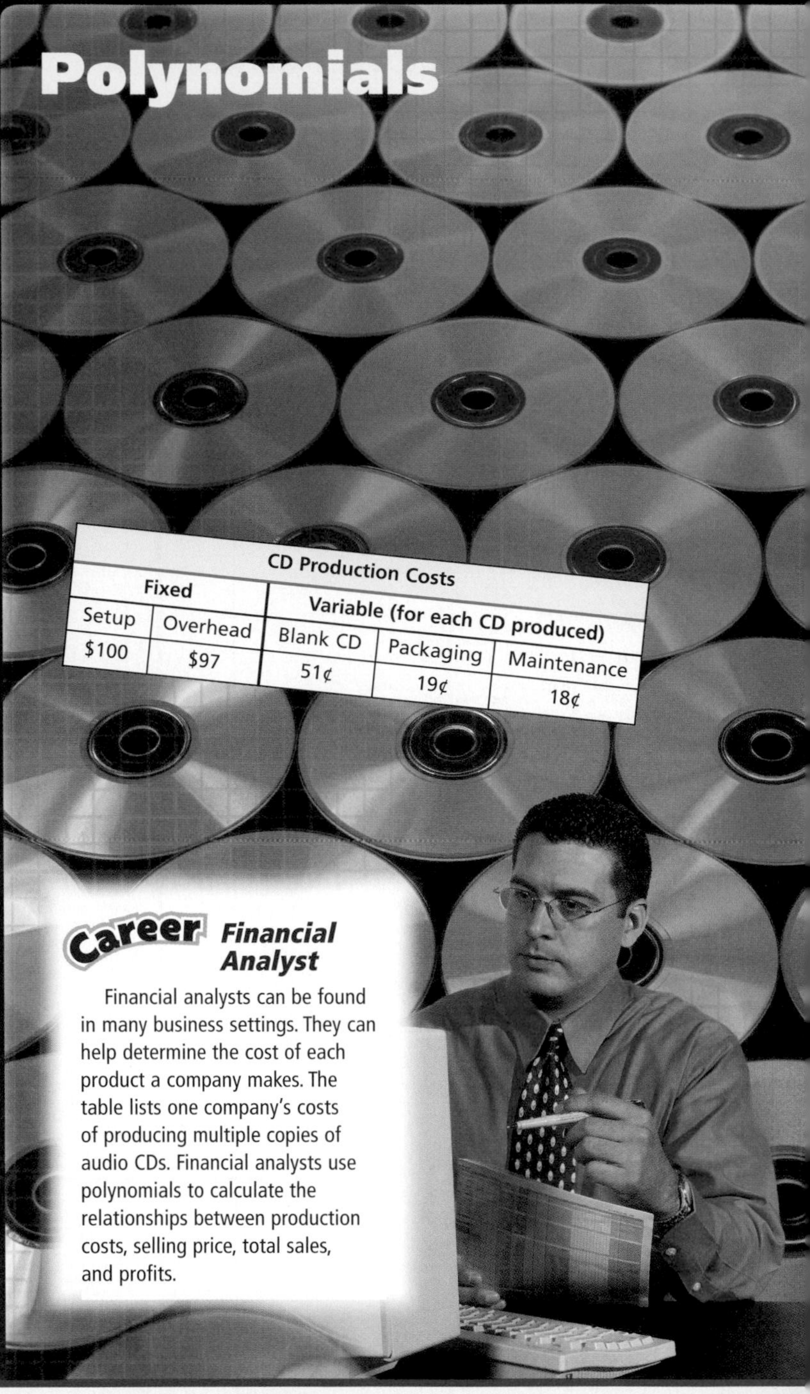

CD Production Costs				
Fixed		**Variable (for each CD produced)**		
Setup	Overhead	Blank CD	Packaging	Maintenance
$100	$97	51¢	19¢	18¢

Career *Financial Analyst*

Financial analysts can be found in many business settings. They can help determine the cost of each product a company makes. The table lists one company's costs of producing multiple copies of audio CDs. Financial analysts use polynomials to calculate the relationships between production costs, selling price, total sales, and profits.

Problem Solving Project

Social Studies Connection

Purpose: To use algebra and polynomials to solve problems

Materials: What Did We Make? worksheet

Understand, Plan, Solve, and Look Back

Have students:

✔ Complete the What Did We Make? worksheet to learn more about algebra and polynomials.

✔ Write this statement as an algebraic equation: Total cost equals fixed costs plus variable costs times the number of units. Does this equation contain any polynomials? Why or why not?

✔ Explain which variables affect the fixed costs of a company.

✔ Explain which aspects are affected by the number of CDs that are produced at one time.

✔ Check students' work.

internet connect

Chapter Project Online: *go.hrw.com*
KEYWORD: MP4 PSProject13

ARE YOU READY?

Choose the best term from the list to complete each sentence.

1. __?__ have the same variables raised to the same powers. **like terms**
2. In the expression $4x^2$, 4 is the __?__. **coefficient**
3. $5 + (4 + 3) = (5 + 4) + 3$ by the __?__. **Associative Property**
4. $3 \cdot 2 + 3 \cdot 4 = 3(2 + 4)$ by the __?__. **Distributive Property**

Associative Property

coefficient

Distributive Property

like terms

Complete these exercises to review skills you will need for this chapter.

✔ **Subtract Integers**

Subtract.

5. $12 - 4$ **8**
6. $8 - 10$ **−2**
7. $14 - (-4)$ **18**
8. $-9 - 5$ **−14**
9. $-9 - (-5)$ **−4**
10. $9 - (-5)$ **14**

✔ **Exponents**

Multiply. Write each product as one power.

11. $3^4 \cdot 3^6$ **3^{10}**
12. $10^2 \cdot 10^3$ **10^5**
13. $x \cdot x^5$ **x^6**
14. $5^5 \cdot 5^5$ **5^{10}**
15. $y^2 \cdot y^6$ **y^8**
16. $z^3 \cdot z^3$ **z^6**
17. $a^2 \cdot a$ **a^3**
18. $b \cdot b$ **b^2**

✔ **Distributive Property**

Rewrite using the Distributive Property.

19. $5(7 + 8)$
   $5 \cdot 7 + 5 \cdot 8$
20. $3(x + y)$
   $3x + 3y$
21. $(a + b)6$
   $6a + 6b$
22. $(r + s)4$
   $4r + 4s$

✔ **Area**

Find the area of the shaded portion in each figure.

23. 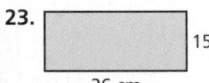 15 cm, 36 cm
24. 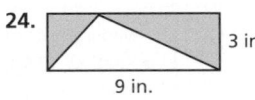 3 in., 9 in.
25. 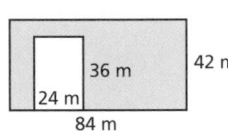 36 m, 24 m, 84 m, 42 m

26. 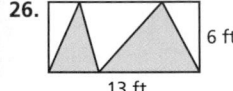 6 ft, 13 ft
27. 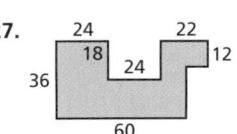 24, 18, 22, 24, 36, 60, 12
28. 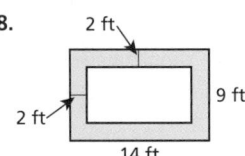 2 ft, 2 ft, 9 ft, 14 ft

ARE YOU READY?
Were students successful with Are You Ready?

NO INTERVENE ⟵ ⟶ **YES** ENRICH

✔ **Subtract Integers** 5–10
Intervention Practice, Skill 47
CD-ROM Intervention Activities, Skill 47

✔ **Distributive Property** 19–22
Intervention Practice, Skill 49
CD-ROM Intervention Activities, Skill 49

Are You Ready? Enrichment, pp. 429–430

✔ **Exponents** 11–18
Intervention Practice, Skill 12
CD-ROM Intervention Activities, Skill 12

✔ **Area** 23–28
Intervention Practice, Skill 85
CD-ROM Intervention Activities, Skill 85

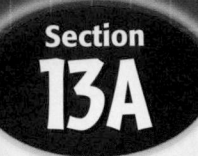

Introduction to Polynomials

One-Minute Section Planner

Lesson	Materials	Resources

Lesson 13-1 Polynomials
NCTM: Algebra, Communication
☐ SAT-9 ☐ SAT-10 ☐ ITBS ☑ CTBS ☐ MAT ☑ CAT
	Optional
	Recording Sheet for
	Reaching All Learners
	(CRB, p. 67)
		• *Chapter 13 Resource Book,* pp. 6–14
		• Daily Transparency T1, CRB
		• Additional Examples Transparencies
		  T2–T4, CRB
		• *Alternate Openers: Explorations,*
		  p. 100

Hands-On Lab 13A Model Polynomials **NCTM:** Algebra, Communication, Representation ☐ SAT-9 ☐ SAT-10 ☐ ITBS ☐ CTBS ☐ MAT ☑ CAT	**Required** Algebra tiles *(MK)*	• *Hands-On Lab Activities,* pp. 85–86

Lesson 13-2 Simplifying Polynomials
NCTM: Algebra, Communication
☐ SAT-9 ☐ SAT-10 ☐ ITBS ☑ CTBS ☐ MAT ☑ CAT
	Optional
	Teaching Transparency T6
	  (CRB)
	Algebra tiles *(MK)*
	Coins *(MK or CRB, p. 68)*
	Monomial Cards for
	  Reaching All Learners
	  (CRB, p. 69)
		• *Chapter 13 Resource Book,* pp. 15–24
		• Daily Transparency T5, CRB
		• Additional Examples Transparencies
		  T7–T9, CRB
		• *Alternate Openers: Explorations,*
		  p. 101

Section 13A Assessment		• Mid-Chapter Quiz, SE p. 654 • Section 13A Quiz, AR p. 29 • *Test and Practice Generator* CD-ROM

SAT = *Stanford Achievement Tests*   **ITBS** = *Iowa Test of Basic Skills*   **CTBS** = *Comprehensive Test of Basic Skills/Terra Nova*
MAT = *Metropolitan Achievement Test*   **CAT** = *California Achievement Test*

NCTM = Complete standards can be found on pages T29–T35.   **NAEP** = Complete standards can be found on pages A54–A58.

SE = *Student Edition*   **TE** = *Teacher's Edition*   **AR** = *Assessment Resources*   **CRB** = *Chapter Resource Book*   **MK** = *Manipulatives Kit*

Section Overview

Polynomials — Lesson 13-1, Hands-On Lab 13A

Why? **Polynomial** expressions are the building blocks of polynomial functions, which are used to model, represent, and analyze many real-world situations.

A **monomial** is a number or a product of numbers and variables with exponents that are whole numbers.

Examples	
Monomials	$2n$, x^3, $4a^4b^3$, 7
Not Monomials	$p^{2.4}$, 2^x, $\sqrt{x}$, $\frac{5}{g^2}$

A **polynomial** is one monomial or the sum or difference of monomials.

Polynomials	
Monomial (1 term)	$10ab^2$
Binomial (2 terms)	$9x^2 + 2$
Trinomial (3 terms)	$2a^2 + 3a - 5$

The **degree** of a polynomial is the degree of the term with the greatest degree.

$$4x^2 \quad + \quad 2x^5 \quad + \quad x \quad + \quad 5$$

Degree 2 Degree 5 Degree 1 Degree 0

Degree 5

Simplifying Polynomials — Lesson 13-2

Why? In order to solve polynomial equations, you need to know how to simplify polynomials.

Like terms have the same variables raised to the same powers.

Example: $\mathbf{5x^2y} + 2xy^2 + \mathbf{6x^2y} + 7y$

Like terms

To **simplify a polynomial,** add or subtract like terms. You may need to use the Distributive Property to simplify a polynomial.

$$2(3ab^2 - 6b) + 2ab^2 + 5$$

$$\mathbf{2} \cdot 3ab^2 - \mathbf{2} \cdot 6b + 2ab^2 + 5 \qquad \textit{Distributive Property}$$

$$\mathbf{6ab^2} - 12b + \mathbf{2ab^2} + 5$$

$$8ab^2 - 12b + 5 \qquad \textit{Combine like terms.}$$

Objective: Students classify polynomials by degree and by the number of terms.

Warm Up

Identify the base and exponent of each power.

1. 3^4 3; 4   **2.** 2^a 2; a **3.** x^5 x; 5

Determine whether each number is a whole number.

4. 0 yes   **5.** -3 no   **6.** 5 yes

Problem of the Day

If you take a whole number n, raise it to the third power, and then divide the result by n, what is the resulting expression? n^2

Available on Daily Transparency in CRB

There are different mathematical meanings of the word *degree*. It is used as a unit of measure for temperature, as a unit of measure for angles, and as a means to classify polynomials.

 **13-1 Polynomials**

Learn to classify polynomials by degree and by the number of terms.

Vocabulary

monomial

polynomial

binomial

trinomial

degree of a polynomial

Some fireworks shows are synchronized to music for dramatic effect. *Polynomials* are used to compute the exact height of each firework when it explodes.

The simplest type of polynomial is called a *monomial*. A **monomial** is a number or a product of numbers and variables with exponents that are whole numbers.

Monomials	$2n$, x^3, $4a^4b^3$, 7
Not monomials	$p^{2.4}$, 2^x, $\sqrt{x}$, $\frac{5}{g^2}$

 EXAMPLE 1 **Identifying Monomials**

Determine whether each expression is a monomial.

A $\frac{1}{2}x^2y^5$

monomial

2 and 5 are whole numbers.

B $12xy^{0.4}$

not a monomial

0.4 is not a whole number.

A **polynomial** is one monomial or the sum or difference of monomials. Polynomials can be classified by the number of terms. A monomial has 1 term, a **binomial** has 2 terms, and a **trinomial** has 3 terms.

 EXAMPLE 2 **Classifying Polynomials by the Number of Terms**

Classify each expression as a monomial, a binomial, a trinomial, or not a polynomial.

A $49.99h + 24.99g$

binomial

Polynomial with 2 terms

B $-3x^4y$

monomial

Polynomial with 1 term

C $4x^2 - 2xy + \frac{3}{x}$

not a polynomial

A variable is in the denominator.

D $5mn + 2m - 3n$

trinomial

Polynomial with 3 terms

---

1 Introduce

Alternate Opener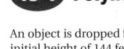

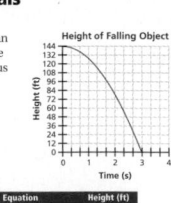

EXPLORATION

13-1 Polynomials

An object is dropped from an initial height of 144 feet. The graph shows its height versus time.

Height of Falling Object

Time (s) x	Equation $y = 144 - 16x^2$	Height (ft) y
0	$y = 144 - 16(0)^2 = 144$	144
1		
2		
3		

1. What does the point (0, 144) represent?
2. When does the object reach the ground? How can you tell?
3. You can use the equation $y = 144 - 16x^2$ to model the object's fall. Complete the table and label the points on the graph.

Think and Discuss
4. **Compare** your solutions with a partner's.
5. **Explain** why the graph of a falling object is not a straight line.

Motivate

Give students a group of words like *monolingual, bilingual,* and *trilingual.* Discuss the meaning of each word and its prefix. able to speak one, two, or three languages, respectively Then introduce the new vocabulary terms *monomial, binomial,* and *trinomial.* Ask students to think about the meaning of each word.

Exploration worksheet and answers on Chapter 13 Resource Book pp. 7 and 71

2 Teach

Lesson Presentation

Guided Instruction

In this lesson, students learn to classify polynomials by degree and by the number of terms. After discussing the meanings of *monomial, binomial,* and *trinomial,* review Examples 1 and 2. You may want to give some additional examples of expressions that are not polynomials, such as $(3x^2 + 2\sqrt{x})$ and $(4x^3 + 5x^{-2})$. Explain why these are not polynomials. Explain that the degree of a monomial with one variable is the exponent of the variable. Point out that the degree of a constant is zero.

A polynomial can also be classified by its degree. The **degree of a polynomial** is the degree of the term with the greatest degree.

$$\underbrace{4x^2}_{\text{Degree 2}} + \underbrace{2x^5}_{\text{Degree 5}} + \underbrace{x}_{\text{Degree 1}} + \underbrace{5}_{\text{Degree 0}}$$
$$\underbrace{}_{\text{Degree 5}}$$

EXAMPLE 3 Classifying Polynomials by Their Degrees

Find the degree of each polynomial.

A $5x^2 + 2x + 3$

$$\underset{\text{Degree 2}}{5x^2} + \underset{\text{Degree 1}}{2x} + \underset{\text{Degree 0}}{3}$$

The degree of $5x^2 + 2x + 3$ is 2.

B $5 + 2m^3 + 3m^6$

$$\underset{\text{Degree 0}}{5} + \underset{\text{Degree 3}}{2m^3} + \underset{\text{Degree 6}}{3m^6}$$

The degree of $5 + 2m^3 + 3m^6$ is 6.

C $h + 2h^3 + h^2$

$$\underset{\text{Degree 1}}{h} + \underset{\text{Degree 3}}{2h^3} + \underset{\text{Degree 2}}{h^2}$$

The degree of $h + 2h^3 + h^2$ is 3.

EXAMPLE 4 *Physics Application*

The height in feet of a firework launched straight up into the air from s feet off the ground at velocity v after t seconds is given by the polynomial $-16t^2 + vt + s$. Find the height of a firework launched from a 10 ft platform at 200 ft/s after 5 seconds.

$-16t^2 + vt + s$	*Write the polynomial expression for height.*
$-16(5)^2 + 200(5) + 10$	*Substitute 5 for t, 200 for v, and 10 for s.*
$-400 + 1000 + 10$	*Simplify.*
610	

The firework is 610 ft high 5 seconds after launching.

These colorfully decorated fireworks are part of a traditional Chinese New Year celebration.

Think and Discuss

1. **Describe** two ways you can classify a polynomial. Give a polynomial with three terms, and classify it two ways.

2. **Explain** why $-5x^2 - 3$ is a polynomial but $-5x^{-2} - 3$ is not.

---

3 Close

Reaching All Learners
Through Critical Thinking

Give each student a recording sheet (Chapter 13 Resource Book p. 67) that contains some related polynomials, such as $3xy$ and $3 + x - y$. Have students classify each polynomial by its number of terms. Then have them identify the operations (+, −, or ×) used in each polynomial. Finally, have students draw a conclusion about how operations are related to types of polynomials.

Summarize

Remind students that a polynomial is a monomial or the sum or difference of monomials. A binomial has two monomial terms, and a trinomial has three monomial terms. The degree of a monomial with one variable is the exponent of the variable. The degree of a polynomial is the degree of the monomial with the greatest degree.

Answers to Think and Discuss

Possible answers:

1. You can classify a polynomial by the number of terms and by the degree. The polynomial $5x^3 + 3x + 1$ is a trinomial with a degree of 3.

2. $-5x^2 - 3$ is a polynomial because it is the difference of monomials. $-5x^{-2} - 3$ is not a polynomial because the exponent of the variable in the term $-5x^{-2}$ is not a whole number.

13-1 Exercises

FOR EXTRA PRACTICE
see page 756

✓ internet connect
Homework Help Online
go.hrw.com Keyword: MP4 13-1

GUIDED PRACTICE

See Example **1** Determine whether each expression is a monomial.

1. $-2x^2y$ **yes**

2. $\frac{3}{2x}$ **no**

3. $4a^{2.4}b^{3.2}$ **no**

4. $3m^2n^2$ **yes**

See Example **2** Classify each expression as a monomial, a binomial, a trinomial, or not a polynomial.

5. $3x^2 - 4x$ **binomial**

6. $5r - 3r^2 + 6$ **trinomial**

7. $\frac{5}{x^2} + 3x$ **not a polynomial**

8. 3 **monomial**

See Example **3** Find the degree of each polynomial.

9. $-5m^4 + 2m^7$ **7**

10. $9w^3 + 4$ **3**

11. $-4b^4 + 5b^6 - 2b$ **6**

12. $x^3 + 2x^2 - 18$ **3**

See Example **4** **13.** The trinomial $-16t^2 + 20t + 50$ describes the height in feet of a ball thrown straight up from a 50 ft platform with a velocity of 20 ft/s after t seconds. What is the ball's height after 2 seconds? **26 feet**

INDEPENDENT PRACTICE

See Example **1** Determine whether each expression is a monomial.

14. $6.7x^4$ **yes**

15. $-2x^{-4}$ **no**

16. $\frac{4y^3}{5x}$ **no**

17. $\frac{4}{7}x^4y^2$ **yes**

See Example **2** Classify each expression as a monomial, a binomial, a trinomial, or not a polynomial.

18. $-8m^3n^5$ **monomial**

19. $4g^{\frac{1}{2}}h^3$ **not a polynomial**

20. $4x^3 + 2x^5 + 3$ **trinomial**

21. $-a + 2$ **binomial**

See Example **3** Find the degree of each polynomial.

22. $2x^2 - 7x + 1$ **2**

23. $-5m^3 + 6m^4 - 3$ **4**

24. $-1 + 2x + 3x^3$ **3**

25. $5p^4 + 7p^3$ **4**

See Example **4** **26.** The volume of a box with height x, length $x + 1$, and width $2x - 4$ is given by the trinomial $2x^3 - 2x^2 - 4x$. What is the volume of the box if its height is 3 inches? **24 in³**

Assignment Guide

If you finished Example **1** assign:
Core 1–4, 14–17, 44–51
Enriched 1–4, 14–17, 44–51

If you finished Example **2** assign:
Core 1–8, 14–21, 44–51
Enriched 1–8, 14–21, 44–51

If you finished Example **3** assign:
Core 1–12, 14–25, 27–37 odd, 44–51
Enriched 14–25, 27–38, 41–42, 44–51

If you finished Example **4** assign:
Core 1–26, 27–39 odd, 44–51
Enriched 14–51

Notes

Math Background

The lesson addresses degree only for polynomials containing one variable. The degree of a monomial is the sum of the exponents of each variable. For example, $5x^2yz^4$ has a degree of 7 because $2 + 1 + 4 = 7$.

The degree of a polynomial is the greatest degree of its terms. For example, $3ab^5 + a^2b^2$ has a degree of 6, because the degree of $3ab^5$ is 6, and the degree of a^2b^2 is 4.

RETEACH 13-1

LESSON **13-1** **Reteach**
Polynomials

You have seen expressions like $2x$ and $4y^2$. These expressions are called **monomials**. A monomial has only one term. Monomials do not have fractional exponents, negative exponents, variable exponents, roots of variables, or variables in a denominator.

Determine whether each expression is a monomial.

1. $3x - 5$ **no**

2. $-9a^4$ **yes**

3. $21m^{0.5}$ **no**

4. $7m^3n^2$ **yes**

A monomial or a sum or difference of monomials is called a **polynomial**. Polynomials can be classified by the number of terms. A monomial has 1 term, a **binomial** has 2 terms, and a **trinomial** has 3 terms.

Classify each expression as a monomial, a binomial, a trinomial, or not a polynomial.

5. $7y + 3x^2 + 5$ **trinomial**

6. $6y + \sqrt{x}$ **not a polynomial**

7. m^2n **monomial**

8. $-6a + 2b^4$ **binomial**

In a polynomial, each monomial is called a **term**. The **degree** of a term that has only one variable is the value of that variable's exponent.

terms

$3x^5 + 5x^3 + 6$
5th degree 3rd degree 0 degree

The degree of a polynomial is the degree of the term with the greatest degree. The above polynomial is a 5th degree trinomial.

Find the degree of each polynomial.

9. $5x + 3x^3 + 2x^2$ **3**

10. $-3m^4 + m^2 + 2$ **4**

11. $4y + 2y^3 + y$ **5**

12. $7a^2 + 8a$ **2**

PRACTICE 13-1

LESSON **13-1** **Practice B**
Polynomials

Determine whether each expression is a monomial.

1. $-135x^5$ **yes**

2. $2.4x^3y^{19}$ **yes**

3. $\frac{(2p^2)}{(q^3)}$ **no**

4. $3r^{\frac{1}{2}}$ **no**

5. $43a^2b^{6.1}$ **no**

6. $\left(\frac{7}{9}\right)x^2yz^5$ **yes**

Classify each expression as a monomial, a binomial, a trinomial, or not a polynomial.

7. $-8.9xy + \frac{6}{y^5}$ **not a polynomial**

8. $\left(\frac{9}{8}\right)ab^8c^2d$ **monomial**

9. $x^8 + x + 1$ **trinomial**

10. $-7pq^{-2}r^4$ **not a polynomial**

11. $5n^{15} - 9n + \left(\frac{1}{3}\right)$ **trinomial**

12. $r^8 - 5.5r^{75}$ **binomial**

Find the degree of each polynomial.

13. $7 - 14x$ **1**

14. $5a + a^2 + \left(\frac{6}{7}\right)a^3$ **3**

15. $7w - 16u + 3v$ **1**

16. $9p - 9q - 9p^3 - 9q^2$ **3**

17. $z^9 + 10y^8 - x$ **9**

18. $100{,}050 + \left(\frac{4}{5}\right)k - k^4$ **4**

19. The volume of a box with height x, length $x - 1$, and width $2x + 2$ is given by the trinomial $2x^3 - 2x$. What is the volume of the box if its height is 4 feet? **120 ft³**

20. The trinomial $-16t^2 + 32t + 32$ describes the height of an upward-flying ball in inches. What is the height of the ball (in feet) after $\frac{5}{8}$ seconds? **45.75 feet**

PRACTICE AND PROBLEM SOLVING

Classify each expression as a monomial, a binomial, a trinomial, or not a polynomial. If it is a polynomial, give its degree.

27. $3x^2$ **monomial; 2**

28. $5x^{0.5} + 2x$
not a polynomial

29. $-\frac{4}{5}x + \frac{2}{3}x^2$

30. $5y^2 - 4y$

31. $3f^4 + 6f^6 - f$

32. $6 - \frac{4}{x}$

33. $5x + 3\sqrt{x}$

34. $5x^{-3}$

35. $2b^2 - 7b - 6b^3$

36. $3 + 4x$ **binomial; 1**

37. $3x^{\frac{2}{3}} - 4x^3 + 6$
not a polynomial

38. 8 **monomial; 0**

39. TRANSPORTATION Gas mileage at speed s can be estimated using the given polynomials. Evaluate the polynomials to complete the table.

		Gas Mileage (mi/gal)		
		40 mi/h	50 mi/h	60 mi/h
Compact	$-0.025s^2 + 2.45s - 30$	28	30	27
Midsize	$-0.015s^2 + 1.45s - 13$	21	22	20
Van	$-0.03s^2 + 2.9s - 53$	15	17	13

40. TRANSPORTATION The distance in feet required for a car traveling at r mi/h to come to a stop can be approximated by the binomial $\frac{r^2}{20} + r$. About how many feet will be required for a car to stop if it is traveling at 60 mi/h? **about 240 feet**

41. WHAT'S THE QUESTION? For the polynomial $4b^5 - 7b^9 + 6b$, the answer is 9. What is the question? **Possible answer: What is the degree of the trinomial?**

42. WRITE ABOUT IT Give some examples of words that start with *mono-*, *bi-*, *tri-*, and *poly-*, and relate the meaning of each to polynomials.

43. CHALLENGE The base of a triangle is described by the binomial $x + 2$, and its height is described by the trinomial $2x^2 + 3x - 7$. What is the area of the triangle if $x = 5$? **203 units^2**

Spiral Review

Write each number or product in scientific notation. (Lesson 2-9)

44. 3,400,000,000 **3.4×10^9**

45. 0.00000045 **4.5×10^{-7}**

46. $(3.2 \times 10^4) \times (2 \times 10^{-5})$
6.4×10^{-1}

Simplify. (Lesson 3-8)

47. $\sqrt{144}$ **12**

48. $\sqrt{64}$ **8**

49. $\sqrt{169}$ **13**

50. $\sqrt{225}$ **15**

51. TEST PREP The length of the base of an isosceles triangle is half the length of a leg. Which expression shows the perimeter of the triangle if the length of the base is x? (Lesson 6-2) **B**

A $\frac{5}{2}x$ **B** $5x$ **C** $6x$ **D** $\frac{3}{2}x$

CHALLENGE 13-1

PROBLEM SOLVING 13-1

Hands-On LAB 13A
Model Polynomials

Pacing: Traditional 1 day
Block $\frac{1}{2}$ day

Objective: To use algebra tiles to model polynomials

Materials: Algebra tiles

Lab Resources

Hands-On Lab Activities . . . . pp. 85–86

Using the Pages

Discuss with students what each algebra tile represents.

Represent each monomial with algebra tiles.

1. $3x$

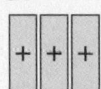

2. $-2x^2$

3. -4

Represent each polynomial with algebra tiles.

1. $2x + 1$

2. $-x^2 + x - 2$

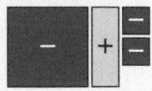

Model Polynomials

Use with Lesson 13-1

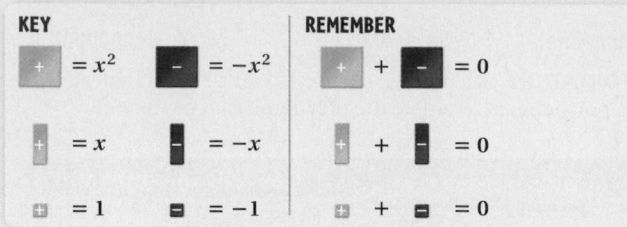

You can use algebra tiles to model polynomials. To model the polynomial $4x^2 + x - 3$, you need four x^2-tiles, one x-tile, and three -1-tiles.

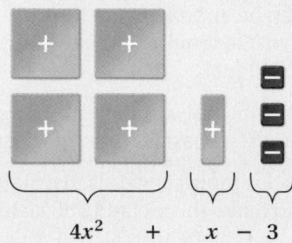

$$4x^2 \quad + \quad x \quad - \quad 3$$

Activity 1

❶ Use algebra tiles to model the polynomial $2x^2 + 4x + 6$.

All signs are positive, so use all yellow tiles.

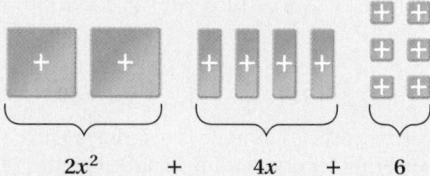

$$2x^2 \quad + \quad 4x \quad + \quad 6$$

Miguel Carrizales
San Antonio, Texas

Teacher to Teacher

As students progress from the concrete to the abstract, it is important for them to be able to visualize abstract concepts such as polynomials. I like to have students create large models of polynomials that can be displayed in the classroom. Students work in pairs to create the models on poster board. After the posters are displayed, each team can identify the different polynomials that are modeled. The students enjoy moving around and seeing other students' work.

❷ Use algebra tiles to model the polynomial $-x^2 + 6x - 4$.

Modeling $-x^2 + 6x - 4$ is similar to modeling $2x^2 + 4x + 6$. Remember to use red tiles for negative values.

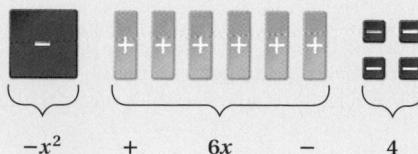

$$-x^2 \qquad + \qquad 6x \qquad - \qquad 4$$

Think and Discuss

1. How do you know when to use red tiles?

Try This

Use algebra tiles to model each polynomial.

1. $3x^2 + 2x - 4$ 　　　　 **2.** $-5x^2 + 4x - 1$ 　　　　 **3.** $4x^2 - x + 7$

Activity 2

❶ Write the polynomial modeled by the tiles below.

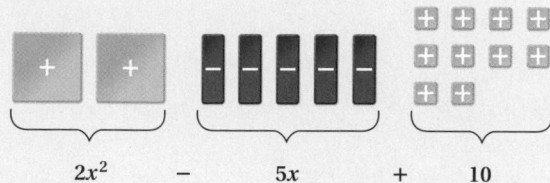

$$2x^2 \qquad - \qquad 5x \qquad + \qquad 10$$

The polynomial modeled by the tiles is $2x^2 - 5x + 10$.

Think and Discuss

1. How do you know the coefficient of the x^2 term in Activity 2?

Try This

Write a polynomial modeled by each group of algebra tiles.

1. 　　 **2.** 　　 **3.**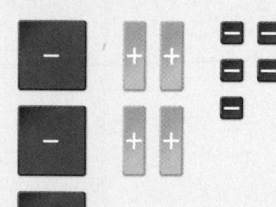

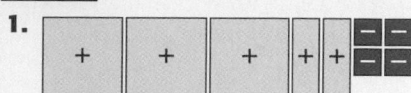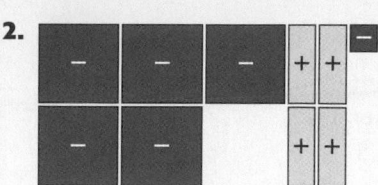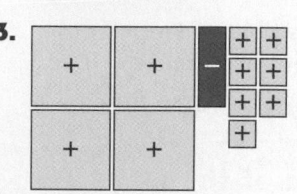

13-2 Organizer

Pacing: Traditional 1 day
Block $\frac{1}{2}$ day

Objective: Students simplify polynomials.

Warm Up

Identify the coefficient of each monomial.

1. $3x^4$ 3   **2.** ab 1

3. $\frac{x}{2}$ $\frac{1}{2}$ **4.** $-cb^3$ -1

Use the Distributive Property to simplify each expression.

5. $9(6+7)$ 117   **6.** $4(10-2)$ 32

Problem of the Day

Warren drank 3.5 gallons of water in one week. Find the average number of *ounces* of water Warren drank each day that week.   64 oz

Available on Daily Transparency in CRB

Math Humor

How do you know when your parrot is a mathematician? It says, "Polly wants a nomial."

13-2 Simplifying Polynomials

Learn to simplify polynomials.

You can simplify a polynomial by adding or subtracting like terms. Remember that like terms have the same variables raised to the same powers.

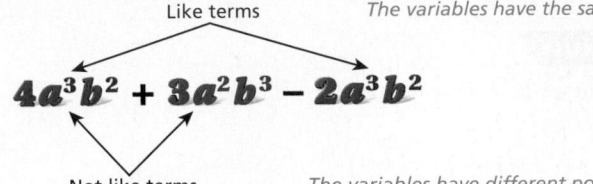

Like terms — *The variables have the same powers.*

$$4a^3b^2 + 3a^2b^3 - 2a^3b^2$$

Not like terms — *The variables have different powers.*

EXAMPLE 1   Identifying Like Terms

Identify the like terms in each polynomial.

A $3a + 2a^2 - 3 + 6a - 4a^2$

$\boxed{3a} + \boxed{2a^2} - 3 + \boxed{6a} - \boxed{4a^2}$ *Identify like terms.*

Like terms: $3a$ and $6a$, $2a^2$ and $-4a^2$

B $-3x^4y^2 + 10x^4y^2 - 3x^2 - 5x^4y^2$

$\boxed{-3x^4y^2} + \boxed{10x^4y^2} - 3x^2 - \boxed{5x^4y^2}$ *Identify like terms.*

Like terms: $-3x^4y^2$, $10x^4y^2$, and $-5x^4y^2$

C $4m^2 - 2mn + 3m$

$4m^2 - 2mn + 3m$ *Identify like terms.*

There are no like terms.

To simplify a polynomial, combine like terms. It may be easier to arrange the terms in *descending* order (highest degree to lowest degree) before combining like terms.

EXAMPLE 2   Simplifying Polynomials by Combining Like Terms

Simplify.

A $x^2 + 6x^4 - 8 + 9x^2 + 2x^4 - 6x^2$

$x^2 + 6x^4 - 8 + 9x^2 + 2x^4 - 6x^2$

$6x^4 + 2x^4 + x^2 + 9x^2 - 6x^2 - 8$ *Arrange in descending order.*

$\boxed{6x^4} + \boxed{2x^4} + \boxed{x^2} + \boxed{9x^2} - \boxed{6x^2} - 8$ *Identify like terms.*

$8x^4 + 4x^2 - 8$ *Combine coefficients:*
        $6 + 2 = 8$ and $1 + 9 - 6 = 4$

---

1 Introduce

Alternate Opener

EXPLORATION

13-2 Simplifying Polynomials

You can use algebra tiles to model polynomials. The polynomial $2x^2 + 2x + 2 + x^2 - x - 3$ is modeled below.

$2x^2$ $2x$ 2 x^2 $-x-3$

1. Use tiles to show that $2x^2 + 2x + 2 + x^2 - x - 3 = 3x^2 + x - 1$.

Use algebra tiles to simplify each expression.

2. $4x^2 - 2x - 5 - 3x^2 + x - 4$

3. $3x^2 - x + 1 - x^2 - x + 3$

Think and Discuss

4. Explain how you can use tiles to simplify polynomials.
5. Explain why you cannot simplify the polynomial $3x^2 + 4x - 9$.

Motivate

Show students a small pile of various coins. You may want to use the play money provided in the Manipulatives Kit. Ask students how they would sort the money. Group like coins together. Explain that like terms in polynomials can be grouped in the same way. You may want to remind students that they have worked with like terms before (Lesson 1-6).

Exploration worksheet and answers on Chapter 13 Resource Book pp. 16 and 73

2 Teach

Lesson Presentation

Guided Instruction

In this lesson, students learn to simplify polynomials. First, be sure that students know how to identify like terms. You may want to use Teaching Transparency T6 (CRB). Explain that like terms are either constants or terms with the same variables raised to the same powers (Lesson 1-6). After students are able to identify like terms, discuss how to simplify polynomials by combining like terms. Then discuss how to use the Distributive Property to simplify expressions.

Simplify.

B $-4a^2b + 10ab^2 - 3a^2b - ab^2 + 2ab$

$\boxed{-4a^2b} + \boxed{10ab^2} - \boxed{3a^2b} - \boxed{ab^2} + 2ab$ *Identify like terms.*

$-7a^2b + 9ab^2 + 2ab$ *Combine coefficients:*

                               $-4 - 3 = -7$ and $10 - 1 = 9$

Sometimes you may need to use the Distributive Property to simplify a polynomial.

EXAMPLE 3   **Simplifying Polynomials by Using the Distributive Property**

Simplify.

A $5(2x^2 + 6x)$

$5(2x^2 + 6x)$ *Distributive Property*

$5 \cdot 2x^2 + 5 \cdot 6x$

$10x^2 + 30x$

B $2(3ab^2 - 6b) + 2ab^2 + 5$

$2(3ab^2 - 6b) + 2ab^2 + 5$ *Distributive Property*

$2 \cdot 3ab^2 - 2 \cdot 6b + 2ab^2 + 5$

$6ab^2 - 12b + 2ab^2 + 5$

$8ab^2 - 12b + 5$ *Combine like terms.*

EXAMPLE 4   **Business Application**

A *board foot* is 1 ft by 1 ft by 1 in. of lumber. The amount of lumber that can be harvested from a tree with diameter d in. is approximately $20 + 0.005(d^3 - 30d^2 + 300d - 1000)$ board feet. Use the Distributive Property to write an equivalent expression.

$20 + 0.005(d^3 - 30d^2 + 300d - 1000) = 20 + 0.005d^3 - 0.15d^2 + 1.5d - 5$
$= 15 + 0.005d^3 - 0.15d^2 + 1.5d$

Think and Discuss

1. **Tell** how you know when you can combine like terms.

2. **Give** an example of an expression that you could simplify by using the Distributive Property and an expression that you could simplify by combining like terms.

Additional Examples

Example 1

Identify the like terms in each polynomial.

A. $5x^3 + y^2 + 2 - 6y^2 + 4x^3$
   $5x^3$ and $4x^3$, y^2 and $-6y^2$

B. $3a^3b^2 + 3a^2b^3 + 2a^3b^2 - a^3b^2$
   $3a^3b^2$, $2a^3b^2$, and $-a^3b^2$

C. $7p^3q^2 + 7p^2q^3 + 7pq^2$
   no like terms

Example 2

Simplify.

A. $4x^2 + 2x^2 + 7 - 6x + 9$
   $6x^2 - 6x + 16$

B. $3n^5m^4 - 6n^3m + n^5m^4 - 8n^3m$
   $4n^5m^4 - 14n^3m$

Example 3

Simplify.

A. $3(x^3 + 5x^2)$
   $3x^3 + 15x^2$

B. $-4(3m^3n + 7m^2n) + m^2n$
   $-12m^3n - 27m^2n$

Example 4

The surface area of a right cylinder can be found by using the expression $2\pi(r^2 + rh)$, where r is the radius and h is the height. Use the Distributive Property to write an equivalent expression.
$2\pi r^2 + 2\pi rh$

3 Close

Reaching All Learners
Through Grouping Strategies

Give each student a card containing a monomial (Chapter 13 Resource Book p. 69). Have students find the other students who have cards that contain like terms. Once all of the students have found their partners, have them combine all of the like terms into one term. Then you may want to have the class write all of the terms as one polynomial on the board.

Summarize

Ask students to define *like terms*. Remind students that like terms can be combined by addition or subtraction. Make sure students understand how to apply the Distributive Property to polynomials.

Possible answer: Like terms are terms that have the same variable(s) raised to the same powers. Constants are also considered like terms.

Answers to Think and Discuss

Possible answers:

1. The terms are either numbers or terms with the same variables raised to the same powers, and they are separated by plus or minus signs.

2. Distributive Property: $8(3x + 2) = 24x + 16$; combine like terms: $8x + 3x = 11x$

13-2 PRACTICE & ASSESS

13-2 Exercises

FOR EXTRA PRACTICE
see page 756

internet connect
Homework Help Online
go.hrw.com Keyword: MP4 13-2

Students may want to refer back to the lesson examples.

GUIDED PRACTICE

See Example ① Identify the like terms in each polynomial.

1. $-2b^2 + 4b + 3b^2 - b + 8$
   $-2b^2$ and $3b^2$, $4b$ and $-b$

2. $5mn - 4m^2n^2 + 6m^2n + 3m^2n^2$
   $-4m^2n^2$ and $3m^2n^2$

See Example ② Simplify.

3. $3x^2 - 4x + 6x^2 + 8x - 6$

4. $7 - 4b + 2b^4 - 6b^2 + 8 + 5b - 4b^2$

See Example ③ 5. $3(2x - 7)$ $6x - 21$

6. $6(4a^2 - 7a) + 3a^2 + 5a$ $27a^2 - 37a$

See Example ④ 7. The level of nitric oxide emissions, in parts per million, from a car engine is approximated by the polynomial $-40,000 + 5x(800 - x^2)$, where x is the air-fuel ratio. Use the Distributive Property to write an equivalent expression. $-40,000 + 4000x - 5x^3$

INDEPENDENT PRACTICE

See Example ① Identify the like terms in each polynomial.

8. $-t + 5t^2 - 6t^2 + 6t - 3$
   $-t$ and $6t$, $5t^2$ and $-6t^2$

9. $9rs - 2r^2s^2 + 4r^2s^2 + 3rs - 7$
   $9rs$ and $3rs$, $-2r^2s^2$ and $4r^2s^2$

See Example ② Simplify.

10. $3p - 4p^2 + 6p + 10p^2$ $9p + 6p^2$

11. $2fg + f^2g - fg^2 - 2fg + 3f^2g + 5fg^2$ $4f^2g + 4fg^2$

See Example ③ 12. $4(x^2 - 4x) + 3x^2 - 6x$ $7x^2 - 22x$ 13. $3(b - 4) + 6b - 4b^2$
   $9b - 12 - 4b^2$

See Example ④ 14. The concentration of a certain medication in an average person's bloodstream h hours after injection can be estimated using the expression $7(0.04h - 0.003h^2 - 0.02h^3)$. Use the Distributive Property to write an equivalent expression. $-0.14h^3 - 0.021h^2 + 0.28h$

PRACTICE AND PROBLEM SOLVING

Simplify.

15. $3s^2 - 4s + 12s^2 + 6s - 2$

16. $4gh^2 + 2g^2h + 3g^2h - g^2h$

17. $3(x^2 - 4x + 3) - 2x + 6$

18. $4(x - x^5 + x^3) - 2x$

19. $2(3m - 4m^2) + 6(2m^2 - 5m)$

20. $8b^4 + 3b^2 + 2(b^2 - 8)$

21. $7mn - 4m^3n^2 + 4(m^3n^2 + 2mn)$

22. $4(x + 2y) + 3(2x - 3y)$

23. **LIFE SCIENCE** The rate of flow in cm/s of blood in an artery at d cm from the center is given by the polynomial $1000(0.04 - d^2)$. Use the Distributive Property to write an equivalent expression.

Assignment Guide

If you finished Example ① assign:
Core 1–2, 8–9, 27–34
Enriched 1–2, 8–9, 27–34

If you finished Example ② assign:
Core 1–4, 8–11, 27–34
Enriched 8–11, 15–16, 27–34

If you finished Example ③ assign:
Core 1–6, 8–13, 15–21 odd, 27–34
Enriched 8–13, 15–22, 24–34

If you finished Example ④ assign:
Core 1–14, 15–25 odd, 27–34
Enriched 8–34

Answers

3. $9x^2 + 4x - 6$
4. $2b^4 - 10b^2 + b + 15$
15. $15s^2 + 2s - 2$
16. $4gh^2 + 4g^2h$
17. $3x^2 - 14x + 15$
18. $-4x^5 + 4x^3 + 2x$
19. $4m^2 - 24m$
20. $8b^4 + 5b^2 - 16$
21. $15mn$
22. $10x - y$
23. $40 - 1000d^2$

Math Background

In the lesson, Example 3 addresses using the Distributive Property to simplify expressions. The process of combining like terms used in Example 2 is also an application of the Distributive Property (e.g., $3x + 2x = (3 + 2)x = 5x$).

The Distributive Property may be used several times in the simplification of a polynomial. An example is shown below.

$4(3x + 2z) + 5x - 6z$
$12x + 8z + 5x - 6z$
$12x + 5x + 8z - 6z$
$(12 + 5)x + (8 - 6)z$
$17x + 2z$

RETEACH 13-2

LESSON Reteach
13-2 Simplifying Polynomials

You simplify a polynomial by combining like terms. Like terms that have variables have the same variables raised to the same powers. All constants are like terms. Look at the terms in the polynomial below.

$9 + 6y^3 - 8 + 7x^2y^3 + 3x^2y^3$

| like terms | like terms |

Notice that $7x^2y^3$ and $3x^2y^3$ both have the variable x raised to the 2nd power and the variable y raised to the 3rd power. Therefore, these are like terms.

Identify the like terms in each polynomial

1. $m + 3m^2 - 2m + 6 + 2m^2$
   m and $-2m$; $3m^2$ and $2m^2$

2. $b - a^2b^2 - 2 + a^2 + 2a^2b^2$
   $2a^2b^2$ and $-a^2b^2$

3. $x^3 + 2 + 4x^3 - 9 + x$
   x^3 and $4x^3$; 2 and -9

4. $9 + 4dg^2 + 4 + 6dg^2 + d^2$
   $4dg^2$ and $6dg^2$, 4 and 9

To simplify a polynomial, combine like terms. To combine like terms, combine the coefficients of like terms. The variables and the exponents do not change.

$7x^2y^3 - 6y^3 + 3x^2y^3$
$(7x^2y^3) + 6y^3 + (3x^2y^3)$ Identify like terms.
$10x^2y^3 + 6y^3$ Combine coefficients of like terms.
$7 + 3 = 10$

Simplify.

5. $8a + 3ab^2 + 3a + 2ab^2$
   $5ab^2 + 11a$

6. $x^3 + 1 + 2x^3 + 3xy^2 - 3$
   $3x^3 + 3xy^2 - 2$

7. $y^4 + 2x^2y^3 - 3x^2 + 2y^4$
   $3y^4 + 2x^2y^3 - 3x^2$

PRACTICE 13-2

LESSON Practice B
13-2 Simplifying Polynomials

Identify the like terms in each polynomial.

1. $x^2 - 8x + 3x^2 + 6x - 1$
   x^2 and $3x^2$, $-8x$ and $6x$

2. $2c^2 + d^3 + 3d^3 - 2c^2 + 6$
   $2c^2$ and $-2c^2$, d^3 and $3d^3$

3. $2x^2 - 2xy - 2y^2 + 3xy + 3x^2$
   $2x^2$ and $3x^2$, $-2xy$ and $3xy$

4. $2 - 9x + x^2 - 3 + x$
   $-9x$ and x, 2 and -3

5. $xy - 5x + y - x + 10y - 3y^2$
   $-5x$ and $-x$, y and $10y$

6. $6p + 2p^2 + pq + 2q^3 - 2p$
   $6p$ and $-2p$

7. $3a + 2b + a^2 - 5b + 7a$
   $3a$ and $7a$, $2b$ and $-5b$

8. $10m - 3m^2 + 9m^2 - 3m - m^3$
   $10m$ and $-3m$,
   $-3m^2$ and $9m^2$

Simplify.

9. $2h - 9hk + 3(2h - 2k)$
   $8h - 9hk - 6k$

10. $9(x^2 + 2xy - y^2) - 2(x^2 + xy)$
    $7x^2 + 16xy - 9y^2$

11. $7qr - q^2r^3 + 2(q^2r^3 - 3qr)$
    $q^2r^3 + qr$

12. $8v^4 + 3v^2 + 2(v^2 - 8)$
    $8v^4 + 5v^2 - 16$

13. $3(x + 2y) + 2(2x - 3y)$
    $7x$

14. $7(1 - x) + 3x^2y + 7x - 7$
    $3x^2y$

15. $6(9y + 1) + 8(2 - 3y)$
    $30y + 22$

16. $a^2b - a^2 + ab^2 - 3a^2b + ab$
    $-2a^2b - a^2 + ab^2 + ab$

17. A student in Tracey's class created the following expression: $y^3 - 3y + 4(y^2 - y^3)$. Use the Distributive Property to write an equivalent expression.
    $-3y^3 + 4y^2 - 3y$

Abstract artists often use geometric shapes, such as cubes, prisms, pyramids, and spheres, to create sculptures.

24. Suppose the volume of a sculpture is approximately $s^3 + 0.52s^3 + 0.18s^3 + 0.33s^3$ cm^3 and the surface area is approximately $6s^2 + 3.14s^2 + 7.62s^2 + 3.24s^2$ cm^2.

   a. Simplify the polynomial expression for the volume of the sculpture, and find the volume of the sculpture for $s = 5$.

   b. Simplify the polynomial expression for the surface area of the sculpture, and find the volume of the sculpture for $s = 5$.

*Balanced/Unbalanced O
by Fletcher Benton*

25. A sculpture features a large ring with an outer lateral surface area of about $44xy$ in^2, an inner lateral surface area of about $38xy$ in^2, and 2 bases, each with an area of about $41y$ in^2. Write and simplify a polynomial that expresses the surface area of the ring. **$82xy + 41y$ in^2**

26. ★ **CHALLENGE** The volume of the ring on the sculpture from Exercise 25 is $49\pi xy^2 - 36\pi xy^2$ in^3. Simplify the polynomial, and find the volume for $x = 12$ and $y = 7.5$. Give your answer both in terms of π and to the nearest tenth. **$13\pi xy^2$; $8775\pi \approx 27{,}567.5$ in^3**

Pyramid Balancing Cube and Sphere, artist unknown

go.hrw.com
KEYWORD: MP4 Art
CNN Student News

Spiral Review

Simplify. (Lessons 2-1 to 2-3)

27. $-5 + -8$ **-13**   28. $4 - (-9)$ **13**   29. $-6 \times (-5)$ **30**   30. $-32 \div 8$ **-4**

Solve for x. (Lessons 3-6 and 3-7)

31. $x - \frac{3}{2} \geq \frac{7}{2}$ **$x \geq 5$**   32. $-\frac{3}{4}x + 6 < 8$ **$x > -\frac{8}{3}$**   33. $\frac{1}{2}x - \frac{2}{3} = 6$ **$x = \frac{40}{3}$**

34. **TEST PREP** The point $A(3, -2)$ is reflected over the x-axis and then translated 3 units up. What are the coordinates of A'? (Lesson 5-7) **A**

   A $(3, 5)$ B $(-3, 1)$ C $(0, -2)$ D $(6, 2)$

Art

Exercises 24–26 involve applying polynomial concepts to finding the volume and surface area of sculptures. Knowing these measurements can help artists determine the amount of materials they need to complete a project.

Answers

24. a. $2.03s^3$; 253.75 cm^3
    b. $20s^2$; 500 cm^2

Journal

Show students the following polynomial expression: $3x^2 + 5x + 2x^2 - 4x$. Explain that you want to evaluate the polynomial for $x = 3$. Ask students to explain whether you should evaluate or simplify the polynomial first. Have them give reasons to support their opinions.

Test Prep Doctor

For Exercise 34, encourage students to sketch a coordinate plane, plot the points, and apply the transformations. They can then find the coordinates of the transformed point, $(3, 5)$, or choice **A**. If students chose **B**, they may have reflected the point over the y-axis.

CHALLENGE 13-2

LESSON 13-2 Challenge
Coming To Terms

For each row, use the equivalent polynomials to find the missing term.

	Polynomial	Simplified Polynomial	Missing Term
1.	$7x^2 - 4x + \underline{} + 3x - 5$	$5x^2 - x - 5$	$-2x^2$
2.	$6(9x + \underline{})$	$54x + 18$	3
3.	$12 + 2m + 3m^4 - 8m^2 + 5 + \underline{} - 7m^2$	$3m^4 - 15m^2 + 6m + 17$	$4m$
4.	$\underline{}(2b^2 - 9b) + 4b^2 + 11b$	$14b^2 - 34b$	5
5.	$7(x^3 + 3x) - 5x^3 + \underline{}$	$2x^3 + 11x$	$-10x$
6.	$4ab + \underline{} + 3a^2b^2 + 2ab - 8$	$2a^2b^2 + 6ab - 8$	$-a^2b^2$
7.	$\underline{} + 7 + 10w^2 - 4w^2 + 5w - 2$	$6w^2 + 4w + 5$	$-w$
8.	$2(t - 7) + \underline{} - 2t^2$	$-2t^2 + 14t - 14$	$12t$
9.	$3hk - h^2k + hk^2 + 4hk + \underline{} + 3hk^2$	$4hk^2 - 3h^2k + 7hk$	$-2h^2k$
10.	$4y^3 + 6y - 7y^2 + 2y^3 + \underline{}$	$6y^3 - 6y^2 + 6y$	y^2
11.	$\underline{} - n^3 - \frac{1}{4}n^4 + \frac{1}{2}n^3 - \frac{1}{3}n^3$	$\frac{1}{4}n^4 - \frac{5}{6}n^3$	$\frac{1}{2}n^4$
12.	$3(12v^3 + \underline{} + 2v^3) + v^3$	$46v^3$	v^3
13.	$1.5pq^3 + 0.7p^2q + \underline{} + 2.4p^2q$	$1.5pq^3 + 3.1p^2q - 3.8pq$	$-3.8pq$
14.	$-9(\underline{} + 8w) + 4(-2w^2 + 18w)$	w^2	$-w^2$

PROBLEM SOLVING 13-2

LESSON 13-2 Problem Solving
Simplifying Polynomials

Write the correct answer.

1. The area of a trapezoid can be found by the expression $\frac{h}{2}(b^1 + b^2)$. Use the Distributive Property to write an equivalent expression.

   $$\frac{hb_1}{2} + \frac{hb_2}{2}$$

2. The sum of the measures of the interior angles of a polygon with n sides is $180(n - 2)$ degrees. Use the Distributive Property to write an equivalent expression, and use the expression to find the sum of the measures of the interior angles of an octagon.

   $180n - 360$; $1{,}080$ degrees

3. The volume of a box of height h is $2h^4 + h^3 + h^2 + h^2 + h$ cubic inches. Simplify the polynomial and then find the volume if the height of the box is 3 inches.

   $$2h^4 + h^3 + 2h^2 + h;$$
   210 cubic inches

4. The height, in feet, of a rocket launched upward from the ground with an initial velocity of 64 feet per second after t seconds is given by $16(4t - t^2)$. Write an equivalent expression for the rocket's height after t seconds. What is the height of the rocket after 4 seconds?

   $$64t - 16t^2;\ 0\ ft$$

Circle the letter of the best answer.

5. The surface area of a square pyramid with base b and slant height l is given by the expression $b(b + 2l)$. What is the surface area of a square pyramid with base 3 inches and slant height 5 inches?
   A 13 square inches
   B 19 square inches
   Ⓒ 39 square inches
   D 55 square inches

6. The volume of a box with a width of $3x$, a height of $4x - 2$, and a length of $3x + 5$ can be found using the expression $3x(12x^2 + 14x - 10)$. What is this expression, simplified by using the Distributive Property?
   F $36x^2 + 42x - 30$
   G $15x^3 + 17x^2 - 7x$
   H $36x^3 + 14x - 10$
   Ⓙ $36x^3 + 42x^2 - 30x$

Lesson Quiz

Identify the like terms in each polynomial.

1. $2x^2 - 3z + 5x^2 + z + 8z^2$
   $2x^2$ and $5x^2$, z and $-3z$

2. $2ab^2 + 4a^2b - 5ab^2 - 4 + a^2b$
   $2ab^2$ and $-5ab^2$, $4a^2b$ and a^2b

Simplify.

3. $5(3x^2 + 2)$ **$15x^2 + 10$**

4. $-2k^2 + 10 + 8k^2 + 8k - 2$
   $6k^2 + 8k + 8$

5. $3(2mn^2 + 3n) + 6mn^2$
   $12mn^2 + 9n$

6. $4h^2 + 3h^3 - 7 - 9h^2 + 8h - 2$
   $3h^3 - 5h^2 + 8h - 9$

Available on Daily Transparency in CRB

Chapter

13

Mid-Chapter Quiz

Purpose: *To assess students' mastery of concepts and skills in Lessons 13-1 through 13-2*

Assessment Resources

Section 13A Quiz
Assessment Resources .... p. 29

Test and Practice Generator CD-ROM

Additional mid-chapter assessment items in both multiple-choice and free-response format may be generated for any objective in Lessons 13-1 through 13-2.

LESSON 13-1 (pp. 644–647)

Determine whether each expression is a monomial.

1. $\frac{1}{3y^3}$ **no**

2. $\frac{1}{2}x^2 - x^3$ **no**

3. 1   **yes**

4. $3a^2b^2$ **yes**

Classify each expression as a monomial, a binomial, a trinomial, or not a polynomial.

5. $\frac{1}{y^2} + y$ **not a polynomial**

6. 17   **monomial**

7. $a^2 + a - 20$ **trinomial**

8. $x + 1$ **binomial**

Find the degree of each polynomial.

9. $w^5 + 3$ **5**

10. $2b^4 + b^6 - b$ **6**

11. $-9r^4 + 3r^7$ **7**

12. 12   **0**

13. The trinomial $-16t^2 + 30t + 40$ describes the height in feet of a ball thrown straight up from a 40 ft platform with a velocity of 30 ft/s after t seconds. What is the ball's height after 2 seconds?   **36 ft**

14. The price of a certain piece of artwork y years after it was painted can be approximated by the polynomial $0.03y^2 + 6y + 240$. Estimate the price of the artwork after 88 years.   **about $1000**

LESSON 13-2 (pp. 650–653)

Identify the like terms in each polynomial. $-t^2$ and $2t^2$, $3t$ and $-t$

15. $-4x^2y^2 + 5xy + 3x^2y^2$ $-4x^2y^2$ and $3x^2y^2$

16. $-t^2 + 3t + 2t^2 - t + 5$

17. $y + 3 - 3y - 4$ y and $-3y$, 3 and -4

18. $7ab + 2ac + 4bc - 3ac + 5ab$
    7ab and 5ab, 2ac and −3ac

Simplify.

19. $7 + 2c^4 - 6c^2 + 8 - 4c^2$ $2c^4 - 10c^2 + 15$

20. $y + 3 - 3y - 4$ $-2y - 1$

21. $2x^2 + x + 3x^2 + 4x - 6$ $5x^2 + 5x - 6$

22. $-2(3x - 4)$ $-6x + 8$

23. $2(4z^2 - 3z) + 5z^2 + 3z$ $13z^2 - 3z$

24. $x + 3 - 3x - 2(3x + 1)$ $-8x + 1$

Solve.

25. The area of one face of a cube is given by the expression $3s^2 + 5s$. Write a polynomial to represent the total surface area of the cube. $18s^2 + 30s$

26. The area of each lateral face of a regular square pyramid is given by the expression $\frac{1}{2}b^2 + 2b$. Write a polynomial to represent the lateral surface area of the pyramid. $2b^2 + 8b$

Mid-Chapter Quiz

Focus on Problem Solving

 Look Back

• Estimate to check that your answer is reasonable

Before you solve a word problem, you can often read through the problem and make an estimate of the correct answer. Make sure your answer is reasonable for the situation in the problem. After you have solved the problem, compare your answer with the original estimate. If your answer is not close to your estimate, check your work again.

Each problem below has an incorrect answer given. Explain why the answer is not reasonable, and give your own estimate of the correct answer.

1 The perimeter of rectangle *ABCD* is 50 cm. What is the value of *x*?

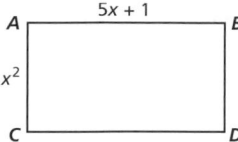

Answer: x = −8

2 A farmer can use $4x + 6y$ ft of fencing material to build three side-by-side enclosures measuring *x* ft long by *y* ft wide. If each enclosure must be at least 15 ft wide and have an area of at least 300 ft^2, what is the minimum amount of fencing needed for the three enclosures?

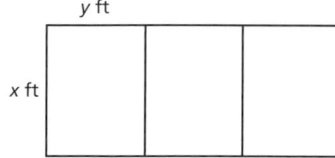

Answer: 70 ft

3 A baseball is thrown straight up from a height of 3 ft at 30 mi/h. The height of the baseball in feet after *t* seconds is $-16t^2 + 44t + 3$. How long will it take the baseball to reach its maximum height?

Answer: 5 minutes

4 Erin deposited $3000 in a savings account that earns 7% simple interest. The amount of money she has in her account after *t* years is $P + Prt$, where *P* is the initial amount of money in the account and *r* is the interest rate expressed as a decimal. How much money will she have in the account after 5 years?

Answer: $2850

Answers

1. 3 cm

2. 170 ft

3. 1.375 s

4. $4050

Polynomial Operations

One-Minute Section Planner

Lesson	Materials	Resources
Lesson 13-3 Adding Polynomials **NCTM:** Algebra, Communication ☐ SAT-9 ☐ SAT-10 ☐ ITBS ☐ CTBS ☐ MAT ☐ CAT	**Optional** Algebra tiles *(MK)* Graphing calculators	• *Chapter 13 Resource Book,* pp. 26–34 • Daily Transparency T10, CRB • Additional Examples Transparencies T11–T13, CRB • *Alternate Openers: Explorations,* p. 102
Lesson 13-4 Subtracting Polynomials **NCTM:** Algebra, Communication ☐ SAT-9 ☐ SAT-10 ☐ ITBS ☐ CTBS ☐ MAT ☐ CAT	**Optional** Algebra tiles *(MK)*	• *Chapter 13 Resource Book,* pp. 35–43 • Daily Transparency T14, CRB • Additional Examples Transparencies T15–T18, CRB • *Alternate Openers: Explorations,* p. 103
Lesson 13-5 Multiplying Polynomials by Monomials **NCTM:** Algebra, Communication ☐ SAT-9 ☐ SAT-10 ☐ ITBS ☑ CTBS ☐ MAT ☐ CAT	**Optional** Algebra tiles *(MK)* Monomial Cards for Reaching All Learners *(CRB, p. 69)*	• *Chapter 13 Resource Book,* pp. 44–52 • Daily Transparency T19, CRB • Additional Examples Transparencies T20–T23, CRB • *Alternate Openers: Explorations,* p. 104
Hands-On Lab 13B Multiply Binomials **NCTM:** Algebra, Communication, Representation ☐ SAT-9 ☐ SAT-10 ☐ ITBS ☐ CTBS ☐ MAT ☐ CAT	**Required** Algebra tiles *(MK)*	• *Hands-On Lab Activities,* pp. 87–88
Lesson 13-6 Multiplying Binomials **NCTM:** Algebra, Communication ☐ SAT-9 ☐ SAT-10 ☐ ITBS ☑ CTBS ☐ MAT ☐ CAT	**Optional** Teaching Transparency T25 *(CRB)* Algebra tiles *(MK)* Graphing calculators Recording Sheet for Reaching All Learners *(CRB, p. 70)*	• *Chapter 13 Resource Book,* pp. 53–61 • Daily Transparency T24, CRB • Additional Examples Transparencies T26–T28, CRB • *Alternate Openers: Explorations,* p. 105
Extension Dividing Polynomials by Monomials **NCTM:** Algebra, Communication ☐ SAT-9 ☐ SAT-10 ☐ ITBS ☐ CTBS ☐ MAT ☐ CAT		• Additional Examples Transparencies T29–T31, CRB
Section 13B Assessment		• Section 13B Quiz, AR p. 30 • *Test and Practice Generator* CD-ROM

SAT = *Stanford Achievement Tests*   **ITBS** = *Iowa Test of Basic Skills*   **CTBS** = *Comprehensive Test of Basic Skills/Terra Nova*
MAT = *Metropolitan Achievement Test*   **CAT** = *California Achievement Test*
NCTM = Complete standards can be found on pages T29–T35.   **NAEP** = Complete standards can be found on pages A54–A58.
SE = *Student Edition*   **TE** = *Teacher's Edition*   **AR** = *Assessment Resources*   **CRB** = *Chapter Resource Book*   **MK** = *Manipulatives Kit*

Section Overview

Why? Sums and differences of polynomials can be used to represent real-world measurements such as perimeters.

Adding Polynomials	Subtracting Polynomials
To add polynomials, combine like terms.	**To subtract polynomials,** add the opposite.
$$\begin{aligned} 3a^2b^2 + 2a^2 &- 5ab \\ a^2 &- 3ab - 2 \\ &+ 6ab + 1 \\ \hline 3a^2b^2 + 3a^2 &- 2ab - 1 \end{aligned}$$	$\begin{aligned}(3x^2y^2 + xy - 5x) \\ - (6x + 4xy - 5)\end{aligned} \longrightarrow \begin{aligned}3x^2y^2 + xy - 5x \\ - 4xy - 6x + 5 \\ \hline 3x^2y^2 - 3xy - 11x + 5\end{aligned}$

Why? Products of polynomials can be used to represent real-world measurements such as areas.

Multiplying Polynomials		
To **multiply two monomials,** multiply the coefficients and add the exponents of the variables that are the same.	To **multiply a polynomial by a monomial,** use the Distributive Property.	To **multiply two binomials,** use the FOIL method.
$(5m^2n^3)(6m^3n^6)$ $5 \cdot 6 \cdot m^{(2+3)}n^{(3+6)}$ $30m^5n^9$	$-4a^2b(2a^4b^3 + 5a^2b^3)$ $-8a^6b^4 - 20a^4b^4$	$(x + y)(x + z)$ $x^2 + xz + yx + yz$ First Outer Inner Last terms terms terms terms

Dividing Polynomials	
To **divide a monomial by a monomial,** divide the coefficients and subtract the exponents of like variables in the denominator from those in the numerator.	To **divide a polynomial by a monomial,** divide each term of the polynomial by the monomial.
$$\begin{aligned}\frac{6x^9y^3}{4x^6y^2} &= \frac{3}{2}x^{(9-6)}y^{(3-2)} \\ &= \frac{3}{2}x^3y^1 \\ &= \frac{3}{2}x^3y\end{aligned}$$	$$\begin{aligned}\frac{x^4 + 5x^3 - 7x^2}{x^2} &= \frac{x^4}{x^2} + \frac{5x^3}{x^2} - \frac{7x^2}{x^2} \\ &= x^{(4-2)} + 5x^{(3-2)} - 7x^{(2-2)} \\ &= x^2 + 5x^1 - 7x^0 \\ &= x^2 + 5x - 7\end{aligned}$$

Pacing: Traditional 1 day
Block $\frac{1}{2}$ day

Objective: Students add polynomials.

Warm Up

Combine like terms.

1. $9x + 4x$ $13x$

2. $-3y + 7y$ $4y$

3. $7n + (-8n) + 12n$ $11n$

Find the perimeter of each rectangle.

4. a 10 ft by 12 ft rectangle   **44 ft**

5. a 5 m by 8 m rectangle   **26 m**

Simplify.

6. $3(2x^2 - x) + x^2 + 1$
    $7x^2 - 3x + 1$

Problem of the Day

Michael has a collection of dimes and quarters worth $6.55. If he has one more quarter than he has dimes, how many of each coin does he have? **18 dimes and 19 quarters**

Available on Daily Transparency in CRB

Math Humor

Why wasn't the monomial reelected? She could have only one term.

13-3 Adding Polynomials

Learn to add polynomials.

Libby wants to to put a mat and a frame on a picture that is 8 inches by 10 inches. If m is the width of the mat and f is the width of the frame, you can add polynomials to find an expression for the amount of framing material Libby needs.

Remember, the Associative Property of Addition states that for any values of a, b, and c, $a + b + c = (a + b) + c = a + (b + c)$. You can use this property to add polynomials.

EXAMPLE 1   Adding Polynomials Horizontally

Add.

A $(8x^2 - 2x + 3) + (9x - 5)$

$(8x^2 - 2x + 3) + (9x - 5)$

$8x^2 - 2x + 3 + 9x - 5$ *Associative Property*

$8x^2 + 7x - 2$ *Combine like terms.*

B $(-3cd^2 - 2cd + 5) + (9cd - 7cd^2 - 5)$

$(-3cd^2 - 2cd + 5) + (9cd - 7cd^2 - 5)$

$-3cd^2 - 2cd + 5 + 9cd - 7cd^2 - 5$ *Associative Property*

$-10cd^2 + 7cd$ *Combine like terms.*

C $(ab^2 + 3a) + (2ab^2 + 3a - 2) + (2a + 4)$

$(ab^2 + 3a) + (2ab^2 + 3a - 2) + (2a + 4)$

$ab^2 + 3a + 2ab^2 + 3a - 2 + 2a + 4$ *Associative Property*

$3ab^2 + 8a + 2$ *Combine like terms.*

You can also add polynomials in a vertical format. Write the second polynomial below the first one, lining up the like terms. If the terms are rearranged, remember to keep the correct sign with each term.

---

1   Introduce

Alternate Opener

EXPLORATION

13-3 Adding Polynomials

You can use algebra tiles to model addition of polynomials. The addition problem $(2x^2 + 2x - 2) + (x^2 - 4x - 3)$ is modeled below.

    $2x^2 + 2x - 2$ $x^2 - 4x - 3$

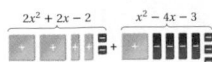

1. Use tiles to show that
$(2x^2 + 2x - 2) + (x^2 - 4x - 3) = 3x^2 - 2x - 5.$

You can use a graphing calculator to check. Enter the left side of the equation in Y1 and the right side in Y2, and compare tables.

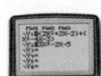 

Use algebra tiles to add each pair of polynomials. Check your answers with a graphing calculator.

2. $(x^2 + 7x) + (3x^2 - 7x - 5)$

3. $(-3x^2 - x - 1) + (4x^2 - 3x - 2)$

Think and Discuss

4. **Explain** how you can use tiles to add polynomials.

Motivate

Tell students to imagine that two students collected coins to donate to a local charity. Louis collected 3 rolls of quarters, 3 rolls of dimes, and 2 rolls of nickels. Lisa collected 2 rolls of quarters, 4 rolls of dimes, and 3 rolls of nickels. Ask students how they could determine the total amount of money collected. Possible answer: Add the number of each type of roll together, and multiply by the value of each roll. Point out that this example models addition of polynomials.

Exploration worksheet and answers on Chapter 13 Resource Book pp. 27 and 75

---

2   Teach

Lesson Presentation

Guided Instruction

In this lesson, students learn to add polynomials. Explain to students that adding polynomials is very similar to the process of simplifying polynomials that they studied in the previous lesson. Show students the horizontal and vertical methods of adding polynomials. Encourage students to be sure to keep the correct sign with each term as they move or reorder the terms in a problem.

EXAMPLE 2 Adding Polynomials Vertically

Add.

A $(4a^2 + 3a + 1) + (5a^2 + 2a + 3)$

$$
\begin{array}{r}
4a^2 + 3a + 1 \\
+\ 5a^2 + 2a + 3 \\
\hline
9a^2 + 5a + 4
\end{array}
$$

Place like terms in columns.
Combine like terms.

B $(3xy^2 + 2x - 3y) + (9xy^2 - x + 2)$

$$
\begin{array}{r}
3xy^2 + 2x - 3y \\
+\ 9xy^2 -\ x\ \ \ \ \ + 2 \\
\hline
12xy^2 +\ x - 3y + 2
\end{array}
$$

Place like terms in columns.
Combine like terms.

C $(3a^2b^2 + 2a^2 - 5ab) + (-3ab + a^2 - 2) + (1 + 6ab)$

$$
\begin{array}{r}
3a^2b^2 + 2a^2 - 5ab \\
a^2 - 3ab - 2 \\
+\ \ \ \ \ \ \ \ \ \ \ \ \ \ 6ab + 1 \\
\hline
3a^2b^2 + 3a^2 - 2ab - 1
\end{array}
$$

Place like terms in columns.
Combine like terms.

EXAMPLE 3 *Art Application*

Libby is putting a mat of width m and a frame of width f around an 8-inch by 10-inch picture. Find an expression for the amount of framing material she needs.

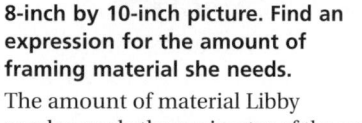

The amount of material Libby needs equals the perimeter of the outside of the frame. Draw a diagram to help you determine the outer dimensions of the frame.

Base $= 10 + m + m + f + f$ Height $= 8 + m + m + f + f$
$\ \ \ \ \ \ = 10 + 2m + 2f$ $\ \ \ \ \ \ \ = 8 + 2m + 2f$

$P = (8 + 2m + 2f) + (10 + 2m + 2f) + (8 + 2m + 2f) + (10 + 2m + 2f)$
$\ \ = 8 + 2m + 2f + 10 + 2m + 2f + 8 + 2m + 2f + 10 + 2m + 2f$
$\ \ = 36 + 8m + 8f$ *Combine like terms.*

She will need $36 + 8m + 8f$ inches of framing material.

Think and Discuss

1. **Compare** adding $(5x^2 + 2x) + (3x^2 - 2x)$ vertically with adding it horizontally.

2. **Explain** why you can remove parentheses from polynomials to add the polynomials.

3 Close

Summarize

Remind students that to add polynomials, they need to identify like terms and combine them. Ask students how many terms they will end up with if they add two trinomials.

Possible answer: It will depend on the number of like terms in the trinomials. If each trinomial is already simplified, the number of terms in the sum will be between 3 and 6.

Answers to Think and Discuss

Possible answers:

1. For both methods, you remove parentheses and then combine like terms. To add horizontally, you apply the Commutative and Associative Properties to reorder and regroup. To add vertically, you line up the like terms in columns (also applications of the same properties).

2. It is an application of the Associative Property of Addition.

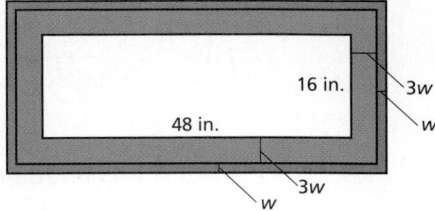

FOR EXTRA PRACTICE
see page 757

internet connect
Homework Help Online
go.hrw.com Keyword: MP4 13-3

> Students may want to refer back to the lesson examples.

Assignment Guide

If you finished Example **1** assign:
Core 1–3, 8–12, 29–35
Enriched 1–3, 8–12, 29–35

If you finished Example **2** assign:
Core 1–6, 8–15, 17–21 odd, 29–35
Enriched 8–15, 17–22, 27–35

If you finished Example **3** assign:
Core 1–16, 17–25 odd, 29–35
Enriched 8–35

Answers

5. $11ab^2 + 3ab + a^2b - 4$
14. $15b^3c^2 - 2b^2c - 7bc + 4$

GUIDED PRACTICE

See Example **1** Add.

1. $(4x^3 + 5x - 1) + (-2x + 6)$ $4x^3 + 3x + 5$

2. $(20x - 8) + (12x - 4)$ $32x - 12$

3. $(m^2n + 2mn) + (3m^2n - 6mn) + (5m^2n + 12mn)$ $9m^2n + 8mn$

See Example **2** 4. $(3b^2 - 4b + 8) + (5b^2 + 6b - 7)$ $8b^2 + 2b + 1$

5. $(7ab^2 - 3ab + 8a^2b) + (6ab - 10a^2b + 8) + (4ab^2 + 3a^2b - 12)$

6. $(h^4j - hj^3 + hj - 4) + (3hj^3 + 3) + (4h^4j - 5hj)$ $5h^4j + 2hj^3 - 4hj - 1$

See Example **3** 7. Colette is putting a mat of width $3w$ and a frame of width w around a 16-inch by 48-inch poster. Find an expression for the amount of frame material she needs.
$128 + 32w$ in.

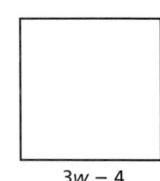

16 in.
48 in.
$3w$
w
$3w$
w

INDEPENDENT PRACTICE

See Example **1** Add.

8. $(4x^2y - 3xy + 2) + (6xy - 2x^2y)$ $2x^2y + 3xy + 2$

9. $(3g - 7) + (5g^2 - 2g + 6)$ $5g^2 + g - 1$

10. $(6bc - 3b^2c^2 + 9bc^2) + (4bc - 2bc^2)$ $10bc - 3b^2c^2 + 7bc^2$

11. $(7h^4 + 3h - 2h^6) + (h^6 - 4h + 2h^4)$ $-h^6 + 9h^4 - h$

12. $(3pq - 4p^2q + 7pq^2) + (5p^2q - 9pq^2) + (2pq^2 - 5pq + 4p^2q)$
  $-2pq + 5p^2q$

See Example **2** 13. $(7t^2 + 3t + 2) + (4t^2 - 7t + 8)$ $11t^2 - 4t + 10$

14. $(6b^3c^2 - 4b^2c + 3bc) + (9b^3c^2 - 4bc + 12) + (2b^2c - 6bc - 8)$

15. $(w^2 - 4w + 6) + (-3w - 4w^2 - 2) + (2w^2 + w - 7)$ $-w^2 - 6w - 3$

See Example **3** 16. Each side of an equilateral triangle has length $w + 2$. Each side of a square has length $3w - 4$. Write an expression for the sum of the perimeter of the equilateral triangle and the perimeter of the square.
$15w - 10$

$w + 2$ $3w - 4$

Math Background

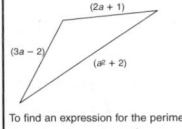

Adding polynomials vertically demonstrates the same principle as using place value to add numbers. Consider the examples below.

$$3x^2 + 4x + 1$$
$$+\ x^2 + 5x + 6$$
$$\overline{4x^2 + 9x + 7}$$

$$341 \longrightarrow 3(10^2) + 4(10) + 1$$
$$+\ 156 \longrightarrow 1(10^2) + 5(10) + 6$$
$$\overline{4(10^2) + 9(10) + 7}$$

RETEACH 13-3

PRACTICE 13-3

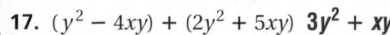

PRACTICE AND PROBLEM SOLVING

Add.

17. $(y^2 - 4xy) + (2y^2 + 5xy)$ **$3y^2 + xy$**

18. $(3x^2 - 2x + 1) + (5x - 4x^2 - 5)$ **$-x^2 + 3x - 4$**

19. $(5s^4t - 6st^3 + 4st^2) + (3st^4 - 8s^4t)$ **$-3s^4t - 6st^3 + 4st^2 + 3st^4$**

20. $(3ab - 5a + 2ab^3) + (3a - 4ab^3)$ **$3ab - 2a - 2ab^3$**

21. $(4w^2y + 2wy^2 - 3wy) + (4wy - 3wy^2 + 8w^2y) + (2wy^2 - 6wy - 4w^2y)$

22. $(4p^2t - 5pt + 7) + (p^2t + 4pt^2 - 5pt) + (3 - 7pt^2 + 2p^2t)$
     $7p^2t - 10pt + 10 - 3pt^2$

23. **BUSINESS** The cost of producing n toys at a factory is given by the polynomial $0.5n^2 + 3n + 12$. The cost of packaging is $0.25n^2 + 5n + 4$. Write and simplify an expression for the total cost of producing and packaging n toys. **$0.75n^2 + 8n + 16$**

24. **GEOMETRY** Write and simplify an expression for the combined volumes of a sphere with volume $\frac{4}{3}\pi r^3$, a cube with volume r^3, and a prism with volume $10r^3 - 5r^2 - 5r$. Use 3.14 for π.
     about $15.19r^3 - 5r^2 - 5r$

25. **TRANSPORTATION** Two airplanes are traveling in opposite directions. After 2 hours, one airplane is $x^2 + 2x + 400$ miles from the airport, and the other airplane is $3x^2 - 50x + 100$ miles from the same airport. How far apart are the two airplanes after 2 hours?
     $4x^2 - 48x + 500$ miles

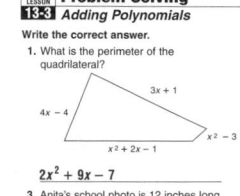

26. **WRITE A PROBLEM** A plane leaves an airport heading north at $x + 3$ mi/h. At the same time, another plane leaves the same airport, heading south at $x + 4$ mi/h. Write a problem using the speeds of both planes.

27. **WRITE ABOUT IT** Explain how to add polynomials.

28. **CHALLENGE** What polynomial would have to be added to $4x^2 - 5x + 6$ so that the sum is $2x^2 + 5x - 8$? **$-2x^2 + 10x - 14$**

Spiral Review

Find the volume of each figure. (Lesson 6-6)

29. a cube with edges 8 cm long **512 cm^3**

30. a cylinder with radius 5 in. and height 10 in.
     $250\pi \approx 785.4$ in^3

Solve for x. (Lesson 7-4)

31. $\frac{4}{5} = \frac{x}{60}$ **48**

32. $\frac{5}{x} = \frac{90}{36}$ **2**

33. $\frac{3}{7} = \frac{x}{30}$ **$\frac{90}{7}$**

34. $\frac{4.8}{2.5} = \frac{x}{17.5}$ **33.6**

35. **TEST PREP** A scale model of a shopping mall is 20 feet long. The actual shopping mall will be 1800 feet long and 45 feet tall. How tall is the model? (Lesson 7-8) **C**

   A 5 feet   B 5 inches   C 6 inches   D 50 feet

Answers

21. $8w^2y + wy^2 - 5wy$

26. Possible answer: How far apart will the planes be after 3 hours? Answer: $6x + 21$ mi

27. Possible answer: First identify like terms. Then add like terms by adding their coefficients.

Journal

Show students the following polynomial addition problem:

$(2x^2 + 3x + 4) + (3x^2 + 5x + 3)$.

Have them compare this problem to the addition problem $234 + 353$.

Test Prep Doctor

For Exercise 35, encourage students to set up a proportion to solve the problem. Remind them to use ratios that compare a dimension of the model with a dimension of the actual object. In this case, the proportion $\frac{20}{1800} = \frac{h}{45}$ gives $h = 0.5$ feet, or 6 in. This is choice **C**.

Lesson Quiz

Add.

1. $(2m^2 - 3m + 7) + (7m^2 - 1)$
   $9m^2 - 3m + 6$

2. $(yz^2 + 5yz + 7) + (2yz^2 - yz)$
   $3yz^2 + 4yz + 7$

3. $(2xy^2 + 2x - 6)$
   $+ (5xy^2 + 3y + 8)$
   _____
   $7xy^2 + 2x + 3y + 2$

4. $(3np^3 + 4n)$
   $(5np^3 - n - 6)$
   $+ \quad (2n - 3)$
   _____
   $8np^3 + 5n - 9$

5. The base of an isosceles triangle has length $x + 4$. The two legs of the triangle have lengths $3x + y$. Write an expression for the perimeter of the triangle.
   $7x + 2y + 4$

Available on Daily Transparency in CRB

Pacing: Traditional 1 day
Block $\frac{1}{2}$ day
Objective: Students subtract polynomials.

Warm Up

Write the opposite of each integer.

1. 10 −10   **2.** −7 7

Subtract.

3. 19 − (−12)   **4.** −16 − 21
31   −37

Add.

5. $(3x^2 + 7) + (x^2 - 3x)$
$4x^2 - 3x + 7$

6. $(2m^2 - 3m) + (-5m^2 + 2)$
$-3m^2 - 3m + 2$

Problem of the Day

Tara has 4 pairs of shorts, 3 tops, and 2 pairs of sandals. If she wants to wear a completely different outfit than she wore yesterday, how many combinations does she have to choose from? **6**

Available on Daily Transparency in CRB

Math Humor

Why was the expression $\frac{0}{3x^5}$ always cold? It was 5 degrees below zero.

13-4 Subtracting Polynomials

Learn to subtract polynomials.

Manufacturers can use polynomials to estimate the cost of making a product and the revenue from sales. To estimate profits, they would subtract these polynomials.

Subtraction is the opposite of addition. To subtract a polynomial, you need to find its opposite.

EXAMPLE 1 Finding the Opposite of a Polynomial

Find the opposite of each polynomial.

A $9x^2y^4z$
$-(9x^2y^4z)$
$-9x^2y^4z$ *The opposite of a is −a.*

B $10x^2 - 3x$
$-(10x^2 - 3x)$
$-10x^2 + 3x$ *Distribute the sign.*

C $-2ab^2 - 3ab + 2$
$-(-2ab^2 - 3ab + 2)$
$2ab^2 + 3ab - 2$ *Distribute the sign.*

To subtract a polynomial, add its opposite.

EXAMPLE 2 Subtracting Polynomials Horizontally

Subtract.

A $(n^3 - n + 4n^2) - (6n - 3n^2 + 8)$
$= (n^3 - n + 4n^2) + (-6n + 3n^2 - 8)$ *Add the opposite.*
$= n^3 - n + 4n^2 - 6n + 3n^2 - 8$ *Associative Property*
$= n^3 + 7n^2 - 7n - 8$ *Combine like terms.*

B $(-3cd^2 + cd + 6) - (-9cd^2 + 2 - 7cd)$
$= (-3cd^2 + cd + 6) + (9cd^2 - 2 + 7cd)$ *Add the opposite.*
$= -3cd^2 + cd + 6 + 9cd^2 - 2 + 7cd$ *Associative Property*
$= 6cd^2 + 8cd + 4$ *Combine like terms.*

---

① Introduce

Alternate Opener

 EXPLORATION

13-4 Subtracting Polynomials

You can use algebra tiles to find the opposite of a polynomial. To do this, replace each tile with its opposite.

The opposite of
$x^2 + 3x - 2$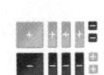
is
$-x^2 - 3x + 2$.

Use algebra tiles to find the opposite of each polynomial.

1. $4x^2 - 2x - 5$
2. $-x^2 - 7x + 3$
3. $2x^2 + 3x - 3$

Think and Discuss

4. Explain how you can use tiles to find opposites.
5. Discuss how opposites of polynomials might be useful if you were subtracting polynomials.

Motivate

Ask students what steps they would take to solve the subtraction problem 8 − (−8). Possible answer: Change subtraction to addition of the opposite. Remind students that they have learned this concept before (Lesson 2-2). Explain that the same principle applies to subtraction of polynomials.

Exploration worksheet and answers on Chapter 13 Resource Book pp. 36 and 77

② Teach

 **Lesson Presentation**

Guided Instruction

In this lesson, students learn to subtract polynomials. First make sure students know how to find and express the opposite of a polynomial. Emphasize the importance of distributing the opposite sign to every term in the polynomial that is being subtracted. Students can do this operation horizontally or vertically.

Teaching Tip It may be helpful to point out that distributing a negative sign results in changing the sign of every term of a polynomial.

You can also subtract polynomials in a vertical format. Write the second polynomial below the first one, lining up the like terms.

EXAMPLE 3 **Subtracting Polynomials Vertically**

Subtract.

A $(x^3 + 3x + 1) - (5x^3 + 2x + 4)$

$$
\begin{array}{ll}
(x^3 + 3x + 1) & x^3 + 3x + 1 \\
- (5x^3 + 2x + 4) \longrightarrow & + -5x^3 - 2x - 4 \\
\hline
& -4x^3 + x - 3
\end{array}
$$
Add the opposite.

B $(3m^2n - 4mn - 3m) - (-9m^2n - 7mn + 2)$

$$
\begin{array}{ll}
(3m^2n - 4mn - 3m) & 3m^2n - 4mn - 3m \\
- (-9m^2n - 7mn + 2) \longrightarrow & + 9m^2n + 7mn \qquad - 2 \\
\hline
& 12m^2n + 3mn - 3m - 2
\end{array}
$$
Add the opposite.

C $(3x^2y^2 + xy - 5x) - (6x + 4xy - 5)$

$$
\begin{array}{ll}
(3x^2y^2 + xy - 5x) & 3x^2y^2 + xy - 5x \\
- (6x + 4xy - 5) \longrightarrow & + \qquad - 4xy - 6x + 5 \\
\hline
& 3x^2y^2 - 3xy - 11x + 5
\end{array}
$$
Rearrange terms as needed.

EXAMPLE 4 **Business Application**

Suppose the cost in dollars of producing *x* model kits is given by the polynomial 500,000 + 2*x* and the revenue generated from sales is given by the polynomial 30*x* − 0.00005*x*². Find a polynomial expression for the profit from making and selling *x* model kits, and evaluate the expression for *x* = 300,000.

$$
\begin{array}{ll}
30x - 0.00005x^2 - (500,000 + 2x) & \text{revenue} - \text{cost} \\
30x - 0.00005x^2 + (-500,000 - 2x) & \text{Add the opposite.} \\
30x - 0.00005x^2 - 500,000 - 2x & \text{Associative Property} \\
28x - 0.00005x^2 - 500,000 & \text{Combine like terms.}
\end{array}
$$

The profit is given by the polynomial 28*x* − 0.00005*x*² − 500,000.
For *x* = 300,000,

$$28(300,000) - 0.00005(300,000)^2 - 500,000 = 3,400,000$$

The profit is $3,400,000, or $3.4 million.

Think and Discuss

1. **Explain** how to find the opposite of a polynomial.

2. **Compare** subtracting polynomials with adding polynomials.

3 Close

Summarize

Remind students that to subtract a polynomial, they must add its opposite. Remind them that the opposite of a polynomial contains the opposite of every monomial in the polynomial.

Answers to Think and Discuss

Possible answers:

1. Change the sign of every term in the polynomial to its opposite.

2. To subtract a polynomial, you first change the polynomial that is being subtracted to its opposite, and then you add.

13-4 **Exercises**

FOR EXTRA PRACTICE
see page 757

internet connect
Homework Help Online
go.hrw.com Keyword: MP4 13-4

Students may want to refer back to the lesson examples.

GUIDED PRACTICE

See Example 1 Find the opposite of each polynomial.

1. $4x^2y$ $-4x^2y$

2. $-4x + 3xy^4$ $4x - 3xy^4$

3. $2x^2 - 7x + 4$ $-2x^2 + 7x - 4$

4. $-6y^2 - 3y + 5$ $6y^2 + 3y - 5$

See Example 2 Subtract.

5. $(2b^3 + 5b^2 - 8) - (4b^3 + b - 12)$ $-2b^3 + 5b^2 - b + 4$

6. $9b - (3b^2 + 5b - 10)$ $-3b^2 + 4b + 10$

7. $(3m^2n - 6mn + 2mn^2) - (-4mn - 3m^2n)$ $6m^2n + 2mn^2 - 2mn$

See Example 3 8. $(7x^2 - 5x + 3) - (4x^2 + 3x + 5)$ $3x^2 - 8x - 2$

9. $(-2x^2y - xy + 3x - 4) - (4xy - 7x + 4)$ $-2x^2y - 5xy + 10x - 8$

10. $(-4ab^2 + 3ab - 2a^2b) - (6 - 4ab + 2ab^2 + 5a^2b)$
$-6ab^2 - 7a^2b + 7ab - 6$

See Example 4 11. The volume of a rectangular prism, in cubic inches, is given by the expression $x^3 + 2x^2 - 4x + 6$. The volume of a smaller rectangular prism is given by the expression $4x^3 - 5x^2 + 6x - 12$. How much greater is the volume of the larger rectangular prism?
$-3x^3 + 7x^2 - 10x + 18$ in^3

INDEPENDENT PRACTICE

See Example 1 Find the opposite of each polynomial.

12. $-3rn^2$ $3rn^2$

13. $2v - 4v^2$ $-2v + 4v^2$

14. $3m^2 - 5m + 1$
$-3m^2 + 5m - 1$

15. $4xy^2 + 2xy$ $-4xy^2 - 2xy$

See Example 2 Subtract.

16. $(4w^2 + 2w + 4) - (2w^2 + 3w - 4)$ $2w^2 - w + 8$

17. $(12a + a^2) - (5 + a^2 + 6a)$ $6a - 5$

18. $(5r^2s^2 - 3rs^2 + 4r^2s + 5rs) - (2rs^2 - 2r^2s + 6rs)$ $5r^2s^2 - 5rs^2 + 6r^2s - rs$

See Example 3 19. $(5x^2 + 7x - 1) - (2x^2 + 8x - 4)$ $3x^2 - x + 3$

20. $(3a^2b^2 - 4ab - 2a - 4) - (4a^2b^2 + 5a - 3b + 6)$

21. $(3pt^2 - 5p^3 + 4p^2t^2) - (4p^2 - 5pt^2 + 6p^2t^2)$

See Example 4 22. The population of a bacteria colony after h hours is $4h^3 - 5h^2 + 2h + 200$. The population of another bacteria colony is $3h^3 - 2h^2 + 5h + 100$. Write an expression to show the difference of the two populations.

Assignment Guide

If you finished Example **1** assign:
Core 1–4, 12–15, 35–38
Enriched 1–4, 12–15, 35–38

If you finished Example **2** assign:
Core 1–7, 12–18, 35–38
Enriched 1–7, 12–18, 35–38

If you finished Example **3** assign:
Core 1–10, 12–21, 35–38
Enriched 12–21, 23–28, 32–38

If you finished Example **4** assign:
Core 1–22, 35–38
Enriched 12–38

Answers

20. $-a^2b^2 - 4ab - 7a + 3b - 10$
21. $-5p^3 - 4p^2 - 2p^2t^2 + 8pt^2$
22. $h^3 - 3h^2 - 3h + 100$

Math Background

The concept of distributing a minus sign through a polynomial can be difficult to grasp. It may be useful to insert a coefficient of 1 in front of the parentheses before distributing. The example below shows one method of finding the opposite of $4x^2 - 2x + 7$.

$$-(4x^2 - 2x + 7)$$
$$-1(4x^2 - 2x + 7)$$
$$(-1)(4x^2) + (-1)(-2x) + (-1)(7)$$
$$-4x^2 + 2x - 7$$

RETEACH 13-4

LESSON 13-4 Reteach
Subtracting Polynomials

When subtracting polynomials, you have to remember to distribute a factor of −1.

Subtract. $(5x^2 + 7x + 3) - (4x^2 + 3x - 5)$.

Rewrite the expression. $(5x^2 + 7x + 3) + (-1)(4x^2 + 3x - 5)$.

Apply the Distributive property.

$-1(4x^2 + 3x - 5) = (-1 \cdot 4x^2) + (-1 \cdot 3x) + (-1 \cdot -5) = -4x^2 - 3x + 5$

Notice how distributing the −1 changes the sign of each term. Now you have the following: $(5x^2 + 7x + 3) + (-4x^2 - 3x + 5)$.

You can now use the Associative Property to remove parentheses and combine like terms.

$5x^2 + 7x + 3 - 4x^2 - 3x + 5 = x^2 + 4x + 8$

Subtract.

1. $(3b^3 + 4b^2 + 6) - (b^3 - 5b - 3)$
$3b^3 + 4b^2 + 6 + -1(b^3 - 5b - 3)$ Rewrite the expression.
$3b^3 + 4b^2 + 6 + (-b^3 + 5b + 3)$ Apply the Distributive property.
$3b^3 + 4b^2 + 6 - b^3 + 5b + 3$ Remove the parentheses.
$\underline{2b^3 + 4b^2 + 5b + 9}$

2. $(3m^2n^2 - 4m^2n + m^2) - (m^2n + 5m^2n - 5)$
$\underline{2m^2n^2 - 9m^2n + m^2 + 5}$

3. $(2x^3y^2 + x^2y - 4) - (x^2y - 8x + 3)$
$\underline{2x^3y^2 + 8x - 7}$

4. $(6y^2 + 3xy - 9x^2) - (-4y^2 + 8xy + x^2)$
$\underline{10y^2 - 5xy - 10x^2}$

PRACTICE 13-4

LESSON 13-4 Practice B
Subtracting Polynomials

Find the opposite of each polynomial.

1. $18xy^3$
$\underline{-18xy^3}$

2. $-9a + 4$
$\underline{9a - 4}$

3. $6d^2 - 2d - 8$
$\underline{-6d^2 + 2d + 8}$

Subtract.

4. $(4n^3 - 4n + 4n^2) - (6n + 3n^2 - 8)$
$\underline{4n^3 + n^2 - 10n + 8}$

5. $(-2h^4 + 3h - 4) - (2h - 3h^4 + 2)$
$\underline{h^4 + h - 6}$

6. $(6m + 2m^2 - 7) - (-6m^2 - m - 7)$
$\underline{8m^2 + 7m}$

7. $(17x^2 - x + 3) - (14x^2 + 3x + 5)$
$\underline{3x^2 - 4x - 2}$

8. $w + 7 - (3w^4 + 5w^3 - 7w^2 + 2w - 10)$
$\underline{-3w^4 - 5w^3 + 7w^2 - w + 17}$

9. $(9r^3s - 3rs + 4rs^3 + 5r^2s^3) - (2rs^2 - 2r^2s^2 + 6rs + 7r^3s - 9)$
$\underline{2r^3s + 7r^2s^2 + 4rs^3 - 2rs^2 - 9rs + 9}$

10. $(3qr^2 - 2 + 14q^2r^2 - 9qr) - (-10qr + 11 - 5qr^2 + 6q^2r^2)$
$\underline{8q^2r^2 + 8qr^2 + qr - 13}$

11. The volume of a rectangular prism, in cubic meters, is given by the expression $x^3 + 7x^2 + 14x + 8$. The volume of a smaller rectangular prism is given by the expression $x^3 + 5x^2 + 6x$. How much greater is the volume of the larger rectangular prism?
$\underline{2x^2 + 8x + 8 \text{ cubic meters}}$

12. Sarah has a table with an area of $y^2 + 30y + 200$ square inches. She has a tablecloth with an area of $y^2 + 18y + 80$ square inches. She wants the tablecloth to cover the top of the table. What is the expression that represents the number of square inches of additional fabric she needs to cover the top of the table?
$\underline{12y + 120 \text{ more square inches of fabric}}$

Subtract.

23. $(3s^2 - 4s + 2) - (5s + 7)$ $3s^2 - 9s - 5$

24. $(2x^3 - 4x + 1) - (3x^2 + 2x - 4)$ $2x^3 - 3x^2 - 6x + 5$

25. $(3g^2h + 2gh) - (5gh + 2g^2h)$ $g^2h - 3gh$

26. $(5a + 2b - 4ab) - (5a + 4b - 6ab)$ $-2b + 2ab$

27. $(3pq^2 - 5p^2q + 2pq) - (6pq^2 + 6p^2q - 2pq)$ $-3pq^2 - 11p^2q + 4pq$

28. $(8y^2 - 4x^2y + x^2) - (2y^2 + 6x^2y - 3x^2)$ $6y^2 - 10x^2y + 4x^2$

29. The area of the rectangle is $2a^2 - 4a + 5$ cm². The area of the square is $a^2 - 2a - 6$ cm². What is the area of the shaded region?
$a^2 - 2a + 11$ cm²

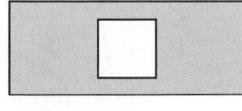

30. The area of the square is $4x^2 - 2x - 6$ in². The area of the triangle is $2x^2 + 4x - 5$ in². What is the area of the shaded region?
$2x^2 - 6x - 1$ in²

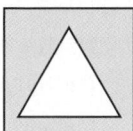

31. *BUSINESS* The price in dollars of one share of stock after y years is modeled by the expression $4y^3 - 5y + 6.25$. The price of one share of another stock is modeled by $4y^3 + 20y + 22.5$. What expression shows the difference in price of the two stocks after y years?
$25y + 16.25$ dollars

 32. *CHOOSE A STRATEGY* Which polynomial has the greatest value when $x = 5$?   **B**

A $x^2 - 2x + 6$ **C** $-x^3 - 40x - 300$

B $3x^4 + 6x + 12$ **D** $x^5 - 120x^4 + 10$

 33. *WRITE ABOUT IT* Explain how to subtract the polynomial $4x^3 - 2x - 8$ from $3x^3 + 8x + 1$.

34. *CHALLENGE* Find the values of $a, b, c,$ and d that make the equation true. $(2t^3 - at^2 - 4bt - 6) - (ct^3 + 4t^2 + 7t + 1) = 4t^3 - 5t^2 - 15t + d$ **$a = 1; b = 2; c = -2; d = -7$**

Spiral Review

Add or subtract. (Lesson 3-5)

35. $\frac{7}{8} + \frac{1}{6}$ $1\frac{1}{24}$ **36.** $4\frac{2}{3} + 5\frac{3}{4}$ $10\frac{5}{12}$ **37.** $6\frac{5}{8} - 2\frac{1}{20}$ $4\frac{23}{40}$

38. **TEST PREP** Which shape **cannot** be used to create a tessellation of a plane? (Lesson 5-9)   **D**

A Square   **B** Equilateral triangle   **C** Regular hexagon   **D** Regular pentagon

Answers

33. Possible answer: Find the opposite of $4x^3 - 2x - 8$, and add it to $3x^3 + 8x + 1$. The result is $-x^3 + 10x + 9$.

Journal

Ask students to consider whether the difference of two polynomials can be zero. Ask them to explain their answer and give an example.

Test Prep Doctor

For Exercise 38, remind students that a regular polygon can tessellate a plane only if the measure of each interior angle is a factor of 360°. This condition is true for a square (90°), an equilateral triangle (60°), and a regular hexagon (120°). However, this is not true for a regular pentagon (108°). The correct choice is **D**.

Lesson Quiz

Find the opposite of each polynomial.

1. $3a^2b^2c^3$
$-3a^2b^2c^3$

2. $-3m^3 + 2m^2n$
$3m^3 - 2m^2n$

Subtract.

3. $(3z^2 - 7z + 6) - (2z^2 + z - 12)$
$z^2 - 8z + 18$

4. $-18h^3 - (4h^3 + h^2 - 12h + 2)$
$-22h^3 - h^2 + 12h - 2$

5. $(3b^2c + 5bc^2 - 8b^2)$
$\ - (4b^2c + 2bc^2 - c^2)$
$-b^2c + 3bc^2 - 8b^2 + c^2$

Available on Daily Transparency in CRB

Warm Up

Multiply. Write each product as one power.

1. $x \cdot x$ x^2 **2.** $6^2 \cdot 6^3$ 6^5

3. $k^2 \cdot k^8$ k^{10} **4.** $19^5 \cdot 19^2$ 19^7

5. $m \cdot m^5$ m^6 **6.** $26^6 \cdot 26^5$ 26^{11}

7. Find the volume of a rectangular prism that measures 5 cm by 2 cm by 6 cm.   **60 cm³**

Problem of the Day

Charlie added 3 binomials, 2 trinomials, and 1 monomial. What is the greatest possible number of terms in the sum?   **13**

Available on Daily Transparency in CRB

Why did the binomial get mad when the mathematician added the term $4x^3$? He didn't like being given the third degree.

Learn to multiply polynomials by monomials.

Carlos is making a stained-glass box with a square base. He wants the box's height to be 2 inches less than the side length of its base. The volume of the box is found by multiplying a polynomial by a monomial.

Remember that when you multiply two powers with the same bases, you add the exponents. To multiply two monomials, multiply the coefficients and add the exponents of the variables that are the same.

$$(5m^2n^3)(6m^3n^6) = 5 \cdot 6 \cdot m^{2+3}n^{3+6} = 30m^5n^9$$

EXAMPLE 1   **Multiplying Monomials**

Multiply.

A $(3r^2s^3)(5r^4s^5)$

    $(3r^2s^3)(5r^4s^5)$

    $15r^6s^8$ *Multiply coefficients and add exponents.*

B $(7x^2y)(-3x^4yz^8)$

    $(7x^2y)(-3x^4yz^8)$

    $-21x^6y^2z^8$ *Multiply coefficients and add exponents.*

To multiply a polynomial by a monomial, use the Distributive Property. Multiply every term of the polynomial by the monomial.

EXAMPLE 2   **Multiplying a Polynomial by a Monomial**

Multiply.

A $\frac{1}{2}h(b_1 + b_2)$

    $\frac{1}{2}h(b_1 + b_2)$ *Multiply each term in the parentheses by $\frac{1}{2}h$.*

    $\frac{1}{2}b_1h + \frac{1}{2}b_2h$

B $-4a^2b\,(2a^4b^3 + 5a^2b^3)$

    $-4a^2b\,(2a^4b^3 + 5a^2b^3)$

    $-8a^6b^4 - 20a^4b^4$ *Multiply each term in the parentheses by $-4a^2b$.*

---

1   Introduce

Alternate Opener

EXPLORATION

13-5 Multiplying Polynomials by Monomials

You can use algebra tiles to model multiplication. The model shows that $2(-3) = -6$.

Find each product.

1. $2(2x - 3)$ **2.** $x(2x - 3)$

Use algebra tiles to model and find each product.

3. $2(3x + 2)$ **4.** $2x(3x - 1)$

Think and Discuss

5. Explain how area applies to modeling multiplication.

6. Write the factors and product modeled by the tiles shown.

Motivate

Ask students to add the following polynomials:

$$\begin{aligned} 2x^2 + 3x - 6 \\ 2x^2 + 3x - 6 \\ 2x^2 + 3x - 6 \\ + \; 2x^2 + 3x - 6 \end{aligned}$$

The sum is $8x^2 + 12x - 24$. Ask students whether they can tell you another way to find the answer. Multiply $4(2x^2 + 3x - 6)$.

Exploration worksheet and answers on Chapter 13 Resource Book pp. 45 and 79

2   Teach

Lesson Presentation

Guided Instruction

In this lesson, students learn to multiply polynomials by monomials. Review the rules for multiplying powers (Lesson 2-7). Show students how to multiply monomials by multiplying coefficients and adding the exponents of powers with the same bases. Then show them how to use the Distributive Property to multiply a polynomial by a monomial. Remind students to pay attention to the sign of each term when they multiply.

Multiply.

C $4rs^2(r^2s^4 + 2rs^3 - 3rst)$

$4rs^2(r^2s^4 + 2rs^3 - 3rst)$
$4r^3s^6 + 8r^2s^5 - 12r^2s^3t$

Multiply each term in the parentheses by $4rs^2$.

EXAMPLE 3 **PROBLEM SOLVING APPLICATION**

Carlos is making a stained-glass box with a square base. He wants the height of the box to be 2 inches less than the side length of the base. If he wants the volume of the box to be 32 in³, what should the side length of the base be?

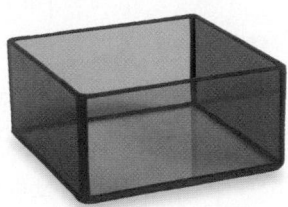

 Understand the Problem

If the side length of the base is s, then the height is $s - 2$. The volume is $s \cdot s \cdot (s - 2) = s^2(s - 2)$. The **answer** will be a value of s that makes the volume of the box equal to 32 in³.

 Make a Plan

You can make a table of values for the polynomial to try to find the value of s. Use the Distributive Property to write the expression $s^2(s - 2)$ another way. Use substitution to complete the table.

3 **Solve**

$s^2(s - 2) = s^3 - 2s^2$ *Distributive Property*

s	1	2	3	4
$s^3 - 2s^2$	$1^3 - 2(1)^2$ $= -1$	$2^3 - 2(2)^2$ $= 0$	$3^3 - 2(3)^2$ $= 9$	$4^3 - 2(4)^2$ $= 32$

The side length of the base should be 4 inches.

 Look Back

If the side length of the base were 4 inches, and the height was 2 inches less, or 2 inches, then the volume would be $4 \cdot 4 \cdot 2 = 32$ inches. The answer is reasonable.

Think and Discuss

1. **Compare** multiplying two monomials with multiplying a polynomial by a monomial.

Additional Examples

Example 1

Multiply.

A. $(2x^3y^2)(6x^5y^3)$ $12x^8y^5$

B. $(9a^5b^7)(-2a^4b^3)$ $-18a^9b^{10}$

Example 2

Multiply.

A. $3m(5m^2 + 2m)$
   $15m^3 + 6m^2$

B. $-6x^2y^3(5xy^4 + 3x^4)$
   $-30x^3y^7 - 18x^6y^3$

C. $-5y^3(y^2 + 6y - 8)$
   $-5y^5 - 30y^4 + 40y^3$

Example 3

The length of a picture in a frame is 8 in. less than three times its width. Find the length and the width if the area is 60 in². $\ell = 10$ in.; $w = 6$ in.

---

3 Close

Reaching All Learners
Through Grouping Strategies

Divide students into groups of three. Give each student a card containing a monomial (Chapter 13 Resource Book p. 69). Have two students in each group add their expressions to create a binomial. (If they happen to have like terms, they can create a monomial.) Then have the group multiply the binomial (or monomial) by the remaining monomial. Repeat with each of the other two pairs in each group.

Summarize

Remind students that the Distributive Property can be used to multiply a polynomial by a monomial. Explain that the Distributive Property works for numbers, expressions, and polynomials. Ask students how many terms result from multiplying a polynomial by a monomial.

Possible answer: The number of terms in the resulting polynomial is the same as in the original polynomial.

Answers to Think and Discuss

1. Possible answer: To multiply two monomials, multiply the coefficients and add exponents of powers with the same bases. To multiply a polynomial by a monomial, apply the Distributive Property and multiply the resulting monomials.

FOR EXTRA PRACTICE
see page 757

internet connect
Homework Help Online
go.hrw.com   Keyword: MP4 13-5

go.hrw.com

Students may want to refer back to the lesson examples.

GUIDED PRACTICE

See Example **1** Multiply.

1. $(-4s^2t^2)(2st^3)$ $-8s^3t^5$

2. $(x^2y^3)(6x^4y^3)$ $6x^6y^6$

3. $(4h^2j^4)(-6h^4j^6)$ $-24h^6j^{10}$

4. $5m(3m^4)$ $15m^5$

See Example **2** 5. $2h(3m - 4h)$ $6hm - 8h^2$

6. $3ab(a^2b - ab^2)$ $3a^3b^2 - 3a^2b^3$

7. $-2x(x^2 - 4x + 12)$
$-2x^3 + 8x^2 - 24x$

8. $5c^2d(2cd^3 - 4c^3d^2 + 3cd)$
$10c^3d^4 - 20c^5d^3 + 15c^3d^2$

See Example **3** 9. The formula for the area of a trapezoid is $A = \frac{1}{2}h(b_1 + b_2)$, where h is the trapezoid's height and b_1 and b_2 are the lengths of its bases. Multiply to write the expression another way. Then use the expression to find the area of a trapezoid with height 10 in. and base lengths 8 in. and 6 in. $A = \frac{1}{2}b_1h + \frac{1}{2}b_2h$; 70 in^2

Assignment Guide

If you finished Example **1** assign:
Core 1–4, 10–13, 37–41
Enriched 10–13, 21–23, 37–41

If you finished Example **2** assign:
Core 1–8, 10–19, 21–31 odd, 37–41
Enriched 10–19, 21–32, 34–35, 37–41

If you finished Example **3** assign:
Core 1–20, 21–33 odd, 37–41
Enriched 10–41

INDEPENDENT PRACTICE

See Example **1** Multiply.

10. $(5x^2y^5)(-2xy^4)$ $-10x^3y^9$

11. $(-gh^3)(-3g^2h^5)$ $3g^3h^8$

12. $(4a^2b)(2b^3)$ $8a^2b^4$

13. $(-s^4t^3)(st)$ $-s^5t^4$

See Example **2** 14. $(2m^2n^3)(1 - 4mn^4)$
$2m^2n^3 - 8m^3n^7$

15. $2z(4z^2 - 3z)$ $8z^3 - 6z^2$

16. $-2h^2(4h + 2h^3)$

17. $-3cd(2c^3d^2 - 4cd^2)$

18. $-3b(5b^4 - 8b + 12)$
$-15b^5 + 24b^2 - 36b$

19. $-4s^2t^2(5s^2t + 6st - 2s^2t^2)$
$-20s^4t^3 - 24s^3t^3 + 8s^4t^4$

See Example **3** 20. A rectangle has a base of length $3x^2y$ and a height of $2x^3 - 4xy - 3$. Write and simplify an expression for the area of the rectangle. Then find the area of the rectangle if $x = 2$ and $y = 1$.
$6x^5y - 12x^3y^2 - 9x^2y$; 60 units^2

Answers

16. $-8h^3 - 4h^5$

17. $-6c^4d^3 + 12c^2d^3$

26. $12a^3b^2 + 12a^2b^3$

27. $x^4 - x^5y^4$

28. $mx + 3m$

29. $2f^2g^2 + f^3g^2 - f^2g^5$

30. $x^4 - 3x^3 + 7x^2$

31. $12m^4p^8 - 6m^3p^7 + 15m^4p^5$

32. $-10w^5z^3 - 6w^2z^3 + 8w^3z^3$

PRACTICE AND PROBLEM SOLVING

Multiply.

21. $(-4b^2)(9b^4)$ $-36b^6$

22. $(5m^2n)(3mn^4)$ $15m^3n^5$

23. $(-2a^2b^2)(-3ab^4)$ $6a^3b^6$

24. $9g(g - 7)$ $9g^2 - 63g$

25. $-2m^2(m^3 - 6m)$ $-2m^5 + 12m^3$ 26. $3ab(4a^2b + 4ab^2)$

27. $x^3(x - x^2y^4)$

28. $m(x + 3)$

29. $f^2g^2(2 + f - g^3)$

30. $x^2(x^2 - 3x + 7)$

31. $(3m^2p^4)(4m^2p^4 - 2mp^3 + 5m^2p)$ 32. $-2wz(5w^4z^2 + 3wz^2 - 4w^2z^2)$

Math Background

Factoring the greatest common factor from a polynomial is an important algebraic skill that is essentially the opposite of multiplying a polynomial by a monomial. (See Chapter 13 Extension.) Understanding how to multiply polynomials is a prerequisite for factoring. When you factor the GCF from a polynomial, you divide out the greatest monomial factor that is common to all terms. Then you write the new expression as a monomial times a polynomial. For example, the greatest common factor of the polynomial $4x^4 + 6x^3 + 8x^2$ is $2x^2$. Factoring the polynomial yields $2x^2(2x^2 + 3x + 4)$.

RETEACH 13-5

LESSON **13-5** Reteach
Multiplying Polynomials by Monomials

Remember that a monomial has only one term. To multiply a monomial by a monomial, follow the steps used in the example below.

Multiply $(7x^2y^3)(3xy^4)$.

1. Multiply the coefficients.
$(7)(3) = 21$

2. Multiply the variables.
To multiply two powers with the same base, you keep the base and **add** the exponents.
$(x^2)(x) = (x^2)(x^1) = x^3$ $(y^3)(y^4) = y^7$

Remember: If there is no exponent on a variable, the exponent is 1.
$x = x^1$

3. Write the monomial product.
$21x^3y^7$

Multiply.

1. $(3x^2)(4x^3y^2)$
$12x^5y^2$

2. $(6a^3b)(2a^3b^4)$
$12a^6b^5$

3. $(2m^4n^2)(-5m^2n^2)$
$-10m^6n^4$

When you multiply a polynomial by a monomial, you multiply each term of the polynomial by the monomial.

$\begin{array}{r} 4a^2 + 2ab + 6b^2 \\ \times \qquad\qquad 3a^3 \\ \hline 12a^5 + 6a^4b + 18a^3b^2 \end{array}$

Multiply.

4. $\begin{array}{r} 3r^2s^3 - 2r^2 + 10 \\ \times \qquad\qquad 2s \\ \hline 6r^2s^4 - 4r^2s + 20s \end{array}$

5. $\begin{array}{r} 5x^5 + x^2 - 3x \\ \times \qquad\qquad 4x^3 \\ \hline 20x^8 + 4x^5 - 12x^4 \end{array}$

6. $\begin{array}{r} m^2n - 3mn^2 - 8n^3 \\ \times \qquad\qquad -3mn \\ \hline -3m^3n^2 + 9m^2n^3 + 24mn^4 \end{array}$

PRACTICE 13-5

LESSON **13-5** Practice B
Multiplying Polynomials by Monomials

Multiply.

1. $(x^2)(-3x^2y^3)$
$-3x^4y^3$

2. $(-9pr^4)(p^2r^2)$
$-9p^3r^6$

3. $(2st^9)(-st^2)$
$-2s^2t^{11}$

4. $(3efg^2)(-3e^2f^2g)$
$-9e^3f^3g^3$

5. $2q(4q^2 - 2)$
$8q^3 - 4q$

6. $-x(x^2 + 2)$
$-x^3 - 2x$

7. $5m(-3m^2 + 2m)$
$-15m^3 + 10m^2$

8. $6x(-x^5 + 2x^3 + x)$
$-6x^6 + 12x^4 + 6x^2$

9. $-4st(st - 12t - 2s)$
$-4s^2t^2 + 48st^2 + 8s^2t$

10. $-9ab(a^2 + 2ab - b^2)$
$-9a^3b - 18a^2b^2 + 9ab^3$

11. $-7v^2w^2(vw^2 + 2vw + 1)$
$-7v^3w^4 - 14v^3w^3 - 7v^2w^2$

12. $8p^4(p^2 - 8p + 17)$
$8p^6 - 64p^5 + 136p^4$

13. $4x(-x^2 - 2xy + 3)$
$-4x^3 - 8x^2y + 12x$

14. $7x^2(3x^2y + 7x^2 - 2x)$
$21x^4y + 49x^4 - 14x^3$

15. $-4t^3r^2(3t^2r - t^5r - 6t^2r^2)$
$-12t^5r^3 + 4t^8r^3 + 24t^5r^4$

16. $h^2k(2hk^2 - hk + 7k)$
$2h^3k^3 - h^3k^2 + 7h^2k^2$

17. A triangle has a base of $4x^2$ and a height of $6x + 3$. Write and simplify an expression for the area of the triangle.
$12x^3 + 6x^2$

33. HEALTH The table gives some formulas for finding the target heart rate for a person of age a exercising at p percent of his or her maximum heart rate.

Target Heart Rate		
	Male	Female
Nonathletic	$p(220 - a)$	$p(226 - a)$
Fit	$\frac{1}{2}p(410 - a)$	$\frac{1}{2}p(422 - a)$

a. Use the Distributive Property to write each expression another way.

b. Use your answer from part **a** to write an expression for the difference between the target heart rate for a fit male and for a fit female, both of age a and exercising at p percent of their maximum heart rates.

34. WHAT'S THE QUESTION? A square prism has a base area of x^2 and a height of $2x + 3$. If the answer is $2x^3 + 3x^2$, what is the question? If the answer is $10x^2 + 12x$, what is the question?

35. WRITE ABOUT IT If a polynomial is multiplied by a monomial, what can you say about the number of terms in the answer? What can you say about the degree of the answer?

36. CHALLENGE On a multiple-choice test, if the probability of guessing each question correctly is p, then the probability of guessing two or more correctly out of four is $6p^2(1 - 2p - p^2) + 4(1 - p) + p^4$. Simplify the expression, and write an expression for the probability of guessing fewer than two out of four correctly.
$6p^2 - 12p^3 - 5p^4 + 4 - 4p; -3 - 6p^2 + 12p^3 + 5p^4 + 4p$

Spiral Review

Classify each triangle by its angle measurements and by the lengths of its sides. (Lesson 5-3)

37. 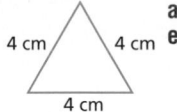 4 cm 4 cm 4 cm **acute equilateral**

38. 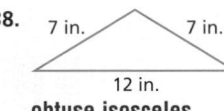 7 in. 7 in. 12 in. **obtuse isosceles**

39. 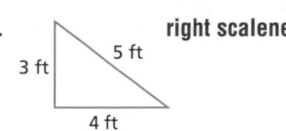 5 ft 3 ft 4 ft **right scalene**

40. TEST PREP Which of the following sets of three lengths could represent the lengths of the sides of a right triangle? (Lesson 6-3) **B**

A 6, 8, 12   **B** 5, 12, 13   **C** 1, 1, 2   **D** 3, 5, 8

41. TEST PREP There are 72 boys in the eighth-grade class at Lincoln Middle School. The other 55% of the class are girls. How many girls are there? (Lesson 8-3) **H**

F 55   **G** 127   **H** 88   **J** 72

33. a.

	Male	Female
Nonathletic	$220p - pa$	$226p - pa$
Fit	$205p - \frac{1}{2}pa$	$211p - \frac{1}{2}pa$

b. $-6p$

34. Possible answers: What is the volume of the prism? What is the surface area of the prism?

35. The number of terms in the answer is the same as the number of terms in the polynomial. The degree of the answer is the degree of the polynomial plus the degree of the monomial.

Journal

Ask students to explain how multiplying 321 by 3 is similar to multiplying a polynomial by a monomial.

Test Prep Doctor

For Exercise 41, students can use logic to eliminate some choices. Because 55% of the class are girls, there must be more than 72 girls in the class. So choices **F** and **J** are incorrect. An estimate should determine that choice **G** is also incorrect. The correct answer is choice **H.**

CHALLENGE 13-5

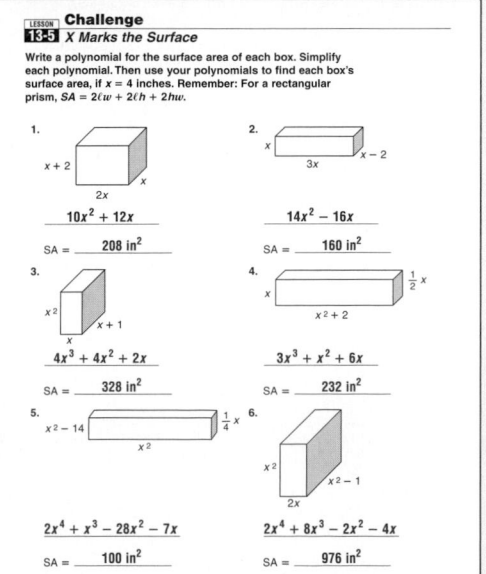

Challenge
13-5 *X Marks the Surface*

Write a polynomial for the surface area of each box. Simplify each polynomial. Then use your polynomials to find each box's surface area, if $x = 4$ inches. Remember: For a rectangular prism, $SA = 2\ell w + 2\ell h + 2hw$.

1. $x + 2$; $2x$; x
$10x^2 + 12x$
$SA = \underline{208 \text{ in}^2}$

2. $3x$; $x - 2$; x
$14x^2 - 16x$
$SA = \underline{160 \text{ in}^2}$

3. x^2 ; $x + 1$; x
$4x^3 + 4x^2 + 2x$
$SA = \underline{328 \text{ in}^2}$

4. x ; $\frac{1}{2}x$; $x^2 + 2$
$3x^3 + x^2 + 6x$
$SA = \underline{232 \text{ in}^2}$

5. $x^2 - 14$; x^2
$2x^4 + x^3 - 28x^2 - 7x$
$SA = \underline{100 \text{ in}^2}$

6. $\frac{1}{4}x$; x^2 ; $2x$; $x^2 - 1$
$2x^4 + 8x^3 - 2x^2 - 4x$
$SA = \underline{976 \text{ in}^2}$

PROBLEM SOLVING 13-5

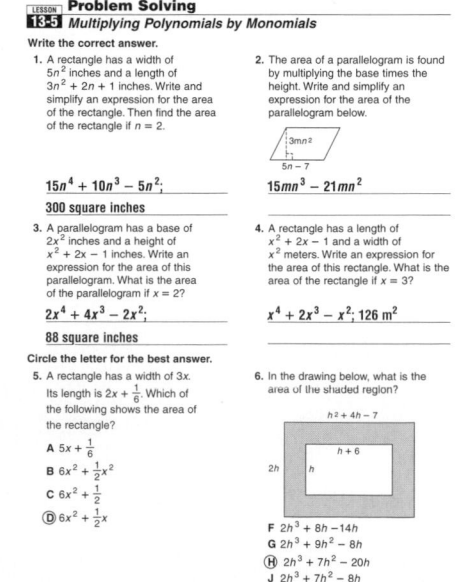

Problem Solving
13-5 *Multiplying Polynomials by Monomials*

Write the correct answer.

1. A rectangle has a width of $5n^2$ inches and a length of $3n^2 + 2n + 1$ inches. Write and simplify an expression for the area of the rectangle. Then find the area of the rectangle if $n = 2$.
$15n^4 + 10n^3 - 5n^2$;
300 square inches

2. The area of a parallelogram is found by multiplying the base times the height. Write and simplify an expression for the area of the parallelogram below.
$3mn2$; $5n - 7$
$15mn^3 - 21mn^2$

3. A parallelogram has a base of $2x^2$ inches and a height of $x^2 + 2x - 1$ inches. Write an expression for the area of this parallelogram. What is the area of the parallelogram if $x = 2$?
$2x^4 + 4x^3 - 2x^2$;
88 square inches

4. A rectangle has a length of $x^2 + 2x - 1$ and a width of x^2 meters. Write an expression for the area of this rectangle. What is the area of the rectangle if $x = 3$?
$x^4 + 2x^3 - x^2$; 126 m²

Circle the letter for the best answer.

5. A rectangle has a width of $3x$. Its length is $2x + \frac{1}{6}$. Which of the following shows the area of the rectangle?

A $5x + \frac{1}{6}$
B $6x^2 + \frac{1}{2}x^2$
C $6x^2 + \frac{1}{2}$
Ⓓ $6x^2 + \frac{1}{2}x$

6. In the drawing below, what is the area of the shaded region?
$h^2 + 4h - 7$; $h + 6$; $2h$; h

F $2h^3 + 8h - 14h$
G $2h^3 + 9h^2 - 8h$
Ⓗ $2h^3 + 7h^2 - 20h$
J $2h^3 + 7h^2 - 8h$

Lesson Quiz

Multiply.

1. $(3a^2b)(2ab^2)$ $6a^3b^3$

2. $(4x^2y^2z)(-5xy^3z^2)$ $-20x^3y^5z^3$

3. $3n(2n^3 - 3n)$ $6n^4 - 9n^2$

4. $-5p^2(3q - 6p)$ $-15p^2q + 30p^3$

5. $-2xy(2x^2 + 2y^2 - 2)$
$-4x^3y - 4xy^3 + 4xy$

6. The width of a garden is 5 feet less than 2 times its length. Find the garden's length and width if its area is 63 ft².
$\ell = 7 \text{ ft}, w = 9 \text{ ft}$

Available on Daily Transparency in CRB

13B Multiply Binomials

Pacing: Traditional 1 day
Block $\frac{1}{2}$ day

Objective: To use algebra tiles to model multiplying binomials

Materials: Algebra tiles

Lab Resources

Hands-On Lab Activities . . . . pp. 87–88

Using the Pages

Discuss with students what each algebra tile represents and how to model the product of two binomials.

Use algebra tiles to model each product.

1. $(x + 2)(x - 4)$

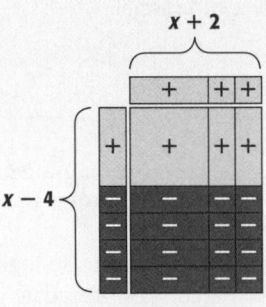

$x^2 - 2x - 8$

2. $(2x - 1)(x + 3)$

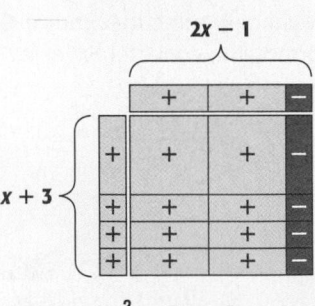

$2x^2 + 5x - 3$

Multiply Binomials

Use with Lesson 13-6

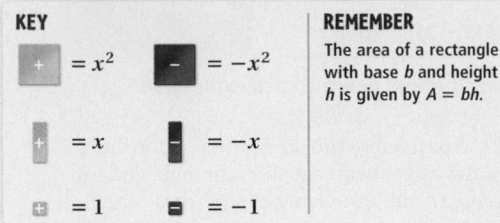

KEY

$\blacksquare = x^2$ $\blacksquare = -x^2$

$\blacksquare = x$ $\blacksquare = -x$

$\blacksquare = 1$ $\blacksquare = -1$

REMEMBER

The area of a rectangle with base b and height h is given by $A = bh$.

You can use algebra tiles to find the product of two binomials.

Activity 1

❶ To model the product of $(x + 3)(2x + 1)$ with algebra tiles, make a rectangle with base $x + 3$ and height $2x + 1$.

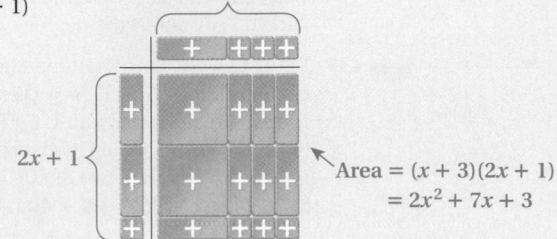

Area $= (x + 3)(2x + 1)$
$= 2x^2 + 7x + 3$

❷ Use algebra tiles to find the product of $(x - 2)(-x + 1)$.

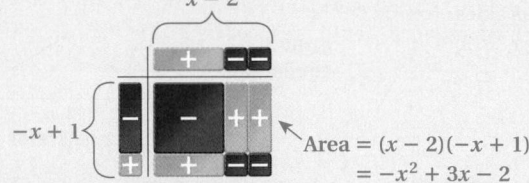

Area $= (x - 2)(-x + 1)$
$= -x^2 + 3x - 2$

Think and Discuss

1. Explain how to determine the signs of each term in the product when you are multiplying $(x - 4)(x - 3)$.

2. How can you use algebra tiles to find $(x + 2)(x - 2)$?

Try This

Use algebra tiles to find each product.

1. $(x + 5)(x - 5)$
2. $(x - 4)(x + 3)$
3. $(x - 6)(-x + 2)$

Activity 2

1 Write two binomials whose product is modeled by the algebra tiles below, and then write the product as a polynomial expression.

The base of the rectangle is $x - 5$ and the height is $x - 2$, so the binomial product is $(x - 5)(x - 2)$.

The model shows one x^2-tile, seven $-x$-tiles, and ten 1-tiles, so the polynomial expression is $x^2 - 7x + 10$.

Think and Discuss

1. Write an expression modeled by the algebra tiles below. How many zero pairs are modeled? Describe them.

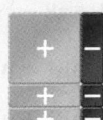

Try This

Write two binomials whose product is modeled by each set of algebra tiles below, and then write the product as a polynomial expression.

1.
2.
3.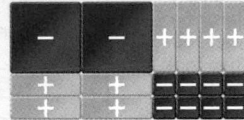

Activity 2

Think and Discuss

1. $x^2 + x - 2$; one zero pair is modeled; the pair is one yellow x-tile and one red x-tile.

Try This

1. $(x + 2)(x + 1)$; $x^2 + 3x + 2$
2. $(x - 3)(x + 3)$; $x^2 - 9$
3. $(-2x + 4)(x - 2)$; $-2x^2 + 8x - 8$

Answers

Activity 1

Think and Discuss

Possible answers:

1. The product of two negatives or two positives is positive, so the x^2 and constant terms are positive. The product of a positive and a negative is negative, so the x-terms are negative.

2. Make a rectangle with base $x + 2$ and height $x - 2$. The product has one yellow x^2-tile, 2 red x-tiles, 2 yellow x-tiles, and 4 red unit tiles. The 2 red and 2 yellow x-tiles form 2 zero pairs. The product is $x^2 - 4$.

Try This

1.

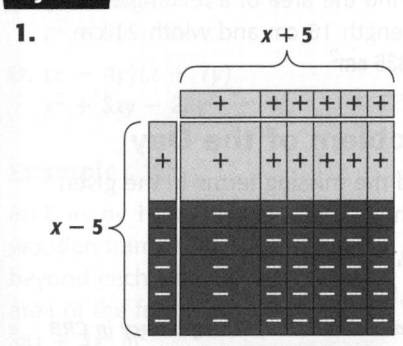

$$(x + 5)(x - 5) = x^2 - 25$$

2.

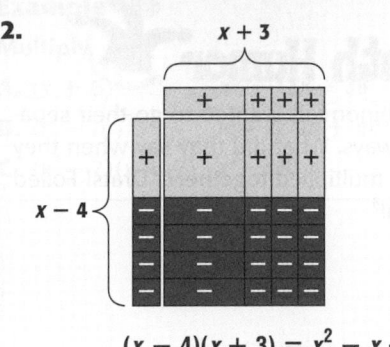

$$(x - 4)(x + 3) = x^2 - x - 12$$

3.

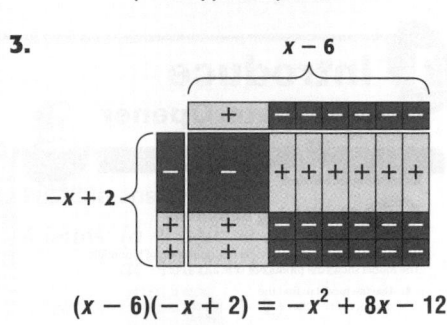

$$(x - 6)(-x + 2) = -x^2 + 8x - 12$$

Purpose: *To provide additional practice for problem-solving skills in Chapters 1–13*

The Brooklyn Bridge

- After problem 1a, have students discuss why *x* is zero at the bridge's center. *x* is the distance from the bridge's center.

- After problem 2, have students discuss the series of steps they need to take to find the length of a cable in the main span. Find the cable heights used in the formula for *n*. Determine *n* by using the formula. Substitute the values for *n* and *K* into the formula $K\left(1 + \frac{8}{3}n^2\right)$, and evaluate.

Extension Have students research suspension bridges. Ask them to compare the Brooklyn Bridge to others, comparing length, maximum allowable weight, and other characteristics. Have them present their findings on a poster or in a presentation. Check students' work.

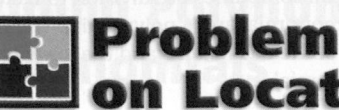

Problem Solving on Location

New York

The Brooklyn Bridge

A *suspension bridge* hangs from cables that are attached to tall towers. The weight of the bridge pulls the cables into a shape that is nearly a parabola. One famous suspension bridge is the Brooklyn Bridge in New York City. The Brooklyn Bridge spans the East River, connecting Brooklyn and Manhattan.

1. The height in feet above the river of the main span of the suspension cables can be approximated by the quadratic function $f(x) = 0.0002x^2 + 140$, where *x* is the horizontal distance from the bridge's center.

   a. Estimate the height of the cable at the bridge's center. **140 ft**

   b. Estimate the height of the cable at the towers. **267 ft**

   Main span: 1596 feet

2. The bridge has four suspension cables. The length of each cable in the main span is approximately $K\left(1 + \frac{8}{3}n^2\right)$, where *K* is the width of the main span and
   $$n = \frac{\text{cable height at towers} - \text{cable height at center}}{K}.$$

   a. Estimate the length of each cable in the main span of the bridge. **1623 ft**

   b. Use the fact that the diameter of each cable is $15\frac{3}{4}$ in. to estimate the volume of each cable in the main span. **2196 ft³**

3. In 1884, P. T. Barnum led a parade of 21 elephants across the Brooklyn Bridge. Suppose the elephants weighed an average of 4.5 tons each. If the four cables can each support 11,200 tons and the suspended part of the bridge weighs 6620 tons, what percent of the maximum allowable weight was the total weight of the elephants? **about 0.2%**

The Finger Lakes

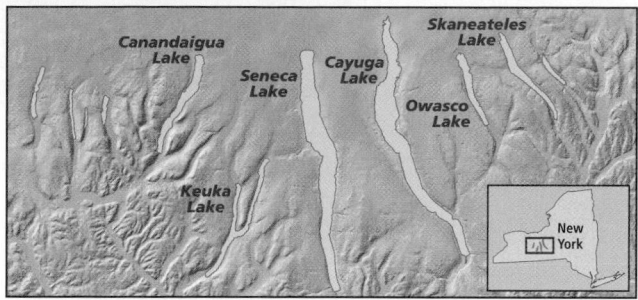

The Finger Lakes are a group of long, narrow lakes in central New York State. Some facts about the six largest Finger Lakes are given in the table.

1. The combined length of the Finger Lakes is 138 miles. Cayuga Lake is twice the length of Keuka Lake, and Canandaigua Lake is 0.8 times the length of Keuka Lake. Find the lengths of Cayuga, Canandaigua, and Keuka Lakes.
   40, 16, 20 miles

2. Because of its greater depth, Seneca Lake holds more water than the next five largest Finger Lakes combined. The sum of the volumes of Cayuga, Canandaigua, Skaneateles, Keuka, and Owasco Lakes is 230 billion gallons less than the volume of Seneca Lake. The volume of Seneca Lake is 10 times the volume of Skaneateles Lake. Find the volumes of Seneca and Skaneateles Lakes.
   4200, 420 billion gallons

Lake	Length (mi)	Surface Area (mi²)	Estimated Volume (billion gallons)
Seneca	36	67.7	▪ 4200
Cayuga	▪ 40	66.4	2500
Canandaigua	▪ 16	16.3	450
Skaneateles	15	13.9	▪ 420
Owasco	11	10.3	210
Keuka	▪ 20	18.1	390

3. The volume of a lake is the surface area of the lake times the lake's average depth. Write this as a formula, and solve the formula for the average depth. Then use your answers from problem **2** and the values given in the table to find the average depth to the nearest foot of each of the six largest Finger Lakes. (*Hint:* There are 5280 feet in a mile and 7.48 gallons in a cubic foot.)

Game Resources

Puzzles, Twisters & Teasers

Chapter 13 Resource Book

Short Cuts

Purpose: *To apply the skill of multiplying binomials to developing a multiplication trick*

Discuss: Ask students to explain how the trick works. What general expression can you write to represent a two-digit number that ends in 5? $10n + 5$, where n is the digit in the tens place What is the square of the expression? $100n^2 + 100n + 25$ or $100n(n + 1) + 25$ How does the square of the expression model a numerical answer? The expression $100n(n + 1)$ provides the hundreds (and thousands) place of the answer; the 25 provides the tens and ones places.

Extend: Challenge students to create a trick for multiplying two-digit numbers with a first digit of 2, and then with first digits of 3, 4, and 5. Ask them to look for a pattern.

$(20 + n)(20 + b) = 400 + 20(n + b) + nb$

$(30 + n)(30 + b) = 900 + 30(n + b) + nb$

$(40 + n)(40 + b) = 1600 + 40(n + b) + nb$

$(50 + n)(50 + b) = 2500 + 50(n + b) + nb$

Rolling for Tiles

Purpose: *To practice operations with monomials in a game format*

Discuss: When a student models an expression that can be added, subtracted, multiplied, or divided to equal a polynomial on the game board, have him or her demonstrate the operation for the class.

Extend: Have students create new polynomial expressions for the game board. Use the new expressions to play again.

Short Cuts

You can use properties of algebra to explain many arithmetic shortcuts. For example, to square a two-digit number that ends in 5, multiply the first digit by one more than the first digit, and then place a 25 at the end.

To find 35^2, multiply the first digit, 3, by one more than the first digit, 4. You get $3 \cdot 4 = 12$. Place a 25 at the end, and you get 1225. So $35^2 = 1225$.

Why does this shortcut work? You can use FOIL to multiply 35 by itself:

$$35^2 = 35 \cdot 35 = (30 + 5)(30 + 5) = 900 + 150 + 150 + 25$$
$$= 900 + 300 + 25$$
$$= 1200 + 25 \qquad \textit{1200 = 30 · 40}$$
$$= 1225$$

First use the shortcut to find each square. Then use FOIL to multiply the number by itself.

1. 15^2 **225**   **2.** 45^2 **2025**   **3.** 85^2 **7225**   **4.** 65^2 **4225**   **5.** 25^2 **625**

6. Can you explain why the shortcut works?

Use FOIL to multiply each pair of numbers.

7. $11 \cdot 14$ **154**   **8.** $12 \cdot 16$ **192**   **9.** $13 \cdot 15$ **195**   **10.** $14 \cdot 17$ **238**   **11.** $18 \cdot 19$ **342**

12. Write a shortcut for multiplying two-digit numbers with a first digit of 1.

Rolling for Tiles

For this game, you will need a number cube, a set of algebra tiles, and a game board. Roll the number cube, and draw an algebra tile:

$1 = $ ⬜ , $2 = $ ◼ , $3 = $ + , $4 = $ ▌ , $5 = $ + , $6 = $ ◼ .

The goal is to model expressions that can be added, subtracted, multiplied, or divided to equal the polynomials on the game board.

✦ internet connect

Go to **go.hrw.com** for a complete set of rules and a game board.
KEYWORD: MP4 Game13

Answers

6. Possible answer: The shortcut is an application of the FOIL method. The expression $(10n + 5)^2$ is equal to $100n^2 + 100n + 25$, or $100n(n + 1) + 25$. The 25 provides the tens and ones places, and $n(n + 1)$ provides the hundreds (and thousands) place(s).

12. Possible answer: The general product can be represented by $(10 + a)(10 + b)$. Applying FOIL yields $100 + 10(a + b) + ab$. So the shortcut is to add 100, 10 times the sum of the last 2 digits, and the product of the last 2 digits together.

Technology
LAB

Evaluate and Compare Polynomials

Use with Lesson 13-6

You can check the result of a polynomial operation by comparing the result to the original expression or expressions.

internet connect
Lab Resources Online
go.hrw.com
KEYWORD: MP4 TechLab13

Activity

1 Multiply $(x + 3)^2$.

Suppose your answer is $x^2 + 9$. Press **Y=**, and enter $(x + 3)^2$ as Y_1 and $x^2 + 9$ as Y_2, as shown.

Press **2nd** **GRAPH** (TABLE). You can see that the values of Y_1 and Y_2 are not equal, so $(x + 3)^2 \neq x^2 + 9$.

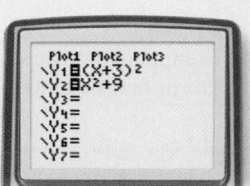

Press **Y=**, and change Y_2 to $x^2 + 6x + 9$. When you press **2nd** **GRAPH** (TABLE), you can see that the values of Y_1 and Y_2 are equal for all values of x shown in the table.

$(x + 3)^2 = x^2 + 6x + 9$

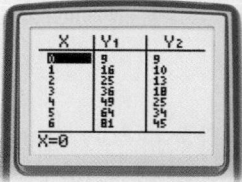

Think and Discuss

1. How could you use a table to subtract $x^2 - 3x + 2$ from $2x^2 + 3x - 1$ and verify that the difference is correct?

Try This

1. Multiply $(x - 4)^2$. Compare each of the following to $(x - 4)^2$: $x^2 - 16$, $x^2 - 8x + 16$, and $x^2 - 8x - 16$. Which expression is the product? $\quad x^2 - 8x + 16$

2. Multiply $(x + 7)(x - 7)$. Compare each of the following to $(x + 7)(x - 7)$: $x^2 - 7$, $x^2 - 49$, and $x^2 + 49$. Which expression is the product? $\quad x^2 - 49$

Answers

Think and Discuss

Possible answer:

1. Enter $2x^2 + 3x - 1 - (x^2 - 3x + 2)$ in **Y1,** and enter the expression you found for the difference in **Y2.** Press **2nd** **GRAPH** and compare.

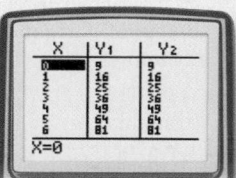

Technology
LAB

Evaluate and Compare Polynomials

Objective: To use a graphing calculator to evaluate and compare polynomials

Materials: Graphing calculator

Lab Resources
Technology Lab Activities . . . . p. 89

Using the Page

This technology activity shows students how to evaluate and compare polynomials, which can be done on a graphing calculator. Specific keystrokes may vary, depending on the make and model of the graphing calculator used. The keystrokes given are for a TI-83 model. For keystrokes to other models, visit go.hrw.com.

The Think and Discuss problem can be used to assess students' understanding of the technology activity. While Try This problems 1–2 can be done without a graphing calculator, they are meant to help students become familiar with using a graphing calculator to evaluate and compare polynomials.

Assessment

1. Explain how to use the graphing calculator to determine whether the expressions $x^2 + x + 1$ and $(x + 1)^2$ are equivalent. Enter $x^2 + x + 1$ in **Y1** and $(x + 1)^2$ in **Y2.** Create the table, and look at the values of **Y1** and **Y2** for each value of x. The values are not equal, so the expressions are not equivalent.

Purpose: *To help students review and practice concepts and skills presented in Chapter 13*

Assessment Resources

Chapter 13 Review
Chapter 13 Resource Book... pp. 62–63

 Test and Practice Generator CD-ROM

Additional review assessment items in both multiple-choice and free-response format may be generated for any objective in Chapter 13.

Answers

1. polynomial; degree

2. FOIL; binomials

3. binomial; trinomial

4. trinomial

5. not a polynomial

6. not a polynomial

7. monomial

8. not a polynomial

9. binomial

10. 8

11. 4

12. 3

13. 5

14. 6

Study Guide and Review

Vocabulary

binomial 644
degree of a polynomial 645
FOIL 670

monomial 644
polynomial 644
trinomial 644

Complete the sentences below with vocabulary words from the list above. Words may be used more than once.

1. $4x^3 - 10x^2 + 4x - 12$ is an example of a ___?___ whose ___?___ is 3.

2. Use the ___?___ method to find the product of two ___?___.

3. A polynomial with 2 terms is called a ___?___. A polynomial with 3 terms is called a ___?___.

13-1 Polynomials (pp. 644–647)

EXAMPLE

Classify each expression as a monomial, a binomial, a trinomial, or not a polynomial.

- $4x^5 - 2x^3 + 7$
  trinomial

- $4xy - \frac{3}{x^4} + 7x^2y^4$
  not a polynomial

Find the degree of each polynomial.

- $x^3 - 2x + 1$
  degree 3

- $n + 3n^4 + 16n^2$
  degree 4

EXERCISES

Classify each expression as a monomial, a binomial, a trinomial, or not a polynomial.

4. $-4t^2 + 6t - 7$

5. $r^{-3} + 2r^{-1} + 6$

6. $10g + 4g^5 - \frac{6}{g^3}$

7. $-4a^2b^3c^5$

8. $\sqrt{x} - 2\sqrt{xy}$

9. $5st - 6s$

Find the degree of each polynomial.

10. $-2x^5 - 7x^8 + 3x$

11. $x^4 - 3x^2 + 4x - 1$

12. $12 + 4r^2 - 6r^3$

13. $\frac{1}{2}m^3 - \frac{1}{4}m^5 + \frac{3}{8}m^2$

14. $-2x^6 + 4x^5 - 8x$

13-2 Simplifying Polynomials (pp. 650–653)

EXAMPLE

Simplify.

■ $5x^2 - 2x + 4 - 5x - 3 + 4x^2$

$\boxed{5x^2} - \boxed{2x} + \boxed{4} - \boxed{5x} - \boxed{3} + \boxed{4x^2}$

$9x^2 - 7x + 1$

■ $4(2x - 7) - 5x + 4$

$\boxed{8x} - \boxed{28} - \boxed{5x} + \boxed{4}$

$3x - 24$

EXERCISES

Simplify.

15. $3t^2 - 7t + 5t - 3t^2 + 6t^2 + 1$
16. $4gh - 5g^2h + 7gh - 4g^2h$
17. $3(4mn - 2m)$
18. $4(2a^2 - 4b) + 6b$
19. $4(3st^2 - 5t) + 14st^2 + 5t$

13-3 Adding Polynomials (pp. 656–659)

EXAMPLE

Add.

■ $(3x^2 - 2x) + (5x^2 + 3x + 2)$

$\boxed{3x^2} - \boxed{2x} + \boxed{5x^2} + \boxed{3x} + 2$ *Identify like terms.*

$8x^2 + x + 2$ *Combine like terms.*

■ $(8t^3 + 4t + 6) + (4t^2 - 7t - 2)$

$$
\begin{array}{l}
8t^3 \qquad\quad + 4t + 6 \\
+ \qquad 4t^2 - 7t - 2 \\
\hline
8t^3 + 4t^2 - 3t + 4
\end{array}
$$
Place like terms in columns.
Combine like terms.

EXERCISES

Add.

20. $(4x^2 + 3x - 7) + (2x^2 - 5x + 12)$
21. $(4x^4 - 2x^2 + 3x - 1) + (3x^2 - 4x + 8)$
22. $(6h + 6) + (3h^2 + 4) + (2h - 1)$
23. $(2xy^2 - 4x^2y - 3xy) + (2x^2y + 5xy - xy^2)$
24. $(4n^2 + 6) + (3n^2 - 2) + (8 + 6n^2)$

13-4 Subtracting Polynomials (pp. 660–663)

EXAMPLE

■ Subtract.

$(5x^2 - 3x + 4) - (6x^2 - 7x + 1)$

$5x^2 - 3x + 4 + (-6x^2 + 7x - 1)$ *Add the opposite.*

$5x^2 - 3x + 4 - 6x^2 + 7x - 1$ *Associative Property*

$-x^2 + 4x + 3$ *Combine like terms.*

EXERCISES

Subtract.

25. $(x^2 - 3) - (3 - 4x^2)$
26. $(w^2 - 4w + 6) - (2w^2 + 8w - 8)$
27. $(2x^2 + 7x - 8) - (6x^2 - 7x + 4)$
28. $(4ab^2 - 5ab + 7a^2b) - (3a^2b + 6ab)$
29. $(4p^3q^2 - 5p^2q^2) - (2pq^2 + 5p^3q^2)$

Answers

15. $6t^2 - 2t + 1$
16. $11gh - 9g^2h$
17. $12mn - 6m$
18. $8a^2 - 10b$
19. $26st^2 - 15t$
20. $6x^2 - 2x + 5$
21. $4x^4 + x^2 - x + 7$
22. $3h^2 + 8h + 9$
23. $xy^2 - 2x^2y + 2xy$
24. $13n^2 + 12$
25. $5x^2 - 6$
26. $-w^2 - 12w + 14$
27. $-4x^2 + 14x - 12$
28. $4ab^2 - 11ab + 4a^2b$
29. $-p^3q^2 - 5p^2q^2 - 2pq^2$

Answers

30. $5s^2t^3 - 10s^2t^4 + 35st^3$

31. $12a^4b^3 + 30a^3b^3 - 36a^3b + 24a^2b^2$

32. $6m^3 - 15m^2 + 3m$

33. $-24gh^5 + 12g^3h^3 - 30h^2 + 12gh$

34. $2j^5k^3 - \frac{3}{2}j^4k^4 + j^6k^5$

35. $-8x^6y^{12} + 10x^7y^{14} - 14x^3y^6 + 6x^3y^7$

36. $p^2 - 8p + 15$

37. $b^2 + 10b + 24$

38. $4r^2 + 19r - 5$

39. $2a^2 - 5ab - 12b^2$

40. $m^2 - 16m + 64$

41. $4t^2 - 25$

42. $8b^2 + 4bt - 40t^2$

43. $100 - 4x^2$

44. $y^2 - 20y + 100$

Study Guide and Review

13-5 Multiplying Polynomials by Monomials (pp. 664–667)

EXAMPLE

Multiply.

■ $(4x^3y^4)(3xy^3)$
Multiply the coefficients, and add the exponents of the variables.
$(4x^3y^4)(3xy^3)$
$4 \cdot 3 \cdot x^{3+1}y^{4+3}$
$12x^4y^7$

■ $(-2ab^2)(4a^2b^2 - 3ab + 6a - 8)$

$(-2ab^2)(4a^2b^2 - 3ab + 6a - 8)$
$-8a^3b^4 + 6a^2b^3 - 12a^2b^2 + 16ab^2$

EXERCISES

Multiply.

30. $(5st^3)(s - 2st + 7)$

31. $-6a^2b(-2a^2b^2 - 5ab^2 + 6a - 4b)$

32. $3m(2m^2 - 5m + 1)$

33. $-6h(4gh^4 - 2g^3h^2 + 5h - 2g)$

34. $\frac{1}{2}j^3k^2(4j^2k - 3jk^2 + 2j^3k^3)$

35. $2x^2y^5(-4x^4y^7 + 5x^5y^9 - 7xy + 3xy^2)$

13-6 Multiplying Binomials (pp. 670–673)

EXAMPLE

Multiply.

■ $(r + 7)(r - 5)$

$(r + 7)(r - 5)$

$r^2 - 5r + 7r - 35$
$r^2 + 2r - 35$

■ $(b + 5)^2$

$(b + 5)(b + 5)$

$b^2 + 5b + 5b + 25$
$b^2 + 10b + 25$

EXERCISES

Multiply.

36. $(p - 5)(p - 3)$

37. $(b + 4)(b + 6)$

38. $(4r - 1)(r + 5)$

39. $(2a + 3b)(a - 4b)$

40. $(m - 8)^2$

41. $(2t - 5)(2t + 5)$

42. $(4b - 8t)(2b + 5t)$

43. $(20 - 4x)(5 + x)$

44. $(y - 10)^2$

Classify each expression as a monomial, a binomial, a trinomial, or not a polynomial.

1. $-2t^4 + 3t - t^{0.5}$

2. $-\frac{2}{3}a^4b^7$

3. $5m^4 - 3t + 4$

4. $4 + n^2$

5. $f^2g^3 - \sqrt{g}$

6. 5

Find the degree of each polynomial.

7. $3b^7 - 8b^{10} + 6b - 12$

8. $5 - 8m + 3m^4$

9. $6 + y$

10. $x^2 - 4x + 6$

11. $6a^3 - \frac{1}{5}a^6 + 11a^2$

12. $7h + 4h^7 - 2h^3$

Simplify.

13. $2a - 4b - 5b + 6a - 2b$

14. $2(x^2 - 7x + 12)$

15. $-2x^2y + 3xy^2 - 4x^2y + 2x^2y$

16. $12m^2 + 4m + 3(2m - 4m^2 + 5)$

17. $5(4a^2b - 3a^3b^2 + ab) - 2ab + 6a^2b$

18. $2(x^2y - 4xy^3 - 3x^2y^2) + 8xy^3 + 4x^2y$

Add.

19. $(2x^2 + 4x - 8) + (6x^2 - 9x - 1)$

20. $(2r^3 - 8r + 2) + (5r^3 - 2r^2 - 8r + 7)$

21. $(5st^3 - 6s^2t^2 + 4st^2) + (2s^2t^2 - 8st^2 + 2st^3)$

22. $(x + y^2) + (2y^2 - 6x + y) + (y^2 - 4y + 4)$

23. Harold is placing a mat of width $w + 4$ around a 16 in. by 20 in. portrait. Find an expression for the perimeter of the outer edge of the mat.

Subtract.

24. $(5x^2 + x - 1) - (2x^2 + 4x - 8)$

25. $(3m^3 - 2m^2 - 4m + 2) - (6m^3 - 8m - 1)$

26. $(3a^2b - 5a^2b^2 + 6ab^2) - (2a^2b^2 - 7a^2b)$

27. $(j^4 + 7j^2 - 4j) - (5j^3 - 2j^2 + 6j + 1)$

28. A circle whose area is $2x^2 + 3x - 4$ is cut from a rectangular piece of plywood with area $4x^2 - 3x - 1$ and discarded. Find an expression for the area of the remaining plywood.

Multiply.

29. $(3x)(5x^4)$

30. $(2x^2y)(-4xy^3)$

31. $(2a^2b^4)(5a^4b^5)$

32. $a(a^2 - 3a + 7)$

33. $3m^3n^4(2m^3n^4 - 5m^2n^2)$

34. $5a^4(ab^3 - 2ab + 6a)$

35. $(x + 2)(x + 12)$

36. $(x + 3)(x - 4)$

37. $(a - 3)(a - 7)$

38. $(x + 4)(2x + 6)$

39. $(x + 3)(x - 3)$

40. $(x - 12)^2$

29. $15x^5$

30. $-8x^3y^4$

31. $10a^6b^9$

32. $a^3 - 3a^2 + 7a$

33. $6m^6n^8 - 15m^5n^6$

34. $5a^5b^3 - 10a^5b + 30a^5$

35. $x^2 + 14x + 24$

36. $x^2 - x - 12$

37. $a^2 - 10a + 21$

38. $2x^2 + 14x + 24$

39. $x^2 - 9$

40. $x^2 - 24x + 144$

Purpose: *To assess students' mastery of concepts and skills in Chapter 13*

Assessment Resources

Chapter 13 Tests (Levels A, B, C)
Assessment Resources.... pp. 105–110

Test and Practice Generator CD-ROM

Additional assessment items in both multiple-choice and free-response format may be generated for any objective in Chapter 13.

Answers

1. not a polynomial

2. monomial

3. trinomial

4. binomial

5. not a polynomial

6. monomial

7. 10

8. 4

9. 1

10. 2

11. 6

12. 7

13. $8a - 11b$

14. $2x^2 - 14x + 24$

15. $-4x^2y + 3xy^2$

16. $10m + 15$

17. $26a^2b - 15a^3b^2 + 3ab$

18. $6x^2y - 6x^2y^2$

19. $8x^2 - 5x - 9$

20. $7r^3 - 16r + 9 - 2r^2$

21. $7st^3 - 4s^2t^2 - 4st^2$

22. $-5x + 4y^2 - 3y + 4$

23. $8w + 104$ in.

24. $3x^2 - 3x + 7$

25. $-3m^3 - 2m^2 + 4m + 3$

26. $10a^2b - 7a^2b^2 + 6ab^2$

27. $j^4 + 9j^2 - 10j - 5j^3 - 1$

28. $2x^2 - 6x + 3$ units^2

Chapter 13 Performance Assessment

Purpose: To assess students' understanding of concepts in Chapter 13 and combined problem-solving skills

Assessment Resources

Performance Assessment
Assessment Resources . . . . p. 142

Performance Assessment
Teacher Support
Assessment Resources . . . . p. 141

Answers

1–4. See page A18.

5. See Level 3 work sample below.

Scoring Rubric for Problem Solving Item 5

Level 3
Accomplishes the purpose of the task.

Student gives clear explanations, shows understanding of mathematical ideas and processes, and computes accurately.

Level 2
Purposes of the task not fully achieved.

Student demonstrates satisfactory but limited understanding of the mathematical ideas and processes.

Level 1
Purposes of the task not accomplished.

Student shows little evidence of understanding the mathematical ideas and processes and makes computational and/or procedural errors.

Performance Assessment

 Show What You Know

Create a portfolio of your work from this chapter. Complete this page and include it with your four best pieces of work from Chapter 13. Choose from your homework or lab assignments, mid-chapter quiz, or any journal entries you have done. Put them together using any design you want. Make your portfolio represent what you consider your best work.

⭐ Short Response

1. Simplify $2a + 3a$. Explain how the Distributive Property is used to simplify the expression.

2. Tell what polynomial you would have to add to $3x - 6y$ so that the sum is $6x + 2y$. Explain how you found the polynomial.

3. Can the product of two binomials be a binomial? Explain.

4. The polynomial $x^2 + x + 41$ was discovered by Leonard Euler in 1772. For integer values of x from 0 to 49, the value of the polynomial is prime. Use this polynomial to find at least 5 prime numbers. Show all your steps.

🧩 Extended Problem Solving

5. A box is made by cutting two squares from a 16 in. by 25 in. piece of cardboard and folding the sides as shown.

   a. Write an expression for the length, width, and height of the box in terms of x.

   b. Multiply the expressions from part a to find a polynomial that gives the volume of the box.

   c. Evaluate the polynomial for $x = 1$, $x = 2$, $x = 3$, and $x = 4$. Which value of x gives the box with the largest volume? Give the dimensions and the volume of the largest box.

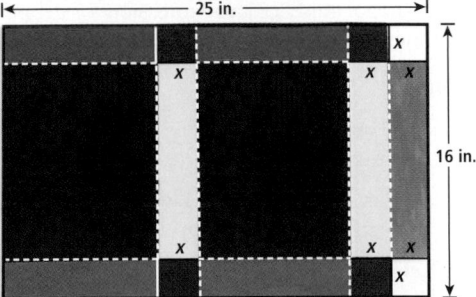

Student Work Samples for Item 5

Level 3	Level 2	Level 1

Level 3

a. length $= \frac{25-3x}{2}$ or $12.5-1.5x$
   width $= 16-2x$
   height $= x$

b. $(12.5-1.5x)(16-2x)(x) =$
   $(200-25x-24x+3x^2)x =$
   $200x - 49x^2 + 3x^3$

c. $x=1, 154$ $x=2, 228$
   $x=3, 240$ $x=4, 208$
   largest
   height $=3$, width$=10$, length$=8$
   volume $= 240$ in^3

Level 2

Ⓐ $\ell = 25-3x$ $w=16-2x$ $h=x$

Ⓑ $(25-3x)(16-2x)(x) =$
   $400x - 98x^2 - 6x^3$

Ⓒ $296,$ $360,$ $156, -352$
   Box is $2 \times 12 \times 19$
   Volume is 456

Level 1

a. $25, 16, x$

b. $400x$

c. 1600 in

The student correctly created the expressions, multiplied them together, and identified the dimensions of the box that maximized volume.

The student applied the proper method at each step but created one expression incorrectly. The student also failed to check the reasonableness of the answer.

The student demonstrated very little understanding of polynomials.

Cumulative Assessment, Chapters 1–13

1. The solution of $12x = -24$ is ___?___. **B**
   (A) $x = -288$
   (C) $x = 2$
   (B) $x = -2$
   (D) $x = 288$

2. If the product of five integers is positive, then at most how many of the five integers could be negative? **H**
   (F) Two
   (H) Four
   (G) Three
   (J) Five

TEST TAKING TIP!
If a problem involves decimals, you may be able to eliminate answer choices that do not have the correct number of places after the decimal point.

3. Find the product of 1.8×0.541. **A**
   (A) 0.9738
   (C) 97.3800
   (B) 9.738
   (D) 9.738×10^4

4. The simplest form of the product of the binomials $(x + 2)$ and $(x - 3)$ is which type of polynomial? **H**
   (F) Monomial
   (G) Binomial
   (H) Trinomial
   (J) Polynomial with four terms

5. Which number is equivalent to 2^{-3}? **C**
   (A) $-\frac{1}{6}$
   (C) $\frac{1}{8}$
   (B) $-\frac{1}{8}$
   (D) $\frac{1}{6}$

6. What is the length of the diagonal of a rectangle with a length of 4 in. and width of 3 in.? **F**
   (F) 5 in.
   (H) 12 in.
   (G) 7 in.
   (J) 14 in.

7. For which set of data are the mean and mode equal? **B**
   (A) 1, 1, 1, 2
   (B) 1, 2, 2, 3
   (C) 2, 3, 4, 5
   (D) 2, 3, 3, 5

8. Point R' is formed by reflecting $R(-3, -2)$ across the y-axis. What are the coordinates of R'? **G**
   (F) (3, 2)
   (H) (−3, 2)
   (G) (3, −2)
   (J) (−2, −3)

9. **SHORT RESPONSE** A fair number cube is rolled twice. What is the probability that the outcomes of the two rolls will have a sum of 4? Explain.

10. **SHORT RESPONSE** What is the area of the shaded region in the figure below? Give your answer in terms of π. Show or explain how you got your answer.

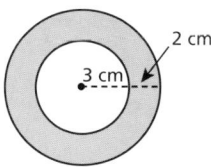
2 cm
3 cm

Standardized Test Prep

Assessment Resources

Cumulative Tests (Levels A, B, C)
Assessment Resources. . . . pp. 289–300

State-Specific Test Practice Online
KEYWORD: MP4 TestPrep

Test Prep Doctor

For item 7, remind students that for the mode to equal the mean, it must be a repeated value near the middle of an ordered list. Choice **A** can be eliminated because the mode is the minimum value, and choice **C** can be eliminated because there is no mode. Finding the modes for choices **B** and **D** will show that **B** is the correct answer.

Expand on the test-taking tip given before item 3 by pointing out that students should work the first step of the multiplication to see whether there will be any zeros at the end of the product before they eliminate any answer choices.

Answers

9. $\frac{1}{12}$; Possible answer: There are 36 possible outcomes, and 3 of those have sums of 4 ($1 + 3$, $2 + 2$, and $3 + 1$). So the probability is $\frac{3}{36}$, or $\frac{1}{12}$.

10. 16π cm^2; Possible answer: The area of the shaded region is the difference of the area of the large circle and the area of the small circle: $A = 25\pi - 9\pi = 16\pi$ cm^2.

Chapter 14 Tests

Three levels (A,B,C) of tests are available for each chapter in the *Assessment Resources*.

LEVEL A

CHAPTER Chapter Test
14 Form A

Use the correct symbol to make each statement true. Choose ∈ or ∉.

1. 13 ☐ {prime numbers}

   ∈

2. pentagon ☐ {parallelograms}

   ∉

Tell whether the first set is a subset of the second set. Use the correct symbol.

3. N = {positive integers};
   R = {whole numbers}

   $N \subset R$

4. E = {circles}; G = {plane figures}

   $E \subset G$

Find the intersection of the sets.

5. A = {0, 1, 2, 3}; B = {2, 3, 4, 5, 6}

   {2, 3}

6. R = {−2, −1, 0}; T = {0, 1, 2, 3, 4}

   {0}

For 16−20, find the union of the sets.

7. M = {0, 1, 2, 3}; N = {4, 5, 6}

   {0, 1, 2, 3, 4, 5, 6}

8. G = {multiples of 4};
   H = {factors of 8}

   {multiples of 4, 1, 2}

For 9−11, use the Venn diagram.

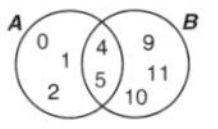

A circle labeled A containing 0, 1, 4, 5, 2 and circle labeled B containing 9, 11, 10, with 4, 5 in overlap.

9. Identify the intersection of sets A and B.

   $A \cap B$ = {4, 5}

10. Identify the union of sets A and B.

   $A \cup B$ = {0, 1, 2, 4, 5, 9, 10, 11}

11. True or False: $A \subset B$.

   False

12. Make a truth table for the conjunction P and Q.
   P: A number is a multiple of 5.
   Q: A number is a multiple of 2.

P	Q	P and Q
T	T	T
T	F	F
F	T	F
F	F	F

13. Complete the truth table.
   P: Today is Monday.
   Q: There is school today.

P	Q	P and Q	P or Q
T	T	T	T
T	F	F	T
F	T	F	T
F	F	F	F

CHAPTER Chapter Test
14 Form A, continued

For 14−15, name the hypothesis and the conclusion in each statement.

14. If you sleep, you will feel rested.

   H: You sleep.

   C: You feel rested.

15. In order to feel well, eat a good breakfast.

   H: You eat a good breakfast.

   C: You feel well.

16. Make a conclusion, if possible, from the deductive argument.
   If a rectangle has 4 equal sides, it is a square. Rectangle ABCD has 4 equal sides.

   Rectangle ABCD is a square.

For 17−19, use the graph.

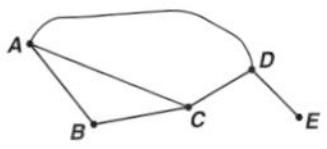

A graph with vertices A, B, C, D, E.

17. Find the degree of each vertex.

   A: 3, B: 2, C: 3,

   D: 3, E: 1

18. Is the graph connected? How do you know?

   The graph is connected because there is a path from each vertex to another vertex.

19. Determine whether the graph can be traversed through an Euler circuit. If your answer is yes, describe an Euler circuit in the graph.

   No; Not all vertices have even degrees.

For 20−21, use the graph.

A graph with vertices W, X, Y, Z with edges labeled 7 mi, 3 mi, 5 mi, 4 mi, 2 mi.

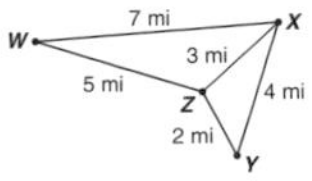

20. A Hamiltonian circuit must pass through every vertex. How is it different from an Euler circuit?

   It is not necessary to traverse every edge.

21. What is the length of the shortest Hamiltonian circuit that begins and ends at W?

   18 mi

LEVEL B

CHAPTER Chapter Test
14 Form B

Insert the correct symbol to make each statement true. Choose ∈ or ∉.

1. $x^2 − 2x = 0$ ☐ {quadratic equations}

   ∈

2. Delaware ☐ {United Nations}

   ∉

Determine whether the first set is a subset of the second set. Use the correct symbol.

3. N = {natural numbers};
   R = {real numbers}

   $N \subset R$

4. P = {polygons}; T = {triangles}

   $P \not\subset T$

Find the intersection of the sets.

5. A = {2, 4, 6, 8}; B = {6, 8, 10, 12}

   {6, 8}

6. J = {negative integers};
   K = {−4, −3, −2, −1, 0, 1}

   {−4, −3, −2, −1}

For 16−20, find the union of the sets.

7. G = {multiples of 2};
   H = {multiples of 4}

   {multiples of 2}

8. X = {quadrilaterals};
   Y = {polygons}

   {polygons}

For 9−11, use the Venn diagram.

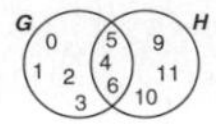

A circle labeled G containing 1, 2, 4, 3, 5, 6 and circle labeled H containing 9, 11, 10, with 5, 6 in overlap.

9. Identify the intersection of sets G and H.

   $G \cap H$ = {4, 5, 6}

10. Identify the union of sets G and H.

   $G \cup H$ = {0, 1, 2, 3, 4, 5, 6, 9, 10, 11}

11. Identify a subset in the diagram.

   no subsets

12. Make a truth table for the conjunction P and Q.
   P: A number is an even number.
   Q: A number is a factor of 90.

P	Q	P and Q
T	T	T
T	F	F
F	T	F
F	F	F

13. Complete the truth table.
   P: Today is Saturday.
   Q: There is practice today.

P	Q	P and Q	P or Q
T	T	T	T
T	F	F	T
F	T	F	T
F	F	F	F

CHAPTER Chapter Test
14 Form B, continued

For 14−15, identify the hypothesis and the conclusion in each conditional.

14. If you study, the test will be easy.

   H: You study.

   C: The test is easy.

15. To be healthy, eat well.

   H: You eat well.

   C: You will be healthy.

16. Make a conclusion, if possible, from the deductive argument.
   If a quadrilateral has 4 right angles, it is a rectangle. Quadrilateral ABCD has angle measures of 90°, 90°, 120°, and 60°.

   No conclusion possible.

For 17−19, use the graph.

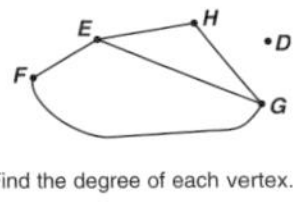

A graph with vertices E, F, G, H, D.

17. Find the degree of each vertex.

   D: 0, E: 3, F: 2,

   G: 3, H: 2

18. Is the graph connected? Explain.

   No; There is no path to vertex D.

19. Determine whether the graph can be traversed through an Euler circuit. If your answer is yes, describe an Euler circuit in the graph.

   No; Not all vertices have even degrees.

For 20−21, use the graph.

A graph with vertices J, K, L, M with edges labeled 8 mi, 10 mi, 12 mi, 9 mi, 15 mi.

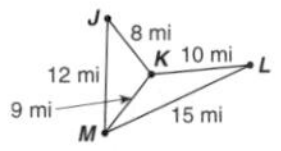

20. Describe how a Hamiltonian circuit differs from an Euler circuit.

   Both pass through every vertex, but an Euler circuit also must pass over every edge.

21. Determine the length of the shortest Hamiltonian circuit beginning at L.

   45 mi

LEVEL C

CHAPTER Chapter Test
14 Form C

Insert the correct symbol to make each statement true. Choose ∈ or ∉.

1. $y = x^3 − 3x$ ☐ {quadratic functions}

   ∉

2. Ohio River ☐ {rivers in the United States}

   ∈

Determine whether the first set is a subset of the second set. Use the correct symbol.

3. P = {perimeter}; A = {area}

   $P \not\subset A$

4. V = {x^2, $3x^3$, $2x$}; M = {monomials}

   $V \subset M$

Find the intersection of the sets.

5. A = {$x | x < 9$}; B = {$x | x > 12$}

   { } (empty set)

6. C = {first 5 prime numbers};
   D = {first 10 positive integers}

   {2, 3, 5, 7}

Find the union of the sets.

7. G = {multiples of 2};
   H = {factors of 12}

   {1, 3, multiples of 2}

8. W = {square roots of 16, 36, 64};
   Z = {squares of 2, 4, 6, 8}

   {4, 6, 8, 16, 36, 64}

For 9−11, use the Venn diagram.

A circle labeled D containing $x > 5$, E ($x > 15$) and circle labeled F containing $x < 2$, with E in overlap.

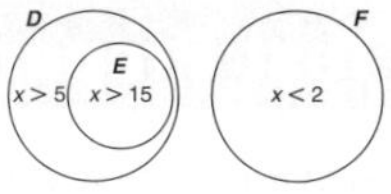

9. Identify the union of D and F.

   $D \cup F$ = {$x | x > 5$ or $x < 2$}

10. Identify the intersections.

   $D \cap E$ = {$x | x > 15$}

   $D \cap F$ = ∅

   $E \cap F$ = ∅

11. Identify any subsets.

   $E \subset D$

12. Make a truth table for the disjunction P or Q.
   P: A figure is a quadrilateral.
   Q: A figure is a regular polygon.

P	Q	P or Q
T	T	T
T	F	T
F	T	T
F	F	F

CHAPTER Chapter Test
14 Form C, continued

13. Complete the truth table.
   P: Ann is on the basketball team.
   Q: Ann plays the violin.

P	Q	P and Q	P or Q
T	T	T	T
T	F	F	T
F	T	F	T
F	F	F	F

For 14−15, identify the hypothesis and the conclusion in each conditional.

14. Work hard, and you will be rewarded.

   H: You work hard.

   C: You will be rewarded.

15. Plants need water to grow.

   H: Don't water your plants.

   C: Your plants won't grow.

16. Make a conclusion, if possible, from the deductive argument.
   If a rectangle has sides of 8 ft and 12 ft, then its area is 96 ft². Rectangle WXYZ has an area of 96 ft².

   No conclusion; The rectangle could be 4 ft by 24 ft.

For 17−19, use the graph.

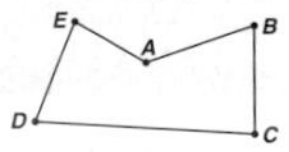

A graph with vertices E, A, B, D, C.

17. Find the degree of each vertex.

   A: 2, B: 2, C: 2, D: 2, E: 2

18. Is the graph connected? Explain.

   Yes; There is a path from each vertex to another vertex.

19. Determine whether the graph can be traversed through an Euler circuit. If your answer is yes, describe an Euler circuit in the graph.

   The graph represents an Euler circuit; D, E, A, B, C, D

For 20−21, use the graph.

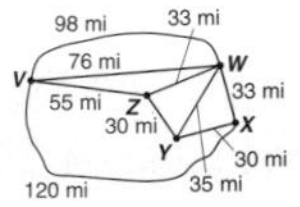

A graph with vertices V, W, X, Y, Z with edges labeled 98 mi, 33 mi, 76 mi, 33 mi, 55 mi, 30 mi, 30 mi, 120 mi, 35 mi.

20. Determine the length of the shortest Hamiltonian circuit beginning at V.

   224 mi

21. Determine the length of the longest Hamiltonian circuit beginning at X.

   311 mi

Test and Practice Generator
CD-ROM

Create and customize multiple versions of the same tests with corresponding answers for any chosen chapter objectives.

Chapter 14 State and Standardized Test Preparation

Test Taking Skill Builder and Standardized Test Practice
are provided for each chapter in the *Test Prep Tool Kit.*

TEST TAKING SKILL BUILDER

Test Taking Strategy   **Answering the Right Question**
Chapter 14

When you first read a question you might answer it too quickly
before you actually understand what is being asked. Be sure to
reread the question slowly so that you are certain that you are
answering the right question.

Example 1

Multiple Choice Shea is creating a Venn Diagram
that shows the factors of two numbers, 55 and 110.
What is the smallest value that can be correctly
placed in the shaded area of the diagram?

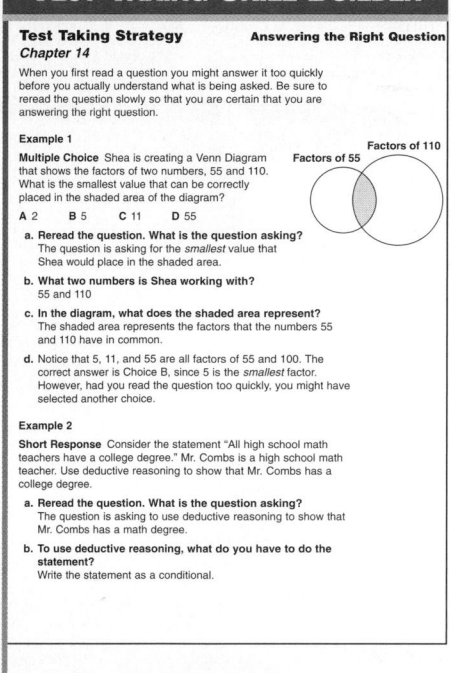

Factors of 55   Factors of 110

A 2   B 5   C 11   D 55

a. **Reread the question. What is the question asking?**
The question is asking for the *smallest* value that
Shea would place in the shaded area.

b. **What two numbers is Shea working with?**
55 and 110

c. **In the diagram, what does the shaded area represent?**
The shaded area represents the factors that the numbers 55
and 110 have in common.

d. Notice that 5, 11, and 55 are all factors of 55 and 100. The
correct answer is Choice B, since 5 is the *smallest* factor.
However, had you read the question too quickly, you might have
selected another choice.

Example 2

Short Response Consider the statement "All high school math
teachers have a college degree." Mr. Combs is a high school math
teacher. Use deductive reasoning to show that Mr. Combs has a
college degree.

a. **Reread the question. What is the question asking?**
The question is asking to use deductive reasoning to show that
Mr. Combs has a math degree.

b. **To use deductive reasoning, what do you have to do in the
statement?**
Write the statement as a conditional.

Test Taking Strategy
Chapter 14, continued

Answer each question. **Possible answers are given.**

1. **Multiple Choice** After school Wally goes to
several different places. If he starts from school,
travels to practice and/or to work, and then goes
home, how many different paths can Wally take?

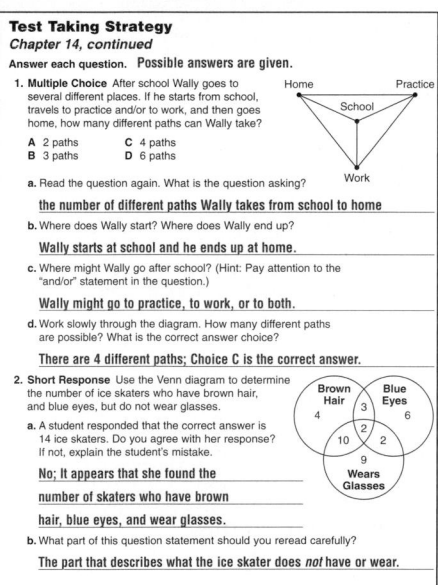

Home   Practice
School
Work

A 2 paths   C 4 paths
B 3 paths   D 6 paths

a. Read the question again. What is the question asking?

the number of different paths Wally takes from school to home

b. Where does Wally start? Where does Wally end up?

Wally starts at school and he ends up at home.

c. Where might Wally go after school? (Hint: Pay attention to the
"and/or" statement in the question.)

Wally might go to practice, to work, or to both.

d. Work slowly through the diagram. How many different paths
are possible? What is the correct answer choice?

There are 4 different paths; Choice C is the correct answer.

2. **Short Response** Use the Venn diagram to determine
the number of ice skaters who have brown hair,
and blue eyes, but do not wear glasses.

Brown Hair   Blue Eyes
4   3
10   2
9
Wears Glasses

a. A student responded that the correct answer is
14 ice skaters. Do you agree with her response?
If not, explain the student's mistake.

No; It appears that she found the

number of skaters who have brown

hair, blue eyes, and wear glasses.

b. What part of this question statement should you reread carefully?

The part that describes what the ice skater does *not* have or wear.

STANDARDIZED TEST PRACTICE

Standardized Test Practice
Chapter 14

**Select the best answer for Questions
1–6.**

1. "If our team wins its next game, they
will go to the state finals." The
statement "they will go to the state
finals" is the _____.
A conjunction
B hypothesis
C disjunction
D conclusion

2. When is the conjunction *P* and *Q*
true?
F When *P* is true, but *Q* is false.
G When *Q* is true, but *P* is false.
H When both *P* and *Q* are true.
I When both *P* and *Q* are false.

3. Which statement is false for the
graph?

N
M   O
P

A The graph is connected.
B *MNOPM* is a circuit.
C *MNOPM* is an Euler circuit.
D *MNOPM* is a Hamiltonian circuit.

4. What is the degree of vertex *N* in the
network shown in Exercise 3?
F 3   H 1
G 2   I 0

5. Andrea is filling numbers in the Venn
diagram. No number may be entered
more than once. What is the greatest
number that can be correctly placed
in the shaded area of the diagram?

Factors of 45   Factors of 60
45   2
9   12   4
6

A 180   C 10
B 15   D 5

6. Which group of statements represent
a valid argument?
F Given: All quadrilaterals have four
sides.
All squares have four sides.
Conclusion: All quadrilaterals are
squares.
G Given: All rectangles have right
angles.
All squares have right
angles.
Conclusion: All rectangles are
squares.
H Given: All quadrilaterals with four
equal sides are rhombuses.
All squares have four equal
sides.
Conclusion: All squares are
rhombuses.
I Given: All parallelograms have
four sides.
All squares have four sides.
Conclusion: All parallelograms are
squares.

Standardized Test Practice
Chapter 14, continued

Gridded Response
Solve the problems. Use the answer
sheet to write and grid-in your answer.

7. The Venn diagram shows the
number of students who have pens,
pencils, or markers in their book bag.
How many students have pens and
pencils, but not markers?

Pencils   Markers
9   4   1
12   3   2
6
Pens

8. As part of his exercise routine, Jerry
takes several different bike routes. If
he starts from home, travels to each
location one time and then returns
home, how many different paths can
he take?

Park   School
Home
Mall

Short Response

9. Write a statement using one of the
symbols $\not\subset$, $\subset$, $\notin$, or $\in$ to show the
relationship between a triangle and
the set of quadrilaterals. Explain why
your statement is true.

10. If *Q* and *R* are two sets, write an
expression that represents the set of
all elements that are in either *Q* or *R*.

11. Explain why the path *ABCDBEA* is
not a Hamiltonian circuit.

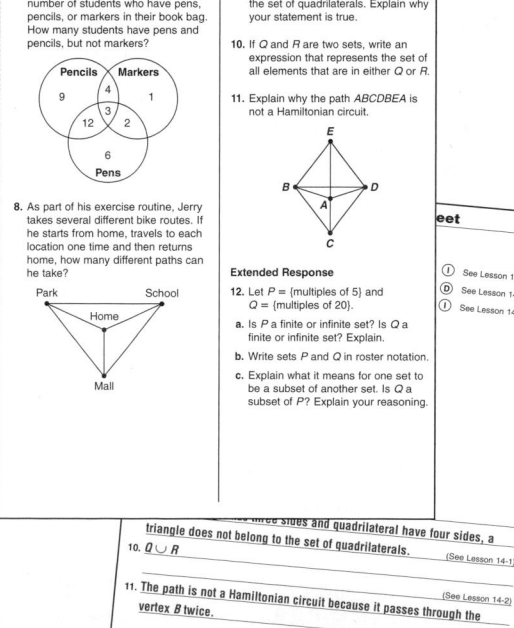

E
B   D
A
C

Extended Response

12. Let *P* = {multiples of 5} and
Q = {multiples of 20}.
a. Is *P* a finite or infinite set? Is *Q* a
finite or infinite set? Explain.
b. Write sets *P* and *Q* in roster notation.
c. Explain what it means for one set to
be a subset of another set. Is *Q* a
subset of *P*? Explain your reasoning.

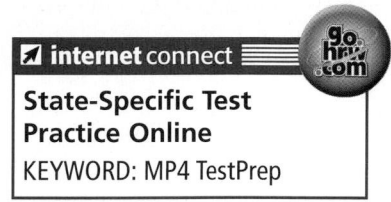

internet connect

**State-Specific Test
Practice Online**
KEYWORD: MP4 TestPrep

Test Prep Tool Kit

- Standardized Test Prep Workbook
- Countdown to Testing transparencies
- State Test Prep CD-ROM
- Standardized Test Prep Video

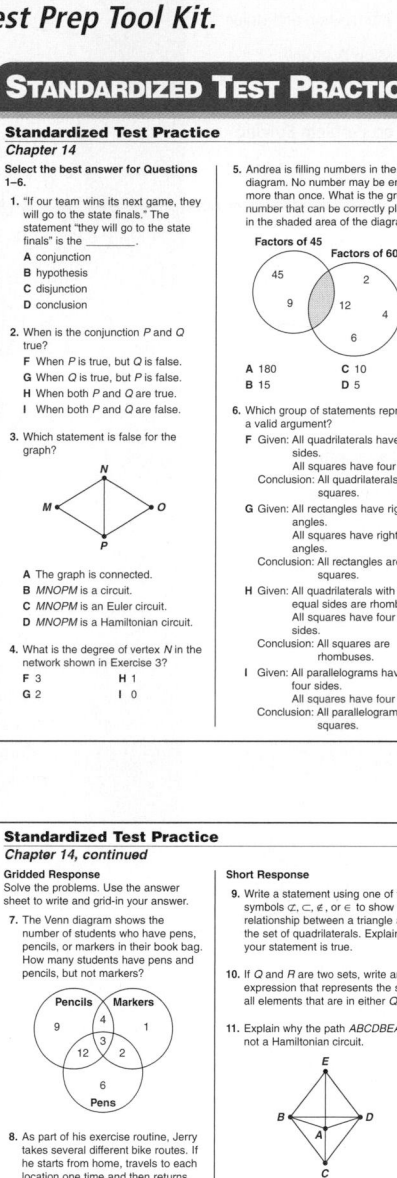

...three sides and quadrilateral have four sides, a
triangle does not belong to the set of quadrilaterals.
(See Lesson 14-1)

10. $Q \cup R$

11. The path is not a Hamiltonian circuit because it passes through the
vertex *B* twice.
(See Lesson 14-2)

Extended Response
Write your answers for Problem 12 on the back of this paper.
See Lesson 14-1.
(See Lesson 14-7)

I See Lesson 14-6.
D See Lesson 14-3.
I See Lesson 14-5.

**Customized answer sheets give
students realistic practice for
actual standardized tests.**

Set Theory and Discrete Math

Why Learn This?

Tell students that discrete math is a branch of mathematics that deals with distinct objects. One application of discrete math is binary code, in which situations are modeled with only the digits 0 and 1. Binary notation is used to represent a variety of situations, including information flow in computer chips. In this application of binary notation, 0 means "closed" and 1 means "open."

Using Data

To begin the study of this chapter, have students:

- Explain why only the last row of the table allows for information flow. Because the table shows data for an "and" circuit, both gates must be open for information to flow.

- Identify the number of permutations of 0 and 1 possible with two gates. **4** Identify the number of permutations of 0 and 1 possible with three gates. **8**

- Determine a rule for the number of permutations in a circuit of n gates. 2^n

Set Theory and Discrete Math

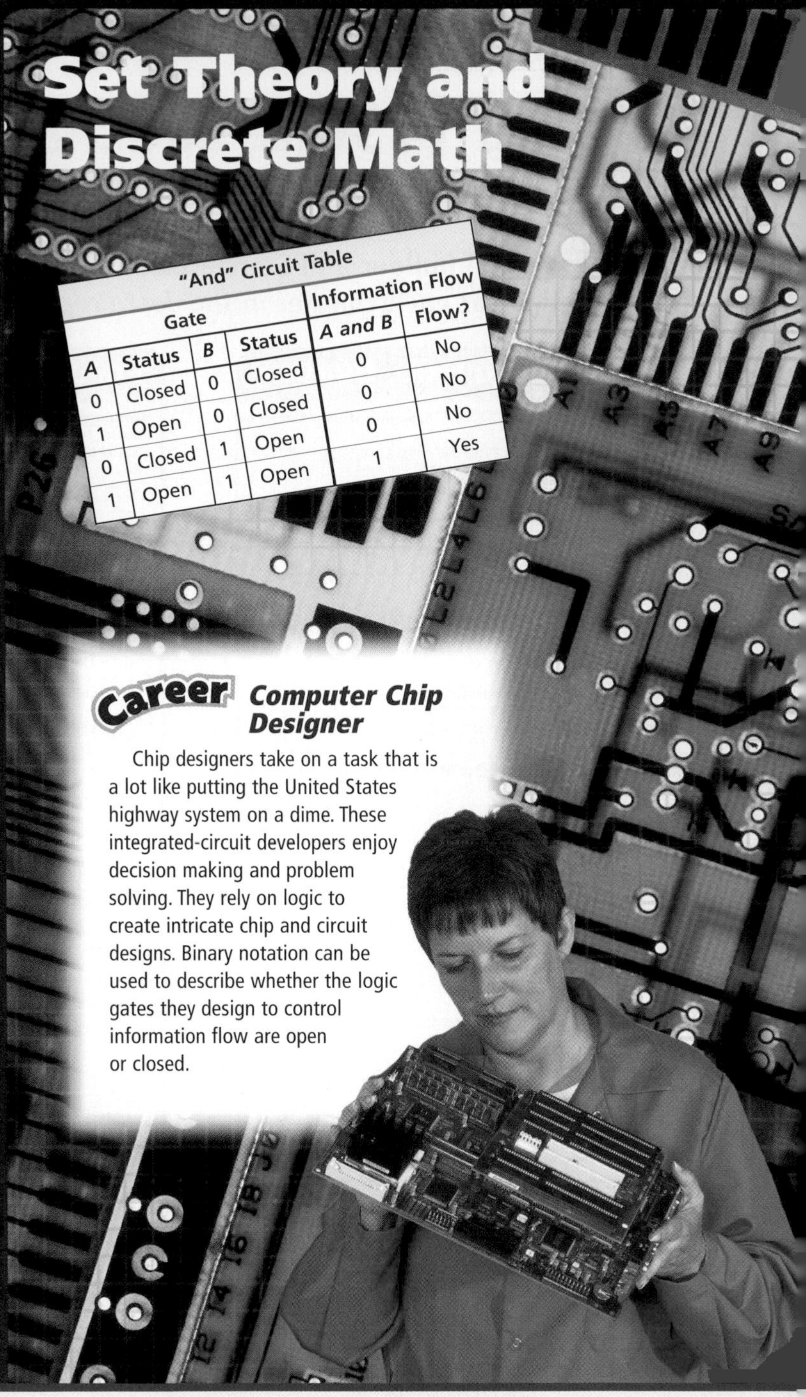

"And" Circuit Table

Gate				Information Flow	
A	Status	B	Status	A and B	Flow?
0	Closed	0	Closed	0	No
1	Open	0	Closed	0	No
0	Closed	1	Open	0	No
1	Open	1	Open	1	Yes

Career Computer Chip Designer

Chip designers take on a task that is a lot like putting the United States highway system on a dime. These integrated-circuit developers enjoy decision making and problem solving. They rely on logic to create intricate chip and circuit designs. Binary notation can be used to describe whether the logic gates they design to control information flow are open or closed.

internet connect
Chapter Opener Online
go.hrw.com
KEYWORD: MP4 Ch14

Problem Solving Project

Physical Science Connection

Purpose: To use logic and the concept of a circuit to solve problems

Materials: What Goes Around Comes Around worksheet, flashlight batteries (cells), insulated bell wire, flashlight bulbs

Understand, Plan, Solve, and Look Back

Have students:

✔ Complete the What Goes Around Comes Around worksheet to learn more about logic, circuits, and networks.

✔ Create a table for an "or" gate.

✔ Decide when an "and" situation or an "or" situation is of value in a circuit.

✔ Check students' work.

internet connect
Chapter Project Online: *go.hrw.com*
KEYWORD: MP4 PSProject14

ARE YOU READY?

Choose the best term from the list to complete each sentence.

1. Numbers divisible by only themselves and 1 are ___?___. **prime numbers**
2. The set of whole numbers consists of the set of ___?___ and 0. **counting numbers**
3. Numbers that cannot be written as decimals that terminate or repeat are called ___?___. **irrational numbers**
4. The set ... $-4, -3, -2, -1, 0, 1, 2, 3, 4, ...$ is the set of ___?___. **integers**

counting numbers

integers

irrational numbers

prime numbers

rational numbers

real numbers

Complete these exercises to review skills you will need for this chapter.

✔ Composite Numbers

List the factors of each number. Tell whether the number is composite.

5. 37 **1, 37; no**
6. 57 **1, 57, 3, 19; yes**
7. 63 **1, 63, 3, 21, 7, 9; yes**
8. 83 **1, 83; no**
9. 103 **1, 103; no**
10. 155 **1, 155, 5, 31; yes**

✔ Identify Sets of Numbers

State whether each number is rational, irrational, or not a real number.

11. $\frac{0}{3}$ **rational**
12. $\sqrt{12}$ **irrational**
13. $\frac{3}{0}$ **not real**
14. $\sqrt{2}$ **irrational**
15. $\sqrt{-5}$ **not real**
16. $-\sqrt{9}$ **rational**
17. $\sqrt{81}$ **rational**
18. π **irrational**

✔ Identify Polygons

Give all of the names that apply to each figure.

19.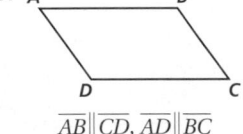
$\overline{AB}\|\overline{CD}, \overline{AD}\|\overline{BC}$

20.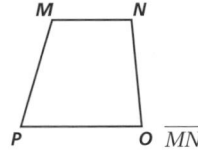
$\overline{MN}\|\overline{OP}$

21.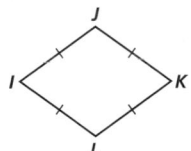

22.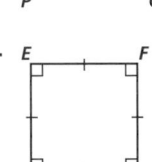

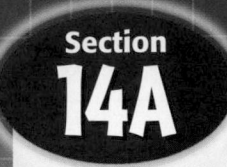

Set Theory

One-Minute Section Planner

Lesson	Materials	Resources
Lesson 14-1 Sets **NCTM:** Reasoning and Proof, Communication, Representation ☐ SAT-9 ☐ SAT-10 ☐ ITBS ☐ CTBS ☐ MAT ☑ CAT	**Optional** Teaching Transparency T2 *(CRB)*	• *Chapter 14 Resource Book,* pp. 6–14 • *Daily Transparency T1, CRB* • *Additional Examples Transparencies T3–T4, CRB* • *Alternate Openers: Explorations,* p. 106
Lesson 14-2 Intersection and Union **NCTM:** Reasoning and Proof, Communication, Representation ☐ SAT-9 ☐ SAT-10 ☐ ITBS ☐ CTBS ☐ MAT ☑ CAT	**Optional** Street map Recording Sheet for Reaching All Learners *(CRB, p. 78)*	• *Chapter 14 Resource Book,* pp. 15–23 • *Daily Transparency T5, CRB* • *Additional Examples Transparencies T6–T7, CRB* • *Alternate Openers: Explorations,* p. 107
Lesson 14-3 Venn Diagrams **NCTM:** Reasoning and Proof, Communication, Representation ☐ SAT-9 ☐ SAT-10 ☐ ITBS ☑ CTBS ☐ MAT ☑ CAT	**Optional** Examples of graphs Teaching Transparency T9 *(CRB)* Construction paper Pattern blocks *(MK or CRB, p. 79)*	• *Chapter 14 Resource Book,* pp. 24–33 • *Daily Transparency T8, CRB* • *Additional Examples Transparencies T10–T12, CRB* • *Alternate Openers: Explorations,* p. 108
Section 14A Assessment		• Mid-Chapter Quiz, SE p. 700 • Section 14A Quiz, AR p. 31 • *Test and Practice Generator* CD-ROM

SAT = *Stanford Achievement Tests*   **ITBS** = *Iowa Test of Basic Skills*   **CTBS** = *Comprehensive Test of Basic Skills/Terra Nova*
MAT = *Metropolitan Achievement Test*   **CAT** = *California Achievement Test*
NCTM = Complete standards can be found on pages T29–T35.   **NAEP** = Complete standards can be found on pages A54–A58.
SE = *Student Edition*   **TE** = *Teacher's Edition*   **AR** = *Assessment Resources*   **CRB** = *Chapter Resource Book*   **MK** = *Manipulatives Kit*

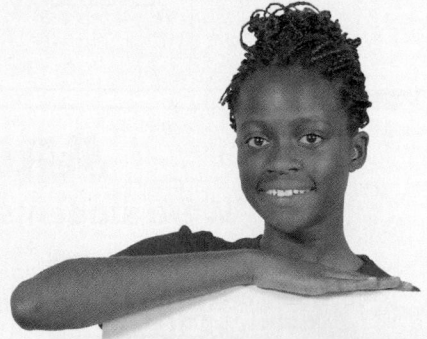

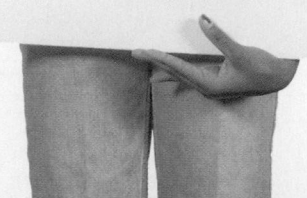

Symbols for Set Theory
$\in$ = is an element of
$\notin$ = is *not* an element of
$\subset$ = is a subset of
$\not\subset$ = is *not* a subset of
$\varnothing$ = empty set
$\cap$ = intersection
$\cup$ = union

Section Overview

Sets, Intersections, and Unions

Lessons 14-1, 14-2

Why? Sets can be used to model many situations that require counting or identifying specific objects. A set is a collection of objects called *elements,* and it is indicated by brackets. The set with no elements is called the *empty set,* which is indicated by the symbol ∅.

The symbol ∈ is read as **"is an element of."** The symbol ∉ is read as **"is *not* an element of."** apple ∈ {fruits} apple ∉ {grains}	The symbol ⊂ is read as **"is a subset of."** The symbol ⊄ is read as **"is not a subset of."** {fruits} ⊂ {foods} {fruits} ⊄ {clothing}

Infinite Set	Finite Set
A set that contains an infinite, or unlimited, number of elements	A set that contains a finite, usually countable, number of elements
Example: {integers greater than 8}	**Example:** {integers between −3 and 7}

The **intersection,** indicated by the symbol ∩, of sets *A* and *B* is the set of all elements that are in both *A* **and** *B.*	The **union,** indicated by the symbol ∪, of sets *A* and *B* is the set of all elements that are in either *A* **or** *B.*

Examples: $A = \{♦, ♣, ♥\}$ and $B = \{♣, ♥, ♠\}$

$A \cap B = \{♣, ♥\}$ $A \cup B = \{♦, ♣, ♥, ♠\}$

Venn Diagrams

Lesson 14-3

Why? A Venn diagram is a diagram used to illustrate relationships between sets.

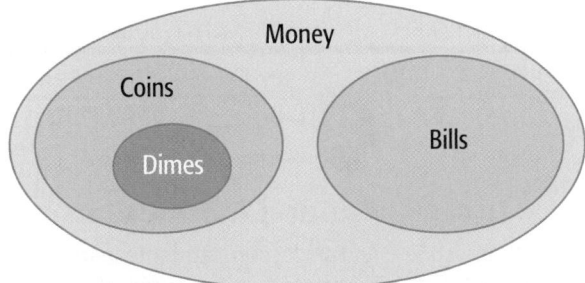

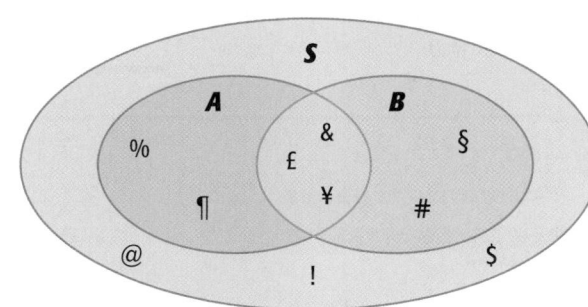

{coins} ⊂ {money}   {bills} ⊄ {coins}
{dimes} ⊂ {coins}   {coins} ⊄ {dimes}
{bills} ⊂ {money}   {coins} ⊄ {bills}
{dimes} ⊂ {money}   {money} ⊄ {bills}

$S = \{@, \%, ¶, £, \&, ¥, \#, §, !, \$\}$
$A \cup B = \{\%, ¶, £, \&, ¥, \#, §\}$
$A \cap B = \{£, \&, ¥\}$
$A \subset S$
$B \subset S$

Pacing: Traditional 1 day
Block $\frac{1}{2}$ day
Objective: Students understand mathematical sets and set notation.

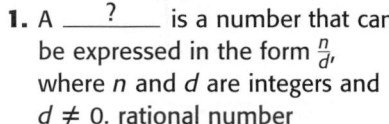

Warm Up

Fill in each blank.

1. A ___?___ is a number that can be expressed in the form $\frac{n}{d}$, where n and d are integers and $d \neq 0$. **rational number**

2. The set of ___?___ consists of the set of rational numbers and the set of irrational numbers. **real numbers**

3. The set of ___?___ consists of the set of counting numbers, their opposites, and 0. **integers**

Problem of the Day

Use a traditional clock face to determine the next three terms in the following sequence:

1, 6, 11, 4, 9, … **2, 7, 12**

Available on Daily Transparency in CRB

Math Humor

What do you call a fleet of submarines that is part of the navy? A sub-set

---

 **14-1 Sets**

Learn to understand mathematical sets and set notation.

Vocabulary
set
element
subset
finite set
infinite set

Shana and Robert are collectors. Shana collects seashells and shell-related objects, and Robert collects owl-related objects.

A **set** is a collection of objects, called **elements**. Elements of a set can be described in two ways: *roster notation* and *set-builder notation*. An owl made from seashells may be a member of Shana's or Robert's sets.

Set	Roster Notation	Set-Builder Notation
Even counting numbers	{2, 4, 6, 8, 10, …}	{x|x is an even counting number} *Read as "the set of all x such that x is an even counting number."*
Great Lakes	{Huron, Ontario, Michigan, Erie, Superior}	{x|x is one of the Great Lakes}

Helpful Hint

Think of the element symbol $\in$ as the letter *e*.

The symbol $\in$ is read as "is an element of." Read the statement $3 \in$ {odd numbers} as "3 is an element of the set of odd numbers." The symbol $\notin$ is read as "is *not* an element of." Read the statement $2 \notin$ {odd numbers} as "2 is *not* an element of the set of odd numbers."

EXAMPLE 1 **Identifying Elements of a Set**

Insert $\in$ or $\notin$ to make each statement true.

A 1 ☐ {numbers that are their own reciprocals}
1 $\in$ {numbers that are their own reciprocals}   *1 is equivalent to $\frac{1}{1}$.*

B broccoli ☐ {red vegetables}
broccoli $\notin$ {red vegetables}   *Broccoli is not a red vegetable.*

C ☐ {polygons}
 $\notin$ {polygons}   *A semicircle is not a polygon.*

---

1 Introduce

Alternate Opener

EXPLORATION

14-1 Sets

Sets of numbers are often displayed in circles. Set *A* contains the multiples of 5 from 1 to 20.

A: 5 10 15 20

For each set description fill in the circle with the appropriate counting numbers from 1 to 20.
1. odd numbers   2. prime numbers   3. multiples of 3

B   C   D

Think and Discuss
4. **Give** an example of a set that does not contain numbers.
5. **Describe** the set *S*.

S: 1 4 9 16

Motivate

Show students a list of symbols with which they are familiar, such as $, %, #, =, and >. Have students identify the meaning of each symbol. Then show them the symbols that they will be learning in this lesson, $\in$, $\notin$, $\subset$, and $\not\subset$, and give students the meanings of these symbols.

Exploration worksheet and answers on Chapter 14 Resource Book pp. 7 and 81

2 Teach

Lesson Presentation

Guided Instruction

In this lesson, students learn to understand mathematical sets and set notation. Define *set,* and give some examples of sets. You may want to display Teaching Transparency T2 (CRB). Explain that each member of a set is called an *element* and that the symbol $\in$ is read as "is an element of." Explain that if a set is formed with elements of a given set, then the new set is a *subset* of the original set. Show students how to read and write statements using $\in$, $\notin$, $\subset$, and $\not\subset$. Then define *finite* and *infinite* sets.

Set *A* is a **subset** of set *B* if every element in *A* is also in *B*. The symbol ⊂ is read as "is a subset of," and the symbol ⊄ is read as "is *not* a subset of."

EXAMPLE 2 Identifying Subsets

Determine whether the first set is a subset of the second set. Use the correct symbol.

A *Q* = {rational numbers}   *R* = {real numbers}
Yes, *Q* ⊂ *R*.   *Every rational number is a real number.*

B *T* = {0, 1, 2, 3}   *N* = {counting numbers}
No, *T* ⊄ *N*.   *0 is not a counting number.*

C *H* = {rhombuses}   *G* = {rectangles}
No, *H* ⊄ *G*.   *Some rhombuses are not rectangles.*

A **finite set** contains a finite number of elements. An **infinite set** contains an infinite number of elements.

EXAMPLE 3 Identifying Finite and Infinite Sets

Tell whether each set is finite or infinite.

A {letters of the alphabet}
finite   *There are exactly 26 elements in the set.*

B {rational numbers between 99 and 100}
infinite   *There are an infinite number of rational numbers between any two rational numbers.*

C {integers with absolute value less than 3}
finite   *Only −2, −1, 0, 1, and 2 have absolute values less than 3.*

Think and Discuss

1. **Describe** the set of whole numbers that are not counting numbers.

2. **Name** three different sets that have {apricots} as a subset.

3. **Give** two examples of finite sets that have 20 as an element. Give two examples of infinite sets that have 20 as an element.

3 Close

Reaching All Learners
Through Grouping Strategies

Explain to students that together they represent a set and each student represents one element of the set. Have students list the elements of various subsets, such as students in the third row, boys, students with brown hair, or students wearing blue. Ask students to tell whether any of these subsets is a subset of any other. For example, if all of the students in the third row are boys, then the students in the third row would be a subset of the set of boys.

Summarize

Remind students that a set is a collection of objects. Each object in the set is called an element. If every element in set *A* is also in set *B*, then *A* is a subset of *B*.

Answers to Think and Discuss

Possible answers:

1. The set of whole numbers is {0, 1, 2, 3, …}. The set of counting numbers is {1, 2, 3, 4, …}. So the set of whole numbers that are not counting numbers is the finite set {0}.

2. {fruits}, {things that are orange}, {things that grow on trees}

3. finite sets: {positive integers less than or equal to 25}, {whole numbers between 19 and 23}; infinite sets: {integers}, {real numbers between 19 and 21}

FOR EXTRA PRACTICE
see page 758

internet connect
Homework Help Online
go.hrw.com   Keyword: MP4 14-1

Students may want to refer back to the lesson examples.

Assignment Guide

If you finished Example **1** assign:
  Core 1–2, 7–8, 37–41
  Enriched 1–2, 7–8, 37–41

If you finished Example **2** assign:
  Core 1–4, 7–12, 17–22, 37–41
  Enriched 7–12, 17–22, 29, 34–41

If you finished Example **3** assign:
  Core 1–28, 29, 31, 37–41
  Enriched 7–41

Notes

GUIDED PRACTICE

See Example **1**   Insert $\in$ or $\notin$ to make each statement true.

1. oak tree ▮ {living things}
   $\in$

2. $x^2 - \frac{4}{x} + 2$ ▮ {trinomials}
   $\notin$

See Example **2**   Determine whether the first set is a subset of the second set. Use the correct symbol.

3. E = {even numbers}   **Yes, $E \subset R$.**
   R = {real numbers}

4. P = {parallelograms}   **No, $P \not\subset S$.**
   S = {squares}

See Example **3**   Tell whether each set is finite or infinite.

5. {letters that are vowels} **finite**

6. {number of radii in a circle} **infinite**

INDEPENDENT PRACTICE

See Example **1**   Insert $\in$ or $\notin$ to make each statement true.

7. Spanish ▮ {world languages}
   $\in$

8. $2\frac{3}{7}$ ▮ {rational numbers}
   $\in$

See Example **2**   Determine whether the first set is a subset of the second set. Use the correct symbol.

9. F = {football players}
   T = {team athletes}   **Yes, $F \subset T$.**

10. C = {counting numbers}
    P = {prime numbers}   **No, $C \not\subset P$.**

11. P = {prime numbers}
    O = {odd numbers}   **No, $P \not\subset O$.**

12. S = {squares}
    P = {parallelograms}   **Yes, $S \subset P$.**

See Example **3**   Tell whether each set is finite or infinite.

13. {composite numbers} **infinite**

14. {rational numbers less than 0} **infinite**

15. {seconds in a year} **finite**

16. {past U.S. presidents} **finite**

PRACTICE AND PROBLEM SOLVING

Choose the symbol that best completes each statement.
Use the symbols $\subset$, $\not\subset$, $\in$, and $\notin$.

17. ▮ {cats} $\in$

18. ▮ {shapes that tessellate} $\in$

19. ▮ {edible items} $\in$

20. 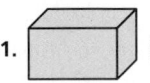 ▮ {flags of South America} $\notin$

21. ▮ {polyhedra} $\in$

22. ▮ {U.S. currency} $\in$

Math Background

A set that contains no elements is called the *empty set* or the *null set* (Lesson 14-2). The symbols used to represent this set are $\varnothing$ and { }. The empty set is a subset of every set.

Subsets are not always sets with fewer elements than the original set. Any set is considered to be a subset of itself. Therefore, there are four subsets of the set {3, 5}: $\varnothing$, {3}, {5}, and {3, 5}.

RETEACH 14-1

LESSON **14-1** Reteach
Sets

A **set** is a collection of objects. Each object in a set is called an **element**. The symbol $\in$ means "is an element of." The symbol $\notin$ means "is not an element of."

For example, $4 \in$ {even counting numbers}, because the even counting numbers are 2, 4, 6, 8, . . .

Insert the correct symbol to make each statement true.

1. soccer $\in$ {ball sports}   2. 20 $\notin$ {multiples of 3}   3. red $\in$ {colors}

All sets can be broken down into **subsets**.
Set A is a **subset** of set B if every element in A is also in B.
The symbol $\subset$ means "is a subset of" a given set.
The symbol $\not\subset$ means "is not a subset of" a given set.

For example, if A = {1, 2, 3, 4, 5, 6, 7, 8, 9, 10} and B = {1, 3, 5, 7, 9}, then $B \subset A$ because all the elements in B are also in A.

Determine whether the first set is a subset of the second set. Use the correct symbol.

4. A = {U.S. state capitals}
   B = {cities in the U.S.} $A \subset B$

5. X = {factors of 15}
   Y = {factors of 25} $X \not\subset Y$

6. P = {natural numbers}
   Q = {composite numbers} $P \not\subset Q$

There are two kinds of sets.
A **finite set** contains a countable number of elements.
An **infinite set** contains elements that CANNOT be counted.

For example, the set of positive odd integers less than 10 is finite, because you can count the numbers in that set.

Tell whether each set is finite or infinite.

7. The set of all factors of 12   finite

8. The set of all multiples of 12   infinite

PRACTICE 14-1

LESSON **14-1** Practice B
Sets

Insert the correct symbol to make each statement true.

1. △ $\notin$ {right triangles}

2. ◡ $\in$ {semicircles}

3. {Jefferson, Washington} $\subset$ {U.S. Presidents}

4. {Pacific Ocean, Lake Erie} $\subset$ {bodies of water}

Determine whether the first set is a subset of the second set. Use the correct symbol.

5. S = {cola, root beer, ginger ale}
   D = {soft drinks}   Yes, $S \subset D$

6. C = {canoe, speedboat, raft}
   A = {water vehicles}   Yes, $C \subset A$

7. Q = {happy, sad, angry}
   M = {feelings}   Yes, $Q \subset M$

8. A = {$\frac{1}{2}, \frac{2}{3}, \frac{8}{10}, \frac{9}{11}$}
   B = {fractions in lowest terms}   No, $A \not\subset B$

Tell whether each set is finite or infinite.

9. {hours in a year}   finite

10. {the length of a line}   infinite

11. {positive integers}   infinite

12. {bones in the human body}   finite

13. Write a statement using one of the symbols $\not\subset$, $\subset$, $\in$, or $\notin$ to show the relationship between the students who like chocolate and the students who do not like chocolate.

   Possible answer: {students who like chocolate} $\not\subset$ {students who do not like chocolate}

14. Write a statement using one of the symbols $\not\subset$, $\subset$, $\in$, or $\notin$ to show the relationship between the state of Virginia and the set of states on the east coast of the United States.

   Possible answer: Virginia $\in$ {states on the east coast of the United States}

Determine whether each set is finite or infinite.

23. {people on Earth} **finite**

24. {counting numbers} **infinite**

25. {trinomials} **infinite**

26. {integers between 0 and 2} **finite**

27. {whole number factors of 20} **finite**

28. {solutions of $x < 0$} **infinite**

29. Set S is composed of the square of every element of the set $\{-5, 5\}$. What is S? **{25}**

Music **LINK**

The first percussion instruments to be used in orchestras were the *timpani*, or kettle drums, in the 1600's.

30. The *closure property* states that a set is *closed* under an operation if performing that operation on any elements of the set always results in an element of the set. The set of integers is closed under multiplication because multiplying integers always results in an integer. Tell whether each set is closed under the given operation.

   a. {0, 1}; multiplication **yes** **b.** {positive numbers}; subtraction **no**

   c. {counting numbers}; division **no** **d.** {even numbers}; addition **yes**

31. **LIFE SCIENCE** Write a statement using one of the symbols $\not\subset$, $\subset$, $\in$, or $\notin$ to show the relationship between the femur (thigh bone) and the set of human bones. **femur $\in$ {human bones}**

32. **MUSIC** Write a statement using one of the symbols $\not\subset$, $\subset$, $\in$, or $\notin$ to show the relationship between the set of percussion instruments and the set of string instruments. **{percussion instruments} $\not\subset$ {string instruments}**

33. **SOCIAL STUDIES** Write a statement using one of the symbols $\not\subset$, $\subset$, $\in$, or $\notin$ to show the relationship between the city of Miami, Florida, and the set of state capitals. **Miami $\notin$ {state capitals}**

 34. **WRITE A PROBLEM** Using facts you find in your social studies or science textbook, show that one set is a subset of another set.

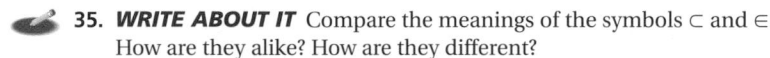 35. **WRITE ABOUT IT** Compare the meanings of the symbols $\subset$ and $\in$. How are they alike? How are they different?

36. **CHALLENGE** If $P = \{2, 4, 6, 8\}$ and $Q = \{$even integers between 0 and 10$\}$, is P a subset of Q? Explain. **Yes; they have the same elements, and a set is a subset of itself.**

Spiral Review

Simplify. (Lesson 13-2)

37. $-4(m^2 - 3m + 6)$ **$-4m^2 + 12m - 24$**

38. $3(a^2b - 4a + 3ab) - 2ab$ **$3a^2b - 12a + 7ab$**

39. $x^2y + 4(xy^2 - 3x^2y + 4xy)$ **$-11x^2y + 4xy^2 + 16xy$**

40. **TEST PREP** Which polynomial shows the result of using the FOIL method to find $(x - 2)(x + 6)$? (Lesson 13-6) **B**

   A $x^2 - 12$ **B** $x^2 + 6x - 2x - 12$ **C** $2x - 2x - 12$ **D** $x^2 + 4$

41. **TEST PREP** Which is equivalent to $x^2 - 16$? (Lesson 13-6) **F**

   F $(x - 4)(x + 4)$ **G** $(x - 4)^2$ **H** $(x + 4)^2$ **J** $(x)(x - 16)$

Journal

Ask students to use each symbol they learned in the lesson in a different statement.

Test Prep Doctor

For Exercise 41, encourage students to multiply out each of the answer choices to find the correct product. By multiplying the polynomials in choice **F**, they will see that it is the correct choice. If students chose **G**, they may have neglected to use the FOIL method properly. If they chose **J**, they may have distributed incorrectly.

CHALLENGE 14-1

Challenge
14-1 *Subset Formulas*

In this activity you will discover two important formulas for finding the number of subsets of any given finite set. To do so, you must know the following facts:

• The order of elements does not affect a set: $\{a, b, c\} = \{c, a, b\}$
• Every set is a subset of itself: $\{a\} \subset \{a\}$.
• The empty set—the set with no elements—is a subset of every set.
• Use the symbol { } for an empty set: { } $\subset \{a\}$ and { } $\subset$ { }.
• Any nonzero number to the zero power equals 1.
   $1^0 = 1$, $2^0 = 1$, $(-\frac{3}{4})^0 = 1$, etc.

Complete the chart below. Then look for patterns in the second and last columns to complete the subset formulas at the bottom of the page.

Set	Number Of Elements	Subsets	Number Of Subsets
{ }	0	{ }	$1 = 2^0$
{a}	1	{a}, { }	$2 = 2^1$
{a, b}	2	{a, b}, {a}, {b}, { }	$4 = 2^2$
{a, b, c}	3	{a, b, c}, {a, b}, {a, c}, {b, c}, {a}, {b}, {c}, { }	$8 = 2^3$
{a, b, c, d}	4	{a, b, c, d}, {a, b, c}, {a, b, d}, {a, c, d}, {b, c, d}, {a, b}, {a, c}, {a, d}, {b, c}, {b, d}, {c, d}, {a}, {b}, {c}, {d}, { }	$16 = 2^4$
{a, b, c, d, e}	5	{a, b, c, d, e}, {a, b, c, d}, {a, b, c, e}, {a, b, d, e}, {a, c, d, e}, {b, c, d, e}, {a, b, c}, {a, b, d}, {a, b, e}, {a, c, d}, {a, c, e}, {a, d, e}, {b, c, d}, {b, c, e}, {b, d, e}, {c, d, e}, {a, b}, {a, c}, {a, d}, {a, e}, {b, c}, {b, d}, {b, e}, {c, d}, {c, e}, {d, e}, {a}, {b}, {c}, {d}, {e}, { }	$32 = 2^5$

Formulas For The Number Of Subsets of a Finite Set:

The number of subsets of a finite set that has n elements is ____ 2^n

The number of non-empty subsets of a finite set that has n elements is ____ $2^n - 1$

PROBLEM SOLVING 14-1

Problem Solving
14-1 *Sets*

Write the correct answer.

1. $A = \{1, 2, 3\}$ and $B = \{2, 3, 4, 5\}$. Use the symbols $\in$ and $\notin$ to explain why $A \not\subset B$.

   Possible answer: $A \not\subset B$ because $1 \in A$ and $1 \notin B$

2. Write the following in set-builder notation: $C = \{1, 3, 5, 7, 9, \ldots\}$

   Possible answer: $C = \{x \mid x$ is a positive odd integer$\}$

3. Set A is **equal** to set B if and only if both sets contain the same elements. Give an example of two sets of numbers that are equal.

   Possible answer: $A = \{1, 2, 3\}$ $B = \{3, 1, 2\}$

4. For finite sets A and B, A is **equivalent** to B if and only if both sets contain the same number of elements. Give an example of two sets of animals that are equivalent, but not equal.

   Possible answer: $A = \{$dog, cat, horse$\}$ $B = \{$lion, tiger, bear$\}$

5. Write a statement using one of the following symbols $\not\subset$, $\subset$, $\in$, or $\notin$ to show the relationship between the set of all quadrilaterals and the set of all parallelograms.

   Possible answer: {parallelograms} $\subset$ {quadrilaterals}

6. Write a statement using one of the following symbols $\not\subset$, $\subset$, $\in$, or $\notin$ to show the relationship between a cylinder and the set of all polyhedrons.

   Possible answer: cylinder $\notin$ {polyhedrons}

7. Describe two different infinite sets of positive integers such that one set is a subset of the other set.

   Possible answer: {even integers} {multiples of 4}

Circle the letter of the best answer.

8. Which of the following statements is NOT true?
   Ⓐ $2 \in \{x \mid x > 2\}$
   B {consonants} $\subset$ {letters of the alphabet}
   C $a \in \{a, b, c, d, e\}$
   D $\sqrt{0.0625} \in \{\frac{1}{2}, \frac{1}{4}, \frac{1}{8}\}$

9. Which of the following is a finite set?
   F $A = \{x \mid x$ is a number less than 9$\}$
   G $A = \{x \mid x$ is a number greater than 3 but less than 5$\}$
   Ⓗ $A = \{x \mid x$ is a positive integer less than 100$\}$
   J $A = \{x \mid x$ is a multiple of 2$\}$

Lesson Quiz

Insert $\in$ or $\notin$ to make each statement true.

1. 0 ▢ {whole numbers} $\in$

2. $-\frac{1}{2}$ ▢ {integers} $\notin$

Determine whether the first set is a subset of the second set. Use the correct symbol.

3. $T = \{$counting numbers$\}$ $P = \{$prime numbers$\}$ No; $T \not\subset P$.

4. $A = \{a, e, o\}$ $Z = \{$vowels$\}$ Yes; $A \subset Z$.

Tell whether each set is finite or infinite.

5. {two-digit whole numbers} finite

6. {polygons} infinite

Available on Daily Transparency in CRB

14-2 Organizer

Pacing: Traditional 1 day
Block $\frac{1}{2}$ day

Objective: Students describe the intersection and union of sets.

Warm Up

Insert the correct symbol to make each statement true. Use the symbols $\in$, $\notin$, $\subset$, and $\not\subset$.

1. {rational numbers} ▓ {real numbers} $\subset$

2. 4.5 ▓ {rational numbers} $\in$

3. π ▓ {rational numbers} $\notin$

4. {real numbers} ▓ {irrational numbers} $\not\subset$

Problem of the Day

You have 10 black socks and 8 blue socks in a drawer. If you reach into the drawer without looking, what is the least number of socks you need to pull out in order to guarantee that you have a matching pair? 3

Available on Daily Transparency in CRB

Math Humor

Where do all the elements of a set go after they graduate? A reunion

---

14-2 Intersection and Union

Learn to describe the intersection and union of sets.

Vocabulary

intersection

empty set

union

The Caspian Sea, surrounded by the countries Azerbaijan, Iran, Kazakhstan, Russia, and Turkmenistan, is one of the world's largest lakes as well as one of the world's deepest.

The **intersection** of sets A and B is the set of all elements that are in both A **and** B. In other words, the intersection of sets A and B is the set of all elements that are common to both A and B.

To indicate the intersection of sets A and B, write $A \cap B$.

If A is the set of the world's five largest lakes by area, then A = {Caspian Sea, Lake Superior, Lake Victoria, Lake Huron, Lake Michigan}.

If B is the set of the world's five deepest lakes, then B = {Lake Baikal, Lake Tanganyika, Caspian Sea, Lake Nyasa, Issyk Kul}.

Reading Math

The empty set may also be represented by empty brackets, { }.

$A \cap B$ = {Caspian Sea} because the Caspian Sea is the only lake in both sets.

The set with no elements is called the **empty set**, or *null set*. The symbol for the empty set is $\varnothing$.

EXAMPLE 1 Finding the Intersection of Two Sets

Find the intersection of the sets.

A Z = {0, 1, 2, 3} T = {2, 4, 6, 8}

The only element that appears in both Z and T is 2.

$Z \cap T$ = {2}

B Q = {rational numbers} I = {irrational numbers}

There are no numbers that are both rational and irrational.

$Q \cap I$ = { } or $\varnothing$

---

1 Introduce

Alternate Opener

EXPLORATION

14-2 Intersection and Union

The diagram shows two sets, A and B. The region where the sets overlap is called the *intersection*. Numbers that belong to both sets are found in the intersection.

1. Describe set A and set B.

Use the counting numbers from 1 to 20 to complete the diagram for each description.

2. Set C contains the multiples of 3. Set D contains the factors of 18.

3. Set E contains the prime numbers. Set F contains the odd numbers.

Think and Discuss

4. Explain what an intersection of two sets is.

5. Give an example of two sets that have no intersection.

Motivate

Show students a small section of a local map. Choose two intersecting streets that are familiar to students. Ask them to locate the intersection of these two streets. Once they identify it, ask them what they think the word *intersection* means. Explain that the intersection on a map is where the two roads overlap.

Exploration worksheet and answers on Chapter 14 Resource Book pp. 16 and 83

2 Teach

Lesson Presentation

Guided Instruction

In this lesson, students learn to describe the intersection and union of sets. Define *empty set*. Explain that the *intersection* of two sets contains all of the elements that are in both sets. Explain that the *union* of two sets contains all of the elements that appear in either set. Show students the symbols for each of these terms. Point out that the symbol for union looks like the letter *U*. Remind students that elements common to both sets should be listed only once in the union.

Find the intersection of the sets.

C $L = \{x \mid x < 10\}$ $G = \{x \mid x > 5\}$

$L \cap G = \{x \mid 5 < x < 10\}$

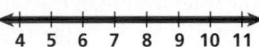

The **union** of sets Q and R is the set of all elements that are in either Q *or* R. To show the union of sets Q and R, write $Q \cup R$.

If $Q = \{-4, 2, 6, 10\}$ and $R = \{-2, 2, 6\}$, then $Q \cup R = \{-4, -2, 2, 6, 10\}$. If an element appears in both sets, represent it only once in the union.

EXAMPLE 2   **Finding the Union of Two Sets**

Find the union of the sets.

A $Q = \{\text{rational numbers}\}$ $I = \{\text{irrational numbers}\}$
Every real number is either rational or irrational.
$Q \cup I = \{\text{real numbers}\}$

B $Z = \{0, 1, 2, 3\}$ $T = \{2, 3, 4, 5\}$
$Z \cup T = \{0, 1, 2, 3, 4, 5\}$

C $N = \{\text{negative integers}\}$ $W = \{\text{whole numbers}\}$
The negative integers are $\{\dots, -3, -2, -1\}$. The whole numbers are $\{0, 1, 2, 3, \dots\}$.
$N \cup W = \{\text{integers}\}$

D $T = \{2, 4, 8, 16\}$ $E = \{\text{even integers}\}$
T is a subset of E, so the union of T and E is E.
$T \cup E = \{\text{even integers}\}$

E $L = \{x \mid x < 10\}$ $G = \{x \mid x > 5\}$
Every real number can be found in either set L or set G.
$L \cup G = \{\text{real numbers}\}$

4 5 6 7 8 9 10 11

Think and Discuss

1. Describe two sets whose intersection is $\{7, 8, 9, 10\}$.

2. Describe two sets whose union is $\{7, 8, 9, 10\}$.

3. Give an example of two sets whose intersection is the empty set.

COMMON ERROR ALERT

Students may confuse the symbols for intersection and union. Emphasize that the symbol for union, $\cup$, looks like the letter *U*.

Additional Examples

Example 1

Find the intersection of the sets.

A. $M = \{1, 3, 5, 7\}$
$N = \{5, 7, 9, 11\}$
$M \cap N = \{5, 7\}$

B. $Q = \{\text{rational numbers}\}$
$R = \{\text{real numbers}\}$
$Q \cap R = \{\text{rational numbers}\}$ or Q

C. $P = \{x \mid x < 4\}$
$T = \{x \mid x > 4\}$
$P \cap T = \{\ \}$ or $\varnothing$

Example 2

Find the union of the sets.

A. $M = \{1, 3, 5, 7\}$
$N = \{5, 7, 9, 11\}$
$M \cup N = \{1, 3, 5, 7, 9, 11\}$

B. $Q = \{\text{rational numbers}\}$
$R = \{\text{real numbers}\}$
$Q \cup R = \{\text{real numbers}\}$ or R

C. $P = \{x \mid x < 4\}$
$T = \{x \mid x > 4\}$
$P \cup T = \{x \mid x \neq 4\}$

D. $E = \{\text{even integers}\}$
$O = \{\text{odd integers}\}$
$E \cup O = \{\text{integers}\}$

E. $Q = \{x \mid x < 4\}$
$R = \{x \mid x < -4\}$
$Q \cup R = \{x \mid x < 4\}$ or Q

3 Close

Reaching All Learners
Through Curriculum Integration

Social Studies Give each student a recording sheet that contains data about the ten largest cities in the United States (Chapter 14 Resource Book p. 78). Have students use the data to list the elements of different sets. Then have them find the intersection or union of these sets.

Summarize

Review the vocabulary terms and symbols from the lesson. Describe some sets of students in the class, such as students who are girls or students who are wearing blue jeans. Ask the class to identify all of the elements of both sets, and write the elements on the board. Then ask students to list all of the elements in the intersection and the union of the two sets.

Answers to Think and Discuss

Possible answers:

1. $\{1, 2, 3, 4, 5, 6, 7, 8, 9, 10\}$, $\{\text{real numbers greater than } 6\}$

2. $\{7, 9\}, \{8, 10\}$

3. $\{\text{odd numbers}\}, \{\text{even numbers}\}$

14-2 Exercises

FOR EXTRA PRACTICE
see page 758

internet connect
Homework Help Online
go.hrw.com Keyword: MP4 14-2

Students may want to refer back to the lesson examples.

GUIDED PRACTICE

See Example 1 Find the intersection of the sets.

1. $B = \{-2, 0, 2, 4, 6\}$
   $D = \{2, 4, 6, 8, 10\}$ **{2, 4, 6}**

2. $A = \{10, 11, 12, 13, 14\}$
   $E = \{$even numbers$\}$ **{10, 12, 14}**

3. $G = \{x \mid x \geq 2\}$
   $H = \{x \mid x \leq 5\}$ **{x | 2 ≤ x ≤ 5}**

4. $M = \{x \mid x \leq 7\}$
   $N = \{x \mid x \geq 0\}$ **{x | 0 ≤ x ≤ 7}**

See Example 2 Find the union of the sets.

5. $R = \{2, 4, 6, 8, 10, 12\}$
   $S = \{1, 2, 3, 4, 5\}$

6. $B = \{x \mid 0 < x < 10\}$
   $C = \{x \mid x \geq 2\}$ **{x | x > 0}**

7. $Q = \{$negative integers$\}$
   $W = \{$whole numbers$\}$
   {integers}

8. $Q = \{$rational numbers$\}$
   $I = \{$integers$\}$
   {rational numbers}

Assignment Guide

If you finished Example **1** assign:
Core 1–4, 9–12, 30–35
Enriched 1–4, 9–12, 30–35

If you finished Example **2** assign:
Core 1–16, 17–27 odd, 30–35
Enriched 9–35

INDEPENDENT PRACTICE

See Example 1 Find the intersection of the sets.

9. $R = \{-10, -8, -6, -4\}$
   $T = \{-4, -2, 0, 2, 4\}$ **{-4}**

10. $L = \{$negative integers$\}$
    $N = \{$natural numbers$\}$ **∅**

11. $O = \{$positive odd integers$\}$
    $X = \{x \mid -10 \leq x \leq 5\}$ **{1, 3, 5}**

12. $K = \{x \mid x < 5\}$
    $R = \{x \mid x < 2\}$ **{x | x < 2}**

See Example 2 Find the union of the sets.

13. $G = \{-12, -10, -8, -6, -4\}$
    $H = \{-12, -8, -4, 0\}$

14. $D = \{1, 2, 3, 4, 5\}$
    $F = \{2, 4, 6\}$ **{1, 2, 3, 4, 5, 6}**

15. $Y = \{x \mid x \leq 0\}$
    $W = \{x \mid x > 0\}$
    {real numbers}

16. $K = \{$positive integers$\}$
    $T = \{$rational numbers$\}$
    {rational numbers}

Answers

5. $\{1, 2, 3, 4, 5, 6, 8, 10, 12\}$

13. $\{-12, -10, -8, -6, -4, 0\}$

17. $F \cup G = \{-2, -1, 0, 1, 2\}$;
    $F \cap G = \{-2, 0, 2\}$

18. $W \cup R = \{3, 5, 7,$ and even integers$\}$; $W \cap R = \{2, 4, 6\}$

19. $R \cup M = \{x \mid x < 6$ or $x \geq 7\}$;
    $R \cap M = \varnothing$

20. $T \cup P = \{x \mid 0 \leq x \leq 15\}$;
    $T \cap P = \{x \mid 5 < x \leq 10\}$

21. $A \cup B = \{$integers$\}$; $A \cap B = \varnothing$

22. $Q \cup T = \{$real numbers$\}$;
    $Q \cap T = \{x \mid 3 < x < 5\}$

23. $P \cup M = \{$even integers$\}$;
    $P \cap M = \{$positive multiples of 2$\}$

24. $J \cup R = \{\frac{1}{4}, \frac{1}{3}, \frac{1}{2}, 1, 4, 9, 16\}$;
    $J \cap R = \{1\}$

PRACTICE AND PROBLEM SOLVING

Find the union and the intersection of the sets.

17. $F = \{-2, -1, 0, 1, 2\}$
    $G = \{-2, 0, 2\}$

18. $W = \{2, 3, 4, 5, 6, 7\}$
    $R = \{$even integers$\}$

19. $R = \{x \mid x \geq 7\}$
    $M = \{x \mid x < 6\}$

20. $T = \{x \mid 0 \leq x \leq 10\}$
    $P = \{x \mid 5 < x \leq 15\}$

21. $A = \{$even integers$\}$
    $B = \{$odd integers$\}$

22. $Q = \{x \mid x < 5\}$
    $T = \{x \mid x > 3\}$

23. $P = \{$positive multiples of 2$\}$
    $M = \{$even integers$\}$

24. $J = \{$reciprocals of 1, 2, 3, and 4$\}$
    $R = \{$squares of 1, 2, 3, and 4$\}$

Math Background

Solving a system of equations graphically (Chapter 11 Extension) involves finding the intersection of two sets. When both equations are graphed on a coordinate plane, the solution is the intersection of the lines. Each line represents an infinite set of points, and there are three possibilities for the intersection of these sets. The intersection is the empty set if the lines are parallel, exactly one point if the lines intersect, and an infinite set of points if the lines are the same.

RETEACH 14-2

Reteach
14-2 Intersection and Union

The **intersection** of two sets is the set of all elements that are common to both sets.

The symbol ∩ indicates the intersection of two sets.

If two sets have no elements in common, then their intersection has no elements—it is the **empty set**. The symbols { } or ∅ indicate the empty set.

Find $A \cap B$.
$A = \{1, 2, 3, 4, 5, 6, 7, 8, 9, 10\}$
$B = \{3, 6, 9, 12, 15\}$

Find all the elements that are common to both given sets.
$A = \{1, 2, 3, 4, 5, 6, 7, 8, 9, 10\}$
$B = \{3, 6, 9, 12, 15\}$
So $A \cap B = \{3, 6, 9\}$

Find the intersection of the sets.

1. $A = \{-2, -1, 3, 4, 5\}$
   $B = \{1, 2, 3, 4, 5\}$
   $A \cap B = \{3, 4, 5\}$

2. $X = \{10, 15, 20, 25, 30\}$
   $Y = \{$odd numbers$\}$
   $X \cap Y = \{15, 25\}$

3. $P = \{2, 3, 4, 5, 6, 7\}$
   $Q = \{7, 8, 9, 10,\}$
   $P \cap Q = \{7\}$

The union of two sets is the set of all elements that are in one set or the other, or both.

The symbol ∪ indicates the union of two sets.

Find $A \cup B$.
$A = \{5, 10, 15, 20, 25, 30\}$
$B = \{10, 20, 30, 40, 50\}$

Combine both sets.
Then cross out any repeated elements.
$\{5, 10, 15, 20, 25, 30, 10, 20, 30, 40, 50\}$
So, $A \cup B = \{5, 10, 15, 20, 25, 30, 40, 50\}$

Find the union of the sets.

4. $A = \{1, 2, 3, 4, 5, 6, 7\}$
   $B = \{2, 4, 6, 8\}$
   $A \cup B = \{1, 2, 3, 4, 5, 6, 7, 8\}$

5. $X = \{$odd integers$\}$
   $Y = \{$even integers$\}$
   $X \cup Y = \{$all integers$\}$

6. $P = \{1, 2, 3, 4, 5, 6\}$
   $Q = \{2, 3, 4, 5, 6, 7\}$
   $P \cup Q = \{1, 2, 3, 4, 5, 6, 7\}$

PRACTICE 14-2

Practice B
14-2 Intersection and Union

Find the intersection of the sets.

1. $N = \{-1, -4, -9, -16, -25\}$
   $S = \{(-1)^2, (-2)^2, (-3)^2, (-4)^2, (-5)^2\}$
   $N \cap S = \varnothing$

2. $K = \{$scalene triangles, isosceles triangles, equilateral triangles, right triangles$\}$
   $T = \{$triangles with at least 2 equal sides$\}$
   $K \cap T = \{$isosceles triangles, equilateral triangles$\}$

3. $M = \{m \mid -20 \leq m < 3\}$
   $N = \{m \mid 3 \leq m\}$
   $M \cap N = \varnothing$

4. $T = \{y \mid y = y^2\}$
   $Y = \{y \mid y = y^3\}$
   $T \cap Y = \{0, 1\}$

Find the union of the sets.

5. $A = \{x \mid |x| \leq 3\}$
   $B = \{x \mid -5 \leq x \leq -3\}$
   $A \cup B = \{x \mid -5 \leq x \leq 3\}$

6. $J = \{$positive real numbers$\}$
   $I = \{$negative real numbers$\}$
   $J \cup I = \{$all real numbers except 0$\}$

7. $M = \{$positive integer multiples of 7$\}$
   $R = \{$positive integer multiples of 14$\}$
   $M \cup R = M$

8. $E = \{$negative even integers$\}$
   $F = \{$positive even integers$\}$
   $E \cup F = \{$all even integers except 0$\}$

Find the intersection and union of the sets.

9. $C = \{2, 4, 6, 8, 10\}$
   $P = \{$prime numbers$\}$
   $C \cap P = \{2\}$
   $C \cup P = \{4, 6, 8, 10,$ the prime numbers$\}$

10. $B = \{\frac{1}{2}, \frac{8}{9}, \frac{1}{3}, \frac{1}{4}, \frac{1}{6}\}$
    $V = \{\frac{1}{2}, \frac{1}{3}, \frac{1}{4}, \frac{1}{5}, \frac{1}{6}\}$
    $B \cap V = \{\frac{1}{2}, \frac{1}{3}, \frac{1}{4}\}$
    $B \cup V = \{\frac{1}{2}, \frac{1}{3}, \frac{1}{4}, \frac{1}{5}, \frac{1}{6}, \frac{1}{10}, \frac{8}{9}\}$

Groups of birds of a species may be known by several names.

Bird	Group Names
Chickens	{flock, run, brood, clutch}
Crows	{clan, murder, hover}
Ducks	{bed, brace, flock, flight, paddling, raft}
Flamingos	{stand, flamboyance}
Geese	{covert, flock, gaggle, plump, skein}
Peacocks	{pride, muster, ostentation}
Pheasants	{nye, brood, nide}
Pigeons	{flock, flight}
Starlings	{chattering, murmeration}
Swans	{bank, bevy, herd, team, wedge}

List the set of group names represented by the following kinds of birds.

25. Find {swans} ∪ {geese}.

26. Find {pheasants} ∩ {chickens}.   **{brood}**

27. Find two sets whose intersection is the empty set.
Possible answer: {crows} and {starlings}

28. Find two sets whose intersection is one of the sets.
{ducks} and {pigeons}

29. 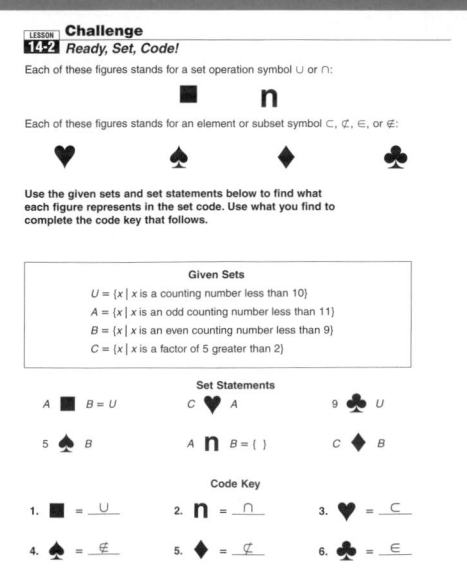 **CHALLENGE** {Quails} ∩ {swans} = {bevy} and {quails} ∪ {swans} = {bank, bevy, herd, covey, team, wedge}. What are all of the names for a group of quails?
{bevy, covey}

Spiral Review

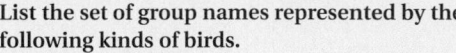

Simplify each expression. (Lesson 3-8)

30. $\sqrt{121} + \sqrt{25}$ **16** **31.** $(4 + 3)^2$ **49** **32.** $\dfrac{\sqrt{441}}{\sqrt{144}}$ **$\frac{7}{4}$** **33.** $\sqrt{5^2 + 12^2}$ **13**

34. TEST PREP Which figure has the fewest lines of symmetry? (Lesson 5-8)   **C**

A    B    C    D

35. TEST PREP The lengths of the sides of a rectangle are whole numbers. If the rectangle's perimeter is 24 units, which of the following could **not** be the rectangle's area? (Lesson 6-1)   **H**

F 27 square units   **G** 20 square units   **H** 24 square units   **J** 11 square units

---

CHALLENGE 14-2

LESSON 14-2 Challenge
Ready, Set, Code!

Each of these figures stands for a set operation symbol ∪ or ∩:

■   n

Each of these figures stands for an element or subset symbol ⊂, ⊄, ∈, or ∉:

♥   ♠   ♦   ♣

Use the given sets and set statements below to find what each figure represents in the set code. Use what you find to complete the code key that follows.

Given Sets

$U = \{x \mid x \text{ is a counting number less than } 10\}$
$A = \{x \mid x \text{ is an odd counting number less than } 11\}$
$B = \{x \mid x \text{ is an even counting number less than } 9\}$
$C = \{x \mid x \text{ is a factor of 5 greater than } 2\}$

Set Statements

$A \;■\; B = U$ $C \;♥\; A$ $9 \;♣\; U$
$5 \;♠\; B$ $A \;n\; B = \{\ \}$ $C \;♦\; B$

Code Key

1. ■ = __∪__   2. n = __∩__   3. ♥ = __⊂__
4. ♠ = __∉__   5. ♦ = __⊄__   6. ♣ = __∈__

PROBLEM SOLVING 14-2

LESSON 14-2 Problem Solving
Intersection and Union

Answer each question using the information in the table below.

Food	Ingredients
Caramel Rolls	{butter, pecans, caramel toppings, crescent rolls}
Cinnamon Buns	{white cake mix, yeast, water, flour, cinnamon, sugar}
Cheese Biscuits	{baking mix, milk, cheese, butter, garlic powder}
Beefy Taco Bake	{ground beef, tomato soup, salsa, milk, cheese}
Cherry Chocolate Brownies	{cherry pie filling, brownie mix, eggs, water, oil}

1. Find Cheese Biscuits ∩ Beefy Taco Bake.

{milk, cheese}

2. Find Cinnamon Buns ∪ Cherry Chocolate Brownies.

{white cake mix, yeast, water, flour, cinnamon, sugar, cherry pie filling, brownie mix, eggs, water, oil}

3. Find two foods that have the empty set as the intersection of their sets of ingredients.

Possible answer: Beefy Taco Bake and Cherry Chocolate Brownies

4. Find two foods that have 10 elements in the union of their sets of ingredients.

Possible answer: Cinnamon Buns and Caramel Rolls

Circle the letter of the best answer.

5. Which of the following is the intersection of the ingredients for Caramel Rolls and Cheese Biscuits?
   Ⓐ butter
   B pecans
   C milk
   D cheese

6. The union of the ingredients for Cinnamon Buns and the ingredients for which of the following is NOT the empty set?
   F Beefy Taco Bake
   G Cheese Biscuits
   Ⓗ Cherry Chocolate Brownies
   J Caramel Rolls

Life Science

Exercises 25–29 involve applying set theory to names of bird groups. Birds are studied in middle school life science courses, such as *Holt Science & Technology.*

Answers

25. {covert, flock, gaggle, plump, skein, bank, bevy, herd, team, wedge}

Journal

Ask students to write about the difference between the intersection and union of two sets. Have them include an example.

Test Prep Doctor

For Exercise 35, remind students that the sum of the length and width is equal to half the perimeter, or 12 units. Because the area of a rectangle is its length times its width, the dimensions must be factors of the answer choices.

By testing the answer choices, students can determine which ones have factors that add to 12. Choice **H** gives possible dimensions of 1 × 24, 2 × 12, 3 × 8, and 4 × 6, but none of these add to 12. So choice **H** is correct.

Lesson Quiz

Find the intersection of the sets.

1. $A = \{1, 2, 3, 4\}$
   $B = \{4, 5, 6\}$ **{4}**

2. $W = \{\text{whole numbers}\}$
   $N = \{\text{counting numbers}\}$
   {counting numbers}

3. $E = \{6, 12, 18, 24\}$
   $F = \{12, 24, 36\}$ **{12, 24}**

Find the union of the sets.

4. $G = \{2, 4, 6, 8\}$
   $H = \{10, 12, 14, 16\}$
   {2, 4, 6, 8, 10, 12, 14, 16}

5. $I = \{\text{even whole numbers}\}$
   $J = \{\text{odd whole numbers}\}$
   {whole numbers}

6. $K = \{x \mid x < 9\}$
   $L = \{x \mid x > -2\}$ **{real numbers}**

Available on Daily Transparency in CRB

Warm Up

Find the intersection and union of the sets in each pair.

1. $A = \{2, 4, 6, 8\}$ $B = \{1, 2, 3, 4\}$
$A \cap B = \{2, 4\}$
$A \cup B = \{1, 2, 3, 4, 6, 8\}$

2. $C = \{x | x > 3\}$ $D = \{x | x < 11\}$
$C \cap D = \{x | 3 < x < 11\}$
$C \cup D = \{real\ numbers\}$

Problem of the Day

A grizzly bear can run up to 30 mi/h. How many feet can the grizzly bear run in 1 second? **44 ft**

Available on Daily Transparency in CRB

Math Humor

Teacher: Use these sets to draw a diagram.

Student: Venn?

Teacher: Now.

14-3 Venn Diagrams

Learn to make and use Venn diagrams.

A computer and a human brain share some characteristics, but they obviously differ in many ways. If you consider their characteristics and abilities as sets, those that they share would be contained in their intersection.

A Venn diagram shows relationships among sets. In a Venn diagram, circles are used to represent sets. When two circles overlap, the region shared by both circles represents the intersection of the two sets.

The intersection of the set of all triangles and the set of all regular polygons for example, is the set of equilateral triangles.

Computer   **Brain**

Nonliving   Living
Must be   Memory   New ideas
programmed   Stores info   Dreams
Unemotional Can be damaged   Creates
Analyzes all   Multitask   Has emotions
possible   Math and logic   Fatigue
outcomes   Needs energy   Sleeps
Hard   Chess   Soft
Dry   Moist

EXAMPLE **1** **Drawing Venn Diagrams**

Draw a Venn diagram to show the relationship between the sets.

A Vowels: $\{A, E, I, O, U\}$
Letters used to represent musical notes: $\{A, B, C, D, E, F, G\}$

To draw the Venn diagram, first determine what is in the intersection of the sets.

The intersection of the sets is $\{A, E\}$.

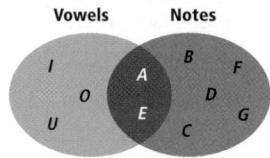

B Factors of 28: $\{1, 2, 4, 7, 14, 28\}$
Factors of 32: $\{1, 2, 4, 8, 16, 32\}$

The intersection of the sets is $\{1, 2, 4\}$.

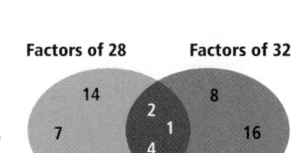

---

1 Introduce

Alternate Opener

Motivate

Show students some examples of graphs, such as a bar graph, a stem-and-leaf plot, and a scatter plot. Ask students how graphic organizers can be helpful. Explain that a type of graphic organizer that is useful for showing the relationship between sets is called a *Venn diagram*.

Exploration worksheet and answers on Chapter 14 Resource Book pp. 25 and 85

2 Teach

Lesson Presentation

Guided Instruction

In this lesson, students learn to make and use Venn diagrams. Show students the sample Venn diagrams in the lesson (Teaching Transparency T9, CRB). Explain that the intersection of sets is illustrated by overlapping circles. Point out that a subset is represented by a circle inside another circle. Show how Venn diagrams can be used to identify the elements of an intersection or a union. Review Example 3, and show students how a Venn diagram can illustrate a logical argument. You may want to give students some other examples of logical arguments.

EXAMPLE 2 **Analyzing Venn Diagrams**

Use each Venn diagram to identify intersections, unions, and subsets.

A

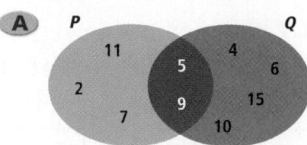

Intersection: $P \cap Q = \{5, 9\}$
Union: $P \cup Q = \{2, 4, 5, 6, 7, 9, 10, 11, 15\}$
Subsets: none

B

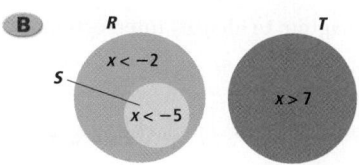

Intersections: $R \cap S = S$, $R \cap T = \varnothing$, $S \cap T = \varnothing$
Unions: $R \cup S = R$, $R \cup T = \{x \mid x < -2 \text{ or } x > 7\}$, and
$S \cup T = \{x \mid x < -5 \text{ or } x > 7\}$
Subsets: $S \subset R$

> **Remember!**
> S is a subset of R if every element of S is also an element of R.

The symbol $\therefore$ means "therefore," and it symbolizes the conclusion of a logical argument.

EXAMPLE 3 **Using Venn Diagrams**

Use a Venn diagram to show the following logical argument.

All frogs are amphibians.
No opossums are amphibians.
$\therefore$ No opossums are frogs.

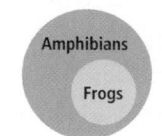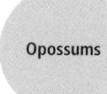

Think and Discuss

1. **Describe** how a subset is shown in a Venn diagram.

2. **Give** an example of a Venn diagram in which the intersection is the empty set.

Additional Examples

Example 1

Draw a Venn diagram to show the relationship between the sets.

primary colors: {red, blue, yellow}; colors in the American flag: {red, white, blue}

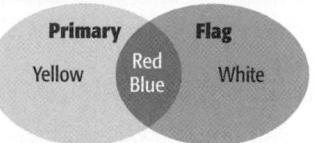

Example 2

Use the Venn diagram to identify intersections, unions, and subsets.

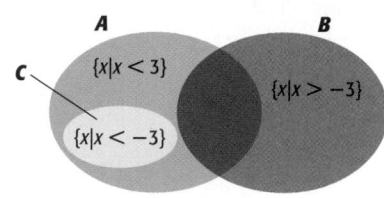

intersections: $A \cap B = \{x \mid -3 < x < 3\}$, $A \cap C = C$, $B \cap C = \varnothing$
unions: $A \cup B = \{$real numbers$\}$, $A \cup C = A$, $B \cup C = \{x \mid x \neq -3\}$
subsets: $C \subset A$

Example 3

Use a Venn diagram to show the following logical argument.

All dogs are animals. All beagles are dogs. $\therefore$ All beagles are animals.

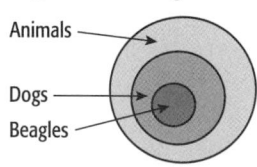

---

3 Close

Reaching All Learners
Through Concrete Manipulatives

Have students work in pairs or small groups. Give each group a large piece of construction paper and a few pattern blocks of geometric figures (Manipulatives Kit). Have students create a Venn diagram with a circle representing polygons, a circle representing quadrilaterals, and a circle representing triangles. Then have them place each pattern block in the appropriate section of the diagram and trace the shape. You may want to display the diagrams around the classroom.

Summarize

Remind students that a Venn diagram is a visual representation of the relationship between sets. Ask students how they would represent each situation below in a Venn diagram.

1. a subset of a subset of a set
2. the intersection of two sets
3. two sets whose intersection is the empty set

1. a circle in a circle in a circle
2. overlapping circles
3. two separate circles

Answers to Think and Discuss

1. A subset is shown as a circle within a circle.

2. Possible answer:

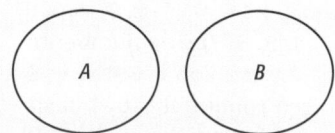

14-3 Venn Diagrams **697**

FOR EXTRA PRACTICE
see page 758

internet connect
Homework Help Online
go.hrw.com Keyword: MP4 14-3

go.hrw.com

> Students may want to refer back to the lesson examples.

Assignment Guide

If you finished Example **1** assign:
  Core 1–2, 6–7, 18–22
Enriched 1–2, 6–7, 18–22

If you finished Example **2** assign:
  Core 1–4, 6–9, 18–22
Enriched 6–9, 15–22

If you finished Example **3** assign:
  Core 1–12, 18–22
Enriched 6–22

Answers

1.
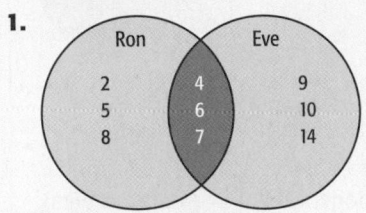

2.
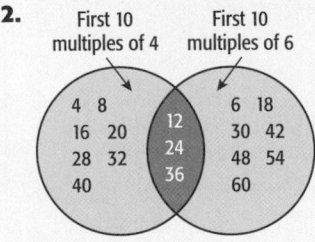

3–10. See p. A18.

GUIDED PRACTICE

See Example **1** Draw a Venn diagram to show the relationship between the sets.

1.

Set	Elements
Ron's favorite TV channels	2, 4, 5, 6, 7, 8
Eve's favorite TV channels	4, 6, 7, 9, 10, 14

2.

Set	Elements
First ten multiples of 4	4, 8, 12, 16, 20, 24, 28, 32, 36, 40
First ten multiples of 6	6, 12, 18, 24, 30, 36, 42, 48, 54, 60

See Example **2** Use each Venn diagram to identify intersections, unions, and subsets.

3.
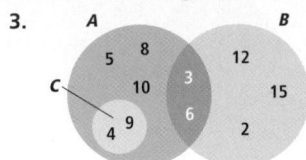

4.
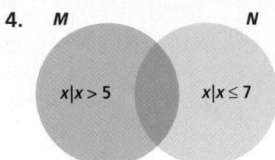

See Example **3** Use a Venn diagram to show the following logical argument.

5. All squares are rectangles. All rectangles are parallelograms.
∴ All squares are parallelograms.

INDEPENDENT PRACTICE

See Example **1** Draw a Venn diagram to show the relationship between the sets.

6.

Set	Elements
Faces on U.S. bills	{Washington, Lincoln, Hamilton, Jackson, Grant, Franklin}
Faces on U.S. coins	{Lincoln, F.D.R., Kennedy, Jefferson, Washington, Sacagawea}

7.

Set	Elements
Integers from −3 to 5	{−3, −2, −1, 0, 1, 2, 3, 4, 5}
Integers from −6 to 0	{−6, −5, −4, −3, −2, −1, 0}

See Example **2** Use each Venn diagram to identify intersections, unions, and subsets.

8.
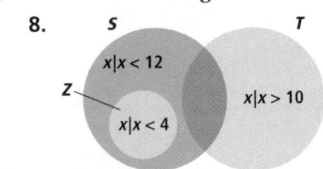

9.
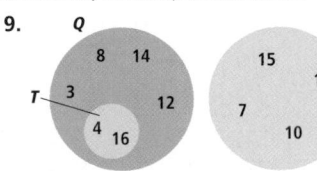

See Example **3** Use a Venn diagram to show the following logical argument.

10. All quadrilaterals are polygons. No circles are polygons.
∴ No circles are quadrilaterals.

Math Background

If A and B are two finite sets, the number of elements in $A \cup B$ is equal to the number of elements in set A plus the number of elements in set B minus the number of elements in $A \cap B$.

For example, let $A = \{1, 2, 3, 4, 5, 6\}$ and $B = \{4, 5, 6, 7\}$. The number of elements in the union is $6 + 4 - 3 = 7$. You can confirm this by listing the elements in the intersection and the union.

$A \cap B = \{4, 5, 6\}$
$A \cup B = \{1, 2, 3, 4, 5, 6, 7\}$

RETEACH 14-3

LESSON **14-3** Reteach
Venn Diagrams

A **Venn diagram** uses circles to show relationships between sets. Follow these steps to draw a Venn diagram for two sets.

Step 1: List the elements in each set

Step 2: Find the intersection of the sets.

Step 3: Draw your Venn diagram.
  • Draw 1 circle for each set. Label each circle with a set letter.
  • If the intersection of the sets has one or more elements, overlap the circles.
  • Don't overlap circles if the intersection of the sets is the empty set.

Step 4: Fill in your Venn diagram with the set elements.
  • Write each element only once in your diagram.
  • The intersection elements go in the overlapping section.
  • Write the rest of the elements inside the correct circle, but outside the overlapping section.

Draw a Venn diagram for sets A and B.
A = {factors of 16}
B = {factors of 20}
A = {1, 2, 4, 8, 16}
B = {1, 2, 4, 5, 10, 20}
A ∩ B = {1, 2, 4}

Draw a Venn diagram for each pair of sets.
1. A = {1, 3, 5, 7, 9, 11, 13}
   B = {factors of 9}

2. X = {integers from 1 to 20}
   Y = {odd integers from 5 to 25}

PRACTICE 14-3

LESSON **14-3** Practice B
Venn Diagrams

Draw a Venn Diagram to show the relationship between the sets.

1.

Set	Elements
A: All integers from −6 to 2, inclusive	−6, −5, −4, −3, −2, −1, 0, 1, 2
B: All integers from −4 to 4, inclusive	−4, −3, −2, −1, 0, 1, 2, 3, 4

Use a Venn Diagram to identify intersections, unions, and subsets.

2.

A ∪ B = {−3, −2, −1, 0, 1, 2, 3, 4}
A ∪ C = {−3, −2, −1, 0, 1, 2, 3}
A ∩ B = {−1, 0, 1}
B ∪ C = B
B ∩ C = C
C ∪ B
A ∩ B ∩ C = {1}
A ∪ B ∪ C = {−3, −2, −1, 0, 1, 2, 3, 4}

Use a Venn diagram to show the following logical argument.

3. All snakes are reptiles.
   The python is a snake.
   ∴ The python is a reptile.

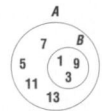

4. All quadrilaterals are polygons.
   No line is a polygon.
   ∴ No line is a quadrilateral.

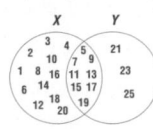

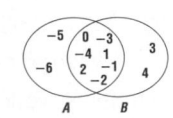

History LINK

Socrates (469–399 B.C.) was an ancient Greek philosopher who taught people to question accepted truths.

go.hrw.com
KEYWORD: MP4 Socrates
CNN Student News

PRACTICE AND PROBLEM SOLVING

11. All prime numbers except 2 are odd. Use a Venn diagram to display the statement.

12. HISTORY A well-known argument states:

"All men are mortal. Socrates was a man. Therefore, Socrates was mortal."

Use a Venn diagram to show the argument.

13. MUSIC All reed instruments are woodwinds. Some reed instruments are double-reed instruments. Therefore, all double-reed instruments are woodwinds. Use a Venn diagram to show the argument.

14. ENTERTAINMENT Use the information shown to write inequalities to show the age limits. Then make a Venn diagram for the situation. Identify the union and intersection of the sets.

Funland Amusement Park
Toddler Land Toddler Coaster
No children over the age of 8 allowed
Children ages 3 and up only

 15. CHOOSE A STRATEGY The following uniform numbers have been retired by the New York Yankees: 1, 3, 4, 5, 7, 8, 9, 10, 15, 16, 23, 32, 37, 42, and 44. The Los Angeles Dodgers have retired 1, 2, 4, 19, 20, 24, 32, 39, 42, and 53. How many uniform numbers from 0–99 are available to be worn by players on either team? **79**

 16. WRITE ABOUT IT Describe how Venn diagrams are useful for finding the union and intersection of two sets.

 17. CHALLENGE In a class, 22 students have been on a plane, 28 on a train, 23 on a boat, 15 on a plane and train, 20 on a train and boat, 14 on a plane and boat, 12 on all three, and 1 on none of them. How many students are in the class? **37**

Spiral Review

Find the volume of each figure. Use 3.14 for π. (Lessons 6-6, 6-7)

18. a 3 ft by 5 ft by 11 ft rectangular prism **165 ft³**

19. a cylinder with radius 3 in. and height 8 in. **226.08 in³**

20. a cone with diameter 7 in. and height 12 in. **153.86 in³**

21. a square pyramid with base length 5 cm and height 10 cm **$83\frac{1}{3}$ cm³**

22. TEST PREP Rachel reached into a bag containing 7 malt energy bars and 4 berry energy bars and pulled out 2 bars. What is the probability that Rachel chose 2 berry bars? (Lesson 9-7) **D**

A $\frac{2}{11}$ B $\frac{12}{121}$ C $\frac{21}{55}$ D $\frac{6}{55}$

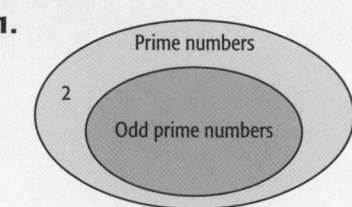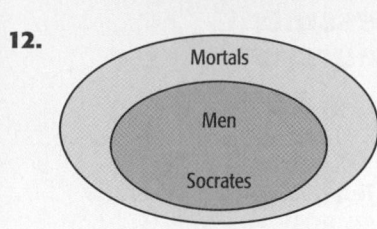
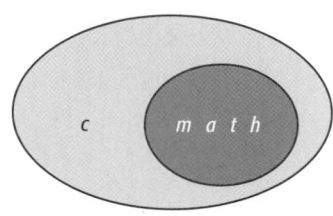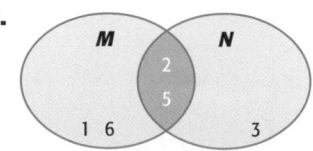
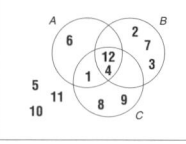
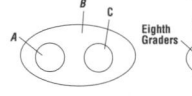
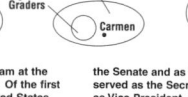

Chapter 14 **Mid-Chapter Quiz**

Purpose: *To assess students' mastery of concepts and skills in Lessons 14-1 through 14-3*

Assessment Resources

Section 14A Quiz
Assessment Resources . . . . p. 31

Test and Practice Generator CD-ROM

Additional mid-chapter assessment items in both multiple-choice and free-response format may be generated for any objective in Lessons 14-1 through 14-3.

Answers

11.

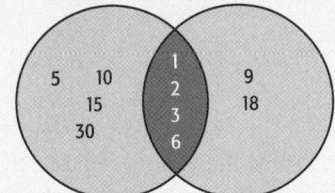

Factors of 30   Factors of 18

12.

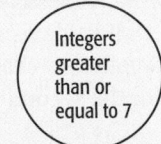

13. intersections: $P \cap C = \varnothing$, $P \cap W = P$, $C \cap W = C$; unions: $P \cup C = \{2, 3, 4, 5, ...\}$ or {whole numbers ≥ 2}, $P \cup W = W$, $C \cup W = W$; subsets: $P \subset W$, $C \subset W$

14. intersections: $P \cap A = P$, $P \cap B = \varnothing$, $B \cap A = \{x \mid 3 < x < 6\}$; unions: $P \cup A = A$, $P \cup B = \{x \mid x > 3 \text{ or } x < 0\}$, $B \cup A = $ {real numbers}; subsets: $P \subset A$

15.

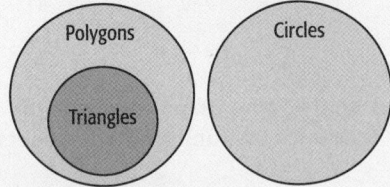

16.

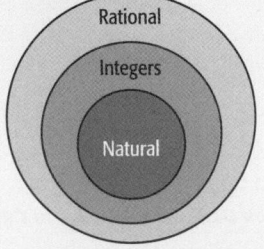

Mid-Chapter Quiz

LESSON 14-1 (pp. 688–691)

Insert $\in$ or $\notin$ to make each statement true.

1. Nevada ▮ {U.S. states} $\in$

2. Mexico ▮ {continents} $\notin$

Determine whether the first set is a subset of the second set. Use the correct symbol.

3. $K = $ {pyramids} $K \not\subset J$
   $J = $ {prisms}

4. $H = \{1, 2, 3, 4, 5\}$ $H \subset S$
   $S = $ {rational numbers}

Tell whether each set is finite or infinite.

5. {integers less than 200}   **infinite**

6. {factors of 1500}   **finite**

LESSON 14-2 (pp. 692–695)

Find the intersection of the sets.

7. $R = \{10, 20, 30, 40, 50\}$
   $S = \{5, 10, 15, 20\}$ **{10, 20}**

8. $W = $ {integers}
   $X = $ {natural numbers}
   {natural numbers}

Find the union of the sets.

9. $G = \{-3, -2, -1, 0\}$
   $H = \{0, 1, 2, 3\}$
   {−3, −2, −1, 0, 1, 2, 3}

10. $P = $ {positive integers}
   $R = $ {factors of 24}
   {positive integers}

LESSON 14-3 (pp. 696–699)

Draw a Venn diagram to show the relationships between the sets.

11.

Set	Elements
Factors of 30	{1, 2, 3, 5, 6, 10, 15, 30}
Factors of 18	{1, 2, 3, 6, 9, 18}

12.

Set	Elements
Integers greater than or equal to 7	{7, 8, 9, 10, ...}
Integers less than or equal to 5	{5, 4, 3, 2, ...}

Use the Venn diagrams to identify intersections, unions, and subsets.

13.

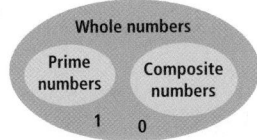

14.

A   B

$x \mid x < 6$ $x \mid x > 3$ P $x \mid x < 0$

Use a Venn diagram to show each logical argument.

15. No circles are polygons.
   All triangles are polygons.
   ∴ No triangles are circles.

16. All natural numbers are integers.
   All integers are rational numbers.
   ∴ All natural numbers are rational numbers.

Focus on Problem Solving

Plan

Make a Plan
• **Prioritize and sequence information**

Some problems contain a lot of information. Read the entire problem carefully to be sure you understand all of the facts. You may need to read it over several times—perhaps aloud so that you can hear yourself say the words.

Then decide which information is most important (prioritize). Is there any information that is absolutely necessary to solve the problem? This information is most important.

Finally, put the information in order (sequence). Use comparison words like *before, after, longer, shorter,* and so on to help you. Write down the sequence before you try to solve the problem.

 Read each problem below, and then answer the questions that follow.

1 Five friends are standing in line for the opening of a movie. They are in line according to their arrival. Tiffany arrived 3 minutes after Cedric. Roy took his place in line at 8:01 P.M. He was 1 minute behind Celeste and 7 minutes ahead of Tiffany. The first person arrived at 8:00 P.M. Blanca showed up 6 minutes after the first person. List the time of each person's arrival.

  a. Whose arrival information helped you determine each arrival time?

  b. Can you determine the order without the time?

  c. List the friends' order from the earliest to arrive to the last to arrive.

2 There are four children in the Putman family. Isabelle is half the age of Maxwell. Joe is 2 years older than Isabelle. Maxwell is 14. Hazel is twice Joe's age and 4 years older than Maxwell. What are the ages of the children?

  a. Whose age must you figure out first before you can find Joe's age?

  b. What are two ways to figure out Hazel's age?

  c. List the Putman children from oldest to youngest.

Answers

1. a. Roy's
  b. yes
  c. Celeste, Roy, Cedric, Blanca, Tiffany

2. a. Isabelle's
  b. Add 4 years to Maxwell's age, or multiply Joe's age by 2.
  c. Hazel, Maxwell, Joe, Isabelle

Focus on Problem Solving

Purpose: *To focus on making a plan to solve a problem by prioritizing and sequencing information*

Problem Solving Resources
Interactive Problem Solving. pp. 106–112
Math: Reading and Writing in the Content Area . . . . pp. 106–112

Problem Solving Process
This page focuses on the second step of the problem-solving process:
Make a Plan

Discuss
Have students discuss the order and priority of the information in each problem.

Possible answers:

1. Cedric is in front of Tiffany. Celeste and Roy are in front of Cedric. The order of these four can be found without knowing the time. The times between arrivals can then be used to find the final order. Celeste is first in line because she arrived 1 minute before Roy. Roy is second at 8:01, and Cedric is third at 8:05. Blanca is fourth at 8:06, and Tiffany is last at 8:08.

2. Maxwell's age, 14, is given in the problem, so you can figure out Isabelle's age, 7, and Hazel's age, 18. You can then use Isabelle's age to find Joe's age, which is 9.

Pacing: Traditional 1 day
Block $\frac{1}{2}$ day

Objective: Students differentiate between conjunctions and disjunctions and make truth tables.

Warm Up

Determine whether each statement is true or false.

1. All rational numbers are real numbers. T

2. All integers are whole numbers. F

3. All rational numbers are integers. F

Problem of the Day

The diameter of Earth is 7926 miles. The diameter of Mercury is 3032 miles. About how many times greater is Earth's surface area than Mercury's? about 6.8 times greater
Available on Daily Transparency in CRB

Teacher: What's the opposite of *disjunction*?

Student: Dat junction

---

1 Introduce
Alternate Opener

The intersection of the Venn diagram shows the counting numbers from 1 to 20 that are prime *and* are factors of 18.

Use a Venn diagram and the counting numbers from 1 to 20 to illustrate each intersection.

1. numbers that are perfect squares *and* are odd
2. numbers that are multiples of 3 *and* multiples of 2

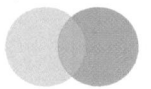 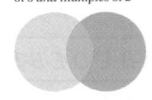

Think and Discuss

3. **Explain** why there is no intersection of perfect-square numbers and prime numbers.

4. **Give** an example of an *and* statement to describe an intersection of objects that are not numbers.

14-4 Compound Statements

14-4

Learn to differentiate between conjunctions and disjunctions and to make truth tables.

Vocabulary

compound statement
conjunction
truth value
truth table
disjunction

To remain on a cheerleading squad, a cheerleader must often meet several conditions, such as the following:

• Maintain a grade point average (GPA) of 2.5 or better.
• Miss no more than two practices.

A **compound statement** is formed by combining two or more simple statements. If *P* and *Q* represent simple statements, then the compound statement *P and Q* is called a **conjunction**.

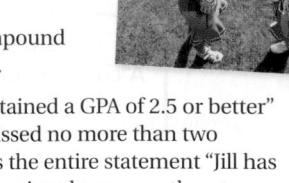

If *P* represents the statement "Jill has maintained a GPA of 2.5 or better" and *Q* represents the statement "Jill has missed no more than two practices," then the conjunction *P and Q* is the entire statement "Jill has maintained a GPA of 2.5 or better *and* has missed no more than two practices." A compound statement can be either true or false.

The **truth value** of a statement is either true or false. A **truth table** is a way to show the truth value of a compound statement, as determined by each different arrangement of the truth values of its simple statements.

EXAMPLE **Making Truth Tables for Conjunctions**

Make a truth table for the conjunction *P and Q*, where *P* is "Jill has maintained a GPA of 2.5 or better" and *Q* is "Jill has missed no more than two practices."

The conjunction *P and Q* is "Jill maintains a GPA of 2.5 or better *and* has missed no more than two practices."

Helpful Hint

To read the first line of the truth table, say, "When *P* is true and *Q* is true, the conjunction *P and Q* is true."

Example	P	Q	P and Q
Jill has a 3.2 GPA and has missed 1 practice.	True	True	True
Jill has a 3.6 GPA and has missed 3 practices.	True	False	False; only one of the conditions is met.
Jill has a 2.25 GPA and has missed no practices.	False	True	False; only one of the conditions is met
Jill has a 2.4 GPA and has missed 3 practices.	False	False	False; neither condition is met.

A conjunction is true only when all of its simple statements are true. If any statement is false, the conjunction is false.

Motivate

Ask students whether the following statement is true or false: "Two is even or two is odd." true Then ask the same of the statement "Two is even and two is odd." false Ask students to describe the difference between the two statements. *or* versus *and* Point out that in the *or* statement, only one of the parts must be true for the entire statement to be true. In the *and* statement, both parts must be true.

Exploration worksheet and answers on Chapter 14 Resource Book pp. 36 and 87

---

2 Teach

Lesson Presentation

Guided Instruction

In this lesson, students learn to differentiate between conjunctions and disjunctions and to make truth tables. Explain that a compound statement is formed by joining two statements that are either true or false. Define *conjunction* and *disjunction*. Provide several examples so that students can become familiar with these terms. Then show students how to create truth tables for conjunctions and disjunctions. Point out that truth tables show all possible outcomes for each compound statement.

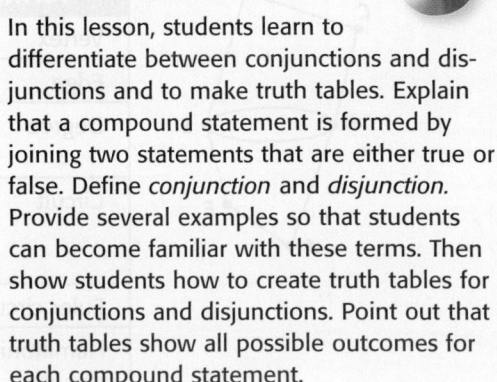

A compound statement of the form *P or Q* is called a **disjunction**. For example, you will be out of school if it is a weekend *or* a holiday.

EXAMPLE 2 **Making Truth Tables for Disjunctions**

Make a truth table for the disjunction *P or Q*, where *P* is "It is a weekend" and *Q* is "It is a holiday."

The disjunction *P or Q* is "It is a weekend *or* a holiday."

Helpful Hint

To read the first line of the truth table, say, "When *P* is true or *Q* is true, the conjunction *P or Q* is true."

Example	P	Q	P or Q
It is Saturday. It is a holiday.	True	True	True
It is Sunday. It is not a holiday.	True	False	True; only one of the conditions needs to be met.
It is Wednesday. It is a holiday.	False	True	True; only one of the conditions needs to be met.
It is Tuesday. It is not a holiday.	False	False	False; neither condition is met.

Note that a disjunction is true when any of its simple statements are true. The disjunction is false only when all of its simple statements are false.

Think and Discuss

1. **Explain** why there are four rows in each truth table in Examples 1 and 2.

2. **Tell** whether:
   a. *P* must be true if the conjunction *P and Q* is true.
   b. *Q* must be true if the disjunction *P or Q* is true.
   c. *P* must be false if the conjunction *P and Q* is false.
   d. *Q* must be false if the disjunction *P or Q* is false.

3. **Consider** the expression "A chain is only as strong as its weakest link." Use a conjunction to describe this situation.

COMMON ERROR ALERT

Students may think that the disjunction *P or Q* is false when both statements are true. Give students a real-world example to show that the disjunction is true. For example, if you have $5 or a free pass, you can see a movie. Explain that if you have $5 and a free pass, you can still see the movie.

Additional Examples

Example 1

Make a truth table for the conjunction *P and Q*, where *P* is "Barry is on the soccer team" and *Q* is "Barry plays the cello."

The conjunction *P and Q* is "Barry is on the soccer team *and* plays the cello."

P	Q	P and Q
T	T	T
T	F	F
F	T	F
F	F	F

Example 2

Make a truth table for the disjunction *P or Q*, where *P* is "The sun is shining" and *Q* is "It is a warm day."

The disjunction *P or Q* is "The sun is shining *or* it is a warm day."

P	Q	P or Q
T	T	T
T	F	T
F	T	T
F	F	F

3 Close

Reaching All Learners
Through Graphic Cues

Give students a recording sheet that contains a blank Venn diagram and some data including prime and odd numbers (Chapter 14 Resource Book p. 80). Have students write elements in their proper places on the Venn diagram and use the diagram to solve problems involving conjunctions and disjunctions. Discuss the connections between conjunction and intersection and between disjunction and union.

Summarize

Remind students that each statement in a compound statement is either true or false. Have students decide on the truth value for each situation shown in the table.

Situation	P and Q	P or Q
Both *P* and *Q* are true.	T	T
Only *P* is true.	F	T
Only *Q* is true.	F	T
P and *Q* are false.	F	F

Answers to Think and Discuss

1. There are two possible truth values for each statement, so there are $2 \cdot 2 = 4$ possible permutations for each compound statement.

2. **a.** Yes; if the conjunction is true, both simple statements are true.
   b. No; it's possible that *P* is true and *Q* is false. **c.** No; it's possible that *Q* is false and *P* is true. **d.** Yes; a disjunction is only false when both statements are false.

3. If the first link is strong and the second link is strong and the third link is strong ... and the last link is strong, then the chain is strong.

Pacing: Traditional 1 day
Block $\frac{1}{2}$ day

Objective: To use a graphing calculator to explore true and false statements

Materials: Graphing calculator

Lab Resources

Technology Lab Activities . . . pp. 90–91

Using the Pages

This technology activity shows students how to use a graphing calculator to determine whether statements are true or false. Specific keystrokes may vary, depending on the make and model of the graphing calculator used. The keystrokes given are for a TI-83 model. For keystrokes to other models, visit go.hrw.com.

The Think and Discuss problems can be used to assess students' understanding of the technology activity. While the Try This problems can be done without a graphing calculator, they are meant to help students become familiar with using a graphing calculator to determine whether statements are true or false.

Assessment

1. What series of keystrokes is needed to make a table of true and false values for the inequality $x + 1 < 4$?

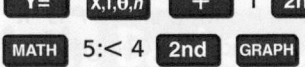

2. What is the difference between the words *and* and *or* when stating restrictions on the variable? *And* means that both conditions must be met; *or* means only one of the conditions must be met.

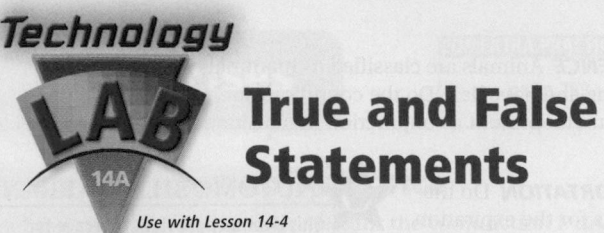

Technology
14A True and False Statements

Use with Lesson 14-4

internet connect
For an interactive lab, visit
go.hrw.com
KEYWORD: MP4 TechLab14A

Your calculator displays the number 1 for a true statement and 0 for a false statement.

Example:

$2 + 2 = 4$ returns a 1.

$10 \div 2 = 6$ returns a 0.

Activity 1

1 Test values of x for the statement $x - 2 \geq 7$.

Press 5 STO▶ X,T,θ,n ENTER and then X,T,θ,n − 2 2nd MATH .

Select **4:≥** and press 7 ENTER .

The statement is false.

Repeat the steps above for $x = 7$ (it will return a 0) and for $x = 10$ (it will return a 1).

2 Make a table of true and false values of x for $x - 2 \geq 7$.

Press Y= and enter X−2≥7. To enter the ≥ symbol, press 2nd MATH , select **4:≥**, and press ENTER .

Press 2nd GRAPH , and use the down arrow key to see the values of x for which the inequality is true.

You can see that the inequality is true for values of x greater than or equal to 9.

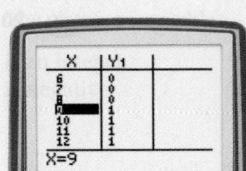

Think and Discuss

1. For what values of x would $|x| = x$ return a value of 0?

Try This

1. Enter a true statement and a false statement for each of the test commands: =, ≠, >, ≥, <, and ≤.

2. Make a table of true and false values of x for $2x - 1 \leq 15$.

Answers

Activity 1

Think and Discuss

1. Possible answer: $\{x \mid x < 0\}$

Try This

1. Check students' work.

2.

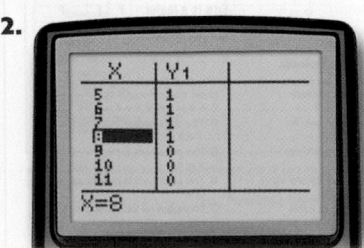

The results of logical calculations can be useful in graphing pieces of functions.

❶ Graph the line $y = x - 2$ for $\{x \mid x \geq 9\}$.

Press **Y=**, and enter **(X−2)/(X≥9)**. To enter the ≥ symbol, press
2nd **MATH** (TEST), select **4:≥**, and press **ENTER**.

Press **ZOOM** **6:Standard**, then **ZOOM** **8:Integer**, and then **ENTER**.
The graph shows integer values of x.

Press **TRACE**. Then press the right arrow key to see ordered pairs.

The first ordered pair that shows a value of y is (9, 7).

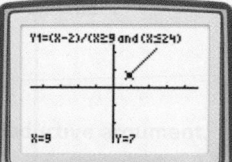

❷ Add the condition $\{x \mid x \leq 24\}$ to the condition in **❶**.

Change the function to the one shown, **(X−2)/(X ≥ 9 and X ≤ 24)**.

To enter the word *and*, use **2nd** **MATH** (TEST) ▶ to go to the **LOGIC** menu, and select **1:and**.

Press **TRACE** to display the values of x that satisfy the conditions.

❸ Change the condition $\{x \mid x \leq 24\}$ to an *or* condition.

Change the function to the one shown, **(X−2)/(X≥9 or X≤24)**.

To enter the word *or*, use **2nd** **MATH** (TEST) ▶ to go to the **LOGIC** menu, and select **2:or**.

Press **TRACE** to display the values of x that satisfy the conditions.

Think and Discuss

1. Why doesn't dividing by $(x \geq 9)$ for a value of x that is less than 9 return a value for y?

Try This

1. Graph the line $y = 2x - 1$ for $\{x \mid x \leq 7\}$.

2. Add the *or* condition $\{x \mid x \geq 12\}$ to Try This problem 1.

Think and Discuss

1. Possible answer: If x is less than 9, then the value of $x \geq 9$ is zero. Because division by zero is undefined, no value is returned for y.

Try This

1.

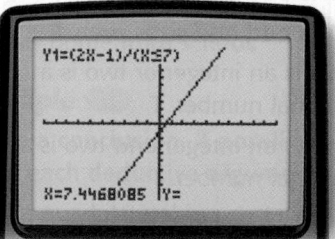

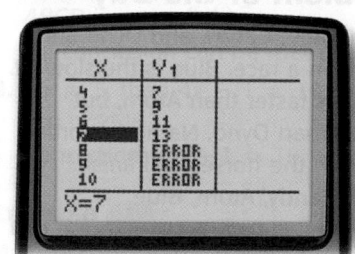

2.

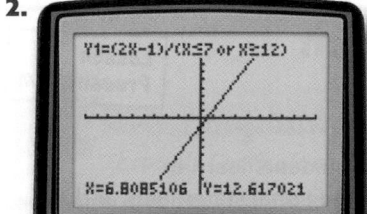

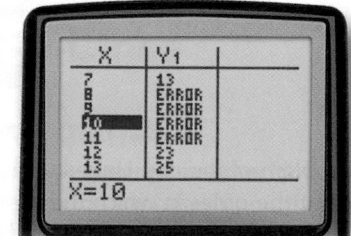

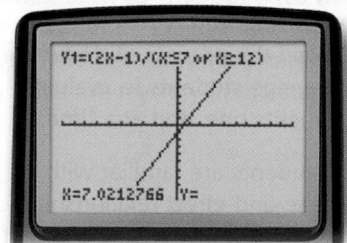

MATH-ABLES

Game Resources

Puzzles, Twisters & Teasers
Chapter 14 Resource Book

Find the Phony!

Purpose: *To apply logic to solving a brainteaser*

Discuss: Have students discuss why there were three groups of three pearls. Because the scale has two trays, dividing the pearls into groups of three ensures that you can compare two groups and evaluate the third by elimination. Because each group has three pearls, you can repeat the process and solve the problem with only two weighings.

Extend: Have students use a balance scale, eight real pennies, and one play-money penny (provided in the Manipulatives Kit) to carry out their solution. Check students' work.

Sprouts

Purpose: *To practice constructing networks by playing a game*

Discuss: Ask students to describe a situation in which no more legal moves are possible. Possible answer: All of the dots have three paths connected, or those that have only two paths are surrounded by others that already have three paths connected.

Extend: Challenge players to consider the shortest or longest games possible. Is it possible for the game to go on indefinitely? **no** What is the least number of moves possible before the game ends? **6** What is the greatest number of moves possible before the game ends? **8**

MATH-ABLES

Find the Phony!

Suppose you have nine identical-looking pearls. Eight are real, and one is fake. Using a balance scale that consists of two pans, you must find the bogus pearl. The real pearls weigh the same, and the fake weighs less. The scale can be used only twice. How can you find the phony?

First you must split the pearls into equal groups. Place any three pearls on one side of the scale and any other three on the other side. If one side weighs less than the other, then the fake pearl is on that side. But you are not done yet! You still need to find the imitation, and you can use the scale only once more. Take any of the two pearls from the lighter pan, and weigh them against each other. If one pan is lighter, then that pan contains the fake pearl. If they balance, then the leftover pearl of the group is the fake.

If the scale balances during the first weighing, then you know the fake is in the third group. Then you can choose two pearls from that group for the second weighing. If the scale balances, the fake is the one left. If it is unbalanced, the false pearl is the lighter one.

You Play Detective

Suppose you have 12 identical gold coins in front of you. One is counterfeit and weighs slightly more than the others. How can you identify the counterfeit in three weighings?

Sprouts

You and a partner play against each other to try to make the last move in the game. You start with three dots. Player one draws a path to join two dots or a path that starts and ends at the same dot. A new dot is then placed somewhere on that path. No dot can have more than three paths drawn from it, and no path can cross another. The last player to make a move is the winner!

☑ internet connect

Go to **go.hrw.com** for a complete set of game rules.
KEYWORD: MP4 Game14

Answers

Possible answer: Arrange the coins in three groups of four. Weigh any two groups against each other. If one is heavier, that group contains the heavy coin. If they balance, the third group contains the heavy coin. Split the group containing the heavy coin into two groups of two, and weigh them. Then split the heavier group into two coins, and weigh them.

Alternatively, arrange the coins in four groups of three. Weigh two of these groups. If they balance, weigh the other two groups. Once you have found the group containing the heavy coin, weigh any two of its coins. If one side is heavier, that is the heavy coin. If they balance, the heavy coin is the one not weighed.

Technology LAB

Logic and Programs

Use with Lesson 14-4

Your calculator has built-in menus for making logical comparisons. These are especially useful when you are writing programs.

internet connect
For an interactive lab, visit
go.hrw.com
KEYWORD: MP4 TechLab14

Activity

❶ Write and run a simple program that tests whether $a < 7$ and $b < 7$ and displays YES if so and NO if not.

Press **PRGM**. Select **NEW** and press **ENTER**. The calculator is in **ALPHA** mode. Type the name LOGIC for your program using the green alphabetic characters L O G I C. Press **ENTER**. Enter the first line of the program:

Press **PRGM** **If**. Press **ALPHA** A **2nd** **MATH** (TEST) < 7 **2nd** **MATH** (TEST) ▶ and **ALPHA** B **2nd** **MATH** (TEST) < 7 **ENTER**.

Enter the next three lines of the program.

Press **PRGM** **THEN** **ENTER**.

Press **PRGM** ▶ **Disp** **2nd** **ALPHA** (A-LOCK) " Y E S " **ENTER**.

Press **PRGM** **Else** **ALPHA** : **PRGM** ▶ **Disp** **2nd** **ALPHA** (A-LOCK) " N O " **ENTER**.

Enter the last line of the program. **PRGM** **End** **2nd** **MODE** (QUIT).

Store values for *A* and *B* and run the program. 2 **STO▶** **ALPHA** A **ENTER** 9 **STO▶** **ALPHA** B **ENTER** **PRGM** **LOGIC** **ENTER**.

You can continue to store values for *a* and *b*, and run the program again.

Think and Discuss

1. How would you modify the program to test whether $a \le 7$ or $b \le 7$? What would be displayed by the program above? Explain.

Try This

1. Write a program to test whether d^2 is greater than cd and displays YES or NO. Test the program by entering values for *c* and *d*.

Answers

Think and Discuss

1. Change the first line of the function from an AND expression to an OR expression. Change both $<$ symbols to $\le$. The calculator will display YES if either of the values entered is less than or equal to 7.

Try This

1. PROGRAM: LOGIC2
If D^2 > C*D
Then:Disp "YES"
Else:Disp "NO"
End

Technology LAB

Logic and Programs

Pacing: Traditional 1 day
Block $\frac{1}{2}$ day

Objective: To use a graphing calculator to write logic programs

Materials: Graphing calculator

Lab Resources

Technology Lab Activities . . . . p. 92

Using the Page

This technology activity shows students how to write logic programs, which can be done on a graphing calculator. Specific keystrokes may vary, depending on the make and model of the graphing calculator used. The keystrokes given are for a TI-83 model. For keystrokes to other models, visit go.hrw.com.

The Think and Discuss problem can be used to assess students' understanding of the technology activity. While the Try This problems can be done without a graphing calculator, they are meant to help students become familiar with writing logic programs on a graphing calculator.

Assessment

1. Explain what the calculator is instructed to do in each line of the progam LOGIC.

   The first line tells the calculator to test whether $a < 7$ and $b < 7$. The second and third lines tell the calculator to display YES if the conditions in line 1 are met. The fourth line tells the calculator to display NO if the conditions in line 1 are not met. The fifth line ends the program.

Chapter
14
**Study Guide
and Review**

Chapter
14
**Study Guide
and Review**

Purpose: *To help students review and practice concepts and skills presented in Chapter 14*

Assessment Resources

Chapter Review
Chapter 14 Resource Book . . pp. 73–74

 Test and Practice Generator CD-ROM

Additional review assessment items in both multiple-choice and free-response format may be generated for any objective in Chapter 14.

Answers

1. Euler circuit, vertex

2. truth table

3. Venn diagram; intersection

4. empty set

5. ∈

6. ∉

7. ⊂

8. finite

9. infinite

Vocabulary

Complete the sentences below with vocabulary words from the list above. Words may be used more than once.

1. A(n) __?__ is a path that begins and ends at the same __?__ and traverses each edge exactly once.

2. A(n) __?__ shows all combinations of the truth or falsity of two statements *P* and *Q*.

3. A(n) __?__ uses circles to represent sets. The area common to both circles shows the __?__ of the sets.

4. If two sets have no elements in common, their intersection is the __?__.

14-1 Sets (pp. 688–691)

EXAMPLE

■ Use ∈, ∉, ⊂, or ⊄ to make the statement true.

John ∈ {male names}; apple ∉ {vegetables}
{6, 8, 10} ⊄ {odd integers}
{1, 3, 5} ⊂ {odd integers}

■ Tell whether each set is finite or infinite.

{whole numbers} infinite
{planets in the solar system} finite

EXERCISES

Use ∈, ∉, ⊂, or ⊄ to make the statement true.

5. vanilla {ice cream flavors}

6. cone ▪ {polygons}

7. *R* = {multiples of 10} ▪ *H* = {even numbers}

Tell whether each set is finite or infinite.

8. {species in the animal kingdom}

9. {rational numbers less than 0}

14-2 Intersection and Union (pp. 692–695)

EXAMPLE

■ Find the intersection and union.

$N = \{1, 3, 5, 7, 9\}$
$M = \{0, 3, 6, 9\}$
$N \cap M = \{3, 9\}$
$N \cup M = \{0, 1, 3, 5, 6, 7, 9\}$

EXERCISES

Find the intersection and union.

10. $P = \{1, 2, 3, 4, 5\}$ $Q = \{0, 2, 4, 6\}$
11. $E = \{\text{even integers}\}$ $O = \{\text{odd integers}\}$
12. $H = \{x \mid x > 3\}$ $R = \{x \mid x < 7\}$

14-3 Venn Diagrams (pp. 696–699)

EXAMPLE

■ Use the Venn diagram to identify intersections, unions, and subsets.

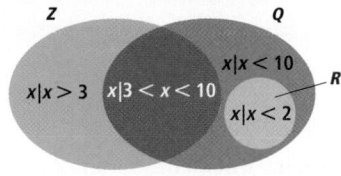

Intersections: $Z \cap Q = \{x \mid 3 < x < 10\}$
$\qquad Q \cap R = R, Z \cap R = \varnothing$
Unions: $Z \cup Q = \{\text{real numbers}\}$
$\qquad R \cup Q = Q$
$\qquad Z \cup R = \{x \mid x > 3 \text{ or } x < 2\}$
Subsets: $R \subset Q$

EXERCISES

Use the Venn diagram to identify intersections, unions, and subsets.

13.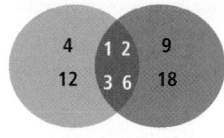

Factors of 12   Factors of 18

Use a Venn diagram to show the following logical argument.

14. All cubes are rectangular prisms.
No cones are rectangular prisms.
∴ No cones are cubes.

14-4 Compound Statements (pp. 702–705)

EXAMPLE

■ Make a truth table for the conjunction *P and Q*, where *P* is "Levon knows CPR" and *Q* is "Levon's only job is teaching."

Example	P	Q	P and Q
Levon knows CPR and is a teacher.	T	T	T
Levon knows CPR and is a judge.	T	F	F
Levon doesn't know CPR and is a teacher.	F	T	F
Levon doesn't know CPR and is a designer.	F	F	F

EXERCISES

Make a truth table for the conjunction *P and Q.*

15. *P*: Carl is less than 6 feet tall.
Q: Carl is more than 12 years old.

16. *P*: A figure is a parallelogram.
Q: A figure is a square.

Make a truth table for the disjunction *P or Q.*

17. *P*: Jill can run a mile in 10 minutes.
Q: Jill can do 50 sit-ups.

18. *P*: John graduated from college.
Q: John's job is to design bridges.

Study Guide and Review

Answers

10. $P \cap Q = \{2, 4\}$;
$P \cup Q = \{0, 1, 2, 3, 4, 5, 6\}$

11. $E \cap O = \varnothing$; $E \cup O = \{\text{integers}\}$

12. $H \cap R = \{x \mid 3 < x < 7\}$;
$H \cup R = \{\text{real numbers}\}$

13. intersection: $\{1, 2, 3, 6\}$;
union: $\{1, 2, 3, 4, 6, 9, 12, 18\}$;
subsets: none

14.

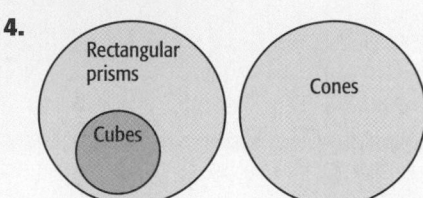

15.

Example	P	Q	P and Q
Carl is 5 feet tall and is 13 years old.	T	T	T
Carl is 5 feet tall and is 10 years old.	T	F	F
Carl is 6 feet 2 inches tall and is 13 years old.	F	T	F
Carl is 6 feet 1 inch tall and is 10 years old.	F	F	F

16.

Example	P	Q	P and Q
ABCD is a parallelogram, and EFGH is a square.	T	T	T
ABCD is a parallelogram, and EFGH is a rhombus.	T	F	F
ABCD is not a parallelogram, and EFGH is a square.	F	T	F
ABCD is not a parallelogram, and EFGH is a trapezoid.	F	F	F

17.

Example	P	Q	P or Q
Jill ran a 9-minute mile and did 50 sit-ups.	T	T	T
Jill ran a 9-minute mile and did 40 sit-ups.	T	F	T
Jill ran an 11-minute mile and did 50 sit-ups.	F	T	T
Jill ran a 12-minute mile and did 35 sit-ups.	F	F	F

18.

Example	P	Q	P or Q
John graduated from college and now designs bridges.	T	T	T
John graduated from college and is now a college professor.	T	F	T
John graduated from high school and now designs bridges.	F	T	T
John graduated from high school and is now the manager of a shoe store.	F	F	F

Answers

19. No conclusion can be made.

20. No conclusion can be made.

21. Figure *ABCD* is a polygon.

22. no

23. The paths *Y-X-Z-W-B-Y, Y-X-B-Z-W-Y,* and their reverses are each 26 in. long.

Study Guide and Review

14-5 Deductive Reasoning (pp. 708–711)

EXAMPLE

■ Make a conclusion, if possible, from the deductive argument.

If a number *n* is greater than 5, then it is greater than 2.
n = 4
No conclusion can be made. Since 4 is not greater than 5, the hypothesis is not true.

EXERCISES

Make a conclusion, if possible, from each deductive argument.

19. If the temperature of water goes above 212°F, then the water is boiling.
The temperature of a pot of water is 210°F.

20. If *x* = 12, then 4*x* = 48.
2*x* = 6

21. If a figure is a parallelogram, then it is a polygon.
Figure *ABCD* is a rectangle.

14-6 Networks and Euler Circuits (pp. 712–715)

EXAMPLE

■ Determine whether the graph can be traversed through an Euler circuit. If so, name one.

Yes. *A-B-C-D-B-A*

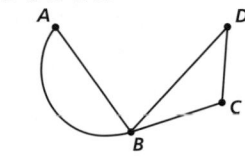

EXERCISES

22. Determine whether the graph can be traversed through an Euler circuit. If so, name one.

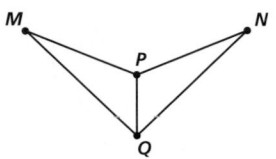

14-7 Hamiltonian Circuits (pp. 716–719)

EXAMPLE

■ Use the information in the graph to find the shortest Hamiltonian circuit starting at vertex *T*.

The paths *T-R-M-N-S-T* and *T-S-N-M-R-T* are each 51 miles long.

EXERCISES

Use the information in the graph to find the shortest Hamiltonian circuit starting at vertex *Y*.

23.

Use $\in$, $\notin$, $\subset$, or $\not\subset$ to make each statement true.

1. 4 ▢ {odd numbers} $\notin$

2. {integers} ▢ {real numbers} $\subset$

3. {−1, −2, −3} ▢ {positive integers} $\not\subset$

4. Triangle *ABC* ▢ {polygons} $\in$

Find the union and intersection of the sets.

5. $A = \{-2, 0, 2, 4, 6, 8\}$ $B = \{2, 4, 6, 8, 10\}$

6. $M = \{x \mid x \le 7\}$ $N = \{x \mid x < 5\}$

Use each Venn diagram to identify the intersections, unions, and subsets.

7.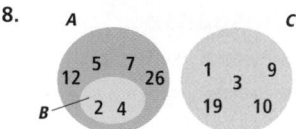

8.

Complete the truth tables for the given statements.

H: Karen is less than 20 years old.

K: Karen is more than 60 inches tall.

D: Sam has driven at least 100 miles.

P: Sam has driven for at least 2 hours.

Statement	H	K	H and K
9. ▢	T	T	**13.** T
10. ▢	T	F	**14.** F
11. ▢	F	T	**15.** F
12. ▢	F	F	**16.** F

Statement	D	P	D or P
17. ▢	T	T	**21.** T
18. ▢	T	F	**22.** T
19. ▢	F	T	**23.** T
20. ▢	F	F	**24.** F

Identify the hypothesis and conclusion in each conditional statement.

25. If gas costs more than $1.75 per gallon, Karl will buy only half a tank.

26. All triangles are polygons.

Make a conclusion, if possible, from the deductive argument.

27. If a polygon is an equilateral triangle, then it has three sides. Polygon *XYZ* has three sides. **No conclusion can be made.**

Use the figure for items 28–32.

28. What is the degree of vertex *N*? **4**

29. Is the graph connected? Explain. **Yes; there is a path from any vertex to any other vertex.**

30. Name an Euler circuit for the graph.

31. Find a Hamiltonian circuit for which the starting vertex is *M*. **Possible answer: *M-P-N-R-Q-T-M***

32. Find the length of the Hamiltonian circuit starting at vertex *M*. **34 ft**

25. Hypothesis: Gas costs more than $1.75 per gallon. Conclusion: Karl will buy only half a tank.

26. Hypothesis: A figure is a triangle. Conclusion: The figure is a polygon.

30. Possible answer: *M-P-N-M-T-Q-R-N-Q-M*

Chapter Test

Chapter **14**

Purpose: *To assess students' mastery of concepts and skills in Chapter 14*

Assessment Resources ✔

Chapter 14 Tests (Levels A, B, C)
Assessment Resources . . . . pp. 111–116

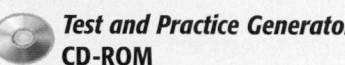 ***Test and Practice Generator*** **CD-ROM**

Additional assessment items in both multiple-choice and free-response format may be generated for any objective in Chapter 14.

Answers

5. $A \cup B = \{-2, 0, 2, 4, 6, 8, 10\}$; $A \cap B = \{2, 4, 6, 8\}$

6. $M \cup N = \{x \mid x \le 7\}$; $M \cap N = \{x \mid x < 5\}$

7. intersection: $R \cap S = \{2, 5\}$; union: $R \cup S = \{1, 2, 3, 5, 6, 7, 9, 12, 14\}$; subsets: none

8. intersections: $A \cap B = B$, $A \cap C = \varnothing$, $B \cap C = \varnothing$; unions: $A \cup B = A$, $A \cup C = \{1, 2, 3, 4, 5, 7, 9, 10, 12, 19, 26\}$, $C \cup B = \{1, 2, 3, 4, 9, 10, 19\}$; subsets: $B \subset A$

9–12. Possible answers:

9. Karen is 10 years old and is 61 inches tall.

10. Karen is 10 years old and is 58 inches tall.

11. Karen is 22 years old and is 61 inches tall.

12. Karen is 25 years old and is 59 inches tall.

17–20. Possible answers:

17. Sam has driven 110 miles in 2.5 hours.

18. Sam has driven 110 miles in 1.75 hours.

19. Sam has driven 90 miles in 3 hours.

20. Sam has driven 85 miles in 1.5 hours.

Performance Assessment

Purpose: To assess students' understanding of concepts in Chapter 14 and combined problem-solving skills

Assessment Resources ✔

Performance Assessment
Assessment Resources . . . . p. 144

Performance Assessment Teacher Support
Assessment Resources . . . . p. 143

Answers

1–5. See pp. A19–A20.

6. See Level 3 work sample below.

Scoring Rubric for Problem Solving Item 6

Level 3
Accomplishes the purpose of the task.

Student gives clear explanations, shows understanding of mathematical ideas and processes, and computes accurately.

Level 2
Purposes of the task not fully achieved.

Student demonstrates satisfactory but limited understanding of the mathematical ideas and processes.

Level 1
Purposes of the task not accomplished.

Student shows little evidence of understanding the mathematical ideas and processes and makes computational and/or procedural errors.

Chapter 14 Performance Assessment

Performance Assessment

🥄 Show What You Know

Create a portfolio of your work from this chapter. Complete this page and include it with your four best pieces of work from Chapter 14. Choose from your homework or lab assignments, mid-chapter quiz, or any journal entries you have done. Put them together using any design you want. Make your portfolio represent what you consider your best work.

⭐ Short Response

1. Is every set a subset of itself? Explain.

2. Let $A = \{2, 4, 6, 8\}$. Let $B = \{6, 7, 8, 9\}$. How can you use a Venn diagram to show $A \cap B$?

3. What is the degree of each vertex of the graph $ABCDE$ formed by rectangle $ABCD$ and its two diagonals? Explain.

4. What is the sum of the degrees of the vertices of a triangle? of a quadrilateral? of a pentagon? of an n-sided polygon?

5. Let P be true and Q be true. What is the truth value of $\sim(P \text{ and } Q)$? What is the truth value of $\sim P \text{ or } \sim Q$?

🧩 Extended Problem Solving

Choose any strategy to solve each problem.

6. **a.** Which of the graphs have Hamiltonian circuits?

   b. Do any of the graphs have Euler circuits? Explain.

   c. Draw a graph with 6 vertices that has an Euler circuit but no Hamiltonian circuit.

   d. Draw a graph with 6 vertices that has a Hamiltonian circuit but no Euler circuit.

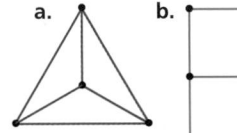

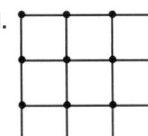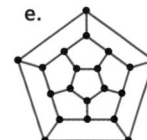

Student Work Samples for Item 6

Level 3

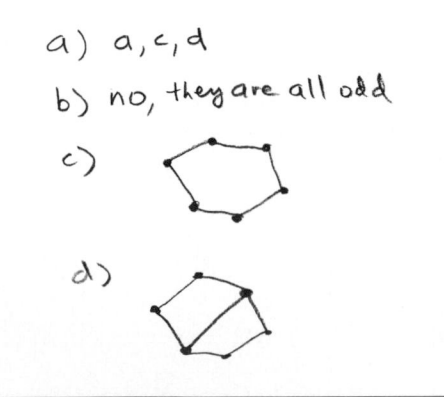

a. a, c, d, e
b. No, they all have some vertices with odd degrees.
c.
d.

The student correctly answered each part and drew accurate diagrams.

Level 2

a) a, c, d
b) no, they are all odd
c)
d)

The student gave an incomplete answer for part **a** and drew an incorrect diagram for part **c**.

Level 1

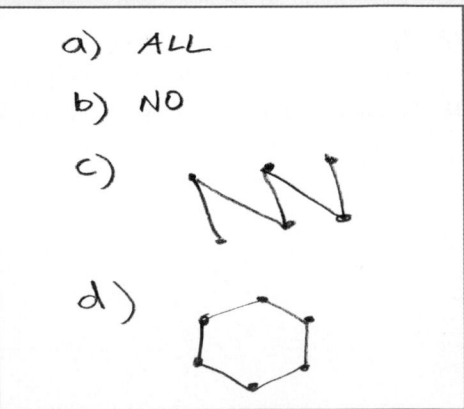

a) ALL
b) NO
c)
d)

The student's answers and diagrams do not show an understanding of Euler or Hamiltonian circuits.

Standardized
Test Prep

Chapter
14

Cumulative Assessment, Chapters 1–14

1. When 3 times a number is decreased by 5, the result is 145. What is the number? **C**

   Ⓐ −150   Ⓒ 50
   Ⓑ −50   Ⓓ 150

2. If $x = -\frac{1}{3}$, which is greatest? **F**

   Ⓕ $1 - x$ Ⓗ $-x$
   Ⓖ $x - 1$ Ⓙ $1 \div x$

3. Which ordered pair is located above and to the left of the origin? **B**

   Ⓐ $(-4, -4)$ Ⓒ $(4, -4)$
   Ⓑ $(-4, 4)$ Ⓓ $(4, 4)$

TEST TAKING TIP!

When assigning test values, try different kinds of numbers, such as negatives and fractions.

4. If x is any real number, then which statement **must** be true? **H**

   Ⓕ $x^2 > x$ Ⓗ $|x| \geq x$
   Ⓖ $-x > x$ Ⓙ No relationship can be determined.

5. What percent of 5 is 4? **B**

   Ⓐ 75%   Ⓒ 125%
   Ⓑ 80%   Ⓓ 150%

6. If $x = 2$, what is $4y(5 - 3x)$ in terms of y? **G**

   Ⓕ $-14y$ Ⓗ $14y$
   Ⓖ $-4y$ Ⓙ $20y - 6$

7. What number is 7.9×10^{-6} in standard notation? **A**

   Ⓐ 0.0000079   Ⓒ 7,900,000
   Ⓑ 0.00000079   Ⓓ 79,000,000

8. Amy paid $3.20 for 20 ounces of fruit. What is the unit price per ounce? **G**

   Ⓕ $0.02   Ⓗ $0.20
   Ⓖ $0.16   Ⓙ $1.60

9. **SHORT RESPONSE** Find the next three numbers in the sequence 4, 9, 18, 31, 48 . . . . Explain how you found your answer.

10. **SHORT RESPONSE** What is the perimeter of the figure below? Explain.

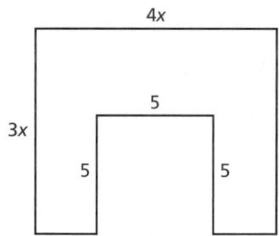

Standardized Test Prep

Purpose: *To provide review and practice for Chapters 1–14 and standardized tests*

Assessment Resources ✓

Cumulative Tests (Levels A, B, C)
Assessment Resources . . . . pp. 301–312

State-Specific Test Practice Online
KEYWORD: MP4 TestPrep

Test Prep Doctor ➕

For item 7, remind students that numbers in scientific notation that have a negative power of 10 are decimal numbers less than 1. This eliminates **C** and **D**. Because the power is −6, the decimal point should be moved 6 spaces to the left. This shows that **A** is the correct answer.

Expand on the test-taking tip given before item 4 by pointing out that in item 4 a fraction will prove that **F** is false and a negative number will prove that **G** is false. Choice **H** is true for all real numbers, including fractions and negative numbers.

Answers

9. 69, 94, 123; possible answer: the first differences are 5, 9, 13, and 17. The common second difference is 4. The pattern of first differences can be continued by adding 4, and then the original sequence can be continued.

10. $14x + 10$ units; possible answer: the perimeter of the figure is the sum of its side lengths. The sum of the lengths of the two unlabeled sides on the bottom is $4x - 5$ units, and the length of the right unlabeled side is $3x$ units. The perimeter is $3x + 4x + 3x + 4x - 5 + 5 + 5 + 5$, which simplifies to $14x + 10$ units.

Student Handbook

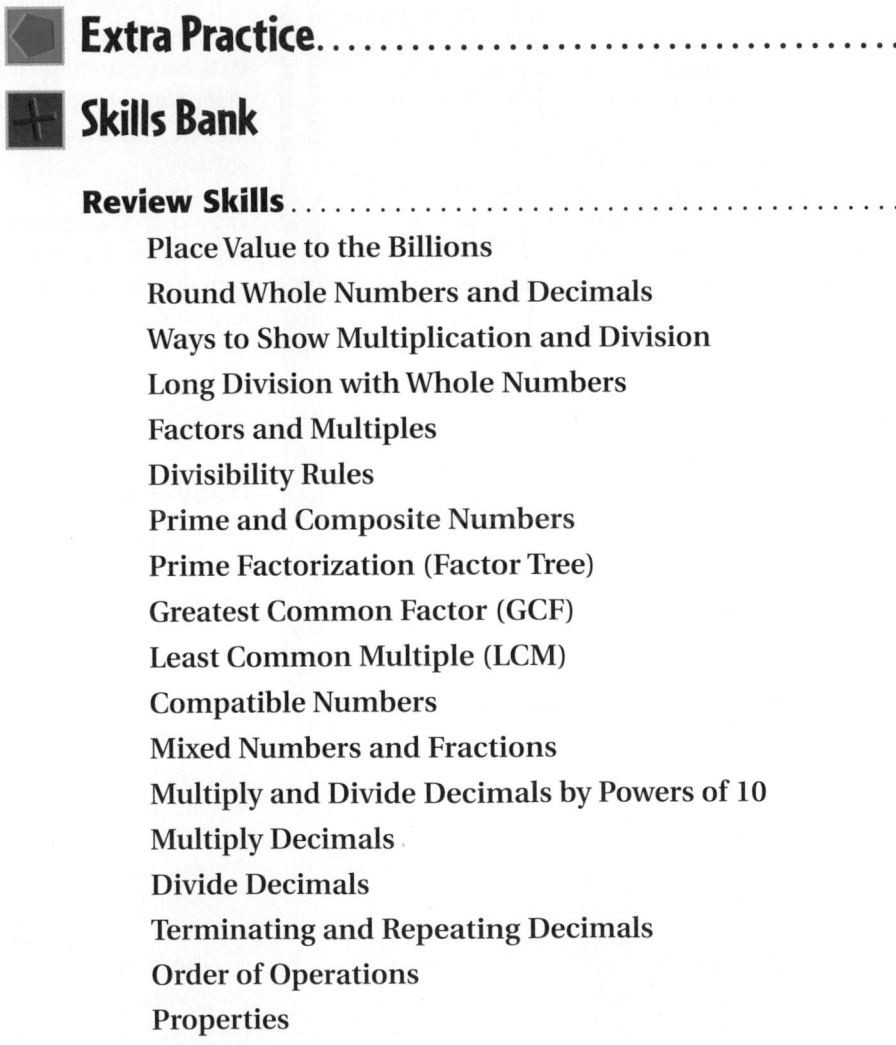

Exponent

Base → 2^4

Extra Practice

1A Equations and Inequalities

LESSON 1-1

Evaluate each expression for the given value of the variable.
1. $2 + x$ for $x = 7$ **9**   2. $4m - 3$ for $m = 2$ **5**   3. $2(p + 3)$ for $p = 8$ **22**

Evaluate each expression for the given values of the variables.
4. $3x + y$ for $x = 2$, $y = 4$ **10**   5. $2y - x$ for $x = 2$, $y = 5$ **8**   6. $5x + 2y$ for $x = 1$, $y = 3$ **11**
7. $3x + 2.5y$ for $x = 1$, $y = 2$ **8**   8. $5.7x + 2y$ for $x = 2$, $y = 1$ **13.4**   9. $4.2x + 3y$ for $x = 2$, $y = 3$ **17.4**

LESSON 1-2

Write an algebraic expression for each word phrase.
10. seven less than a number b $b - 7$ 11. eight more than the product of 7 and a $7a + 8$
12. a quotient of 8 and a number m $8 \div m$ 13. five times the sum of c and 18 $5(c + 18)$

Solve.
14. The formula for converting a temperature in C degrees Celsius (°C) to degrees Fahrenheit (°F) is $F = 1.8C + 32$. Convert the temperature 28°C to degrees Fahrenheit. **82.4°F**

LESSON 1-3

Solve.
15. $4 + x = 13$ $x = 9$ 16. $t - 3 = 8$ $t = 11$ 17. $17 = m + 11$ $m = 6$ 18. $5 + a = 7$ $a = 2$
19. $p - 5 = 23$ $p = 28$ 20. $31 + y = 50$ $y = 19$ 21. $18 + k = 34$ $k = 16$ 22. $g - 16 = 23$ $g = 39$

LESSON 1-4

Solve.
23. $5x = 30$ $x = 6$ 24. $\frac{m}{4} = 13$ $m = 52$ 25. $9a = 54$ $a = 6$ 26. $\frac{n}{7} = 7$ $n = 49$
27. $3p = 96$ $p = 32$ 28. $\frac{s}{6} = 3$ $s = 18$ 29. $3k + 2 = 20$ $k = 6$ 30. $\frac{r}{4} - 5 = 3$ $r = 32$
31. Four friends split the cost of a $16.68 pizza. How much did each friend pay? **$4.17**

LESSON 1-5

Compare. Write < or >.
32. $15 - 8$ ▊ 6 >   33. $3(7)$ ▊ 23 <   34. $51 - 18$ ▊ 34 <   35. $4(16)$ ▊ 62 >

Solve and graph. For 36–39, see p. A20.
36. $x - 3.5 \geq 7$ $x \geq 10.5$ 37. $5p < 40$ $p < 8$ 38. $2 \leq \frac{a}{3}$ $6 \leq a$ 39. $h - 5 \leq 13$ $h \leq 18$

LESSON 1-6

Combine like terms.
40. $3x + 2x + 5x$ $10x$ 41. $4x - 2x + 8 + 3x + 5$ $5x + 13$ 42. $5a - 3b + 4 + 6b - 2a$ $3a + 4 + 3b$

Solve. For 43–46, see p. A20.
43. $3x + 9 = 84$ 44. $2a - 3 = 41$ 45. $7b + 5 = 61$ 46. $6h - 12 = 78$

1B Graphing

LESSON 1-7

Determine whether each ordered pair is a solution of $3x + 5y = 25$.
1. $(4, 3)$ **no**   2. $(5, 2)$ **yes**   3. $(6, 1)$ **no**   4. $(3, 4)$ **no**

Use the given values to make a table of solutions. For 5–6, see p. A20.
5. $y = x - 3$ for $x = -2, -1, 0, 1, 2$ 6. $y = 2x + 1$ for $x = -2, -1, 0, 1, 2$
7. If sales tax is 6%, the equation for the total cost c of an item is $c = 1.06p$, where p is the price of the item before tax. What is the total cost of a $20 shirt, including sales tax? **$21.20**

LESSON 1-8

Graph each point on a coordinate plane.
8. $(4, 3)$ **right 4, up 3**   9. $(3, 0)$ **right 3**   10. $(-1, 3)$ **left 1, up 3**
11. $(0, -5)$ **down 5**   12. $(-2, -4)$ **left 2, down 4**   13. $(4, -2)$ **right 4, down 2**

Complete each table of ordered pairs. Graph each equation on a coordinate plane. For 14–15, see p. A20.
14. $x + 3 = y$

x	$x + 3$	y	(x, y)
1	4	4	(1, 4)
2	5	5	(2, 5)
3	6	6	(3, 6)
4	7	7	(4, 7)

15. $3x = y$

x	$3x$	y	(x, y)
2	6	6	(2, 6)
4	12	12	(4, 12)
6	18	18	(6, 18)
8	24	24	(8, 24)

LESSON 1-9

Match each situation to the correct graph.

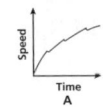 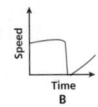 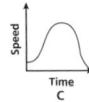

A B C

16. A skier increases speed going down a hill, and then comes to a stop. **C**
17. A skier travels cross-country, stopping only to rest for a minute before going up a hill. **B**
18. A skier accelerates going downhill, decreases speed slightly before sharp turns, and then accelerates again. **A**

2A Integers

LESSON 2-1

Add.
1. $-4 + 6$ **2**   2. $3 + (-8)$ **−5**   3. $-6 + (-2)$ **−8**   4. $7 + (-11)$ **−4**
5. $-6 + 3$ **−3**   6. $7 + (-2)$ **5**   7. $-4 + (-1)$ **−5**   8. $9 + (-5)$ **4**

Evaluate each expression for the given value of the variable.
9. $x + 9$ for $x = -8$ **1**   10. $x + 3$ for $x = -3$ **0**   11. $x + 5$ for $x = -7$ **−2**
12. $x + 1$ for $x = -5$ **−4**   13. $x + 6$ for $x = -9$ **−3**   14. $x + 2$ for $x = -8$ **−6**

LESSON 2-2

Subtract.
15. $-5 - 3$ **−8**   16. $4 - (-1)$ **5**   17. $-9 - (-4)$ **−5**   18. $-4 - 7$ **−11**
19. $-2 - 5$ **−7**   20. $3 - (-8)$ **11**   21. $-6 - (-12)$ **6**   22. $-1 - 6$ **−7**
23. An elevator rises to 281 feet above ground level and then drops 314 feet to the basement. What is the position of the elevator relative to ground level? **−33 ft**

Evaluate each expression for the given value of the variable.
24. $4 - x$ for $x = -7$ **11**   25. $-7 - s$ for $s = -5$ **−2**   26. $-5 - b$ for $b = 9$ **−14**
27. $12 - y$ for $y = -8$ **20**   28. $-13 - f$ for $f = -8$ **−5**   29. $-2 - c$ for $c = 5$ **−7**

LESSON 2-3

Multiply or divide.
30. $5(-8)$ **−40**   31. $\frac{-81}{9}$ **−9**   32. $-6(-4)$ **24**   33. $\frac{24}{-3}$ **−8**
34. $7(-3)$ **−21**   35. $\frac{-36}{6}$ **−6**   36. $-8(-4)$ **32**   37. $\frac{48}{-8}$ **−6**
38. $-9(-12)$ **108**   39. $\frac{-54}{9}$ **−6**   40. $13(-5)$ **−65**   41. $\frac{96}{-12}$ **−8**

LESSON 2-4

Solve.
42. $x + 13 = 8$ $x = -5$ 43. $-7 + t = -15$ $t = -8$ 44. $h = -8 + 17$ $h = 9$ 45. $g + 15 = 3$ $g = -12$
46. $-8 + p = -20$ $p = -12$ 47. $n = -4 + 31$ $n = 27$ 48. $m + 4 = 9$ $m = 5$ 49. $d = -8 + 2$ $d = -6$
50. $\frac{a}{-4} = -2$ $a = 8$ 51. $-49 = 7d$ $d = -7$ 52. $\frac{c}{-2} = -8$ $c = 16$ 53. $-57 = 3p$ $p = -19$

LESSON 2-5

Solve and graph. For 54–65, see p. A20.
54. $w - 1 < -4$ $w < -3$ 55. $x - 3 \geq -2$ $x \geq 1$ 56. $h - 2 \leq -5$ $h \leq -3$ 57. $g - 6 > -1$ $g > 5$
58. $k - 3 > -9$ $k > -6$ 59. $m - 5 > -8$ $m > -3$ 60. $f - 9 < -2$ $f < 7$ 61. $m - 2 \leq -1$ $m \leq 1$
62. $-3a > 15$ $a < -5$ 63. $\frac{x}{-4} < 6$ $x > -24$ 64. $-5b \leq 65$ $b \geq -13$ 65. $\frac{a}{-8} \geq 4$ $a \leq -32$

2B Exponents and Scientific Notation

LESSON 2-6

Write using exponents.
1. $2 \times 2 \times 2 \times 2$ 2^4 2. $5 \times 5 \times 5 \times 5 \times 5 \times 5 \times 5$ 5^7 3. $4 \cdot 4 \cdot 4 \cdot 4 \cdot 4$ 4^5
4. $9 \cdot 9 \cdot 9 \cdot 9 \cdot 9 \cdot 9 \cdot 9 \cdot 9$ 9^8 5. $a \cdot a \cdot a \cdot a \cdot a \cdot a \cdot a$ a^7 6. p p^1

Evaluate.
7. 2^4 **16**   8. 3^3 **27**   9. $(-5)^2$ **25**   10. $(-3)^5$ **−243**
11. 8^3 **512**   12. 6^5 **7776**   13. $(-2)^8$ **256**   14. $(-4)^3$ **−64**

Simplify.
15. $20 + 3(2^3)$ **44**   16. $14 + 5(3^4)$ **419**   17. $19 + 3(2 \cdot 4^2)$ **115**   18. $22 + 5(8 + 2^4)$ **142**
19. $8 + 2(3 \cdot 4^3)$ **392**   20. $17 + 2(4 + 5^3)$ **275**   21. $32 + 4(5 + 2^5)$ **180**   22. $58 + 3(9 + 6^3)$ **733**

LESSON 2-7

Multiply or divide. Write as one power.
23. $5^4 \cdot 5^3$ 5^7 24. $2^6 \cdot 2^3$ 2^9 25. $4^4 \cdot 4^8$ 4^{12} 26. $7^3 \cdot 7^9$ 7^{12}
27. $12^8 \cdot 12^5$ 12^{13} 28. $a^8 \cdot a^5$ a^{13} 29. $b^6 \cdot b^{12}$ b^{18} 30. $w^7 \cdot w^7$ w^{14}
31. $\frac{16^4}{16^2}$ 16^2 32. $\frac{8^9}{8^3}$ 8^6 33. $\frac{7^{12}}{7^5}$ 7^7 34. $\frac{15^{12}}{15^{11}}$ 15^1 or 15
35. $\frac{a^7}{a^4}$ a^3 36. $\frac{w^{11}}{w^4}$ w^7 37. $\frac{c^6}{c^2}$ c^4 38. $\frac{z^{16}}{z^9}$ z^7

LESSON 2-8

Evaluate each power of 10.
39. 10^{-2} **0.01**   40. 10^{-3} **0.001**   41. 10^{-4} **0.0001**   42. 10^{-5} **0.00001**
43. 10^{-6} **0.000001**   44. 10^{-7} **0.0000001**   45. 10^{-8} **0.00000001**   46. 10^{-9} **0.000000001**

Evaluate.
47. $(-3)^{-2}$ **0.1111**   48. 4^{-3} **0.015625**   49. $(-6)^{-4}$ **0.0007716**   50. 7^{-3} **0.002915**
51. $10^4 \cdot 10^{-2}$ **100**   52. $\frac{3^2}{3^4}$ **0.1111**   53. $2^5 \cdot 2^{-2}$ **8**   54. $\frac{4^3}{4^5}$ **0.0625**

LESSON 2-9

Write each number in standard notation.
55. 3.6×10^3 **3600**   56. 5.62×10^5 **562,000**   57. 7.13×10^{-4} **0.000713**   58. 8.39×10^{-7} **0.000000839**
59. 1.6×10^2 **160**   60. 3.12×10^7 **31,200,000**   61. 1.13×10^{-5} **0.0000113**   62. 5.92×10^{-8} **0.0000000592**

Write each number in scientific notation.
63. 0.000483 4.83×10^{-4} 64. 5,410,000,000 5.41×10^9 65. 0.00328 3.28×10^{-3}
66. 12,600,000 1.26×10^7 67. 0.0000000000912 9.12×10^{-11} 68. 432,000,000,000,000 4.32×10^{14}

3A Rational Numbers and Operations

LESSON 3-1

Write each decimal as a fraction in simplest form.

1. 0.4 $\frac{2}{5}$
2. 0.05 $\frac{1}{20}$
3. 0.12 $\frac{3}{25}$
4. 0.625 $\frac{5}{8}$

Write each fraction as a decimal.

5. $\frac{3}{8}$ 0.375
6. $\frac{1}{4}$ 0.25
7. $\frac{9}{4}$ 2.25
8. $\frac{3}{5}$ 0.6

LESSON 3-2

Add or subtract.

9. $\frac{2}{3} - \frac{5}{3}$ −1
10. $\frac{17}{4} + \frac{13}{4}$ $\frac{15}{2}$, or $7\frac{1}{2}$
11. $\frac{5}{8} - \frac{5}{4}$, or $-1\frac{1}{4}$
12. $-\frac{8}{3} + \frac{11}{3}$ 1
13. $\frac{9}{2} - \frac{15}{2}$ −3
14. $\frac{19}{3} + \frac{27}{3}$ $\frac{46}{3}$, or $15\frac{1}{3}$
15. $\frac{9}{4} - \frac{22}{4}$ $-\frac{13}{4}$, or $-3\frac{1}{4}$
16. $-\frac{31}{5} + \frac{24}{5}$ $-\frac{7}{5}$, or $-1\frac{2}{5}$

Evaluate each expression for the given value of the variable.

17. $32.9 + x$ for $x = -15.8$ **17.1**
18. $21.3 + a$ for $a = -37.6$ **−16.3**
19. $-\frac{3}{5} + z$ for $z = 3\frac{1}{5}$ $\frac{13}{5}$, or $2\frac{3}{5}$

LESSON 3-3

Multiply. Write each answer in simplest form.

20. $-\frac{2}{3}\left(-\frac{5}{8}\right)$ $\frac{5}{12}$
21. $\frac{7}{10}\left(-\frac{2}{3}\right)$ $-\frac{7}{15}$
22. $-\frac{4}{5}\left(-\frac{9}{10}\right)$ $\frac{18}{25}$
23. $-\frac{5}{8}\left(\frac{11}{12}\right)$ $-\frac{55}{96}$
24. $-3.9(-9)$ **35.1**
25. $-4.1(8.6)$ **−35.26**
26. $-0.08(3.1)$ **−0.248**
27. $-0.004(-1.9)$ **0.0076**

LESSON 3-4

Divide. Write each answer in simplest form.

28. $3\frac{1}{2} \div \frac{1}{4}$ 14, or $14\frac{2}{3}$
29. $5\frac{1}{5} \div \frac{7}{35}$ $\frac{208}{35}$, or $5\frac{33}{35}$
30. $6\frac{3}{5} \div \frac{2}{16}$ $\frac{159}{16}$, or $9\frac{15}{16}$
31. $4\frac{1}{7} \div \frac{3}{7}$ $\frac{259}{27}$, or $9\frac{16}{27}$
32. $5.68 \div 0.2$ **28.4**
33. $9.45 \div 0.05$ **189**
34. $2.31 \div 0.7$ **3.3**
35. $0.522 \div 6$ **0.087**

LESSON 3-5

Add or subtract.

36. $\frac{9}{10} + \frac{3}{8}$ $\frac{51}{40}$, or $1\frac{11}{40}$
37. $\frac{2}{7} - \frac{3}{4}$ $-\frac{13}{36}$
38. $\frac{3}{4} + \frac{1}{9}$ $\frac{31}{36}$
39. $\frac{5}{8} - \frac{3}{10}$ $\frac{13}{40}$
40. $5\frac{1}{3} + \left(-2\frac{1}{7}\right)$ $\frac{77}{24}$, or $3\frac{5}{24}$
41. $3\frac{3}{8} + \left(-1\frac{7}{9}\right)$ $\frac{43}{24}$, or $1\frac{19}{24}$
42. $4\frac{1}{8} + \left(-1\frac{3}{5}\right)$ $\frac{101}{40}$, or $2\frac{21}{40}$
43. $9\frac{1}{9} + \left(-5\frac{2}{11}\right)$ $\frac{389}{99}$, or $3\frac{92}{99}$

LESSON 3-6

Solve.

44. $x - 3.2 = 5.1$ **x = 8.3**
45. $-3.1p = 15.5$ **p = −5**
46. $\frac{a}{-2.3} = 7.9$ **a = −18.17**
47. $-4.3x = 34.4$ **x = −8**
48. $m - \frac{1}{3} = \frac{5}{8}$ $m = \frac{23}{24}$
49. $x - \frac{3}{7} = \frac{1}{9}$ $x = \frac{34}{63}$
50. $\frac{4}{5}w = \frac{2}{3}$ $w = \frac{5}{6}$
51. $\frac{9}{10}z = \frac{5}{8}$ $z = \frac{25}{36}$

LESSON 3-7

Solve. For 52–59, see p. A20.

52. $1.2x > 7.2$
53. $a - 3.8 < 5.4$
54. $3.8b \geq 26.6$
55. $d - 5.3 \leq 7.9$
56. $w + \frac{2}{3} > \frac{2}{5}$
57. $-2\frac{1}{4}b < 9$
58. $b + \frac{3}{8} \geq \frac{9}{10}$
59. $4\frac{2}{5}x \leq 39\frac{3}{5}$

3B Real Numbers

LESSON 3-8

Find the two square roots of each number.

1. 25 **5, −5**
2. 81 **9, −9**
3. 144 **12, −12**
4. 169 **13, −13**
5. 100 **10, −10**
6. 225 **15, −15**
7. 36 **6, −6**
8. 400 **20, −20**

Evaluate each expression.

9. $3\sqrt{9}$ **9**
10. $5\sqrt{36}$ **30**
11. $7\sqrt{16}$ **28**
12. $3\sqrt{49}$ **21**
13. $\sqrt{97 + 24}$ **11**
14. $\sqrt{111 + 85}$ **14**
15. $\sqrt{231 + 253}$ **22**
16. $\sqrt{45 - 9}$ **6**

Solve.

17. The area of a square room is 729 square feet. What are the dimensions of the room? **27 ft by 27 ft**
18. The area of a square garden is 1,444 square feet. What are the dimensions of the garden? **38 ft by 38 ft**

LESSON 3-9

Each square root is between two integers. Name the integers.

19. $\sqrt{29}$ **5 and 6**
20. $\sqrt{51}$ **7 and 8**
21. $\sqrt{93}$ **9 and 10**
22. $\sqrt{74}$ **8 and 9**
23. $\sqrt{32}$ **5 and 6**
24. $\sqrt{12}$ **3 and 4**
25. $\sqrt{48}$ **6 and 7**
26. $\sqrt{128}$ **11 and 12**

Use a calculator to find the square root of each number. Round to the nearest tenth.

27. $\sqrt{212}$ **14.6**
28. $\sqrt{186}$ **13.6**
29. $\sqrt{542}$ **23.3**
30. $\sqrt{219}$ **14.8**
31. $\sqrt{384}$ **19.6**
32. $\sqrt{410}$ **20.2**
33. $\sqrt{334}$ **18.3**
34. $\sqrt{96}$ **9.8**
35. $\sqrt{54}$ **7.3**
36. $\sqrt{683}$ **26.1**
37. $\sqrt{614}$ **24.8**
38. $\sqrt{304}$ **17.4**

LESSON 3-10

Write the names that apply to each number.

39. $\sqrt{7}$ **irrational, real**
40. -61.2 **rational, real**
41. $\frac{\sqrt{16}}{2}$ **whole, integer, rational, real**
42. -8 **whole, integer, rational, real**
43. 4.168 **rational, real**
44. $\frac{\sqrt{25}}{\sqrt{1}}$ **whole, integer, rational, real**
45. $\sqrt{11}$ **irrational, real**
46. $\sqrt{13}$ **irrational, real**

State whether the number is rational, irrational, or not a real number.

47. $\sqrt{\frac{9}{16}}$ **rational**
48. $\sqrt{-4}$ **not real**
49. $\sqrt{19}$ **irrational**
50. $\sqrt{-13}$ **not real**
51. 12 **rational**
52. $\frac{8}{0}$ **not real**
53. $\sqrt{\frac{36}{49}}$ **rational**
54. $\frac{13}{0}$ **not real**

Find a real number between the two given numbers. **Possible answer:**

55. $5\frac{1}{8}$ and $5\frac{2}{8}$ $5\frac{3}{16}$
56. $2\frac{1}{3}$ and $2\frac{2}{3}$ $2\frac{1}{2}$
57. $4\frac{4}{9}$ and $4\frac{5}{9}$ $4\frac{1}{2}$
58. $1\frac{5}{7}$ and $1\frac{6}{7}$ $1\frac{11}{14}$
59. $3\frac{1}{8}$ and $3\frac{1}{4}$ $3\frac{3}{16}$
60. $9\frac{4}{7}$ and $9\frac{5}{7}$ $9\frac{9}{14}$

4A Collecting and Describing Data

LESSON 4-1

Identify the population and the sample. Give a reason why the sample could be biased.

1. A company chooses 2000 veterinarians who belong to the same veterinary association for a survey on their opinion about a new dog medicine. **Population: veterinarians who belong to the association; Sample: 2,000 veterinarians; Possible bias: not all veterinarians belong to the association**

Identify the sampling method used.

2. In a nationwide survey, 7 states are chosen at random, and 150 people are chosen from each state. **stratified**
3. A questionnaire is distributed to every fifth adult shopper at a grocery store. **systematic**

LESSON 4-2 For 4–5, see p. A20.

4. Use the given data to make a stem-and-leaf plot.

Number of Floors in Selected Major Buildings					
Promenade	40	One Park Tower	32	Commerce Plaza	31
One Financial Center	46	One Post Office Square	40	Water Tower Place	74
Park Tower Condos	54	City Plaza	40	Harbour Point	54
Park Millennium	53	Energy Plaza	49	San Felipe Plaza	45
The Spires	41	Santa Maria	51	Cityspire	72

5. Use the given data to make a back-to-back stem-and-leaf plot.

World Series Win/Loss Records of Selected Teams (through 2001)							
Team	Yankees	Pirates	Giants	Tigers	Cardinals	Dodgers	Orioles
Wins	26	5	5	4	9	6	3
Losses	12	2	11	5	6	12	4

LESSON 4-3

Find the mean, median, and mode of each data set.

6. 8, 3, 9, 10, 8, 4, 5, 7, 6, 7, 8, 5 **mean: 6.67; median: 7; mode: 8**
7. 31, 28, 25, 41, 52, 40, 38, 24, 43, 27, 24, 35 **mean: 34; median: 33; mode: 24**

LESSON 4-4

Find the range and first and third quartiles for each data set.

8. 18, 20, 15, 13, 13, 20, 17, 20, 15, 13 **range: 7; first quartile: 15; third quartile: 20**
9. 82, 77, 74, 71, 85, 89, 81, 85, 80, 91, 72, 81, 88, 86, 75 **range: 20; first quartile: 75; third quartile: 86**

Use the given data to make a box-and-whisker plot. For 10–11, see p. A20.

10. 3, 12, 17, 9, 8, 4, 13, 24, 17, 19, 5
11. 57, 53, 52, 31, 48, 59, 64, 86, 56, 54, 55

4B Displaying Data

LESSON 4-5 For 1–2, see p. A20.

Organize the data into a frequency table, and make a bar graph.

1. The following are the ages at which a randomly chosen group of 20 students graduated from college: 20, 21, 23, 19, 20, 21, 21, 21, 19, 22, 21, 21, 21, 20, 21, 20, 21, 21, 22, 21

Make a line graph of the given data. Use the graph to estimate the population density in 1975.

2.

Year	Population Density (people per square mile)
1950	42.6
1960	50.6
1970	57.5
1980	64.0
1990	70.3
2000	79.6

The population density would be about 87–88 in 2010.

LESSON 4-6

Explain why each graph or statistic is misleading.

3.

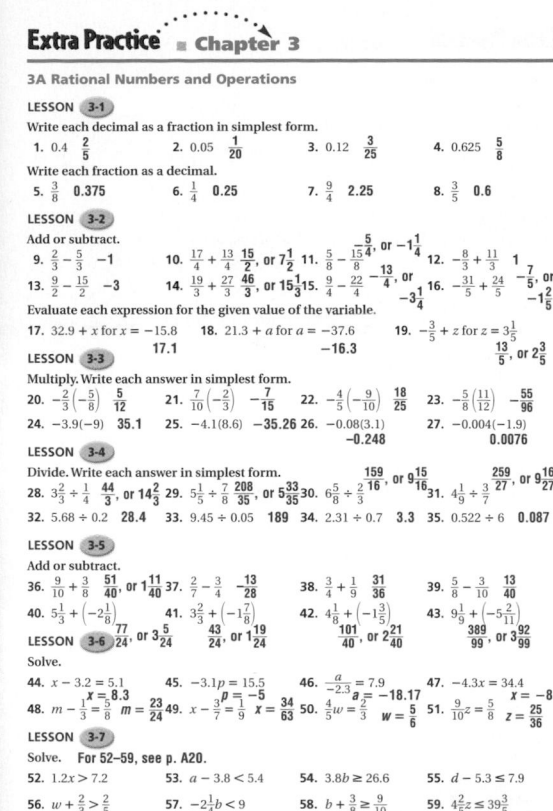

Trucks Sold in the U.S.

(Sales (millions) vs. Manufacturer A, B, C, D)

The intervals on the scale are unequal.

4. A market researcher randomly selects 12 shoppers to sample 3 brands of sausage labeled A, B, and C. Of the shoppers, 8 selected B, 2 selected A, and 2 selected C. An ad for brand B reads, "Preferred 4 to 1 over other brands." **The sample is too small.**

5. A real-estate agency sold houses for $75,000, $420,000, $88,000, $80,000, and $82,000. Its ads boast an average selling price of $149,000. **One extreme outlier distorts the average, giving a false impression.**

LESSON 4-7

Use the given data to make a scatter plot. For 6, see p. A20.

6. The table shows the relationship between the number of years of post-high-school education and salary.

1	$18,000	4	$51,000	6	$64,000
1	$20,500	4	$43,000	6	$58,000
3	$28,000	5	$48,000	8	$75,000
4	$35,000	5	$52,000	8	$73,500

Do the data sets have a positive, a negative, or no correlation?

7. the temperature of an oven and the amount of time it takes a roast to brown **negative**
8. the lengths of the pencils used on a test and the test scores **no correlation**

5A Plane Figures

LESSON 5-1

Classify each angle as acute, obtuse, or right.

1. **right**   2. **acute**   3. **obtuse**

In the figure, ∠1 and ∠3 are vertical angles, and ∠2 and ∠4 are vertical angles.

4. If m∠1 = 83°, find m∠3. **83°**

5. If m∠2 = 136°, find m∠4. **136°**

LESSON 5-2

In the figure, $d \parallel f$. Find the measure of each angle.

6. ∠1 **140°**   7. ∠2 **40°**   8. ∠3 **140°**

LESSON 5-3

Find the missing measures in each triangle.

9. **70°**   10. **56°**   11. **35°**

12. The first angle of a triangle is 3 times as large as the second angle. The third angle is twice as large as the second angle. Find the angle measures.
    first: 90°; second: 30°; third: 60°

LESSON 5-4

Find the angle measures in each regular polygon.

13. hexagon (6 sides) **120°**   14. nonagon (9 sides) **140°**   15. decagon (10 sides) **144°**

Write all the names that apply to each figure.

16. **quadrilateral; trapezoid** $\overline{AB} \parallel \overline{CD}$

17. **quadrilateral; parallelogram; rectangle; rhombus; square**

LESSON 5-5

Determine whether the slope of each line is positive, negative, 0, or undefined. Then find the slope of each line.

18. line a **negative; −2**   19. line b **positive; 2**

20. line c **positive; $\frac{2}{3}$**   21. line d **negative; $-\frac{1}{2}$**

22. Which lines are perpendicular? **b and d**

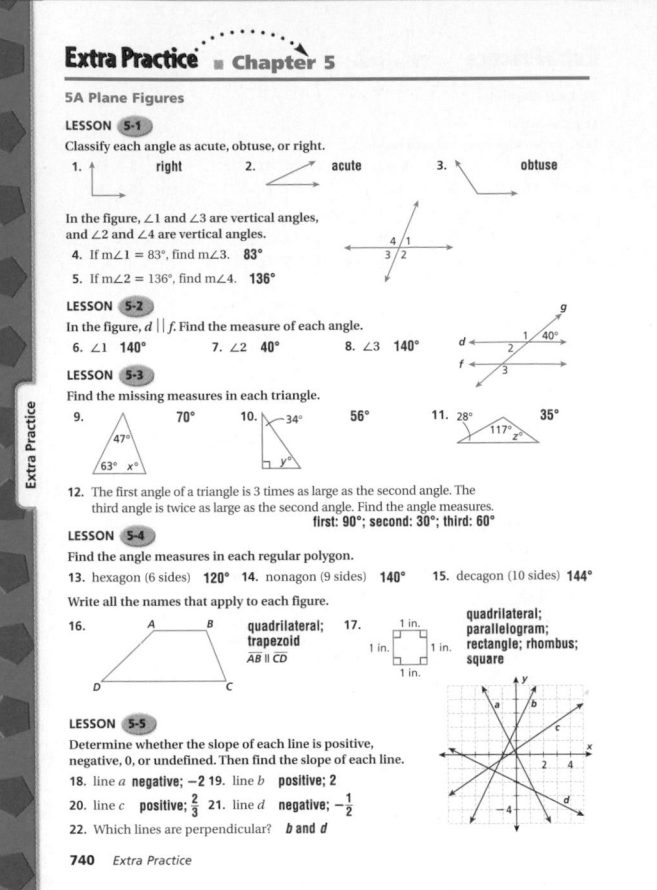

5B Patterns in Geometry

LESSON 5-6

Quadrilateral $ABCD \cong$ quadrilateral $KLMN$. Find each value.

1. x **5**

2. y **6**

3. z **60**

LESSON 5-7

Identify each as a translation, rotation, reflection, or none of these.

4. **translation**   5. **rotation**   6. **none**   7. **reflection**

Draw the image of a triangle with vertices (1, 1), (4, 2), and (4, 4) after each transformation.   **For 8–15, see pp. A20–A21.**

8. reflection across the y-axis

9. rotation 180° around the origin

10. reflection across the x-axis

LESSON 5-8

Complete each figure. The dashed line is the line of symmetry.

11.   12.

Complete each figure. The point is the center of rotation.

13. 4-fold   14. 6-fold

LESSON 5-9

Create a tessellation with the figure.

15.

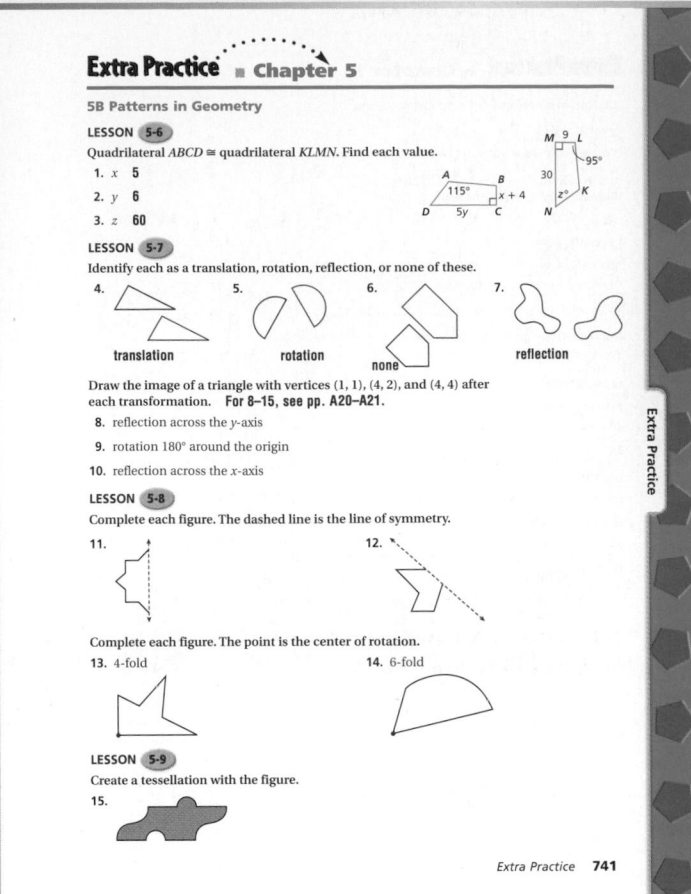

6A Perimeter and Area

LESSON 6-1

Find the perimeter of each figure.

1. **40 m**   2. **48 in.**   3. **52 units**

For 4–6, see p. A21.
Graph each figure with the given vertices. Then find the area of each figure.

4. (−2, 4), (4, 4), (−2, 8), (4, 8) **24 sq. units**

5. (1, 2), (2, −1), (5, 2), (6, −1) **12 sq. units**

6. (3, 3), (1, −2), (−3, 3), (−5, −2) **30 sq. units**

LESSON 6-2

Find the perimeter of each figure.

7. **46 cm**   8. **85 ft**   9. **16x units**

Graph and find the area of each figure with the given vertices.   **For 10–12, see p. A21.**

10. (5, 1), (5, 4), (−1, 4) **9 sq. units**

11. (2, 1), (−2, 1), (5, −3), (−4, −3) **26 sq. units**

12. (2, −1), (5, 3), (0, −1), (−3, 3) **20 sq. units**

LESSON 6-3

Find the missing measure in each triangle.

13. **x = 15 mm**   14. **y = 4 in.**   15. **z = 19.6214**

LESSON 6-4

Find the circumference and area of each circle both in terms of π and to the nearest tenth of a unit. Use 3.14 for π.

16. **$8\pi \approx$ 25.1 cm**
     $16\pi \approx$ 50.2 cm²

17. **$24\pi \approx$ 75.4 in.**
     $144\pi \approx$ 452.2 in.²

18. **$15\pi \approx$ 47.1 ft**
     $56.25\pi \approx$ 176.6 ft²

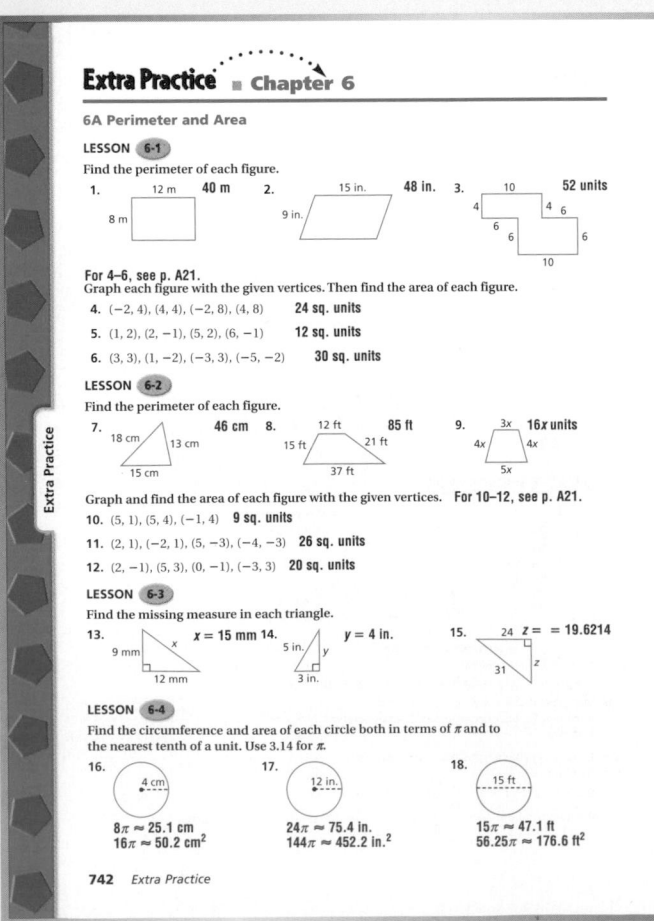

6B Three-Dimensional Geometry

LESSON 6-5   **For 1–4, see p. A21.**

Use isometric dot paper to draw rectangular boxes with the given dimensions.

1. 3 units long, 2 units wide, 5 units high   2. 5 units long, 3 units wide, 4 units high

3. Sketch a one-point perspective drawing of a cube.

4. Sketch a two-point perspective drawing of a cube.

LESSON 6-6

Find the volume of each figure to the nearest tenth of a unit. Use 3.14 for π.

5. **120 ft³**   6. **113.0 m³**   7. a cylinder 16 units tall with a radius of 2 units **201 cubic units**

8. **339 cubic units**   9. **1978.2 cubic units**   10. **1221.3 cm³**

LESSON 6-7

Find the volume of each figure to the nearest tenth of a unit. Use 3.14 for π.

11. **1512 m³**   12. **90.5 cubic units**   13. **14735.3 mm³**

LESSON 6-8

Find the surface area of each figure to the nearest tenth. Use 3.14 for π.

14. a cylinder with radius 5 cm and height 3 cm **251.2 cm²**   15. **102 m²**   16. **3220 ft²**

LESSON 6-9   **For 17–19, see p. A21.**

Find the surface area of each figure to the nearest tenth. Use 3.14 for π.

17.   18. a square pyramid with a 6 in. by 6 in. base and a height of 4 in.   19. a pyramid with an equilateral triangle base with side length 10 units and all lateral faces equilateral triangles

LESSON 6-10   **For 20–21, see p. A21.**

Find the volume and surface area of each figure to the nearest tenth. Use 3.14 for π.

20. a sphere with radius 6 ft   21. a sphere with diameter 80 cm

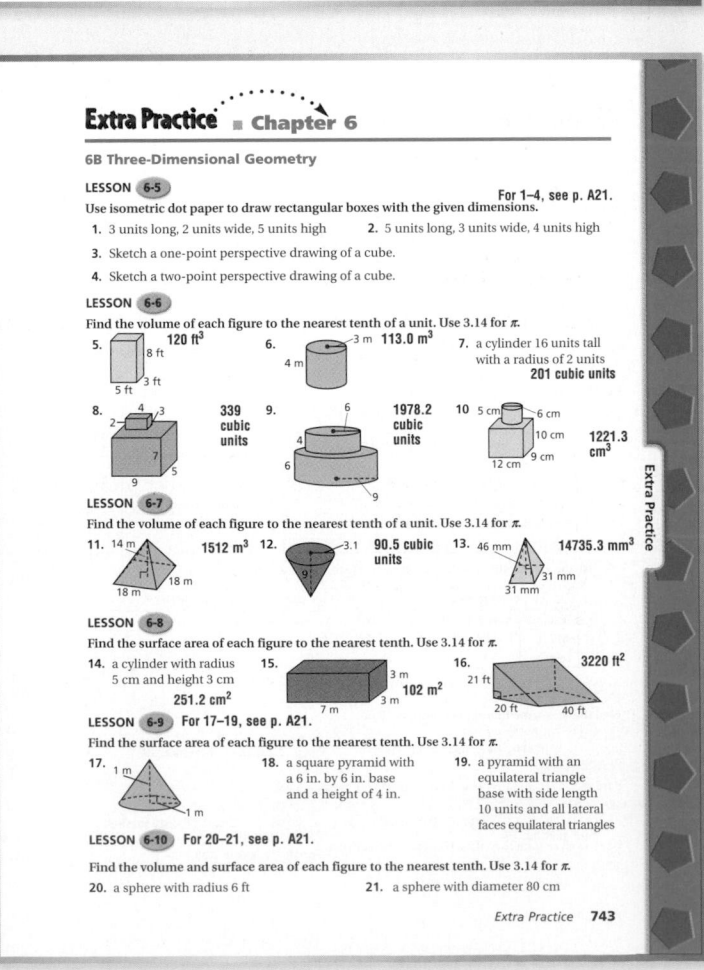

7A Ratios, Rates, and Proportions

LESSON 7-1

Find two ratios that are equivalent to each given ratio. **Possible answer:**

1. $\frac{7}{14}$ $\frac{14}{28}, \frac{1}{2}$
2. $\frac{9}{12}$ $\frac{18}{24}, \frac{3}{4}$
3. $\frac{21}{35}$ $\frac{42}{70}, \frac{3}{5}$
4. $\frac{42}{49}$ $\frac{84}{98}, \frac{6}{7}$

Simplify to tell whether the ratios form a proportion.

5. $\frac{6}{30}$ and $\frac{4}{20}$ **yes**
6. $\frac{10}{16}$ and $\frac{15}{24}$ **yes**
7. $\frac{21}{24}$ and $\frac{14}{18}$ **no**
8. $\frac{52}{64}$ and $\frac{91}{112}$ **yes**

LESSON 7-2

9. Find the unit rate for each brand of detergent, and determine which brand is the best buy. **Bubbling detergent, at 2.3¢ per ounce**

10. A computer monitor has a viewable screen area that is 15 inches wide and 12 inches tall. What is the aspect ratio of this monitor? **5:4**

Product	Size	Price
Pizzazz detergent	128 oz	$3.08
Spring Clean detergent	64 oz	$1.60
Bubbling detergent	196 oz	$4.51

LESSON 7-3

Find the appropriate conversion factor for each conversion.

11. pint to quart $\frac{1\ quart}{2\ pints}$
12. mile to foot $\frac{5280\ ft}{1\ mi}$
13. kilogram to gram $\frac{1000g}{1\ kg}$
14. milliliter to liter $\frac{1\ L}{1000\ mL}$

Solve.

15. In 1911, the first year of the Indianapolis 500 auto race, the winning car had an average speed of 109.416 feet per second. What is that speed in miles per hour? **74.6018 mi/h**

16. A three-toed sloth has a top speed of 0.22 feet per second. A giant tortoise has a top speed of 2.992 inches per second. Convert both speeds to miles per hour, and determine which animal is faster.
Sloth: 0.15 mi/h; tortoise: 0.17 mi/h; the tortoise is faster.

LESSON 7-4

Tell whether the ratios are proportional.

17. $\frac{7}{9}$ and $\frac{3}{4}$ **no**
18. $\frac{2}{3}$ and $\frac{16}{24}$ **yes**
19. $\frac{32}{48}$ and $\frac{18}{27}$ **yes**
20. $\frac{14}{52}$ and $\frac{31}{52}$ **no**

Solve each proportion.

21. $\frac{3}{8} = \frac{n}{12}$ $n = 4.5$
22. $\frac{c}{15} = \frac{3}{45}$ $c = 1$
23. $\frac{7}{18} = \frac{3}{m}$ $m = 7.7$
24. $\frac{8}{p} = \frac{15}{9}$ $p = 4.8$
25. $\frac{5}{f} = \frac{8}{12}$ $f = 7.5$
26. $\frac{12}{15} = \frac{z}{24}$ $z = 19.2$
27. $\frac{a}{32} = \frac{6}{12}$ $a = 16$
28. $\frac{30}{b} = \frac{6}{17}$ $b = 85$

7B Similarity and Scale

LESSON 7-5

Tell whether each transformation is a dilation.

1. 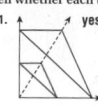 **yes**
2. 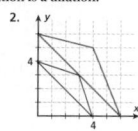 **no**
3. 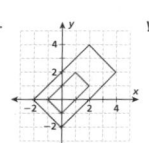 **yes**

4. A figure has vertices at (1, 2), (2, 5), (5, 6), and (6, 1). The figure is dilated by a scale factor of 2.5. What are the coordinates of the image?
(2.5, 5), (5, 12.5), (12.5, 15) and (15, 2.5)

LESSON 7-6

5. Find the missing dimensions of $\triangle XYZ$. $\triangle ABC \sim \triangle XYZ$.
m = 2.5 units; n = 4 units

6. A rectangle is 15 cm long and 8 cm tall. Another rectangle is 20 cm long and 12 cm tall. Are the rectangles similar? **no**

LESSON 7-7

7. On a scale drawing of a house plan, the master bathroom is $1\frac{1}{2}$ inches wide and $2\frac{5}{8}$ inches long. If the scale of the drawing is $\frac{3}{16}$ inches = 1 foot, what are the actual dimensions of the bathroom? **8 ft × 14 ft**

8. Julio uses a scale of $\frac{1}{8}$ inch = 1 foot when he paints landscapes. In one painting, a giant sequoia tree is 34.375 inches tall. How tall is the real tree? **275 ft**

LESSON 7-8

Tell whether each scale reduces, enlarges, or preserves the size of the actual object.

9. 10 cm:1 dm **preserves**
10. 3 ft:3 yd **reduces**
11. 5 km:5 m **enlarges**
12. 1760 yd:5280 ft **preserves**

13. A model of a skyscraper was made using a scale of 0.5 in:5 ft. If the actual skyscraper is 570 feet tall, how many feet tall is the model? **4.75 ft**

LESSON 7-9 For 14–16, see p. A21.

A 9 cm cube and a 2 cm cube are both part of a demonstration kit for architects. Compare the following values of the two cubes.

14. side length
15. surface area
16. volume

17. A popcorn machine makes enough popcorn to fill a rectangular box that measures 5 in. × 8 in. × 2 in. in 45 seconds. How long would it take the same machine to fill a rectangular box that measures 10 in. × 16 in. × 4 in.? **6 minutes**

8A Numbers and Percents

LESSON 8-1

Find the equivalent values missing from the table for each value given on the circle graph.

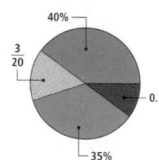

Fraction	Decimal	Percent
$\frac{3}{20}$	1. **0.15**	2. **15%**
3. **$\frac{1}{10}$**	0.1	4. **10%**
5. **$\frac{2}{5}$**	6. **0.40**	40%
7. **$\frac{7}{20}$**	8. **0.35**	35%

LESSON 8-2

Find each percent or number. Round to the nearest tenth if necessary.

9. What percent of 264 is 93? **35.2%**
10. What number is to 100 as 4 is to 78? **5.1**
11. What percent of 68 is 5? **7.4%**
12. What number is to 100 as 13 is to 107? **12.1**
13. What percent of 144 is 24? **16.7%**
14. What number is to 100 as 57 is to 72? **79.2**
15. What percent of 318 is 156? **49.1%**
16. What number is to 100 as 31 is to 148? **20.9**
17. What percent of 984 is 593? **60.3%**
18. What number is to 100 as 264 is to 985? **26.8**

19. Mt. McKinley, in Alaska is 20,320 feet tall. The height of Mt. Everest is about 143% of the height of Mt. McKinley. Estimate the height of Mt. Everest. Round to the nearest thousand. **29,000 ft**

20. Adelaide Island, in Antarctica has an area of 1400 square miles. The area of Alexander Island is 1193% as great as that of Adelaide Island. Estimate the area of Alexander Island. Round to the nearest hundred. **16,700 square miles**

LESSON 8-3

Find each number. Round to the nearest tenth if necessary.

21. 26 is 53% of what number? **49.1**
22. 42 is 86% of what number? **48.8**
23. 17 is 8% of what number? **212.5**
24. 93 is 62% of what number? **150**
25. 215 is 94% of what number? **228.7**
26. 370 is 44% of what number? **840.9**
27. 73 is 18% of what number? **405.6**
28. 61 is 77% of what number? **79.2**

29. A certain rock is a compound of several minerals. Tests show that the sample contains 20.2 grams of quartz. If 37.5% of the rock is quartz, find the mass in grams of the entire rock. **53.9 g**

30. The Alabama River is 729 miles in length, or about 31% of the length of the Mississippi River. Estimate the length of the Mississippi River. Round to the nearest mile. **2352 mi**

8B Applying Percents

LESSON 8-4

Find the percent increase or decrease to the nearest percent.

1. 15 to 27 **80%**
2. 41 to 75 **83%**
3. 91 to 44 **52%**
4. 7 to 31 **343%**
5. 94 to 53 **44%**
6. 38 to 46 **21%**
7. 24 to 80 **233%**
8. 85 to 22 **74%**

9. A computer that sells for $1295 is on sale for 30% off the regular price. What is the sale price of the computer? **$906.50**

LESSON 8-5

Estimate.

10. 26% of 37 **10**
11. 16% of 51 **8**
12. 48% of 19 **10**
13. 75% of 88 **66**
14. 52% of 64 **32**
15. 9% of 31 **3**
16. 81% of 77 **60**
17. 32% of 61 **20**

Estimate to solve.

18. The highest point in Australia is Mount Kosciusko. This mountain is 32% as high as the highest point in South America, Mount Aconcagua at 22,834 feet. Estimate the height of Mount Kosciusko. **about 6900 feet**

LESSON 8-6

19. A furniture salesperson sold $8759 worth of furniture last month. If he makes 4% commission on all sales and earns a monthly salary of $1500, what was his total pay last month? **$1850.36**

20. Simon bought a set of speakers for $279 and a new tuner for $549. Sales tax on these items was 7.5%. What is Simon's total bill for these items? **$890.10**

21. Antwaan earns $1250 per month. Of that, $89.38 is withheld for Social Security and Medicare taxes. What percent of Antwaan's earnings are withheld? **7.15%**

22. In her shop, Ashley earns 22% on all the glassware she sells. This month she earned $2750. What were the total sales of glassware? **$12,500.00**

LESSON 8-7

23. Nigel borrowed $7500 to make home repairs and to put in a new bathroom. The bank charges $6\frac{1}{2}$% simple interest over 3 years. What is the total Nigel will repay the bank? **$8962.50**

24. Gwen invested $10,000 in a mutual fund at a yearly rate of 7%. She earned $5600 in simple interest. How long was the money invested? **8 years**

25. Ray earned $5000, which he used to buy a 5-year certificate of deposit (CD). The CD paid simple interest at 6%. What will the CD be worth at the end of the 5 years? **$6500**

26. Rich borrowed $16,000 for 12 years at simple interest to help pay for his schooling. If he repaid a total of $31,360, at what interest rate did he borrow the money? **8%**

9A Experimental Probability

LESSON 9-1 For 7, see p. A21.

Refer to the spinner at right. Give the probability of each outcome.

1. red $\frac{3}{8}$
2. blue $\frac{3}{8}$
3. yellow $\frac{1}{4}$
4. not red $\frac{5}{8}$
5. not blue $\frac{5}{8}$
6. not yellow $\frac{3}{4}$

7. The probability that Kara will win a game is $\frac{1}{5}$. Kevin and Cheryl have half as much chance of winning as Kara does. Sherry and Jameel are both three times as likely to win the game as Kevin is. Create a table of probabilities for the sample space.

LESSON 9-2

A utensil is drawn from a drawer and replaced. The table shows the results after 100 draws.

Outcomes	Draws
Spoon	37
Knife	32
Fork	31

8. Estimate the probability of drawing a spoon. **P(spoon) = 0.37, or 37%**
9. Estimate the probability of not drawing a spoon. **P(not a spoon) = 0.63, or 63%**

A sales assistant tracks the sales of a particular sweater. The table shows the data after 1000 sales.

Outcomes	Sales
Turquoise	361
Lavender	207
Pink	189
Green	243

10. Estimate the probability that the next customer will buy a pink sweater. **P(pink) = 0.189, or 18.9%**
11. Estimate the probability of the next sweater sold not being pink or lavender. **P(not pink and not lavender) = 0.604, or 60.4%**

LESSON 9-3

Use the table of random numbers to simulate each situation. Use at least 10 trials for each simulation.

```
53736  85815  87649  31119  16635  65161  27919  86585  32848  94425  61378  41256
11632  46278  38783  87649  13325  60848  74681  54238  94228  82794  23426  46498
46278  65264  13906  24794  85976  98713  51876  25847  65972  41973  58927  16842
58147  52697  28467  21358  20650  59731  20587  20648  91845  27364  59421  18579
```

12. A golfer has an 81% chance of making a putt on the first try. Estimate the probability that he will make the putt on the first try at least 8 of his next 10 times. **Possible answer: 0.5, or 50%**

13. A field-goal kicker has a 94% chance of making successful field goals. Estimate the probability that he will make at least 9 of his next 10 field goal attempts. **Possible answer: 0.9, or 90%**

9B Theoretical Probability and Counting

LESSON 9-4

An experiment consists of rolling a fair number cube. There are 6 possible outcomes: 1, 2, 3, 4, 5, and 6. Find each probability.

1. P(rolling an even number) $\frac{1}{2}$
2. P(rolling a 3) $\frac{1}{6}$
3. P(rolling a number greater than 4) $\frac{1}{3}$
4. P(rolling a 7) 0

An experiment consists of rolling two fair number cubes. Find each probability.

5. P(rolling a total of 8) $\frac{5}{36}$
6. P(rolling a total less than 3) $\frac{1}{36}$
7. P(rolling a 2 on at least one number cube) $\frac{11}{36}$
8. P(rolling a total of 6) $\frac{5}{36}$
9. P(rolling a total greater than 5) $\frac{13}{18}$
10. P(rolling a total of 12) $\frac{1}{36}$

LESSON 9-5 For 13, see p. A21.

A computer randomly generates a 4-character computer password of 2 digits followed by 2 letters.

11. Find the number of possible passwords. **67,600**
12. Find the probability that an assigned password does not contain a K. **P(no K) = 0.9246, or 92.5%**
13. A dancer has a choice of 2 dresses, 4 scarves, and 4 pairs of shoes. Draw a tree diagram to show all the possible outcomes.

LESSON 9-6

Evaluate each expression.

14. 8! **40,320**
15. $\frac{7!}{2!}$ **2520**
16. $\frac{5!}{11!}$ **0.000003**
17. $\frac{6!}{(14-6)!}$ **0.018**
18. There are 12 college football teams in the conference. Find the number of orders in which all 12 teams can finish the season. **479,001,600**
19. Find the number of ways the 12 teams can finish first, second, and third in the conference. **1320**

LESSON 9-7

20. An experiment consists of rolling a fair number cube 4 times. For each toss, all outcomes are equally likely. What is the probability of rolling a 3 four times in a row? **P(3, 3, 3, 3) = 0.00077, or 0.077%**
21. A jar contains 8 black marbles and 5 white marbles. What is the probability of drawing 2 white marbles at the same time? **P(white, white) = $\frac{5}{39}$**

LESSON 9-8

22. At a track meet, 250 participants competed for 15 trophies. Estimate the odds of winning a trophy. **3:47**
23. If the odds against winning a contest are 2999:1, what is the probability of winning the contest? **P(winning) = 0.000333, or 0.0333%**

10A Solving Linear Equations

LESSON 10-1

Solve.

1. $\frac{a}{2} - 3 = 8$ **a = 22**
2. $2.4 = -0.8x + 3.2$ **x = 1**
3. $\frac{6+z}{3} = 4$ **z = 6**
4. $\frac{c}{6} + 2 = 5$ **c = 18**
5. $0.9m - 1.6 = -5.2$ **m = -4**
6. $\frac{x-4}{3} = 7$ **x = 25**
7. $\frac{b}{2} + 2 = -3$ **b = -25**
8. $2.1d + 0.7 = 7$ **d = 3**
9. $\frac{p+5}{3} = 6$ **p = 13**
10. $\frac{c}{6} - 8 = 3$ **c = 66**
11. $-8.6 = 3.4k - 1.82$ **k = -2**
12. $\frac{r-6}{9} = 5$ **r = 51**

13. A bill from the plumber was $383. The plumber charged $175 for parts and $52 per hour for labor. How long did the plumber work at this job? **4 hours**

14. Alicia bought $116 worth of flowers and some bushes for around her house. The bushes cost $28 each, and the bill totaled $340. How many bushes did she buy? **8 bushes**

LESSON 10-2

Solve.

15. $4a - 3 + 2a + 7 = 34$ **a = 5**
16. $7 - 6b + 4 - 3b = 74$ **b = -7**
17. $5x - 8 - 7x - 9 = 5$ **x = -11**
18. $g - 9 + 4g + 6 = 12$ **g = 3**
19. $3 - 5f - 7 + 3f = -6.5$ **f = 5**
20. $2r - 6 + 9 - 4r = -7$ **r = 5**
21. $\frac{2a}{4} - \frac{4}{3} = -\frac{2}{3}$ **a = 1**
22. $\frac{2}{5} - \frac{3b}{5} = \frac{8}{5}$ **b = -2**
23. $\frac{4z}{13} + \frac{3}{13} = -1$ **z = -4**
24. $\frac{8}{9} - \frac{5m}{9} = \frac{23}{9}$ **m = -3**
25. $\frac{9}{4} - \frac{3s}{4} = \frac{3}{4}$ **s = 2**
26. $\frac{5p}{3} - \frac{2}{3} = 9$ **p = 6**
27. $\frac{2f}{2} - 4 = -\frac{24}{4}$ **f = -2**
28. $\frac{10c}{4} - \frac{32}{4} = \frac{56}{4}$ **c = 3**
29. $\frac{6x}{3} - \frac{54}{4} + \frac{30x}{4} = -\frac{180}{4}$ **x = -2**
30. $\frac{42y}{2} - \frac{9}{4} + \frac{16y}{4} = \frac{396}{12}$ **y = 431**
31. $\frac{18a}{9} + \frac{12}{4} - \frac{6a}{4} = \frac{30}{4}$ **a = -132**
32. $\frac{2b}{8} + \frac{4}{8} - \frac{4}{8} = \frac{34}{8}$ **b = 5**

33. Jack had a $5 coupon for a CD by his favorite group. After the CD was rung up and 8% sales tax was added, the $5 was subtracted. Jack paid a total of $11.20. What was the original price of the CD? **$15.00**

LESSON 10-3

Solve.

34. $4x - 7 = 3x$ **x = 7**
35. $3w + 4 = 24 - w$ **w = 5**
36. $2y + 6 = 4y$ **y = 3**
37. $2b + 8 = -b + 2$ **b = -2**
38. $5z - 3 = z + 1$ **z = 1**
39. $-2a - 6 = a + 3$ **a = -3**
40. $2p - 3 = 3 + p$ **no solution**
41. $4 + 3c = 7c - 4$ **c = 2**
42. $7d - 3 + 2d = 5d - 8 + 1$ **d = -1**
43. $5f - 2 - 3f = 2f + f$ **f = -4**
44. $7k - 6 + 2k = 3k - 8 + 3k$ **k = $\frac{2}{3}$**
45. $\frac{w}{4} + \frac{5}{8} = \frac{2w}{2} - \frac{9}{8}$ **w = -1**
46. $\frac{2a}{3} - \frac{11}{6} + \frac{3a}{4} = \frac{a}{6} + \frac{7}{3}$ **a = 4**
47. $\frac{4q}{9} + \frac{7}{9} - \frac{3q}{6} = \frac{2q}{6} - \frac{13}{18}$ **q = -3**

48. A cafeteria charges a fixed price per ounce for the salad bar. A sandwich costs $2.10, and a drink costs $1.30. If a 6 ounce salad and a drink costs the same as a 4 ounce salad and a sandwich, how much does the salad cost per ounce? **$0.40**

10B Solving Equations and Inequalities

LESSON 10-4

Solve and graph. For 1-14, see pp. A21–A22.

1. $4a + 3 < 11$ **a < 2**
2. $-12 \le 5x + 3$ **x ≥ -3**
3. $2b + 8 > 16$ **b > 4**
4. $5c + 6 \ge -4$ **c ≥ -2**
5. $4 > 3d - 2$ **d < 2**
6. $-6f + 4 \le 10$ **f ≥ -1**
7. $-3g + 2 \ge -4$ **g ≤ 2**
8. $-3 < 5h - 8$ **1 < h or h > 1**
9. $4z + 8 - z \le -1$ **z ≤ -3**
10. $\frac{6a}{4} + \frac{3}{7} \ge \frac{3}{7}$ **a > $\frac{1}{8}$**
11. $2x + 3 - 6x > -5x + 1$ **-2 < x or x > -2**
12. $5k - 3 + k \ge 9$ **k ≥ 2**
13. $\frac{5d}{4} - \frac{15}{8} \le \frac{15}{8}$ **d < $\frac{12}{5}$**
14. $4p - 9 + 3p < 5p - 3$ **p < 3**

15. Shelly sews doll dresses and sells them for $12 each. The unit cost of the material is $4 each, and the cost of the sewing machine is $360. How many dresses does Shelly have to sell to make a profit? **> 45**

LESSON 10-5

Solve for the indicated variable.

16. Solve $P = s_1 + s_2 + s_3$ for s_2. **$s_2 = P - s_1 - s_3$**
17. Solve $P = s_1 + s_2 + s_3$ for s_3. **$s_3 = P - s_1 - s_2$**
18. Solve $A = s^2$ for s. **$s = \pm\sqrt{A}$**
19. Solve $A = \frac{1}{2}h(b_1 + b_2)$ for b_1. **$b_1 = \frac{2A}{h} - b_2$**
20. Solve $a^2 + b^2 = c^2$ for a. **$a = \pm\sqrt{c^2 - b^2}$**
21. Solve $V = \frac{1}{3}\pi r^2 h$ for h. **$h = \frac{3V}{\pi r^2}$**

Solve for y and graph. For 22-29, see p. A22.

22. $3y + 6x = 6$ **y = 2 - 2x**
23. $5y + 2x = 5$ **y = 1 - $\frac{2}{5}$x**
24. $2x - 2y = 0$ **y = x**
25. $3y + x = 7$ **y = $\frac{7}{3}$ - $\frac{1}{3}$x**
26. $3y - x = 5$ **y = $\frac{1}{3}$x - $\frac{5}{3}$**
27. $2x + 4y = 6$ **y = $\frac{3}{2}$ - $\frac{1}{2}$x**
28. $4y - 2x = 4$ **y = $\frac{1}{2}$x + 1**
29. $2x - 3y = 2$ **y = $\frac{2}{3}$x + $\frac{3}{3}$**

LESSON 10-6

Determine whether each ordered pair is a solution of the given system of equations.

30. $(2, -2)$ $3y - 2x = -2$; $-3x + 2y = -10$ **yes**
31. $(4, 3)$ $y - x = -1$; $3y - 2x = 1$ **yes**
32. $(3, 1)$ $3y - x = 0$; $4x - y = 11$ **yes**
33. $(-1, -3)$ $-3x + y = 3$; $2y - 2x = -2$ **no**
34. $(-1, 5)$ $y - 4x = 7$; $2x + 2y = 4$ **no**
35. $(5, 7)$ $3y - 4x = 1$; $y + x = 12$ **yes**

Solve each system of equations.

36. $y = x - 1$; $y = -2x + 5$ **(2, 1)**
37. $-y = x + 1$; $y = -2x - 4$ **(-3, 2)**
38. $y = 2x - 3$; $y = -2x + 13$ **(4, 5)**
39. $x + y = -5$; $x - 2y = 7$ **(-1, -4)**
40. $x + y = 1$; $x - 3y = -11$ **(-2, 3)**
41. $x - y = 6$; $x + 2y = 3$ **(3, -3)**
42. $x - 2y = 11$; $3y + 5x = 3$ **(3, -4)**
43. $y - 2x = 7$; $4y + x = 10$ **(-2, 3)**
44. $3y - 2x = -2$; $y + 2x = -6$ **(-2, -2)**
45. $y - x = 4$; $3x + 2y = 3$ **(-1, 3)**
46. $-3y - x = 2$; $2y + 2x = 4$ **(4, -2)**
47. $2y - 2x = 4$; $x + y = 8$ **(3, 5)**

11A Linear Equations

LESSON 11-1

Graph each equation and tell whether it is linear. For 1–5, see p. A22.

1. $y = 3x - 4$ **yes**
2. $y = -2x + 1$ **yes**
3. $y = x^2 - 3$ **no**
4. $y = -x - 2$ **yes**

5. A limousine company charges a base fee of $200, plus $50 for each hour of rental. The cost C for h hours is given by $C = 50h + 200$. Find the cost for 2, 3, 4, 5, and 6 hours. Is this a linear equation? Draw a graph that represents the relationship between the cost and the number of hours of rental. **2 hr = $300; 3 hr = $350; 4 hr = $400; 5 hr = $450; 6 hr = $500; yes, it is a linear equation**

LESSON 11-2

Find the slope of the line that passes through each pair of given points.

6. (2, 4) and (−3, 1) $\frac{3}{5}$
7. (5, 1) and (−1, −5) **1**
8. (3, 3) and $\left(1, -4\right)$ $\frac{7}{2}$
9. (−3, 5) and $\left(-1, \frac{-1}{3}\right)$

Tell whether the lines passing through the given points are parallel or perpendicular.

10. A: (−2, −6) and (2, −4)   B: (−4, 1) and (4, 5) **parallel**
11. A: (−1, −7) and (5, 2)   B: (−1, 1) and (−4, 3) **perpendicular**
12. A: (2, 1) and (1, −4)   B: (−2, 2) and (−1, 7) **parallel**

13. Graph the line passing through (4, −2) with slope $\frac{1}{2}$. For 13–14, see p. A22.

14. Graph the line passing through (−3, 1) with slope −2.

LESSON 11-3

Find the x-intercept and y-intercept of each line, and use the intercepts to graph the equation. For 15–18, see p. A22.

15. $4x - 3y = 7$
16. $2y - x = 4$
17. $5x + 3 = 4y$
18. $3y + x = 5$

Write each equation in slope-intercept form, and then find the slope and the y-intercept. For 19–22, see p. A22.

19. $2x = y$ $y = 2x$; $m = 2, b = 0$
20. $3y = 5x$ $y = \frac{5}{3}x$; $m = \frac{5}{3}, b = 0$
21. $4x - y = 7$ $y = 4x - 7$; $m = 4, b = -7$
22. $4y + 5 = 2x$ $y = \frac{1}{2}x - \frac{5}{4}$; $m = \frac{1}{2}, b = -\frac{5}{4}$

Write the equation of the line in slope-intercept form that passes through the given points.

23. (2, −3) and (−4, −5) $y = \frac{1}{3}x - \frac{11}{3}$
24. (4, 1) and (−1, −4) $y = x - 3$
25. (3, 8) and (−5, 2) $y = \frac{3}{4}x + \frac{23}{4}$

LESSON 11-4

Identify a point that the line passes through and the slope of the line.

26. $y - 3 = \frac{1}{2}(x + 2)$ **(−2, 3); $m = \frac{1}{2}$**
27. $y + 2 = -2(x - 1)$ **(1, −2); $m = -2$**
28. $y - 4 = -\frac{1}{3}(x - 5)$ **(5, 4); $m = -\frac{1}{3}$**
29. $y + 5 = 2(x - 1)$ **(1, −5); $m = 2$**
30. $y - 1 = \frac{3}{5}(x + 4)$ **(−4, 1); $m = \frac{3}{5}$**
31. $y = -\frac{2}{3}(x - 3)$ **(3, 0); $m = -\frac{2}{3}$**

Write the point-slope form of the equation of each line.

32. the line with slope 2 passing through (1, 4) $y - 4 = 2(x - 1)$

33. the line with slope $-\frac{1}{3}$ passing through (−2, 1) $y - 1 = -\frac{1}{3}(x + 2)$

11B Linear Relationships

LESSON 11-5

Determine whether the data set shows direct variation.

1.

Weight of Patient	Medication Prescribed (mg)
100	50
120	60
140	70
160	80

yes

2.

Cost of Item	Shipping and Handling
$12.50	$3
$34.97	$5
$52.10	$6
$64.00	$7

no

Find each equation of direct variation, given that y varies directly with x.

3. y is 36 when x is 9. $y = 4x$
4. y is 15 when x is 10. $y = \frac{3}{2}x$
5. y is 84 when x is 2. $y = 42x$
6. y is 6 when x is 3. $y = 2x$
7. y is 90 when x is 18. $y = 5x$
8. y is 13 when x is 8. $y = \frac{13}{8}x$

9. Instructions for a cleaning fluid concentrate state that 3 ounces of concentrate should be added to every $2\frac{1}{2}$ gallons of water used. How many ounces of concentrate should be added to 20 gallons of water? **24 ounces**

10. The distance d an object falls varies directly with the square of the time t of the fall. This is expressed by the formula $d = k \cdot t^2$. An object falls 90 feet in 3 seconds. How far will the object fall in 15 seconds? **2250 ft**

LESSON 11-6

Graph each inequality. For 11–19, see p. A23.

11. $y \le x - 4$
12. $y > x + 3$
13. $4x - 2y \ge 8$
14. $6y - 12 < 3x$
15. $5y - 10x < 20$
16. $3y + 9 > 5x$
17. $x - 4y \le 2$
18. $-2y \ge x - 3$

19. A golf cart gets at most 3 miles per gallon and has a 2.3-gallon gas tank. Graph the relationship between the distance the cart can travel and the number of gallons of gas used. Will the driver be able to make two full trips around the golf course without refueling if one trip is 3.8 miles? **No, the cart will need to be refueled for the second round.**

LESSON 11-7

Plot the data and find the line of best fit. For 20, see p. A23.

20.

x	6	4	8	5	1	7	2	3
y	5	3	6	2	2	5	1	4

12A Sequences

LESSON 12-1

Determine whether each sequence could be arithmetic. If so, give the common difference.

1. 203, 195, 187, 179, 171, 163, . . . **yes; −8**
2. 13, 24, 36, 49, 63, 78, . . . **no**
3. 18.3, 18.8, 19.3, 19.8, 20.3, 20.8, . . . **yes; 0.5**
4. 151, 156, 162, 167, 173, 178, . . . **no**

Find the given term in each arithmetic sequence.

5. 17th term: 7, 14, 21, 28, . . . **119**
6. 25th term: 100, 97, 94, 91, . . . **28**
7. 19th term: 52, 41, 30, 19, . . . **−146**
8. 31st term: 761, 748, 735, 722, . . . **371**

9. Courtney received 200 bonus points when she signed up for a savings card at the grocery store. For every $100 she spends, she will receive 50 more points. How much does she have to spend to collect 1500 points? **$n = 27$; she will have to spend $(n - 1)\$100$, or $26 \times \$100$ ($2600), to reach 1500 points.**

LESSON 12-2

Determine whether each sequence could be geometric. If so, give the common ratio.

10. 6561, 2187, 729, 243, 81, 27, . . . **yes; $\frac{1}{3}$**
11. 1, 7, 49, 343, 2401, 16,807, . . . **yes; 7**
12. 4, 8, 24, 120, 720, 5040, . . . **no**
13. 18, 54, 162, 486, 1458, 4374, . . . **yes; 3**

Find the given term in each geometric sequence.

14. 13th term: 4, −4, 4, −4, . . . **4**
15. 44th term: 2, 4, 8, 16, . . . **17,592,186,044,416**
16. 9th term: 212, 106, 53, 26.5, . . . **0.828125**
17. 23rd term: 3, 6, 12, 24, . . . **12,582,912**

18. The water in a 16,000-gallon swimming pool evaporates at 2% per week during the hot summer months. If the pool is not refilled, how many gallons of water would be left after 8 weeks? **13,890 gallons**

LESSON 12-3

Use the first and second differences to find the next three terms in each sequence.

19. 13, 22, 36, 55, 79, 108, . . . **142, 181, 225**
20. 17, 23, 32, 44, 59, 77, . . . **98, 122, 149**
21. 10.5, 15.25, 20.75, 27, 34, 41.75, . . . **50.25, 59.5, 69.5**
22. 8, 15, 23, 33, 46, 63, . . . **85, 113, 148**
23. 214, 230, 247, 265, 284, 304, . . . **325, 347, 370**
24. 51, 57, 63.5, 71, 80, 91, . . . **104.5, 121, 141**

Give the next three terms in each sequence using the simplest rule you can find.

25. 1, 3, 5, 7, 9, . . . **11, 13, 15**
26. $1, \frac{1}{2}, \frac{1}{4}, \frac{1}{6}, \frac{1}{8}, \frac{1}{10}, \dots$ **$\frac{1}{12}, \frac{1}{14}, \frac{1}{16}$**
27. 3, 7, 11, 15, 19, . . . **23, 27, 31**

Find the first five terms of the sequence defined by the given rule.

28. $a_n = \frac{n}{n + 2}$ **$\frac{1}{3}, \frac{2}{4}, \frac{3}{5}, \frac{4}{6}, \frac{5}{7}$**
29. $a_n = n(n + 1)$ **2, 6, 12, 20, 30**
30. $a_n = n(n - 1) + 3n$ **3, 8, 15, 24, 35**
31. $a_n = 2n\left(\frac{1}{n}\right)$ **2, 2, 2, 2, 2**
32. $a_n = 4n$ **4, 8, 12, 16, 20**
33. $a_n = \left(\frac{n}{n + 1}\right)n$ **$\frac{1}{2}, \frac{4}{3}, \frac{9}{4}, \frac{16}{5}, \frac{25}{6}$**

12B Functions

LESSON 12-4

Determine whether each relationship represents a function.

1. 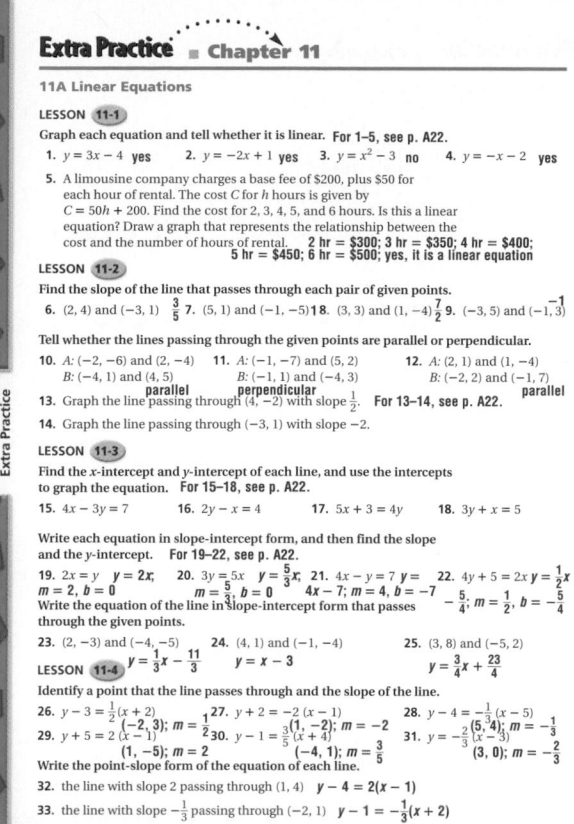 **no**

2.

x	y
−3	1
−1	−1
0	−2
2	0
4	2

yes

3. **yes**

For each function, find $f(-1)$, $f(1)$, and $f(3)$.

4. $f(x) = x^2 + 1$ **2, 2, 10**
5. $f(x) = |x| - 2$ **−1, −1, 1**
6. $f(x) = 3x + 1$ **−2, 4, 10**
7. $f(x) = \frac{x^2}{x - 2}$ **$-\frac{1}{3}$, −1, 9**

LESSON 12-5

Write the rule for each linear function.

8. 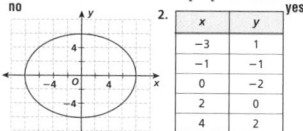 $y = \frac{x}{2} + 3$

9.

x	y
−2	−7
−1	−5
0	−3
1	−1
2	1

$y = 2x - 3$

10.

x	y
−2	3
−1	2
0	1
1	0
2	−1

$y = 1 - x$

LESSON 12-6 For 11–14, see p. A23.

Create a table for each exponential function, and use it to graph the function.

11. $f(x) = 3 \cdot 4^x$
12. $f(x) = \frac{1}{2} \cdot 3^x$
13. $f(x) = 0.75 \cdot 2^x$
14. $f(x) = 2 \cdot 10^x$

15. The isotope cobalt-60, found in radioactive waste, has a half-life of 5 years. How much of a 150 g sample of cobalt-60 would remain after 35 years? **1.17188 g**

LESSON 12-7 For 16–18, see p. A23.

Create a table for each quadratic function, and use it to make a graph.

16. $f(x) = x^2 - 3$
17. $f(x) = x^2 - x + 6$
18. $f(x) = (x - 1)(x + 2)$

LESSON 12-8

Tell whether the relationship is an inverse variation.

19.

Outdoor Temperature (°F)	40°	25°	20°	10°	5°
Cups of Coffee Sold	200	320	400	800	1600

yes

Graph each inverse variation function. For 20–22, see p. A23.

20. $f(x) = \frac{3}{x}$
21. $f(x) = \frac{-0.5}{x}$
22. $f(x) = \frac{3}{2x}$

13A Introduction to Polynomials

LESSON 13-1

Determine whether each expression is a monomial.

1. $\frac{2}{3}r^2st^3$ **yes**   2. $-4p^5q$ **yes**   3. 5^xy^2 **no**   4. $\frac{4m^2}{n^4}$ **no**

Classify each expression as a monomial, a binomial, a trinomial, or not a polynomial.

5. $6x^2 + 3x + \frac{1}{2}$ **trinomial**
6. $-3a^4bc^4$ **monomial**
7. $\frac{3}{4}m^3n^2 + m^2$ **binomial**
8. $5f + 3f^{\frac{1}{2}}g^2$ **not a polynomial**
9. $-mn^3 - 109$ **binomial**
10. $-\frac{2}{z^3}$ **not a polynomial**
11. $-9h^3 + h^2 - 2$ **trinomial**
12. $3xy^2$ **monomial**

Find the degree of each polynomial.

13. $2x^2 + 3x^4 + 7$ **4**   14. $8r + r^3 + 3r^2$ **3**   15. $-10y^4 + 4 + 5y^5$ **5**   16. $6m^3 + 11m^4 - 3m$ **4**

17. The trinomial $-16t^2 + vt + 3$ describes the height in feet of a model rocket launched straight up from a 3-foot platform with a velocity of v ft/s after t seconds. Find the height of the rocket after 4 seconds if $v = 70$ ft/s. **27 feet**

18. The trinomial $-16t^2 + vt + 10$ describes the height in feet of a model rocket launched straight up from a 10-foot platform with a velocity of v ft/s after t seconds. Find the height of the rocket after 3 seconds if $v = 50$ ft/s. **16 feet**

LESSON 13-2

Identify the like terms in each polynomial.

19. $5s - 2rs^2 + 3rs^2 + 2rs - s$ **5s and −s; −2rs² and 3rs²**

20. $-2x^3y^2 + 2x^2y^2 - x^3y + 4x^3y^2$ **−2x³y² and 4x³y²**

21. $6b + 4b^2 - 3b^3 + 5b - b^2$ **6b and 5b; 4b² and −b²**

Simplify.

22. $8r^3 - 2r + 6(r^2 - 3r)$ **8r³ + 6r² − 20r**
23. $5(a^2b^2 + 3ab) + 3(ab^2 - 5ab)$ **5a²b² + 3ab**
24. $7x - 3x^3 + 4x + 12x^2$ **−3x³ + 12x² + 11x**
25. $2s^2t^2 + st^2 + 5s^2t^2 - 7s^2t - 3s^3t^2 + s^2t$ **7s²t² − 2st² − 6s²t**

26. A rectangle has a width of 13 cm and a length of $(4x^2 + 18)$ cm. The area is given by the expression $13(4x^2 + 18)$ cm². Use the Distributive Property to write an equivalent expression. **52x² + 234**

27. A parallelogram has a base of $(3x^2 - 4)$ in. and a height of 4 in. The area is given by the expression $4(3x^2 - 4)$ in². Use the Distributive Property to write an equivalent expression. **12x² − 16**

13B Polynomial Operations

LESSON 13-3

Add.

1. $(5x^2y^2 - 3xy^2 + 2y^2) + (3x^2y^2 + 5y^2)$ **8x²y² − 3xy + 7y²**
2. $(4a^2 + 3ab^2) + (2ab^2 + b^2) + (-5a^2 - 2b^2)$ **−a² + 5ab² − b²**
3. $(m^3 + 3m^2n^2 + 4) + (6m^2n^2 - 9)$ **m³ + 9m²n² − 5**
4. $(10r^3s^2 - 7r^2s + 4r) + (-4r^3s^2 + 3r)$ **6r³s² − 7r²s + 7r**

5. A rectangle has a width of $(x + 5)$ in. and a length of $(4x - 3)$ in. A square has sides of length $(x^2 + 2x - 3)$ in. Write an expression for the sum of the perimeter of the rectangle and the perimeter of the square. **4x² + 18x − 8**

LESSON 13-4

Find the opposite of each polynomial.

6. $-6xy - 2y^3$ **6xy + 2y³**
7. $5a^2b^2 + 3ab - 2$ **−5a²b² − 3ab + 2**
8. $-4x^4 - 5x + x^3$ **4x⁴ + 5x − x³**
9. $9m^3 + mn^2$ **−9m³ − mn²**

Subtract.

10. $(5x^2 + 2xy - 3y^2) - (3x^2 + 2y^2 - 8)$ **2x² + 2xy − 5y² + 8**
11. $12a - (4a^3 - 2a + 7)$ **−4a³ + 14a − 7**
12. $(8r^2s^2 + 4r^2s + rs) - (-2r^2s - 6rs + 3r^2)$ **8r²s² + 6r²s + 7rs − 3r²**
13. $(12y^3 - 6xy + 1) - (8xy - 2x + 1)$ **12y³ − 14xy + 2x**

14. The area of the larger rectangle is $15x^2 + 11x - 14$ cm². The area of the smaller rectangle is $6x^2 + 8x$ cm². What is the area of the shaded region? **9x² + 3x − 14**

LESSON 13-5

Multiply.

15. $(3x^2y)(4x^3y)$ **12x⁵y³**
16. $(2a^2bc^2)(-5a^3b^2)$ **−10a⁵b³c²**
17. $(6m^3n^4)(2mn)$ **12m⁴n⁵**
18. $3s(5t - 8s)$ **15st − 24s²**
19. $-p(3p^2 + 2pq - 9)$ **−3p³ − 2p²q + 9p**
20. $2x^2y(3x^2y^3 + 5x^2y - xy + 12y)$ **6x⁴y⁴ + 10x⁴y² − 2x³y² + 24x²y²**

21. A rectangle has a width of $3x^2y$ ft and a length of $2x^2 + 4xy + 7$ ft. Write and simplify an expression for the area of the rectangle. Then find the area of the rectangle if $x = 2$ and $y = 3$. **Area of the rectangle: 6x⁴y + 12x³y² + 21x²y; If x = 2 and y = 3, then area = 1,404**

LESSON 13-6

Multiply.

22. $(y + 5)(y - 3)$ **y² + 2y − 15**
23. $(t + 1)(t - 6)$ **t² − 5t − 6**
24. $(3m + 2)(4m - 3)$ **12m² − m − 6**
25. $(y + 2)^2$ **y² + 4y + 4**
26. $(a - 4)^2$ **a² − 8a + 16**
27. $(c - 2)(c + 2)$ **c² − 4**

14A Set Theory

LESSON 14-1

Insert the correct symbol to make each statement true.

1. pear ___ {fruit} **∈**   2. $\sqrt{4}$ ___ {prime numbers} **∈**

Determine whether the first set is a subset of the second set. Use the correct symbol.

3. $T = $ {trapezoids}
$P = $ {parallelograms} **No; T ⊄ P**
4. $N = \{(x + 3), x^2y^2, \frac{1}{2}x\}$
$P = $ {polynomials} **Yes; N ⊂ P**

Tell whether each set is finite or infinite.

5. {points on a line} **infinite**   6. {prime numbers less than 100} **finite**

LESSON 14-2

Find the intersection of the sets.

7. $A = \{-3, -1, 3, 5, 7\}$
$B = \{1, 3, 5, 7, 9\}$
$A \cap B = $ **{3, 5, 7}**
8. $N = \{-2, \frac{1}{3}, 0, 1.5, 2\frac{1}{3}, 8\}$
$I = $ {integers}
$N \cap I = $ **{−2, 0, 8}**
9. $P = $ {prime numbers}
$E = $ {even numbers}
$P \cap E = $ **{2}**

Find the union of the sets.

10. $E = \{0, 2, 4, 6, 8\}$
$F = \{-4, -2, 0, 2\}$
$E \cup F = $ **{−4, −2, 0, 2, 4, 6, 8}**
11. $O = $ {odd numbers}
$M = \{1, 3, 7, 11\}$
$O \cup M = $ **{odd numbers}**
12. $X = \{-3, -2, 0, 2, 3\}$
$Y = \{1, 3, 7, 13\}$
$X \cup Y = $ **{−3, −2, 0, 1, 2, 3}**

LESSON 14-3

Draw a Venn diagram to show the relationship between the sets. **13–14. See p. A23.**

13. Set	Elements
First 10 multiples of 3	{3, 6, 9, 12, 15, 18 21, 24, 27, 30}
Factors of 24	{1, 2, 3, 4, 6, 8, 12, 24}

14. Set	Elements
Students in the science club	{Mark, Tina, Maria, Jacob, Patty, Lucas, Vivian, Bob, Missy, Ariana, Cindy, Dan}
Students in the jazz band	{Nick, Jacob, Rob, Missy, Cathy, Natalie, Cindy}

Use each Venn diagram to identify intersections, unions, and subsets.

15.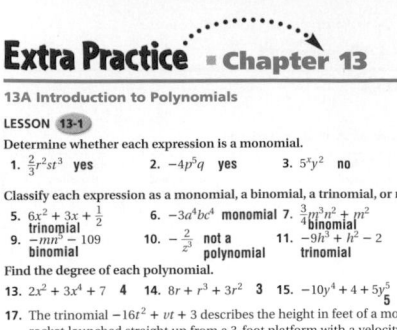
$T \cap S = $ {2, 4}
$T \cup S = $ {1, 2, 3, 4, 5, 6, 9, 10, 12}
There are no subsets.

16. **See p. A23.**

Use a Venn diagram to show the following logical argument. **See p. A24.**

17. All squares are rhombuses.
All rhombuses are quadrilaterals.
∴ All squares are quadrilaterals.

14B Logic and Discrete Math

LESSON 14-4 **1–2. See p. A24.**

Make a truth table for P and Q.

1. P: Greg is in eighth grade.
Q: Greg has a GPA greater than 3.0.

Make a truth table for P or Q.

2. P: The number x is a multiple of 5.
Q: The number x is even.

LESSON 14-5

Identify the hypothesis and the conclusion in each conditional.

3. The baseball games are canceled when it rains. **Hypothesis: It rains. Conclusion: The baseball games are canceled.**
4. If $x + 4 = 10$, then $x = 6$. **Hypothesis: x + 4 = 10. Conclusion: x = 6.**

5. If a polygon has four sides, it is a quadrilateral. **Hypothesis: A polygon has four sides. Conclusion: The polygon is a quadrilateral.**

Make a conclusion, if possible, from each deductive argument.

6. A polygon with eight sides is an octagon. Figure D is a polygon with eight sides. **Figure D is an octagon.**
7. If x is a multiple of 8, it is a multiple of 4. $x = 4^2 + (3)(7)$ **No conclusion can be made.**

8. If a triangle has a base of 8 cm and a height of 5 cm, it has an area of 20 cm². Triangle ABC has an area of 20 cm². **No conclusion can be made.**

LESSON 14-6

Find the degree of each vertex, and determine whether the graph is connected.

9. **A:1, B:3; C:4; D:3; E:2; F:3; Yes.**

10. **R:2; S:2; T:3; U:2; V:2; W:3; Yes**

Determine whether each graph above can be traversed through an Euler circuit. If your answer is yes, describe an Euler circuit in the graph.

11. The graph in Exercise 9. **No**   12. The graph in Exercise 10. **No**

LESSON 14-7

Find a Hamiltonian circuit in each graph.

13. **Possible answer: B-C-D-E-F-A-B**
14. **Possible answer: L-M-O-P-N-Q-L**

Determine the shortest Hamiltonian circuit beginning at A.

15. **Possible answer: A-D-E-B-C-A; 44 m**

Skills Bank

Skills Bank → Review Skills

Place Value to the Billions

A place-value chart can help you read and write numbers. The number 345,012,678,912.5784 (three hundred forty-five billion, twelve million, six hundred seventy-eight thousand, nine hundred twelve and five thousand seven hundred eighty-four ten-thousandths) is shown.

Billions	Millions	Thousands	Ones	Tenths	Hundredths	Thousandths	Ten-Thousandths
345,	012,	678,	912	5	7	8	4

EXAMPLE

Name the place value of the digit.

A the 7 in the thousands column
7 ⟶ ten thousands place

B the 0 in the millions column
0 ⟶ hundred millions place

C the 5 in the billions column
5 ⟶ one billion, or billions, place

D the 8 to the right of the decimal point
8 ⟶ thousandths

PRACTICE

Name the place value of the underlined digit.

1. 123,456,789,123.0594
2. 123,456,789,123.0594
3. 123,456,789,123.0594
4. 123,456,789,123.0594
5. 123,456,789,123.0594
6. 123,456,789,123.0594

1. hundred millions
2. ten billions
3. thousandths
4. ones
5. tenths
6. millions

Round Whole Numbers and Decimals

To round to a certain place, follow these steps.
1. Locate the digit in that place, and consider the next digit to the right.
2. If the digit to the right is 5 or greater, round up. Otherwise, round down.
3. Change each digit to the right of the rounding place to zero.

EXAMPLE

A Round 125,439.378 to the nearest thousand.
125,439.378   Locate digit
The digit to the right is less than 5, so round down.
125,000.000 = 125,000

B Round 125,439.378 to the nearest tenth.
125,439.378   Locate digit.
The digit to the right is greater than 5, so round up.
125,439.400 = 125,539.4

PRACTICE

Round 259,345.278 to the place indicated.

1. hundred thousand
2. ten thousand
3. thousand
4. hundred

1. 300,000
2. 260,000
3. 259,000
4. 259,300

760   Skills Bank

Ways to Show Multiplication and Division

Multiplication and division can be shown in several ways.

EXAMPLE

1. Show the product of 7 and 8 in several ways.

7×8 $7 \cdot 8$ $(7)(8)$ $(7)(8)$

When a variable is used in an expression with multiplication, the multiplication sign is usually omitted. An expression such as $5 \times n$ can be written as $5n$.

2. Show the quotient 15 divided by 3 in several ways.

$15 \div 3$ $15/3$ $\frac{15}{3}$ $3\overline{)15}$

PRACTICE

Write each expression in two other ways.   Possible answers are given.

1. 4×8 $4(8); 4 \cdot 8$
2. 9×10 $9(10); 9 \cdot 10$
3. $18 \div 3$ $\frac{18}{3}; 3\overline{)18}$
4. 2×11 $2(11); 2 \cdot 11$
5. $(9)(2)(5)$ $9(2)(5); 9 \times 2 \times 5$
6. $7 \div n$ $\frac{7}{n}; n\overline{)7}$
7. $\frac{b}{2}$ $b \div 2; 2\overline{)b}$
8. $7 \cdot y$ $7y; 7(y)$
9. $4(c)$ $4c; 4 \times c$
10. $(3)(b)(f)$ $3bf; 3 \cdot b \cdot f$
11. $24/6$ $6\overline{)24}; 24 \div 6$
12. $11\overline{)55}$ $55 \div 11; \frac{55}{11}$

Long Division with Whole Numbers

You can use long division to divide large numbers.

EXAMPLE

Divide 8208 by 72.

```
     114
72)8208      Place the first number under the long division symbol.
   72         Subtract.
   100        Bring down the next digit.
    72        Subtract.
   288        Bring down the next digit.
   288        Subtract.
     0
```

PRACTICE

Divide.

1. $125\overline{)4125}$ 33
2. $158\overline{)20,698}$ 131
3. $268\overline{)4556}$ 17
4. $39\overline{)3471}$ 89
5. $99\overline{)4653}$ 47
6. $321\overline{)38,841}$ 121
7. $120\overline{)5040}$ 42
8. $108\overline{)10,476}$ 97
9. $741\overline{)107,445}$ 145

Skills Bank   761

Factors and Multiples

When two numbers are multiplied to form a third, the two numbers are said to be **factors** of the third number. **Multiples** of a number can be found by multiplying the number by 1, 2, 3, 4, and so on.

EXAMPLE

A List all the factors of 48.
$1 \cdot 48 = 48, 2 \cdot 24 = 48, 3 \cdot 16 = 48$,
$4 \cdot 12 = 48$, and $6 \cdot 8 = 48$
So the factors of 48 are
1, 2, 3, 4, 6, 8, 12, 16, 24, and 48.

B Find the first five multiples of 3.
$3 \cdot 1 = 3, 3 \cdot 2 = 6, 3 \cdot 3 = 9$,
$3 \cdot 4 = 12$, and $3 \cdot 5 = 15$
So the first five multiples of 3 are
3, 6, 9, 12, and 15.

PRACTICE

List all the factors of each number.

1. 8   1, 2, 4, 8
2. 20   1, 2, 4, 5, 10, 20
3. 9   1, 3, 9
4. 51   1, 3, 17, 51
5. 16   1, 2, 4, 8, 16
6. 27   1, 3, 9, 27

Write the first five multiples of each number.

7. 9   9, 18, 27, 36, 45
8. 10   10, 20, 30, 40, 50
9. 20   20, 40, 60, 80, 100
10. 15   15, 30, 45, 60, 75
11. 7   7, 14, 21, 28, 35
12. 18   18, 36, 54, 72, 90

Divisibility Rules

A number is divisible by another number if the division results in a remainder of 0. Some divisibility rules are shown below.

A number is divisible by . . .	Divisible	Not Divisible
2 if the last digit is an even number.	11,994	2,175
3 if the sum of the digits is divisible by 3.	216	79
4 if the last two digits form a number divisible by 4.	1,028	621
5 if the last digit is 0 or 5.	15,195	10,007
6 if the number is even and divisible by 3.	1,332	44
8 if the last three digits form a number divisible by 8.	25,016	14,100
9 if the sum of the digits is divisible by 9.	144	33
10 if the last digit is 0.	2,790	9,325

PRACTICE

Determine which of these numbers each number is divisible by: 2, 3, 4, 5, 6, 8, 9, 10

1. 56   2, 4, 8
2. 200   2, 4, 5, 8, 10
3. 75   3, 5
4. 324   2, 3, 6, 9
5. 42   2, 3, 6
6. 812   2, 4
7. 784   2, 4, 8
8. 501   3
9. 2345   5
10. 555,555   3, 5
11. 3009   3
12. 2001   3

762   Skills Bank

Prime and Composite Numbers

A **prime number** has exactly two factors, 1 and the number itself.

A **composite number** has more than two factors.

2   Factors: 1 and 2; prime
11   Factors: 1 and 11; prime
47   Factors: 1 and 47; prime

4   Factors: 1, 2, and 4; composite
12   Factors: 1, 2, 3, 4, 6, and 12; composite
63   Factors: 1, 3, 7, 9, 21, and 63; composite

EXAMPLE

Determine whether each number is prime or composite.

A 17
Factors
1, 17 ⟶ prime

B 16
Factors
1, 2, 4, 8, 16 ⟶ composite

C 51
Factors
1, 3, 17, 51 ⟶ composite

PRACTICE

Determine whether each number is prime or composite.

1. 5   prime
2. 14   composite
3. 18   composite
4. 2   prime
5. 23   prime
6. 27   composite
7. 13   prime
8. 39   composite
9. 72   composite
10. 49   composite
11. 9   composite
12. 89   prime

Prime Factorization (Factor Tree)

A composite number can be expressed as a product of prime numbers. This is the **prime factorization** of the number. To find the prime factorization of a number, you can use a factor tree.

EXAMPLE

Find the prime factorization of 24 by using a factor tree.

24	24	24
2 · 12	3 · 8	4 · 6
2 · 3 · 4	3 · 2 · 4	2 · 2 · 2 · 3
2 · 3 · 2 · 2	3 · 2 · 2 · 2	

The prime factorization of 24 is $2 \cdot 2 \cdot 2 \cdot 3$, or $2^3 \cdot 3$.

PRACTICE

Find the prime factorization of each number by using a factor tree.

1. 25 5×5, or 5^2
2. 16 $2 \times 2 \times 2 \times 2$, or 2^4
3. 56 $2 \times 2 \times 2 \times 7$, or $2^3 \times 7$
4. 18 $2 \times 3 \times 3$, or 2×3^2
5. 72 $2 \times 2 \times 2 \times 3 \times 3$, or $2^3 \times 3^2$
6. 40 $2 \times 2 \times 2 \times 5$, or $2^3 \times 5$

Skills Bank   763

Greatest Common Factor (GCF)

The **greatest common factor (GCF)** of two whole numbers is the greatest factor the numbers have in common.

EXAMPLE

Find the GCF of 24 and 32.

Method 1: List all the factors of both numbers.

Find all the common factors.

24: **1**, **2**, 3, **4**, 6, **8**, 12, 24
32: **1**, **2**, **4**, **8**, 16, 32

The common factors are 1, 2, 4, and 8.
So the GCF is 8.

Method 2: Find the prime factorizations. Then find the common prime factors.

24: **2 · 2** · 2 · 3
32: **2 · 2** · 2 · 2 · 2

The common prime factors are 2, 2, and 2.
The product of these is the GCF.
So the GCF is 2 · 2 · 2 = 8.

PRACTICE

Find the GCF of each pair of numbers by either method.

1. 9, 15 **3** 2. 25, 75 **25** 3. 18, 30 **6** 4. 4, 10 **2** 5. 12, 17 **1** 6. 30, 96 **6**
7. 54, 72 **18** 8. 15, 20 **5** 9. 40, 60 **20** 10. 40, 50 **10** 11. 14, 21 **7** 12. 14, 28 **14**

Least Common Multiple (LCM)

The **least common multiple (LCM)** of two numbers is the smallest common multiple the numbers share.

EXAMPLE

Find the least common multiple of 8 and 10.

Method 1: List multiples of both numbers.

8: 8, 16, 24, 32, **40**, 48, 56, 64, 72, **80**
10: 10, 20, 30, **40**, 50, 60, 70, **80**, 90

The smallest common multiple is 40.
So the LCM is 40.

Method 2: Find the prime factorizations. Then find the most occurrences of each factor.

8: **2 · 2 · 2**
10: 2 · **5**

The LCM is the product of the factors.

2 · 2 · 2 · 5 = 40 So the LCM is 40.

PRACTICE

Find the LCM of each pair of numbers by either method.

1. 2, 4 **4** 2. 3, 15 **15** 3. 10, 25 **50** 4. 10, 15 **30** 5. 3, 7 **21** 6. 18, 27 **54**
7. 12, 21 **84** 8. 9, 21 **63** 9. 24, 30 **120** 10. 9, 18 **18** 11. 16, 24 **48** 12. 8, 36 **72**

Compatible Numbers

Compatible numbers are close to the numbers in a problem and divide without a remainder. You can use compatible numbers to estimate quotients.

EXAMPLE

Use compatible numbers to estimate each quotient.

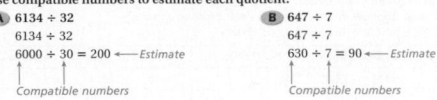

A 6134 ÷ 32
6134 ÷ 32
6000 ÷ 30 = 200 ←— *Estimate*
‾‾‾‾‾‾‾‾‾‾‾
Compatible numbers

B 647 ÷ 7
647 ÷ 7
630 ÷ 7 = 90 ←— *Estimate*
‾‾‾‾‾‾‾‾‾‾
Compatible numbers

PRACTICE

Estimate the quotient by using compatible numbers.   **Possible answers are given.**

1. 345 ÷ 5 **70** 2. 5474 ÷ 23 **220** 3. 46,170 ÷ 18 **2,500** 4. 749 ÷ 7 **100**
5. 861 ÷ 41 **20** 6. 1225 ÷ 2 **600** 7. 968 ÷ 47 **20** 8. 3456 ÷ 432 **8**
9. 5765 ÷ 26 **200** 10. 25,012 ÷ 64 **500** 11. 99,170 ÷ 105 **100** 12. 868 ÷ 8 **100**

Mixed Numbers and Fractions

Mixed numbers can be written as fractions greater than 1, and fractions greater than 1 can be written as mixed numbers.

EXAMPLE

A Write $\frac{23}{5}$ as a mixed number.

$\frac{23}{5}$ *Divide the numerator by the denominator.*

$5\overline{)23}$ → $4\frac{3}{5}$ ←— *Write the remainder as the numerator of a fraction.*
$\underline{20}$
3

B Write $6\frac{2}{7}$ as a fraction.

Multiply the denominator by the whole number.   *Add the product to the numerator.*

$6\frac{2}{7}$ → 7 · 6 = 42 → 42 + 2 = 44

Write the sum over the denominator. → $\frac{44}{7}$

PRACTICE

Write each mixed number as a fraction. Write each fraction as a mixed number.

1. $\frac{22}{5}$ **$4\frac{2}{5}$** 2. $9\frac{1}{7}$ **$\frac{64}{7}$** 3. $\frac{41}{8}$ **$5\frac{1}{8}$** 4. $5\frac{7}{9}$ **$\frac{52}{9}$**
5. $\frac{7}{3}$ **$2\frac{1}{3}$** 6. $4\frac{9}{11}$ **$\frac{53}{11}$** 7. $\frac{47}{16}$ **$2\frac{15}{16}$** 8. $3\frac{3}{8}$ **$\frac{27}{8}$**
9. $\frac{31}{9}$ **$3\frac{4}{9}$** 10. $8\frac{2}{3}$ **$\frac{26}{3}$** 11. $\frac{33}{5}$ **$6\frac{3}{5}$** 12. $12\frac{1}{9}$ **$\frac{109}{9}$**

Multiply and Divide Decimals by Powers of 10

Notice the pattern below.

0.24 · 10   = 2.4
0.24 · 100   = 24
0.24 · 1000   = 240
0.24 · 10,000 = 2400

10   = 10^1
100   = 10^2
1000   = 10^3
10,000 = 10^4

Think: When multiplying decimals by powers of 10, move the decimal point one place to the right for each power of 10, or for each zero.

Notice the pattern below.

0.24 ÷ 10   = 0.024
0.24 ÷ 100   = 0.0024
0.24 ÷ 1000   = 0.00024
0.24 ÷ 10,000 = 0.000024

Think: When dividing decimals by powers of 10, move the decimal point one place to the left for each power of 10, or for each zero.

PRACTICE

Find each product or quotient.

1. 10 · 9.26 **92.6** 2. 0.642 · 100 **64.2** 3. 10^3 · 84.2 **84,200** 4. 0.44 · 10^4 **4400**
5. 69.7 · 1000 **69,700** 6. 11.32 ÷ 10 **1.132** 7. 678 · 10^8 **67,800,000,000** 8. 1.276 ÷ 1000 **0.001276**
9. 536.5 ÷ 10^2 **5.365** 10. 5.92 ÷ 10^3 **0.00592** 11. 25 ÷ 10,000 **0.0025** 12. 6.519 · 10^2 **651.9**

Multiply Decimals

When multiplying decimals, multiply as you would with whole numbers. The sum of the number of decimal places in the factors equals the number of decimal places in the product.

EXAMPLE

Find each product.

A 81.2 · 6.547

  6.547 ←— *3 decimal places*
× 81.2 ←— *1 decimal place*
‾‾‾‾‾‾
  1 3094
  6 5470
523 7600
‾‾‾‾‾‾‾‾
531.6164 ←— *4 decimal places*

B 0.376 · 0.12

0.376 ←— *3 decimal places*
× 0.12 ←— *2 decimal places*
‾‾‾‾‾‾
  752
 3760
‾‾‾‾‾
0.04512 ←— *5 decimal places*

PRACTICE

Find each product.

1. 6.8 · 3.4 **23.12** 2. 2.56 · 4.6 **11.776** 3. 6.787 · 7.6 **51.5812** 4. 0.98 · 4.6 **4.508**
5. 0.97 · 0.76 **0.7372** 6. 0.5 · 3.761 **1.8805** 7. 42 · 17.654 **741.468** 8. 7.005 · 32.1 **224.8605**
9. 9.76 · 16.254 **158.63904** 10. 296.5 · 2.4 **711.60 or 711.6** 11. 7.7 · 6.5 **50.05** 12. 8.92 · 2.8 **24.976**
13. 3.65 · 4.2 **15.33** 14. 0.002 · 8.1 **0.0162** 15. 0.03 · 0.204 **0.00612** 16. 98.6 · 4.9 **483.14**

Divide Decimals

When dividing with decimals, set up the division as you would with whole numbers. Pay attention to the decimal places, as shown below.

EXAMPLE

Find each quotient.

A 89.6 ÷ 16

   5.6
16$\overline{)89.6}$ *Place decimal point.*
 80
‾‾
 96
 96
‾‾
  0

B 3.4 ÷ 4

   0.85   *Place decimal point.*
4$\overline{)3.40}$ ←— *Insert zeros if necessary.*
 3 2
‾‾‾
  20
  20
‾‾
   0

PRACTICE

Find each quotient.

1. 242.76 ÷ 68 **3.57** 2. 40.5 ÷ 18 **2.25** 3. 121.03 ÷ 98 **1.235** 4. 3.6 ÷ 4 **0.9**
5. 1.58 ÷ 5 **0.316** 6. 0.2835 ÷ 2.7 **0.105** 7. 8.1 ÷ 0.09 **90** 8. 0.42 ÷ 0.28 **1.5**
9. 480.48 ÷ 7.7 **62.4** 10. 36.9 ÷ 0.003 **12,300** 11. 0.784 ÷ 0.04 **19.6** 12. 15.12 ÷ 0.063 **240**

Terminating and Repeating Decimals

You can change a fraction to a decimal by dividing. If the resulting decimal has a finite number of digits, it is **terminating**. Otherwise, it is **repeating**.

EXAMPLE

Write $\frac{4}{5}$ and $\frac{2}{3}$ as decimals. Are the decimals terminating or repeating?

$\frac{4}{5}$ = 4 ÷ 5

   0.8
5$\overline{)4.0}$ → $\frac{4}{5}$ = 0.8
 4 0
‾‾‾
   0

$\frac{2}{3}$ = 2 ÷ 3

   0.6666
3$\overline{)2.0000}$ → $\frac{2}{3}$ = 0.6666...
 1 8   ←— *This pattern will repeat.*
‾‾‾
  20

The number 0.8 is a terminating decimal.   The number 0.6666 . . . is a repeating decimal.

PRACTICE

For 1–18, see p. A24.

Write as a decimal. Is the decimal terminating or repeating?

1. $\frac{1}{5}$ 2. $\frac{1}{3}$ 3. $\frac{3}{11}$ 4. $\frac{3}{5}$ 5. $\frac{7}{9}$ 6. $\frac{7}{15}$
7. $\frac{1}{4}$ 8. $\frac{5}{6}$ 9. $\frac{4}{11}$ 10. $\frac{5}{10}$ 11. $\frac{1}{9}$ 12. $\frac{11}{12}$
13. $\frac{5}{9}$ 14. $\frac{8}{11}$ 15. $\frac{7}{8}$ 16. $\frac{23}{25}$ 17. $\frac{3}{20}$ 18. $\frac{5}{11}$

Order of Operations

When simplifying expressions, follow the order of operations.

1. Simplify within parentheses.
2. Evaluate exponents and roots.
3. Multiply and divide from left to right.
4. Add and subtract from left to right.

EXAMPLE

A Simplify the expression $3^2 \times (11 - 4)$.

$3^2 \times (11 - 4)$

$3^2 \times 7$ *Simplify within parentheses.*

9×7 *Evaluate the exponent.*

63 *Multiply.*

B Use a calculator to simplify the expression $19 - 100 \div 5^2$.

If your calculator follows the order of operations, enter the following keystrokes:

$19 - 100 \div$ [x²] [ENTER] The result is 15.

If your calculator does not follow the order of operations, insert parentheses so that the expression is simplified correctly.

$19 - (100 \div 5$ [x²]) [ENTER] The result is 15.

PRACTICE

Simplify each expression.

1. $45 - 15 \div 3$ **40**
2. $51 + 48 \div 8$ **57**
3. $35 \div (15 - 8)$ **5**
4. $\sqrt{9} \times 5 - 15$ **0**
5. $24 \div 3 - 6 + 12$ **14**
6. $(6 \times 8) \div 2^2$ **12**
7. $20 - 3 \times 4 + 30 \div 6$ **13**
8. $3^2 - 10 \div 2 + 4 \times 2$ **12**
9. $27 \div (3 + 6) + 6^2$ **39**
10. $4 \div 2 + 8 \times 2^3 - 4$ **62**
11. $33 - \sqrt{64} \times 3 - 5$ **4**
12. $(8^2 \times 4) - 12 \times 13 + 5$ **105**

Use a calculator to simplify each expression.

13. $6 + 20 \div 4$ **11**
14. $37 - 21 + 7$ **34**
15. $9^2 - 32 \div 8$ **77**
16. $10 \div 2 + 8 \times 2$ **21**
17. $\sqrt{25} + 4 \times 6$ **29**
18. $4 \times 12 - 4 + 8 \div 2$ **48**
19. $28 - 3^2 + 27 \div 3$ **28**
20. $9 + (50 - 16) \div 2$ **26**
21. $4^2 - (10 \times 8) \div 5$ **0**
22. $30 + 22 \div 11 - 7 - 3^2$ **16**
23. $3 + 7 \times 5 - 1$ **37**
24. $38 \div 2 + \sqrt{81} \times 4 - 31$ **24**

768   *Skills Bank*

Properties

The following are basic properties of addition and multiplication when a, b, and c are real numbers.

	Addition		Multiplication
Closure:	$a + b$ is a real number.	Closure:	$a \cdot b$ is a real number.
Commutative:	$a + b = b + a$	Commutative:	$a \cdot b = b \cdot a$
Associative:	$(a + b) + c = a + (b + c)$	Associative:	$(a \cdot b) \cdot c = a \cdot (b \cdot c)$
Identity Property of Zero:	$a + 0 = a$ and $0 + a = a$	Identity Property of One:	$a \cdot 1 = a$ and $1 \cdot a = a$
		Multiplication Property of Zero:	$a \cdot 0 = 0$ and $0 \cdot a = 0$

The following properties are true when a, b, and c are real numbers.

Distributive: $a \cdot (b + c) = a \cdot b + a \cdot c$ **Transitive:** If $a = b$ and $b = c$, then $a = c$.

EXAMPLE

Name the property shown.

A $4 \cdot (7 \cdot 2) = (4 \cdot 7) \cdot 2$
Associative Property of Multiplication

B $4 \cdot (7 + 2) = (4 \cdot 7) + (4 \cdot 2)$
Distributive Property

PRACTICE

Give an example of each of the following properties, using real numbers. **Answers will vary.**

1. Associative Property of Addition
2. Commutative Property of Multiplication
3. Closure Property of Multiplication
4. Distributive Property
5. Multiplication Property of Zero
6. Identity Property of Addition
7. Transitive Property
8. Closure Property of Addition

Name the property shown.

9. $4 + 0 = 4$ **Identity of Addition**
10. $(6 + 3) + 1 = 6 + (3 + 1)$ **Associative of Addition**
11. $7 \cdot 51 = 51 \cdot 7$ **Commutative of Multiplication**
12. $5 \cdot 456 = 456 \cdot 5$ **Commutative of Multiplication**
13. $17 \cdot (1 + 3) = 17 \cdot 1 + 17 \cdot 3$ **Distributive**
14. $1 \cdot 5 = 5$ **Identity of Multiplication**
15. $(8 \cdot 2) \cdot 5 = 8 \cdot (2 \cdot 5)$ **Associative of Multiplication**
16. $72 + 1234 = 1234 + 72$ **Commutative of Addition**
17. $0 \cdot 12 = 0$ **Zero Property of Multiplication**
18. $15.7 \cdot 1.3 = 1.3 \cdot 15.7$ **Commutative of Multiplication**
19. $8.2 + (9.3 + 7) = (8.2 + 9.3) + 7$ **Associative of Addition**
20. $85.98 \cdot 0 = 0$ **Zero Property of Multiplication**
21. If $x = 3.5$ and $3.5 = y$, then $x = y$. **Transitive**
22. $12a \cdot 15b = 15b \cdot 12a$ **Commutative of Multiplication**
23. $(2x + 3y) + 8z = 2x + (3y + 8z)$ **Associative of Addition**
24. $0 \cdot 6m^2n = 0$ **Zero Property of Multiplication**
25. $8j + 32k = 32k + 8j$ **Commutative of Addition**
26. If $3 + 8 = 11$ and $11 = x$, then $3 + 8 = x$. **Transitive**

Skills Bank   769

Compare and Order Rational Numbers

A number line is helpful when you compare and order rational numbers.

EXAMPLE

A Compare. Write $<$ or $>$.

$-\frac{1}{2}$ ▮ -2.5

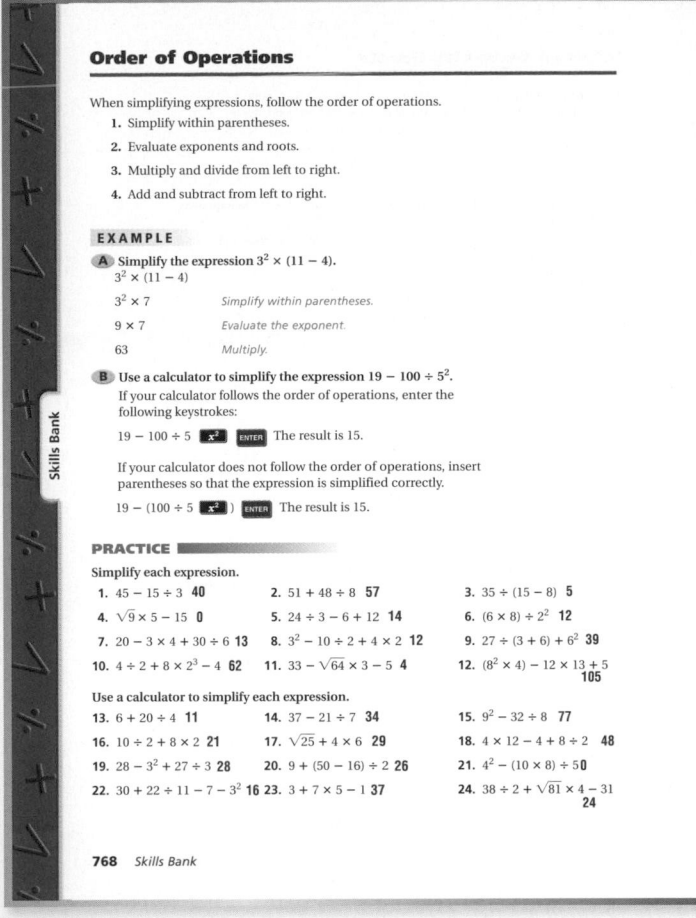

Graph both numbers on a number line.

$-\frac{1}{2}$ is to the right of -2.5.

$-\frac{1}{2} > -2.5$

B Order 40%, 70%, and 10% in order from least to greatest. Use $<$ between numbers.

Graph all three percents on a number line.

10% is to the left of 40%, which is to the left of 70%.

$10\% < 40\% < 70\%$

PRACTICE

Compare. Write $<$ or $>$.

1. -0.3 ▮ -0.1 **<**
2. $-\frac{3}{4}$ ▮ $-\frac{5}{8}$ **<**
3. 35% ▮ 6% **>**
4. -8.65 ▮ -9.97 **>**
5. 0.25 ▮ $\frac{2}{5}$ **<**
6. 6.05 ▮ 6.31 **<**
7. $-\frac{4}{5}$ ▮ -0.5 **<**
8. 75% ▮ 0.80 **<**
9. -0.07 ▮ -0.7 **>**
10. 4.5 ▮ 445% **>**
11. 0.43 ▮ 4.3% **>**
12. $-9\frac{1}{3}$ ▮ -9.03 **<**

Order the numbers from least to greatest. Use $<$ between numbers. **For 13–24, see p. A24.**

13. $1.5, 0.15, 1.05$
14. $34\%, 76\%, 9.8\%$
15. $0.4, -\frac{3}{5}, -1\frac{1}{2}$
16. $-2.6, -1.3, -6.3$
17. $-7.1, 0, -2.4$
18. $2.5\%, 105\%, 53\%$
19. $-0.25, -\frac{2}{5}, -1.2$
20. $0.65, 61\%, 3$
21. $13\%, 8.3\%, 6.7\%$
22. $5\frac{3}{4}, 5\frac{4}{25}, 5\frac{2}{5}$
23. $-0.1003, -0.018, -0.008$
24. $2.7, \frac{28}{100}, 0.029$

770   *Skills Bank*

Absolute Value and Opposites

The **absolute value** of a number is the number's distance from zero on a number line. The symbol for absolute value is $|\ |$. Integers that are the same distance from 0 on a number line and are on opposite sides of 0 are **opposites**.

EXAMPLE

A Name the opposite of 24.
The opposite of 24 is -24.

B Name the opposite of -8.
The opposite of -8 is 8.

C Evaluate $|-5|$ and $|3|$.

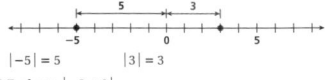

$|-5| = 5$ $|3| = 3$

D Evaluate $|-8 + 6|$.

$|-2|$ *Simplify within the absolute value bars.*

2

PRACTICE

Name the opposite.

1. 13 **−13**
2. 9 **−9**
3. -28 **28**
4. -54 **54**
5. 85 **−85**
6. 1 **−1**
7. -16 **16**
8. -125 **125**
9. a **−a**
10. $-2x$ **2x**
11. $18x^2y$ **−18x²y**
12. $-20mn$ **20mn**

Evaluate.

13. $|-6|$ **6**
14. $|-12|$ **12**
15. $|2.5|$ **2.5**
16. $|18|$ **18**
17. $|-120|$ **120**
18. $|-4.4|$ **4.4**
19. $\left|\frac{1}{2}\right|$ **$\frac{1}{2}$**
20. $|0|$ **0**
21. $\left|-3\frac{2}{5}\right|$ **$3\frac{2}{5}$**
22. $|-100,100|$ **100,100**
23. $|15.75|$ **15.75**
24. $|-52|$ **52**
25. $|8 + 6|$ **14**
26. $|19 - 3|$ **16**
27. $|2 - 6|$ **4**
28. $|-3 + 10|$ **7**
29. $|27 - 28|$ **1**
30. $|-107 + 120|$ **13**
31. $|-3| + |12|$ **15**
32. $|6| + |-4|$ **10**
33. $|-33| + |-17|$ **50**
34. $|25| - |30|$ **−5**
35. $|15| - |-11|$ **4**
36. $|-7| + |7|$ **14**

Use $<$ or $>$ to compare.

37. $|-6|$ ▮ $|5|$ **>**
38. $|-10|$ ▮ $|-17|$ **<**
39. $|3.5|$ ▮ $|-3.7|$ **<**
40. $\left|-\frac{1}{2}\right|$ ▮ $\left|\frac{2}{3}\right|$ **<**

Skills Bank   771

Measure Angles

You can use a protractor to measure angles. To measure an angle, place the base of the protractor on one of the rays of the angle and center the base on the vertex. Look at the protractor scale that has zero on the first ray. Read the scale where the second ray crosses it. Extend the rays, if necessary.

EXAMPLE

A Measure ∠ABC.

The measure of ∠ABC, or m∠ABC, equals 120°.

B Measure ∠XYZ.

The measure of ∠XYZ, or m∠XYZ, equals 50°.

PRACTICE

Use a protractor to measure each angle.

1.   **90°**   2.   **135°**

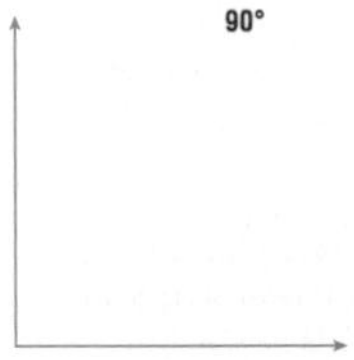

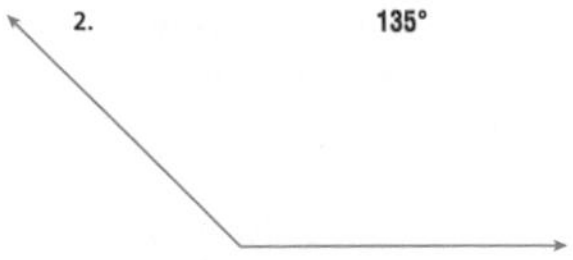

3.   **60°** 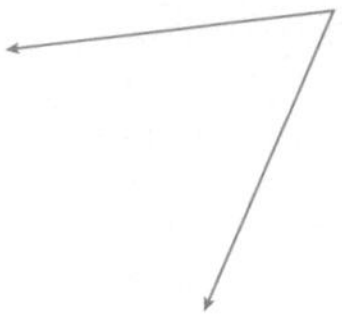   4.   **45°**

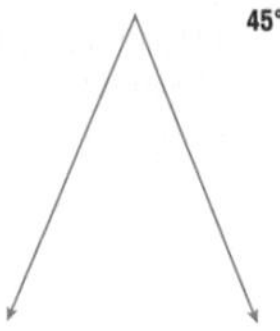

Informal Geometry Proofs

Inductive reasoning involves examining a set of data to determine a pattern and then making a conjecture about the data. In **deductive reasoning**, you reach a conclusion by using logical reasoning based on given statements or premises that you assume to be true.

EXAMPLE

A Use inductive reasoning to determine the 30th number of the sequence.
3, 5, 7, 9, 11, . . .
Examine the pattern to determine the relationship between each term in the sequence and its value.

Term	1st	2nd	3rd	4th	5th
Value	3	5	7	9	11

$1 \cdot 2 + 1 = 2 + 1 = 3$ $4 \cdot 2 + 1 = 8 + 1 = 9$
$2 \cdot 2 + 1 = 4 + 1 = 5$ $5 \cdot 2 + 1 = 10 + 1 = 11$
$3 \cdot 2 + 1 = 6 + 1 = 7$
To obtain each value, multiply the term by 2 and add 1. So the 30th term is
$30 \cdot 2 + 1 = 60 + 1 = 61$.

B Use deductive reasoning to make a conclusion from the given premises.
Premise: Makayla needs at least an 89 on her exam to get a B for the quarter in math class.
Premise: Makayla got a B for the quarter in math class.
Conclusion: Makayla got at least an 89 on her exam.

PRACTICE

Use inductive reasoning to determine the 100th number in each pattern.

1. $\frac{1}{2}$, 1, $1\frac{1}{2}$, 2, $2\frac{1}{2}$, . . .   **50**   2. 1, 4, 9, 16, 25, . . .   **10,000**

3. 4, 6, 8, 10, 12, . . .   **202**   4. 0, 3, 6, 9, 12, 15, . . .   **297**

Use deductive reasoning to make a conclusion from the given premises.

5. Premise: If it is raining, then there must be a cloud in the sky.
   Premise: It is raining.   **There is a cloud in the sky.**

6. Premise: A quadrilateral with four congruent sides and four right angles is a square.
   Premise: Quadrilateral ABCD has four right angles.
   Premise: Quadrilateral ABCD has four congruent sides.   **Quadrilateral ABCD is a square.**

7. Premise: Darnell is 3 years younger than half his father's age.
   Premise: Darnell's father is 40 years old.   **Darnell is 17 years old.**

Iteration

An **iteration** is a step in the process of repeating something over and over again. You can show the steps of the process in an **iteration diagram**.

EXAMPLE

A Use the iteration diagram below, and complete the process three times.

Start with 4. → Add 8.

4 → 12 → 20 → 28
Start   Stage 1 Stage 2 Stage 3

B For the pattern below, state the iteration and give the next three numbers in the pattern.
1, 5, 25, 125, . . .
To get from one stage to the next, the iteration is to multiply by 5.
   $125 \cdot 5 = 625$ $625 \cdot 5 = 3125$ $3125 \cdot 5 = 15{,}625$
The next three numbers in the pattern are 625, 3125, and 15,625.

PRACTICE

Use the diagram at right. Write the results of the first three iterations.

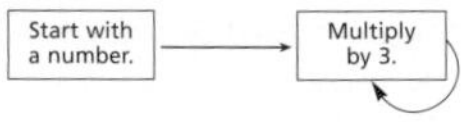

1. Start with 1.   2. Start with 8.
   3, 9, 27   **24, 72, 216**
3. Start with 2.   4. Start with 25.
   6, 18, 54   **75, 225, 675**
5. Start with −3.   6. Start with −7.
   −9, −27, −81   **−21, −63, −189**

Start with a number. → Multiply by 3.

For each pattern, state the iteration and give the next three numbers in the pattern.

7. 11, 17, 23, 29, . . .   8. 5, 10, 20, 40, . . .   9. 345, 323, 301, 279, . . .
   add 6; 35, 41, 47   **multiply by 2; 80, 160, 320**   **subtract 22; 257, 235, 213**
10. 30, 75, 120, 165, . . .   11. 15, 7, −1, −9, . . .   12. 1, $1\frac{1}{3}$, $2\frac{2}{3}$, 3, . . .
    add 45; 210, 255, 300   **subtract 8; −17, −25, −33**   **add $\frac{2}{3}$; $3\frac{2}{3}$, $4\frac{1}{3}$, 5**

A **fractal** is a geometric pattern that is *self similar*, so each stage of the pattern is similar to a portion of another stage of the pattern. For example, the Koch snowflake is a fractal formed by beginning with a triangle and then adding an equilateral triangle to each segment of the triangle.

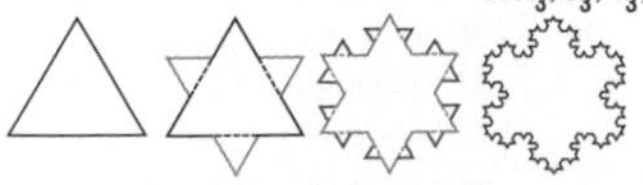

Draw the next two stages of each fractal.   For 13–14, see p. A24.

13.   14.

Stage 0   Stage 1   Stage 0   Stage 1

Skills Bank  Preview Skills

Relative, Cumulative, and Relative Cumulative Frequency

A **frequency table** lists each value or range of values of the data set followed by its **frequency**, or number of times it occurs.

Relative frequency is the frequency of a value or range of values divided by the total number of data values.

Cumulative frequency is the frequency of all data values that are less than a given value.

Relative cumulative frequency is the cumulative frequency divided by the total number of values.

Test Score	Frequency
66–70	3
71–75	1
76–80	4
81–85	7
86–90	5
91–95	6
96–100	2

EXAMPLE

The frequency table above shows a range of test scores and the frequency, or the number of students who scored in that range.

A Find the relative frequency of test scores in the range 76–80.

$3 + 1 + 4 + 7 + 5 + 6 + 2 = 28$ *Find the total number of test scores.*

There are 4 test scores in the range 76–80. The relative frequency is $\frac{4}{28} \approx 0.14$.

B Find the cumulative frequency of test scores less than 86.

$7 + 4 + 1 + 3 = 15$ *Add the frequencies of all test scores less than 86.*

The cumulative frequency of test scores less than 86 is 15.

C Find the relative cumulative frequency of test scores less than 86.

$\frac{15}{28} \approx 0.54$ *Divide the cumulative frequency by the total number of values.*

The relative cumulative frequency of test scores less than 86 is 0.54.

PRACTICE

The frequency table shows the frequency of each range of heights among Mrs. Dawkin's students.

Height	Frequency
4 ft–4 ft 5 in.	2
4 ft 6 in–4 ft 11 in.	8
5 ft–5 ft 5 in.	10
5 ft 6 in–5 ft 11 in.	6
6 ft–6 ft 5 in.	1

1. What is the relative frequency of heights in the range 5 ft–5 ft 5 in.? **0.37**

2. What is the relative frequency of heights in the range 4 ft–4 ft 5 in.? **0.07**

3. What is the cumulative frequency of heights less than 6 ft? **26**

4. What is the cumulative frequency of heights less than 5 ft? **10**

5. What is the relative cumulative frequency of heights less than 5 ft 6 in.? **0.74**

6. What is the relative cumulative frequency of heights less than 5 ft? **0.37**

Frequency Polygons

A **histogram** is a common way to represent frequency tables. A histogram is a bar graph with no space between the bars. Each bar can represent a range of values of a data set.

A **frequency polygon** is made by connecting the midpoints of the tops of all of the bars of a histogram.

EXAMPLE

A The frequency table shows the frequency of the number of push-ups done by the students in a gym class. Draw a histogram and frequency polygon of the data.
Label the horizontal axis with the number of push-ups.
Label the vertical axis with the frequency.

Push-ups Done in 1 Minute	
Number of Push-ups	Frequency
0–9	3
10–19	6
20–29	11
30–39	10
40–49	4
50–59	2

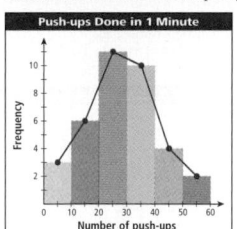

The frequency polygon is made up of the red points and red segments connecting the points.

PRACTICE

For 1–2, see p. A24.

Use each frequency table to draw a histogram and frequency polygon of the data.

1.

Books Read over the Summer	
Number of Books	Frequency
0–2	5
3–5	8
6–8	12
9–11	6
12–14	4
15–17	2

2.

Miles Driven One Way to Work	
Number of Miles	Frequency
0–4	6
5–9	5
10–14	13
15–19	9
20–24	4
25–29	1

Exponential Growth and Quadratic Behavior

An **exponential growth function** is in the form $y = C(1 + r)^t$, where C is the starting amount, r is the percent increase, and t is the time.

EXAMPLE

A Patrick invested $2000 for 5 years at a 3% annual interest rate. Write an exponential growth function to represent this situation.

C = starting amount = $2000

r = percent increase = 3% = 0.03

t = time = 5 years

$y = 2000(1 + 0.03)^5$

$y = 2000(1.03)^5$

A function of the form $y = ax^2 + bx + c$ is called a **quadratic function**. The graph of a quadratic function is called a **parabola**. The most basic quadratic function is $y = x^2$. The graph of $y = x^2$ is shown at right. By examining the value of a in $y = ax^2$, you can determine the effect it will have on the graph of $y = x^2$.

- If a is positive, the graph opens upward.
- If a is negative, the graph opens downward.
- If $|a| < 1$, the graph is wider than the graph of $y = x^2$.
- If $|a| > 1$, the graph is narrower than the graph of $y = x^2$.

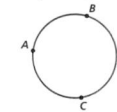

EXAMPLE

B Compare the graph of $y = -2x^2$ with the graph of $y = x^2$.

Since a is negative, the graph will open downward. Since $|a| = 2 (2 > 1)$, the graph will be narrower than the graph of $y = x^2$.

PRACTICE

Write an exponential growth function to represent each situation.

1. The population of a small town in 1997 was 25,500. Over a 5-year period, the population of the town increased at a rate of 2% each year. **$y = 25,500(1.02)^5$**

2. Shante invested $1800 at a 4.5% annual interest rate for 10 years. **$y = 1,800(1.045)^{10}$**

3. Tyler took a job that paid $30,000 annually with a 4% salary increase each year. He stayed at that job for 8 years. **$y = 30,000(1.04)^8$**

For 4–11, see p. A24.

Compare the graph of each quadratic function with the graph of $y = x^2$.

4. $y = -x^2$ 5. $y = \frac{1}{2}x^2$ 6. $y = 3x^2$ 7. $y = -\frac{1}{4}x^2$

8. $y = -5x^2$ 9. $y = 0.2x^2$ 10. $y = -\frac{3}{2}x^2$ 11. $6x^2 = y$

Circles

A circle can be named by its center, using the $\odot$ symbol. A circle with a center labeled C would be named $\odot C$. An unbroken part of a circle is called an **arc**. There are major arcs and minor arcs.

A **minor arc** of a circle is an arc that is shorter than half the circle and named by its endpoints. A **major arc** of a circle is an arc that is longer than half the circle and named by its endpoints and one other point on the arc.

$\overset{\frown}{AB}$ is a minor arc.

$\overset{\frown}{BAC}$ is a major arc.

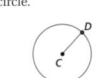

A **radius** connects the center with a point on a circle.

radius $\overline{CD}$

A **chord** connects two points point on a circle. A **diameter** is a chord that passes through the center of a circle.

chord $\overline{AB}$

A **secant** is a line that intersects a circle at two points.

secant $\overleftrightarrow{EF}$

A **tangent** is a line that intersects a circle at one point.

tangent $\overleftrightarrow{GH}$

A **central angle** has its vertex at the center of the circle.

central angle $\angle JKL$

An **inscribed angle** has its vertex on the circle.

inscribed angle $\angle MNP$

PRACTICE

Use the given diagram of $\odot A$ for exercises 1–6.

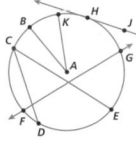

1. Name a radius. **$\overline{AB}$ or $\overline{AK}$**
2. What two chords make up the inscribed angle?
3. Name a secant. **$\overleftrightarrow{FG}$**   **$\overline{CD}$ and $\overline{CE}$**
4. Give the tangent line. **$\overleftrightarrow{HJ}$**
5. Name the central angle. **$\angle BAK$**
6. Name the inscribed angle. **$\angle DCE$**

Matrices

A **matrix** is a rectangular arrangement of data enclosed in brackets. Matrices are used to list, organize, and sort data.

The **dimensions** of a matrix are given by the number of horizontal **rows** and vertical columns in the matrix. For example, Matrix A below is an example of a 3×2 ("3-by-2") matrix because it has 3 rows and 2 columns, for a total of 6 **elements**. The number of rows is always given first. So a 3×2 matrix is not the same as a 2×3 matrix.

$$A = \begin{bmatrix} 86 & 137 \\ 103 & 0 \\ 115 & 78 \end{bmatrix} \begin{matrix} \leftarrow \text{Row 1} \\ \leftarrow \text{Row 2} \\ \leftarrow \text{Row 3} \end{matrix}$$

Column 1 Column 2

Each matrix element is identified by its row and column. The element in row 2 column 1 is 103. You can use the notation $a_{21} = 103$ to express this.

EXAMPLE

Use the data shown in the bar graph to create a matrix.

The matrix can be organized with the votes in each year

as the columns: $\begin{bmatrix} 12 & 5 \\ 6 & 11 \\ 2 & 4 \end{bmatrix}$

Number of Votes

or with the votes in each year as the rows:

$\begin{bmatrix} 12 & 6 & 2 \\ 5 & 11 & 4 \end{bmatrix}$

First place Second place Third place
■ 2003 ■ 2004

For 4–6, see p. A24.

PRACTICE

Use matrix B for Exercises 1–3. $\qquad B = \begin{bmatrix} 1 & 0 & 7 & 4 \\ 0 & 1 & 3 & 8 \\ 6 & 5 & 2 & 9 \end{bmatrix}$

1. B is a ▦ × ▦ matrix.   **3 4**
2. Name the element with a value of 5. b_{32}
3. What is the value of b_{13}? **7**

4. A football team scored 24, 13, and 35 points in three playoff games. Use this data to write a 3×1 matrix.

5. The greatest length and average weight of some whale species are as follows: finback whale—50 ft, 82 tons; humpback whale—33 ft, 49 tons; bowhead whale—50 ft, 59 tons; blue whale—84 ft, 98 tons; right whale—50 ft, 56 tons. Organize this data in a matrix.

6. The second matrix in the example is called the *transpose* of the first matrix. Write the transpose of matrix B above. What are its dimensions?

Skills Bank **779**

Skills Bank Science Skills

Conversion of Units in 1, 2, and 3 Dimensions

When converting between the metric and customary system, use **conversion factors**.

Common Metric to Customary Conversions		
Length	Area	Volume
1 cm ≈ 0.394 in.	1 cm² ≈ 0.155 in²	1 cm³ ≈ 0.061 in³
1 m ≈ 3.281 ft	1 m² ≈ 10.764 ft²	1 m³ ≈ 35.315 ft³
1 m ≈ 1.094 yd	1 m² ≈ 1.196 yd²	1 m³ ≈ 1.308 yd³
1 km ≈ 0.621 mi	1 km² ≈ 0.386 mi²	1 km³ ≈ 0.239 mi³

Common Customary to Metric Conversions		
Length	Area	Volume
1 in. ≈ 2.54 cm	1 in² ≈ 6.452 cm²	1 in³ ≈ 16.387 cm³
1 ft ≈ 0.305 m	1 ft² ≈ 0.093 m²	1 ft³ ≈ 0.028 m³
1 yd ≈ 0.914 m	1 yd² ≈ 0.836 m²	1 yd³ ≈ 0.765 m³
1 mi ≈ 1.609 km	1 mi² ≈ 2.590 km²	1 mi³ ≈ 4.168 km³

EXAMPLES

A 8 cm ≈ ▦ in.

1 cm ≈ 0.394 in.
8 cm ≈ 8(0.394) in.
8 cm ≈ 3.152 in.

B 45 mi² ≈ ▦ km²

1 mi² ≈ 2.590 km²
45 mi² ≈ 45(2.590) km²
45 mi² ≈ 116.550 km²

PRACTICE

Complete each conversion.

1. 2 in. ≈ ▦ cm **5.08**
2. 3 km³ ≈ ▦ mi³ **0.717**
3. 4.2 m² ≈ ▦ ft² **45.2088**
4. 5 ft² ≈ ▦ m² **0.465**
5. 10 mi ≈ ▦ km **16.09**
6. 1.1 m³ ≈ ▦ yd³ **1.4388**
7. 4 yd ≈ ▦ m **3.656**
8. 15 in² ≈ ▦ cm² **96.78**
9. 12 yd ≈ ▦ m **10.968**
10. 1 cm³ ≈ ▦ in³ **0.061**
11. 9 m³ ≈ ▦ ft³ **317.835**
12. 2 mi ≈ ▦ km **3.218**

13. Approximately how many meters are in a mile? **1609 m**

Temperature Conversion

In the United States, the Fahrenheit (°F) temperature scale is the common scale used. For example, weather reports and body temperatures are given in degrees Fahrenheit. The metric temperature scale is Celsius (°C) and is commonly used in science applications. Temperatures given in one scale can be converted to the other system using one of the formulas below.

Formulas

Fahrenheit to Celsius (°F to °C) $\frac{5}{9}(F - 32) = C$

Celsius to Fahrenheit (°C to °F) $\frac{9}{5}C + 32 = F$

EXAMPLES

A Convert 77°F to degrees Celsius.

$\frac{5}{9}(F - 32) = C$
$\frac{5}{9}(77 - 32) = C$
$\frac{5}{9}(45) = C$
$25 = C$

B Convert 103°C to degrees Fahrenheit.

$\frac{9}{5}C + 32 = F$
$\frac{9}{5}(103) + 32 = F$
$185.4 + 32 = F$
$217.4 = F$

PRACTICE

Convert each temperature to degrees Celsius. Give the temperature to the nearest tenth of a degree.

1. 7°F **−13.9°C**
2. 0°F **−17.8°C**
3. 12°F **−11.1°C**
4. 40°F **4.4°C**
5. 100°F **37.8°C**
6. 32°F **0°C**
7. 25°F **−3.9°C**
8. 212°F **100°C**
9. −50°F **−45.6°C**
10. −8°F **−22.2°C**

Convert each temperature to degrees Fahrenheit. Give the temperature to the nearest tenth of a degree.

11. 0°C **32°F**
12. 10°C **50°F**
13. 22°C **71.6°F**
14. 55°C **131°F**
15. 212°C **413.6°F**
16. 1°C **33.8°F**
17. 100°C **212°F**
18. 80°C **176°F**
19. 95°C **203°F**
20. 32°C **89.6°F**
21. 31°C **87.8°F**
22. 42°C **107.6°F**
23. −6°C **21.2°F**
24. −40°C **−40°F**

Skills Bank **781**

Customary and Metric Rulers

A metric ruler is divided into centimeter units, and each centimeter is divided into 10 millimeter units. A metric ruler that is 1 meter long is a *meter stick*.

1 m = 100 cm
1 cm = 10 mm

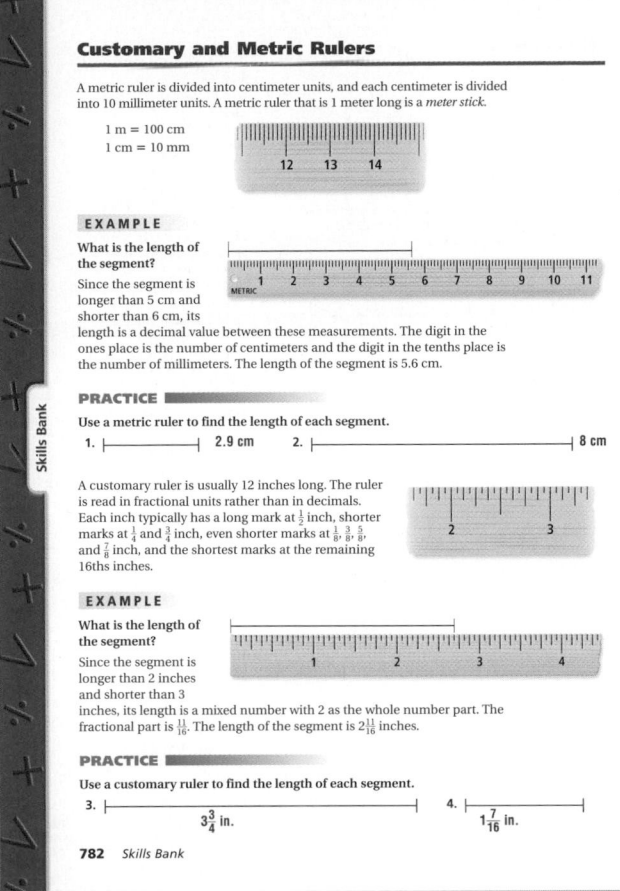

EXAMPLE

What is the length of the segment?

Since the segment is longer than 5 cm and shorter than 6 cm, its length is a decimal value between these measurements. The digit in the ones place is the number of centimeters and the digit in the tenths place is the number of millimeters. The length of the segment is 5.6 cm.

PRACTICE

Use a metric ruler to find the length of each segment.

1. |———————| **2.9 cm**   2. |————————————————| **8 cm**

A customary ruler is usually 12 inches long. The ruler is read in fractional units rather than in decimals. Each inch typically has a long mark at $\frac{1}{2}$ inch, shorter marks at $\frac{1}{4}$ and $\frac{3}{4}$ inch, even shorter marks at $\frac{1}{8}, \frac{3}{8}, \frac{5}{8},$ and $\frac{7}{8}$ inch, and the shortest marks at the remaining 16ths inches.

EXAMPLE

What is the length of the segment?

Since the segment is longer than 2 inches and shorter than 3 inches, its length is a mixed number with 2 as the whole number part. The fractional part is $\frac{11}{16}$. The length of the segment is $2\frac{11}{16}$ inches.

PRACTICE

Use a customary ruler to find the length of each segment.

3. |———————————| $3\frac{3}{4}$ in.   4. |————| $1\frac{7}{16}$ in.

Precision and Significant Digits

In a measurement, all digits that are known with certainty are called **significant digits** . The more precise a measurement is, the more significant digits there are in the measurement. The table shows some rules for identifying significant digits.

Rule	Example	Number of Significant Digits
All nonzero digits	15.32	All 4
Zeros beween significant digits	43,001	All 5
Zeros after the last nonzero digit that are to the right of the decimal point	0.0070	2; 0.0070

Zeros at the end of a whole number are assumed to be nonsignificant. (Example: 500)

EXAMPLE

A **Which is a more precise measurement, 14 ft or 14.2 ft?**

Because 14.2 ft has three significant digits and 14 has only two, 14.2 ft is more precise. In the measurement 14.2 ft, each 0.1 ft is measured.

B **Determine the number of significant digits in 20.04 m, 200 m, and 200.0 m.**

20.04 All 4 digits are significant.
200 There is 1 significant digit.
200.0 All 4 digits are significant.

When calculating with measurements, the answer can only be as precise as the least precise measurement.

C **Multiply 16.3 m by 2.5 m. Use the correct number of significant digits in your answer.**

When muliplying or dividing, use the least number of significant digits of the numbers.

16.3 m · 2.5 m = 40.75
Round to 2 significant digits. ⟶ **41 m²**

D **Add 4500 in. and 70 in. Use the correct number of significant digits in your answer.**

When adding or subtracting, line up the numbers. Round the answer to the last significant digit that is farthest to the left.

4500 in. *5 is farthest left. Round to*
+ 70 in. *hundreds.*
4570 Round to the hundreds. ⟶ 4600 in.

PRACTICE

Tell which is more precise.

1. 31.8 g or 32 g **31.8 g**   2. 496.5 mi or 496.50 mi **496.50 mi**   3. 3.0 ft or 3.001 ft **3.001 ft**

Determine the number of significant digits in each measurement.

4. 12 lb **2**   5. 14.00 mm **4**   6. 1.009 yd **4**   7. 20.87 s   **4**

Perform the indicated operation. Use the correct number of significant digits in your answer.

8. 210 m + 43 m **250 m**   9. 4.7 ft · 1.04 ft **4.9 ft²**   10. 6.7 s − 0.08 s **6.6 s**

Greatest Possible Error

The smaller the units used to measure something, the greater the precision of the measurement. The **greatest possible error** of a measurement is half the smallest unit. This is written as ± 0.5 unit, which is read as "plus or minus 0.5 unit."

EXAMPLES

A **Which is a more precise measurement, 292 cm or 3 m?**

The more precise measurement is 292 cm because its unit of measurement, 1 cm, is smaller than 1 m.

B **Find the greatest possible error for a measurement of 2.4 cm.**

The smallest unit is 0.1 cm.
0.5 × 0.1 = 0.05
The greatest possible error is ± 0.05 cm.

|←|→|←|→|←|→|←|→|
2.3 cm   2.35 cm   2.4 cm   2.45 cm   2.5 cm

PRACTICE

Tell which is a more precise measurement.

1. 40 cm or 412 mm **412 mm** 2. 3.2 ft or 1 yd **3.2 ft**   3. 7 ft or 87 in. **87 in.**
4. 3116 m or 3 km **3,116 m** 5. 1 mi or 5281 ft **5,281 ft**   6. 0.04 m or 4.2 cm **4.2 cm**

Find the greatest possible error of each measurement.

7. 5 ft **±0.5 ft**   8. 22 mm   **±0.5 mm**   9. 12.5 mi   **±0.05 mi**
10. 60 km **±0.5 km**   11. 2.06 cm   **±0.005 cm**   12. 0.08 g   **±0.005 g**

pH (Logarithmic Scale)

pH is a measure of the concentration of hydrogen ions in a solution. pH ranges from 0 to 14. An *acid* has a pH below 7 and a *base* has a pH above 7. A pH of 7 is *neutral* and a hydrogen ion concentration of 1×10^{-7} mol/L. The exponent is the opposite of the pH.

0 Strong acids   Weak acids 7 Weak bases   Strong bases 14

EXAMPLES

A **Write the pH of the solution, given the hydrogen ion concentration.**

coffee: 1×10^{-5} mol/L
The coffee is acidic, with a pH of 5.

B **Write the hydrogen ion concentration of the solution in mol/L.**

antacid solution: pH = 10.0
1×10^{-10} mol/L in the antacid solution

PRACTICE

Write the pH of each solution, given the hydrogen ion concentration.

1. seawater: 1×10^{-8} mol/L **8**   2. lye: 1×10^{-13} mol/L **13**   3. borax: 1×10^{-9} mol/L **9**

Write the hydrogen ion concentration in mol/L.

4. drain cleaner: pH = 14.0   5. lemon juice: pH = 2.0   6. milk: pH = 7.0
1×10^{-14}   **1×10^{-2}**   **1×10^{-7}**

Richter Scale

An earthquake is classified according to its magnitude. The Richter scale is a mathematical system that compares the sizes and magnitudes of earthquakes.

The magnitude is related to the height, or *amplitude*, of seismic waves as recorded by a seismograph during an earthquake. The higher the number is on the Richter scale, the greater the amplitude of the earthquake's waves.

Earthquakes per Year	Magnitude on the Richter Scale	Severity
1	8.0 and higher	Great
18	7.0–7.9	Major
120	6.0–6.9	Strong
800	5.0–5.9	Moderate
6200	4.0–4.9	Light
49,000	3.0–3.9	Minor
≈ 3,300,000	below 3.0	Very minor

The Richter scale is a *logarithmic scale*, which means that the numbers in the scale measure factors of 10. An earthquake that measures 6.0 on the Richter scale is 10 times as great as one that measures 5.0.

The largest earthquake ever measured registered 8.9 on the Richter scale.

EXAMPLE

How many times greater is an earthquake that measures 5.0 on the Richter scale than one that measures 3.0?

You can divide powers of 10, with the magnitudes as the exponents.

$$\frac{10^5}{10^3} = 10^2$$

A 5.0 quake is 100 times greater than a 3.0 quake.

PRACTICE

Describe the severity of an earthquake with each given Richter scale reading.

1. 7.6 **major**   2. 4.2 **light**   3. 5.0 **moderate**
4. 2.0 **very minor**   5. 3.6 **minor**   6. 8.4 **great**

Each pair of numbers repesents two earthquake magnitudes on the Richter scale. How many times greater is the first earthquake in each pair? (Use a calculator for 10–12.)

7. 6.0 and 4.0 **100**   8. 8.0 and 5.0 **1000**   9. 7.0 and 3.0 **10,000**
10. 7.5 and 5.5 **100**   11. 5.7 and 5.3 **2.5**   12. 8.6 and 7.1 **31.6**

Selected Answers

Selected Answers

Chapter 1

1-1 Exercises
1. 17 2. 23 3. 3 4. 44 5. 1.8
6. 5 tbsp 7. 8 tbsp 8. 11.5 tbsp
9. 17 tbsp 11. 33 13. 67
15. 4 gal 17. 2 gal 19. 0 21. 22
23. 9 25. 6 27. 10 29. 16 31. 11
33. 20 35. 34 37. 12.6 39. 18
41. 105 43. 17 45. 30.5 47. 24
49. 0 51. Possible range: 204 to 208 beats per minute
53. b. 165,600 frames
57. 15, 21, 71 59. 49, 81 61. C

1-2 Exercises
1. $6 + t$ 2. $y - 25$ 3. $7(m + 6)$
4. $7m + 6$ 5. a. $8n$ b. $8(23) = \$184$
6. $\$15 + d$; $\$17.50$ 7. $k + 34$
9. $5 + 5z$ 11. a. $42 \div p$
b. 7 students 13. $\$1.75n$; $\$14.00$
15. $6(4 + y)$ 17. $\frac{1}{2}(m + 5)$
19. $13y - 6$ 21. $2\left(\frac{m}{45}\right)$
25. $2(r - 1)$; $2(2.50 - 1) = \$3$
27.

$24 + 4(2 - 2)$	24
$24 + 4(3 - 2)$	28
$24 + 4(4 - 2)$	32
$24 + 4(5 - 2)$	36
$24 + 4(6 - 2)$	40

31. 202 33. 400 35. 200.2 37. 40
39. C

1-3 Exercises
1. 5 2. 21 3. $m = 32$ 4. $t = 5$
5. $w = 17$ 6. 15,635 feet 7. 22
9. $w = 1$ 11. $t = 12$ 13. 20
15. 30 17. 7 19. 0 21. $t = 5$
23. $m = 24$ 25. $h = 3$
27. $t = 2621$ 29. $x = 110$
31. $n = 45$ 33. $t = 0.5$
35. $w = 1.9$ 37. a. $497 + m = 1696$;
1199 miles b. $1278 + m = 1696$;
418 miles 39. a. $0.24 + c = 4.23$;
$\$3.99$ b. $c - 3.82 = 0.53$; $\$4.35$
43. 22 45. 26

1-4 Exercises
1. $x = 7$ 2. $t = 7$ 3. $y = 14$
4. $w = 13$ 5. $l = 60$ 6. $k = 72$
7. $h = 57$ 8. $m = 6$ 9. $8n = 32$; $n = 4$ servings 10. $\frac{1}{4}c = \$60$ or $\frac{c}{4} = \$60$; $c = \$240$ 11. $x = 7$
12. $k = 40$ 13. $y = 3$ 14. $m = 36$
15. $d = 19$ 17. $g = 10$
19. $n = 567$ 21. $a = 612$
23. $10n = 80$; $n = 8$ mg 25. $x = 2$
27. $y = 2$ 29. $x = 7$ 31. $y = 2$
33. $k = 56$ 35. $b = 72$ 37. $x = 17$
39. $y = 3$ 41. $b = 48$ 43. $n = 35$
45. $16m = 42{,}000$; $m = 2625$ miles
47. $\frac{1}{6}m = 22$ or $\frac{m}{6} = 22$; $m = 132$ miles 49. $x = 8$ 51. $w = 2$ 53. A

1-5 Exercises
1. $<$ 2. $>$ 3. $>$ 4. $>$ 5. $>$ 6. $<$
7. $>$ 8. $>$ 9. $x < 1$ 10. $b \geq 5$
11. $m \leq 32$ 12. $15 > x$ 13. $y \geq 17$
14. $f < 5$ 15. $z > 21$ 16. $14 \leq x$
17. $m > 40$; more than 40 members 19. $<$ 21. $>$ 23. $<$
25. $<$ 27. $x \geq 7$ 29. $4 < t$
31. $x \geq 4$ 33. $6 < a$ 35. $x < 6$
37. $x > 4$ 39. $x < 1$ 41. $x \geq 5$
43. $50(50) > 2200$; $2500 > 2200$;
no 45. $x \geq 53$ 51. 22; 19; 16; 13
53. 13; 21; 29; 37 55. 15; 13; 11; 9
57. H

1-6 Exercises
1. $4x$ 2. $5z + 5$ 3. $8f + 8$ 4. $17g$
5. $4p - 8$ 6. $4x + 12$ 7. $3x$ I $5y$
8. $9x + y$ 9. $5x + y$ 10. $9p + 3z$
11. $7g + 5h - 12$ 12. $10h$
13. $r + 6$ 14. $10 + 8x$ 15. $2t + 56$
16. $n = 42$ 17. $y = 24$ 18. $p = 17$
19. $13y$ 21. $7a + 11$ 23. $3x + 2$
25. $5p$ 27. $9x + 3$ 29. $5a + z$
31. $7x + 5q + 3$ 33. $9a + 7c + 3$
35. $20y - 18$ 37. $6y + 17$
39. $11x - 9$ 41. $p = 5$ 43. $y = 8$
45. $x = 14$ 47. $8d + 1$ 49. $x = 2$
51. $52g + 41s + 49b$ 57. $x = 13$

1-7 Exercises
1. no 2. yes 3. yes 4. no
5.

x	y	(x, y)
2	4	(2, 4)
3	6	(3, 6)
4	8	(4, 8)
5	10	(5, 10)
6	12	(6, 12)

6.

x	y	(x, y)
1	1	(1, 1)
2	4	(2, 4)
3	9	(3, 9)
4	10	(4, 10)
5	15	(5, 15)
6	16	(6, 16)

7. $\$1.29$ 9. no 11. no
13.

x	y	(x, y)
1	10	(1, 10)
2	12	(2, 12)
3	14	(3, 14)
4	16	(4, 16)
5	18	(5, 18)
6	20	(6, 20)

15.

x	y	(x, y)
2	2	(2, 2)
4	8	(4, 8)
6	14	(6, 14)
8	20	(8, 20)
10	26	(10, 26)

17. yes 19. yes 21. no 23. yes
25.

x	y	(x, y)
1	1	(1, 1)
2	2.5	(2, 2.5)
3	3.9	(3, 3.9)
4	4.15	(4, 4.15)
5	5	(5, 5)
6	20	(6, 20)

27.

x	y	(x, y)
1	9	(1, 9)
2	10	(2, 10)
3	11	(3, 11)
4	13	(4, 13)
5	13	(5, 13)
6	14	(6, 14)

29.

x	y	(x, y)
2	8	(2, 8)
4	16	(4, 16)
6	16	(6, 16)
8	20	(8, 20)
10	24	(10, 24)

31. Possible answer: $x = y$ 33. no; (13, 52) or (12.75, 51)
35. a. (1980, 74) b. (2020, 81)
39. 7 41. 4 43. 12 45. B

59. $x = 8$ 61. $x = 32$ 63. $x = 16$
65. B

786 Selected Answers

1-8 Exercises
1. $(-2, 3)$ 2. $(3, 5)$ 3. $(2, -3)$
4. $(5, -1)$ 5. $(5, 6)$ 6. $(-3, -4)$
7–10.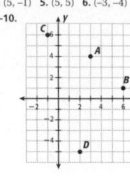

11.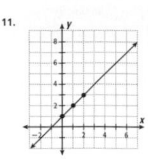

12.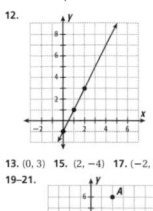

13. $(0, 3)$ 15. $(2, -4)$ 17. $(-2, 5)$
19–21.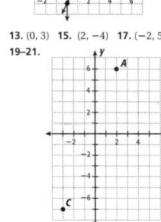

23.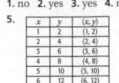

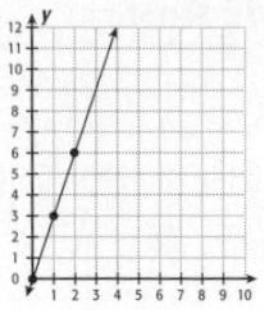

25–31. Possible answers given.
25. $(1, 0)$, $(2, 0)$ 27. $(2, 7)$, $(4, 7)$
29. $(4, 3)$, $(4, 5)$ 31. $(0, 4)$, $(0, 5)$
33. 75 beats
35.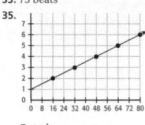

7 studs
39. $x - 13$ 41. $x + 31$ 43. C

1-9 Exercises
1. table 2 2. table 2; table 1; table 3; none
3.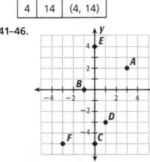

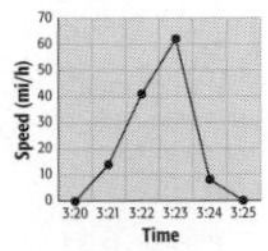

5. table 1; table 3; table 2
7.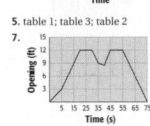

9. a. Old Faithful b. Riverside
11. $x = 9$ 13. $x = 11$ 15. D

Chapter 1 Study Guide and Review
1. ordered pair; x-coordinate;
y-coordinate 2. solution set; inequality 3. 147 4. 152 5. 278
6. $2(k + 4)$ 7. $4r + 5$ 8. $z = 23$
9. $t = 8$ 10. $k = 15$ 11. $x = 11$
12. 1300 lb 13. 3300 mi^2
14. $g = 8$ 15. $k = 9$ 16. $p = 80$
17. $w = 48$ 18. $y = 40$
19. $z = 19.2$ 20. 352.5 mi
21. 24 months 22. $h < 4$
23. $y > 7$ 24. $x \geq 4$ 25. $p < \frac{1}{2}$
26. $m > 2.3$ 27. $q \leq 0$ 28. $w \geq 8$
29. $x \leq 3$ 30. $y > 16$ 31. $x > 3$
32. $y > 6$ 33. $x \leq 2$ 34. $11m - 4$
35. $14w + 6$ 36. $y = 5$ 37. $z = 8$
38. yes 39. no
40.

x	y	(x, y)
0	2	(0, 2)
1	5	(1, 5)
2	8	(2, 8)
3	11	(3, 11)
4	14	(4, 14)

41–46.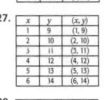

47. 5 48. 8 49. 20 50. Oven E

Chapter 2

2-1 Exercises
1. 5 2. 3 3. 4 4. -6 5. -8 6. 6
7. 3 8. -16 9. 11 10. 4 11. -8
12. $\$297$ 13. -2 15. -3 17. 21
19. -18 21. 22 23. 9
25. $-6 + (-2) = -8$ 27. -13
29. -18 31. -3 33. 43 35. 0
37. -19 39. 8 41. -20 43. -15
45. 5 51. $f = 6$ 53. $q = 6$

Selected Answers **787**

2-2 Exercises
1. -15 2. -3 3. 14 4. -7 5. 13
6. -6 7. -15 8. 49°F 9. -11
11. 17 13. 15 14. 17. 16
19. -17 21. -14 23. 40 m below
sea level, or -40 m 25. $5 - 8 = -3$
27. 51 29. -62 31. -16
33. 13 35. 2 37. -42
39. Great Pyramid to Cleopatra; about 500 years
41. Cleopatra takes throne and Napoleon invades Egypt.
45. no like terms 47. C

2-3 Exercises
1. -27 2. -8 3. 30 4. -4 5. 49
6. -77 7. -24 8. -72
9.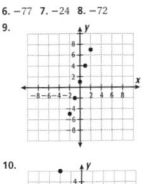

10.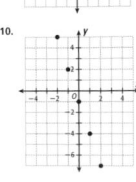

11.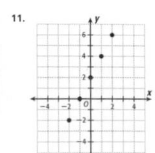

13. -11 15. -7 17. 130 19. -2

21.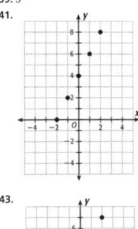

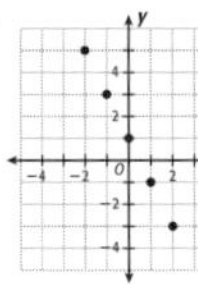

23. -45 25. 36 27. 24 29. -72
31. -80 33. 63 35. -19 37. 14
39. 3
41.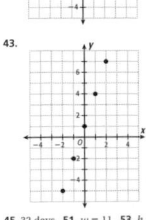

43.

45. 32 days 51. $w = 11$ 53. $h = 0$
55. G

2-4 Exercises
1. $y = 6$ 2. $d = 12$ 3. $x = -11$
4. $b = -7$ 5. $t = -16$ 6. $g = -4$
7. $a = 12$ 8. -9 9. 9 427°C
11. $a = 13$ 13. $b = -3$
15. $y = -37$ 17. $h = -31$
19. $n = -39$ 21. $c = 84$
23. $a = 45$ 25. $r = -64$
27. $s = -11$ 29. $x = 4$
31. $m = -27$ 33. $z = 16$
35. $h = -4$ 37. $y = -105$

39. $x = 24$ 41. $p = -6$
43. a. $-4t = d$, t is time in minutes and d is depth. b. -68 m
c. $-4t = -24$; $t = 6$ minutes
49. $w = 2$ 51. C

2-5 Exercises
1. $x \geq -5$ 2. $y < 2$ 3. $b \leq -7$
4. $h < 1$ 5. $f > 4$ 6. $k \leq 5$
7. $x < -3$ 8. $y < -2$ 9. $w \leq 3$
10. $x \geq -3$ 11. $z > -8$ 12. $n \leq 6$
13. $k > -3$ 15. $x < -1$ 17. $r \geq 2$
19. $n > 5$ 21. $x \geq -4$ 23. $y > -5$
25. $x > -2$ 27. $k \geq 10$
29. $a \leq -12$ 31. $r \leq -1$ 33. $t = 2$
35. $b > 0$ 37. $f = -18$ 39. $c \leq 2$
41. $n < -6$ 43. $g = 8$ 45. $p = -9$
47. $3x + (-7x) > -12$; $x < 3$
49. $-1 + x < -7$; $x < -6$;
less than 6 under par 55. 9
57. -254 59. -16 61. 3 63. H

2-6 Exercises
1. 14^1 2. 15^2 3. b^4 4. $(-1)^3$ 5. 81
6. 25 7. -243 8. 2401 9. -33
10. 90 11. -117 12. -47 13. 78
15. $(-7)^3$ 17. c^5 19. 256
21. -512 23. 77 25. -360
27. $(-2)^3$ 29. 4^4 31. 343
33. -1728 35. 729 37. 4
39. -1 41. -166 43. -4
45. -1 47. 216 49. 257
51. $2^{18} = 262{,}144$ bacteria 59. 9
61. 104 63. C

2-7 Exercises
1. 3^{11} 2. 12^5 3. m^6 4. cannot combine 5. 8^2 6. a^8 7. $12^0 = 1$
8. 7^{12} 9. 10^2 plants 11. 2^6
13. 16^4 15. cannot combine
17. $10^0 = 1$ 19. 6^3 21. a^{10}
25. 6^5 27. cannot combine
29. $y^0 = 1$ 31. x^8 33. 4^6
35. 10^{14} 37. h^{16} 39. 4^2 41. 6^9
43. 26^2, or 676 more ways
45. 12^2; 12^1 47. 22^3 trips 51. 3
53. -12 55. -16 57. -12 59. D

788 Selected Answers

2-8 Exercises
1. 0.0000001 2. 0.001 3. 0.000001
4. 0.1 5. $\frac{1}{16}$ 6. $\frac{1}{9}$ 7. $\frac{1}{8}$ 8. $-\frac{1}{32}$
9. 1000 10. 0.5 11. 216 12. -25
13. 0.01 15. 0.00001 17. $-\frac{1}{64}$
19. 0.0001 21. 10.000 23. 1
25. $\frac{1}{8}$ 27. 0.001 29. 128 31. m^7
33. $\frac{1}{8}$ 35. 1024 37. $\frac{3}{5}$ 39. $\frac{1}{4}$
41. $\frac{1}{144}$ 43. 4 45. 1 kilometer
47. a. $10^{-5} \cdot 10^3 = 10^{-2}$ g
b. $10^{-2} \cdot 10^7 = 10^5$ g
c. $10^3 + 10^1 = 10^{5-1} = 10^4$;
10^4 decagrams 51. 30 53. 85

2-9 Exercises
1. 3150 2. 0.000000125
3. 410.000 4. 0.00039
5. 5.7×10^{-5} 6. 3×10^{-4}
7. 4.89×10^6 8. 1.4×10^{-7}
9. (1.485×10^8)°C 11. 0.00067
13. 63,700,000 15. 7.8×10^6
17. 3×10^{-8} 19. 13,000 21. 56
23. 0.00000053 25. 8,580,000
27. 9,112,000 29. 0.00029
31. 4.67×10^{-3} 33. 5.6×10^7
35. 7.6×10^{-3} 37. 3.5×10^3
39. 9×10^2 41. 6×10^6
43. a. $\approx 2.21 \times 10^7$; $\approx 1.4 \times 10^4$ mi^2 b. 6.35×10^{-4} mi^2/person
45. 0.000078 51. -20 53. 21
55. $t = -9$ 57. $b = -27$

Chapter 2 Study Guide and Review
1. opposite 2. scientific notation; power 3. exponent; base 4. -2
5. -12 6. -3 7. 8 8. -24 9. 8
10. -8 11. -16 12. 17 13. 3
14. 15 15. -22 16. -14 17. 5
18. -5 19. -35 20. -18 21. 52
22. 25 23. -6 24. -3 25. 9
26. $t = 3$ 27. $k = 3$ 28. $g = -6$
29. $w = -80$ 30. $b = -20$
31. $a = -4$ 32. $h = -91$
33. $S = 38$ 34. $b < -2$ 35. $t < 5$
36. $m \geq 37$ 37. $p < -2$ 38. $<$
-5 39. $q \geq 3$ 40. $m \geq 4$ 41. $x >$
-3 42. $y < 4$ 43. $x > -3$ 44. b
45. a. 3.63 quadrillion Btu

≤ 0 45. $y < 6$ 46. 7^3 47. $(-3)^2$
48. K^4 49. 625 50. -32 51. -1
52. 4^7 53. 9^6 54. p^4 55. 8^3
56. m^9 57. $58. 5^3$ 59. x^7
60. k^0 61. $\frac{1}{125}$ 62. $-\frac{1}{64}$ 63. $\frac{1}{11}$
64. 1 65. 1 66. 1 67. $\frac{1}{8}$ 68. $-\frac{1}{27}$
69. 1620 70. 0.00162 71. 910,000
72. 0.000091 73. 8.0×10^{-9}
74. 7.3×10^7 75. 9.6×10^{-6}
76. 5.64×10^{10}

Chapter 3

3-1 Exercises
1. $\frac{4}{5}$ 2. $\frac{2}{3}$ 3. $-\frac{2}{5}$ 4. $\frac{11}{27}$ 5. $\frac{19}{23}$
6. $-\frac{5}{27}$ 7. $-\frac{2}{7}$ 8. $\frac{9}{16}$ 9. $\frac{3}{10}$ 10. $1\frac{1}{8}$
11. $\frac{431}{1000}$ 12. $\frac{4}{5}$ 13. $-2\frac{5}{16}$ 14. $\frac{1}{5}$
15. $3\frac{21}{100}$ 16. $-\frac{1939}{5000}$ 17. 0.875
18. 0.6 19. 0.416 20. 0.75
21. 4.0 22. 0.125 23. 2.4
24. 2.25 25. $\frac{3}{4}$ 22. $\frac{1}{4}$ 29. $\frac{13}{17}$
31. $\frac{16}{19}$ 33. $\frac{5}{6}$ 35. $\frac{71}{100}$ 37. $1\frac{377}{1000}$
39. $-1\frac{5}{9}$ 41. 0.375 43. 1.4
45. 0.68 47. 1.16 49. Possible answer: $\frac{25}{36}$ 51. a. $\frac{3}{4}$; $\frac{1}{2}$; $\frac{5}{20}$
$\frac{11}{25}$; $\frac{13}{24}$; $\frac{13}{15}$ b. 2×2; 2×3;
3×3; 2×3; 5; $2 \times 2 \times 2 \times 2$;
5×5; $2 \times 2 \times 2 \times 3$; 3×5
c. 0.75 terminating;
0.16 repeating; 0.5 repeating;
0.85 terminating; 0.40625
terminating; 0.44 terminating;
0.7196 repeating; 0.53 repeating
53. GCF = 4; $\frac{10}{18}$, No 59. 28; 48
61. 35; 14 63. H

3-2 Exercises
1. 9.693 seconds 2. 1.4 3. -2
4. -0.4 5. $-2\frac{1}{2}$ 6. -1.5 7. $-\frac{5}{9}$
12. $\frac{5}{12}$ 13. $\frac{5}{12}$ 14. $\frac{1}{12}$ 15. $\frac{3}{11}$ 16. $4\frac{5}{12}$
21. $\frac{1}{2}$ 23. -1.6 25. 1.6 27. 1.9
29. -2.7 31. $\frac{1}{12}$ 35. $\frac{2}{12}$ 35. $\frac{1}{2}$
37. $1\frac{7}{20}$ 39. 28.7 41. -16.34
43. a. $\frac{29}{32}$ in. b. $1\frac{7}{32}$ in. c. $\frac{9}{32}$ in.
45. a. 3.63 quadrillion Btu

b. 2.717 quadrillion Btu
49. $7x - 5y + 18$
51. $16x + 22y + 11$ 53. A

3-3 Exercises
1. $1\frac{1}{3}$ 2. $-14\frac{2}{3}$ 3. $1\frac{7}{12}$ 4. $-3\frac{4}{5}$
5. $3\frac{1}{3}$ 6. $-8\frac{7}{12}$ 7. $6\frac{1}{2}$ 8. $6\frac{3}{8}$
9. $\frac{4}{21}$ 10. $-\frac{11}{80}$ 11. $3\frac{5}{9}$ 12. $\frac{7}{4}$
13. $-\frac{25}{78}$ 14. $2\frac{3}{32}$ 15. $\frac{7}{12}$
16. $-\frac{25}{192}$ 17. 12.4 18. 0.144
19. 36.5 20. -0.42 21. 41.3
22. 3.65 23. 14.1 24. -0.416
25. $13\frac{1}{2}$ 26. $5\frac{3}{4}$ 27. $-6\frac{1}{2}$
28. $-1\frac{20}{21}$ 29. 23 30. $2\frac{5}{7}$ 31. $-\frac{5}{7}$
32. $-\frac{69}{70}$ 33. $\frac{5}{3}$ 35. 1 38. $8\frac{5}{9}$
39. 4 43. $\frac{5}{9}$ 49. 8.7 51. 43.4
53. 33.6 55. 28.8 57. $16\frac{1}{5}$
59. -11 61. $8\frac{1}{4}$ 63. $-19\frac{1}{4}$
65. $72\frac{1}{2}$ ounces 67. a. $1\frac{1}{2}$ tsp
b. $1\frac{1}{2}$ tsp c. 2 tsp 73. $x = 12$
75. $x = 34$ 77. $x = 44$ 79. F

3-4 Exercises
1. $\frac{4}{5}$ 2. $\frac{45}{4}$ 3. $-\frac{2}{3}$ 4. $2\frac{11}{12}$ 5. $1\frac{3}{14}$
6. $-\frac{5}{54}$ 7. $1\frac{1}{8}$ 8. $2\frac{9}{10}$ 9. 12.4
10. 68 11. 15.3 12. 8.6 13. 3.84
14. 17.6 15. 1310 16. 9.2
17. 22.5 18. 21 19. 45 20. 4
21. 13 22. 22 270 23. $\frac{4}{9}$ serving
25. $1\frac{13}{15}$ 27. $3\frac{7}{8}$ 29. $-\frac{8}{11}$ 31. $2\frac{1}{28}$
33. $\frac{1}{3}$ 35. $-\frac{4}{5}$ 37. 97
39. 17.1 41. 27.4 43. 25.4 45. 32
47. 5.76 49. 13 51. 11
53. 370 55. 0.7 57. 6 chairs
59. $2\frac{5}{9}$ tiles 61. Yes 65. $x = 6.5$
67. $x = 9$ 69. $x = 4.5$ 71. C

3-5 Exercises
1. $\frac{1}{24}$ 2. $\frac{67}{112}$ 3. $\frac{4}{49}$ 4. $-\frac{7}{12}$
5. $\frac{7}{15}$ 6. $-\frac{11}{24}$ 7. $-\frac{7}{60}$ 8. $\frac{29}{40}$
9. $-\frac{11}{112}$ 10. $\frac{7}{10}$ 11. $-\frac{11}{112}$
13. $6\frac{5}{6}$ ft 15. $\frac{45}{112}$ 17. $1\frac{1}{2}$ 19. $\frac{11}{112}$
21. $1\frac{5}{12}$ 23. $-\frac{3}{48}$ 25. $\frac{7}{60}$
27. 660 $\frac{729}{800}$ in. 29. $18\frac{21}{50}$ in.

Selected Answers **789**

31. $47\frac{2}{25}$ meters 35. −27 37. 88 39. 18 41. H

3-6 Exercises

1. $y = -75.4$ 2. $f = -7$ 3. $m = -19.2$ 4. $r = 54.7$ 5. $s = 68.692$ 6. $g = 6.3$ 7. $x = -\frac{4}{7}$ 8. $k = -\frac{1}{3}$ 9. $w = -\frac{7}{9}$ 10. $m = 0$ 11. $y = -9$ 12. $t = 0$ 13. $17\frac{24}{25}$ mm 15. $m = -9$ 17. $k = -2.4$ 19. $c = 5.16$ 21. $d = \frac{8}{15}$ 23. $x = \frac{1}{2}$ 25. $c = \frac{7}{20}$ 27. $z = \frac{2}{3}$ 29. $j = -32.4$ 31. $g = 9$ 33. $v = -30.25$ 35. $y = -5.4$ 37. $c = -\frac{1}{24}$ 39. $y = 64.1$ 41. $m = -2.8$ 43. a. 15 tiles b. 9 tiles c. 5 boxes 49. 21 51. 5.24×10^{-6} 53. 6.4×10^{10}

3-7 Exercises

1. $x \geq 2$ 2. $k > 9.3$ 3. $g \leq 7$ 4. $h < 0.79$ 5. $w \leq 0.24$ 6. $z > 0$ 7. $k > \frac{3}{5}$ 8. $y \geq 0$ 9. $q \leq -\frac{1}{169}$ 10. $x < 1\frac{2}{3}$ 11. $f > \frac{4}{15}$ 12. $m \geq 4$ 13. between 6.7 and 8.1 hours 15. $m \leq -.07$ 17. $g \leq -24.3$ 19. $w \leq -1.5$ 21. $k \geq \frac{25}{36}$ 23. $x \geq 4\frac{3}{5}$ 25. $m \leq -1\frac{1}{7}$ 27. $d \leq -3$ 29. $g \geq -2$ 31. $t > \frac{3}{13}$ 33. $y \geq -8$ 35. $w \leq -\frac{1}{3}$ 37. $c > 3.1$ 39. $c < 3\frac{1}{3}$ 41. $t \leq 6$ 43. at least 12.5 in., but not more than 3600 in. 47. 0.3 49. −0.26 51. 16.8 53. −0.258 55. C

3-8 Exercises

1. ± 5 2. ±12 3. ±2 4. ±20 5. ±1 6. ±9 7. ±3 8. ±4 9. 16 ft 10. 5 11. 2 12. −55 13. −1 15. ±15 17. ±13 19. ±21 21. ±19 23. −3 25. −20 27. ±7 29. ±17 31. ±30 33. ±23 35. $\pm\frac{1}{2}$ 37. $\pm\frac{5}{2}$ 39. $\pm\frac{3}{2}$ 41. $\pm\frac{1}{10}$ 43. 26 ft 45. 327 47. a. 81; 1 b. 18 51. $t = 9$ 53. $t = 22$ 55. $\frac{1}{9}$ 57. 1 59. D

3-9 Exercises

1. 6 and 7 2. −8 and −9 3. 14 and 15 4. −18 and −19 5. ≈ 13.27 ft 6. 9.1 7. 6.5 8. 50 9. 13.8 11. 1 and 2 13. −31 and −32 15. 8.3 17. 25.5 19. B 21. E 23. F 25. 7.14 27. 11.62 29. 42.85 31. −11.62 33. −32.83 35. ±5.20 37. ±317.02 39. 800 ft/s 43. $y = -4.4$ 45. $m = -25.6$ 47. $x < 5\frac{2}{3}$ 49. $m \geq 8$ 51. 4 and −4 53. 10 and −10 55. D

3-10 Exercises

1. irrational, real 2. whole, integer, rational, real 3. rational, real 4. rational, real 5. rational 6. rational 7. irrational 8. not real 9. rational 10. not real 11. not real 12. not real 13–15. Possible answers given. 13. $5\frac{1}{4}$ 14. $\frac{2199}{700}$ 15. $\frac{3}{16}$ 17. rational, real 19. integer, rational, real 21. rational 23. irrational 25. irrational 27. not real 29. $-\frac{1}{200}$ 31. whole, integer, rational, real 33. irrational, real 35. rational, real 37. rational, real 39. rational, real 41. integer, rational, real 43–51. Possible answers given. 43. $-\sqrt{50}$ 45. $\frac{11}{18}$ 47. $\frac{3}{4}$ 49. 3 51. −4.25 53. $x \geq 0$ 55. $x \geq -3$ 57. $x \geq -\frac{2}{5}$ 63. 6.32 65. 7.75 67. −4.12 69. 3.46 71. 2.5×10^6 73. 5.68×10^{15} 75. J

Chapter 3 Study Guide and Review

1. rational number 2. real numbers; irrational numbers 3. relatively prime 4. principal square root 5. perfect square 6. $\frac{3}{5}$ 7. $\frac{1}{4}$ 8. $\frac{21}{40}$ 9. $\frac{2}{3}$ 10. $\frac{2}{3}$ 11. $\frac{3}{4}$ 12. $\frac{-6}{13}$ 13. $1\frac{2}{5}$ 14. $\frac{5}{9}$ 15. $\frac{1}{6}$ 16. $-1\frac{1}{5}$ 17. $7\frac{3}{5}$ 18. $\frac{8}{15}$ 19. −4 20. $2\frac{1}{4}$ 21. $3\frac{1}{4}$ 22. 6 23. $\frac{3}{8}$ 24. $\frac{2}{9}$ 25. −16 26. $1\frac{1}{4}$ 27. 2 28. $1\frac{1}{6}$ 29. $\frac{5}{18}$ 30. $11\frac{3}{10}$ 31. $4\frac{7}{20}$ 32. $y = -21.8$ 33. $z = -18$ 34. $w = -\frac{5}{8}$ 35. $p = 2$ 36. $m > -\frac{1}{12}$ 37. $t \geq -12$ 38. $y \leq -3\frac{1}{4}$ 39. $x > -\frac{1}{2}$ 40. ±4 41. ±30 42. ±26 43. 5 44. $\frac{1}{2}$ 45. 9 46. 89.4 in. 47. 167.3 cm 48. rational 49. irrational 50. not real 51. irrational 52. rational 53. not real

Chapter 4

4-1 Exercises

1. Population: pet store customers; sample: 100 customers; possible bias: not all customers have dogs. 2. systematic 3. random 5. systematic 7. Population: students; sample: students who buy the entrée; possible bias: the students who buy the entrée may be the people who like the food in the cafeteria. 9. Population: restaurant customers; sample: first four customers who order the cheese sauce; possible bias: if the customers ordered cheese sauce, then they probably like cheese. 11. systematic 13. stratified 15. systematic 17 a. Possible answer: Randomly select visitors leaving the zoo. b. Possible answer: Select every tenth visitor leaving the zoo. c. Possible answer: People visiting with children might visit the zoo only because they have children. 23. $y = -7.2$ 25. $c = -\frac{2}{7}$ 27. $x > 25.6$

4-2 Exercises

1.

Nutrition in Potatoes	Baked Potato (100 g)	French Fries (100 g)	Potato Chips (100 g)
Fiber	2.4 g	3.2 g	4.5 g
Ca	10 mg	10 mg	24 mg
Mg	27 mg	22 mg	67 mg

2. 2, 3, 3, 7, 11, 13, 17, 17, 18, 20, 20, 27, 34, 34, 35, 35 3. 63, 66, 68, 73, 73, 75, 77, 80, 80, 81, 81, 90, 94, 95, 99

4.

Tens	Ones
0	1 6 7
1	8
2	0 2 6
3	5 6
4	7
5	3 6

Key: 1|8 means 18

5.

Democrats		Republicans
	3	2 6 7 8
6 6	4	1 2 3 4
8 7 6 4	5	3 4
8 4 1 1	6	

Key: |4|1 means 41; 6|4| means 46

7. 50, 51, 54, 58, 62, 66, 67, 71, 74, 75, 76, 76, 82

9.

Dollars	Cents
0.9	3 5 5
1.0	2 6
1.1	1 1 3 4 7
1.2	1 3 3 4
1.3	0 8

Key: 1.1|1 means \$1.11

11.

Tens	Ones
4	3
5	7
6	5 8
7	2 2 3 5 6
8	1 2 4 8
9	1

Key: 5|7 means 57

13.

Energy Use in U.S.	1980	1990	2000
Fossil Fuels	89%	86%	85%
Nuclear Power	3%	7%	8%
Renewable Resources	7%	7%	7%

15.

Numbers		Time	
One	9	Night	12
Two	3	Day	4
Three	6	Supper-time	1
Ten	2	Bed-time	1
Twelve	1	Evening	1
Fourteen	1		

19. 5^{11} 21. cannot combine 23. population: students; sample: students on every other bus

4-3 Exercises

1. ≈ 34.43; 35; no mode 2. 4.4; 4.4; 4.4 and 6.2 3. 5; 5; 5 4. ≈ 55.67; 56; no mode 5. 2.39 million 6. approximately 1.43 million 7. 3.35 million 9. 87.6; 88; 88 11. 5.85; 4.4; no mode 13. approximately 74.33 million 15. 25; 26; no mode; no outlier 17. 11; 12; 10 and 13; 3 19. 4; 2; 2; 29 21. 1105 million miles; 484 million miles; no mode 29. $14x - 45$ 31. $x = 13$ 33. $m = 100$ 35. J

4-4 Exercises

1. 56; 42; 66 2. 6; 1.5; 4.5

3.
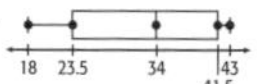

18
23.5
34
43
41.5

4.
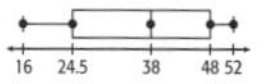

16
24.5
38
48 52

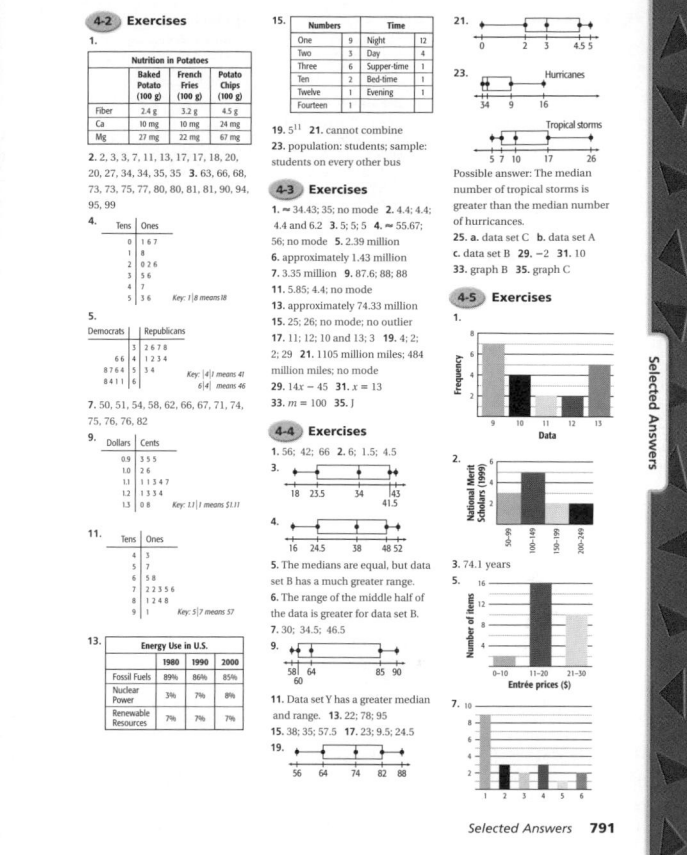

5. The medians are equal, but data set B has a much greater range. 6. The range of the middle half of the data is greater for data set B. 7. 30; 34.5; 46.5

9.
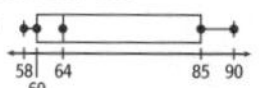

58
60
64
85 90

11. Data set Y has a greater median and range. 13. 22; 78; 95 15. 38; 35; 57.5 17. 23; 9.5; 24.5

19.
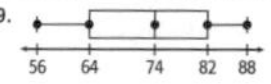

56
64
74
82
88

21.
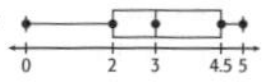

0
2
3
4.5 5

23.
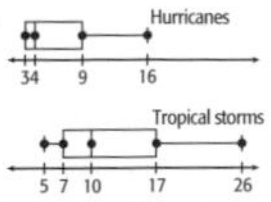

Hurricanes
34
9
16
Tropical storms
5 7 10
17
26

Possible answer: The median number of tropical storms is greater than the median number of hurricances. 25. a. data set C b. data set A c. data set B 29. −2 31. 10 33. graph B 35. graph C

4-5 Exercises

1.
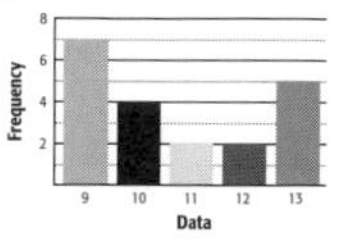

Frequency
Data

2.
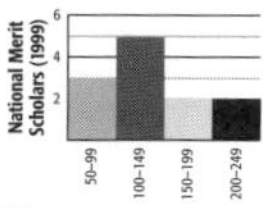

National Merit Scholars (1999)
50–99
100–149
150–199
200–249

3. 74.1 years

5.
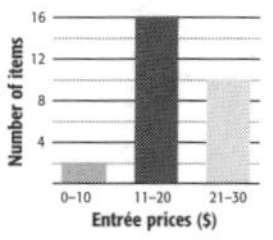

Number of items
0–10
11–20
21–30
Entrée prices ($)

7.
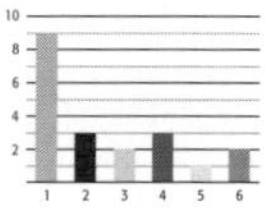

9. a. 34.9 hours b. \$11.88 13. $x < 5$ 15. $x \leq 2$ 17. $x > 6$ 19. $6 \geq x$ 21. B

4-6 Exercises

1–9. Possible answers given. 1. The scale does not start at zero, so changes appear exaggerated. 2. The intervals used in the histogram are not equal. 3. The fruits are all different sizes. A better comparison would be the same serving size of each fruit. 4. The sales are for different lengths of time. 5. The graph has no scale, so it's impossible to compare the money earned. 7. The difference between the two groups' responses is only 3 people out of 1000. 9. The areas of the sails distort the comparison. Your graph should use bars or pictures that are the same width. 15. $b = 6$ 17. $a = 21$ 19. $1.5 = h$ 21. $f = 1.5$

4-7 Exercises

1.
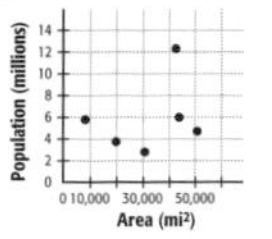

Population (millions)
Area (mi²)

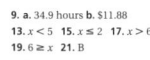

2. positive 3. no correlation 4. 66°F

5.
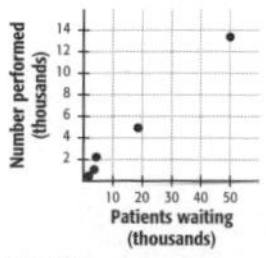

Number performed (thousands)
Patients waiting (thousands)

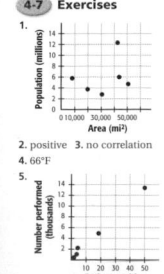

7. positive 9. There is a positive correlation between the pollen levels. 11. negative 15. $x = 5$ 17. $x = 6$ 19. $x = 18$

Chapter 4 Extension

1. 2.4 3. 12.9 5. 2.3 7. 0 9. data set B 11. a. week 1: 1.7; week 2: 3.1 b. week 2 13. Zero; the sum for the differences of the data values would be zero.

Chapter 4 Study Guide and Review

1. median; mode 2. variability; variability; range 3. line of best fit; scatter plot; correlation 4. population: moviegoers; sample: 25 people in line for a Star Wars movie; possible bias: people in line for Star Wars might have a preference for science fiction movies. 5. population: community members; sample: 50 parents of middle-school-age children; possible bias: parents of middle-school-age children may support the field more than other community members. 6. population: constituents; sample: 75 constituents who visited the office; possible bias: constituents who visit the senator probably are strong supporters of the senator.

7.

Inaugural Age		Age at Death
3	4	6
7 7 2 1	5	6
	6	3 7
	7	
	8	3

Key: 3|4| means 43; |4|6 means 46

8. 760; 570; 500 9. 9.25; 9; 8, 9, and 10 10. 6; 6; 6 11. 3.1; 3.1; 3.1 12. 10; 80; 90 13. 32; 68; 99

14.
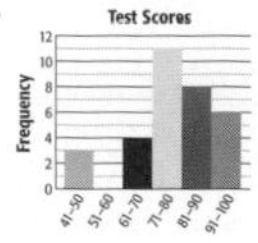

Test Scores
Frequency
41–50
51–60
61–70
71–80
81–90
91–100

15.
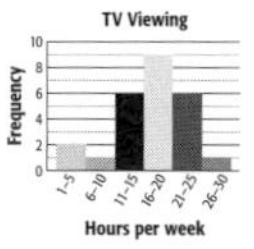

TV Viewing
Frequency
1–5
6–10
11–15
16–20
21–25
26–30
Hours per week

16. Possible answer: The symbols are different sizes even though they represent the same number of sightings. 17. positive 18. no correlation

Chapter 5

5-1 Exercises

1. points A, B, C 2. $\overrightarrow{BC}$ 3. plane Z or plane ABC 4. $\overline{AB}, \overline{BC}, \overline{AC}$ 5. $\overrightarrow{BA}, \overrightarrow{BC}, \overrightarrow{CB}$ 7. $\angle LJM$, $\angle MJK$ 9. $\angle LJM$ and $\angle MJK$ 11. 115° 13. points V, W, X, Y 15. plane N or plane VWX 17. $\overrightarrow{WV}, \overrightarrow{VW}, \overrightarrow{WY}, \overrightarrow{YW}, \overrightarrow{WX}$ 19. $\angle DEH$, $\angle GEF$ 21. $\angle FEG$ and $\angle HED$ 23. 117° 25. False 27. False 29. False 31. False 33. False 35. a. 145° b. They are supplementary angles. 41. 18; 18; 29 43. B

5-2 Exercises

1. $\angle 1 \cong \angle 4 \cong \angle 5 \cong \angle 8$ (45°); $\angle 2 \cong \angle 3 \cong \angle 6 \cong \angle 7$ (135°) 2. 59° 3. 59° 4. 121° 5. 59° 7. 60° 9. 120° 11. $\angle 4$, $\angle 5$, $\angle 8$ 13. Possible answers: $\angle 1$ and $\angle 2$, $\angle 1$ and $\angle 3$, $\angle 3$ and $\angle 4$. 15. 51° 17. 90° 19. Possible answer:

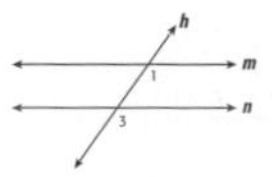

h
m
n
1
3

21. a. $\overline{AB}$ b. $m\angle 2 = m\angle 3 = m\angle 4 = 45°$ 27. 32 29. 0.00000001 31. 128 33. m^{13} 35. Population: shoppers; sample: paid shoppers at a mall; possible bias: The people may answer favorably because they are being paid.

5-3 Exercises

1. $q° = 78°$ 2. $r° = 51°$ 3. $s° = 120°$ 4. $a° = 60°$ 5. $c° = 66°$ 6. $d° = 18°, 3d° = 54°, 6d° = 108°$ 7. 60°, 30°, 90° 9. $s° = 69°$ 11. $w° = 60°$ 13. $g° = 15°$, $4g° = 60°, 7g° = 105°$ 15. $x° = 56°$ 17. $w° = 40°$ 19. $y° = 18°$ 27. always 29. sometimes 31. never 33. never 35. a. $w° = 75°$; $y° = 75°$; two right angles b. $x° = 30°$; $z° = 75°$; $m° = 75°$ c. The two blue triangles are right scalene triangles, and the white triangle is an acute isosceles triangle. 39. 11 41. C

5-4 Exercises

1. 360° 2. 720° 3. $t° = 90°$ 4. $v° = 144°$ 5. quadrilateral, trapezoid 6. quadrilateral, parallelogram, rhombus 7. 540° 9. $m° = 120°$ 11. quadrilateral, parallelogram, rhombus, rectangle, square 13. 3240°; 162° 15. 12,600°; 175° 17. 2880°; 160° 19. $x° = 110°$ 21. $w° = 123°$ 23. $x° = 130°$ 25. hexagon 27. 13-gon 29. pentagon 35. a. $x° = 98°$ b. $y° = 145°$ 39. 6.4×10^{-7} 41. -1.6×10^{-6} 43. C

5-5 Exercises

1. 0 2. Slope is undefined. 3. positive slope; 1 4. negative slope; $-\frac{1}{2}$ 5. $\overleftrightarrow{AB} \parallel \overleftrightarrow{CD}$ 6. $\overleftrightarrow{MN} \perp \overleftrightarrow{AB}$, $\overleftrightarrow{MN} \perp \overleftrightarrow{CD}$, and $\overleftrightarrow{AD} \perp \overleftrightarrow{BE}$ 7. parallelogram, rhombus, rectangle, square 8. trapezoid 9. positive slope, 1 11. 0 13. $\overleftrightarrow{CD} \parallel \overleftrightarrow{AB}$ 15. parallelogram, rhombus, rectangle, square 17. 3 19. 0 27. 90° 29. 33°

5-6 Exercises

1. triangle $ABC \cong$ triangle FED 2. quadrilateral $LMNO \cong$ quadrilateral $STQR$ 3. $q = 13$ 4. $r = 4$ 5. $s = 4$ 7. trapezoid $PQRS \cong$ trapezoid $ZYXW$ 9. $n = 5$ 11. $x = 16, y = 25, z = 14.2$ 13. $s = 120, t = 33, r = 33$ 19. $16 = x$ 21. $-15 = m$ 23. $b = -6$ 25. $a = -32$

5-7 Exercises

1. reflection 2. rotation

3.
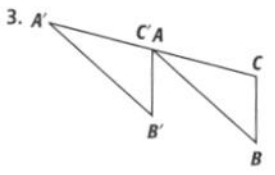

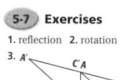

A′
C′A
C
B′
B

4.
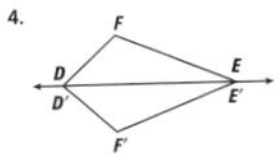

F
D
D′
E
E′
F′

5.
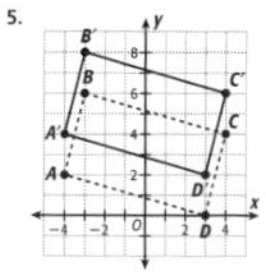

6.
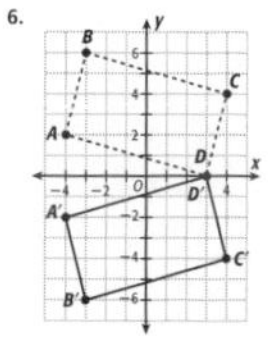

7.
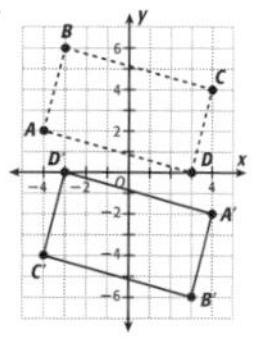

9. translation

11.
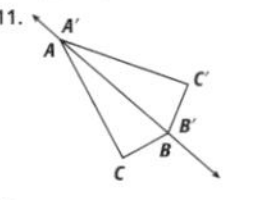

A′
A
C′
B′
B
C

13.
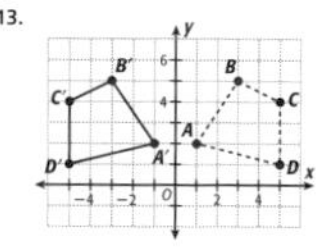

15.
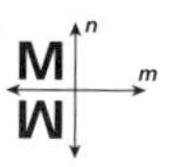

17.
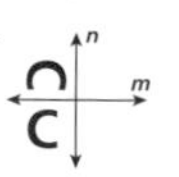

19. (−2, −1) 21. (−4, −3) 23. $(-m, n)$ 25. (6, −1)

27.
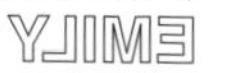

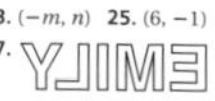

reflection across a vertical line 31. 32 33. −343 35. −128 37. 16 39. A

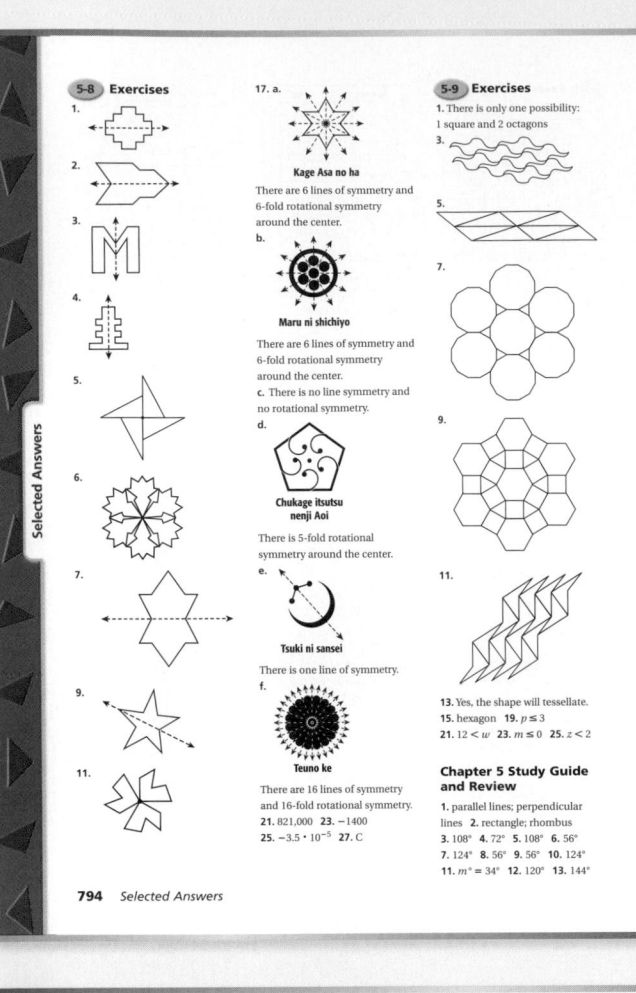

5-8 Exercises

1.
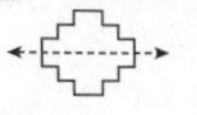
2.
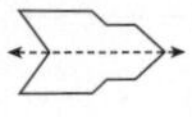
3.

4.
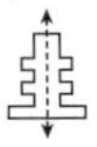
5.
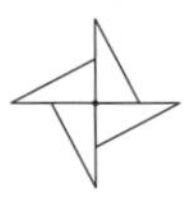
6.

7.
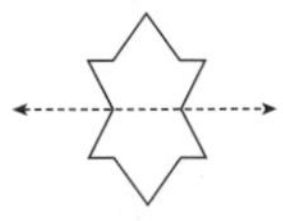
9.
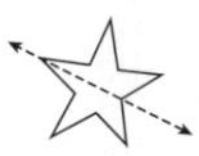
11.
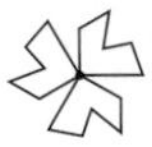

17. a.
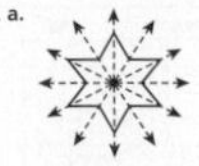

Kage Asa no ha

There are 6 lines of symmetry and 6-fold rotational symmetry around the center.

b.
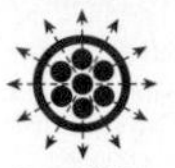

Maru ni shichiyo

There are 6 lines of symmetry and 6-fold rotational symmetry around the center.

c. There is no line symmetry and no rotational symmetry.

d.

Chukage itsutsu nenji Aoi

There is 5-fold rotational symmetry around the center.

e.
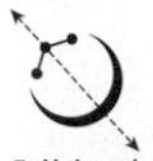

Tsuki ni sansei

There is one line of symmetry.

f.
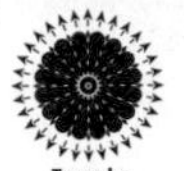

Teuno ke

There are 16 lines of symmetry and 16-fold rotational symmetry.

21. 821,000 **23.** −1400 **25.** $-3.5 \cdot 10^{-5}$ **27.** C

5-9 Exercises

1. There is only one possibility: 1 square and 2 octagons

3.
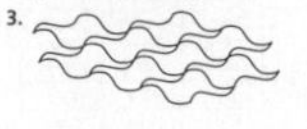
5.
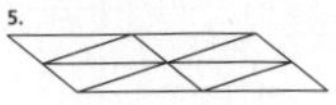
7.
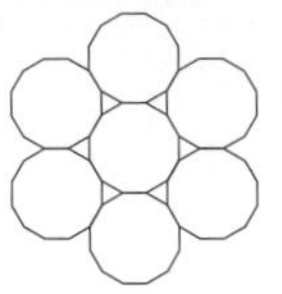
9.
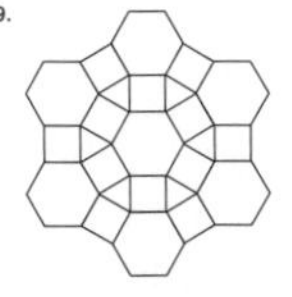
11.
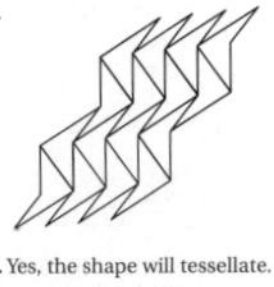

13. Yes, the shape will tessellate. **15.** hexagon **19.** $p \leq 3$ **21.** $12 < w$ **23.** $m \leq 0$ **25.** $z < 2$

Chapter 5 Study Guide and Review

1. parallel lines; perpendicular lines **2.** rectangle; rhombus **3.** 108° **4.** 72° **5.** 108° **6.** 56° **7.** 124° **8.** 56° **9.** 56° **10.** 124° **11.** $m° = 34°$ **12.** 120° **13.** 144°

794 Selected Answers

14. trapezoid **15.** parallelogram, rhombus **16.** parallelogram **17.** $x = 23$ **18.** $t = 3.2$ **19.** $q = 5$

20.
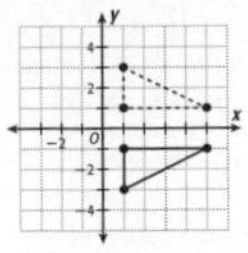
21.
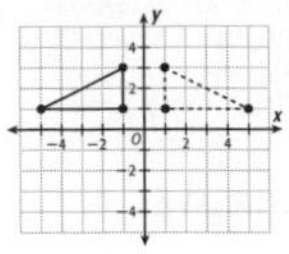
22.
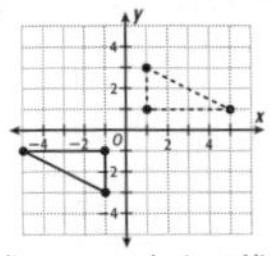

23. line symmetry: horizontal line of symmetry **24.** 2-fold rotational symmetry **25.** line symmetry: horizontal and vertical lines of symmetry; 2-fold rotational symmetry

26.
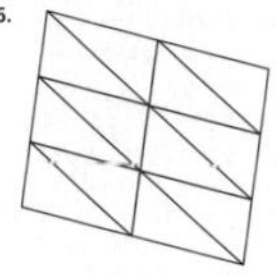
27.
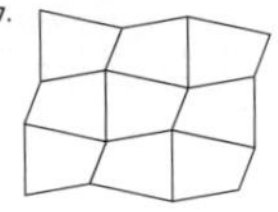

Chapter 6

6-1 Exercises

1. 20 units **2.** 36 units **3.** 19.4x units **4.** 15 units^2 **5.** 28 units^2 **6.** 32 units^2 **7.** 14 units^2 **8.** 44 units; 53 units^2 **9.** 34 units **11.** 26x units **13.** 24 units^2 **15.** 18 units^2 **17.** 64 units **19.** 46 units; 72 units^2 **21.** 46 units; 84 units^2 **23.** 33 in.; 792 in^2 **25. a.** \$1125 **b.** 375 people **31.** $y < -2$ **33.** $w > 3$

6-2 Exercises

1. 22 units **2.** $11\frac{1}{4}$ units **3.** 30 units **4.** 34.5 units **5.** 84 units **6.** $(4x + 1)$ units **7.** 15 units^2 **8.** 28 units^2 **9.** 12 units^2 **10.** 25 units^2 **11.** 29 units **13.** 70 units **15.** $(30a + 8)$ units **17.** 20 units^2 **19.** 12 units^2 **21.** 49.5 units^2 **23.** 21x units^2 **25.** 9.1 ft **27. a.** 1929.5 ft^2 **b.** 466.6 ft **29.** 49.8 ft **31.** 874.6 ft^2; 160.4 ft **33.** 0.75 **35.** 2.5 **37.** negative

6-3 Exercises

1. 5 **2.** 10.6 **3.** 7.8 **4.** 5 **5.** 5.3 **6.** 20 **7.** $\sqrt{24} \approx 4.9$ units; 19.6 units^2 **9.** 17 **11.** 8.9 **13.** 9.2 **15.** $\sqrt{80} \approx 8.9$ units; 71.2 units^2 **17.** 7 **19.** $\sqrt{1716} \approx 41.4$ **21.** 72 **23.** yes **25.** yes **27.** no **29.** yes **31.** 139 km **33.** 475 mi **37.** $x = 9$ **39.** $y = 5$ **41.** A

6-4 Exercises

1. 8π cm; 25.1 cm **2.** 6.4π in.; 20.1 in. **3.** 2.25π ft^2; 7.1 ft^2 **4.** 56.25π cm^2; 176.6 cm^2 **5.** $A = 9\pi$ units$^2 \approx 28.3$ units^2; $C = 6\pi$ units ≈ 18.8 units **6.** $\frac{175}{99} \approx 1.8$ ft **7.** 14π in.; 44.0 in. **9.** 40.4π cm; 126.9 cm

11. 144π cm^2; 452.2 cm^2 **13.** 324π in^2; 1017.4 in^2 **15.** $A = 36\pi$ units$^2 \approx 113.0$ units^2; $C = 12\pi$ units ≈ 37.7 units **17.** $C \approx 7.5$ m; $A \approx 4.5$ m^2 **19.** $C \approx 25.1$ in.; $A \approx 50.2$ in^2 **21.** 6.4 cm **23.** 4 cm **25.** 11.7 m **27.** 297.7 m^2 **29.** $C = 12\pi \approx 37.7$ ft; $A = 36\pi \approx 113.1$ ft^2 **35.** $\frac{6}{19}$ **37.** $-\frac{21}{40}$

6-5 Exercises

1. Possible answer:
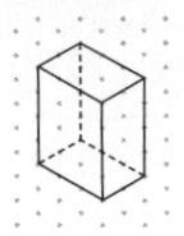

2. Possible answer:
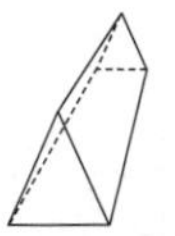

3. Possible answer:
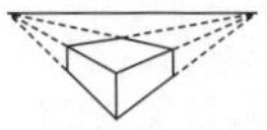

5. Possible answer:
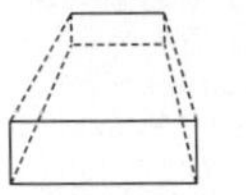

7. rectangles *JKLM, PQRN, JMNR, KLPQ, JKQR,* and *LMNP* **9.** triangles *SVW* and *TUX* and rectangles *STUV, UVWX,* and *STXW* **15.** *A* **17.** *PQSR* **19.** $\overline{RY}$, $\overline{WY}$, and $\overline{YZ}$ **21.** $\overline{UV} \parallel \overline{PQ} \parallel \overline{ST}$ **23.** $\overline{PQ}$ **25.** one-point

Selected Answers 795

27.
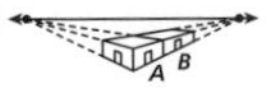

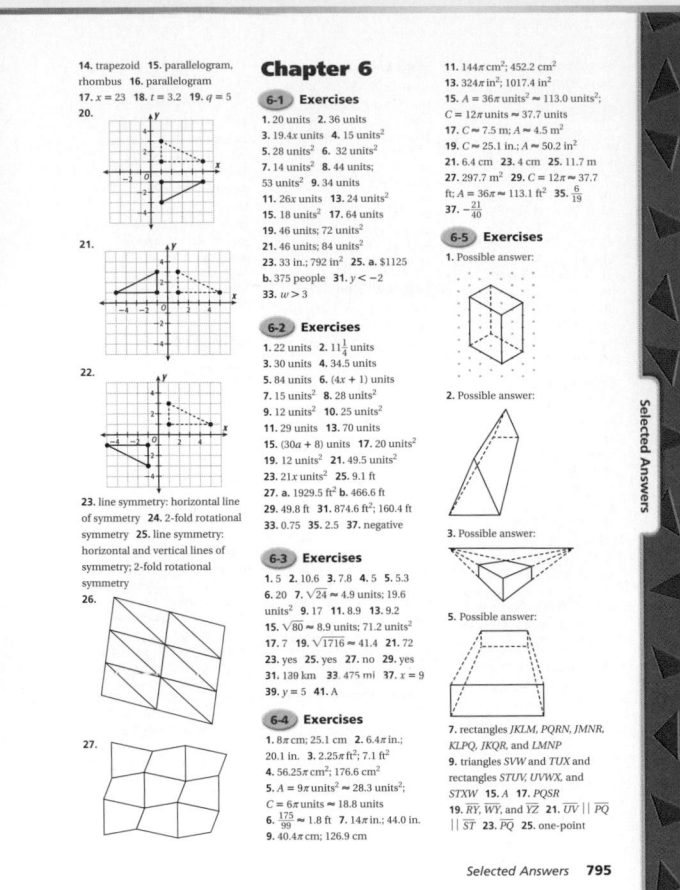

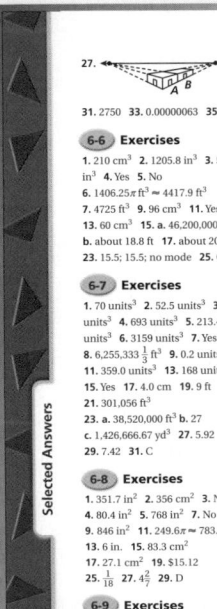

31. 2750 **33.** 0.00000063 **35.** B

6-6 Exercises

1. 210 cm^3 **2.** 1205.8 in^3 **3.** 556 in^3 **4.** Yes **5.** No **6.** 1406.25π ft$^3 \approx 4417.9$ ft^3 **7.** 4725 ft^3 **9.** 96 cm^3 **11.** Yes **13.** 60 cm^3 **15. a.** 46,200,000 in^3 **b.** about 18.8 ft **17.** about 20.5 in. **23.** 15.5; 15.5; no mode **25.** C

6-7 Exercises

1. 70 units^3 **2.** 52.5 units^3 **3.** 14.8 units^3 **4.** 693 units^3 **5.** 213.4 units^3 **6.** 3159 units^3 **7.** Yes **8.** $6{,}255{,}333\frac{1}{3}$ ft^3 **9.** 0.2 units^3 **11.** 359.0 units^3 **13.** 168 units^3 **15.** Yes **17.** 4.0 cm **19.** 9 ft **21.** 301,056 ft^3 **23. a.** 38,520,000 ft^3 **b.** 27 **c.** 1,426,666.67 yd^3 **27.** 5.92 **29.** 7.42 **31.** C

6-8 Exercises

1. 351.7 in^2 **2.** 356 cm^2 **3.** No **4.** 80.4 in^2 **5.** 768 in^2 **7.** No **9.** 846 in^2 **11.** $249.6\pi \approx 783.7$ cm^2 **13.** 6 in. **15.** 83.3 cm^2 **17.** 27.1 cm^2 **19.** \$15.12 **25.** $\frac{1}{18}$ **27.** $4\frac{2}{7}$ **29.** D

6-9 Exercises

1. 144 m^2 **2.** 74.6 ft^2 **3.** No **4.** ≈ 702.5 ft^2 **5.** 24.1 in^2 **7.** No **9.** 765 cm^2 **11.** $1368\pi \approx 4295.5$ ft^2 **13.** $\approx 877{,}201{,}312$ mi^2 **15. a.** ≈ 588; ≈ 216 **b.** Khufu; 925,344 ft^2 **c.** Menkaure; $\approx 8{,}619{,}552$ ft^3 **19.** 0.6 **21.** −1.4 **23.** $\sqrt{709} \approx 26.63$ m

6-10 Exercises

1. 10.7π cm^3; 33.6 cm^3 **2.** 1333.3π ft^3; 4186.6 ft^3 **3.** 6.6π m^3; 20.7 m^3 **4.** 85.3π mi^3; 267.8 mi^3 **5.** 4π in^2; 12.6 in^2 **6.** 174.2π mm^2; 547.0 mm^2 **7.** 324π cm^2; 1017.4 cm^2 **8.** 225π yd^2; 706.5 yd^2 **9.** The volume of the sphere and the cube are about equal (≈ 268 in^3). The surface area of the sphere is about 201 in^2, and the surface area of the cube is about 250 in^2. **11.** 147.5π cm^3; 463.2 cm^3 **13.** 0.17π in^3; 0.5 in^3 **15.** 207.4π m^2; 651.2 m^2 **17.** 2500π cm^2; 7850 cm^2 **19.** 221.83π in^3; 696.55 in^3 **21.** $V = 39.72\pi \approx 124.72$ yd^3; $S = 38.44\pi \approx 120.70$ yd^2 **23.** 30 km; $36{,}000\pi \approx 113{,}040$ km^3 **25.** ≈ 5392 cm^3 **27.** ≈ 0.0314 mm^2 **31.** $\frac{3}{10}$ **33.** $3\frac{17}{75}$ **35.** G

Chapter 6 Extension

1. rotational and bilateral **3.** rotational and bilateral

5.
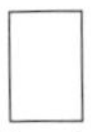
line and rotational

7. rotational **9.** rotational and bilateral

11.
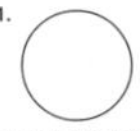
line and rotational

13. square; smaller **15.** circles

Chapter 6 Study Guide and Review

1. perimeter; area **2.** edge; vertex **3.** great circle; hemispheres **4.** $13\frac{2}{9}$ in^2; 16 in. **5.** 208 m^2; 80 m **6.** 16 cm^2; 20.2 cm **7.** 21 in^2; 34 in. **8.** $c = 10$ **9.** $a = 10$ **10.** $A = 225\pi \approx 706.5$ in^2; $C = 30\pi \approx 94.2$ in. **11.** $A = 5.8\pi \approx 18.2$ cm^2; $C = 4.8\pi \approx 15.1$ cm **12.** $A = 16\pi \approx 50.2$ m^2; $C = 8\pi \approx 25.1$ m **13.** $A = 0.4\pi \approx 1.3$ ft^2; $C = 1.2\pi \approx 3.8$ ft

14.
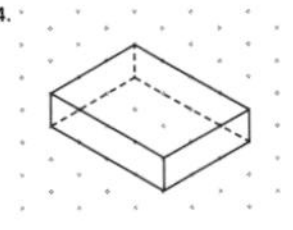
15.
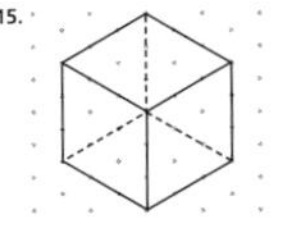
16.
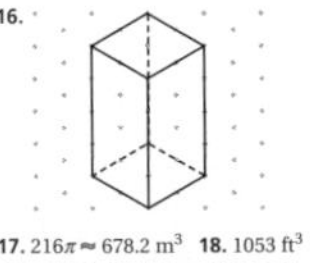

17. $216\pi \approx 678.2$ m^3 **18.** 1053 ft^3 **19.** 320 ft^3 **20.** $120\pi \approx 376.8$ in^3 **21.** 90 cm^2 **22.** 95 cm^2 **23.** 340 in^2 **24.** 972π in$^3 \approx 3052.1$ in^3 **25.** 4500π m$^3 \approx 14{,}130$ m^3

Chapter 7

7-1 Exercises

1–4. Possible answers given. **1.** $\frac{2}{5}$, $\frac{8}{20}$ **2.** $\frac{1}{3}$, $\frac{6}{18}$ **3.** $\frac{3}{1}$, $\frac{42}{14}$ **4.** $\frac{20}{16}$, $\frac{10}{8}$ **5.** yes **6.** no **7.** yes **8.** No; $2\frac{1}{4}$ cups are needed. **9.** Possible answers: $\frac{2}{14}$, $\frac{3}{21}$ **11.** Possible answers: $\frac{8}{7}$, $\frac{32}{28}$ **13.** no **15.** yes **17–25.** Possible answers given. **17.** no; $\frac{4}{7}$ **19.** no; $\frac{8}{14}$ **21.** yes **23.** no; $\frac{8}{42}$ **25.** yes **27.** no; 4 gallons **29.** $\frac{39}{18}$ **35.** $-1\frac{11}{36}$ **37.** $1\frac{5}{99}$ **39.** 3 and −3 **41.** 13 and −13

7-2 Exercises

1. 1:5 **2.** 35 wpm **3.** 42 wpm **4.** 22 oz can **5.** dozen golf balls **7.** 171.6 gal/h **9.** 4 boxes

796 Selected Answers

11. \$26.25 per hour **13.** \$0.77 per slice **15.** \$2.49/yard; \$2.26/yard; 5 yards **17.** \$1.37/gal; \$1.42/gal; 10 gal **19. a.** Super-Cell: \$0.10/min; Easy Phone: \$0.11/min **b.** Super-Cell offers a better rate. **21. a.** Tom: $25\frac{3}{8}$ frames per hour; Cherise: 27 frames per hour; Tina: $28\frac{3}{8}$ frames per hour **b.** Tina **c.** $1\frac{5}{8}$ **d.** 24 **25.** −4 **27.** −5 **29.** −4.4 **31.** D

7-3 Exercises

1. 12 in./1 ft **2.** 8 pt/1 gal **3.** 1 m/100 cm **4.** 91.25 gal **5.** 7.5 mi/h **6.** 0.09 m/s **7.** ≈ 1.14 g **9.** 1 yd/36 in. **11.** 585 ft **13.** 57,600 bricks **15.** 900 radios **17.** 4 hot dogs **19.** 4.98 mi **21.** A ≈ 22.88 mi/h; B ≈ 23.16 mi/h; C ≈ 21.76 mi/h **23.** 200 times **29.** 14 units^2 **31.** 226.9 in^2 **33.** 3.8 mi^2

7-4 Exercises

1. yes **2.** yes **3.** no **4.** yes **5.** no; $\frac{1}{8} \neq \frac{8}{56}$ **6.** $x = 1$ **7.** $n = 8$ **8.** $d = 2$ **9.** $h = 6$ **10.** $f = 9.75$ **11.** $t = 2$ **12.** $s = 9$ **13.** $q = 12.5$ **14.** ≈ 3.3 cm **15.** no **17.** no **19.** yes; $\frac{18}{12} = \frac{15}{10}$ **21.** $b = 3$ **23.** $y = 6$ **25.** $n = 4$ **27.** $d = 0.5$ **29.** $\frac{6}{3}$, $\frac{18}{9}$ **31.** $\frac{66}{21}$, $\frac{22}{7}$ **33.** $\frac{0.25}{4}$, $\frac{1}{16}$ **35.** 12 molecules **37. a.** about 1.53:1 **b.** about 134.6 mm Hg **41.** $-1\frac{1}{4}$ **43.** $11\frac{17}{100}$

7-5 Exercises

1. no **2.** yes

3.
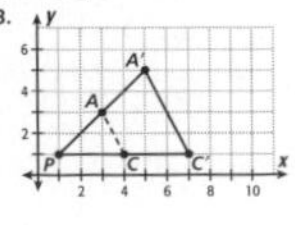

4.
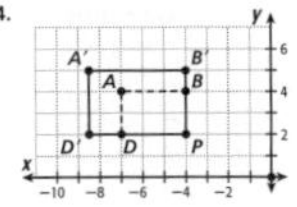

5. $A'(1.5, -1)$; $B'(1, -2.5)$; $C'(4, -3)$; $D'(5, -0.5)$ **6.** $A'(16, 4)$; $B'(28, 4)$; $C'(20, 12)$ **7.** no

9.
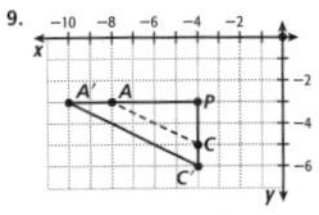

11. $A'(-9, 6)$; $B'(15, 12)$; $C'(-6, -9)$ **13.** 3 **15.** Yes **21.** 24 units^2 **23.** 21 units^2

7-6 Exercises

1. ≈ 5.4 in. **2.** ≈ 14.7 cm **3.** *A* and *C* are similar. **5.** ≈ 22.9 ft **7.** similar **9.** similar **11.** $x = 6$ ft **13.** $x = 24$ ft **15.** yes; $\frac{1}{15}$ or $\frac{4\text{ in.}}{5\text{ ft}}$ **17.** 24 ft **21.** 1256 mm^3 **23.** 2044.3 cm^3 **25.** D

7-7 Exercises

1. 1 in:1.25 ft **2.** 20.25 m **3.** 0.0085 in. **4.** 7.5 mm **5.** 52 ft **6.** 27 in. **7.** 1 cm = 1.5 m **9.** 0.023 mm **11.** 20 in. **13.** 2 in. **15.** 0.5 in. **17.** 18 ft **19.** 58.5 ft **21.** about 580 mi **23–27.** The scale is 1.2 cm:36 in. **23.** ≈ 18 in. **25.** No; each wall is only ≈ 45 in. wide. **27.** ≈ 298 ft^2 **31.** no **33.** no

7-8 Exercises

1. reduces **2.** enlarges **3.** preserves **4.** preserves **5.** reduces **6.** enlarges **7.** $\frac{1}{24}$ **8.** 14 in. **9.** 0.000028 mm **11.** preserves **13.** reduces **15.** reduces **17.** 7.5 ft **19.** $\frac{12}{1}$ **21.** $\frac{1}{45}$ **23.** $\frac{1}{12.5}$ **25.** $\frac{1}{28}$ **27.** 630 ft **33.** ≈ 1869.4 ft^2 **35.** 1256 cm^2 **37.** \$0.17 per apple **39.** A

7-9 Exercises

1. 4:1 **2.** 16:1 **3.** 64:1 **4.** width: 30 in.; height: 10 in. **5.** 72 min **6.** 7:1 **7.** 49:1 **9.** 32 cm **11.** 2 cm; 8 cubes **13.** 4 cm; 64 cubes **15.** 5 cm; 125 cubes **17.** 1,000,000 cm^3 **19.** 256,000 **21.** 14.58 oz **25.** Possible answers: $\frac{6}{10}$, $\frac{9}{15}$ **27.** Possible answers: $\frac{8}{22}$, $\frac{12}{33}$ **29.** 1.5 ft **31.** 18 ft **33.** D

Chapter 7 Extension

1. 0.777 **3.** 0.017 **5.** 45 ft **7.** 16.7 m **9.** 137.7 m **11.** 11.7 yd **13.** 10 ft **15.** 45°

Chapter 7 Study Guide and Review

1. ratio; proportion **2.** rate; unit rate **3.** similar; scale factor **4.** dilation; enlargement; reduction **5–7.** Possible answers given. **5.** $\frac{1}{2}$, $\frac{2}{4}$ **6.** $\frac{3}{6}$, $\frac{4}{8}$ **7.** $\frac{7}{12}$, $\frac{14}{24}$ **8.** yes **9.** no **10.** yes **11.** no **12.** \$0.30 per disk; \$0.29 per disk; 75 disks **13.** \$3.75 per box; \$3.75 per box; unit prices are the same. **14.** \$2.89 per divider; \$4.00 per divider; 8-pack **15.** 90,000 m/h **16.** 4500 ft/min **17.** $583\frac{1}{3}$ m/min **18.** $80\frac{2}{3}$ ft/s **19.** 2160 m/h **20.** $x = 15$ **21.** $h = 6$ **22.** $w = 21$ **23.** $y = 29\frac{1}{3}$

24.
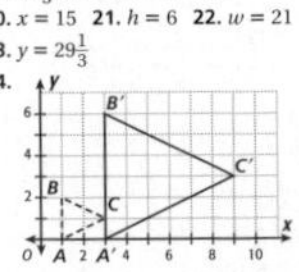
25.
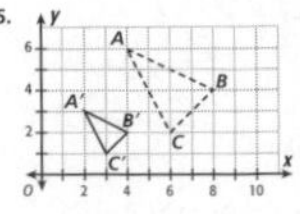

Selected Answers 797

748 Selected Answers pages 794-797

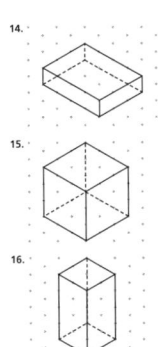

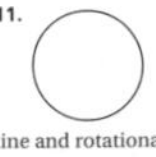

26.
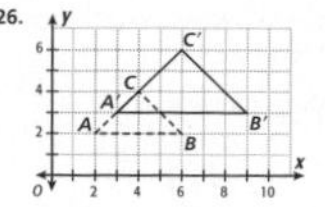

27. 12.5 in. 28. 3.125 in. 29. 64.8 m 30. 6.6 in. 31. $2.\overline{7}$:1; enlarges 32. 2.5:1; enlarges 33. 1:100; reduces 34. 1:1; preserves 35. 3:1 36. 9:1 37. 27:1

Chapter 8

8-1 Exercises

1. $\frac{3}{10}$ 2. 46% 3. 62.5% 4. $\frac{17}{20}$ 5. 40% 6. $\frac{8}{25}$ 7. 0.875 8. $33\frac{1}{3}$% 9. 10% 11. $\frac{3}{5}$ 13. 0.32 15. $\frac{109}{200}$ 17. 40%, 30%, 20%, 10% 19. 40%, 30%, 25%, 5% 21. 85% 23. a. $\frac{4}{25}$; 0.16 b. 23% 27. perpendicular 29. parallel 31. D

8-2 Exercises

1. 49.3% 2. 19.9% 3. 70.6% 4. 31.5 pages 5. 300% 7. 1% 9. 1.0% 11. 30 13. 2.6 15. 266 17. a. 30 b. 45 c. 150 19. a. 100 b. 50 c. 25 21. 21.5% 23. 16 29. irrational 31. not real

8-3 Exercises

1. 34.4 2. 168 3. 166.7 4. 320 5. ≈ 1.7 oz 6. 28 ft 7. 315 9. 850 11. 16.7 13. 570 15. 336 17. a. 300 b. 150 c. 75 19. a. 40 b. 20 c. 10 21. 48.2% 23. a. 49.5% b. 50.5% c. 49.1% d. 50.9% 27. 882 29. 45, 65

8-4 Exercises

1. 38% increase 2. 65% decrease 3. 100% increase 4. 64% increase 5. 25% decrease 6. 34% increase 7. 96.6% increase 8. $9773.60 9. 9% increase 11. 44% decrease 13. 20% decrease 15. ≈ 8.6% 17. 23% decrease 19. 30% decrease 21. 17% decrease 23. $600 25. 200 27. 50 29. 40% 31. a. $84 b. $156 c. $104 d. $56\frac{2}{3}$% 33. about 8361 ft 37. 364 m^2 39. 84 yd^2

8-5 Exercises

Note: All answers are estimates.

1. 100 2. 24 3. 25% 4. 21 5. 50% 6. 900 7. $4.50 9. 50% 11. 440 13. 10% 15. B 17. A 19. C 21. 150 23. 250 25. 1600 27. 33% 29. 400 31. 750 33. 50% 35. 50% 39. 120 ft^3 41. 132 in^3 43. 0.48 ft^3 45. D

8-6 Exercises

1. $510 2. $5.18 3. 22.5% 4. $499 5. $389.50 7. 18% 9. $330 11. $2.16 13. $1963.75 15. $2800 plus 3% of sales: $3100 to $3400 a month 17. a. $64,208 b. $14,275.95 c. ≈ 20.0% d. ≈ 22.2% 21. 40:3 23. 10,000:1

8-7 Exercises

1. $1794.38; $10,044.38 2. 5 years 3. $1635.30 4. 5.5% 5. $23,032.50 7. $1846.50 9. $33.75, $258.75 11. $446.25, $4696.25 13. $14.89, $411.89 15. $87.50, $787.50 17. $270, $1770 19. 6% 25. 14.4 27. $16\frac{2}{3}$% 29. 5

Chapter 8 Extension

1. $12,597.12 3. $14,802.44 5. $15,208.16 7. $2462.88 9. $6744.25

Chapter 8 Study Guide and Review

1. percent 2. percent change 3. commission 4. simple interest; principal; rate of interest 5. 0.4375 6. 43.75% 7. $1\frac{1}{8}$ 8. 112.5% 9. $\frac{7}{10}$ 10. 0.7 11. $\frac{1}{250}$ 12. 0.4% 13. 39% 14. 4200 ft 15. 3030 mi 16. 5 lb 7 oz 17. 20% 18. 472,750% 19. ≈ 12.38% 20. ≈ 25% 21. ≈ 25% 22. ≈ 13 23. ≈ 16 24. ≈ 6 25. ≈ 4.5 26. $10,990 27. $3.04 28. $1796.88 29. $500 30. 7% 31. $\frac{1}{2}$ yr, or 6 mo 32. 2-year loan; $50

Chapter 9

9-1 Exercises

1. 0.55; 0.45 2. 0.262 3. 0.738

4.

Team	A	B	C	D
Prob.	0.25	0.3	0.15	0.3

5. $\frac{1}{3}$; $\frac{1}{3}$; $\frac{1}{6}$; $\frac{1}{6}$ 7. 0.155 9. 0.319 11. 1 13. 0.465

15.

Person	Probability
Jamal	0.1
Elroy	0.2
Tina	0.2
Mel	0.2
Gina	0.3

17. 0.56 21. 59 in^2 23. ≈ 354.43 yd^2 25. B

9-2 Exercises

1. 0.34 2. 0.11 3. ≈ 0.186; ≈ 0.281; more likely to listen to a rock station 5. 0.433 7. 0.26 9. 0.06 11. 0.36 13. 0.308 17. $x = 2$ 19. $b = 5$ 21. D

9-3 Exercises

1–9. Possible answers are given.

1. 90% 2. 50% 3. 60% 5. 30% 7. 30% 9. 50% 13. 84 ft tall 15. 42 ft tall 17. B

9-4 Exercises

1. $\frac{1}{2}$ 2. $\frac{1}{3}$ 3. $\frac{1}{6}$ 4. $\frac{1}{36}$ 5. $\frac{1}{4}$ 6. $\frac{5}{18}$ 7. $\frac{5}{18}$ 9. $\frac{5}{6}$ 11. $\frac{1}{36}$ 13. 1 15. $\frac{1}{4}$ 17. $\frac{1}{8}$ 19. $\frac{3}{8}$ 21. $\frac{7}{8}$ 23. $\frac{1}{4}$ 25. a. $\frac{1}{4}$ b. $\frac{3}{4}$ 27. 90% 29. 0.375 31. $\frac{39}{50}$

9-5 Exercises

1. 1,757,600 2. ≈ 0.000000569 3. 0.81 4. ≈ 0.107 5. 6 ways 6. 9 combinations 7. 17,576,000 9. ≈ 0.5269 11. 18 chairs 13. 6 15. 12 17. 1920 23. 140 25. 20 27. 40

9-6 Exercises

1. 5040 2. 360 3. 20,160 4. 20 5. 3,628,800 6. 720 7. 56 8. 56 9. 6 11. 24 13. 5040 15. 3,838,380 17. 72 19. 39,916,800 21. 1 23. 10 25. n 27. $n!$ 29. n 31. 1 33. 504 35. 720 37. 21 39. a. 5040 b. 210 45. $135, $885 47. $15.99, $425.99 49. $21.60, $111.60

9-7 Exercises

1. dependent 2. independent 3. $\frac{1}{32}$ 4. $\frac{1}{8}$ 5. $\frac{28}{435}$ 6. $\frac{8}{203}$ 7. dependent 9. $\frac{1}{4}$ 11. $\frac{1}{20}$ 13. $\frac{1}{14}$ 15. a. $\frac{9}{100} = 0.09$ b. $\frac{3}{275} \approx 0.01$ c. $\frac{19}{825} \approx 0.02$ 19. 50% decrease 21. 440% increase 23. 28% decrease 25. B

9-8 Exercises

1. 1:20 2. 20:1 3. $\frac{1}{1000}$ 4. $\frac{1}{2250}$ 5. 1:74 6. 22,749:1 7. 1:17 9. $\frac{1}{10{,}000}$ 11. 1:844 13. 1:35, 35:1 15. 1:17, 17:1 17. 1:3, 3:1 19. 1:2 21. $\frac{1}{1275}$ 23. a. 237:1 b. $\frac{237}{238}$ 27. $\frac{1}{7}$ 29. 0 31. $\frac{22}{35}$ 33. A

Chapter 9 Study Guide and Review

1. probability; impossible; certain 2. sample space 3. permutation; combination 4. 0.75; 0.25 5. 0.17, or 17% 6. 0.28, or 28% 7. Possible answer: 40% 8. $\frac{1}{2}$ 9. 36 10. 30 11. 210 12. 126 13. $\frac{1}{216}$ 14. $\frac{13}{51}$ 15. 5:21

Chapter 10

10-1 Exercises

1. 4 hr 2. $t = 7$ 3. $x = 3.5$ 4. $r = 98$ 5. $b = 2$ 6. $q = 4$ 7. $a = 87$ 9. $m = -30$ 11. $g = 3\frac{1}{3}$ 13. $y = -5$ 15. $w = 2.5$ 17. $m = 15$ 19. $q = 3$ 21. $z = \frac{1}{3}$ 23. $k = 5$ 25. $n = 23$ 27. $y = 1.3$ 29. $b = -2$ 31. $\frac{x+5}{7} = 12; x = 79$ 33. 110,000 35. 25 in. 37. $12x + 3$ 39. $w - 15$ 41. C

10-2 Exercises

1. $d = 2$ 2. $y = 3$ 3. $e = 6$ 4. $c = 4$ 5. $h = 9$ 6. $x = -7$ 7. $x = -1$ 8. $y = 2$ 9. $p = -1$ 10. $z = \frac{2}{3}$ 11. $24.49 13. $k = -10$ 15. $w = 3$ 17. $y = 5$ 19. $h = 3$ 21. $m = -2$ 23. $x = -4$ 25. $n = \frac{3}{4}$ 27. $b = -13$ 29. $x = 17$ 31. $y = -11$ 33. $h = 4$ 35. $b = \frac{10}{13}$ 37. $12.20 per hr 39. 212°F 45. −2.5 47. $-2\frac{1}{7}$ 49. $\frac{17}{19}$

10-3 Exercises

1. $x = 1$ 2. $a = 7$ 3. $x = 2$ 4. $y = -4$ 5. $x = 1$ 6. $n = 2$ 7. $d = 3$ 8. $x = 7$ 9. 32 chairs 11. $x = 1$ 13. $y = 7$ 15. no solution 17. $x = 22.9$ 19. $a = 11$ 21. $y = 2$ 23. $n = 2$ 25. $x = 5$ 27. $m = 3.5$ 29. $x = 7$ 31. 360 units 33. 24, 25 35. a. 17 b. 11 41. ≈ $0.175 per oz; ≈ $0.187 per oz; 20 oz 43. ≈ $0.199 per oz; $0.184 per oz; 20 oz 45. $249 per monitor; $275 per monitor; 3 monitors

10-4 Exercises

1. $k > 3$ 2. $z \leq 20$ 3. $y < -7$ 4. $x \leq -2$ 5. $y \geq 3$ 6. $k > 5$ 7. $x < 3$ 8. $b \geq -\frac{1}{4}$ 9. $h \leq 1$ 10. $c > 2$ 11. $d < \frac{7}{18}$ 12. $m \leq 1\frac{3}{4}$ 13. at least 16 caps 15. $x > 4$ 17. $q \leq 2$ 19. $x \leq -8$ 21. $a \geq -3$ 23. $k \geq 3$ 25. $r < 3$ 27. $p \leq \frac{22}{3}$ 29. $w > -1$ 31. $a > \frac{1}{2}$ 33. $q < 6$ 35. $b < 2.7$ 37. $f \leq -27$ 39. at most 26 chairs 41. at least 38 games 47. 650 49. 1.44

10-5 Exercises

1. $\ell_2 = P - \ell_1 - \ell_3$ 2. $\ell_1 = P - \ell_2 - \ell_3$ 3. $A = C + B - 2$ 4. $B = A - C + 2$ 5. $d_1 = \frac{2A}{d_2}$ 6. $b = \sqrt{c^2 - a^2}$ 7. $n = \frac{S}{180} + 2$ 8. $C = \frac{5}{9}(F - 32)$ 9. $y = -3x + 15$ 10. $y = \frac{9}{2}x + 7$ 11. $y = 2x - 1$ 13. $A_3 = 180 - A_1 - A_2$ 15. $c = p - a - 100$ 17. $c = \sqrt{\frac{E}{m}}$ 19. $b_1 = \frac{2A}{h} - b_2$ 21. $y = -\frac{9}{2}x - 5$ 23. $x = 2y + 4$ 25. $k = \frac{10\ell^2}{3}$ 27. $y = \frac{x}{22}$ 29. $z = \sqrt{4y}$ 31. $m = \frac{y-b}{x}$ 33. $b = y - mx$ 35. $y = -6x + 8$ 37. a. $T = \frac{E}{P}$ b. 1.5 hr 39. a. $T_1 = \frac{P_2 \cdot T_2}{P_1}$ b. 2.5 hr 43. $x = 0.5$ 45. B

10-6 Exercises

1. yes 2. yes 3. yes 4. no 5. (2, 3) 6. (1, −1) 7. (3, 12) 8. (4, 13) 9. (3, 0) 10. (0, 7) 11. (5, 3) 12. (8, 12) 13. (5, 2) 14. (3, −7) 15. (0, −2) 16. (−9, 3) 17. no 19. no 21. (−1, −1) 23. (2, −1) 25. (−6, 0) 27. (2, 3) 29. (1, 3) 31. (4, 1) 33. (2, 4) 35. (2, −3) 37. (0, 0) 39. (2, 1) 41. $\left(\frac{1}{10}, \frac{2}{15}\right)$ 43. (12, 8) 45. (6, −3) 47. $\left(\frac{1}{5}, \frac{2}{3}\right)$ 49. 14 and 4 51. a. $m + u = 2000$ b. $40m + 25u = 62{,}000$ c. 800 main-floor and 1200 upper-level 55. 12 57. 9 59. D

Chapter 10 Study Guide and Review

1. system of equations 2. solution of a system of equations 3. $m = 10$ 4. $y = -8$ 5. $c = -16$ 6. $r = -3$ 7. $t = 16$ 8. $w = 64$ 9. $r = -6$ 10. $h = -25$ 11. $x = 52$ 12. $d = -33$ 13. $a = 67$ 14. $c = 90$ 15. $y = 2$ 16. $h = 3$ 17. $t = -1$ 18. $r = 3$ 19. $z = 4$ 20. $a = 12$ 21. $s = 4$ 22. $c = 24$ 23. $x = \frac{3}{4}$ 24. $y = \frac{3}{5}$ 25. $z > 1$ 26. $h \geq 6$ 27. $a < 24$ 28. $x \geq -6$ 29. $\ell = \frac{P - 2w}{2}$ 30. $r = \frac{A - P}{Pt}$ 31. $C = \frac{5}{9}(F - 32)$ 32. $y = -\frac{2}{3}x + 3$ 33. $y = \frac{7 - x}{3}$ 34. $x = 3y + 2$ 35. (2, 9) 36. (8, 10) 37. (3, 5) 38. (3, −2) 39. (6, −2) 40. $\left(-\frac{1}{2}, -2\right)$

Chapter 11

11-1 Exercises

1. linear 2. linear 3. not linear 4. 130 ft, 133 ft, 136 ft; linear 5. linear 7. not linear 9. not linear 11. $900, $1275, $1650, $2025, $2400; linear 13. (−1, 3), (0, 5), (1, 7) 15. (−1, −11), (0, −10), (1, −9) 17. (−1, −1), (0, 3), (1, 7) 19. (−1, 6), (0, 7), (1, 8) 21. 509.6 N 23. linear 25. a. $w = \$7.50$ b. $E = 7.50h$

c.
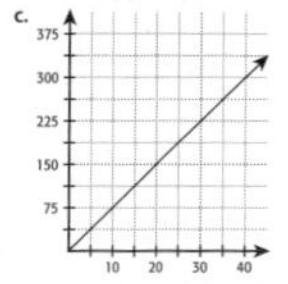

d. yes 29. $\frac{1}{4}$ 31. $\frac{1}{36}$ 33. B

11-2 Exercises

1. 1 2. 2 3. $\frac{1}{2}$ 4. $\frac{1}{2}$ 5. −2 6. perpendicular 7. parallel

8.
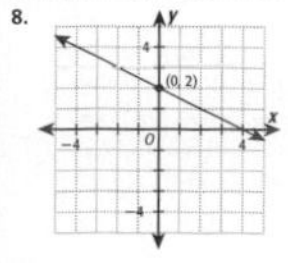

9.
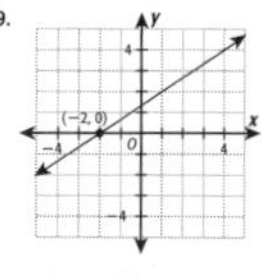

11. $-\frac{1}{2}$ 13. $-\frac{2}{3}$ 15. $-\frac{5}{6}$ 17. 0 19. perpendicular

21.
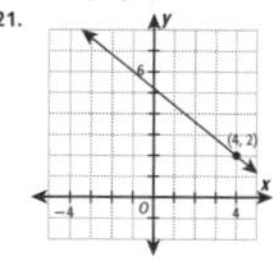

23. line 1: 1; line 2: −2; neither 25. line 1: −1; line 2: slope = $-\frac{1}{2}$; neither

27.
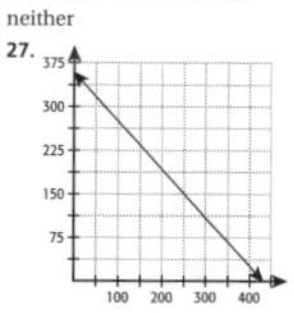

31. 12 $units^2$ 33. 120 $units^2$ 35. D

11-3 Exercises

1. x-intercept: 5, y-intercept: −5 2. x-intercept: 6, y-intercept: 4 3. x-intercept: −5, y-intercept: −3 4. x-intercept: 2, y-intercept: −5 5. $y = \frac{1}{2}x$; $m = \frac{1}{2}$; $b = 0$ 6. $y = 3x - 14$; $m = 3$, $b = -14$ 7. $y = \frac{1}{3}x - 3$; $m = \frac{1}{3}$; $b = -3$ 8. $y = -\frac{1}{2}x + 4$; $m = -\frac{1}{2}$; $b = 4$ 9. $m = 3.5$; $b = 22$ 10. $y = 4x - 2$ 11. $y = -2x + 5$ 12. $y = \frac{1}{3}x + 4$ 13. x-intercept: 5, y-intercept: 10 15. x-intercept: 3, y-intercept: −18 17. $y = -2x$; $m = -2$; $b = 0$ 19. $y = -2x - 2$; $m = -2$; $b = -2$ 21. $m = 15$; $b = 300$ 23. $y = -x$

25.
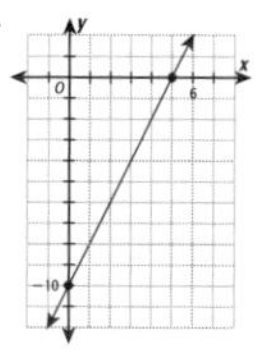

27.
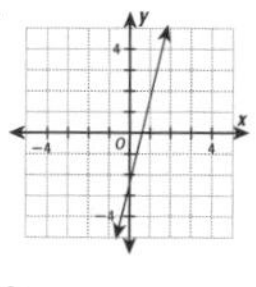

29. a.
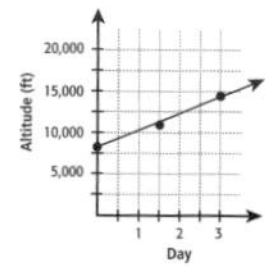

b. 8255 ft c. $y = 2000x + 8255$ d. yes 31. Slope: 955 ft per day; y-intercept: 16,500 ft 33. 100 35. 40% 37. A

11-4 Exercises

1–6. Possible answers given.

1. (−7, 4), −2 2. (12, 9), 5 3. (1.8, −2.4), 2.1 4. (1, −1), 11 5. (9, −8), −6 6. (−3, 7), 4 7. $y - 4 = 3x$ 8. $y - 8 = -10(x + 13)$ 9. $y - 2{,}450 = -12.5(x - 44)$, (240, 0) or 240 minutes

11–15. Possible answers given.

11. (−4, −7), 3 13. (8, 11), 14 15. (5, −7), 1 17. $y = 4(x + 1)$ 19. $y - 4 = 3(x + 1)$ 21. $y + 8 = -1(x + 6)$ 23. Possible answer: $y - 315 = -0.6(x - 50)$

25. $\ell - 15 = d$ 29. $x > 4$ 31. $x < -13$ 33. C

11-5 Exercises

1. yes 2. $y = 5x$ 3. $y = 4x$ 4. $y = \frac{4}{5}x$ 5. $y = \frac{1}{2}x$ 6. $y = 110x$ 7. $y = \frac{1}{8}x$ 8. no direct variation 9. no 11. $y = \frac{1}{4}x$ 13. $y = \frac{4}{11}x$ 15. $y = \frac{1}{10}x$ 17. yes 19. no 21. a. yes b. 156 lb 23. yes 27. $x = -4$ 29. $a = -37$ 31. B

11-6 Exercises

1.
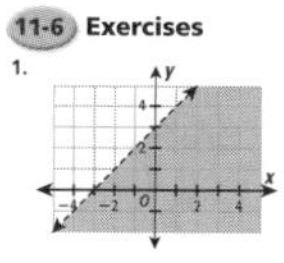

2.
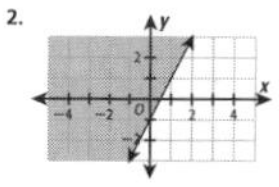

3.
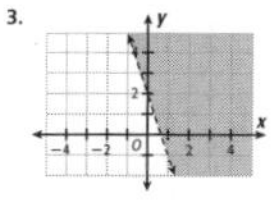

4.
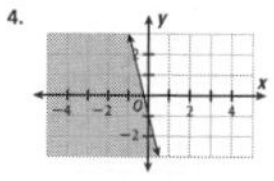

5.
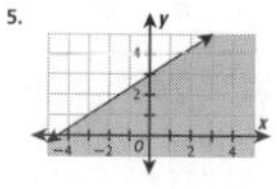

6.
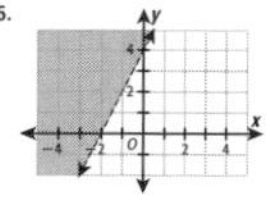

7. a. $10g + 12h \leq 150$ b. yes

9.
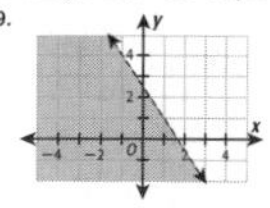

11.
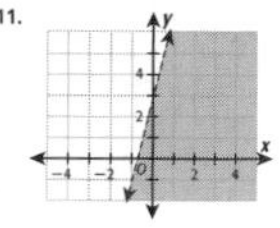

13.
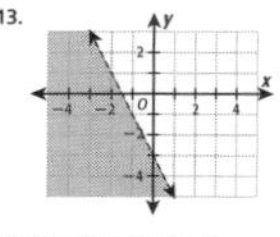

15. yes 17. yes 19. yes 21. b. Possible answer: (1, 6) c. No d. the upper side e. Possible answer: (0, 4) 23. $2x + 3y \leq 18$ 25. a. $d \leq 800t$ c. yes 29. $h = \frac{2A}{b}$ 31. $b_2 = \frac{2A}{h} - b_1$ 33. A

11-7 Exercises

1. $y = 1.9x - 0.26$ 2. $y = -\frac{13}{30}x + 20.8$ 3. $y = 0.8x + 72.2$ 5. $y = -10.7x + 9.6$ 7. negative 9. positive 11. a. 5.75 b. ≈ 31.25% 13. about 42% 17. perpendicular 19. neither

Chapter 11 Extension Systems of Equations

1. yes 3. no 5. (0, 3) 7. yes

Chapter 11 Study Guide and Review

1. x-intercept; y-intercept 2. slope-intercept form; point-slope form 3. direct variation 4. linear 5. linear 6. not linear 7. not linear 8. not linear 9. linear 10. not linear 11. not linear 12. $\frac{2}{3}$ 13. −4 14. $\frac{6}{5}$ 15. −1 16. −1 17. $\frac{7}{5}$ 18. $-\frac{9}{4}$ 19. $y = \frac{3}{2}x + 4$; $m = \frac{3}{2}$; $b = 4$ 20. $y = \frac{5}{3}x - 3$; $m = \frac{5}{3}$; $b = -3$ 21. $y = -\frac{4}{5}x + 2$; $m = -\frac{4}{5}$; $b = 2$ 22. $y = \frac{7}{4}x + 3$; $m = \frac{7}{4}$; $b = 3$ 23. $y = 3x + 4$ 24. $y = -2x + 1$ 25. $y = -\frac{1}{2}x + 5$ 26. $y = \frac{1}{2}x - \frac{5}{2}$ 27. $y - 3 = 4(x - 1)$ 28. $y - 4 = -2(x + 3)$ 29. $y + 2 = -\frac{3}{5}x$ 30. $y = \frac{2}{7}x$ 31. $y = 6x$ 32. $y = 12x$ 33. $y = \frac{1}{7}x$

34.
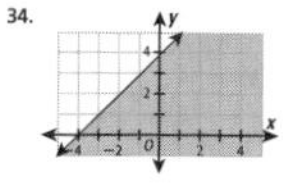

35.
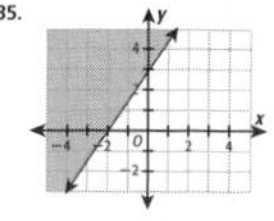

36.
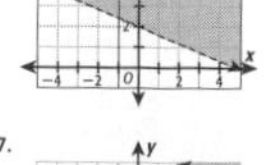

37.
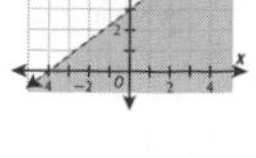

38.

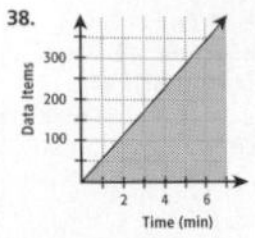

39–42. Possible answers given.
39. $y = 1.5x$ **40.** $y = 1.25x - 0.7$
41. $y = 1.1x - 3.5$
42. $y = -0.6x + 70$

Chapter 12

12-1 Exercises

1. yes; 2 **2.** no **3.** yes; $\frac{1}{9}$ **4.** yes; −7 **5.** no **6.** yes; −3 **7.** 37 **8.** 94 **9.** −84 **10.** 156 **11.** 6 oz **13.** yes; −1 **15.** yes; 23 **17.** yes; 0.3 **19.** 1.2 **21.** −15 **23.** 23, 26, 29 **25.** 57, 46, 35 **27.** −54, −66, −78 **29.** 1, 2, 3, 4, 5 **31.** 0, 0.25, 0.5, 0.75, 1 **33.** 32, $33\frac{4}{5}$, $35\frac{3}{5}$, $37\frac{2}{5}$, $39\frac{1}{5}$ **35.** 78, 92, 106, 120 **37.** 11:55, 11:50, 11:45, 11:40 **39. a.** \$127.50, \$180, \$232.50, \$285 **b.** 3 hr **43.** $x > 2$ **45.** $p \geq \frac{13}{2}$ **47.** $c > 5$ **49.** A

12-2 Exercises

1. no **2.** yes; 3 **3.** yes; −1 **4.** yes; 1.5 **5.** yes; 2 **6.** yes; 2 **7.** 6144 **8.** $\frac{1}{3}$ **9.** $\frac{1}{8}$ **10.** 16,384 **11.** \$7.62 per hour **13.** no **15.** yes; $\frac{1}{2}$ **17.** yes; $\frac{1}{3}$ **19.** 2401 **21.** 192 **23.** 7.59375 **25.** 12, 6, 3 **27.** $-\frac{1}{27}$, $\frac{1}{9}$, $-\frac{1}{3}$ **29.** 1, 1, 1, 1, 1 **31.** 100, 110, 121, 133.1, 146.41 **33.** 10, 2.5, 0.625, 0.15625, 0.0390625 **35.** $\frac{2}{5}$ **37.** $\frac{3}{5}$ **39.** 18 **41.** 162 **43.** 4096 cells **49.** $\frac{1 \text{ gal}}{4 \text{ qt}}$ **51.** $\frac{100 \text{ cg}}{1 \text{ g}}$ **53.** $\frac{36 \text{ in}}{1 \text{ yd}}$

12-3 Exercises

1. 232, 301, 379 **2.** 260, 355, 465 **3.** 180, 264, 372 **4.** 524, 778, 1104 **5.** $\frac{7}{8}, \frac{8}{9}, \frac{9}{10}$ **6.** −12, 13, −14 **7.** 5, 6, 4 **8.** 216, 343, 512 **9.** $\frac{4}{3}$, 2, $\frac{12}{5}$, $\frac{8}{3}$, $\frac{20}{7}$ **10.** 12, 20, 30, 42, 56 **11.** 2, 1, $\frac{2}{3}$, $\frac{1}{2}$, $\frac{2}{5}$

12.

2	1
24	25
442	441
7920	7921

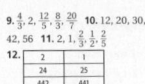

13. 92, 109, 127 **15.** 109, 145, 190 **17.** 5, 5, 5 **19.** 1.00001, 1.000001, 1.0000001 **21.** 0, $\frac{1}{3}$, $\frac{1}{2}$, $\frac{3}{5}$, $\frac{2}{3}$ **23.** $\frac{3}{2}$, 2, $\frac{9}{4}$, $\frac{12}{5}$, $\frac{5}{2}$ **25.** 3rd, 6th, 9th, 12th terms; 15th, 18th, 21st, 24th terms **27.** geometric; $a_n = 55 \cdot 2^{(n-1)}$ **29.** arithmetic; $a_n = 55n$ **33.** 3, −9 **35.** 9, 3

12-4 Exercises

1.

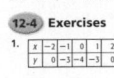

x	−2	−1	0	1	2
y	0	−3	−4	−3	0

2.

x	−2	−1	0	1	2
y	−2	1	4	7	10

3.

x	−2	−1	0	1	2
y	5	−1	−3	−1	5

4.

x	−2	−1	0	1	2
y	3	2	1	0	−1

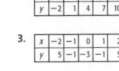

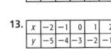

5. yes **6.** yes **7.** yes **8.** 1.2, 11.4, −2.2 **9.** 2, 1, 1 **10.** 5, 7, 3

11.

x	−2	−1	0	1	2
y	−8	−6	−4	−2	0

13.

x	−2	−1	0	1	2
y	−5	−4	−3	−2	−1

15. no **17.** yes **19.** 1, 19, 64 **21.** D = 1, 4, 8, 14; R = 27, 39, 50, 62 **23.** D = 30, 40, 55, 75; R = 20, 30, 45, 65 **25. a.** \$0.20 **b.** any non-negative number of hours ($x \geq 0$) **c.** 550 hr **26.** 55 **27. a.** yes **b.** D = 0, 20, 40, 60, 80, 100; R = 0, 200, 400, 600, 800, 1000 **31.** 50% **33.** 311.75 **35.** 64% **37.** A

12-5 Exercises

1. $f(x) = x$ **2.** $f(x) = \frac{5}{3}x - 2$ **3.** $y = 3x + 2$ **4.** $y = -2x + 4$ **5.** $f(x) = 15x + 400$; \$505 **7.** $f(x) = \frac{1}{2}x + 2$ **9.** $y = 6x - 5$ **11. a.** $f(x) = 5x + 1245$ **b.** 2745 ft; 1500 ft **13.** $f(x) = 3x + 4$; 28 lb **19.** $y + 6 = -2(x - 6)$ **21.** $y + 3 = 1.4(x - 1)$

12-6 Exercises

1.

x	$f(x)$
−2	$\frac{1}{9}$
−1	$\frac{1}{3}$
0	1
1	3
2	9

2.

x	$f(x)$
−2	450
−1	150
0	50
1	$\frac{50}{3}$
2	$\frac{50}{9}$

3.

x	$f(x)$
−2	$\frac{3}{4}$
−1	$\frac{3}{2}$
0	3
1	6
2	12

4.

x	$f(x)$
−2	0.0004
−1	0.002
0	0.01
1	0.05
2	0.25

5. 5.12×10^{-4} g **6.** ≈ 0.0046 mg

7.

x	$f(x)$
−2	$\frac{2}{9}$
−1	$\frac{2}{3}$
0	2
1	6
2	18

9.

x	$f(x)$
−2	$\frac{9}{4}$
−1	$\frac{3}{2}$
0	1
1	$\frac{2}{3}$
2	$\frac{4}{9}$

11. $f(x) = 500 \cdot 2^x$; \$8000 **13.** $\frac{1}{32}$, 1, 32 **15.** $\frac{1}{100{,}000}$, 1, 100,000 **17.** $f(x) = 3 \cdot 2^x$ **19.** $f(x) = 1 \cdot 3^x$

21.

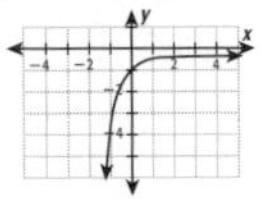

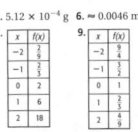

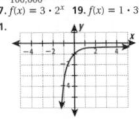

23. 1.5625% **25.** 3 hr; $f(x) = 160 \cdot \left(\frac{1}{2}\right)^x$ where x is the number of 3-hour intervals. **a.** 10 mg **b.** 0.625 mg **27.** 15 mg **31.** yes; 2 **33.** no **35.** yes; 0.1

Selected Answers

12-7 Exercises

1.

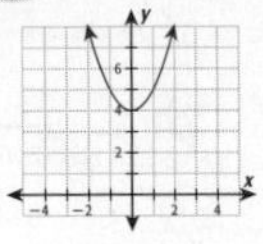

2.

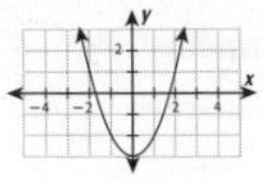

3.

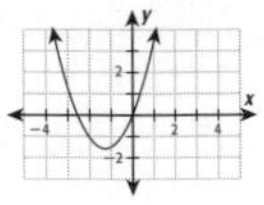

4.

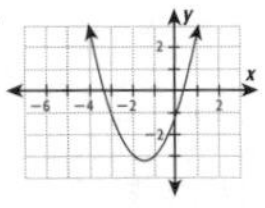

5.

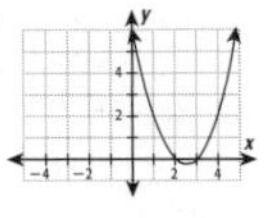

6.

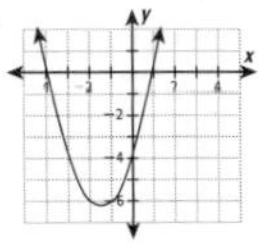

7.

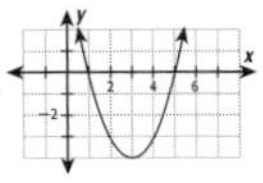

8.

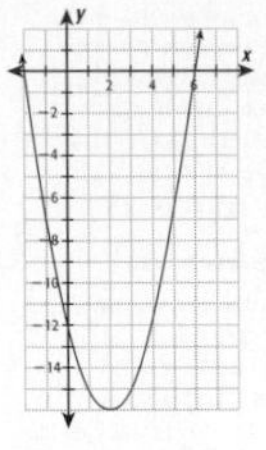

9. 5.1 ft

11.

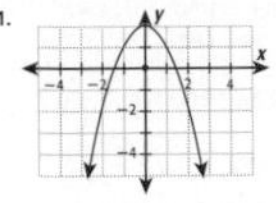

13.

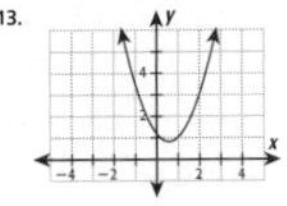

15.

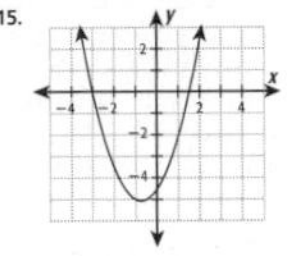

17.

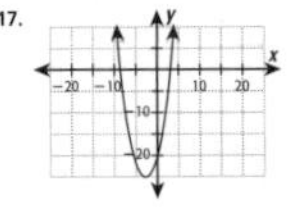

19. 14, 5, 14 **21.** 3, 0, 15 **23.** 26, 5, 20 **25.** $x = 5$, $x = -11$ **27.** $x = 2$, $x = -1$ **29.** $x = 1.8$, $x = -2.6$ **31.** 5 and 5; 50 **33. a.** \$4860, \$4940, \$5000, \$5040, \$5060; 7 **b.** \$115 **39.** −2, 12 **41.** −1, −4 **43.** −4, 1 **45.** −2, 4 **47.** H

12-8 Exercises

1. no **2.** yes

3.

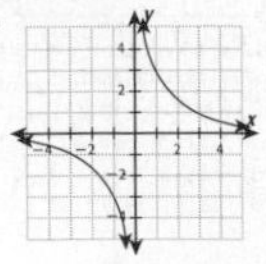

4.

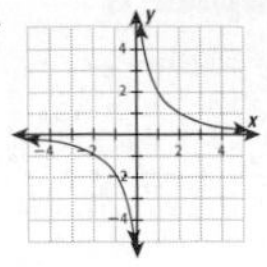

5.

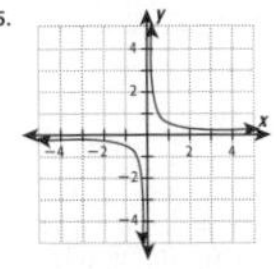

6. $y = \frac{12}{x}$; $1\frac{1}{3}$ amps **7.** yes

9.

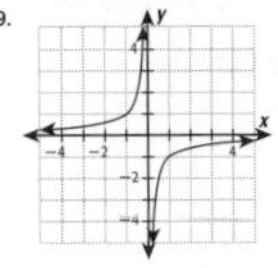

11.

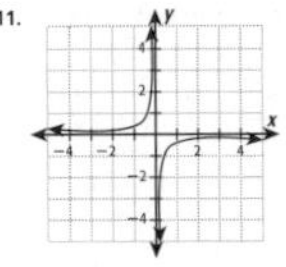

Selected Answers

13. $y = \frac{4}{x}$ **15.** $y = \frac{32}{x}$ **17.** 10 cm **19.** \$1250 **23.** 9, 1, −1 **25.** 300, 243, 192 **27.** −36, −35, −32 **29.** C

Chapter 12 Study Guide and Review

1. sequence **2.** arithmetic sequence; geometric sequence **3.** Fibonacci sequence **4.** function; domain; range **5.** 31 **6.** 0.65 **7.** $\frac{14}{3}$ **8.** −640 **9.** $\frac{32}{729}$, or ≈ 0.0439 **10.** −1 **11.** 4, 7, 10, 13 **12.** 2, 5, 10, 17 **13.** −6, 12, −2, 16 **14.** 3, 4, 8, 26 **15.** −4, 10, −11 **16.** 1, 17, −1 **17.** 0, 2, −4 **18.** 0, 0, 3 **19.** 5, 15, 9 **20.** 1, −1, −1 **21.** $f(x) = x - 1$ **22.** $f(x) = \frac{1}{2}x + 4$

23.

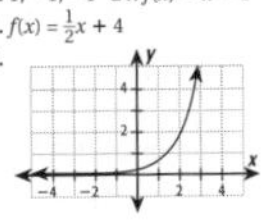

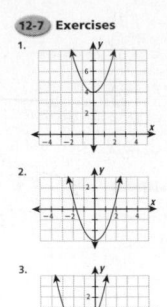

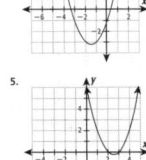

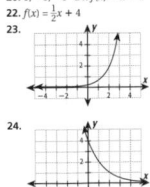

24.

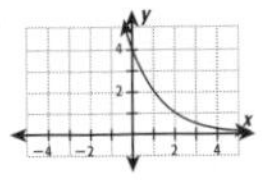

25.

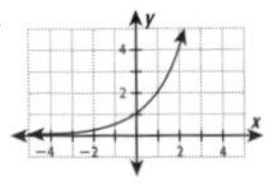

26.

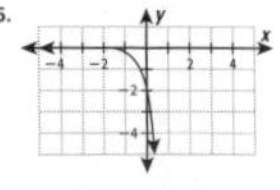

27.

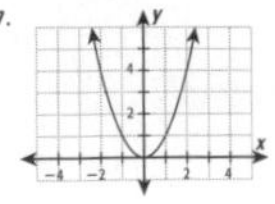

28.

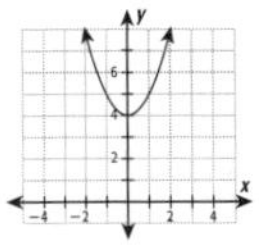

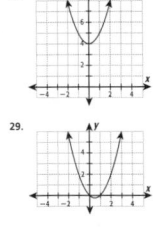

29.

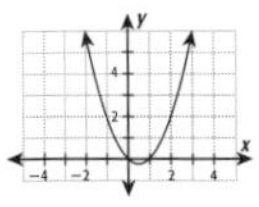

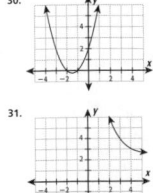

30.

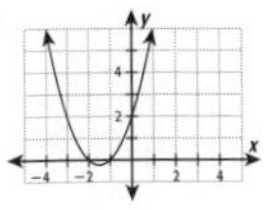

31.

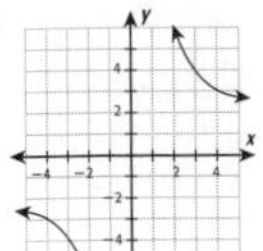

32.

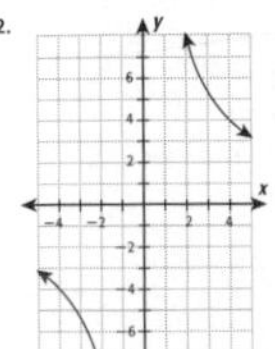

33.

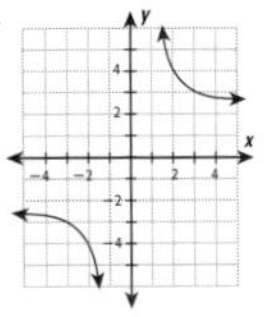

34.

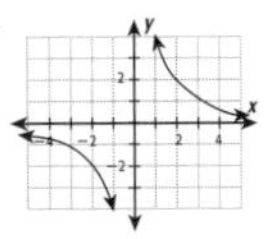

Chapter 13

13-1 Exercises

1. yes **2.** no **3.** no **4.** yes **5.** binomial **6.** trinomial **7.** not a polynomial **8.** monomial **9.** 7 **10.** 3 **11.** 6 **12.** 3 **13.** 26 feet **15.** no **17.** yes **19.** not a polynomial **21.** binomial **23.** 4 **25.** 4 **27.** monomial; 2 **29.** binomial; 2 **31.** trinomial; 6 **33.** not a polynomial **35.** trinomial; 3 **37.** not a polynomial

39.

Gas Mileage (mi/gal)			
	40 mi/h	50 mi/h	60 mi/h
Compact	28	30	27
Midsize	21	22	20
Van	15	17	13

45. 4.5×10^{-7} **47.** 12 **49.** 13 **51.** B

13-2 Exercises

1. $-2b^2$ and $3b^2$, $4b$ and $-b$ **2.** $-4m^2n^2$ and $3m^2n^2$ **3.** $9x^2 + 4x - 6$ **4.** $2b^4 - 10b^2 + b + 15$ **5.** $6x - 21$ **6.** $27a^2 - 37a$ **7.** $-40{,}000 + 4000x - 5x^3$ **9.** $9rs$ and $3rs$, $-2r^2s^2$ and $4r^2s^2$ **11.** $4f^2g + 4fg^2$ **13.** $9b - 12 - 4b^2$ **15.** $15s^2 + 2s - 2$ **17.** $3x^2 - 14x + 15$ **19.** $4m^2 - 24m$ **21.** $15mn$ **23.** $40 - 1000d^2$ **25.** $82xy + 41y$ in^2 **27.** −13 **29.** 30 **31.** $x \geq 5$ **33.** $x = \frac{40}{3}$

13-3 Exercises

1. $4x^3 + 3x + 5$ **2.** $32x - 12$ **3.** $9m^2n + 8mn$ **4.** $8b^2 + 2b + 1$ **5.** $11ab^2 + 3ab + ab^2 - 4$ **6.** $5h^4j + 2hj^3 - 4hj - 1$ **7.** $128 + 32w$ in. **9.** $5g^2 + g - 1$ **11.** $-h^6 + 9h^4 - h$ **13.** $11t^2 - 4t + 10$ **15.** $-w^2 - 6w - 3$ **17.** $3y^2 + xy$ **19.** $-3s^4t - 6st^3 + 4st^2 + 3st^4$ **21.** $8w^2y + wy^2 - 5wy$ **23.** $0.75n^2 + 8n + 16$ **25.** $4x^2 - 48x + 500$ miles **29.** 512 cm^3 **31.** 48 **33.** $\frac{90}{7}$ **35.** C

13-4 Exercises

1. $-4x^2y$ **2.** $4x - 3xy^4$ **3.** $-2x^2 + 7x - 4$ **4.** $6y^2 + 3y - 5$ **5.** $-2b^3 + 5b^2 - b + 4$ **6.** $-3b^2 + 4b + 10$ **7.** $6m^2n + 2mn^2 - 2mn$ **8.** $3x^2 - 8x - 2$ **9.** $-2x^2y - 5xy + 10x - 8$ **10.** $-6ab^2 - 7a^2b + 7ab - 6$ **11.** $-3x^3 + 7x^2 - 10x + 18$ in^3 **13.** $-2v + 4v^2$ **15.** $-4xy^2 - 2xy$ **17.** $6a - 5$ **19.** $3x^2 - x + 3$ **21.** $-5p^3 - 4p^2 - 2p^2t^2 + 8pt^2$ **23.** $3s^2 - 9s - 5$ **25.** $g^2h - 3gh$ **27.** $-3pq^2 - 11p^2q + 4pq$ **29.** $a^2 - 2a + 11$ cm^2 **31.** $25y + 16.25$ dollars **35.** $1\frac{1}{24}$ **37.** $4\frac{23}{40}$

13-5 Exercises

1. $-8s^3t^5$ **2.** $6x^6y^6$ **3.** $-24h^6j^{10}$ **4.** $15m^5$ **5.** $6hm - 8h^2$ **6.** $3a^3b^2 - 3a^2b^3$ **7.** $-2x^3 + 8x^2 - 24x$ **8.** $10c^3d^4 - 20c^5d^3 + 15c^3d^2$ **9.** $A = \frac{1}{2}b_1h + \frac{1}{2}b_2h$; 70 in^2 **11.** $3g^3h^8$ **13.** $-s^5t^4$ **15.** $8z^3 - 6z^2$ **17.** $-6c^4d^3 + 12c^2d^3$ **19.** $-20s^4t^3 - 24s^3t^3 + 8s^4t^4$ **21.** $-36b^6$ **23.** $6a^3b^6$ **25.** $-2m^5 + 12m^3$ **27.** $x^4 - x^5y^4$ **29.** $2f^2g^2 + f^3g^2 - f^2g^5$ **31.** $12m^4p^8 - 6m^3p^7 + 15m^4p^5$

33. a.

$220p - pa$	$226p - pa$
$205p - \frac{1}{2}pa$	$211p - \frac{1}{2}pa$

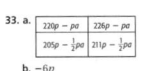

b. $-6p$

37. acute equilateral **39.** right scalene **41.** H

13-6 Exercises

1. $xy + 3x - 4y - 12$ **2.** $x^2 + 4x - 12$ **3.** $6m^2 + 4m - 32$ **4.** $2h^2 + 11h + 15$ **5.** $m^2 - 8m + 15$ **6.** $3b^2 + 7bc + 2c^2$ **7.** $200 - 60x + 4x^2$ ft^2 **8.** $x^2 + 6x + 9$ **9.** $b^2 - 16$ **10.** $x^2 - 10x + 25$ **11.** $x^2 + 3x - 10$ **13.** $w^2 + 8w + 15$ **15.** $6m^2 + 7m - 3$ **17.** $8t^2 + 2t - 1$ **19.** $6n^2 + 14bn - 12b^2$ **21.** $x^2 - 8x + 16$ **23.** $x^2 - 9$ **25.** $9x^2 - 6x + 1$ **27.** $m^2 - 25$ **29.** $q^2 + 9q + 20$ **31.** $g^2 - 9$ **33.** $10t^2 + 7t - 6$ **35.** $4a^2 + 20ab + 25b^2$ **37.** $w^2 - 25$ **39.** $15r^2 - 22rs + 8s^2$ **41.** $p^2 + 20p + 100$ **43.** $(M + \frac{1}{4}M)(V + b) = c$; $\frac{5}{4}MV + \frac{5}{4}Mb = c$ **47.** $\frac{2}{9}$ **49.** 0

Chapter 13 Extension

1. $3a^3$ **3.** $6a^2b$ **5.** $3ab^4c^4$ **7.** $2x^3 + 3x^2$ **9.** $p^4q^3 - 4p^5q$ **11.** $3a^3b^5 - 2a^9$ **13.** $2m^2n^2(3n^2 - 4m)$ **15.** $5z^3(3 + 5z^3)$ **17.** $6a^2(4 + 3a + a^5)$

Chapter 13 Study Guide and Review

1. polynomial; degree **2.** FOIL; binomials **3.** binomial; trinomial **4.** trinomial **5.** not a polynomial **6.** not a polynomial **7.** monomial **8.** not a polynomial **9.** binomial **10.** 8 **11.** 4 **12.** 3 **13.** 5 **14.** 6 **15.** $6t^2 - 2t + 1$ **16.** $11gh - 9g^2h$ **17.** $12mn - 6m$ **18.** $8a^2 - 10b$ **19.** $26st^2 - 15t$ **20.** $6x^2 - 2x + 5$ **21.** $4x^4 + x^2 - x + 7$ **22.** $3h^2 + 8h + 9$ **23.** $xy^2 - 2x^2y + 2xy$ **24.** $13n^2 + 12$ **25.** $5x^2 - 6$ **26.** $-w^2 - 12w + 14$ **27.** $-4x^2 + 14x - 12$ **28.** $4ab^2 - 11ab + 4a^2b$ **29.** $-p^3q^2 - 5p^2q^2 - 2pq^2$ **30.** $5s^2t^3 - 10s^2t^4 + 35st^3$ **31.** $12a^4b^3 + 30a^3b^3 - 36a^3b + 24a^2b^2$ **32.** $6m^3 - 15m^2 + 3m$ **33.** $-24gh^5 + 12g^3h^3 - 30h^2 + 12gh$ **34.** $2j^5k^3 - \frac{3}{2}j^4k^4 + j^6k^5$ **35.** $-8x^6y^{12} + 10x^7y^{14} - 14x^3y^6 + 6x^3y^7$ **36.** $p^2 - 8p + 15$ **37.** $b^2 + 10b + 24$ **38.** $4r^2 + 19r + 5$ **39.** $2a^2 - 5ab - 12b^2$ **40.** $m^2 - 16m + 64$ **41.** $4t^2 - 25$ **42.** $8b^2 + 4bt - 40t^2$ **43.** $100 - 4x^2$ **44.** $y^2 + 20y + 100$

Chapter 14

14-1 Exercises

1. ∈ **2.** ∉ **3.** Yes, $E \subset R$. **4.** No, $P \not\subset S$. **5.** finite **6.** infinite **7.** ∈ **9.** Yes, $F \subset T$. **11.** No, $P \not\subset O$. **13.** infinite **15.** finite **17.** ∈ **19.** ∈ **21.** ∈ **23.** finite **25.** infinite **27.** finite **29.** {25} **31.** femur ∈ {human bones} **33.** Miami ∈ {state capitals} **37.** $-4m^2 + 12m - 24$ **39.** $-11x^2y + 4xy^2 + 16xy$ **41.** F

14-2 Exercises

1. {2, 4, 6} **2.** {10, 12, 14} **3.** $\{x \mid 2 \leq x \leq 5\}$ **4.** $\{x \mid 0 \leq x \leq 7\}$ **5.** {1, 2, 3, 4, 5, 6, 8, 10, 12} **6.** $\{x \mid x > 0\}$ **7.** {integers} **8.** {rational numbers} **9.** {−4} **11.** {1, 3, 5} **13.** {−12, −10, −8, −6, −4, 0} **15.** {real numbers} **17.** $F \cup G$ = {−2, −1, 0, 1, 2}; $F \cap G$ = {−2, 0, 2} **19.** $R \cup M = \{x \mid x < 6$ or $x \geq 7\}$; $R \cap M = \varnothing$ **21.** $A \cup B$ = {integers}; $A \cap B = \varnothing$ **23.** $P \cup M$ = {even integers}; $P \cap M$ = {positive multiples of 2} **25.** {covert, flock, gaggle, plump, skein, bank, bevy, herd, team, wedge} **27.** Possible answer: {crows} and {starlings} **31.** 49 **33.** 13 **35.** H

14-3 Exercises

1.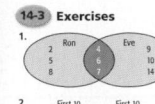

2. First 10 multiples of 4 / First 10 multiples of 6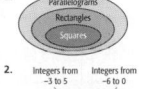

3. $A \cap C = C$, $A \cap B = \{3, 6\}$, $B \cap C = \emptyset$; $A \cup C = A$, $A \cup B = \{2, 3, 4, 5, 6, 8, 9, 10, 12, 15\}$, $B \cup C = \{2, 3, 4, 6, 9, 12, 15\}$; $C \subset A$

4. $M \cap N = \{x|5 < x \leq 7\}$; $A \cup B = \{$all real numbers$\}$; none

5. Parallelograms / Rectangles / Squares

2. Integers from −5 to 5 / Integers from −6 to 0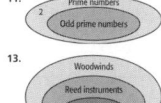

9. $Q \cap T = T$, $Q \cap Z = \emptyset$, $Z \cap T = \emptyset$; $Q \cup T = Q$, $Q \cup Z = \{1, 3, 4, 7, 8, 10, 12, 14, 15, 16\}$, $T \cup Z = \{1, 4, 7, 10, 15, 16\}$; $T \subset Q$

11. Prime numbers / Odd prime numbers

13. Woodwinds / Reed instruments / Double-reed instruments

19. 226.08 in³ **21.** $83\frac{1}{3}$ cm³

14-4 Exercises

1–13. Possible examples given.

1.

Example	P	Q	P and Q
58 in. tall, 11 years old	T	T	T
40 in. tall, 7 years old	T	F	F
62 in. tall, 12 years old	F	T	F
63 in. tall, 8 years old	F	F	F

2.

Example	P	Q	P and Q
$x = 6$	T	T	T
$x = 8$	T	F	F
$x = 9$	F	T	F
$x = 7$	F	F	F

3.

Example	P	Q	P or Q
7 A.M., 65° outside	T	T	T
3 A.M., 82° outside	T	F	T
4 P.M., 30° outside	F	T	T
1 P.M., 90° outside	F	F	F

4.

Example	P	Q	P or Q
Live in AL, vacation in FL	T	T	T
Live in FL, are home	T	F	T
Live in TX, vacation in Mexico	F	T	T
Live in MI, are home	F	F	F

5.

Example	P	Q	P and Q
Blond hair, size 9	T	T	T
Blond hair, size 10	T	F	F
Red hair, size 9	F	T	F
Brown hair, size 11	F	F	F

7.

Example	P	Q	P and Q
ABCD is a rectangle, perimeter 25 cm	T	T	T
ABCD is a rectangle, perimeter 22 cm	T	F	F
ABCD is a trapezoid, perimeter 25 cm	F	T	F
ABCD is a trapezoid, perimeter 20 cm	F	F	F

9.

Example	P	Q	P or Q
10 A.M., math	T	T	T
10 A.M., science	T	F	T
3 P.M., math	F	T	T
5 P.M., at the movies	F	F	F

11.

Example	P	Q	P or Q
The word is *strong*.	T	T	T
The word is *wide*.	T	F	T
The word is *wisdom*.	F	T	T
The word is *smile*.	F	F	F

13.

Example	P	Q	P and Q	P or Q
20 yrs old, no drivers ed	T	F	F	T
14 yrs old, no drivers ed	F	F	F	F
16 yrs old, drivers ed	T	F	F	T
17 yrs old, drivers ed	T	T	T	T

15. Disjunction; if either condition is met, the warranty expires.

17. conjunction; possible answer:

Example	P	Q	P and Q
37 yrs old, lived in U.S. all his life	T	T	T
42 yrs old, lived in U.S. 12 yrs	T	F	F
21 yrs old, lived in U.S. 20 yrs	F	T	F
5 yrs old, lived in U.S. all his life	F	F	F

21. 0.625 **23.** 0.71 **25.** $\frac{11}{10}$, or $1\frac{1}{10}$
27. $\frac{9}{25}$ **29.** B

14-5 Exercises

1. Ron eats peanuts. Ron has an allergic reaction. **2.** A number is divisible by 4. The number is even. **3.** A pot is watched. The pot never boils. **4.** Figure A has 5 sides. **5.** $x + 2 = 9$ **6.** No conclusion can be made. **7.** $x - 1 = 6$; $x = 7$ **9.** It is the first Friday of the month. The garden club will hold a meeting. **11.** The expression $x^3 - 4x + 2$ is a trinomial. **13.** Quadrilateral XYWZ is a square. **25.** 60 **27.** D

14-6 Exercises

1. A: 2; B: 3; C: 2; D: 5; E: 2; F: 2; G: 0; no **2.** A: 2; B: 4; C: 2; yes **3.** no **4.** yes; possible answer: A-B-C-B-A **5.** M: 2; R: 2; S: 4; T: 2; yes **7.** yes; possible answer: M-R-S-T-S-M **9.** connected; A: 3; B: 2; C: 2; D: 3; E: 4; no **11.** land masses; bridges and tunnels **13.** 13 **15.** Yes; there is a path from each vertex to any other. **19.** $x = -1$ **21.** $z = \frac{11}{12}$ **23.** $x = 9$ **25.** $r = 3$ **27.** B

14-7 Exercises

1.–9. Possible answers given.
1. A-B-C-D-A **2.** W-S-V-R-T-W
3. A-C-B-D-A; 24 mi
5. A-D-F-E-C-B-A **7.** B-T-N-M-R-B; 43 mi **9.** J-K-M-L-N-J; 226 mi **11.** S-A-C-B-S; S-A-B-C-S; S-B-C-A-S; S-B-A-C-S; S-C-A-B-S; S-C-B-A-S
17. $-3x^3y^2 - 2x^2y$
19. $18x^2 - 36x - 6$

Chapter 14 Study Guide Review

1. Euler circuit, vertex **2.** truth table **3.** Venn diagram; intersection **4.** empty set **5.** ∈ **6.** ∉ **7.** ⊂ **8.** finite **9.** infinite **10.** $P \cap Q = \{2, 4\}$; $P \cup Q = \{0, 1, 2, 3, 4, 5, 6\}$ **11.** $E \cap O = \emptyset$; $E \cup O = \{$integers$\}$ **12.** $H \cap R = \{x|3 < x < 6\}$; $H \cup R = \{$real numbers$\}$ **13.** intersection: $\{1, 2, 3, 6\}$; union: $\{1, 2, 3, 4, 6, 9, 12, 18\}$; subsets: none

14. Rectangular prisms / Cubes / Cones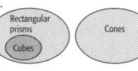

15.

Example	P	Q	P and Q
5 ft tall, 13 yrs old	T	T	T
5 ft tall, 10 yrs old	T	F	F
6 ft 2 in. tall, 13 yrs old	F	T	F
6 ft 1 in. tall, 10 yrs old	F	F	F

16.

Example	P	Q	P and Q
ABCD parallelogram, EFGH square	T	T	T
ABCD parallelogram, EFGH rhombus.	T	F	F
ABCD not a parallelogram, EFGH square	F	T	F
ABCD not a parallelogram, EFGH trapezoid	F	F	F

17.

Example	P	Q	P or Q
9-min mile, 50 sit-ups	T	T	T
9-min mile, 40 sit-ups	T	F	T
11-min mile, 50 sit-ups	F	T	T
12-min mile, 35 sit-ups	F	F	F

18.

Example	P	Q	P or Q
Graduated college, designs bridges	T	T	T
Graduated college, college professor	T	F	T
Graduated high school, designs bridges	F	T	T
Graduated high school, manager of a shoe store	F	F	F

19. No conclusion can be made.
20. No conclusion can be made.
21. Figure ABCD is a polygon.
22. no **23.** Y-X-Z-W-B-Y or Y-X-B-Z-W-Y, (or reverses); 26 in.

Notes

Additional Answers

Chapter 1

Lesson 1-5

47. Possible answer: $185x \leq 2500$; $x \leq 13.51$ (rounded); since a person cannot be split into parts, the elevator can safely carry a maximum of 13 people.

48. Possible answer: Switch the variable and the constant, and then reverse the inequality sign. $4 \leq x$ becomes $x \geq 4$.

Lesson 1-7

5.

x	2x	y	(x, y)
1	2(1)	2	(1, 2)
2	2(2)	4	(2, 4)
3	2(3)	6	(3, 6)
4	2(4)	8	(4, 8)
5	2(5)	10	(5, 10)
6	2(6)	12	(6, 12)

6.

x	3x − 2	y	(x, y)
1	3(1) − 2	1	(1, 1)
2	3(2) − 2	4	(2, 4)
3	3(3) − 2	7	(3, 7)
4	3(4) − 2	10	(4, 10)
5	3(5) − 2	13	(5, 13)
6	3(6) − 2	16	(6, 16)

12.

x	4x − 1	y	(x, y)
1	4(1) − 1	3	(1, 3)
2	4(2) − 1	7	(2, 7)
3	4(3) − 1	11	(3, 11)
4	4(4) − 1	15	(4, 15)
5	4(5) − 1	19	(5, 19)
6	4(6) − 1	23	(6, 23)

13.

x	2x + 8	y	(x, y)
1	2(1) + 8	10	(1, 10)
2	2(2) + 8	12	(2, 12)
3	2(3) + 8	14	(3, 14)
4	2(4) + 8	16	(4, 16)
5	2(5) + 8	18	(5, 18)
6	2(6) + 8	20	(6, 20)

14.

x	2x − 3	y	(x, y)
2	2(2) − 3	1	(2, 1)
4	2(4) − 3	5	(4, 5)
6	2(6) − 3	9	(6, 9)
8	2(8) − 3	13	(8, 13)
10	2(10) − 3	17	(10, 17)

15.

x	3x−4	y	(x, y)
2	3(2)−4	2	(2, 2)
4	3(4)−4	8	(4, 8)
6	3(6)−4	14	(6, 14)
8	3(8)−4	20	(8, 20)
10	3(10)−4	26	(10, 26)

25.

x	4x − 3	y	(x, y)
1	4(1) − 3	1	(1, 1)
2	4(2) − 3	5	(2, 5)
3	4(3) − 3	9	(3, 9)
4	4(4) − 3	13	(4, 13)
5	4(5) − 3	17	(5, 17)
6	4(6) − 3	21	(6, 21)

26.

x	3x − 1	y	(x, y)
1	3(1) − 1	2	(1, 2)
2	3(2) − 1	5	(2, 5)
3	3(3) − 1	8	(3, 8)
4	3(4) − 1	11	(4, 11)
5	3(5) − 1	14	(5, 14)
6	3(6) − 1	17	(6, 17)

27.

x	x + 8	y	(x, y)
1	1 + 8	9	(1, 9)
2	2 + 8	10	(2, 10)
3	3 + 8	11	(3, 11)
4	4 + 8	12	(4, 12)
5	5 + 8	13	(5, 13)
6	6 + 8	14	(6, 14)

28.

x	2x + 1	y	(x, y)
2	2(2) + 1	5	(2, 5)
4	2(4) + 1	9	(4, 9)
6	2(6) + 1	13	(6, 13)
8	2(8) + 1	17	(8, 17)
10	2(10) + 1	21	(10, 21)

29.

x	2x + 4	y	(x, y)
2	2(2) + 4	8	(2, 8)
4	2(4) + 4	12	(4, 12)
6	2(6) + 4	16	(6, 16)
8	2(8) + 4	20	(8, 20)
10	2(10) + 4	24	(10, 24)

30.

x	2x − 3	y	(x, y)
3	2(3) − 3	3	(3, 3)
6	2(6) − 3	9	(6, 9)
9	2(9) − 3	15	(9, 15)
12	2(12) − 3	21	(12, 21)
15	2(15) − 3	27	(15, 27)

36. Possible answer: The x- and y-coordinates have been reversed. (10, 4) is a solution of the equation.

37. Possible answer: $y = 2x - 1$; I multiplied the x-coordinate by 2 and found the number I needed to subtract 1 from to get the y-coordinate.

Lesson 1-8

11.

x	x + 1	y	(x, y)
0	0 + 1	1	(0, 1)
1	1 + 1	2	(1, 2)
2	2 + 1	3	(2, 3)

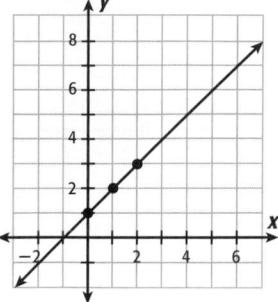

12.

x	2x − 1	y	(x, y)
0	2(0) − 1	−1	(0, −1)
1	2(1) − 1	1	(1, 1)
2	2(2) − 1	3	(2, 3)

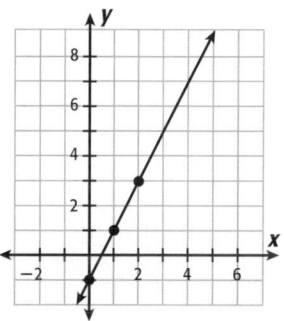

19–22.

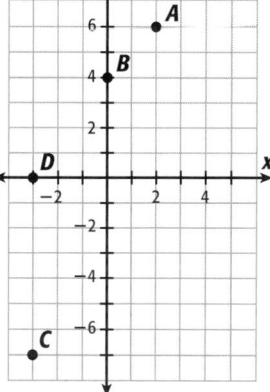

23.

x	3x	y	(x, y)
0	3(0)	0	(0, 0)
1	3(1)	3	(1, 3)
2	3(2)	6	(2, 6)

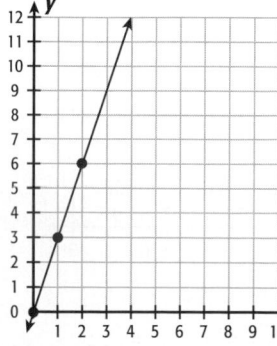

24.

x	2x + 1	y	(x, y)
0	2(0) + 1	1	(0, 1)
1	2(1) + 1	3	(1, 3)
2	2(2) + 1	5	(2, 5)

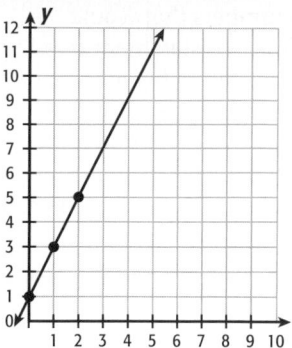

34. 210 miles

x	60x	y	(x, y)
1	60(1)	60	(1, 60)
2	60(2)	120	(2, 120)
3	60(3)	180	(3, 180)
4	60(4)	240	(4, 240)
5	60(5)	300	(5, 300)

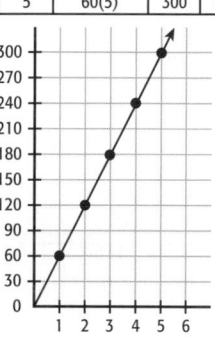

35. 7 studs

x	$\frac{x}{16}$ + 1	y	(x, y)
16	$\frac{16}{16}$ + 1	2	(16, 2)
32	$\frac{32}{16}$ + 1	3	(32, 3)
48	$\frac{48}{16}$ + 1	4	(48, 4)
64	$\frac{64}{16}$ + 1	5	(64, 5)
80	$\frac{80}{16}$ + 1	6	(80, 6)

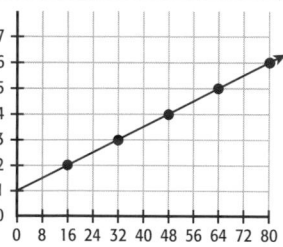
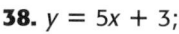

38. $y = 5x + 3$;

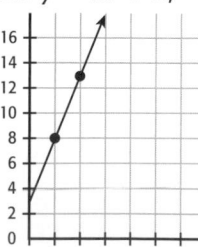

Performance Assessment

1. $x = 4$; $x \geq 4$; The number 4 is in both sets, and both sets describe all numbers that would make a statement true. They are different because the set $x = 4$ contains only one number, and $x \geq 4$ contains infinitely many numbers.

2. Let e be the measurement for Eastport and p be the measurement for Philadelphia. Convert the measurements into inches: 5 ft 10 in. = 70 in., and 6 ft 9 in. = 81 in. Then $e = 70 + 2(81)$, so $e = 232$ in., or 19 ft 4 in.

Chapter 2

Lesson 2-1

1.

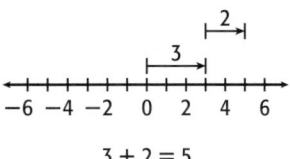

$3 + 2 = 5$

2.

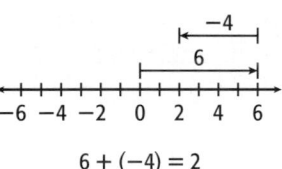

$6 + (-4) = 2$

3.

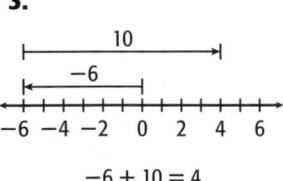

$-6 + 10 = 4$

4.

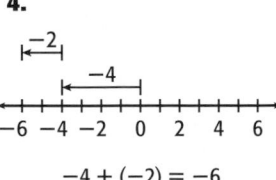

$-4 + (-2) = -6$

13.

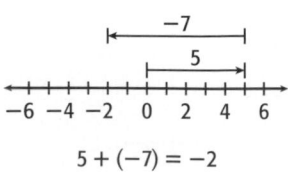

$5 + (-7) = -2$

14.

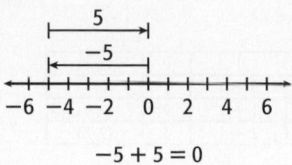

$-5 + 5 = 0$

15.

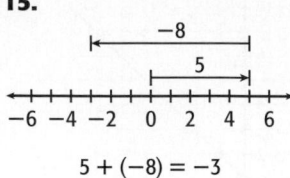

$5 + (-8) = -3$

16.

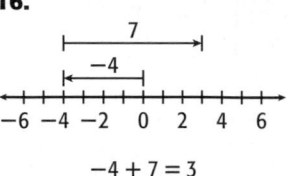

$-4 + 7 = 3$

27. -13

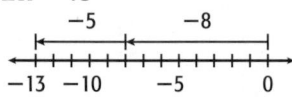

28. -6

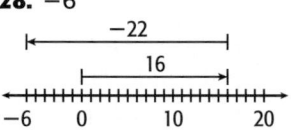

29. -18

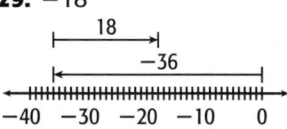

30. 82

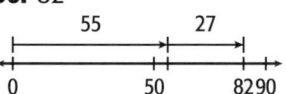

31. -2

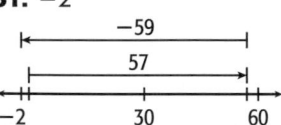

32. 71

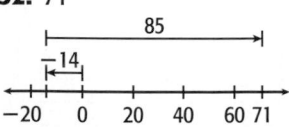

33. 43

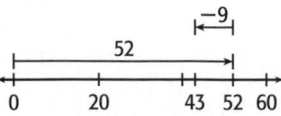

34. -52

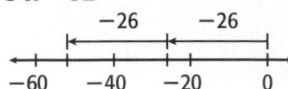

35. 0

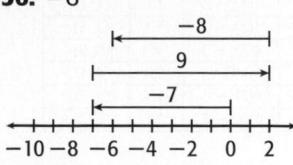

36. -6

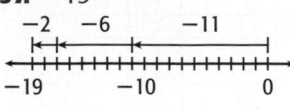

37. -19

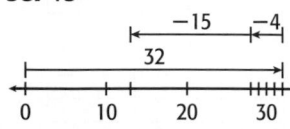

38. 13

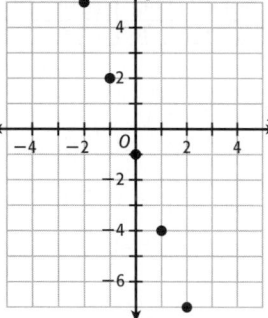

Lesson 2-3

10.

x	$-3x - 1$	y	(x, y)
-2	$-3(-2) - 1$	5	$(-2, 5)$
-1	$-3(-1) - 1$	2	$(-1, 2)$
0	$-3(0) - 1$	-1	$(0, -1)$
1	$-3(1) - 1$	-4	$(1, -4)$
2	$-3(2) - 1$	-7	$(2, -7)$

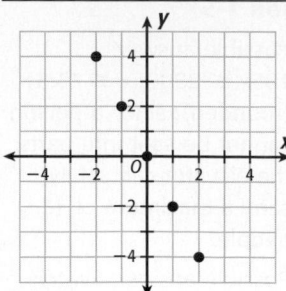

11.

x	$2x + 2$	y	(x, y)
-2	$2(-2) + 2$	-2	$(-2, -2)$
-1	$2(-1) + 2$	0	$(-1, 0)$
0	$2(0) + 2$	2	$(0, 2)$
1	$2(1) + 2$	4	$(1, 4)$
2	$2(2) + 2$	6	$(2, 6)$

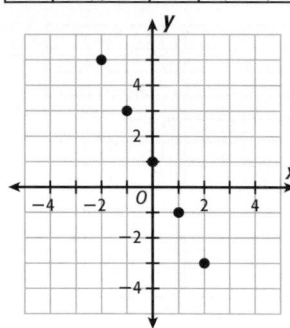

20.

x	$-2x$	y	(x, y)
-2	$-2(-2)$	4	$(-2, 4)$
-1	$-2(-1)$	2	$(-1, 2)$
0	$-2(0)$	0	$(0, 0)$
1	$-2(1)$	-2	$(1, -2)$
2	$-2(2)$	-4	$(2, -4)$

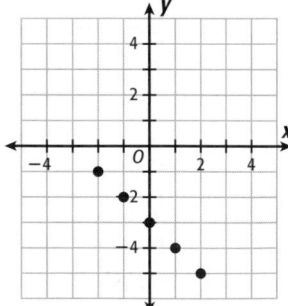

21.

x	$-2x + 1$	y	(x, y)
-2	$-2(-2) + 1$	5	$(-2, 5)$
-1	$-2(-1) + 1$	3	$(-1, 3)$
0	$-2(0) + 1$	1	$(0, 1)$
1	$-2(1) + 1$	-1	$(1, -1)$
2	$-2(2) + 1$	-3	$(2, -3)$

22.

x	$-x - 3$	y	(x, y)
-2	$-(-2) - 3$	-1	$(-2, -1)$
-1	$-(-1) - 3$	-2	$(-1, -2)$
0	$-(0) - 3$	-3	$(0, -3)$
1	$-(1) - 3$	-4	$(1, -4)$
2	$-(2) - 3$	-5	$(2, -5)$

41.

x	2x + 4	y	(x, y)
−2	2(−2) + 4	0	(−2, 0)
−1	2(−1) + 4	2	(−1, 2)
0	2(0) + 4	4	(0, 4)
1	2(1) + 4	6	(1, 6)
2	2(2) + 4	8	(2, 8)

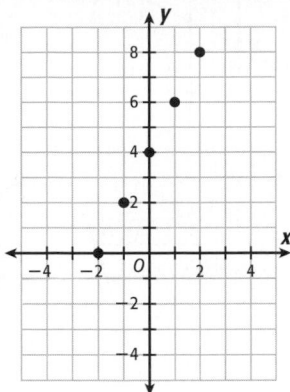

42.

x	5 − 4x	y	(x, y)
−2	5 − 4(−2)	13	(−2, 13)
−1	5 − 4(−1)	9	(−1, 9)
0	5 − 4(0)	5	(0, 5)
1	5 − 4(1)	1	(1, 1)
2	5 − 4(2)	−3	(2, −3)

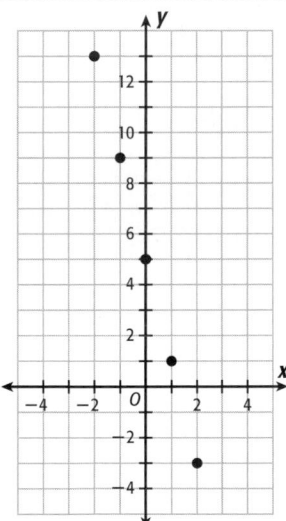

43.

x	1 + 3x	y	(x, y)
−2	1 + 3(−2)	−5	(−2, −5)
−1	1 + 3(−1)	−2	(−1, −2)
0	1 + 3(0)	1	(0, 1)
1	1 + 3(1)	4	(1, 4)
2	1 + 3(2)	7	(2, 7)

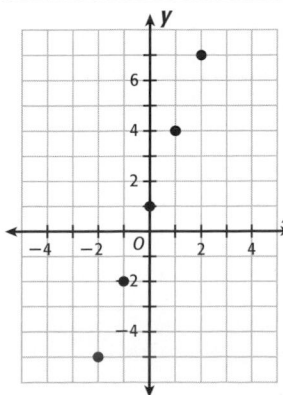

Lesson 2-5

13. $k > -3$

14. $z \le 9$

15. $x < -1$

16. $b \le -4$

17. $r \ge 2$

18. $p > 6$

19. $n > 5$

20. $g \le 4$

21. $x \ge -4$

22. $h > -7$

23. $x > -5$

24. $p \le 5$

25. $x > -2$

26. $y \ge 5$

27. $k \ge 10$

28. $b < 8$

29. $a \le -12$

30. $z \le -8$

31. $r \le -1$

32. $x < -3$

33. $t = 2$

34. $s \ge 3$

35. $b > 0$

36. $a \le 4$

37. $f = -18$

38. $h \ge -4$

39. $c \le 2$

40. $y < -4$

41. $n < -6$

42. $k \ge 0$

43. $g = 8$

44. $f > -4$

45. $p = -9$

Performance Assessment

1. a. even; even; odd; even; odd; even

b. If two integers have the same sign, their product is positive. This is similar to the rule for adding odd and even numbers. If two numbers are both even or both odd, the sum is even. Also, if two integers have a different sign, their product is negative, and if one number is even and the other is odd, their sum is odd.

2. $-2 + 6 = 4$

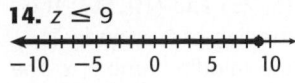

3. $\frac{1}{2}n \le -2$

$2 \cdot \frac{1}{2}n \le 2 \cdot (-2)$

$n \le -4$

Chapter 3

Performance Assessment

1. Each side of a chessboard has $\sqrt{64} = 8$ squares. If you multiply 8 by the number of sides, you get $8 \cdot 4 = 32$. This would count each corner square twice, so you subtract 4. $32 - 4 = 28$ pieces.

2. The length of 26 dashes is $26 \cdot \frac{1}{8} = \frac{26}{8} = \frac{13}{4}$ in. There is a space after every dash except the last, so there are 25 spaces. The length of 25 spaces is $25 \cdot \frac{1}{32} = \frac{25}{32}$ in.

The total length is

$\frac{13}{4} + \frac{25}{32} = \frac{129}{32} = 4\frac{1}{32}$ in.

3. $\frac{15}{28} \div \frac{5}{7} = \frac{3}{4}$; Because $\frac{\overset{3}{\cancel{15}}}{\cancel{28}_4} \cdot \frac{\overset{1}{\cancel{7}}}{\cancel{5}_1} = \frac{3}{4}$, the quotient of $\frac{15}{28} \div \frac{5}{7}$ is the same as the product $\frac{15}{28} \cdot \frac{7}{5}$.

Chapter 4

Lesson 4-5

2.

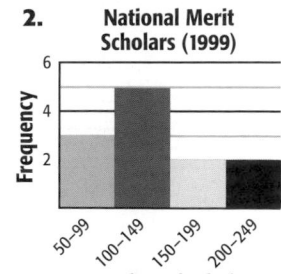

National Merit Scholars (1999)

3.

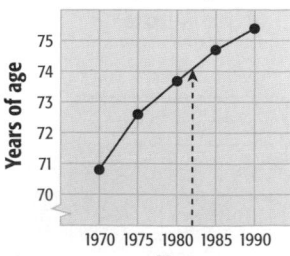

Life Expectancy in U.S.

estimated life expectancy for 1982: 74.1

4.

Data	Frequency
−46	2
−34	4
−32	2
−25	1
−20	1
−17	2
−2	1

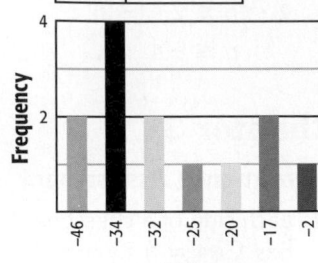

5.

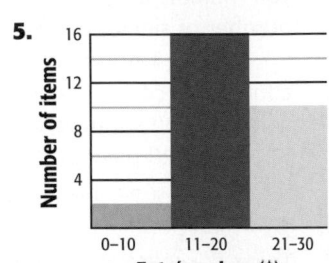

6.

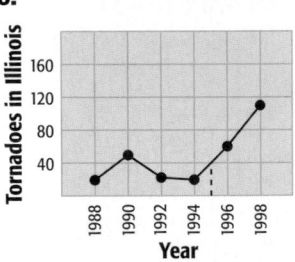

estimate for 1995: 40 tornadoes

7.

Data	Frequency
1	9
2	3
3	2
4	3
5	1
6	2

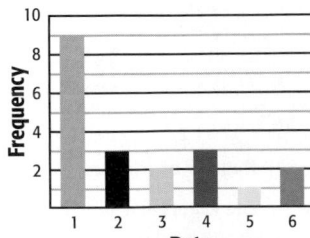

8.

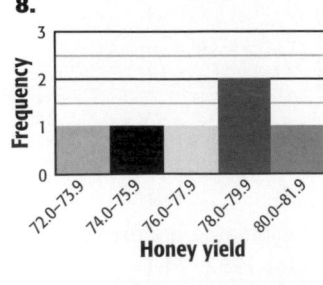

---

9. a.

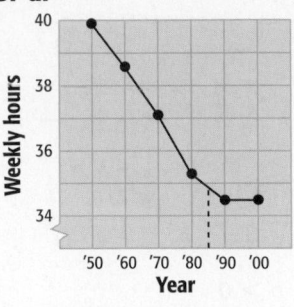

estimate for 1985: 34.9 hours

b.

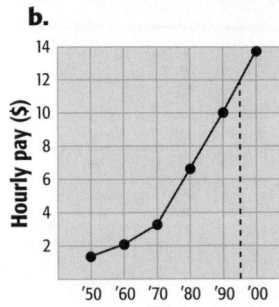

estimate for 1995: $11.88

13. $x < 5$

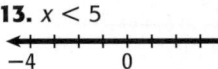

14. $x \geq 2$

15. $x \leq 2$

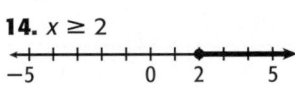

16. $x < 8$

17. $x > 6$

18. $x > 6$

19. $6 \geq x$

20. $32 < x$

Performance Assessment

1. mean 74; median 3; mode 1 and 3; You would use the mean to show the greatest measure of central tendency, the median to reduce the effect of the outlier 500, and the mode to show the value or values that occur most often.

2.

$$\frac{1 \cdot 4 + 2 \cdot 7 + 3 \cdot 1 + 4 \cdot 6 + 5 \cdot 2}{4 + 7 + 1 + 6 + 2}$$

$$= \frac{55}{20} = 2.75$$

3. (5, 25) and (10, 50); The median of 0, x, and y is the middle number, x. The mean of 0, x, and y is $\frac{0 + x + y}{3} = \frac{x + y}{3}$. The mean is twice the median, so $\frac{x + y}{3} = 2x$. Multiply both sides by 3 to get $x + y = 6x$. Solve for y to find $y = 5x$. So $n = 5$. If $x = 5$, then $y = 25$, so the ordered pair is (5, 25). If $x = 10$, then $y = 50$, so the ordered pair is (10, 50).

Chapter 5

Lesson 5-5

7. parallelogram, rhombus, rectangle, square

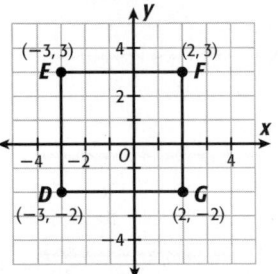

8. trapezoid

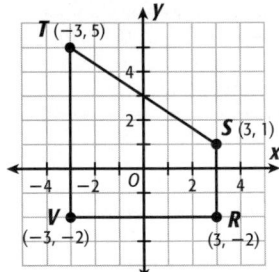

15. parallelogram, rhombus, rectangle, square

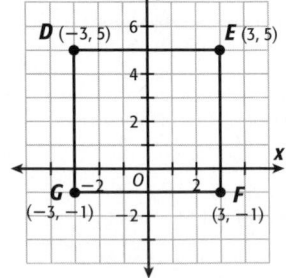

---

16. trapezoid

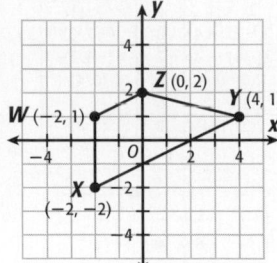

17. 3

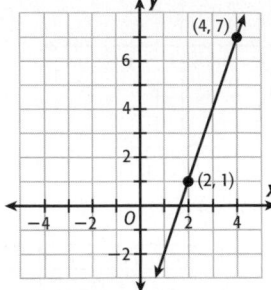

18. undefined

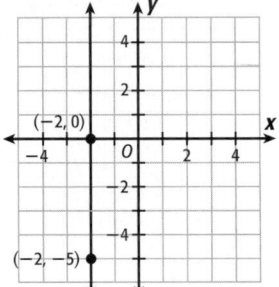

19. 0

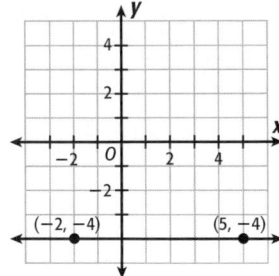

20. $-\frac{3}{7}$

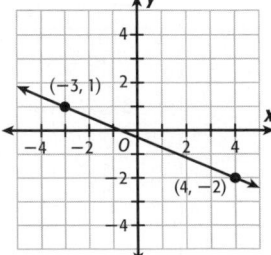

21. Possible answer:

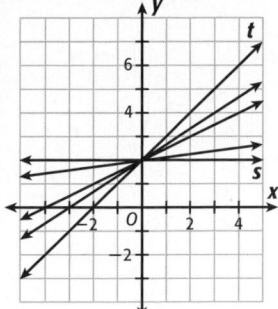

22. Possible answer:

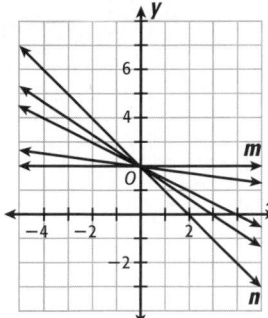

23. Possible answer:
Segments $\overline{PQ}$ and $\overline{RS}$ have slopes of $-\frac{5}{2}$ and are parallel. Segments $\overline{QR}$ and $\overline{PS}$ have slopes of $\frac{5}{2}$ and are parallel. So *PQRS* is a parallelogram. However, none of these segments are perpendicular, so *PQRS* is not a rectangle and, therefore, not a square.

24. Possible answer: The slope of a line will have the same value, regardless of which two points on a line you use to determine the slope.

25. Possible answer: Draw a square with vertices (0, 0), (1, 0), (1, 1), and (0, 1). A line through the diagonal has a slope of 1 and cuts the square into two congruent right triangles that have angle measures 45°, 45°, and 90°.

Lesson 5-7

5.

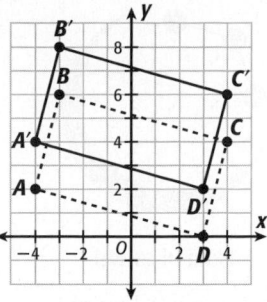

6.

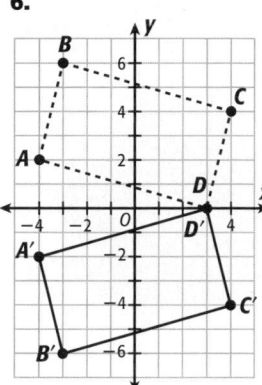

7.

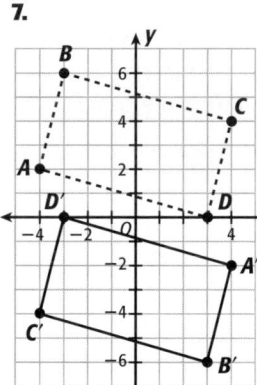

10.

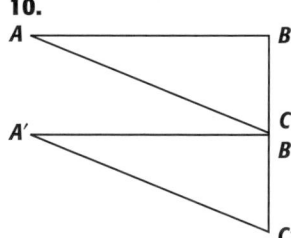

11.

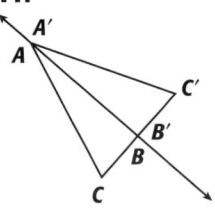

12.

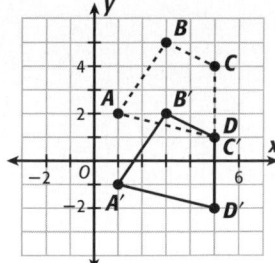

13.

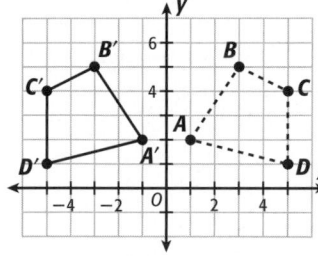

14.

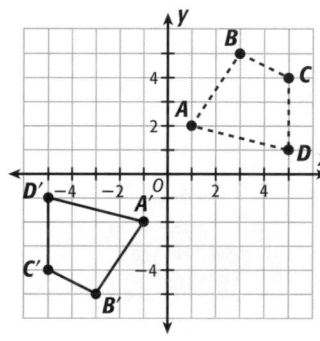

15.

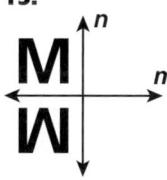

16.

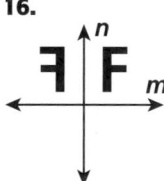

17.

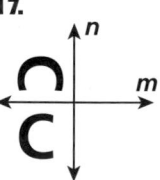

27.

reflection across a vertical line

28. Possible answer: Draw a triangle. Translate it down 5 units and up 5 units. Translate all three figures left 5 units and right 5 units.

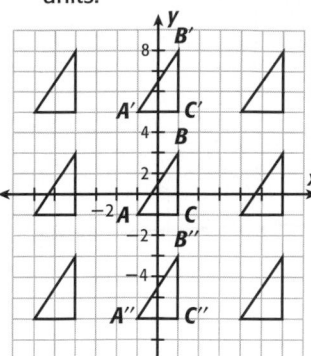

Lesson 5-8

5.

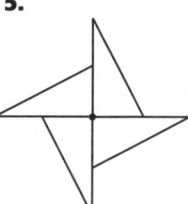

6.

7.

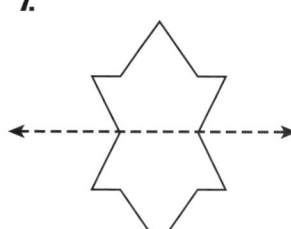

8.

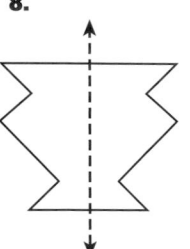

9.

10.

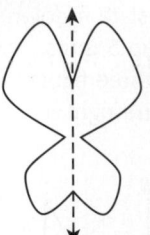

11.

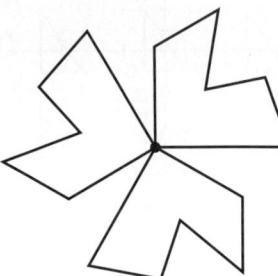

12.

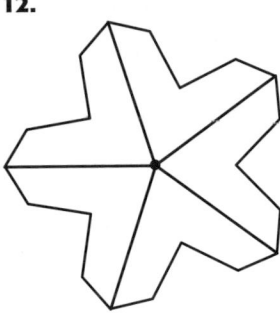

13. Possible answer:

14. Possible answer:

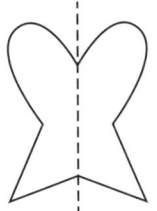

15. Possible answer:

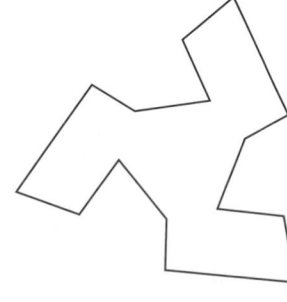

16. Possible answer:

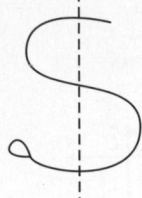

17. a.

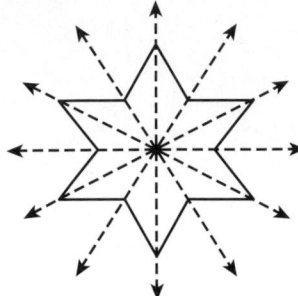

Kage Asa no ha

There are 6 lines of symmetry and 6-fold rotational symmetry around the center.

b.

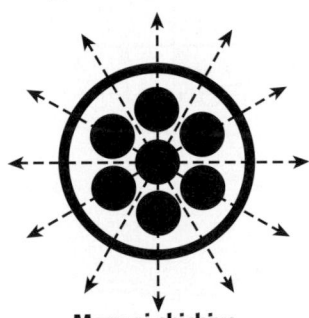

Maru ni shichiyo

There are 6 lines of symmetry and 6-fold rotational symmetry around the center.

c. There is no line symmetry and no rotational symmetry.

d.

Chukage itsutsu nenji Aoi

There is 5-fold rotational symmetry around the center.

e.

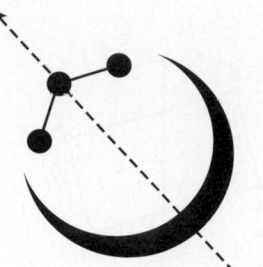

Tsuki ni sansei

There is 1 line of symmetry.

f.

Teuno ke

There are 16 lines of symmetry and 16-fold rotational symmetry around the center.

Lesson 5-9

4. The angles in a square all measure 90°; the angles in an equilateral triangle measure 60°. There are two arrangements of 3 triangles and 2 squares: 3(60°) + 2(90°) = 360°.

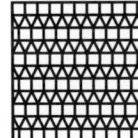 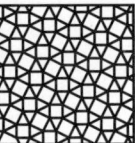

5. Possible answer:

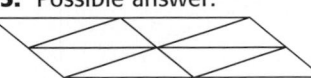

6. Possible answer:

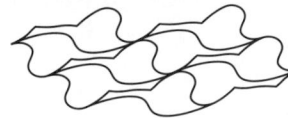

7. Possible answer:

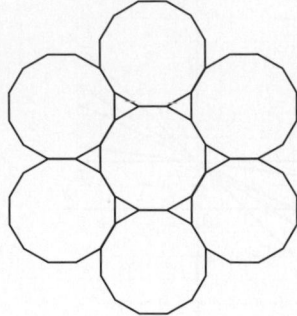

8. Possible answer:

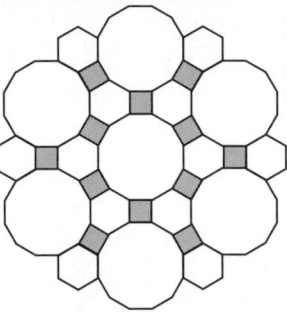

9. Possible answer:

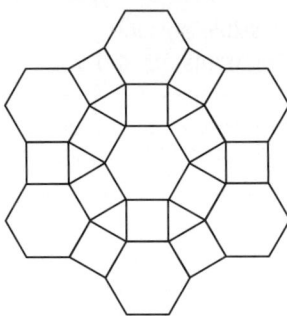

10. Possible answer:

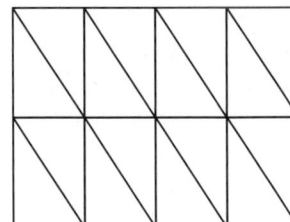

11. Possible answer:

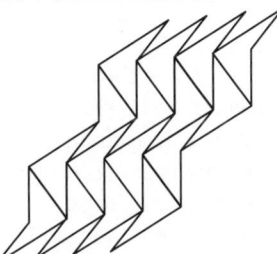

12. Possible answer:

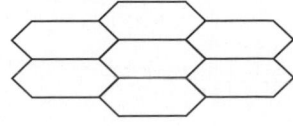

13. Yes, the shape will tesselate.

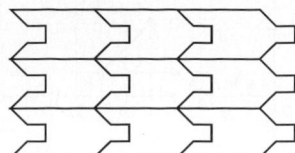

14.

18. $y > -3$

19. $p \leq 3$

20. $f \geq 5$

21. $12 < w$

22. $p \geq 5$

23. $m \leq 0$

24. $6 > n$

25. $z < 2$

Performance Assessment

1. 90°; The sum of all the angles around point A must be 360°, so subtract the given measures of the angles from 360°: 360° − 30° − 60° − 70° − 20° − 30° − 60° = 90°

2. ∠BAC and ∠CAF;
∠BAC and ∠BAG;
∠CAD and ∠CAH;
∠CAD and ∠DAG;
∠BAE and ∠EAF;
∠GAF and ∠BAG;
∠GAF and ∠FAC;
∠GAH and ∠CAH;
∠GAH and ∠DAG;
∠BAD and ∠DAF;
∠CAE and ∠EAG;
∠DAE and ∠EAH;
∠DAF and ∠FAH;
∠BAH and ∠FAH;

The measures of supplementary angles must add to 180°. For each angle, subtract its measure from 180° and look for angles with that measure.

3. 2; 5; 9; 14

Chapter 6

Lesson 6-1

6.

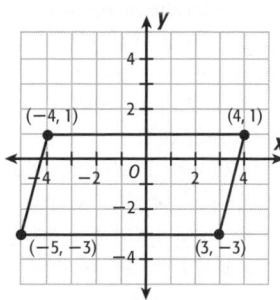

32 units²

7.

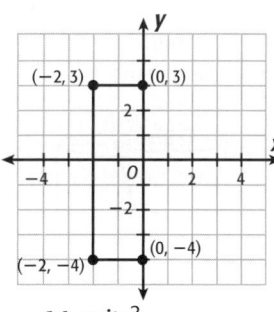

14 units²

12.

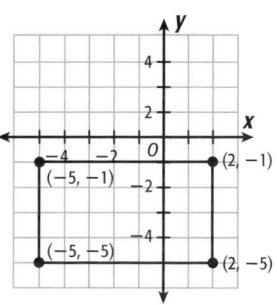

28 units²

13.

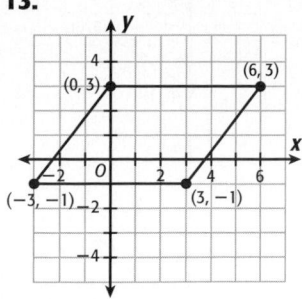

24 units²

14.

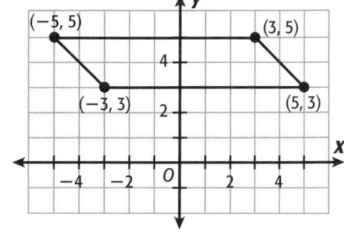

16 units²

15.

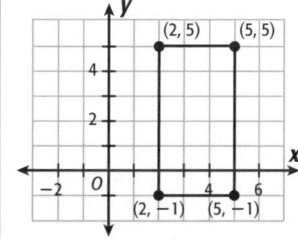

18 units²

Lesson 6-2

7.

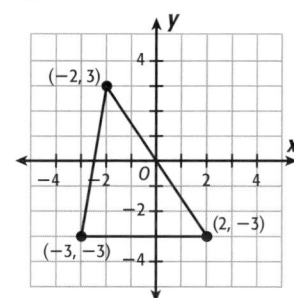

15 units²

8.

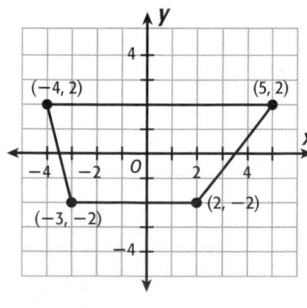

28 units²

9.

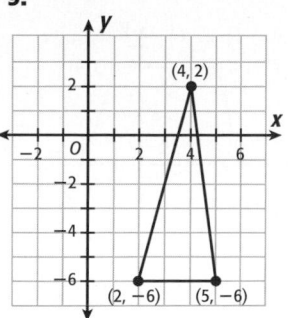

12 units²

10.

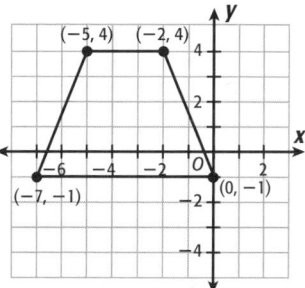

25 units²

17.

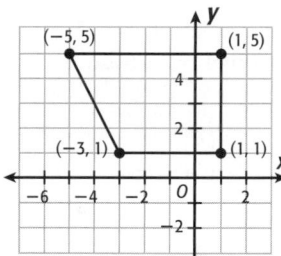

20 units²

18.

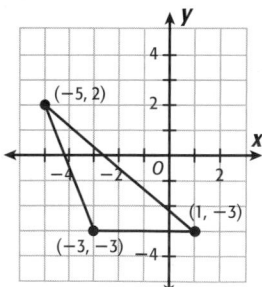

10 units²

19.

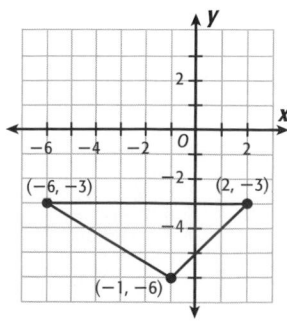

12 units²

20.

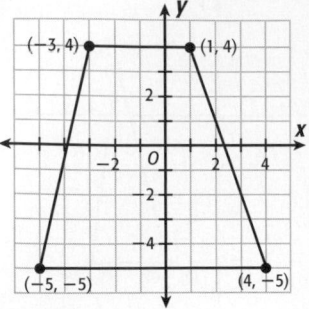

58.5 units^2

Lesson 6-4

5.

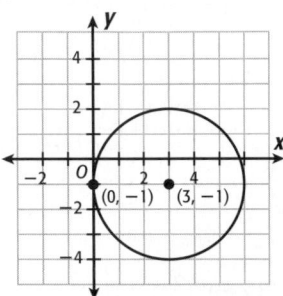

$A = 9\pi$ units$^2 \approx$ 28.3 units^2; $C = 6\pi$ units ≈ 18.8 units

15.

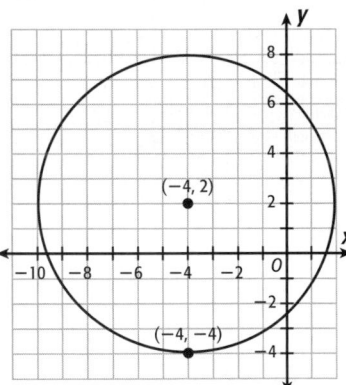

$A = 36\pi$ units$^2 \approx$ 113.0 units^2; $C = 12\pi$ units ≈ 37.7 units

Mid-Chapter Quiz

11.

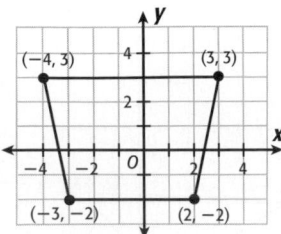

30 units^2

12.

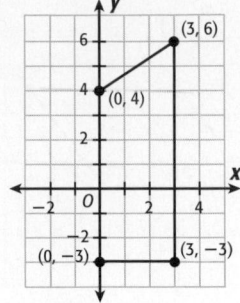

24 units^2

Lesson 6-5

10. Possible answer:

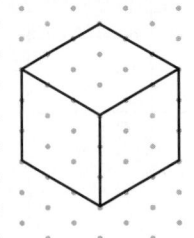

11. Possible answer:

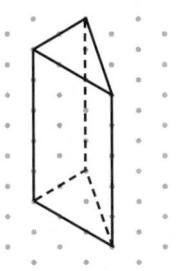

12. Possible answer:

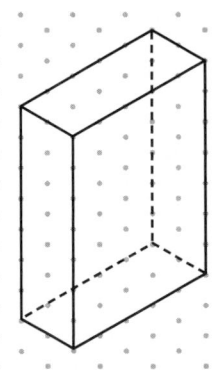

13. Possible answer:

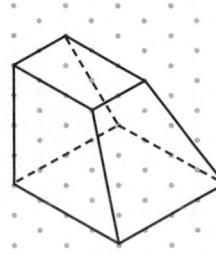

14. Possible answer:

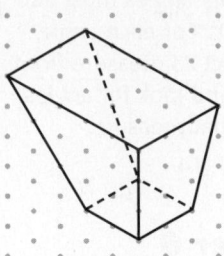

Performance Assessment

1. Check students' drawings. The rectangle with base length 7 cm and height 4 cm has an area of $7 \cdot 4 = 28$ cm^2 and a perimeter of $2 \cdot 7 + 2 \cdot 4 = 22$ cm. The rectangle with base length 14 cm and height 1 cm has an area of $14 \cdot 1 = 14$ cm^2 and a perimeter of $2 \cdot 14 + 2 \cdot 1 = 30$ cm. The first rectangle has the larger area, and the second rectangle has the larger perimeter.

2.

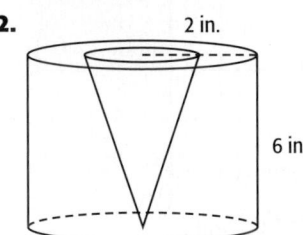

2 in.

6 in.

The amount of water left is the volume of the cylinder minus the volume of the cone:

Volume of the cylinder: $V = \pi r^2 h = \pi \cdot 2^2 \cdot 6 = 24\pi$.

Volume of the cone: $V = \frac{1}{3}\pi r^2 h = \frac{1}{3}\pi \cdot 1^2 \cdot 6 = 2\pi$

Volume of the water: $24\pi - 2\pi = 22\pi \approx 69.1$ in^3

Chapter 7

Lesson 7-5

5.

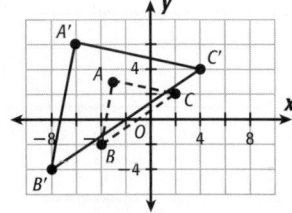

6.

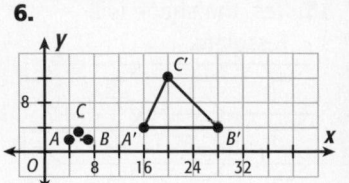

11.

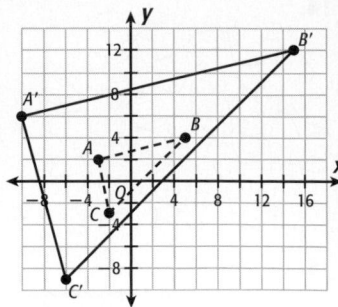

12.

15. Yes, the aperture is a dilation of the image as the light flows into the camera.

16. $A'(10, 10)$; $B'(22.5, 10)$; $C'(22.5, 0)$; $D'(10, 0)$

18. Possible answer: Use a straightedge to make sure segments connecting to corresponding vertices intersect at the center of dilation, or measure all of the sides to make sure they are in proportion.

Performance Assessment

1. The ratio is 4 pints of chocolate milk for every 7 pints of plain milk.
$\frac{4 \text{ chocolate}}{7 \text{ plain}} = \frac{x \text{ chocolate}}{168 \text{ plain}}$
$672 = 7x$
$96 = x$
There would be 96 pints of chocolate milk sold.

2. a. The 12 count box is the better buy because it has a lower unit cost. Unit costs:
$\frac{0.89}{8} \approx 0.11$, $\frac{1.25}{12} \approx 0.10$

b. 8 count box: $6(0.89) = \$5.34$; 12 count box: $4(1.25) = \$5.00$; you would save 34 cents.

Chapter 8

Performance Assessment

1. The total weight of the solution will be $10 + 15 = 25$ kg, and the amount of pure acid is 10 kg, so the percent of the solution that is acid is $\frac{10}{25} = 0.4$, or 40%.

2. Each quantity Jim pours in the 10 L jar is 1 L. He continues this process until there is only 1 L remaining. Since he started with 5 L, this means he repeats the process 4 times so there are 4 L in the 10 L jar, so it is 40% full.

Chapter 9

Performance Assessment

1. Suppose the side length of the yellow square is 1. Then the circle has a diameter of 1, which is the diagonal of the blue square. A square with a diagonal of 1 has a side length of $\frac{1}{\sqrt{2}}$. The area of the yellow square is 1, and the area of the blue square is $\left(\frac{1}{\sqrt{2}}\right)^2 = \frac{1}{2}$, so the probability of landing in the blue square is $\frac{1}{2}$.

2. The area of the field is $2^2 = 4$ km². The area of the four quarter-circles around the trees is $\pi\left(\frac{1}{7}\right)^2 = \frac{\pi}{49} \approx 0.064$. The probability of getting caught in a tree is $\frac{0.064}{4} = 0.016$, or about 1.6%, so the probability of not getting caught is $100\% - 1.6\% = 98.4\%$.

Chapter 10

Lesson 10-4

1. $k > 3$

2. $z \le 20$

---

3. $y < -7$

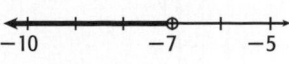

4. $x \le -2$

5. $y \ge 3$

6. $k > 5$

7. $x < 3$

8. $b \ge -\frac{1}{4}$

9. $h \le 1$

10. $c > 2$

11. $d < \frac{7}{18}$

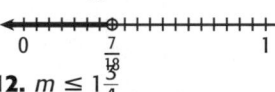

12. $m \le 1\frac{3}{4}$

14. $k > 5$

15. $x > 4$

16. $p \le 10$

17. $q \le 2$

18. $n < 3$

19. $x \le -8$

20. $p < -17$

21. $a \ge -3$

22. $n > \frac{3}{13}$

23. $k \ge 3$

---

24. $n \le -14$

25. $r < 3$

27. $p \le \frac{22}{3}$

28. $n > -3$

29. $w > -1$

30. $x \le -4$

31. $a > \frac{1}{2}$

32. $y \le -\frac{7}{2}$

33. $q < 6$

34. $m > \frac{5}{6}$

35. $b < 2.7$

36. $k > -\frac{2}{3}$

37. $f \le -27$

38. $v \le \frac{3}{5}$

Lesson 10-5

9. $y = -3x + 15$

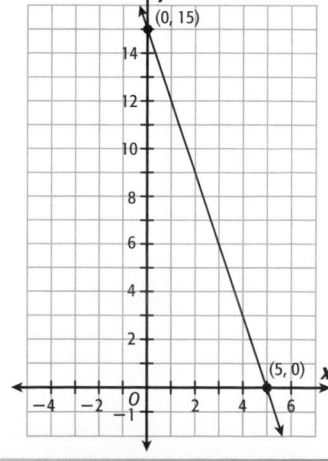

---

10. $y = \frac{9}{2}x + 7$

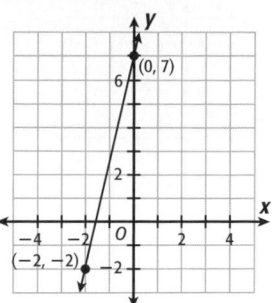

11. $y = 2x - 1$

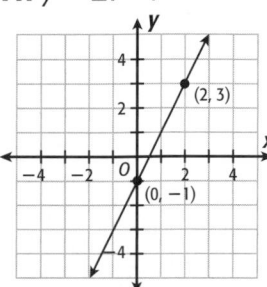

20. $y = -2x + 8$

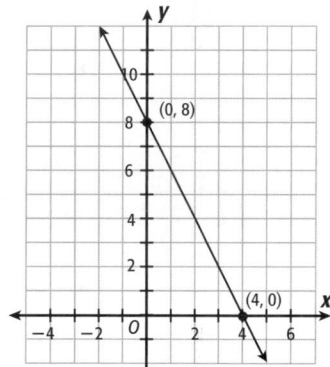

21. $y = -\frac{9}{2}x - 5$

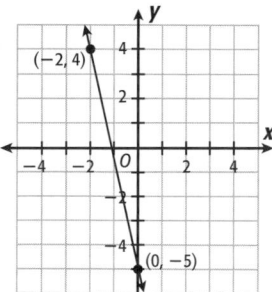

22. $y = \frac{3}{4}x + \frac{5}{4}$

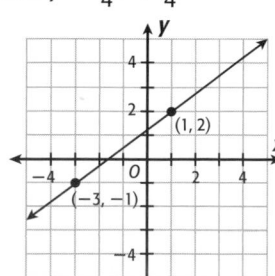

34. $y = 2x + 3$

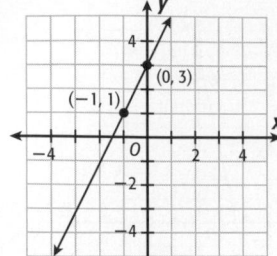

35. $y = -6x + 8$

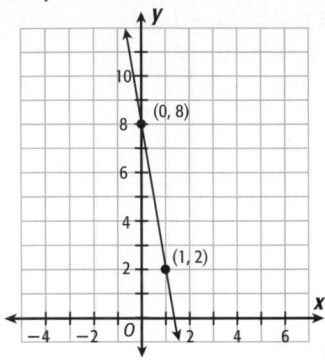

36. $y = 6x + 16$

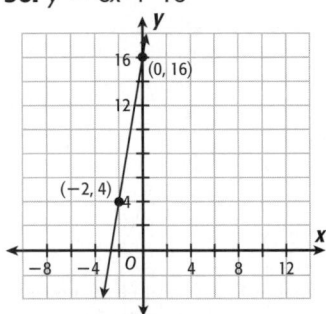

Performance Assessment

1. $7x - 4 < 9x + 14$

$\qquad 7x < 9x + 18$

$\qquad -2x < 18$

$\qquad x > -9$

2. $x - y = -3$

$2x - 4y = 22$

Solve both equations for x.

$x = -3 + y$

$x = 11 + 2y$

Set them equal and solve.

$-3 + y = 11 + 2y$

$-14 = y$

Substitute to solve for x.

$x = -3 + -14$

$x = -17$

3. $4c + 30 = 62$ or

$2c + 46 = 62$; $8

Chapter 11

Lesson 11-1

1.

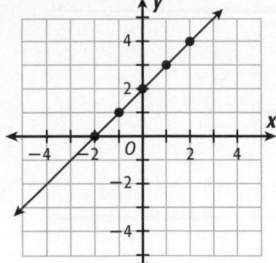

linear

2.

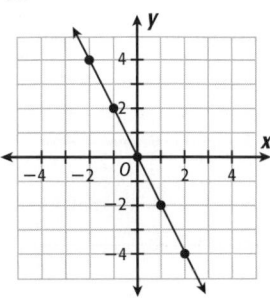

linear

3.

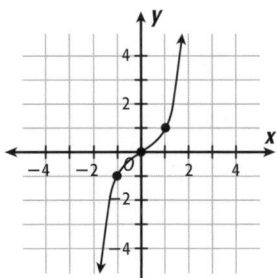

not linear

4.

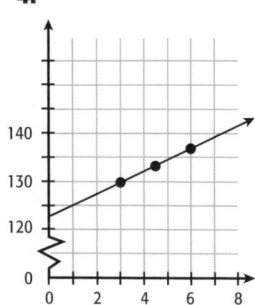

130 ft, 133 ft, 136 ft; linear

5.

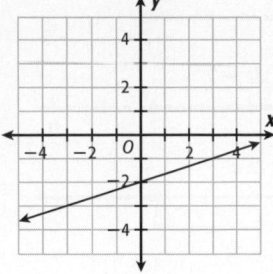

linear

6.

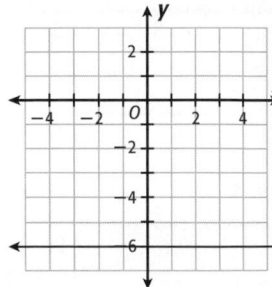

linear

7.

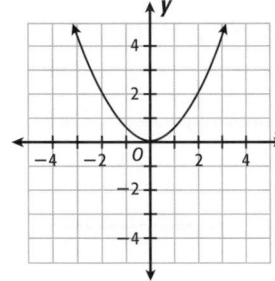

not linear

8.

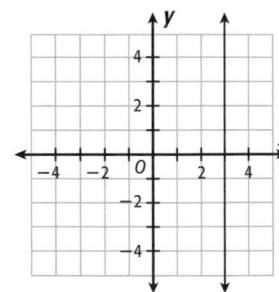

linear

9.

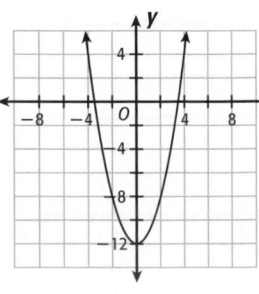

not linear

10.

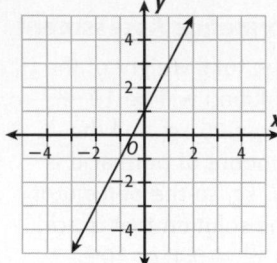

linear

11. $900, $1275, $1650, $2025, $2400; linear

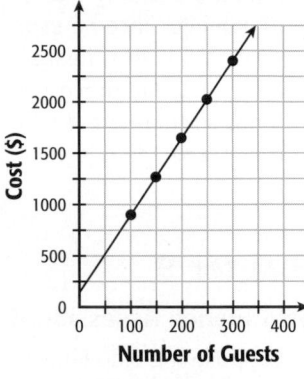
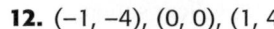

12. $(-1, -4), (0, 0), (1, 4)$

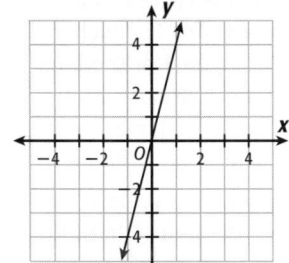

13. $(-1, 3), (0, 5), (1, 7)$

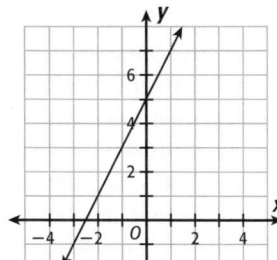

14. $(-1, -9), (0, -3), (1, 3)$

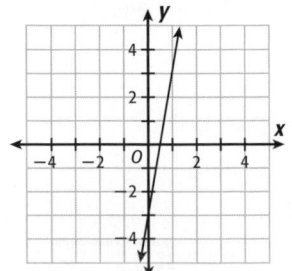

15. $(-1, -11)$, $(0, -10)$, $(1, -9)$

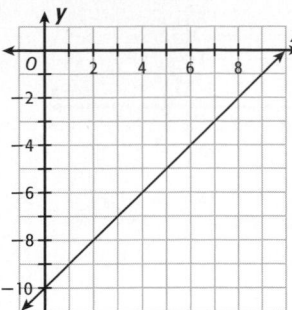

16. $(-1, -6)$, $(0, -2)$, $(1, 2)$

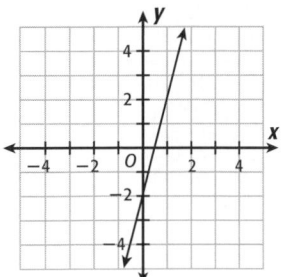

17. $(-1, -1)$, $(0, 3)$, $(1, 7)$

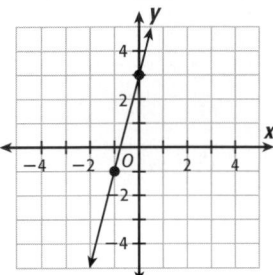

18. $(-1, -6)$, $(0, -4)$, $(1, -2)$

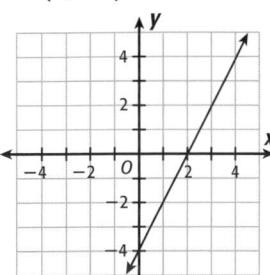

19. $(-1, 6)$, $(0, 7)$, $(1, 8)$

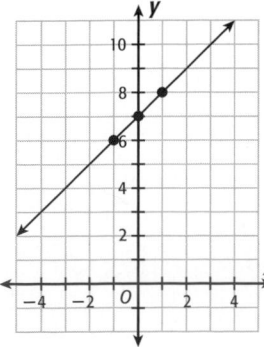

20. $(-1, -0.5)$, $(0, 2.5)$, $(1, 5.5)$

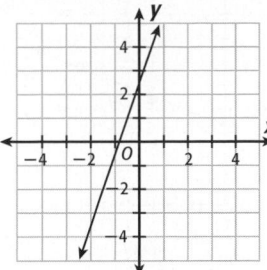

22. $43.20, $46.40, $49.60, $52.80, $56, $59.20

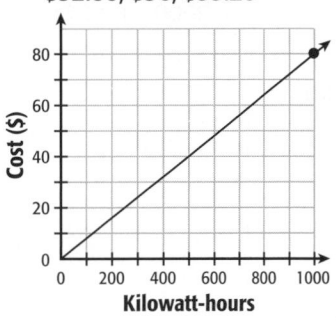

23.

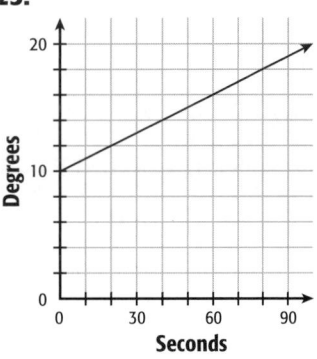

linear

24. $C = 1.75g + 4$;

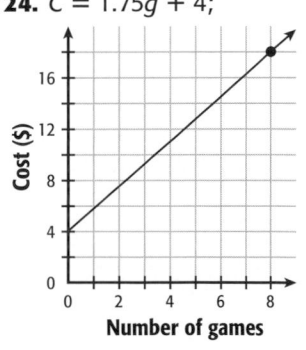

linear

28. Possible answer:

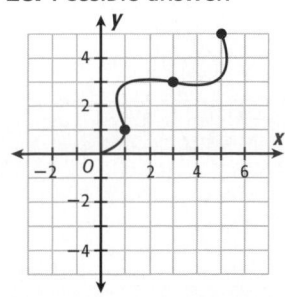

Lesson 11-2

27. $y = -\frac{4}{5}x + 350$

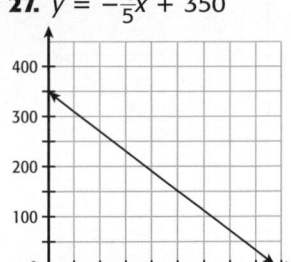

28. Possible answer: The formula for slope is $\frac{y_2 - y_1}{x_2 - x_1}$, so the y-values should be in the numerator instead of the x-values.

29. Possible answer: For the line $x = 2$, find the slope by using any two points on the line. Using the points $(2, 4)$ and $(2, 6)$, the slope is $\frac{6 - 4}{2 - 2} = \frac{2}{0}$, which is undefined.

30.

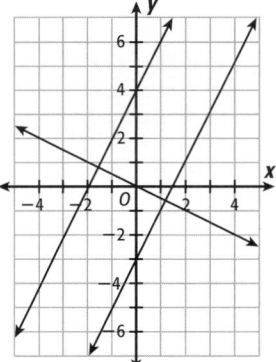

Slopes of $y = 2x - 3$ and $y = 2x + 4$ are 2. Slope of $y = -\frac{1}{2}x$ is $-\frac{1}{2}$. The line $y = -\frac{1}{2}x$ is perpendicular to $y = 2x - 3$ and $y = 2x + 4$, which are parallel; to tell whether two lines are parallel, perpendicular, or neither by their equations, write the equations in the form $y = mx + b$, where m is the slope, and determine whether the slopes are the same, their product is -1, or neither.

Lesson 11-3

1. x-intercept: 5, y-intercept: -5

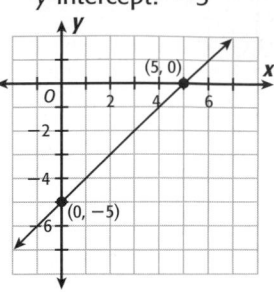

2. x-intercept: 6, y-intercept: 4

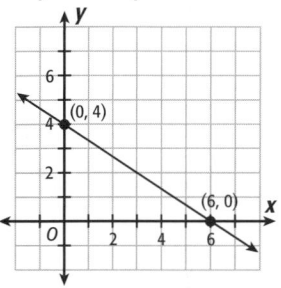

3. x-intercept: -5, y-intercept: -3

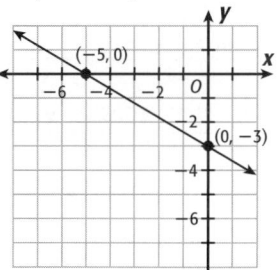

4. x-intercept: 2, y-intercept: -5

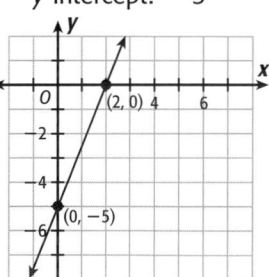

9. $m = 3.5$; $b = 22$

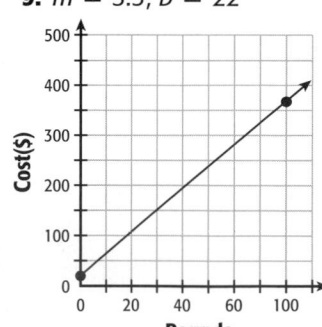

13. x-intercept: 5,
y-intercept: 10

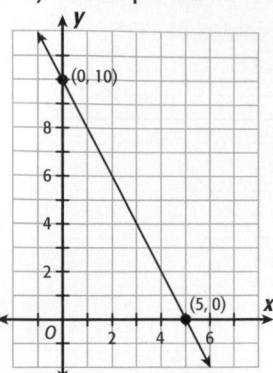

14. x-intercept: 3,
y-intercept: −4

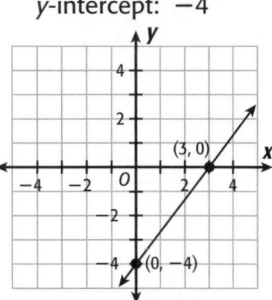

15. x-intercept: 3,
y-intercept: −18

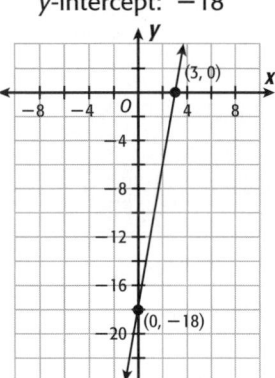

16. x-intercept: 3.5,
y-intercept: 7

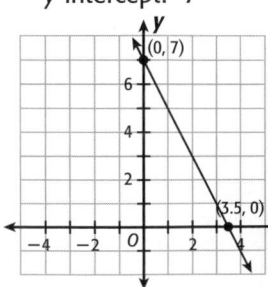

21. m = 15; b = 300

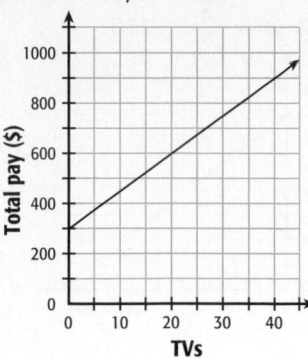

25.

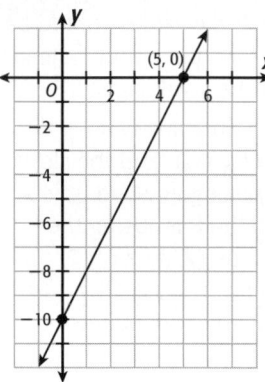

26.

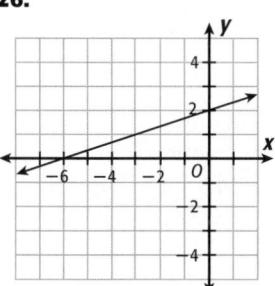

27.

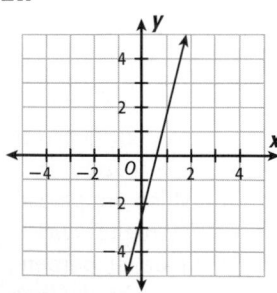

28.

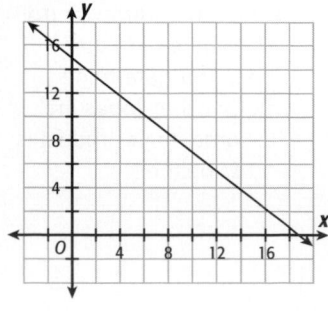

29. a. slope = 2000

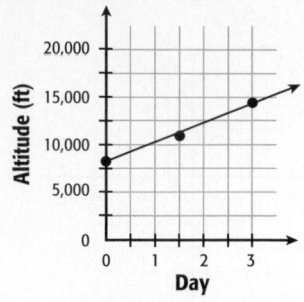

b. 8255 ft, the altitude of
the base camp

c. y = 2000x + 8255

d. yes

32.

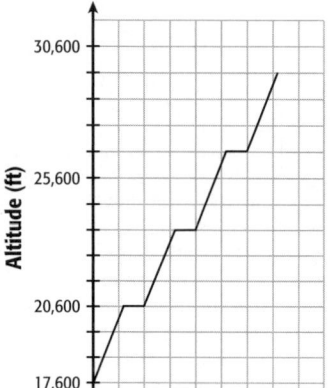

No, the graph is not linear,
because on some days
the team does not ascend
at all.

Lesson 11-6

1.

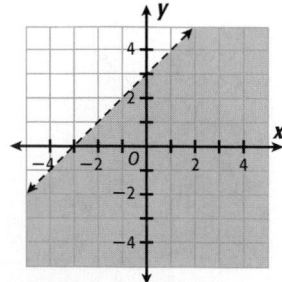

2.

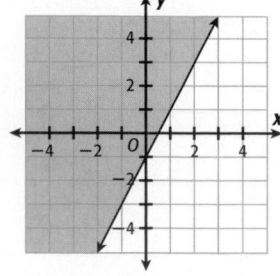

3.

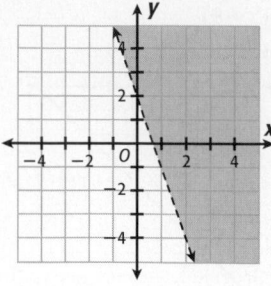

4.

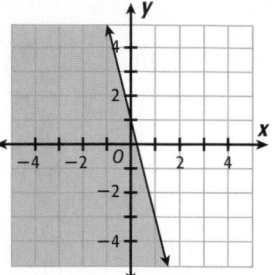

5.

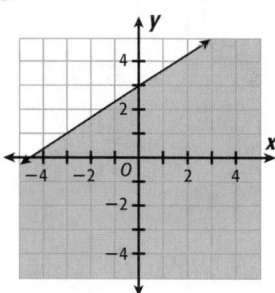

6.

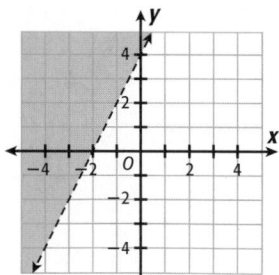

7. a. 10g + 12h ≤ 150

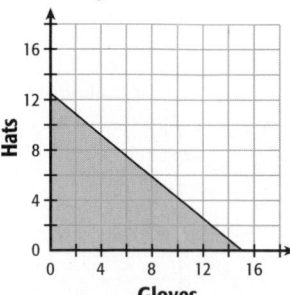

b. yes

8.

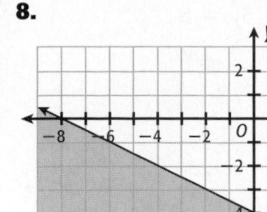

9.

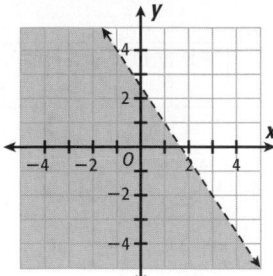

10.

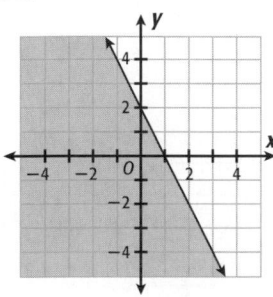

11.

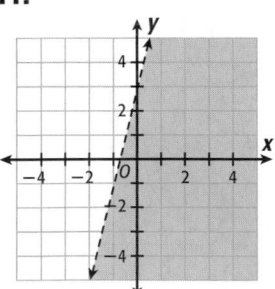

12.

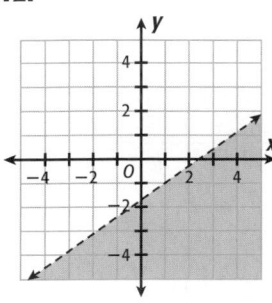

13.

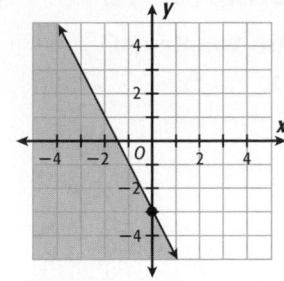

14. a. $d \leq 30t$

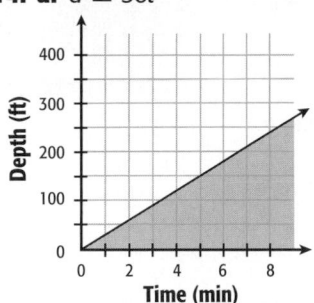

b. no

22. $3x + 5y \leq 30$

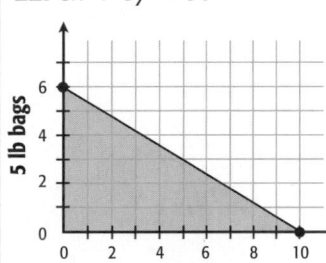

23. $2x + 3y \leq 18$

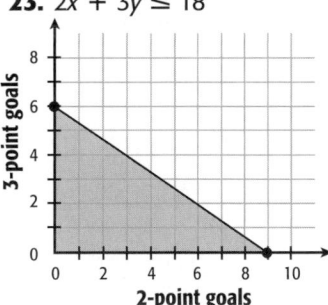

24. a. $35a + 25b \leq 250$

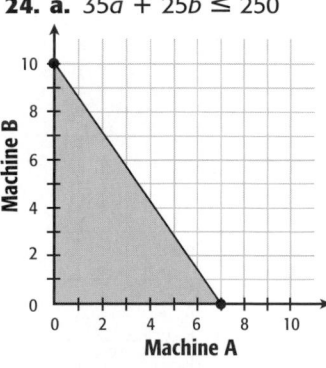

b. 4.4 hr

25. a. $d \leq 800t$

b.

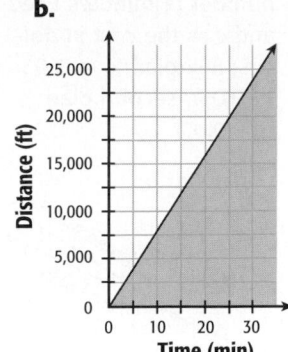

c. yes

28.

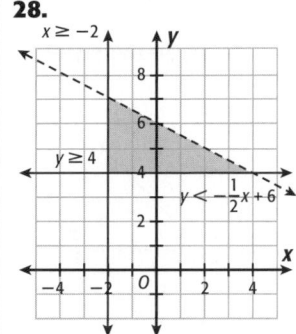

Lesson 11-7

1–6. Possible answers given.

1.

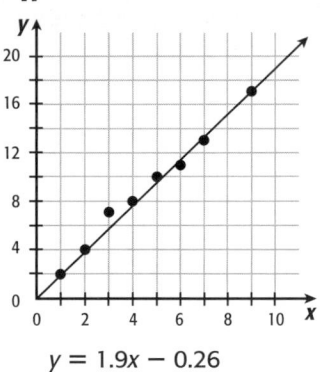

$y = 1.9x - 0.26$

2.

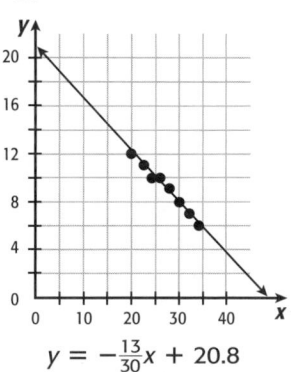

$y = -\frac{13}{30}x + 20.8$

3.

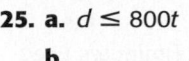

$y = 0.8x + 72.2$; as exercise increases, so does heart rate.

4.

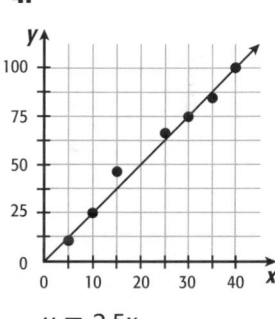

$y = 2.5x$

5.

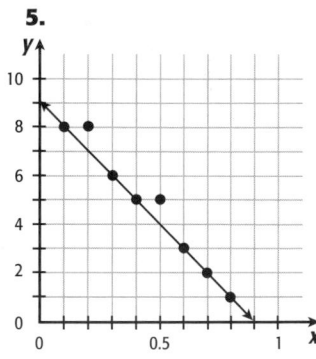

$y = -10.7x + 9.6$

6.

$y = 0.3x + 10.4$ (1990 represents year 0); $14.30

Extension

4.

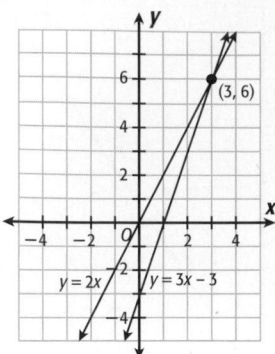

(3, 6)

5.

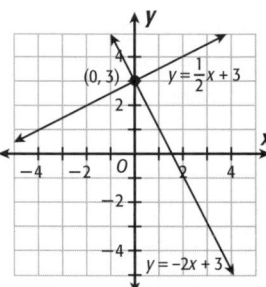

(0, 3)

6.

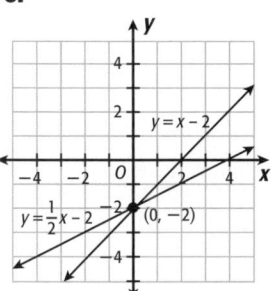

(0, −2)

7. Yes; let t = the number of minutes and d = the distance in meters; lioness: $d = 660t$; cub: $d = 480t + 450$.

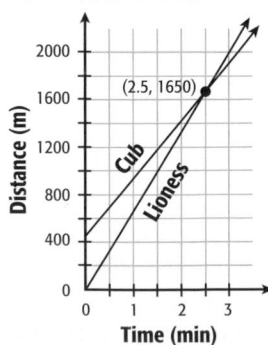

8. First plan; let t = the number of minutes used and c = the cost in dollars; first plan: $c = 3.95 + 0.05t$; second plan: $c = 0.07t$.

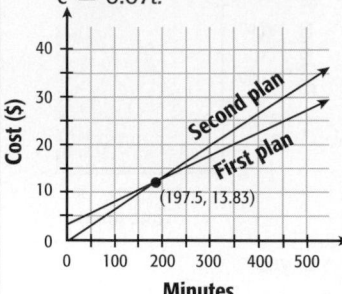

Performance Assessment

1.

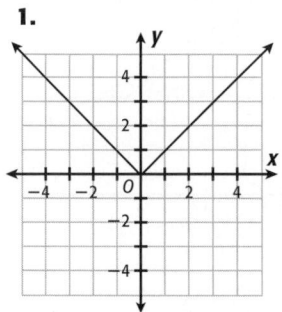

not linear

2. Use (63, 100) and (77, 178) to find a linear equation.

Slope $= \dfrac{178 - 100}{77 - 63} = \dfrac{78}{14} \approx 5.6$

$y - \mathbf{100} = \mathbf{5.6}(x - \mathbf{63})$
$y - 100 = 5.6x - 352.8$
$ y = 5.6x - 252.8$

Substitute 126 for y.
$\mathbf{126} = 5.6x - 252.8$
$378.8 = 5.6x$
$67.6 \approx x$

The temperature is about 68°F.

3.

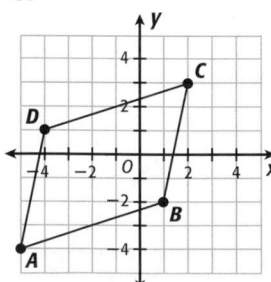

The slopes of $\overline{AB}$ and $\overline{CD}$ are both $\frac{1}{3}$, and the slopes of $\overline{BC}$ and $\overline{AD}$ are both 5. Opposite sides have equal slopes, so they are parallel; parallelogram

Chapter 12

Are You Ready?

12. $y = \frac{2}{3}x + 4$

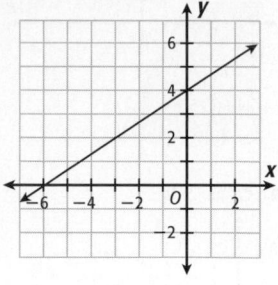

13. $y = -\frac{1}{2}x - 2$

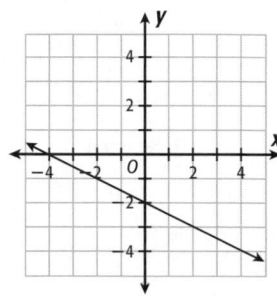

14. $y = 3x + 1$

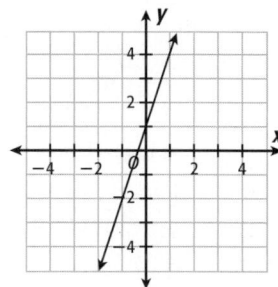

15. $2y = 3x - 8$

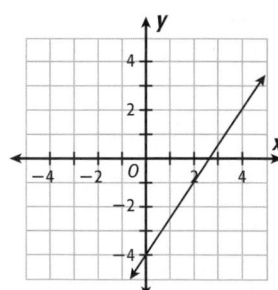

16. $3y + 2x = 6$

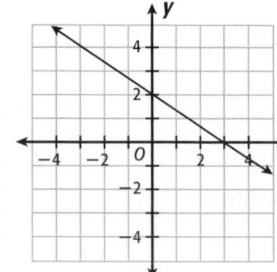

17. $x - 5y = 5$

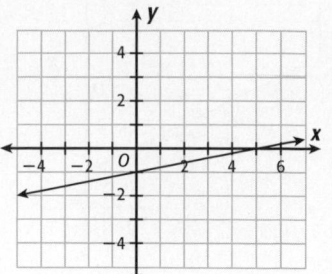

Lesson 12-4

1.

x	−2	−1	0	1	2
y	0	−3	−4	−3	0

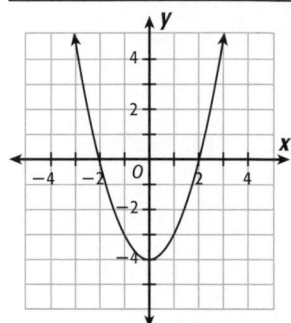

2.

x	−2	−1	0	1	2
y	−2	1	4	7	10

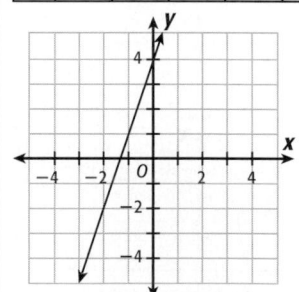

3.

x	−2	−1	0	1	2
y	5	−1	−3	−1	5

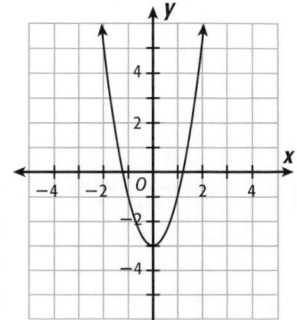

4.

x	−2	−1	0	1	2
y	3	2	1	0	−1

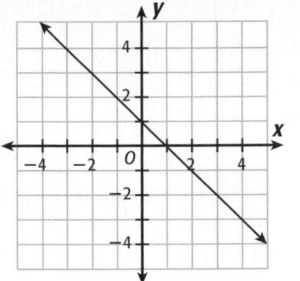

13.

x	−2	−1	0	1	2
y	−5	−4	−3	−2	−1

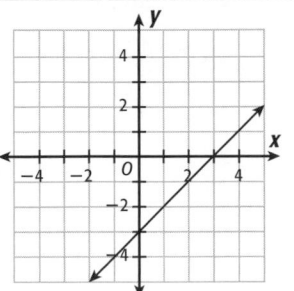

14.

x	−2	−1	0	1	2
y	−6	0	2	0	−6

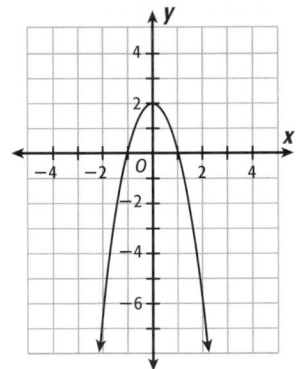

30.

x	y
−3	$-\frac{1}{3}$
−2	$-\frac{1}{2}$
−1	−1
−0.5	−2
−0.25	−4
0.25	4
0.5	2
1	1
2	$\frac{1}{2}$
3	$\frac{1}{3}$

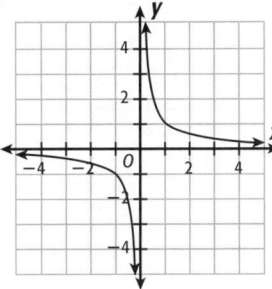

When x = 0, the function is not defined.

Lesson 12-6

1.

x	f(x)
−2	$\frac{1}{9}$
−1	$\frac{1}{3}$
0	1
1	3
2	9

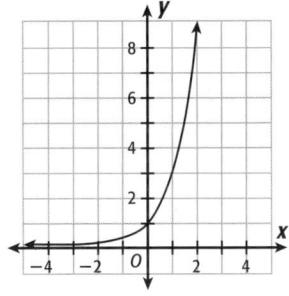

2.

x	f(x)
0	50
1	$\frac{50}{3}$
2	$\frac{50}{9}$
3	$\frac{50}{27}$
4	$\frac{50}{81}$

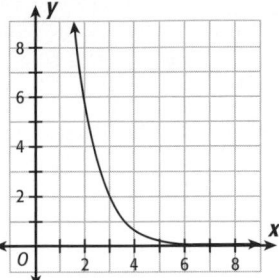

3.

x	f(x)
−2	$\frac{3}{4}$
−1	$\frac{3}{2}$
0	3
1	6
2	12

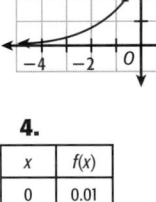

4.

x	f(x)
0	0.01
1	0.05
2	0.25
3	1.25
4	6.25

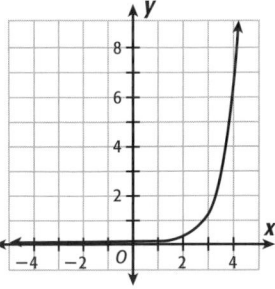

7.

x	f(x)
−2	$\frac{2}{9}$
−1	$\frac{2}{3}$
0	2
1	6
2	18

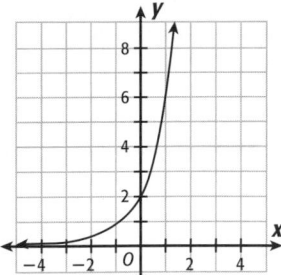

8.

x	f(x)
−2	−50
−1	−10
0	−2
1	−0.4
2	−0.08

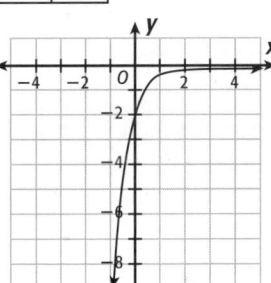

9.

x	f(x)
−2	$\frac{9}{4}$
−1	$\frac{3}{2}$
0	1
1	$\frac{2}{3}$
2	$\frac{4}{9}$

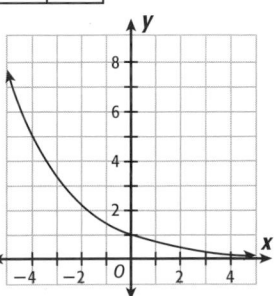

10.

x	f(x)
−2	250
−1	50
0	10
1	2
2	$\frac{2}{5}$

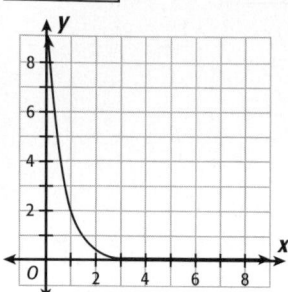

20. $f(x) = 6 \cdot 5^x$

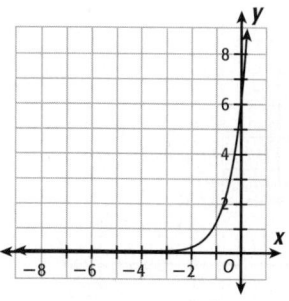

21. $f(x) = -1 \cdot \left(\frac{1}{4}\right)^x$

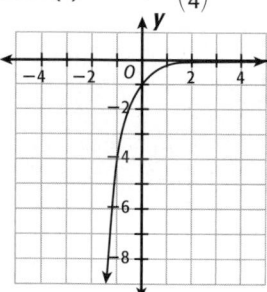

22. $f(x) = 100 \cdot (0.01)^x$

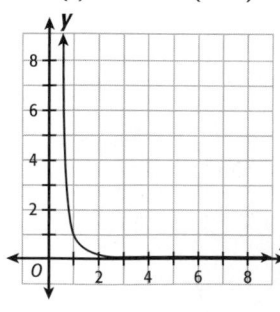

Lesson 12-7

1.

x	f(x)
−3	13
−2	8
−1	5
0	4
1	5
2	8
3	13

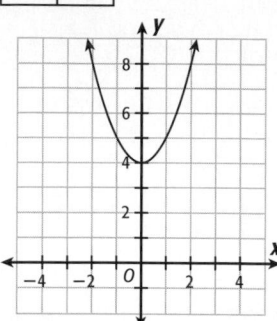

2.

x	f(x)
−3	6
−2	1
−1	−2
0	−3
1	−2
2	1
3	6

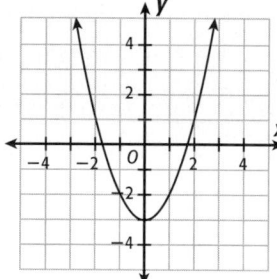

3.

x	f(x)
−3	1.5
−2	−1
−1	−1.5
0	0
1	3.5
2	9
3	16.5

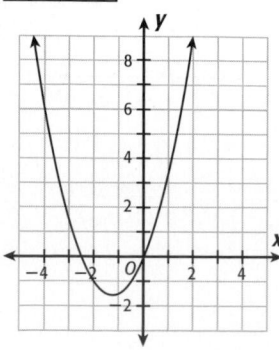

4.

x	f(x)
−3	−1
−2	−3
−1	−3
0	−1
1	3
2	9
3	17

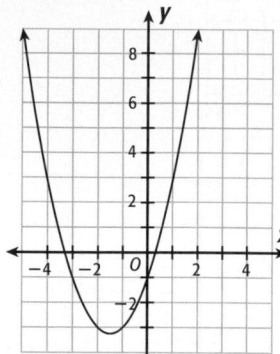

5.

x	f(x)
−1	12
0	6
1	2
2	0
3	0
4	2
5	6

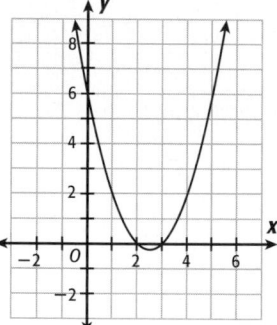

6.

x	f(x)
−4	0
−3	−4
−2	−6
−1	−6
0	−4
1	0
2	6

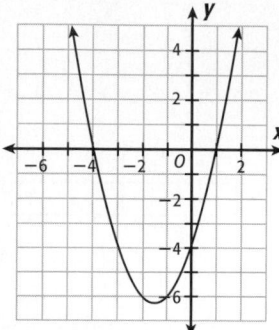

7.

x	f(x)
0	5
1	0
2	−3
3	−4
4	−3
5	0
6	5

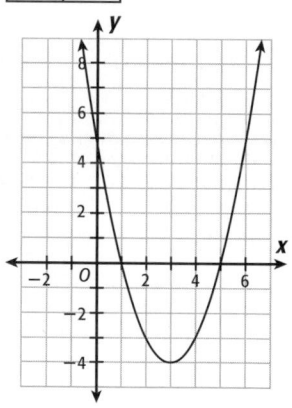

8.

x	f(x)
−3	9
−2	0
−1	−7
0	−12
1	−15
2	−16
3	−15

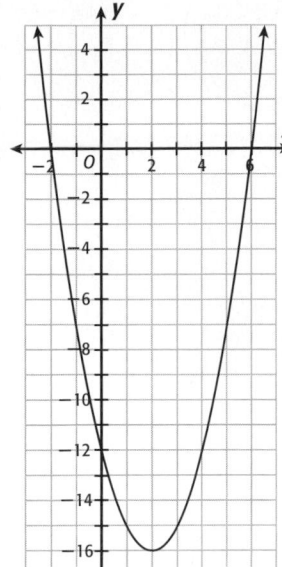

10.

x	f(x)
−3	9
−2	5
−1	3
0	3
1	5
2	9
3	15

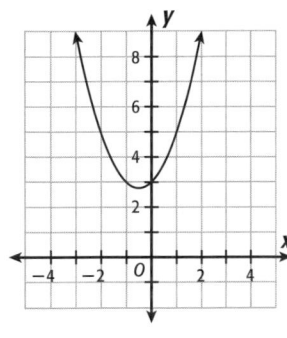

11.

x	f(x)
−3	−7
−2	−2
−1	1
0	2
1	1
2	−2
3	−7

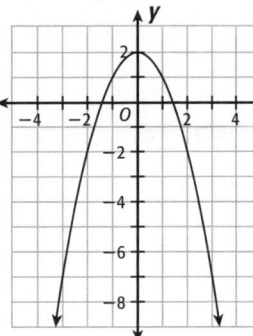

12.

x	f(x)
−3	17
−2	7
−1	1
0	−1
1	1
2	7
3	17

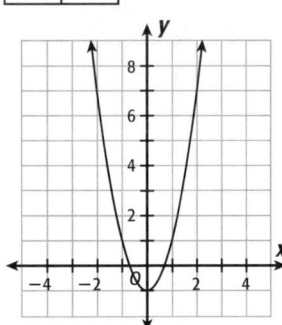

13.

x	f(x)
−3	13
−2	7
−1	3
0	1
1	1
2	3
3	7

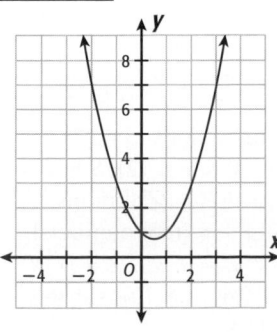

14.

x	f(x)
−3	8
−2	3
−1	0
0	−1
1	0
2	3
3	8

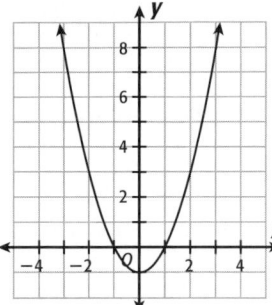

15.

x	f(x)
−4	5.5
−3	0
−2	−3.5
−1	−5
0	−4.5
1	−2
2	2.5

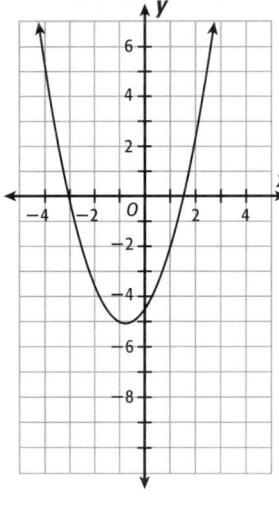

16.

x	f(x)
−1	9
0	4
1	1
2	0
3	1
4	4
5	9

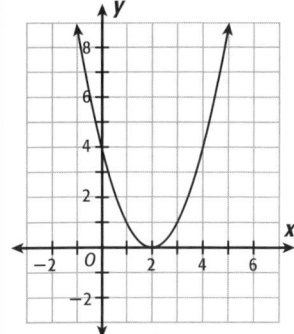

17.

x	f(x)
−3	−24
−2	−25
−1	−24
0	−21
1	−16
2	−9
3	0

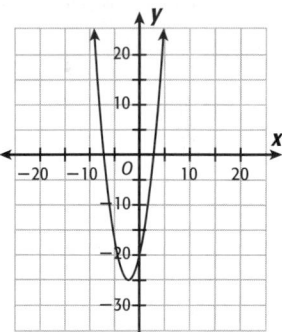

32. a.

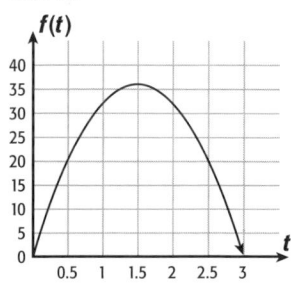

b. $t = 1.5$ s; 36 ft

c. 3 s

37.

x	-2	-1	0	1	2
$f(x)$	-10	-4	-2	-4	-10

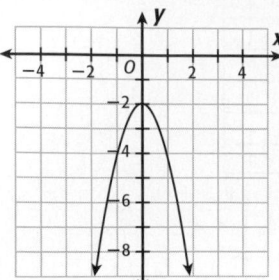

It has no x-intercepts.

Lesson 12-8

3.

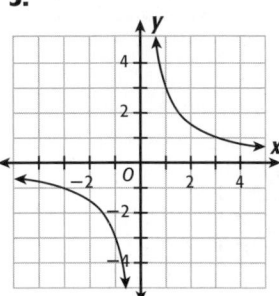

4.

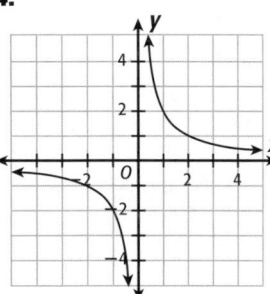

5.

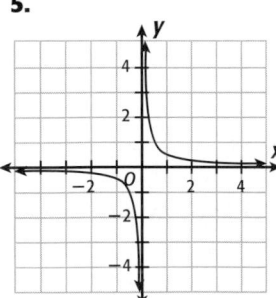

9.

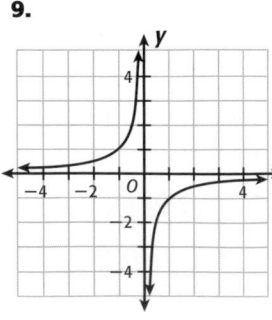

10.

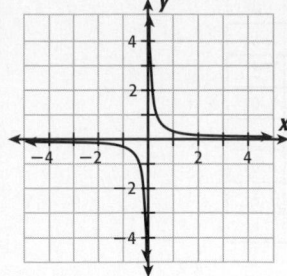

11.

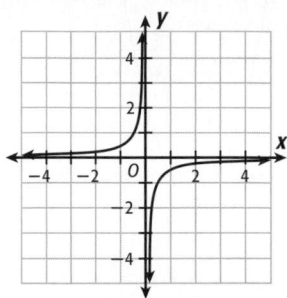

Performance Assessment

1. Check students' work.
Next term is
$$\sqrt{2+\sqrt{2+\sqrt{2+\sqrt{2+\sqrt{2}}}}}$$
$1.414\ldots$, $1.847\ldots$,
$1.962\ldots$, $1.990\ldots$
They are getting closer to 2.

2.

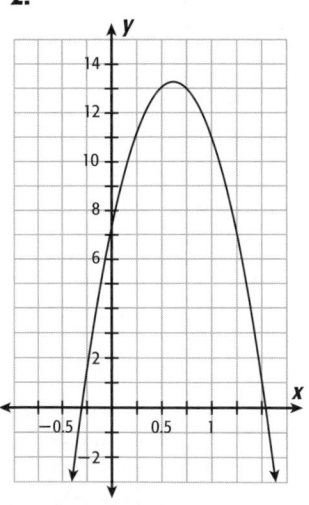

About 0.6 seconds

3. $\dfrac{119.3}{p} = \dfrac{p}{134}$

$p^2 = 15986.2$

$p \approx 126.4$ in.

$r \approx \dfrac{126.4}{119.3} \approx 1.06$

$t_0 = \dfrac{119.3}{1.06} \approx 112.5$

$t_n = 112.5 \cdot 1.06^{n-1}$

Chapter 13

Performance Assessment

1. $5a$; $2a + 3a = (2 + 3)a$
$= 5a$

2. $3x + 8y$; subtract
$(6x + 2y) - (3x - 6y)$.

3. yes; possible answer:
$(a + b)(a - b) = a^2 - b^2$

4. Possible answer: 41, 43,
47, 53, 61

Chapter 14

Lesson 14-3

3. intersections: $A \cap C = C$,
$A \cap B = \{3, 6\}$, $B \cap C =$
$\varnothing$; unions: $A \cup C = A$, A
$\cup B = \{2, 3, 4, 5, 6, 8, 9,$
$10, 12, 15\}$, $B \cup C = \{2, 3,$
$4, 6, 9, 12, 15\}$; subsets:
$C \subset A$

4. intersection: $M \cap N =$
$\{x \mid 5 < x \leq 7\}$; union: M
$\cup N = \{$all real numbers$\}$;
subsets: none

5.

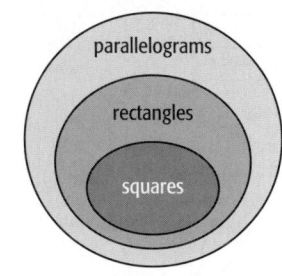

6.

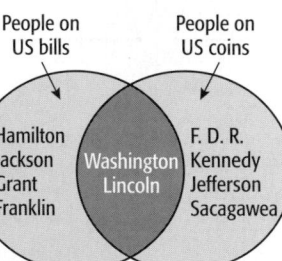

7.

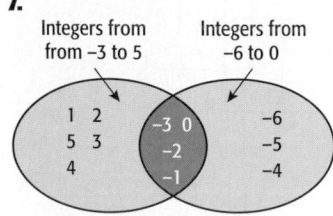

8. intersections: $S \cap T =$
$\{x \mid 10 < x < 12\}$, $S \cap Z =$
Z, $T \cap Z = \varnothing$; unions:
$S \cup T = \{$real numbers$\}$,
$S \cup Z = S$, $T \cup Z = \{x \mid x$
< 4 or $x > 10\}$; subsets:
$Z \subset S$

---

9. intersections: $Q \cap T = T$,
$Q \cap Z = \varnothing$, $Z \cap T = \varnothing$;
unions: $Q \cup T = Q$, $Q \cup$
$Z = \{1, 3, 4, 7, 8, 10, 12,$
$14, 15, 16\}$, $T \cup Z = \{1, 4,$
$7, 10, 15, 16\}$; subsets: T
$\subset Q$

10.

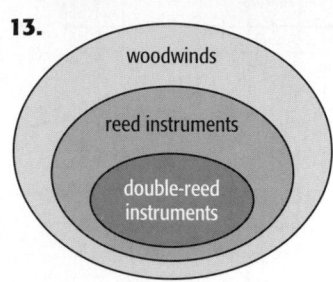

13.

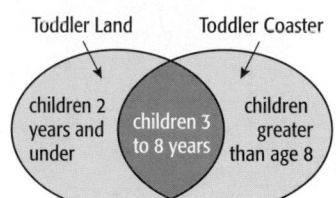

14. $x \leq 8$; $x \geq 3$;

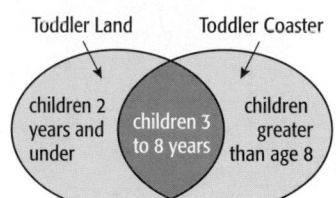

union: {all children}; intersection: {children ages 3
to 8 years}

16. Venn Diagrams make it
easy to see the intersec-
tion of two sets. The overlapping portion of the
circles is the intersection.
To find the union, make
a set whose elements
include every number
inside the circles.

Lesson 14-4

2.

Example	P	Q	P and Q
$x = 6$	T	T	T
$x = 8$	T	F	F
$x = 9$	F	T	F
$x = 7$	F	F	F

3.

Example	P	Q	P or Q
It is 7 A.M. and 65° outside.	T	T	T
It is 3 A.M. and 82° outside.	T	F	T
It is 4 P.M. and 30° outside.	F	T	T
It is 1 P.M. and 90° outside.	F	F	F

4.

Example	P	Q	P or Q
You live in Alabama and are on vacation in Florida.	T	T	T
You live in Florida and are home.	T	F	T
You live in Texas and are on vacation in Mexico.	F	T	T
You live in Michigan and are home.	F	F	F

5.

Example	P	Q	P and Q
Matt has blond hair and wears size 9 shoes.	T	T	T
Matt has blond hair and wears size 10 shoes.	T	F	F
Matt has red hair and wears size 9 shoes.	F	T	F
Matt has brown hair and wears size 11 shoes.	F	F	F

6.

Example	P	Q	P and Q
Harrison has a dog named Bear and has no other pets.	T	T	T
Harrison has a dog named Rick and has no other pets.	T	F	F
Harrison has a cat named Bear and has no other pets.	F	T	F
Harrison has a fish named Shark.	F	F	F

7.

Example	P	Q	P and Q
Polygon *ABCD* is a rectangle with a perimeter of 25 cm.	T	T	T
Polygon *ABCD* is a rectangle with a perimeter of 22 cm.	T	F	F
Polygon *ABCD* is a trapezoid with a perimeter of 25 cm.	F	T	F
Polygon *ABCD* is a trapezoid with a perimeter of 20 cm.	F	F	F

8.

Example	P	Q	P and Q
$n = 37$	T	T	T
$n = 2$	T	F	F
$n = 15$	F	T	F
$n = 12$	F	F	F

9.

Example	P	Q	P or Q
It is 10 A.M. and you are in math class.	T	T	T
It is 10 A.M. and you are in science class.	T	F	T
It is 3 P.M. and you are in math class.	F	T	T
It is 5 P.M. and you are at the movies.	F	F	F

10.

Example	P	Q	P or Q
The food on the plate is a red pepper.	T	T	T
The food on the plate is raspberries.	T	F	T
The food on the plate is broccoli.	F	T	T
The food on the plate is an orange.	F	F	F

11.

Example	P	Q	P or Q
The word is *strong*.	T	T	T
The word is *wide*.	T	F	T
The word is *wisdom*.	F	T	T
The word is *smile*.	F	F	F

12.

Example	P	Q	P or Q
$y = -3$	T	T	T
$y = 5$	T	F	T
$y = -0.5$	F	T	T
$y = 4.8$	F	F	F

17. conjunction; possible answer:

Example	P	Q	P and Q
John is 37 years old and has lived in the U.S. all his life.	T	T	T
John is 42 years old and has lived in the U.S. for 12 years.	T	F	F
John is 21 years old and has lived in the U.S. for 20 years.	F	T	F
John is 5 years old and has lived in the U.S. all his life.	F	F	F

18. Possible answer: What is an example of a true statement of the form *P or Q* in which *P* is true and *Q* is false?

19. Possible answer: A truth table shows all of the combinations of the truth values of the statements *P* and *Q*. Each row of the table shows one possible combination. You can determine whether a conjunction or disjunction is true or false using the truth values of *P* and *Q*.

20. Possible answer: Roger is not at least 35 years old. If *P* is true, then ~*P* is false. If *P* is false, then ~*P* is true. Examples: *P*: 4 is an even number (true); ~*P*: 4 is not an even number (false); *Q*: 12 is not greater than 8 (false); ~*Q*: 12 is greater than 8 (true).

Lesson 14-5

14–16. Possible answers:

14. If you are a 10th grade student, then you are called a *sophomore*. Hypothesis: You are a 10th grade student. Conclusion: You are called a *sophomore*. Raji is a 10th grade student.

15. If there are four objects, they can be chosen two at a time in six different ways. Hypothesis: There are four objects. Conclusion: The objects can be chosen two at a time in six different ways. Carl has four baseball cards.

16. If a figure is a pentagon, then the sum of its interior angle measures is 540°. Hypothesis: A figure is a pentagon. Conclusion: The sum of the interior angle measures is 540°. Figure *ABCDE* is a pentagon.

19. If you use our shampoo, then your hair will be beautiful. No conclusion can be drawn. A conclusion can be drawn only when the hypothesis "you use our shampoo" is true.

20. If a tornado has winds that are between 113 mi/h and 157 mi/h, then the storm is an F2. The tornado is an F2 because the hypothesis "winds are between 113 and 157" is true.

21. Possible answer: The conditional statement is "If a figure is a square, then it is a rectangle." In order to draw a conclusion from this statement, the hypothesis "a figure is a square" must be true. Rectangle *GHJK* is not necessarily a square.

22. Possible answer: A conditional statement can be used to make a conclusion when the hypothesis is true. Example: If it is raining, there are clouds in the sky. You can only make a conclusion, if it is raining.

23. If a figure is a pyramid, then it has a square base. The conjecture is false; a counterexample is a pyramid with a triangular base.

Lesson 14-6

15. Yes; there is a path from each vertex to any other.

16. No; the vertices representing New Jersey and Manhattan each have an odd degree.

17. Possible answer:

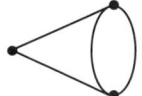

Two vertices have odd degrees.

18. Yes; NJ-M-NJ-M-BX-M-BX-Q-M-Q-M-BR-M-BR-M

Performance Assessment

1. Yes, any set *A* is a subset of set *A* because every element in *A* is also in *A*.

2. Possible answer: Draw two overlapping circles. Label one circle A and one circle B. In the overlapping area of the two circles, write the numbers 6 and 8. In the region of circle A that does not overlap, write 2 and 4. In the region of circle B that does not overlap, write 7 and 9. $A \cap B = \{6, 8\}$

3. A, B, C, and D are each degree three because each is the intersection of two sides of the rectangle and a diagonal. The degree of E is four because it is the intersection of the four segments formed by the intersecting diagonals.

4. 6; 8; 10; $2n$

5. false; false

Extra Practice

1A Equations and Inequalities

36.
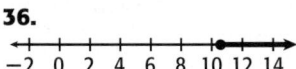
$-2\ 0\ 2\ 4\ 6\ 8\ 10\ 12\ 14$

37.
$-2\ 0\ 2\ 4\ 6\ 8\ 10\ 12\ 14$

38.
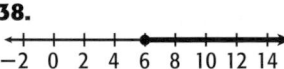
$-2\ 0\ 2\ 4\ 6\ 8\ 10\ 12\ 14$

39.

$10\ 12\ 14\ 16\ 18\ 20\ 22\ 24$

43. $x = 25$

44. $a = 22$

45. $b = 8$

46. $h = 15$

1B Graphing

5.

x	-2	-1	0	1	2
y	-5	-4	-3	-2	-1

6.

x	-2	-1	0	1	2
y	-3	-1	1	3	5

14.
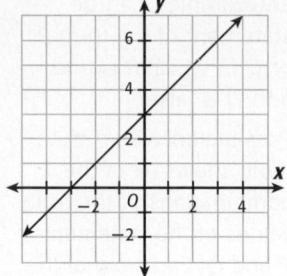

15.
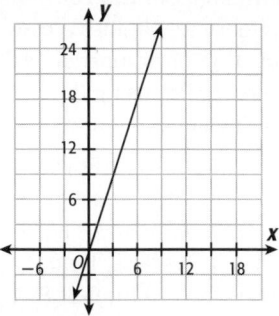

2A Integers

54. $w < -3$;
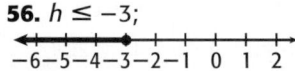
$-5\ -4\ -3\ -2\ -1\ \ 0\ \ 1\ \ 2$

55. $x \geq 1$;
$-4\ -3\ -2\ -1\ 0\ 1\ 2\ 3\ 4$

56. $h \leq -3$;
$-6\ -5\ -4\ -3\ -2\ -1\ 0\ 1\ 2$

57. $g > 5$;
$-1\ 0\ 1\ 2\ 3\ 4\ 5\ 6\ 7$

58. $k > -6$;
$-8\ -7\ -6\ -5\ -4\ -3\ -2\ -1\ \ 0$

59. $m > -3$;
$-6\ -5\ -4\ -3\ -2\ -1\ 0\ 1\ 2$

60. $f < 7$;
$-1\ 0\ 1\ 2\ 3\ 4\ 5\ 6\ 7\ 8\ 9$

61. $m \leq 1$;
$-4\ -3\ -2\ -1\ 0\ 1\ 2\ 3\ 4$

62. $a < -5$;
$-7\ -6\ -5\ -4\ -3\ -2\ -1\ 0\ 1$

63. $x > -24$;
$-30\ -25\ -20\ -15\ -10\ -5\ \ \ 0$

64. $b \geq -13$;
$-14\ -12\ -10\ -8\ -6\ -4\ -2$

65. $a \leq -32$;
$-38\ -36\ -34\ -32\ -30\ -28\ -26$

3A Rational Numbers

52. $x > 6$

53. $a < 9.2$

54. $b \geq 7$

55. $d \leq 13.2$

56. $w > -\dfrac{4}{15}$

57. $b > 4$

58. $b \geq \dfrac{21}{40}$

59. $x \leq 9$

4A Collecting and Describing Data

4.

Stems	Leaves
3	1 2
4	0 0 0 1 5 6 9
5	1 3 4 4
6	
7	2 4

Key: 3|1 means 31

5.

Wins	Stems	Losses
9 6 5 5 4 3	0	2 4 5 6
	1	1 2 2
6	2	

Key: 3|0 means 3 and 0|2 means 2

10.
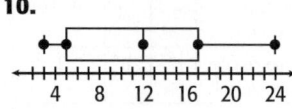
$4\ \ 8\ \ 12\ \ 16\ \ 20\ \ 24$

11.
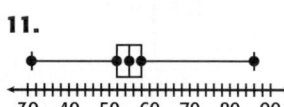
$30\ \ 40\ \ 50\ \ 60\ \ 70\ \ 80\ \ 90$

4B Displaying Data

1.

Age	Frequency
19	2
20	4
21	11
22	2
23	1

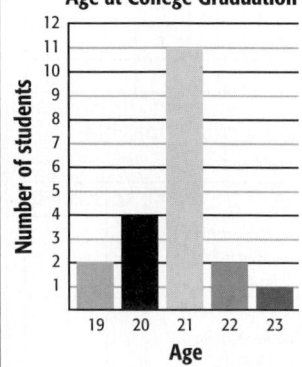

Age at College Graduation

2.
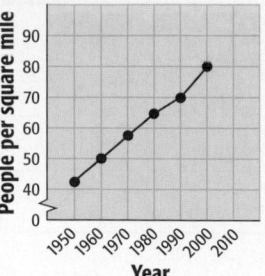
Population Density in U.S.

6.
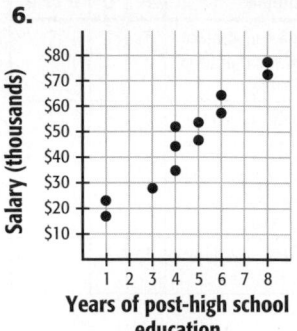

5B Patterns in Geometry

8.
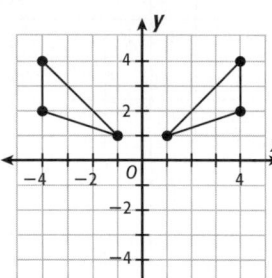

9.
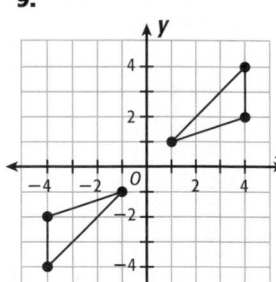

10.
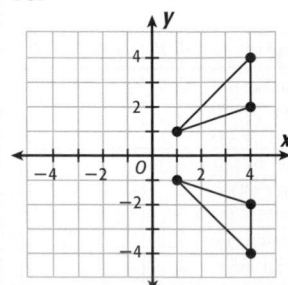

11.

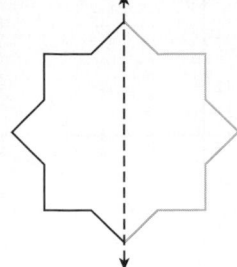

12.

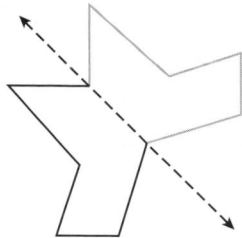

13.

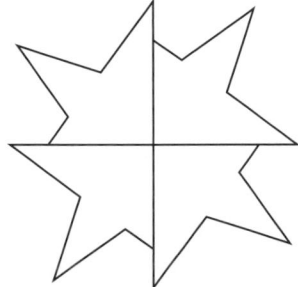

14.

15.

6A Perimeter and Area

4. 24 units²

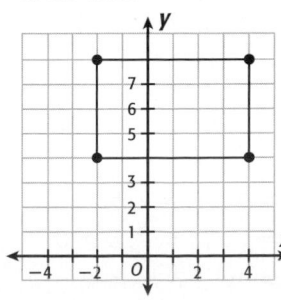

5. 12 units²

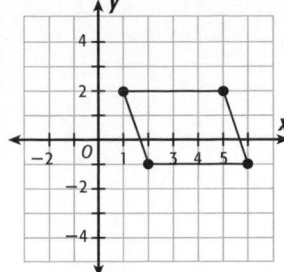

6. 30 units²

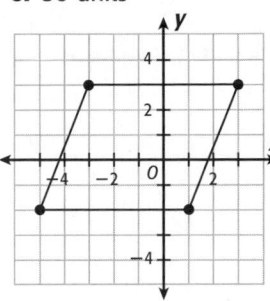

10. 9 units²

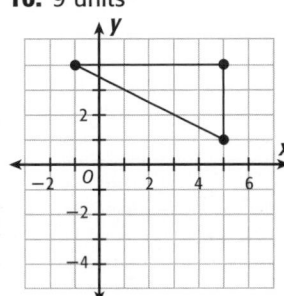

11. 26 units²

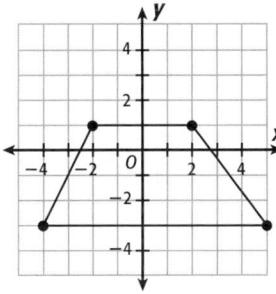

12. 20 units²

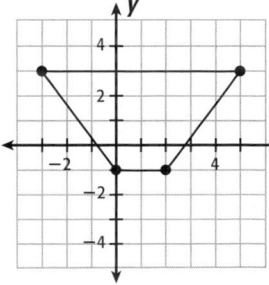

6B Three-Dimensional Geometry

1.

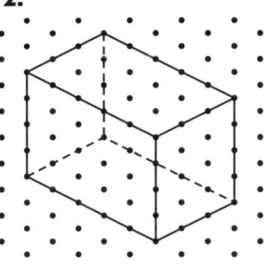

2.

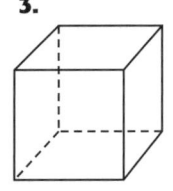

3.

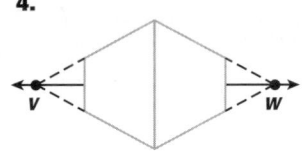

4.

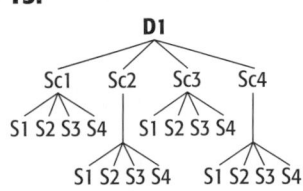

17. 7.6 m²

18. 96 in²

19. 173.2 units²

20. $V = 904.3$ ft³;
$SA = 452.2$ ft²

21. $V = 267,946.7$ cm³;
$SA = 20,096$ cm²

7B Similarity and Scale

14. 9 cm:2 cm; the sides of the larger cube are $\frac{9}{2}$, or 4.5 times, longer than those of the smaller cube.

15. 486 cm²:24 cm²; the surface area of the larger cube is 20.25 times greater than the surface area of the smaller cube.

16. 729 cm³:8 cm³; the volume of the larger cube is 91.125 times greater than that of the smaller cube.

9A Experimental Probability

7.

Outcomes	Probability
Kara	$\frac{1}{5}$
Kevin	$\frac{1}{10}$
Cheryl	$\frac{1}{10}$
Sherry	$\frac{3}{10}$
Jameel	$\frac{3}{10}$

9B Probability and Counting

13.

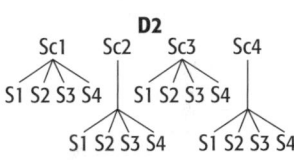

[D1 = dress 1, etc.; Sc1 = scarf 1, etc.; S1 = shoes 1, etc.]

10B Solving Equations and Inequalities

1.

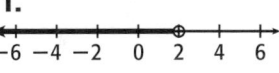

2.

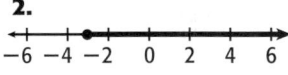

3.

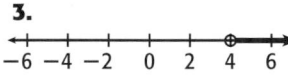

4.

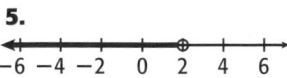

5.

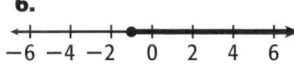

6.

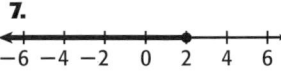

7.

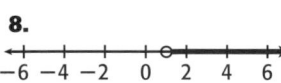

8.

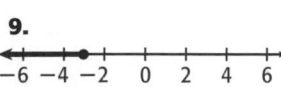

9.

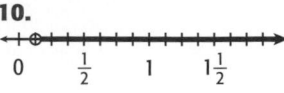

10.

11.

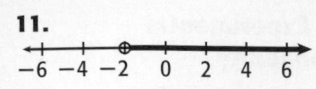

12.

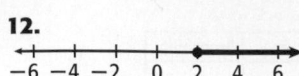

13.

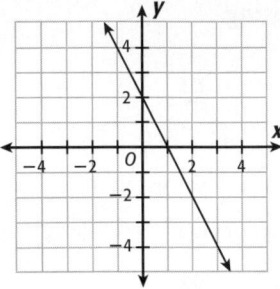

14.

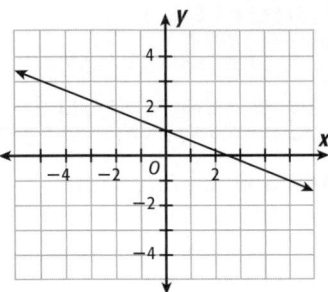

22.

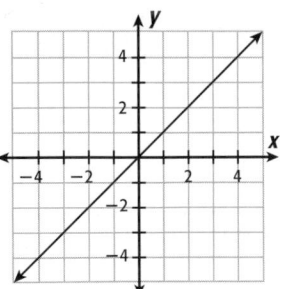

23.

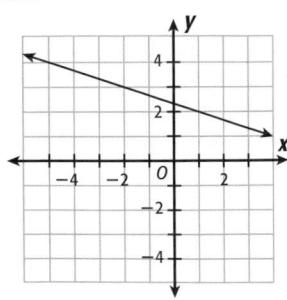

24.

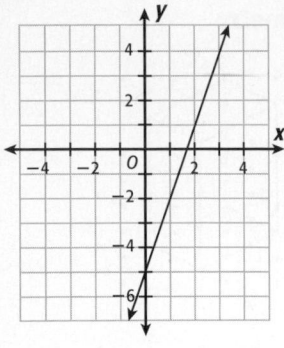

25.

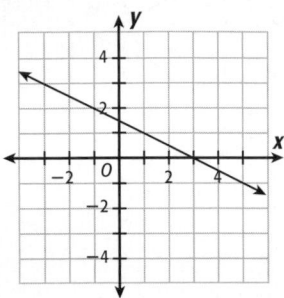

26.

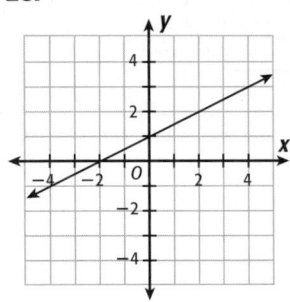

27.

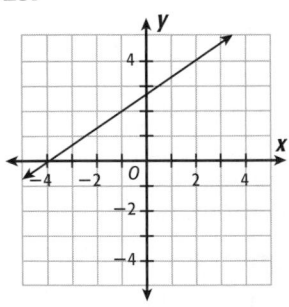

28.

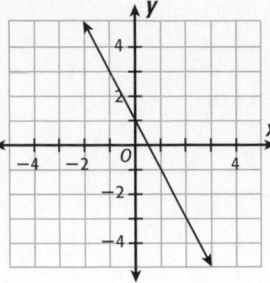

29.

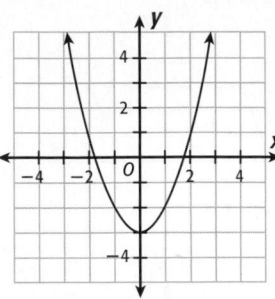

11A Linear Equations

1.

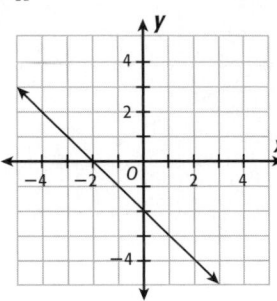

2.

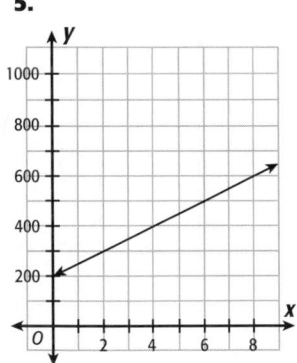

3.

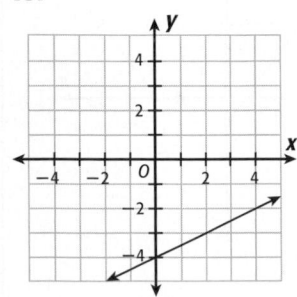

4.

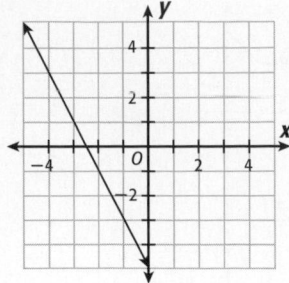

5.

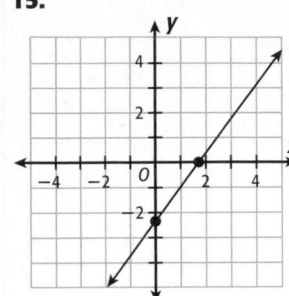

13.

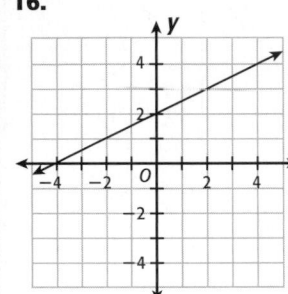

14.

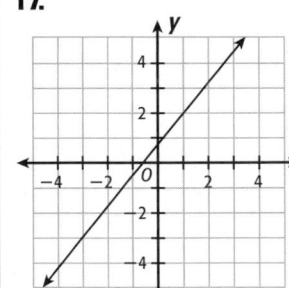

15.

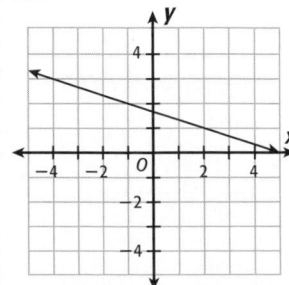

16.

17.

18.

19. $y = 2x$; $m = 2$, $b = 0$

20. $y = \frac{5}{3}x$; $m = \frac{5}{3}$, $b = 0$

21. $y = 4x - 7$; $m = 4$, $b = -7$

22. $y = \frac{1}{2}x - \frac{5}{4}$; $m = \frac{1}{2}$, $b = -\frac{5}{4}$

11B Linear Relationships

11.

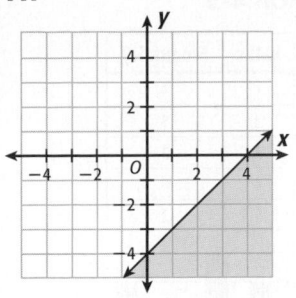

12.

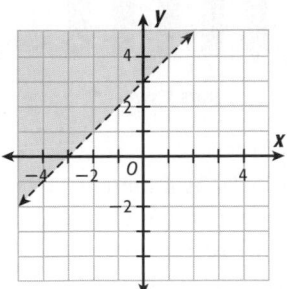

13.

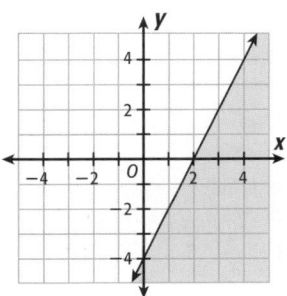

14.

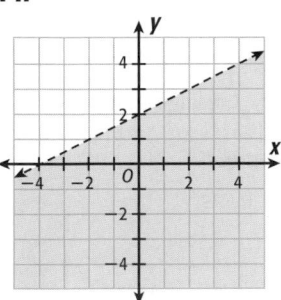

15.

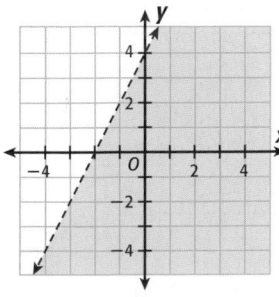

16.

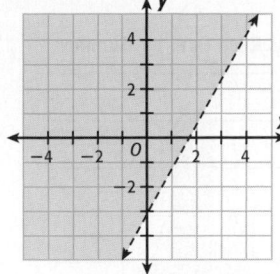

17.

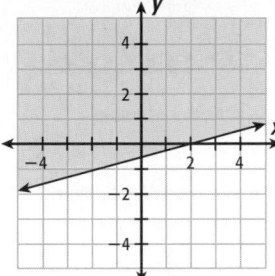

18.

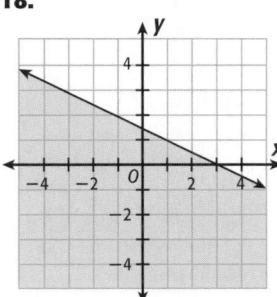

19.

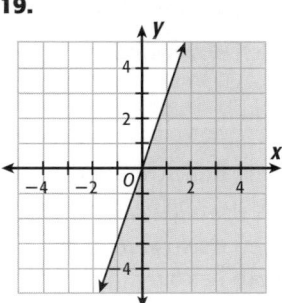

No, the cart will need to be refueled for the second round.

20.

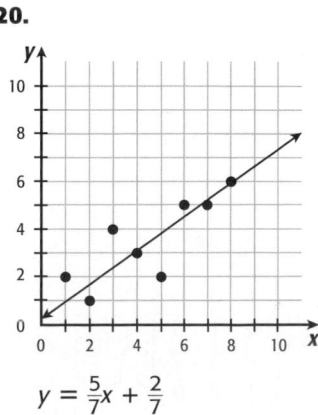

$$y = \frac{5}{7}x + \frac{2}{7}$$

12B Functions

11.

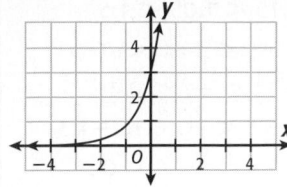

12.

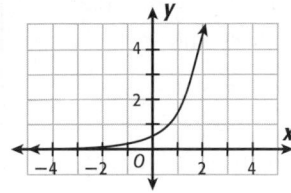

13.

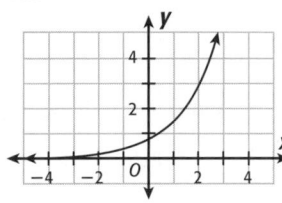

14.

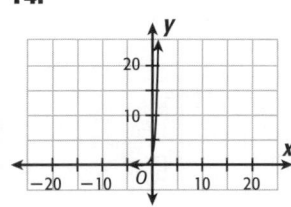

16.

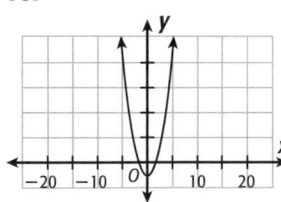

17.

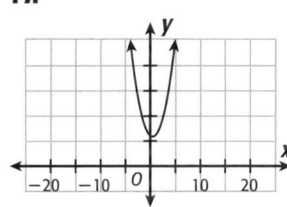

18.

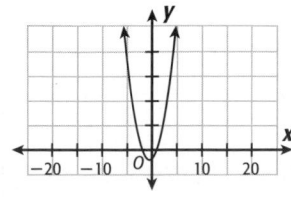

20.

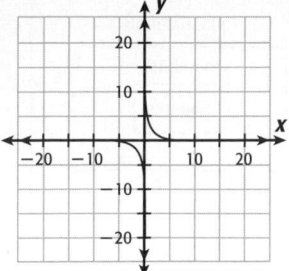

21.

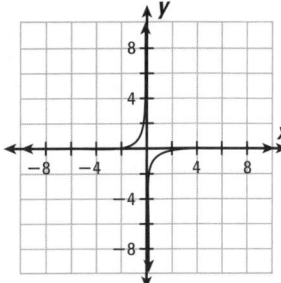

22.

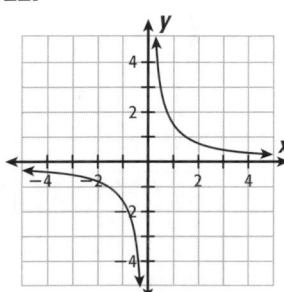

14A Set Theory

13.

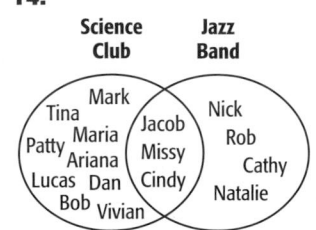

14.

Science Club	Jazz Band

Tina, Mark, Patty, Maria, Ariana, Lucas, Dan, Bob, Vivian, Jacob, Missy, Cindy, Nick, Rob, Cathy, Natalie

16. $X \cap Y = \{6\}$, $X \cap Z = Z$, $Y \cap Z = \varnothing$, $X \cup Y = \{1, 2, 3, 4, 5, 6, 8, 10\}$, $X \cup Z = X$, $Y \cup Z = \{1, 2, 3, 4, 6, 8\}$, $Z \subset X$

17.

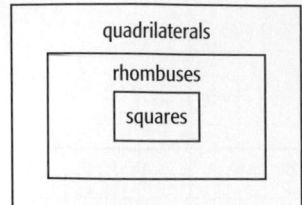

14B Logic and Discrete Math

Possible answers:

1.

Example	P	Q	$P \cup Q$
Greg is in 8th grade and has a 3.2 GPA.	T	T	T
Greg is in 8th grade and has a 2.9 GPA.	T	F	F
Greg is in 7th grade and has a 3.2 GPA.	F	T	F
Greg is in 7th grade and has a 2.9 GPA.	F	F	F

2.

Example	P	Q	$P \cap Q$
$x = 20$	T	T	T
$x = 15$	T	F	T
$x = 14$	F	T	T
$x = 9$	F	F	F

Skills Bank

Terminating and Repeating Decimals

1. 0.2; terminating
2. $0.\overline{3}$; repeating
3. $0.\overline{27}$; repeating
4. 0.375; terminating
5. $0.\overline{7}$; repeating
6. $0.4\overline{6}$; repeating
7. 0.75; terminating
8. $0.8\overline{3}$; repeating
9. $0.\overline{36}$; repeating
10. 0.5; terminating
11. $0.\overline{1}$; repeating
12. $0.91\overline{6}$; repeating
13. $0.\overline{5}$; repeating
14. $0.\overline{72}$; repeating
15. 0.875; terminating
16. 0.92; terminating
17. 0.15; terminating
18. $0.\overline{45}$; repeating

Compare and Order Rational Numbers

13. $0.15 < 1.05 < 1.5$
14. $9.8\% < 34\% < 76\%$
15. $-1\frac{1}{2} < -\frac{3}{5} < 0.4$
16. $-6.3 < -2.6 < -1.3$
17. $-7.1 < -2.4 < 0$
18. $2.5\% < 53\% < 105\%$
19. $-1.2 < -\frac{2}{5} < -0.25$
20. $61\% < 0.65 < 3$
21. $6.7\% < 8.3\% < 13\%$
22. $5\frac{4}{25} < 5\frac{2}{5} < 5\frac{3}{4}$
23. $-0.1003 < -0.018 < -0.008$
24. $0.029 < \frac{28}{100} < 2.7$

Iteration

13.

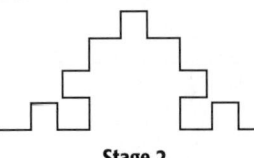

Stage 2

Stage 3

14.

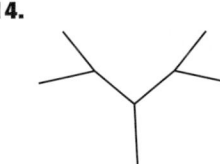

Stage 2

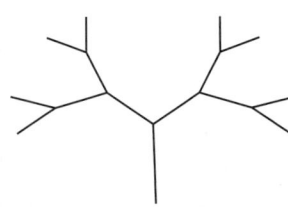

Stage 3

Frequency Polygons

1.
Books Read Over the Summer

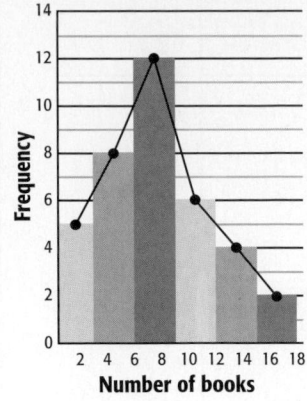

2.
Miles Driven One Way to Work

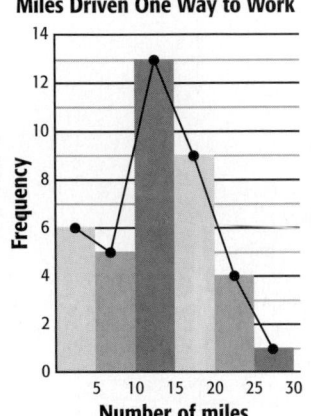

Exponential Growth and Quadratic Behavior

4. open downward
5. wider
6. more narrow
7. open downward and wider
8. open downward and more narrow
9. wider
10. open downward and more narrow
11. more narrow

Matrices

4. $\begin{bmatrix} 24 \\ 13 \\ 35 \end{bmatrix}$

5. Possible answer:
$\begin{bmatrix} 50 & 33 & 50 & 84 & 50 \\ 82 & 49 & 59 & 98 & 56 \end{bmatrix}$

6. $\begin{bmatrix} 1 & 0 & 6 \\ 0 & 1 & 5 \\ 7 & 3 & 2 \\ 4 & 8 & 9 \end{bmatrix}$; 4×3

Lesson Quizzes

Lesson 4-5

1.

Value	Frequency
1	4
2	5
3	3
4	4
5	2

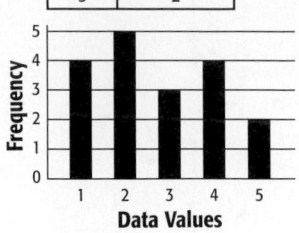

Lesson 4-6

1. It looks as if there is an equal number of cars and semi-trucks due to the different sized icons.
2. The price of the wrecked car skews the average.

Lesson 4-7

1.

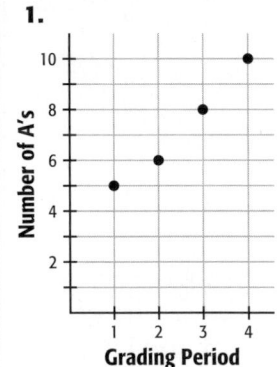

Lesson 6-1

3. 50 units^2

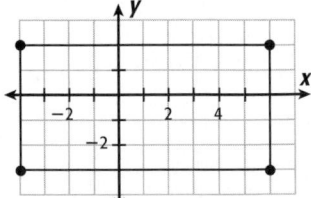

4. 42 units^2

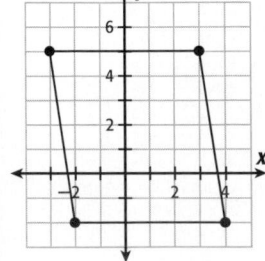

Lesson 6-5

1.

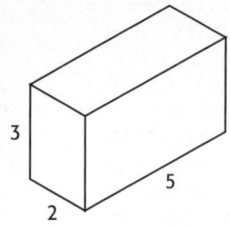

2.

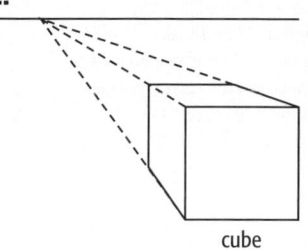

cube

3.

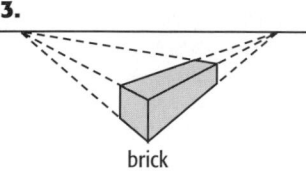

brick

Lesson 11-7

1.

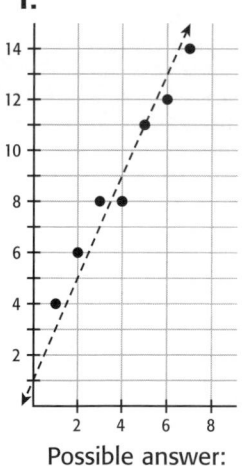

Possible answer:
$y = 2x + 1$

2.

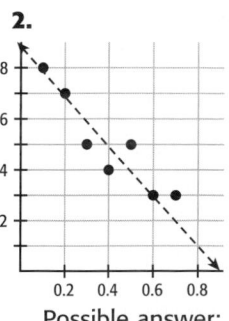

Possible answer:
$y = -10x + 9$

Lesson 12-4

1.

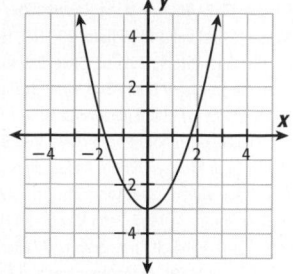

Lesson 12-7

1.

x	−2	−1	0	1	2
y	−1	−2	−1	2	7

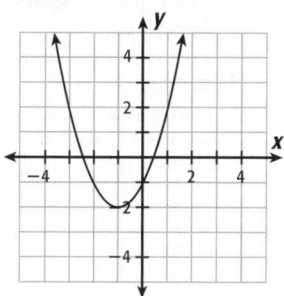

2.

x	−2	−1	0	1	2
y	0	−3	−4	−3	0

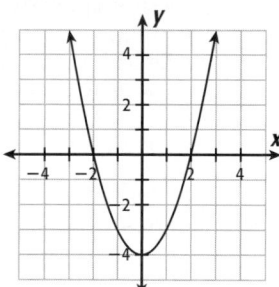

Credits

■ Photo

Cover (all), Pronk & Associates.; **Title Page** (all), Pronk & Associates.; *Master Icons* — teens (all), Sam Dudgeon/HRW.

Problem Solving Handbook: xix, Thomas Wiewandt/Visions of America, LLC/PictureQuest; xxi, xxii, xxiii, Victoria Smith/HRW; xxvi, xxvii, Sam Dudgeon/HRW; xxix, Digital Image ©2004 EyeWire.

All author photos by Sam Dudgeon/HRW. Jan Scheer photo by Ron Shipper.

Chapter One: 2-3 (bkgd), Peter Skinner/Photo Researchers, Inc.; 2 (b), Tom Tracy/Getty Images/FPG International; 4 (tl), Roy King/SuperStock; 4 (tr), Douglas Faulkner/Photo Researchers, Inc.; 7, The Kobal Collection; 8, Robert Landau/CORBIS; 10, ©(2002) PhotoDisc, Inc./HRW; 13, Robert Llewellyn/SuperStock; 15, Mark Lewis/Getty Images/Stone; 18 (tr), Danny Lehman/CORBIS; 18 (tc), Peter Van Steen/HRW; 19, ©2004 PhotoDisc, Inc./HRW; 27, Stephen Munday/Allsport/Getty Images; 33 (tr), Sam Dudgeon/HRW; 33 (tr), Sam Dudgeon/HRW; 34, Peter Van Steen/HRW; 37, Bettmann/CORBIS; 41, Laurence Fleury/Photo Researchers, Inc.; 43, Peter Van Steen/HRW; 47, Alec Pytlowany/Masterfile; 48 (c), Jack Olson; 48 (b), Jack Olson; 49 (t), Mark Segal/Getty Images/Stone; 49 (b), James Blank/Photophile; 50, Randall Hyman; 56, Peter Van Steen/HRW; **Chapter Two**: 58-59 (bkgrd), Science Photo Library/Photo Researchers, Inc.; 58 (b), Dean Conger/CORBIS; 60, Peter Van Steen/HRW; 63, Peter Van Steen/HRW; 64, Lloyd Sutton/Masterfile; 67 (tl), Steve Vidler/SuperStock; 67 (cl), Araldo de Luca/CORBIS; 67 (cr), The Art Archive/Napoleonic Museum Rome/Dagli Orti; 67 (tc), Bettmann/CORBIS; 68, Jeopardy Productions Inc.; 71, Peter David/Masterfile; 75, Sam Dudgeon/HRW; 78, Peter Van Steen/HRW; 81, Luke Frazza/AFP/CORBIS; 83 (b), Dean Conger/CORBIS; 87, S. Lowry/Univ.Ulster/Getty Images/Stone; 92, Courtesy Cornell University; 95 (tc), Francois Gohier/Photo Researchers, Inc.; 95 (bc), Flip Nicklin/Minden Pictures; 96, Sam Dudgeon/HRW; 97, Peter Van Steen/HRW; 99, Joe McDonald/CORBIS; 100 (br), Joseph Sohm; ChromoSohm Inc./CORBIS; 101 (tl), Sam Dudgeon/HRW; 101, John Belliveau; 102 (b), Randall Hyman; 108, Seth Carter/SuperStock; **Chapter Three**: 110-111 (bkgd), Bohemian Nomad Picturemakers/CORBIS; 110 (br), Sam Dudgeon/HRW; 112 (tr), Allsport/Getty Images; 117 (tr), AFP/CORBIS; 125 (tl), John Giustina/Bruce Coleman, Inc.; 130 (tr), Mark Tomalty/Masterfile; 131 (tr), Joe Viesti/Viesti Collection, Inc.; 132 (cl), Lindsay Hebberd/CORBIS; 134 (tr), Wofgang Kaehler/CORBIS; 134 (cr), Sam Dudgeon/HRW; 137 (cl), National Museum of Natural History © 2002 Smithsonian Institution; 140 (tr), Peter Van Steen/HRW Photo; 141, Bettmann/CORBIS; 141, Leonard de Selva/CORBIS;143 (cr), Stuart Westmorland/Getty Images/The Image Bank; 146 (tr), Roman Soumar/CORBIS; 147 (tr), Roberto Rivera; 149 (tr), Peter Van Steen/HRW Photo; 149 (tl), Uimonen Ilkka/CORBIS SYGMA; 150 (tr), Dave Bartruff/Index Stock Imagery, Inc.; 153 (tr), Chris Butler/Photo Researchers, Inc.; 156 (tr), John Garrett/CORBIS; 162 (b), Cosmo Condina/Getty Images/Stone; 163 (tr,br), Morton Beebe/CORBIS; 163 (t), Gail Mooney/CORBIS;164 (br), Jenny Thomas/HRW; 170 (br), Peter Van Steen/HRW;

Chapter Four: 172-173 (bkgrd), David Joel/Getty Images/Stone; 172 (br), Sam Dudgeon/HRW; 179 (tr), Aaron Weithoff; 183 (tr), Richard Schultz; 185 (cl), Corbis Images; 188 (tr), Peter Van Steen/HRW/Kittens courtesy of Austin Humane Society/SPCA; 195 (br), Richard Cummins/CORBIS; 204 (tr), Custom Medical Stock Photo; 207 (c), Peter Van Steen/HRW; 210 (b), Bruce Schulman/Reuters/TimePix; 210 (cl), Michael Clevenger/AP/Wide World Photos; 211 (t), Layne Kennedy/CORBIS; 213 (br), ; 218 (cr), Michal Heron/Corbis Stock Market; **Chapter Five**: 220-211 (bkgrd), Richard T. Nowitz/CORBIS; 220 (br), Victoria Smith/HRW; 226 (tr), Stephen Dalton/Photo Researchers, Inc.; 226 (cr), Roberto Rivera; 228 (tr), Daryl Benson/Masterfile; 231 (tr), Hulton-Deutsch Collection/CORBIS; 243 (cl), Johnathan Blair/CORBIS; 244 , Lucasfilm, Ltd.; 250 (tr), Seth Kushner/Getty Images/Stone; 254 (tr), Angelo Cavalli/Getty Images/The Image Bank; 259 (butterfly), Bob Jensen/Bruce Coleman, Inc.; 259, Jeff Lepore/Photo Researchers, Inc.; 259, R.N. Mariscal/Bruce Coleman, Inc.; 259, Jeff Rotman/International Stock Photography; 259 (shells), SuperStock; 260 (tc) ; 260 (tr), Garry Black/Masterfile; 262 (tl), Grant V. Faint/Getty Images/The Image Bank; 263 (tr), SuperStock; 263 (cr), Adam Woolfitt/CORBIS; 267 (tr), Hand With Reflecting Sphere by M.C. Escher , Cordon Art - Baarn - Holland. All rights reserved.; 267 (cr), Reptiles by M.C. Escher. © 2004 Cordon Art - Baarn - Holland. All rights reserved; 268 (br), ©2004 EyeWire/Getty; 269 (t), Paul A. Souders/CORBIS; 269 (br), Mae Scanlan; 270 (br), Jenny Thomas/HRW; **Chapter Six**: 278-279 (bkgrd), UHB Trust/Getty Images/Stone; 278 (br), Rob

Crandall/Alamy Photos; 280 (tr), Sam Dudgeon/HRW, Woodwork by Carl Childs; 288 (tr), Benelux/ZEFA/H. Armstrong Roberts; 290 (tr), Loukas Hapsis/On Location; 294 (tr), Michelle Bridwell/HRW Photo; 294 (tc), Peter Van Steen/HRW Photo; 297 (cl), Steve vidler/SuperStock; 299 (b), Dave G. Houser/Houserstock; 302 (tr, cr), Jeremy Boon; 306 (cr), Jeremy Boon; 306 (tl), Prat Thierry/Corbis/Sygma; 307 (tr), SuperStock; 309 (tr), Reuters/NewsCom; 311 (cr), Dallas and John Heaton/CORBIS; 311 (tl), G. Leavens/Photo Researchers, Inc.; 312 (tr), Steve Vidler/SuperStock; 313 (cr), Will & Deni McIntyre/Photo Researchers, Inc.; 315 (cl), Owen Franken/CORBIS; 315 (cr), Steve Vidler/SuperStock; 316 (tr), ©2004 Kelly Houle; 317 (cr), Peter Van Steen/HRW Photo; 319 (tr), Peter Van Steen/HRW; 319 (tl), Todd Patrick; 321 (cl), Robert & Linda Mitchell Photography; 322 (cr), Baldwin H. Ward & Kathryn C. Ward/CORBIS; 324 (tr), Imtek Imagineering/Masterfile; 327 (fossil eggs), Sinclair Stammers/Science Photo Library/Photo Researchers, Inc.; 327 (turtle eggs), Dwight Kuhn Photography; 327 (cr), Bob Gossington/Bruce Coleman, Inc.; 327 (c), Frank Lane Picture Agency/CORBIS; 327 (tr), Darryl Torckler/Getty Images/Stone; 328 (tl), Sam Dudgeon/HRW; 328 (tc), Art Stein/Photo Researchers, Inc.; 328 (tr), Neil Rabinowitz/CORBIS; 329 (br), John Elk III; 330 (br, cr), Waverly Traylor; 331 (tr), Courtesy of Great Lakes Aquarium; 331 (bl), Gary Meszaros/Photo Researchers, Inc.; 338 (br), Gunter Marx/CORBIS; **Chapter Seven:** 340-341 (bkgd), Galen Rowell/CORBIS; 340 (br), Michael S. Yamashita/CORBIS; 343 (cl), Biophoto Associates/Photo Researchers, Inc.; 345 (c), Sam Dudgeon/HRW; 346 (tr), Peter Van Steen/HRW Photo; 350 (tr), Stephen Dalton/Photo Researchers, Inc.; 352 (tl), Dr. Harold E. Edgerton/The Harold E. EdgertonTrust ©2004/courtesy Palm Press, Inc.; 356 (tr), Art on File/CORBIS; 359 (tr), Eyewire collection; 359 (tc), Andrew Syred / Microscopix Photolibrary; 359 (bc), Ed Reschke/PA; 361 (b), Nik Wheeler/CORBIS; 362 (tr), Phil Jude/Science Photo Library/Photo Researchers, Inc.; 365 (cl), Peter Van Steen/HRW, 368 (tr), Joseph Sohm; ChromoSohm Inc./CORBIS; 371 (tl), Layne Kennedy/CORBIS; 372 (tr), "Iowa Countryside Outside of Cedar Rapids Iowa" by Stan Herd, photo Jon Blumb; 373 (tr), Eric Grave/Photo Researchers, Inc.; 375 (c), Jeremy Boon, Sam Dudgeon/HRW Photo; 375 (tr), David Young-Wolff/PhotoEdit; 376 (tr), Jonathan Blair/CORBIS; 376 (cr), Peter Van Steen/HRW; 377 (cr), Digital Art/CORBIS; 379 (tl), SuperStock; 379 (cr), Michael S. Yamashita/CORBIS; 380 (cr), Lee Snider/CORBIS; 381 (cr), Peter Van Steen/HRW; 383 (tr), Gail Mooney/CORBIS; 385 (tr), Chris Lisle/CORBIS; 386 (br), Craig Aurness/CORBIS; 387 (tl), Bill Ross/CORBIS; 387 (tr), Robert Holmes/CORBIS; 388 (tr), Isaac Menashe/Zuma Press/NewsCom; 388 (br), AP/Wide World Photos; 388 (bc), William Manning/CORBIS; 389 (tr), Waverly Traylor; 389 (br), Lynda Richardson/CORBIS; 390 (br), Ken Karp/HRW; 390 (tr), Digital Image © 2004 PhotoDisc; 396 (cr), Bettmann/CORBIS; **Chapter Eight:** 398-399 (bkgrd), Photo File/TimePix; 398 (br), Clive Mason/Allsport/Getty Images; 400 (tr), SuperStock; 405 (tr), Ric Ergenbright/CORBIS; 410 (tr), Robert Jensen/Getty Images/Stone; 411 (cl), Hans Reinhard/Bruce Coleman, Inc.; 415 (insects), HRW Photo/Royalty Free; 420 (tr), John Langford/HRW; 421 (tr), Ken Fisher/Getty Images/Stone; 425 (cr), Peter Van Steen; 427, Sam Dudgeon/HRW; 431 (tl), AFP/CORBIS; 434 (cr), Reuters NewMedia Inc./Jacon Cohn/CORBIS; 435 (t), Bob Krist/CORBIS; 435 (br), © 2004 Conrad Gloos c/o MIRA; 436 (br), Victoria Smith/HRW; 442 (br), Sam Dudgeon/HRW; **Chapter Nine:** 444-445 (bkgrd), Erlendur Berg/SuperStock; 444 (br), Bettmann/CORBIS; 446 (tr), Peter Van Steen/HRW Photo; 446 (cr), Sam Dudgeon/HRW; 451 (tr), Joe Richard/AP/Wide World Photos; 454 (tc), Reuters NewMedia Inc./CORBIS; 454 (tr), David Weintraub/Photo Researchers, Inc.; 456 (tr), Duomo/CORBIS; 459 (tl), Raymond Gehman/CORBIS; 461 (br), Susan Marie Anderson/FoodPix; 462 (tr), Sam Dudgeon/HRW; 462 (tr), Sam Dudgeon/HRW; 462 (tr), Sam Dudgeon/HRW; 463 (bl), Peter Van Steen/HRW; 464 (cr), Peter Van Steen/HRW; 466 (tr), Sam Dudgeon/HRW; 467 (tr), ; 470 (tl), Steve Kahn/Getty Images/FPG International; 471 (tr), Peter Van Steen/HRW; 471 (tr), Peter Van Steen/HRW; 475 (tl), The Newark Museum/Art Resource, NY; 477 (tr), Jeffrey Cable/SuperStock; 481 (tl), Corbis/Sygma; 485 (tr), Jeff Greenberg/Photo Researchers, Inc.; 486 (br), From the U.S. Senate Collection, Center for Legislative Archives/Clifford Berryman/Cartoon A-24/May 21, 1912, Washington Evening Star, Washington, D.C.; 486 (cr), CORBIS; 487 (t), Bruce Burkhardt/CORBIS; 487 (br), Paul Sakuma/AP/Wide World Photos; 488 (br), Jenny Thomas/HRW; 494 (cl), Peter Van Steen/HRW Photo; 494 (cr), Peter Van Steen/HRW Photo; **Chapter Ten:** 496-497 (bkgrd), Tom Bean/Getty Images/Stone; 496 (br), David Edwards Photography; 498 (tr), Peter Van Steen; 501 (tr), Karl H. Switak/Photo Researchers, Inc.; 501 (cr), AFP/CORBIS; 503 (cl), Peter Van Steen/HRW; 503 (cl), Sam Dudgeon/HRW; 503 (cl), © 2004 EyeWire, Inc. All rights reserved.; 505 (tr), Peter Van Steen/HRW; 505 (tl), Buddy Mays/CORBIS; 511 (tl), Andrew Syred/Science Photo Library/Photo Researchers, Inc.; 513 (b), Sam Dudgeon/HRW; 514 (tr), Sam Dudgeon/HRW; 516 (cl), Sam Dudgeon/HRW; 518 (tc), Peter Van Steen/HRW; 523 (tr), Kelly-Mooney

Photography/CORBIS; 527 (tl), Rafael Macia/Photo Researchers, Inc.; 528 (b), Tony Arruza/CORBIS; 529 (t), Ric Ergenbright/CORBIS; 529 (br), Raymond Gehman/CORBIS; 529 (cr), Erwin Nielsen/Painet; 530 (br), Jenny Thomas; **Chapter Eleven:** 538-539 (bkgrd), Tom Stack/Painet; 538 (br), Gary Braasch; 540 (tr), Dick Reed/Corbis Stock Market; 540 (cr), Courtesy of Peabody Advertising; www.peabody-adv.com; 542 (cr), HRW Photo Research Library; 545 (tr), ; 554 (tr), (artist)/AlaskaStock Images; 556 (tr), John Greim/Science Photo Library/Photo Researchers, Inc.; 559 (tl), Art Wolfe/Getty Images/The Image Bank; 561 (b), Sam Dudgeon/HRW; 562 (tr), NASA/Science Photo Library/Photo Researchers, Inc.; 566 (tl), E.R. Degginger/Bruce Coleman, Inc.; 567 (tr), NASA/Science Photo Library/Photo Researchers, Inc.; 569 (tr), ; 572 (tr), Duomo/CORBIS; 578 (br), Craig Aurness/CORBIS; 578 (cr), Nat Farbman/TimePix; 579 (tr), Bettmann/CORBIS; 579 (b), Robert Holmes/CORBIS; 580 (br), Jenny Thomas; **Chapter Twelve:** 588-589 (bkgrd), C.N.R.I./Phototake; 588 (br), Stevie Grand/Science Photo Library/Photo Researchers, Inc.; 590 (tr), Getty Images/The Image Bank; 592 (cl), © 2004 PhotoDisc ; 599 (c), Peter Van Steen/HRW; 607 (bl), George McCarthy/CORBIS; 612 (tl), Schenectady Museum; Hall of Electrical History Foundation/CORBIS; 613 (tr), G. C. Kelley/Photo Researchers, Inc.; 616 (tl), Liz Hymans/CORBIS; 618 (bl), GJLP/Science Photo Library/Photo Researchers, Inc.; 620 (tr), John Langford/HRW; 621 (tr), Chip Simons Photography; 625 (tr), Sam Dudgeon/HRW; 628 (tr), Getty Images/Stone; 632 (cr), James A. Sugar/CORBIS; 633 (tr,b), Butch Dill; 633 (cr), Courtesy, New Deal Network; newdeal.feri.org; 634 (br), Randall Hyman; 640 (br), Sam Dudgeon/HRW. **Chapter Thirteen:** 642-643 (bkgd), © W. Cody/CORBIS; 642 (br), Victoria Smith/HRW; 644 © Otto Rogge/CORBIS; 645 © Dave G. Houser/CORBIS; 653 (b), © Paul Eekhoff/Masterfile; 653 (t), Private Collection/Bridgeman Art Library/© 2002 Fletcher Benton/Artists Rights Society (ARS), New York; 655 © Steve Gottlieb/ Stock Connection/ PictureQuest; 656 (l), Victoria Smith/HRW; 656 (r), Sam Dudgeon/HRW; 656 (frame), Victoria Smith/HRW; 657 Sam Dudgeon/HRW; 659 © Getty Images/The Image Bank; 660 Victoria Smith/HRW/Image of assembled plastic model kit used courtesy of Revell-Monogram, LLC © 2004; 661 Victoria Smith/HRW/Images of parts trees used courtesy of Revell-Monogram, LLC © 2004; 664 Sam Dudgeon/HRW; 665 Victoria Smith/HRW; 667 Sam Dudgeon/HRW; 670 © Robert Harding World Imagery/Alamy Photos; 673 (b), Sam Dudgeon/HRW; 676 (b), © RIchard Berenholtz/CORBIS; 677 (br), Bill Banaszewski/New England Stock Photos; 677 (t), © James Schwabel/Panoramic Images; 678 (b), Sam Dudgeon/HRW; 684 (t), Victoria Smith/HRW; 684 (b), Victoria Smith/HRW. **Chapter Fourteen:** 686 (br), Victoria Smith/HRW; 686-687 (bkgd), Corbis Images; 688 (tr, tc), Victoria Smith/HRW; 690 (br), Sam Dudgeon/HRW; 691 (tl), Tony Freeman/PhotoEdit; 692 (tr), © Jeremy Homer/CORBIS; 695 (flamingos), © Royalty-Free/CORBIS; 695 (swans), © Renee Lynn/CORBIS; 695 (geese), Paul J. Fusco/Photo Researchers, Inc.; 695 (chickens), Peter Cade/Getty Images/The Image Bank; 699 (tl), Silvio Fiore/SuperStock; 701 (b), Victoria Smith/HRW; 702 (tr), Lisette Le Bon/SuperStock; 708 (tr), Andrew Toos/CartoonResource.com; 712 (tr), Corbis Images; 715 (tr), © Owaki - Kulla/CORBIS; 716 (tr), Getty Images/The Image Bank; 717 (cl), Sam Dudgeon/HRW; 719 (r), © Roger Ressmeyer/CORBIS; 720 (br), © Sheldon Schafer/Lakeview Museum of Arts & Sciences 2002; 721 (t), © David Muench/CORBIS; 721 (br), AP Photo/The Daily Times, Tom Sistak; 722 (br), Victoria Smith/HRW.

■ Illustrations

All work, unless otherwise noted, contributed by Holt, Rinehart & Winston.

Table of Contents: Page xx (tr), Gary Otteson; xxv (tr), Rosie Sanders; xxvi (c), HRW; xxvi (c), HRW; xxvi (c), HRW; xxviii (cr), Cindy Jeftovic; xxiv (tr), Lori Bilter.

Chapter One: Page 4 (tl), Greg Geisler; 8 (cl), Greg Geisler; 8 (cl), Greg Geisler; 8 (cl), Greg Geisler; 8 (cl), Greg Geisler; 12 (c), Jeffrey Oh; 17 (t), Ortelius Design; 22 (tr), Mark Betcher; 23 (tr), Jeffrey Oh; 23 (c), Greg Geisler; 27 (tr), Ortelius Design; 28 (t), Jeffrey Oh; 28 (c), Greg Geisler; 31 (t), Argosy; 33 (tr), Argosy; 33 (t), Ortelius

Design; 38 (tr), Mark Heine; 38 (c), Greg Geisler; 38 (c), Greg Geisler; 38 (c), Greg Geisler; 48 (tc), Ortelius Design; 50 (tr), Ted Williams. **Chapter Two:** Page 61 (br), Argosy; 65 (br), Argosy; 67 (t), Stephen Durke/Washington Artists; 71 (tr), Argosy; 77 (cr), Argosy; 77 (tr), Argosy; 81 (cr), Argosy; 84 (tr), Greg Geisler; 87 (cr), Argosy; 87 (cr), Stephen Durke/Washington Artists; 96 (c), Greg Geisler; 99 (tr), Argosy; 100 (tr), Ortelius Design; 100 (tr), Ortelius Design; 102 (tr), Jeffrey Oh; 108 (br), Stephen Durke/Washington Artists; 109 (br), Argosy. **Chapter Three:** Page 112 (tr), Greg Geisler; 113 (c), Argosy; 114 (t), Greg Geisler; 116 (cr), Argosy; 121 (tr), Kim Malek; 125 (tr), Argosy; 128 (tr), Mark Heine; 136 (tr), Jeffrey Oh; 139 (cr), Mark Heine; 143 (tr), Argosy; 145 (b), Argosy; 153 (cr), Argosy; 160 (tr), Robert Salinas; 162 (tr), Ortelius Design; 164 (tr), Nenad Jakesevic. **Chapter Four:** Page 174 (tr), Gary Otteson; 181 (c), Argosy; 182 (c), Ortelius Design; 186 (t), Ortelius Design; 187 (t), Argosy; 191 Argosy; 196 (tr), Jeffrey Oh; 196 (bc), Argosy; 197 (br), Argosy; 198 (b), Argosy; 199 (t), Argosy; 199 (br), Argosy; 200 (tr), Jeffrey Oh; 200 (c), Argosy; 200 (b), Argosy; 201 (t), Argosy; 202 (tl), Argosy; 202 (bl), Argosy; 202 (tr), Argosy; 202 (br), Argosy; 203 (tl), Argosy; 203 (tr), Argosy; 203 (br), Argosy; 206 (tr), Ortelius Design; 207 (l), Argosy; 207 (r), Argosy; 210 (tc), Ortelius Design; 216 (tl), Argosy; 216 (bl), Argosy; 216 (br), Argosy; 218 (br), Argosy. **Chapter Five:** Page 226 (t), Argosy; 228 (c), Argosy; 231 (tr), Jeffrey Oh; 234 (tr), Argosy; 238 (cr), Argosy; 239 (tr), Argosy; 243 (cr), Argosy; 260 (tl), Argosy; 262 (c), Argosy; 262 (c), Argosy; 262 (c), Argosy; 262 (c), Argosy; 262 (c), Argosy; 262 (c), Argosy; 262 (br), Argosy; 267 (cl), Argosy; 268 (cr), Argosy; 268 (tc), Ortelius Design; 276 (br) Jeffrey Oh.

Chapter Six: Page 281 (t), Argosy; 284 (cr), Ortelius Design; 285 (t), Argosy; 288 (r), Argosy; 293 (cl), Argosy; 293 (br), Ortelius Design; 294 (cr), Greg Geisler; 297 (cr), Jeffrey Oh; 306 (tr), Mark Heine; 307 (tr), Ortelius Design; 310 (cr), Mark Heine; 310 (br), John White/The Neis Group; 311 (tr), Argosy; 321 (cr), Argosy; 323 (tr), Don Dixon; 324 (cr), Argosy; 328 (cr), Argosy; 329 (tl), Argosy; 329 (tc), Argosy; 329 (tr), Argosy; 329 (cl), Argosy; 329 (c), Argosy; 330 (tc), Ortelius Design; 332 (cl), Argosy; 332 (c), Argosy; 332 (cr), Argosy; 338 (cr), Argosy. **Chapter Seven:** Page 345 (tr), Argosy; 349 (cr), Argosy; 351 (tr), Argosy; 354 (tr), Argosy 374 (br), Ortelius Design; 375 (c), Mark Heine; 381 (tr), Argosy; 388 (tc), Ortelius Design; 397 (br), Ortelius Design. **Chapter Eight:** Page 403 (tr), Jane Sanders; 406 (tr), Gary Otteson; 408 (tr), Doug Bowles; 414 (br), Nenad Jakesevic; 419 (tr), John Bindon; 424 (t), Greg Geisler; 427 (cr), Stephen Durke/Washington Artists; 428 (t), Greg Geisler; 431 (cr), Argosy; 432 (tr), Greg Geisler; 434 (br), Jeffrey Oh; 434 (tc), Ortelius Design; 436 (tr), Gary Otteson. **Chapter Nine:** Page 453 (br), Argosy; 468 (b), Argosy; 470 (tr), Jeffrey Oh; 471 (tl), Greg Geisler; 478 (br), Jeffrey Oh; 482 (tr), Polly Powell; 482 (c), Greg Geisler; 486 (tr), Ortelius Design; 488 (tr), Gary Otteson; 494 (br), Bruno Paciulli. **Chapter Ten:** Page 507 (tc), Greg Geisler; 507 (tr), Mark Heine; 519 (tr), Nenad Jakesevic; 519 (tl), Greg Geisler; 522 (r), Gary Otteson; 528 (tr), Ortelius Design; 530 (tr), John Etheridge; 536 (br), John White/The Neis Group. **Chapter Eleven:** Page 540 (tr), HRW; 544 (c), Argosy; 545 (bc), Greg Geisler; 550 (tr), Cindy Jeftovic; 551 (tc), Greg Geisler; 554 (tr), Nenad Jakesevic; 559 (tr), Patrick Gnan; 559; 566 (tr), Christy Krames; 571 (cr), HRW; 575 (tr), Gary Otteson; 578 (tr), Ortelius Design; 580 (tc), Argosy; 580 (tr), Lance Lekander; 586 (br), Jeffrey Oh. **Chapter Twelve:** Page 594 (tr), Gary Otteson; 595 (tr), Fian Arroyo; 601 (tr), HRW; 603 (cr), HRW; 605 (r), Argosy; 608 (tr), Jeffrey Oh; 608 (cl), Jeffrey Oh; 608 (cr), Jeffrey Oh; 617 (tr), Argosy; 623 (tl), Argosy; 632 (br), HRW; 632 (tr), Ortelius Design; 634 (tr), Gary Otteson; 682 (b), Argosy; 688 (b), Argosy; 688 (c), Argosy; 688 (cr), Argosy; 688 (t), Argosy. **Chapter Thirteen:** Page 647 (t), Argosy; 650 (t), Greg Geisler; 670 (c), Greg Geisler; 671 (tr), Argosy; 673 (tr), Gary Otteson; 673 (cr), Leslie Kell; 676 (cr), Nenad Jakesevic; 677 (tl), Ortelius Design; 678 (tr), Gary Otteson. **Chapter Fourteen:** Page 690 (bl), Argosy; 690 (bl), Kim Malek; 690 (br), Argosy; 692 (cr), Ortelius Design; 696 (tr), John Etheridge; 699 (cr), Argosy; 703 (tr), Dan Vasconcellos; 705 (tr), Uhl Studios, Inc.; 712 (tr), Ortelius Design; 713 (cl), Nenad Jakesevic; 715 (c), Argosy; 720 (r), Mark Betcher; 721 (r), Ortelius Design; 722 (tr), Cindy Jeftovic.

■ Teacher's Edition Credits

All One-Minute Section Planner teens: HRW Photo

T24 (lizard), Digital Image copyright © 2004 PhotoDisc ; T25 (earth), Corbis Images; T26 (beaker), Charlie Winters; T34 (girl), Sam Dudgeon/HRW Photo; T39 (apple), Digital Image copyright © 2004 PhotoDisc; T41 (girl with backpack), Sam Dudgeon/HRW Photo.

Glossary

☑ internet connect
Glossary Online: *go.hrw.com*
Keyword: MT4 Glossary

A

absolute value The distance of a number from zero on a number line; shown by | |. (pp. 60, 679)

Example: $|-5| = 5$

accuracy The closeness of a given measurement or value to the actual measurement or value.

acute angle An angle that measures less than 90°. (p. 223)

acute triangle A triangle with all angles measuring less than 90°. (p. 234)

Addition Property of Equality The property that states that if you add the same number to both sides of an equation, the new equation will have the same solution. (p. 14)

Addition Property of Opposites The property that states that the sum of a number and its opposite equals zero. (p. 60)

Example: $12 + (-12) = 0$

additive inverse The opposite of a number.

Example: The additive inverse of 6 is -6.

adjacent angles Angles in the same plane that have a common vertex and a common side.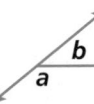

algebraic expression An expression that contains at least one variable. (p. 4)

Example: $x + 8, 4(m - b)$

algebraic inequality An inequality that contains at least one variable. (p. 23)

Example: $x + 3 > 10; 5a > b + 3$

alternate exterior angles A pair of angles on the outer sides of two lines cut by a transversal that are on opposite sides of the transversal. The pairs of alternate exterior angles are $\angle a$ and $\angle d$, and $\angle b$ and $\angle c$. (p. 229)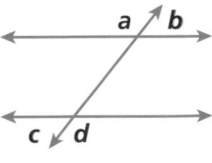

alternate interior angles A pair of angles on the inner sides of two lines cut by a transversal that are on opposite sides of the transversal. The pairs of alternate interior angles are $\angle r$ and $\angle v$, and $\angle s$ and $\angle t$. (p. 229)

angle A figure formed by two rays with a common endpoint called the vertex. (p. 222)

angle bisector A line, segment, or ray that divides an angle into two congruent angles. (p. 227)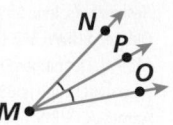

arc An unbroken part of a circle. (p. 686)

area The number of square units needed to cover a given surface. (p. 281)

arithmetic sequence An ordered list of numbers in which the difference between consecutive terms is always the same. (p. 590)

Associative Property
Addition: The property that states that for all real numbers a, b, and c, the sum is always the same, regardless of their grouping: $a + b + c = (a + b) + c = a + (b + c)$. (p. 677)

Multiplication: The property that states that for all real numbers a, b, and c, their product is always the same, regardless of their grouping: $a \cdot b \cdot c = (a \cdot b) \cdot c = a \cdot (b \cdot c)$. (p. 677)

average The sum of a set of data divided by the number of items in the data set; also called *mean*. (p. 184)

average deviation The average distance a data value is from the mean. (p. 208)

axes The two perpendicular lines of a coordinate plane that intersect at the origin. (p. 38)

B

back-to-back stem-and-leaf plot A stem-and-leaf plot that compares two sets of data by displaying one set of data to the left of the stem and the other to the right. (p. 180)

bar graph A graph that uses vertical or horizontal bars to display data. (p. 196)

base-10 system A number system in which all numbers are expressed using the digits 0–9. (p. 160)

base (in numeration) When a number is raised to a power, the number that is used as a factor is the base. (p. 84)

Example: $3^5 = 3 \cdot 3 \cdot 3 \cdot 3 \cdot 3$

base (of a polygon or three-dimensional figure) A side of a polygon; a face of a three-dimensional figure by which the figure is measured or classified. (p. 307)

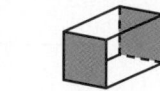  

| Bases of a cylinder | Bases of a prism | Base of a cone | Base of a pyramid |

biased sample A sample that does not fairly represent the population. (p. 174)

binary number system A number system in which all numbers are expressed using only two digits, 0 and 1. (p. 160)

binomial A polynomial with two terms. (p. 644)

bisect To divide into two congruent parts. (p. 227)

boundary line The set of points where the two sides of a two-variable linear inequality are equal. (p. 567)

box-and-whisker plot A graph that displays the highest and lowest quarters of data as whiskers, the middle two quarters of the data as a box, and the median. (p. 189)

break (graph) A zigzag on a horizontal or vertical scale of a graph that indicates that some of the numbers on the scale have been omitted.

capacity The amount a container can hold when filled. (p. 382)

Celsius A metric scale for measuring temperature in which 0°C is the freezing point of water and 100°C is the boiling point of water; also called *centigrade*.

center (of a circle) The point inside a circle that is the same distance from all the points on the circle. (p. 294)

center (of dilation) The point of intersection of lines from all the angles in a dilation. (p. 362)

center (of rotation) The point about which a figure is rotated. (p. 254)

central angle An angle formed by two radii with its vertex at the center of a circle. (p. 778)

certain (probability) Sure to happen; having a probability of 1. (p. 446)

chord A segment with its endpoints on a circle. (p. 778)

circle The set of all points in a plane that are the same distance from a given point called the center. (p. 294)

circle graph A graph that uses sectors of a circle to compare parts to the whole and parts to other parts.

circuit A path in a graph that begins and ends at the same vertex. (p. 713)

circumference The distance around a circle. (p. 294)

clockwise A circular movement to the right in the direction shown.

coefficient The number that is multiplied by the variable in an algebraic expression. (p. 4)

Example: 5 is the coefficient in 5b.

combination An arrangement of items or events in which order does not matter. (p. 472)

commission A fee paid to a person for making a sale. (p. 424)

commission rate The fee paid to a person who makes a sale expressed as a percent of the selling price. (p. 424)

common denominator A denominator that is the same in two or more fractions.

Example: The common denominator of $\frac{5}{8}$ and $\frac{2}{8}$ is 8.

common difference The difference between any two successive terms in an arithmetic sequence. (p. 590)

common factor A number that is a factor of two or more numbers. (p. 764)

Example: 8 is a common factor of 16 and 40.

common multiple A number that is a multiple of each of two or more numbers. (p. 764)

Example: 15 is a common multiple of 3 and 5.

common ratio The ratio each term is multiplied by to produce the next term in a geometric sequence. (p. 595)

Commutative Property
Addition: The property that states that two or more numbers can be added in any order without changing the sum. (p. 769)

Example: $8 + 20 = 20 + 8; a + b = b + a$

Multiplication: The property that states that two or more numbers can be multiplied in any order without changing the product.

Example: $6 \cdot 12 = 12 \cdot 6; a \cdot b = b \cdot a$ (p. 769)

compatible numbers Numbers that are close to the given numbers that make estimation or mental calculation easier. (pp. 420, 765)

complementary angles Two angles whose measures add to 90°. (p. 223)

composite number A number greater than 1 that has more than two whole-number factors. (p. 763)

compound inequality A combination of more than one inequality.

Example: $x \geq -2$ or $x < 10$, or $-2 \leq x < 10$. x is greater than or equal to -2 and less than 10.

compound interest Interest earned or paid on principal and previously earned or paid interest. (p. 432)

compound statement A statement formed by combining two or more simple statements. (p. 702)

conclusion The second statement in a conditional statement. (p. 708)

conditional A compound statement of the form "If P, then Q." Also called an *if-then* statement. (p. 708)

cone A three-dimensional figure with one vertex and one circular base. (p. 312)

congruent Having the same size and shape. (p. 223)

congruent angles Angles that have the same measure. (p. 223)

congruent segments Segments that have the same length. (p. 223)

conjunction A compound statement of the form "*P and Q*." (p. 702)

connected graph A graph in which a path exists from every vertex to every other vertex. (p. 712)

constant A value that does not change. (p. 4)

constant of proportionality A constant ratio of two variables related proportionally. (p. 562)

conversion factor A fraction whose numerator and denominator represent the same quantity but use different units; the fraction is equal to 1 because the numerator and denominator are equal. (pp. 350, 780)

Example: $\dfrac{24 \text{ hours}}{1 \text{ day}}$ and $\dfrac{1 \text{ day}}{24 \text{ hours}}$

coordinate plane (coordinate grid) A plane formed by the intersection of a horizontal number line called the x-axis and a vertical number line called the y-axis. (p. 38)

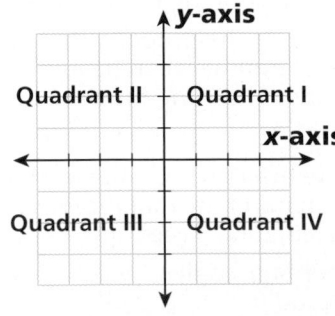

coordinate One of the numbers of an ordered pair that locate a point on a coordinate graph. (p. 38)

correlation The description of the relationship between two data sets. (p. 204)

correspondence The relationship between two or more objects that are matched. (p. 250)

corresponding angles (for lines) Angles formed by a transversal cutting two or more lines and that are in the same relative position.

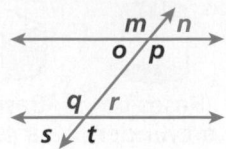

When a transversal cuts two lines as shown in the diagram, the pairs of corresponding angles are $\angle m$ and $\angle q$, $\angle n$ and $\angle r$, $\angle o$ and $\angle s$, and $\angle p$ and $\angle t$. (p. 229)

corresponding angles (in polygons) Matching angles of two or more polygons. (p. 250)

corresponding sides Matching sides of two or more polygons. (p. 250)

cosine (cos) In a right triangle, the ratio of the length of the side adjacent to an acute angle to the length of the hypotenuse. (p. 386)

counterclockwise A circular movement to the left in the direction shown.

cross product The product of numbers on the diagonal when comparing two ratios. (p. 356)

Example:  $2 \cdot 6 = 12$
                                 $3 \cdot 4 = 12$

cube (geometric figure) A rectangular prism with six congruent square faces. (pp. 154, 300)

cube (in numeration) A number raised to the third power. (p. 154)

cumulative frequency The sum of successive data items. (p. 775)

customary system of measurement The measurement system often used in the United States.

Example: inches, feet, miles, ounces, pounds, tons, cups, quarts, gallons

cylinder A three-dimensional figure with two parallel, congruent circular bases connected by a curved lateral surface. (p. 307)

D

decagon A polygon with ten sides.

decimal system A base-10 place value system. (p. 160)

deductive reasoning A form of argument using conditional statements. (p. 709)

degree The unit of measure for angles or temperature. (p. 222)

degree of a polynomial The highest power of the variable in a polynomial. (p. 645)

degree (of a vertex) The number of edges touching a vertex. (p. 712)

horizon line A horizontal line that represents the viewer's eye level. (p. 303)

hypotenuse In a right triangle, the side opposite the right angle. (p. 290)

hypothesis The first statement in a conditional statement. (p. 708)

icosahedron A polyhedron with 20 faces. (p. 300)

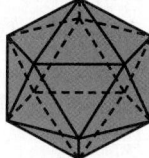

Identity Property of One The property that states that the product of 1 and any number is that number. (p. 769)

Identity Property of Zero The property that states the sum of zero and any number is that number. (p. 769)

if-then statement A compound statement of the form "If P, then Q." Also called a conditional statement. (p. 708)

image A figure resulting from a transformation. (p. 254)

impossible (probability) Can never happen; having a probability of 0. (p. 446)

improper fraction A fraction in which the numerator is greater than or equal to the denominator. (p. 765)

independent events Events for which the outcome of one event does not affect the probability of the other. (p. 477)

indirect measurement The technique of using similar figures and proportions to find a measure.

inductive reasoning Using a pattern to make a conclusion. (p. 773)

inequality A mathematical sentence that shows the relationship between quantities that are not equivalent. (p. 23)

Example: $5 < 8$; $5x + 2 \geq 12$

infinite set A set that contains an infinite number of elements. (p. 689)

input The value substituted into an expression or function. (p. 608)

inscribed angle An angle formed by two chords with its vertex on a circle. (p. 778)

integers The set of whole numbers and their opposites. (p. 60)

interest The amount of money charged for borrowing or using money. (p. 428)

interior angles Angles on the inner sides of two lines cut by a transversal. In the diagram, $\angle c$, $\angle d$, $\angle e$, and $\angle f$ are interior angles.

intersecting lines Lines that cross at exactly one point.

intersection (sets) The set of elements common to two or more sets. (p. 692)

interval The space between marked values on a number line or the scale of a graph.

inverse operations Operations that undo each other: addition and subtraction, or multiplication and division. (p. 14)

inverse variation A relationship in which one variable quantity increases as another variable quantity decreases; the product of the variables is a constant. (p. 628)

irrational number A number that cannot be expressed as a ratio of two integers or as a repeating or terminating decimal. (p. 156)

isolate the variable To get a variable alone on one side of an equation or inequality in order to solve the equation or inequality. (p. 14)

isometric drawing A representation of a three-dimensional figure that is drawn on a grid of equilateral triangles. (p. 302)

isosceles triangle A triangle with at least two congruent sides. (p. 235)

L

lateral face In a prism or a pyramid, a face that is not a base. (p. 316)

lateral surface In a cylinder, the curved surface connecting the circular bases; in a cone, the curved surface that is not a base. (p. 316)

least common denominator (LCD) The least common multiple of two or more denominators.

least common multiple (LCM) The smallest number, other than zero, that is a multiple of two or more given numbers. (p. 764)

legs In a right triangle, the sides that include the right angle; in an isoceles triangle, the pair of congruent sides. (p. 290)

like fractions Fractions that have the same denominator.

like terms Two or more terms that have the same variable raised to the same power. (p. 28)

Example: In the expression $3a + 5b + 12a$, $3a$ and $12a$ are like terms.

line A straight path that extends without end in opposite directions. (p. 222)

line graph A graph that uses line segments to show how data changes. (p. 197)

line of best fit A straight line that comes closest to the points on a scatter plot. (p. 204)

line of reflection A line that a figure is flipped across to create a mirror image of the original figure. (p. 254)

line of symmetry The imaginary "mirror" in line symmetry. (p. 259)

line segment A part of a line between two endpoints. (p. 222)

line symmetry A figure has line symmetry if one half is a mirror-image of the other half. (p. 259)

linear equation An equation whose solutions form a straight line on a coordinate plane. (p. 540)

linear function A function whose graph is a straight line. (p. 613)

linear inequality A mathematical sentence using <, >, ≤, or ≥ whose graph is a region with a straight-line boundary. (p. 567)

major arc An arc that is more than half of a circle. (p. 778)

matrix A rectangular arrangement of data enclosed in brackets. (p. 779)

mean The sum of a set of data divided by the number of items in the data set; also called *average*. (p. 184)

measure of central tendency A measure used to describe the middle of a data set; the mean, median, and mode are measures of central tendency. (p. 184)

median The middle number, or the mean (average) of the two middle numbers, in an ordered set of data. (p. 184)

metric system of measurement A decimal system of weights and measures that is used universally in science and commonly throughout the world.

Example: centimeters, meters, kilometers, gram, kilograms, milliliters, liters

midpoint The point that divides a line segment into two congruent line segments.

minor arc An arc that is less than half of a circle. (p. 778)

mixed number A number made up of a whole number that is not zero and a fraction. (p. 765)

mode The number or numbers that occur most frequently in a set of data; when all numbers occur with the same frequency, we say there is no mode. (p. 184)

monomial A number or a product of numbers and variables with exponents that are whole numbers. (p. 644)

Multiplication Property of Equality The property that states that if you multiply both sides of an equation by the same number, the new equation will have the same solution. (p. 19)

Multiplication Property of Zero The property that states that for all real numbers a, $a \cdot 0 = 0$ and $0 \cdot a = 0$. (p. 769)

multiplicative inverse A number times its multiplicative inverse is equal to 1; also called *reciprocal*. (p. 126)

Example: The multiplicative inverse of $\frac{4}{5}$ is $\frac{5}{4}$.

multiple The product of any number and a whole number is a multiple of that number. (p. 762)

mutually exclusive Two events are mutually exclusive if they cannot occur in the same trial of an experiment. (p. 464)

negative correlation Two data sets have a negative correlation if one set of data values increases while the other decreases. (p. 205)

negative integer An integer less than zero. (p. 60)

net An arrangement of two-dimensional figures that can be folded to form a polyhedron. (p. 300)

network A set of points and line segments or arcs that connect the points. Also called a graph. (p. 712)

no correlation Two data sets have no correlation when there is no relationship between their data values. (p. 205)

nonlinear function A function whose graph is not a straight line.

nonterminating decimal A decimal that never ends. (p. 156)

numerator The top number of a fraction that tells how many parts of a whole are being considered. (p. 112)

numerical expression An expression that contains only numbers and operations.

obtuse angle An angle whose measure is greater than 90° but less than 180°. (p. 223)

obtuse triangle A triangle containing one obtuse angle. (p. 234)

octagon An eight-sided polygon. (p. 239)

Glossary

octahedron A polyhedron with eight faces. (p. 300)

odd number A whole number that is not divisible by two.

odds A comparison of favorable outcomes and unfavorable outcomes. (p. 482)

odds against The ratio of the number of unfavorable outcomes to the number of favorable outcomes. (p. 482)

odds in favor The ratio of the number of favorable outcomes to the number of unfavorable outcomes. (p. 482)

opposites Two numbers that are an equal distance from zero on a number line; also called *additive inverse*. (p. 60)

order of operations A rule for evaluating expressions: first perform the operations in parentheses, then compute powers and roots, then perform all multiplication and division from left to right, and then perform all addition and subtraction from left to right. (p. 768)

ordered pair A pair of numbers that can be used to locate a point on a coordinate plane. (p. 34)

origin The point where the *x*-axis and *y*-axis intersect on the coordinate plane; (0, 0). (p. 38)

outcome (probability) A possible result of a probability experiment. (p. 446)

outlier A value much greater or much less than the others in a data set. (p. 185)

output The value that results from the substitution of a given input into an expression or function. (p. 608)

overestimate An estimate that is greater than the exact answer.

parabola The graph of a quadratic function. (p. 621)

parallel lines Lines in a plane that do not intersect. (p. 228)

parallelogram A quadrilateral with two pairs of parallel sides. (p. 240)

Pascal's triangle A triangular arrangement of numbers in which each row starts and ends with 1 and each other number is the sum of the two numbers above it. (p. 476)

path A way to get from one vertex of a graph to another along one or more edges. (p. 712)

pentagon A five-sided polygon. (p. 239)

percent A ratio comparing a number to 100. (p. 400)

percent change The amount stated as a percent that a number increases or decreases. (p. 416)

percent decrease A percent change describing a decrease in a quantity. (p. 416)

percent increase A percent change describing an increase in a quantity. (p. 416)

perfect square A square of a whole number. (p. 146)

perimeter The distance around a polygon. (p. 280)

permutation An arrangement of items or events in which order is important. (p. 471)

perpendicular bisector A line that intersects a segment at its midpoint and is perpendicular to the segment. (p. 227)

perpendicular lines Lines that intersect to form right angles. (p. 228)

perspective A technique used to make three-dimensional objects appear to have depth and distance on a flat surface. (p. 303)

pi (π) The ratio of the circumference of a circle to the length of its diameter; $\pi \approx 3.14$ or $\frac{22}{7}$. (p. 294)

plane A flat surface that extends forever. (p. 222)

point An exact location in space. (p. 222)

point-slope form The equation of a line in the form of $y - y_1 = m(x - x_1)$, where m is the slope and (x_1, y_1) is a specific point on the line. (p. 556)

point symmetry A figure has point symmetry if it coincides with itself after a 180° rotation.

polygon A closed plane figure formed by three or more line segments that intersect only at their endpoints (vertices). (p. 239)

polyhedron A three-dimensional figure in which all the surfaces or faces are polygons.

polynomial One monomial of the sum or difference of monomials. (p. 644)

population The entire group of objects or individuals considered for a survey. (p. 174)

positive correlation Two data sets have a positive correlation when their data values increase or decrease together.

positive integer An integer greater than zero. (p. 60)

power A number produced by raising a base to an exponent. (p. 84)

Example: $2^3 = 8$, so 8 is the 3rd power of 2.

precision The level of detail of a measurement, determined by the unit of measure. (p. 783)

premise A conditional statement used in deductive reasoning. (p. 709)

prime factorization A number written as the product of its prime factors. (p. 763)

prime number A whole number greater than 1 that has exactly two factors, itself and 1. (p. 763)

principal The initial amount of money borrowed or saved. (p. 428)

principal square root The nonnegative square root of a number. (p. 146)

Example: $\sqrt{25} = 5$. 5 is the principal square root.

prism A polyhedron that has two congruent, polygon-shaped bases and other faces that are all parallelograms. (p. 307)

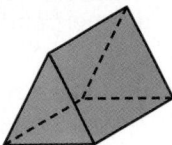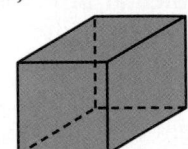

probability A number from 0 to 1 (or 0% to 100%) that describes how likely an event is to occur. (p. 446)

product The result when two or more numbers are multiplied.

proper fraction A fraction in which the numerator is less than the denominator.

Example: $\frac{3}{4}, \frac{1}{12}, \frac{7}{8}$

proportion An equation that states that two ratios are equivalent. (p. 343)

protractor A tool for measuring angles. (pp. 228, 772)

pyramid A polyhedron with a polygon base and triangular sides that all meet at a common vertex. (p. 312)

Pythagorean Theorem In a right triangle, the square of the length of the hypotenuse is equal to the sum of the squares of the lengths of the legs. (p. 290)

Q

quadrant The x- and y-axes divide the coordinate plane into four regions. Each region is called a quadrant.

quadratic function A function of the form $y = ax^2 + bx + c$, where $a \neq 0$. (p. 621)

Example: $y = 2x^2 - 12x + 10$, $y = -3x^2$

quadrilateral A four-sided polygon. (p. 239)

quarterly Four times a year. (p. 432)

quartile Three values, one of which is the median, that divide a data set into fourths. See also *first quartile, third quartile.* (p. 188)

quotient The result when one number is divided by another.

R

radical symbol The symbol $\sqrt{}$ used to represent the nonnegative square root of a numbers. (p. 146)

radius A line segment with one endpoint at the center of the circle and the other endpoint on the circle, or the length of that segment. (p. 294)

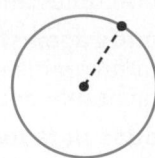

random numbers In a set of random numbers, each number has an equal chance of being selected. (p. 456)

random sample A sample in which each individual or object in the entire population has an equal chance of being selected. (p. 175)

range (in statistics) The difference between the greatest and least values in a data set. (p. 188)

range (in a function) The set of all possible output values of a function. (p. 608)

rate A ratio that compares two quantities measured in different units. (p. 346)

Example: The speed limit is 55 miles per hour, or 55 mi/h.

rate of interest The percent charged or earned on an amount of money; see *simple interest.* (p. 428)

ratio A comparison of two quantities by division. (p. 342)

Example: 12 to 25, 12:25, $\frac{12}{25}$

rational number Any number that can be expressed as a ratio of two integers. (p. 112)

Example: 6 can be expressed as $\frac{6}{1}$, and 0.5 as $\frac{1}{2}$.

ray A part of a line that starts at one endpoint and extends forever. (p. 222)

real number A rational or irrational number. (p. 156)

reciprocal One of two numbers whose product is 1; also called *multiplicative inverse.* (p. 126)

Example: The reciprocal of $\frac{2}{3}$ is $\frac{3}{2}$. The reciprocal of n is $\frac{1}{n}$.

rectangle A parallelogram with four right angles. (p. 240)

rectangular prism A polyhedron whose bases are rectangles and whose other faces are parallelograms. (p. 307)

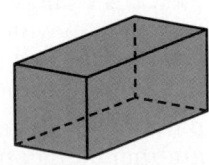

reduction A decrease in size of all dimensions. (p. 373)

reflection A transformation of a figure that flips the figure across a line. (p. 254)

regular polygon A polygon with congruent sides and angles. (p. 240)

regular pyramid A pyramid whose base is a regular polygon and whose lateral faces are all congruent. (p. 320)

regular tessellation A tessellation formed by using regular polygons. (p. 263)

relatively prime Two numbers are relatively prime if their greatest common factor (GCF) is 1. (p. 112)

Example: 7 and 15 are relatively prime.

repeating decimal A decimal in which one or more digits repeat infinitely. (pp. 156, 767)

Example: $0.757575... = 0.\overline{75}$

rhombus A parallelogram with all sides congruent. (p. 240)

right angle An angle that measures 90°. (p. 223)

right cone A cone in which a perpendicular line drawn from the base to the tip (vertex) passes through the center of the base. (p. 320)

right triangle A triangle containing a right angle. (p. 234)

rise The vertical change when the slope of a line is expressed as the ratio $\frac{rise}{run}$, or "rise over run." (p. 244)

rotation A transformation in which a figure is turned around a point. (p. 254)

rotational symmetry A figure has rotational symmetry if it can be rotated less than 360° around a central point and coincide with the original figure. (p. 260)

rounding Replacing a number with an estimate of that number to a given place value. (p. 760)

Example: 2354 rounded to the nearest thousand is 2000, and 2354 rounded to the nearest 100 is 2400.

run The horizontal change when the slope of a line is expressed as the ratio $\frac{rise}{run}$, or "rise over run." (p. 244)

sales tax A percent of the cost of an item, which is charged by governments to raise money. (p. 424)

sample A part of the population. (p. 174)

sample space All possible outcomes of an experiment. (p. 446)

scale The ratio between two sets of measurements. (p. 372)

scale drawing A drawing that uses a scale to make an object smaller than (a reduction) or larger than (an enlargement) the real object. (p. 372)

scale factor The ratio used to enlarge or reduce similar figures. (p. 362)

scale model A proportional model of a three-dimensional object. (p. 376)

scalene triangle A triangle with no congruent sides. (p. 235)

scatter plot A graph with points plotted to show a possible relationship between two sets of data. (p. 204)

scientific notation A method of writing very large or very small numbers by using powers of 10. (p. 96)

secant A line that intersects a circle at two points. (p. 778)

second differences A sequence formed from differences of differences between terms of a sequence. (p. 601)

second quartile The median of a set of data. (p. 188)

sector (data) A section of a circle graph representing part of the data set. (p. 404)

segment A part of a line between two endpoints. (p. 222)

semiregular tessellation A tessellation formed with two or more regular polygons in which every vertex is identical. (p. 263)

sequence An ordered list of numbers. (p. 590)

set A group of items. (p. 688)

side A line bounding a geometric figure; one of the faces forming the outside of an object. (p. 280)

significant digits The digits used to express the precision of a measurement. (p. 783)

similar Figures with the same shape but not necessarily the same size are similar. (p. 367)

simple interest A fixed percent of the principal. It is found using the formula $I = Prt$, where P represents the principal, r the rate of interest, and t the time. (p. 428)

simplest form A fraction is in simplest form when the numerator and denominator have no common factors other than 1. (p. 112)

simplify To write a fraction or expression in simplest form. (p. 29)

simulation A model of an experiment, often one that would be too difficult or too time-consuming to actually perform. (p. 456)

sine (sin) In a right triangle, the ratio of the length of the side opposite an acute angle to the length of the hypotenuse. (p. 386)

skew lines Lines that lie in different planes that are neither parallel nor intersecting.

slant height The distance from the base of a cone to its vertex, measured along the lateral surface. (p. 320)

slope A measure of the steepness of a line on a graph; the rise divided by the run. (p. 244)

slope-intercept form A linear equation written in the form $y = mx + b$, where m represents slope and b represents the y-intercept. (p. 551)

solution of an equation A value or values that make an equation true. (p. 13)

solution of an inequality A value or values that make an inequality true. (p. 23)

solution of a system of equations A set of values that make all equations in a system true. (p. 523)

solution set The set of values that make a statement true. (p. 23)

solve To find an answer or a solution. (p. 13)

sphere A three-dimensional figure with all points the same distance from the center. (p. 324)

square (geometry) A rectangle with four congruent sides. (p. 240)

square (numeration) A number raised to the second power. (p. 146)

Example: In 5^2, the number 5 is squared.

square root One of the two equal factors of a number. (p. 146)

Example: $16 = 4 \cdot 4$, or $16 = -4 \cdot -4$, so 4 and -4 are square roots of 16.

standard form (in numeration) A way to write numbers by using digits.

Example: Five thousand, two hundred ten in standard form is 5210.

stem-and-leaf plot A graph used to organize and display data so that the frequencies can be compared. (p. 179)

stratified sample A sample of a population that has been divided into subgroups. (p. 175)

subset A set contained within another set. (p. 689)

substitute To replace a variable with a number or another expression in an algebraic expression. (p. 4)

Subtraction Property of Equality The property that states that if you subtract the same number from both sides of an equation, the new equation will have the same solution. (p. 14)

sum The result when two or more numbers are added.

supplementary angles Two angles whose measures have a sum of 180°. (p. 223)

surface area The sum of the areas of the faces, or surfaces, of a three-dimensional figure. (p. 316)

system of equations A set of two or more equations that contain two or more variables. (p. 523)

system of linear equations Two or more linear equations graphed in the same coordinate plane. (p. 576)

systematic sample A sample of a population that has been selected using a pattern. (p. 175)

tangent (geometry) A line that intersects a circle at one point. (p. 778)

tangent (tan) In a right triangle, the ratio of the length of the side opposite an acute angle to the length of the side adjacent to that acute angle. (p. 386)

term (in an expression) The parts of an expression that are added or subtracted. (p. 28)

Example: $5x^2$ is an expression with one term, -10 is an expression with one term, and $x + 1$ is an expression with two terms.

term (in a sequence) An element or number in a sequence. (p. 590)

terminating decimal A decimal number that ends or terminates. (pp. 156, 767)

Example: 6.75

tessellation A repeating pattern of plane figures that completely cover a plane with no gaps or overlaps. (p. 263)

tetrahedron A polyhedron with four faces. (p. 300)

theoretical probability The ratio of the number of equally likely outcomes in an event to the total number of possible outcomes. (p. 462)

third quartile The median of the upper half of a set of data; also called *upper quartile*. (p. 188)

tip The amount of money added to a bill for service; usually a percent of the bill. (p. 420)

transformation A change in the size or position of a figure. (p. 254)

translation A movement (slide) of a figure along a straight line. (p. 254)

transversal A line that intersects two or more lines. (p. 228)

trapezoid A quadrilateral with exactly one pair of parallel sides. (p. 240)

tree diagram A branching diagram that shows all possible combinations or outcomes of an event. (p. 468)

trial In probability, a single repetition or observation of an experiment. (p. 446)

triangle A three-sided polygon.

Triangle Sum Theorem The theorem that states that the measures of the angles in a triangle add up to 180°. (p. 234)

triangular prism A polyhedron whose bases are triangles and whose other faces are parallelograms. (p. 307)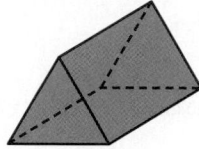

trigonometric ratios Ratios that compare the lengths of the sides of a right triangle; the common ratios are tangent, sine, and cosine. (p. 386)

trinomial A polynomial with three terms. (p. 644)

truth table A way to show the truth value of a compound statement. (p. 702)

truth value Either true or false. (p. 702)

unbiased sample A sample is unbiased if every individual in the population has an equal chance of being selected. (p. 174)

underestimate An estimate that is less than the exact answer.

union The set of all elements that belong to two or more sets. (p. 693)

unit conversion The process of changing one unit of measure to another.

unit conversion factor A fraction used in unit conversion in which the numerator and denominator represent the same amount but are in different units. (p. 350)

Example: $\frac{60 \text{ min}}{1 \text{ h}}$ or $\frac{1 \text{ h}}{60 \text{ min}}$

unit price A unit rate used to compare prices. (p. 347)

unit rate A rate in which the second quantity in the comparison is one unit. (p. 346)

Example: 10 centimeters per minute

unlike fractions Fractions with different denominators. (p. 131)

vanishing point In a perspective drawing, a point where lines running away from the viewer meet. (p. 303)

variability The spread of values in a set of data. (p. 188)

variable A symbol used to represent a quantity that can change. (p. 4)

Venn diagram A diagram that is used to show relationships between sets. (p. 696)

vertex On an angle or polygon, the point where two sides intersect; on a polyhedron, the intersection of three or more faces; on a cone or pyramid, the top point. (p. 302)

vertex (of a graph) The points in a graph. (p. 712)

vertical angles A pair of opposite congruent angles formed by intersecting lines; in the diagram ∠*a* and ∠*c* are congruent and ∠*b* and ∠*d* are congruent. (p. 223)

volume The number of cubic units needed to fill a given space. (p. 307)

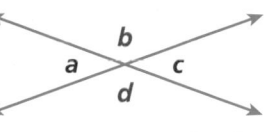

withholding tax A deduction from earnings as an advance payment on income tax. (p. 425)

x-axis The horizontal axis on a coordinate plane. (p. 38)

x-coordinate The first number in an ordered pair; it tells the distance to move right or left from the origin (0, 0). (p. 38)

Example: 5 is the *x*-coordinate in (5, 3).

x-intercept The *x*-coordinate of the point where the graph of a line crosses the *x*-axis. (p. 550)

y-axis The vertical axis on a coordinate plane. (p. 38)

y-coordinate The second number in an ordered pair; it tells the distance to move up or down from the origin (0, 0). (p. 38)

Example: 3 is the *y*-coordinate in (5, 3).

y-intercept The *y*-coordinate of the point where the graph of a line crosses the *y*-axis. (p. 550)

zero pair A number and its opposite, which add to 0.

Index

A

Absolute value, 60, 771
  using, for addition of integers, 61
Across the Curriculum, *2B, 58B, 110B, 172B,*
  220B, 278B, 340B, 398B, 444B, 496B, 538B,
  588B, 642B, 686B
Across the Series, *2B, 58B, 110B, 172B, 220B,*
  278B, 340B, 398B, 444B, 496B, 538B, 588B,
  642B, 686B
Accuracy, 808
Acute angles, 223
Acute triangles, 234
Addition, *see also* Sums
  of fractions
    with like denominators, 118
    with unlike denominators, 131
  of integers, 60–61
    using a number line for, 60
    using absolute value for, 61
  of rational numbers, 117–118
  repeated, multiplication as, 121
  solving equations with, 13–15, 74
  solving for variables with, 519
  solving inequalities with, 78
  with unlike denominators, 131–132
  word phrases for, 8
Addition Property of Equality, 14
Addition Property of Opposites, 60
Additive inverse, 60, 808
Algebra
The development of algebra skills and concepts
is a central focus of this course and is found
throughout this book.
  absolute value, 60, 771
    using, for addition of integers, 61
  arithmetic sequences, 590–592
  coefficient, 4
  combining like terms, 28–29
  equations, 13
    of direct variation, 563
    graphs of, 39
    linear, *see* Linear equations
    literal, 519–520
    point-slope form of, 557
    simple two-step, solving, 20
    solutions of, 13
    solving word problems using, 137
    systems of, *see* Systems
    writing, for linear functions, 613–614
  expressions, 4
    and combining like terms, 29
    evaluating, 4–5
    simplifying, by combining like terms, 29
    translating word phrases into, 8–9
    in word problems, 10
    writing, 8–10
  functions, 608–610
    exponential, 617
    inverse variation, 629

    linear, *see* Linear functions
    quadratic, *see* Quadratic functions
  graphs
    of equations, 39
    finding slopes from, 546
    line, *see* Line graphs
    using, to solve systems of linear
      equations, 576
    writing equations for linear functions
      from, 613
  inequalities, 23
    algebraic, 23, *see also* Inequalities
    equations and, 496–537
    graphing, 24, 78, 567–568
    linear, 567
    multistep, 514–516
    simple, solving, 23–25
    solutions of, 23
    in two variables, graphing, 567–569
  intercepts, 550–552
  isolating variables, 14, 520
  like terms, 28
    solving multistep equations that contain, 502
  linear equations, 540
    graphing, 540–542
    finding intercepts for, 550
  linear functions, 613–614
  linear inequalities, 567
  lines, 222
    of best fit, 204, 572–573
    graphing, 538–587
    using points and slopes, 547
    point-slope form of, 556
    slope-intercept form of, 551
  multistep equations,
    502–503, 507–509
  multistep inequalities, 514–516
  point-slope form, 556–557
  points, 222
    and slopes, graphing lines
      using, 547
    two, finding slopes, given, 545
  polynomials, 642–685
    addition of, 656–657
    classifying, 644–645
    division of, by monomials, 674–675
    factoring, 675
    multiplication of, by monomials, 664–665
    opposites of, 660
    simplifying, 650–651
    subtraction of, 660–661
  quadratic functions, 621–623
  simple inequalities, 23–25
  slope-intercept form, 551
  slopes, 244, 545–547
    and intercepts, using, 550–552
    negative, 545
    of parallel and perpendicular lines, 245
    points and graphing lines
      using, 547
    positive, 545
    undefined, 545
    zero, 545
  solution set, 23
  solutions, 13
    of equations, 13, 34
    of inequalities, 23

    ordered pair, 35
    of systems of equations, 523
  solving inequalities, 24
    with addition, 78
    with decimals, 140
    with division, 79
    with fractions, 140
    with integers, 78–79
    with multiplication, 79
    multistep, 514–516
    with rational numbers, 140–141
    with subtraction, 78
    two-step, 514–515
  systems of equations, 523–525
    using graphs to solve, 576–577
  tiles, 72–73, 506
  toolbox, 2–57
  two-step equations, 498–499
  variables
    on both sides, solving equations with,
      507–509
    expressions and, 4–5
    isolating, 14, 520
    solving for, 519–520
  variation
    direct, 562–564
    inverse, 628–629
  word problems
    interpreting which operation to use in, 9
    solving, using equations, 137
    writing algebraic expressions in, 10
Algebra tiles, 72–73, *502, 506, 507, 657, 661*
Algebra toolbox, 2–57
Algebraic expressions, 4
  combining like terms, 29
  with one variable, evaluating, 4
  simplifying, by combining like terms, 29
  translating word phrases into, 8–9
  with two variables, evaluating, 5
  writing, 8–10
    in word problems, 10
Alternate Opener, *4, 8, 13, 18, 23, 28, 34, 38,*
  43, 60, 64, 68, 74, 78, 84, 88, 92, 96, 112, 117,
  121, 126, 131, 136, 140, 146, 150, 156, 174,
  179, 184, 188, 196, 200, 204, 222, 228, 234,
  239, 244, 250, 254, 259, 263, 280, 285, 290,
  294, 302, 307, 312, 316, 320, 324, 342, 346,
  350, 356, 362, 368, 372, 376, 382, 400, 405,
  410, 416, 420, 424, 428, 446, 451, 456, 462,
  467, 471, 477, 482, 498, 502, 507, 514, 519,
  523, 540, 545, 550, 556, 562, 567, 572, 590,
  595, 601, 608, 613, 617, 621, 628, 644, 650,
  656, 660, 664, 670, 688, 692, 696, 702, 708,
  712, 716
Analyze units, 350–352
Analyzing
  data, 172–219
  scale factors, 376
Anamorphic image, 316
Angle measures
  of parallel lines cut by transversals, 229
  in polygons, finding sums of the, 239
Angle bisector, 227
Angles, 222–224
  acute, 223
  alternate exterior, 229
  alternate interior, 229

Deductive reasoning, 711, 773
Degree,
 of a polynomial, 645
 of a vertex, 712
Denominators, 112, 813
 like, *see* Like denominators
 unlike, *see* Unlike denominators
Density Property of real numbers, 157
Dependent events, 477–479
 probabilities of, 478–479
Dependent variables, 34
Design, 120
Deviation, 208–209
Diagonals, 813
Diagrams, tree, *see* Tree diagrams
Diameter, 294, 778
Difference, common, 590
Differences, *see also* Subtraction
 first and second, 601
Dilating figures, 363
Dilation, center of, *see* Center of dilation
Dilations, 362–363
Dimensional analysis, 350
Dimensions
 changing, exploring effects of, 308, 313,
  317, 321
 of matrices, 779
 missing
  using equivalent ratios to find, 369
  using scale factors to find, 368
 and scale factors, 377
 three, symmetry in, 328–329
Direct variation, 562–564
 equations of, 563
Disjunctions, 703
Displaying data, 172–219, 196–197
 in bar graphs, 196
 in histograms, 197
 in line graphs, 197
Dividend, 813
Divisibility rules, 762
Distributive Property, 29
 and polynomials, 651, 670
Division, *see also* Quotients
 of decimals, by decimals, 127
 of fractions, by fractions, 126–127
 of integers, 68
 long, 761
 of monomials by monomials, 674
 and multiplication as inverse operations, 126
 of polynomials by monomials, 674–675
 of powers with the same base, 88–89, 93
 of rational numbers, 126–128
 solving equations with, 18–20, 74
 solving for variables by, 520
 solving inequalities with, 79
 word phrases for, 8
Division Property of Equality, 18
Divisor, 813
Dodecahedron, 300
Domain, 608
Drawings
 perspective, 303–304
 scale, 372–373

E

Earth science, 47, 71, 134, 139, 191, 243, 323,
 343, 408, 423, 454, *496,* 511, 559, 571, *588B,* 611
Economics, 63, 427, 518, 575
Edges, 302
 of a graph, 712
Element,
 of a matrix, 779
 of a set, 688
Empty set, 692
Endpoint, 222
Energy, 120
Enlargement, 373
Enrichment, 3, 59, 111, 173, 221, 279, 341, 399,
 445, 497, 539, 589, 643, 687
Entertainment, 7, 12, 17, 297, 311, 345, 346,
 349, 379, 450, 517, 527, 544, 552, 699
Equality
 Addition Property of, 14
 Division Property of, 18
 Multiplication Property of, 19
 Subtraction Property of, 14
Equally likely outcomes, 462
 theoretical probability for, 462
Equations, 13
 determining whether numbers are solutions
  of, 13
 of direct variation, 563
 graphing, 39
 graphs of, 39
 inequalities and, 496–537
 linear, *see* Linear equations
 literal, 519–520
 point-slope form of, 557
 simple two-step, solving, 20
 solutions of, 13
 solving, *see* Solving equations
 solving word problems using, 137
 systems of, *see* Systems of equations
 writing, for linear functions
  from graphs, 613
  from tables, 614
Equilateral triangles, 235
Equivalent expressions, 28
Equivalent ratios, 342
 finding, 342, 400
 using, to find missing dimensions, 369
Eratosthenes, sieve of, 212
Error, greatest possible, 784
Escher, M. C., 267
Estimating
 odds from experiments, 482
 with percents, 420–421
 probabilities of events, 451–452
 square roots of numbers, 150–151
Euler circuits, 713
Euler path, 715
Evaluating
 algebraic expressions
  with one variable, 4
  with two variables, 5
 expressions
  containing factorials, 471

 with fractions and decimals, 127
 with integers, 61, 65
 with rational numbers, 118, 123, 132
 functions, 609–610
 negative exponents, 93
 powers, 84–85
 products and quotients of negative
  exponents, 93
Events, 446
 classifying, as independent or dependent, 477
 compound, 446, 464
 dependent, *see* Dependent events
 finding probabilities of, 447
 independent, *see* Independent events
 mutually exclusive, 464
 probabilities of, estimating, 451–452
Experimental probability, 451–452
Experiments, 446
 estimating odds from, 482
Exploration, 4, 8, 13, 18, 23, 28, 34, 38, 43, 60,
 64, 68, 74, 78, 84, 88, 92, 96, 112, 117, 121,
 126, 131, 136, 140, 146, 150, 156, 174, 179, 184,
 188, 196, 200, 204, 222, 228, 234, 239, 244,
 250, 254, 259, 263, 280, 285, 290, 294, 302,
 307, 312, 316, 320, 324, 342, 346, 350, 356,
 362, 368, 372, 376, 382, 400, 405, 410, 416,
 420, 424, 428, 446, 451, 456, 462, 467, 471,
 477, 482, 498, 502, 507, 514, 519, 523, 540,
 545, 550, 556, 562, 567, 572, 590, 595, 601,
 608, 613, 617, 621, 628, 644, 650, 656, 660,
 664, 670, 688, 692, 696, 702, 708, 712, 716
Exponential decay functions, 618
Exponential form, 84
Exponential functions, 617–618
 graphing, 617
Exponential growth functions, 618, 777
Exponents, 84–85
 integer, looking for patterns in, 92–93
 integers and, 58–109
 negative, *see* Negative exponents
 properties of, 88–89
Expressions
 algebraic, *see* Algebraic expressions
 containing factorials, evaluating, 471
 containing powers, simplifying, 85
 with decimals, evaluating, 127
 equivalent, 28
 with fractions, evaluating, 127
 with integers, evaluating, 61, 65
 with rational numbers, evaluating, 118,
  123, 132
 two-variable, combining like terms in, 29
 variables and, 4–5
Extension
 Average Deviation, 208–209
 Compound Interest, 432–433
 Dividing Polynomials by Monomials, 674–675
 games, 50, 102, 164, 212, 270, 332, 390, 436,
  488, 530, 580, 634
 Other Number Systems, 160–161
 problem solving, 48, 49, 100, 101, 162, 163,
  210, 211, 268, 269, 330, 331, 388, 389, 434,
  435, 486, 487, 528, 529, 578, 579, 632, 633
 Symmetry in Three Dimensions, 328–329
 Systems of Equations, 576–577
 Trigonometric Ratios, 386–387
Exterior angles, 229, 271

F

Faces
 of figures, 302
 lateral, of prisms, 316
Factoring polynomials, 675
Factors, 762
Factor tree, 763
Factorials, 471
 evaluating expressions containing, 471
Fahrenheit, 781
Fair, 462
Fibonacci sequence, 603
Figures
 composite, *see* Composite figures
 congruent, 223
 dilating, 363
 drawing
  with line symmetry, 259
  with rotational symmetry, 260
 similar, *see* Similar figures
 solid, *see* Solid figures
 three-dimensional
  drawing, 302–304
  scaling, 382–383
Finance, 599, 616, 631
Finite sets, 689
Firefighter, 2
First differences
 using, to find terms of sequences, 601
First quartile, 188
FOIL, 670
Flip, *see* Reflection
Food, 19, 297, 570
Force, 75
Formula,
 area of a circle, 295
 area of a parallelogram, 281–282
 area of a rectangle, 281
 area of a triangle, 286
 area of a trapezoid, 286
 arithmetic sequence, 591
 circumference of a circle, 294–295
 compound interest, 433
 Fahrenheit to Celsius, 781
 geometric sequence, 596
 Pythagorean Theorem, 290–291
 simple interest, 428
 surface area of a cylinder, 316–317
 surface area of a rectangular prism, 316–317
 surface area of a sphere, 325
 volume of a cone, 312–313
 volume of a cylinder, 307–308
 volume of a prism, 307–309
 volume of a pyramid, 312–313
 volume of a sphere, 324
Foster, Don, 183
Fractal, 285
Fraction bars, 112, 132
Fractions
 addition of, with unlike denominators, 131
 and decimals and percents, relating, 400–401
 division of, by fractions, 126–127
 expressions with, evaluating, 127

improper
 writing as mixed numbers, 765
 writing mixed numbers as, 765
with like denominators
 addition of, 118
 subtraction of, 118
in lowest terms, 122
multiplication of
 by fractions, 122
 by integers, 121
simplifying, 112–113
solving equations with, 136–137
solving inequalities with, 140
solving multistep equations that contain, 502–503
subtraction of, with unlike denominators, 131
unit, 164
with unlike denominators, addition and subtraction of, 131
writing, as decimals, 114
writing decimals as, 113
Frequency tables, 196, 775
Frequency polygon, 776
Function notation, 609
Functions, 608–610
 evaluating, 609–610
 exponential, *see* Exponential functions
 exponential decay, 618
 exponential growth, 618, 685
 finding different representations of, 608
 identifying, 609
 linear, *see* Linear functions
 quadratic, *see* Quadratic functions
 sequences and, 588–641
Fundamental Counting Principle, 467–468
Fundraising, 485

G

Games, 481
 Crazy Cubes, 50
 Egg Fractions, 164
 Equation Bingo, 102
 Line Solitaire, 580
 Math in the Middle, 212
 Percent Tiles, 436
 Permutations, 488
 Polygon Rummy, 270
 Tic-Frac-Toe, 390
 Triple Concentration, 332
 24 Points, 530
 What's Your Function?, 634
Games Extension, *50, 102, 164, 212, 270, 332, 390, 436, 488, 530, 580, 634*
Geography, 15, 192, 715
Geometric sequences, 595–597
 *n*th term of, 596
Geometry
 The development of Geometry skills and concepts is a central focus of this course and is found throughout this book.
 acute angles, 223
 acute triangles, 234
 anamorphic image, 316

angle bisector, 227
angle measures
 of parallel lines, 229
 in polygons, 239
angles, 222–224
 acute, 223
 alternate exterior, 229
 alternate interior, 229
 classifying, 223
 complementary, 223
 congruent, *see* Congruent angles
 corresponding, 229, 250
 exterior, 229, 271
 finding, 234–236
 interior, 229
 obtuse, 223
 in regular polygons, 240
 right, 223
 supplementary, 223
 vertical, 223–224
axes, 38
bilateral symmetry, 328
circles, 294–295, 778
complementary angles, 223
cones
 right, 320
 surface area of, 320–321
 volume of, 312–313
congruent angles, 228
congruent figures, 223
coordinate geometry, 244–246
coordinate plane, 38
 area on a, 295
 circumference on a, 295
 graphing on a, 38–39
corresponding sides, 250
cube, 154
cylinders, 307
 surface area of, 316–317
 volume of, 307–309
dilations, 362–363
dimensional analysis, 350
dimensions
 changing, exploring effects of, 308, 313, 317, 321
 and scale factors, 368, 377
 three, symmetry in, 328–329
dodecahedron, 300
equilateral triangles, 235
Escher, M. C., 267
exterior angles, 271
faces, 302
 lateral, of prisms, 316
flip, *see* Reflection
heptagons, 239
hexagons, 239
hypotenuse, 290
icosahedron, 300
image, 254
 anamorphic, 316
informal proof, 773
isometric drawing, 250
isosceles triangles, 235
lateral faces of prisms, 316
lateral surface, 316
legs, 290
 in right triangles, 291

Helpful Hint, 9, 10, 14, 23, 28, 34, 35, 38, 60, 75, 79, 84, 89, 96, 121, 122, 146, 156, 161, 197, 205, 241, 245, 255, 259, 280, 282, 290, 307, 350, 356, 362, 382, 406, 420, 457, 472, 508, 509, 520, 524, 525, 546, 552, 562, 567, 568, 569, 590, 618, 628, 670, 688, 702, 703, 708

Hemispheres, 324

Heptagons, 239

Hexagons, 239

Hill, A. V, 673

    Hill's equation, 673

Histograms, 196, 776

History, 37, *44*, 699

Hobbies, 345, 625

Home Economics, 612

Homework Help

Homework Help Online is available for every lesson. Refer to the Internet Connect box at the beginning of each exercise set. Some examples: 6, 11, 16, 21, 26

Horizon line, 303

Horticulturist, 340

Hydrologist, 496

Hypotenuse, 290

Hypothesis, 708

Icosahedron, 301

Identity Property of One, 769

Identity Property of Zero, 769

If-then **statements,** 708

Image, 254

    anamorphic, 316

Impossible event, 446

Independent events, 477–479

Independent variables, 34

Index cards, 24, 462, 508

Indirect measurement, 386

Inductive reasoning, 773

Industrial Arts, 149

Inequalities, 23

    algebraic, 23

    equations and, 496–537

    graphing, 24, 78, 567–568

    linear, 567

    simple, solving, 23–25

    solutions of, 23

    in two variables, graphing, 567–569

Infinite sets, 689

Informal geometry proofs, 773

Input, 608

Inscribed angle, 778

Integer exponents, 92–93

Integers, 60

    addition of, 60–61

    division of, 68–69

    exponents and, 58–109

    expressions with, evaluating, 61, 65

    multiplication of, 68–69

        by fractions, 121

    negative, 60

    positive, 60

    solving equations with, 74–75

    solving inequalities with, 78–79

    subtraction of, 64–65

    using order of operations with, 69

Intercepts, slopes and, using, 550–552

Interdisciplinary Bulletin Board

    astronomy, 642B

    Earth science, 2B, 588B

    life science, 340B, 444B

    physical education, 110B

    physical science, 2B, 278B, 398B, 496B

    social studies, 58B, 172B, 686B

    sports, 538B

    structural engineering, 220B

Interest, 428

    compound, 432–433

    on a loan, 428

    rate of, 428–429

    simple, 428

Intersection of sets, 692–693

    and Venn diagrams, 696–697

Intervention, *3, 59, 111, 173, 221, 279, 341, 399, 445, 497, 539, 589, 643, 687*

Inverse operations, 14

    multiplication and division as, 126

Inverse variation, 628–629

    graphing, 629

Irrational numbers, 156

Isometric drawing, 302

Isolating variables, 14, 520

Isosceles triangles, 235

Iteration, 774

Journal, *7, 12, 17, 22, 27, 31, 37, 41, 47, 63, 67, 71, 77, 81, 87, 91, 95, 99, 116, 120, 125, 130, 134, 139, 143, 149, 153, 159, 177, 183, 187, 192, 199, 203, 207, 226, 231, 238, 243, 247, 253, 257, 262, 267, 284, 288, 293, 297, 306, 311, 315, 319, 323, 327, 345, 349, 354, 359, 365, 371, 375, 379, 385, 403, 408, 413, 419, 423, 427, 431, 450, 454, 459, 466, 470, 475, 481, 485, 501, 505, 511, 518, 522, 527, 544, 549, 554, 559, 566, 571, 575, 594, 599, 605, 612, 616, 620, 625, 631, 647, 653, 659, 663, 667, 673, 691, 695, 699, 705, 711, 715, 719*

Lab Resources Online, 42, 51, 72, 103, 135, 154, 165, 178, 193, 213, 227, 232, 258, 271, 289, 300, 333, 355, 366, 380, 391, 404, 409, 437, 455, 476, 489, 506, 531, 555, 600, 626

Language arts, 90, 149, 183, 229, 408, 711

Lateral faces of prisms, 316

Lateral surface, 316

Least common denominator (LCD), 131

Least common multiple (LCM), 131, 764

Legs, 290

    in right triangles, finding length of, 291

Lengths

    using proportions to find, 372

    using trigonometric ratios to find, 386

Lesson Quiz, *7, 12, 17, 22, 27, 31, 37, 41, 47, 63, 67, 71, 77, 81, 87, 91, 95, 99, 116, 120, 125, 130, 134, 139, 143, 149, 153, 159, 177, 183, 187, 192, 199, 203, 207, 226, 231, 238, 243, 247, 253, 257, 262, 267, 284, 288, 293, 297, 306, 311, 315, 319, 323, 327, 345, 349, 354, 359, 365, 371, 375, 379, 385, 403, 408, 413, 419, 423, 427, 431, 450, 454, 459, 466, 470, 475, 481, 485, 501, 505, 511, 518, 522, 527, 544, 549, 554, 559, 566, 571, 575, 594, 599, 605, 612, 616, 620, 625, 631, 647, 653, 659, 663, 667, 673, 691, 695, 699, 705, 711, 715, 719*

Life science, 7, 37, 87, 99, 110, 139, 143, 207, 278, 311, 321, 327, *340B*, 340, 354, 373, 377, 378, 403, 408, 411, 416–417, 419, *444B*, 459, 466, 475, 501, *538*, 554, 559, 566, *588*, 599, 614, 616, 652, 673, 691, 695, 705

Like denominators, fractions with

    addition of, 118

    subtraction of, 118

Like terms, 28, 650

    combining, *see* Combining like terms

    solving multistep equations that contain, 502

Line, *see also* Lines

    boundary, 567

    horizon, 303

    of reflection, 259

    of symmetry, 259

Line graphs, 197

Line segments, 222

Line symmetry, 259–260

    drawing figures with, 259

Linear equations, 540

    graphing, 540–542

    finding *x*-intercepts and *y*-intercepts for, 550

Linear functions, 613–614

    writing equations for

        from graphs, 613

        from tables, 614

Linear inequalities, 567

Lines, 222, *see also* Line

    of best fit, 204, 572–573

    graphing, 538–587

        using points and slopes, 547

    naming, 222

    parallel, *see* Parallel lines

    perpendicular, *see* Perpendicular lines

    point-slope form of, 556

    slope-intercept form of, 551

    slopes of, *see* Slopes

Link

    architecture, 375, 379

    art, 267, 371, 475, 653

    astronomy, 185

    business, 659

    career, 125, 315

    Earth science, 47, 134, 139, 243, 343, 454, 559

Multiplicative inverse, 126
Music, 41, 605, 629, 691, 699
Mutually exclusive events, 464

n-gons, 239
Naming
    points, lines, planes, segments, and rays, 222
Negative correlation, 205
Negative exponents
    evaluating, 93
    products and quotients of, evaluating, 93
    using patterns to evaluate, 92
Negative integers, 60
Negative slopes, 545
Negative square roots of numbers, 146
Nets, 300, *313, 316, 321*
Networks, 712–713
Newtons (N), 75
No correlation, 205
Nonterminating decimal, 156
Notation
    function, 609
    scientific, 96–97
    standard, 96–97
*n*th term
    of arithmetic sequences, 591–592
    of geometric sequences, 596
Nuclear physicist, 58
Null set, 692
Number cards, 147, 151, 401
Number cubes, 429, 452, 717
Number line, 60, 117
Number systems, 160–161
Numbers
    bases of, 84, 160–161
    classification of, 157
    compatible, 420
    composite, 763
    irrational, 156
    percents of, 406
    positive and negative square roots of, 146
    prime, 763
    random, 456
    rational, *see* Rational numbers
    real, *see* Real numbers
    relatively prime, 112
    square roots of, *see* Square roots
    triangular, 601
Numerator, 112
Nutritionist, 110

Obtuse angles, 223
Obtuse triangles, 234–235
Octagons, 239
Octahedrons, 300
Octal number system, 160–161
Odds, 482–483
    against, 482

converting, to probabilities, 482–483
converting probabilities to, 483
estimating, from experiments, 482
in favor, 482
One-Minute Section Planner, 4A, 34A, 60A,
    84A, 112A, 146A, 174A, 196A, 222A, 250A,
    280A, 300A, 342A, 362A, 400A, 416A, 446A,
    462A, 498A, 514A, 540A, 562A, 590A, 608A,
    644A, 656A, 688A, 702A
One-point perspective drawings, 303
Operations, 9
    inverse, *see* Inverse operations
    order of, *see* Order of operations
Opposites, 60, 771
    of polynomials, 660
Order of operations, 4, 69, 676 *see also*
    Parentheses
Ordered pairs, 34–35
    as solutions of equations, 34
Organizing data, 179–180
    in back-to-back stem-and-leaf
        plots, 180
    in histograms, 196
    in stem-and-leaf plots, 180
    in tables, 179
Origin, 38
    as the center of dilation, 363
Outcomes, 446
    equally likely, 462
    in sample spaces, finding probabilities of,
        446–447
Outlier, 185
Output, 608

Pacing Guide, 2A, 58A, 110A, 172A, 220A,
    278A, 340A, 398A, 444A, 496A, 538A, 588A,
    642A, 686A
Parabola, 621, 777
Parallel lines, 228–229
    cut by transversals, finding angle
        measures of, 229
    finding, 245
    identifying, by slopes, 547
    slopes of, 245
Parallelograms, 240, 280
    area of, 281–282
    height of, 282
    perimeter of, 280
Parentheses, 84, 676 *see also* Order of
    Operations
Pascal's Triangle, 476
Path, 712
    Euler, 715
Pattern blocks, 255, 264, 697
Patterns
    looking for, in integer exponents, 92–93
    using, to evaluate negative
        exponents, 92
Payment, finding total, on a loan, 428
Pei, I. M., 315
Pentagons, 239
Percent change, 416–417

Percent decrease, 416–417
Percent increase, 416–417
Percent problems, types of, 411
Percents, 398–443
    applications of, 424–425, 428–429
    and decimals and fractions, relating,
        400–401
    division by, to find total sales, 425
    estimating with, 420–421
    finding, 400, 405–406
    finding numbers when percents are known,
        410–411
    multiplication by,
        to find commission amounts, 424
        to find sales tax amounts, 424
    of tax withheld, using proportions to
        find the, 425
Perfect squares, 146
Performance Assessment, 56, 108, 170, 218,
    276, 338, 396, 442, 494, 536, 586, 640,
    684, 728
    rubric for, 56, 108, 170, 218, 276, 338, 396,
        442, 494, 536, 586, 640, 684, 728
Perimeter, 280
    and area and volume, 278–339
    of composite figures, 282
    of rectangles and parallelograms, 280
    of trapezoids, 285
    of triangles, 285
Permutations, 471–472
Perpendicular bisector, 227
Perpendicular lines, 228–229
    identifying, by slopes, 547
    slopes of, 245
    symbol for, 229
Perspective, 303
Perspective drawings
    one-point, 303
    two-point, 304
pH scale, 784
Photography, 365
Physical education, 110B
Physical science, *2, 5, 41, 58, 81, 89, 99, 220,*
    231, 278B, 278, 288, 352, 357, 358, 371, 385,
    398B, 401, 403, 410, 423, 496B, 505, 511, 522,
    543, 564, 566, 599, 614, 618, 625, 631, 645
Pi, 294
Pixels, 147
Place-value table, 113, 760
Plane geometry, 220–277
Planes, 222
    coordinate, *see* Coordinate plane
    naming, 222
Playground equipment designer, 220
Playing cards, 463
Point symmetry, *see* Rotational symmetry
Point-slope form, 556–557
    of equations, writing, 557
    using, to identify information about lines, 556
Points, 222
    coordinates of, on a coordinate plane, 38
    graphing, on a coordinate plane, 39
    naming geometric, 222
    and slopes, graphing lines using, 547
    two, finding slopes given, 545

Use a simulation, 456–457

Vanishing point, 303
Variability, 188–190
Variables
  on both sides, solving equations with, 507–509
  expressions and, 4–5
  isolating, 14, 520
  one, evaluating expressions with, 4
  solving for, 519–520
    by addition or subtraction, 519
    by division or square roots, 520
  two, expressions with,
    evaluating, 5
    inequalities in, graphing, 567–569
Variation
  direct, 562–564
  inverse, 628–629
Venn diagrams, 690–697
Vertex of figure, 302
Vertex (networks), 712
Vertical angles, 223–224
Volume
  and area and perimeter, 278–339
  of composite figures, 309
  of cones, 312–313
  in cubic units, 307
  of cylinders, 307–309
  of cylinders, finding, 307–308
  of prisms, 307–309
  of pyramids, 312–313
  of spheres, 324
Volumes, comparing, 325

Warm Up, 4, 8, 13, 18, 23, 28, 34, 38, 43, 60, 64, 68, 74, 78, 84, 88, 92, 96, 112, 117, 121, 126, 131, 136, 140, 146, 150, 156, 174, 179, 184, 188, 196, 200, 204, 222, 228, 234, 239, 244, 250, 254, 259, 263, 280, 285, 290, 294, 302, 307, 312, 316, 320, 324, 342, 346, 350, 356, 362, 368, 372, 376, 382, 400, 405, 410, 416, 420, 424, 428, 446, 451, 456, 462, 467, 471, 477, 482, 498, 502, 507, 514, 519, 523, 540, 545, 550, 556, 562, 567, 572, 590, 595, 601, 608, 613, 617, 621, 628, 644, 650, 656, 660, 664, 670, 688, 692, 696, 702, 708, 712, 716
Weak correlation, 205
What's the Error?, 12, 31, 37, 63, 77, 81, 91, 116, 125, 130, 139, 149, 153, 159, 161, 177, 192, 203, 231, 238, 243, 247, 253, 297, 306, 311, 315, 323, 349, 354, 379, 403, 431, 450, 459, 470, 485, 505, 549, 599, 711
What's the Question?, 143, 284, 475, 544, 612, 616, 647, 667, 705
Whole numbers
  rounding, 760
Wildlife ecologist, 538
Williams, Venus and Serena, 112
Withholding tax, 425
Word phrases, translating, into algebraic expressions, 8–9
Word problems
  interpreting operations to use in, 9
  solving, using equations, 137
  writing algebraic expressions in, 10
Write a Problem, 17, 27, 41, 99, 120, 187, 199, 257, 262, 293, 345, 371, 481, 518, 559, 566, 594, 631, 659, 691, 719

Write About It, 12, 17, 27, 31, 37, 41, 63, 67, 71, 77, 81, 87, 91, 99, 116, 120, 125, 130, 139, 143, 149, 159, 177, 187, 192, 199, 203, 226, 231, 238, 243, 247, 253, 257, 262, 284, 293, 297, 306, 311, 315, 319, 323, 345, 349, 354, 365, 371, 379, 385, 403, 408, 419, 423, 431, 450, 454, 459, 470, 475, 481, 485, 505, 511, 518, 527, 544, 549, 559, 566, 571, 594, 599, 612, 616, 625, 631, 647, 659, 663, 667, 673, 691, 699, 705, 711, 719
Writing Math, 229, 591

x-**axis,** 38
x-**coordinate,** 38
x-**intercepts,** 550
  finding, to graph linear equations, 550

y, **solving for,** and graphing, 520
y-**axis,** 38
y-**coordinate,** 38
y-**intercepts,** 550
  finding, to graph linear equations, 550
  using slope-intercept form to find slopes and, 551

Zero, as its own opposite, 60
Zero power, 89
Zero slopes, 545

2005 NAEP Mathematics Framework
Grade 8 Assessment

I. Number Properties

1. Number sense

a. Use place value to model and describe integers and decimals.

b. Model or describe rational numbers or numerical relationships using number lines and diagrams.

d. Write or rename rational numbers.

e. Recognize, translate between, or apply multiple representations of rational numbers (fractions, decimals, and percents) in meaningful contexts.

f. Express or interpret numbers using scientific notation from real life contexts.

g. Find or model absolute value or apply to problem situations.

i. Order or compare rational numbers (fractions, decimals, percents, or integers) using various models and representations (e.g., number line).

j. Order or compare rational numbers including very large and small integers, and decimals and frations close to zero.

2. Estimation

a. Establish or apply benchmarks for rational numbers and common irrational numbers (e.g., π) in contexts.

b. Make estimates appropriate to a given situation by:
   • identifying when estimation is appropriate,
   • determining the level of accuracy needed,
   • selecting the appropriate method of estimation, or
   • analyzing the effect of an estimation method on the accuracy of results.

c. Verify solutions or determine the reasonableness of results in a variety of situations including calculator and computer results.

d. Estimate square or cube roots of numbers less than 1,000 between two whole numbers.

3. Number operations

a. Perform computations with rational numbers.

d. Describe the effect of multiplying and dividing by numbers, including the effect of multiplying or dividing a rational number by:
   • zero, or
   • a number less than zero, or
   • a number between zero and one, or
   • one, or
   • a number greater than one.

e. Provide a mathematical argument to explain operations with two or more fractions.

f. Interpret rational number operations and the relationship between them.

g. Solve application problems involving rational numbers and operations using exact answers or estimates as appropriate.

4. Ratios and proportional reasoning

a. Use ratios to describe problem situations.

b. Use fractions to represent and express ratios and proportions.

c. Use proportional reasoning to model and solve problems (including rates, scaling, and similarity).

d. Solve problems involving percentages (including percent increase and decrease, interest rates, tax, discount, tips, or part/whole relationships).

5. Properties of number and operations

a. Describe odd and even integers and how they behave under different operations.

b. Recognize, find, or use factors, multiples, or prime factorization.

c. Recognize or use prime and composite numbers to solve problems.

d. Use divisibility or remainders in problem settings.

e. Apply basic properties of operations.

f. Explain or justify a mathematical concept or relationship (e.g., explain why 17 is prime).

2005 NAEP Mathematics Framework
Grade 8 Assessment

II. Measurement

1. Measuring physical attributes

b. Compare objects with respect to length, area, volume, angle measurement, weight, or mass.

c. Estimate the size of an object with respect to a given measurement attribute (e.g., area).

g. Select or use appropriate measurement instrument to determine or create a given length, area, volume, angle, weight, or mass.

h. Solve mathematical or real-world problems involving perimeter or area of plane figures such as triangles, rectangles, circles, or composite figures.

j. Solve problems involving volume or surface area of rectangular solids, cylinders, prisms, or composite shapes.

k. Solve problems involving indirect measurement such as finding the height of a building by comparing its shadow with the height and shadow of a known object.

l. Solve problems involving rates such as speed or population density.

2. Systems of measurement

a. Select or use appropriate type of unit for the attribute being measured such as length, area, angle, time, or volume.

b. Solve problems involving conversions within the same measurement system such as conversions involving square inches and square feet.

c. Estimate the measure of an object in one system given the measure of that object in another system and the approximate conversion factor.
For example:
- Distance Conversion:
  1 kilometer is approximately $\frac{5}{8}$ of a mile.
- Money Conversion:
  US dollar is approximately 1.5 Canadian dollars.
- Temperature Conversion:
  Fahrenheit to Celsius

d. Determine appropriate size of unit of measurement in problem situation involving such attributes as length, area, or volume.

e. Determine appropriate accuracy of measurement in problem situations (e.g., the accuracy of each of several lengths needed to obtain a specified accuracy of total length) and find the measure to that degree of accuracy.

f. Construct or solve problems (e.g., floor area of a room) involving scale drawings.

2005 NAEP Mathematics Framework
Grade 8 Assessment

III. Geometry

1. Dimension and shape

a. Draw or describe a path of shortest length between points to solve problems in context.

b. Identify a geometric object given written description of its properties.

c. Identify, define, or describe geometric shapes in the plane and in 3-dimensional space given a visual representation.

d. Draw or sketch from a written description polygons, circles, or semicircles.

e. Represent or describe a three-dimensional situation in a two-dimensional drawing using perspective.

f. Demonstrate an understanding about the two- and three-dimensional shapes in our world through identifying, drawing, modeling, building, or taking apart.

2. Transformation of shapes and preservation of properties

a. Identify lines of symmetry in plane figures or recognize and classify types of symmetries of plane figures.

c. Recognize or informally describe the effect of a transformation on two-dimensional geometric shapes (reflections across lines of symmetry, rotations, translations, magnifications, and contractions).

d. Predict results of combining, subdividing, and changing shapes of plane figures and solids (e.g., paper folding, tiling, and cutting up and rearranging pieces).

e. Justify relationships of congruence and similarity, and apply these relationships using scaling and proportional reasoning.

f. For similar figures, identify and use the relationships of conservation of angles and of proportionality of side length and perimeter.

3. Relationships between geometric figures

b. Apply geometric properties and relationships in solving simple problems in two- and three-dimensions.

c. Represent problem situations with simple geometric models to solve mathematical or real-world problems.

d. Use the Pythagorean Theorem to solve problems.

f. Describe or analyze simple properties of, or relationships between, triangles, quadrilaterals, and other polygonal plane figures.

g. Describe or analyze properties and relationships of parallel or intersecting lines.

4. Position and direction

a. Describe relative positions of points and lines using the geometric ideas of midpoint, points on common line through a common point, parallelism, or perpendicularity.

b. Describe the intersection of two or more geometric figures in the plane (e.g., intersection of a circle and a line).

c. Visualize or describe the cross-section of a solid.

d. Represent geometric figures using rectangular coordinates on a plane.

5. Mathematical reasoning

a. Make and test a geometric conjecture about regular polygons.

2005 NAEP Mathematics Framework
Grade 8 Assessment

IV. Data Analysis and Probability

1. Data representation

a. Read or interpret data, including interpolating or extrapolating from data.

b. Given a set of data, complete a graph and then solve a problem using the data in the graph (circle graphs, histograms, bar graphs, line graphs, scatter plots).

c. Solve problems by estimating and computing with data from a single set or across sets of data.

d. Given a graph or a set of data, determine whether information is represented effectively and appropriately (circle graphs, histograms, bar graphs, line graphs, scatter plots).

e. Compare and contrast the effectiveness of different representations of the same data.

2. Characteristics of data sets

a. Calculate, use, or interpret mean, median, mode, or range.

b. Describe how mean, median, mode, range, or interquartile ranges relate to the shape of the distribution.

c. Identify outliers and determine their effect on mean, median, mode, or range.

d. Using appropriate statistical measures, compare two or more data sets describing the same characteristic for two different populations or subsets of the same population.

e. Visually choose the line that best fits given a scatter plot and informally explain the meaning of the line. Use the line to make predictions.

3. Experiments and samples

a. Given a sample, identify possible sources of bias in sampling.

b. Distinguish between a random and non-random sample.

d. Evaluate the design of an experiment.

4. Probability

a. Analyze a situation that involves probability of an independent event.

b. Determine the theoretical probability of simple and compound events in familiar contexts.

c. Estimate the probability of simple and compound events through experimentation or simulation.

d. Distinguish between experimental and theoretical probability.

e. Determine the sample space for a given situation.

f. Use a sample space to determine the probability of the possible outcomes of an event.

g. Represent probability using fractions, decimals, and percents.

h. Determine the probability of independent and dependent events. (Dependent events should be limited to linear functions with a small sample size.)

j. Interpret probabilities within a given context.

2005 NAEP Mathematics Framework
Grade 8 Assessment

V. Algebra

1. Patterns, relations, and functions

a. Recognize, describe, or extend numerical and geometric patterns using tables, graphs, words, or symbols.

b. Generalize a pattern appearing in a numerical sequence or table or graph using words or symbols.

c. Analyze or create patterns, sequences, or linear functions given a rule.

e. Identify functions as linear or non-linear or contrast distinguishing properties of functions from tables, graphs, or equations.

f. Interpret the meaning of slope or intercepts in linear functions.

2. Algebraic representations

a. Translate between different representations of linear expressions using symbols, graphs, tables, diagrams, or written descriptions.

b. Analyze or interpret linear relationships expressed in symbols, graphs, tables, diagrams, or written descriptions.

c. Graph or interpret points that are represented by ordered pairs of numbers on a rectangular coordinate system.

d. Solve problems involving coordinate pairs on the rectangular coordinate system.

e. Make, validate, and justify conclusions and generalizations about linear relationships.

g. Identify or represent functional relationships in meaningful contexts including proportional, linear, and common non-linear (e.g., compound interest, bacterial growth) in tables, graphs, words, or symbols.

3. Variables, expressions, and operations

a. Write algebraic expressions, equations, or inequalities to represent a situation.

b. Perform basic operations, using appropriate tools, on linear algebraic expressions (including grouping and order of multiple operations involving basic operations, exponents, roots, simplifying, and expanding).

4. Equations and inequalities

a. Solve linear equations or inequalities (e.g., $ax + b = c$ or $ax + b = cx + d$ or $ax + b > c$).

b. Interpret "=" as an equivalence between two expressions and use this interpretation to solve problems.

c. Analyze situations or solve problems using linear equations and inequalities with rational coefficients symbolically or graphically (e.g., $ax + b = c$ or $ax + b = cx + d$).

d. Interpret relationships between symbolic linear expressions and graphs of lines by identifying and computing slope and intercepts (e.g., know in $y = ax + b$, that a is the rate of change and b is the vertical intercept of the graph).

e. Use and evaluate common formulas e.g., relationship between a circle's circumference and diameter ($C = \pi d$), distance and time under constant speed].

Formulas

Perimeter

Polygon	P = sum of the lengths of the sides
Rectangle	$P = 2(b + h)$
Square	$P = 4s$

Circumference

Circle	$C = 2\pi r$, or $C = \pi d$ $d = 2r$

Volume

Prism	$V = Bh$
Rectangular prism	$V = \ell wh$
Cube	$V = s^3$
Cylinder	$V = \pi r^2 h$
Pyramid	$V = \frac{1}{3}Bh$
Cone	$V = \frac{1}{3}\pi r^2 h$
Sphere	$V = \frac{4}{3}\pi r^3$

Area

Circle	$A = \pi r^2$
Parallelogram	$A = bh$
Rectangle	$A = bh$
Square	$A = s^2$
Triangle	$A = \frac{1}{2}bh$
Trapezoid	$A = \frac{1}{2}h(b_1 + b_2)$

Surface Area

Prism	$S = 2B + ph$
Rectangular prism	$S = 2\ell w + 2\ell h + 2wh$
Cube	$S = 6s^2$
Cylinder	$S = 2B + 2\pi rh$
Regular pyramid	$S = B + \frac{1}{2}p\ell$
Cone	$S = \pi r^2 + \pi r\ell$
Sphere	$S = 4\pi r^2$

Trigonometry

Sine	$\sin A = \dfrac{\text{length of side opposite } \angle A}{\text{length of hypotenuse}}$
Cosine	$\cos A = \dfrac{\text{length of side adjacent to } \angle A}{\text{length of hypotenuse}}$
Tangent	$\tan A = \dfrac{\text{length of side opposite } \angle A}{\text{length of side adjacent to } \angle A}$

Probability

Experimental	$\text{probability} \approx \dfrac{\text{number of times event occurs}}{\text{total number of trials}}$
Theoretical	$\text{probability} = \dfrac{\text{number of outcomes in the event}}{\text{number of outcomes in sample space}}$
Permutations	$_nP_r = \dfrac{n!}{(n - r)!}$
Combinations	$_nC_r = \dfrac{_nP_r}{r!} = \dfrac{n!}{r!(n - r)!}$
Dependent events	$P(A \text{ and } B) = P(A) \cdot P(B \text{ after } A)$
Independent events	$P(A \text{ and } B) = P(A) \cdot P(B)$